W9-AWU-766

Glencoe

ALGEBRA 2

Aligned to the COMMON CORE STATE STANDARDS

Bothell, WA • Chicago, IL • Columbus, OH • New York, NY

connectED.mcgraw-hill.com Your Digital Math Portal

Animation 
Vocabulary 
eGlossary 
Personal Tutor 
Virtual Manipulatives 
Graphing Calculator 
Audio 
Foldables 
Self-Check Practice 
Worksheets 

In this Common Core State Standards edition of *Glencoe Algebra 2,* students are challenged to develop 21st century skills such as critical thinking and creative problem solving while engaging with exciting careers within Science, Technology, Engineering, and Mathematics (STEM) related fields.

Photo Credits:
Cover Dodecahedral Knot. By Paul Nylander (http://bugman123.com)

The McGraw·Hill Companies

connectED.mcgraw-hill.com

Education

Send all inquiries to:
McGraw-Hill Education
STEM Learning Solutions Center
8787 Orion Place
Columbus, OH 43240-4027

ISBN: 978-0-07-895265-4
MHID: 0-07-895265-4

...inted in the United States of America.

10 11 12 QVR 20 19 18 17 16 15 14 13 12

McGraw-Hill is committed to providing instructional materials in Science, Technology, Engineering, and Mathematics (STEM) that give students a solid foundation, one that prepares them for college and careers in the 21st Century.

Reviewers

Corey Andreasen
Mathematics Teacher
North High School
Sheboygan, Wisconsin

Mark B. Baetz
Mathematics Coordinating Teacher
Salem City Schools
Salem, Virginia

Kathryn Ballin
Mathematics Supervisor
Newark Public Schools
Newark, New Jersey

Kevin C. Barhorst
Mathematics Department Chair
Independence High School
Columbus, Ohio

Brenda S. Berg
Mathematics Teacher
Carbondale Community High School
Carbondale, Illinois

Dawn Brown
Mathematics Department Chair
Kenmore West High School
Buffalo, New York

Sheryl Pernell Clayton
Mathematics Teacher
Hume Fogg Magnet School
Nashville, Tennessee

Bob Coleman
Mathematics Teacher
Cobb Middle School
Tallahassee, Florida

Jane E. Cotts
Mathematics Teacher
O'Fallon Township High School
O'Fallon, Illinois

Michael D. Cuddy
Mathematics Instructor
Zypherhills High School
Zypherhills, Florida

Melissa M. Dalton, NBCT
Mathematics Instructor
Rural Retreat High School
Rural Retreat, Virginia

Tina S. Dohm
Mathematics Teacher
Naperville Central High School
Naperville, Illinois

Laurie L.E. Ferrari
Mathematics Teacher
L'Anse Creuse High School–North
Macomb, Michigan

Steve Freshour
Mathematics Teacher
Parkersburg South High School
Parkersburg, West Virginia

Shirley D. Glover
Mathematics Teacher
TC Roberson High School
Asheville, North Carolina

Caroline W. Greenough
Mathematics Teacher
Cape Fear Academy
Wilmington, North Carolina

Susan Hack, NBCT
Mathematics Teacher
Oldham County High School
Buckner, Kentucky

Michelle Hanneman
Mathematics Teacher
Moore High School
Moore, Oklahoma

Theresalynn Haynes
Mathematics Teacher
Glenbard East High School
Lombard, Illinois

Sandra Hester
Mathematics Teacher/AIG Specialist
North Henderson High School
Hendersonville, North Carolina

Jacob K. Holloway
Mathematics Teacher
Capitol Heights Junior High School
Montgomery, Alabama

Robert Hopp
Mathematics Teacher
Harrison High School
Harrison, Michigan

Eileen Howanitz
Mathematics Teacher/Department Chairperson
Valley View High School
Archbald, Pennsylvania

Charles R. Howard, NBCT
Mathematics Teacher
Tuscola High School
Waynesville, North Carolina

Sue Hvizdos
Mathematics Department Chairperson
Wheeling Park High School
Wheeling, West Virginia

Elaine Keller
Mathematics Teacher
Mathematics Curriculum Director K-12
Northwest Local Schools
Canal Fulton, Ohio

Sheila A. Kotter
Mathematics Educator
River Ridge High School
New Port Richey, Florida

Frank Lear
Mathematics Department Chair
Cleveland High School
Cleveland, Tennessee

Jennifer Lewis
Mathematics Teacher
Triad High School
Troy, Illinois

Catherine McCarthy
Mathematics Teacher
Glen Ridge High School
Glen Ridge, New Jersey

Jacqueline Palmquist
Mathematics Department Chair
Waubonsie Valley High School
Aurora, Illinois

Thom Schacher
Mathematics Teacher
Otsego High School
Otsego, Michigan

Laurie Shappee
Teacher/Mathematics Coordinator
Larson Middle School
Troy, Michigan

Jennifer J. Southers
Mathematics Teacher
Hillcrest High School
Simpsonville, South Carolina

Sue Steinbeck
Mathematics Department Chair
Parkersburg High School
Parkersburg, West Virginia

Kathleen D. Van Sise
Mathematics Teacher
Mandarin High School
Jacksonville, Florida

Karen Wiedman
Mathematics Teacher
Taylorville High School
Taylorville, Illinois

Online Guide

connectED.mcgraw-hill.com

The eStudentEdition allows you to access your math curriculum anytime, anywhere.

The icons found throughout your textbook provide you with the opportunity to connect the print textbook with online interactive learning.

Investigate

Animations illustrate math concepts through movement and audio.

Vocabulary tools include fun Vocabulary Review Games.

Multilingual eGlossary presents key vocabulary in 13 languages.

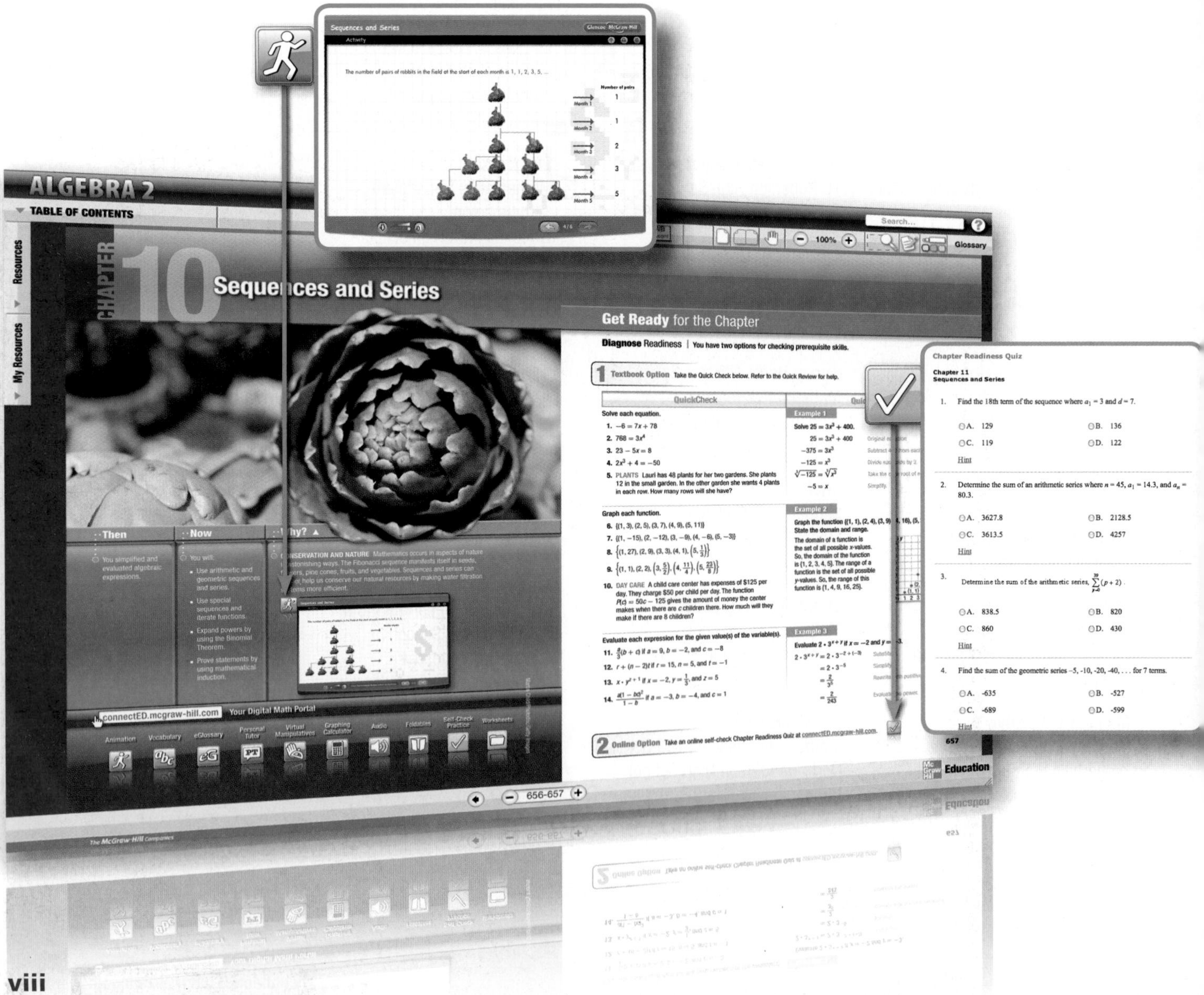

Learn

Personal Tutor presents an experienced educator explaining step-by-step solutions to problems.

Virtual Manipulatives are outstanding tools for enhancing understanding.

Graphing Calculator provides other calculator keystrokes for each Graphing Technology Lab.

Audio is provided to enhance accessibility.

Foldables provide a unique way to enhance study skills.

Practice

Self-Check Practice allows you to check your understanding and send results to your teacher.

Worksheets provide additional practice.

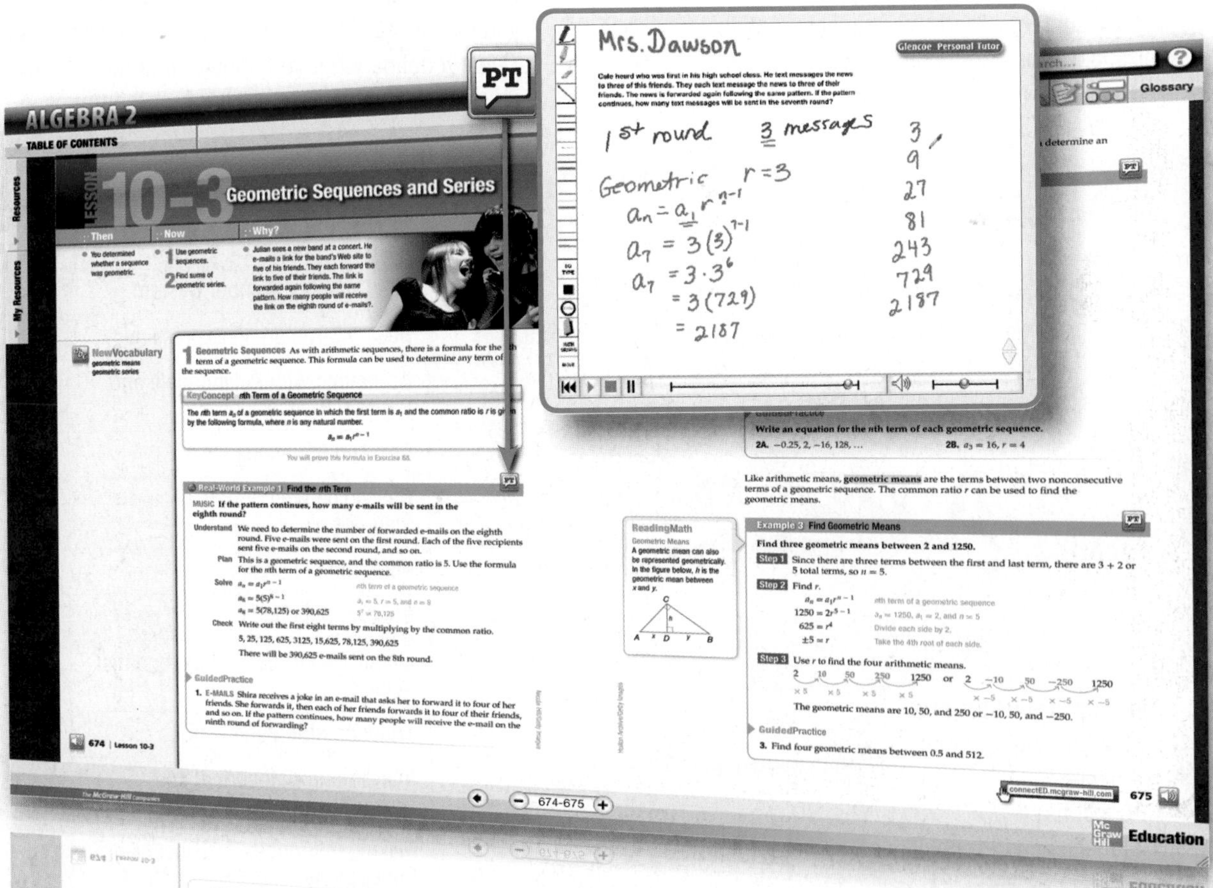

Common Core State Standards

With American students fully prepared for the future, our communities will be best positioned to compete successfully in the global economy. — Common Core State Standards Initiative

What is the goal of the Common Core State Standards?

The mission of the *Common Core State Standards* is to provide a consistent, clear understanding of what you are expected to learn, so teachers and parents know what they need to do to help you. The standards are designed to be robust and relevant to the real world, reflecting the knowledge and skills needed for success in college and careers.

Who wrote the standards?

The National Governors Association Center for Best Practices and the Council of Chief State School Officers worked with representatives from participating states, a wide range of educators, content experts, researchers, national organizations, and community groups.

What are the major points of the standards?

The standards seek to develop both mathematical understanding and procedural skill. The *Standards for Mathematical Practice* describe varieties of expertise that you will develop in yourself. The *Standards for Mathematical Content* define what you should understand and be able to do at each level in your study of mathematics.

How do I decode the standards?

This diagram provides clarity for decoding the standard identifiers.

A.REI.2

Conceptual Category
N = Number and Quantity
A = Algebra
F = Functions
S = Statistics and Probability

Domain

Standard

Domain Names	Abbreviations
The **C**omplex **N**umber System	**CN**
Seeing **S**tructure in **E**xpressions	**SSE**
Arithmetic with **P**olynomials and **R**ational Expressions	**APR**
Creating **E**quations	**CED**
Reasoning with **E**quations and **I**nequalities	**REI**
Interpreting **F**unctions	**IF**
Building **F**unctions	**BF**
Linear, Quadratic, and **E**xponential Models	**LE**
Trigonometric **F**unctions	**TF**
Interpreting Categorical and Quantitative **D**ata	**ID**
Making Inferences and Justifying **C**onclusions	**IC**
Using Probability to **M**ake **D**ecisions	**MD**

CHAPTER 0

Preparing for Advanced Algebra

connectED.mcgraw-hill.com **Your Digital Math Portal**

Vocabulary p. P4

Multilingual eGlossary p. P2

Personal Tutor p. P6

Foldables p. P2

CHAPTER 1 Equations and Inequalities

connectED.mcgraw-hill.com Your Digital Math Portal

Animation pp. 12, 84

Vocabulary pp. 5, 69

Multilingual eGlossary pp. 4, 60

Personal Tutor pp. 6, 63

CHAPTER 2

Linear Relations and Functions

Roine Magnusson/Photodisc/Getty Images

Virtual Manipulatives pp. 28, 61

Graphing Calculator p. 90

Foldables pp. 4, 60

Self-Check Practice pp. 7, 127

CHAPTER 3

Systems of Equations and Inequalities

DreamPictures/Blend Images/Getty Images

connectED.mcgraw-hill.com Your Digital Math Portal

Animation pp. 146, 247

Vocabulary pp. 206, 290

Multilingual eGlossary pp. 134, 218

Personal Tutor pp. 136, 222

CHAPTER 4

Quadratic Functions and Relations

Pete Saloutos/UpperCut Images/Getty Images

Virtual Manipulatives pp. 172, 219

Graphing Calculator pp. 205, 237

Foldables pp. 134, 218

Self-Check Practice pp. 183, 295

CHAPTER 5 Polynomials and Polynomial Functions

connectED.mcgraw-hill.com **Your Digital Math Portal**

Animation pp. 306, 387

Vocabulary pp. 303, 400

Multilingual eGlossary pp. 302, 384

Personal Tutor pp. 304, 386

Photodisc/Getty Images

CHAPTER 6

Inverses and Radical Functions and Relations

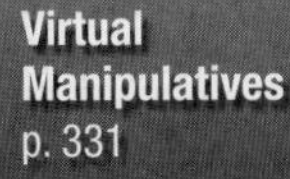
Virtual Manipulatives p. 331

Graphing Calculator pp. 338, 436

Foldables pp. 302, 384

Self-Check Practice pp. 326, 443

CHAPTER 7

Exponential and Logarithmic Functions and Relations

connectED.mcgraw-hill.com **Your Digital Math Portal**

Animation pp. 453, 532

Vocabulary pp. 517, 581

Multilingual eGlossary pp. 450, 528

Personal Tutor pp. 455, 531

Purestock/Getty Images

CHAPTER 8 Rational Functions and Relations

Image Source/Masterfile

Virtual Manipulatives pp. 451, 547

Graphing Calculator pp. 516, 579

Foldables pp. 450, 528

Self-Check Practice pp. 464, 585

CHAPTER 9 Conic Sections

connectED.mcgraw-hill.com **Your Digital Math Portal**

Animation pp. 641, 659

Vocabulary pp. 599, 710

Multilingual eGlossary pp. 592, 658

Personal Tutor pp. 600, 660

CHAPTER 10

Sequences and Series

Martin Child/Photodisc/Getty Images

Virtual Manipulatives pp. 601, 681

Graphing Calculator pp. 639, 690

Foldables pp. 592, 658

Self-Check Practice pp. 603, 715

CHAPTER 11

Statistics and Probability

connectED.mcgraw-hill.com **Your Digital Math Portal**

Animation pp. 720, 799

Vocabulary pp. 742, 859

Multilingual eGlossary pp. 722, 788

Personal Tutor pp. 724, 789

MIXA/Getty Images

CHAPTER 12 Trigonometric Functions

Virtual Manipulatives pp. 737, 822

Graphing Calculator pp. 731, 844

Foldables pp. 722, 788

Self-Check Practice pp. 738, 865

CHAPTER 13 Trigonometric Identities and Equations

CCSS Mathematical Content

connectED.mcgraw-hill.com Your Digital Math Portal

Animation p. 870

Vocabulary p. 873

Multilingual eGlossary p. 872

Personal Tutor p. 875

Graphing Calculator p. 900

Foldables p. 872

Self-Check Practice p. 897

Student Handbook

Contents

How to Use the Student Handbook

Getty Images

CHAPTER 0

Preparing for Advanced Algebra

Now

Chapter 0 contains lessons on topics from previous courses. You can use this chapter in various ways.

- Begin the school year by taking the Pretest. If you need additional review, complete the lessons in this chapter. To verify that you have successfully reviewed the topics, take the Posttest.
- As you work through the text, you may find that there are topics you need to review. When this happens, complete the individual lessons that you need.
- Use this chapter for reference. When you have questions about any of these topics, flip back to this chapter to review definitions or key concepts.

connectED.mcgraw-hill.com **Your Digital Math Portal**

Animation  Vocabulary   eGlossary  Personal Tutor  Virtual Manipulatives   Graphing Calculator  Audio  Foldables  Self-Check Practice  Worksheets 

Get Started on the Chapter

You will review several concepts, skills, and vocabulary terms as you study Chapter 0. To get ready, identify important terms and organize your resources.

FOLDABLES® StudyOrganizer

Throughout this text, you will be invited to use Foldables to organize your notes.

Why should you use them?

- They help you organize, display, and arrange information.
- They make great study guides, specifically designed for you.
- You can use them as your math journal for recording main ideas, problem-solving strategies, examples, or questions you may have.
- They give you a chance to improve your math vocabulary.

How should you use them?

- Write general information—titles, vocabulary terms, concepts, questions, and main ideas—on the front tabs of your Foldable.
- Write specific information—ideas, your thoughts, answers to questions, steps, notes, and definitions—under the tabs.
- Use the tabs for:
    - math concepts in parts, like types of triangles,
    - steps to follow, or
    - parts of a problem, like *compare and contrast* (2 parts) or *what, where, when, why,* and *how* (5 parts).
- You may want to store your Foldables in a plastic zipper bag that you have three-hole punched to fit in your notebook.

When should you use them?

- Set up your Foldable as you begin a chapter, or when you start learning a new concept.
- Write in your Foldable every day.
- Use your Foldable to review for homework, quizzes, and tests.

ReviewVocabulary

English		Español
domain	p. P4	dominio
range	p. P4	rango
quadrants	p. P4	cuadrantes
mapping	p. P4	transformaciones
function	p. P4	función
outcome	p. P9	resultados
sample space	p. P9	espacio muestral
permutation	p. P9	permutación
factorial	p. P10	factorial
combination	p. P11	combinación
theoretical probability	p. P13	probabilidad teórica
experimental probability	p. P13	probabilidad experimental
simple event	p. P14	evento simple
compound event	p. P14	suceso compuesto
mutually exclusive	p. P14	mutuamente exclusiva
independent events	p. P16	eventos independientes
dependent events	p. P16	eventos dependientes
conditional probability	p. P16	probabilidad condicional
two-way frequency table	p. P18	tabla de doble entrada o de contingencia
population	p. P24	población
sample	p. P24	muestra
mean	p. P24	media
median	p. P24	mediana
mode	p. P24	moda
range	p. P25	rango
variance	p. P25	varianza
standard deviation	p. P25	desviación estándar
five-number summary	p. P26	resumen de cinco números
outlier	p. P27	valor atípico

CHAPTER 0 Pretest

State the domain and range of each relation. Then determine whether each relation is a function. Write *yes* or *no*.

1. $\{(14, 1), (-3, 6), (8, 4)\}$

2.

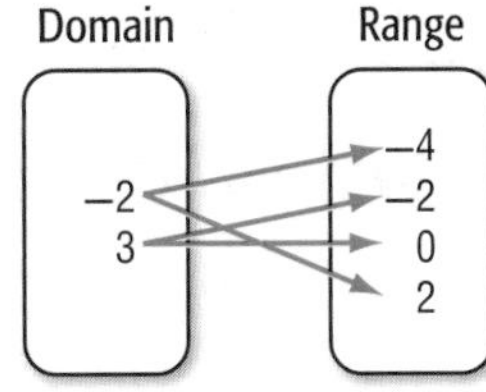

Name the quadrant in which each point is located.

3. $(-6, -2)$

4. $(4, -3)$

5. $(-5, 7)$

Find each product.

6. $(x + 1)(x + 4)$

7. $(a - 3)(a + 6)$

8. $(m - 2)(m - 5)$

9. $(c + 8)(c - 8)$

10. NUMBER THEORY There are two integers. One is 5 more than a number, and the other is 1 less than the same number.

a. Write expressions for the two numbers.

b. Write a polynomial expression for the product of the numbers.

Factor each polynomial.

11. $10ab^2 + 5b$

12. $15d - 12cd^2$

13. $y^2 + 6y - 7$

14. $a^2 - 13a + 36$

15. ICE CREAM How many different 1-scoop ice cream cones can you order if you have a choice of two cones types and 15 flavors of ice cream?

Determine whether each situation involves *permutations* or *combinations*. Then solve.

16. How many ways are there to place an algebra book, a geometry book, a chemistry book, an English book, and a health book on a shelf?

17. How many ways are there to select 3 of 15 flavors of juice at the grocery store?

A card is drawn at random from a standard deck. Determine whether the events are *mutually exclusive* or *not mutually exclusive*. Then find each probability.

18. P(2 or black card)

19. P(jack or 4)

Determine whether the events are *independent* or *dependent*. Then find the probability.

20. Two cards are selected at random one after the other from a standard deck without replacement. What is the probability that both cards are aces?

21. A coin is tossed and a die is rolled. What is the probability that the coin lands on heads and the number rolled is greater than 4?

22. SHOPPING During one month, 18% of *INET Clothing* customers purchased item #345A. If 5% of customers order this item and item #345B, find the probability that someone who ordered item #345A also ordered item #345B.

23. Determine whether the triangles are *similar*, *congruent*, or *neither*.

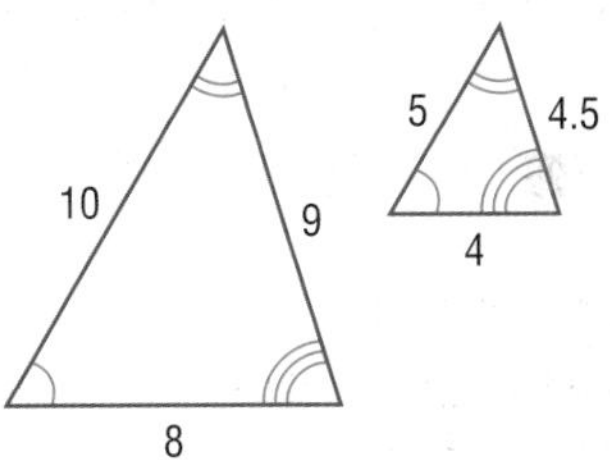

24. PHOTOGRAPHS A photo that is 3 inches wide by 5 inches long is being enlarged so that it is 12 inches long. How wide will the enlarged photo be?

Find each missing measure. Round to the nearest tenth, if necessary.

25.

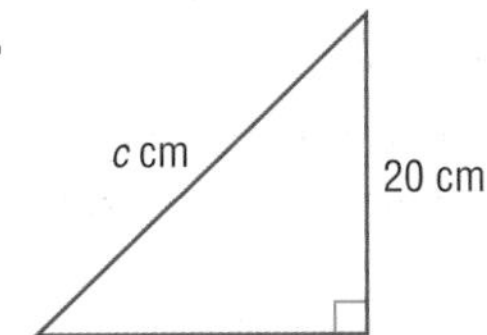

26. $a = 6$ yd,
$b = 9$ yd,
$c = ?$

The lengths of three sides of a triangle are given. Determine whether each triangle is a right triangle.

27. 12 yd, 14 yd, 16 yd

28. 15 km, 20 km, 25 km

Find the mean, median, mode, range, and standard deviation of each data set. Then identify any outliers.

29. scores for 16 students on a math test: 85, 100, 92, 36, 74, 88, 92, 86, 88, 82, 98, 70, 78, 92, 84, 100

30. weight to the nearest pound of 17 cats: 10, 15, 9, 8, 11, 10, 9, 8, 13, 9, 12, 10, 13, 11, 9, 12, 10

LESSON 0-1

Representing Functions

Objective

- Identify the domain and range of functions.

NewVocabulary
domain
range
quadrants
mapping
function

Recall that a *relation* is a set of ordered pairs. The **domain** of a relation is the set of all first coordinates (x-coordinates) from the ordered pairs, and the **range** is the set of all second coordinates (y-coordinates) from the ordered pairs.

Example 1 Domain and Range

State the domain and the range of the relation.
{(−3, 3), (0, −7), (1, −5), (2, 4)}

The domain is the set of x-coordinates.
D = {−3, 0, 1, 2}

The range is the set of y-coordinates.
R = {−7, −5, 3, 4}

A relation can be graphed on a coordinate plane. A coordinate plane is formed by the intersection of the horizontal axis, or x-axis, and the vertical axis, or y-axis. The axes meet at the origin (0, 0) and divide the plane into four **quadrants**. Any ordered pair in the coordinate plane can be written in the form (x, y).

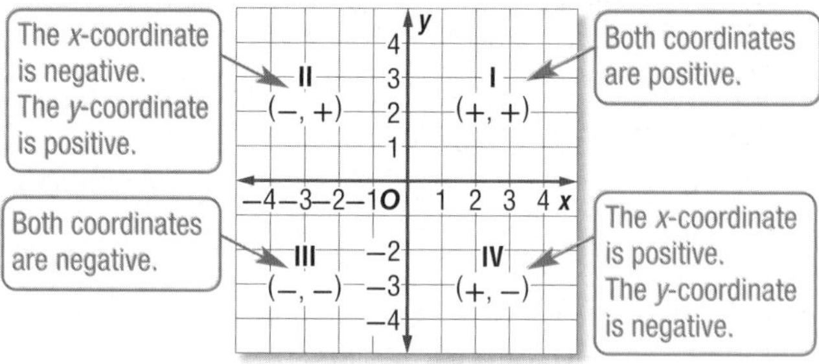

Example 2 Locate Coordinates

Name the quadrant in which $T(-8, 5)$ is located.

Point T has a negative x-coordinate and a positive y-coordinate. The point is located in Quadrant II.

A relation can also be represented by a table or a mapping. A **mapping** illustrates how each element of the domain is paired with an element in the range.

Ordered Pairs

(1, 2)
(−2, 3)
(0, −3)

Table

x	y
1	2
−2	3
0	−3

Graph

Mapping

Domain: 1, −2, 0
Range: 2, 3, −3

A **function** is a relation in which each element of the domain is paired with *exactly one* element of the range.

Example 3 Identify Domain and Range

WatchOut!

Functions Remember that in a function, an element of the range can be paired with more than one element of the domain. But an element of the domain cannot be paired with more than one element of the range.

State the domain and range of each relation. Then determine whether each relation is a function.

a. **{(10, 3), (6, −2), (7, 4), (−8, −9)}**

D = {−8, 6, 7, 10}
R = {−9, −2, 3, 4}
For each element of the domain, there is only one corresponding element in the range. So, this relation is a function.

b.

x	y
1	3, 4
2	7
3	4

D = {1, 2, 3}
R = {3, 4, 7}
Because 1 is paired with 3 and 4, this is not a function.

c.

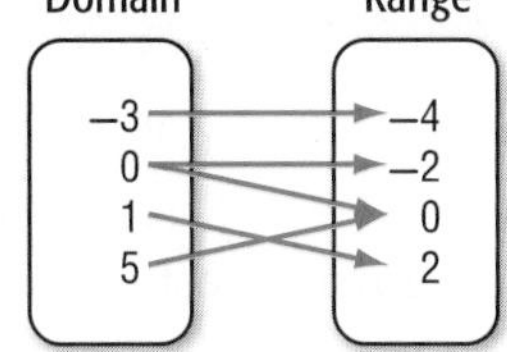

D = {−3, 0, 1, 5}
R = {−4, −2, 0, 2}
Because 0 is paired with −2 and 0, this is not a function.

d.

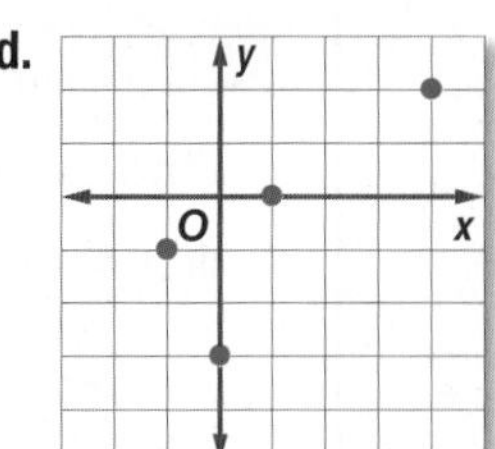

D = {−1, 0, 1, 4}
R = {−3, −1, 0, 2}
This is a function.

Exercises

State the domain and range of each relation. Then determine whether each relation is a function. Write *yes* or *no*.

1. {(2, 7), (3, 10), (1, 6)}

2. {(−6, 0), (5, 5), (9, −2), (−2, −9)}

3.

x	y
1	5
2	7
1	9

4.

x	y
−12	0
−10	1
−8	2
−6	4

5.

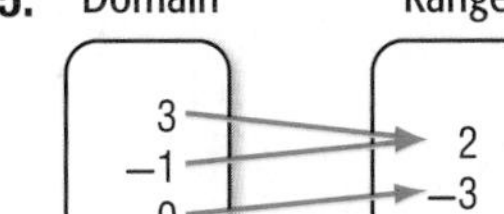

6.

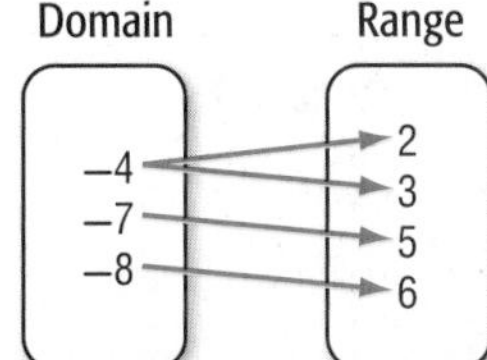

7.

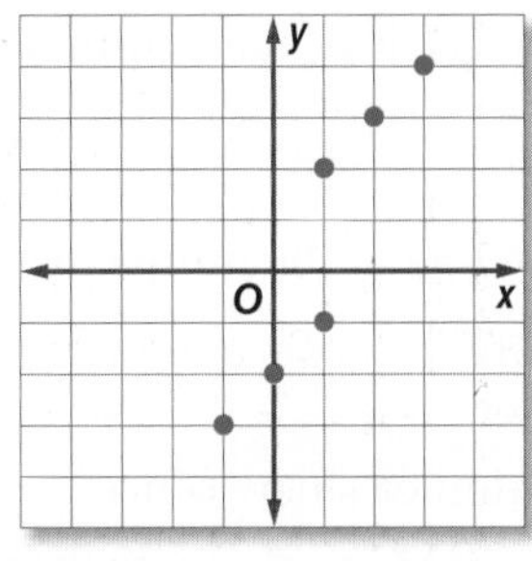

8.

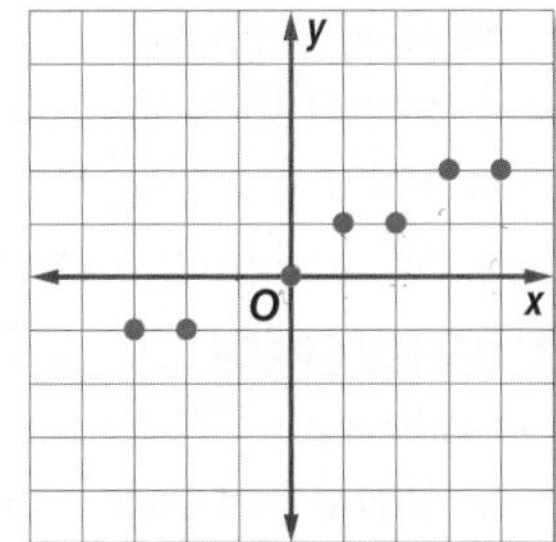

Name the quadrant in which each point is located.

9. (5, 3) **10.** (8, −6) **11.** (2, 0) **12.** (−7, −1)

LESSON 0-2 FOIL

Objective

- Use the FOIL method to multiply binomials.

The product of two binomials is the sum of the products of the *first* terms, the *outer* terms, the *inner* terms, and the *last* terms.

F **O** **I** **L**

PT

Example 1 Use the FOIL Method

Find each product.

a. $(x + 3)(x - 5)$

$$(x + 3)(x - 5) = \overset{\text{First}}{x \cdot x} + \overset{\text{Outer}}{x \cdot (-5)} + \overset{\text{Inner}}{3 \cdot x} + \overset{\text{Last}}{3 \cdot (-5)}$$

$$= x^2 - 5x + 3x - 15$$
$$= x^2 - 2x - 15$$

b. $(3y + 2)(5y + 4)$

$$(3y + 2)(5y + 4) = 3y \cdot 5y + 3y \cdot 4 + 2 \cdot 5y + 2 \cdot 4$$
$$= 15y^2 + 12y + 10y + 8$$
$$= 15y^2 + 22y + 8$$

Exercises

Find each product.

1. $(a + 2)(a + 4)$

2. $(v - 7)(v - 1)$

3. $(h + 4)(h - 4)$

4. $(d - 1)(d + 1)$

5. $(b + 4)(b - 3)$

6. $(t - 9)(t + 11)$

7. $(r + 3)(r - 8)$

8. $(k - 2)(k + 5)$

9. $(p + 8)(p + 8)$

10. $(x - 15)(x - 15)$

11. $(2c + 1)(c - 5)$

12. $(7n - 2)(n + 3)$

13. $(3m + 4)(2m - 5)$

14. $(5g + 1)(6g + 9)$

15. $(2q - 17)(q + 2)$

16. $(4t - 7)(3t - 12)$

17. **NUMBERS** I am thinking of two integers. One is 7 less than a number, and the other is 2 greater than the same number.

a. Write expressions for the two numbers.

b. Write a polynomial expression for the product of the numbers.

18. **OFFICE SPACE** Monica's current office is square. Her office in the company's new building will be 3 feet wider and 5 feet longer.

a. Write expressions for the dimensions of Monica's new office.

b. Write a polynomial expression for the area of Monica's new office.

c. Suppose Monica's current office is 7 feet by 7 feet. How much larger will her new office be?

LESSON 0-3 Factoring Polynomials

Objective

- Use various techniques to factor polynomials.

Some polynomials can be factored using the Distributive Property.

Example 1 Use the Distributive Property

Factor $4a^2 + 8a$.

Find the GCF of $4a^2$ and $8a$.

$4a^2 = 2 \cdot 2 \cdot a \cdot a \qquad 8a = 2 \cdot 2 \cdot 2 \cdot a \quad \rightarrow \quad$ GCF: $2 \cdot 2 \cdot a$ or $4a$

$4a^2 + 8a = 4a(a) + 4a(2)$ Rewrite each term using the GCF.

$= 4a(a + 2)$ Distributive Property

Thus, the completely factored form of $4a^2 + 8a$ is $4a(a + 2)$.

To factor trinomials of the form $x^2 + bx + c$, find two integers m and p with a product equal to c and a sum equal to b. Then write $x^2 + bx + c$ using the pattern $(x + m)(x + p)$.

Example 2 Use Factors and Sums

Factor each polynomial.

a. $x^2 + 5x + 6$ ← Both b and c are positive.

b is 5 and c is 6. Find two numbers with a product of 6 and a sum of 5.

Factors of 6	Sum of factors
1, 6	7
2, 3	5

The correct factors are 2 and 3.

$x^2 + 5x + 6 = (x + m)(x + p)$ Write the pattern.

$= (x + 2)(x + 3)$ $m = 2$ and $p = 3$

b. $x^2 - 8x + 12$ ← b is negative, and c is positive.

$b = -8$ and $c = 12$. This means that $m + p$ is negative and mp is positive. So m and p must both be negative.

Factors of 12	Sum of factors
−1, −12	−13
−2, −6	−8

The correct factors are −2 and −6.

$x^2 - 8x + 12 = (x + m)(x + p)$ Write the pattern.

$= [x + (-2)][x + (-6)]$ $m = -2$ and $p = -6$

$= (x - 2)(x - 6)$ Simplify.

c. $x^2 + 14x - 15$ ← b is positive, and c is negative.

$b = 14$ and $c = -15$. This means that $m + p$ is positive and mp is negative. So either m or p must be negative, but not both.

Factors of −15	Sum of factors
1, −15	−14
−1, 15	14

The correct factors are −1 and 15.

$x^2 + 14x - 15 = (x + m)(x + p)$ Write the pattern.

$= [x + (-1)](x + 15)$ $m = -1$ and $p = 15$

$= (x - 1)(x + 15)$ Simplify.

To factor quadratic trinomials of the form $ax^2 + bx + c$, find two integers m and p with a product equal to ac and a sum equal to b. Write $ax^2 + bx + c$ using the pattern $ax^2 + mx + px + c$. Then factor by grouping.

Example 3 Use Factors and Sums

Factor $6x^2 + 7x - 3$.

$a = 6$, $b = 7$, and $c = -3$. This means that $m + p$ is positive and mp is negative. So either m or p must be negative, but not both.

Factors of −18	Sum of factors	
1, −18	−17	
−1, 18	17	
2, −9	−7	
−2, 9	7	The correct factors are −2 and 9.

$6x^2 + 7x - 3 = 6x^2 + mx + px - 3$ Write the pattern.

$= 6x^2 + (-2)x + 9x - 3$ $m = -2$ and $p = 9$

$= (6x^2 - 2x) + (9x - 3)$ Group terms with common factors.

$= 2x(3x - 1) + 3(3x - 1)$ Factor the GCF from each group.

$= (2x + 3)(3x - 1)$ Distributive Property

StudyTip

Checking Solutions You can check to see that you have factored correctly by multiplying the factors and comparing the product to the original polynomial.

Here are some special products.

Perfect Square Trinomials

$(a + b)^2 = (a + b)(a + b) = a^2 + 2ab + b^2$

$(a - b)^2 = (a - b)(a - b) = a^2 - 2ab + b^2$

Difference of Squares

$a^2 - b^2 = (a + b)(a - b)$

Example 4 Use Special Products

Factor each polynomial.

a. $4x^2 + 20x + 25$

The first and last terms are perfect squares.
The middle term is equal to $2(2x)(5)$.
This is a perfect square trinomial of the form $(a + b)^2$.

$4x^2 + 20x + 25 = (2x)^2 + 2(2x)(5) + 5^2$ Write as $a^2 + 2ab + b^2$.

$= (2x + 5)^2$ Factor using the pattern.

b. $x^2 - 4$

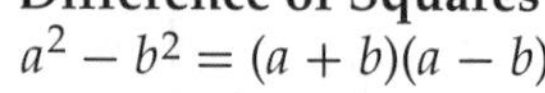

$x^2 - 4 = x^2 - (2)^2$ Write in the form $a^2 - b^2$.

$= (x + 2)(x - 2)$ Factor the difference of squares.

Exercises

Factor each polynomial.

1. $12x^2 + 4x$
2. $6x^2y + 2x$
3. $8ab^2 - 12ab$
4. $x^2 + 5x + 4$
5. $y^2 + 12y + 27$
6. $x^2 + 6x + 8$
7. $3y^2 + 13y + 4$
8. $7x^2 + 51x + 14$
9. $3x^2 + 28x + 32$
10. $x^2 - 5x + 6$
11. $y^2 - 5y + 4$
12. $6x^2 - 13x + 5$
13. $6a^2 - 50ab + 16b^2$
14. $11x^2 - 78x + 7$
15. $18x^2 - 31xy + 6y^2$
16. $x^2 + 4xy + 4y^2$
17. $9x^2 - 24x + 16$
18. $4a^2 + 12ab + 9b^2$
19. $x^2 - 144$
20. $4c^2 - 9$
21. $16y^2 - 1$
22. $25x^2 - 4y^2$
23. $36y^2 - 16$
24. $9a^2 - 49b^2$

LESSON 0-4 Counting Techniques

Objective

- Find the total number of outcomes using a variety of methods.

NewVocabulary

outcome
probability experiment
sample space
tree diagram
permutation
factorial
combination

Recall that an **outcome** is the result of a single trial of a process involving chance, called a **probability experiment**. The set of all possible outcomes of such an experiment is called a **sample space**. A **tree diagram**, like the one shown, can be used to systematically list the outcomes in a sample space. Notice that there are 8 total outcomes in this sample space.

Blood Type	Rh Factor	Sample Space
A	+	A+
	–	A–
B	+	B+
	–	B–
AB	+	AB+
	–	AB–
O	+	O+
	–	O–

When you only need to know the total number of outcomes, you can use the **Fundamental Counting Principle**.

KeyConcept Fundamental Counting Principle

Words The number of possible outcomes in a sample space can be found by multiplying the number of possible outcomes for each event.

Symbols If event M can occur in m ways and is followed by event N that can occur in n ways, then the event M followed by N can occur in $m \cdot n$ ways.

This rule can be extended to three or more events.

If there are 4 possible blood types and 2 possible Rh-factors, there are 4 • 2 or 8 possible ways to label blood donations.

Example 1 Fundamental Counting Principle

BICYCLES A bicycle manufacturer makes five- and ten-speed bikes in seven different colors and four different frame sizes. How many different bicycles does the manufacturer make?

There are 2 gear choices, 7 color choices, and 4 frame choices. Apply the Fundamental Counting Principle.

gear		color		frame		number of bicycles
2	•	7	•	4	=	56

56 different bicycles can be made.

Counting problems can also involve determining the number of different arrangements of objects. An arrangement of a group of distinct objects in a certain order is called a **permutation**. For example, there are 6 permutations of the letters A, B, and C.

ABC ACB BAC BCA CAB CBA

To determine this number mathematically, apply the Fundamental Counting Principle. There are 3 choices for the first letter. Once this letter is chosen, 2 choices remain for the second letter. After the first two letters are chosen, only one choice remains.

first letter		second letter		third letter		permutations
3	•	2	•	1	=	6

The product 3 • 2 • 1 can also be written as 3!, which is read *3 factorial*.

KeyConcept Factorial

Words *n* **factorial** is the product of all counting numbers beginning with *n* and counting backward to 1. Zero factorial is defined to equal 1.

Symbols $n! = n \cdot (n-1) \cdot (n-2) \cdot (n-3) \cdot \cdots \cdot 3 \cdot 2 \cdot 1$ and $0! = 1$

The number of permutations of n distinct objects is $n!$.

Example 2 Permutations of *n* Objects

BAND There are 8 finalists in a band competition. In how many different ways can the bands be ranked if they cannot receive the same ranking?

$8! = 8 \cdot 7 \cdot 6 \cdot 5 \cdot 4 \cdot 3 \cdot 2 \cdot 1$ or 40,320 — Permutations of *n* distinct objects

The bands can be ranked in 40,320 different ways.

Suppose in Example 2 you wanted to know how many different ways the first-, second-, and third-place rankings in the competition could be awarded. This also involves permutations. It is the number of permutations of 8 finalists taken 3 at a time.

first place		second place		third place		permutations
8	·	7	·	6	=	336

The number of permutations can also be found using factorials by finding the number of permutations of all 8 bands and dividing out the number of arrangements of bands that finished with rankings other than first, second, or third.

$$\frac{8!}{(8-3)!} = \frac{8!}{5!} = \frac{8 \cdot 7 \cdot 6 \cdot 5 \cdot 4 \cdot 3 \cdot 2 \cdot 1}{5 \cdot 4 \cdot 3 \cdot 2 \cdot 1} = 8 \cdot 7 \cdot 6$$

This result is generalized below.

ReadingMath

Permutations The number of permutations of *n* objects taken *r* at a time can also be denoted as $P(n, r)$.

KeyConcept Permutations of *n* Objects Taken *r* at a Time

The number of permutations of *r* objects taken from a group of *n* distinct objects is given by

$$_nP_r = \frac{n!}{(n-r)!}.$$

Example 3 Permutation of *n* Objects Taken *r* at a Time

TUTORING How many different ways can two students be assigned to five tutors if only one student is assigned to each tutor?

Since the order in which the students are arranged determines the tutor to which a student is assigned, this situation involves permutations.

$_5P_2 = \frac{5!}{(5-2)!}$ — Permutations of 5 things taken 2 at a time

$= \frac{5!}{3!}$ — Subtract.

$= \frac{5 \cdot 4 \cdot \cancel{3 \cdot 2 \cdot 1}}{\cancel{3 \cdot 2 \cdot 1}}$ or 20 — Divide out common factors and simplify.

The students can be assigned in 20 different ways.

StudyTip

With Repetition The number of permutations of *n* objects in which one object is repeated *p* times and another object is repeated *q* times is $\frac{n!}{p!q!}$. This rule can be extended to permutations involving three or more repeated objects.

A selection of distinct objects in which the order of the objects selected is *not* important is called a **combination**. Suppose you select 3 out of 4 books, A, B, C, and D, for a book report. Selecting books A, B, and C is not different from selecting books C, B, A, or any other arrangements of these three books.

To find the number of combinations of n objects taken r at a time, divide the number of permutations by the number of ways to arrange each selection of r objects.

KeyConcept Combinations of n Objects Taken r at a Time

The number of combinations of r objects taken from a group of n distinct objects is given by

$${}_nC_r = \frac{{}_nP_r}{r!} = \frac{n!}{(n-r)!\,r!}.$$

PT

Example 4 Combinations of n Objects Taken r at a Time

CARDS How many ways are there to choose 5 cards from a standard deck of 52 playing cards?

Since the order in which the cards are chosen is not important, this situation involves combinations.

$${}_{52}C_5 = \frac{52!}{(52-5)!\,5!}$$ Combinations of 52 things taken 5 at a time

$$= \frac{52!}{47!\,5!}$$ Subtract.

$$= 2{,}598{,}960$$ Use a calculator.

There are 2,598,960 ways to choose 5 cards from a standard deck of playing cards.

StudyTip

Evaluating Permutations and Combinations You can also use the **nPr** and **nCr** features on your graphing calculator to evaluate permutations and combinations, respectively.

A critical first step in solving a counting problem is deciding whether order is important. If the order of objects selected is important, then the problem involves permutations. If objects are selected without regard to order, the problem involves combinations.

PT

Example 5 Permutations or Combinations?

Twenty-five students write their names on slips of paper. Then three different names are chosen at random to receive prizes. Determine whether each situation involves *permutations* or *combinations*.

a. choosing 3 people to each receive a "no homework" coupon

Since the prizes are identical, the order in which the students' names are drawn *is not* important. This situation involves combinations.

b. choosing 3 people to each receive one of the following prizes: 1st prize, a new graphing calculator; 2nd prize, a "no homework" coupon; 3rd prize, a new pencil

Because the prizes are different, the order in which the students' names are drawn *is* important. This situation involves permutations.

Exercises

Use the Fundamental Counting Principle to determine the number of outcomes.

1. **FOOD** How many different combinations of sandwich, side, and beverage are possible?

Sandwiches	Sides	Beverages
hot dog	chips	bottled water
hamburger	apple	soda
veggie burger	pasta salad	juice
bratwurst		milk
grilled chicken		

2. **QUIZZES** Each question on a five-question multiple-choice quiz has answer choices labeled A, B, C, and D. How many different ways can a student answer the five questions?

3. **DANCES** Dane is renting a tuxedo for prom. Once he has chosen his jacket, he must choose from three types of pants and six colors of vests. How many different ways can he select his attire for prom?

4. **MANUFACTURING** A baseball glove manufacturer makes a glove with the different options shown in the table. How many different gloves are possible?

Option	Number of Choices
sizes	4
types by position	3
materials	2
levels of quality	2

Evaluate each permutation or combination.

5. ${}_6P_3$
6. ${}_7P_5$
7. ${}_4C_2$
8. ${}_{12}C_7$
9. ${}_6C_1$
10. ${}_9P_5$

Determine whether each situation involves *permutations* or *combinations*. Then solve the problem.

11. **SCHOOL** Charlita wants to take 6 different classes next year. Assuming that each class is offered each period, how many different schedules could she have?

12. **BALLOONS** How many 4-colored groups can be selected from 13 different colored balloons?

13. **CONTEST** How many ways are there to choose the winner and first, second, and third runners-up in a contest with 10 finalists?

14. **BANDS** A band is choosing 3 new backup singers from a group of 18 who try out. How many ways can they choose the new singers?

15. **PIZZA** How many different two-topping pizzas can be made if there are 6 options for toppings?

16. **SOFTBALL** How many ways can the manager of a softball team choose players for the top 4 spots in the lineup if she has 7 possible players in mind?

17. **NEWSPAPERS** A newspaper has 9 reporters available to cover 4 different stories. How many ways can the reporters be assigned?

18. **READING** Jack has a reading list of 12 books. How many ways can he select 9 books from the list to check out of the library?

19. **CHALLENGE** Abby is registering at a Web site and must select a six-character password. The password can contain either letters or digits.

    a. How many passwords are possible if characters can be repeated? if no characters can be repeated?

    b. How many passwords are possible if all characters are letters that can be repeated? if the password must contain exactly one digit? Which type of password is more secure? Explain.

LESSON 0-5

Adding Probabilities

Objectives

- Compute theoretical and experimental probabilities.
- Compute probabilities of compound events.

NewVocabulary

probability
probability model
uniform or simple probability model
theoretical probability
experimental probability
simple event
compound event
mutually exclusive events
odds

Probability is a measure of the chance that a given event E will occur. The probability $P(E)$ that an event will occur is always given as a ratio between 0 and 1, inclusive.

E will not occur.	E is equally likely to occur or not occur.	E is certain to occur.
$P(E) = 0$	$P(E) = \frac{1}{2}$	$P(E) = 1$

A **probability model** is a mathematical model used to represent the outcomes of an experiment. A **uniform** or **simple probability model** is used to describe an experiment for which all outcomes are equally likely or have the same probability of occurring.

If we assume that all outcomes of an experiment are equally likely, then we can calculate the **theoretical probability** that an event will occur using the sample space of possible outcomes. We can also calculate the empirical or **experimental probability** using outcomes obtained by actually performing trials of the experiment.

KeyConcept Theoretical and Experimental Probability

If each outcome is assumed to be equally likely, the theoretical probability P of an event E is given by

$$P(E) = \frac{\text{number of favorable outcomes}}{\text{number of possible outcomes}}.$$

Given the frequency of outcomes from a certain number of trials of an experiment, the experimental probability P of an event E is given by

$$P(E) = \frac{\text{number of favorable trials}}{\text{number of trials}}.$$

Example 1 Theoretical and Experimental Probability

The graph shows the results of several trials of an experiment in which a single die is rolled.

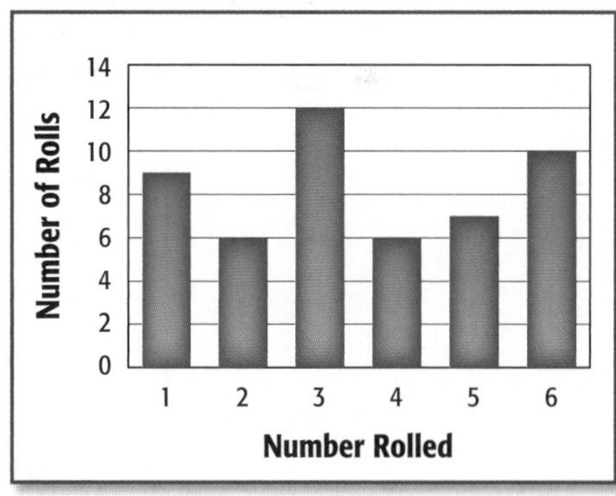

a. What is the experimental probability of rolling a 6?

From the graph we can see that there were 10 trials that resulted in 6. The total number of trials was $9 + 6 + 12 + 6 + 7 + 10$ or 50.

$$P(6) = \frac{\text{number of favorable trials}}{\text{number of trials}}$$
$$= \frac{10}{50} \text{ or } \frac{1}{5}, \text{ which is } 20\%$$

b. What is the theoretical probability of rolling a 6?

List the sample space. The possible outcomes for this experiment are 1, 2, 3, 4, 5, or 6. Of these 6 possible outcomes, only one is a 6.

$$P(6) = \frac{\text{number of favorable outcomes}}{\text{number of possible outcomes}} = \frac{1}{6} \text{ or about } 17\%$$

An event that has a single outcome, such as rolling a 6 on a die, is called a **simple event**. Many problems involve finding the probability of a composite or **compound event**, which consists of two or more simple events. To find the probability of a compound event, you must consider whether the events can occur at the same time.

StudyTip

Mutually and Not Mutually Exclusive Events that are mutually exclusive are sometimes said to be *disjoint*. Events that are not mutually exclusive are sometimes said to be *inclusive* events.

Events that cannot occur at the same time are said to be **mutually exclusive**. Mutually exclusive events have no outcomes in common. For example, because it is not possible to draw a card from a standard deck that is both a king and a queen, these two events are mutually exclusive. It is possible to draw a card that is both a king and a spade, so these events are *not* mutually exclusive.

Mutually Exclusive

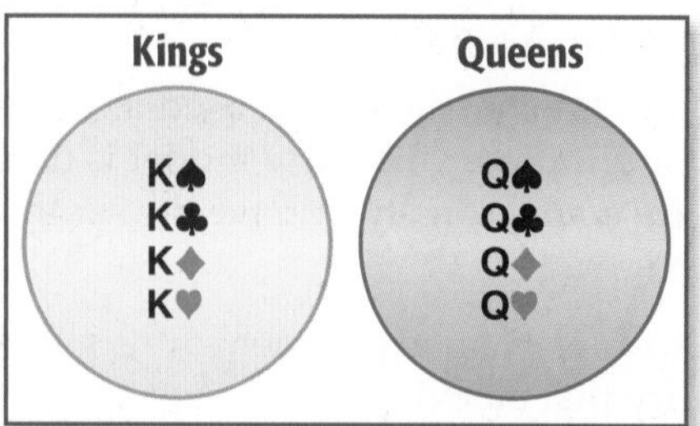

Not Mutually Exclusive

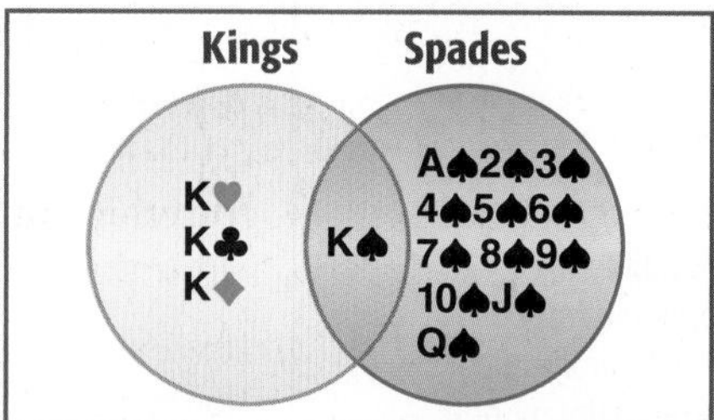

StudyTip

Why Subtract? Notice that when two events are not mutually exclusive, you add the probabilities and then *subtract* the probabilities of the outcomes that are common to both. If these common probabilities are not subtracted, you have actually counted them twice.

KeyConcept Addition Rules for Probability

If two events A and B are **mutually exclusive**, the probability that A or B will occur is	If two events A and B are **not mutually exclusive**, the probability that A or B will occur is
$P(A \text{ or } B) = P(A) + P(B)$.	$P(A \text{ or } B) = P(A) + P(B) - P(A \text{ and } B)$.

PT

Example 2 Add Probabilities

Determine whether the events are *mutually exclusive* or *not mutually exclusive*. Then find the probability.

a. Keisha has a stack of 8 baseball cards, 5 basketball cards, and 6 hockey cards. If she selects a card at random from the stack, what is the probability that it is a baseball or a hockey card?

These events are mutually exclusive, because the card cannot be both a baseball card *and* a hockey card. There is a total of 8 + 5 + 6 or 19 cards.

$P(\text{baseball or hockey}) = P(\text{baseball}) + P(\text{hockey})$ — Mutually exclusive events

$= \frac{8}{19} + \frac{6}{19} \text{ or } \frac{14}{19}$ — Substitute and add.

b. Suppose that of 1400 students, 550 take Spanish, 700 take biology, and 400 take both Spanish and biology. What is the probability that a student selected at random takes Spanish or biology?

Because some students take both Spanish S and biology B, the events are not mutually exclusive.

$P(S) = \frac{550}{1400} \text{ or } \frac{11}{28}$ $\quad P(B) = \frac{700}{1400} \text{ or } \frac{14}{28}$ $\quad P(S \text{ and } B) = \frac{400}{1400} \text{ or } \frac{8}{28}$

$P(S \text{ or } B) = P(S) + P(B) - P(S \text{ and } B)$ — Not mutually exclusive events

$= \frac{11}{28} + \frac{14}{28} - \frac{8}{28} \text{ or } \frac{17}{28}$ — Substitute and simplify.

Exercises

1. **CARNIVAL GAMES** A spinner has sections of equal size. The table shows the results of several spins.

Color	Frequency
red	6
blue	7
yellow	9
orange	12
purple	5
green	11

   a. Copy the table and add a column to show the experimental probability of the spinner landing on each of the colors with the next spin.

   b. Create a bar graph that shows these experimental probabilities.

   c. Add a column to your table that shows the theoretical probability of the spinner landing on each of the colors with the next spin.

   d. Create a bar graph that shows these theoretical probabilities.

   e. Interpret and compare the graphs you created in parts **b** and **d.**

Determine whether the events are *mutually exclusive* or *not mutually exclusive.* Then find the probability.

2. Two dice are rolled.

   a. P(sum of 10 or doubles) **b.** P(sum of 6 or 7) **c.** $P(\text{sum} < 3 \text{ or sum} > 10)$

3. A card is drawn at random from a standard deck of cards.

   a. P(club or diamond) **b.** P(ace or spade) **c.** P(jack or red card)

4. In a French class, there are 10 freshmen, 8 sophomores, and 2 juniors. Of these students, 9 freshmen, 2 sophomores, and 1 junior are female. A student is selected at random.

   a. P(freshman or female) **b.** P(sophomore or male) **c.** P(freshman or sophomore)

5. There are 40 vehicles on a rental car lot. All are either sedans or SUVs. There are 18 red vehicles, and 3 of them are sedans. There are 15 blue vehicles, and 9 of them are SUVs. Of the remaining vehicles, all are black and 2 are SUVs. A vehicle is selected at random.

   a. P(blue or black) **b.** P(red or SUV) **c.** P(black or sedan)

6. **DRIVING** A survey of Longview High School students found that the probability of a student driving while texting was 0.16, the probability of a student getting into an accident while driving was 0.07, and the probability of a student getting into an accident while driving and texting was 0.05. What is the probability of a student driving while texting or getting into an accident while driving?

7. **REASONING** Explain why the rule $P(A \text{ or } B) = P(A) + P(B) - P(A \text{ and } B)$ can be used for both mutually exclusive and not mutually exclusive events.

ODDS Another measure of the chance that an event will occur is called odds. The odds of an event occurring is a ratio that compares the number of ways an event can occur *s* (successes) to the number of ways it cannot occur *f* (failure), or *s* to *f*. The sum of the number of success and failures equals the number of possible outcomes.

8. A card is drawn from a standard deck of 52 cards. Find the odds in favor of drawing a heart. Then find the odds against drawing an ace.

9. Two fair coins are tossed. Find the odds in favor of both landing on heads. Then find the odds in favor of exactly one landing on tails.

10. The results of rolling a die 120 times are shown.

Roll	1	2	3	4	5	6
Frequency	16	24	17	25	30	8

   Find the experimental odds against rolling a 1 or a 6. Then find the experimental odds in favor of rolling a number less than 3.

LESSON 0-6 Multiplying Probabilities

Objectives

- Find probabilities of independent and dependent events.
- Use two-way frequency tables to find conditional probabilities.

NewVocabulary

independent events
dependent events
conditional probability
two-way frequency table

If the occurrence of one event does not affect the probability of a second event occurring, then the two events are **independent events**.

KeyConcept Probability of Independent Events

If two events A and B are independent, then the probability that A and B will occur is

$$P(A \text{ and } B) = P(A) \cdot P(B).$$

This rule can be extended to three or more independent events.

Example 1 Probability of Independent Events

A coin is tossed and a die is rolled. What is the probability of the coin landing on tails and rolling a 3?

Since the outcome of tossing the coin does not affect the outcome of rolling the die, these events are independent.

$P(\text{tails and } 3) = P(\text{tails}) \cdot P(3)$ — Probability of independent events

$= \frac{1}{2} \cdot \frac{1}{6}$ or $\frac{1}{12}$ — $P(\text{tails}) = \frac{1}{2}$ and $P(3) = \frac{1}{6}$

CHECK Use a listing of the possible outcomes from this compound event.

H1, H2, H3, H4, H5, H6, T1, T2, T3, T4, T5, T6

There is only 1 outcome out of the 12 possible outcomes that indicates landing on tails and rolling a 3, so $P(\text{tails and } 3)$ is $\frac{1}{12}$. ✓

If the occurrence of the first event *does* affect the probability of the second event occurring, then the events are **dependent events**. An example of dependent events is drawing a card from a standard deck of cards, not putting it back, and then drawing a second card. To find the probability of drawing a jack and then a queen in this situation, we modify the multiplication rule.

$$P(\text{jack and queen}) = \frac{4}{52} \cdot \frac{4}{51}$$

- $\frac{4}{52}$: Probability of drawing a jack on the first draw. Out of 52 cards, 4 are jacks.
- $\frac{4}{51}$: Probability of drawing a queen on second draw given that a jack was drawn first. Out of the 51 cards left, 4 are queens.

The probability of an event A occurring given that event B has already occurred is called a **conditional probability** and is represented by $P(B \mid A)$, read *the probability of B given A*. This notation is used in the rule for the probability of two dependent events.

KeyConcept Probability of Dependent Events

If two events A and B are dependent, then the probability that A and B will occur is

$$P(A \text{ and } B) = P(A) \cdot P(B \mid A).$$

This rule can be extended to three or more dependent events.

PT

Example 2 Probability of Dependent Events

A bag contains 12 red, 9 blue, 11 yellow, and 8 green marbles. If two marbles are drawn at random and not replaced, what is the probability that a red and then a blue marble are drawn?

The event of drawing the first marble affects the probability of drawing the second marble, because there is one fewer marble from which to choose. So, the events are dependent.

$P(\text{red}) = \frac{12}{40}$ ← number of red marbles / ← total number of marbles

$= \frac{3}{10}$

$P(\text{blue} \mid \text{red}) = \frac{9}{39}$ ← number of blue marbles / ← number of marbles remaining

$= \frac{3}{13}$

$P(\text{red and blue}) = P(\text{red}) \cdot P(\text{blue} \mid \text{red})$ Probability of dependent events

$= \frac{3}{10} \cdot \frac{3}{13}$ Substitute.

$= \frac{9}{130}$ Multiply.

The probability that a red and then a blue marble are drawn is $\frac{9}{130}$.

Solving the equation for the probability of two dependent events for $P(A \mid B)$ gives a rule for computing the conditional probability of an event.

$P(A \text{ and } B) = P(A) \cdot P(B \mid A)$ Probability of dependent events

$\frac{P(A \text{ and } B)}{P(A)} = P(B \mid A)$ Divide each side by $P(A)$.

StudyTip

Independent Events If A and B are independent events, then $P(B \mid A) = P(B)$.

KeyConcept Conditional Probability

If A and B are dependent events, then the conditional probability of event B occurring, given that event A has already occurred, is

$$P(B \mid A) = \frac{P(A \text{ and } B)}{P(A)}, \text{ where } P(A) \neq 0.$$

PT

Example 3 Conditional Probability

FOOD At a restaurant, 25% of customers order chili. If 4% of customers order chili and a baked potato, find the probability that someone who orders chili also orders a baked potato.

$P(\text{baked potato} \mid \text{chili}) = \frac{P(\text{chili and baked potato})}{P(\text{chili})}$ Conditional probability

$= \frac{0.04}{0.25}$ $P(\text{chili and baked potato}) = 0.04$ and $P(\text{chili}) = 0.25$

$= 0.16$ Simplify.

The probability that someone who orders chili also orders a baked potato is 16%.

Review Vocabulary

relative frequency the ratio of the number of observations in a statistical category to the total number of observations

A *contingency* or **two-way frequency table** is often used to show the observed or relative frequencies of data from an experiment classified according to two variables, with the rows indicating one variable and the columns indicating the other. These tables can be used to find conditional probabilities.

Real-World Example 4 Two-Way Frequency Table

MEDICINE A drug company conducted an experiment to determine the effectiveness of a certain new drug. Test subjects were randomly assigned to one of two groups: a treatment group, which received the drug, or a control group, which received a placebo instead of the drug. The contingency table below shows the results.

Group	Condition Improves (*Y*)	Condition Does Not Improve (N)
Treatment (*T*)	1600	400
Control (*C*)	1200	800

a. Find the probability that a test subject's condition improved given that he or she was in the treatment group.

Step 1 Add a row and a column to the table, and calculate the totals.

Group	Condition Improves (*Y*)	Condition Does Not Improve (N)	Totals
Treatment (*T*)	1600	400	2000
Control (*C*)	1200	800	2000
Totals	1800	1200	4000

Study Tip

Joint and Marginal Frequencies The totals row and totals column in a two-way table report the *marginal frequencies*, while the cells in the interior of the table report the *joint frequencies.* The total number of observations is reported in the bottom right corner of the table.

Step 2 Divide each observed frequency by the total number of people in the study, 4000, to obtain a table of relative frequencies.

Group	Condition Improves (*Y*)	Condition Does Not Improve (N)	Totals
Treatment (*T*)	**0.4**	0.1	**0.5**
Control (*C*)	0.3	0.2	0.5
Totals	0.45	0.3	1

Step 3 Find the conditional probability $P(Y \mid T)$.

$P(Y \mid T) = \frac{P(T \text{ and } Y)}{P(T)}$ Conditional probability

$= \frac{0.4}{0.5}$ or 0.8 $P(T$ and $Y) = 0.4$ and $P(T) = 0.5$

The probability that a subject's condition improved given that he or she was in the treatment group is 0.8 or 80%.

b. Find the probability that a test subject was in the control group given that his or her condition did not improve.

$P(C \mid N) = \frac{P(N \text{ and } C)}{P(N)}$ Conditional probability

$= \frac{0.2}{0.3}$ or about 0.667 $P(N$ and $C) = 0.2$ and $P(N) = 0.3$

The probability that a test subject was in the control group given that his or her condition did not improve is about 0.667 or about 66.7%.

Exercises

Determine whether the events are *independent* or *dependent*. Then find the probability.

1. A red die and a blue die are rolled. What is the probability of getting the result shown?

2. Yana has 4 black socks, 6 blue socks, and 8 white socks in his drawer. If he selects three socks at random with no replacement, what is the probability that he will first select a blue sock, then a black sock, and then another blue sock?

A die is rolled twice. Find each probability.

3. P(2 and 3)
4. P(two 4s)
5. P(no 6s)
6. P(two of the same number)

A bag contains 8 blue marbles, 6 red marbles, and 5 green marbles. Three marbles are drawn one at a time. Find each probability.

7. The second marble is green, given that the first marble is blue and not replaced.
8. The second marble is red, given that the first marble is green and is replaced.
9. The third marble is red, given that the first two are red and blue and not replaced.
10. The third marble is green, given that the first two are red and are replaced.

DVDS There are 8 action, 3 comedy, and 5 drama DVDs on a shelf. Suppose three DVDs are selected at random from the shelf. Find each probability.

11. P(3 action), with replacement
12. P(2 action, then a comedy), without replacement

13. **CARDS** You draw a card from a standard deck of cards and show it to a friend. The friend tells you that the card is red. What is the probability that you correctly guess that the card is the ace of diamonds?

14. **HONOR ROLL** Suppose the probability that a student takes AP Calculus and is on the honor roll is 0.0035, and the probability that a student is on the honor roll is 0.23. Find the probability that a student takes AP Calculus given that he or she is on the honor roll.

15. **DRIVING TESTS** The table shows how students in Mr. Diaz's class fared on their first driving test. Some took a class to prepare, while others did not. Find each probability.

Status	Class	No Class
passed	64	48
failed	18	32

   a. Paige passed, given that she took the class.
   b. Madison failed, given that she did not take the class.
   c. Jamal did not take the class, given that he passed.

16. **SCHOOL CLUBS** King High School tallied the number of males and females that were members of at least one after school club. Find each probability.

Gender	Clubs	No Clubs
male	156	242
female	312	108

   a. A student is a member of a club given that he is male.
   b. A student is not a member of a club given that she is female.
   c. A student is a male given that he is not a member of a club.

17. **FOOTBALL ATTENDANCE** The number of students who have attended a football game at North Coast High School is shown. Find each probability.

Class	Freshman	Sophomore	Junior	Senior
attended	48	90	224	254
not attended	182	141	36	8

   a. A student is a freshman and has not attended a game.
   b. A student has attended a game and is an upperclassman (a junior or senior).

LESSON 0-7 Congruent and Similar Figures

:: **Objective**

- Identify and use congruent and similar figures.

NewVocabulary
congruent
similar

Congruent figures have the same size and the same shape. Two polygons are congruent if their corresponding sides and their corresponding angles are congruent.

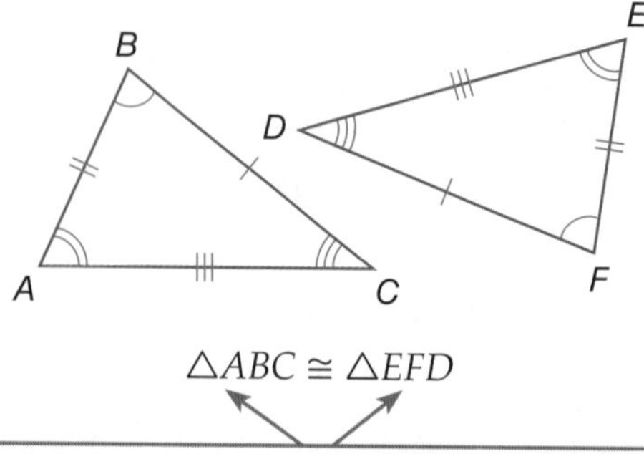

$\triangle ABC \cong \triangle EFD$

The order of the vertices indicates the corresponding parts.

Congruent Angles	Congruent Sides
$\angle A \cong \angle E$	$\overline{AB} \cong \overline{EF}$
$\angle B \cong \angle F$	$\overline{BC} \cong \overline{FD}$
$\angle C \cong \angle D$	$\overline{AC} \cong \overline{ED}$

Read the symbol $\cong$ as *is congruent to.*

Example 1 Congruence Statements

The corresponding parts of two congruent triangles are marked on the figure. Write a congruence statement for the two triangles.

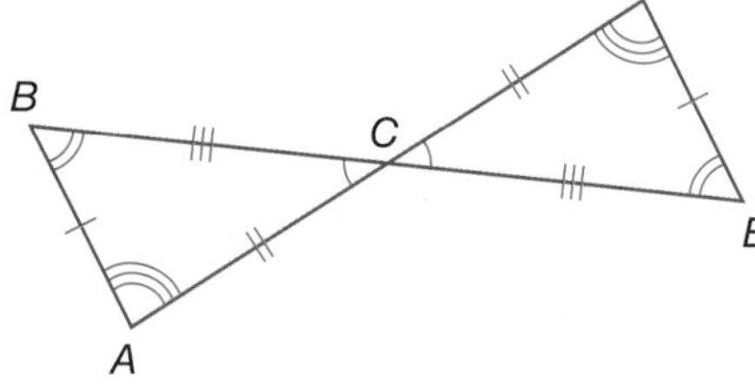

List the congruent angles and sides.

$\angle A \cong \angle D$ $\angle B \cong \angle E$ $\angle ACB \cong \angle DCE$

$\overline{AB} \cong \overline{DE}$ $\overline{AC} \cong \overline{DC}$ $\overline{BC} \cong \overline{EC}$

Match the vertices of the congruent angles. Therefore, $\triangle ABC \cong \triangle DEC$.

Similar figures have the same shape, but not necessarily the same size.

In similar figures, corresponding angles are congruent, and the measures of corresponding sides are proportional. (They have equivalent ratios.)

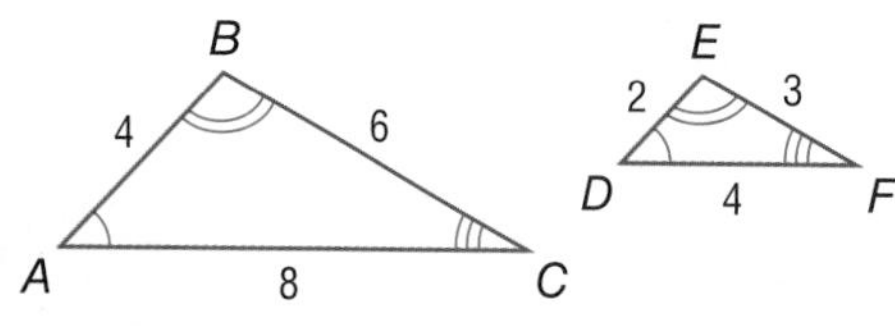

Congruent Angles
$\angle A \cong \angle D$, $\angle B \cong \angle E$, $\angle C \cong \angle F$

Proportional Sides
$\frac{AB}{DE} = \frac{BC}{EF} = \frac{AC}{DF}$

$\triangle ABC \sim \triangle DEF$

Read the symbol $\sim$ as *is similar to.*

Example 2 Determine Similarity

Determine whether the polygons are similar. Justify your answer.

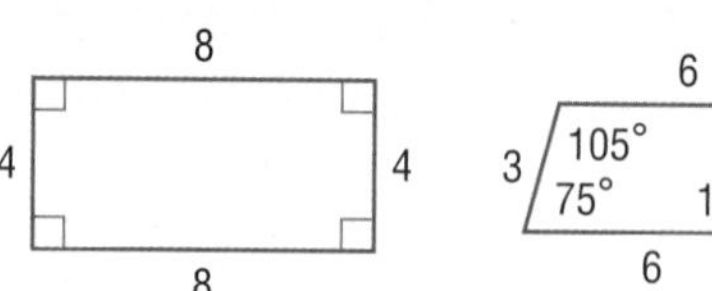

Because $\frac{4}{3} = \frac{8}{6} = \frac{4}{3} = \frac{8}{6}$, the measures of the sides are proportional. However, the corresponding angles are not congruent. The polygons are not similar.

Example 3 Solve a Problem Involving Similarity

CIVIL ENGINEERING The city of Mansfield plans to build a bridge across Pine Lake. Use the information in the diagram at the right to find the distance across Pine Lake.

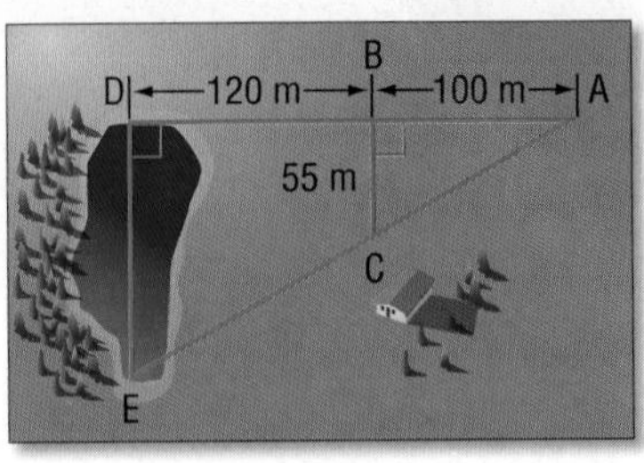

$\triangle ABC \sim \triangle ADE$

$\frac{AB}{AD} = \frac{BC}{DE}$ Definition of similar polygons

$\frac{100}{220} = \frac{55}{DE}$ $AB = 100$, $AD = 100 + 120$ or 220, $BC = 55$

$100DE = 220(55)$ Cross products

$100DE = 12{,}100$ Simplify.

$DE = 121$ Divide each side by 100.

The distance across the lake is 121 meters.

StudyTip

Reasonableness When solving a problem using a proportion, examine your solution for reasonableness. In Example 3, *DA* is more than twice *BA*, so *DE* should be more than twice *BC*. The solution is reasonable.

Exercises

Determine whether each pair of figures is *similar*, *congruent*, or *neither*.

1. 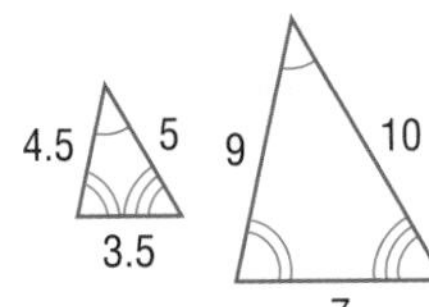

2. 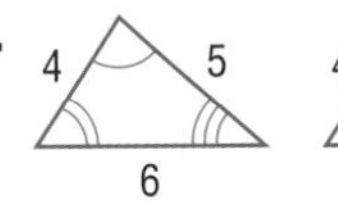

3. 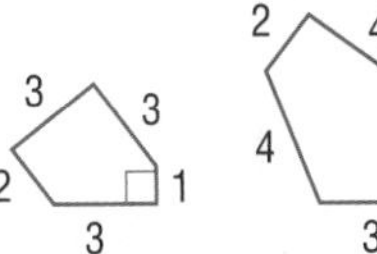

4. 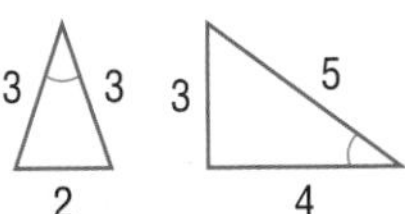

5. 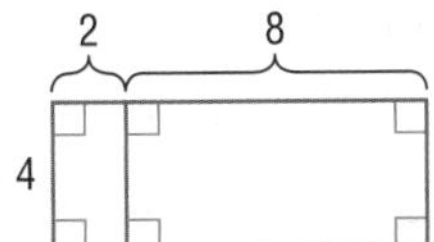

6. 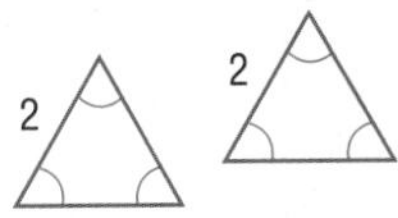

Each pair of polygons is similar. Find the values of *x* and *y*.

7. 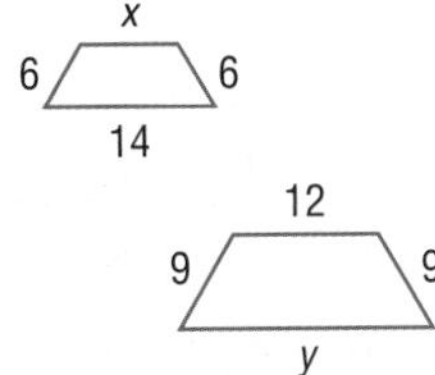

8. 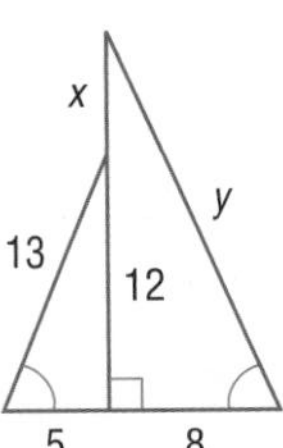

9. 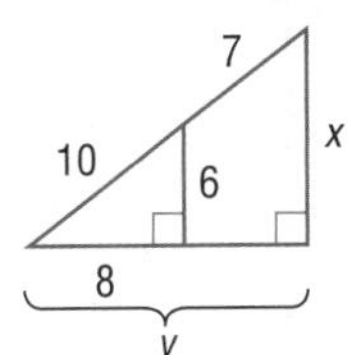

10. **SHADOWS** On a sunny day, Jason measures the length of his shadow and the length of a tree's shadow. Use the figures at the right to find the height of the tree.

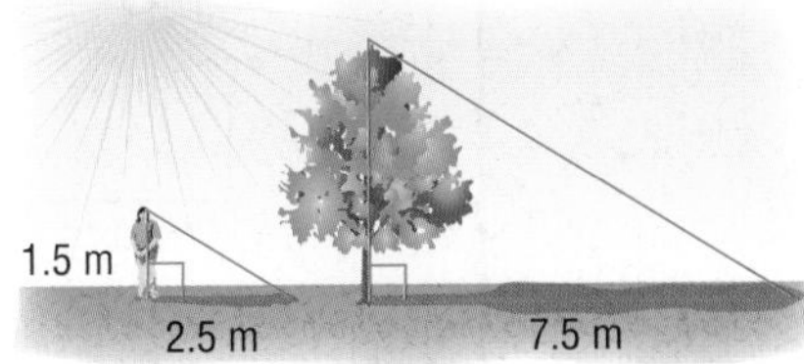

11. **PHOTOGRAPHY** A photo that is 4 inches wide by 6 inches long must be reduced to fit in a space 3 inches wide. How long will the reduced photo be?

12. **SURVEYING** Surveyors use instruments to measure objects that are too large or too far away to measure by hand. They can use the shadows that objects cast to find the height of the objects without measuring them. A surveyor finds that a telephone pole that is 25 feet tall is casting a shadow 20 feet long. A nearby building is casting a shadow 52 feet long. What is the height of the building?

LESSON 0-8 The Pythagorean Theorem

Objective

- Use the Pythagorean Theorem and its converse.

The **Pythagorean Theorem** states that in a right triangle, the square of the length of the hypotenuse c is equal to the sum of the squares of the lengths of the legs a and b.

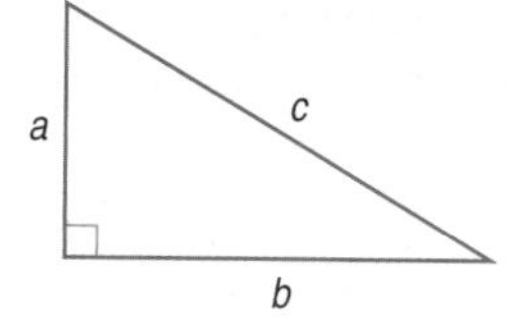

That is, in any right triangle, $c^2 = a^2 + b^2$.

Example 1 Find Hypotenuse Measures

Find the length of the hypotenuse of each right triangle.

a.

5 in.
c in.
12 in.

$c^2 = a^2 + b^2$	Pythagorean Theorem
$c^2 = 5^2 + 12^2$	$a = 5$ and $b = 12$
$c^2 = 25 + 144$	Simplify.
$c^2 = 169$	Add.
$c = \sqrt{169}$	Take the positive square root of each side.
$c = 13$	The length of the hypotenuse is 13 inches.

b.

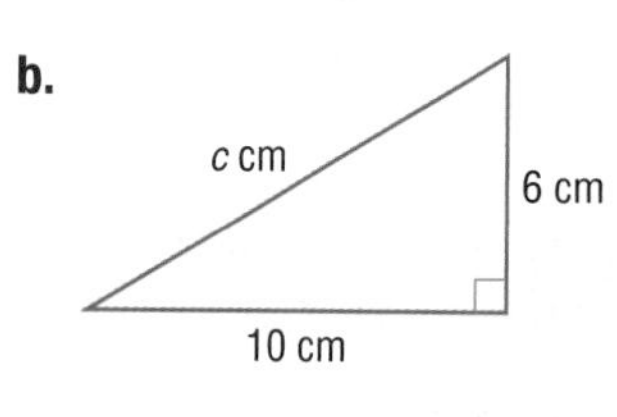

$c^2 = a^2 + b^2$	Pythagorean Theorem
$c^2 = 6^2 + 10^2$	$a = 6$ and $b = 10$
$c^2 = 36 + 100$	Simplify.
$c^2 = 136$	Add.
$c = \sqrt{136}$	Take the positive square root of each side.
$c \approx 11.7$	Use a calculator.

To the nearest tenth, the length of the hypotenuse is 11.7 centimeters.

You can also find the length of a leg of a right triangle given the lengths of the hypotenuse and the other leg.

Example 2 Find Leg Measures

Find the length of the missing leg in each right triangle.

a.

7 ft
25 ft
a ft

$c^2 = a^2 + b^2$	Pythagorean Theorem
$25^2 = a^2 + 7^2$	$c = 25$ and $b = 7$
$625 = a^2 + 49$	Simplify.
$625 - 49 = a^2 + 49 - 49$	Subtract 49 from each side.
$576 = a^2$	Simplify.
$\sqrt{576} = a$	Take the positive square root of each side.
$24 = a$	The length of the leg is 24 feet.

WatchOut!

Positive Square Root When finding the length of a side using the Pythagorean Theorem, use only the positive and not the negative square root, since length cannot be negative.

b.

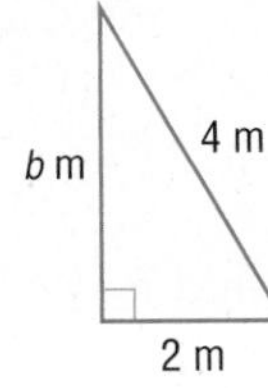

$c^2 = a^2 + b^2$	Pythagorean Theorem
$4^2 = 2^2 + b^2$	$c = 4$ and $a = 2$
$16 = 4 + b^2$	Simplify.
$16 - 4 = 4 - 4 + b^2$	Subtract 4 from each side.
$12 = b^2$	Simplify.
$\sqrt{12} = b$	Take the positive square root of each side.
$3.5 \approx b$	Use a calculator.

To the nearest tenth, the length of the leg is 3.5 meters.

The **converse of the Pythagorean Theorem** states that if the sides of a triangle have lengths a, b, and c, and $c^2 = a^2 + b^2$, then the triangle is a right triangle.

Example 3 Identify a Right Triangle

The lengths of the three sides of a triangle are 5, 7, and 9 inches. Determine whether this triangle is a right triangle.

Because the longest side is 9 inches, use 9 as c, the measure of the hypotenuse.

$c^2 = a^2 + b^2$	Pythagorean Theorem
$9^2 \stackrel{?}{=} 5^2 + 7^2$	$c = 9$, $a = 5$, and $b = 7$
$81 \stackrel{?}{=} 25 + 49$	Evaluate 9^2, 5^2, and 7^2.
$81 \neq 74$	Simplify.

Because $c^2 \neq a^2 + b^2$, the triangle is *not* a right triangle.

Exercises

Find each missing measure. Round to the nearest tenth, if necessary.

1.

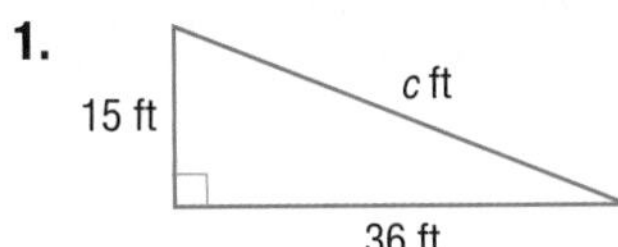

2.

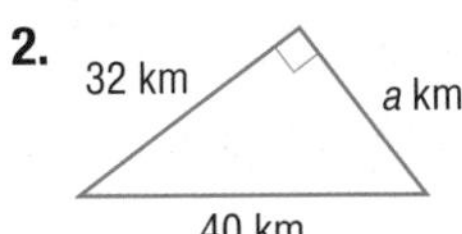

3.

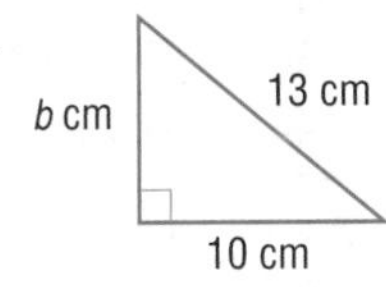

4. $a = 3, b = 4, c = ?$

5. $a = ?, b = 12, c = 13$

6. $a = 14, b = ?, c = 50$

7. $a = 2, b = 9, c = ?$

8. $a = 6, b = ?, c = 13$

9. $a = ?, b = 7, c = 11$

The lengths of three sides of a triangle are given. Determine whether each triangle is a right triangle.

10. 5 in., 7 in., 8 in.

11. 9 m, 12 m, 15 m

12. 6 cm, 7 cm, 12 cm

13. 11 ft, 12 ft, 16 ft

14. 10 yd, 24 yd, 26 yd

15. 11 km, 60 km, 61 km

16. **FLAGPOLES** If a flagpole is 30 feet tall and Mai-Lin is standing a distance of 15 feet from the flagpole, what is the distance from her feet to the top of the flagpole?

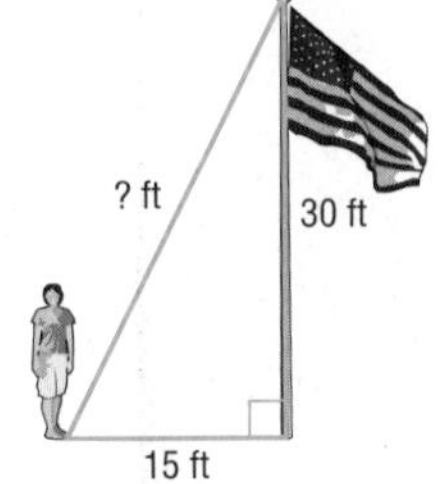

17. **CONSTRUCTION** The walls of a recreation center are being covered with paneling. A doorway is 0.9 meter wide and 2.5 meters high. What is the length of the widest rectangular panel that can be taken through this doorway?

18. **OPEN ENDED** Create an application problem involving right triangles and the Pythagorean Theorem. Then solve your problem, drawing diagrams if necessary.

LESSON 0-9 Measures of Center, Spread, and Position

Objective

- Find measures of center, spread, and position.

NewVocabulary

statistics
descriptive statistics
population
variable
data
univariate data
sample
measures of center or central tendency
mean
median
mode
measures of spread or variation
range
variance
standard deviation
quartile
lower quartile
upper quartile
five-number summary
interquartile range
outlier

Statistics is the science of collecting, organizing, displaying, and analyzing data in order to draw conclusions and make predictions. The branch of statistics that focuses on collecting, summarizing, and displaying data is called **descriptive statistics**.

The entire group of interest to a statistician is called a **population**. A **variable** is a characteristic of a population that can assume different values called **data**. Data that include only one variable are called **univariate data**. When it is not possible to obtain data about every member of a population, a representative **sample** or subset of the population is selected.

Population

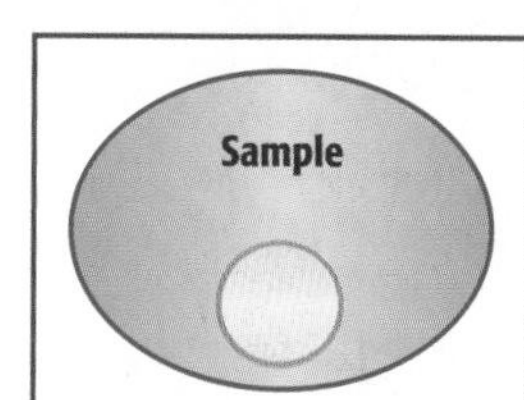

Univariate data are often summarized using a single number to represent what is average or typical. Measures of average are also called **measures of center** or **central tendency**. The most common measures of center are mean, median, and mode.

KeyConcept Measures of Center

- The **mean** is the sum of the values in a set of data $x_1, x_2, \ldots, x_n$ divided by the total number of values n in the set. The formula for population mean μ is $\mu = \frac{x_1 + x_2 + \ldots + x_n}{n}$.
- The **median** is the middle value or the mean of the two middle values in a set of data when the data are arranged in numerical order.
- The **mode** is the value or values that appear most often in a set of data. A set of data can have no mode, one mode, or more than one mode.

For sample mean, replace μ with $\bar{x}$.

Example 1 Measures of Center

SOFT DRINKS **The number of milligrams of sodium in a 12-ounce can of ten different brands of regular cola are shown below. Find the mean, median, and mode.**

50, 30, 25, 20, 40, 35, 35, 10, 15, 35

Mean
$$\bar{x} = \frac{50 + 30 + 25 + 20 + 40 + 35 + 35 + 10 + 15 + 35}{10}$$
$$= \frac{295}{10} \text{ or } 29.5 \text{ milligrams}$$

Find the sum of the data values and divide by the number of values, 10.

Median 10, 15, 20, 25, 30, 35, 35, 35, 40, 50

Arrange the data in order.

$\frac{30 + 35}{2}$ or 32.5 milligrams

Find the mean of the middle two values.

Mode The value that occurs most often in the set is 35, so the mode of the data set is 35 milligrams.

Because two very different data sets can have the same mean, statisticians also use **measures of spread** or **variation** to describe how widely the data values vary and how much the values differ from what is typical. The most common of these measures are listed below.

KeyConcept Measures of Spread

- The **range** is the difference between the greatest and least values in a set of data.
- The **variance** in a set of data $x_1, x_2, \ldots, x_n$ is the mean of the squares of the deviations or differences from the mean. The formula for population variance σ^2 is

$$\sigma^2 = \frac{(x_1 - \mu)^2 + (x_2 - \mu)^2 + \ldots + (x_n - \mu)^2}{n}.$$

- The **standard deviation** in a set of data $x_1, x_2, \ldots, x_n$ is the average amount by which each individual value deviates or differs from the mean. It is the square root of the variance. The formula for population standard deviation σ is

$$\sigma = \sqrt{\sigma^2} \text{ or } \sqrt{\frac{(x_1 - \mu)^2 + (x_2 - \mu)^2 + \ldots + (x_n - \mu)^2}{n}}.$$

For sample variance and standard deviation, replace σ^2 with s^2, μ with $\bar{x}$, and n with $n - 1$.

StudyTip

Unbiased Estimator Dividing by $n - 1$ instead of n when finding a sample variance and standard deviation gives a slightly larger value, and therefore an unbiased estimate, of the population standard deviation or variance.

PT

Example 2 Measures of Spread

MIDTERM EXAMS Two classes took the same midterm exam. The scores of five students from each class are shown. Both sets of scores have a mean of 84.2.

Class A
85, 76, 92, 88, 80

Class B
75, 85, 95, 98, 68

a. Find the range, variance, and standard deviation for the sample scores from Class A.

Range 92 − 76 or 16 range = greatest value − least value

Variance Subtract the mean of the data, 84.2, from each value. Then square each result and find the sum of these squares. Finally, since these data were obtained from a sample of all scores in Class A, divide the sum by $n - 1$.

X	$X - \bar{X}$	$(X - \bar{X})^2$
85	85 − 84.2 = 0.8	$0.8^2 = 0.64$
76	76 − 84.2 = −8.2	$(-8.2)^2 = 67.24$
92	92 − 84.2 = 7.8	$7.8^2 = 60.84$
88	88 − 84.2 = 3.8	$3.8^2 = 14.44$
80	80 − 84.2 = −4.2	$(-4.2)^2 = 17.64$
		Sum = 160.8

StudyTip

Squared Differences The sum of the differences of the data values from the mean of the set will always be zero. For this reason, variance is calculated using the sum of the *squares* of the differences.

$$s^2 = \frac{(x_1 - \bar{x})^2 + \ldots + (x_n - \bar{x})^2}{n - 1}$$ Sample variance

$$= \frac{160.8}{5 - 1} \text{ or about } 40.2$$ $(x_1 - \bar{x})^2 + \ldots + (x_5 - \bar{x})^2 = 160.8$ and $n = 5$

Standard Deviation $s = \sqrt{s^2} = \sqrt{\frac{160.8}{5 - 1}}$ or about 6.3 Sample standard deviation

(continued on the next page)

b. Use a calculator to find the range, variance, and standard deviation for the sample scores from Class B.

To find measures of center and spread using a graphing calculator, first clear all lists. Press STAT ENTER to enter each data value. Then press STAT and select **1-Var Stats** from the **CALC** menu. Scroll to see all one-variable statistics.

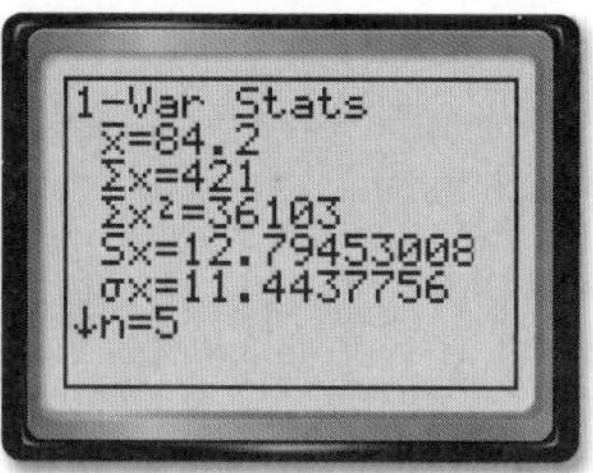

The range **maxX** – **minX** for Class B is 98 – 68 or 30. The sample standard deviation **Sx** is about 12.8, so the variance is about 12.8^2 or 163.8.

StudyTip

Comparing Standard Deviations Standard deviation can only be compared in this way when the units of both data sets are the same.

c. Compare the sample standard deviations of Class A and Class B.

Since the sample standard deviation of Class A is 6.3 and the sample standard deviation of Class B is 12.8, there is more variability in the sample scores from Class B than from Class A.

Statisticians use measures of position to describe where specific values fall within a data set. **Quartiles** are three position measures that divide a data set arranged in ascending order into four groups, each containing about 25% of the data. The median marks the second quartile Q_2 and separates the data into upper and lower halves.
The first or **lower quartile** Q_1 is the median of the lower half, while the third or **upper quartile** Q_3 is the median of the upper half.

StudyTip

Calculating Quartiles When the number of values in a set of data is odd, the median is not included in either half of the data when calculating Q_1 or Q_3.

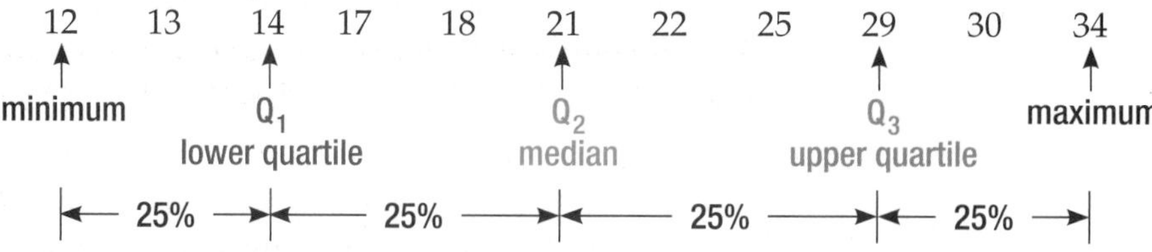

The three quartiles, along with the minimum and maximum values, are called a **five-number summary** of a data set.

PT

Example 3 Five-Number Summary

PART-TIME JOB The number of hours Liana worked each week for the last 12 weeks were 21, 10, 18, 12, 15, 13, 20, 20, 19, 16, 18, and 14.

a. Find the minimum, lower quartile, median, upper quartile, and maximum of the data set.

Use a calculator to find the one-variable statistics for the data set. The minimum **minX** of the data set is 10, the lower quartile $\mathbf{Q_1}$ is 13.5, the median **Med** is 17, the upper quartile $\mathbf{Q_3}$ is 19.5, and the maximum **maxX** is 21.

1-Var Stats
↑n=12
minX=10
Q1=13.5
Med=17
Q3=19.5
maxX=21

b. Interpret this five-number summary.

Over the last 12 weeks, Liana worked a minimum of 10 hours and a maximum of 21 hours. She worked less than 13.5 hours 25% of the time, less than 17 hours 50% of the time, and less than 19.5 hours 75% of the time.

StudyTip

Interquartile Range A large interquartile range means that the data are spread out.

The difference between Q_3 and Q_1, is called the **interquartile range** IQR. The interquartile range contains about 50% of the values. Before deciding on which measures best describe a set of data, check the data set for outliers. An **outlier** is an extremely high or extremely low value when compared with the rest of the values in the set. To check for outliers, look for data values that are beyond the upper or lower quartiles by more than 1.5 times the interquartile range.

Example 4 Effect of an Outlier

HOMEWORK The number of minutes each of the 22 students in a class spent working on the same algebra assignment is shown below.

15, 12, 25, 15, 27, 10, 16, 18, 30, 35, 22, 25, 65, 20, 18, 25, 15, 13, 25, 22, 15, 28

a. Identify any outliers in the data.

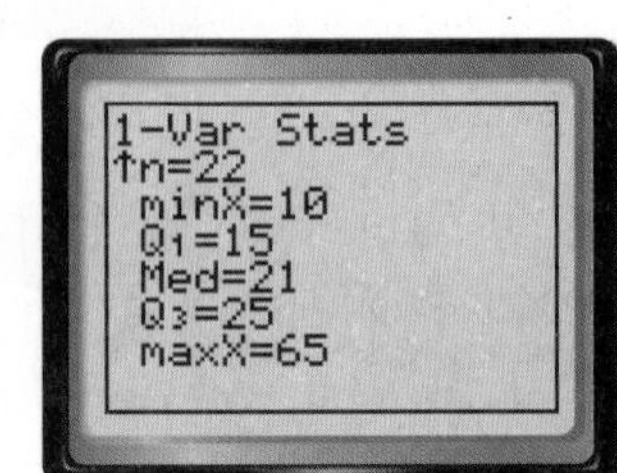

Calculate the interquartile range IQR by using a graphing calculator to find Q_1 and Q_3.

$$\text{IQR} = Q_3 - Q_1 = 25 - 15 \text{ or } 10$$

Use the interquartile range to find the values beyond which any outliers would lie.

$Q_1 - 1.5(\text{IQR})$ and	$Q_3 + 1.5(\text{IQR})$	Values beyond which outliers lie
$15 - 1.5(10)$	$25 + 1.5(10)$	$Q_1 = 15, Q_3 = 25, \text{IQR} = 10$
0	40	Simplify.

There are no values less than 0, but there is one value greater than 40. The value 65 can be considered an outlier for this data set.

b. Find the mean, median, mode, range, and standard deviation of the data set with and without the outlier. Describe the effect on each measure.

You can use a graphing calculator. The data set represents the entire population of the class, so use the population standard deviation σ.

Data Set	Mean	Median	Mode	Range	Standard Deviation
with outlier	≈22.5	21	15	55	≈11.2
without outlier	≈20.5	20	15	25	≈6.5

Removal of the outlier causes the mean, median, range, and standard deviation to decrease. Notice that the mean is affected more by the outlier than the median.

Exercises

Find the mean, median, and mode for each set of data.

1. number of pages in each novel assigned for summer reading:
224, 272, 374, 478, 960, 394, 404, 308, 480, 624

2. height in centimeters of bean plants at the end of an experiment:
14.5, 12, 16, 11, 14, 11, 10.5, 14, 11.5, 15, 13.5

3. number of text messages sent each day during the last two weeks:
18, 35, 53, 44, 26, 57, 23, 27, 47, 33, 4, 35, 39, 41

State whether the data in sets A and B represent *sample* or *population* data. Then find the range, variance, and standard deviation of each set. Use the standard deviations to compare the variability between the data sets.

4.

Wait Times (min)					
Ride A			Ride B		
45	22	40	35	50	32
48	11	51	31	35	45
36	55	60	45	49	40
32	24	37	43	37	45

5.

Number of Sponsors Obtained by Participants					
Charity Walk A			Charity Walk B		
44	14	61	8	28	15
22	27	25	100	42	19
38	50	49	25	75	82

6.

Number of Days Each Student Missed This Year														
Class A														
10	8	5	9	7	3	6	8	14	11	8	4	7	8	2
5	13	0	15	9	7	9	10	9	11	14	8	12	10	1
Class B														
5	8	13	7	9	4	10	2	12	9	6	11	3	8	5
12	6	7	8	11	12	8	9	3	10	5	13	9	1	8

Find the minimum, lower quartile, median, upper quartile, and maximum of each data set. Then interpret this five-number summary.

7.

Number of Students in Each Math Class at Central High														
25	27	26	26	19	27	24	23	19	28	25	24	20	22	22
24	26	18	28	29	29	26	24	24	23	23	25	25	29	28

8.

State Mean ACT Scores									
20.2	21.3	21.5	20.4	21.6	20.3	22.5	21.5	17.8	20.5
20.0	21.7	21.3	20.2	21.6	22.0	21.6	20.3	19.8	22.6
20.8	22.4	21.4	22.2	18.8	21.5	21.7	21.2	22.5	21.2
20.1	22.3	20.3	21.2	21.4	20.6	22.5	21.8	21.9	19.3
21.5	20.5	20.3	21.5	22.7	20.9	22.5	22.2	21.4	20.7

Identify any outliers in each data set, and explain your reasoning. Then find the mean, median, mode, range, and standard deviation of the data set with and without the outlier. Describe the effect on each measure.

9. fuel efficiency in miles per gallon of 15 randomly selected automobiles: 40, 36, 29, 45, 51, 36, 48, 34, 36, 22, 13, 42, 31, 44, 32, 34

10. number of posts to a certain blog each month during a particular year: 25, 23, 21, 27, 29, 19, 10, 21, 20, 18, 26, 23

11. **CEREAL** The weights, in ounces, of 20 randomly selected boxes of a certain brand of cereal are shown.
16.7, 16.8, 15.9, 16.1, 16.5, 16.6, 16.5, 15.9, 16.7, 16.5,
16.6, 14.9, 16.5, 16.1, 15.8, 16.7, 16.2, 16.5, 16.4, 16.6

   a. Identify any outliers in the data set, and explain your reasoning.

   b. If the outlier was removed and an additional cereal box that was 17.35 ounces was added, would this value be an outlier of the new data set? Explain.

   c. What are some possible causes of outliers in this situation?

CHAPTER 0 Posttest

State the domain and range of each relation. Then determine whether each relation is a function. Write *yes* or *no*.

1. $\{(4, 5), (5, -1), (0, 12), (0, -2), (7, 9)\}$

2.

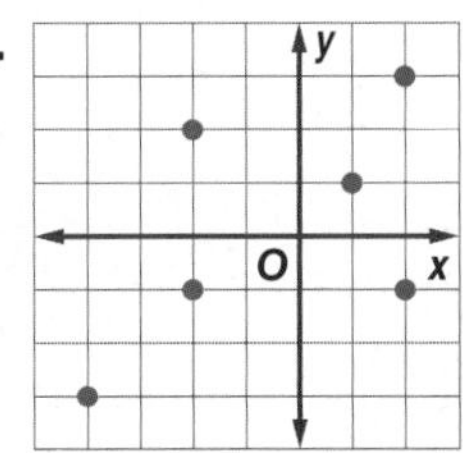

Name the quadrant in which each point is located.

3. $(-3, 7)$ **4.** $(10, -11)$ **5.** $(-15, -3)$

Find each product.

6. $(4n - 3)(2n + 2)$ **7.** $(5p - 1)(6p - 10)$

8. $(7x + 4)(7x + 4)$ **9.** $(3k - 2)(6k + 9)$

10. GEOMETRY The height of a rectangle is 3 millimeters less than twice the width.

a. Write an expression for each measure.

b. Write a polynomial expression for the area of the rectangle.

Factor each polynomial.

11. $4x^2 + 4xy + y^2$ **12.** $25a^2 - 20a + 4$

13. $4a^2 + 16ab + 16b^2$ **14.** $81t^2 - 36$

15. STUDENT COUNCIL A student council has 6 seniors, 5 juniors, and 1 sophomore as members. How many ways can a 3-member committee be formed that includes one member from each class?

Determine whether each situation involves *permutations* or *combinations*. Then solve.

16. How many ways are there to select one competitor and one alternate out of 8 students?

17. How many ways are there to form a team of 7 athletes from a group of 15 who try out?

RESTAURANT There are 24 male and 36 female patrons in a restaurant. Of the 11 patrons under 10 years old, 6 are male. Of the 14 patrons over 55 years old, 9 are female. A patron is selected at random. Determine whether the events are *mutually exclusive* or *not mutually exclusive*. Then find each probability.

18. P(female or under 10) **19.** P(under 10 or over 55)

Determine whether the events are *independent* or *dependent*. Then find the probability.

20. Slips of paper numbered 1 through 10 are placed into a bag. What is the probability of drawing the number 10 three times in a row if a slip is drawn at random and then replaced?

21. Two students are selected at random from a class that consists of 13 males and 7 females. What is the probability that both students are female?

22. TESTING Of the students who took both the Mid-Chapter 4 Quiz and the Chapter 4 Test, 56% passed the quiz and 48% passed both the quiz and the test. If a student passed the quiz, find the probability that he or she also passed the test.

23. Determine whether the rectangles are *similar*, *congruent*, or *neither*.

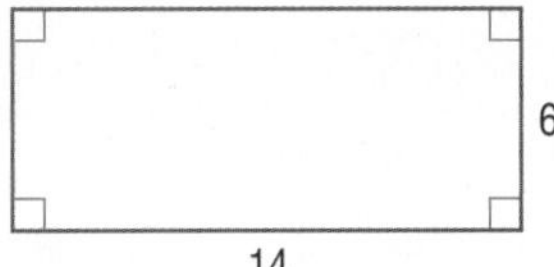

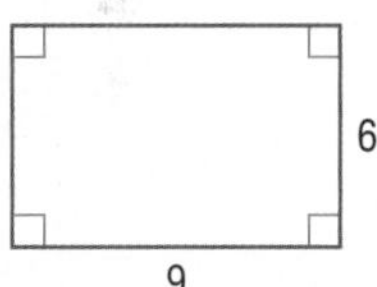

24. COMPUTERS A computer image of a painting 320 pixels wide by 240 pixels high. If the actual painting is 42 inches wide, how high is it?

Find each missing measure. Round to the nearest tenth, if necessary.

25.

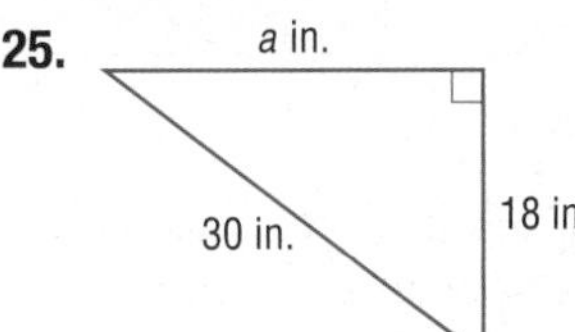

26. $a = 33$ cm, $b = ?$ cm, $c = 45$ cm

The lengths of three sides of a triangle are given. Determine whether each triangle is a right triangle.

27. 6 in., 8 in., 12 in. **28.** 30 m, 34 m, 16 m

Find the mean, median, mode, range, and standard deviation of each data set. Then identify any outliers.

29. number of students present at 8 student council meetings: 23, 45, 16, 75, 32, 35, 28, 35

30. running time in minutes for 17 movies: 95, 102, 148, 140, 110, 103, 107, 104, 99, 111, 109, 124, 109, 90, 92, 110, 129

CHAPTER 1

Equations and Inequalities

Then

You wrote expressions with variables.

Now

You will:

- Simplify and evaluate algebraic expressions.
- Solve linear and absolute value equations.
- Solve and graph inequalities.

Why?

MONEY Connecting money to mathematics is one of the most practical skills you can learn. As long as you use money, you will be using mathematics. In this chapter, you will explore money topics such as sales tax, income, and budgeting for your first apartment.

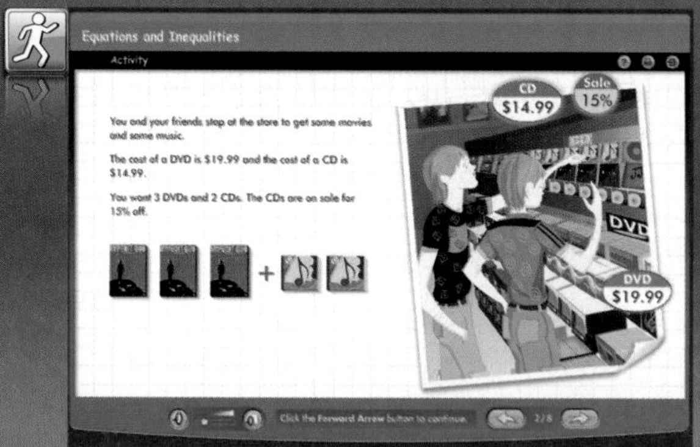

Jon Boyes/Photographer's Choice/Getty Ima

connectED.mcgraw-hill.com Your Digital Math Portal

Animation  Vocabulary  eGlossary  Personal Tutor  Virtual Manipulatives  Graphing Calculator  Audio  Foldables  Self-Check Practice  Worksheets

Get Ready for the Chapter

Diagnose Readiness | You have two options for checking prerequisite skills.

Textbook Option Take the Quick Check below. Refer to the Quick Review for help.

QuickCheck

Simplify.

1. $15.7 + (-3.45)$
2. $-18.54 - (-32.05)$
3. $-9.8 \cdot 6.75$
4. $4 \div (-0.5)$
5. $3\frac{2}{3} + \left(-1\frac{4}{5}\right)$
6. $\frac{54}{7} - \frac{26}{6}$
7. $\left(\frac{6}{5}\right)\left(-\frac{10}{9}\right)$
8. $-3 \div \frac{7}{8}$
9. **CRAFTS** Felisa needs $\frac{7}{8}$ yard of one type of material to make a quilt. How much of this material will she need to make 12 quilts?

Evaluate each power.

10. 6^3
11. $(-4)^3$
12. $-(0.6)^2$
13. $-(-2.5)^3$
14. $\left(\frac{4}{5}\right)^2$
15. $\left(\frac{7}{3}\right)^4$
16. $\left(-\frac{7}{10}\right)^2$
17. $-\left(\frac{15}{2}\right)^3$
18. **FOOD** Nate's Deli offers 3 types of bread, 3 types of meat, and 3 types of cheese. How many different sandwiches can be made with 1 type each of bread, meat, and cheese?

Identify each statement as *true* or *false*.

19. $-6 \geq -7$
20. $8 > -5$
21. $\frac{1}{7} \leq \frac{1}{9}$
22. $\frac{5}{6} \leq \frac{25}{30}$
23. **MEASUREMENT** Christy has a board that is 0.6 yard long. Marissa has a board that is $\frac{2}{3}$ yard long. Marissa states that $\frac{2}{3} > 0.6$. Is she correct?

QuickReview

Example 1

Simplify $\left(\frac{3}{16}\right)\left(-\frac{4}{5}\right)$.

$\left(\frac{3}{16}\right)\left(-\frac{4}{5}\right) = -\frac{3(4)}{16(5)}$ Multiply the numerators and the denominators.

$= -\frac{12}{80}$ Simplify.

$= -\frac{12 \div 4}{80 \div 4}$ Divide the numerator and denominator by the GCF, 4.

$= -\frac{3}{20}$ Simplify.

Example 2

Evaluate $(-1.5)^3$.

$(-1.5)^3 = (-1.5)(-1.5)(-1.5)$ $(-1.5)^3$ means 1.5 is a factor 3 times.

$= -3.375$ Simplify.

Example 3

Identify $\frac{3}{8} > \frac{12}{24}$ as *true* or *false*.

$\frac{3}{8} \overset{?}{>} \frac{12 \div 3}{24 \div 3}$ Divide 12 and 24 by 3 to get a denominator of 8.

$\frac{3}{8} \not> \frac{4}{8}$ Simplify.

False; $\frac{3}{8} \not> \frac{4}{8}$ because $\frac{3}{8} < \frac{4}{8}$.

2 Online Option Take an online self-check Chapter Readiness Quiz at connectED.mcgraw-hill.com.

Get Started on the Chapter

You will learn several new concepts, skills, and vocabulary terms as you study Chapter 1. To get ready, identify important terms and organize your resources. You may wish to refer to Chapter 0 to review prerequisite skills.

FOLDABLES® StudyOrganizer

Equations and Inequalities Make this Foldable to help you organize your Chapter 1 notes about equations and inequalities. Begin with one sheet of 11" × 17" paper.

1 **Fold** 2" tabs on each of the short sides.

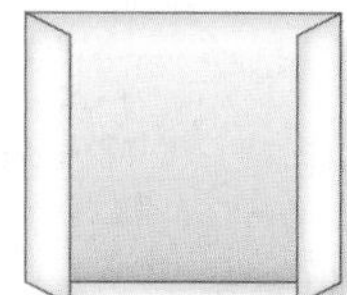

2 **Then** fold in half in both directions. Open and cut as shown.

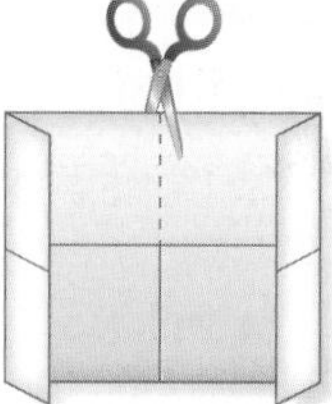

3 **Refold** along the width. Staple each pocket. Label pockets as *Algebraic Expressions, Properties of Real Numbers, Solving Equations,* and *Solve and Graph Inequalities*. Place index cards for notes in each pocket.

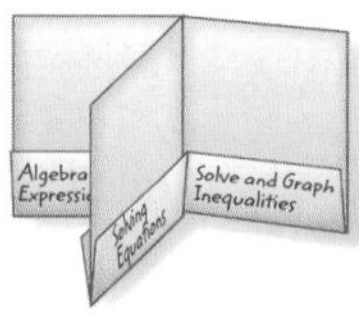

NewVocabulary

English		Español
variable	p. 5	variable
algebraic expression	p. 5	expressión algebraica
order of operations	p. 5	orden de las operaciones
formula	p. 6	fórmula
real numbers	p. 11	números reales
rational numbers	p. 11	números racional
irrational numbers	p. 11	números irracional
integers	p. 11	enteros
whole numbers	p. 11	números enteros
natural numbers	p. 11	números naturales
open sentence	p. 18	enunciado abierto
equation	p. 18	ecuación
solution	p. 18	solución
absolute value	p. 27	valor absoluto
empty set	p. 28	conjunto vacío
set-builder notation	p. 35	notación de construcción de conjuntos
compound inequality	p. 41	desigualdad compuesta
intersection	p. 41	intersección
union	p. 42	unión

ReviewVocabulary

evaluate evaluar to find the value of an expression

inequality desigualdad an open sentence that contains the symbol $<$, $\leq$, $>$, or $\geq$

power potencia an expression of the form x^n, read *x to the nth power*

The *base* is the number that is multiplied.

x^n

The *exponent* tells how many times the base is used as a factor.

The number that can be expressed using an exponent is called a *power.*

LESSON 1-1 Expressions and Formulas

Then

- You used the rules of exponents.

Now

1. Use the order of operations to evaluate expressions.
2. Use formulas.

Why?

- The following formula can be used to calculate a baseball player's on-base percentage x.

$$x = \frac{h + w + p}{b + w + p + s}$$

- h is the number of hits.
- w is the number of walks.
- p is the number of times the player has been hit by a pitch.
- b is the number of times at bat.
- s is the number of sacrifice flies and sacrifice bunts.

During the first twenty games of a season, Ian has 9 hits, 2 walks, 38 at bats, 5 sacrifice flies, and he is hit by 1 pitch. The expression $\frac{9 + 2 + 1}{38 + 2 + 1 + 5}$ gives Ian's on-base percentage.

NewVocabulary

variables
algebraic expressions
order of operations
formula

Common Core State Standards

Content Standards

A.SSE.1.a Interpret parts of an expression, such as terms, factors, and coefficients.

A.SSE.1.b Interpret complicated expressions by viewing one or more of their parts as a single entity.

Mathematical Practices

1 Make sense of problems and persevere in solving them.

1 Order of Operations

Variables are letters used to represent unknown quantities. Expressions that contain at least one variable are called **algebraic expressions**. You can evaluate an algebraic expression by replacing each variable with a number and then applying the **order of operations**.

KeyConcept Order of Operations

Step 1 Evaluate the expressions inside grouping symbols.

Step 2 Evaluate all powers.

Step 3 Multiply and/or divide from left to right.

Step 4 Add and/or subtract from left to right.

Example 1 Evaluate Algebraic Expressions

Evaluate $m + (p - 1)^2$ if $m = 3$ and $p = -4$.

$m + (p - 1)^2 = 3 + (-4 - 1)^2$	Replace m with 3 and p with -4.
$= 3 + (-5)^2$	Add -4 and -1.
$= 3 + 25$	Evaluate $(-5)^2$.
$= 28$	Add 3 and 25.

GuidedPractice

Evaluate each expression if $m = 12$ and $q = -1$.

1A. $m + (3 - q)^2$

1B. $m \div 2q + 4$

Taxi/Getty Images

Example 2 Evaluate Algebraic Expressions

a. Evaluate $a + b^2(b - a)$ if $a = 5$ and $b = -3.2$.

$a + b^2(b - a) = 5 + (-3.2)^2(-3.2 - 5)$	$a = 5$ and $b = -3.2$
$= 5 + (-3.2)^2(-8.2)$	Subtract 5 from -3.2.
$= 5 + 10.24(-8.2)$	Evaluate $(-3.2)^2$.
$= 5 + (-83.968)$	Multiply 10.24 and -8.2.
$= -78.968$	Add 5 and -83.968.

b. Evaluate $\frac{x^4 - 3wy}{y^3 + 2w}$ if $w = 4$, $x = -3$, and $y = -5$.

$\frac{x^4 - 3wy}{y^3 + 2w} = \frac{(-3)^4 - 3(4)(-5)}{(-5)^3 + 2(4)}$	$w = 4$, $x = -3$, and $y = -5$
$= \frac{81 - 3(4)(-5)}{-125 + 2(4)}$	Evaluate the numerator and denominator separately.
$= \frac{81 - (-60)}{-125 + 8}$	Multiply in the numerator and denominator.
$= \frac{141}{-117}$ or $-\frac{47}{39}$	Simplify the numerator and denominator. Then simplify the fraction.

StudyTip

CCSS Structure Remember that a fraction bar is a type of grouping symbol. Evaluate the expressions in the numerator and denominator separately before dividing.

GuidedPractice

Evaluate each expression if $h = 4$, $j = -1$, and $k = 0.5$.

2A. $h^2k + h(h - k)$

2B. $j + (3 - h)^2$

2C. $\frac{j^2 - 3h^2k}{j^3 + 2}$

2 Formulas

A **formula** is a mathematical sentence that expresses the relationship between certain quantities. If you know the value of every variable in the formula except one, you can find the value of the remaining variable.

ReadingMath

Exponents x^2 may be read as *x squared* or *x to the second power*. x^3 may be read as *x cubed* or *x to the third power.*

Real-World Example 3 Use a Formula

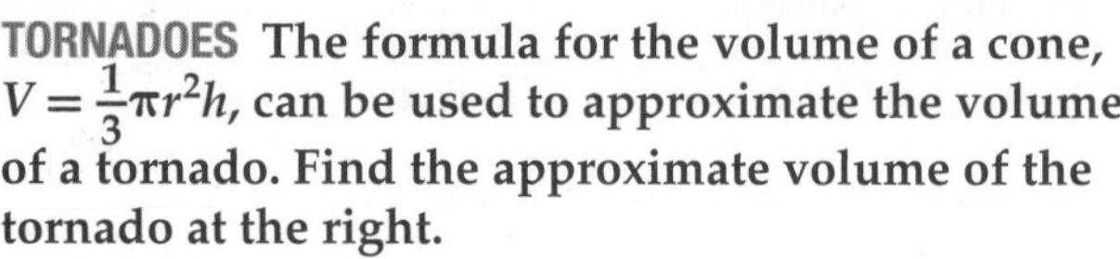

TORNADOES The formula for the volume of a cone, $V = \frac{1}{3}\pi r^2 h$, can be used to approximate the volume of a tornado. Find the approximate volume of the tornado at the right.

$V = \frac{1}{3}\pi r^2 h$	Volume of a cone
$= \frac{1}{3}\pi(75)^2(225)$	$r = 75$ and $h = 225$
$= \frac{1}{3}\pi(5625)(225)$	Evaluate 75^2.
$\approx 1{,}325{,}359$	Multiply.

The approximate volume of the tornado is about 1,325,359 cubic meters.

GuidedPractice

3. GEOMETRY The formula for the volume V of a rectangular prism is $V = \ell wh$, where ℓ represents the length, w represents the width, and h represents the height. Find the volume of a rectangular prism with a length of 4 feet, a width of 2 feet, and a height of 3.5 feet.

Check Your Understanding

= Step-by-Step Solutions begin on page R14.

Example 1 **Evaluate each expression if $a = -2$, $b = 3$, and $c = 4.2$.**

1. $a - 2b + 3c$

2. $2a + (b + 3)^2$

3. $a + 3[b^2 - (a + c)]$

Example 2

4. $5c - 2[(b - a) + c]$

5. $4(2a + 3b) - 2c$

6. $\frac{a^2 + 4c}{3b + 2a}$

7. $\frac{b^3 + ac}{ab + 2bc}$

8. $\frac{3b + 2a}{5 - c}$

9. $\frac{3a - 2c}{4ab}$

Example 3 **10. VOLLEYBALL** A player's attack percentage A is calculated using the formula $A = \frac{k - e}{t}$, where k represents the number of kills, e represents the number of attack errors including blocks, and t represents the total attacks attempted. Find the attack percentage given each set of values.

a. $k = 22, e = 11, t = 35$

b. $k = 33, e = 9, t = 50$

Practice and Problem Solving

Extra Practice is on page R1.

Example 1 **Evaluate each expression if $w = -3$, $x = 4$, $y = 2.6$, and $z = \frac{1}{3}$.**

11. $y + x - z$

12. $w - 2x + y \div 2$

13. $4(x - w)$

14. $6(y + x)$

15. $9z - 4y + 2w$

16. $3y - 4z + x$

17. GAS MILEAGE The gasoline used by a car is measured in miles per gallon and is related to the distance traveled by the following formula.

miles per gallon × number of gallons = distance traveled

a. During a trip your car used a total of 46.2 gallons of gasoline. If your car gets 33 miles to the gallon, how far did you travel?

b. Your friend has decided to buy a hybrid car that gets 60 miles per gallon. The gasoline tank holds 12 gallons. How far can the car go on one tank of gasoline?

Example 2 **Evaluate each expression if $a = -4$, $b = -0.8$, $c = 5$, and $d = \frac{1}{5}$.**

18. $\frac{a + b}{c - d}$

19. $\frac{a - b}{bd}$

20. $\frac{ac}{d + b}$

21 $\frac{b^2c^2}{ad}$

22. $\frac{b + 6}{4(d + c)}$

23. $\frac{5(d + a)}{2ab^2}$

24. CCSS **SENSE-MAKING** The formula $C = \frac{5(F - 32)}{9}$ can be used to convert temperatures in degrees Fahrenheit to degrees Celsius.

a. Room temperature commonly ranges from 64°F to 73°F. Determine the room temperature range in degrees Celsius.

b. The normal average human body temperature is 98.6°F. A temperature above this indicates a fever. If your temperature is 42°C, do you have a fever? Explain your reasoning.

Example 3 **25. GEOMETRY** The formula for the area A of a triangle with height h and base b is $A = \frac{1}{2}bh$. Write an expression to represent the area of the triangle.

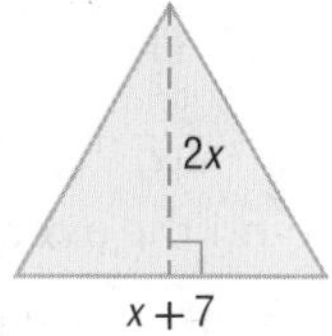

26. FINANCIAL LITERACY The profit that a business made during a year is $536,897,000. If the business divides the profit evenly for each share, estimate how much each share made if there are 10,995,000 shares.

27. CCSS **REASONING** The radius of Earth's orbit is 93,000,000 miles.

a. Find the circumference of Earth's orbit assuming that the orbit is a circle. The formula for the circumference of a circle is $2\pi r$.

b. Earth travels at a speed of 66,698 miles per hour around the Sun. Use the formula $T = \frac{C}{V}$, where T is time in hours, C is circumference, and V is velocity to find the number of hours it takes Earth to revolve around the Sun.

c. Did you prove that it takes 1 year for Earth to go around the Sun? Explain.

28. **ANCIENT PYRAMID** The Great Pyramid in Cairo, Egypt, is approximately 146.7 meters high, and each side of its base is approximately 230 meters.

a. Find the area of the base of the pyramid. Remember $A = \ell w$.

b. The volume of a pyramid is $\frac{1}{3}Bh$, where B is the area of the base and h is the height. What is the volume of the Great Pyramid?

Evaluate each expression if $w = \frac{3}{4}$, $x = 8$, $y = -2$, and $z = 0.4$.

29. $x^3 + 2y^4$

30. $(x - 6z)^2$

31. $2(6w - 2y) - 8z$

32. $\frac{(y + z)^2}{xw}$

33. $\frac{12w - 6y}{z^2}$

34. $\frac{wx + yz}{wx - yz}$

35. **GEOMETRY** The formula for the volume V of a cone with radius r and height h is $V = \frac{1}{3}\pi r^2 h$. Write an expression for the volume of the cone at the right.

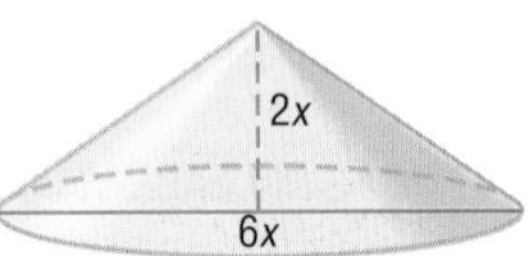

36. **SEARCH ENGINES** Page rank is a numerical value that represents how important a page is on the Web. One formula used to calculate the page rank for a page is $PR = 0.15 + 0.85L$, where L is the page rank of the linking page divided by the number of outbound links on the page. Determine the page rank of a page in which $L = 10$.

37. **WEATHER** In 1898, A.E. Dolbear studied various species of crickets to determine their "chirp rate" based on temperatures. He determined that the formula $t = 50 + \frac{n - 40}{4}$, where n is the number of chirps per minute, could be used to find the temperature t in degrees Fahrenheit. What is the temperature if the number of chirps is 120?

38. **FOOTBALL** The following formula can be used to calculate a quarterback efficiency rating.

$$\left(\frac{\frac{C}{A} - 0.3}{0.2} + \frac{\frac{Y}{A} - 3}{4} + \frac{\frac{T}{A}}{0.05} + \frac{0.095 - \frac{I}{A}}{0.04}\right) \cdot \frac{100}{6}$$

- C is the number of passes completed.
- A is the number of passes attempted.
- Y is passing yardage.
- T is the number of touchdown passes.
- I is the number of interceptions.

Find Peyton Manning's efficiency rating to the nearest tenth for the season statistics shown.

39. **MOVIES** The average price for a movie ticket can be represented by $P = \frac{y^2}{400} + \frac{7y}{100} + 2.96$ where y is the number of years since 1980.

a. Find the average price of a ticket in 1990, 2000, and 2010.

b. Another equation that can be used to represent ticket prices is $P = \frac{y^3}{2500} - \frac{y^2}{100} + \frac{6y}{25} + 2.62$. Find the price of a ticket in 1990, 2000, and 2010. How do these values compare to those you found in part **a**?

40. **GEOMETRY** The area of a triangle can be found using Heron's Formula, $A = \sqrt{s(s-a)(s-b)(s-c)}$, where a, b, and c are the lengths of the three sides of the triangle, and $s = \frac{a+b+c}{2}$. Find the area of the triangle at the right.

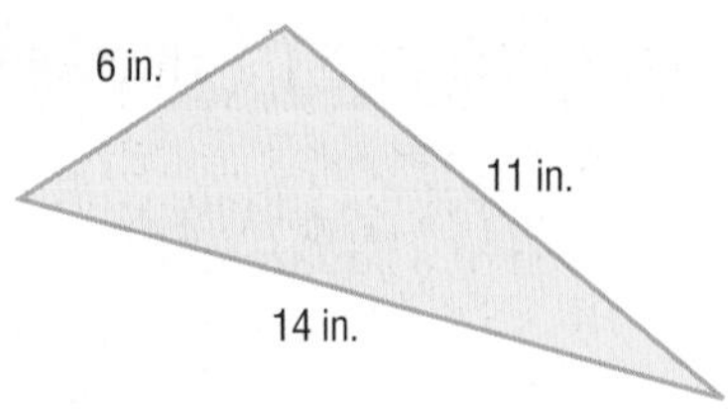

41. Evaluate $y = \sqrt{b^2\left(1 - \frac{x^2}{a^2}\right)}$ if $a = 6$, $b = 8$, and $x = 3$. Round to the nearest tenth.

42. **MULTIPLE REPRESENTATIONS** You will write expressions using the formula for the volume of a cylinder. Recall that the volume of a cylinder can be found using the formula $v = \pi r^2 h$, in which v = volume, r = radius, and h = height.

   a. **Geometric** Draw two cylinders of different sizes.

   b. **Tabular** Use a ruler to measure the radius and height of each cylinder. Organize the measures for each cylinder into a table. Include a column in your table to calculate the volume of each cylinder.

   c. **Verbal** Write a verbal expression for the difference in volume of the two cylinders.

   d. **Algebraic** Write and solve an algebraic expression for the difference in volume of the two cylinders.

H.O.T. Problems Use Higher-Order Thinking Skills

43. **CCSS CRITIQUE** Lauren and Rico are evaluating $\frac{-3d - 4c}{2ab}$ for $a = -2$, $b = -3$, $c = 5$, and $d = 4$. Is either of them correct? Explain your reasoning.

Lauren

$$\frac{-3d - 4c}{2ab} = \frac{-3(4) - 4(5)}{2(-2)(-3)}$$
$$= \frac{-12 - 20}{12} = \frac{-32}{12} = -\frac{8}{3}$$

Rico

$$\frac{-3d - 4c}{2ab} = \frac{-3(4) - 4(5)}{2(-2)(-3)}$$
$$= \frac{-12 - 20}{12} = \frac{8}{12} = \frac{2}{3}$$

44. **CHALLENGE** For any three distinct numbers a, b, and c, $a\$b\$c$ is defined as $a\$b\$c = \frac{-a - b - c}{c - b - a}$. Find $-2\$(-4)\$5$.

45. **REASONING** The following equivalent expressions represent the height in feet of a stone thrown downward off a bridge where t is the time in seconds after release. Which do you find most useful for finding the maximum height of the stone? Explain.

   a. $-4t^2 - 2t + 6$

   b. $-2t(2t + 1) + 6$

   c. $-2(t - 1)(2t + 3)$

46. **CHALLENGE** Let m, n, p, and q represent nonzero positive integers. Find a number in terms of m, n, p, and q that is halfway between $\frac{m}{n}$ and $\frac{p}{q}$.

47. **OPEN ENDED** Write an algebraic expression using $x = -2$, $y = -3$, and $z = 4$ and all four operations for which the value of the expression is 10.

48. **WRITING IN MATH** Provide an example of a formula used in everyday situations. Explain its usefulness and what happens if the formula is not used correctly.

49. **WRITING IN MATH** Use the information for on-base percentage given at the beginning of the lesson to explain why a formula for on-base percentage is more useful than a table of specific percentages.

Standardized Test Practice

50. SAT/ACT If the area of a square with side x is 9, what is the area of a square of side $4x$?

A 36
B 144
C 212
D 324
E 1296

51. SHORT RESPONSE A coffee shop owner wants to open a second shop when his daily customer average reaches 800 people. He has calculated the daily customer average in the table below for each month since he has opened.

Month	Daily Customer Average
1	225
2	298
3	371
4	444

If the trend continues, during what month can he open a second shop?

52. GEOMETRY In $\triangle DFG$, $\overline{FH}$ and $\overline{HG}$ are angle bisectors and $m\angle D = 84°$. How many degrees are in $\angle FHG$?

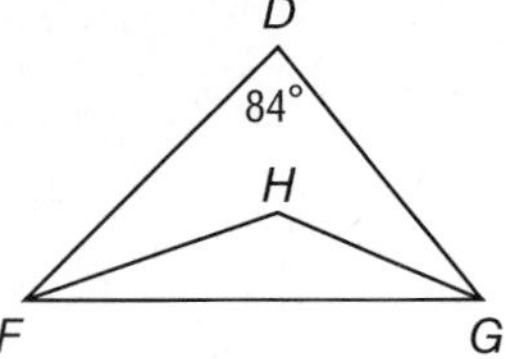

F 96
G 132
H 145
J 192

53. A skydiver in a computer game free-falls from a height of 3000 meters at a rate of 55 meters per second. Which equation can be used to find h, the height of the skydiver after t seconds of free fall?

A $h = -55t - 3000$
B $h = -55t + 3000$
C $h = 3000t - 55$
D $h = 3000t + 55$

Spiral Review

54. The lengths of the three sides of a triangle are 10, 14, and 18 inches. Determine whether this triangle is a right triangle. (Lesson 0-8)

55. The legs of a right triangle measure 6 centimeters and 8 centimeters. Find the length of the hypotenuse. (Lesson 0-8)

56. MAPS On a map of the U.S., the cities of Milwaukee, Wisconsin, and Charlotte, North Carolina, are $6\frac{1}{2}$ inches apart. The actual distance between Milwaukee and Charlotte is 670 miles. If Birmingham, Alabama, and St. Petersburg, Florida, are 465 miles apart, how far apart are they on the map? (Lesson 0-7)

57. Factor $6x^2 + 12x$. (Lesson 0-3)

58. Find the product of $(a + 2)(a - 4)$. (Lesson 0-2)

59. NUMBER An integer is 2 less than a number, and another integer is 1 greater than double that same number. What are the two integers if their sum is 14? (Lesson 0-2)

Skills Review

Evaluate each expression.

60. $\sqrt{4}$

61. $\sqrt{25}$

62. $\sqrt{81}$

63. $\sqrt{121}$

64. $-\sqrt{9}$

65. $-\sqrt{16}$

66. $\sqrt{\frac{49}{100}}$

67. $\sqrt{\frac{25}{64}}$

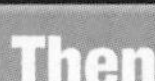

Lesson 1-2 Properties of Real Numbers

Then	Now	Why?
You identified and used the arithmetic properties of real numbers.	**1** Classify real numbers. **2** Use the properties of real numbers to evaluate expressions.	The Central High School Boosters sell snacks and beverages at school functions. The items are priced the same to make determining the total cost easy. You can use the Distributive Property to calculate the total cost when multiple items are purchased.

NewVocabulary
- real numbers
- rational numbers
- irrational numbers
- integers
- whole numbers
- natural numbers

Common Core State Standards

Content Standards
A.SSE.2 Use the structure of an expression to identify ways to rewrite it.

Mathematical Practices
2 Reason abstractly and quantitatively.
7 Look for and make use of structure.

1 Real Numbers

Real numbers consist of several different kinds of numbers.

- **Rational numbers** can be expressed as a ratio $\frac{a}{b}$, where a and b are integers and b is not zero. The decimal form of a rational number is either a terminating or repeating decimal.
- The decimal form of an **irrational number** neither terminates nor repeats. Square roots of numbers that are not perfect squares are irrational numbers.
- The sets of **integers**, $\{\ldots, -3, -2, -1, 0, 1, 2, 3, \ldots\}$, **whole numbers**, $\{0, 1, 2, 3, 4, \ldots\}$, and **natural numbers**, $\{1, 2, 3, 4, 5, \ldots\}$, are subsets of the rational numbers. These numbers are subsets of the rational numbers because every integer n is equal to $\frac{n}{1}$.

KeyConcept Real Numbers (R)

Letter	Set	Examples
Q	rationals	$0.125, -\frac{7}{8}, \frac{2}{3} = 0.66\ldots$
I	irrationals	$\pi = 3.14159\ldots$ $\sqrt{3} = 1.73205\ldots$
Z	integers	$-5, 17, -23, 8$
W	wholes	$2, 96, 0, \sqrt{36}$
N	naturals	$3, 17, 6, 86$

Example 1 Classify Numbers

Name the sets of numbers to which each number belongs.

a. -23 integers (Z), rationals (Q), reals (R)

b. $\sqrt{50}$ irrationals (I), reals (R)

c. $-\frac{4}{9}$ rationals (Q), reals (R)

GuidedPractice

1A. -185 **1B.** $-\sqrt{49}$ **1C.** $\sqrt{95}$ **1D.** $-\frac{7}{8}$

2 Properties of Real Numbers

Some of the properties of real numbers are summarized below.

StudyTip

Real Numbers A number can belong to more than one set of numbers. For example, if a number is natural, it is also whole, an integer, rational, and real.

ConceptSummary Real Number Properties

For any real numbers a, b, and c:

Property	Addition	Multiplication
Commutative	$a + b = b + a$	$a \cdot b = b \cdot a$
Associative	$(a + b) + c = a + (b + c)$	$(a \cdot b) \cdot c = a \cdot (b \cdot c)$
Identity	$a + 0 = a = 0 + a$	$a \cdot 1 = a = 1 \cdot a$
Inverse	$a + (-a) = 0 = (-a) + a$	$a \cdot \frac{1}{a} = 1 = \frac{1}{a} \cdot a, a \neq 0$
Closure	$a + b$ is a real number.	$a \cdot b$ is a real number.
Distributive	$a(b + c) = ab + ac$ and $(b + c)a = ba + ca$	

Example 2 Name Properties of Real Numbers

Name the property illustrated by $5 \cdot (4 \cdot 13) = (5 \cdot 4) \cdot 13$.

Associative Property of Multiplication

The Associative Property of Multiplication states that the way in which you group factors does not affect the product.

GuidedPractice

2. Name the property illustrated by $2(x + 3) = 2x + 6$.

You can use the properties of real numbers to identify related values.

StudyTip

Additive and Multiplicative Inverses The additive inverse of a number has the opposite sign as the number. The multiplicative inverse of a number has the same sign as the number.

Example 3 Additive and Multiplicative Inverses

Find the additive inverse and multiplicative inverse for $-\frac{5}{8}$.

Since $-\frac{5}{8} + \frac{5}{8} = 0$, the additive inverse of $-\frac{5}{8}$ is $\frac{5}{8}$.

Since $\left(-\frac{5}{8}\right)\left(-\frac{8}{5}\right) = 1$, the multiplicative inverse of $-\frac{5}{8}$ is $-\frac{8}{5}$.

GuidedPractice

Find the additive and multiplicative inverse for each number.

3A. 1.25

3B. $2\frac{1}{2}$

Many real-world applications involve working with real numbers.

Real-WorldCareer

Retail Store Manager Store managers are responsible for the day-to-day operations of a retail store. A store manager may range from being a high school graduate to having a 4-year degree, depending on the business.

Real-World Example 4 Distributive Property

MONEY The prices of the components of a computer package offered by Computer Depot are shown in the table. If a 6% sales tax is added to the purchase price, how much sales tax is charged for this computer package?

Component	Price ($)
Computer	359.95
Monitor	219.99
Printer	79.00
Digital Camera	149.50
Software Bundle	99.00

There are two ways to determine the total sales tax.

Method 1 Multiply, then add.

Multiply each dollar amount by 6% or 0.06 and then add.

$T = 0.06(359.95) + 0.06(219.99) + 0.06(79.00) + 0.06(149.50) + 0.06(99.00)$

$= 21.60 + 13.20 + 4.74 + 8.97 + 5.94$

$= 54.45$

Method 2 Add, then multiply.

Find the total cost of the computer package, and then multiply the total by 0.06.

$T = 0.06(359.95 + 219.99 + 79.00 + 149.50 + 99.00)$

$= 0.06(907.44)$

$= 54.45$

The sales tax charged is $54.45. Notice that both methods result in the same answer.

GuidedPractice

4. JOBS Kayla makes $8 per hour working at a grocery store. The number of hours Kayla worked each day in one week are 3, 2.5, 2, 1, and 4. How much money did Kayla earn this week?

The properties of real numbers can be used to simplify algebraic expressions.

Example 5 Simplify an Expression

Simplify $3(2q + r) + 5(4q - 7r)$.

$3(2q + r) + 5(4q - 7r)$

$= 3(2q) + 3(r) + 5(4q) - 5(7r)$	Distributive Property
$= 6q + 3r + 20q - 35r$	Multiply.
$= 6q + 20q + 3r - 35r$	Commutative Property (+)
$= (6 + 20)q + (3 - 35)r$	Distributive Property
$= 26q - 32r$	Simplify.

GuidedPractice

5. Simplify $3(4x - 2y) - 2(3x + y)$.

Jon Riley/Stone/Getty Images

Check Your Understanding

= Step-by-Step Solutions begin on page R14.

Example 1 **Name the sets of numbers to which each number belongs.**

1. 62
2. $\frac{5}{4}$
3. $\sqrt{11}$
4. -12

Example 2 **Name the property illustrated by each equation.**

5. $(6 \cdot 8) \cdot 5 = 6 \cdot (8 \cdot 5)$
6. $7(9 - 5) = 7 \cdot 9 - 7 \cdot 5$
7. $84 + 16 = 16 + 84$
8. $(12 + 5)6 = 12 \cdot 6 + 5 \cdot 6$

Example 3 **Find the additive inverse and multiplicative inverse for each number.**

9. -7
10. $\frac{4}{9}$
11. 3.8
12. $\sqrt{5}$

Example 4 13. **CCSS REASONING** Melba is mowing lawns for \$22 each to earn money for a video game console that costs \$550.

a. Write an expression to represent the total amount of money Melba earned during this week.

b. Evaluate the expression from part **a** by using the Distributive Property.

c. When do you think Melba will earn enough for the video game console? Is this reasonable? Explain.

Lawns Mowed in One Week

Day	Lawns Mowed
Monday	2
Tuesday	4
Wednesday	3
Thursday	1
Friday	5
Saturday	6
Sunday	7

Example 5 **Simplify each expression.**

14. $5(3x + 6y) + 4(2x - 9y)$
15. $6(6a + 5b) - 3(4a + 7b)$
16. $-4(6c - 3d) - 5(-2c - 4d)$
17. $-5(8x - 2y) - 4(-6x - 3y)$

Practice and Problem Solving

Extra Practice is on page R1.

Example 1 **Name the sets of numbers to which each number belongs.**

18. $-\frac{4}{3}$
19. -8.13
20. $\sqrt{25}$
21. $0.\overline{61}$
22. $\frac{9}{3}$
23. $-\sqrt{144}$
24. $\frac{21}{7}$
25. $\sqrt{17}$

Example 2 **Name the property illustrated by each equation.**

26. $-7y + 7y = 0$
27. $8\sqrt{11} + 5\sqrt{11} = (8 + 5)\sqrt{11}$
28. $(16 + 7) + 23 = 16 + (7 + 23)$
29. $\left(\frac{22}{7}\right)\left(\frac{7}{22}\right) = 1$

Example 3 **Find the additive inverse and multiplicative inverse for each number.**

30. -8
31. 12.1
32. -0.25
33. $\frac{6}{13}$
34. $-\frac{3}{8}$
35. $\sqrt{15}$

Example 4 36. **CONSTRUCTION** Jorge needs two different kinds of concrete: quick drying and slow drying. The quick-drying concrete mix calls for $2\frac{1}{2}$ pounds of dry cement, and the slow-drying concrete mix calls for $1\frac{1}{4}$ pounds of dry cement. He needs 5 times more quick-drying concrete and 3 times more slow-drying concrete than the mixes make.

a. How many pounds of dry cement mix will he need?

b. Use the properties of real numbers to show how Jorge could compute this amount mentally. Justify each step.

Stockbyte/Getty Images

Example 5 **Simplify each expression.**

37. $8b - 3c + 4b + 9c$

38. $-2a + 9d - 5a - 6d$

39. $4(4x - 9y) + 8(3x + 2y)$

40. $6(9a - 3b) - 8(2a + 4b)$

41. $-2(-5g + 6k) - 9(-2g + 4k)$

42. $-5(10x + 8z) - 6(4x - 7z)$

43. **FOOTBALL** Illustrate the Distributive Property by writing two expressions for the area of a college football field. Then find the area of the football field.

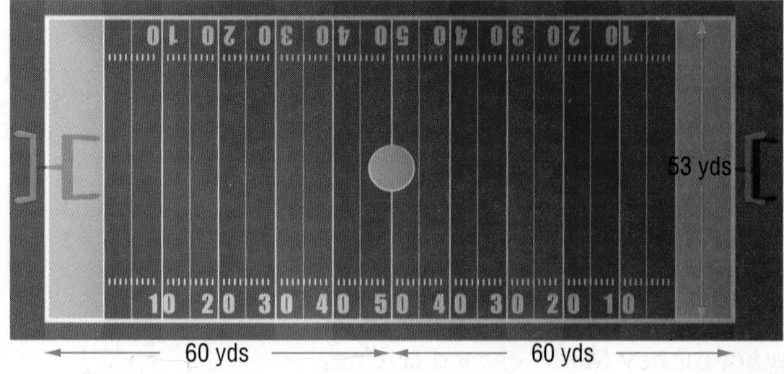

44. **PETS** The chart shows the percent of dogs registered with the American Kennel Club that are of the eight most popular breeds.

a. Illustrate the Distributive Property by writing two expressions to represent the number of registered dogs of the top four breeds.

b. Evaluate the expressions you wrote to find the number of registered dogs of the top four breeds.

Top Dogs

Breed	Percent of Registered Dogs
Labrador Retrievers	14.2
Yorkshire Terriers	5.6
German Shepherds	5.0
Golden Retrievers	4.9
Beagles	4.5
Dachshunds	4.1
Boxers	4.1
Poodles	3.4
Total Registered Dogs	**870,192**

Source: American Kennel Club

45. **FINANCIAL LITERACY** Billie is given \$20 in lunch money by her parents once every two weeks. On some days, she packs her lunch, and on other days, she buys her lunch. A hot lunch from the cafeteria costs \$4.50, and a cold sandwich from the lunch line costs \$2.

a. Billie decides that she wants to buy a hot lunch on Thursday and Friday of the first week and on Wednesday of the second week. Use the Distributive Property to determine how much that will cost.

b. How many cold sandwiches can Billie buy with the amount left over?

c. Assuming that both weeks are Monday through Friday, how many times will Billie have to pack her lunch?

Simplify each expression.

46. $\frac{1}{3}(5x + 8y) + \frac{1}{4}(6x - 2y)$

47. $\frac{2}{5}(6c - 8d) + \frac{3}{4}(4c - 9d)$

48. $-6(3a + 5b) - 3(6a - 8c)$

49. $-9(3x + 8y) - 3(5x + 10z)$

50. CCSS **MODELING** Mary is making curtains out of the same fabric for 5 windows. The two larger windows are the same size, and the three smaller windows are the same size.

One larger window requires $3\frac{3}{4}$ yards of fabric, and one smaller window needs $2\frac{1}{3}$ yards of fabric.

a. How many yards of material will Mary need?

b. Use the properties of real numbers to show how Mary could compute this amount mentally.

51 **MULTIPLE REPRESENTATIONS** Consider the following real numbers.

$$-\sqrt{6}, 3, \frac{-15}{3}, 4.1, \pi, 0, \frac{3}{8}, \sqrt{36}$$

a. **Tabular** Organize the numbers into a table according to the sets of numbers to which each belongs.

b. **Algebraic** Convert each number to decimal form. Then list the numbers from least to greatest.

c. **Graphical** Graph the numbers on a number line.

d. **Verbal** Make a conjecture about using decimal form to list real numbers in order.

52. **CLOTHING** A department store sells shirts for $12.50 each. Dalila buys 2, Latisha buys 3, and Pilar buys 1.

a. Illustrate the Distributive Property by writing two expressions to represent the cost of these shirts.

b. Use the Distributive Property to find how much money the store received from selling these shirts.

H.O.T. Problems Use Higher-Order Thinking Skills

53. **WHICH ONE DOESN'T BELONG?** Identify the number that does not belong with the other three. Explain your reasoning.

$\sqrt{21}$	$\sqrt{35}$	$\sqrt{67}$	$\sqrt{81}$

54. **CHALLENGE** If $12(5r + 6t) = w$, then in terms of w, what is $48(30r + 36t)$?

55. **CCSS CRITIQUE** Luna and Sophia are simplifying $4(14a - 10b) - 6(b + 4a)$. Is either of them correct? Explain your reasoning.

Luna	Sophia
$4(14a - 10b) - 6(b + 4a)$	$4(14a - 10b) - 6(b + 4a)$
$56a - 40b - 6b + 24a$	$56a - 40b - 6a - 24b$
$80a - 46b$	$50a - 64b$

56. **REASONING** Determine whether the following statement is *sometimes, always,* or *never* true. Explain your reasoning.

An irrational number is a real number within a radical sign.

57. **OPEN ENDED** Determine whether the Closure Property of Multiplication applies to irrational numbers. If not, provide a counterexample.

OPEN ENDED The set of all real numbers is *dense,* meaning between any two distinct members of the set there lies infinitely many other members of the set. Find an example of (a) a rational number, and (b) an irrational number between the given numbers.

58. 2.45 and 2.5

59. π and $\frac{10}{3}$

60. $1.\overline{9}$ and 2.01

61. **WRITING IN MATH** Explain and provide examples to show why the Commutative Property does not hold true for subtraction or division.

Standardized Test Practice

62. EXTENDED RESPONSE Lenora bought several pounds of cashews and several pounds of almonds for a party. The cashews cost \$8 per pound, and the almonds cost \$6 per pound. Lenora bought a total of 7 pounds and paid a total of \$48. Write and solve a system of equations to determine the pounds of cashews and the pounds of almonds that Lenora purchased.

63. SAT/ACT Find the 10th term in the series 2, 4, 7, 11, 16, … .

A 41
B 46
C 56
D 67
E 72

64. GEOMETRY What are the coordinates of point A in the parallelogram?

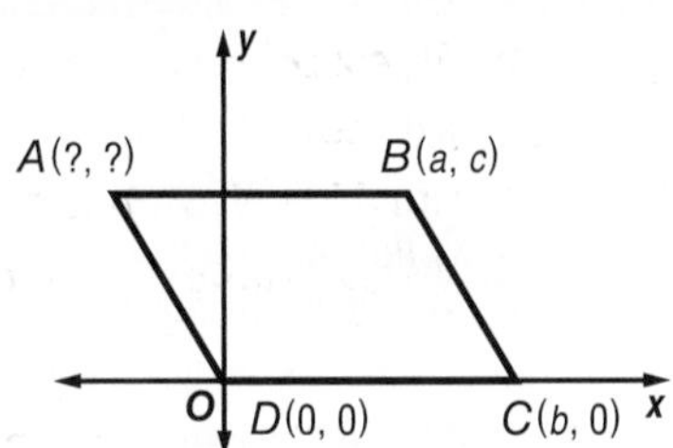

F $(b - a, c)$
G $(a - b, c)$
H (b, c)
J (c, c)

65. What is the domain of the function that contains the points $(-3, 0)$, $(0, 4)$, $(-2, 5)$, and $(6, 4)$?

A $\{-3, 6\}$
B $\{-3, -2, 0, 6\}$
C $\{0, 4, 5, 6\}$
D $\{-3, -2, 0, 4, 5, 6\}$

Spiral Review

66. Evaluate $8(4 - 2)^3$. (Lesson 1-1)

67. Evaluate $a + 3(b + c) - d$, if $a = 5$, $b = 4$, $c = 3$, and $d = 2$. (Lesson 1-1)

68. GEOMETRY The formula for the area A of a circle with diameter d is $A = \pi\left(\frac{d}{2}\right)^2$. Write an expression to represent the area of the circle. (Lesson 1-1)

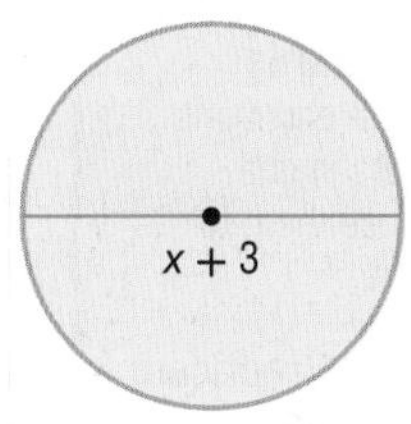

69. CONSTRUCTION A 10-meter ladder leans against a building so that the top is 9.64 meters above the ground. How far from the base of the wall is the bottom of the ladder? (Lesson 0-8)

Factor each polynomial. (Lesson 0-3)

70. $14x^2 + 10x - 8$
71. $9x^2 - 3x + 18$
72. $8x^2 + 16x + 12$
73. $10x^2 - 20x$
74. $7x^2 - 14x - 21$
75. $12x^2 - 18x - 24$

Find each product. (Lesson 0-2)

76. $(x + 2)(x - 3)$
77. $(y + 2)(y - 1)$
78. $(a - 5)(a + 4)$
79. $(b - 7)(b - 3)$
80. $(n + 6)(n + 8)$
81. $(p - 9)(p + 1)$

Skills Review

Evaluate each expression if $a = 3$, $b = \frac{2}{3}$, and $c = -1.7$.

82. $6b - 5$
83. $\frac{1}{6}b + 1$
84. $2.3c - 7$
85. $-8(a - 4)$
86. $a + b + c$
87. $\frac{a \cdot b}{c}$
88. $a^2 - c$
89. $\frac{a \cdot c}{a}$

LESSON 1-3

Solving Equations

Then

- You used properties of real numbers to evaluate expressions.

Now

1. Translate verbal expressions into algebraic expressions and equations, and vice versa.
2. Solve equations using the properties of equality.

Why?

- The United States is one of the few countries in the world that measures distances in miles. When traveling by car in different countries, it is often useful to convert miles to kilometers. To find the approximate number of kilometers k in miles m, divide the number of miles by 0.62137.

$$m \text{ miles} \times \frac{1 \text{ kilometer}}{0.62137 \text{ mile}} \approx k \text{ kilometers}$$

$$\frac{m}{0.62137} \approx k \text{ kilometers}$$

NewVocabulary

open sentence
equation
solution

Common Core State Standards

Content Standards

A.CED.1 Create equations and inequalities in one variable and use them to solve problems.

Mathematical Practices

3 Construct viable arguments and critique the reasoning of others.

8 Look for and express regularity in repeated reasoning.

1 Verbal Expressions and Algebraic Expressions

Verbal expressions can be translated into algebraic expressions by using the language of algebra.

Example 1 Verbal to Algebraic Expression

Write an algebraic expression to represent each verbal expression.

a. **2 more than 4 times the cube of a number** $4x^3 + 2$

b. **the quotient of 5 less than a number and 12** $\frac{n-5}{12}$

GuidedPractice

1A. the cube of a number increased by 4 times the same number

1B. three times the difference of a number and 8

A mathematical sentence containing one or more variables is called an **open sentence**. A mathematical sentence stating that two mathematical expressions are equal is called an **equation**.

Example 2 Algebraic to Verbal Sentence

Write a verbal sentence to represent each equation.

a. $6x = 72$ The product of 6 and a number is 72.

b. $n + 15 = 91$ The sum of a number and 15 is ninety-one.

GuidedPractice

2A. $g - 5 = -2$

2B. $2c = c^2 - 4$

Open sentences are neither true nor false until the variables have been replaced by numbers. Each replacement that results in a true sentence is called a **solution** of the open sentence.

2 Properties of Equality

Properties of Equality To solve equations, we can use properties of equality. Some of these properties are listed below.

Math HistoryLink

Diophantus of Alexandria (c. 200–284)

Diophantus was famous for his work in algebra. His main work was titled *Arithmetica* and introduced symbolism to Greek algebra as well as propositions in number theory and polygonal numbers.

KeyConcept Properties of Equality

Property	Symbols	Examples
Reflexive	For any real number a, $a = a$.	$b + 12 = b + 12$
Symmetric	For all real numbers a and b, if $a = b$, then $b = a$.	If $18 = -2n + 4$, then $-2n + 4 = 18$.
Transitive	For all real numbers a, b, and c, if $a = b$ and $b = c$, then $a = c$.	If $5p + 3 = 48$ and $48 = 7p - 15$, then $5p + 3 = 7p - 15$.
Substitution	If $a = b$, then a may be replaced by b and b may be replaced by a.	If $(6 + 1)x = 21$, then $7x = 21$.

PT

Example 3 Identify Properties of Equality

Name the property illustrated by each statement.

a. If $3a - 4 = b$, and $b = a + 17$, then $3a - 4 = a + 17$.

Transitive Property of Equality

b. If $2g - h = 62$, and $h = 24$, then $2g - 24 = 62$.

Substitution Property of Equality

GuidedPractice

3. If $-11a + 2 = -3a$, then $-3a = -11a + 2$.

Solving most equations requires assuming that the original equation has a solution, and performing the same operations on each side of the equals sign. The properties of equality allow for the equation to be solved in this way.

KeyConcept

Addition and Subtraction Properties of Equality	
Symbols	For any real numbers, a, b, and c, if $a = b$, then $a + c = b + c$ and $a - c = b - c$.
Examples	If $x - 6 = 14$, then $x - 6 + 6 = 14 + 6$. If $n + 5 = -32$, then $n + 5 - 5 = -32 - 5$.

Multiplication and Division Properties of Equality	
Symbols	For any real numbers, a, b, and c, $c \neq 0$, if $a = b$, then $a \cdot c = b \cdot c$ and $\frac{a}{c} = \frac{b}{c}$.
Examples	If $\frac{m}{8} = -7$, then $8 \cdot \frac{m}{8} = 8 \cdot (-7)$. If $-2y = 12$, then $\frac{-2y}{-2} = \frac{12}{-2}$.

Example 4 Solve One-Step Equations

Solve each equation. Check your solution.

a. $n - 3.24 = 42.1$

$n - 3.24 = 42.1$	Original equation
$n - 3.24 + 3.24 = 42.1 + 3.24$	Add 3.24 to each side.
$n = 45.34$	Simplify.

The solution is 45.34.

CHECK	
$n - 3.24 = 42.1$	Original equation
$45.34 - 3.24 \stackrel{?}{=} 42.1$	Substitute 45.34 for n.
$42.1 = 42.1$ ✓	Simplify.

StudyTip

CCSS Regularity In Example 4b, notice that multiplying both sides of the equation by $-\frac{8}{5}$ is the same as dividing both sides by $-\frac{5}{8}$.

b. $-\frac{5}{8}x = 20$

$-\frac{5}{8}x = 20$	Original equation
$-\frac{8}{5}\left(-\frac{5}{8}\right)x = -\frac{8}{5}(20)$	Multiply each side by $-\frac{8}{5}$.
$x = -32$	Simplify.

The solution is −32.

CHECK	
$-\frac{5}{8}x = 20$	Original equation
$-\frac{5}{8}(-32) \stackrel{?}{=} 20$	Replace x with −32.
$20 = 20$ ✓	Simplify.

GuidedPractice

4A. $x - 14.29 = 25$

4B. $\frac{2}{3}y = -18$

To solve an equation with more than one operation, undo operations by working backward.

Example 5 Solve a Multi-Step Equation

Solve $5(x + 3) + 2(1 - x) = 14$.

$5(x + 3) + 2(1 - x) = 14$	Original equation
$5x + 15 + 2 - 2x = 14$	Apply the Distributive Property.
$3x + 17 = 14$	Simplify the left side.
$3x = -3$	Subtract 17 from each side.
$x = -1$	Divide each side by 3.

StudyTip

Checking Answers When solving for a variable, you can use substitution to check your answer by replacing the variable in the original equation with your answer.

GuidedPractice

Solve each equation.

5A. $-10x + 3(4x - 2) = 6$

5B. $2(2x - 1) - 4(3x + 1) = 2$

You can use properties to solve an equation for a variable.

PT

Example 6 Solve for a Variable

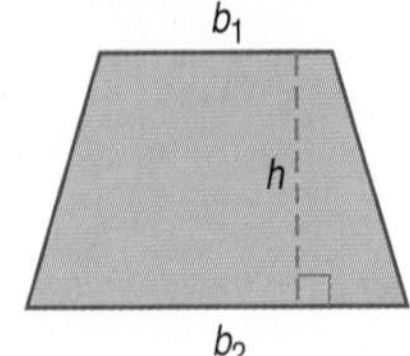

GEOMETRY The formula for the area A of a trapezoid is $A = \frac{1}{2}h(b_1 + b_2)$, where h represents the height, and b_1 and b_2 represent the measures of the bases. Solve the formula for b_2.

$A = \frac{1}{2}h(b_1 + b_2)$ Area formula

$2A = 2\left[\frac{1}{2}h(b_1 + b_2)\right]$ Multiply each side by 2.

$2A = h(b_1 + b_2)$ Simplify.

$\frac{2A}{h} = \frac{h(b_1 + b_2)}{h}$ Divide each side by h.

$\frac{2A}{h} = b_1 + b_2$ Simplify.

$\frac{2A}{h} - b_1 = b_1 + b_2 - b_1$ Subtract b_1 from each side.

$\frac{2A}{h} - b_1 = b_2$ Simplify.

GuidedPractice

6. The formula for the surface area S of a cylinder is $S = 2\pi r^2 + 2\pi rh$, where r is the radius of the base and h is the height of the cylinder. Solve the formula for h.

There are often many ways to solve a problem. Using the properties of equality can help you find a simpler way.

PT

Standardized Test Example 7 Use Properties of Equality

If $6x - 12 = 18$, what is the value of $6x + 5$?

A 5 **B** 11 **C** 35 **D** 41

Test-TakingTip

Read the Question Read the question carefully before solving the equation. In Example 7, you are to find the value of $6x + 5$, not the value of x.

Read the Test Item

You are asked to find the value of $6x + 5$. Note that you do not have to find the value of x. Instead, you can use the Addition Property of Equality to make the left side of the equation $6x + 5$.

Solve the Test Item

$6x - 12 = 18$ Original equation

$6x - 12 + 17 = 18 + 17$ Add 17 to each side because $-12 + 17 = 5$.

$6x + 5 = 35$ Simplify.

The answer is C.

GuidedPractice

7. If $5y + 2 = \frac{8}{3}$, what is the value of $5y - 6$?

F $\frac{-20}{3}$ **G** $\frac{-16}{3}$ **H** $\frac{16}{3}$ **J** $\frac{32}{3}$

Check Your Understanding

= Step-by-Step Solutions begin on page R14.

Example 1 **Write an algebraic expression to represent each verbal expression.**

1. the product of 12 and the sum of a number and negative 3
2. the difference between the product of 4 and a number and the square of the number

Example 2 **Write a verbal sentence to represent each equation.**

3. $5x + 7 = 18$
4. $x^2 - 9 = 27$
5. $5y - y^3 = 12$
6. $\frac{x}{4} + 8 = -16$

Example 3 **Name the property illustrated by each statement.**

7. $(8x - 3) + 12 = (8x - 3) + 12$
8. If $a = -3$ and $-3 = d$, then $a = d$.

Examples 4–5 CCSS PRECISION **Solve each equation. Check your solution.**

9. $z + 19 = 34$
10. $x + 13 = 7$
11. $-y = 8$
12. $-6x = 42$
13. $5x - 3 = -33$
14. $-6y - 8 = 16$
15. $3(2a + 3) - 4(3a - 6) = 15$
16. $5(c - 8) - 3(2c + 12) = -84$
17. $-3(-2x + 20) + 8(x + 12) = 92$
18. $-4(3m - 10) - 6(-7m - 6) = -74$

Example 6 **Solve each equation or formula for the specified variable.**

19. $8r - 5q = 3$, for q
20. $Pv = nrt$, for n

Example 7

21. **MULTIPLE CHOICE** If $\frac{y}{5} + 8 = 7$, what is the value of $\frac{y}{5} - 2$?

A -10 B -3 C 1 D 5

Practice and Problem Solving

Extra Practice is on page R1.

Example 1 **Write an algebraic expression to represent each verbal expression.**

22. the difference between the product of four and a number and 6
23. the product of the square of a number and 8
24. fifteen less than the cube of a number
25. five more than the quotient of a number and 4

Example 2 **Write a verbal sentence to represent each equation.**

26. $8x - 4 = 16$
27. $\frac{x + 3}{4} = 5$
28. $4y^2 - 3 = 13$

29. **BASEBALL** During a recent season, Miguel Cabrera and Mike Jacobs of the Florida Marlins hit a combined total of 46 home runs. Cabrera hit 6 more home runs than Jacobs. How many home runs did each player hit? Define a variable, write an equation, and solve the problem.

Example 3 **Name the property illustrated by each statement.**

30. If $x + 9 = 2$, then $x + 9 - 9 = 2 - 9$
31. If $y = -3$, then $7y = 7(-3)$
32. If $g = 3h$ and $3h = 16$, then $g = 16$
33. If $-y = 13$, then $-(-y) = -13$

34. MONEY Aiko and Kendra arrive at the state fair with $32.50. What is the total number of rides they can go on if they each pay the entrance fee?

Examples 4–5 **CCSS PRECISION** **Solve each equation. Check your solution.**

35. $3y + 4 = 19$

36. $-9x - 8 = 55$

37. $7y - 2y + 4 + 3y = -20$

38. $5g + 18 - 7g + 4g = 8$

39. $5(-2x - 4) - 3(4x + 5) = 97$

40. $-2(3y - 6) + 4(5y - 8) = 92$

41. $\frac{2}{3}(6c - 18) + \frac{3}{4}(8c + 32) = -18$

42. $\frac{3}{5}(15d + 20) - \frac{1}{6}(18d - 12) = 38$

43. GEOMETRY The perimeter of a regular pentagon is 100 inches. Find the length of each side.

44. MEDICINE For Nina's illness her doctor gives her a prescription for 28 pills. The doctor says that she should take 4 pills the first day and then 2 pills each day until her prescription runs out. For how many days does she take 2 pills?

Example 6 **Solve each equation or formula for the specified variable.**

45. $E = mc^2$, for m

46. $c(a + b) - d = f$, for a

47. $z = \pi q^3 h$, for h

48. $\frac{x + y}{z} - a = b$, for y

49. $y = ax^2 + bx + c$, for a

50. $wx + yz = bc$, for z

51. GEOMETRY The formula for the volume of a cylinder with radius r and height h is π times the radius times the radius times the height.

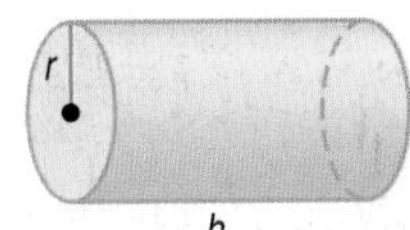

a. Write this as an algebraic expression.

b. Solve the expression in part **a** for h.

52. AWARDS BANQUET A banquet room can seat a maximum of 69 people. The coach, principal, and vice principal have invited the award-winning girls' tennis team to the banquet. If the tennis team consists of 22 girls, how many guests can each student bring?

CCSS PRECISION **Solve each equation. Check your solution.**

53. $5x - 9 = 11x + 3$

54. $\frac{1}{x} + \frac{1}{4} = \frac{7}{12}$

55. $5.4(3k - 12) + 3.2(2k + 6) = -136$

56. $8.2p - 33.4 = 1.7 - 3.5p$

57. $\frac{4}{9}y + 5 = -\frac{7}{9}y - 8$

58. $\frac{3}{4}z - \frac{1}{3} = \frac{2}{3}z + \frac{1}{5}$

59. FINANCIAL LITERACY Benjamin spent $10,734 on his living expenses last year. Most of these expenses are listed at the right. Benjamin's only other expense last year was rent. If he paid rent 12 times last year, how much was Benjamin's rent each month?

Expense	Annual Cost
Electric	$622
Gas	$428
Water	$240
Renter's Insurance	$144

60 CCSS **SENSE-MAKING** The Sunshine Skyway Bridge spans Tampa Bay, Florida. Suppose one crew began building south from St. Petersburg, and another crew began building north from Bradenton. The two crews met 10,560 feet south of St. Petersburg approximately 5 years after construction began.

a. Suppose the St. Petersburg crew built an average of 176 feet per month. Together the two crews built 21,120 feet of bridge. Determine the average number of feet built per month by the Bradenton crew.

b. About how many miles of bridge did each crew build?

c. Is this answer reasonable? Explain.

61 **MULTIPLE REPRESENTATIONS** The absolute value of a number describes the distance of the number from zero.

a. **Geometric** Draw a number line. Label the integers from −5 to 5.

b. **Tabular** Create a table of the integers on the number line and their distance from zero.

c. **Graphical** Make a graph of each integer x and its distance from zero y using the data points in the table.

d. **Verbal** Make a conjecture about the integer and its distance from zero. Explain the reason for any changes in sign.

H.O.T. Problems Use Higher-Order Thinking Skills

62. ERROR ANALYSIS Steven and Jade are solving $A = \frac{1}{2}h(b_1 + b_2)$ for b_2. Is either of them correct? Explain your reasoning.

Steven	Jade
$A = \frac{1}{2}h(b_1 + b_2)$	$A = \frac{1}{2}h(b_1 + b_2)$
$\frac{2A}{h} = (b_1 + b_2)$	$\frac{2A}{h} = (b_1 + b_2)$
$\frac{2A - b_1}{h} = b_2$	$\frac{2A}{h} - b_1 = b_2$

63. CHALLENGE Solve $d = \sqrt{(x_2 - x_1)^2 + (y_2 - y_1)^2}$ for y_1.

64. REASONING Use what you have learned in this lesson to explain why the following number trick works.

- Take any number.
- Multiply it by ten.
- Subtract 30 from the result.
- Divide the new result by 5.
- Add 6 to the result.
- Your new number is twice your original.

65. OPEN ENDED Provide one example of an equation involving the Distributive Property that has no solution and another example that has infinitely many solutions.

66. WRITING IN MATH Compare and contrast the Substitution Property of Equality and the Transitive Property of Equality.

Standardized Test Practice

67. The graph shows the solution of which inequality?

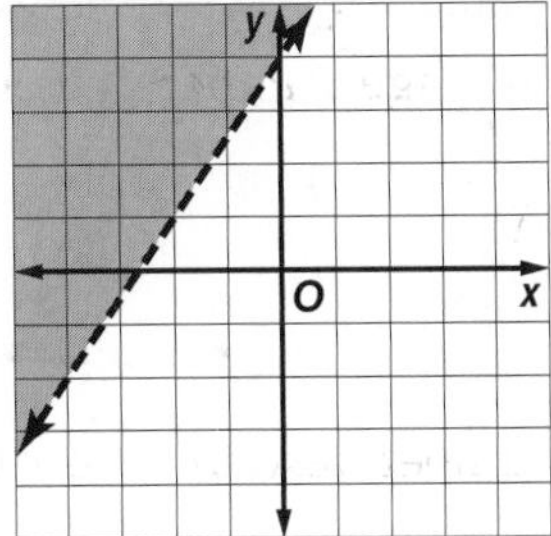

A $y < \frac{2}{3}x + 4$

B $y > \frac{2}{3}x + 4$

C $y < \frac{3}{2}x + 4$

D $y > \frac{3}{2}x + 4$

68. SAT/ACT What is $1\frac{1}{3}$ subtracted from its reciprocal?

F $-2\frac{2}{3}$

G $-\frac{7}{12}$

H $-\frac{1}{12}$

J $\frac{1}{4}$

K $\frac{3}{4}$

69. GEOMETRY Which of the following describes the transformation of $\triangle ABC$ to $\triangle A'B'C'$?

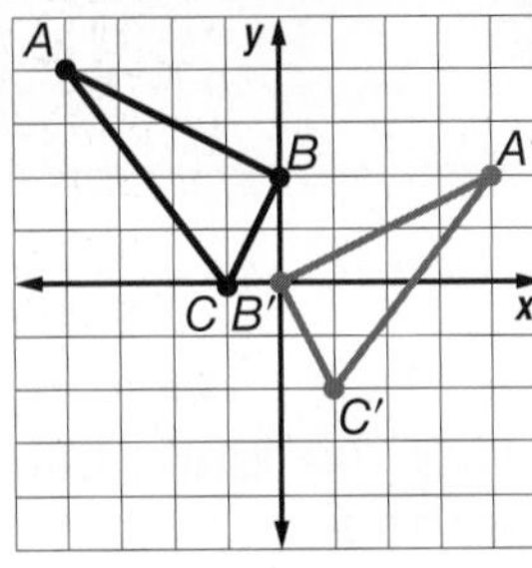

A a reflection across the y-axis and a translation down 2 units

B a reflection across the x-axis and a translation down 2 units

C a rotation 90° to the right and a translation down 2 units

D a rotation 90° to the right and a translation right 2 units

70. SHORT RESPONSE A local theater sold 1200 tickets during the opening weekend of a movie. On the following weekend, 840 tickets were sold. What was the percent decrease of tickets sold?

Spiral Review

71. Simplify $3x + 8y + 5z - 2y - 6x + z$. (Lesson 1-2)

72. BAKING Tamera is making two types of bread. The first type of bread needs $2\frac{1}{2}$ cups of flour, and the second needs $1\frac{3}{4}$ cups of flour. Tamera wants to make 2 loaves of the first recipe and 3 loaves of the second recipe. How many cups of flour does she need? (Lesson 1-2)

73. LANDMARKS Suppose the Space Needle in Seattle, Washington, casts a 220-foot shadow at the same time a nearby tourist casts a 2-foot shadow. If the tourist is $5\frac{1}{2}$ feet tall, how tall is the Space Needle? (Lesson 0-7)

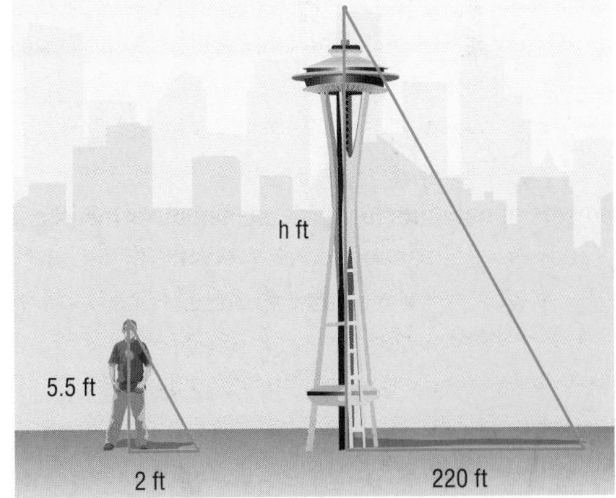

74. Evaluate $a - [c(b - a)]$, if $a = 5$, $b = 7$, and $c = 2$. (Lesson 1-1)

Skills Review

Identify the additive inverse for each number or expression.

75. $-4\frac{1}{5}$

76. 3.5

77. $-2x$

78. $6 - 7y$

79. $3\frac{2}{3}$

80. -1.25

81. $5x$

82. $4 - 9x$

CHAPTER 1

Mid-Chapter Quiz

Lessons 1-1 through 1-3

1. Evaluate $3c - 4(a + b)$ if $a = -1$, $b = 2$ and $c = \frac{1}{3}$. (Lesson 1-1)

2. **TRAVEL** The distance that Maurice traveled in 2.5 hours riding his bicycle can be found by using the formula $d = rt$, where d is the distance traveled, r is the rate, and t is the time. How far did Maurice travel if he traveled at a rate of 16 miles per hour? (Lesson 1-1)

3. Evaluate $(5 - m)^3 + n(m - n)$ if $m = 6$ and $n = -3$. (Lesson 1-1)

4. **GEOMETRY** The formula for the surface area of the rectangular prism below is given by the formula $S = 2xy + 2yz + 2xz$. What is the surface area of the prism if $x = 2.2$, $y = 3.5$, and $z = 5.1$? (Lesson 1-1)

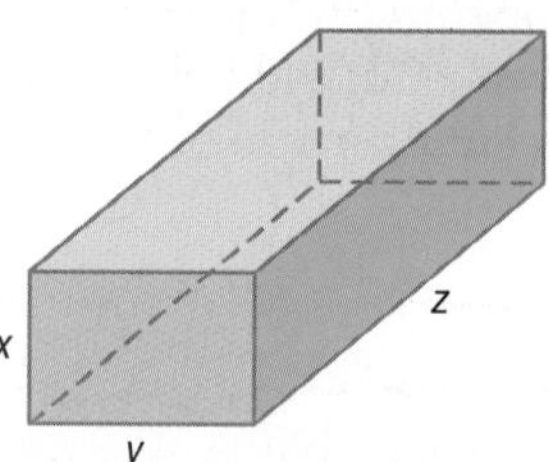

5. **MULTIPLE CHOICE** What is the value of $\frac{q^2 + rt}{qr - 2t}$ if $q = -4$, $r = 3$, and $t = 8$? (Lesson 1-1)

   A $-\frac{17}{6}$

   B $-\frac{10}{7}$

   C $-\frac{2}{7}$

   D $-\frac{1}{6}$

Name the sets of numbers to which each number belongs. (Lesson 1-2)

6. $\frac{25}{11}$

7. $-\frac{128}{32}$

8. $\sqrt{50}$

9. -32.4

10. What is the property illustrated by the equation $(4 + 15)7 = 4 \cdot 7 + 15 \cdot 7$? (Lesson 1-2)

11. Simplify $-3(7a - 4b) + 2(-3a + b)$. (Lesson 1-2)

12. **CLOTHES** Brittany is buying T-shirts and jeans for her new job. T-shirts cost \$10.50, and jeans cost \$26.50. She buys 3 T-shirts and 3 pairs of jeans. Illustrate the Distributive Property by writing two expressions representing how much Brittany spent. (Lesson 1-2)

13. **MULTIPLE CHOICE** Which expression is equivalent to $\frac{2}{3}(4m - 5n) + \frac{1}{5}(2m + n)$? (Lesson 1-2)

   F $\frac{46}{15}m - \frac{47}{15}n$

   G $46m - 47n$

   H $-\frac{mn}{15}$

   J $\frac{5}{4}m - \frac{9}{8}n$

14. Identify the additive inverse and the multiplicative inverse for $\frac{7}{6}$. (Lesson 1-2)

15. Write a verbal sentence to represent the equation $\frac{a}{a - 3} = 1$. (Lesson 1-3)

16. Solve $6x + 4y = -1$ for x. (Lesson 1-3)

17. **MULTIPLE CHOICE** Which algebraic expression represents the verbal expression, the product of 4 and the difference of a number and 13? (Lesson 1-3)

   A $4n - 13$

   B $4(n - 13)$

   C $\frac{4}{n - 13}$

   D $\frac{4n}{13}$

18. Solve $-3(6x + 5) + 2(4x) = 20$. (Lesson 1-3)

19. What is the height of the trapezoid below? (Lesson 1-3)

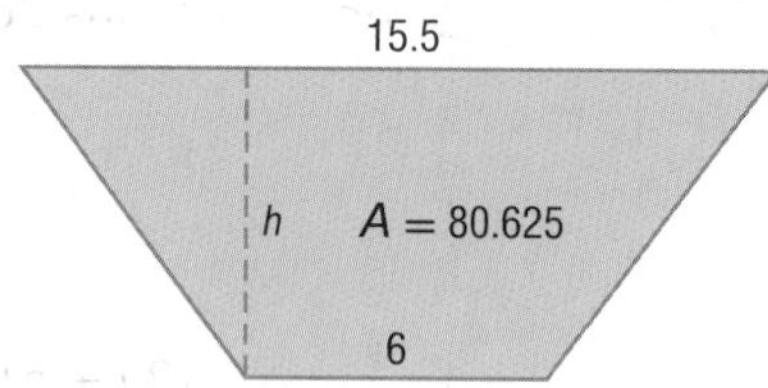

20. **GEOMETRY** The formula for the surface area of a sphere is $SA = 4\pi r^2$, and the formula for the volume of a sphere is $V = \frac{4}{3}\pi r^3$. (Lesson 1-3)

   a. Find the volume and surface area of a sphere with radius 2 inches. Write your answers in terms of π.

   b. Is it possible for a sphere to have the same numerical value for the surface area and volume? If so, find the radius of such a sphere.

LESSON 1-4

Solving Absolute Value Equations

Then

- You solved equations using properties of equality.

Now

1. Evaluate expressions involving absolute values.
2. Solve absolute value equations.

Why?

- Sailors sometimes use a laser range finder to determine distances. Suppose one such range finder is accurate to within ± 0.5 yard. This means that if a sailor estimating the distance to shore reads 323.1 yards on the laser range finder, the distance to shore might actually be as close as 322.6 or as far away as 323.6 yards. These extremes can be described by the equation $|E - 323.1| = 0.5$.

New Vocabulary

absolute value
empty set
constraint
extraneous solution

Common Core State Standards

Content Standards

A.SSE.1.b Interpret complicated expressions by viewing one or more of their parts as a single entity.

A.CED.1 Create equations and inequalities in one variable and use them to solve problems.

Mathematical Practices

6 Attend to precision.

1 Absolute Value Expressions

The **absolute value** of a number is its distance from 0 on a number line. Since distance is nonnegative, the absolute value of a number is always nonnegative. The symbol $|x|$ is used to represent the absolute value of a number x.

Key Concept Absolute Value

Words	For any real number a, if a is positive or zero, the absolute value of a is a. If a is negative, the absolute value of a is the opposite of a.
Symbols	For any real number a, $\lvert a\rvert = a$ if $a \geq 0$, and $\lvert a\rvert = -a$ if $a < 0$.
Model	$\lvert -4\rvert = 4$ and $\lvert 4\rvert = 4$

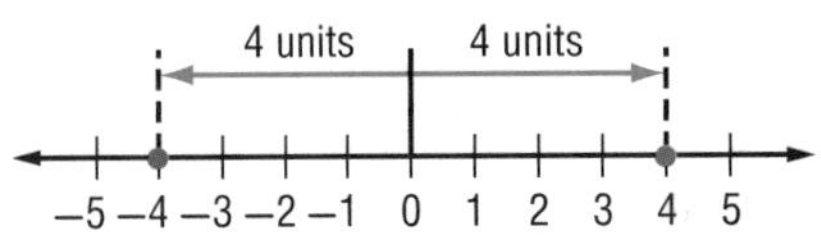

When evaluating expressions, absolute value bars act as a grouping symbol. Perform any operations inside the absolute value bars first.

Example 1 Evaluate an Expression with Absolute Value

Evaluate $8.4 - |2n + 5|$ if $n = -7.5$.

$8.4 - |2n + 5| = 8.4 - |2(-7.5) + 5|$ Replace n with -7.5.

$= 8.4 - |-15 + 5|$ Multiply 2 and -7.5.

$= 8.4 - |-10|$ Add -15 and 5.

$= 8.4 - 10$ $|-10| = 10$

$= -1.6$ Subtract 10 from 8.4.

Guided Practice

1A. Evaluate $|4x + 3| - 3\frac{1}{2}$ if $x = -2$.

1B. Evaluate $1\frac{1}{3} - |2y + 1|$ if $y = -\frac{2}{3}$.

OLIVIER MORIN/AFP/Getty Images

2 Absolute Value Equations

Some equations contain absolute value expressions. The definition of absolute value is used in solving these equations. For any real numbers a and b, where $b \geq 0$, if $|a| = b$, then $a = b$ or $-a = b$. This second case is often written as $a = -b$.

Real-World Example 2 Solve an Absolute Value Equation

TENNIS A standard adult tennis racket has a 100-square-inch head, plus or minus 20 square inches. Write and solve an absolute value equation to determine the least and greatest possible sizes for the head of an adult tennis racket.

Understand We need to determine the greatest and least possible sizes for the head of a tennis racket given the middle size and the range in sizes.

Plan When writing an absolute value equation, the middle or *central* value is always placed inside the absolute value symbols. The *range* is always placed on the other side of the equality symbol.

central value → $|x - c| = r$ ← range

Solve

$|x - c| = r$ Absolute value equation

$|x - 100| = 20$ $c = 100$, and $r = 20$

Case 1 $a = b$

$x - 100 = 20$

$x - 100 + 100 = 20 + 100$

$x = 120$

Case 2 $a = -b$

$x - 100 = -20$

$x - 100 + 100 = -20 + 100$

$x = 80$

Check

$|x - 100| = 20$

$|120 - 100| \stackrel{?}{=} 20$

$|20| \stackrel{?}{=} 20$

$20 = 20$ ✓

$|x - 100| = 20$

$|80 - 100| \stackrel{?}{=} 20$

$|-20| \stackrel{?}{=} 20$

$20 = 20$ ✓

On a number line, you can see that both solutions are 20 units away from 100.

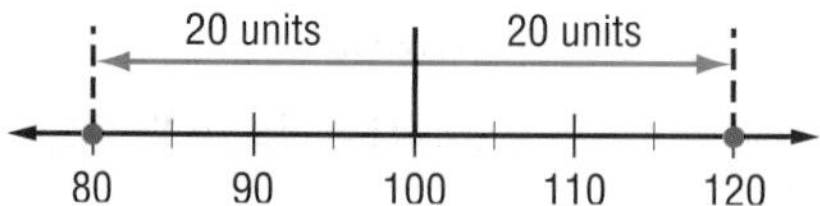

The solutions are 120 and 80. The greatest size is 120 square inches and the least is 80 square inches.

Real-WorldLink

Originally, players used leather gloves to hit tennis balls. Soon after, the glove was placed at the end of a stick to extend the reach of the "hand."

Source: The Cliff Richard Tennis Foundation

Problem-SolvingTip

Write an Equation Frequently, the best way to solve a problem is to use the given information to write and solve an equation.

GuidedPractice

Solve each equation. Check your solutions.

2A. $9 = |x + 12|$

2B. $8 = |y + 5|$

Because the absolute value of a number is always positive or zero, an equation like $|x| = -4$ is never true. Thus, it has no solution. The solution set for this type of equation is the **empty set**, symbolized by { } or ∅.

Example 3 No Solution

Solve $|3x - 2| + 8 = 1$.

$$|3x - 2| + 8 = 1 \quad \text{Original equation}$$

$$|3x - 2| + 8 - 8 = 1 - 8 \quad \text{Subtract 8 from each side.}$$

$$|3x - 2| = -7 \quad \text{Simplify.}$$

This sentence is *never* true. The solution set is $\varnothing$.

GuidedPractice **Solve each equation. Check your solutions.**

3A. $-2|3a| = 6$

3B. $|4b + 1| + 8 = 0$

In mathematics, a **constraint** is a condition that a solution must satisfy. Equations can be viewed as constraints in a problem situation. The solutions of the equation meet the constraints of the problem.

Even if the correct procedure for solving the equation is used, the answers may not be actual solutions to the original equation. Such a number is called an **extraneous solution**.

StudyTip

CCSS Precision It is possible for an absolute value equation to have only one solution. Remember to set up two cases. Then check your solutions.

Example 4 One Solution

Solve $|x + 10| = 4x - 8$. Check your solutions.

Case 1

$$a = b$$
$$x + 10 = 4x - 8$$
$$10 = 3x - 8$$
$$18 = 3x$$
$$6 = x$$

Case 2

$$a = -b$$
$$x + 10 = -(4x - 8)$$
$$x + 10 = -4x + 8$$
$$5x + 10 = 8$$
$$5x = -2$$
$$x = -\frac{2}{5}$$

There appear to be two solutions, 6 and $-\frac{2}{5}$.

CHECK Substitute each value in the original equation.

$$|x + 10| = 4x - 8$$
$$|6 + 10| \stackrel{?}{=} 4(6) - 8$$
$$|16| \stackrel{?}{=} 24 - 8$$
$$16 = 16 \checkmark$$

$$|x + 10| = 4x - 8$$
$$\left|-\frac{2}{5} + 10\right| \stackrel{?}{=} 4\left(-\frac{2}{5}\right) - 8$$
$$\left|9\frac{3}{5}\right| \stackrel{?}{=} -1\frac{3}{5} - 8$$
$$9\frac{3}{5} \neq -9\frac{3}{5} \ \times$$

Because $9\frac{3}{5} \neq -9\frac{3}{5}$, the only solution is 6. The solution set is $\{6\}$.

GuidedPractice **Solve each equation. Check your solutions.**

4A. $2|x + 1| - x = 3x - 4$

4B. $3|2x + 2| - 2x = x + 3$

Check Your Understanding

= Step-by-Step Solutions begin on page R14.

Example 1 **Evaluate each expression if $x = -4$ and $y = -9$.**

1. $|x - 8|$ **2.** $|7y|$ **3.** $-3|xy|$ **4.** $-2|3x + 8| - 4$

5. **CCSS MODELING** Most freshwater tropical fish thrive if the water is within 2°F of 78°F.

a. Write an equation to determine the least and greatest optimal temperatures.

b. Solve the equation you wrote in part **a**.

c. If your aquarium's thermometer is accurate to within plus or minus 1°F, what should the temperature of the water be to ensure that it reaches the minimum temperature? Explain.

Examples 2–4 **Solve each equation. Check your solutions.**

6. $|x + 8| = 12$ **7.** $|y - 4| = 11$

8. $|a - 5| + 4 = 9$ **9.** $|b - 3| + 8 = 3$

10. $3|2x - 3| - 5 = 4$ **11.** $-2|5y - 1| = -10$

12. $|a - 4| = 3a - 6$ **13.** $|b + 5| = 2b + 3$

Practice and Problem Solving

Extra Practice is on page R1.

Example 1 **Evaluate each expression if $a = -3$, $b = -5$, and $c = 4.2$.**

14. $|-3c|$ **15.** $|5b|$ **16.** $|a - b|$ **17.** $|b - c|$

18. $|3b - 4a|$ **19.** $2|4a - 3c|$ **20.** $-|3c - a|$ **21.** $-|abc|$

22. **FOOD** To make cocoa powder, cocoa beans are roasted. The ideal temperature for roasting is 300°F, plus or minus 25°. Write and solve an equation describing the maximum and minimum roasting temperatures for cocoa beans.

Examples 2–4 **Solve each equation. Check your solutions.**

23. $|z - 13| = 21$ **24.** $|w + 9| = 17$

25. $9 = |d + 5|$ **26.** $35 = |x - 6|$

27. $5|q + 6| = 20$ **28.** $-3|r + 4| = -21$

29. $3|2a - 4| = 0$ **30.** $8|5w - 1| = 0$

31. $2|3x - 4| + 8 = 6$ **32.** $4|7y + 2| - 8 = -7$

33. $-3|3t - 2| - 12 = -6$ **34.** $-5|3z + 8| - 5 = -20$

35. **MONEY** The U.S. Mint produces quarters that weigh about 5.67 grams each. After the quarters are produced, a machine weighs them. If the quarter weighs 0.02 gram more or less than the desired weight, the quarter is rejected. Write and solve an equation to find the heaviest and lightest quarters the machine will approve.

Evaluate each expression if $q = -8$, $r = -6$, and $t = 3$.

36. $12 - t|3r + 2|$ **37.** $2q + |2rt + q|$ **38.** $-5t - q|8r - t|$

Solve each equation. Check your solutions.

39. $8x = 2|6x - 2|$

40. $-6y + 4 = |4y + 12|$

41. $8z + 20 = -|2z + 4|$

42. $-3y - 2 = |6y + 25|$

43 **SEA LEVEL** Florida is on average 100 feet above sea level. This level varies by as much as 245 feet depending on precipitation and your location. Write and solve an equation describing the maximum and minimum sea levels for Florida. Is this solution reasonable? Explain.

44. **MULTIPLE REPRESENTATIONS** Draw a number line.

a. Geometric Label any 5 integers on the number line points *A*, *B*, *C*, *D*, and *F*.

b. Tabular Fill in each blank in the table with either > or < using the points from the number line.

A ___ B	A + C ___ B + C A + D ___ B + D A + F ___ B + F
B ___ A	B + C ___ A + C B + D ___ A + D B + F ___ A + F

A ___ B	A − C ___ B − C A − D ___ B − D A − F ___ B − F
B ___ A	B − C ___ A − C B − D ___ A − D B − F ___ A − F

c. Verbal Describe the patterns in the table.

d. Algebraic Describe the patterns algebraically, using the variable *x* to replace *C*, *D*, and *F*.

H.O.T. Problems Use Higher-Order Thinking Skills

45. **CCSS CRITIQUE** Ana and Ling are solving $|3x + 14| = -6x$. Is either of them correct? Explain your reasoning.

Ana

$|3x + 14| = -6x$

$3x + 14 = -6x$ or $3x + 14 = 6x$

$9x = -14$ $\quad$ $14 = 3x$

$x = -\frac{14}{9}$ ✓ $\quad$ $x = \frac{14}{3}$ ✓

Ling

$|3x + 14| = -6x$

$3x + 14 = -6x$ or $3x + 14 = 6x$

$9x = -14$ $\quad$ $14 = 3x$

$x = -\frac{14}{9}$ ✓ $\quad$ $x = \frac{14}{3}$ X

46. **CHALLENGE** Solve $|2x - 1| + 3 = |5 - x|$. List all cases and resulting equations. (*Hint*: There are four possible cases to examine as potential solutions.)

REASONING **If *a*, *x*, and *y* are real numbers, determine whether each statement is *sometimes*, *always*, or *never* true. Explain your reasoning.**

47. If $|a| > 7$, then $|a + 3| > 10$.

48. If $|x| < 3$, then $|x| + 3 > 0$.

49. If y is between 1 and 5, then $|y - 3| \leq 2$.

50. **OPEN ENDED** Write an absolute value equation of the form $|ax + b| = cx + d$ that has no solution. Assume that a, b, c, and $d \neq 0$.

51. **WRITING IN MATH** How are symbols used to represent mathematical ideas? Use an example to justify your reasoning.

Standardized Test Practice

52. If $4x - y = 3$ and $2x + 3y = 19$, what is the value of y?

A 2
B 3
C 4
D 5

53. GRIDDED RESPONSE Two male and 2 female students from each of the 9th, 10th, 11th, and 12th grades comprise the Student Council. If a Student Council representative is chosen at random to attend a board meeting, what is the probability that the student will be either an 11th grader or male?

54. Which equation is equivalent to $4(9 - 3x) = 7 - 2(6 - 5x)$?

F $8x = 41$
G $22x = 41$
H $8x = 24$
J $22x = 24$

55. SAT/ACT A square with side length 4 units has one vertex at the point (1, 2). Which of the following points *cannot* be diagonally opposite that vertex?

A $(-3, -2)$
B $(-3, 6)$
C $(5, -2)$
D $(5, 6)$
E $(1, 6)$

Spiral Review

Solve each equation. Check your solution. (Lesson 1-3)

56. $4x + 6 = 30$

57. $5p - 10 = 4(7 + 6p)$

58. $\frac{3}{5}y - 7 = \frac{2}{5}y + 3$

59. MONEY Nhu is saving to buy a car. In the first 6 months, his savings were \$80 less than $\frac{3}{4}$ the price of the car. In the second six months, Nhu saved \$50 more than $\frac{1}{5}$ the price of the car. He still needs \$370. (Lesson 1-3)

a. What is the price of the car?

b. What is the average amount of money Nhu saved each month?

c. If Nhu continues to save the average amount each month, in how many months will he be able to afford the car?

Name the property illustrated by each equation. (Lesson 1-2)

60. $(1 + 8) + 11 = 11 + (1 + 8)$

61. $z(9 - 4) = z \cdot 9 - z \cdot 4$

Simplify each expression. (Lesson 1-2)

62. $7a + 3b - 4a - 5b$

63. $3x + 5y + 7x - 3y$

64. $3(15x - 9y) + 5(4y - x)$

65. $2(10m - 7a) + 3(8a - 3m)$

66. $8(r + 7t) - 4(13t + 5r)$

67. $4(14c - 10d) - 6(d + 4c)$

68. GEOMETRY The formula for the surface area of a rectangular prism is $SA = 2\ell w + 2\ell h + 2wh$, where ℓ represents the length, w represents the width, and h represents the height. Find the surface area of the rectangular prism at the right. (Lesson 1-1)

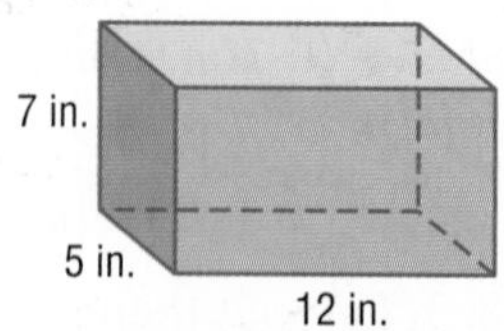

Skills Review

Solve each equation.

69. $15x + 5 = 35$

70. $2.4y + 4.6 = 20$

71. $8a + 9 = 6a - 7$

72. $3(w - 1) = 2w - 6$

73. $\frac{1}{2}(2b - 4) = 2 + 8b$

74. $\frac{1}{3}(6p - 24) = 18 + 3p$

LESSON 1-5 Solving Inequalities

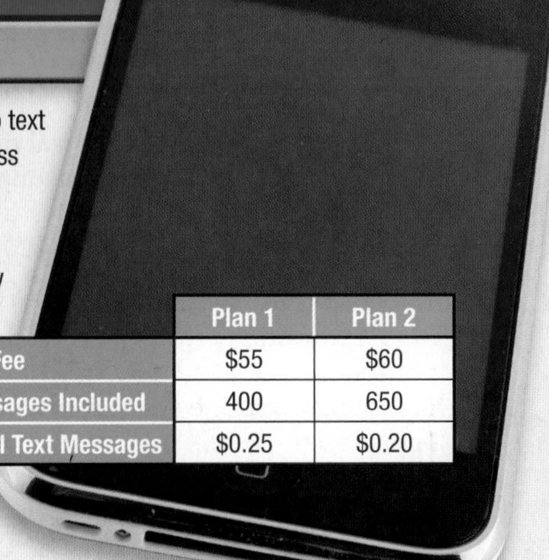

Then	Now	Why?
You solved equations involving absolute values.	**1** Solve one-step inequalities. **2** Solve multi-step inequalities.	Josh is trying to decide between two text messaging plans offered by a wireless telephone company. To compare these two rate plans, we can use inequalities. The monthly access fee for Plan 1 is less than the fee for Plan 2, $55 < $60. However, the additional text messaging fee for Plan 1 is greater than that of Plan 2, $0.25 > $0.20.

	Plan 1	Plan 2
Monthly Fee	$55	$60
Text Messages Included	400	650
Additional Text Messages	$0.25	$0.20

NewVocabulary
set-builder notation

Common Core State Standards

Content Standards

A.CED.1 Create equations and inequalities in one variable and use them to solve problems.

A.CED.3 Represent constraints by equations or inequalities, and by systems of equations and/or inequalities, and interpret solutions as viable or nonviable options in a modeling context.

Mathematical Practices

4 Model with mathematics.

1 One-Step Inequalities

For any two real numbers, a and b, exactly one of the following statements is true.

$$a < b \qquad a = b \qquad a > b$$

Adding the same number to, or subtracting the same number from, each side of an inequality does not change the truth of the inequality.

KeyConcept

Addition Property of Inequality

Words

For any real numbers, a, b, and c:

If $a > b$, then $a + c > b + c$.

If $a < b$, then $a + c < b + c$.

Models

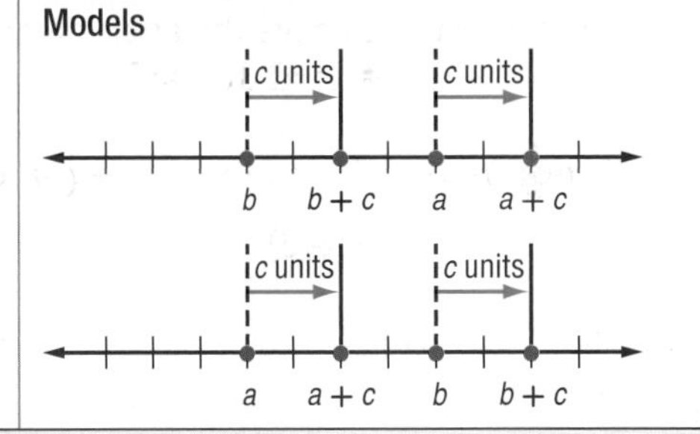

Subtraction Property of Inequality

Words

For any real numbers, a, b, and c:

If $a > b$, then $a - c > b - c$.

If $a < b$, then $a - c < b - c$.

Models

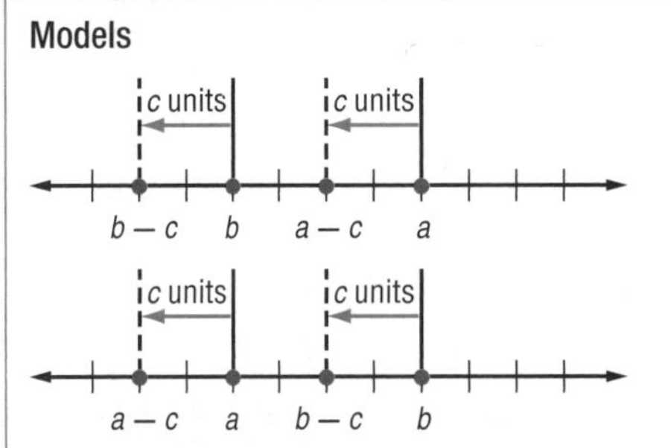

These properties are also true for $\leq$, $\geq$, and $\neq$.

These properties can be used to solve inequalities. The solution sets of inequalities in one variable can then be graphed on number lines.

Review Vocabulary

Inequality Symbols

- $>$ greater than; is more than
- $<$ less than; is fewer than
- $\geq$ greater than or equal to; is at least; is no less than
- $\leq$ less than or equal to; is at most; is no more than

Example 1 Solve an Inequality Using Addition or Subtraction

Solve $y - 6 < 3$. Graph the solution set on a number line.

$y - 6 < 3$ Original inequality

$y - 6 + 6 < 3 + 6$ Add 6 to each side.

$y < 9$ Simplify.

Any real number less than 9 is a solution of this inequality. The graph of the solution set is shown at the right.

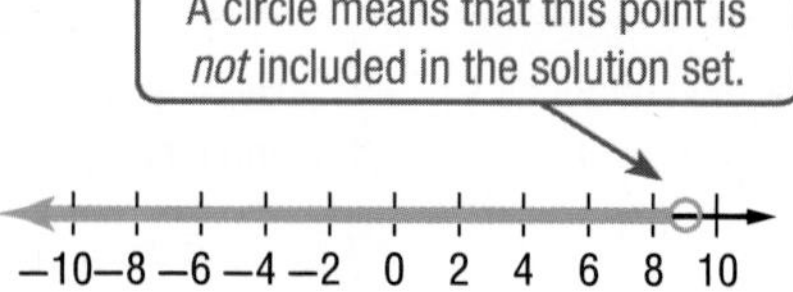

CHECK Substitute 8 and then 10 for y in $y - 6 < 3$. The inequality should be true for $y = 8$ and false for $y = 10$. ✓

Guided Practice

Solve each inequality. Graph the solution set on a number line.

1A. $5w + 3 > 4w + 9$

1B. $5x - 3 > 4x + 2$

Study Tip

Graphing Inequalities A circle is used for $<$ and $>$. A dot is used for $\leq$ and $\geq$.

Multiplying or dividing each side of an inequality by a positive number does not change the truth of the inequality. However, multiplying or dividing each side of an inequality by a *negative* number requires that the order of the inequality be *reversed*. For example, to reverse $\leq$, replace it with $\geq$.

Key Concept

Multiplication Property of Inequality

Words	Examples
For any real numbers, a, b, and c,	$-5 < -3$
where c is positive:	$-5(6) < -3(6)$
If $a > b$, then $ac > bc$.	$-30 < -18$
If $a < b$, then $ac < bc$.	
where c is negative:	$12 > -7$
If $a > b$, then $ac < bc$.	$12(-4) < -7(-4)$
If $a < b$, then $ac > bc$.	$-48 < 28$

Division Property of Inequality

Words	Examples
For any real numbers, a, b, and c,	$-12 < -8$
where c is positive:	$\frac{-12}{4} < \frac{-8}{4}$
If $a > b$, then $\frac{a}{c} > \frac{b}{c}$.	$-3 < -2$
If $a < b$, then $\frac{a}{c} < \frac{b}{c}$.	
where c is negative:	$-21 > -14$
If $a > b$, then $\frac{a}{c} < \frac{b}{c}$.	$\frac{-21}{-7} < \frac{-14}{-7}$
If $a < b$, then $\frac{a}{c} > \frac{b}{c}$.	$3 > 2$

These properties are also true for $\leq$, $\geq$, and $\neq$.

Reading Math

Set-Builder Notation

$\{y \mid y < 9\}$ is read *the set of all numbers y such that y is less than 9.*

The solution set of an inequality can be expressed by using **set-builder notation**. For example, the solution set in Example 1 can be expressed as $\{y \mid y < 9\}$.

Example 2 Solve an Inequality Using Multiplication or Division

Solve $-4.2x \leq -29.4$. Graph the solution set on a number line.

$-4.2x \leq -29.4$	Original inequality
$\frac{-4.2x}{-4.2} \geq \frac{-29.4}{-4.2}$	Divide each side by −4.2, reversing the inequality symbol.
$x \geq 7$	Simplify.

The solution set is $\{x \mid x \geq 7\}$. The graph of the solution is shown below.

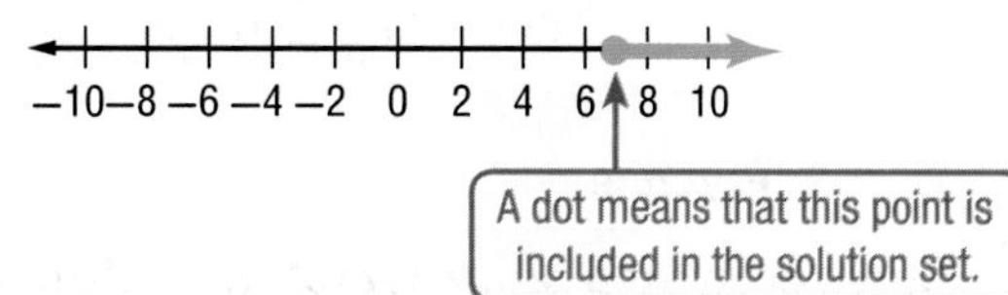

CHECK Substitute 6 and then 8 for x in $-4.2x \leq -29.4$. The inequality should be true for $x = 8$ and false for $x = 6$. ✓

Guided Practice

Solve each inequality. Graph the solution set on a number line.

2A. $-4x \geq -24$

2B. $-9.2y < 23$

2 Multi-Step Inequalities

Solving multi-step inequalities is similar to solving multi-step equations.

Example 3 Solve Multi-Step Inequalities

Solve $-4c \leq \frac{5c + 58}{6}$. Graph the solution set on a number line.

$-4c \leq \frac{5c + 58}{6}$	Original inequality
$-24c \leq 5c + 58$	Multiply each side by 6.
$-29c \leq 58$	Add −5c to each side.
$c \geq -2$	Divide each side by −29, reversing the inequality symbol.

Study Tip

When multiplying or dividing by a negative number, remember to reverse the inequality symbol.

The solution set is $\{c \mid c \geq -2\}$ and is graphed below.

−5 −4 −3 −2 −1 0 1 2 3 4 5

CHECK Substitute −3 and then −1 for x in $-4c \leq \frac{5c + 58}{6}$. The inequality should be true for $x = -1$ and false for $x = -3$. ✓

Guided Practice

Solve each inequality. Graph the solution set on a number line.

3A. $-3x \leq \frac{-4x + 22}{5}$

3B. $8y \geq \frac{-5y + 9}{-4}$

3C. $-6(-4v + 3) \leq 2(10v + 3)$

3D. $-5(3d - 7) > 3(2d + 14)$

Real-WorldLink

In 2007, the Netcraft Web Server Survey found over 108,000,000 distinct Web sites.

Source: Netcraft

Real-World Example 4 Write and Solve an Inequality

WEB SITES Enrique's company pays Salim to advertise on Salim's Web site. Salim's Web site earns \$15 per month plus \$0.05 every time a visitor clicks on the advertisement. What is the least number of clicks per month that Salim needs in order to earn \$50 per month or more?

Understand Let $c =$ the number of clicks on the advertisement. Salim earns \$15 per month and \$0.05 per click, and he wants to earn a minimum of \$50 for the advertisement.

Plan Write an inequality.

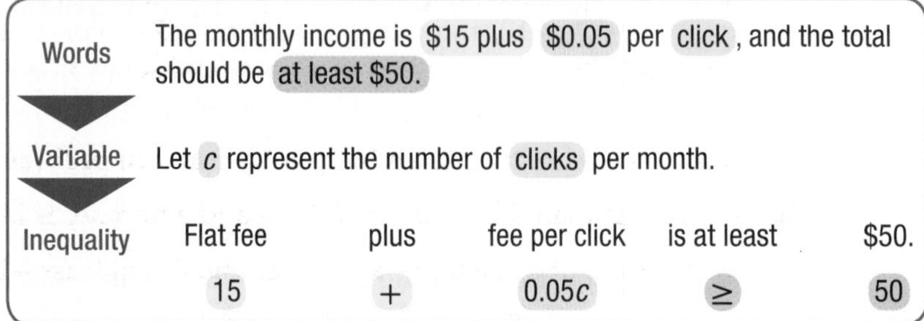

Solve

$15 + 0.05c \geq 50$	Original inequality
$0.05c \geq 35$	Subtract 15 from each side.
$c \geq 700$	Divide each side by 0.05.

Check

$15 + 0.05c \geq 50$	Original inequality
$5 + 0.05(700) \overset{?}{\geq} 50$	Replace c with 700.
$15 + 35 \overset{?}{\geq} 50$	Multiply.
$50 \geq 50$ ✓	Add.

Visitors to Salim's Web site need to click on Enrique's advertisement at least 700 times per month in order for Salim to earn \$50 or more from Enrique's company.

GuidedPractice

4. Rosa's cell phone plan costs her \$50 per month plus \$0.25 for each minute she goes beyond her free minutes. How many minutes can she go beyond her free minutes and still pay less than a total of \$70?

Check Your Understanding

 = Step-by-Step Solutions begin on page R14. 

Examples 1–3 **Solve each inequality. Then graph the solution set on a number line.**

1. $b + 6 < 14$

2. $12 - d > -8$

3. $18 \leq -3x$

4. $-5y \geq -35$

5. $-4w - 13 > -21$

6. $8z - 9 \geq -15$

7. $s \geq \frac{s + 6}{5}$

8. $\frac{2x - 9}{4} \leq x + 2$

Example 4

9. **CCSS MODELING** Tara is delivering bags of mulch. Each bag weighs 48 pounds, and the push cart weighs 65 pounds. If her flat-bed truck is capable of hauling 2000 pounds, how many bags of mulch can Tara safely take on each trip?

Ed Kashi/Terra/CORBIS

Practice and Problem Solving

Extra Practice is on page R1.

Examples 1–3 **Solve each inequality. Then graph the solution set on a number line.**

10. $m - 8 > -12$

11. $n + 6 \leq 3$

12. $6r < -36$

13. $-12t \geq -6$

14. $-\frac{w}{4} \leq -7$

15. $\frac{k}{3} - 14 < -5$

16. $4x - 15 \leq 21$

17. $-6z - 14 > -32$

18. $-16 \geq 5(2z - 11)$

19. $12 < -4(3c - 6)$

20. $\frac{3y - 4}{0.2} - 8 > 12$

21. $\frac{9z + 5}{4} + 18 < 26$

Example 4

22. GYMNASTICS In a gymnastics competition, an athlete's final score is calculated by taking 75% of the average technical score and adding 25% of the artistic score. All scores are out of 10, and one gymnast has a 7.6 average technical score. What artistic score does the gymnast need to have a final score of at least 8.0?

Define a variable and write an inequality for each problem. Then solve.

23. Twelve less than the product of three and a number is less than 21.

24. The quotient of three times a number and 4 is at least −16.

25. The difference of 5 times a number and 6 is greater than the number.

26. The quotient of the sum of 3 and a number and 6 is less than −2.

27. HIKING Danielle can hike 3 miles in an hour, but she has to take a one-hour break for lunch and a one-hour break for dinner. If Danielle wants to hike at least 18 miles, solve $3(x - 2) \geq 18$ to determine how many hours the hike should take.

Solve each inequality. Then graph the solution set on a number line.

28. $18 - 3x < 12$

29. $-8(4x + 6) < -24$

30. $\frac{1}{4}n + 12 \geq \frac{3}{4}n - 4$

31. $0.24y - 0.64 > 3.86$

32. $10x - 6 \leq 4x + 42$

33. $-6v + 8 > -14v - 28$

34. $n > \frac{-3n - 15}{8}$

35. $-2r < \frac{6 - 2r}{5}$

36. $\frac{9z - 4}{5} \leq \frac{7z + 2}{4}$

37. MONEY Jin is selling advertising space in *Central City Magazine* to local businesses. Jin earns 3% commission for every advertisement he sells plus a salary of $250 a week. If the average amount of money that a business spends on an advertisement is $500, how many advertisements must he sell each week to make a salary of at least $700 that week?

a. Write an inequality to describe this situation.

b. Solve the inequality and interpret the solution.

Define a variable and write an inequality for each problem. Then solve.

38. One third of the sum of 5 times a number and 3 is less than one fourth the sum of six times that number and 5.

39. The sum of one third a number and 4 is at most the sum of twice that number and 12.

40. CCSS **SENSE-MAKING** The sides of square *ABCD* are extended to form rectangle *DEFG*. If the perimeter of the rectangle is at least twice the perimeter of the square, what is the maximum length of a side of square *ABCD*?

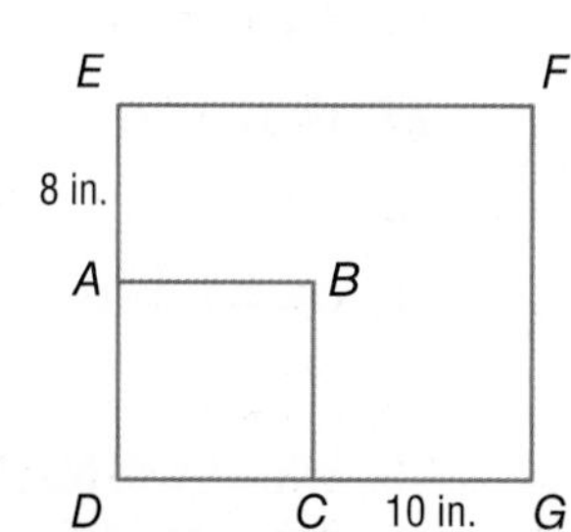

41 **MARATHONS** Jamie wants to be able to run at least the standard marathon distance of 26.2 miles. A good rule for training is that runners generally have enough endurance to finish a race that is up to 3 times his or her average daily distance.

a. If the length of her current daily run is 5 miles, write an inequality to find the amount by which she needs to increase her daily run to have enough endurance to finish a marathon.

b. Solve the inequality and interpret the solution.

42. **CCSS MODELING** The costs for renting a car from Ace Car Rental and from Basic Car Rental are shown in the table. For what mileage does Basic have the better deal? Use the inequality $38 + 0.1x > 42 + 0.05x$. Explain why this inequality works.

Rental Car Costs		
Company	Cost per Day	Cost per Mile
Ace	\$38	\$0.10
Basic	\$42	\$0.05

43. **MULTIPLE REPRESENTATIONS** In this exercise, you will explore graphing inequalities on a coordinate plane.

a. **Tabular** Organize the following into a table. Substitute 5 points into the inequality $y \geq -\frac{1}{2}x + 3$. State whether the resulting statement is *true* or *false*.

b. **Graphical** Graph $y = -\frac{1}{2}x + 3$. Also graph the 5 points from the table. Label all points that resulted in a true statement with a T. Label all points that resulted in a false statement with an F.

c. **Verbal** Describe the pattern produced by the points you have labeled. Make a conjecture about which points on the coordinate plane would result in true and false statements.

H.O.T. Problems Use Higher-Order Thinking Skills

44. **CHALLENGE** If $-4 < x < 5$ and $0.25 < y < 4$, then $a < \frac{x}{y} < b$. What is $a + b$?

45. **ERROR ANALYSIS** Madlynn and Emilie were comparing their homework. Is either of them correct? Explain your reasoning.

Madlynn

$$\frac{4x+5}{-2} - 1 > -3$$
$$\frac{4x+5}{-2} < -2$$
$$4x + 5 > 4$$
$$4x > -1$$
$$x > -\frac{1}{4}$$

Emilie

$$\frac{4x+5}{-2} - 1 > -3$$
$$\frac{4x+5}{-2} > -2$$
$$4x + 5 > 4$$
$$4x > -1$$
$$x > -\frac{1}{4}$$

46. **REASONING** Determine whether the following statement is *sometimes*, *always*, or *never* true. Explain your reasoning.

The opposite of the absolute value of a negative number is less than the opposite of that number.

47. **CHALLENGE** Given $\triangle ABC$ with sides $AB = 3x + 4$, $BC = 2x + 5$, and $AC = 4x$, determine the values of x such that $\triangle ABC$ exists.

48. **OPEN ENDED** Write an inequality for which the solution is all real numbers in the form $ax + b > c(x + d)$. Explain how you know this.

49. **WRITING IN MATH** Why does the inequality symbol need to be reversed when multiplying or dividing by a negative number?

Standardized Test Practice

50. SHORT RESPONSE Rogelio found a cookie recipe that requires $\frac{3}{4}$ cup of sugar and 2 cups of flour. How many cups of sugar would he need if he used 6 cups of flour?

51. STATISTICS The mean score for Samantha's first six algebra quizzes was 88. If she scored a 95 on her next quiz, what will her mean score be for all 7 quizzes?

A 89
B 90
C 91
D 92

52. SAT/ACT The average of five numbers is 9. The average of 7 other numbers is 8. What is the average of all 12 numbers?

F $8\frac{5}{12}$
G $8\frac{1}{2}$
H $8\frac{7}{12}$
J $8\frac{3}{4}$
K $8\frac{11}{12}$

53. What is the complete solution of the equation $|8 - 4x| = 40$?

A $x = 8; x = 12$
B $x = 8; x = -12$
C $x = -8; x = -12$
D $x = -8; x = 12$

Spiral Review

Solve each equation. Check your solutions. (Lesson 1-4)

54. $|x - 5| = 12$

55. $7|3y - 4| = 35$

56. $|a + 6| = a$

57. ASTRONOMY Pluto travels in a path that is not circular. Pluto's farthest distance from the Sun is 4539 million miles, and its closest distance is 2756 million miles. Write an equation that can be solved to find the minimum and maximum distances from the Sun to Pluto. (Lesson 1-4)

58. POPULATION In 2005, the population of Bay City was 19,611. For each of the next five years, the population decreased by an average of 715 people per year. (Lesson 1-3)

a. What was the population in 2010?

b. If the population continues to decline at the same rate as from 2005 to 2010, what would you expect the population to be in 2025?

59. GEOMETRY The formula for the surface area of a cylinder is $SA = 2\pi r^2 + 2\pi rh$. (Lesson 1-2)

a. Use the Distributive Property to rewrite the formula by factoring out the greatest common factor of the two terms.

b. Find the surface area for a cylinder with radius 3 centimeters and height 10 centimeters using both formulas. Leave the answer in terms of π.

c. Which formula do you prefer? Explain your reasoning.

60. CONSTRUCTION The Sawyers are adding a family room to their house. The dimensions of the room are 26 feet by 28 feet. Show how to use the Distributive Property to mentally calculate the area of the room. (Lesson 1-2)

Skills Review

Solve each equation. Check your solutions.

61. $|x| = 9$

62. $|x + 3| = 10$

63. $|4y - 15| = 13$

64. $18 = |3x - 9|$

65. $16 = 4|w + 2|$

66. $|y + 3| + 4 = 20$

EXPLORE

1-6 Algebra Lab
Interval Notation

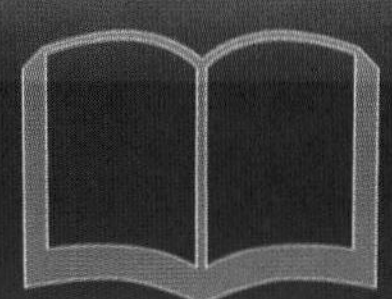

The solution set of an inequality can be described by using **interval notation**. The **infinity** symbols below are used to indicate that a set is unbounded in the positive or negative direction, respectively.

Common Core State Standards
Content Standards
A.CED.1 Create equations and inequalities in one variable and use them to solve problems.
A.CED.3 Represent constraints by equations or inequalities, and by systems of equations and/or inequalities, and interpret solutions as viable or nonviable options in a modeling context.

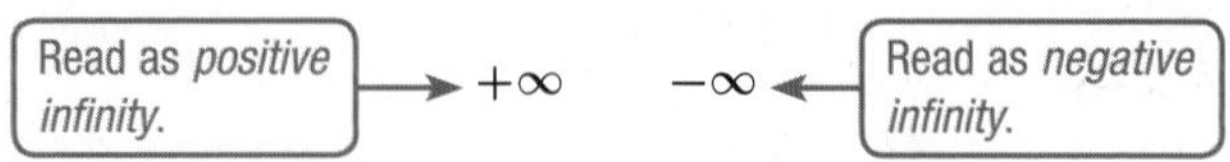

To indicate that an endpoint is *not* included in the set, a parenthesis, (or), is used. Parentheses are always used with the symbols $+\infty$ and $-\infty$, because they do not include endpoints.

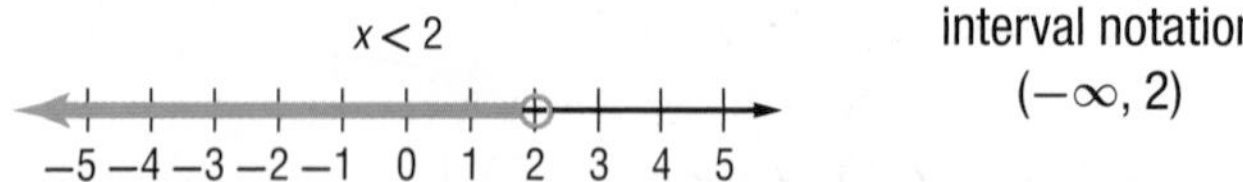

interval notation
$(-\infty, 2)$

A bracket is used to indicate that the endpoint, -2, *is* included in the solution set below.

$x \geq -2$

–5 –4 –3 –2 –1 0 1 2 3 4 5

interval notation
$[-2, +\infty)$

In interval notation, the symbol for the union of the two sets is $\cup$. The compound inequality $y \leq -7$ or $y > -1$ is written as $(-\infty, -7] \cup (-1, +\infty)$.

Exercises

Write each inequality using interval notation.

1. $\{a \mid a \leq -3\}$

2. $\{n \mid n > -8\}$

3. $\{y \mid y < 2 \text{ or } y \geq 14\}$

4. $\{b \mid b \leq -9 \text{ or } b > 1\}$

5. $\{t \mid 1 < t < 3\}$

6. $\{m \mid m \geq 4 \text{ or } m \leq -7\}$

7. $\{x \mid x \geq 0\}$

8. $\{r \mid -3 < r < 4\}$

9. –5–4–3–2–1 0 1 2 3 4 5

10. –5–4–3–2–1 0 1 2 3 4 5

11. –8 –4 0 4 8

12. –2 –1 0 1 2

13. 5 6 7 8 9 10 11 12 13

14. –15–10–5 0 5 10 15 20 25 30 35

Graph each solution set on a number line.

15. $(-1, -\infty)$

16. $(-\infty, 4]$

17. $(-\infty, 5] \cup (7, +\infty)$

18. **WRITING IN MATH** Write in words the meaning of $(-\infty, 3) \cup [10, +\infty)$. Then write the compound inequality that this notation represents.

19. **WRITING IN MATH** How are symbols used to write solution sets for inequalities? Explain.

LESSON 1-6

Solving Compound and Absolute Value Inequalities

Then	Now	Why?
You solved one-step and multi-step inequalities.	**1** Solve compound inequalities. **2** Solve absolute value inequalities.	Marine biologists often have to transplant a dolphin from its natural habitat to a pool. Dolphins prefer the temperature of water to be at least 22°C but no more than 29°C. The acceptable temperature of water t for dolphins can be described by the following compound inequality. $t \geq 22$ and $t \leq 29$

NewVocabulary
compound inequality
intersection
union

Common Core State Standards

Mathematical Practices
5 Use appropriate tools strategically.

1 Compound Inequalities

A **compound inequality** consists of two inequalities joined by the word *and* or the word *or*. To solve a compound inequality, you must solve each part of the inequality. The graph of a compound inequality containing *and* is the **intersection** of the solution sets of the two inequalities.

KeyConcept "And" Compound Inequalities

Words A compound inequality containing the word *and* is true if and only if *both* inequalities are true.

Example

$x \geq -4$

$x < 3$

$x \geq -4$ and $x < 3$

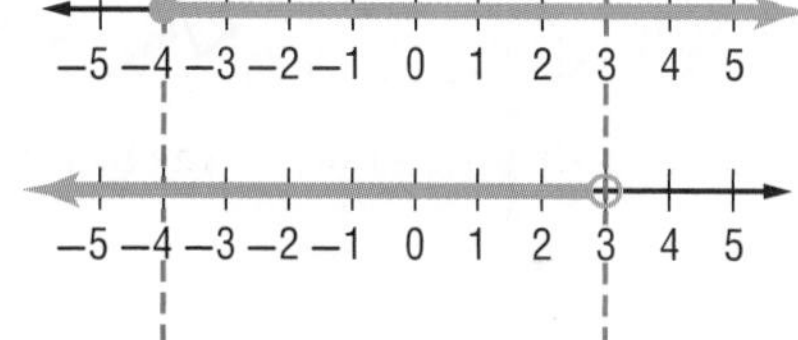

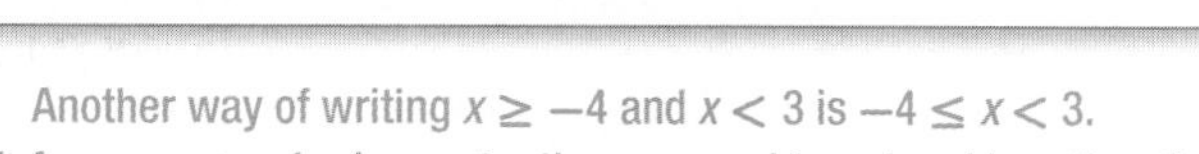

Another way of writing $x \geq -4$ and $x < 3$ is $-4 \leq x < 3$.
Both forms are read *x is greater than or equal to −4 and less than 3.*

Example 1 Solve an "And" Compound Inequality

Solve $8 < 3y - 7 \leq 23$. Graph the solution set on a number line.

Method 1 Solve separately.

Write the compound inequality using the word *and*. Then solve each inequality.

$$8 < 3y - 7 \quad \text{and} \quad 3y - 7 \leq 23$$
$$15 < 3y \qquad\qquad 3y \leq 30$$
$$5 < y \qquad\qquad y \leq 10$$
$$5 < y \leq 10$$

Method 2 Solve both together.

Solve both parts at the same time by adding 7 to each part. Then divide each part by 3.

$$8 < 3y - 7 \leq 23$$
$$15 < 3y \leq 30$$
$$5 < y \leq 10$$
$$5 < y \leq 10$$

(continued on the next page)

Graph the solution set for each inequality and find their intersection.

$5 < y$

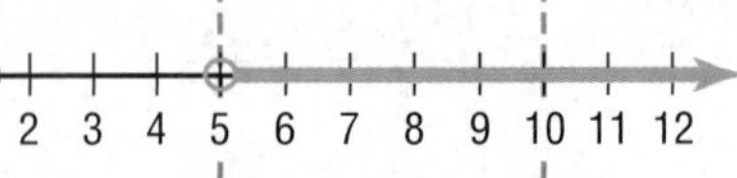

$y \leq 10$

$5 < y \leq 10$

The solution set is $\{y \mid 5 < y \leq 10\}$ or $(5, 10]$.

GuidedPractice

Solve each inequality. Graph the solution set on a number line.

1A. $-12 \leq 4x + 8 \leq 32$

1B. $-5 \geq 3z - 2 > -14$

The graph of a compound inequality containing *or* is the **union** of the solution sets of the two inequalities.

KeyConcept "Or" Compound Inequalities

Words A compound inequality containing the word *or* is true if one or more of the inequalities is true.

Example

$x \geq 5$

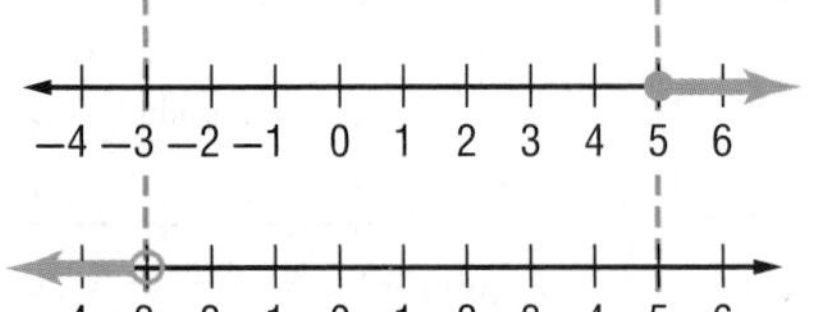

$x < -3$

$x \geq 5$ or $x < -3$

Example 2 Solve an "Or" Compound Inequality

Solve $k + 6 < -4$ or $3k \geq 14$. Graph the solution set.

Solve each inequality separately.

$k + 6 < -4$ or $3k \geq 14$

$k < -10$ $\quad k \geq \frac{14}{3}$

$k < -10$

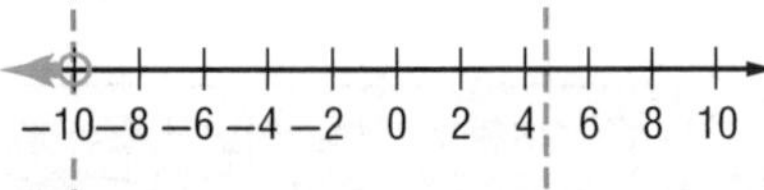

$k \geq \frac{14}{3}$

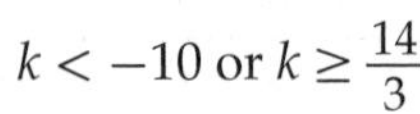

$k < -10$ or $k \geq \frac{14}{3}$

GuidedPractice

Solve each inequality. Graph the solution set on a number line.

2A. $5j \geq 15$ or $-3j \geq 21$

2B. $g - 6 > -11$ or $2g + 4 < -15$

2 Absolute Value Inequalities

In Lesson 1-4, you learned that the absolute value of a number is its distance from 0 on the number line. You can use this definition to solve inequalities involving absolute value.

Example 3 Solve Absolute Value Inequalities

Solve each inequality. Graph the solution set on a number line.

a. $|x| < 3$

$|x| < 3$ means that the distance between x and 0 on a number line is less than 3 units. To make $|x| < 3$ true, substitute numbers for x that are fewer than 3 units from 0.

All of the numbers between −3 and 3 are less than 3 units from 0. The solution set is $\{x \mid -3 < x < 3\}$ or $(-3, 3)$.

b. $|x| > 5$

$|x| > 5$ means that the distance between x and 0 on a number line is more than 5 units. To make $|x| > 5$ true, substitute numbers for x that are more than 5 units from 0.

All of the numbers between and including −5 and 5 are no more than 5 units from 0. The solution set is $\{x \mid -5 > x \text{ or } x > 5\}$ or $(-\infty, -5) \cup (5, \infty)$.

ReadingMath

within and *between* When solving problems involving inequalities, *within* is meant to be inclusive. Use $\le$ or $\ge$.

Between is meant to be exclusive. Use $<$ or $>$.

GuidedPractice

Solve each inequality. Graph the solution set on a number line.

3A. $|t| < 6$

3B. $|u| < -3$

3C. $|t| > 3$

3D. $|u| > -2$

StudyTip

Absolute Value Inequalities Because the absolute value of a number is never negative, solutions involving negative numbers are as follows.
$|x| < -5$ is the empty set.
$|x| > -5$ is infinite solutions.

An absolute value inequality can be solved by rewriting it as a compound inequality.

KeyConcept Absolute Value Inequalities

For all real numbers a, b, c, and x, $c > 0$, the following statements are true.

Absolute Value Inequality	Compound Inequality	Example
$\lvert ax + b\rvert > c$	$ax + b > c$ or $ax + b < -c$	If $\lvert 4x + 5\rvert > 7$, then $4x + 5 > 7$ or $4x + 5 < -7$.
$\lvert ax + b\rvert < c$	$-c < ax + b < c$	If $\lvert 4x + 5\rvert < 7$, then $-7 < 4x + 5 < 7$.

These statements are also true for $\le$ and $\ge$, respectively.

Example 4 Solve a Multi-Step Absolute Value Inequality

Solve $|6y - 5| \geq 13$. Graph the solution set on a number line.

$|6y - 5| \geq 13$ is equivalent to $6y - 5 \geq 13$ or $6y - 5 \leq -13$. Solve the inequality.

$6y - 5 \geq 13$ or $6y - 5 \leq -13$ Rewrite the inequality.

$6y \geq 18$ $\quad 6y \leq -8$ Add 5 to each side.

$y \geq 3$ $\quad y \leq -\frac{8}{6}$ or $-\frac{4}{3}$ Divide each side by 6.

The solution set is $\left\{y \mid y \leq -\frac{4}{3} \text{ or } y \geq 3\right\}$ or $\left(-\infty, -\frac{4}{3}\right] \cup [3, \infty)$.

−5 −4 −3 −2 −1 0 1 2 3 4 5

GuidedPractice

Solve each inequality. Graph the solution set on a number line.

4A. $|4x - 7| > 13$

4B. $|5z + 2| \leq 17$

An inequality can be viewed as a constraint in a problem situation. Each solution of the inequality represents a combination that meets the constraint.

In real-world problems, the domain and range are often restricted to nonnegative or whole numbers.

Real-World Example 5 Write and Solve an Absolute Value Inequality

MONEY Amanda is apartment hunting in a specific area. She discovers that the average monthly rent for a 2-bedroom apartment is \$600 a month, but the actual price could differ from the average as much as \$225 a month.

a. Write an absolute value inequality to describe this situation.

Let r = average monthly rent. $\quad |600 - r| \leq 225$

b. Solve the inequality to find the range of monthly rent.

Rewrite the absolute value inequality as a compound inequality. Then solve for r.

$$-225 \leq 600 - r \leq 225$$

$$-225 - 600 \leq 600 - r - 600 \leq 225 - 600$$

$$-825 \leq -r \leq -375$$

$$825 \geq r \geq 375$$

The solution set is $\{r \mid 375 \leq r \leq 825\}$ or $[375, 825]$. Thus, monthly rent could fall between \$375 and \$825, inclusive.

GuidedPractice

5. TUITION Rachel is considering colleges to attend and determines that the average tuition among her choices is \$3725 per year, but the tuition at a school could differ by as much as \$1650 from the average. Write and solve an absolute value inequality to find the range of tuition.

Apartment For Rent

Studio

Real-WorldLink

Apartment costs vary greatly depending on location. Of the major U.S. cities, New York has the highest average monthly rent of \$2400, while Oklahoma City is lowest at \$543.

Source: MSN

Check Your Understanding

● = Step-by-Step Solutions begin on page R14.

Examples 1–4 **Solve each inequality. Graph the solution set on a number line.**

1. $-4 < g + 8 < 6$

2. $-9 \leq 4y - 3 \leq 13$

3. $z + 6 > 3$ or $2z < -12$

4. $m - 7 \geq -3$ or $-2m + 1 \geq 11$

5. $|c| \geq 8$

6. $|q| \geq -1$

7. $|z| < 6$

8. $|x| \leq -4$

9. $|3v + 5| > 14$

10. $|4t - 3| \leq 7$

Example 5

11. MONEY Khalid is considering several types of paint for his bedroom. He estimates that he will need between 2 and 3 gallons. The table at the right shows the price per gallon for each type of paint Khalid is considering. Write a compound inequality and determine how much he could be spending.

Paint Type	Price per Gallon
Flat	$21.98
Satin	$23.98
Semi-Gloss	$24.98
Gloss	$25.98

Practice and Problem Solving

Extra Practice is on page R1.

Examples 1–4 **Solve each inequality. Graph the solution set on a number line.**

12. $8 < 2v - 4 < 16$

13. $-7 \leq 4d - 3 \leq -1$

14. $4r + 3 < -6$ or $3r - 7 > 2$

15. $6y - 3 < -27$ or $-4y + 2 < -26$

16. $|6h| < 12$

17. $|-4k| > 16$

18. $|3x - 4| > 10$

(19) $|8t + 3| \leq 4$

20. $|-9n - 3| < 6$

21. $|-5j - 4| \geq 12$

Example 5

22. CCSS **MODELING** Forensic scientists use the equation $h = 2.6f + 47.2$ to estimate the height h of a woman given the length in centimeters f of her femur bone.

a. Suppose the equation has a margin of error of ±3 centimeters. Write an inequality to represent the height of a woman given the length of her femur bone.

b. If the length of a female skeleton's femur is 50 centimeters, write and solve an absolute value inequality that describes the woman's height in centimeters.

Write an absolute value inequality for each graph.

23.

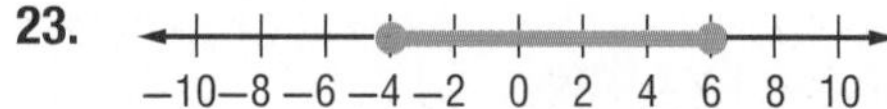

24.

−10 −8 −6 −4 −2 0 2 4 6 8 10

25.

26.

−5 −4 −3 −2 −1 0 1 2 3 4 5

27.

−20 −16 −12 −8 −4 0 4 8 12 16 20

28.

−10 −8 −6 −4 −2 0 2 4 6 8 10

29.

−5 −4 −3 −2 −1 0 1 2 3 4 5

30.

−10 −8 −6 −4 −2 0 2 4 6 8 10

31 **DOGS** The Labrador retriever is one of the most recognized and popular dogs kept as a pet. Using the information given, write a compound inequality to describe the range of healthy weights for a fully grown female Labrador retriever.

Healthy Heights and Weights for Labrador Retrievers		
Gender	Height (in.)	Weight (lb)
Male	22.5–24.5	65–80
Female	21.5–23.5	55–70

32. **GEOMETRY** The *Exterior Angle Inequality Theorem* states that an exterior angle measure is greater than the measure of either of its corresponding remote interior angles. Write two inequalities to express the relationships among the measures of the angles of $\triangle ABC$.

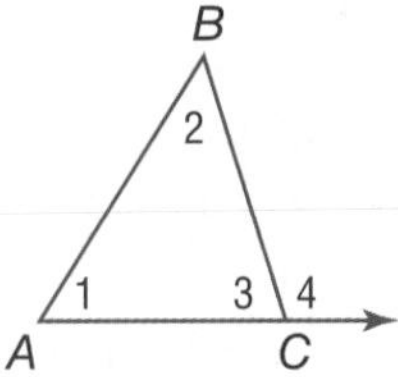

Solve each inequality. Graph the solution set on a number line.

33. $28 > 6k + 4 > 16$

34. $m - 7 > -12$ or $-3m + 2 > 38$

35. $|-6h| > 90$

36. $-|-5k| > 15$

37. $3|2z - 4| - 6 > 12$

38. $6|4p + 2| - 8 < 34$

39. $\frac{|5f - 2|}{6} > 4$

40. $\frac{|2w + 8|}{5} \geq 3$

Write an algebraic expression to represent each verbal expression.

41. numbers that are at least 4 units from -5

42. numbers that are no more than $\frac{3}{8}$ unit from 1

43. numbers that are at least 6 units but no more than 10 units from 2

44. CCSS **REASONING** NASCAR rules stipulate that a car must conform to a set of 32 templates, each shaped to fit a different contour of the car. When a template is placed on a car, the gap between it and the car cannot exceed the specified tolerance. Each template is marked on its edge with a colored line that indicates the tolerance for the template.

a. Suppose a certain template is 24.42 inches long. Use the information in the table at the right to write an absolute value inequality for templates with each line color.

b. Find the acceptable lengths for that part of a car if the template has each line color.

c. Graph the solution set for each line color on a number line.

d. The tolerance of which line color includes the tolerances of the other line colors? Explain your reasoning.

Line Color	Tolerance (in.)
Red	0.07
Blue	0.25
Green	0.5

Solve each inequality. Graph the solution set on a number line.

45. $n + 6 > 2n + 5 > n - 2$

46. $y + 7 < 2y + 2 < 0$

47. $2x + 6 < 3(x - 1) \leq 2(x + 3)$

48. $a - 16 \leq 2(a - 4) < a + 2$

49. $4g + 8 \geq g + 6$ or $7g - 14 \geq 2g - 4$

50. $5t + 7 > 2t + 4$ and $3t + 3 < 24 - 4t$

51. **HEALTH** Hypoglycemia (low blood sugar) and hyperglycemia (high blood sugar) are potentially dangerous and occur when a person's blood sugar fluctuates by more than 38 mg from the normal blood sugar level of 88 mg. Write and solve an absolute value inequality to describe blood sugar levels that are considered potentially dangerous.

52. **AIR TRAVEL** The airline on which Drew is flying has weight restrictions for checked baggage. Drew is checking one bag.

a. Describe the ranges of weights that would classify Drew's bag as free, \$25, \$50, and unacceptable.

b. If Drew's bag weighs 68 pounds, how much will he pay to take it on the plane?

Cost for Checked Baggage	
Weight	**Cost**
Up to 50 lb limit	free
20 lb over limit	\$25
More than 20, but less than 50 lb over limit	\$50
More than 50 lb over limit	not accepted

H.O.T. Problems Use Higher-Order Thinking Skills

53. CCSS **ARGUMENTS** David and Sarah are solving $4|-5x-3|-6 \geq 34$. Is either of them correct? Explain your reasoning.

David	Sarah
$4\|-5x-3\|-6 \geq 34$	$4\|-5x-3\|-6 \geq 34$
$\|-5x-3\| \geq 10$	$\|-5x-3\| \geq 10$
$-5x-3 \geq 10$ or $-5x-3 \leq -10$	$-5x-3 \leq 10$ or $-5x-3 \geq -10$
$-5x \geq 13$ $-5x \leq -7$	$-5x \leq 13$ $-5x \geq -7$
$x \leq -\frac{13}{5}$ $x \geq \frac{7}{5}$	$x \geq -\frac{13}{5}$ $x \leq \frac{7}{5}$

54. **CHALLENGE** Solve $|x-2|-|x+2|>x$.

REASONING Determine whether each statement is *true* or *false*. If false, provide a counterexample.

55. The graph of a compound inequality involving an *and* statement is bounded on the left and right by two values of x.

56. The graph of a compound inequality involving an *or* statement contains a region of values that are not solutions.

57. The graph of a compound inequality involving an *and* statement includes values that make all parts of the given statement true.

58. **WRITING IN MATH** An alternate definition of absolute value is to define $|a-b|$ as the distance between a and b on the number line. Explain how this definition can be used to solve inequalities of the form $|x-c|<r$.

59. **REASONING** The graphs of the solutions of two different absolute value inequalities are shown. Compare and contrast the absolute value inequalities.

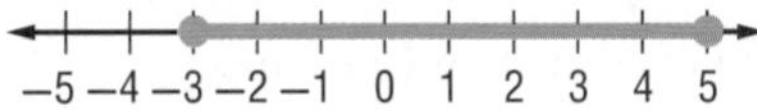

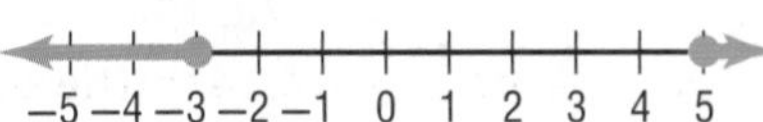

60. **OPEN ENDED** Write an absolute value inequality with a solution of $a \leq x \leq b$.

61. **WHICH ONE DOESN'T BELONG?** Identify the compound inequality that is not the same as the other three. Explain your reasoning.

$-3 < x < 5$	$x > 2$ and $x < 3$	$x > 5$ and $x < 1$	$x > -4$ and $x > -2$

62. **WRITING IN MATH** Summarize the difference between *and* compound inequalities and *or* compound inequalities.

Standardized Test Practice

63. Which of the following best describes the graph of the equations below?

$$24y = 8x + 11$$
$$36y = 12x + 11$$

A The lines have the same x-intercept.
B The lines have the same y-intercept.
C The lines are parallel.
D The lines are perpendicular.

64. SAT/ACT Find an expression equivalent to $\left(\frac{3x^3}{y^3}\right)^3$.

F $\frac{9x^6}{3y}$

G $\frac{9x^9}{y^3}$

H $\frac{9x^6}{y^3}$

J $\frac{27x^6}{3y}$

K $\frac{27x^9}{y^3}$

65. GRIDDED RESPONSE How many cubes that measure 4 centimeters on each side can be placed completely inside the box below?

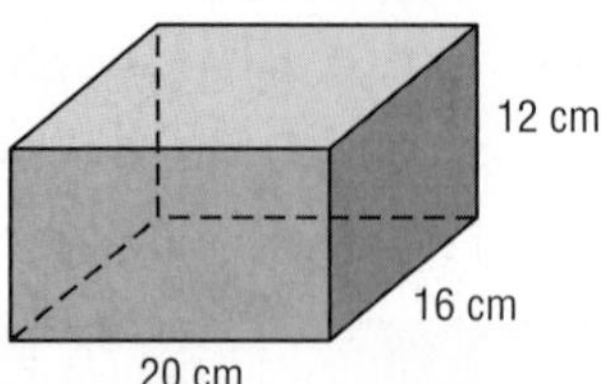

66. Which graph represents the solution set for $|3x - 6| + 8 \geq 17$?

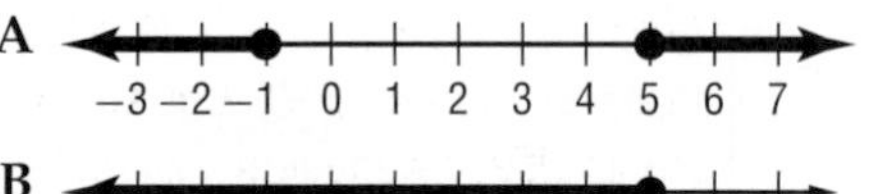

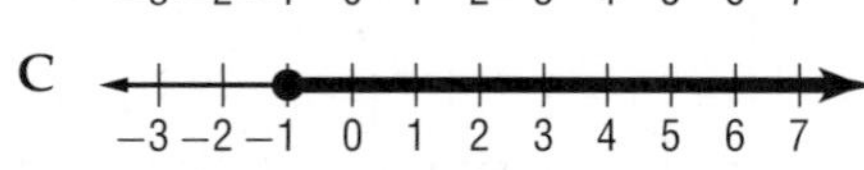

Spiral Review

67. HEALTH The National Heart Association recommends that less than 30% of a person's total daily caloric intake come from fat. One gram of fat yields nine Calories. Consider a healthy 21-year-old whose average caloric intake is between 2500 and 3300 Calories. (Lesson 1-5)

a. Write an inequality that represents the suggested fat intake for the person.

b. What is the greatest suggested fat intake for the person?

68. TRAVEL Maggie is planning a 5-day, 5-night, trip to a family reunion. She wants to spend no more than \$1000. Her plane ticket is \$375, and the hotel is \$85 per night. (Lesson 1-5)

a. Let f represent the cost of food for one day. Write an inequality to represent this situation.

b. Solve the inequality and interpret the solution.

Solve each equation. Check your solutions. (Lesson 1-4)

69. $4|x - 5| = 20$

70. $|3y + 10| = 25$

71. $|7z + 8| = -9$

Skills Review

Name the property illustrated by each statement.

72. If $5x = 7$, then $5x + 3 = 7 + 3$.

73. If $-3x + 9 = 11$ and $6x + 2 = 11$, then $-3x + 9 = 6x + 2$.

74. If $[x + (-2)] + (-4) = 5$, then $x + [-2 + (-4)] = 5$.

Study Guide and Review

Study Guide

KeyConcepts

Expressions and Formulas (Lesson 1-1)

- Use the order of operations to solve equations.

Properties of Real Numbers (Lesson 1-2)

- Real numbers can be classified as rational (Q) or irrational (I). Rational numbers can be classified as integers (Z), whole numbers (W), natural numbers (N), and/or quotients of these.

Solving Equations (Lessons 1-3 and 1-4)

- Verbal expressions can be translated into algebraic expressions.
- The absolute value of a number is the number of units it is from 0 on a number line.
- For any real numbers a and b, where $b \geq 0$, if $|a| = b$, then $a = b$ or $a = -b$.

Solving Inequalities (Lessons 1-5 and 1-6)

- Adding or subtracting the same number from each side of an inequality does not change the truth of the inequality.
- When you multiply or divide each side of an inequality by a negative number, the direction of the inequality symbol must be reversed.
- The graph of an *and* compound inequality is the intersection of the solution sets of the two inequalities. The graph of an *or* compound inequality is the union of the solution sets of the two inequalities.
- An *and* compound inequality can be expressed in two different ways. For example, $-2 \leq x \leq 3$ is equivalent to $x \geq -2$ and $x \leq 3$.
- For all real numbers a and b, where $b > 0$, the following statements are true.
  1. If $|a| < b$ then $-b < a < b$.
  2. If $|a| > b$ then $a > b$ or $a < -b$.

FOLDABLES StudyOrganizer

Be sure the Key Concepts are noted in your Foldable.

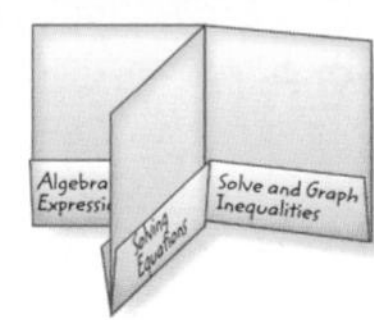

KeyVocabulary

absolute value (p. 27)	irrational numbers (p. 11)
algebraic expressions (p. 5)	natural numbers (p. 11)
compound inequality (p. 41)	open sentence (p. 18)
constraint (p. 29)	order of operations (p. 5)
empty set (p. 28)	rational numbers (p. 11)
equation (p. 18)	real numbers (p. 11)
extraneous solution (p. 29)	set-builder notation (p. 35)
formula (p. 6)	solution (p. 18)
infinity (p. 40)	union (p. 42)
integers (p. 11)	variables (p. 5)
intersection (p. 41)	whole numbers (p. 11)
interval notation (p. 40)	

VocabularyCheck

State whether each sentence is *true* or *false*. If *false*, replace the underlined term to make a true sentence.

1. The absolute value of a number is always <u>negative</u>.
2. $\sqrt{12}$ belongs to the set of <u>rational</u> numbers.
3. An <u>equation</u> is a statement that two expressions have the same value.
4. A solution of an equation is a value that makes the equation <u>false</u>.
5. The empty set contains <u>no</u> elements.
6. A mathematical sentence containing one or more variables is called an <u>open sentence</u>.
7. The graph of a compound inequality containing *and* is the union of the solution sets of the two inequalities.
8. Variables are used to represent <u>unknown</u> quantities.
9. The set of <u>rational</u> numbers includes terminating and repeating decimals.
10. Expressions that contain at least one variable are called <u>algebraic expressions</u>.

CHAPTER 1

Study Guide and Review *Continued*

Lesson-by-Lesson Review

1-1 Expressions and Formulas

Evaluate each expression.

11. $[28 - (16 + 3)] \div 3$

12. $\frac{2}{3}(3^3 + 12)$

13. $\frac{15(9 - 7)}{3}$

Evaluate each expression if $w = 0.2$, $x = 10$, $y = \frac{1}{2}$, and $z = -4$.

14. $4w - 8y$

15. $z^2 + xy$

16. $\frac{5w - xy}{z}$

17. **GEOMETRY** The formula for the volume of a cylinder is $V = \pi r^2 h$, where V is the volume, r is the radius, and h is the height. What is the volume of a cylinder that is 6 inches high and has a radius of 3 inches?

Example 1

Evaluate $(12 - 15) \div 3^2$.

$(12 - 15) \div 3^2 = -3 \div 3^2$ Subtract.

$= -3 \div 9$ $3^2 = 9$

$= -\frac{1}{3}$ Divide.

Example 2

Evaluate $\frac{a^2}{2ac - b}$ if $a = -6$, $b = 5$, and $c = 0.25$.

$\frac{a^2}{2ac - b} = \frac{(-6)^2}{2(-6)(0.25) - 5}$ $a = -6$, $b = 5$, and $c = 0.25$

$= \frac{36}{2(-1.5) - 5}$ Evaluate the numerator and denominator separately.

$= \frac{36}{-8}$ or $-\frac{9}{2}$ Simplify.

1-2 Properties of Real Numbers

Name the sets of numbers to which each value belongs.

18. $1.\overline{3}$ **19.** $\sqrt{4}$ **20.** $-\frac{3}{4}$

Simplify each expression.

21. $4x - 3y + 7x + 5y$

22. $2(a + 3) - 4a + 8b$

23. $4(2m + 5n) - 3(m - 7n)$

24. **MONEY** At Fun City Amusement Park, hot dogs sell for \$3.50 and sodas sell for \$2.50. Dion bought 3 hot dogs and 3 sodas during one day at the park.

a. Illustrate the Distributive Property by writing two expressions to represent the cost of the hot dogs and the sodas.

b. Use the Distributive Property to find how much money Dion spent on food and drinks.

Example 3

Name the sets of numbers to which $\sqrt{50}$ belongs.

$\sqrt{50} = 5\sqrt{2}$ Irrationals (I), and reals (R)

Example 4

Simplify $-4(a + 3b) + 5b$.

$-4(a + 3b) + 5b$ Original expression

$= -4(a) + -4(3b) + 5b$ Distributive Property

$= -4a - 12b + 5b$ Multiply.

$= -4a - 7b$ Simplify.

1-3 Solving Equations

Solve each equation. Check your solution.

25. $8 + 5r = -27$

26. $4w + 10 = 6w - 13$

27. $\frac{x}{6} + \frac{x}{3} = \frac{3}{4}$

28. $6b - 5 = 3(b + 2)$

29. **MONEY** It cost Lori \$14 to go to the movies. She bought popcorn for \$3.50 and a soda for \$2.50. How much was her ticket?

Solve each equation or formula for the specified variable.

30. $2k - 3m = 16$ for k

31. $\frac{r + 5}{mn} = p$ for m

32. $A = \frac{1}{2}h(a + b)$ for h

33. **GEOMETRY** Yu-Jun wants to fill the water container at the right. He knows that the radius is 2 inches and the volume is 100.48 cubic inches. What is the height of the water bottle? Use the formula for the volume of a cylinder, $V = \pi r^2 h$, to find the height of the bottle.

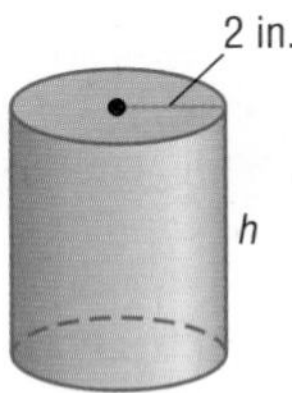

Example 5

Solve $-3(a - 3) + 2(3a - 2) = 14$.

$-3(a - 3) + 2(3a - 2) = 14$	Original equation
$-3a + 9 + 6a - 4 = 14$	Distributive Property
$-3a + 6a + 9 - 4 = 14$	Commutative Property
$3a + 5 = 14$	Substitution Property
$3a = 9$	Subtraction Property
$a = 3$	Division Property

Example 6

Solve each equation or formula for the specified variable.

a. $y = 2x + 3z$ for x

$y = 2x + 3z$	Original equation
$y - 3z = 2x$	Subtract $3z$ from each side.
$\frac{y - 3z}{2} = x$	Divide each side by 2.

b. $V = \frac{\pi r^2 h}{3}$ for h

$V = \frac{\pi r^2 h}{3}$	Original equation
$3V = \pi r^2 h$	Multiply each side by 3.
$\frac{3V}{\pi r^2} = h$	Divide each side by πr^2.

1-4 Solving Absolute Value Equations

Solve each equation. Check your solution.

34. $|r + 5| = 12$

35. $4|a - 6| = 16$

36. $|3x + 7| = -15$

37. $|b + 5| = 2b - 9$

38. **MEASUREMENT** Marcos is cutting ribbons for a craft project. Each ribbon needs to be $\frac{3}{4}$ yard long. If each piece is always within plus or minus $\frac{1}{16}$ yard, how long are the shortest and longest pieces of ribbon?

Example 7

Solve $|3m + 7| = 13$.

Case 1	Case 2
$a = b$	$a = -b$
$3m + 7 = 13$	$3m + 7 = -13$
$3m = 6$	$3m = -20$
$m = 2$	$m = -\frac{20}{3}$

The solutions are 2 and $-\frac{20}{3}$.

CHAPTER 1

Study Guide and Review *Continued*

1-5 Solving Inequalities

Solve each inequality. Then graph the solution set on a number line.

39. $-4a \leq 24$

40. $\frac{r}{5} - 8 > 3$

41. $4 - 7x \geq 2(x + 3)$

42. $-p - 13 < 3(5 + 4p) - 2$

43. **MONEY** Ms. Hawkins is taking her science class on a field trip to a museum. She has \$572 to spend on the trip. There are 52 students that will go to the museum. The museum charges \$5 per student, and Ms. Hawkins gets in for free. If the students will have slices of pizza for lunch that cost \$2 each, how many slices can each student have?

Example 8

Solve $2m - 7 < -11$. Graph the solution set on a number line.

$2m - 7 < -11$	Original inequality
$2m < -4$	Add 7 to each side.
$m < -2$	Divide each side by 2.

The solution set is $\{m \mid m < -2\}$.

The graph of the solution set is shown below.

−5 −4 −3 −2 −1 0 1 2 3 4 5

1-6 Solving Compound and Absolute Value Inequalities

Solve each inequality. Graph the solution set on a number line.

44. $2m + 4 < 7$ or $3m + 5 > 14$

45. $-5 < 4x + 3 < 19$

46. $6y - 1 > 17$ or $8y - 6 \leq -10$

47. $-2 \leq 5(m - 3) < 9$

48. $|a| + 2 < 15$

49. $|p - 14| \leq 19$

50. $|6k - 1| < 15$

51. $|2r + 7| < -1$

52. $\frac{1}{3}|8q + 5| \geq 7$

53. **MONEY** Cara is making a beaded necklace for a gift. She wants to spend between \$20 and \$30 on the necklace. The bead store charges \$2.50 for large beads and \$1.25 for small beads. If she buys 3 large beads, how many small beads can she buy to stay within her budget? Write and solve a compound inequality to describe the range of possible beads.

Example 9

Solve each inequality. Graph the solution set on a number line.

a. $-14 \leq 3x - 8 < 16$

$-14 \leq 3x - 8 < 16$	Original inequality
$-6 \leq 3x < 24$	Add 8 to each part.
$-2 \leq x < 8$	Divide each part by 3.

The solution set is $\{x \mid -2 \leq x < 8\}$.

−4 −3 −2 −1 0 1 2 3 4 5 6 7 8 9 10

b. $|3a - 5| > 13$

$|3a - 5| > 13$ is equivalent to $3a - 5 > 13$ or $3a - 5 < -13$.

$3a - 5 > 13$	or	$3a - 5 < -13$	
$3a > 18$		$3a < -8$	Subtract.
$a > 6$		$a < -\frac{8}{3}$	Divide.

The solution set is $\left\{a \middle| a > 6 \text{ or } a < -\frac{8}{3}\right\}$.

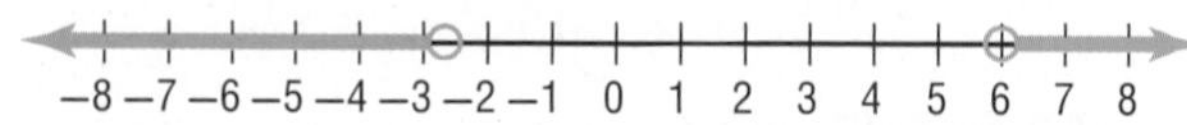

CHAPTER 1

Practice Test

1. Evaluate $x + y^2(2 + x)$ if $x = 3$ and $y = -1$.
2. Simplify $-4(3a + b) - 2(a - 5b)$.
3. **MULTIPLE CHOICE** If $3m + 5 = 23$, what is the value of $2m - 3$?

   A 105

   B 9

   C $\frac{47}{3}$

   D 6

4. Solve $r = \frac{1}{2}m^2p$ for p.

Write an algebraic expression to represent each verbal expression.

5. twice the difference of a number and 11
6. the product of the square of a number and 5
7. Evaluate $2|3y - 8| + y$ if $y = 2.5$.
8. Solve $-2b > \frac{18 - b}{5}$. Graph the solution set on a number line.
9. **MONEY** Carson has $35 to spend at the water park. The admission price is $25 and each soda is $2.50. Write an inequality to show how many sodas he can buy.
10. Solve $r - 3 < -5$ or $4r + 1 > 15$. Graph the solution set.
11. Solve $|p - 4| \leq 11$. Graph the solution set on a number line.
12. **MULTIPLE CHOICE** Which graph represents the solution set for $4 < 6t + 1 \leq 43$?

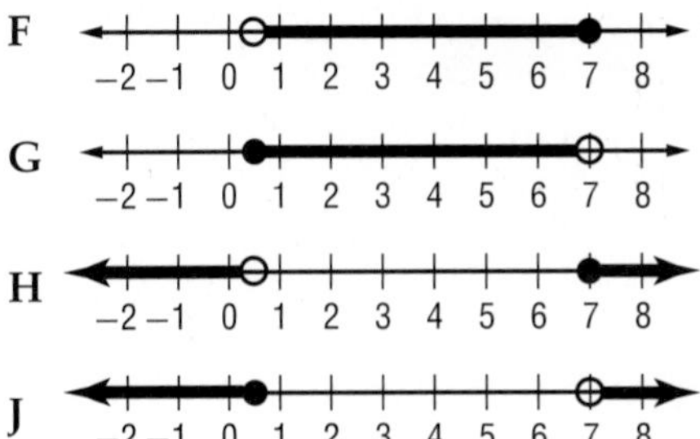

13. **MONEY** Sofia is buying new skis. She finds that the average price of skis is $500 but the actual price could differ from the average by as much as $250. Write and solve an absolute value inequality to describe this situation.
14. **GARDENING** Andy is making 3 trapezoidal garden boxes for his backyard. Each trapezoid will be the size of the trapezoid below. He will place stone blocks around the borders of the boxes. How many feet of stones will Andy need?

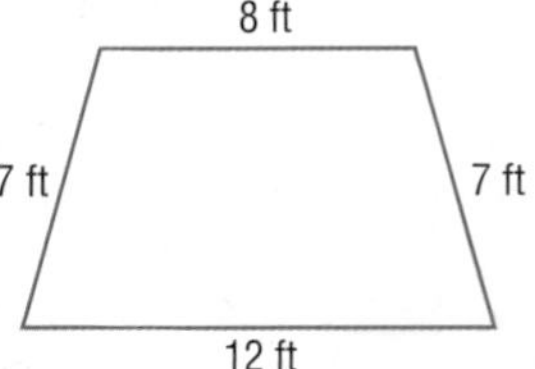

Solve each equation.

15. $|x + 4| = 3$
16. $|3m + 2| = 1$
17. $|3a + 2| = -4$
18. $|2t + 5| - 7 = 4$
19. $|5n - 2| - 6 = -3$
20. $|p + 6| + 9 = 8$
21. **GEOMETRY** The volume of a cylinder is given by the formula $V = \pi r^2h$. What is the volume of the cylinder below?

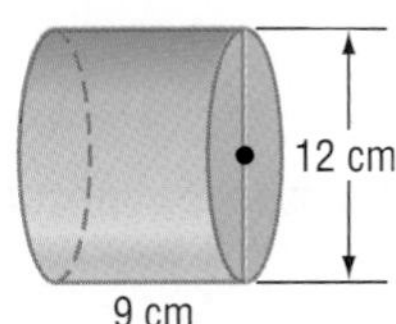

22. Solve $-3b - 5 \geq -6b - 13$. Graph the solution set on a number line.
23. Evaluate $\frac{3(x + y)}{4xy^2}$ if $x = \frac{2}{3}$ and $y = -2$.
24. Name the set(s) of numbers to which $-\frac{1}{3}$ belongs.
25. **MONEY** The costs for making necklaces at two craft stores are shown in the table. For what quantity of beads does The Accessory Store have a better deal? Use the inequality $15 + 3.25b < 20 + 2.50b$.

Shop	Cost per Chain	Cost per Bead
The Accessory Store	$15	$3.25
Finishing Touch	$20	$2.50

Preparing for Standardized Tests

Eliminate Unreasonable Answers

You can eliminate unreasonable answers to help you find the correct answer when solving multiple-choice test items.

BananaStock/PictureQuest

Strategies for Eliminating Unreasonable Answers

Step 1

Read the problem statement carefully to determine exactly what you are being asked to find.

Ask yourself:

- What am I being asked to solve?
- In what format (that is, fraction, number, decimal, percent, type of graph) will the correct answer be?
- What units (if any) will the correct answer have?

Step 2

Carefully look over each possible answer choice and evaluate for reasonableness.

- Identify any answer choices that are clearly incorrect and eliminate them.
- Eliminate any answer choices that are not in the proper format.
- Eliminate any answer choices that do not have the correct units.

Step 3

Solve the problem and choose the correct answer from those remaining. Check your answer.

Standardized Test Example

Read the problem. Identify what you need to know. Then use the information in the problem to solve.

The formula for the area A of a trapezoid with height h and bases b_1 and b_2 is $A = \frac{h}{2}(b_1 + b_2)$. Write an expression to represent the area of the trapezoid at the right.

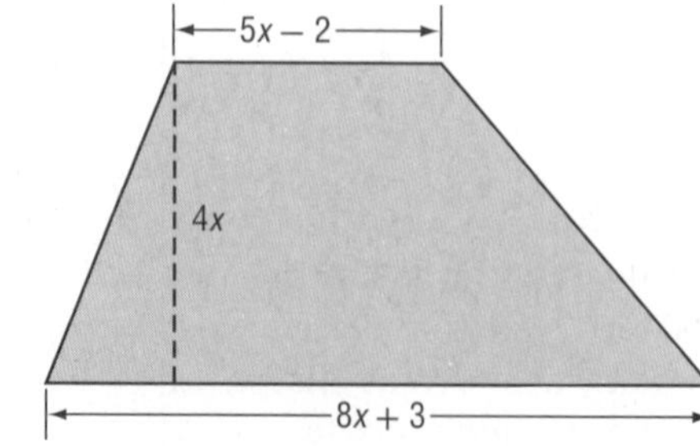

A $26x^2 + 2x$ **C** $13x + 1$

B $52x^2 + 4x$ **D** $28x + 10$

To compute the area of the trapezoid, you need to multiply half the height, $2x$, by another linear factor in x. So, the correct answer will contain an x^2 term. Since choices C and D are both linear, they can be eliminated. The correct answer is either A or B. Multiply to find the expression for the area.

$$A = \frac{h}{2}(b_1 + b_2)$$

$$= \frac{4x}{2}(8x + 3 + 5x - 2)$$

$$= 2x(13x + 1)$$

$$= 26x^2 + 2x$$

The correct answer is A.

Exercises

Read each problem. Eliminate any unreasonable answers. Then use the information in the problem to solve.

1. The graph below shows the solution to which inequality?

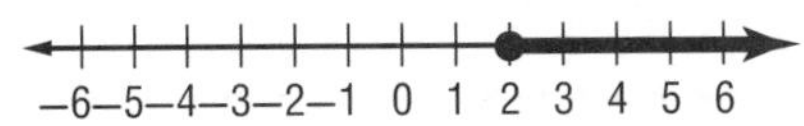

A $8x - 9 \leq 5x - 3$

B $8x - 9 < 5x - 3$

C $8x - 9 \geq 5x - 3$

D $8x - 9 > 5x - 3$

2. Einstein's theory of relativity relates the energy E of an object to its mass m and the speed of light c. This relationship can be represented by the formula $E = mc^2$. Solve the formula for m.

F $m = \frac{c^2}{E}$

G $m = \frac{E}{c^2}$

H $m = \frac{c}{E^2}$

J $m = \frac{E^2}{c}$

3. A rectangle has a width of 8 inches and a perimeter of 30 inches. What is the perimeter, in inches, of a similar rectangle with a width of 12 inches?

A 40

B 45

C 48

D 360

4. The rectangular prism below has a volume of 82 cubic inches. What will the volume be if the length, width, and height of the prism are all doubled?

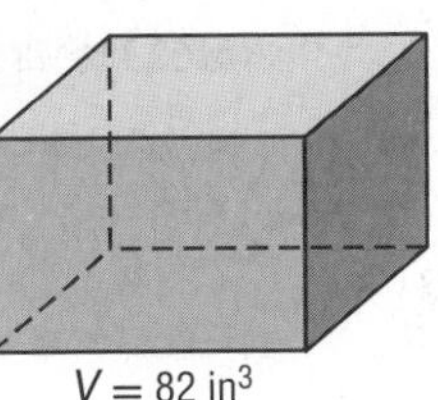

F 41 in^3

G 164 in^3

H 482 in^3

J 656 in^3

5. Evaluate $a + (b + 1)^2$ if $a = 3$ and $b = 2$.

A −6

B −1

C 12

D 15

6. At a veterinarian's office, 2 cats and 4 dogs are seen in a random order. What is the probability that the 2 cats are seen in a row?

F $\frac{1}{3}$

G $\frac{2}{3}$

H $\frac{1}{2}$

J $\frac{3}{5}$

CHAPTER 1

Standardized Test Practice

Cumulative

Multiple Choice

Read each question. Then fill in the correct answer on the answer document provided by your teacher or on a sheet of paper.

1. Evaluate $\frac{m^2 + 2mn}{n^2 - 1}$ if $m = -3$ and $n = 2$.

A -3

B -1

C 2

D 4

2. The volume of a cone with height h and radius r can be found by multiplying one-third π by the product of the height and the square of the radius. Which equation represents the volume of a cone?

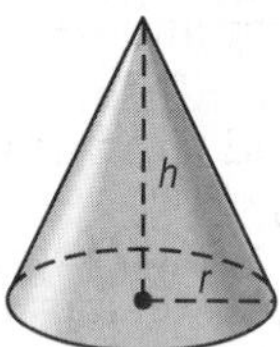

F $V = \frac{1}{3}\pi r^2 h$

G $V = 3\pi r^2 h$

H $V = \frac{1}{3}\pi r h$

J $V = \frac{1}{3}\pi r h^2$

3. Which property of equality is illustrated by the equation below?

$$a + 2 = 4 \quad \rightarrow \quad 4 = a + 2$$

A Reflexive

B Substitution

C Symmetric

D Transitive

Test-Taking Tip

Question 1 Substitute -3 for m and 2 for n in the expression. Then use the order of operations to evaluate the expression.

4. Suppose a thermometer is accurate to within plus or minus 0.2°F. If the thermometer reads 81.5°F, which absolute value inequality represents the actual temperature T?

F $|T - 81.5| < 0.2$

G $|T - 81.5| \leq 0.2$

H $|T - 0.2| < 81.5$

J $|T - 0.2| \leq 81.5$

5. To which set of numbers does -25 *not* belong?

A integers

B rationals

C reals

D wholes

6. Which number line shows the solution of the inequality $2n - 3 \geq 5n - 6$?

F

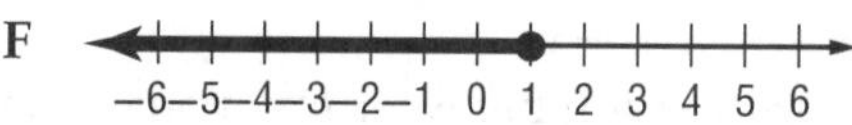

G

−6 −5 −4 −3 −2 −1 0 1 2 3 4 5 6

H

−6 −5 −4 −3 −2 −1 0 1 2 3 4 5 6

J

−6 −5 −4 −3 −2 −1 0 1 2 3 4 5 6

7. Write an algebraic expression to represent the verbal expression below.

two more than the product of a number and 5

A $\frac{n}{5} + 2$

B $2n + 5$

C $5n + 2$

D $\frac{n}{2} + 5$

Short Response/Gridded Response

Record your answers on the answer sheet provided by your teacher or on a sheet of paper.

8. Use the absolute value equation below to answer each question.

$$|x - 3| - 2 = 0$$

a. How many solutions are there of the absolute value equation?

b. Solve the equation.

9. GRIDDED RESPONSE The table below shows the fill amounts and tolerances of different size soft drinks at a fountain drink vending machine. What is the maximum acceptable fill amount, in fluid ounces, for a medium drink?

Size	Amount (fl. oz)	Tolerance (fl. oz)
small	16	0.25
medium	21	0.35
large	32	0.4

10. Simplify the expression below. Show your work.

$$-4(3a - b) + 3(-2a + 5b)$$

11. While grilling steaks, Washington likes to keep the grill temperature at 425°, plus or minus 15°.

a. Write an absolute value inequality to model this situation. Let t represent the temperature of the grill.

b. Within what range of temperatures does Washington like the grill to be when he cooks his steaks?

12. GRIDDED RESPONSE Cameron uses a laser range finder to determine distances on the golf course. Her range finder is accurate to within 0.5 yard. If Cameron measures the distance from the tee to the flag on a par 3 to be 136 yards, what is the minimum number of yards that the distance could actually be?

Extended Response

Record your answers on a sheet of paper. Show your work.

13. Cindy is evaluating the expression $\frac{-5m - 3n}{-2p + r}$ for $m = 1, n = -4, p = -3$, and $r = -2$. Her work is shown below.

$$\frac{-5m - 3n}{-2p + r} = \frac{-5(1) - 3(-4)}{-2(-3) + (-2)}$$

$$= \frac{-5 - 12}{6 - 2} = -\frac{17}{4} = -4\frac{1}{4}$$

a. What error did Cindy make in her computation?

b. What is the correct answer?

14. The table at the right shows Ricardo's scores on the first 5 math quizzes this quarter. Each quiz is worth 100 points. There will be 1 more quiz this quarter.

Quiz	Score
1	86
2	79
3	80
4	85
5	77

a. In order to receive a B, Ricardo must have a quiz average of 82 or better. Write an inequality that can be solved to find the minimum score he must earn on Quiz 6.

b. Solve the inequality you wrote in part **a**.

c. What does the solution mean?

Need ExtraHelp?

If you missed Question...	1	2	3	4	5	6	7	8	9	10	11	12	13	14
Go to Lesson...	1-1	1-3	1-3	1-6	1-2	1-5	1-3	1-4	1-6	1-2	1-6	1-4	1-1	1-5

CHAPTER 2

Linear Relations and Functions

Then

You solved equations and inequalities.

Now

You will:

- Use equations of relations and functions.
- Determine the slope of a line.
- Use scatter plots and prediction equations.
- Graph linear inequalities.

Why? ▲

RECREATION **Linear functions** can be used to model many aspects of recreational activities such as distance ridden on a bicycle, the amount of money a group of people would spend at a state fair, the height of a water slide at various points, or the amount of money you could earn from a hobby.

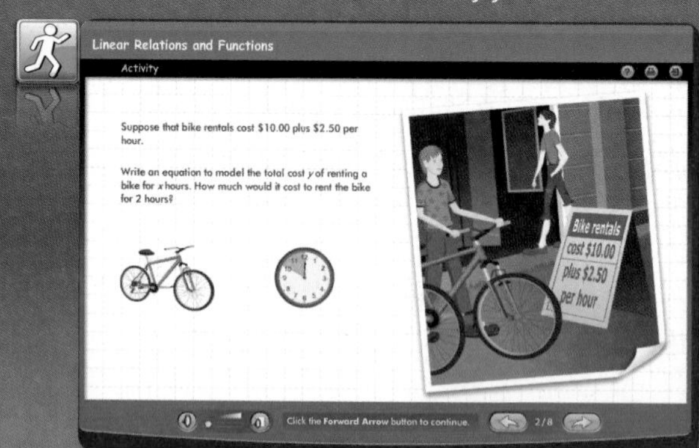

connectED.mcgraw-hill.com **Your Digital Math Portal**

Animation 
Vocabulary 
eGlossary 
Personal Tutor  
Virtual Manipulatives 
Graphing Calculator  
Audio 
Foldables 
Self-Check Practice  
Worksheets 

Get Ready for the Chapter

Diagnose Readiness | You have two options for checking prerequisite skills.

1 Textbook Option Take the Quick Check below. Refer to the Quick Review for help.

QuickCheck

Write the ordered pair for each point. Then name the quadrant in which it is located.

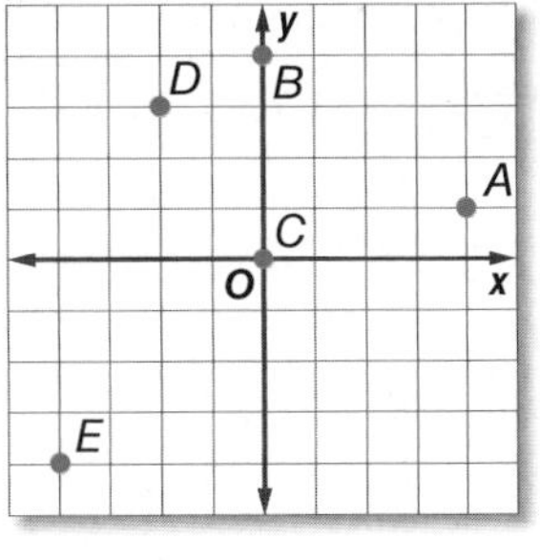

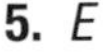

1. A
2. B
3. C
4. D
5. E

6. **BABYSITTING** Aliza earns \$6 per hour babysitting. Make a table in which the x-coordinate represents the number of hours Aliza babysits, and the y-coordinate represents the amount of money she earns.

Evaluate each expression if $a = -3$, $b = 4$, and $c = -2$.

7. $4a - 3$
8. $2b - 5c$
9. $b^2 - 3b + 6$
10. $\frac{2a + 4b}{c}$

11. **PHONE SERVICE** A cell phone company uses the expression $20 + 0.25m$ to determine the monthly charge for m minutes of air time. Find the monthly charge for 80 minutes of air time.

Solve each equation for the given variable.

12. $4x + 2y = 12$ for y
13. $a = 3b + 9$ for b
14. $15w - 10 = 5v$ for v
15. $3x - 4y = 8$ for x
16. $\frac{d}{6} + \frac{f}{3} = 4$ for d

QuickReview

Example 1

Write the ordered pair for point M. Then name the quadrant in which it is located.

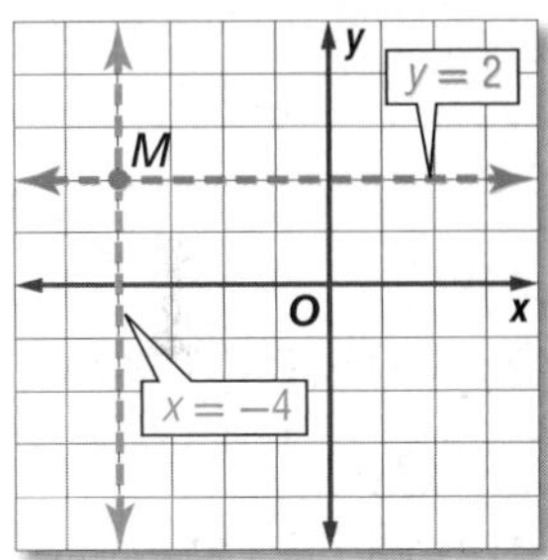

Step 1 Follow a vertical line through the point to find the x-coordinate on the x-axis.

Step 2 Follow a horizontal line through the point to find the y-coordinate on the y-axis.

Step 3 The ordered pair for point M is $(-4, 2)$. It can also be written as $M(-4, 2)$.

The x-coordinate of M is negative, while the y-coordinate is positive. So M lies in Quadrant II.

Example 2

Evaluate $3a^2 - 2ab + b^2$ if $a = 4$ and $b = -3$.

$$\begin{aligned} 3a^2 - 2ab + b^2 &= 3(4^2) - 2(4)(-3) + (-3)^2 \\ &= 3(16) - 2(4)(-3) + 9 \\ &= 48 - (-24) + 9 \\ &= 48 + 24 + 9 \\ &= 81 \end{aligned}$$

Example 3

Solve $3x + 6y = 24$ for y.

$3x + 6y = 24$	Original equation
$3x + 6y - 3x = 24 - 3x$	Subtract $3x$ from each side.
$6y = 24 - 3x$	Simplify.
$\frac{6y}{6} = \frac{24}{6} - \frac{3x}{6}$	Divide each side by 6.
$y = 4 - \frac{1}{2}x$	Simplify.

2 Online Option Take an online self-check Chapter Readiness Quiz at connectED.mcgraw-hill.com.

Get Started on the Chapter

You will learn several new concepts, skills, and vocabulary terms as you study Chapter 2. To get ready, identify important terms and organize your resources. You may wish to refer to Chapter 0 to review prerequisite skills.

FOLDABLES® StudyOrganizer

Linear Relations and Functions Make this Foldable to help you organize your Chapter 2 notes about linear relations and functions. Begin with four sheets of notebook paper.

1 **Fold** each sheet of paper in half from top to bottom.

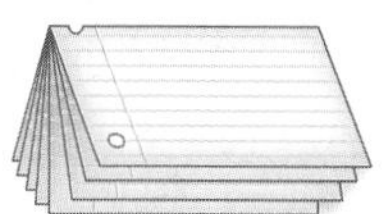

2 **Cut** along the fold. Staple the eight half-sheets together to form a booklet.

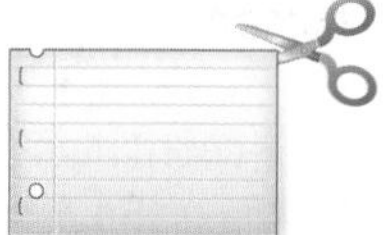

3 **Cut** tabs into the margin. The top tab is 2 lines deep, the next tab is 6 lines deep, and so on.

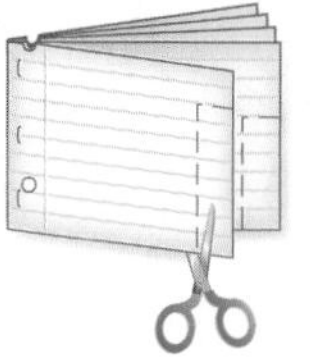

4 **Label** each of the tabs with a lesson number.

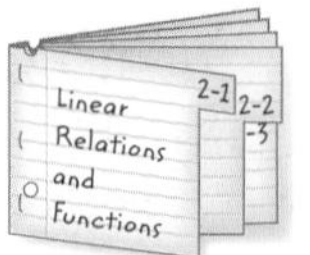

NewVocabulary

English		Español
one-to-one function	p. 61	función biunívoca
onto function	p. 61	función
discrete relation	p. 62	relación discreta
continuous relation	p. 62	relación continua
vertical line test	p. 62	prueba de la recta vertical
independent variable	p. 64	variable independiente
dependent variable	p. 64	variable dependiente
linear equation	p. 69	ecuación lineal
linear function	p. 69	función lineal
rate of change	p. 76	tasa de cambio
bivariate data	p. 92	datos bivariados
positive correlation	p. 92	correlación positiva
negative correlation	p. 92	correlación negativa
line of fit	p. 92	recta de ajuste
regression line	p. 94	línea de regresión
piecewise-linear function	p. 102	función a intervalos lineal
absolute value function	p. 103	función del valor absoluto
parent function	p. 109	función madre
quadratic function	p. 109	función cuadrática
linear inequality	p. 117	desigualdad lineal

ReviewVocabulary

equation ecuación a mathematical sentence stating that two mathematical expressions are equal

function función a relation in which each x-coordinate is paired with exactly one y-coordinate

relation relación a set of ordered pairs

Relation

Domain: −1, 1, 3, 5 — Range: 0, 2, 4, 6

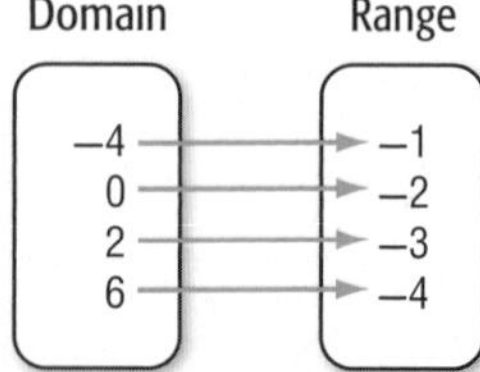

LESSON 2-1 Relations and Functions

:: Then	:: Now	:: Why?
You identified domains and ranges for given situations.	**1** Analyze relations and functions. **2** Use equations of relations and functions.	The table shows the monthly average low and high temperatures for Charlotte, North Carolina. Each month's average temperatures can be represented by the ordered pair (average low, average high). For example, January's average temperatures can be expressed as (32, 51).

Monthly Average Temperature (°F) Charlotte, NC						
Month	Jan	Feb	Mar	Apr	May	Jun
Low	32	34	42	49	58	66
High	51	56	64	73	80	87
Month	Jul	Aug	Sep	Oct	Nov	Dec
Low	71	69	63	51	42	35
High	90	88	82	73	63	54

Source: The Weather Channel

NewVocabulary
one-to-one function
onto function
discrete relation
continuous relation
vertical line test
independent variable
dependent variable
function notation

Common Core State Standards

Content Standards

F.IF.4 For a function that models a relationship between two quantities, interpret key features of graphs and tables in terms of the quantities, and sketch graphs showing key features given a verbal description of the relationship.

F.IF.5 Relate the domain of a function to its graph and, where applicable, to the quantitative relationship it describes.

Mathematical Practices

1 Make sense of problems and persevere in solving them.

7 Look for and make use of structure.

1 Relations and Functions

Recall that a function is a relation in which each element of the domain is paired with exactly one element in the range. All functions map elements of the domain to elements of the range, but they may differ in the way the elements of the domain and range are paired.

KeyConcept Functions

one-to-one function	onto function	both one-to-one and onto
Each element of the domain pairs to exactly one unique element of the range.	Each element of the range corresponds to an element of the domain.	Each element of the domain is paired to exactly one element of the range, and each element of the range corresponds to a unique element of the domain.

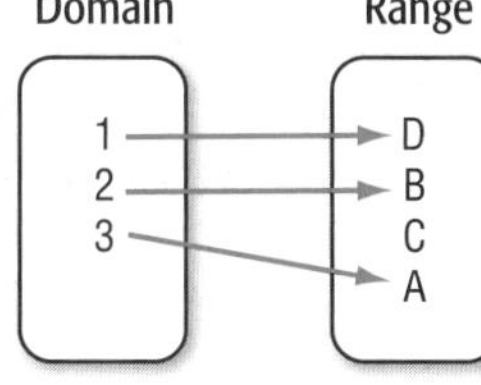

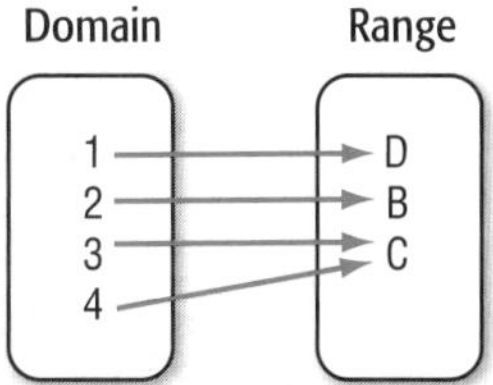

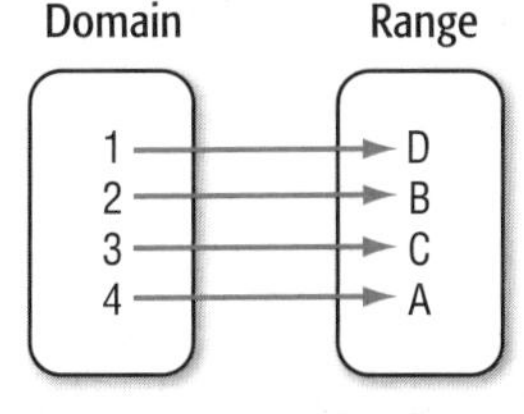

Example 1 Domain and Range

State the domain and range of each relation. Then determine whether each relation is a *function*. If it is a function, determine if it is *one-to-one*, *onto*, *both*, or *neither*.

a. $\{(-6, -1), (-5, -9), (-3, -7), (-1, 7), (6, -9)\}$

Domain: $\{-6, -5, -3, -1, 6\}$ Range: $\{-9, -7, -1, 7\}$

function: Yes, because each element of the domain is paired with one element of the range.

one-to-one: No, because each element of the domain is not paired with a unique element of the range.

onto: Yes, because each element of the range corresponds to an element of the domain.

b.

x	2	−1	−2	−1	2
y	−2	−1	0	1	2

Domain: {−2, −1, 2} Range: {−2, −1, 0, 1, 2}

The relation is not a function because 2 is mapped to both −2 and 2, and −1 is mapped to both −1 and 1.

Guided Practice

State the domain and range of each relation. Then determine whether each relation is a *function*. If it is a function, determine if it is *one-to-one*, *onto*, *both*, or *neither*.

1A.

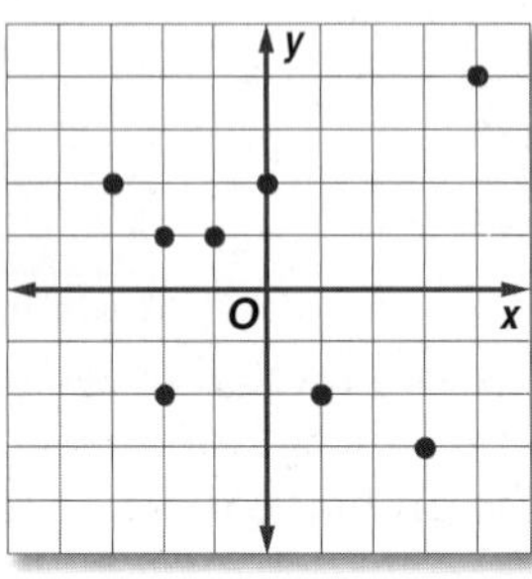

1B.

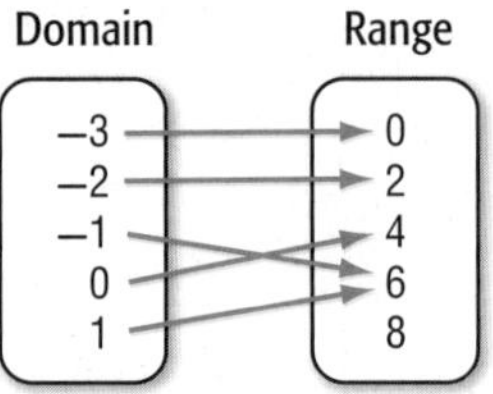

Study Tip

CCSS Structure

Notice that Graph A is composed of individual, or discrete, points while Graph B continues from one point to the next with no gaps.

A relation in which the domain is a set of individual points, like the relation in Graph A, is said to be a **discrete relation**. Notice that its graph consists of points that are not connected. When the domain of a relation has an infinite number of elements and the relation can be graphed with a line or smooth curve, the relation is a **continuous relation**.

Graph A

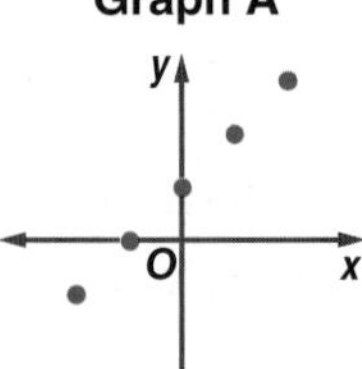

discrete relation

Graph B

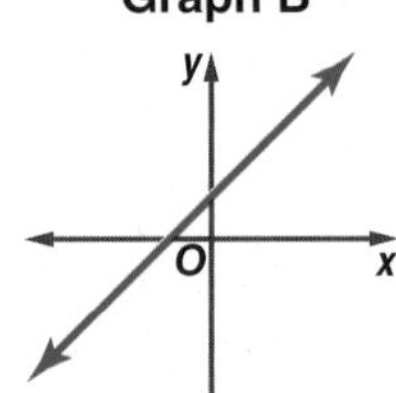

continuous relation

With both discrete and continuous graphs, you can use the **vertical line test** to determine whether the relation is a function.

Key Concept Vertical Line Test

Words

If no vertical line intersects a graph in more than one point, the graph represents a function.

If a vertical line intersects a graph in two or more points, the graph does not represent a function.

Models

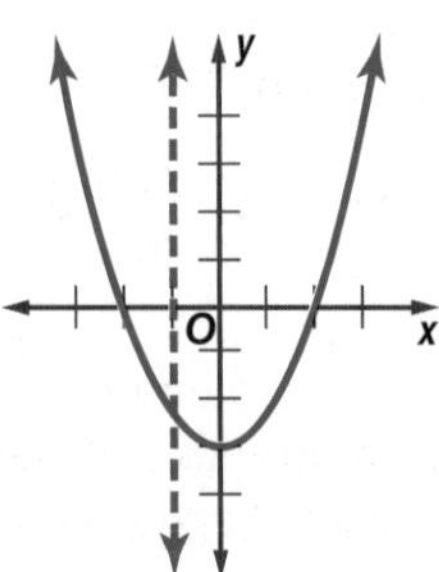

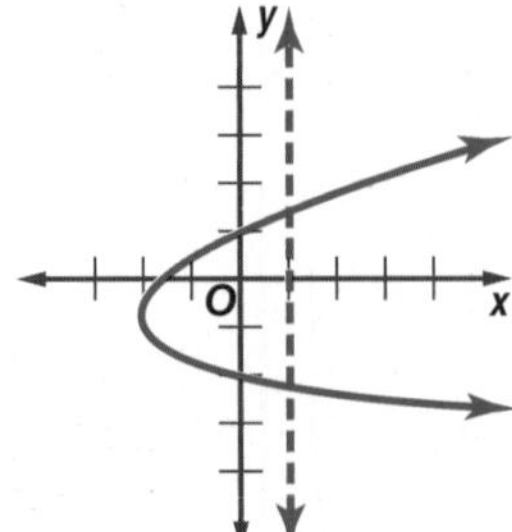

Real-WorldLink

Lance Armstrong has won the Tour de France more than any other cyclist, having won 7 consecutive races from 1999 through 2005.

Source: *USA Cycling*

Real-World Example 2

BICYCLING The graph shows the length of the Tour de France in kilometers each year from 2000 through 2009. Is the relation *discrete* or *continuous*? Does the graph represent a function?

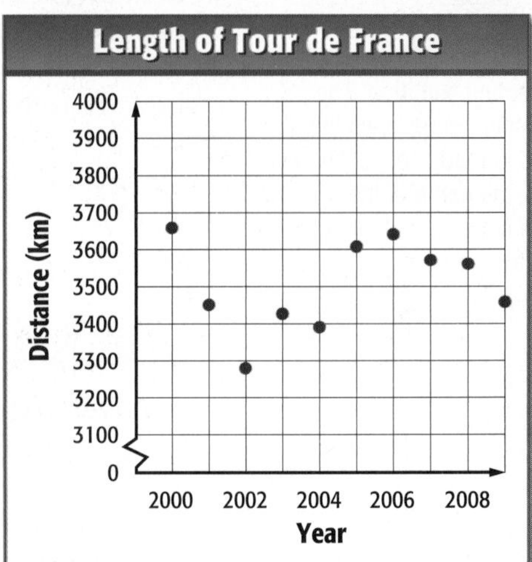

Because the graph consists of distinct points, the function is discrete. Use the vertical line test. No vertical line can be drawn that contains more than one of the data points. Therefore, the relation is a function.

GuidedPractice

2. The number of employees a company had in each year from 2004 to 2009 were 25, 28, 34, 31, 27, and 29. Graph this information and determine whether the relation is *discrete* or *continuous*. Does the graph represent a function?

2 Equations of Relations and Functions

Relations and functions can also be represented by equations. The solutions of an equation in x and y are the set of ordered pairs (x, y) that make the equation true. To determine whether an equation represents a function, it is often simplest to look at the graph of the relation.

Example 3 Graph a Relation

Graph $y = \frac{1}{2}x - 3$, and determine the domain and range. Then determine whether the equation is a *function*, is *one-to-one*, *onto*, *both*, or *neither*. State whether it is *discrete* or *continuous*.

Make a table of values that satisfy the equation. Then graph the equation.

x	y
−4	−5
−2	−4
0	−3
2	−2
4	−1

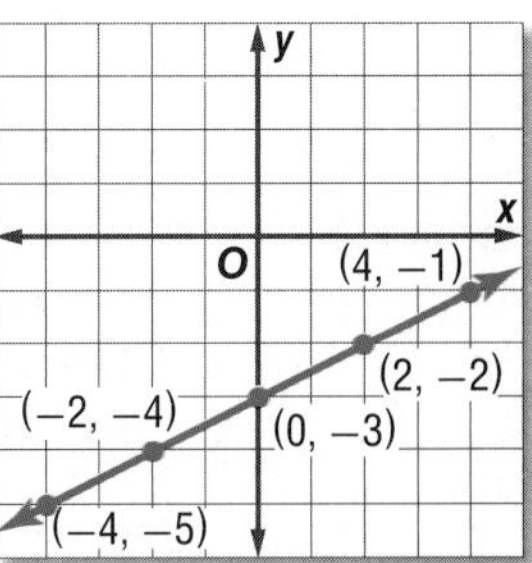

Every real number is the x-coordinate of some point on the line, and every real number is the y-coordinate of some point on the line. So the domain and range are both all real numbers.

The graph passes the vertical line test, so the equation is a function. Every x-value is paired with exactly one unique y-value, and every y-value corresponds to an x-value. Thus, the function is both one-to-one and onto.

Because the graph is a solid line without breaks, the function is continuous.

GuidedPractice

3. Graph $y = x^2 + 1$, and determine the domain and range. Then determine whether the equation is a *function*, is *one-to-one*, *onto*, *both*, or *neither*. State whether it is *discrete* or *continuous*.

BERND THISSEN/DPA/epa/Corbis

When an equation represents a function, the variable, often x, with values making up the domain is called the **independent variable**. The other variable, often y, is called the **dependent variable** because its values depend on x.

ReadingMath

Function Notation The symbol $f(x)$ replaces the y and is read "f of x." The f is just the name of the function. It is not a variable that is multiplied by x.

Equations that represent functions are often written in **function notation**. The equation $y = 5x - 1$ can be written as $f(x) = 5x - 1$. Suppose you want to find the value in the range that corresponds to the element -6 in the domain of the function. The value $f(-6)$ is found by substituting -6 for each x in the equation. Therefore, $f(-6) = 5(-6) - 1$ or -31.

PT

Example 4 Evaluate a Function

Given $f(x) = 2x^2 - 8$, find each value.

a. $f(6)$

$f(x) = 2x^2 - 8$	Original function
$f(6) = 2(6)^2 - 8$	Substitute.
$= 2(36) - 8$	Evaluate 6^2.
$= 72 - 8$ or 64	Simplify.

b. $f(2y)$

$f(x) = 2x^2 - 8$	Original function
$f(2y) = 2(2y)^2 - 8$	Substitute.
$= 2(4y^2) - 8$	$(2y)^2 = 2^2y^2$
$= 8y^2 - 8$	Simplify.

GuidedPractice

Given $g(x) = 0.5x^2 - 5x + 3.5$, find each value.

4A. $g(2.8)$ **4B.** $g(4a)$

Check Your Understanding

 = Step-by-Step Solutions begin on page R14.

Example 1

CCSS STRUCTURE State the domain and range of each relation. Then determine whether each relation is a *function*. If it is a function, determine if it is *one-to-one, onto, both,* or *neither*.

1.

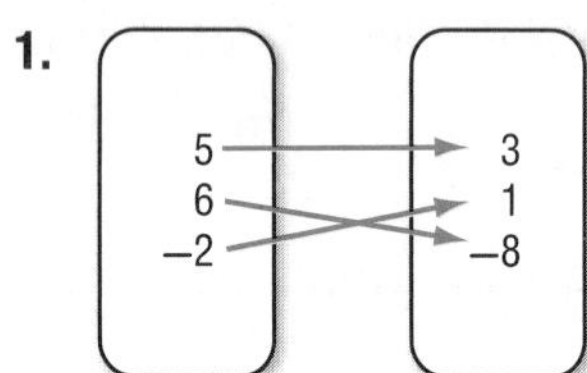

2.

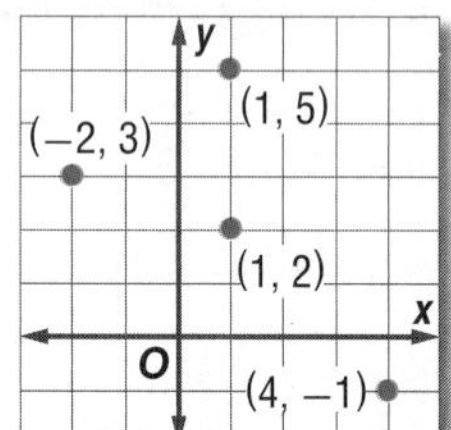

3.

x	y
−2	−4
1	−4
4	−2
8	6

Example 2

4. BASKETBALL The table shows the average points per game for Dwayne Wade of the Miami Heat for four seasons.

Season	Dwayne Wade's Age	Average Points Per Game
2005–2006	24	27.2
2006–2007	25	27.4
2007–2008	26	24.6
2008–2009	27	30.2

Source: *Basketball-Reference*

a. Assume that the ages are the domain. Identify the domain and range.

b. Write a relation of ordered pairs for the data.

c. State whether the relation is *discrete* or *continuous*.

d. Graph the relation. Is this relation a function?

Jed Jacobsohn/Getty Images Sport/Getty Images

Example 3

Graph each equation, and determine the domain and range. Determine whether the equation is a *function*, is *one-to-one, onto, both,* or *neither*. Then state whether it is *discrete* or *continuous*.

5. $y = 5x + 4$ **6.** $y = -4x - 2$ **7.** $y = 3x^2$ **8.** $x = 7$

Example 4 **Evaluate each function.**

9. $f(-3)$ if $f(x) = -4x - 8$

10. $g(5)$ if $g(x) = -2x^2 - 4x + 1$

Practice and Problem Solving

Extra Practice is on page R2.

Example 1 **State the domain and range of each relation. Then determine whether each relation is a *function*. If it is a function, determine if it is *one-to-one*, *onto*, *both*, or *neither*.**

11.

x	y
−0.3	−6
0.4	−3
1.2	−1
1.2	4

12.

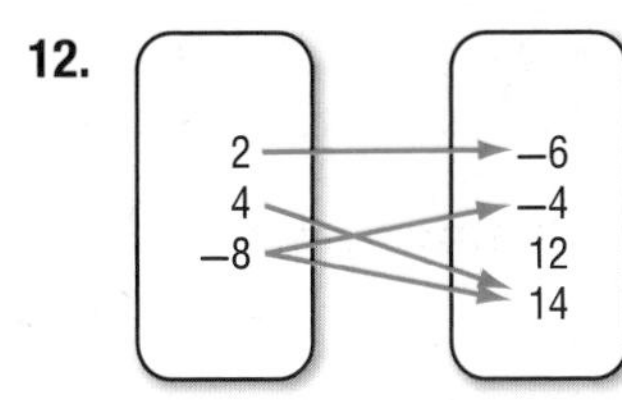

13. $\{(-3, -4), (-1, 0), (3, 0), (5, 3)\}$

Example 2 **14. POLITICS** The table below shows the population of several states and the number of U.S. representatives from those states.

a. Make a graph of the data with population on the horizontal axis and representatives on the vertical axis.

b. Identify the domain and range.

c. Is the relation *discrete* or *continuous*?

d. Does the graph represent a function? Explain your reasoning.

State	Population (millions)	Number of Representatives
California	33.93	53
Florida	16.03	25
Illinois	12.44	19
New York	19.00	29
North Carolina	8.07	13
Texas	20.90	32

Source: U.S. Bureau of the Census

Example 3 **CCSS STRUCTURE** **Graph each equation, and determine the domain and range. Determine whether the equation is a *function*, is *one-to-one*, *onto*, *both*, or *neither*. Then state whether it is *discrete* or *continuous*.**

15. $y = -3x + 2$

16. $y = 0.5x - 3$

17. $y = 2x^2$

18. $y = -5x^2$

19. $y = 4x^2 - 8$

20. $y = -3x^3 - 1$

Example 4 **Evaluate each function.**

21. $f(-8)$ if $f(x) = 5x^3 + 1$

22. $f(2.5)$ if $f(x) = 16x^2$

23. DIVING The table below shows the pressure on a diver at various depths.

Depth (ft)	0	20	40	60	80	100
Pressure (atm)	1	1.6	2.2	2.8	3.4	4

a. Write a relation to represent the data.

b. Graph the relation.

c. Identify the domain and range. Is the relation *discrete* or *continuous*?

d. Is the relation a function? Explain your reasoning.

Find each value if $f(x) = 3x + 2$, $g(x) = -2x^2$, and $h(x) = -4x^2 - 2x + 5$.

24. $f(-5)$

25. $f(9)$

26. $g(-3)$

27. $g(-6)$

28. $h(3)$

29. $h(8)$

30. $f\left(\frac{2}{3}\right)$

31. $g\left(\frac{3}{2}\right)$

32. $h\left(\frac{1}{5}\right)$

33 **PODCASTS** Chaz has a collection of 15 podcasts downloaded on his digital audio player. He decides to download 3 more podcasts each month. The function $P(t) = 15 + 3t$ counts the number of podcasts $P(t)$ he has after t months. How many podcasts will he have after 8 months?

34. **MULTIPLE REPRESENTATIONS** In this problem you will investigate one-to-one and onto functions.

a. Graphical Graph each function on a separate graphing calculator screen.

$f(x) = x^2$ $\quad g(x) = 2^x$ $\quad h(x) = x^3 - 3x^2 - 5x + 6$ $\quad j(x) = x^3$

b. Tabular Use the graphs to create a table showing the number of times a horizontal line could intersect the graph of each function. List all possibilities.

c. Analytical For a function to be one-to-one, a horizontal line on the graph of the function can intersect the function at most once. Which functions meet this condition? Which do not? Explain your reasoning.

d. Analytical For a function to be onto, every possible horizontal line on the graph of the function must intersect the function at least once. Which functions meet this condition? Which do not? Explain your reasoning.

e. Graphical Create a table showing whether each function is one-to-one and/or onto.

H.O.T. Problems Use Higher-Order Thinking Skills

35. **CCSS CRITIQUE** Omar and Madison are finding $f(3d)$ for the function $f(x) = -4x^2 - 2x + 1$. Is either of them correct? Explain your reasoning.

Omar	Madison
$f(3d) = -4(3d)^2 - 2(3d) + 1$	$f(3d) = -4(3d)^2 - 2(3d) + 1$
$= -4(9d^2) - 6d + 1$	$= 12d^2 - 6d + 1$
$= -36d^2 - 6d + 1$	

36. **CHALLENGE** Consider the functions $f(x)$ and $g(x)$. $f(a) = 19$ and $g(a) = 33$, while $f(b) = 31$ and $g(b) = 51$. If $a = 5$ and $b = 8$, find two possible functions to represent $f(x)$ and $g(x)$.

37. **REASONING** If the graph of a relation crosses the y-axis at more than one point, is the relation *sometimes, always,* or *never* a function? Explain your reasoning.

38. **OPEN ENDED** Graph a relation that can be used to represent each of the following.

a. the height of a baseball that is hit into the outfield

b. the speed of a car that travels to the store, stopping at two lights along the way

c. the height of a person from age 5 to age 80

d. the temperature on a typical day from 6 A.M. to 11 P.M.

39. **REASONING** Determine whether the following statement is *true* or *false*. Explain your reasoning.

If a function is onto, then it must be one-to-one as well.

40. **WRITING IN MATH** Explain why the vertical line test can determine if a relation is a function.

Standardized Test Practice

41. Patricia's swimming pool contains 19,500 gallons of water. She drains the pool at a rate of 6 gallons per minute. Which of these equations represents the number of gallons of water g remaining in the pool after m minutes?

A $g = 19{,}500 - 6m$

B $g = 19{,}500 + 6m$

C $g = \frac{19{,}500}{6m}$

D $g = \frac{6m}{19{,}500}$

42. SHORT RESPONSE Look at the pattern below.

$$-\frac{5}{2}, -2, -\frac{3}{2}, -1, \ldots$$

If the pattern continues, what will the next term be?

43. GEOMETRY Which set of dimensions represents a triangle similar to the triangle shown below?

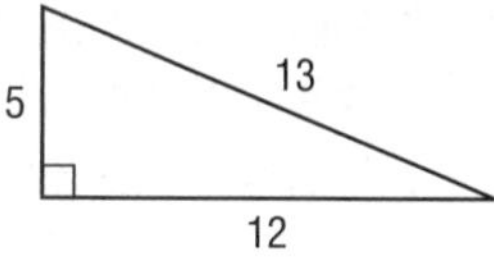

F 1 unit, 2 units, 3 units

G 7 units, 11 units, 12 units

H 10 units, 23 units, 24 units

J 20 units, 48 units, 52 units

44. SAT/ACT If $g(x) = x^2$, which expression is equal to $g(x + 1)$?

A 1

B $x^2 + 1$

C $x^2 + 2x + 1$

D $x^2 - x$

E $x^2 + x + 1$

Spiral Review

Solve each inequality. (Lesson 1-6)

45. $48 > 7y + 6 > 20$

46. $z + 12 > 18$ or $-2z + 16 > 12$

47. $2|4x + 2| + 3 > 21$

48. CLUBS Mr. Willis is starting a chess club at his high school. He sent the advertisement at the right to all of the homerooms. Write an absolute value inequality representing the situation. (Lesson 1-6)

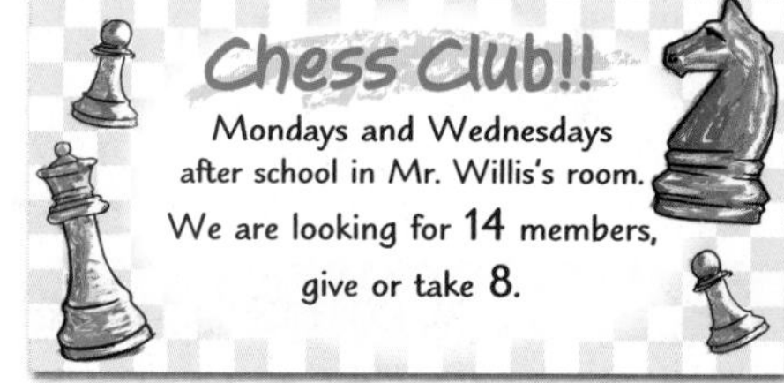

49. SALES Ling can spend no more than \$120 at the summer sale of a department store. She wants to buy shirts on sale for \$15 each. Write and solve an inequality to determine the number of shirts she can buy. (Lesson 1-5)

Solve each equation. Check your solutions. (Lesson 1-4)

50. $18 = 2|2a + 6| - 2$

51. $2 = -3|4c - 5| + 8$

52. $-5 = 2|3b + 4| - 9$

Simplify each expression. (Lesson 1-2)

53. $6(3a - 2b) + 3(5a + 4b)$

54. $-4(5x - 3y) + 2(y + 3x)$

55. $-7(2c - 4d) + 8(3c + d)$

Skills Review

Solve each equation. Check your solutions.

56. $5x + 2 = 32$

57. $6a - 3 = 21$

58. $-2x + 5 = 5x + 19$

59. $6b + 4 = -2b - 28$

60. $2(x + 5) - 3(x - 4) = 19$

61. $4(2y - 3) + 5(3y + 1) = -99$

62. $5c - 8 + 2c = 4c + 10$

63. $8d - 4 + 3d = 2d - 100 - 7d$

64. $10y - 5 - 3y = 4(2y + 3) - 20$

EXTEND 2-1 Algebra Lab
Discrete and Continuous Functions

A cup of frozen yogurt costs \$2 at the Yogurt Shack. We might describe the cost of x cups of yogurt using the continuous function $y = 2x$, where y is the total cost in dollars. The graph of that function is shown at the right.

Common Core State Standards
Content Standards
F.IF.4 For a function that models a relationship between two quantities, interpret key features of graphs and tables in terms of the quantities, and sketch graphs showing key features given a verbal description of the relationship.

From the graph, you can see that 2 cups of yogurt cost \$4, 3 cups cost \$6, and so on. The graph also shows that 1.5 cups of yogurt cost 2(1.5) or \$3. However, the Yogurt Shack probably will not sell partial cups of yogurt. This function is more accurately modeled with a discrete function.

The graph of the discrete function at the right also models the cost of buying cups of frozen yogurt. The domain in this graph makes sense in this situation.

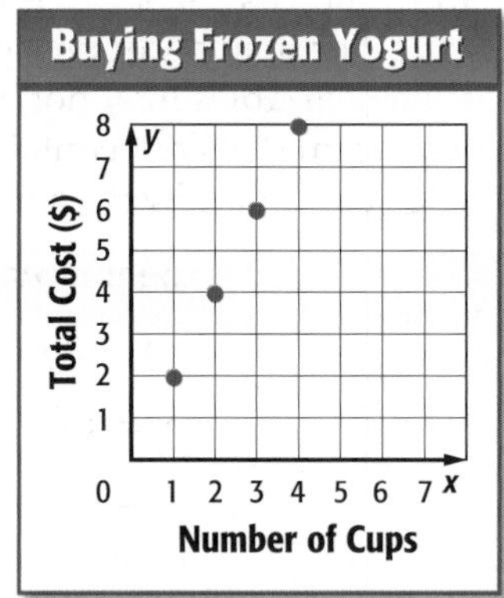

When choosing a discrete function or a continuous function to model a real-world situation, consider whether all real numbers make sense as part of the domain.

Exercises

Determine whether each function is correctly modeled using a discrete or continuous function. Explain your reasoning.

1. 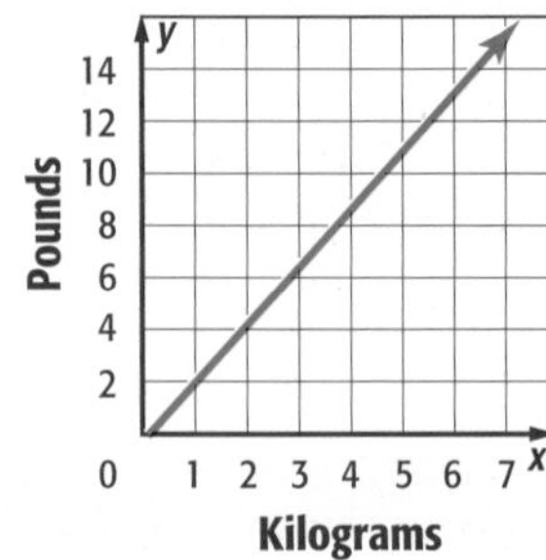

2. 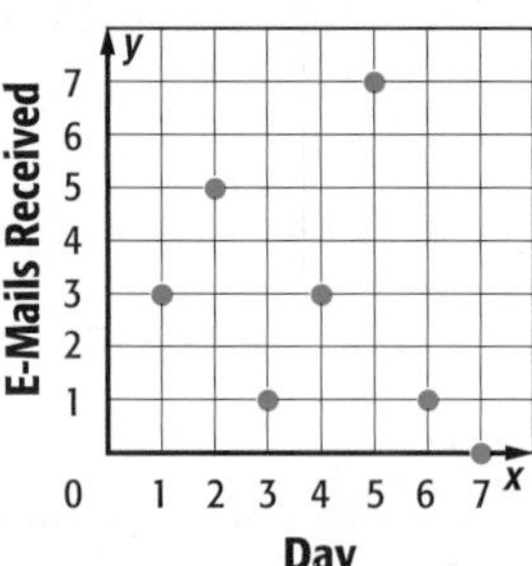

3. y represents the distance a car travels in x hours.

4. y represents the total number of riders who have ridden on a roller coaster after x rides.

5. **WRITING IN MATH** Give an example of a real-world function that is discrete and a real-world function that is continuous. Explain your reasoning.

LESSON 2-2 Linear Relations and Functions

Then	Now	Why?
• You analyzed relations and functions.	1 Identify linear relations and functions. 2 Write linear equations in standard form.	• Laura does yard work to earn money during the summer. She either cuts grass x or does general gardening y, and she schedules 5 jobs per day. The equation $x + y = 5$ can be used to relate how many of each task Laura can do in a day.

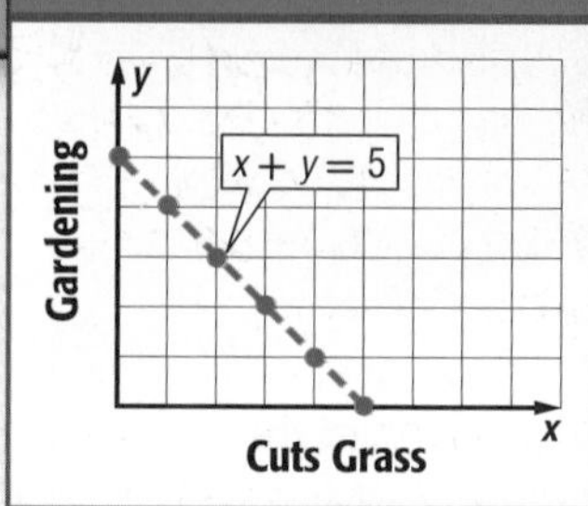

NewVocabulary
linear relation
nonlinear relation
linear equation
linear function
standard form
y-intercept
x-intercept

Common Core State Standards

Content Standards
F.IF.4 For a function that models a relationship between two quantities, interpret key features of graphs and tables in terms of the quantities, and sketch graphs showing key features given a verbal description of the relationship.
F.IF.9 Compare properties of two functions each represented in a different way (algebraically, graphically, numerically in tables, or by verbal descriptions).

Mathematical Practices
3 Construct viable arguments and critique the reasoning of others.

1 Linear Relations and Functions

The points on the graph above lie along a straight line. Relations that have straight line graphs are called **linear relations**. Relations that are not linear are called **nonlinear relations**.

An equation such as $x + y = 5$ is called a linear equation. A **linear equation** has no operations other than addition, subtraction, and multiplication of a variable by a constant. The variables may not be multiplied together or appear in a denominator. A linear equation does not contain variables with exponents other than 1. The graph of a linear equation is always a line.

Linear equations	Nonlinear equations
$4x - 5y = 16$	$2x + 6y^2 = -25$
$x = 10$	$y = \sqrt{x} + 2$
$y = -\frac{2}{3}x - 1$	$x + xy = -\frac{5}{8}$
$y = \frac{1}{2}x$	$y = \frac{1}{x}$

A **linear function** is a function with ordered pairs that satisfy a linear equation. Any linear function can be written in the form $f(x) = mx + b$, where m and b are real numbers.

Example 1 Identify Linear Functions

State whether each function is a linear function. Write *yes* or *no*. Explain.

a. $f(x) = 8 - \frac{3}{4}x$

Yes; it can be written as $f(x) = -\frac{3}{4}x + 8$.

$m = -\frac{3}{4}, b = 8$

b. $f(x) = \frac{2}{x}$

No; the expression includes division by the variable.

c. $g(x, y) = 3xy - 4$

No; the two variables are multiplied together.

GuidedPractice

1A. $f(x) = \frac{5}{x + 6}$

1B. $g(x) = -\frac{3}{2}x + \frac{1}{3}$

You can evaluate linear functions by substituting values for x or $f(x)$.

Real-WorldLink

The largest member of the grass family, bamboo, is capable of growing from 1 to 4 feet per day.

Source: Infoplease

PT

Real-World Example 2 Evaluate a Linear Function

PLANTS The growth rate of a sample of Bermuda grass is given by the function $f(x) = 5.9x + 3.25$, where $f(x)$ is the total height in inches x days after an initial measurement.

a. How tall is the sample after 3 days?

$f(x) = 5.9x + 3.25$ Original function

$f(3) = 5.9(3) + 3.25$ Substitute 3 for x.

$= 20.95$ Simplify.

The height of the sample after 3 days is 20.95 inches.

b. The term 3.25 in the function represents the height of the grass when it was initially measured. The sample is how many times as tall after 3 days?

Divide the height after 3 days by the initial height. $\frac{20.95}{3.25} \approx 6.4$

The height after 3 days is about 6.4 times as great as the initial height.

GuidedPractice

2A. If the Bermuda grass is 50.45 inches tall, how many days has it been since it was last cut?

2B. Is it reasonable to think that this rate of growth can be maintained for long periods of time? Explain.

2 Standard Form

Any linear equation can be written in **standard form**, $Ax + By = C$, where A, B, and C are integers with a greatest common factor of 1.

KeyConcept Standard Form of a Linear Equation

Words	The standard form of a linear equation is $Ax + By = C$, where A, B, and C are integers with a greatest common factor of 1, $A \geq 0$, and A and B are not both zero.
Example	$3x + 5y = 12$; $A = 3$, $B = 5$, and $C = 12$

PT

Example 3 Standard Form

Write $-\frac{3}{10}x = 8y - 15$ in standard form. Identify A, B, and C.

$-\frac{3}{10}x = 8y - 15$ Original equation

$-\frac{3}{10}x - 8y = -15$ Subtract $8y$ from each side.

$3x + 80y = 150$ Multiply each side by -10.

$A = 3$, $B = 80$, and $C = 150$

GuidedPractice

Write each equation in standard form. Identify A, B, and C.

3A. $2y = 4x + 5$

3B. $3x - 6y - 9 = 0$

Since two points determine a line, one way to graph a linear function is to find the points at which the graph intersects each axis and connect them with a line. The y-coordinate of the point at which a graph crosses the y-axis is called the **y-intercept**. Likewise, the x-coordinate of the point at which it crosses the x-axis is called the **x-intercept**.

StudyTip

Vertical and Horizontal Lines When C represents a constant, an equation of the form $x = C$ represents a vertical line with only an x-intercept. The equation $y = C$ represents a horizontal line with only a y-intercept.

Example 4 Use Intercepts to Graph a Line

Find the x-intercept and the y-intercept of the graph of $2x - 3y + 8 = 0$. Then graph the equation.

The x-intercept is the value of x when $y = 0$.

$2x - 3y + 8 = 0$ Original equation

$2x - 3(0) + 8 = 0$ Substitute 0 for y.

$2x = -8$ Subtract 8 from each side.

$x = -4$ Divide each side by 2.

The x-intercept is -4.

Likewise, the y-intercept is the value of y when $x = 0$.

$2x - 3y + 8 = 0$ Original equation

$2(0) - 3y + 8 = 0$ Substitute 0 for x.

$-3y = -8$ Subtract 8 from each side.

$y = \frac{8}{3}$ Divide each side by 3.

The y-intercept is $\frac{8}{3}$.

Use these ordered pairs to graph the equation.

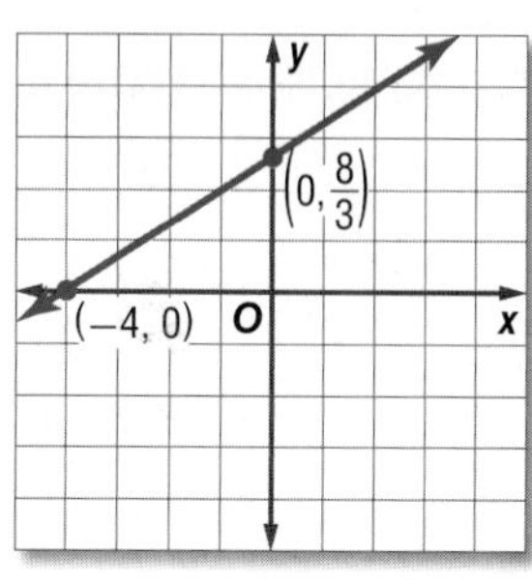

GuidedPractice

4. Find the x-intercept and the y-intercept of the graph of $2x + 5y - 10 = 0$. Then graph the equation.

Check Your Understanding

= Step-by-Step Solutions begin on page R14.

Example 1 **State whether each function is a linear function. Write *yes* or *no*. Explain.**

1. $f(x) = \frac{x + 12}{5}$ **2.** $g(x) = \frac{7 - x}{x}$ **3.** $p(x) = 3x^2 - 4$ **4.** $q(x) = -8x - 21$

Example 2 **5** **RECREATION** You want to make sure that you have enough music for a car trip. If each CD is an average of 45 minutes long, the linear function $m(x) = 0.75x$ could be used to find out how many CDs you need to bring.

a. How many hours of music are there on 4 CDs?

b. If the trip you are taking is 6 hours, how many CDs should you bring?

Example 3 **CCSS STRUCTURE** **Write each equation in standard form. Identify A, B, and C.**

6. $y = -4x - 7$ **7.** $y = 6x + 5$ **8.** $3x = -2y - 1$

9. $-8x = 9y - 6$ **10.** $12y = 4x + 8$ **11.** $4x - 6y = 24$

Example 4 **Find the x-intercept and the y-intercept of the graph of each equation. Then graph the equation using the intercepts.**

12. $y = 5x + 12$ **13.** $y = 4x - 10$ **14.** $2x + 3y = 12$ **15.** $3x - 4y - 6 = 15$

Practice and Problem Solving

Extra Practice is on page R2.

Example 1

State whether each equation or function is a linear function. Write *yes* or *no*. Explain.

16. $3y - 4x = 20$
17. $y = x^2 - 6$
18. $h(x) = 6$
19. $j(x) = 2x^2 + 4x + 1$
20. $g(x) = 5 + \frac{6}{x}$
21. $f(x) = \sqrt{7 - x}$
22. $4x + \sqrt{y} = 12$
23. $\frac{1}{x} + \frac{1}{y} = 1$
24. $f(x) = \frac{4x}{5} + \frac{8}{3}$

Example 2

25. **ROLLER COASTERS** The speed of the Steel Dragon 2000 roller coaster in Mie Prefecture, Japan, can be modeled by $y = 10.4x$, where y is the distance traveled in meters in x seconds.
    a. How far does the coaster travel in 25 seconds?
    b. The speed of the Kingda Ka roller coaster in Jackson, New Jersey, can be described by $y = 33.9x$. Which coaster travels faster? Explain your reasoning.

Example 3

Write each equation in standard form. Identify *A*, *B*, and *C*.

26. $-7x - 5y = 35$
27. $8x + 3y + 6 = 0$
28. $10y - 3x + 6 = 11$
29. $-6x - 3y - 12 = 21$
30. $3y = 9x - 12$
31. $2.4y = -14.4x$
32. $\frac{2}{3}y - \frac{3}{4}x + \frac{1}{6} = 0$
33. $\frac{4}{5}y + \frac{1}{8}x = 4$
34. $-0.08x = 1.24y - 3.12$

Example 4

Find the *x*-intercept and the *y*-intercept of the graph of each equation. Then graph the equation using the intercepts.

35. $y = -8x - 4$
36. $5y = 15x - 90$
37. $-4y + 6x = -42$
38. $-9x - 7y = -30$
39. $\frac{1}{3}x - \frac{2}{9}y = 4$
40. $\frac{3}{4}y - \frac{2}{3}x = 12$

41. **CCSS MODELING** Latonya earns a commission of \$1.75 for each magazine subscription that she sells and \$1.50 for each newspaper subscription that she sells. Her goal is to earn a total of \$525 in commissions in the next two weeks.
    a. Write an equation that is a model for the different numbers of magazine and newspaper subscriptions that can be sold to meet the goal.
    b. Graph the equation. Does this equation represent a function? Explain.
    c. If Latonya sells 100 magazine subscriptions and 200 newspaper subscriptions, will she meet her goal? Explain.

42. **SNAKES** Suppose the body length L in inches of a baby snake is given by $L(m) = 1.5 + 2m$, where m is the age of the snake in months until it becomes 12 months old.
    a. Find the length of an 8-month-old snake.
    b. Find the snake's age if the length of the snake is 25.5 inches.

43. **STATE FAIR** The Ohio State Fair charges \$8 for admission and \$5 for parking. After Joey pays for admission and parking, he plans to spend all of his remaining money at the ring game, which costs \$3 per game.
    a. Write an equation representing the situation.
    b. How much did Joey spend at the fair if he paid \$6 for food and drinks and played the ring game 4 times?

Write each equation in standard form. Identify *A*, *B*, and *C*.

44. $\frac{x + 5}{3} = -2y + 4$

45. $\frac{4x - 1}{5} = 8y - 12$

46. $\frac{-2x - 8}{3} = -12y + 18$

Find the *x*-intercept and the *y*-intercept of the graph of each equation.

47 $\frac{6x + 15}{4} = 3y - 12$

48. $\frac{-8x + 12}{3} = 16y + 24$

49. $\frac{15x + 20}{4} = \frac{3y + 6}{5}$

50. FUNDRAISING The Freshman Class Student Council wanted to raise money by giving car washes. The students spent \$10 on supplies and charged \$2 per car wash.

a. Write an equation to model the situation.

b. Graph the equation.

c. How much money did they earn after 20 car washes?

d. How many car washes are needed for them to earn \$100?

51. MULTIPLE REPRESENTATIONS Consider the following linear functions.

$$f(x) = -2x + 4 \qquad g(x) = 6 \qquad h(x) = \frac{1}{3}x + 5$$

a. Graphical Graph the linear functions on separate graphs.

b. Tabular Use the graphs to complete the table.

Function	One-to-One	Onto
$f(x) = -2x + 4$		
$g(x) = 6$		
$h(x) = \frac{1}{3}x + 5$		

c. Verbal Are all linear functions one-to-one and/or onto? Explain your reasoning.

H.O.T. Problems Use Higher-Order Thinking Skills

52. CHALLENGE Write a function with an *x*-intercept of $(a, 0)$ and a *y*-intercept of $(0, b)$.

53. OPEN ENDED Write an equation of a line with an *x*-intercept of 3.

54. REASONING Determine whether an equation of the form $x = a$, where a is a constant, is *sometimes*, *always*, or *never* a function. Explain your reasoning.

55. CCSS ARGUMENTS Of the four equations shown, identify the one that does not belong. Explain your reasoning.

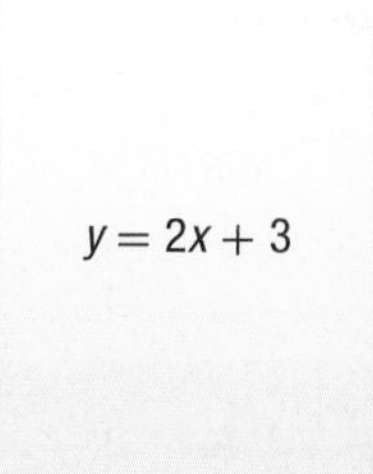

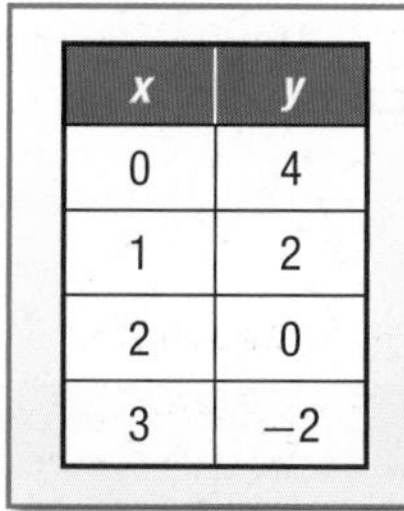

x	y
0	4
1	2
2	0
3	−2

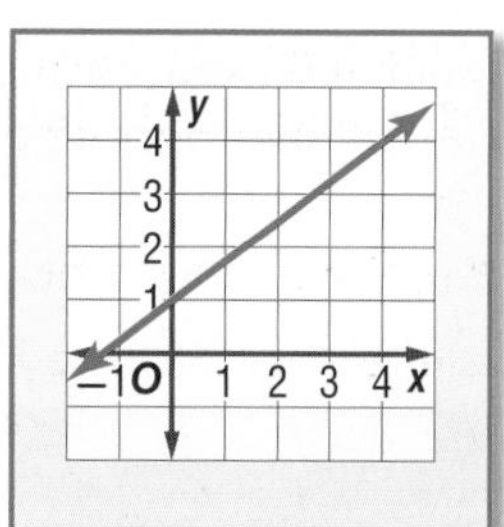

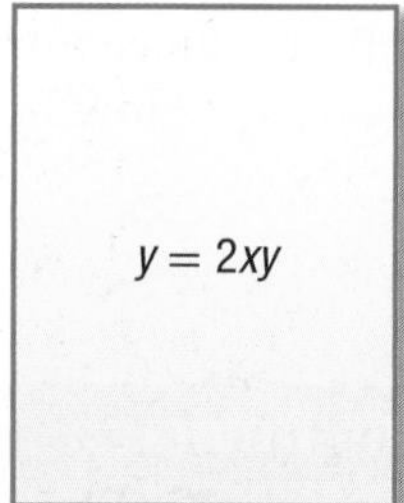

56. WRITING IN MATH Consider the graph of the relationship between hours worked and earnings.

a. When would this graph represent a linear relationship? Explain your reasoning.

b. Provide another example of a linear relationship in a real-world situation.

Standardized Test Practice

57. Tom bought n DVDs for a total cost of $15n - 2$ dollars. Which expression represents the cost of each DVD?

A $n(15n - 2)$

B $n + (15n - 2)$

C $(15n - 2) \div n; n \neq 0$

D $(15n - 2) - n$

58. SHORT RESPONSE What is the complete solution of the equation?

$$|9 - 3x| = 18$$

59. NUMBER THEORY If a, b, c, and d are consecutive odd integers and $a < b < c < d$, how much greater is $c + d$ than $a + b$?

F 2 **H** 6

G 4 **J** 8

60. SAT/ACT Which function is linear?

A $f(x) = x^2$

B $g(x) = \sqrt{x - 1}$

C $f(x) = \sqrt{9 - x^2}$

D $g(x) = \frac{2.7}{x}$

E $f(x) = 2x$

Spiral Review

State the domain and range of each relation. Then determine whether each relation is a *function*. If it is a function, determine if it is *one-to-one*, *onto*, *both*, or *neither*. (Lesson 2-1)

61.

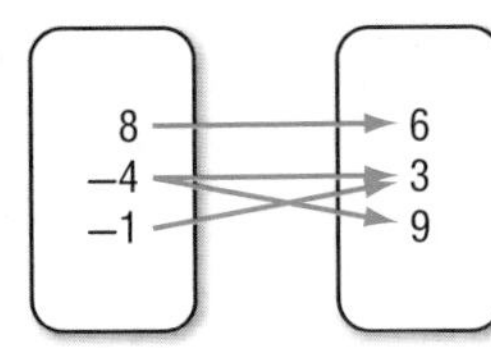

62.

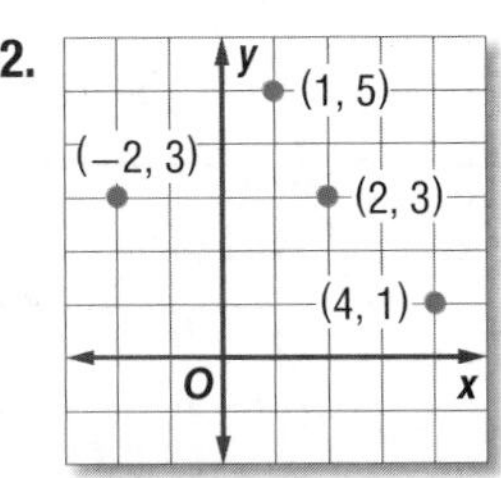

63.

x	y
−4	−2
−3	−1
−3	−1
7	9

64. SHOPPING Claudio is shopping for a new television. The average price of the televisions he likes is $800, and the actual prices differ from the average by up to $350. Write and solve an absolute value inequality to determine the price range of the televisions. (Lesson 1-6)

Evaluate each expression if $a = -6$, $b = 5$, and $c = 3.6$. (Lesson 1-1)

65. $\frac{6a - 3c}{2ab}$

66. $\frac{a + 7b}{4bc}$

67. $\frac{b - c}{a + c}$

68. FOOD Brandi can order a small, medium, or large pizza with pepperoni, mushrooms, or sausage. How many different one-topping pizzas can she order? (Lesson 0-4)

Skills Review

Evaluate each expression.

69. $\frac{12 - 8}{4 - (-2)}$

70. $\frac{5 - 9}{-3 - (-6)}$

71. $\frac{-2 - 8}{3 - (-5)}$

72. $\frac{-2 - (-6)}{-1 - (-8)}$

73. $\frac{-7 - (-11)}{-3 - 9}$

74. $\frac{-1 - 8}{7 - (-3)}$

75. $\frac{-12 - (-3)}{-6 - (-5)}$

76. $\frac{4 - 3}{2 - 5}$

EXTEND 2-2

Algebra Lab
Roots of Equations and Zeros of Functions

The *solution* of an equation is called the *root* of the equation.

CCSS Common Core State Standards
Content Standards
F.IF.4 For a function that models a relationship between two quantities, interpret key features of graphs and tables in terms of the quantities, and sketch graphs showing key features given a verbal description of the relationship.

Example Determine Roots

Find the root of $0 = 5x - 10$.

$0 = 5x - 10$ Original equation

$10 = 5x$ Add 10 to each side.

$2 = x$ Divide each side by 5.

The root of the equation is 2.

You can also find the root of an equation by finding the *zero* of its related function. Values of x for which $f(x) = 0$ are called *zeros* of the function f.

Linear Equation	Related Linear Function
$0 = 5x - 10$	$f(x) = 5x - 10$ or $y = 5x - 10$

The zero of a function is the *x-intercept* of its graph. Since the graph of $y = 5x - 10$ intersects the x-axis at 2, the zero of the function is 2.

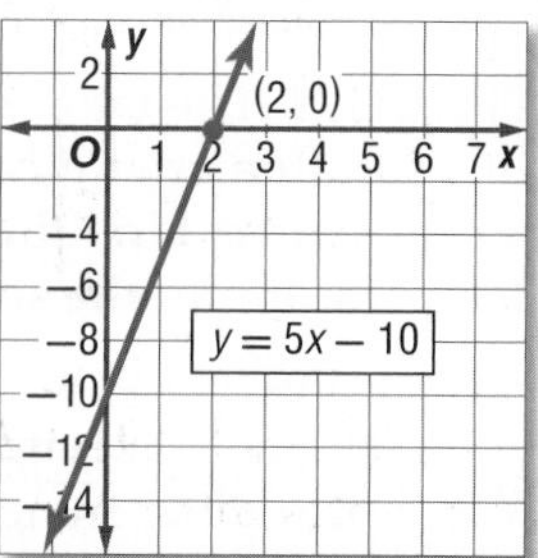

Exercises

1. Use $0 = 4x + 10$ and $f(x) = 4x + 10$ to distinguish among roots, solutions, and zeros.

2. Relate solutions of equations and x-intercepts of graphs.

Determine whether each statement is *true* or *false*. Explain your reasoning.

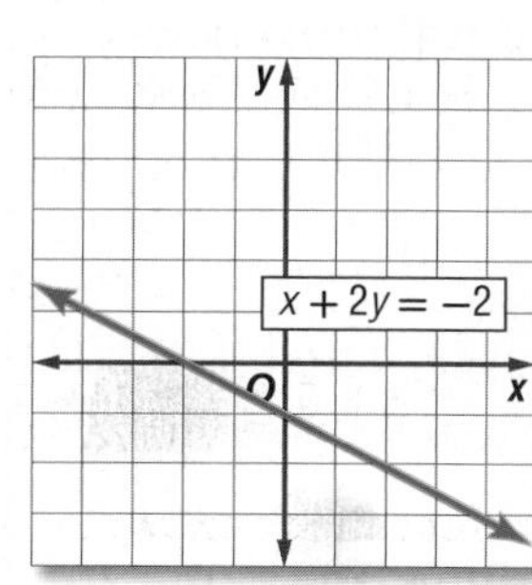

3. The function graphed at the right has two zeros, −2 and −1.

4. The root of $6x + 9 = 0$ is −1.5.

5. $f(0)$ is a zero of the function $f(x) = -\frac{2}{3}x + 12$.

6. FUNDRAISERS The function $y = 2x - 150$ represents the money raised y when the Boosters sell x soft drinks at a basketball game. Find the zero and describe what it means in the context of this situation. Make a connection between the zero of the function and the root of $0 = 2x - 150$.

LESSON

2-3 Rate of Change and Slope

Then

- You graphed linear relations.

Now

1. Find rate of change.
2. Determine the slope of a line.

Why?

- The table shows the total distance a car traveled over various time intervals. The distance formula, $rt = d$ or $r = \frac{d}{t}$, relates time and distance.

Time (h)	Distance (mi)
1	68
2.5	170
3	204
4.5	306
5	340

NewVocabulary
rate of change
slope

Common Core State Standards

Content Standards

F.IF.4 For a function that models a relationship between two quantities, interpret key features of graphs and tables in terms of the quantities, and sketch graphs showing key features given a verbal description of the relationship.

F.IF.6 Calculate and interpret the average rate of change of a function (presented symbolically or as a table) over a specified interval. Estimate the rate of change from a graph.

Mathematical Practices

8 Look for and express regularity in repeated reasoning.

1 Rate of Change

Rate of change is a ratio that compares how much one quantity changes, on average, relative to the change in another quantity. If x is the independent variable and y is the dependent variable, then rate of change $= \frac{\text{change in } y}{\text{change in } x}$. This is sometimes referred to as $\frac{\Delta y}{\Delta x}$.

Real-World Example 1 Constant Rate of Change

CHEMISTRY **The table shows the temperature of a solution after it has been removed from a heat source. Find the rate of change in temperature for the solution.**

Time (min)	Temperature (°C)
0	143.6
2	139.4
5	133.1
8	126.8
12	118.4

Use the ordered pairs (2, 139.4) and (5, 133.1).

$$\begin{aligned}\text{rate of change} &= \frac{\text{change in } y}{\text{change in } x} \\ &= \frac{\text{change in temperature}}{\text{change in time}} \\ &= \frac{133.1 - 139.4}{5 - 2} \\ &= \frac{-6.3}{3} \text{ or } \frac{-2.1}{1}\end{aligned}$$

The rate of change is −2.1. This means that the temperature is decreasing by 2.1°C each minute.

GuidedPractice

1. RECREATION The graph at the right shows the number of gallons of water in a swimming pool as it is being filled. At what rate is the pool being filled?

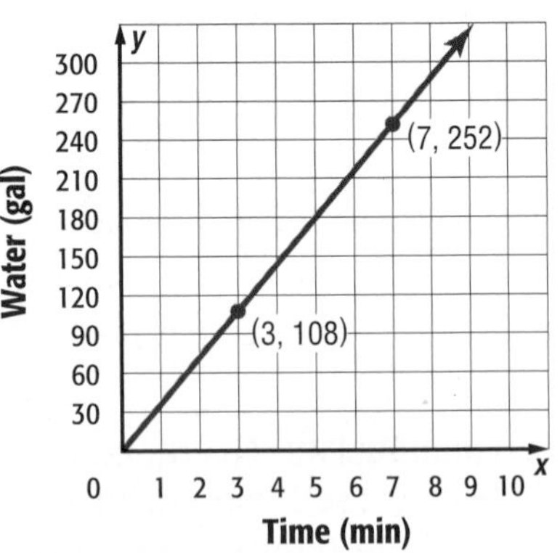

Glen Allison/Photodisc/Getty Images

StudyTip

Independent Quantities Rates of change often include a measure of *time* as the independent variable.

Up to this point, you have used rates of change that are constant. Many real-world situations involve rates of change that are not constant. These situations are often described using an average rate of change over a specified interval.

Real-World Example 2 Average Rate of Change

MUSIC Refer to the graph at the right. Find the average rate of change of the percent of total music sales for both CDs and downloads from 2001 to 2008. Compare the rates.

Percent of Total Music Sales

Percent: 100, 80, 60, 40, 20

Year: 2000, 2002, 2004, 2006, 2008

CDs: 89.2%, 77.8%

Downloads: 0.2%, 12.8%

Source: Recording Industry Association of America

CDs:

$$\text{rate of change} = \frac{\text{change in } y}{\text{change in } x}$$

$$= \frac{\text{change in percent}}{\text{change in time}}$$

$$= \frac{77.8 - 89.2}{2008 - 2001}$$

$$= \frac{-11.4}{7} \text{ or } \frac{-1.63}{1}$$

Downloads:

$$\text{rate of change} = \frac{\text{change in } y}{\text{change in } x}$$

$$= \frac{\text{change in percent}}{\text{change in time}}$$

$$= \frac{12.8 - 0.2}{2008 - 2001}$$

$$= \frac{12.6}{7} \text{ or } \frac{1.8}{1}$$

The percent of CD music sales declined at an average rate of 1.63% per year, while the percent of downloaded music sales increased at an average rate of 1.8% per year.

Real-WorldLink

As of April 2007, 36% of teens preferred to purchase CDs and 64% preferred to download music online.

Source: CNN

GuidedPractice

2. **EDUCATION** In 2002, 23,142 students applied to State College, and 34,689 students applied to Central University. In 2010, 29,563 students applied to State College, and 36,107 applied to Central University. Determine the average rate of change in applicants for both schools from 2002 to 2010.

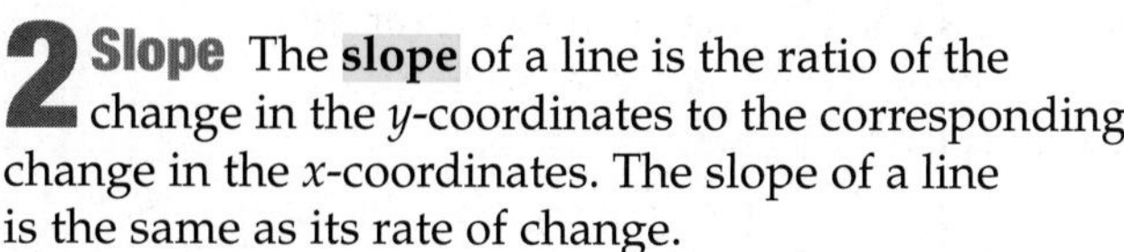

2 Slope

The **slope** of a line is the ratio of the change in the y-coordinates to the corresponding change in the x-coordinates. The slope of a line is the same as its rate of change.

Suppose a line passes through points at (x_1, y_1) and (x_2, y_2).

$$\text{Slope} = \frac{\text{change in } y\text{-coordinates}}{\text{change in } x\text{-coordinates}} = \frac{y_2 - y_1}{x_2 - x_1}$$

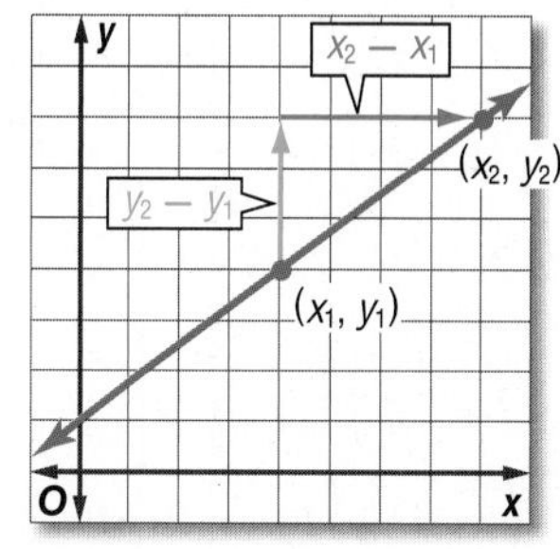

Rubberball/Getty Images

StudyTip

Slope The formula for slope is often remembered as *rise over run*, where the rise is the difference in *y*-coordinates and the run is the difference in *x*-coordinates.

KeyConcept Slope of a Line

Words	The slope of a line is the ratio of the change in *y*-coordinates to the change in *x*-coordinates.
Symbols	The slope m of a line passing through (x_1, y_1) and (x_2, y_2) is given by $m = \frac{y_2 - y_1}{x_2 - x_1}$, where $x_1 \neq x_2$.

Example 3 Find Slope Using Coordinates

Find the slope of the line that passes through (−4, 3) and (2, 5).

$m = \frac{y_2 - y_1}{x_2 - x_1}$ Slope Formula

$= \frac{5 - 3}{2 - (-4)}$ $(x_1, y_1) = (-4, 3), (x_2, y_2) = (2, 5)$

$= \frac{2}{6}$ or $\frac{1}{3}$ Simplify.

GuidedPractice

Find the slope of the line that passes through each pair of points.

3A. (1, −3) and (3, 5)

3B. (−8, 11) and (24, −9)

StudyTip

Slope is Constant The slope of a line is the same, no matter what two points on the line are used.

You can choose any two points from the graph of a line to find the slope.

Example 4 Find Slope Using a Graph

Find the slope of the line shown at the right.

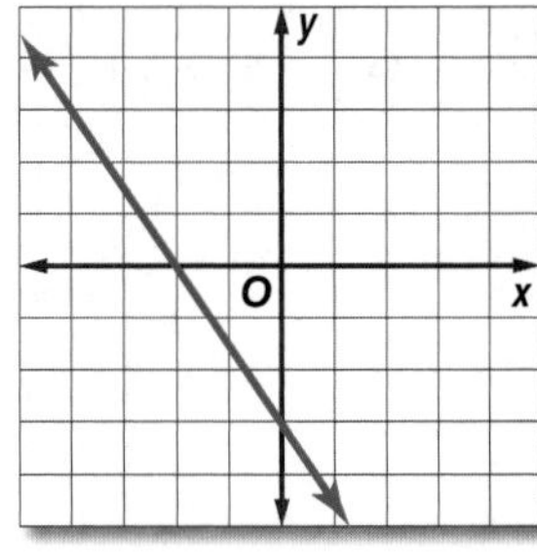

The line passes through (−2, 0) and (0, −3).

$m = \frac{y_2 - y_1}{x_2 - x_1}$ Slope Formula

$= \frac{-3 - 0}{0 - (-2)}$ $(x_1, y_1) = (-2, 0), (x_2, y_2) = (0, -3)$

$= \frac{-3}{2}$ or $-\frac{3}{2}$ Simplify.

GuidedPractice

Find the slope of each line.

4A.

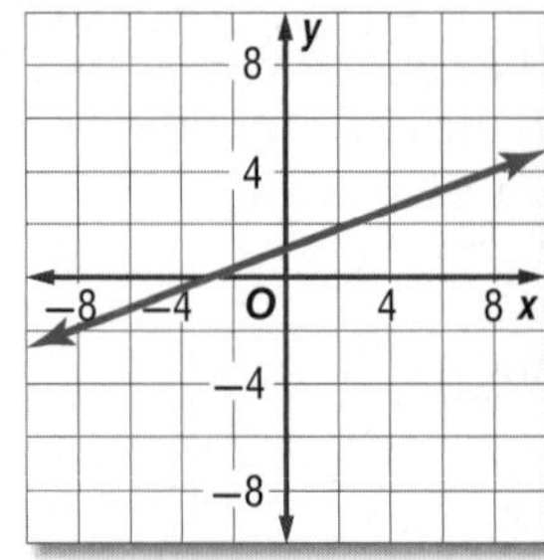

4B.

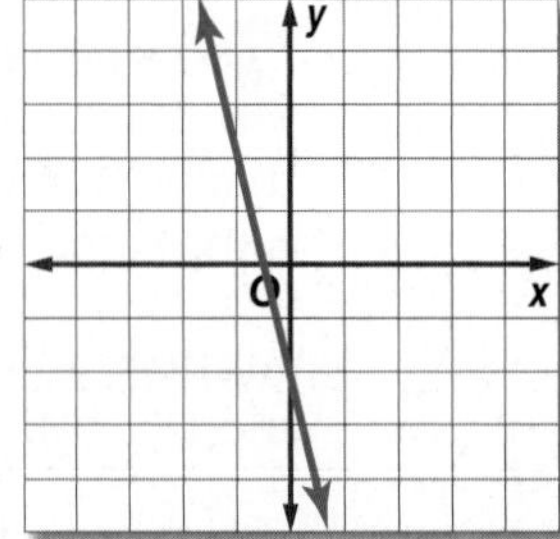

Check Your Understanding

= Step-by-Step Solutions begin on page R14.

Example 1 **CCSS REGULARITY** **Find the rate of change for each set of data.**

1.

Time (min)	2	4	6	8	10
Distance (ft)	12	24	36	48	60

2.

Time (sec)	5	10	15	20	25
Volume (cm^3)	16	32	48	64	80

Example 2

3. **CAMERAS** The graph shows the number of digital still cameras and film cameras sold by Yellow Camera Stores in recent years.

   a. Find the average rate of change of the number of digital cameras sold from 2004 to 2009.

   b. Find the average rate of change of the number of film cameras sold from 2004 to 2009.

   c. What do the signs of each rate of change represent?

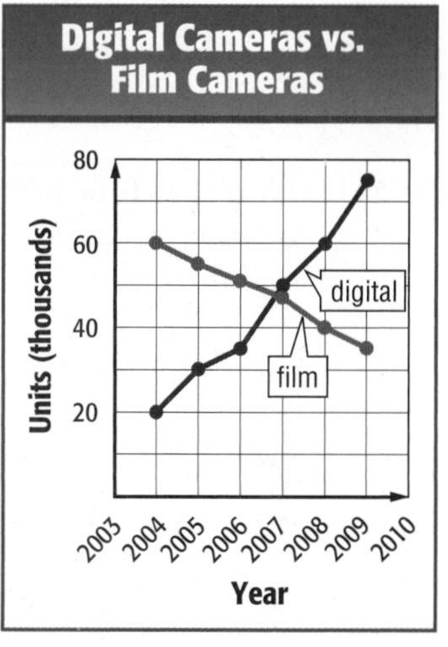

Example 3 **Find the slope of the line that passes through each pair of points.**

4. (3, 2), (8, 12)

5. (−1, 4), (3, −8)

6. (−2, −5), (−7, 10)

Example 4 **Determine the rate of change of each graph.**

7.

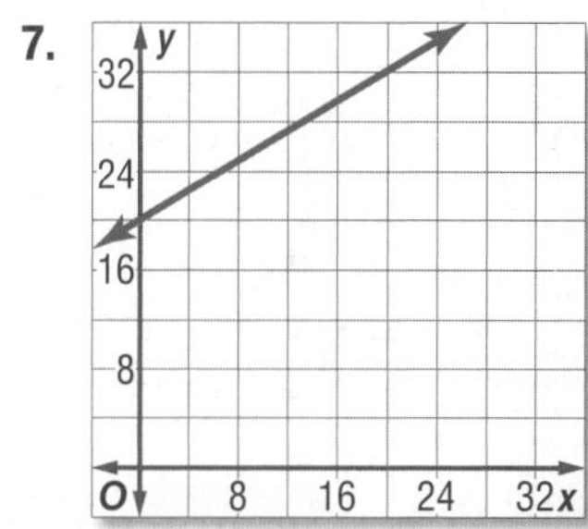

8.

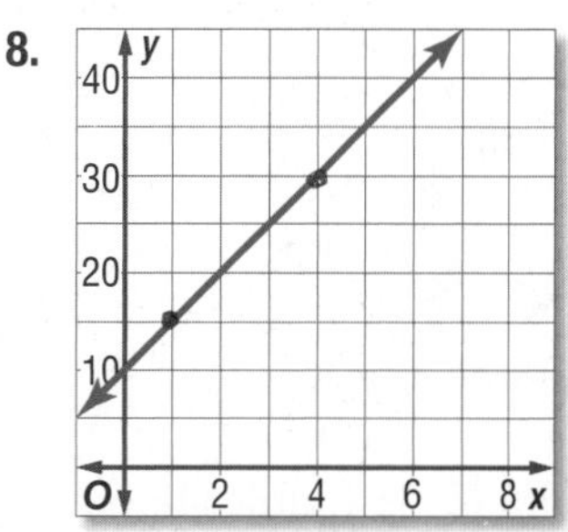

Practice and Problem Solving

Extra Practice is on page R2.

Example 1 **Find the rate of change for each set of data.**

9.

Time (day)	3	6	9	12	15
Height (mm)	20	40	60	80	100

10.

Weight (lb)	11	22	33	44	55
Cost ($)	8	16	24	32	40

Example 2

11. **HEALTH** The table below shows Lisa's temperature during an illness over a 3-day period.

Day	Monday		Tuesday		Wednesday	
Time	8:00 A.M.	8:00 P.M.	8:00 A.M.	8:00 P.M.	8:00 A.M.	8:00 P.M.
Temp (°F)	100.5	102.3	103.1	100.7	99.9	98.6

a. What was the average rate of change in Lisa's temperature from 8:00 A.M. on Monday to 8:00 P.M. on Monday?

b. What was the average rate of change in Lisa's temperature from 8:00 A.M. on Tuesday to 8:00 P.M. on Wednesday? Is your answer reasonable? What does the sign of the rate mean?

c. During which 12-hour period was the average rate of change in Lisa's temperature the greatest?

Example 3

Find the slope of the line that passes through each pair of points. Express as a fraction in simplest form.

12. $(-2, 11), (5, 6)$
13. $(-9, -11), (6, 3)$
14. $(-1.5, 3.5), (4.5, 6)$
15. $(-4.5, 9.5), (-1, 2.5)$
16. $(-8, -0.5), (-4, 5)$
17. $(-6, -2), (-1.5, 5.5)$

Example 4

Determine the rate of change of each graph.

18.
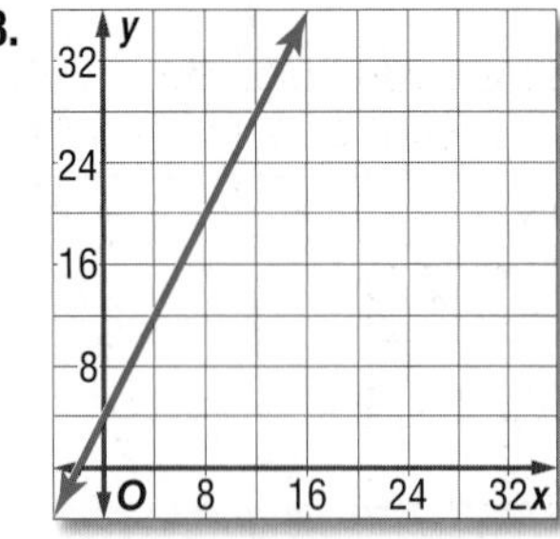

19.
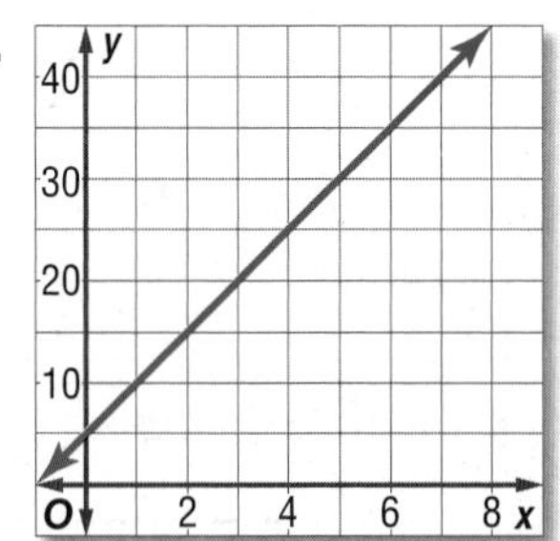

20.
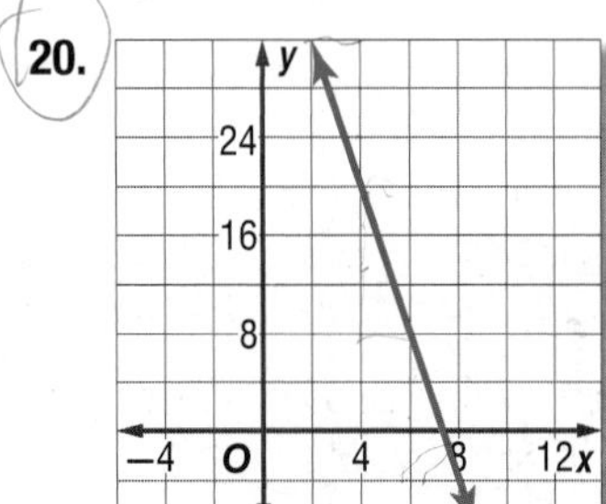

21.
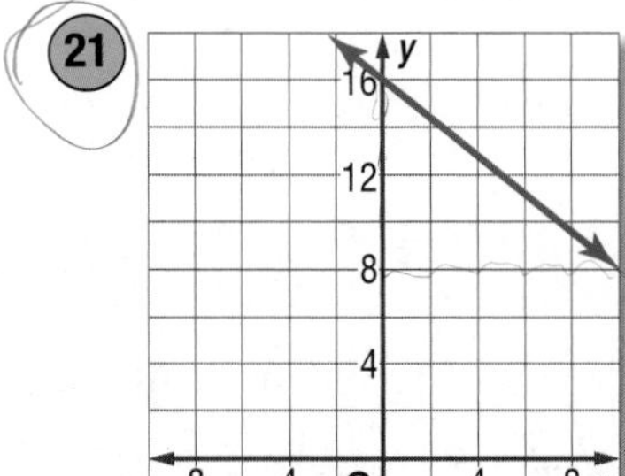

22. **CCSS REASONING** The table shows your height on a water slide at various time intervals.

a. Graph the height versus the time on the water slide.

b. Find the average rate of change in a rider's height between 1 and 3 seconds.

c. Find the average rate of change in a rider's height between 0 and 5 seconds.

d. What is another word for *rate of change* in this situation?

Time (s)	Height (ft)
0	120
1	90
2	60
3	30
4	0
5	0

Determine the rate of change for each equation.

23. $6y = 8x - 40$
24. $-2y - 16x = 41$
25. $12x - 4y + 5 = 18$
26. $20x + 85y = 120$
27. $\frac{3}{2}x - \frac{5}{4}y = 15$
28. $\frac{1}{6}y + \frac{3}{8}x = 24$

29. **WASHINGTON MONUMENT** The Washington Monument is 555 feet $5\frac{1}{8}$ inches tall and weighs 90,854 tons. The monument is topped by an aluminum square pyramid. The sides of the pyramid's base measure 5.6 inches, and the pyramid is 8.9 inches tall. Estimate the slope that a face of the pyramid makes with its base.

30. **MARINE LIFE** The illustrations show the growth of a starfish over time.

a. Find the average rate of change in the measure over time.

b. Predict the size of the starfish in 2014.

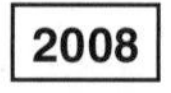

Find the value of *r* so that the line that passes through each pair of points has the given slope.

31. $(6, r), (3, 3), m = 2$

32. $(8, 1), (5, r), m = \frac{1}{3}$

33. $(10, r), (4, -3), m = \frac{4}{3}$

34. $(8, -2), (r, -6), m = -4$

35. **MULTIPLE REPRESENTATIONS** In this problem, you will explore the rate of change for the function $f(x) = x^2$.

a. **Graphical** Graph $f(x) = x^2$.

b. **Tabular** Copy and complete the table. To complete the slope row, find the slope of the line containing two consecutive points such as (−4, 16) and (−3, 9). The first one is completed for you.

x	−4	−3	−2	−1	0	1	2	3	4
f(x)	16	9							
slope		−7							

c. **Verbal** Describe what happens to the rate of change for $f(x) = x^2$ as x increases.

H.O.T. Problems Use Higher-Order Thinking Skills

36. **CCSS CRITIQUE** Patty and Tim are asked to find the slope of the line passing through the points (4, 3) and (7, 9). Is either of them correct? Explain.

Patty	Tim
$m = \frac{9-3}{7-4}$ $= \frac{6}{3}$ or 2	$m = \frac{7-4}{9-3}$ $= \frac{3}{6}$ or $\frac{1}{2}$

37. **CHALLENGE** The graph of a line passes through the points (2, 3) and (5, 8). Explain how you would find the *y*-coordinate of the point (11, *y*) on the same line. Then find *y*.

38. **WRITING IN MATH** In what ways can change be represented mathematically?

39. **REASONING** Determine whether the statement *A line has a slope that is a real number* is *sometimes, always,* or *never* true. Explain your reasoning.

40. **WRITING IN MATH** Describe the process of finding the rate of change for each.

a. a table of values **b.** a graph **c.** an equation

Standardized Test Practice

41. GRIDDED RESPONSE What is the slope of the line shown in the graph?

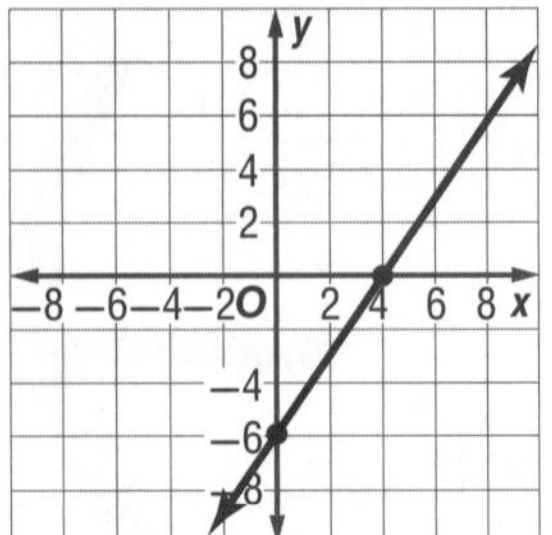

42. SAT/ACT In the figure below, the large square contains two smaller squares. If the areas of the two smaller squares are 4 and 25, what is the sum of the perimeters of the two shaded rectangles?

A 14
B 20
C 24
D 28
E 49

43. GEOMETRY In $\triangle ABC$ shown, $AC = 16$ and $m\angle DAB = 60°$. What is the measure of $\overline{BD}$?

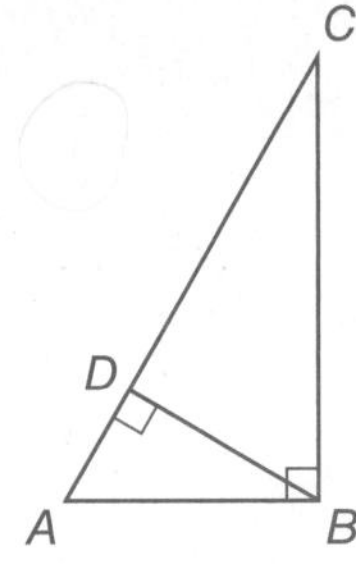

F $9\sqrt{2}$

G 9

H $4\sqrt{3}$

J 4

44. The table shows the cost of bananas depending on the amount purchased. Which conclusion can be made based on information in the table?

Cost of Bananas	
Number of Pounds	Cost ($)
5	1.45
20	4.60
50	10.50
100	19.00

A The cost of 10 pounds of bananas would be more than $4.

B The cost of 200 pounds of bananas would be at most $38.

C The cost of bananas is always more than $0.20 per pound.

D The cost of bananas is always less than $0.28 per pound.

Spiral Review

State whether each equation or function is a linear function. Write *yes* or *no*. Explain. (Lesson 2-2)

45. $6y - 8x = 19$

46. $4x^2 = 2y - 9$

47. $18 = 2xy + 6$

Evaluate each function. (Lesson 2-1)

48. $f(-9)$ if $f(x) = -7x + 8$

49. $g(-4)$ if $g(x) = -3x^2 + 2$

50. $h(12)$ if $h(x) = 4x^2 - 10x$

51. RACING There are 8 contestants in a 400-meter race. In how many different ways can the top three runners finish? (Lesson 0-4)

Determine the quadrant of the coordinate plane where each point is located. (Lesson 0-1)

52. $(-4, -8)$

53. $(-2, 6)$

54. $(3, -1)$

Skills Review

Solve each equation.

55. $8 = 4m - 6$

56. $-6 = 3(8) + b$

57. $-2 = -3x + 5$

LESSON

2-4 Writing Linear Equations

Then	Now	Why?
You determined slopes of lines.	**1** Write an equation of a line given the slope and a point on the line. **2** Write an equation of a line parallel or perpendicular to a given line.	Medical insurance companies often require their customers to make a co-payment for every doctor's office visit in addition to an annual insurance premium. If an insurance company charges \$2280 annually and requires a copayment of \$35 per doctor's office visit, then the linear equation $y = 35x + 2280$ represents the total annual cost y for x doctor's office visits.

NewVocabulary
slope-intercept form
point-slope form
parallel
perpendicular

CCSS Common Core State Standards

Content Standards
A.SSE.1.b Interpret complicated expressions by viewing one or more of their parts as a single entity.
A.CED.2 Create equations in two or more variables to represent relationships between quantities; graph equations on coordinate axes with labels and scales.

Mathematical Practices
2 Reason abstractly and quantitatively.

1 Forms of Equations

Consider the line through $A(0, b)$ and $C(x, y)$. Notice that b is the y-intercept. You can use these two points to find the slope of $\overleftrightarrow{AC}$.

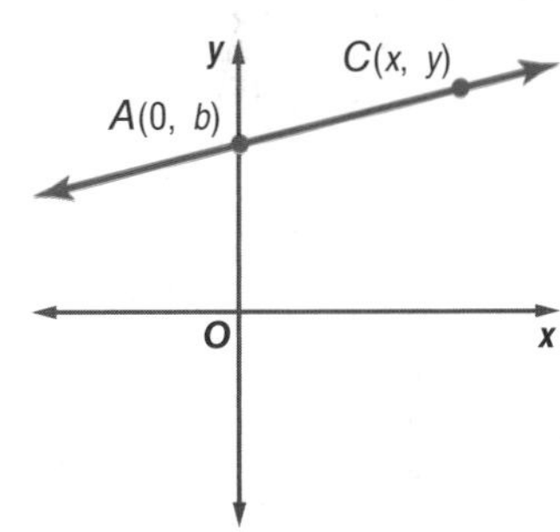

$m = \frac{y_2 - y_1}{x_2 - x_1}$ Slope Formula

$= \frac{y - b}{x - 0}$ $(x_1, y_1) = (0, b), (x_2, y_2) = (x, y)$

$= \frac{y - b}{x}$ Simplify.

Now solve the equation for y.

$mx = y - b$ Multiply each side by x.

$mx + b = y$ Add b to each side.

$y = mx + b$ Symmetric Property of Equality

Equations written in this format are in **slope-intercept form**.

KeyConcept Slope-Intercept Form

Words The slope-intercept form of the equation of a line is $y = mx + b$, where m is the slope and b is the y-intercept.

Model

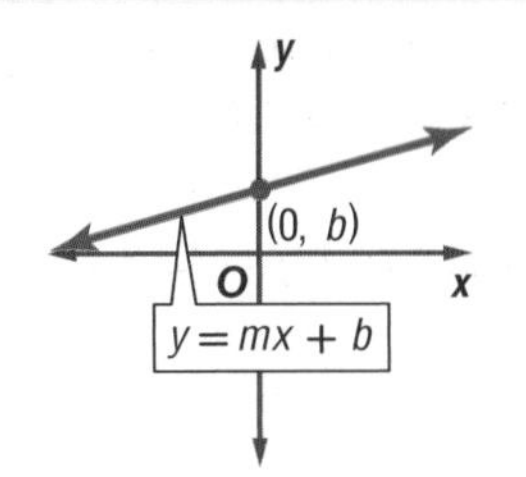

Symbols $y = mx + b$

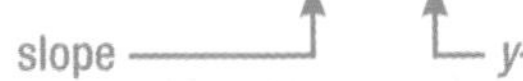
slope → m; y-intercept → b

If you are given the slope and y-intercept of a line, you can find an equation of the line by substituting the values of m and b into the slope-intercept form.

WatchOut!

CCSS Reasoning The equation of a vertical line cannot be written in slope-intercept form because its slope is undefined.

Sometimes it is necessary to calculate the slope before you can write an equation.

Example 1 Write an Equation in Slope-Intercept Form

Write an equation in slope-intercept form for the line.

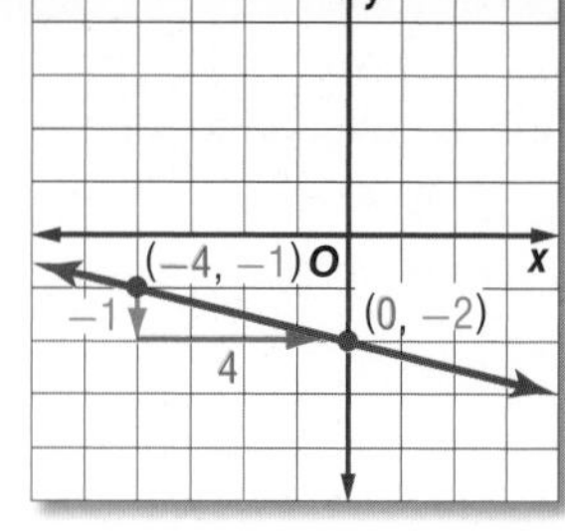

The graph intersects the y-axis at −2. So $b = -2$.

Step 1 Find the slope.

$m = \frac{y_2 - y_1}{x_2 - x_1}$ Slope Formula

$= \frac{-2 - (-1)}{0 - (-4)}$ $(x_1, y_1) = (-4, -1), (x_2, y_2) = (0, -2)$

$= \frac{-1}{4}$ or $-\frac{1}{4}$ Simplify.

Step 2 Substitute the values into the slope-intercept equation.

$y = mx + b$ Slope-intercept form

$y = -\frac{1}{4}x - 2$ $m = -\frac{1}{4}, b = -2$

StudyTip

Check Your Results You can check that your equation satisfies the conditions by graphing it.

GuidedPractice

Write an equation in slope-intercept form for the line described.

1A. slope $\frac{4}{3}$, passes through (0, 4)

1B. passes through (0, −6) and (−4, 10)

If you know the slope of a line and the coordinates of a point on the line, you can use the **point-slope form** to find an equation of the line.

KeyConcept Point-Slope Form

Words The point-slope form of the equation of a line is $y - y_1 = m(x - x_1)$, where (x_1, y_1) are the coordinates of a point on the line and m is the slope of the line.

Symbols

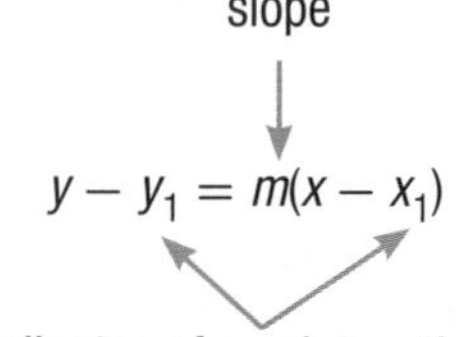

Example 2 Write an Equation Given Slope and One Point

Write an equation of the line through (6, −2) with a slope of −4.

$y - y_1 = m(x - x_1)$ Point-slope form

$y - (-2) = -4(x - 6)$ $(x_1, y_1) = (6, -2), m = -4$

$y + 2 = -4x + 24$ Simplify.

$y = -4x + 22$ Subtract 2 from each side.

GuidedPractice

Write an equation in slope-intercept form for the line described.

2A. passes through (2, 3); $m = \frac{1}{2}$

2B. passes through (−2, −1); $m = -3$

You can use any two points on a line to write an equation.

PT

Standardized Test Example 3 Write an Equation Given Two Points

Which is an equation of the line that passes through (–2, 7) and (3, –3)?

A $y = -\frac{1}{2}x - \frac{3}{2}$

B $y = -2x + 3$

C $y = \frac{1}{2}x + 8$

D $y = 2x + 11$

Test-TakingTip

Definitions Be certain to review key vocabulary, such as *y*-intercept, so that you understand what is being asked in a question.

Read the Test Item

You are given the coordinates of two points on the line.

Solve the Test Item

Step 1 Find the slope of the line.

$m = \frac{y_2 - y_1}{x_2 - x_1}$ Slope Formula

$= \frac{-3 - 7}{3 - (-2)}$ $(x_1, y_1) = (-2, 7)$, $(x_2, y_2) = (3, -3)$

$= -\frac{10}{5}$ or -2 Simplify.

Step 2 Write an equation. Use either ordered pair for (x_1, y_1).

$y - y_1 = m(x - x_1)$ Point-slope form

$y - (-3) = -2(x - 3)$ $(x_1, y_1) = (3, -3)$ and $m = -2$

$y + 3 = -2x + 6$ Simplify.

$y = -2x + 3$ Subtract 3 from each side.

The answer is B.

GuidedPractice

3. Which is an equation of the line that passes through (4, –9) and (2, –4)?

F $y = -\frac{5}{2}x + 1$

G $y = -\frac{5}{2}x - 1$

H $y = -\frac{2}{5}x + \frac{37}{5}$

J $y = -\frac{2}{5}x - \frac{37}{5}$

2 Parallel and Perpendicular Lines

Parallel and Perpendicular Lines Slopes can help you determine whether two lines are parallel or perpendicular.

KeyConcept Parallel and Perpendicular Lines

Parallel Lines	Perpendicular Lines
Two nonvertical lines are **parallel** if and only if they have the same slope. All vertical lines are parallel.	Two nonvertical lines are **perpendicular** if and only if the product of the slopes is –1. Vertical lines and horizontal lines are perpendicular.
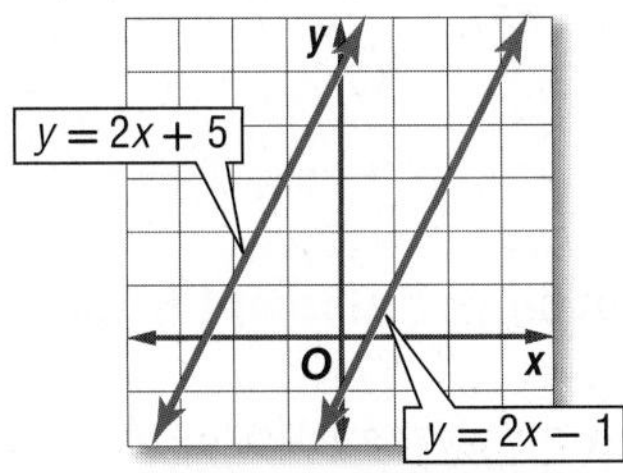	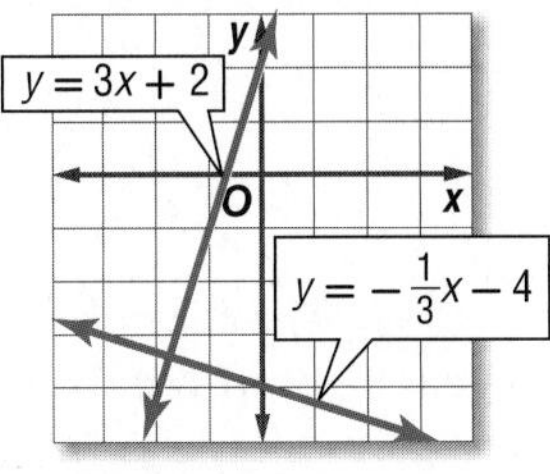
$y = 2x + 5$ and $y = 2x - 1$	$y = 3x + 2$ and $y = -\frac{1}{3}x - 4$

Example 4 Write an Equation of a Parallel or Perpendicular Line

Write an equation in slope-intercept form for the line that passes through (5, −6) and is perpendicular to the line with equation $y = -\frac{3}{2}x + 7$.

The slope of the given line is $-\frac{3}{2}$. Because the slopes of perpendicular lines are opposite reciprocals, the slope of the line perpendicular to the given line is $\frac{2}{3}$.

Use the point-slope form and the ordered pair (5, −6).

$y - y_1 = m(x - x_1)$ Point-slope form

$y - (-6) = \frac{2}{3}(x - 5)$ $(x_1, y_1) = (5, -6)$ and $m = \frac{2}{3}$

$y + 6 = \frac{2}{3}x - \frac{10}{3}$ Distributive Property

$y = \frac{2}{3}x - \frac{28}{3}$ Subtract 6 from each side and simplify.

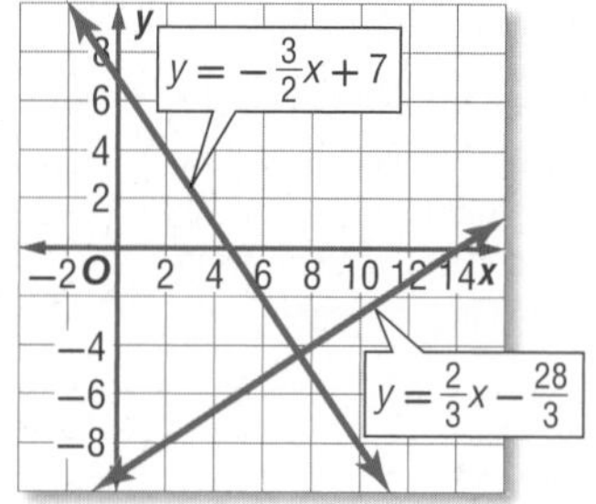

CHECK Graph both equations to verify the solution.

GuidedPractice

4. Write an equation in slope-intercept form for the line that passes through (3, 7) and is parallel to the line with equation $y = \frac{3}{4}x - 5$.

Check Your Understanding

= Step-by-Step Solutions begin on page R14.

Example 1 **Write an equation in slope-intercept form for the line described.**

1. slope 1.5, passes through (0, 5)

2. passes through (−2, 3) and (0, 1)

Example 2

3. passes through (3, 5); $m = -2$

4. passes through (−8, −2); $m = \frac{5}{2}$

Example 3

5. MULTIPLE CHOICE Which is an equation of the line?

A $y = -4x - 25$

B $y = -\frac{2}{3}x - 5$

C $y = \frac{4}{5}x + \frac{29}{25}$

D $y = 6x + 35$

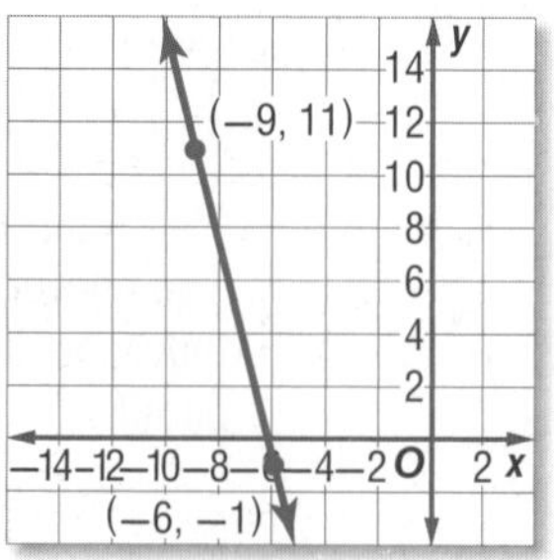

Example 4 **CCSS PERSEVERANCE** **Write an equation in slope-intercept form for the line that satisfies each set of conditions.**

6. passes through (−9, −3), perpendicular to $y = -\frac{5}{3}x - 8$

7. passes through (4, −10), parallel to $y = \frac{7}{8}x - 3$

Practice and Problem Solving

Extra Practice is on page R2.

Example 1 **Write an equation in slope-intercept form for the line described.**

8. slope 3, passes through (0, −2)
9. slope $-\frac{1}{2}$, passes through (0, 5)
10. slope $-\frac{6}{5}$, passes through (0, 8)
11. slope $\frac{9}{2}$, passes through $\left(0, -\frac{13}{2}\right)$

Example 2

12. slope −2, passes through (−3, 14)
13. slope 4, passes through (6, 9)
14. slope $\frac{3}{5}$, passes through (−6, −8)
15. slope $-\frac{1}{4}$, passes through (12, −4)

16. **PART-TIME JOB** Each week, Carmen earns a base pay of $15 plus $0.17 for every pamphlet that she delivers. Write an equation that can be used to find how much Carmen earns each week. How much will she earn the week that she delivers 300 pamphlets?

Example 3 **Write an equation of the line passing through each pair of points.**

17. (−2, −6), (4, 6)
18. (−8, −5), (−3, 10)
19. (−4, 12), (−2, −4)
20. (4.6, 3.4), (2.2, 2.8)
21. (5.5, 0.6), (1.1, 2.8)
22. (−25, −16), (−29, 12)

Example 4 **CCSS PERSEVERANCE Write an equation in slope-intercept form for the line that satisfies each set of conditions.**

23. passes through (4, 2), perpendicular to $y = -2x + 3$
24. passes through (−6, −6), parallel to $y = \frac{4}{3}x + 8$
25. passes through (12, 0), parallel to $y = -\frac{1}{2}x - 3$
26. passes through (10, 2), perpendicular to $y = 4x + 6$

27. **FINANCIAL LITERACY** Julio buys a used car for $5900. Monthly expenses for the car—which include insurance, maintenance, and gas—average $180 per month. Write an equation that represents the total cost of buying and owning the car for x months.

28. **DELI** The sales of a sandwich store increased approximately linearly from $52,000 to $116,000 during the first five years of business. Write an equation that models the sales y after x years. Determine what the sales will be at the end of 12 years if the pattern continues.

29. **WHALES** In 2009, it was estimated that there were 300 northern right whales in existence. The population of northern right whales is expected to decline by at least 25 whales each generation. Write an equation that represents the number of northern right whales that will be in existence in x generations.

Write an equation in slope-intercept form for each graph.

30.

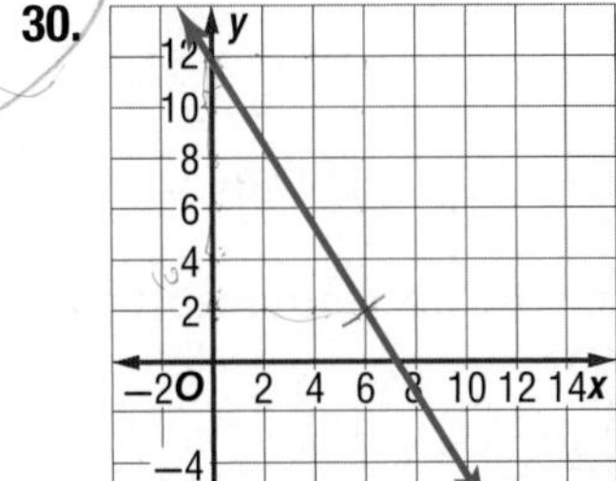

31.

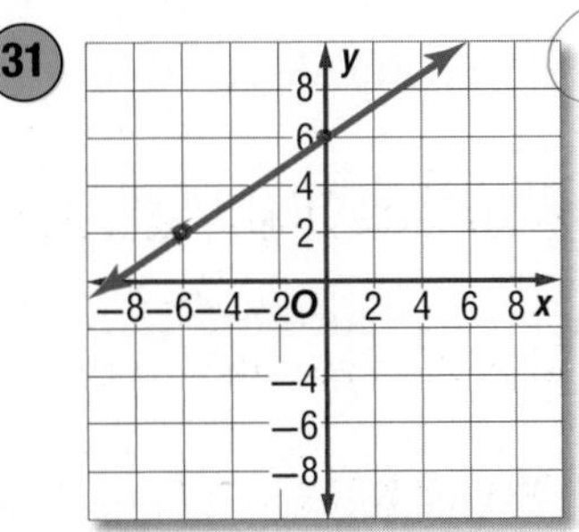

32.

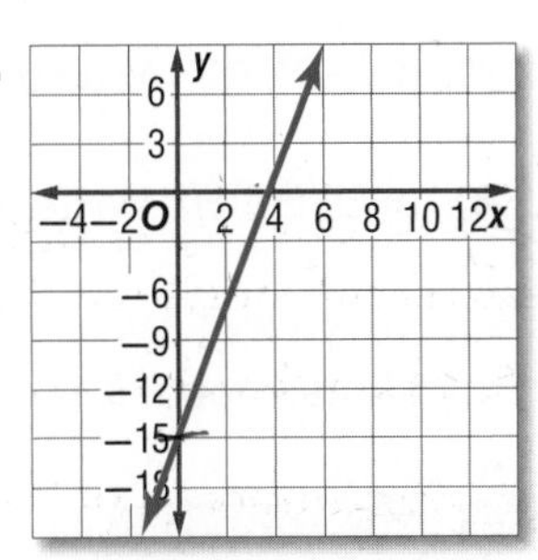

33. ROSES Brad wants to send his girlfriend Kelli a dozen roses. He visits two stores. For what distance do the two stores charge the same amount to deliver a dozen roses?

Full Bloom	Flowers R US
Dozen roses \$30 Delivery: \$3 per mile	Dozen roses \$40 Delivery: \$2 per mile

34. TYPING The equation $y = 55(23 - x)$ can be used to model the number of words y you have left to type after x minutes.

a. Write this equation in slope-intercept form.

b. Identify the slope and y-intercept.

c. Find the number of words you have left to type after 20 minutes.

35. RECRUITING As an army recruiter, Ms. Cooper is paid a daily salary plus commission. When she recruits 10 people, she earns \$100. When she recruits 14 people, she earns \$120.

a. Write a linear equation to model this situation.

b. What is Ms. Cooper's daily salary?

c. How much would Ms. Cooper earn in a day if she recruits 20 people?

36. CCSS MODELING Refer to the table at the right.

Miles	Kilometers
100	161
50	80.5

a. Write and graph the linear equation that gives the distance y in kilometers in terms of the number x in miles.

b. What distance in kilometers corresponds to 20 miles?

c. What number is the same in kilometers and miles? Explain your reasoning.

H.O.T. Problems Use Higher-Order Thinking Skills

37. REASONING Determine whether the following statement is *always*, *sometimes*, or *never* true. Explain your reasoning.

The quadrilateral formed by any two parallel lines and two lines perpendicular to those lines is a square.

38. CHALLENGE Given $\square ABCD$ with vertices $A(a, b)$, $B(c - a, d)$, $C(c + a, d)$, and $D(c, b)$, write an equation of a line perpendicular to diagonal $\overline{BD}$ that contains A.

39. REASONING Write $y = ax + b$ in point-slope form.

40. OPEN ENDED Write the equations of two parallel lines with negative slopes.

41. REASONING Write an equation in point-slope form of a line with an x-intercept of c and y-intercept of d.

42. WRITING IN MATH Why do we represent linear equations in more than one form?

Todd Gipstein/National Geographic/Getty Images

Standardized Test Practice

43. The total cost c in dollars to go to a water park and ride n water rides is given by the equation

$$c = 15 + 3n.$$

If the total cost was \$33, how many water rides were ridden?

A 6 B 7 C 8 D 9

44. SHORT RESPONSE To raise money, the service club bought 1000 candy bars for \$0.60 each. If the club sells all of the candy bars for \$1 each, what will be their total profit?

45. PROBABILITY A fair six-sided die is tossed. What is the probability that a number less than 3 will show on the face of the die?

F $\frac{1}{6}$ G $\frac{1}{3}$ H $\frac{1}{2}$ J $\frac{2}{3}$

46. SAT/ACT What is an equation of the line through $\left(\frac{1}{2}, -\frac{3}{2}\right)$ and $\left(-\frac{1}{2}, \frac{1}{2}\right)$?

A $y = -2x - \frac{1}{2}$

B $y = -3x$

C $y = 2x - 5$

D $y = \frac{1}{2}x + 1$

E $y = -2x - \frac{5}{2}$

Spiral Review

Determine the rate of change of each graph. (Lesson 2-3)

47.

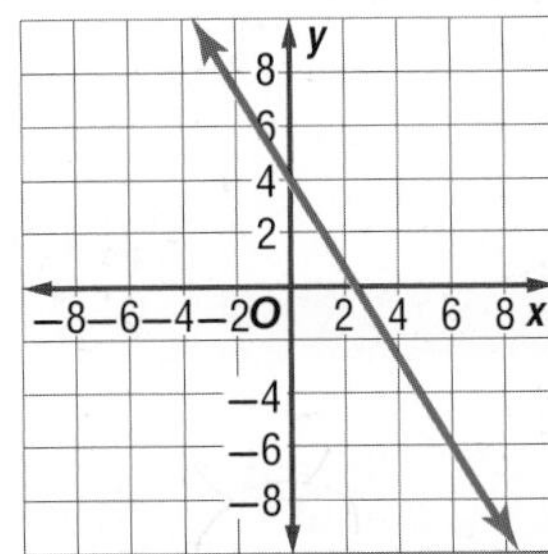

48.

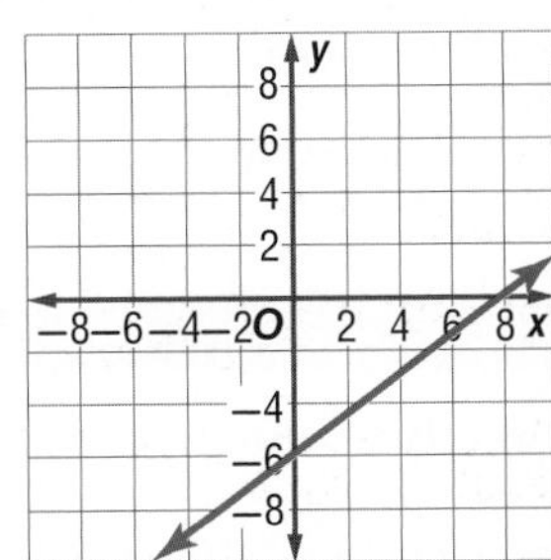

49.

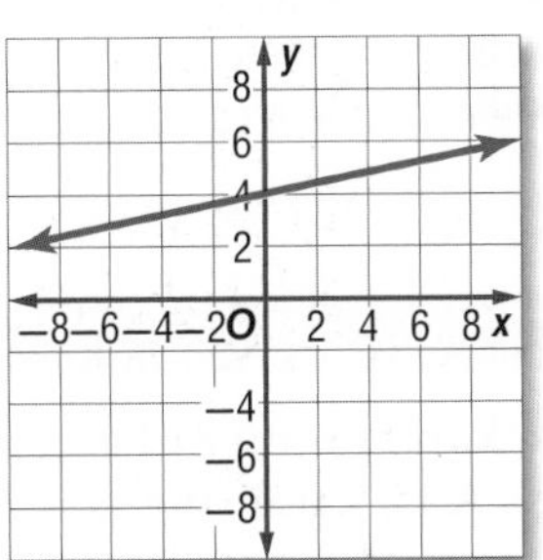

50. RECREATION Scott is currently on page 210 of an epic novel that is 980 pages long. He plans to read 30 pages per day until he finishes the novel. Write and solve a linear equation to determine how many days it will take Scott to complete the novel. (Lesson 2-2)

Solve each inequality. (Lesson 1-5)

51. $-6x - 4 \leq 12 - 2x$

52. $\frac{x + 2}{5} > -3x + 1$

53. $\frac{5x + 3}{3} \geq \frac{4x - 2}{5}$

Determine if the triangles with the following lengths are right triangles. (Lesson 0-8)

54. 5, 12, 13

55. 36, 48, 60

56. 7, 23, 25

Multiply. (Lesson 0-2)

57. $(4c - 6)(2c + 5)$

58. $(-3b + 2)(b + 3)$

59. $(2a - 5)(-3a - 4)$

Skills Review

Find the slope of the line that passes through each pair of points. Express as a fraction in simplest form.

60. (4, 8), (−2, −6)

61. (−6, 3), (−2, 9)

62. (−4, −1), (−8, −8)

63. (12, 4), (42, 10)

64. (10.5, −3), (18, −8)

65. (3.5, −2.5), (−1, −2)

EXTEND 2-4

Graphing Technology Lab
Direct Variation

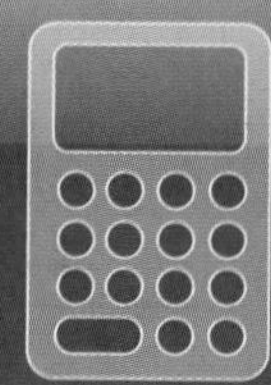

An equation of a direct variation is a special case of a linear equation. A **direct variation** can be expressed in the form $y = kx$. This means that y is a multiple of x. The k in this equation is a constant and is called the **constant of variation**.

CCSS Common Core State Standards
Content Standards
F.IF.4 For a function that models a relationship between two quantities, interpret key features of graphs and tables in terms of the quantities, and sketch graphs showing key features given a verbal description of the relationship.

Notice that the graph of $y = 4x$ is a straight line through the origin. An equation of a direct variation is a special case of an equation written in slope-intercept form, $y = mx + b$. When $m = k$ and $b = 0$, $y = mx + b$ becomes $y = kx$. So the slope of a direct variation equation is its constant of variation.

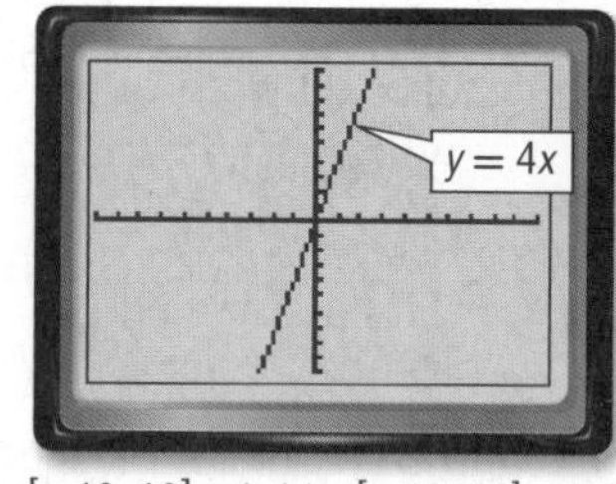

[−10, 10] scl: 1 by [−10, 10] scl: 1

To express a direct variation, we say that y varies directly as x. In other words, as x increases, y increases or decreases at a constant rate.

KeyConcept Direct Variation

y varies directly as x if there is some nonzero constant k such that $y = kx$. k is called the *constant of variation.*

Activity

GOLD The karat rating r of a gold object varies directly as the percentage p of gold in the object. A 14-karat ring is 58.25% gold.

a. Write and graph a direct variation equation relating r and p.

Use the point (0.5825, 14) to find the constant of variation.

$y = kx$	Direct variation equation
$14 = k(0.5825)$	$x = 0.5825$, $y = 14$
$24.03 \approx k$	Divide each side by 0.5825.

The direct variation equation is $r = 24.03p$.

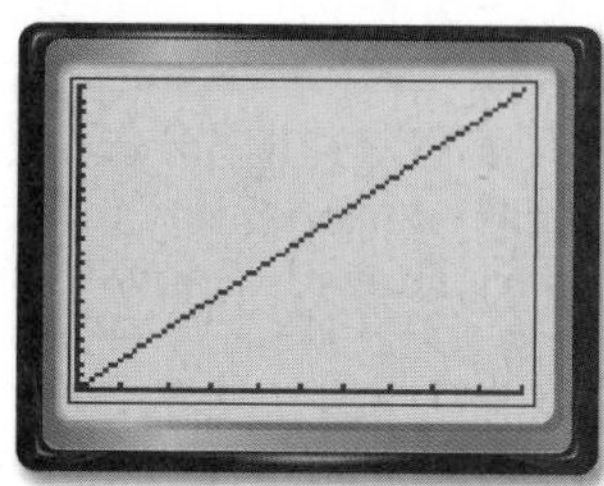

[0, 1] scl: 0.1 by [0, 24] scl: 1

b. Find the karat rating of a ring that is 75% gold.

Use the calculator to find the karat rating.

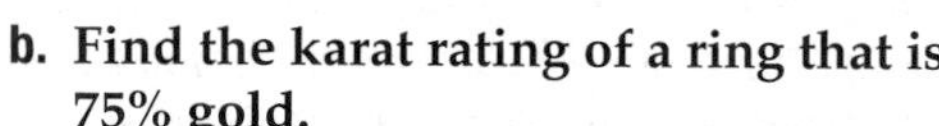

KEYSTROKES: 2nd [CALC] 0.75 ENTER 18.0225

The karat rating of a ring that is 75% gold is 18 karats.

MENTAL CHECK 75% of $24 = \frac{3}{4}$ of 24   Think $75\% = \frac{3}{4}$.
$= 18$ ✓   Think $\frac{3}{4}$ of 24 is 18.

Exercises

1. **SWIMMING** When you swim under water, the pressure on your ears varies directly with the depth at which you are swimming. If you are swimming in 8 feet of water, the pressure on your ears is 3.44 pounds per square inch. Write and graph a direct variation equation relating pressure and depth. Then find the pressure at a depth of 65 feet.

2. Graph the direct variation equations $y = -4x$, $y = -2x$, $y = 4x$, and $y = 2x$. Compare and contrast the graphs of the equations.

CHAPTER 2

Mid-Chapter Quiz

Lessons 2-1 through 2-4

1. State the domain and range of the relation {(−3, 2), (4, 1), (0, 3), (5, −2), (2, 7)}. Then determine whether the relation is a function.

2. Graph $y = 2x - 3$ and determine whether the equation is a *function*, is *one-to-one*, *onto*, *both*, or *neither*. State whether it is *discrete* or *continuous*.

Given $f(x) = 3x^3 - 2x + 7$, find each value.

3. $f(-2)$
4. $f(2y)$
5. $f(1.4)$

6. State whether $f(x) = 2x^2 - 9$ is a linear function. Explain.

7. **MULTIPLE CHOICE** The daily pricing for renting a mid-sized car is given by the function $f(x) = 0.35x + 49$, where $f(x)$ is the total rental price for a car driven x miles. Find the rental cost for a car driven 250 miles.

   A $84

   B $112.50

   C $136.50

   D $215

Write each equation in standard form. Identify *A*, *B*, and *C*.

8. $y = -6x + 5$
9. $y = 10x$
10. $-\frac{5}{8}x = 2y + 11$
11. $0.5x = 3$

Find the *x*-intercept and the *y*-intercept of the graph of each equation. Then graph the equation using the intercepts.

12. $4x - 3y + 12 = 0$
13. $10 - x = 2y$

14. **SPEED** The table shows the distance traveled by a car after each time given in minutes. Find the rate of change in distance for the car.

Time (min)	Distance (mi)
15	20
30	40
45	60
60	80
75	100

Find the slope of the line that passes through each pair of points. Express as a fraction in simplest form.

15. (−2, 6), (1, 15)
16. (3, 5), (7, 15)
17. (4, 8), (4, −3)
18. (−2.5, 4), (1.5, −2)

19. Find the slope of the line shown.

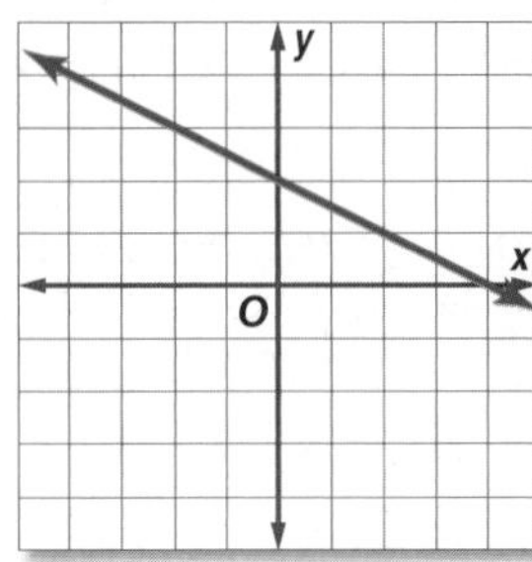

Write an equation for the line that satisfies each set of conditions.

20. slope $\frac{2}{3}$, passes through (3, −4)
21. slope −2.5, passes through (1, 2)

Write an equation of the line through each set of points.

22. (−2, 3), (4, 1)
23. (4.2, 3.6), (1.8, −1.2)

24. **MULTIPLE CHOICE** Each week, Jaya earns $32 plus $0.25 for each newspaper she delivers. Write an equation that can be used to determine how much Jaya earns each week. How much will she earn during a week in which she delivers 240 papers?

   F $75

   G $92

   H $148

   J $212

25. **PART-TIME JOB** Jesse is a pizza delivery driver. Each day his employer gives him $20 plus $0.50 for every pizza that he delivers.

    a. Write an equation that can be used to determine how much Jesse earns each day if he delivers x pizzas.

    b. How much will he earn the day he delivers 20 pizzas?

LESSON 2-5

Scatter Plots and Lines of Regression

Then	Now	Why?
You wrote linear equations.	1 Use scatter plots and prediction equations. 2 Model data using lines of regression.	The scatter plot shows the number of visitors to Isle Royale National Park in Michigan per year.

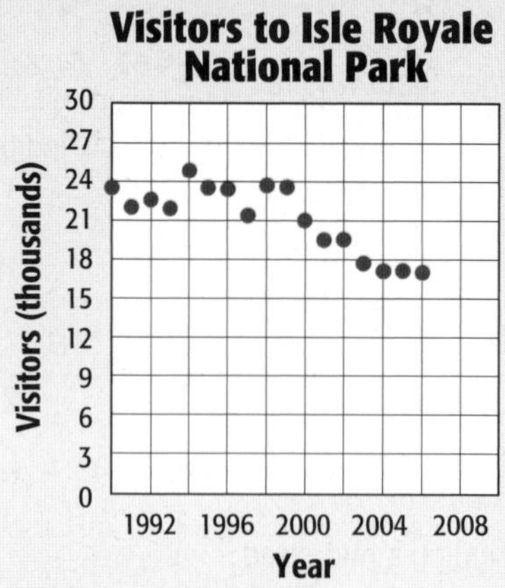

Source: National Park Service

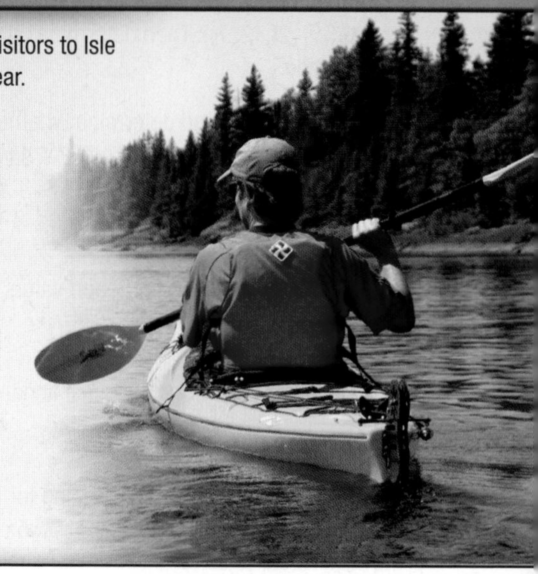

NewVocabulary

bivariate data
scatter plot
dot plot
positive correlation
negative correlation
line of fit
prediction equation
regression line
correlation coefficient

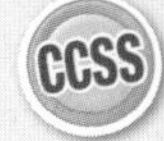

Common Core State Standards

Content Standards

F.IF.4 For a function that models a relationship between two quantities, interpret key features of graphs and tables in terms of the quantities, and sketch graphs showing key features given a verbal description of the relationship.

Mathematical Practices

4 Model with mathematics.
5 Use appropriate tools strategically.

1 Scatter Plots and Prediction Equations

Data with two variables, such as year and number of visitors, are called **bivariate data**. A set of bivariate data graphed as ordered pairs in a coordinate plane is called a **scatter plot** or **dot plot**.

A scatter plot can show whether there is a positive, negative, or no correlation between two variables. Correlations are usually described as *strong* or *weak*. In a strong correlation, the points of the scatterplot are closer to the graph of a line than the points representing a weak correlation.

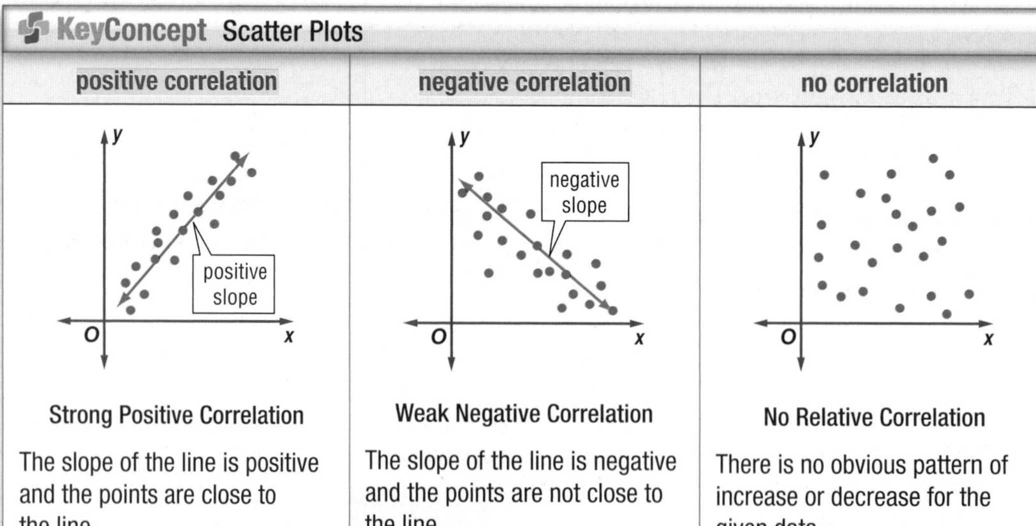

KeyConcept Scatter Plots

positive correlation	negative correlation	no correlation
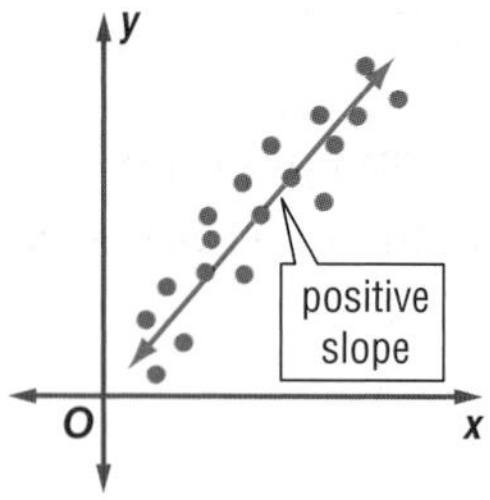	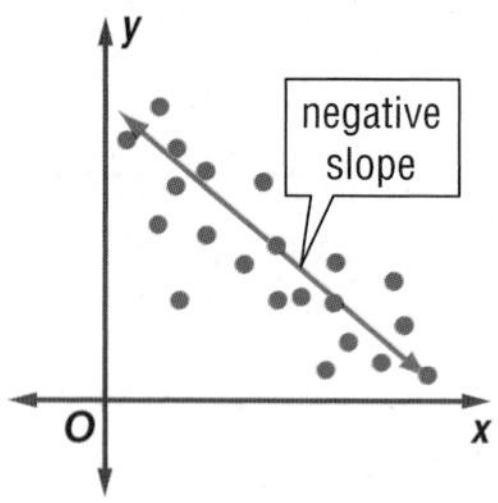	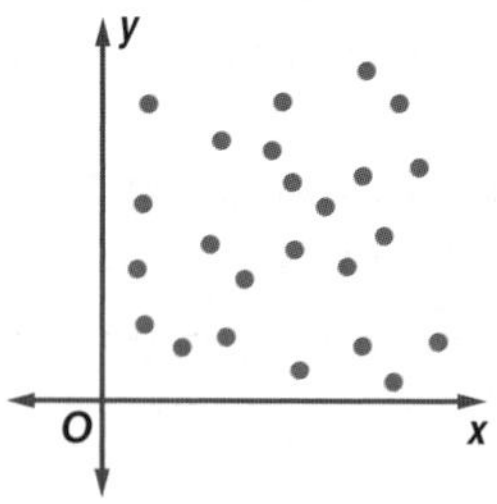
Strong Positive Correlation	Weak Negative Correlation	No Relative Correlation
The slope of the line is positive and the points are close to the line.	The slope of the line is negative and the points are not close to the line.	There is no obvious pattern of increase or decrease for the given data.

When you find a line that closely approximates a set of data, you are finding a **line of fit** for the data. An equation of such a line is often called a **prediction equation** because it can be used to predict one of the variables given the other variable.

To find a line of fit and a prediction equation for a set of data, select two points that appear to represent the data well. This is a matter of personal judgment, so your line and prediction equation may differ from those of others.

Real-WorldLink

According to the National Technology Scan, about 70% of U.S. households have computers connected to the Internet.

Source: *The National Technology Scan*

Real-World Example 1 Use a Scatter Plot and Prediction Equation

TECHNOLOGY The table shows the percent of U.S. households with Internet access.

Year	1997	2000	2001	2003	2007
Percent	18.0	41.5	50.4	54.7	61.7

Source: U.S. Census Bureau

a. Make a scatter plot and a line of fit, and describe the correlation. Let x be the number of years since 1995.

Graph the data as ordered pairs with the number of years since 1995 on the horizontal axis and the percent of households on the vertical axis.

The points (5, 41.5) and (12, 61.7) appear to represent the data well. Draw a line through these two points. The data show a strong positive correlation.

Percent of Households with Internet Access

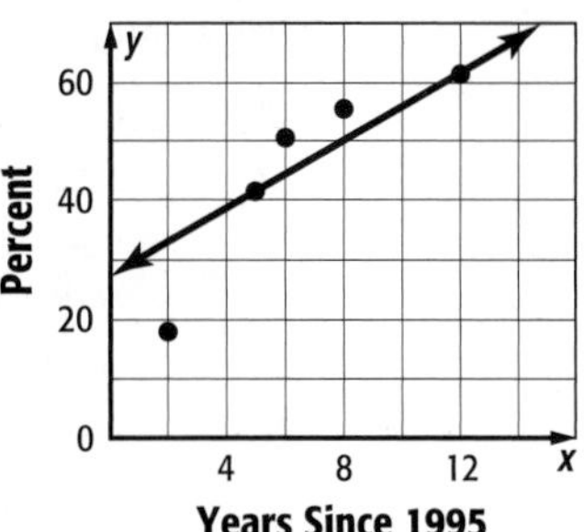

b. Use two ordered pairs to write a prediction equation.

Find an equation of the line through (5, 41.5) and (12, 61.7).

$m = \frac{y_2 - y_1}{x_2 - x_1}$ Slope Formula

$= \frac{61.7 - 41.5}{12 - 5}$ Substitute.

≈ 2.89 Simplify.

$y - y_1 = m(x - x_1)$ Point-Slope form

$y - 41.5 \approx 2.89(x - 5)$ Substitute.

$y - 41.5 \approx 2.89x - 14.45$ Distributive Property

$y \approx 2.89x + 27.05$ Simplify.

One prediction equation is $y = 2.89x + 27.05$.

c. Predict the percent of households with Internet access in 2020.

The year 2020 is 25 years after 1995, so find y when $x = 25$.

$y \approx 2.89x + 27.05$ Prediction equation

$\approx 2.89(25) + 27.05$ $x = 25$

≈ 99.3 Simplify.

The model predicts that 99.3% of U.S. households will have Internet access in 2020.

ReviewVocabulary

outlier a data point that does not appear to belong to the rest of the set.

d. How accurate does your prediction appear to be?

Except for the outlier at (2, 18.0), the line fits the data well, so the prediction value should be fairly accurate.

GuidedPractice

1. HOUSING The table shows the mean selling price of new, privately-owned, single-family homes for six consecutive years.

Year	0	1	2	3	4	5
Price($1000)	154.5	166.4	181.9	207.0	228.7	273.5

A. Make a scatter plot and a line of fit, and describe the correlation.

B. Write a prediction equation.

C. Predict the selling price of a new home for year 8.

D. How accurate does your prediction appear to be?

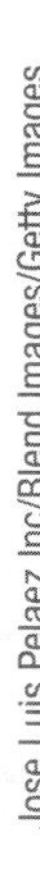

2 Lines of Regression

Lines of Regression Another method for writing a line of fit is to use a line of regression. A **regression line** is determined through complex calculations to ensure that the distance of all data points to the line of fit are at a minimum. Most graphing calculators and spreadsheets can perform these calculations easily.

The **correlation coefficient** r, $-1 \leq r \leq 1$, is a measure that shows how well data are modeled by a linear equation.

- When r is close to -1, the data have a negative correlation.
- When $r = 0$, the data have no correlation.
- When r is close to 1, the data have a positive correlation.

Real-World Example 2 Regression Line

The table shows the life expectancy for people born in the United States.

Year of Birth	1980	1983	1990	1995	2000	2006
Life Expectancy (yr)	73.7	74.6	75.4	75.8	76.8	77.7

Source: U.S. CDC

Use a graphing calculator to make a scatter plot of the data. Find an equation for and graph a line of regression. Then use the equation to predict the life expectancy of a person born in 2025.

Step 1 Make a scatter plot.

- Enter the years of birth in **L1** and the ages in **L2**.

KEYSTROKES: STAT ENTER 1980 ENTER 1983 ENTER 1990 ENTER ...

- Set the viewing window to fit the data.

KEYSTROKES: WINDOW 1975 ENTER 2010 ENTER 5 ENTER 70 ENTER 90 ENTER 2

- Use **STAT PLOT** to graph the scatter plot.

KEYSTROKES: 2nd [STAT PLOT] ENTER ENTER GRAPH

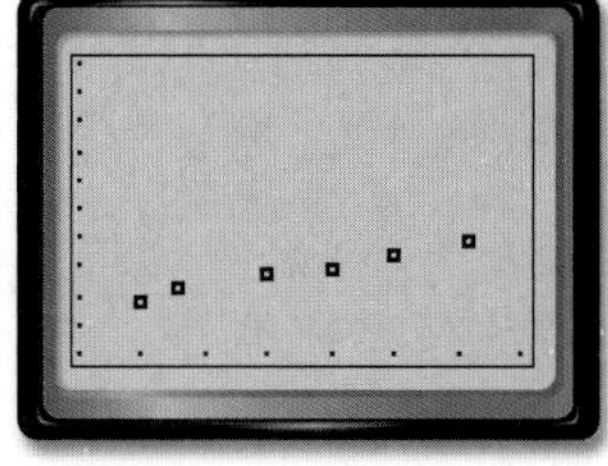

[1975, 2010] scl: 5 by [70, 90] scl: 2

> **StudyTip**
>
> **CCSS Modeling** An alternate method for creating a model of the data is to let x represent the number of years since the first year in the data set. This simplifies the math involved in finding a function to model the data. But the rate of change and fit of the regression equation are the same. Only the y-intercept of the equation is changed.

Step 2 Find the equation of the line of regression.

- Find the regression equation by selecting **LinReg(**$ax + b$**)** on the **STAT CALC** menu.

KEYSTROKES: STAT ▶ 4 ENTER

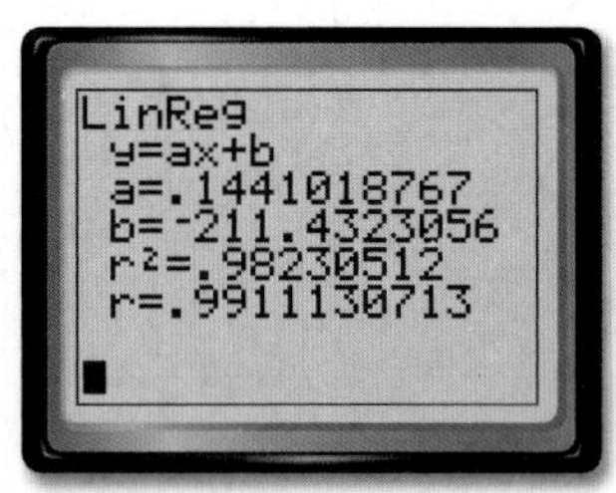

The regression equation is about $y = 0.14x - 211.43$. The slope indicates that the life expectancy increases at a rate of about 0.14 per year. The correlation coefficient r is about 0.99, which is very close to 1. So, the data fit the regression line very well.

Step 3 Graph the regression equation.

- Copy the equation to the **Y=** list and graph.

KEYSTROKES: Y= VARS 5 ▶ ▶ 1 GRAPH

Notice that the regression line comes close to all of the data points. As the correlation coefficient indicated, the line fits the data very well.

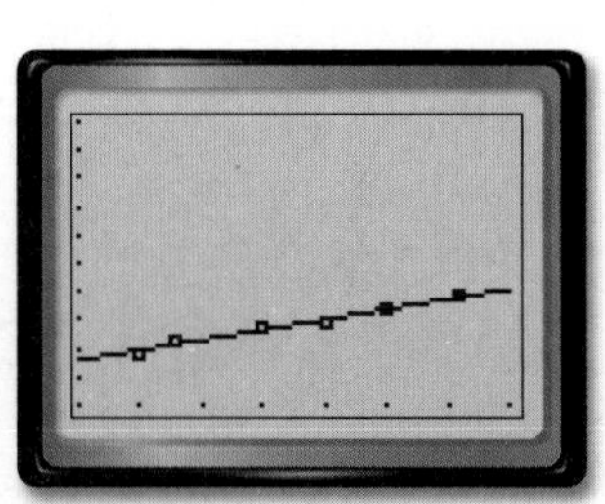

[1975, 2010] scl: 5 by [70, 90] scl: 2

ReadingMath

Predictions
When you are predicting an x-value greater than or less than any in the data set, the process is known as **extrapolation**.
When you are predicting an x-value between the least and greatest in the data set, the process is known as **interpolation**.

Step 4 Predict using the function.

- Find y when $x = 2025$. Use **VALUE** on the **CALC** menu. Reset the window size to accommodate the x-value of 2025.

KEYSTROKES: 2nd [CALC] 1 2025 ENTER

According to the function, the life expectancy of a person born in 2025 will be about 80.4 years.

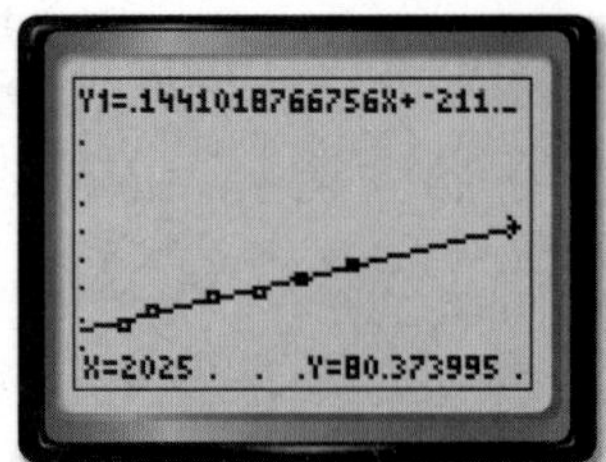

[1975, 2025] scl: 5 by [70, 90] scl: 2

GuidedPractice

2. MUSIC The table at the right shows the percent of sales that were made in music stores in the United States for the period 1999–2008. Use a graphing calculator to make a scatter plot of the data. Find and graph a line of regression. Then use the function to predict the percent of sales made in a music store in 2018.

Music Store Sales	
Year	Sales (percent)
1999	44.5
2000	42.4
2001	42.5
2002	36.8
2003	33.2
2004	32.5
2005	39.4
2006	35.4
2007	31.1
2008	30.0

Source: Recording Industry Association of America

Check Your Understanding

○ = Step-by-Step Solutions begin on page R14.

Example 1

1 OCEANS The table shows the temperature in the ocean at various depths.

Depth (in meters)	0	300	500	1000	2000	2500
Temp (°C)	22	20	13	7	6	?

Source: NDAA

a. Make a scatter plot and a line of fit, and describe the correlation.

b. Use two ordered pairs to write a prediction equation.

c. Use your prediction equation to predict the missing value.

Example 2

2. CCSS TOOLS The table shows the median income of families in North Carolina by family size in a recent year. Use a graphing calculator to make a scatter plot of the data. Find an equation for and graph a line of regression. Then use the equation to predict the median income of a North Carolina family of 9.

Family Size	Income ($)
1	33,265
2	44,625
3	50,528
4	59,481

Source: U.S. Department of Justice

Practice and Problem Solving

Extra Practice is on page R2.

Example 1

For Exercises 3–6, complete parts *a–c*.

a. Make a scatter plot and a line of fit, and describe the correlation.

b. Use two ordered pairs to write a prediction equation.

c. Use your prediction equation to predict the missing value.

3. **COMPACT DISC SALES** The table shows the number of CDs sold in recent years at Jerome's House of Music. Let x be the number of years since 2000.

Year	2004	2005	2006	2007	2008	2017
Number of CDs sold	49,300	47,280	43,450	40,125	35,792	?

4. **BASKETBALL** The table shows the number of field goals and assists for some of the members of the Miami Heat in a recent NBA season.

Field Goals	472	353	278	283	238	265	186	162	144
Assists	384	97	81	79	18	130	94	95	?

Source: NBA

5. **ICE CREAM** The table shows the amount of ice cream Sunee's Homemade Ice Creams sold for eight months. Let $x = 1$ for January.

Month	Jan	Feb	Mar	Apr	May	June	July	Aug	Sept
Gallons sold	37	44	72	80	105	110	119	131	?

6. **DRAMA CLUB** The table shows the total revenue of all of Central High School's plays in recent school years. Let x be the number of years since 2003.

School Year	2005	2006	2007	2008	2009	2016
Revenue ($)	603	666	643	721	771	?

Example 2

7. **SALES** The table shows the sales of Chayton's Computers. Let x be the number of years since 2002 and use a graphing calculator to make a scatter plot of the data. Find an equation for and graph a line of regression. Then use the function to predict the sales in 2018.

Year	Sales ($ thousands)
2004	640
2005	715
2006	791
2007	852
2008	910
2009	944

8. CCSS **TOOLS** The table shows the number of employees of a small company. Let x be the number of years since 2000 and use a graphing calculator to make a scatter plot of the data. Find an equation for and graph a line of regression. Then use the function to predict the number of employees in 2025.

Year	Number of Employees
2002	4
2003	7
2004	11
2005	14
2006	20
2007	23

9. BASEBALL The table at the right shows the total attendance and wins for the Cleveland Indians in some recent years.

Year	Wins	Attendance
2009	65	1,776,904
2008	81	2,182,087
2007	96	2,275,916
2006	78	1,997,936
2005	93	2,014,220
2004	80	1,814,401
2003	68	1,730,001

a. Make a scatter plot of the years and attendance.

b. If a linear shape is apparent, find a regression equation to summarize the trend.

c. Interpret the slope of the regression line in the context of the data.

d. Is the relationship between years and attendance stronger than the one between wins and attendance? Explain.

10. CLASS SIZE The table at the right shows the relationship between the number of students in a mathematics class and the average grade for each class.

Class Size	Class Average
16	81.2
19	80.5
24	82.5
25	79.9
27	78.6
29	79.3
32	77.7

a. Make a scatter plot and find a regression equation for the data. Then graph the regression line.

b. What is the correlation coefficient of the data?

c. Describe the correlation. How accurate is the regression equation?

11. CCSS TOOLS Jocelyn is analyzing the sales of her company. The table at the right shows the total sales for each of six years.

Year	Sales ($ millions)
2003	31.2
2004	34.6
2005	18.6
2006	37.7
2007	41.3
2008	45.1

a. Find a regression equation and correlation coefficient for the data. Let x be the year.

b. Use the regression equation to predict the 2020 sales.

c. Remove the outlier from the data set and find a new regression equation and correlation coefficient.

d. Use the new regression equation to predict the sales in 2020.

e. Compare the correlation coefficients for the two equations. Which function fits the data better? Which prediction should Jocelyn expect to be more accurate?

H.O.T. Problems Use Higher-Order Thinking Skills

12. REASONING What is the relevance of the correlation coefficient of a linear regression line? Explain your reasoning.

13. CHALLENGE If statements a and b have a positive correlation, b and c have a negative correlation, and c and d have a positive correlation, what can you determine about the correlation between statements a and d? Explain your reasoning.

14. OPEN ENDED Provide real-world quantities that represent each of the following.

a. positive correlation **b.** negative correlation **c.** no correlation

15. CHALLENGE Draw a scatter plot for the following data set.

x	1.0	1.5	2.0	2.8	3.2	4.0	4.8	5.8
y	3.5	4.7	5.1	6.8	7.1	7.5	8.8	10.3

Which of the following best represents the correlation coefficient r for the data? Justify your answer.

a. 0.99 **b.** −0.98 **c.** 0.62 **d.** 0.08

16. WRITING IN MATH What are the strengths and weaknesses of using a regression equation to approximate data?

Standardized Test Practice

17. SHORT RESPONSE What is the value of the expression below?

$$17 - 3[-1 + 2(7 - 4)]$$

18. Anna took brownies to a club meeting. She gave half of her brownies to Selena. Selena gave a third of her brownies to Randall. Randall gave a fourth of his brownies to Trina. If Trina has 3 brownies, how many brownies did Anna bring to the meeting?

A 12

B 36

C 72

D 144

19. GEOMETRY Which is always true?

F A parallelogram is a square.

G A parallelogram is a rectangle.

H A quadrilateral is a trapezoid.

J A square is a rectangle.

20. SAT/ACT Which line best fits the data in the graph?

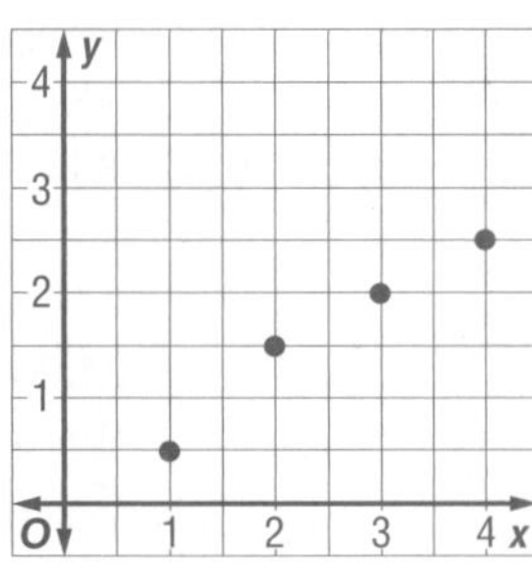

A $y = x$

B $y = -0.5x + 4$

C $y = -0.5x - 4$

D $y = 0.5x + 0.5$

E $y = 1.5x - 1.5$

Spiral Review

Write an equation in slope-intercept form for each graph. (Lesson 2-4)

21.

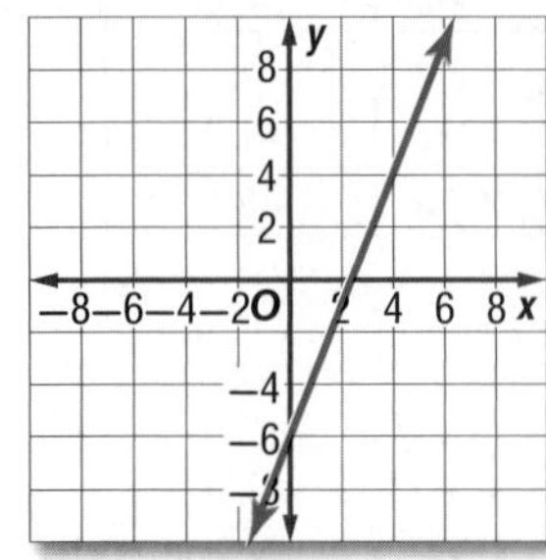

22.

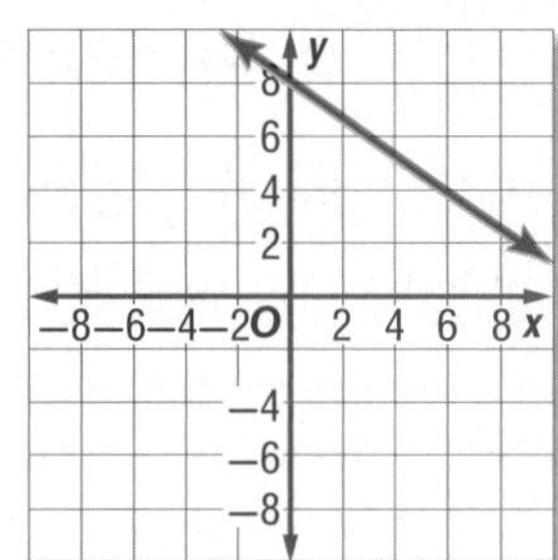

23.

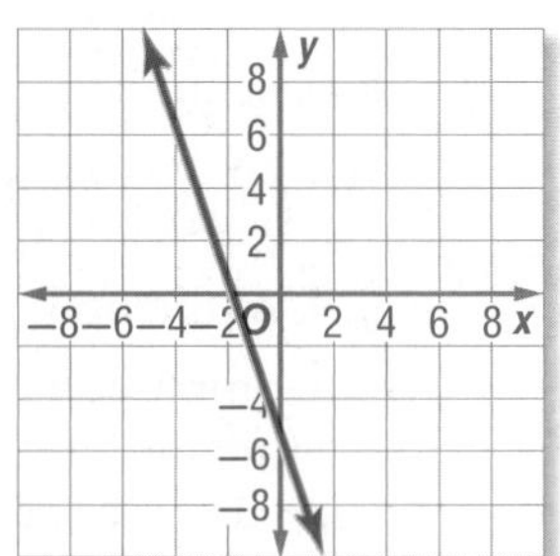

Find the rate of change for each set of data. (Lesson 2-3)

24.

Time (day)	3	6	9	12	15
Height (mm)	12	24	36	48	60

25.

Time (h)	2	4	6	8
Distance (mi)	35	70	105	140

26.

Time (s)	12	16	20	24	28
Volume (cm^3)	45	60	75	90	105

27.

Force (*N*)	32	40	48	56	64
Work (*J*)	48	60	72	84	96

28. RECREATION Ramona estimates that she will need 50 tennis balls for every player that signs up for the tennis club and at least 150 more just in case. Write an inequality to express the situation. (Lesson 1-5)

29. DODGEBALL Six teams played in a dodgeball tournament. In how many ways can the top three teams finish? (Lesson 0-4)

Skills Review

Solve each equation.

30. $-4|x - 2| = -12$

31. $|3x + 4| = 21$

32. $2|4x - 1| + 3 = 9$

EXTEND 2-5

Algebra Lab
Correlation and Causation

You have learned that the correlation coefficient measures how well an equation fits a set of data. A correlation coefficient close to 1 or −1 indicates a high correlation. However, this does not imply causation. When there is **causation**, one data set is the direct cause of the other data set. When there is a correlation between two sets of data, the data sets are related.

NEWS

Cell Phones Cause Brain Tumors!

Studies have shown that brain tumors are linked to cell phone use. One particular study showed a very strong correlation between the two events.

Common Core State Standards
Content Standards
F.IF.4 For a function that models a relationship between two quantities, interpret key features of graphs and tables in terms of the quantities, and sketch graphs showing key features given a verbal description of the relationship.

The news clip to the right uses correlation to imply causation. While a study may have found a high correlation between these two variables, this does not mean that cell phone use causes brain tumors. Other variables, such as an individual's family history, diet, and environment could also have an effect on the formation of brain tumors.

A **lurking variable** affects the relationship between two other variables, but is not included in the study that compares them.

EXAMPLE 1

Determine whether the following correlations possibly show causation. Write *yes* or *no*. If not, identify other lurking variables.

a. Studies have shown that students are less energetic after they eat lunch.

No; students could have had gym class before lunch, or the class after lunch could be one that they do not find interesting.

b. If Kevin lifts weights, he will make the football team.

No; Kevin may need to maintain a certain grade-point average to make the team. He also needs to be talented enough to make the team.

c. When the Sun is visible, we have daylight.

Yes; no other variables cause daylight.

Exercises

Determine whether the following correlations possibly show causation. Write *yes* or *no*. If not, identify other lurking variables.

1. If Allison studies, she will get an A.
2. When Lisa exercises, she is in a better mood.
3. The gun control laws have reduced violent crime.
4. Smoking causes lung cancer.
5. If we have a Level 2 snow emergency, we do not have school.
6. Reading more increases one's intelligence.

Think About It

7. What do you think must be done to show that a correlation between two variables actually shows causation?

(continued on the next page)

Algebra Lab
Correlation and Causation *Continued*

Statistics are often provided that show a correlation between two events and causation is implied, but not confirmed. The best method to find evidence that x causes y is to actually perform multiple experiments with large sample sizes. When an experiment is not available, the only option is to determine if causation is *possible.*

EXAMPLE 2

The table shows IQ scores for ten high school students and their corresponding grade-point averages.

IQ	GPA	IQ	GPA
106	3.4	118	3.9
110	3.6	107	3.3
109	3.5	111	3.6
98	2.9	109	3.7
115	3.8	102	3.1

a. Calculate the correlation coefficient.
The value of r is about 0.96.

b. Describe the correlation.
There is a strong positive correlation.

c. Is there causation? Explain your reasoning.
There is no causation. Many other factors determine grade-point averages, including effort, study habits, knowledge of the material in the particular classes, and extracurricular activities.

EXAMPLE 3

The table shows the approximate boiling temperatures of water at different altitudes.

Altitude (ft)	Temp. (°F)
0	212
984	210
2000	208
3000	206
5000	203
7500	198
10,000	194
20,000	178
26,000	168

a. Calculate the correlation coefficient.
The value of r is about -0.99.

b. Describe the correlation.
There is a strong negative correlation.

c. Is there causation? Explain your reasoning.
There appears to be causation. Altitude is a direct cause for the boiling temperature of water to decrease.

Exercises

Calculate the correlation coefficient and describe the correlation. Determine if causation is possible. Explain your reasoning.

8.

Car Value ($)	Miles per Gallon	Car Value ($)	Miles per Gallon
14,000	55	25,000	33
16,000	48	29,000	29
18,500	37	32,000	19
20,000	35	35,000	26
22,500	29	40,000	33

9.

Temp. (°F)	Ice Cream Sales ($)	Temp. (°F)	Ice Cream Sales ($)
76	850	85	905
78	885	86	1005
80	875	87	1060
82	920	88	1105
84	955	89	1165

10.

Year	Sales via Internet ($)	Year	Sales via Internet ($)
2001	1923	2006	7067
2002	2806	2007	8568
2003	3467	2008	9991
2004	4848	2009	10,587
2005	5943	2010	12,695

11.

Number of Rings from Pith to Bark	Age of Tree (yr.)	Number of Rings from Pith to Bark	Age of Tree (yr.)
5	5	30	30
10	10	40	40
15	15	50	50
20	20	60	60

LESSON

2-6 Special Functions

Then

- You modeled data using lines of regression.

Now

1. Write and graph piecewise-defined functions.
2. Write and graph step and absolute value functions.

Why?

- The table shows a recent federal income tax rate schedule. The amount of federal income tax an individual is required to pay is a function of income.

Federal Tax Rate Schedule – Filing Single		
If taxable income is over	But not over	The tax is:
\$0	\$7,825	10% of the amount over \$0
\$7,825	\$31,850	\$782.50 plus 15% of the amount over \$7,825
\$31,850	\$77,100	\$4,386.25 plus 25% of the amount over \$31,850
\$77,100	\$160,850	\$15,698.75 plus 28% of the amount over \$77,100
\$160,850	\$349,700	\$39,148.75 plus 33% of the amount over \$160,850
\$349,700	no limit	\$101,469.25 plus 35% of the amount over \$349,700

Source: Internal Revenue Service

NewVocabulary

piecewise-defined function
piecewise-linear function
step function
greatest integer function
absolute value function

Common Core State Standards

Content Standards

F.IF.4 For a function that models a relationship between two quantities, interpret key features of graphs and tables in terms of the quantities, and sketch graphs showing key features given a verbal description of the relationship.

F.IF.7.b Graph square root, cube root, and piecewise-defined functions, including step functions and absolute value functions.

Mathematical Practices

1 Make sense of problems and persevere in solving them.

1 Piecewise-Defined Functions

The function relating income and tax is not a linear function because each interval, or piece, of the function is defined by a different expression. A function that is written using two or more expressions is called a **piecewise-defined function.** On the graph of a piecewise-defined function, a dot indicates that the point is included in the graph. A circle indicates that the point is not included in the graph.

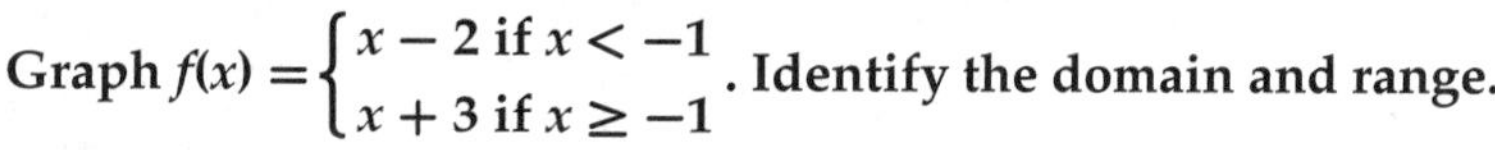

Example 1 Piecewise-Defined Function

Graph $f(x) = \begin{cases} x - 2 \text{ if } x < -1 \\ x + 3 \text{ if } x \geq -1 \end{cases}$. Identify the domain and range.

Step 1 Graph $f(x) = x - 2$ for $x < -1$.

$$f(x) = x - 2$$
$$= (-1) - 2$$
$$= -3$$

Because -1 does not satisfy the inequality, begin with a circle at $(-1, -3)$.

Step 2 Graph $f(x) = x + 3$ for $x \geq -1$.

$$f(x) = x + 3$$
$$= (-1) + 3$$
$$= 2$$

Because -1 satisfies the inequality, begin with a dot at $(-1, 2)$.

The function is defined for all values of x, so the domain is all real numbers.

The $f(x)$-coordinates of points on the graph are all real numbers less than -3 and all real numbers greater than or equal to 2, so the range is $\{f(x) \mid f(x) < -3 \text{ or } f(x) \geq 2\}$.

GuidedPractice

1. Graph $f(x) = \begin{cases} x + 2 \text{ if } x < 0 \\ x \quad\quad \text{if } x \geq 0 \end{cases}$. Identify the domain and range.

Piecewise-defined functions are often defined by several linear functions.

StudyTip

Graphs of Piecewise-Defined Functions The graphs of different parts of a piecewise-defined function may or may not connect. A graph may also stop at a given x-value and then begin at a different y-value for the same value of x.

Example 2 Write a Piecewise-Defined Function

Write the piecewise-defined function shown in the graph.

Examine and write a function for each portion of the graph.

The left portion of the graph is the graph of $f(x) = 2x + 3$. There is a circle at (1, 5), so the linear function is defined for $\{x|x < 1\}$.

The center portion of the graph is the graph of $f(x) = -x + 2$. There are dots at (1, 1) and (2, 0), so the linear function is defined for $\{x|1 \leq x \leq 2\}$.

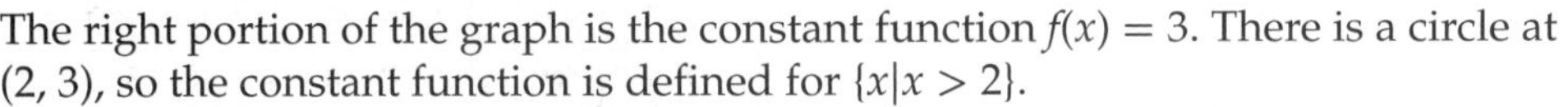

The right portion of the graph is the constant function $f(x) = 3$. There is a circle at (2, 3), so the constant function is defined for $\{x|x > 2\}$.

Write the piecewise-defined function.

$$f(x) = \begin{cases} 2x + 3 \text{ if } x < 1 \\ -x + 2 \text{ if } 1 \leq x \leq 2 \\ 3 \text{ if } x > 2 \end{cases}$$

CHECK The graph shows a portion of a line with positive slope for $x < 1$. The graph has negative slope for $1 \leq x \leq 2$ and constant slope for $x > 2$. The function is reasonable for the graph.

GuidedPractice

Write the piecewise-defined function shown in each graph.

2A.

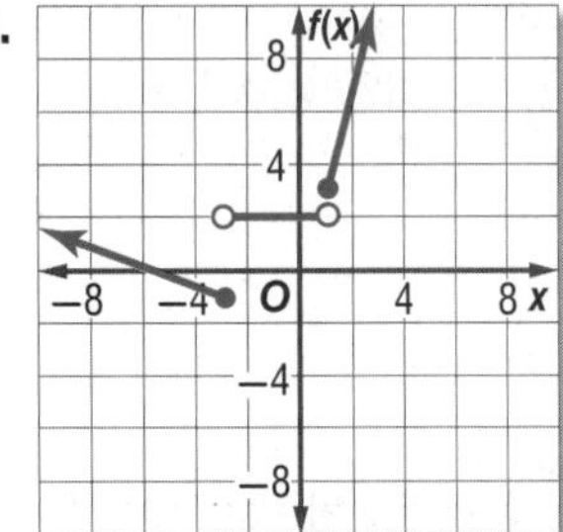

2B.

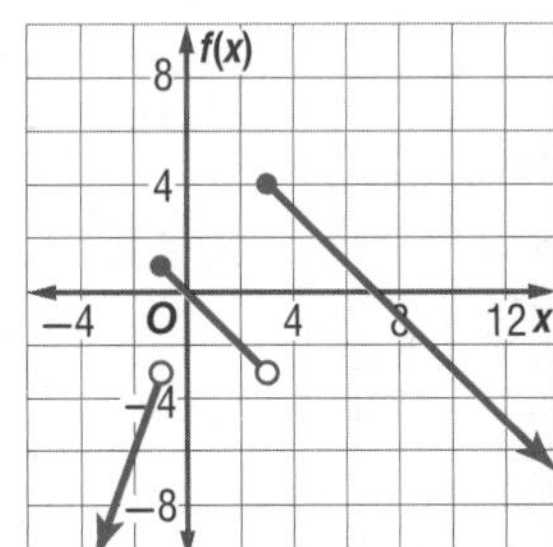

2 Step Functions and Absolute Value Functions

Unlike a piecewise-defined function, a **piecewise-linear function** contains a single expression. A common piecewise-linear function is the step function. The graph of a **step function** consists of line segments.

StudyTip

Greatest Integer Function Notice that the domain of this step function is all real numbers and the range is all integers.

The **greatest integer function**, written $f(x) = [\![x]\!]$, is one kind of step function. The symbol $[\![x]\!]$ means *the greatest integer less than or equal to x*. For example, $[\![3.25]\!] = 3$ and $[\![-4.6]\!] = -5$.

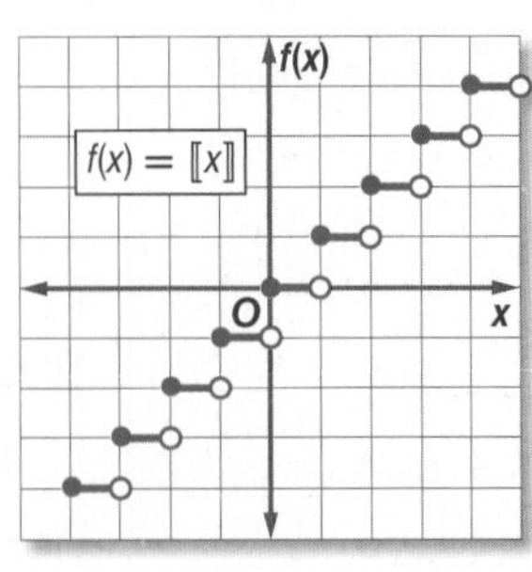

Real-World Example 3 Use a Step Function

BUSINESS An automotive repair center charges \$50 for any part of the first hour of labor, and \$35 for any part of each additional hour. Draw a graph that represents this situation.

Understand The total labor charge is \$50 for the first hour plus \$35 for each additional fraction of an hour, so the graph will be a step function.

Plan If the time spent on labor is greater than 0 hours, but less than or equal to 1 hour, then the labor charge is \$50. If the time is greater than 1 hour but less than 2 hours, then the labor charge is \$85, and so on.

Solve Use the pattern of times and costs to make a table, where x is the number of hours of labor and $T(x)$ is the total labor charge. Then graph.

x	$T(x)$
$0 < x \leq 1$	\$50
$1 < x \leq 2$	\$85
$2 < x \leq 3$	\$120
$3 < x \leq 4$	\$155
$4 < x \leq 5$	\$190

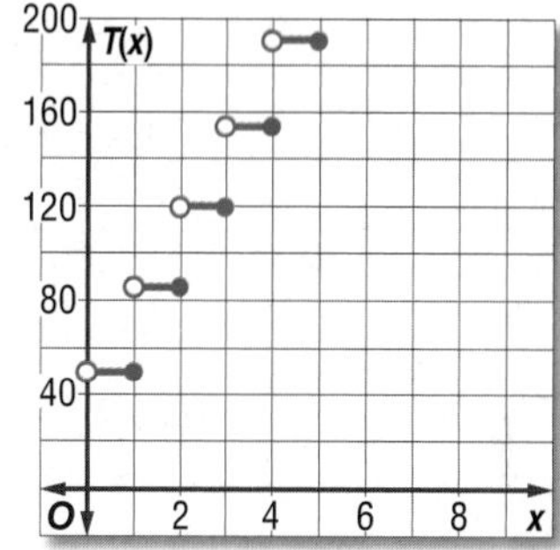

Check Since the repair center rounds any fraction of an hour up to the next whole number, each segment of the graph has a circle at the left endpoint and a dot at the right endpoint.

Guided Practice

3. RECYCLING A recycling company pays \$5 for every full box of newspaper. They do not give any money for partial boxes. Draw a graph that shows the amount of money $P(x)$ for the number of boxes x brought to the recycling center.

Math History Link

Karl Weierstrass (1815–1897) At the wishes of his father, Weierstrass studied law, economics, and finance at the University of Bonn, but then dropped out to study his true interest, mathematics, at the University of Münster. In an 1841 essay, Weierstrass first used | | to denote absolute value.

Photo: The Granger Collection

Another piecewise-linear function is the absolute value function. An **absolute value function** is a function that contains an algebraic expression within absolute value symbols.

Key Concept Parent Function of Absolute Value Functions

Parent function: $f(x) = |x|$, defined as

$$f(x) = \begin{cases} x \text{ if } x > 0 \\ 0 \text{ if } x = 0 \\ -x \text{ if } x < 0 \end{cases}$$

Type of graph: V-shaped

Domain: all real numbers

Range: all nonnegative real numbers

Intercepts: $x = 0$, $f(x) = 0$

Not defined: $f(x) < 0$

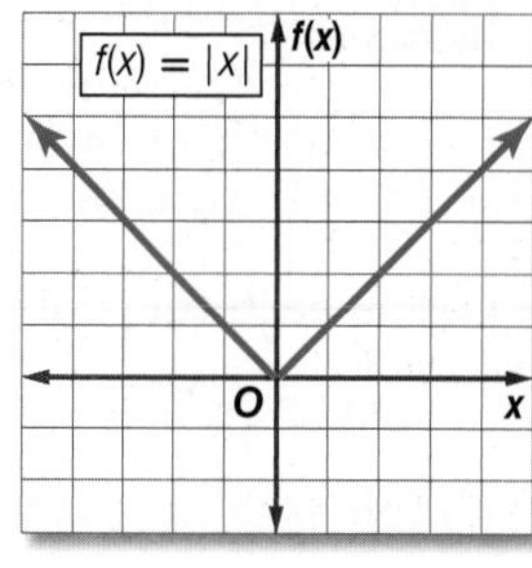

Example 4 Absolute Value Functions

Graph $f(x) = |2x| - 4$. Identify the domain and range.

Create a table of values.

x	$\|2x\| - 4$
−3	2
−2	0
−1	−2
0	−4
1	−2
2	0
3	2

Graph the points and connect them.

The domain is the set of all real numbers. The range is $\{f(x) | f(x) \geq -4\}$.

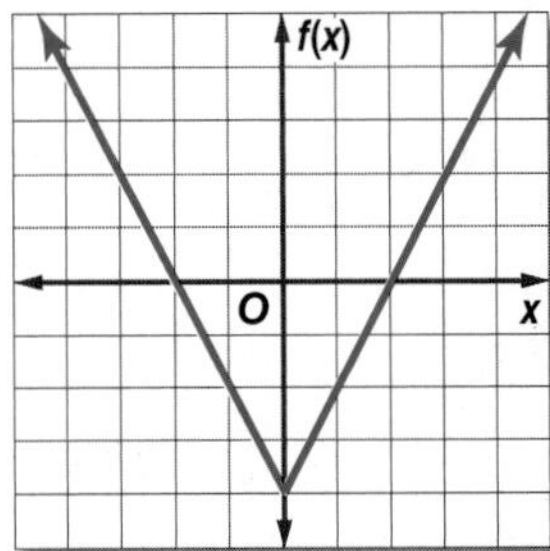

GuidedPractice

Graph each function. Identify the domain and range.

4A. $f(x) = |x - 2|$

4B. $f(x) = -|x| + 1$

Check Your Understanding

= Step-by-Step Solutions begin on page R14.

Example 1 **Graph each function. Identify the domain and range.**

1. $g(x) = \begin{cases} -3 \text{ if } x \leq -4 \\ x \text{ if } -4 < x < 2 \\ -x + 6 \text{ if } x \geq 2 \end{cases}$

2. $f(x) = \begin{cases} 8 \text{ if } x \leq -1 \\ 2x \text{ if } -1 < x < 4 \\ -4 - x \text{ if } x \geq 4 \end{cases}$

Example 2 **Write the piecewise-defined function shown in each graph.**

3.

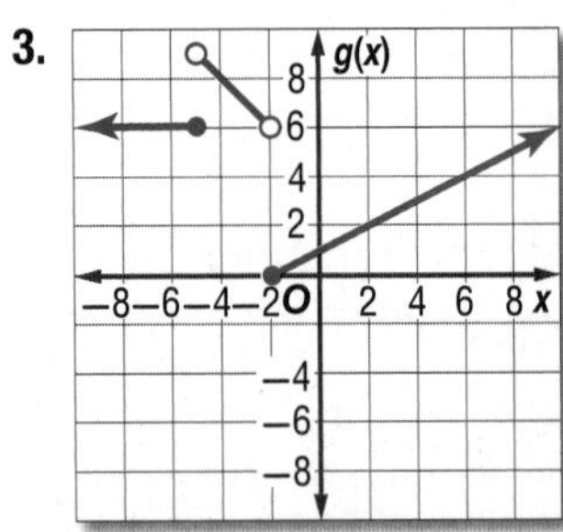

4.

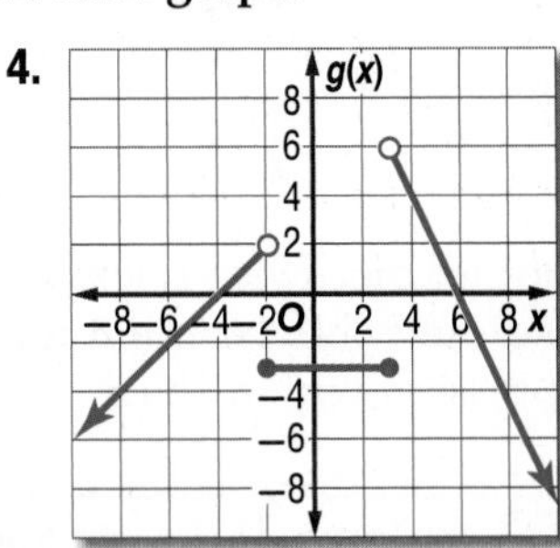

Example 3

5. **CCSS REASONING** Springfield High School's theater can hold 250 students. The drama club is performing a play in the theater. Draw a graph of a step function that shows the relationship between the number of tickets sold x and the minimum number of plays y that the drama club must perform.

Graph each function. Identify the domain and range.

6. $g(x) = -2[\![x]\!]$

7. $h(x) = [\![x - 5]\!]$

Example 4 **Graph each function. Identify the domain and range.**

8. $g(x) = |-3x|$

9. $f(x) = 2|x|$

10. $h(x) = |x + 4|$

11. $s(x) = |-2x| + 6$

Practice and Problem Solving

Extra Practice is on page R2.

Example 1 **Graph each function. Identify the domain and range.**

12. $f(x) = \begin{cases} -3x \text{ if } x \leq -4 \\ x \text{ if } 0 < x \leq 3 \\ 8 \text{ if } x > 3 \end{cases}$

13. $f(x) = \begin{cases} 2x \text{ if } x \leq -6 \\ 5 \text{ if } -6 < x \leq 2 \\ -2x + 1 \text{ if } x > 4 \end{cases}$

14. $g(x) = \begin{cases} 2x + 2 \text{ if } x < -6 \\ x \text{ if } -6 \leq x \leq 2 \\ -3 \text{ if } x > 2 \end{cases}$

15. $g(x) = \begin{cases} -2 \text{ if } x < -4 \\ x - 3 \text{ if } -1 \leq x \leq 5 \\ 2x - 15 \text{ if } x > 7 \end{cases}$

Example 2 **Write the piecewise-defined function shown in each graph.**

16.

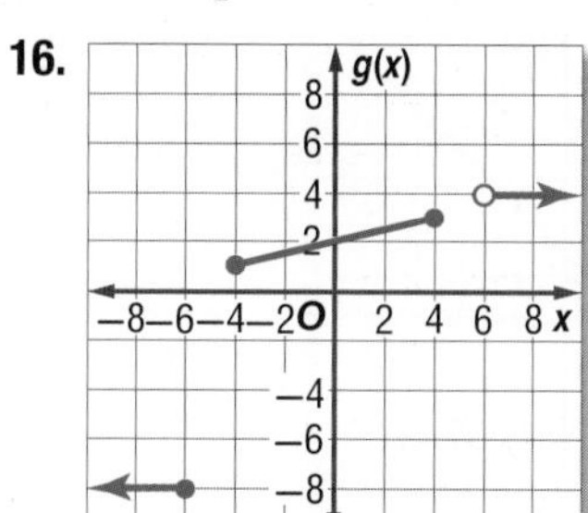

17.

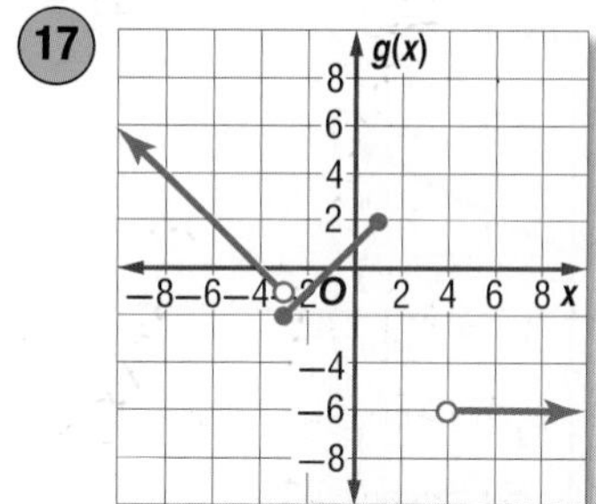

18.

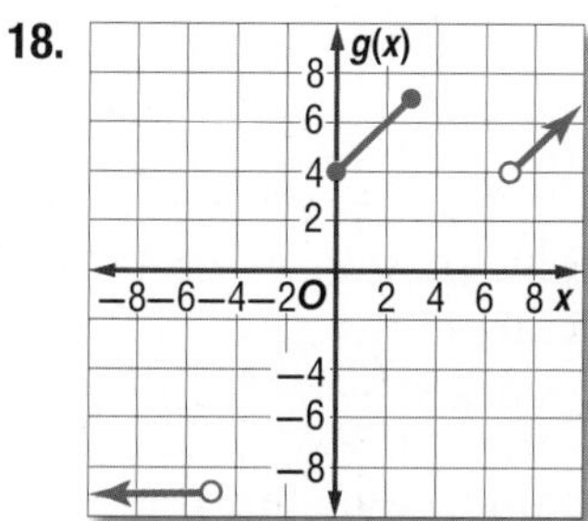

19.

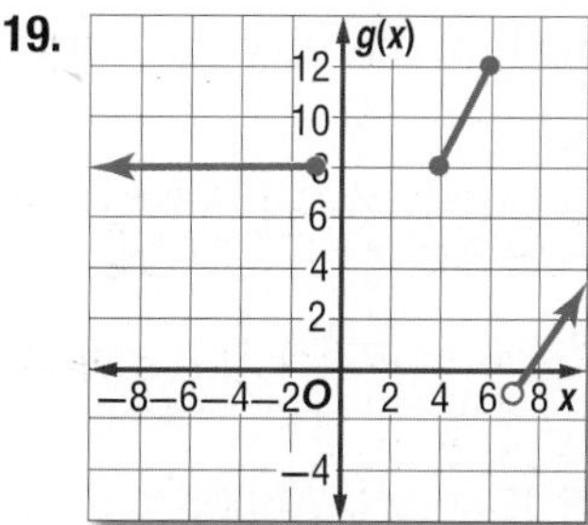

Example 3 **Graph each function. Identify the domain and range.**

20. $f(x) = [\![x]\!] - 6$

21. $h(x) = [\![3x]\!] - 8$

22. $f(x) = [\![3x + 2]\!]$

23. $g(x) = 2[\![0.5x + 4]\!]$

Example 4 **Graph each function. Identify the domain and range.**

24. $f(x) = |x - 5|$

25. $g(x) = |x + 2|$

26. $h(x) = |2x| - 8$

27. $k(x) = |-3x| + 3$

28. $f(x) = 2|x - 4| + 6$

29. $h(x) = -3|0.5x + 1| - 2$

30. **GIVING** Patrick is donating money and volunteering his time to an organization that restores homes for the needy. His employer will match his monetary donations up to $100.

a. Identify the type of function that models the total amount of money received by the charity when Patrick donates x dollars.

b. Write and graph a function for the situation.

31. **CCSS SENSE-MAKING** A car's speedometer reads 60 miles an hour.

a. Write an absolute value function for the difference between the car's actual speed a and the reading on the speedometer.

b. What is an appropriate domain for the function? Explain your reasoning.

c. Use the domain to graph the function.

32. RECREATION The charge for renting a bicycle from a rental shop for different amounts of time is shown at the right.

a. Identify the type of function that models this situation.

b. Write and graph a function for the situation.

Time	Price
$\frac{1}{2}$ hour	\$6
1 hour	\$10
2 hours	\$16
Daily	\$24

Use each graph to write the absolute value function.

33.

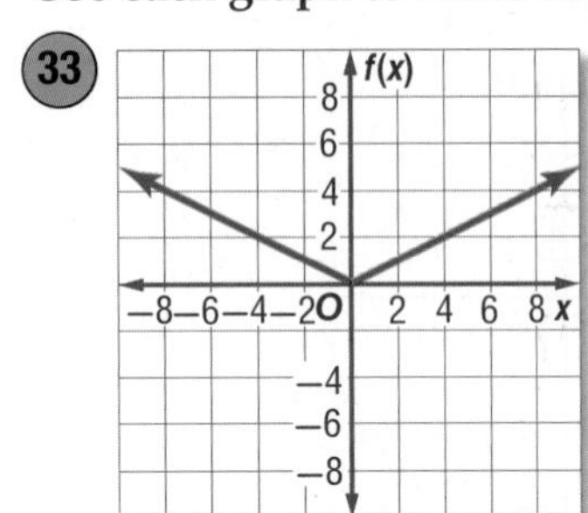

34.

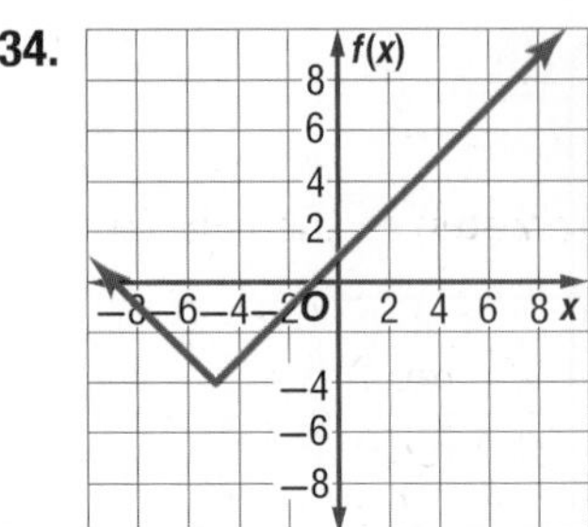

Graph each function. Identify the domain and range.

35. $f(x) = [\![\,|0.5x|\,]\!]$

36. $g(x) = |\,[\![2x]\!]\,|$

37. $g(x) = \begin{cases} [\![x]\!] \text{ if } x < -4 \\ x + 1 \text{ if } -4 \leq x \leq 5 \\ -|x| \text{ if } x > 3 \end{cases}$

38. $h(x) = \begin{cases} -|x| \text{ if } x < -6 \\ |x| \text{ if } -6 \leq x \leq 2 \\ |-x| \text{ if } x > 2 \end{cases}$

39. MULTIPLE REPRESENTATIONS Consider the following absolute value functions.

$$f(x) = |x| - 4 \qquad g(x) = |3x|$$

a. Tabular Use a graphing calculator to create a table of $f(x)$ and $g(x)$ values for $x = -4$ to $x = 4$.

b. Graphical Graph the functions on separate graphs.

c. Numerical Determine the slope between each two consecutive points in the table.

d. Verbal Describe how the slopes of the two sections of an absolute value graph are related.

H.O.T. Problems Use Higher-Order Thinking Skills

40. OPEN ENDED Write an absolute value relation in which the domain is all nonnegative numbers and the range is all real numbers.

41. CHALLENGE Graph $|y| = 2|x + 3| - 5$.

42. CCSS ARGUMENTS Find a counterexample to the statement and explain your reasoning.

In order to find the greatest integer function of x when x is not an integer, round x to the nearest integer.

43. OPEN ENDED Write an absolute value function in which $f(5) = -3$.

44. WRITING IN MATH Explain how piecewise functions can be used to accurately represent real-world problems.

Ben Blankenburg/CORBIS

Standardized Test Practice

45. SHORT RESPONSE What expression gives the nth term of the linear pattern defined by the table?

2	4	6	8	n
7	13	19	25	?

46. Solve: $5(x + 4) = x + 4$

Step 1: $5x + 20 = x + 4$

Step 2: $4x + 20 = 4$

Step 3: $4x = 24$

Step 4: $x = 6$

Which is the first *incorrect* step in the solution shown above?

A Step 4
B Step 3
C Step 2
D Step 1

47. NUMBER THEORY Twelve consecutive integers are arranged in order from least to greatest. If the sum of the first six integers is 381, what is the sum of the last six integers?

F 345
G 381
H 387
J 417

48. SAT/ACT For which function does $f\left(-\frac{1}{2}\right) \neq -1$?

A $f(x) = 2x$
B $f(x) = |-2x|$
C $f(x) = [\![x]\!]$
D $f(x) = [\![2x]\!]$
E $f(x) = -|2x|$

Spiral Review

49. FOOTBALL The table shows the relationship between the total number of male students per school and the number of students who tried out for the football team. (Lesson 2-5)

a. Find a regression equation for the data.

b. Determine the correlation coefficient.

c. Predict how many students will try out for football at a school with 800 male students.

Number of Male Students	Number of Tryouts
180	46
212	51
274	62
401	75
513	81
589	90

Write an equation in slope-intercept form for the line described. (Lesson 2-4)

50. passes through $(-3, -6)$, perpendicular to $y = -2x + 1$

51. passes through $(4, 0)$, parallel to $3x + 2y = 6$

52. passes through the origin, perpendicular to $4x - 3y = 12$

Find each value if $f(x) = -4x + 6$, $g(x) = -x^2$, and $h(x) = -2x^2 - 6x + 9$. (Lesson 2-1)

53. $f(2c)$

54. $g(a + 1)$

55. $h(6)$

56. Determine whether the figures below are similar. (Lesson 0-7)

33

26.4

9.6

Skills Review

Graph each equation.

57. $y = -0.25x + 8$

58. $y = \frac{4}{3}x + 2$

59. $8x + 4y = 32$

EXPLORE

2-7 Graphing Technology Lab
Families of Lines

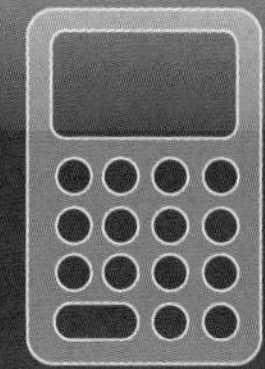

The parent function of the family of linear functions is $f(x) = x$. You can use a graphing calculator to investigate how changing the parameters m and b in $f(x) = mx + b$ affects the graphs as compared to the parent function.

CCSS Common Core State Standards
Content Standards
F.BF.3 Identify the effect on the graph of replacing $f(x)$ by $f(x) + k$, $k\,f(x)$, $f(kx)$, and $f(x + k)$ for specific values of k (both positive and negative); find the value of k given the graphs. Experiment with cases and illustrate an explanation of the effects on the graph using technology.

PT

Activity 1 b in $f(x) = mx + b$

Graph $f(x) = x$, $f(x) = x + 3$, and $f(x) = x - 5$ in the standard viewing window.

Enter the equations in the **Y=** list as **Y1**, **Y2**, and **Y3**. Then graph the equations.

KEYSTROKES: [Y=] [X,T,θ,n] [ENTER] [X,T,θ,n] [+] 3 [ENTER] [X,T,θ,n] [−] 5 [ENTER] [ZOOM] 6

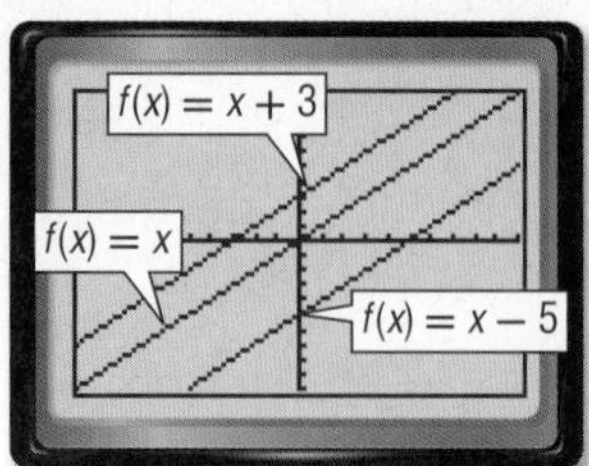

[−10, 10] scl: 1 by [−10, 10] scl: 1

1A. Compare and contrast the graphs.

1B. How would you obtain the graphs of $f(x) = x + 3$ and $f(x) = x - 5$ from the graph of $f(x) = x$?

The parameter m in $f(x) = mx + b$ affects the graphs in a different way than b.

Activity 2 m in $f(x) = mx + b$

Graph $f(x) = x$, $f(x) = 3x$, and $f(x) = \frac{1}{2}x$ in the standard viewing window.

Enter the equations in the **Y=** list and graph.

2A. How do the graphs compare?

2B. Which graph is steepest? Which graph is the least steep?

2C. Graph $f(x) = -x$, $f(x) = -3x$, and $f(x) = -\frac{1}{2}x$ in the standard viewing window. How do these graphs compare?

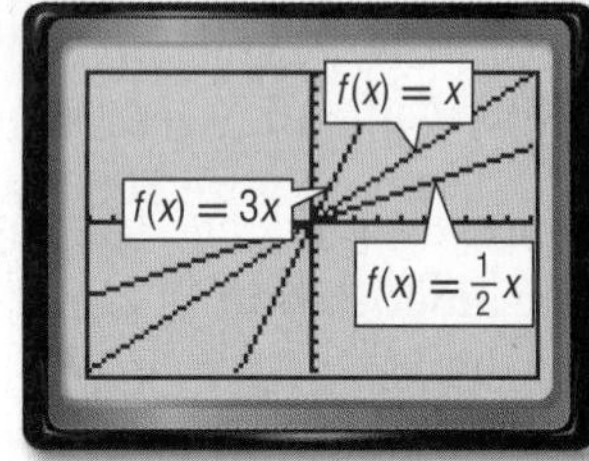

[−10, 10] scl: 1 by [−10, 10] scl: 1

Analyze the Results

Graph each set of equations on the same screen. Describe the similarities or differences among the graphs.

1. $f(x) = 3x$
$f(x) = 3x + 1$
$f(x) = 3x - 2$

2. $f(x) = x + 2$
$f(x) = 5x + 2$
$f(x) = \frac{1}{2}x + 2$

3. $f(x) = x - 3$
$f(x) = 2x - 3$
$f(x) = 0.75x - 3$

4. What do the graphs of equations of the form $f(x) = mx + b$ have in common?

5. What are the domain and range of functions of the form $f(x) = mx + b$, where $m \neq 0$?

6. How do the values of b and m affect the graph of $f(x) = mx + b$ as compared to the parent function $f(x) = x$?

7. Summarize your results. How can knowing about the effects of m and b help you sketch the graph of a function?

LESSON

2-7 Parent Functions and Transformations

Then	Now	Why?
You analyzed and used relations and functions.	**1** Identify and use parent functions. **2** Describe transformations of functions.	Nick makes \$8 an hour working at a pizza shop. The red line represents his wages. If he also delivers the pizzas, he is paid \$2 more per hour. The blue line represents Nick's wages when he delivers the pizzas. These graphs are examples of transformations.

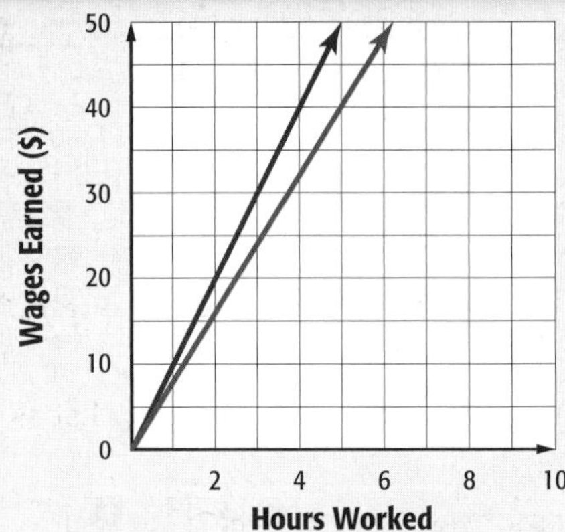

NewVocabulary

family of graphs
parent graph
parent function
constant function
identity function
quadratic function
translation
reflection
line of reflection
dilation

Common Core State Standards

Content Standards

F.IF.4 For a function that models a relationship between two quantities, interpret key features of graphs and tables in terms of the quantities, and sketch graphs showing key features given a verbal description of the relationship.

F.BF.3 Identify the effect on the graph of replacing $f(x)$ by $f(x) + k$, $k\,f(x)$, $f(kx)$, and $f(x + k)$ for specific values of k (both positive and negative); find the value of k given the graphs. Experiment with cases and illustrate an explanation of the effects on the graph using technology.

Mathematical Practices

6 Attend to precision.

1 Parent Graphs A **family of graphs** is a group of graphs that display one or more similar characteristics. The **parent graph**, which is the graph of the **parent function**, is the simplest of the graphs in a family. This is the graph that is transformed to create other members in a family of graphs.

KeyConcept Parent Functions

Constant Function

y
$y = 1$
O
x

The general equation of a **constant function** is $f(x) = a$, where a is any number. The domain is all real numbers, and the range consists of a single real number a.

Identity Function

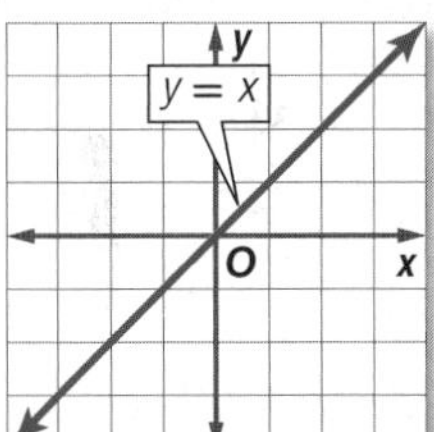

The **identity function** $f(x) = x$ passes through all points with coordinates (a, a). It is the parent function of most linear functions. Its domain and range are all real numbers.

Absolute Value Function

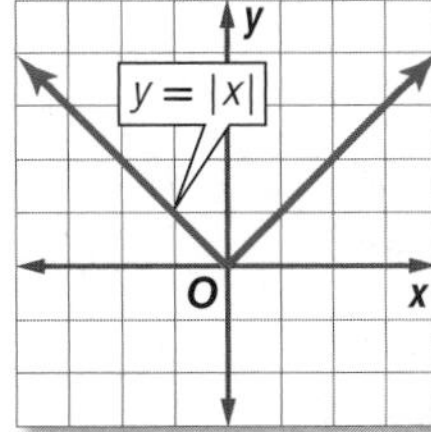

Recall that the parent function of absolute value functions is $f(x) = |x|$. The domain of $f(x) = |x|$ is the set of real numbers, and the range is the set of real numbers greater than or equal to 0.

Quadratic Function

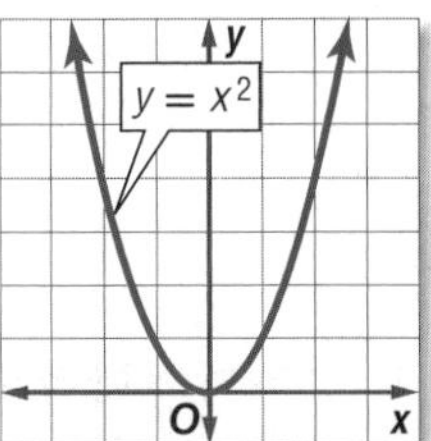

The parent function of **quadratic functions** is $f(x) = x^2$. The domain of $f(x) = x^2$ is the set of real numbers, and the range is the set of real numbers greater than or equal to 0.

Example 1 Identify a Function Given the Graph

Identify the type of function represented by each graph.

a.

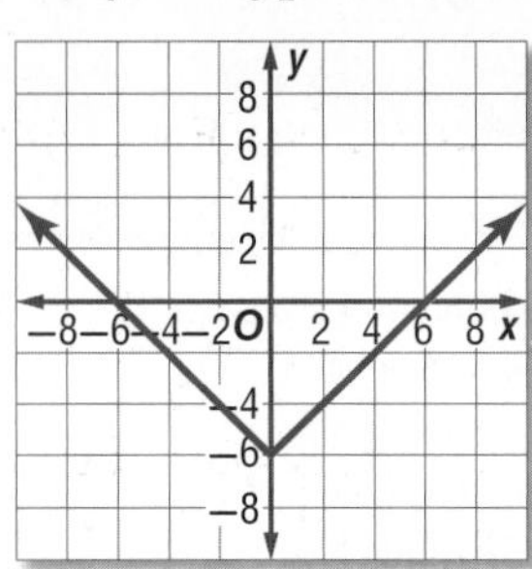

The graph is in the shape of a V. The graph represents an absolute value function.

b.

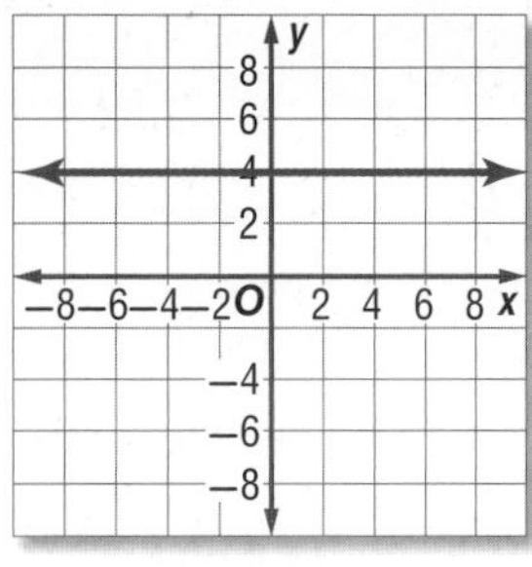

The graph is a horizontal line that crosses the y-axis at 4. The graph represents a constant function.

Guided Practice

1A.

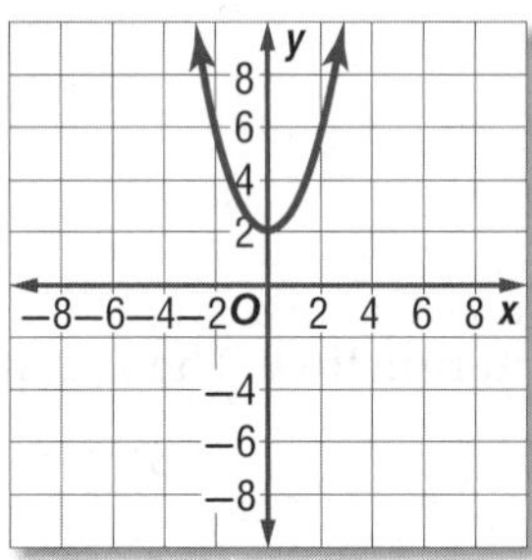

1B.

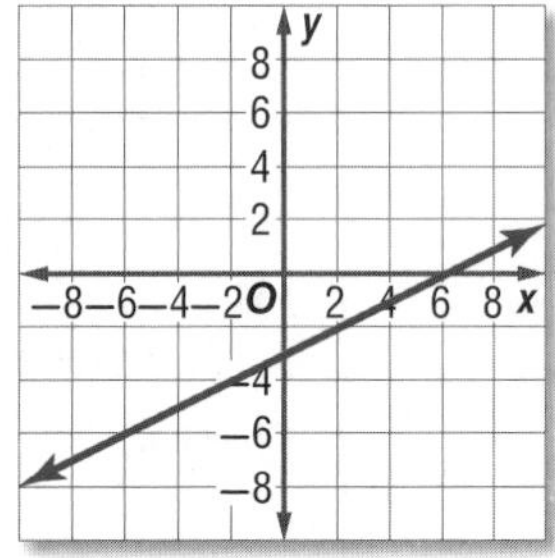

2 Transformations

Transformations Transformations of a parent graph may appear in a different location, flip over an axis, or appear to have been stretched or compressed. The transformed graph may resemble the parent graph, or it may not.

Reading Math

translation A translation is also called a *slide*, a *shift*, or a *glide*.

A **translation** moves a figure up, down, left, or right.

- When a constant k is added to or subtracted from a parent function, the result $f(x) \pm k$ is a translation of the graph up or down.
- When a constant h is added to or subtracted from x before evaluating a parent function, the result, $f(x \pm h)$, is a translation left or right.

Example 2 Describe and Graph Translations

Describe the translation in $y = |x| + 2$. Then graph the function.

The graph of $y = |x| + 2$ is a translation of the graph of $y = |x|$ up 2 units.

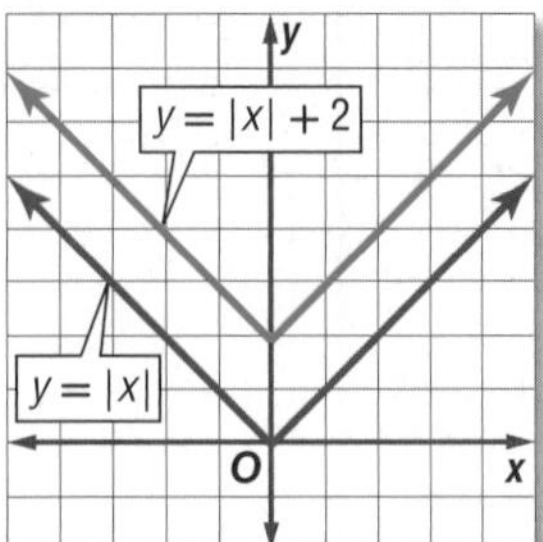

Guided Practice

Describe the translation in each function. Then graph the function.

2A. $y = |x + 3|$

2B. $y = x^2 - 4$

A **reflection** flips a figure over a line called the **line of reflection**.

- When a parent function is multiplied by -1, the result $-f(x)$ is a reflection of the graph in the x-axis.
- When only the variable is multiplied by -1, the result $f(-x)$ is a reflection of the graph in the y-axis.

Example 3 Describe and Graph Reflections

Describe the reflection in $y = -x^2$. Then graph the function.

The graph of $y = -x^2$ is a reflection of the graph of $y = x^2$ in the x-axis.

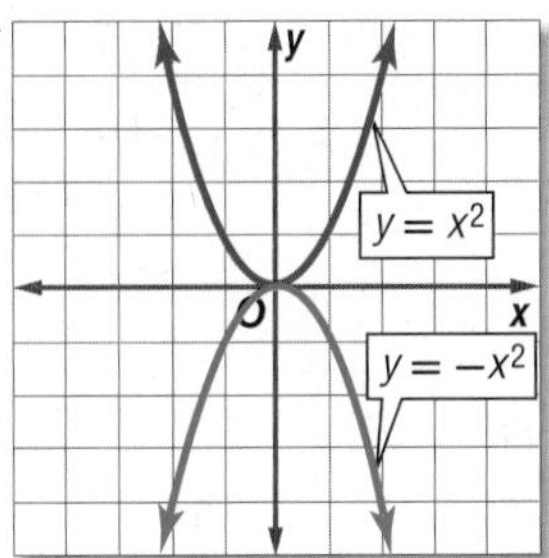

GuidedPractice

Describe the reflection in each function. Then graph the function.

3A. $y = -|x|$

3B. $y = -x$

A **dilation** shrinks or enlarges a figure proportionally. When the variable in a linear parent function is multiplied by a nonzero number, the slope of the graph changes.

- When a nonlinear parent function is multiplied by a nonzero number, the function is stretched or compressed vertically.
- Coefficients greater than 1 cause the graph to be stretched vertically, and coefficients between 0 and 1 cause the graph to be compressed vertically.

Example 4 Describe and Graph Dilations

Describe the dilation in $y = 4x$. Then graph the function.

The graph of $y = 4x$ is a dilation of the graph of $y = x$. The slope of the graph of $y = 4x$ is steeper than that of the graph of $y = x$.

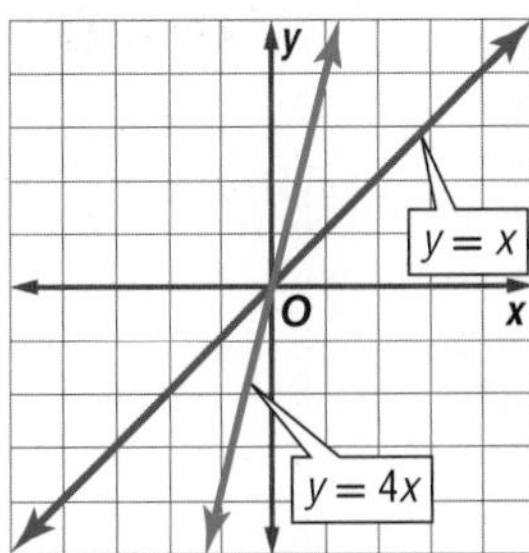

GuidedPractice

Describe the dilation in each function. Then graph the function.

4A. $y = 2x^2$

4B. $y = \left|\frac{1}{3}x\right|$

Real-WorldLink

The average American homeowner spends over \$350 each year on indoor and outdoor plants.

Source: American Nursery & Landscape Association

Real-World Example 5 Identify Transformations

LANDSCAPING Ethan is going to add a brick walkway around the perimeter of his vegetable garden. The area of the walkway can be represented by the function $f(x) = 4(x + 2.5)^2 - 25$. Describe the transformations in the function. Then graph the function.

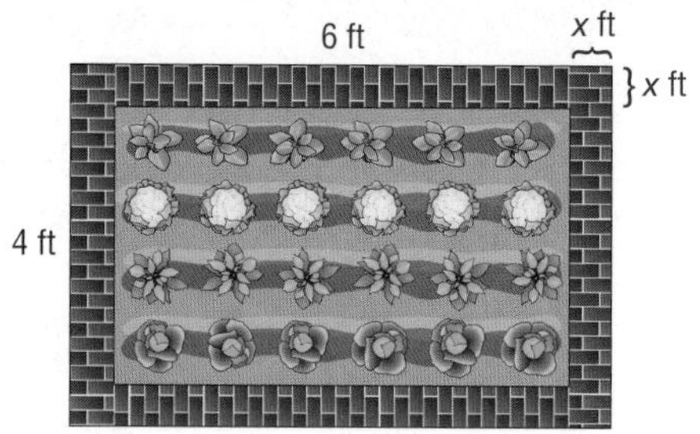

The graph of $f(x) = 4(x + 2.5)^2 - 25$ is a combination of transformations of the parent graph $f(x) = x^2$. Determine how each transformation affects the parent graph.

$f(x) = 4(x + 2.5)^2 - 25$

$+ 2.5$ translates $f(x) = x^2$ left 2.5 units.

$- 25$ translates $f(x) = x^2$ down 25 units.

4 stretches $f(x) = x^2$ vertically.

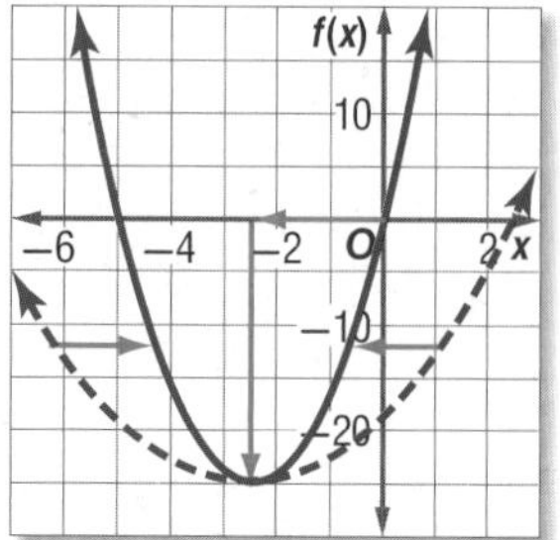

GuidedPractice

5. **SCIENCE** The function $C(x) = \frac{5}{9}(x - 32)$ can be used to determine the temperature in degrees Celsius when given the temperature in degrees Fahrenheit. Describe the transformations in the function. Then graph the function.

The table summarizes the changes to the parent graph under different transformations.

StudyTip

 Regularity

Ask yourself these questions to help you identify transformations.

1. What type of function is it?
2. Does the graph open up or down?
3. Does the vertex lie on an axis?

ConceptSummary Transformations of Functions

Transformation	Change to Parent Graph
Translation	
$f(x + h), h > 0$	Translates graph h units left.
$f(x - h), h > 0$	Translates graph h units right.
$f(x) + k, k > 0$	Translates graph k units up.
$f(x) - k, k > 0$	Translates graph k units down.
Reflection	
$-f(x)$	Reflects graph in the x-axis.
$f(-x)$	Reflects graph in the y-axis.
Dilation	
$a \cdot f(x), \|a\| > 1$	Stretches graph vertically.
$a \cdot f(x), 0 < \|a\| < 1$	Compresses graph vertically
$f(bx), \|b\| > 1$	Compresses graph horizontally.
$f(bx), 0 < \|b\| < 1$	Stretches graph horizontally.

Check Your Understanding

= Step-by-Step Solutions begin on page R14.

Example 1 **Identify the type of function represented by each graph.**

1. 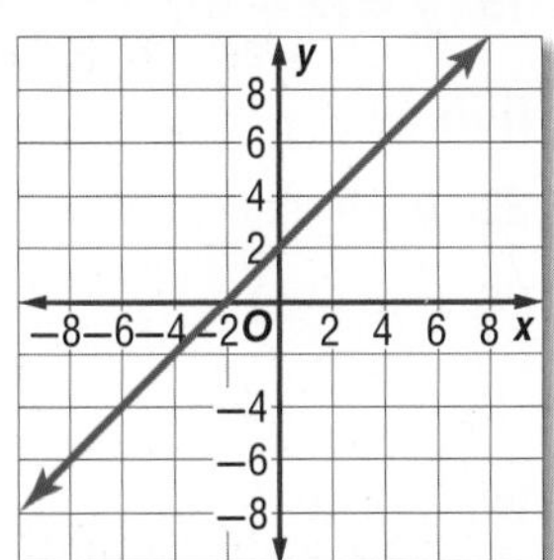

2. 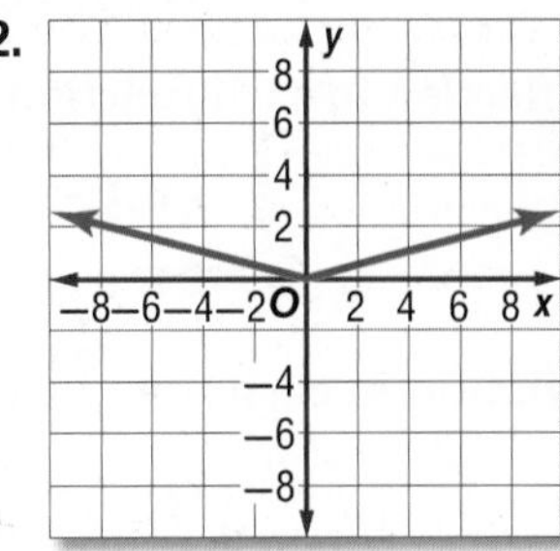

Example 2 CCSS **SENSE-MAKING** **Describe the translation in each function. Then graph the function.**

3. $y = x^2 - 4$

4. $y = |x + 1|$

Example 3 **Describe the reflection in each function. Then graph the function.**

5. $y = -|x|$

6. $y = (-x)^2$

Example 4 **Describe the dilation in each function. Then graph the function.**

7. $y = \frac{3}{5}x$

8. $y = 3x^2$

Example 5

9. **FOOD** The manager of a coffee shop is randomly checking coffee drinks prepared by employees to ensure that the correct amount of coffee is in each cup. Each 12-ounce drink should contain half coffee and half steamed milk. The amount of coffee by which each drink varies can be represented by $f(x) = \frac{1}{2}|x - 12|$. Describe the transformations in the function. Then graph the function.

Practice and Problem Solving

Extra Practice is on page R2.

Example 1 **Identify the type of function represented by each graph.**

10. 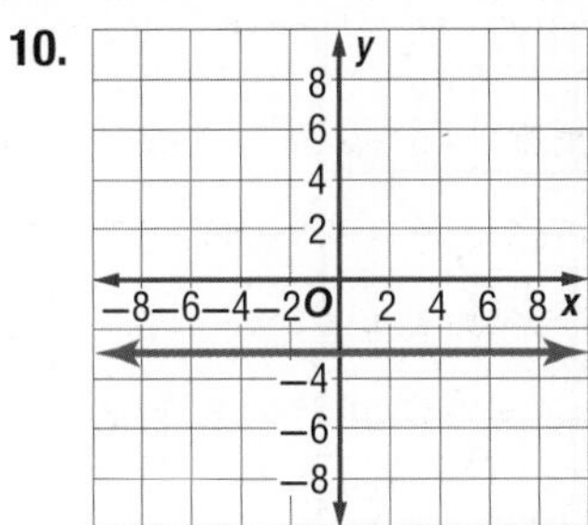

11. 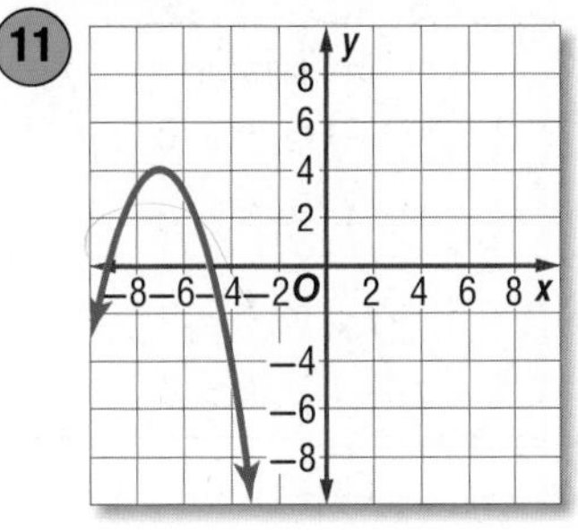

12. 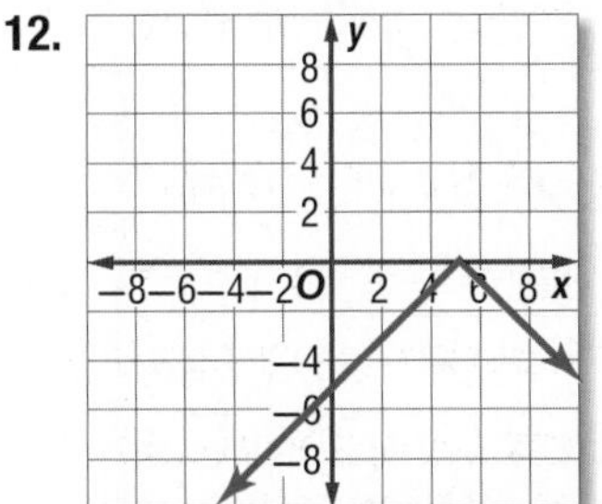

13. 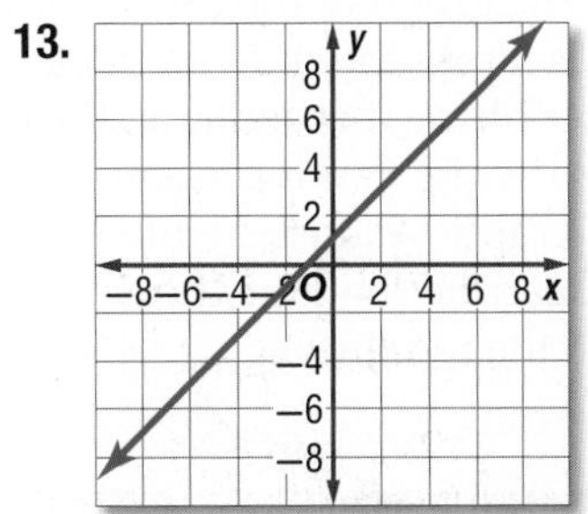

Example 2 **Describe the translation in each function. Then graph the function.**

14. $y = x^2 + 4$ **15.** $y = |x| - 3$ **16.** $y = x - 1$

17. $y = x + 2$ **18.** $y = (x - 5)^2$ **19.** $y = |x + 6|$

Example 3 **Describe the reflection in each function. Then graph the function.**

20. $y = -x$ **21.** $y = -x^2$ **22.** $y = (-x)^2$

23. $y = |-x|$ **24.** $y = -|x|$ **25.** $y = (-x)$

Example 4 **Describe the dilation in each function. Then graph the function.**

26. $y = (3x)^2$ **27.** $y = 6x$ **28.** $y = 4|x|$

29. $y = |2x|$ **30.** $y = \frac{2}{3}x$ **31.** $y = \frac{1}{2}x^2$

Example 5 **32.** **CCSS SENSE-MAKING** A non-impact workout can burn up to 7.5 Calories per minute. The equation to represent how many Calories a person burns after m minutes of the workout is $C(m) = 7.5m$. Identify the transformation in the function. Then graph the function.

Write an equation for each function.

33.

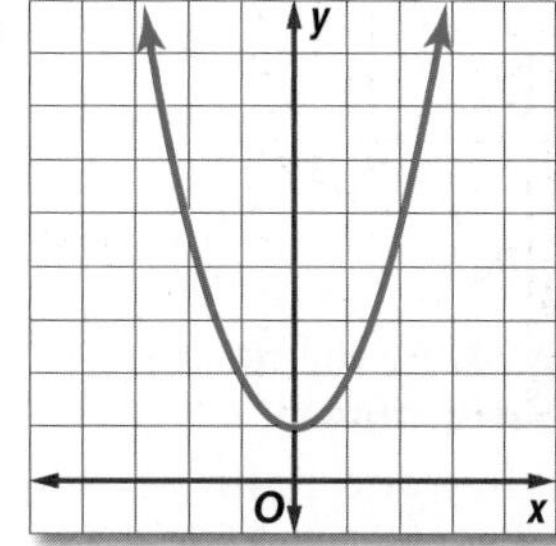

34.

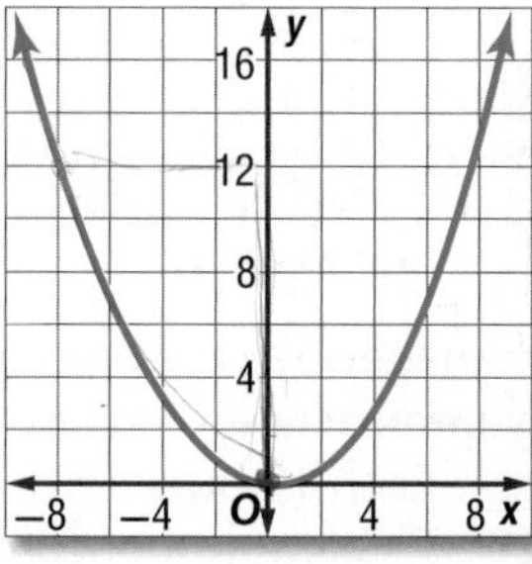

35.

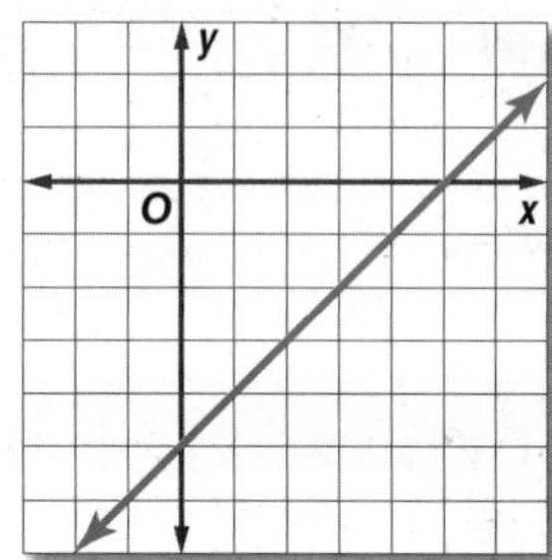

36.

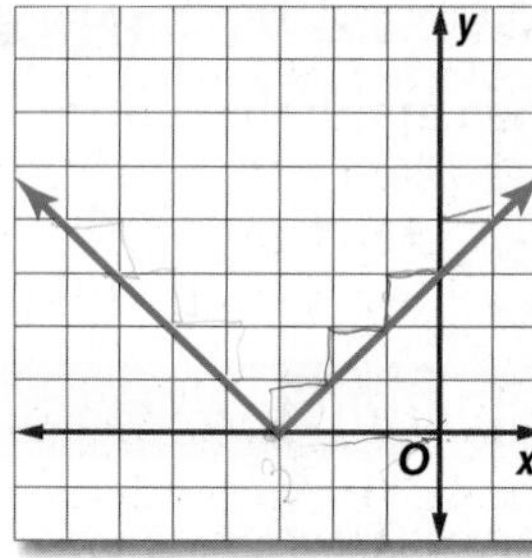

37.

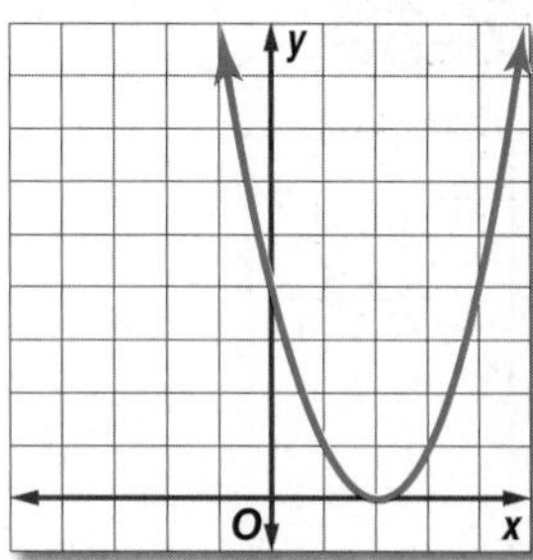

38.

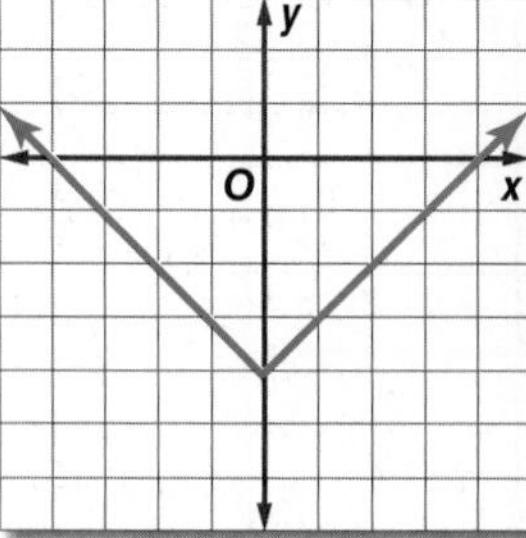

39 **BUSINESS** The graph of the cost of producing x widgets is represented by the blue line in the graph. After hiring a consultant, the cost of producing x widgets is represented by the red line in the graph. Write the equations of both lines and describe the transformation from the blue line to the red line.

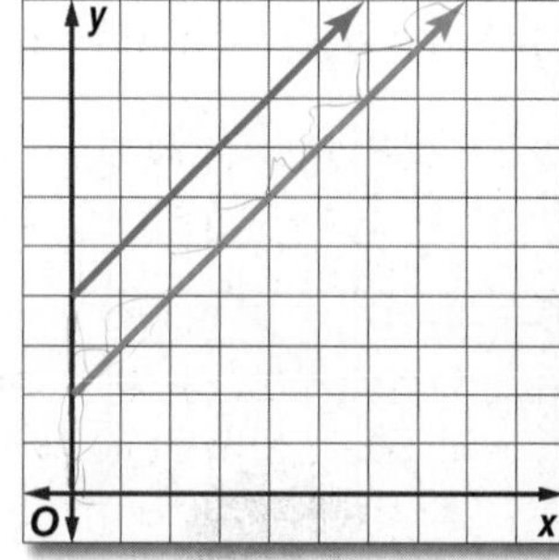

40. **ROCKETRY** Kenji launched a toy rocket from ground level. The height $h(t)$ of Kenji's rocket after t seconds is shown in blue. Emily believed that her rocket could fly higher and longer than Kenji's. The flight of Emily's rocket is shown in red.

a. Identify the type of function shown.

b. How much longer than Kenji's rocket did Emily's rocket stay in the air?

c. How much higher than Kenji's rocket did Emily's rocket go?

d. Describe the type of transformation between the two graphs.

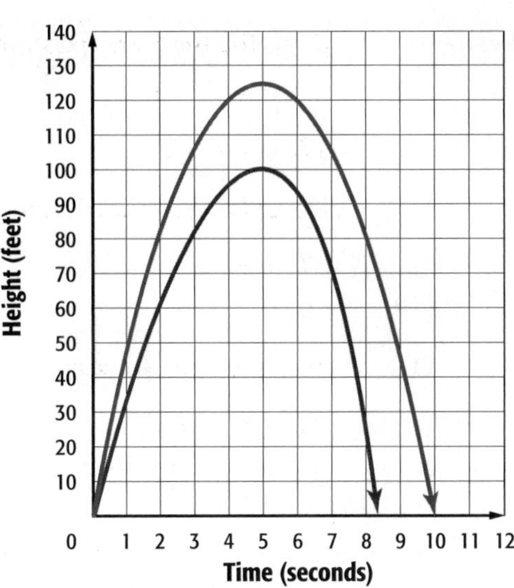

Write an equation for each function.

41.

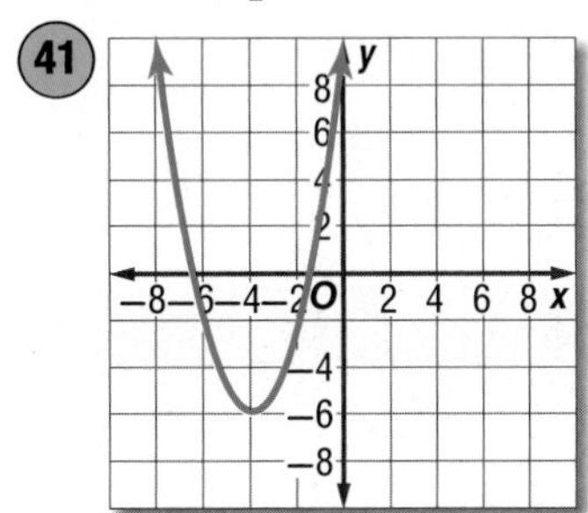

42.

H.O.T. Problems Use Higher-Order Thinking Skills

43. **CHALLENGE** Explain why performing a horizontal translation followed by a vertical translation ends up being the same transformation as performing a vertical translation followed by a horizontal translation.

44. **CCSS CRITIQUE** Kimi thinks that the graph and table below are representations of the same linear relation. Carla disagrees. Who is correct? Explain your reasoning.

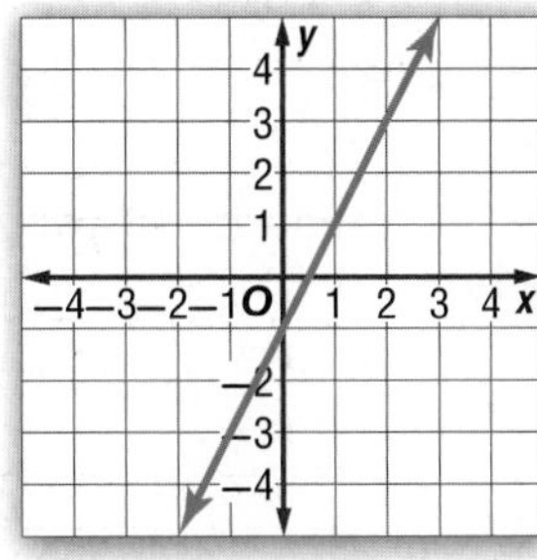

x	y
0	−1
1	1
2	3
3	5

45. **OPEN ENDED** Draw a figure in Quadrant II. Use any of the transformations you learned in this lesson to move your figure to Quadrant IV. Describe your transformation.

46. **REASONING** Study the parent graphs at the beginning of this lesson. Select a parent graph with positive y-values at its leftmost points and positive y-values at its rightmost points.

47. **WRITING IN MATH** Explain why the reflection of the graph of $f(x) = x^2$ in the y-axis is the same as the graph of $f(x) = x^2$. Is this true for all reflections of quadratic equations? If not, describe a case when it is false.

Standardized Test Practice

48. What is the solution set of the inequality?

$$6 - |x + 7| \leq -2$$

A $\{x \mid -15 \leq x \leq 1\}$

B $\{x \mid x \leq -1 \text{ or } x \geq 3\}$

C $\{x \mid -1 \leq x \leq 3\}$

D $\{x \mid x \leq -15 \text{ or } x \geq 1\}$

49. GEOMETRY The measures of two angles of a triangle are x and $4x$. Which of these expressions represents the measure of the third angle?

F $180 + x + 4x$

G $180 - x - 4x$

H $180 - x + 4x$

J $180 + x - 4x$

50. GRIDDED RESPONSE Find the value of x that makes $\frac{1}{2} = \frac{x-2}{x+2}$ true.

51. SAT/ACT Which could be the equation for the graph?

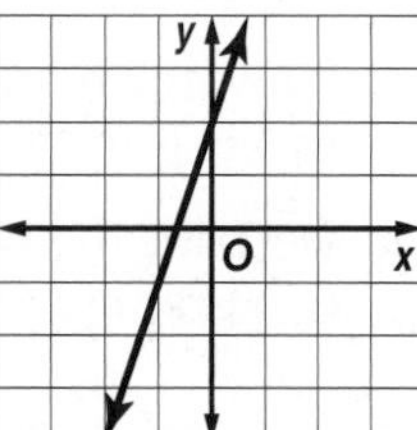

A $y = 3x + 2$

B $y = 3x - 2$

C $y = -3x + 2$

D $y = -\frac{1}{3}x + 2$

E $y = \frac{1}{3}x + 2$

Spiral Review

Graph each function. Identify the domain and range. (Lesson 2-6)

52. $f(x) = |x - 3|$

53. $h(x) = [\![x]\!] - 5$

54. $f(x) = \begin{cases} -2x \text{ if } x \leq -2 \\ x \text{ if } -2 < x \leq 1 \\ 4 \text{ if } x > 1 \end{cases}$

55. ATTENDANCE The table shows the annual attendance to West High School's Summer Celebration. (Lesson 2-5)

a. Find a regression equation for the data.

b. Determine the correlation coefficient.

c. Predict how many people will attend the Summer Celebration in 2014.

Year	Attendance
2008	61
2009	83
2010	85
2011	92
2012	97
2013	106

Solve each inequality. (Lesson 1-6)

56. $-12 \leq 2x + 4 \leq 8$

57. $-4 < -3y + 2 < 11$

58. $|x - 3| > 7$

59. CARS Loren is buying her first car. She is considering 4 different models and 5 different colors. How many different cars could she buy? (Lesson 0-4)

Determine if each relation is a function. (Lesson 0-1)

60.

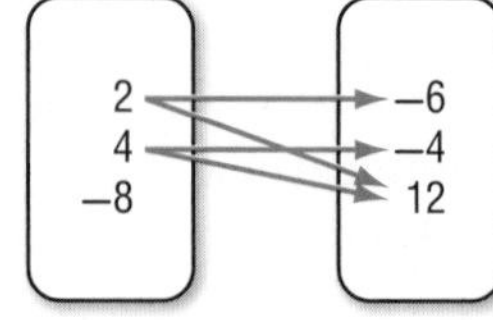

61.

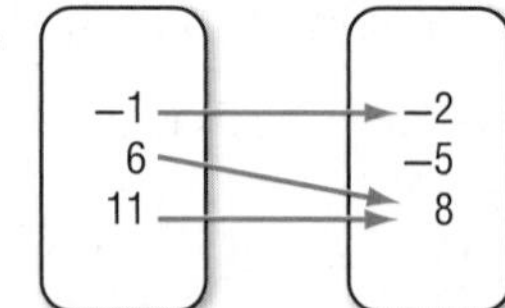

62.

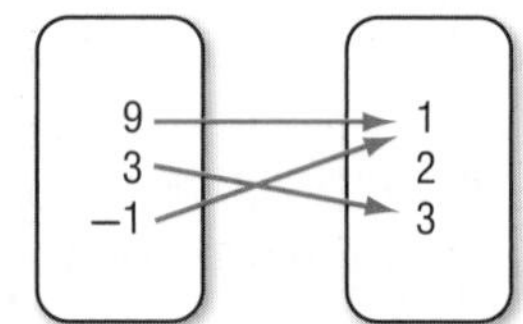

Skills Review

Evaluate each expression if $x = -4$ and $y = 6$.

63. $4x - 8y + 12$

64. $5y + 3x - 8$

65. $-12x + 10y - 24$

LESSON 2-8

Graphing Linear and Absolute Value Inequalities

∴ Then	∴ Now	∴ Why?
● You described transformations of functions.	● 1 Graph linear inequalities. 2 Graph absolute value inequalities.	● Randy is planning to treat his lacrosse team to a pizza party after the championship game, but he does not want to spend more than \$200. Randy can use the inequality $11p + 2.25d \leq 200$, where p represents the number of pizzas and d represents the number of soft drinks, to check whether certain combinations of pizzas and drinks will fall within his budget.

NewVocabulary
linear inequality
boundary

Common Core State Standards

Content Standards
A.CED.3 Represent constraints by equations or inequalities, and by systems of equations and/or inequalities, and interpret solutions as viable or nonviable options in a modeling context.

Mathematical Practices
1 Make sense of problems and persevere in solving them.

1 Graph Linear Inequalities

A **linear inequality** resembles a linear equation, but with an inequality symbol instead of an equals symbol. For example, $y > -3x - 2$ is a linear inequality and $y = -3x - 2$ is the related linear equation.

The graph of the inequality $y > -3x - 2$ is shown at the right as a shaded region. Every point in the shaded region satisfies the inequality. The graph of $y = -3x - 2$ is the **boundary** of the region. It is drawn as a dashed line to show that points on the line do not satisfy the inequality. If the symbol was $\leq$ or $\geq$, then points on the boundary would satisfy the inequality, so the boundary would be drawn as a solid line.

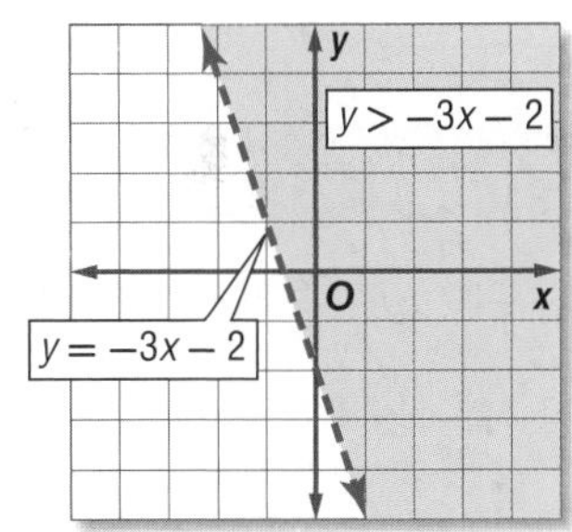

Example 1 Dashed Boundary

Graph $x + 4y > 2$.

Step 1 The boundary of the graph is the graph of $x + 4y = 2$. Since the inequality symbol is $>$, the boundary will be dashed.

Step 2 Test the point (0, 0) because it is not on the boundary.

$x + 4y > 2$	Original inequality
$0 + 4(0) \overset{?}{>} 2$	$(x, y) = (0, 0)$
$0 > 2$ ✗	False

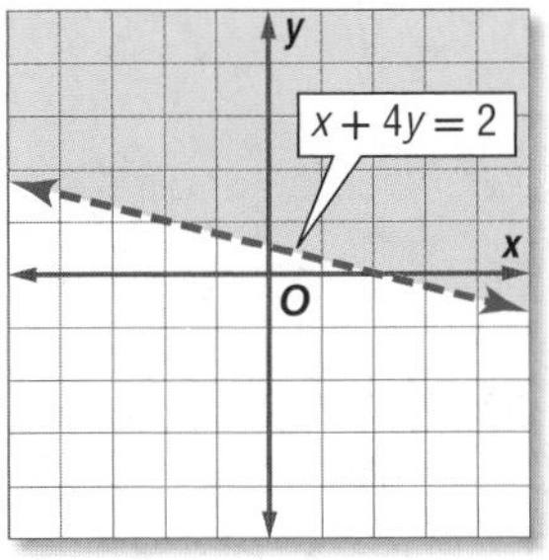

The region that does *not* contain (0, 0) is shaded.

x + 4y = 2 (graph: y-axis, x-axis, O)

CHECK The graph indicates that (0, 3) is a solution.

$x + 4y > 2$	Original inequality
$0 + 4(3) \overset{?}{>} 2$	$(x, y) = (0, 3)$
$12 > 2$ ✓	True

The solution checks.

GuidedPractice

1A. Graph $3x + \frac{1}{2}y < 2$.

1B. Graph $-x + 2y > 4$.

Real-World Example 2 Solid Boundary

RECREATION **A recreation center offers various 30-minute and 60-minute art classes. The recreation director has allotted up to 20 hours per week for art classes.**

a. Write an inequality to represent the number of classes that can be offered per week. Graph the inequality.

Let x represent the number of 30-minute or $\frac{1}{2}$-hour art classes, and let y represent the number of 60-minute or 1-hour art classes. Because the sum can equal the maximum, the inequality symbol is $\leq$, and the boundary is solid. The inequality is $\frac{1}{2}x + y \leq 20$.

Step 1 Graph the boundary $\frac{1}{2}x + y = 20$.

Step 2 Test the point (0, 0).

$\frac{1}{2}x + y \leq 20$	Original inequality
$\frac{1}{2}(0) + (0) \overset{?}{\leq} 20$	$(x, y) = (0, 0)$
$0 \leq 20$ ✓	True

The region that contains (0, 0) is shaded.

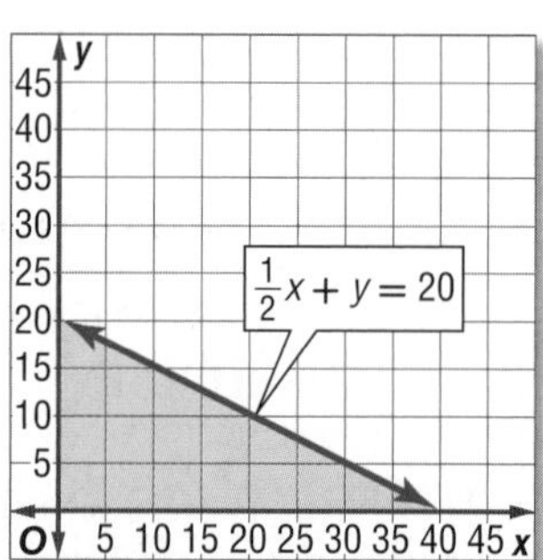

b. Can the recreation director schedule 25 of the 30-minute classes and 15 of the 60-minute classes during a given week? Explain your reasoning.

The point (25, 15) lies outside the shaded region, so it does not satisfy the inequality. Thus, the recreation director cannot schedule 25 30-minute and 15 60-minute classes.

Guided Practice

2. Manuel has \$15 to spend at the county fair. The fair costs \$5 for admission, \$0.75 for each ride ticket, and \$0.25 for each game ticket. Write an inequality, and draw a graph that represents the number of r ride and g game tickets that Manuel can buy.

Real-World Career

Recreation Workers Recreation workers plan, organize, and manage recreational activities to meet the needs of a variety of people. Educational requirements range from a high school diploma to a graduate degree.

2 Graph Absolute Value Inequalities

Graphing absolute value inequalities is similar to graphing linear inequalities. First you graph the absolute value equation. Then you determine whether the boundary is dashed or solid and which region should be shaded.

Example 3 Absolute Value Inequality

Graph $y \geq |x| - 4$.

Since the inequality symbol is $\geq$, the boundary is solid. Graph the equation. Then test (0, 0).

$y \geq |x| - 4$	Original inequality
$0 \overset{?}{\geq} |0| - 4$	$(x, y) = (0, 0)$
$0 \geq -4$ ✓	True

The region that includes (0, 0) is shaded.

Guided Practice

3A. Graph $y \leq 2|x| + 3$.

3B. Graph $y \geq 3|x + 1|$.

Check Your Understanding

Example 1 **Graph each inequality.**

1. $y \leq 4$
2. $x \geq -6$
3. $x + 4y \leq 2$
4. $3x + y > -8$

Example 2

5. **CCSS MODELING** Gregg needs to buy gas and oil for his car. Gas costs \$3.45 a gallon, and oil costs \$2.41 a quart. He has \$50 to spend.
   - **a.** Write an inequality to represent the situation, where g is the number of gallons of gas he buys and q is the number of quarts of oil.
   - **b.** Graph the inequality.
   - **c.** Can Gregg buy 10 gallons of gasoline and 8 quarts of oil? Explain.

Example 3 **Graph each inequality.**

6. $y \geq |x + 3|$
7. $y - 6 < |x|$

Practice and Problem Solving

Extra Practice is on page R2.

Example 1 **Graph each inequality.**

8. $x + 2y > 6$
9. $y \geq -3x - 2$
10. $2y + 3 \leq 11$
11. $4x - 3y > 12$
12. $6x + 4y \leq -24$
13. $y \geq \frac{3}{4}x + 6$

Example 2

14. **COLLEGE** April's guidance counselor says that she needs a combined score of at least 1700 on her college entrance exams to be eligible for the college of her choice. The highest possible score is 2400. There are 1200 possible points on the math portion and 1200 on the verbal portion.
    - **a.** The inequality $x + y \geq 1700$ represents this situation, where x is the verbal score and y is the math score. Graph this inequality.
    - **b.** Refer to your graph. If she scores a 680 on the math portion of the test and 910 on the verbal portion of the test, will April be eligible for the college of her choice?

Example 3 **Graph each inequality.**

15. $y > |3x|$
16. $y + 4 \leq |x - 2|$
17. $y - 6 < |-2x|$
18. $y + 8 < 2\left|\frac{2}{3}x + 6\right|$
19. $2y > |4x - 5|$
20. $-y \leq |3x - 4|$

21. **SCHOOL DANCE** Carlos estimates that he will need to earn at least \$700 to take his girlfriend to the prom. Carlos works two jobs as shown in the table.
    - **a.** Write an inequality to represent this situation.
    - **b.** Graph the inequality.
    - **c.** Will he make enough money if he works 50 hours at each job?

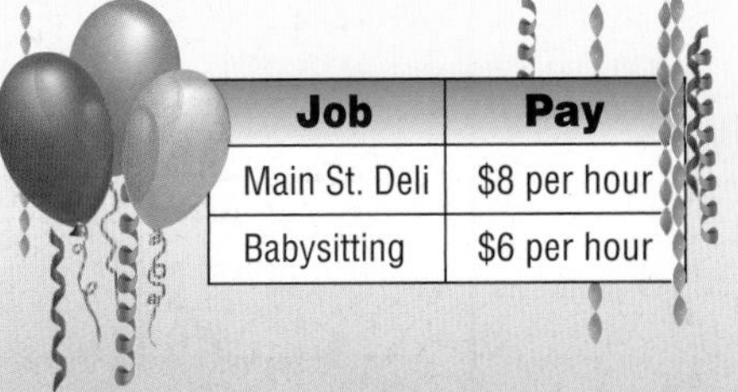

Job	Pay
Main St. Deli	\$8 per hour
Babysitting	\$6 per hour

Graph each inequality.

22. $y \geq |-2x - 6|$
23. $y \leq |x - 3| + 4$
24. $y - 3 > -2|x + 4|$
25. $|y| > |x|$
26. $|x - y| > 5$
27. $|x + 3y| \geq -2$

28. **CCSS MODELING** Mei is making necklaces and bracelets to sell at a craft show. She has enough beads to make 50 pieces. Let x represent the number of bracelets and y represent the number of necklaces.

a. Write an inequality that shows the possible number of necklaces and bracelets Mei can make.

b. Graph the inequality.

c. Give three possible solutions for the number of necklaces and bracelets that can be made.

29. **GIFT CARDS** Susan received a gift card from an electronics store for $400. She wants to spend the money on DVDs, which cost $20 each, and CDs, which cost $15 each.

a. Let d equal the number of DVDs, and let c equal the number of CDs. Write an inequality that shows the possible combinations of DVDs and CDs that Susan can purchase.

b. Graph the inequality.

c. Give three possible solutions for the number of DVDs and CDs she can buy.

Graph each inequality.

30. $y \geq [\![x]\!]$

31. $y < [\![x + 2]\!]$

32. $y \geq |[\![x]\!]|$

H.O.T. Problems Use Higher-Order Thinking Skills

33. **OPEN ENDED** Create an absolute value inequality in which none of the possible solutions fall in the second or third quadrant.

34. **CHALLENGE** Graph the following inequality.

$$g(x) > \begin{cases} |x + 1| \text{ if } x \leq -4 \\ -|x| \text{ if } -4 < x < 2 \\ |x - 4| \text{ if } x \geq 2 \end{cases}$$

35. **ERROR ANALYSIS** Paulo and Janette are graphing $x - y \geq 2$. Is either of them correct? Explain your reasoning.

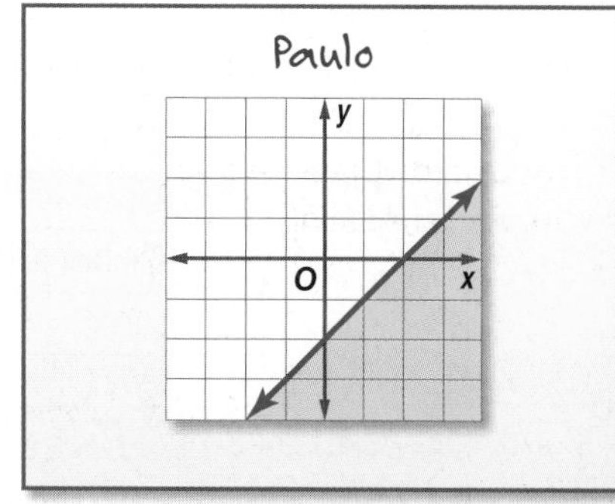

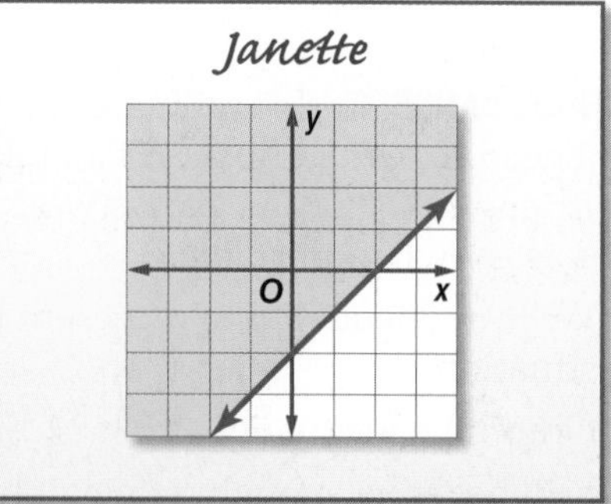

36. **REASONING** When will it be possible to shade two different areas when graphing a linear absolute value inequality? Explain your reasoning.

37. **WRITING IN MATH** Describe a situation in which there are no solutions to an absolute value inequality. Explain your reasoning.

Standardized Test Practice

38. EXTENDED RESPONSE Craig scored 85%, 96%, 79%, and 81% on his first four math tests. He hopes to score high enough on the final test to earn a 90% average. If the final test is worth twice as much as one of the other tests, determine if Craig can earn a 90% average. If so, what score does Craig need to get on the final test to accomplish this? Explain how you found your answer.

39. Which of the following sets of numbers represents an infinite set?

A {2, 4, 6}

B {whole numbers between −50 and 50}

C {integers}

D $\left\{\frac{1}{2}, \frac{3}{4}, \frac{4}{5}, \frac{5}{6}\right\}$

40. SHORT RESPONSE Which theorem of congruence should be used to prove $\triangle ABC \cong \triangle XYZ$?

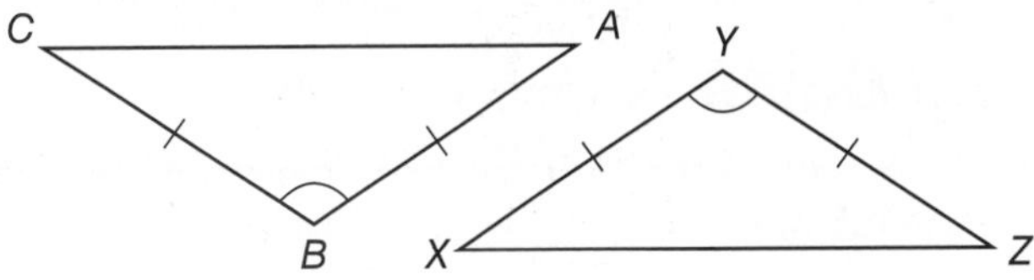

41. SAT/ACT For which function is the range $\{f(x) \mid f(x) \leq 0\}$?

F $f(x) = -x$

G $f(x) = [\![x]\!]$

H $f(x) = [\![-x]\!]$

J $f(x) = |x|$

K $f(x) = -|x|$

Spiral Review

Write an equation for each graph. (Lesson 2-7)

42.

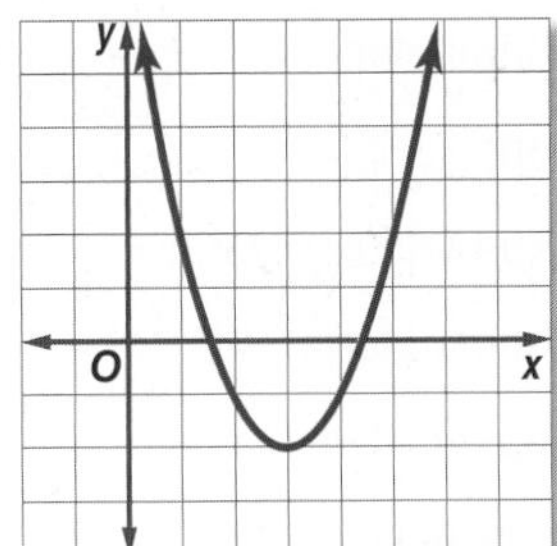

43.

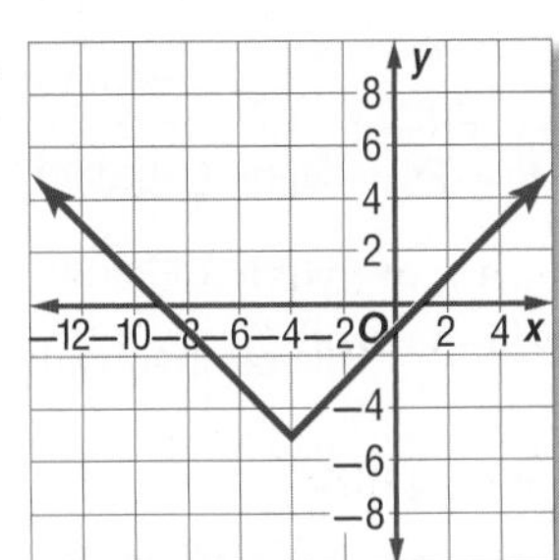

44.

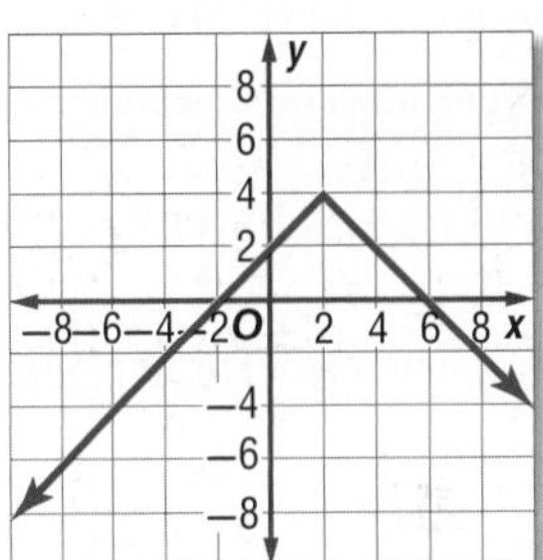

Graph each function. (Lesson 2-6)

45. $f(x) = \begin{cases} x \text{ if } x < 1 \\ 3 \text{ if } 1 \leq x \leq 3 \\ -2x \text{ if } x > 3 \end{cases}$

46. $f(x) = \begin{cases} x + 3 \text{ if } x < -2 \\ 2x \text{ if } -2 \leq x \leq 2 \\ -3x \text{ if } x > 2 \end{cases}$

47. $f(x) = \begin{cases} -2x \text{ if } x \leq -2 \\ x + 1 \text{ if } 0 < x \leq 6 \\ x - 5 \text{ if } x > 6 \end{cases}$

Write each equation in standard form. Identify *A*, *B*, and *C*. (Lesson 2-2)

48. $-6y = 8x - 3$

49. $12y + x = -3y + 5x - 6$

50. $\frac{x+3}{4} + \frac{y-1}{2} = 3$

51. TENNIS Sixteen players signed up for tennis lessons. The instructor plans to use 50 tennis balls for every player and have 200 extra. How many tennis balls are needed for the lessons? (Lesson 1-3)

Multiply. (Lesson 0-2)

52. $(3x - 4)(2x + 1)$

53. $(6x + 5)(-x - 3)$

54. $(5x + 2)(-2x + 3)$

Skills Review

Graph each linear equation.

55. $y = 2x - 8$

56. $y = -\frac{3}{4}x + 2$

57. $3y - 4x = 24$

CHAPTER 2

Study Guide and Review

Study Guide

KeyConcepts

Relations and Functions (Lesson 2-1)

- A function is a relation where each member of the domain is paired with exactly one member of the range.

Linear Equations and Slope (Lessons 2-2 to 2-4)

- Standard Form: $Ax + By = C$, where A, B, and C are integers whose greatest common factor is 1, $A \geq 0$, and A and B are not both zero.
- Slope-Intercept Form: $y = mx + b$
- Point-Slope Form: $y - y_1 = m(x - x_1)$

Scatter Plots and Lines of Regression (Lesson 2-5)

- A prediction equation can be used to predict the value of one of the variables given the value of the other variable.
- A line of regression can be used to model data.

Special Functions and Parent Functions (Lessons 2-6 and 2-7)

- A piecewise-defined function is made up of two or more expressions.
- Translations, reflections, and dilations of a parent graph form a family of graphs.

Graphing Linear and Absolute Value Inequalities (Lesson 2-8)

- You can graph an inequality by following these steps.

  Step 1 Determine whether the boundary is solid or dashed. Graph the boundary.

  Step 2 Choose a point not on the boundary and test it in the inequality.

  Step 3 If a true inequality results, shade the region containing your test point. If a false inequality results, shade the other region.

FOLDABLES StudyOrganizer

Be sure the Key Concepts are noted in your Foldable.

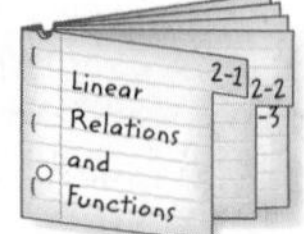

KeyVocabulary

absolute value function (p. 103)
bivariate data (p. 92)
continuous relation (p. 62)
correlation coefficient (p. 94)
dependent variable (p. 64)
dilation (p. 111)
direct variation (p. 90)
discrete relation (p. 62)
family of graphs (p. 109)
greatest integer function (p. 102)
independent variable (p. 64)
linear equation (p. 69)
linear function (p. 69)
linear inequality (p. 117)
line of fit (p. 92)
negative correlation (p. 92)
nonlinear relation (p. 69)
parent function (p. 109)
piecewise-defined function (p. 101)
point-slope form (p. 84)
positive correlation (p. 92)
prediction equation (p. 92)
quadratic function (p. 109)
rate of change (p. 76)
reflection (p. 111)
regression line (p. 94)
scatter plot (p. 92)
slope (p. 77)
slope-intercept form (p. 83)
standard form (p. 70)
step function (p. 102)
translation (p. 110)
vertical line test (p. 62)

VocabularyCheck

Choose the correct term to complete each sentence.

1. A function is (discrete, one-to-one) if each element of the domain is paired to exactly one unique element of the range.
2. The (domain, range) of a relation is the set of all first coordinates from the ordered pairs which determine the relation.
3. The (constant, identity) function is a linear function described by $f(x) = x$.
4. If you are given the coordinates of two points on a line, you can use the (slope-intercept, point-slope) form to find the equation of the line that passes through them.
5. A set of bivariate data graphed as ordered pairs in a coordinate plane is called a (scatter plot, line of fit).
6. A function that is written using two or more expressions is called a (linear, piecewise-defined) function.

Lesson-by-Lesson Review

2-1 Relations and Functions

State the domain and range of each relation. Then determine whether the relation is a function. If it is a function, determine if it is *one-to-one, onto, both,* or *neither.*

7. {(1, 2), (3, 4), (5, 6), (7, 8)}

8. {(−3, 0), (0, 2), (2, 4), (4, 5), (5, 2)}

9. {(−4, 1), (3, 3), (1, 1), (−2, 5), (3, −4)}

10. {(7, −4), (5, −2), (3, 0), (1, 2), (−1, 4)}

Find each value if $f(x) = -3x + 2$.

11. $f(4)$

12. $f(-3)$

13. $f(0)$

14. $f(y)$

15. $f(-a)$

16. $f(2w)$

17. BOWLING A bowling alley charges $2.50 for shoe rental and $3.25 per game bowled. The amount a bowler is charged can be expressed as $y = 2.50 + 3.25x$, when $x \geq 1$, and is an integer. Find the domain and range. Then determine whether the equation is a function. Is the equation discrete or continuous?

Example 1

State the domain and range of the relation {(−4, 3), (−1, 0), (−2, 4), (3, −1), (2, 6)}. Then determine whether the relation is a function. If it is a function, determine if it is *one-to-one, onto, both,* or *neither.*

Domain: {−4, −2, −1, 2, 3}
Range: {−1, 0, 3, 4, 6}

Each element of the domain is paired with one element of the range, so the relation is a function. The function is both because each element of the domain is paired with a unique element of the range and each element of the range is paired with a unique element of the domain.

Example 2

Find $f(-2)$ if $f(x) = 4x - 3$.

$f(-2) = 4(-2) - 3$	Substitute −3 for *x*.
$= -8 - 3$	Multiply.
$= -11$	Simplify.

2-2 Linear Relations and Functions

State whether each function is a linear function. Write *yes* or *no*. Explain.

18. $3x + 4y = 12$

19. $x^2 + y^2 = 4$

20. $y = x^3 - 6$

21. $y = 6x - 19$

22. $f(x) = -2x + 9$

23. $\frac{1}{x} + 3y = -5$

Write each equation in standard form. Identify *A*, *B*, and *C*.

24. $2x + 5y = 10$

25. $y = 12x$

26. $-4y = 3x - 24$

27. $4x = 8y - 12$

28. TRAVEL The distance the Green family traveled during their family vacation is given by the equation $y = 65x$, where x represents the number of hours spent driving. How far does the Green family travel in 8 hours?

Example 3

State whether $f(x) = 3x^2$ is a linear function. Write *yes* or *no*. Explain.

No, because the expression includes a variable raised to the second power.

Example 4

Write the equation $y = -5x + 8$ in standard form. Identify *A*, *B*, and *C*.

$y = -5x + 8$	Original equation
$5x + y = 8$	Add $5x$ to each side.

$A = 5$, $B = 1$, and $C = 8$

Study Guide and Review *Continued*

2-3 Rate of Change and Slope

29. RETAIL The table shows the number of DVDs sold each week at the Super Movie Store. Find the average rate of change of the number of DVDs sold from week 2 to week 5.

Week	1	2	3	4	5
DVDs Sold	76	58	94	83	112

Find the slope of the line that passes through each pair of points.

30. (2, 5), (6, −3)

31. (8, 2), (2, 8)

32. Determine the rate of change of the graph.

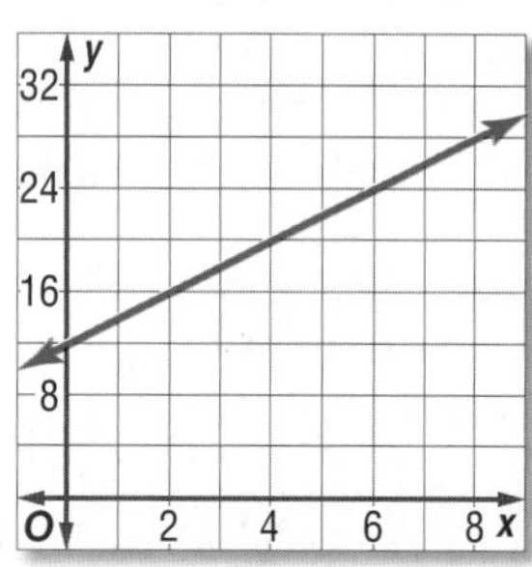

Example 5

Find the slope of the line that passes through each pair of points.

a. (−2, 9), (1, 4)

$m = \frac{y_2 - y_1}{x_2 - x_1}$ Slope Formula

$= \frac{4 - 9}{1 - (-2)}$ $(x_1, y_1) = (-2, 9), (x_2, y_2) = (1, 4)$

$= -\frac{5}{3}$ Simplify.

b. (−3, 6), (4, 6)

$m = \frac{y_2 - y_1}{x_2 - x_1}$ Slope Formula

$= \frac{6 - 6}{4 - (-3)}$ $(x_1, y_1) = (-3, 6), (x_2, y_2) = (4, 6)$

$= \frac{0}{7}$ or 0 Simplify.

2-4 Writing Linear Equations

Write an equation in slope-intercept form for the line that satisfies each set of conditions.

33. slope −2, passes through (−3, −5)

34. slope $\frac{2}{3}$, passes through (4, −1)

35. passes through (−2, 4) and (0, 8)

36. passes through (3, 5) and (−1, 5)

Write an equation of the line passing through each pair of points.

37. (6, 1), (4, 9)

38. (−4, 2), (6, 8)

Write an equation in slope-intercept form for the line that satisfies each set of conditions.

39. through (1, 2), parallel to $y = 4x - 3$

40. through (−3, 5), perpendicular to $y = \frac{2}{3}x - 8$

41. PETS Drew paid a $250 fee when he adopted a puppy. The average monthly cost of feeding and caring for the puppy is $32. Write an equation that represents the total cost of adopting and caring for the puppy for *x* months.

Example 6

Write an equation of the line through (−2, 5) and (0, −9).

Find the slope of the line.

$m = \frac{y_2 - y_1}{x_2 - x_1}$ Slope Formula

$= \frac{-9 - 5}{0 - (-2)}$ $(x_1, y_1) = (-2, 5), (x_2, y_2) = (0, -9)$

$= \frac{-14}{2}$ or −7 Simplify.

Write an equation.

$y - y_1 = m(x - x_1)$ Point-slope form

$y - 5 = -7(x - (-2))$ Substitute.

$y - 5 = -7(x + 2)$ Simplify.

$y - 5 = -7x - 14$ Distributive Property

$y = -7x - 9$ Add 5 to each side.

The equation is $y = -7x - 9$.

2-5 Scatter Plots and Lines of Regression

Make a scatter plot and a line of fit and describe the correlation for each set of data. Then, use two ordered pairs to write a prediction equation.

42. HEATING The table shows the monthly heating cost for a large home.

Month	Sep	Oct	Nov	Dec	Jan	Feb
Bill ($)	72	114	164	198	224	185

43. AMUSEMENT PARK The table shows the annual attendance in thousands at an amusement park during the last 5 years.

Year	1	2	3	4	5
People (thousands)	44	42	39	31	24

Example 7

SCHOOL ENROLLMENT The table shows the number of students each year at a school.

Year	'07	'08	'09	'10	'11	'12
Students	125	116	142	154	146	175

Use (2007, 125) and (2012, 175) to find a prediction equation.

$m = \frac{y_2 - y_1}{x_2 - x_1}$ Slope Formula

$= \frac{175 - 125}{2012 - 2007}$ Substitution

$= \frac{50}{5}$ or 10 Simplify.

$y - y_1 = m(x - x_1)$ Point-slope form

$y - 125 = 10(x - 2007)$ Substitution

$y - 125 = 10x - 20{,}070$ Distributive Property

$y = 10x - 20{,}195$ Add 125 to each side.

2-6 Special Functions

Graph each function. Identify the domain and range.

44. $f(x) = \begin{cases} -2x \text{ if } x \leq -1 \\ x + 1 \text{ if } -1 < x < 3 \\ x \text{ if } x \geq 3 \end{cases}$

45. $f(x) = \begin{cases} -3 \text{ if } x < -1 \\ 4x - 3 \text{ if } -1 \leq x \leq 3 \\ x \text{ if } x > 3 \end{cases}$

46. Write the piecewise-defined function shown in the graph.

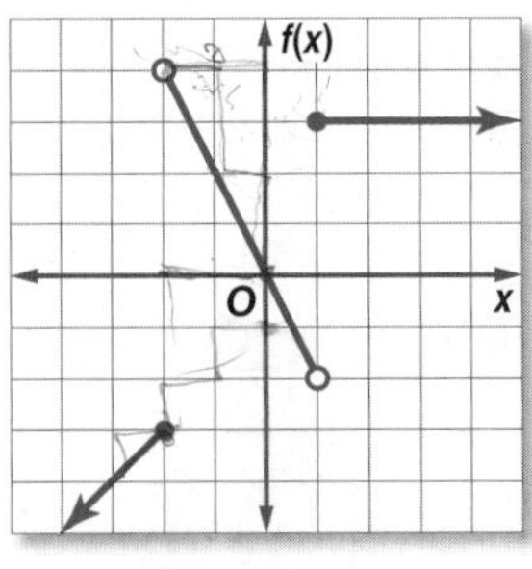

Graph each function. Identify the domain and range.

47. $f(x) = [\![x]\!] + 2$

48. $f(x) = [\![x + 3]\!]$

Example 8

Write the piecewise-defined function shown in the graph.

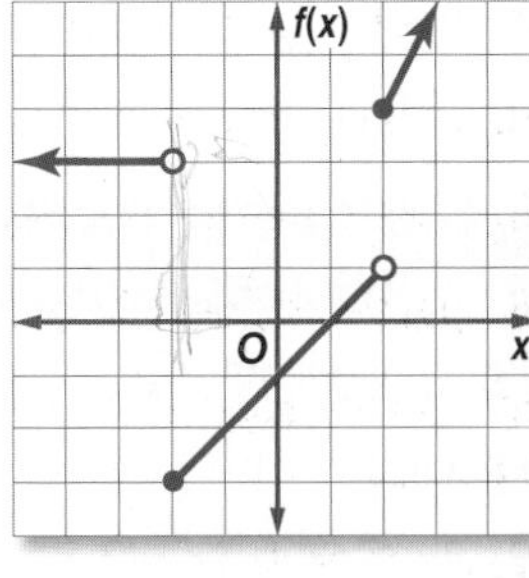

The left portion of the graph is the graph of $f(x) = 3$. There is a circle at $(-2, 3)$, so the linear function is defined for $x < -2$.

The center portion of the graph is the graph of $f(x) = x - 1$. There is a dot at $(-2, -3)$ and a circle at $(2, 1)$, so the linear function is defined for $-2 \leq x < 2$.

The right portion of the graph is the graph of $f(x) = 2x$. There is a dot at $(2, 4)$, so the linear function is defined for $x \geq 2$.

$$f(x) = \begin{cases} 3 \text{ if } x < -2 \\ x - 1 \text{ if } -2 \leq x < 2 \\ 2x \text{ if } x \geq 2 \end{cases}$$

Study Guide and Review *Continued*

2-7 Parent Functions and Transformations

Identify the type of function represented by each graph.

49.

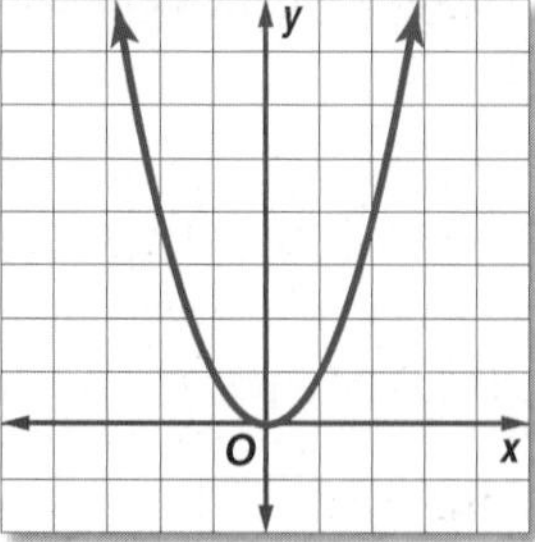

50.

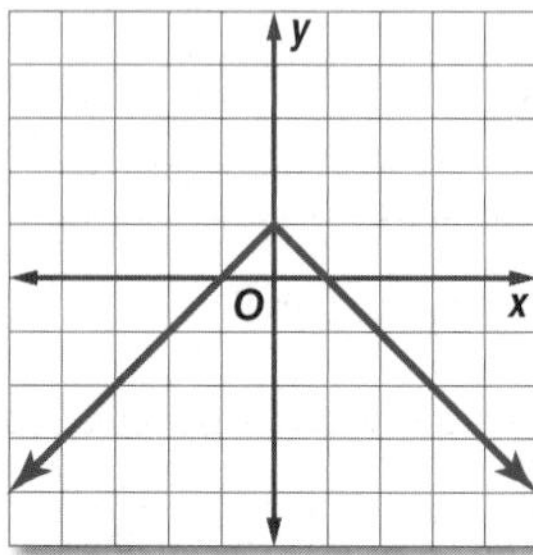

51. Describe the translation in $y = x^2 - 3$.

52. Describe the reflection in $y = -x^2$.

53. **CONSTRUCTION** A large arch is being constructed at the entrance of a new city hall building. The shape of the arch resembles the graph of the function $f(x) = -0.025x^2 + 3.64x - 0.038$. Describe the shape of the arch.

Example 9

Identify the type of function represented by the graph.

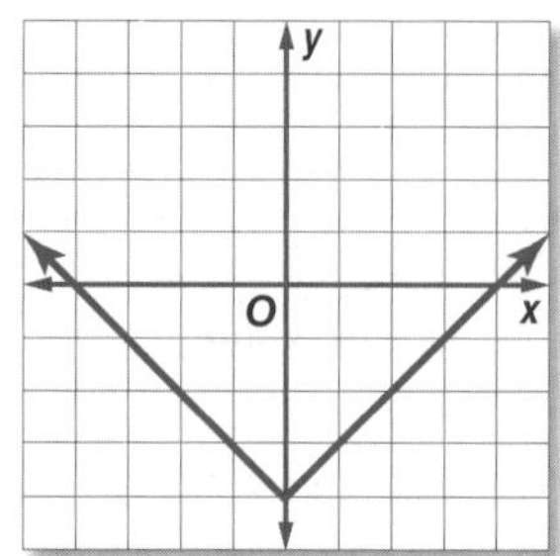

The graph is in the shape of a V. The graph represents an absolute value function.

Example 10

Describe the translation in $y = |x + 6|$.

The graph of $y = |x + 6|$ is a translation of the graph of $y = |x|$ shifted left 6 units.

2-8 Graphing Linear and Absolute Value Inequalities

Graph each inequality.

54. $x - 3y < 6$

55. $y \geq 2x + 1$

56. $2x + 4y \leq 12$

57. $y < -3x - 5$

58. $y > |2x|$

59. $y \geq |2x - 2|$

60. $y + 3 < |x + 1|$

61. $2y \leq |x - 3|$

62. **BOOKS** Spencer has saved $96 for a trip to his favorite bookstore. Each paperback book costs $8 and each hardback book costs $12. Write and graph an inequality that shows the number of paperback books and hardback books Spencer can purchase.

Example 11

Graph $x - 2y > 6$.

Since the inequality symbol is >, the graph of the boundary line should be dashed. Graph $x - 2y = 6$.

Test $x - 2y > 6$ at (0, 0).

$$x - 2y > 6$$
$$0 - 2(0) \overset{?}{>} 6$$
$$0 > 6 \quad ✗$$

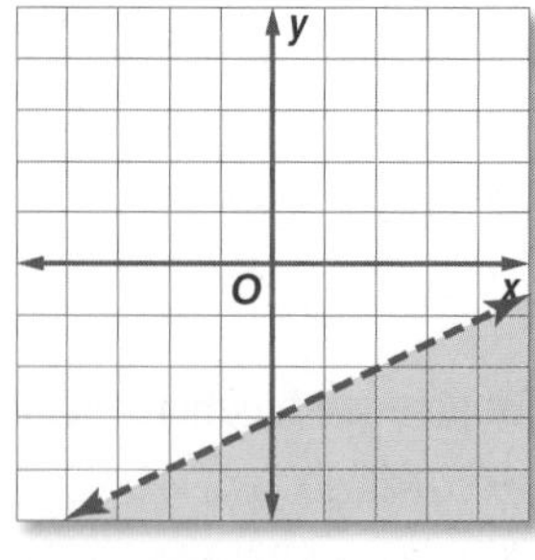

CHAPTER 2

Practice Test

1. State the domain and range of the relation shown in the table. Then determine if it is a function. If it is a function, determine if it is *one-to-one*, *onto*, *both*, or *neither*.

x	y
−2	3
4	−1
3	2
6	3

Find each value if $f(x) = -2x + 3$.

2. $f(-4)$
3. $f(3y)$

4. Write $2y = -6x + 4$ in standard form. Identify A, B and C.

5. Find the x-intercept and the y-intercept for $3x - 4y = -24$.

6. **MULTIPLE CHOICE** The cost of producing x pumpkin pies at a small bakery is given by $C(x) = 49 + 1.75x$. Find the cost of producing 25 pies.

   A $74.00

   B $81.50

   C $92.75

   D $108.25

Find the slope of the line that passes through each pair of points.

7. (1, 6), (3, 10)
8. (−2, 7), (3, −1)

9. **MULTIPLE CHOICE** Find the equation of the line that passes through (0, −3) and (4, 1).

   F $y = -x + 3$

   G $y = -x - 3$

   H $y = x - 3$

   J $y = x + 3$

10. Write an equation in slope-intercept form for the line that has slope −2 and passes through the point (3, −4).

11. Write an equation of the line that passes through the points (2, −4) and (1, 6).

12. Write an equation in slope-intercept form for the line that passes through (−3, 5) and is parallel to $y = -6x + 1$.

13. **EMERGENCY ROOM** A hospital tracks the number of emergency room visits during the fall and winter months.

Month	Oct	Nov	Dec	Jan	Feb
Visits	124	163	155	171	192

   a. Make a scatter plot and describe the correlation.

   b. Use two ordered pairs to write a prediction equation.

   c. Use your prediction equation to predict the number of emergency room visits for March.

14. Graph $f(x) = \begin{cases} -x \text{ if } x < -2 \\ x + 2 \text{ if } -2 \le x \le 2 \\ 5 \text{ if } x > 2 \end{cases}$.

15. Write the piecewise-defined function shown.

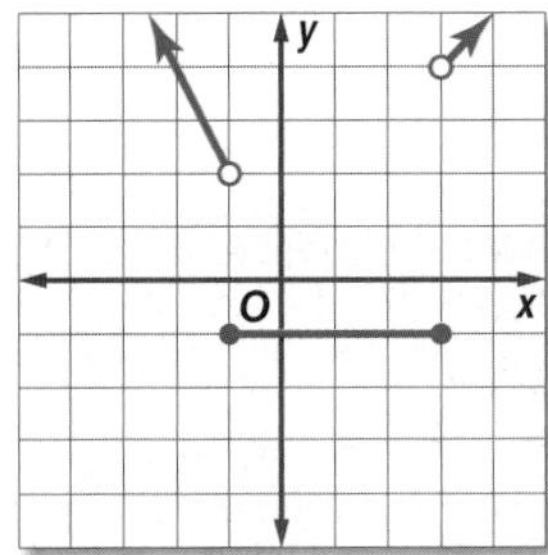

16. Identify the domain and range of $y = [\![x]\!] + 2$.

17. Describe the translation to $y = x^2 + 5$.

18. Describe the reflection in $y = -|x|$.

Graph each inequality.

19. $y \ge 4x - 1$

20. $2x + 6y < -12$

CHAPTER 2

Preparing for Standardized Tests

Reading Math Problems

The first step to solving any math problem is to read the problem. When reading a math problem to get the information you need to solve, it is helpful to use special reading strategies.

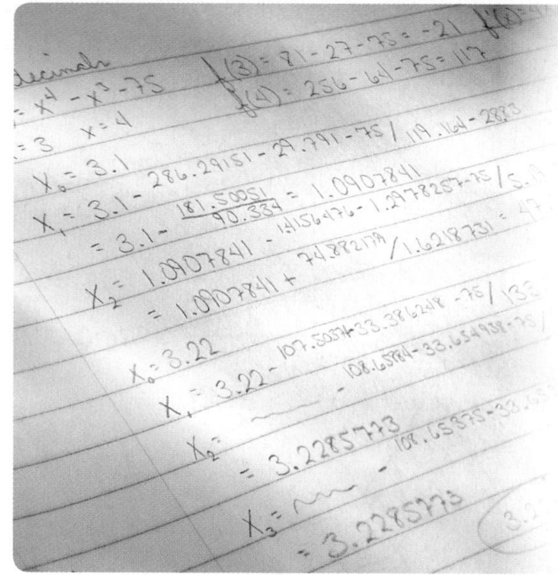

CORBIS

Strategies for Reading Math Problems

Step 1

Read the problem quickly to gain a general understanding of it.

- **Ask yourself:** "What do I know?" "What do I need to find out?"
- **Think:** "Is there enough information to solve the problem? Is there extra information?"
- **Highlight:** If you are allowed to write in your test booklet, underline or highlight important information. Cross out any information you do not need.

Step 2

Reread the problem to identify relevant facts.

- **Analyze:** Determine how the facts are related.
- **Key Words:** Look for key words to solve the problem.
- **Vocabulary:** Identify mathematical terms. Think about the concepts and how they are related.
- **Plan:** Make a plan to solve the problem.
- **Estimate:** Quickly estimate the answer.

Step 3

Identify any obvious wrong answers.

- **Eliminate:** Eliminate any choices that are very different from your estimate.
- **Units of Measure:** Identify choices that are possible answers based on the units of measure in the question. For example, if the question asks for area, eliminate all the answer choices that are not in square units.

Step 4

Look back after solving the problem.

Check: Make sure you have answered the question.

Standardized Test Example

Read the problem. Identify what you need to know. Then use the information in the problem to solve.

Sandy heated a solution over a burner and then removed it from the heat source. The temperature of the solution decreased linearly as it cooled. The temperatures after 0, 2, 5, and 9 minutes are shown in the table. What is the rate of change in the temperature of the solution as it cools?

Time (min)	Temperature (°C)
0	133.2
2	130.4
5	126.2
9	120.6

A −1.4 degrees per minute

B −0.8 degrees per minute

C 0.8 degrees per minute

D 1.4 degrees per minute

Read the problem carefully. There is extra information in the problem. To determine the slope, you only need information from two points on the linear function. Use two of the points to find the slope.

$$m = \frac{130.4 - 133.2}{2 - 0} = -1.4$$

The correct answer is A.

Exercises

Read each question. Then fill in the correct answer on the answer document provided by your teacher or on a sheet of paper.

1. The graph shows the cost of shipping packages. How much would it cost to ship a package that weighs 2 pounds 8 ounces?

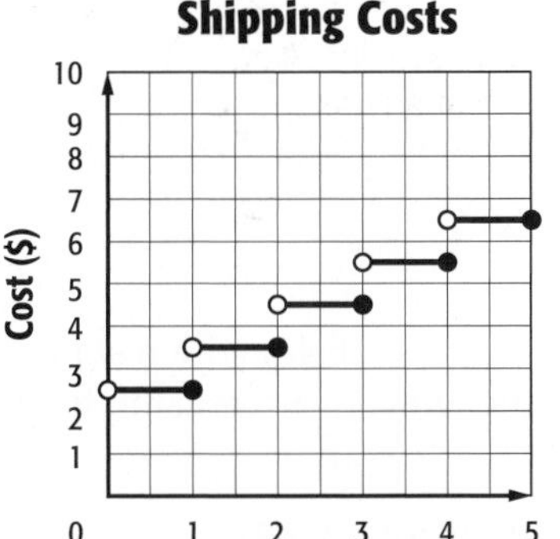

A $3.50

B $4.50

C $5.00

D $5.50

2. What is the slope of the line shown in the graph?

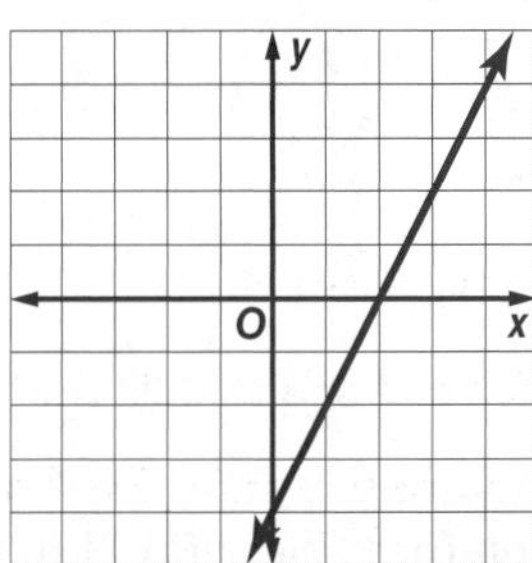

F −2

G $-\frac{1}{2}$

H $\frac{1}{2}$

J 2

Standardized Test Practice

Cumulative, Chapters 1 through 2

Multiple Choice

Read each question. Then fill in the correct answer on the answer document provided by your teacher or on a sheet of paper.

1. What is the domain of the relation shown below?

x	y
−3	4
1	−1
2	0
6	−3

A {0, 1, 2, 4, 6}

B {−3, −1, 0, 4}

C {−3, 1, 2, 6}

D {−3, −1}

2. What is the slope of the line?

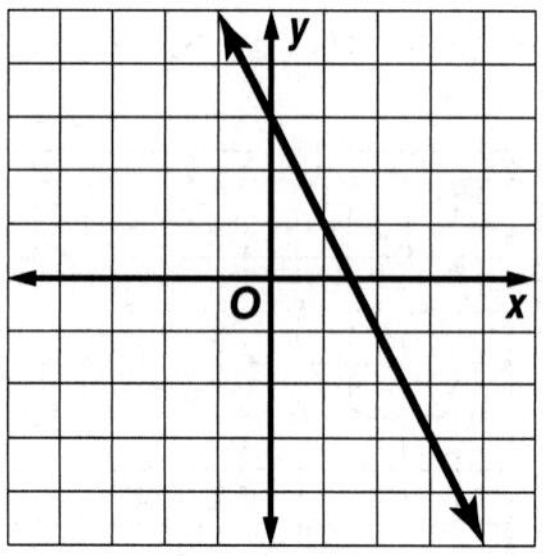

F −2

G $-\frac{1}{2}$

H $\frac{1}{2}$

J 2

3. The Robinson family bought their home in 1998 for $152,400. When they sold it in 2010, the value was $174,900. What was the annual rate of change in the value of the home?

A $1225

B $1875

C $22,500

D $27,275

Test-TakingTip

Question 2 Since the graph slopes downward from left to right, you know the slope is negative. So, answer choices H and J can be eliminated.

4. Carmen works for an electronics retailer. She earns a weekly salary of $450 plus a commission of 4.5% on her weekly sales. Write a linear equation for Carmen's weekly earnings E if she has d dollars in sales.

F $E = (450 + 4.5)d$

G $E = (450 + 0.045)d$

H $E = 450 + 0.045d$

J $E = 450 + 4.5d$

5. Which of the graphs represents the solution set for $|x - 3| - 4 = 0$?

A 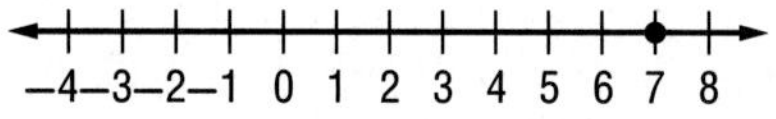

B 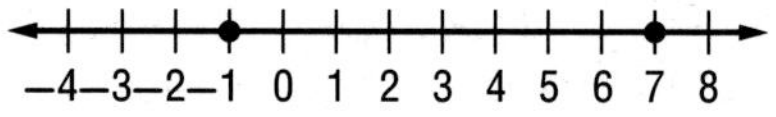

C 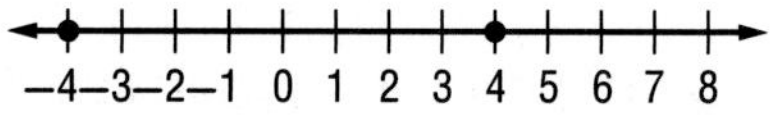

D 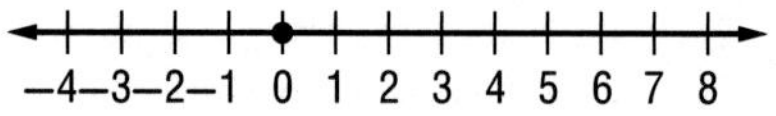

6. Karissa has $10 per month to spend text messaging on her cell phone. The phone company charges $4.95 for the first 100 messages and $0.10 for each additional message. How many text messages can Karissa afford to send and receive each month?

F 50

G 100

H 150

J 151

7. Given $y = 2.24x + 16.45$, which statement best describes the effect of decreasing the y-intercept by 20.25?

A The x-intercept increases.

B The y-intercept increases.

C The new line has a greater rate of change.

D The new line is perpendicular to the original.

Short Response/Gridded Response

Record your answers on the answer sheet provided by your teacher or on a sheet of paper.

8. Write an equation for the piecewise-defined function shown in the graph below.

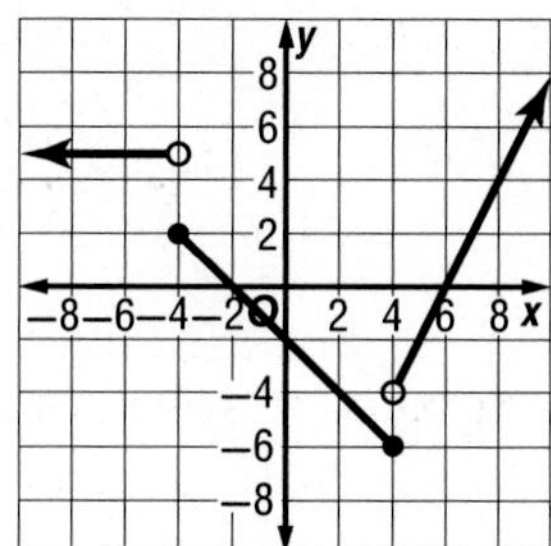

9. GRIDDED RESPONSE Evaluate the piecewise-defined function shown in Exercise 8 for $x = -3$.

10. The graph of the absolute value parent function is shown below.

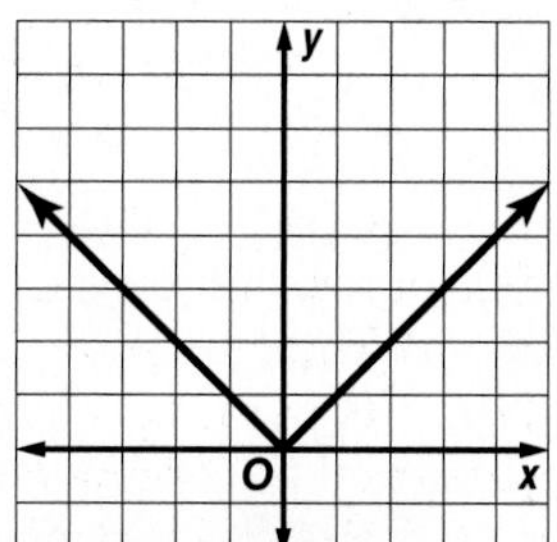

a. What is the equation of this parent function?

b. What equation would result in the graph of the parent function being reflected over the x-axis and shifted up 2 units?

c. What equation would result in the graph of the parent function being shifted left 3 units and down 1 unit?

Extended Response

Record your answers on a sheet of paper. Show your work.

11. The soccer team is having a bake sale this week to raise money for the program. For each cookie sold, the profit is \$0.45, and for each brownie sold, the profit is \$0.50.

a. The team hopes to earn \$150 in profits from the bake sale. Let x represent the number of cookies sold and y the number of brownies sold. Write an inequality to model the situation.

b. Graph the inequality.

c. If the team sells 180 cookies and 160 brownies this week, will they meet their goal? Explain.

12. Use the scatter plot to answer each question.

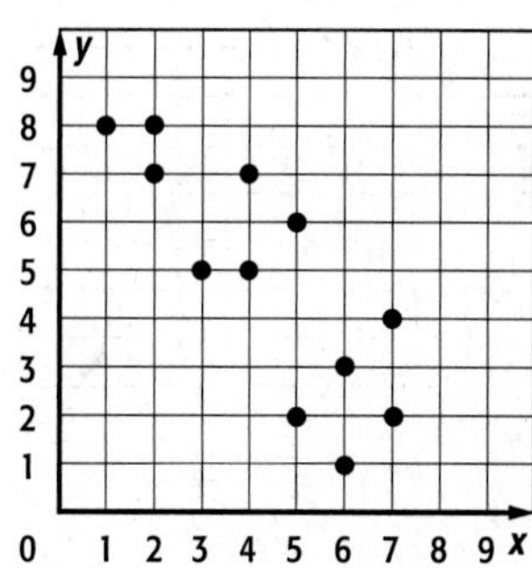

a. What type of correlation is shown by the data in the plot?

b. Determine a regression line for the data.

c. Use your regression line to predict the value of y when $x = 12$.

Need ExtraHelp?

If you missed Question...	1	2	3	4	5	6	7	8	9	10	11	12
Go to Lesson...	2-1	2-3	2-3	2-4	1-4	1-3	2-7	2-6	2-6	2-7	2-8	2-5

CHAPTER 3

Systems of Equations and Inequalities

Then

You graphed equations of lines, transformed functions, and solved equations.

Now

You will:

- Solve systems of linear equations and linear inequalities.
- Solve problems by using linear programming.
- Perform operations with matrices and determinants.

Why? ▲

BUSINESS Most of the time, being successful in business means that you have to have good math skills. In this chapter, you will learn how to maximize your profits and minimize your costs. By doing this you will earn the most money possible.

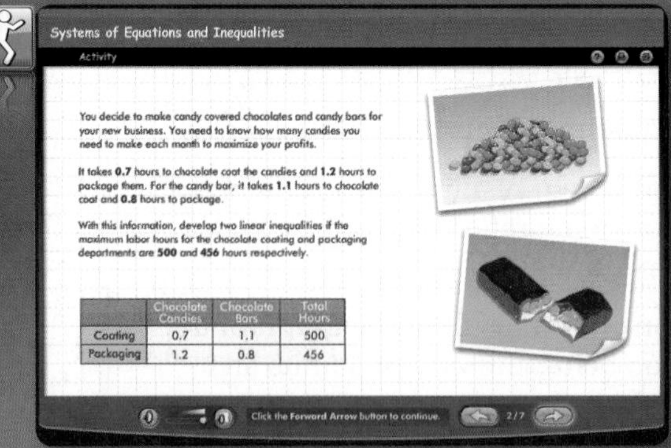

	Chocolate Candies	Chocolate Bars	Total Hours
Coating	0.7	1.1	500
Packaging	1.2	0.8	456

connectED.mcgraw-hill.com **Your Digital Math Portal**

Animation 
Vocabulary 
eGlossary  
Personal Tutor 
Virtual Manipulatives 
Graphing Calculator 
Audio  
Foldables 
Self-Check Practice 
Worksheets 

Get Ready for the Chapter

Diagnose Readiness | **You have two options for checking prerequisite skills.**

1 Textbook Option Take the Quick Check below. Refer to the Quick Review for help.

QuickCheck

Graph each equation.

1. $x = 4y$

2. $y = \frac{1}{3}x + 5$

3. $x + 2y = 4$

4. $y = -x + 6$

5. $3x + 5y = 15$

6. $3y - 2x = -12$

7. BUSINESS A museum charges $8.50 for adult tickets and $5.25 for children's tickets. On Friday they made $650.

a. Write an equation that can be used to model the ticket sales.

b. Graph the equation.

QuickReview

Example 1

Graph $2y + 5x = -10$.

Find the *x*- and *y*-intercepts.

$$\begin{aligned} 2(0) + 5x &= -10 \\ 5x &= -10 \\ x &= -2 \end{aligned} \qquad \begin{aligned} 2y + 5(0) &= -10 \\ 2y &= -10 \\ y &= -5 \end{aligned}$$

The graph crosses the *x*-axis at $(-2, 0)$ and the *y*-axis at $(0, -5)$. Use these ordered pairs to graph the equation.

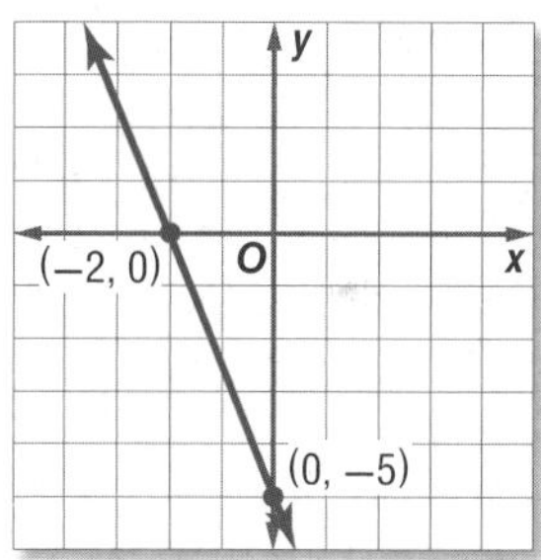

QuickCheck

Graph each inequality.

8. $y < 3$

9. $x + y \geq 1$

10. $3x - y > 6$

11. $x + 2y \leq 5$

12. $y > 4x - 1$

13. $5x - 4y < 12$

14. FUNDRAISER The student council is selling T-shirts for $15 and sweatshirts for $25. They must make $2500 to cover expenses. Write and graph an inequality to show the number of T-shirts and sweatshirts that they must sell.

QuickReview

Example 2

Graph $y \geq 3x - 2$.

The boundary is the graph of $y = 3x - 2$. Since the inequality symbol is $\geq$, the boundary will be solid.

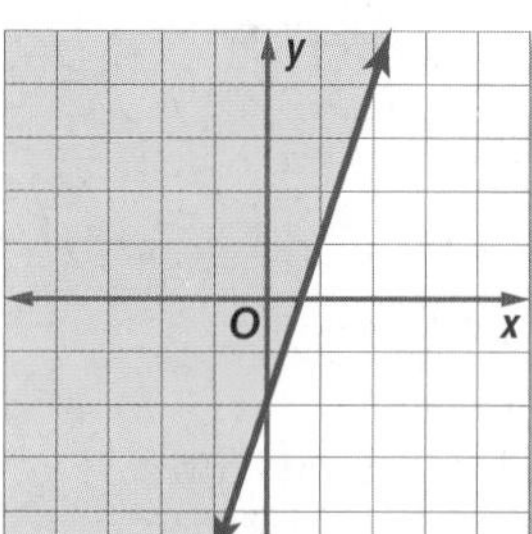

Test the point (0, 0).

$0 \overset{?}{\geq} 3(0) - 2 \qquad (x, y) = (0, 0)$

$0 \geq -2$ ✓

Shade the region that includes (0, 0).

2 Online Option Take an online self-check Chapter Readiness Quiz at connectED.mcgraw-hill.com.

Get Started on the Chapter

You will learn several new concepts, skills, and vocabulary terms as you study Chapter 3. To get ready, identify important terms and organize your resources. You may refer to Chapter 0 to review prerequisite skills.

FOLDABLES StudyOrganizer

Systems of Equations and Inequalities Make this Foldable to help you organize your Chapter 3 notes about systems of equations and inequalities. Begin with a sheet of $8\frac{1}{2}''$ by $11''$ paper.

1 **Fold** in half along the height.

2 **Cut** along the fold.

3 **Fold** each sheet along the width into fourths.

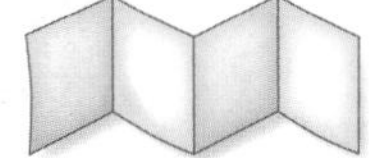

4 **Tape** the ends of two sheets together.

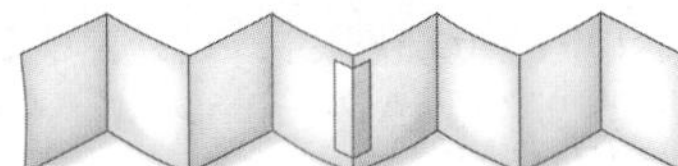

5 **Label** the tabs with *Solve by Graphing, Substitution Method, Elimination Method, Optimization, Systems in Three Variables, Matrix Multiplication, Cramer's Rule,* and *Inverse Matrices.*

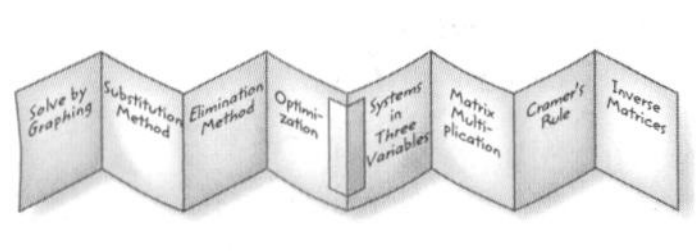

NewVocabulary

English		Español
break-even point	p. 136	punto de equilibrio
consistent	p. 137	consistente
inconsistent	p. 137	inconsistente
independent	p. 137	independiente
dependent	p. 137	dependiente
substitution method	p. 138	método de substitución
elimination method	p. 139	método de eliminación
feasible region	p. 154	región viable
bounded	p. 154	acotada
unbounded	p. 154	no acotado
optimize	p. 156	optimizer
dimensions	p. 169	tamaño
scalar	p. 173	escalar
determinant	p. 189	determinante
Cramer's Rule	p. 192	regula de Crámer
coefficient matrix	p. 192	matriz coefficiente
identity matrix	p. 198	matriz identidad
square matrix	p. 198	matriz cuadrada
inverse matrix	p. 198	matriz inversa
variable matrix	p. 200	matriz variables
constant matrix	p. 200	matriz constante

ReviewVocabulary

inequality desigualdad
an open sentence that contains the symbol $<, \leq, >,$ or $\geq$

linear equation ecuación lineal
an equation that has no operations other than addition, subtraction, and multiplication of a variable by a constant

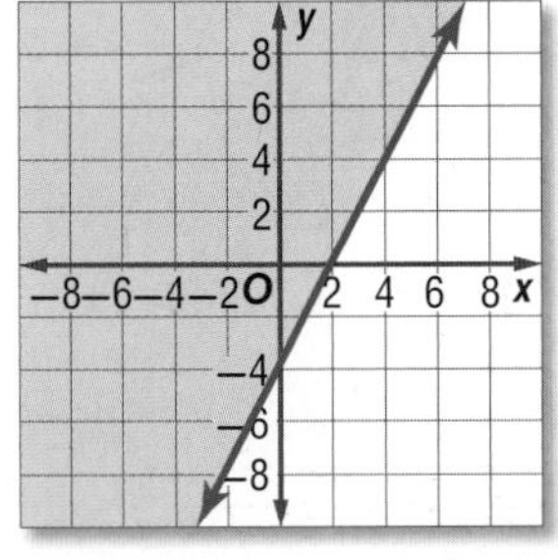

EXPLORE 3-1

Graphing Technology Lab
Intersections of Graphs

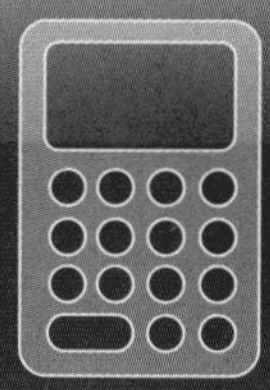

You can use a TI-83/84 Plus graphing calculator to find where two graphs intersect. You can use the Y= menu to graph each equation on the same set of axes.

CCSS Common Core State Standards
Content Standards
A.CED.3 Represent constraints by equations or inequalities, and by systems of equations and/or inequalities, and interpret solutions as viable or nonviable options in a modeling context.
A.REI.11 Explain why the *x*-coordinates of the points where the graphs of the equations $y = f(x)$ and $y = g(x)$ intersect are the solutions of the equation $f(x) = g(x)$; find the solutions approximately, e.g., using technology to graph the functions, make tables of values, or find successive approximations. Include cases where $f(x)$ and/or $g(x)$ are linear, polynomial, rational, absolute value, exponential, and logarithmic functions.

Example Intersection of Two Graphs

Graph both equations in the standard viewing window.

$3x + y = 9$
$x - y = -1$

Step 1 Write each equation in the form $y = mx + b$.

$$3x + y = 9 \qquad y = -3x + 9$$

$$x - y = 1 \qquad -y = -x - 1 \qquad y = x + 1$$

Step 2 Enter $y = -3x + 9$ as **Y1** and $y = x + 1$ as **Y2**. Then graph the lines.

KEYSTROKES: [Y=] [(–)] 3 [X,T,θ,n] [+] 9 [ENTER] [X,T,θ,n] [+] 1 [ENTER] [ZOOM] 6

Step 3 Find the intersection of the lines.

KEYSTROKES: [2nd] [CALC] 5 [ENTER] [ENTER] [ENTER]

The intersection is at (2, 3).

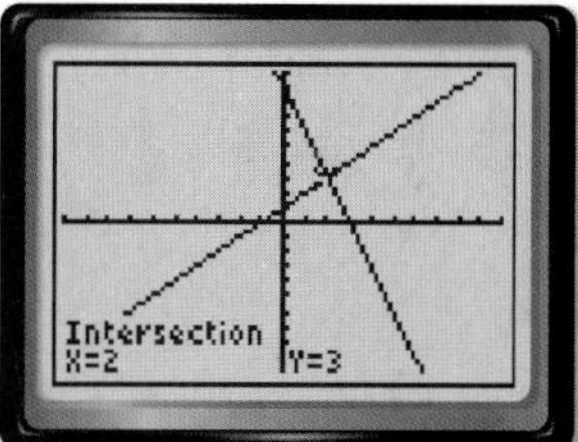

[−10, 10] scl: 1 by [−10, 10] scl: 1

Exercises

Use a graphing calculator to find where each pair of graphs intersect.

1. $2x + 4y = 36$
   $10y - 5x = 0$
2. $2y - 3x = 7$
   $5x = 4y - 12$
3. $4x - 2y = 16$
   $7x + 3y = 15$
4. $2x + 4y = 4$
   $x + 3y = 13$
5. $5x + y = 13$
   $3x = 15 - 3y$
6. $4y - 5 = 20 - 3x$
   $4x - 7y + 16 = 0$
7. $\frac{1}{4}x + y = \frac{11}{4}$
   $x - \frac{1}{2}y = 2$
8. $3x + 2y = -3$
   $x + \frac{1}{3}y = -4$
9. $3x - 6y = 6$
   $2x - 4y = 4$
10. $6x + 8y = -16$
    $3x + 4y = 12$

LESSON 3-1 Solving Systems of Equations

Then

- You graphed and solved linear equations.

Now

1. Solve systems of linear equations graphically.
2. Solve systems of linear equations algebraically.

Why?

- Libby borrowed $450 to start a lawn-mowing business. She charges $35 per lawn and incurs $8 in operating costs per lawn. A system of equations can be used to determine the break-even point. The **break-even point** is the point at which income equals cost.

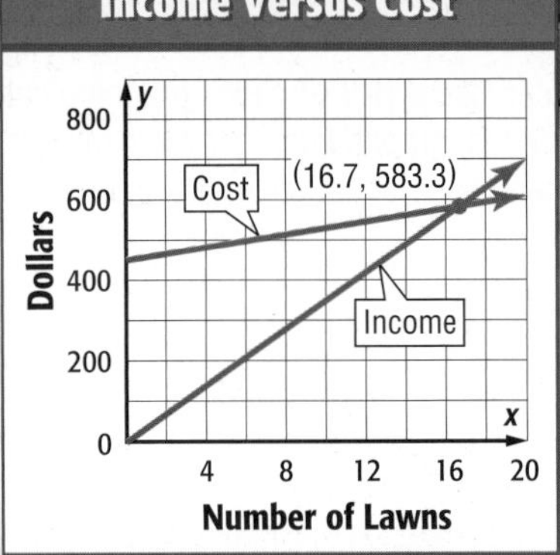

NewVocabulary
break-even point
system of equations
consistent
inconsistent
independent
dependent
substitution method
elimination method

Common Core State Standards

Mathematical Practices
2 Reason abstractly and quantitatively.
6 Attend to precision.

1 Solve Systems Graphically

A **system of equations** is two or more equations with the same variables. To solve a system of equations with two variables, find the ordered pair that satisfies all of the equations.

To solve a system of equations by using a table, first write each equation in slope-intercept form. Then substitute different values for x and solve for the corresponding y-values. For ease of use, choose 0 and 1 as your first x-values.

$y_1 = -2x + 8$
$y_2 = 4x - 7$

x	y_1	y_2	Difference
0	8	−7	15
1	6	−3	9
2	4	1	3
3	2	5	−3

Because the difference between the y-values is closer to 0 for $x = 1$ than $x = 0$, a value greater than 1 should be tried next.

Because the difference between the y-values changed signs from $x = 2$ to $x = 3$, a value between these should be tried next.

The solution is between 2 and 3.

Example 1 Solve by Using a Table

PT

Solve the system of equations.

$$3x + 2y = -2$$
$$-4x + 5y = -28$$

Write each equation in slope-intercept form.

$3x + 2y = -2 \rightarrow y = -1.5x - 1$
$-4x + 5y = -28 \rightarrow y = 0.8x - 5.6$

Use a table to find the solution that satisfies both equations.

x	y_1	y_2	(x, y_1)	(x, y_2)
0	−1	−5.6	(0, −1)	(0, −5.6)
1	−2.5	−4.8	(1, −2.5)	(1, −4.8)
2	**−4**	**−4**	**(2, −4)**	**(2, −4)**

The solution of the system is (2, −4).

GuidedPractice

1A. $2x - 5y = 11$
$-3x + 4y = -13$

1B. $4x + 3y = -17$
$-7x - 2y = 20$

Another method for solving a system of equations is to graph the equations on the same coordinate plane. The point of intersection represents the solution.

Example 2 Solve by Graphing

Solve the system of equations by graphing.

$\mathbf{2x - y = -1}$ $\qquad$ $\mathbf{2y + 5x = -16}$

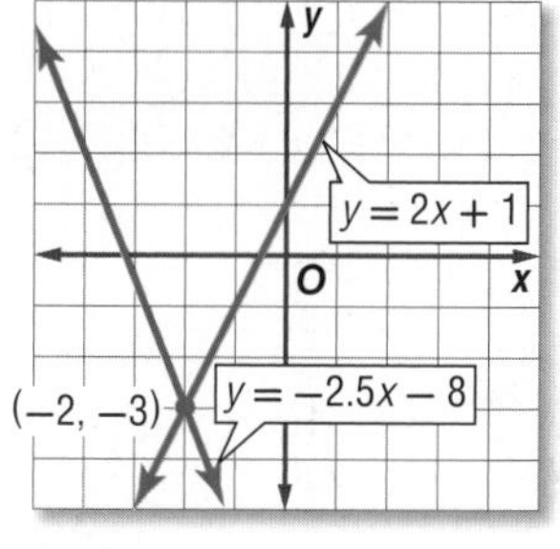

Write each equation in slope-intercept form.

$2x - y = -1 \quad \rightarrow \quad y = 2x + 1$

$2y + 5x = -16 \quad \rightarrow \quad y = -2.5x - 8$

The graphs of the lines appear to intersect at (−2, −3).

CHECK Substitute the coordinates into each original equation.

$2x - y = -1$	$2y + 5x = -16$	Original equations
$2(-2) - (-3) \stackrel{?}{=} -1$	$2(-3) + 5(-2) \stackrel{?}{=} -16$	$x = -2$ and $y = -3$
$-1 = -1$ ✓	$-16 = -16$ ✓	Simplify.

The solution of the system is (−2, −3).

StudyTip

Checking Solutions Always check to see if the values work for **both** of the original equations.

GuidedPractice

2A. $4x + 3y = 12$
$-6x + 4y = -1$

2B. $-3y + 8x = 36$
$6x + y = -21$

Systems of equations can be classified by the number of solutions. A system is **consistent** if it has at least one solution and **inconsistent** if it has no solutions. If it has exactly one solution, it is **independent**, and if it has an infinite number of solutions, it is **dependent**.

StudyTip

Slope and Classifying Systems If the equations have different slopes, then the system is consistent and independent.

Example 3 Classify Systems

Graph each system of equations and describe them as *consistent and independent*, *consistent and dependent*, or *inconsistent*.

a. $\mathbf{4x + 3y = 24}$
$\mathbf{-3x + 5y = 30}$

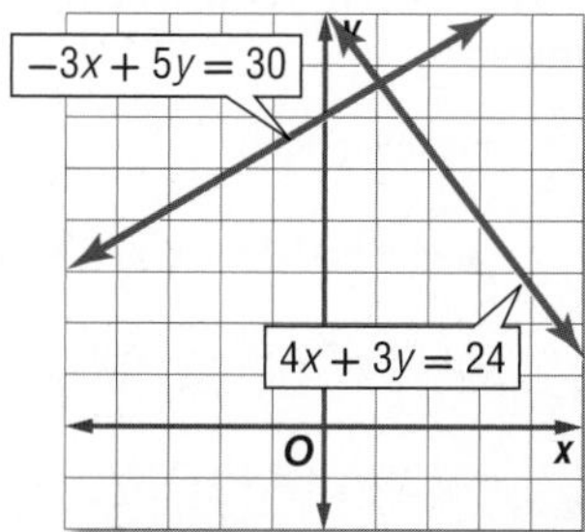

The graphs of the lines intersect at one point, so there is one solution. The system is consistent and independent.

b. $\mathbf{-2x + 5y = 10}$
$\mathbf{4x - 10y = -20}$

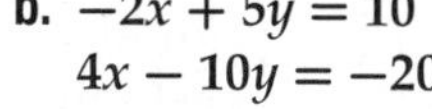

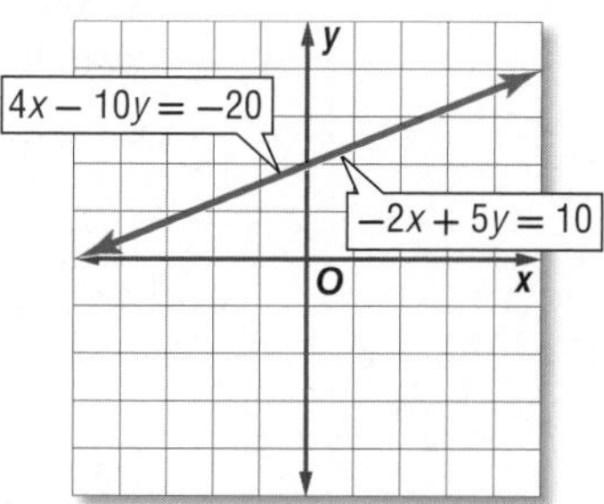

Because the equations are equivalent, their graphs are the same line. The system is consistent and dependent.

GuidedPractice

3A. $6x - 4y = 15$
$-6x + 4y = 18$

3B. $-4x + 5y = -17$
$-4x - 2y = 15$

The relationship between the graph and the solutions of a system is summarized below.

ConceptSummary Characteristics of Linear Systems

Consistent and Independent	Consistent and Dependent	Inconsistent
intersecting lines; one solution	same line; infinitely many solutions	parallel lines; no solution

2 Solve Systems Algebraically

Algebraic methods are used to find exact solutions of systems of equations. One algebraic method is called the **substitution method**.

KeyConcept Substitution Method

Step 1 Solve one equation for one of the variables.

Step 2 Substitute the resulting expression into the other equation to replace the variable. Then solve the equation.

Step 3 Substitute to solve for the other variable.

Systems of equations can be used to solve many real-world problems involving constraints modeled by two or more different functions.

Real-World Example 4 Use the Substitution Method

PT

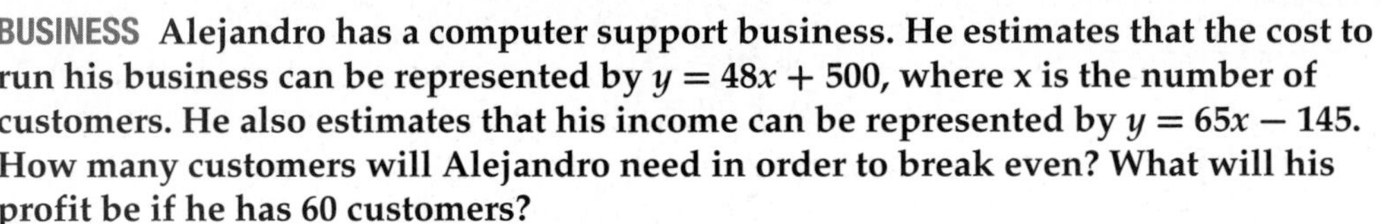

BUSINESS Alejandro has a computer support business. He estimates that the cost to run his business can be represented by $y = 48x + 500$, where x is the number of customers. He also estimates that his income can be represented by $y = 65x - 145$. How many customers will Alejandro need in order to break even? What will his profit be if he has 60 customers?

$y = 65x - 145$	Income equation
$48x + 500 = 65x - 145$	Substitute $48x + 500$ for y.
$500 = 17x - 145$	Subtract $48x$ from each side.
$645 = 17x$	Add 145 to each side.
$37.9 \approx x$	Divide each side by 17.

Alejandro needs 38 customers to break even. If he has 60 customers, his income will be 65(60) − 145 or \$3755, and his costs will be 48(60) + 500 or \$3380, so his profit will be 3755 − 3380 or \$375.

CHECK You can use a graphing calculator to check this solution. The break-even point is near (37.9, 2321.2). Use the **CALC** function to find the cost and income for 60 customers.

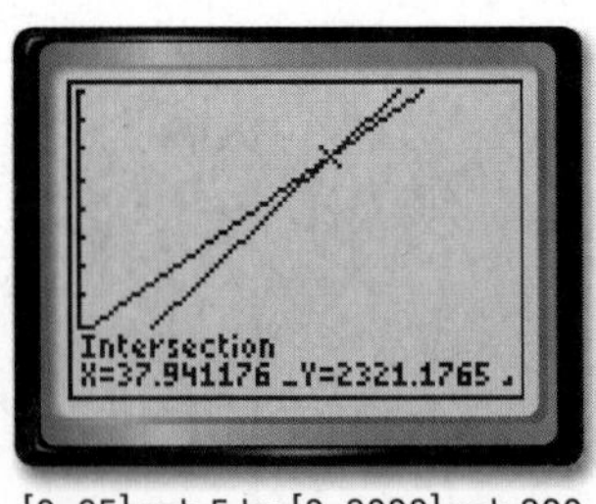

[0, 65] scl: 5 by [0, 3000] scl: 300

GuidedPractice

Use substitution to solve each system of equations.

4A. $5x - 3y = 23$
$2x + y = 7$

4B. $x - 7y = 11$
$5x + 4y = -23$

4C. $-6x - y = 27$
$3x + 8y = 9$

You can use the **elimination method** to solve a system when one of the variables has the same coefficient in both equations.

KeyConcept Elimination Method

Step 1 Multiply one or both equations by a number to result in two equations that contain opposite terms.

Step 2 Add the equations, eliminating one variable. Then solve the equation.

Step 3 Substitute to solve for the other variable.

Variables can be eliminated by addition or subtraction.

StudyTip

Perseverance

Remember when you add or subtract one equation from another to add or subtract every term, including the constant on the other side of the equal sign.

Example 5 Solve by Using Elimination

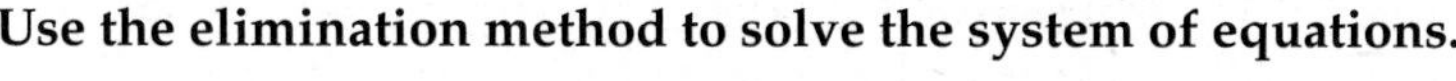

Use the elimination method to solve the system of equations.

$$\mathbf{5x + 3y = -19}$$
$$\mathbf{8x + 3y = -25}$$

Notice that solving by substitution would involve fractions.

Step 1 Multiply one equation by -1 so the equations contain $3y$ and $-3y$.

$8x + 3y = -25$ → Multiply by -1. → $-8x - 3y = 25$

Step 2 Add the equations to eliminate one variable.

$5x + 3y = -19$	Equation 1
$(+)\ -8x - 3y = 25$	Equation 2 × (−1)
$-3x = 6$	Add the equations.
$x = -2$	Divide each side by −3.

Step 3 Substitute -2 for x into either original equation.

$8x + 3y = -25$	Equation 2
$8(-2) + 3y = -25$	$x = -2$
$-16 + 3y = -25$	Multiply.
$3y = -9$	Add 16 to each side.
$y = -3$	Divide each side by 3.

The solution is $(-2, -3)$.

GuidedPractice

5A. $4x - 3y = -22$
$2x + 3y = 16$

5B. $6x - 5y = -8$
$4x - 5y = -12$

5C. $2x - 9y = 34$
$-2x + 6y = -28$

ReviewVocabulary

Least Common Multiple
the least number that is a common multiple of two or more numbers

Sometimes, adding or subtracting equations will not eliminate either variable. You can use multiplication and least common multiples to find a common coefficient.

Standardized Test Example 6 No Solution and Infinite Solutions

Solve the system of equations.

$5x + 3y = 52$

$15x + 9y = 54$

A (3, 1) **B** (8, 4) **C** no solution **D** infinite solutions

Read the Test Item

You are given a system of two linear equations and are asked to find the solution.

Solve the Test Item

Neither variable has a common coefficient. The coefficients of the y-variables are 3 and 9 and their least common multiple is 9, so multiply each equation by the value that will make the y-coefficient 9.

$5x + 3y = 52$ → Multiply by 4. → $15x + 9y = 156$

$15x + 9y = 54$ → $\underline{(-)\ 15x + 9y = 54}$

$0 = 102$ Subtract the equations.

Because $0 = 102$ is not true, this system has no solution.

The correct answer is C.

StudyTip

Adding and Subtracting Equations If you add or subtract two equations in a system and the result is an equation that is never true, then the system is inconsistent. When you add or subtract two equations in a system and the result is an equation that is always true, then the system is dependent.

GuidedPractice

6. Solve the system of equations. $2x + 3y = 5$
$6x + 9y = 15$

F (−2, 3) **G** (7, 3) **H** no solution **J** infinite solutions

The following summarizes the various methods for solving systems.

Math HistoryLink

Nina Karlovna Bari (1901–1961) Russian mathematician Nina Karlovna Bari was considered the principal leader of mathematics at Moscow State University, shown above. She is best known for her textbooks *Higher Algebra* and *The Theory of Series.*

ConceptSummary Solving Systems of Equations

Method	The Best Time to Use
Table	to estimate the solution, since a table may not provide an exact solution
Graphing	to estimate the solution, since graphing usually does not give an exact solution
Substitution	if one of the variables in either equation has a coefficient of 1 or −1
Elimination Using Addition	if one of the variables has opposite coefficients in the two equations
Elimination Using Subtraction	if one of the variables has the same coefficient in the two equations
Elimination Using Multiplication	if none of the coefficients are 1 or −1 and neither of the variables can be eliminated by simply adding or subtracting the equations

Brendan Hoffman/Alamy

Check Your Understanding

= Step-by-Step Solutions begin on page R14.

Example 1 **Solve each system of equations by using a table.**

1. $y = 3x - 4$
   $y = -2x + 11$

2. $4x - y = 1$
   $5x + 2y = 24$

Example 2 **Solve each system of equations by graphing.**

3. $y = -3x + 6$
   $2y = 10x - 36$

4. $y = -x - 9$
   $3y = 5x + 5$

5. $y = 0.5x + 4$
   $3y = 4x - 3$

6. $-3y = 4x + 11$
   $2x + 3y = -7$

7. $4x + 5y = -41$
   $3y - 5x = 5$

8. $8x - y = 50$
   $x + 4y = -2$

9. **CCSS MODELING** Refer to the table at the right.

   a. Write equations that represent the cost of printing digital photos at each lab.

   b. Under what conditions is the cost to print digital photos the same at both stores?

   c. When is it best to use EZ Online Digital Photos and when is it best to use the local store?

Digital Photos
Online Store
\$0.15 per photo + \$2.70 shipping
Local Store
\$0.25 per photo

Example 3 **Graph each system of equations and describe it as *consistent and independent*, *consistent and dependent*, or *inconsistent*.**

10. $y + 4x = 12$
    $3y = 8 - 12x$

11. $-2x - 3y = 9$
    $4x + 6y = -18$

12. $9x - 2y = 11$
    $5x + 4y = 13$

Example 4 **Solve each system of equations by using substitution.**

13. $x + 5y = 3$
    $3x - 2y = -8$

14. $y = 2x - 10$
    $y = -4x + 8$

15. $2a + 8b = -8$
    $3a - 5b = 22$

16. $a - 3b = -22$
    $4a + 2b = -4$

17. $6x - 7y = 23$
    $8x + 4y = 44$

18. $9c - 3d = -33$
    $6c + 5d = -8$

Examples 5–6 **Solve each system of equations by using elimination.**

19. $-6w - 8z = -44$
    $3w + 6z = 36$

20. $4x - 3y = 29$
    $4x + 3y = 35$

21. $3a + 5b = -27$
    $4a + 10b = -46$

22. $8a - 3b = -11$
    $5a + 2b = -3$

23. $5a + 15b = -24$
    $-2a - 6b = 28$

24. $6x - 4y = 30$
    $12x + 5y = -18$

25. **MULTIPLE CHOICE** What is the solution of the linear system?

$$4x + 3y = 2$$
$$4x - 2y = 12$$

A (8, −10) **B** (2, −2) **C** (−10, 14) **D** no solution

Practice and Problem Solving

Extra Practice is on page R3.

Example 1 **Solve each system of equations by using a table.**

26. $y = 5x + 3$
$y = x - 9$

27. $3x - 4y = 16$
$-6x + 5y = -29$

28. $2x - 5 = y$
$-3x + 4y = 0$

29. **FUNDRAISER** To raise money for new uniforms, the band boosters sell T-shirts and hats. The cost and sale price of each item is shown. The boosters spend a total of $2000 on T-shirts and hats. They sell all of the merchandise, and make $3375. How many T-shirts did they sell?

Item	Cost	Sale Price
T-Shirt	$6	$10
Hat	$4	$7

Example 2 **Solve each system of equations by graphing.**

30. $-3x + 2y = -6$
$-5x + 10y = 30$

31. $4x + 3y = -24$
$8x - 2y = -16$

32. $6x - 5y = 17$
$6x + 2y = 31$

33. $-3x - 8y = 12$
$12x + 32y = -48$

34. $y - 3x = -29$
$9x - 6y = 102$

35. $-10x + 4y = 7$
$2x - 5y = 7$

36. **CCSS MODELING** Jerilyn has a $10 coupon and a 15% discount coupon for her favorite store. The store has a policy that only one coupon may be used per purchase. When is it best for Jerilyn to use the $10 coupon, and when is it best for her to use the 15% discount coupon?

Example 3 **Graph each system of equations and describe it as *consistent and independent, consistent and dependent,* or *inconsistent.***

37. $y = 3x - 4$
$y = 6x - 8$

38. $y = 2x - 1$
$y = 2x + 6$

39. $2x + 5y = 10$
$-4x - 10y = 20$

40. $x - 6y = 12$
$3x + 18y = 14$

41. $-5x - 6y = 13$
$12y + 10x = -26$

42. $8y - 3x = 15$
$-16y + 6x = -30$

Example 4 **Solve each system of equations by using substitution.**

43. $9y + 3x = 18$
$-3y - x = -6$

44. $5x - 20y = 70$
$6x + 5y = -32$

45. $-4x - 16y = -96$
$7x + 3y = 68$

46. $-4a - 5b = 14$
$9a + 3b = -48$

47. $-9c - 4d = 31$
$6c + 6d = -24$

48. $8f + 3g = 12$
$-32f - 12g = 48$

49. **TENNIS** At a park, there are 38 people playing tennis. Some are playing doubles, and some are playing singles. There are 13 matches in progress. A doubles match requires 4 players, and a singles match requires 2 players.

a. Write a system of two equations that represents the number of singles and doubles matches going on.

b. How many matches of each kind are in progress?

Examples 5–6 **Solve each system of equations by using elimination.**

50. $8x + y = 27$
$-3x + 4y = 3$

51. $2a - 5b = -20$
$2a + 5b = 20$

52. $6j + 4k = -46$
$2j + 4k = -26$

53. $3x - 8y = 24$
$-12x + 32y = 96$

54. $5a - 2b = -19$
$8a + 5b = -55$

55. $r - 6t = 44$
$9r + 12t = 0$

56. $6d + 5f = -32$
$5d - 9f = 26$

57. $11u = 5v + 35$
$8v = -6u + 62$

58. $-1.2c + 3.4d = 6$
$6c = -30 + 17d$

Use a graphing calculator to solve each system of equations. Round the coordinates of the intersection to the nearest hundredth.

59. $12y = 5x - 15$
$4.2y + 6.1x = 11$

60. $-3.8x + 2.9y = 19$
$6.6x - 5.4y = -23$

61. $5.8x - 6.3y = 18$
$-4.3x + 8.8y = 32$

Solve each system of equations.

62. $11p + 3q = 6$
$-0.75q - 2.75p = -1.5$

63. $8r - 5t = -60$
$6r + 3t = -18$

64. $10t + 4v = 13$
$-4t - 7v = 11$

65. $6w = 12 - 4x$
$6x = -9w + 18$

66. $\frac{3}{2}y + z = 3$
$-y - \frac{2}{3}z = -2$

67. $\frac{5}{2}a - \frac{3}{4}b = 46$
$-\frac{7}{8}a - 3b = 10$

68. ROWING Allison can row a boat 1 mile upstream (against the current) in 24 minutes. She can row the same distance downstream in 13 minutes. Assume that both the rowing speed and the speed of the current are constant.

a. Find the speed at which Allison is rowing and the speed of the current.

b. If Allison plans to meet her friends 3 miles upstream one hour from now, will she be on time? Explain.

69. CCSS **MODELING** The table shows the winning times in seconds for the 100-meter dash at the Olympics between 1964 and 2008.

Years Since 1964, x	Men's Gold Medal Time	Women's Gold Medal Time
0	10.0	11.4
4	9.90	11.0
8	10.14	11.07
12	10.06	11.08
16	10.25	11.06
20	9.99	10.97
24	9.92	10.54
28	9.96	10.82
32	9.84	10.94
36	9.87	10.75
40	9.85	10.93
44	9.69	10.78

a. Write equations that represent the winning times for men and women since 1964. Assume that both times continue along the same trend.

b. Graph both equations. Estimate when the women's performance will catch up to the men's performance. Do you think that your prediction is reasonable? Explain.

70. JOBS Levi has a job offer in which he will receive \$800 per month plus a commission of 2% of the total price of the cars that he sells. At his current job, he receives \$1200 per month plus a commission of 1.5% of his total sales. How much must he sell per month to make the new job a better deal?

71. TRAVEL A youth group went on a trip to an amusement park, travelling in two vans. The number of people in each van and the total cost of admission are shown in the table. Find the adult price and student price of admission.

Van	Adults	Students	Total Cost
A	2	5	\$77
B	2	7	\$95

TICKETS

GEOMETRY **Find the point at which the diagonals of the quadrilaterals intersect.**

72.

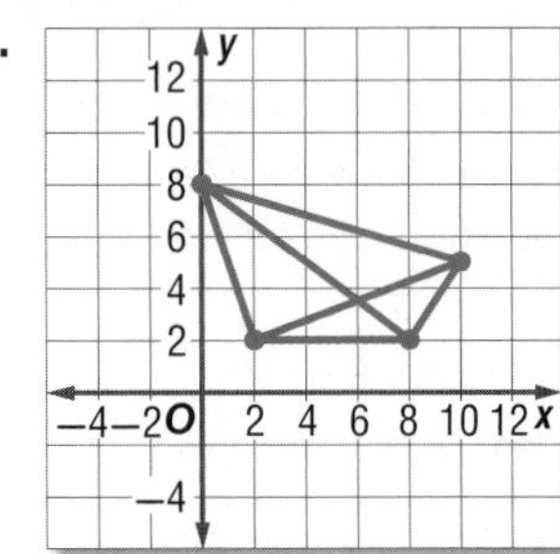

73.

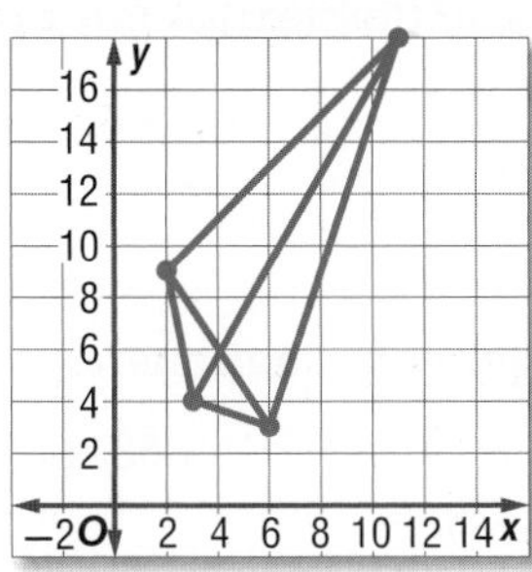

74. **ELECTIONS** In the election for student council, Candidate A received 55% of the total votes, while Candidate B received 1541 votes. If Candidate C received 40% of the votes that Candidate A received, how many total votes were cast?

75. **MULTIPLE REPRESENTATIONS** In this problem, you will explore systems of three linear equations and two variables.

$$3y + x = 16$$
$$y - 2x = -4$$
$$y + 5x = 10$$

a. Tabular Make a table of x- and y-values for each equation.

b. Analytical Which values from the table indicate intersections? Is there a solution that satisfies all three equations?

c. Graphical Graph the three equations on a single coordinate plane.

d. Verbal What conditions must be met for a system of three equations with two variables to have a solution? What conditions result in no solution?

H.O.T. Problems Use Higher-Order Thinking Skills

76. **CCSS CRITIQUE** Gloria and Syreeta are solving the system $6x - 4y = 26$ and $-3x + 4y = -17$. Is either of them correct? Explain your reasoning.

Gloria

$6x - 4y = 26$ $\quad$ $6(3) - 4y = 26$
$-3x + 4y = -17$ $\quad$ $18 - 4y = 26$
$3x = 9$ $\quad$ $-4y = 8$
$x = 3$ $\quad$ $y = -2$

The solution is $(3, -2)$.

Syreeta

$6x - 4y = 26$ $\quad$ $6(-3) - 4y = 26$
$-3x + 4y = -17$ $\quad$ $-18 - 4y = 26$
$3x = -9$ $\quad$ $-4y = 44$
$x = -3$ $\quad$ $y = -11$

The solution is $(-3, -11)$.

77. **CHALLENGE** Find values of a and b for which the following system has a solution of $(b - 1, b - 2)$.

$$-8ax + 4ay = -12a$$
$$2bx - by = 9$$

78. **REASONING** If a is consistent and dependent with b, b is inconsistent with c, and c is consistent and independent with d, then a will *sometimes*, *always*, or *never* be consistent and independent with d. Explain your reasoning.

79. **OPEN ENDED** Write a system of equations in which one equation needs to be multiplied by 3 and the other needs to be multiplied by 4 in order to solve the system with elimination. Then solve your system.

80. **WRITING IN MATH** Why is substitution sometimes more helpful than elimination, and vice versa?

Standardized Test Practice

81. SHORT RESPONSE Simplify $3y(4x + 6y - 5)$.

82. SAT/ACT Which of the following best describes the graph of the equations?

$$4y = 3x + 8$$
$$-6x = -8y + 24$$

A The lines are parallel.

B The lines are perpendicular.

C The lines have the same x-intercept.

D The lines have the same y-intercept.

E The lines are the same.

83. GEOMETRY Which set of dimensions corresponds to a triangle similar to the one shown at the right?

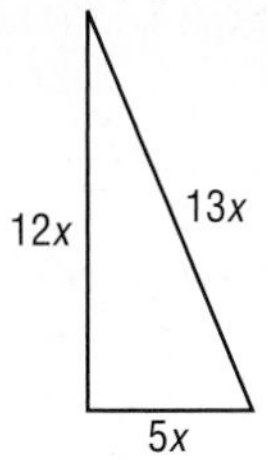

F 1 unit, 2 units, 3 units

G 7 units, 11 units, 12 units

H 10 units, 23 units, 24 units

J 20 units, 48 units, 52 units

84. Move-A-Lot Rentals will rent a moving truck for \$100 plus \$0.10 for every mile it is driven. Which equation can be used to find C, the cost of renting a moving truck, and driving it for m miles?

A $C = 0.1(100 + m)$

B $C = 100 + 0.1m$

C $C = 100m + 0.1$

D $C = 100(m + 0.1)$

Spiral Review

85. CRAFTS Priscilla sells stuffed animals at a local craft show. She charges \$10 for the small ones and \$15 for the large ones. To cover her expenses, she needs to sell at least \$350 worth of animals. (Lesson 2-8)

a. Write an inequality for this situation.

b. Graph the inequality.

c. If she sells 10 small and 15 large animals, will she cover her expenses?

Write an equation for each function. (Lesson 2-7)

86.

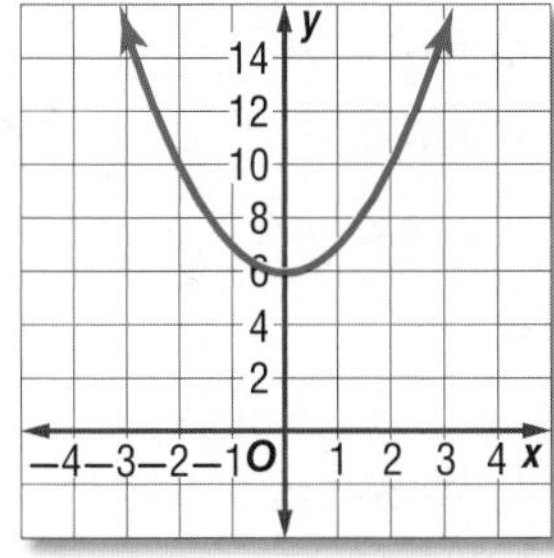

87.

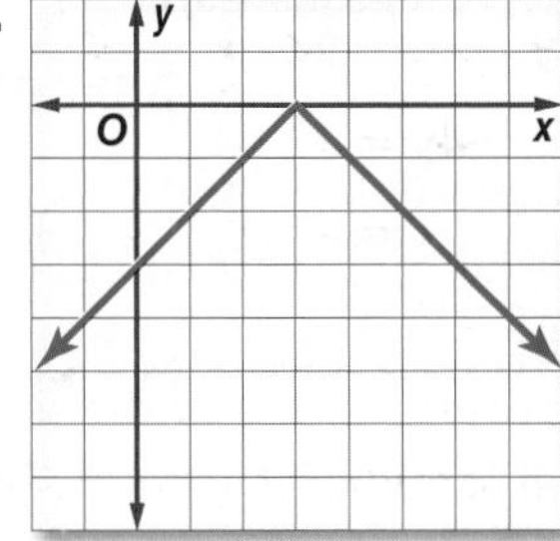

88.

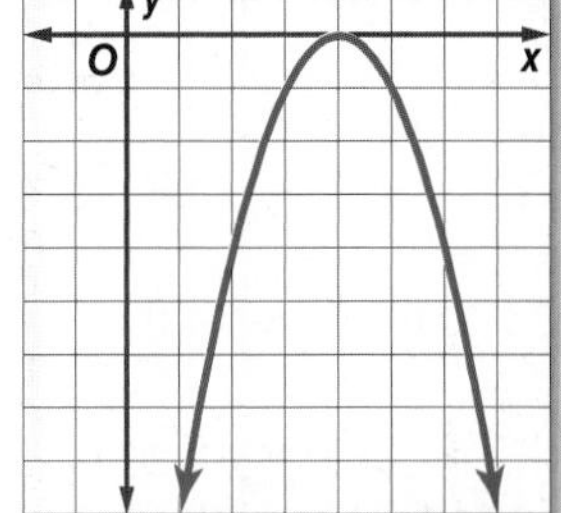

Solve each equation. Check your solution. (Lesson 1-3)

89. $2p = 14$

90. $-14 + n = -6$

91. $7a - 3a + 2a - a = 16$

92. $x + 9x - 6x + 4x = 20$

93. $27 = -9(y + 5) + 6(y + 8)$

94. $-7(p + 7) + 3(p - 4) = -17$

Skills Review

Determine whether the given point satisfies each inequality.

95. $4x + 5y \leq 15$; $(2, -2)$

96. $3x + 5y \geq 8$; $(1, 1)$

97. $6x + 9y < -1$; $(0, 0)$

LESSON 3-2 Solving Systems of Inequalities by Graphing

Then

- You solved systems of linear equations graphically and algebraically.

Now

1. Solve systems of inequalities by graphing.
2. Determine the coordinates of the vertices of a region formed by the graph of a system of inequalities.

Why?

- Many weather conditions need to be met before a space shuttle can launch. The temperature must be greater than 35°F and less than 100°F, and the wind speed cannot exceed 30 knots. A system of inequalities can be used to show these three conditions.

NewVocabulary
system of inequalities

Common Core State Standards

Content Standards
A.CED.3 Represent constraints by equations or inequalities, and by systems of equations and/or inequalities, and interpret solutions as viable or nonviable options in a modeling context.

Mathematical Practices
1 Make sense of problems and persevere in solving them.

1 Systems of Inequalities

Solving a **system of inequalities** means finding the ordered pairs that satisfy all of the inequalities in the system.

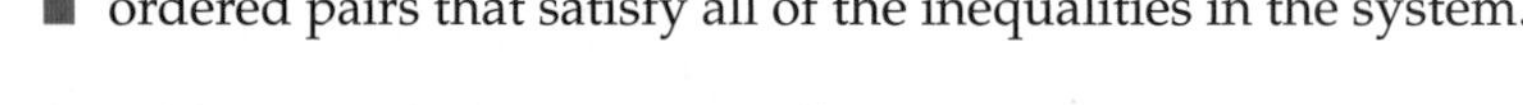

KeyConcept Solving Systems of Inequalities

Step 1 Graph each inequality, shading the correct area.

Step 2 Identify the region that is shaded for all of the inequalities. This is the solution of the system.

Example 1 Intersecting Regions

Solve the system of inequalities.
$y > 2x - 4$
$y \leq -0.5x + 3$

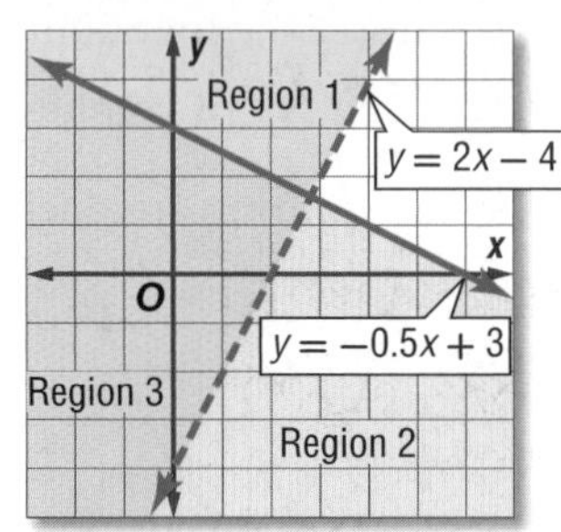

Solution of $y > 2x - 4 \rightarrow$ Regions 1 and 3

Solution of $y \leq -0.5x + 3 \rightarrow$ Regions 2 and 3

Region 3 is part of the solution of both inequalities, so it is the solution of the system.

CHECK Notice that the origin is part of the solution of the system. The origin can be used as a test point. You can test the solution by substituting (0, 0) for x and y in each equation.

$y > 2x - 4$	$y \leq -0.5x + 3$
$0 \overset{?}{>} 2(0) - 4$	$0 \overset{?}{\leq} -0.5(0) + 3$
$0 \overset{?}{>} 0 - 4$	$0 \overset{?}{\leq} 0 + 3$
$0 > -4$ ✓	$0 \leq 3$ ✓

GuidedPractice

1A. $y \leq -2x + 5$
$y > -\frac{1}{4}x - 6$

1B. $y \geq |x|$
$y < \frac{4}{3}x + 5$

Imagestate Pictor/Photolibrary

ReadingMath

Empty Set The empty set is also called the *null set*. It can be represented by Ø or { }.

It is possible that the regions do not intersect. When this occurs, there is no solution of the system or the solution set is the *empty set*.

PT

Example 2 Separate Regions

Solve the system of inequalities by graphing.

$y \geq x + 5$
$y < x - 4$

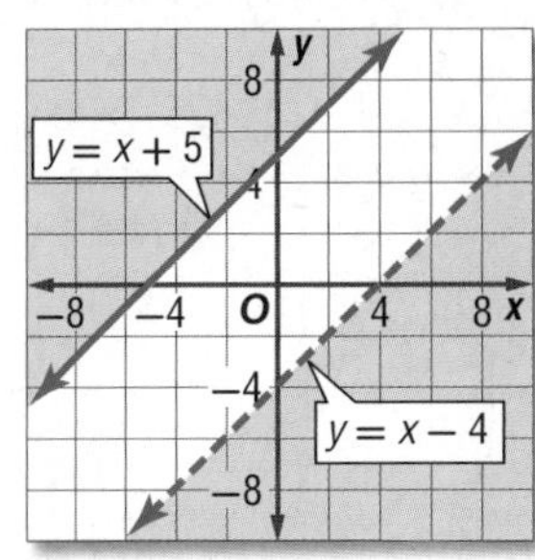

Graph both inequalities.

Since the graphs of the inequalities do not overlap, there are no points in common and there is no solution to the system.

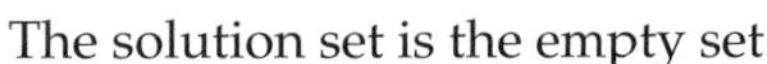

The solution set is the empty set.

GuidedPractice

2A. $y \geq -4x + 8$
$y < -4x + 4$

2B. $y \geq |x|$
$y < 2x - 24$

Real-WorldLink

Typical incoming freshmen will spend more than 3 times as many hours studying in college as in high school.

Source: *National Survey of Student Engagement*

Real-World Example 3 Write and Use a System of Inequalities

TIME MANAGEMENT **Chelsea has final exams in calculus, physics, and history. She has up to 25 hours to study for the exams. She plans to study history for 2 hours. She needs to spend at least 7 hours studying for calculus, but over 14 is too much. She hopes to spend between 8 and 12 hours on physics. Write and graph a system of inequalities to represent the situation.**

Calculus: at least 7 hours, but no more than 14

$$7 \leq c \leq 14$$

Physics: at least 8 hours, but no more than 12

$$8 \leq p \leq 12$$

Chelsea has 25 hours available, and 2 of those will be spent on history. She has up to 23 hours left for calculus and physics.

$$c + p \leq 23$$

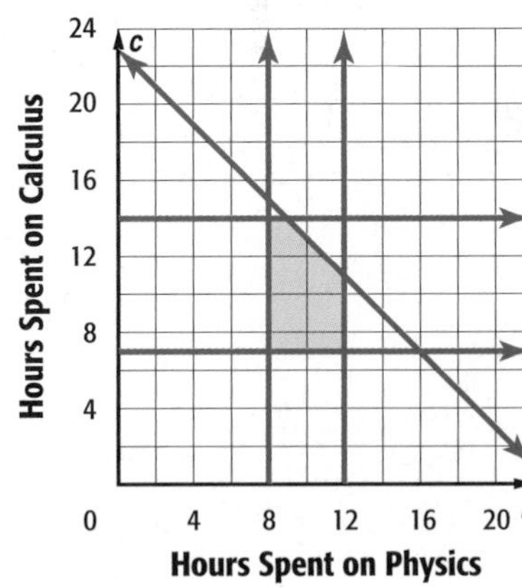

Graph all of the inequalities. Any ordered pair in the intersection is a solution of the system. One solution is 10 hours on physics and 12 hours on calculus.

GuidedPractice

3. TRAVEL Mr. and Mrs. Rodriguez are driving across the country with their two children. They plan on driving a maximum of 10 hours each day. Mr. Rodriguez wants to drive at least 4 hours a day but no more than 8 hours a day. Mrs. Rodriguez can drive between 2 and 5 hours per day. Write and graph a system of inequalities that represents this information.

BananaStock/Jupiterimages

StudyTip

Boundaries If the inequality that forms the boundary is < or >, then the boundary is not included in the solution, and the line should be dashed.

2 Find Vertices of an Enclosed Region

Sometimes the graph of a system of inequalities produces an enclosed region in the form of a polygon. To find the vertices of the region, determine the coordinates of the points at which the boundaries intersect.

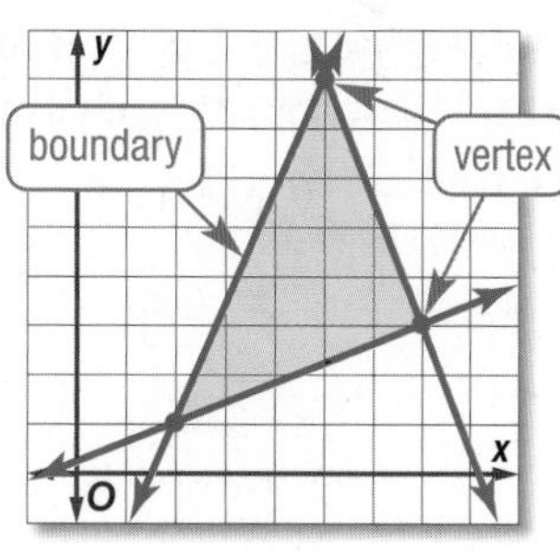

Example 4 Find Vertices

Find the coordinates of the vertices of the triangle formed by $y \geq 2x - 8$, $y \leq -\frac{1}{4}x + 6$, and $4y \geq -15x - 32$.

Step 1 Graph each inequality.

The coordinates $(-4, 7)$ and $(0, -8)$ can be determined from the graph. To find the coordinates of the third vertex, solve the system of equations $y = 2x - 8$ and $y = -\frac{1}{4}x + 6$.

Step 2 Substitute for y in the second equation.

$2x - 8 = -\frac{1}{4}x + 6$ — Replace y with $2x - 8$.

$2x = -\frac{1}{4}x + 14$ — Add 8 to each side.

$\frac{9}{4}x = 14$ — Add $\frac{1}{4}x$ to each side.

$x = \frac{56}{9}$ or $6\frac{2}{9}$ — Divide each side by $\frac{9}{4}$.

Step 3 Find y.

$y = 2\left(6\frac{2}{9}\right) - 8$ — Replace x with $6\frac{2}{9}$.

$= 12\frac{4}{9} - 8$ — Distributive Property

$= 4\frac{4}{9}$ — Simplify.

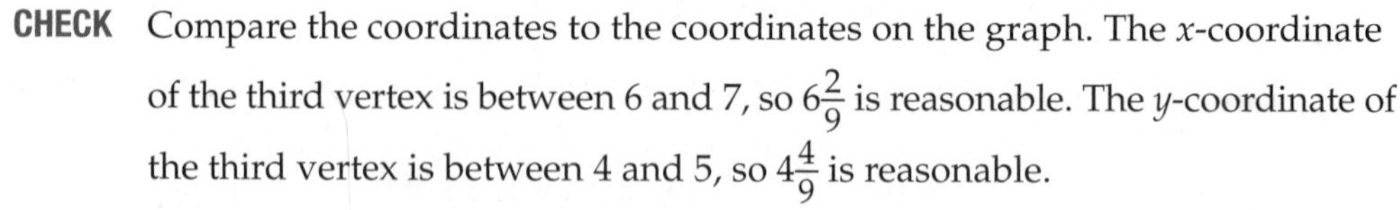

CHECK Compare the coordinates to the coordinates on the graph. The x-coordinate of the third vertex is between 6 and 7, so $6\frac{2}{9}$ is reasonable. The y-coordinate of the third vertex is between 4 and 5, so $4\frac{4}{9}$ is reasonable.

The vertices of the triangle are at $(-4, 7)$, $(0, -8)$, and $\left(6\frac{2}{9}, 4\frac{4}{9}\right)$.

GuidedPractice

Find the coordinates of the vertices of the triangle formed by each system of inequalities.

4A. $y \geq -3x - 6$
$2y \geq x - 16$
$11y + 7x \leq 12$

4B. $5y \leq 2x + 9$
$y \leq -x + 6$
$9y \geq -2x + 5$

Check Your Understanding

= Step-by-Step Solutions begin on page R14.

Examples 1–2 **Solve each system of inequalities by graphing.**

1. $y \leq 6$
$y > -3 + x$

2. $y \leq -3x + 4$
$y \geq 2x - 1$

3. $y > -2x + 4$
$y \leq -3x - 3$

Example 3

4. **CCSS REASONING** The most Kala can spend on hot dogs and buns for her cookout is $35. A package of 10 hot dogs costs $3.50. A package of buns costs $2.50 and contains 8 buns. She needs to buy at least 40 hot dogs and 40 buns.

a. Graph the region that shows how many packages of each item she can purchase.

b. Give an example of three different purchases she can make.

Example 4 **Find the coordinates of the vertices of the triangle formed by each system of inequalities.**

5. $y \geq 2x + 1$
$y \leq 8$
$4x + 3y \geq 8$

6. $y \geq -2x - 4$
$6y \leq x + 28$
$y \geq 13x - 34$

Practice and Problem Solving

Extra Practice is on page R3.

Examples 1–2 **Solve each system of inequalities by graphing.**

7. $x < 3$
$y \geq -4$

8. $y > 3x - 5$
$y \leq 4$

9. $y < -3x + 4$
$3y + x > -6$

10. $y \geq 0$
$y < x$

11. $6x - 2y \geq 12$
$3x + 4y > 12$

12. $-8x > -2y - 1$
$-4y \geq 2x - 5$

13. $5y < 2x + 10$
$y - 4x > 8$

14. $3y - 2x \leq -24$
$y \geq \frac{2}{3}x - 1$

15. $y > -\frac{2}{5}x + 2$
$5y \leq -2x - 15$

Example 3

16. **RECORDING** Jane's band wants to spend no more than $575 recording their first CD. The studio charges at least $35 per hour to record. Graph a system of inequalities to represent this situation.

17. **SUMMER TRIP** Rondell has to save at least $925 to go to Rome with his Latin class in 8 weeks. He earns $9 an hour working at the Pizza Palace and $12 an hour working at a car wash. By law, he cannot work more than 25 hours per week. Graph two inequalities that Rondell can use to determine the number of hours he needs to work at each job if he wants to make the trip.

Example 4 **Find the coordinates of the vertices of the triangle formed by each system of inequalities.**

18. $x \geq 0$
$y \geq 0$
$x + 2y < 4$

19. $y \geq 3x - 7$
$y \leq 8$
$x + y > 1$

20. $x \leq 4$
$y > -3x + 12$
$y \leq 9$

21. $-3x + 4y \leq 15$
$2y + 5x > -12$
$10y + 60 \geq 27x$

22. $8y - 19x < 74$
$38y + 26x \leq 119$
$54y - 12x \geq -198$

23. $6y - 24x \geq -168$
$8y + 7x > 10$
$20y - 2x \leq 64$

24. **BAKING** Rebecca wants to bake cookies and cupcakes for a bake sale. She can bake 15 cookies at a time and 12 cupcakes at a time. She needs to make at least 120 baked goods, but no more than 360, and she wants to have at least three times as many cookies as cupcakes. What combination of batches of each could Rebecca make?

25. **CELL PHONES** Dale has a maximum of 800 minutes on his cell phone plan that he can use each month. Daytime minutes cost \$0.15, and nighttime minutes cost \$0.10. Dale plans to use at least twice as many daytime minutes as nighttime minutes. If Dale uses at least 200 nighttime minutes and does not go over his limit, what is his maximum bill? his minimum bill?

26. **TREES** Trees are divided into four categories according to height and trunk circumference. Data for the trees in a forest are described in the table.

Crown Class	dominant	co-dominant	intermediate	suppressed
Height (in feet)	over 72	56–72	40–55	under 39
Trunk Circumference (in inches)	over 60	48–60	34–48	under 33

Source: USDA Forest Service

a. Write and graph the system of inequalities that represents the range of heights h and circumferences c for a co-dominant tree.

b. Determine the crown class of a basswood that is 48 feet tall. Find the expected trunk circumference.

27. **CCSS REASONING** On a camping trip, Jessica needs at least 3 pounds of food and 0.5 gallon of water per day. Marc needs at least 5 pounds of food and 0.5 gallon of water per day. Jessica's equipment weighs 10 pounds, and Marc's equipment weighs 20 pounds. A gallon of water weighs approximately 8 pounds. Each of them carries their own supplies, and Jessica is capable of carrying 35 pounds while Marc can carry 50 pounds.

a. Graph the inequalities that represent how much they can carry.

b. How many days can they camp, assuming that they bring all their supplies in at once?

c. Who will run out of supplies first?

Solve each system of inequalities by graphing.

28. $y \geq |2x + 4| - 2$
$3y + x \leq 15$

29. $y \geq |6 - x|$
$|y| \leq 4$

30. $|y| \geq x$
$y < 2x$

31. $y > -3x + 1$
$4y \leq x - 8$
$3x - 5y < 20$

32. $6y + 2x \leq 9$
$2y + 18 \geq 5x$
$y > -4x - 9$

33. $|x| > y$
$y \leq 6$
$y \geq -2$

34. $2x + 3y \geq 6$
$y \leq |x - 6|$

35. $8x + 4y < 10$
$y > |2x - 1|$

36. $y \geq |x - 2| + 4$
$y \leq [\![x]\!] - 3$

37. **MUSIC** Steve is trying to decide what to put on his MP3 player. Audio books are 3 hours long and songs are 2.5 minutes long. Steve wants no more than 4 audio books on his MP3 player, but at least ten songs and one audio book. Each book costs \$15.00 and each song costs \$0.95. Steve has \$63 to spend on books and music. Graph the inequalities to show possible combinations of books and songs that Steve can have.

38. **JOBS** Louie has two jobs and can work no more than 25 total hours per week. He wants to earn at least \$150 per week. Graph the inequalities to show possible combinations of hours worked at each job that will help him reach his goal.

Job	Pay
Busboy	\$6.50
Clerk	\$8.00

39. **TIME MANAGEMENT** Ramir uses his spare time to write a novel and to exercise. He has budgeted 35 hours per week. He wants to exercise at least 7 hours a week but no more than 15. He also hopes to write between 20 and 25 hours per week. Write and graph a system of inequalities that represents this situation.

Find the coordinates of the vertices of the figure formed by each system of inequalities.

40. $y \geq 2x - 12$
$y \leq -4x + 20$
$4y - x \leq 8$
$y \geq -3x + 2$

41. $y \geq -x - 8$
$2y \geq 3x - 20$
$4y + x \leq 24$
$y \leq 4x + 22$

42. $2y - x \geq -20$
$y \geq -3x - 6$
$y \leq -2x + 2$
$y \leq 2x + 14$

43. **FINANCIAL LITERACY** Mr. Hoffman is investing $10,000 in two funds. One fund will pay 6% interest, and a riskier second fund will pay 10% interest. What is the least amount he can invest in the risky fund and still earn at least $740 after one year?

44. **DODGEBALL** A high school is selecting a dodgeball team to play in a fund-raising exhibition against their rival. There can be between 10 and 15 players on the team and there must be more girls than boys on the team.

a. Write and graph a system of inequalities to represent the situation.

b. List all of the possible combinations of boys and girls for the team.

c. Explain why there is not an infinite number of possibilities.

H.O.T. Problems Use Higher-Order Thinking Skills

45. **CHALLENGE** Find the area of the region defined by the following inequalities.

$$y \geq -4x - 16$$
$$4y \leq 26 - x$$
$$3y + 6x \leq 30$$
$$4y - 2x \geq -10$$

46. **OPEN ENDED** Write a system of two inequalities in which the solution:

a. lies only in the third quadrant.

b. does not exist.

c. lies only on a line.

d. lies on exactly one point.

47. **CHALLENGE** Write a system of inequalities to represent the solution shown at the right. How many points with integer coordinates are solutions of the system?

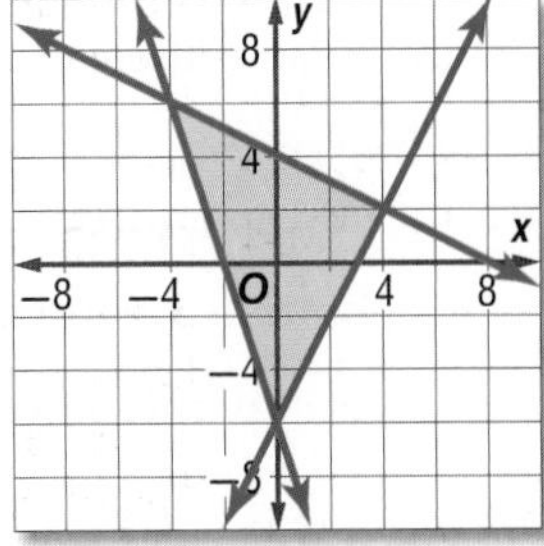

48. **CCSS ARGUMENTS** Determine whether the statement is *true* or *false*. If false, give a counterexample.

A system of two linear inequalities has either no points or infinitely many points in its solution.

49. **WRITING IN MATH** Write a how-to manual for determining where to shade when graphing a system of inequalities.

50. **WRITING IN MATH** Explain how you would test to see whether (−4, 6) is a solution of a system of inequalities.

Standardized Test Practice

51. To be a member of the marching band, a student must have a grade-point average of at least 2.0 and must have attended at least five after-school practices. Choose the system of inequalities that best represents this situation.

A $x \geq 2$, $y \geq 5$

B $x \leq 2$, $y \leq 5$

C $x < 2$, $y < 5$

D $x > 2$, $y > 5$

52. SAT/ACT The table at the right shows a relationship between x and y. Which equation represents this relationship?

x	y
1	5
2	8
3	11
4	14
5	17
6	20

F $y = 3x - 2$

G $y = 3x + 2$

H $y = 4x + 1$

J $y = 4x + 2$

K $y = 4x - 1$

53. SHORT RESPONSE If $3x = 2y$ and $5y = 6z$, what is the value of x in terms of z?

54. GEOMETRY Look at the graph below. Which of these statements describes the relationship between the two lines?

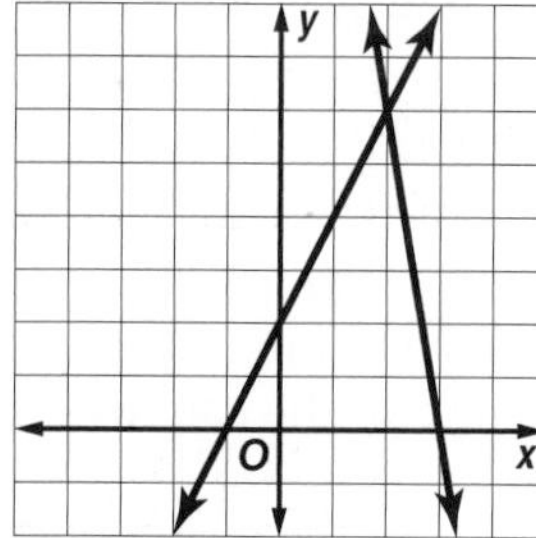

A They intersect at (6, 2).

B They intersect at (0, 2).

C They intersect at (3.5, 0).

D They intersect at (2, 6).

Spiral Review

55. GEOMETRY Find the coordinates of the vertices of the parallelogram with sides that are contained in the lines with equations $y = 3$, $y = 7$, $y = 2x$, and $y = 2x - 13$. (Lesson 3-1)

Graph each inequality. (Lesson 2-8)

56. $x + y \geq 6$

57. $4x - 3y < 10$

58. $5x + 7y \geq -20$

Graph each function. Identify the domain and range. (Lesson 2-6)

59. $g(x) = \begin{cases} 0 \text{ if } x < 0 \\ -x + 2 \text{ if } x \geq 0 \end{cases}$

60. $h(x) = \begin{cases} x + 3 \text{ if } x \leq -1 \\ 2x \text{ if } x > -1 \end{cases}$

61. $h(x) = \begin{cases} -1 \text{ if } x < -2 \\ 1 \text{ if } x > 2 \end{cases}$

62. BOOK CLUB For each meeting of the Putnam High School book club, \$25 is taken from the activities account to buy snacks and materials. After their sixth meeting, there will be \$350 left in the activities account. (Lesson 2-4)

a. If no money is put back into the account, what equation can be used to show how much money is left in the activities account after having x number of meetings?

b. How much money was originally in the account?

c. After how many meetings will there be no money left in the activities account?

Skills Review

Find each value if $f(x) = 2x + 5$ and $g(x) = 3x - 4$.

63. $f(-3)$

64. $g(-2)$

65. $f(-1)$

66. $g(-0.5)$

67. $f(-0.25)$

68. $g(-0.75)$

EXTEND 3-2

Graphing Technology Lab
Systems of Linear Inequalities

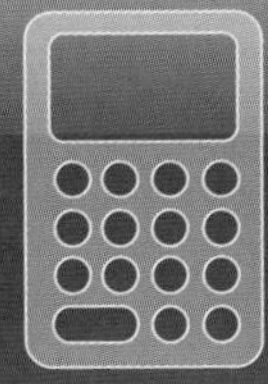

You can graph systems of linear inequalities with a graphing calculator by using the Y= menu. You can choose different graphing styles to shade above or below a line.

Example Intersection of Two Graphs

Graph the system of inequalities in the standard viewing window.
$y \geq -3x + 4$
$y \leq 2x - 1$

Step 1 Enter $-3x + 4$ as **Y1**. Because y is greater than $-3x + 4$, shade above the line.

KEYSTROKES: Y= ◀ ◀ ENTER ENTER ▶ ▶ (–) 3 X,T,θ,n + 4 ENTER

Plot1 Plot2 Plot3
Y1 = -3X+4
Y2=
Y3=
Y4=
Y5=
Y6=
Y7=

Step 2 Enter $2x - 1$ as **Y2**. Because y is less than $2x - 1$, shade below the line.

KEYSTROKES: ◀ ◀ ENTER ENTER ENTER ▶ ▶ 2 X,T,θ,n – 1 ENTER

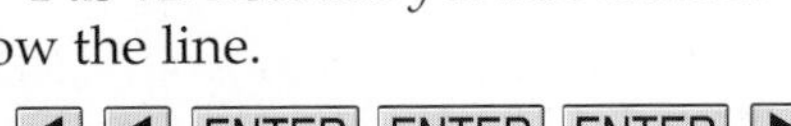
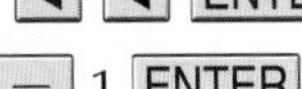

Step 3 Display the graphs in the standard viewing window.

KEYSTROKES: ZOOM 6

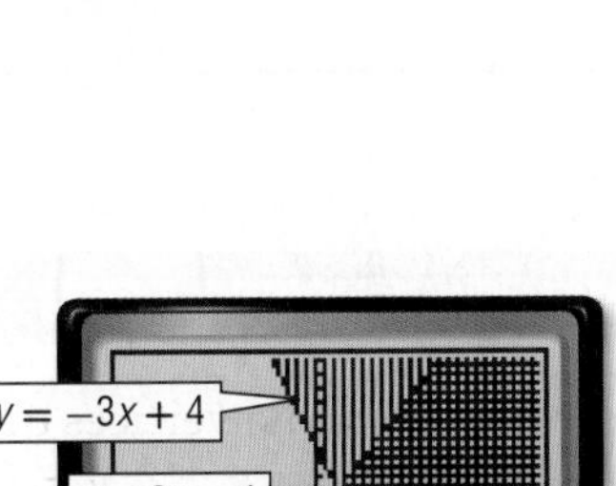

[−10, 10] scl: 1 by [−10, 10] scl: 1

Notice the shading pattern above the line $y = -3x + 4$ and the shading pattern below the line $y = 2x - 1$. The intersection of the graphs is the region where the patterns overlap. This region includes all the points that satisfy the system $y \geq -3x + 4$ and $y \leq 2x - 1$.

Exercises

Use a graphing calculator to solve each system of inequalities.

1. $y \geq 3$
$y \leq -x + 1$

2. $y \geq -4x$
$y \leq -5$

3. $y \geq 2 - x$
$y \leq x + 3$

4. $y \geq 2x + 1$
$y \leq -x - 1$

5. $2y \geq 3x - 1$
$3y \leq -x + 7$

6. $y + 5x \geq 12$
$y - 3 \leq 10$

7. $5y + 3x \geq 11$
$3y - x \leq -8$

8. $10y - 7x \geq -19$
$7y - 5x \leq 11$

9. $\frac{1}{6}y - x \geq -3$
$\frac{1}{5}y + x \leq 7$

LESSON 3-3 Optimization with Linear Programming

Then

- You solved systems of linear inequalities by graphing.

Now

1. Find the maximum and minimum values of a function over a region.
2. Solve real-world optimization problems using linear programming.

Why?

- An electronics company produces digital audio players and phones. A sign on the company bulletin board is shown.

If at least 2000 items must be produced per shift, how many of each type should be made to minimize costs?

The company is experiencing limitations, or constraints, on production caused by customer demand, shipping, and the productivity of their factory. A system of inequalities can be used to represent these constraints.

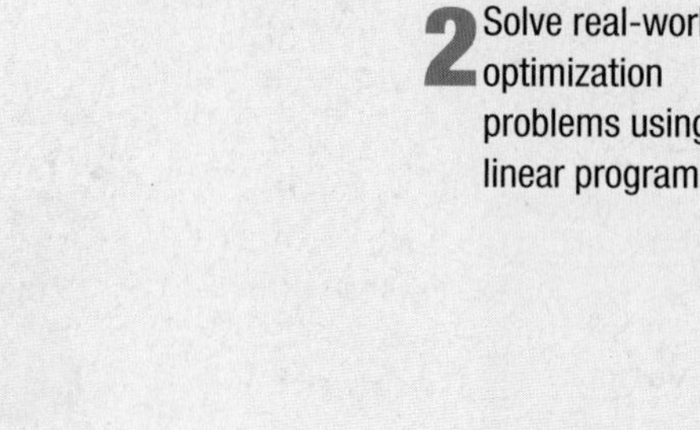

Keeping Costs Down: We Can Do It!

Our Goal: Production per Shift			
Unit	Minimum	Maximum	Cost per Unit
audio	600	1500	\$55
phone	800	1700	\$95

NewVocabulary
linear programming
feasible region
bounded
unbounded
optimize

Common Core State Standards

Content Standards
A.CED.3 Represent constraints by equations or inequalities, and by systems of equations and/or inequalities, and interpret solutions as viable or nonviable options in a modeling context.

Mathematical Practices
4 Model with mathematics.
8 Look for and express regularity in repeated reasoning.

1 Maximum and Minimum Values

Situations often occur in business in which a company hopes to either maximize profits or minimize costs, and many constraints need to be considered. These issues can often be addressed by the use of systems of inequalities in linear programming.

Linear programming is a method for finding maximum or minimum values of a function over a given system of inequalities with each inequality representing a constraint. After the system is graphed and the vertices of the solution set, called the **feasible region**, are substituted into the function, you can determine the maximum or minimum value.

KeyConcept Feasible Regions

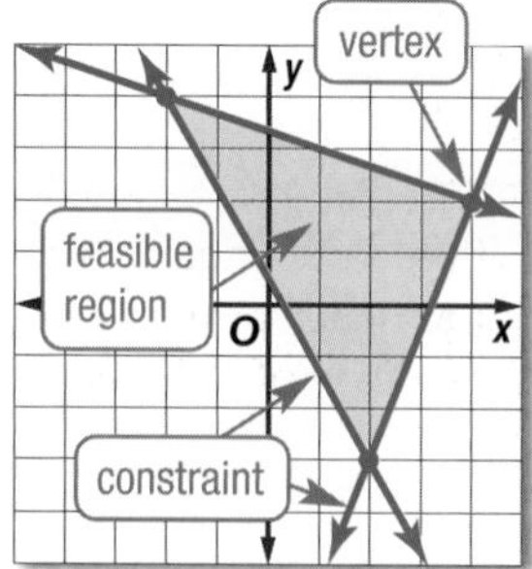

The feasible region is enclosed, or **bounded**, by the constraints. The maximum or minimum value of the related function *always* occurs at a vertex of the feasible region.

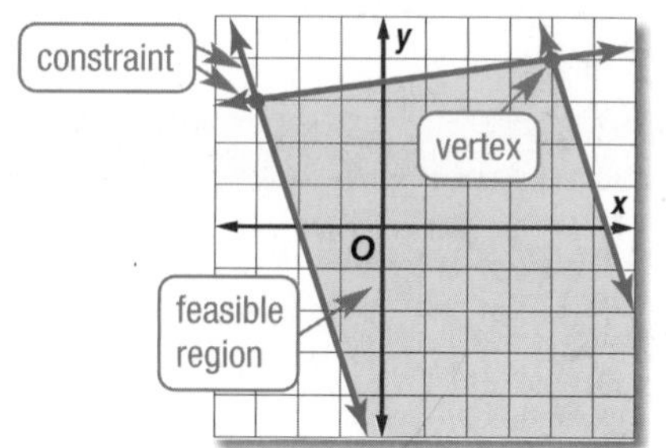

The feasible region is open and can go on forever. It is **unbounded**. Unbounded regions have either a maximum or a minimum.

Bloomberg/Bloomberg Contributor/Getty Images

Example 1 Bounded Region

Graph the system of inequalities. Name the coordinates of the vertices of the feasible region. Find the maximum and minimum values of the function for this region.

$3 \le y \le 6$
$y \le 3x + 12$
$y \le -2x + 6$
$f(x, y) = 4x - 2y$

> **ReadingMath**
> **Function Notation** The notation $f(x, y)$ is used to represent a function with two variables, x and y. It is read *f of x and y*.

Step 1 Graph the inequalities and locate the vertices.

Step 2 Evaluate the function at each vertex.

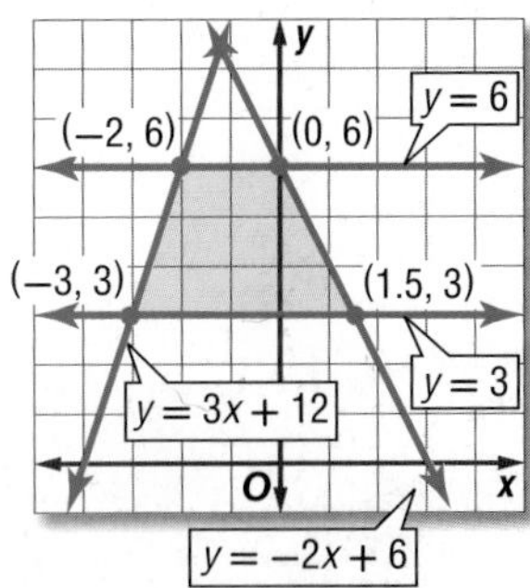

(x, y)	$4x - 2y$	$f(x, y)$	
$(-3, 3)$	$4(-3) - 2(3)$	-18	
$(1.5, 3)$	$4(1.5) - 2(3)$	0	← maximum
$(0, 6)$	$4(0) - 2(6)$	-12	
$(-2, 6)$	$4(-2) - 2(6)$	-20	← minimum

The maximum value is 0 at (1.5, 3). The minimum value is −20 at (−2, 6).

GuidedPractice

1A. $-2 \le x \le 6$
$1 \le y \le 5$
$y \le x + 3$
$f(x, y) = -5x + 2y$

1B. $-6 \le y \le -2$
$y \le -x + 2$
$y \le 2x + 2$
$f(x, y) = 6x + 4y$

When a system of inequalities does not form a closed region, it is unbounded.

Example 2 Unbounded Region

> **WatchOut!**
> **CCSS Precision** Do not assume that there is no maximum if the feasible region is unbounded above the vertices. Test points are needed to determine if there is a minimum or maximum.

Graph the system of inequalities. Name the coordinates of the vertices of the feasible region. Find the maximum and minimum values of the function for this region.

$2y + 3x \ge -12$
$y \le 3x + 12$
$y \ge 3x - 6$
$f(x, y) = 9x - 6y$

Evaluate the function at each vertex.

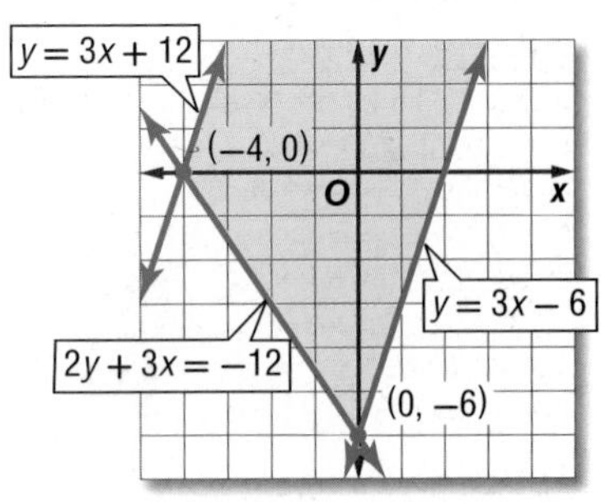

(x, y)	$9x - 6y$	$f(x, y)$
$(-4, 0)$	$9(-4) - 6(0)$	-36
$(0, -6)$	$9(0) - 6(-6)$	36

The maximum value is 36 at (0, −6). There is no minimum value. Notice that another point in the feasible region, (0, 8), yields a value of −48, which is less than −36.

GuidedPractice

2A. $y \le 8$
$y \ge -x + 4$
$y \le -x + 10$
$f(x, y) = -6x + 8y$

2B. $y \ge x - 9$
$y \le -4x + 16$
$y \ge -4x - 4$
$f(x, y) = 10x + 7y$

2 Optimization

To **optimize** means to seek the best price or amount to minimize costs or maximize profits. This is often obtained with the use of linear programming.

KeyConcept Optimization with Linear Programming

Step 1 Define the variables.

Step 2 Write a system of inequalities.

Step 3 Graph the system of inequalities.

Step 4 Find the coordinates of the vertices of the feasible region.

Step 5 Write a linear function to be maximized or minimized.

Step 6 Substitute the coordinates of the vertices into the function.

Step 7 Select the greatest or least result. Answer the problem.

When using a system of inequalities to describe constraints in real-world problems, often only whole-number solutions will make sense.

Real-World Example 3 Optimization with Linear Programming

BUSINESS Refer to the application at the beginning of the lesson. Determine how many of each type of device should be made per shift.

Step 1 Let a = number of audio players produced. Let p = number of phones produced.

Step 2

$600 \leq a \leq 1500$

$800 \leq p \leq 1700$

$a + p \geq 2000$

Steps 3 and 4 The system is graphed at the right. Note the vertices of the feasible region.

Number of Phones; 2000; 1600; 1200; 800; 400; 0; 400; 800; 1200; 1600; 2000; Number of Audio Players; p; a; (600, 1700); (1500, 1700); (600, 1400); (1500, 800); (1200, 800)

Step 5 The function to be minimized is $f(a, p) = 55a + 95p$.

Step 6

(a, p)	$55a + 95p$	$f(a, p)$	
(600, 1700)	55(600) + 95(1700)	194,500	
(600, 1400)	55(600) + 95(1400)	166,000	
(1500, 1700)	55(1500) + 95(1700)	244,000	← maximum
(1500, 800)	55(1500) + 95(800)	158,500	
(1200, 800)	55(1200) + 95(800)	142,000	← minimum

Step 7 Produce 1200 audio players and 800 phones to minimize costs.

GuidedPractice

3. **JEWELRY** Each week, Mackenzie can make 10 to 25 necklaces and 15 to 40 pairs of earrings. If she earns profits of $3 on each pair of earrings and $5 on each necklace, and she plans to sell at least 30 pieces of jewelry, how can she maximize profit?

Real-WorldCareer

Operations Manager Operations management is an area of business that is concerned with the production of goods and services, and involves the responsibility of ensuring that business operations are efficient and effective. A master's degree in business and experience in operations are preferred.

StudyTip

Reasonableness Check your solutions for reasonableness by thinking of the context of the problem.

Check Your Understanding

⬤ = Step-by-Step Solutions begin on page R14.

Examples 1–2 **Graph each system of inequalities. Name the coordinates of the vertices of the feasible region. Find the maximum and minimum values of the given function for this region.**

1. $y \leq 5$
$x \leq 4$
$y \geq -x$
$f(x, y) = 5x - 2y$

2. $y \leq -3x + 6$
$-y \leq x$
$y \leq 3$
$f(x, y) = 8x + 4y$

3. $y \geq -3x + 2$
$9x + 3y \leq 24$
$y \geq -4$
$f(x, y) = 2x + 14y$

4. $-2 \leq y \leq 6$
$3y \leq 4x + 26$
$y \leq -2x + 2$
$f(x, y) = -3x - 6y$

5. $-3 \leq y \leq 7$
$4y \geq 4x - 8$
$6y + 3x \leq 24$
$f(x, y) = -12x + 9y$

6. $y \leq 2x + 6$
$y \geq 2x - 8$
$y \geq -2x - 18$
$f(x, y) = 5x - 4y$

Example 3

7. **CCSS PRECISION** The total number of workers' hours per day available for production in a skateboard factory is 85 hours. There are 40 hours available for finishing decks and quality control each day. The table shows the number of hours needed in each department for two different types of skateboards.

Skateboard Manufacturing Time		
Board Type	**Production Time**	**Deck Finishing/Quality control**
Pro Boards	1.5 hours	2 hours
Specialty Boards	1 hour	0.5 hour

a. Write a system of inequalities to represent the situation.

b. Draw the graph showing the feasible region.

c. List the coordinates of the vertices of the feasible region.

d. If the profit on a pro board is \$50 and the profit on a specialty board is \$65, write a function for the total profit on the skateboards.

e. Determine the number of each type of skateboard that needs to be made to have a maximum profit. What is the maximum profit?

Practice and Problem Solving

Extra Practice is on page R3.

Examples 1–2 **Graph each system of inequalities. Name the coordinates of the vertices of the feasible region. Find the maximum and minimum values of the given function for this region.**

8. $1 \leq y \leq 4$
$4y - 6x \geq -32$
$2y \geq -x + 4$
$f(x, y) = -6x + 3y$

9. $2 \geq x \geq -3$
$y \geq -2x - 6$
$4y \leq 2x + 32$
$f(x, y) = -4x - 9y$

10. $-2 \leq x \leq 4$
$5 \leq y \leq 8$
$2x + 3y \leq 26$
$f(x, y) = 8x - 10y$

11. $-8 \leq y \leq -2$
$y \leq x$
$y \leq -3x + 10$
$f(x, y) = 5x + 14y$

12. $x + 4y \geq 2$
$2x + 4y \leq 24$
$2 \leq x \leq 6$
$f(x, y) = 6x + 7y$

13. $3 \leq y \leq 7$
$2y + x \leq 8$
$y - 2x \leq 23$
$f(x, y) = -3x + 5y$

Examples 1–2 **Graph each system of inequalities. Name the coordinates of the vertices of the feasible region. Find the maximum and minimum values of the given function for this region.**

14. $-9 \leq x \leq -3$
$-9 \leq y \leq -5$
$3y + 12x \leq -75$
$f(x, y) = 20x + 8y$

15. $x \geq -8$
$3x + 6y \leq 36$
$2y + 12 \geq 3x$
$f(x, y) = 10x - 6y$

16. $y \geq |x - 2|$
$y \leq 8$
$8y + 5x \leq 49$
$f(x, y) = -5x - 15y$

17. $x \geq -6$
$y + x \leq -1$
$2x + 3y \geq -9$
$f(x, y) = -10x - 12y$

18. $-5 \geq y \geq -17$
$y \leq 3x + 19$
$y \leq -4x + 15$
$f(x, y) = 8x - 3y$

19. $-8 \leq x \leq 16$
$y \geq 2x - 10$
$2y + x \leq 80$
$f(x, y) = 12x + 15y$

20. $y \leq x + 4$
$y \geq x - 4$
$y \leq -x + 10$
$y \geq -x - 10$
$f(x, y) = -10x + 9y$

21. $-4 \leq x \leq 8$
$-8 \leq y \leq 6$
$y \geq x - 6$
$4y + 7x \leq 31$
$f(x, y) = 12x + 8y$

22. $y \geq |x + 1| - 2$
$0 \leq y \leq 6$
$-6 \leq x \leq 2$
$x + 3y \leq 14$
$f(x, y) = 5x + 4y$

Example 3

23. COOKING Jenny's Bakery makes two types of birthday cakes: yellow cake, which sells for \$25, and strawberry cake, which sells for \$35. Both cakes are the same size, but the decorating and assembly time required for the yellow cake is 2 hours, while the time is 3 hours for the strawberry cake. There are 450 hours of labor available for production. How many of each type of cake should be made to maximize revenue?

24. BUSINESS The manager of a travel agency is printing brochures and fliers to advertise special discounts on vacation spots during the summer months. Each brochure costs \$0.08 to print, and each flier costs \$0.04 to print. A brochure requires 3 pages, and a flier requires 2 pages. The manager does not want to use more than 600 pages, and she needs at least 50 brochures and 150 fliers. How many of each should she print to minimize the cost?

25. CCSS PRECISION Sean has 20 days to paint as many play houses and sheds as he is able. The sheds can be painted at a rate of 2.5 per day, and the play houses can be painted at a rate of 2 per day. He has 45 structures that need to be painted.

a. Write a system of inequalities to represent the possible ways Sean can paint the structures.

b. Draw a graph showing the feasible region and list the coordinates of the vertices of the feasible region.

c. If the profit is \$26 per shed and \$30 per play house, how many of each should he paint?

d. What is the maximum profit?

26. MOVIES Employees at a local movie theater work 8-hour shifts from noon to 8 P.M. or from 4 P.M. to midnight. The table below shows the number of employees needed and their corresponding pay. Find the numbers of day-shift workers and night-shift workers that should be scheduled to minimize the cost. What is the minimal cost?

Time	noon to 4 P.M.	4 P.M. to 8 P.M.	8 P.M. to midnight
Number of Employees Needed	at least 5	at least 14	6
Rate per Hour	\$5.50	\$7.50	\$7.50

27. BUSINESS Each car on a freight train can hold 4200 pounds of cargo and has a capacity of 480 cubic feet. The freight service handles two types of packages: small—which weigh 25 pounds and are 3 cubic feet each, and large—which are 50 pounds and are 5 cubic feet each. The freight service charges \$5 for each small package and \$8 for each large package.

a. Find the number of each type of package that should be placed on a train car to maximize revenue.

b. What is the maximum revenue per train car?

c. In this situation, is maximizing the revenue necessarily the best thing for the company to do? Explain.

28. RECYCLING A recycling plant processes used plastic into food or drink containers. The plant processes up to 1200 tons of plastic per week. At least 300 tons must be processed for food containers, while at least 450 tons must be processed for drink containers. The profit is \$17.50 per ton for processing food containers and \$20 per ton for processing drink containers. What is the profit if the plant maximizes processing?

H.O.T. Problems Use Higher-Order Thinking Skills

29. OPEN ENDED Create a set of inequalities that forms a bounded region with an area of 20 units^2 and lies only in the fourth quadrant.

30. CHALLENGE Find the area of the bounded region formed by the following constraints: $y \geq |x| - 3$, $y \leq -|x| + 3$, and $x \geq |y|$.

31. CCSS ARGUMENTS Identify the system of inequalities that is not the same as the other three. Explain your reasoning.

a.

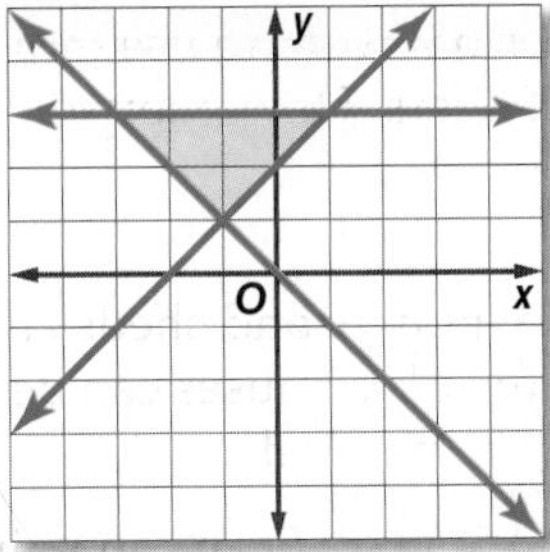

c.

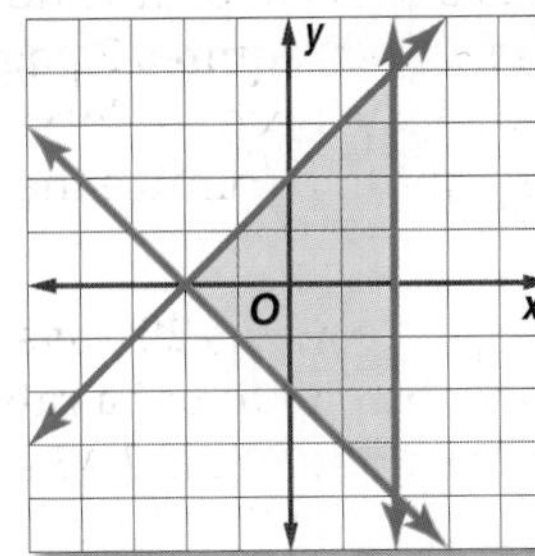

b.

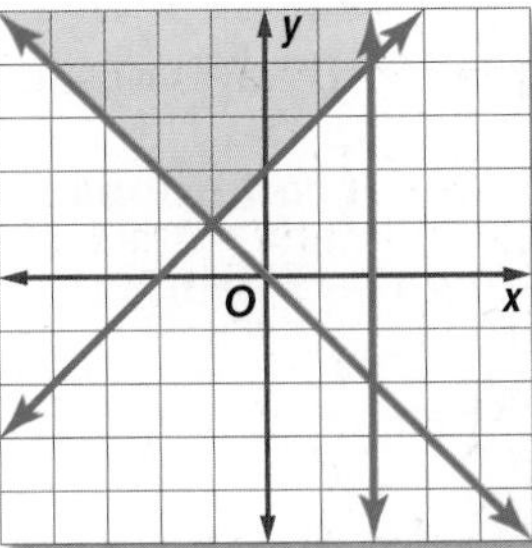

d.

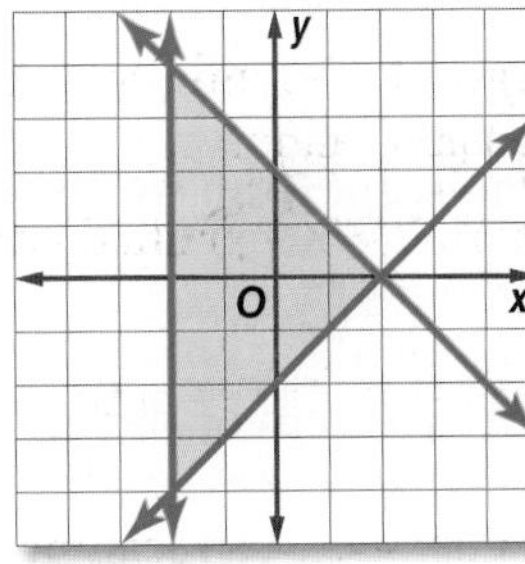

32. REASONING Determine whether the following statement is *sometimes*, *always*, or *never* true. Explain your reasoning.

An unbounded region will not have both a maximum and minimum value.

33. WRITING IN MATH Upon determining a bounded feasible region, Ayumi noticed that vertices $A(-3, 4)$ and $B(5, 2)$ yielded the same maximum value for $f(x, y) = 16y + 4x$. Kelvin confirmed that his constraints were graphed correctly and his vertices were correct. Then he said that those two points were not the only maximum values in the feasible region. Explain how this could have happened.

Standardized Test Practice

34. Kelsey worked 350 hours during the summer and earned $2978.50. She earned $6.85 per hour when she worked at a video store and $11 per hour as an architectural intern. Let x represent the number of hours she worked at the video store and y represent the number of hours that she interned. Which system of equations represents this situation?

A $x + y = 350$
$11x + 6.85y = 2978.50$

B $x + y = 350$
$6.85x + 11y = 2978.50$

C $x + y = 2978.50$
$6.85x + 11y = 350$

D $x + y = 2978.50$
$11x + 6.85y = 350$

35. **SHORT RESPONSE** A family of four went out to dinner. Their bill, including tax, was $60. They left a 17% tip on the total cost of their bill. What is the total cost of the dinner including tip?

36. **SAT/ACT** For a game she is playing, Liz must draw a card from a deck of 26 cards, one with each letter of the alphabet on it, and roll a die. What is the probability that Liz will draw a letter in her name and roll an odd number?

F $\frac{2}{3}$

G $\frac{1}{13}$

H $\frac{3}{52}$

J $\frac{1}{26}$

K $\frac{1}{52}$

37. **GEOMETRY** Which of the following best describes the graphs of $y = 3x - 5$ and $4y = 12x + 16$?

A The lines have the same y-intercept.
B The lines have the same x-intercept.
C The lines are perpendicular.
D The lines are parallel.

Spiral Review

Solve each system of inequalities by graphing. (Lesson 3-2)

38. $3x + 2y \geq 6$
$4x - y \geq 2$

39. $4x - 3y < 7$
$2y - x < -6$

40. $3y \leq 2x - 8$
$y \geq \frac{2}{3}x - 1$

41. **BUSINESS** Last year the chess team paid $7 per hat and $15 per shirt for a total purchase of $330. This year they spent $360 to buy the same number of shirts and hats because the hats now cost $8 and the shirts cost $16. Write and solve a system of two equations that represents the number of hats and shirts bought each year. (Lesson 3-1)

Write an equation in slope-intercept form for the line that satisfies each set of conditions. (Lesson 2-4)

42. passes through (5, 1) and (8, −4)

43. passes through (−3, 5) and (3, 2)

Find the x-intercept and the y-intercept of the graph of each equation. Then graph the equation. (Lesson 2-2)

44. $5x + 3y = 15$

45. $2x - 6y = 12$

46. $3x - 4y - 10 = 0$

47. $2x + 5y - 10 = 0$

48. $y = x$

49. $y = 4x - 2$

Skills Review

Evaluate each expression if $x = -1$, $y = 3$, and $z = 7$.

50. $x + y + z$

51. $2x - y + 2z$

52. $-x + 4y - 3z$

53. $4x + 2y - z$

54. $5x - y + 4z$

55. $-3x - 3y + 3z$

LESSON 3-4 Systems of Equations in Three Variables

Then

- You solved systems of linear equations in two variables.

Now

1. Solve systems of linear equations in three variables.
2. Solve real-world problems using systems of linear equations in three variables.

Why?

- Seats closest to an amphitheater stage cost \$30. The seats in the next section cost \$25, and lawn seats are \$20. There are twice as many seats in section B as in section A. When all 19,200 seats are sold, the amphitheater makes \$456,000.

A system of equations in three variables can be used to determine the number of seats in each section.

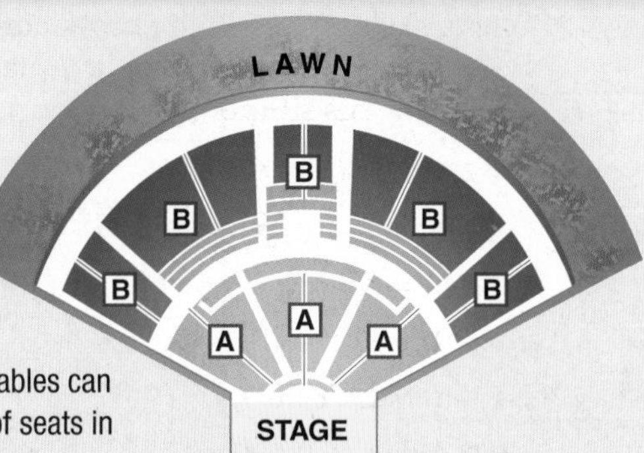

NewVocabulary
ordered triple

Common Core State Standards

Content Standards
A.CED.3 Represent constraints by equations or inequalities, and by systems of equations and/or inequalities, and interpret solutions as viable or nonviable options in a modeling context.

Mathematical Practices
3 Construct viable arguments and critique the reasoning of others.

1 Systems in Three Variables

Like systems of equations in two variables, systems in three variables can have one solution, infinite solutions, or no solution. A solution of such a system is an **ordered triple** (x, y, z).

The graph of an equation in three variables is a three-dimensional graph in the shape of a plane. The graphs of a system of equations in three variables form a system of planes.

One Solution

The three individual planes intersect at a specific point.

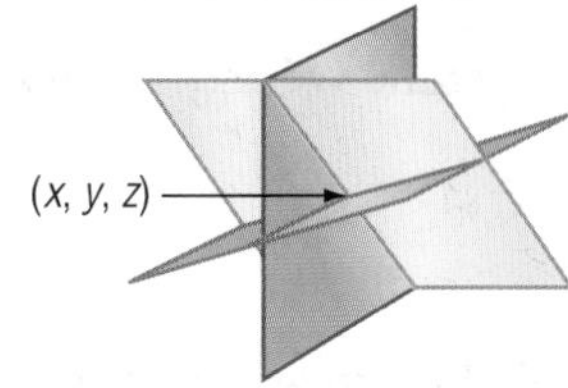

Infinitely Many Solutions

The planes intersect in a line.

Every coordinate on the line represents a solution of the system.

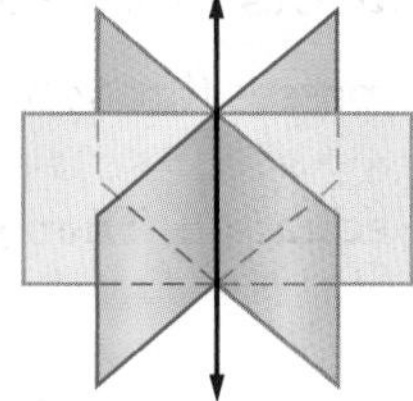

The planes intersect in the same plane.

Every equation is equivalent.
Every coordinate in the plane represents a solution of the system.

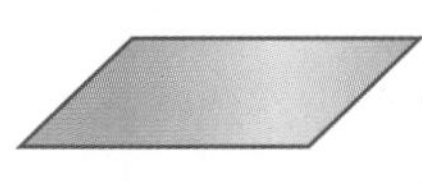

No Solution There are no points in common with all three planes.

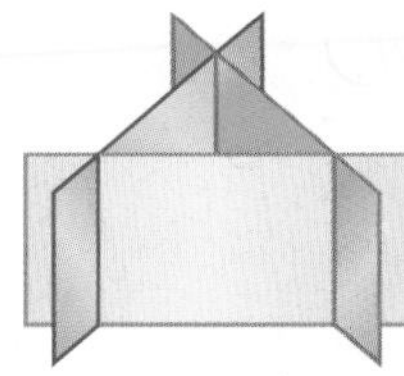

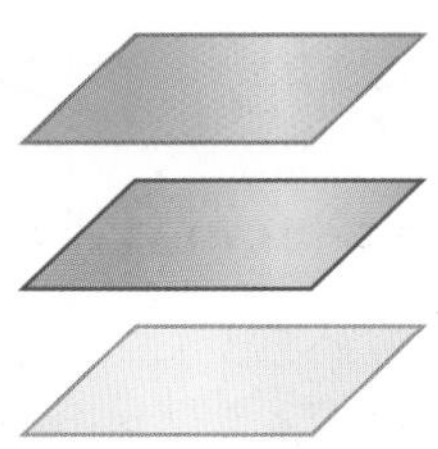

Solving systems of equations in three variables is similar to solving systems of equations in two variables. Use the strategies of substitution and elimination to find the ordered triple that represents the solution of the system.

Example 1 A System with One Solution

Solve the system of equations.

$$\begin{aligned} 3x - 2y + 4z &= 35 \\ -4x + y - 5z &= -36 \\ 5x - 3y + 3z &= 31 \end{aligned}$$

The coefficient of 1 in the second equation makes y a good choice for elimination.

Step 1 Eliminate one variable by using two pairs of equations.

$3x - 2y + 4z = 35$
$-4x + y - 5z = -36$

Multiply by 2.

$$\begin{array}{rl} 3x - 2y + 4z = 35 & \text{Equation 1} \\ (+)\ -8x + 2y - 10z = -72 & \text{Equation 2} \times 2 \\ \hline -5x \quad - 6z = -37 & \end{array}$$

$-4x + y - 5z = -36$
$5x - 3y + 3z = 31$

Multiply by 3.

$$\begin{array}{rl} -12x + 3y - 15z = -108 & \text{Equation 2} \times 3 \\ (+)\ 5x - 3y + 3z = 31 & \text{Equation 1} \\ \hline -7x \quad - 12z = -77 & \end{array}$$

The y-terms in each equation have been eliminated. We now have a system of two equations and two variables, x and z.

Step 2 Solve the system of two equations.

$-5x - 6z = -37$
$-7x - 12z = -77$

Multiply by −2.

$$\begin{array}{rl} 10x + 12z = 74 & \\ (+)\ -7x - 12z = -77 & \\ \hline 3x \quad = -3 & \text{Eliminate } z. \\ x = -1 & \text{Divide by 3.} \end{array}$$

Use substitution to solve for z.

$$\begin{array}{rl} -5x - 6z = -37 & \text{Equation with two variables} \\ -5(-1) - 6z = -37 & \text{Substitution} \\ 5 - 6z = -37 & \text{Multiply.} \\ -6z = -42 & \text{Subtract 5 from each side.} \\ z = 7 & \text{Divide each side by } -6. \end{array}$$

The result is $x = -1$ and $z = 7$.

Step 3 Substitute the two values into one of the original equations to find y.

$$\begin{array}{rl} -4x + y - 5z = -36 & \text{Equation 1} \\ -4(-1) + y - 5(7) = -36 & \text{Substitution} \\ 4 + y - 35 = -36 & \text{Multiply.} \\ y = -5 & \text{Add 31 to each side.} \end{array}$$

CHECK

$$\begin{array}{rl} -4x + y - 5z = -36 & \text{Equation 2} \\ -4(-1) + (-5) - 5(7) \stackrel{?}{=} -36 & x = -1, y = -5, z = 7 \\ 4 + (-5) - 35 \stackrel{?}{=} -36 & \text{Simplify.} \\ -36 = -36\ \checkmark & \end{array}$$

The solution is $(-1, -5, 7)$.

StudyTip

Checking Solutions Always substitute your answer into all of the original equations to confirm your answer.

GuidedPractice

1A. $2x + 4y - 5z = 18$
$-3x + 5y + 2z = -27$
$-5x + 3y - z = -17$

1B. $4x - 3y + 6z = 18$
$-x + 5y + 4z = 48$
$6x - 2y + 5z = 0$

When solving a system of three linear equations with three variables, it is important to check your answer using all three of the original equations. This is necessary because it is possible for a solution to work for two of the equations but not the third.

Example 2 No Solution and Infinite Solutions

Solve each system of equations.

a. $5x + 4y - 5z = -10$
$-4x - 10y - 8z = -16$
$6x + 15y + 12z = 24$

Eliminate x in the second two equations.

$$-4x - 10y - 8z = -16 \xrightarrow{\text{Multiply by 3.}} -12x - 30y - 24z = -48$$
$$6x + 15y + 12z = 24 \xrightarrow{\text{Multiply by 2.}} (+)\ 12x + 30y + 24z = 48$$
$$0 = 0$$

The equation $0 = 0$ is always true. This indicates that the last two equations represent the same plane. Check to see if this plane intersects the first plane.

$$5x + 4y - 5z = -10 \xrightarrow{\text{Multiply by 4.}} 20x + 16y - 20z = -40$$
$$-4x - 10y - 8z = -16 \xrightarrow{\text{Multiply by 5.}} (+)\ -20x - 50y - 40z = -80$$
$$-34y - 60z = -120$$

The planes intersect in a line. So, there are an infinite number of solutions.

b. $-6a + 9b - 12c = 21$
$-2a + 3b - 4c = 7$
$10a - 15b + 20c = -30$

Eliminate a in the first two equations.

$$-6a + 9b - 12c = 21 \qquad -6a + 9b - 12c = 21$$
$$-2a + 3b - 4c = 7 \xrightarrow{\text{Multiply by }-3.} (+)\ 6a - 9b + 12c = -21$$
$$0 = 0$$

The equation $0 = 0$ is always true. This indicates that the first two equations represent the same plane. Check to see if this plane intersects the last plane.

$$-2a + 3b - 4c = 7 \xrightarrow{\text{Multiply by 5.}} -10a + 15b - 20c = 35$$
$$10a - 15b + 20c = -30 \qquad (+)\ 10a - 15b + 20c = -30$$
$$0 = 5$$

The equation $0 = 5$ is never true. So, there is no solution of this system.

Study Tip

Infinite and No Solutions When solving a system of more than two equations, $0 = 5$ will always yield no solution, while $0 = 0$ may not always yield infinite solutions.

Guided Practice

2A. $-4x - 2y - z = 15$
$12x + 6y + 3z = 45$
$2x + 5y + 7z = -29$

2B. $3x + 5y - 2z = 13$
$-5x - 2y - 4z = 20$
$-14x - 17y + 2z = -19$

2 Real-World Problems

When solving problems involving three variables, use the four-step plan to help organize the information. Identify the three variables and what they represent. Then use the information in the problem to form equations using the variables. Once you have three equations and all three variables are represented, you can solve the problem.

Real-WorldCareer

Music Management Music management involves acting as the talent manager to the artists. Other duties include negotiating with record labels, music promoters, and tour promoters. Managers usually earn a percentage of the artist's income. A bachelor's degree is usually required.

Real-World Example 3 Write and Solve a System of Equations

CONCERTS Refer to the beginning of the lesson. Write and solve a system of equations to determine how many seats are in each section of the amphitheater.

Understand Define the variables.

x = seats in section A
y = seats in section B
z = lawn seats

Plan There are 19,200 seats.

$x + y + z = 19{,}200$

The total revenue is $456,000. — Equation 1

$30x + 25y + 20z = 456{,}000$ — Equation 2

There are twice as many seats in section B as in section A.

$y = 2x$ — Equation 3

Solve Solve the system.

Step 1 Substitute $y = 2x$ in the first two equations.

$x + y + z = 19{,}200$	Equation 1
$x + 2x + z = 19{,}200$	$y = 2x$
$3x + z = 19{,}200$	Add.
$30x + 25y + 20z = 456{,}000$	Equation 2
$30x + 25(2x) + 20z = 456{,}000$	$y = 2x$
$80x + 20z = 456{,}000$	Simplify.

Step 2 Solve the system of two equations in two variables.

$3x + z = 19{,}200$ — Multiply by −20. → $-60x - 20z = -384{,}000$

$80x + 20z = 456{,}000$ → $(+)\ 80x + 20z = 456{,}000$

$20x = 72{,}000$

$x = 3600$

Step 3 Substitute to find z.

$3x + z = 19{,}200$	Remaining equation in two variables
$3(3600) + z = 19{,}200$	$x = 3600$
$10{,}800 + z = 19{,}200$	Distribute.
$z = 8400$	Simplify.

Step 4 Substitute to find y.

$y = 2x$	Equation 3
$y = 2(3600)$ or 7200	$x = 3600$

The solution is (3600, 7200, 8400). There are 3600 seats in section A, 7200 in section B, and 8400 lawn seats.

Check Substitute the values into either of the first two equations.

GuidedPractice

3. Ms. Garza invested $50,000 in three different accounts. She invested three times as much money in an account that paid 8% interest than an account that paid 10% interest. The third account earned 12% interest. If she earned a total of $5160 in interest in a year, how much did she invest in each account?

Check Your Understanding

= Step-by-Step Solutions begin on page R14.

Examples 1–2 **Solve each system of equations.**

1. $-3a - 4b + 2c = 28$
$a + 3b - 4c = -31$
$2a + 3c = 11$

2. $3y - 5z = -23$
$4x + 2y + 3z = 7$
$-2x - y - z = -3$

3. $3x + 6y - 2z = -6$
$2x + y + 4z = 19$
$-5x - 2y + 8z = 62$

4. $-4r - s + 3t = -9$
$3r + 2s - t = 3$
$r + 3s - 5t = 29$

5. $3x + 5y - z = 12$
$-2x - 3y + 5z = 14$
$4x + 7y + 3z = 38$

6. $2a - 3b + 5c = 58$
$-5a + b - 4c = -51$
$-6a - 8b + c = 22$

Example 3

7. DOWNLOADING Heather downloaded some television shows. A sitcom uses 0.3 gigabyte of memory; a drama, 0.6 gigabyte; and a talk show, 0.6 gigabyte. She downloaded 7 programs totaling 3.6 gigabytes. There were twice as many episodes of the drama as the sitcom.

a. Write a system of equations for the number of episodes of each type of show.

b. How many episodes of each show did she download?

Practice and Problem Solving

Extra Practice is on page R3.

Examples 1–2 **Solve each system of equations.**

8. $-5x + y - 4z = 60$
$2x + 4y + 3z = -12$
$6x - 3y - 2z = -52$

9. $4a + 5b - 6c = 2$
$-3a - 2b + 7c = -15$
$-a + 4b + 2c = -13$

10. $-2x + 5y + 3z = -25$
$-4x - 3y - 8z = -39$
$6x + 8y - 5z = 14$

11. $4r + 6s - t = -18$
$3r + 2s - 4t = -24$
$-5r + 3s + 2t = 15$

12. $-2x + 15y + z = 44$
$4x + 3y + 3z = 18$
$-3x + 6y - z = 8$

13. $4x + 2y + 6z = 13$
$-12x + 3y - 5z = 8$
$-4x + 7y + 7z = 34$

14. $8x + 3y + 6z = 43$
$-3x + 5y + 2z = 32$
$5x - 2y + 5z = 24$

15. $-6x - 5y + 4z = 53$
$5x + 3y + 2z = -11$
$8x - 6y + 5z = 4$

16. $-9a + 3b - 2c = 61$
$8a + 7b + 5c = -138$
$5a - 5b + 8c = -45$

17. $2x - y + z = 1$
$x + 2y - 4z = 3$
$4x + 3y - 7z = -8$

18. $x + 2y = 12$
$3y - 4z = 25$
$x + 6y + z = 20$

19. $r - 3s + t = 4$
$3r - 6s + 9t = 5$
$4r - 9s + 10t = 9$

Example 3

20. CCSS **SENSE-MAKING** A friend e-mails you the results of a recent high school swim meet. The e-mail states that 24 individuals placed, earning a combined total of 53 points. First place earned 3 points, second place earned 2 points, and third place earned 1 point. There were as many first-place finishers as second- and third-place finishers combined.

a. Write a system of three equations that represents how many people finished in each place.

b. How many swimmers finished in first place, in second place, and in third place?

c. Suppose the e-mail had said that the athletes scored a combined total of 47 points. Explain why this statement is false and the solution is unreasonable.

21. AMUSEMENT PARKS Nick goes to the amusement park to ride roller coasters, bumper cars, and water slides. The wait for the roller coasters is 1 hour, the wait for the bumper cars is 20 minutes long, and the wait for the water slides is only 15 minutes long. Nick rode 10 total rides during his visit. Because he enjoys roller coasters the most, the number of times he rode the roller coasters was the sum of the times he rode the other two rides. If Nick waited in line for a total of 6 hours and 20 minutes, how many of each ride did he go on?

22. BUSINESS Ramón usually gets one of the routine maintenance options at Annie's Garage. Today however, he needs a different combination of work than what is listed.

a. Assume that the price of an option is the same price as purchasing each item separately. Find the prices for an oil change, a radiator flush, and a brake pad replacement.

b. If Ramón wants his brake pads replaced and his radiator flushed, how much should he plan to spend?

23. FINANCIAL LITERACY Kate invested $100,000 in three different accounts. If she invested $30,000 more in account A than account C and is expected to earn $6300 in interest, how much did she invest in each account?

Account	Expected Interest
A	4%
B	8%
C	10%

H.O.T. Problems Use Higher-Order Thinking Skills

24. CCSS REASONING Write a system of equations to represent the three rows of figures below. Use the system to find the number of red triangles that will balance one green circle.

25. CHALLENGE The general form of an equation for a parabola is $y = ax^2 + bx + c$, where (x, y) is a point on the parabola. If three points on a parabola are $(2, -10)$, $(-5, -101)$, and $(6, -90)$, determine the values of a, b, and c and write the general form of the equation.

26. PROOF Consider the following system and prove that if $b = c = -a$, then $ty = a$.

$$rx + ty + vz = a$$
$$rx - ty + vz = b$$
$$rx + ty - vz = c$$

27. OPEN ENDED Write a system of three linear equations that has a solution of $(-5, -2, 6)$. Show that the ordered triple satisfies all three equations.

28. REASONING Use the diagrams of solutions of systems of equations on page 167 to consider a system of inequalities in three variables. Describe the solution of such a system.

29. WRITING IN MATH Use your knowledge of solving a system of three linear equations with three variables to explain how to solve a system of four equations with four variables.

EXPLORE 3-5

Spreadsheet Lab: Organizing Data with Matrices

A **matrix** is a rectangular array of variables or constants in rows and columns, usually enclosed in brackets. In a matrix, the numbers or data are organized so that each position in the matrix has a purpose. Each value in the matrix is called an **element**. A matrix is usually named using an uppercase letter.

$$A = \begin{bmatrix} 8 & -2 & 5 & 6 \\ -1 & 3 & -3 & 6 \\ 7 & -8 & 1 & 4 \end{bmatrix}$$

The element -1 is in Row 2, Column 1, depicted by a_{21}.

The element -8 is in Row 3, Column 2, depicted by a_{32}.

3 rows

4 columns

A matrix can be described by its **dimensions**. A matrix with m rows and n columns is an $m \times n$ matrix (read "m by n"). Matrix A above is a 3×4 matrix because it has 3 rows and 4 columns. a_{12} refers to an element of A, whereas b_{12} refers to an element of B.

Activity 1 Organize Data in a Matrix

FOOTBALL The West High School football team used five running backs throughout its season. Coach Williams wanted to compare the statistics of each player.

Joey: 11 games, 72 attempts, 439 yards, 6.10 average, 8 TDs

DeShawn: 9 games, 143 attempts, 1024 yards, 7.16 average, 12 TDs

Dario: 11 games, 164 attempts, 885 yards, 5.40 average, 15 TDs

Leo: 11 games, 84 attempts, 542 yards, 6.45 average, 7 TDs

Alex: 10 games, 151 attempts, 966 yards, 6.40 average, 11 TDs

a. Organize the data in a matrix from greatest to least attempts.

Player	Games	Attempts	Yards	Average	TDs
Dario	11	164	885	5.40	15
Alex	10	151	966	6.40	11
DeShawn	9	143	1024	7.16	12
Leo	11	84	542	6.45	7
Joey	11	72	439	6.10	8

b. What are the dimensions of the matrix? What value is a_{34}?

There are five rows and five columns, so the dimensions are 5×5. The value a_{34}, which is in the third row and fourth column, is 7.16.

$$\begin{bmatrix} 11 & 164 & 885 & 5.40 & 15 \\ 10 & 151 & 966 & 6.40 & 11 \\ 9 & 143 & 1024 & \boxed{7.16} & 12 \\ 11 & 84 & 542 & 6.45 & 7 \\ 11 & 72 & 439 & 6.10 & 8 \end{bmatrix}$$

Exercise

1. Julie is shopping for a new smartphone and discovers that many different options are available.
   a. Organize the data in a matrix. List the options in descending order, and label the columns options, price, memory, color, and interface respectively.
   b. What are the dimensions of the matrix? What is the value of a_{23}?

Option 1
$420
Memory: 512, Color: 24, Interface: infrared

Option 2
$399
Memory: 512, Color: 24, Interface: Bluetooth

Option 3
$315
Memory: 256, Color: 24, Interface: infrared

Option 4
$289
Memory: 128, Color: 18, Interface: wi-fi

(continued on the next page)

Spreadsheet Lab

Organizing Data with Matrices *Continued*

People in the workforce often use computer **spreadsheets** to organize, display, and analyze data. Similar to a matrix, data in a spreadsheet are entered into rows and columns. Then the data can be used to create graphs or perform calculations.

Activity 2 Organize Data in a Spreadsheet

The manager of a gourmet food store has gathered data on the number of pounds of bulk coffees they have sold each week in January. Enter the data into a spreadsheet.

Weekly Sales for January				
Coffee	1/5	1/12	1/19	1/26
Hawaiian Kona	17	22	11	23
Mocha Java	31	34	22	29
House Blend	55	61	44	71
Espresso	41	36	60	77
Decaf Espresso	23	29	19	44
Breakfast Blend	8	18	19	31
Decaf Breakfast Blend	22	18	30	32
Organic Italian Roast	26	16	31	39

Use Column A for the types of coffee, Column B for the sales in the week starting 1/5, Column C for sales in the week starting 1/12, and Columns D and E for the sales in the weeks starting 1/19 and 1/26.

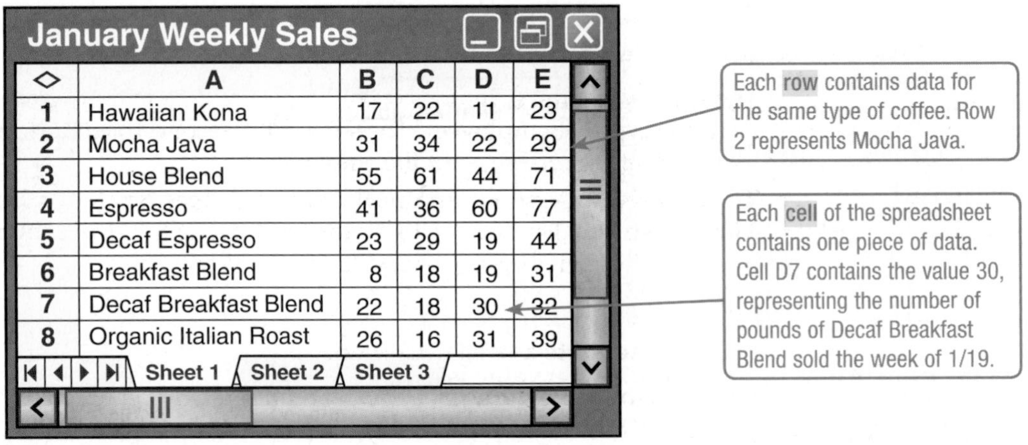

Exercises

2. Refer to Activity 2. A SUM formula allows you to find the sum of the entries in a column or row.

a. The formula =SUM(B1:B8) finds the sum of column B. Enter formulas in cells B9, C9, D9, and E9 to find the sums of those columns. What do the sums of the columns represent in the situation?

b. Enter formulas in cells F1 through F8 to find the sums of rows 1 through 8. What do the sums of the rows represent in the situation?

c. Find the sum of row 9 and the sum of column F. What do you observe? Explain.

3. Enter the data on smartphones from Exercise 1 into a spreadsheet.

4. Compare and contrast how data are organized in a spreadsheet and in a matrix.

LESSON 3-5

Operations with Matrices

Then	Now	Why?
You organized data into matrices.	**1** Analyze data in matrices. **2** Perform algebraic operations with matrices.	Coastal Sales Company has three locations in Florida. The matrices below show the average daily wages and annual sales of all of the representatives.

$$\begin{array}{l} \\ \\ \text{Entry} \\ \text{Assistant} \\ \text{Associate} \end{array} \begin{array}{c} \text{Miami} \\ \text{Wages} \quad \text{Sales} \\ \begin{bmatrix} 900 & 145{,}000 \\ 2400 & 225{,}000 \\ 2700 & 290{,}000 \end{bmatrix} \end{array} \begin{array}{c} \text{Tampa} \\ \text{Wages} \quad \text{Sales} \\ \begin{bmatrix} 900 & 122{,}000 \\ 1800 & 145{,}500 \\ 1800 & 160{,}000 \end{bmatrix} \end{array} \begin{array}{c} \text{Tallahassee} \\ \text{Wages} \quad \text{Sales} \\ \begin{bmatrix} 1050 & 109{,}500 \\ 1800 & 135{,}000 \\ 1800 & 150{,}500 \end{bmatrix} \end{array}$$

NewVocabulary
scalar
scalar multiplication

1 Analyze Data

Data that are organized in a matrix can be analyzed and interpreted. Sometimes, further analysis is needed. Other times, the data are meaningless.

Real-World Example 1 Analyze Data with Matrices

BUSINESS The manager at the Miami location would like to use their matrix to further analyze the representatives.

$$\begin{array}{l} \\ \text{Entry} \\ \text{Assistant} \\ \text{Associate} \end{array} \begin{array}{c} \text{Wages} \quad \text{Sales} \\ \begin{bmatrix} 900 & 145{,}000 \\ 2400 & 225{,}000 \\ 2700 & 290{,}000 \end{bmatrix} \end{array}$$

a. Add the elements in each column and interpret the results.

The sum of the first column is 6000. This is the total average daily wages of the three types of employees. The sum of the second column is 660,000. This is the total average annual sales of the employees.

b. The manager wants to determine the average wages for all of the employees at the Miami location. He decides to divide the total of the first column by three, the number of different positions. What is the average?

The average is $6000 \div 3$ or 2000.

c. Is this an accurate average? Explain.

If there is the same number of each type of representative, then the average is accurate. If there is more of one type of representative than the others, then the average is not accurate and would need to be weighted accordingly.

d. Would adding the rows provide any meaningful data for the manger?

No. The sum of a row includes two different forms of data, wages and sales.

GuidedPractice

1. **POPULATION** The table displays U.S. Census data.
   A. Organize the data in a matrix.
   B. Add the elements in the columns and interpret the results.
   C. Add the elements in the rows and interpret the results.
   D. Would finding the average of the rows or columns provide any meaningful data?

Latino Population in the U.S. (millions)

Age	Male	Female
0–19	7.1	6.6
20–39	6.8	5.9
40–59	3.2	2.2
60+	1.1	1.4

2 Algebraic Operations Several algebraic operations can be performed on data that are organized in matrices. Matrices can be added or subtracted if and only if they have the same dimensions.

KeyConcept Adding and Subtracting Matrices

Words To add or subtract two matrices with the same dimensions, add or subtract their corresponding elements.

Symbols

$$\underset{A}{\begin{bmatrix} a & b \\ c & d \end{bmatrix}} + \underset{B}{\begin{bmatrix} e & f \\ g & h \end{bmatrix}} = \underset{A + B}{\begin{bmatrix} a + e & b + f \\ c + g & d + h \end{bmatrix}}$$

$$\underset{A}{\begin{bmatrix} a & b \\ c & d \end{bmatrix}} - \underset{B}{\begin{bmatrix} e & f \\ g & h \end{bmatrix}} = \underset{A - B}{\begin{bmatrix} a - e & b - f \\ c - g & d - h \end{bmatrix}}$$

Example

$$\begin{bmatrix} 3 & -5 \\ 1 & 7 \end{bmatrix} + \begin{bmatrix} 2 & 0 \\ -9 & 10 \end{bmatrix} = \begin{bmatrix} 3 + 2 & -5 + 0 \\ 1 + (-9) & 7 + 10 \end{bmatrix}$$

Example 2 Add and Subtract Matrices

StudyTip

Corresponding Elements Elements are *corresponding* if they are in the exact same position in each matrix.

Find each of the following for $A = \begin{bmatrix} 16 & 2 \\ -9 & 8 \end{bmatrix}$, $B = \begin{bmatrix} -4 & -1 \\ -3 & -7 \end{bmatrix}$, and $C = \begin{bmatrix} 8 \\ 6 \end{bmatrix}$.

a. $A + B$

$A + B = \begin{bmatrix} 16 & 2 \\ -9 & 8 \end{bmatrix} + \begin{bmatrix} -4 & -1 \\ -3 & -7 \end{bmatrix}$ Substitution

$= \begin{bmatrix} 16 + (-4) & 2 + (-1) \\ -9 + (-3) & 8 + (-7) \end{bmatrix}$ Add corresponding elements.

$= \begin{bmatrix} 12 & 1 \\ -12 & 1 \end{bmatrix}$ Simplify.

b. $B - C$

$B - C = \begin{bmatrix} -4 & -1 \\ -3 & -7 \end{bmatrix} - \begin{bmatrix} 8 \\ 6 \end{bmatrix}$ Substitution

Since the dimensions of B and C are different, you cannot subtract the matrices.

c. $B - A$

$B - A = \begin{bmatrix} -4 & -1 \\ -3 & -7 \end{bmatrix} - \begin{bmatrix} 16 & 2 \\ -9 & 8 \end{bmatrix}$ Substitution

$= \begin{bmatrix} -4 - 16 & -1 - 2 \\ -3 - (-9) & -7 - 8 \end{bmatrix}$ Subtract corresponding elements.

$= \begin{bmatrix} -20 & -3 \\ 6 & -15 \end{bmatrix}$ Simplify.

GuidedPractice

2A. $\begin{bmatrix} -3 & 4 \\ -9 & -5 \end{bmatrix} - \begin{bmatrix} -4 & 12 \\ 8 & -7 \end{bmatrix}$

2B. $\begin{bmatrix} -9 & 8 & 3 \\ -2 & 4 & -7 \end{bmatrix} + \begin{bmatrix} -4 & -3 & 6 \\ -9 & -5 & 18 \end{bmatrix}$

2C. $\begin{bmatrix} 8 & -3 \\ -2 & 0 \\ 1 & 7 \end{bmatrix} - \begin{bmatrix} 5 & 1 & -4 & 2 \\ 10 & -6 & 9 & 0 \end{bmatrix}$

You can multiply any matrix by a constant called a **scalar**. When you do this, you multiply each individual element by the value of the scalar. This operation is called **scalar multiplication.**

ReadingMath

Scalar Think of a scalar as a coefficient for a variable, but instead it is for a matrix.

KeyConcept Multiplying by a Scalar

Words To multiply a matrix by a scalar k, multiply each element by k.

$$k \cdot A = kA$$

Symbols $k\begin{bmatrix} a & b \\ c & d \end{bmatrix} = \begin{bmatrix} ka & kb \\ kc & kd \end{bmatrix}$

Example $-3\begin{bmatrix} 4 & 1 \\ 7 & -2 \end{bmatrix} = \begin{bmatrix} -3(4) & -3(1) \\ -3(7) & -3(-2) \end{bmatrix}$

PT

Example 3 Multiply a Matrix by a Scalar

If $R = \begin{bmatrix} -12 & 8 & 6 \\ -16 & 4 & 19 \end{bmatrix}$, find $5R$.

$5R = 5\begin{bmatrix} -12 & 8 & 6 \\ -16 & 4 & 19 \end{bmatrix}$ Substitution

$= \begin{bmatrix} 5(-12) & 5(8) & 5(6) \\ 5(-16) & 5(4) & 5(19) \end{bmatrix}$ Distribute the scalar.

$= \begin{bmatrix} -60 & 40 & 30 \\ -80 & 20 & 95 \end{bmatrix}$ Multiply.

StudyTip

Scalar Multiplication The bracket of a matrix is treated just like a regular grouping symbol. So when multiplying by a scalar, distribute the same way as with a grouping symbol.

GuidedPractice

3. If $T = \begin{bmatrix} 8 & 0 & 3 & -2 \\ -1 & -4 & -2 & 9 \end{bmatrix}$, find $-4T$.

Many properties of real numbers also hold true for matrices. A summary of these properties is listed below.

KeyConcept Properties of Matrix Operations

For any matrices A, B, and C for which the matrix sum and product are defined and any scalar k, the following properties are true.

Commutative Property of Addition	$A + B = B + A$
Associative Property of Addition	$(A + B) + C = A + (B + C)$
Left Scalar Distributive Property	$k(A + B) = kA + kB$
Right Scalar Distributive Property	$(A + B)k = kA + kB$

Multi-step operations can be performed on matrices. The order of these operations is the same as with real numbers.

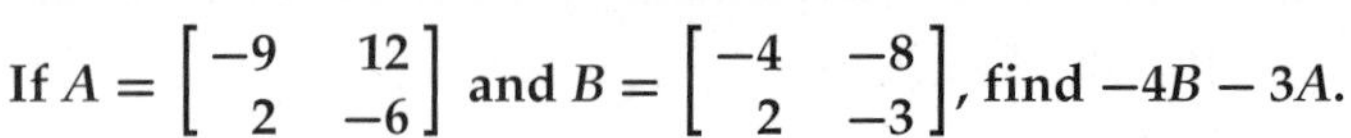

Example 4 Multi-Step Operations

If $A = \begin{bmatrix} -9 & 12 \\ 2 & -6 \end{bmatrix}$ and $B = \begin{bmatrix} -4 & -8 \\ 2 & -3 \end{bmatrix}$, find $-4B - 3A$.

$$-4B - 3A = -4\begin{bmatrix} -4 & -8 \\ 2 & -3 \end{bmatrix} - 3\begin{bmatrix} -9 & 12 \\ 2 & -6 \end{bmatrix}$$ Substitution

$$= \begin{bmatrix} -4(-4) & -4(-8) \\ -4(2) & -4(-3) \end{bmatrix} - \begin{bmatrix} 3(-9) & 3(12) \\ 3(2) & 3(-6) \end{bmatrix}$$ Distribute the scalars in each matrix.

$$= \begin{bmatrix} 16 & 32 \\ -8 & 12 \end{bmatrix} - \begin{bmatrix} -27 & 36 \\ 6 & -18 \end{bmatrix}$$ Multiply.

$$= \begin{bmatrix} 16-(-27) & 32-36 \\ -8-6 & 12-(-18) \end{bmatrix}$$ Subtract corresponding elements.

$$= \begin{bmatrix} 43 & -4 \\ -14 & 30 \end{bmatrix}$$ Simplify.

GuidedPractice

4. If $A = \begin{bmatrix} -5 & 3 \\ 6 & -8 \\ 2 & 9 \end{bmatrix}$ and $B = \begin{bmatrix} 12 & 5 \\ 5 & -4 \\ 4 & -7 \end{bmatrix}$, find $-6B + 7A$.

Real-WorldCareer

Financial Planner Financial planners often use matrices to organize and describe the data they use. Financial planners need a bachelor's degree. Usually they have degrees in accounting, finance, economics, business, marketing, or commerce.

Matrices can be used in many business applications.

Real-World Example 5 Use Multi-Step Operations with Matrices

BUSINESS Refer to the application at the beginning of the lesson. Express the average wages and sales for the entire company for a 5-day week.

To calculate the 5-day sales for the entire company, each matrix needs to be multiplied by 5 and the totals added together.

$$5\begin{bmatrix} 900 & 145{,}000 \\ 2400 & 225{,}000 \\ 2700 & 290{,}000 \end{bmatrix} + 5\begin{bmatrix} 900 & 122{,}000 \\ 1800 & 145{,}500 \\ 1800 & 160{,}000 \end{bmatrix} + 5\begin{bmatrix} 1050 & 109{,}500 \\ 1800 & 135{,}000 \\ 1800 & 150{,}500 \end{bmatrix}$$ Write matrices.

$$= \begin{bmatrix} 4500 & 725{,}000 \\ 12{,}000 & 1{,}125{,}000 \\ 13{,}500 & 1{,}450{,}000 \end{bmatrix} + \begin{bmatrix} 4500 & 610{,}000 \\ 9000 & 727{,}500 \\ 9000 & 800{,}000 \end{bmatrix} + \begin{bmatrix} 5250 & 547{,}500 \\ 9000 & 675{,}000 \\ 9000 & 752{,}500 \end{bmatrix}$$ Multiply scalars.

$$= \begin{array}{r} \\ \text{Entry} \\ \text{Assistant} \\ \text{Associate} \end{array} \begin{array}{c} \begin{array}{cc} \text{Wages} & \text{Sales} \end{array} \\ \begin{bmatrix} 14{,}250 & 1{,}882{,}500 \\ 30{,}000 & 2{,}527{,}500 \\ 31{,}500 & 3{,}002{,}500 \end{bmatrix} \end{array}$$ Add matrices.

The final matrix indicates the average weekly sales and wages for all of the representatives of the company.

StudyTip

Corresponding Elements When representing quantities with multiple matrices, make sure the corresponding elements represent corresponding quantities.

GuidedPractice

5. Use the data above to calculate the average yearly sales and wages for the company, assuming 260 working days.

moodboard/CORBIS

Check Your Understanding

● = Step-by-Step Solutions begin on page R14.

Example 1

1. **CCSS MODELING** Use the table that shows the city and highway gas mileage of five different types of vehicles.

Vehicle	SUV	Mini-van	Sedan	Compact	APV
City	23	21	21	42	61
Highway	25	24	32	49	70

Source: Auto Hoppers

a. Organize the gas mileages in a matrix.
b. Add the elements of each row and interpret the results.
c. Add the elements of each column and interpret the results.

Example 2

Perform the indicated operations. If the matrix does not exist, write *impossible*.

2. $\begin{bmatrix} -8 & 2 & 6 \end{bmatrix} + \begin{bmatrix} 11 & -7 & 1 \end{bmatrix}$

3. $\begin{bmatrix} 9 & -8 & 4 \end{bmatrix} + \begin{bmatrix} 12 & 2 \end{bmatrix}$

4. $\begin{bmatrix} 7 & -12 \\ 15 & 4 \end{bmatrix} - \begin{bmatrix} 9 & 6 \\ 4 & -9 \end{bmatrix}$

5. $\begin{bmatrix} 5 & 13 & -6 \\ 3 & -17 & 2 \end{bmatrix} - \begin{bmatrix} -2 & -18 & 8 \\ 2 & -11 & 0 \end{bmatrix}$

Example 3

Perform the indicated operations. If the matrix does not exist, write *impossible*.

6. $3\begin{bmatrix} 6 & 4 & 0 \\ -2 & 14 & -8 \\ -4 & -6 & 7 \end{bmatrix}$

7. $-6\begin{bmatrix} 15 & -9 & 2 & 3 \\ 6 & -11 & 14 & -2 \\ 4 & -8 & -10 & 27 \end{bmatrix}$

Example 4

Use matrices *A*, *B*, *C*, and *D* to find the following.

$A = \begin{bmatrix} 6 & -4 \\ 3 & -5 \end{bmatrix}$ $B = \begin{bmatrix} 8 & -1 \\ -2 & 7 \end{bmatrix}$ $C = \begin{bmatrix} -4 & -6 \\ 12 & -7 \end{bmatrix}$ $D = \begin{bmatrix} 9 & 6 & 0 \\ -2 & 8 & 0 \end{bmatrix}$

8. $4B + 2A$

9. $-8C + 3A$

10. $-5B - 2D$

11. $-4C - 5B$

Example 5

12. **GRADES** Geraldo, Olivia, and Nikki have had two tests in their math class. The table shows the test grades for each student.

a. Write a matrix for the information from each test.
b. Find the sum of the scores from the two tests expressed as a matrix.
c. Express the difference in scores from test 1 to test 2 as a matrix.

Student	Test 1	Test 2
Geraldo	85	72
Olivia	75	74
Nikki	96	83

Practice and Problem Solving

Extra Practice is on page R3.

Example 1

13. **SHOES** A consumer service company rated several pairs of shoes by cost, level of comfort, look, and longevity using a scale of 1–5, with 1 being low and 5 being high.

a. Write a 4 × 4 matrix to organize this information.
b. Which shoe would you buy based on this information, and why?
c. Would finding the sum of the rows or columns provide any useful information? Explain your reasoning.

Brand	Cost	Comfort	Look	Longevity
A	3	2	2	1
B	4	3	2	3
C	5	5	4	4
D	1	5	5	2

Example 2

Perform the indicated operations. If the matrix does not exist, write *impossible*.

14. $\begin{bmatrix} 12 & -5 \\ -8 & -3 \end{bmatrix} + \begin{bmatrix} -6 & 11 \\ -7 & 2 \end{bmatrix}$

15. $\begin{bmatrix} 9 & 5 \\ -2 & 16 \end{bmatrix} + \begin{bmatrix} -6 & -3 & 7 \\ 12 & 2 & -4 \end{bmatrix}$

Examples 3–4

16. **BUSINESS** The drink menu from a fast-food restaurant is shown at the right. The store owner has decided that all of the prices must be increased by 10%.

Drink	Small	Medium	Large
Soda	$0.95	$1.00	$1.05
Iced tea	$0.75	$0.80	$0.85
Lemonade	$0.75	$0.80	$0.85
Coffee	$1.00	$1.10	$1.20

a. Write matrix C to represent the current prices.

b. What scalar can be used to determine a matrix N to represent the new prices?

c. Find N.

d. What is $N - C$? What does this represent in this situation?

Use matrices A, B, C, and D to find the following.

$$A = \begin{bmatrix} 0 & -7 \\ 8 & 12 \end{bmatrix} \quad B = \begin{bmatrix} 11 & 4 \\ -3 & -17 \end{bmatrix} \quad C = \begin{bmatrix} 8 & 2 & -2 \\ 1 & -9 & 15 \end{bmatrix} \quad D = \begin{bmatrix} -2 & -8 & 0 \\ 4 & 13 & 1 \end{bmatrix}$$

17. $-3B + 2A$

18. $9C - 4D$

19. $2C + 11A$

20. $7A - 2B$

Example 5

21. **CCSS MODELING** Library A has 10,000 novels, 5000 biographies, and 5000 children's books. Library B has 15,000 novels, 10,000 biographies, and 2500 children's books. Library C has 4000 novels, 700 biographies, and 800 children's books.

a. Express each library's number of books as a matrix. Label the matrices A, B, and C.

b. Find the total number of each type of book in all 3 libraries. Express as a matrix.

c. How many more books of each type does Library A have than Library C?

d. Find $A + B$. Does the matrix have meaning in this situation? Explain.

Perform the indicated operations. If the matrix does not exist, write *impossible*.

22. $-2\begin{bmatrix} -9.2 & -8.4 \\ 5.6 & -4.3 \end{bmatrix} - 4\begin{bmatrix} 4.1 & -2.9 \\ 7.2 & -8.2 \end{bmatrix}$

23. $-\frac{3}{4}\begin{bmatrix} 12 & -16 \\ 15 & 8 \end{bmatrix} + \frac{2}{3}\begin{bmatrix} 21 & 18 \\ -4 & -6 \end{bmatrix}$

24. $-3\begin{bmatrix} 18 & -6 & -8 \\ -5 & -3 & 12 \\ 0 & 3x & -y \end{bmatrix}$

25. $8\begin{bmatrix} -a & 4b & c - b \\ -13 & 10 & -5c \end{bmatrix}$

26. $-4\begin{bmatrix} -7 \\ 4 \\ -3 \end{bmatrix} + 3\begin{bmatrix} -8 \\ 3x \\ -9 \end{bmatrix} - 5\begin{bmatrix} 4 \\ x - 6 \\ 12 \end{bmatrix}$

27. $-5\left(\begin{bmatrix} 4 & -8 \\ 8 & -9 \end{bmatrix} + \begin{bmatrix} 4 & -2 \\ -3 & -6 \end{bmatrix}\right)$

28. **WEATHER** The table shows snowfall in inches.

City	Normal Snowfall			2007 Snowfall		
	Jan	Feb	Mar	Jan	Feb	Mar
Grand Rapids, MI	21.1	12.2	9.0	15.4	33.6	13.6
Boston, MA	13.3	11.3	8.3	1.0	4.6	10.2
Buffalo, NY	26.1	17.8	12.4	15.5	33.5	5.4
Pittsburgh, PA	12.3	8.5	7.9	11.3	14.0	9.3

Source: National Weather Service

a. Express the normal snowfall data and the 2007 data in two 4×3 matrices.

b. Subtract the matrix of normal data from the matrix of 2007 data. What does the difference represent in the context of the situation?

c. Explain the meaning of positive and negative numbers in the difference matrix. What trends do you see in the data?

29 **CCSS MODELING** The table shows some of the world, Olympic, and American women's freestyle swimming records.

Distance (m)	World	Olympic	American
50	24.13 s	24.13 s	24.63 s
100	53.52 s	53.52 s	53.99 s
200	1:56.54 min	1:57.65 min	1:57.41 min
800	8:16.22 min	8:19.67 min	8:16.22 min

Source: USA Swimming

a. Find the difference between the American and World records expressed as a matrix.

b. What is the meaning of each row in the column?

c. In which events were the fastest times set at the Olympics?

30. **MULTIPLE REPRESENTATIONS** In this problem, you will investigate using matrices to represent transformations.

a. Algebraic The matrix $\begin{bmatrix} -3 & -4 & 1 \\ 8 & 6 & 0 \end{bmatrix}$ represents a triangle with vertices at $(-3, 8)$, $(-4, 6)$, and $(1, 0)$. Write a matrix to represent $\triangle ABC$.

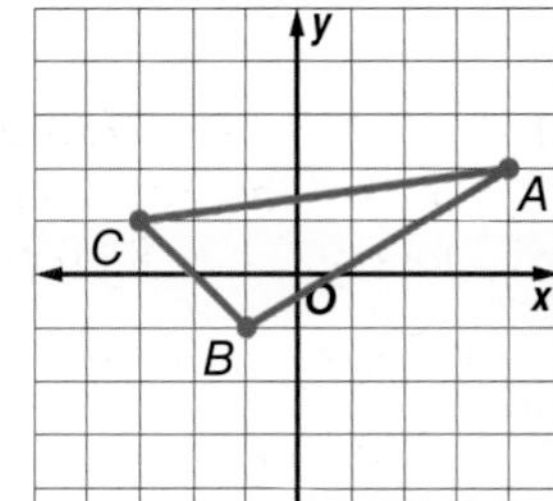

b. Geometric Multiply the matrix you wrote by 2. Then graph the figure represented by the new matrix.

c. Analytical How do the figures compare? Make a conjecture about the result of multiplying the matrix by 0.5. Verify your conjecture.

H.O.T. Problems Use Higher-Order Thinking Skills

31. **PROOF** Prove that matrix addition is commutative for 2×2 matrices.

32. **PROOF** Prove that matrix addition is associative for 2×2 matrices.

33. **CHALLENGE** Find the elements of C if:

$A = \begin{bmatrix} -3 & -4 \\ 8 & 6 \end{bmatrix}$, $B = \begin{bmatrix} 5 & -1 \\ 2 & -4 \end{bmatrix}$, and $3A - 4B + 6C = \begin{bmatrix} 13 & 22 \\ 10 & 4 \end{bmatrix}$.

34. **REASONING** Determine whether each statement is *sometimes, always,* or *never* true for matrices A and B. Explain your reasoning.

a. If $A + B$ exists, then $A - B$ exists.

b. If k is a real number, then kA and kB exist.

c. If $A - B$ does not exist, then $B - A$ does not exist.

d. If A and B have the same number of elements, then $A + B$ exists.

e. If kA exists and kB exists, then $kA + kB$ exists.

35. **OPEN ENDED** Give an example of matrices A and B if $4B - 3A = \begin{bmatrix} -6 & 5 \\ -2 & -1 \end{bmatrix}$.

36. **WRITING IN MATH** Explain how to find $4D - 3C$ for two given matrices, C and D with the same dimensions.

Standardized Test Practice

37. What is the solution of the system of equations?

$$0.06p + 4q = 0.88$$
$$p - q = -2.25$$

A (−0.912, −1.338)
B (0.912, −3.162)
C (−2, 0.25)
D (−2, −4.25)

38. SHORT RESPONSE Find $A + B$ if $A = \begin{bmatrix} -7 & 3 \\ 2 & 6 \end{bmatrix}$ and $B = \begin{bmatrix} 4 & 2 \\ 0 & 1 \end{bmatrix}$.

39. SAT/ACT Solve for x and y.

$$x + 3y = 16$$
$$7 - x = 12$$

F $x = -5, y = 7$
G $x = 7, y = 3$
H $x = 7, y = 5$
J $x = 5, y = 7$
K $x = -5, y = 3$

40. PROBABILITY A local pizzeria offers 5 different meat toppings and 6 different vegetable toppings. You decide to get two vegetable toppings and one meat topping. How many different types of pizzas can you order?

A 60
B 75
C 120
D 150

Spiral Review

Solve each system of equations. (Lesson 3-4)

41. $2x + 3y - z = -1$
$5x + y + 4z = 30$
$-8x - 2y + 5z = -2$

42. $3x - 4y + 6z = 26$
$5x + 3y + 2z = 5$
$-2x + 5y - 3z = -9$

43. $5x + 2y - 4z = 22$
$6x + 3y + 5z = 5$
$-2x - 4y + z = 2$

44. PACKAGING The Cookie Factory sells chocolate chip and peanut butter cookies in combination packages that contain between six and twelve cookies. At least three of each type of cookie should be in each package. How many of each type of cookie should be in each package to maximize the profit? (Lesson 3-3)

Cookie	chocolate chip	peanut butter
Cost	\$0.19	\$0.13
Price	\$0.44	\$0.39

Solve each system of inequalities by graphing. (Lesson 3-2)

45. $x - 2y > -4$
$y < -2x - 3$

46. $y \geq -4x + 6$
$3y < 2x + 9$

47. $4x + 2y > 8$
$4y - 3x \leq 12$

48. RAKING LEAVES A student can earn \$20 plus an extra \$5 for each trash bag he or she completely fills with leaves. Write and solve an equation to determine how many bags the student will need to fill in order to earn \$100. (Lesson 2-4)

49. SPORTS There are 15,991 more student athletes in New York than in Illinois. Write and solve an equation to find the number of student athletes in Illinois. (Lesson 1-3)

Skills Review

Simplify each expression.

50. $4(2x - 3y) + 2(5x - 6y)$

51. $-3(2a - 5b) - 4(4b + a)$

52. $-7(x - y) + 5(y - x)$

LESSON 3-6

Multiplying Matrices

Then

- You multiplied matrices by a scalar.

Now

1. Multiply matrices.
2. Use the properties of matrix multiplication.

Why?

- The table shows the scoring summary for Lisa Leslie, the WNBA's all-time scoring leader, during her highest scoring seasons. Her total baskets can be summarized in the baskets matrix B. The point values for each type of basket made can be organized in the point value matrix P.

Lisa Leslie Regular Season Scoring

Type	2005	2006	2008	2009
Field Goal	197	249	184	143
3-Point Field Goal	7	8	4	1
Free Throw	102	158	117	65

Source: WNBA

You can use matrix multiplication to find the points scored during each season.

Baskets **Point Values**

$$B = \begin{bmatrix} 197 & 249 & 184 & 143 \\ 7 & 8 & 4 & 1 \\ 102 & 158 & 117 & 65 \end{bmatrix} \qquad P = \begin{bmatrix} 2 & 3 & 1 \end{bmatrix}$$

1 Multiply Matrices

You can multiply two matrices A and B if and only if the number of columns in A is equal to the number of rows in B. When you multiply two matrices $A_{m \times r}$ and $B_{r \times t}$, the resulting matrix AB is an $m \times t$ matrix.

Example 1 Dimensions of Matrix Products

Determine whether each matrix product is defined. If so, state the dimensions of the product.

a. $A_{3 \times 4}$ **and** $B_{4 \times 2}$

$A \cdot B = AB$

$3 \times 4 \quad 4 \times 2 \quad 3 \times 2$

The inner dimensions are equal, so the product is defined. Its dimensions are 3×2.

b. $A_{5 \times 3}$ **and** $B_{5 \times 4}$

$A \cdot B$

$5 \times 3 \quad 5 \times 4$

The inner dimensions are not equal, so the matrix product is not defined.

GuidedPractice

1A. $A_{4 \times 6}$ and $B_{6 \times 2}$

1B. $A_{3 \times 2}$ and $B_{3 \times 2}$

C Squared Studios/Getty Images

The product of two matrices is found by multiplying columns and rows.

KeyConcept Multiplying Matrices

Words The element in the mth row and rth column of matrix AB is the sum of the products of the corresponding elements in row m of matrix A and column r of matrix B.

Symbols

$$\underset{A}{\begin{bmatrix} a & b \\ c & d \end{bmatrix}} \cdot \underset{B}{\begin{bmatrix} e & f \\ g & h \end{bmatrix}} = \underset{AB}{\begin{bmatrix} ae + bg & af + bh \\ ce + dg & cf + dh \end{bmatrix}}$$

Example 2 Multiply Square Matrices

Find XY if $X = \begin{bmatrix} 6 & -3 \\ -10 & -2 \end{bmatrix}$ and $Y = \begin{bmatrix} -5 & -4 \\ 3 & 3 \end{bmatrix}$.

$$XY = \begin{bmatrix} 6 & -3 \\ -10 & -2 \end{bmatrix} \cdot \begin{bmatrix} -5 & -4 \\ 3 & 3 \end{bmatrix}$$

> **WatchOut!**
>
> **Saving Your Place** It is easy to lose your place as you multiply matrices. It may help to cover rows or columns not being multiplied as you find elements of the product matrix.

Step 1 Multiply the numbers in the first row of X by the numbers in the first column of Y, add the products, and put the result in the first row, first column of XY.

$$\begin{bmatrix} 6 & -3 \\ -10 & -2 \end{bmatrix} \cdot \begin{bmatrix} -5 & -4 \\ 3 & 3 \end{bmatrix} = \begin{bmatrix} 6(-5) + (-3)(3) & \\ & \end{bmatrix}$$

Step 2 Follow the same procedure as in Step 1 using the first row and the second column numbers. Write the result in the first row, second column.

$$\begin{bmatrix} 6 & -3 \\ -10 & -2 \end{bmatrix} \cdot \begin{bmatrix} -5 & -4 \\ 3 & 3 \end{bmatrix} = \begin{bmatrix} 6(-5) + (-3)(3) & 6(-4) + (-3)(3) \\ & \end{bmatrix}$$

Step 3 Follow the same procedure with the second row and the first column numbers. Write the result in the second row, first column.

$$\begin{bmatrix} 6 & -3 \\ -10 & -2 \end{bmatrix} \cdot \begin{bmatrix} -5 & -4 \\ 3 & 3 \end{bmatrix} = \begin{bmatrix} 6(-5) + (-3)(3) & 6(-4) + (-3)(3) \\ -10(-5) + (-2)(3) & \end{bmatrix}$$

Step 4 The procedure is the same for the numbers in the second row, second column.

$$\begin{bmatrix} 6 & -3 \\ -10 & -2 \end{bmatrix} \cdot \begin{bmatrix} -5 & -4 \\ 3 & 3 \end{bmatrix} = \begin{bmatrix} 6(-5) + (-3)(3) & 6(-4) + (-3)(3) \\ -10(-5) + (-2)(3) & -10(-4) + (-2)(3) \end{bmatrix}$$

Step 5 Simplify the product matrix.

$$\begin{bmatrix} 6(-5) + (-3)(3) & 6(-4) + (-3)(3) \\ -10(-5) + (-2)(3) & -10(-4) + (-2)(3) \end{bmatrix} = \begin{bmatrix} -39 & -33 \\ 44 & 34 \end{bmatrix}$$

GuidedPractice

2. Find UV if $U = \begin{bmatrix} 5 & 9 \\ -3 & -2 \end{bmatrix}$ and $V = \begin{bmatrix} 2 & -1 \\ 6 & -5 \end{bmatrix}$.

Real-World Example 3 Multiply Matrices

SWIM MEET **At a particular swim meet, 7 points were awarded for each first-place finish, 4 points for second, and 2 points for third. Find the total number of points for each school. Which school won the meet?**

School	First Place	Second Place	Third Place
Central	4	7	3
Franklin	8	9	1
Hayes	10	5	3
Lincoln	3	3	6

Real-WorldLink

Swim meets consist of racing and diving competitions. There are more than 241,000 high schools that participate each year.

Source: National Federation of State High School Associations

Understand The final scores can be found by multiplying the swim results for each school by the points awarded for each first-, second-, and third-place finish.

Plan Write the results of the races and the points awarded in matrix form. Set up the matrices so that the number of rows in the points matrix equals the number of columns in the results matrix.

Results **Points**

$$R = \begin{bmatrix} 4 & 7 & 3 \\ 8 & 9 & 1 \\ 10 & 5 & 3 \\ 3 & 3 & 6 \end{bmatrix} \quad P = \begin{bmatrix} 7 \\ 4 \\ 2 \end{bmatrix}$$

Solve Multiply the matrices.

$$RP = \begin{bmatrix} 4 & 7 & 3 \\ 8 & 9 & 1 \\ 10 & 5 & 3 \\ 3 & 3 & 6 \end{bmatrix} \cdot \begin{bmatrix} 7 \\ 4 \\ 2 \end{bmatrix}$$ Write an equation.

$$= \begin{bmatrix} 4(7) + 7(4) + 3(2) \\ 8(7) + 9(4) + 1(2) \\ 10(7) + 5(4) + 3(2) \\ 3(7) + 3(4) + 6(2) \end{bmatrix}$$ Multiply columns by rows.

$$= \begin{bmatrix} 62 \\ 94 \\ 96 \\ 45 \end{bmatrix}$$ Simplify.

The product matrix shows the scores for Central, Franklin, Hayes, and Lincoln, respectively. Hayes won the swim meet with a total of 96 points.

Check R is a 4×3 matrix and P is a 3×1 matrix, so their product should be a 4×1 matrix.

GuidedPractice

3. BASKETBALL Refer to the beginning of the lesson. Use matrix multiplication to determine in which season Lisa Leslie scored the most points. How many points did she score that season?

2 Multiplicative Properties

Recall that the properties of real numbers also held true for matrix addition. However, some of these properties do *not* always hold true for matrix multiplication.

Example 4 Test of the Commutative Property

Find each product if $G = \begin{bmatrix} 1 & 3 & -5 \\ 4 & -2 & 0 \end{bmatrix}$ **and** $H = \begin{bmatrix} 2 & 3 \\ -2 & -8 \\ 1 & 7 \end{bmatrix}$.

a. GH

$$GH = \begin{bmatrix} 1 & 3 & -5 \\ 4 & -2 & 0 \end{bmatrix} \cdot \begin{bmatrix} 2 & 3 \\ -2 & -8 \\ 1 & 7 \end{bmatrix}$$ Substitution

$$= \begin{bmatrix} 2-6-5 & 3-24-35 \\ 8+4+0 & 12+16+0 \end{bmatrix} \text{ or } \begin{bmatrix} -9 & -56 \\ 12 & 28 \end{bmatrix}$$

b. HG

$$HG = \begin{bmatrix} 2 & 3 \\ -2 & -8 \\ 1 & 7 \end{bmatrix} \cdot \begin{bmatrix} 1 & 3 & -5 \\ 4 & -2 & 0 \end{bmatrix}$$ Substitution

$$= \begin{bmatrix} 2+12 & 6-6 & -10+0 \\ -2-32 & -6+16 & 10+0 \\ 1+28 & 3-14 & -5+0 \end{bmatrix} \text{ or } \begin{bmatrix} 14 & 0 & -10 \\ -34 & 10 & 10 \\ 29 & -11 & -5 \end{bmatrix}$$ Notice that $GH \neq HG$.

GuidedPractice

4. Determine if $AB = BA$ is true for $A = \begin{bmatrix} 4 & -1 \\ 5 & -2 \end{bmatrix}$ and $B = \begin{bmatrix} -3 & 6 \\ -4 & 5 \end{bmatrix}$.

StudyTip

Proof and Counterexamples To show that a property is *not* always true, you need to find only one counterexample.

Example 4 demonstrates that the Commutative Property of Multiplication does not hold for matrix multiplication. The order in which you multiply matrices is very important.

Example 5 Test of the Distributive Property

Find each product if $J = \begin{bmatrix} 2 & 4 \\ -5 & -2 \end{bmatrix}$, $K = \begin{bmatrix} 3 & 2 \\ -1 & 3 \end{bmatrix}$, **and** $L = \begin{bmatrix} -4 & -1 \\ 3 & 0 \end{bmatrix}$.

a. $J(K + L)$

$$J(K+L) = \begin{bmatrix} 2 & 4 \\ -5 & -2 \end{bmatrix} \cdot \left(\begin{bmatrix} 3 & 2 \\ -1 & 3 \end{bmatrix} + \begin{bmatrix} -4 & -1 \\ 3 & 0 \end{bmatrix} \right)$$ Substitution

$$= \begin{bmatrix} 2 & 4 \\ -5 & -2 \end{bmatrix} \cdot \begin{bmatrix} -1 & 1 \\ 2 & 3 \end{bmatrix}$$ Add.

$$= \begin{bmatrix} -2+8 & 2+12 \\ 5-4 & -5-6 \end{bmatrix} \text{ or } \begin{bmatrix} 6 & 14 \\ 1 & -11 \end{bmatrix}$$ Multiply.

b. $JK + JL$

$$JK + JL = \begin{bmatrix} 2 & 4 \\ -5 & -2 \end{bmatrix} \cdot \begin{bmatrix} 3 & 2 \\ -1 & 3 \end{bmatrix} + \begin{bmatrix} 2 & 4 \\ -5 & -2 \end{bmatrix} \cdot \begin{bmatrix} -4 & -1 \\ 3 & 0 \end{bmatrix}$$

$$= \begin{bmatrix} 2(3)+4(-1) & 2(2)+4(3) \\ -5(3)+(-2)(-1) & -5(2)+(-2)(3) \end{bmatrix} + \begin{bmatrix} 2(-4)+4(3) & 2(-1)+4(0) \\ -5(-4)+(-2)(3) & -5(-1)+(-2)(0) \end{bmatrix}$$

$$= \begin{bmatrix} 2 & 16 \\ -13 & -16 \end{bmatrix} + \begin{bmatrix} 4 & -2 \\ 14 & 5 \end{bmatrix} \text{ or } \begin{bmatrix} 6 & 14 \\ 1 & -11 \end{bmatrix}$$ Notice that $J(K + L) = JK + JL$.

GuidedPractice

5. Use the matrices $R = \begin{bmatrix} 2 & -1 \\ 1 & 3 \end{bmatrix}$, $S = \begin{bmatrix} 4 & 6 \\ -2 & 5 \end{bmatrix}$, and $T = \begin{bmatrix} -3 & 7 \\ -4 & 8 \end{bmatrix}$ to determine if $(S + T)R = SR + TR$.

The previous example suggests that the Distributive Property is true for matrix multiplication. Some properties of matrix multiplication are shown below.

KeyConcept Properties of Matrix Multiplication

For any matrices A, B, and C for which the matrix product is defined and any scalar k, the following properties are true.

Associative Property of Matrix Multiplication	$(AB)C = A(BC)$
Associative Property of Scalar Multiplication	$k(AB) = (kA)B = A(kB)$
Left Distributive Property	$C(A + B) = CA + CB$
Right Distributive Property	$(A + B)C = AC + BC$

Check Your Understanding

Example 1 **Determine whether each matrix product is defined. If so, state the dimensions of the product.**

1. $A_{2 \times 4} \cdot B_{4 \times 3}$

2. $C_{5 \times 4} \cdot D_{5 \times 4}$

3. $E_{8 \times 6} \cdot F_{6 \times 10}$

Examples 2–3 **Find each product, if possible.**

4. $\begin{bmatrix} 2 & 1 \\ 7 & -5 \end{bmatrix} \cdot \begin{bmatrix} -6 & 3 \\ -2 & -4 \end{bmatrix}$

5. $\begin{bmatrix} 10 & -2 \\ -7 & 3 \end{bmatrix} \cdot \begin{bmatrix} 1 & 4 \\ 5 & -2 \end{bmatrix}$

6. $\begin{bmatrix} 9 & -2 \end{bmatrix} \cdot \begin{bmatrix} -2 & 4 \\ 6 & -7 \end{bmatrix}$

7. $\begin{bmatrix} -9 \\ 6 \end{bmatrix} \cdot \begin{bmatrix} -1 & -10 & 1 \end{bmatrix}$

8. $\begin{bmatrix} -8 & 7 & 4 \\ -5 & -3 & 8 \end{bmatrix} \cdot \begin{bmatrix} 10 & 6 \\ 8 & 4 \end{bmatrix}$

9. $\begin{bmatrix} 2 & 8 \\ 3 & -1 \end{bmatrix} \cdot \begin{bmatrix} 6 \\ -7 \end{bmatrix}$

10. $\begin{bmatrix} -4 & 3 & 2 \\ -1 & -5 & 4 \end{bmatrix} \cdot \begin{bmatrix} 2 & 1 & 6 \\ 8 & 4 & -1 \\ 5 & 3 & -2 \end{bmatrix}$

11. $\begin{bmatrix} 2 & 5 & 3 & -1 \\ -3 & 1 & 8 & -3 \end{bmatrix} \cdot \begin{bmatrix} 6 & -3 \\ -7 & 1 \\ 2 & 0 \\ -1 & 0 \end{bmatrix}$

12. **CCSS SENSE-MAKING** The table shows the number of people registered for aerobics for the first quarter.

Quinn's Gym charges the following registration fees: class-by-class, \$165; 11-class pass, \$110; unlimited pass, \$239.

Quinn's Gym		
Payment	Aerobics	Step Aerobics
class-by-class	35	28
11-class pass	32	17
unlimited pass	18	12

a. Write a matrix for the registration fees and a matrix for the number of students.

b. Find the total amount of money the gym received from aerobics and step aerobic registrations.

Examples 4–5 **Use** $X = \begin{bmatrix} -10 & -3 \\ 2 & -8 \end{bmatrix}$, $Y = \begin{bmatrix} -5 & 6 \\ -1 & 9 \end{bmatrix}$, **and** $Z = \begin{bmatrix} -5 & -1 \\ -8 & -4 \end{bmatrix}$ **to determine whether the following equations are true for the given matrices.**

13. $XY = YX$

14. $X(YZ) = (XY)Z$

Practice and Problem Solving

Extra Practice is on page R3.

Example 1 **Determine whether each matrix product is defined. If so, state the dimensions of the product.**

15. $P_{2 \times 3} \cdot Q_{3 \times 4}$

16. $A_{5 \times 5} \cdot B_{5 \times 5}$

17. $M_{3 \times 1} \cdot N_{2 \times 3}$

18. $X_{2 \times 6} \cdot Y_{6 \times 3}$

19. $J_{2 \times 1} \cdot K_{2 \times 1}$

20. $S_{5 \times 2} \cdot T_{2 \times 4}$

Examples 2–3 **Find each product, if possible.**

21. $[1 \quad 6] \cdot \begin{bmatrix} -10 \\ 6 \end{bmatrix}$

22. $\begin{bmatrix} 6 \\ -3 \end{bmatrix} \cdot [2 \quad -7]$

23. $\begin{bmatrix} -3 & -7 \\ -2 & -1 \end{bmatrix} \cdot \begin{bmatrix} 4 & 4 \\ 9 & -3 \end{bmatrix}$

24. $\begin{bmatrix} -1 & 0 \\ 5 & 2 \end{bmatrix} \cdot \begin{bmatrix} 6 & -3 \\ 7 & -2 \end{bmatrix}$

25. $\begin{bmatrix} -1 & 0 & 6 \\ -4 & -10 & 4 \end{bmatrix} \cdot \begin{bmatrix} 5 & -7 \\ -2 & -9 \end{bmatrix}$

26. $\begin{bmatrix} -6 & 4 & -9 \\ 2 & 8 & 7 \end{bmatrix} \cdot \begin{bmatrix} 7 \\ 2 \\ 4 \end{bmatrix}$

27. $\begin{bmatrix} 2 & 9 & -3 \\ 4 & -1 & 0 \end{bmatrix} \cdot \begin{bmatrix} 4 & 2 \\ -6 & 7 \\ -2 & 1 \end{bmatrix}$

28. $\begin{bmatrix} -4 \\ 8 \end{bmatrix} \cdot [-3 \quad -1]$

29. TRAVEL The Wolf family owns three bed and breakfasts in a vacation spot. A room with a single bed is \$220 per night, a room with two beds is \$250 per night, and a suite is \$360.

a. Write a matrix for the number of each type of room at each bed and breakfast. Then write a room-cost matrix.

b. Write a matrix for total daily income, assuming that all the rooms are rented.

Available Rooms at a Wolf Bed and Breakfast

B & B	Single	Double	Suite
1	3	2	2
2	2	3	1
3	4	3	0

c. What is the total daily income from all three bed and breakfasts, assuming that all the rooms are rented?

Examples 4–5 **Use** $P = \begin{bmatrix} 4 & -1 \\ 1 & 2 \end{bmatrix}$, $Q = \begin{bmatrix} 6 & 4 \\ -2 & -5 \end{bmatrix}$, $R = \begin{bmatrix} 4 & 6 \\ -6 & 4 \end{bmatrix}$, **and** $k = 2$ **to determine whether the following equations are true for the given matrices.**

30. $k(PQ) = P(kQ)$

31. $PQR = RQP$

32. $PR + QR = (P + Q)R$

33. $R(P + Q) = PR + QR$

34. CCSS SENSE-MAKING Student Council is selling flowers for Mother's Day. They bought 200 roses, 150 daffodils, and 100 orchids for the purchase prices shown. They sold all of the flowers for the sales prices shown.

Flower	Purchase Price	Sales Price
rose	\$1.67	\$3.00
daffodil	\$1.03	\$2.25
orchid	\$2.59	\$4.50

a. Organize the data in two matrices, and use matrix multiplication to find the total amount that was spent on the flowers.

b. Write two matrices, and use matrix multiplication to find the total amount the student council received for the flower sale.

c. Use matrix operations to find how much money the student council made on their project.

EXTEND 3-6

Graphing Technology Lab
Operations with Matrices

A graphing calculator can be used to perform operations with matrices.

Activity 1 Perform Operations

Use A, B, and C to find the following.

$$A = \begin{bmatrix} -3.2 & 1.7 \\ 0.4 & -5.8 \end{bmatrix} \quad B = \begin{bmatrix} 4.9 & 0.3 \\ -7.1 & 2.6 \end{bmatrix} \quad C = \begin{bmatrix} 5.6 & -6.1 & 2.1 \\ -8.2 & 7.6 & 0.2 \end{bmatrix}$$

a. $3A + 2B$

Begin by entering matrix A into a graphing calculator.

KEYSTROKES: 2nd [MATRIX] ▶ ▶ ENTER 2 ENTER 2 ENTER (–) 3.2 ENTER 1.7 ENTER 0.4 ENTER (–) 5.8 ENTER 2nd [QUIT]

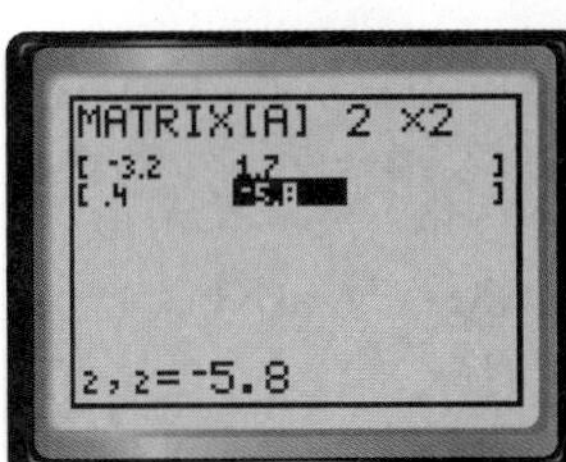

Enter matrix B into the graphing calculator using similar keystrokes. Then, perform the indicated operations.

KEYSTROKES: 3 2nd [MATRIX] ENTER + 2 2nd [MATRIX] ▼ ENTER ENTER

$3A + 2B$ is equal to $\begin{bmatrix} 0.2 & 5.7 \\ -13 & -12.2 \end{bmatrix}$.

b. $4C + 3A$

Enter matrix C into the graphing calculator. Perform the indicated operations. Notice that the calculator displays an error message when the dimensions do not allow the operations to be performed.

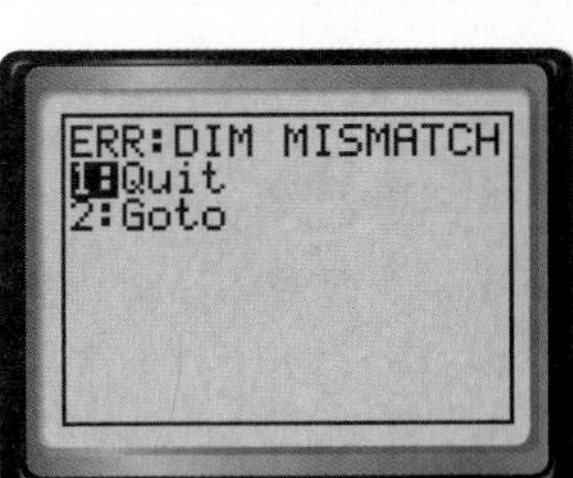

Exercises

Use A, B, C, and D to find the following. If the matrix does not exist, write *impossible*.

$$A = \begin{bmatrix} 4.5 & -9.0 \\ -7.4 & 9.4 \end{bmatrix} \quad B = \begin{bmatrix} 1.9 & -5.9 \\ 2.9 & 5.0 \end{bmatrix} \quad C = \begin{bmatrix} 7.0 & 5.5 & -1.9 \\ 7.6 & -9.9 & 0.5 \end{bmatrix} \quad D = \begin{bmatrix} 8.5 & 8.0 \\ -1.2 & -5.9 \\ 0.7 & 8.9 \end{bmatrix}$$

1. $CD + 4A$

2. $-2B + 7A$

3. $4(DC)$

4. $6B + DC$

5. $2(AB) - 3B + 5A$

6. $-3(CD) + 4(BA) + 7A$

(continued on the next page)

Activity 2 Explore Properties of Matrix Operations

Use A, B, and C to find the following.

$A = \begin{bmatrix} -6 & 2 \\ 7 & 3 \end{bmatrix}$ $B = \begin{bmatrix} 5 & 1 \\ -8 & 4 \end{bmatrix}$ $C = \begin{bmatrix} 11 & -3 \\ 4 & -1 \end{bmatrix}$

a. $A + B$ **and** $B + A$

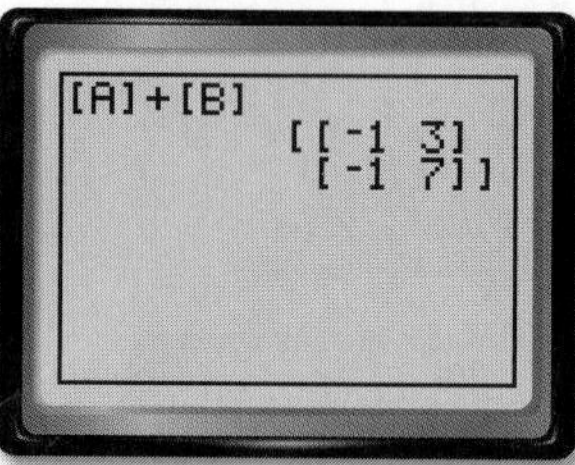

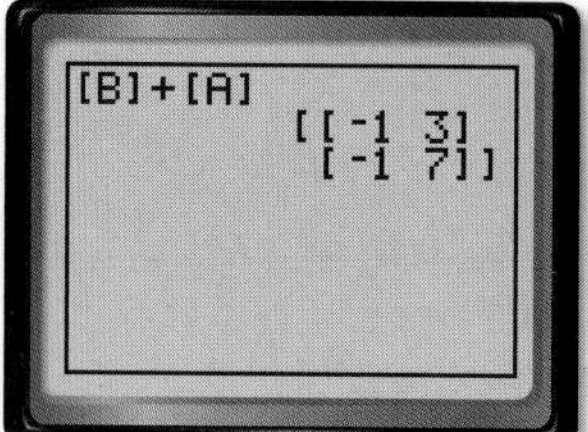

Find $A + B$.
Then find $B + A$.

$A + B$ and $B + A$ are both equal to $\begin{bmatrix} -1 & 3 \\ -1 & 7 \end{bmatrix}$.

b. $(A + B) + C$ **and** $A + (B + C)$

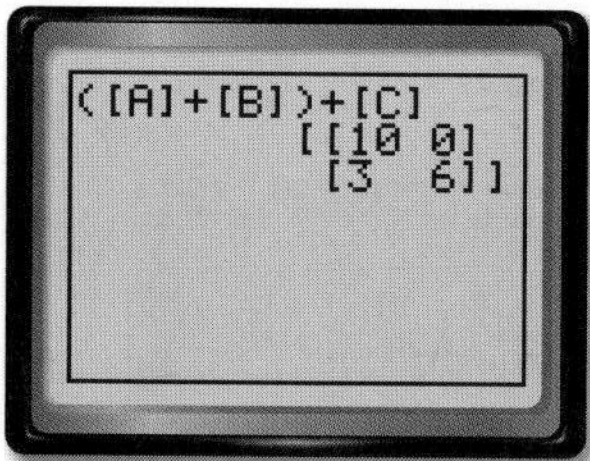

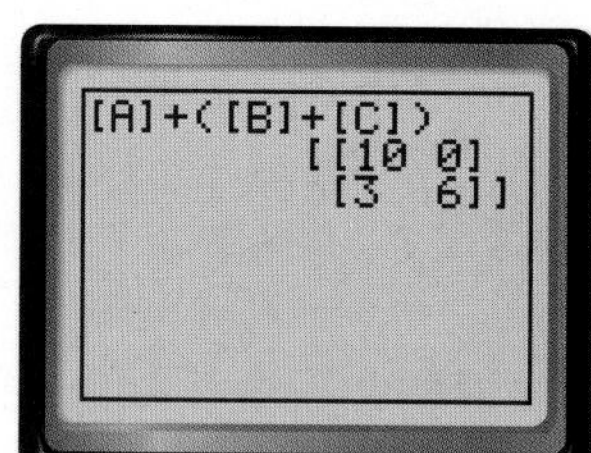

Find $(A + B) + C$.
Then find $A + (B + C)$.

$(A + B) + C$ and $A + (B + C)$ are both equal to $\begin{bmatrix} 10 & 0 \\ 3 & 6 \end{bmatrix}$.

Analyze the Results

What property is illustrated in each part of Activity 2?

7. part **a**

8. part **b**

Find each set of products. How are they similar or different? What property is illustrated?

9. $(AB)C$, $A(BC)$

10. $3(AB)$, $(3A)B$, $A(3B)$

11. The zero matrix O is an $m \times n$ matrix with all elements equal to 0. If $A + O$ is defined, verify the Additive Identity Property for Matrices, $A + O = A$.

12. Two matrices are additive inverses if their sum is the zero matrix. Find matrix E so that $A + E$ are additive inverses. Then verify that $A + E = O$.

13. Find the additive inverses for matrix B and matrix C.

14. **CHALLENGE** What observations can be made about a matrix and its additive inverse? Find the additive inverse for $\begin{bmatrix} w & -x \\ -y & z \end{bmatrix}$.

LESSON 3-7

Solving Systems of Equations Using Cramer's Rule

Then	Now	Why?
You solved systems of equations algebraically.	1 Evaluate determinants. 2 Solve systems of linear equations by using Cramer's Rule.	A zoologist tagged a tiger with a GPS tracker so that she could determine the tiger's territory. After several days, the zoologist determined that the tiger's territory was a triangular region. By using the coordinates of the vertices of this triangle, she could use matrices and determinants to determine the size of the tiger's territory.

NewVocabulary
determinant
second-order determinant
third-order determinant
diagonal rule
Cramer's Rule
coefficient matrix

Common Core State Standards

Content Standards
A.CED.3 Represent constraints by equations or inequalities, and by systems of equations and/or inequalities, and interpret solutions as viable or nonviable options in a modeling context

Mathematical Practices
7 Look for and make use of structure.

MORITSCH, MARC (VIA HORNOCKER, MAURICE)/National Geographic Stock

1 Determinants

Every square matrix has a **determinant**. The determinant of a 2×2 matrix is called a **second-order determinant**.

KeyConcept Second-Order Determinant

Words The value of a second-order determinant is the difference of the products of the two diagonals.

Symbols $\det \begin{bmatrix} a & b \\ c & d \end{bmatrix} = \begin{vmatrix} a & b \\ c & d \end{vmatrix} = ad - bc$

Example $\begin{vmatrix} 4 & 5 \\ -3 & 6 \end{vmatrix} = 4(6) - (5)(-3) = 39$

Example 1 Second-Order Determinant

Evaluate each determinant.

a. $\begin{vmatrix} 5 & -4 \\ 8 & 9 \end{vmatrix}$

$$\begin{vmatrix} 5 & -4 \\ 8 & 9 \end{vmatrix} = 5(9) - (-4)(8) \quad \text{Definition of determinant}$$
$$= 45 + 32 \quad \text{Simplify.}$$
$$= 77$$

b. $\begin{vmatrix} 0 & 6 \\ 4 & -11 \end{vmatrix}$

$$\begin{vmatrix} 0 & 6 \\ 4 & -11 \end{vmatrix} = 0(-11) - 6(4) \quad \text{Definition of determinant}$$
$$= 0 - 24 \quad \text{Simplify.}$$
$$= -24$$

GuidedPractice

1A. $\begin{vmatrix} -6 & -7 \\ 10 & 8 \end{vmatrix}$

1B. $\begin{vmatrix} 7 & 5 \\ 9 & -4 \end{vmatrix}$

StudyTip

Diagonal Rule The diagonal rule can only be used for 3×3 matrices.

Determinants of 3×3 matrices are called **third-order determinants**. They can be evaluated by using the **diagonal rule**.

KeyConcept Diagonal Rule

Step 1 Rewrite the first two columns to the right of the determinant.

Step 2 Draw diagonals, beginning with the upper left-hand element. Multiply the elements in each diagonal. Repeat the process, beginning with the upper right-hand element.

Step 3 Find the sum of the products of the elements in each set of diagonals.

Step 4 Subtract the second sum from the first sum.

PT

Example 2 Use Diagonals

Evaluate $\begin{vmatrix} 4 & -8 & 3 \\ -3 & 2 & 6 \\ -4 & 5 & 9 \end{vmatrix}$ **using diagonals.**

Step 1 Rewrite the first two columns to the right of the determinant.

$$\left|\begin{matrix} 4 & -8 & 3 \\ -3 & 2 & 6 \\ -4 & 5 & 9 \end{matrix}\right|\begin{matrix} 4 & -8 \\ -3 & 2 \\ -4 & 5 \end{matrix}$$

Step 2 Find the products of the elements of the diagonals.

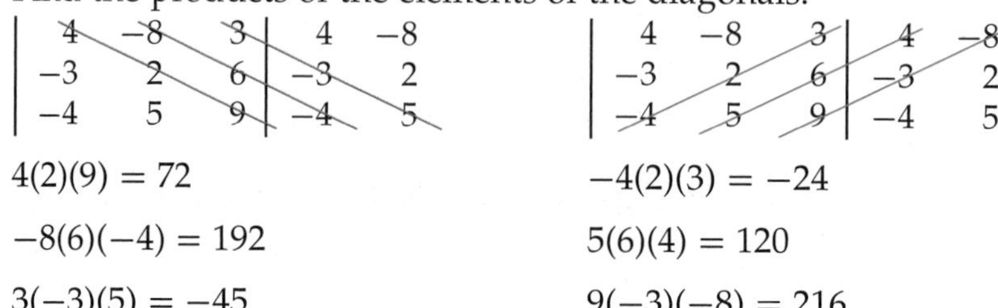

$4(2)(9) = 72$ $\qquad -4(2)(3) = -24$

$-8(6)(-4) = 192$ $\qquad 5(6)(4) = 120$

$3(-3)(5) = -45$ $\qquad 9(-3)(-8) = 216$

Step 3 Find the sum of each group.

$72 + 192 + (-45) = 219$ $\qquad -24 + 120 + 216 = 312$

Step 4 Subtract the sum of the second group from the sum of the first group.

$219 - 312 = -93$

The value of the determinant is -93.

GuidedPractice

Evaluate each determinant.

2A. $\begin{vmatrix} -5 & 9 & 4 \\ -2 & -1 & 5 \\ -4 & 6 & 2 \end{vmatrix}$ **2B.** $\begin{vmatrix} -8 & -4 & 4 \\ 0 & -5 & -8 \\ 3 & 4 & 1 \end{vmatrix}$

Determinants can also be used to find the areas of triangles. If the coordinates of the vertices of the triangle are known, the formula below can be used to calculate the area of the triangle.

KeyConcept Area of a Triangle

Words The area of a triangle with vertices (a, b), (c, d), and (e, f) is $|A|$, where

$$A = \frac{1}{2}\begin{vmatrix} a & b & 1 \\ c & d & 1 \\ e & f & 1 \end{vmatrix}.$$

Example

$$A = \frac{1}{2}\begin{vmatrix} -4 & 3 & 1 \\ 3 & 1 & 1 \\ -2 & -2 & 1 \end{vmatrix}$$

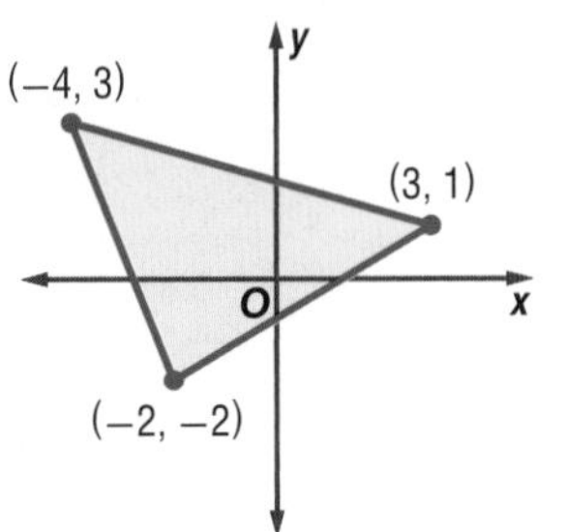

Real-WorldLink

Tigers are very territorial, solitary animals. Their territories can measure up to 100 square kilometers.

Source: *National Geographic*

Real-World Example 3 Use Determinants

ZOOLOGY Refer to the application at the beginning of the lesson. The coordinates of the vertices of the tiger's territory are shown to the right. Use determinants to find the area of the tiger's territory.

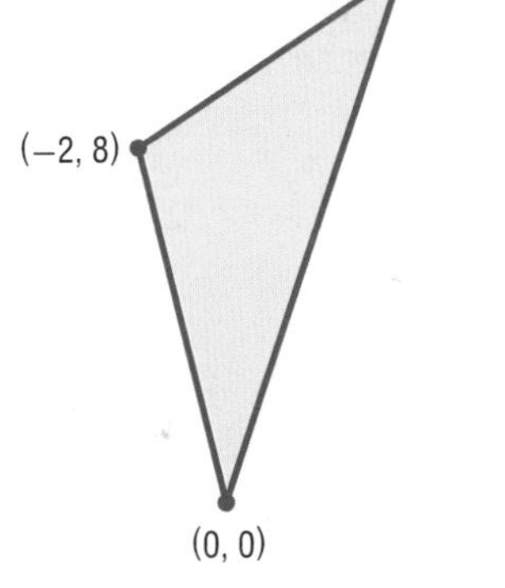

$$A = \frac{1}{2}\begin{vmatrix} a & b & 1 \\ c & d & 1 \\ e & f & 1 \end{vmatrix}$$

$$= \frac{1}{2}\begin{vmatrix} 0 & 0 & 1 \\ 4 & 12 & 1 \\ -2 & 8 & 1 \end{vmatrix} \quad \begin{aligned} (a, b) &= (0, 0) \\ (c, d) &= (4, 12) \\ (e, f) &= (-2, 8) \end{aligned}$$

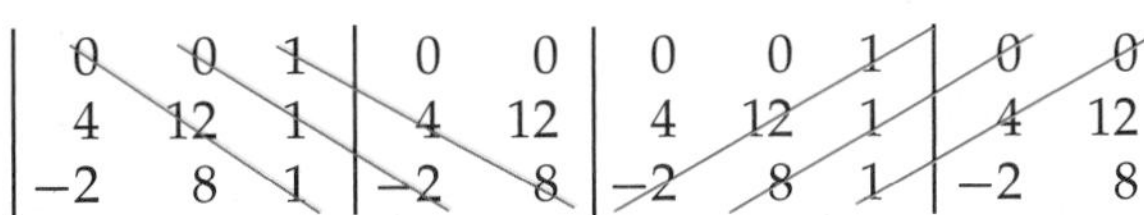

Diagonal Rule

$0 + 0 + 32 = 32$ $\qquad -24 + 0 + 0 = -24$ Sum of products of diagonals

$$A = \frac{1}{2}\begin{vmatrix} 0 & 0 & 1 \\ 4 & 12 & 1 \\ -2 & 8 & 1 \end{vmatrix}$$ Area of a Triangle

$$= \left(\frac{1}{2}\right)[32 - (-24)] \text{ or } 28$$ Simplify.

The area of the tiger's territory is 28 square kilometers.

GuidedPractice

3. CAR WASH To raise money for their rowing club, Hannah, Christina, and Dario are advertising a car wash at three different street corners in a neighborhood. On a map, the coordinates for the corners are (3, 15), (6, 4), and (11, 9). Each unit represents 0.5 kilometer. What is the area of the neighborhood in which they are advertising?

ReadingMath

Determinants The determinant is used to *determine* whether a system has a unique solution.

2 Cramer's Rule

You can use determinants to solve systems of equations. If a determinant is nonzero, then the system has a unique solution. If a determinant is 0, then the system either has no solution or infinite solutions. A method called **Cramer's Rule** uses the coefficient matrix. The **coefficient matrix** is a matrix that contains only the coefficients of the system.

KeyConcept Cramer's Rule

Let C be the coefficient matrix of the system $\begin{aligned} ax + by &= m \\ fx + gy &= n \end{aligned} \rightarrow \begin{bmatrix} a & b \\ f & g \end{bmatrix}$.

The solution of this system is $x = \dfrac{\begin{vmatrix} m & b \\ n & g \end{vmatrix}}{|C|}$ and $y = \dfrac{\begin{vmatrix} a & m \\ f & n \end{vmatrix}}{|C|}$, if $|C| \neq 0$.

Test-TakingTip

Cramer's Rule When the determinant of the coefficient matrix C is 0, the system does not have a unique solution.

Example 4 Solve a System of Two Equations

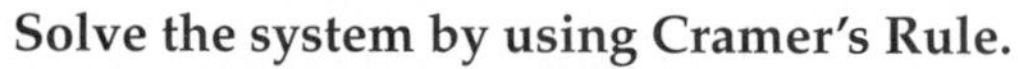

Solve the system by using Cramer's Rule.

$\mathbf{5x - 6y = 15}$

$\mathbf{3x + 4y = -29}$

$x = \dfrac{\begin{vmatrix} m & b \\ n & g \end{vmatrix}}{\lvert C \rvert}$	Cramer's Rule	$y = \dfrac{\begin{vmatrix} a & m \\ f & n \end{vmatrix}}{\lvert C \rvert}$
$= \dfrac{\begin{vmatrix} 15 & -6 \\ -29 & 4 \end{vmatrix}}{\begin{vmatrix} 5 & -6 \\ 3 & 4 \end{vmatrix}}$	Substitute values.	$= \dfrac{\begin{vmatrix} 5 & 15 \\ 3 & -29 \end{vmatrix}}{\begin{vmatrix} 5 & -6 \\ 3 & 4 \end{vmatrix}}$
$= \dfrac{15(4) - (-29)(-6)}{5(4) - (3)(-6)}$	Evaluate.	$= \dfrac{5(-29) - 3(15)}{5(4) - (3)(-6)}$
$= \dfrac{60 - 174}{20 + 18}$	Multiply.	$= \dfrac{-145 - 45}{20 + 18}$
$= -\dfrac{114}{38}$	Add and subtract.	$= -\dfrac{190}{38}$
$= -3$	Simplify.	$= -5$

The solution of the system is $(-3, -5)$.

CHECK

$5(-3) - 6(-5) \stackrel{?}{=} 15$ $x = -3, y = -5$

$-15 + 30 \stackrel{?}{=} 15$ Simplify.

$15 = 15$ ✓

$3(-3) + 4(-5) \stackrel{?}{=} -29$ $x = -3, y = -5$

$-9 - 20 \stackrel{?}{=} -29$ Simplify.

$-29 = -29$ ✓

GuidedPractice

Solve each system using Cramer's Rule.

4A. $7x + 3y = 37$
$-5x - 7y = -41$

4B. $8x - 5y = 70$
$9x + 7y = 3$

Cramer's Rule can also be used for systems of three equations.

KeyConcept Cramer's Rule for a System of Three Equations

Let C be the coefficient matrix of the system

$$\begin{aligned} ax + by + cz &= m \\ fx + gy + hz &= n \\ jx + ky + \ell z &= p \end{aligned} \rightarrow \begin{bmatrix} a & b & c \\ f & g & h \\ j & k & \ell \end{bmatrix}.$$

The solution of this system is $x = \dfrac{\begin{vmatrix} m & b & c \\ n & g & h \\ p & k & \ell \end{vmatrix}}{|C|}$, $y = \dfrac{\begin{vmatrix} a & m & c \\ f & n & h \\ j & p & \ell \end{vmatrix}}{|C|}$, and $z = \dfrac{\begin{vmatrix} a & b & m \\ f & g & n \\ j & k & p \end{vmatrix}}{|C|}$, if $|C| \neq 0$.

Example 5 Solve a System of Three Equations

Solve the system by using Cramer's Rule.

$$\begin{aligned} 4x + 5y - 6z &= -14 \\ 3x - 2y + 7z &= 47 \\ 7x - 6y - 8z &= 15 \end{aligned}$$

$$x = \frac{\begin{vmatrix} m & b & c \\ n & g & h \\ p & k & \ell \end{vmatrix}}{|C|} \qquad y = \frac{\begin{vmatrix} a & m & c \\ f & n & h \\ j & p & \ell \end{vmatrix}}{|C|} \qquad z = \frac{\begin{vmatrix} a & b & m \\ f & g & n \\ j & k & p \end{vmatrix}}{|C|}$$

$$= \frac{\begin{vmatrix} -14 & 5 & -6 \\ 47 & -2 & 7 \\ 15 & -6 & -8 \end{vmatrix}}{\begin{vmatrix} 4 & 5 & -6 \\ 3 & -2 & 7 \\ 7 & -6 & -8 \end{vmatrix}} \qquad = \frac{\begin{vmatrix} 4 & -14 & -6 \\ 3 & 47 & 7 \\ 7 & 15 & -8 \end{vmatrix}}{\begin{vmatrix} 4 & 5 & -6 \\ 3 & -2 & 7 \\ 7 & -6 & -8 \end{vmatrix}} \qquad = \frac{\begin{vmatrix} 4 & 5 & -14 \\ 3 & -2 & 47 \\ 7 & -6 & 15 \end{vmatrix}}{\begin{vmatrix} 4 & 5 & -6 \\ 3 & -2 & 7 \\ 7 & -6 & -8 \end{vmatrix}}$$

$$= \frac{3105}{621} \text{ or } 5 \qquad = -\frac{1242}{621} \text{ or } -2 \qquad = \frac{2484}{621} \text{ or } 4$$

The solution of the system is (5, −2, 4).

StudyTip

Check for Accuracy Always substitute your answers into the initial equations to confirm accuracy.

CHECK

$$\begin{aligned} 4(5) + 5(-2) - 6(4) &\stackrel{?}{=} -14 \\ 20 - 10 - 24 &\stackrel{?}{=} -14 \\ -14 &= -14 \checkmark \end{aligned} \qquad \begin{aligned} 3(5) - 2(-2) + 7(4) &\stackrel{?}{=} 47 \\ 15 + 4 + 28 &\stackrel{?}{=} 47 \\ 47 &= 47 \checkmark \end{aligned}$$

$$\begin{aligned} 7(5) - 6(-2) - 8(4) &\stackrel{?}{=} 15 \\ 35 + 12 - 32 &\stackrel{?}{=} 15 \\ 15 &= 15 \checkmark \end{aligned}$$

GuidedPractice

Solve each system using Cramer's Rule.

5A. $3x + 5y + 2z = -7$
$-4x + 3y - 5z = -19$
$5x + 4y - 7z = -15$

5B. $6x + 5y + 2z = -1$
$-x + 3y + 7z = 12$
$5x - 7y - 3z = -52$

Check Your Understanding

= Step-by-Step Solutions begin on page R14.

Example 1 **Evaluate each determinant.**

1. $\begin{vmatrix} 8 & 6 \\ 5 & 7 \end{vmatrix}$

2. $\begin{vmatrix} -6 & -6 \\ 8 & 10 \end{vmatrix}$

3. $\begin{vmatrix} -4 & 12 \\ 9 & 5 \end{vmatrix}$

4. $\begin{vmatrix} 16 & -10 \\ -8 & 5 \end{vmatrix}$

Example 2 **Evaluate each determinant using diagonals.**

5. $\begin{vmatrix} 3 & -2 & 2 \\ -4 & 2 & -5 \\ -3 & 1 & 4 \end{vmatrix}$

6. $\begin{vmatrix} 2 & -3 & 5 \\ -4 & 6 & -2 \\ 4 & -1 & -6 \end{vmatrix}$

7. $\begin{vmatrix} 8 & 4 & 0 \\ -2 & -6 & -1 \\ 5 & -3 & 6 \end{vmatrix}$

8. $\begin{vmatrix} -5 & -3 & 4 \\ -2 & -4 & -3 \\ 8 & -2 & 4 \end{vmatrix}$

9. $\begin{vmatrix} 8 & 3 & 4 \\ 2 & 4 & 2 \\ 1 & 6 & 5 \end{vmatrix}$

10. $\begin{vmatrix} -4 & 3 & 0 \\ 1 & 5 & -2 \\ -1 & -8 & -3 \end{vmatrix}$

11. $\begin{vmatrix} 2 & -6 & -3 \\ 7 & 9 & -4 \\ -6 & 4 & 9 \end{vmatrix}$

12. $\begin{vmatrix} -5 & -6 & 7 \\ 4 & 0 & 5 \\ -3 & 8 & 2 \end{vmatrix}$

Example 4 **Use Cramer's Rule to solve each system of equations.**

13. $4x - 5y = 39$
$3x + 8y = -6$

14. $5x + 6y = 20$
$-3x - 7y = -29$

15. $-8a - 5b = -27$
$7a + 6b = 22$

16. $10c - 7d = -59$
$6c + 5d = -63$

Examples 3–5 17. **CCSS PERSEVERANCE** The "Bermuda Triangle" is an area located off the southeastern Atlantic coast of the United States, and is noted for reports of unexplained losses of ships, small boats, and aircraft.

a. Find the area of the triangle on the map.

b. Suppose each grid represents 175 miles. What is the area of the Bermuda Triangle?

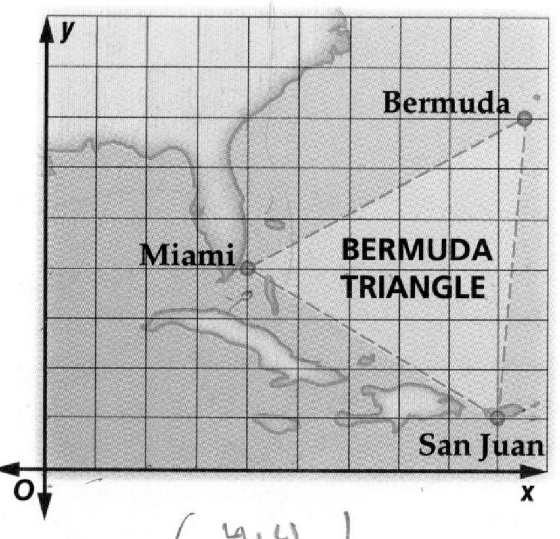

(4, 4)
(7, 9.5)
(1, 9)

Use Cramer's Rule to solve each system of equations.

18. $4x - 2y + 7z = 26$
$5x + 3y - 5z = -50$
$-7x - 8y - 3z = 49$

19. $-3x - 5y + 10z = -4$
$-8x + 2y - 3z = -91$
$6x + 8y - 7z = -35$

20. $6x - 5y + 2z = -49$
$-5x - 3y - 8z = -22$
$-3x + 8y - 5z = 55$

21. $-9x + 5y + 3z = 50$
$7x + 8y - 2z = -60$
$-5x + 7y + 5z = 46$

22. $x + 2y = 12$
$3y - 4z = 25$
$x + 6y + z = 20$

23. $9a + 7b = -30$
$8b + 5c = 11$
$-3a + 10c = 73$

24. $2n + 3p - 4w = 20$
$4n - p + 5w = 13$
$3n + 2p + 4w = 15$

25. $x + y + z = 12$
$6x - 2y - z = 16$
$3x + 4y + 2z = 28$

Practice and Problem Solving

Extra Practice is on page R3.

Examples 1–2 **Evaluate each determinant.**

26. $\begin{vmatrix} -7 & 12 \\ 5 & 6 \end{vmatrix}$

27. $\begin{vmatrix} -8 & -9 \\ 11 & 12 \end{vmatrix}$

28. $\begin{vmatrix} -5 & 8 \\ -6 & -7 \end{vmatrix}$

29. $\begin{vmatrix} 3 & 5 & -2 \\ -1 & -4 & 6 \\ -6 & -2 & 5 \end{vmatrix}$

30. $\begin{vmatrix} 2 & 0 & -6 \\ -3 & -4 & -5 \\ -2 & 5 & 8 \end{vmatrix}$

31. $\begin{vmatrix} -5 & -1 & -2 \\ 1 & 8 & 4 \\ 0 & -6 & 9 \end{vmatrix}$

32. $\begin{vmatrix} 6 & -3 & -5 \\ 0 & -7 & 0 \\ 3 & -6 & -4 \end{vmatrix}$

33. $\begin{vmatrix} -8 & -3 & -9 \\ 0 & 0 & 0 \\ 8 & -2 & -4 \end{vmatrix}$

34. $\begin{vmatrix} 1 & 6 & 7 \\ -2 & -5 & -8 \\ 4 & 4 & 9 \end{vmatrix}$

35. $\begin{vmatrix} 1 & -8 & -9 \\ 6 & 5 & -6 \\ -2 & -8 & 10 \end{vmatrix}$

36. $\begin{vmatrix} 5 & -5 & -5 \\ -8 & -3 & -2 \\ -2 & 4 & 6 \end{vmatrix}$

37. $\begin{vmatrix} -4 & 1 & -2 \\ 10 & 12 & 9 \\ -6 & 0 & 13 \end{vmatrix}$

38. **TRAVEL** Mr. Smith's art class took a bus trip to an art museum. The bus averaged 65 miles per hour on the highway and 25 miles per hour in the city. The art museum is 375 miles away from the school, and it took the class 7 hours to get there. Use Cramer's Rule to find how many hours the bus was on the highway and how many hours it was driving in the city.

Examples 4–5 **Use Cramer's Rule to solve each system of equations.**

39. $6x - 5y = 73$
$-7x + 3y = -71$

40. $10a - 3b = -34$
$3a + 8b = -28$

41. $-4c - 5d = -39$
$5c + 8d = 54$

42. $-6f - 8g = -22$
$-11f + 5g = -60$

43. $9r + 4s = -55$
$-5r - 3s = 36$

44. $-11u - 7v = 4$
$9u + 4v = -24$

45. $5x - 4y + 6z = 58$
$-4x + 6y + 3z = -13$
$6x + 3y + 7z = 53$

46. $8x - 4y + 7z = 34$
$5x + 6y + 3z = -21$
$3x + 7y - 8z = -85$

47. **DOUGHNUTS** Mi-Ling is ordering doughnuts for a class party. The box contains 2 dozen doughnuts, some of which are plain and some of which are jelly-filled. The plain doughnuts each cost $0.50, and the jelly-filled cost $0.60. If the total cost is $12.60, use Cramer's Rule to find how many jelly-filled doughnuts Mi-Ling ordered.

Examples 3–4 48. **CCSS PERSEVERANCE** The salary for each of the stars of a new movie is $5 million, and the supporting actors each receive $1 million. The total amount spent for the salaries of the actors and actresses is $19 million. If the cast has 7 members, use Cramer's Rule to find the number of stars in the movie.

49. **ARCHAEOLOGY** Archaeologists found whale bones at coordinates (0, 3), (4, 7), and (5, 9). If the units of the coordinates are meters, find the area of the triangle formed by these finds.

Use Cramer's Rule to solve each system of equations.

50. $6a - 7b = -55$
$2a + 4b - 3c = 35$
$-5a - 3b + 7c = -37$

51. $3a - 5b - 9c = 17$
$4a - 3c = 31$
$-5a - 4b - 2c = -42$

52. $4x - 5y = -2$
$7x + 3z = -47$
$8y - 5z = -63$

53. $7x + 8y + 9z = -149$
$-6x + 7y - 5z = 54$
$4x + 5y - 2z = -44$

54. GARDENING Rob wants to build a triangular flower garden. To plan out his garden he uses a coordinate grid where each of the squares represents one square foot. The coordinates for the vertices of his garden are (−1, 7), (2, 6), and (4, −3). Find the area of the garden.

55. FINANCIAL LITERACY A vendor sells small drinks for $1.15, medium drinks for $1.75, and large drinks for $2.25. During a week in which he sold twice as many small drinks as medium drinks, his total sales were $2,238.75 for 1385 drinks.

a. Use Cramer's Rule to determine how many of each drink were sold.

b. The vendor decided to increase the price for small drinks to $1.25 the next week. The next week, he sold 140 fewer small drinks, 125 more medium drinks, and 35 more large drinks. Calculate his sales for that week.

c. Was raising the price of the small drink a good business move? Explain.

H.O.T. Problems Use Higher-Order Thinking Skills

56. REASONING Some systems of equations cannot be solved by using Cramer's Rule.

a. Find the value of $\begin{vmatrix} a & b \\ f & g \end{vmatrix}$. When is the value 0?

b. Choose values for a, b, f, and g to make the determinant of the coefficient matrix 0. What type of system is formed?

57. REASONING What can you determine about the solution of a system of linear equations if the determinant of the coefficients is 0?

58. CCSS CRITIQUE James and Amber are finding the value of $\begin{vmatrix} 8 & 3 \\ -5 & 2 \end{vmatrix}$.

James

$\begin{vmatrix} 8 & 3 \\ -5 & 2 \end{vmatrix} = 16 - (-15)$
$= 31$

Amber

$\begin{vmatrix} 8 & 3 \\ -5 & 2 \end{vmatrix} = 16 - 15$
$= 1$

Is either of them correct? Explain your reasoning.

59. CHALLENGE Find the determinant of a 3 × 3 matrix defined by

$$a_{mn} = \begin{cases} 0 & \text{if } m + n \text{ is even} \\ m + n & \text{if } m + n \text{ is odd} \end{cases}.$$

60. OPEN ENDED Write a 2 × 2 matrix with each of the following characteristics.

a. The determinant equals 0.

b. The determinant equals 25.

c. The elements are all negative numbers and the determinant equals −32.

61. WRITING IN MATH Describe the possible graphical representations of a 2 × 2 system of linear equations if the determinant of the matrix of coefficients is 0.

Standardized Test Practice

62. Tyler paid \$25.25 to play three games of miniature golf and two rides on go-karts. Brent paid \$25.75 for four games of miniature golf and one ride on the go-karts. How much does one game of miniature golf cost?

A \$4.25 **C** \$5.25

B \$4.75 **D** \$5.75

63. Use the table to determine the expression that best represents the number of faces of any prism having a base with n sides.

Base	Sides of Base	Faces of Prisms
triangle	3	5
quadrilateral	4	6
pentagon	5	7
hexagon	6	8
heptagon	7	9
octagon	8	10

F $2(n - 1)$ **H** $n + 2$

G $2(n + 1)$ **J** $2n$

64. SHORT RESPONSE A right circular cone has radius 4 inches and height 6 inches.

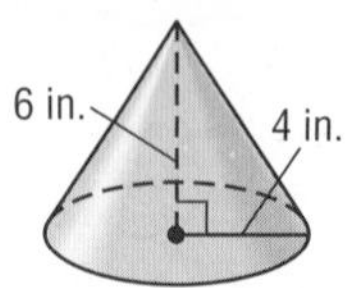

What is the lateral area of the cone? (lateral area of cone $= \pi r \ell$, where $\ell =$ slant height)

65. SAT/ACT Find the area of $\triangle ABC$.

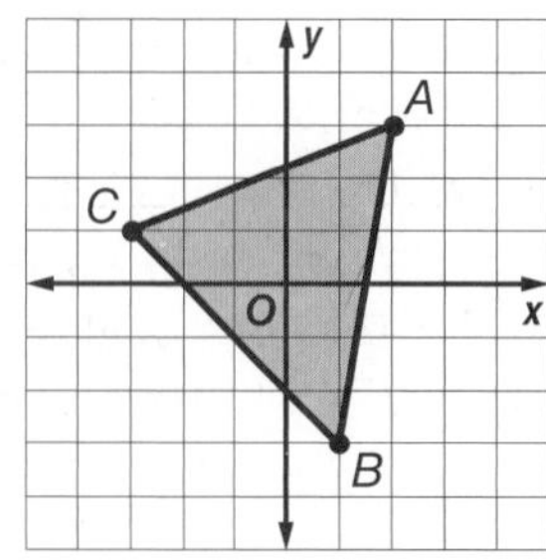

A 10 units^2 **D** 14 units^2

B 12 units^2 **E** 16 units^2

C 13 units^2

Spiral Review

Determine whether each matrix product is defined. If so, state the dimensions of the product. (Lesson 3-6)

66. $A_{4 \times 2} \cdot B_{2 \times 6}$

67. $C_{5 \times 4} \cdot D_{5 \times 3}$

68. $E_{2 \times 7} \cdot F_{7 \times 1}$

69. BUSINESS The table lists the prices at the Sandwich Shoppe. (Lesson 3-5)

a. List the prices in a 4×3 matrix.

b. The manager decides to cut the prices of every item by 20%. List this new set of data in a 4×3 matrix.

c. Subtract the second matrix from the first and determine the savings to the customer for each sandwich.

Sandwich	Small	Medium	Large
ham	\$4.50	\$6.75	\$9.50
salami	\$4.50	\$6.75	\$9.50
veggie	\$4.00	\$6.25	\$8.75
meatball	\$4.75	\$7.50	\$10.25

Graph each function. (Lesson 2-6)

70. $f(x) = 2|x - 3| - 4$

71. $f(x) = -3|2x| + 4$

72. $f(x) = |3x - 1| + 2$

Skills Review

Solve each system of equations.

73. $2x - 5y = -26$
$5x + 3y = -34$

74. $4y + 6x = 10$
$2x - 7y = 22$

75. $-3x - 2y = 17$
$-4x + 5y = -8$

LESSON 3-8 Solving Systems of Equations Using Inverse Matrices

Then	Now	Why?
You solved systems of linear equations algebraically.	**1** Find the inverse of a 2×2 matrix. **2** Write and solve matrix equations for a system of equations.	Maria's Sandwich Shop offers three lunch options as shown at the right. To determine how much each individual item costs, you can solve the following matrix equation in which w represents the cost of a sandwich, s the cost of a side, and d the cost of a drink.

$$\begin{bmatrix} 1 & 2 & 0 \\ 2 & 2 & 2 \\ 4 & 3 & 4 \end{bmatrix}\begin{bmatrix} w \\ s \\ d \end{bmatrix} = \begin{bmatrix} 9 \\ 16.50 \\ 30.75 \end{bmatrix}$$

NewVocabulary
identity matrix
square matrix
inverse matrix
matrix equation
variable matrix
constant matrix

Common Core State Standards

Content Standards
A.CED.3 Represent constraints by equations or inequalities, and by systems of equations and/or inequalities, and interpret solutions as viable or nonviable options in a modeling context.

Mathematical Practices
5 Use appropriate tools strategically.

1 Identity and Inverse Matrices

Recall that in real numbers, two numbers are multiplicative inverses if their product is the multiplicative identity, 1. Similarly, for matrices, the **identity matrix** is a *square matrix* that, when multiplied by another matrix, equals that same matrix. A **square matrix** is a matrix with the same number of rows and columns.

2 × 2 Identity Matrix

$$\begin{bmatrix} 1 & 0 \\ 0 & 1 \end{bmatrix}$$

3 × 3 Identity Matrix

$$\begin{bmatrix} 1 & 0 & 0 \\ 0 & 1 & 0 \\ 0 & 0 & 1 \end{bmatrix}$$

KeyConcept Identity Matrix for Multiplication

Words The identity matrix for multiplication I is a square matrix with 1 for every element of the main diagonal, from upper left to lower right, and 0 in all other positions. For any square matrix A of the same dimension as I, $A \cdot I = I \cdot A = A$.

Symbols If $A = \begin{bmatrix} a & b \\ c & d \end{bmatrix}$, then $I = \begin{bmatrix} 1 & 0 \\ 0 & 1 \end{bmatrix}$ such that

$$\begin{bmatrix} a & b \\ c & d \end{bmatrix} \cdot \begin{bmatrix} 1 & 0 \\ 0 & 1 \end{bmatrix} = \begin{bmatrix} 1 & 0 \\ 0 & 1 \end{bmatrix} \cdot \begin{bmatrix} a & b \\ c & d \end{bmatrix} = \begin{bmatrix} a & b \\ c & d \end{bmatrix}.$$

Two $n \times n$ matrices are **inverses** of each other if their product is the identity matrix. If matrix A has an inverse symbolized by A^{-1}, then $A \cdot A^{-1} = A^{-1} \cdot A = I$.

Example 1 Verify Inverse Matrices

Determine whether the matrices in each pair are inverses.

a. $A = \begin{bmatrix} -4 & 2 \\ -2 & 1 \end{bmatrix}$ **and** $B = \begin{bmatrix} \frac{1}{4} & -\frac{1}{2} \\ \frac{1}{2} & -1 \end{bmatrix}$

If A and B are inverses, then $A \cdot B = B \cdot A = I$.

$$A \cdot B = \begin{bmatrix} -4 & 2 \\ -2 & 1 \end{bmatrix} \cdot \begin{bmatrix} \frac{1}{4} & -\frac{1}{2} \\ \frac{1}{2} & -1 \end{bmatrix}$$ Write an equation.

$$= \begin{bmatrix} -1 + 1 & 2 - 2 \\ -\frac{1}{2} + \frac{1}{2} & 1 - 1 \end{bmatrix} \text{ or } \begin{bmatrix} 0 & 0 \\ 0 & 0 \end{bmatrix}$$ Matrix multiplication

Since $A \cdot B \neq I$, they are *not* inverses.

b. $F = \begin{bmatrix} 3 & -5 \\ -2 & 6 \end{bmatrix}$ **and** $G = \begin{bmatrix} \frac{3}{4} & \frac{5}{8} \\ \frac{1}{4} & \frac{3}{8} \end{bmatrix}$

If F and G are inverses, then $F \cdot G = G \cdot F = I$.

$$F \cdot G = \begin{bmatrix} 3 & -5 \\ -2 & 6 \end{bmatrix} \cdot \begin{bmatrix} \frac{3}{4} & \frac{5}{8} \\ \frac{1}{4} & \frac{3}{8} \end{bmatrix}$$ Write an equation.

$$= \begin{bmatrix} \frac{9}{4} - \frac{5}{4} & \frac{15}{8} - \frac{15}{8} \\ -\frac{6}{4} + \frac{6}{4} & -\frac{10}{8} + \frac{18}{8} \end{bmatrix} \text{ or } \begin{bmatrix} 1 & 0 \\ 0 & 1 \end{bmatrix}$$ Matrix multiplication

$$G \cdot F = \begin{bmatrix} \frac{3}{4} & \frac{5}{8} \\ \frac{1}{4} & \frac{3}{8} \end{bmatrix} \cdot \begin{bmatrix} 3 & -5 \\ -2 & 6 \end{bmatrix}$$ Write an equation.

$$= \begin{bmatrix} \frac{9}{4} - \frac{10}{8} & -\frac{15}{4} + \frac{30}{8} \\ \frac{3}{4} - \frac{6}{8} & -\frac{5}{4} + \frac{18}{8} \end{bmatrix} \text{ or } \begin{bmatrix} 1 & 0 \\ 0 & 1 \end{bmatrix}$$ Matrix multiplication

Since $F \cdot G = G \cdot F = I$, F and G are inverses.

StudyTip

CCSS Structure Since multiplication of matrices is not commutative, it is necessary to check the product in both orders.

GuidedPractice

1. Determine whether $X = \begin{bmatrix} 4 & -1 \\ 2 & -2 \end{bmatrix}$ and $Y = \begin{bmatrix} \frac{1}{3} & -\frac{1}{6} \\ \frac{1}{3} & -\frac{2}{3} \end{bmatrix}$ are inverses of each other.

Some matrices do not have inverses. You can determine whether a matrix has an inverse by using the determinant.

KeyConcept Inverse of a 2 × 2 Matrix

The inverse of matrix $A = \begin{bmatrix} a & b \\ c & d \end{bmatrix}$ is $A^{-1} = \frac{1}{ad - bc}\begin{bmatrix} d & -b \\ -c & a \end{bmatrix}$, where $ad - bc \neq 0$.

Notice that $ad - bc$ is the value of det A. Therefore, if the value of the determinant of a matrix is 0, the matrix cannot have an inverse.

Example 2 Find the Inverse of a Matrix

Find the inverse of each matrix, if it exists.

a. $P = \begin{bmatrix} 7 & -5 \\ 2 & -1 \end{bmatrix}$

$\begin{vmatrix} 7 & -5 \\ 2 & -1 \end{vmatrix} = -7 - (-10) \text{ or } 3$ Find the determinant.

Since the determinant does not equal 0, P^{-1} exists.

$P^{-1} = \frac{1}{ad - bc}\begin{bmatrix} d & -b \\ -c & a \end{bmatrix}$ Definition of inverse

$= \frac{1}{7(-1) - (-5)(2)}\begin{bmatrix} -1 & 5 \\ -2 & 7 \end{bmatrix}$ $a = 7, b = -5, c = 2, d = -1$

$= \frac{1}{3}\begin{bmatrix} -1 & 5 \\ -2 & 7 \end{bmatrix} \text{ or } \begin{bmatrix} -\frac{1}{3} & \frac{5}{3} \\ -\frac{2}{3} & \frac{7}{3} \end{bmatrix}$ Simplify.

CHECK Find the product of the matrices. If the product is I, then they are inverses.

$$\begin{bmatrix} 7 & -5 \\ 2 & -1 \end{bmatrix} \cdot \begin{bmatrix} -\frac{1}{3} & \frac{5}{3} \\ -\frac{2}{3} & \frac{7}{3} \end{bmatrix} = \begin{bmatrix} -\frac{7}{3} + \frac{10}{3} & \frac{35}{3} - \frac{35}{3} \\ -\frac{2}{3} + \frac{2}{3} & \frac{10}{3} - \frac{7}{3} \end{bmatrix} \text{ or } \begin{bmatrix} 1 & 0 \\ 0 & 1 \end{bmatrix} \checkmark$$

b. $Q = \begin{bmatrix} -8 & -6 \\ 12 & 9 \end{bmatrix}$

$\begin{vmatrix} -8 & -6 \\ 12 & 9 \end{vmatrix} = -72 - (-72) = 0$ Find the determinant.

Since the determinant equals 0, Q^{-1} does not exist.

Guided Practice

2A. $\begin{bmatrix} 3 & 7 \\ 1 & -4 \end{bmatrix}$

2B. $\begin{bmatrix} 2 & 1 \\ -4 & 3 \end{bmatrix}$

Math History Link

Seki Kowa **(1642–1708)** Known as The Arithmetical Sage, Seki Kowa was the first to develop the theory of determinants.

2 Matrix Equations

Matrices can be used to represent and solve systems of equations. You can write a **matrix equation** to solve the system of equations below.

$$\begin{aligned} x + 2y &= 9 \\ 3x - 6y &= 3 \end{aligned} \rightarrow \begin{bmatrix} x + 2y \\ 3x - 6y \end{bmatrix} = \begin{bmatrix} 9 \\ 3 \end{bmatrix}$$

Write the left side of the matrix equation as the product of the coefficient matrix and the variable matrix. Write the right side as a constant matrix.

$$\begin{matrix} A & \cdot & X & = & B \\ \begin{bmatrix} 1 & 2 \\ 3 & -6 \end{bmatrix} & \cdot & \begin{bmatrix} x \\ y \end{bmatrix} & = & \begin{bmatrix} 9 \\ 3 \end{bmatrix} \end{matrix}$$

coefficient matrix

variable matrix only the variables of a system

constant matrix only the constants of a system

Courtesy Ichinoseki City Museum, Japan

StudyTip

Inverse You can use this method to solve systems of equations only if A has an inverse. If A does not have an inverse, then the system either has no solution or infinitely many solutions.

Then solve the matrix equation in the same way that you would solve any other equation.

$ax = b$	Write the equation.	$AX = B$
$\left(\frac{1}{a}\right)ax = \left(\frac{1}{a}\right)b$	Multiply each side by the inverse of the coefficient, if it exists.	$A^{-1}AX = A^{-1}B$
$1x = \frac{b}{a}$	$\left(\frac{1}{a}\right)a = 1$, $A^{-1}A = I$	$IX = A^{-1}B$
$x = \frac{b}{a}$	$1x = x$, $IX = X$	$X = A^{-1}B$

Notice that the solution of the matrix equation is the product of the inverse of the coefficient matrix and the constant matrix.

Real-WorldLink

Average gas prices increased fivefold from \$0.70 per gallon in 1977 to \$3.50 per gallon in 2007.

Source: U.S. Department of Energy

Real-World Example 3 Solve a System of Equations

TRAVEL Helena stopped for gasoline twice during a car trip. The price of gasoline at the first station where she stopped was \$3.75 per gallon. At the second station, the price was \$3.50 per gallon. Helena bought a total of 24.2 gallons of gasoline and spent \$88.05. How much gasoline did Helena buy at each gas station?

A system of equations to represent the situation is as follows.

$x + y = 24.2$
$3.75x + 3.50y = 88.05$

The matrix equation is $\begin{bmatrix} 1 & 1 \\ 3.75 & 3.50 \end{bmatrix} \cdot \begin{bmatrix} x \\ y \end{bmatrix} = \begin{bmatrix} 24.2 \\ 88.05 \end{bmatrix}$.

Step 1 Find the inverse of the coefficient matrix.

$$A^{-1} = \frac{1}{3.50 - 3.75}\begin{bmatrix} 3.50 & -1 \\ -3.75 & 1 \end{bmatrix} \text{ or } -\frac{1}{0.25}\begin{bmatrix} 3.50 & -1 \\ -3.75 & 1 \end{bmatrix}$$

Step 2 Multiply each side of the matrix equation by the inverse matrix.

$$-\frac{1}{0.25}\begin{bmatrix} 3.50 & -1 \\ -3.75 & 1 \end{bmatrix} \cdot \begin{bmatrix} 1 & 1 \\ 3.75 & 3.50 \end{bmatrix} \cdot \begin{bmatrix} x \\ y \end{bmatrix} = -\frac{1}{0.25}\begin{bmatrix} 3.50 & -1 \\ -3.75 & 1 \end{bmatrix} \cdot \begin{bmatrix} 24.2 \\ 88.05 \end{bmatrix}$$

$$\begin{bmatrix} 1 & 0 \\ 0 & 1 \end{bmatrix} \cdot \begin{bmatrix} x \\ y \end{bmatrix} = -\frac{1}{0.25}\begin{bmatrix} -3.35 \\ -2.70 \end{bmatrix}$$

$$\begin{bmatrix} x \\ y \end{bmatrix} = \begin{bmatrix} 13.4 \\ 10.8 \end{bmatrix}$$

The solution is (13.4, 10.8), where x represents the amount of gas Helena purchased at the first gas station, and y represents the amount purchased at the second gas station.

CHECK You can check your answer by using inverses.

Enter $\begin{bmatrix} 1 & 1 \\ 3.75 & 3.50 \end{bmatrix}$ as matrix A.

Enter $\begin{bmatrix} 24.2 \\ 88.05 \end{bmatrix}$ as matrix B.

Multiply the inverse of A by B.

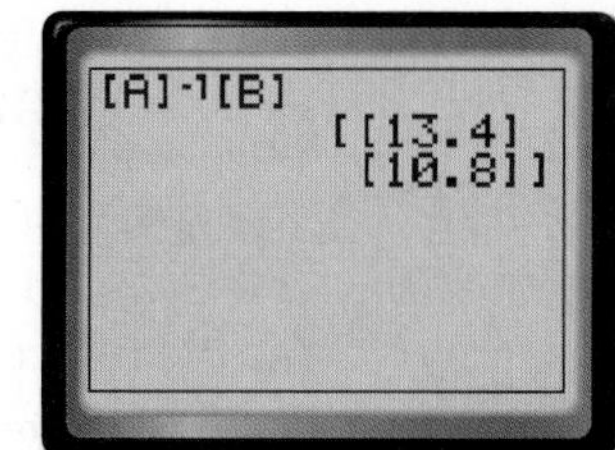

GuidedPractice

3. COMIC BOOKS Dante and Erica just returned from a comic book store that sells new and used comic books. Dante spent \$11.25 on 3 new and 4 old books, and Erica spent \$15.75 on 10 used and 3 new ones. If comic books of one type are sold at the same price, what is the price in dollars of a new comic book?

Tetra Images/Alamy

Check Your Understanding

= Step-by-Step Solutions begin on page R14.

Example 1 **Determine whether the matrices in each pair are inverses.**

1. $A = \begin{bmatrix} 2 & 1 \\ -1 & 0 \end{bmatrix}, B = \begin{bmatrix} 1 & 2 \\ 2 & 1 \end{bmatrix}$

2. $C = \begin{bmatrix} 2 & 1 \\ 5 & 3 \end{bmatrix}, D = \begin{bmatrix} 2 & 1 \\ 5 & -3 \end{bmatrix}$

3. $F = \begin{bmatrix} -1 & 1 \\ 0 & -1 \end{bmatrix}, G = \begin{bmatrix} -1 & -1 \\ 0 & -1 \end{bmatrix}$

4. $H = \begin{bmatrix} -3 & -1 \\ -4 & -2 \end{bmatrix}, J = \begin{bmatrix} -1 & 2 \\ 3 & -4 \end{bmatrix}$

Example 2 **Find the inverse of each matrix, if it exists.**

5. $\begin{bmatrix} 6 & -3 \\ -1 & 0 \end{bmatrix}$

6. $\begin{bmatrix} 2 & -4 \\ -3 & 0 \end{bmatrix}$

7. $\begin{bmatrix} -3 & 0 \\ 5 & 2 \end{bmatrix}$

8. $\begin{bmatrix} 2 & 4 \\ 1 & 2 \end{bmatrix}$

Example 3 **Use a matrix equation to solve each system of equations.**

9. $-2x + y = 9$
$x + y = 3$

10. $4x - 2y = 22$
$6x + 9y = -3$

11. $-2x + y = -4$
$3x + y = 1$

12. MONEY Kevin had 25 quarters and dimes. The total value of all the coins was $4. How many quarters and dimes did Kevin have?

Practice and Problem Solving

Extra Practice is on page R3.

Example 1 **Determine whether each pair of matrices are inverses of each other.**

13. $K = \begin{bmatrix} 1 & 2 \\ 3 & 0 \end{bmatrix}, L = \begin{bmatrix} 0 & 1 \\ 2 & -1 \end{bmatrix}$

14. $M = \begin{bmatrix} 0 & 2 \\ 4 & 5 \end{bmatrix}, N = \begin{bmatrix} 1 & 1 \\ 0 & 0 \end{bmatrix}$

15. $P = \begin{bmatrix} 4 & 0 \\ 3 & 0 \end{bmatrix}, Q = \begin{bmatrix} -1 & -1 \\ \frac{2}{3} & 5 \end{bmatrix}$

16. $R = \begin{bmatrix} \frac{1}{2} & -\frac{1}{4} \\ \frac{1}{4} & -\frac{1}{2} \end{bmatrix}, S = \begin{bmatrix} 2 & 4 \\ 4 & 2 \end{bmatrix}$

Example 2 **Find the inverse of each matrix, if it exists.**

17. $\begin{bmatrix} 3 & 0 \\ 0 & 2 \end{bmatrix}$

18. $\begin{bmatrix} 2 & 3 \\ 3 & 2 \end{bmatrix}$

19. $\begin{bmatrix} 3 & 0 \\ 5 & 1 \end{bmatrix}$

20. $\begin{bmatrix} 1 & -1 \\ -6 & -1 \end{bmatrix}$

21. $\begin{bmatrix} -5 & -4 \\ 4 & 2 \end{bmatrix}$

22. $\begin{bmatrix} -5 & 9 \\ 4 & -8 \end{bmatrix}$

23. $\begin{bmatrix} 6 & -5 \\ 4 & 9 \end{bmatrix}$

24. $\begin{bmatrix} -4 & -2 \\ 7 & 8 \end{bmatrix}$

25. $\begin{bmatrix} -6 & 8 \\ 8 & -7 \end{bmatrix}$

Example 3 **26. BAKING** Peggy is preparing a colored frosting for a cake. For the right shade of purple, she needs 25 milliliters of a 44% concentration food coloring. The store has a 25% red and a 50% blue concentration of food coloring. How many milliliters each of blue food coloring and red food coloring should be mixed to make the necessary amount of purple food coloring?

CCSS PERSEVERANCE **Use a matrix equation to solve each system of equations.**

27. $-x + y = 4$
$-x + y = -4$

28. $-x + y = 3$
$-2x + y = 6$

29. $x + y = 4$
$-4x + y = 9$

30. $3x + y = 3$
$5x + 3y = 6$

31. $y - x = 5$
$2y - 2x = 8$

32. $4x + 2y = 6$
$6x - 3y = 9$

33. $1.6y - 0.2x = 1$
$0.4y - 0.1x = 0.5$

34. $4y - x = -2$
$3y - x = 6$

35. $2y - 4x = 3$
$4x - 3y = -6$

36. POPULATIONS The diagram shows the annual percent migration between a city and its suburbs.

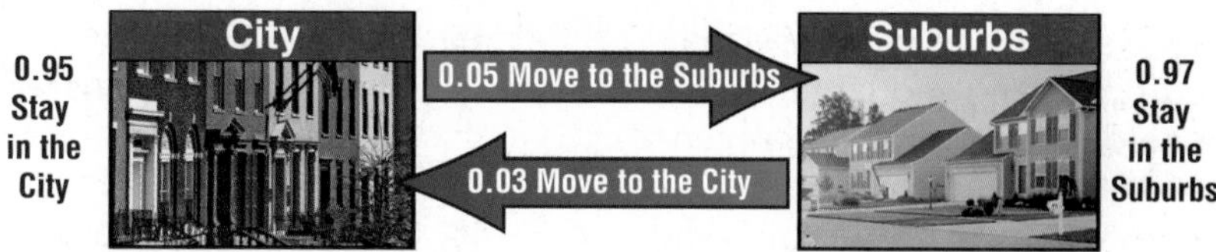

a. Write a matrix to represent the transitions in city population and suburb population.

b. There are currently 16,275 people living in the city and 17,552 people living in the suburbs. Assuming that the trends continue, predict the number of people who will live in the suburbs next year.

c. Use the inverse of the matrix from part b to find the number of people who lived in the city last year.

37 **MUSIC** The diagram shows the trends in digital audio player and portable CD player ownership over the past five years for Central City. Every person in Central City has either a digital audio player or a portable CD player. Central City has a stable population of 25,000 people, of whom 17,252 own digital audio players and 7748 own portable CD players.

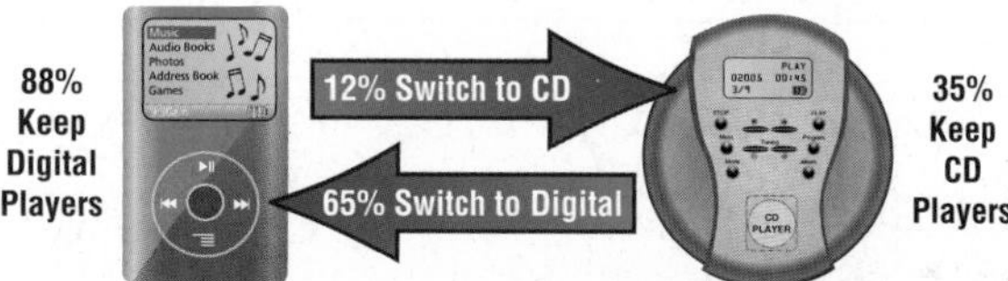

a. Write a matrix to represent the transitions in player ownership.

b. Assume that the trends continue. Predict the number of people who will own digital audio players next year.

c. Use the inverse of the matrix from part **b** to find the number of people who owned digital audio players last year.

H.O.T. Problems Use Higher-Order Thinking Skills

38. CCSS **CRITIQUE** Cody and Megan are setting up matrix equations for the system $5x + 7y = 19$ and $3y + 4x = 10$. Is either of them correct? Explain your reasoning.

Cody

$\begin{bmatrix} 5 & 7 \\ 3 & 4 \end{bmatrix}\begin{bmatrix} x \\ y \end{bmatrix} = \begin{bmatrix} 19 \\ 10 \end{bmatrix}$

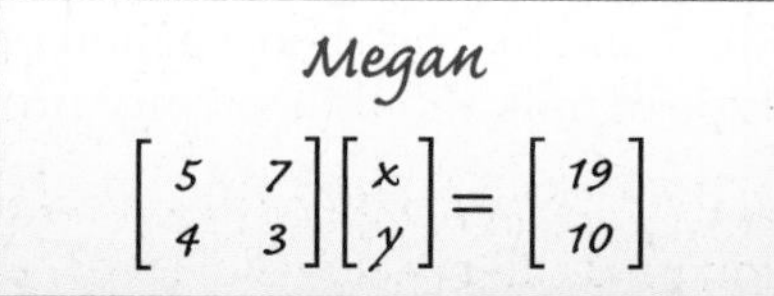

39. CHALLENGE Describe what a matrix equation with infinite solutions looks like.

40. REASONING Determine whether the following statement is *always, sometimes,* or *never* true. Explain your reasoning.

A square matrix has a multiplicative inverse.

41. OPEN ENDED Write a matrix equation that does not have a solution.

42. **WRITING IN MATH** When would you prefer to solve a system of equations using algebraic methods, and when would you prefer to use matrices? Explain.

Standardized Test Practice

43. The Yogurt Shoppe sells cones in three sizes: small, \$0.89; medium, \$1.19; and large, \$1.39. One day Santos sold 52 cones. He sold seven more medium cones than small cones. If he sold \$58.98 in cones, how many medium cones did he sell?

A 11 B 17 C 24 D 36

44. The chart shows an expression evaluated for different values of x. A student concludes that for all values of x, $x^2 + x + 1$ produces a prime number. Which value of x serves as a counterexample to prove this conclusion false?

x	$x^2 + x + 1$
1	3
2	7
3	13
4	21

F −4 G −3 H 2 J 4

45. SHORT RESPONSE What is the solution of the system of equations $6a + 8b = 5$ and $10a - 12b = 2$?

46. SAT/ACT Each year at Capital High School the students vote to choose the theme of the homecoming dance. The theme "A Night Under the Stars" received 225 votes, and "The Time of My Life" received 480 votes. If 40% of girls voted for "A Night Under the Stars" and 75% of boys voted for "The Time of My Life," how many girls and boys voted?

A 176 boys and 351 girls
B 395 boys and 310 girls
C 380 boys and 325 girls
D 705 boys and 325 girls
E 854 boys and 176 girls

Spiral Review

Evaluate each determinant. (Lesson 3-7)

47. $\begin{vmatrix} 8 & -3 \\ 6 & -9 \end{vmatrix}$

48. $\begin{vmatrix} 9 & -7 \\ -5 & -3 \end{vmatrix}$

49. $\begin{vmatrix} 8 & 6 & -1 \\ -4 & 5 & 1 \\ -3 & -2 & 9 \end{vmatrix}$

Find each product, if possible. (Lesson 3-6)

50. $\begin{bmatrix} 4 & 2 \\ -1 & -3 \end{bmatrix} \cdot \begin{bmatrix} 6 & 2 \\ 5 & 1 \end{bmatrix}$

51. $\begin{bmatrix} 8 & -2 \\ -4 & -5 \end{bmatrix} \cdot \begin{bmatrix} -2 \\ 3 \end{bmatrix}$

52. $\begin{bmatrix} -3 \\ -4 \end{bmatrix} \cdot \begin{bmatrix} -6 & -8 \\ -4 & 5 \end{bmatrix}$

53. MILK The Yoder Family Dairy produces at most 200 gallons of skim and whole milk each day for delivery to large bakeries and restaurants. Regular customers require at least 15 gallons of skim and 21 gallons of whole milk each day. If the profit on a gallon of skim milk is \$0.82 and the profit on a gallon of whole milk is \$0.75, how many gallons of each type of milk should the dairy produce each day to maximize profits? (Lesson 3-4)

Skills Review

Identify the type of the function represented by each graph.

54.

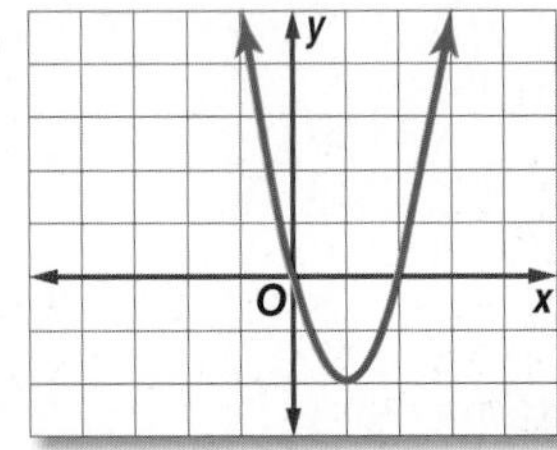

55.

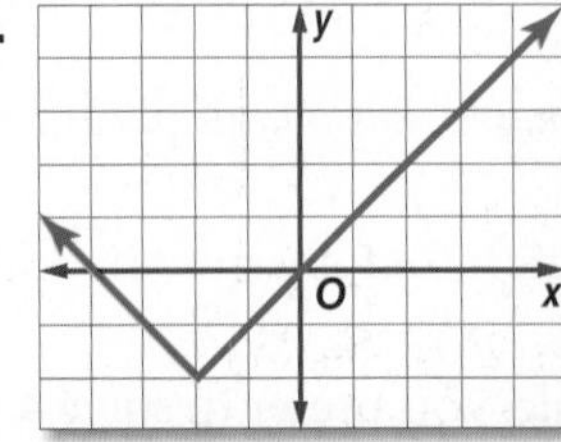

56.

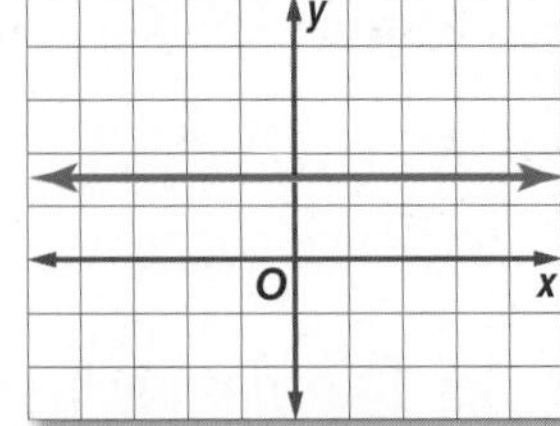

EXTEND 3-8

Graphing Technology Lab
Augmented Matrices

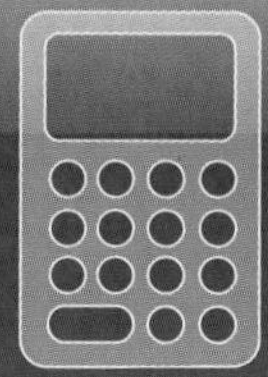

Using a TI-83/84 Plus graphing calculator, you can solve a system of linear equations using the MATRIX function. An **augmented matrix** contains the coefficient matrix with an extra column containing the constant terms. You can use a graphing calculator to reduce the augmented matrix so that the solution of the system of equations can be easily determined.

Example

Write an augmented matrix for the following system of equations. Then solve the system by using a graphing calculator.

$2x + y + z = 1$
$3x + 2y + 3z = 12$
$4x + y + 2z = -1$

Step 1 Write the augmented matrix and enter it into a calculator.

The augmented matrix $B = \left[\begin{array}{ccc:c} 2 & 1 & 1 & 1 \\ 3 & 2 & 3 & 12 \\ 4 & 1 & 2 & -1 \end{array}\right]$.

Begin by entering the matrix.

KEYSTROKES: 2nd [MATRIX] ▶ ▶ ENTER 3 ENTER 4 ENTER 2 ENTER 1 ENTER 1 ENTER 1 ENTER 3 ENTER 2 ENTER 3 ENTER 12 ENTER 4 ENTER 1 ENTER 2 ENTER (–) 1 ENTER

Step 2 Find the reduced row echelon form (rref) using the graphing calculator.

KEYSTROKES: 2nd [QUIT] 2nd [MATRIX] ▶ ALPHA [B] 2nd [MATRIX] ENTER) ENTER

Study the reduced echelon matrix. The first three columns are the same as a 3 × 3 identity matrix. The first row represents $x = -4$, the second row represents $y = 3$, and the third row represents $z = 6$. The solution is $(-4, 3, 6)$.

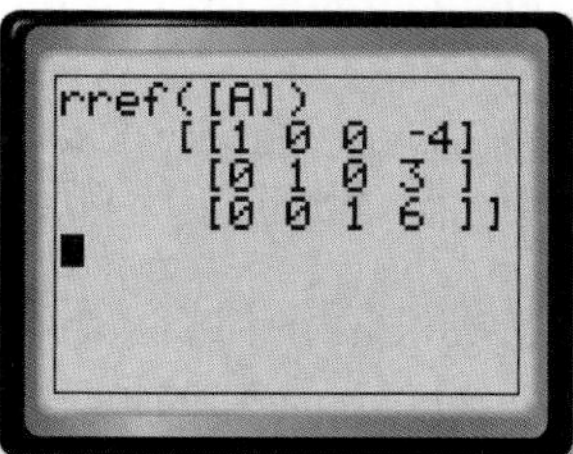

Exercises

Write an augmented matrix for each system of equations. Then solve with a graphing calculator.

1. $3x + 2y = -4$
$4x + 7y = 13$

2. $2x + y = 6$
$6x - 2y = 0$

3. $2x + 2y = -4$
$7x + 3y = 10$

4. $4x + 6y = 0$
$8x - 2y = 7$

5. $6x - 4y + 2z = -4$
$2x - 2y + 6z = 10$
$2x + 2y + 2z = -2$

6. $5x - 5y + 5z = 10$
$5x - 5z = 5$
$5y + 10z = 0$

CHAPTER 3

Study Guide and Review

Study Guide

KeyConcepts

Systems of Equations and Inequalities (Lessons 3-1 and 3-2)

- In the substitution method, one equation is solved for a variable and substituted to find the value of another variable. In the elimination method, one variable is eliminated by adding or subtracting the equations.
- The solution of a system of inequalities is found by graphing the inequalities and determining the intersection of the graphs.

Linear Programming (Lesson 3-3)

- Linear programming is a method for finding maximum or minimum values of a function over a given system of inequalities with each inequality representing a constraint.

Systems of Equations in Three Variables (Lesson 3-4)

- A system of equations in three variables can be solved algebraically by using the substitution method or the elimination method.

Operations with Matrices (Lessons 3-5 and 3-6)

- Matrices can be added or subtracted if they have the same dimensions. Add or subtract corresponding elements.
- Two matrices can be multiplied if and only if the number of columns in the first matrix is equal to the number of rows in the second matrix.

Solving Systems Using Cramer's Rule (Lesson 3-7)

- If a determinant is nonzero, then the system has a unique solution. If the determinant is 0, then the system either has no solution or infinite solutions.

Solving Systems with Inverse Matrices (Lesson 3-8)

- An identity matrix is a square matrix with ones on the main diagonal and zeros in the other positions.
- Two matrices are inverses of each other if their product is the identity matrix.
- To solve a matrix equation, find the inverse of the coefficient matrix. Then multiply each side of the equation by the inverse of the coefficient matrix.

FOLDABLES StudyOrganizer

Be sure the Key Concepts are noted in your Foldable.

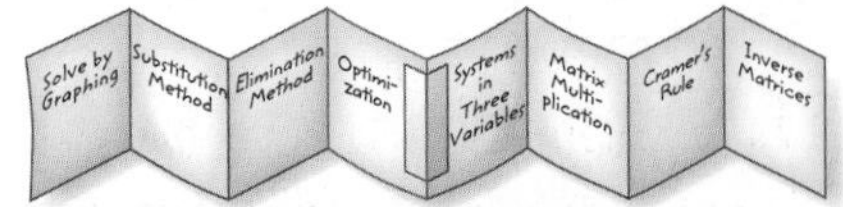

KeyVocabulary

bounded (p. 154)
break-even point (p. 136)
coefficient matrix (p. 192)
consistent (p. 137)
constant matrix (p. 200)
Cramer's Rule (p. 192)
dependent (p. 137)
determinant (p. 189)
diagonal rule (p. 190)
dimensions (p. 169)
elimination method (p. 139)
feasible region (p.154)
identity matrix (p. 198)
inconsistent (p. 137)
independent (p. 137)
inverse matrices (p. 198)
matrix (p. 169)
matrix equation (p. 200)
optimize (p. 156)
ordered triple (p. 161)
scalar (p. 173)
scalar multiplication (p. 173)
substitution method (p. 138)
unbounded (p. 154)
variable matrix (p. 200)

VocabularyCheck

Choose the term from above to complete each sentence.

1. A feasible region that is open and can go on forever is called __________.
2. To __________ means to seek the best price or profit using linear programming.
3. A matrix that contains the constants in a system of equations is called a(n) __________.
4. A matrix can be multiplied by a constant called a(n) __________.
5. The __________ of a matrix with 4 rows and 3 columns are 4×3.
6. A system of equations is __________ if it has at least one solution.
7. The __________ matrix is a square matrix that, when multiplied by another matrix, equals that same matrix.
8. The __________ is the point at which income equals cost.
9. A system of equations is __________ if it has no solutions.
10. If the product of two matrices is the identity matrix, they are __________.

Lesson-by-Lesson Review

3-1 Solving Systems of Equations

Solve each system of equations by graphing.

11. $3x + 4y = 8$
$x - 3y = -6$

12. $x + \frac{8}{3}y = 12$
$\frac{1}{2}x + \frac{4}{3}y = 6$

13. $y - 3x = 13$
$y = \frac{1}{3}x + 5$

14. $6x - 14y = 5$
$3x - 7y = 5$

15. **LAWN CARE** André and Paul each mow lawns. André charges a \$30 service fee and \$10 per hour. Paul charges a \$10 service fee and \$15 per hour. After how many hours will André and Paul charge the same amount?

Solve each system of equations by using either substitution or elimination.

16. $x + y = 6$
$3x - 2y = -2$

17. $5x - 2y = 4$
$-2y + x = 12$

18. $x + y = 3.5$
$x - y = 7$

19. $3y - 5x = 0$
$2y - 4x = -2$

20. **SCHOOL SUPPLIES** At an office supply store, Emilio bought 3 notebooks and 5 pens for \$13.75. If a notebook costs \$1.25 more than a pen, how much does a notebook cost? How much does a pen cost?

Example 1

Solve the system of equations by graphing.

$x + y = 4$

$x + 2y = 5$

Graph both equations on the coordinate plane.

The solution of the system is (3, 1).

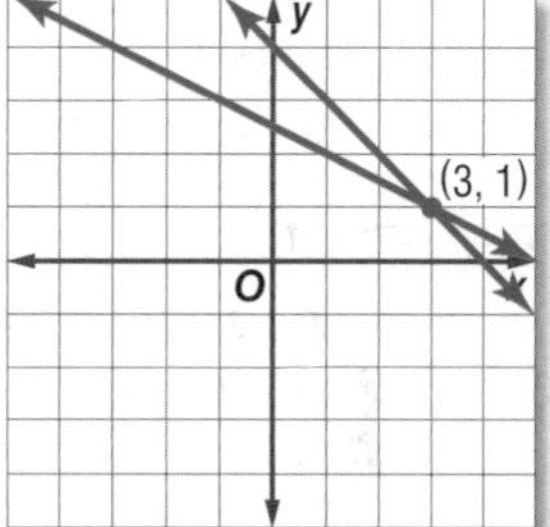

Example 2

Solve the system of equations by using either substitution or elimination.

$3x + 2y = 1$

$y = -x + 1$

Substitute $-x + 1$ for y in the first equation. Then solve for y.

$$\begin{aligned} 3x + 2y &= 1 \\ 3x + 2(-x + 1) &= 1 \\ 3x - 2x + 2 &= 1 \\ x + 2 &= 1 \\ x &= -1 \end{aligned}$$

$$\begin{aligned} y &= -x + 1 \\ &= -(-1) + 1 \text{ or } 2 \end{aligned}$$

The solution is (−1, 2).

3-2 Solving Systems of Inequalities by Graphing

Solve each system of inequalities by graphing.

21. $y < 2x - 3$
$y \geq 4$

22. $|y| > 2$
$x > 3$

23. $y \geq x + 3$
$2y \leq x - 5$

24. $y > x + 1$
$x < -2$

25. **JEWELRY** Payton makes jewelry to sell at her mother's clothing store. She spends no more than 3 hours making jewelry on Saturdays. It takes her 15 minutes to set up her supplies and 25 minutes to make each bracelet. Draw a graph that represents this.

Example 3

Solve the system of inequalities by graphing.

$y \geq \frac{3}{2}x - 3$

$y < 4 - 2x$

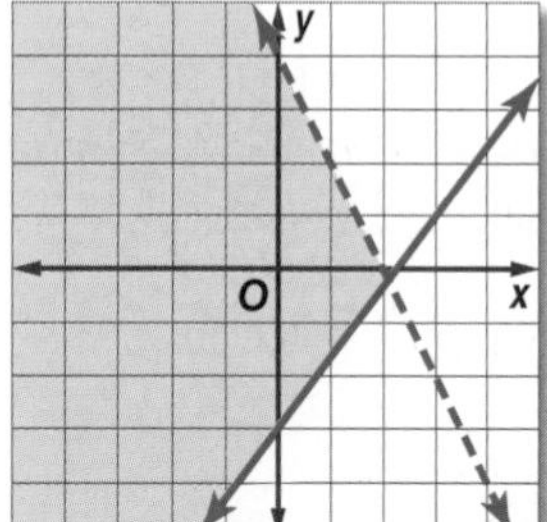

The solution of the system is the region that satisfies both inequalities. The solution of this system is the shaded region.

Study Guide and Review *Continued*

3-3 Optimization with Linear Programming

26. FLOWERS A florist can make a grand arrangement in 18 minutes or a simple arrangement in 10 minutes. The florist makes at least twice as many of the simple arrangements as the grand arrangements. The florist can work only 40 hours per week. The profit on the simple arrangements is \$10 and the profit on the grand arrangements is \$25. Find the number and type of arrangements that the florist should produce to maximize profit.

27. MANUFACTURING A shoe manufacturer makes outdoor and indoor soccer shoes. There is a two-step process for both kinds of shoes. Each pair of outdoor shoes requires 2 hours in step one and 1 hour in step two, and produces a profit of \$20. Each pair of indoor shoes requires 1 hour in step one and 3 hours in step two, and produces a profit of \$15. The company has 40 hours of labor available per day for step one and 60 hours available for step two. What is the manufacturer's maximum profit? What is the combination of shoes for this profit?

Example 4

A gardener is planting two types of herbs in a 5184-square-inch garden. Herb A requires 6 square inches of space, and herb B requires 24 square inches of space. The gardener will plant no more than 300 plants. If herb A can be sold for \$8 and herb B can be sold for \$12, how many of each herb should be sold to maximize income?

Let $a =$ the number of herb A and $b =$ the number of herb B.

$a \geq 0$, $b \geq 0$, $6a + 24b \leq 5184$, and $a + b \leq 300$

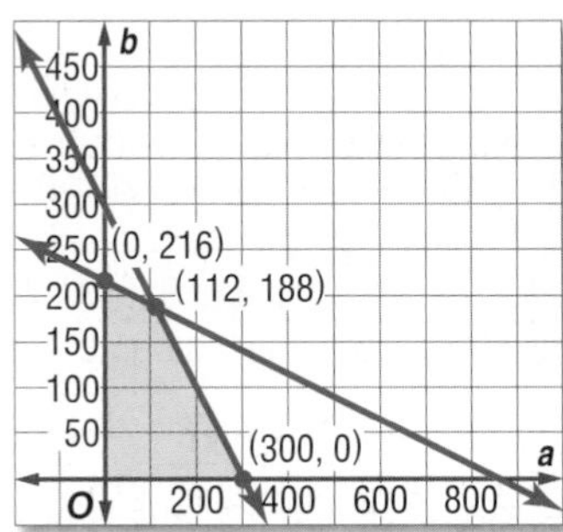

Graph the inequalities. The vertices of the feasible region are (0, 0), (300, 0), (0, 216), and (112, 188).

The profit function is $f(a, b) = 8a + 12b$.

The maximum value of \$3152 occurs at (112, 188). So the gardener should plant 112 of herb A and 188 of herb B.

3-4 Systems of Equations in Three Variables

Solve each system of equations.

28. $a - 4b + c = 3$
$b - 3c = 10$
$3b - 8c = 24$

29. $2x - z = 14$
$3x - y + 5z = 0$
$4x + 2y + 3z = -2$

30. AMUSEMENT PARKS Dustin, Luis, and Marci went to an amusement park. They purchased snacks from the same vendor. Their snacks and how much they paid are listed in the table. How much did each snack cost?

Name	Hot Dogs	Popcorn	Soda	Price
Dustin	1	2	3	\$15.25
Luis	2	0	3	\$14.00
Marci	1	2	1	\$10.25

Example 5

Solve the system of equations.

$$x + y + 2z = 6$$
$$2x + 5z = 12$$
$$x + 2y + 3z = 9$$

$$\begin{array}{rl} 2x + 2y + 4z = 12 & \text{Equation 1} \times 2 \\ (-)\ x + 2y + 3z = 9 & \text{Equation 3} \\ \hline x + z = 3 & \text{Subtract.} \end{array}$$

Solve the system of two equations.

$$\begin{array}{rl} 2x + 5z = 12 & \text{Equation 1} \\ (-)\ 2x + 2z = 6 & 2 \times (x + z = 3) \\ \hline 3z = 6 & \text{Subtract.} \\ z = 2 & \text{Divide each side by 2.} \end{array}$$

Substitute 2 for z in one of the equations with two variables, and solve for y. Then, substitute 2 for z and the value you got for y into an equation from the original system to solve for x.

The solution is (1, 1, 2).

3-5 Operations with Matrices

Perform the indicated operations. If the matrix does not exist, write *impossible.*

31. $3\left(\begin{bmatrix} -2 & 0 \\ 6 & 8 \end{bmatrix} + \begin{bmatrix} 1 & 9 \\ -3 & -4 \end{bmatrix}\right)$

32. $\begin{bmatrix} 2 \\ -6 \end{bmatrix} - \begin{bmatrix} -3 \\ 2 \end{bmatrix} + \begin{bmatrix} 6 \\ 0 \end{bmatrix}$

33. RETAIL Current Fashions buys shirts, jeans, and shoes from a manufacturer, marks them up, and then sells them. The table shows the purchase price and the selling price.

Item	Purchase Price	Selling Price
shirts	$15	$35
jeans	$25	$55
shoes	$30	$85

a. Write a matrix for the purchase price.

b. Write a matrix for the selling price.

c. Use matrix operations to find the profit on 1 shirt, 1 pair of jeans, and 1 pair of shoes.

Example 6

Find $2B + 3B$ if $A = \begin{bmatrix} 9 & 1 \\ 1 & 2 \end{bmatrix}$ and $B = \begin{bmatrix} 1 & 4 \\ 3 & 7 \end{bmatrix}$.

$2B = 2\begin{bmatrix} 1 & 4 \\ 3 & 7 \end{bmatrix}$ or $\begin{bmatrix} 2 & 8 \\ 6 & 14 \end{bmatrix}$

$3A = 3\begin{bmatrix} 9 & 1 \\ 1 & 2 \end{bmatrix}$ or $\begin{bmatrix} 27 & 3 \\ 3 & 6 \end{bmatrix}$

$2B + 3A = \begin{bmatrix} 2 & 8 \\ 6 & 14 \end{bmatrix} + \begin{bmatrix} 27 & 3 \\ 3 & 6 \end{bmatrix}$ or $\begin{bmatrix} 29 & 11 \\ 9 & 20 \end{bmatrix}$

Example 7

Find $3C - 5D$ if $C = \begin{bmatrix} 3 \\ -7 \end{bmatrix}$ and $D = [9 \quad 8]$.

$3C - 5D = 3\begin{bmatrix} 3 \\ -7 \end{bmatrix} - 5\,[9 \quad 8]$.

Because the dimensions are different, you cannot subtract the matrices.

3-6 Multiplying Matrices

Find each product, if possible.

34. $[3 \quad -7] \cdot \begin{bmatrix} 9 \\ -5 \end{bmatrix}$

35. $\begin{bmatrix} -3 & 0 & 2 \\ 6 & -1 & 5 \end{bmatrix} \cdot \begin{bmatrix} 8 & -1 \\ -4 & 3 \\ 6 & 7 \end{bmatrix}$

36. $\begin{bmatrix} 2 & 11 \\ 0 & -3 \\ -6 & 7 \end{bmatrix} \cdot \begin{bmatrix} 0 & 8 & -5 \\ 12 & 0 & 9 \\ 4 & -6 & 7 \end{bmatrix}$

37. GROCERIES Martin bought 1 gallon of milk, 2 apples, 4 frozen dinners, and 1 box of cereal. The following matrix shows the prices for each item, respectively.

[$2.59 $0.49 $5.25 $3.99]

Use matrix multiplication to find the total amount of money Martin spent at the grocery store.

Example 8

Find XY if $X = \begin{bmatrix} 0 & -6 \\ 3 & 5 \end{bmatrix}$ and $Y = \begin{bmatrix} 8 \\ -1 \end{bmatrix}$.

$XY = \begin{bmatrix} 0 & -6 \\ 3 & 5 \end{bmatrix} \cdot \begin{bmatrix} 8 \\ -1 \end{bmatrix}$ Write an equation.

$= \begin{bmatrix} 0(8) + (-6)(-1) \\ 3(8) + 5(-1) \end{bmatrix}$ Multiply columns by rows.

$= \begin{bmatrix} 6 \\ 19 \end{bmatrix}$ Simplify.

CHAPTER 3

Study Guide and Review *Continued*

3-7 Solving Systems of Equations Using Cramer's Rule

Evaluate each determinant.

38. $\begin{vmatrix} 2 & 4 \\ 7 & -3 \end{vmatrix}$

39. $\begin{vmatrix} 2 & 3 & -1 \\ 0 & 2 & 4 \\ -2 & 5 & 6 \end{vmatrix}$

Use Cramer's Rule to solve each system of equations.

40. $3x - y = 0$
$5x + 2y = 22$

41. $5x + 2y = 4$
$3x + 4y + 2z = 6$
$7x + 3y + 4z = 29$

42. **JEWELRY** Alana paid $98.25 for 3 necklaces and 2 pairs of earrings. Petra paid $133.50 for 2 necklaces and 4 pairs of earrings. Use Cramer's Rule to find out how much 1 necklace costs and how much 1 pair of earrings costs.

Example 9

Evaluate $\begin{vmatrix} 4 & -6 \\ 2 & 5 \end{vmatrix}$.

$\begin{vmatrix} 4 & -6 \\ 2 & 5 \end{vmatrix} = 4(5) - (-6)(2)$ Definition of determinant

$= 20 + 12 \text{ or } 32$ Simplify.

Example 10

Use Cramer's Rule to solve $2a + 6b = -1$ and $a + 8b = 2$.

$a = \dfrac{\begin{vmatrix} -1 & 6 \\ 2 & 8 \end{vmatrix}}{\begin{vmatrix} 2 & 6 \\ 1 & 8 \end{vmatrix}}$ Cramer's Rule $b = \dfrac{\begin{vmatrix} 2 & -1 \\ 1 & 2 \end{vmatrix}}{\begin{vmatrix} 2 & 6 \\ 1 & 8 \end{vmatrix}}$

$= \dfrac{-8 - 12}{16 - 6}$ Evaluate each determinant. $= \dfrac{4 + 1}{16 - 6}$

$= \dfrac{-20}{10} \text{ or } -2$ Simplify. $= \dfrac{5}{10} \text{ or } \dfrac{1}{2}$

The solution is $\left(-2, \frac{1}{2}\right)$.

3-8 Solving Systems of Equations Using Inverse Matrices

Find the inverse of each matrix, if it exists.

43. $\begin{bmatrix} 7 & 4 \\ 3 & 2 \end{bmatrix}$

44. $\begin{bmatrix} 2 & 5 \\ -5 & -13 \end{bmatrix}$

45. $\begin{bmatrix} 6 & -3 \\ -8 & 4 \end{bmatrix}$

Use a matrix equation to solve each system of equations.

46. $\begin{bmatrix} 5 & 3 \\ 3 & 2 \end{bmatrix} \cdot \begin{bmatrix} x \\ y \end{bmatrix} = \begin{bmatrix} 4 \\ 0 \end{bmatrix}$

47. $\begin{bmatrix} 3 & -1 \\ 1 & 2 \end{bmatrix} \cdot \begin{bmatrix} a \\ b \end{bmatrix} = \begin{bmatrix} 5 \\ 4 \end{bmatrix}$

48. **HEALTH FOOD** Heath sells nuts and raisins by the pound. Sonia bought 2 pounds of nuts and 2 pounds of raisins for $23.50. Drew bought 3 pounds of nuts and 1 pound of raisins for $22.25. What is the cost of 1 pound of nuts and 1 pound of raisins?

Example 11

Solve $\begin{bmatrix} 2 & -5 \\ 3 & -6 \end{bmatrix} \cdot \begin{bmatrix} x \\ y \end{bmatrix} = \begin{bmatrix} 15 \\ 36 \end{bmatrix}$.

Step 1 Find the inverse of the coefficient matrix.

$A^{-1} = \dfrac{1}{-12 - (-15)} \begin{bmatrix} -6 & 5 \\ -3 & 2 \end{bmatrix} \text{ or } \dfrac{1}{3}\begin{bmatrix} -6 & 5 \\ -3 & 2 \end{bmatrix}$

Step 2 Multiply each side by the inverse matrix.

$$\frac{1}{3}\begin{bmatrix} -6 & 5 \\ -3 & 2 \end{bmatrix} \cdot \begin{bmatrix} 2 & -5 \\ 3 & -6 \end{bmatrix} \cdot \begin{bmatrix} x \\ y \end{bmatrix} = \frac{1}{3}\begin{bmatrix} -6 & 5 \\ -3 & 2 \end{bmatrix} \cdot \begin{bmatrix} 15 \\ 36 \end{bmatrix}$$

$$\begin{bmatrix} 1 & 0 \\ 0 & 1 \end{bmatrix} \cdot \begin{bmatrix} x \\ y \end{bmatrix} = \frac{1}{3}\begin{bmatrix} 90 \\ 27 \end{bmatrix}$$

$$\begin{bmatrix} x \\ y \end{bmatrix} = \begin{bmatrix} 30 \\ 9 \end{bmatrix}$$

CHAPTER 3 Practice Test

Solve each system of equations by using either substitution or elimination.

1. $y = x + 4$
   $x + y = -12$

2. $3x + 5y = -7$
   $6x - 4y = 0$

3. $5x + 2y = 4$
   $3y - 4x = -40$

4. $8x - 3y = -13$
   $-3x + 5y = 1$

5. **MULTIPLE CHOICE** Which graph shows the solution of the system of inequalities?

$$y \le 2x + 3$$
$$y < \frac{1}{3}x + 5$$

A
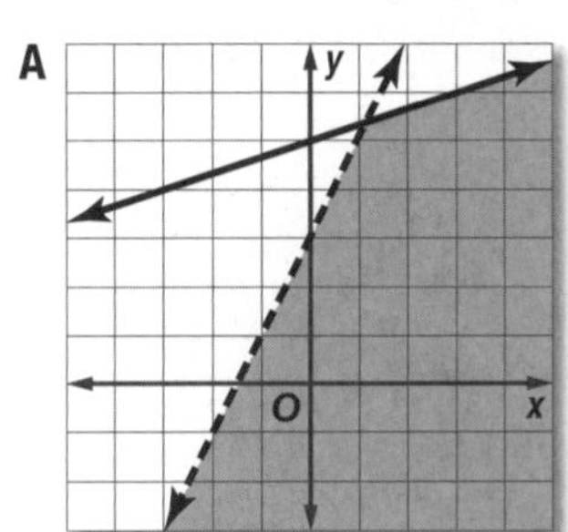

C
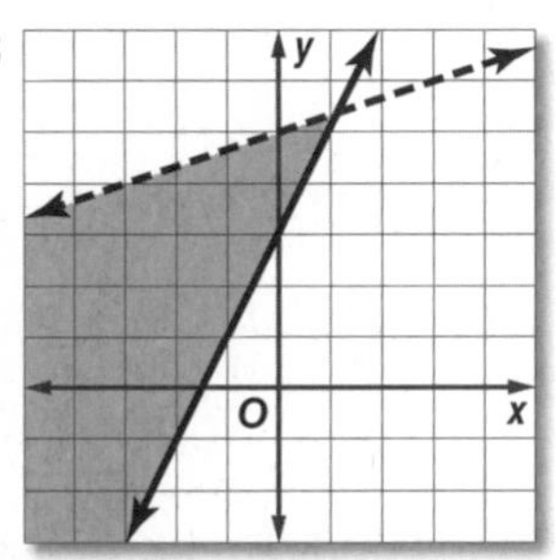

B
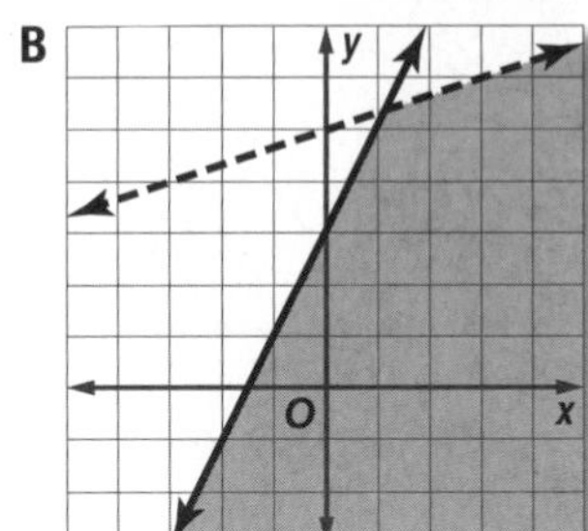

D
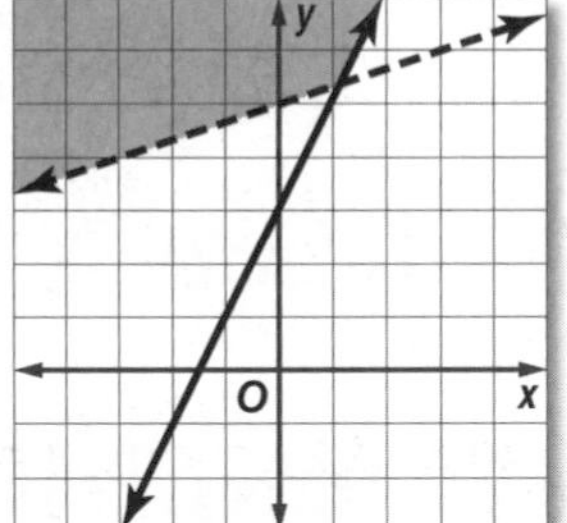

Solve each system of inequalities by graphing.

6. $x + y > 6$
   $x - y < 0$

7. $y \ge 2x - 5$
   $y \le x + 4$

8. $3x + 4y \le 12$
   $6x - 3y \ge 18$

9. $5y + 2x \le 20$
   $4x + 3y > 12$

10. **SALONS** Sierra King is a nail technician. She allots 20 minutes for a manicure and 45 minutes for a pedicure in her 7-hour work day. No more than 5 pedicures can be scheduled each day. The prices are \$18 for a manicure and \$45 for a pedicure. How many manicures and pedicures should Ms. King schedule to maximize her daily income? What is her maximum daily income?

11. **COLLEGE FOOTBALL** Darren McFadden of Arkansas placed second overall in the Heisman Trophy voting. Players are given 3 points for every first-place vote, 2 points for every second-place vote, and 1 point for every third-place vote. McFadden received 490 total votes for first, second, and third place, for a total of 878 points. If he had 4 more than twice as many second-place votes as third-place votes, how many votes did he receive for each place?

Perform the indicated operations. If the matrix does not exist, write *impossible*.

12. $-3\begin{bmatrix} 4a \\ 0 \\ -3 \end{bmatrix} + 4\begin{bmatrix} -2 \\ 3 \\ -1 \end{bmatrix}$

13. $\begin{bmatrix} -3 & 0 \\ 1 & 5 \end{bmatrix} \cdot \begin{bmatrix} 2 & 4 \\ -6 & 0 \end{bmatrix}$

14. $\begin{bmatrix} 2 & 0 \\ -3 & 5 \\ 1 & 4 \end{bmatrix} \cdot \begin{bmatrix} 3 \\ -2 \end{bmatrix}$

15. $\begin{bmatrix} -5 & 7 \\ 6 & 8 \end{bmatrix} - \begin{bmatrix} 4 & 0 & -2 \\ 9 & 0 & 1 \end{bmatrix}$

16. **MULTIPLE CHOICE** What is the value of

$$\begin{vmatrix} 2 & 3 & -1 \\ 0 & 2 & 4 \\ -2 & 5 & 6 \end{vmatrix}?$$

F -44

G $-\frac{1}{44}$

H $\frac{1}{44}$

J 44

Find the inverse of each matrix, if it exists.

17. $\begin{bmatrix} 5 & 0 \\ 0 & 1 \end{bmatrix}$

18. $\begin{bmatrix} 1 & 2 \\ 2 & 1 \end{bmatrix}$

19. $\begin{bmatrix} 6 & 3 \\ 8 & 4 \end{bmatrix}$

20. $\begin{bmatrix} -3 & -2 \\ 6 & 4 \end{bmatrix}$

Use Cramer's Rule to solve each system of equations.

21. $2x - y = -9$
    $x + 2y = 8$

22. $x - y + 2z = 0$
    $3x + z = 11$
    $-x + 2y = 0$

CHAPTER 3 Preparing for Standardized Tests

Short Answer Questions

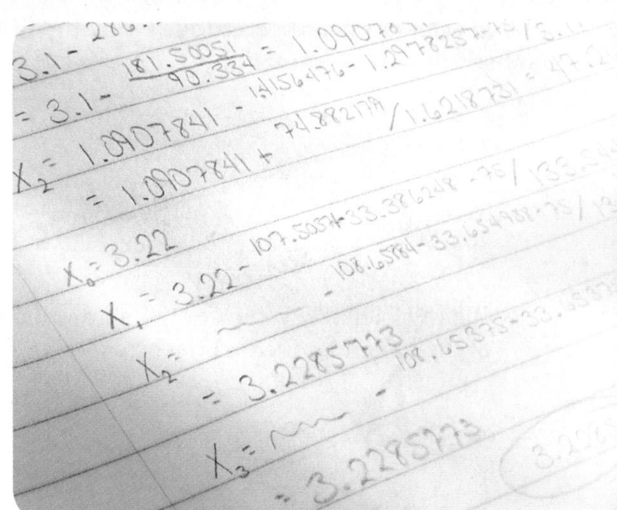

Short answer questions require you to provide a solution to the problem, along with a method, explanation, and/or justification used to arrive at the solution.

Strategies for Solving Short Answer Questions

Short answer questions are typically graded using a **rubric**, or a scoring guide. The following is an example of a short answer question scoring rubric.

Scoring Rubric	
Criteria	Score
Full Credit: The answer is correct and a full explanation is provided that shows each step.	2
Partial Credit: • The answer is correct but the explanation is incomplete. • The answer is incorrect but the explanation is correct.	1
No Credit: Either an answer is not provided or the answer does not make sense.	0

In solving short answer questions, remember to...

- explain your reasoning or state your approach to solving the problem.
- show all of your work or steps.
- check your answer if time permits.

Standardized Test Example

Read the problem. Identify what you need to know. Then use the information in the problem to solve.

Company A charges a monthly fee of \$14.50 plus \$0.05 per minute for cell phone service. Company B charges \$20.00 per month plus \$0.04 per minute. For what number of minutes would the total monthly charge be the same with each company?

Read the problem carefully. You are given information about two different cell phone companies and their monthly charges. Since the situation involves a fixed amount and a variable rate, you can set up and solve a system of equations.

Example of a 2-point response:

Set up and solve a system of equations.

flat fee + rate × minutes = total charges
y = total charges, x = minutes used

$y = 14.5 + 0.05x$ (Company A)

$y = 20 + 0.04x$ (Company B)

CORBIS

Solve the system by graphing.

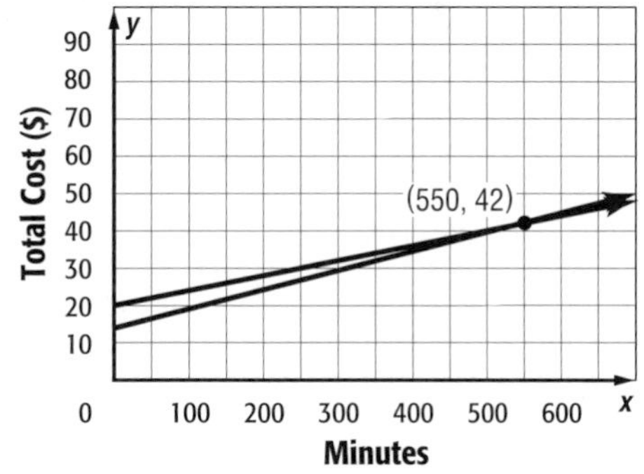

The solution is (550, 42). So, with each company, if the customer uses 550 minutes the total monthly charge is $42.

The steps, calculations, and reasoning are clearly stated. The student also arrives at the correct answer. So, this response is worth the full 2 points.

Exercises

Read each problem. Then use the information in the problem to solve.

1. Shawn and Jerome borrowed $1400 to start a lawn mowing business. They charge their customers $45 per lawn, and with each lawn that they mow, they incur $10.50 in operating expenses. How many lawns must they mow in order to start earning a profit?

2. A circle of radius r is circumscribed about a square. What is the exact ratio of the area of the circle to the area of the square?

3. Mr. Williams can spend no more than $50 on art supplies. Packages of paint brushes cost $4.75 each, and boxes of colored pencils cost $6.50 each. He wants to buy at least 2 packages of each supply. Write a system of inequalities and plot the feasible region on a coordinate grid. Give three different solutions to the system.

4. Marla sells engraved necklaces over the Internet. She purchases 50 necklaces for $400, and it costs her an additional $3 for each personalized engraving. If she charges $20 for each necklace, how many will she need to sell in order to make a profit of at least $225?

5. An auto dealership sold 7378 cars during 2011. This was an 8.5% increase in the number of cars sold during 2010. What was the increase in the number of cars sold in 2011?

6. The sides of two similar triangles are in a ratio of 3:5. If the area of the larger triangle is 600 square centimeters, what is the area of the smaller triangle?

7. Raul had $35 in a savings account and started adding $25 a week. At the same time, his sister Tina had $365 in her account and began spending $30 a week. After how many weeks will Raul and Tina have the same amount in their savings accounts?

8. A city planner wishes to build a sidewalk diagonally across a rectangular-shaped park. The park measures 140 feet by 225 feet. It will cost $30 per foot to construct the sidewalk. What will be the total cost of the sidewalk?

CHAPTER 3

Standardized Test Practice

Cumulative, Chapters 1 through 3

Multiple Choice

Read each question. Then fill in the correct answer on the answer document provided by your teacher or on a sheet of paper.

1. Matrix L shows the average low temperature, in degrees Fahrenheit, each month where Terrance lives. Matrix H shows the monthly average high temperature.

$$L = \begin{bmatrix} 24.1 & 27.7 & 35.9 \\ 44.1 & 53.6 & 62.2 \\ 66.4 & 64.9 & 57.9 \\ 46.4 & 37.3 & 28.4 \end{bmatrix}$$

$$H = \begin{bmatrix} 39.9 & 45.2 & 55.3 \\ 65.1 & 74.0 & 82.3 \\ 85.9 & 84.6 & 78.1 \\ 66.9 & 54.5 & 44.3 \end{bmatrix}$$

Which operation would you use to find the difference between the average high temperature and the average low temperature each month?

A $L + H$
B $H - L$
C $H \times L$
D $L - H$

2. Find $[3 \quad 1] \cdot \begin{bmatrix} 2 \\ 5 \end{bmatrix}$, if possible.

F $[-3]$
G $[11]$
H $\begin{bmatrix} 8 & -4 \\ 12 & 6 \end{bmatrix}$
J undefined

3. Which equation is equivalent to $4x - 3(2x + 7) = 5x$?

A $-2x - 21 = 5x$
B $-2x + 7 = 5x$
C $-2x + 21 = 5x$
D $6x - 7 = 5x$

Test-Taking Tip

Question 2 The product of a 1-by-2 matrix and a 2-by-1 matrix is a 1-by-1 matrix. So, answer choices H and J can be eliminated.

4. Triangle DEF has vertices $D(-6, 2)$, $E(3, 5)$, and $F(8, -7)$. Evaluate the determinant below to find the area of the triangle.

$$A = \frac{1}{2}\begin{vmatrix} -6 & 2 & 1 \\ 3 & 5 & 1 \\ 8 & -7 & 1 \end{vmatrix}$$

F 54.5 square units
G 58 square units
H 60 square units
J 61.5 square units

5. Suppose Kendall sells apples and tomatoes at a farmer's market. If he sold 280 items one morning and earned \$65.20, how many apples did he sell?

Item	Cost
apple	\$0.25
tomato	\$0.20

A 96
B 126
C 168
D 184

6. What are the dimensions of $D = \begin{bmatrix} 4 & -6 \\ 9 & 2 \\ 1 & 0 \\ -3 & -5 \end{bmatrix}$?

F 4×2
G 2×4
H 4×8
J 8×4

7. What is the solution set of $6 - |x + 7| \leq -2$?

A $\{x \mid -15 \leq x \leq 1\}$
B $\{x \mid -1 \leq x \leq 3\}$
C $\{x \mid x \leq -1 \text{ or } x \geq 3\}$
D $\{x \mid x \leq -15 \text{ or } x \geq 1\}$

Short Response/Gridded Response

Record your answers on the answer sheet provided by your teacher or on a sheet of paper.

8. Does matrix B have an inverse? Explain why or why not.

$$B = \begin{bmatrix} 3 & -2 \\ -9 & 6 \end{bmatrix}$$

9. GRIDDED RESPONSE Evaluate the determinant of
$$W = \begin{bmatrix} 3 & 1 & 0 \\ 2 & 5 & -4 \\ 0 & -1 & 1 \end{bmatrix}.$$

10. Valeria has 14 quarters and dimes. The total value of all the coins is $2.75. Use this information to answer each question.

a. Let d represent the number of dimes that Valeria has, and let q represent the number of quarters. Write a system of equations to model the situation.

b. Write a matrix equation that can be used to solve for d and q.

c. Solve your matrix equation using inverses. How many dimes and quarters does Valeria have?

11. GRIDDED RESPONSE Andrea is using a coordinate grid to design a new deck for her backyard. The deck is represented by the intersection of $y \leq 20$, $x \leq 16$, $y \geq 0$, $x \geq 0$ and $y \leq -x + 32$. If each unit of the coordinate grid represents 1 foot, what is the area of the deck? Express your answer in square feet.

12. What are the coordinates of the x- and y-intercepts of the graph of $2y = 4x + 3$?

Extended Response

Record your answers on a sheet of paper. Show your work.

13. Suppose Tonya is baking cookies and muffins for a bake sale. Each tray of cookies uses 5 cups of flour and 2 cups of sugar. Each tray of muffins uses 5 cups of flour and 1 cup of sugar. She has 40 cups of flour and 15 cups of sugar available for baking. Tonya will make $12 profit for each tray of cookies that is sold and $8 profit for each tray of muffins sold.

a. Let x represent the number of trays of cookies baked, and let y represent the number of trays of muffins baked. Write a system of inequalities to model the different number of trays Tonya can bake.

b. Graph the system of inequalities to show the feasible region. List the coordinates of the vertices of the feasible region.

c. Write a profit function for selling x trays of cookies and y trays of muffins.

d. How any trays of cookies and muffins should Tonya bake to maximize the profit? What will the total profit be?

Need ExtraHelp?

If you missed Question...	1	2	3	4	5	6	7	8	9	10	11	12	13
Go to Lesson...	3-5	3-6	1-2	3-7	3-1	3-5A	2-8	3-8	3-7	3-8	3-2	2-2	3-3

CHAPTER 4

Quadratic Functions and Relations

Then

You graphed linear equations and inequalities.

Now

You will:

- Graph quadratic functions.
- Solve quadratic equations.
- Perform operations with complex numbers.
- Graph and solve quadratic inequalities.

Why? ▲

MOTION The path that a soccer ball or a firework takes can be modeled by a quadratic function. Quadratic functions can map an object in motion. In this chapter you will look at a pumpkin catapult, an amusement park ride, and a diver in motion.

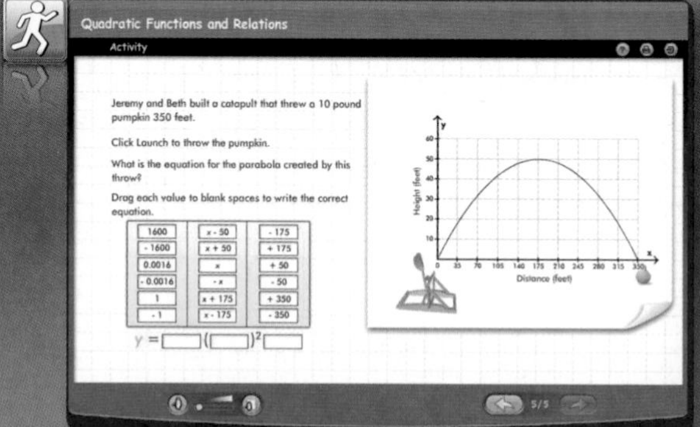

connectED.mcgraw-hill.com Your Digital Math Portal

Animation 
Vocabulary 
eGlossary 
Personal Tutor  
Virtual Manipulatives 
Graphing Calculator 
Audio 
Foldables 
Self-Check Practice 
Worksheets 

Pete Saloutos/UpperCut Images/Getty Ima

Get Ready for the Chapter

Diagnose Readiness | You have two options for checking prerequisite skills.

1 Textbook Option Take the Quick Check below. Refer to the Quick Review for help.

QuickCheck

Given $f(x) = 2x^2 + 4$ and $g(x) = -x^2 - 2x + 3$, find each value.

1. $f(-1)$

2. $f(3)$

3. $f(0)$

4. $g(4)$

5. $g(0)$

6. $g(-3)$

7. FISH Tuna swim at a steady rate of 9 miles per hour until they die, and they never stop moving.

a. Write a function that is a model for the situation.

b. Evaluate the function to estimate how far a 2-year old tuna has traveled.

8. BUDGET Marla has budgeted \$65 per day on food during a business trip. Write a function that is a model for the situation and evaluate what she would spend on a 2 week business trip.

Factor completely. If the polynomial is not factorable, write *prime*.

9. $x^2 + 13x + 40$

10. $x^2 - 10x + 21$

11. $2x^2 + 7x - 4$

12. $2x^2 - 7x - 15$

13. $x^2 - 11x + 15$

14. $x^2 + 12x + 36$

15. FLOOR PLAN The rectangular room pictured below has an area of $x^2 + 14x + 48$ square feet. If the width of the room is $(x + 6)$ feet, what is the length?

$A = (x^2 + 14x + 48)$ ft^2 $(x + 6)$ ft

QuickReview

Example 1

Given $f(x) = -2x^2 + 3x - 1$ and $g(x) = 3x^2 - 5$, find each value.

a. $f(2)$

$f(x) = -2x^2 + 3x - 1$	Original function
$f(2) = -2(2)^2 + 3(2) - 1$	Substitute 2 for x.
$= -8 + 6 - 1$ or -3	Simplify.

b. $g(-2)$

$g(x) = 3x^2 - 5$	Original function
$g(-2) = 3(-2)^2 - 5$	Substitute -2 for x.
$= 12 - 5$ or 7	Simplify.

Example 2

Factor $2x^2 - x - 3$ completely. If the polynomial is not factorable, write *prime*.

To find the coefficients of the x-terms, you must find two numbers whose product is $2(-3)$ or -6, and whose sum is -1. The two coefficients must be 2 and -3 since $2(-3) = -6$ and $2 + (-3) = -1$. Rewrite the expression and factor by grouping.

$2x^2 - x - 3$

$= 2x^2 + 2x - 3x - 3$	Substitute $2x - 3x$ for $-x$.
$= (2x^2 + 2x) + (-3x - 3)$	Associative Property
$= 2x(x + 1) + -3(x + 1)$	Factor out the GCF.
$= (2x - 3)(x + 1)$	Distributive Property

2 Online Option Take an online self-check Chapter Readiness Quiz at connectED.mcgraw-hill.com.

Get Started on the Chapter

You will learn several new concepts, skills, and vocabulary terms as you study Chapter 4. To get ready, identify important terms and organize your resources. You may wish to refer to Chapter 0 to review prerequisite skills.

FOLDABLES® Study Organizer

Quadratic Functions Make this Foldable to help you organize your Chapter 4 notes about quadratic functions and relations. Begin with one sheet of 11″ by 17″ paper.

1 **Fold** in half lengthwise.

2 **Fold** in fourths crosswise.

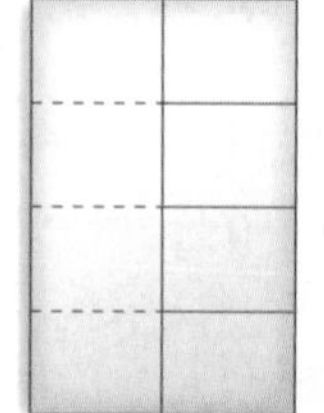

3 **Cut** along the middle fold from the edge to the last crease as shown.

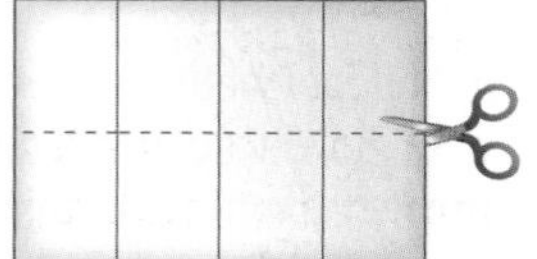

4 **Refold** along the lengthwise fold and tape the uncut section at the top. Label each section with a lesson number and close to form a booklet.

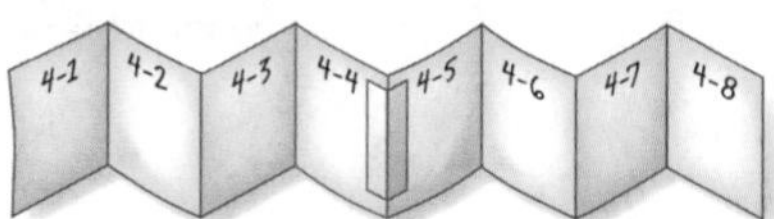

NewVocabulary

English		Español
quadratic term	p. 219	término cuadrático
linear term	p. 219	término lineal
constant term	p. 219	término constante
vertex	p. 220	vértice
maximum value	p. 222	valor máximo
minimum value	p. 222	valor mínimo
quadratic equation	p. 229	ecuación cuadrática
standard form	p. 229	forma estándar
root	p. 229	raíz
zero	p. 229	cero
imaginary unit	p. 246	unidad imaginaria
pure imaginary number	p. 246	número imaginario puro
complex number	p. 247	número complejo
complex conjugates	p. 249	conjugados complejos
completing the square	p. 257	completar el cuadrado
Quadratic Formula	p. 264	fórmula cuadrática
discriminant	p. 267	discriminante
vertex form	p. 275	forma de vértice
quadratic inequality	p. 282	desigualdad cuadrática

ReviewVocabulary

domain dominio the set of all x-coordinates of the ordered pairs of a relation

function función a relation in which each x-coordinate is paired with exactly one y-coordinate

range rango the set of all y-coordinates of the ordered pairs of a relation

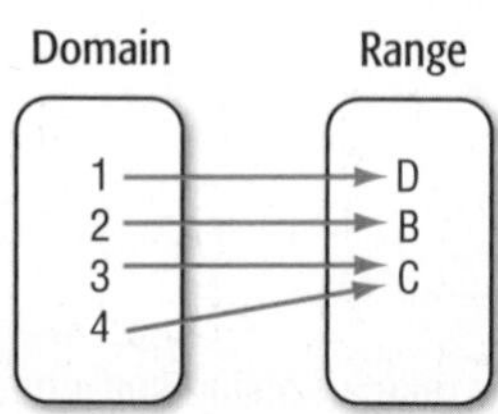

LESSON 4-1 Graphing Quadratic Functions

Then

- You identified and manipulated graphs of functions.

Now

1. Graph quadratic functions.
2. Find and interpret the maximum and minimum values of a quadratic function.

Why?

- Eddie is organizing a charity tournament. He plans to charge a $20 entry fee for each of the 80 players. He recently decided to raise the entry fee by $5, and 5 fewer players entered with the increase. He used this information to determine how many fee increases will maximize the money raised.

The quadratic function at the right represents this situation. The tournament prize pool increases when he first increases the fee, but eventually the pool starts to decrease as the fee gets even higher.

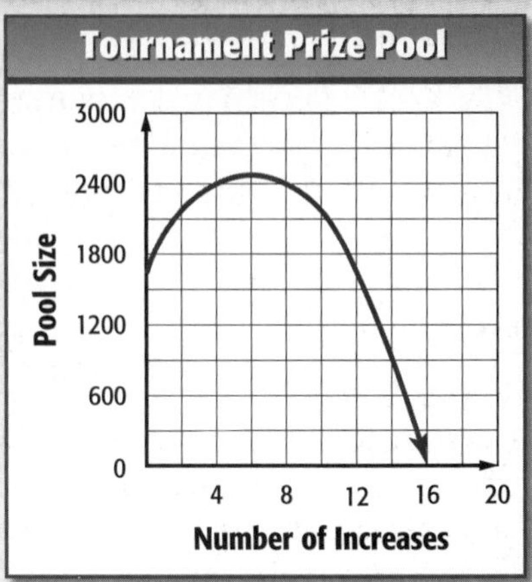

NewVocabulary

quadratic function
quadratic term
linear term
constant term
parabola
axis of symmetry
vertex
maximum value
minimum value

Common Core State Standards

Content Standards

A.SSE.1.a Interpret parts of an expression, such as terms, factors, and coefficients.

F.IF.9 Compare properties of two functions each represented in a different way (algebraically, graphically, numerically in tables, or by verbal descriptions).

Mathematical Practices

1 Make sense of problems and persevere in solving them.

1 Graph Quadratic Functions

In a **quadratic function**, the greatest exponent is 2. These functions can have a **quadratic term**, a **linear term**, and a **constant term**. The general quadratic function is shown below.

$$f(x) = ax^2 + bx + c, \text{ where } a \neq 0$$

quadratic term (ax^2), linear term (bx), constant term (c)

The graph of a quadratic function is called a **parabola**. To graph a quadratic function, graph ordered pairs that satisfy the function.

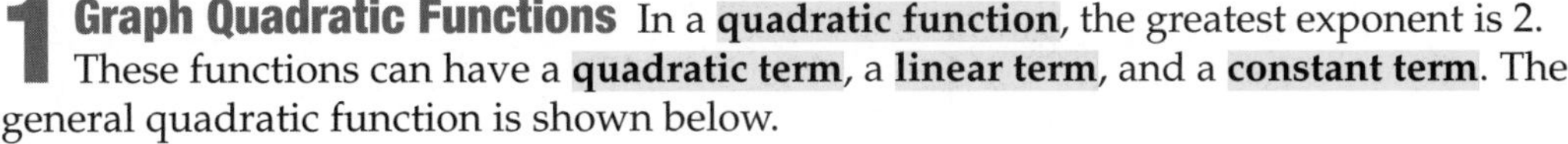

Example 1 Graph a Quadratic Function by Using a Table

Graph $f(x) = 3x^2 - 12x + 6$ by making a table of values.

Choose integer values for x, and evaluate the function for each value. Graph the resulting coordinate pairs, and connect the points with a smooth curve.

x	$3x^2 - 12x + 6$	$f(x)$	$(x, f(x))$
0	$3(0)^2 - 12(0) + 6$	6	(0, 6)
1	$3(1)^2 - 12(1) + 6$	−3	(1, −3)
2	$3(2)^2 - 12(2) + 6$	−6	(2, −6)
3	$3(3)^2 - 12(3) + 6$	−3	(3, −3)
4	$3(4)^2 - 12(4) + 6$	6	(4, 6)

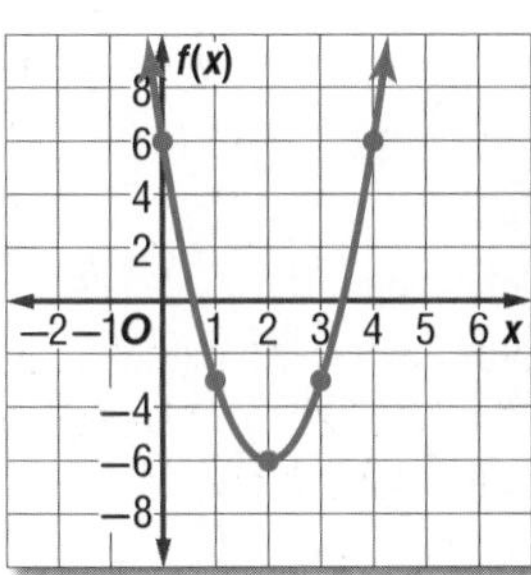

GuidedPractice

Graph each function by making a table of values.

1A. $g(x) = -2x^2 + 8x - 3$

1B. $h(x) = 4x^2 - 8x + 1$

Notice in Example 1 that there seemed to be a pattern in the values for $f(x)$. This is due to the axis of symmetry of parabolas. The **axis of symmetry** is a line through the graph of a parabola that divides the graph into two congruent halves. Each side of the parabola is a reflection of the other side.

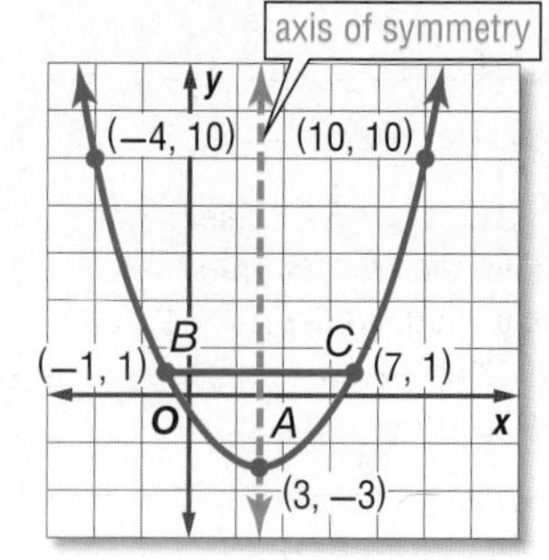

ReviewVocabulary

Symmetry When something is symmetrical, its opposite sides are mirror images of each other.

The axis of symmetry will intersect a parabola at only one point, called the **vertex**. The vertex of the graph at the right is $A(3, -3)$.

Notice that the x-coordinates of points B and C are both 4 units away from the x-coordinate of the vertex, and they have the same y-coordinate. This is due to the symmetrical nature of the graph.

KeyConcept Graph of a Quadratic Function—Parabola

Words Consider the graph of $y = ax^2 + bx + c$, where $a \neq 0$.

- The y-intercept is $a(0)^2 + b(0) + c$ or c.
- The equation of the axis of symmetry is $x = -\frac{b}{2a}$.
- The x-coordinate of the vertex is $-\frac{b}{2a}$.

Model

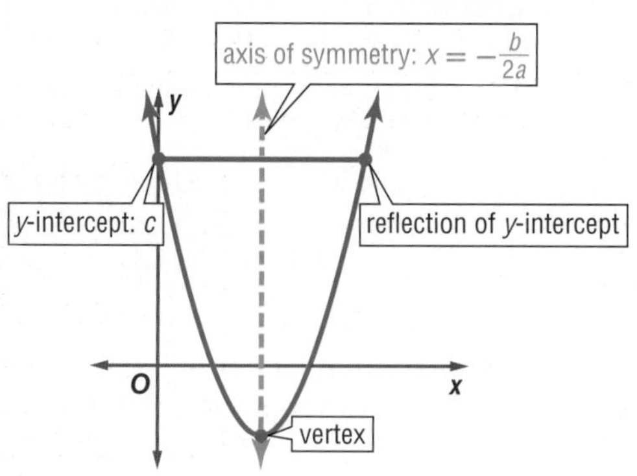

Now you can use the axis of symmetry to help plot points and graph a parabola. For $y = x^2 + 6x - 2$ below, the axis of symmetry is $x = -\frac{b}{2a} = -\frac{6}{2(1)}$ or $x = -3$.

StudyTip

Plotting Reflections The reflection of a point is its mirror image on the other side of the axis of symmetry.

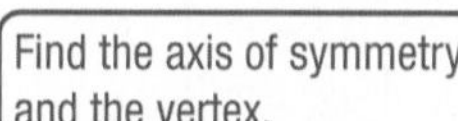

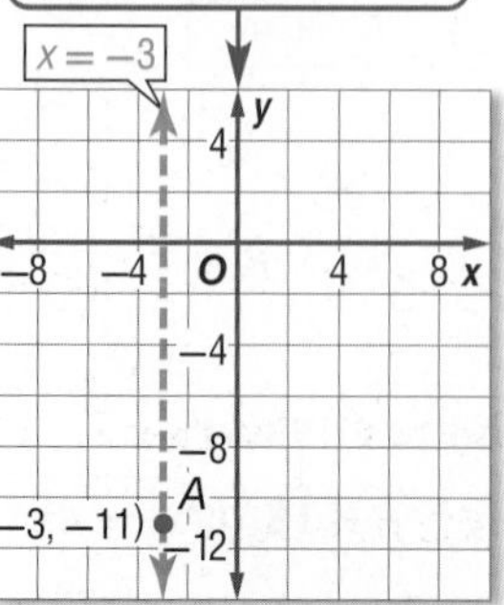

Find the y-intercept and its reflection.

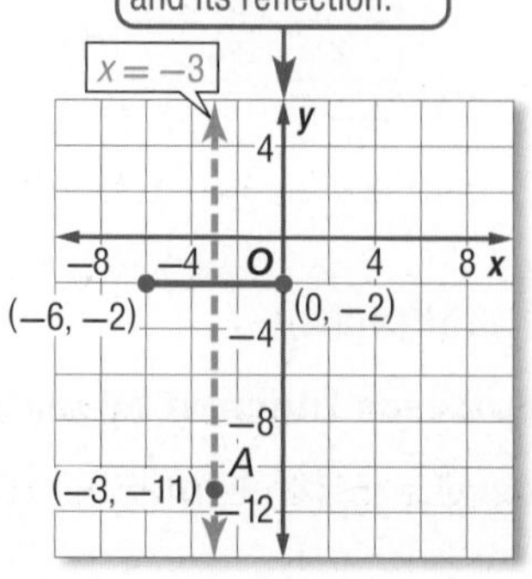

Connect the points with a smooth curve.

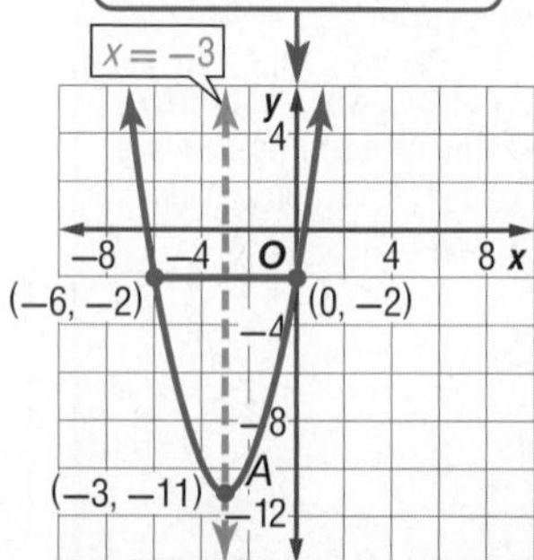

Example 2 Axis of Symmetry, y-intercept, and Vertex

StudyTip

Quadratic Form Make sure the function is in standard quadratic form, $y = ax^2 + bx + c$, before graphing.

Consider $f(x) = x^2 + 4x - 3$.

a. Find the y-intercept, the equation of the axis of symmetry, and the x-coordinate of the vertex.

The function is of the form $f(x) = ax^2 + bx + c$, so we can identify a, b, and c.

$$f(x) = ax^2 + bx + c$$
$$\downarrow \quad \downarrow \quad \downarrow$$
$$f(x) = 1x^2 + 4x - 3 \quad \rightarrow \quad a = 1, b = 4, \text{ and } c = -3$$

The y-intercept is $c = -3$.

Use a and b to find the equation of the axis of symmetry.

$x = -\frac{b}{2a}$ Equation of the axis of symmetry

$= -\frac{4}{2(1)}$ $a = 1$ and $b = 4$

$= -2$ Simplify.

The equation of the axis of symmetry is $x = -2$. Therefore, the x-coordinate of the vertex is -2.

StudyTip

Fractions When the x-coordinate of the vertex is a fraction, select the nearest integer for the next point to avoid using fractions and simplify the calculations.

b. Make a table of values that includes the vertex.

Select five specific points, with the vertex in the middle and two points on either side of the vertex, including the y-intercept and its reflection. Use symmetry to determine the y-values of the reflections.

x	$x^2 + 4x - 3$	$f(x)$	$(x, f(x))$
-6	$(-6)^2 + 4(-6) - 3$	9	$(-6, 9)$
-4	$(-4)^2 + 4(-4) - 3$	-3	$(-4, -3)$
-2	$(-2)^2 + 4(-2) - 3$	-7	$(-2, -7)$
0	$(0)^2 + 4(0) - 3$	-3	$(0, -3)$
2	$(2)^2 + 4(2) - 3$	9	$(2, 9)$

c. Use this information to graph the function.

Graph the points from the table and connect them with a smooth curve.

Draw the axis of symmetry, $x = -2$, as a dashed line. The graph should be symmetrical about this line.

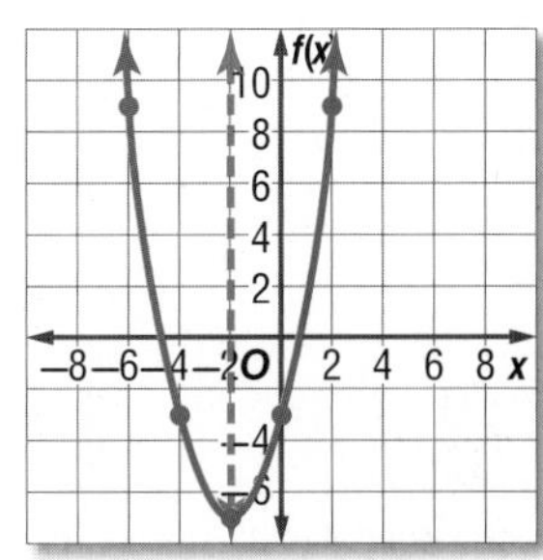

GuidedPractice

2. Consider $f(x) = -5x^2 - 10x + 6$.

A. Find the y-intercept, the equation of the axis of symmetry, and the x-coordinate of the vertex.

B. Make a table of values that includes the vertex.

C. Use this information to graph the function.

WatchOut!

Maxima and Minima The terms *minimum point* and *minimum value* are not interchangeable. The minimum point on the graph of a quadratic function is the ordered pair that describes the location of the vertex. The minimum value of a function is the y-coordinate of the minimum point. It is the smallest value obtained when $f(x)$ is evaluated for all values of x.

2 Maximum and Minimum Values

The y-coordinate of the vertex of a quadratic function is the **maximum value** or the **minimum value** of the function. These values represent the greatest or lowest possible value the function can reach.

KeyConcept Maximum and Minimum Value

Words The graph of $f(x) = ax^2 + bx + c$, where $a \neq 0$,

- opens up and has a minimum value when $a > 0$, and
- opens down and has a maximum value when $a < 0$.

Model

a is positive.

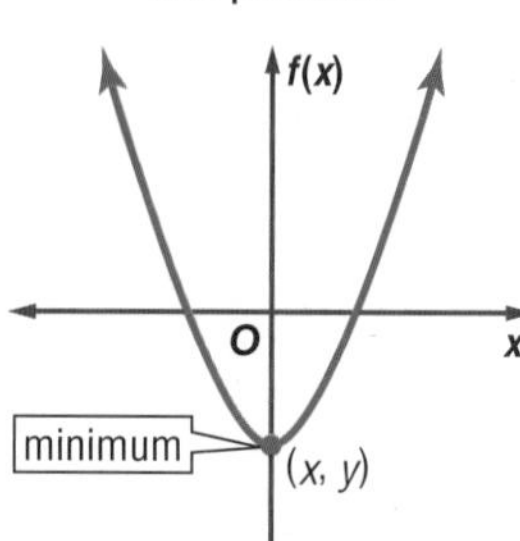

The y-coordinate is the minimum value.

a is negative.

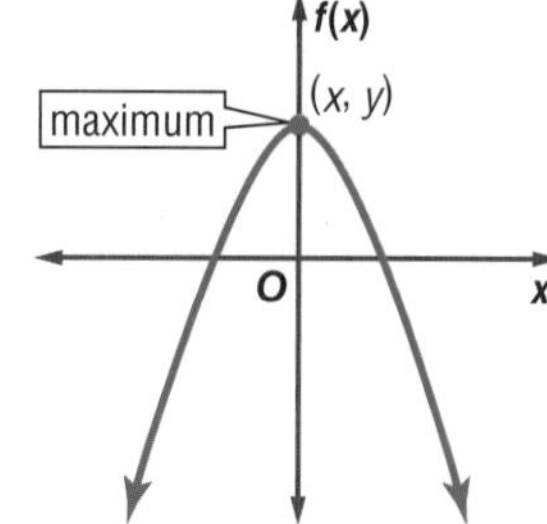

The y-coordinate is the maximum value.

Example 3 Maximum or Minimum Values

PT

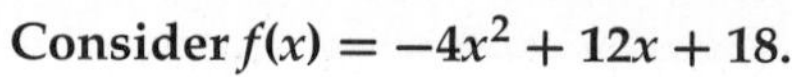
Consider $f(x) = -4x^2 + 12x + 18$.

a. Determine whether the function has a *maximum* or *minimum* value.

For this function, $a = -4$, so the graph opens down and the function has a maximum value.

b. State the maximum or minimum value of the function.

The maximum value of the function is the y-coordinate of the vertex.

The x-coordinate of the vertex is $-\frac{12}{2(-4)}$ or 1.5.

Find the y-coordinate of the vertex by evaluating the function for $x = 1.5$.

$f(x) = -4x^2 + 12x + 18$ Original function

$= -4(1.5)^2 + 12(1.5) + 18$ $x = 1.5$

$= -9 + 18 + 18$ or 27 The maximum value of the function is 27.

c. State the domain and range of the function.

The domain is all real numbers. The range is all real numbers less than or equal to the maximum value, or $\{f(x) \mid f(x) \leq 27\}$.

StudyTip

Domain and Range The domain of a quadratic function will always be all real numbers. The range will either be all real numbers less than or equal to the maximum or all real numbers greater than or equal to the minimum.

GuidedPractice

3. Consider $f(x) = 4x^2 - 24x + 11$.
   - **A.** Determine whether the function has a maximum or minimum value.
   - **B.** State the maximum or minimum value of the function.
   - **C.** State the domain and range of the function.

Real-WorldLink

As of 2006, there were approximately 900,000 public charities in the United States.

Source: National Center for Charitable Statistics

Real-World Example 4 Quadratic Equations in the Real World

PT

CHARITY **Refer to the beginning of the lesson.**

a. How much should Eddie charge in order to maximize charity income?

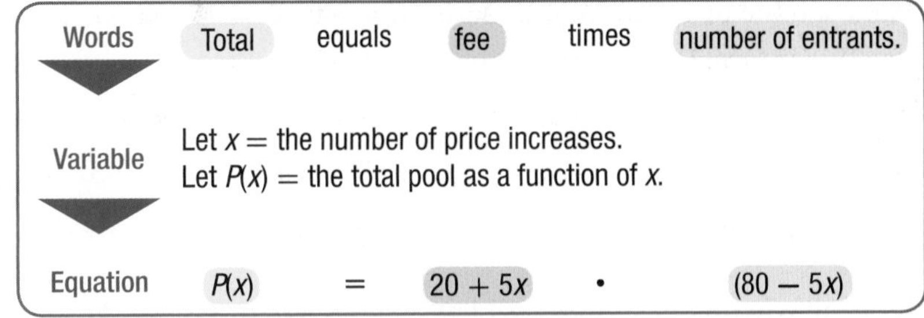

Solve for the x-value of the vertex.

$$
\begin{aligned}
P(x) &= (20 + 5x) \cdot (80 - 5x) \\
&= 20(80) + 20(-5x) + 5x(80) + 5x(-5x) && \text{Distribute.} \\
&= 1600 - 100x + 400x - 25x^2 && \text{Multiply.} \\
&= 1600 + 300x - 25x^2 && \text{Simplify.} \\
&= -25x^2 + 300x + 1600 && ax^2 + bx + c \text{ form}
\end{aligned}
$$

Use the formula for the axis of symmetry, $x = -\frac{b}{2a}$, to find the x-coordinate.

$x = -\frac{300}{2(-25)}$ or 6 $\qquad a = -25$ and $b = 300$

Eddie needs to have 6 price increases, so he should charge 20 + 6(5) or \$50.

b. What will be the maximum value of the pool?

Find the maximum value of the quadratic function $P(x)$ by evaluating $P(6)$.

$$
\begin{aligned}
P(x) &= -25x^2 + 300x + 1600 && \text{Total pool function} \\
P(6) &= -25(6)^2 + 300(6) + 1600 && x = 6 \\
&= -900 + 1800 + 1600 \text{ or } 2500 && \text{Simplify.}
\end{aligned}
$$

Thus, the maximum prize pool is \$2500 after 6 price increases.

CHECK Graph the function on a graphing calculator and use the **CALC:Maximum** function to confirm the solution.

Select a left bound of 0 and a right bound of 10. The calculator will display the coordinates of the maximum at the bottom of the screen.

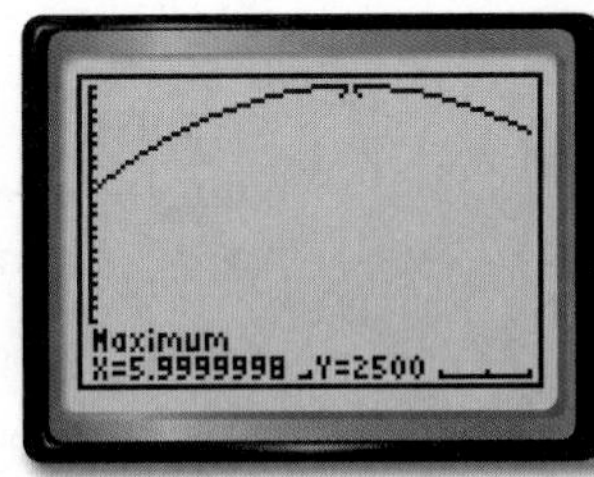

[0, 10] scl: 1 by [0, 2500] scl: 100

The domain is $\{x \mid x \geq 0\}$ because there can be no negative increases in price. The range is $\{y \mid 0 \leq y \leq 2500\}$ because the prize pool cannot have a negative monetary value.

StudyTip

CCSS Modeling Use logic and the information from the problem to determine the domain and range that are reasonable in the situation.

GuidedPractice

4. Suppose a different tournament that Eddie organizes has 120 players and the entry fee is \$40. Each time he increases the fee by \$5, he loses 10 players. Determine what the entry fee should be to maximize the value of the pool.

Peter Dazeley/Getty Images

Check Your Understanding

= Step-by-Step Solutions begin on page R14.

Examples 1–2 **Complete parts a–c for each quadratic function.**

a. **Find the y-intercept, the equation of the axis of symmetry, and the x-coordinate of the vertex.**

b. **Make a table of values that includes the vertex.**

c. **Use this information to graph the function.**

1. $f(x) = 3x^2$

2. $f(x) = -6x^2$

3. $f(x) = x^2 - 4x$

4. $f(x) = -x^2 - 3x + 4$

5. $f(x) = 4x^2 - 6x - 3$

6. $f(x) = 2x^2 - 8x + 5$

Example 3 **Determine whether each function has a *maximum* or *minimum* value, and find that value. Then state the domain and range of the function.**

7. $f(x) = -x^2 + 6x - 1$

8. $f(x) = x^2 + 3x - 12$

9. $f(x) = 3x^2 + 8x + 5$

10. $f(x) = -4x^2 + 10x - 6$

Example 4 **11.** **BUSINESS** A store rents 1400 videos per week at \$2.25 per video. The owner estimates that they will rent 100 fewer videos for each \$0.25 increase in price. What price will maximize the income of the store?

Practice and Problem Solving

Extra Practice is on page R4.

Examples 1–2 **Complete parts a–c for each quadratic function.**

a. **Find the y-intercept, the equation of the axis of symmetry, and the x-coordinate of the vertex.**

b. **Make a table of values that includes the vertex.**

c. **Use this information to graph the function.**

12. $f(x) = 4x^2$

13. $f(x) = -2x^2$

14. $f(x) = x^2 - 5$

15. $f(x) = x^2 + 3$

16. $f(x) = 4x^2 - 3$

17. $f(x) = -3x^2 + 5$

18. $f(x) = x^2 - 6x + 8$

19. $f(x) = x^2 - 3x - 10$

20. $f(x) = -x^2 + 4x - 6$

21. $f(x) = -2x^2 + 3x + 9$

Example 3 **Determine whether each function has a *maximum* or *minimum* value, and find that value. Then state the domain and range of the function.**

22. $f(x) = 5x^2$

23. $f(x) = -x^2 - 12$

24. $f(x) = x^2 - 6x + 9$

25. $f(x) = -x^2 - 7x + 1$

26. $f(x) = 8x - 3x^2 + 2$

27. $f(x) = 5 - 4x - 2x^2$

28. $f(x) = 15 - 5x^2$

29. $f(x) = x^2 + 12x + 27$

30. $f(x) = -x^2 + 10x + 30$

31. $f(x) = 2x^2 - 16x - 42$

Example 4 **32.** **CCSS MODELING** A financial analyst determined that the cost, in thousands of dollars, of producing bicycle frames is $C = 0.000025f^2 - 0.04f + 40$, where f is the number of frames produced.

a. Find the number of frames that minimizes cost.

b. What is the total cost for that number of frames?

Complete parts a–c for each quadratic function.

a. Find the y-intercept, the equation of the axis of symmetry, and the x-coordinate of the vertex.

b. Make a table of values that includes the vertex.

c. Use this information to graph the function.

33. $f(x) = 2x^2 - 6x - 9$

34. $f(x) = -3x^2 - 9x + 2$

35. $f(x) = -4x^2 + 5x$

36. $f(x) = 2x^2 + 11x$

37. $f(x) = 0.25x^2 + 3x + 4$

38. $f(x) = -0.75x^2 + 4x + 6$

39. $f(x) = \frac{3}{2}x^2 + 4x - \frac{5}{2}$

40. $f(x) = \frac{2}{3}x^2 - \frac{7}{3}x + 9$

41. FINANCIAL LITERACY A babysitting club sits for 50 different families. They would like to increase their current rate of \$9.50 per hour. After surveying the families, the club finds that the number of families will decrease by about 2 for each \$0.50 increase in the hourly rate.

a. Write a quadratic function that models this situation.

b. State the domain and range of this function as it applies to the situation.

c. What hourly rate will maximize the club's income? Is this reasonable?

d. What is the maximum income the club can expect to make?

42. ACTIVITIES Last year, 300 people attended the Franklin High School Drama Club's winter play. The ticket price was \$8. The advisor estimates that 20 fewer people would attend for each \$1 increase in ticket price.

a. What ticket price would give the greatest income for the Drama Club?

b. If the Drama Club raised its tickets to this price, how much income should it expect to bring in?

CCSS TOOLS Use a calculator to find the maximum or minimum of each function. Round to the nearest hundredth if necessary.

43. $f(x) = -9x^2 - 12x + 19$

44. $f(x) = 12x^2 - 21x + 8$

45. $f(x) = -8.3x^2 + 14x - 6$

46. $f(x) = 9.7x^2 - 13x - 9$

47. $f(x) = 28x - 15 - 18x^2$

48. $f(x) = -16 - 14x - 12x^2$

Determine whether each function has a *maximum* or *minimum* value, and find that value. Then state the domain and range of the function.

49. $f(x) = -5x^2 + 4x - 8$

50. $f(x) = -4x^2 - 3x + 2$

51. $f(x) = -9 + 3x + 6x^2$

52. $f(x) = 2x - 5 - 4x^2$

53. $f(x) = \frac{2}{3}x^2 + 6x - 10$

54. $f(x) = -\frac{3}{5}x^2 + 4x - 8$

Determine the function represented by each graph.

55.

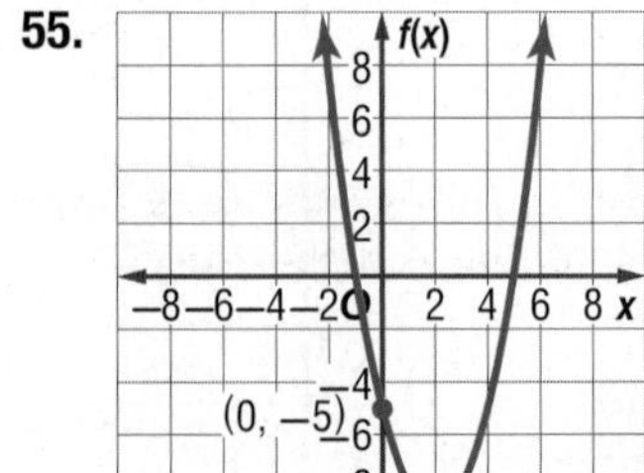

56.

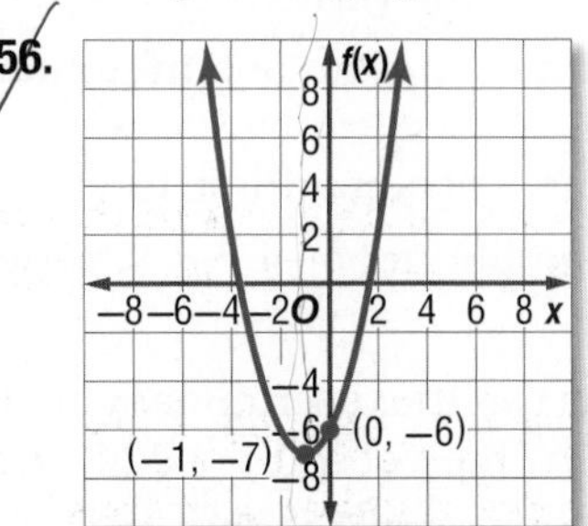

57.

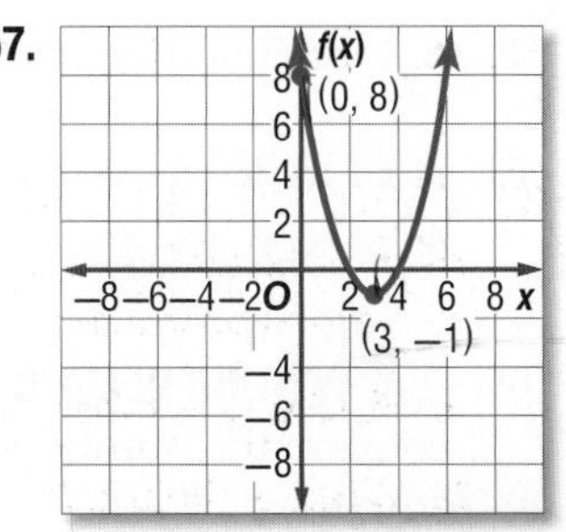

58. **MULTIPLE REPRESENTATIONS** Consider $f(x) = x^2 - 4x + 8$ and $g(x) = 4x^2 - 4x + 8$.

a. Tabular Make a table of values for $f(x)$ and $g(x)$ if $-4 \leq x \leq 4$.

b. Graphical Graph $f(x)$ and $g(x)$.

c. Verbal Explain the difference in the shapes of the graphs of $f(x)$ and $g(x)$. What value was changed to cause this difference?

d. Analytical Predict the appearance of the graph of $h(x) = 0.25x^2 - 4x + 8$. Confirm your prediction by graphing all three functions if $-10 \leq x \leq 10$.

59. **VENDING MACHINES** Omar owns a vending machine in a bowling alley. He currently sells 600 cans of soda per week at \$0.65 per can. He estimates that he will lose 100 customers for every \$0.05 increase in price and gain 100 customers for every \$0.05 decrease in price. (*Hint*: The charge *must* be a multiple of 5.)

a. Write and graph the related quadratic equation for a price increase.

b. If Omar lowers the price, what should it be to maximize his income?

c. What will be his income per week from the vending machine?

60. **BASEBALL** Lolita throws a baseball into the air and the height h of the ball in feet at a given time t in seconds after she releases the ball is given by the function $h(t) = -16t^2 + 30t + 5$.

a. State the domain and range for this situation.

b. Find the maximum height the ball will reach.

H.O.T. Problems Use Higher-Order Thinking Skills

61. **CCSS CRITIQUE** Trent thinks that the function $f(x)$ graphed below, and the function $g(x)$ described next to it have the same maximum. Madison thinks that $g(x)$ has a greater maximum. Is either of them correct? Explain your reasoning.

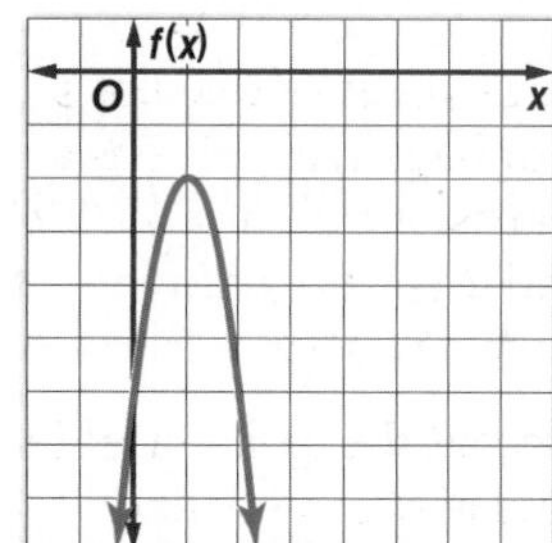

$g(x)$ is a quadratic function with roots of 4 and 2 and a y-intercept of -8.

62. **REASONING** Determine whether the following statement is *sometimes*, *always*, or *never* true. Explain your reasoning.

In a quadratic function, if two x-coordinates are equidistant from the axis of symmetry, then they will have the same y-coordinate.

63. **CHALLENGE** The table at the right represents some points on the graph of a quadratic function.

a. Find the values of a, b, c, and d.

b. What is the x-coordinate of the vertex?

c. Does the function have a maximum or a minimum?

x	y
-20	-377
c	-13
-5	-2
-1	22
$d - 1$	a
5	$a - 24$
7	$-b$
15	-202
$14 - c$	-377

64. **OPEN ENDED** Give an example of a quadratic function with a

a. maximum of 8. **b.** minimum of -4. **c.** vertex of $(-2, 6)$.

65. **WRITING IN MATH** Why can the discriminant be used to confirm the number and the type of solutions of a quadratic equation?

Standardized Test Practice

66. Which expression is equivalent to $\frac{8!}{5!}$?

A $\frac{8}{5}$

B $8 \cdot 7 \cdot 6$

C 3!

D $8 \cdot 7 \cdot 6 \cdot 5$

67. SAT/ACT The price of coffee beans is d dollars for 6 ounces, and each ounce makes c cups of coffee. In terms of c and d, what is the cost of the coffee beans required to make 1 cup of coffee?

F $\frac{cd}{6}$

G $\frac{6c}{d}$

H $\frac{6}{cd}$

J $6cd$

K $\frac{d}{6c}$

68. SHORT RESPONSE Each side of the square base of a pyramid is 20 feet, and the pyramid's height is 90 feet. What is the volume of the pyramid?

69. Which ordered pair is the solution of the following system of equations?

$$3x - 5y = 11$$
$$3x - 8y = 5$$

A (2, 1)

B (7, −2)

C (7, 2)

D $\left(\frac{1}{3}, -2\right)$

Spiral Review

Find the inverse of each matrix, if it exists. (Lesson 3-8)

70. $\begin{bmatrix} 3 & -4 \\ 2 & -1 \end{bmatrix}$

71. $\begin{bmatrix} -4 & -1 \\ 0 & 6 \end{bmatrix}$

72. $\begin{bmatrix} 2 & 8 \\ -3 & -5 \end{bmatrix}$

Evaluate each determinant. (Lesson 3-7)

73. $\begin{vmatrix} 6 & -3 \\ -1 & 8 \end{vmatrix}$

74. $\begin{vmatrix} -3 & -5 \\ -1 & -9 \end{vmatrix}$

75. $\begin{vmatrix} 8 & 6 \\ 4 & 3 \end{vmatrix}$

76. MANUFACTURING The Community Service Committee is making canvas tote bags and leather tote bags for a fundraiser. They will line both types of bags with canvas and use leather handles on both. For the canvas bags, they need 4 yards of canvas and 1 yard of leather. For the leather bags, they need 3 yards of leather and 2 yards of canvas. The committee leader purchased 56 yards of leather and 104 yards of canvas. (Lesson 3-3)

a. Let c represent the number of canvas bags, and let ℓ represent the number of leather bags. Write a system of inequalities for the number of bags that can be made.

b. Draw the graph showing the feasible region.

c. List the coordinates of the vertices of the feasible region.

d. If the club plans to sell the canvas bags at a profit of $20 each and the leather bags at a profit of $35 each, write a function for the total profit on the bags.

e. How can the club make the maximum profit?

f. What is the maximum profit?

State whether each function is a linear function. Write *yes* or *no*. Explain. (Lesson 2-2)

77. $y = 4x^2 - 3x$

78. $y = -2x - 4$

79. $y = 4$

Skills Review

Evaluate each function for the given value.

80. $f(x) = 3x^2 - 4x + 6, x = -2$

81. $f(x) = -2x^2 + 6x - 5, x = 4$

82. $f(x) = 6x^2 + 18, x = -5$

EXTEND 4-1

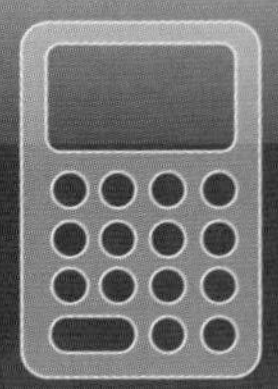

Graphing Technology Lab
Modeling Real-World Data

You can use a TI-83/84 Plus graphing calculator to model data points for which a curve of best fit is a quadratic function.

CCSS Common Core State Standards
Content Standards
F.IF.4 For a function that models a relationship between two quantities, interpret key features of graphs and tables in terms of the quantities, and sketch graphs showing key features given a verbal description of the relationship.
Mathematical Practices
5 Use appropriate tools strategically.

WATER A bottle is filled with water. The water is allowed to drain from a hole made near the bottom of the bottle. The table shows the level of the water y measured in centimeters from the bottom of the bottle after x seconds.

Time (s)	0	20	40	60	80	100	120	140	160	180	200	220
Water level (cm)	42.6	40.7	38.9	37.2	35.8	34.3	33.3	32.3	31.5	30.8	30.4	30.1

Find and graph a linear regression equation and a quadratic regression equation. Determine which equation is a better fit for the data.

Activity

Step 1 **Find and graph a linear regression equation.**

- Enter the times in **L1** and the water levels in **L2**. Then find a linear regression equation.

  KEYSTROKES: *Refer to Lesson 2-5.*

- Use **STAT PLOT** to graph a scatter plot. Copy the equation to the **Y=** list and graph.

  KEYSTROKES: *Review statistical plots and graphing a regression equation in Lesson 2-5.*

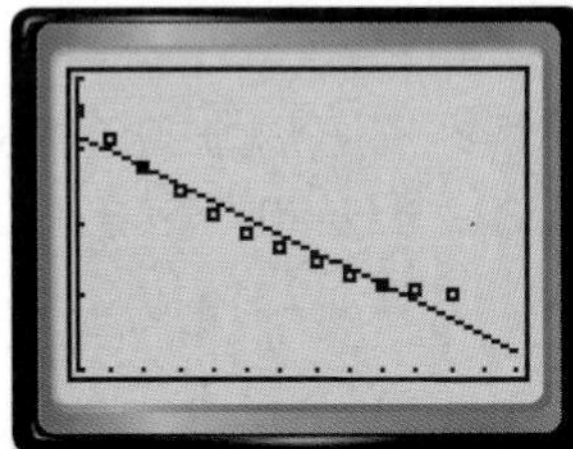

[0, 260] scl: 20 by [25, 45] scl: 5

Step 2 **Find and graph a quadratic regression equation.**

- Find the quadratic regression equation. Then copy the equation to the **Y=** list and graph.

  KEYSTROKES: STAT ▶ 5 ENTER Y= VARS 5 ▶ ▶ ENTER GRAPH

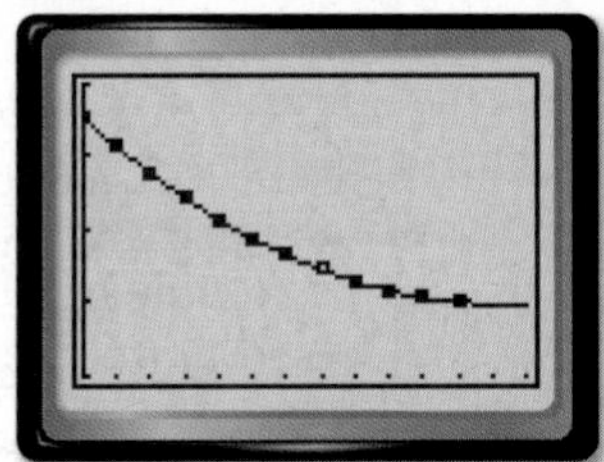

[0, 260] scl: 20 by [25, 45] scl: 5

Notice that the graph of the linear regression equation appears to pass through just two data points. However, the graph of the quadratic regression equation fits the data very well.

Exercises

Refer to the table.

Height of Player's Feet above Floor

Time (s)	Height (in.)
0.1	3.04
0.2	5.76
0.3	8.16
0.4	10.24
0.5	12
0.6	13.44
0.7	14.56

1. Find and graph a linear regression equation and a quadratic regression equation for the data. Determine which equation is a better fit for the data.
2. Estimate the height of the player's feet after 1 second and 1.5 seconds. Use mental math to check the reasonableness of your estimates.
3. Compare and contrast the estimates you found in Exercise 2.
4. How might choosing a regression equation that does not fit the data well affect predictions made by using the equation?

LESSON 4-2

Solving Quadratic Equations by Graphing

Then

- You solved systems of equations by graphing.

Now

1. Solve quadratic equations by graphing.
2. Estimate solutions of quadratic equations by graphing.

Why?

- Arielle works in the marketing department of a major retailer. Her job is to set prices for new products sold in the stores. Arielle determined that for a certain product, the function $f(p) = -6p^2 + 192p - 1440$ tells the profit $f(p)$ made at price p.

Arielle can determine the price range by finding the prices for which the profit is equal to \$0. This can be done by finding the solutions of the related quadratic equation $0 = -6p^2 + 192p - 1440$.

The graph of the function indicates that the profit is zero at 12 and 20, so the profitable price range of the item is between \$12 and \$20.

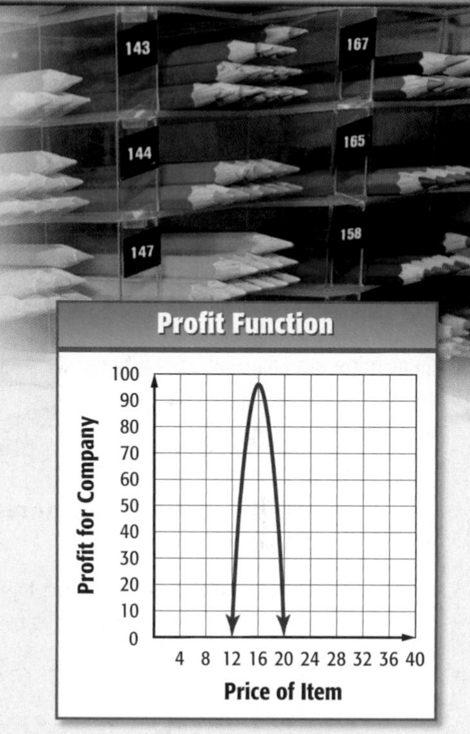

NewVocabulary
quadratic equation
standard form
root
zero

Common Core State Standards

Content Standards

A.CED.2 Create equations in two or more variables to represent relationships between quantities; graph equations on coordinate axes with labels and scales.

F.IF.4 For a function that models a relationship between two quantities, interpret key features of graphs and tables in terms of the quantities, and sketch graphs showing key features given a verbal description of the relationship.

Mathematical Practices

3 Construct viable arguments and critique the reasoning of others.

1 Solve Quadratic Equations

Quadratic equations are quadratic functions that are set equal to a value. The **standard form** of a quadratic equation is $ax^2 + bx + c = 0$, where $a \neq 0$ and a, b, and c are integers.

The solutions of a quadratic equation are called the **roots** of the equation. One method for finding the roots of a quadratic equation is to find the **zeros** of the related quadratic function.

The zeros of the function are the x-intercepts of its graph.

Quadratic Function

$$f(x) = x^2 - x - 6$$

$$f(-2) = (-2)^2 - (-2) - 6 \text{ or } 0$$
$$f(3) = 3^2 - 3 - 6 \text{ or } 0$$

−2 and 3 are zeros of the function.

Quadratic Equation

$$x^2 - x - 6 = 0$$

$$(-2)^2 - (-2) - 6 \text{ or } 0$$
$$3^2 - 3 - 6 \text{ or } 0$$

−2 and 3 are roots of the equation.

Graph of Function

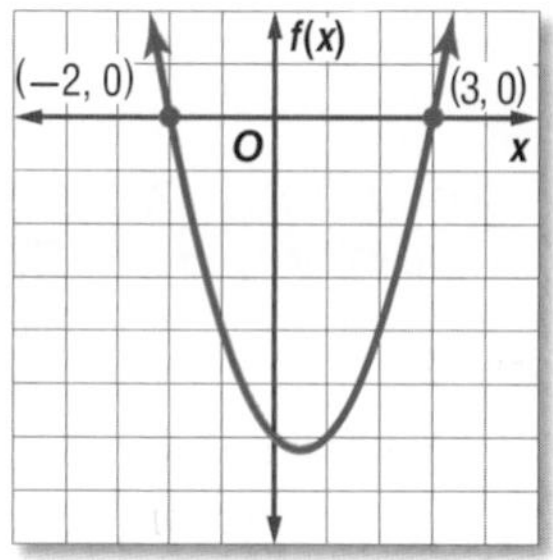

The x-intercepts are −2 and 3.

Jack Sullivan/Alamy

Example 1 Two Real Solutions

Solve $x^2 - 3x - 4 = 0$ by graphing.

Graph the related function, $f(x) = x^2 - 3x - 4$. The equation of the axis of symmetry is $x = -\frac{-3}{2(1)}$ or 1.5. Make a table using x-values around 1.5. Then graph each point.

x	−1	0	1	1.5	2	3	4
$f(x)$	0	−4	−6	−6.25	−6	−4	0

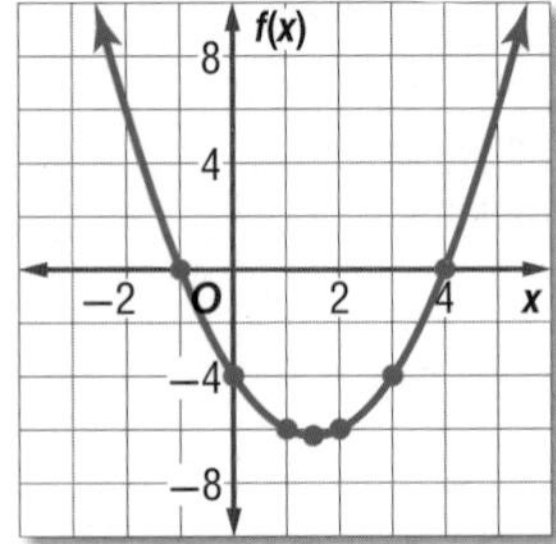

The zeros of the function are −1 and 4. Therefore, the solutions of the equation are −1 and 4 or $\{x \mid x = -1, 4\}$.

StudyTip

Set-Builder Notation In Lesson 1-5, you learned how to express the solution set of an inequality using set-builder notation. The solutions of an equation can also be expressed in set-builder notation. For example, the solutions of $x^2 = 25$ can be expressed as $\{x \mid x = -5, 5\}$.

GuidedPractice

Solve each equation by graphing.

1A. $x^2 + 2x - 15 = 0$

1B. $x^2 - 8x = -12$

The graph of the related function in Example 1 has two zeros; therefore, the quadratic equation has two real solutions. This is one of the three possible outcomes when solving a quadratic equation.

KeyConcept Solutions of a Quadratic Equation

Words A quadratic equation can have one real solution, two real solutions, or no real solutions.

Models

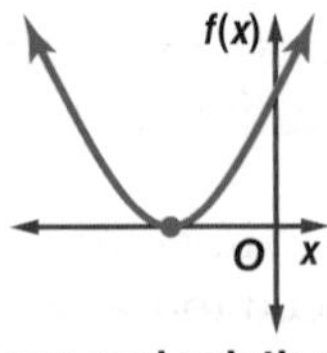

one real solution

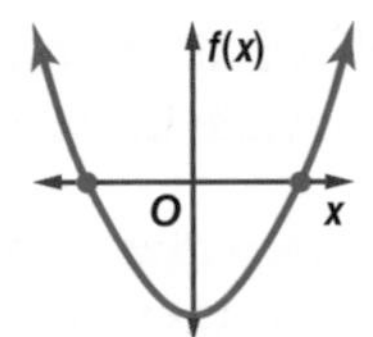

two real solutions

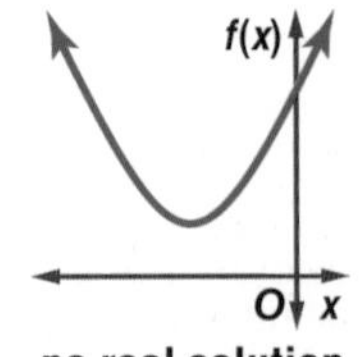

no real solution

Example 2 One Real Solution

Solve $14 - x^2 = -6x + 23$ by graphing.

$14 - x^2 = -6x + 23$ Original equation

$14 = x^2 - 6x + 23$ Add x^2 to each side.

$0 = x^2 - 6x + 9$ Subtract 14.

Graph the related function $f(x) = x^2 - 6x + 9$.

x	1	2	3	4	5
$f(x)$	4	1	0	1	4

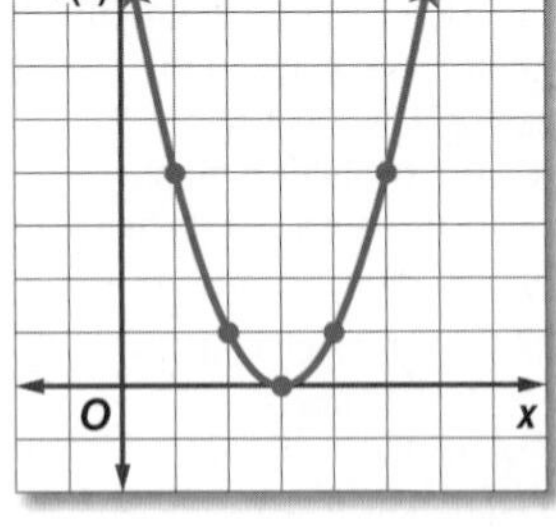

The function has only one zero, 3. Therefore, the solution is 3 or $\{x \mid x = 3\}$.

StudyTip

Optional Graph $f(x) = -x^2 + 6x - 9$ could also have been graphed for this example. The graph would appear different, but it would have the same solution.

GuidedPractice

Solve each equation by graphing.

2A. $x^2 + 5 = -8x - 11$

2B. $12 - x^2 = 48 - 12x$

Example 3 No Real Solution

NUMBER THEORY **Use a quadratic equation to find two real numbers with a sum of 15 and a product of 63.**

Understand Let x represent one of the numbers. Then $15 - x$ is the other number.

Plan

$x(15 - x) = 63$ — The product of the numbers is 63.

$15x - x^2 = 63$ — Distributive Property

$-x^2 + 15x - 63 = 0$ — Subtract 63.

Solve Graph the related function.

The graph has no x-intercepts. This means the original equation has no real solution. Thus, it is not possible for two real numbers to have a sum of 15 and a product of 63.

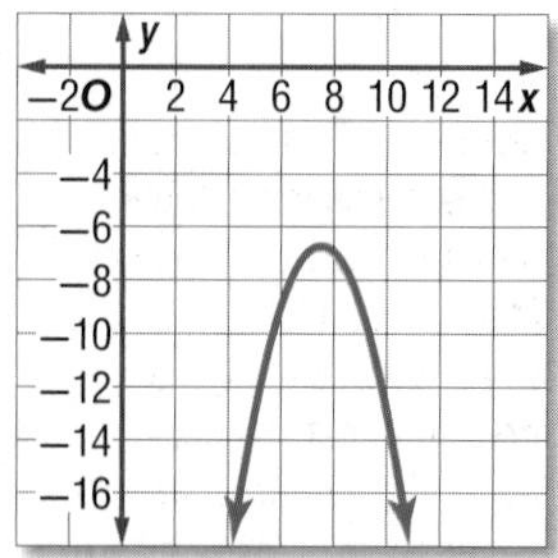

Check Try finding the product of several pairs of numbers with sums of 15. Is each product less than 63 as the graph suggests?

GuidedPractice

3. Find two real numbers with a sum of 6 and a product of −55, or show that no such numbers exist.

2 Estimate Solutions

Often exact roots cannot be found by graphing. You can estimate the solutions by stating the integers between which the roots are located.

> **WatchOut!**
> **Zeros** You will see in later chapters that many zeros can appear within small intervals.

When the value of the function is positive for one value and negative for a second value, then there is at least one zero between those two values.

x	−3	−2	−1	0	1	2	3
$f(x)$	12	3	−6	−2	4	8	14

zero (between −2 and −1) zero (between 0 and 1)

Example 4 Estimate Roots

Solve $x^2 - 6x + 4 = 0$ by graphing. If exact roots cannot be found, state the consecutive integers between which the roots are located.

x	0	1	2	3	4	5	6
$f(x)$	4	−1	−4	−5	−4	−1	4

The x-intercepts of the graph indicate that one solution is between 0 and 1, and the other solution is between 5 and 6.

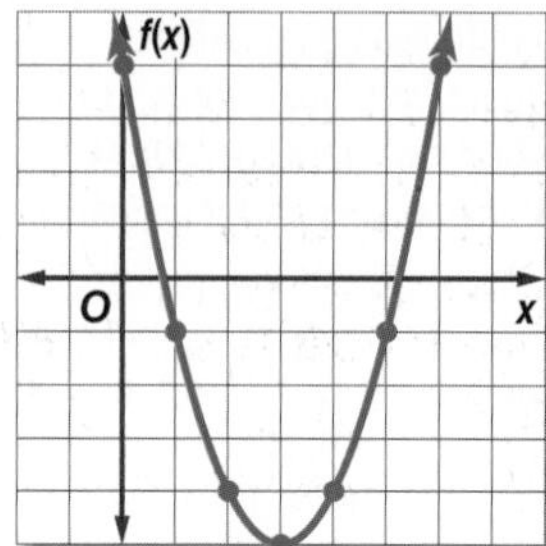

GuidedPractice

4. Solve $x^2 - x - 10 = 0$ by graphing. If exact roots cannot be found, state the consecutive integers between which the roots are located.

You can also use tables to solve quadratic equations. After entering the equation in your calculator, scroll through the table to locate the solutions.

PT

Example 5 Solve by Using a Table

Solve $x^2 - 6x + 2 = 0$.

Enter $y_1 = x^2 - 6x + 2$ in your graphing calculator. Use the **TABLE** window to find where the sign of **Y1** changes. Change **△Tbl** to 0.1 and look again for the sign change. Repeat the process with 0.01 and 0.001 to get a more accurate location of the zero.

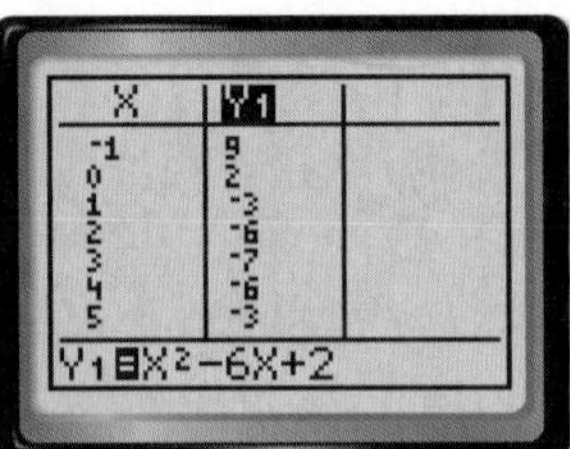

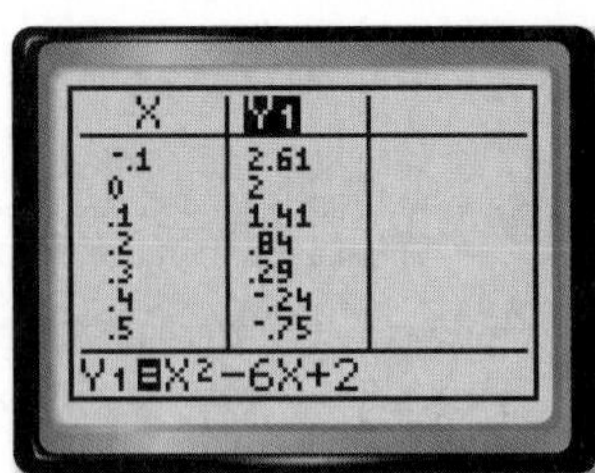

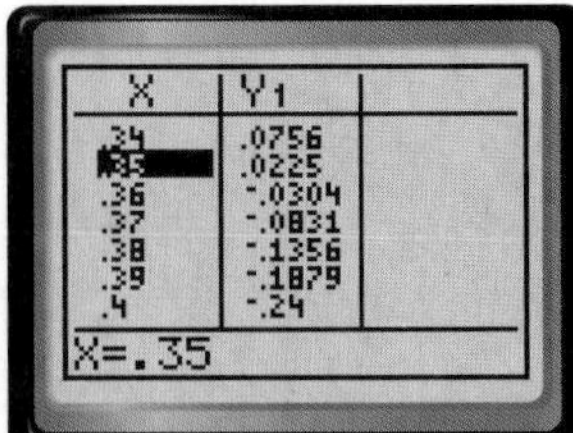

One solution is approximately 0.354.

GuidedPractice

5. Locate the second zero in the function above to the nearest thousandth.

Quadratic equations can be solved with a calculator as well. After entering the equation, use the **ZERO** operation in the **CALC** menu.

PT

Real-World Example 6 Solve by Using a Calculator

RELIEF A package of supplies is tossed from a helicopter at an altitude of 200 feet. The package's height above the ground is modeled by $h(t) = -16t^2 + 28t + 200$, where t is the time in seconds after it is tossed. How long will it take the package to reach the ground?

We need to find t when $h(t)$ is 0. Solve $0 = -16t^2 + 28t + 200$. Then graph the related function $f(t) = -16t^2 + 28t + 200$ on a graphing calculator.

- Use the **ZERO** feature in the **CALC** menu to find the positive zero of the function, since time cannot be negative.
- Use the arrow keys to select a left bound and press ENTER.
- Locate a right bound and press ENTER twice.
- The positive zero of the function is about 4.52. The package would take about 4.52 seconds to reach the ground.

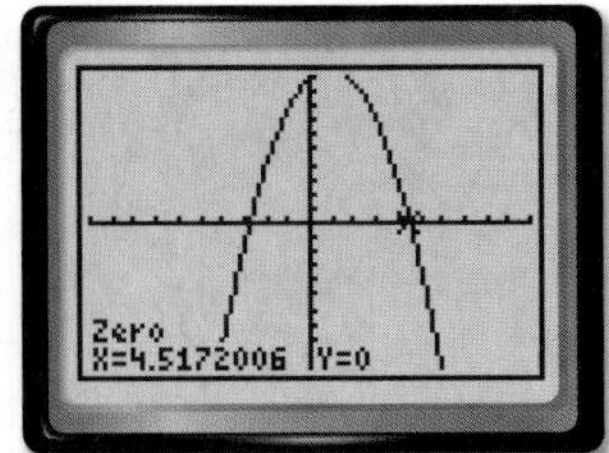

[−10, 10] scl: 1 by [−200, 200] scl: 20

GuidedPractice

6. How long would it take to reach the ground if the height was modeled by $h(t) = -16t^2 + 48t + 400$?

Real-WorldLink

Each year, the American Red Cross responds to over 70,000 disaster situations ranging from house fires to natural disasters such as hurricanes and earthquakes.

Check Your Understanding

= Step-by-Step Solutions begin on page R14.

Example 1 **Use the related graph of each equation to determine its solutions.**

1. $x^2 + 2x + 3 = 0$

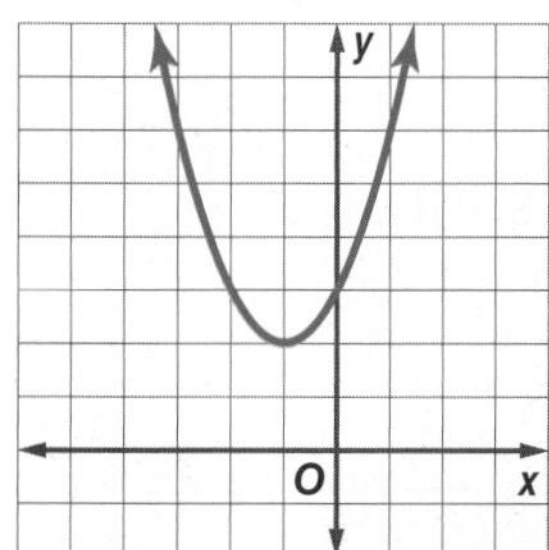

2. $x^2 - 3x - 10 = 0$

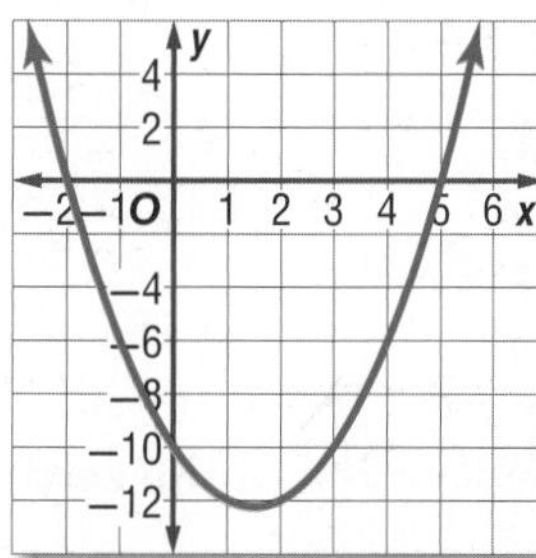

3. $-x^2 - 8x - 16 = 0$

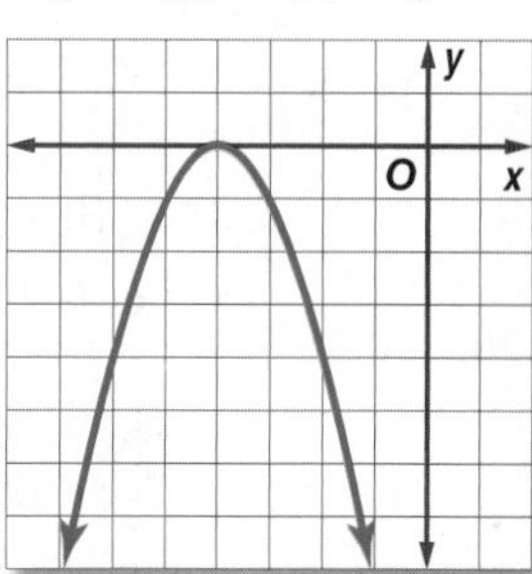

Examples 2-5 CCSS **PRECISION** **Solve each equation. If exact roots cannot be found, state the consecutive integers between which the roots are located.**

4. $x^2 + 8x = 0$
5. $x^2 - 3x - 18 = 0$
6. $4x - x^2 + 8 = 0$
7. $-12 - 5x + 3x^2 = 0$
8. $x^2 - 6x + 4 = -8$
9. $9 - x^2 = 12$
10. $5x^2 + 10x - 4 = -6$
11. $x^2 - 20 = 2 + x$
12. **NUMBER THEORY** Use a quadratic equation to find two real numbers with a sum of 2 and a product of -24.

Example 6

13. **PHYSICS** How long will it take an object to fall from the roof of a building 400 feet above the ground? Use the formula $h(x) = -16t^2 + h_0$, where t is the time in seconds and the initial height h_0 is in feet.

Practice and Problem Solving

Extra Practice is on page R4.

Example 1 **Use the related graph of each equation to determine its solutions.**

14. $x^2 + 4x = 0$

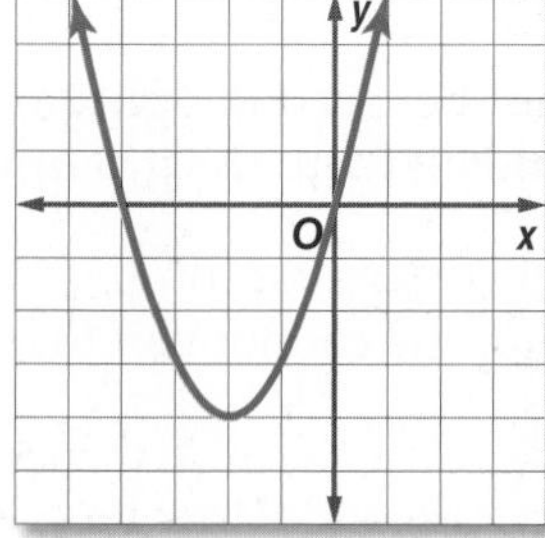

15. $-2x^2 - 4x - 5 = 0$

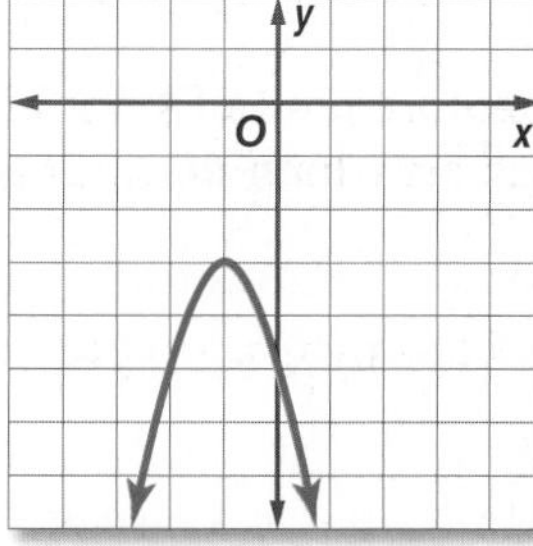

16. $0.5x^2 - 2x + 2 = 0$

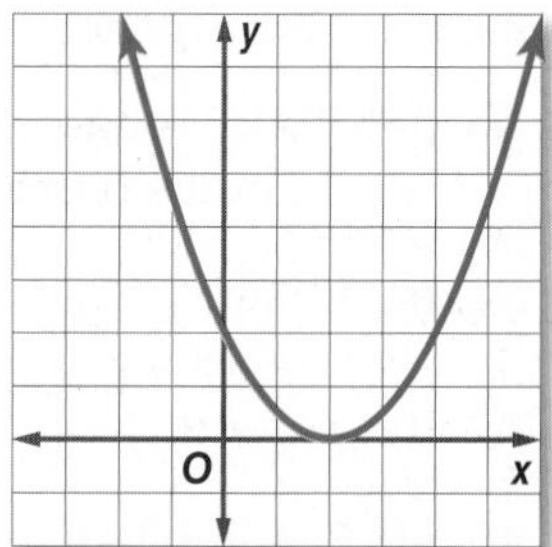

17. $-0.25x^2 - x - 1 = 0$

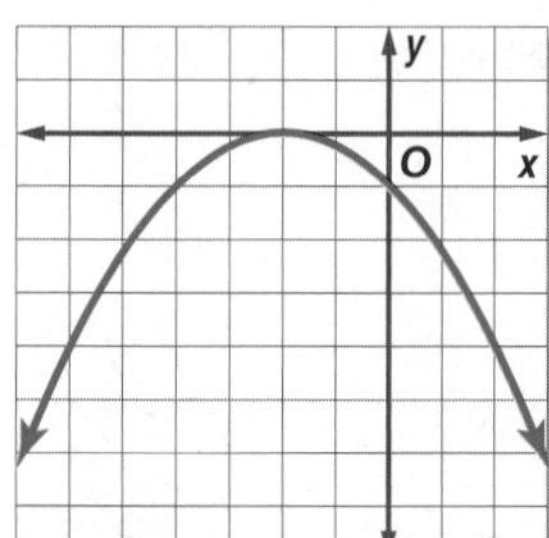

18. $x^2 - 6x + 11 = 0$

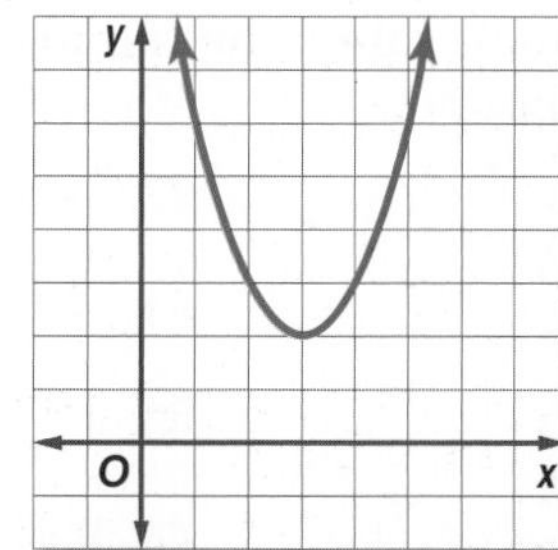

19. $-0.5x^2 + 0.5x + 6 = 0$

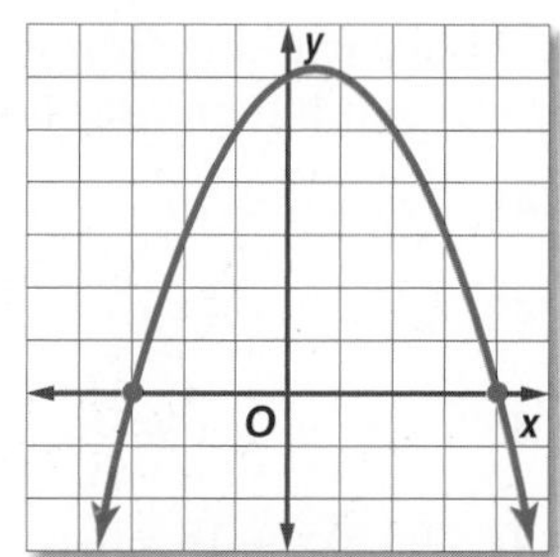

Examples 2–4 **Solve each equation. If exact roots cannot be found, state the consecutive integers between which the roots are located.**

20. $x^2 = 5x$

21. $-2x^2 - 4x = 0$

22. $x^2 - 5x - 14 = 0$

23. $-x^2 + 2x + 24 = 0$

24. $x^2 - 18x = -81$

25. $2x^2 - 8x = -32$

26. $2x^2 - 3x - 15 = 4$

27. $-3x^2 - 7 + 2x = -11$

28. $-0.5x^2 + 3 = -5x - 2$

29. $-2x + 12 = x^2 + 16$

Example 5 **Use the tables to determine the location of the zeros of each quadratic function.**

30.

x	−7	−6	−5	−4	−3	−2	−1	0
$f(x)$	−8	−1	4	4	−1	−8	−22	−48

31.

x	−2	−1	0	1	2	3	4	5
$f(x)$	32	14	2	−3	−3	2	14	32

32.

x	−6	−3	0	3	6	9	12	15
$f(x)$	−6	−1	3	5	3	−1	−6	−14

Example 6 **NUMBER THEORY** **Use a quadratic equation to find two real numbers that satisfy each situation, or show that no such numbers exist.**

33. Their sum is −15, and their product is −54.

34. Their sum is 4, and their product is −117.

35. Their sum is 12, and their product is −84.

36. Their sum is −13, and their product is 42.

37. Their sum is −8 and their product is −209.

CCSS MODELING **For Exercises 38–40, use the formula $h(t) = v_0t - 16t^2$, where $h(t)$ is the height of an object in feet, v_0 is the object's initial velocity in feet per second, and t is the time in seconds.**

38. **BASEBALL** A baseball is hit with an initial velocity of 80 feet per second. Ignoring the height of the baseball player, how long does it take for the ball to hit the ground?

39. **CANNONS** A cannonball is shot with an initial velocity of 55 feet per second. Ignoring the height of the cannon, how long does it take for the cannonball to hit the ground?

40. **GOLF** A golf ball is hit with an initial velocity of 100 feet per second. How long will it take for it to hit the ground?

Solve each equation. If exact roots cannot be found, state the consecutive integers between which the roots are located.

41. $2x^2 + x = 15$

42. $-5x - 12 = -2x^2$

43. $4x^2 - 15 = -4x$

44. $-35 = -3x - 2x^2$

45. $-3x^2 + 11x + 9 = 1$

46. $13 - 4x^2 = -3x$

47. $-0.5x^2 + 18 = -6x + 33$

48. $0.5x^2 + 0.75 = 0.25x$

49 **WATER BALLOONS** Tony wants to drop a water balloon so that it splashes on his brother. Use the formula $h(t) = -16t^2 + h_0$, where t is the time in seconds and the initial height h_0 is in feet, to determine how far his brother should be from the target when Tony lets go of the balloon.

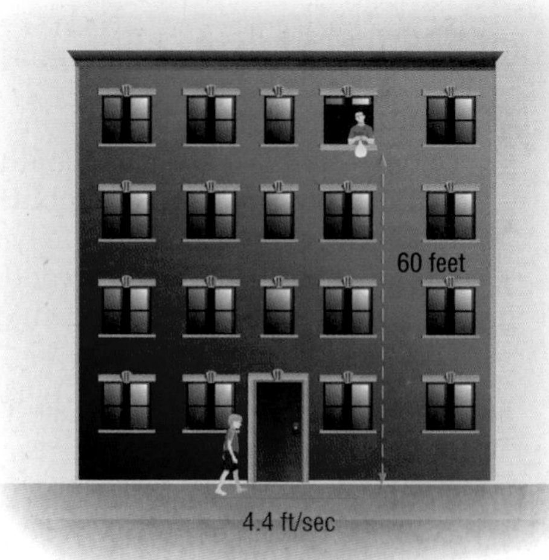

50. **WATER HOSES** A water hose can spray water at an initial velocity of 40 feet per second. Use the formula $h(t) = v_0t - 16t^2$, where $h(t)$ is the height of the water in feet, v_0 is the initial velocity in feet per second, and t is the time in seconds.

a. How long will it take the water to hit the nozzle on the way down?

b. Assuming the nozzle is 5 feet up, what is the maximum height of the water?

51. **SKYDIVING** In 2003, John Fleming and Dan Rossi became the first two blind skydivers to be in free fall together. They jumped from an altitude of 14,000 feet and free fell to an altitude of 4000 feet before their parachutes opened. Ignoring air resistance and using the formula $h(t) = -16t^2 + h_0$, where t is the time in seconds and the initial height h_0 is in feet, determine how long they were in free fall.

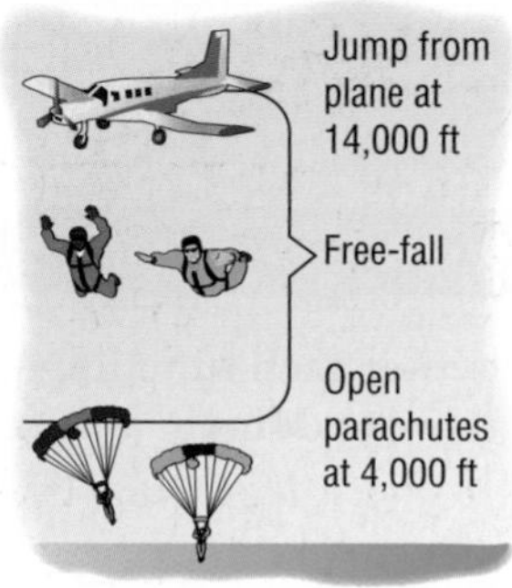

H.O.T. Problems Use Higher-Order Thinking Skills

52. **CCSS CRITIQUE** Hakeem and Tanya were asked to find the location of the roots of the quadratic function represented by the table. Is either of them correct? Explain.

x	−4	−2	0	2	4	6	8	10
$f(x)$	52	26	8	−2	−4	2	16	38

Hakeem

The roots are between 4 and 6 because f(x) stops decreasing and begins to increase between x = 4 and x = 6.

Tanya

The roots are between −2 and 0 because x changes signs at that location.

53. **CHALLENGE** Find the value of a positive integer k such that $f(x) = x^2 - 2kx + 55$ has roots at $k + 3$ and $k - 3$.

54. **REASONING** If a quadratic function has a minimum at $(-6, -14)$ and a root at $x = -17$, what is the other root? Explain your reasoning.

55. **OPEN ENDED** Write a quadratic function with a maximum at $(3, 125)$ and roots at -2 and 8.

56. **WRITING IN MATH** Explain how to solve a quadratic equation by graphing its related quadratic function.

Standardized Test Practice

57. SHORT RESPONSE A bag contains five different colored marbles. The colors of the marbles are black, silver, red, green, and blue. A student randomly chooses a marble. Then, without replacing it, he chooses a second marble. What is the probability that the student chooses the red and then the green marble?

58. Which number would be closest to zero on the number line?

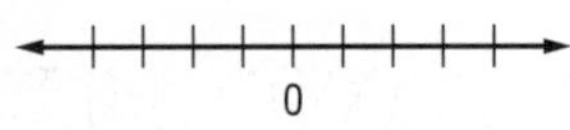

A -0.6 C 0.5

B $\frac{2}{5}$ D $\frac{\sqrt{2}}{2}$

59. SAT/ACT A salesman's monthly gross pay consists of $3500 plus 20 percent of the dollar amount of his sales. If his gross pay for one month was $15,500, what was the dollar amount of his sales for that month?

F $12,000 J $70,000

G $16,000 K $77,500

H $60,000

60. Find the next term in the sequence below.

$$\frac{2x}{5}, \frac{3x}{5}, \frac{4x}{5}, \ldots$$

A x C $\frac{x}{5}$

B $5x$ D $\frac{5x}{4}$

Spiral Review

Determine whether each function has a *maximum* or *minimum* value, and find that value. Then state the domain and range of the function. (Lesson 4-1)

61. $f(x) = -4x^2 + 8x - 16$ **62.** $f(x) = 3x^2 + 12x - 18$ **63.** $f(x) = 4x + 13 - 2x^2$

Determine whether each pair of matrices are inverses of each other. (Lesson 3-8)

64. $\begin{bmatrix} 4 & -3 \\ -1 & -6 \end{bmatrix}$ and $\begin{bmatrix} \frac{3}{13} & -\frac{1}{18} \\ -\frac{1}{26} & -\frac{2}{13} \end{bmatrix}$ **65.** $\begin{bmatrix} 6 & -3 \\ 4 & 8 \end{bmatrix}$ and $\begin{bmatrix} \frac{1}{10} & \frac{1}{20} \\ -\frac{1}{15} & \frac{2}{15} \end{bmatrix}$ **66.** $\begin{bmatrix} 2 & 4 \\ -3 & -2 \end{bmatrix}$ and $\begin{bmatrix} -\frac{1}{4} & -\frac{1}{2} \\ \frac{3}{8} & \frac{1}{4} \end{bmatrix}$

67. SALES Alex is in charge of stocking shirts for the concession stand at the high school football game. The numbers of shirts needed for a regular season game are listed in the matrix. Alex plans to double the number of shirts stocked for a playoff game. (Lesson 3-6)

Size	small	medium	large
Child	10	10	15
Adult	25	35	45

a. Write a matrix A to represent the regular season stock.

b. What scalar can be used to determine a matrix M to represent the new numbers? Find M.

c. What is $M - A$? What does this represent in this situation?

Solve each system of equations. (Lesson 3-1)

68. $4x - 7y = -9$
$5x + 2y = -22$

69. $8y - 2x = 38$
$5x - 3y = -27$

70. $3x + 8y = 24$
$-16y - 6x = 48$

Solve each inequality. (Lesson 1-5)

71. $3x - 6 \leq -14$ **72.** $6 - 4x \leq 2$ **73.** $-6x + 3 \geq 3x - 16$

Skills Review

Find the GCF of each set of numbers.

74. 16, 48, 128 **75.** 15, 21, 49 **76.** 12, 28, 36

EXTEND 4-2

Graphing Technology Lab
Solving Quadratic Equations by Graphing

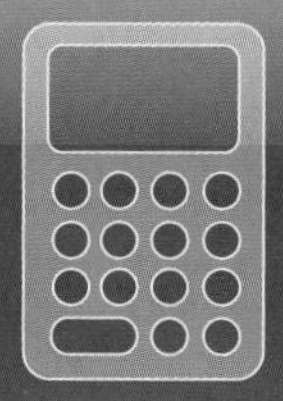

You can use a TI-83/84 Plus graphing calculator to solve quadratic equations.

Common Core State Standards
Content Standards
A.REI.11 Explain why the *x*-coordinates of the points where the graphs of the equations $y = f(x)$ and $y = g(x)$ intersect are the solutions of the equation $f(x) = g(x)$; find the solutions approximately, e.g., using technology to graph the functions, make tables of values, or find successive approximations. Include cases where $f(x)$ and/or $g(x)$ are linear, polynomial, rational, absolute value, exponential, and logarithmic functions.

Activity Solving Quadratic Equations

Solve $x^2 - 8x + 15 = 0$.

Step 1 Let **Y1** $= x^2 - 8x + 15$ and **Y2** $= 0$.

Step 2 Graph **Y1** and **Y2** in the standard viewing window.

KEYSTROKES: [Y=] [X,T,θ,n] [x^2] [(–)] 8 [X,T,θ,n] [+] 15 [ENTER] 0 [ZOOM] 6

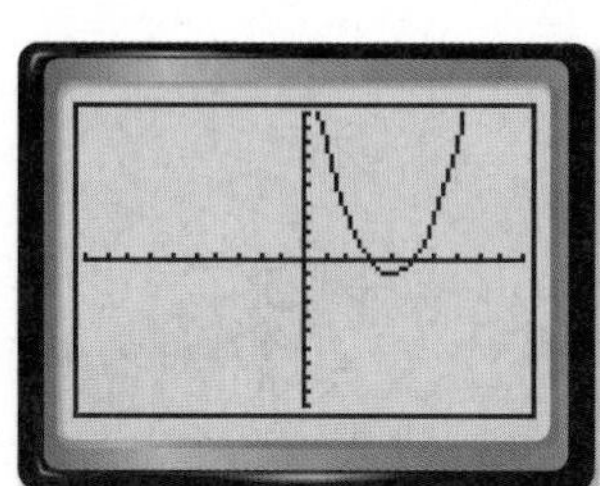

[–10, 10] scl: 1 by [–10, 10] scl: 1

Step 3 To find the *x*-intercepts, determine the points where **Y1** = **Y2**.

KEYSTROKES: [2nd] [CALC] 5

Press [ENTER] for the first equation. Press [ENTER] for the second equation.

Move the cursor as close to the left *x*-intercept and press [ENTER].

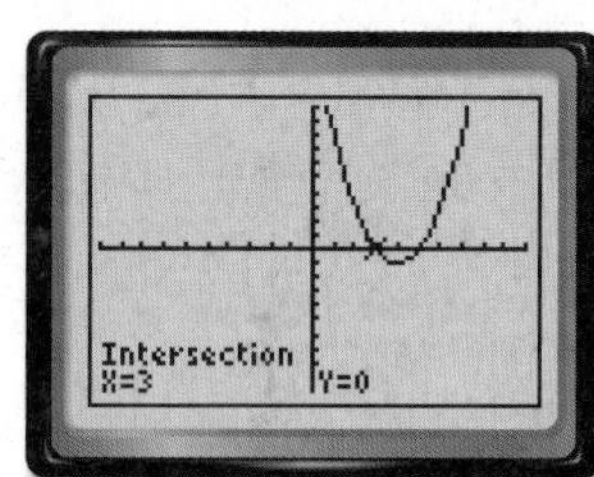

[–10, 10] scl: 1 by [–10, 10] scl: 1

Find the second *x*-intercept.

KEYSTROKES: [2nd] [CALC] 5

Press [ENTER] for the first equation. Press [ENTER] for the second equation.

Move the cursor as close to the right *x*-intercept and press [ENTER].

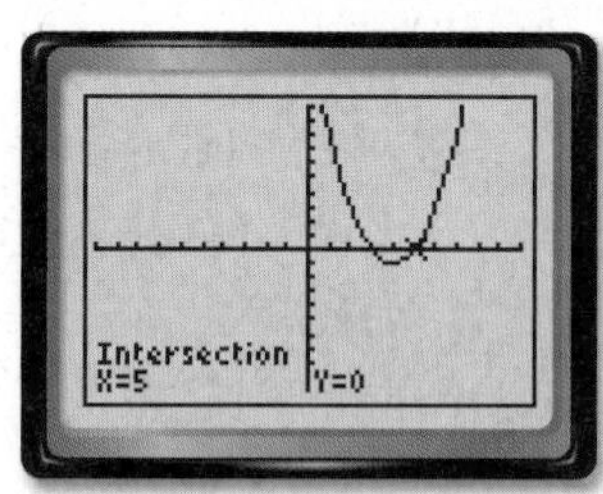

[–10, 10] scl: 1 by [–10, 10] scl: 1

The *x*-intercepts are 3 and 5, so $x = 3$ and $x = 5$.

Exercises

Solve each equation. Round to the nearest tenth if necessary.

1. $x^2 - 7x + 12 = 0$

2. $x^2 + 5x + 6 = 0$

3. $x^2 - 3 = 2x$

4. $x^2 + 5x + 6 = 12$

5. $x^2 + 5x = 0$

6. $x^2 - 4 = 0$

7. $x^2 + 8x + 16 = 0$

8. $x^2 - 10x = -25$

9. $9x^2 + 48x + 64 = 0$

10. $2x^2 + 3x - 1 = 0$

11. $5x^2 - 7x = -2$

12. $6x^2 + 2x + 1 = 0$

LESSON

4-3 Solving Quadratic Equations by Factoring

Then

- You found the greatest common factors of sets of numbers.

Now

1. Write quadratic equations in standard form.
2. Solve quadratic equations by factoring.

Why?

- The **factored form** of a quadratic equation is $0 = a(x - p)(x - q)$. In the equation, p and q represent the x-intercepts of the graph of the equation.

The x-intercepts of the graph at the right are 2 and 6. In this lesson, you will learn how to change a quadratic equation in factored form into standard form and vice versa.

Standard Form	Factored Form	Related Graph
$0 = x^2 - 8x + 12$	$0 = (x - 6)(x - 2)$ (Factors)	2 and 6 are x-intercepts.

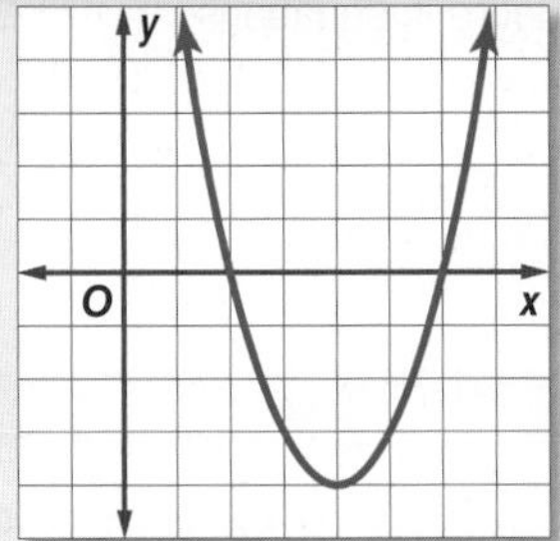

NewVocabulary
factored form
FOIL method

Common Core State Standards

Content Standards

A.SSE.2 Use the structure of an expression to identify ways to rewrite it.

F.IF.8.a Use the process of factoring and completing the square in a quadratic function to show zeros, extreme values, and symmetry of the graph, and interpret these in terms of a context.

Mathematical Practices

2 Reason abstractly and quantitatively.

1 Standard Form

You can use the FOIL method to write a quadratic equation that is in factored form in standard form. The **FOIL method** uses the Distributive Property to multiply binomials.

KeyConcept FOIL Method for Multiplying Binomials

Words To multiply two binomials, find the sum of the products of ***F*** the *First terms*, ***O*** the *Outer terms*, ***I*** the *Inner terms*, and ***L*** the *Last terms*.

Examples

		Product of First Terms		Product of Outer Terms		Product of Inner Terms		Product of Last Terms
$(x - 6)(x - 2)$ (F O I L)	=	$(x)(x)$	+	$(x)(-2)$	+	$(-6)(x)$	+	$(-6)(-2)$

$$= x^2 - 2x - 6x + 12 \text{ or } x^2 - 8x + 12$$

Example 1 Translate Sentences into Equations

Write a quadratic equation in standard form with $-\frac{1}{3}$ and 6 as its roots.

$(x - p)(x - q) = 0$ Write the pattern.

$\left[x - \left(-\frac{1}{3}\right)\right](x - 6) = 0$ Replace p with $-\frac{1}{3}$ and q with 6.

$\left(x + \frac{1}{3}\right)(x - 6) = 0$ Simplify.

$x^2 - \frac{17}{3}x - 2 = 0$ Multiply.

$3x^2 - 17x - 6 = 0$ Multiply each side by 3 so that b and c are integers.

GuidedPractice

1. Write a quadratic equation in standard form with $\frac{3}{4}$ and -5 as its roots.

2 Solve Equations by Factoring

Solving quadratic equations by factoring is an application of the Zero Product Property.

KeyConcept Zero Product Property

Words For any real numbers a and b, if $ab = 0$, then either $a = 0$, $b = 0$, or both a and b equal zero.

Example If $(x + 3)(x - 5) = 0$, then $x + 3 = 0$ or $x - 5 = 0$.

PT

Example 2 Factor the GCF

Solve $16x^2 + 8x = 0$.

$16x^2 + 8x = 0$	Original equation.
$8x(2x) + 8x^2(1) = 0$	Factor the GCF.
$8x(2x + 1) = 0$	Distributive Property
$8x = 0$ or $2x + 1 = 0$	Zero Product Property
$x = 0$ $\quad 2x = -1$	Solve both equations.
$x = -\frac{1}{2}$	

GuidedPractice **Solve each equation.**

2A. $20x^2 + 15x = 0$ **2B.** $4y^2 + 16y = 0$ **2C.** $6a^5 + 18a^4 = 0$

ReviewVocabulary

perfect square a number with a positive square root that is a whole number

Trinomials and binomials that are perfect squares have special factoring rules. In order to use these rules, the first and last terms need to be perfect squares and the middle term needs to be twice the product of the square roots of the first and last terms.

Example 3 Perfect Squares and Differences of Squares

Solve each equation.

a. $x^2 + 16x + 64 = 0$

$x^2 = (x)^2$; $64 = (8)^2$	First and last terms are perfect squares.
$16x = 2(x)(8)$	Middle term equals $2ab$.

$x^2 + 16x + 64$ is a perfect square trinomial.

$x^2 + 16x + 64 = 0$	Original equation
$(x + 8)^2 = 0$	Factor using the pattern.
$x + 8 = 0$	Take the square root of each side.
$x = -8$	Solve.

b. $x^2 = 64$

$x^2 = 64$	Original equation
$x^2 - 64 = 0$	Subtract 64 from each side.
$x^2 - (8)^2 = 0$	Write in the form $a^2 - b^2$.
$(x + 8)(x - 8) = 0$	Factor the difference of squares.
$x + 8 = 0$ or $x - 8 = 0$	Zero Product Property
$x = -8$ $\quad x = 8$	Solve.

StudyTip

Square Roots By inspection, notice that the square root of 64 is −8 and 8. Also, for $x^2 = 4$, the solutions would be −2 and 2.

GuidedPractice

3A. $4x^2 - 12x + 9 = 0$ **3B.** $81x^2 - 9x = 0$ **3C.** $6a^2 - 3a = 0$

StudyTip

CCSS Structure If values for m and p exist, then the trinomial can always be factored.

A special pattern is used when factoring trinomials of the form $ax^2 + bx + c$. First, multiply the values of a and c. Then, find two values, m and p, such that their product equals ac and their sum equals b.

Consider $6x^2 + 13x - 5$: $ac = 6(-5) = -30$.

Factors of −30	Sum	Factors of −30	Sum
1, −30	−29	−1, 30	29
2, −15	−13	**−2, 15**	**13**
3, −10	−7	−3, 10	7
5, −6	−1	−5, 6	1

Now the middle term, $13x$, can be rewritten as $-2x + 15x$.

This polynomial can now be factored by grouping.

$6x^2 + 13x - 5 = 6x^2 + mx + px - 5$ Write the pattern.

$= 6x^2 - 2x + 15x - 5$ $m = -2$ and $p = 15$

$= (6x^2 - 2x) + (15x - 5)$ Group terms.

$= 2x(3x - 1) + 5(3x - 1)$ Factor the GCF.

$= (2x + 5)(3x - 1)$ Distributive Property

PT

Example 4 Factor Trinomials

Solve each equation.

a. $x^2 + 9x + 20 = 0$

$ac = 20$ $a = 1, c = 20$

Factors of 20	Sum	Factors of 20	Sum
1, 20	21	−1, −20	−21
2, 10	12	−2, −10	−12
4, 5	**9**	−4, −5	−9

StudyTip

Trinomials It does not matter if the values of m and p are switched when grouping.

$x^2 + 9x + 20 = 0$ Original expression

$x^2 + mx + px + 20 = 0$ Write the pattern.

$x^2 + 4x + 5x + 20 = 0$ $m = 4$ and $p = 5$

$(x^2 + 4x) + (5x + 20) = 0$ Group terms with common factors.

$x(x + 4) + 5(x + 4) = 0$ Factor the GCF from each grouping.

$(x + 5)(x + 4) = 0$ Distributive Property

$x + 5 = 0$ or $x + 4 = 0$ Zero Product Property

$x = -5$ $\quad x = -4$ Solve each equation.

b. $6y^2 - 23y + 20 = 0$

$ac = 120$ $a = 6, c = 20$

$m = -8, p = -15$ $-8(-15) = 120; -8 + (-15) = -23$

$6y^2 - 23y + 20 = 0$ Original equation

$6y^2 + my + py + 20 = 0$ Write the pattern.

$6y^2 - 8y - 15y + 20 = 0$ $m = -8$ and $p = -15$

$(6y^2 - 8y) + (-15y + 20) = 0$ Group terms with common factors.

$2y(3y - 4) - 5(3y - 4) = 0$ Factor the GCF from each grouping.

$(2y - 5)(3y - 4) = 0$ Distributive Property

$2y - 5 = 0$ or $3y - 4 = 0$ Zero Product Property

$2y = 5$ $\quad 3y = 4$ Solve both equations.

$y = \frac{5}{2}$ $\quad y = \frac{4}{3}$

GuidedPractice

4A. $x^2 - 11x + 30 = 0$

4B. $x^2 - 4x - 21 = 0$

4C. $15x^2 - 8x + 1 = 0$

4D. $-12x^2 + 8x + 15 = 0$

Real-WorldLink

Cuba's Osleidys Menendez broke the javelin world record in 2002 with a distance of 234 feet 8 inches.

Source: *New York Times*

Real-World Example 5 Solve Equations by Factoring

TRACK AND FIELD The height of a javelin in feet is modeled by $h(t) = -16t^2 + 79t + 5$, where t is the time in seconds after the javelin is thrown. How long is it in the air?

To determine how long the javelin is in the air, we need to find when the height equals 0. We can do this by solving $-16t^2 + 79t + 5 = 0$.

$-16t^2 + 79t + 5 = 0$	Original equation
$m = 80; p = -1$	$-16(5) = -80, 80 \cdot (-1) = -80, 80 + (-1) = 79$
$-16t^2 + 80t - t + 5 = 0$	Write the pattern.
$(-16t^2 + 80t) + (-t + 5) = 0$	Group terms with common factors.
$16t(-t + 5) + 1(-t + 5) = 0$	Factor GCF from each group.
$(16t + 1)(-t + 5) = 0$	Distributive Property
$16t + 1 = 0$ or $-t + 5 = 0$	Zero Product Property
$16t = -1$ $\quad -t = -5$	Solve both equations.
$t = -\frac{1}{16}$ $\quad t = 5$	Solve.

CHECK We have two solutions.

- The first solution is negative and since time cannot be negative, this solution can be eliminated.
- The second solution of 5 seconds seems reasonable for the time a javelin spends in the air.
- The answer can be confirmed by substituting back into the original equation.

$$-16t^2 + 79t + 5 = 0$$
$$-16(5)^2 + 79(5) + 5 \stackrel{?}{=} 0$$
$$-400 + 395 + 5 \stackrel{?}{=} 0$$
$$0 = 0 \checkmark$$

The javelin is in the air for 5 seconds.

GuidedPractice

5. BUNGEE JUMPING Juan recorded his brother bungee jumping from a height of 1100 feet. At the time the cord lifted his brother back up, he was 76 feet above the ground. If Juan started recording as soon as his brother fell, how much time elapsed when the cord snapped back? Use $f(t) = -16t^2 + c$, where c is the height in feet.

Check Your Understanding

= Step-by-Step Solutions begin on page R14.

Example 1 **Write a quadratic equation in standard form with the given root(s).**

1. $-8, 5$ **2.** $\frac{3}{2}, \frac{1}{4}$ **3.** $-\frac{2}{3}, \frac{5}{2}$

Examples 2–4 **Factor each polynomial.**

4. $35x^2 - 15x$ **5.** $18x^2 - 3x + 24x - 4$ **6.** $x^2 - 12x + 32$

7. $x^2 - 4x - 21$ **8.** $2x^2 + 7x - 30$ **9.** $16x^2 - 16x + 3$

Example 5 **Solve each equation.**

10. $x^2 - 36 = 0$ **11.** $12x^2 - 18x = 0$ **12.** $12x^2 - 2x - 2 = 0$

13. $x^2 - 9x = 0$ **14.** $x^2 - 3x - 28 = 0$ **15.** $2x^2 - 24x = -72$

16. CCSS **SENSE-MAKING** Tamika wants to double the area of her garden by increasing the length and width by the same amount. What will be the dimensions of her garden then?

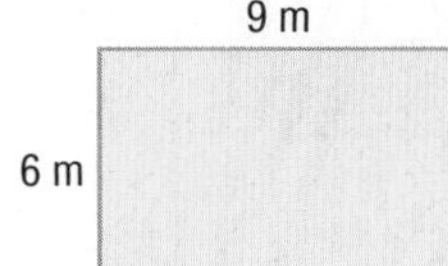

Practice and Problem Solving

Extra Practice is on page R4.

Example 1 **Write a quadratic equation in standard form with the given root(s).**

17. 7 **18.** $-5, \frac{1}{2}$ **19.** $\frac{1}{5}, 6$

Examples 2–4 **Factor each polynomial.**

20. $40a^2 - 32a$ **21.** $51c^3 - 34c$ **22.** $32xy + 40bx - 12ay - 15ab$

23. $3x^2 - 12$ **24.** $15y^2 - 240$ **25.** $48cg + 36cf - 4dg - 3df$

26. $x^2 + 13x + 40$ **27.** $x^2 - 9x - 22$ **28.** $3x^2 + 12x - 36$

29. $15x^2 + 7x - 2$ **30.** $4x^2 + 29x + 30$ **31.** $18x^2 + 15x - 12$

32. $8x^2z^2 - 4xz^2 - 12z^2$ **33.** $9x^2 - 25$ **34.** $18x^2y^2 - 24xy^2 + 36y^2$

Example 5 **Solve each equation.**

35. $15x^2 - 84x - 36 = 0$ **36.** $12x^2 + 13x - 14 = 0$ **37.** $12x^2 - 108x = 0$

38. $x^2 + 4x - 45 = 0$ **39.** $x^2 - 5x - 24 = 0$ **40.** $x^2 = 121$

41. $x^2 + 13 = 17$ **42.** $-3x^2 - 10x + 8 = 0$ **43.** $-8x^2 + 46x - 30 = 0$

44. **GEOMETRY** The hypotenuse of a right triangle is 1 centimeter longer than one side and 4 centimeters longer than three times the other side. Find the dimensions of the triangle.

45. **NUMBER THEORY** Find two consecutive even integers with a product of 624.

GEOMETRY **Find x and the dimensions of each rectangle.**

46. $A = 96$ ft^2; sides: $x - 2$ ft, $x + 2$ ft

47. $A = 432$ in^2; sides: $x - 2$ in., $x + 4$ in.

48. $A = 448$ ft^2; sides: $3x - 4$ ft, $x + 2$ ft

Solve each equation by factoring.

49. $12x^2 - 4x = 5$

50. $5x^2 = 15x$

51. $16x^2 + 36 = -48x$

52. $75x^2 - 60x = -12$

53. $4x^2 - 144 = 0$

54. $-7x + 6 = 20x^2$

55. **MOVIE THEATER** A company plans to build a large multiplex theater. The financial analyst told her manager that the profit function for their theater was $P(x) = -x^2 + 48x - 512$, where x is the number of movie screens, and $P(x)$ is the profit earned in thousands of dollars. Determine the range of production of movie screens that will guarantee that the company will not lose money.

Write a quadratic equation in standard form with the given root(s).

56. $-\frac{4}{7}, \frac{3}{8}$

57. 3.4, 0.6

58. $\frac{2}{11}, \frac{5}{9}$

Solve each equation by factoring.

59. $10x^2 + 25x = 15$

60. $27x^2 + 5 = 48x$

61. $x^2 + 0.25x = 1.25$

62. $48x^2 - 15 = -22x$

63. $3x^2 + 2x = 3.75$

64. $-32x^2 + 56x = 12$

65. **DESIGN** A square is cut out of the figure at the right. Write an expression for the area of the figure that remains, and then factor the expression.

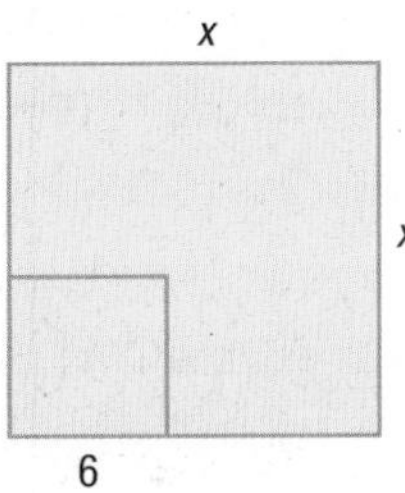

66. **CCSS PERSEVERANCE** After analyzing the market, a company that sells Web sites determined the profitability of their product was modeled by $P(x) = -16x^2 + 368x - 2035$, where x is the price of each Web site and $P(x)$ is the company's profit. Determine the price range of the Web sites that will be profitable for the company.

67. **PAINTINGS** Enola wants to add a border to her painting, distributed evenly, that has the same area as the painting itself. What are the dimensions of the painting with the border included?

15 in.

10 in.

68. **MULTIPLE REPRESENTATIONS** In this problem, you will consider $a(x - p)(x - q) = 0$.

a. Graphical Graph the related function for $a = 1$, $p = 2$, and $q = -3$.

b. Analytical What are the solutions of the equation?

c. Graphical Graph the related functions for $a = 4, -3$, and $\frac{1}{2}$ on the same graph.

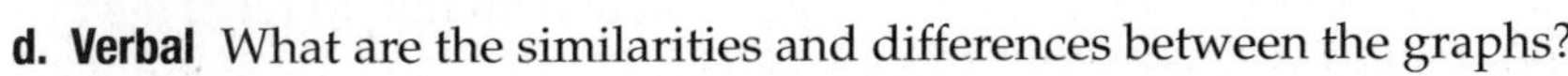

d. Verbal What are the similarities and differences between the graphs?

e. Verbal What conclusion can you make about the relationship between the factored form of a quadratic equation and its solutions?

69. **GEOMETRY** The area of the triangle is 26 square centimeters. Find the length of the base.

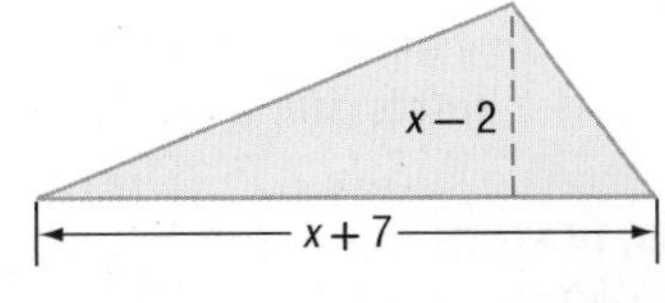

70. SOCCER When a ball is kicked in the air, its height in meters above the ground can be modeled by $h(t) = -4.9t^2 + 14.7t$ and the distance it travels can be modeled by $d(t) = 16t$, where t is the time in seconds.

a. How long is the ball in the air?

b. How far does it travel before it hits the ground? (*Hint*: Ignore air resistance.)

c. What is the maximum height of the ball?

Factor each polynomial.

71. $18a - 24ay + 48b - 64by$

72. $3x^2 + 2xy + 10y + 15x$

73. $6a^2b^2 - 12ab^2 - 18b^3$

74. $12a^2 - 18ab + 30ab^3$

75. $32ax + 12bx - 48ay - 18by$

76. $30ac + 80bd + 40ad + 60bc$

77. $5ax^2 - 2by^2 - 5ay^2 + 2bx^2$

78. $12c^2x + 4d^2y - 3d^2x - 16c^2y$

H.O.T. Problems Use Higher-Order Thinking Skills

79. ERROR ANALYSIS Gwen and Morgan are solving $-12x^2 + 5x + 2 = 0$. Is either of them correct? Explain your reasoning.

Gwen

$-12x^2 + 5x + 2 = 0$

$-12x^2 + 8x - 3x + 2 = 0$

$4x(-3x + 2) - (3x + 2) = 0$

$(4x - 1)(3x + 2) = 0$

$x = \frac{1}{4}$ or $-\frac{2}{3}$

Morgan

$-12x^2 + 5x + 2 = 0$

$-12x^2 + 8x - 3x + 2 = 0$

$4x(-3x + 2) + (-3x + 2) = 0$

$(4x + 1)(-3x + 2) = 0$

$x = -\frac{1}{4}$ or $\frac{2}{3}$

80. CHALLENGE Solve $3x^6 - 39x^4 + 108x^2 = 0$ by factoring.

81. CHALLENGE The rule for factoring a difference of cubes is shown below. Use this rule to factor $40x^5 - 135x^2y^3$.

$$a^3 - b^3 = (a - b)(a^2 + ab + b^2)$$

82. OPEN ENDED Choose two integers. Then write an equation in standard form with those roots. How would the equation change if the signs of the two roots were switched?

83. CHALLENGE For a quadratic equation of the form $(x - p)(x - q) = 0$, show that the axis of symmetry of the related quadratic function is located halfway between the x-intercepts p and q.

84. WRITE A QUESTION A classmate is using the guess-and-check strategy to factor trinomials of the form $x^2 + bx + c$. Write a question to help him think of a way to use that strategy for $ax^2 + bx + c$.

85. CCSS ARGUMENTS Determine whether the following statement is *sometimes*, *always*, or *never* true. Explain your reasoning.

In a quadratic equation in standard form where a, b, and c are integers, if b is odd, then the quadratic cannot be a perfect square trinomial.

86. WRITING IN MATH Explain how to factor a trinomial in standard form with $a > 1$.

Standardized Test Practice

87. SHORT RESPONSE If $ABCD$ is transformed by $(x, y) \rightarrow (3x, 4y)$, determine the area of $A'B'C'D'$.

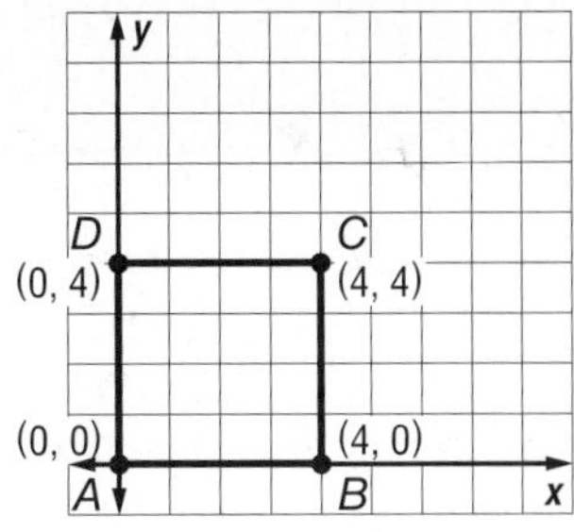

88. For $y = 2|6 - 3x| + 4$, which set describes x when $y < 6$?

A $\left\{x \middle| \frac{5}{3} < x < \frac{7}{3}\right\}$

B $\left\{x \middle| x < \frac{5}{3} \text{ or } x > \frac{7}{3}\right\}$

C $\left\{x \middle| x < \frac{5}{3}\right\}$

D $\left\{x \middle| x > \frac{7}{3}\right\}$

89. PROBABILITY A 5-character password can contain the numbers 0 through 9 and 26 letters of the alphabet. None of the characters can be repeated. What is the probability that the password begins with a consonant?

F $\frac{21}{26}$

G $\frac{21}{35}$

H $\frac{21}{36}$

J $\frac{5}{36}$

90. SAT/ACT If $c = \frac{8a^3}{b}$, what happens to the value of c when both a and b are doubled?

A c is unchanged.

B c is halved.

C c is doubled.

D c is multiplied by 4.

E c is multiplied by 8.

Spiral Review

Use the related graph of each equation to determine its solutions. (Lesson 4-2)

91. $x^2 - 2x - 8 = 0$

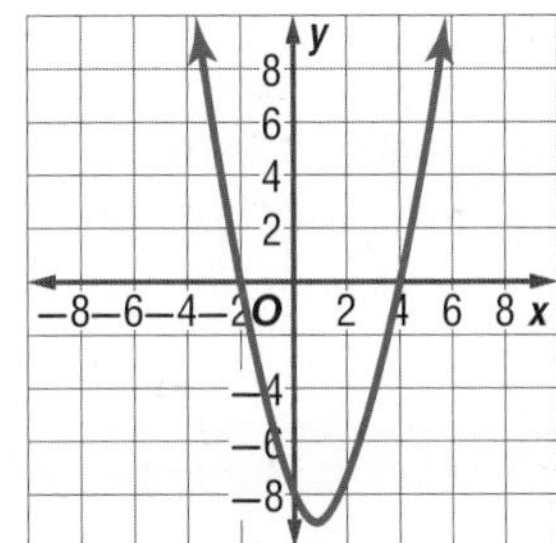

92. $x^2 + 4x = 12$

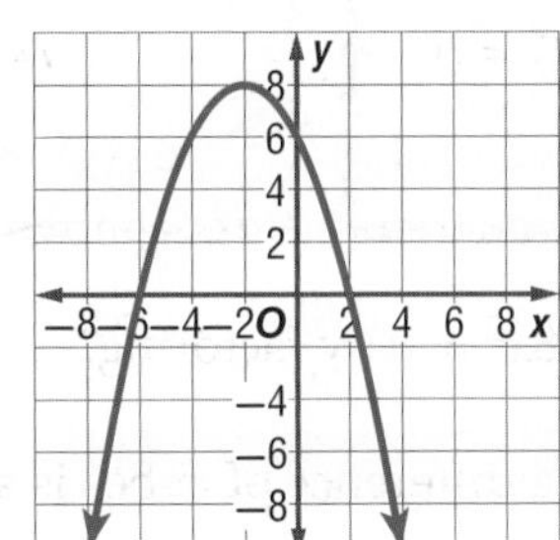

93. $x^2 + 4x + 4 = 0$

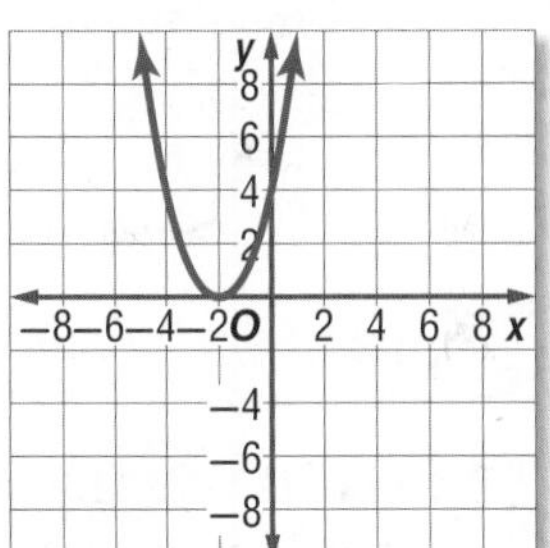

Graph each function. (Lesson 4-1)

94. $f(x) = x^2 - 6x + 2$

95. $f(x) = -2x^2 + 4x + 1$

96. $f(x) = (x - 3)(x + 4)$

97. FUNDRAISING Lawrence High School sold wrapping paper and boxed cards for their fundraising event. The school gets \$1.00 for each roll of wrapping paper sold and \$0.50 for each box of cards sold. (Lesson 3-6)

a. Write a matrix that represents the amounts sold for each class and a matrix that represents the amount of money the school earns for each item sold.

b. Write a matrix that shows how much each class earned.

c. Which class earned the most money?

d. What is the total amount of money the school made from the fundraiser?

Total Amounts for Each Class

Class	Wrapping Paper	Cards
freshmen	72	49
sophomores	68	63
juniors	90	56
seniors	86	62

Skills Review

Simplify.

98. $\sqrt{5} \cdot \sqrt{15}$

99. $\sqrt{8} \cdot \sqrt{32}$

100. $2\sqrt{3} \cdot \sqrt{27}$

LESSON 4-4 Complex Numbers

Then

- You simplified square roots.

Now

1. Perform operations with pure imaginary numbers.
2. Perform operations with complex numbers.

Why?

- Consider the graph of $y = x^2 + 2x + 4$ at the right. Notice how this graph has no x-intercepts and therefore does not have any roots. Does this mean there are no solutions to $0 = x^2 + 2x + 4$?

Use the **Solver** function located in the MATH menu of a graphing calculator. Enter the equation and select $x = 2$ as your guess to a solution.

Press ALPHA ENTER and the calculator will attempt to solve the equation. The calculator indicates there is no solution with the error message. So there are no real solutions. However, there are imaginary solutions.

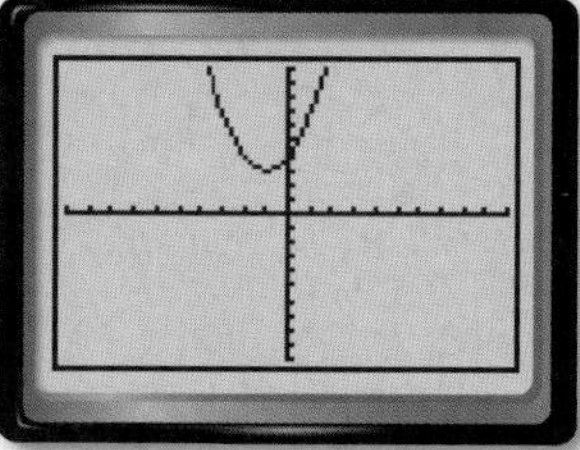

[−10, 10] scl: 1 by [−10, 10] scl: 1

NewVocabulary
imaginary unit
pure imaginary number
complex number
complex conjugates

Common Core State Standards

Content Standards
N.CN.1 Know there is a complex number i such that $i^2 = -1$, and every complex number has the form $a + bi$ with a and b real.
N.CN.2 Use the relation $i^2 = -1$ and the commutative, associative, and distributive properties to add, subtract, and multiply complex numbers.

Mathematical Practices
6 Attend to precision.

1 Pure Imaginary Numbers

In your math studies so far, you have worked with real numbers. Equations like the one above led mathematicians to define imaginary numbers. The **imaginary unit i** is defined to be $i^2 = -1$. The number i is the principal square root of -1; that is, $i = \sqrt{-1}$.

Numbers of the form $6i$, $-2i$, and $i\sqrt{3}$ are called **pure imaginary numbers**. Pure imaginary numbers are square roots of negative real numbers. For any positive real number b, $\sqrt{-b^2} = \sqrt{b^2} \cdot \sqrt{-1}$ or bi.

Example 1 Square Roots of Negative Numbers

Simplify.

a. $\sqrt{-27}$

$$\begin{aligned}\sqrt{-27} &= \sqrt{-1 \cdot 3^2 \cdot 3}\\ &= \sqrt{-1} \cdot \sqrt{3^2} \cdot \sqrt{3}\\ &= i \cdot 3 \cdot \sqrt{3} \text{ or } 3i\sqrt{3}\end{aligned}$$

b. $\sqrt{-216}$

$$\begin{aligned}\sqrt{-216} &= \sqrt{-1 \cdot 6^2 \cdot 6}\\ &= \sqrt{-1} \cdot \sqrt{6^2 \cdot 6}\\ &= i \cdot 6 \cdot \sqrt{6} \text{ or } 6i\sqrt{6}\end{aligned}$$

GuidedPractice

1A. $\sqrt{-18}$

1B. $\sqrt{-125}$

The Commutative and Associative Properties of Multiplication hold true for pure imaginary numbers. The first few powers of i are shown below.

$i^1 = i$	$i^2 = -1$	$i^3 = i^2 \cdot i$ or $-i$	$i^4 = (i^2)^2$ or 1
$i^5 = i^4 \cdot i$ or i	$i^6 = i^4 \cdot i^2$ or -1	$i^7 = i^4 \cdot i^3$ or $-i$	$i^8 = (i^2)^4$ or 1

Example 2 Products of Pure Imaginary Numbers

Simplify.

a. $-5i \cdot 3i$

$$\begin{aligned} -5i \cdot 3i &= -15i^2 && \text{Multiply.} \\ &= -15(-1) && i^2 = -1 \\ &= 15 && \text{Simplify.} \end{aligned}$$

b. $\sqrt{-6} \cdot \sqrt{-15}$

$$\begin{aligned} \sqrt{-6} \cdot \sqrt{-15} &= i\sqrt{6} \cdot i\sqrt{15} && i = \sqrt{-1} \\ &= i^2\sqrt{90} && \text{Multiply.} \\ &= -1 \cdot \sqrt{9} \cdot \sqrt{10} && \text{Simplify.} \\ &= -3\sqrt{10} && \text{Multiply.} \end{aligned}$$

GuidedPractice

2A. $3i \cdot 4i$ **2B.** $\sqrt{-20} \cdot \sqrt{-12}$ **2C.** i^{31}

You can solve some quadratic equations by using the **Square Root Property**. Similar to a difference of squares, the sum of squares can be factored over the complex numbers.

Example 3 Equation with Pure Imaginary Solutions

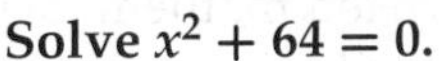

Solve $x^2 + 64 = 0$.

Method 1 Square Root Property

$$\begin{aligned} x^2 + 64 &= 0 \\ x^2 &= -64 \\ x &= \pm\sqrt{-64} \\ x &= \pm 8i \end{aligned}$$

Method 2 Factoring

$$\begin{aligned} x^2 + 64 &= 0 \\ x^2 + 8^2 &= 0 \\ x^2 - (-8^2) &= 0 \\ (x + 8i)(x - 8i) &= 0 \end{aligned}$$

$(x + 8i) = 0$ or $(x - 8i) = 0$

$x = -8i$ $\quad$ $x = 8i$

GuidedPractice

Solve each equation.

3A. $4x^2 + 100 = 0$ **3B.** $x^2 + 4 = 0$

Real-WorldCareer

Electrical Engineer Electrical engineers design, develop, test, and supervise the making of electrical equipment such as digital music players, electric motors, lighting, and radar and navigation systems. A bachelor's degree in engineering is required for almost all entry-level engineering jobs.

FABRICE COFFRINI/AFP/Getty Images

2 Operations with Complex Numbers

Consider $2 + 3i$. Since 2 is a real number and $3i$ is a pure imaginary number, the terms are not like terms and cannot be combined. This type of expression is called a **complex number**.

KeyConcept Complex Numbers

Words	A complex number is any number that can be written in the form $a + bi$, where a and b are real numbers and i is the imaginary unit. a is called the real part, and b is called the imaginary part.
Examples	$5 + 2i$ $\quad$ $1 - 3i = 1 + (-3)i$

The Venn diagram shows the set of complex numbers.

- If $b = 0$, the complex number is a real number.
- If $b \neq 0$, the complex number is imaginary.
- If $a = 0$, the complex number is a pure imaginary number.

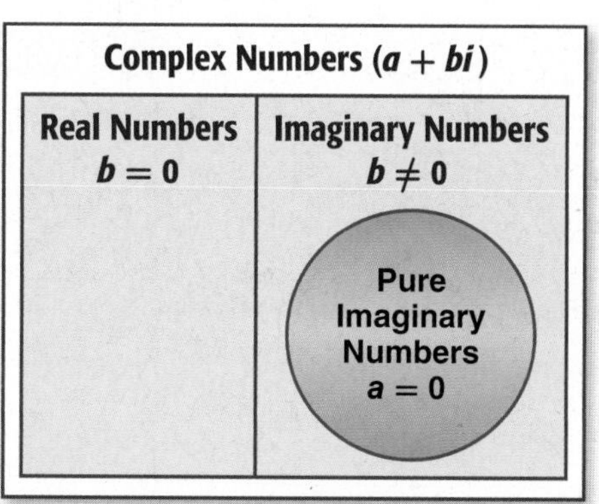

Two complex numbers are equal if and only if their real parts are equal and their imaginary parts are equal. That is, $a + bi = c + di$ if and only if $a = c$ and $b = d$.

StudyTip

Complex Numbers Whereas all real numbers are also complex, the term *complex number* usually refers to a number that is not real.

Example 4 Equate Complex Numbers

Find the values of x and y that make $3x - 5 + (y - 3)i = 7 + 6i$ true.

Set the real parts equal to each other and the imaginary parts equal to each other.

$3x - 5 = 7$	Real parts	$y - 3 = 6$	Imaginary parts
$3x = 12$	Add 5 to each side.	$y = 9$	Add 3 to each side.
$x = 4$	Divide each side by 3.		

GuidedPractice

4. Find the values of x and y that make $5x + 1 + (3 + 2y)\boldsymbol{i} = 2x - 2 + (y - 6)\boldsymbol{i}$ true.

The Commutative, Associative, and Distributive Properties of Multiplication and Addition hold true for complex numbers. To add or subtract complex numbers, combine like terms. That is, combine the real parts, and combine the imaginary parts.

Example 5 Add and Subtract Complex Numbers

Simplify.

a. $(5 - 7i) + (2 + 4i)$

$(5 - 7i) + (2 + 4i) = (5 + 2) + (-7 + 4)i$ — Commutative and Associative Properties

$= 7 - 3i$ — Simplify.

b. $(4 - 8i) - (3 - 6i)$

$(4 - 8i) - (3 - 6i) = (4 - 3) + [-8 - (-6)]i$ — Commutative and Associative Properties

$= 1 - 2i$ — Simplify.

GuidedPractice

5A. $(-2 + 5i) + (1 - 7i)$

5B. $(4 + 6i) - (-1 + 2i)$

StudyTip

Reading Math Electrical engineers use ***j*** as the imaginary unit to avoid confusion with the ***i*** for current.

Complex numbers are used with electricity. In these problems, ***j*** usually represents the imaginary unit. In a circuit with alternating current, the voltage, current, and impedance, or hindrance to current, can be represented by complex numbers. To multiply these numbers, use the FOIL method.

Real-World Example 6 Multiply Complex Numbers

ELECTRICITY In an AC circuit, the voltage V, current C, and impedance I are related by the formula $V = C \cdot I$. Find the voltage in a circuit with current $2 + 4j$ amps and impedance $9 - 3j$ ohms.

$V = C \cdot I$	Electricity formula
$= (2 + 4j) \cdot (9 - 3j)$	$C = 2 + 4j$ and $I = 9 - 3j$
$= 2(9) + 2(-3j) + 4j(9) + 4j(-3j)$	FOIL Method
$= 18 - 6j + 36j - 12j^2$	Multiply.
$= 18 + 30j - 12(-1)$	$j^2 = -1$
$= 30 + 30j$	Add.

The voltage is $30 + 30j$ volts.

GuidedPractice

6. Find the voltage in a circuit with current $2 - 4j$ amps and impedance $3 - 2j$ ohms.

Real-WorldLink

An example of a series circuit is a string of holiday lights. The number of bulbs on a circuit affects the strength of the current, which in turn affects the brightness of the lights.

Source: *Popular Science*

Two complex numbers of the form $a + bi$ and $a - bi$ are called **complex conjugates**. The product of complex conjugates is always a real number. You can use this fact to simplify the quotient of two complex numbers.

Example 7 Divide Complex Numbers

Simplify.

a. $\frac{2i}{3 + 6i}$

$\frac{2i}{3 + 6i} = \frac{2i}{3 + 6i} \cdot \frac{3 - 6i}{3 - 6i}$	$3 + 6i$ and $3 - 6i$ are complex conjugates.
$= \frac{6i - 12i^2}{9 - 36i^2}$	Multiply.
$= \frac{6i - 12(-1)}{9 - 36(-1)}$	$i^2 = -1$
$= \frac{6i + 12}{45}$	Simplify.
$= \frac{4}{15} + \frac{2}{15}i$	$a + bi$ form

b. $\frac{4 + i}{5i}$

$\frac{4 + i}{5i} = \frac{4 + i}{5i} \cdot \frac{i}{i}$	Multiply by $\frac{i}{i}$.
$= \frac{4i + i^2}{5i^2}$	Multiply.
$= \frac{4i - 1}{-5}$	$i^2 = -1$
$= \frac{1}{5} - \frac{4}{5}i$	$a + bi$ form

GuidedPractice

7A. $\frac{-2i}{3 + 5i}$

7B. $\frac{2 + i}{1 - i}$

StudyTip

Technology Operations with complex numbers can be preformed with a TI-83/84 Plus graphing calculator. Use the [2nd] [i] function to enter the expression. Then press [MATH] [ENTER] [ENTER] to view the answer.

Check Your Understanding

= Step-by-Step Solutions begin on page R14.

Examples 1–2 **Simplify.**

1. $\sqrt{-81}$
2. $\sqrt{-32}$
3. $(4i)(-3i)$
4. $3\sqrt{-24} \cdot 2\sqrt{-18}$
5. i^{40}
6. i^{63}

Example 3 **Solve each equation.**

7. $4x^2 + 32 = 0$
8. $x^2 + 1 = 0$

Example 4 **Find the values of a and b that make each equation true.**

9. $3a + (4b + 2)i = 9 - 6i$
10. $4b - 5 + (-a - 3)i = 7 - 8i$

Examples 5 and 7 **Simplify.**

11. $(-1 + 5i) + (-2 - 3i)$
12. $(7 + 4i) - (1 + 2i)$
13. $(6 - 8i)(9 + 2i)$
14. $(3 + 2i)(-2 + 4i)$
15. $\frac{3 - i}{4 + 2i}$
16. $\frac{2 + i}{5 + 6i}$

Example 6

17. **ELECTRICITY** The current in one part of a series circuit is $5 - 3j$ amps. The current in another part of the circuit is $7 + 9j$ amps. Add these complex numbers to find the total current in the circuit.

Practice and Problem Solving

Extra Practice is on page R4.

Examples 1–2 CCSS **STRUCTURE** **Simplify.**

18. $\sqrt{-121}$
19. $\sqrt{-169}$
20. $\sqrt{-100}$
21. $\sqrt{-81}$
22. $(-3i)(-7i)(2i)$
23. $4i(-6i)^2$
24. i^{11}
25. i^{25}
26. $(10 - 7i) + (6 + 9i)$
27. $(-3 + i) + (-4 - i)$
28. $(12 + 5i) - (9 - 2i)$
29. $(11 - 8i) - (2 - 8i)$
30. $(1 + 2i)(1 - 2i)$
31. $(3 + 5i)(5 - 3i)$
32. $(4 - i)(6 - 6i)$
33. $\frac{2i}{1 + i}$
34. $\frac{5}{2 + 4i}$
35. $\frac{5 + i}{3i}$

Example 3 **Solve each equation.**

36. $4x^2 + 4 = 0$
37. $3x^2 + 48 = 0$
38. $2x^2 + 50 = 0$
39. $2x^2 + 10 = 0$
40. $6x^2 + 108 = 0$
41. $8x^2 + 128 = 0$

Example 4 **Find the values of x and y that make each equation true.**

42. $9 + 12i = 3x + 4yi$
43. $x + 1 + 2yi = 3 - 6i$
44. $2x + 7 + (3 - y)i = -4 + 6i$
45. $5 + y + (3x - 7)i = 9 - 3i$
46. $a + 3b + (3a - b)i = 6 + 6i$
47. $(2a - 4b)i + a + 5b = 15 + 58i$

Examples 5 and 7 **Simplify.**

48. $\sqrt{-10} \cdot \sqrt{-24}$

49. $4i\left(\frac{1}{2}i\right)^2(-2i)^2$

50. i^{41}

51. $(4 - 6i) + (4 + 6i)$

52. $(8 - 5i) - (7 + i)$

53. $(-6 - i)(3 - 3i)$

54. $\frac{(5 + i)^2}{3 - i}$

55. $\frac{6 - i}{2 - 3i}$

56. $(-4 + 6i)(2 - i)(3 + 7i)$

57. $(1 + i)(2 + 3i)(4 - 3i)$

58. $\frac{4 - i\sqrt{2}}{4 + i\sqrt{2}}$

59. $\frac{2 - i\sqrt{3}}{2 + i\sqrt{3}}$

Example 6

60. **ELECTRICITY** The impedance in one part of a series circuit is $7 + 8j$ ohms, and the impedance in another part of the circuit is $13 - 4j$ ohms. Add these complex numbers to find the total impedance in the circuit.

ELECTRICITY **Use the formula $V = C \cdot I$.**

61. The current in a circuit is $3 + 6j$ amps, and the impedance is $5 - j$ ohms. What is the voltage?

62. The voltage in a circuit is $20 - 12j$ volts, and the impedance is $6 - 4j$ ohms. What is the current?

63. Find the sum of $ix^2 - (4 + 5i)x + 7$ and $3x^2 + (2 + 6i)x - 8i$.

64. Simplify $[(2 + i)x^2 - ix + 5 + i] - [(-3 + 4i)x^2 + (5 - 5i)x - 6]$.

65. **MULTIPLE REPRESENTATIONS** In this problem, you will explore quadratic equations that have complex roots.

a. **Algebraic** Write a quadratic equation in standard form with $3i$ and $-3i$ as its roots.

b. **Graphical** Graph the quadratic equation found in part **a** by graphing its related function.

c. **Algebraic** Write a quadratic equation in standard form with $2 + i$ and $2 - i$ as its roots.

d. **Graphical** Graph the quadratic equation found in part **c** by graphing its related function.

e. **Analytical** How do you know when a quadratic equation will have only complex solutions?

H.O.T. Problems Use Higher-Order Thinking Skills

66. **CCSS CRITIQUE** Joe and Sue are simplifying $(2i)(3i)(4i)$. Is either of them correct? Explain your reasoning.

Joe	Sue
$24i^3 = -24$	$24i^3 = -24i$

67. **CHALLENGE** Simplify $(1 + 2i)^3$.

68. **REASONING** Determine whether the following statement is *always, sometimes,* or *never* true. Explain your reasoning.

Every complex number has both a real part and an imaginary part.

69. **OPEN ENDED** Write two complex numbers with a product of 20.

70. **WRITING IN MATH** Explain how complex numbers are related to quadratic equations.

Standardized Test Practice

71. EXTENDED RESPONSE Refer to the figure to answer the following.

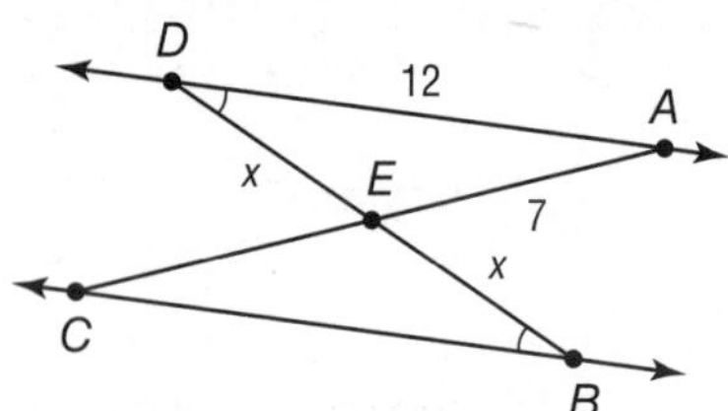

a. Name two congruent triangles with vertices in correct order.

b. Explain why the triangles are congruent.

c. What is the length of $\overline{EC}$? Explain your procedure.

72. $(3 + 6)^2 =$

A $2 \times 3 + 2 \times 6$

B 9^2

C $3^2 + 6^2$

D $3^2 \times 6^2$

73. SAT/ACT A store charges $49 for a pair of pants. This price is 40% more than the amount it costs the store to buy the pants. After a sale, any employee is allowed to purchase any remaining pairs of pants at 30% off the store's cost. How much would it cost an employee to purchase the pants after the sale?

F $10.50

G $12.50

H $13.72

J $24.50

K $35.00

74. What are the values of x and y when $(5 + 4i) - (x + yi) = (-1 - 3i)$?

A $x = 6, y = 7$

B $x = 4, y = i$

C $x = 6, y = i$

D $x = 4, y = 7$

Spiral Review

Solve each equation by factoring. (Lesson 4-3)

75. $2x^2 + 7x = 15$

76. $4x^2 - 12 = 22x$

77. $6x^2 = 5x + 4$

NUMBER THEORY Use a quadratic equation to find two real numbers that satisfy each situation, or show that no such numbers exist. (Lesson 4-2)

78. Their sum is -3, and their product is -40.

79. Their sum is 19, and their product is 48.

80. Their sum is -15, and their product is 56.

81. Their sum is -21, and their product is 108.

82. RECREATION Refer to the table. (Lesson 3-5)

a. Write a matrix that represents the cost of admission for residents and a matrix that represents the cost of admission for nonresidents.

b. Write the matrix that represents the additional cost for nonresidents.

c. Write a matrix that represents the difference in cost if a child or adult goes after 6:00 P.M. instead of before 6:00 P.M.

Daily Admission Fees		
Residents Time of day	**Child**	**Adult**
Before 6:00 p.m.	$3.00	$4.50
After 6:00 p.m.	$2.00	$3.50
Nonresidents Time of day	**Child**	**Adult**
Before 6:00 p.m.	$4.50	$6.75
After 6:00 p.m.	$3.00	$5.25

83. PART-TIME JOBS Terrell makes $10 per hour cutting grass and $12 per hour for raking leaves. He cannot work more than 15 hours per week. Graph two inequalities that Terrell can use to determine how many hours he needs to work at each job if he wants to earn at least $120 per week. (Lesson 3-2)

Skills Review

Determine whether each trinomial is a perfect square trinomial. Write *yes* or *no*.

84. $x^2 + 16x + 64$

85. $x^2 - 12x + 36$

86. $x^2 + 8x - 16$

87. $x^2 - 14x - 49$

88. $x^2 + x + 0.25$

89. $x^2 + 5x + 6.25$

EXTEND 4-4 Algebra Lab
The Complex Plane

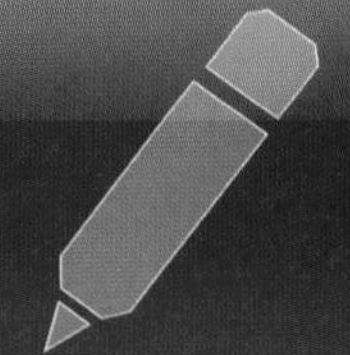

A complex number $a + bi$ can be graphed in the **complex plane** by representing it with the point (a, b). Similar to a coordinate plane, the complex plane is comprised of two axes. The real component is plotted on the **real axis**, which is horizontal. The imaginary component is plotted on the **imaginary axis**, which is vertical. The complex plane may also be referred to as the **Argand (ar GON) plane**.

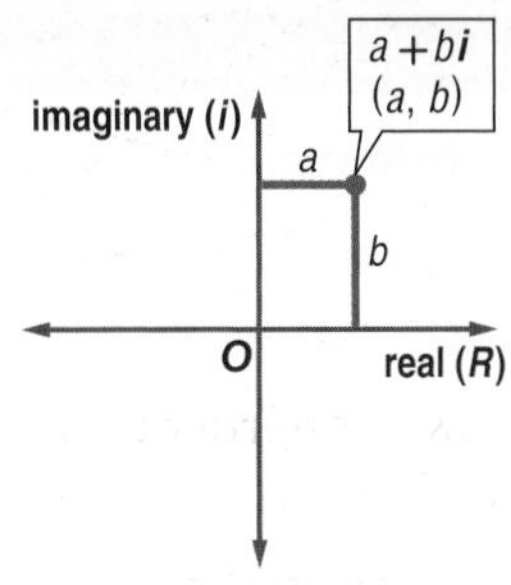

Example 1 Graph in the Complex Plane

Graph $z = 3 + 4i$ in the complex plane.

Step 1 Represent z with the point (a, b).

The real component a of z is 3.

The imaginary component bi of z is $4i$.

z can be represented by the point (a, b) or $(3, 4)$.

Step 2 Graph z in the complex plane.

Construct the complex plane and plot the point $(3, 4)$.

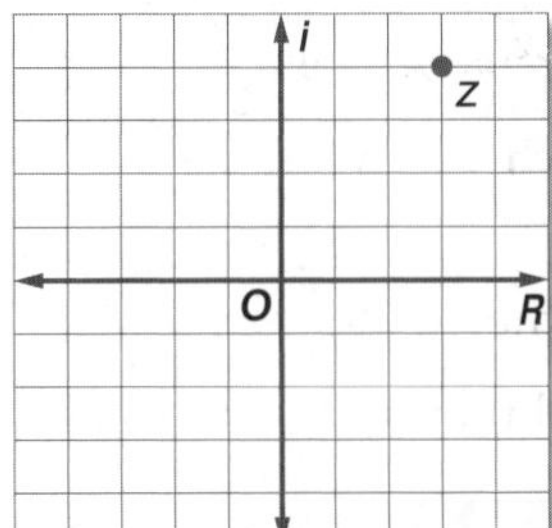

Recall that for a real number, the absolute value is its distance from zero on the number line. Similarly, the **absolute value of a complex number** is its distance from the origin in the complex plane. When $a + bi$ is graphed in the complex plane, the absolute value of $a + bi$ is the distance from (a, b) to the origin. This can be found by using the Distance Formula.

$$\sqrt{(a - 0)^2 + (b - 0)^2} \text{ or } \sqrt{a^2 + b^2}$$

KeyConcept Absolute Value of a Complex Number

The absolute value of the complex number $z = a + bi$ is

$$|z| = |a + bi| = \sqrt{a^2 + b^2}.$$

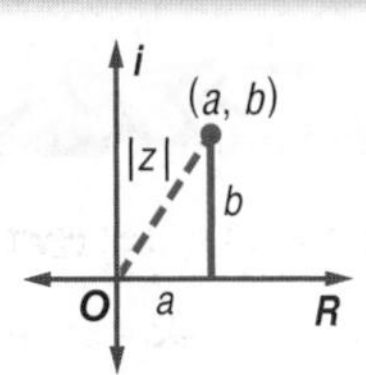

Algebra Lab
The Complex Plane *Continued*

Example 2 Absolute Value of a Complex Number

Find the absolute value of $z = -5 + 12i$.

Step 1 Determine values for a and b.

The real component a of z is -5. The imaginary component bi of z is $12i$.

Thus, $a = -5$ and $b = 12$.

Step 2 Find the absolute value of z.

$|z| = \sqrt{a^2 + b^2}$ Absolute value of a complex number

$= \sqrt{(-5)^2 + 12^2}$ $a = -5$ and $b = 12$

$= \sqrt{169}$ or 13 Simplify.

The absolute value of $z = -5 + 12i$ is 13.

Addition and subtraction of complex numbers can be performed graphically.

Example 3 Simplify by Graphing

Simplify $(1 - 2i) - (-2 - 5i)$ by graphing.

Step 1 Write $(1 - 2i) - (-2 - 5i)$ as $(1 - 2i) + (2 + 5i)$.

Step 2 Graph $1 - 2i$ and $2 + 5i$ on the same complex plane. Connect each point with the origin using a dashed segment.

Step 3 Complete the parallelogram that has the two segments as two of its sides. Plot a point where the two additional sides meet. The solution of $(1 - 2i) - (-2 - 5i)$ is $3 + 3i$.

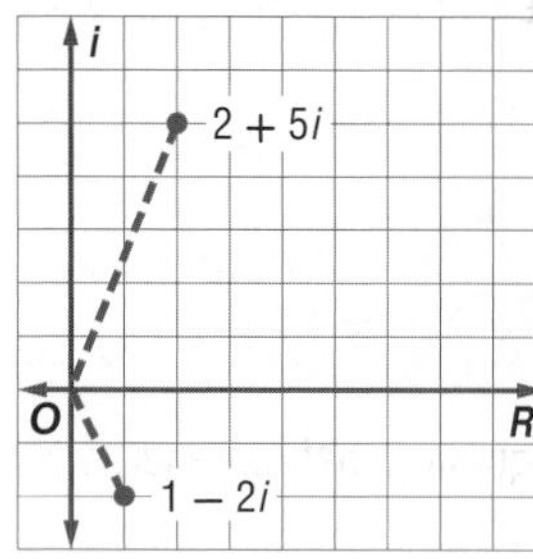

Step 2

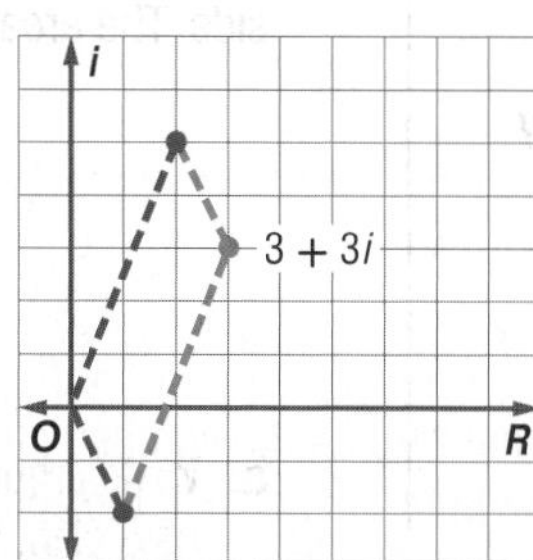

Step 3

Exercises

Graph each number in the complex plane.

1. $z = 3 + i$
2. $z = -4 - 2i$
3. $z = 2 - 2i$

Find the absolute value of each complex number.

4. $z = -4 - 3i$
5. $z = 7 - 2i$
6. $z = -6 - i$

Simplify by graphing.

7. $(6 + 5i) + (-2 - 3i)$
8. $(8 - 2i) - (4 + 7i)$
9. $(5 + 6i) + (-4 + 3i)$

CHAPTER 4

Mid-Chapter Quiz

Lessons 4-1 through 4-4

1. Find the y-intercept, the equation of the axis of symmetry, and the x-coordinate of the vertex for $f(x) = 2x^2 + 8x - 3$. Then graph the function by making a table of values. (Lesson 4-1)

2. **MULTIPLE CHOICE** For which equation is the axis of symmetry $x = 5$? (Lesson 4-1)

   A $f(x) = x^2 - 5x + 3$

   B $f(x) = x^2 - 10x + 7$

   C $f(x) = x^2 + 10x - 3$

   D $f(x) = x^2 + 5x + 2$

3. Determine whether $f(x) = 5 - x^2 + 2x$ has a maximum or a minimum value. Then find this maximum or minimum value and state the domain and range of the function. (Lesson 4-1)

4. **PHYSICAL SCIENCE** From 4 feet above the ground, Maya throws a ball upward with a velocity of 18 feet per second. The height $h(t)$ of the ball t seconds after Maya throws the ball is given by $h(t) = -16t^2 + 18t + 4$. Find the maximum height reached by the ball and the time that this height is reached. (Lesson 4-1)

5. Solve $3x^2 - 17x + 5 = 0$ by graphing. If exact roots cannot be found, state the consecutive integers between which the roots are located. (Lesson 4-2)

Use a quadratic equation to find two real numbers that satisfy each situation, or show that no such numbers exist. (Lesson 4-2)

6. Their sum is 15, and their product is 36.

7. Their sum is 7, and their product is 15.

8. **MULTIPLE CHOICE** Using the graph of the function $f(x) = x^2 + 6x - 7$, what are the solutions to the equation $x^2 + 6x - 7 = 0$? (Lesson 4-2)

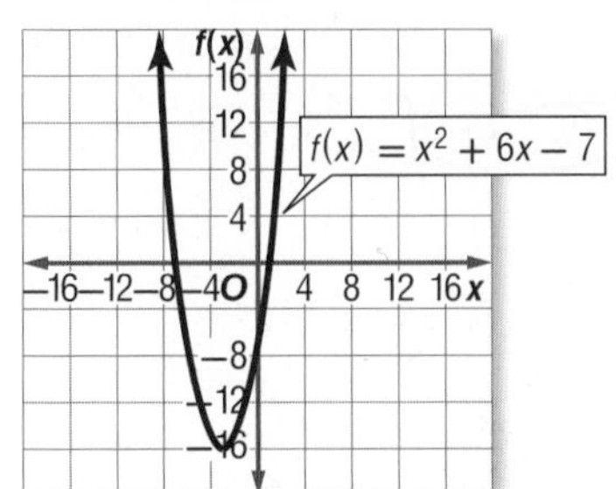

   F $-1, 6$ 　　 **H** $-1, 7$

   G $1, -6$ 　　 **J** $1, -7$

9. **BASEBALL** A baseball is hit upward with a velocity of 40 feet per second. Ignoring the height of the baseball player, how long does it take for the ball to fall to the ground? Use the formula $h(t) = v_0t - 16t^2$ where $h(t)$ is the height of an object in feet, v_0 is the object's initial velocity in feet per second, and t is the time in seconds. (Lesson 4-2)

Solve each equation by factoring. (Lesson 4-3)

10. $x^2 - x - 12 = 0$

11. $3x^2 + 7x + 2 = 0$

12. $x^2 - 2x - 15 = 0$

13. $2x^2 + 5x - 3 = 0$

14. Write a quadratic equation in standard form with roots -6 and $\frac{1}{4}$. (Lesson 4-3)

15. **TRIANGLES** Find the dimensions of a triangle if the base is $\frac{2}{3}$ the measure of the height and the area is 12 square centimeters. (Lesson 4-3)

16. **PATIO** Eli is putting a cement slab in his backyard. The original slab was going to have dimensions of 8 feet by 6 feet. He decided to make the slab larger by adding x feet to each side. The area of the new slab is 120 square feet. (Lesson 4-3)

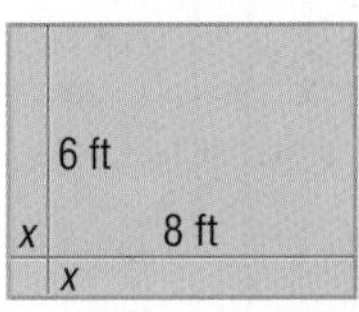

   a. Write a quadratic equation that represents the area of the new slab.

   b. Find the new dimensions of the slab.

Simplify. (Lesson 4-4)

17. $\sqrt{-81}$

18. $\sqrt{-25x^4y^5}$

19. $(15 - 3i) - (4 - 12i)$

20. i^{37}

21. $(5 - 3i)(5 + 3i)$

22. $\frac{3 - i}{2 + 5i}$

23. The impedance in one part of a series circuit is $3 + 4j$ ohms and the impedance in another part of the circuit is $6 - 7j$ ohms. Add these complex numbers to find the total impedance in the circuit. (Lesson 4-4)

LESSON 4-5

Completing the Square

Then

- You factored perfect square trinomials.

Now

1. Solve quadratic equations by using the Square Root Property.
2. Solve quadratic equations by completing the square.

Why?

- When going through a school zone, drivers must slow to a speed of 20 miles per hour. Once they are out of the school zone, the drivers can increase their speed.

Suppose Arturo is leaving school to go home for lunch, and he lives 5000 feet from the school zone. If Arturo accelerates at a constant rate of 8 feet per second squared, the equation $t^2 + 2t + 8 = 16$ represents the time t it takes him to reach home.

To solve this equation, you can use the Square Root Property.

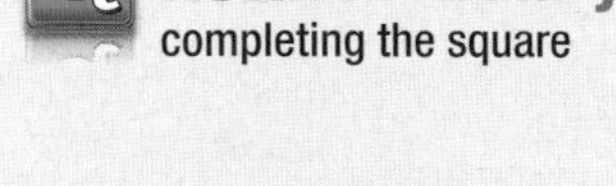

NewVocabulary
completing the square

Common Core State Standards

Content Standards
N.CN.7 Solve quadratic equations with real coefficients that have complex solutions.
F.IF.8.a Use the process of factoring and completing the square in a quadratic function to show zeros, extreme values, and symmetry of the graph, and interpret these in terms of a context.

Mathematical Practices
7 Look for and make use of structure.

1 Square Root Property

You have solved equations like $x^2 - 25 = 0$ by factoring. You have also used the Square Root Property to solve such equations. This method can be useful with equations like the one above that describes the car's speed.

Example 1 Equation with Rational Roots

Solve $x^2 + 6x + 9 = 36$ by using the Square Root Property.

$x^2 + 6x + 9 = 36$	Original equation
$(x + 3)^2 = 36$	Factor the perfect square trinomial.
$x + 3 = \pm\sqrt{36}$	Square Root Property
$x + 3 = \pm 6$	$\sqrt{36} = 6$
$x = -3 \pm 6$	Subtract 3 from each side.
$x = -3 + 6$ or $x = -3 - 6$	Write as two equations.
$= 3$ $\quad$ $= -9$	Simplify.

The solution set is $\{-9, 3\}$ or $\{x \mid x = -9, 3\}$.

CHECK Substitute both values into the original equation.

$x^2 + 6x + 9 = 36$	Original equation	$x^2 + 6x + 9 = 36$
$3^2 + 6(3) + 9 \stackrel{?}{=} 36$	Substitute 3 and −9.	$(-9)^2 + 6(-9) + 9 \stackrel{?}{=} 36$
$9 + 18 + 9 \stackrel{?}{=} 36$	Simplify.	$81 - 54 + 9 \stackrel{?}{=} 36$
$36 = 36$ ✓	Both solutions are correct.	$36 = 36$ ✓

GuidedPractice

Solve each equation by using the Square Root Property.

1A. $x^2 - 12x + 36 = 25$ $\qquad$ **1B.** $x^2 - 16x + 64 = 49$

Arthur S. Aubry/Getty Images

Roots that are irrational numbers may be written as exact answers in radical form or as *approximate* answers in decimal form when a calculator is used.

Example 2 Equation with Irrational Roots

Solve $x^2 - 10x + 25 = 27$ by using the Square Root Property.

$x^2 - 10x + 25 = 27$	Original equation
$(x - 5)^2 = 27$	Factor the perfect square trinomial.
$x - 5 = \pm\sqrt{27}$	Square Root Property
$x = 5 \pm 3\sqrt{3}$	Add 5 to each side; $\sqrt{27} = 3\sqrt{3}$.
$x = 5 + 3\sqrt{3}$ or $x = 5 - 3\sqrt{3}$	Write as two equations.
≈ 10.2 $\quad\approx -0.2$	Use a calculator.

StudyTip

Plus or Minus When using the Square Root Property, remember to put a $\pm$ sign before the radical.

The exact solutions of this equation are $5 + 3\sqrt{3}$ and $5 - 3\sqrt{3}$. The approximate solutions are -0.2 and 10.2. Check these results by finding and graphing the related quadratic function.

$x^2 - 10x + 25 = 27$	Original equation
$x^2 - 10x - 2 = 0$	Subtract 27 from each side.
$y = x^2 - 10x - 2$	Related quadratic function

CHECK Use the **ZERO** function of a graphing calculator. The approximate zeros of the related function are -0.2 and 10.2.

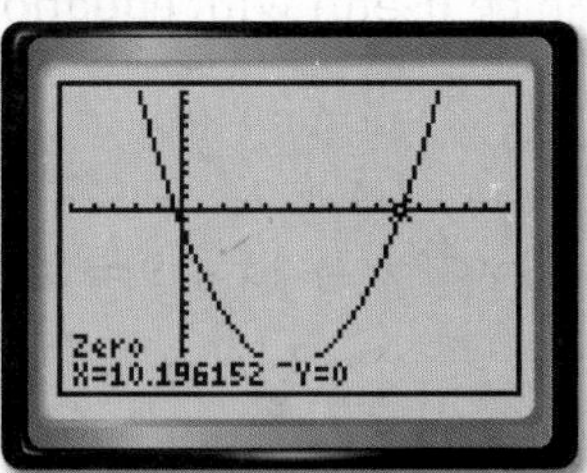

[−5, 15] scl: 1 by [−30, 20] scl: 2

GuidedPractice

Solve each equation by using the Square Root Property.

2A. $x^2 + 8x + 16 = 20$

2B. $x^2 - 6x + 9 = 32$

2 Complete the Square

All quadratic equations can be solved using the Square Root Property by manipulating the equation until one side is a perfect square. This method is called **completing the square**.

Consider $x^2 + 16x = 9$. Remember to perform each operation on each side of the equation.

$x^2 + 16x + \blacksquare = 9$	What value is needed for the perfect square?
$x^2 + 16x + \mathbf{64} = 9 + \mathbf{64}$	$\left(\frac{16}{2}\right)^2 = 64$; add 64 to each side.
$x^2 + 16x + 64 = 73$	Simplify.
$(x + 8)^2 = 73$	We can now use the Square Root Property.

Use this pattern of coefficients to complete the square of a quadratic expression.

KeyConcept Completing the Square

Words To complete the square for any quadratic expression of the form $x^2 + bx$, follow the steps below.

Step 1 Find one half of b, the coefficient of x.

Step 2 Square the result in Step 1.

Step 3 Add the result of Step 2 to $x^2 + bx$.

Symbols $x^2 + bx + \left(\frac{b}{2}\right)^2 = \left(x + \frac{b}{2}\right)^2$

PT

Example 3 Complete the Square

Find the value of c that makes $x^2 + 16x + c$ a perfect square. Then write the trinomial as a perfect square.

Step 1 Find one half of 16. $\frac{16}{2} = 8$

Step 2 Square the result in Step 1. $8^2 = 64$

Step 3 Add the result of Step 2 to $x^2 + 16x$. $x^2 + 16x + 64$

The trinomial $x^2 + 16x + 64$ can be written as $(x + 8)^2$.

GuidedPractice

3. Find the value of c that makes $x^2 - 14x + c$ a perfect square. Then write the trinomial as a perfect square.

You can solve any quadratic equation by completing the square. Because you are solving an equation, add the value you use to complete the square to each side.

Example 4 Solve an Equation by Completing the Square

Solve $x^2 + 10x - 11 = 0$ by completing the square.

$x^2 + 10x - 11 = 0$	Notice that $x^2 + 10x - 11$ is not a perfect square.
$x^2 + 10x = 11$	Rewrite so the left side is of the form $x^2 + bx$.
$x^2 + 10x + 25 = 11 + 25$	Since $\left(\frac{10}{2}\right)^2 = 25$, add 25 to each side.
$(x + 5)^2 = 36$	Write the left side as a perfect square.
$x + 5 = \pm 6$	Square Root Property
$x = -5 \pm 6$	Subtract 5 from each side.
$x = -5 + 6$ or $x = -5 - 6$	Write as two equations.
$= 1$ $\quad$ $= -11$	Simplify.

The solution set is $\{-11, 1\}$ or $\{x \mid x = -11, 1\}$. Check the result by using factoring.

WatchOut!

Each Side When solving equations by completing the square, don't forget to add $\left(\frac{b}{2}\right)^2$ to *each* side of the equation.

GuidedPractice

Solve each equation by completing the square.

4A. $x^2 - 10x + 24 = 0$ **4B.** $x^2 + 10x + 9 = 0$

When the coefficient of the quadratic term is not 1, you must divide the equation by that coefficient before completing the square.

Example 5 Equation with $a \neq 1$

Solve $2x^2 - 7x + 5 = 0$ by completing the square.

$2x^2 - 7x + 5 = 0$	Notice that $2x^2 - 7x + 5$ is not a perfect square.
$x^2 - \frac{7}{2}x + \frac{5}{2} = 0$	Divide by the coefficient of the quadratic term, 2.
$x^2 - \frac{7}{2}x = -\frac{5}{2}$	Subtract $\frac{5}{2}$ from each side.
$x^2 - \frac{7}{2}x + \frac{49}{16} = -\frac{5}{2} + \frac{49}{16}$	Since $\left(-\frac{7}{2} \div 2\right)^2 = \frac{49}{16}$, add $\frac{49}{16}$ to each side.
$\left(x - \frac{7}{4}\right)^2 = \frac{9}{16}$	Write the left side as a perfect square by factoring. Simplify the right side.
$x - \frac{7}{4} = \pm\frac{3}{4}$	Square Root Property
$x = \frac{7}{4} \pm \frac{3}{4}$	Add $\frac{7}{4}$ to each side.
$x = \frac{7}{4} + \frac{3}{4}$ or $x = \frac{7}{4} - \frac{3}{4}$	Write as two equations.
$= \frac{5}{2}$ $\qquad = 1$	

The solution set is $\left\{1, \frac{5}{2}\right\}$ or $\left\{x \mid x = 1, \frac{5}{2}\right\}$.

GuidedPractice

Solve each equation by completing the square.

5A. $3x^2 + 10x - 8 = 0$ **5B.** $3x^2 + 14x - 16 = 0$

Not all solutions of quadratic equations are real numbers. In some cases, the solutions are complex numbers of the form $a + bi$, where $b \neq 0$.

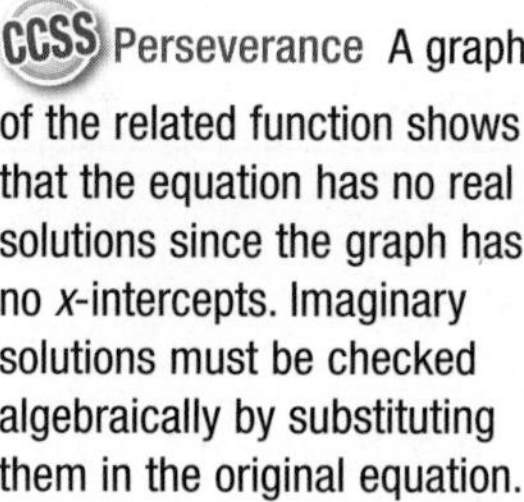

StudyTip

CCSS Perseverance A graph of the related function shows that the equation has no real solutions since the graph has no x-intercepts. Imaginary solutions must be checked algebraically by substituting them in the original equation.

Example 6 Equation with Imaginary Solutions

Solve $x^2 + 8x + 22 = 0$ by completing the square.

$x^2 + 8x + 22 = 0$	Notice that $x^2 + 8x + 22$ is not a perfect square.
$x^2 + 8x = -22$	Rewrite so the left side is of the form $x^2 + bx$.
$x^2 + 8x + 16 = -22 + 16$	Since $\left(\frac{8}{2}\right)^2 = 16$, add 16 to each side.
$(x + 4)^2 = -6$	Write the left side as a perfect square.
$x + 4 = \pm\sqrt{-6}$	Square Root Property
$x + 4 = \pm i\sqrt{6}$	$\sqrt{-1} = i$
$x = -4 \pm i\sqrt{6}$	Subtract 4 from each side.

The solution set is $\{-4 + i\sqrt{6}, -4 - i\sqrt{6}\}$ or $\{x \mid x = -4 + i\sqrt{6}, -4 - i\sqrt{6}\}$.

GuidedPractice

Solve each equation by completing the square.

6A. $x^2 + 2x + 2 = 0$ **6B.** $x^2 - 6x + 25 = 0$

Check Your Understanding

= Step-by-Step Solutions begin on page R14.

Examples 1–2 **Solve each equation by using the Square Root Property. Round to the nearest hundredth if necessary.**

1. $x^2 + 12x + 36 = 6$
2. $x^2 - 8x + 16 = 13$
3. $x^2 + 18x + 81 = 15$
4. $9x^2 + 30x + 25 = 11$

5. **LASER LIGHT SHOW** The area A in square feet of a projected laser light show is given by $A = 0.16d^2$, where d is the distance from the laser to the screen in feet. At what distance will the projected laser light show have an area of 100 square feet?

Example 3 **Find the value of c that makes each trinomial a perfect square. Then write the trinomial as a perfect square.**

6. $x^2 - 10x + c$
7. $x^2 - 5x + c$

Examples 4–6 **Solve each equation by completing the square.**

8. $x^2 + 2x - 8 = 0$
9. $x^2 - 4x + 9 = 0$
10. $2x^2 - 3x - 3 = 0$
11. $2x^2 + 6x - 12 = 0$
12. $x^2 + 4x + 6 = 0$
13. $x^2 + 8x + 10 = 0$

Practice and Problem Solving

Extra Practice is on page R4.

Examples 1–2 **Solve each equation by using the Square Root Property. Round to the nearest hundredth if necessary.**

14. $x^2 + 4x + 4 = 10$
15. $x^2 - 6x + 9 = 20$
16. $x^2 + 8x + 16 = 18$
17. $x^2 + 10x + 25 = 7$
18. $x^2 + 12x + 36 = 5$
19. $x^2 - 2x + 1 = 4$
20. $x^2 - 5x + 6.25 = 4$
21. $x^2 - 15x + 56.25 = 8$
22. $x^2 + 32x + 256 = 1$
23. $x^2 - 3x + \frac{9}{4} = 6$
24. $x^2 + 7x + \frac{49}{4} = 4$
25. $x^2 - 9x + \frac{81}{4} = \frac{1}{4}$

Example 3 **Find the value of c that makes each trinomial a perfect square. Then write the trinomial as a perfect square.**

26. $x^2 + 8x + c$
27. $x^2 + 16x + c$
28. $x^2 - 11x + c$
29. $x^2 + 9x + c$

Examples 4–6 **Solve each equation by completing the square.**

30. $x^2 - 4x + 12 = 0$
31. $x^2 + 2x - 12 = 0$
32. $x^2 + 6x + 8 = 0$
33. $x^2 - 4x + 3 = 0$
34. $2x^2 + x - 3 = 0$
35. $2x^2 - 3x + 5 = 0$
36. $2x^2 + 5x + 7 = 0$
37. $3x^2 - 6x - 9 = 0$
38. $x^2 - 2x + 3 = 0$
39. $x^2 + 4x + 11 = 0$
40. $x^2 - 6x + 18 = 0$
41. $x^2 - 10x + 29 = 0$
42. $3x^2 - 4x = 2$
43. $2x^2 - 7x = -12$
44. $x^2 - 2.4x = 2.2$
45. $x^2 - 5.3x = -8.6$
46. $x^2 - \frac{1}{5}x - \frac{11}{5} = 0$
47. $x^2 - \frac{9}{2}x - \frac{24}{5} = 0$

48. **CCSS MODELING** An architect's blueprints call for a dining room measuring 13 feet by 13 feet. The customer would like the dining room to be a square, but with an area of 250 square feet. How much will this add to the dimensions of the room?

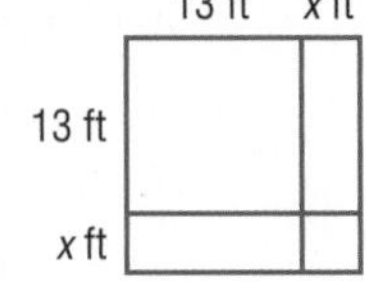

Solve each equation. Round to the nearest hundredth if necessary.

49. $4x^2 - 28x + 49 = 5$
50. $9x^2 + 30x + 25 = 11$
51. $x^2 + x + \frac{1}{3} = \frac{2}{3}$
52. $x^2 + 1.2x + 0.56 = 0.91$

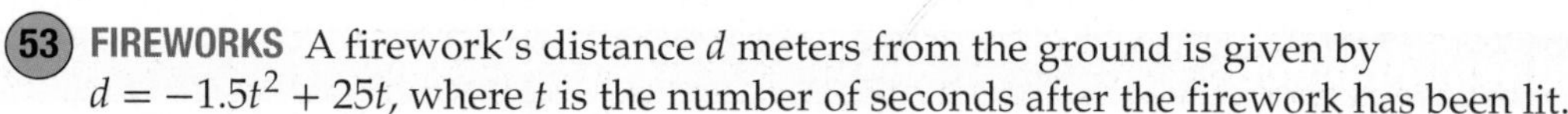

53 **FIREWORKS** A firework's distance d meters from the ground is given by $d = -1.5t^2 + 25t$, where t is the number of seconds after the firework has been lit.

a. How many seconds have passed since the firework was lit when the firework explodes if it explodes at the maximum height of its path?

b. What is the height of the firework when it explodes?

Find the value of c that makes each trinomial a perfect square. Then write the trinomial as a perfect square.

54. $x^2 + 0.7x + c$ **55.** $x^2 - 3.2x + c$ **56.** $x^2 - 1.8x + c$

57. **MULTIPLE REPRESENTATIONS** In this problem, you will use quadratic equations to investigate golden rectangles and the golden ratio.

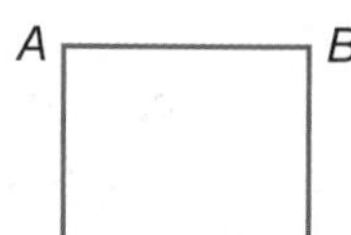

a. Geometric

- Draw square $ABCD$.
- Locate the midpoint of $\overline{CD}$. Label the midpoint P. Draw $\overline{PB}$.
- Construct an arc with a radius of $\overline{PB}$ from B clockwise past the bottom of the square.
- Extend $\overline{CD}$ until it intersects the arc. Label this point Q.
- Construct rectangle $ARQD$.

b. Algebraic Let $AD = x$ and $CQ = 1$. Use completing the square to solve $\frac{DQ}{AD} = \frac{QR}{CQ}$ for x.

c. Tabular Make a table of x and values for $CQ = 2, 3,$ and 4.

d. Verbal What do you notice about the x-values? Write an equation you could use to determine x for $CQ = n$, where n is a nonzero real number.

H.O.T. Problems Use Higher-Order Thinking Skills

58. **ERROR ANALYSIS** Alonso and Aida are solving $x^2 + 8x - 20 = 0$ by completing the square. Is either of them correct? Explain your reasoning

Alonso

$x^2 + 8x - 20 = 0$

$x^2 + 8x = 20$

$x^2 + 8x + 16 = 20 + 16$

$(x + 4)^2 = 36$

$x + 4 = \pm 6$

$x = -4 \pm 6$

Aida

$x^2 + 8x - 20 = 0$

$x^2 + 8x = 20$

$x^2 + 8x + 16 = 20$

$(x + 4)^2 = 20$

$x + 4 = \pm\sqrt{20}$

$x = -4 \pm \sqrt{20}$

59. **CHALLENGE** Solve $x^2 + bx + c = 0$ by completing the square. Your answer will be an expression for x in terms of b and c.

60. CCSS **ARGUMENTS** Without solving, determine how many unique solutions there are for each equation. Are they rational, real, or complex? Justify your reasoning.

a. $(x + 2)^2 = 16$ **b.** $(x - 2)^2 = 16$ **c.** $-(x - 2)^2 = 16$

d. $36 - (x - 2)^2 = 16$ **e.** $16(x + 2)^2 = 0$ **f.** $(x + 4)^2 = (x + 6)^2$

61. **OPEN ENDED** Write a perfect square trinomial equation in which the linear coefficient is negative and the constant term is a fraction. Then solve the equation.

62. **WRITING IN MATH** Explain what it means to complete the square. Describe each step.

Standardized Test Practice

63. SAT/ACT If $x^2 + y^2 = 2xy$, then y must equal

A -1 C 1   E x

B 0   D $-x$

64. GEOMETRY Find the area of the shaded region.

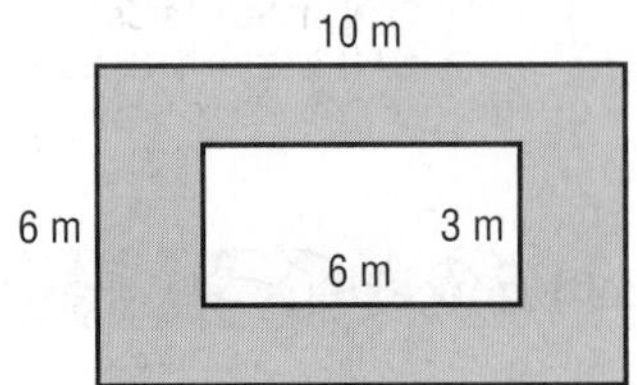

F 14 m^2 G 18 m^2 H 42 m^2 J 60 m^2

65. SHORT RESPONSE What value of c should be used to solve the following equation by completing the square?

$$5x^2 - 50x + c = 12 + c$$

66. If $5 - 3i$ is a solution for $x^2 + ax + b = 0$, where a and b are real numbers, what is the value of b?

A 10   C 34

B 14   D 40

Spiral Review

Simplify. (Lesson 4-4)

67. $(8 + 5i)^2$

68. $4(3 - i) + 6(2 - 5i)$

69. $\frac{5 - 2i}{6 + 9i}$

Write a quadratic equation in standard form with the given root(s). (Lesson 4-3)

70. $\frac{4}{5}, \frac{3}{4}$

71. $-\frac{2}{5}, 6$

72. $-\frac{1}{4}, -\frac{6}{7}$

73. TRAVEL Yoko is going with the Spanish Club to Costa Rica. She buys 10 traveler's checks in denominations of \$20, \$50, and \$100, totaling \$370. She has twice as many \$20 checks as \$50 checks. How many of each denomination of traveler's checks does she have? (Lesson 3-4)

74. SHOPPING Main St. Media sells all DVDs for one price and all books for another price. Alex bought 4 DVDs and 6 books for \$170, while Matt bought 3 DVDs and 8 books for \$180. What is the cost of a DVD and the cost of a book? (Lesson 3-1)

Graph each inequality. (Lesson 2-8)

75. $y \geq 4x - 3$

76. $2x - 3y < 6$

77. $5x + 2y + 3 \leq 0$

Write the piecewise function shown in each graph. (Lesson 2-6)

78.

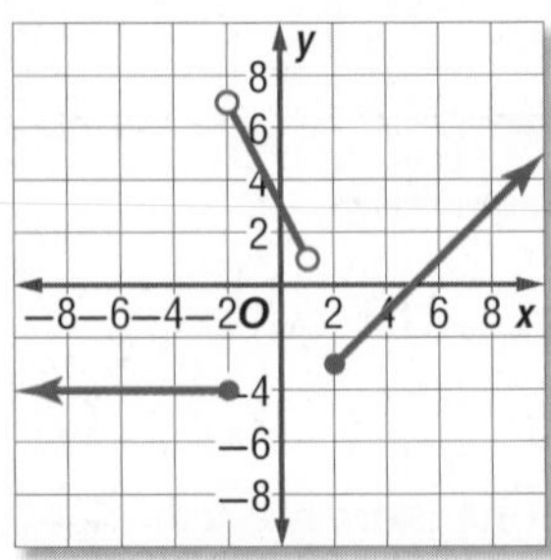

79.

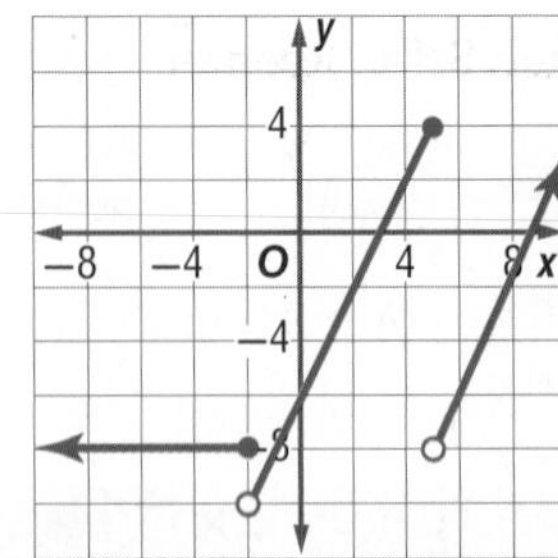

80.

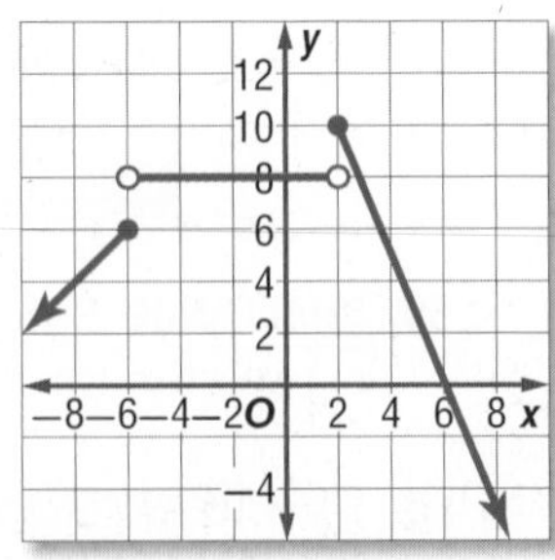

Skills Review

Evaluate $b^2 - 4ac$ for the given values of a, b, and c.

81. $a = 5, b = 6, c = 2$

82. $a = -2, b = -7, c = 3$

83. $a = -5, b = -8, c = -10$

EXTEND 4-5

Graphing Technology Lab
Solving Quadratic Equations

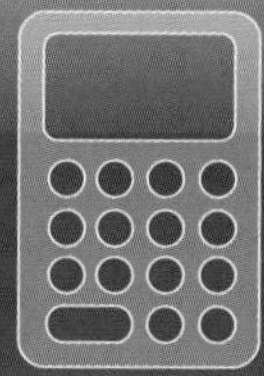

You can use a TI-Nspire™ CAS Technology to solve quadratic equations.

CCSS Common Core State Standards
Content Standards
N.CN.7 Solve quadratic equations with real coefficients that have complex solutions.

Activity Finding Roots

Solve each equation.

a. $3x^2 - 4x + 1 = 0$

Step 1 Add a new **Calculator** page.

Step 2 Select the **Solve** tool from the **Algebra** menu.

Step 3 Type $3x^2 - 4x + 1 = 0$ followed by a comma, x, and then **enter**.

The solutions are $x = \frac{1}{3}$ or $x = 1$.

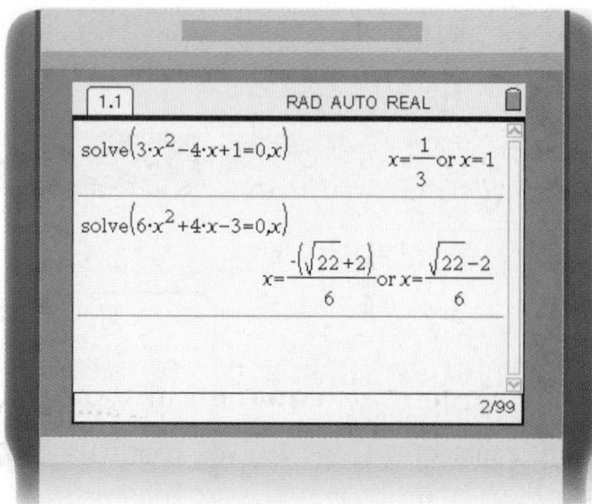

b. $6x^2 + 4x - 3 = 0$

Step 1 Select the **Solve** tool from the **Algebra** menu.

Step 2 Type $6x^2 - 4x - 3 = 0$ followed by a comma, x, and then **enter**.

The solutions are $x = \frac{-2 \pm \sqrt{22}}{6}$.

c. $x^2 - 6x + 10 = 0$.

Step 1 Select the **Solve** tool from the **Algebra** menu.

Step 2 Type $x^2 - 6x + 10 = 0$ followed by a comma, x, and then **enter**.

The calculator returns a value of *false*, meaning that there are no real solutions.

Step 3 Under menu, select **Algebra**, **Complex**, then **Solve**. Reenter the equation.

The solutions are $x = 3 \pm i$.

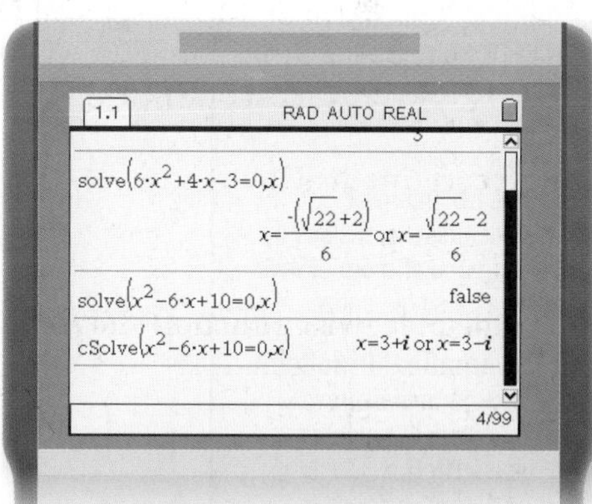

Exercises

Solve each equation.

1. $x^2 - 2x - 24 = 0$

2. $-x^2 + 4x - 1 = 0$

3. $0 = -3x^2 - 6x + 9$

4. $x^2 - 2x + 5 = 0$

5. $0 = 4x^2 - 8$

6. $0 = 2x^2 - 4x + 1$

7. $x^2 + 3x + 8 = 5$

8. $25 + 4x^2 = -20x$

9. $x^2 - x = -6$

LESSON 4-6 The Quadratic Formula and the Discriminant

Then

- You solved equations by completing the square.

Now

1. Solve quadratic equations by using the Quadratic Formula.
2. Use the discriminant to determine the number and type of roots of a quadratic equation.

Why?

- Pumpkin catapult is an event in which a contestant builds a catapult and launches a pumpkin at a target.

The path of the pumpkin can be modeled by the quadratic function $h = -4.9t^2 + 117t + 42$, where h is the height of the pumpkin and t is the number of seconds.

To predict when the pumpkin will hit the target, you can solve the equation $0 = -4.9t^2 + 117t + 42$. This equation would be difficult to solve using factoring, graphing, or completing the square.

Izzy Schwartz/Photodisc/Getty Images

NewVocabulary
Quadratic Formula
discriminant

Common Core State Standards

Content Standards
N.CN.7 Solve quadratic equations with real coefficients that have complex solutions.
A.SSE.1.b Interpret complicated expressions by viewing one or more of their parts as a single entity.

Mathematical Practices
8 Look for and express regularity in repeated reasoning.

1 Quadratic Formula

You have found solutions of some quadratic equations by graphing, by factoring, and by using the Square Root Property. There is also a formula that can be used to solve any quadratic equation. This formula can be derived by solving the standard form of a quadratic equation.

General Case		Specific Case
$ax^2 + bx + c = 0$	Standard quadratic equation	$2x^2 + 8x + 1 = 0$
$x^2 + \frac{b}{a}x + \frac{c}{a} = 0$	Divide each side by a.	$x^2 + 4x + \frac{1}{2} = 0$
$x^2 + \frac{b}{a}x = -\frac{c}{a}$	Subtract $\frac{c}{a}$ from each side.	$x^2 + 4x = -\frac{1}{2}$
$x^2 + \frac{b}{a}x + \frac{b^2}{4a^2} = -\frac{c}{a} + \frac{b^2}{4a^2}$	Complete the square.	$x^2 + 4x + \left(\frac{4}{2}\right)^2 = -\frac{1}{2} + \left(\frac{4}{2}\right)^2$
$\left(x + \frac{b}{2a}\right)^2 = -\frac{c}{a} + \frac{b^2}{4a^2}$	Factor the left side.	$(x + 2)^2 = -\frac{1}{2} + \left(\frac{4}{2}\right)^2$
$\left(x + \frac{b}{2a}\right)^2 = \frac{b^2 - 4ac}{4a^2}$	Simplify the right side.	$(x + 2)^2 = \frac{7}{2}$
$x + \frac{b}{2a} = \pm\frac{\sqrt{b^2 - 4ac}}{2a}$	Square Root Property	$x + 2 = \pm\sqrt{\frac{7}{2}}$
$x = -\frac{b}{2a} \pm \frac{\sqrt{b^2 - 4ac}}{2a}$	Subtract $\frac{b}{2a}$ from each side.	$x = -2 \pm \sqrt{\frac{7}{2}}$
$x = \frac{-b \pm \sqrt{b^2 - 4ac}}{2a}$	Simplify.	$x = \frac{-4 \pm \sqrt{14}}{2}$

The equation $x = \frac{-b \pm \sqrt{b^2 - 4ac}}{2a}$ is known as the **Quadratic Formula**.

StudyTip

Quadratic Formula Although factoring may be an easier method to solve some of the equations, the Quadratic Formula can be used to solve any quadratic equation.

KeyConcept Quadratic Formula

Words The solutions of a quadratic equation of the form $ax^2 + bx + c = 0$, where $a \neq 0$, are given by the following formula.

$$x = \frac{-b \pm \sqrt{b^2 - 4ac}}{2a}$$

Example $x^2 + 5x + 6 = 0 \rightarrow x = \frac{-5 \pm \sqrt{5^2 - 4(1)(6)}}{2(1)}$

Example 1 Two Rational Roots

Solve $x^2 - 10x = 11$ by using the Quadratic Formula.

First, write the equation in the form $ax^2 + bx + c = 0$ and identify a, b, and c.

$$ax^2 + bx + c = 0$$

$$x^2 - 10x = 11 \quad \rightarrow \quad 1x^2 - 10x - 11 = 0$$

Then, substitute these values into the Quadratic Formula.

$x = \frac{-b \pm \sqrt{b^2 - 4ac}}{2a}$ Quadratic Formula

$= \frac{-(-10) \pm \sqrt{(-10)^2 - 4(1)(-11)}}{2(1)}$ Replace a with 1, b with -10, and c with -11.

$= \frac{10 \pm \sqrt{100 + 44}}{2}$ Multiply.

$= \frac{10 \pm \sqrt{144}}{2}$ Simplify.

$= \frac{10 \pm 12}{2}$ $\sqrt{144} = 12$

$x = \frac{10 + 12}{2}$ or $x = \frac{10 - 12}{2}$ Write as two equations.

$= 11$ $\quad = -1$ Simplify.

The solutions are -1 and 11.

CHECK Substitute both values into the original equation.

$x^2 - 10x = 11$	$x^2 - 10x = 11$
$(-1)^2 - 10(-1) \stackrel{?}{=} 11$	$(11)^2 - 10(11) \stackrel{?}{=} 11$
$1 + 10 \stackrel{?}{=} 11$	$121 - 110 \stackrel{?}{=} 11$
$11 = 11$ ✓	$11 = 11$ ✓

GuidedPractice

Solve each equation by using the Quadratic Formula.

1A. $x^2 + 6x = 16$

1B. $2x^2 + 25x + 33 = 0$

ReviewVocabulary

radicand the value underneath the radical symbol

When the value of the radicand in the Quadratic Formula is 0, the quadratic equation has exactly one rational root.

Example 2 One Rational Root

Solve $x^2 + 8x + 16 = 0$ by using the Quadratic Formula.

Identify a, b, and c. Then, substitute these values into the Quadratic Formula.

$x = \frac{-b \pm \sqrt{b^2 - 4ac}}{2a}$ Quadratic Formula

$= \frac{-(8) \pm \sqrt{(8)^2 - 4(1)(16)}}{2(1)}$ Replace a with 1, b with 8, and c with 16.

$= \frac{-8 \pm \sqrt{0}}{2}$ Simplify.

$= \frac{-8}{2}$ or -4 $\sqrt{0} = 0$

The solution is -4.

CHECK A graph of the related function shows that there is one solution at $x = -4$.

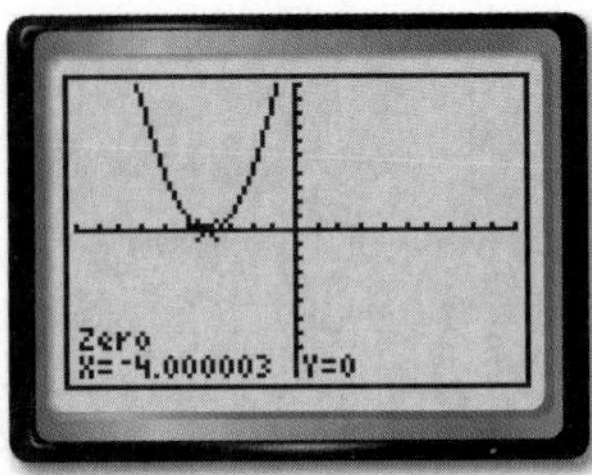

[−10, 10] scl: 1 by [−10, 10] scl: 1

Guided Practice

Solve each equation by using the Quadratic Formula.

2A. $x^2 - 16x + 64 = 0$

2B. $x^2 + 34x + 289 = 0$

Math History Link

Brahmagupta (598–668) Indian mathematician Brahmagupta offered the first general solution of the quadratic equation $ax^2 + bx = c$, now known as the Quadratic Formula.

You can express irrational roots exactly by writing them in radical form.

Example 3 Irrational Roots

Solve $2x^2 + 6x - 7 = 0$ by using the Quadratic Formula.

$x = \frac{-b \pm \sqrt{b^2 - 4ac}}{2a}$ Quadratic Formula

$= \frac{-(6) \pm \sqrt{(6)^2 - 4(2)(-7)}}{2(2)}$ Replace a with 2, b with 6, and c with -7.

$= \frac{-6 \pm \sqrt{92}}{4}$ Simplify.

$= \frac{-6 \pm 2\sqrt{23}}{4}$ or $\frac{-3 \pm \sqrt{23}}{2}$ $\sqrt{92} = \sqrt{4 \cdot 23}$ or $2\sqrt{23}$

The approximate solutions are -3.9 and 0.9.

CHECK Check these results by graphing the related quadratic function, $y = 2x^2 + 6x - 7$. Using the **ZERO** function of a graphing calculator, the approximate zeros of the related function are -3.9 and 0.9.

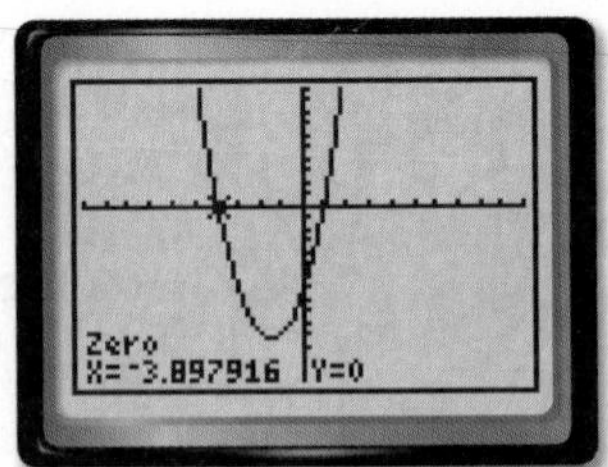

[−10, 10] scl: 1 by [−10, 10] scl: 1

Guided Practice

Solve each equation by using the Quadratic Formula.

3A. $3x^2 + 5x + 1 = 0$

3B. $x^2 - 8x + 9 = 0$

StudyTip

Complex Numbers Remember to write your solutions in the form $a + bi$, sometimes called the *standard form* of a complex number.

When using the Quadratic Formula, if the value of the radicand is negative, the solutions will be complex. Complex solutions always appear in conjugate pairs.

PT

Example 4 Complex Roots

Solve $x^2 - 6x = -10$ by using the Quadratic Formula.

$x = \frac{-b \pm \sqrt{b^2 - 4ac}}{2a}$	Quadratic Formula
$= \frac{-(-6) \pm \sqrt{(-6)^2 - 4(1)(10)}}{2(1)}$	Replace a with 1, b with −6, and c with 10.
$= \frac{6 \pm \sqrt{-4}}{2}$	Simplify.
$= \frac{6 \pm 2i}{2}$	$\sqrt{-4} = \sqrt{4 \cdot (-1)}$ or $2i$
$= 3 \pm i$	Simplify.

The solutions are the complex numbers $3 + i$ and $3 - i$.

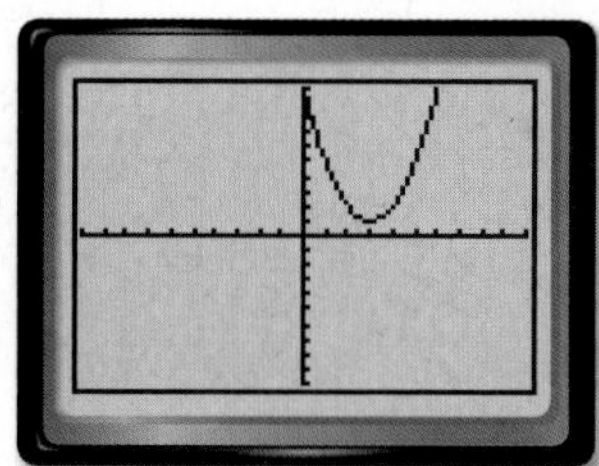

[−10, 10] scl: 1 by [−10, 10] scl: 1

CHECK A graph of the related function shows that the solutions are complex, but it cannot help you find them. To check complex solutions, substitute them into the original equation.

$x^2 - 6x = -10$	Original equation
$(3 + i)^2 - 6(3 + i) \stackrel{?}{=} -10$	$x = 3 + i$
$9 + 6i + i^2 - 18 - 6i \stackrel{?}{=} -10$	Square of a sum; Distributive Property
$-9 + i^2 \stackrel{?}{=} -10$	Simplify.
$-9 - 1 = -10$ ✓	$i^2 = -1$

$x^2 - 6x = -10$	Original equation
$(3 - i)^2 - 6(3 - i) \stackrel{?}{=} -10$	$x = 3 - i$
$9 - 6i + i^2 - 18 + 6i \stackrel{?}{=} -10$	Square of a sum; Distributive Property
$-9 + i^2 \stackrel{?}{=} -10$	Simplify.
$-9 - 1 = -10$ ✓	$i^2 = -1$

GuidedPractice

Solve each equation by using the Quadratic Formula.

4A. $3x^2 + 5x + 4 = 0$

4B. $x^2 - 4x = -13$

2 Roots and the Discriminant

In the previous examples, observe the relationship between the value of the expression under the radical and the roots of the quadratic equation. The expression $b^2 - 4ac$ is called the **discriminant**.

$$x = \frac{-b \pm \sqrt{b^2 - 4ac}}{2a} \quad \leftarrow \text{discriminant}$$

The value of the discriminant can be used to determine the number and type of roots of a quadratic equation. The table on the following page summarizes the possible types of roots.

The discriminant can also be used to confirm the number and type of solutions after you solve the quadratic equation.

StudyTip

Roots Remember that the solutions of an equation are called *roots* or *zeros* and are the value(s) where the graph crosses the x-axis.

KeyConcept Discriminant

Consider $ax^2 + bx + c = 0$, where a, b, and c are rational numbers and $a \neq 0$.

Value of Discriminant	Type and Number of Roots	Example of Graph of Related Function
$b^2 - 4ac > 0$; $b^2 - 4ac$ is a perfect square.	2 real, rational roots	
$b^2 - 4ac > 0$; $b^2 - 4ac$ is *not* a perfect square.	2 real, irrational roots	
$b^2 - 4ac = 0$	1 real rational root	
$b^2 - 4ac < 0$	2 complex roots	

Example 5 Describe Roots

Find the value of the discriminant for each quadratic equation. Then describe the number and type of roots for the equation.

a. $7x^2 - 11x + 5 = 0$

$a = 7, b = -11, c = 5$

$b^2 - 4ac = (-11)^2 - 4(7)(5)$

$= 121 - 140$

$= -19$

The discriminant is negative, so there are two complex roots.

b. $x^2 + 22x + 121 = 0$

$a = 1, b = 22, c = 121$

$b^2 - 4ac = (22)^2 - 4(1)(121)$

$= 484 - 484$

$= 0$

The discriminant is 0, so there is one rational root.

GuidedPractice

5A. $-5x^2 + 8x - 1 = 0$

5B. $-7x + 15x^2 - 4 = 0$

You have studied a variety of methods for solving quadratic equations. The table below summarizes these methods.

StudyTip

Study Notebook You may wish to copy this list of methods to your math notebook or Foldable to keep as a reference as you study.

ConceptSummary Solving Quadratic Equations

Method	Can be Used	When to Use
graphing	sometimes	Use only if an exact answer is not required. Best used to check the reasonableness of solutions found algebraically.
factoring	sometimes	Use if the constant term is 0 or if the factors are easily determined. **Example** $x^2 - 7x = 0$
Square Root Property	sometimes	Use for equations in which a perfect square is equal to a constant. **Example** $(x - 5)^2 = 18$
completing the square	always	Useful for equations of the form $x^2 + bx + c = 0$, where b is even. **Example** $x^2 + 6x - 14 = 0$
Quadratic Formula	always	Useful when other methods fail or are too tedious. **Example** $2.3x^2 - 1.8x + 9.7 = 0$

Check Your Understanding

 = Step-by-Step Solutions begin on page R14. 

Examples 1–4 **Solve each equation by using the Quadratic Formula.**

1. $x^2 + 12x - 9 = 0$
2. $x^2 + 8x + 5 = 0$
3. $4x^2 - 5x - 2 = 0$
4. $9x^2 + 6x - 4 = 0$
5. $10x^2 - 3 = 13x$
6. $22x = 12x^2 + 6$
7. $-3x^2 + 4x = -8$
8. $x^2 + 3 = -6x + 8$

Examples 3–4 9. **CCSS MODELING** An amusement park ride takes riders to the top of a tower and drops them at speeds reaching 80 feet per second. A function that models this ride is $h = -16t^2 - 64t + 60$, where h is the height in feet and t is the time in seconds. About how many seconds does it take for riders to drop from 60 feet to 0 feet?

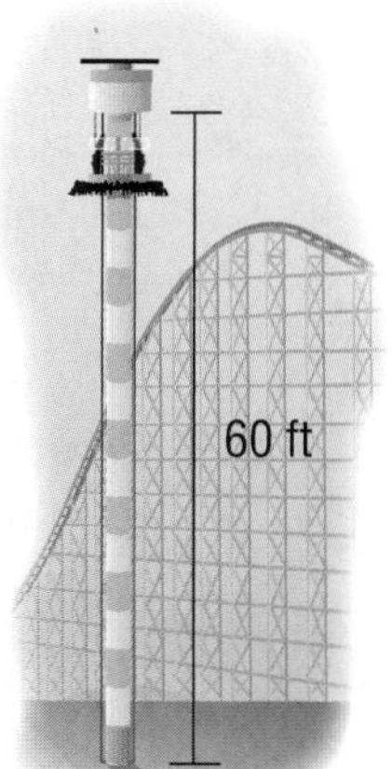

Example 5 **Complete parts a and b for each quadratic equation.**

a. Find the value of the discriminant.

b. Describe the number and type of roots.

10. $3x^2 + 8x + 2 = 0$
11. $2x^2 - 6x + 9 = 0$
12. $-16x^2 + 8x - 1 = 0$
13. $5x^2 + 2x + 4 = 0$

Practice and Problem Solving

Extra Practice is on page R4.

Examples 1–4 **Solve each equation by using the Quadratic Formula.**

14. $x^2 + 45x = -200$

15. $4x^2 - 6 = -12x$

16. $3x^2 - 4x - 8 = -6$

17. $4x^2 - 9 = -7x - 4$

18. $5x^2 - 9 = 11x$

19. $12x^2 + 9x - 2 = -17$

20. **DIVING** Competitors in the 10-meter platform diving competition jump upward and outward before diving into the pool below. The height h of a diver in meters above the pool after t seconds can be approximated by the equation $h = -4.9t^2 + 3t + 10$.

a. Determine a domain and range for which this function makes sense.

b. When will the diver hit the water?

Example 5 **Complete parts a–c for each quadratic equation.**

a. Find the value of the discriminant.

b. Describe the number and type of roots.

c. Find the exact solutions by using the Quadratic Formula.

21. $2x^2 + 3x - 3 = 0$

22. $4x^2 - 6x + 2 = 0$

23. $6x^2 + 5x - 1 = 0$

24. $6x^2 - x - 5 = 0$

25. $3x^2 - 3x + 8 = 0$

26. $2x^2 + 4x + 7 = 0$

27. $-5x^2 + 4x + 1 = 0$

28. $x^2 - 6x = -9$

29. $-3x^2 - 7x + 2 = 6$

30. $-8x^2 + 5 = -4x$

31. $x^2 + 2x - 4 = -9$

32. $-6x^2 + 5 = -4x + 8$

33. **VIDEO GAMES** While Darnell is grounded his friend Jack brings him a video game. Darnell stands at his bedroom window, and Jack stands directly below the window. If Jack tosses a game cartridge to Darnell with an initial velocity of 35 feet per second, an equation for the height h feet of the cartridge after t seconds is $h = -16t^2 + 35t + 5$.

a. If the window is 25 feet above the ground, will Darnell have 0, 1, or 2 chances to catch the video game cartridge?

b. If Darnell is unable to catch the video game cartridge, when will it hit the ground?

34. CCSS **SENSE-MAKING** Civil engineers are designing a section of road that is going to dip below sea level. The road's curve can be modeled by the equation $y = 0.00005x^2 - 0.06x$, where x is the horizontal distance in feet between the points where the road is at sea level and y is the elevation. The engineers want to put stop signs at the locations where the elevation of the road is equal to sea level. At what horizontal distances will they place the stop signs?

Complete parts a–c for each quadratic equation.

a. Find the value of the discriminant.

b. Describe the number and type of roots.

c. Find the exact solutions by using the Quadratic Formula.

35. $5x^2 + 8x = 0$

36. $8x^2 = -2x + 1$

37. $4x - 3 = -12x^2$

38. $0.8x^2 + 2.6x = -3.2$

39. $0.6x^2 + 1.4x = 4.8$

40. $-4x^2 + 12 = -6x - 8$

41. **SMOKING** A decrease in smoking in the United States has resulted in lower death rates caused by lung cancer. The number of deaths per 100,000 people y can be approximated by $y = -0.26x^2 - 0.55x + 91.81$, where x represents the number of years after 2000.

Year	Deaths per 100,000
2000	91.8
2002	89.7
2004	85.5
2010	60.3
2015	?
2017	?

a. Calculate the number of deaths per 100,000 people for 2015 and 2017.

b. Use the Quadratic Formula to solve for x when $y = 50$.

c. According to the quadratic function, when will the death rate be 0 per 100,000? Do you think that this prediction is reasonable? Why or why not?

42. **NUMBER THEORY** The sum S of consecutive integers 1, 2, 3, …, n is given by the formula $S = \frac{1}{2}n(n + 1)$. How many consecutive integers, starting with 1, must be added to get a sum of 666?

H.O.T. Problems Use Higher-Order Thinking Skills

43. **CCSS CRITIQUE** Tama and Jonathan are determining the number of solutions of $3x^2 - 5x = 7$. Is either of them correct? Explain your reasoning.

Tama

$3x^2 - 5x = 7$

$b^2 - 4ac = (-5)^2 - 4(3)(7)$

$= -59$

Since the discriminant is negative, there are no real solutions.

Jonathan

$3x^2 - 5x = 7$

$3x^2 - 5x - 7 = 0$

$b^2 - 4ac = (-5)^2 - 4(3)(-7)$

$= 109$

Since the discriminant is positive, there are two real roots.

44. **CHALLENGE** Find the solutions of $4ix^2 - 4ix + 5i = 0$ by using the Quadratic Formula.

45. **REASONING** Determine whether each statement is *sometimes, always,* or *never* true. Explain your reasoning.

a. In a quadratic equation in standard form, if a and c are different signs, then the solutions will be real.

b. If the discriminant of a quadratic equation is greater than 1, the two roots are real irrational numbers.

46. **OPEN ENDED** Sketch the corresponding graph and state the number and type of roots for each of the following.

a. $b^2 - 4ac = 0$

b. A quadratic function in which $f(x)$ never equals zero.

c. A quadratic function in which $f(a) = 0$ and $f(b) = 0$; $a \neq b$.

d. The discriminant is less than zero.

e. a and b are both solutions and can be represented as fractions.

47. **CHALLENGE** Find the value(s) of m in the quadratic equation $x^2 + x + m + 1 = 0$ such that it has one solution.

48. **WRITING IN MATH** Describe three different ways to solve $x^2 - 2x - 15 = 0$. Which method do you prefer, and why?

Standardized Test Practice

49. A company determined that its monthly profit P is given by $P = -8x^2 + 165x - 100$, where x is the selling price for each unit of product. Which of the following is the best estimate of the maximum price per unit that the company can charge without losing money?

A $10 B $20 C $30 D $40

50. SAT/ACT For which of the following sets of numbers is the mean greater than the median?

F {4, 5, 6, 7, 8} J {3, 5, 6, 7, 8}
G {4, 6, 6, 6, 8} K {2, 6, 6, 6, 6}
H {4, 5, 6, 7, 9}

51. SHORT RESPONSE In the figure below, P is the center of the circle with radius 15 inches. What is the area of $\triangle APB$?

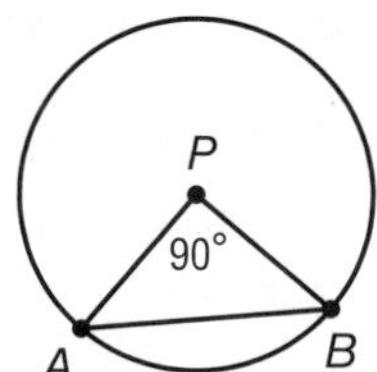

52. 75% of 88 is the same as 60% of what number?

A 100 B 101 C 108 D 110

Spiral Review

Find the value of c that makes each trinomial a perfect square. Then write the trinomial as a perfect square. (Lesson 4-5)

53. $x^2 + 13x + c$ **54.** $x^2 + 2.4x + c$ **55.** $x^2 + \frac{4}{5}x + c$

Simplify. (Lesson 4-4)

56. i^{26} **57.** $\sqrt{-16}$ **58.** $4\sqrt{-9} \cdot 2\sqrt{-25}$

59. PILOT TRAINING Evita is training for her pilot's license. Flight instruction costs $105 per hour, and the simulator costs $45 per hour. She spent 4 more hours in airplane training than in the simulator. If Evita spent $3870, how much time did she spend training in an airplane and in a simulator? (Lesson 3-8)

60. BUSINESS Ms. Larson owns three fruit farms on which she grows apples, peaches, and apricots. She sells apples for $22 a case, peaches for $25 a case, and apricots for $18 a case. (Lesson 3-6)

a. Write an inventory matrix for the number of cases for each type of fruit for each farm and a cost matrix for the price per case for each type of fruit.

b. Find the total income of the three fruit farms expressed as a matrix.

c. What is the total income from all three fruit farms?

Number of Cases in Stock of Each Type of Fruit

Fruit	Farm 1	Farm 2	Farm 3
apples	290	175	110
peaches	165	240	75
apricots	210	190	0

Skills Review

Write an equation for each graph.

61.

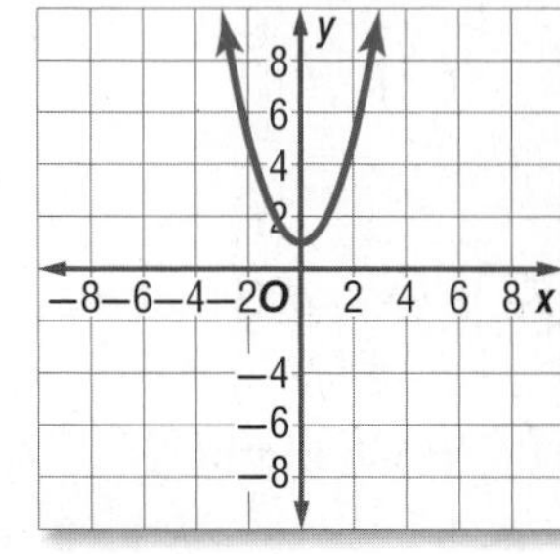

62.

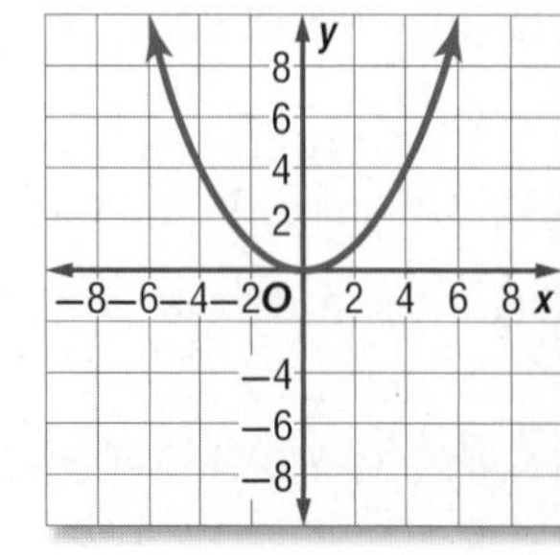

63.

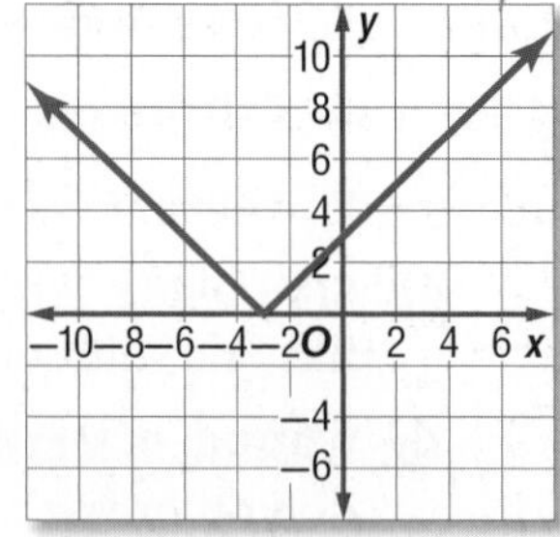

EXPLORE

4-7 Graphing Technology Lab
Families of Parabolas

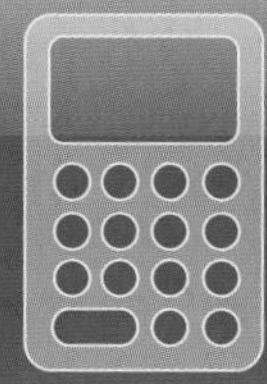

The general form of a quadratic function is $y = a(x - h)^2 + k$. Changing the values of a, h, and k results in a different parabola in the family of quadratic functions. You can use a TI-83/84 Plus graphing calculator to analyze the effects that result from changing each of these parameters.

CCSS Common Core State Standards
Content Standards
F.IF.4 For a function that models a relationship between two quantities, interpret key features of graphs and tables in terms of the quantities, and sketch graphs showing key features given a verbal description of the relationship.
F.BF.3 Identify the effect on the graph of replacing $f(x)$ by $f(x) + k$, $k\,f(x)$, $f(kx)$, and $f(x + k)$ for specific values of k (both positive and negative); find the value of k given the graphs. Experiment with cases and illustrate an explanation of the effects on the graph using technology.

Activity 1 Change in k

Graph each set of functions on the same screen in the standard viewing window. Describe any similarities and differences among the graphs.

$y = x^2$, $y = x^2 + 4$, $y = x^2 - 3$

The graphs have the same shape, and all open up. The vertex of each graph is on the y-axis. However, the graphs have different vertical positions.

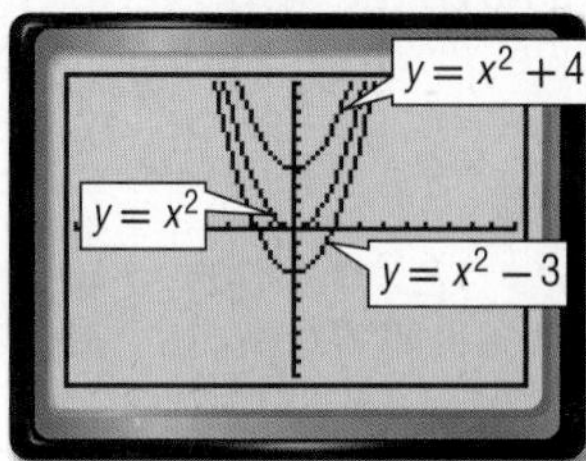

[−10, 10] scl: 1 by [−10, 10] scl: 1

Activity 1 shows how changing the value of k in the function $y = a(x - h)^2 + k$ *translates* the parabola along the y-axis. If $k > 0$, the parabola is translated k units up, and if $k < 0$, it is translated $|k|$ units down.

How do you think changing the value of h will affect the graph of $y = a(x - h)^2 + k$?

Activity 2 Change in h

Graph each set of functions on the same screen in the standard viewing window. Describe any similarities and differences among the graphs.

$y = x^2$, $y = (x + 4)^2$, $y = (x - 3)^2$

These three graphs all open up and have the same shape. The vertex of each graph is on the x-axis. However, the graphs have different horizontal positions.

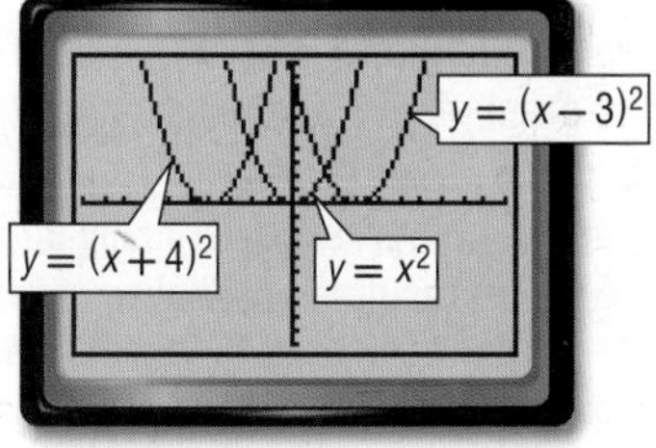

[−10, 10] scl: 1 by [−10, 10] scl: 1

Activity 2 shows how changing the value of h in the equation $y = a(x - h)^2 + k$ *translates* the graph horizontally. If $h > 0$, the graph translates to the right h units. If $h < 0$, the graph translates to the left $|h|$ units.

Activity 3 Change in h and k

Graph each set of functions on the same screen in the standard viewing window. Describe any similarities and differences among the graphs.

$y = x^2$, $y = (x + 6)^2 - 5$, $y = (x - 4)^2 + 6$

These three graphs all open up and have the same shape. However, the graphs have different horizontal and vertical positions.

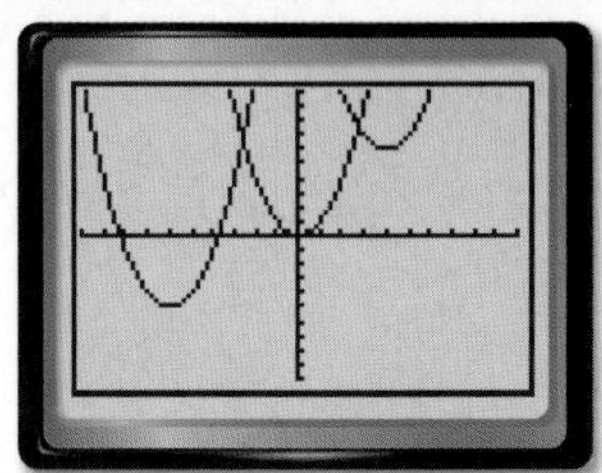

[−10, 10] scl: 1 by [−10, 10] scl: 1

(continued on the next page)

Families of Parabolas *Continued*

How does the value of a affect the graph of $y = a(x - h)^2 + k$?

Activity 4 Change in a

Graph each set of functions on the same screen in the standard viewing window. Describe any similarities and differences among the graphs.

a. $y = x^2, y = -x^2$

The graphs have the same vertex and the same shape. However, the graph of $y = x^2$ opens up and the graph of $y = -x^2$ opens down.

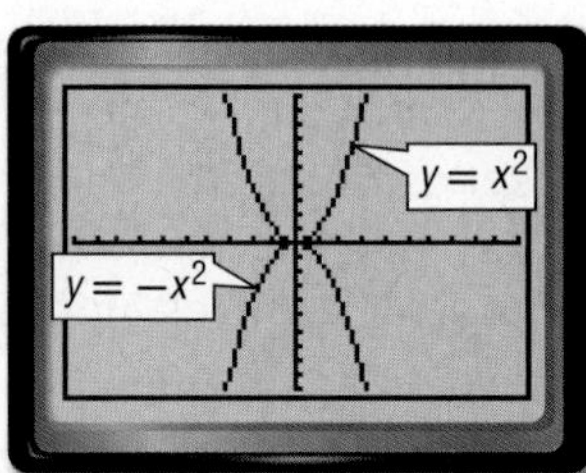

[−10, 10] scl: 1 by [−10, 10] scl: 1

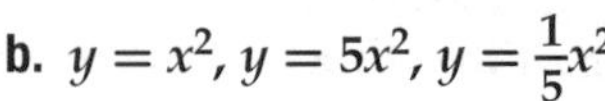

b. $y = x^2, y = 5x^2, y = \frac{1}{5}x^2$

The graphs have the same vertex, (0, 0), but each has a different shape. The graph of $y = 5x^2$ is narrower than the graph of $y = x^2$. The graph of $y = \frac{1}{5}x^2$ is wider than the graph of $y = x^2$.

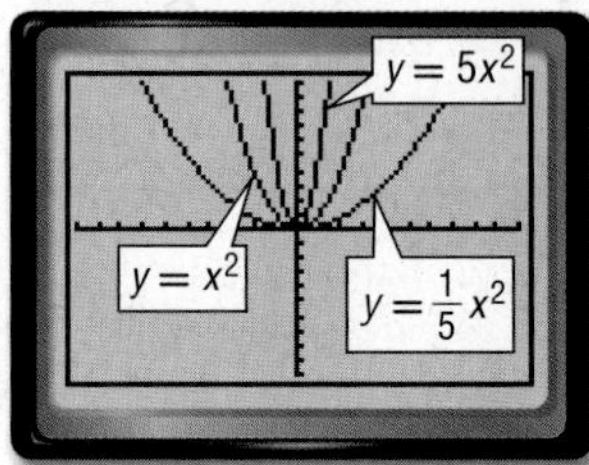

[−10, 10] scl: 1 by [−10, 10] scl: 1

Changing the value of a in the function $y = a(x - h)^2 + k$ can affect the direction of the opening and the shape of the graph. If $a > 0$, the graph opens up, and if $a < 0$, the graph opens down or is *reflected* over the x-axis. If $|a| > 1$, the graph is expanded vertically and is narrower than the graph of $y = x^2$. If $|a| < 1$, the graph is compressed vertically and is wider than the graph of $y = x^2$. Thus, a change in the absolute value of a results in a *dilation* of the graph of $y = x^2$.

Analyze the Results

1. How does changing the value of h in $y = a(x - h)^2 + k$ affect the graph? Give an example.
2. How does changing the value of k in $y = a(x - h)^2 + k$ affect the graph? Give an example.
3. How does using $-a$ instead of a in $y = a(x - h)^2 + k$ affect the graph? Give an example.

Examine each pair of functions and predict the similarities and differences in their graphs. Use a graphing calculator to confirm your predictions. Write a sentence or two comparing the two graphs.

4. $y = x^2, y = x^2 + 3.5$
5. $y = -x^2, y = x^2 - 7$
6. $y = x^2, y = 4x^2$
7. $y = x^2, y = -8x^2$
8. $y = x^2, y = (x + 2)^2$
9. $y = -\frac{1}{6}x^2, y = -\frac{1}{6}x^2 + 2$
10. $y = x^2, y = (x - 5)^2$
11. $y = x^2, y = 2(x + 3)^2 - 6$
12. $y = x^2, y = -\frac{1}{8}x^2 + 1$
13. $y = (x + 5)^2 - 4, y = (x + 5)^2 + 7$
14. $y = 2(x + 1)^2 - 4, y = 5(x + 3)^2 - 1$
15. $y = 5(x - 2)^2 - 3, y = \frac{1}{4}(x - 5)^2 - 6$

Transformations of Quadratic Graphs

Then

- You transformed graphs of functions.

Now

1. Write a quadratic function in the form $y = a(x - h)^2 + k$.
2. Transform graphs of quadratic functions of the form $y = a(x - h)^2 + k$.

Why?

- Recall that a family of graphs is a group of graphs that display one or more similar characteristics. The parent graph is the simplest graph in the family. For the family of quadratic functions, $y = x^2$ is the parent graph.

Other graphs in the family of quadratic functions, such as $y = (x - 2)^2$ and $y = x^2 - 4$, can be drawn by transforming the graph of $y = x^2$.

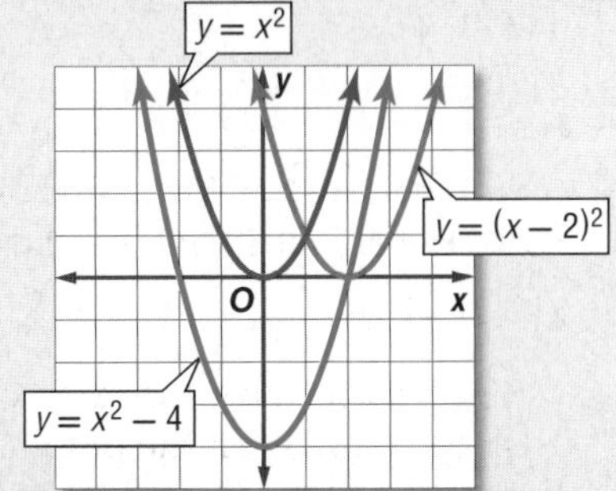

NewVocabulary
vertex form

Common Core State Standards

Content Standards

F.IF.8.a Use the process of factoring and completing the square in a quadratic function to show zeros, extreme values, and symmetry of the graph, and interpret these in terms of a context.

F.BF.3 Identify the effect on the graph of replacing $f(x)$ by $f(x) + k$, $k\,f(x)$, $f(kx)$, and $f(x + k)$ for specific values of k (both positive and negative); find the value of k given the graphs. Experiment with cases and illustrate an explanation of the effects on the graph using technology.

Mathematical Practices

7 Look for and make use of structure.

1 Write Quadratic Functions in Vertex Form

Each function above is written in **vertex form**, $y = a(x - h)^2 + k$, where (h, k) is the vertex of the parabola, $x = h$ is the axis of symmetry, and a determines the shape of the parabola and the direction in which it opens.

When a quadratic function is in the form $y = ax^2 + bx + c$, you can complete the square to write the function in vertex form. If the coefficient of the quadratic term is not 1, then factor the coefficient from the quadratic and linear terms *before* completing the square. After completing the square and writing the function in vertex form, the value of k indicated a minimum value if $a < 0$ or a maximum value if $a > 0$.

Example 1 Write Functions in Vertex Form

Write each function in vertex form.

a. $y = x^2 + 6x - 5$

$y = x^2 + 6x - 5$	Original function
$y = (x^2 + 6x + 9) - 5 - 9$	Complete the square.
$y = (x + 3)^2 - 14$	Simplify.

b. $y = -2x^2 + 8x - 3$

$y = -2x^2 + 8x - 3$	Original function
$y = -2(x^2 - 4x) - 3$	Group $ax^2 + bx$ and factor, dividing by a.
$y = -2(x^2 - 4x + 4) - 3 - (-2)(4)$	Complete the square.
$y = -2(x - 2)^2 + 5$	Simplify.

GuidedPractice

1A. $y = x^2 + 4x + 6$

1B. $y = 2x^2 - 12x + 17$

If the vertex and one additional point on the graph of a parabola are known, you can write the equation of the parabola in vertex form.

Standardized Test Example 2 Write an Equation Given a Graph

Which is an equation of the function shown in the graph?

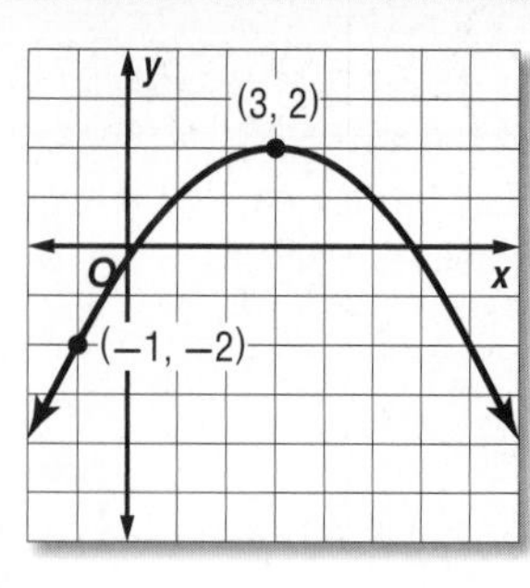

A $y = -4(x - 3)^2 + 2$

B $y = -\frac{1}{4}(x - 3)^2 + 2$

C $y = \frac{1}{4}(x + 3)^2 - 2$

D $y = 4(x + 3)^2 - 2$

Read the Test Item

You are given a graph of a parabola with the vertex and a point on the graph labeled. You need to find an equation of the parabola.

Solve the Test Item

The vertex of the parabola is at (3, 2), so $h = 3$ and $k = 2$. Since (−1, −2) is a point on the graph, let $x = -1$ and $y = -2$. Substitute these values into the vertex form of the equation and solve for a.

$y = a(x - h)^2 + k$	Vertex form
$-2 = a(-1 - 3)^2 + 2$	Substitute −2 for *y*, −1 for *x*, 3 for *h* and 2 for *k*.
$-2 = a(16) + 2$	Simplify.
$-4 = 16a$	Subtract 2 from each side.
$-\frac{1}{4} = a$	Divide each side by 16.

The equation of the parabola in vertex form is $y = -\frac{1}{4}(x - 3)^2 + 2$.

The answer is B.

Test-Taking Tip

The Meaning of *a* The sign of *a* in vertex form does not determine the width of the parabola. The sign indicates whether the parabola opens up or down. The width of a parabola is determined by the absolute value of *a*.

Guided Practice

2. Which is an equation of the function shown in the graph?

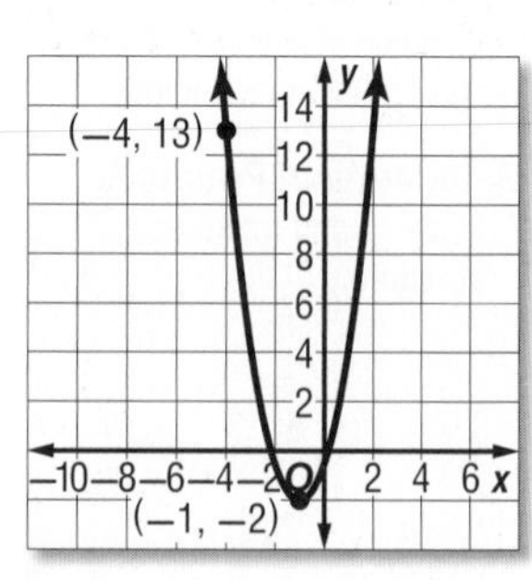

F $y = \frac{9}{25}(x - 1)^2 + 2$

G $y = \frac{3}{5}(x + 1)^2 - 2$

H $y = \frac{5}{3}(x + 1)^2 - 2$

J $y = \frac{25}{9}(x - 1)^2 + 2$

2 Transformations of Quadratic Graphs In Lesson 2-7, you learned how different transformations affect the graphs of parent functions. The following summarizes these transformations for quadratic functions.

ConceptSummary Transformations of Quadratic Functions

$f(x) = a(x - h)^2 + k$

h, Horizontal Translation
h units to the right if h is positive
$|h|$ units to the left if h is negative

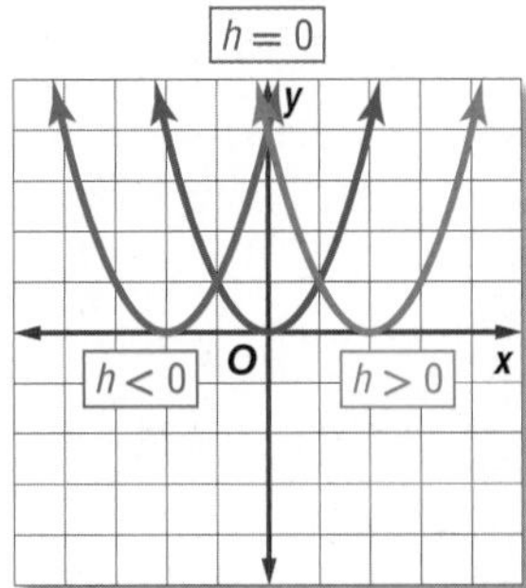

k, Vertical Translation
k units up if k is positive
$|k|$ units down if k is negative

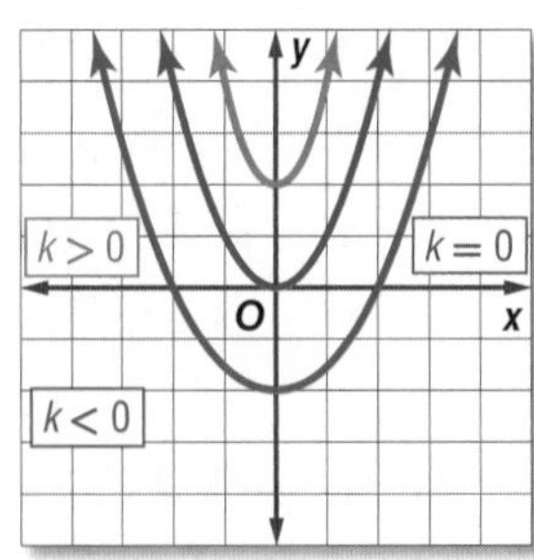

a, Reflection
If $a > 0$, the graph opens up.
If $a < 0$, the graph opens down.

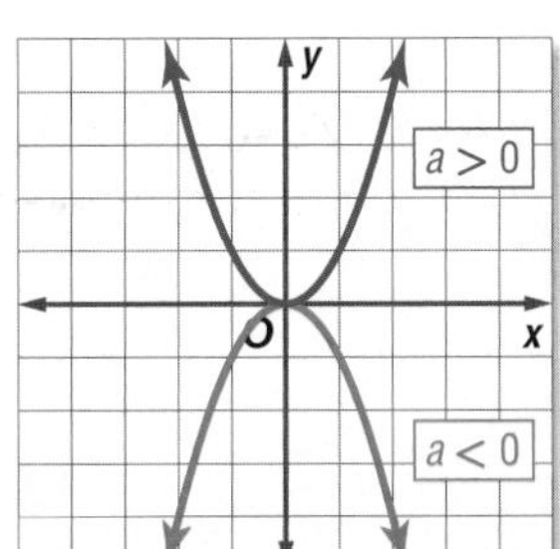

a, Dilation
If $|a| > 1$, the graph is stretched vertically. If $0 < |a| < 1$, the graph is compressed vertically.

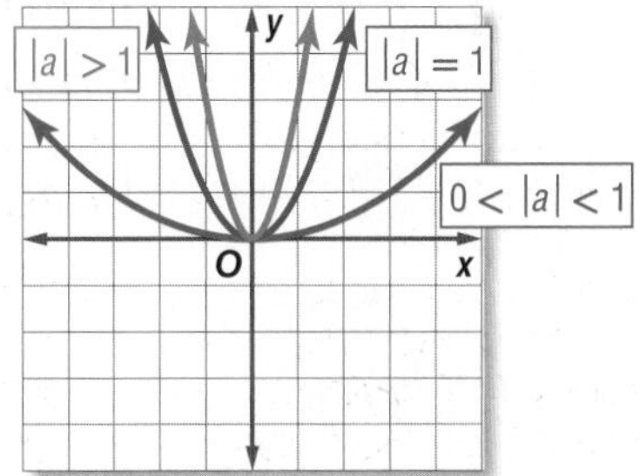

StudyTip

Absolute Value
$0 < |a| < 1$ means that a is a rational number between 0 and 1, such as $\frac{3}{4}$, or a rational number between −1 and 0, such as −0.3.

Example 3 Graph Equations in Vertex Form

Graph $y = 4x^2 - 16x - 40$.

Step 1 Rewrite the equation in vertex form.

$y = 4x^2 - 16x - 40$	Original equation
$y = 4(x^2 - 4x) - 40$	Distributive Property
$y = 4(x - 4x + 4) - 40 - 4(4)$	Complete the square.
$y = 4(x - 2)^2 - 56$	Simplify.

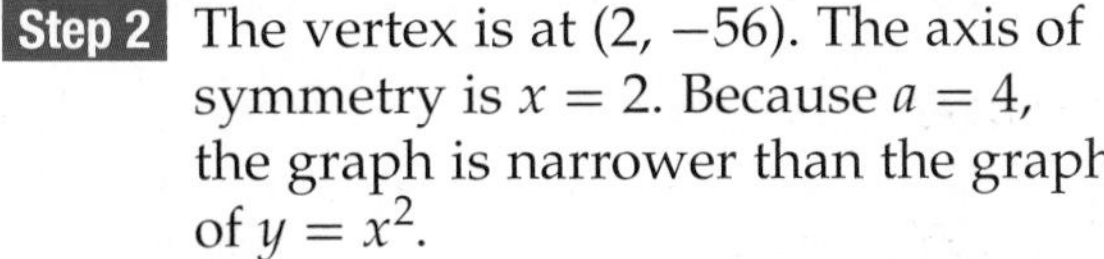

Step 2 The vertex is at (2, −56). The axis of symmetry is $x = 2$. Because $a = 4$, the graph is narrower than the graph of $y = x^2$.

Step 3 Plot additional points to help you complete the graph.

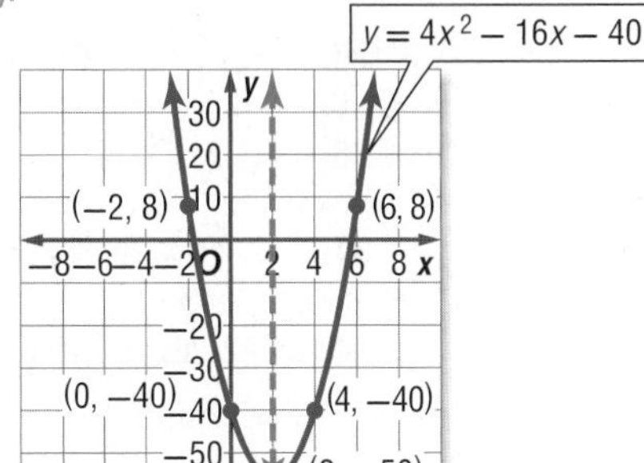

GuidedPractice

3A. $y = (x - 3)^2 - 2$

3B. $y = 0.25(x + 1)^2$

Check Your Understanding

= Step-by-Step Solutions begin on page R14.

Example 1 **Write each function in vertex form.**

1. $y = x^2 + 6x + 2$
2. $y = -2x^2 + 8x - 5$
3. $y = 4x^2 + 24x + 24$

Example 2

4. **MULTIPLE CHOICE** Which function is shown in the graph?

   A $y = -(x + 3)^2 + 6$

   B $y = -(x - 3)^2 - 6$

   C $y = -2(x + 3)^2 + 6$

   D $y = -2(x - 3)^2 - 6$

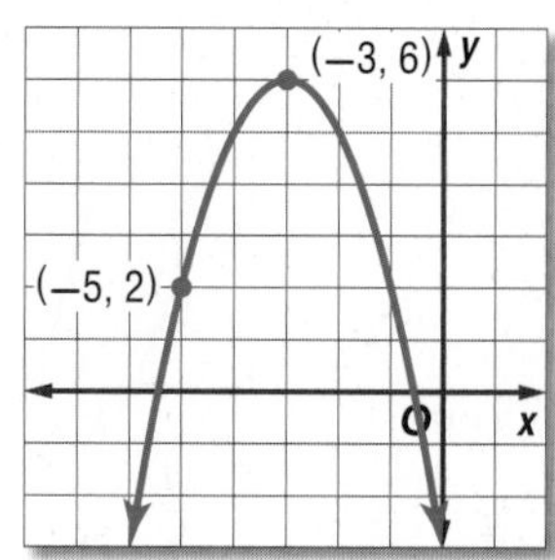

Example 3 **Graph each function.**

5. $y = (x - 3)^2 - 4$
6. $y = -2x^2 + 5$
7. $y = \frac{1}{2}(x + 6)^2 - 8$

Practice and Problem Solving

Extra Practice is on page R4.

Example 1 **Write each function in vertex form.**

8. $y = x^2 + 9x + 8$
9. $y = x^2 - 6x + 3$
10. $y = -2x^2 + 5x$
11. $y = x^2 + 2x + 7$
12. $y = -3x^2 + 12x - 10$
13. $y = x^2 + 8x + 16$
14. $y = 2x^2 - 4x - 3$
15. $y = 3x^2 + 10x$
16. $y = x^2 - 4x + 9$
17. $y = -4x^2 - 24x - 15$
18. $y = x^2 - 12x + 36$
19. $y = -x^2 - 4x - 1$

Example 2

20. **FIREWORKS** During an Independence Day fireworks show, the height h in meters of a specific rocket after t seconds can be modeled by $h = -4.9(t - 4)^2 + 80$. Graph the function.

21. **FINANCIAL LITERACY** A bicycle rental shop rents an average of 120 bicycles per week and charges \$25 per day. The manager estimates that there will be 15 additional bicycles rented for each \$1 reduction in the rental price. The maximum income the manager can expect can be modeled by $y = -15x^2 + 255x + 3000$, where y is the weekly income and x is the number of bicycles rented. Write this function in vertex form. Then graph.

Example 3 **Graph each function.**

22. $y = (x - 5)^2 + 3$
23. $y = 9x^2 - 8$
24. $y = -2(x - 5)^2$
25. $y = \frac{1}{10}(x + 6)^2 + 6$
26. $y = -3(x - 5)^2 - 2$
27. $y = -\frac{1}{4}x^2 - 5$
28. $y = 2x^2 + 10$
29. $y = -(x + 3)^2$
30. $y = \frac{1}{6}(x - 3)^2 - 10$
31. $y = (x - 9)^2 - 7$
32. $y = -\frac{5}{8}x^2 - 8$
33. $y = -4(x - 10)^2 - 10$

34. **CCSS MODELING** A sailboard manufacturer uses an automated process to manufacture the masts for its sailboards. The function $f(x) = \frac{1}{250}x^2 + \frac{3}{5}x$ is programmed into a computer to make one such mast.

    a. Write the quadratic function in vertex form. Then graph the function.

    b. Describe how the manufacturer can adjust the function to make its masts with a greater or smaller curve.

Write an equation in vertex form for each parabola.

35. 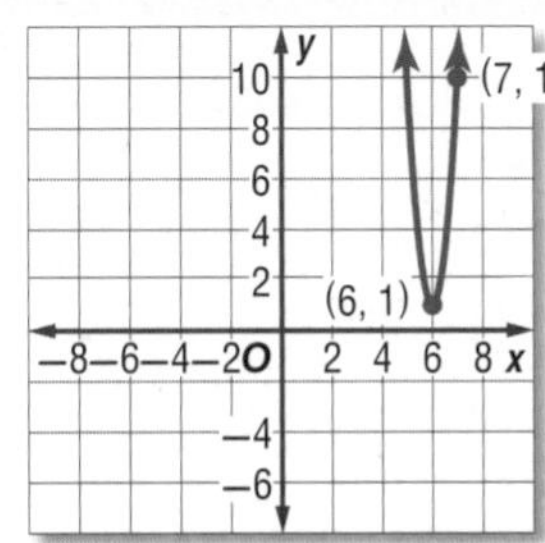

36. 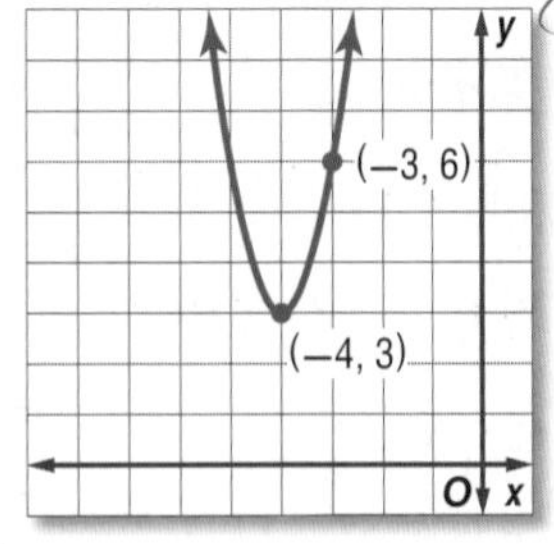

37. 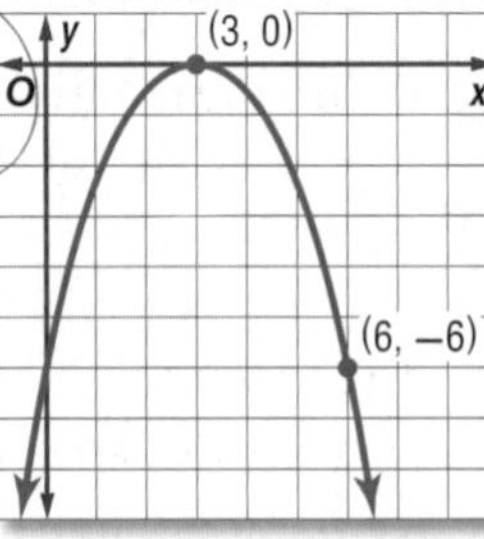

38. 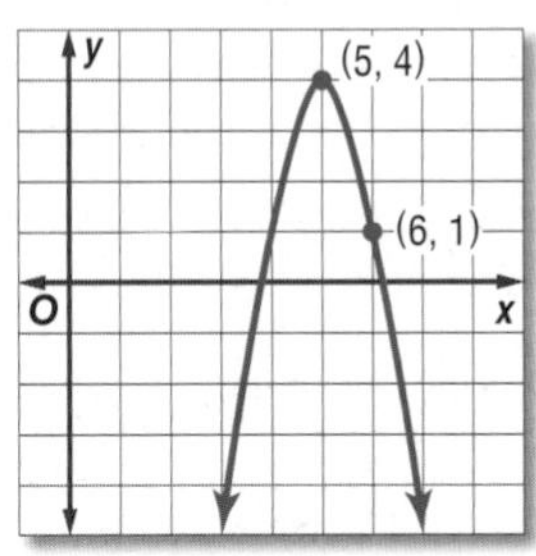

39. 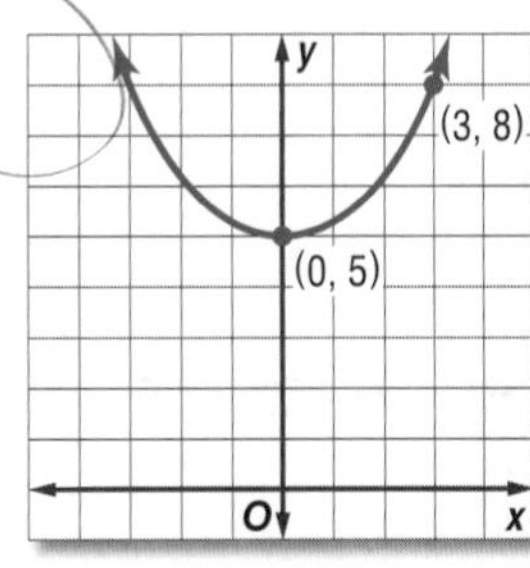

40. 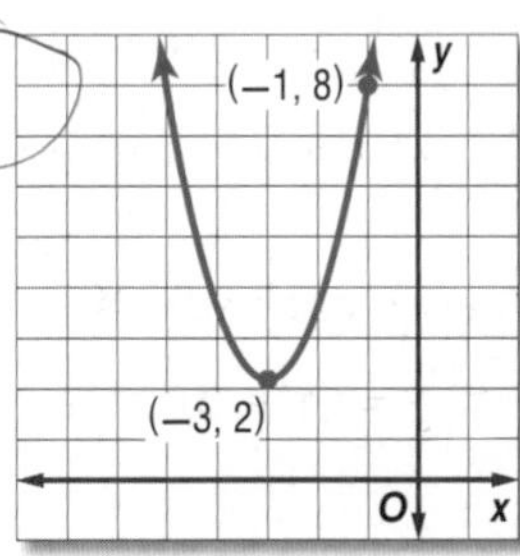

Write each function in vertex form. Then identify the vertex, axis of symmetry, and direction of opening.

41. $3x^2 - 4x = 2 + y$

42. $-2x^2 + 7x = y - 12$

43. $-x^2 - 4.7x = y - 2.8$

44. $x^2 + 1.4x - 1.2 = y$

45. $x^2 - \frac{2}{3}x - \frac{26}{9} = y$

46. $x^2 + 7x + \frac{49}{4} = y$

47. **CARS** The formula $S(t) = \frac{1}{2}at^2 + v_0t$ can be used to determine the position $S(t)$ of an object after t seconds at a rate of acceleration a with initial velocity v_0. Valerie's car can accelerate 0.002 miles per second squared.

a. Express $S(t)$ in vertex form as she accelerates from 35 miles per hour to enter highway traffic.

b. How long will it take Valerie to match the average speed of highway traffic of 68 miles per hour? (*Hint*: Use *acceleration* • *time* = *velocity*.)

c. If the entrance ramp is $\frac{1}{8}$-mile long, will Valerie have sufficient time to match the average highway speed? Explain.

H.O.T. Problems Use Higher-Order Thinking Skills

48. **OPEN ENDED** Write an equation for a parabola that has been translated, compressed, and reflected in the x-axis.

49. **CHALLENGE** Explain how you can find an equation of a parabola using the coordinates of three points on the graph.

50. **CHALLENGE** Write the standard form of a quadratic function $ax^2 + bx + c = y$ in vertex form. Identify the vertex and the axis of symmetry.

51. **REASONING** Describe the graph of $f(x) = a(x - h)^2 + k$ when $a = 0$. Is the graph the same as that of $g(x) = ax^2 + bx + c$ when $a = 0$? Explain.

52. CCSS **ARGUMENTS** Explain how the graph of $y = x^2$ can be used to graph any quadratic function. Include a description of the effects produced by changing a, h, and k in the equation $y = a(x - h)^2 + k$, and a comparison of the graph of $y = x^2$ and the graph of $y = a(x - h)^2 + k$ using values you choose for a, h, and k.

Standardized Test Practice

53. Flowering bushes need a mixture of 70% soil and 30% vermiculite. About how many buckets of vermiculite should you add to 20 buckets of soil?

A 6.0
C 14.0
B 8.0
D 24.0

54. SAT/ACT The sum of the integers x and y is 495. The units digit of x is 0. If x is divided by 10, the result is equal to y. What is the value of x?

F 40
J 250
G 45
K 450
H 245

55. What is the solution set of the inequality $|4x - 1| < 9$?

A $\{x \mid 2.5 < x \text{ or } x < -2\}$
B $\{x \mid x < 2.5\}$
C $\{x \mid x > -2\}$
D $\{x \mid -2 < x < 2.5\}$

56. SHORT RESPONSE At your store, you buy wrenches for \$30.00 a dozen and sell them for \$3.50 each. What is the percent markup for the wrenches?

Spiral Review

Solve each equation by using the method of your choice. Find exact solutions. (Lesson 4-6)

57. $4x^2 + 15x = 21$
58. $-3x^2 + 19 = 5x$
59. $6x - 5x^2 + 9 = 3$

Find the value of c that makes each trinomial a perfect square. (Lesson 4-5)

60. $x^2 - 12x + c$
61. $x^2 + 0.1x + c$
62. $x^2 - 0.45x + c$

Determine whether each function has a maximum or minimum value, and find that value. (Lesson 4-1)

63. $f(x) = 6x^2 - 8x + 12$
64. $f(x) = -4x^2 + x - 18$
65. $f(x) = 3x^2 - 9 + 6x$

66. ARCHAEOLOGY A coordinate grid is laid over an archaeology dig to identify the location of artifacts. Three corners of a building have been partially unearthed at $(-1, 6)$, $(4, 5)$, and $(-1, -2)$. If each square on the grid measures one square foot, estimate the area of the floor of the building. (Lesson 3-7)

67. ART Rai can spend no more than \$225 on the art club's supply of brushes and paint. She needs at least 20 brushes and 56 tubes of paint. Graph the region that shows how many boxes of each item can be purchased. (Lesson 3-2)

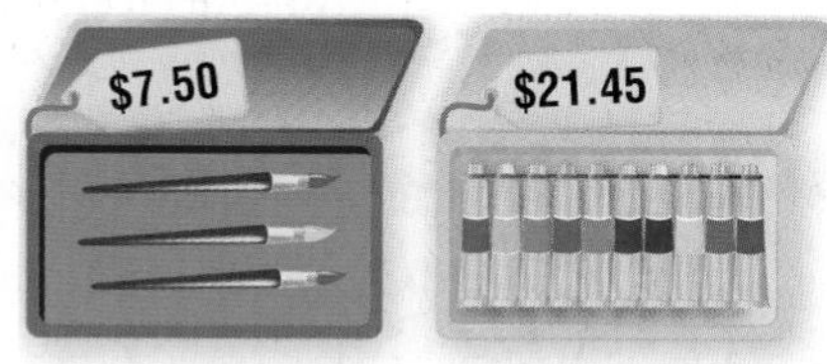

Solve each system of equations by graphing. (Lesson 3-1)

68. $y = 3x - 4$
$y = -2x + 16$

69. $2x + 5y = 1$
$6y - 5x = 16$

70. $4x + 3y = -30$
$3x - 2y = 3$

Evaluate each function. (Lesson 2-1)

71. $f(3)$ if $f(x) = x^2 - 4x + 12$
72. $f(-2)$ if $f(x) = -4x^2 + x - 8$
73. $f(4)$ if $f(x) = 3x^2 + x$

Skills Review

Determine whether the given value satisfies the inequality.

74. $3x^2 - 5 > 6;\ x = 2$
75. $-2x^2 + x - 1 < 4;\ x = -2$
76. $4x^2 + x - 3 \leq 36;\ x = 3$

EXTEND 4-7

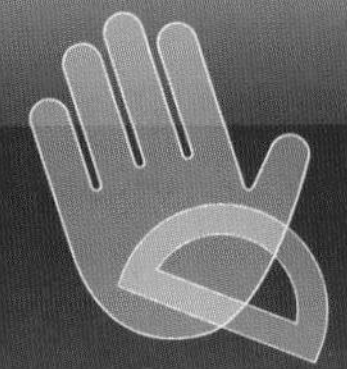

Algebra Lab: Quadratics and Rate of Change

You have learned that a linear function has a constant rate of change. In this lab, you will investigate the rate of change for quadratic functions.

Common Core State Standards
Content Standards
F.IF.4 For a function that models a relationship between two quantities, interpret key features of graphs and tables in terms of the quantities, and sketch graphs showing key features given a verbal description of the relationship.
F.IF.6 Calculate and interpret the average rate of change of a function (presented symbolically or as a table) over a specified interval. Estimate the rate of change from a graph.

Activity: Determine Rate of Change

Consider $f(x) = 0.1875x^2 - 3x + 12$.

Step 1 Make a table like the one below. Use values from 0 through 16 for x.

x	0	1	2	3	...	16
y	12	9.1875	6.75			
First-Order Differences						
Second-Order Differences						

Step 2 Find each y-value. For example, when $x = 1$, $y = 0.1875(1)^2 - 3(1) + 12$ or 9.1875.

Step 3 Graph the ordered pairs (x, y). Then connect the points with a smooth curve. Notice that the function *decreases* when $0 < x < 8$ and *increases* when $8 < x < 16$.

Step 4 The rate of change from one point to the next can be found by using the slope formula. From (0, 12) to (1, 9.1875), the slope is $\frac{9.1875 - 12}{1 - 0}$ or -2.8125. This is the first-order difference at $x = 1$. Complete the table for all the first-order differences. Describe any patterns in the differences.

Step 5 The second-order differences can be found by subtracting consecutive first-order differences. For example, the second-order difference at $x = 2$ is found by subtracting the first order difference at $x = 1$ from the first-order difference at $x = 2$. Describe any patterns in the differences.

Exercises

For each function make a table of values for the given x-values. Graph the function. Then determine the first-order and second-order differences.

1. $y = -x^2 + 2x - 1$ for $x = -3, -2, -1, 0, 1, 2, 3$
2. $y = 0.5x^2 + 2x - 2$ for $x = -5, -4, -3, -2, -1, 0, 1$
3. $y = -3x^2 - 18x - 26$ for $x = -6, -5, -4, -3, -2, -1, 0$
4. **MAKE A CONJECTURE** Repeat the activity for a cubic function. At what order difference would you expect $g(x) = x^4$ to be constant? $h(x) = x^n$?

LESSON 4-8 Quadratic Inequalities

Then

- You solved linear inequalities.

Now

1. Graph quadratic inequalities in two variables.
2. Solve quadratic inequalities in one variable.

Why?

- A water balloon launched from a slingshot can be represented by several different quadratic equations and inequalities.

Suppose the height of a water balloon $h(t)$ in meters above the ground t seconds after being launched is modeled by the quadratic function $h(t) = -4.9t^2 + 32t + 1.2$. You can solve a quadratic inequality to determine how long the balloon will be a certain distance above the ground.

NewVocabulary
quadratic inequality

Common Core State Standards

Content Standards

A.CED.1 Create equations and inequalities in one variable and use them to solve problems.

A.CED.3 Represent constraints by equations or inequalities, and by systems of equations and/or inequalities, and interpret solutions as viable or nonviable options in a modeling context.

Mathematical Practices

1 Make sense of problems and persevere in solving them.

1 Graph Quadratic Inequalities

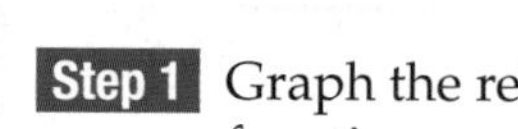

You can graph **quadratic inequalities** in two variables by using the same techniques used to graph linear inequalities in two variables.

Step 1 Graph the related function.

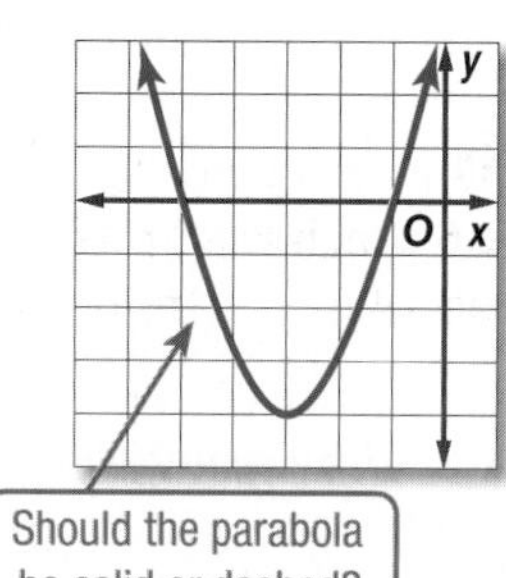

Step 2 Test a point not on the parabola.

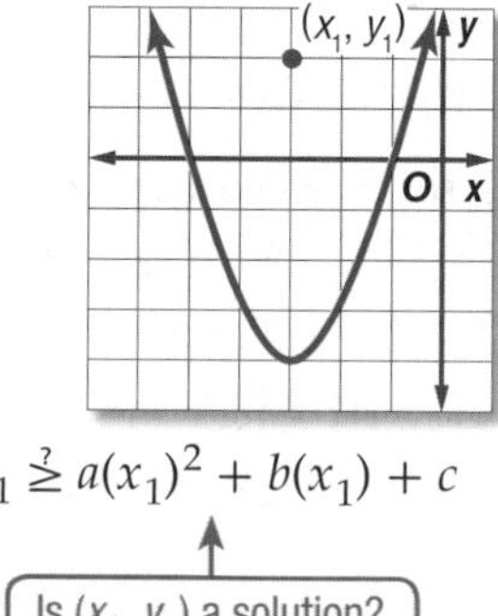

$y_1 \overset{?}{\geq} a(x_1)^2 + b(x_1) + c$

Is (x_1, y_1) a solution?

Step 3 Shade accordingly.

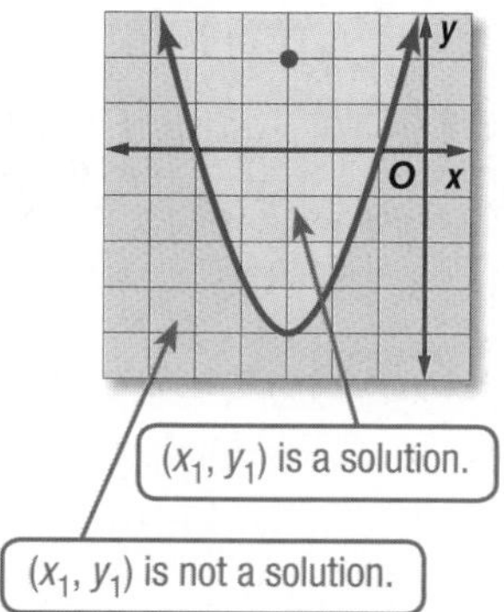

Example 1 Graph a Quadratic Inequality

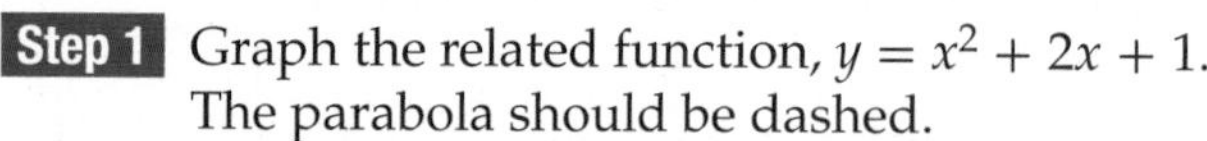

Graph $y > x^2 + 2x + 1$.

Step 1 Graph the related function, $y = x^2 + 2x + 1$. The parabola should be dashed.

Step 2 Test a point not on the graph of the parabola.

$y > x^2 + 2x + 1$

$-1 \overset{?}{>} 0^2 + 2(0) + 1$

$-1 \not> 1$ So, $(0, -1)$ is *not* a solution of the inequality.

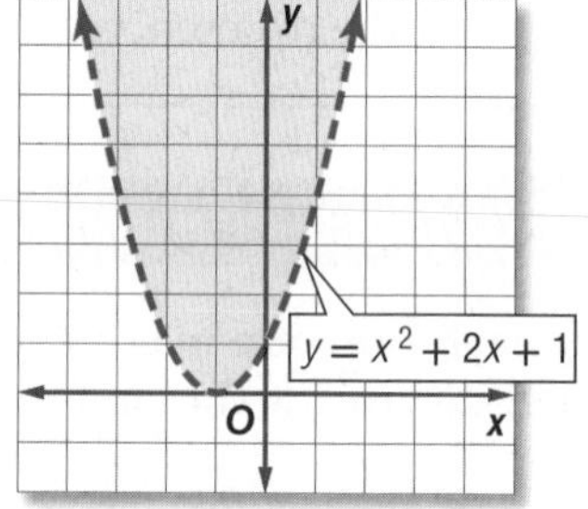

Step 3 Shade the region that does not contain the point $(0, -1)$.

GuidedPractice

Graph each inequality.

1A. $y \leq x^2 + 2x + 4$

1B. $y < -2x^2 + 3x + 5$

madelyn mulvaney/Flickr/Getty Images

2 Solve Quadratic Inequalities

Quadratic inequalities in one variable can be solved using the graphs of the related quadratic functions.

$ax^2 + bx + c < 0$

Graph $y = ax^2 + bx + c$ and identify the x-values for which the graph lies *below* the x-axis.

For $\leq$, include the x-intercepts in the solution.

$a > 0$

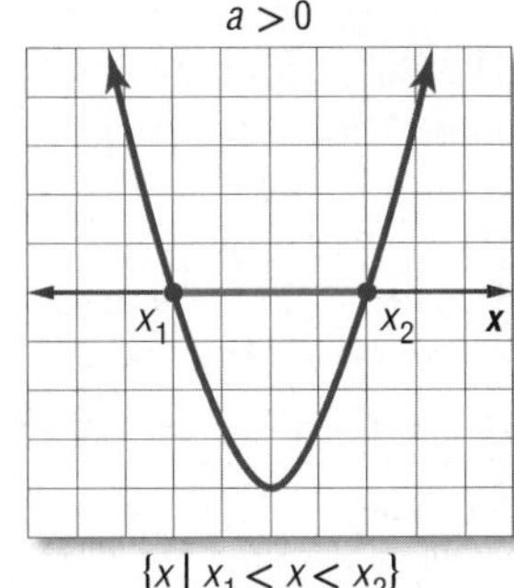

$\{x \mid x_1 < x < x_2\}$

$a < 0$

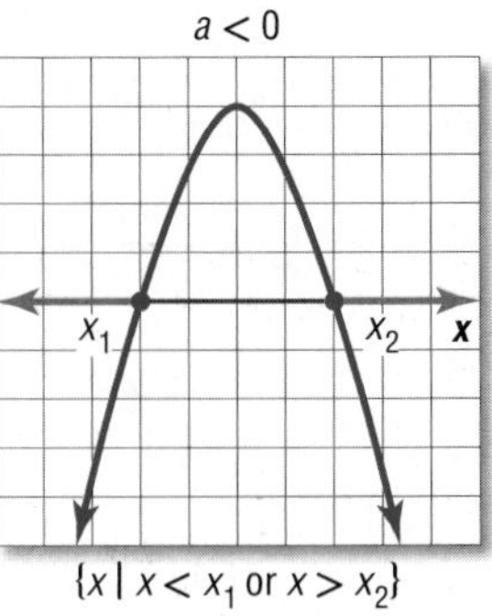

$\{x \mid x < x_1 \text{ or } x > x_2\}$

$ax^2 + bx + c > 0$

Graph $y = ax^2 + bx + c$ and identify the x-values for which the graph lies *above* the x-axis.

For $\geq$, include the x-intercepts in the solution.

$a > 0$

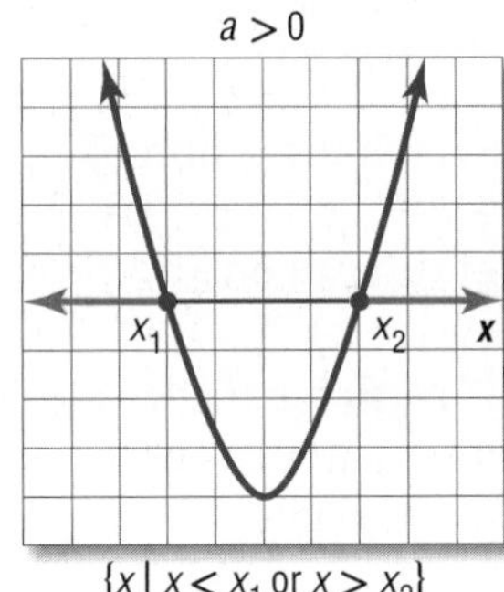

$\{x \mid x < x_1 \text{ or } x > x_2\}$

$a < 0$

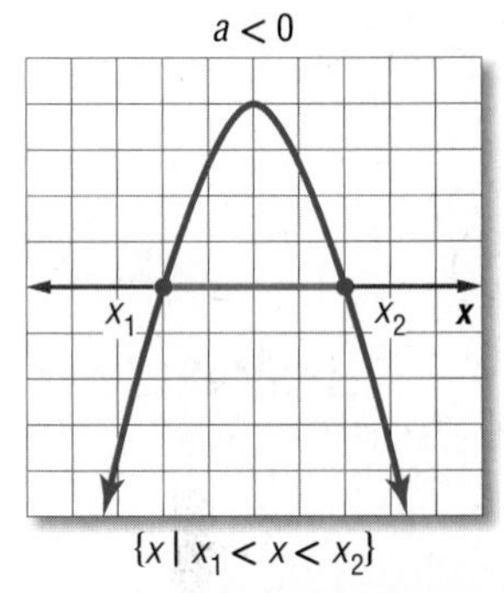

$\{x \mid x_1 < x < x_2\}$

StudyTip

Solving Quadratic Inequalities by Graphing A precise graph of the related quadratic function is not necessary since the zeros of the function were found algebraically.

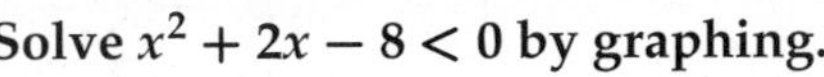

Example 2 Solve $ax^2 + bx + c < 0$ by Graphing

Solve $x^2 + 2x - 8 < 0$ by graphing.

The solution consists of x-values for which the graph of the related function lies *below* the x-axis. Begin by finding the roots of the related function.

$x^2 + 2x - 8 = 0$	Related equation
$(x - 2)(x + 4) = 0$	Factor.
$x - 2 = 0$ or $x + 4 = 0$	Zero Product Property
$x = 2$ $\quad$ $x = -4$	Solve each equation.

Sketch the graph of a parabola that has x-intercepts at -4 and 2. The graph should open up because $a > 0$.

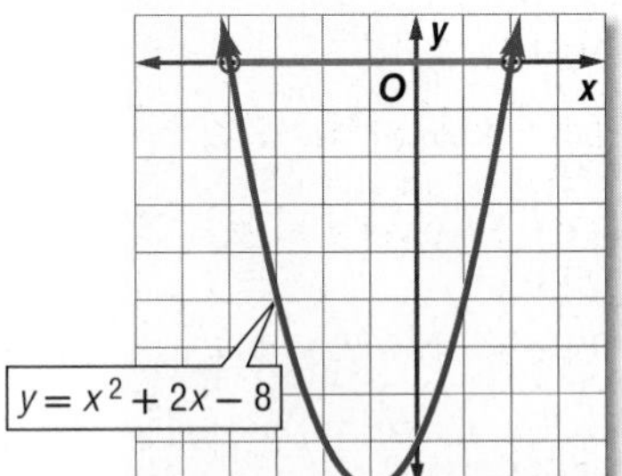

The graph lies below the x-axis between $x = -4$ and $x = 2$. Thus, the solution set is $\{x \mid -4 < x < 2\}$ or $(-4, 2)$.

CHECK Test one value of x less than -4, one between -4 and 2, and one greater than 2 in the original inequality.

Test $x = -6$.	**Test $x = 0$.**	**Test $x = 5$.**
$x^2 + 2x - 8 < 0$	$x^2 + 2x - 8 < 0$	$x^2 + 2x - 8 < 0$
$(-6)^2 + 2(-6) - 8 \overset{?}{<} 0$	$0^2 + 2(0) - 8 \overset{?}{<} 0$	$5^2 + 2(5) - 8 \overset{?}{<} 0$
$16 < 0$ ✗	$-8 < 0$ ✓	$27 < 0$ ✗

GuidedPractice

Solve each inequality by graphing.

2A. $0 > x^2 + 5x - 6$

2B. $-x^2 + 3x + 10 \leq 0$

Example 3 Solve $ax^2 + bx + c \geq 0$ by Graphing

Solve $2x^2 + 4x - 5 \geq 0$ by graphing.

The solution consists of x-values for which the graph of the related function lies *on and above* the x-axis. Begin by finding the roots of the related function.

$2x^2 + 4x - 5 = 0$ — Related equation

$x = \frac{-b \pm \sqrt{b^2 - 4ac}}{2a}$ — Use the Quadratic Formula

$x = \frac{-4 \pm \sqrt{4^2 - 4(2)(-5)}}{2(2)}$ — Replace a with 4, b with 2, and c with −5.

$x = \frac{-4 + \sqrt{56}}{4}$ or $x = \frac{-4 - \sqrt{56}}{4}$ — Simplify and write as two equations.

≈ 0.87 ≈ -2.87 — Simplify.

Sketch the graph of a parabola with x-intercepts at -2.87 and 0.87. The graph opens up since $a > 0$. The graph lies on and above the x-axis at about $x \leq -2.87$ and $x \geq 0.87$. Therefore, the solution is approximately $\{x \mid x \leq -2.87 \text{ or } x \geq 0.87\}$ or $(-\infty, -2.87] \cup [0.87, \infty)$.

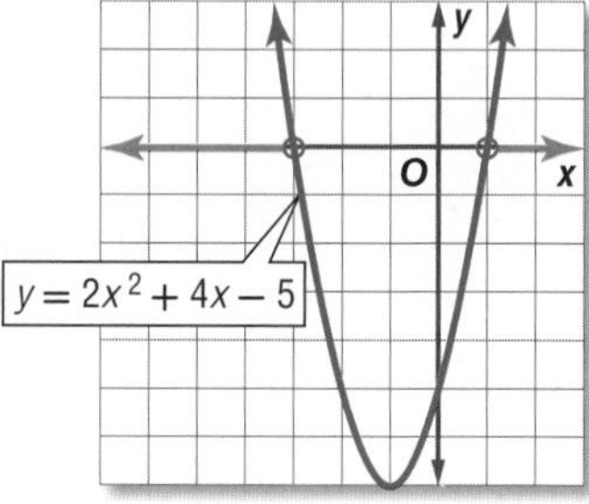

GuidedPractice

Solve each inequality by graphing.

3A. $x^2 - 6x + 2 > 0$

3B. $-4x^2 + 5x + 7 \geq 0$

Real-world problems can be solved by graphing quadratic inequalities.

Real-World Example 4 Solve a Quadratic Inequality

WATER BALLOONS Refer to the beginning of the lesson. At what time will a water balloon be within 3 meters of the ground after it has been launched?

The function $h(t) = -4.9t^2 + 32t + 1.2$ describes the height of the water balloon. Therefore, you want to find the values of t for which $h(t) \leq 3$.

$h(t) \leq 3$ — Original inequality

$-4.9t^2 + 32t + 1.2 \leq 3$ — $h(t) = -4.9t^2 + 32t + 1.2$

$-4.9t^2 + 32t - 1.8 \leq 0$ — Subtract 3 from each side.

Graph the related function $y = -4.9x^2 + 32x - 1.8$ using a graphing calculator. The zeros of the function are about 0.06 and 6.47, and the graph lies below the x-axis when $x < 0.06$ and $x > 6.47$.

So, the water balloon is within 3 meters of the ground during the first 0.06 second after being launched and again after about 6.47 seconds until it hits the ground.

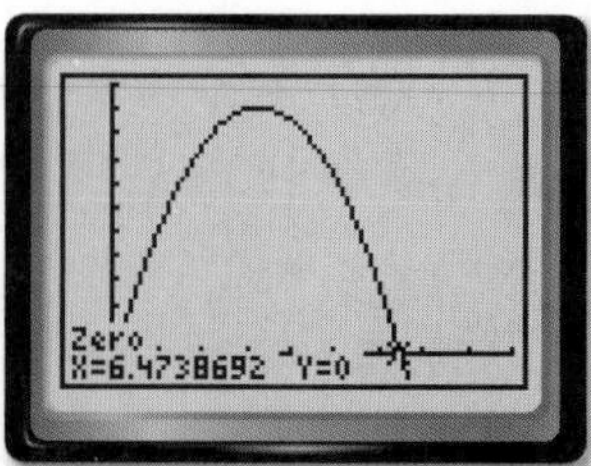

[−1, 9] scl: 1 by [−5, 55] scl: 5

Real-WorldLink

It takes just milliseconds for a water balloon to break. A high-speed camera can capture the impact on the fluid before gravity makes it fall.

Source: NASA

GuidedPractice

4. ROCKETS The height $h(t)$ of a model rocket in feet t seconds after its launch can be represented by the function $h(t) = -16t^2 + 82t + 0.25$. During what interval is the rocket at least 100 feet above the ground?

RubberBall/Alamy

StudyTip

Solving Quadratic Inequalities Algebraically The solution set of a quadratic inequality is all real numbers when all three test points satisfy the inequality. It is the empty set when none of the test points satisfy the inequality.

Example 5 Solve a Quadratic Inequality Algebraically

Solve $x^2 - 3x \leq 18$ algebraically.

Step 1 Solve the related quadratic equation $x^2 - 3x = 18$.

$x^2 - 3x = 18$	Related quadratic equation
$x^2 - 3x - 18 = 0$	Subtract 18 from each side.
$(x + 3)(x - 6) = 0$	Factor.
$x + 3 = 0$ or $x - 6 = 0$	Zero Product Property
$x = -3$ $\quad$ $x = 6$	Solve each equation.

Step 2 Plot −3 and 6 on a number line. Use dots since these values are solutions of the original inequality. Notice that the number line is divided into three intervals.

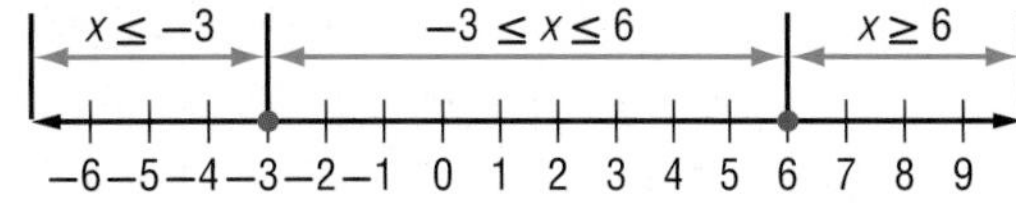

Step 3 Test a value from each interval to see if it satisfies the original inequality.

$x \leq -3$	$-3 \leq x \leq 6$	$x \geq 6$
Test $x = -5$.	Test $x = 0$.	Test $x = 8$.
$x^2 - 3x \leq 18$	$x^2 - 3x \leq 18$	$x^2 - 3x \leq 18$
$(-5)^2 - 3(-5) \overset{?}{\leq} 18$	$(0)^2 - 3(0) \overset{?}{\leq} 18$	$(8)^2 - 3(8) \overset{?}{\leq} 18$
$40 \not\leq 18$	$0 \leq 18$	$40 \not\leq 18$

The solution set is $\{x \mid -3 \leq x \leq 6\}$ or $[-3, 6]$.

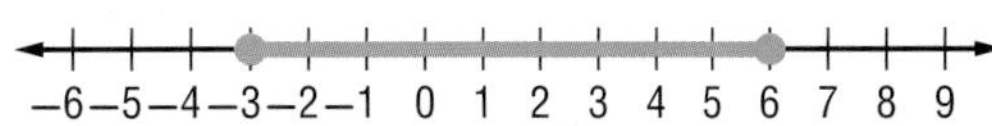

GuidedPractice

Solve each inequality algebraically.

5A. $x^2 + 5x < -6$

5B. $x^2 + 11x + 30 \geq 0$

Check Your Understanding

= Step-by-Step Solutions begin on page R14.

Example 1 **Graph each inequality.**

1. $y \leq x^2 - 8x + 2$

2. $y > x^2 + 6x - 2$

3. $y \geq -x^2 + 4x + 1$

Examples 2–3 CCSS **SENSE-MAKING** **Solve each inequality by graphing.**

4. $0 < x^2 - 5x + 4$

5. $x^2 + 8x + 15 < 0$

6. $-2x^2 - 2x + 12 \geq 0$

7. $0 \geq 2x^2 - 4x + 1$

Example 4

8. **SOCCER** A midfielder kicks a ball toward the goal during a match. The height of the ball in feet above the ground $h(t)$ at time t can be represented by $h(t) = -0.1t^2 + 2.4t + 1.5$. If the height of the goal is 8 feet, at what time during the kick will the ball be able to enter the goal?

Example 5 **Solve each inequality algebraically.**

9. $x^2 + 6x - 16 < 0$

10. $x^2 - 14x > -49$

11 $-x^2 + 12x \geq 28$

12. $x^2 - 4x \leq 21$

Practice and Problem Solving

Extra Practice is on page R4.

Example 1 **Graph each inequality.**

13. $y \geq x^2 + 5x + 6$ **14.** $x^2 - 2x - 8 < y$ **15.** $y \leq -x^2 - 7x + 8$

16. $-x^2 + 12x - 36 > y$ **17.** $y > 2x^2 - 2x - 3$ **18.** $y \geq -4x^2 + 12x - 7$

Examples 2–3 **Solve each inequality by graphing.**

19. $x^2 - 9x + 9 < 0$ **20.** $x^2 - 2x - 24 \leq 0$ **21.** $x^2 + 8x + 16 \geq 0$

22. $x^2 + 6x + 3 > 0$ **23.** $0 > -x^2 + 7x + 12$ **24.** $-x^2 + 2x - 15 < 0$

25. $4x^2 + 12x + 10 \leq 0$ **26.** $-3x^2 - 3x + 9 > 0$ **27.** $0 > -2x^2 + 4x + 4$

28. $3x^2 + 12x + 36 \leq 0$ **29.** $0 \leq -4x^2 + 8x + 5$ **30.** $-2x^2 + 3x + 3 \leq 0$

Example 4

31. **ARCHITECTURE** An arched entry of a room is shaped like a parabola that can be represented by the equation $f(x) = -x^2 + 6x + 1$. How far from the sides of the arch is its height at least 7 feet?

32. **MANUFACTURING** A box is formed by cutting 4-inch squares from each corner of a square piece of cardboard and then folding the sides. If $V(x) = 4x^2 - 64x + 256$ represents the volume of the box, what should the dimensions of the original piece of cardboard be if the volume of the box cannot exceed 750 cubic inches?

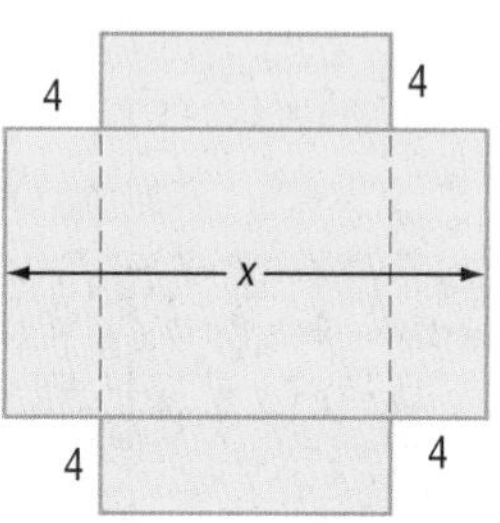

Example 5 **Solve each inequality algebraically.**

33. $x^2 - 9x < -20$ **34.** $x^2 + 7x \geq -10$ **35.** $2 > x^2 - x$

36. $-3 \leq -x^2 - 4x$ **37.** $-x^2 + 2x \leq -10$ **38.** $-6 > x^2 + 4x$

39. $2x^2 + 4 \geq 9$ **40.** $3x^2 + x \geq -3$ **41.** $-4x^2 + 2x < 3$

42. $-11 \geq -2x^2 - 5x$ **43.** $-12 < -5x^2 - 10x$ **44.** $-3x^2 - 10x > -1$

45. CCSS **PERSEVERANCE** The Sanchez family is adding a deck along two sides of their swimming pool. The deck width will be the same on both sides and the total area of the pool and deck cannot exceed 750 square feet.

a. Graph the quadratic inequality.

b. Determine the possible widths of the deck.

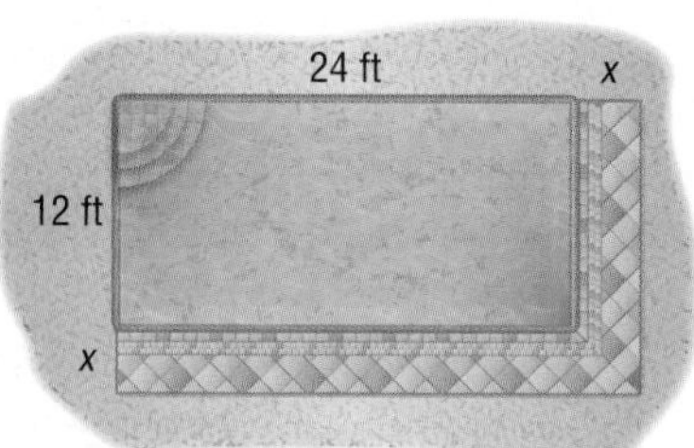

Write a quadratic inequality for each graph.

46. (−2, 6) (6, 6) (0, −6) (4, −6) (2, −10)

47.

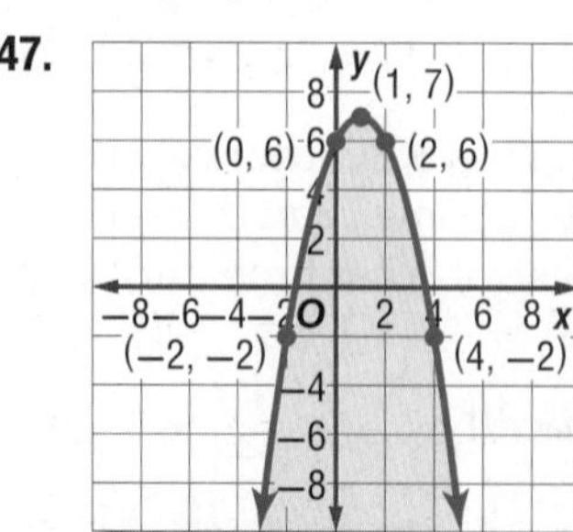

48.

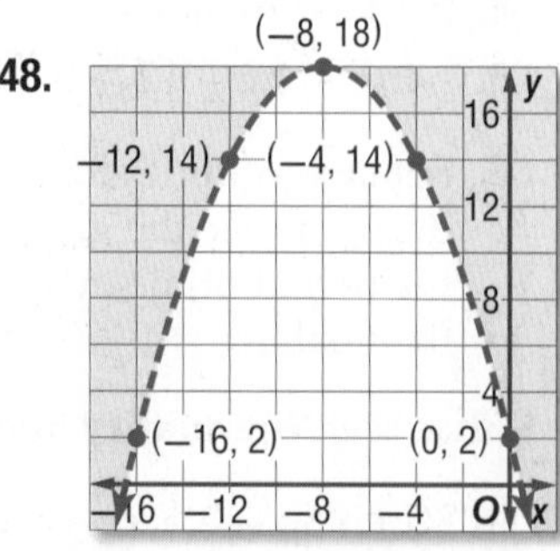

Solve each quadratic inequality by using a graph, a table, or algebraically.

49. $-2x^2 + 12x < -15$ **50.** $5x^2 + x + 3 \geq 0$ **51** $11 \leq 4x^2 + 7x$

52. $x^2 - 4x \leq -7$ **53.** $-3x^2 + 10x < 5$ **54.** $-1 \geq -x^2 - 5x$

55. BUSINESS An electronics manufacturer uses the function $P(x) = x(-27.5x + 3520) + 20{,}000$ to model their monthly profits when selling x thousand digital audio players.

a. Graph the quadratic inequality for a monthly profit of at least \$100,000.

b. How many digital audio players must the manufacturer sell to earn a profit of at least \$100,000 in a month?

c. Suppose the manufacturer has an additional monthly expense of \$25,000. Explain how this affects the graph of the profit function. Then determine how many digital audio players the manufacturer needs to sell to have at least \$100,000 in profits.

56. UTILITIES A contractor is installing drain pipes for a shopping center's parking lot. The outer diameter of the pipe is to be 10 inches. The cross sectional area of the pipe must be at least 35 square inches and should not be more than 42 square inches.

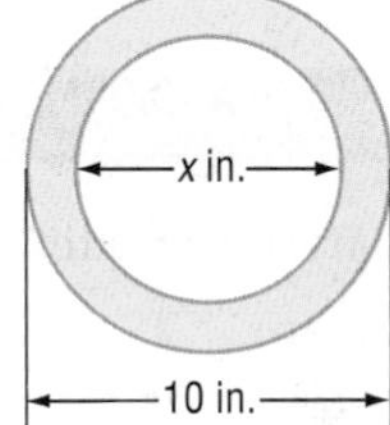

a. Graph the quadratic inequalities.

b. What thickness of drain pipe can the contractor use?

H.O.T. Problems Use Higher-Order Thinking Skills

57. OPEN ENDED Write a quadratic inequality for each condition.

a. The solution set is all real numbers.

b. The solution set is the empty set.

58. CCSS CRITIQUE Don and Diego used a graph to solve the quadratic inequality $x^2 - 2x - 8 > 0$. Is either of them correct? Explain.

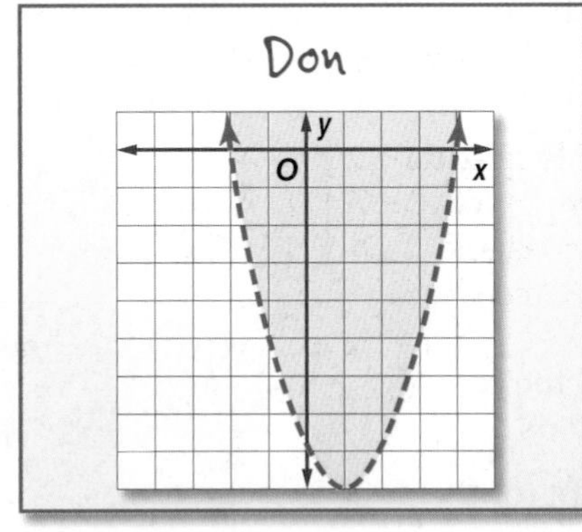

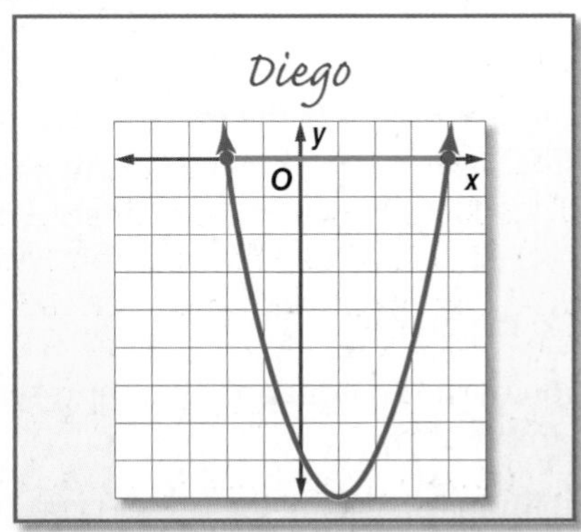

59. REASONING Are the boundaries of the solution set of $x^2 + 4x - 12 \leq 0$ twice the value of the boundaries of $\frac{1}{2}x^2 + 2x - 6 \leq 0$? Explain.

60. REASONING Determine if the following statement is *sometimes*, *always*, or *never* true. Explain your reasoning.

The intersection of $y \leq -ax^2 + c$ and $y \geq ax^2 - c$ is the empty set.

61. CHALLENGE Graph the intersection of the graphs of $y \leq -x^2 + 4$ and $y \geq x^2 - 4$.

62. WRITING IN MATH How are the techniques used when solving quadratic inequalities and quadratic equations similar? different?

Standardized Test Practice

63. **GRIDDED RESPONSE** You need to seed an area that is 80 feet by 40 feet. Each bag of seed can cover 25 square yards of land. How many bags of seed will you need?

64. **SAT/ACT** The product of two integers is between 107 and 116. Which of the following *cannot* be one of the integers?

A 5
B 10
C 12
D 15
E 23

65 **PROBABILITY** Five students are to be arranged side by side with the tallest student in the center and the two shortest students on the ends. If no two students are the same height, how many different arrangements are possible?

F 2
G 4
H 5
J 6

66. **SHORT RESPONSE** Simplify $\frac{5+i}{6-3i}$.

Spiral Review

Write an equation in vertex form for each parabola. (Lesson 4-7)

67.

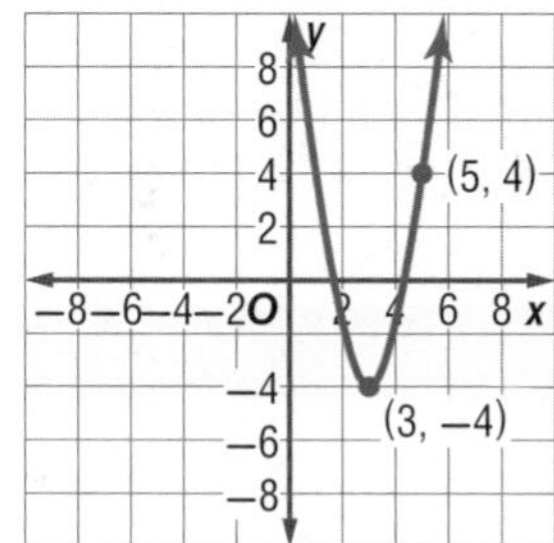

68.

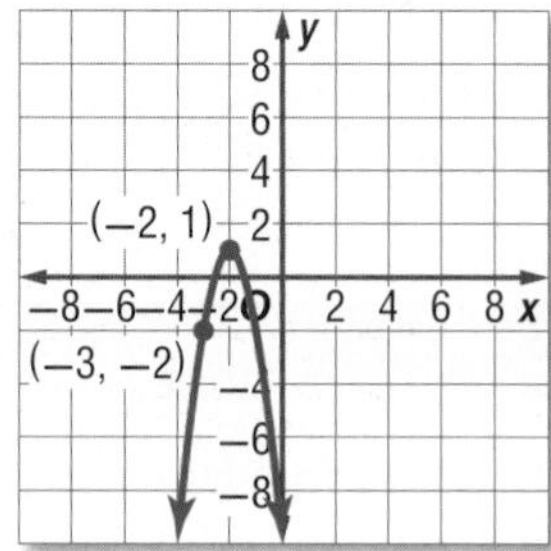

69.

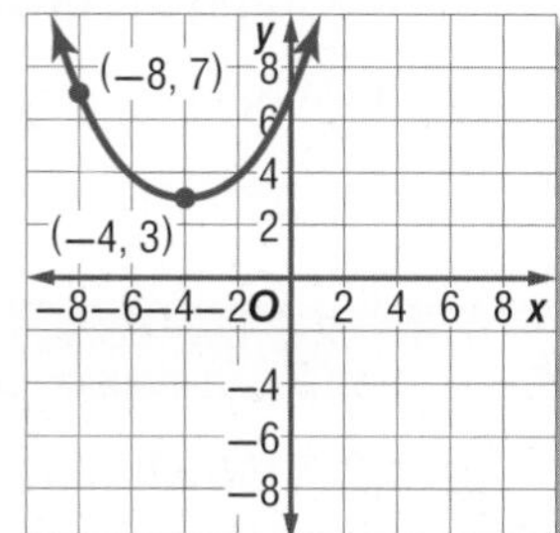

Complete parts a and b for each quadratic equation.

a. Find the value of the discriminant.

b. Describe the number and type of roots. (Lesson 4-6)

70. $4x^2 + 7x - 3 = 0$

71. $-3x^2 + 2x - 4 = 9$

72. $6x^2 + x - 4 = 12$

Perform the indicated operation. If the matrix does not exist, write *impossible*. (Lesson 3-5)

73. $4\begin{bmatrix} 3 & -6 \\ -5 & 2 \end{bmatrix} - 3\begin{bmatrix} 4 & -1 \\ -2 & 8 \end{bmatrix}$

74. $-2\begin{bmatrix} 5 & -9 \\ 5 & 11 \end{bmatrix} - 6\begin{bmatrix} 3 & -7 \\ -5 & 8 \end{bmatrix}$

75. $\begin{bmatrix} 2 & -6 \\ -4 & 6 \end{bmatrix} \cdot \begin{bmatrix} 2 & -1 & 1 \\ -1 & 6 & 4 \end{bmatrix}$

76. **EXERCISE** Refer to the graphic. (Lesson 3-1)

a. For each option, write an equation that represents the cost of belonging to the gym.

b. Graph the equations. Estimate the break-even point for the gym memberships.

c. Explain what the break-even point means.

d. If you plan to visit the gym at least once per week during the year, which option should you chose?

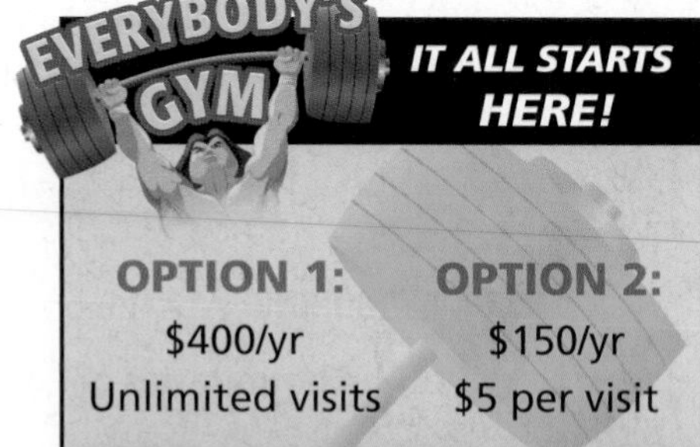

Skills Review

Use the Distributive Property to find each product.

77. $-6(x - 4)$

78. $8(w + 3x)$

79. $-4(-2y + 3z)$

80. $-1(c - d)$

81. $0.5(5x + 6y)$

82. $-3(-6y - 4z)$

EXTEND 4-8

Graphing Technology Lab
Modeling Motion

Set Up the Lab

- Place a board on a stack of books to create a ramp.
- Connect the data collection device to the graphing calculator. Place at the top of the ramp so that the data collection device can read the motion of the car on the ramp.
- Hold the car still about 6 inches up from the bottom of the ramp and zero the collection device.

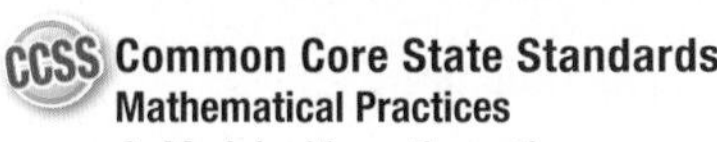

Common Core State Standards
Mathematical Practices
4 Model with mathematics.

Activity

Step 1 One group member should press the button to start collecting data.

Step 2 Another group member places the car at the bottom of the ramp. After data collection begins, gently but quickly push the car so it travels up the ramp toward the motion detector.

Step 3 Stop collecting data when the car returns to the bottom of the ramp. Save the data as Trial 1.

Step 4 Remove one book from the stack. Then repeat the experiment. Save the data as Trial 2. For Trial 3, create a steeper ramp and repeat the experiment.

Analyze the Results

1. What type of function could be used to represent the data? Justify your answer.
2. Use the **CALC** menu to find the vertex of the graph. Record the coordinates in a table like the one at the right.
3. Use the **TRACE** feature of the calculator to find the coordinates of another point on the graph. Then use the coordinates of the vertex and the point to find an equation of the graph.
4. Find an equation for each of the graphs of Trials 2 and 3.
5. How do the equations for Trials 1, 2, and 3 compare? Which graph is widest and which is most narrow? Explain what this represents in the context of the situation. How is this represented in the equations?
6. What do the x-intercepts and vertex of each graph represent?
7. Why were the values of h and k different in each trial?

Trial	Vertex (h, k)	Point (x, y)	Equation
1			
2			
3			

Ed-Imaging

CHAPTER 4

Study Guide and Review

Study Guide

KeyConcepts

Graphing Quadratic Functions (Lesson 4-1)

- The graph of $y = ax^2 + bx + c$, $a \neq 0$, opens up, and the function has a minimum value when $a > 0$. The graph opens down, and the function has a maximum value when $a < 0$.

Solving Quadratic Equations (Lessons 4-2 and 4-3)

- Roots of a quadratic equation are the zeros of the related quadratic function. You can find the zeros of a quadratic function by finding the x-intercepts of the graph.

Complex Numbers (Lesson 4-4)

- i is the imaginary unit; $i^2 = -1$ and $i = \sqrt{-1}$.

Solving Quadratic Equations (Lessons 4-5 and 4-6)

- Completing the square: **Step 1** Find one half of b, the coefficient of x. **Step 2** Square the result in Step 1. **Step 3** Add the result of Step 2 to $x^2 + bx$.
- Quadratic Formula: $x = \frac{-b \pm \sqrt{b^2 - 4ac}}{2a}$

Transformations of Quadratic Graphs (Lesson 4-7)

- The graph of $y = (x - h)^2 + k$ is the graph of $y = x^2$ translated $|h|$ units left if h is negative or h units right if h is positive and k units up if k is positive or $|k|$ units down if k is negative.
- Consider $y = a(x - h)^2 + k$, $a \neq 0$. If $a > 0$, the graph opens up; if $a < 0$ the graph opens down. If $|a| > 1$, the graph is narrower than the graph of $y = x^2$. If $|a| < 1$, the graph is wider than the graph of $y = x^2$.

Quadratic Inequalities (Lesson 4-8)

- Graph the related function, test a point *not* on the parabola and determine if it is a solution, and shade accordingly.

FOLDABLES StudyOrganizer

Be sure the Key Concepts are noted in your Foldable.

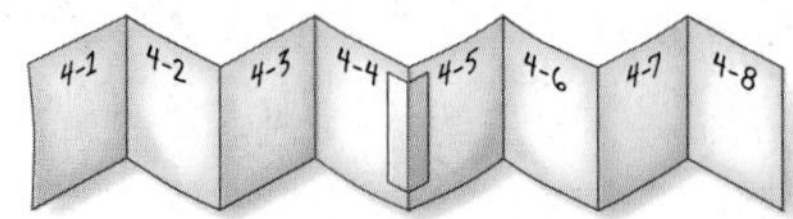

KeyVocabulary

axis of symmetry (p. 220)
completing the square (p. 257)
complex conjugates (p. 249)
complex number (p. 247)
constant term (p. 219)
discriminant (p. 267)
factored form (p. 238)
FOIL method (p. 238)
imaginary unit (p. 246)
linear term (p. 219)
maximum value (p. 222)
minimum value (p. 222)
parabola (p. 219)
pure imaginary number (p. 246)
quadratic equation (p. 229)
Quadratic Formula (p. 264)
quadratic function (p. 219)
quadratic inequality (p. 282)
quadratic term (p. 219)
root (p. 229)
Square Root Property (p. 247)
standard form (p. 229)
vertex (p. 220)
vertex form (p. 275)
zero (p. 229)

VocabularyCheck

State whether each sentence is *true* or *false*. If *false*, replace the underlined term to make a true sentence.

1. The <u>factored form</u> of a quadratic equation is $ax^2 + bx + c = 0$, where $a \neq 0$ and a, b, and c are integers.

2. The graph of a quadratic function is called a <u>parabola</u>.

3. The <u>vertex form</u> of a quadratic function is $y = a(x - p)(x - q)$.

4. The axis of symmetry will intersect a parabola in one point called the <u>vertex</u>.

5. A method called <u>FOIL method</u> is used to make a quadratic expression a perfect square in order to solve the related equation.

6. The equation $x = \frac{-b \pm \sqrt{b^2 - 4ac}}{2a}$ is known as the <u>discriminant</u>.

7. The number $6i$ is called a <u>pure imaginary number</u>.

8. The two numbers $2 + 3i$ and $2 - 3i$ are called <u>complex conjugates</u>.

Lesson-by-Lesson Review

4-1 Graphing Quadratic Functions

Complete parts a–c for each quadratic function.

a. Find the y-intercept, the equation of the axis of symmetry, and the x-coordinate of the vertex.

b. Make a table of values that includes the vertex.

c. Use this information to graph the function.

9. $f(x) = x^2 + 5x + 12$ **10.** $f(x) = x^2 - 7x + 15$

11. $f(x) = -2x^2 + 9x - 5$ **12.** $f(x) = -3x^2 + 12x - 1$

Determine whether each function has a maximum or minimum value and find the maximum or minimum value. Then state the domain and range of the function.

13. $f(x) = -x^2 + 3x - 1$ **14.** $f(x) = -3x^2 - 4x + 5$

15. BUSINESS Sal's Shirt Store sells 100 T-shirts per week at a rate of $10 per shirt. Sal estimates that he will sell 5 fewer shirts for each $1 increase in price. What price will maximize Sal's T-shirt income?

Example 1

Consider the quadratic function $f(x) = x^2 - 4x + 11$. Find the y-intercept, the equation for the axis of symmetry, and the x-coordinate of the vertex.

In the function, $a = 1$, $b = -4$, and $c = 11$. The y-intercept is $c = 11$.

Use a and b to find the equation of the axis of symmetry.

$x = -\frac{b}{2a}$ Equation of the axis of symmetry

$= -\frac{-4}{2(1)}$ $a = 1$ and $b = -4$

$= 2$ Simplify.

The equation of the axis of symmetry is $x = 2$. Therefore, the x-coordinate of the vertex is 2.

4-2 Solving Quadratic Equations by Graphing

Solve each equation by graphing. If exact roots cannot be found, state the consecutive integers between which the roots are located.

16. $x^2 - x - 20 = 0$

17. $2x^2 - x - 3 = 0$

18. $4x^2 - 6x - 15 = 0$

19. BASEBALL A baseball is hit upward at 120 feet per second. Use the formula $h(t) = v_0t - 16t^2$, where $h(t)$ is the height of an object in feet, v_0 is the object's initial velocity in feet per second, and t is the time in seconds. Ignoring the height of the ball when it was hit, how long does it take for the ball to hit the ground?

Example 2

Solve $2x^2 - 7x + 3 = 0$ by graphing.

The equation of the axis of symmetry is $-\frac{-7}{2(2)}$ or $x = \frac{7}{4}$.

x	0	1	$\frac{7}{4}$	2	3
$f(x)$	3	−2	$-2\frac{5}{8}$	−3	0

The zeros of the related function are $\frac{1}{2}$ and 3. Therefore, the solutions of the equation are $\frac{1}{2}$ and 3.

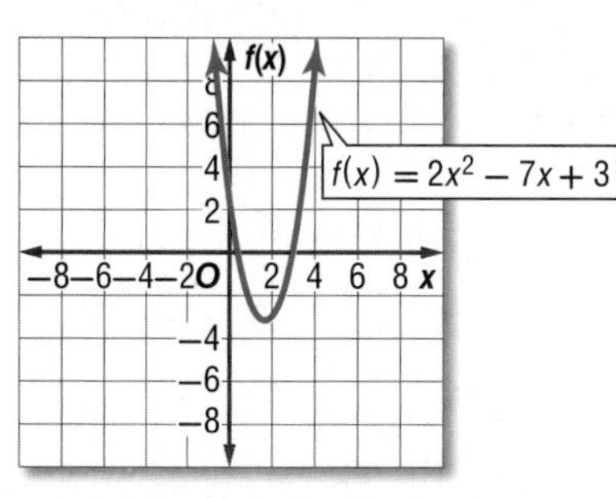

Study Guide and Review *Continued*

4-3 Solving Quadratic Equations by Factoring

Write a quadratic equation in standard form with the given roots.

20. 5, 6

21. $-3, -7$

22. $-4, 2$

23. $-\frac{2}{3}, 1$

24. $\frac{1}{6}, 5$

25. $-\frac{1}{4}, -1$

Solve each equation by factoring.

26. $2x^2 - 2x - 24 = 0$

27. $2x^2 - 5x - 3 = 0$

28. $3x^2 - 16x + 5 = 0$

29. Find x and the dimensions of the rectangle below.

$A = 126$ ft^2

$x - 3$

$x + 2$

Example 3

Write a quadratic equation in standard form with $-\frac{1}{2}$ and 4 as its roots.

$(x - p)(x - q) = 0$ Write the pattern.

$\left[x - \left(-\frac{1}{2}\right)\right](x - 4) = 0$ Replace p with $-\frac{1}{2}$ and q with 4.

$\left(x + \frac{1}{2}\right)(x - 4) = 0$ Simplify.

$x^2 - \frac{7}{2}x - 2 = 0$ Multiply.

$2x^2 - 7x - 4 = 0$ Multiply each side by 2 so that b and c are integers.

Example 4

Solve $2x^2 - 3x - 5 = 0$ by factoring.

$2x^2 - 3x - 5 = 0$ Original equation

$(2x - 5)(x + 1) = 0$ Factor the trinomial.

$2x - 5 = 0$ or $x + 1 = 0$ Zero Product Property

$x = \frac{5}{2}$ $\quad x = -1$

The solution set is $\left\{-1, \frac{5}{2}\right\}$ or $\left\{x \mid x = -1, \frac{5}{2}\right\}$.

4-4 Complex Numbers

Simplify.

30. $\sqrt{-8}$

31. $(2 - i) + (13 + 4i)$

32. $(6 + 2i) - (4 - 3i)$

33. $(6 + 5i)(3 - 2i)$

34. **ELECTRICITY** The impedance in one part of a series circuit is $3 + 2j$ ohms, and the impedance in the other part of the circuit is $4 - 3j$ ohms. Add these complex numbers to find the total impedance in the circuit.

Solve each equation.

35. $2x^2 + 50 = 0$

36. $4x^2 + 16 = 0$

37. $3x^2 + 15 = 0$

38. $8x^2 + 16 = 0$

39. $4x^2 + 1 = 0$

Example 5

Simplify $(12 + 3i) - (-5 + 2i)$.

$(12 + 3i) - (-5 + 2i)$

$= [12 - (-5)] + (3 - 2)i$ Group the real and imaginary parts.

$= 17 + i$ Simplify.

Example 6

Solve $3x^2 + 12 = 0$.

$3x^2 + 12 = 0$ Original equation

$3x^2 = -12$ Subtract 12 from each side.

$x^2 = -4$ Divide each side by 3.

$x = \pm\sqrt{-4}$ Square Root Property

$x = \pm 2i$ $\quad \sqrt{-4} = \sqrt{4} \cdot \sqrt{-1}$

4-5 Completing the Square

Find the value of c that makes each trinomial a perfect square. Then write the trinomial as a perfect square.

40. $x^2 + 18x + c$

41. $x^2 - 4x + c$

42. $x^2 - 7x + c$

43. $x^2 + 2.4x + c$

44. $x^2 - \frac{1}{2}x + c$

45. $x^2 + \frac{6}{5}x + c$

Solve each equation by completing the square.

46. $x^2 - 6x - 7 = 0$

47. $x^2 - 2x + 8 = 0$

48. $2x^2 + 4x - 3 = 0$

49. $2x^2 + 3x - 5 = 0$

50. **FLOOR PLAN** Mario's living room has a length 6 feet wider than the width. The area of the living room is 280 square feet. What are the dimensions of his living room?

Example 7

Find the value of c that makes $x^2 + 14x + c$ a perfect square. Then write the trinomial as a perfect square.

Step 1 Find one half of 14.

Step 2 Square the result of Step 1.

Step 3 Add the result of Step 2 to $x^2 + 14x$.

The trinomial $x^2 + 14x + 49$ can be written as $(x + 7)^2$.

Example 8

Solve $x^2 + 12x - 13 = 0$ by completing the square.

$$x^2 + 12x - 13 = 0$$
$$x^2 + 12x = 13$$
$$x^2 + 12x + 36 = 13 + 36$$
$$(x + 6)^2 = 49$$
$$x + 6 = \pm 7$$
$$x + 6 = 7 \quad \text{or} \quad x + 6 = -7$$
$$x = 1 \qquad\qquad x = -13$$

The solution set is $\{-13, 1\}$ or $\{x \mid x = -13, 1\}$.

4-6 The Quadratic Formula and the Discriminant

Complete parts **a–c** for each quadratic equation.

a. Find the value of the discriminant.

b. Describe the number and type of roots.

c. Find the exact solutions by using the Quadratic Formula.

51. $x^2 - 10x + 25 = 0$

52. $x^2 + 4x - 32 = 0$

53. $2x^2 + 3x - 18 = 0$

54. $2x^2 + 19x - 33 = 0$

55. $x^2 - 2x + 9 = 0$

56. $4x^2 - 4x + 1 = 0$

57. $2x^2 + 5x + 9 = 0$

58. **PHYSICAL SCIENCE** Lauren throws a ball with an initial velocity of 40 feet per second. The equation for the height of the ball is $h = -16t^2 + 40t + 5$, where h represents the height in feet and t represents the time in seconds. When will the ball hit the ground?

Example 9

Solve $x^2 - 4x - 45 = 0$ by using the Quadratic Formula.

In $x^2 - 4x - 45 = 0$, $a = 1$, $b = -4$, and $c = -45$.

$$x = \frac{-b \pm \sqrt{b^2 - 4ac}}{2a} \qquad \text{Quadratic Formula}$$
$$= \frac{-(-4) \pm \sqrt{(-4)^2 - 4(1)(-45)}}{2(1)}$$
$$= \frac{4 \pm 14}{2}$$

Write as two equations.

$$x = \frac{4 + 14}{2} \quad \text{or} \quad x = \frac{4 - 14}{2}$$
$$= 9 \qquad\qquad = -5$$

The solution set is $\{-5, 9\}$ or $\{x \mid x = -5, 9\}$.

CHAPTER 4 Study Guide and Review *Continued*

4-7 Transformations of Quadratic Graphs

Write each quadratic function in vertex form, if not already in that form. Then identify the vertex, axis of symmetry, and direction of opening. Then graph the function.

59. $y = -3(x - 1)^2 + 5$

60. $y = 2x^2 + 12x - 8$

61. $y = -\frac{1}{2}x^2 - 2x + 12$

62. $y = 3x^2 + 36x + 25$

63. The graph at the right shows a product of 2 numbers with a sum of 10. Find a function that models this product and use it to determine the two numbers that would give a maximum product.

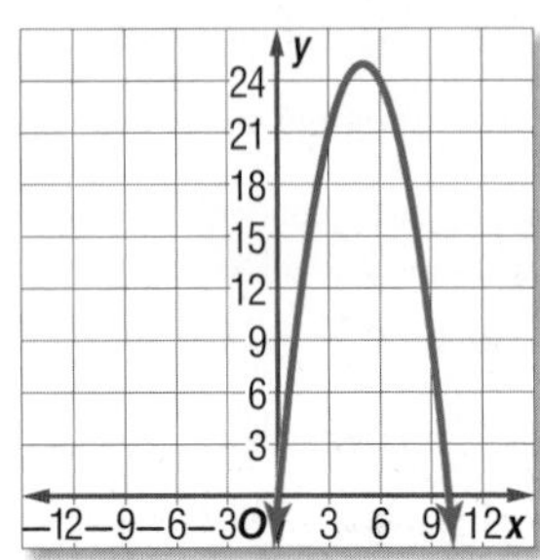

Example 10

Write the quadratic function $y = 3x^2 + 24x + 15$ in vertex form. Then identify the vertex, axis of symmetry, and direction of opening.

$y = 3x^2 + 24x + 15$ — Original equation

$y = 3(x^2 + 8x) + 15$ — Group and factor.

$y = 3(x^2 + 8x + 16) + 15 - 3(16)$ — Complete the square.

$y = 3(x + 4)^2 - 33$ — Rewrite $x^2 + 8x + 16$ as a perfect square.

So, $a = 3$, $h = -4$, and $k = -33$. The vertex is at $(-4, -33)$ and the axis of symmetry is $x = -4$. Since a is positive, the graph opens up.

4-8 Quadratic Inequalities

Graph each quadratic inequality.

64. $y \geq x^2 + 5x + 4$

65. $y < -x^2 + 5x - 6$

66. $y > x^2 - 6x + 8$

67. $y \leq x^2 + 10x - 4$

68. Solomon wants to put a deck along two sides of his garden. The deck width will be the same on both sides and the total area of the garden and deck cannot exceed 500 square feet. How wide can the deck be?

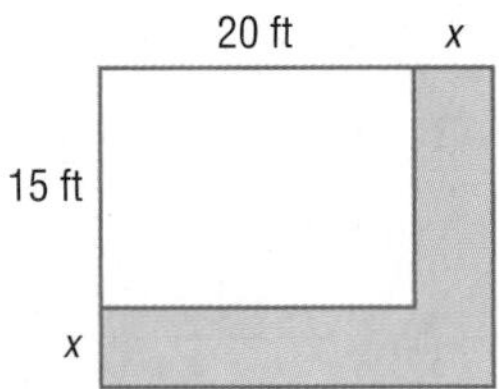

Solve each inequality using a graph or algebraically.

69. $x^2 + 8x + 12 > 0$

70. $6x + x^2 \geq -9$

71. $2x^2 + 3x - 20 > 0$

72. $4x^2 - 3 < -5x$

73. $3x^2 + 4 > 8x$

Example 11

Graph $y > x^2 + 3x + 2$.

Step 1 Graph the related function, $y > x^2 + 3x + 2$. Because the inequality symbol $>$ is used, the parabola should be dashed.

Step 2 Test a point not on the graph of the parabola such as (0, 0).

$y > x^2 + 3x + 2$

$(0) \stackrel{?}{>} (0)^2 + 3(0) + 2$

$0 \not> 2$

So, (0, 0) is not a solution of the inequality.

Step 3 Shade the region that does not contain the point (0, 0).

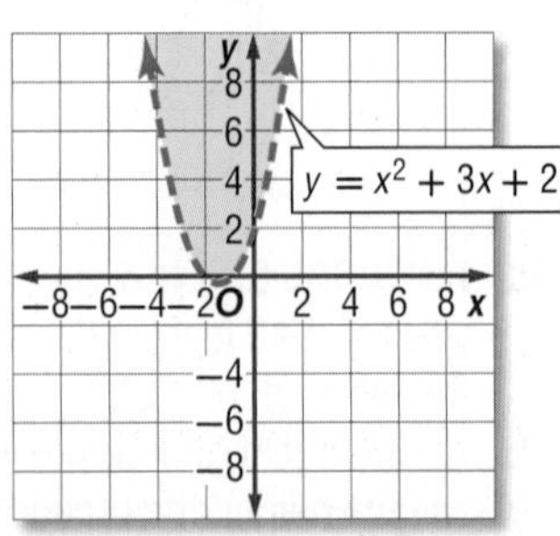

CHAPTER 4 Practice Test

Complete parts *a–c* for each quadratic function.

a. **Find the *y*-intercept, the equation of the axis of symmetry, and the *x*-coordinate of the vertex.**

b. **Make a table of values that includes the vertex.**

c. **Use this information to graph the function.**

1. $f(x) = x^2 + 4x - 7$

2. $f(x) = -2x^2 + 5x$

3. $f(x) = -x^2 - 6x - 9$

Determine whether each function has a maximum or minimum value. State the maximum or minimum value of each function.

4. $f(x) = x^2 + 10x + 25$ **5.** $f(x) = -x^2 + 6x$

Solve each equation using the method of your choice. Find exact solutions.

6. $x^2 - 8x - 9 = 0$

7. $-4.8x^2 + 1.6x + 24 = 0$

8. $12x^2 + 15x - 4 = 0$

9. $x^2 - 7x - \frac{17}{4} = 0$

10. $4x^2 + x = 3$

11. $-9x^2 + 40x + 84 = 0$

12. **PHYSICAL SCIENCE** Parker throws a ball off the top of a building. The building is 350 feet high and the initial velocity of the ball is 30 feet per second. Find out how long it will take the ball to hit the ground by solving the equation $-16t^2 - 30t + 350 = 0$.

13. **MULTIPLE CHOICE** Which equation below has roots at -6 and $\frac{1}{5}$?

A $0 = 5x^2 - 29x - 6$

B $0 = 5x^2 + 31x + 6$

C $0 = 5x^2 + 29x - 6$

D $0 = 5x^2 - 31x + 6$

14. **PHYSICS** A ball is thrown into the air vertically with a velocity of 112 feet per second. The ball was released 6 feet above the ground. The height above the ground t seconds after release is modeled by $h(t) = -16t^2 + 112t + 6$.

a. When will the ball reach 130 feet?

b. Will the ball ever reach 250 feet? Explain.

c. In how many seconds after its release will the ball hit the ground?

15. The rectangle below has an area of 104 square inches. Find the value of x and the dimensions of the rectangle.

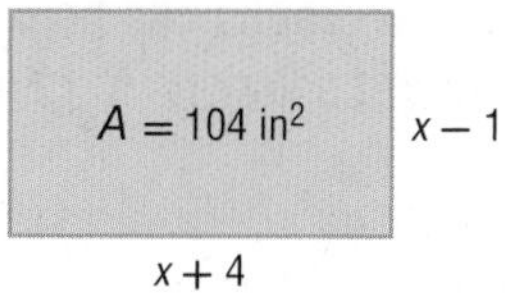

Simplify.

16. $(3 - 4i) - (9 - 5i)$

17. $\frac{4i}{4 - i}$

18. **MULTIPLE CHOICE** Which value of c makes the trinomial $x^2 - 12x + c$ a perfect square trinomial?

F 6

G 12

H 36

J 144

Complete parts *a–c* for each quadratic equation.

a. **Find the value of the discriminant.**

b. **Describe the number and type of roots.**

c. **Find the exact solution by using the Quadratic Formula.**

19. $6x^2 + 7x = 0$

20. $5x^2 = -6x + 1$

21. $2x^2 + 5x - 8 = -13$

Write each quadratic function in vertex form. Then identify the vertex, axis of symmetry, and direction of opening.

22. $3x^2 + 6x = 2 + y$

23. $x^2 + 9x + \frac{81}{4} = y$

24. Graph the quadratic inequality $0 < -3x^2 + 4x + 10$.

Solve each inequality by using a graph or algebraically.

25. $x^2 + 6x > -5$

26. $4x^2 - 19x \leq -12$

Preparing for Standardized Tests

Use a Graph

Using a graph can help you solve many different kinds of problems on standardized tests. Graphs can help you solve equations, evaluate functions, and interpret solutions to real-world problems.

Strategies for Using a Graph

Step 1

Read the problem statement carefully.

Ask yourself:

- What am I being asked to solve?
- What information is given in the problem?
- How could a graph help me solve the problem?

Step 2

Create your graph.

- Sketch your graph on scrap paper if appropriate.
- If allowed, you can also use a graphing calculator to create the graph.

Step 3

Solve the problem.

- Use your graph to help you model and solve the problem.
- Check to be sure your answer makes sense.

Standardized Test Example

Read the problem. Identify what you need to know. Then use the information in the problem to solve.

The students in Mr. Himebaugh's physics class built a model rocket. The rocket is launched in a large field with an initial upward velocity of 128 feet per second. The function $h(t) = -16t^2 + 128t$ models the height of the rocket above the ground (in feet) t seconds after it is launched. How long will it take for the rocket to reach its maximum height?

A 4 seconds

B 5 seconds

C 6 seconds

D 8 seconds

JUPITERIMAGES/Polka Dot/Alamy

Graphing the quadratic function will allow you to determine the peak height of the rocket and when it occurs. A graphing calculator can help you quickly graph the function and analyze it.

KEYSTROKES: [Y=] [(–)] 16 [X,T,θ,n] [x^2] [+] 128 [X,T,θ,n] [GRAPH]

After graphing the equation, use **maximum** under the **CALC** menu.

Press [2nd] **[CALC]** 4. Then use [◀] to place the cursor to the left of the maximum point and press [ENTER]. Use [▶] to place the cursor to the right of the maximum point and press [ENTER] [ENTER].

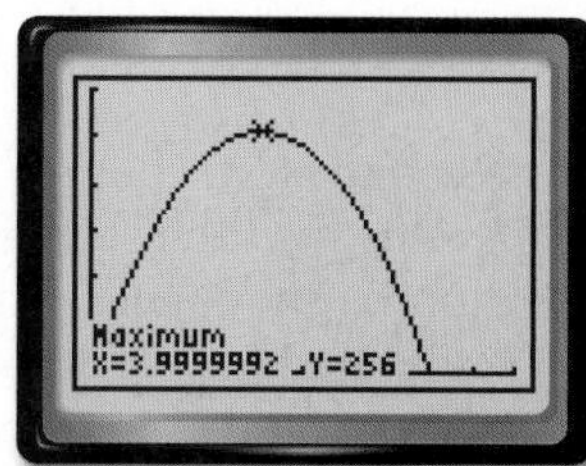

[0, 10] scl: 1 by [0, 300] scl: 50

The graph shows that the rocket takes 4 seconds to reach its maximum height of 256 feet. The correct answer is A.

Exercises

Read each problem. Identify what you need to know. Then use the information in the problem to solve.

1. What are the roots of $y = 2x^2 + 10x - 48$?

A $-5, 4$

B $-6, 1$

C $-8, 3$

D $2, 3$

2. How many times does the graph of $f(x) = 2x^2 - 3x + 2$ cross the x-axis?

F 0 **H** 2

G 1 **J** 3

3. Which statement best describes the graphs of the two equations?

$$16x - 2y = 24$$
$$12x = 3y - 36$$

A The lines are parallel.

B The lines are the same.

C The lines intersect in only one point.

D The lines intersect in more than one point, but are not the same.

4. Adrian is using 120 feet of fencing to enclose a rectangular area for her puppy. One side of the enclosure will be her house.

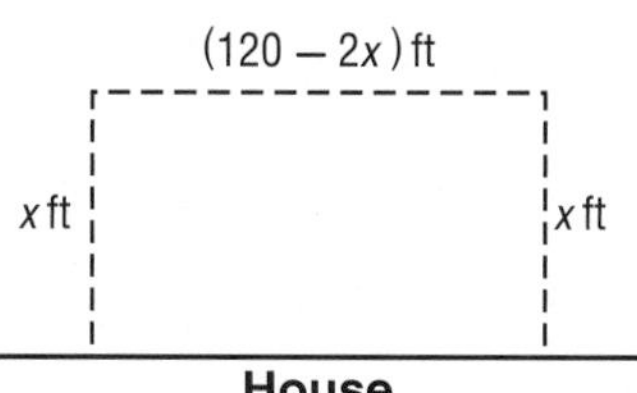

The function $f(x) = x(120 - 2x)$ represents the area of the enclosure. What is the greatest area Adrian can enclose with the fencing?

F 1650 ft^2 **H** 1980 ft^2

G 1800 ft^2 **J** 2140 ft^2

5. For which equation is the x-coordinate of the vertex at 4?

A $f(x) = x^2 - 8x + 15$ **C** $f(x) = x^2 + 6x + 8$

B $f(x) = -x^2 - 4x + 12$ **D** $f(x) = -x^2 - 2x + 2$

6. For what value of x does $f(x) = x^2 + 5x + 6$ reach its minimum value?

F -5 **H** $-\frac{5}{2}$

G -3 **J** -2

CHAPTER 4

Standardized Test Practice

Cumulative, Chapters 1 through 4

Multiple Choice

Read each question. Then fill in the correct answer on the answer document provided by your teacher or on a sheet of paper.

1. A rental store charges \$32 per day to rent a bicycle. At this rate, the store rents about 200 bikes per month. The owner of the store estimates that they will rent 5 fewer bikes per month for each \$2 increase in the rental price. What price will maximize the income of the store?

A \$46 **C** \$55

B \$51 **D** \$56

2. What are the solutions of the quadratic equation graphed below?

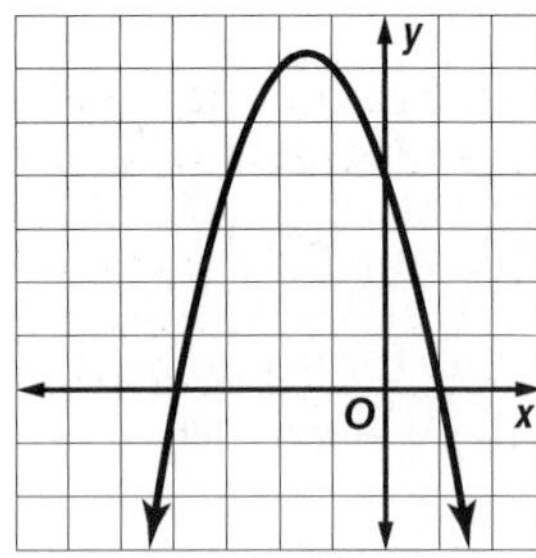

F $-4, -1$ **H** $-1, 4$

G $1, 4$ **J** $-4, 1$

3. Peyton works as a nanny. She charges at least \$10 to drive to a home and \$10.50 per hour. Which inequality best represents the relationship between the number of hours working n and the total charge c?

A $c \geq 10 + 10.50n$

B $c \geq 10.50 + 10n$

C $c \leq 10.50 + 10$

D $c \leq 10n + 10.50n$

> **Test-TakingTip**
>
> Question 1 Multiply the expressions for the new price and new number of customers after x price increases to write a quadratic equation.

4. Leo sells T-shirts at a local swim meet. It costs him \$250 to set up the stand and rent the machine. It costs him an additional \$5 to make each T-shirt. If he sells each T-shirt for \$15, how many T-shirts does he have to sell before he can make a profit?

F 10 **H** 25

G 15 **J** 50

5. Solve $x^2 - 2x = 15$ by completing the square.

A $-4, -1$ **C** $-2, 3$

B $-3, 5$ **D** $5, 7$

6. The graph of $g(x) = \frac{2}{5}x^2 - 4x + 2$ is translated down 5 units to produce the graph of the function $h(x)$. Which of the following could be $h(x)$?

F $h(x) = \frac{2}{5}x^2 - 4x + 7$

G $h(x) = \frac{2}{5}x^2 - 4x - 3$

H $h(x) = \frac{2}{5}x^2 - 9x + 2$

J $h(x) = \frac{2}{5}x^2 + x + 2$

7. Which of the following points is ***not*** a vertex of the feasible region for the system of linear inequalities below?

$$x \geq 0, y \geq 0$$
$$y \leq -2x + 6$$

A $(0, 0)$ **B** $(0, 3)$ **C** $(0, 6)$ **D** $(3, 0)$

8. The function $P(t) = -0.068t^2 + 7.85t + 56$ can be used to approximate the population, in thousands, of Clarksville between 1960 and 2000. The domain t of the function is the number of years since 1960. According to the model, in what year did the population of Clarksville reach 200,000 people?

F 1974 **H** 1981

G 1977 **J** 1983

Short Response/Gridded Response

Record your answers on the answer sheet provided by your teacher or on a sheet of paper.

9. GRIDDED RESPONSE What is the y-coordinate of the solution of the system of equations below?

$$y = 4x - 7$$
$$y = -\frac{1}{2}x + 2$$

10. GRIDDED RESPONSE Simplify $-2i \cdot 5i$.

11. Use the quadratic function below to answer each question.

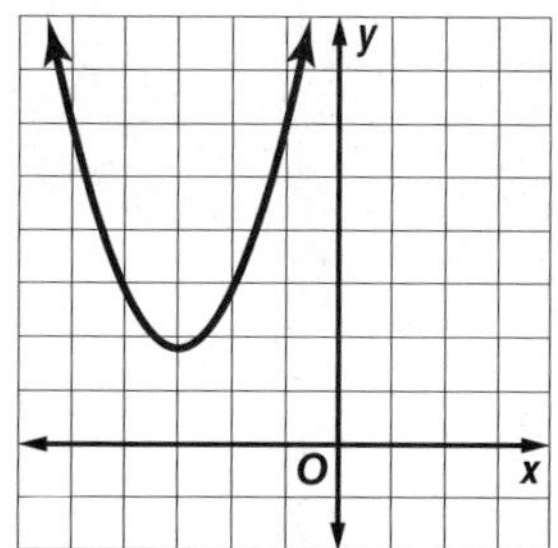

a. How many real roots does the quadratic function have?

b. How many complex roots does the function have?

c. What do you know about the discriminant of the quadratic equation? Explain.

12. If one of the roots of $x^2 + kx - 12 = 0$ is 4, what is the value of k?

13. Describe the translation of the graph of $y = (x + 5)^2 - 1$ to the graph of $y = (x - 1)^2 + 3$.

Extended Response

Record your answers on a sheet of paper. Show your work.

14. For a given quadratic equation $y = ax^2 + bx + c$, describe what the discriminant $b^2 - 4ac$ tells you about the roots of the equation.

15. Craig is checking in a shipment of ink jet printers that cost \$200 each and laser printers that cost \$500 each. There are 40 boxes in the shipment, and the invoice total is \$11,600.

a. Write a system of equations to model the situation. Let x represent the number of ink jet printers, and let y represent the number of laser printers.

b. Write a matrix equation that can be used to solve the system of equations you wrote in part **a**.

c. Find the inverse of the coefficient matrix and solve the matrix equation. How many ink jet printers and laser printers were included in the shipment?

16. If Robert kicks a football straight up into the air with an initial velocity of 100 feet per second, the function $h(t) = -16t^2 + 100t$ gives the height, in feet, of the ball after t seconds.

a. Graph the quadratic function on a coordinate grid.

b. What is the maximum height reached by the football? Round to the nearest foot.

c. How long is the football in the air before it hits the ground?

Need ExtraHelp?

If you missed Question...	1	2	3	4	5	6	7	8	9	10	11	12	13	14	15	16
Go to Lesson...	4-1	4-2	2-4	1-3	4-5	4-7	3-3	4-6	3-1	4-4	4-2	4-2	4-7	4-6	3-8	4-1

CHAPTER 5

Polynomials and Polynomial Functions

Then

You graphed quadratic functions and solved quadratic equations.

Now

You will:

- Add, subtract, multiply, divide, and factor polynomials.
- Analyze and graph polynomial functions.
- Evaluate polynomial functions and solve polynomial equations.
- Find factors and zeros of polynomial functions.

Why? ▲

TRANSPORTATION Polynomial functions can be used to determine bus schedules, highway capacity, traffic patterns, average fuel costs, and the prices of new and used cars.

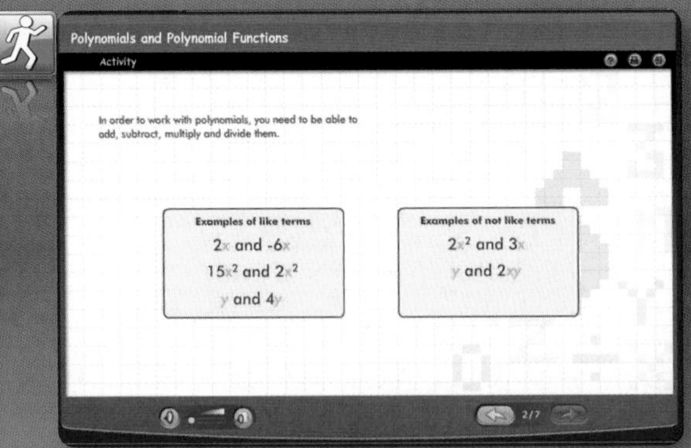

connectED.mcgraw-hill.com Your Digital Math Portal

Animation 
Vocabulary 
eGlossary 
Personal Tutor 
Virtual Manipulatives 
Graphing Calculator 
Audio 
Foldables 
Self-Check Practice 
Worksheets 

Get Ready for the Chapter

Diagnose Readiness | You have two options for checking prerequisite skills.

1 Textbook Option Take the Quick Check below. Refer to the Quick Review for help.

QuickCheck

Rewrite each difference as a sum.

1. $-5 - 13$
2. $5 - 3y$
3. $5mr - 7mp$
4. $3x^2y - 14xy^2$
5. **PARTIES** Twenty people attended a going away party for Zach. The guests left in groups of 2. By 9:00, x groups had left. Rewrite the number of guests remaining at 9:00 as a sum.

Use the Distributive Property to rewrite each expression without parentheses.

6. $-4(a + 5)$
7. $-1(3b^2 + 2b - 1)$
8. $-\frac{1}{2}(2m - 5)$
9. $-\frac{3}{4}(3z + 5)$
10. **MONEY** Mr. Chávez is buying pizza and soda for the members of the science club. A slice of pizza costs \$2.25, and a soda costs \$1.25. Write an expression to represent the amount that Mr. Chávez will spend on 15 students. Evaluate the expression by using the Distributive Property.

Solve each equation.

11. $x^2 + 2x - 8 = 0$
12. $2x^2 + 7x + 3 = 0$
13. $6x^2 + 5x - 4 = 0$
14. $4x^2 - 2x - 1 = 0$
15. **PHYSICS** If an object is dropped from a height of 50 feet above the ground, then its height after t seconds is given by $h = -16t^2 + 50$. Use the equation $0 = -16t^2 + 50$ to find how long it will take until the ball reaches the ground.

QuickReview

Example 1

Rewrite $2xy - 3 - z$ as a sum.

$2xy - 3 - z$ Original expression

$= 2xy + (-3) + (-z)$ Rewrite using addition.

Example 2

Use the Distributive Property to rewrite $-3(a + b - c)$.

$-3(a + b - c)$ Original expression

$= -3(a) + (-3)(b) + (-3)(-c)$ Distributive Property

$= -3a - 3b + 3c$ Simplify.

Example 3

Solve $2x^2 + 8x + 1 = 0$.

$x = \frac{-b \pm \sqrt{b^2 - 4ac}}{2a}$ Quadratic Formula

$= \frac{-8 \pm \sqrt{8^2 - 4(2)(1)}}{2(2)}$ $a = 2, b = 8, c = 1$

$= \frac{-8 \pm \sqrt{56}}{4}$ Simplify.

$= -2 \pm \frac{\sqrt{14}}{2}$ $\sqrt{56} = \sqrt{4 \cdot 14}$ or $2\sqrt{14}$

The exact solutions are $-2 + \frac{\sqrt{14}}{2}$ and $-2 - \frac{\sqrt{14}}{2}$.

The approximate solutions are -0.13 and -3.87.

2 Online Option Take an online self-check Chapter Readiness Quiz at connectED.mcgraw-hill.com.

Get Started on the Chapter

You will learn several new concepts, skills, and vocabulary terms as you study Chapter 5. To get ready, identify important terms and organize your resources. You may wish to refer to Chapter 0 to review prerequisite skills.

FOLDABLES® StudyOrganizer

Polynomials and Polynomial Functions Make this Foldable to help you organize your Chapter 5 notes about polynomials and polynomial functions. Begin with one sheet of $8\frac{1}{2}$" by 14" paper.

1. **Fold** a 2" tab along the bottom of a long side.

2. **Fold** along the width into thirds.

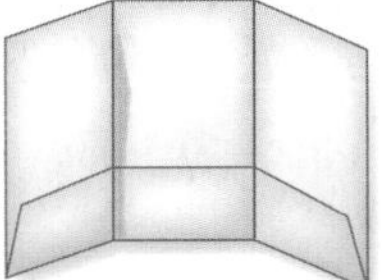

3. **Staple** the outer edges of the tab.

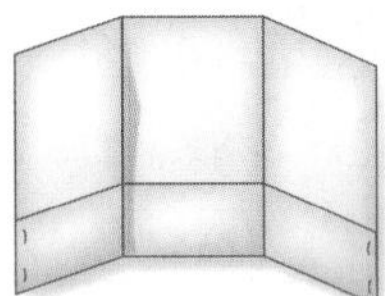

4. **Label** the tabs *Polynomials, Polynomial Functions and Graphs,* and *Solving Polynomial Equations.*

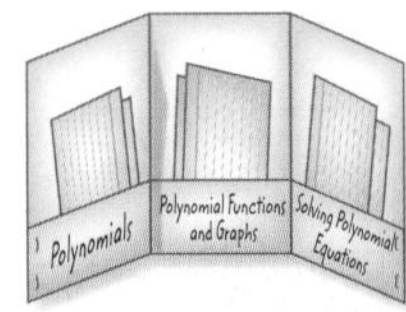

NewVocabulary

English		Español
simplify	p. 303	reducer
degree of a polynomial	p. 305	grado de un polinomio
synthetic division	p. 312	división sintética
polynomial in one variable	p. 322	polinomio de una variable
leading coefficient	p. 322	coeficiente líder
polynomial function	p. 323	función polinomial
power function	p. 323	función potencia
end behavior	p. 324	comportamiento final
relative maximum	p. 331	máximo relativo
relative minimum	p. 331	mínimo relativo
extrema	p. 331	extrema
turning points	p. 331	momentos cruciales
prime polynomials	p. 342	polinomios primeros
quadratic form	p. 345	forma de ecuación cuadrática
synthetic substitution	p. 352	sustitución sintética
depressed polynomial	p. 354	polinomio reducido

ReviewVocabulary

factoring factorización to express a polynomial as the product of monomials and polynomials

function función a relation in which each element of the domain is paired with exactly one element in the range

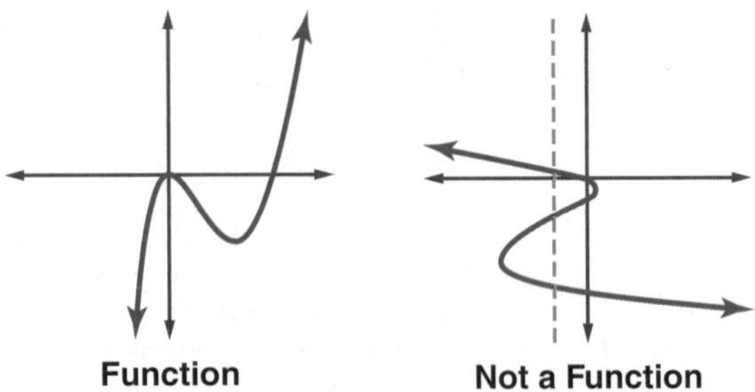

polynomial polinomio a monomial or sum of monomials

LESSON 5-1 Operations with Polynomials

Then	Now	Why?
You evaluated powers.	**1** Multiply, divide, and simplify monomials and expressions involving powers. **2** Add, subtract, and multiply polynomials.	The light from the Sun takes approximately 8 minutes to reach Earth. So if you are outside right now you are basking in sunlight that the Sun emitted approximately 8 minutes ago. Light travels very fast, at a speed of about 3×10^8 meters per second. How long would it take light to get here from the Andromeda galaxy, which is approximately 2.367×10^{21} meters away?

NewVocabulary
simplify
degree of a polynomial

Common Core State Standards

Content Standards
A.APR.1 Understand that polynomials form a system analogous to the integers, namely, they are closed under the operations of addition, subtraction, and multiplication; add, subtract, and multiply polynomials.

Mathematical Practices
2 Reason abstractly and quantitatively.

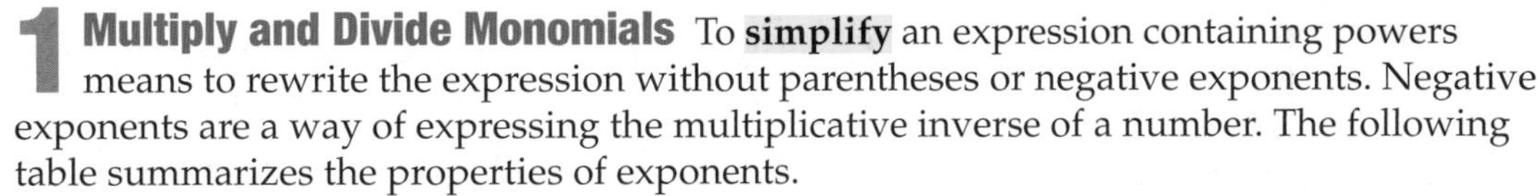

1 Multiply and Divide Monomials

To **simplify** an expression containing powers means to rewrite the expression without parentheses or negative exponents. Negative exponents are a way of expressing the multiplicative inverse of a number. The following table summarizes the properties of exponents.

ConceptSummary Properties of Exponents

For any real numbers x and y, integers a and b:

Property	Definition	Examples
Product of Powers	$x^a \cdot x^b = x^{a+b}$	$3^2 \cdot 3^4 = 3^{2+4}$ or 3^6 $p^2 \cdot p^9 = p^{2+9}$ or p^{11}
Quotient of Powers	$\frac{x^a}{x^b} = x^{a-b}, x \neq 0$	$\frac{9^5}{9^2} = 9^{5-2}$ or 9^3 $\frac{b^6}{b^4} = b^{6-4}$ or b^2
Negative Exponent	$x^{-a} = \frac{1}{x^a}$ and $\frac{1}{x^{-a}} = x^a, x \neq 0$	$3^{-5} = \frac{1}{3^5}$ $\frac{1}{b^{-7}} = b^7$
Power of a Power	$(x^a)^b = x^{ab}$	$(3^3)^2 = 3^{3 \cdot 2}$ or 3^6 $(d^2)^4 = d^{2 \cdot 4}$ or d^8
Power of a Product	$(xy)^a = x^a y^a$	$(2k)^4 = 2^4k^4$ or $16k^4$ $(ab)^3 = a^3b^3$
Power of a Quotient	$\left(\frac{x}{y}\right)^a = \frac{x^a}{y^a}, y \neq 0$, and $\left(\frac{x}{y}\right)^{-a} = \left(\frac{y}{x}\right)^a$ or $\frac{y^a}{x^a}, x \neq 0, y \neq 0$	$\left(\frac{x}{y}\right)^2 = \frac{x^2}{y^2}$ $\left(\frac{a}{b}\right)^{-5} = \frac{b^5}{a^5}$
Zero Power	$x^0 = 1, x \neq 0$	$7^0 = 1$

Recall that a *monomial* is a number, a variable, or an expression that is the product of one or more variables with nonnegative integer exponents.

When simplifying a monomial, check to be sure that it has been simplified fully.

KeyConcept Simplifying Monomials

A monomial expression is in simplified form when:

- there are no powers of powers,
- each base appears exactly once,
- all fractions are in simplest form, and
- there are no negative exponents.

PT

Example 1 Simplify Expressions

Simplify each expression. Assume that no variable equals 0.

a. $(2a^{-2})(3a^3b^2)(c^{-2})$

$(2a^{-2})(3a^3b^2)(c^{-2})$	Original expression
$= 2\left(\frac{1}{a^2}\right)(3a^3b^2)\left(\frac{1}{c^2}\right)$	Definition of negative exponents
$= \left(\frac{2}{a \cdot a}\right)(3 \cdot a \cdot a \cdot a \cdot b \cdot b)\left(\frac{1}{c \cdot c}\right)$	Definition of exponents
$= \left(\frac{2}{\not{a} \cdot \not{a}}\right)(3 \cdot \not{a} \cdot \not{a} \cdot a \cdot b \cdot b)\left(\frac{1}{c \cdot c}\right)$	Divide out common factors.
$= \frac{6ab^2}{c^2}$	Simplify.

b. $\frac{q^2r^4}{q^7r^3}$

$\frac{q^2r^4}{q^7r^3} = q^{2-7} \cdot r^{4-3}$	Quotient of powers
$= q^{-5}r$	Subtract powers.
$= \frac{r}{q^5}$	Simplify.

Problem-SolvingTip

Check You can always check your answer using the definition of exponents.

$$\frac{q^2}{q^7} = \frac{q \cdot q}{q \cdot q \cdot q \cdot q \cdot q \cdot q \cdot q}$$
$$= \frac{1}{q^5}$$

c. $\left(\frac{-2a^4}{b^2}\right)^3$

$\left(\frac{-2a^4}{b^2}\right)^3 = \frac{(-2a^4)^3}{(b^2)^3}$	Power of a quotient
$= \frac{(-2)^3(a^4)^3}{(b^2)^3}$	Power of a product
$= \frac{-8a^{12}}{b^6}$	Power of a power

GuidedPractice

1A. $(2x^{-3}y^3)(-7x^5y^{-6})$

1B. $\frac{15c^5d^3}{-3c^2d^7}$

1C. $\left(\frac{a}{4}\right)^{-3}$

1D. $(-2x^3y^2)^5$

StudyTip

Power of 1 Remember that a variable with no exponent indicated can be written as a power of 1.

2 Operations With Polynomials

The **degree of a polynomial** is the degree of the monomial with the greatest degree.

Example 2 Degree of a Polynomial

Determine whether each expression is a polynomial. If it is a polynomial, state the degree of the polynomial.

a. $\frac{1}{4}x^4y^3 - 8x^5$

This expression is a polynomial because each term is a monomial. The degree of the first term is 4 + 3 or 7, and the degree of the second term is 5. The degree of the polynomial is 7.

b. $\sqrt{x} + x + 4$

This expression is not a polynomial because $\sqrt{x}$ is not a monomial.

c. $x^{-3} + 2x^{-2} + 6$

This expression is not a polynomial because x^{-3} and x^{-2} are not monomials: $x^{-3} = \frac{1}{x^3}$ and $x^{-2} = \frac{1}{x^2}$. Monomials cannot contain variables in the denominator.

GuidedPractice

2A. $\frac{x}{y} + 3x^2$

2B. $x^5y + 9x^4y^3 - 2xy$

You can simplify a polynomial just like you simplify a monomial. Perform the operations indicated, and combine like terms.

StudyTip

Alternative Methods Notice that Example 3a uses a horizontal method, and Example 3b uses a vertical method to simplify. Either method will yield a correct solution.

Example 3 Simplify Polynomial Expressions

Simplify each expression.

a. $(4x^2 - 5x + 6) - (2x^2 + 3x - 1)$

Remove parentheses, and group like terms together.

$(4x^2 - 5x + 6) - (2x^2 + 3x - 1)$

$= 4x^2 - 5x + 6 - 2x^2 - 3x + 1$ — Distribute the −1.

$= (4x^2 - 2x^2) + (-5x - 3x) + (6 + 1)$ — Group like terms.

$= 2x^2 - 8x + 7$ — Combine like terms.

b. $(6x^2 - 7x + 8) + (-4x^2 + 9x - 5)$

Align like terms vertically and add.

$$\begin{array}{r} 6x^2 - 7x + 8 \\ \underline{(+)\ -4x^2 + 9x - 5} \\ 2x^2 + 2x + 3 \end{array}$$

GuidedPractice

3A. $(-x^2 - 3x + 4) - (x^2 + 2x + 5)$

3B. $(3x^2 - 6) + (-x + 1)$

Adding or subtracting integers results in an integer, so the set of integers is closed under addition and subtraction. Similarly, because adding or subtracting polynomials results in a polynomial, the set of polynomials is closed under addition and subtraction.

You can use the Distributive Property to multiply polynomials.

Example 4 Simplify by Using the Distributive Property

Find $3x(2x^2 - 4x + 6)$.

$3x(2x^2 - 4x + 6) = 3x(2x^2) + 3x(-4x) + 3x(6)$ — Distributive Property

$= 6x^3 - 12x^2 + 18x$ — Multiply the monomials.

Guided Practice

Find each product.

4A. $\frac{4}{3}x^2(6x^2 + 9x - 12)$

4B. $-2a(-3a^2 - 11a + 20)$

Real-World Example 5 Write a Polynomial Expression

DRIVING The U.S. Department of Transportation limits the time a truck driver can work between periods of rest to ten hours. For the first part of his shift, Tom drives at a speed of 60 miles per hour, and for the second part of the shift, he drives at a speed of 70 miles per hour. Write a polynomial to represent the distance driven.

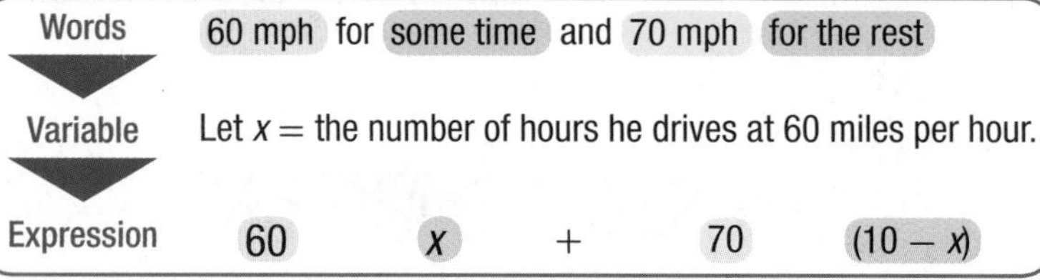

$60x + 70(10 - x)$ — Original expression

$= 60x + 700 - 70x$ — Distributive Property

$= 700 - 10x$ — Combine like terms.

The polynomial is $700 - 10x$.

Guided Practice

5. Paul has $900 to invest in a savings account that has an annual interest rate of 1.8%, and a money market account that pays 4.2% per year. Write a polynomial for the interest he will earn in one year if he invests x dollars in the savings account.

Real-World Career

Truck Driver Truck drivers are considered technical professionals because they are required to obtain specialized education and professional licensure. Although state motor vehicle departments administer the Commercial Driver's License program, federal law spells out the requirements to obtain one.

Like addition and subtraction, polynomials are closed under multiplication.

Example 6 Multiply Polynomials

Find $(n^2 + 4n - 6)(n + 2)$.

$(n^2 + 4n - 6)(n + 2)$

$= n^2(n + 2) + 4n(n + 2) + (-6)(n + 2)$ — Distributive Property

$= n^2 \cdot n + n^2 \cdot 2 + 4n \cdot n + 4n \cdot 2 + (-6) \cdot n + (-6) \cdot 2$ — Distributive Property

$= n^3 + 2n^2 + 4n^2 + 8n - 6n - 12$ — Multiply monomials.

$= n^3 + 6n^2 + 2n - 12$ — Combine like terms.

Guided Practice

Find each product.

6A. $(x^2 + 4x + 16)(x - 4)$

6B. $(2x^2 - 4x + 5)(3x - 1)$

Check Your Understanding

● = Step-by-Step Solutions begin on page R14.

Example 1 **Simplify. Assume that no variable equals 0.**

1. $(2a^3b^{-2})(-4a^2b^4)$
2. $\frac{12x^4y^2}{2xy^5}$
3. $\left(\frac{2a^2}{3b}\right)^3$
4. $(6g^5h^{-4})^3$

Example 2 **Determine whether each expression is a polynomial. If it is a polynomial, state the degree of the polynomial.**

5. $3x + 4y$
6. $\frac{1}{2}x^2 - 7y$
7. $x^2 + \sqrt{x}$
8. $\frac{ab^3 - 1}{az^4 + 3}$

Examples 3–4, and 6 **Simplify.**

9. $(x^2 - 5x + 2) - (3x^2 + x - 1)$
10. $(3a + 4b) + (6a - 6b)$
11. $2a(4b + 5)$
12. $3x^2(2xy - 3xy^2 + 4x^2y^3)$
13. $(n - 9)(n + 7)$
14. $(a + 4)(a - 6)$

Example 5

15. **EXERCISE** Tara exercises 75 minutes a day. She does cardio, which burns an average of 10 Calories per minute, and weight training, which burns an average of 7.5 Calories per minute. Write a polynomial to represent the amount of Calories Tara burns in one day if she does x minutes of weight training.

Practice and Problem Solving

Extra Practice is on page R5

Example 1 **Simplify. Assume that no variable equals 0.**

16. $(5x^3y^{-5})(4xy^3)$
17. $(-2b^3c)(4b^2c^2)$
18. $\frac{a^3n^7}{an^4}$
19. $\frac{-y^3z^5}{y^2z^3}$
20. $\frac{-7x^5y^5z^4}{21x^7y^5z^2}$
21. $\frac{9a^7b^5c^5}{18a^5b^9c^3}$
22. $(n^5)^4$
23. $(z^3)^6$

Example 2 **Determine whether each expression is a polynomial. If it is a polynomial, state the degree of the polynomial.**

24. $2x^2 - 3x + 5$
25. $a^3 - 11$
26. $\frac{5np}{n^2} - \frac{2g}{h}$
27. $\sqrt{m - 7}$

Examples 3–4, and 6 **CCSS REGULARITY Simplify.**

28. $(6a^2 + 5a + 10) - (4a^2 + 6a + 12)$
29. $(7b^2 + 6b - 7) - (4b^2 - 2)$
30. $3p(np - z)$
31. $4x(2x^2 + y)$
32. $(x - y)(x^2 + 2xy + y^2)$
33. $(a + b)(a^3 - 3ab - b^2)$
34. $4(a^2 + 5a - 6) - 3(2a^3 + 4a - 5)$
35. $5c(2c^2 - 3c + 4) + 2c(7c - 8)$
36. $5xy(2x - y) + 6y^2(x^2 + 6)$
37. $3ab(4a - 5b) + 4b^2(2a^2 + 1)$
38. $(x - y)(x + y)(2x + y)$
39. $(a + b)(2a + 3b)(2x - y)$

Example 5

40. **PAINTING** Connor has hired two painters to paint his house. The first painter charges \$12 per hour and the second painter charges \$11 per hour. It will take 15 hours of labor to paint the house.
    a. Write a polynomial to represent the total cost of the job if the first painter does x hours of the labor.
    b. Write a polynomial to represent the total cost of the job if the second painter does y hours of the labor.

Simplify. Assume that no variable equals 0.

41. $\left(\frac{8x^2y^3}{24x^3y^2}\right)^4$ **42.** $\left(\frac{12a^3b^5}{4a^6b^3}\right)^3$ **43.** $\left(\frac{4x^{-2}y^3}{xy^{-4}}\right)^{-2}$ **44.** $\left(\frac{5a^{-7}b^2}{ab^{-6}}\right)^{-3}$

45. $(a^2b^3)(ab)^{-2}$ **46.** $(-3x^3y)^2(4xy^2)$ **47.** $\frac{3c^2d(2c^3d^5)}{15c^4d^2}$

48. $\frac{-10g^6h^9(g^2h^3)}{30g^3h^3}$ **49.** $\frac{5x^4y^2(2x^5y^6)}{20x^3y^5}$ **50.** $\frac{-12n^7p^5(n^2p^4)}{36n^6p^7}$

51 **ASTRONOMY** Refer to the beginning of the lesson.

a. How long does it take light from Andromeda to reach Earth?

b. The average distance from the Sun to Mars is approximately 2.28×10^{11} meters. How long does it take light from the Sun to reach Mars?

Simplify.

52. $\frac{1}{4}g^2(8g + 12h - 16gh^2)$ **53.** $\frac{1}{3}n^3(6n - 9p + 18np^4)$ **54.** $x^{-2}(x^4 - 3x^3 + x^{-1})$

55. $a^{-3}b^2(ba^3 + b^{-1}a^2 + b^{-2}a)$ **56.** $(g^3 - h)(g^3 + h)$ **57.** $(n^2 - 7)(2n^3 + 4)$

58. $(2x - 2y)^3$ **59.** $(4n - 5)^3$ **60.** $(3z - 2)^3$

61. **CCSS MODELING** The polynomials $0.108x^2 - 0.876x + 474.1$ and $0.047x^2 + 9.694x + 361.7$ approximate the number of bachelor's degrees, in thousands, earned by males and females, respectively, where x is the number of years after 1971.

a. Find the polynomial that represents the total number of bachelor's degrees (in thousands) earned by both men and women.

b. Find the polynomial that represents the difference between bachelor's degrees earned by men and by women.

62. If $5^{k+7} = 5^{2k-3}$, what is the value of k?

63. What value of k makes $q^{41} = q^{4k} \cdot q^5$ true?

64. **MULTIPLE REPRESENTATIONS** Use the model at the right that represents the product of $x + 3$ and $x + 4$.

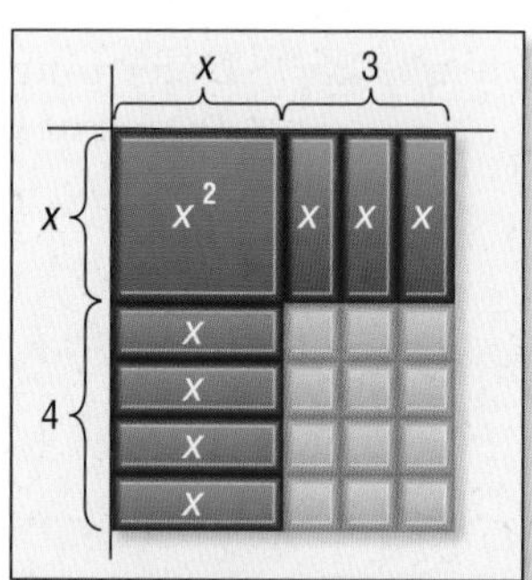

a. Geometric The area of each rectangle is the product of its length and width. Use the model to find the product of $x + 3$ and $x + 4$.

b. Algebraic Use FOIL to find the product of $x + 3$ and $x + 4$.

c. Verbal Explain how each term of the product is represented in the model.

H.O.T. Problems Use Higher-Order Thinking Skills

65. PROOF Show how the property of negative exponents can be proven using the Quotient of Powers Property and the Zero Power Property.

66. CHALLENGE What happens to the quantity of x^{-y} as y increases, for $y > 0$ and $x > 1$?

67. REASONING Explain why the expression 0^{-2} is undefined.

68. OPEN ENDED Write three different expressions that are equivalent to x^{12}.

69. WRITING IN MATH Explain why properties of exponents are useful in astronomy. Include an explanation of how to find the amount of time it takes for light from a source to reach a planet.

Standardized Test Practice

70. SHORT RESPONSE Simplify $\frac{(2x^2)^3}{12x^4}$.

71. STATISTICS For the numbers a, b, and c, the average (arithmetic mean) is twice the median. If $a = 0$ and $a < b < c$, what is the value of $\frac{c}{b}$?

A 2
B 3
C 4
D 5

72. Which is not a factor of $x^3 - x^2 - 2x$?

F x
G $x + 1$
H $x - 1$
J $x - 2$

73. SAT/ACT The expression $(-6 + i)^2$ is equivalent to which of the following expressions?

A 35
B $-12i$
C $-12 + i$
D $35 - 12i$
E $37 - 12i$

Spiral Review

Solve each inequality algebraically. (Lesson 4-8)

74. $x^2 - 6x \leq 16$

75. $x^2 + 3x > 40$

76. $2x^2 - 12 \leq -5x$

Graph each function. (Lesson 4-7)

77. $y = 3(x - 2)^2 - 4$

78. $y = -2(x + 4)^2 + 3$

79. $y = \frac{1}{3}(x + 1)^2 + 6$

80. BASEBALL A baseball player hits a high pop-up with an initial upward velocity of 30 meters per second, 1.4 meters above the ground. The height $h(t)$ of the ball in meters t seconds after being hit is modeled by $h(t) = -4.9t^2 + 30t + 1.4$. How long does an opposing player have to get under the ball if he catches it 1.7 meters above the ground? Does your answer seem reasonable? Explain. (Lesson 4-3)

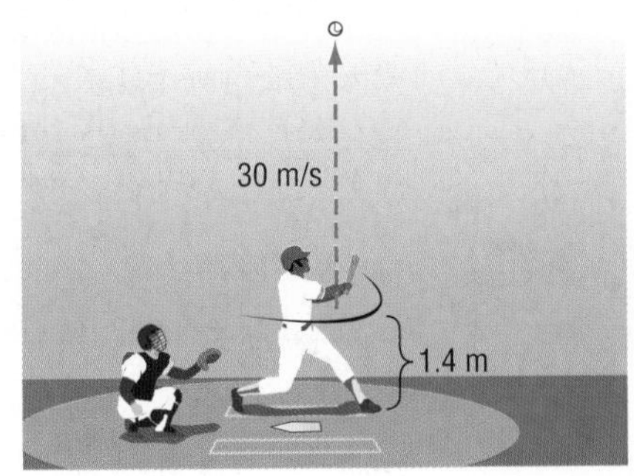

Evaluate each determinant. (Lesson 3-7)

81. $\begin{vmatrix} 3 & 0 & -2 \\ -1 & 4 & 3 \\ 5 & -2 & -1 \end{vmatrix}$

82. $\begin{vmatrix} -2 & -4 & -6 \\ 0 & 6 & -5 \\ -1 & 3 & -1 \end{vmatrix}$

83. $\begin{vmatrix} -3 & -1 & -2 \\ -2 & 3 & 4 \\ 6 & 1 & 0 \end{vmatrix}$

84. FINANCIAL LITERACY A couple is planning to invest \$15,000 in certificates of deposit (CDs). For tax purposes, they want their total interest the first year to be \$800. They want to put \$1000 more in a 2-year CD than in a 1-year CD and then invest the rest in a 3-year CD. How much should they invest in each type of CD? (Lesson 3-4)

Years	1	2	3
Rate	3.4%	5.0%	6.0%

Find the slope of the line that passes through each pair of points. (Lesson 2-3)

85. $(6, -2)$ and $(-2, -9)$

86. $(-4, -1)$ and $(3, 8)$

87. $(3, 0)$ and $(-7, -5)$

88. $\left(\frac{1}{2}, \frac{2}{3}\right)$ and $\left(\frac{1}{4}, \frac{1}{3}\right)$

89. $\left(\frac{2}{5}, \frac{1}{4}\right)$ and $\left(\frac{1}{10}, \frac{1}{12}\right)$

90. $(-4.5, 2.5)$ and $(-3, -1)$

Skills Review

Factor each polynomial.

91. $12ax^3 + 20bx^2 + 32cx$

92. $x^2 + 2x + 6 + 3x$

93. $12y^2 + 9y + 8y + 6$

94. $2my + 7x + 7m + 2xy$

95. $8ax - 6x - 12a + 9$

96. $10x^2 - 14xy - 15x + 21y$

EXTEND 5-1

Algebra Lab
Dimensional Analysis

Real-world problems often involve units of measure. Performing operations with units is called **dimensional analysis** or **unit analysis**. You can use dimensional analysis to convert units or to perform calculations.

Example

A car is traveling at 65 miles per hour. How fast is the car traveling in meters per second?

You want to find the speed in meters per second, so you need to change the unit of distance from miles to meters and the unit of time from hours to seconds. To make the conversion, use fractions that you can multiply.

Step 1 Change the units of length from miles to meters.
Use the relationships of miles to feet and feet to meters.

$$\frac{65 \text{ miles}}{1 \text{ hour}} \cdot \frac{5280 \text{ feet}}{1 \text{ mile}} \cdot \frac{1 \text{ meter}}{3.3 \text{ feet}}$$

Step 2 Change the units of time from hours to seconds.
Write fractions relating hours to minutes and minutes to seconds.

$$\frac{65 \text{ miles}}{1 \text{ hour}} \cdot \frac{5280 \text{ feet}}{1 \text{ mile}} \cdot \frac{1 \text{ meter}}{3.3 \text{ feet}} \cdot \frac{1 \text{ hour}}{60 \text{ minutes}} \cdot \frac{1 \text{ minute}}{60 \text{ seconds}}$$

Step 3 Simplify and check by canceling the units.

$$\frac{65 \cancel{\text{ miles}}}{1 \cancel{\text{ hour}}} \cdot \frac{5280 \cancel{\text{ feet}}}{1 \cancel{\text{ mile}}} \cdot \frac{1 \text{ meter}}{3.3 \cancel{\text{ feet}}} \cdot \frac{1 \cancel{\text{ hour}}}{60 \cancel{\text{ minutes}}} \cdot \frac{1 \cancel{\text{ minute}}}{60 \text{ seconds}}$$

$= \frac{65 \cdot 5280}{3.3 \cdot 60 \cdot 60}$ m/s Simplify.

≈ 28.9 m/s Use a calculator.

So, 65 miles per hour is about 28.9 meters per second. This answer is reasonable because the final units are m/s, not m/hr, ft/s, or mi/hr.

Exercises

Solve each problem by using dimensional analysis. Include the appropriate units with your answer.

1. A zebra can run 40 miles per hour. How far can a zebra run in 3 minutes?
2. A cyclist traveled 43.2 miles at an average speed of 12 miles per hour. How long did the cyclist ride?
3. If you are driving 50 miles per hour, how many feet per second are you traveling?
4. The equation $d = \frac{1}{2}(9.8 \text{ m/s}^2)(3.5 \text{ s})^2$ represents the distance d that a ball falls 3.5 seconds after it is dropped from a tower. Find the distance.
5. **WRITING IN MATH** Explain how dimensional analysis can be useful in checking the reasonableness of your answer.

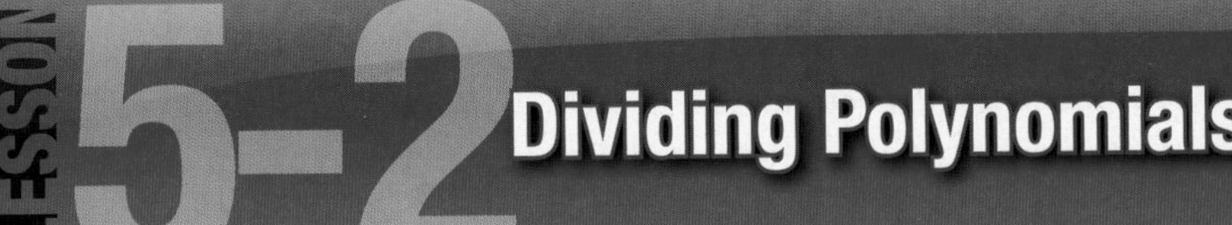

Dividing Polynomials

Then	Now	Why?
• You divided monomials.	**1** Divide polynomials using long division. **2** Divide polynomials using synthetic division.	• Arianna needed $140x^2 + 60x$ square inches of paper to make a book jacket $10x$ inches tall. In figuring the area, she allowed for a front and back flap. If the spine is $2x$ inches wide, and the front and back are $6x$ inches wide, how wide are the front and back flaps? You can use a quotient of polynomials to help you find the answer.

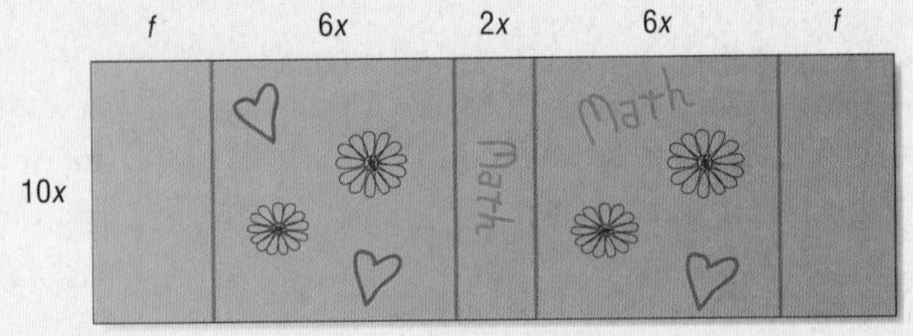

f = flap width

NewVocabulary
synthetic division

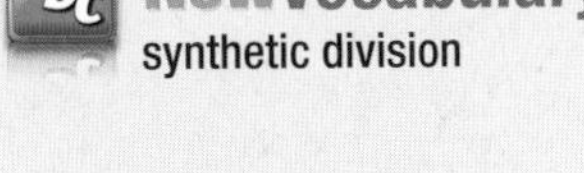

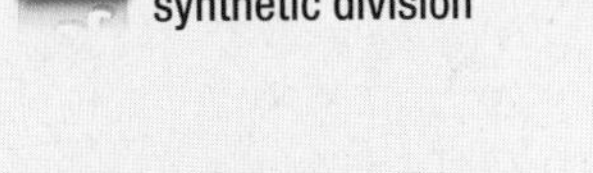

Common Core State Standards

Content Standards
A.APR.6 Rewrite simple rational expressions in different forms; write $a(x)/b(x)$ in the form $q(x) + r(x)/b(x)$, where $a(x)$, $b(x)$, $q(x)$, and $r(x)$ are polynomials with the degree of $r(x)$ less than the degree of $b(x)$, using inspection, long division, or, for the more complicated examples, a computer algebra system.

Mathematical Practices
6 Attend to precision.

1 Long Division

In Lesson 5-1, you learned how to divide monomials. You can divide a polynomial by a monomial by using those same skills.

Example 1 Divide a Polynomial by a Monomial

Simplify $\frac{6x^4y^3 + 12x^3y^2 - 18x^2y}{3xy}$.

$$\frac{6x^4y^3 + 12x^3y^2 - 18x^2y}{3xy} = \frac{6x^4y^3}{3xy} + \frac{12x^3y^2}{3xy} - \frac{18x^2y}{3xy} \quad \text{Sum of quotients}$$

$$= \frac{6}{3} \cdot x^{4-1}y^{3-1} + \frac{12}{3} \cdot x^{3-1}y^{2-1} - \frac{18}{3} \cdot x^{2-1}y^{1-1} \quad \text{Divide.}$$

$$= 2x^3y^2 + 4x^2y - 6x \quad y^{1-1} = y^0 \text{ or } 1$$

GuidedPractice **Simplify.**

1A. $(20c^4d^2f - 16cdf^2 + 4cdf) \div (4cdf)$

1B. $(18x^2y + 27x^3y^2z)(3xy)^{-1}$

You can use a process similar to long division to divide a polynomial by a polynomial with more than one term. The process is known as the *division algorithm*.

Example 2 Division Algorithm

Use long division to find $(x^2 + 3x - 40) \div (x - 5)$.

$$\begin{array}{r l}
x + 8 & \\
x - 5\overline{)x^2 + 3x - 40} & \\
\underline{(-)\ x^2 - 5x} & \text{Multiply the divisor by } x \text{ since } \frac{x^2}{x} = x. \\
8x - 40 & \text{Subtract. Bring down the next term.} \\
\underline{(-)\ 8x - 40} & \text{Multiply the divisor by 8 since } \frac{8x}{x} = 8. \\
0 & \text{Subtract.}
\end{array}$$

The quotient is $x + 8$. The remainder is 0.

GuidedPractice **Use long division to find each quotient.**

2A. $(x^2 + 7x - 30) \div (x - 3)$

2B. $(x^2 - 13x + 12) \div (x - 1)$

Just as with the division of whole numbers, the division of two polynomials may result in a quotient with a remainder. Remember that $11 \div 3 = 3 + R2$, which is often written as $3\frac{2}{3}$. The result of a division of polynomials with a remainder can be written in a similar manner.

Standardized Test Example 3 Divide Polynomials

Which expression is equal to $(a^2 + 7a - 11)(3 - a)^{-1}$?

A $a + 10 - \frac{19}{3 - a}$

B $-a + 10$

C $-a - 10 + \frac{19}{3 - a}$

D $-a - 10 - \frac{19}{3 - a}$

Test-Taking Tip

Multiple Choice You may be able to eliminate some of the answer choices by substituting the same value for a in the original expression and the answer choices, and then evaluate.

Read the Test Item

Since the second factor has an exponent of −1, this is a division problem.

$$(a^2 + 7a - 11)(3 - a)^{-1} = \frac{a^2 + 7a - 11}{3 - a}$$

Solve the Test Item

$$\begin{array}{r|l} -a - 10 & \\ -a + 3\overline{)a^2 + 7a - 11} & \text{For ease in dividing, rewrite } 3 - a \text{ as } -a + 3. \\ (-)\ a^2 - 3a & -a(-a + 3) = a^2 - 3a \\ 10a - 11 & 7a - (-3a) = 10a \\ (-)\ 10a - 30 & -10(-a + 3) = 10a - 30 \\ 19 & -11 - (-30) = 19 \end{array}$$

The quotient is $-a - 10$, and the remainder is 19.

Therefore, $(a^2 + 7a - 11)(3 - a)^{-1} = -a - 10 + \frac{19}{3 - a}$. The answer is **C**.

Guided Practice

3. Which expression is equal to $(r^2 + 5r + 7)(1 - r)^{-1}$?

F $-r - 6 + \frac{13}{1 - r}$

G $r + 6$

H $r - 6 + \frac{13}{1 - r}$

J $r + 6 - \frac{13}{1 - r}$

2 Synthetic Division

Synthetic division is a simpler process for dividing a polynomial by a binomial. Suppose you want to divide $2x^3 - 13x^2 + 26x - 24$ by $x - 4$ using long division. Compare the coefficients in this division with those in Example 4.

$$\begin{array}{r} 2x^2 - 5x + 6 \\ x - 4\overline{)2x^3 - 13x^2 + 26x - 24} \\ (-)\ 2x^3 - 8x^2 \qquad\qquad \\ -5x^2 + 26x \qquad \\ (-)\ -5x^2 + 20x \qquad \\ 6x - 24 \\ (-)\ 6x - 24 \\ 0 \end{array}$$

When the polynomial in the dividend is missing a term, a zero must be used to represent the missing term. So, with a dividend of $2x^3 - 4x^2 + 6$, a 0 will be used as a placeholder for the x-term.

$$\overline{)2x^3 - 4x^2 + 0x + 6}$$

KeyConcept Synthetic Division

Step 1 Write the coefficients of the dividend so that the degrees of the terms are in descending order. Write the constant r of the divisor $x - r$ in the box. Bring the first coefficient down.

Step 2 Multiply the first coefficient by r, and write the product under the second coefficient.

Step 3 Add the product and the second coefficient.

Step 4 Repeat Steps 2 and 3 until you reach a sum in the last column. The numbers along the bottom row are the coefficients of the quotient. The power of the first term is one less than the degree of the dividend. The final number is the remainder.

Example 4 Synthetic Division

Use synthetic division to find $(2x^3 - 13x^2 + 26x - 24) \div (x - 4)$.

Step 1 Write the coefficients of the dividend. Write the constant r in the box. In this case, $r = 4$. Bring the first coefficient, 2, down.

$$\begin{array}{r|rrrr} 4 & 2 & -13 & 26 & -24 \\ & \downarrow & & & \\ \hline & 2 & & & \end{array}$$

Step 2 Multiply the first coefficient by r: $2 \cdot 4 = 8$. Write the product under the second coefficient.

$$\begin{array}{r|rrrr} 4 & 2 & -13 & 26 & -24 \\ & & 8 & & \\ \hline & 2 \nearrow & & & \end{array}$$

Step 3 Add the product and the second coefficient: $-13 + 8 = -5$.

$$\begin{array}{r|rrrr} 4 & 2 & -13 & 26 & -24 \\ & & 8 & & \\ \hline & 2 & -5 & & \end{array}$$

Step 4 Multiply the sum, -5, by r: $-5 \times 4 = -20$. Write the product under the next coefficient, and add: $26 + (-20) = 6$. Multiply the sum, 6, by r: $6 \cdot 4 = 24$. Write the product under the next coefficient and add: $-24 + 24 = 0$.

$$\begin{array}{r|rrrr} 4 & 2 & -13 & 26 & -24 \\ & & 8 & -20 & 24 \\ \hline & 2 & -5 \nearrow & 6 \nearrow & \vert\, 0 \end{array}$$

CHECK Multiply the quotient by the divisor. The answer should be the dividend.

$$\begin{array}{rr} & 2x^2 - 5x + 6 \\ (\times) & x - 4 \\ \hline & -8x^2 + 20x - 24 \\ 2x^3 - 5x^2 + 6x & \\ \hline 2x^3 - 13x^2 + 26x - 24 & \end{array}$$

The quotient is $2x^2 - 5x + 6$. The remainder is 0.

GuidedPractice

Use synthetic division to find each quotient.

4A. $(2x^3 + 3x^2 - 4x + 15) \div (x + 3)$

4B. $(3x^3 - 8x^2 + 11x - 14) \div (x - 2)$

4C. $(4a^4 + 2a^2 - 4a + 12) \div (a + 2)$

4D. $(6b^4 - 8b^3 + 12b - 14) \div (b - 2)$

WatchOut!
Synthetic Division Remember to *add* terms when performing synthetic division.

To use synthetic division, the divisor must be of the form $x - r$. If the coefficient of x in a divisor is not 1, you can rewrite the division expression so that you can use synthetic division.

PT

Example 5 Divisor with First Coefficient Other than 1

WatchOut!

CCSS Precision Remember to divide *all* terms in the numerator and denominator.

Use synthetic division to find $(3x^4 - 5x^3 + x^2 + 7x) \div (3x + 1)$.

$$\frac{3x^4 - 5x^3 + x^2 + 7x}{3x + 1} = \frac{(3x^4 - 5x^3 + x^2 + 7x) \div 3}{(3x + 1) \div 3}$$

Rewrite the divisor with a leading coefficient of 1. Then divide the numerator and denominator by 3.

$$= \frac{x^4 - \frac{5}{3}x^3 + \frac{1}{3}x^2 + \frac{7}{3}x}{x + \frac{1}{3}}$$

Simplify the numerator and the denominator.

Since the numerator does not have a constant term, use a coefficient of 0 for the constant term.

$x - r = x + \frac{1}{3}$, so $r = -\frac{1}{3}$. →

$$\begin{array}{r|rrrrr} -\frac{1}{3} & 1 & -\frac{5}{3} & \frac{1}{3} & \frac{7}{3} & 0 \\ & & -\frac{1}{3} & \frac{2}{3} & -\frac{1}{3} & -\frac{2}{3} \\ \hline & 1 & -2 & 1 & 2 & \big| -\frac{2}{3} \end{array}$$

The result is $x^3 - 2x^2 + x + 2 - \frac{\frac{2}{3}}{x + \frac{1}{3}}$. Now simplify the fraction.

$$\frac{\frac{2}{3}}{x + \frac{1}{3}} = \frac{2}{3} \div \left(x + \frac{1}{3}\right)$$

Rewrite as a division expression.

$$= \frac{2}{3} \div \frac{3x + 1}{3}$$

$x + \frac{1}{3} = \frac{3x}{3} + \frac{1}{3} = \frac{3x + 1}{3}$

$$= \frac{2}{3} \cdot \frac{3}{3x + 1}$$

Multiply by the reciprocal.

$$= \frac{2}{3x + 1}$$

Simplify.

The solution is $x^3 - 2x^2 + x + 2 - \frac{2}{3x + 1}$.

CHECK Divide using long division.

$$\begin{array}{r} x^3 - 2x^2 + x + 2 \\ 3x + 1 \overline{)\, 3x^4 - 5x^3 + x^2 + 7x} \\ \underline{(-)\, 3x^4 + x^3 } \\ -6x^3 + x^2 \\ \underline{(-)\, -6x^3 - 2x^2 } \\ 3x^2 + 7x \\ \underline{(-)\, 3x^2 + x} \\ 6x + 0 \\ \underline{(-)\, 6x + 2} \\ -2 \end{array}$$

The result is $x^3 - 2x^2 + x + 2 - \frac{2}{3x + 1}$. ✓

GuidedPractice

Use synthetic division to find each quotient.

5A. $(8x^4 - 4x^2 + x + 4) \div (2x + 1)$

5B. $(8y^5 - 2y^4 - 16y^2 + 4) \div (4y - 1)$

5C. $(15b^3 + 8b^2 - 21b + 6) \div (5b - 4)$

5D. $(6c^3 - 17c^2 + 6c + 8) \div (3c - 4)$

Check Your Understanding

= Step-by-Step Solutions begin on page R14.

Examples 1, 2, and 4

Simplify.

1. $\dfrac{4xy^2 - 2xy + 2x^2y}{xy}$
2. $(3a^2b - 6ab + 5ab^2)(ab)^{-1}$
3. $(x^2 - 6x - 20) \div (x + 2)$
4. $(2a^2 - 4a - 8) \div (a + 1)$
5. $(3z^4 - 6z^3 - 9z^2 + 3z - 6) \div (z + 3)$
6. $(y^5 - 3y^2 - 20) \div (y - 2)$

Example 3

7. **MULTIPLE CHOICE** Which expression is equal to $(x^2 + 3x - 9)(4 - x)^{-1}$?

A $-x - 7 + \dfrac{19}{4 - x}$ B $-x - 7$ C $x + 7 - \dfrac{19}{4 - x}$ D $-x - 7 - \dfrac{19}{4 - x}$

Example 5

Simplify.

8. $(10x^2 + 15x + 20) \div (5x + 5)$
9. $(18a^2 + 6a + 9) \div (3a - 2)$
10. $\dfrac{12b^2 + 23b + 15}{3b + 8}$
11. $\dfrac{27y^2 + 27y - 30}{9y - 6}$

Practice and Problem Solving

Extra Practice is on page R5.

Example 1

Simplify.

12. $\dfrac{24a^3b^2 - 16a^2b^3}{8ab}$
13. $\dfrac{5x^2y - 10xy + 15xy^2}{5xy}$
14. $\dfrac{7g^3h^2 + 3g^2h - 2gh^3}{gh}$
15. $\dfrac{4a^3b - 6ab + 2ab^2}{2ab}$
16. $\dfrac{16c^4d^4 - 24c^2d^2}{4c^2d^2}$
17. $\dfrac{9n^3p^3 - 18n^2p^2 + 21n^2p^3}{3n^2p^2}$

18. **ENERGY** Compact fluorescent light (CFL) bulbs reduce energy waste. The amount of energy waste that is reduced each day in a certain community can be estimated by $-b^2 + 8b$, where b is the number of bulbs. Divide by b to find the average amount of energy saved per CFL bulb.

19. **BAKING** The number of cookies produced in a factory each day can be estimated by $-w^2 + 16w + 1000$, where w is the number of workers. Divide by w to find the average number of cookies produced per worker.

Examples 2, 4, and 5

Simplify.

20. $(a^2 - 8a - 26) \div (a + 2)$
21. $(b^3 - 4b^2 + b - 2) \div (b + 1)$
22. $(z^4 - 3z^3 + 2z^2 - 4z + 4)(z - 1)^{-1}$
23. $(x^5 - 4x^3 + 4x^2) \div (x - 4)$
24. $\dfrac{y^3 + 11y^2 - 10y + 6}{y + 2}$
25. $(g^4 - 3g^2 - 18) \div (g - 2)$
26. $(6a^2 - 3a + 9) \div (3a - 2)$
27. $\dfrac{6x^5 + 5x^4 + x^3 - 3x^2 + x}{3x + 1}$
28. $\dfrac{4g^4 - 6g^3 + 3g^2 - g + 12}{4g - 4}$
29. $(2b^3 - 6b^2 + 8b) \div (2b + 2)$
30. $(6z^6 + 3z^4 - 9z^2)(3z - 6)^{-1}$
31. $(10y^6 + 5y^5 + 10y^3 - 20y - 15)(5y + 5)^{-1}$

32. CCSS **REASONING** A rectangular box for a new product is designed in such a way that the three dimensions always have a particular relationship defined by the variable x. The volume of the box can be written as $6x^3 + 31x^2 + 53x + 30$, and the height is always $x + 2$. What are the width and length of the box?

33. **PHYSICS** The voltage V is related to current I and power P by the equation $V = \dfrac{P}{I}$. The power of a generator is modeled by $P(t) = t^3 + 9t^2 + 26t + 24$. If the current of the generator is $I = t + 4$, write an expression that represents the voltage.

34. **ENTERTAINMENT** A magician gives these instructions to a volunteer.

- Choose a number and multiply it by 4.
- Then add the sum of your number and 15 to the product you found.
- Now divide by the sum of your number and 3.

a. What number will the volunteer always have at the end?

b. Explain the process you used to discover the answer.

35. **BUSINESS** The number of magazine subscriptions sold can be estimated by $n = \frac{3500a^2}{a^2 + 100}$, where a is the amount of money the company spent on advertising in hundreds of dollars and n is the number of subscriptions sold.

a. Perform the division indicated by $\frac{3500a^2}{a^2 + 100}$.

b. About how many subscriptions will be sold if \$1500 is spent on advertising?

Simplify.

36. $(x^4 - y^4) \div (x - y)$

37. $(28c^3d^2 - 21cd^2) \div (14cd)$

38. $(a^3b^2 - a^2b + 2b)(-ab)^{-1}$

39. $\frac{n^3 + 3n^2 - 5n - 4}{n + 4}$

40. $\frac{p^3 + 2p^2 - 7p - 21}{p + 3}$

41. $\frac{3z^5 + 5z^4 + z + 5}{z + 2}$

42. **MULTIPLE REPRESENTATIONS** Consider a rectangle with area $2x^2 + 7x + 3$ and length $2x + 1$.

a. Concrete Use algebra tiles to represent this situation. Use the model to find the width.

b. Symbolic Write an expression to represent the model.

c. Numerical Solve this problem algebraically using synthetic or long division. Does your concrete model check with your algebraic model?

H.O.T. Problems Use Higher-Order Thinking Skills

43. **ERROR ANALYSIS** Sharon and Jamal are dividing $2x^3 - 4x^2 + 3x - 1$ by $x - 3$. Sharon claims that the remainder is 26. Jamal argues that the remainder is −100. Is either of them correct? Explain your reasoning.

44. **CHALLENGE** If a polynomial is divided by a binomial and the remainder is 0, what does this tell you about the relationship between the binomial and the polynomial?

45. **REASONING** Review any of the division problems in this lesson. What is the relationship between the degrees of the dividend, the divisor, and the quotient?

46. **OPEN ENDED** Write a quotient of two polynomials for which the remainder is 3.

47. **CCSS ARGUMENTS** Identify the expression that does not belong with the other three. Explain your reasoning.

$3xy + 6x^2$	$\frac{5}{x^2}$	$x + 5$	$5b + 11c - 9ad^2$

48. **WRITING IN MATH** Use the information at the beginning of the lesson to write assembly instructions using the division of polynomials to make a paper cover for your textbook.

Standardized Test Practice

49. An office employs x women and 3 men. What is the ratio of the total number of employees to the number of women?

A $\frac{x+3}{x}$ C $\frac{3}{x}$

B $\frac{x}{x+3}$ D $\frac{x}{3}$

50. SAT/ACT Which polynomial has degree 3?

F $x^3 + x^2 - 2x^4$ J $x^2 + x + 12^3$

G $-2x^2 - 3x + 4$ K $1 + x + x^3$

H $3x - 3$

51. GRIDDED RESPONSE In the figure below, $m + n + p = ?$

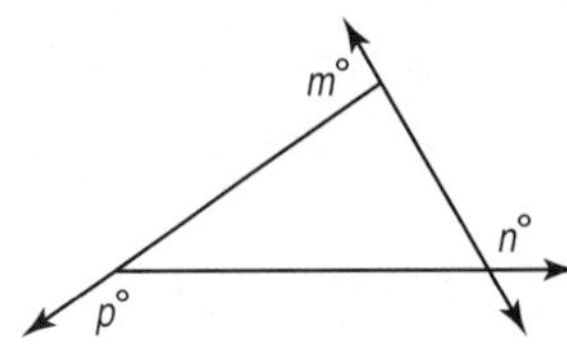

52. $(-4x^2 + 2x + 3) - 3(2x^2 - 5x + 1) =$

A $2x^2$ C $-10x^2 + 17x$

B $-10x^2$ D $2x^2 + 17x$

Spiral Review

Simplify. (Lesson 5-1)

53. $(5x^3 + 2x^2 - 3x + 4) - (2x^3 - 4x)$

54. $(2y^3 - 3y + 8) + (3y^2 - 6y)$

55. $4a(2a - 3) + 3a(5a - 4)$

56. $(c + d)(c - d)(2c - 3d)$

57. $(xy)^2(2xy^2z)^3$

58. $(3ab^2)^{-2}(2a^2b)^2$

59. LANDSCAPING Amado wants to plant a garden and surround it with decorative stones. He has enough stones to enclose a rectangular garden with a perimeter of 68 feet, but he wants the garden to cover no more than 240 square feet. What could the width of his garden be? (Lesson 4-8)

Solve each equation by completing the square. (Lesson 4-5)

60. $x^2 + 6x + 2 = 0$

61. $x^2 - 8x - 3 = 0$

62. $2x^2 + 6x + 5 = 0$

State the consecutive integers between which the zeros of each quadratic function are located. (Lesson 4-2)

63.

x	−7	−6	−5	−4	−3	−2	−1	0
$f(x)$	4	1	−3	−8	−1	2	8	16

64.

x	−2	−1	0	1	2	3	4	5
$f(x)$	−16	−7	−4	3	3	−4	−7	−16

65.

x	−2	−1	0	1	2	3	4	5
$f(x)$	6	1	−3	−5	−3	1	6	14

66. BUSINESS A landscaper can mow a lawn in 30 minutes and perform a small landscape job in 90 minutes. He works at most 10 hours per day, 5 days per week. He earns $35 per lawn and $125 per landscape job. He cannot do more than 3 landscape jobs per day. Find the combination of lawns mowed and completed landscape jobs per week that will maximize income. Then find the maximum income. (Lesson 3-3)

Skills Review

Find each value if $f(x) = 4x + 3$, $g(x) = -x^2$, and $h(x) = -2x^2 - 2x + 4$.

67. $f(-6)$

68. $g(-8)$

69. $h(3)$

70. $f(c)$

71. $g(3d)$

72. $h(2b + 1)$

EXTEND 5-2

Graphing Technology Lab
Dividing Polynomials

Long division and synthetic division are two alternatives for dividing polynomials with linear divisors. You can use a graphing calculator with a computer algebra system (CAS) to divide polynomials with any divisor.

CCSS Common Core State Standards
Content Standards
A.APR.6 Rewrite simple rational expressions in different forms; write $a(x)/b(x)$ in the form $q(x) + r(x)/b(x)$, where $a(x)$, $b(x)$, $q(x)$, and $r(x)$ are polynomials with the degree of $r(x)$ less than the degree of $b(x)$, using inspection, long division, or, for the more complicated examples, a computer algebra system.

Activity 1 Divide Polynomials Without Remainders

Use CAS to find $(x^4 + 3x^3 - x^2 - 5x + 2) \div (x^2 + 2x - 1)$.

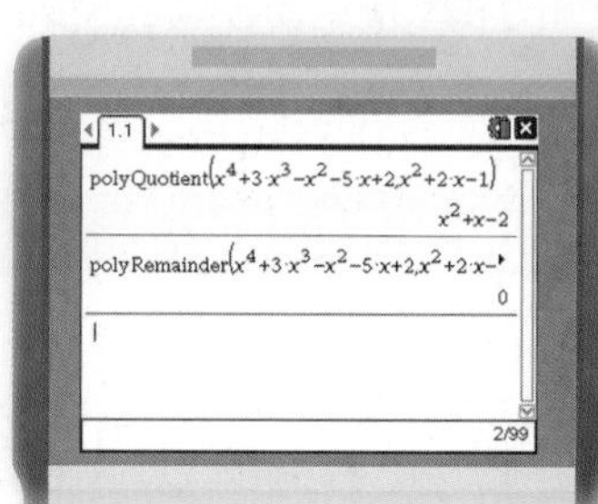

Step 1 Add a new **Calculator** page on the TI-Nspire.

Step 2 From the menu, select **Algebra**, then **Polynomial Tools** and **Quotient of Polynomial**.

Step 3 Type the dividend, a comma, and the divisor.

The CAS indicates that $(x^4 + 3x^3 - x^2 - 5x + 2) \div (x^2 + 2x - 1)$ is $x^2 + x - 2$.

Step 4 To verify that there is no remainder, select **Remainder of a Polynomial** from the **Algebra, Polynomial Tools** menu then type the dividend, a comma, and the divisor.

In Activity 1, there was no remainder. But in many cases, there will be a remainder.

Activity 2 Divide Polynomials With Remainders

Use CAS to find $(4x^5 - 12x^4 - 7x^3 + 32x^2 + 3x + 20) \div (x^2 - 2x + 4)$.

Step 1 Add a new **Calculator** page on the TI-Nspire.

Step 2 From the menu, select **Algebra, Polynomial Tools,** and **Quotient of Polynomial**.

Step 3 Type the dividend, a comma, and the divisor.

The CAS indicates that $(4x^5 - 12x^4 - 7x^3 + 32x^2 + 3x + 20) \div (x^2 - 2x + 4)$ is $4x^3 - 4x^2 - 31x - 14$.

We need to determine whether there is a remainder.

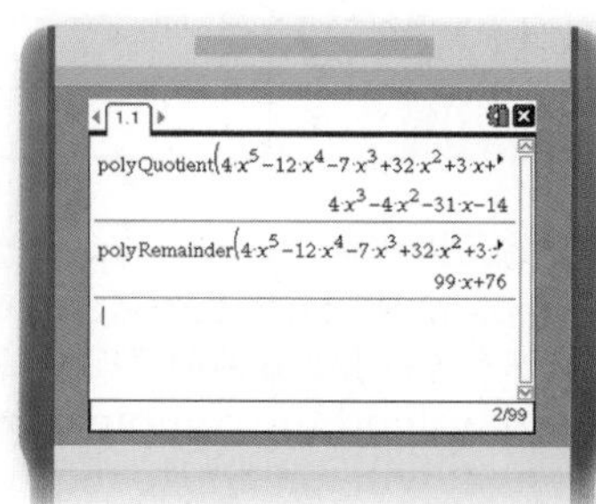

Step 4 Use the **Remainder of a Polynomial** option from the **Algebra, Polynomial Tools** menu to determine the remainder. Then type the dividend, a comma, and the divisor.

The remainder is $99x + 76$.
Therefore, $(4x^5 - 12x^4 - 7x^3 + 32x^2 + 3x + 20) \div (x^2 - 2x + 4)$ is $4x^3 - 4x^2 - 31x - 14 + \frac{99x + 76}{x^2 - 2x + 4}$.

You can also use a graphing calculator to determine roots of a polynomial so you can divide with synthetic division.

Activity 3 Divide with Synthetic Division

Use synthetic division to find $(x^6 - 28x^4 + 14x^3 + 147x^2 - 14x - 120) \div (x^3 + 3x^2 - 18x - 40)$.

Step 1 Graph the dividend as $f1(x)$ and the divisor as $f2(x)$ on the same calculator page. Use the **intersection points** tool from **Points & Lines** menu to find where the graphs have the same x-intercepts.

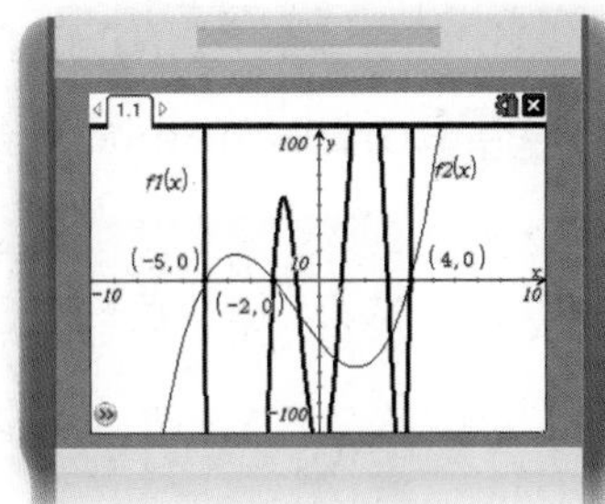

[−10, 10] scl: 1 by [−100, 100] scl: 10

Step 2 Use the roots from Step 1 as the divisors for synthetic division.

−5	1	0	−28	14	147	−14	−120
		−5	25	15	−145	−10	120
	1	−5	−3	29	2	−24	0

−2	1	−5	−3	29	2	−24
		−2	14	−22	−14	24
	1	−7	11	7	−12	0

4	1	−7	11	7	−12
		4	−12	−4	12
	1	−3	−1	3	0

Step 3 Use the **Expand** function to verify that $x^3 - 3x^2 - x + 3$ is the quotient when −5, −2, and 4 are roots.

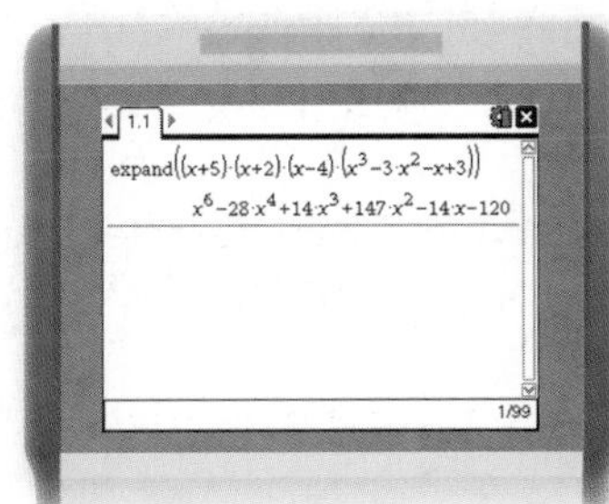

Thus, $(x^6 - 28x^4 + 14x^3 + 147x^2 - 14x - 120) \div (x^3 + 3x^2 - 18x - 40)$ is $x^3 - 3x^2 - x + 3$.

Exercises

Find each quotient.

1. $(2x^4 + x^3 - 8x^2 + 17x - 12) \div (x^2 + 2x - 3)$
2. $(x^4 + 7x^3 + 8x^2 + x - 12) \div (x^2 + 3x - 4)$
3. $(9x^5 - 9x^3 - 5x^2 + 5) \div (9x^3 - 5)$
4. $(x^5 - 8x^4 + 10x^3 + 14x^2 + 61x - 30) \div (x^2 - 5x + 3)$
5. $(2x^6 + 2x^5 - 4x^4 - 18x^3 - 16x^2 + 8x + 16) \div (2x^3 + 2x^2 - 4x - 2)$
6. $(6x^6 - 2x^5 - 14x^4 + 10x^3 - 4x^2 - 28x - 5) \div (3x^3 - x^2 - 7x - 1)$
7. Use synthetic division to find $(x^6 - 7x^5 - 21x^4 + 175x^3 + 56x^2 - 924x + 720) \div (x^3 - 5x^2 - 12x + 36)$.

EXPLORE 5-3

Graphing Technology Lab
Power Functions

A **power function** is any function of the form $f(x) = ax^n$, where a and n are nonzero constant real numbers. A power function in which n is a positive integer is called a **monomial function**.

CCSS Common Core State Standards
Content Standards
F.IF.4 For a function that models a relationship between two quantities, interpret key features of graphs and tables in terms of the quantities, and sketch graphs showing key features given a verbal description of the relationship.

Activity 1 $y = ax^4$

Graph $f(x) = 2x^4$. Analyze the graph.

Step 1 Graph the function.

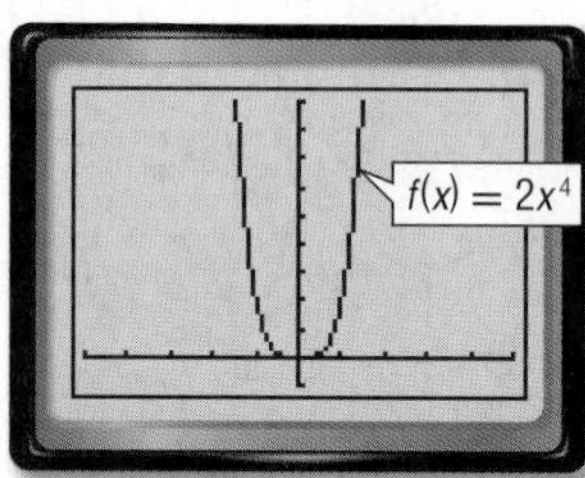

[−5, 5] scl: 1 by [−1, 9] scl: 1

Step 2 Analyze the graph.

The graph resembles the graph of $g(x) = x^2$, but it flattens out more at its vertex.

Analyze the Results

1. What assumptions can you make about the graph of $f(x) = 2x^4$ as x becomes more positive or more negative? Use the **TABLE** feature to evaluate the function for values of x to confirm this assumption, if necessary.

2. State the domain and range of $f(x) = 2x^4$. Compare these values with the domain and range of $g(x) = x^2$.

3. What other characteristics does the graph of $f(x) = 2x^4$ share with the graph of $g(x) = x^2$?

Activity 2 $y = ax^5$

Graph $f(x) = 3x^5$. Analyze the graph.

Step 1 Graph the function.

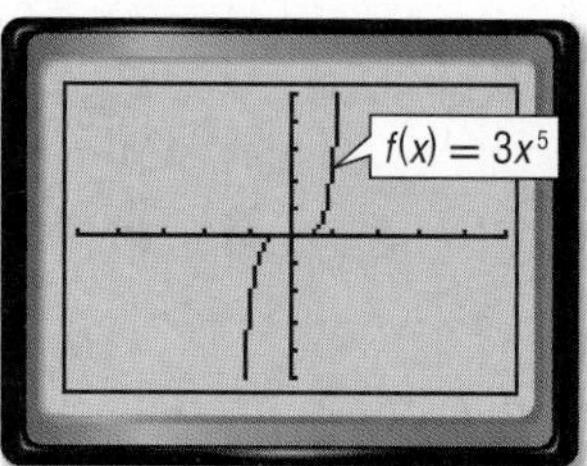

[−5, 5] scl: 1 by [−5, 5] scl: 1

Step 2 Analyze the graph.

The graph resembles the graph of $f(x) = x^3$, but it flattens out more as the graph approaches the origin.

Analyze the Results

4. What assumptions can you make about the graph of $f(x) = 3x^5$ as x becomes more positive or more negative? Use the **TABLE** feature to evaluate the function for values of x to confirm this assumption, if necessary.

5. State the domain and range of $f(x) = 3x^5$. Compare these values with the domain and range of $g(x) = x^3$.

6. What other characteristics does the graph of $f(x) = 3x^5$ share with the graph of $g(x) = x^3$?

7. Compare the characteristics of the graph $h(x) = -3x^5$ with the graph of $f(x) = 3x^5$. What conclusions can you make about $f(x) = ax^5$ when a is positive and when a is negative?

Activity 3 $y = x^n$, where n is even

Graph $f(x) = x^2$, $g(x) = x^4$, and $h(x) = x^6$ on the same screen. Analyze the graphs.

Step 1 Graph the function.

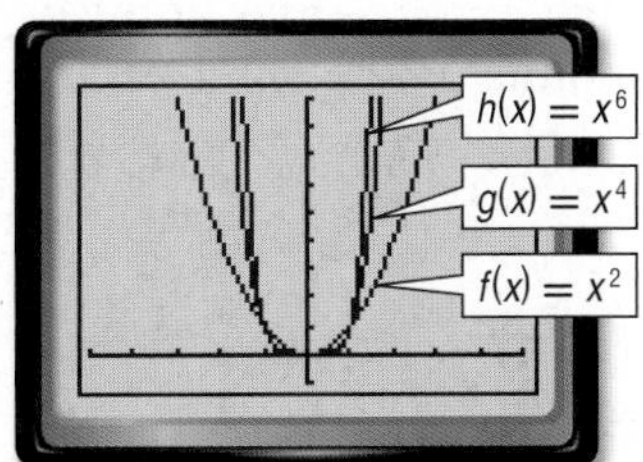

[−5, 5] scl: 1 by [−1, 9] scl: 1

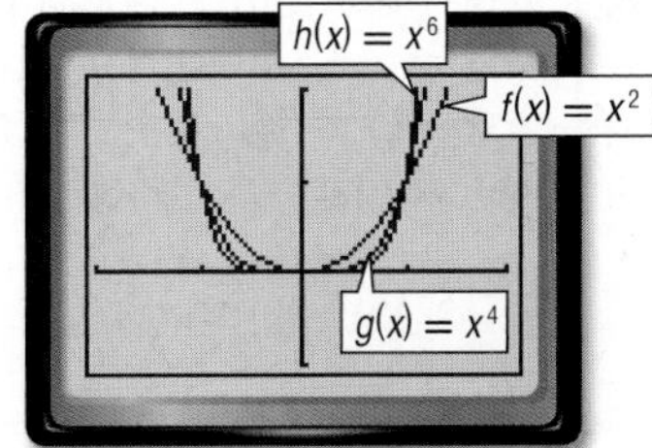

[−2, 2] scl: 1 by [−1, 2] scl: 1

Step 2 The graphs are all U-shaped, but the widths are different. The graphs also differ around the origin. If you zoom in around the origin, you can see this more clearly. You can do this by using the **ZOOM** feature or by adjusting the window manually.

Analyze the Results

8. What assumptions can you make about the graphs of $a(x) = x^8$, $b(x) = x^{10}$, and so on?

9. Identify the common characteristics of the graphs of power functions in which the power is an even number.

10. Graph $f(x) = x^3$, $g(x) = x^5$, and $h(x) = x^7$ on the same graph.

11. What assumptions can you make about the graphs of $a(x) = x^9$, $b(x) = x^{11}$, and so on?

12. Identify the common characteristics of the graphs of power functions in which the power is an odd number.

13. Graph $f(x) = x^4$ and $g(x) = -x^4$ on the same graph.

14. Graph $f(x) = x^5$ and $g(x) = -x^5$ on the same graph.

15. What assumptions can you make about the effects of a negative value of a in $f(x) = ax^n$ when n is even? when n is odd?

LESSON 5-3 Polynomial Functions

Then	Now	Why?
You analyzed graphs of quadratic functions.	**1** Evaluate polynomial functions. **2** Identify general shapes of graphs of polynomial functions.	The volume of air in the lungs during a 5-second respiratory cycle can be modeled by $v(t) = -0.037t^3 + 0.152t^2 + 0.173t$, where v is the volume in liters and t is the time in seconds. This model is an example of a polynomial function.

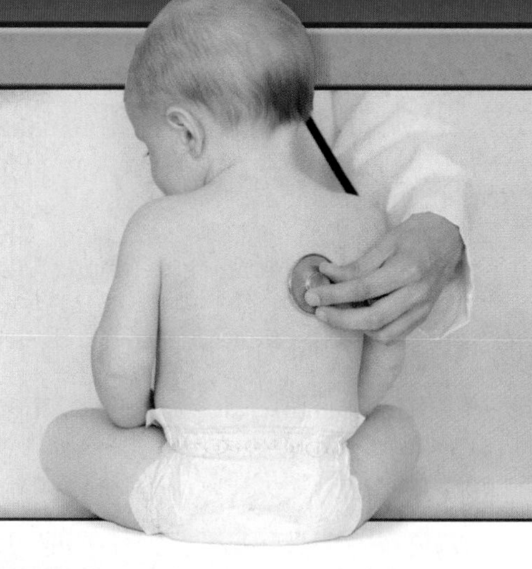

NewVocabulary
- polynomial in one variable
- leading coefficient
- polynomial function
- power function
- quartic function
- quintic function
- end behavior

Common Core State Standards

Content Standards

F.IF.4 For a function that models a relationship between two quantities, interpret key features of graphs and tables in terms of the quantities, and sketch graphs showing key features given a verbal description of the relationship.

F.IF.7.c Graph polynomial functions, identifying zeros when suitable factorizations are available, and showing end behavior.

Mathematical Practices

1 Make sense of problems and persevere in solving them.

1 Polynomial Functions

A **polynomial in one variable** is an expression of the form $a_nx^n + a_{n-1}x^{n-1} + \cdots + a_2x^2 + a_1x + a_0$, where $a_n \neq 0$, a_{n-1}, a_2, a_1, and a_0 are real numbers, and n is a nonnegative integer.

The polynomial is written in standard form when the values of the exponents are in descending order. The degree of the polynomial is the value of the greatest exponent. The coefficient of the first term of a polynomial in standard form is called the **leading coefficient**.

Polynomial	Expression	Degree	Leading Coefficient
Constant	12	0	12
Linear	$4x - 9$	1	4
Quadratic	$5x^2 - 6x - 9$	2	5
Cubic	$8x^3 + 12x^2 - 3x + 1$	3	8
General	$a_nx^n + a_{n-1}x^{n-1} + \cdots + a_1x + a_0$	n	a_n

Example 1 Degrees and Leading Coefficients

State the degree and leading coefficient of each polynomial in one variable. If it is not a polynomial in one variable, explain why.

a. $8x^5 - 4x^3 + 2x^2 - x - 3$

This is a polynomial in one variable. The greatest exponent is 5, so the degree is 5 and the leading coefficient is 8.

b. $12x^2 - 3xy + 8x$

This is not a polynomial in one variable. There are two variables, x and y.

c. $3x^4 + 6x^3 - 4x^8 + 2x$

This is a polynomial in one variable. The greatest exponent is 8, so the degree is 8 and the leading coefficient is -4.

GuidedPractice

1A. $5x^3 - 4x^2 - 8x + \frac{4}{x}$ **1B.** $5x^6 - 3x^4 + 12x^3 - 14$ **1C.** $8x^4 - 2x^3 - x^6 + 3$

A **polynomial function** is a continuous function that can be described by a polynomial equation in one variable. For example, $f(x) = 3x^3 - 4x + 6$ is a cubic polynomial function. The simplest polynomial functions of the form $f(x) = ax^b$ where a and b are real numbers are called **power functions**.

If you know an element in the domain of any polynomial function, you can find the corresponding value in the range.

Real-WorldLink

Total lung capacity is approximately 6 liters in a healthy young adult.

Source: *Family Practice Notebook*

Real-World Example 2 Evaluate a Polynomial Function

RESPIRATION Refer to the beginning of the lesson. Find the volume of air in the lungs 2 seconds into the respiratory cycle.

By substituting 2 into the function we can find $v(2)$, the volume of air in the lungs 2 seconds into the respiratory cycle.

$v(t) = -0.037t^3 + 0.152t^2 + 0.173t$	Original function
$v(2) = -0.037(2)^3 + 0.152(2)^2 + 0.173(2)$	Replace t with 2.
$= -0.296 + 0.608 + 0.346$	Simplify.
$= 0.658$ L	Add.

GuidedPractice

2. Find the volume of air in the lungs 4 seconds into the respiratory cycle.

You can also evaluate functions for variables and algebraic expressions.

Example 3 Function Values of Variables

Find $f(3c - 4) - 5f(c)$ if $f(x) = x^2 + 2x - 3$.

To evaluate $f(3c - 4)$, replace the x in $f(x)$ with $3c - 4$.

$f(x) = x^2 + 2x - 3$	Original function
$f(3c - 4) = (3c - 4)^2 + 2(3c - 4) - 3$	Replace x with $3c - 4$.
$= 9c^2 - 24c + 16 + 6c - 8 - 3$	Multiply.
$= 9c^2 - 18c + 5$	Simplify.

To evaluate $5f(c)$, replace x with c in $f(x)$, then multiply by 5.

$f(x) = x^2 + 2x - 3$	Original function
$5f(c) = 5(c^2 + 2c - 3)$	Replace x with c.
$= 5c^2 + 10c - 15$	Distributive Property

Now evaluate $f(3c - 4) - 5f(c)$.

$f(3c - 4) - 5f(c) = (9c^2 - 18c + 5) - (5c^2 + 10c - 15)$	
$= 9c^2 - 18c + 5 - 5c^2 - 10c + 15$	Distribute.
$= 4c^2 - 28c + 20$	Simplify.

GuidedPractice

3A. Find $g(5a - 2) + 3g(2a)$ if $g(x) = x^2 - 5x + 8$.

3B. Find $h(-4d + 3) - 0.5h(d)$ if $h(x) = 2x^2 + 5x + 3$.

Purestock/Alamy

2 Graphs of Polynomial Functions

The general shapes of the graphs of several polynomial functions show the *maximum* number of times the graph of each function may intersect the x-axis. This is the same number as the degree of the polynomial.

Constant function
Degree 0

Linear function
Degree 1

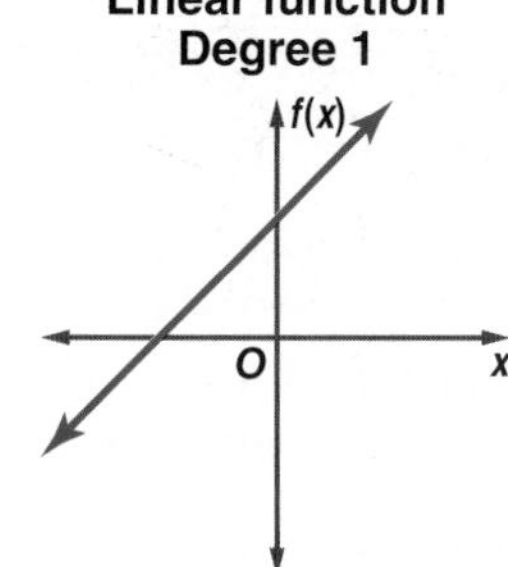

Quadratic function
Degree 2

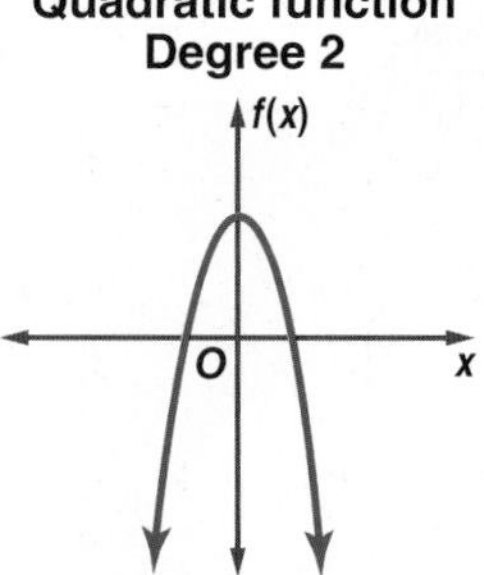

Cubic function
Degree 3

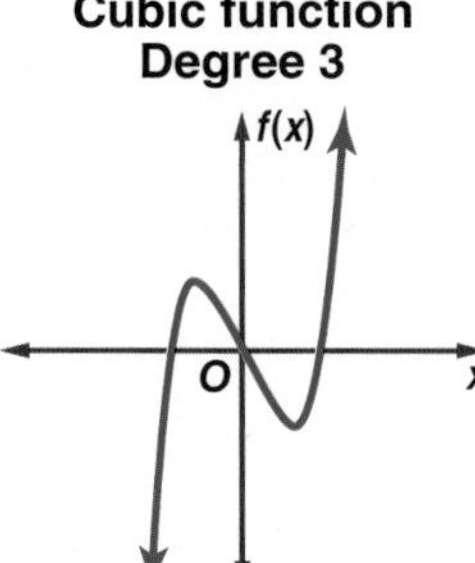

Quartic function
Degree 4

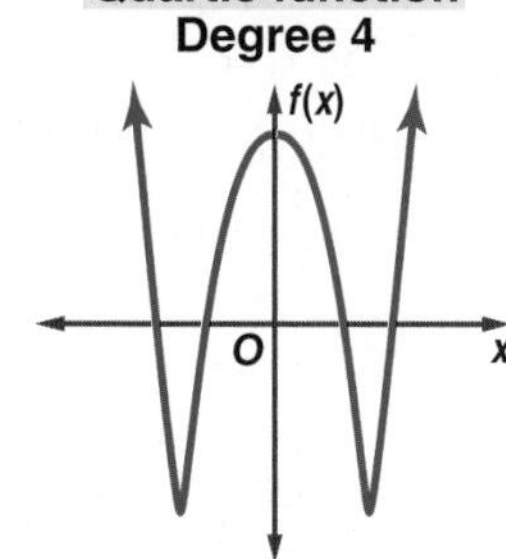

Quintic function
Degree 5

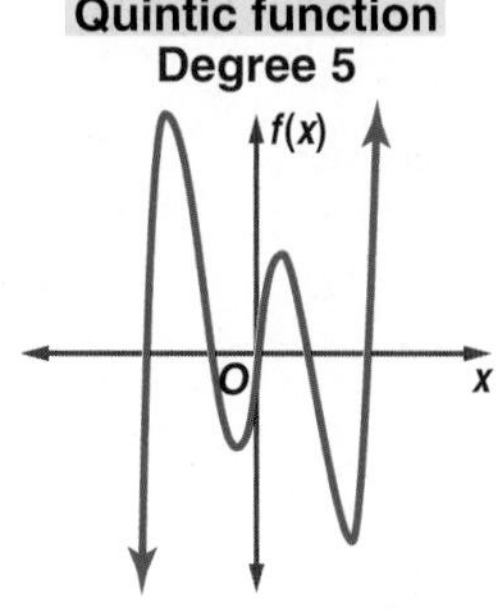

StudyTip

CCSS Sense-Making The leading coefficient and degree are the sole determining factors for the end behavior of a polynomial function. With very large or very small numbers, the rest of the polynomial is insignificant in the appearance of the graph.

The domain of any polynomial function is all real numbers. The **end behavior** is the behavior of the graph of $f(x)$ as x approaches positive infinity ($x \to +\infty$) or negative infinity ($x \to -\infty$). The degree and leading coefficient of a polynomial function determine the end behavior of the graph and the range of the function.

KeyConcept End Behavior of a Polynomial Function

Degree: even
Leading Coefficient: positive
End Behavior:

$f(x) \to +\infty$ as $x \to -\infty$

$f(x) \to +\infty$ as $x \to +\infty$

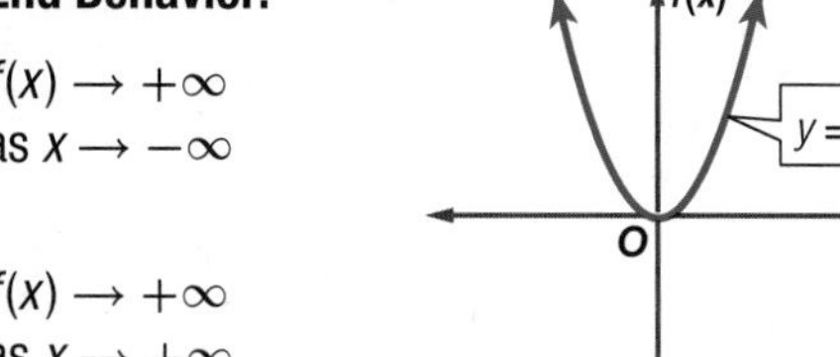

Domain: all real numbers
Range: all real numbers $\geq$ minimum

Degree: odd
Leading Coefficient: positive
End Behavior:

$f(x) \to -\infty$ as $x \to -\infty$

$f(x) \to +\infty$ as $x \to +\infty$

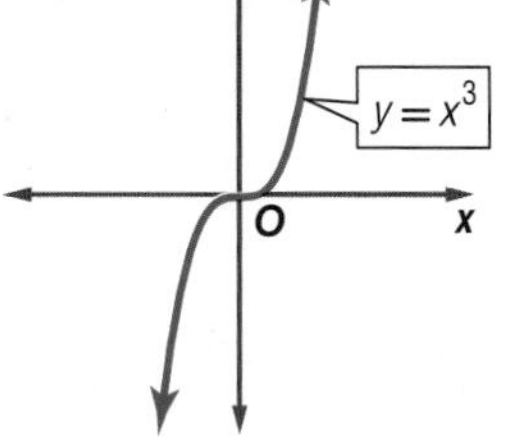

Domain: all real numbers
Range: all real numbers

Degree: even
Leading Coefficient: negative
End Behavior:

$f(x) \to -\infty$ as $x \to -\infty$

$f(x) \to -\infty$ as $x \to +\infty$

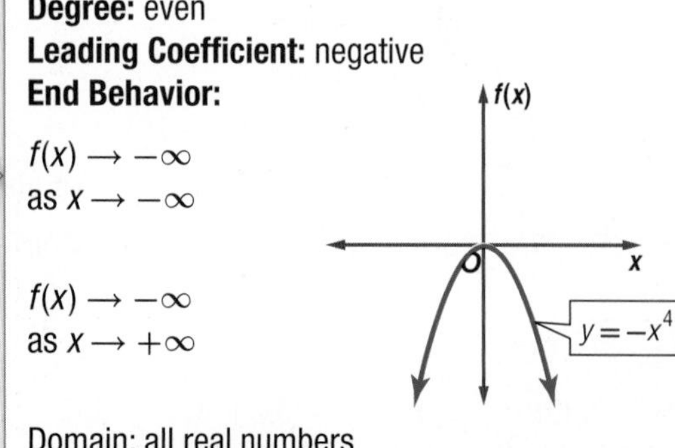

Domain: all real numbers
Range: all real numbers $\leq$ maximum

Degree: odd
Leading Coefficient: negative
End Behavior:

$f(x) \to +\infty$ as $x \to -\infty$

$f(x) \to -\infty$ as $x \to +\infty$

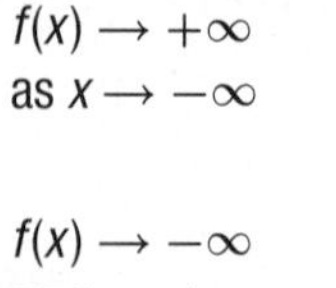

Domain: all real numbers
Range: all real numbers

ReviewVocabulary

infinity endless or boundless

ReviewVocabulary

zero the x-coordinate of the point at which a graph intersects the x-axis

The number of real zeros of a polynomial function can be determined by examining its graph. Recall that real zeros occur at x-intercepts, so the number of times a graph crosses the x-axis equals the number of real zeros.

KeyConcept Zeros of Even- and Odd-Degree Functions

Odd-degree functions will always have an odd number of real zeros. Even-degree functions will always have an even number of real zeros or no real zeros at all.

Even-Degree Polynomials

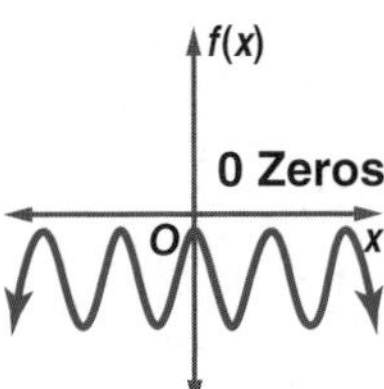

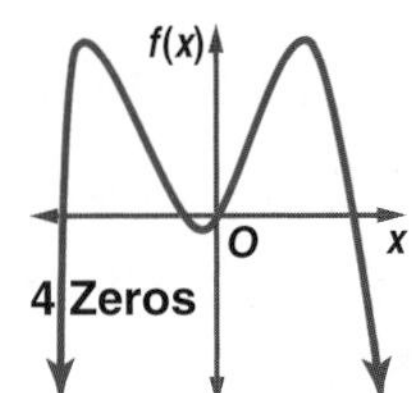

Odd-Degree Polynomials

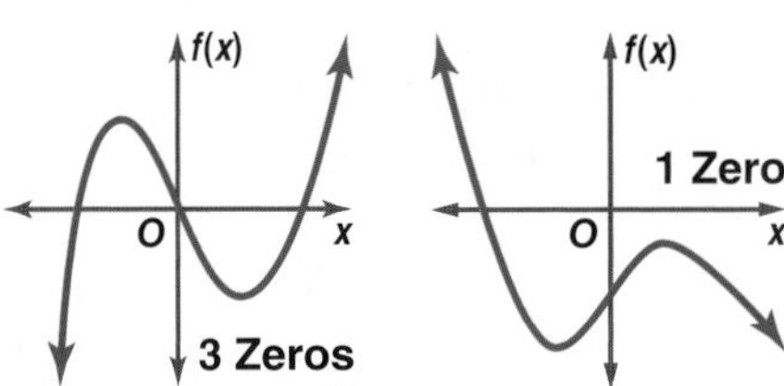

StudyTip

Double roots When a graph is tangent to the x-axis, there is a *double root*, which represents two of the same root.

Example 4 Graphs of Polynomial Functions

For each graph,

- **describe the end behavior,**
- **determine whether it represents an odd-degree or an even-degree polynomial function, and**
- **state the number of real zeros.**

a.

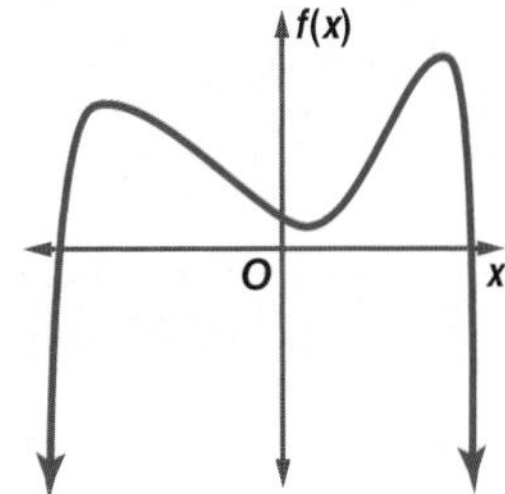

$f(x) \to -\infty$ as $x \to -\infty$.
$f(x) \to -\infty$ as $x \to +\infty$.

Since the end behavior is in the same direction, it is an even-degree function. The graph intersects the x-axis at two points, so there are two real zeros.

b.

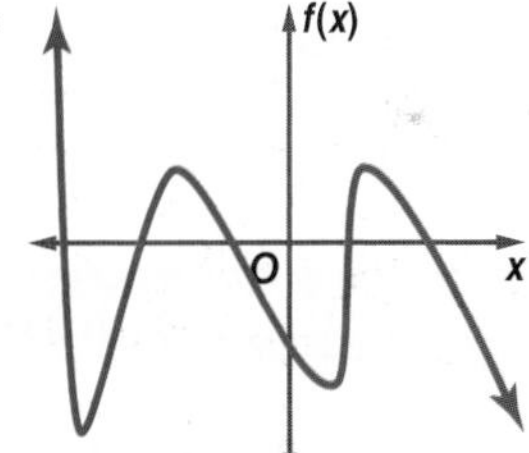

$f(x) \to +\infty$ as $x \to -\infty$.
$f(x) \to -\infty$ as $x \to +\infty$.

Since the end behavior is in opposite directions, it is an odd-degree function. The graph intersects the x-axis at five points, so there are five real zeros.

GuidedPractice

4A.

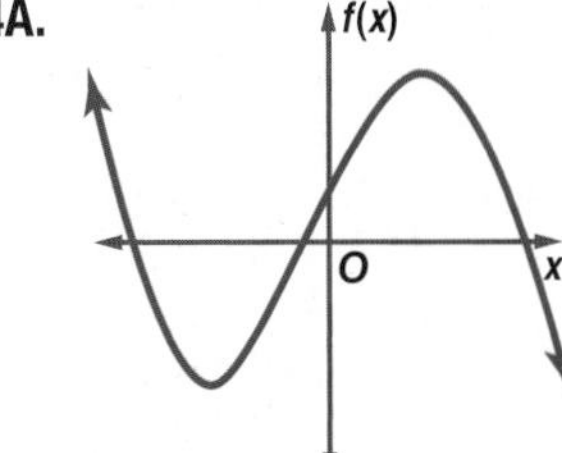

4B.

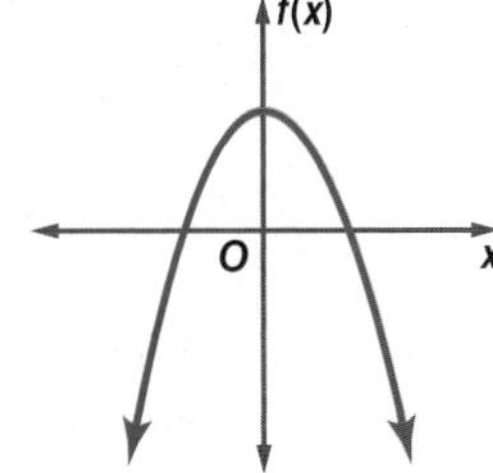

Check Your Understanding

= Step-by-Step Solutions begin on page R14.

Example 1 State the degree and leading coefficient of each polynomial in one variable. If it is not a polynomial in one variable, explain why.

1. $11x^6 - 5x^5 + 4x^2$
2. $-10x^7 - 5x^3 + 4x - 22$
3. $14x^4 - 9x^3 + 3x - 4y$
4. $8x^5 - 3x^2 + 4xy - 5$

Example 2 Find $w(5)$ and $w(-4)$ for each function.

5. $w(x) = -2x^3 + 3x - 12$
6. $w(x) = 2x^4 - 5x^3 + 3x^2 - 2x + 8$

Example 3 If $c(x) = 4x^3 - 5x^2 + 2$ and $d(x) = 3x^2 + 6x - 10$, find each value.

7. $c(y^3)$
8. $-4[d(3z)]$
9. $6c(4a) + 2d(3a - 5)$
10. $-3c(2b) + 6d(4b - 3)$

Example 4 For each graph,

a. describe the end behavior,

b. determine whether it represents an odd-degree or an even-degree function, and

c. state the number of real zeros.

11.

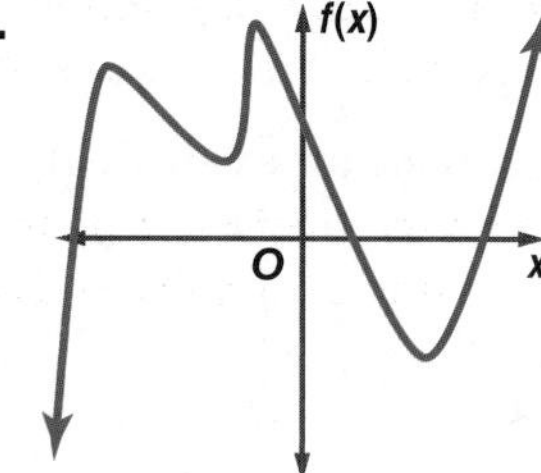

12.

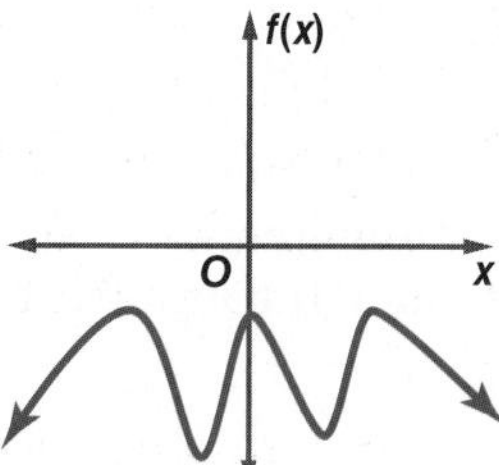

Practice and Problem Solving

Extra Practice is on page R5.

Example 1 CCSS PERSEVERANCE State the degree and leading coefficient of each polynomial in one variable. If it is not a polynomial in one variable, explain why.

13. $-6x^6 - 4x^5 + 13xy$
14. $3a^7 - 4a^4 + \frac{3}{a}$
15. $8x^5 - 12x^6 + 14x^3 - 9$
16. $-12 - 8x^2 + 5x - 21x^7$
17. $15x - 4x^3 + 3x^2 - 5x^4$
18. $13b^3 - 9b + 3b^5 - 18$
19. $(d + 5)(3d - 4)$
20. $(5 - 2y)(4 + 3y)$
21. $6x^5 - 5x^4 + 2x^9 - 3x^2$
22. $7x^4 + 3x^7 - 2x^8 + 7$

Example 2 Find $p(-6)$ and $p(3)$ for each function.

23. $p(x) = x^4 - 2x^2 + 3$
24. $p(x) = -3x^3 - 2x^2 + 4x - 6$
25. $p(x) = 2x^3 + 6x^2 - 10x$
26. $p(x) = x^4 - 4x^3 + 3x^2 - 5x + 24$
27. $p(x) = -x^3 + 3x^2 - 5$
28. $p(x) = 2x^4 + x^3 - 4x^2$

Example 3 If $c(x) = 2x^2 - 4x + 3$ and $d(x) = -x^3 + x + 1$, find each value.

29. $c(3a)$
30. $5d(2a)$
31. $c(b^2)$
32. $d(4a^2)$
33. $d(4y - 3)$
34. $c(y^2 - 1)$

Example 4 **For each graph,**

a. **describe the end behavior,**

b. **determine whether it represents an odd-degree or an even-degree function, and**

c. **state the number of real zeros.**

35.

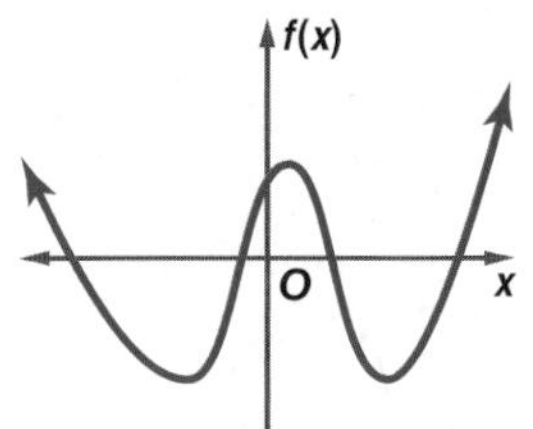

36.

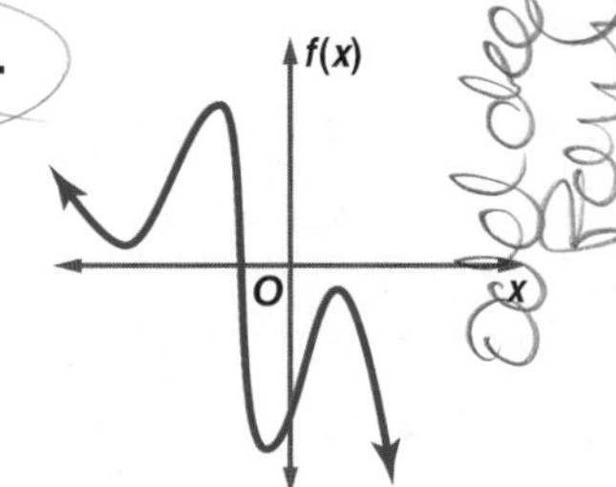

37.

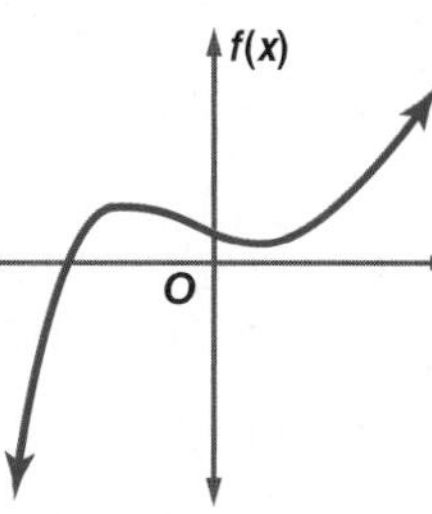

38.

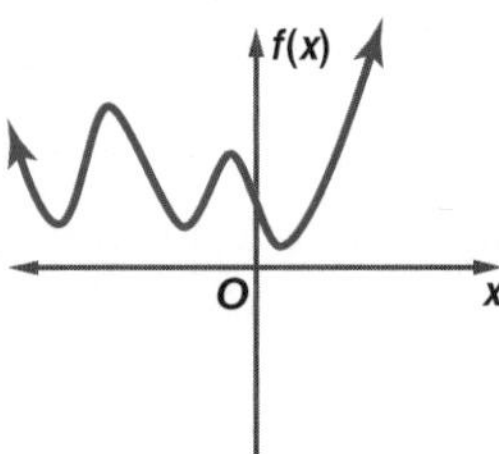

39.

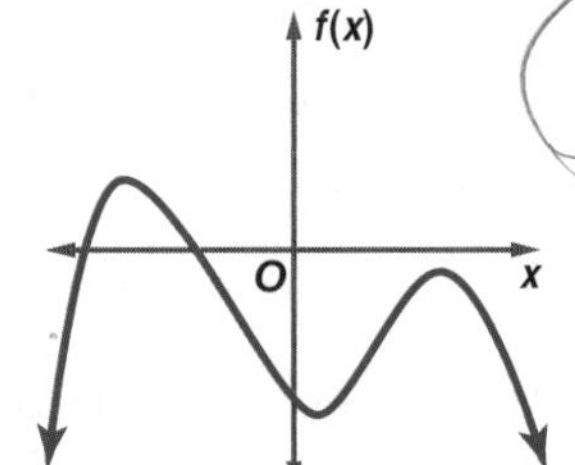

40.

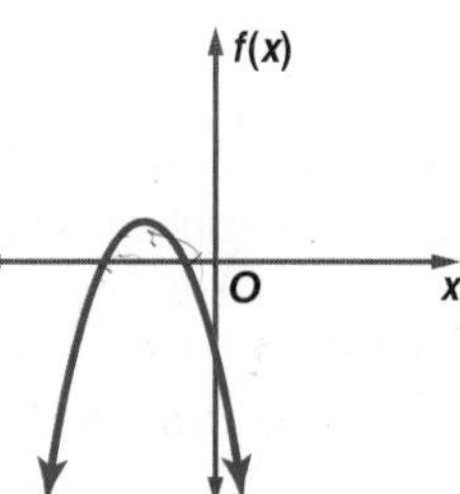

41 **PHYSICS** For a moving object with mass m in kilograms, the kinetic energy KE in joules is given by the function $KE(v) = 0.5mv^2$, where v represents the speed of the object in meters per second. Find the kinetic energy of an all-terrain vehicle with a mass of 171 kilograms moving at a speed of 11 meters/second.

42. **CCSS MODELING** A microwave manufacturing firm has determined that their profit function is $P(x) = -0.0014x^3 + 0.3x^2 + 6x - 355$, where x is the number of microwaves sold annually.

a. Graph the profit function using a calculator.

b. Determine a reasonable viewing window for the function.

c. Approximate all of the zeros of the function using the **CALC** menu.

d. What must be the range of microwaves sold in order for the firm to have a profit?

Find $p(-2)$ and $p(8)$ for each function.

43. $p(x) = \frac{1}{4}x^4 + \frac{1}{2}x^3 - 4x^2$

44. $p(x) = \frac{1}{8}x^4 - \frac{3}{2}x^3 + 12x - 18$

45. $p(x) = \frac{3}{4}x^4 - \frac{1}{8}x^2 + 6x$

46. $p(x) = \frac{5}{8}x^3 - \frac{1}{2}x^2 + \frac{3}{4}x + 10$

Use the degree and end behavior to match each polynomial to its graph.

A

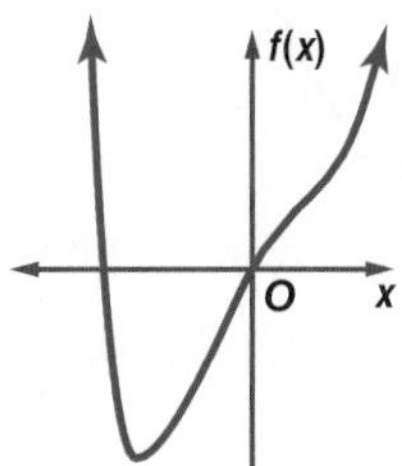

B

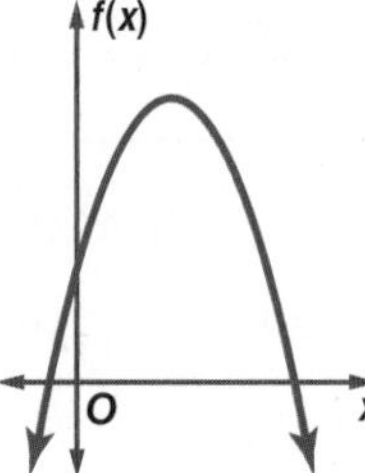

C

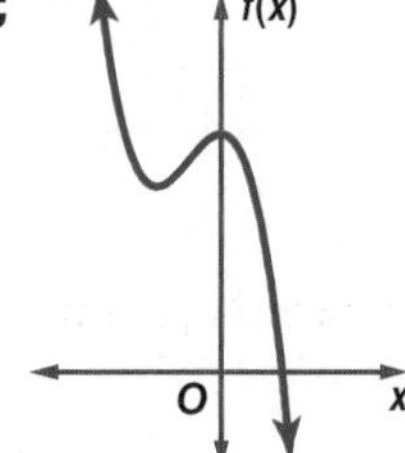

D

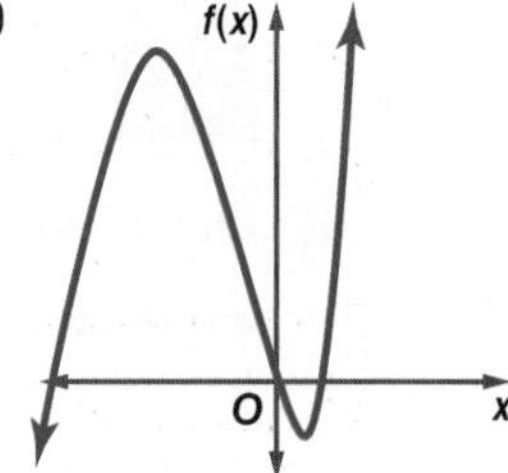

47. $f(x) = x^3 + 3x^2 - 4x$

48. $f(x) = -2x^2 + 8x + 5$

49. $f(x) = x^4 - 3x^2 + 6x$

50. $f(x) = -4x^3 - 4x^2 + 8$

If $c(x) = x^3 - 2x$ and $d(x) = 4x^2 - 6x + 8$, find each value.

51. $3c(a - 4) + 3d(a + 5)$

52. $-2d(2a + 3) - 4c(a^2 + 1)$

53. $5c(a^2) - 8d(6 - 3a)$

54. $-7d(a^3) + 6c(a^4 + 1)$

55. **BUSINESS** A clothing manufacturer's profitability can be modeled by $p(x) = -x^4 + 40x^2 - 144$, where x is the number of items sold in thousands and $p(x)$ is the company's profit in thousands of dollars.

a. Use a table of values to sketch the function.

b. Determine the zeros of the function.

c. Between what two values should the company sell in order to be profitable?

d. Explain why only two of the zeros are considered in part **c**.

56. **MULTIPLE REPRESENTATIONS** Consider $g(x) = (x - 2)(x + 1)(x - 3)(x + 4)$.

a. Analytical Determine the x- and y-intercepts, roots, degree, and end behavior of $g(x)$.

b. Algebraic Write the function in standard form

c. Tabular Make a table of values for the function.

d. Graphical Sketch a graph of the function by plotting points and connecting them with a smooth curve.

Describe the end behavior of the graph of each function.

57. $f(x) = -5x^4 + 3x^2$

58. $g(x) = 2x^5 + 6x^4$

59. $h(x) = -4x^7 + 8x^6 - 4x$

60. $f(x) = 6x - 7x^2$

61. $g(x) = 8x^4 + 5x^5$

62. $h(x) = 9x^6 - 5x^7 + 3x^2$

H.O.T. Problems Use Higher-Order Thinking Skills

63. **CCSS CRITIQUE** Shenequa and Virginia are determining the number of zeros of the graph at the right. Is either of them correct? Explain your reasoning.

Shenequa	Virginia
There are 7 zeros because the graph crosses the x-axis 7 times.	There are 8 zeros because the graph crosses the x-axis 7 times, and there is a double root.

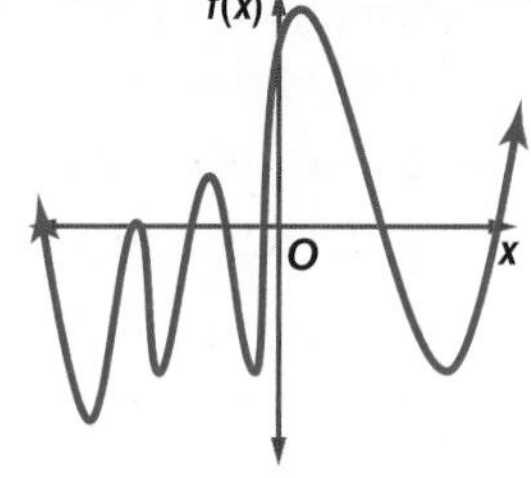

64. **CHALLENGE** Of $f(x)$ and $g(x)$, which function has more potential real roots? What is the degree of that function?

x	−24	−18	−12	−6	0	6	12	18	24
f(x)	−8	−1	3	−2	4	7	−1	−8	5

$g(x) = x^4 + x^3 - 13x^2 + x + 4$

65. **CHALLENGE** If $f(x)$ has a degree of 5 and a positive leading coefficient and $g(x)$ has a degree of 3 and a positive leading coefficient, determine the end behavior of $\frac{f(x)}{g(x)}$. Explain your reasoning.

66. **OPEN ENDED** Sketch the graph of an even-degree polynomial with 8 real roots, one of them a double root.

67. **REASONING** Determine whether the following statement is *always*, *sometimes*, or *never* true. Explain.

A polynomial function that has four real roots is a fourth-degree polynomial.

68. **WRITING IN MATH** Describe what the end behavior of a polynomial function is and how to determine it.

Standardized Test Practice

69. SHORT RESPONSE Four students solved the same math problem. Each student's work is shown below. Who is correct?

Student A

$x^2x^{-5} = \frac{x^2}{x^5}$

$= \frac{1}{x^3}, x \neq 0$

Student C

$x^2x^{-5} = \frac{x^2}{x^{-5}}$

$= x^7, x \neq 0$

Student B

$x^2x^{-5} = \frac{x^2}{x^{-5}}$

$= x^{-7}, x \neq 0$

Student D

$x^2x^{-5} = \frac{x^2}{x^5}$

$= x^3, x \neq 0$

70. SAT/ACT What is the remainder when $x^3 - 7x + 5$ is divided by $x + 3$?

A −11

B −1

C 1

D 11

E 35

71. EXTENDED RESPONSE A company manufactures tables and chairs. It costs $40 to make each table and $25 to make each chair. There is $1440 available to spend on manufacturing each week. Let t = the number of tables produced and c = the number of chairs produced.

a. The manufacturing equation is $40t + 25c = 1500$. Construct a graph of this equation.

b. The company always produces two chairs with each table. Write and graph an equation to represent this situation on the same graph as the one in part **a**.

c. Determine the number of tables and chairs that the company can produce each week.

d. Explain how to determine this answer using the graph.

72. If $i = \sqrt{-1}$, then $5i(7i) =$

F 70

G 35

H −35

J −70

Spiral Review

Simplify. (Lesson 5-2)

73. $\frac{16x^4y^3 + 32x^6y^5z^2}{8x^2y}$

74. $\frac{18ab^4c^5 - 30a^4b^3c^2 + 12a^5bc^3}{6abc^2}$

75. $\frac{18c^5d^2 - 3c^2d^2 + 12a^5c^3d^4}{3c^2d^2}$

Determine whether each expression is a polynomial. If it is a polynomial, state the degree of the polynomial. (Lesson 5-1)

76. $8x^2 + 5xy^3 - 6x + 4$

77. $9x^4 + 12x^6 - 16$

78. $3x^4 + 2x^2 - x^{-1}$

79. FOUNTAINS The height of a fountain's water stream can be modeled by a quadratic function. Suppose the water from a jet reaches a maximum height of 8 feet at a distance 1 foot away from the jet. (Lesson 4-7)

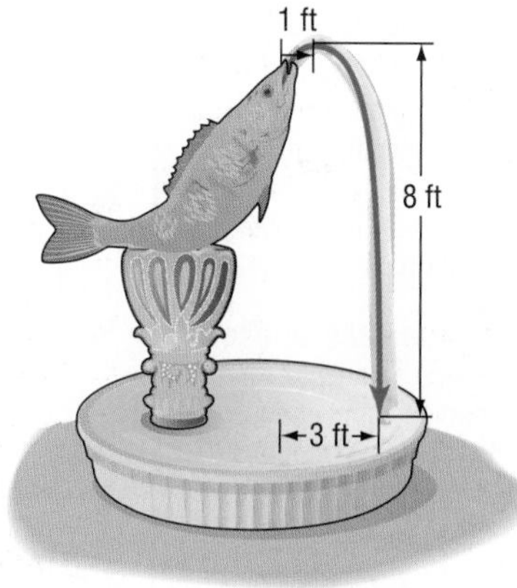

a. If the water lands 3 feet away from the jet, find a quadratic function that models the height $h(d)$ of the water at any given distance d feet from the jet. Then compare the graph of the function to the parent function.

b. Suppose a worker increases the water pressure so that the stream reaches a maximum height of 12.5 feet at a distance of 15 inches from the jet. The water now lands 3.75 feet from the jet. Write a new quadratic function for $h(d)$. How do the changes in h and k affect the shape of the graph?

Solve each inequality. (Lesson 1-6)

80. $|2x + 4| \leq 8$

81. $|-3x + 2| \geq 4$

82. $|2x - 8| - 4 \leq -6$

Skills Review

Determine whether each function has a maximum or minimum value, and find that value.

83. $f(x) = 3x^2 - 8x + 4$

84. $f(x) = -4x^2 + 2x - 10$

85. $f(x) = -0.25x^2 + 4x - 5$

LESSON 5-4 Analyzing Graphs of Polynomial Functions

Then	Now	Why?
You used maxima and minima and graphs of polynomials.	**1** Graph polynomial functions and locate their zeros. **2** Find the relative maxima and minima of polynomial functions.	Annual attendance at the movies has fluctuated since the first movie theater, the Nickelodeon, opened in Pittsburgh in 1906. Overall attendance peaked during the 1920s, and it was at its lowest during the 1970s. A graph of the annual attendance to the movies can be represented by a polynomial function.

NewVocabulary
Location Principle
relative maximum
relative minimum
extrema
turning points

Common Core State Standards

Content Standards
F.IF.4 For a function that models a relationship between two quantities, interpret key features of graphs and tables in terms of the quantities, and sketch graphs showing key features given a verbal description of the relationship.
F.IF.7.c Graph polynomial functions, identifying zeros when suitable factorizations are available, and showing end behavior.

Mathematical Practices
3 Construct viable arguments and critique the reasoning of others.

1 Graphs of Polynomial Functions

To graph a polynomial function, make a table of values to find several points and then connect them to make a smooth continuous curve. Knowing the end behavior of the graph will assist you in completing the graph.

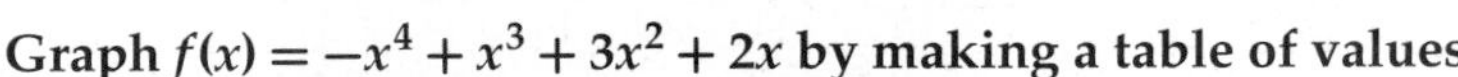

Example 1 Graph of a Polynomial Function

Graph $f(x) = -x^4 + x^3 + 3x^2 + 2x$ by making a table of values.

x	$f(x)$	x	$f(x)$
−2.5	≈ −41	0.5	≈ 1.8
−2.0	−16	1.0	5.0
−1.5	≈ −4.7	1.5	≈ 8.1
−1.0	−1.0	2.0	8.0
−0.5	≈ −0.4	2.5	≈ 0.3
0.0	0.0	3.0	−21

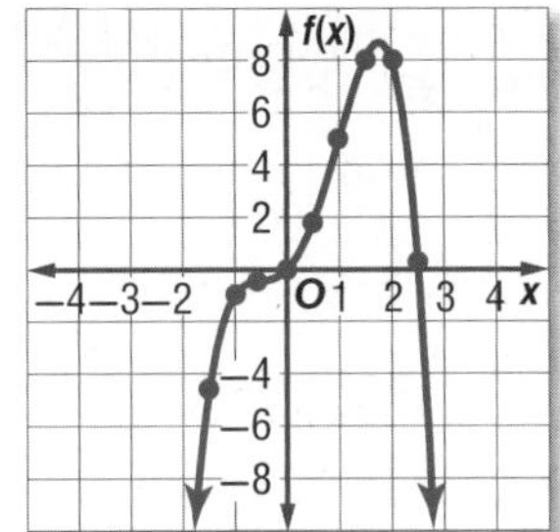

This is an even-degree polynomial with a negative leading coefficient, so $f(x) \to -\infty$ as $x \to -\infty$ and $f(x) \to -\infty$ as $x \to +\infty$. Notice that the graph intersects the x-axis at two points, indicating there are two zeros for this function.

GuidedPractice

1. Graph $f(x) = x^4 - x^3 - 2x^2 + 4x - 6$ by making a table of values.

In Example 1, one of the zeros occurred at $x = 0$. Another zero occurred between $x = 2.5$ and $x = 3.0$. Because $f(x)$ is positive for $x = 2.5$ and negative for $x = 3.0$ and all polynomial functions are continuous, we know there is a zero between these two values.

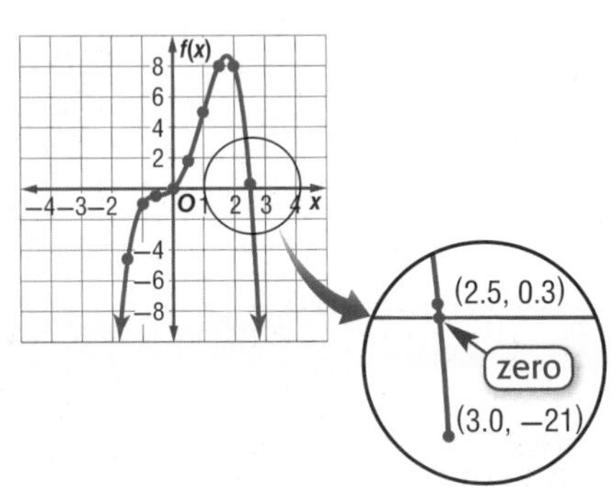

So, if the value of $f(x)$ *changes signs* from one value of x to the next, then there is a zero between those two x-values. This idea is called the **Location Principle**.

StudyTip

Degree Recall that the degree of the function is also the maximum number of zeros the function can have.

KeyConcept Location Principle

Words Suppose $y = f(x)$ represents a polynomial function and a and b are two real numbers such that $f(a) < 0$ and $f(b) > 0$. Then the function has at least one real zero between a and b.

Model

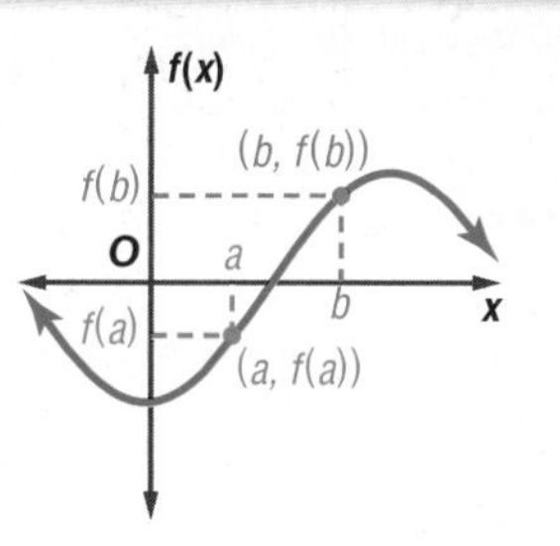

Example 2 Locate Zeros of a Function

Determine consecutive integer values of x between which each real zero of $f(x) = x^3 - 4x^2 + 3x + 1$ is located. Then draw the graph.

Make a table of values. Since $f(x)$ is a third-degree polynomial function, it will have either 3 or 1 real zeros. Look at the values of $f(x)$ to locate the zeros. Then use the points to sketch a graph of the function.

x	$f(x)$	
−2	−29	
−1	−7	
0	1	← change in sign
1	1	
2	−1	← change in sign
3	1	← change in sign
4	13	

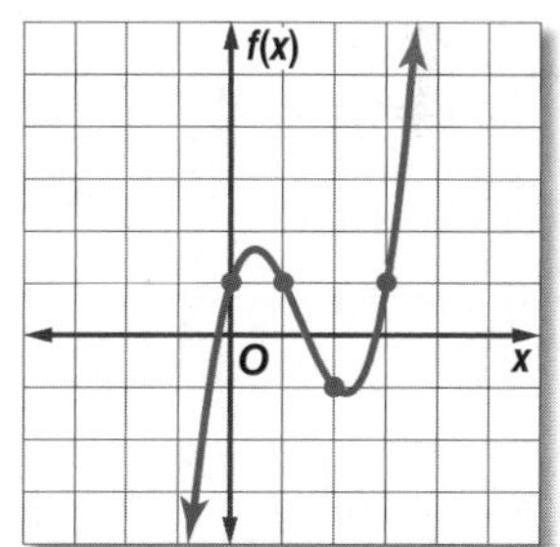

The changes in sign indicate that there are zeros between $x = -1$ and $x = 0$, between $x = 1$ and $x = 2$, and between $x = 2$ and $x = 3$.

GuidedPractice

2. Determine consecutive integer values of x between which each real zero of the function $f(x) = x^4 - 3x^3 - 2x^2 + x + 1$ is located. Then draw the graph.

2 Maximum and Minimum Points

The graph below shows the general shape of a third-degree polynomial function.

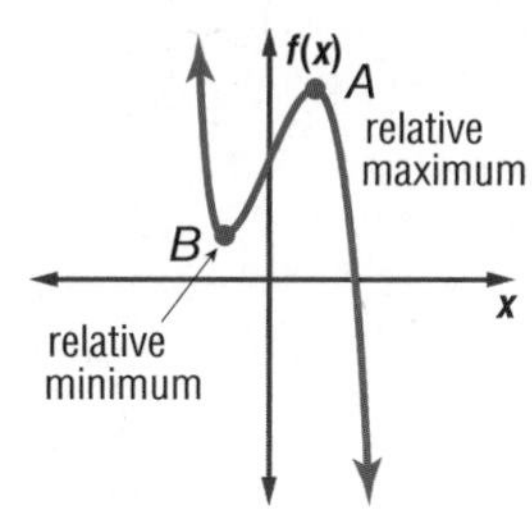

Point A on the graph is a **relative maximum** of the function since no other nearby points have a greater y-coordinate. The graph is increasing as it approaches A and decreasing as it moves from A.

Likewise, point B is a **relative minimum** since no other nearby points have a lesser y-coordinate. The graph is decreasing as it approaches B and increasing as it moves from B. The maximum and minimum values of a function are called the **extrema**.

StudyTip

Odd Functions Some odd functions, like $f(x) = x^3$, have no turning points.

These points are often referred to as **turning points**. The graph of a polynomial function of degree n has at most $n - 1$ turning points.

> **StudyTip**
> **Maximum and Minimum** A polynomial with a degree greater than 3 may have more than one relative maximum or relative minimum.

Example 3 Maximum and Minimum Points

Graph $f(x) = x^3 - 4x^2 - 2x + 3$. Estimate the x-coordinates at which the relative maxima and relative minima occur.

Make a table of values and graph the function.

x	$f(x)$
−2	−17
−1	0
0	3
1	−2
2	−9
3	−12
4	−4
5	18

zero

indicates a relative maximum

indicates a relative minimum

zero between 4 and 5

Look at the table of values and the graph.

The value of $f(x)$ changes signs between $x = 4$ and $x = 5$, indicating a zero of the function.

The value of $f(x)$ at $x = 0$ is greater than the surrounding points, so there must be a relative maximum *near* $x = 0$.

The value of $f(x)$ at $x = 3$ is less than the surrounding points, so there must be a relative minimum *near* $x = 3$.

> **StudyTip**
> **Rational Values** Zeros and turning points will not always occur at integral values of x.

CHECK You can use a graphing calculator to find the relative maximum and relative minimum of a function and confirm your estimates.

Enter $y = x^3 - 4x^2 - 2x + 3$ in the **Y=** list and graph the function.

Use the **CALC** menu to find each maximum and minimum.

When selecting the left bound, move the cursor to the left of the maximum or minimum. When selecting the right bound, move the cursor to the right of the maximum or minimum.

Press ENTER twice.

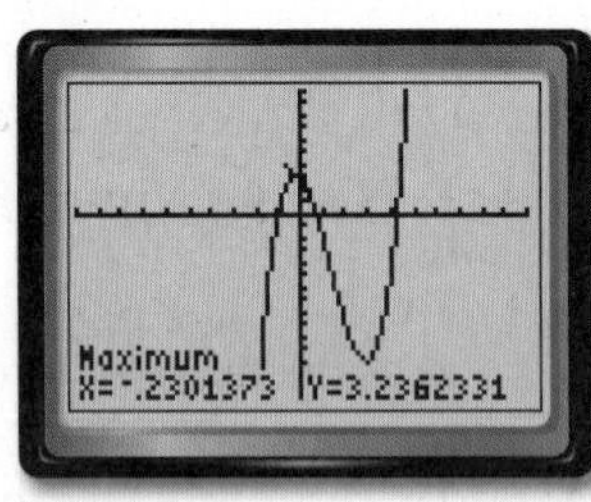

[−10, 10] scl: 1 by [−15, 10] scl: 1

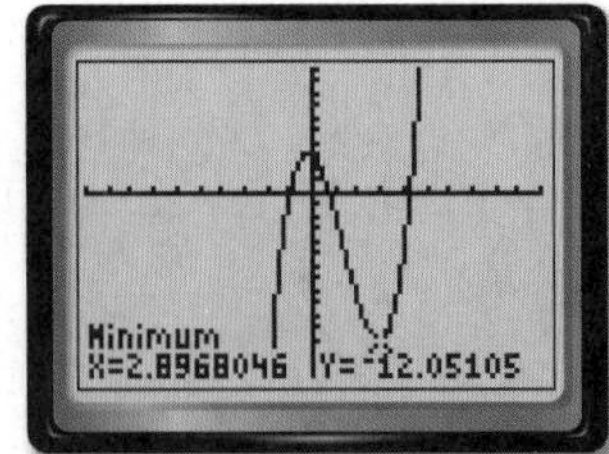

[−10, 10] scl: 1 by [−15, 10] scl: 1

The estimates for a relative maximum near $x = 0$ and a relative minimum near $x = 3$ are accurate.

GuidedPractice

3. Graph $f(x) = 2x^3 + x^2 - 4x - 2$. Estimate the x-coordinates at which the relative maxima and relative minima occur.

The graph of a polynomial function can reveal trends in real-world data. It is often helpful to note when the graph is increasing or decreasing.

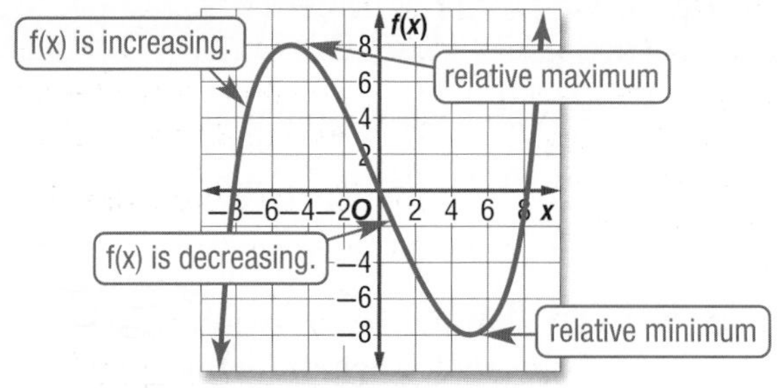

Real-WorldLink

Over 1.4 billion movie tickets were sold in the United States in 2006.

Source: CNN

Real-World Example 4 Graph a Polynomial Model

MOVIES Refer to the beginning of the lesson. Annual admissions to movies in the United States can be modeled by the function $f(x) = -0.0017x^4 + 0.31x^3 - 17.66x^2 + 277x + 3005$, where x is the number of years since 1926 and $f(x)$ is the annual admissions in millions.

a. Graph the function.

Make a table of values for the years 1926–2006. Plot the points and connect with a smooth curve. Finding and plotting the points for every tenth year gives a good approximation of the graph.

x	$f(x)$
0	3005
10	4302
20	3689
30	2414
40	1317
50	830
60	977
70	1374
80	1229

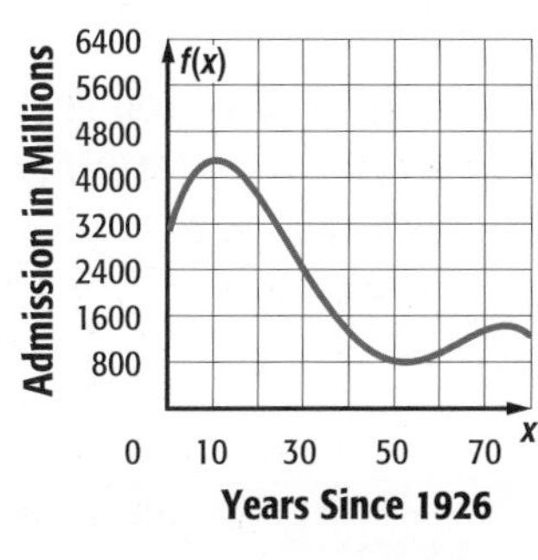

b. Describe the turning points of the graph and its end behavior.

There are relative maxima near 1936 and 2000 and a relative minimum between 1976 and 1981. $f(x) \to -\infty$ as $x \to -\infty$ and $f(x) \to -\infty$ as $x \to \infty$.

c. What trends in movie admissions does the graph suggest? Is it reasonable that the trend will continue indefinitely?

Movie attendance peaked around 1936 and declined until about 1978. It then increased until 2000 and began a decline.

d. Is it reasonable that the trend will continue indefinitely?

This trend may continue for a couple of years, but the graph will soon become unreasonable as it predicts negative attendance for the future.

GuidedPractice

4. **FAX MACHINES** The annual sales of fax machines for home use can be modeled by $f(x) = -0.17x^4 + 6.29x^3 - 77.65x^2 + 251x + 1100$, where x is the number of years after 1990 and $f(x)$ is the annual sales in millions of dollars.
   - **A.** Graph the function.
   - **B.** Describe the turning points of the graph and its end behavior.
   - **C.** What trends in fax machine sales does the graph suggest?
   - **D.** Is it reasonable that the trend will continue indefinitely?

Check Your Understanding

= Step-by-Step Solutions begin on page R14.

Example 1 **Graph each polynomial equation by making a table of values.**

1. $f(x) = 2x^4 - 5x^3 + x^2 - 2x + 4$
2. $f(x) = -2x^4 + 4x^3 + 2x^2 + x - 3$
3. $f(x) = 3x^4 - 4x^3 - 2x^2 + x - 4$
4. $f(x) = -4x^4 + 5x^3 + 2x^2 + 3x + 1$

Example 2 **Determine the consecutive integer values of x between which each real zero of each function is located. Then draw the graph.**

5. $f(x) = x^3 - 2x^2 + 5$
6. $f(x) = -x^4 + x^3 + 2x^2 + x + 1$
7. $f(x) = -3x^4 + 5x^3 + 4x^2 + 4x - 8$
8. $f(x) = 2x^4 - x^3 - 3x^2 + 2x - 4$

Example 3 **Graph each polynomial function. Estimate the x-coordinates at which the relative maxima and relative minima occur. State the domain and range for each function.**

9. $f(x) = x^3 + x^2 - 6x - 3$
10. $f(x) = 3x^3 - 6x^2 - 2x + 2$
11. $f(x) = -x^3 + 4x^2 - 2x - 1$
12. $f(x) = -x^3 + 2x^2 - 3x + 4$

Example 4

13. **CCSS SENSE-MAKING** Annual compact disc sales can be modeled by the quartic function $f(x) = 0.48x^4 - 9.6x^3 + 53x^2 - 49x + 599$, where x is the number of years after 1995 and $f(x)$ is annual sales in millions.
    a. Graph the function for $0 \le x \le 10$.
    b. Describe the turning points of the graph, its end behavior, and the intervals on which the graph is increasing or decreasing.
    c. Continue the graph for $x = 11$ and $x = 12$. What trends in compact disc sales does the graph suggest?
    d. Is it reasonable that the trend will continue indefinitely? Explain.

Practice and Problem Solving

Extra Practice is on page R5.

Examples 1–3 **Complete each of the following.**

a. **Graph each function by making a table of values.**
b. **Determine the consecutive integer values of x between which each real zero is located.**
c. **Estimate the x-coordinates at which the relative maxima and minima occur.**

14. $f(x) = x^3 + 3x^2$
15. $f(x) = -x^3 + 2x^2 - 4$
16. $f(x) = x^3 + 4x^2 - 5x$
17. $f(x) = x^3 - 5x^2 + 3x + 1$
18. $f(x) = -2x^3 + 12x^2 - 8x$
19. $f(x) = 2x^3 - 4x^2 - 3x + 4$
20. $f(x) = x^4 + 2x - 1$
21. $f(x) = x^4 + 8x^2 - 12$

Example 4

22. **FINANCIAL LITERACY** The average annual price of gasoline can be modeled by the cubic function $f(x) = 0.0007x^3 - 0.014x^2 + 0.08x + 0.96$, where x is the number of years after 1987 and $f(x)$ is the price in dollars.
    a. Graph the function for $0 \le x \le 30$.
    b. Describe the turning points of the graph and its end behavior.
    c. What trends in gasoline prices does the graph suggest?
    d. Is it reasonable that the trend will continue indefinitely? Explain.

Use a graphing calculator to estimate the x-coordinates at which the maxima and minima of each function occur. Round to the nearest hundredth.

23. $f(x) = x^3 + 3x^2 - 6x - 6$
24. $f(x) = -2x^3 + 4x^2 - 5x + 8$
25. $f(x) = -2x^4 + 5x^3 - 4x^2 + 3x - 7$
26. $f(x) = x^5 - 4x^3 + 3x^2 - 8x - 6$

Sketch the graph of polynomial functions with the following characteristics.

27. an odd function with zeros at $-5, -3, 0, 2$ and 4

28. an even function with zeros at $-2, 1, 3$, and 5

29. a 4-degree function with a zero at -3, maximum at $x = 2$, and minimum at $x = -1$

30. a 5-degree function with zeros at $-4, -1$, and 3, maximum at $x = -2$

31. an odd function with zeros at $-1, 2$, and 5 and a negative leading coefficient

32. an even function with a minimum at $x = 3$ and a positive leading coefficient

33. **DIVING** The deflection d of a 10-foot-long diving board can be calculated using the function $d(x) = 0.015x^2 - 0.0005x^3$, where x is the distance between the diver and the stationary end of the board in feet.

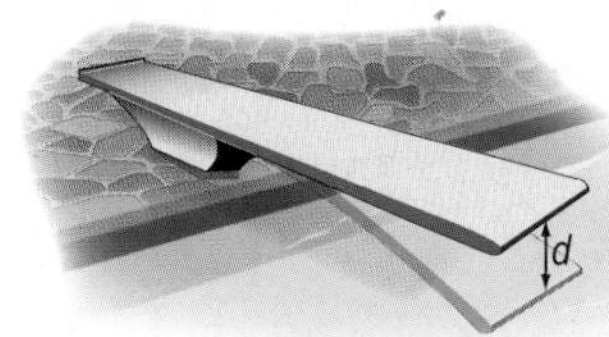

a. Make a table of values of the function for $0 \le x \le 10$.

b. Graph the function.

c. What does the end behavior of the graph suggest as x increases?

d. Will this trend continue indefinitely? Explain your reasoning.

Complete each of the following.

a. Estimate the x-coordinate of every turning point and determine if those coordinates are relative maxima or relative minima.

b. Estimate the x-coordinate of every zero.

c. Determine the smallest possible degree of the function.

d. Determine the domain and range of the function.

34.

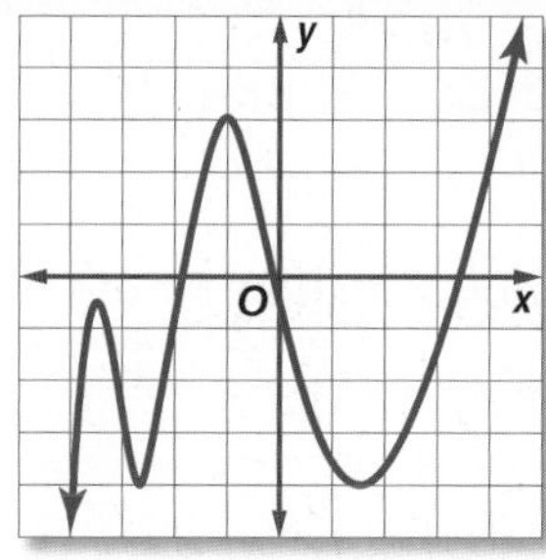

35.

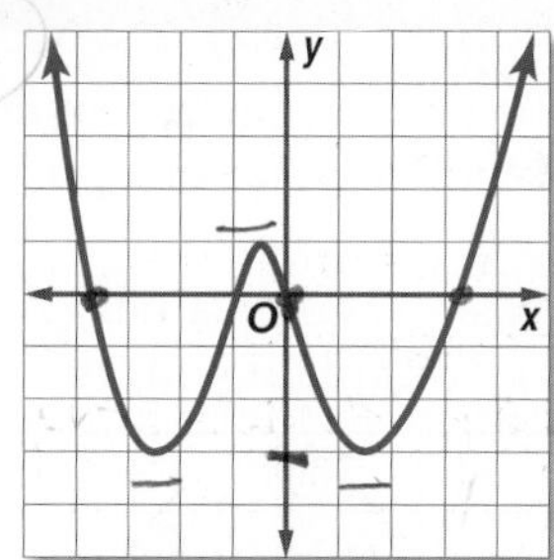

36.

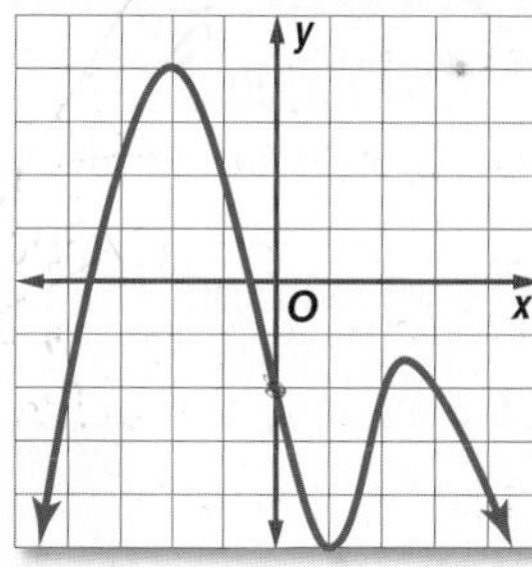

37.

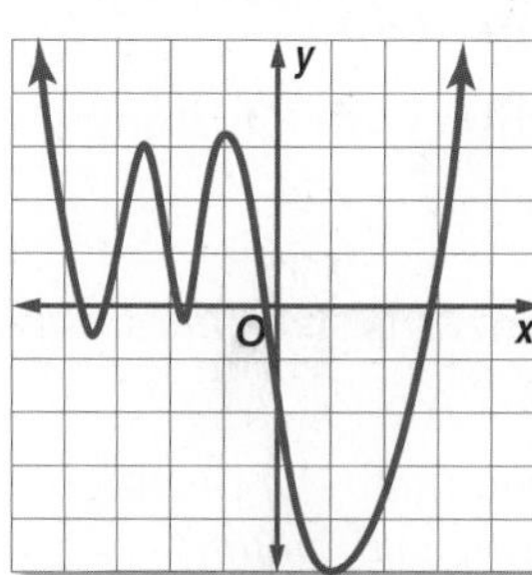

38.

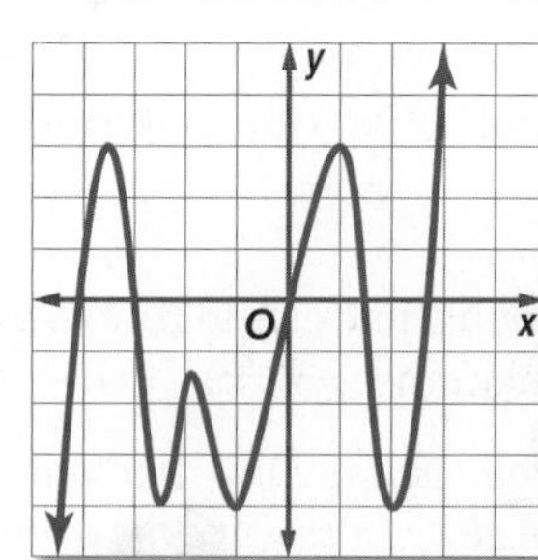

39.

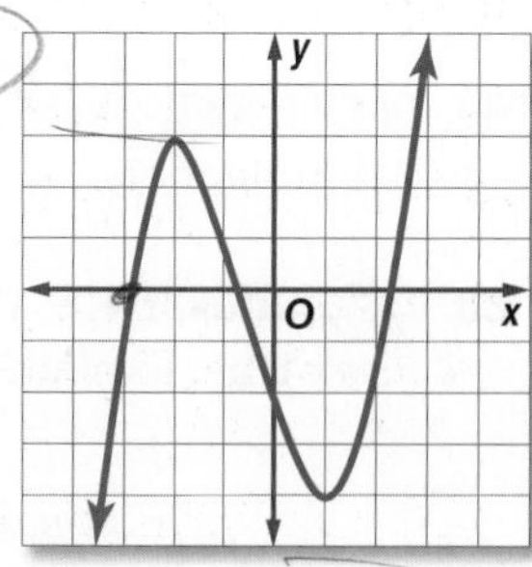

40. **CCSS REASONING** The number of subscribers using pagers in the United States can be modeled by $f(x) = 0.015x^4 - 0.44x^3 + 3.46x^2 - 2.7x + 9.68$, where x is the number of years after 1990 and $f(x)$ is the number of subscribers in millions.

a. Graph the function.

b. Describe the end behavior of the graph.

c. What does the end behavior suggest about the number of pager subscribers?

d. Will this trend continue indefinitely? Explain your reasoning.

41. **PRICING** Jin's vending machines currently sell an average of 3500 beverages per week at a rate of \$0.75 per can. She is considering increasing the price. Her weekly earnings can be represented by $f(x) = -5x^2 + 100x + 2625$, where x is the number of \$0.05 increases. Graph the function and determine the most profitable price for Jin.

For each function,

a. determine the zeros, x- and y-intercepts, and turning points,

b. determine the axis of symmetry, and

c. determine the intervals for which it is increasing, decreasing, or constant.

42. $y = x^4 - 8x^2 + 16$

43. $y = x^5 - 3x^3 + 2x - 4$

44. $y = -2x^4 + 4x^3 - 5x$

45. $y = \begin{cases} x^2 \text{ if } x \leq -4 \\ 5 \text{ if } -4 < x \leq 0 \\ x^3 \text{ if } x > 0 \end{cases}$

46. **MULTIPLE REPRESENTATIONS** Consider the following function.

$$f(x) = x^4 - 8.65x^3 + 27.34x^2 - 37.2285x + 18.27$$

a. **Analytical** What are the degree, leading coefficient, and end behavior?

b. **Tabular** Make a table of integer values $f(x)$ if $-4 \leq x \leq 4$. How many zeros does the function appear to have from the table?

c. **Graphical** Graph the function by using a graphing calculator.

d. **Graphical** Change the viewing window to [0, 4] scl: 1 by [−0.4, 0.4] scl: 0.2. What conclusions can you make from this new view of the graph?

H.O.T. Problems Use Higher-Order Thinking Skills

47. **REASONING** Explain why the leading coefficient and the degree are the only determining factors in the end behavior of a polynomial function.

48. **REASONING** The table below shows the values of $g(x)$, a cubic function. Could there be a zero between $x = 2$ and $x = 3$? Explain your reasoning.

x	−2	−1	0	1	2	3
$g(x)$	4	−2	−1	1	−2	−2

49. **OPEN ENDED** Sketch the graph of an odd polynomial function with 6 turning points and 2 double roots.

50. **CCSS ARGUMENTS** Determine whether the following statement is *sometimes*, *always*, or *never* true. Explain your reasoning.

For any continuous polynomial function, the y-coordinate of a turning point is also either a relative maximum or relative minimum.

51. **REASONING** A function is said to be even if for every x in the domain of f, $f(x) = f(-x)$. Is every even-degree polynomial function also an even function? Explain.

52. **REASONING** A function is said to be *odd* if for every x in the domain, $-f(x) = f(-x)$. Is every odd-degree polynomial function also an odd function? Explain.

53. **WRITING IN MATH** How can you use the characteristics of a polynomial function to sketch its graph?

Standardized Test Practice

54. Which of the following is the factorization of $2x - 15 + x^2$?

A $(x - 3)(x - 5)$
B $(x - 3)(x + 5)$
C $(x + 3)(x - 5)$
D $(x + 3)(x + 5)$

55. SHORT RESPONSE In the figure below, if $x = 35$ and $z = 50$, what is the value of y?

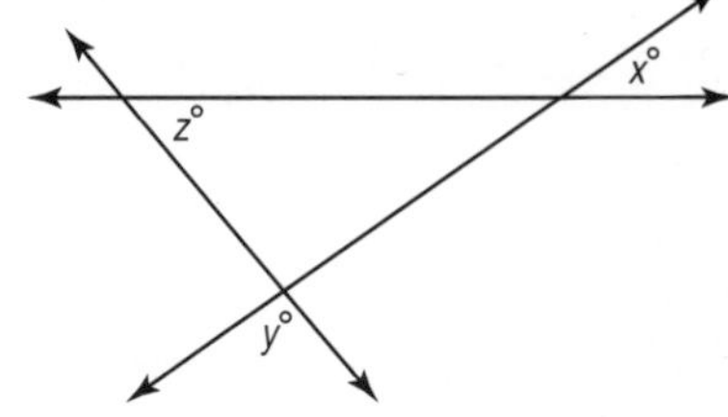

56. Which polynomial represents $(4x^2 + 5x - 3)(2x - 7)$?

F $8x^3 - 18x^2 - 41x - 21$
G $8x^3 + 18x^2 + 29x - 21$
H $8x^3 - 18x^2 - 41x + 21$
J $8x^3 + 18x^2 - 29x + 21$

57. SAT/ACT The figure at the right shows the graph of a polynomial function $f(x)$. Which of the following could be the degree of $f(x)$?

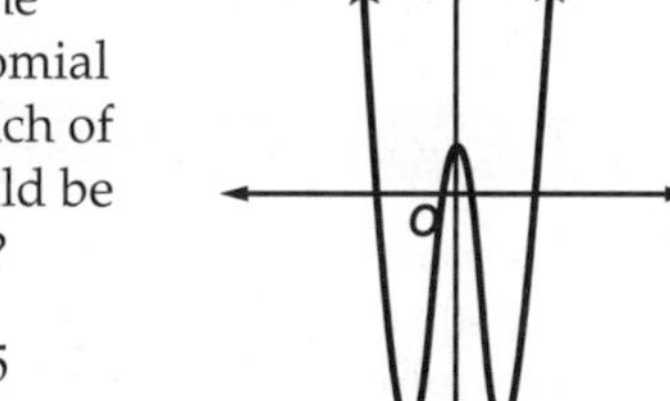

A 2
B 3
C 4
D 5
E 6

Spiral Review

For each graph,

a. describe the end behavior,

b. determine whether it represents an odd-degree or an even-degree function, and

c. state the number of real zeros. (Lesson 5-3)

58.

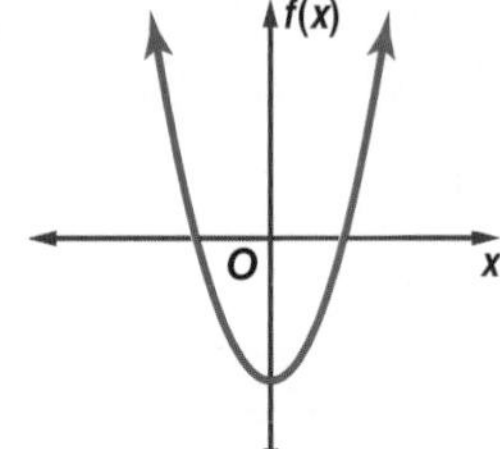

59.

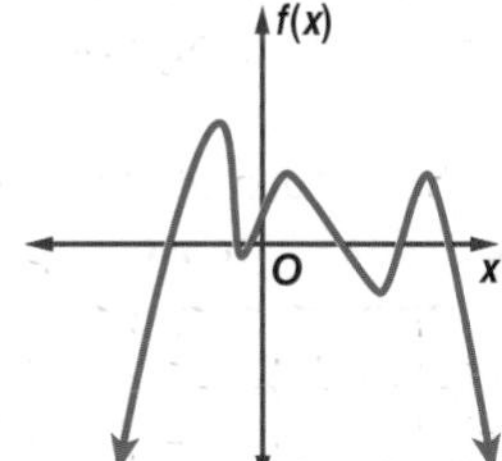

60.

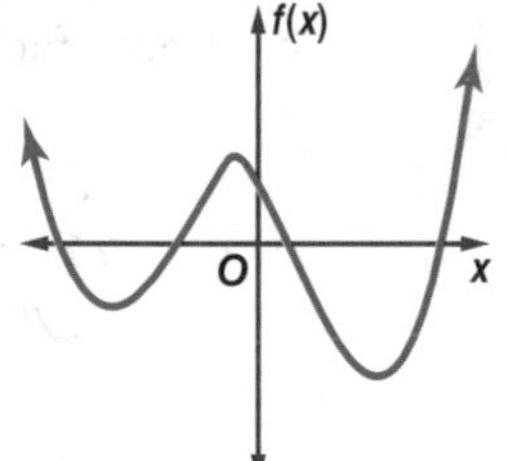

Simplify. (Lesson 5-2)

61. $(x^3 + 2x^2 - 5x - 6) \div (x + 1)$

62. $(4y^3 + 18y^2 + 5y - 12) \div (y + 4)$

63. $(2a^3 - a^2 - 4a) \div (a - 1)$

64. CHEMISTRY Tanisha needs 200 milliliters of a 48% concentration acid solution. She has 60% and 40% concentration solutions in her lab. How many milliliters of 40% acid solution should be mixed with 60% acid solution to make the required amount of 48% acid solution? (Lesson 3-8)

Skills Review

Factor.

65. $x^2 + 6x + 3x + 18$

66. $y^2 - 5y - 8y + 40$

67. $a^2 + 6a - 16$

68. $b^2 - 4b - 21$

69. $6x^2 - 5x - 4$

70. $4x^2 - 7x - 15$

EXTEND 5-4 Graphing Technology Lab
Modeling Data Using Polynomial Functions

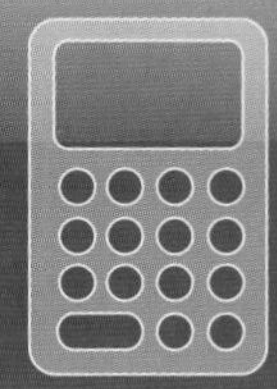

You can use a TI-83/84 Plus graphing calculator to model data points when a curve of best fit is a polynomial function.

CCSS Common Core State Standards
Content Standards
F.IF.4 For a function that models a relationship between two quantities, interpret key features of graphs and tables in terms of the quantities, and sketch graphs showing key features given a verbal description of the relationship.
F.IF.7.c Graph polynomial functions, identifying zeros when suitable factorizations are available, and showing end behavior.
Mathematical Practices
5 Use appropriate tools strategically.

Example

The table shows the distance a seismic wave produced by an earthquake travels from the epicenter. Draw a scatter plot and a curve of best fit to show how the distance is related to time. Then determine approximately how far away from the epicenter a seismic wave will be felt 8.5 minutes after an earthquake occurs.

Travel Time (min)	1	2	5	7	10	12	13
Distance (km)	400	800	2500	3900	6250	8400	10,000

Source: University of Arizona

Step 1 Enter time in **L1** and distance in **L2**.

KEYSTROKES: STAT 1 1 ENTER 2 ENTER 5 ENTER 7 ENTER 10 ENTER 12 ENTER 13 ENTER ▶ 400 ENTER 800 ENTER 2500 ENTER 3900 ENTER 6250 ENTER 8400 ENTER 10000 ENTER

Step 2 Graph the scatter plot.

KEYSTROKES: 2nd [STAT PLOT] 1 ENTER ▼ ENTER ZOOM 9

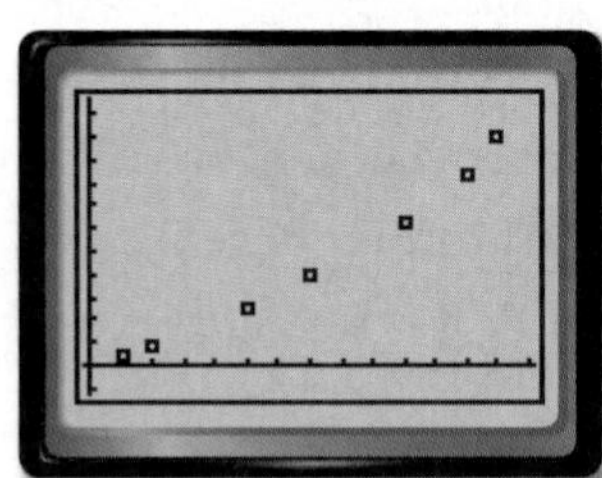

[−0.2, 14.2] scl: 1 by [−1232, 11632] scl: 1000

Step 3 Determine and graph the equation for a curve of best fit. Use a quartic regression for the data.

KEYSTROKES: STAT ▶ 7 ENTER Y= VARS 5 ▶ ▶ 1 GRAPH

The equation is shown in the Y= screen. If rounded, the regression equation shown on the calculator can be written as the algebraic equation $y = 0.7x^4 - 17x^3 + 161x^2 - 21x + 293$.

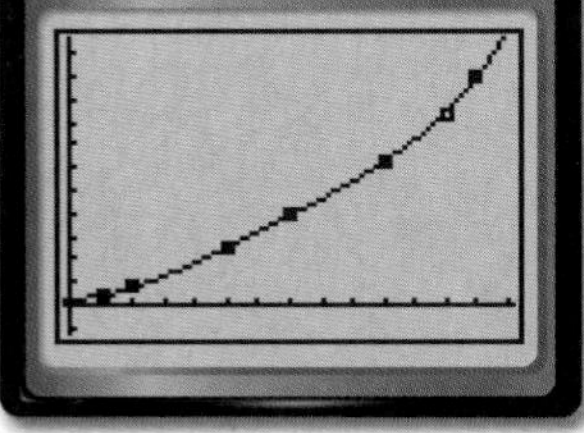

Step 4 Use the **[CALC]** feature to find the value of the function for $x = 8.5$.

KEYSTROKES: 2nd [CALC] 1 8.5 ENTER

After 8.5 minutes, the wave could be expected to be felt approximately 4980 kilometers from the epicenter.

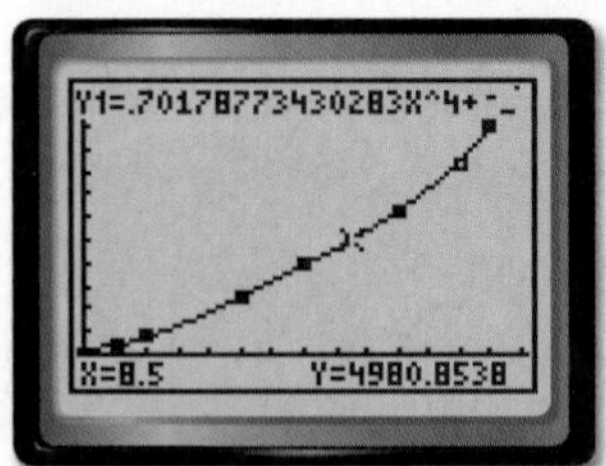

MENTAL CHECK The table gives the distance for 7 minutes as 3900 and the distance for 10 minutes as 6250. Since 8.5 is halfway between 7 and 10 a reasonable estimate for the distance is halfway between 3900 and 6250. ✓

(continued on the next page)

Exercises

The table shows how many minutes out of each eight-hour work day are used to pay one day's worth of taxes.

Year	Minutes
1930	56
1940	86
1950	119
1960	134
1970	144
1980	147
1990	148
2000	163
2005	151

Source: Tax Foundation

1. Draw a scatter plot of the data. Then graph several curves of best fit that relate the number of minutes to the number of years. Try LinReg, QuadReg, and CubicReg.
2. Write the equation for the curve that best fits the data.
3. Based on this equation, how many minutes should you expect to work each day in the year 2020 to pay one day's taxes? Use mental math to check the reasonableness of your estimate.

The table shows the estimated number of alternative-fueled vehicles in use in the United States per year from 1998 to 2007.

Year	Number of Vehicles	Year	Number of Vehicles
1998	295,030	2003	533,999
1999	322,302	2004	565,492
2000	394,664	2005	592,122
2001	425,457	2006	634,559
2002	471,098	2007	695,763

Source: U.S. Department of Energy

4. Draw a scatter plot of the data. Then graph several curves of best fit that relate the distance to the month.
5. Which curve best fits the data? Is that curve best for predicting future values?
6. Use the best-fit equation you think will give the most accurate prediction for how many alternative-fuel vehicles will be in use in 2018. Use mental math to check the reasonableness of your estimate.

The table shows the average distance from the Sun to Earth during each month of the year.

Month	Distance (astronomical units)
January	0.9840
February	0.9888
March	0.9962
April	1.0050
May	1.0122
June	1.0163
July	1.0161
August	1.0116
September	1.0039
October	0.9954
November	0.9878
December	0.9837

Source: The Astronomy Cafe

7. Draw a scatter plot of the data. Then graph several curves of best fit that relate the distance to the month.
8. Write the equation for the curve that best fits the data.
9. Based on your regression equation, what is the distance from the Sun to Earth halfway through September?
10. Would you use this model to find the distance from the Sun to Earth in subsequent years? Explain your reasoning.

Extension

11. Write a question that could be answered by examining data. For example, you might estimate the number of people living in your town 5 years from now or predict the future cost of a car.
12. Collect and organize the data you need to answer the question you wrote. You may need to research your topic on the Internet or conduct a survey to collect the data you need.
13. Make a scatter plot and find a regression equation for your data. Then use the regression equation to answer the question.

CHAPTER 5

Mid-Chapter Quiz

Lessons 5-1 through 5-4

Simplify. Assume that no variable equals 0. (Lesson 5-1)

1. $(3x^2y^{-3})(-2x^3y^5)$

2. $4t(3rt - r)$

3. $\frac{3a^4b^3c}{6a^2b^5c^3}$

4. $\left(\frac{p^2r^3}{pr^4}\right)^2$

5. $(4m^2 - 6m + 5) - (6m^2 + 3m - 1)$

6. $(x + y)(x^2 + 2xy - y^2)$

7. MULTIPLE CHOICE The volume of the rectangular prism is $6x^3 + 19x^2 + 2x - 3$. Which polynomial expression represents the area of the base? (Lesson 5-1)

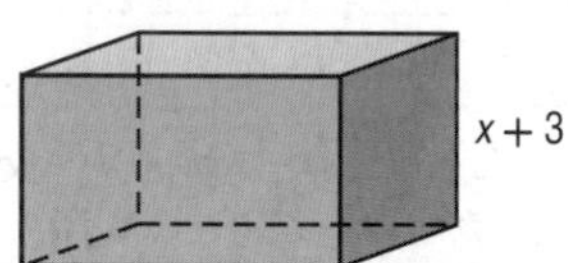

A $6x^4 + 37x^3 + 59x^2 + 3x - 9$

B $6x^2 + x + 1$

C $6x^2 + x - 1$

D $6x + 1$

Simplify. (Lesson 5-2)

8. $(4r^3 - 8r^2 - 13r + 20) \div (2r - 5)$

9. $\frac{3x^3 - 16x^2 + 9x - 24}{x - 5}$

10. Describe the end behavior of the graph. Then determine whether it represents an odd-degree or an even-degree polynomial function and state the number of real zeros. (Lesson 5-3)

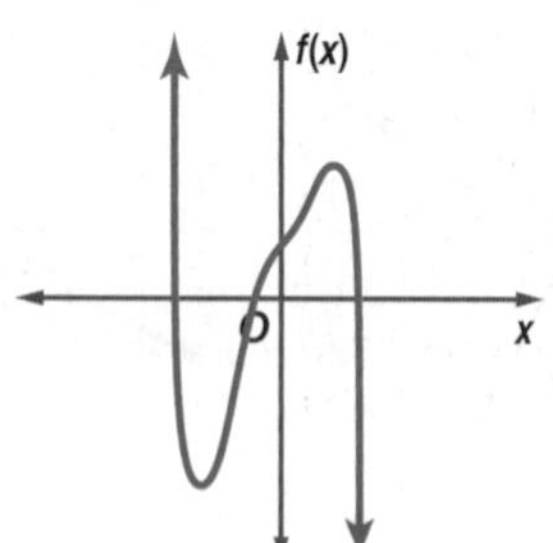

11. MULTIPLE CHOICE Find $p(-3)$ if $p(x) = \frac{2}{3}x^3 + \frac{1}{3}x^2 - 5x$. (Lesson 5-3)

F 0

G 11

H 30

J 36

12. PENDULUMS The formula $L(t) = \frac{8t^2}{\pi^2}$ can be used to find the length of a pendulum in feet when it swings back and forth in t seconds. Find the length of a pendulum that makes one complete swing in 4 seconds. (Lesson 5-3)

13. MULTIPLE CHOICE Find $3f(a - 4) - 2h(a)$ if $f(x) = x^2 + 3x$ and $h(x) = 2x^2 - 3x + 5$. (Lesson 5-3)

A $-a^2 + 15a - 74$

B $-a^2 - 2a - 1$

C $a^2 + 9a - 2$

D $-a^2 - 9a + 2$

14. ENERGY The power generated by a windmill is a function of the speed of the wind. The approximate power is given by the function $P(s) = \frac{s^3}{1000}$, where s represents the speed of the wind in kilometers per hour. Find the units of power $P(s)$ generated by a windmill when the wind speed is 18 kilometers per hour. (Lesson 5-3)

Use $f(x) = x^3 - 2x^2 - 3x$ for Exercises 15–17. (Lesson 5-4)

15. Graph the function.

16. Estimate the x-coordinates at which the relative maxima and relative minima occur.

17. State the domain and range of the function.

18. Determine the consecutive integer values of x between which each real zero is located for $f(x) = 3x^2 - 3x - 1$. (Lesson 5-4)

Refer to the graph. (Lesson 5-4)

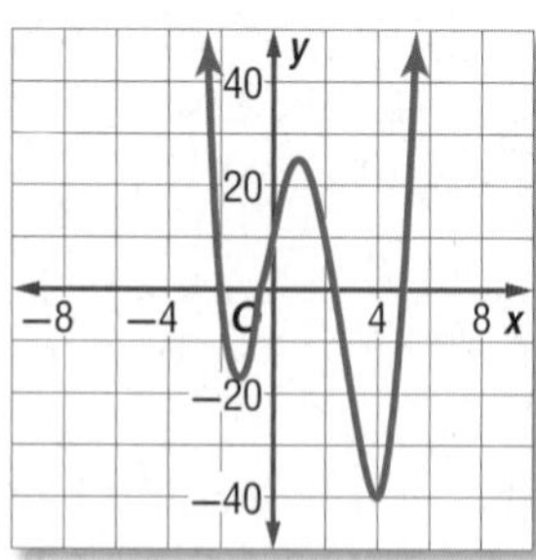

19. Estimate the x-coordinate of every turning point, and determine if those coordinates are relative maxima or relative minima.

20. Estimate the x-coordinate of every zero.

21. What is the least possible degree of the function?

EXPLORE 5-5

Graphing Technology Lab
Solving Polynomial Equations by Graphing

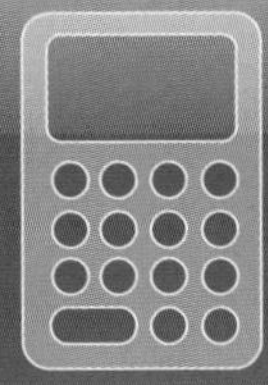

You can use a TI-83/84 Plus graphing calculator to solve polynomial equations.

Activity

Solve $x^4 + 2x^3 = 7$.

Method 1

Step 1 Graph each side of the equation separately in a standard viewing window.

Let **Y1** $= x^4 + 2x^3$ and **Y2** $= 7$.

KEYSTROKES: [Y=] [X,T,θ,n] [^] 4 [+] 2 [X,T,θ,n] [^] 3 [ENTER] 7 [ZOOM] 6

Step 2 Find the points of intersection.

KEYSTROKES: [2nd] [CALC] 5

Use [◄] or [►] to position the cursor on **Y1** near the first point of intersection.
Press [ENTER] [ENTER] [ENTER].
Then use [►] to position the cursor near the second intersection point.
Press [ENTER] [ENTER] [ENTER].

Step 3 Examine the graphs.

Determine where the graph of $y = x^4 + 2x^3$ intersects $y = 7$.

The intersections of the graphs of **Y1** and **Y2** are approximately −2.47 and 1.29, so the solution is approximately −2.47 and 1.29

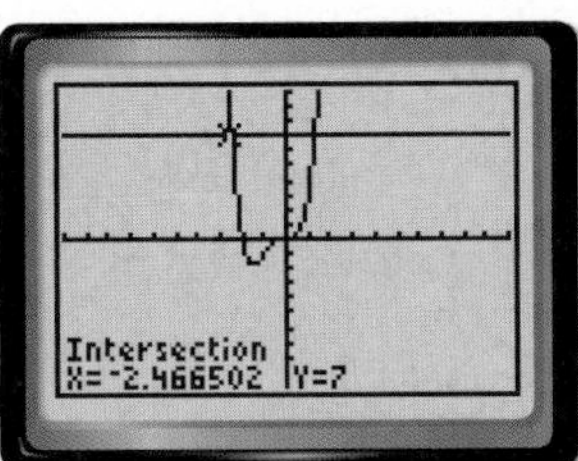

[−10, 10] scl: 1 by [−10, 10] scl: 1

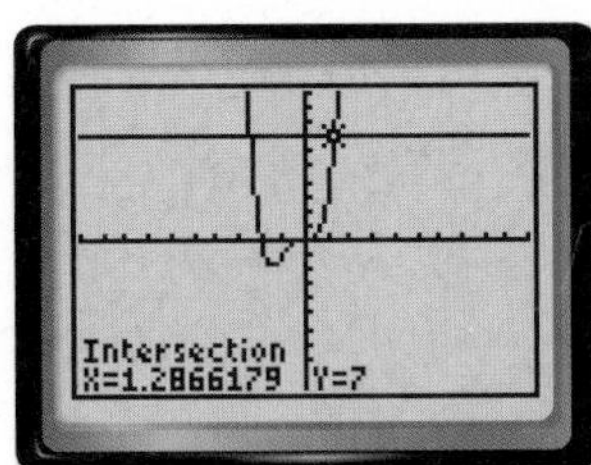

[−10, 10] scl: 1 by [−10, 10] scl: 1

Method 2

Step 1 Rewrite the related equation so it is equal to 0.

$$x^4 + 2x^3 = 7$$
$$x^4 + 2x^3 - 7 = 0$$

Let **Y1** $= x^4 + 2x^3 - 7$ and **Y2** $= 0$.

KEYSTROKES: [Y=] [X,T,θ,n] [^] 4 [+] 2 [X,T,θ,n] [^] 3 [−] 7 [ENTER] 0 [ZOOM] 6

Step 2 Because **Y2** $= 0$, to find the intersection points of **Y1** and **Y2**, find where **Y1** crosses the x-axis.

KEYSTROKES: [2nd] [CALC] 5

Use [◄] or [►] to position the cursor on **Y1** near the first point of intersection.
Press [ENTER] [ENTER] [ENTER].
Then use [►] to position the cursor near the second intersection point.
Press [ENTER] [ENTER] [ENTER].

Step 3 Examine the graphs.

Determine where the graph of $y = x^4 + 2x^3 - 7$ crosses the x-axis.

The intersections of the graphs of **Y1** and **Y2** are approximately −2.47 and 1.29, so the solution is approximately −2.47 and 1.29

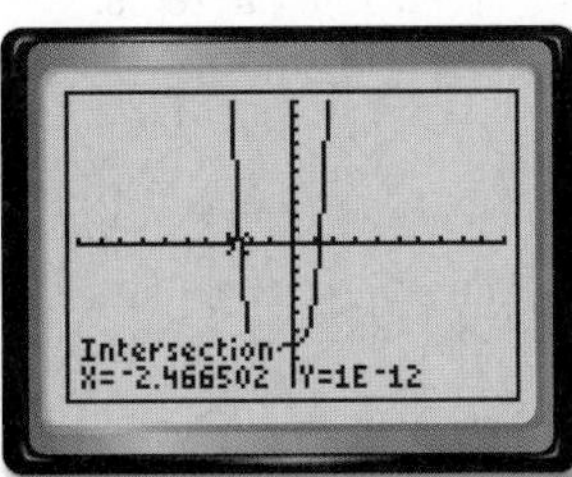

[−10, 10] scl: 1 by [−10, 10] scl: 1

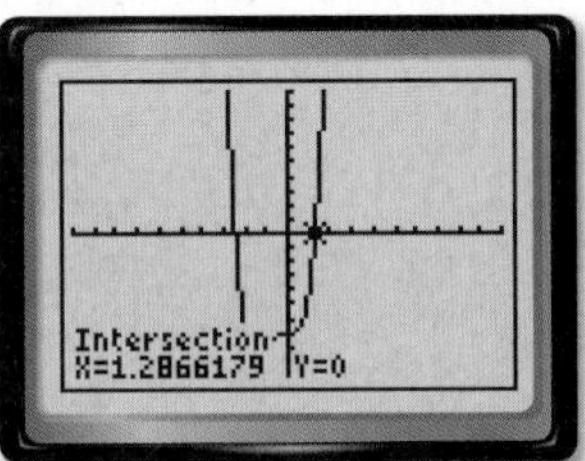

[−10, 10] scl: 1 by [−10, 10] scl: 1

Exercises

Solve each equation. Round to the nearest hundredth.

1. $\frac{2}{3}x^3 + x^2 - 5x = -9$
2. $x^3 - 9x^2 + 27x = 20$
3. $x^3 + 1 = 4x^2$
4. $x^6 - 15 = 5x^4 - x^2$
5. $\frac{1}{2}x^5 = \frac{1}{5}x^2 - 2$
6. $x^8 = -x^7 + 3$
7. $x^4 - 15x^2 = -24$
8. $x^3 - 6x^2 + 4x = -6$
9. $x^4 - 15x^2 + x + 65 = 0$

LESSON

5-5 Solving Polynomial Equations

Then

- You solved quadratic functions by factoring.

Now

1. Factor polynomials.
2. Solve polynomial equations by factoring.

Why?

- A small cube is cut out of a larger cube. The volume of the remaining figure is given and the dimensions of each cube need to be determined.

This can be accomplished by factoring the cubic polynomial $x^3 - y^3$.

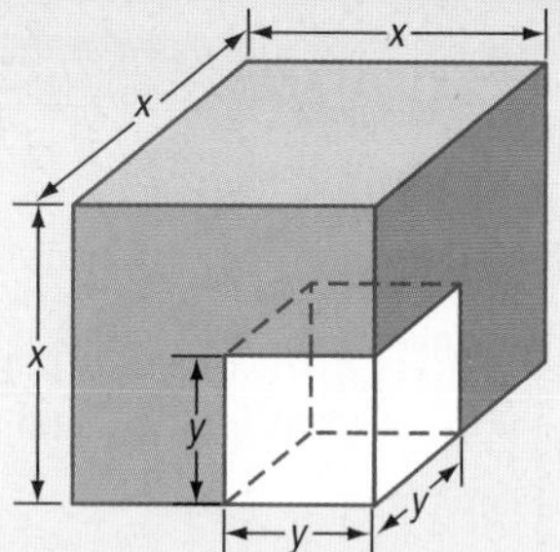

NewVocabulary
prime polynomials
quadratic form

Common Core State Standards

Content Standards
A.CED.1 Create equations and inequalities in one variable and use them to solve problems.

Mathematical Practices
4 Model with mathematics.

1 Factor Polynomials

In Lesson 4-3, you learned that quadratics can be factored just like whole numbers. Their factors, however, are other polynomials. Like quadratics, some cubic polynomials can also be factored with special rules.

KeyConcept Sum and Difference of Cubes

Factoring Technique	General Case
Sum of Two Cubes	$a^3 + b^3 = (a + b)(a^2 - ab + b^2)$
Difference of Two Cubes	$a^3 - b^3 = (a - b)(a^2 + ab + b^2)$

Polynomials that cannot be factored are called **prime polynomials**.

Example 1 Sum and Difference of Cubes

Factor each polynomial. If the polynomial cannot be factored, write *prime*.

a. $16x^4 + 54xy^3$

$16x^4 + 54xy^3 = 2x(8x^3 + 27y^3)$ Factor out the GCF.

$8x^3$ and $27y^3$ are both perfect cubes, so we can factor the sum of two cubes.

$8x^3 + 27y^3 = (2x)^3 + (3y)^3$ $(2x)^3 = 8x^3$; $(3y)^3 = 27y^3$

$= (2x + 3y)[(2x)^2 - (2x)(3y) + (3y)^2]$ Sum of two cubes

$= (2x + 3y)(4x^2 - 6xy + 9y^2)$ Simplify.

$16x^4 + 54xy^3 = 2x(2x + 3y)(4x^2 - 6xy + 9y^2)$ Replace the GCF.

b. $9y^3 + 5x^3$

The first term is a perfect cube, but the second term is not. So, the polynomial cannot be factored using the sum of two cubes pattern. The polynomial also cannot be factored using quadratic methods or the GCF. Therefore, it is a prime polynomial.

GuidedPractice

1A. $5y^4 - 320yz^3$ **1B.** $-54w^4 - 250wz^3$

Math HistoryLink

Sophie Germain (1776–1831)
Sophie Germain taught herself mathematics with books from her father's library during the French Revolution when she was confined for safety. Germain discovered the identity $x^4 + 4y^4 = (x^2 + 2y^2 + 2xy)(x^2 + 2y^2 - 2xy)$, which is named for her.

The table below summarizes the most common factoring techniques used with polynomials. Whenever you factor a polynomial, always look for a common factor first. Then determine whether the resulting polynomial factors can be factored again using one or more of the methods below.

ConceptSummary Factoring Techniques

Number of Terms	Factoring Technique	General Case
any number	Greatest Common Factor (GCF)	$4a^3b^2 - 8ab = 4ab(a^2b - 2)$
two	Difference of Two Squares Sum of Two Cubes Difference of Two Cubes	$a^2 - b^2 = (a + b)(a - b)$ $a^3 + b^3 = (a + b)(a^2 - ab + b^2)$ $a^3 - b^3 = (a - b)(a^2 + ab + b^2)$
three	Perfect Square Trinomials	$a^2 + 2ab + b^2 = (a + b)^2$ $a^2 - 2ab + b^2 = (a - b)^2$
	General Trinomials	$acx^2 + (ad + bc)x + bd$ $= (ax + b)(cx + d)$
four or more	Grouping	$ax + bx + ay + by$ $= x(a + b) + y(a + b)$ $= (a + b)(x + y)$

Example 2 Factoring by Grouping

Factor each polynomial. If the polynomial cannot be factored, write *prime*.

a. $8ax + 4bx + 4cx + 6ay + 3by + 3cy$

$8ax + 4bx + 4cx + 6ay + 3by + 3cy$ — Original expression
$= (8ax + 4bx + 4cx) + (6ay + 3by + 3cy)$ — Group to find a GCF.
$= 4x(2a + b + c) + 3y(2a + b + c)$ — Factor the GCF.
$= (4x + 3y)(2a + b + c)$ — Distributive Property

StudyTip
Answer Checks
Multiply the factors to check your answer.

b. $20fy - 16fz + 15gy + 8hz - 10hy - 12gz$

$20fy - 16fz + 15gy + 8hz - 10hy - 12gz$ — Original expression
$= (20fy + 15gy - 10hy) + (-16fz - 12gz + 8hz)$ — Group to find a GCF.
$= 5y(4f + 3g - 2h) - 4z(4f + 3g - 2h)$ — Factor the GCF.
$= (5y - 4z)(4f + 3g - 2h)$ — Distributive Property

GuidedPractice

2A. $30ax - 24bx + 6cx - 5ay^2 + 4by^2 - cy^2$

2B. $13ax + 18bz - 15by - 14az - 32bx + 9ay$

Factoring by grouping is the only method that can be used to factor polynomials with four or more terms. For polynomials with two or three terms, it may be possible to factor according to one of the patterns listed above.

When factoring two terms in which the exponents are 6 or greater, look to factor perfect squares before factoring perfect cubes.

Example 3 Combine Cubes and Squares

Factor each polynomial. If the polynomial cannot be factored, write *prime*.

a. $x^6 - y^6$

This polynomial could be considered the difference of two squares or the difference of two cubes. The difference of two squares should always be done before the difference of two cubes for easier factoring.

$x^6 - y^6 = (x^3 + y^3)(x^3 - y^3)$ Difference of two squares

$= (x + y)(x^2 - xy + y^2)(x - y)(x^2 + xy + y^2)$ Sum and difference of two cubes

StudyTip

Grouping 6 or more terms Group the terms that have the *most* common values.

b. $a^3x^2 - 6a^3x + 9a^3 - b^3x^2 + 6b^3x - 9b^3$

With six terms, factor by grouping first.

$a^3x^2 - 6a^3x + 9a^3 - b^3x^2 + 6b^3x - 9b^3$

$= (a^3x^2 - 6a^3x + 9a^3) + (-b^3x^2 + 6b^3x - 9b^3)$ Group to find a GCF.

$= a^3(x^2 - 6x + 9) - b^3(x^2 - 6x + 9)$ Factor the GCF.

$= (a^3 - b^3)(x^2 - 6x + 9)$ Distributive Property

$= (a - b)(a^2 + ab + b^2)(x^2 - 6x + 9)$ Difference of cubes

$= (a - b)(a^2 + ab + b^2)(x - 3)^2$ Perfect squares

GuidedPractice

3A. $a^6 + b^6$

3B. $x^5 + 4x^4 + 4x^3 + x^2y^3 + 4xy^3 + 4y^3$

2 Solve Polynomial Equations

In Chapter 4, you learned to solve quadratic equations by factoring and using the Zero Product Property. You can extend these techniques to solve higher-degree polynomial equations.

Real-World Example 4 Solve Polynomial Functions by Factoring

GEOMETRY Refer to the beginning of the lesson. If the small cube is half the length of the larger cube and the figure is 7000 cubic centimeters, what should be the dimensions of the cubes?

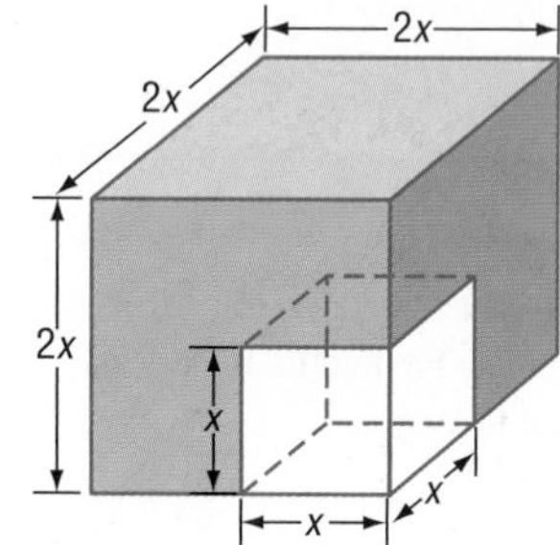

Since the length of the smaller cube is half the length of the larger cube, then their lengths can be represented by x and $2x$, respectively. The volume of the object equals the volume of the larger cube minus the volume of the smaller cube.

$(2x)^3 - x^3 = 7000$ Volume of object

$8x^3 - x^3 = 7000$ $(2x)^3 = 8x^3$

$7x^3 = 7000$ Subtract.

$x^3 = 1000$ Divide.

$x^3 - 1000 = 0$ Subtract 1000 from each side.

$(x - 10)(x^2 + 10x + 100) = 0$ Difference of cubes

$x - 10 = 0$ or $x^2 + 10x + 100 = 0$ Zero Product Property

$x = 10$ $x = -5 \pm 5i\sqrt{3}$

Since 10 is the only real solution, the lengths of the cubes are 10 cm and 20 cm.

GuidedPractice

4. Determine the dimensions of the cubes if the length of the smaller cube is one third of the length of the larger cube, and the volume of the object is 3250 cubic centimeters.

In some cases, you can rewrite a polynomial in x in the form $au^2 + bu + c$. For example, by letting $u = x^2$, the expression $x^4 + 12x^2 + 32$ can be written as $(x^2)^2 + 12(x^2) + 32$ or $u^2 + 12u + 32$. This new, but equivalent, expression is said to be in **quadratic form**.

KeyConcept Quadratic Form

Words An expression that is in quadratic form can be written as $au^2 + bu + c$ for any numbers a, b, and c, $a \neq 0$, where u is some expression in x. The expression $au^2 + bu + c$ is called the quadratic form of the original expression.

Example $12x^6 + 8x^6 + 1 = 3(2x^3)^2 + 2(2x^3)^2 + 1$

StudyTip

Quadratic Form When writing a polynomial in quadratic form, choose the expression equal to u by examining the terms with variables. Pay special attention to the exponents in those terms. Not every polynomial can be written in quadratic form.

PT

Example 5 Quadratic Form

Write each expression in quadratic form, if possible.

a. $150n^8 + 40n^4 - 15$

$150n^8 + 40n^4 - 15 = 6(5n^4)^2 + 8(5n^4) - 15$ $\quad$ $(5n^4)^2 = 25n^8$

b. $y^8 + 12y^3 + 8$

This cannot be written in quadratic form since $y^8 \neq (y^3)^2$.

GuidedPractice

5A. $x^4 + 5x + 6$

5B. $8x^4 + 12x^2 + 18$

You can use quadratic form to solve equations with larger degrees.

PT

Example 6 Solve Equations in Quadratic Form

Solve $18x^4 - 21x^2 + 3 = 0$.

$18x^4 - 21x^2 + 3 = 0$		Original equation
$2(3x^2)^2 - 7(3x^2) + 3 = 0$		$2(3x^2)^2 = 18x^4$
$2u^2 - 7u + 3 = 0$		Let $u = 3x^2$.
$(2u - 1)(u - 3) = 0$		Factor.
$u = \frac{1}{2}$	or $\quad u = 3$	Zero Product Property
$3x^2 = \frac{1}{2}$	$3x^2 = 3$	Replace u with $3x^2$.
$x^2 = \frac{1}{6}$	$x^2 = 1$	Divide by 3.
$x = \pm\frac{\sqrt{6}}{6}$	$x = \pm 1$	Take the square root.

The solutions of the equation are $1, -1, \frac{\sqrt{6}}{6}$, and $-\frac{\sqrt{6}}{6}$.

GuidedPractice

6A. $4x^4 - 8x^2 + 3 = 0$

6B. $8x^4 + 10x^2 - 12 = 0$

Check Your Understanding

= Step-by-Step Solutions begin on page R14.

Examples 1–3 **Factor completely. If the polynomial is not factorable, write *prime*.**

1. $3ax + 2ay - az + 3bx + 2by - bz$

2. $2kx + 4mx - 2nx - 3ky - 6my + 3ny$

3. $2x^3 + 5y^3$

4. $16g^3 + 2h^3$

5. $12qw^3 - 12q^4$

6. $3w^2 + 5x^2 - 6y^2 + 2z^2 + 7a^2 - 9b^2$

7. $a^6x^2 - b^6x^2$

8. $x^3y^2 - 8x^3y + 16x^3 + y^5 - 8y^4 + 16y^3$

9. $8c^3 - 125d^3$

10. $6bx + 12cx + 18dx - by - 2cy - 3dy$

Example 4 **Solve each equation.**

11. $x^4 - 19x^2 + 48 = 0$

12. $x^3 - 64 = 0$

13. $x^3 + 27 = 0$

14. $x^4 - 33x^2 + 200 = 0$

15. CCSS **PERSEVERANCE** A boardwalk that is x feet wide is built around a rectangular pond. The pond is 30 feet wide and 40 feet long. The combined area of the pond and the boardwalk is 2000 square feet. What is the width of the boardwalk?

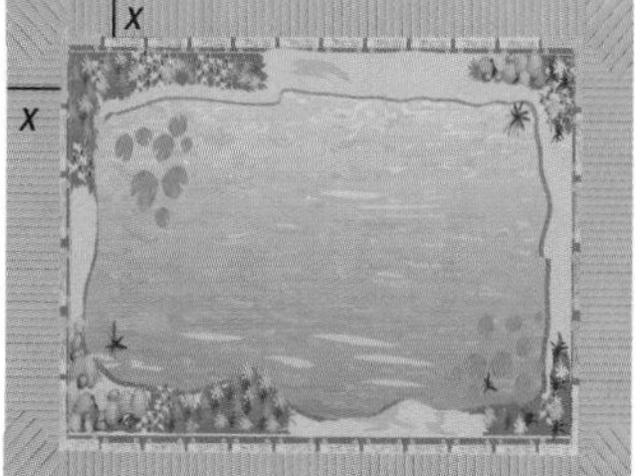

Example 5 **Write each expression in quadratic form, if possible.**

16. $4x^6 - 2x^3 + 8$

17. $25y^6 - 5y^2 + 20$

Example 6 **Solve each equation.**

18. $x^4 - 6x^2 + 8 = 0$

19. $y^4 - 18y^2 + 72 = 0$

Practice and Problem Solving

Extra Practice is on page R5.

Examples 1–3 **Factor completely. If the polynomial is not factorable, write *prime*.**

20. $8c^3 - 27d^3$

21. $64x^4 + xy^3$

22. $a^8 - a^2b^6$

23. $x^6y^3 + y^9$

24. $18x^6 + 5y^6$

25. $w^3 - 2y^3$

26. $gx^2 - 3hx^2 - 6fy^2 - gy^2 + 6fx^2 + 3hy^2$

27. $12ax^2 - 20cy^2 - 18bx^2 - 10ay^2 + 15by^2 + 24cx^2$

28. $a^3x^2 - 16a^3x + 64a^3 - b^3x^2 + 16b^3x - 64b^3$

29 $8x^5 - 25y^3 + 80x^4 - x^2y^3 + 200x^3 - 10xy^3$

Example 4 **Solve each equation.**

30. $x^4 + x^2 - 90 = 0$

31. $x^4 - 16x^2 - 720 = 0$

32. $x^4 - 7x^2 - 44 = 0$

33. $x^4 + 6x^2 - 91 = 0$

34. $x^3 + 216 = 0$

35. $64x^3 + 1 = 0$

Example 5 **Write each expression in quadratic form, if possible.**

36. $x^4 + 12x^2 - 8$

37. $-15x^4 + 18x^2 - 4$

38. $8x^6 + 6x^3 + 7$

39. $5x^6 - 2x^2 + 8$

40. $9x^8 - 21x^4 + 12$

41. $16x^{10} + 2x^5 + 6$

Example 6 **Solve each equation.**

42. $x^4 + 6x^2 + 5 = 0$

43. $x^4 - 3x^2 - 10 = 0$

44. $4x^4 - 14x^2 + 12 = 0$

45. $9x^4 - 27x^2 + 20 = 0$

46. $4x^4 - 5x^2 - 6 = 0$

47. $24x^4 + 14x^2 - 3 = 0$

48. **ZOOLOGY** A species of animal is introduced to a small island. Suppose the population of the species is represented by $P(t) = -t^4 + 9t^2 + 400$, where t is the time in years. Determine when the population becomes zero.

Factor completely. If the polynomial is not factorable, write *prime*.

49. $x^4 - 625$
50. $x^6 - 64$
51. $x^5 - 16x$
52. $8x^5y^2 - 27x^2y^5$
53. $15ax - 10bx + 5cx + 12ay - 8by + 4cy + 15az - 10bz + 5cz$
54. $6a^2x^2 - 24b^2x^2 + 18c^2x^2 - 5a^2y^3 + 20b^2y^3 - 15c^2y^3 + 2a^2z^2 - 8b^2z^2 + 6c^2z^2$
55. $6x^5 - 11x^4 - 10x^3 - 54x^3 + 99x^2 + 90x$
56. $20x^6 - 7x^5 - 6x^4 - 500x^4 + 175x^3 + 150x^2$

57. **GEOMETRY** The volume of the figure at the right is 440 cubic centimeters. Find the value of x and the length, height, and width.

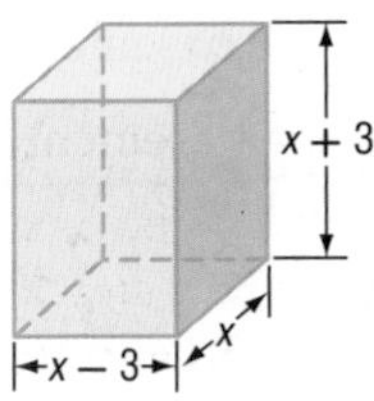

Solve each equation.

58. $8x^4 + 10x^2 - 3 = 0$
59. $6x^4 - 5x^2 - 4 = 0$
60. $20x^4 - 53x^2 + 18 = 0$
61. $18x^4 + 43x^2 - 5 = 0$
62. $8x^4 - 18x^2 + 4 = 0$
63. $3x^4 - 22x^2 - 45 = 0$
64. $x^6 + 7x^3 - 8 = 0$
65. $x^6 - 26x^3 - 27 = 0$
66. $8x^6 + 999x^3 = 125$
67. $4x^4 - 4x^2 - x^2 + 1 = 0$
68. $x^6 - 9x^4 - x^2 + 9 = 0$
69. $x^4 + 8x^2 + 15 = 0$

70. **CCSS SENSE-MAKING** A rectangular prism with dimensions $x - 2$, $x - 4$, and $x - 6$ has a volume equal to $40x$ cubic units.
    a. Write a polynomial equation using the formula for volume.
    b. Use factoring to solve for x.
    c. Are any values for x unreasonable? Explain.
    d. What are the dimensions of the prism?

71. **POOL DESIGN** Andrea wants to build a pool following the diagram at the right. The pool will be surrounded by a sidewalk of a constant width.
    a. If the total area of the pool itself is to be 336 ft^2, what is x?
    b. If the value of x were doubled, what would be the new area of the pool?
    c. If the value of x were halved, what would be the new area of the pool?

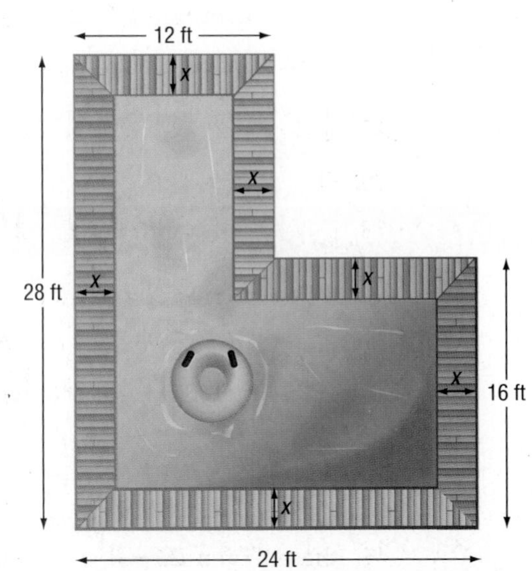

72. **BIOLOGY** During an experiment, the number of cells of a virus can be modeled by $P(t) = -0.012t^3 - 0.24t^2 + 6.3t + 8000$, where t is the time in hours and P is the number of cells. Jack wants to determine the times at which there are 8000 cells.

a. Solve for t by factoring.

b. What method did you use to factor?

c. Which values for t are reasonable and which are unreasonable? Explain.

d. Graph the function for $0 \leq t \leq 20$ using your calculator.

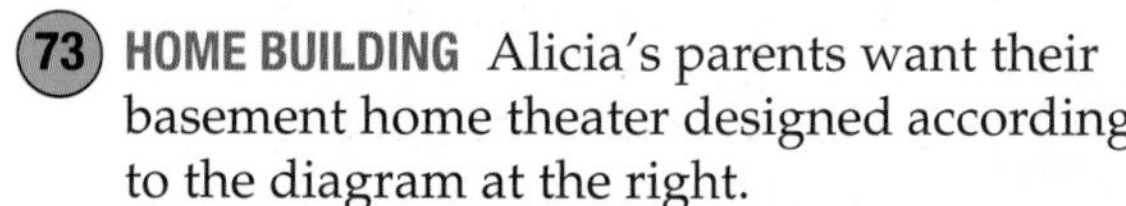

73. **HOME BUILDING** Alicia's parents want their basement home theater designed according to the diagram at the right.

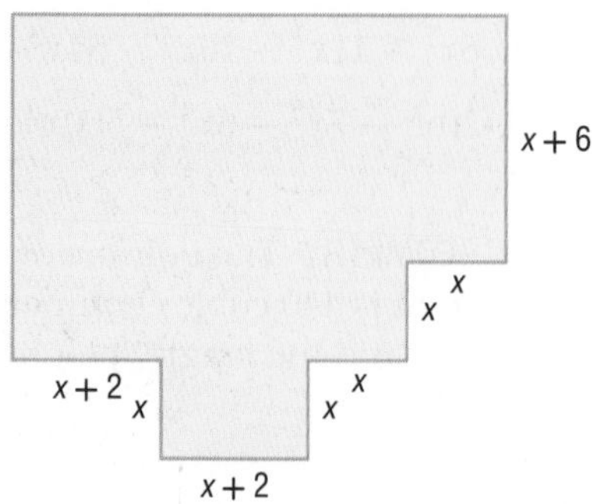

a. Write a function in terms of x for the area of the basement.

b. If the basement is to be 1366 square feet, what is x?

74. **BIOLOGY** A population of parasites in an experiment can be modeled by $f(t) = t^3 + 5t^2 - 4t - 20$, where t is the time in days.

a. Use factoring by grouping to determine the values of t for which $f(t) = 0$.

b. At what times does the population reach zero?

c. Are any of the values of t unreasonable? Explain.

Factor completely. If the polynomial is not factorable, write *prime*.

75. $x^6 - 4x^4 - 8x^4 + 32x^2 + 16x^2 - 64$

76. $y^9 - y^6 - 2y^6 + 2y^3 + y^3 - 1$

77. $x^6 - 3x^4y^2 + 3x^2y^4 - y^6$

78. **CCSS SENSE-MAKING** Fredo's corral, an enclosure for livestock, is currently 32 feet by 40 feet. He wants to enlarge the area to 4.5 times its current area by increasing the length and width by the same amount.

a. Draw a diagram to represent the situation.

b. Write a polynomial equation for the area of the new corral. Then solve the equation by factoring.

c. Graph the function.

d. Which solution is irrelevant? Explain.

H.O.T. Problems Use Higher-Order Thinking Skills

79. **CHALLENGE** Factor $36x^{2n} + 12x^n + 1$.

80. **CHALLENGE** Solve $6x - 11\sqrt{3x} + 12 = 0$.

81. **REASONING** Find a counterexample to the statement $a^2 + b^2 = (a + b)^2$.

82. **OPEN ENDED** The cubic form of an equation is $ax^3 + bx^2 + cx + d = 0$. Write an equation with degree 6 that can be written in *cubic* form.

83. **WRITING IN MATH** Explain how the graph of a polynomial function can help you factor the polynomial.

Standardized Test Practice

84. SHORT RESPONSE Tiles numbered from 1 to 6 are placed in a bag and are drawn to determine which of six tasks will be assigned to six people. What is the probability that the tiles numbered 5 and 6 are the last two drawn?

85. STATISTICS Which of the following represents a negative correlation?

A 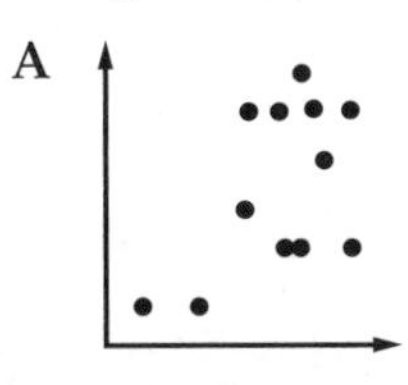

C 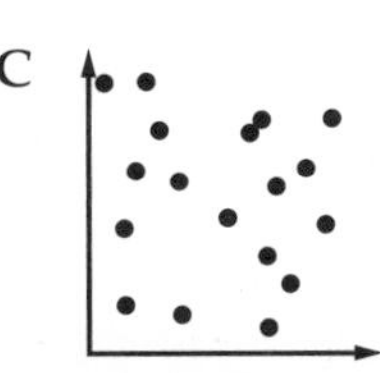

B 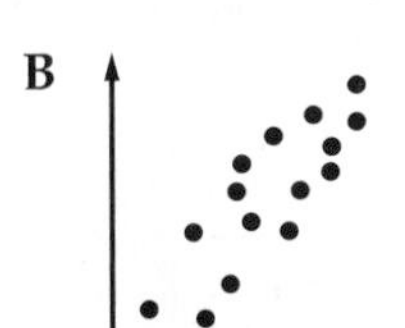

D 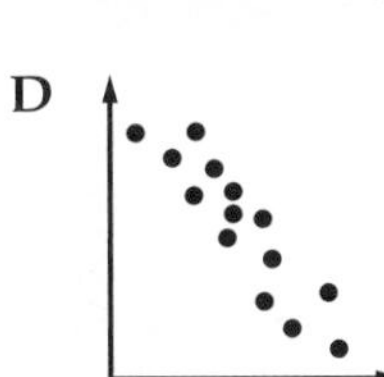

86. Which of the following most accurately describes the translation of the graph $y = (x + 4)^2 - 3$ to the graph of $y = (x - 1)^2 + 3$?

F down 1 and to the right 3

G down 6 and to the left 5

H up 1 and to the left 3

J up 6 and to the right 5

87. SAT/ACT The positive difference between k and $\frac{1}{12}$ is the same as the positive difference between $\frac{1}{3}$ and $\frac{1}{5}$. Which of the following is the value of k?

A $\frac{1}{60}$

B $\frac{1}{20}$

C $\frac{1}{15}$

D $\frac{13}{60}$

E $\frac{37}{60}$

Spiral Review

Graph each polynomial function. Estimate the x-coordinates at which the relative maxima and relative minima occur. (Lesson 5-4)

88. $f(x) = 2x^3 - 4x^2 + x + 8$

89. $f(x) = -3x^3 + 6x^2 + 2x - 1$

90. $f(x) = -x^3 + 3x^2 + 4x - 6$

State the degree and leading coefficient of each polynomial in one variable. If it is not a polynomial in one variable, explain why. (Lesson 5-3)

91. $f(x) = 4x^3 - 6x^2 + 5x^4 - 8x$

92. $f(x) = -2x^5 + 5x^4 + 3x^2 + 9$

93. $f(x) = -x^4 - 3x^3 + 2x^6 - x^7$

94. ELECTRICITY The impedance in one part of a series circuit is $3 + 4j$ ohms, and the impedance in another part of the circuit is $2 - 6j$ ohms. Add these complex numbers to find the total impedance of the circuit. (Lesson 4-4)

95. SKIING All 28 members of a ski club went on a trip. The club paid a total of \$478 for the equipment. How many skis and snowboards did they rent? (Lesson 3-2)

96. GEOMETRY The sides of an angle are parts of two lines whose equations are $2y + 3x = -7$ and $3y - 2x = 9$. The angle's vertex is the point where the two sides meet. Find the coordinates of the vertex of the angle. (Lesson 3-1)

Skills Review

Divide.

97. $(x^2 + 6x - 2) \div (x + 4)$

98. $(2x^2 + 8x - 10) \div (2x + 1)$

99. $(8x^3 + 4x^2 + 6) \div (x + 2)$

EXTEND 5-5

Graphing Technology Lab
Polynomial Identities

An **identity** is an equation that is satisfied by any numbers that replace the variables. Thus, a **polynomial identity** is a polynomial equation that is true for any values that are substituted for the variables.

CCSS Common Core State Standards
Content Standards
A.REI.11 Explain why the x-coordinates of the points where the graphs of the equations $y = f(x)$ and $y = g(x)$ intersect are the solutions of the equation $f(x) = g(x)$; find the solutions approximately, e.g., using technology to graph the functions, make tables of values, or find successive approximations. Include cases where $f(x)$ and/or $g(x)$ are linear, polynomial, rational, absolute value, exponential, and logarithmic functions.

You can use a table or a spreadsheet on your graphing calculator to determine whether a polynomial equation may be an identity.

Activity 1 Use a Table

Determine whether $x^3 - y^3 = (x - y)(x^2 + xy + y^2)$ may be an identity.

Step 1 Add a new **Lists & Spreadsheet** page on the TI-Nspire. Label column A x and column B y. Type any values in columns A and B.

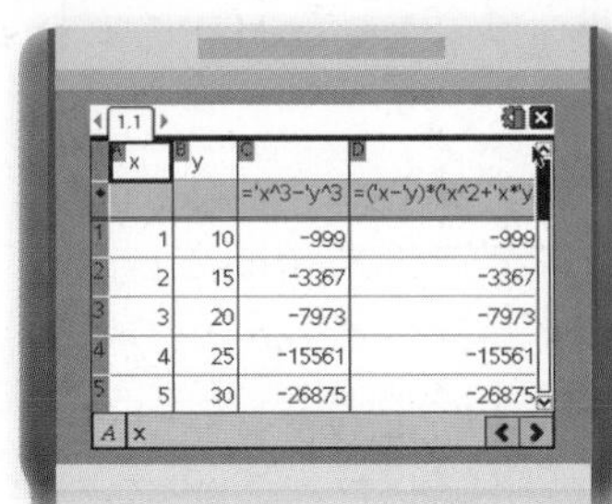

Step 2 Move the cursor to the formula row in column C and type $= x^3 - y^3$. In column D, type $= (x - y)(x^2 + xy + y^2)$.

No matter what values are entered for x and y in columns A and B, the values in columns C and D are the same. Thus, the equation may be an identity.

If you want to prove that an equation is an identity, you need to show that it is true for all values of the variables.

KeyConcept Verifying Identities by Transforming One Side

Step 1 Simplify one side of an equation until the two sides of the equation are the same. It is often easier to work with the more complicated side of the equation.

Step 2 Transform that expression into the form of the simpler side.

Activity 2 Transform One Side

Prove that $(x + y)^2 = x^2 + 2xy + y^2$ is an identity.

$(x + y)^2 \stackrel{?}{=} x^2 + 2xy + y^2$	Original equation
$(x + y)(x + y) \stackrel{?}{=} x^2 + 2xy + y^2$	Write $(x + y)^2$ as two factors.
$x^2 + xy + xy + y^2 \stackrel{?}{=} x^2 + 2xy + y^2$	FOIL Method
$x^2 + 2xy + y^2 = x^2 + 2xy + y^2$ ✓	Simplify.

Thus, the identity $(x + y)^2 = x^2 + 2xy + y^2$ is verified.

You can also use a TI-Nspire with a computer algebra system (CAS) to prove an identity.

Activity 3 Use CAS

Prove that $x^3 - y^3 = (x - y)(x^2 + xy + y^2)$ is an identity.

Step 1 Add a new **Calculator** page on the TI-Nspire CAS. Simplify the right side of the equation one step at a time.

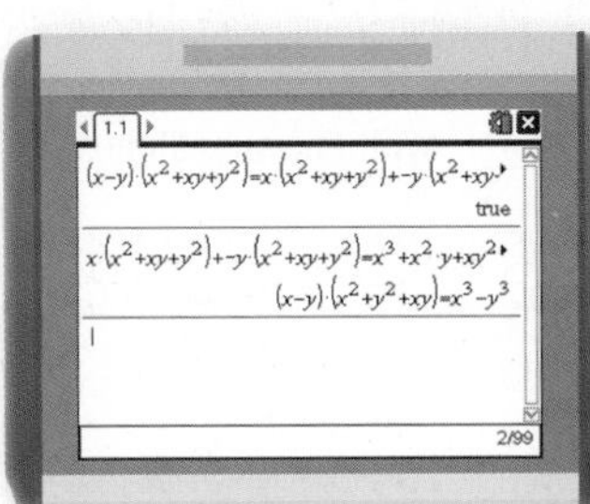

Step 2 Enter the right side of the equation and then distribute.

Step 3 Multiply next. The CAS system will do the final simplification step.

The final step shown on the CAS screen is the results in $x^3 - y^3$. Thus, the identity has been proved.

You can also prove identities by transforming each side of the equation.

Activity 4 Use CAS to Transform Each Side

Prove that $(x + 2)^3(x - 1)^3 = (x^2 + x - 2)(x^4 + 2x^3 - 3x^2 - 4x + 4)$ is an identity.

Add a new **Calculator** page on the TI-Nspire. Simplify the left and the right sides of the equation simultaneously.

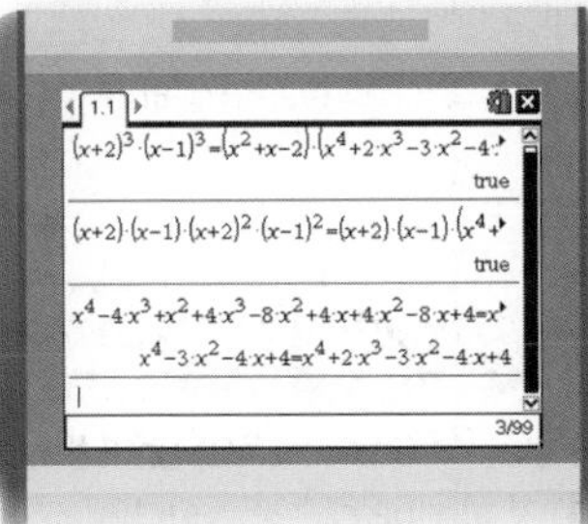

The CAS will indicate if the changes are true, otherwise it will simplify for you.

The CAS system will do the final simplification step. The identity $(x + 2)^3 (x - 1)^3 = (x^2 + x - 2)(x^4 + 2x^3 - 3x^2 - 4x + 4)$ has been proved.

Analyze

1. Determine whether $a^2 - b^2 = (a + b)(a - b)$ may be an identity.

Exercises

Use CAS to prove each identity.

2. $x^3 + y^3 = (x + y)(x^2 - xy + y^2)$
3. $p^4 - q^4 = (p - q)(p + q)(p^2 + q^2)$
4. $a^5 - b^5 = (a - b)(a^4 + a^3b + a^2b^2 + ab^3 + b^4)$
5. $g^6 + h^6 = (g^2 + h^2)(g^4 - g^2h^2 + h^4)$
6. $a^5 + b^5 = (a + b)(a^4 - a^3b + a^2b^2 - ab^3 + b^4)$
7. $u^6 - w^6 = (u + w)(u - w)(u^2 + vw + w^2)(u^2 - vw + w^2)$
8. $(x + 1)^2(x - 4)^3 = (x^2 - 3x - 4)(x^3 - 7x^2 + 8x + 16)$

LESSON 5-6 The Remainder and Factor Theorems

Then

- You used the Distributive Property and factoring to simplify algebraic expressions.

Now

1. Evaluate functions by using synthetic substitution.
2. Determine whether a binomial is a factor of a polynomial by using synthetic substitution.

Why?

- The number of college students from the United States who study abroad can be modeled by the function $S(x) = 0.02x^4 - 0.52x^3 + 4.03x^2 + 0.09x + 77.54$, where x is the number of years since 1993 and $S(x)$ is the number of students in thousands.

You can use this function to estimate the number of U.S. college students studying abroad in 2018 by evaluating the function for $x = 25$. Another method you can use is *synthetic substitution*.

NewVocabulary
synthetic substitution
depressed polynomial

Common Core State Standards

Content Standards
A.APR.2 Know and apply the Remainder Theorem: For a polynomial $p(x)$ and a number a, the remainder on division by $x - a$ is $p(a)$, so $p(a) = 0$ if and only if $(x - a)$ is a factor of $p(x)$.
F.IF.7.c Graph polynomial functions, identifying zeros when suitable factorizations are available, and showing end behavior.

Mathematical Practices
7 Look for and make use of structure.

1 Synthetic Substitution

Synthetic division can be used to find the value of a function. Consider the polynomial function $f(x) = -3x^2 + 5x + 4$. Divide the polynomial by $x - 3$.

Method 1 Long Division

$$\begin{array}{r} -3x - 4 \\ x - 3 \overline{) -3x^2 + 5x + 4} \\ \underline{-3x^2 + 9x} \\ -4x + 4 \\ \underline{-4x + 12} \\ -8 \end{array}$$

Method 2 Synthetic Division

$$\begin{array}{r|rrr} 3 & -3 & 5 & 4 \\ & & -9 & -12 \\ \hline & -3 & -4 & -8 \end{array}$$

Compare the remainder of -8 to $f(3)$.

$f(3) = -3(3)^2 + 5(3) + 4$ Replace x with 3.
$= -27 + 15 + 4$ Multiply.
$= -8$ Simplify.

Notice that the value of $f(3)$ is the same as the remainder when the polynomial is divided by $x - 3$. This illustrates the **Remainder Theorem**.

KeyConcept Remainder Theorem

Words If a polynomial $P(x)$ is divided by $x - r$, the remainder is a constant $P(r)$, and

Dividend	equals	quotient	times	divisor	plus	remainder.
$P(x)$	$=$	$Q(x)$	$\cdot$	$(x - r)$	$+$	$P(r)$,

where $Q(x)$ is a polynomial with degree one less than $P(x)$.

Example $x^2 + 6x + 2 = (x - 4) \cdot (x + 10) + 42$

Applying the Remainder Theorem using synthetic division to evaluate a function is called **synthetic substitution**. It is a convenient way to find the value of a function, especially when the degree of the polynomial is greater than 2.

Example 1 Synthetic Substitution

If $f(x) = 3x^4 - 2x^3 + 5x + 2$, find $f(4)$.

Method 1 **Synthetic Substitution**

By the Remainder Theorem, $f(4)$ should be the remainder when the polynomial is divided by $x - 4$.

4	3	−2	0	5	2
		12	40	160	660
	3	10	40	165	662

Because there is no x^2 term, a zero is placed in this position as a placeholder.

The remainder is 662. Therefore, by using synthetic substitution, $f(4) = 662$.

Method 2 **Direct Substitution**

Replace x with 4.

$f(x) = 3x^4 - 2x^3 + 5x + 2$ Original function

$f(4) = 3(4)^4 - 2(4)^3 + 5(4) + 2$ Replace x with 4.

$= 768 - 128 + 20 + 2$ or 662 Simplify.

By using direct substitution, $f(4) = 662$. Both methods give the same result.

Guided Practice

1A. If $f(x) = 3x^3 - 6x^2 + x - 11$, find $f(3)$.

1B. If $g(x) = 4x^5 + 2x^3 + x^2 - 1$, find $f(-1)$.

Slow Images/Photographer's Choice/Getty Images

Real-WorldLink

Some benefits of studying abroad include learning a new language, becoming more independent, and improving communication skills. Studying abroad also gives students a chance to experience different cultures and customs as well as try new foods.

Source: StudyAbroad

Synthetic substitution can be used in situations in which direct substitution would involve cumbersome calculations.

Real-World Example 2 Find Function Values

COLLEGE Refer to the beginning of the lesson. How many U.S. college students will study abroad in 2018?

Use synthetic substitution to divide $0.02x^4 - 0.52x^3 + 4.03x^2 + 0.09x + 77.54$ by $x - 20$.

25	0.02	−0.52	4.03	0.09	77.54
		0.5	−0.5	88.25	2208.5
	0.02	−0.02	3.53	88.34	2286.04

In 2018, there will be about 2,286,040 U.S. college students studying abroad.

Guided Practice

2. COLLEGE The function $C(x) = 2.46x^3 - 22.37x^2 + 53.81x + 548.24$ can be used to approximate the number, in thousands, of international college students studying in the United States x years since 2000. How many international college students can be expected to study in the U.S. in 2015?

WatchOut!

Synthetic Substitution Remember that synthetic substitution is used to divide a polynomial by $(x - a)$. If the binomial is $(x - a)$, use a. If the binomial is $(x + a)$, use $-a$.

2 Factors of Polynomials

The synthetic division below shows that the quotient of $2x^3 - 3x^2 - 17x + 30$ and $x + 3$ is $2x^2 - 9x + 10$.

$$\begin{array}{r|rrrr} -3 & 2 & -3 & -17 & 30 \\ & & -6 & 27 & -30 \\ \hline & 2 & -9 & 10 & 0 \end{array}$$

When you divide a polynomial by one of its binomial factors, the quotient is called a depressed polynomial. A **depressed polynomial** has a degree that is one less than the original polynomial. From the results of the division, and by using the Remainder Theorem, we can make the following statement.

$$\underbrace{2x^3 - 3x^2 - 17x + 30}_{\text{Dividend}} \underbrace{=}_{\text{equals}} \underbrace{(2x^2 - 9x + 10)}_{\text{quotient}} \underbrace{\cdot}_{\text{times}} \underbrace{(x + 3)}_{\text{divisor}} \underbrace{+}_{\text{plus}} \underbrace{0}_{\text{remainder.}}$$

Since the remainder is 0, $f(-3) = 0$. This means that $x + 3$ is a factor of $2x^3 - 3x^2 - 17x + 30$. This illustrates the **Factor Theorem**, which is a special case of the Remainder Theorem.

KeyConcept Factor Theorem

The binomial $x - r$ is a factor of the polynomial $P(x)$ if and only if $P(r) = 0$.

The Factor Theorem can be used to determine whether a binomial is a factor of a polynomial. It can also be used to determine all of the factors of a polynomial.

Example 3 Use the Factor Theorem

Determine whether $x - 5$ is a factor of $x^3 - 7x^2 + 7x + 15$. Then find the remaining factors of the polynomial.

The binomial $x - 5$ is a factor of the polynomial if 5 is a zero of the related polynomial function. Use the Factor Theorem and synthetic division.

$$\begin{array}{r|rrrr} 5 & 1 & -7 & 7 & 15 \\ & & 5 & -10 & -15 \\ \hline & 1 & -2 & -3 & 0 \end{array}$$

Because the remainder is 0, $x - 5$ is a factor of the polynomial. The polynomial $x^3 - 7x^2 + 7x + 15$ can be factored as $(x - 5)(x^2 - 2x - 3)$. The polynomial $x^2 - 2x - 3$ is the depressed polynomial. Check to see if this polynomial can be factored.

$x^2 - 2x - 3 = (x + 1)(x - 3)$ Factor the trinomial.

So, $x^3 - 7x^2 + 7x + 15 = (x - 5)(x + 1)(x - 3)$.

You can check your answer by multiplying out the factors and seeing if you come up with the initial polynomial.

StudyTip

Factoring The factors of a polynomial do not have to be binomials. For example, the factors of $x^3 + x^2 - x + 15$ are $x + 3$ and $x^2 - 2x + 5$.

GuidedPractice

3. Show that $x - 2$ is a factor of $x^3 - 7x^2 + 4x + 12$. Then find the remaining factors of the polynomial.

Check Your Understanding

= Step-by-Step Solutions begin on page R14.

Example 1 **Use synthetic substitution to find $f(4)$ and $f(-2)$ for each function.**

1. $f(x) = 2x^3 - 5x^2 - x + 14$

2. $f(x) = x^4 + 8x^3 + x^2 - 4x - 10$

Example 2

3. NATURE The approximate number of bald eagle nesting pairs in the United States can be modeled by the function $P(x) = -0.16x^3 + 15.83x^2 - 154.15x + 1147.97$, where x is the number of years since 1970. About how many nesting pairs of bald eagles can be expected in 2018?

Example 3 **Given a polynomial and one of its factors, find the remaining factors of the polynomial.**

4. $x^3 - 6x^2 + 11x - 6; x - 1$

5. $x^3 + x^2 - 16x - 16; x + 1$

6. $3x^3 + 10x^2 - x - 12; x - 1$

7. $2x^3 - 5x^2 - 28x + 15; x + 3$

Practice and Problem Solving

Extra Practice is on page R5.

Example 1 **Use synthetic substitution to find $f(-5)$ and $f(2)$ for each function.**

8. $f(x) = x^3 + 2x^2 - 3x + 1$

9. $f(x) = x^2 - 8x + 6$

10. $f(x) = 3x^4 + x^3 - 2x^2 + x + 12$

11. $f(x) = 2x^3 - 8x^2 - 2x + 5$

12. $f(x) = x^3 - 5x + 2$

13. $f(x) = x^5 + 8x^3 + 2x - 15$

14. $f(x) = x^6 - 4x^4 + 3x^2 - 10$

15. $f(x) = x^4 - 6x - 8$

Example 2

16. FINANCIAL LITERACY A specific car's fuel economy in miles per gallon can be approximated by $f(x) = 0.00000056x^4 - 0.000018x^3 - 0.016x^2 + 1.38x - 0.38$, where x represents the car's speed in miles per hour. Determine the fuel economy when the car is traveling 40, 50 and 60 miles per hour.

Example 3 **Given a polynomial and one of its factors, find the remaining factors of the polynomial.**

17. $x^3 - 3x + 2; x + 2$

18. $x^4 + 2x^3 - 8x - 16; x + 2$

19. $x^3 - x^2 - 10x - 8; x + 2$

20. $x^3 - x^2 - 5x - 3; x - 3$

21. $2x^3 + 17x^2 + 23x - 42; x - 1$

22. $2x^3 + 7x^2 - 53x - 28; x - 4$

23. $x^4 + 2x^3 + 2x^2 - 2x - 3; x - 1$

24. $x^3 + 2x^2 - x - 2; x + 2$

25. $6x^3 - 25x^2 + 2x + 8; 2x + 1$

26. $16x^5 - 32x^4 - 81x + 162; 2x - 3$

27. BOATING A motor boat traveling against waves accelerates from a resting position. Suppose the speed of the boat in feet per second is given by the function $f(t) = -0.04t^4 + 0.8t^3 + 0.5t^2 - t$, where t is the time in seconds.

a. Find the speed of the boat at 1, 2, and 3 seconds.

b. It takes 6 seconds for the boat to travel between two buoys while it is accelerating. Use synthetic substitution to find $f(6)$ and explain what this means.

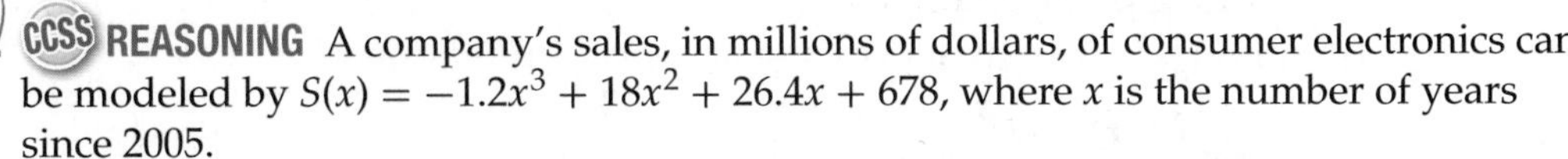

28. CCSS REASONING A company's sales, in millions of dollars, of consumer electronics can be modeled by $S(x) = -1.2x^3 + 18x^2 + 26.4x + 678$, where x is the number of years since 2005.

a. Use synthetic substitution to estimate the sales for 2017 and 2020.

b. Do you think this model is useful in estimating future sales? Explain.

Use the graph to find all of the factors for each polynomial function.

29.

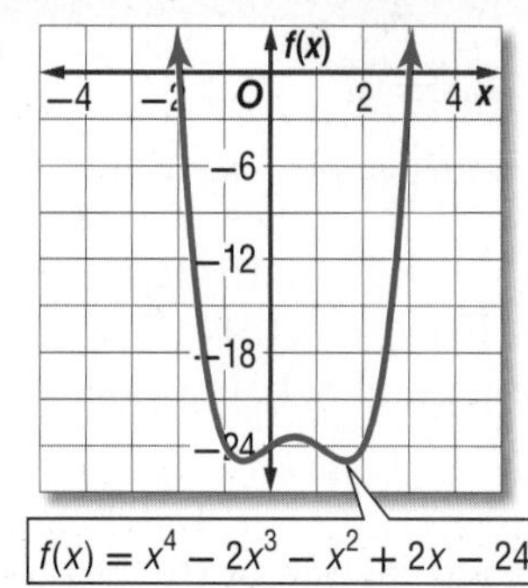

30.

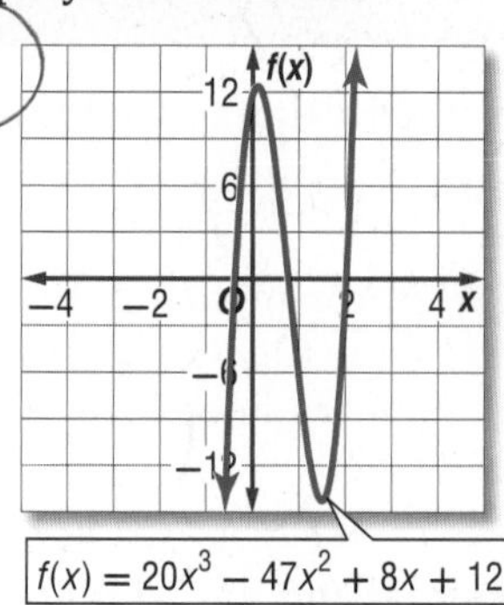

31. **MULTIPLE REPRESENTATIONS** In this problem, you will consider the function $f(x) = -9x^5 + 104x^4 - 249x^3 - 456x^2 + 828x + 432$.

a. Algebraic If $x - 6$ is a factor of the function, find the depressed polynomial.

b. Tabular Make a table of values for $-5 \leq x \leq 6$ for the depressed polynomial.

c. Analytical What conclusions can you make about the locations of the other zeros based on the table? Explain your reasoning.

d. Graphical Graph the original function to confirm your conclusions.

CCSS PERSEVERANCE Find values of k so that each remainder is 3.

32. $(x^2 - x + k) \div (x - 1)$

33. $(x^2 + kx - 17) \div (x - 2)$

34. $(x^2 + 5x + 7) \div (x + k)$

35. $(x^3 + 4x^2 + x + k) \div (x + 2)$

H.O.T. Problems Use Higher-Order Thinking Skills

36. OPEN ENDED Write a polynomial function that has a double root of 1 and a double root of −5. Graph the function.

CHALLENGE Find the solutions of each polynomial function.

37. $(x^2 - 4)^2 - (x^2 - 4) - 2 = 0$

38. $(x^2 + 3)^2 - 7(x^2 + 3) + 12 = 0$

39. REASONING Polynomial $f(x)$ is divided by $x - c$. What can you conclude if:

a. the remainder is 0?

b. the remainder is 1?

c. the quotient is 1, and the remainder is 0?

40. CHALLENGE Review the definition for the Factor Theorem. Provide a proof of the theorem.

41. OPEN ENDED Write a cubic function that has a remainder of 8 for $f(2)$ and a remainder of −5 for $f(3)$.

42. CHALLENGE Show that the quartic function $f(x) = ax^4 + bx^3 + cx^2 + dx + e$ will always have a rational zero when the numbers 1, −2, 3, 4, and −6 are randomly assigned to replace a through e, and all of the numbers are used.

43. WRITING IN MATH Explain how the zeros of a function can be located by using the Remainder Theorem and making a table of values for different input values and then comparing the remainders.

Standardized Test Practice

44. $27x^3 + y^3 =$

A $(3x + y)(3x + y)(3x + y)$

B $(3x + y)(9x^2 - 3xy + y^2)$

C $(3x - y)(9x^2 + 3xy + y^2)$

D $(3x - y)(9x^2 + 9xy + y^2)$

45. GRIDDED RESPONSE In the figure, a square with side length $2\sqrt{2}$ is inscribed in a circle. The area of the circle is $k\pi$. What is the exact value of k?

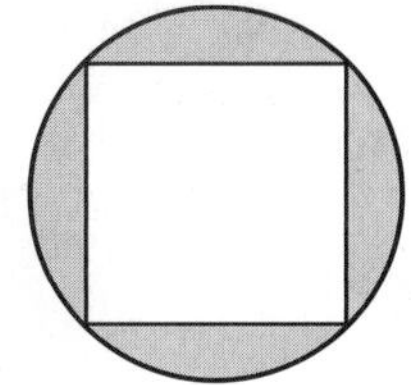

46. What is the product of the complex numbers $(4 + i)(4 - i)$?

F 15 **H** 17

G $16 - i$ **J** $17 - 8i$

47. SAT/ACT The measure of the largest angle of a triangle is 14 less than twice the measure of the smallest angle. The third angle measure is 2 more than the measure of the smallest angle. What is the measure of the smallest angle?

A 46 **D** 52

B 48 **E** 82

C 50

Spiral Review

Solve each equation. (Lesson 5-5)

48. $x^4 - 4x^2 - 21 = 0$

49. $x^4 - 6x^2 = 27$

50. $4x^4 - 8x^2 - 96 = 0$

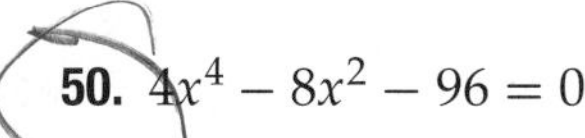

Complete each of the following. (Lesson 5-4)

a. **Estimate the x-coordinate of every turning point and determine if those coordinates are relative maxima or relative minima.**

b. **Estimate the x-coordinate of every zero.**

c. **Determine the smallest possible degree of the function.**

d. **Determine the domain and range of the function.**

51.

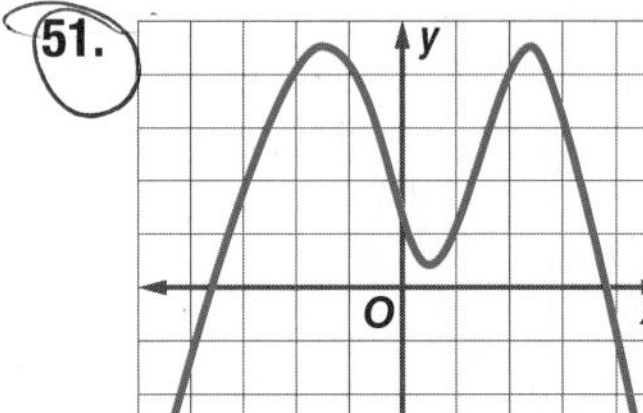

52.

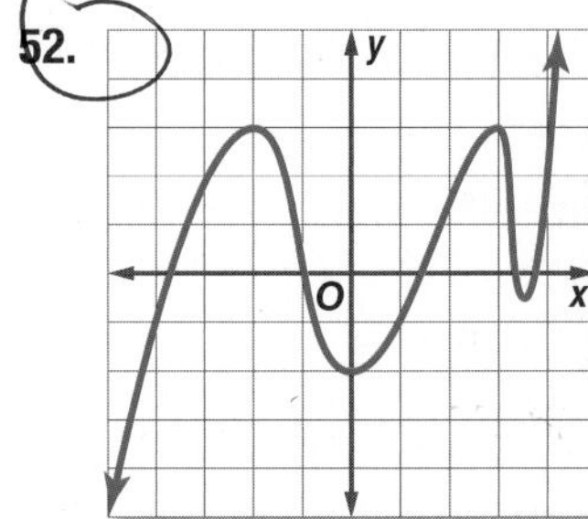

53.

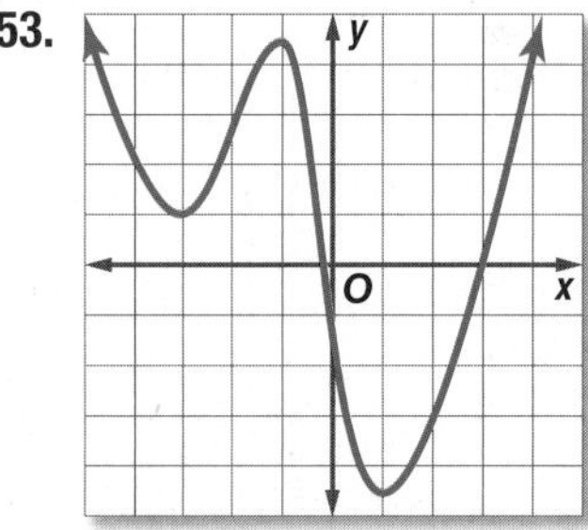

54. HIGHWAY SAFETY Engineers can use the formula $d = 0.05v^2 + 1.1v$ to estimate the minimum stopping distance d in feet for a vehicle traveling v miles per hour. If a car is able to stop after 125 feet, what is the fastest it could have been traveling when the driver first applied the brakes? (Lesson 4-6)

Solve by graphing. (Lesson 3-1)

55. $y = 3x - 1$
$y = -2x + 4$

56. $3x + 2y = 8$
$-4x + 6y = 11$

57. $5x - 2y = 6$
$3x - 2y = 2$

Skills Review

If $c(x) = x^2 - 2x$ and $d(x) = 3x^2 - 6x + 4$, find each value.

58. $c(a + 2) - d(a - 4)$

59. $c(a - 3) + d(a + 1)$

60. $c(-3a) + d(a + 4)$

61. $3d(3a) - 2c(-a)$

62. $c(a) + 5d(2a)$

63. $-2d(2a + 3) - 4c(a^2 + 1)$

LESSON 5-7 Roots and Zeros

Then

- You used complex numbers to describe solutions of quadratic equations.

Now

1. Determine the number and type of roots for a polynomial equation.
2. Find the zeros of a polynomial function.

Why?

- The function $g(x) = 1.384x^4 - 0.003x^3 + 0.28x^2 - 0.078x + 1.365$ can be used to model the average price of a gallon of gasoline in a given year if x is the number of years since 1990. To find the average price of gasoline in a specific year, you can use the roots of the related polynomial equation.

Common Core State Standards

Content Standards

N.CN.9 Know the Fundamental Theorem of Algebra; show that it is true for quadratic polynomials.

A.APR.3 Identify zeros of polynomials when suitable factorizations are available, and use the zeros to construct a rough graph of the function defined by the polynomial.

Mathematical Practices

6 Attend to precision.

1 Synthetic Types of Roots

Previously, you learned that a zero of a function $f(x)$ is any value c such that $f(c) = 0$. When the function is graphed, the real zeros of the function are the x-intercepts of the graph.

ConceptSummary Zeros, Factors, Roots, and Intercepts

Words Let $P(x) = a_nx^n + \cdots + a_1x + a_0$ be a polynomial function. Then the following statements are equivalent.

- c is a zero of $P(x)$.
- c is a root or solution of $P(x) = 0$.
- $x - c$ is a factor of $a_nx^n + \cdots + a_1x + a_0$.
- If c is a real number, then $(c, 0)$ is an x-intercept of the graph of $P(x)$.

Example Consider the polynomial function $P(x) = x^4 + 2x^3 - 7x^2 - 8x + 12$.

The zeros of $P(x) = x^4 + 2x^3 - 7x^2 - 8x + 12$ are -3, -2, 1, and 2.

The roots of $x^4 + 2x^3 - 7x^2 - 8x + 12 = 0$ are -3, -2, 1, and 2.

The factors of $x^4 + 2x^3 - 7x^2 - 8x + 12$ are $(x + 3)$, $(x + 2)$, $(x - 1)$, and $(x - 2)$.

The x-intercepts of the graph of $P(x) = x^4 + 2x^3 - 7x^2 - 8x + 12$ are $(-3, 0)$, $(-2, 0)$, $(1, 0)$, and $(2, 0)$.

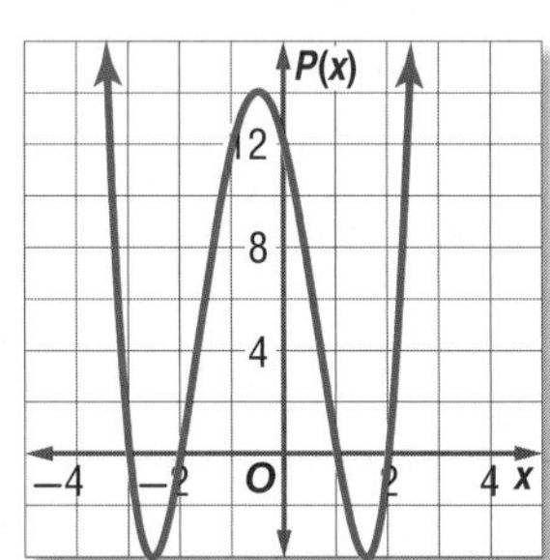

When solving a polynomial equation with degree greater than zero, there may be one or more real roots or no real roots (the roots are imaginary numbers). Since real numbers and imaginary numbers both belong to the set of complex numbers, all polynomial equations with degree greater than zero will have at least one root in the set of complex numbers. This is the **Fundamental Theorem of Algebra**.

KeyConcept Fundamental Theorem of Algebra

Every polynomial equation with degree greater than zero has at least one root in the set of complex numbers.

TORU YAMANAKA/AFP/Getty Images

Example 1 Determine Number and Type of Roots

Solve each equation. State the number and type of roots.

a. $x^2 + 6x + 9 = 0$

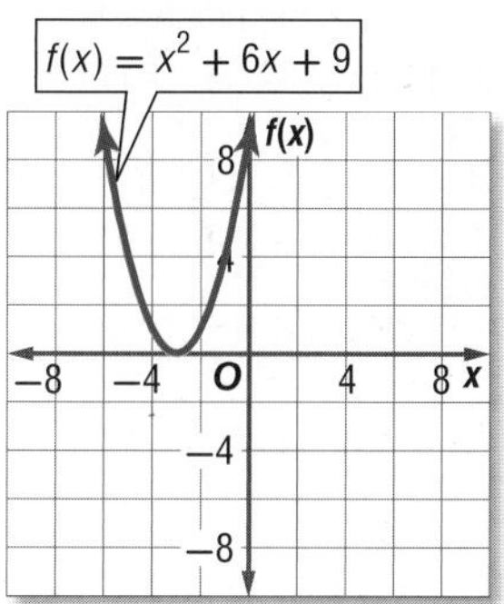

$x^2 + 6x + 9 = 0$	Original equation
$(x + 3)^2 = 0$	Factor.
$x + 3 = 0$	Take the root of each side.
$x = -3$	Solve for x.

Because $(x + 3)$ is twice a factor of $x^2 + 6x + 9$, -3 is a double root. Thus, the equation has one real repeated root, -3.

CHECK The graph of the equation touches the x-axis at $x = -3$. Since -3 is a double root, the graph does not cross the axis. ✓

ReadingMath

Repeated Roots Polynomial equations can have double roots, triple roots, quadruple roots, and so on. In general, these are referred to as *multiple roots*.

b. $x^3 + 25x = 0$

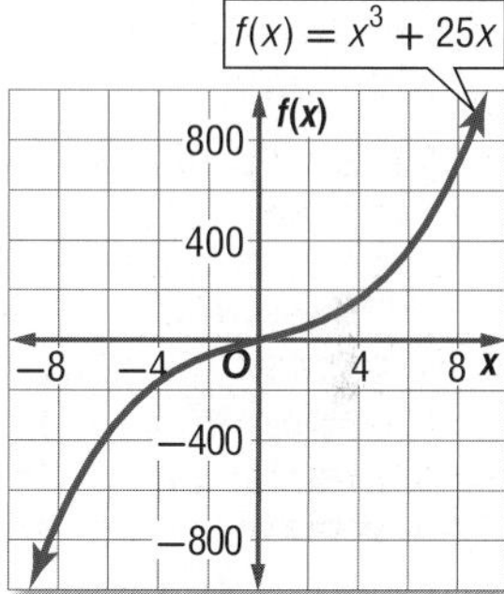

$x^3 + 25x = 0$	Original equation
$x(x^2 + 25) = 0$	Factor.

$x = 0$ or $x^2 + 25 = 0$

$x^2 = -25$

$x = \pm\sqrt{-25}$ or $\pm 5i$

This equation has one real root, 0, and two imaginary roots, $5i$ and $-5i$.

CHECK The graph of this equation crosses the x-axis at only one place, $x = 0$. ✓

GuidedPractice

1A. $x^3 + 2x = 0$

1B. $x^4 - 16 = 0$

1C. $x^3 + 4x^2 - 7x - 10 = 0$

1D. $3x^3 - x^2 + 9x - 3 = 0$

Examine the solutions for each equation in Example 1. Notice that the number of solutions for each equation is the same as the degree of each polynomial. The following corollary to the Fundamental Theorem of Algebra describes this relationship between the degree and the number of roots of a polynomial equation.

KeyConcept Corollary to the Fundamental Theorem of Algebra

Words A polynomial equation of degree n has exactly n roots in the set of complex numbers, including repeated roots.

Example	$x^3 + 2x^2 + 6$	$4x^4 - 3x^3 + 5x - 6$	$-2x^5 - 3x^2 + 8$
	3 roots	4 roots	5 roots

Similarly, an nth degree polynomial function has exactly n zeros.

Additionally, French mathematician René Descartes discovered a relationship between the signs of the coefficients of a polynomial function and the number of positive and negative real zeros.

StudyTip

Zero at the Origin If a zero of a function is at the origin, the sum of the number of positive real zeros, negative real zeros, and imaginary zeros is reduced by how many times 0 is a zero of the function.

KeyConcept Descartes' Rule of Signs

Let $P(x) = a_n x^n + \cdots + a_1 x + a_0$ be a polynomial function with real coefficients. Then

- the number of positive real zeros of $P(x)$ is the same as the number of changes in sign of the coefficients of the terms, or is less than this by an even number, and
- the number of negative real zeros of $P(x)$ is the same as the number of changes in sign of the coefficients of the terms of $P(-x)$, or is less than this by an even number.

Example 2 Find Numbers of Positive and Negative Zeros

State the possible number of positive real zeros, negative real zeros, and imaginary zeros of $f(x) = x^6 + 3x^5 - 4x^4 - 6x^3 + x^2 - 8x + 5$.

Because $f(x)$ has degree 6, it has six zeros, either real or imaginary. Use Descartes' Rule of Signs to determine the possible number and type of *real* zeros.

Count the number of changes in sign for the coefficients of $f(x)$.

$$f(x) = x^6 + 3x^5 - 4x^4 - 6x^3 + x^2 - 8x + 5$$

x^6 to $3x^5$	$3x^5$ to $4x^4$	$4x^4$ to $6x^3$	$6x^3$ to x^2	x^2 to $8x$	$8x$ to 5
no	yes	no	yes	yes	yes
+ to +	+ to −	− to −	− to +	+ to −	− to +

There are 4 sign changes, so there are 4, 2, or 0 positive real zeros.

Count the number of changes in sign for the coefficients of $f(-x)$.

$$f(-x) = (-x)^6 + 3(-x)^5 - 4(-x)^4 - 6(-x)^3 + (-x)^2 - 8(-x) + 5$$
$$= x^6 - 3x^5 - 4x^4 + 6x^3 + x^2 + 8x + 5$$

x^6 to $3x^5$	$3x^5$ to $4x^4$	$4x^4$ to $6x^3$	$6x^3$ to x^2	x^2 to $8x$	$8x$ to 5
yes	no	yes	no	no	no
+ to −	− to −	− to +	+ to +	+ to +	+ to +

There are 2 sign changes, so there are 2, or 0 negative real zeros.
Make a chart of the possible combinations of real and imaginary zeros.

Number of Positive Real Zeros	Number of Negative Real Zeros	Number of Imaginary Zeros	Total Number of Zeros
4	2	0	$4 + 2 + 0 = 6$
4	0	2	$4 + 0 + 2 = 6$
2	2	2	$2 + 2 + 2 = 6$
2	0	4	$2 + 0 + 4 = 6$
0	2	4	$0 + 2 + 4 = 6$
0	0	6	$0 + 0 + 6 = 6$

GuidedPractice

2. State the possible number of positive real zeros, negative real zeros, and imaginary zeros of $h(x) = 2x^5 + x^4 + 3x^3 - 4x^2 - x + 9$.

2 Find Zeros

You can use the various strategies and theorems you have learned to find all of the zeros of a function.

Example 3 Use Synthetic Substitution to Find Zeros

Find all of the zeros of $f(x) = x^4 - 18x^2 + 12x + 80$.

Step 1 Determine the total number of zeros.

Since $f(x)$ has degree 4, the function has 4 zeros.

Step 2 Determine the type of zeros.

Examine the number of sign changes for $f(x)$ and $f(-x)$.

$f(x) = x^4 - 18x^2 + 12x + 80$ — yes, yes, no

$f(-x) = x^4 - 18x^2 - 12x + 80$ — yes, no, yes

Because there are 2 sign changes for the coefficients of $f(x)$, the function has 2 or 0 positive real zeros. Because there are 2 sign changes for the coefficients of $f(-x)$, $f(x)$ has 2 or 0 negative real zeros. Thus, $f(x)$ has 4 real zeros, 2 real zeros and 2 imaginary zeros, or 4 imaginary zeros.

StudyTip

Testing for Zeros If a value is not a zero for a polynomial, then it will not be a zero for the depressed polynomial either, so it does not need to be checked again.

Step 3 Determine the real zeros.

List some possible values, and then use synthetic substitution to evaluate $f(x)$ for real values of x.

Each row shows the coefficients of the depressed polynomial and the remainder.

x	1	0	−18	12	80
−3	1	−3	−9	39	−37
−2	1	−2	−14	40	0
−1	1	−1	−17	29	51
0	1	0	−18	12	80
1	1	1	−17	−5	75
2	1	2	−14	−2	76

From the table, we can see that one zero occurs at $x = -2$. Since there are 2 negative real zeros, use synthetic substitution with the depressed polynomial function $f(x) = x^3 - 2x^2 - 14x + 40$ to find a second negative zero.

A second negative zero is at $x = -4$. Since the depressed polynomial $x^2 - 6x + 10$ is quadratic, use the Quadratic Formula to find the remaining zeros of $f(x) = x^2 - 6x + 10$.

x	1	−2	−14	40
−4	1	−6	10	0
−5	1	−7	21	−65
−6	1	−8	34	−164

$$x = \frac{-b \pm \sqrt{b^2 - 4ac}}{2a}$$ Quadratic Formula

$$= \frac{-(-6) \pm \sqrt{(-6)^2 - 4(1)(10)}}{2(1)}$$ Replace a with 1, b with −6, and c with 10.

$$= 3 \pm i$$ Simplify.

The function has zeros at −4, −2, $3 + i$, and $3 - i$.

StudyTip

Locating Zeros Refer to Lesson 4-2 on how to use the CALC menu to locate a zero on your calculator.

CHECK Graph the function on a graphing calculator. The graph crosses the x-axis two times, so there are two real zeros. Use the zero function under the **CALC** menu to locate each zero. The two real zeros are −4 and −2.

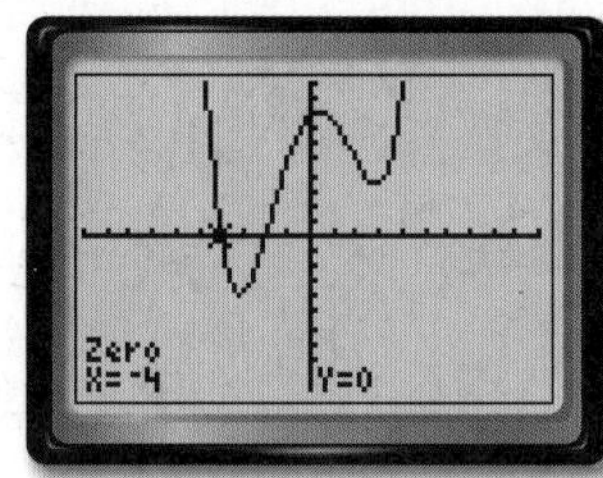

[−10, 10] scl: 1 by [−100, 100] scl: 10

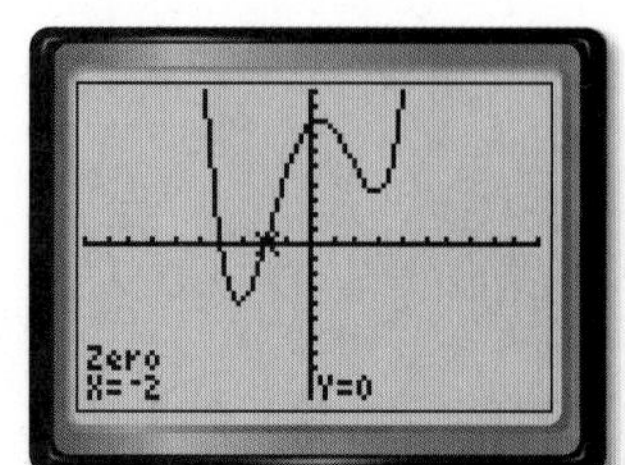

[−10, 10] scl: 1 by [−100, 100] scl: 10

GuidedPractice

3. Find all of the zeros of $h(x) = x^3 + 2x^2 + 9x + 18$.

ReviewVocabulary
complex conjugates two complex numbers of the form $a + bi$ and $a - bi$

In Chapter 4, you learned that the product of complex conjugates is always a real number and that complex roots always come in conjugate pairs. For example, if one root of $x^2 - 8x + 52 = 0$ is $4 + 6i$, then the other root is $4 - 6i$.

This applies to the zeros of polynomial functions as well. For any polynomial function with real coefficients, if an imaginary number is a zero of that function, its conjugate is also a zero. This is called the **Complex Conjugates Theorem**.

KeyConcept Complex Conjugates Theorem

Words Let a and b be real numbers, and $b \neq 0$. If $a + bi$ is a zero of a polynomial function with real coefficients, then $a - bi$ is also a zero of the function.

Example If $3 + 4i$ is a zero of $f(x) = x^3 - 4x^2 + 13x + 50$, then $3 - 4i$ is also a zero of the function.

When you are given all of the zeros of a polynomial function and are asked to determine the function, convert the zeros to factors and then multiply all of the factors together. The result is the polynomial function.

Example 4 Use Zeros to Write a Polynomial Function

Write a polynomial function of least degree with integral coefficients, the zeros of which include −1 and $5 - i$.

Understand If $5 - i$ is a zero, then $5 + i$ is also a zero according to the Complex Conjugates Theorem. So, $x + 1$, $x - (5 - i)$, and $x - (5 + i)$ are factors of the polynomial.

Plan Write the polynomial function as a product of its factors.

$$P(x) = (x + 1)[x - (5 - i)][x - (5 + i)]$$

Solve Multiply the factors to find the polynomial function.

$P(x) = (x + 1)\,[x - (5 - i)][x - (5 + i)]$	Write the equation.
$= (x + 1)\,[(x - 5) + i][(x - 5) - i]$	Regroup terms.
$= (x + 1)\,[(x - 5)^2 - i^2]$	Difference of squares
$= (x + 1)\,[(x^2 - 10x + 25 - (-1)]$	Square terms.
$= (x + 1)\,(x^2 - 10x + 26)$	Simplify.
$= x^3 - 10x^2 + 26x + x^2 - 10x + 26$	Multiply.
$= x^3 - 9x^2 + 16x + 26$	Combine like terms.

Check Because there are 3 zeros, the degree of the polynomial function must be 3, so $P(x) = x^3 - 9x^2 + 16x + 26$ is a polynomial function of least degree with integral coefficients and zeros of -1, $5 - i$, and $5 + i$.

GuidedPractice

4. Write a polynomial function of least degree with integral coefficients having zeros that include -1 and $1 + 2i$.

Check Your Understanding

= Step-by-Step Solutions begin on page R14.

Example 1 **Solve each equation. State the number and type of roots.**

1. $x^2 - 3x - 10 = 0$
2. $x^3 + 12x^2 + 32x = 0$
3. $16x^4 - 81 = 0$
4. $0 = x^3 - 8$

Example 2 **State the possible number of positive real zeros, negative real zeros, and imaginary zeros of each function.**

5. $f(x) = x^3 - 2x^2 + 2x - 6$
6. $f(x) = 6x^4 + 4x^3 - x^2 - 5x - 7$
7. $f(x) = 3x^5 - 8x^3 + 2x - 4$
8. $f(x) = -2x^4 - 3x^3 - 2x - 5$

Example 3 **Find all of the zeros of each function.**

9. $f(x) = x^3 + 9x^2 + 6x - 16$
10. $f(x) = x^3 + 7x^2 + 4x + 28$
11. $f(x) = x^4 - 2x^3 - 8x^2 - 32x - 384$
12. $f(x) = x^4 - 6x^3 + 9x^2 + 6x - 10$

Example 4 **Write a polynomial function of least degree with integral coefficients that have the given zeros.**

13. 4, −1, 6
14. 3, −1, 1, 2
15. −2, 5, $-3i$
16. −4, $4 + i$

Practice and Problem Solving

Extra Practice is on page R5.

Example 1 **Solve each equation. State the number and type of roots.**

17. $2x^2 + x - 6 = 0$
18. $4x^2 + 1 = 0$
19. $x^3 + 1 = 0$
20. $2x^2 - 5x + 14 = 0$
21. $-3x^2 - 5x + 8 = 0$
22. $8x^3 - 27 = 0$
23. $16x^4 - 625 = 0$
24. $x^3 - 6x^2 + 7x = 0$
25. $x^5 - 8x^3 + 16x = 0$
26. $x^5 + 2x^3 + x = 0$

Example 2 **State the possible number of positive real zeros, negative real zeros, and imaginary zeros of each function.**

27. $f(x) = x^4 - 5x^3 + 2x^2 + 5x + 7$
28. $f(x) = 2x^3 - 7x^2 - 2x + 12$
29. $f(x) = -3x^5 + 5x^4 + 4x^2 - 8$
30. $f(x) = x^4 - 2x^2 - 5x + 19$
31. $f(x) = 4x^6 - 5x^4 - x^2 + 24$
32. $f(x) = -x^5 + 14x^3 + 18x - 36$

Example 3 **Find all of the zeros of each function.**

33. $f(x) = x^3 + 7x^2 + 4x - 12$
34. $f(x) = x^3 + x^2 - 17x + 15$
35. $f(x) = x^4 - 3x^3 - 3x^2 - 75x - 700$
36. $f(x) = x^4 + 6x^3 + 73x^2 + 384x + 576$
37. $f(x) = x^4 - 8x^3 + 20x^2 - 32x + 64$
38. $f(x) = x^5 - 8x^3 - 9x$

Example 4 **Write a polynomial function of least degree with integral coefficients that have the given zeros.**

39. 5, −2, −1
40. −4, −3, 5
41. −1, −1, $2i$
42. −3, 1, $-3i$
43. 0, −5, $3 + i$
44. −2, −3, $4 - 3i$

45. **CCSS REASONING** A computer manufacturer determines that the profit for producing x computers per day is $P(x) = -0.006x^4 + 0.15x^3 - 0.05x^2 - 1.8x$.
    a. How many positive real zeros, negative real zeros, and imaginary zeros exist?
    b. What is the meaning of the zeros in this situation?

Sketch the graph of each function using its zeros.

46. $f(x) = x^3 - 5x^2 - 2x + 24$

47. $f(x) = 4x^3 + 2x^2 - 4x - 2$

48. $f(x) = x^4 - 6x^3 + 7x^2 + 6x - 8$

49. $f(x) = x^4 - 6x^3 + 9x^2 + 4x - 12$

Match each graph to the given zeros.

a. $-3, 4, i, -i$

b. $-4, 3$

c. $-4, 3, i, -i$

50.
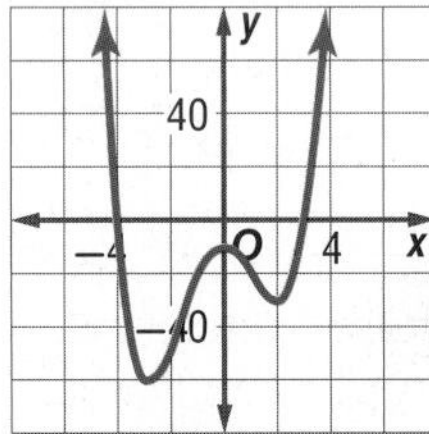

51.
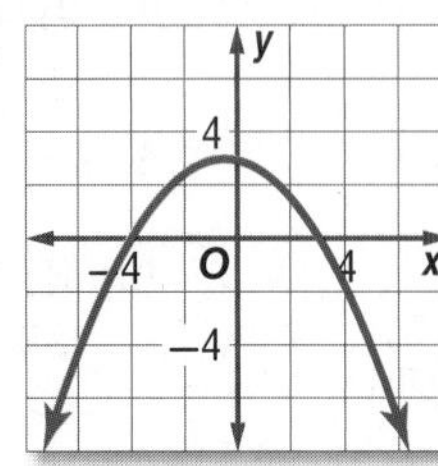

52.
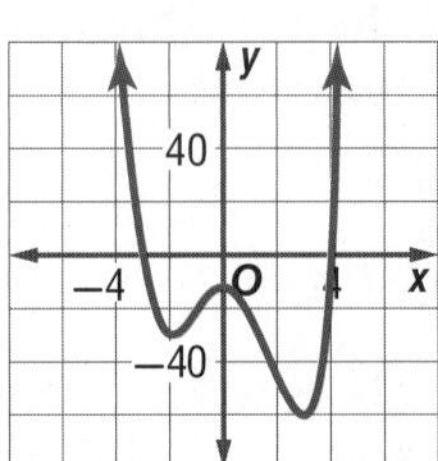

53. **CONCERTS** The amount of money Hoshi's Music Hall took in from 2003 to 2010 can be modeled by $M(x) = -2.03x^3 + 50.1x^2 - 214x + 4020$, where x is the years since 2003.

a. How many positive real zeros, negative real zeros, and imaginary zeros exist?

b. Graph the function using your calculator.

c. Approximate all real zeros to the nearest tenth. What is the significance of each zero in the context of the situation?

Determine the number of positive real zeros, negative real zeros, and imaginary zeros for each function. Explain your reasoning.

54.
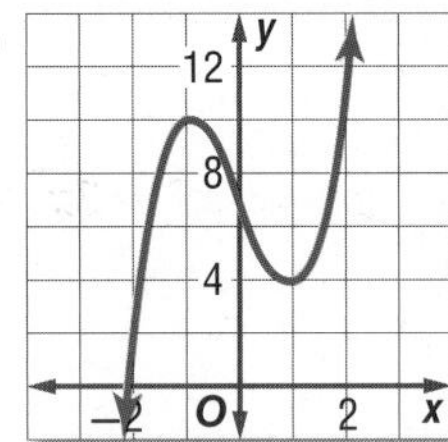

degree: 3

55.

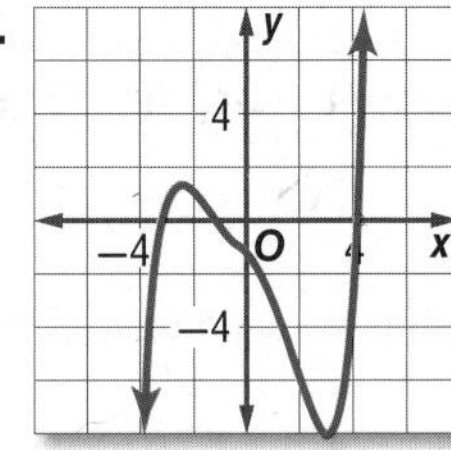

degree: 5

H.O.T. Problems Use Higher-Order Thinking Skills

56. **OPEN ENDED** Sketch the graph of a polynomial function with:

a. 3 real, 2 imaginary zeros
b. 4 real zeros
c. 2 imaginary zeros

57. **CHALLENGE** Write an equation in factored form of a polynomial function of degree 5 with 2 imaginary zeros, 1 nonintegral zero, and 2 irrational zeros. Explain.

58. **CCSS ARGUMENTS** Determine which equation is not like the others. Explain.

$r^4 + 1 = 0$	$r^3 + 1 = 0$	$r^2 - 1 = 0$	$r^3 - 8 = 0$

59. **REASONING** Provide a counterexample for each statement.

a. All polynomial functions of degree greater than 2 have at least 1 negative real root.

b. All polynomial functions of degree greater than 2 have at least 1 positive real root.

60. **WRITING IN MATH** Explain to a friend how you would use Descartes' Rule of Signs to determine the number of possible positive real roots and the number of possible negative roots of the polynomial function $f(x) = x^4 - 2x^3 + 6x^2 + 5x - 12$.

Standardized Test Practice

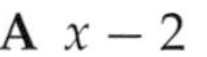

61. Use the graph of the polynomial function below. Which is not a factor of the polynomial $x^5 + x^4 - 3x^3 - 3x^2 - 4x - 4$?

A $x - 2$
B $x + 2$
C $x - 1$
D $x + 1$

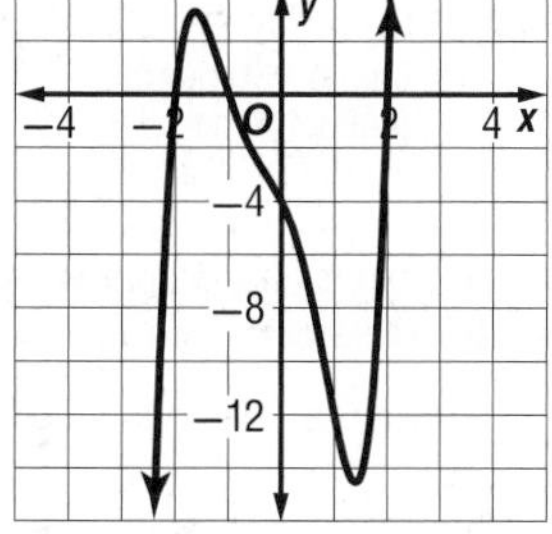

62. SHORT RESPONSE A window is in the shape of an equilateral triangle. Each side of the triangle is 8 feet long. The window is divided in half by a support from one vertex to the midpoint of the side of the triangle opposite the vertex. Approximately how long is the support?

63. GEOMETRY In rectangle $ABCD$, $\overline{AD}$ is 8 units long. What is the length of $\overline{AB}$?

F 4 units
G 8 units
H $8\sqrt{3}$ units
J 16 units

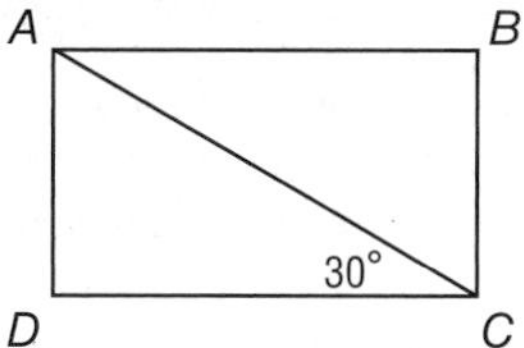

64. SAT/ACT The total area of a rectangle is $25a^4 - 16b^2$ square units. Which factors could represent the length and width?

A $(5a^2 + 4b)$ units and $(5a^2 + 4b)$ units
B $(5a^2 + 4b)$ units and $(5a^2 - 4b)$ units
C $(5a^2 - 4b)$ units and $(5a^2 - 4b)$ units
D $(5a - 4b)$ units and $(5a - 4b)$ units
E $(5a + 4b)$ units and $(5a - 4b)$ units

Spiral Review

Use synthetic substitution to find $f(-8)$ and $f(4)$ for each function. (Lesson 5-6)

65. $f(x) = 4x^3 + 6x^2 - 3x + 2$

66. $f(x) = 5x^4 - 2x^3 + 4x^2 - 6x$

67. $f(x) = 2x^5 - 3x^3 + x^2 - 4$

Factor completely. If the polynomial is not factorable, write *prime*. (Lesson 5-5)

68. $x^6 - y^6$

69. $a^6 + b^6$

70. $4x^2y + 8xy + 16y - 3x^2z - 6xz - 12z$

71. $5a^3 - 30a^2 + 40a + 2a^2b - 12ab + 16b$

72. BUSINESS A mall owner has determined that the relationship between monthly rent charged for store space r (in dollars per square foot) and monthly profit $P(r)$ (in thousands of dollars) can be approximated by $P(r) = -8.1r^2 + 46.9r - 38.2$. Solve each quadratic equation or inequality. Explain what each answer tells about the relationship between monthly rent and profit for this mall. (Lesson 4-8)

a. $-8.1r^2 + 46.9r - 38.2 = 0$

b. $-8.1r^2 + 46.9r - 38.2 > 0$

c. $-8.1r^2 + 46.9r - 38.2 > 10$

d. $-8.1r^2 + 46.9r - 38.2 < 10$

73. DIVING To avoid hitting any rocks below, a cliff diver jumps up and out. The equation $h = -16t^2 + 4t + 26$ describes her height h in feet t seconds after jumping. Find the time at which she returns to a height of 26 feet. (Lesson 4-3)

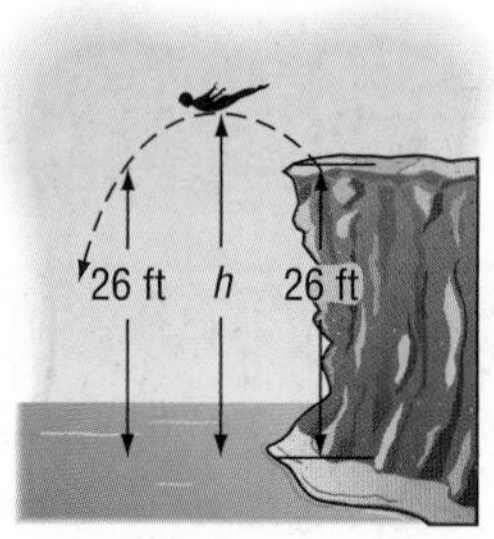

Skills Review

Find all of the possible values of $\pm\frac{b}{a}$ for each replacement set.

74. $a = \{1, 2, 4\}$; $b = \{1, 2, 3, 6\}$

75. $a = \{1, 5\}$; $b = \{1, 2, 4, 8\}$

76. $a = \{1, 2, 3, 6\}$; $b = \{1, 7\}$

EXTEND 5-7

Graphing Technology Lab
Analyzing Polynomial Functions

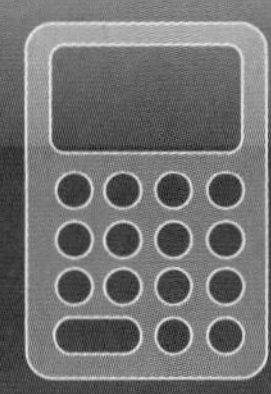

You can use graphing technology to help you identify real zeros, maximum and minimum points, number and type of zeros, y-intercepts, and symmetry of polynomial functions.

CCSS Common Core State Standards
Content Standards
A.APR.4 Prove polynomial identities and use them to describe numerical relationships.

Activity Identify Polynomial Characteristics

Graph each function. Identify the real zeros, maximum and minimum points, number and type of zeros, y-intercepts, and symmetry.

a. $g(x) = 3x^4 - 15x^3 + 87x^2 - 375x + 300$

Step 1 Graph the equation.

Step 2 Use 2nd [CALC] **zero** to find the real zeros at $x = 1$ and $x = 4$.

Step 3 Use 2nd [CALC] **minimum** to find the relative minimum at (2.68, −214.11). There is no relative maximum point.

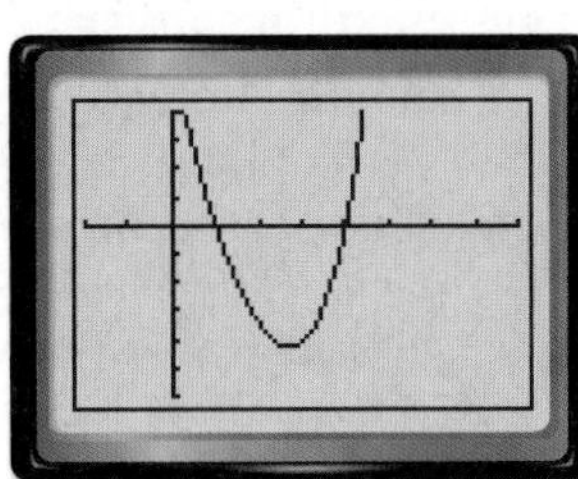

[−2, 8] scl: 1 by [−300, 200] scl: 50

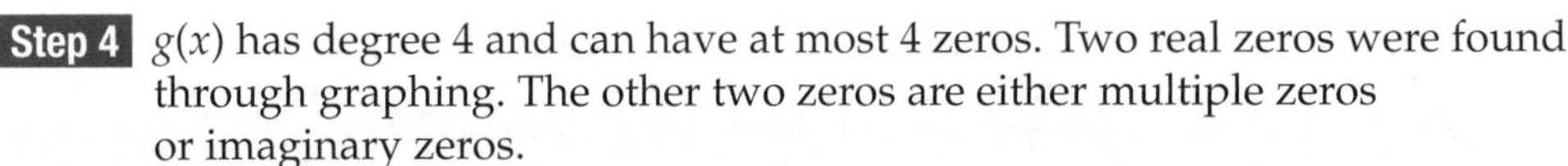

Step 4 $g(x)$ has degree 4 and can have at most 4 zeros. Two real zeros were found through graphing. The other two zeros are either multiple zeros or imaginary zeros.

Step 5 Use 2nd [CALC] **value** 0 to find the y-intercept, 300.

Step 6 The line of symmetry passes through the vertex. Its equation is $x = 2.68$.

b. $f(x) = 2x^5 - 5x^4 - 3x^3 + 8x^2 + 4x$

Step 1 Graph the equation.

Step 2 Locate the real zeros at $x = -1$, $x = -\frac{1}{2}$, $x = 0$ and $x = 2$.

Step 3 Find the relative maxima at (−0.81, 0.75) and (1.04, 6.02) and the relative minima at (−0.24, −0.48) and (2, 0).

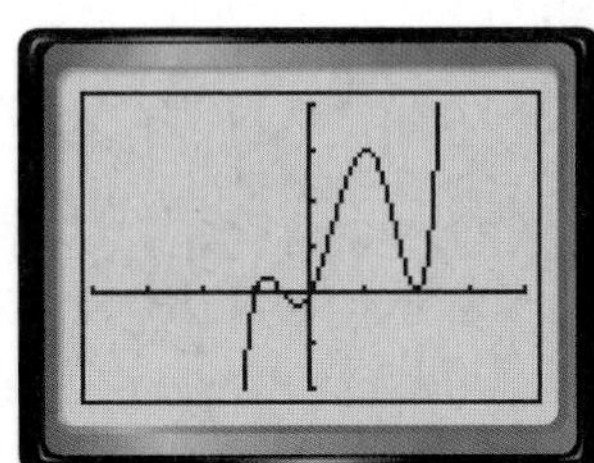

[−4, 4] scl: 1 by [−4, 8] scl: 2

Step 4 $f(x)$ has degree 5 and can have at most 5 zeros. Four real zeros were found through graphing. The other zero is either a multiple zero or an imaginary zero. In this case, there is a double zero at $x = 2$.

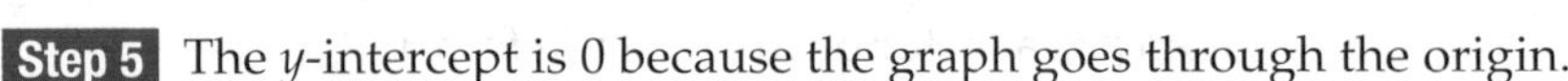

Step 5 The y-intercept is 0 because the graph goes through the origin.

Step 6 There is no symmetry.

Exercises

Graph each function. Identify the real zeros, maximum and minimum points, number and type of zeros, y-intercepts, and symmetry.

1. $f(x) = x^3 - 5x^2 + 6x$

2. $g(x) = x^4 - 3x^2 - 4$

3. $k(x) = -x^4 - x^3 + 2x^2$

4. $f(x) = -2x^3 - 4x^2 + 16x$

5. $g(x) = 3x^5 - 18x^4 + 27x^3$

6. $k(x) = x^4 - 8x^2 + 15$

7. $f(x) = -x^3 + 2x^2 + 8x$

8. $g(x) = x^5 + 3x^4 - 10x^2$

LESSON

5-8 Rational Zero Theorem

Then

- You found zeros of quadratic functions of the form $f(x) = ax^2 + bx + c$.

Now

1. Identify possible rational zeros of a polynomial function.
2. Find all of the rational zeros of a polynomial function.

Why?

- Annual sales of recorded music in the United States can be approximated by $d(t) = 30x^3 - 478x^2 + 1758x + 10{,}092$, where $d(t)$ is the total sales in millions of dollars and t is the number of years since 2005. You can use this function to estimate when music sales will be $9 billion.

Common Core State Standards

Mathematical Practices

8 Look for and express regularity in repeated reasoning.

1 Identify Rational Zeros

Usually it is not practical to test all possible zeros of a polynomial function using synthetic substitution. The **Rational Zero Theorem** can help you choose some possible zeros to test. If the leading coefficient is 1, the corollary applies.

KeyConcept Rational Zero Theorem

Words If $P(x)$ is a polynomial function with integral coefficients, then every rational zero of $P(x) = 0$ is of the form $\frac{p}{q}$, a rational number in simplest form, where p is a factor of the constant term and q is a factor of the leading coefficient.

Example Let $f(x) = 6x^4 + 22x^3 + 11x^2 - 80x - 40$. If $\frac{4}{3}$ is a zero of $f(x)$, then 4 is a factor of -40, and 3 is a factor of 6.

Corollary to the Rational Zero Theorem

If $P(x)$ is a polynomial function with integral coefficients, a leading coefficient of 1, and a nonzero constant term, then any rational zeros of $P(x)$ must be factors of the constant term.

Example 1 Identify Possible Zeros

List all of the possible rational zeros of each function.

a. $f(x) = 4x^5 + x^4 - 2x^3 - 5x^2 + 8x + 16$

If $\frac{p}{q}$ is a rational zero, then p is a factor of 16 and q is a factor of 4.

p: ±1, ±2, ±4, ±8, ±16 q: ±1, ±2, ±4

Write the possible values of $\frac{p}{q}$ in simplest form.

$\frac{p}{q} = \pm1, \pm2, \pm4, \pm8, \pm16, \pm\frac{1}{2}, \pm\frac{1}{4}$

b. $f(x) = x^3 - 2x^2 + 5x + 12$

If $\frac{p}{q}$ is a rational zero, then p is a factor of 12 and q is a factor of 1.

p: ±1, ±2, ±3, ±4, ±6, ±12 q: ±1

So, $\frac{p}{q} = \pm1, \pm2, \pm3, \pm4, \pm6$, and ±12

GuidedPractice

1A. $g(x) = 3x^3 - 4x + 10$

1B. $h(x) = x^3 + 11x^2 + 24$

Javier Larrea/age fotostock

2 Find Rational Zeros

Find Rational Zeros Once you have written the possible rational zeros, you can test each number using synthetic substitution and use the other tools you have learned to determine the zeros of a function.

Real-World Example 2 Find Rational Zeros

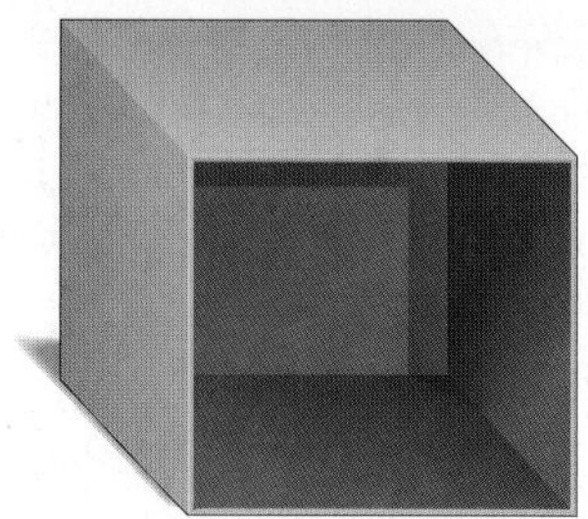

WOODWORKING Adam is building a computer desk with a separate compartment for the computer. The compartment for the computer is a rectangular prism and will be 8019 cubic inches. The compartment will be 24 inches longer than it is wide and the height will be 18 inches greater than the width. Find the dimensions of the computer compartment.

Let x = width, $x + 24$ = length, and $x + 18$ = height.

Write an equation for the volume.

$\ell wh = V$ Formula for volume

$(x + 24)(x)(x + 18) = 8019$ Substitute.

$x^3 + 42x^2 + 432x = 8019$ Multiply.

$x^3 + 42x^2 + 432x - 8019 = 0$ Subtract 8019 from each side.

The leading coefficient is 1, so the possible rational zeros are factors of 8019.

$\pm1, \pm3, \pm9, \pm11, \pm27, \pm33, \pm81, \pm99, \pm243, \pm297, \pm729, \pm891, \pm2673$, and ±8019

StudyTip

CCSS Structure Examine the signs of the coefficients of the equation. In this case, there is only one change of sign, so there is only one positive real zero.

Since length can only be positive, we only need to check positive values.

There is one change of sign of the coefficients, so by Descartes' Rule of Signs, there is only one positive real zero. Make a table for synthetic division and test possible values.

p	1	42	432	−8019
1	1	43	475	−7544
2	1	45	567	−6318
9	1	51	891	0

One zero is 9. Since there is only one positive real zero, we do not have to test the other numbers. The other dimensions are 9 + 24 or 33 inches, and 9 + 18 or 27 inches.

CHECK Multiply the dimensions and see if they equal the volume of 8019 cubic inches.
$9 \times 33 \times 27 = 8019$ ✓

GuidedPractice

2. GEOMETRY The volume of a rectangular prism is 1056 cubic centimeters. The length is 1 centimeter more than the width, and the height is 3 centimeters less than the width. Find the dimensions of the prism.

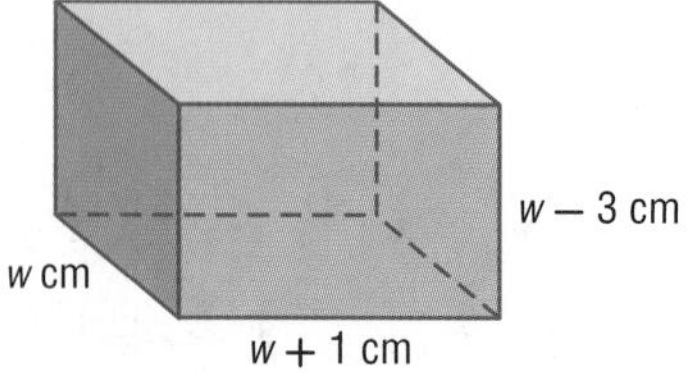

You usually do not need to test all of the possible zeros. Once you find a zero, you can try to factor the depressed polynomial to find any other zeros.

Example 3 Find All Zeros

Find all of the zeros of $f(x) = 5x^4 - 8x^3 + 41x^2 - 72x - 36$.

From the corollary to the Fundamental Theorem of Algebra, there are exactly 4 complex zeros. According to Descartes' Rule of Signs, there are 3 or 1 positive real zeros and exactly 1 negative real zero. The possible rational zeros are $\pm1, \pm2, \pm3, \pm4, \pm6, \pm9, \pm12, \pm18, \pm36, \pm\frac{1}{5}, \pm\frac{2}{5}, \pm\frac{3}{5}, \pm\frac{4}{5}, \pm\frac{6}{5}, \pm\frac{9}{5}, \pm\frac{12}{5}, \pm\frac{18}{5}$, and $\pm\frac{36}{5}$.

Make a table and test some possible rational zeros.

$\frac{p}{q}$	5	−8	41	−72	−36
−1	5	−13	54	−126	90
1	5	−3	38	−34	−70
2	5	2	45	18	0

Because $f(2) = 0$, there is a zero at $x = 2$. Factor the depressed polynomial $5x^3 + 2x^2 + 45x + 18$.

$5x^3 + 2x^2 + 45x + 18 = 0$ — Write the depressed polynomial.

$(5x^3 + 2x^2) + (45x + 18) = 0$ — Group terms.

$x^2(5x + 2) + 9(5x + 2) = 0$ — Factor.

$(x^2 + 9)(5x + 2) = 0$ — Distributive Property

$x^2 + 9 = 0$ or $5x + 2 = 0$ — Zero Product Property

$x^2 = -9$ $\qquad 5x = -2$

$x = \pm 3i$ $\qquad x = -\frac{2}{5}$

There is another real zero at $x = -\frac{2}{5}$ and two imaginary zeros at $x = 3i$ and $x = -3i$.

The zeros of the function are $-\frac{2}{5}$, 2, $3i$, and $-3i$.

Guided Practice

Find all of the zeros of each function.

3A. $h(x) = 9x^4 + 5x^2 - 4$

3B. $k(x) = 2x^4 - 5x^3 + 20x^2 - 45x + 18$

Check Your Understanding

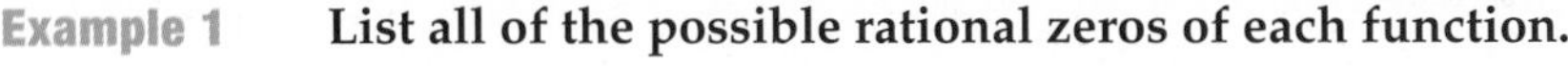

= Step-by-Step Solutions begin on page R14.

Example 1 **List all of the possible rational zeros of each function.**

1. $f(x) = x^3 - 6x^2 - 8x + 24$

2. $f(x) = 2x^4 + 3x^2 - x + 15$

Example 2 **3.** **CCSS REASONING** The volume of the triangular pyramid is 210 cubic inches. Find the dimensions of the solid.

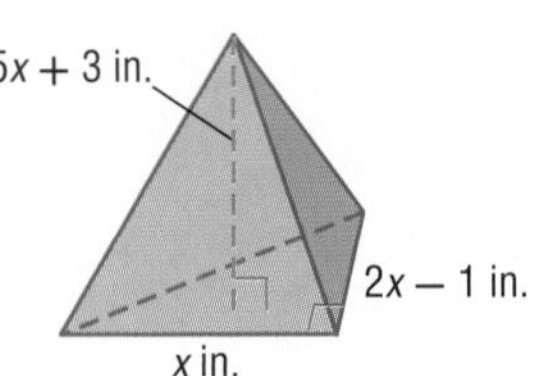

Find all of the rational zeros of each function.

4. $f(x) = x^3 - 6x^2 - 13x + 42$

5. $f(x) = 2x^4 + 11x^3 + 26x^2 + 29x + 12$

Example 3 **Find all of the zeros of each function.**

6. $f(x) = 3x^3 - 2x^2 - 8x + 5$

7. $f(x) = 8x^3 + 14x^2 + 11x + 3$

8. $f(x) = 4x^4 + 13x^3 - 8x^2 + 13x - 12$

9. $f(x) = 4x^4 - 12x^3 + 25x^2 - 14x - 15$

Practice and Problem Solving

Extra Practice is on page R5.

Example 1 **List all of the possible rational zeros of each function.**

10. $f(x) = x^4 + 8x - 32$
11. $f(x) = x^3 + x^2 - x - 56$
12. $f(x) = 2x^3 + 5x^2 - 8x - 10$
13. $f(x) = 3x^6 - 4x^4 - x^2 - 35$
14. $f(x) = 6x^5 - x^4 + 2x^3 - 3x^2 + 2x - 18$
15. $f(x) = 8x^4 - 4x^3 - 4x^2 + x + 42$
16. $f(x) = 15x^3 + 6x^2 + x + 90$
17. $f(x) = 16x^4 - 5x^2 + 128$

Example 2

18. **MANUFACTURING** A box is to be constructed by cutting out equal squares from the corners of a square piece of cardboard and turning up the sides.
    a. Write a function $V(x)$ for the volume of the box.
    b. For what value of x will the volume of the box equal 1152 cubic centimeters?
    c. What will be the volume of the box if $x = 6$ centimeters?

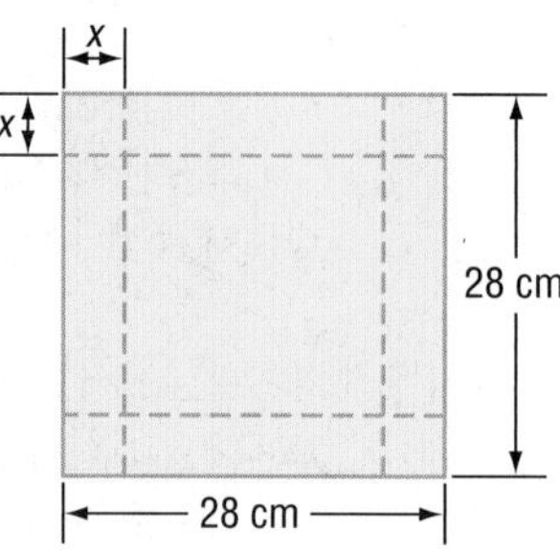

Find all of the rational zeros of each function.

19. $f(x) = x^3 + 10x^2 + 31x + 30$
20. $f(x) = x^3 - 2x^2 - 56x + 192$
21. $f(x) = 4x^3 - 3x^2 - 100x + 75$
22. $f(x) = 4x^4 + 12x^3 - 5x^2 - 21x + 10$
23. $f(x) = x^4 + x^3 - 8x - 8$
24. $f(x) = 2x^4 - 3x^3 - 24x^2 + 4x + 48$
25. $f(x) = 4x^3 + x^2 + 16x + 4$
26. $f(x) = 81x^4 - 256$

Example 3 **Find all of the zeros of each function.**

27. $f(x) = x^3 + 3x^2 - 25x + 21$
28. $f(x) = 6x^3 + 5x^2 - 9x + 2$
29. $f(x) = x^4 - x^3 - x^2 - x - 2$
30. $f(x) = 10x^3 - 17x^2 - 7x + 2$
31. $f(x) = x^4 - 3x^3 + x^2 - 3x$
32. $f(x) = 6x^3 + 11x^2 - 3x - 2$
33. $f(x) = 6x^4 + 22x^3 + 11x^2 - 38x - 40$
34. $f(x) = 2x^3 - 7x^2 - 8x + 28$
35. $f(x) = 9x^5 - 94x^3 + 27x^2 + 40x - 12$
36. $f(x) = x^5 - 2x^4 - 12x^3 - 12x^2 - 13x - 10$
37. $f(x) = 48x^4 - 52x^3 + 13x - 3$
38. $f(x) = 5x^4 - 29x^3 + 55x^2 - 28x$

39. **SWIMMING POOLS** A diagram of the swimming pool at the Midtown Community Center is shown below. The pool can hold 9175 cubic feet of water.

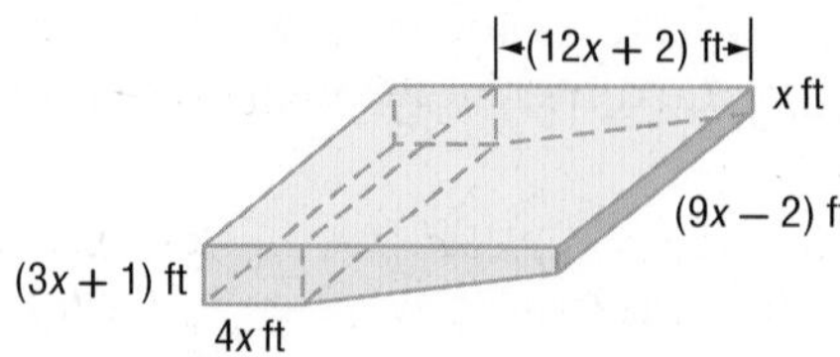

    a. Write a polynomial function that represents the volume of the swimming pool.
    b. What are the possible values of x? Which of these values are reasonable?

40. **CCSS MODELING** A portion of the path of a certain roller coaster can be modeled by $f(t) = t^4 - 31t^3 + 308t^2 - 1100t + 1200$ where t represents the time in seconds and $f(t)$ represents the height of the roller coaster. Use the Rational Zero Theorem to determine the four times at which the roller coaster is at ground level.

41 **FOOD** A restaurant orders spaghetti sauce in cylindrical metal cans. The volume of each can is about 160π cubic inches, and the height of the can is 6 inches more than the radius.

a. Write a polynomial equation that represents the volume of a can. Use the formula for the volume of a cylinder, $V = \pi r^2 h$.

b. What are the possible values of r? Which of these values are reasonable for this situation?

c. Find the dimensions of the can.

42. **Refer to the graph at the right.**

a. Find all of the zeros of $f(x) = 2x^3 + 7x^2 + 2x - 3$ and $g(x) = 2x^3 - 7x^2 + 2x + 3$.

b. Determine which function, f or g, is shown in the graph at the right.

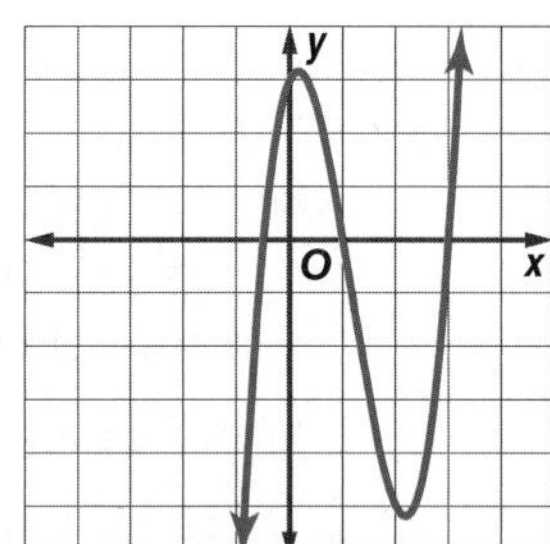

43. **MUSIC SALES** Refer to the beginning of the lesson.

a. Write a polynomial equation that could be used to determine the year in which music sales would be about $9,000,000,000.

b. List the possible whole number solutions for your equation in part **a**.

c. Determine the approximate year in which music sales will be 9,000,000,000.

d. Does the model represent a realistic estimate for all future music sales? Explain your reasoning.

Find all of the zeros of each function.

44. $f(x) = x^5 + 3x^4 - 19x^3 - 43x^2 + 18x + 40$

45. $f(x) = x^5 - x^4 - 23x^3 + 33x^2 + 126x - 216$

H.O.T. Problems Use Higher-Order Thinking Skills

46. **CCSS CRITIQUE** Doug and Mika are listing all of the possible rational zeros for $f(x) = 4x^4 + 8x^5 + 10x^2 + 3x + 16$. Is either of them correct? Explain your reasoning.

Doug	Mika
$\pm1, \pm2, \pm4, \pm8, \pm16, \pm\frac{1}{2}, \pm\frac{1}{4}$	$\pm1, \pm2, \pm4, \pm8, \pm16, \pm\frac{1}{2}, \pm\frac{1}{4}, \pm\frac{1}{8}$

47. **CHALLENGE** Give a polynomial function that has zeros at $1 + \sqrt{3}$ and $5 + 2i$.

48. **REASONING** Determine if the following statement is *sometimes*, *always*, or *never* true. Explain your reasoning.

If all of the possible zeros of a polynomial function are integers, then the leading coefficient of the function is 1 or −1.

49. **OPEN ENDED** Write a function that has possible zeros of $\pm18, \pm9, \pm6, \pm3, \pm2, \pm1, \pm\frac{9}{4}, \pm\frac{9}{2}, \pm\frac{3}{2}, \pm\frac{3}{4}, \pm\frac{1}{2}$, and $\pm\frac{1}{4}$.

50. **CHALLENGE** The roots of $x^2 + bx + c = 0$ are M and N. If $|M - N| = 1$, express c in terms of b.

51. **WRITING IN MATH** How can you find the zeros of a polynomial function?

Standardized Test Practice

52. ALGEBRA Which of the following is a zero of the function $f(x) = 12x^5 - 5x^3 + 2x - 9$?

A -6

B $-\frac{2}{3}$

C $\frac{3}{8}$

D 1

53. SAT/ACT How many negative real zeros does $f(x) = x^5 - 2x^4 - 4x^3 + 4x^2 - 5x + 6$ have?

F 5

G 3

H 2

J 1

K 0

54. ALGEBRA For all nonnegative numbers n, let $\boxed{n}$ be defined by $\boxed{n} = \frac{\sqrt{n}}{2}$. If $\boxed{n} = 4$, what is the value of n?

A 2

B 4

C 16

D 64

55. GRIDDED RESPONSE What is the y-intercept of a line that contains the point $(-1, 4)$ and has the same x-intercept as $x + 2y = -3$?

Spiral Review

Write a polynomial function of least degree with integral coefficients that has the given zeros. (Lesson 5-7)

56. $6, -3, \sqrt{2}$

57. $5, -1, 4i$

58. $-4, -2, i\sqrt{2}$

Given a polynomial and one of its factors, find the remaining factors of the polynomial. (Lesson 5-6)

59. $x^4 + 5x^3 + 5x^2 - 5x - 6; x + 3$

60. $a^4 - 2a^3 - 17a^2 + 18a + 72; a - 3$

61. $x^4 + x^3 - 11x^2 + x - 12; x + i$

62. BRIDGES The supporting cables of the Golden Gate Bridge approximate the shape of a parabola. The parabola can be modeled by the quadratic function $y = 0.00012x^2 + 6$, where x represents the distance from the axis of symmetry and y represents the height of the cables. The related quadratic equation is $0.00012x^2 + 6 = 0$. (Lesson 4-6)

a. Calculate the value of the discriminant.

b. What does the discriminant tell you about the supporting cables of the Golden Gate Bridge?

63. RIDES An amusement park ride carries riders to the top of a 225-foot tower. The riders then free-fall in their seats until they reach 30 feet above the ground. (Lesson 4-2)

a. Use the formula $h(t) = -16t^2 + h_0$, where the time t is in seconds and the initial height h_0, is in feet, to find how long the riders are in free-fall.

b. Suppose the designer of the ride wants the riders to experience free-fall for 5 seconds before stopping 30 feet above the ground. What should be the height of the tower?

Skills Review

Simplify.

64. $(x - 4)(x + 3)$

65. $3x(x^2 + 4)$

66. $x^2(x - 2)(x + 1)$

Find each value if $f(x) = 6x + 2$ and $g(x) = -4x^2$.

67. $f(5)$

68. $g(-3)$

69. $f(3c)$

Study Guide and Review

Study Guide

KeyConcepts

Operations with Polynomials (Lessons 5-1 and 5-2)

- To add or subtract: Combine like terms.
- To multiply: Use the Distributive Property.
- To divide: Use long division or synthetic division.

Polynomial Functions and Graphs (Lessons 5-3 and 5-4)

- Turning points of a function are called *relative maxima* and *relative minima.*

Solving Polynomial Equations (Lesson 5-5)

- You can factor polynomials by using the GCF, grouping, or quadratic techniques.

The Remainder and Factor Theorems (Lesson 5-6)

- Factor Theorem: The binomial $x - a$ is a factor of the polynomial $f(x)$ if and only if $f(a) = 0$.

Roots, Zeros, and the Rational Zero Theorem (Lessons 5-7 and 5-8)

- Complex Conjugates Theorem: If $a + bi$ is a zero of a function, then $a - bi$ is also a zero.
- Integral Zero Theorem: If the coefficients of a polynomial function are integers such that $a_0 = 1$ and $a_n = 0$, any rational zeros of the function must be factors of a_n.
- Rational Zero Theorem: If $P(x)$ is a polynomial function with integral coefficients, then every rational zero of $P(x) = 0$ is of the form $\frac{p}{q}$, a rational number in simplest form, where p is a factor of the constant term and q is a factor of the leading coefficient.

FOLDABLES® StudyOrganizer

Be sure the Key Concepts are noted in your Foldable.

KeyVocabulary

degree of a polynomial (p. 305)
depressed polynomial (p. 354)
end behavior (p. 324)
extrema (p. 331)
leading coefficient (p. 322)
Location principle (p. 330)
polynomial function (p. 323)
polynomial in one variable (p. 322)
power function (p. 323)
prime polynomials (p. 342)
quadratic form (p. 345)
relative maximum (p. 331)
relative minimum (p. 331)
simplify (p. 303)
synthetic division (p. 312)
synthetic substitution (p. 352)
turning points (p. 331)

VocabularyCheck

State whether each sentence is *true* or *false*. If *false*, replace the underlined term to make a true sentence.

1. The coefficient of the first term of a polynomial in standard form is called the <u>leading coefficient</u>.
2. Polynomials that cannot be factored are called <u>polynomials in one variable</u>.
3. A <u>prime polynomial</u> has a degree that is one less than the original polynomial.
4. A point on the graph of a function where no other nearby point has a greater *y*-coordinate is called a <u>relative maximum</u>.
5. A <u>polynomial function</u> is a continuous function that can be described by a polynomial equation in one variable.
6. To <u>simplify</u> an expression containing powers means to rewrite the expression without parentheses or negative exponents.
7. <u>Synthetic division</u> is a shortcut method for dividing a polynomial by a binomial.
8. The relative maximum and relative minimum of a function are often referred to as <u>end behavior</u>.
9. When a polynomial is divided by one of its binomial factors, the quotient is called a <u>depressed polynomial</u>.
10. $(x^3)^2 + 3x^3 - 8 = 0$ is a <u>power function</u>.

CHAPTER 5

Study Guide and Review *Continued*

Lesson-by-Lesson Review

5-1 Operations with Polynomials

Simplify. Assume that no variable equals 0.

11. $\frac{14x^4y}{2x^3y^5}$

12. $3t(tn - 5)$

13. $(4r^2 + 3r - 1) - (3r^2 - 5r + 4)$

14. $(x^4)^3$

15. $(m + p)(m^2 - 2mp + p^2)$

16. $3b(2b - 1) + 2b(b + 3)$

Example 1

Simplify each expression.

a. $(-4a^3b^5)(5ab^3)$

$(-4a^3b^5)(5ab^3) = (-4)(5)(a^3 \cdot a)(b^5 \cdot b^3)$ Product of Powers

$= -20a^4b^8$ Simplify.

b. $(2x^2 + 3x - 8) + (3x^2 - 5x - 7)$

$(2x^2 + 3x - 8) + (3x^2 - 5x - 7)$

$= (2x^2 + 3x^2) + (3x - 5x) + [-8 + (-7)]$

$= 5x^2 - 2x - 15$

5-2 Dividing Polynomials

Simplify.

17. $\frac{12x^4y^5 + 8x^3y^7 - 16x^2y^6}{4xy^5}$

18. $(6y^3 + 13y^2 - 10y - 24) \div (y + 2)$

19. $(a^4 + 5a^3 + 2a^2 - 6a + 4)(a + 2)^{-1}$

20. $(4a^6 - 5a^4 + 3a^2 - a) \div (2a + 1)$

21. **GEOMETRY** The volume of the rectangular prism is $3x^3 + 11x^2 - 114x - 80$ cubic units. What is the area of the base?

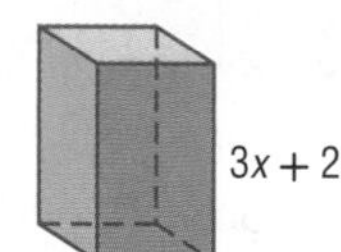

Example 2

Simplify $(6x^3 - 31x^2 - 34x + 22) \div (2x - 1)$.

$$
\begin{array}{r}
3x^2 - 14x - 24 \\
2x - 1 \overline{)\,6x^3 - 31x^2 - 34x + 22} \\
(-)\ 6x^3 - 3x^2 \qquad\qquad \\
\hline
-28x^2 - 34x \qquad \\
(-)\ -28x^2 + 14x \qquad \\
\hline
-48x + 22 \\
(-)\ -48x + 24 \\
\hline
-2
\end{array}
$$

The result is $3x^2 - 14x - 24 - \frac{2}{2x - 1}$.

5-3 Polynomial Functions

State the degree and leading coefficient of each polynomial in one variable. If it is not a polynomial in one variable, explain why.

22. $5x^6 - 3x^4 + x^3 - 9x^2 + 1$

23. $6xy^2 - xy + y^2$

24. $12x^3 - 5x^4 + 6x^8 - 3x - 3$

Find $p(-2)$ and $p(x + h)$ for each function.

25. $p(x) = x^2 + 2x - 3$

26. $p(x) = 3x^2 - x$

27. $p(x) = 3 - 5x^2 + x^3$

Example 3

What are the degree and leading coefficient of $4x^3 + 3x^2 - 7x^7 + 4x - 1$?

The greatest exponent is 7, so the degree is 7. The leading coefficient is -7.

Example 4

Find $p(a - 2)$ if $p(x) = 3x + 2x^2 - x^3$.

$p(a - 2) = 3(a - 2) + 2(a - 2)^2 - (a - 2)^3$

$= 3a - 6 + 2a^2 - 8a + 8 - (a^3 - 6a^2 + 12a - 8)$

$= -a^3 + 8a^2 - 17a + 10$

5-4 Analyzing Graphs of Polynomial Functions

Complete each of the following.

a. Graph each function by making a table of values.

b. Determine the consecutive integer values of x between which each real zero is located.

c. Estimate the x-coordinates at which the relative maxima and minima occur.

28. $h(x) = x^3 - 4x^2 - 7x + 10$

29. $g(x) = 4x^4 - 21x^2 + 5$

30. $f(x) = x^3 - 3x^2 - 4x + 12$

31. $h(x) = 4x^3 - 6x^2 + 1$

32. $p(x) = x^5 - x^4 + 1$

33. **BUSINESS** Milo tracked the monthly profits for his sports store business for the first six months of the year. They can be modeled by using the following six points: (1, 675), (2, 950), (3, 550), (4, 250), (5, 600), and (6, 400). How many turning points would the graph of a polynomial function through these points have? Describe them.

Example 5

Graph $f(x) = x^3 + 3x^2 - 4$ by making a table of values.

Make a table of values for several values of x.

x	−3	−2	−1	0	1	2
$f(x)$	−4	0	−2	−4	0	16

Plot the points and connect the points with a smooth curve.

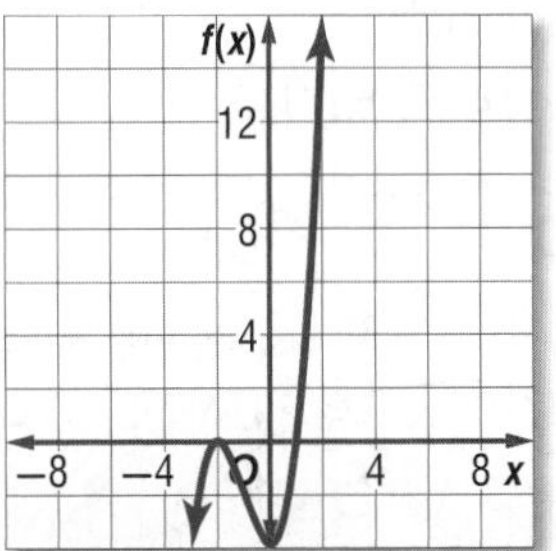

5-5 Solving Polynomial Equations

Factor completely. If the polynomial is not factorable, write *prime.*

34. $a^4 - 16$

35. $x^3 + 6y^3$

36. $54x^3y - 16y^4$

37. $6ay + 4by - 2cy + 3az + 2bz - cz$

Solve each equation.

38. $x^3 + 2x^2 - 35x = 0$

39. $8x^4 - 10x^2 + 3 = 0$

40. **GEOMETRY** The volume of the prism is 315 cubic inches. Find the value of x and the length, height, and width.

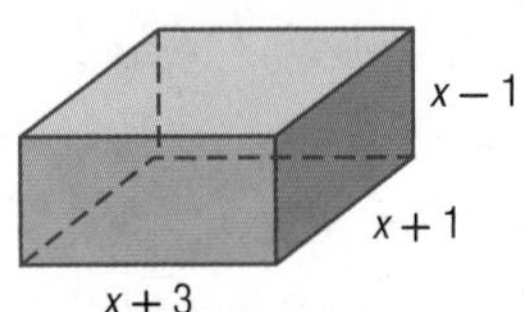

Example 6

Factor $r^7 + 64r$.

$r^7 + 64r = r(r^6 + 64)$ Factor by GCF.

$= r[(r^2)^3 + 4^3]$ Write as cubes.

$= r(r^2 + 4)(r^4 - 4r^2 + 16)$

Example 7

Solve $4x^4 - 25x^2 + 36 = 0$.

$(x^2 - 4)(4x^2 - 9) = 0$

$x^2 - 4 = 0$ or $4x^2 - 9 = 0$

$x^2 = 4$ $\qquad x^2 = \frac{9}{4}$

$x = \pm 2$ $\qquad x = \pm\frac{3}{2}$

The solutions are $-2, 2, -\frac{3}{2}$, and $\frac{3}{2}$.

Study Guide and Review *Continued*

5-6 The Remainder and Factor Theorems

Use synthetic substitution to find $f(-2)$ and $f(4)$ for each function.

41. $f(x) = x^2 - 3$

42. $f(x) = x^2 - 5x + 4$

43. $f(x) = x^3 + 4x^2 - 3x + 2$

44. $f(x) = 2x^4 - 3x^3 + 1$

Given a polynomial and one of its factors, find the remaining factors of the polynomial.

45. $3x^3 + 20x^2 + 23x - 10$; $x + 5$

46. $2x^3 + 11x^2 + 17x + 5$; $2x + 5$

47. $x^3 + 2x^2 - 23x - 60$; $x - 5$

Example 8

Determine whether $x - 6$ is a factor of $x^3 - 2x^2 - 21x - 18$.

6	1	−2	−21	−18
		6	24	18
	1	4	3	0

$x - 6$ is a factor because $r = 0$.

$x^3 - 2x^2 - 21x - 18 = (x - 6)(x^2 + 4x + 3)$

5-7 Roots and Zeros

State the possible number of positive real zeros, negative real zeros, and imaginary zeros of each function.

48. $f(x) = -2x^3 + 11x^2 - 3x + 2$

49. $f(x) = -4x^4 - 2x^3 - 12x^2 - x - 23$

50. $f(x) = x^6 - 5x^3 + x^2 + x - 6$

51. $f(x) = -2x^5 + 4x^4 + x^2 - 3$

52. $f(x) = -2x^6 + 4x^4 + x^2 - 3x - 3$

Example 9

State the possible number of positive real zeros, negative real zeros, and imaginary zeros of $f(x) = 3x^4 + 2x^3 - 2x^2 - 26x - 48$.

$f(x)$ has one sign change, so there is 1 positive real zero.

$f(-x)$ has 3 sign changes, so there are 3 or 1 negative real zeros.

There are 0 or 2 imaginary zeros.

5-8 Rational Zero Theorem

Find all of the zeros of each function.

53. $f(x) = x^3 + 4x^2 + 3x - 2$

54. $f(x) = 4x^3 + 4x^2 - x - 1$

55. $f(x) = x^3 + 2x^2 + 4x + 8$

56. **STORAGE** Melissa is building a storage box that is shaped like a rectangular prism. It will have a volume of 96 cubic feet. Using the diagram below, find the dimensions of the box.

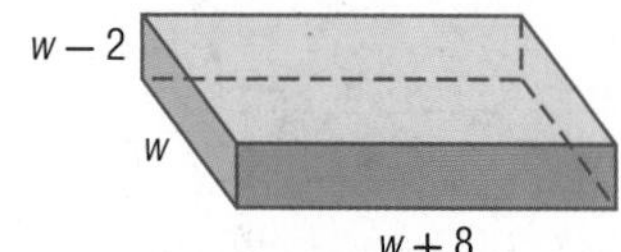

Example 10

Find all of the zeros of $f(x) = x^3 + 4x^2 - 11x - 30$.

There are exactly 3 zeros.

There are 1 positive real zero and 2 negative real zeros. The possible rational zeros are ±1, ±2, ±3, ±5, ±6, ±10, ±15, ±30.

3	1	4	−11	−30
		3	21	30
	1	7	10	0

$$x^3 + 4x^2 - 11x - 30 = (x - 3)(x^2 + 7x + 10)$$
$$= (x - 3)(x + 2)(x + 5)$$

Thus, the zeros are 3, −2, and −5.

CHAPTER 5 Practice Test

Simplify.

1. $(3a)^2(7b)^4$
2. $(7x - 2)(2x + 5)$
3. $(2x^2 + 3x - 4) - (4x^2 - 7x + 1)$
4. $(4x^3 - x^2 + 5x - 4) + (5x - 10)$
5. $(x^4 + 5x^3 + 3x^2 - 8x + 3) \div (x + 3)$
6. $(3x^3 - 5x^2 - 23x + 24) \div (x - 3)$

7. **MULTIPLE CHOICE** How many unique real zeros does the graph have?

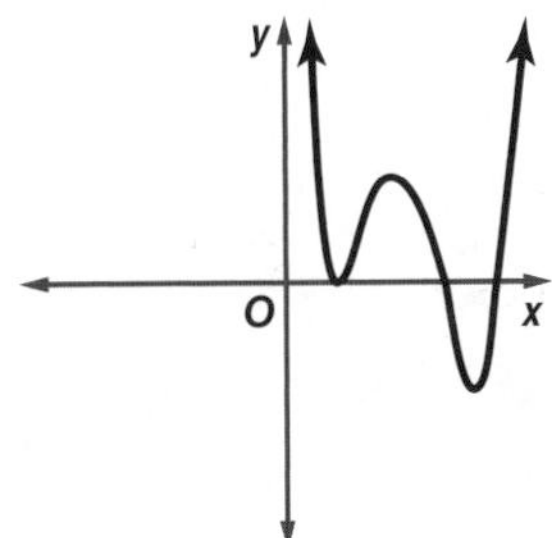

A 0 C 3

B 2 D 5

8. If $c(x) = 3x^3 + 5x^2 - 4$, what is the value of $4c(3b)$?

Complete each of the following.

a. Graph each function by making a table of values.

b. Determine consecutive integer values of x between which each real zero is located.

c. Estimate the x-coordinates at which the relative maxima and relative minima occur.

9. $g(x) = x^3 + 4x^2 - 3x + 1$
10. $h(x) = x^4 - 4x^3 - 3x^2 + 6x + 2$

Factor completely. If the polynomial is not factorable, write *prime*.

11. $8y^4 + x^3y$
12. $2x^2 + 2x + 1$
13. $a^2x + 3ax + 2x - a^2y - 3ay - 2y$

Solve each equation.

14. $8x^3 + 1 = 0$
15. $x^4 - 11x^2 + 28 = 0$

16. **FRAMING** The area of the picture and frame shown below is 168 square inches. What is the width of the frame?

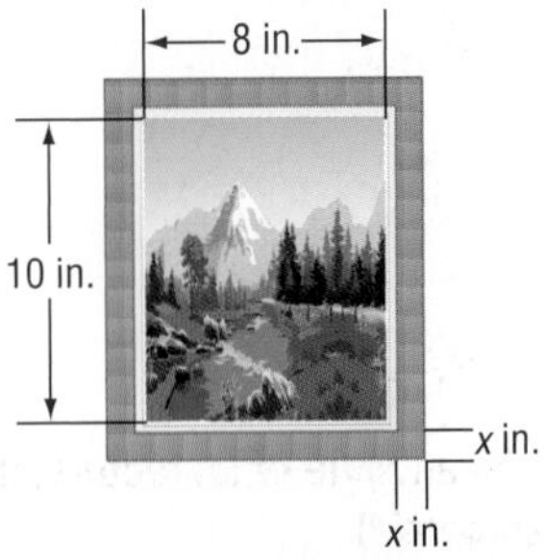

17. **MULTIPLE CHOICE** Let $f(x) = x^4 - 3x^3 + 5x - 3$. Use synthetic substitution to find $f(-2)$.

F 37 H −21

G 27 J −33

Given a polynomial and one of its factors, find the remaining factors of the polynomial.

18. $2x^3 + 15x^2 + 22x - 15$; $x + 5$
19. $x^3 - 4x^2 + 10x - 12$; $x - 2$

State the possible number of positive real zeros, negative real zeros, and imaginary zeros of each function.

20. $p(x) = x^3 - x^2 - x - 3$
21. $p(x) = 2x^6 + 5x^4 - x^3 - 5x - 1$

Find all zeros of each function.

22. $p(x) = x^3 - 4x^2 + x + 6$
23. $p(x) = x^3 + 2x^2 + 4x + 8$

24. **GEOMETRY** The volume of the rectangular prism shown is 612 cubic centimeters. Find the dimensions of the prism.

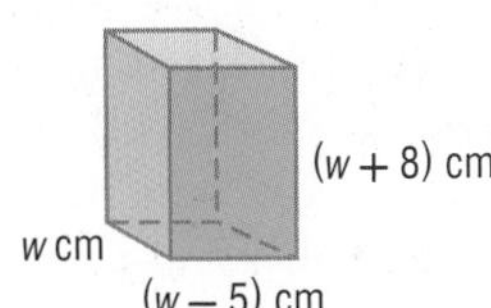

25. List all possible rational zeros of $f(x) = 2x^4 + 3x^2 - 12x + 8$.

Preparing for Standardized Tests

Draw a Picture

Drawing a picture can be a helpful way for you to visualize how to solve a problem. Sketch your picture on scrap paper or in your test booklet (if allowed). Do not make any marks on your answer sheet other than your answers.

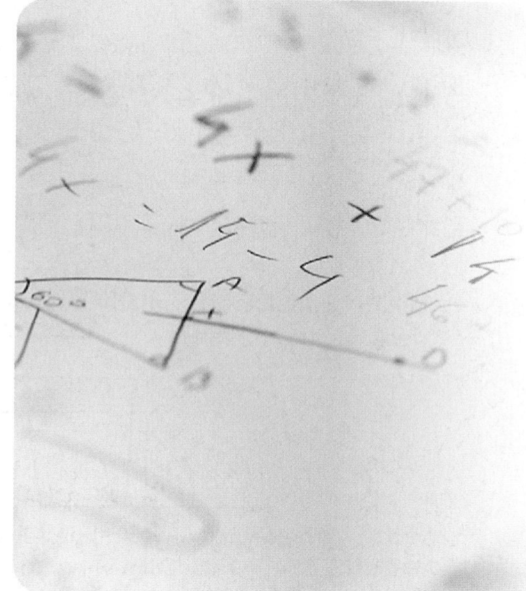

Strategies for Drawing a Picture

Step 1

Read the problem statement carefully.

Ask yourself:

- What am I being asked to solve?
- What information is given in the problem?
- What are the unknowns that I need to model and solve for?

Step 2

Sketch and label your picture.

- Draw your picture as clearly and accurately as possible.
- Label the picture carefully. Be sure to include all of the information given in the problem statement.

Step 3

Solve the problem.

- Use your picture to help you model the problem situation with an equation. Solve the equation.
- Check to be sure your answer makes sense.

Standardized Test Example

Read the problem. Identify what you need to know. Then use the information in the problem to solve.

Mr. Nolan has a rectangular swimming pool that measures 25 feet by 14 feet. He wants to have a cement walkway installed around the perimeter of the pool. The combined area of the pool and walkway will be 672 square feet. What will be the width of the walkway?

A 2.75 ft **C** 3.25 ft

B 3 ft **D** 3.5 ft

Draw a picture to help you visualize the problem situation. Let x represent the unknown width of the cement walkway.

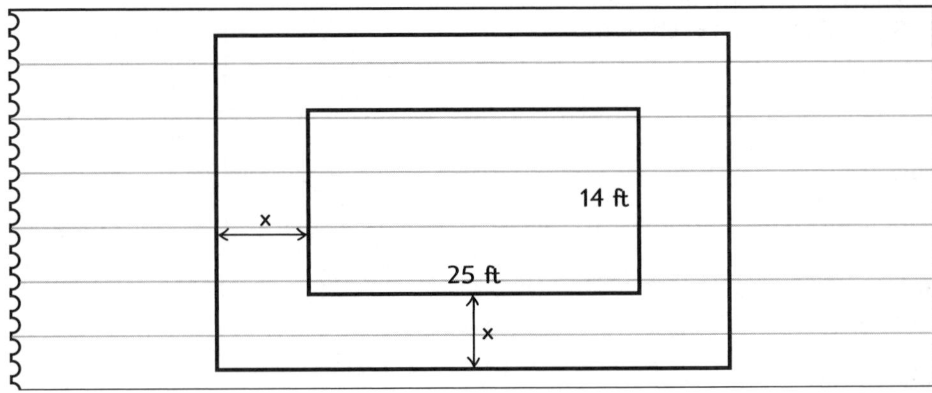

The width of the pool and walkway is $14 + 2x$, and the length is $25 + 2x$. Multiply these polynomial expressions and set the result equal to the combined area, 672 square feet. Then solve for x.

$$(14 + 2x)(25 + 2x) = 672$$
$$350 + 78x + 4x^2 = 672$$
$$4x^2 + 78x - 322 = 0$$
$$x = -23 \text{ or } 3.5$$

Since the width cannot be negative, the width of the walkway will be 3.5 feet. The correct answer is D.

Exercises

Read each problem. Identify what you need to know. Then use the information in the problem to solve.

1. A farmer has 240 feet of fencing that he wants to use to enclose a rectangular area for his chickens. He plans to build the enclosure using the wall of his barn as one of the walls. What is the maximum amount of area he can enclose?

   A 7200 ft^2

   B 4960 ft^2

   C 3600 ft^2

   D 3280 ft^2

2. Metal washers are made by cutting a hole in a circular piece of metal. Suppose a washer is made by removing the center of a piece of metal with a 1.8-inch diameter. What is the radius of the hole if the washer has an area of 0.65π square inches?

   F 0.35 in.

   G 0.38 in.

   H 0.40 in.

   J 0.42 in.

CHAPTER 5

Standardized Test Practice

Cumulative, Chapters 1 through 5

Multiple Choice

Read each question. Then fill in the correct answer on the answer document provided by your teacher or on a sheet of paper.

1. Simplify the following expression.

$$(5n^2 + 11n - 6) - (2n^2 - 5)$$

A $3n^2 + 11n - 11$

B $3n^2 + 11n - 1$

C $7n^2 + 11n - 11$

D $7n^2 + 11n - 1$

2. What is the effect on the graph of the equation $y = x^2 + 4$ when it is changed to $y = x^2 - 3$?

F The slope of the graph changes.

G The graph widens.

H The graph is the same shape, and the vertex of the graph is moved down.

J The graph is the same shape, and the vertex of the graph is shifted to the left.

3. Let p represent the price that Ella charges for a necklace. Let $f(x)$ represent the total amount of money that Ella makes for selling x necklaces. The function $f(x)$ is best represented by

A $f(x) = x + p$

B $f(x) = xp^2$

C $f(x) = px$

D $f(x) = x^2 + p$

4. Which of the following is *not* a solution to the cubic equation below?

$$x^3 - 37x - 84 = 0$$

F -4

G -3

H 6

J 7

5. What is the solution set for the equation $3(2x + 1)^2 = 27$?

A $\{-5, 4\}$

B $\{-2, 1\}$

C $\{2, -1\}$

D $\{-3, 3\}$

Test-Taking Tip

Question 5 You can use substitution to check each possible solution and identify the one that does not result in a true number sentence.

6. How many real zeros does the polynomial function graphed below have?

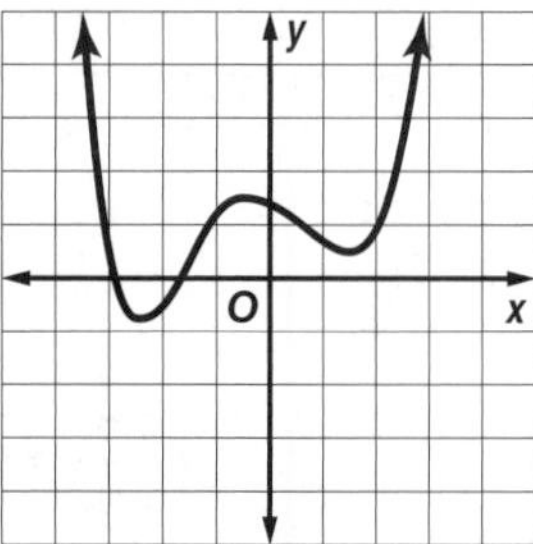

F 2

G 3

H 4

J 5

7. For Marla's vacation, it will cost \$100 to drive her car, plus between \$0.50 and \$0.75 per mile. If she will drive her car 400 miles, what is a reasonable conclusion about c, the total cost to drive her car on the vacation?

A $300 < c < 400$

B $200 < c \leq 400$

C $100 < c < 400$

D $200 \leq c \leq 300$

8. The function $P(x) = -0.000047x^2 + 0.027x + 3$ can be used to approximate the population of Ling's home country between 1960 and 2000. The domain of the function x represents the number of years since 1960, and P is given in millions of people. Evaluate $P(20)$ to estimate the population of the country in 1980.

F about 2 million people

G about 2.5 million people

H about 3 million people

J about 3.5 million people

9. Solve $4x - 5 = 2x + 5 - 3x$ for x.

A -2

B -1

C 1

D 2

Short Response/Gridded Response

Record your answers on the answer sheet provided by your teacher or on a sheet of paper.

10. A stone path that is x feet wide is built around a rectangular flower garden. The garden is 12 feet wide and 25 feet long as shown below.

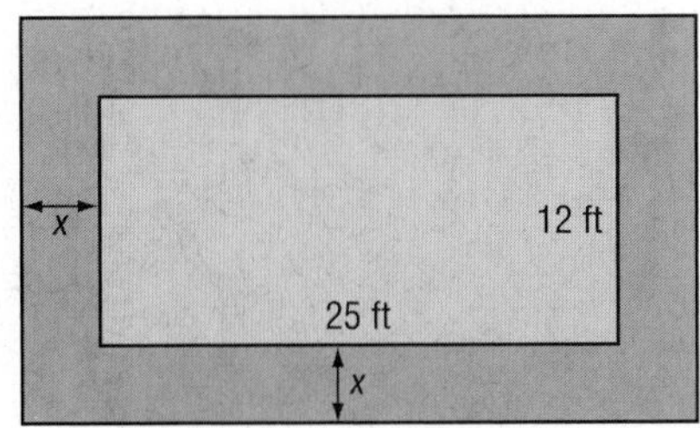

If the combined area of the garden and the stone path is 558 square feet, what is the width of the walkway? Express your answer in feet.

11. Factor $64a^4 + ab^3$ completely. Show your work.

12. Simplify $\frac{3x^3 - 4x^2 - 28x - 16}{x + 2}$. Give your answer in factored form. Show your work.

13. GRIDDED RESPONSE What is the value of a in the matrix equation below?

$$\begin{bmatrix} 4 & 3 \\ 2 & 2 \end{bmatrix} \cdot \begin{bmatrix} a \\ b \end{bmatrix} = \begin{bmatrix} 21 \\ 9 \end{bmatrix}$$

14. GRIDDED RESPONSE Matt has a cubic aquarium as shown below. He plans to fill it by emptying cans of water with the dimensions as shown.

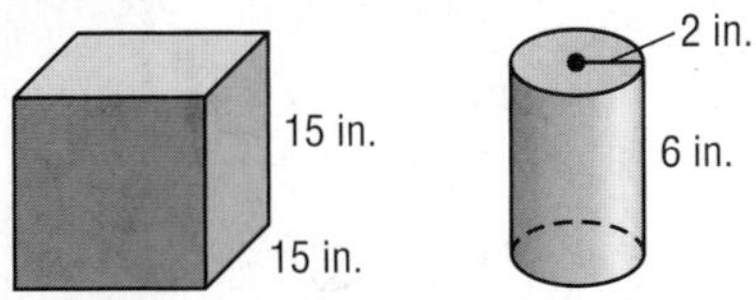

About how many cylindrical cans will it take to fill the aquarium?

15. What are the slope and y-intercept of the equation of the line graphed at the right?

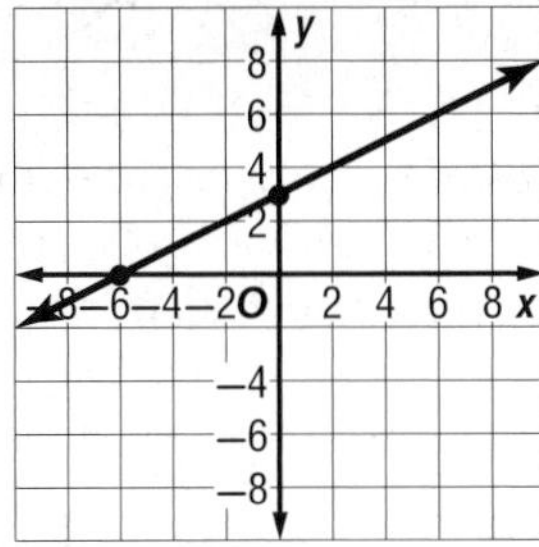

Extended Response

Record your answers on a sheet of paper. Show your work.

16. The volume of a rectangular prism is 864 cubic centimeters. The length is 1 centimeter less than the height, and the width is 3 centimeters more than the height.

a. Write a polynomial equation that can be used to solve for the height of the prism h.

b. How many possible roots are there for h in the polynomial equation you wrote? Explain.

c. Solve the equation from part **a** for all real roots h. What are the dimensions of the prism?

17. Scott launches a model rocket from ground level. The rocket's height h in meters is given by the equation $h = -4.9t^2 + 56t$, where t is the time in seconds after the launch.

a. What is the maximum height the rocket will reach? Round to the nearest tenth of a meter. Show each step and explain your method.

b. How long after it is launched will the rocket reach its maximum height? Round to the nearest tenth of a second.

Need ExtraHelp?

If you missed Question...	1	2	3	4	5	6	7	8	9	10	11	12	13	14	15	16	17
Go to Lesson...	5-2	4-7	2-4	5-5	4-5	5-7	3-2	5-3	1-3	5-5	5-5	5-2	3-8	5-7	2-3	5-8	4-7

CHAPTER 6

Inverses and Radical Functions and Relations

Then

You simplified polynomial expressions.

Now

You will:

- Find compositions and inverses of functions.
- Graph and analyze square root functions and inequalities.
- Simplify and solve equations involving roots, radicals, and rational exponents.

Why? ▲

FINANCE Connecting finances to mathematics is a skill that, once mastered, you will use your entire life. Learning to manage your finances entails creating a budget and living within that budget. In this chapter, you will explore financial topics such as saving for college, income, profit, inflation, and converting money when traveling.

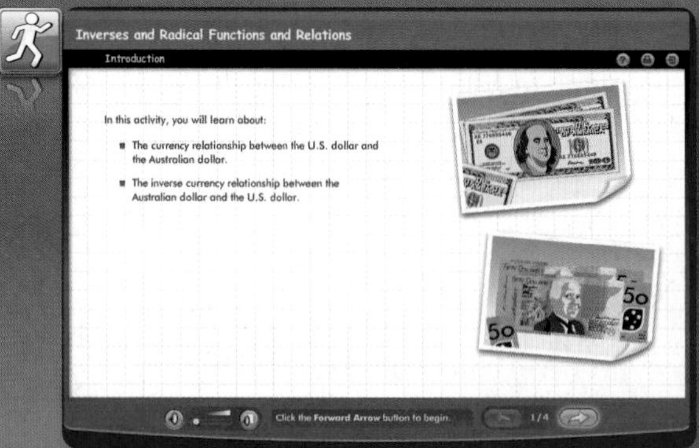

connectED.mcgraw-hill.com Your Digital Math Portal

Animation 
Vocabulary  
eGlossary  
Personal Tutor 
Virtual Manipulatives  
Graphing Calculator 
Audio  
Foldables 
Self-Check Practice 
Worksheets

Stockbyte/Getty Images

Get Ready for the Chapter

Diagnose Readiness | **You have two options for checking prerequisite skills.**

1 Textbook Option Take the Quick Check below. Refer to the Quick Review for help.

QuickCheck

Use the related graph of each equation to determine its roots. If exact roots cannot be found, state the consecutive integers between which the roots are located.

1. $x^2 - 4x + 1 = 0$
2. $2x^2 + x - 6 = 0$
3. **PHYSICS** Allie drops a ball from the top of a 30-foot building. How long does it take for the ball to reach the ground, assuming there is no air resistance? Use the formula $h(t) = -16t^2 + h_0$, where t is the time in seconds and the initial height h_0 is in feet.

Simplify each expression by using synthetic division.

4. $(5x^2 - 22x - 15) \div (x - 5)$
5. $(3x^2 + 14x - 12) \div (x + 4)$
6. $(2x^3 - 7x^2 - 36x + 36) \div (x - 6)$
7. $(3x^4 - 13x^3 + 17x^2 - 18x + 15) \div (x - 3)$
8. **FINANCE** The number of specialty coffee mugs sold at a coffee shop can be estimated by $n = \frac{4000x^2}{x^2 + 50}$, where x is the amount of money spent on advertising in hundreds of dollars and n is the number of mugs sold.
   a. Perform the division indicated by $\frac{4000x^2}{x^2 + 50}$.
   b. About how many mugs will be sold if $1000 is spent on advertising?

QuickReview

Example 1

Use the related graph of $0 = 3x^2 - 4x + 1$ to determine its roots. If exact roots cannot be found, state the consecutive integers between which the roots are located.

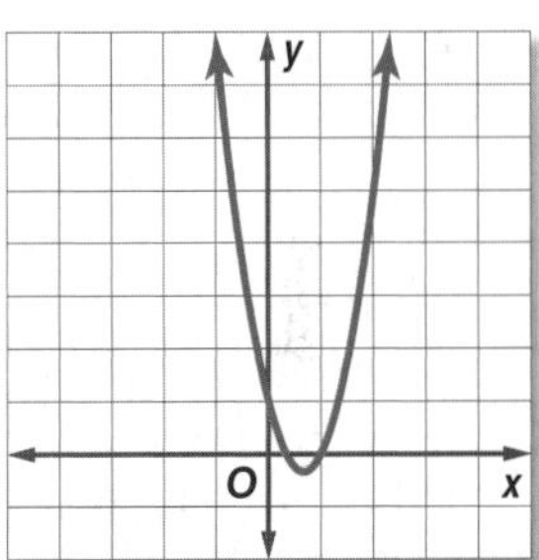

The roots are the x-coordinates where the graph crosses the x-axis.

The graph crosses the x-axis between 0 and 1 and at 1.

Example 2

Simplify $(3x^4 + 4x^3 + x^2 + 9x - 6) \div (x + 2)$ by using synthetic division.

$x - r = x + 2$, so $r = -2$.

−2	3	4	1	9	−6
	↓	−6	4	−10	2
	3	−2	5	−1	−4

The result is $3x^3 - 2x^2 + 5x - 1 - \frac{4}{x + 2}$.

2 Online Option Take an online self-check Chapter Readiness Quiz at connectED.mcgraw-hill.com.

Get Started on the Chapter

You will learn several new concepts, skills, and vocabulary terms as you study Chapter 6. To get ready, identify important terms and organize your resources. You may wish to refer to Chapter 0 to review prerequisite skills.

FOLDABLES® StudyOrganizer

Inverses and Radical Functions Make this Foldable to help you organize your Chapter 6 notes about radical equations and inequalities. Begin with four sheets of notebook paper.

1 **Stack** four sheets of notebook paper so that each sheet is one inch higher than the sheet in front of it.

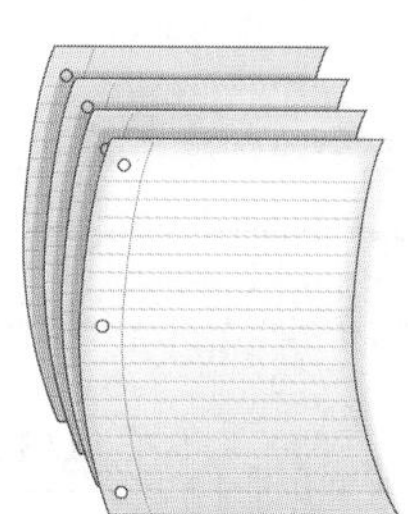

2 **Bring** the bottom of all the sheets upward and align the edges so that all of the layers or tabs are the same distance apart.

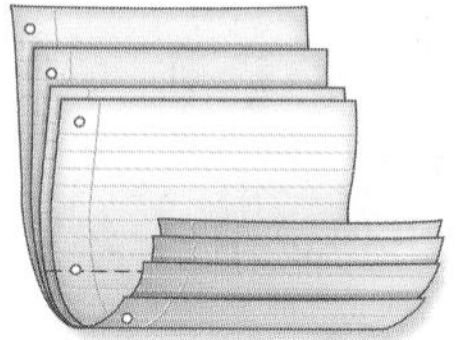

3 **When** all the tabs are an equal distance apart, fold the papers and crease well. Open the papers and staple them together along the valley or inner center fold. Label the pages with lesson titles.

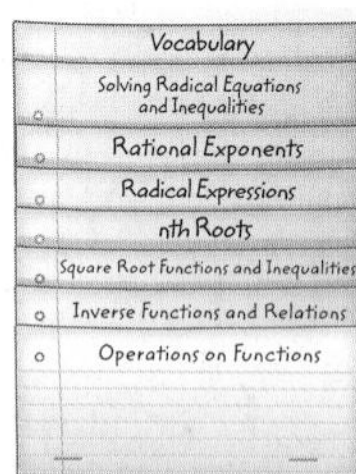

NewVocabulary

English		Español
composition of functions	p. 387	composición de funciones
inverse relation	p. 393	relaciones inversas
inverse function	p. 393	función inversa
square root function	p. 400	función raíz cuadrada
radical function	p. 400	función radical
square root inequality	p. 402	desigualdad raíz cuadrada
*n*th root	p. 407	raíz enésima
radical sign	p. 407	signo radical
index	p. 407	índice
radicand	p. 407	radicando
principal root	p. 407	raíz principal
rationalizing the denominator	p. 416	racionalizar el denominador
conjugates	p. 418	conjugados
radical equation	p. 429	ecuación radical
extraneous solution	p. 429	solución extraña
radical inequality	p. 431	desigualdad radical

ReviewVocabulary

absolute value valor absoluto a number's distance from zero on the number line, represented by $|x|$

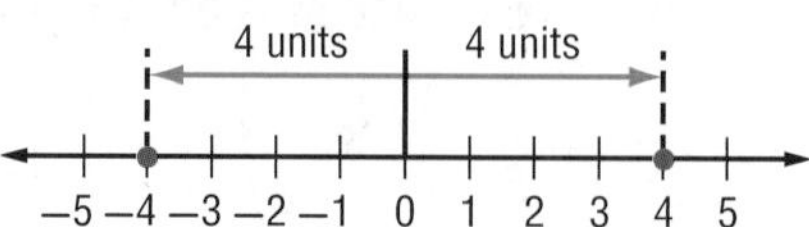

rational number número racional Any number $\frac{m}{n}$, where m and n are integers and n is not zero; the decimal form is either a terminating or repeating decimal.

relation relación a set of ordered pairs

LESSON 6-1 Operations on Functions

Then

- You performed operations on polynomials.

Now

1. Find the sum, difference, product, and quotient of functions.
2. Find the composition of functions.

Why?

- The graphs model the income for the Brooks family since 2000, where $m(x)$ represents Mr. Brooks' income and $f(x)$ represents Mrs. Brooks' income.

The total household income for the Brooks household can be represented by $f(x) + m(x)$.

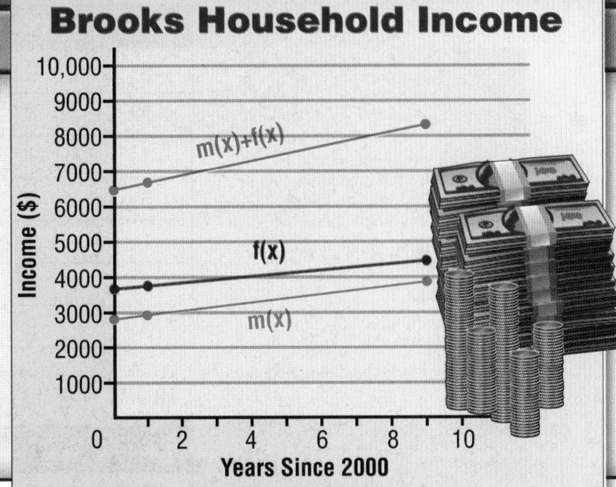

NewVocabulary
composition of functions

Common Core State Standards

Content Standards
F.IF.9 Compare properties of two functions each represented in a different way (algebraically, graphically, numerically in tables, or by verbal descriptions).
F.BF.1.b Combine standard function types using arithmetic operations.

Mathematical Practices
2 Reason abstractly and quantitatively.

1 Arithmetic Operations

In Chapter 6, you performed arithmetic operations with polynomials. You can also use addition, subtraction, multiplication, and division with functions.

You can perform arithmetic operations according to the following rules.

KeyConcept Operations on Functions

Operation	Definition	Example Let $f(x) = 2x$ and $g(x) = -x + 5$.
Addition	$(f + g)(x) = f(x) + g(x)$	$2x + (-x + 5) = x + 5$
Subtraction	$(f - g)(x) = f(x) - g(x)$	$2x - (-x + 5) = 3x - 5$
Multiplication	$(f \cdot g)(x) = f(x) \cdot g(x)$	$2x(-x + 5) = -2x^2 + 10x$
Division	$\left(\frac{f}{g}\right)(x) = \frac{f(x)}{g(x)}, g(x) \neq 0$	$\frac{2x}{-x + 5}, x \neq 5$

PT

Example 1 Add and Subtract Functions

Given $f(x) = x^2 - 4$ and $g(x) = 2x + 1$, find each function. Indicate any restrictions in the domain or range.

a. $(f + g)(x)$

$(f + g)(x) = f(x) + g(x)$ — Addition of functions

$= (x^2 - 4) + (2x + 1)$ — $f(x) = x^2 - 4$ and $g(x) = 2x + 1$

$= x^2 + 2x - 3$ — Simplify.

b. $(f - g)(x)$

$(f - g)(x) = f(x) - g(x)$ — Subtraction of functions

$= (x^2 - 4) - (2x + 1)$ — $f(x) = x^2 - 4$ and $g(x) = 2x + 1$

$= x^2 - 2x - 5$ — Simplify.

GuidedPractice

Given $f(x) = x^2 + 5x - 2$ and $g(x) = 3x - 2$, find each function.

1A. $(f + g)(x)$ **1B.** $(f - g)(x)$

> **Review Vocabulary**
> **intersection** the intersection of two sets is the set of elements common to both

You can graph sum and difference functions by graphing each function involved separately, then adding their corresponding functional values. Let $f(x) = x^2$ and $g(x) = x$. Examine the graphs of $f(x)$, $g(x)$, and their sum and difference.

Find $(f + g)(x)$.

x	$f(x) = x^2$	$g(x) = x$	$(f + g)(x) = x^2 + x$
−3	9	−3	$9 + (-3) = 6$
−2	4	−2	$4 + (-2) = 2$
−1	1	−1	$1 + (-1) = 0$
0	0	0	$0 + 0 = 0$
1	1	1	$1 + 1 = 2$
2	4	2	$4 + 2 = 6$
3	9	3	$9 + 3 = 12$

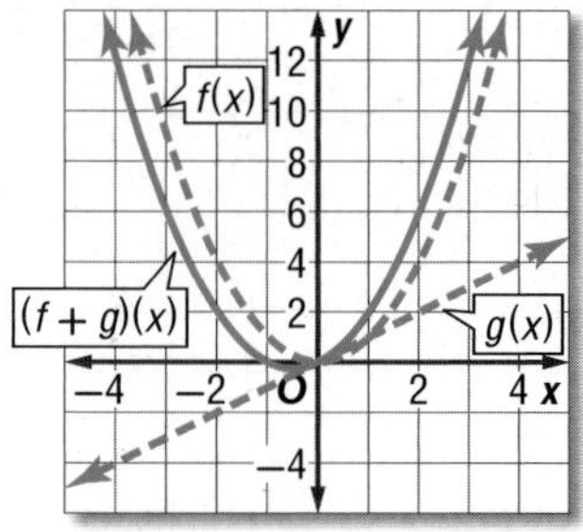

Find $(f - g)(x)$.

x	$f(x) = x^2$	$g(x) = x$	$(f - g)(x) = x^2 - x$
−3	9	−3	$9 - (-3) = 12$
−2	4	−2	$4 - (-2) = 6$
−1	1	−1	$1 - (-1) = 2$
0	0	0	$0 - 0 = 0$
1	1	1	$1 - 1 = 0$
2	4	2	$4 - 2 = 2$
3	9	3	$9 - 3 = 6$

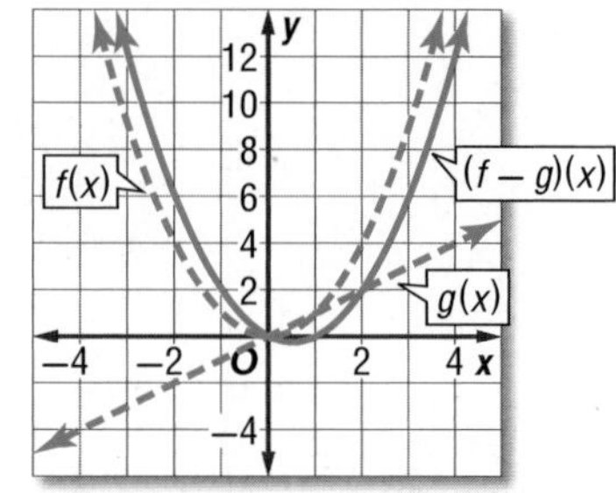

In Example 1, the functions $f(x)$ and $g(x)$ have the same domain of all real numbers. The functions $(f + g)(x)$ and $(f - g)(x)$ also have domains that include all real numbers. For each new function, the domain consists of the intersection of the domains of $f(x)$ and $g(x)$. Under division, the domain of the new function is restricted by excluded values that cause the denominator to equal zero.

PT

Example 2 Multiply and Divide Functions

Given $f(x) = x^2 + 7x + 12$ and $g(x) = 3x - 4$, find each function. Indicate any restrictions in the domain or range.

a. $(f \cdot g)(x)$

$$\begin{aligned}(f \cdot g)(x) &= f(x) \cdot g(x) && \text{Multiplication of functions}\\ &= (x^2 + 7x + 12)(3x - 4) && \text{Substitution}\\ &= 3x^3 + 21x^2 + 36x - 4x^2 - 28x - 48 && \text{Distributive Property}\\ &= 3x^3 + 17x^2 + 8x - 48 && \text{Simplify.}\end{aligned}$$

b. $\left(\frac{f}{g}\right)(x)$

$$\begin{aligned}\left(\frac{f}{g}\right)(x) &= \frac{f(x)}{g(x)} && \text{Division of functions}\\ &= \frac{x^2 + 7x + 12}{3x - 4}, x \neq \frac{4}{3} && \text{Substitution}\end{aligned}$$

Because $x = \frac{4}{3}$ makes the denominator $3x - 4 = 0$, $\frac{4}{3}$ is excluded from the domain of $\left(\frac{f}{g}\right)(x)$.

Guided Practice **Given $f(x) = x^2 - 7x + 2$ and $g(x) = x + 4$, find each function.**

2A. $(f \cdot g)(x)$

2B. $\left(\frac{f}{g}\right)(x)$

2 Composition of Functions

Another method used to c[illegible] a composition of functions. In a **composition of functio[illegible]** function are used to evaluate a second function.

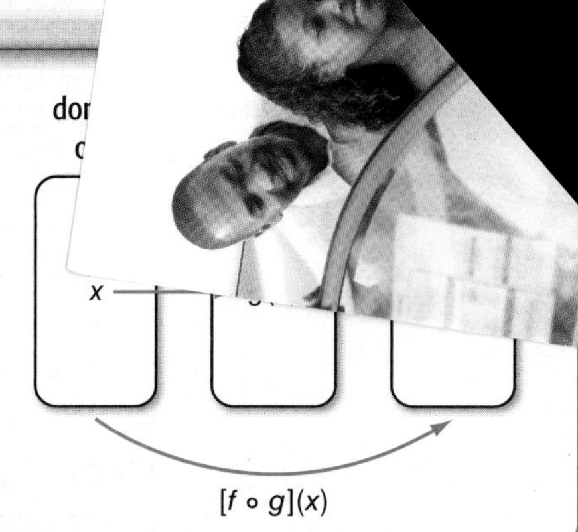

> **ReadingMath**
> **Composition of Functions** The composition of f and g, denoted by $f \circ g$ or $f[g(x)]$, is read f *of* g.

KeyConcept Composition of Functions

Words Suppose f and g are functions such that the range of g is a subset of the domain of f. Then the composition function $f \circ g$ can be described by

$$[f \circ g](x) = f[g(x)].$$

Model

The composition of two functions may not exist. Given two functions f and g, $[f \circ g](x)$ is defined only if the range of $g(x)$ is a subset of the domain of f. Likewise, $[g \circ f](x)$ is defined only if the range of $f(x)$ is a subset of the domain of g.

> **StudyTip**
> **Composition** Be careful not to confuse a composition $f[g(x)]$ with multiplication of functions $(f \cdot g)(x)$.

Example 3 Compose Functions

For each pair of functions, find $[f \circ g](x)$ and $[g \circ f](x)$, if they exist. State the domain and range for each composed function.

a. $f = \{(1, 8), (0, 13), (15, 11), (14, 9)\}$, $g = \{(8, 15), (5, 1), (10, 14), (9, 0)\}$

To find $f \circ g$, evaluate $g(x)$ first. Then use the range to evaluate $f(x)$.

$f[g(8)] = f(15)$ or 11 $g(8) = 15$ $f[g(10)] = f(14)$ or 9 $g(10) = 14$

$f[g(5)] = f(1)$ or 8 $g(5) = 1$ $f[g(9)] = f(0)$ or 13 $g(9) = 0$

$f \circ g = \{(8, 11), (5, 8), (10, 9), (9, 13)\}$, D = {5, 8, 9, 10}, R = {8, 9, 11, 13}

To find $g \circ f$, evaluate $f(x)$ first. Then use the range to evaluate $g(x)$.

$g[f(1)] = g(8)$ or 15 $f(1) = 8$ $g[f(15)] = g(11)$ $g(11)$ is undefined.

$g[f(0)] = g(13)$ $g(13)$ is undefined. $g[f(14)] = g(9)$ or 0 $f(14) = 0$

Because 11 and 13 are not in the domain of g, $g \circ f$ is undefined for $x = 11$ and $x = 13$. However, $g[f(1)] = 15$ and $g[f(14)] = 0$, so $g \circ f = \{(1, 15), (14, 0)\}$, D = {5, 8, 9, 10}, and R = {8, 9, 11, 13}.

b. $f(x) = 2a - 5$, $g(x) = 4a$

$[f \circ g](x) = f[g(x)]$	Composition of functions	$[g \circ f](x) = g[f(x)]$
$= f(4a)$	Substitute.	$= g(2a - 5)$
$= 2(4a) - 5$	Substitute again.	$= 4(2a - 5)$
$= 8a - 5$	Simplify.	$= 8a - 20$

For $[f \circ g](x)$, D = {all real numbers} and R = {all real numbers}, and for $[g \circ f](x)$, D = {all real numbers} and R = {all real numbers}.

GuidedPractice

3A. $f(x) = \{(3, -2), (-1, -5), (4, 7), (10, 8)\}$, $g(x) = \{(4, 3), (2, -1), (9, 4), (3, 10)\}$

3B. $f(x) = x^2 + 2$ and $g(x) = x - 6$

Notice that in most cases, $f \circ g \neq g \circ f$. Therefore, the order in which two functions are composed is important.

Real-WorldLink

Adjusted for inflation, the average price of a new car declined from $23,014 in 1995 to $22,013 in 2005.

Source: U.S. Department of Commerce

PT

Real-World Example 4 Use Composition of Functions

SHOPPING A new car dealer is discounting all new cars by 12%. At the same time, the manufacturer is offering a $1500 rebate on all new cars. Mr. Navarro is buying a car that is priced $24,500. Will the final price be lower if the discount is applied before the rebate or if the rebate is applied before the discount?

Understand Let x represent the original price of a new car, $d(x)$ represent the price of a car after the discount, and $r(x)$ the price of the car after the rebate.

Plan Write equations for $d(x)$ and $r(x)$.

The original price is discounted by 12%. $d(x) = x - 0.12x$

There is a $1500 rebate on all new cars. $r(x) = x - 1500$

Solve If the discount is applied *before* the rebate, then the final price of Mr. Navarro's new car is represented by $[r \circ d](24{,}500)$.

$[r \circ d](x) = r[d(x)]$

$$\begin{aligned}[r \circ d](24{,}500) &= r[24{,}500 - 0.12(24{,}500)] \\ &= r(24{,}500 - 2940) \\ &= r(21{,}560) \\ &= 21{,}560 - 1500 \\ &= 20{,}060\end{aligned}$$

If the rebate is given *before* the discount is applied, then the final price of Mr. Navarro's car is represented by $[d \circ r](24{,}500)$.

$[d \circ r](x) = d[r(x)]$

$$\begin{aligned}[d \circ r](24{,}500) &= d(24{,}500 - 1500) \\ &= d(23{,}000) \\ &= 23{,}000 - 0.12(23{,}000) \\ &= 23{,}000 - 2760 \\ &= 20{,}240\end{aligned}$$

$[r \circ d](24{,}500) = 20{,}060$ and $[d \circ r](24{,}500) = 20{,}240$. So, the final price of the car is less when the discount is applied before the rebate.

Check The answer seems reasonable because the 12% discount is being applied to a greater amount. Thus, the dollar amount of the discount is greater.

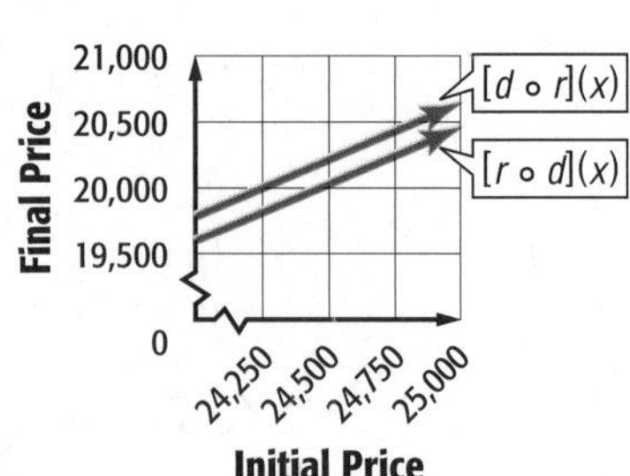

GuidedPractice

4. **SHOPPING** Sounds-to-Go offers both an in-store $35 rebate and a 15% discount on a digital audio player that normally sells for $300. Which provides the better price: taking the discount before the rebate or taking the discount after the rebate?

Randy Faris/Flame/CORBIS

Check Your Understanding

= Step-by-Step Solutions begin on page R14.

Examples 1–2 **Find $(f + g)(x)$, $(f - g)(x)$, $(f \cdot g)(x)$, and $\left(\frac{f}{g}\right)(x)$ for each $f(x)$ and $g(x)$. Indicate any restrictions in domain or range.**

1. $f(x) = x + 2$
   $g(x) = 3x - 1$

2. $f(x) = x^2 - 5$
   $g(x) = -x + 8$

Example 3 **For each pair of functions, find $f \circ g$ and $g \circ f$, if they exist. State the domain and range for each composed function.**

3. $f = \{(2, 5), (6, 10), (12, 9), (7, 6)\}$
   $g = \{(9, 11), (6, 15), (10, 13), (5, 8)\}$

4. $f = \{(-5, 4), (14, 8), (12, 1), (0, -3)\}$
   $g = \{(-2, -4), (-3, 2), (-1, 4), (5, -6)\}$

Find $[f \circ g](x)$ and $[g \circ f](x)$, if they exist. State the domain and range for each composed function.

5. $f(x) = -3x$
   $g(x) = 5x - 6$

6. $f(x) = x + 4$
   $g(x) = x^2 + 3x - 10$

Example 4

7. **CCSS MODELING** Dora has 8% of her earnings deducted from her paycheck for a college savings plan. She can choose to take the deduction either before taxes are withheld, which reduces her taxable income, or after taxes are withheld. Dora's tax rate is 17.5%. If her pay before taxes and deductions is \$950, will she save more money if the deductions are taken before or after taxes are withheld? Explain.

Practice and Problem Solving

Extra Practice is on page R6.

Examples 1–2 **Find $(f + g)(x)$, $(f - g)(x)$, $(f \cdot g)(x)$, and $\left(\frac{f}{g}\right)(x)$ for each $f(x)$ and $g(x)$. Indicate any restrictions in domain or range.**

8. $f(x) = 2x$
   $g(x) = -4x + 5$

9. $f(x) = x - 1$
   $g(x) = 5x - 2$

10. $f(x) = x^2$
    $g(x) = -x + 1$

11. $f(x) = 3x$
    $g(x) = -2x + 6$

12. $f(x) = x - 2$
    $g(x) = 2x - 7$

13. $f(x) = x^2$
    $g(x) = x - 5$

14. $f(x) = -x^2 + 6$
    $g(x) = 2x^2 + 3x - 5$

15. $f(x) = 3x^2 - 4$
    $g(x) = x^2 - 8x + 4$

16. **POPULATION** In a particular county, the population of the two largest cities can be modeled by $f(x) = 200x + 25$ and $g(x) = 175x - 15$, where x is the number of years since 2000 and the population is in thousands.
    a. What is the population of the two cities combined after any number of years?
    b. What is the difference in the populations of the two cities?

Example 3 **For each pair of functions, find $f \circ g$ and $g \circ f$, if they exist. State the domain and range for each composed function.**

17. $f = \{(-8, -4), (0, 4), (2, 6), (-6, -2)\}$
    $g = \{(4, -4), (-2, -1), (-4, 0), (6, -5)\}$

18. $f = \{(-7, 0), (4, 5), (8, 12), (-3, 6)\}$
    $g = \{(6, 8), (-12, -5), (0, 5), (5, 1)\}$

19. $f = \{(5, 13), (-4, -2), (-8, -11), (3, 1)\}$
    $g = \{(-8, 2), (-4, 1), (3, -3), (5, 7)\}$

20. $f = \{(-4, -14), (0, -6), (-6, -18), (2, -2)\}$
    $g = \{(-6, 1), (-18, 13), (-14, 9), (-2, -3)\}$

For each pair of functions, find $f \circ g$ and $g \circ f$, if they exist. State the domain and range for each composed function.

21. $f = \{(-15, -5), (-4, 12), (1, 7), (3, 9)\}$
$g = \{(3, -9), (7, 2), (8, -6), (12, 0)\}$

22. $f = \{(-1, 11), (2, -2), (5, -7), (4, -4)\}$
$g = \{(5, -4), (4, -3), (-1, 2), (2, 3)\}$

23. $f = \{(7, -3), (-10, -3), (-7, -8), (-3, 6)\}$
$g = \{(4, -3), (3, -7), (9, 8), (-4, -4)\}$

24. $f = \{(1, -1), (2, -2), (3, -3), (4, -4)\}$
$g = \{(1, -4), (2, -3), (3, -2), (4, -1)\}$

25. $f = \{(-4, -1), (-2, 6), (-1, 10), (4, 11)\}$
$g = \{(-1, 5), (3, -4), (6, 4), (10, 8)\}$

26. $f = \{(12, -3), (9, -2), (8, -1), (6, 3)\}$
$g = \{(-1, 5), (-2, 6), (-3, -1), (-4, 8)\}$

Find $[f \circ g](x)$ and $[g \circ f](x)$, if they exist. State the domain and range for each composed function.

27. $f(x) = 2x$
$g(x) = x + 5$

28. $f(x) = -3x$
$g(x) = -x + 8$

29. $f(x) = x + 5$
$g(x) = 3x - 7$

30. $f(x) = x - 4$
$g(x) = x^2 - 10$

31. $f(x) = x^2 + 6x - 2$
$g(x) = x - 6$

32. $f(x) = 2x^2 - x + 1$
$g(x) = 4x + 3$

33. $f(x) = 4x - 1$
$g(x) = x^3 + 2$

34. $f(x) = x^2 + 3x + 1$
$g(x) = x^2$

35. $f(x) = 2x^2$
$g(x) = 8x^2 + 3x$

36. FINANCE A ceramics store manufactures and sells coffee mugs. The revenue $r(x)$ from the sale of x coffee mugs is given by $r(x) = 6.5x$. Suppose the function for the cost of manufacturing x coffee mugs is $c(x) = 0.75x + 1850$.

a. Write the profit function.

b. Find the profit on 500, 1000, and 5000 coffee mugs.

37. CCSS SENSE-MAKING Ms. Smith wants to buy an HDTV, which is on sale for 35% off the original price of $2299. The sales tax is 6.25%.

a. Write two functions representing the price after the discount $p(x)$ and the price after sales tax $t(x)$.

b. Which composition of functions represents the price of the HDTV, $[p \circ t](x)$ or $[t \circ p](x)$? Explain your reasoning.

c. How much will Ms. Smith pay for the HDTV?

Perform each operation if $f(x) = x^2 + x - 12$ and $g(x) = x - 3$. State the domain of the resulting function.

38. $(f - g)(x)$

39. $2(g \cdot f)(x)$

40. $\left(\frac{f}{g}\right)(x)$

If $f(x) = 5x$, $g(x) = -2x + 1$, and $h(x) = x^2 + 6x + 8$, find each value.

41. $f[g(-2)]$

42. $g[h(3)]$

43. $h[f(-5)]$

44. $h[g(2)]$

45. $f[h(-3)]$

46. $h[f(9)]$

47. $f[g(3a)]$

48. $f[h(a + 4)]$

49. $g[f(a^2 - a)]$

50. MULTIPLE REPRESENTATIONS Let $f(x) = x^2$ and $g(x) = x$.

a. Tabular Make a table showing values for $f(x)$, $g(x)$, $(f + g)(x)$, and $(f - g)(x)$.

b. Graphical Graph $f(x)$, $g(x)$, and $(f + g)(x)$ on the same coordinate grid.

c. Graphical Graph $f(x)$, $g(x)$, and $(f - g)(x)$ on the same coordinate grid.

d. Verbal Describe the relationship among the graphs of $f(x)$, $g(x)$, $(f + g)(x)$, and $(f - g)(x)$.

51 **EMPLOYMENT** The number of women and men age 16 and over employed each year in the United States can be modeled by the following equations, where x is the number of years since 1994 and y is the number of people in thousands.

women: $y = 1086.4x + 56{,}610$ men: $y = 999.2x + 66{,}450$

a. Write a function that models the total number of men and women employed in the United States during this time.

b. If f is the function for the number of men, and g is the function for the number of women, what does $(f - g)(x)$ represent?

If $f(x) = x + 2$, $g(x) = -4x + 3$, and $h(x) = x^2 - 2x + 1$, find each value.

52. $(f \cdot g \cdot h)(3)$ **53.** $[(f + g) \cdot h](1)$ **54.** $\left(\frac{h}{fg}\right)(-6)$

55. $[f \circ (g \circ h)](2)$ **56.** $[g \circ (h \circ f)](-4)$ **57.** $[h \circ (f \circ g)](5)$

58. **MULTIPLE REPRESENTATIONS** You will explore $(f \cdot g)(x)$, $\left(\frac{f}{g}\right)(x)$, $[f \circ g](x)$, and $[g \circ f](x)$ if $f(x) = x^2 + 1$ and $g(x) = x - 3$.

a. Tabular Make a table showing values for $(f \cdot g)(x)$, $\left(\frac{f}{g}\right)(x)$, $[f \circ g](x)$, and $[g \circ f](x)$.

b. Graphical Use a graphing calculator to graph $(f \cdot g)(x)$ and $\left(\frac{f}{g}\right)(x)$ on the same coordinate plane.

c. Verbal Explain the relationship between $(f \cdot g)(x)$ and $\left(\frac{f}{g}\right)(x)$.

d. Graphical Use a graphing calculator to graph $[f \circ g](x)$, and $[g \circ f](x)$ on the same coordinate plane.

e. Verbal Explain the relationship between $[f \circ g](x)$, and $[g \circ f](x)$.

H.O.T. Problems Use Higher-Order Thinking Skills

59. **OPEN ENDED** Write two functions $f(x)$ and $g(x)$ such that $(f \circ g)(4) = 0$.

60. CCSS **CRITIQUE** Chris and Tobias are finding $(f \circ g)(x)$, where $f(x) = x^2 + 2x - 8$ and $g(x) = x^2 + 8$. Is either of them correct? Explain your reasoning.

Chris	Tobias
$(f \circ g)(x) = f[g(x)]$	$(f \circ g)(x) = f[g(x)]$
$= (x^2 + 8)^2 + 2x - 8$	$= (x^2 + 8)^2 + 2(x^2 + 8) - 8$
$= x^4 + 16x^2 + 64 + 2x - 8$	$= x^4 + 16x^2 + 64 + 2x^2 + 16 - 8$
$= x^4 + 16x^2 + 2x + 58$	$= x^4 + 18x^2 + 72$

61. **CHALLENGE** Given $f(x) = \sqrt{x^3}$ and $g(x) = \sqrt{x^6}$, determine the domain for each of the following.

a. $g(x) \cdot g(x)$ **b.** $f(x) \cdot f(x)$

62. **REASONING** State whether each statement is *sometimes*, *always*, or *never* true. Explain.

a. The domain of two functions $f(x)$ and $g(x)$ that are composed $g[f(x)]$ is restricted by the domain of $f(x)$.

b. The domain of two functions $f(x)$ and $g(x)$ that are composed $g[f(x)]$ is restricted by the domain of $g(x)$.

63. **WRITING IN MATH** In the real world, why would you ever perform a composition of functions?

Standardized Test Practice

64. What is the value of x in the equation $7(x - 4) = 44 - 11x$?

A 1
B 2
C 3
D 4

65. If $g(x) = x^2 + 9x + 21$ and $h(x) = 2(x + 5)^2$, which is an equivalent form of $h(x) - g(x)$?

F $k(x) = -x^2 - 11x - 29$
G $k(x) = x^2 + 11x + 29$
H $k(x) = x + 4$
J $k(x) = x^2 + 7x + 11$

66. GRIDDED RESPONSE In his first three years of coaching basketball at North High School, Coach Lucas' team won 8 games the first year, 17 games the second year, and 6 games the third year. How many games does the team need to win in the fourth year so the coach's average will be 10 wins per year?

67. SAT/ACT What is the value of $f[g(6)]$ if $f(x) = 2x + 4$ and $g(x) = x^2 + 5$?

A 38
B 43
C 57
D 86
E 261

Spiral Review

Find all of the rational zeros of each function. (Lesson 5-8)

68. $f(x) = 2x^3 - 13x^2 + 17x + 12$

69. $f(x) = x^3 - 3x^2 - 10x + 24$

70. $f(x) = x^4 - 4x^3 - 7x^2 + 34x - 24$

71. $f(x) = 2x^3 - 5x^2 - 28x + 15$

State the possible number of positive real zeros, negative real zeros, and imaginary zeros of each function. (Lesson 5-7)

72. $f(x) = 2x^4 - x^3 + 5x^2 + 3x - 9$

73. $f(x) = -4x^4 - x^2 - x + 1$

74. $f(x) = 3x^4 - x^3 + 8x^2 + x - 7$

75. $f(x) = 2x^4 - 3x^3 - 2x^2 + 3$

76. MANUFACTURING A box measures 12 inches by 16 inches by 18 inches. The manufacturer will increase each dimension of the box by the same number of inches and have a new volume of 5985 cubic inches. How much should be added to each dimension? (Lesson 5-7)

Solve each system of equations. (Lesson 3-4)

77. $x + 4y - z = 6$
$3x + 2y + 3z = 16$
$2x - y + z = 3$

78. $2a + b - c = 5$
$a - b + 3c = 9$
$3a - 6c = 6$

79. $y + z = 4$
$2x + 4y - z = -3$
$3y = -3$

80. INTERNET A webmaster estimates that the time, in seconds, to connect to the server when n people are connecting is given by $t(n) = 0.005n + 0.3$. Estimate the time to connect when 50 people are connecting. (Lesson 2-2)

Skills Review

Solve each equation or formula for the specified variable.

81. $5x - 7y = 12$, for x

82. $3x^2 - 6xy + 1 = 4$, for y

83. $4x + 8yz = 15$, for x

84. $D = mv$, for m

85. $A = k^2 + b$, for k

86. $(x + 2)^2 - (y + 5)^2 = 4$, for y

LESSON

6-2 Inverse Functions and Relations

Then

- You transformed and solved equations for a specific variable.

Now

1. Find the inverse of a function or relation.
2. Determine whether two functions or relations are inverses.

Why?

- The table shows the value of \$1 (U.S.) compared to Canadian dollars and Mexican pesos.

The equation $p = 10.75d$ represents the number of pesos p you can receive for every U.S. dollar d. To determine how many U.S. dollars you can receive for one Mexican peso, solve the equation $p = 10.75d$ for d. The result, $d \approx 0.09p$, is the inverse function.

	U.S.	Canada	Mexico
U.S.		1.05	10.75
Canada	0.95		10.26
Mexico	0.09	0.10	

NewVocabulary
inverse relation
inverse function

Common Core State Standards

Content Standards

F.IF.4 For a function that models a relationship between two quantities, interpret key features of graphs and tables in terms of the quantities, and sketch graphs showing key features given a verbal description of the relationship.

F.BF.4.a Find inverse functions. - Solve an equation of the form $f(x) = c$ for a simple function f that has an inverse and write an expression for the inverse.

Mathematical Practices

7 Look for and make use of structure.

8 Look for and express regularity in repeated reasoning.

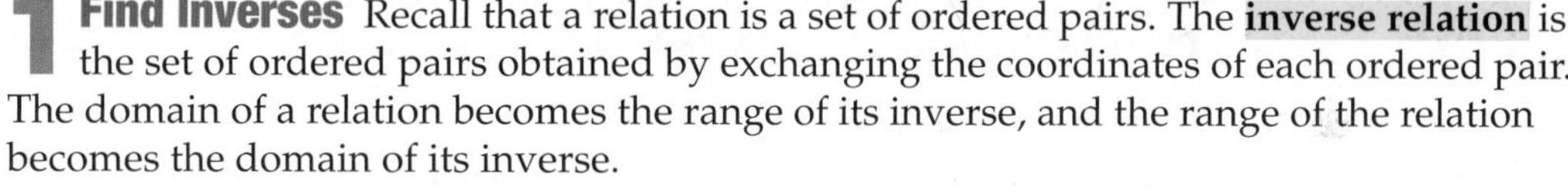

1 Find Inverses

Recall that a relation is a set of ordered pairs. The **inverse relation** is the set of ordered pairs obtained by exchanging the coordinates of each ordered pair. The domain of a relation becomes the range of its inverse, and the range of the relation becomes the domain of its inverse.

KeyConcept Inverse Relations

Words	Two relations are inverse relations if and only if whenever one relation contains the element (a, b), the other relation contains the element (b, a).
Example	A and B are inverse relations.
	$A = \{(1, 5), (2, 6), (3, 7)\}$ $B = \{(5, 1), (6, 2), (7, 3)\}$

Example 1 Find an Inverse Relation

GEOMETRY The vertices of $\triangle ABC$ can be represented by the relation $\{(1, -2), (2, 5), (4, -1)\}$. Find the inverse of this relation. Describe the graph of the inverse.

Graph the relation. To find the inverse, exchange the coordinates of the ordered pairs. The inverse of the relation is $\{(-2, 1), (5, 2), (-1, 4)\}$.

Plotting these points shows that the ordered pairs describe the vertices of $\triangle A'B'C'$ as a reflection of $\triangle ABC$ in the line $y = x$.

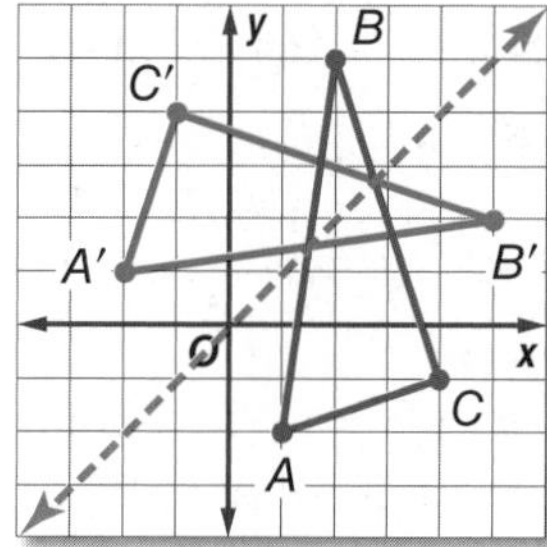

GuidedPractice

1. **GEOMETRY** The ordered pairs of the relation $\{(-8, -3), (-8, -6), (-3, -6)\}$ are the coordinates of the vertices of a right triangle. Find the inverse of this relation. Describe the graph of the inverse.

As with relations, the ordered pairs of **inverse functions** are also related. We can write the inverse of the function $f(x)$ as $f^{-1}(x)$.

ReadingMath

CCSS Sense-Making f^{-1} is read *f inverse* or *the inverse of f*. Note that -1 is *not* an exponent.

KeyConcept Property of Inverses

Words If f and f^{-1} are inverses, then $f(a) = b$ if and only if $f^{-1}(b) = a$.

Example Let $f(x) = x - 4$ and represent its inverse as $f^{-1}(x) = x + 4$.

Evaluate $f(6)$.	Evaluate $f^{-1}(2)$.
$f(x) = x - 4$	$f^{-1}(x) = x + 4$
$f(6) = 6 - 4$ or 2	$f^{-1}(2) = 2 + 4$ or 6

Because $f(x)$ and $f^{-1}(x)$ are inverses, $f(6) = 2$ and $f^{-1}(2) = 6$.

When the inverse of a function is a function, the original function is one-to-one. Recall that the vertical line test can be used to determine whether a relation is a function. Similarly, the *horizontal line test* can be used to determine whether the inverse of a function is also a function.

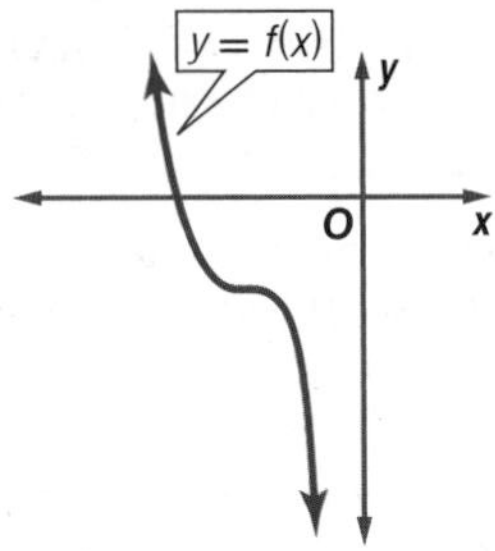

No horizontal line can be drawn so that it passes through more than one point. The inverse of $y = f(x)$ is a function.

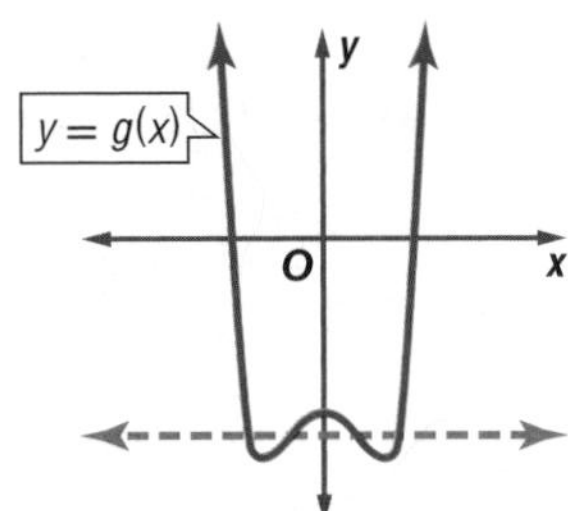

A horizontal line can be drawn that passes through more than one point. The inverse of $y = g(x)$ is not a function.

The inverse of a function can be found by exchanging the domain and the range.

Example 2 Find and Graph an Inverse

Find the inverse of each function. Then graph the function and its inverse.

a. $f(x) = 2x - 5$

Step 1 Rewrite the function as an equation relating x and y.

$f(x) = 2x - 5 \rightarrow y = 2x - 5$

Step 2 Exchange x and y in the equation. $x = 2y - 5$

Step 3 Solve the equation for y.

$x = 2y - 5$	Inverse of $y = 2x - 5$
$x + 5 = 2y$	Add 5 to each side.
$\frac{x + 5}{2} = y$	Divide each side by 2.

Step 4 Replace y with $f^{-1}(x)$.

$y = \frac{x + 5}{2} \rightarrow f^{-1}(x) = \frac{x + 5}{2}$

The inverse of $f(x) = 2x - 5$ is $f^{-1}(x) = \frac{x + 5}{2}$. The graph of $f^{-1}(x) = \frac{x + 5}{2}$ is the reflection of the graph of $f(x) = 2x - 5$ in the line $y = x$.

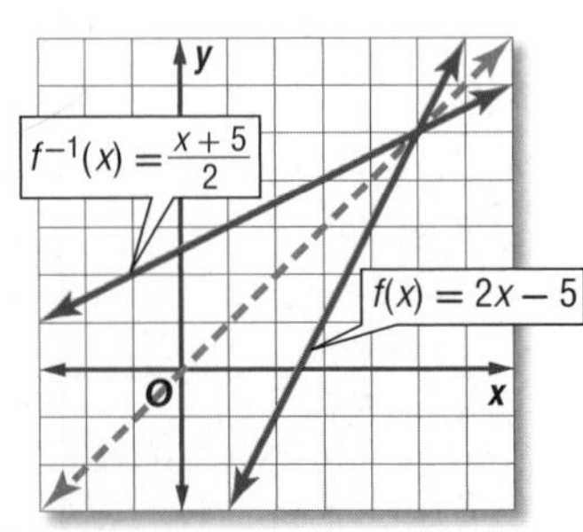

b. $f(x) = x^2 + 1$

Step 1 $f(x) = x^2 + 1 \rightarrow y = x^2 + 1$

Step 2 $x = y^2 + 1$

Step 3
$$x = y^2 + 1$$
$$x - 1 = y^2$$
$$\pm\sqrt{x - 1} = y$$
Take the square root of each side.

Step 4 $y = \pm\sqrt{x - 1}$

Graph $y = \pm\sqrt{x - 1}$ by reflecting the graph of $f(x) = x^2 + 1$ in the line $y = x$.

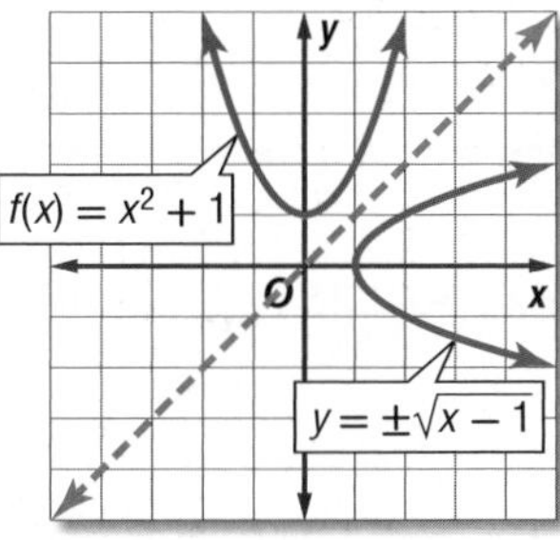

StudyTip

Functions The inverse of the function in part **b** is not a function since it does not pass the vertical line test.

GuidedPractice

Find the inverse of each function. Then graph the function and its inverse.

2A. $f(x) = \frac{x - 3}{5}$

2B. $f(x) = 3x^2$

2 Verifying Inverses

You can determine whether two functions are inverses by finding both of their compositions. If both compositions equal the identity function $I(x) = x$, then the functions are inverse functions.

KeyConcept Inverse Functions

Words	Two functions f and g are inverse functions if and only if both of their compositions are the identity function.
Symbols	$f(x)$ and $g(x)$ are inverses if and only if $[f \circ g](x) = x$ and $[g \circ f](x) = x$.

Example 3 Verify that Two Functions are Inverses

Determine whether each pair of functions are inverse functions. Explain your reasoning.

a. $f(x) = 3x + 9$ **and** $g(x) = \frac{1}{3}x - 3$

Verify that the compositions of $f(x)$ and $g(x)$ are identity functions.

$$\begin{aligned}[f \circ g](x) &= f[g(x)] \\ &= f\left(\tfrac{1}{3}x - 3\right) \\ &= 3\left(\tfrac{1}{3}x - 3\right) + 9 \\ &= x - 9 + 9 \text{ or } x\end{aligned}$$

$$\begin{aligned}[g \circ f](x) &= g[f(x)] \\ &= g(3x + 9) \\ &= \tfrac{1}{3}(3x + 9) - 3 \\ &= x + 3 - 3 \text{ or } x\end{aligned}$$

The functions are inverses because $[f \circ g](x) = [g \circ f](x) = x$.

b. $f(x) = 4x^2$ **and** $g(x) = 2\sqrt{x}$

$$\begin{aligned}[f \circ g](x) &= f\left(2\sqrt{x}\right) \\ &= 4\left(2\sqrt{x}\right)^2 \\ &= 4(4x) \text{ or } 16x\end{aligned}$$

Because $[f \circ g](x) \neq x$, $f(x)$ and $g(x)$ are not inverses.

WatchOut!

Inverse Functions Be sure to check both $[f \circ g](x)$ and $[g \circ f](x)$ to verify that functions are inverses. By definition, both compositions must be the identity function.

GuidedPractice

3A. $f(x) = 3x - 3$, $g(x) = \frac{1}{3}x + 4$

3B. $f(x) = 2x^2 - 1$, $g(x) = \sqrt{\frac{x + 1}{2}}$

Check Your Understanding

= Step-by-Step Solutions begin on page R14.

Example 1 **Find the inverse of each relation.**

1. $\{(-9, 10), (1, -3), (8, -5)\}$
2. $\{(-2, 9), (4, -1), (-7, 9), (7, 0)\}$

Example 2 **Find the inverse of each function. Then graph the function and its inverse.**

3. $f(x) = -3x$
4. $g(x) = 4x - 6$
5. $h(x) = x^2 - 3$

Example 3 **Determine whether each pair of functions are inverse functions. Write *yes* or *no*.**

6. $f(x) = x - 7$
   $g(x) = x + 7$
7. $f(x) = \frac{1}{2}x + \frac{3}{4}$
   $g(x) = 2x - \frac{4}{3}$
8. $f(x) = 2x^3$
   $g(x) = \frac{1}{3}\sqrt{x}$

Practice and Problem Solving

Extra Practice is on page R6.

Example 1 **Find the inverse of each relation.**

9. $\{(-8, 6), (6, -2), (7, -3)\}$
10. $\{(7, 7), (4, 9), (3, -7)\}$
11. $\{(8, -1), (-8, -1), (-2, -8), (2, 8)\}$
12. $\{(4, 3), (-4, -4), (-3, -5), (5, 2)\}$
13. $\{(1, -5), (2, 6), (3, -7), (4, 8), (5, -9)\}$
14. $\{(3, 0), (5, 4), (7, -8), (9, 12), (11, 16)\}$

Example 2 CCSS **SENSE-MAKING** **Find the inverse of each function. Then graph the function and its inverse.**

15. $f(x) = x + 2$
16. $g(x) = 5x$
17. $f(x) = -2x + 1$
18. $h(x) = \frac{x - 4}{3}$
19. $f(x) = -\frac{5}{3}x - 8$
20. $g(x) = x + 4$
21. $f(x) = 4x$
22. $f(x) = -8x + 9$
23. $f(x) = 5x^2$
24. $h(x) = x^2 + 4$
25. $f(x) = \frac{1}{2}x^2 - 1$
26. $f(x) = (x + 1)^2 + 3$

Example 3 **Determine whether each pair of functions are inverse functions. Write *yes* or *no*.**

27. $f(x) = 2x + 3$
    $g(x) = 2x - 3$
28. $f(x) = 4x + 6$
    $g(x) = \frac{x - 6}{4}$
29. $f(x) = -\frac{1}{3}x + 3$
    $g(x) = -3x + 9$
30. $f(x) = -6x$
    $g(x) = \frac{1}{6}x$
31. $f(x) = \frac{1}{2}x + 5$
    $g(x) = 2x - 10$
32. $f(x) = \frac{x + 10}{8}$
    $g(x) = 8x - 10$
33. $f(x) = 4x^2$
    $g(x) = \frac{1}{2}\sqrt{x}$
34. $f(x) = \frac{1}{3}x^2 + 1$
    $g(x) = \sqrt{3x - 3}$
35. $f(x) = x^2 - 9$
    $g(x) = x + 3$
36. $f(x) = \frac{2}{3}x^3$
    $g(x) = \sqrt{\frac{2}{3}x}$
37. $f(x) = (x + 6)^2$
    $g(x) = \sqrt{x} - 6$
38. $f(x) = 2\sqrt{x - 5}$
    $g(x) = \frac{1}{4}x^2 - 5$

39. **FUEL** The average miles traveled for every gallon g of gas consumed by Leroy's car is represented by the function $m(g) = 28g$.
    a. Find a function $c(g)$ to represent the cost per gallon of gasoline.
    b. Use inverses to determine the function used to represent the cost per mile traveled in Leroy's car.

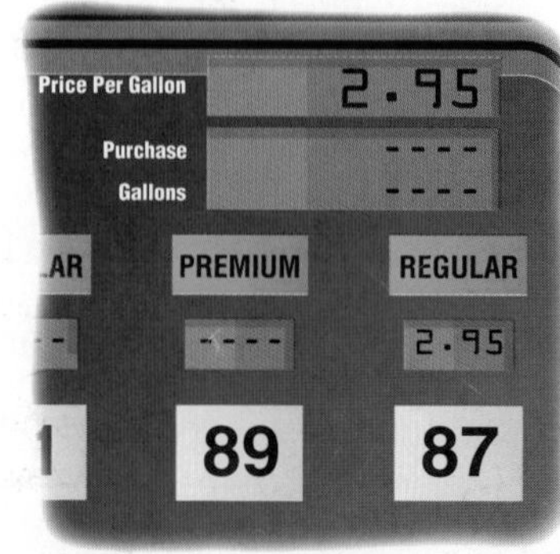

40. **SHOES** The shoe size for the average U.S. teen or adult male can be determined using the formula $M(x) = 3x - 22$, where x is length of a foot in measured inches. The shoe size for the average U.S. teen or adult female can be found by using the formula $F(x) = 3x - 21$.

a. Find the inverse of each function.

b. If Lucy wears a size $7\frac{1}{2}$ shoe, how long are her feet?

41. **GEOMETRY** The formula for the area of a circle is $A = \pi r^2$.

a. Find the inverse of the function.

b. Use the inverse to find the radius of a circle with an area of 36 square centimeters.

Use the horizontal line test to determine whether the inverse of each function is also a function.

42. $f(x) = 2x^2$

43. $f(x) = x^3 - 8$

44. $g(x) = x^4 - 6x^2 + 1$

45. $h(x) = -2x^4 - x - 2$

46. $g(x) = x^5 + x^2 - 4x$

47. $h(x) = x^3 + x^2 - 6x + 12$

48. **SHOPPING** Felipe bought a used car. The sales tax rate was 7.25% of the selling price, and he paid $350 in processing and registration fees. Find the selling price if Felipe paid a total of $8395.75.

49. **TEMPERATURE** A formula for converting degrees Celsius to Fahrenheit is $F(x) = \frac{9}{5}x + 32$.

a. Find the inverse $F^{-1}(x)$. Show that $F(x)$ and $F^{-1}(x)$ are inverses.

b. Explain what purpose $F^{-1}(x)$ serves.

50. **MEASUREMENT** There are approximately 1.852 kilometers in a nautical mile.

a. Write a function that converts nautical miles to kilometers.

b. Find the inverse of the function that converts kilometers back to nautical miles.

c. Using composition of functions, verify that these two functions are inverses.

51. **MULTIPLE REPRESENTATIONS** Consider the functions $y = x^n$ for $n = 0, 1, 2, \ldots$.

a. **Graphing** Use a graphing calculator to graph $y = x^n$ for $n = 0, 1, 2, 3$, and 4.

b. **Tabular** For which values of n is the inverse a function? Record your results in a table.

c. **Analytical** Make a conjecture about the values of n for which the inverse of $f(x) = x^n$ is a function. Assume that n is a whole number.

H.O.T. Problems Use Higher-Order Thinking Skills

52. **REASONING** If a relation is *not* a function, then its inverse is *sometimes*, *always*, or *never* a function. Explain your reasoning.

53. **OPEN ENDED** Give an example of a function and its inverse. Verify that the two functions are inverses.

54. **CHALLENGE** Give an example of a function that is its own inverse.

55. **CCSS ARGUMENTS** Show that the inverse of a linear function $y = mx + b$, where $m \neq 0$ and $x \neq b$, is also a linear function.

56. **WRITING IN MATH** Suppose you have a composition of two functions that are inverses. When you put in a value of 5 for x, why is the result always 5?

Standardized Test Practice

57. SHORT RESPONSE If the length of a rectangular television screen is 24 inches and its height is 18 inches, what is the length of its diagonal in inches?

58. GEOMETRY If the base of a triangle is represented by $2x + 5$ and the height is represented by $4x$, which expression represents the area of the triangle?

A $(2x + 5) + (4x)$

B $(2x + 5)(4x)$

C $\frac{1}{2}(2x + 5) + (4x)$

D $\frac{1}{2}(2x + 5)(4x)$

59. Which expression represents $f[g(x)]$ if $f(x) = x^2 + 3$ and $g(x) = -x + 1$?

F $x^2 - x + 2$

G $-x^2 - 2$

H $-x^3 + x^2 - 3x + 3$

J $x^2 - 2x + 4$

60. SAT/ACT Which of the following is the inverse of $f(x) = \frac{3x - 5}{2}$?

A $g(x) = \frac{2x + 5}{3}$

B $g(x) = \frac{2x - 5}{3}$

C $g(x) = \frac{3x + 5}{2}$

D $g(x) = 2x + 5$

E $g(x) = \frac{3x - 5}{2}$

Spiral Review

If $f(x) = 3x + 5$, $g(x) = x - 2$, and $h(x) = x^2 - 1$, find each value. (Lesson 6-1)

61. $g[f(3)]$

62. $f[h(-2)]$

63. $h[g(1)]$

64. CONSTRUCTION A picnic area has the shape of a trapezoid. The longer base is 8 more than 3 times the length of the shorter base, and the height is 1 more than 3 times the shorter base. What are the dimensions if the area is 4104 square feet? (Lesson 5-8)

Find the value of c that makes each trinomial a perfect square. Then write the trinomial as a perfect square. (Lesson 4-5)

65. $x^2 + 34x + c$

66. $x^2 - 11x + c$

Simplify. (Lesson 4-4)

67. $(3 + 4i)(5 - 2i)$

68. $(\sqrt{6} + i)(\sqrt{6} - i)$

69. $\frac{1 + i}{1 - i}$

70. $\frac{4 - 3i}{1 + 2i}$

Determine the rate of change of each graph. (Lesson 2-3)

71.

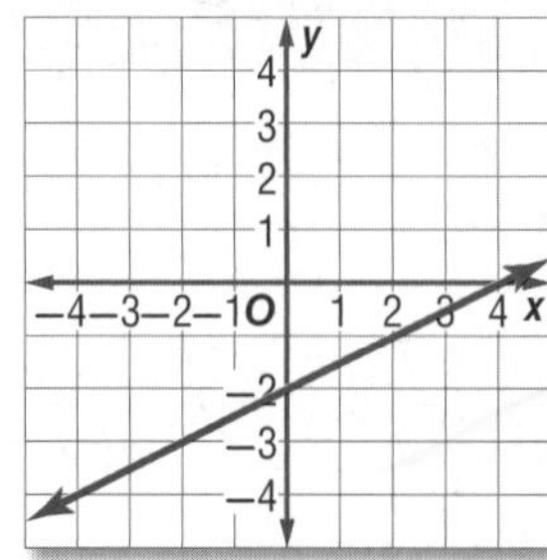

72.

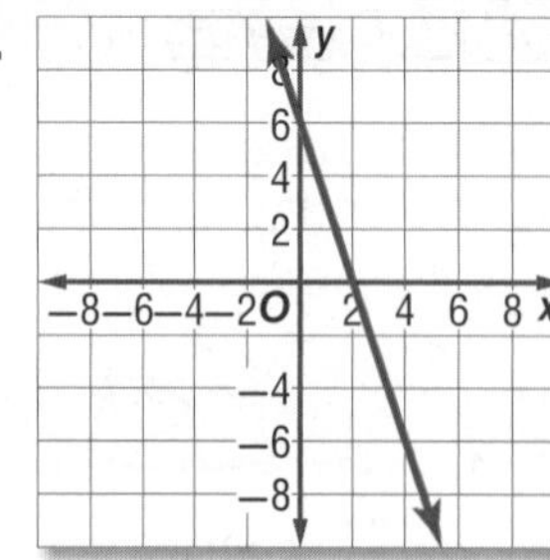

73.

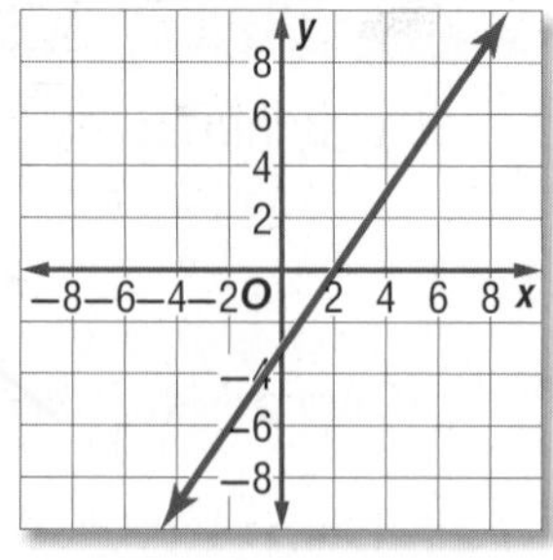

Skills Review

Graph each inequality.

74. $y > \frac{3}{4}x - 2$

75. $y \leq -3x + 2$

76. $y < -x - 4$

EXTEND 6-2

Graphing Technology Lab
Inverse Functions and Relations

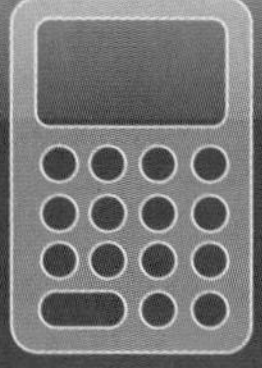

You can use a TI-83/84 Plus graphing calculator to compare a function and its inverse using tables and graphs. Note that before you enter any values in the calculator, you should clear all lists.

Activity 1 Graph Inverses with Ordered Pairs

Graph $f(x) = \{(1, 2), (2, 4), (3, 6), (4, 8), (5, 10), (6, 12)\}$ and its inverse.

Step 1 Enter the x-values in **L1** and the y-values in **L2**. Then graph the function.

KEYSTROKES: [STAT] [ENTER] 1 [ENTER] 2 [ENTER] 3 [ENTER] 4 [ENTER] 5 [ENTER] 6 [ENTER] [▶] 2 [ENTER] 4 [ENTER] 6 [ENTER] 8 [ENTER] 10 [ENTER] 12 [ENTER] [2nd] [STAT PLOT] [ENTER] [ENTER] [GRAPH]

Adjust the window to reflect the domain and range.

Step 2:

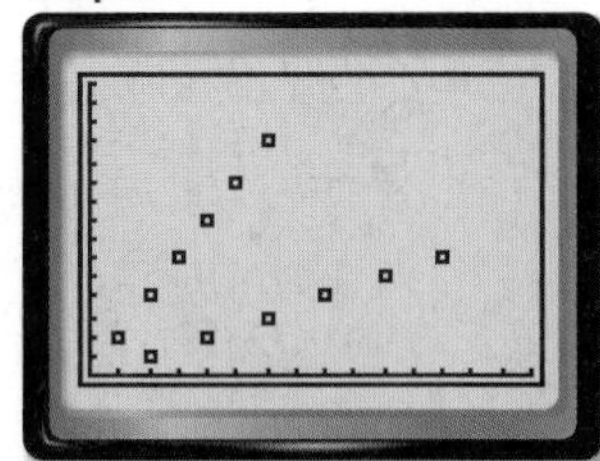

[0, 14] scl: 1 by [0, 14] scl: 1

Step 2 Define the inverse function by setting **Xlist** to **L2** and **Ylist** to **L1**. Then graph the inverse function.

KEYSTROKES: [2nd] [STAT PLOT] [▼] [ENTER] [ENTER] [▼] [▼] [2nd] [L2] [▼] [2nd] [L1] [GRAPH]

Step 3 Graph the line $y = x$.

KEYSTROKES: [Y=] [X,T,θ,n] [GRAPH]

Step 3:

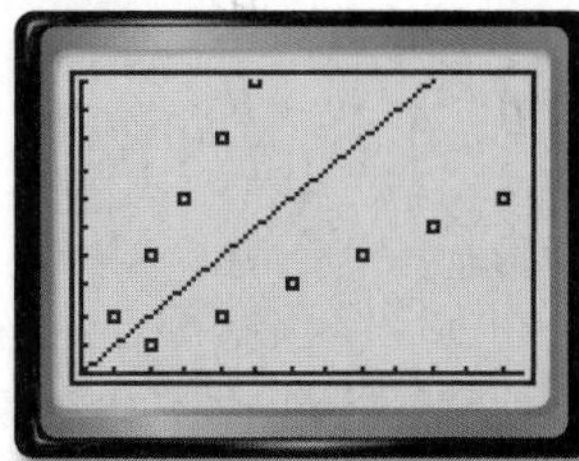

[0, 14] scl: 1 by [0, 14] scl: 1

Activity 2 Graph Inverses with Function Notation

Graph $f(x) = 3x$ and its inverse $g(x) = \frac{x}{3}$.

Step 1 Clear the data from Activity 1.

KEYSTROKES: [2nd] [STAT PLOT] [ENTER] [▶] [ENTER] [▲] [▶] [ENTER] [▶] [ENTER] [2nd] [QUIT]

Step 2 Enter $f(x)$ as **Y1**, $g(x)$ as **Y2**, and $y = x$ as **Y3**. Then graph.

KEYSTROKES: [Y=] 3 [X,T,θ,n] [ENTER] [X,T,θ,n] [÷] 3 [ENTER] [X,T,θ,n] [GRAPH]

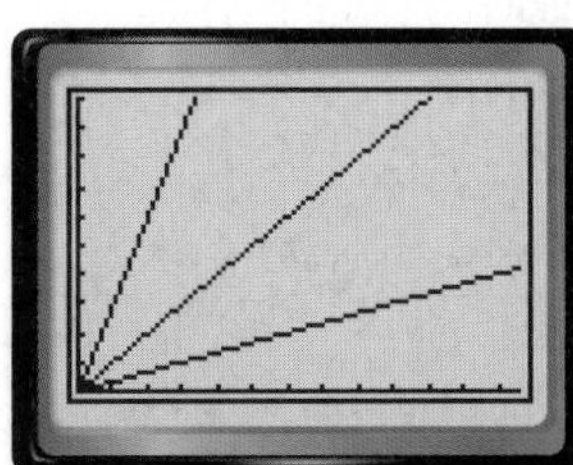

[0, 14] scl: 1 by [0, 14] scl: 1

Exercises

Graph each function $f(x)$ and its inverse $g(x)$. Then graph $(f \circ g)(x)$.

1. $f(x) = 5x$

2. $f(x) = x - 3$

3. $f(x) = 2x + 1$

4. $f(x) = \frac{1}{2}x + 3$

5. $f(x) = x^2$

6. $f(x) = x^2 - 3$

7. What is the relationship between the graphs of a function and its inverse?

8. **MAKE A CONJECTURE** For any function $f(x)$ and its inverse $g(x)$, what is $(f \circ g)(x)$?

LESSON 6-3 Square Root Functions and Inequalities

Then

- You simplified expressions with square roots.

Now

1. Graph and analyze square root functions.
2. Graph square root inequalities.

Why?

- With guitars, pitch is dependent on string length and string tension. The longer the string, the higher the tension needed to produce a desired pitch. Likewise, the heavier the string, the higher the tension needed to reach a desired pitch.

  This can be modeled by the square root function $f = \frac{1}{2L}\sqrt{\frac{T}{P}}$, where T is the tension, P is the mass of the string, L is the length of the string, and f is the pitch.

NewVocabulary
square root function
radical function
square root inequality

Common Core State Standards

Content Standards

F.IF.7.b Graph square root, cube root, and piecewise-defined functions, including step functions and absolute value functions.

F.BF.3 Identify the effect on the graph of replacing $f(x)$ by $f(x) + k$, $k\,f(x)$, $f(kx)$, and $f(x + k)$ for specific values of k (both positive and negative); find the value of k given the graphs. Experiment with cases and illustrate an explanation of the effects on the graph using technology.

Mathematical Practices

3 Construct viable arguments and critique the reasoning of others.

1 Square Root Functions

If a function contains the square root of a variable, it is called a **square root function**. The square root function is a type of **radical function**.

KeyConcept Parent Function of Square Root Functions

Parent function: $f(x) = \sqrt{x}$

Domain: $\{x \mid x \geq 0\}$

Range: $\{f(x) \mid f(x) \geq 0\}$

Intercepts: $x = 0, f(x) = 0$

Not defined: $x < 0$

End behavior: $x \to 0, f(x) \to 0$

$x \to +\infty, f(x) \to +\infty$

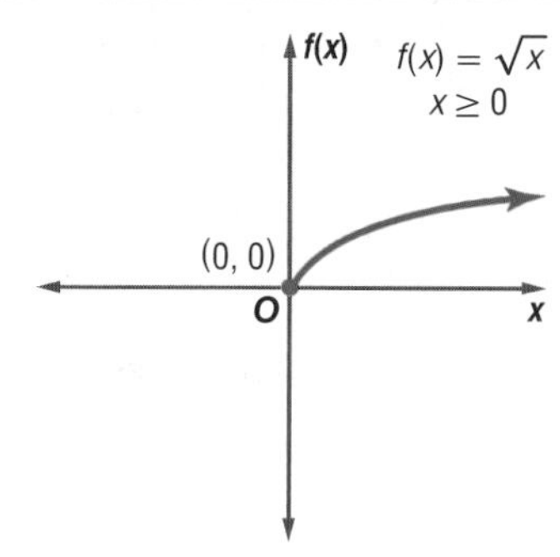

The domain of a square root function is limited to values for which the function is defined.

Example 1 Identify Domain and Range

Identify the domain and range of $f(x) = \sqrt{x + 4}$.

The domain only includes values for which the radicand is nonnegative.

$x + 4 \geq 0$ Write an inequality.

$x \geq -4$ Subtract 4 from each side.

Thus, the domain is $\{x \mid x \geq -4\}$.

Find $f(-4)$ to determine the lower limit of the range.

$f(-4) = \sqrt{-4 + 4}$ or 0

So, the range is $\{f(x) \mid f(x) \geq 0\}$.

GuidedPractice

Identify the domain and range of each function.

1A. $f(x) = \sqrt{x - 3}$

1B. $f(x) = \sqrt{x + 6} + 2$

Ryan McVay/Lifesize/Getty Images

The same techniques used to transform the graph of other functions you have studied can be applied to the graphs of square root functions.

KeyConcept Transformations of Square Root Functions

$$f(x) = a\sqrt{x - h} + k$$

***h*—Horizontal Translation**	***k*—Vertical Translation**
h units right if h is positive $\lvert h \rvert$ units left if h is negative The domain is $\{x \mid x \geq h\}$.	k units up if k is positive $\lvert k \rvert$ units down if k is negative If $a > 0$, then the range is $\{f(x) \mid f(x) \geq k\}$. If $a < 0$, then the range is $\{f(x) \mid f(x) \leq k\}$.

***a*—Orientation and Shape**

- If $a < 0$, the graph is reflected across the x-axis.
- If $\lvert a \rvert > 1$, the graph is stretched vertically.
- If $0 < \lvert a \rvert < 1$, the graph is compressed vertically.

StudyTip

Domain and Range The limits on the domain and range also represent the initial point of the graph of a square root function.

PT

Example 2 Graph Square Root Functions

Graph each function. State the domain and range.

a. $y = \sqrt{x - 2} + 5$

The minimum point is at $(h, k) = (2, 5)$. Make a table of values for $x \geq 2$, and graph the function. The graph is the same shape as $f(x) = \sqrt{x}$, but is translated 2 units right and 5 units up. Notice the end behavior. As x increases, y increases.

The domain is $\{x \mid x \geq 2\}$ and the range is $\{y \mid y \geq 5\}$.

x	y
2	5
3	6
4	6.4
5	7
6	7.2
7	7.2
8	7.4

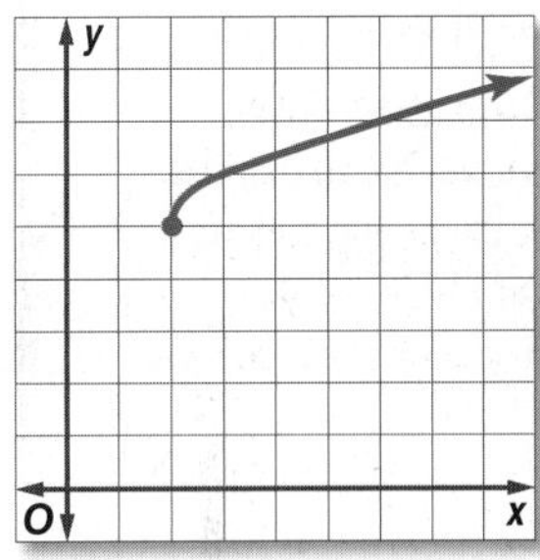

b. $y = -2\sqrt{x + 3} - 1$

The minimum domain value is at h or -3. Make a table of values for $x \geq -3$, and graph the function. Because a is negative, the graph is similar to the graph of $f(x) = \sqrt{x}$, but is reflected in the x-axis. Because $\lvert a \rvert > 1$, the graph is vertically stretched. It is also translated 3 units left and 1 unit down.

x	y
-3	-1
-2	-3
-1	-3.8
0	-4.5
1	-5
2	-5.5
3	-5.9

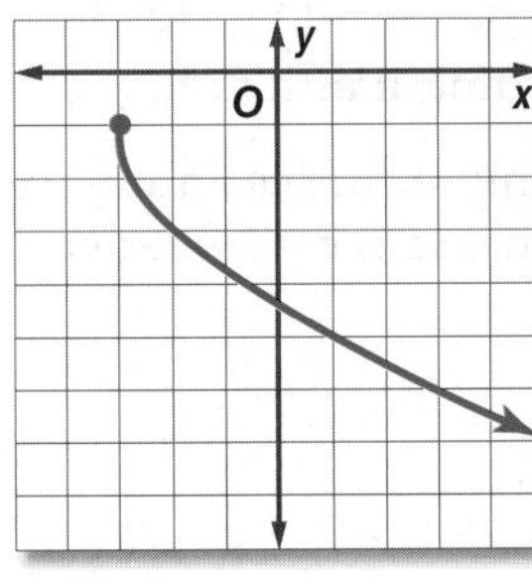

The domain is $\{x \mid x \geq -3\}$ and the range is $\{y \mid y \leq -1\}$.

GuidedPractice

2A. $f(x) = 2\sqrt{x + 4}$

2B. $f(x) = \frac{1}{4}\sqrt{x - 5} + 3$

Real-World Example 3 Use Graphs to Analyze Square Root Functions

MUSIC Refer to the application at the beginning of the lesson. The pitch, or frequency, measured in hertz (Hz) of a certain string can be modeled by $f(T) = \frac{1}{1.28}\sqrt{\frac{T}{0.0000708}}$, where T is tension in kilograms.

a. Graph the function for tension in the domain $\{T \mid 0 \leq T \leq 10\}$.

Make a table of values for $0 \leq T \leq 10$ and graph.

T	$y(T)$
0	0
1	92.8
2	131.3
3	160.8
4	185.7
5	207.6

T	$f(T)$
6	227.4
7	245.7
8	262.6
9	278.5
10	293.6

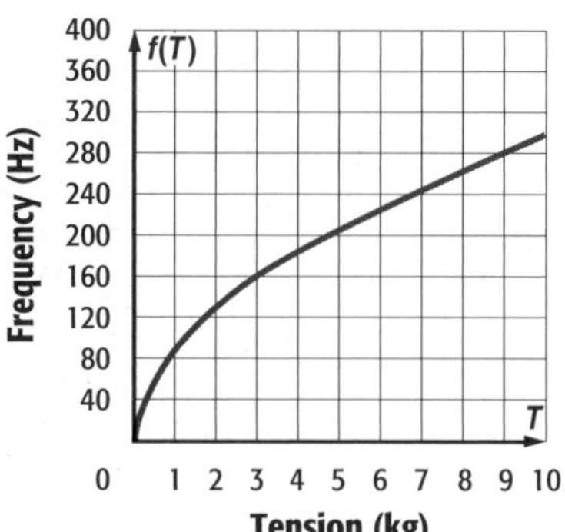

b. How much tension is needed for a pitch of over 200 Hz?

According to the graph and the table, more than 4.5 kilograms of tension is needed for a pitch of more than 200 hertz.

Real-WorldLink

On every string, the guitar player has an option of decreasing the length of the string in about 24 different ways. This will produce 24 different frequencies on each string.

Source: *Guitar World*

Problem-SolvingTip

CCSS Modeling Making a table is a good way to organize ordered pairs in order to see the general behavior of a graph.

GuidedPractice

3. MUSIC The frequency of vibrations for a certain guitar string when it is plucked can be determined by $F = 200\sqrt{T}$, where F is the number of vibrations per second and T is the tension measured in pounds. Graph the function for $0 \leq T \leq 10$. Then determine the frequency for $T = 3$, 6, and 9 pounds.

2 Square Root Inequalities

A **square root inequality** is an inequality involving square roots. They are graphed using the same method as other inequalities.

Example 4 Graph a Square Root Inequality

Graph $y < \sqrt{x - 4} - 6$.

Graph the boundary $y = \sqrt{x - 4} - 6$.

The domain is $\{x \mid x \geq 4\}$. Because y is *less than*, the shaded region should be *below* the boundary and within the domain.

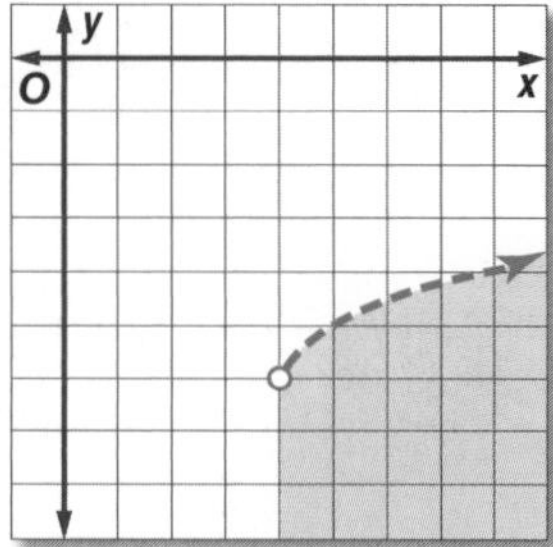

CHECK Select a point in the shaded region, and verify that it is a solution of the inequality.

Test (7, –5): $-5 \overset{?}{<} \sqrt{7 - 4} - 6$

$-5 \overset{?}{<} \sqrt{3} - 6$

$-5 < -4.27$ ✓

GuidedPractice

4A. $f(x) \geq \sqrt{2x + 1}$

4B. $f(x) < -\sqrt{x + 2} - 4$

Check Your Understanding

= Step-by-Step Solutions begin on page R14.

Example 1 **Identify the domain and range of each function.**

1. $f(x) = \sqrt{4x}$ **2.** $f(x) = \sqrt{x - 5}$ **3.** $f(x) = \sqrt{x + 8} - 2$

Example 2 **Graph each function. State the domain and range.**

4. $f(x) = \sqrt{x} - 2$ **5.** $f(x) = 3\sqrt{x - 1}$

6. $f(x) = \frac{1}{2}\sqrt{x + 4} - 1$ **7.** $f(x) = -\sqrt{3x - 5} + 5$

Example 3 **8. OCEAN** The speed that a tsunami, or tidal wave, can travel is modeled by the equation $v = 356\sqrt{d}$, where v is the speed in kilometers per hour and d is the average depth of the water in kilometers. A tsunami is found to be traveling at 145 kilometers per hour. What is the average depth of the water? Round to the nearest hundredth of a kilometer.

Example 4 **Graph each inequality.**

9. $f(x) \geq \sqrt{x} + 4$ **10.** $f(x) \leq \sqrt{x - 6} + 2$

11. $f(x) < -2\sqrt{x + 3}$ **12.** $f(x) > \sqrt{2x - 1} - 3$

Practice and Problem Solving

Extra Practice is on page R6.

Example 1 **Identify the domain and range of each function.**

13. $f(x) = -\sqrt{2x} + 2$ **14.** $f(x) = \sqrt{x} - 6$ **15.** $f(x) = 4\sqrt{x - 2} - 8$

16. $f(x) = \sqrt{x + 2} + 5$ **17.** $f(x) = \sqrt{x - 4} - 6$ **18.** $f(x) = -\sqrt{x - 6} + 5$

Example 2 **Graph each function. State the domain and range.**

19. $f(x) = \sqrt{6x}$ **20.** $f(x) = -\sqrt{5x}$

21. $f(x) = \sqrt{x - 8}$ **22.** $f(x) = \sqrt{x + 1}$

23. $f(x) = \sqrt{x + 3} + 2$ **24.** $f(x) = \sqrt{x - 4} - 10$

25. $f(x) = 2\sqrt{x - 5} - 6$ **26.** $f(x) = \frac{3}{4}\sqrt{x + 12} + 3$

27. $f(x) = -\frac{1}{5}\sqrt{x - 1} - 4$ **28.** $f(x) = -3\sqrt{x + 7} + 9$

Example 3 **29. SKYDIVING** The approximate time t in seconds that it takes an object to fall a distance of d feet is given by $t = \sqrt{\frac{d}{16}}$. Suppose a parachutist falls 11 seconds before the parachute opens. How far does the parachutist fall during this time?

30. CCSS **MODELING** The velocity of a roller coaster as it moves down a hill is $V = \sqrt{v^2 + 64h}$, where v is the initial velocity in feet per second and h is the vertical drop in feet. The designer wants the coaster to have a velocity of 90 feet per second when it reaches the bottom of the hill.

a. If the initial velocity of the coaster at the top of the hill is 10 feet per second, write an equation that models the situation.

b. How high should the designer make the hill?

Example 4 **Graph each inequality.**

31. $y < \sqrt{x - 5}$

32. $y > \sqrt{x + 6}$

33. $y \geq -4\sqrt{x + 3}$

34. $y \leq -2\sqrt{x - 6}$

35. $y > 2\sqrt{x + 7} - 5$

36. $y \geq 4\sqrt{x - 2} - 12$

37. $y \leq 6 - 3\sqrt{x - 4}$

38. $y < \sqrt{4x - 12} + 8$

39. **PHYSICS** The kinetic energy of an object is the energy produced due to its motion and mass. The formula for kinetic energy, measured in joules j, is $E = 0.5mv^2$, where m is the mass in kilograms and v is the velocity of the object in meters per second.

 a. Solve the above formula for v.

 b. If a 1500-kilogram vehicle is generating 1 million joules of kinetic energy, how fast is it traveling?

 c. *Escape velocity* is the minimum velocity at which an object must travel to escape the gravitational field of a planet or other object. Suppose a 100,000-kilogram ship must have a kinetic energy of 3.624×10^{14} joules to escape the gravitational field of Jupiter. Estimate the escape velocity of Jupiter.

40.  **REASONING** After an accident, police can determine how fast a car was traveling before the driver put on his or her brakes by using the equation $v = \sqrt{30fd}$. In this equation, v represents the speed in miles per hour, f represents the coefficient of friction, and d represents the length of the skid marks in feet. The coefficient of friction varies depending on road conditions. Assume that $f = 0.6$.

 a. Find the speed of a car that skids 25 feet.

 b. If your car is going 35 miles per hour, how many feet would it take you to stop?

 c. If the speed of a car is doubled, will the skid be twice as long? Explain.

Write the square root function represented by each graph.

41. 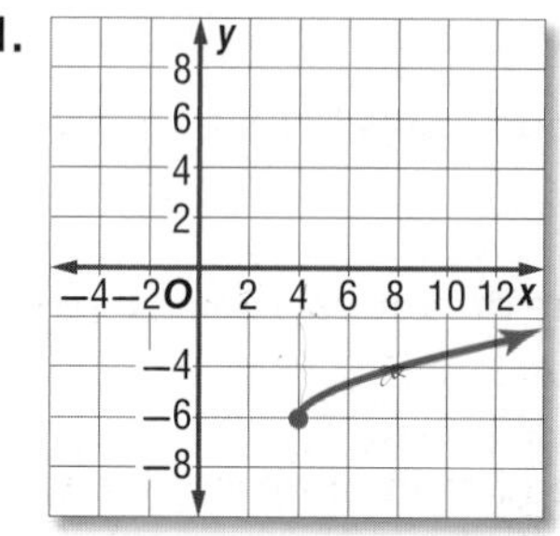

42. 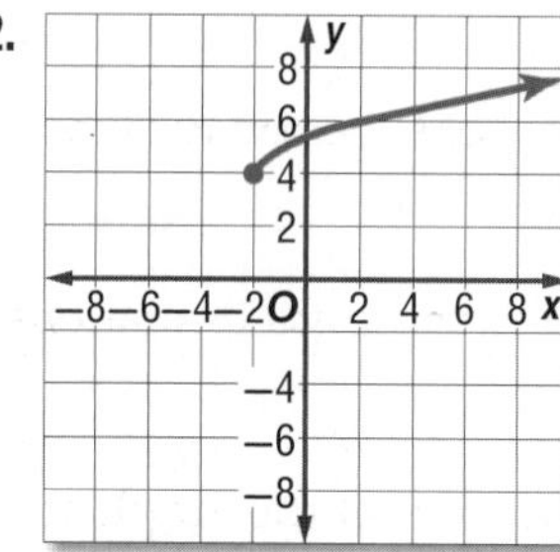

43. 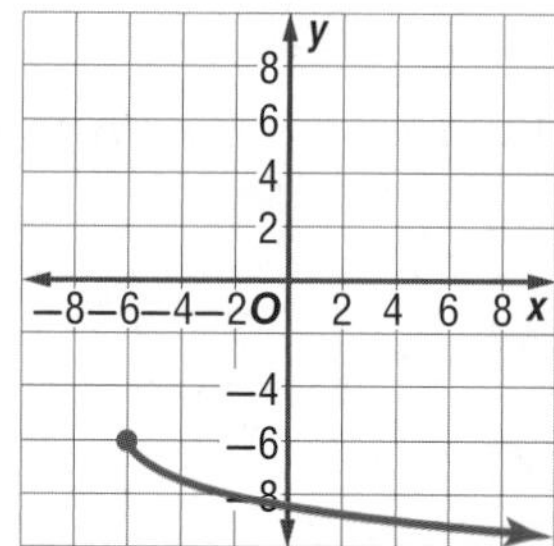

44. **MULTIPLE REPRESENTATIONS** In this problem, you will use the following functions to investigate transformations of square root functions.

$f(x) = 4\sqrt{x - 6} + 3$ | $g(x) = \sqrt{16x + 1} - 6$ | $h(x) = \sqrt{x + 3} + 2$

 a. **Graphical** Graph each function on the same set of axes.

 b. **Analytical** Identify the transformation on the graph of the parent function. What values caused each transformation?

 c. **Analytical** Which functions appear to be stretched or compressed vertically? Explain your reasoning.

 d. **Verbal** The two functions that are stretched appear to be stretched by the same magnitude. How is this possible?

 e. **Tabular** Make a table of the rate of change for all three functions between 8 and 12 as compared to 12 and 16. What generalization about rate of change in square root functions can be made as a result of your findings?

45 **PENDULUMS** The period of a pendulum can be represented by $T = 2\pi\sqrt{\frac{L}{g}}$, where T is the time in seconds, L is the length in feet, and g is gravity, 32 feet per second squared.

a. Graph the function for $0 \leq L \leq 10$.

b. What is the period for lengths of 2, 5, and 8 feet?

46. PHYSICS Using the function $m = \frac{m_0}{\sqrt{1 - \left(\frac{v^2}{c^2}\right)}}$, Einstein's theory of relativity states that the apparent mass m of a particle depends on its velocity v. An object that is traveling extremely fast, close to the speed of light c, will *appear* to have more mass compared to its mass at rest, m_0.

a. Use a graphing calculator to graph the function for a 10,000-kilogram ship for the domain $0 \leq v \leq 300{,}000{,}000$. Use 300 million meters per second for the speed of light.

b. What viewing window did you use to view the graph?

c. Determine the apparent mass m of the ship for speeds of 100 million, 200 million, and 299 million meters per second.

H.O.T. Problems Use Higher-Order Thinking Skills

47. CHALLENGE Write an equation for a square root function with a domain of $\{x \mid x \geq -4\}$, a range of $\{y \mid y \leq 6\}$, and that passes through (5, 3).

48. REASONING For what positive values of a are the domain and range of $f(x) = \sqrt[a]{x}$ the set of real numbers?

49. OPEN ENDED Write a square root function for which the domain is $\{x \mid x \geq 8\}$ and the range is $\{y \mid y \leq 14\}$.

50. WRITING IN MATH Explain why there are limitations on the domain and range of square root functions.

51. CCSS CRITIQUE Cleveland thinks that the graph and the equation represent the same function. Molly disagrees. Who is correct? Explain your reasoning.

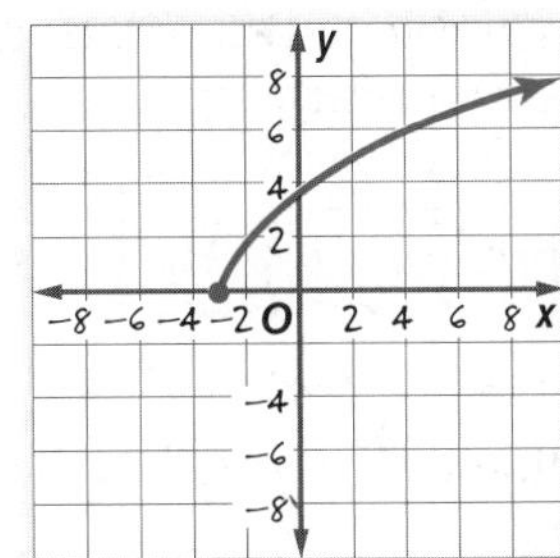

$y = \sqrt{5x + 10}$

52. WRITING IN MATH Explain why $y = \pm\sqrt{x}$ is not a function.

53. OPEN ENDED Write an equation of a relation that contains a radical and its inverse such that:

a. the original relation is a function, and its inverse is not a function.

b. the original relation is not a function, and its inverse is a function.

Standardized Test Practice

54. The expression $-\frac{64x^6}{8x^3}$, $x \neq 0$, is equivalent to

A $8x^2$　　C $-8x^2$

B $8x^3$　　D $-8x^3$

55. PROBABILITY For a game, Patricia must roll a standard die and draw a card from a deck of 26 cards, each card having a letter of the alphabet on it. What is the probability that Patricia will roll an odd number and draw a letter in her name?

F $\frac{2}{3}$　　H $\frac{1}{13}$

G $\frac{3}{26}$　　J $\frac{1}{26}$

56. SHORT RESPONSE What is the product of $(d + 6)$ and $(d - 3)$?

57. SAT/ACT Given the graph of the square root function below, which must be true?

I. The domain is all real numbers.

II. The function is $y = \sqrt{x} + 3.5$.

III. The range is $\{y \mid y \geq 3.5\}$.

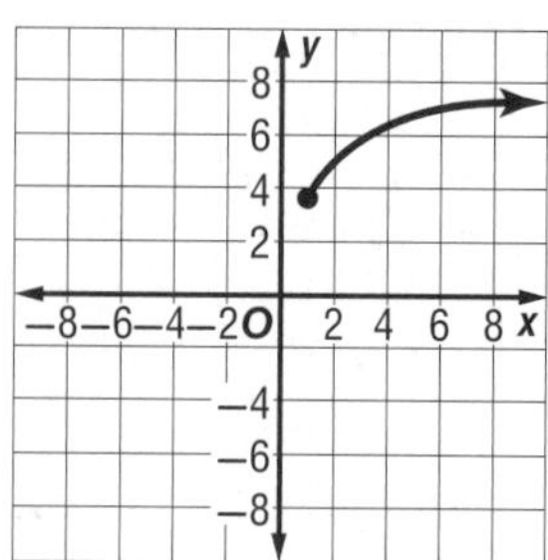

A I only　　D II only

B I, II, and III　　E III only

C II and III only

Spiral Review

Determine whether each pair of functions are inverse functions. Write *yes* or *no*. (Lesson 6-2)

58. $f(x) = 2x$
$g(x) = \frac{1}{2}x$

59. $f(x) = 3x - 7$
$g(x) = \frac{1}{3}x - \frac{7}{16}$

60. $f(x) = \frac{3x + 2}{5}$
$g(x) = \frac{5x - 2}{3}$

61. TIME The formula $h = \frac{m}{60}$ converts minutes m to hours h, and $d = \frac{h}{24}$ converts hours h to days d. Write a function that converts minutes to days. (Lesson 6-1)

62. CABLE TV The number of households in the United States with cable TV after 1985 can be modeled by the function $C(t) = -43.2t^2 + 1343t + 790$, where t represents the number of years since 1985. (Lesson 5-4)

a. Graph this equation for the years 1985 to 2005.

b. Describe the turning points of the graph and its end behavior.

c. What is the domain of the function? Use the graph to estimate the range for the function.

d. What trends in households with cable TV does the graph suggest? Is it reasonable to assume that the trend will continue indefinitely?

Skills Review

Determine whether each number is *rational* or *irrational*.

63. 6.34　　**64.** 3.787887888…　　**65.** 5.333…　　**66.** 1.25

LESSON 6-4 nth Roots

Then	Now	Why?
You worked with square root functions.	1 Simplify radicals. 2 Use a calculator to approximate radicals.	According to a world-wide injury prevention study, the number of collisions between bicycles and automobiles increased as the number of bicycles per intersection increased. The relationship can be expressed using the equation $c = \sqrt[5]{b^2}$, where b is the number of bicycles and c is the number of collisions.

NewVocabulary
nth root
radical sign
index
radicand
principal root

Common Core State Standards

Content Standards
A.SSE.2 Use the structure of an expression to identify ways to rewrite it.

Mathematical Practices
6 Attend to precision.

1 Simplify Radicals

Finding the square root of a number and squaring a number are inverse operations. To find the square root of a number a, you must find a number with a square of a. Similarly, the inverse of raising a number to the nth power is finding the **nth root** of a number.

Powers	Factors	Words	Roots
$x^3 = 64$	$4 \cdot 4 \cdot 4 = 64$	4 is a cube root of 64.	$\sqrt[3]{64} = 4$
$x^4 = 625$	$5 \cdot 5 \cdot 5 \cdot 5 = 625$	5 is a fourth root of 625.	$\sqrt[4]{625} = 5$
$x^5 = 32$	$2 \cdot 2 \cdot 2 \cdot 2 \cdot 2 = 32$	2 is a fifth root of 32.	$\sqrt[5]{32} = 2$
$a^n = b$	$\underbrace{a \cdot a \cdot a \cdot \ldots \cdot a}_{n \text{ factors of } a} = b$	a is an nth root of b.	$\sqrt[n]{b} = a$

This pattern suggests the following formal definition of an nth root.

KeyConcept Definition of nth Root

Words For any real numbers a and b, and any positive integer n, if $a^n = b$, then a is an nth root of b.

Example Because $(-3)^4 = 81$, -3 is a fourth root of 81 and 3 is a principal root.

The symbol $\sqrt[n]{}$ indicates an nth root.

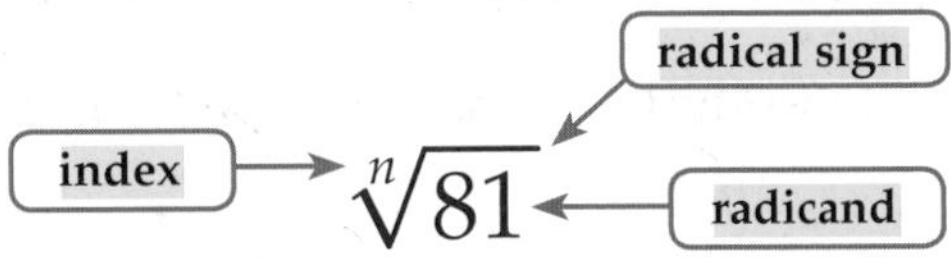

Some numbers have more than one real nth root. For example, 64 has two square roots, 8 and -8, since 8^2 and $(-8)^2$ both equal 64. When there is more than one real root and n is even, the nonnegative root is called the **principal root**.

Some examples of nth roots are listed below.

$\sqrt{25} = 5$ — $\sqrt{25}$ indicates the principal square root of 25.

$-\sqrt{25} = -5$ — $-\sqrt{25}$ indicates the opposite of the principal square root of 25.

$\pm\sqrt{25} = \pm 5$ — $\pm\sqrt{25}$ indicates both square roots of 25.

Ingram Publishing/age fotostock

KeyConcept Real nth Roots

Suppose n is an integer greater than 1, and a is a real number.

a	n is even.	n is odd.
$a > 0$	1 unique positive and 1 unique negative real root: $\pm\sqrt[n]{a}$; positive root is principal root	1 unique positive and 0 negative real roots: $\sqrt[n]{a}$
$a < 0$	0 real roots	0 positive and 1 negative real root: $\sqrt[n]{a}$
$a = 0$	1 real root: $\sqrt[n]{0} = 0$	1 real root: $\sqrt[n]{0} = 0$

ReviewVocabulary

pure imaginary numbers Square roots of negative real numbers; for any positive real number b, $\sqrt{-b^2} = \sqrt{b^2} \cdot \sqrt{-1}$, or bi, where i is the imaginary unit.

Example 1 Find Roots

PT

Simplify.

a. $\pm\sqrt{16y^4}$

$$\pm\sqrt{16y^4} = \pm\sqrt{(4y^2)^2}$$
$$= \pm 4y^2$$

The square roots of $16y^4$ are $\pm 4y^2$.

b. $-\sqrt{(x^2 - 6)^8}$

$$-\sqrt{(x^2 - 6)^8} = -\sqrt{[(x^2 - 6)^4]^2}$$
$$= -(x^2 - 6)^4$$

The opposite of the principal square root of $(x^2 - 6)^8$ is $-(x^2 - 6)^4$.

c. $\sqrt[5]{243a^{20}b^{25}}$

$$\sqrt[5]{243a^{20}b^{25}} = \sqrt[5]{(3a^4b^5)^5}$$
$$= 3a^4b^5$$

The fifth root of $243a^{20}b^{25}$ is $3a^4b^5$

d. $\sqrt{-16x^4y^8}$

$$\sqrt[2]{-16x^4y^8}$$

b is negative.

n is even.

There are no real roots since $\sqrt{-16}$ is not a real number. However, there are two imaginary roots, $4ix^2y^4$ and $-4ix^2y^4$.

StudyTip

Odd Index If n is odd, there is only one real root. Therefore, there is no principal root when n is odd, and absolute value symbols are never needed.

GuidedPractice

1A. $\pm\sqrt{36x^{10}}$

1B. $-\sqrt{(y + 7)^{16}}$

When you find an even root of an even power and the result is an odd power, you must use the absolute value of the result to ensure that the answer is nonnegative.

Example 2 Simplify Using Absolute Value

Simplify.

a. $\sqrt[4]{y^4}$

$$\sqrt[4]{y^4} = |y|$$

Since y could be negative, you must take the absolute value of y to identify the principal root.

b. $\sqrt[6]{64(x^2 - 3)^{18}}$

$$\sqrt[6]{64(x^2 - 3)^{18}} = 2\left|(x^2 - 3)^3\right|$$

Since the index 6 is even and the exponent 3 is odd, you must use absolute value.

GuidedPractice

2A. $\sqrt{36y^6}$

2B. $\sqrt[4]{16(x - 3)^{12}}$

2 Approximate Radicals with a Calculator

Recall that real numbers that cannot be expressed as terminating or repeating decimals are irrational numbers. Approximations for irrational numbers are often used in real-world problems.

Real-WorldLink

77% of the employees in China commute by bicycle.

Source: International Bicycle Fund

PT

Real-World Example 3 Approximate Radicals

INJURY PREVENTION Refer to the beginning of the lesson.

a. If $c = \sqrt[5]{b^2}$ represents the number of collisions and b represents the number of bicycle riders per intersection, estimate the number of collisions at an intersection that has 1000 bicycle riders per week.

Understand You want to find out how many collisions there are.

Plan Let 1000 be the number of bicycle riders. The number of collisions is c.

Solve

$c = \sqrt[5]{b^2}$	Original formula
$= \sqrt[5]{1000^2}$	$b = 1000$
≈ 15.85	Use a calculator.

There are about 16 collisions per week at the intersection.

Check

$15.85 \stackrel{?}{=} \sqrt[5]{b^2}$	$c = 15.85$
$15.85^5 \stackrel{?}{=} b^2$	Raise each side to the fifth power.
$1{,}000{,}337 \stackrel{?}{=} b^2$	Simplify.
$1000 \approx b$ ✓	Take the positive square root of each side.

b. If the total number of collisions reported in one week is 21, estimate the number of bicycle riders that passed through that intersection.

$c = \sqrt[5]{b^2}$	Original formula
$21 = \sqrt[5]{b^2}$	$c = 27$
$21^5 = b^2$	Raise each side to the fifth power.
$4{,}084{,}101 = b^2$	Simplify.
$2021 \approx b$	Take the positive square root of each side.

GuidedPractice

3A. The surface area of a sphere can be determined from the volume of the sphere using the formula $S = \sqrt[3]{36\pi V^2}$, where V is the volume. Determine the surface area of a sphere with a volume of 200 cubic inches.

3B. If the surface area of a sphere is about 214.5 square inches, determine the volume.

Check Your Understanding

= Step-by-Step Solutions begin on page R14.

Examples 1–2 **Simplify.**

1. $\pm\sqrt{100y^8}$
2. $-\sqrt{49u^8v^{12}}$
3. $\sqrt{(y-6)^8}$
4. $\sqrt[4]{16g^{16}h^{24}}$
5. $\sqrt{-16y^4}$
6. $\sqrt[6]{64(2y+1)^{18}}$

Example 3 **Use a calculator to approximate each value to three decimal places.**

7. $\sqrt{58}$
8. $-\sqrt{76}$
9. $\sqrt[5]{-43}$
10. $\sqrt[4]{71}$

11. **CCSS PERSEVERANCE** The radius r of the orbit of a television satellite is given by $\sqrt[3]{\dfrac{GMt^2}{4\pi^2}}$, where G is the universal gravitational constant, M is the mass of Earth, and t is the time it takes the satellite to complete one orbit. Find the radius of the satellite's orbit if G is $6.67 \times 10^{-11} N \cdot m^2/kg^2$, M is 5.98×10^{24} kg, and t is 2.6×10^6 seconds.

Practice and Problem Solving

Extra Practice is on page R6.

Examples 1–2 **Simplify.**

12. $\pm\sqrt{121x^4y^{16}}$
13. $\pm\sqrt{225a^{16}b^{36}}$
14. $\pm\sqrt{49x^4}$
15. $-\sqrt{16c^4d^2}$
16. $-\sqrt{81a^{16}b^{20}c^{12}}$
17. $-\sqrt{400x^{32}y^{40}}$
18. $\sqrt{(x+15)^4}$
19. $\sqrt{(x^2+6)^{16}}$
20. $\sqrt{(a^2+4a)^{12}}$
21. $\sqrt[3]{8a^6b^{12}}$
22. $\sqrt[6]{d^{24}x^{36}}$
23. $\sqrt[3]{27b^{18}c^{12}}$
24. $-\sqrt{(2x+1)^6}$
25. $\sqrt{-(x+2)^8}$
26. $\sqrt[3]{-(y-9)^9}$
27. $\sqrt[6]{x^{18}}$
28. $\sqrt[4]{a^{12}}$
29. $\sqrt[3]{a^{12}}$
30. $\sqrt[4]{81(x+4)^4}$
31. $\sqrt[3]{(4x-7)^{24}}$
32. $\sqrt[3]{(y^3+5)^{18}}$
33. $\sqrt[4]{256(5x-2)^{12}}$
34. $\sqrt[8]{x^{16}y^8}$
35. $\sqrt[5]{32a^{15}b^{10}}$

Example 3

36. **SHIPPING** An online book store wants to increase the size of the boxes it uses to ship orders. The new volume N is equal to the old volume V times the scale factor F cubed, or $N = V \cdot F^3$. What is the scale factor if the old volume was 0.8 cubic feet and the new volume is 21.6 cubic feet?

37. **GEOMETRY** The side length of a cube is determined by $r = \sqrt[3]{V}$, where V is the volume in cubic units. Determine the side length of a cube with a volume of 512 cm^3.

Use a calculator to approximate each value to three decimal places.

38. $\sqrt{92}$
39. $-\sqrt{150}$
40. $\sqrt{0.43}$
41. $\sqrt{0.62}$
42. $\sqrt[3]{168}$
43. $\sqrt[5]{-4382}$
44. $\sqrt[6]{(8912)^2}$
45. $\sqrt[5]{(4756)^2}$

46. **GEOMETRY** The radius r of a sphere with volume V can be found using the formula $r = \sqrt[3]{\frac{3V}{4\pi}}$.
    a. Determine the radius for volumes of 1000 cm^3, 8000 cm^3, and 64,000 cm^3.
    b. How does the volume of the sphere change if the radius is doubled? Explain.

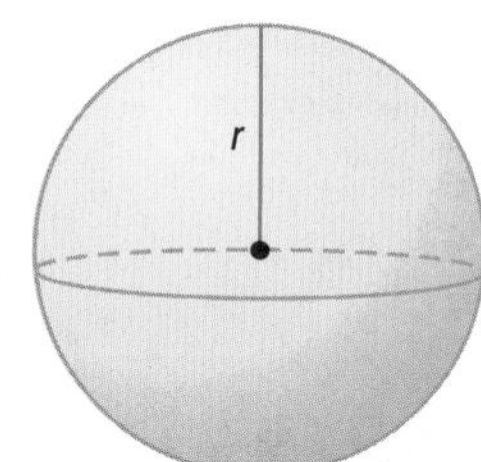

Simplify.

47. $\sqrt{196c^6d^4}$
48. $\sqrt{-64y^8z^6}$
49. $\sqrt[3]{-27a^{15}b^9}$
50. $\sqrt[4]{-16x^{16}y^8}$
51. $\sqrt{400x^{16}y^6}$
52. $\sqrt[3]{8c^3d^{12}}$
53. $\sqrt[3]{64(x+y)^6}$
54. $\sqrt[5]{-(y-z)^{15}}$

55. **PHYSICS** Johannes Kepler developed the formula $d = \sqrt[3]{6t^2}$, where d is the distance of a planet from the Sun in millions of miles and t is the number of Earth-days that it takes for the planet to orbit the Sun. If the length of a year on Mars is 687 Earth-days, how far from the Sun is Mars?

56. CCSS **SENSE-MAKING** All matter is composed of atoms. The nucleus of an atom is the center portion of the atom that contains most of the mass of the atom. A theoretical formula for the radius r of the nucleus of an atom is $r = (1.3 \times 10^{-15})\sqrt[3]{A}$ meters, where A is the mass number of the nucleus. Find the radius of the nucleus for each atom in the table.

Atom	Mass Number
carbon	6
oxygen	8
sodium	11
aluminum	13
chlorine	17

57. **BIOLOGY** Kleiber's Law, $P = 73.3\sqrt[4]{m^3}$, shows the relationship between the mass m in kilograms of an organism and its metabolism P in Calories per day. Determine the metabolism for each of the animals listed at the right.

Animal	Mass (kg)
bald eagle	4.5
golden retriever	30
komodo dragon	72
bottlenose dolphin	156
Asian elephant	2300

58. **MULTIPLE REPRESENTATIONS** In this problem, you will use $f(x) = x^n$ and $g(x) = \sqrt[n]{x}$ to explore inverses.

a. **Tabular** Make tables for $f(x)$ and $g(x)$ using $n = 3$ and $n = 4$.

b. **Graphical** Graph the equations.

c. **Analytical** Which equations are functions? Which functions are one-to-one?

d. **Analytical** For what values of n are $g(x)$ and $f(x)$ inverses of each other?

e. **Verbal** What conclusions can you make about $g(x) = \sqrt[n]{x}$ and $f(x) = x^n$ for all positive even values of n? for odd values of n?

H.O.T. Problems Use Higher-Order Thinking Skills

59. **CCSS CRITIQUE** Ashley and Kimi are simplifying $\sqrt[4]{16x^4y^8}$. Is either of them correct? Explain your reasoning.

Ashley

$\sqrt[4]{16x^4y^8} = \sqrt[4]{(2xy^2)^4}$

$= 2|xy^2|$

Kimi

$\sqrt[4]{16x^4y^8} = \sqrt[4]{(2xy^2)^4}$

$= 2y^2|x|$

60. **CHALLENGE** Under what conditions is $\sqrt{x^2 + y^2} = x + y$ true?

61. **REASONING** Determine whether the statement $\sqrt[4]{(-x)^4} = x$ is *sometimes*, *always*, or *never* true.

62. **CHALLENGE** For what real values of x is $\sqrt[3]{x} > x$?

63. **OPEN ENDED** Write a number for which the principal square root and cube root are both integers.

64. **WRITING IN MATH** Explain when and why absolute value symbols are needed when taking an nth root.

65. **CHALLENGE** Write an equivalent expression for $\sqrt[3]{2x} \cdot \sqrt[3]{8y}$. Simplify the radical.

CHALLENGE **Simplify each expression.**

66. $\sqrt[4]{0.0016}$

67. $\sqrt[7]{-0.0000001}$

68. $\dfrac{\sqrt[5]{-0.00032}}{\sqrt[3]{-0.027}}$

69. **CHALLENGE** Solve $-\dfrac{5}{\sqrt{a}} = -125$ for a.

Standardized Test Practice

70. What is the value of w in the equation $\frac{1}{2}(4w + 36) = 3(4w - 3)$?

A 2

B 2.7

C 27

D 36

71. What is the product of the complex numbers $(5 + i)$ and $(5 - i)$?

F 24

G 26

H $25 - i$

J $26 - 10i$

72. EXTENDED RESPONSE A cylindrical cooler has a diameter of 9 inches and a height of 11 inches. Tate plans to use it for soda cans that have a diameter of 2.5 inches and a height of 4.75 inches.

a. Tate plans to place two layers consisting of 9 cans each into the cooler. What is the volume of the space that will not be filled with the cans?

b. Find the ratio of the volume of the cooler to the volume of the cans in part **a**.

73. SAT/ACT Which of the following is closest to $\sqrt[3]{7.32}$?

A 1.8

B 1.9

C 2.0

D 2.1

E 2.2

Spiral Review

Graph each function. (Lesson 6-3)

74. $y = \sqrt{x - 5}$

75. $y = \sqrt{x} - 2$

76. $y = 3\sqrt{x} + 4$

77. HEALTH The average weight of a baby born at a certain hospital is $7\frac{1}{2}$ pounds and the average length is 19.5 inches. One kilogram is about 2.2 pounds and 1 centimeter is about 0.3937 inches. Find the average weight in kilograms and the length in centimeters. (Lesson 6-2)

Simplify. (Lesson 5-1)

78. $(4c - 5) - (c + 11) + (-6c + 17)$

79. $(11x^2 + 13x - 15) - (7x^2 - 9x + 19)$

80. $(d - 5)(d + 3)$

81. $(2a^2 + 6)^2$

82. GAS MILEAGE The gas mileage y in miles per gallon for a certain vehicle is given by the equation $y = 10 + 0.9x - 0.01x^2$, where x is the speed of the vehicle between 10 and 75 miles per hour. Find the range of speeds that would give a gas mileage of at least 25 miles per gallon. (Lesson 4-8)

Write each equation in vertex form, if not already in that form. Identify the vertex, axis of symmetry, and direction of opening. Then graph the function. (Lesson 4-7)

83. $y = -6(x + 2)^2 + 3$

84. $y = -\frac{1}{3}x^2 + 8x$

85. $y = (x - 2)^2 - 2$

86. $y = 2x^2 + 8x + 10$

Skills Review

Find each product.

87. $(x + 4)(x + 5)$

88. $(y - 3)(y + 4)$

89. $(a + 2)(a - 9)$

90. $(a - b)(a - 3b)$

91. $(x + 2y)(x - y)$

92. $2(w + z)(w - 4z)$

EXTEND 6-4

Graphing Technology Lab
Graphing *n*th Root Functions

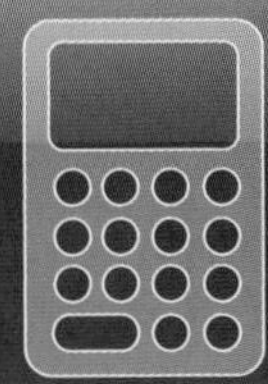

You can use a graphing calculator to graph *n*th root functions.

CCSS Common Core State Standards
Content Standards
F.IF.7.b Graph square root, cube root, and piecewise-defined functions, including step functions and absolute value functions.
F.BF.3 Identify the effect on the graph of replacing $f(x)$ by $f(x) + k$, $k\,f(x)$, $f(kx)$, and $f(x + k)$ for specific values of k (both positive and negative); find the value of k given the graphs. Experiment with cases and illustrate an explanation of the effects on the graph using technology.
Mathematical Practices
5 Use appropriate tools strategically.

Example 1 Graph an *n*th Root Function

Graph $y = \sqrt[5]{x}$.

Enter the equation as **Y1** and graph.

KEYSTROKES: [Y=] 5 [MATH] 5 [X,T,θ,*n*] [GRAPH]

Another way to enter the equation is to use $y = x^{\frac{1}{5}}$. You will learn about this later in Chapter 6.

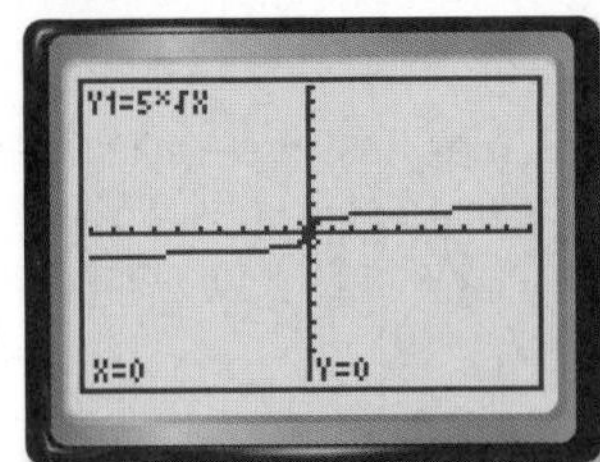

[−10, 10] scl: 1 by [−10, 10] scl: 1

Example 2 *n*th Root Functions with Different Roots

Graph and compare $y = \sqrt{x}$ and $y = \sqrt[4]{x}$.

Enter $y = \sqrt{x}$ as **Y1** and $y = \sqrt[4]{x}$ as **Y2**. Change the viewing window. Then graph.

KEYSTROKES: [Y=] [CLEAR] [2nd] [√] [X,T,θ,*n*] [ENTER] 4
[MATH] 5 [X,T,θ,*n*] [GRAPH]
[WINDOW] [(−)] 2 [ENTER] 10 [ENTER] 1 [ENTER]
[(−)] 2 [ENTER] 10 [ENTER] 1

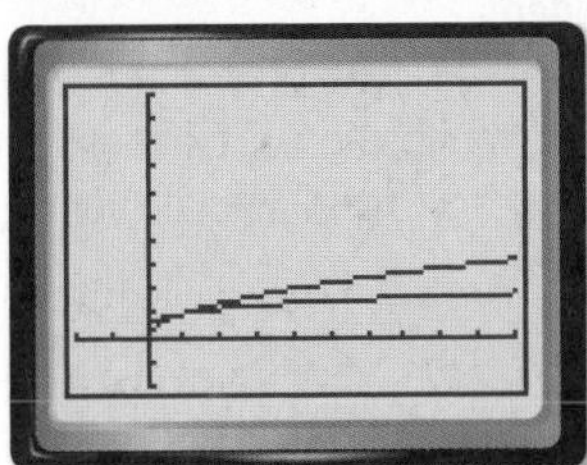

[−2, 10] scl: 1 by [−2, 10] scl: 1

Example 3 *n*th Root Functions with Different Radicands

Graph and compare $y = \sqrt[3]{x}$, $y = \sqrt[3]{x + 4}$, and $y = \sqrt[3]{x} + 4$.

Enter $y = \sqrt[3]{x}$ as **Y1**, $y = \sqrt[3]{x + 4}$ as **Y2**, and $y = \sqrt[3]{x} + 4$ as **Y3**. Then graph in the standard viewing window.

KEYSTROKES: [Y=] [CLEAR] 3 [MATH] 5 [X,T,θ,*n*] [ENTER] [CLEAR] 3
[MATH] 5 [(] [X,T,θ,*n*] [+] 4 [)] [ENTER] 3 [MATH]
5 [X,T,θ,*n*] [+] 4 [ENTER] [ZOOM] 6

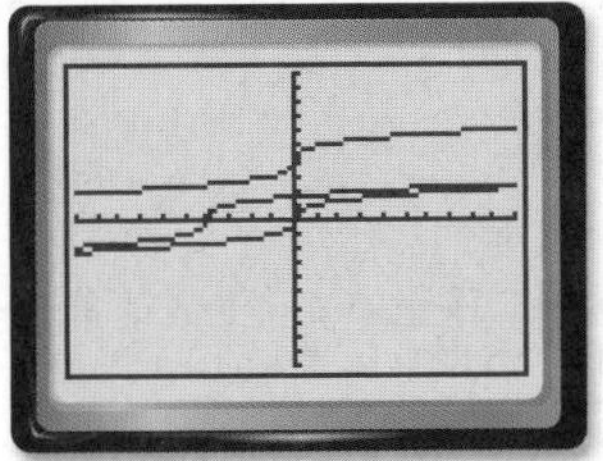

[−10, 10] scl: 1 by [−10, 10] scl: 1

Exercises

Graph each function.

1. $y = \sqrt[4]{x}$
2. $y = \sqrt[4]{x + 2}$
3. $y = \sqrt[4]{x} + 2$
4. $y = \sqrt[5]{x}$
5. $y = \sqrt[5]{x - 5}$
6. $y = \sqrt[5]{x} - 5$
7. What is the effect of adding or subtracting a constant under the radical sign?
8. What is the effect of adding or subtracting a constant outside the radical sign?

CHAPTER 6

Mid-Chapter Quiz

Lessons 6-1 through 6-4

Given $f(x) = 2x^2 + 4x - 3$ and $g(x) = 5x - 2$, find each function. (Lesson 6-1)

1. $(f + g)(x)$
2. $(f - g)(x)$
3. $(f \cdot g)(x)$
4. $\left(\frac{f}{g}\right)(x)$
5. $[f \circ g](x)$
6. $[g \circ f](x)$

7. **SHOPPING** Mrs. Ross is shopping for her children's school clothes. She has a coupon for 25% off her total. The sales tax of 6% is added to the total after the coupon is applied. (Lesson 6-1)

   a. Express the total price after the discount and the total price after the tax using function notation. Let x represent the price of the clothing, $p(x)$ represent the price after the 25% discount, and $g(x)$ represent the price after the tax is added.

   b. Which composition of functions represents the final price, $p[g(x)]$ or $g[p(x)]$? Explain your reasoning.

Determine whether each pair of functions are inverse functions. Write *yes* or *no*. (Lesson 6-2)

8. $f(x) = 2x + 16$
   $g(x) = \frac{1}{2}x - 8$
9. $g(x) = 4x + 15$
   $h(x) = \frac{1}{4}x - 15$
10. $f(x) = x^2 - 5$
    $g(x) = 5 + x^{-2}$
11. $g(x) = -6x + 8$
    $h(x) = \frac{8 - x}{6}$

Find the inverse of each function, if it exists. (Lesson 6-2)

12. $h(x) = \frac{2}{5}x + 8$
13. $f(x) = \frac{4}{9}(x - 3)$
14. $h(x) = -\frac{10}{3}(x + 5)$
15. $f(x) = \frac{x + 12}{7}$

16. **JOBS** Louise runs a lawn care service. She charges $25 for supplies plus $15 per hour. The function $f(h) = 15h + 25$ gives the cost $f(h)$ for h hours of work. (Lesson 6-2)

    a. Find $f^{-1}(h)$. What is the significance of $f^{-1}(h)$?

    b. If Louise charges a customer $85, how many hours did she work?

Graph each inequality. (Lesson 6-3)

17. $y < \sqrt{x - 5}$
18. $y \leq -2\sqrt{x}$
19. $y > \sqrt{x + 9} + 3$
20. $y \geq \sqrt{x + 4} - 5$

Graph each function. State the domain and range of each function. (Lesson 6-3)

21. $y = 2 + \sqrt{x}$
22. $y = \sqrt{x + 4} - 1$

23. **MULTIPLE CHOICE** What is the domain of $f(x) = \sqrt{2x + 5}$? (Lesson 6-3)

    A $\left\{x \middle| x > \frac{5}{2}\right\}$
    B $\left\{x \middle| x > -\frac{5}{2}\right\}$
    C $\left\{x \middle| x \geq \frac{5}{2}\right\}$
    D $\left\{x \middle| x \geq -\frac{5}{2}\right\}$

Simplify. (Lesson 6-4)

24. $\pm\sqrt{121a^4b^{18}}$
25. $\sqrt{(x^4 + 3)^{12}}$
26. $\sqrt[3]{27(2x - 5)^{15}}$
27. $\sqrt[5]{-(y - 6)^{20}}$
28. $\sqrt[3]{8(x + 4)^6}$
29. $\sqrt[4]{16(y + x)^8}$

30. **MULTIPLE CHOICE** The radius of the cylinder below is equal to the height of the cylinder. The radius r can be found using the formula $r = \sqrt[3]{\frac{V}{\pi}}$. Find the radius of the cylinder if the volume is 500 cubic inches. (Lesson 6-4)

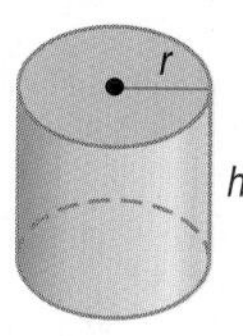

    F 2.53 inches
    G 5.42 inches
    H 7.94 inches
    J 24.92 inches

31. **PRODUCTION** The cost in dollars of producing p cell phones in a factory is represented by $C(p) = 5p + 60$. The number of cell phones produced in h hours is represented by $P(h) = 40h$. (Lesson 6-1)

    a. Find the composition function.

    b. Determine the cost of producing cell phones for 8 hours.

LESSON

6-5 Operations with Radical Expressions

Then

- You simplified expressions with nth roots.

Now

1. Simplify radical expressions.
2. Add, subtract, multiply, and divide radical expressions.

Why?

- Golden rectangles have been used by artists and architects to create beautiful designs. Many golden rectangles appear in the Parthenon in Athens, Greece. The ratio of the lengths of the sides of a golden rectangle is $\frac{2}{\sqrt{5}-1}$. In this lesson, you will learn to simplify radical expressions like $\frac{2}{\sqrt{5}-1}$.

NewVocabulary

rationalizing the denominator
like radical expressions
conjugate

Common Core State Standards

Content Standards

A.SSE.2 Use the structure of an expression to identify ways to rewrite it.

Mathematical Practices

1 Make sense of problems and persevere in solving them.

1 Simplify Radicals

The properties you have used to simplify radical expressions involving square roots also hold true for expressions involving nth roots.

KeyConcept Product Property of Radicals

Words	For any real numbers a and b and any integer $n > 1$, $\sqrt[n]{ab} = \sqrt[n]{a} \cdot \sqrt[n]{b}$, if n is even and a and b are both nonnegative or if n is odd.
Examples	$\sqrt{2} \cdot \sqrt{8} = \sqrt{16}$ or 4 and $\sqrt[3]{3} \cdot \sqrt[3]{9} = \sqrt[3]{27}$ or 3

In order for a radical to be in simplest form, the radicand must contain no factors that are nth powers of an integer or polynomial.

Example 1 Simplify Expressions with the Product Property

Simplify.

a. $\sqrt{32x^8}$

$\sqrt{32x^8} = \sqrt{4^2 \cdot 2 \cdot (x^4)^2}$ Factor into squares.

$= \sqrt{4^2} \cdot \sqrt{(x^4)^2} \cdot \sqrt{2}$ Product Property of Radicals

$= 4x^4\sqrt{2}$ Simplify.

b. $\sqrt[4]{16a^{24}b^{13}}$

$\sqrt[4]{16a^{24}b^{13}} = \sqrt[4]{2^4 \cdot (a^6)^4(b^3)^4 \cdot b}$ Factor into squares.

$= \sqrt[4]{2^4} \cdot \sqrt[4]{(a^6)^4} \cdot \sqrt[4]{(b^3)^4} \cdot \sqrt[4]{b}$ Product Property of Radicals

$= 2a^6|b^3|\sqrt[4]{b}$ Simplify.

In this case, the absolute value symbols are not necessary because in order for $\sqrt[4]{16a^{24}b^{13}}$ to be defined, b must be nonnegative.

Thus, $\sqrt[4]{16a^{24}b^{13}} = 2a^6b^3\sqrt[4]{b}$.

GuidedPractice

1A. $\sqrt{12c^6d^3}$

1B. $\sqrt[3]{27y^{12}z^7}$

Photodisc/PunchStock

The Quotient Property of Radicals is another property used to simplify radicals.

KeyConcept Quotient Property of Radicals

Words For any real numbers a and $b \neq 0$ and any integer $n > 1$,
$\sqrt[n]{\frac{a}{b}} = \frac{\sqrt[n]{a}}{\sqrt[n]{b}}$, if all roots are defined.

Examples $\frac{\sqrt{27}}{\sqrt{3}} = \sqrt{9}$ or 3 $\quad \sqrt[3]{\frac{x^6}{8}} = \frac{\sqrt[3]{x^6}}{\sqrt[3]{8}} = \frac{x^2}{2}$ or $\frac{1}{2}x^2$

To eliminate radicals from a denominator or fractions from a radicand, you can use a process called **rationalizing the denominator**. To rationalize a denominator, multiply the numerator and denominator by a quantity so that the radicand has an exact root.

If the denominator is:	Multiply the numerator and denominator by:	Examples
$\sqrt{b}$	$\sqrt{b}$	$\frac{2}{\sqrt{3}} = \frac{2}{\sqrt{3}} \cdot \frac{\sqrt{3}}{\sqrt{3}}$ or $\frac{2\sqrt{3}}{3}$
$\sqrt[n]{b^x}$	$\sqrt[n]{b^{n-x}}$	$\frac{5}{\sqrt[3]{2}} = \frac{5}{\sqrt[3]{2}} \cdot \frac{\sqrt[3]{2^2}}{\sqrt[3]{2^2}}$ or $\frac{5\sqrt[3]{4}}{2}$

Example 2 Simplify Expressions with the Quotient Property

Simplify.

a. $\sqrt{\frac{x^6}{y^7}}$

$\sqrt{\frac{x^6}{y^7}} = \frac{\sqrt{x^6}}{\sqrt{y^7}}$ Quotient Property

$= \frac{\sqrt{(x^3)^2}}{\sqrt{(y^3)^2 \cdot y}}$ Factor into squares.

$= \frac{\sqrt{(x^3)^2}}{\sqrt{(y^3)^2} \cdot \sqrt{y}}$ Product Property

$= \frac{|x^3|}{y^3\sqrt{y}}$ Simplify.

$= \frac{|x^3|}{y^3\sqrt{y}} \cdot \frac{\sqrt{y}}{\sqrt{y}}$ Rationalize the denominator.

$= \frac{|x^3|\sqrt{y}}{y^4}$ $\sqrt{y} \cdot \sqrt{y} = y$

b. $\sqrt[4]{\frac{6}{5x}}$

$\sqrt[4]{\frac{6}{5x}} = \frac{\sqrt[4]{6}}{\sqrt[4]{5x}}$ Quotient Property

$= \frac{\sqrt[4]{6}}{\sqrt[4]{5x}} \cdot \frac{\sqrt[4]{5^3x^3}}{\sqrt[4]{5^3x^3}}$ Rationalize the denominator.

$= \frac{\sqrt[4]{6 \cdot 5^3x^3}}{\sqrt[4]{5x \cdot 5^3x^3}}$ Product Property

$= \frac{\sqrt[4]{750x^3}}{\sqrt[4]{5^4x^4}}$ Multiply.

$= \frac{\sqrt[4]{750x^3}}{5x}$ $\sqrt[4]{5^4x^4} = 5x$

GuidedPractice

2A. $\frac{\sqrt{a^9}}{\sqrt{b^5}}$

2B. $\sqrt[5]{\frac{3}{4y}}$

Here is a summary of the rules used to simplify radicals.

ConceptSummary Simplifying Radical Expressions

A radical expression is in simplified form when the following conditions are met.

- The index n is as small as possible.
- The radicand contains no factors (other than 1) that are nth powers of an integer or polynomial.
- The radicand contains no fractions.
- No radicals appear in a denominator.

2 Operations with Radicals

Operations with Radicals You can use the Product and Quotient Properties to multiply and divide some radicals.

StudyTip

CCSS Regularity Exact roots occur when the powers of the constants and variables are all identical to or multiples of the index. For example, $\sqrt[3]{2} \cdot \sqrt[3]{2^2} = \sqrt[3]{2^3}$ or 2.

Example 3 Multiply Radicals

Simplify $5\sqrt[3]{-12ab^4} \cdot 3\sqrt[3]{18a^2b^2}$.

$5\sqrt[3]{-12ab^4} \cdot 3\sqrt[3]{18a^2b^2} = 5 \cdot 3 \cdot \sqrt[3]{-12ab^4 \cdot 18a^2b^2}$	Product Property of Radicals
$= 15 \cdot \sqrt[3]{-2^2 \cdot 3 \cdot ab^4 \cdot 2 \cdot 3^2 \cdot a^2b^2}$	Factor constants.
$= 15 \cdot \sqrt[3]{-2^3 \cdot 3^3 \cdot a^3b^6}$	Group into cubes if possible.
$= 15 \cdot \sqrt[3]{-2^3} \cdot \sqrt[3]{3^3} \cdot \sqrt[3]{a^3} \cdot \sqrt[3]{b^6}$	Product Property of Radicals
$= 15 \cdot (-2) \cdot 3 \cdot a \cdot b^2$	Simplify.
$= -90ab^2$	Multiply.

GuidedPractice

Simplify.

3A. $6\sqrt{8c^3d^5} \cdot 4\sqrt{2cd^3}$

3B. $2\sqrt[4]{8x^3y^2} \cdot 3\sqrt[4]{2x^5y^2}$

Radicals can be added and subtracted in the same manner as monomials. In order to add or subtract, the radicals must be like terms. Radicals are **like radical expressions** if *both* the index and the radicand are identical.

Like: $\sqrt{3b}$ and $4\sqrt{3b}$ | Unlike: $\sqrt{3b}$ and $\sqrt[3]{3b}$ | Unlike: $\sqrt{2b}$ and $\sqrt{3b}$

StudyTip

Adding and Subtracting Radicals Simplify the individual radicals before attempting to combine like terms.

Example 4 Add and Subtract Radicals

Simplify $\sqrt{98} - 2\sqrt{32}$.

$\sqrt{98} - 2\sqrt{32} = \sqrt{2 \cdot 7^2} - 2\sqrt{4^2 \cdot 2}$	Factor using squares.
$= \sqrt{7^2} \cdot \sqrt{2} - 2 \cdot \sqrt{4^2} \cdot \sqrt{2}$	Product Property
$= 7\sqrt{2} - 2 \cdot 4 \cdot \sqrt{2}$	Simplify radicals.
$= 7\sqrt{2} - 8\sqrt{2}$	Multiply.
$= -\sqrt{2}$	$(7 - 8)\sqrt{2} = (-1)(\sqrt{2})$

GuidedPractice

4A. $4\sqrt{8} + 3\sqrt{50}$

4B. $5\sqrt{12} + 2\sqrt{27} - \sqrt{128}$

Just as you can add and subtract radicals like monomials, you can multiply radicals using the FOIL method as you do when multiplying binomials.

Example 5 Multiply Radicals

Simplify $(4\sqrt{3} + 5\sqrt{2})(3\sqrt{2} - 6)$.

$$(4\sqrt{3} + 5\sqrt{2})(3\sqrt{2} - 6) = \overset{F}{4\sqrt{3} \cdot 3\sqrt{2}} + \overset{O}{4\sqrt{3} \cdot (-6)} + \overset{I}{5\sqrt{2} \cdot 3\sqrt{2}} + \overset{L}{5\sqrt{2} \cdot (-6)}$$

$$= 12\sqrt{3 \cdot 2} - 24\sqrt{3} + 15\sqrt{2^2} - 30\sqrt{2} \quad \text{Product Property}$$

$$= 12\sqrt{6} - 24\sqrt{3} + 30 - 30\sqrt{2} \quad \text{Simplify.}$$

GuidedPractice

Simplify.

5A. $(6\sqrt{3} - 5)(2\sqrt{5} + 4\sqrt{2})$

5B. $(7\sqrt{2} - 3\sqrt{3})(7\sqrt{2} + 3\sqrt{3})$

StudyTip

Conjugates The product of conjugates is always a rational number.

Binomials of the form $a\sqrt{b} + c\sqrt{d}$ and $a\sqrt{b} - c\sqrt{d}$, where a, b, c, and d are rational numbers, are called **conjugates** of each other. You can use conjugates to rationalize denominators.

Real-World Example 6 Use a Conjugate to Rationalize a Denominator

ARCHITECTURE Refer to the beginning of the lesson. Use a conjugate to rationalize the denominator and simplify $\frac{2}{\sqrt{5} - 1}$.

$$\frac{2}{\sqrt{5} - 1} = \frac{2}{\sqrt{5} - 1} \cdot \frac{\sqrt{5} + 1}{\sqrt{5} + 1} \quad \sqrt{5} + 1 \text{ is the conjugate of } \sqrt{5} - 1.$$

$$= \frac{2\sqrt{5} + 2(1)}{(\sqrt{5})^2 + 1(\sqrt{5}) - 1(\sqrt{5}) - 1(1)} \quad \text{Multiply.}$$

$$= \frac{2\sqrt{5} + 2}{5 + \sqrt{5} - \sqrt{5} - 1} \quad \text{Simplify.}$$

$$= \frac{2\sqrt{5} + 2}{4} \quad \text{Subtract.}$$

$$= \frac{\sqrt{5} + 1}{2} \quad \text{Simplify.}$$

GuidedPractice

6. GEOMETRY The area of the rectangle at the right is 900 ft^2. Write and simplify an equation for L in terms of x.

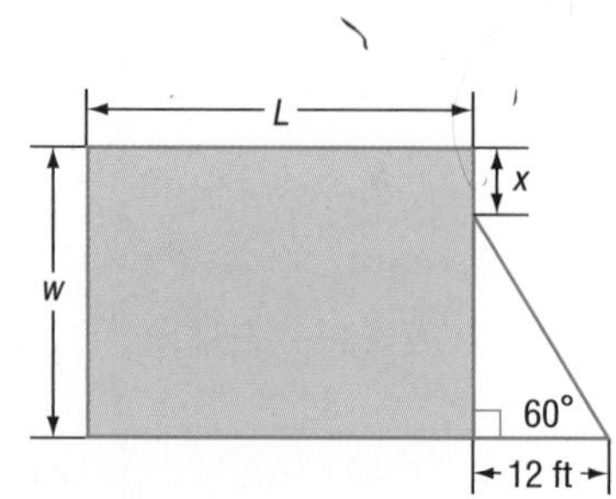

Math HistoryLink

Theano (c. 5th century B.C.) Theano is believed to have been the wife of Pythagoras. It is also believed that she directed a famous mathematics academy and carried on the work of Pythagoras after his death. Her most important work was on the idea of the golden mean, which is the irrational number $\frac{1 + \sqrt{5}}{2}$.

Source: The Granger Collection, New York

SuperStock

Check Your Understanding

● = Step-by-Step Solutions begin on page R14.

Examples 1–5 CCSS **PRECISION** Simplify.

1. $\sqrt{36ab^4c^5}$
2. $\sqrt{144x^7y^5}$
3. $\dfrac{\sqrt{c^5}}{\sqrt{d^9}}$
4. $\sqrt[4]{\dfrac{5x}{8y}}$
5. $5\sqrt{2x} \cdot 3\sqrt{8x}$
6. $4\sqrt{5a^5} \cdot \sqrt{125a^3}$
7. $3\sqrt[3]{36xy} \cdot 2\sqrt[3]{6x^2y^2}$
8. $\sqrt[4]{3x^3y^2} \cdot \sqrt[4]{27xy^2}$
9. $5\sqrt{32} + \sqrt{27} + 2\sqrt{75}$
10. $4\sqrt{40} + 3\sqrt{28} - \sqrt{200}$
11. $(4 + 2\sqrt{5})(3\sqrt{3} + 4\sqrt{5})$
12. $(8\sqrt{3} - 2\sqrt{2})(8\sqrt{3} + 2\sqrt{2})$
13. $\dfrac{5}{\sqrt{2} + 3}$
14. $\dfrac{8}{\sqrt{6} - 5}$
15. $\dfrac{4 + \sqrt{2}}{\sqrt{2} - 3}$
16. $\dfrac{6 - \sqrt{3}}{\sqrt{3} + 4}$

Example 6

17. **GEOMETRY** Find the altitude of the triangle if the area is $189 + 4\sqrt{3}$ square centimeters.

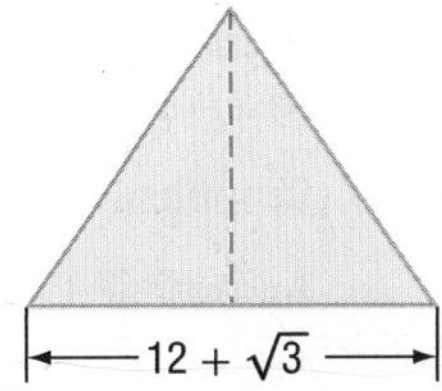

Practice and Problem Solving

Extra Practice is on page R6.

Examples 1–4 Simplify.

18. $\sqrt{72a^8b^5}$
19. $\sqrt{9a^{15}b^3}$
20. $\sqrt{24a^{16}b^8c}$
21. $\sqrt{18a^6b^3c^5}$
22. $\dfrac{\sqrt{5a^5}}{\sqrt{b^{13}}}$
23. $\sqrt{\dfrac{7x}{10y^3}}$
24. $\dfrac{\sqrt[3]{6x^2}}{\sqrt[3]{5y}}$
25. $\sqrt[4]{\dfrac{7x^3}{4b^2}}$
26. $3\sqrt{5y} \cdot 8\sqrt{10yz}$
27. $2\sqrt{32a^3b^5} \cdot \sqrt{8a^7b^2}$
28. $6\sqrt{3ab} \cdot 4\sqrt{24ab^3}$
29. $5\sqrt{x^8y^3} \cdot 5\sqrt{2x^5y^4}$
30. $3\sqrt{90} + 4\sqrt{20} + \sqrt{162}$
31. $9\sqrt{12} + 5\sqrt{32} - \sqrt{72}$
32. $4\sqrt{28} - 8\sqrt{810} + \sqrt{44}$
33. $3\sqrt{54} + 6\sqrt{288} - \sqrt{147}$
34. **GEOMETRY** Find the perimeter of the rectangle.
35. **GEOMETRY** Find the area of the rectangle.
36. **GEOMETRY** Find the exact surface area of a sphere with radius of $4 + \sqrt{5}$ inches.

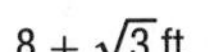

Examples 5–6 Simplify.

37. $(7\sqrt{2} - 3\sqrt{3})(4\sqrt{6} + 3\sqrt{12})$
38. $(8\sqrt{5} - 6\sqrt{3})(8\sqrt{5} + 6\sqrt{3})$
39. $(12\sqrt{10} - 6\sqrt{5})(12\sqrt{10} + 6\sqrt{5})$
40. $(6\sqrt{3} + 5\sqrt{2})(2\sqrt{6} + 3\sqrt{8})$
41. $\dfrac{6}{\sqrt{3} - \sqrt{2}}$
42. $\dfrac{\sqrt{2}}{\sqrt{5} - \sqrt{3}}$
43. $\dfrac{9 - 2\sqrt{3}}{\sqrt{3} + 6}$
44. $\dfrac{2\sqrt{2} + 2\sqrt{5}}{\sqrt{5} + \sqrt{2}}$

Simplify.

45. $\sqrt[3]{16y^4z^{12}}$

46. $\sqrt[3]{-54x^6y^{11}}$

47. $\sqrt[4]{162a^6b^{13}c}$

48. $\sqrt[4]{48a^9b^3c^{16}}$

49. $\sqrt[4]{\dfrac{12x^3y^2}{5a^2b}}$

50. $\dfrac{\sqrt[3]{36xy^2}}{\sqrt[3]{10xz}}$

51. $\dfrac{x+1}{\sqrt{x}-1}$

52. $\dfrac{x-2}{\sqrt{x^2-4}}$

53. $\dfrac{\sqrt{x}}{\sqrt{x^2-1}}$

54. **APPLES** The diameter of an apple is related to its weight and can be modeled by the formula $d = \sqrt[3]{3w}$, where d is the diameter in inches and w is the weight in ounces. Find the diameter of an apple that weighs 6.47 ounces.

Simplify each expression if b is an even number.

55. $\sqrt[b]{a^b}$

56. $\sqrt[b]{a^{4b}}$

57. $\sqrt[b]{a^{2b}}$

58. $\sqrt[b]{a^{3b}}$

59. **MULTIPLE REPRESENTATIONS** In this problem, you will explore operations with like radicals.

   a. **Numerical** Copy the diagram at the right on dot paper. Use the Pythagorean Theorem to prove that the length of the red segment is $\sqrt{2}$ units.

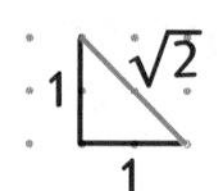

   b. **Graphical** Extend the segment to represent $\sqrt{2} + \sqrt{2}$.

   c. **Analytical** Use your drawing to show that $\sqrt{2} + \sqrt{2} \neq \sqrt{2+2}$ or 2.

   d. **Graphical** Use the dot paper to draw a square with side lengths $\sqrt{2}$ units.

   e. **Numerical** Prove that the area of the square is $\sqrt{2} \cdot \sqrt{2} = 2$ square units.

H.O.T. Problems Use Higher-Order Thinking Skills

60. **ERROR ANALYSIS** Twyla and Ben are simplifying $4\sqrt{32} + 6\sqrt{18}$. Is either of them correct? Explain your reasoning.

Twyla

$4\sqrt{32} + 6\sqrt{18}$

$= 4\sqrt{4^2 \cdot 2} + 6\sqrt{3^2 \cdot 2}$

$= 16\sqrt{2} + 18\sqrt{2}$

$= 34\sqrt{2}$

Ben

$4\sqrt{32} + 6\sqrt{18}$

$= 4\sqrt{16 \cdot 2} + 6\sqrt{9 \cdot 2}$

$= 64\sqrt{2} + 54\sqrt{2}$

$= 118\sqrt{2}$

61. **CHALLENGE** Show that $\dfrac{-1 - i\sqrt{3}}{2}$ is a cube root of 1.

62. **CCSS ARGUMENTS** For what values of a is $\sqrt{a} \cdot \sqrt{-a}$ a real number? Explain.

63. **CHALLENGE** Find four combinations of whole numbers that satisfy $\sqrt[a]{256} = b$.

64. **OPEN ENDED** Find a number other than 1 that has a positive whole number for a square root, cube root, and fourth root.

65. **WRITING IN MATH** Explain why absolute values may be unnecessary when an nth root of an even power results in an odd power.

Standardized Test Practice

66. PROBABILITY A six-sided number cube has faces with the numbers 1 though 6 marked on it. What is the probability that a number less than 4 will occur on one toss of the number cube?

A $\frac{1}{2}$ C $\frac{1}{4}$

B $\frac{1}{3}$ D $\frac{1}{5}$

67. When the number of a year is divisible by 4, the year is a leap year. However, when the year is divisible by 100, the year is not a leap year, unless the year is divisible by 400. Which is *not* a leap year?

F 1884 H 1904

G 1900 J 1940

68. SHORT RESPONSE Which property is illustrated by $4x + 0 = 4x$?

69. SAT/ACT The expression $\sqrt{180a^2b^8}$ is equivalent to which of the following?

A $3\sqrt{10}\,|a|b^4$

B $5\sqrt{6}\,|a|b^4$

C $6\sqrt{5}\,|a|b^4$

D $18\sqrt{10}\,|a|b^4$

E $36\sqrt{5}\,|a|b^4$

Spiral Review

Simplify. (Lesson 6-4)

70. $\sqrt{81x^6}$ **71.** $\sqrt[3]{729a^3b^9}$ **72.** $\sqrt{(g+5)^2}$

73. Graph $y \leq \sqrt{x-2}$. (Lesson 6-3)

Solve each equation. (Lesson 5-5)

74. $x^4 - 34x^2 + 225 = 0$ **75.** $x^4 - 15x^2 - 16 = 0$ **76.** $x^4 + 6x^2 - 27 = 0$

77. $x^3 + 64 = 0$ **78.** $27x^3 + 1 = 0$ **79.** $8x^3 - 27 = 0$

80. MODELS A model car builder is building a display table for model cars. He wants the perimeter of the table to be 26 feet, but he wants the area of the table to be no more than 30 square feet. What could be the width of the table? (Lesson 4-8)

81. CONSTRUCTION Cho charges $1500 to build a small deck and $2500 to build a large deck. During the spring and summer, she built 5 more small decks than large decks. If she earned $23,500 how many of each type of deck did she build? (Lesson 3-8)

82. FOOD The Hot Dog Grille offers the lunch combinations shown. Assume that the price of a combo meal is the same price as purchasing each item separately. Find the prices for a hot dog, a soda, and a bag of potato chips. (Lesson 3-4)

Lunch Combo Meals	
1. Two hot dogs, one soda	$5.40
2. One hot dog, potato chips, one soda	$4.35
3. Two hot dogs, two bags of chips	$5.70

Skills Review

Evaluate each expression.

83. $2\left(\frac{1}{6}\right)$ **84.** $3\left(\frac{1}{8}\right)$ **85.** $\frac{1}{4} + \frac{1}{3}$

86. $\frac{1}{2} + \frac{3}{8}$ **87.** $\frac{2}{3} - \frac{1}{4}$ **88.** $\frac{5}{6} - \frac{2}{5}$

LESSON 6-6 Rational Exponents

Then

- You used properties of exponents.

Now

1. Write expressions with rational exponents in radical form and vice versa.
2. Simplify expressions in exponential or radical form.

Why?

- The formula $C = c(1 + r)^n$ can be used to estimate the future cost of an item due to inflation. C represents the future cost, c represents the current cost, r is the rate of inflation, and n is the number of years for the projection.

  For example, $C = c(1 + r)^{\frac{1}{2}}$ can be used to estimate the cost of a video game system in six months.

Common Core State Standards

Mathematical Practices

1 Make sense of problems and persevere in solving them.

1 Rational Exponents and Radicals

You know that squaring a number and taking the square root of a number are inverse operations. But how would you evaluate an expression that contains a fractional exponent such as the one above? You can investigate such an expression by assuming that fractional exponents behave as integral exponents.

$\left(b^{\frac{1}{2}}\right)^2 = b^{\frac{1}{2}} \cdot b^{\frac{1}{2}}$ Write as a multiplication expression.

$= b^{\frac{1}{2} + \frac{1}{2}}$ Add the exponents.

$= b^1$ or b Simplify.

Thus, $b^{\frac{1}{2}}$ is a number with a square equal to b. So $b^{\frac{1}{2}} = \sqrt{b}$.

KeyConcept $b^{\frac{1}{n}}$

Words For any real number b and any positive integer n, $b^{\frac{1}{n}} = \sqrt[n]{b}$, except when $b < 0$ and n is even. When $b < 0$ and n is even, a complex root may exist.

Examples $27^{\frac{1}{3}} = \sqrt[3]{27}$ or 3 $\quad (-16)^{\frac{1}{2}} = \sqrt{-16}$ or $4i$

Example 1 Radical and Exponential Forms

Simplify.

a. Write $x^{\frac{1}{6}}$ in radical form.

$x^{\frac{1}{6}} = \sqrt[6]{x}$ Definition of $b^{\frac{1}{n}}$

b. Write $\sqrt[4]{z}$ in exponential form.

$\sqrt[4]{z} = z^{\frac{1}{4}}$ Definition of $b^{\frac{1}{n}}$

GuidedPractice

1A. Write $a^{\frac{1}{5}}$ in radical form.

1B. Write $\sqrt[8]{c}$ in exponential form.

1C. Write $d^{\frac{7}{4}}$ in radical form.

1D. Write $\sqrt[3]{c^{-5}}$ in exponential form.

ReadingMath

Power Functions Square root functions are also *power functions* when the exponent is a fraction. In the case of square roots, the exponent is $\frac{1}{2}$.

The rules for negative exponents also apply to negative rational exponents.

PT

Example 2 Evaluate Expressions with Rational Exponents

Evaluate each expression.

a. $81^{-\frac{1}{4}}$

$81^{-\frac{1}{4}} = \frac{1}{81^{\frac{1}{4}}}$ — $b^{-n} = \frac{1}{b^n}$

$= \frac{1}{\sqrt[4]{81}}$ — $81^{\frac{1}{4}} = \sqrt[4]{81}$

$= \frac{1}{\sqrt[4]{3^4}}$ — $81 = 3^4$

$= \frac{1}{3}$ — Simplify.

b. $216^{\frac{2}{3}}$

$216^{\frac{2}{3}} = (6^3)^{\frac{2}{3}}$ — $216 = 6^3$

$= 6^{3 \cdot \frac{2}{3}}$ — Power of a Power

$= 6^2$ — Multiply exponents.

$= 36$ — Simplify.

GuidedPractice

2A. $-3125^{-\frac{1}{5}}$

2B. $256^{\frac{3}{8}}$

Examples 2a and 2b use the definition of $b^{\frac{1}{n}}$ and the properties of powers to evaluate an expression. Both methods suggest the following general definition of rational exponents

KeyConcept Rational Exponents

Words For any real nonzero number b, and any integers x and y, with $y > 1$, $b^{\frac{x}{y}} = \sqrt[y]{b^x} = \left(\sqrt[y]{b}\right)^x$, except when $b < 0$ and y is even. When $b < 0$ and y is even, a complex root may exist.

Examples $27^{\frac{2}{3}} = \left(\sqrt[3]{27}\right)^2 = 3^2$ or 9 $\qquad (-16)^{\frac{3}{2}} = \left(\sqrt{-16}\right)^3 = (4\boldsymbol{i})^3$ or $-64\boldsymbol{i}$

Real-World Example 3 Solve Equations with Rational Exponents

FINANCIAL LITERACY Refer to the beginning of the lesson. Suppose a video game system costs \$390 now. How much would the price increase in six months with an annual inflation rate of 5.3%?

$C = c(1 + r)^n$ — Original formula

$= 390(1 + 0.053)^{\frac{1}{2}}$ — $c = 390$, $r = 0.053$, and $n = \frac{6 \text{ months}}{12 \text{ months}}$ or $\frac{1}{2}$

≈ 400.20 — Use a calculator.

In six months the price of the video game system will be \$400.20 − \$390.00 or \$10.20 more than its current price.

GuidedPractice

3. Suppose a gallon of milk costs \$2.99 now. How much would the price increase in 9 months with an annual inflation rate of 5.3%?

Real-WorldCareer

Economist Economists are employed in banking, finance, accounting, commerce, marketing, and business administration. Politicians often consult economists before enacting policy. Economists use shopping as one way of judging the economy. A bachelor's or master's degree in economics is required.

Ariel Skelley/Blend Images/Jupiterimages

2 Simplify Expressions

Simplify Expressions All of the properties of powers you learned in Lesson 6-1 apply to rational exponents. Write each expression with all positive exponents. Also, any exponents in the denominator of a fraction must be positive *integers*. So, it may be necessary to rationalize a denominator.

Example 4 Simplify Expressions with Rational Exponents

StudyTip

Simplifying Expressions When simplifying expressions containing rational exponents, leave the exponent in rational form rather than writing the expression as a radical.

Simplify each expression.

a. $a^{\frac{2}{7}} \cdot a^{\frac{4}{7}}$

$a^{\frac{2}{7}} \cdot a^{\frac{4}{7}} = a^{\frac{2}{7}+\frac{4}{7}}$ Add powers.

$= a^{\frac{6}{7}}$ Add exponents.

b. $b^{-\frac{5}{6}}$

$b^{-\frac{5}{6}} = \frac{1}{b^{\frac{5}{6}}}$ $b^{-n} = \frac{1}{b^n}$

$= \frac{1}{b^{\frac{5}{6}}} \cdot \frac{b^{\frac{1}{6}}}{b^{\frac{1}{6}}}$ Why use $\frac{b^{\frac{1}{6}}}{b^{\frac{1}{6}}}$?

$= \frac{b^{\frac{1}{6}}}{b^{\frac{6}{6}}}$ $b^{\frac{5}{6}} \cdot b^{\frac{1}{6}} = b^{\frac{5}{6}+\frac{1}{6}}$

$= \frac{b^{\frac{1}{6}}}{b}$ $b^{\frac{6}{6}} = b^1$ or b

GuidedPractice

4A. $p^{\frac{1}{4}} \cdot p^{\frac{9}{4}}$

4B. $r^{-\frac{4}{5}}$

When simplifying a radical expression, always use the least index possible. Using rational exponents makes this process easier, but the answer should be written in radical form.

Example 5 Simplify Radical Expressions

Simplify each expression.

a. $\frac{\sqrt[4]{27}}{\sqrt{3}}$

$\frac{\sqrt[4]{27}}{\sqrt{3}} = \frac{27^{\frac{1}{4}}}{3^{\frac{1}{2}}}$ Rational exponents

$= \frac{(3^3)^{\frac{1}{4}}}{3^{\frac{1}{2}}}$ $27 = 3^3$

$= \frac{3^{\frac{3}{4}}}{3^{\frac{1}{2}}}$ Power of a Power

$= 3^{\frac{3}{4}-\frac{1}{2}}$ Quotient of Powers

$= 3^{\frac{1}{4}}$ Simplify.

$= \sqrt[4]{3}$ Rewrite in radical form.

b. $\sqrt[3]{64z^6}$

$\sqrt[3]{64z^6} = (64z^6)^{\frac{1}{3}}$ Rational exponents

$= (8^2 \cdot z^6)^{\frac{1}{3}}$ $64 = 8^2$

$= 8^{\frac{2}{3}} \cdot z^{\frac{6}{3}}$ Power of a Power

$= 4z^2$ $8^{\frac{2}{3}} = 4$

c. $\dfrac{x^{\frac{1}{2}} - 2}{3x^{\frac{1}{2}} + 2}$

$\dfrac{x^{\frac{1}{2}} - 2}{3x^{\frac{1}{2}} + 2} = \dfrac{x^{\frac{1}{2}} - 2}{3x^{\frac{1}{2}} + 2} \cdot \dfrac{3x^{\frac{1}{2}} - 2}{3x^{\frac{1}{2}} - 2}$ — $3x^{\frac{1}{2}} - 2$ is the conjugate of $3x^{\frac{1}{2}} + 2$.

$= \dfrac{3x^{\frac{2}{2}} - 8x^{\frac{1}{2}} + 4}{9x^{\frac{2}{2}} - 4}$ — Multiply.

$= \dfrac{3x - 8x^{\frac{1}{2}} + 4}{9x - 4}$ — Simplify.

GuidedPractice

5A. $\dfrac{\sqrt[4]{32}}{\sqrt[3]{2}}$ **5B.** $\sqrt[3]{16x^4}$ **5C.** $\dfrac{y^{\frac{1}{2}} + 2}{y^{\frac{1}{2}} - 2}$

StudyTip

Radical Expressions Write the simplified expression in the same form as the beginning expression. When you start with a radical expression, end with a radical expression. When you start with an expression with rational exponents, end with an expression with rational exponents.

ConceptSummary Expressions with Rational Exponents

An expression with rational exponents is simplified when all of the following conditions are met.

- It has no negative exponents.
- It has no exponents that are not positive integers in the denominator.
- It is not a complex fraction.
- The index of any remaining radical is the least number possible.

Check Your Understanding

◯ = Step-by-Step Solutions begin on page R14.

Example 1 **Write each expression in radical form, or write each radical in exponential form.**

1. $10^{\frac{1}{4}}$ **2.** $x^{\frac{3}{5}}$ **3.** $\sqrt[3]{15}$ **4.** $\sqrt[4]{7x^6y^9}$

Example 2 **Evaluate each expression.**

5. $343^{\frac{1}{3}}$ **6.** $32^{-\frac{1}{5}}$ **7.** $125^{\frac{2}{3}}$ **8.** $\dfrac{24}{4^{\frac{3}{2}}}$

Example 3 **9** **GARDENING** If the area A of a square is known, then the lengths of its sides ℓ can be computed using $\ell = A^{\frac{1}{2}}$. You have purchased a 169 ft² share in a community garden for the season. What is the length of one side of your square garden?

Examples 4–5 **CCSS PRECISION** **Simplify each expression.**

10. $a^{\frac{3}{4}} \cdot a^{\frac{1}{2}}$ **11.** $\dfrac{x^{\frac{4}{5}}}{x^{\frac{1}{5}}}$ **12.** $\dfrac{b^3}{c^{\frac{1}{2}}} \cdot \dfrac{c}{b^{\frac{1}{3}}}$

13. $\sqrt[4]{9g^2}$ **14.** $\dfrac{\sqrt[5]{64}}{\sqrt[5]{4}}$ **15.** $\dfrac{g^{\frac{1}{2}} - 1}{g^{\frac{1}{2}} + 1}$

Practice and Problem Solving

Extra Practice is on page R6.

Example 1 **Write each expression in radical form, or write each radical in exponential form.**

16. $8^{\frac{1}{5}}$
17. $4^{\frac{2}{7}}$
18. $a^{\frac{3}{4}}$
19. $(x^3)^{\frac{3}{2}}$
20. $\sqrt{17}$
21. $\sqrt[4]{63}$
22. $\sqrt[3]{5xy^2}$
23. $\sqrt[4]{625x^2}$

Example 2 **Evaluate each expression.**

24. $27^{\frac{1}{3}}$
25. $256^{\frac{1}{4}}$
26. $16^{-\frac{1}{2}}$
27. $81^{-\frac{1}{4}}$

Example 3

28. **CCSS SENSE-MAKING** A women's regulation-sized basketball is slightly smaller than a men's basketball. The radius r of the ball that holds V cubic units of air is $\left(\frac{3V}{4\pi}\right)^{\frac{1}{3}}$.
    a. Find the radius of a women's basketball.
    b. Find the radius of a men's basketball.

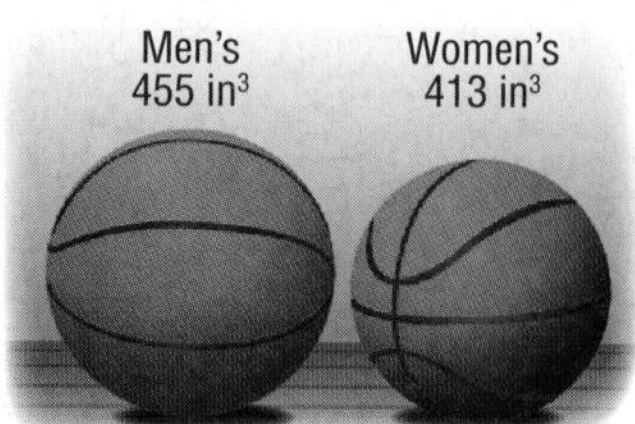

29. **GEOMETRY** The radius r of a sphere with volume V is given by $r = \left(\frac{3V}{4\pi}\right)^{\frac{1}{3}}$. Find the radius of a ball with a volume of 77 cm^3.

Examples 4–5 **Simplify each expression.**

30. $x^{\frac{1}{3}} \cdot x^{\frac{2}{5}}$
31. $a^{\frac{4}{9}} \cdot a^{\frac{1}{4}}$
32. $b^{-\frac{3}{4}}$
33. $y^{-\frac{4}{5}}$
34. $\frac{\sqrt[8]{81}}{\sqrt[6]{3}}$
35. $\frac{\sqrt[4]{27}}{\sqrt[4]{3}}$
36. $\sqrt[4]{25x^2}$
37. $\sqrt[6]{81g^3}$
38. $\frac{h^{\frac{1}{2}} + 1}{h^{\frac{1}{2}} - 1}$
39. $\frac{x^{\frac{1}{4}} + 2}{x^{\frac{1}{4}} - 2}$

GEOMETRY **Find the area of each figure.**

40.

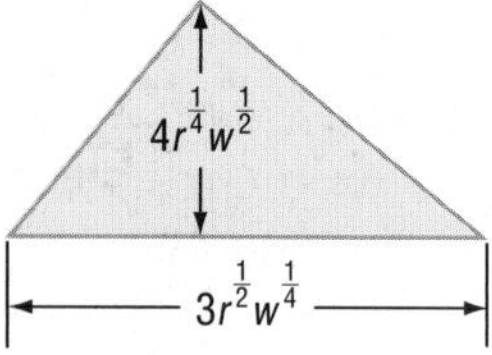

41.

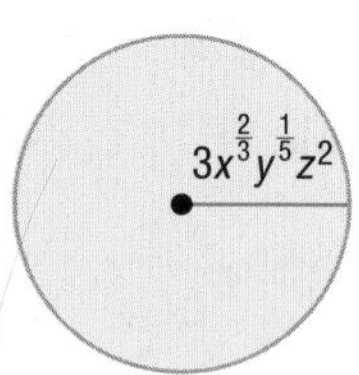

42. Find the simplified form of $18^{\frac{1}{2}} + 2^{\frac{1}{2}} - 32^{\frac{1}{2}}$.

43. What is the simplified form of $64^{\frac{1}{3}} - 32^{\frac{1}{3}} + 8^{\frac{1}{3}}$?

Simplify each expression.

44. $a^{\frac{7}{4}} \cdot a^{\frac{5}{4}}$
45. $x^{\frac{2}{3}} \cdot x^{\frac{8}{3}}$
46. $\left(b^{\frac{3}{4}}\right)^{\frac{1}{3}}$
47. $\left(y^{-\frac{3}{5}}\right)^{-\frac{1}{4}}$
48. $\sqrt[4]{64}$
49. $\sqrt[6]{216}$
50. $d^{-\frac{5}{6}}$
51. $w^{-\frac{7}{8}}$

52. **WILDLIFE** A population of 100 deer is reintroduced to a wildlife preserve. Suppose the population does extremely well and the deer population doubles in two years. Then the number D of deer after t years is given by $D = 100 \cdot 2^{\frac{t}{2}}$.

a. How many deer will there be after $4\frac{1}{2}$ years?

b. Make a table that charts the population of deer every year for the next five years.

c. Make a graph using your table.

d. Using your table and graph, decide whether this is a reasonable trend over the long term. Explain.

Simplify each expression.

53. $\dfrac{f^{-\frac{1}{4}}}{4f^{\frac{1}{2}} \cdot f^{-\frac{1}{3}}}$

54. $\dfrac{g^{\frac{5}{2}}}{g^{\frac{1}{2}} + 2}$

55. $\dfrac{c^{\frac{2}{3}}}{c^{\frac{1}{6}}}$

56. $\dfrac{z^{\frac{4}{5}}}{z^{\frac{1}{2}}}$

57. $\sqrt{23} \cdot \sqrt[3]{23^2}$

58. $\sqrt[8]{36h^4j^4}$

59. $\sqrt{\sqrt{81}}$

60. $\sqrt[4]{\sqrt{256}}$

61. $\dfrac{ab}{\sqrt{c}}$

62. $\dfrac{xy}{\sqrt[3]{z}}$

63. $\dfrac{8^{\frac{1}{6}} - 9^{\frac{1}{4}}}{\sqrt{3} + \sqrt{2}}$

64. $\dfrac{x^{\frac{5}{3}} - x^{\frac{1}{3}}z^{\frac{4}{3}}}{x^{\frac{2}{3}} + z^{\frac{2}{3}}}$

65. **MULTIPLE REPRESENTATIONS** In this problem, you will explore the functions $f(x) = x^3$ and $g(x) = x^{\frac{1}{3}}$.

a. **Tabular** Copy and complete the table to the right.

b. **Graphical** Graph $f(x)$ and $g(x)$.

c. **Verbal** Explain the transformation between $f(x)$ and $g(x)$.

x	$f(x)$	$g(x)$
−2		
−1		
0		
1		
2		

H.O.T. Problems Use Higher-Order Thinking Skills

66. **REASONING** Determine whether $-x^{-2} = (-x)^{-2}$ is *always*, *sometimes*, or *never* true. Explain your reasoning.

67. **CHALLENGE** Consider $\sqrt[4]{(-16)^3}$.

a. Explain why the expression is not a real number.

b. Find n such that $n\sqrt[4]{(-16)^3}$ is a real number.

68. **OPEN ENDED** Find two different expressions that equal 2 in the form $x^{\frac{1}{a}}$.

69. **WRITING IN MATH** Explain how it might be easier to simplify an expression using rational exponents rather than using radicals.

70. **CCSS CRITIQUE** Ayana and Kenji are simplifying $\dfrac{x^{\frac{3}{4}}}{x^{\frac{1}{2}}}$. Is either of them correct? Explain your reasoning.

Ayana

$\dfrac{x^{\frac{3}{4}}}{x^{\frac{1}{2}}} = x^{\frac{3}{4} + \frac{1}{2}}$

$= x^{\frac{3}{4} + \frac{2}{4}}$

$= x^{\frac{5}{4}}$

Kenji

$\dfrac{x^{\frac{3}{4}}}{x^{\frac{1}{2}}} = x^{\frac{3}{4} \div \frac{1}{2}}$

$= x^{\frac{3}{4} \cdot \frac{2}{1}}$

$= x^{\frac{3}{2}}$

Standardized Test Practice

71. The expression $\sqrt{56 - c}$ is equivalent to a positive integer when c is equal to

A -10 B -8 C 36 D 56

72. SAT/ACT Which of the following sentences is true about the graphs of $y = 2(x - 3)^2 + 1$ and $y = 2(x + 3)^2 + 1$?

F Their vertices are maximums.
G The graphs have the same shape with different vertices.
H The graphs have different shapes with the same vertices.
J The graphs have different shapes with different vertices.
K One graph has a vertex that is a maximum; the other has a vertex that is a minimum.

73. GEOMETRY What is the converse of the statement?
If it is summer, then it is hot outside.

A If it is not hot outside, then it is not summer.
B If it is not summer, then it is not hot outside.
C If it is hot outside, then it is summer.
D If it is hot outside, it is not summer.

74. SHORT RESPONSE If $3^5 \cdot p = 3^3$, then find p.

Spiral Review

Simplify. (Lesson 6-5)

75. $\sqrt{243}$ **76.** $\sqrt[3]{16y^3}$ **77.** $3\sqrt[3]{56y^6z^3}$

78. PHYSICS The speed of sound in a liquid is $s = \sqrt{\frac{B}{d}}$, where B is the bulk modulus of the liquid and d is its density. For water, $B = 2.1 \times 10^9$ N/m^2 and $d = 10^3$ kg/m^3. Find the speed of sound in water to the nearest meter per second. (Lesson 6-4)

Find $p(-4)$ and $p(x + h)$ for each function. (Lesson 5-3)

79. $p(x) = x - 2$ **80.** $p(x) = -x + 4$ **81.** $p(x) = 6x + 3$

82. $p(x) = x^2 + 5$ **83.** $p(x) = x^2 - x$ **84.** $p(x) = 2x^3 - 1$

Solve each equation by factoring. (Lesson 4-3)

85. $x^2 - 11x = 0$ **86.** $x^2 + 6x - 16 = 0$ **87.** $4x^2 - 13x = 12$

88. $x^2 - 14x = -49$ **89.** $x^2 + 9 = 6x$ **90.** $x^2 - 3x = -\frac{9}{4}$

91. GEOMETRY A rectangle is inscribed in an isosceles triangle as shown. Find the dimensions of the inscribed rectangle with maximum area. (*Hint*: Use similar triangles.) (Lesson 4-1)

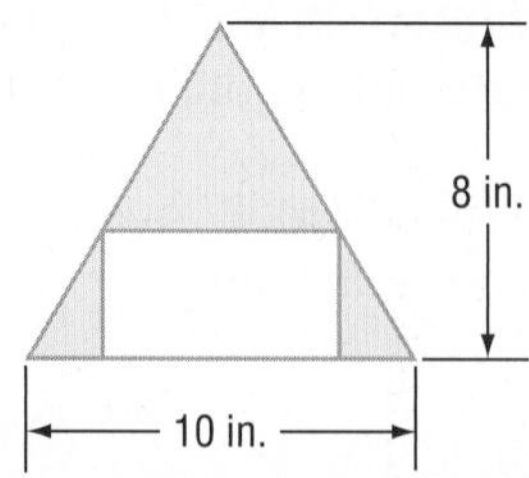

Skills Review

Find each power.

92. $(\sqrt{x - 3})^2$ **93.** $(\sqrt[3]{3x - 4})^3$ **94.** $(\sqrt[4]{7x - 1})^4$

95. $(\sqrt{x} - 4)^2$ **96.** $(2\sqrt{x} - 5)^2$ **97.** $(3\sqrt{x} + 1)^2$

LESSON 6-7

Solving Radical Equations and Inequalities

Then

- You solved polynomial equations.

Now

1. Solve equations containing radicals.
2. Solve inequalities containing radicals.

Why?

- When you jump, the time that you are in the air is your hang time. Hang time can be calculated in seconds t if you know the height h of the jump in feet. The formula for hang time is $t = 0.5\sqrt{h}$.

  Michael Jordan had a hang time of about 0.98 second. How would you calculate the height of Jordan's jump?

NewVocabulary
radical equation
extraneous solution
radical inequality

Common Core State Standards

Content Standards
A.SSE.2 Use the structure of an expression to identify ways to rewrite it.

Mathematical Practices
4 Model with mathematics.

1 Solve Radical Equations

Radical equations include radical expressions. You can solve a radical equation by raising each side of the equation to a power.

KeyConcept Solving Radical Equations

Step 1 Isolate the radical on one side of the equation.

Step 2 Raise each side of the equation to a power equal to the index of the radical to eliminate the radical.

Step 3 Solve the resulting polynomial equation. Check your results.

When solving radical equations, the result may be a number that does not satisfy the original equation. Such a number is called an **extraneous solution**.

Example 1 Solve Radical Equations

Solve each equation.

a. $\sqrt{x+2} + 4 = 7$

$\sqrt{x+2} + 4 = 7$	Original equation
$\sqrt{x+2} = 3$	Subtract 4 from each side to isolate the radical.
$(\sqrt{x+2})^2 = 3^2$	Square each side to eliminate the radical.
$x + 2 = 9$	Find the squares.
$x = 7$	Subtract 2 from each side.

CHECK

$\sqrt{x+2} + 4 = 7$	Original equation
$\sqrt{7+2} + 4 \stackrel{?}{=} 7$	Replace x with 7.
$7 = 7$ ✓	Simplify.

b. $\sqrt{x-12} = 2 - \sqrt{x}$

$\sqrt{x-12} = 2 - \sqrt{x}$	Original equation
$(\sqrt{x-12})^2 = (2-\sqrt{x})^2$	Square each side.
$x - 12 = 4 - 4\sqrt{x} + x$	Find the squares.
$-16 = -4\sqrt{x}$	Isolate the radical.
$4 = \sqrt{x}$	Divide each side by −4.
$16 = x$	Square each side.

Streeter Lecka/Getty Images Sport/Getty Images

StudyTip

Solution Check You can use a graphing calculator to check solutions of equations. Graph each side of the original equation, and examine the intersection.

CHECK $\sqrt{x-12} = 2 - \sqrt{x}$

$\sqrt{16-12} \stackrel{?}{=} 2 - \sqrt{16}$

$\sqrt{4} \stackrel{?}{=} 2 - 4$

$2 \neq -2$ ✗

The solution does not check, so the equation has an extraneous solution. The graphs of $y = \sqrt{x-12}$ and $y = 2 - \sqrt{x}$ do not intersect, which confirms that there is no real solution.

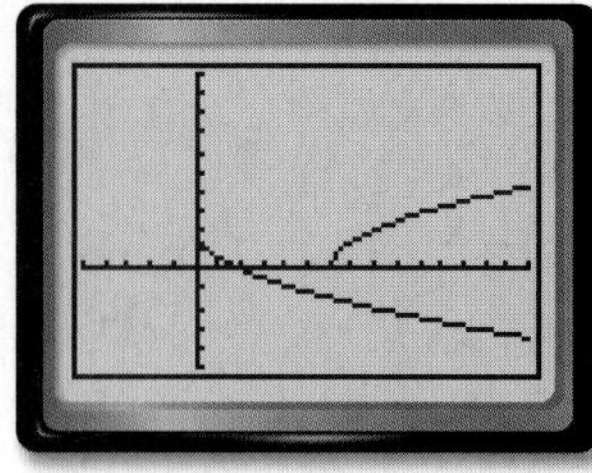

[−10, 30] scl: 2 by [−5, 10] scl: 1

GuidedPractice

1A. $5 = \sqrt{x-2} - 1$

1B. $\sqrt{x+15} = 5 + \sqrt{x}$

To undo a square root, you square the expression. To undo a cube root, you must raise the expression to the third power.

Example 2 Solve a Cube Root Equation

Solve $2(6x-3)^{\frac{1}{3}} - 4 = 0$.

In order to remove the $\frac{1}{3}$ power, or cube root, you must first isolate it and then raise each side of the equation to the third power.

$2(6x-3)^{\frac{1}{3}} - 4 = 0$	Original equation
$2(6x-3)^{\frac{1}{3}} = 4$	Add 4 to each side.
$(6x-3)^{\frac{1}{3}} = 2$	Divide each side by 2.
$\left[(6x-3)^{\frac{1}{3}}\right]^3 = 2^3$	Cube each side.
$6x - 3 = 8$	Evaluate the cubes.
$6x = 11$	Add 3 to each side.
$x = \frac{11}{6}$	Divide each side by 6.

CHECK

$2(6x-3)^{\frac{1}{3}} - 4 = 0$	Original equation
$2\left(6 \cdot \frac{11}{6} - 3\right)^{\frac{1}{3}} - 4 \stackrel{?}{=} 0$	Replace x with $\frac{11}{6}$.
$2(8)^{\frac{1}{3}} - 4 \stackrel{?}{=} 0$	Simplify.
$2(2) - 4 \stackrel{?}{=} 0$	The cube root of 8 is 2.
$0 = 0$ ✓	Subtract.

GuidedPractice

Solve each equation.

2A. $(3n+2)^{\frac{1}{3}} + 1 = 0$

2B. $3(5y-1)^{\frac{1}{3}} - 2 = 0$

You can apply the methods used to solve square and cube root equations to solving equations with roots of any index. To undo an nth root, raise to the nth power.

Standardized Test Example 3 Solve a Radical Equation

Test-Taking Tip

Substitute Values You could also solve the test question by substituting each answer for n in the equation to see if the solution is correct.

What is the solution of $3(\sqrt[4]{2n + 6}) - 6 = 0$?

A -1 **B** 1 **C** 5 **D** 11

$3(\sqrt[4]{2n + 6}) - 6 = 0$	Original equation
$3(\sqrt[4]{2n + 6}) = 6$	Add 6 to each side.
$\sqrt[4]{2n + 6} = 2$	Divide each side by 3.
$(\sqrt[4]{2n + 6})^4 = 2^4$	Raise each side to the fourth power.
$2n + 6 = 16$	Evaluate each side.
$2n = 10$	Subtract 6 from each side.
$n = 5$	The answer is C.

Guided Practice

3. What is the solution of $4(3x + 6)^{\frac{1}{4}} - 12 = 0$?

F $x = 7$ **G** $x = 25$ **H** $x = 29$ **J** $x = 37$

2 Solve Radical Inequalities

A 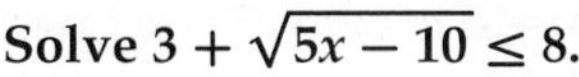 **radical inequality** has a variable in the radicand. To solve radical inequalities, complete the following steps.

Key Concept Solving Radical Inequalities

Step 1 If the index of the root is even, identify the values of the variable for which the radicand is nonnegative.

Step 2 Solve the inequality algebraically.

Step 3 Test values to check your solution.

Example 4 Solve a Radical Inequality

Study Tip

Radical Inequalities Since a principal square root is never negative, inequalities that simplify to the form $\sqrt{ax + b} \leq c$, where c is a negative number, have no solutions.

Solve $3 + \sqrt{5x - 10} \leq 8$.

Step 1 Since the radicand of a square root must be greater than or equal to zero, first solve $5x - 10 \geq 0$ to identify the values of x for which the left side of the inequality is defined.

$5x - 10 \geq 0$	Set the radicand ≥ 0.
$5x \geq 10$	Add 10 to each side.
$x \geq 2$	Divide each side by 5.

Step 2 Solve $3 + \sqrt{5x - 10} \leq 8$.

$3 + \sqrt{5x - 10} \leq 8$	Original inequality
$\sqrt{5x - 10} \leq 5$	Isolate the radical.
$5x - 10 \leq 25$	Eliminate the radical.
$5x \leq 35$	Add 10 to each side.
$x \leq 7$	Divide each side by 5.

Step 3 It appears that $2 \leq x \leq 7$. You can test some x-values to confirm the solution. Use three test values: one less than 2, one between 2 and 7, and one greater than 7. Organize the test values in a table.

$x = 0$	$x = 4$	$x = 9$
$3 + \sqrt{5(0) - 10} \stackrel{?}{\leq} 8$ $3 + \sqrt{-10} \leq 8$ ✗ Since $\sqrt{-10}$ is not a real number, the inequality is not satisfied.	$3 + \sqrt{5(4) - 10} \stackrel{?}{\leq} 8$ $6.16 \leq 8$ ✓ Since $6.16 \leq 8$, the inequality is satisfied.	$3 + \sqrt{5(9) - 10} \stackrel{?}{\leq} 8$ $8.92 \leq 8$ ✗ Since $8.92 \not\leq 8$, the inequality is not satisfied.

The solution checks. Only values in the interval $2 \leq x \leq 7$ satisfy the inequality. You can summarize the solution with a number line.

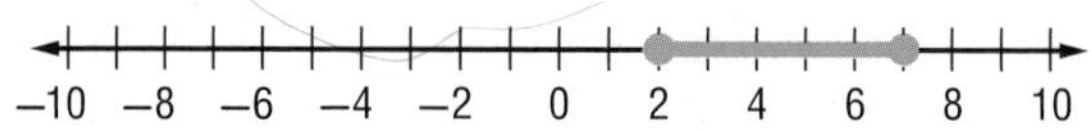

GuidedPractice

Solve each inequality.

4A. $\sqrt{2x + 2} + 1 \geq 5$

4B. $\sqrt{4x - 4} - 2 < 4$

Check Your Understanding

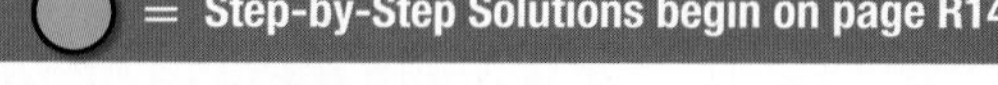
= Step-by-Step Solutions begin on page R14.

Examples 1–2 **Solve each equation.**

1. $\sqrt{x - 4} + 6 = 10$

2. $\sqrt{x + 13} - 8 = -2$

3. $8 - \sqrt{x + 12} = 3$

4. $\sqrt{x - 8} + 5 = 7$

5. $\sqrt[3]{x - 2} = 3$

6. $(x - 5)^{\frac{1}{3}} - 4 = -2$

7. $(4y)^{\frac{1}{3}} + 3 = 5$

8. $\sqrt[3]{n + 8} - 6 = -3$

9. $\sqrt{y} - 7 = 0$

10. $2 + 4z^{\frac{1}{2}} = 0$

11. $5 + \sqrt{4y - 5} = 12$

12. $\sqrt{2t - 7} = \sqrt{t + 2}$

13. **CCSS REASONING** The time T in seconds that it takes a pendulum to make a complete swing back and forth is given by the formula $T = 2\pi\sqrt{\frac{L}{g}}$, where L is the length of the pendulum in feet and g is the acceleration due to gravity, 32 feet per second squared.

a. In Tokyo, Japan, a huge pendulum in the Shinjuku building measures 73 feet 9.75 inches. How long does it take for the pendulum to make a complete swing?

b. A clockmaker wants to build a pendulum that takes 20 seconds to swing back and forth. How long should the pendulum be?

Example 3

14. **MULTIPLE CHOICE** Solve $(2y + 6)^{\frac{1}{4}} - 2 = 0$.

A $y = 1$ **B** $y = 5$ **C** $y = 11$ **D** $y = 15$

Example 4 **Solve each inequality.**

15. $\sqrt{3x + 4} - 5 \leq 4$

16. $\sqrt{b - 7} + 6 \leq 12$

17. $2 + \sqrt{4y - 4} \leq 6$

18. $\sqrt{3a + 3} - 1 \leq 2$

19. $1 + \sqrt{7x - 3} > 3$

20. $\sqrt{3x + 6} + 2 \leq 5$

21. $-2 + \sqrt{9 - 5x} \geq 6$

22. $6 - \sqrt{2y + 1} < 3$

Practice and Problem Solving

Extra Practice is on page R6.

Example 1 **Solve each equation. Confirm by using a graphing calculator.**

23. $\sqrt{2x+5} - 4 = 3$
24. $6 + \sqrt{3x+1} = 11$
25. $\sqrt{x+6} = 5 - \sqrt{x+1}$
26. $\sqrt{x-3} = \sqrt{x+4} - 1$
27. $\sqrt{x-15} = 3 - \sqrt{x}$
28. $\sqrt{x-10} = 1 - \sqrt{x}$
29. $6 + \sqrt{4x+8} = 9$
30. $2 + \sqrt{3y-5} = 10$
31. $\sqrt{x-4} = \sqrt{2x-13}$
32. $\sqrt{7a-2} = \sqrt{a+3}$
33. $\sqrt{x-5} - \sqrt{x} = -2$
34. $\sqrt{b-6} + \sqrt{b} = 3$

35. **CCSS SENSE-MAKING** Isabel accidentally dropped her keys from the top of a Ferris wheel. The formula $t = \frac{1}{4}\sqrt{d-h}$ describes the time t in seconds at which the keys are h meters above the ground and Isabel is d meters above the ground. If Isabel was 65 meters high when she dropped the keys, how many meters above the ground will the keys be after 2 seconds?

Example 2 **Solve each equation.**

36. $(5n-6)^{\frac{1}{3}} + 3 = 4$
37. $(5p-7)^{\frac{1}{3}} + 3 = 5$
38. $(6q+1)^{\frac{1}{4}} + 2 = 5$
39. $(3x+7)^{\frac{1}{4}} - 3 = 1$
40. $(3y-2)^{\frac{1}{5}} + 5 = 6$
41. $(4z-1)^{\frac{1}{5}} - 1 = 2$
42. $2(x-10)^{\frac{1}{3}} + 4 = 0$
43. $3(x+5)^{\frac{1}{3}} - 6 = 0$
44. $\sqrt[3]{5x+10} - 5 = 0$
45. $\sqrt[3]{4n-8} - 4 = 0$
46. $\frac{1}{7}(14a)^{\frac{1}{3}} = 1$
47. $\frac{1}{4}(32b)^{\frac{1}{3}} = 1$

Example 3

48. **MULTIPLE CHOICE** Solve $\sqrt[4]{y+2} + 9 = 14$.

A 23 **B** 53 **C** 123 **D** 623

49. **MULTIPLE CHOICE** Solve $(2x-1)^{\frac{1}{4}} - 2 = 1$.

F 41 **G** 28 **H** 13 **J** 1

Example 4 **Solve each inequality.**

50. $1 + \sqrt{5x-2} > 4$
51. $\sqrt{2x+14} - 6 \geq 4$
52. $10 - \sqrt{2x+7} \leq 3$
53. $6 + \sqrt{3y+4} < 6$
54. $\sqrt{2x+5} - \sqrt{9+x} > 0$
55. $\sqrt{d+3} + \sqrt{d+7} > 4$
56. $\sqrt{3x+9} - 2 < 7$
57. $\sqrt{2y+5} + 3 \leq 6$
58. $-2 + \sqrt{8-4z} \geq 8$
59. $-3 + \sqrt{6a+1} > 4$
60. $\sqrt{2} - \sqrt{b+6} \leq -\sqrt{b}$
61. $\sqrt{c+9} - \sqrt{c} > \sqrt{3}$

62. **PENDULUMS** The formula $s = 2\pi\sqrt{\frac{\ell}{32}}$ represents the swing of a pendulum, where s is the time in seconds to swing back and forth, and ℓ is the length of the pendulum in feet. Find the length of a pendulum that makes one swing in 1.5 seconds.

63. **FISH** The relationship between the length and mass of certain fish can be approximated by the equation $L = 0.46\sqrt[3]{M}$, where L is the length in meters and M is the mass in kilograms. Solve this equation for M.

64. HANG TIME Refer to the information at the beginning of the lesson regarding hang time. Describe how the height of a jump is related to the amount of time in the air. Write a step-by-step explanation of how to determine the height of Jordan's 0.98-second jump.

65. CONCERTS The organizers of a concert are preparing for the arrival of 50,000 people in the open field where the concert will take place. Each person is allotted 5 square feet of space, so the organizers rope off a circular area of 250,000 square feet. Using the formula $A = \pi r^2$, where A represents the area of the circular region and r represents the radius of the region, find the radius of this region.

66. WEIGHTLIFTING The formula $M = 512 - 146{,}230B^{-\frac{8}{5}}$ can be used to estimate the maximum total mass that a weightlifter of mass B kilograms can lift using the snatch and the clean and jerk. According to the formula, how much does a person weigh who can lift at most 470 kilograms?

H.O.T. Problems Use Higher-Order Thinking Skills

67. CCSS ARGUMENTS Which equation does not have a solution?

$\sqrt{x-1} + 3 = 4$

$\sqrt{x+1} + 3 = 4$

$\sqrt{x-2} + 7 = 10$

$\sqrt{x+2} - 7 = -10$

68. CHALLENGE Lola is working to solve $(x + 5)^{\frac{1}{4}} = -4$. She said that she could tell there was no real solution without even working the problem. Is Lola correct? Explain your reasoning.

69. REASONING Determine whether $\frac{\sqrt{(x^2)^2}}{-x} = x$ is *sometimes, always,* or *never* true when x is a real number. Explain your reasoning.

70. OPEN ENDED Select a whole number. Now work backward to write two radical equations that have that whole number as solutions. Write one square root equation and one cube root equation. You may need to experiment until you find a whole number you can easily use.

71. WRITING IN MATH Explain the relationship between the index of the root of a variable in an equation and the power to which you raise each side of the equation to solve the equation.

72. OPEN ENDED Write an equation that can be solved by raising each side of the equation to the given power.

a. $\frac{3}{2}$ power **b.** $\frac{5}{4}$ power **c.** $\frac{7}{8}$ power

73. CHALLENGE Solve $7^{3x-1} = 49^{x+1}$ for x. (*Hint:* $b^x = b^y$ if and only if $x = y$.)

REASONING Determine whether the following statements are *sometimes, always,* or *never* true for $x^{\frac{1}{n}} = a$. Explain your reasoning.

74. If n is odd, there will be extraneous solutions.

75. If n is even, there will be extraneous solutions.

Standardized Test Practice

76. What is an equivalent form of $\frac{4}{5+i}$?

A $\frac{10-2i}{13}$

B $\frac{5-i}{6}$

C $\frac{6-i}{6}$

D $\frac{6-i}{13}$

77. Which set of points describes a function?

F {(3, 0), (−2, 5), (2, −1), (2, 9)}

G {(−3, 5), (−2, 3), (−1, 5), (0, 7)}

H {(2, 5), (2, 4), (2, 3), (2, 2)}

J {(3, 1), (−3, 2), (3, 3), (−3, 4)}

78. SHORT RESPONSE The perimeter of an isosceles triangle is 56 inches. If one leg is 20 inches long, what is the measure of the base of the triangle?

79. SAT/ACT If $\sqrt{x+5}+1=4$, what is the value of x?

A 4

B 10

C 11

D 12

E 20

Spiral Review

Evaluate. (Lesson 6-6)

80. $27^{-\frac{2}{3}}$

81. $9^{\frac{1}{3}} \cdot 9^{\frac{5}{3}}$

82. $\left(\frac{8}{27}\right)^{-\frac{2}{3}}$

83. GEOMETRY The measures of the legs of a right triangle can be represented by the expressions $4x^2y^2$ and $8x^2y^2$. Use the Pythagorean Theorem to find a simplified expression for the measure of the hypotenuse. (Lesson 6-5)

Find the inverse of each function. (Lesson 6-2)

84. $y = 3x - 4$

85. $y = -2x - 3$

86. $y = x^2$

87. $y = (2x + 3)^2$

For each graph,

a. describe the end behavior,

b. determine whether it represents an odd-degree or an even-degree polynomial function, and

c. state the number of real zeros. (Lesson 5-3)

88.

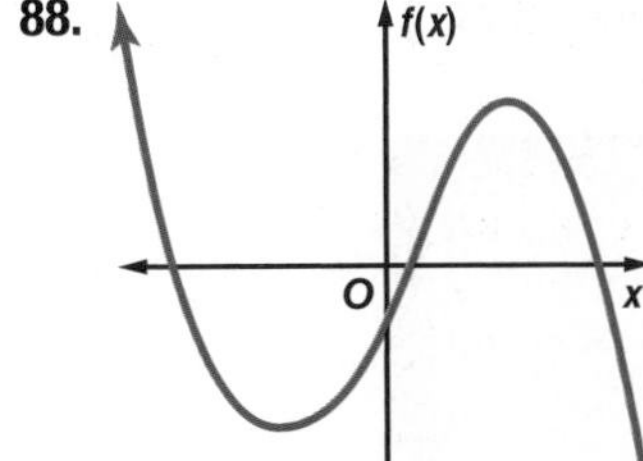

89.

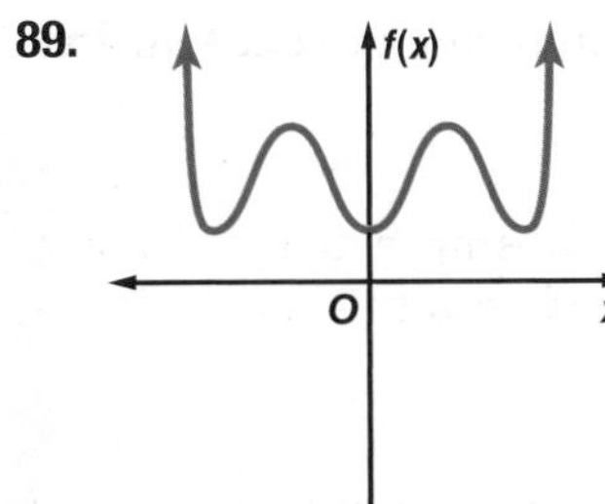

90.

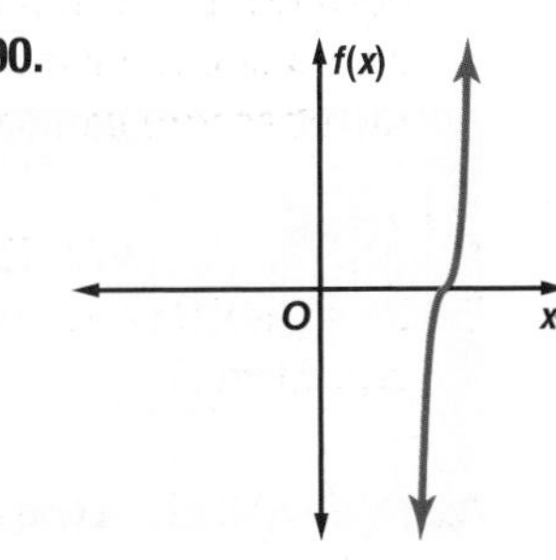

Skills Review

Solve each equation. Write in simplest form.

91. $\frac{8}{5}x = \frac{4}{15}$

92. $\frac{27}{14}y = \frac{6}{7}$

93. $\frac{3}{10} = \frac{12}{25}a$

94. $\frac{6}{7} = 9m$

95. $\frac{9}{8}b = 18$

96. $\frac{6}{7}n = \frac{3}{4}$

97. $\frac{1}{3}p = \frac{5}{6}$

98. $\frac{2}{3}q = 7$

EXTEND 6-7 Graphing Technology Lab
Solving Radical Equations and Inequalities

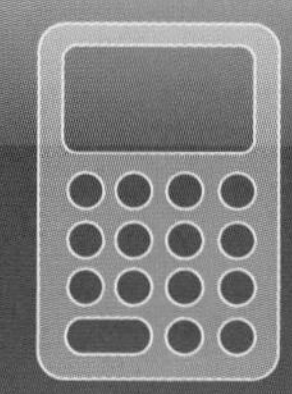

You can use a TI-83/84 Plus graphing calculator to solve radical equations and inequalities. One way to do this is to rewrite the equation or inequality so that one side is 0. Then use the zero feature on the calculator.

CCSS Common Core State Standards
Content Standards
A.REI.2 Solve simple rational and radical equations in one variable, and give examples showing how extraneous solutions may arise.
A.REI.11 Explain why the x-coordinates of the points where the graphs of the equations $y = f(x)$ and $y = g(x)$ intersect are the solutions of the equation $f(x) = g(x)$; find the solutions approximately, e.g., using technology to graph the functions, make tables of values, or find successive approximations. Include cases where $f(x)$ and/or $g(x)$ are linear, polynomial, rational, absolute value, exponential, and logarithmic functions.

Example 1 Radical Equation

Solve $\sqrt{x} + \sqrt{x+2} = 3$.

Step 1 Rewrite the equation.

- Subtract 3 from each side of the equation to get $\sqrt{x} + \sqrt{x+2} - 3 = 0$.
- Enter the function $y = \sqrt{x} + \sqrt{x+2} - 3$ in the **Y=** list.

KEYSTROKES: Y= 2nd [√] X,T,θ,n) + 2nd [√] X,T,θ,n + 2) − 3 ENTER

Step 2 Use a table.

- You can use the **TABLE** function to locate intervals where the solution(s) lie. First, enter the starting value and the interval for the table.

KEYSTROKES: 2nd [TBLSET] 0 ENTER 1 ENTER

Step 3 Estimate the solution.

- Complete the table and estimate the solution(s).

KEYSTROKES: 2nd [TABLE]

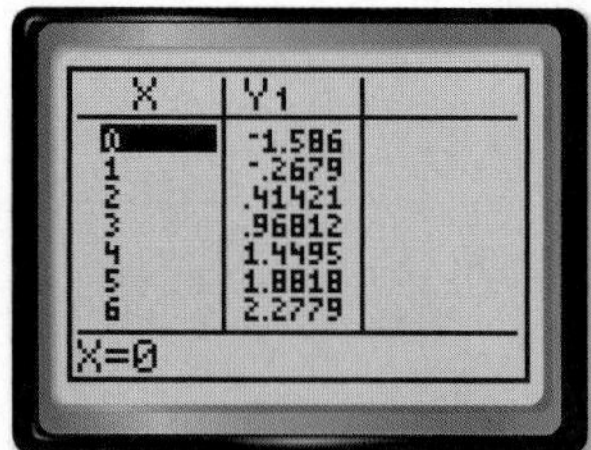

Since the function changes sign from negative to positive between $x = 1$ and $x = 2$, there is a solution between 1 and 2.

Step 4 Use the ZERO feature.

- Graph the function in the standard viewing window; then select **ZERO** from the **CALC** menu.

KEYSTROKES: ZOOM 6 2nd [CALC] 2

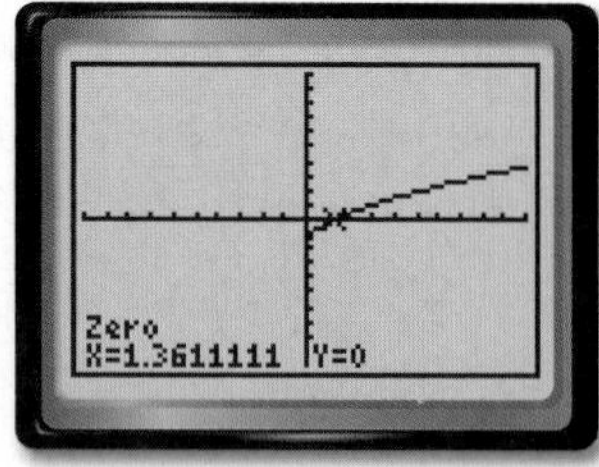

[−10, 10] scl: 1 by [−10, 10] scl: 1

Place the cursor on a point at which $y < 0$ and press ENTER for the **LEFT BOUND**. Then place the cursor on a point at which $y > 0$ and press ENTER for the **RIGHT BOUND**. You can use the same point for the **GUESS**.The solution is about 1.36. This is consistent with the estimate made by using the **TABLE**.

Example 2 Radical Inequality

Solve $2\sqrt{x} > \sqrt{x+2} + 1$.

Step 1 **Graph each side of the inequality and use the TRACE feature.**

- In the **Y=** list, enter $y_1 = 2\sqrt{x}$ and $y_2 = \sqrt{x+2} + 1$. Then press GRAPH.

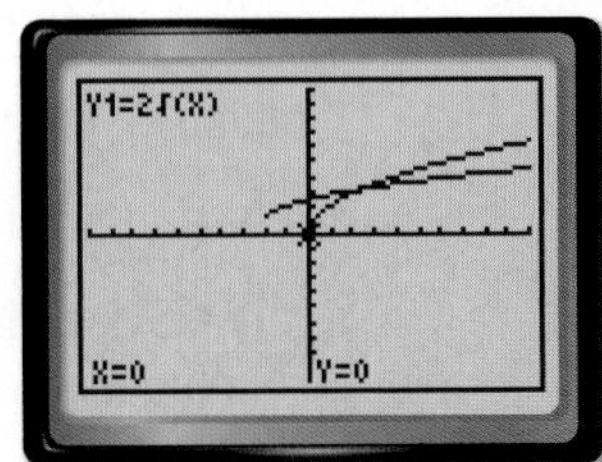

[−10, 10] scl: 1 by [−10, 10] scl: 1

- Press TRACE. You can use ▲ or ▼ to switch the cursor between the two curves.

The calculator screen above shows that, for points to the left of where the curves cross, **Y1** < **Y2** or $2\sqrt{x} < \sqrt{x+2} + 1$. To solve the original inequality, you must find points for which **Y1** > **Y2**. These are the points to the right of where the curves cross.

Step 2 **Use the intersect feature.**

- You can use the **intersect** feature on the **CALC** menu to approximate the x-coordinate of the point at which the curves cross.

KEYSTROKES: 2nd [CALC] 5

- Press ENTER for each of **FIRST CURVE?**, **SECOND CURVE?**, and **GUESS?**.

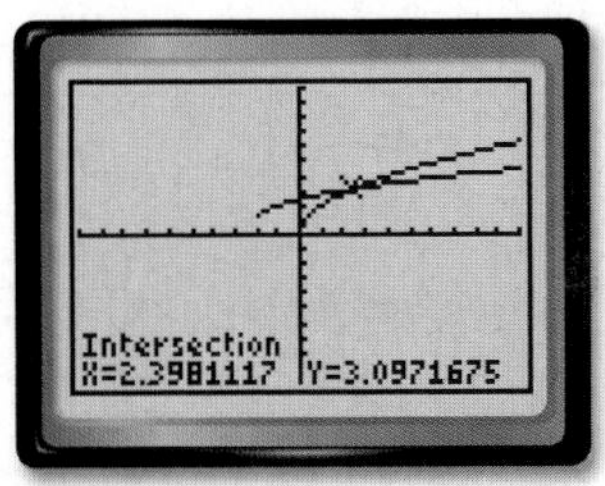

[−10, 10] scl: 1 by [−10, 10] scl: 1

The calculator screen shows that the x-coordinate of the point at which the curves cross is about 2.40. Therefore, the solution of the inequality is about $x > 2.40$. *Use the symbol > in the solution because the symbol in the original inequality is >.*

Step 3 **Use the TABLE feature to check your solution.**

- Start the table at 2 and show x-values in increments of 0.1. Scroll through the table.

KEYSTROKES: 2nd [TBLSET] 2 ENTER .1 ENTER 2nd [TABLE]

Notice that when x is less than or equal to 2.4, **Y1** < **Y2**. This verifies the solution $\{x \mid x > 2.40\}$.

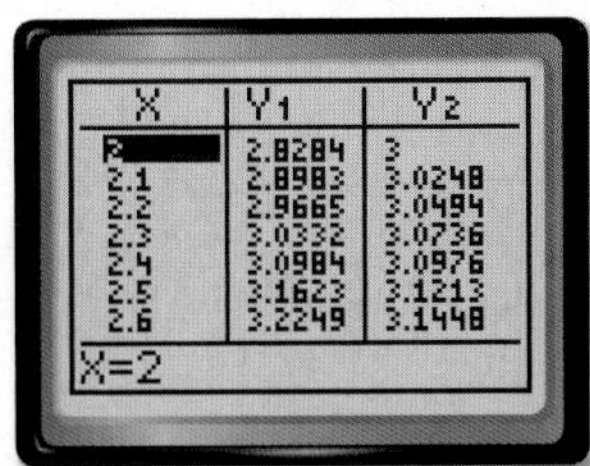

X	Y1	Y2
2	2.8284	3
2.1	2.8983	3.0248
2.2	2.9665	3.0494
2.3	3.0332	3.0736
2.4	3.0984	3.0976
2.5	3.1623	3.1213
2.6	3.2249	3.1448

X=2

Exercises

Use a graphical method to solve each equation or inequality.

1. $\sqrt{x+4} = 3$

2. $\sqrt{3x-5} = 1$

3. $\sqrt{x+5} = \sqrt{3x+4}$

4. $\sqrt{x+3} + \sqrt{x-2} = 4$

5. $\sqrt{3x-7} = \sqrt{2x-2} - 1$

6. $\sqrt{x+8} - 1 = \sqrt{x+2}$

7. $\sqrt{x-3} \geq 2$

8. $\sqrt{x+3} > 2\sqrt{x}$

9. $\sqrt{x} + \sqrt{x-1} < 4$

10. **WRITING IN MATH** Explain how you could apply the technique in the first example to solving an inequality.

CHAPTER 6

Study Guide and Review

Study Guide

KeyConcepts

Operations on Functions (Lesson 6-1)

Operation	Definition
Sum	$(f + g)(x) = f(x) + g(x)$
Difference	$(f - g)(x) = f(x) - g(x)$
Product	$(f \cdot g)(x) = f(x) \cdot g(x)$
Quotient	$\left(\frac{f}{g}\right)(x) = \frac{f(x)}{g(x)}, g(x) \neq 0$
Composition	$[f \circ g](x) = f[g(x)]$

Inverse and Square Root Functions (Lessons 6-2 and 6-3)

- Two functions are inverses if and only if both their compositions are the identity function.

***n*th Roots** (Lesson 6-4)

Real *n*th roots of b, $\sqrt[n]{b}$, or $-\sqrt[n]{b}$			
n	$\sqrt[n]{b}$ if $b > 0$	$\sqrt[n]{b}$ if $b < 0$	$\sqrt[n]{b}$ if $b = 0$
even	one positive root one negative root	no real roots	one real root, 0
odd	one positive root no negative roots	no positive roots one negative root	one real root, 0

Radicals (Lessons 6-5 through 6-7)

For any real numbers a and b and any integers n, x, and y, with $b \neq 0$, $n > 1$, and $y > 1$, the following are true.

- Product Property: $\sqrt[n]{ab} = \sqrt[n]{a} \cdot \sqrt[n]{b}$
- Quotient Property: $\sqrt[n]{\frac{a}{b}} = \frac{\sqrt[n]{a}}{\sqrt[n]{b}}$
- Rational Exponents: $b^{\frac{x}{y}} = \sqrt[y]{b^x} = \left(\sqrt[y]{b}\right)^x, b \geq 0$

FOLDABLES StudyOrganizer

Be sure the Key Concepts are noted in your Foldable.

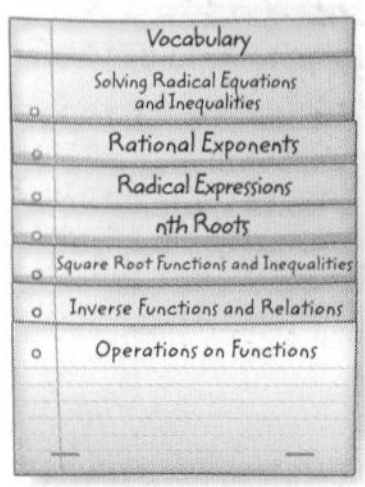

KeyVocabulary

composition of functions (p. 387)
conjugates (p. 418)
extraneous solution (p. 429)
index (p. 407)
inverse function (p. 393)
inverse relation (p. 393)
like radical expressions (p. 417)
*n*th root (p. 407)
principal root (p. 407)
radical equation (p. 429)
radical function (p. 400)
radical inequality (p. 431)
radical sign (p. 407)
radicand (p. 407)
rationalizing the denominator (p. 416)
square root function (p. 400)
square root inequality (p. 402)

VocabularyCheck

Choose a word or term that best completes each statement.

1. If both compositions result in the ______________, then the functions are inverse functions.
2. Radicals are ______________ if *both* the index and the radicand are identical.
3. In a(n) ______________, the results of one function are used to evaluate a second function.
4. When there is more than one real root, the nonnegative root is called the ______________.
5. To eliminate radicals from a denominator or fractions from a radicand, you use a process called ______________.
6. Equations with radicals that have variables in the radicands are called ______________.
7. Two relations are ______________ if and only if one relation contains the element (b, a) when the other relation contains the element (a, b).
8. When solving a radical equation, sometimes you will obtain a number that does not satisfy the original equation. Such a number is called a(n) ______________.
9. The square root function is a type of ______________.

Lesson-by-Lesson Review

6-1 Operations on Functions

Find $[f \circ g](x)$ and $[g \circ f](x)$.

10. $f(x) = 2x + 1$
    $g(x) = 4x - 5$
11. $f(x) = x^2 + 1$
    $g(x) = x - 7$
12. $f(x) = x^2 + 4$
    $g(x) = -2x + 1$
13. $f(x) = 4x$
    $g(x) = 5x - 1$
14. $f(x) = x^3$
    $g(x) = x - 1$
15. $f(x) = x^2 + 2x - 3$
    $g(x) = x + 1$
16. **MEASUREMENT** The formula $f = 3y$ converts yards y to feet f and $f = \frac{n}{12}$ converts inches n to feet f. Write a composition of functions that converts yards to inches.

Example 1

If $f(x) = x^2 + 3$ and $g(x) = 3x - 2$, find $g[f(x)]$ and $f[g(x)]$.

$g[f(x)] = 3(x^2 + 3) - 2$	Replace $f(x)$ with $x^2 + 3$.
$= 3x^2 + 9 - 2$	Distributive Property
$= 3x^2 + 7$	Simplify.
$f[g(x)] = (3x - 2)^2 + 3$	Replace $g(x)$ with $3x - 2$.
$= 9x^2 - 12x + 4 + 3$	Multiply.
$= 9x^2 - 12x + 7$	Simplify.

6-2 Inverse Functions and Relations

Find the inverse of each function. Then graph the function and its inverse.

17. $f(x) = 5x - 6$
18. $f(x) = -3x - 5$
19. $f(x) = \frac{1}{2}x + 3$
20. $f(x) = \frac{4x + 1}{5}$
21. $f(x) = x^2$
22. $f(x) = (2x + 1)^2$
23. **SHOPPING** Samuel bought a computer. The sales tax rate was 6% of the sale price, and he paid \$50 for shipping. Find the sale price if Samuel paid a total of \$1322.

Use the horizontal line test to determine whether the inverse of each function is also a function.

24. $f(x) = 3x^2$
25. $h(x) = x^3 - 3$
26. $g(x) = -3x^4 + 2x - 1$
27. $g(x) = 4x^3 - 5x$
28. $f(x) = -3x^5 + x^2 - 3$
29. $h(x) = 4x^4 + 7x$
30. **FINANCIAL LITERACY** During the last month, Jonathan has made two deposits of \$45, made a deposit of double his original balance, and has withdrawn \$35 five times. His balance is now \$189. Write an equation that models this problem. How much money did Jonathan have in his account at the beginning of the month?

Example 2

Find the inverse of $f(x) = -2x + 7$.

Rewrite $f(x)$ as $y = -2x + 7$. Then interchange the variables and solve for y.

$x = -2y + 7$	Interchange the variables.
$2y = -x + 7$	Solve for y.
$y = \frac{-x + 7}{2}$	Divide each side by 2.
$f^{-1}(x) = \frac{-x + 7}{2}$	Rewrite using function notation.

Example 3

Use the horizontal line test to determine whether the inverse of $f(x) = 2x^3 + 1$ is also a function.

Graph the function.

No horizontal line can be drawn so that it passes through more than one point. The inverse of this function is a function.

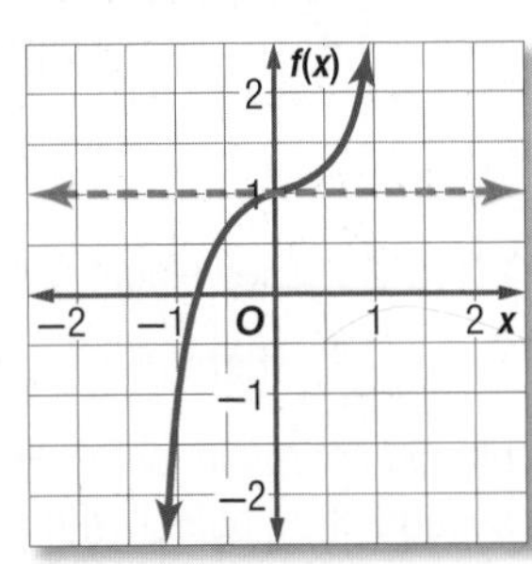

CHAPTER 6 Study Guide and Review *Continued*

6-3 Square Root Functions and Inequalities

Graph each function. State the domain and range.

31. $f(x) = \sqrt{3x}$

32. $f(x) = -\sqrt{6x}$

33. $f(x) = \sqrt{x - 7}$

34. $f(x) = \sqrt{x + 5} - 3$

35. $f(x) = \frac{3}{4}\sqrt{x - 1} + 5$

36. $f(x) = -\frac{1}{3}\sqrt{x + 4} - 1$

37. GEOMETRY The area of a circle is given by the formula $A = \pi r^2$. What is the radius of a circle with an area of 300 square inches?

Graph each inequality.

38. $y \geq \sqrt{x} + 3$

39. $y < 2\sqrt{x - 5}$

40. $y > -\sqrt{x - 1} + 2$

Example 4

Graph $f(x) = \sqrt{x + 1} - 2$. State the domain and range.

Identify the domain.

$x + 1 \geq 0$ — Write the radicand as greater than or equal to 0.

$x \geq -1$ — Subtract 1 from each side.

Make a table of values for $x \geq -1$ and graph the function.

x	$f(x)$
−1	−2
0	−1
1	−0.59
2	−0.27
3	0
4	0.24
5	0.45

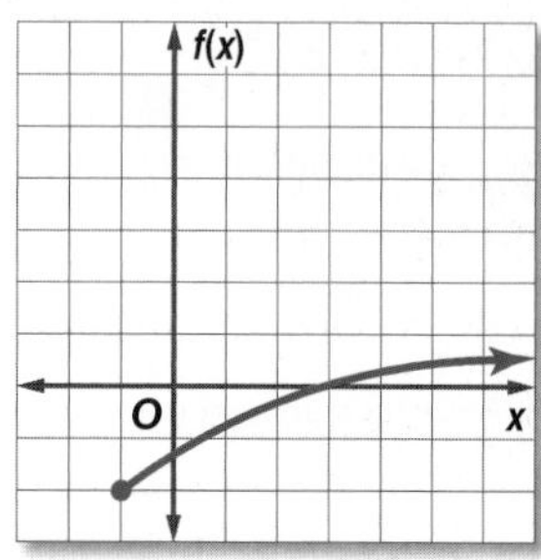

The domain is $\{x | x \geq -1\}$, and the range is $\{f(x) | f(x) \geq -2\}$.

6-4 *n*th Roots

Simplify.

41. $\pm\sqrt{121}$

42. $\sqrt[3]{-125}$

43. $\sqrt{(-6)^2}$

44. $\sqrt{-(x + 3)^4}$

45. $\sqrt[6]{(x^2 + 2)^{18}}$

46. $\sqrt[3]{27(x + 3)^3}$

47. $\sqrt[4]{a^8 b^{12}}$

48. $\sqrt[5]{243x^{10}y^{25}}$

49. PHYSICS The velocity v of an object can be defined as $v = \sqrt{\frac{2K}{m}}$, where m is the mass of an object and K is the kinetic energy in joules. Find the velocity in meters per second of an object with a mass of 17 grams and a kinetic energy of 850 joules.

Example 5

Simplify $\sqrt{64x^6}$.

$\sqrt{64x^6} = \sqrt{(8x^3)^2}$ — $64x^6 = (8x^3)^2$

$= 8|x^3|$ — Simplify.

Use absolute value symbols because x could be negative.

Example 6

Simplify $\sqrt[6]{4096x^{12}y^{24}}$.

$\sqrt[6]{4096x^{12}y^{24}} = \sqrt[6]{(4x^2y^4)^6}$ — $4096x^{12}y^{24} = (4x^2y^4)^6$

$= 4x^2y^4$ — Simplify.

6-5 Operations with Radical Expressions

Simplify.

50. $\sqrt[3]{54}$

51. $\sqrt{144a^3b^5}$

52. $4\sqrt{6y} \cdot 3\sqrt{7x^2y}$

53. $6\sqrt{72} + 7\sqrt{98} - \sqrt{50}$

54. $(6\sqrt{5} - 2\sqrt{2})(3\sqrt{5} + 4\sqrt{2})$

55. $\dfrac{\sqrt{6m^5}}{\sqrt{p^{11}}}$

56. $\dfrac{3}{5 + \sqrt{2}}$

57. $\dfrac{\sqrt{3}}{\sqrt{5} - \sqrt{6}}$

58. **GEOMETRY** What are the perimeter and the area of the rectangle?

$6 - \sqrt{2}$

$8 + \sqrt{3}$

Example 7

Simplify $2\sqrt[3]{18a^2b} \cdot 3\sqrt[3]{12ab^5}$.

$2\sqrt[3]{18a^2b} \cdot 3\sqrt[3]{12ab^5}$

$= (2 \cdot 3)\sqrt[3]{18a^2b \cdot 12ab^5}$ — Product Property

$= 6\sqrt[3]{2^3 3^3 a^3 b^6}$ — Factor.

$= 6 \cdot \sqrt[3]{2^3} \cdot \sqrt[3]{3^3} \cdot \sqrt[3]{a^3} \cdot \sqrt[3]{b^6}$ — Product Property

$= 6 \cdot 2 \cdot 3 \cdot a \cdot b^2$ — Find cube roots.

$= 36ab^2$ — Simplify.

Example 8

Simplify $\sqrt{\dfrac{x^4}{y^5}}$.

$\sqrt{\dfrac{x^4}{y^5}} = \dfrac{\sqrt{x^4}}{\sqrt{y^5}}$ — Quotient Property

$= \dfrac{\sqrt{(x^2)^2}}{\sqrt{(y^2)^2} \cdot \sqrt{y}}$ — Factor into squares.

$= \dfrac{x^2}{y^2\sqrt{y}} \cdot \dfrac{\sqrt{y}}{\sqrt{y}}$ — Rationalize the denominator.

$= \dfrac{x^2\sqrt{y}}{y^3}$ — $\sqrt{y} \cdot \sqrt{y} = y$

6-6 Rational Exponents

Simplify each expression.

59. $x^{\frac{1}{2}} \cdot x^{\frac{2}{3}}$

60. $m^{-\frac{3}{4}}$

61. $\dfrac{d^{\frac{1}{6}}}{d^{\frac{3}{4}}}$

Simplify each expression.

62. $\dfrac{1}{y^{\frac{1}{4}}}$

63. $\sqrt[3]{\sqrt{729}}$

64. $\dfrac{x^{\frac{2}{3}} - x^{\frac{1}{3}}y^{\frac{2}{3}}}{x^{\frac{1}{3}}}$

65. **GEOMETRY** What is the area of the circle?

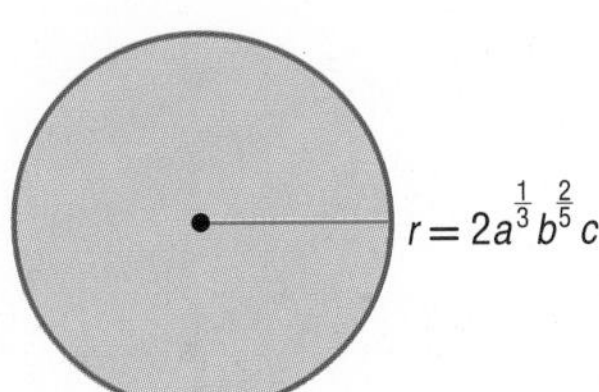

Example 9

Simplify $a^{\frac{2}{3}} \cdot a^{\frac{1}{5}}$.

$a^{\frac{2}{3}} \cdot a^{\frac{1}{5}} = a^{\frac{2}{3} + \frac{1}{5}}$ — Product of Powers

$= a^{\frac{13}{15}}$ — Add.

Example 10

Simplify $\dfrac{2a}{\sqrt[3]{b}}$.

$\dfrac{2a}{\sqrt[3]{b}} = \dfrac{2a}{b^{\frac{1}{3}}}$ — Rational exponents

$= \dfrac{2a}{b^{\frac{1}{3}}} \cdot \dfrac{b^{\frac{2}{3}}}{b^{\frac{2}{3}}}$ — Rationalize the denominator.

$= \dfrac{2ab^{\frac{2}{3}}}{b}$ or $\dfrac{2a\sqrt[3]{b^2}}{b}$ — Rewrite in radical form.

6-7 Solving Radical Equations and Inequalities

Solve each equation.

66. $\sqrt{x-3}+5=15$

67. $-\sqrt{x-11}=3-\sqrt{x}$

68. $4+\sqrt{3x-1}=8$

69. $\sqrt{m+3}=\sqrt{2m+1}$

70. $\sqrt{2x+3}=3$

71. $(x+1)^{\frac{1}{4}}=-3$

72. $a^{\frac{1}{3}}-4=0$

73. $3(3x-1)^{\frac{1}{3}}-6=0$

74. PHYSICS The formula $t=2\pi\sqrt{\frac{\ell}{32}}$ represents the swing of a pendulum, where t is the time in seconds for the pendulum to swing back and forth and ℓ is the length of the pendulum in feet. Find the length of a pendulum that makes one swing in 2.75 seconds.

Solve each inequality.

75. $2+\sqrt{3x-1}<5$

76. $\sqrt{3x+13}-5\geq 5$

77. $6-\sqrt{3x+5}\leq 3$

78. $\sqrt{-3x+4}-5\geq 3$

79. $5+\sqrt{2y-7}<5$

80. $3+\sqrt{2x-3}\geq 3$

81. $\sqrt{3x+1}-\sqrt{6+x}>0$

Example 11

Solve $\sqrt{2x+9}-2=5$.

$\sqrt{2x+9}-2=5$	Original equation
$\sqrt{2x+9}=7$	Add 2 to each side.
$(\sqrt{2x+9})^2=7^2$	Square each side.
$2x+9=49$	Evaluate the squares.
$2x=40$	Subtract 9 from each side.
$x=20$	Divide each side by 2.

Example 12

Solve $\sqrt{2x-5}+2>5$.

$\sqrt{2x-5}\geq 0$	Radicand must be ≥ 0.
$2x-5\geq 0$	Square each side.
$2x\geq 5$	Add 5 to each side.
$x\geq 2.5$	Divide each side by 2.

The solution must be greater than or equal to 2.5 to satisfy the domain restriction.

$\sqrt{2x-5}+2>5$	Original inequality
$\sqrt{2x-5}>3$	Subtract 2 from each side.
$(\sqrt{2x-5})^2>3^2$	Square each side.
$2x-5>9$	Evaluate the squares.
$2x>14$	Add 5 to each side.
$x>7$	Divide each side by 2.

Since $x\geq 2.5$ contains $x>7$, the solution of the inequality is $x>7$.

CHAPTER 6 Practice Test

Determine whether each pair of functions are inverse functions. Write *yes* or *no*. Explain your reasoning.

1. $f(x) = 3x + 8, g(x) = \frac{x-8}{3}$
2. $f(x) = \frac{1}{3}x + 5, g(x) = 3x - 15$
3. $f(x) = x + 7, g(x) = x - 7$
4. $g(x) = 3x - 2, f(x) = \frac{x-2}{3}$

5. **MULTIPLE CHOICE** Which inequality represents the graph below?

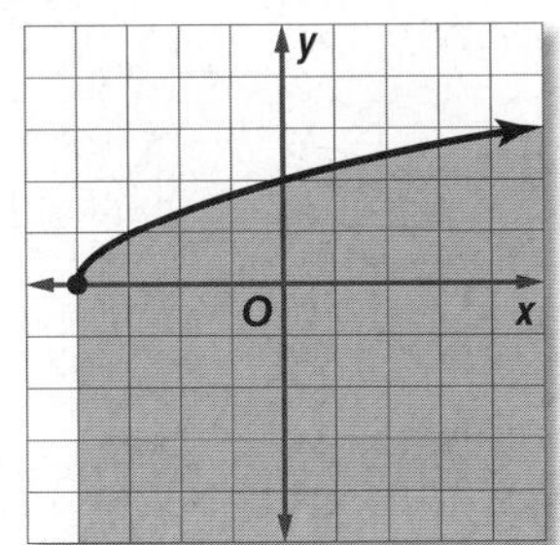

A $y \geq \sqrt{x+4}$
B $y \leq \sqrt{x+4}$
C $y \geq \sqrt{x-4}$
D $y \leq \sqrt{x-4}$

If $f(x) = 3x + 2$ and $g(x) = x^2 - 2x + 1$, find each function.

6. $(f + g)(x)$
7. $(f \cdot g)(x)$
8. $(f - g)(x)$
9. $\left(\frac{f}{g}\right)(x)$

Solve each equation.

10. $\sqrt{a+12} = \sqrt{5a-4}$
11. $\sqrt{3x} = \sqrt{x-2}$
12. $4(\sqrt[4]{3x+1}) - 8 = 0$
13. $\sqrt[3]{5m+6} + 15 = 21$
14. $\sqrt{3x+21} = \sqrt{5x+27}$
15. $1 + \sqrt{x+11} = \sqrt{2x+15}$
16. $\sqrt{x-5} = \sqrt{2x-4}$
17. $\sqrt{x-6} - \sqrt{x} = 3$

18. **MULTIPLE CHOICE** Which expression is equivalent to $125^{-\frac{1}{3}}$?

F -5
G $-\frac{1}{5}$
H $\frac{1}{5}$
J 5

Simplify.

19. $(2+\sqrt{5})(6-3\sqrt{5})$
20. $(3-2\sqrt{2})(-7+\sqrt{2})$
21. $\frac{12}{2-\sqrt{3}}$
22. $\frac{m^{\frac{1}{2}}-1}{2m^{\frac{1}{2}}+1}$
23. $4\sqrt{3} - 8\sqrt{48}$
24. $5^{\frac{2}{3}} \cdot 5^{\frac{1}{2}} \cdot 5^{\frac{5}{6}}$
25. $\sqrt[6]{729a^9b^{24}}$
26. $\sqrt[5]{32x^{15}y^{10}}$
27. $w^{-\frac{4}{5}}$
28. $\frac{r^{\frac{2}{3}}}{r^{\frac{1}{6}}}$
29. $\frac{a^{-\frac{1}{2}}}{6a^{\frac{1}{3}} \cdot a^{-\frac{1}{4}}}$
30. $\frac{y^{\frac{3}{2}}}{y^{\frac{1}{2}}+2}$

31. **MULTIPLE CHOICE** What is the area of the rectangle?

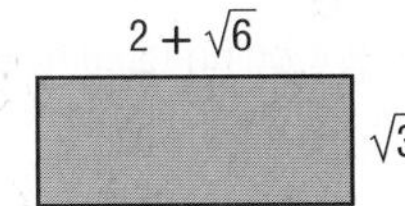

A $2\sqrt{3} + 3\sqrt{2}$ units^2
B $4 + 2\sqrt{6} + 2\sqrt{3}$ units^2
C $2\sqrt{3} + \sqrt{6}$ units^2
D $2\sqrt{3} + 3$ units^2

Solve each inequality.

32. $\sqrt{4x-3} < 5$
33. $-2 + \sqrt{3m-1} < 4$
34. $2 + \sqrt{4x-4} \leq 6$
35. $\sqrt{2x+3} - 4 \leq 5$
36. $\sqrt{b+12} - \sqrt{b} > 2$
37. $\sqrt{y-7} + 5 \geq 10$
38. $\sqrt{a-5} - \sqrt{a+7} \leq 4$
39. $\sqrt{c+5} + \sqrt{c+10} > 2$

40. **GEOMETRY** The area of a triangle with sides of length a, b, and c is given by $A = \sqrt{s(s-a)(s-b)(s-c)}$, where $s = \frac{1}{2}(a+b+c)$. What is the area of the triangle expressed in radical form?

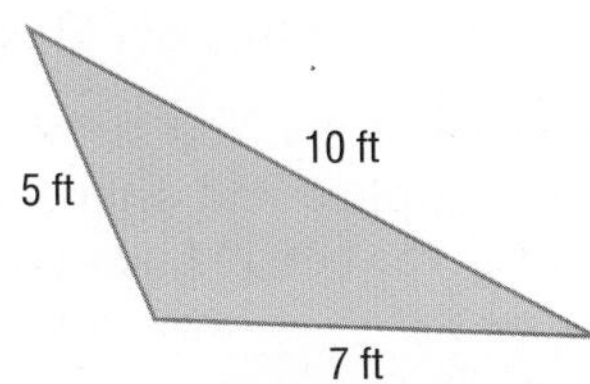

CHAPTER 6 Preparing for Standardized Tests

Work Backward

In certain math problems, you are given information about an end result, but you need to find out something that happened earlier. You can work backward to solve problems like this.

Sean Justice/Stone/Getty Images

Strategies for Working Backward

Step 1

Read the problem statement carefully.

Ask yourself:

- What information am I given?
- What am I being asked to solve?
- Does any of the information given relate to an end result?
- Am I being asked to solve for a quantity that occurred "earlier" in the problem statement?
- What operations are being used in the problem?

Step 2

Model the problem situation with an equation, an inequality, or a graph as appropriate. Then work backward to solve the problem.

- If needed, sketch a flow of events to show the sequence described in the problem statement.
- Use inverse operations to undo any operations while working backward until you arrive at your answer.

Step 3

Check by beginning with your answer and seeing if you arrive at the same result given in the problem statement.

Standardized Test Example

Read the problem. Identify what you need to know. Then use the information in the problem to solve.

Maria bought a used car. The sales tax rate was 6.75% of the selling price, and she had to pay $450 in processing, title, and registration fees. If Maria paid a total of $15,768.63, what was the sale price of the car? Show your work.

Read the problem carefully. You know the total amount that Maria paid for the car after sales tax was applied and after she paid all of the other fees. You need to find the sale price of the car before taxes and fees.

Let x represent the sale price of the car and set up an equation. Use the work backward strategy to solve the problem.

Words	The sale price of the car plus the sales tax and other fees is equal to the final price.
Variable	Let x = sale price.
Equation	$1.0675x + 450 = 15{,}768.63$

Using the work backward strategy results in a simple equation. Use inverse operations to solve for x.

$1.0675x + 450 = 15{,}768.63$	Original equation
$1.0675x = 15{,}318.63$	Subtract 450 from each side.
$x \approx 14{,}350$	Divide each side by 1.0675.

Check your answer by working the problem forward. Begin with your answer and see if you get the same result as in the problem statement.

$14{,}350(1.0675) \approx 15{,}318.63$	Compute the sales tax.
$15{,}318.63 + 450 = 15{,}768.63$	Add the other fees.
$15{,}768.63 = 15{,}768.63$	The result is the same.

So, the sale price of the car was $14,350.

Exercises

Read the problem. Identify what you need to know. Then use the information in the problem to solve.

1. The equation $d = \frac{s^2}{30f}$ can be used to model the length of the skid marks left by a car when a driver applies the brakes to come to a sudden stop. In the equation, d is the length (in feet) of the skid marks left on the road, s is the speed of the car in miles per hour, and f is a coefficient of friction that describes the condition of the road. Suppose a car left skid marks that are 120 feet long.
   a. Solve the equation for s, the speed of the car.
   b. If the coefficient of friction for the road is 0.75, about how fast was the car traveling?
   c. How fast was the car traveling if the coefficient of friction for the road is 1.1?

2. An object is shot straight upward into the air with an initial speed of 800 feet per second. The height h that the object will be after t seconds is given by the equation $h = -16t^2 + 800t$. When will the object reach a height of 10,000 feet?
   A 10 seconds
   B 25 seconds
   C 100 seconds
   D 625 seconds

3. Pedro is creating a scale drawing of a car. He finds that the height of the car in the drawing is $\frac{1}{32}$ of the actual height of the car x. Which equation best represents this relationship?
   F $y = x - \frac{1}{32}$
   G $y = -\frac{1}{32}x$
   H $y = \frac{1}{32}x$
   J $y = x + \frac{1}{32}$

CHAPTER 6

Standardized Test Practice

Cumulative, Chapters 1 through 6

Multiple Choice

Read each question. Then fill in the correct answer on the answer document provided by your teacher or on a sheet of paper.

1. A sporting goods store is discounting all camping equipment by 20% during the off-season. Charles also has a coupon good for $5.00 off his next purchase from the store. If the coupon is applied *after* the store discount, which of the following functions can be used to find the final price of a tent that originally cost d dollars?

A $P(d) = 0.8 \times (d + 5)$

B $P(d) = (0.8 \times d) - 5$

C $P(d) = 0.2 \times (d - 5)$

D $P(d) = 0.8 \times (d - 5)$

2. Find the equation that can be used to determine the total area of the composite figure below.

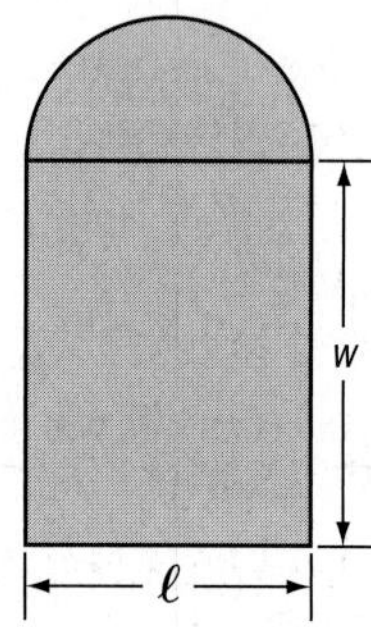

F $A = \ell w + \frac{1}{2}\ell w$

G $A = \ell w + \pi\left(\frac{1}{2}\ell\right)^2$

H $A = \ell w + \frac{1}{2}\pi\ell^2$

J $A = \ell w + \pi\left(\frac{1}{2}\ell\right)^2\left(\frac{1}{2}\right)$

3. Which expression is equivalent to $3a(2a + 1) - (2a - 2)(a + 3)$?

A $2a^2 + 6a + 7$

B $4a^2 - a + 6$

C $4a^2 + 6a - 6$

D $4a^2 - 3a + 7$

4. Kay bought a used car. The sales tax rate was 6.5% of the selling price, and she also had to pay $325 in registration fees. Find the selling price if Kay spent a total of $15,501.25.

F $13,850

G $14,120

H $14,250

J $14,650

5. Simplify $\sqrt[3]{-27b^6c^{12}}$.

A $-3b^3c^6$

B $-3b^2c^4$

C $3b^2c^4$

D $3b^3c^6$

6. What is the equation of the square root function graphed at the right?

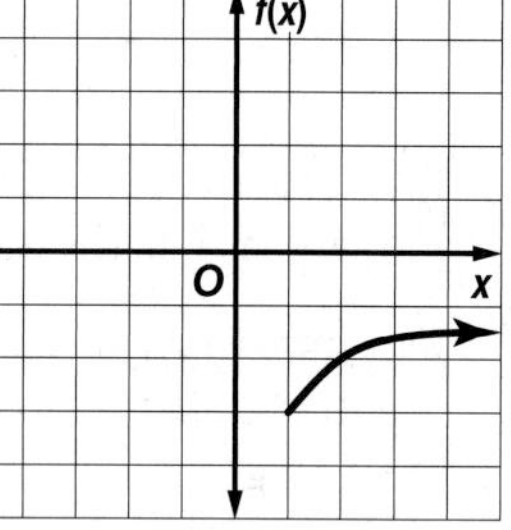

F $f(x) = \sqrt{x - 3} - 1$

G $f(x) = \sqrt{x + 1} - 3$

H $f(x) = \sqrt{x + 3} + 1$

J $f(x) = \sqrt{x - 1} - 3$

7. Which equation will produce the narrowest parabola when graphed?

A $y = 3x^2$

B $y = \frac{3}{4}x^2$

C $y = -6x^2$

D $y = -\frac{3}{4}x^2$

8. Find the inverse of $f(x) = x - 5$.

F $f(x) = x + 5$

G $f(x) = 5x$

H $f(x) = \frac{x}{5}$

J $f(x) = 5 - x$

9. The equations of two lines are $2x - y = 6$ and $4x - y = -2$. Which of the following describes their point of intersection?

A $(2, -2)$

B $(-8, -38)$

C $(-4, -14)$

D no intersection

Test-TakingTip

Question 4 You know the final price but need to know the sale price. Work backward to find the solution.

Short Response/Gridded Response

Record your answers on the answer sheet provided by your teacher or on a sheet of paper.

10. Which of the following terms does *not* describe the system of equations graphed below: consistent, dependent, independent, or intersecting?

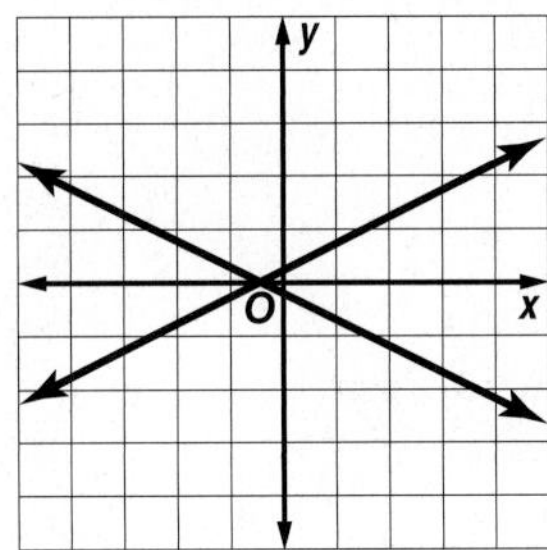

11. Suppose a projectile is launched into the air from a platform. The formula $h = -16t^2 + 40t + 70$ relates the height h of the object (in feet) and the time t since it was launched (in seconds). What is the maximum height the object reaches?

12. The radius of a sphere with volume V can be found using the formula $r = \sqrt[3]{\frac{3V}{4\pi}}$.

a. What is the radius of the sphere at the right? Round to the nearest tenth.

V = 8580 cu. in.

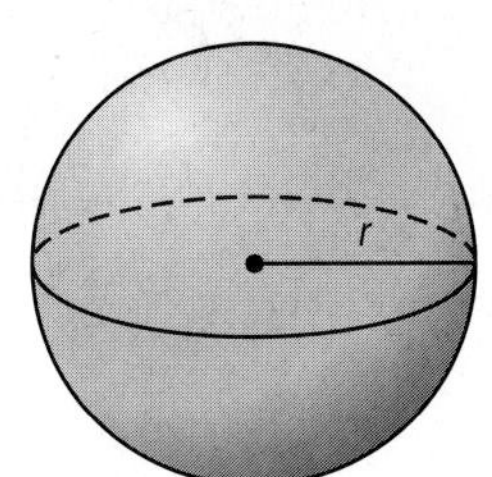

b. Solve the formula for V to find the formula for the volume of a sphere, given its radius.

c. What is the volume of a basketball that has a diameter of 9 inches? Round to the nearest tenth.

13. GRIDDED RESPONSE The perimeter of the quadrilateral below is 160. What is the value of m?

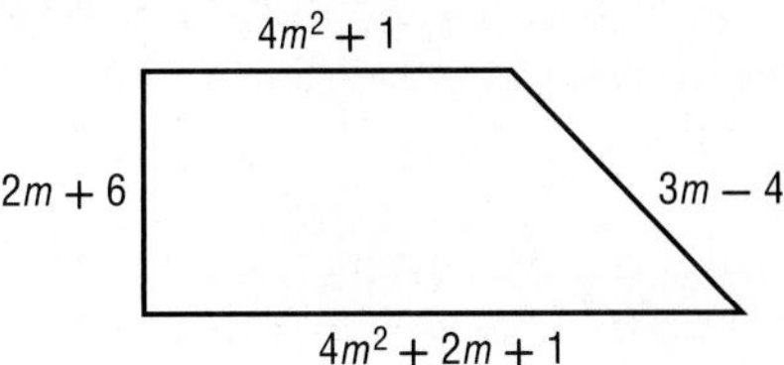

Extended Response

Record your answers on a sheet of paper. Show your work.

14. The amount that a retailer charges for shipping an electronics purchase is determined by the weight of the package. The charges for several different weights are given in the table.

a. Find the rate of change of the shipping charge per pound.

b. Write an equation that could be used to find the shipping charge y for a package that weighs x pounds.

c. Find the shipping charge for a package that weighs 19 pounds.

Electronics Shipping Charges

Weight (lb)	Shipping ($)
1	5.58
3	6.76
4	7.35
7	9.12
10	10.89
13	12.66
15	13.84

15. Suppose $f(x)$ and $g(x)$ are inverse functions.

a. Describe how the graphs of $f(x)$ and $g(x)$ would appear on a coordinate grid.

b. What is the value of the composition $f[g(2)]$? Explain.

Need ExtraHelp?

If you missed Question...	1	2	3	4	5	6	7	8	9	10	11	12	13	14	15
Go to Lesson...	6-1	1-1	5-1	6-2	6-4	6-3	4-7	6-2	3-1	3-1	6-7	6-4	5-1	2-4	6-2

CHAPTER 7

Exponential and Logarithmic Functions and Relations

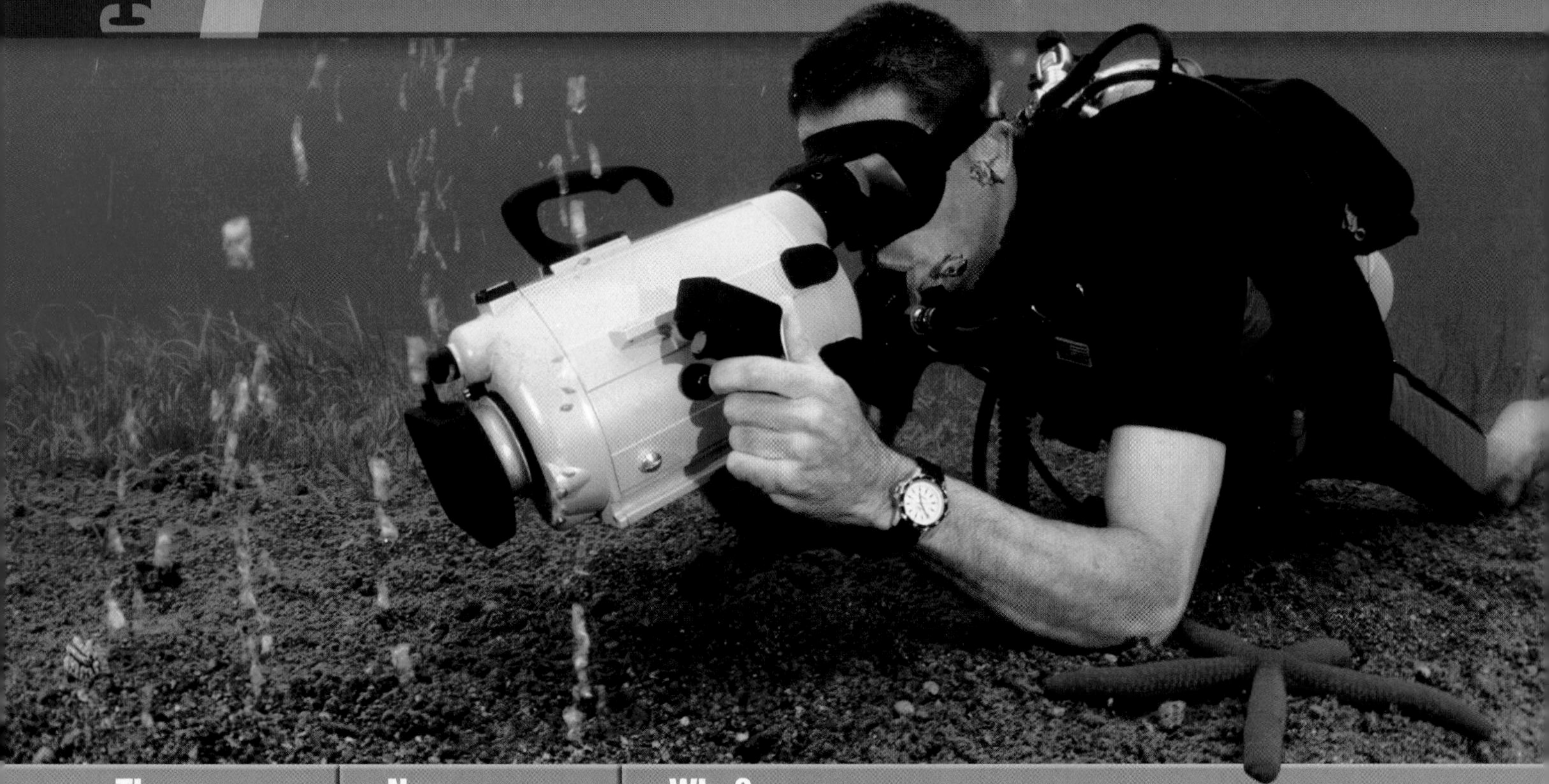

Then

- You graphed functions and transformations of functions.

Now

- You will:
  - Graph exponential and logarithmic functions.
  - Solve exponential and logarithmic equations and inequalities.
  - Solve problems involving exponential growth and decay.

Why? ▲

- **SCIENCE** Mathematics and science go hand in hand. Whether it is chemistry, biology, paleontology, zoology, or anthropology, you will need strong math skills. In this chapter, you will learn mathematical aspects of science such as computer viruses, populations of insects, bacteria growth, cell division, astronomy, tornados, and earthquakes.

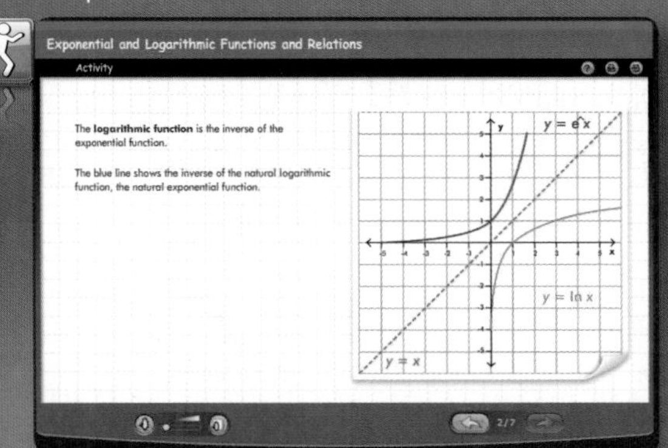

connectED.mcgraw-hill.com Your Digital Math Portal

Animation  Vocabulary  eGlossary  Personal Tutor  Virtual Manipulatives  Graphing Calculator  Audio  Foldables  Self-Check Practice  Worksheets  

Get Ready for the Chapter

Diagnose Readiness | You have two options for checking prerequisite skills.

1 Textbook Option
Take the Quick Check below. Refer to the Quick Review for help.

QuickCheck

Simplify. Assume that no variable equals zero.

1. $a^4a^3a^5$
2. $(2xy^3z^2)^3$
3. $\frac{-24x^8y^5z}{16x^2y^8z^6}$
4. $\left(\frac{-8r^2n}{36n^3t}\right)^2$
5. **DENSITY** The density of an object is equal to the mass divided by the volume. An object has a mass of 7.5×10^3 grams and a volume of 1.5×10^3 cubic centimeters. What is the density of the object?

QuickReview

Example 1

Simplify $\frac{(a^3bc^2)^2}{a^4a^2b^2bc^5c^3}$. Assume that no variable equals zero.

$\frac{(a^3bc^2)^2}{a^4a^2b^2bc^5c^3}$

$= \frac{a^6b^2c^4}{a^6b^3c^8}$ Simplify the numerator by using the Power of a Power Rule and the denominator by using the Product of Powers Rule.

$= \frac{1}{bc^4}$ or $b^{-1}c^{-4}$ Simplify by using the Quotient of Powers Rule.

QuickCheck

Find the inverse of each function. Then graph the function and its inverse.

6. $f(x) = 2x + 5$
7. $f(x) = x - 3$
8. $f(x) = -4x$
9. $f(x) = \frac{1}{4}x - 3$
10. $f(x) = \frac{x-1}{2}$
11. $y = \frac{1}{3}x + 4$

Determine whether each pair of functions are inverse functions.

12. $f(x) = x - 6$
    $g(x) = x + 6$
13. $f(x) = 2x + 5$
    $g(x) = 2x - 5$
14. **FOOD** A pizzeria charges $12 for a medium cheese pizza and $2 for each additional topping. If $f(x) = 2x + 12$ represents the cost of a medium pizza with x toppings, find $f^{-1}(x)$ and explain its meaning.

QuickReview

Example 2

Find the inverse of $f(x) = 3x - 1$.

Step 1 Replace $f(x)$ with y in the original equation:
$f(x) = 3x - 1 \rightarrow y = 3x - 1$.

Step 2 Interchange x and y: $x = 3y - 1$.

Step 3 Solve for y.

$x = 3y - 1$ Inverse

$x + 1 = 3y$ Add 1 to each side.

$\frac{x+1}{3} = y$ Divide each side by 3.

$\frac{1}{3}x + \frac{1}{3} = y$ Simplify.

Step 4 Replace y with $f^{-1}(x)$.

$y = \frac{1}{3}x + \frac{1}{3} \rightarrow f^{-1}(x) = \frac{1}{3}x + \frac{1}{3}$

2 Online Option
Take an online self-check Chapter Readiness Quiz at connectED.mcgraw-hill.com.

Get Started on the Chapter

You will learn several new concepts, skills, and vocabulary terms as you study Chapter 7. To get ready, identify important terms and organize your resources. You may wish to refer to Chapter 0 to review prerequisite skills.

FOLDABLES® StudyOrganizer

Exponential and Logarithmic Functions and Relations
Make this Foldable to help you organize your Chapter 7 notes about exponential and logarithmic functions. Begin with two sheets of grid paper.

1 **Fold** in half along the width.

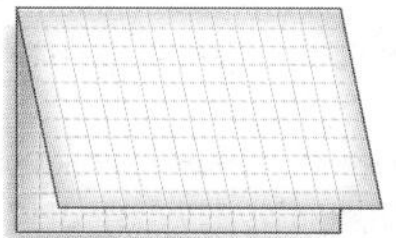

2 **On** the first sheet, cut 5 cm along the fold at the ends.

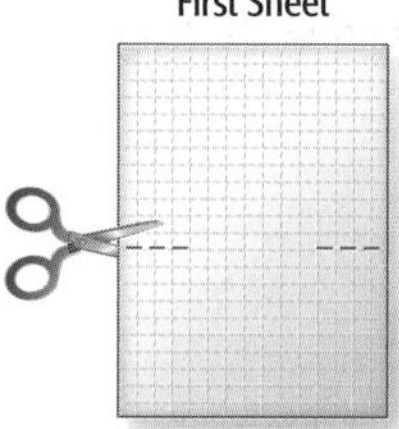

3 **On** the second sheet, cut in the center, stopping 5 cm from the ends.

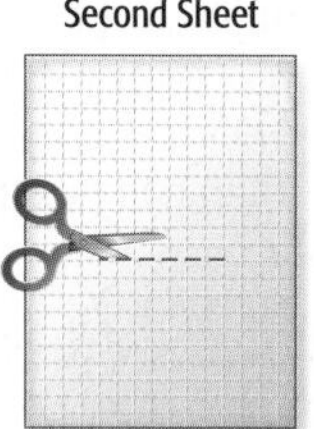

4 **Insert** the first sheet through the second sheet and align the folds. Label the pages with lesson numbers.

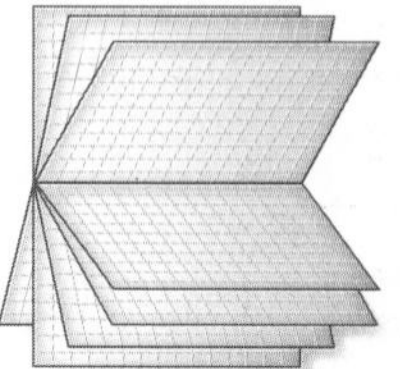

NewVocabulary

English		Español
exponential function	p. 451	función exponencial
exponential growth	p. 451	crecimiento exponencial
asymptote	p. 451	asíntota
growth factor	p. 453	factor de crecimiento
exponential decay	p. 453	desintegración exponencial
decay factor	p. 454	factor de desintegración
exponential equation	p. 461	ecuación exponencial
compound interest	p. 462	interés compuesto
exponential inequality	p. 463	desigualdad exponencial
logarithm	p. 468	logaritmo
logarithmic function	p. 469	función logarítmica
logarithmic equation	p. 478	ecuación logarítmica
logarithmic inequality	p. 479	desigualdad logarítmica
common logarithm	p. 492	logaritmos communes
Change of Base Formula	p. 494	fórmula del cambio de base
natural base, *e*	p. 501	*e* base natural
natural base exponential function	p. 501	base natural función exponencial
natural logarithm	p. 501	logaritmo natural

ReviewVocabulary

domain dominio the set of all *x*-coordinates of the ordered pairs of a relation

function función a relation in which each element of the domain is paired with exactly one element in the range

range rango the set of all *y*-coordinates of the ordered pairs of a relation

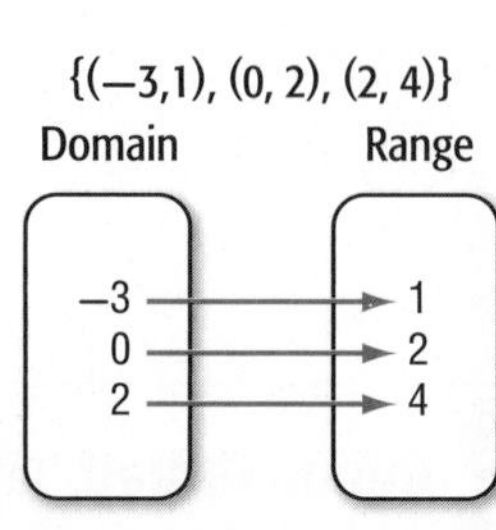

LESSON 7-1

Graphing Exponential Functions

Then

- You graphed polynomial functions.

Now

1. Graph exponential growth functions.
2. Graph exponential decay functions.

Why?

- Have you ever received an e-mail that tells you to forward it to 5 friends? If each of those 5 friends then forwards it to 5 of their friends, who each forward it to 5 of their friends, the number of people receiving the e-mail is growing exponentially.

The equation $y = 5^x$ can be used to represent this situation, where x is the number of rounds that the e-mail has been forwarded.

NewVocabulary

exponential function
exponential growth
asymptote
growth factor
exponential decay
decay factor

Common Core State Standards

Content Standards

F.IF.7.e Graph exponential and logarithmic functions, showing intercepts and end behavior, and trigonometric functions, showing period, midline, and amplitude.

F.IF.8.b Use the properties of exponents to interpret expressions for exponential functions.

Mathematical Practices

3 Construct viable arguments and critique the reasoning of others.

1 Exponential Growth

A function like $y = 5^x$, where the base is a constant and the exponent is the independent variable, is an **exponential function**. One type of exponential function is exponential growth. An **exponential growth** function is a function of the form $f(x) = b^x$, where $b > 1$. The graph of an exponential function has an **asymptote**, which is a line that the graph of the function approaches.

KeyConcept Parent Function of Exponential Growth Functions

Parent Functions: $f(x) = b^x, b > 1$

Type of graph: continuous, one-to-one, and increasing

Domain: all real numbers

Range: all positive real numbers

Asymptote: x-axis

Intercept: (0, 1)

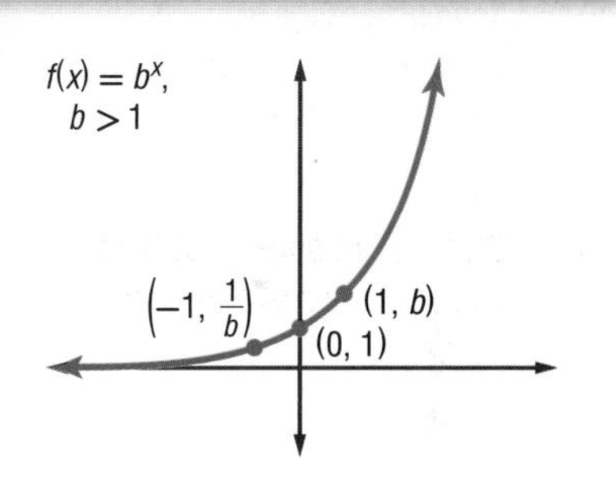

Example 1 Graph Exponential Growth Functions

Graph $y = 3^x$. State the domain and range.

Make a table of values. Then plot the points and sketch the graph.

x	-3	-2	$-\frac{1}{2}$	0
$y = 3^x$	$3^{-3} = \frac{1}{27}$	$3^{-2} = \frac{1}{9}$	$3^{-\frac{1}{2}} = \frac{\sqrt{3}}{3}$	$3^0 = 1$

x	1	$\frac{3}{2}$	2
$y = 3^x$	$3^1 = 3$	$3^{\frac{3}{2}} = \sqrt{27}$	$3^2 = 9$

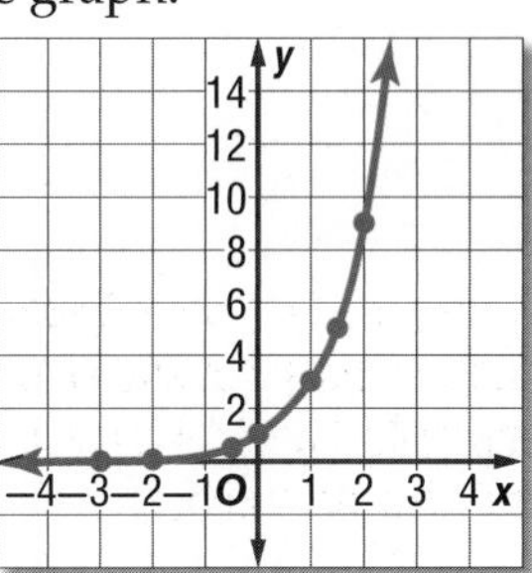

The domain is all real numbers, and the range is all positive real numbers.

GuidedPractice

1. Graph $y = 4^x$. State the domain and range.

Stockbyte/Alamy

The graph of $f(x) = b^x$ represents a parent graph of the exponential functions. The same techniques used to transform the graphs of other functions you have studied can be applied to the graphs of exponential functions.

KeyConcept Transformations of Exponential Functions

$$f(x) = ab^{x-h} + k$$

h – Horizontal Translation	k – Vertical Translation
h units right if h is positive $\|h\|$ units left if h is negative	k units up if k is positive $\|k\|$ units down if k is negative

a – Orientation and Shape

If $a < 0$, the graph is reflected in the x-axis.

If $|a| > 1$, the graph is stretched vertically.
If $0 < |a| < 1$, the graph is compressed vertically.

StudyTip

Precision Remember that end behavior is the action of the graph as x approaches positive infinity or negative infinity. In Example 2a, as x approaches infinity, y approaches infinity. In Example 2b, as x approaches infinity, y approaches negative infinity.

Example 2 Graph Transformations

Graph each function. State the domain and range.

a. $y = 2^x + 1$

The equation represents a translation of the graph of $y = 2^x$ one unit up.

x	$y = 2^x + 1$
−3	$2^{-3} + 1 = 1.125$
−2	$2^{-2} + 1 = 1.25$
−1	$2^{-1} + 1 = 1.5$
0	$2^0 + 1 = 2$
1	$2^1 + 1 = 3$
2	$2^2 + 1 = 5$
3	$2^3 + 1 = 9$

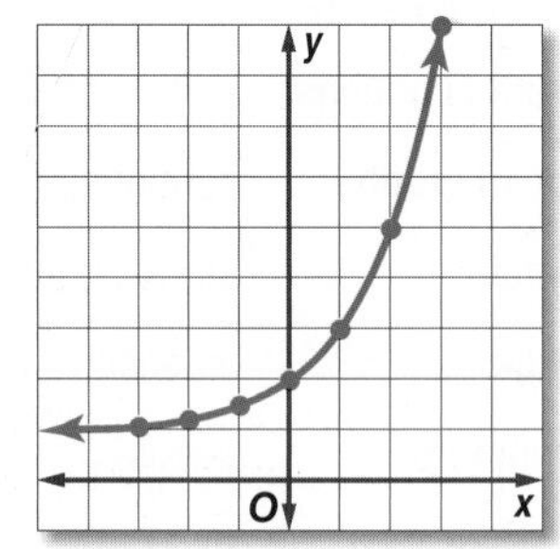

Domain = {all real numbers}; Range = $\{y \mid y > 1\}$

b. $y = -\frac{1}{2} \cdot 5^{x-2}$

The equation represents a transformation of the graph of $y = 5^x$.

Graph $y = 5^x$ and transform the graph.

- $a = -\frac{1}{2}$: The graph is reflected in the x-axis and compressed vertically.
- $h = 2$: The graph is translated 2 units right.
- $k = 0$: The graph is not translated vertically.

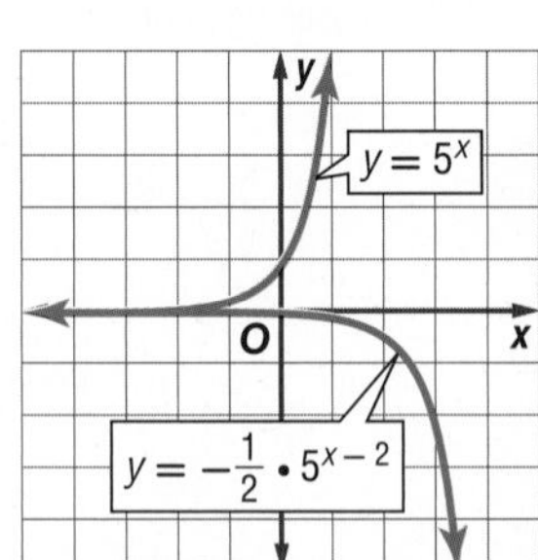

Domain = {all real numbers}

Range = $\{y \mid y < 0\}$

GuidedPractice

2A. $y = 2^{x+3} - 5$

2B. $y = 0.1(6)^x - 3$

You can model exponential growth with a constant percent increase over specific time periods using the following function.

$$A(t) = a(1 + r)^t$$

The function can be used to find the amount $A(t)$ after t time periods, where a is the initial amount and r is the percent of increase per time period. Note that the base of the exponential expression, $1 + r$, is called the **growth factor**.

The exponential growth function is often used to model population growth.

Real-WorldLink

The U.S. Census Bureau's American Community Survey is mailed to approximately 1 out of every 480 households.

Source: Census Bureau

Real-World Example 3 Graph Exponential Growth Functions

CENSUS The first U.S. census was conducted in 1790. At that time, the population was 3,929,214. Since then, the U.S. population has grown by approximately 2.03% annually. Draw a graph showing the population growth of the U.S. since 1790.

First, write an equation using $a = 3{,}929{,}214$, and $r = 0.0203$.

$y = 3{,}929{,}214(1.0203)^t$

Then graph the equation.

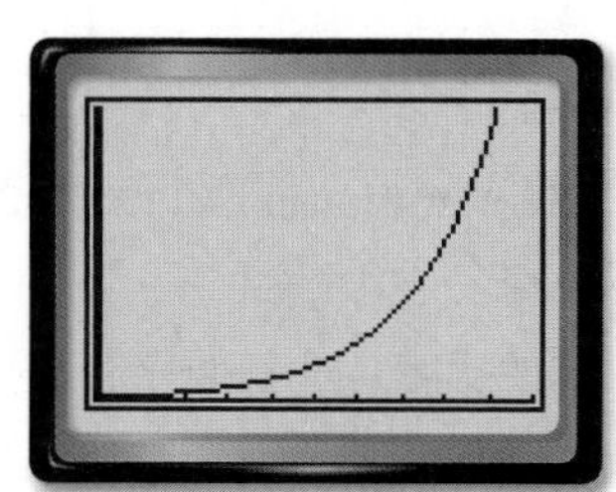

[0, 250] scl: 25 by [0, 400,000,000] scl: 40,000,000

GuidedPractice

3. FINANCIAL LITERACY Teen spending is expected to grow 3.5% annually from \$79.7 billion in 2006. Draw a graph to show the spending growth.

StudyTip

Interest The formula for simple interest, $i = prt$, illustrates linear growth over time. However, the formula for compound interest, $A(t) = a(1 + r)^t$, illustrates exponential growth over time. This is why investments with compound interest make more money.

2 Exponential Decay

The second type of exponential function is **exponential decay**.

KeyConcept Parent Function of Exponential Decay Functions

Parent Functions:	$f(x) = b^x$, $0 < b < 1$
Type of graph:	continuous, one-to-one, and decreasing
Domain:	all real numbers
Range:	positive real numbers
Asymptote:	x-axis
Intercept:	(0, 1)

Model

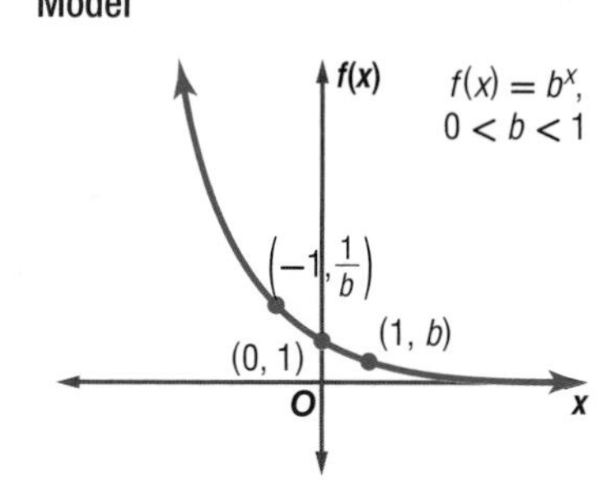

The graphs of exponential decay functions can be transformed in the same manner as those of exponential growth.

Creatas/SuperStock

StudyTip

Exponential Decay Be sure not to confuse a dilation in which $|a| < 1$ with exponential decay in which $0 < b < 1$.

Example 4 Graph Exponential Decay Functions

Graph each function. State the domain and range.

a. $y = \left(\frac{1}{3}\right)^x$

x	$y = \left(\frac{1}{3}\right)^x$
-3	$\left(\frac{1}{3}\right)^{-3} = 27$
-2	$\left(\frac{1}{3}\right)^{-2} = 9$
$-\frac{1}{2}$	$\left(\frac{1}{3}\right)^{-\frac{1}{2}} = \sqrt{3}$
0	$\left(\frac{1}{3}\right)^{0} = 1$
1	$\left(\frac{1}{3}\right)^{1} = \frac{1}{3}$
$\frac{3}{2}$	$\left(\frac{1}{3}\right)^{\frac{3}{2}} = \sqrt{\frac{1}{27}}$
2	$\left(\frac{1}{3}\right)^{2} = \frac{1}{9}$

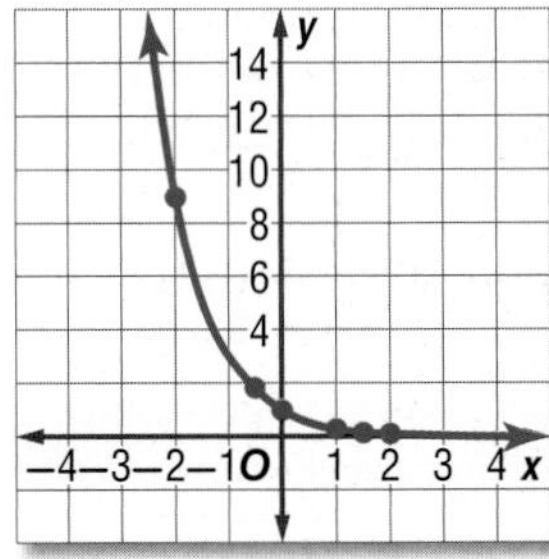

The domain is all real numbers, and the range is all positive real numbers.

b. $y = 2\left(\frac{1}{4}\right)^{x+2} - 3$

The equation represents a transformation of the graph of $y = \left(\frac{1}{4}\right)^x$.

Examine each parameter.

- $a = 2$: The graph is stretched vertically.
- $h = -2$: The graph is translated 2 units left.
- $k = -3$: The graph is translated 3 units down.

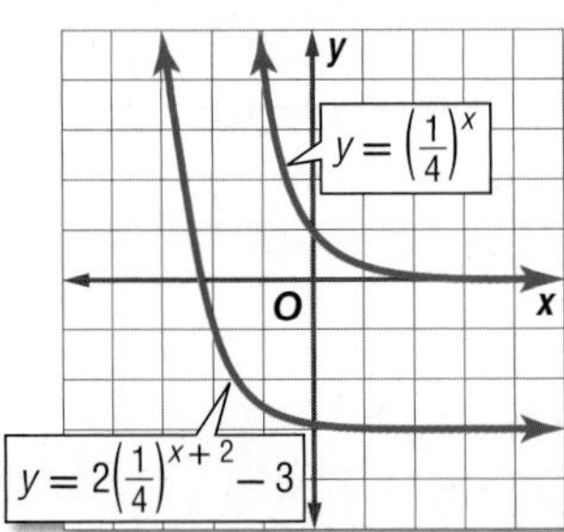

The domain is all real numbers, and the range is all real numbers greater than -3.

GuidedPractice

4A. $y = -3\left(\frac{2}{5}\right)^{x-4} + 2$

4B. $y = \frac{3}{8}\left(\frac{5}{6}\right)^{x-1} + 1$

Similar to exponential growth, you can model exponential decay with a constant percent of decrease over specific time periods using the following function.

$$A(t) = a(1 - r)^t$$

The base of the exponential expression, $1 - r$, is called the **decay factor**.

Real-WorldLink

After water, tea is the most consumed beverage in the U.S. It can be found in over 80% of American households. Just over half the American population drinks tea daily.

Source: The Tea Association of the USA

Real-World Example 5 Graph Exponential Decay Functions

TEA A cup of green tea contains 35 milligrams of caffeine. The average teen can eliminate approximately 12.5% of the caffeine from their system per hour.

a. Draw a graph to represent the amount of caffeine remaining after drinking a cup of green tea.

$y = a(1 - r)^t$

$= 35(1 - 0.125)^t$

$= 35(0.875)^t$

Graph the equation.

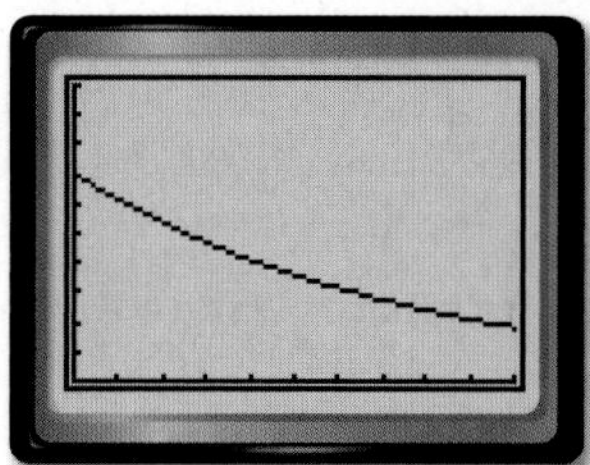

[0, 10] scl: 1 by [0, 50] scl: 5

b. Estimate the amount of caffeine in a teenager's body 3 hours after drinking a cup of green tea.

$y = 35(0.875)^t$ Equation from part a

$= 35(0.875)^3$ Replace t with 3.

≈ 23.45 Use a calculator.

The caffeine in a teenager will be about 23.45 milligrams after 3 hours.

GuidedPractice

5. A cup of black tea contains about 68 milligrams of caffeine. Draw a graph to represent the amount of caffeine remaining in the body of an average teen after drinking a cup of black tea. Estimate the amount of caffeine in the body 2 hours after drinking a cup of black tea.

Check Your Understanding

= Step-by-Step Solutions begin on page R14.

Examples 1–2 **Graph each function. State the domain and range.**

1. $f(x) = 2^x$

2. $f(x) = 5^x$

3 $f(x) = 3^{x-2} + 4$

4. $f(x) = 2^{x+1} + 3$

5. $f(x) = 0.25(4)^x - 6$

6. $f(x) = 3(2)^x + 8$

Example 3

7. CCSS **SENSE-MAKING** A virus spreads through a network of computers such that each minute, 25% more computers are infected. If the virus began at only one computer, graph the function for the first hour of the spread of the virus.

Example 4 **Graph each function. State the domain and range.**

8. $f(x) = 2\left(\frac{2}{3}\right)^{x-3} - 4$

9. $f(x) = -\frac{1}{2}\left(\frac{3}{4}\right)^{x+1} + 5$

10. $f(x) = -\frac{1}{3}\left(\frac{4}{5}\right)^{x-4} + 3$

11. $f(x) = \frac{1}{8}\left(\frac{1}{4}\right)^{x+6} + 7$

Example 5

12. FINANCIAL LITERACY A new SUV depreciates in value each year by a factor of 15%. Draw a graph of the SUV's value for the first 20 years after the initial purchase.

Practice and Problem Solving

Extra Practice is on page R7.

Examples 1–2 **Graph each function. State the domain and range.**

13. $f(x) = 2(3)^x$
14. $f(x) = -2(4)^x$
15. $f(x) = 4^{x+1} - 5$
16. $f(x) = 3^{2x} + 1$
17. $f(x) = -0.4(3)^{x+2} + 4$
18. $f(x) = 1.5(2)^x + 6$

Example 3

19. **SCIENCE** The population of a colony of beetles grows 30% each week for 10 weeks. If the initial population is 65 beetles, graph the function that represents the situation.

Example 4 **Graph each function. State the domain and range.**

20. $f(x) = -4\left(\frac{3}{5}\right)^{x+4} + 3$
21. $f(x) = 3\left(\frac{2}{5}\right)^{x-3} - 6$
22. $f(x) = \frac{1}{2}\left(\frac{1}{5}\right)^{x+5} + 8$
23. $f(x) = \frac{3}{4}\left(\frac{2}{3}\right)^{x+4} - 2$
24. $f(x) = -\frac{1}{2}\left(\frac{3}{8}\right)^{x+2} + 9$
25. $f(x) = -\frac{5}{4}\left(\frac{4}{5}\right)^{x+4} + 2$

Example 5

26. **ATTENDANCE** The attendance for a basketball team declined at a rate of 5% per game throughout a losing season. Graph the function modeling the attendance if 15 home games were played and 23,500 people were at the first game.

27. **PHONES** The function $P(x) = 2.28(0.9^x)$ can be used to model the number of pay phones in millions x years since 1999.
    a. Classify the function as either exponential *growth* or *decay*, and identify the growth or decay factor. Then graph the function.
    b. Explain what the $P(x)$-intercept and the asymptote represent in this situation.

28. **HEALTH** Each day, 10% of a certain drug dissipates from the system.
    a. Classify the function representing this situation as either exponential *growth* or *decay*, and identify the growth or decay factor. Then graph the function.
    b. How much of the original amount remains in the system after 9 days?
    c. If a second dose should not be taken if more than 50% of the original amount is in the system, when should the label say it is safe to redose? Design the label and explain your reasoning.

29. **CCSS REASONING** A sequence of numbers follows a pattern in which the next number is 125% of the previous number. The first number in the pattern is 18.
    a. Write the function that represents the situation.
    b. Classify the function as either exponential *growth* or *decay*, and identify the growth or decay factor. Then graph the function for the first 10 numbers.
    c. What is the value of the tenth number? Round to the nearest whole number.

For each graph, $f(x)$ is the parent function and $g(x)$ is a transformation of $f(x)$. Use the graph to determine the equation of $g(x)$.

30. $f(x) = 3^x$

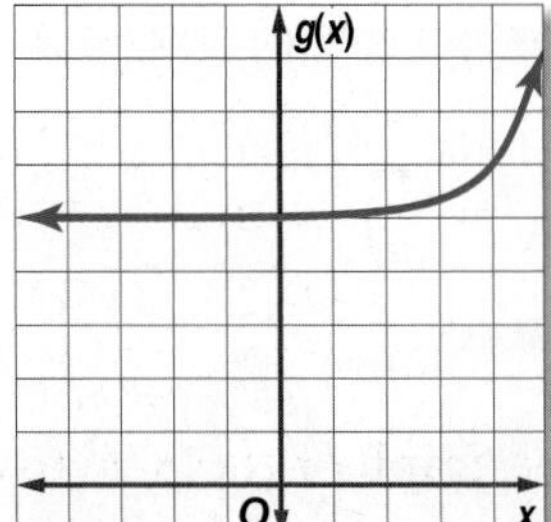

31. $f(x) = 2^x$

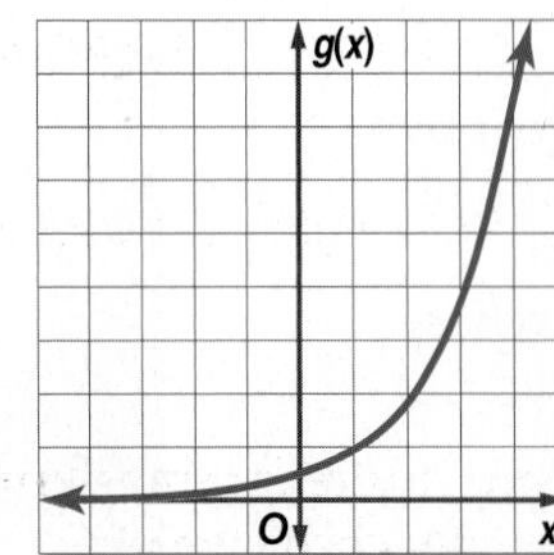

32. $f(x) = 4^x$

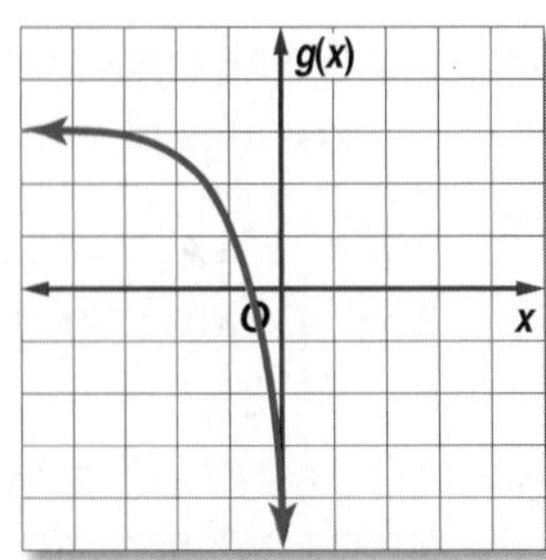

33. **MULTIPLE REPRESENTATIONS** In this problem, you will use the tables below for exponential functions $f(x)$, $g(x)$, and $h(x)$.

x	−1	0	1	2	3	4	5
$f(x)$	2.5	2	1	−1	−5	−13	−29

x	−1	0	1	2	3	4	5
$g(x)$	5	11	23	47	95	191	383

x	−1	0	1	2	3	4	5
$h(x)$	3	2.5	2.25	2.125	2.0625	2.0313	2.0156

a. Graphical Graph the functions for $-1 \le x \le 5$ on separate graphs.

b. Verbal List any function with a negative coefficient. Explain your reasoning.

c. Analytical List any function with a graph that is translated to the left.

d. Analytical Determine which functions are growth models and which are decay models.

H.O.T. Problems Use Higher-Order Thinking Skills

34. REASONING Determine whether each statement is *sometimes*, *always*, or *never* true. Explain your reasoning.

a. An exponential function of the form $y = ab^{x-h} + k$ has a y-intercept.

b. An exponential function of the form $y = ab^{x-h} + k$ has an x-intercept.

c. The function $f(x) = |b|^x$ is an exponential growth function if b is an integer.

35. CCSS CRITIQUE Vince and Grady were asked to graph the following functions. Vince thinks they are the same, but Grady disagrees. Who is correct? Explain your reasoning.

x	y
0	2
1	1
2	0.5
3	0.25
4	0.125
5	0.0625
6	0.03125

an exponential function with rate of decay of $\frac{1}{2}$ and an initial amount of 2

36. CHALLENGE A substance decays 35% each day. After 8 days, there are 8 milligrams of the substance remaining. How many milligrams were there initially?

37. OPEN ENDED Give an example of a value of b for which $f(x) = \left(\frac{8}{b}\right)^x$ represents exponential decay.

38. WRITING IN MATH Write the procedure for transforming the graph of $g(x) = b^x$ to the graph of $f(x) = ab^{x-h} + k$. Justify each step.

Standardized Test Practice

39. GRIDDED RESPONSE In the figure, $\overline{PO} \parallel \overline{RN}$, $ON = 12$, $MN = 6$, and $RN = 4$. What is the length of $\overline{PO}$?

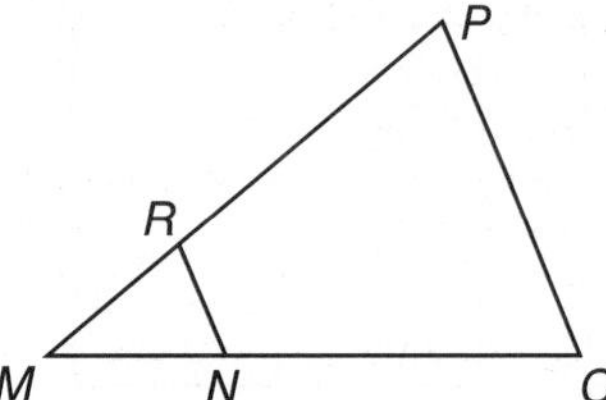

40. Ivan has enough money to buy 12 used CDs. If the cost of each CD was \$0.20 less, Ivan could buy 2 more CDs. How much money does Ivan have to spend on CDs?

A \$16.80
B \$16.40
C \$15.80
D \$15.40

41. One hundred students will attend the fall dance if tickets cost \$30 each. For each \$5 increase in price, 10 fewer students will attend. What price will deliver the maximum dollar sales?

F \$30
G \$35
H \$40
J \$45

42. SAT/ACT Javier mows a lawn in 2 hours. Tonya mows the same lawn in 1.5 hours. About how many minutes will it take to mow the lawn if Javier and Tonya work together?

A 28 minutes
B 42 minutes
C 51 minutes
D 1.2 hours
E 1.4 hours

Spiral Review

Solve each equation or inequality. (Lesson 6-7)

43. $\sqrt{y+5} = \sqrt{2y-3}$

44. $\sqrt{y+1} + \sqrt{y-4} = 5$

45. $10 - \sqrt{2x+7} \leq 3$

46. $6 + \sqrt{3y+4} < 6$

47. $\sqrt{d+3} + \sqrt{d+7} > 4$

48. $\sqrt{2x+5} - \sqrt{9+x} > 0$

Simplify. (Lesson 6-6)

49. $\frac{1}{y^{\frac{2}{5}}}$

50. $\frac{xy}{\sqrt[3]{z}}$

51. $\frac{3x + 4x^2}{x^{-\frac{2}{3}}}$

52. $\sqrt[6]{27x^3}$

53. $\frac{\sqrt[4]{27}}{\sqrt[4]{3}}$

54. $\frac{a^{-\frac{1}{2}}}{6a^{\frac{1}{3}} \cdot a^{-\frac{1}{4}}}$

55. FOOTBALL The path of a football thrown across a field is given by the equation $y = -0.005x^2 + x + 5$, where x represents the distance, in feet, the ball has traveled horizontally and y represents the height, in feet, of the ball above ground level. About how far has the ball traveled horizontally when it returns to ground level? (Lesson 4-6)

56. COMMUNITY SERVICE A drug awareness program is being presented at a theater that seats 300 people. Proceeds will be donated to a local drug information center. If every two adults must bring at least one student, what is the maximum amount of money that can be raised? (Lesson 3-3)

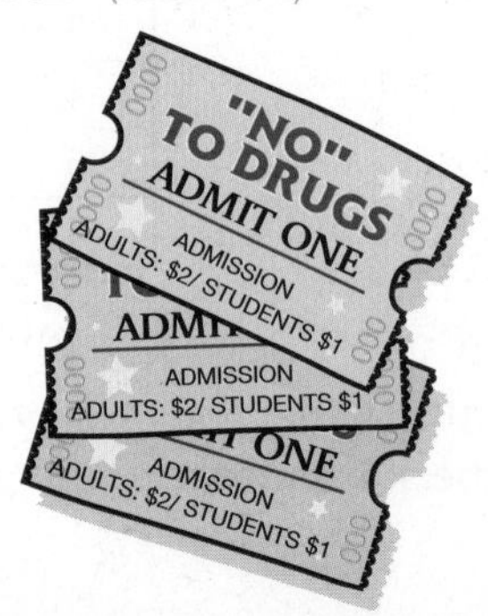

Skills Review

Simplify. Assume that no variable equals 0.

57. $f^{-7} \cdot f^4$

58. $(3x^2)^3$

59. $(2y)(4xy^3)$

60. $\left(\frac{3}{5}c^2f\right)\left(\frac{4}{3}cd\right)^2$

EXPLORE

7-2 Graphing Technology Lab
Solving Exponential Equations and Inequalities

You can use a TI-83/84 Plus graphing calculator to solve exponential equations by graphing or by using the table feature. To do this, you will write the equations as systems of equations.

CCSS Common Core State Standards
Content Standards
A.REI.11 Explain why the x-coordinates of the points where the graphs of the equations $y = f(x)$ and $y = g(x)$ intersect are the solutions of the equation $f(x) = g(x)$; find the solutions approximately, e.g., using technology to graph the functions, make tables of values, or find successive approximations. Include cases where $f(x)$ and/or $g(x)$ are linear, polynomial, rational, absolute value, exponential, and logarithmic functions.

Activity 1

Solve $3^{x-4} = \frac{1}{9}$.

Step 1 Graph each side of the equation as a separate function. Enter 3^{x-4} as **Y1**. Be sure to include parentheses around the exponent. Enter $\frac{1}{9}$ as **Y2**. Then graph the two equations.

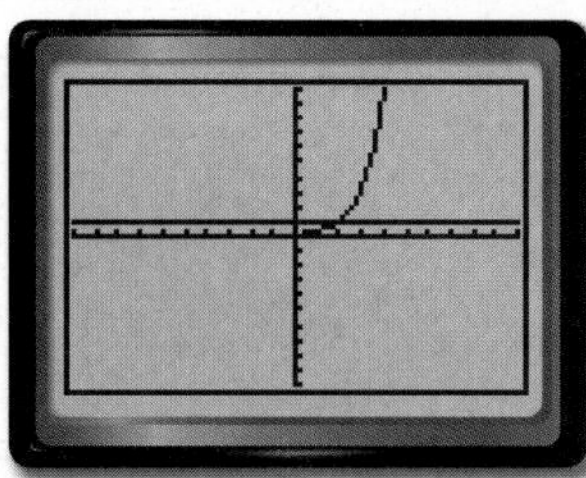

[−10, 10] scl: 1 by [−1, 1] scl: 0.1

Step 2 Use the **intersect** feature.

You can use the **intersect** feature on the **CALC** menu to approximate the ordered pair of the point at which the graphs cross.

The calculator screen shows that the x-coordinate of the point at which the curves cross is 2. Therefore, the solution of the equation is 2.

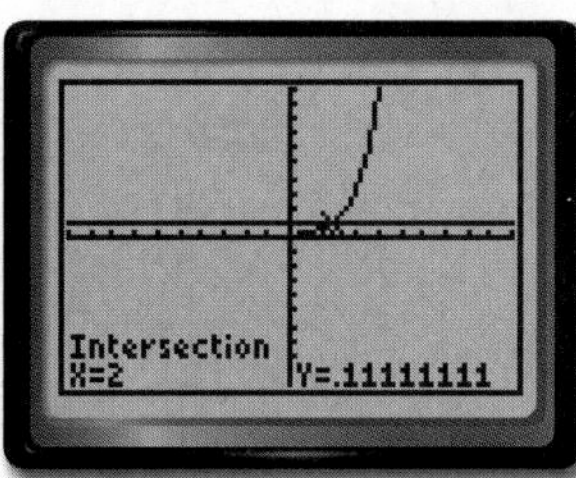

[−10, 10] scl: 1 by [−1, 1] scl: 0.1

Step 3 Use the **TABLE** feature.

You can also use the **TABLE** feature to locate the point at which the curves intersect.

The table displays x-values and corresponding y-values for each graph. Examine the table to find the x-value for which the y-values of the graphs are equal.

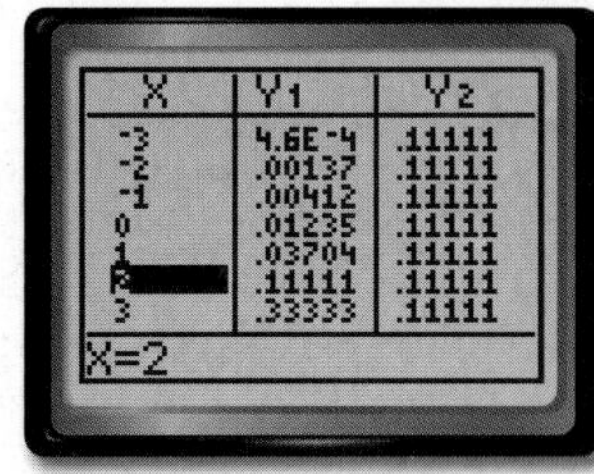

X	Y1	Y2
-3	4.6E-4	.11111
-2	.00137	.11111
-1	.00412	.11111
0	.01235	.11111
1	.03704	.11111
2	.11111	.11111
3	.33333	.11111

X=2

At $x = 2$, both functions have a y-value of $0.\overline{1}$ or $\frac{1}{9}$. Thus, the solution of the equation is 2.

CHECK Substitute 2 for x in the original equation.

$3^{x-4} \stackrel{?}{=} \frac{1}{9}$ Original equation

$3^{2-4} \stackrel{?}{=} \frac{1}{9}$ Substitute 2 for x.

$3^{-2} \stackrel{?}{=} \frac{1}{9}$ Simplify.

$\frac{1}{9} = \frac{1}{9}$ ✓ The solution checks.

A similar procedure can be used to solve exponential inequalities.

(continued on the next page)

Graphing Technology Lab

Solving Exponential Equations and Inequalities *Continued*

Activity 2 Description

Solve $2^{x-2} \geq 0.5^{x-3}$.

Step 1 Enter the related inequalities.

Rewrite the problem as a system of inequalities.

The first inequality is $2^{x-2} \geq y$ or $y \leq 2^{x-2}$. Since this inequality includes the *less than or equal to* symbol, shade below the curve.

First enter the boundary, and then use the arrow and ENTER keys to choose the shade below icon, ▙.

The second inequality is $y \geq 0.5^{x-3}$. Shade above the curve since this inequality contains *greater than or equal to.*

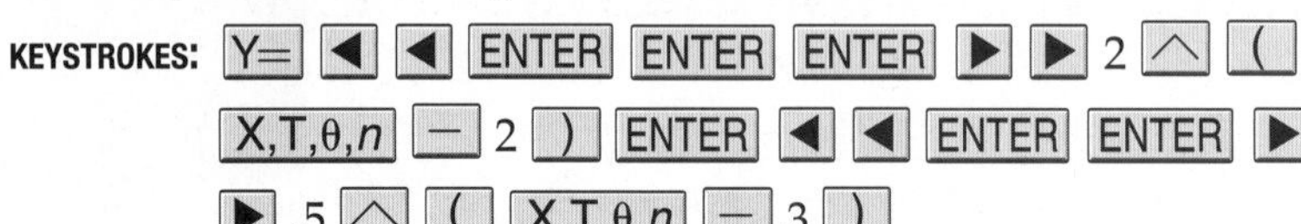

Step 2 Graph the system.

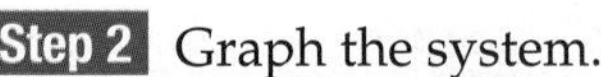

The x-values of the points in the region where the shadings overlap is the solution set of the original inequality. Using the **intersect** feature, you can conclude that the solution set is $\{x \mid x \geq 2.5\}$.

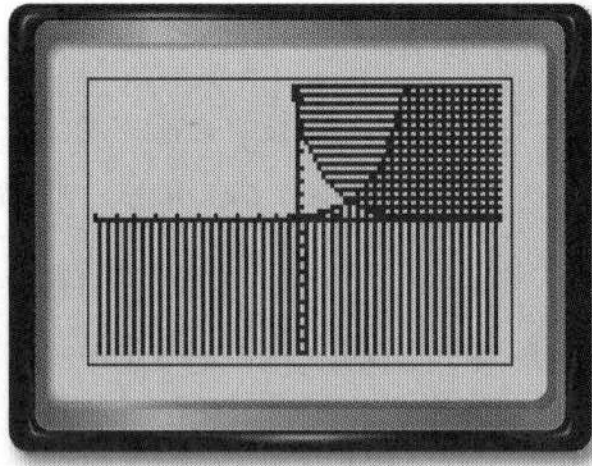

[−10, 10] scl: 1 by [−10, 10] scl: 1

Step 3 Use the **TABLE** feature.

Verify using the **TABLE** feature. Set up the table to show x-values in increments of 0.5.

KEYSTROKES: 2nd [TBLSET] 0 ENTER .5 ENTER 2nd [TABLE]

Notice that for x-values greater than $x = 2.5$, **Y1** > **Y2**. This confirms that the solution of the inequality is $\{x \mid x \geq 2.5\}$.

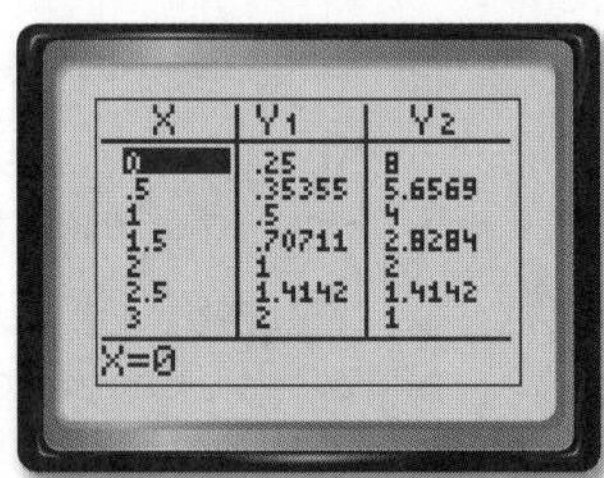

X	Y1	Y2
0	.25	8
.5	.35355	5.6569
1	.5	4
1.5	.70711	2.8284
2	1	2
2.5	1.4142	1.4142
3	2	1

X=0

Exercises

Solve each equation or inequality.

1. $9^{x-1} = \frac{1}{81}$
2. $4^{x+3} = 2^{5x}$
3. $5^{x-1} = 2^x$
4. $3.5^{x+2} = 1.75^{x+3}$
5. $-3^{x+4} = -0.5^{2x+3}$
6. $6^{2-x} - 4 < -0.25^{x-2.5}$
7. $16^{x-1} > 2^{2x+2}$
8. $3^x - 4 \leq 5^{\frac{x}{2}}$
9. $5^{x+3} \leq 2^{x+4}$
10. **WRITING IN MATH** Explain why this technique of graphing a system of equations or inequalities works to solve exponential equations and inequalities.

LESSON 7-2

Solving Exponential Equations and Inequalities

Then

- You graphed exponential functions.

Now

1. Solve exponential equations.
2. Solve exponential inequalities.

Why?

- Membership on Internet social networking sites tends to increase exponentially. The membership growth of one Web site can be modeled by the equation $y = 2.2(1.37)^x$, where x is the number of years since 2004 and y is the number of members in millions.

You can use $y = 2.2(1.37)^x$ to determine how many members there will be in a given year, or to determine the year in which membership was at a certain level.

NewVocabulary
exponential equation
compound interest
exponential inequality

Common Core State Standards

Content Standards
A.CED.1 Create equations and inequalities in one variable and use them to solve problems.
F.LE.4 For exponential models, express as a logarithm the solution to $ab^{ct} = d$ where a, c, and d are numbers and the base b is 2, 10, or e; evaluate the logarithm using technology.

Mathematical Practices
2 Reason abstractly and quantitatively.

1 Solve Exponential Equations

In an **exponential equation**, variables occur as exponents.

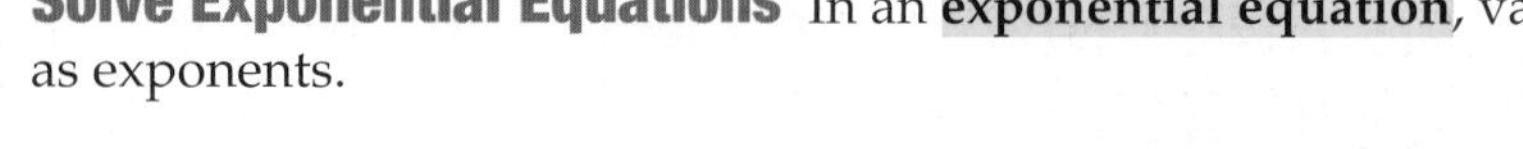

KeyConcept Property of Equality for Exponential Functions

Words	Let $b > 0$ and $b \neq 1$. Then $b^x = b^y$ if and only if $x = y$.
Example	If $3^x = 3^5$, then $x = 5$. If $x = 5$, then $3^x = 3^5$.

The Property of Equality can be used to solve exponential equations.

Example 1 Solve Exponential Equations

Solve each equation.

a. $2^x = 8^3$

$2^x = 8^3$	Original equation
$2^x = (2^3)^3$	Rewrite 8 as 2^3.
$2^x = 2^9$	Power of a Power
$x = 9$	Property of Equality for Exponential Functions

b. $9^{2x-1} = 3^{6x}$

$9^{2x-1} = 3^{6x}$	Original equation
$(3^2)^{2x-1} = 3^{6x}$	Rewrite 9 as 3^2.
$3^{4x-2} = 3^{6x}$	Power of a Power
$4x - 2 = 6x$	Property of Equality for Exponential Functions
$-2 = 2x$	Subtract $4x$ from each side.
$-1 = x$	Divide each side by 2.

GuidedPractice

1A. $4^{2n-1} = 64$

1B. $5^{5x} = 125^{x+2}$

You can use information about growth or decay to write the equation of an exponential function.

PT

Real-World Example 2 Write an Exponential Function

SCIENCE Kristin starts an experiment with 7500 bacteria cells. After 4 hours, there are 23,000 cells.

a. Write an exponential function that could be used to model the number of bacteria after x hours if the number of bacteria changes at the same rate.

At the beginning of the experiment, the time is 0 hours and there are 7500 bacteria cells. Thus, the y-intercept, and the value of a, is 7500.

When $x = 4$, the number of bacteria cells is 23,000. Substitute these values into an exponential function to determine the value of b.

$y = ab^x$	Exponential function
$23{,}000 = 7500 \cdot b^4$	Replace x with 4, y with 23,000, and a with 7500.
$3.067 \approx b^4$	Divide each side by 7500.
$\sqrt[4]{3.067} \approx b$	Take the 4th root of each side.
$1.323 \approx b$	Use a calculator.

An equation that models the number of bacteria is $y \approx 7500(1.323)^x$.

b. How many bacteria cells can be expected in the sample after 12 hours?

$y \approx 7500(1.323)^x$	Modeling equation
$\approx 7500(1.323)^{12}$	Replace x with 12.
$\approx 215{,}665$	Use a calculator.

There will be approximately 215,665 bacteria cells after 12 hours.

Guided Practice

2. RECYCLING A manufacturer distributed 3.2 million aluminum cans in 2005.

A. In 2010, the manufacturer distributed 420,000 cans made from the recycled cans it had previously distributed. Assuming that the recycling rate continues, write an equation to model the distribution each year of cans that are made from recycled aluminum.

B. How many cans made from recycled aluminum can be expected in the year 2050?

Real-World Link

In 2008, the U.S. recycling rate for metals of 35% prevented the release of approximately 25 million metric tons of carbon into the air—roughly the amount emitted annually by 4.5 million cars.

Source: Environmental Protection Agency

Exponential functions are used in situations involving compound interest. **Compound interest** is interest paid on the principal of an investment and any previously earned interest.

Key Concept Compound Interest

You can calculate compound interest using the following formula.

$$A = P\left(1 + \frac{r}{n}\right)^{nt},$$

where A is the amount in the account after t years, P is the principal amount invested, r is the annual interest rate, and n is the number of compounding periods each year.

PT

Example 3 Compound Interest

An investment account pays 4.2% annual interest compounded monthly. If $2500 is invested in this account, what will be the balance after 15 years?

Understand Find the total amount in the account after 15 years.

Plan Use the compound interest formula.

$P = 2500$, $r = 0.042$, $n = 12$, and $t = 15$

Solve

$A = P\left(1 + \frac{r}{n}\right)^{nt}$ — Compound Interest Formula

$= 2500\left(1 + \frac{0.042}{12}\right)^{12 \cdot 15}$ — $P = 2500$, $r = 0.042$, $n = 12$, $t = 15$

≈ 4688.87 — Use a calculator.

Check Graph the corresponding equation $y = 2500(1.0035)^{12t}$. Use **CALC: value** to find y when $x = 15$.

The y-value 4688.8662 is very close to 4688.87, so the answer is reasonable.

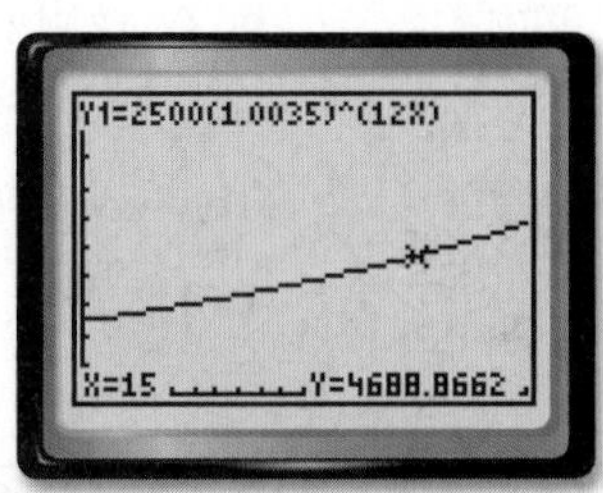

[0, 20] scl: 1 by [0, 10,000] scl: 1000

WatchOut!

Percents Remember to convert all percents to decimal form; 4.2% is 0.042.

GuidedPractice

3. Find the account balance after 20 years if $100 is placed in an account that pays 1.2% interest compounded twice a month.

2 Solve Exponential Inequalities

An **exponential inequality** is an inequality involving exponential functions.

KeyConcept Property of Inequality for Exponential Functions

Words	Let $b > 1$. Then $b^x > b^y$ if and only if $x > y$, and $b^x < b^y$ if and only if $x < y$.
Example	If $2^x > 2^6$, then $x > 6$. If $x > 6$, then $2^x > 2^6$.

This property also holds true for $\leq$ and $\geq$.

PT

Example 4 Solve Exponential Inequalities

Solve $16^{2x-3} < 8$.

$16^{2x-3} < 8$ — Original inequality

$(2^4)^{2x-3} < 2^3$ — Rewrite 16 as 2^4 and 8 as 2^3.

$2^{8x-12} < 2^3$ — Power of a Power

$8x - 12 < 3$ — Property of Inequality for Exponential Functions

$8x < 15$ — Add 12 to each side.

$x < \frac{15}{8}$ — Divide each side by 8.

GuidedPractice

Solve each inequality.

4A. $3^{2x-1} \geq \frac{1}{243}$

4B. $2^{x+2} > \frac{1}{32}$

Check Your Understanding

= Step-by-Step Solutions begin on page R14.

Example 1 **Solve each equation.**

1. $3^{5x} = 27^{2x - 4}$
2. $16^{2y - 3} = 4^{y + 1}$
3. $2^{6x} = 32^{x - 2}$
4. $49^{x + 5} = 7^{8x - 6}$

Example 2

5. **SCIENCE** Mitosis is a process in which one cell divides into two. The *Escherichia coli* is one of the fastest growing bacteria. It can reproduce itself in 15 minutes.
   a. Write an exponential function to represent the number of cells c after t minutes.
   b. If you begin with one *Escherichia coli* cell, how many cells will there be in one hour?

Example 3

6. A certificate of deposit (CD) pays 2.25% annual interest compounded biweekly. If you deposit \$500 into this CD, what will the balance be after 6 years?

Example 4 **Solve each inequality.**

7. $4^{2x + 6} \leq 64^{2x - 4}$
8. $25^{y - 3} \leq \left(\frac{1}{125}\right)^{y + 2}$

Practice and Problem Solving

Extra Practice is on page R7.

Example 1 **Solve each equation.**

9. $8^{4x + 2} = 64$
10. $5^{x - 6} = 125$
11. $81^{a + 2} = 3^{3a + 1}$
12. $256^{b + 2} = 4^{2 - 2b}$
13. $9^{3c + 1} = 27^{3c - 1}$
14. $8^{2y + 4} = 16^{y + 1}$

Example 2

15. **CCSS MODELING** In 2009, My-Lien received \$10,000 from her grandmother. Her parents invested all of the money, and by 2021, the amount will have grown to \$16,960.
    a. Write an exponential function that could be used to model the money y. Write the function in terms of x, the number of years since 2009.
    b. Assume that the amount of money continues to grow at the same rate. What would be the balance in the account in 2031?

Write an exponential function for the graph that passes through the given points.

16. (0, 6.4) and (3, 100)
17. (0, 256) and (4, 81)
18. (0, 128) and (5, 371,293)
19. (0, 144), and (4, 21,609)

Example 3

20. Find the balance of an account after 7 years if \$700 is deposited into an account paying 4.3% interest compounded monthly.
21. Determine how much is in a retirement account after 20 years if \$5000 was invested at 6.05% interest compounded weekly.
22. A savings account offers 0.7% interest compounded bimonthly. If \$110 is deposited in this account, what will the balance be after 15 years?
23. A college savings account pays 13.2% annual interest compounded semiannually. What is the balance of an account after 12 years if \$21,000 was initially deposited?

Example 4 **Solve each inequality.**

24. $625 \geq 5^{a + 8}$
25. $10^{5b + 2} > 1000$
26. $\left(\frac{1}{64}\right)^{c - 2} < 32^{2c}$
27. $\left(\frac{1}{27}\right)^{2d - 2} \leq 81^{d + 4}$
28. $\left(\frac{1}{9}\right)^{3t + 5} \geq \left(\frac{1}{243}\right)^{t - 6}$
29. $\left(\frac{1}{36}\right)^{w + 2} < \left(\frac{1}{216}\right)^{4w}$

30. **SCIENCE** A mug of hot chocolate is 90°C at time $t = 0$. It is surrounded by air at a constant temperature of 20°C. If stirred steadily, its temperature in Celsius after t minutes will be $y(t) = 20 + 70(1.071)^{-t}$.

 a. Find the temperature of the hot chocolate after 15 minutes.

 b. Find the temperature of the hot chocolate after 30 minutes.

 c. The optimum drinking temperature is 60°C. Will the mug of hot chocolate be at or below this temperature after 10 minutes?

31. **ANIMALS** Studies show that an animal will defend a territory, with area in square yards, that is directly proportional to the 1.31 power of the animal's weight in pounds.

 a. If a 45-pound beaver will defend 170 square yards, write an equation for the area a defended by a beaver weighing w pounds.

 b. Scientists believe that thousands of years ago, the beaver's ancestors were 11 feet long and weighed 430 pounds. Use your equation to determine the area defended by these animals.

Solve each equation.

32. $\left(\frac{1}{2}\right)^{4x+1} = 8^{2x+1}$

33. $\left(\frac{1}{5}\right)^{x-5} = 25^{3x+2}$

34. $216 = \left(\frac{1}{6}\right)^{x+3}$

35. $\left(\frac{1}{8}\right)^{3x+4} = \left(\frac{1}{4}\right)^{-2x+4}$

36. $\left(\frac{2}{3}\right)^{5x+1} = \left(\frac{27}{8}\right)^{x-4}$

37. $\left(\frac{25}{81}\right)^{2x+1} = \left(\frac{729}{125}\right)^{-3x+1}$

38. **CCSS MODELING** In 1950, the world population was about 2.556 billion. By 1980, it had increased to about 4.458 billion.

 a. Write an exponential function of the form $y = abx$ that could be used to model the world population y in billions for 1950 to 1980. Write the equation in terms of x, the number of years since 1950. (Round the value of b to the nearest ten-thousandth.)

 b. Suppose the population continued to grow at that rate. Estimate the population in 2000.

 c. In 2000, the population of the world was about 6.08 billion. Compare your estimate to the actual population.

 d. Use the equation you wrote in Part **a** to estimate the world population in the year 2020. How accurate do you think the estimate is? Explain your reasoning.

39. **TREES** The diameter of the base of a tree trunk in centimeters varies directly with the $\frac{3}{2}$ power of its height in meters.

 a. A young sequoia tree is 6 meters tall, and the diameter of its base is 19.1 centimeters. Use this information to write an equation for the diameter d of the base of a sequoia tree if its height is h meters high.

 b. The General Sherman Tree in Sequoia National Park, California, is approximately 84 meters tall. Find the diameter of the General Sherman Tree at its base.

40. **FINANCIAL LITERACY** Mrs. Jackson has two different retirement investment plans from which to choose.

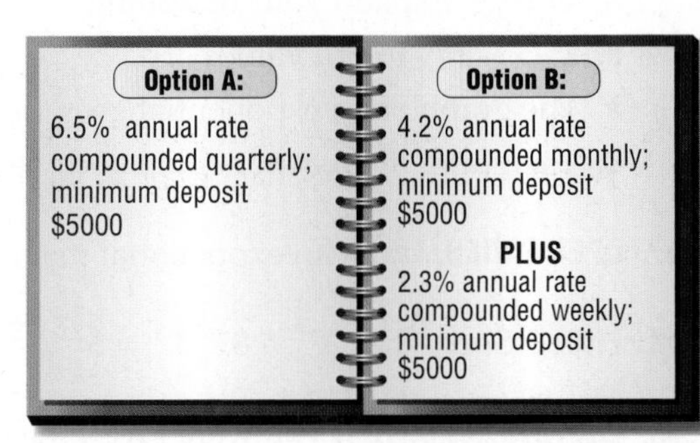

 a. Write equations for Option A and Option B given the minimum deposits.

 b. Draw a graph to show the balances for each investment option after t years.

 c. Explain whether Option A or Option B is the better investment choice.

41. **MULTIPLE REPRESENTATIONS** In this problem, you will explore the rapid increase of an exponential function. A large sheet of paper is cut in half, and one of the resulting pieces is placed on top of the other. Then the pieces in the stack are cut in half and placed on top of each other. Suppose this procedure is repeated several times.

a. **Concrete** Perform this activity and count the number of sheets in the stack after the first cut. How many pieces will there be after the second cut? How many pieces after the third cut? How many pieces after the fourth cut?

b. **Tabular** Record your results in a table.

c. **Symbolic** Use the pattern in the table to write an equation for the number of pieces in the stack after x cuts.

d. **Analytical** The thickness of ordinary paper is about 0.003 inch. Write an equation for the thickness of the stack of paper after x cuts.

e. **Analytical** How thick will the stack of paper be after 30 cuts?

H.O.T. Problems Use Higher-Order Thinking Skills

42. **WRITING IN MATH** In a problem about compound interest, describe what happens as the compounding period becomes more frequent while the principal and overall time remain the same.

43. **ERROR ANALYSIS** Beth and Liz are solving $6^{x-3} > 36^{-x-1}$. Is either of them correct? Explain your reasoning.

Beth	Liz
$6^{x-3} > 36^{-x-1}$	$6^{x-3} > 36^{-x-1}$
$6^{x-3} > (6^2)^{-x-1}$	$6^{x-3} > (6^2)^{-x-1}$
$6^{x-3} > 6^{-2x-2}$	$6^{x-3} > 6^{-x+1}$
$x - 3 > -2x - 2$	$x - 3 > -x + 1$
$3x > 1$	$2x > 4$
$x > \frac{1}{3}$	$x > 2$

44. **CHALLENGE** Solve for x: $16^{18} + 16^{18} + 16^{18} + 16^{18} + 16^{18} = 4^x$.

45. **OPEN ENDED** What would be a more beneficial change to a 5-year loan at 8% interest compounded monthly: reducing the term to 4 years or reducing the interest rate to 6.5%?

46. **CCSS ARGUMENTS** Determine whether the following statements are *sometimes*, *always*, or *never* true. Explain your reasoning.

a. $2^x > -8^{20x}$ for all values of x.

b. The graph of an exponential growth equation is increasing.

c. The graph of an exponential decay equation is increasing.

47. **OPEN ENDED** Write an exponential inequality with a solution of $x \leq 2$.

48. **PROOF** Show that $27^{2x} \cdot 81^{x+1} = 3^{2x+2} \cdot 9^{4x+1}$.

49. **WRITING IN MATH** If you were given the initial and final amounts of a radioactive substance and the amount of time that passes, how would you determine the rate at which the amount was increasing or decreasing in order to write an equation?

Standardized Test Practice

50. $3 \times 10^{-4} =$

A 0.003
B 0.0003
C 0.00003
D 0.000003

51. Which of the following could *not* be a solution to $5 - 3x < -3$?

F 2.5
G 3
H 3.5
J 4

52. GRIDDED RESPONSE The three angles of a triangle are $3x$, $x + 10$, and $2x - 40$. Find the measure of the smallest angle in the triangle.

53. SAT/ACT Which of the following is equivalent to $(x)(x)(x)(x)$ for all x?

A $x + 4$
B $4x$
C $2x^2$
D $4x^2$
E x^4

Spiral Review

Graph each function. (Lesson 7-1)

54. $y = 2(3)^x$

55. $y = 5(2)^x$

56. $y = 4\left(\frac{1}{3}\right)^x$

Solve each equation. (Lesson 6-7)

57. $\sqrt{x + 5} - 3 = 0$

58. $\sqrt{3t - 5} - 3 = 4$

59. $\sqrt[4]{2x - 1} = 2$

60. $\sqrt{x - 6} - \sqrt{x} = 3$

61. $\sqrt[3]{5m + 2} = 3$

62. $(6n - 5)^{\frac{1}{3}} + 3 = -2$

63. $(5x + 7)^{\frac{1}{5}} + 3 = 5$

64. $(3x - 2)^{\frac{1}{5}} + 6 = 5$

65. $(7x - 1)^{\frac{1}{3}} + 4 = 2$

66. SALES A salesperson earns \$10 per hour plus a 10% commission on sales. Write a function to describe the salesperson's income. If the salesperson wants to earn \$1000 in a 40-hour week, what should his sales be? (Lesson 6-2)

67. STATE FAIR A dairy makes three types of cheese—cheddar, Monterey Jack, and Swiss—and sells the cheese in three booths at the state fair. At the beginning of one day, the first booth received x pounds of each type of cheese. The second booth received y pounds of each type of cheese, and the third booth received z pounds of each type of cheese. By the end of the day, the dairy had sold 131 pounds of cheddar, 291 pounds of Monterey Jack, and 232 pounds of Swiss. The table below shows the percent of the cheese delivered in the morning that was sold at each booth. How many pounds of cheddar cheese did each booth receive in the morning? (Lesson 3-4)

Type	Booth 1	Booth 2	Booth 3
Cheddar	40%	30%	10%
Monterey Jack	40%	90%	80%
Swiss	30%	70%	70%

Skills Review

Find $[g \circ h](x)$ and $[h \circ g](x)$.

68. $h(x) = 2x - 1$
$g(x) = 3x + 4$

69. $h(x) = x^2 + 2$
$g(x) = x - 3$

70. $h(x) = x^2 + 1$
$g(x) = -2x + 1$

71. $h(x) = -5x$
$g(x) = 3x - 5$

72. $h(x) = x^3$
$g(x) = x - 2$

73. $h(x) = x + 4$
$g(x) = |x|$

LESSON 7-3 Logarithms and Logarithmic Functions

Then

- You found the inverse of a function.

Now

1. Evaluate logarithmic expressions.
2. Graph logarithmic functions.

Why?

- Many scientists believe the extinction of the dinosaurs was caused by an asteroid striking Earth. Astronomers use the Palermo scale to classify objects near Earth based on the likelihood of impact. To make comparing several objects easier, the scale was developed using *logarithms*. The Palermo scale value of any object can be found using the equation $PS = \log_{10} R$, where R is the relative risk posed by the object.

NewVocabulary

logarithm
logarithmic function

Common Core State Standards

Content Standards

F.IF.7.e Graph exponential and logarithmic functions, showing intercepts and end behavior, and trigonometric functions, showing period, midline, and amplitude.

F.BF.3 Identify the effect on the graph of replacing $f(x)$ by $f(x) + k$, $k f(x)$, $f(kx)$, and $f(x + k)$ for specific values of k (both positive and negative); find the value of k given the graphs. Experiment with cases and illustrate an explanation of the effects on the graph using technology.

Mathematical Practices

6 Attend to precision.

1 Logarithmic Functions and Expressions

Consider the exponential function $f(x) = 2^x$ and its inverse. Recall that you can graph an inverse function by interchanging the x- and y-values in the ordered pairs of the function.

$y = 2^x$	
x	y
-3	$\frac{1}{8}$
-2	$\frac{1}{4}$
-1	$\frac{1}{2}$
0	1
1	2
2	4
3	8

$x = 2^y$	
x	y
$\frac{1}{8}$	-3
$\frac{1}{4}$	-2
$\frac{1}{2}$	-1
1	0
2	1
4	2
8	3

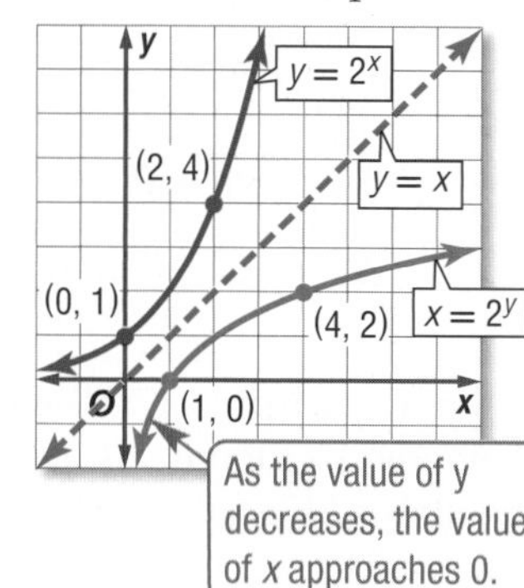

The inverse of $y = 2^x$ can be defined as $x = 2^y$. In general, the inverse of $y = b^x$ is $x = b^y$. In $x = b^y$, the variable y is called the **logarithm** of x. This is usually written as $y = \log_b x$, which is read *y equals log base b of x*.

KeyConcept Logarithm with Base b

Words Let b and x be positive numbers, $b \neq 1$. The *logarithm of x with base b* is denoted $\log_b x$ and is defined as the exponent y that makes the equation $b^y = x$ true.

Symbols Suppose $b > 0$ and $b \neq 1$. For $x > 0$, there is a number y such that

$$\log_b x = y \text{ if and only if } b^y = x.$$

Example If $\log_3 27 = y$, then $3^y = 27$.

The definition of logarithms can be used to express logarithms in exponential form.

Example 1 Logarithmic to Exponential Form

Write each equation in exponential form.

a. $\log_2 8 = 3$

$\log_2 8 = 3 \rightarrow 8 = 2^3$

b. $\log_4 \frac{1}{256} = -4$

$\log_4 \frac{1}{256} = -4 \rightarrow \frac{1}{256} = 4^{-4}$

GuidedPractice

1A. $\log_4 16 = 2$

1B. $\log_3 729 = 6$

The definition of logarithms can also be used to write exponential equations in logarithmic form.

Example 2 Exponential to Logarithmic Form

Write each equation in logarithmic form.

a. $15^3 = 3375$

$15^3 = 3375 \rightarrow \log_{15} 3375 = 3$

b. $4^{\frac{1}{2}} = 2$

$4^{\frac{1}{2}} = 2 \rightarrow \log_4 2 = \frac{1}{2}$

GuidedPractice

2A. $4^3 = 64$

2B. $125^{\frac{1}{3}} = 5$

You can use the definition of a logarithm to evaluate a logarithmic expression.

WatchOut!

Logarithmic Base It is easy to get confused about which number is the base and which is the exponent in logarithmic equations. Consider highlighting each number as you solve to help organize your calculations.

Example 3 Evaluate Logarithmic Expressions

Evaluate $\log_{16} 4$.

$\log_{16} 4 = y$	Let the logarithm equal y.
$4 = 16^y$	Definition of logarithm
$4^1 = (4^2)^y$	$16 = 4^2$
$4^1 = 4^{2y}$	Power of a Power
$1 = 2y$	Property of Equality for Exponential Functions
$\frac{1}{2} = y$	Divide each side by 2.

Thus, $\log_{16} 4 = \frac{1}{2}$.

GuidedPractice

Evaluate each expression.

3A. $\log_3 81$

3B. $\log_{\frac{1}{2}} 256$

2 Graphing Logarithmic Functions

The function $y = \log_b x$, where $b \neq 1$, is called a **logarithmic function**. The graph of $f(x) = \log_b x$ represents a parent graph of the logarithmic functions.

KeyConcept Parent Function of Logarithmic Functions

Parent function:	$f(x) = \log_b x$	**Type of graph:**	continuous, one-to-one
Domain:	all positive real numbers	**Range:**	all real numbers
Asymptote:	$f(x)$-axis	**Intercept:**	(1, 0)

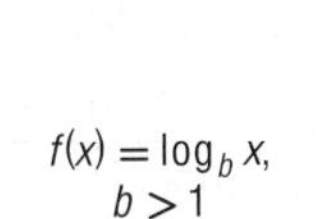

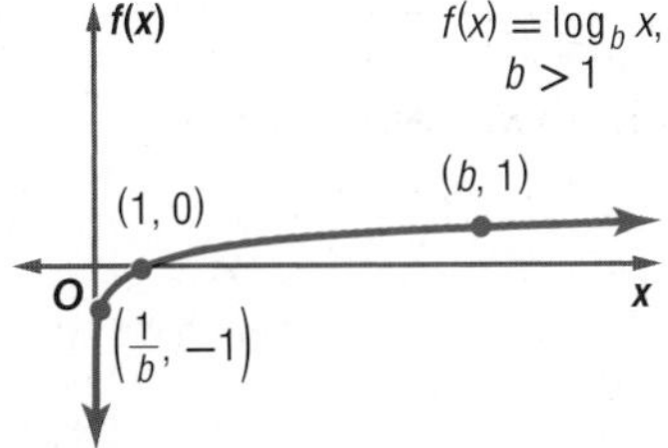

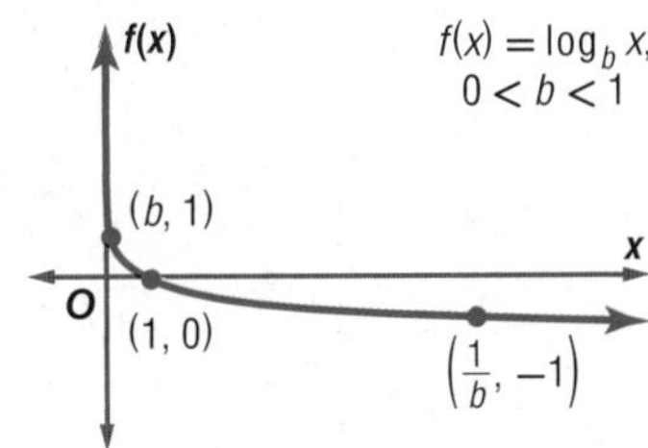

Example 4 Graph Logarithmic Functions

Graph each function.

a. $f(x) = \log_5 x$

Step 1 Identify the base.

$b = 5$

> **StudyTip**
>
> **Zero Exponent** Recall that for any $b \neq 0$, $b^0 = 1$. Therefore, $\log_2 0$ is undefined because $2^x \neq 0$ for any x-value.

Step 2 Determine points on the graph.

Because $5 > 1$, use the points $\left(\frac{1}{b}, -1\right)$, $(1, 0)$, and $(b, 1)$.

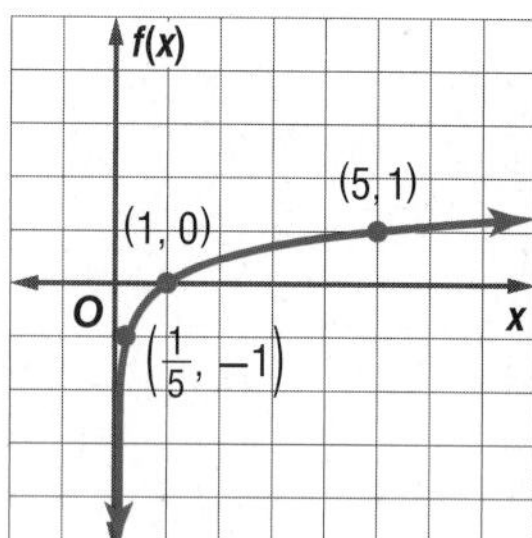

Step 3 Plot the points and sketch the graph.

$\left(\frac{1}{b}, -1\right) \rightarrow \left(\frac{1}{5}, -1\right)$

$(1, 0)$

$(b, 1) \rightarrow (5, 1)$

b. $f(x) = \log_{\frac{1}{3}} x$

Step 1 $b = \frac{1}{3}$

Step 2 $0 < \frac{1}{3} < 1$, so use the points $\left(\frac{1}{3}, 1\right)$, $(1, 0)$ and $(3, -1)$.

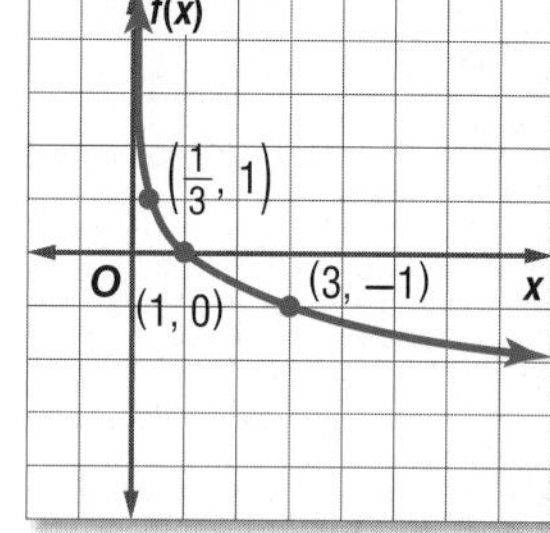

Step 3 Sketch the graph.

GuidedPractice

4A. $f(x) = \log_2 x$

4B. $f(x) = \log_{\frac{1}{8}} x$

The same techniques used to transform the graphs of other functions you have studied can be applied to the graphs of logarithmic functions.

KeyConcept Transformations of Logarithmic Functions

$f(x) = a \log_b (x - h) + k$	
h – Horizontal Translation	**k – Vertical Translation**
h units right if h is positive $\lvert h \rvert$ units left if h is negative	k units up if k is positive $\lvert k \rvert$ units down if k is negative
a – Orientation and Shape	
If $a < 0$, the graph is reflected across the x-axis.	If $\lvert a \rvert > 1$, the graph is stretched vertically. If $0 < \lvert a \rvert < 1$, the graph is compressed vertically.

Example 5 Graph Logarithmic Functions

StudyTip

End Behavior In Example 5a, as x approaches infinity, $f(x)$ approaches infinity.

Graph each function.

a. $f(x) = 3 \log_{10} x + 1$

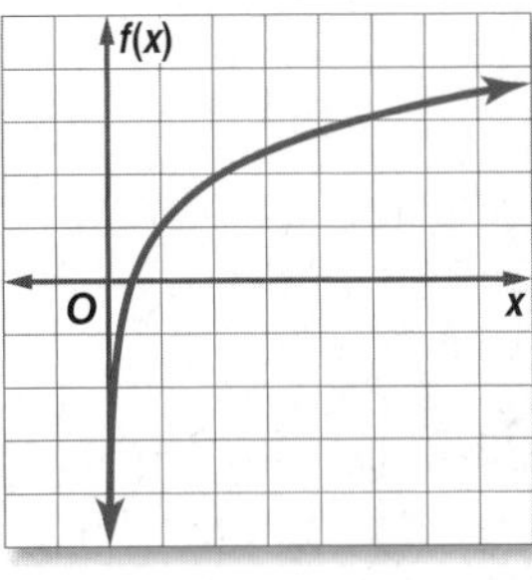

This represents a transformation of the graph of $f(x) = \log_{10} x$.

- $|a| = 3$: The graph stretches vertically
- $h = 0$: There is no horizontal shift.
- $k = 1$: The graph is translated 1 unit up.

b. $f(x) = \frac{1}{2} \log_{\frac{1}{4}} (x - 3)$

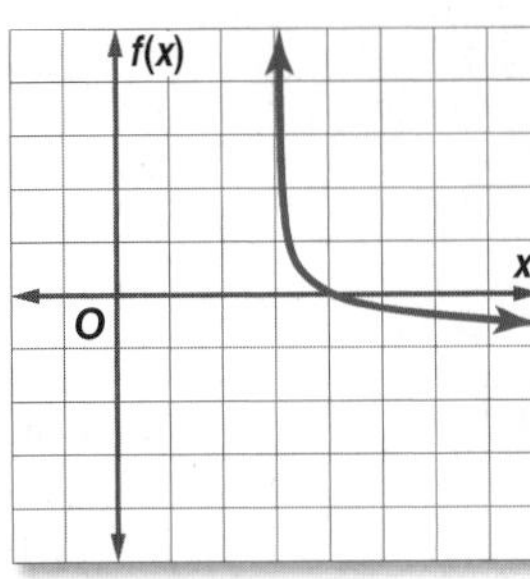

This is a transformation of the graph of $f(x) = \log_{\frac{1}{4}} x$.

- $|a| = \frac{1}{2}$: The graph is compressed vertically.
- $h = 3$: The graph is translated 3 units to the right.
- $k = 0$: There is no vertical shift.

GuidedPractice

Graph each function.

5A. $f(x) = 2 \log_3 (x - 2)$

5B. $f(x) = \frac{1}{4} \log_{\frac{1}{2}} (x + 1) - 5$

Real-World Example 6 Find Inverses of Exponential Functions

Real-WorldLink

The largest recorded earthquake in the United States was a magnitude 9.2 that struck Prince William Sound, Alaska, on Good Friday, March 28, 1964.

Source: United States Geological Survey

Winfield Parks/National Geographic/Getty Images

EARTHQUAKES The Richter scale measures earthquake intensity. The increase in intensity between each number is 10 times. For example, an earthquake with a rating of 7 is 10 times more intense than one measuring 6. The intensity of an earthquake can be modeled by $y = 10^{x-1}$, where x is the Richter scale rating.

a. Use the information at the left to find the intensity of the strongest recorded earthquake in the United States.

$y = 10^{x-1}$	Original equation
$= 10^{9.2-1}$	Substitute 9.2 for x.
$= 10^{8.2}$	Simplify.
$= 158{,}489{,}319.2$	Use a calculator.

b. Write an equation of the form $y = \log_{10} x + c$ for the inverse of the function.

$y = 10^{x-1}$	Original equation
$x = 10^{y-1}$	Replace x with y, replace y with x, and solve for y.
$y - 1 = \log_{10} x$	Definition of logarithm
$y = \log_{10} x + 1$	Add 1 to each side.

GuidedPractice

6. Write an equation for the inverse of the function $y = 0.5^x$.

Check Your Understanding

= Step-by-Step Solutions begin on page R14.

Example 1 Write each equation in exponential form.

1. $\log_8 512 = 3$
2. $\log_5 625 = 4$

Example 2 Write each equation in logarithmic form.

3. $11^3 = 1331$
4. $16^{\frac{3}{4}} = 8$

Example 3 Evaluate each expression.

5. $\log_{13} 169$
6. $\log_2 \frac{1}{128}$
7. $\log_6 1$

Examples 4–5 Graph each function.

8. $f(x) = \log_3 x$
9. $f(x) = \log_{\frac{1}{6}} x$
10. $f(x) = 4 \log_4 (x - 6)$
11. $f(x) = 2 \log_{\frac{1}{10}} x - 5$

Example 6

12. **SCIENCE** Use the information at the beginning of the lesson. The Palermo scale value of any object can be found using the equation $PS = \log_{10} R$, where R is the relative risk posed by the object. Write an equation in exponential form for the inverse of the function.

Practice and Problem Solving

Extra Practice is on page R7.

Example 1 Write each equation in exponential form.

13. $\log_2 16 = 4$
14. $\log_7 343 = 3$
15. $\log_9 \frac{1}{81} = -2$
16. $\log_3 \frac{1}{27} = -3$
17. $\log_{12} 144 = 2$
18. $\log_9 1 = 0$

Example 2 Write each equation in logarithmic form.

19. $9^{-1} = \frac{1}{9}$
20. $6^{-3} = \frac{1}{216}$
21. $2^8 = 256$
22. $4^6 = 4096$
23. $27^{\frac{2}{3}} = 9$
24. $25^{\frac{3}{2}} = 125$

Example 3 Evaluate each expression.

25. $\log_3 \frac{1}{9}$
26. $\log_4 \frac{1}{64}$
27. $\log_8 512$
28. $\log_6 216$
29. $\log_{27} 3$
30. $\log_{32} 2$
31. $\log_9 3$
32. $\log_{121} 11$
33. $\log_{\frac{1}{5}} 3125$
34. $\log_{\frac{1}{8}} 512$
35. $\log_{\frac{1}{3}} \frac{1}{81}$
36. $\log_{\frac{1}{6}} \frac{1}{216}$

Examples 4–5 **CCSS PRECISION** Graph each function.

37. $f(x) = \log_6 x$
38. $f(x) = \log_{\frac{1}{5}} x$
39. $f(x) = 4 \log_2 x + 6$
40. $f(x) = \log_{\frac{1}{9}} x$
41. $f(x) = \log_{10} x$
42. $f(x) = -3 \log_{\frac{1}{12}} x + 2$
43. $f(x) = 6 \log_{\frac{1}{8}} (x + 2)$
44. $f(x) = -8 \log_3 (x - 4)$
45. $f(x) = \log_{\frac{1}{4}} (x + 1) - 9$
46. $f(x) = \log_5 (x - 4) - 5$
47. $f(x) = \frac{1}{6} \log_8 (x - 3) + 4$
48. $f(x) = -\frac{1}{3} \log_{\frac{1}{6}} (x + 2) - 5$

Example 6

49. PHOTOGRAPHY The formula $n = \log_2 \frac{1}{p}$ represents the change in the f-stop setting n to use in less light where p is the fraction of sunlight.

a. Benito's camera is set up to take pictures in direct sunlight, but it is a cloudy day. If the amount of sunlight on a cloudy day is $\frac{1}{4}$ as bright as direct sunlight, how many f-stop settings should he move to accommodate less light?

b. Graph the function.

c. Use the graph in part b to predict what fraction of daylight Benito is accommodating if he moves down 3 f-stop settings. Is he allowing more or less light into the camera?

50. EDUCATION To measure a student's retention of knowledge, the student is tested after a given amount of time. A student's score on an Algebra 2 test t months after the school year is over can be approximated by $y(t) = 85 - 6 \log_2 (t + 1)$, where $y(t)$ is the student's score as a percent.

a. What was the student's score at the time the school year ended ($t = 0$)?

b. What was the student's score after 3 months?

c. What was the student's score after 15 months?

Graph each function.

51. $f(x) = 4 \log_2 (2x - 4) + 6$

52. $f(x) = -3 \log_{12} (4x + 3) + 2$

53. $f(x) = 15 \log_{14} (x + 1) - 9$

54. $f(x) = 10 \log_5 (x - 4) - 5$

55. $f(x) = -\frac{1}{6} \log_8 (x - 3) + 4$

56. $f(x) = -\frac{1}{3} \log_6 (6x + 2) - 5$

57. CCSS MODELING In general, the more money a company spends on advertising, the higher the sales. The amount of money in sales for a company, in thousands, can be modeled by the equation $S(a) = 10 + 20 \log_4(a + 1)$, where a is the amount of money spent on advertising in thousands, when $a \geq 0$.

a. The value of $S(0) \approx 10$, which means that if \$10 is spent on advertising, \$10,000 is returned in sales. Find the values of $S(3)$, $S(15)$, and $S(63)$.

b. Interpret the meaning of each function value in the context of the problem.

c. Graph the function.

d. Use the graph in part c and your answers from part a to explain why the money spent in advertising becomes less "efficient" as it is used in larger amounts.

58. BIOLOGY The generation time for bacteria is the time that it takes for the population to double. The generation time G for a specific type of bacteria can be found using experimental data and the formula $G = \frac{t}{3.3 \log_b f}$, where t is the time period, b is the number of bacteria at the beginning of the experiment, and f is the number of bacteria at the end of the experiment.

a. The generation time for mycobacterium tuberculosis is 16 hours. How long will it take four of these bacteria to multiply into 1024 bacteria?

b. An experiment involving rats that had been exposed to salmonella showed that the generation time for the salmonella was 5 hours. After how long would 20 of these bacteria multiply into 8000?

c. *E. coli* are fast growing bacteria. If 6 *E. coli* can grow to 1296 in 4.4 hours, what is the generation time of *E. coli*?

59 FINANCIAL LITERACY Jacy has spent \$2000 on a credit card. The credit card company charges 24% interest, compounded monthly. The credit card company uses $\log_{\left(1+\frac{0.24}{12}\right)} \frac{A}{2000} = 12t$ to determine how much time it will be until Jacy's debt reaches a certain amount, if A is the amount of debt after a period of time, and t is time in years.

a. Graph the function for Jacy's debt.

b. Approximately how long will it take Jacy's debt to double?

c. Approximately how long will it be until Jacy's debt triples?

H.O.T. Problems Use Higher-Order Thinking Skills

60. **WRITING IN MATH** What should you consider when using exponential and logarithmic models to make decisions?

61. **CCSS ARGUMENTS** Consider $y = \log_b x$ in which b, x, and y are real numbers. Zero can be in the domain *sometimes*, *always* or *never*. Justify your answer.

62. **ERROR ANALYSIS** Betsy says that the graphs of all logarithmic functions cross the y-axis at (0, 1) because any number to the zero power equals 1. Tyrone disagrees. Is either of them correct? Explain your reasoning.

63. **REASONING** Without using a calculator, compare $\log_7 51$, $\log_8 61$, and $\log_9 71$. Which of these is the greatest? Explain your reasoning.

64. **OPEN ENDED** Write a logarithmic equation of the form $y = \log_b x$ for each of the following conditions.

a. y is equal to 25.

b. y is negative.

c. y is between 0 and 1.

d. x is 1.

e. x is 0.

65. **ERROR ANALYSIS** Elisa and Matthew are evaluating $\log_{\frac{1}{7}} 49$. Is either of them correct? Explain your reasoning.

Elisa	Matthew
$\log_{\frac{1}{7}} 49 = y$	$\log_{\frac{1}{7}} 49 = y$
$\frac{1}{7}^y = 49$	$49^y = \frac{1}{7}$
$(7^{-1})^y = 7^2$	$(7^2)^y = (7)^{-1}$
$(7)^{-y} = 7^2$	$7^{2y} = (7)^{-1}$
$y = 2$	$2y = -1$
	$y = -\frac{1}{2}$

66. **WRITING IN MATH** A transformation of $\log_{10} x$ is $g(x) = a \log_{10}(x - h) + k$. Explain the process of graphing this transformation.

Standardized Test Practice

67. A rectangle is twice as long as it is wide. If the width of the rectangle is 3 inches, what is the area of the rectangle in square inches?

A 9
B 12
C 15
D 18

68. SAT/ACT Ichiro has some pizza. He sold 40% more slices than he ate. If he sold 70 slices of pizza, how many did he eat?

F 25
G 50
H 75
J 98
K 100

69. SHORT RESPONSE In the figure $AB = BC$, $CD = BD$, and $m\angle CAD = 70°$. What is the measure of angle ADC?

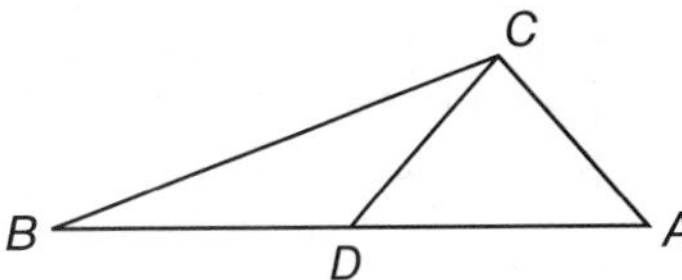

70. If $6x - 3y = 30$ and $4x = 2 - y$ then find $x + y$.

A -4
B -2
C 2
D 4

Spiral Review

Solve each inequality. Check your solution. (Lesson 7-2)

71. $3^{n-2} > 27$

72. $2^{2n} \leq \frac{1}{16}$

73. $16^n < 8^{n+1}$

74. $32^{5p} + 2 \geq 16^{5p}$

Graph each function. (Lesson 7-1)

75. $y = -\left(\frac{1}{5}\right)^x$

76. $y = -2.5(5)^x$

77. $y = 30^{-x}$

78. $y = 0.2(5)^{-x}$

79. GEOMETRY The area of a triangle with sides of length a, b, and c is given by $\sqrt{s(s-a)(s-b)(s-c)}$, where $s = \frac{1}{2}(a + b + c)$. If the lengths of the sides of a triangle are 6, 9, and 12 feet, what is the area of the triangle expressed in radical form? (Lesson 6-5)

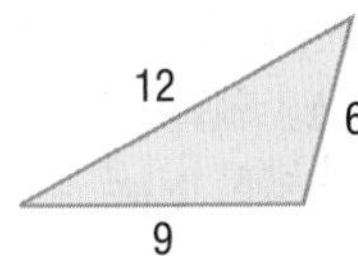

80. GEOMETRY The volume of a rectangular box can be written as $6x^3 + 31x^2 + 53x + 30$ when the height is $x + 2$. (Lesson 5-5)

a. What are the width and length of the box?

b. Will the ratio of the dimensions of the box always be the same regardless of the value of x? Explain.

81. AUTO MECHANICS Shandra is inventory manager for a local repair shop. She orders 6 batteries, 5 cases of spark plugs, and two dozen pairs of wiper blades and pays \$830. She orders 3 batteries, 7 cases of spark plugs, and four dozen pairs of wiper blades and pays \$820. The batteries are \$22 less than twice the price of a dozen wiper blades. Use augmented matrices to determine what the cost of each item on her order is. (Lesson 3-8)

Skills Review

Solve each equation or inequality. Check your solution.

82. $9^x = \frac{1}{81}$

83. $2^{6x} = 4^{5x+2}$

84. $49^{3p+1} = 7^{2p-5}$

85. $9^{x^2} \leq 27^{x^2-2}$

EXTEND

7-3 Graphing Technology Lab: Choosing the Best Model

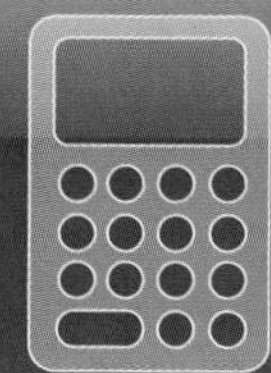

We can find exponential and logarithmic functions of best fit using a TI-83/84 Plus graphing calculator.

CCSS **Common Core State Standards**
Mathematical Practices
4 Model with mathematics.

Activity

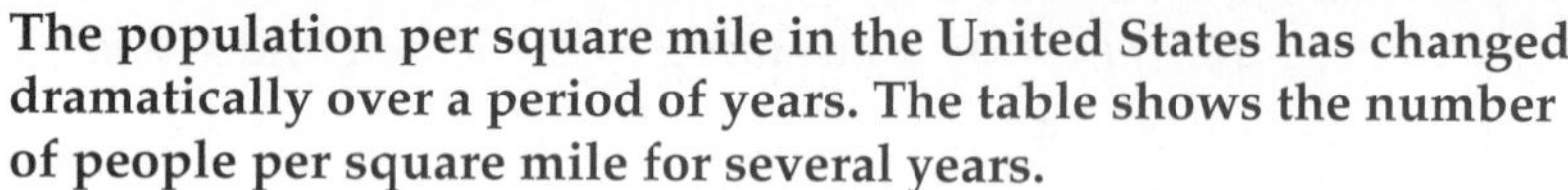

The population per square mile in the United States has changed dramatically over a period of years. The table shows the number of people per square mile for several years.

U.S. Population Density			
Year	People per square mile	Year	People per square mile
1790	4.5	1900	21.5
1800	6.1	1910	26.0
1810	4.3	1920	29.9
1820	5.5	1930	34.7
1830	7.4	1940	37.2
1840	9.8	1950	42.6
1850	7.9	1960	50.6
1860	10.6	1970	57.5
1870	10.9	1980	64.0
1880	14.2	1990	70.3
1890	17.8	2000	80.0

Source: Northeast-Midwest Institute

a. Use a graphing calculator to enter the data. Then draw a scatter plot that shows how the number of people per square mile is related to the year.

Step 1 Enter the year into **L1** and the people per square mile into **L2**.

KEYSTROKES: *See pages 94 and 95 to review how to enter lists.*

Be sure to clear the **Y=** list. Use the [▶] key to move the cursor from **L1** to **L2**.

Step 2 Draw the scatter plot.

KEYSTROKES: *See pages 94 and 95 to review how to graph a scatter plot.*

Make sure that **Plot 1** is on, the scatter plot is chosen, **Xlist** is **L1**, and **Ylist** is **L2**.

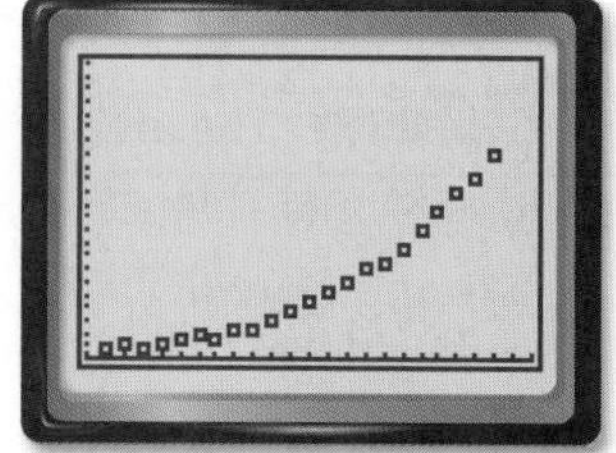

[1780, 2020] scl: 10 by [0, 115] scl: 5

Step 3 Find a regression equation.

To find an equation that best fits the data, use the regression feature of the calculator. Examine various regressions to determine the best model.

Recall that the calculator returns the correlation coefficient r, which is used to indicate how well the model fits the data. The closer r is to 1 or -1, the better the fit.

Linear regression

KEYSTROKES: [STAT] [▶] 4 [ENTER]

Quadratic regression

KEYSTROKES: [STAT] [▶] 5 [ENTER]

$r^2 = 0.9974003374$

$r = \sqrt{0.9974003374}$

$r \approx 0.9986993228$

Exponential regression

KEYSTROKES: STAT ▶ 0 ENTER

Power regression

KEYSTROKES: STAT ▶ ALPHA [A] ENTER

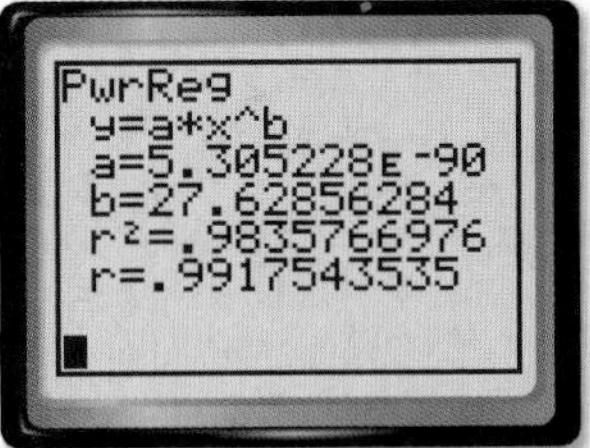

Compare the r-values.

Linear: 0.945411996

Exponential: 0.991887235

Quadratic: 0.9986993228

Power: 0.9917543535

The r-value of the quadratic regression is closest to 1, so it appears to model the data best. You can examine the equation visually by graphing the regression equation with the scatter plot.

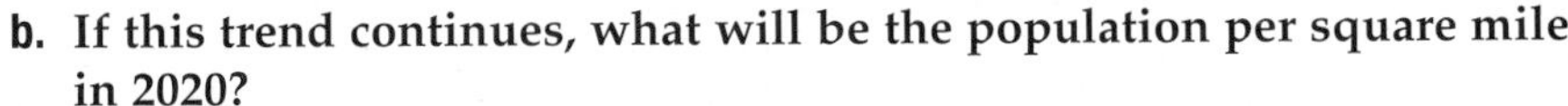

KEYSTROKES: STAT ▶ 5 ENTER Y= VARS 5 ▶ ▶ 1 GRAPH

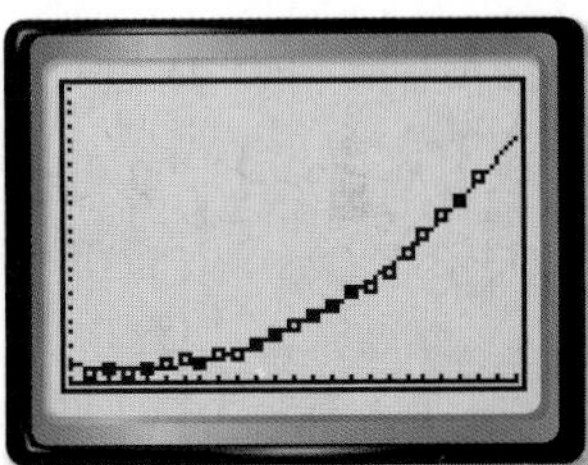

[1780, 2020] scl: 10 by [0, 115] scl: 5

b. If this trend continues, what will be the population per square mile in 2020?

To determine the population per square mile in 2020, find the value of y when $x = 2020$.

KEYSTROKES: 2nd [CALC] ENTER 2020

If this trend continues, there will be approximately 94.9 people per square mile.

[1780, 2020] scl: 10 by [0, 115] scl: 5

Exercises

For Exercises 1–5, Jewel deposited \$50 into an account, then forgot about it and made no further deposits or withdrawals. The table shows the account balance for several years.

Elapsed Time (years)	Balance
0	\$50.00
2	\$55.80
4	\$64.80
6	\$83.09
8	\$101.40
10	\$123.14
12	\$162.67

1. Use a graphing calculator to draw a scatter plot of the data.
2. Calculate and graph a curve of fit using an exponential regression.
3. Write the equation of best fit. What are the domain and range?
4. Based on the model, what will the account balance be after 25 years?
5. Is an exponential model the best fit for the data? Explain.
6. **YOUR TURN** Write a question that can be answered by examining the data of a logarithmic model. First choose a topic and then collect relevant data through Internet research or a survey. Next, make a scatter plot and find a regression equation for your data. Then answer your question.

LESSON 7-4

Solving Logarithmic Equations and Inequalities

Then

- You evaluated logarithmic expressions.

Now

1. Solve logarithmic equations.
2. Solve logarithmic inequalities.

Why?

- Each year the National Weather Service documents about 1000 tornado touchdowns in the United States. The intensity of a tornado is measured on the Fujita scale. Tornados are divided into six categories according to their wind speed, path length, path width, and damage caused.

F-Scale	Wind Speed (mph)	Type of Damage
F-0 Gale	40-72	chimneys, branches
F-1 Moderate	73-112	mobile homes overturned
F-2 Significant	113-157	roof torn off
F-3 Severe	158-206	tree uprooted
F-4 Devastating	207-260	homes leveled, cars thrown
F-5 Incredible	261-318	homes thrown
F-6 Inconceivable	319-379	level has never been achieved

NewVocabulary
logarithmic equation
logarithmic inequality

Common Core State Standards

Content Standards
A.SSE.2 Use the structure of an expression to identify ways to rewrite it.
A.CED.1 Create equations and inequalities in one variable and use them to solve problems.

Mathematical Practices
4 Model with mathematics.

1 Solve Logarithmic Equations

A **logarithmic equation** contains one or more logarithms. You can use the definition of a logarithm to help you solve logarithmic equations.

Example 1 Solve a Logarithmic Equation

Solve $\log_{36} x = \frac{3}{2}$.

$\log_{36} x = \frac{3}{2}$	Original equation
$x = 36^{\frac{3}{2}}$	Definition of logarithm
$x = (6^2)^{\frac{3}{2}}$	$36 = 6^2$
$x = 6^3$ or 216	Power of a Power

GuidedPractice

Solve each equation.

1A. $\log_9 x = \frac{3}{2}$

1B. $\log_{16} x = \frac{5}{2}$

Use the following property to solve logarithmic equations that have logarithms with the same base on each side.

KeyConcept Property of Equality for Logarithmic Functions

Symbols If b is a positive number other than 1, then $\log_b x = \log_b y$ if and only if $x = y$.

Example If $\log_5 x = \log_5 8$, then $x = 8$. If $x = 8$, then $\log_5 x = \log_5 8$.

PT

Standardized Test Example 2 Solve a Logarithmic Equation

Solve $\log_2 (x^2 - 4) = \log_2 3x$.

A -2 B -1 C 2 D 4

Test-TakingTip

Substitution To save time, you can substitute each answer choice in the original equation to find the one that results in a true statement.

Read the Test Item

You need to find x for the logarithmic equation.

Solve the Test Item

$\log_2 (x^2 - 4) = \log_2 3x$	Original equation
$x^2 - 4 = 3x$	Property of Equality for Logarithmic Functions
$x^2 - 3x - 4 = 0$	Subtract $3x$ from each side.
$(x - 4)(x + 1) = 0$	Factor.
$x - 4 = 0$ or $x + 1 = 0$	Zero Product Property
$x = 4$ $\quad x = -1$	Solve each equation.

CHECK Substitute each value into the original equation.

$x = 4$	$x = -1$
$\log_2 (4^2 - 4) \stackrel{?}{=} \log_2 3(4)$	$\log_2 [(-1)^2 - 4] \stackrel{?}{=} \log_2 3(-1)$
$\log_2 12 = \log_2 12$ ✓	$\log_2 (-3) \stackrel{?}{=} \log_2 (-3)$ ✗

The domain of a logarithmic function cannot be 0, so $\log_2 (-3)$ is undefined and -1 is an extraneous solution. The answer is D.

GuidedPractice

2. Solve $\log_3 (x^2 - 15) = \log_3 2x$.

F -3 G -1 H 5 J 15

Math HistoryLink

Zhang Heng (A.D. 78–139) The earliest known seismograph was invented by Zhang Heng in China in 132 B.C. It was a large brass vessel with a heavy pendulum and several arms that tripped when an earthquake tremor was felt. This helped determine the direction of the quake.

Science & Society Picture Library/Contributor/SSPL/Getty Images

2 Solve Logarithmic Inequalities

A **logarithmic inequality** is an inequality that involves logarithms. The following property can be used to solve logarithmic inequalities.

KeyConcept Property of Inequality for Logarithmic Functions

If $b > 1$, $x > 0$, and $\log_b x > y$, then $x > b^y$.

If $b > 1$, $x > 0$, and $\log_b x < y$, then $0 < x < b^y$.

This property also holds true for $\leq$ and $\geq$.

PT

Example 3 Solve a Logarithmic Inequality

Solve $\log_3 x > 4$.

$\log_3 x > 4$	Original inequality
$x > 3^4$	Property of Inequality for Logarithmic Functions
$x > 81$	Simplify.

GuidedPractice

Solve each inequality.

3A. $\log_4 x \geq 3$ **3B.** $\log_2 x < 4$

The following property can be used to solve logarithmic inequalities that have logarithms with the same base on each side. Exclude from your solution set values that would result in taking the logarithm of a number less than or equal to zero in the original inequality.

KeyConcept Property of Inequality for Logarithmic Functions

Symbols	If $b > 1$, then $\log_b x > \log_b y$ if and only if $x > y$, and $\log_b x < \log_b y$ if and only if $x < y$.
Example	If $\log_6 x > \log_6 35$, then $x > 35$

This property also holds true for $\leq$ and $\geq$.

Example 4 Solve Inequalities with Logarithms on Each Side

Solve $\log_4 (x + 3) > \log_4 (2x + 1)$.

$\log_4 (x + 3) > \log_4 (2x + 1)$ Original inequality

$x + 3 > 2x + 1$ Property of Inequality for Logarithmic Functions

$2 > x$ Subtract $x + 1$ from each side.

Exclude all values of x for which $x + 3 \leq 0$ or $2x + 1 \leq 0$. So, $x > -3$, $x > -\frac{1}{2}$, and $x < 2$. The solution set is $\left\{x \mid -\frac{1}{2} < x < 2\right\}$ or $\left(-\frac{1}{2}, 2\right)$.

GuidedPractice

4. Solve $\log_5 (2x + 1) \leq \log_5 (x + 4)$. Check your solution.

Check Your Understanding

= Step-by-Step Solutions begin on page R14.

Example 1 Solve each equation.

1. $\log_8 x = \frac{4}{3}$

2. $\log_{16} x = \frac{3}{4}$

Example 2 **3. MULTIPLE CHOICE** Solve $\log_5 (x^2 - 10) = \log_5 3x$.

A 10 B 2 C 5 D 2, 5

Example 3 Solve each inequality.

4. $\log_5 x > 3$

5. $\log_8 x \leq -2$

6. $\log_4 (2x + 5) \leq \log_4 (4x - 3)$

7. $\log_8 (2x) > \log_8 (6x - 8)$

Practice and Problem Solving

Extra Practice is on page R7.

Examples 1–2 CCSS **STRUCTURE** Solve each equation.

8. $\log_{81} x = \frac{3}{4}$

9. $\log_{25} x = \frac{5}{2}$

10. $\log_8 \frac{1}{2} = x$

11. $\log_6 \frac{1}{36} = x$

12. $\log_x 32 = \frac{5}{2}$

13. $\log_x 27 = \frac{3}{2}$

14. $\log_3 (3x + 8) = \log_3 (x^2 + x)$

15. $\log_{12} (x^2 - 7) = \log_{12} (x + 5)$

16. $\log_6 (x^2 - 6x) = \log_6 (-8)$

17. $\log_9 (x^2 - 4x) = \log_9 (3x - 10)$

18. $\log_4 (2x^2 + 1) = \log_4 (10x - 7)$

19. $\log_7 (x^2 - 4) = \log_7 (-x + 2)$

SCIENCE The equation for wind speed w, in miles per hour, near the center of a tornado is $w = 93 \log_{10} d + 65$, where d is the distance in miles that the tornado travels.

20. Write this equation in exponential form.

21. In May of 1999, a tornado devastated Oklahoma City with the fastest wind speed ever recorded. If the tornado traveled 525 miles, estimate the wind speed near the center of the tornado.

Solve each inequality.

Examples 3–4

22. $\log_6 x < -3$

23. $\log_4 x \geq 4$

24. $\log_3 x \geq -4$

25 $\log_2 x \leq -2$

26. $\log_5 x > 2$

27. $\log_7 x < -1$

28. $\log_2 (4x - 6) > \log_2 (2x + 8)$

29. $\log_7 (x + 2) \geq \log_7 (6x - 3)$

30. $\log_3 (7x - 6) < \log_3 (4x + 9)$

31. $\log_5 (12x + 5) \leq \log_5 (8x + 9)$

32. $\log_{11} (3x - 24) \geq \log_{11} (-5x - 8)$

33. $\log_9 (9x + 4) \leq \log_9 (11x - 12)$

34. **CCSS MODELING** The magnitude of an earthquake is measured on a logarithmic scale called the Richter scale. The magnitude M is given by $M = \log_{10} x$, where x represents the amplitude of the seismic wave causing ground motion.

a. How many times as great is the amplitude caused by an earthquake with a Richter scale rating of 8 as an aftershock with a Richter scale rating of 5?

b. In 1906, San Francisco was almost completely destroyed by a 7.8 magnitude earthquake. In 1911, an earthquake estimated at magnitude 8.1 occurred along the New Madrid fault in the Mississippi River Valley. How many times greater was the New Madrid earthquake than the San Francisco earthquake?

35. **MUSIC** The first key on a piano keyboard corresponds to a pitch with a frequency of 27.5 cycles per second. With every successive key, going up the black and white keys, the pitch multiplies by a constant. The formula for the frequency of the pitch sounded when the nth note up the keyboard is played is given by $n = 1 + 12 \log_2 \frac{f}{27.5}$.

a. A note has a frequency of 220 cycles per second. How many notes up the piano keyboard is this?

b. Another pitch on the keyboard has a frequency of 880 cycles per second. After how many notes up the keyboard will this be found?

36. **MULTIPLE REPRESENTATIONS** In this problem, you will explore the graphs shown: $y = \log_4 x$ and $y = \log_{\frac{1}{4}} x$.

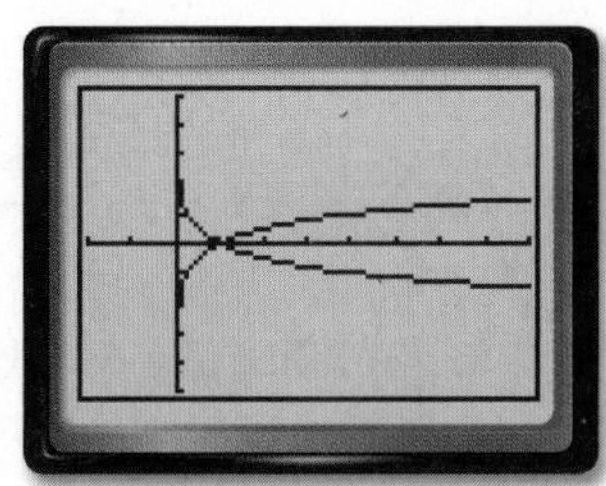

[−2, 8] scl: 1 by [−5, 5] scl: 1

a. **Analytical** How do the shapes of the graphs compare? How do the asymptotes and the x-intercepts of the graphs compare?

b. **Verbal** Describe the relationship between the graphs.

c. **Graphical** Use what you know about transformations of graphs to compare and contrast the graph of each function and the graph of $y = \log_4 x$.

1. $y = \log_4 x + 2$ 2. $y = \log_4 (x + 2)$ 3. $y = 3 \log_4 x$

d. **Analytical** Describe the relationship between $y = \log_4 x$ and $y = -1(\log_4 x)$. What are a reasonable domain and range for each function?

e. **Analytical** Write an equation for a function for which the graph is the graph of $y = \log_3 x$ translated 4 units left and 1 unit up.

37 **SOUND** The relationship between the intensity of sound I and the number of decibels β is $\beta = 10 \log_{10}\left(\frac{I}{10^{-12}}\right)$, where I is the intensity of sound in watts per square meter.

a. Find the number of decibels of a sound with an intensity of 1 watt per square meter.

b. Find the number of decibels of sound with an intensity of 10^{-2} watts per square meter.

c. The intensity of the sound of 1 watt per square meter is 100 times as much as the intensity of 10^{-2} watts per square meter. Why are the decibels of sound not 100 times as great?

Sound	Intensity	Decibels
pin drop	100	0
normal breathing	101	1
clothes dryer	106	6
subway train	1010	10
firecracker	1012	12

H.O.T. Problems Use Higher-Order Thinking Skills

38. **CCSS CRITIQUE** Ryan and Heather are solving $\log_3 x \geq -3$. Is either of them correct? Explain your reasoning.

Ryan

$\log_3 x \geq -3$

$x \geq 3^{-3}$

$x \geq \frac{1}{27}$

Heather

$\log_3 x \geq -3$

$x \geq 3^{-3}$

$0 < x \leq \frac{1}{27}$

39. **CHALLENGE** Find $\log_3 27 + \log_9 27 + \log_{27} 27 + \log_{81} 27 + \log_{243} 27$.

40. **REASONING** The Property of Inequality for Logarithmic Functions states that when $b > 1$, $\log_b x > \log_b y$ if and only if $x > y$. What is the case for when $0 < b < 1$? Explain your reasoning.

41. **WRITING IN MATH** Explain how the domain and range of logarithmic functions are related to the domain and range of exponential functions.

42. **OPEN ENDED** Give an example of a logarithmic equation that has no solution.

43. **REASONING** Choose the appropriate term. Explain your reasoning. All logarithmic equations are of the form $y = \log_b x$.

a. If the base of a logarithmic equation is greater than 1 and the value of x is between 0 and 1, then the value for y is (*less than, greater than, equal to*) 0.

b. If the base of a logarithmic equation is between 0 and 1 and the value of x is greater than 1, then the value of y is (*less than, greater than, equal to*) 0.

c. There is/are (*no, one, infinitely many*) solution(s) for b in the equation $y = \log_b 0$.

d. There is/are (*no, one, infinitely many*) solution(s) for b in the equation $y = \log_b 1$.

44. **WRITING IN MATH** Explain why any logarithmic function of the form $y = \log_b x$ has an x-intercept of (1, 0) and no y-intercept.

Standardized Test Practice

45. Find x if $\frac{6.4}{x} = \frac{4}{7}$.

A 3.4
B 9.4
C 11.2
D 44.8

46. The monthly precipitation in Houston for part of a year is shown.

Month	Precipitation (in.)
April	3.60
May	5.15
June	5.35
July	3.18
August	3.83

Find the median precipitation.

F 4.25 in.
G 4.22 in.
H 3.83 in.
J 3.60 in.

47. Clara received a 10% raise each year for 3 consecutive years. What was her salary after the three raises if her starting salary was $12,000 per year?

A $14,520
B $15,972
C $16,248
D $16,410

48. SAT/ACT A vendor has 14 helium balloons for sale: 9 are yellow, 3 are red, and 2 are green. A balloon is selected at random and sold. If the balloon sold is yellow, what is the probability that the next balloon, selected at random, is also yellow?

F $\frac{1}{9}$
G $\frac{1}{8}$
H $\frac{36}{91}$
J $\frac{8}{13}$
K $\frac{9}{14}$

Spiral Review

Evaluate each expression. (Lesson 7-3)

49. $\log_4 256$

50. $\log_2 \frac{1}{8}$

51. $\log_6 216$

52. $\log_3 27$

53. $\log_5 \frac{1}{125}$

54. $\log_7 2401$

Solve each equation or inequality. Check your solution. (Lesson 7-2)

55. $5^{2x+3} \leq 125$

56. $3^{3x-2} > 81$

57. $4^{4a+6} \leq 16^a$

58. $11^{2x+1} = 121^{3x}$

59. $3^{4x-7} = 27^{2x+3}$

60. $8^{x-4} \leq 2^{4-x}$

61. SHIPPING The height of a shipping cylinder is 4 feet more than the radius. If the volume of the cylinder is 5π cubic feet, how tall is it? Use the formula $V = \pi \cdot r^2 \cdot h$. (Lesson 5-8)

62. NUMBER THEORY Two complex conjugate numbers have a sum of 12 and a product of 40. Find the two numbers. (Lesson 4-4)

Skills Review

Simplify. Assume that no variable equals zero.

63. $x^5 \cdot x^3$

64. $a^2 \cdot a^6$

65. $(2p^2n)^3$

66. $(3b^3c^2)^2$

67. $\frac{x^4y^6}{xy^2}$

68. $\left(\frac{c^9}{d^7}\right)^0$

CHAPTER 7

Mid-Chapter Quiz

Lessons 7-1 through 7-4

Graph each function. State the domain and range. (Lesson 7-1)

1. $f(x) = 3(4)^x$
2. $f(x) = -(2)^x + 5$
3. $f(x) = -0.5(3)^{x+2} + 4$
4. $f(x) = -3\left(\frac{2}{3}\right)^{x-1} + 8$

5. **SCIENCE** You are studying a bacteria population. The population originally started with 6000 bacteria cells. After 2 hours, there were 28,000 bacteria cells. (Lesson 7-1)

   a. Write an exponential function that could be used to model the number of bacteria after x hours if the number of bacteria changes at the same rate.

   b. How many bacteria cells can be expected after 4 hours?

6. **MULTIPLE CHOICE** Which exponential function has a graph that passes through the points at (0, 125) and (3, 1000)? (Lesson 7-1)

   A $f(x) = 125(3)^x$

   B $f(x) = 1000(3)^x$

   C $f(x) = 125(1000)^x$

   D $f(x) = 125(2)^x$

7. **POPULATION** In 1995, a certain city had a population of 45,000. It increased to 68,000 by 2007. (Lesson 7-2)

   a. What is an exponential function that could be used to model the population of this city x years after 1995?

   b. Use your model to estimate the population in 2020.

8. **MULTIPLE CHOICE** Find the value of x for $\log_3 (x^2 + 2x) = \log_3 (x + 2)$. (Lesson 7-3)

   F $x = -2, 1$

   G $x = -2$

   H $x = 1$

   J no solution

Graph each function. (Lesson 7-3)

9. $f(x) = 3 \log_2 (x - 1)$
10. $f(x) = -4 \log_3 (x - 2) + 5$

11. **MULTIPLE CHOICE** Which graph below is the graph of the function $f(x) = \log_3 (x + 5) + 3$? (Lesson 7-3)

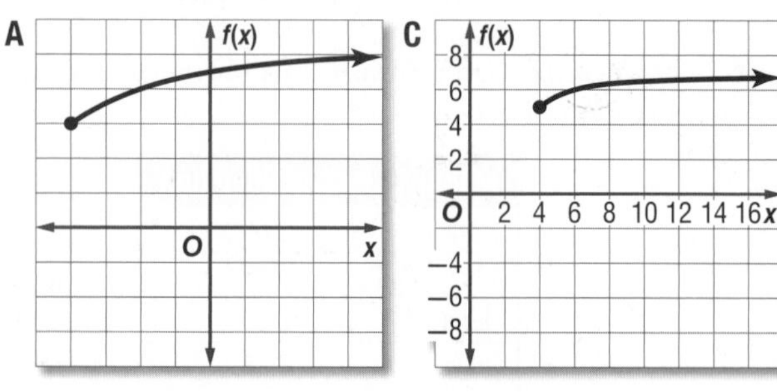

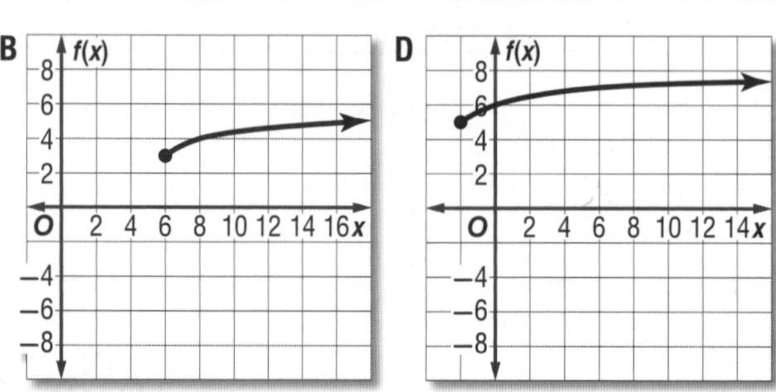

Evaluate each expression. (Lesson 7-3)

12. $\log_4 32$
13. $\log_5 5^{12}$
14. $\log_{16} 4$
15. Write $\log_9 729 = 3$ in exponential form. (Lesson 7-3)

Solve each equation or inequality. Check your solution. (Lessons 7-2 and 7-4)

16. $3^x = 27^2$
17. $4^{3x-1} = 16^x$
18. $\frac{1}{9} = 243^{2x+1}$
19. $16^{2x+3} < 64$
20. $\left(\frac{1}{32}\right)^{x+3} \geq 16^{3x}$
21. $\log_4 x = \frac{3}{2}$
22. $\log_7 (-x + 3) = \log_7 (6x + 5)$
23. $\log_2 x < -3$
24. $\log_8 (3x + 7) = \log_8 (2x - 5)$

LESSON 7-5 Properties of Logarithms

Product	pH Level
lemon juice	2.1
sauerkraut	3.5
tomatoes	4.2
black coffee	5.0
milk	6.4
pure water	7.0
eggs	7.8
milk of magnesia	10.0

:: Then

- You evaluated logarithmic expressions and solved logarithmic equations.

:: Now

1. Simplify and evaluate expressions using the properties of logarithms.
2. Solve logarithmic equations using the properties of logarithms.

:: Why?

- The level of acidity in food is important to some consumers with sensitive stomachs. Most of the foods that we consume are more acidic than basic. The pH scale measures acidity; a low pH indicates an acidic solution, and a high pH indicates a basic solution. It is another example of a logarithmic scale based on powers of ten. Black coffee has a pH of 5, while neutral water has a pH of 7. Black coffee is one hundred times as acidic as neutral water, because $10^{7-5} = 10^2$ or 100.

Common Core State Standards

Content Standards
A.CED.1 Create equations and inequalities in one variable and use them to solve problems.

Mathematical Practices
8 Look for and express regularity in repeated reasoning.

1 Properties of Logarithms

Because logarithms are exponents, the properties of logarithms can be derived from the properties of exponents. The Product Property of Logarithms can be derived from the Product of Powers Property of Exponents.

KeyConcept Product Property of Logarithms

Words	The logarithm of a product is the sum of the logarithms of its factors.
Symbols	For all positive numbers a, b, and x, where $x \neq 1$, $\log_x ab = \log_x a + \log_x b$.
Example	$\log_2 [(5)(6)] = \log_2 5 + \log_2 6$

To show that this property is true, let $b^x = a$ and $b^y = c$. Then, using the definition of logarithm, $x = \log_b a$ and $y = \log_b c$.

$b^x b^y = ac$	Substitution
$b^{x+y} = ac$	Product of Powers
$\log_b b^{x+y} = \log_b ac$	Property of Equality for Logarithmic Functions
$x + y = \log_b ac$	Inverse Property of Exponents and Logarithms
$\log_b a + \log_b c = \log_b ac$	Replace x with $\log_b a$ and y with $\log_b c$.

You can use the Product Property of Logarithms to approximate logarithmic expressions.

Example 1 Use the Product Property

Use $\log_4 3 \approx 0.7925$ to approximate the value of $\log_4 192$.

$\log_4 192 = \log_4 (4^3 \cdot 3)$	Replace 192 with $64 \cdot 3$ or $4^3 \cdot 3$.
$= \log_4 4^3 + \log_4 3$	Product Property
$= 3 + \log_4 3$	Inverse Property of Exponents and Logarithms
$\approx 3 + 0.7925$ or 3.7925	Replace $\log_4 3$ with 0.7925.

GuidedPractice

1. Use $\log_4 2 = 0.5$ to approximate the value of $\log_4 32$.

Recall that the quotient of powers is found by subtracting exponents. The property for the logarithm of a quotient is similar. Let $b^x = a$ and $b^y = c$. Then $\log_b a = x$ and $\log_b c = y$

$\frac{b^x}{b^y} = \frac{a}{c}$	
$b^{x-y} = \frac{a}{c}$	Quotient Property
$\log_b b^{x-y} = \log_b \frac{a}{c}$	Property of Equality for Logarithmic Equations
$x - y = \log_b \frac{a}{c}$	Inverse Property of Exponents and Logarithms
$\log_b a - \log_b c = \log_b \frac{a}{c}$	Replace x with $\log_b a$ and y with $\log_b c$.

KeyConcept Quotient Property of Logarithms

Words The logarithm of a quotient is the difference of the logarithms of the numerator and the denominator.

Symbols For all positive numbers a, b, and x, where $x \neq 1$,
$\log_x \frac{a}{b} = \log_x a - \log_x b$.

Example $\log_2 \frac{5}{6} = \log_2 5 - \log_2 6$

Real-WorldLink

Acid rain is more acidic than normal rain. Smoke and fumes from burning fossil fuels rise into the atmosphere and combine with the moisture in the air to form acid rain. Acid rain can be responsible for the erosion of statues, as in the photo above.

PT

Real-World Example 2 Quotient Property

SCIENCE The pH of a substance is defined as the concentration of hydrogen ions $[H^+]$ in moles. It is given by the formula $\text{pH} = \log_{10} \frac{1}{H^+}$. Find the amount of hydrogen in a liter of acid rain that has a pH of 4.2.

Understand The formula for finding pH and the pH of the rain is given. You want to find the amount of hydrogen in a liter of this rain.

Plan Write the equation. Then, solve for $[H^+]$.

Solve

$\text{pH} = \log_{10} \frac{1}{H^+}$	Original equation
$4.2 = \log_{10} \frac{1}{H^+}$	Substitute 4.2 for pH.
$4.2 = \log_{10} 1 - \log_{10} H^+$	Quotient Property
$4.2 = 0 - \log_{10} H^+$	$\log_{10} 1 = 0$
$4.2 = -\log_{10} H^+$	Simplify.
$-4.2 = \log_{10} H^+$	Multiply each side by −1.
$10^{-4.2} = H^+$	Definition of logarithm

There are $10^{-4.2}$, or about 0.000063, mole of hydrogen in a liter of this rain.

Check

$4.2 = \log_{10} \frac{1}{H^+}$	pH = 4.2
$4.2 \stackrel{?}{=} \log_{10} \frac{1}{10^{-4.2}}$	$H^+ = 10^{-4.2}$
$4.2 \stackrel{?}{=} \log_{10} 1 - \log_{10} 10^{-4.2}$	Quotient Property
$4.2 \stackrel{?}{=} 0 - (-4.2)$	Simplify.
$4.2 = 4.2$ ✓	

GuidedPractice

2. SOUND The loudness L of a sound, measured in decibels, is given by $L = 10 \log_{10} R$, where R is the sound's relative intensity. Suppose one person talks with a relative intensity of 10^6 or 60 decibels. How much louder would 100 people be, talking at the same intensity?

Recall that the power of a power is found by multiplying exponents. The property for the logarithm of a power is similar.

KeyConcept Power Property of Logarithms

Words	The logarithm of a power is the product of the logarithm and the exponent.
Symbols	For any real number p, and positive numbers m and b, where $b \neq 1$, $\log_b m^p = p \log_b m$.
Example	$\log_2 6^5 = 5 \log_2 6$

StudyTip

CCSS Tools You can check this answer by evaluating $2^{4.6438}$ on a calculator. The calculator should give a result of about 25, since $\log_2 25 \approx 4.6438$ means $2^{4.6438} \approx 25$.

Example 3 Power Property of Logarithms

Given $\log_2 5 \approx 2.3219$, approximate the value of $\log_2 25$.

$\log_2 25 = \log_2 5^2$ — Replace 25 with 5^2.

$= 2 \log_2 5$ — Power Property

$\approx 2(2.3219)$ or 4.6438 — Replace $\log_2 5$ with 2.3219.

GuidedPractice

3. Given $\log_3 7 \approx 1.7712$, approximate the value of $\log_3 49$.

2 Solve Logarithmic Equations

Solve Logarithmic Equations You can use the properties of logarithms to solve equations involving logarithms.

PT

Example 4 Solve Equations Using Properties of Logarithms

Solve $\log_6 x + \log_6 (x - 9) = 2$.

$\log_6 x + \log_6 (x - 9) = 2$ — Original equation

$\log_6 x(x - 9) = 2$ — Product Property

$x(x - 9) = 6^2$ — Definition of logarithm

$x^2 - 9x - 36 = 0$ — Subtract 36 from each side.

$(x - 12)(x + 3) = 0$ — Factor.

$x - 12 = 0$ or $x + 3 = 0$ — Zero Product Property

$x = 12$ $\quad x = -3$ — Solve each equation.

CHECK

$$\begin{aligned} \log_6 x + \log_6 (x - 9) &= 2 \\ \log_6 12 + \log_6 (12 - 9) &\stackrel{?}{=} 2 \\ \log_6 12 + \log_6 3 &\stackrel{?}{=} 2 \\ \log_6 (12 \cdot 3) &\stackrel{?}{=} 2 \\ \log_6 36 &\stackrel{?}{=} 2 \\ 2 &= 2 \checkmark \end{aligned}$$

$$\begin{aligned} \log_6 x + \log_6 (x - 9) &= 2 \\ \log_6 (-3) + \log_6 (-3 - 9) &\stackrel{?}{=} 2 \\ \log_6 (-3) + \log_6 (-12) &\stackrel{?}{=} 2 \end{aligned}$$

Because $\log_6 (-3)$ and $\log_6 (-12)$ are undefined, -3 is an extraneous solution.

The solution is $x = 12$.

GuidedPractice

4A. $2 \log_7 x = \log_7 27 + \log_7 3$

4B. $\log_6 x + \log_6 (x + 5) = 2$

Check Your Understanding

= Step-by-Step Solutions begin on page R14.

Examples 1–2 **Use $\log_4 3 \approx 0.7925$ and $\log_4 5 \approx 1.1610$ to approximate the value of each expression.**

1. $\log_4 18$

2. $\log_4 15$

3. $\log_4 \frac{5}{3}$

4. $\log_4 \frac{3}{4}$

Example 2

5. MOUNTAIN CLIMBING As elevation increases, the atmospheric air pressure decreases. The formula for pressure based on elevation is $a = 15{,}500(5 - \log_{10} P)$, where a is the altitude in meters and P is the pressure in pascals (1 psi ≈ 6900 pascals). What is the air pressure at the summit in pascals for each mountain listed in the table at the right?

Mountain	Country	Height (m)
Everest	Nepal/Tibet	8850
Trisuli	India	7074
Bonete	Argentina/Chile	6872
McKinley	United States	6194
Logan	Canada	5959

Example 3 **Given $\log_3 5 \approx 1.465$ and $\log_5 7 \approx 1.2091$, approximate the value of each expression.**

6. $\log_3 25$

7. $\log_5 49$

Example 4 **Solve each equation. Check your solutions.**

8. $\log_4 48 - \log_4 n = \log_4 6$

9. $\log_3 2x + \log_3 7 = \log_3 28$

10. $3 \log_2 x = \log_2 8$

11. $\log_{10} a + \log_{10} (a - 6) = 2$

Practice and Problem Solving

Extra Practice is on page R7.

Examples 1–2 **Use $\log_4 2 = 0.5$, $\log_4 3 \approx 0.7925$, and $\log_4 5 \approx 1.1610$ to approximate the value of each expression.**

12. $\log_4 30$

13. $\log_4 20$

14. $\log_4 \frac{2}{3}$

15 $\log_4 \frac{4}{3}$

16. $\log_4 9$

17. $\log_4 8$

Example 2

18. SCIENCE In 2007, an earthquake near San Francisco registered approximately 5.6 on the Richter scale. The famous San Francisco earthquake of 1906 measured 8.3 in magnitude.

a. How much more intense was the 1906 earthquake than the 2007 earthquake?

b. Richter himself classified the 1906 earthquake as having a magnitude of 8.3. More recent research indicates it was most likely a 7.9. What is the difference in intensities?

Year	Location	Magnitude
1906	San Francisco	8.3
1923	Tokyo, Japan	8.3
1932	Gansu, China	7.6
1960	Chile	9.5
1964	Alaska	9.2
2007	San Francisco	5.6

Source: TLC

Example 3 **Given $\log_6 8 \approx 1.1606$ and $\log_7 9 \approx 1.1292$, approximate the value of each expression.**

19. $\log_6 48$

20. $\log_7 81$

21. $\log_6 512$

22. $\log_7 729$

Example 4 **CCSS PERSEVERANCE** **Solve each equation. Check your solutions.**

23. $\log_3 56 - \log_3 n = \log_3 7$

24. $\log_2 (4x) + \log_2 5 = \log_2 40$

25. $5 \log_2 x = \log_2 32$

26. $\log_{10} a + \log_{10} (a + 21) = 2$

27. **PROBABILITY** In the 1930s, Dr. Frank Benford demonstrated a way to determine whether a set of numbers has been randomly chosen or manually chosen. If the sets of numbers were not randomly chosen, then the Benford formula, $P = \log_{10}\left(1 + \frac{1}{d}\right)$, predicts the probability of a digit d being the first digit of the set. For example, there is a 4.6% probability that the first digit is 9.

   a. Rewrite the formula to solve for the digit if given the probability.

   b. Find the digit that has a 9.7% probability of being selected.

   c. Find the probability that the first digit is 1 ($\log_{10} 2 \approx 0.30103$).

Use $\log_5 3 \approx 0.6826$ and $\log_5 4 \approx 0.8614$ to approximate the value of each expression.

28. $\log_5 40$
29. $\log_5 30$
30. $\log_5 \frac{3}{4}$
31. $\log_5 \frac{4}{3}$
32. $\log_5 9$
33. $\log_5 16$
34. $\log_5 12$
35. $\log_5 27$

Solve each equation. Check your solutions.

36. $\log_3 6 + \log_3 x = \log_3 12$
37. $\log_4 a + \log_4 8 = \log_4 24$
38. $\log_{10} 18 - \log_{10} 3x = \log_{10} 2$
39. $\log_7 100 - \log_7 (y + 5) = \log_7 10$
40. $\log_2 n = \frac{1}{3}\log_2 27 + \log_2 36$
41. $3 \log_{10} 8 - \frac{1}{2}\log_{10} 36 = \log_{10} x$

Solve for n.

42. $\log_a 6n - 3 \log_a x = \log_a x$
43. $2 \log_b 16 + 6 \log_b n = \log_b (x - 2)$

Solve each equation. Check your solutions.

44. $\log_{10} z + \log_{10} (z + 9) = 1$
45. $\log_3 (a^2 + 3) + \log_3 3 = 3$
46. $\log_2 (15b - 15) - \log_2 (-b^2 + 1) = 1$
47. $\log_4 (2y + 2) - \log_4 (y - 2) = 1$
48. $\log_6 0.1 + 2 \log_6 x = \log_6 2 + \log_6 5$
49. $\log_7 64 - \log_7 \frac{8}{3} + \log_7 2 = \log_7 4p$

50. **CCSS REASONING** The humpback whale is an endangered species. Suppose there are 5000 humpback whales in existence today, and the population decreases at a rate of 4% per year.

   a. Write a logarithmic function for the time in years based upon population.

   b. After how long will the population drop below 1000? Round your answer to the nearest year.

State whether each equation is *true* or *false*.

51. $\log_8 (x - 3) = \log_8 x - \log_8 3$
52. $\log_5 22x = \log_5 22 + \log_5 x$
53. $\log_{10} 19k = 19 \log_{10} k$
54. $\log_2 y^5 = 5 \log_2 y$
55. $\log_7 \frac{x}{3} = \log_7 x - \log_7 3$
56. $\log_4 (z + 2) = \log_4 z + \log_4 2$
57. $\log_8 p^4 = (\log_8 p)^4$
58. $\log_9 \frac{x^2y^3}{z^4} = 2 \log_9 x + 3 \log_9 y - 4 \log_9 z$

59. **PARADE** An equation for loudness L, in decibels, is $L = 10 \log_{10} R$, where R is the relative intensity of the sound.

   a. Solve $120 = 10 \log_{10} R$ to find the relative intensity of the Macy's Thanksgiving Day Parade with a loudness of 120 decibels depending on how close you are.

   b. Some parents with young children want the decibel level lowered to 80. How many times less intense would this be? In other words, find the ratio of their intensities.

60. **FINANCIAL LITERACY** The average American carries a credit card debt of approximately \$8600 with an annual percentage rate (APR) of 18.3%. The formula $m = \frac{b\left(\frac{r}{n}\right)}{1 - \left(1 + \frac{r}{n}\right)^{-nt}}$ can be used to compute the monthly payment m that is necessary to pay off a credit card balance b in a given number of years t, where r is the annual percentage rate and n is the number of payments per year.

a. What monthly payment should be made in order to pay off the debt in exactly three years? What is the total amount paid?

b. The equation $t = \frac{\log\left(1 - \frac{br}{mn}\right)}{-n \log\left(1 + \frac{r}{n}\right)}$ can be used to calculate the number of years necessary for a given payment schedule. Copy and complete the table.

Payment (m)	Years (t)
\$50	
\$100	
\$150	
\$200	
\$250	
\$300	

c. Graph the information in the table from part b.

d. If you could only afford to pay \$100 a month, will you be able to pay off the debt? If so, how long will it take? If not, why not?

e. What is the minimum monthly payment that will work toward paying off the debt?

H.O.T. Problems Use Higher-Order Thinking Skills

61. **OPEN ENDED** Write a logarithmic expression for each condition. Then write the expanded expression.

a. a product and a quotient

b. a product and a power

c. a product, a quotient, and a power

62. **CCSS ARGUMENTS** Use the properties of exponents to prove the Power Property of Logarithms.

63. **WRITING IN MATH** Explain why the following are true.

a. $\log_b 1 = 0$ **b.** $\log_b b = 1$ **c.** $\log_b b^x = x$

64. **CHALLENGE** Simplify $\log_{\sqrt{a}} (a^2)$ to find an exact numerical value.

65. **WHICH ONE DOESN'T BELONG?** Find the expression that does not belong. Explain.

$\log_b 24 = \log_b 2 + \log_b 12$

$\log_b 24 = \log_b 20 + \log_b 4$

$\log_b 24 = \log_b 8 + \log_b 3$

$\log_b 24 = \log_b 4 + \log_b 6$

66. **REASONING** Use the properties of logarithms to prove that $\log_a \frac{1}{x} = -\log_a x$.

67. **CHALLENGE** Simplify $x^{3 \log_x 2 - \log_x 5}$ to find an exact numerical value.

68. **WRITING IN MATH** Explain how the properties of exponents and logarithms are related. Include examples like the one shown at the beginning of the lesson illustrating the Product Property, but with the Quotient Property and Power Property of Logarithms.

Standardized Test Practice

69. Find the mode of the data.

22, 11, 12, 23, 7, 6, 17, 15, 21, 19

A 11
B 15
C 16
D There is no mode.

70. SAT/ACT What is the effect on the graph of $y = 4x^2$ when the equation is changed to $y = 2x^2$?

F The graph is rotated 90 degrees about the origin.
G The graph is narrower.
H The graph is wider.
J The graph of $y = 2x^2$ is a reflection of the graph $y = 4x^2$ across the x-axis.
K The graph is unchanged.

71. SHORT RESPONSE In $y = 6.5(1.07)^x$, x represents the number of years since 2000, and y represents the approximate number of millions of Americans 7 years of age and older who went camping two or more times that year. Describe how the number of millions of Americans who go camping is changing over time.

72. What are the x-intercepts of the graph of $y = 4x^2 - 3x - 1$?

A $-\frac{1}{4}$ and $\frac{1}{4}$
B -1 and $\frac{1}{4}$
C -1 and 1
D 1 and $-\frac{1}{4}$

Spiral Review

Solve each equation. Check your solutions. (Lesson 7-4)

73. $\log_5 (3x - 1) = \log_5 (2x^2)$

74. $\log_{10} (x^2 + 1) = 1$

75. $\log_{10} (x^2 - 10x) = \log_{10} (-21)$

Evaluate each expression. (Lesson 7-3)

76. $\log_{10} 0.001$

77. $\log_4 16^x$

78. $\log_3 27^x$

79. ELECTRICITY The amount of current in amperes I that an appliance uses can be calculated using the formula $I = \left(\frac{P}{R}\right)^{\frac{1}{2}}$, where P is the power in watts and R is the resistance in ohms. How much current does an appliance use if $P = 120$ watts and $R = 3$ ohms? Round to the nearest tenth. (Lesson 6-6)

Determine whether each pair of functions are inverse functions. Write *yes* or *no*. (Lesson 6-2)

80. $f(x) = x + 73$
$g(x) = x - 73$

81. $g(x) = 7x - 11$
$h(x) = \frac{1}{7}x + 11$

82. SCULPTING Antonio is preparing to make an ice sculpture. He has a block of ice that he wants to reduce in size by shaving off the same amount from the length, width, and height. He wants to reduce the volume of the ice block to 24 cubic feet. (Lesson 5-7)

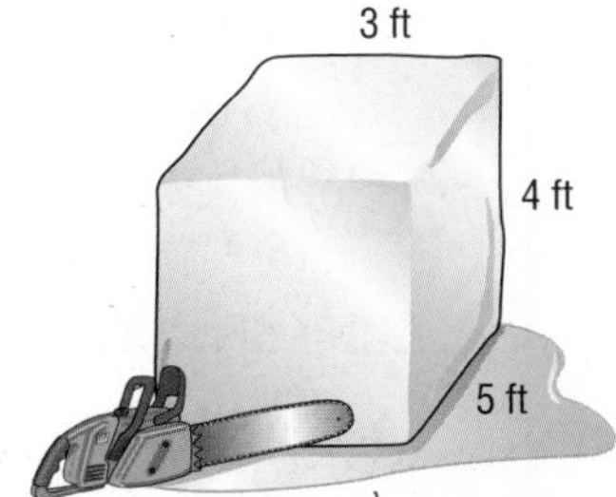

a. Write a polynomial equation to model this situation.

b. How much should he take from each dimension?

Skills Review

Solve each equation or inequality. Check your solution.

83. $3^{4x} = 3^{3 - x}$

84. $3^{2n} \leq \frac{1}{9}$

85. $3^{5x} \cdot 81^{1 - x} = 9^{x - 3}$

86. $49^x = 7^{x^2 - 15}$

87. $\log_2 (x + 6) > 5$

88. $\log_5 (4x - 1) = \log_5 (3x + 2)$

LESSON 7-6 Common Logarithms

Then

- You simplified expressions and solved equations using properties of logarithms.

Now

1. Solve exponential equations and inequalities using common logarithms.
2. Evaluate logarithmic expressions using the Change of Base Formula.

Why?

- Seismologists use the Richter scale to measure the strength or magnitude of earthquakes. The magnitude of an earthquake is determined using the logarithm of the amplitude of waves recorded by seismographs.

The logarithmic scale used by the Richter scale is based on the powers of 10. For example, a magnitude 6.4 earthquake can be represented by $6.4 = \log_{10} x$.

Richter Number	Intensity
1	10^1 micro
2	10^2 minor
3	10^3 minor
4	10^4 light
5	10^5 moderate
6	10^6 strong
7	10^7 major
8	10^8 great

NewVocabulary
common logarithm
Change of Base Formula

Common Core State Standards

Content Standards
A.CED.1 Create equations and inequalities in one variable and use them to solve problems.

Mathematical Practices
4 Model with mathematics.

1 Common Logarithms

You have seen that the base 10 logarithm function, $y = \log_{10} x$, is used in many applications. Base 10 logarithms are called **common logarithms**. Common logarithms are usually written without the subscript 10.

$$\log_{10} x = \log x, x > 0$$

Most scientific calculators have a LOG key for evaluating common logarithms.

Example 1 Find Common Logarithms

Use a calculator to evaluate each expression to the nearest ten-thousandth.

a. log 5

KEYSTROKES: LOG 5 ENTER .6989700043

$\log 5 \approx 0.6990$

b. log 0.3

KEYSTROKES: LOG 0.3 ENTER −.5228787453

$\log 0.3 \approx -0.5229$

GuidedPractice

1A. log 7

1B. log 0.5

The common logarithms of numbers that differ by integral powers of ten are closely related. Remember that a logarithm is an exponent. For example, in the equation $y = \log x$, y is the power to which 10 is raised to obtain the value of x.

$\log x = y$	$\rightarrow$	means	$\rightarrow$	$10^y = x$
$\log 1 = 0$		since		$10^0 = 1$
$\log 10 = 1$		since		$10^1 = 10$
$\log 10^m = m$		since		$10^m = 10^m$

D. Falconer/PhotoLink/Getty Images

Common logarithms are used in the measurement of sound. Soft recorded music is about 36 decibels (dB).

Real-WorldLink

Acoustical Engineer Acoustical engineers are concerned with reducing unwanted sounds, noise control, and making useful sounds. Examples of useful sounds are ultrasound, sonar, and sound reproduction. Employment in this field requires a minimum of a bachelor's degree.

Real-World Example 2 Solve Logarithmic Equations

ROCK CONCERT The loudness L, in decibels, of a sound is $L = 10 \log \frac{I}{m}$, where I is the intensity of the sound and m is the minimum intensity of sound detectable by the human ear. Residents living several miles from a concert venue can hear the music at an intensity of 66.6 decibels. How many times the minimum intensity of sound detectable by the human ear was this sound, if m is defined to be 1?

$L = 10 \log \frac{I}{m}$	Original equation
$66.6 = 10 \log \frac{I}{1}$	Replace L with 66.6 and m with 1.
$6.66 = \log I$	Divide each side by 10 and simplify.
$I = 10^{6.66}$	Exponential form
$I \approx 4{,}570{,}882$	Use a calculator.

The sound heard by the residents was approximately 4,570,000 times the minimum intensity of sound detectable by the human ear.

GuidedPractice

2. EARTHQUAKES The amount of energy E in ergs that an earthquake releases is related to its Richter scale magnitude M by the equation $\log E = 11.8 + 1.5M$. Use the equation to find the amount of energy released by the 2004 Sumatran earthquake, which measured 9.0 on the Richter scale and led to a tsunami.

If both sides of an exponential equation cannot easily be written as powers of the same base, you can solve by taking the logarithm of each side.

Example 3 Solve Exponential Equations Using Logarithms

Solve $4^x = 19$. Round to the nearest ten-thousandth.

$4^x = 19$	Original equation
$\log 4^x = \log 19$	Property of Equality for Logarithmic Functions
$x \log 4 = \log 19$	Power Property of Logarithms
$x = \frac{\log 19}{\log 4}$	Divide each side by log 4.
$x \approx 2.1240$	Use a calculator.

The solution is approximately 2.1240.

CHECK You can check this answer graphically by using a graphing calculator. Graph the line $y = 4^x$ and the line $y = 19$. Then use the **CALC** menu to find the intersection of the two graphs. The intersection is very close to the answer that was obtained algebraically. ✓

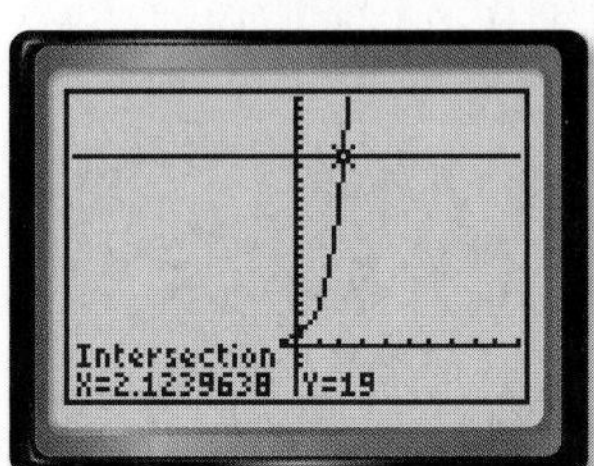

[−10, 10] scl: 1 by [−5, 25] scl: 1

GuidedPractice

3A. $3^x = 15$

3B. $6^x = 42$

UpperCut Images/Getty Images

The same strategies that are used to solve exponential equations can be used to solve exponential inequalities.

StudyTip

Solving Inequalities Remember that the direction of an inequality must be switched if each side is multiplied or divided by a negative number. Since $5 \log 3 - \log 7 > 0$, the inequality does not change.

Example 4 Solve Exponential Inequalities Using Logarithms

Solve $3^{5y} < 7^{y-2}$. Round to the nearest ten-thousandth.

$3^{5y} < 7^{y-2}$	Original inequality
$\log 3^{5y} < \log 7^{y-2}$	Property of Inequality for Logarithmic Functions
$5y \log 3 < (y-2) \log 7$	Power Property of Logarithms
$5y \log 3 < y \log 7 - 2 \log 7$	Distributive Property
$5y \log 3 - y \log 7 < -2 \log 7$	Subtract $y \log 7$ from each side.
$y(5 \log 3 - \log 7) < -2 \log 7$	Distributive Property
$y < \frac{-2 \log 7}{5 \log 3 - \log 7}$	Divide each side by $5 \log 3 - \log 7$.
$\{y \mid y < -1.0972\}$	Use a calculator.

CHECK Test $y = -2$.

$3^{5y} < 7^{y-2}$	Original inequality
$3^{5(-2)} \overset{?}{<} 7^{(-2)-2}$	Replace y with -2.
$3^{-10} \overset{?}{<} 7^{-4}$	Simplify.
$\frac{1}{59,049} < \frac{1}{2401}$ ✓	Negative Exponent Property

GuidedPractice

Solve each inequality. Round to the nearest ten-thousandth.

4A. $3^{2x} \geq 6^{x+1}$

4B. $4^{y} < 5^{2y+1}$

2 Change of Base Formula

The **Change of Base Formula** allows you to write equivalent logarithmic expressions that have different bases.

KeyConcept Change of Base Formula

Symbols For all positive numbers a, b, and n, where $a \neq 1$ and $b \neq 1$,

$$\log_a n = \frac{\log_b n}{\log_b a}.$$

← log base b of original number (numerator)
← log base b of old base (denominator)

Example $\log_3 11 = \frac{\log_{10} 11}{\log_{10} 3}$

To prove this formula, let $\log_a n = x$.

$a^x = n$	Definition of logarithm
$\log_b a^x = \log_b n$	Property of Equality for Logarithmic Functions
$x \log_b a = \log_b n$	Power Property of Logarithms
$x = \frac{\log_b n}{\log_b a}$	Divide each side by $\log_b a$.
$\log_a n = \frac{\log_b n}{\log_b a}$	Replace x with $\log_a n$.

Math HistoryLink

John Napier (1550–1617) John Napier was a Scottish mathematician and theologian who began the use of logarithms to aid in calculations. He is also known for popularizing the use of the decimal point.

Hulton Archive/Getty Images

The Change of Base Formula makes it possible to evaluate a logarithmic expression of any base by translating the expression into one that involves common logarithms.

Example 5 Change of Base Formula

Express $\log_3 20$ in terms of common logarithms. Then round to the nearest ten-thousandth.

$$\log_3 20 = \frac{\log_{10} 20}{\log_{10} 3}$$ Change of Base Formula

$$\approx 2.7268$$ Use a calculator.

GuidedPractice

5. Express $\log_6 8$ in terms of common logarithms. Then round to the nearest ten-thousandth.

Check Your Understanding

● = Step-by-Step Solutions begin on page R14.

Example 1 **Use a calculator to evaluate each expression to the nearest ten-thousandth.**

1. $\log 5$ **2.** $\log 21$ **3.** $\log 0.4$ **4.** $\log 0.7$

Example 2

5. **SCIENCE** The amount of energy E in ergs that an earthquake releases is related to its Richter scale magnitude M by the equation $\log E = 11.8 + 1.5M$. Use the equation to find the amount of energy released by the 1960 Chilean earthquake, which measured 8.5 on the Richter scale.

Example 3 **Solve each equation. Round to the nearest ten-thousandth.**

6. $6^x = 40$ **7.** $2.1^{a+2} = 8.25$ **8.** $7^{x^2} = 20.42$ **9** $11^{b-3} = 5^b$

Example 4 **Solve each inequality. Round to the nearest ten-thousandth.**

10. $5^{4n} > 33$ **11.** $6^{p-1} \le 4^p$

Example 5 **Express each logarithm in terms of common logarithms. Then approximate its value to the nearest ten-thousandth.**

12. $\log_3 7$ **13.** $\log_4 23$ **14.** $\log_9 13$ **15.** $\log_2 5$

Practice and Problem Solving

Extra Practice is on page R7.

Example 1 **Use a calculator to evaluate each expression to the nearest ten-thousandth.**

16. $\log 3$ **17.** $\log 11$ **18.** $\log 3.2$

19. $\log 8.2$ **20.** $\log 0.9$ **21.** $\log 0.04$

Example 2

22. CCSS **SENSE-MAKING** Loretta had a new muffler installed on her car. The noise level of the engine dropped from 85 decibels to 73 decibels.

a. How many times the minimum intensity of sound detectable by the human ear was the car with the old muffler, if m is defined to be 1?

b. How many times the minimum intensity of sound detectable by the human ear is the car with the new muffler? Find the percent of decrease of the intensity of the sound with the new muffler.

Example 3 **Solve each equation. Round to the nearest ten-thousandth.**

23. $8^x = 40$

24. $5^x = 55$

25. $2.9^{a-4} = 8.1$

26. $9^{b-1} = 7^b$

27. $13^{x^2} = 33.3$

28. $15^{x^2} = 110$

Example 4 **Solve each inequality. Round to the nearest ten-thousandth.**

29. $6^{3n} > 36$

30. $2^{4x} \leq 20$

31. $3^{y-1} \leq 4^y$

32. $5^{p-2} \geq 2^p$

Example 5 **Express each logarithm in terms of common logarithms. Then approximate its value to the nearest ten-thousandth.**

33. $\log_7 18$

34. $\log_5 31$

35. $\log_2 16$

36. $\log_4 9$

37. $\log_3 11$

38. $\log_6 33$

39 **PETS** The number n of pet owners in thousands after t years can be modeled by $n = 35[\log_4 (t + 2)]$. Let $t = 0$ represent 2000. Use the Change of Base Formula to answer the following questions.

a. How many pet owners were there in 2010?

b. How long until there are 80,000 pet owners? When will this occur?

40. **CCSS PRECISION** Five years ago the grizzly bear population in a certain national park was 325. Today it is 450. Studies show that the park can support a population of 750.

a. What is the average annual rate of growth in the population if the grizzly bears reproduce once a year?

b. How many years will it take to reach the maximum population if the population growth continues at the same average rate?

Solve each equation or inequality. Round to the nearest ten-thousandth.

41. $3^x = 40$

42. $5^{3p} = 15$

43. $4^{n+2} = 14.5$

44. $8^{z-4} = 6.3$

45. $7.4^{n-3} = 32.5$

46. $3.1^{y-5} = 9.2$

47. $5^x \geq 42$

48. $9^{2a} < 120$

49. $3^{4x} \leq 72$

50. $7^{2n} > 52^{4n+3}$

51. $6^p \leq 13^{5-p}$

52. $2^{y+3} \geq 8^{3y}$

Express each logarithm in terms of common logarithms. Then approximate its value to the nearest ten-thousandth.

53. $\log_4 12$

54. $\log_3 21$

55. $\log_8 2$

56. $\log_6 7$

57. $\log_5 (2.7)^2$

58. $\log_7 \sqrt{5}$

59. **MUSIC** A musical cent is a unit in a logarithmic scale of relative pitch or intervals. One octave is equal to 1200 cents. The formula $n = 1200\left(\log_2 \frac{a}{b}\right)$ can be used to determine the difference in cents between two notes with frequencies a and b.

a. Find the interval in cents when the frequency changes from 443 Hertz (Hz) to 415 Hz.

b. If the interval is 55 cents and the beginning frequency is 225 Hz, find the final frequency.

Solve each equation. Round to the nearest ten-thousandth.

60. $10^{x^2} = 60$

61 $4^{x^2 - 3} = 16$

62. $9^{6y - 2} = 3^{3y + 1}$

63. $8^{2x - 4} = 4^{x + 1}$

64. $16^x = \sqrt{4^{x + 3}}$

65. $2^y = \sqrt{3^{y - 1}}$

66. ENVIRONMENTAL SCIENCE An environmental engineer is testing drinking water wells in coastal communities for pollution, specifically unsafe levels of arsenic. The safe standard for arsenic is 0.025 parts per million (ppm). Also, the pH of the arsenic level should be less than 9.5. The formula for hydrogen ion concentration is pH = $-\log H$. (*Hint*: 1 kilogram of water occupies approximately 1 liter. 1 ppm = 1 mg/kg.)

a. Suppose the hydrogen ion concentration of a well is 1.25×10^{-11}. Should the environmental engineer be worried about too high an arsenic content?

b. The environmental engineer finds 1 milligram of arsenic in a 3 liter sample, is the well safe?

c. What is the hydrogen ion concentration that meets the troublesome pH level of 9.5?

67. MULTIPLE REPRESENTATIONS In this problem, you will solve the exponential equation $4^x = 13$.

a. Tabular Enter the function $y = 4^x$ into a graphing calculator, create a table of values for the function, and scroll through the table to find x when $y = 13$.

b. Graphical Graph $y = 4^x$ and $y = 13$ on the same screen. Use the **intersect** feature to find the point of intersection.

c. Numerical Solve the equation algebraically. Do all of the methods produce the same result? Explain why or why not.

H.O.T. Problems Use Higher-Order Thinking Skills

68. CCSS CRITIQUE Sam and Rosamaria are solving $4^{3p} = 10$. Is either of them correct? Explain your reasoning.

Sam	Rosamaria
$4^{3p} = 10$	$4^{3p} = 10$
$\log 4^{3p} = \log 10$	$\log 4^{3p} = \log 10$
$p \log 4 = \log 10$	$3p \log 4 = \log 10$
$p = \dfrac{\log 10}{\log 4}$	$p = \dfrac{\log 10}{3 \log 4}$

69. CHALLENGE Solve $\log_{\sqrt{a}} 3 = \log_a x$ for x and explain each step.

70. REASONING Write $\dfrac{\log_5 9}{\log_5 3}$ as a single logarithm.

71. PROOF Find the values of $\log_3 27$ and $\log_{27} 3$. Make and prove a conjecture about the relationship between $\log_a b$ and $\log_b a$.

72. WRITING IN MATH Explain how exponents and logarithms are related. Include examples like how to solve a logarithmic equation using exponents and how to solve an exponential equation using logarithms.

Standardized Test Practice

73. Which expression represents $f[g(x)]$ if $f(x) = x^2 + 4x + 3$ and $g(x) = x - 5$?

A $x^2 + 4x - 2$

B $x^2 - 6x + 8$

C $x^2 - 9x + 23$

D $x^2 - 14x + 6$

74. EXTENDED RESPONSE Colleen rented 3 documentaries, 2 video games, and 2 movies. The charge was \$16.29. The next week, she rented 1 documentary, 3 video games, and 4 movies for a total charge of \$19.84. The third week she rented 2 documentaries, 1 video game, and 1 movie for a total charge of \$9.14.

a. Write a system of equations to determine the cost to rent each item.

b. What is the cost to rent each item?

75. GEOMETRY If the surface area of a cube is increased by a factor of 9, what is the change in the length of the sides of the cube?

F The length is 2 times the original length.

G The length is 3 times the original length.

H The length is 6 times the original length.

J The length is 9 times the original length.

76. SAT/ACT Which of the following *most* accurately describes the translation of the graph $y = (x + 4)^2 - 3$ to the graph of $y = (x - 1)^2 + 3$?

A down 1 and to the right 3

B down 6 and to the left 5

C up 1 and to the left 3

D up 1 and to the right 3

E up 6 and to the right 5

Spiral Review

Solve each equation. Check your solutions. (Lesson 7-5)

77. $\log_5 7 + \frac{1}{2}\log_5 4 = \log_5 x$

78. $2 \log_2 x - \log_2 (x + 3) = 2$

79. $\log_6 48 - \log_6 \frac{16}{5} + \log_6 5 = \log_6 5x$

80. $\log_{10} a + \log_{10} (a + 21) = 2$

Solve each equation or inequality. (Lesson 7-4)

81. $\log_4 x = \frac{1}{2}$

82. $\log_{81} 729 = x$

83. $\log_8 (x^2 + x) = \log_8 12$

84. $\log_8 (3y - 1) < \log_8 (y + 5)$

85. SAILING The area of a triangular sail is $16x^4 - 60x^3 - 28x^2 + 56x - 32$ square meters. The base of the triangle is $x - 4$ meters. What is the height of the sail? (Lesson 5-2)

86. HOME REPAIR Mr. Turner is getting new locks installed. The locksmith charges \$85 for the service call, \$25 for each door, and \$30 for each lock. (Lesson 2-4)

a. Write an equation that represents the cost for x number of doors.

b. Mr. Turner wants the front, side, back, and garage door locks changed. How much will this cost?

Skills Review

Write an equivalent exponential equation.

87. $\log_2 5 = x$

88. $\log_4 x = 3$

89. $\log_5 25 = 2$

90. $\log_7 10 = x$

91. $\log_6 x = 4$

92. $\log_4 64 = 3$

EXTEND 7-6

Graphing Technology Lab: Solving Logarithmic Equations and Inequalities

You have solved logarithmic equations algebraically. You can also solve logarithmic equations by graphing or by using a table. The TI-83/84 Plus has $y = \log_{10} x$ as a built-in function. Enter [Y=] [LOG] [X,T,θ,n] [GRAPH] to view this graph.

To graph logarithmic functions with bases other than 10, you must use the Change of Base Formula, $\log_a n = \frac{\log_b n}{\log_b a}$.

Common Core State Standards

Content Standards

A.REI.11 Explain why the x-coordinates of the points where the graphs of the equations $y = f(x)$ and $y = g(x)$ intersect are the solutions of the equation $f(x) = g(x)$; find the solutions approximately, e.g., using technology to graph the functions, make tables of values, or find successive approximations. Include cases where $f(x)$ and/or $g(x)$ are linear, polynomial, rational, absolute value, exponential, and logarithmic functions.

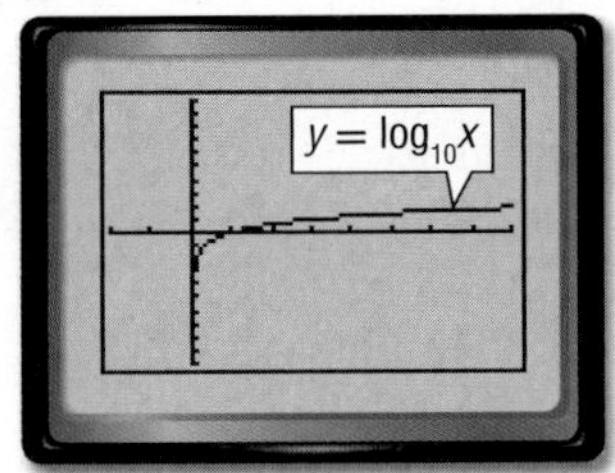

[−2, 8] scl: 1 by [−10, 10] scl: 1

Activity 1

Solve $\log_2 (6x - 8) = \log_3 (20x + 1)$.

Step 1 Graph each side of the equation.

Graph each side of the equation as a separate function. Enter $\log_2 (6x - 8)$ as **Y1** and $\log_3 (20x + 1)$ as **Y2**. Then graph the two equations.

KEYSTROKES: [Y=] [LOG] 6 [X,T,θ,n] [−] 8 [)] [÷] [LOG] 2 [)] [ENTER] [LOG] 20 [X,T,θ,n] [+] 1 [)] [÷] [LOG] 3 [)] [GRAPH]

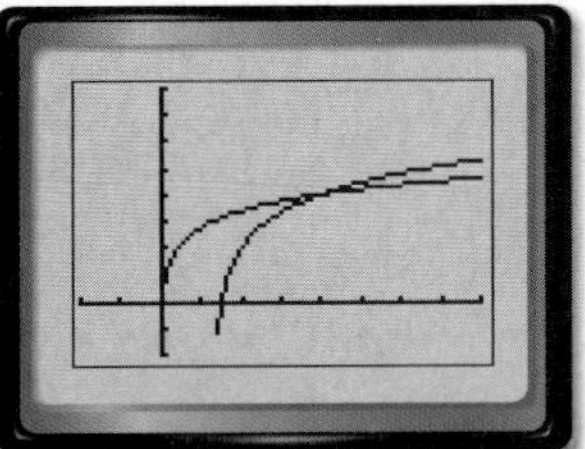

[−2, 8] scl: 1 by [−2, 8] scl: 1

Step 2 Use the **intersect** feature.

Use the **intersect** feature on the **CALC** menu to approximate the ordered pair of the point at which the curves intersect.

The calculator screen shows that the x-coordinate of the point at which the curves intersect is 4. Therefore, the solution of the equation is 4.

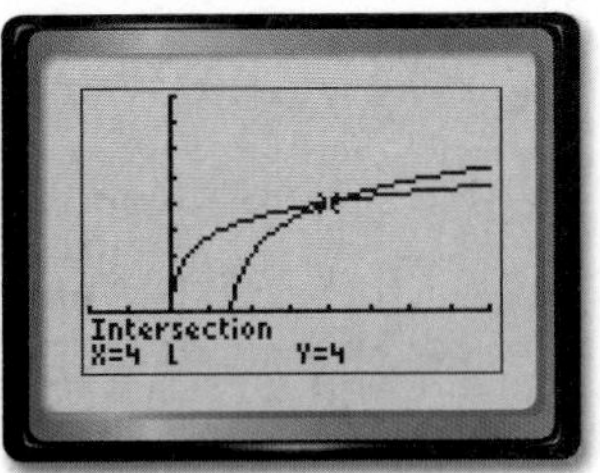

[−2, 8] scl: 1 by [−2, 8] scl: 1

Step 3 Use the **TABLE** feature.

Examine the table to find the x-value for which the y-values for the graphs are equal. At $x = 4$, both functions have a y-value of 4. Thus, the solution of the equation is 4.

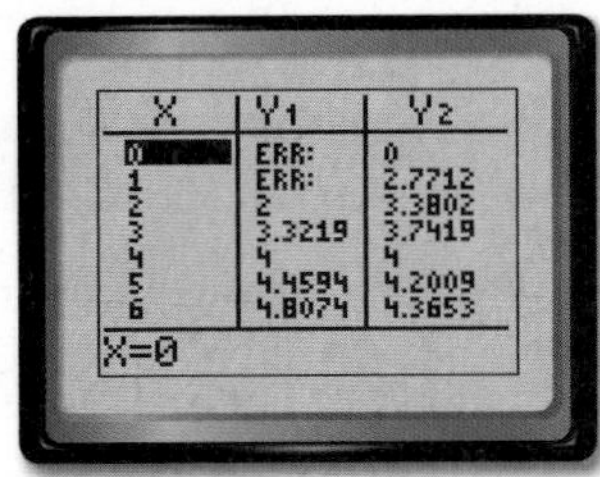

You can use a similar procedure to solve logarithmic inequalities using a graphing calculator.

(continued on the next page)

Graphing Technology Lab
Solving Logarithmic Equations and Inequalities *Continued*

Activity 2

Solve $\log_4 (10x + 1) < \log_5 (16 + 6x)$.

Step 1 Enter the inequalities.

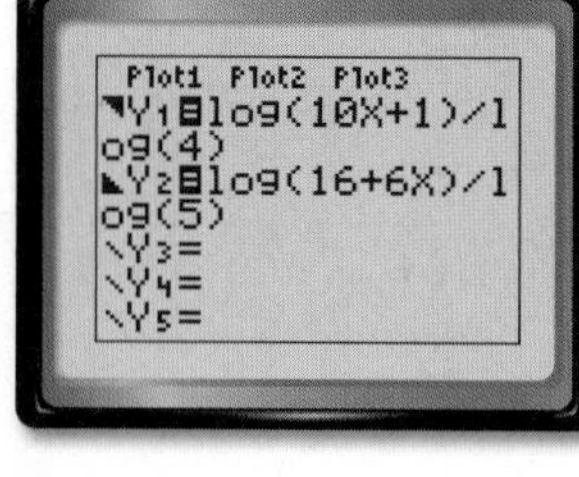

Rewrite the problem as a system of inequalities.

The first inequality is $\log_4 (10x + 1) < y$ or $y > \log_4 (10x + 1)$. Since this inequality includes the *greater than* symbol, shade above the curve.

First enter the boundary and then use the arrow and ENTER keys to choose the shade above icon, ◣.

The second inequality is $y < \log_5 (16 + 6x)$. Shade below the curve since this inequality contains *less than*.

KEYSTROKES: Y= ◀ ◀ ENTER ENTER ▶ ▶ LOG 10 X,T,θ,n + 1) ÷ LOG 4) ENTER ◀ ◀ ENTER ENTER ENTER ▶ ▶ LOG 16 + 6 X,T,θ,n) ÷ LOG 5)

Step 2 Graph the system.

KEYSTROKES: GRAPH

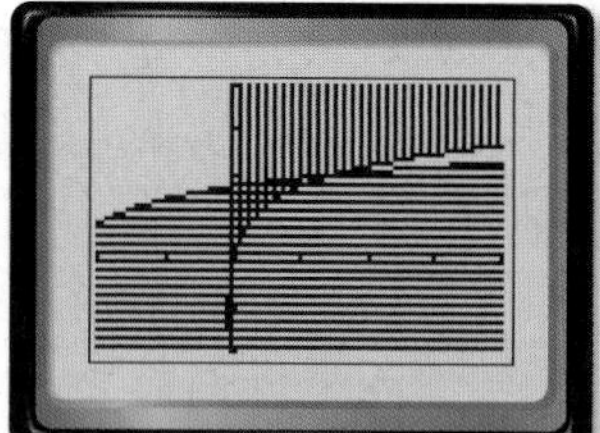

[−2, 4] scl: 1 by [−2, 4] scl: 1

The left boundary of the solution set is where the first inequality is undefined. It is undefined for $10x + 1 \leq 0$.

$$10x + 1 \leq 0$$
$$10x \leq -1$$
$$x \leq -\frac{1}{10}$$

Use the calculator's **intersect** feature to find the right boundary. You can conclude that the solution set is $\{x \mid -0.1 < x < 1.5\}$.

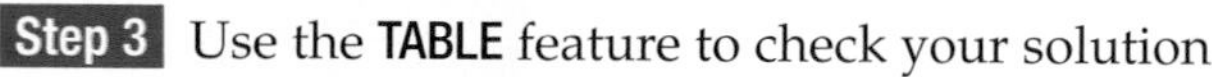

Step 3 Use the **TABLE** feature to check your solution.

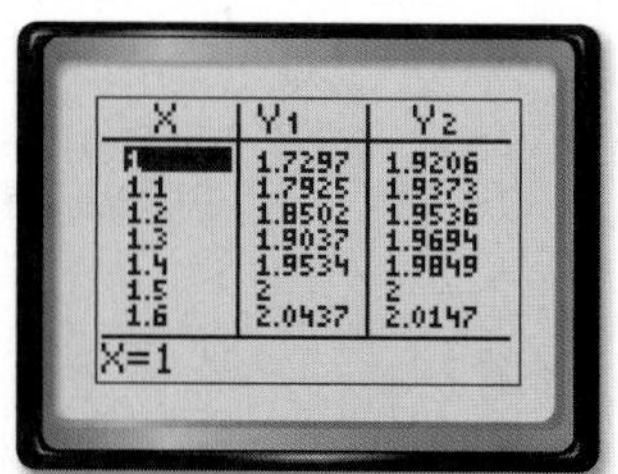

Start the table at −0.1 and show x-values in increments of 0.1. Scroll through the table.

KEYSTROKES: 2nd [TBLSET] −0.1 ENTER 0.1 ENTER 2nd [TABLE]

The table confirms the solution of the inequality is $\{x \mid -0.1 < x < 1.5\}$.

Exercises

Solve each equation or inequality. Check your solution.

1. $\log_2 (3x + 2) = \log_3 (12x + 3)$
2. $\log_6 (7x + 1) = \log_4 (4x - 4)$
3. $\log_2 3x = \log_3 (2x + 2)$
4. $\log_{10} (1 - x) = \log_5 (2x + 5)$
5. $\log_4 (9x + 1) > \log_3 (18x - 1)$
6. $\log_3 (3x - 5) \geq \log_3 (x + 7)$
7. $\log_5 (2x + 1) < \log_4 (3x - 2)$
8. $\log_2 2x \leq \log_4 (x + 3)$

LESSON 7-7 Base e and Natural Logarithms

Then

- You worked with common logarithms.

Now

1. Evaluate expressions involving the natural base and natural logarithm.
2. Solve exponential equations and inequalities using natural logarithms.

Why?

- The St. Louis Gateway Arch in Missouri is in the form of an inverted catenary curve. A catenary curve directs the force of its weight along itself, so that:
  - if a rope or chain is hanging, it is pulled into that shape, and,
  - if a catenary is standing upright, it can support itself.

  The equation for the catenary curve involves e, a special number that appears throughout mathematics and science.

NewVocabulary
natural base, e
natural base exponential function
natural logarithm

Common Core State Standards

Content Standards
A.SSE.2 Use the structure of an expression to identify ways to rewrite it.

Mathematical Practices
7 Look for and make use of structure.

1 Base e and Natural Logarithms

Like π and $\sqrt{2}$, the number e is an irrational number. The value of e is 2.71828... . It is referred to as the **natural base, e**. An exponential function with base e is called a **natural base exponential function**.

KeyConcept Natural Base Functions

The function $f(x) = e^x$ is used to model continuous exponential growth.
The function $f(x) = e^{-x}$ is used to model continuous exponential decay.

The inverse of a natural base exponential function is called the **natural logarithm**. This logarithm can be written as $\log_e x$, but is more often abbreviated as $\ln x$.

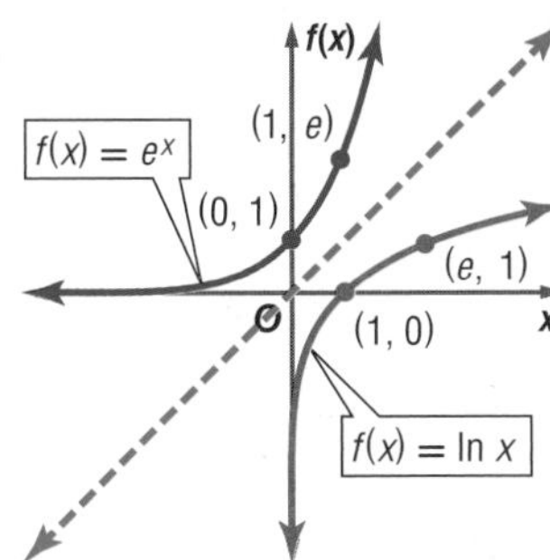

Exponential Growth

Exponential Decay

You can write an equivalent base e exponential equation for a natural logarithmic equation by using the fact that $\ln x = \log_e x$.

$$\ln 4 = x \quad \rightarrow \quad \log_e 4 = x \quad \rightarrow \quad e^x = 4$$

Example 1 Write Equivalent Expressions

Write each exponential equation in logarithmic form.

a. $e^x = 8$

$e^x = 8 \quad \rightarrow \quad \log_e 8 = x$

$\ln 8 = x$

b. $e^5 = x$

$e^5 = x \quad \rightarrow \quad \log_e x = 5$

$\ln x = 5$

GuidedPractice

1A. $e^x = 9$

1B. $e^7 = x$

CORBIS

You can also write an equivalent natural logarithm equation for a natural base *e* exponential equation.

$$e^x = 12 \rightarrow \log_e 12 = x \rightarrow \ln 12 = x$$

PT

Example 2 Write Equivalent Expressions

Write each logarithmic equation in exponential form.

a. $\ln x \approx 0.7741$

$\ln x \approx 0.7741 \rightarrow \log_e x = 0.7741$

$x \approx e^{0.7741}$

b. $\ln 10 = x$

$\ln 10 = x \rightarrow \log_e 10 = x$

$10 = e^x$

GuidedPractice

2A. $\ln x \approx 2.1438$

2B. $\ln 18 = x$

The properties of logarithms you learned in Lesson 7-5 also apply to the natural logarithms. The logarithmic expressions below can be simplified into a single logarithmic term.

PT

Example 3 Simplify Expressions with *e* and the Natural Log

StudyTip

Simplifying When you simplify logarithmic expressions, verify that the logarithm contains no operations and no powers.

Write each expression as a single logarithm.

a. $3 \ln 10 - \ln 8$

$3 \ln 10 - \ln 8 = \ln 10^3 - \ln 8$	Power Property of Logarithms
$= \ln \frac{10^3}{8}$	Quotient Property of Logarithms
$= \ln 125$	Simplify.
$= \ln 5^3$	$5^3 = 125$
$= 3 \ln 5$	Power Property of Logarithms

CHECK Use a calculator to verify the solution.

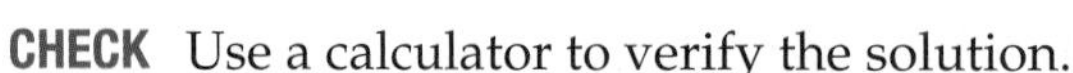

KEYSTROKES: 3 [LN] 10 [ )] [–] [LN] 8 [ )] [ENTER] 4.828313737

KEYSTROKES: 3 [LN] 5 [ )] [ENTER] 4.828313737 ✓

b. $\ln 40 + 2 \ln \frac{1}{2} + \ln x$

$\ln 40 + 2 \ln \frac{1}{2} + \ln x = \ln 40 + \ln \frac{1}{4} + \ln x$	Power Property of Logarithms
$= \ln \left(40 \cdot \frac{1}{4} \cdot x\right)$	Product Property of Logarithms
$= \ln 10x$	Simplify.

GuidedPractice

3A. $6 \ln 8 - 2 \ln 4$

3B. $2 \ln 5 + 4 \ln 2 + \ln 5y$

Because the natural base and natural log are inverse functions, they can be used to *undo* or eliminate each other.

$$e^{\ln x} = x \qquad \ln e^x = x$$

2 Equations and Inequalities with e and ln

Equations and inequalities involving base e are easier to solve by using natural logarithms rather than by using common logarithms, because $\ln e = 1$.

Example 4 Solve Base e Equations

Solve $4e^{-2x} - 5 = 3$. Round to the nearest ten-thousandth.

$4e^{-2x} - 5 = 3$	Original equation
$4e^{-2x} = 8$	Add 5 to each side.
$e^{-2x} = 2$	Divide each side by 4.
$\ln e^{-2x} = \ln 2$	Property of Equality for Logarithms
$-2x = \ln 2$	$\ln e^x = x$
$x = \frac{\ln 2}{-2}$	Divide each side by −2.
$x \approx -0.3466$	Use a calculator.

KEYSTROKES: [LN] 2 [ )] [÷] −2 [ENTER] −.34657359

StudyTip

Calculators Most calculators have an e^x and LN key for evaluating natural base and natural log expressions.

GuidedPractice

Solve each equation. Round to the nearest ten-thousandth.

4A. $3e^{4x} - 12 = 15$

4B. $4e^{-x} + 8 = 17$

Just like the natural logarithm can be used to eliminate e^x, the natural base exponential function can eliminate $\ln x$.

Example 5 Solve Natural Log Equations and Inequalities

Solve each equation or inequality. Round to the nearest ten-thousandth.

a. $3 \ln 4x = 24$

$3 \ln 4x = 24$	Original equation
$\ln 4x = 8$	Divide each side by 3.
$e^{\ln 4x} = e^8$	Property of Equality for Exponential Functions
$4x = e^8$	$e^{\ln x} = x$
$x = \frac{e^8}{4}$	Divide each side by 4.
$x \approx 745.2395$	Use a calculator.

b. $\ln (x - 8)^4 < 4$

$\ln (x - 8)^4 < 4$	Original equation
$e^{\ln (x-8)^4} < e^4$	Write each side using exponents and base e.
$(x - 8)^4 < e^4$	$e^{\ln x} = x$
$x - 8 < e$	Property of Equality for Exponential Functions
$x < e + 8$	Add 8 to each side.
$x < 10.7183$	Use a calculator.

GuidedPractice

Solve each equation or inequality. Round to the nearest ten-thousandth.

5A. $5 \ln 6x = 8$

5B. $\ln (2x - 3)^3 > 6$

Interest that is compounded continuously can be found using e.

KeyConcept Continuously Compounded Interest

Calculate continuously compounded interest using the following formula:

$$A = Pe^{rt},$$

where A is the amount in the account after t years, P is the principal amount invested, and r is the annual interest rate.

Real-WorldLink

The average cost of tuition at four-year public colleges is about $6185 per year.

Source: CNN

Real-World Example 6 Solve Base e Inequalities

FINANCIAL LITERACY When Angelina was born, her grandparents deposited $3000 into a college savings account paying 4% interest compounded continuously.

a. Assuming there are no deposits or withdrawals from the account, what will the balance be after 10 years?

$A = Pe^{rt}$	Continuous Compounding Formula
$= 3000e^{(0.04)(10)}$	$P = 3000$, $r = 0.04$, and $t = 10$
$= 3000e^{0.4}$	Simplify.
≈ 4475.47	Use a calculator.

The balance will be $4475.47.

b. How long will it take the balance to reach at least $10,000?

$A < Pe^{rt}$	Continuous Compounding Formula
$10{,}000 < 3000e^{(0.04)t}$	$P = 3000$, $r = 0.04$, and $A = 10{,}000$
$\frac{10}{3} < e^{0.04t}$	Divide each side by 3000.
$\ln \frac{10}{3} < \ln e^{0.04t}$	Property of Equality of Logarithms
$\ln \frac{10}{3} < 0.04t$	$\ln e^x = x$
$\frac{\ln \frac{10}{3}}{0.04} < t$	Divide each side by 0.04.
$30.099 < t$	Use a calculator.

It will take about 30 years to reach at least $10,000.

c. If her grandparents want Angelina to have $10,000 after 18 years, how much would they need to invest?

$10{,}000 = Pe^{(0.04)18}$	$A = 10{,}000$, $r = 0.04$, and $t = 18$
$\frac{10{,}000}{e^{0.72}} = P$	Divide each side by $e^{0.72}$.
$4867.52 \approx P$	Use a calculator.

They need to invest $4867.52.

StudyTip

CCSS Reasoning In order to avoid any errors due to rounding, do not round until the very end of your calculations.

GuidedPractice

6. Use the information in Example 6 to answer the following.

   A. If they invested $8000 at 3.75% interest compounded continuously, how much money would be in the account in 30 years?

   B. If they could only deposit $10,000 in the account above, at what rate would the account need to grow in order for Angelina to have $30,000 in 18 years?

   C. If Angelina's grandparents found an account that paid 5% compounded continuously and wanted her to have $30,000 after 18 years, how much would they need to deposit?

Cornstock Images/Getty Images

Check Your Understanding

⬤ = Step-by-Step Solutions begin on page R14.

Examples 1–2 **Write an equivalent exponential or logarithmic function.**

1. $e^x = 30$

2. $\ln x = 42$

3. $e^3 = x$

4. $\ln 18 = x$

Example 3 **Write each as a single logarithm.**

5. $3 \ln 2 + 2 \ln 4$

6. $5 \ln 3 - 2 \ln 9$

7. $3 \ln 6 + 2 \ln 9$

8. $3 \ln 5 + 4 \ln x$

Example 4 **Solve each equation. Round to the nearest ten-thousandth.**

9. $5e^x - 24 = 16$

10. $-3e^x + 9 = 4$

11. $3e^{-3x} + 4 = 6$

12. $2e^{-x} - 3 = 8$

Example 5 **Solve each equation or inequality. Round to the nearest ten-thousandth.**

13. $\ln 3x = 8$

14. $-4 \ln 2x = -26$

15. $\ln (x + 5)^2 < 6$

16. $\ln (x - 2)^3 > 15$

17. $e^x > 29$

18. $5 + e^{-x} > 14$

Example 6 **19.** **SCIENCE** A virus is spreading through a computer network according to the formula $v(t) = 30e^{0.1t}$, where v is the number of computers infected and t is the time in minutes. How long will it take the virus to infect 10,000 computers?

Practice and Problem Solving

Extra Practice is on page R7.

Examples 1–2 **Write an equivalent exponential or logarithmic function.**

20. $e^{-x} = 8$

21. $e^{-5x} = 0.1$

22. $\ln 0.25 = x$

23. $\ln 5.4 = x$

24. $e^{x-3} = 2$

25. $\ln (x + 4) = 36$

26. $e^{-2} = x^6$

27. $\ln e^x = 7$

Example 3 **Write each as a single logarithm.**

28. $\ln 125 - 2 \ln 5$

29. $3 \ln 10 + 2 \ln 100$

30. $4 \ln \frac{1}{3} - 6 \ln \frac{1}{9}$

(31) $7 \ln \frac{1}{2} + 5 \ln 2$

32. $8 \ln x - 4 \ln 5$

33. $3 \ln x^2 + 4 \ln 3$

Example 4 **Solve each equation. Round to the nearest ten-thousandth.**

34. $6e^x - 3 = 35$

35. $4e^x + 2 = 180$

36. $3e^{2x} - 5 = -4$

37. $-2e^{3x} + 19 = 3$

38. $6e^{4x} + 7 = 4$

39. $-4e^{-x} + 9 = 2$

Examples 5–6 **40.** **CCSS SENSE-MAKING** The value of a certain car depreciates according to $v(t) = 18500e^{-0.186t}$, where t is the number of years after the car is purchased new.

a. What will the car be worth in 18 months?

b. When will the car be worth half of its original value?

c. When will the car be worth less than $1000?

Solve each inequality. Round to the nearest ten-thousandth.

41. $e^x \leq 8.7$

42. $e^x \geq 42.1$

43. $\ln (3x + 4)^3 > 10$

44. $4 \ln x^2 < 72$

45. $\ln (8x^4) > 24$

46. $-2 [\ln (x - 6)^{-1}] \leq 6$

47 FINANCIAL LITERACY Use the formula for continuously compounded interest.

a. If you deposited \$800 in an account paying 4.5% interest compounded continuously, how much money would be in the account in 5 years?

b. How long would it take you to double your money?

c. If you want to double your money in 9 years, what rate would you need?

d. If you want to open an account that pays 4.75% interest compounded continuously and have \$10,000 in the account 12 years after your deposit, how much would you need to deposit?

Write the expression as a sum or difference of logarithms or multiples of logarithms.

48. $\ln 12x^2$ **49.** $\ln \frac{16}{125}$ **50.** $\ln \sqrt[5]{x^3}$ **51.** $\ln xy^4z^{-3}$

Use the natural logarithm to solve each equation.

52. $8^x = 24$ **53.** $3^x = 0.4$ **54.** $2^{3x} = 18$ **55.** $5^{2x} = 38$

56. CCSS MODELING Newton's Law of Cooling, which can be used to determine how fast an object will cool in given surroundings, is represented by $T(t) = T_s + (T_0 - T_s)e^{-kt}$, where T_0 is the initial temperature of the object, T_s is the temperature of the surroundings, t is the time in minutes, and k is a constant value that depends on the type of object.

a. If a cup of coffee with an initial temperature of 180° is placed in a room with a temperature of 70° and the coffee cools to 140° after 10 minutes, find the value of k.

b. Use this value of k to determine the temperature of the coffee after 20 minutes.

c. When will the temperature of the coffee reach 75°?

57. MULTIPLE REPRESENTATIONS In this problem, you will use $f(x) = e^x$ and $g(x) = \ln x$.

a. Graphical Graph both functions and their axis of symmetry, $y = x$, for $-5 \leq x \leq 5$. Then graph $a(x) = e^{-x}$ on the same graph.

b. Analytical The graphs of $a(x)$ and $f(x)$ are reflections along which axis? What function would be a reflection of $f(x)$ along the other axis?

c. Graphical Determine the two functions that are reflections of $g(x)$. Graph these new functions.

d. Verbal We know that $f(x)$ and $g(x)$ are inverses. Are any of the other functions that we have graphed inverses as well? Explain your reasoning.

H.O.T. Problems Use Higher-Order Thinking Skills

58. CHALLENGE Solve $4^x - 2^{x+1} = 15$ for x.

59. PROOF Prove $\ln ab = \ln a + \ln b$ for natural logarithms.

60. REASONING Determine whether $x > \ln x$ is *sometimes*, *always*, or *never* true. Explain your reasoning.

61. OPEN ENDED Express the value 3 using e^x and the natural log.

62. WRITING IN MATH Explain how the natural log can be used to solve a natural base exponential function.

Standardized Test Practice

63. Given the function $y = 2.34x + 11.33$, which statement best describes the effect of moving the graph down two units?

A The x-intercept decreases.
B The y-intercept decreases.
C The x-intercept remains the same.
D The y-intercept remains the same.

64. GRIDDED RESPONSE Aidan sells wooden picture frames over the Internet. He purchases materials for $85 and pays $19.95 for his website. If he charges $15 each, how many frames will he need to sell in order to make a profit of at least $270?

65. Solve $|2x - 5| = 17$.

F $-6, -11$
G $-6, 11$
H $6, -11$
J $6, 11$

66. A local pet store sells rabbit food. The cost of two 5-pound bags is $7.99. The total cost c of purchasing n bags can be found by—

A multiplying n by c.
B multiplying n by 5.
C multiplying n by the cost of 1 bag.
D dividing n by c.

Spiral Review

Solve each equation or inequality. Round to the nearest ten-thousandth. (Lesson 7-6)

67. $2^x = 53$ **68.** $2.3^{x^2} = 66.6$ **69.** $3^{4x-7} < 4^{2x+3}$

70. $6^{3y} = 8^{y-1}$ **71.** $12^{x-5} \geq 9.32$ **72.** $2.1^{x-5} = 9.32$

73. SOUND Use the formula $L = 10 \log_{10} R$, where L is the loudness of a sound and R is the sound's relative intensity. Suppose the sound of one alarm clock is 80 decibels. Find out how much louder 10 alarm clocks would be than one alarm clock. (Lesson 7-5)

Given a polynomial and one of its factors, find the remaining factors of the polynomial. Some factors may not be binomials. (Lesson 5-6)

74. $x^3 + 5x^2 + 8x + 4;\ x + 1$ **75.** $x^3 + 4x^2 + 7x + 6;\ x + 2$

76. CRAFTS Mrs. Hall is selling crocheted items. She sells large afghans for $60, baby blankets for $40, doilies for $25, and pot holders for $5. She takes the following number of items to the fair: 12 afghans, 25 baby blankets, 45 doilies, and 50 pot holders. (Lesson 3-6)

a. Write an inventory matrix for the number of each item and a cost matrix for the price of each item.

b. Suppose Mrs. Hall sells all of the items. Find her total income as a matrix.

Skills Review

Solve each equation.

77. $2^{3x+5} = 128$ **78.** $5^{n-3} = \frac{1}{25}$ **79.** $\left(\frac{1}{9}\right)^m = 81^{m+4}$

80. $\left(\frac{1}{7}\right)^{y-3} = 343$ **81.** $10^{x-1} = 100^{2x-3}$ **82.** $36^{2p} = 216^{p-1}$

EXPLORE

7-8 Spreadsheet Lab: Compound Interest

You can use a spreadsheet to organize and display data. A spreadsheet is an easy way to track the amount of interest earned over a period of time.

Compound interest is earned not only on the original amount, but also on any interest that has been added to the principal.

CCSS Common Core State Standards
Mathematical Practices
5 Use appropriate tools strategically.

Activity

Find the total amount of money after 5 years if you deposit $100 at 7% compounded annually.

Step 1 Label your columns as shown. The period is one year. Enter the starting values and the rate.

Step 2 Each row will be generated using formulas. Enter the formulas as shown.

Savings Account

	A	B	C	D	E
1	End of Period	Principal	Interest	Balance	Rate per Period
2	0			$100.00	7%
3	=A2+1	=D2	=B3*E2	=C3+D2	
4					

Sheet 1 | Sheet 2 | Sheet 3

Step 3 Use the **FILL DOWN** function to fill 4 additional rows.

Savings Account

	A	B	C	D	E
1	End of Period	Principal	Interest	Balance	Rate per Period
2	0			$100.00	7%
3	1	$100.00	$7.00	$107.00	
4	2	$107.00	$7.49	$114.49	
5	3	$114.49	$8.01	$122.50	
6	4	$122.50	$8.58	$131.08	
7	5	$131.08	$9.18	$140.26	
8					

Sheet 1 | Sheet 2 | Sheet 3

If you deposit $100 at 7% annual interest for 5 years, you will have $140.26 at the end of the 5 years.

Exercises

Find the total balance for each situation.

1. deposit $500 for 7 years at 5%
2. deposit $1000 for 5 years at 6%
3. deposit $200 for 2 years at 10%
4. deposit $800 for 3 years at 8%
5. borrow $10,000 for 5 years at 5.05%
6. borrow $25,000 for 30 years at 8%

LESSON 7-8 Using Exponential and Logarithmic Functions

Then	Now	Why?
You used exponential growth and decay formulas.	**1** Use logarithms to solve problems involving exponential growth and decay. **2** Use logarithms to solve problems involving logistic growth.	The ancient footprints of Acahualinca, discovered in Managua, Nicaragua, are believed to be the oldest human footprints in the world. Using carbon dating, scientists estimate that these footprints are 6000 years old.

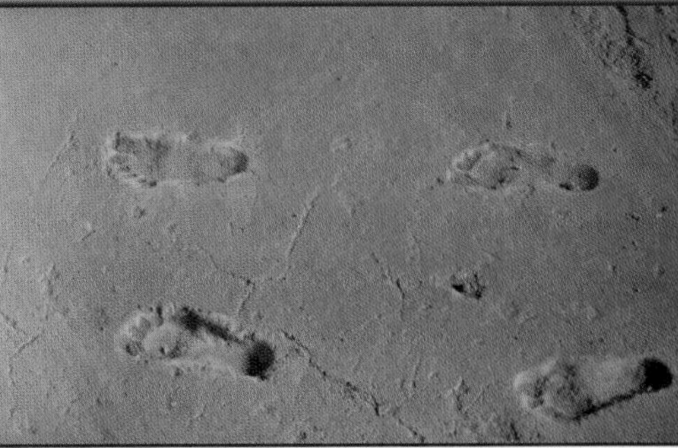

NewVocabulary
rate of continuous growth
rate of continuous decay
logistic growth model

Common Core State Standards

Content Standards

F.IF.8.b Use the properties of exponents to interpret expressions for exponential functions.

F.LE.4 For exponential models, express as a logarithm the solution to $ab^{ct} = d$ where a, c, and d are numbers and the base b is 2, 10, or e; evaluate the logarithm using technology.

Mathematical Practices

1 Make sense of problems and persevere in solving them.

1 Exponential Growth and Decay Scientists and researchers frequently use alternate forms of the growth and decay formulas that you learned in Lesson 7-1.

KeyConcept Exponential Growth and Decay

Exponential Growth	Exponential Decay
Exponential growth can be modeled by the function $f(x) = ae^{kt}$, where a is the initial value, t is time in years, and k is a constant representing the **rate of continuous growth.**	Exponential decay can be modeled by the function $f(x) = ae^{-kt}$, where a is the initial value, t is time in years, and k is a constant representing the **rate of continuous decay.**

Real-World Example 1 Exponential Decay

PT

SCIENCE The half-life of a radioactive substance is the time it takes for half of the atoms of the substance to disintegrate. The half-life of Carbon-14 is 5730 years. Determine the value of k and the equation of decay for Carbon-14.

If a is the initial amount of the substance, then the amount y that remains after 5730 years can be represented by $\frac{1}{2}a$ or $0.5a$.

$y = ae^{-kt}$	Exponential Decay Formula
$0.5a = ae^{-k(5730)}$	$y = 0.5a$ and $t = 5730$
$0.5 = e^{-5730k}$	Divide each side by a.
$\ln 0.5 = \ln e^{-5730k}$	Property of Equality for Logarithmic Functions
$\ln 0.5 = -5730k$	$\ln e^x = x$
$\frac{\ln 0.5}{-5730} = k$	Divide each side by −5730.
$0.00012 \approx k$	Use a calculator.

Thus, the equation for the decay of Carbon-14 is $y = ae^{-0.00012t}$.

GuidedPractice

1. The half-life of Plutonium-239 is 24,000 years. Determine the value of k.

Nik Wheeler/Documentary Value/CORBIS

Now that the value of k for Carbon-14 is known, it can be used to date fossils.

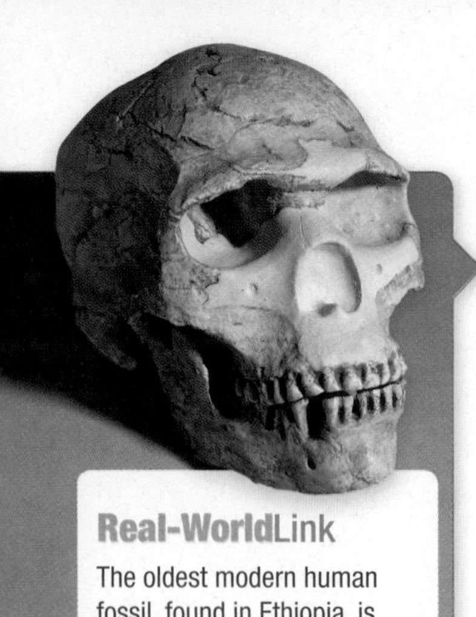

Real-WorldLink

The oldest modern human fossil, found in Ethiopia, is approximately 160,000 years old.

Source: National Public Radio

PT

Real-World Example 2 Carbon Dating

SCIENCE A paleontologist examining the bones of a prehistoric animal estimates that they contain 2% as much Carbon-14 as they would have contained when the animal was alive.

a. How long ago did the animal live?

Understand The formula for the decay of Carbon-14 is $y = ae^{-0.00012t}$. You want to find out how long ago the animal lived.

Plan Let a be the initial amount of Carbon-14 in the animal's body. The amount y that remains after t years is 2% of a or $0.02a$.

Solve

$y = ae^{-0.00012t}$	Formula for the decay of Carbon-14
$0.02a = ae^{-0.00012t}$	$y = 0.02a$
$0.02 = e^{-0.00012t}$	Divide each side by a.
$\ln 0.02 = \ln e^{-0.00012t}$	Property of Equality for Logarithmic Functions
$\ln 0.02 = -0.00012t$	$\ln e^x = x$
$\frac{\ln 0.02}{-0.00012} = t$	Divide each side by -0.00012.
$32{,}600 \approx t$	Use a calculator.

The animal lived about 32,600 years ago.

Check Use the formula to find the amount of a sample remaining after 32,600 years. Use an original amount of 1.

$y = ae^{-0.00012t}$	Original equation
$= 1e^{-0.00012(32{,}600)}$	$a = 1$ and $t = 32{,}600$
≈ 0.02 or 2% ✓	Use a calculator.

StudyTip

Carbon Dating When given a percent or fraction of decay, use an original amount of 1 for a.

b. If prior research points to the animal being around 20,000 years old, how much Carbon-14 should be in the animal?

$y = ae^{-0.00012t}$	Formula for the decay of Carbon-14
$= 1e^{-0.00012(20{,}000)}$	$a = 1$ and $t = 20{,}000$
$= e^{-2.4}$	Simplify.
$= 0.09$ or 9%	Use a calculator.

GuidedPractice

2. Use the information in Example 2 to answer the following questions.

A. A specimen that originally contained 42 milligrams of Carbon-14 now contains 8 milligrams. How old is the fossil?

B. A wooly mammoth specimen was thought to be about 12,000 years old. How much Carbon-14 should be in the animal?

The exponential growth equation $y = ae^{kt}$ is identical to the continuously compounded interest formula you learned in Lesson 7-7.

Continuous Compounding	Population Growth
$A = Pe^{rt}$	$y = ae^{kt}$
P = initial amount	a = initial population
A = amount at time t	y = population at time t
r = interest rate	k = rate of continuous growth

Prehistoric/Getty Images

PT

Real-World Example 3 Continuous Exponential Growth

POPULATION In 2007, the population of the state of Georgia was 9.36 million people. In 2000, it was 8.18 million.

> **Problem-Solving Tip**
>
> **Use a Formula** When dealing with population, it is almost always necessary to use an exponential growth or decay formula.

a. Determine the value of k, Georgia's relative rate of growth.

$y = ae^{kt}$	Formula for continuous exponential growth
$9.36 = 8.18e^{k(7)}$	$y = 9.36$, $a = 8.18$, and $t = 2007 - 2000$ or 7
$\frac{9.36}{8.18} = e^{7k}$	Divide each side by 8.18.
$\ln \frac{9.36}{8.18} = \ln e^{7k}$	Property of Equality for Logarithmic Functions
$\ln \frac{9.36}{8.18} = 7k$	$\ln e^x = x$
$\frac{\ln \frac{9.36}{8.18}}{7} = k$	Divide each side by 7.
$0.01925 = k$	Use a calculator.

Georgia's relative rate of growth is about 0.01925 or about 2%.

b. When will Georgia's population reach 12 million people?

$y = ae^{kt}$	Formula for continuous exponential growth
$12 = 8.18e^{0.01925t}$	$y = 10$, $a = 8.18$, and $k = 0.01925$
$1.4670 = e^{0.01925t}$	Divide each side by 8.18.
$\ln 1.4670 = \ln e^{0.01925t}$	Property of Equality for Logarithmic Functions
$\ln 1.4670 = 0.01925t$	$\ln e^x = x$
$\frac{\ln 1.4670}{0.01925} = t$	Divide each side by 0.01925.
$19.907 \approx t$	Use a calculator.

Georgia's population will reach 12 million people by 2020.

c. Michigan's population in 2000 was 9.9 million and can be modeled by $y = 9.9e^{0.0028t}$. Determine when Georgia's population will surpass Michigan's.

$8.18e^{0.01925t} > 9.9e^{0.0028t}$	Formula for exponential growth
$\ln 8.18e^{0.01925t} > \ln 9.9e^{0.0028t}$	Property of Inequality for Logarithms
$\ln 8.18 + \ln e^{0.01925t} > \ln 9.9 + \ln e^{0.0028t}$	Product Property of Logarithms
$\ln 8.18 + 0.01925t > \ln 9.9 + 0.0028t$	$\ln e^x = x$
$0.01645t > \ln 9.9 - \ln 8.18$	Subtract $(0.0028t + \ln 8.18)$ from each side.
$t > \frac{\ln 9.9 - \ln 8.18}{0.01645}$	Divide each side by 0.01645.
$t > 11.6$	Use a calculator.

Georgia's population will surpass Michigan's during 2012.

Guided Practice

3. BIOLOGY A type of bacteria is growing exponentially according to the model $y = 1000e^{kt}$, where t is the time in minutes.

A. If there are 1000 cells initially and 1650 cells after 40 minutes, find the value of k for the bacteria.

B. Suppose a second type of bacteria is growing exponentially according to the model $y = 50e^{0.0432t}$. Determine how long it will be before the number of cells of this bacteria exceed the number of cells in the other bacteria.

2 Logistic Growth

Refer to the equation representing Georgia's population in Example 3. According to the graph at the right, Georgia's population will be about one billion by the year 2130. Does this seem logical?

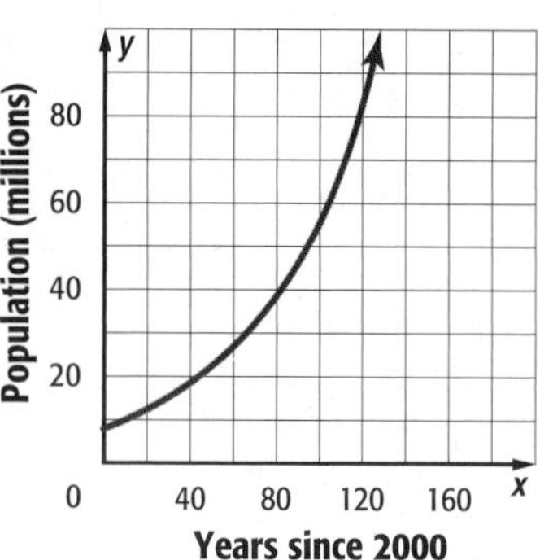

Populations cannot grow infinitely large. There are limitations, such as food supplies, war, living space, diseases, available resources, and so on.

Exponential growth is unrestricted, meaning it will increase without bound. A **logistic growth model**, however, represents growth that has a limiting factor. Logistic models are the most accurate models for representing population growth.

KeyConcept Logistic Growth Function

Let a, b, and c be positive constants where $b < 1$. The logistic growth function is represented by $f(t) = \frac{c}{1 + ae^{-bt}}$, where t represents time.

Real-WorldLink

Phoenix is the fifth largest city in the country and has a population of 1.5 million.

Real-World Example 4 Logistic Growth

The population of Phoenix, Arizona, in millions can be modeled by the logistic function $f(t) = \frac{2.0666}{1 + 1.66e^{-0.048t}}$, where t is the number of years after 1980.

a. Graph the function for $0 \leq t \leq 500$.

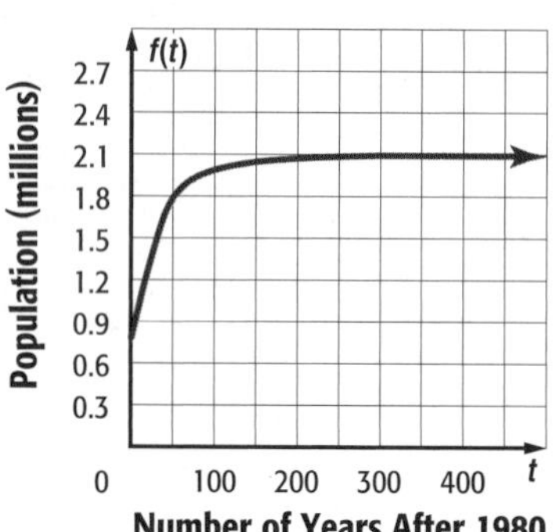

b. What is the horizontal asymptote?

The horizontal asymptote is at $y = 2.0666$.

c. Will the population of Phoenix increase indefinitely? If not, what will be their maximum population?

No. The population will reach a maximum of a little less than 2.0666 million people.

d. According to the function, when will the population of Phoenix reach 1.8 million people?

The graph indicates the population will reach 1.8 million people at $t \approx 50$. Replacing $f(t)$ with 1.8 and solving for t in the equation yields $t = 50.35$ years. So, the population of Phoenix will reach 1.8 million people by 2031.

StudyTip

Intersections To determine where the graph intersects 1.8 on the calculator, graph $y = 1.8$ on the same graph and select *intersection* in the CALC menu.

GuidedPractice

4. The population of a certain species of fish in a lake after t years can be modeled by the function $P(t) = \frac{1880}{1 + 1.42e^{-0.037t}}$, where $t \geq 0$.

A. Graph the function for $0 \leq t \leq 500$.

B. What is the horizontal asymptote?

C. What is the maximum population of the fish in the lake?

D. When will the population reach 1875?

Check Your Understanding

= Step-by-Step Solutions begin on page R14.

Examples 1–2

1. **PALEONTOLOGY** The half-life of Potassium-40 is about 1.25 billion years.
   a. Determine the value of k and the equation of decay for Potassium-40.
   b. A specimen currently contains 36 milligrams of Potassium-40. How long will it take the specimen to decay to only 15 milligrams of Potassium-40?
   c. How many milligrams of Potassium-40 will be left after 300 million years?
   d. How long will it take Potassium-40 to decay to one eighth of its original amount?

Example 3

2. **SCIENCE** A certain food is dropped on the floor and is growing bacteria exponentially according to the model $y = 2e^{kt}$, where t is the time in seconds.
   a. If there are 2 cells initially and 8 cells after 20 seconds, find the value of k for the bacteria.
   b. The "5-second rule" says that if a person who drops food on the floor eats it within 5 seconds, there will be no harm. How much bacteria is on the food after 5 seconds?
   c. Would you eat food that had been on the floor for 5 seconds? Why or why not? Do you think that the information you obtained in this exercise is reasonable? Explain.

Example 4

3. **ZOOLOGY** Suppose the red fox population in a restricted habitat follows the function $P(t) = \frac{16{,}500}{1 + 18e^{-0.085t}}$, where t represents the time in years.
   a. Graph the function for $0 \le t \le 200$.
   b. What is the horizontal asymptote?
   c. What is the maximum population?
   d. When does the population reach 16,450?

Practice and Problem Solving

Extra Practice is on page R7.

Examples 1–2

4. **CCSS PERSEVERANCE** The half-life of Rubidium-87 is about 48.8 billion years.
   a. Determine the value of k and the equation of decay for Rubidium-87.
   b. A specimen currently contains 50 milligrams of Rubidium-87. How long will it take the specimen to decay to only 18 milligrams of Rubidium-87?
   c. How many milligrams of Rubidium-87 will be left after 800 million years?
   d. How long will it take Rubidium-87 to decay to one-sixteenth its original amount?

Example 3

5. **BIOLOGY** A certain bacteria is growing exponentially according to the model $y = 80e^{kt}$, where t is the time in minutes.
   a. If there are 80 cells initially and 675 cells after 30 minutes, find the value of k for the bacteria.
   b. When will the bacteria reach a population of 6000 cells?
   c. If a second type of bacteria is growing exponentially according to the model $y = 35e^{0.0978t}$, determine how long it will be before the number of cells of this bacteria exceed the number of cells in the other bacteria.

Example 4

6. **FORESTRY** The population of trees in a certain forest follows the function $f(t) = \frac{18000}{1 + 16e^{-0.084t}}$, where t is the time in years.
   a. Graph the function for $0 \le t \le 100$.
   b. When does the population reach 17500 trees?

7. **PALEONTOLOGY** A paleontologist finds a human bone and determines that the Carbon-14 found in the bone is 85% of that found in living bone tissue. How old is the bone?

8. **ANTHROPOLOGY** An anthropologist has determined that a newly discovered human bone is 8000 years old. How much of the original amount of Carbon-14 is in the bone?

9. **RADIOACTIVE DECAY** 100 milligrams of Uranium-238 are stored in a container. If Uranium-238 has a half-life of about 4.47 billion years, after how many years will only 10 milligrams be present?

10. **POPULATION GROWTH** The population of the state of Oregon has grown from 3.4 million in 2000 to 3.7 million in 2006.

    a. Write an exponential growth equation of the form $y = ae^{kt}$ for Oregon, where t is the number of years after 2000.

    b. Use your equation to predict the population of Oregon in 2020.

    c. According to the equation, when will Oregon reach 6 million people?

11. **HALF-LIFE** A substance decays 99.9% of its total mass after 200 years. Determine the half-life of the substance.

12. **LOGISTIC GROWTH** The population in millions of the state of Ohio after 1900 can be modeled by $P(t) = \frac{12.95}{1 + 2.4e^{-kt}}$, where t is the number of years after 1900 and k is a constant.

    a. If Ohio had a population of 10 million in 1970, find the value of k.

    b. According to the equation, when will the population of Ohio reach 12 million?

13. **MULTIPLE REPRESENTATIONS** In this problem, you will explore population growth. The population growth of a country follows the exponential function $f(t) = 8e^{0.075t}$ or the logistic function $g(t) = \frac{400}{1 + 16e^{-0.025t}}$. The population is measured in millions and t is time in years.

    a. **Graphical** Graph both functions for $0 \leq t \leq 100$.

    b. **Analytical** Determine the intersection of the graphs. What is the significance of this intersection?

    c. **Analytical** Which function is a more accurate estimate of the country's population 100 years from now? Explain your reasoning.

H.O.T. Problems Use Higher-Order Thinking Skills

14. **OPEN ENDED** Give an example of a quantity that grows or decays at a fixed rate. Write a real-world problem involving the rate and solve by using logarithms.

15. **CHALLENGE** Solve $\frac{120{,}000}{1 + 48e^{-0.015t}} = 24e^{0.055t}$ for t.

16. **CCSS ARGUMENTS** Explain mathematically why $f(t) = \frac{c}{1 + 60e^{-0.5t}}$ approaches, but never reaches the value of c as $t \to +\infty$.

17. **OPEN ENDED** Give an example of a quantity that grows logistically and has limitations to growth. Explain why the quantity grows in this manner.

18. **WRITING IN MATH** How are exponential, continuous exponential, and logistic functions used to model different real-world situations?

Standardized Test Practice

19. Kareem is making a circle graph showing the favorite ice cream flavors of customers at his store. The table summarizes the data. What central angle should Kareem use for the section representing chocolate?

Flavor	Customers
chocolate	35
vanilla	42
strawberry	7
mint chip	12
butter pecan	4

A 35° **C** 126°

B 63° **D** 150°

20. PROBABILITY Lydia has 6 books on her bookshelf. Two are literature books, one is a science book, two are math books, and one is a dictionary. What is the probability that she randomly chooses a science book and the dictionary?

F $\frac{1}{3}$ **H** $\frac{1}{12}$

G $\frac{1}{4}$ **J** $\frac{1}{15}$

21. SAT/ACT Peter has made a game for his daughter's birthday party. The playing board is a circle divided evenly into 8 sectors. If the circle has a radius of 18 inches, what is the approximate area of one of the sectors?

A 4 in^2 **D** 127 in^2

B 14 in^2 **E** 254 in^2

C 32 in^2

22. STATISTICS In a survey of 90 physical trainers, 15 said they went for a run at least 5 times per week. Of that group, 5 said they also swim during the week, and at least 25% of all trainers run and swim every week. Which conclusion is valid based on the information given?

F The report is accurate because 15 out of 90 is 25%.

G The report is accurate because 5 out of 15 is 33%, which is at least 25%.

H The report is inaccurate because 5 out of 90 is only 5.6%.

J The report is inaccurate because no one knows if swimming is really exercising.

Spiral Review

Write an equivalent exponential or logarithmic equation. (Lesson 7-7)

23. $e^7 = y$

24. $e^{2n-4} = 36$

25. $\ln 5 + 4 \ln x = 9$

26. EARTHQUAKES The table shows the magnitudes of some major earthquakes. (*Hint:* A magnitude x earthquake has an intensity of 10^x.) (Lessons 7-5 and 7-6)

Year	Location	Magnitude
1963	Yugoslavia	6.0
1970	Peru	7.8
1988	Armenia	7.0
2004	Morocco	6.4
2007	Indonesia	8.4
2010	Haiti	7.0

a. For which two earthquakes was the intensity of one 10 times that of the other? For which two was the intensity of one 100 times that of the other?

b. What would be the magnitude of an earthquake that is 1000 times as intense as the 1963 earthquake in Yugoslavia?

c. Suppose you know that $\log_7 2 \approx 0.3562$ and $\log_7 3 \approx 0.5646$. Describe two different methods that you could use to approximate $\log_7 2.5$. (You may use a calculator, of course.) Then describe how you can check your result.

Skills Review

Solve each equation. Write in simplest form.

27. $\frac{8}{5}x = \frac{4}{15}$

28. $\frac{27}{14}n = \frac{6}{7}$

29. $\frac{3}{10} = \frac{12}{25}a$

30. $\frac{6}{7} = 9p$

31. $\frac{9}{8}b = 18$

32. $\frac{6}{7}y = \frac{3}{4}$

33. $\frac{1}{3}z = \frac{5}{6}$

34. $\frac{2}{3}q = 7$

EXTEND

7-8 Graphing Technology Lab Cooling

In this lab, you will explore the type of equation that models the change in the temperature of water as it cools under various conditions.

Common Core State Standards
Content Standards
F.BF.1.b Combine standard function types using arithmetic operations.

Set Up the Lab

- Collect a variety of containers, such as a foam cup, a ceramic coffee mug, and an insulated cup.
- Boil water or collect hot water from a tap.
- Choose a container to test and fill with hot water. Place the temperature probe in the cup.
- Connect the temperature probe to your data collection device.

Activity

Step 1 Program the device to collect 20 or more samples in 1 minute intervals.

Step 2 Wait a few seconds for the probe to warm to the temperature of the water.

Step 3 Press the button to begin collecting data.

Analyze the Results

1. When the data collection is complete, graph the data in a scatter plot. Use time as the independent variable and temperature as the dependent variable. Write a sentence that describes the points on the graph.
2. Use the **STAT** menu to find an equation to model the data you collected. Try linear, quadratic, and exponential models. Which model appears to fit the data best? Explain.
3. Would you expect the temperature of the water to drop below the temperature of the room? Explain your reasoning.
4. Use the data collection device to find the temperature of the air in the room. Graph the function $y = t$, where t is the temperature of the room, along with the scatter plot and the model equation. Describe the relationship among the graphs. What is the meaning of the relationship in the context of the experiment?

Make a Conjecture

5. Do you think the results of the experiment would change if you used an insulated container for the water? What part of the function will change, the constant or the rate of decay? Repeat the experiment to verify your conjecture.
6. How might the results of the experiment change if you added ice to the water? What part of the function will change, the constant or the rate of decay? Repeat the experiment to verify your conjecture.

Ed-Imaging

CHAPTER 7

Study Guide and Review

Study Guide

KeyConcepts

Exponential Functions (Lessons 7-1 and 7-2)

- An exponential function is in the form $y = ab^x$, where $a \neq 0$, $b > 0$ and $b \neq 1$.
- Property of Equality for Exponential Functions: If b is a positive number other than 1, then $b^x = b^y$ if and only if $x = y$.
- Property of Inequality for Exponential Functions: If $b > 1$, then $b^x > b^y$ if and only if $x > y$, and $b^x < b^y$ if and only if $x < y$.

Logarithms and Logarithmic Functions
(Lessons 7-3 through 7-6)

- Suppose $b > 0$ and $b \neq 1$. For $x > 0$, there is a number y such that $\log_b x = y$ if and only if $b^y = x$.
- The logarithm of a product is the sum of the logarithms of its factors.
- The logarithm of a quotient is the difference of the logarithms of the numerator and the denominator.
- The logarithm of a power is the product of the logarithm and the exponent.
- The Change of Base Formula: $\log_a n = \dfrac{\log_b n}{\log_b a}$

Base *e* and Natural Logarithms (Lesson 7-7)

- Since the natural base function and the natural logarithmic function are inverses, these two can be used to "undo" each other.

Using Exponential and Logarithmic Functions (Lesson 7-8)

- Exponential growth can be modeled by the function $f(x) = ae^{kt}$, where k is a constant representing the rate of continuous growth.
- Exponential decay can be modeled by the function $f(x) = ae^{-kt}$, where k is a constant representing the rate of continuous decay.

FOLDABLES StudyOrganizer

Be sure the Key Concepts are noted in your Foldable.

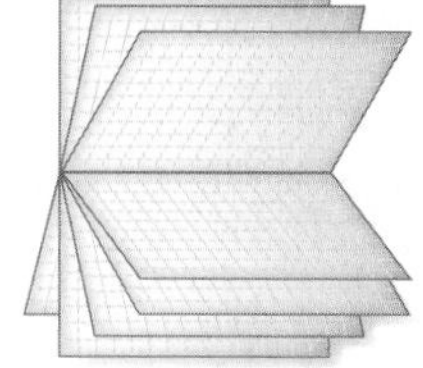

KeyVocabulary

asymptote (p. 451)
Change of Base Formula (p. 494)
common logarithm (p. 492)
compound interest (p. 462)
decay factor (p. 454)
exponential decay (p. 453)
exponential equation (p. 461)
exponential function (p. 451)
exponential growth (p. 451)
exponential inequality (p. 463)
growth factor (p. 453)
logarithm (p. 468)
logarithmic equation (p. 478)
logarithmic function (p. 469)
logarithmic inequality (p. 479)
logistic growth model (p. 512)
natural base, *e* (p. 501)
natural base exponential function (p. 501)
natural logarithm (p. 501)
rate of continuous decay (p. 509)
rate of continuous growth (p. 509)

VocabularyCheck

Choose a word or term from the list above that best completes each statement or phrase.

1. A function of the form $f(x) = b^x$ where $b > 1$ is a(n) ____________ function.
2. In $x = b^y$, the variable y is called the ____________ of x.
3. Base 10 logarithms are called ____________.
4. A(n) ____________ is an equation in which variables occur as exponents.
5. The ____________ allows you to write equivalent logarithmic expressions that have different bases.
6. The base of the exponential function, $A(t) = a(1 - r)^t$, $1 - r$ is called the ____________.
7. The function $y = \log_b x$, where $b > 0$ and $b \neq 1$, is called a(n) ____________.
8. An exponential function with base e is called the ____________.
9. The logarithm with base e is called the ____________.
10. The number e is referred to as the ____________.

Lesson-by-Lesson Review

7-1 Graphing Exponential Functions

Graph each function. State the domain and range.

11. $f(x) = 3^x$

12. $f(x) = -5(2)^x$

13. $f(x) = 3(4)^x - 6$

14. $f(x) = 3^{2x} + 5$

15. $f(x) = 3\left(\frac{1}{4}\right)^{x+3} - 1$

16. $f(x) = \frac{3}{5}\left(\frac{2}{3}\right)^{x-2} + 3$

17. POPULATION A city with a population of 120,000 decreases at a rate of 3% annually.

a. Write the function that represents this situation.

b. What will the population be in 10 years?

Example 1

Graph $f(x) = -2(3)^x + 1$. State the domain and range.

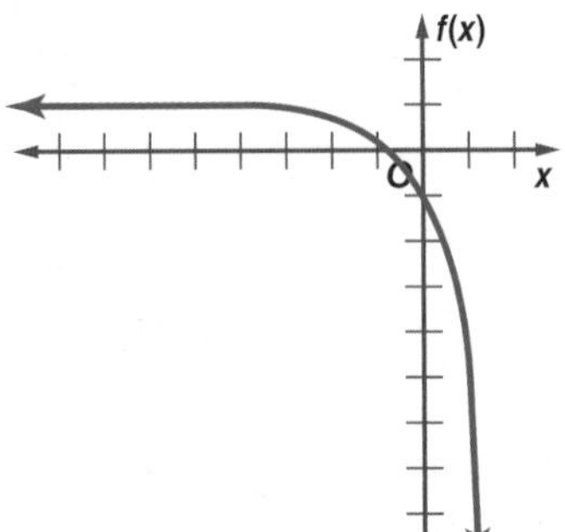

The domain is all real numbers, and the range is all real numbers less than 1.

7-2 Solving Exponential Equations and Inequalities

Solve each equation or inequality.

18. $16^x = \frac{1}{64}$

19. $3^{4x} = 9^{3x+7}$

20. $64^{3n} = 8^{2n-3}$

21. $8^{3-3y} = 256^{4y}$

22. $9^{x-2} > \left(\frac{1}{81}\right)^{x+2}$

23. $27^{3x} \leq 9^{2x-1}$

24. BACTERIA A bacteria population started with 5000 bacteria. After 8 hours there were 28,000 in the sample.

a. Write an exponential function that could be used to model the number of bacteria after x hours if the number of bacteria changes at the same rate.

b. How many bacteria can be expected in the sample after 32 hours?

Example 2

Solve $4^{3x} = 32^{x-1}$ for x.

$4^{3x} = 32^{x-1}$	Original equation
$(2^2)^{3x} = (2^5)^{x-1}$	Rewrite so each side has the same base.
$2^{6x} = 2^{5x-5}$	Power of a Power
$6x = 5x - 5$	Property of Equality for Exponential Functions
$x = -5$	Subtract $5x$ from each side.

The solution is -5.

7-3 Logarithms and Logarithmic Functions

25. Write $\log_2 \frac{1}{16} = -4$ in exponential form.

26. Write $10^2 = 100$ in logarithmic form.

Evaluate each expression.

27. $\log_4 256$

28. $\log_2 \frac{1}{8}$

Graph each function.

29. $f(x) = 2\log_{10} x + 4$

30. $f(x) = \frac{1}{6}\log_{\frac{1}{3}}(x - 2)$

Example 3

Evaluate $\log_2 64$.

$\log_2 64 = y$	Let the logarithm equal y.
$64 = 2^y$	Definition of logarithm
$2^6 = 2^y$	$64 = 2^6$
$6 = y$	Property of Equality for Exponential Functions

7-4 Solving Logarithmic Equations and Inequalities

Solve each equation or inequality.

31. $\log_4 x = \frac{3}{2}$

32. $\log_2 \frac{1}{64} = x$

33. $\log_4 x < 3$

34. $\log_5 x < -3$

35. $\log_9 (3x - 1) = \log_9 (4x)$

36. $\log_2 (x^2 - 18) = \log_2 (-3x)$

37. $\log_3 (3x + 4) \leq \log_3 (x - 2)$

38. **EARTHQUAKE** The magnitude of an earthquake is measured on a logarithmic scale called the Richter scale. The magnitude M is given by $M = \log_{10} x$, where x represents the amplitude of the seismic wave causing ground motion. How many times as great is the amplitude caused by an earthquake with a Richter scale rating of 10 as an aftershock with a Richter scale rating of 7?

Example 4

Solve $\log_{27} x < \frac{2}{3}$.

$\log_{27} x < \frac{2}{3}$	Original inequality
$x < 27^{\frac{2}{3}}$	Logarithmic to Exponential Inequality
$x < 9$	Simplify.

Example 5

Solve $\log_5 (p^2 - 2) = \log_5 p$.

$\log_5 (p^2 - 2) = \log_5 p$	Original equation
$p^2 - 2 = p$	Property of Equality
$p^2 - p - 2 = 0$	Subtract p from each side.
$(p - 2)(p + 1) = 0$	Factor.
$p - 2 = 0$ or $p + 1 = 0$	Zero Product Property
$p = 2$ $\quad p = -1$	Solve each equation.

The solution is $p = 2$, since $\log_5 p$ is undefined for $p = -1$.

7-5 Properties of Logarithms

Use $\log_5 16 \approx 1.7227$ and $\log_5 2 \approx 0.4307$ to approximate the value of each expression.

39. $\log_5 8$

40. $\log_5 64$

41. $\log_5 4$

42. $\log_5 \frac{1}{8}$

43. $\log_5 \frac{1}{2}$

Solve each equation. Check your solution.

44. $\log_5 x - \log_5 2 = \log_5 15$

45. $3 \log_4 a = \log_4 27$

46. $2 \log_3 x + \log_3 3 = \log_3 36$

47. $\log_4 n + \log_4 (n - 4) = \log_4 5$

48. **SOUND** Use the formula $L = 10 \log_{10} R$, where L is the loudness of a sound and R is the sound's relative intensity, to find out how much louder 20 people talking would be than one person talking. Suppose the sound of one person talking has a relative intensity of 80 decibels.

Example 6

Use $\log_5 16 \approx 1.7227$ and $\log_5 2 \approx 0.4307$ to approximate $\log_5 32$.

$\log_5 32 = \log_5 (16 \cdot 2)$	Replace 32 with 16.
$= \log_5 16 + \log_5 2$	Product Property
$\approx 1.7227 + 0.4307$	Use a calculator.
≈ 2.1534	

Example 7

Solve $\log_3 3x + \log_3 4 = \log_3 36$.

$\log_3 3x + \log_3 4 = \log_3 36$	Original equation
$\log_3 3x(4) = \log_3 36$	Product Property
$3x(4) = 36$	Definition of logarithm
$12x = 36$	Multiply.
$x = 3$	Divide each side by 12.

CHAPTER 7

Study Guide and Review *Continued*

7-6 Common Logarithms

Solve each equation or inequality. Round to the nearest ten-thousandth.

49. $3^x = 15$

50. $6^{x^2} = 28$

51. $8^{m+1} = 30$

52. $12^{r-1} = 7r$

53. $3^{5n} > 24$

54. $5^{x+2} \leq 3^x$

55. SAVINGS You deposited $1000 into an account that pays an annual interest rate r of 5% compounded quarterly. Use $A = P\left(1 + \frac{r}{n}\right)^{nt}$.

a. How long will it take until you have $1500 in your account?

b. How long it will take for your money to double?

Example 8

Solve $5^{3x} > 7^{x+1}$.

$5^{3x} > 7^{x+1}$	Original inequality
$\log 5^{3x} > \log 7^{x+1}$	Property of Inequality
$3x \log 5 > (x+1) \log 7$	Power Property
$3x \log 5 > x \log 7 + \log 7$	Distributive Property
$3x \log 5 - x \log 7 > \log 7$	Subtract $x \log 7$.
$x(3 \log 5 - \log 7) > \log 7$	Distributive Property
$x > \frac{\log 7}{3 \log 5 - \log 7}$	Divide by $3 \log 5 - \log 7$.
$x > 0.6751$	Use a calculator.

The solution set is $\{x \mid x > 0.6751\}$.

7-7 Base *e* and Natural Logarithms

Solve each equation or inequality. Round to the nearest ten-thousandth.

56. $4e^x - 11 = 17$

57. $2e^{-x} + 1 = 15$

58. $\ln 2x = 6$

59. $2 + e^x > 9$

60. $\ln (x+3)^5 < 5$

61. $e^{-x} > 18$

62. SAVINGS If you deposit $2000 in an account paying 6.4% interest compounded continuously, how long will it take for your money to triple? Use $A = Pe^{rt}$.

Example 9

Solve $3e^{5x} + 1 = 10$. Round to the nearest ten-thousandth.

$3e^{5x} + 1 = 10$	Original equation
$3e^{5x} = 9$	Subtract 1 from each side.
$e^{5x} = 3$	Divide each side by 3.
$\ln e^{5x} = \ln 3$	Property of Equality
$5x = \ln 3$	$\ln e^x = x$
$x = \frac{\ln 3}{5}$	Divide each side by 5.
$x \approx 0.2197$	Use a calculator.

7-8 Using Exponential and Logarithmic Functions

63. CARS Abe bought a used car for $2500. It is expected to depreciate at a rate of 25% per year. What will be the value of the car in 3 years?

64. BIOLOGY For a certain strain of bacteria, k is 0.728 when t is measured in days. Using the formula $y = ae^{kt}$, how long will it take 10 bacteria to increase to 675 bacteria?

65. POPULATION The population of a city 20 years ago was 24,330. Since then, the population has increased at a steady rate each year. If the population is currently 55,250, find the annual rate of growth for this city.

Example 10

A certain culture of bacteria will grow from 250 to 2000 bacteria in 1.5 hours. Find the constant k for the growth formula. Use $y = ae^{kt}$.

$y = ae^{kt}$	Exponential Growth Formula
$2000 = 250e^{k(1.5)}$	Replace y with 2000, a with 250, and t with 1.5.
$8 = e^{1.5k}$	Divide each side by 250.
$\ln 8 = \ln e^{1.5k}$	Property of Equality
$\ln 8 = 1.5k$	Inverse Property
$\frac{\ln 8}{1.5} = k$	Divide each side by 1.5.
$1.3863 \approx k$	Use a calculator.

CHAPTER 7

Practice Test

Graph each function. State the domain and range.

1. $f(x) = 3^{x-3} + 2$
2. $f(x) = 2\left(\frac{3}{4}\right)^{x+1} - 3$

Solve each equation or inequality. Round to the nearest ten-thousandth if necessary.

3. $8^{c+1} = 16^{2c+3}$
4. $9^{x-2} > \left(\frac{1}{27}\right)^x$
5. $2^{a+3} = 3^{2a-1}$
6. $\log_2 (x^2 - 7) = \log_2 6x$
7. $\log_5 x > 2$
8. $\log_3 x + \log_3 (x - 3) = \log_3 4$
9. $6^{n-1} \leq 11^n$
10. $4e^{2x} - 1 = 5$
11. $\ln (x + 2)^2 > 2$

Use $\log_5 11 \approx 1.4899$ and $\log_5 2 \approx 0.4307$ to approximate the value of each expression.

12. $\log_5 44$
13. $\log_5 \frac{11}{2}$

14. **POPULATION** The population of a city 10 years ago was 150,000. Since then, the population has increased at a steady rate each year. The population is currently 185,000.
    a. Write an exponential function that could be used to model the population after x years if the population changes at the same rate.
    b. What will the population be in 25 years?

15. Write $\log_9 27 = \frac{3}{2}$ in exponential form.

16. **AGRICULTURE** An equation that models the decline in the number of U.S. farms is $y = 3{,}962{,}520(0.98)^x$, where x is the number of years since 1960 and y is the number of farms.
    a. How can you tell that the number is declining?
    b. By what annual rate is the number declining?
    c. Predict when the number of farms will be less than 1 million.

17. **MULTIPLE CHOICE** What is the value of $\log_4 \frac{1}{64}$?
    A -3
    B $-\frac{1}{3}$
    C $\frac{1}{3}$
    D 3

18. **SAVINGS** You put $7500 in a savings account paying 3% interest compounded continuously.
    a. Assuming there are no deposits or withdrawals from the account, what is the balance after 5 years?
    b. How long will it take your savings to double?
    c. In how many years will you have $10,000 in your account?

19. **MULTIPLE CHOICE** What is the solution of $\log_4 16 - \log_4 x = \log_4 8$?
    F $\frac{1}{2}$
    G 2
    H 4
    J 8

20. **MULTIPLE CHOICE** Which function is graphed below?

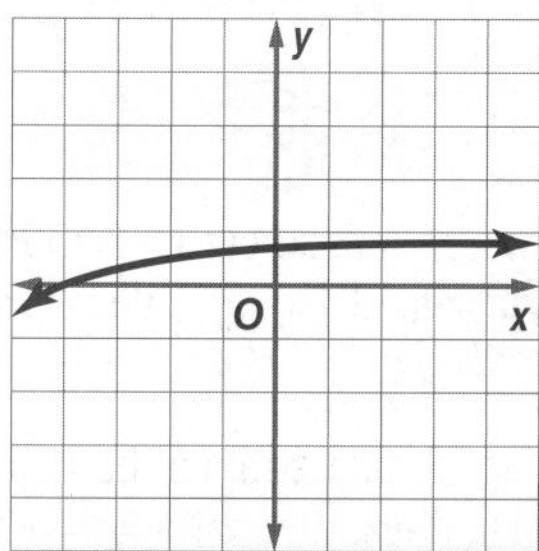

    A $y = \log_{10} (x - 5)$
    B $y = 5 \log_{10} x$
    C $y = \log_{10} (x + 5)$
    D $y = -5 \log_{10} x$

21. Write $2 \ln 6 + 3 \ln 4 - 5 \ln \left(\frac{1}{3}\right)$ as a single logarithm.

Preparing for Standardized Tests

Using Technology

Your calculator can be a useful tool in taking standardized tests. Some problems that you encounter might have steps or computations that require the use of a calculator. A calculator may also help you solve a problem more quickly.

Strategies for Using Technology

Step 1

A calculator is a useful tool, but typically it should be used sparingly. Standardized tests are designed to measure your ability to reason and solve problems, not to measure your ability to punch keys on a calculator.

Before using a calculator, ask yourself:

- How would I normally solve this type of problem?
- Are there any steps that I cannot perform mentally or by using paper and pencil?
- Is a calculator absolutely necessary to solve this problem?
- Would a calculator help me solve this problem more quickly or efficiently?

Step 2

When might a calculator come in handy?

- solving problems that involve large, complex computations
- solving certain problems that involve graphing functions, evaluating functions, solving equations, and so on
- checking solutions of problems

Alamy as Creativeact - Education series/Alamy

Standardized Test Example

Read the problem. Identify what you need to know. Then use the information in the problem to solve.

A certain can of soda contains 60 milligrams of caffeine. The caffeine is eliminated from the body at a rate of 15% per hour. What is the *half-life* of the caffeine? That is, how many hours does it take for half of the caffeine to be eliminated from the body?

A 4 hours

B 4.25 hours

C 4.5 hours

D 4.75 hours

Read the problem carefully. The problem can be solved using an exponential function. Use the exponential decay formula to model the problem and solve for the half-life of caffeine.

$$y = a(1 - r)^t$$

$$y = 60(1 - 0.15)^t$$

Half of 60 milligrams is 30. So, let $y = 30$ and solve for t.

$$30 = 60(1 - 0.15)^t$$

$$0.5 = (0.85)^t$$

Take the log of each side and use the power property.

$$\log 0.5 = \log (0.85)^t$$

$$\log 0.5 = t \log 0.85$$

$$\frac{\log 0.5}{\log 0.85} = t$$

At this point, it is necessary to use a calculator to evaluate the logarithms and solve the problem. Doing so shows that $t \approx 4.265$. So, the half-life of caffeine is about 4.25 hours. The correct answer is B.

Exercises

Read each problem. Identify what you need to know. Then use the information in the problem to solve.

1. Jason recently purchased a new truck for $34,750. The value of the truck decreases by 12% each year. What will the approximate value of the truck be 7 years after Jason purchased it?

   A $13,775

   B $13,890

   C $14,125

   D $14,200

2. A baseball is thrown upward at a velocity of 105 feet per second, and is released when it is 5 feet above the ground. The height of the baseball t seconds after being thrown is given by the formula $h(t) = -16t^2 + 105t + 5$. Find the time at which the baseball reaches its maximum height.

   F 1.0 s **H** 6.6 s

   G 3.3 s **J** 177.3 s

3. Lucinda deposited $2500 in a CD with the terms described below.

   Use the formula below to solve for t, the number of years needed to earn $250 in interest with the CD.

   $$2750 = 2500\left(1 + \frac{0.0425}{365}\right)^{365t}$$

   A about 2.15 years

   B about 2.24 years

   C about 2.35 years

   D about 2.46 years

CHAPTER 7

Standardized Test Practice

Cumulative, Chapters 1 through 7

Multiple Choice

Read each question. Then fill in the correct answer on the answer document provided by your teacher or on a sheet of paper.

1. What is the y-intercept of the exponential function below?

$$y = 4^x - 1$$

A 0 B 1 C 2 D 3

2. Suppose there are only 3500 birds of a particular endangered species left on an island and the population decreases at a rate of about 5% each year. The logarithmic function $t = \log_{0.95} \frac{p}{3500}$ predicts how many years t it will be for the population to decease to a number p. About how long will it take for the population to reach 3000 birds?

F 2 years H 5 years
G 3 years J 8 years

3. Suppose a certain bacteria duplicates to reproduce itself every 20 minutes. If you begin with one cell of the bacteria, how many will there be after 2 hours?

A 2 B 6 C 32 D 64

4. Lucas determined that the total cost C to rent a car could be represented by the equation $C = 0.35m + 125$, where m is the number of miles that he drives. If the total cost to rent the car was $363, how many miles did he drive?

F 125 H 520
G 238 J 680

5. Which of the following best describes the graph of $3y = 4x - 3$ and $8y = -6x - 5$?

A The lines have the same y-intercept.

B The lines have the same x-intercept.

C The lines are perpendicular.

D The lines are parallel

> **Test-Taking Tip**
>
> **Question 2** Use technology and the properties of logarithms to find t when $p = 3000$.

6. Graph $y = \log_5 x$.

F

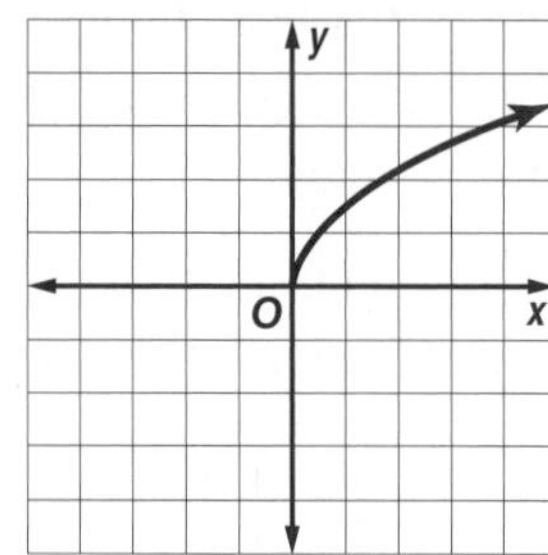

G

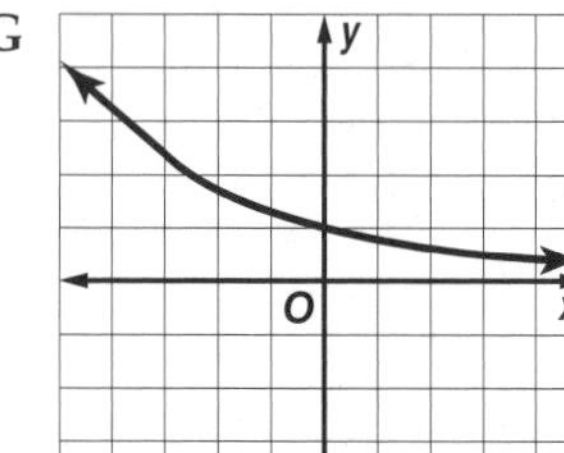

H

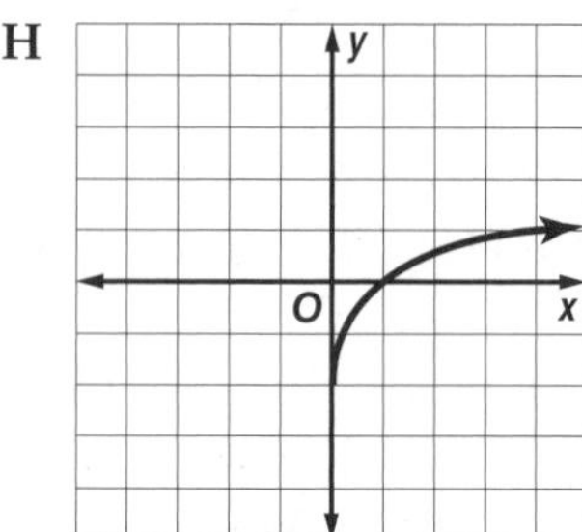

J

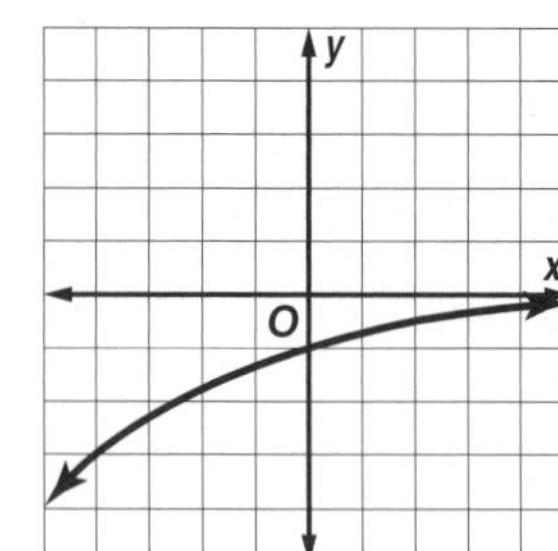

7. Ray's Book Store sells two used books for $7.99. The total cost c of purchasing n books can be found by—

A multiplying n by c.

B multiplying n by 5.

C multiplying n by the cost of 1 book.

D dividing n by c

Short Response/Gridded Response

Record your answers on the answer sheet provided by your teacher or on a sheet of paper.

8. The function $y = \left(\frac{1}{2}\right)^x$ is graphed below. What is the domain of the function?

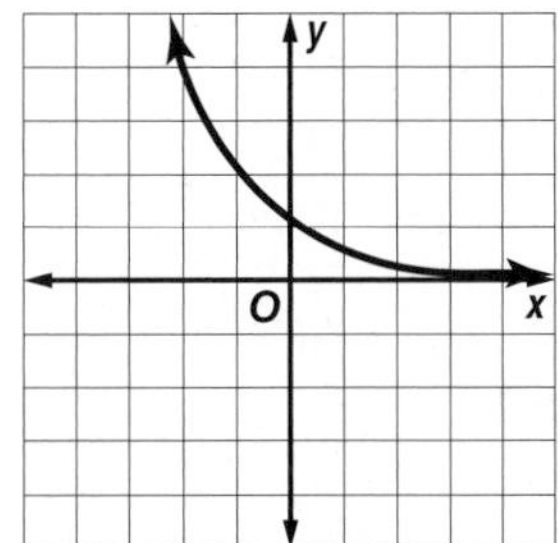

9. Suzanne bought a new car this year for $33,750. The value of the car is expected to decrease by 10.5% per year. What will be the ***approximate*** value of the car 6 years after Suzanne purchased it? Show your work.

10. Julio is solving the system of equations $8x - 2y = 12$ and $-15x + 2y = -19$ by using elimination. His work is shown below.

$$\begin{array}{rl} 8x - 2y = & 12 \\ \underline{-15x + 2y =} & \underline{-19} \\ -7x = & 7 \\ x = & -1 \end{array} \qquad \begin{array}{r} 8x - 2y = 12 \\ 8(-1) - 2y = 12 \\ -8 - 2y = 12 \\ -2y = 20 \\ y = -10 \end{array}$$

The solution is $(-1, -10)$.

a. What error does Julio make?

b. What is the correct solution of the system of equations? Show your work.

11. GRIDDED RESPONSE If $f(x) = 3x$ and $g(x) = x^2 - 1$, what is the value of $f(g(-3))$?

12. Simplify $(-2a^{-2}b^{-6})(-3a^{-1}b^8)$.

13. GRIDDED RESPONSE For what value of x would the rectangle below have an area of 48 square units?

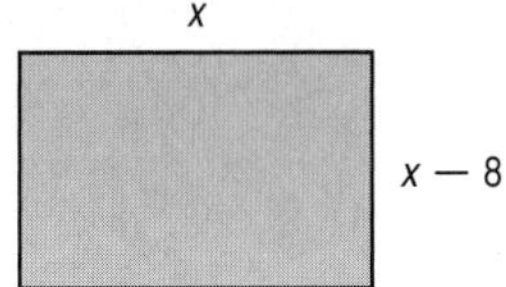

Extended Response

Record your answers on a sheet of paper. Show your work.

14. Suppose the number of whitetail deer in a particular region has increased at an annual rate of about 10% since 1995. There were 135,000 deer in 1995.

a. Write a function to model the number of whitetail deer after t years.

b. About how many whitetail deer inhabited the region in 2000? Round your answer to the nearest hundred deer.

15. Sandy inherited $250,000 from her aunt in 1998. She invested the money and increased it as shown in the table below.

Year	Amount
1998	$250,000
2006	$329,202
2011	$390,989

a. Write an exponential function that could be used to predict the amount of money A after investing for t years.

b. If the money continues to grow at the same rate, in what year will it be worth $500,000?

Need ExtraHelp?

If you missed Question...	1	2	3	4	5	6	7	8	9	10	11	12	13	14	15
Go to Lesson...	7-1	7-4	7-2	1-3	3-1	7-3	2-4	7-1	7-6	3-1	6-1	5-1	4-3	7-6	7-2

CHAPTER 8 Rational Functions and Relations

Then

You used factoring to solve quadratic equations and you graphed quadratic equations.

Now

You will:

- Simplify rational expressions.
- Graph rational functions.
- Solve direct, joint, and inverse variation problems.
- Solve rational equations and inequalities.

Why? ▲

TRAVEL Whether you travel by boat, car, bicycle, or airplane, rational functions can be used to find distance traveled, time spent traveling, and speed. If you want to arrive at a destination on time, rational relations can tell you at what speed you need to travel to reach your goal. When graphing rational functions you see clearly how the speed at which you travel affects the time it takes to get there.

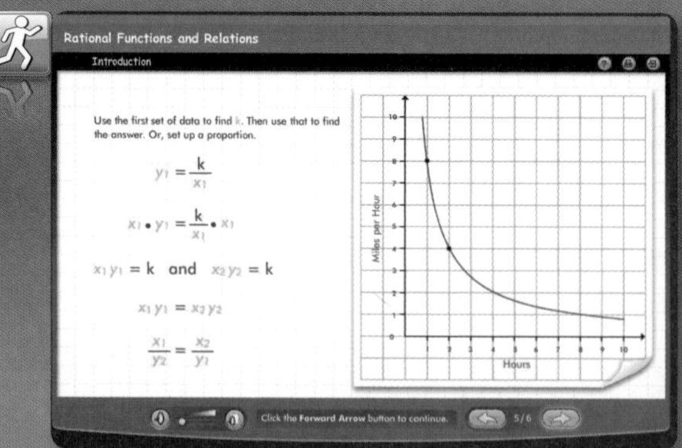

connectED.mcgraw-hill.com **Your Digital Math Portal**

Animation  Vocabulary  eGlossary   Personal Tutor  Virtual Manipulatives  Graphing Calculator  Audio  Foldables  Self-Check Practice  Worksheets 

Get Ready for the Chapter

Diagnose Readiness | You have two options for checking prerequisite skills.

1 Textbook Option Take the Quick Check below. Refer to the Quick Review for help.

QuickCheck

Solve each equation. Write in simplest form.

1. $\frac{5}{14} = \frac{1}{3}x$
2. $\frac{1}{8}m = \frac{7}{3}$
3. $\frac{8}{5} = \frac{1}{4}k$
4. $\frac{10}{9}p = 7$
5. **TRUCKS** Martin used $\frac{1}{3}$ of a tank of gas in his truck to get to work. He began with a full tank of gas. If he has 18 gallons of gas left, how many gallons does his tank hold?

Simplify each expression.

6. $\frac{3}{4} - \frac{7}{8}$
7. $\frac{8}{9} - \frac{7}{6} + \frac{1}{3}$
8. $\frac{9}{10} - \frac{4}{15} + \frac{1}{3}$
9. $\frac{10}{3} + \frac{5}{6} + 3$
10. **BAKING** Annie baked cookies for a bake sale. She used $\frac{2}{3}$ cup of flour for one recipe and $4\frac{1}{2}$ cups of flour for the other recipe. How many cups did she use in all?

Solve each proportion.

11. $\frac{9}{12} = \frac{p}{36}$
12. $\frac{9}{18} = \frac{6}{m}$
13. $\frac{2}{7} = \frac{5}{k}$
14. **SALES TAX** Kirsten pays \$4.40 tax on \$55 worth of clothes. What amount of tax will she pay on \$35 worth of clothes?

QuickReview

Example 1

Solve $\frac{9}{11} = \frac{7}{8}r$. Write in simplest form.

$\frac{9}{11} = \frac{7}{8}r$

$\frac{72}{11} = 7r$ Multiply each side by 8.

$\frac{72}{77} = r$ Divide each side by 7.

Since the GCF of 72 and 77 is 1, the solution is in simplest form.

Example 2

Simplify $\frac{1}{3} + \frac{3}{4} - \frac{5}{6}$.

$\frac{1}{3} + \frac{3}{4} - \frac{5}{6}$

$= \frac{1}{3}\left(\frac{4}{4}\right) + \frac{3}{4}\left(\frac{3}{3}\right) - \frac{5}{6}\left(\frac{2}{2}\right)$ The GCF of 3, 4, and 6 is 12.

$= \frac{4}{12} + \frac{9}{12} - \frac{10}{12}$ Simplify.

$= \frac{3}{12}$ Add and subtract.

$= \frac{3 \div 3}{12 \div 3}$ or $\frac{1}{4}$ Simplify.

Example 3

Solve $\frac{5}{8} = \frac{u}{11}$.

$\frac{5}{8} = \frac{u}{11}$ Write the equation.

$5(11) = 8u$ Find the cross products.

$55 = 8u$ Simplify.

$\frac{55}{8} = u$ Divide each side by 8.

Since the GCF of 55 and 8 is 1, the answer is in simplified form. $u = \frac{55}{8}$ or $6\frac{7}{8}$.

2 Online Option Take an online self-check Chapter Readiness Quiz at connectED.mcgraw-hill.com.

Get Started on the Chapter

You will learn several new concepts, skills, and vocabulary terms as you study Chapter 8. To get ready, identify important terms and organize your resources. You may wish to refer to Chapter 0 to review prerequisite skills.

FOLDABLES® StudyOrganizer

Rational Functions and Relations Make this Foldable to help you organize your Chapter 8 notes about rational functions and relations. Begin with an $8\frac{1}{2}'' \times 11''$ sheet of grid paper.

1 **Fold** in thirds along the height.

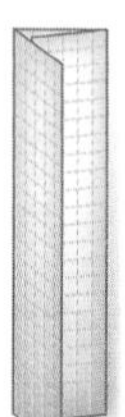

2 **Fold** the top edge down making a 2″ tab at the top. Cut along the folds.

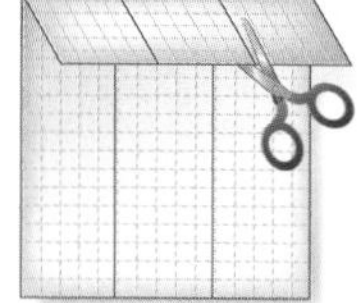

3 **Label** the outside tabs *Expressions*, *Functions*, and *Equations*. Use the inside tabs for definitions and notes.

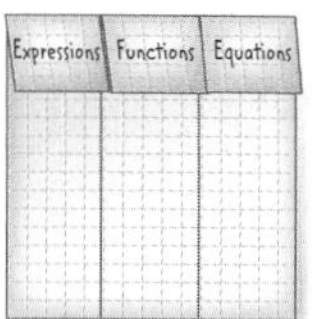

4 **Write** examples of each topic in the space below each tab.

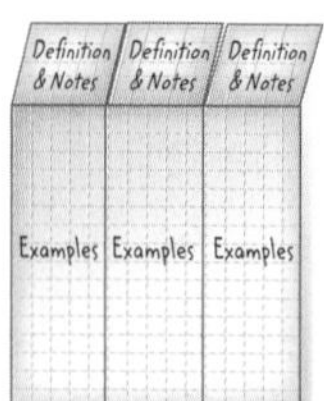

NewVocabulary

English		Español
rational expression	p. 529	expresión racional
complex fraction	p. 532	fracción compleja
reciprocal function	p. 545	función recíproco
hyperbola	p. 545	hipérbola
rational function	p. 553	función racional
vertical asymptote	p. 553	asíntota vertical
horizontal asymptote	p. 553	asíntota horizontal
oblique asymptote	p. 555	asíntota oblicua
point discontinuity	p. 556	discontinuidad evitable
direct variation	p. 562	variación directa
constant of variation	p. 562	constante de variación
joint variation	p. 563	variación conjunta
inverse variation	p. 564	variación inversa
combined variation	p. 565	variación combinada
rational equation	p. 570	ecuación racional
weighted average	p. 572	media ponderada
rational inequality	p. 575	desigualdad racional

ReviewVocabulary

function función a relation in which each element of the domain is paired with exactly one element of the range

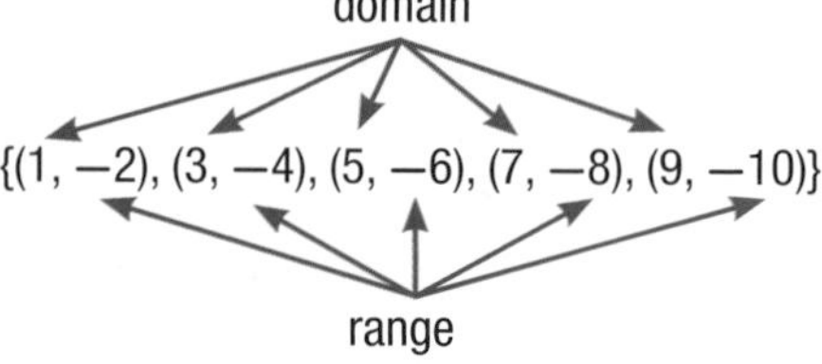

least common multiple mínimo común múltiplo the least number that is a common multiple of two or more numbers

rational number número racional a number expressed in the form $\frac{a}{b}$, where a and b are integers and $b \neq 0$

LESSON 8-1

Multiplying and Dividing Rational Expressions

Then	Now	Why?
You factored polynomials.	1 Simplify rational expressions. 2 Simplify complex fractions.	If a scuba diver goes to depths greater than 33 feet, the rational function $T(d) = \frac{1700}{d - 33}$ gives the maximum time a diver can remain at those depths and still surface at a steady rate with no stops. $T(d)$ represents the dive time in minutes, and d represents the depth in feet.

NewVocabulary

rational expression
complex fraction

Common Core State Standards

Content Standards

A.APR.7 Understand that rational expressions form a system analogous to the rational numbers, closed under addition, subtraction, multiplication, and division by a nonzero rational expression; add, subtract, multiply, and divide rational expressions.

Mathematical Practices

8 Look for and express regularity in repeated reasoning.

1 Simplify Rational Expressions

A ratio of two polynomial expressions such as $\frac{1700}{d - 33}$ is called a **rational expression**.

Because variables in algebra often represent real numbers, operations with rational numbers and rational expressions are similar. Just as with reducing fractions, to simplify a rational expression, you divide the numerator and denominator by their greatest common factor (GCF).

$$\frac{8}{12} = \frac{2 \cdot \cancel{4}^{1}}{3 \cdot \cancel{4}_{1}} = \frac{2}{3} \qquad \text{GCF} = 4$$

$$\frac{x^2 - 4x + 3}{x^2 - 6x + 5} = \frac{(x-3)\cancel{(x-1)}^{1}}{(x-5)\cancel{(x-1)}_{1}} = \frac{(x-3)}{(x-5)} \qquad \text{GCF} = (x - 1)$$

PT

Example 1 Simplify a Rational Expression

a. Simplify $\frac{5x(x^2 + 4x + 3)}{(x - 6)(x^2 - 9)}$.

$$\frac{5x(x^2 + 4x + 3)}{(x - 6)(x^2 - 9)} = \frac{5x(x + 3)(x + 1)}{(x - 6)(x + 3)(x - 3)}$$ Factor numerator and denominator.

$$= \frac{5x(x + 1)}{(x - 6)(x - 3)} \cdot \frac{\cancel{(x + 3)}^{1}}{\cancel{(x + 3)}_{1}}$$ Eliminate common factors.

$$= \frac{5x(x + 1)}{(x - 6)(x - 3)}$$ Simplify.

b. Under what conditions is this expression undefined?

The original factored denominator is $(x - 6)(x + 3)(x - 3)$.
Determine the values that would make the denominator equal to 0.
These values are 6, −3, or 3, so the expression is undefined when $x = 6, -3$ or 3.

GuidedPractice

Simplify each expression. Under what conditions is the expression undefined?

1A. $\frac{4y(y - 3)(y + 4)}{y(y^2 - y - 6)}$

1B. $\frac{2z(z + 5)(z^2 + 2z - 8)}{(z - 1)(z + 5)(z - 2)}$

Standardized Test Example 2 Undefined Values

For what value(s) is $\frac{x^2(x^2 - 5x - 14)}{4x(x^2 + 6x + 8)}$ undefined?

A $-2, -4$

B $-2, 7$

C $0, -2, -4$

D $0, -2, -4, 7$

Read the Test Item

You want to determine which values of x make the denominator equal to 0.

Solve the Test Item

With $4x$ in the denominator, x cannot equal 0. So, choices A and B can be eliminated. Next, factor the denominator.

$x^2 + 6x + 8 = (x + 2)(x + 4)$, so the denominator is $4x(x + 2)(x + 4)$.

Because the denominator equals 0 when $x = 0, -2$, and -4, the answer is C.

Test-Taking Tip

Eliminating Choices Sometimes you can save time by looking at the possible answers and eliminating choices.

GuidedPractice

2. For what value(s) of x is $\frac{x(x^2 + 8x + 12)}{-6(x^2 - 3x - 10)}$ undefined?

F $0, 5, -2$ **G** $5, -2$ **H** $0, -2, -6$ **J** $5, -2, -6$

Sometimes you can factor out -1 in the numerator or denominator to help simplify a rational expression.

Example 3 Simplify Using −1

Simplify $\frac{(4w^2 - 3wy)(w + y)}{(3y - 4w)(5w + y)}$.

$$\frac{(4w^2 - 3wy)(w + y)}{(3y - 4w)(5w + y)} = \frac{w(4w - 3y)(w + y)}{(3y - 4w)(5w + y)} \quad \text{Factor.}$$

$$= \frac{w(-1)\overset{1}{\cancel{(3y - 4w)}}(w + y)}{\underset{1}{\cancel{(3y - 4w)}}(5w + y)} \quad 4w - 3y = -1(3y - 4w)$$

$$= \frac{(-w)(w + y)}{5w + y} \quad \text{Simplify.}$$

GuidedPractice

Simplify each expression.

3A. $\frac{(xz - 4z)}{z^2(4 - x)}$

3B. $\frac{ab^2 - 5ab}{(5 + b)(5 - b)}$

The method for multiplying and dividing fractions also works with rational expressions. Remember that to multiply two fractions, you multiply the numerators and multiply the denominators. To divide two fractions, you multiply by the multiplicative inverse, or the reciprocal, of the divisor.

Multiplication

$$\frac{2}{9} \cdot \frac{15}{4} = \frac{\overset{1}{\cancel{2}} \cdot \overset{1}{\cancel{3}} \cdot 5}{\underset{1}{\cancel{3}} \cdot 3 \cdot \underset{1}{\cancel{2}} \cdot 2} = \frac{5}{3 \cdot 2} = \frac{5}{6}$$

Division

$$\frac{3}{5} \div \frac{6}{35} = \frac{3}{5} \cdot \frac{35}{6} = \frac{\overset{1}{\cancel{3}} \cdot \overset{1}{\cancel{5}} \cdot 7}{\underset{1}{\cancel{5}} \cdot 2 \cdot \underset{1}{\cancel{3}}} = \frac{7}{2}$$

The following table summarizes the rules for multiplying and dividing rational expressions.

KeyConcept

Multiplying Rational Expressions

Words	To multiply rational expressions, multiply the numerators and multiply the denominators.
Symbols	For all rational expressions $\frac{a}{b}$ and $\frac{c}{d}$ with $b \neq 0$ and $d \neq 0$, $\frac{a}{b} \cdot \frac{c}{d} = \frac{ac}{bd}$.

Dividing Rational Expressions

Words	To divide rational expressions, multiply by the reciprocal of the divisor.
Symbols	For all rational expressions $\frac{a}{b}$ and $\frac{c}{d}$ with $b \neq 0$, $c \neq 0$, and $d \neq 0$, $\frac{a}{b} \div \frac{c}{d} = \frac{a}{b} \cdot \frac{d}{c} = \frac{ad}{bc}$.

StudyTip

Eliminating Common Factors Be sure to eliminate factors from both the numerator and denominator.

Example 4 Multiply and Divide Rational Expressions

Simplify each expression.

a. $\frac{6c}{5d} \cdot \frac{15cd^2}{8a}$

$\frac{6c}{5d} \cdot \frac{15cd^2}{8a} = \frac{2 \cdot 3 \cdot c \cdot 5 \cdot 3 \cdot c \cdot d \cdot d}{5 \cdot d \cdot 2 \cdot 2 \cdot 2 \cdot a}$ Factor.

$= \frac{\overset{1}{\cancel{2}} \cdot 3 \cdot c \cdot \overset{1}{\cancel{5}} \cdot 3 \cdot c \cdot \overset{1}{\cancel{d}} \cdot d}{\underset{1}{\cancel{5}} \cdot \underset{1}{\cancel{d}} \cdot \underset{1}{\cancel{2}} \cdot 2 \cdot 2 \cdot a}$ Eliminate common factors.

$= \frac{3 \cdot 3 \cdot c \cdot c \cdot d}{2 \cdot 2 \cdot a}$ Simplify.

$= \frac{9c^2d}{4a}$ Simplify.

b. $\frac{18xy^3}{7a^2b^2} \div \frac{12x^2y}{35a^2b}$

$\frac{18xy^3}{7a^2b^2} \div \frac{12x^2y}{35a^2b} = \frac{18xy^3}{7a^2b^2} \cdot \frac{35a^2b}{12x^2y}$ Multiply by reciprocal of the divisor.

$= \frac{2 \cdot 3 \cdot 3 \cdot x \cdot y \cdot y \cdot y \cdot 5 \cdot 7 \cdot a \cdot a \cdot b}{7 \cdot a \cdot a \cdot b \cdot b \cdot 2 \cdot 2 \cdot 3 \cdot x \cdot x \cdot y}$ Factor.

$= \frac{\overset{1}{\cancel{2}} \cdot \overset{1}{\cancel{3}} \cdot 3 \cdot \overset{1}{\cancel{x}} \cdot \overset{1}{\cancel{y}} \cdot y \cdot y \cdot 5 \cdot \overset{1}{\cancel{7}} \cdot \overset{1}{\cancel{a}} \cdot \overset{1}{\cancel{a}} \cdot \overset{1}{\cancel{b}}}{\underset{1}{\cancel{7}} \cdot \underset{1}{\cancel{a}} \cdot \underset{1}{\cancel{a}} \cdot \underset{1}{\cancel{b}} \cdot b \cdot \underset{1}{\cancel{2}} \cdot 2 \cdot \underset{1}{\cancel{3}} \cdot \underset{1}{\cancel{x}} \cdot x \cdot \underset{1}{\cancel{y}}}$ Eliminate common factors.

$= \frac{3 \cdot 5 \cdot y \cdot y}{2 \cdot b \cdot x}$ Simplify.

$= \frac{15y^2}{2bx}$ Simplify.

GuidedPractice

4A. $\frac{12c^3d^2}{21ab} \cdot \frac{14a^2b}{8c^2d}$

4B. $\frac{6xy}{15ab^2} \cdot \frac{21a^3}{18x^4y}$

4C. $\frac{16mt^2}{21a^4b^3} \div \frac{24m^3}{7a^2b^2}$

4D. $\frac{12x^4y^2}{40a^4b^4} \div \frac{6x^2y^4}{16a^2x}$

Sometimes you must factor the numerator and/or the denominator first before you can simplify a product or a quotient of rational expressions.

StudyTip

CCSS Regularity When simplifying rational expressions, factors in one polynomial will often reappear in other polynomials. In Example 5a, $x - 8$ appears four times. Use this as a guide when factoring challenging polynomials.

Example 5 Polynomials in the Numerator and Denominator

Simplify each expression.

a. $\dfrac{x^2 - 6x - 16}{x^2 - 16x + 64} \cdot \dfrac{x - 8}{x^2 + 5x + 6}$

$\dfrac{x^2 - 6x - 16}{x^2 - 16x + 64} \cdot \dfrac{x - 8}{x^2 + 5x + 6} = \dfrac{(x-8)(x+2)}{(x-8)(x-8)} \cdot \dfrac{x-8}{(x+3)(x+2)}$ Factor.

$= \dfrac{\overset{1}{\cancel{(x-8)}}\overset{1}{\cancel{(x+2)}}}{\underset{1}{\cancel{(x-8)}}\underset{1}{\cancel{(x-8)}}} \cdot \dfrac{\overset{1}{\cancel{x-8}}}{(x+3)\underset{1}{\cancel{(x+2)}}}$ Eliminate common factors.

$= \dfrac{1}{x+3}$ Simplify.

b. $\dfrac{x^2 - 16}{12y + 36} \div \dfrac{x^2 - 12x + 32}{y^2 - 3y - 18}$

$\dfrac{x^2 - 16}{12y + 36} \div \dfrac{x^2 - 12x + 32}{y^2 - 3y - 18} = \dfrac{x^2 - 16}{12y + 36} \cdot \dfrac{y^2 - 3y - 18}{x^2 - 12x + 32}$ Multiply by reciprocal.

$= \dfrac{(x+4)(x-4)}{12(y+3)} \cdot \dfrac{(y-6)(y+3)}{(x-4)(x-8)}$ Factor.

$= \dfrac{(x+4)\overset{1}{\cancel{(x-4)}}}{12\underset{1}{\cancel{(y+3)}}} \cdot \dfrac{(y-6)\overset{1}{\cancel{(y+3)}}}{\underset{1}{\cancel{(x-4)}}(x-8)}$ Eliminate common factors.

$= \dfrac{(x+4)(y-6)}{12(x-8)}$ Simplify.

GuidedPractice

5A. $\dfrac{8x - 20}{x^2 + 2x - 35} \cdot \dfrac{x^2 - 7x + 10}{4x^2 - 16}$

5B. $\dfrac{x^2 - 9x + 20}{x^2 + 10x + 21} \div \dfrac{x^2 - x - 12}{6x + 42}$

2 Simplify Complex Fractions

A **complex fraction** is a rational expression with a numerator and/or denominator that is also a rational expression. The following expressions are complex fractions.

$$\dfrac{\frac{c}{6}}{5d} \qquad \dfrac{\frac{8}{x}}{x-2} \qquad \dfrac{\frac{x-3}{8}}{\frac{x-2}{x+4}} \qquad \dfrac{\frac{4}{a}+6}{\frac{12}{a}-3}$$

To simplify a complex fraction, first rewrite it as a division expression.

Example 6 Simplify Complex Fractions

Simplify each expression.

a. $\dfrac{\frac{a+b}{4}}{\frac{a^2+b^2}{4}}$

$\dfrac{\frac{a+b}{4}}{\frac{a^2+b^2}{4}} = \dfrac{a+b}{4} \div \dfrac{a^2+b^2}{4}$ Express as a division expression.

$= \dfrac{a+b}{4} \cdot \dfrac{4}{a^2+b^2}$ Multiply by the reciprocal.

$= \dfrac{a+b}{\underset{1}{\cancel{4}}} \cdot \dfrac{\overset{1}{\cancel{4}}}{a^2+b^2}$ or $\dfrac{a+b}{a^2+b^2}$ Simplify.

b. $\dfrac{\frac{x^2}{x^2 - y^2}}{\frac{4x}{y - x}}$

$\dfrac{\frac{x^2}{x^2 - y^2}}{\frac{4x}{y - x}} = \dfrac{x^2}{x^2 - y^2} \div \dfrac{4x}{y - x}$ Express as a division expression.

$= \dfrac{x^2}{x^2 - y^2} \cdot \dfrac{y - x}{4x}$ Multiply by the reciprocal.

$= \dfrac{x \cdot x}{(x + y)(x - y)} \cdot \dfrac{(-1)(x - y)}{4x}$ Factor.

$= \dfrac{x \cdot \overset{1}{\cancel{x}}}{(x + y)\underset{1}{\cancel{(x - y)}}} \cdot \dfrac{(-1)\overset{1}{\cancel{(x - y)}}}{\underset{1}{4\cancel{x}}}$ Eliminate Factors.

$= \dfrac{-x}{4(x + y)}$ Simplify.

GuidedPractice

Simplify each expression.

6A. $\dfrac{\frac{(x - 2)^2}{2(x^2 - 5x + 4)}}{\frac{x^2 - 4}{4x - 10}}$

6B. $\dfrac{\frac{x^2 - y^2}{y^2 - 49}}{\frac{y - x}{y + 7}}$

Check Your Understanding

Example 1 **Simplify each expression.**

1. $\dfrac{x^2 - 5x - 24}{x^2 - 64}$

2. $\dfrac{c + d}{3c^2 - 3d^2}$

Example 2

3. MULTIPLE CHOICE Identify all values of x for which $\dfrac{x + 7}{x^2 - 3x - 28}$ is undefined.

A $-7, 4$ **B** $7, 4$ **C** $4, -7, 7$ **D** $-4, 7$

Examples 3–6 **Simplify each expression.**

4. $\dfrac{y^2 + 3y - 40}{25 - y^2}$

5 $\dfrac{a^2x - b^2x}{by - ay}$

6. $\dfrac{27x^2y^4}{16yz^3} \cdot \dfrac{8z}{9xy^3}$

7. $\dfrac{12x^3y}{13ab^2} \div \dfrac{36xy^3}{26b}$

8. $\dfrac{x^2 - 4x - 21}{x^2 - 6x + 8} \cdot \dfrac{x - 4}{x^2 - 2x - 35}$

9. $\dfrac{a^2 - b^2}{3a^2 - 6a + 3} \div \dfrac{4a + 4b}{a^2 - 1}$

10. $\dfrac{\frac{a^3b^3}{xy^4}}{\frac{a^2b}{x^2y}}$

11. $\dfrac{\frac{4x}{x + 6}}{\frac{x^2 - 3x}{x^2 + 3x - 18}}$

12. CCSS **SENSE-MAKING** The volume of a shipping container in the shape of a rectangular prism can be represented by the polynomial $6x^3 + 11x^2 + 4x$, where the height is x.

a. Find the length and width of the container.

b. Find the ratio of the three dimensions of the container when $x = 2$.

c. Will the ratio of the three dimensions be the same for all values of x?

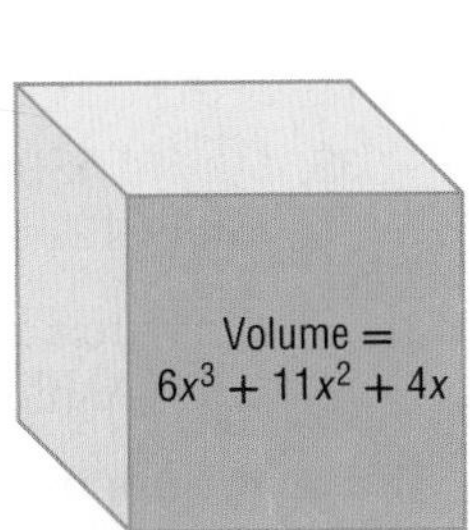

Practice and Problem Solving

Extra Practice is on page R8.

Example 1 **Simplify each expression.**

13. $\frac{x(x-3)(x+6)}{x^2+x-12}$

14. $\frac{y^2(y^2+3y+2)}{2y(y-4)(y+2)}$

15. $\frac{(x^2-9)(x^2-z^2)}{4(x+z)(x-3)}$

16. $\frac{(x^2-16x+64)(x+2)}{(x^2-64)(x^2-6x-16)}$

17. $\frac{x^2(x+2)(x-4)}{6x(x^2+x-20)}$

18. $\frac{3y(y-8)(y^2+2y-24)}{15y^2(y^2-12y+32)}$

Example 2

19. **MULTIPLE CHOICE** Identify all values of x for which $\frac{(x-3)(x+6)}{(x^2-7x+12)(x^2-36)}$ is undefined.

F 3, −6 **G** 4, 6 **H** −6, 6 **J** −6, 3, 4, 6

Example 3 **Simplify each expression.**

20. $\frac{x^2-5x-14}{28+3x-x^2}$

21. $\frac{x^3-9x^2}{x^2-3x-54}$

22. $\frac{(x-4)(x^2+2x-48)}{(36-x^2)(x^2+4x-32)}$

23. $\frac{16-c^2}{c^2+c-20}$

24. **GEOMETRY** The cylinder at the right has a volume of $(x+3)(x^2-3x-18)\pi$ cubic centimeters. Find the height of the cylinder.

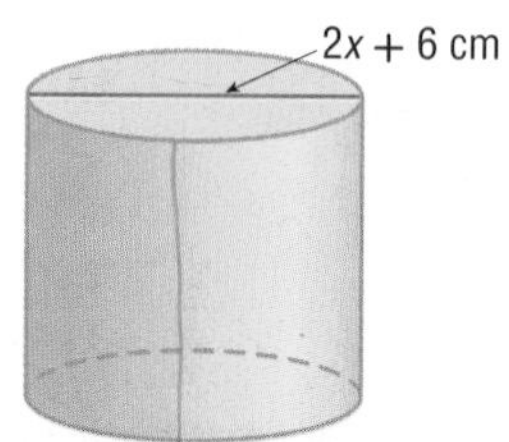

Examples 4–6 **Simplify each expression.**

25. $\frac{3ac^3f^3}{8a^2bcf^4} \cdot \frac{12ab^2c}{18ab^3c^2f}$

26. $\frac{14xy^2z^3}{21w^4x^2yz} \cdot \frac{7wxyz}{12w^2y^3z}$

27. $\frac{64a^2b^5}{35b^2c^3f^4} \div \frac{12a^4b^3c}{70abcf^2}$

28. $\frac{9x^2yz}{5z^4} \div \frac{12x^4y^2}{50xy^4z^2}$

29. $\frac{15a^2b^2}{21ac} \cdot \frac{14a^4c^2}{6ab^3}$

30. $\frac{14c^2f^5}{9a^2} \div \frac{35cf^4}{18ab^3}$

31. $\frac{y^2+8y+15}{y-6} \cdot \frac{y^2-9y+18}{y^2-9}$

32. $\frac{c^2-6c-16}{c^2-d^2} \div \frac{c^2-8c}{c+d}$

33. $\frac{x^2+9x+20}{8x+16} \cdot \frac{4x^2+16x+16}{x^2-25}$

34. $\frac{3a^2+6a+3}{a^2-3a-10} \div \frac{12a^2-12}{a^2-4}$

35. $\frac{\frac{x^2-9}{6x-12}}{\frac{x^2+10x+21}{x^2-x-2}}$

36. $\frac{\frac{y-x}{z^3}}{\frac{x-y}{6z^2}}$

37. $\frac{\frac{a^2-b^2}{b^3}}{\frac{b^2-ab}{a^2}}$

38. $\frac{\frac{x-y}{a+b}}{\frac{x^2-y^2}{b^2-a^2}}$

39. **CCSS REASONING** At the end of her high school soccer career, Ashley had made 33 goals out of 121 attempts.

   a. Write a ratio to represent the ratio of the number of goals made to goals attempted by Ashley at the end of her high school career.

   b. Suppose Ashley attempted a goals and made m goals during her first year at college. Write a rational expression to represent the ratio of the number of career goals made to the number of career goals attempted at the end of her first year in college.

40. GEOMETRY Parallelogram F has an area of $8x^2 + 10x - 3$ square meters and a height of $2x + 3$ meters. Parallelogram G has an area of $6x^2 + 13x - 5$ square meters and a height of $3x - 1$ meters. Find the area of right triangle H.

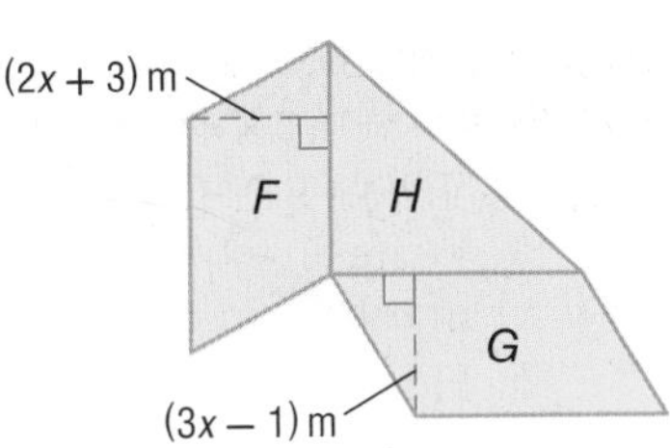

41. POLLUTION The thickness of an oil spill from a ruptured pipe on a rig is modeled by the function $T(x) = \frac{0.4(x^2 - 2x)}{x^3 + x^2 - 6x}$, where T is the thickness of the oil slick in meters and x is the distance from the rupture in meters.

a. Simplify the function.

b. How thick is the slick 100 meters from the rupture?

Simplify each expression.

42. $\frac{x^2 - 16}{3x^3 + 18x^2 + 24x} \cdot \frac{x^3 - 4x}{2x^2 - 7x - 4}$

43. $\frac{3x^2 - 17x - 6}{4x^2 - 20x - 24} \div \frac{6x^2 - 7x - 3}{2x^2 - x - 3}$

44. $\frac{9 - x^2}{x^2 - 4x - 21} \cdot \left(\frac{2x^2 + 7x + 3}{2x^2 - 15x + 7}\right)^{-1}$

45. $\left(\frac{2x^2 + 2x - 12}{x^2 + 4x - 5}\right)^{-1} \cdot \frac{2x^3 - 8x}{x^2 - 2x - 35}$

46. $\left(\frac{3xy^3z}{2a^2bc^2}\right)^3 \cdot \frac{16a^4b^3c^5}{15x^7yz^3}$

47. $\frac{20x^2y^6z^{-2}}{3a^3c^2} \cdot \left(\frac{16x^3y^3}{9acz}\right)^{-1}$

48. $\left(\frac{2xy^3}{3abc}\right)^{-2} \div \frac{6a^2b}{x^2y^4}$

49. $\dfrac{\frac{8x^2 - 10x - 3}{10x^2 + 35x - 20}}{\frac{2x^2 + x - 6}{4x^2 + 18x + 8}}$

50. $\dfrac{\frac{2x^2 + 7x - 30}{-6x^2 + 13x + 5}}{\frac{4x^2 + 12x - 72}{3x^2 - 11x - 4}}$

51. $\dfrac{\frac{4x^2 - 1}{3x^3 - 6x^2 - 24x}}{\frac{12x^2 + 12x - 9}{-2x^2 + 5x + 12}}$

52. GEOMETRY The area of the base of the rectangular prism at the right is 20 square centimeters.

a. Find the length of $\overline{BC}$ in terms of x.

b. If $DC = 3BC$, determine the area of the shaded region in terms of x.

c. Determine the volume of the prism in terms of x.

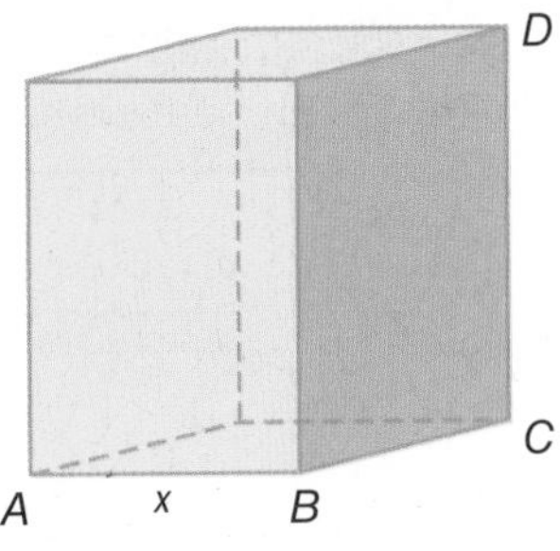

Simplify each expression.

53. $\frac{x^2 + 4x - 32}{2x^2 + 9x - 5} \cdot \frac{3x^2 - 75}{3x^2 - 11x - 4} \div \frac{6x^2 - 18x - 60}{x^3 - 4x}$

54. $\frac{8x^2 + 10x - 3}{3x^2 - 12x - 36} \div \frac{2x^2 - 5x - 12}{3x^2 - 17x - 6} \cdot \frac{4x^2 + 3x - 1}{4x^2 - 40x + 24}$

55. $\frac{4x^2 - 9x - 9}{3x^2 + 6x - 18} \div \frac{-2x^2 + 5x + 3}{x^2 - 4x - 32} \div \frac{8x^2 + 10x + 3}{6x^2 - 6x - 12}$

56. CCSS PERSEVERANCE Use the formula $d = rt$ and the following information. An airplane is traveling at a rate r of 500 miles per hour for a time t of $(6 + x)$ hours. A second airplane travels at the rate r of $(540 + 90x)$ miles per hour for a time t of 6 hours.

a. Write a rational expression to represent the ratio of the distance d traveled by the first airplane to the distance d traveled by the second airplane.

b. Simplify the rational expression. What does this expression tell you about the distances traveled by the two airplanes?

c. Under what condition is the rational expression undefined? Describe what this condition would tell you about the two airplanes.

57. **TRAINS** Trying to get into a train yard one evening, all of the trains are backed up for 2 miles along a system of tracks. Assume that each car occupies an average of 75 feet of space on a track and that the train yard has 5 tracks.

a. Write an expression that could be used to determine the number of train cars involved in the backup.

b. How many train cars are involved in the backup?

c. Suppose that there are 8 attendants doing safety checks on each car, and it takes each vehicle an average of 45 seconds for each check. Approximately how many hours will it take for all the vehicles in the backup to exit?

58. **MULTIPLE REPRESENTATIONS** In this problem, you will investigate the graph of a rational function.

a. Algebraic Simplify $\frac{x^2 - 5x + 4}{x - 4}$.

b. Tabular Let $f(x) = \frac{x^2 - 5x + 4}{x - 4}$. Use the expression you wrote in part **a** to write the related function $g(x)$. Use a graphing calculator to make a table for both functions for $0 \leq x \leq 10$.

c. Analytical What are $f(4)$ and $g(4)$? Explain the significance of these values.

d. Graphical Graph the functions on the graphing calculator. Use the **TRACE** function to investigate each graph, using the ▲ and ▼ keys to switch from one graph to the other. Compare and contrast the graphs.

e. Verbal What conclusions can you draw about the expressions and the functions?

H.O.T. Problems Use Higher-Order Thinking Skills

59. **REASONING** Compare and contrast $\frac{(x - 6)(x + 2)(x + 3)}{x + 3}$ and $(x - 6)(x + 2)$.

60. **CCSS CRITIQUE** Troy and Beverly are simplifying $\frac{x + y}{x - y} \div \frac{4}{y - x}$. Is either of them correct? Explain your reasoning.

Troy

$$\frac{x+y}{x-y} \div \frac{4}{y-x} = \frac{x-y}{x+y} \cdot \frac{4}{y-x} = \frac{-4}{x+y}$$

Beverly

$$\frac{x+y}{x-y} \div \frac{4}{y-x} = \frac{x+y}{x-y} \cdot \frac{y-x}{4} = -\frac{x+y}{4}$$

61. **CHALLENGE** Find the expression that makes the following statement true.

$$\frac{x - 6}{x + 3} \cdot \frac{?}{x - 6} = x - 2$$

62. **WHICH ONE DOESN'T BELONG?** Identify the expression that does not belong with the other three. Explain your reasoning.

$\frac{1}{x - 1}$ $\frac{x^2 + 3x + 2}{x - 5}$ $\frac{x + 1}{\sqrt{x + 3}}$ $\frac{x^2 + 1}{3}$

63. **REASONING** Determine whether the following statement is *sometimes*, *always*, or *never* true. Explain your reasoning.

A rational function that has a variable in the denominator is defined for all real values of x.

64. **OPEN ENDED** Write a rational expression that simplifies to $\frac{x - 1}{x + 4}$.

65. **WRITING IN MATH** The rational expression $\frac{x^2 + 3x}{4x}$ is simplified to $\frac{x + 3}{4}$. Explain why this new expression is not defined for all values of x.

Standardized Test Practice

66. SAT/ACT The Mason family wants to drive an average of 250 miles per day on their vacation. On the first five days, they travel 220 miles, 300 miles, 210 miles, 275 miles, and 240 miles. How many miles must they travel on the sixth day to meet their goal?

A 235 miles
B 251 miles
C 255 miles
D 275 miles
E 315 miles

67. Which of the following equations gives the relationship between N and T in the table?

N	1	2	3	4	5	6
T	1	4	7	10	13	16

F $T = 2 - N$
G $T = 4 - 3N$
H $T = 3N + 1$
J $T = 3N - 2$

68. A monthly cell phone plan costs $39.99 for up to 300 minutes and 20 cents per minute thereafter. Which of the following represents the total monthly bill (in dollars) to talk for x minutes if x is greater than 300?

A $39.99 + 0.20(300 - x)$
B $39.99 + 0.20(x - 300)$
C $39.99 + 0.20x$
D $39.99 + 20x$

69. SHORT RESPONSE The area of a circle 6 meters in diameter exceeds the combined areas of a circle 4 meters in diameter and a circle 2 meters in diameter by how many square meters?

Spiral Review

70. ANTHROPOLOGY An anthropologist studying the bones of a prehistoric person finds there is so little remaining Carbon-14 in the bones that instruments cannot measure it. This means that there is less than 0.5% of the amount of Carbon-14 the bones would have contained when the person was alive. The half-life of Carbon-14 is 5760 years. How long ago did the person die? (Lesson 7-8)

Solve each equation. Round to the nearest ten-thousandth. (Lesson 7-7)

71. $3e^x + 1 = 5$
72. $2e^x - 1 = 0$
73. $-3e^{4x} + 11 = 2$
74. $8 + 3e^{3x} = 26$

75. NOISE ORDINANCE A proposed city ordinance will make it illegal in a residential area to create sound that exceeds 72 decibels during the day and 55 decibels during the night. How many times as intense is the noise level allowed during the day than at night? (Lesson 7-3)

Simplify. (Lesson 6-5)

76. $\sqrt{50x^4}$
77. $\sqrt[3]{16y^3}$
78. $\sqrt{18x^2y^3}$
79. $\sqrt{40a^3b^4}$

80. AUTOMOBILES The length of the cargo space in a sport-utility vehicle is 4 inches greater than the height of the space. The width is 16 inches less than twice the height. The cargo space has a total volume of 55,296 cubic inches. (Lesson 5-8)

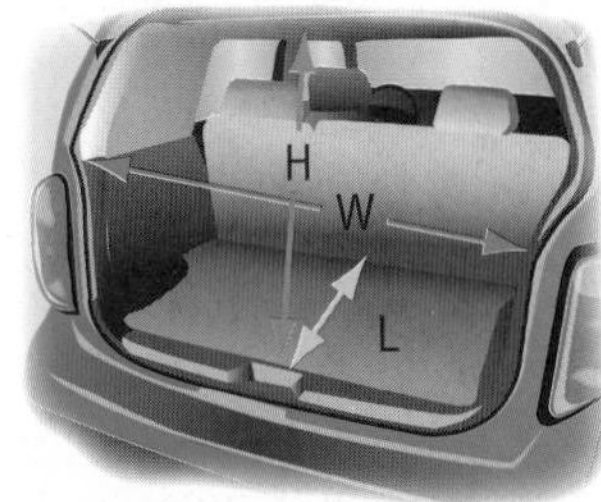

a. Write a polynomial function that represents the volume of the cargo space.

b. Will a package 34 inches long, 44 inches wide, and 34 inches tall fit in the cargo space? Explain.

Skills Review

Simplify.

81. $(2a + 3b) + (8a - 5b)$
82. $(x^2 - 4x + 3) - (4x^2 + 3x - 5)$
83. $(5y + 3y^2) + (-8y - 6y^2)$
84. $2x(3y + 9)$
85. $(x + 6)(x + 3)$
86. $(x + 1)(x^2 - 2x + 3)$

LESSON 8-2 Adding and Subtracting Rational Expressions

Then	Now	Why?
You added and subtracted polynomial expressions.	1 Determine the LCM of polynomials. 2 Add and subtract rational expressions.	As a fire engine moves toward a person, the pitch of the siren sounds higher to that person than it would if the fire engine were at rest. This is because the sound waves are compressed closer together, referred to as the *Doppler effect.* The Doppler effect can be represented by the rational expression $P_0\left(\frac{s_0}{s_0 - v}\right)$, where P_0 is the actual pitch of the siren, v is the speed of the fire truck, and s_0 is the speed of sound in air.

Common Core State Standards

Content Standards

A.APR.7 Understand that rational expressions form a system analogous to the rational numbers, closed under addition, subtraction, multiplication, and division by a nonzero rational expression; add, subtract, multiply, and divide rational expressions.

Mathematical Practices

3 Construct viable arguments and critique the reasoning of others.

1 LCM of Polynomials

Just as with rational numbers in fractional form, to add or subtract two rational expressions that have unlike denominators, you must first find the least common denominator (LCD). The LCD is the least common multiple (LCM) of the denominators.

To find the LCM of two or more numbers or polynomials, factor them. The LCM contains each factor the greatest number of times it appears as a factor.

Numbers	Polynomials
$\frac{5}{6} + \frac{4}{9}$	$\frac{3}{x^2 - 3x + 2} + \frac{5}{2x^2 - 2}$
LCM of 6 and 9	**LCM of $x^2 - 3x + 2$ and $2x^2 - 2$**
$6 = 2 \cdot 3$	$x^2 - 3x + 2 = (x - 1)(x - 2)$
$9 = 3 \cdot 3$	$2x^2 - 2 = 2 \cdot (x - 1)(x + 1)$
$\text{LCM} = 2 \cdot 3 \cdot 3$ or 18	$\text{LCM} = 2(x - 1)(x - 2)(x + 1)$

Example 1 LCM of Monomials and Polynomials

Find the LCM of each set of polynomials.

a. $6xy$, $15x^2$, and $9xy^4$

$6xy = 2 \cdot 3 \cdot x \cdot y$ — Factor the first monomial.

$15x^2 = 3 \cdot 5 \cdot x^2$ — Factor the second monomial.

$9xy^4 = 3 \cdot 3 \cdot x \cdot y^4$ — Factor the third monomial.

$\text{LCM} = 2 \cdot 3 \cdot 3 \cdot 5 \cdot x^2 \cdot y^4$ — Use each factor the greatest number of times it appears.

$= 90x^2y^4$ — Then simplify.

b. $y^4 + 8y^3 + 15y^2$ and $y^2 - 3y - 40$

$y^4 + 8y^3 + 15y^2 = y^2(y + 5)(y + 3)$ — Factor the first polynomial.

$y^2 - 3y - 40 = (y + 5)(y - 8)$ — Factor the second polynomial.

$\text{LCM} = y^2(y + 5)(y + 3)(y - 8)$ — Use each factor the greatest number of times it appears as a factor.

Guided Practice

1A. $12a^2b$, $15abc$, $8b^3c^4$

1B. $4a^2 - 12a - 16$ and $a^3 - 9a^2 + 20a$

Philip Rostron/Masterfile

2 Add and Subtract Rational Expressions

As with fractions, rational expressions must have common denominators in order to be added or subtracted.

KeyConcept

Adding Rational Expressions

Words To add rational expressions, find the least common denominator (LCD). Rewrite each expression with the LCD. Then add.

Symbols For all $\frac{a}{b}$ and $\frac{c}{d}$, with $b \neq 0$ and $d \neq 0$, $\frac{a}{b} + \frac{c}{d} = \frac{ad}{bd} + \frac{bc}{bd} = \frac{ad + bc}{bd}$.

Subtracting Rational Expressions

Words To subtract rational expressions, find the least common denominator (LCD). Rewrite each expression with the LCD. Then subtract.

Symbols For all $\frac{a}{b}$ and $\frac{c}{d}$, with $b \neq 0$ and $d \neq 0$, $\frac{a}{b} - \frac{c}{d} = \frac{ad}{bd} - \frac{bc}{bd} = \frac{ad - bc}{bd}$.

PT

Example 2 Monomial Denominators

Simplify $\frac{3y}{2x^3} + \frac{5z}{8xy^2}$.

$$\frac{3y}{2x^3} + \frac{5z}{8xy^2} = \frac{3y}{2x^3} \cdot \frac{4y^2}{4y^2} + \frac{5z}{8xy^2} \cdot \frac{x^2}{x^2}$$ The LCD is $8x^3y^2$.

$$= \frac{12y^3}{8x^3y^2} + \frac{5x^2z}{8x^3y^2}$$ Multiply fractions.

$$= \frac{12y^3 + 5x^2z}{8x^3y^2}$$ Add the numerators.

GuidedPractice

Simplify each expression.

2A. $\frac{4}{5a^3b^2} + \frac{9c}{10ab}$

2B. $\frac{3a^2}{16b^2} - \frac{8x}{5a^3b}$

The LCD is also used to combine rational expressions with polynomial denominators.

PT

Example 3 Polynomial Denominators

Simplify $\frac{5}{6x - 18} - \frac{x - 1}{4x^2 - 14x + 6}$.

$$\frac{5}{6x - 18} - \frac{x - 1}{4x^2 - 14x + 6} = \frac{5}{6(x - 3)} - \frac{x - 1}{2(2x - 1)(x - 3)}$$ Factor denominators.

$$= \frac{5(2x - 1)}{6(x - 3)(2x - 1)} - \frac{(x - 1)(3)}{2(2x - 1)(x - 3)(3)}$$ Multiply by missing factors.

$$= \frac{10x - 5 - 3x + 3}{6(x - 3)(2x - 1)}$$ Subtract numerators.

$$= \frac{7x - 2}{6(x - 3)(2x - 1)}$$ Simplify.

StudyTip

Simplifying Rational Expressions After you add or subtract rational expressions, it is possible that the resulting expression can be further simplified.

GuidedPractice

Simplify each expression.

3A. $\frac{x - 1}{x^2 - x - 6} - \frac{4}{5x + 10}$

3B. $\frac{x - 8}{4x^2 + 21x + 5} + \frac{6}{12x + 3}$

One way to simplify a complex fraction is to simplify the numerator and the denominator separately, and then simplify the resulting expressions.

StudyTip

Undefined Terms Remember that there are restrictions on variables in the denominator.

Example 4 Complex Fractions with Different LCDs

Simplify $\dfrac{1+\frac{1}{x}}{1-\frac{x}{y}}$.

$$\frac{1+\frac{1}{x}}{1-\frac{x}{y}} = \frac{\frac{x}{x}+\frac{1}{x}}{\frac{y}{y}-\frac{x}{y}}$$ The LCD of the numerator is x. The LCD of the denominator is y.

$$= \frac{\frac{x+1}{x}}{\frac{y-x}{y}}$$ Simplify the numerator and denominator.

$$= \frac{x+1}{x} \div \frac{y-x}{y}$$ Write as a division expression.

$$= \frac{x+1}{x} \cdot \frac{y}{y-x}$$ Multiply by the reciprocal of the divisor.

$$= \frac{xy+y}{xy-x^2}$$ Simplify.

Guided Practice

Simplify each expression.

4A. $\dfrac{1-\frac{y}{x}}{\frac{1}{y}+\frac{1}{x}}$

4B. $\dfrac{\frac{c}{d}-\frac{d}{c}}{\frac{d}{c}+2}$

Another method of simplifying complex fractions is to find the LCD of all of the denominators. Then, the denominators are all eliminated by multiplying by the LCD.

Example 5 Complex Fractions with Same LCDs

Simplify $\dfrac{1+\frac{1}{x}}{1-\frac{x}{y}}$.

$$\frac{1+\frac{1}{x}}{1-\frac{x}{y}} = \frac{\left(1+\frac{1}{x}\right)}{\left(1-\frac{x}{y}\right)} \cdot \frac{xy}{xy}$$ The LCD of all of the denominators is xy. Multiply by $\frac{xy}{xy}$.

$$= \frac{xy+y}{xy-x^2}$$ Distribute xy.

Notice that the same problem is solved in Examples 4 and 5 using different methods, but both produce the same answer. So, how you solve problems similar to these is left up to your own discretion.

Guided Practice

Simplify each expression.

5A. $\dfrac{1+\frac{2}{x}}{\frac{3}{y}-\frac{4}{x}}$

5B. $\dfrac{\frac{1}{d}-\frac{d}{c}}{\frac{1}{c}+6}$

5C. $\dfrac{\frac{1}{y}+\frac{1}{x}}{\frac{1}{y}-\frac{1}{x}}$

5D. $\dfrac{\frac{a}{b}+1}{1-\frac{b}{a}}$

Check Your Understanding

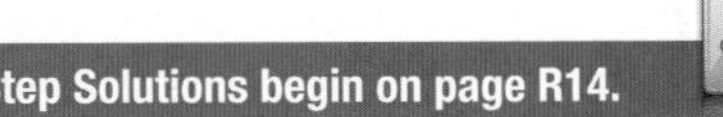

= Step-by-Step Solutions begin on page R14.

Example 1 **Find the LCM of each set of polynomials.**

1. $16x, 8x^2y^3, 5x^3y$
2. $7a^2, 9ab^3, 21abc^4$
3. $3y^2 - 9y, y^2 - 8y + 15$
4. $x^3 - 6x^2 - 16x, x^2 - 4$

Examples 2–3 **Simplify each expression.**

5. $\frac{12y}{5x} + \frac{5x}{4y^3}$
6. $\frac{5}{6ab} + \frac{3b^2}{14a^3}$
7. $\frac{7b}{12a} - \frac{1}{18ab^3}$
8. $\frac{y^2}{8c^2d^2} - \frac{3x}{14c^4d}$
9. $\frac{4x}{x^2 + 9x + 18} + \frac{5}{x + 6}$
10. $\frac{8}{y - 3} + \frac{2y - 5}{y^2 - 12y + 27}$
11. $\frac{4}{3x + 6} - \frac{x + 1}{x^2 - 4}$
12. $\frac{3a + 2}{a^2 - 16} - \frac{7}{6a + 24}$

13. **GEOMETRY** Find the perimeter of the rectangle.

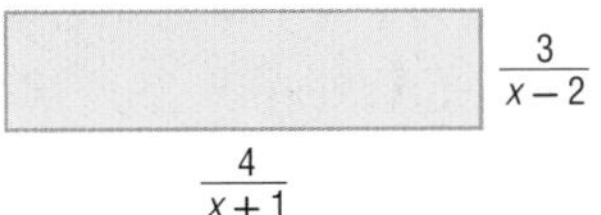

Examples 4–5 **Simplify each expression.**

14. $\frac{4 + \frac{2}{x}}{3 - \frac{2}{x}}$
15. $\frac{6 + \frac{4}{y}}{2 + \frac{6}{y}}$
16. $\frac{\frac{3}{x} + \frac{2}{y}}{1 + \frac{4}{y}}$
17. $\frac{\frac{2}{b} + \frac{5}{a}}{\frac{3}{a} - \frac{8}{b}}$

Practice and Problem Solving

Extra Practice is on page R8.

Example 1 **Find the LCM of each set of polynomials.**

18. $24cd, 40a^2c^3d^4, 15abd^3$
19. $4x^2y^3, 18xy^4, 10xz^2$
20. $x^2 - 9x + 20, x^2 + x - 30$
21. $6x^2 + 21x - 12, 4x^2 + 22x + 24$

Examples 2–3 **CCSS PERSEVERANCE** **Simplify each expression.**

22. $\frac{5a}{24cf^4} + \frac{a}{36bc^4f^3}$
23. $\frac{4b}{15x^3y^2} - \frac{3b}{35x^2y^4z}$
24. $\frac{5b}{6a} + \frac{3b}{10a^2} + \frac{2}{ab^2}$
25. $\frac{4}{3x} + \frac{8}{x^3} + \frac{2}{5xy}$
26. $\frac{8}{3y} + \frac{2}{9} - \frac{3}{10y^2}$
27. $\frac{1}{16a} + \frac{5}{12b} - \frac{9}{10b^3}$
28. $\frac{8}{x^2 - 6x - 16} + \frac{9}{x^2 - 3x - 40}$
29. $\frac{6}{y^2 - 2y - 35} + \frac{4}{y^2 + 9y + 20}$
30. $\frac{12}{3y^2 - 10y - 8} - \frac{3}{y^2 - 6y + 8}$
31. $\frac{6}{2x^2 + 11x - 6} - \frac{8}{x^2 + 3x - 18}$
32. $\frac{2x}{4x^2 + 9x + 2} + \frac{3}{2x^2 - 8x - 24}$
33. $\frac{4x}{3x^2 + 3x - 18} - \frac{2x}{2x^2 + 11x + 15}$

34. **BIOLOGY** After a person eats something, the pH or acid level A of his or her mouth can be determined by the formula $A = \frac{20.4t}{t^2 + 36} + 6.5$, where t is the number of minutes that have elapsed since the food was eaten.
    a. Simplify the equation.
    b. What would the acid level be after 30 minutes?

35. **GEOMETRY** Both triangles in the figure at the right are equilateral. If the area of the smaller triangle is 200 square centimeters and the area of the larger triangle is 300 square centimeters, find the minimum distance from A to B in terms of x and y and simplify.

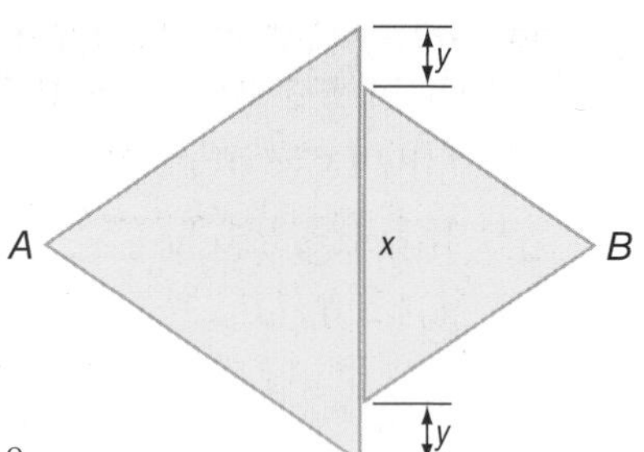

Examples 4–5 **Simplify each expression.**

36. $\dfrac{\frac{2}{x-3}+\frac{3x}{x^2-9}}{\frac{3}{x+3}-\frac{4x}{x^2-9}}$

37. $\dfrac{\frac{4}{x+5}+\frac{9}{x-6}}{\frac{5}{x-6}-\frac{8}{x+5}}$

38. $\dfrac{\frac{5}{x+6}-\frac{2x}{2x-1}}{\frac{x}{2x-1}+\frac{4}{x+6}}$

39. $\dfrac{\frac{8}{x-9}-\frac{x}{3x+2}}{\frac{3}{3x+2}+\frac{4x}{x-9}}$

40. **OIL PRODUCTION** Managers of an oil company have estimated that oil will be pumped from a certain well at a rate based on the function $R(x) = \frac{20}{x} + \frac{200x}{3x^2+20}$, where $R(x)$ is the rate of production in thousands of barrels per year x years after pumping begins.

   a. Simplify $R(x)$.

   b. At what rate will oil be pumping from the well in 50 years?

Find the LCM of each set of polynomials.

41. $12xy^4, 14x^4y^2, 5xyz^3, 15x^5y^3$

42. $-6abc^2, 18a^2b^2, 15a^4c, 8b^3$

43. $x^2 - 3x - 28, 2x^2 + 9x + 4, x^2 - 16$

44. $x^2 - 5x - 24, x^2 - 9, 3x^2 + 8x - 3$

Simplify each expression.

45. $\frac{1}{12a} + 6 - \frac{3}{5a^2}$

46. $\frac{5}{16y^2} - 4 - \frac{8}{3x^2y}$

47. $\frac{5}{6x^2+46x-16} + \frac{2}{6x^2+57x+72}$

48. $\frac{1}{8x^2-20x-12} + \frac{4}{6x^2+27x+12}$

49. $\frac{x^2+y^2}{x^2-y^2} + \frac{y}{x+y} - \frac{x}{x-y}$

50. $\frac{x^2+x}{x^2-9x+8} + \frac{4}{x-1} - \frac{3}{x-8}$

51. $\dfrac{\frac{2}{a-1}+\frac{3}{a-4}}{\frac{6}{a^2-5a+4}}$

52. $\dfrac{\frac{1}{x}+\frac{1}{y}}{\left(\frac{1}{x}-\frac{1}{y}\right)(x+y)}$

53. **GEOMETRY** An expression for the length of one rectangle is $\frac{x^2-9}{x-2}$. The length of a similar rectangle is expressed as $\frac{x+3}{x^2-4}$. What is the scale factor of the lengths of the two rectangles? Write in simplest form.

54. **CCSS MODELING** Cameron is taking a 20-mile kayaking trip. He travels half the distance at one rate. The rest of the distance he travels 2 miles per hour slower.

   a. If x represents the faster pace in miles per hour, write an expression that represents the time spent at that pace.

   b. Write an expression for the amount of time spent at the slower pace.

   c. Write an expression for the amount of time Cameron needed to complete the trip.

Find the slope of the line that passes through each pair of points.

55. $A\left(\frac{2}{p}, \frac{1}{2}\right)$ and $B\left(\frac{1}{3}, \frac{3}{p}\right)$

56. $C\left(\frac{1}{4}, \frac{4}{q}\right)$ and $D\left(\frac{5}{q}, \frac{1}{5}\right)$

57. $E\left(\frac{7}{w}, \frac{1}{7}\right)$ and $F\left(\frac{1}{7}, \frac{7}{w}\right)$

58. $G\left(\frac{6}{n}, \frac{1}{6}\right)$ and $H\left(\frac{1}{6}, \frac{6}{n}\right)$

59. **PHOTOGRAPHY** The focal length of a lens establishes the field of view of the camera. The shorter the focal length is, the larger the field of view. For a camera with a fixed focal length of 70 mm to focus on an object x mm from the lens, the film must be placed a distance y from the lens. This is represented by $\frac{1}{x} + \frac{1}{y} = \frac{1}{70}$.

a. Express y as a function of x.

b. What happens to the focusing distance when the object is 70 mm away?

60. **PHARMACOLOGY** Two drugs are administered to a patient. The concentrations in the bloodstream of each are given by $f(t) = \frac{2t}{3t^2 + 9t + 6}$ and $g(t) = \frac{3t}{2t^2 + 6t + 4}$ where t is the time, in hours, after the drugs are administered.

a. Add the two functions together to determine a function for the total concentration of drugs in the patient's bloodstream.

b. What is the concentration of drugs after 8 hours?

61. **DOPPLER EFFECT** Refer to the application at the beginning of the lesson. George is equidistant from two fire engines traveling toward him from opposite directions.

a. Let x be the speed of the faster fire engine and y be the speed of the slower fire engine. Write and simplify a rational expression representing the difference in pitch between the two sirens according to George.

b. If one is traveling at 45 meters per second and the other is traveling at 70 meters per second, what is the difference in their pitches according to George? The speed of sound in air is 332 meters per second, and both engines have a siren with a pitch of 500 Hz.

62. **RESEARCH** A student studying learning behavior performed an experiment in which a rat was repeatedly sent through a maze. It was determined that the time it took the rat to complete the maze followed the rational function $T(x) = 4 + \frac{10}{x}$, where x represented the number of trials.

a. What is the domain of the function?

b. Graph the function for $0 \leq x \leq 10$.

c. Make a table of the function for $x = 20, 50, 100, 200$, and 400.

d. If it were possible to have an infinite number of trials, what do you think would be the rat's best time? Explain your reasoning.

H.O.T. Problems Use Higher-Order Thinking Skills

63. **CHALLENGE** Simplify $\dfrac{5x^{-2} - \frac{x+1}{x}}{\frac{4}{3 - x^{-1}} + 6x^{-1}}$.

64. **CCSS ARGUMENTS** The sum of any two rational numbers is always a rational number. So, the set of rational numbers is said to be closed under addition. Determine whether the set of rational expressions is closed under addition, subtraction, multiplication, and division by a nonzero rational expression. Justify your reasoning.

65. **OPEN ENDED** Write three monomials with an LCM of $180a^4b^6c$.

66. **WRITING IN MATH** Write a how-to manual for adding rational expressions that have unlike denominators. How does this compare to adding rational numbers?

Standardized Test Practice

67. PROBABILITY A drawing is to be held to select the winner of a new bike. There are 100 seniors, 150 juniors, and 200 sophomores who had correct entries. The drawing will contain 3 tickets for each senior name, 2 for each junior, and 1 for each sophomore. What is the probability that a senior's ticket will be chosen?

A $\frac{1}{8}$ C $\frac{2}{7}$

B $\frac{2}{9}$ D $\frac{3}{8}$

68. SHORT RESPONSE Find the area of the figure.

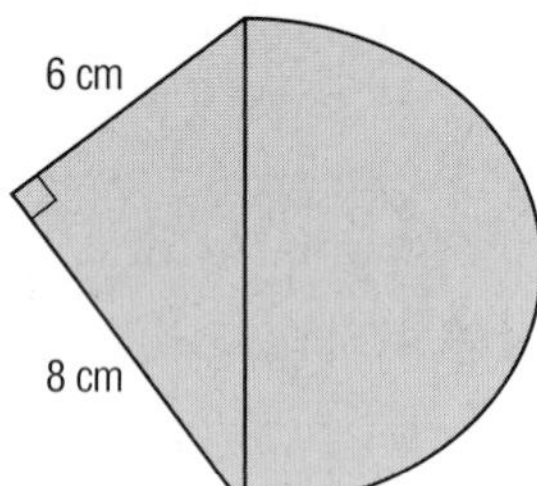

69. SAT/ACT If Mauricio receives b books in addition to the number of books he had, he will have t times as many as he had originally. In terms of b and t, how many books did Mauricio have at the beginning?

F $\frac{b}{t-1}$ J $\frac{b}{t}$

G $\frac{b}{t+1}$ K $\frac{t}{b}$

H $\frac{t+1}{b}$

70. If $\frac{2a}{a} + \frac{1}{a} = 4$, then $a =$ ___.

A 2 C $\frac{1}{8}$

B $\frac{1}{2}$ D $-\frac{1}{8}$

Spiral Review

Simplify each expression. (Lesson 8-1)

71. $\frac{-4ab}{21c} \cdot \frac{14c^2}{22a^2}$ **72.** $\frac{x^2 - y^2}{6y} \div \frac{x + y}{36y^2}$ **73.** $\frac{n^2 - n - 12}{n + 2} \div \frac{n - 4}{n^2 - 4n - 12}$

74. BIOLOGY Bacteria usually reproduce by a process known as *binary fission*. In this type of reproduction, one bacterium divides, forming two bacteria. Under ideal conditions, some bacteria reproduce every 20 minutes. (Lesson 7-8)

a. Find the constant k for this type of bacteria under ideal conditions.

b. Write the equation for modeling the exponential growth of this bacterium.

Graph each function. State the domain and range of each function. (Lesson 6-3)

75. $y = -\sqrt{2x + 1}$ **76.** $y = \sqrt{5x - 3}$ **77.** $y = \sqrt{x + 6} - 3$

78. $y = 5 - \sqrt{x + 4}$ **79.** $y = \sqrt{3x - 6} + 4$ **80.** $y = 2\sqrt{3 - 4x} + 3$

Solve each equation. State the number and type of roots. (Lesson 5-7)

81. $3x + 8 = 0$ **82.** $2x^2 - 5x + 12 = 0$ **83.** $x^3 + 9x = 0$ **84.** $x^4 - 81 = 0$

Skills Review

Graph each function.

85. $y = 4(x + 3)^2 + 1$ **86.** $y = -(x - 5)^2 - 3$ **87.** $y = \frac{1}{4}(x - 2)^2 + 4$

88. $y = \frac{1}{2}(x - 3)^2 - 5$ **89.** $y = x^2 + 6x + 2$ **90.** $y = x^2 - 8x + 18$

LESSON

8-3 Graphing Reciprocal Functions

Then

- You graphed polynomial functions.

Now

1. Determine properties of reciprocal functions.
2. Graph transformations of reciprocal functions.

Why?

- The East High School Chorale wants to raise $5000 to fund a trip to a national competition in Tampa. They have decided to sell candy bars. They will make a $1 profit on each candy bar they sell, so they need to sell 5000 candy bars.

  If c represents the number of candy bars each student has to sell and n represents the number of students, then $c = \frac{5000}{n}$.

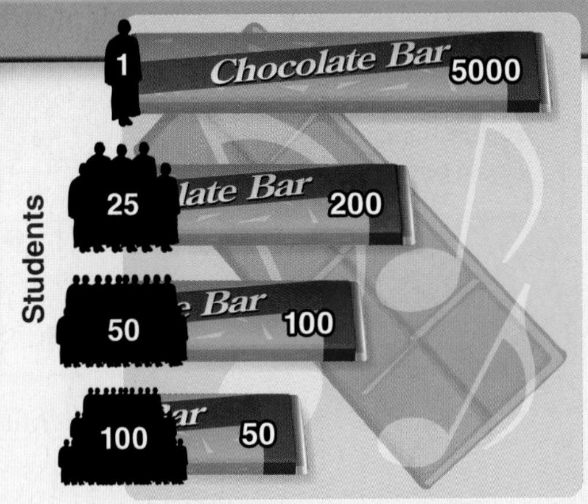

NewVocabulary
reciprocal function
hyperbola

Common Core State Standards

Content Standards

A.CED.2 Create equations in two or more variables to represent relationships between quantities; graph equations on coordinate axes with labels and scales.

F.BF.3 Identify the effect on the graph of replacing $f(x)$ by $f(x) + k$, $kf(x)$, $f(kx)$, and $f(x + k)$ for specific values of k (both positive and negative); find the value of k given the graphs. Experiment with cases and illustrate an explanation of the effects on the graph using technology.

Mathematical Practices

2 Reason abstractly and quantitatively.

1 Vertical and Horizontal Asymptotes

The function $c = \frac{5000}{n}$ is a reciprocal function. A **reciprocal function** has an equation of the form $f(x) = \frac{1}{a(x)}$, where $a(x)$ is a linear function and $a(x) \neq 0$.

KeyConcept Parent Function of Reciprocal Functions

Parent function:	$f(x) = \frac{1}{x}$
Type of graph:	**hyperbola**
Domain and range:	all nonzero real numbers
Asymptotes:	$x = 0$ and $f(x) = 0$
Intercepts:	none
Not defined:	$x = 0$

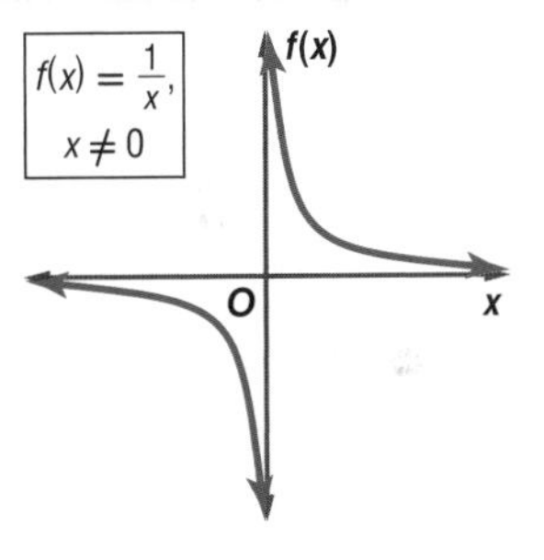

The domain of a reciprocal function is limited to values for which the function is defined.

Functions:	$f(x) = \frac{-3}{x + 2}$	$g(x) = \frac{4}{x - 5}$	$h(x) = \frac{3}{x}$
Not defined at:	$x = -2$	$x = 5$	$x = 0$

PT

Example 1 Limitations on Domain

Determine the value of x for which $f(x) = \frac{3}{2x + 5}$ is not defined.

Find the value for which the denominator of the expression equals 0.

$\frac{3}{2x + 5} \rightarrow 2x + 5 = 0$

$x = -\frac{5}{2}$ The function is undefined for $x = -\frac{5}{2}$.

GuidedPractice

Determine the value of x for which each function is not defined.

1A. $f(x) = \frac{2}{x - 1}$

1B. $f(x) = \frac{7}{3x + 2}$

The graphs of reciprocal functions may have breaks in continuity for excluded values. Some may have an asymptote, which is a line that the graph of the function approaches.

Study Tip

Structure Vertical asymptotes show where a function is undefined, while horizontal asymptotes show the end behavior of a graph.

Example 2 Determine Properties of Reciprocal Functions

Identify the asymptotes, domain, and range of each function.

a.

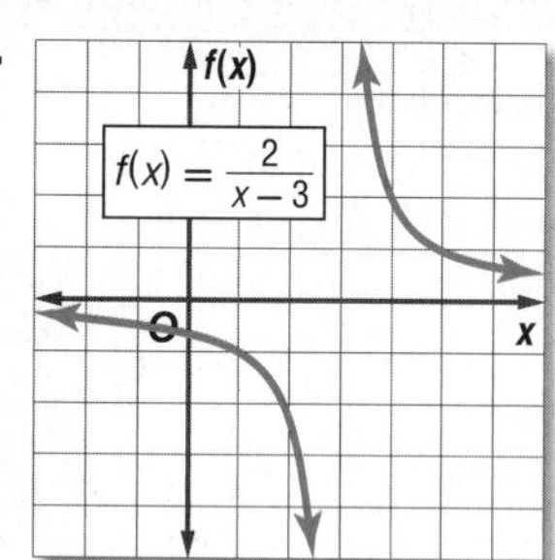

Identify x-values for which $f(x)$ is undefined.

$x - 3 = 0$

$x = 3$

$f(x)$ is not defined when $x = 3$. So there is an asymptote at $x = 3$.

From $x = 3$, as x-values decrease, $f(x)$-values approach 0, and as x-values increase, $f(x)$-values approach 0. So there is an asymptote at $f(x) = 0$.

The domain is all real numbers not equal to 3 and the range is all real numbers not equal to 0.

b.

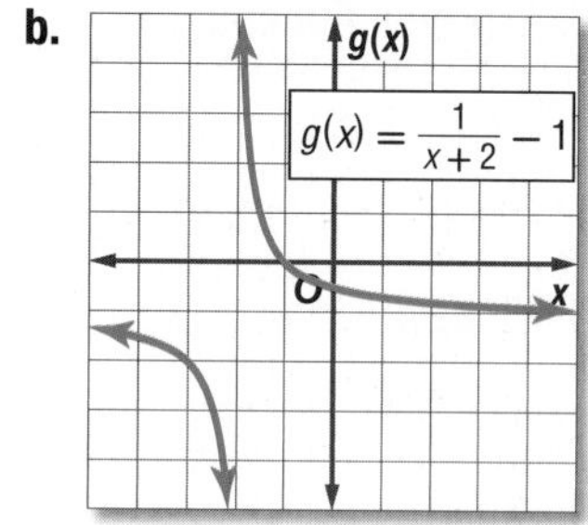

Identify x-values for which $g(x)$ is undefined.

$x + 2 = 0$

$x = -2$

$g(x)$ is not defined when $x = -2$. So there is an asymptote at $x = -2$.

From $x = -2$, as x-values decrease, $g(x)$-values approach -1, and as x-values increase, $g(x)$-values approach -1. So there is an asymptote at $g(x) = -1$.

The domain is all real numbers not equal to -2 and the range is all real numbers not equal to -1.

Guided Practice

2A.

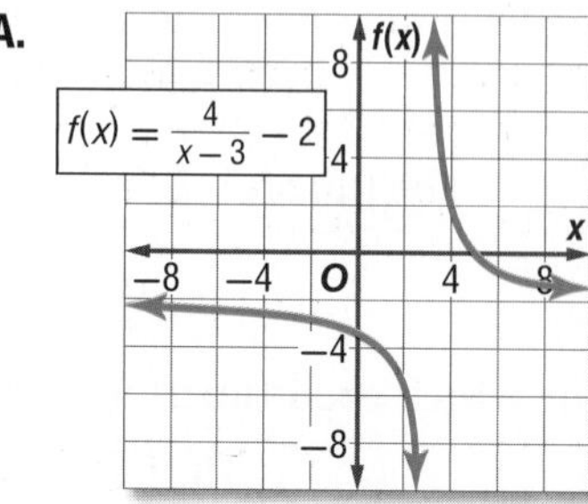

2B.

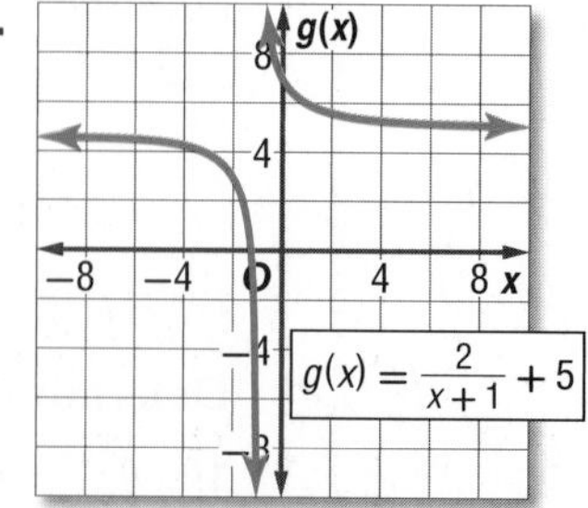

2 Transformations of Reciprocal Functions

The same techniques used to transform the graphs of other functions you have studied can be applied to the graphs of reciprocal functions. In Example 2, note that the asymptotes have been moved along with the graphs of the functions.

StudyTip

Asymptotes The asymptotes of a reciprocal function move with the graph of the function and intersect at (h, k).

KeyConcept Transformations of Reciprocal Functions

$$f(x) = \frac{a}{x - h} + k$$

h – Horizontal Translation	*k* – Vertical Translation
h units right if *h* is positive	*k* units up if *k* is positive
$\lvert h \rvert$ units left if *h* is negative	$\lvert k \rvert$ units down if *k* is negative
The *vertical* asymptote is at $x = h$.	The *horizontal* asymptote is at $f(x) = k$.

a – Orientation and Shape

If $a < 0$, the graph is reflected across the *x*-axis.

If $\lvert a \rvert > 1$, the graph is stretched vertically. If $0 < \lvert a \rvert < 1$, the graph is compressed vertically.

Example 3 Graph Transformations

Graph each function. State the domain and range.

a. $f(x) = \frac{2}{x - 4} + 2$

This represents a transformation of the graph of $f(x) = \frac{1}{x}$.

$a = 2$: The graph is stretched vertically.

$h = 4$: The graph is translated 4 units right. There is an asymptote at $x = 4$.

$k = 2$: The graph is translated 2 units up. There is an asymptote at $f(x) = 2$.

Domain: $\{x \mid x \neq 4\}$ Range: $\{f(x) \mid f(x) \neq 2\}$

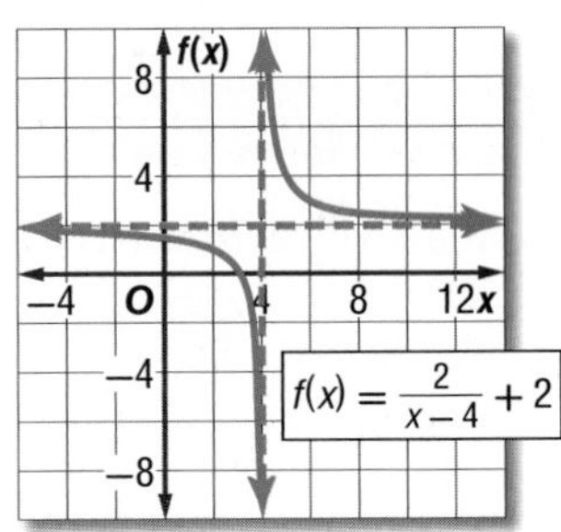

b. $f(x) = \frac{-3}{x + 1} - 4$

This represents a transformation of the graph of $f(x) = \frac{1}{x}$.

$a = -3$: The graph is stretched vertically and reflected across the *x*-axis.

$h = -1$: The graph is translated 1 unit left. There is an asymptote at $x = -1$.

$k = -4$: The graph is translated 4 units down. There is an asymptote at $f(x) = -4$.

Domain: $\{x \mid x \neq -1\}$ Range: $\{f(x) \mid f(x) \neq -4\}$

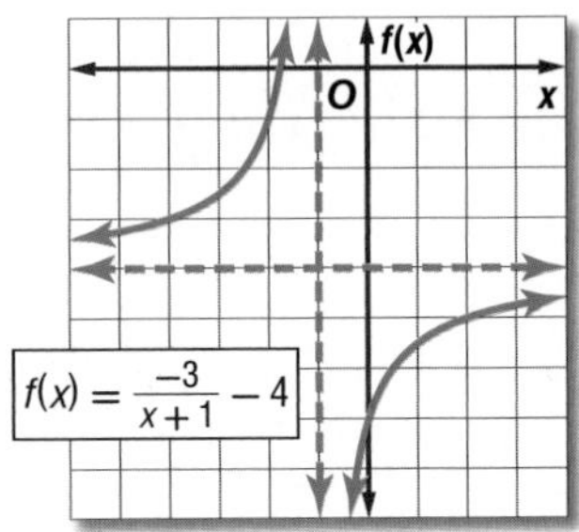

GuidedPractice

3A. $f(x) = \frac{-2}{x + 4} + 1$

3B. $g(x) = \frac{1}{3(x - 1)} - 2$

Real-World Example 4 Write Equations

TRAVEL An airline has a daily nonstop flight between Los Angeles, California, and Sydney, Australia. A one-way trip is about 7500 miles.

a. Write an equation to represent the travel time from Los Angeles to Sydney as a function of flight speed. Then graph the equation.

Solve $rt = d$ for t.

$rt = d$ Original formula

$t = \frac{d}{r}$ Divide each side by r.

$t = \frac{7500}{r}$ $d = 7500$

Graph $t = \frac{7500}{r}$.

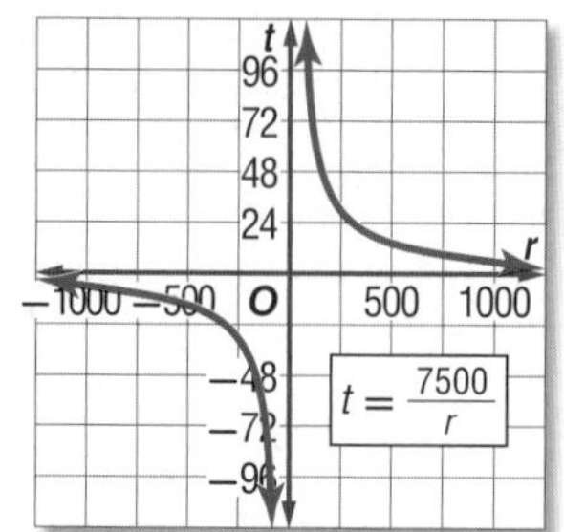

b. Explain any limitations to the range or domain in this situation.

In this situation, the range and domain are limited to all real numbers greater than zero because negative values do not make sense. There will be further restrictions to the domain because the aircraft has minimum and maximum speeds at which it can travel.

Real-World Career

Travel Agent Travel agents assess individual and business needs to help make the best possible travel arrangements. They may specialize by type of travel, such as leisure or business, or by destination, such as Europe or Africa. A high school diploma is required, and vocational training is preferred.

Guided Practice

4. **HOMECOMING DANCE** The junior and senior class officers are sponsoring a homecoming dance. The total cost for the facilities and catering is \$45 per person plus a \$2500 deposit. Write and graph an equation to represent the average cost per person. Then explain any limitations to the domain and range.

Check Your Understanding

○ = Step-by-Step Solutions begin on page R14.

Examples 1–2 **Identify the asymptotes, domain, and range of each function.**

1.

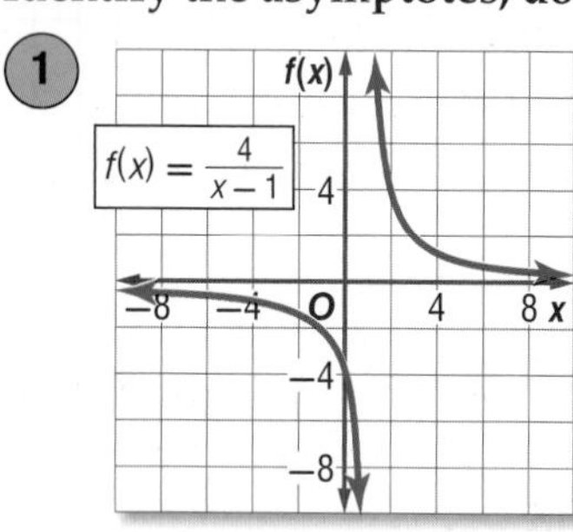

2.

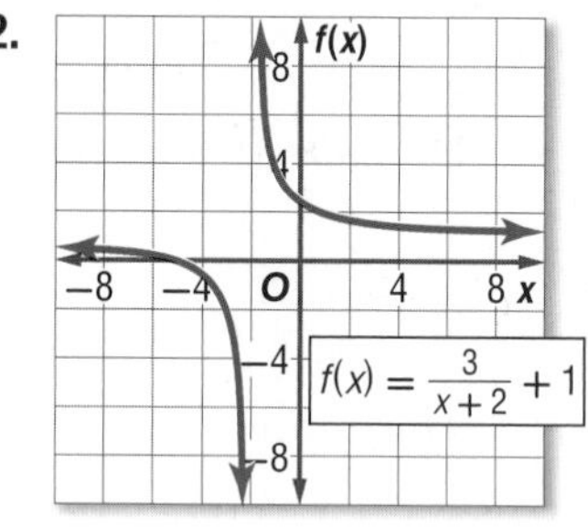

Example 3 **Graph each function. State the domain and range.**

3. $f(x) = \frac{5}{x}$

4. $f(x) = \frac{2}{x+3}$

5. $f(x) = \frac{-1}{x-2} + 4$

Example 4

6. **CCSS SENSE-MAKING** A group of friends plans to get their youth group leader a gift certificate for a day at a spa. The certificate costs \$150.

a. If c represents the cost for each friend and f represents the number of friends, write an equation to represent the cost to each friend as a function of how many friends give.

b. Graph the function.

c. Explain any limitations to the range or domain in this situation.

Practice and Problem Solving

Extra Practice is on page R8.

Examples 1–2 **Identify the asymptotes, domain, and range of each function.**

7.

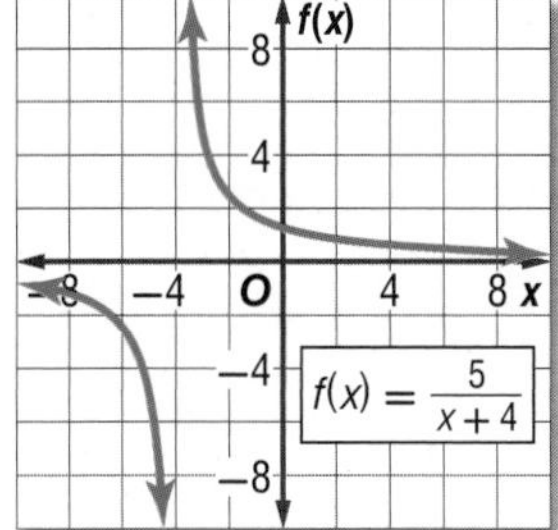

8.

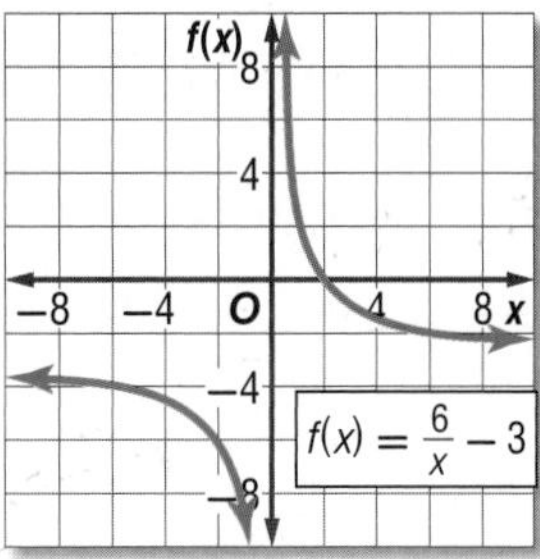

9.

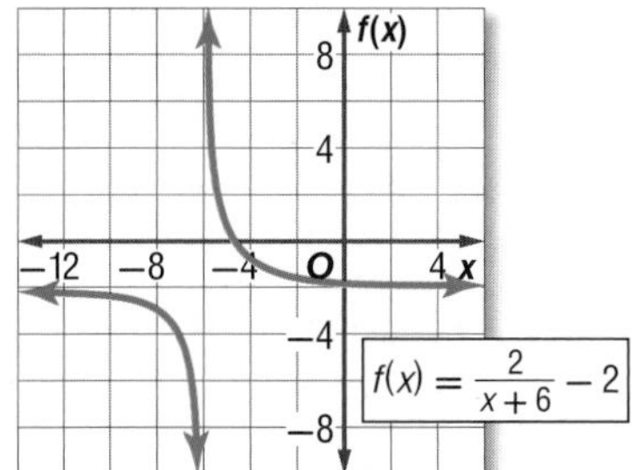

10.

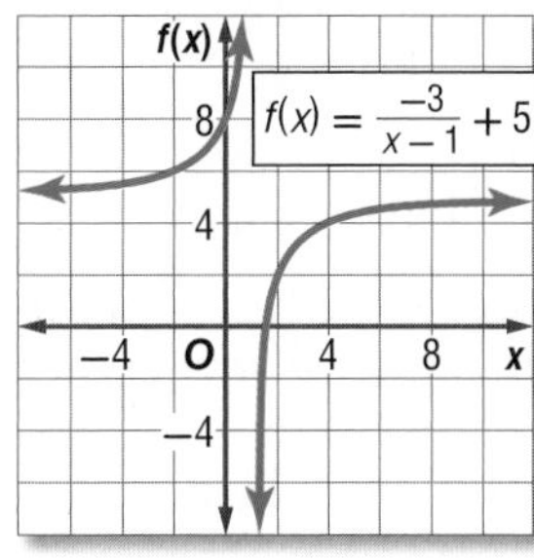

Example 3 **Graph each function. State the domain and range.**

11. $f(x) = \frac{3}{x}$

12. $f(x) = \frac{-4}{x + 2}$

13. $f(x) = \frac{2}{x - 6}$

14. $f(x) = \frac{6}{x} - 5$

15. $f(x) = \frac{2}{x} + 3$

16. $f(x) = \frac{8}{x}$

17. $f(x) = \frac{-2}{x - 5}$

18. $f(x) = \frac{3}{x - 7} - 8$

19. $f(x) = \frac{9}{x + 3} + 6$

20. $f(x) = \frac{8}{x + 3}$

21. $f(x) = \frac{-6}{x + 4} - 2$

22. $f(x) = \frac{-5}{x - 2} + 2$

Example 4

23. CYCLING Marina's New Year's resolution is to ride her bike 5000 miles.

a. If m represents the mileage Marina rides each day and d represents the number of days, write an equation to represent the mileage each day as a function of the number of days that she rides.

b. Graph the function.

c. If she rides her bike every day of the year, how many miles should she ride each day to meet her goal?

24. CCSS MODELING Parker has 200 grams of an unknown liquid. Knowing the density will help him discover what type of liquid this is.

a. Density of a liquid is found by dividing the mass by the volume. Write an equation to represent the density of this unknown as a function of volume.

b. Graph the function.

c. From the graph, identify the asymptotes, domain, and range of the function.

Graph each function. State the domain and range.

25. $f(x) = \frac{3}{2x - 4}$

26. $f(x) = \frac{5}{3x}$

27. $f(x) = \frac{2}{4x + 1}$

28. $f(x) = \frac{1}{2x + 3}$

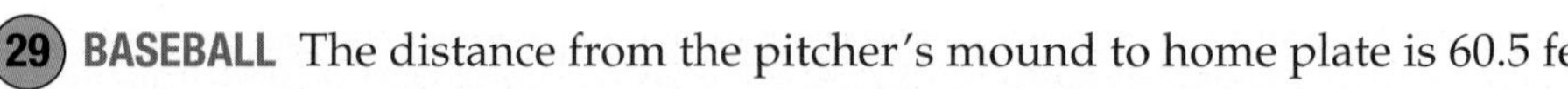

29 **BASEBALL** The distance from the pitcher's mound to home plate is 60.5 feet.

a. If r represents the speed of the pitch and t represents the time it takes the ball to get to the plate, write an equation to represent the speed as a function of time.

b. Graph the function.

c. If a two-seam fastball reaches the plate in 0.48 second, what was its speed?

Graph each function. State the domain and range, and identify the asymptotes.

30. $f(x) = \frac{-3}{x+7} - 1$

31. $f(x) = \frac{-4}{x+2} - 5$

32. $f(x) = \frac{6}{x-1} + 2$

33. $f(x) = \frac{2}{x-4} + 3$

34. $f(x) = \frac{-7}{x-8} - 9$

35. $f(x) = \frac{-6}{x-7} - 8$

36. **FINANCIAL LITERACY** Lawanda's car went 440 miles on one tank of gas.

a. If g represents the number of miles to the gallon that the car gets and t represents the size of the gas tank, write an equation to represent the miles to the gallon as a function of tank size.

b. Graph the function.

c. How many miles does the car get per gallon if it has a 15-gallon tank?

37. **MULTIPLE REPRESENTATIONS** In this problem you will investigate the similarities and differences between power functions with positive and negative exponents.

a. **TABULAR** Make a table of values for $a(x) = x^2$, $b(x) = x^{-2}$, $c(x) = x^3$, and $d(x) = x^{-3}$.

b. **GRAPHICAL** Graph $a(x)$ and $b(x)$ on the same coordinate plane.

c. **VERBAL** Compare the domain, range, end behavior, and behavior at $x = 0$ for $a(x)$ and $b(x)$.

d. **GRAPHICAL** Graph $c(x)$ and $d(x)$ on the same coordinate plane.

e. **VERBAL** Compare the domain, range, end behavior, and behavior at $x = 0$ for $c(x)$ and $d(x)$.

f. **ANALYTICAL** What conclusions can you make about the similarities and differences between power functions with positive and negative exponents?

H.O.T. Problems Use Higher-Order Thinking Skills

38. **OPEN ENDED** Write a reciprocal function for which the graph has a vertical asymptote at $x = -4$ and a horizontal asymptote at $f(x) = 6$.

39. **REASONING** Compare and contrast the graphs of each pair of equations.

a. $y = \frac{1}{x}$ and $y - 7 = \frac{1}{x}$

b. $y = \frac{1}{x}$ and $y = 4\left(\frac{1}{x}\right)$

c. $y = \frac{1}{x}$ and $y = \frac{1}{x+5}$

d. Without making a table of values, use what you observed in parts **a–c** to sketch a graph of $y - 7 = 4\left(\frac{1}{x+5}\right)$.

40. **CCSS ARGUMENTS** Find the function that does not belong. Explain.

$f(x) = \frac{3}{x+1}$

$g(x) = \frac{x+2}{x^2+1}$

$h(x) = \frac{5}{x^2+2x+1}$

$j(x) = \frac{20}{x-7}$

41. **CHALLENGE** Write two different reciprocal functions with graphs having the same vertical and horizontal asymptotes. Then graph the functions.

42. **WRITING IN MATH** Refer to the beginning of the lesson. Explain how rational functions can be used in fundraising. Explain why only part of the graph is meaningful in the context of the problem.

Standardized Test Practice

43. SHORT RESPONSE What is the value of $(x + y)(x + y)$ if $xy = -3$ and $x^2 + y^2 = 10$?

44. GRIDDED RESPONSE If $x = 2y$, $y = 4z$, $2z = w$, and $w \neq 0$, then $\frac{x}{w} =$ ___.

45. If $c = 1 + \frac{1}{d}$ and $d > 1$, then c could equal ___.

A $\frac{5}{7}$ C $\frac{15}{7}$

B $\frac{9}{7}$ D $\frac{19}{7}$

46. SAT/ACT A car travels m miles at the rate of t miles per hour. How many hours does the trip take?

F $\frac{m}{t}$ J $\frac{t}{m}$

G $m - t$ K $t - m$

H mt

47. If $-1 < a < b < 0$, then which of the following has the greatest value?

A $a - b$ C $a + b$

B $b - a$ D $2b - a$

Spiral Review

48. BUSINESS A small corporation decides that 8% of its profits will be divided among its six managers. There are two sales managers and four nonsales managers. Fifty percent will be split equally among all six managers. The other 50% will be split among the four nonsales managers. Let p represent the profits. (Lesson 8-2)

a. Write an expression to represent the share of the profits each nonsales manager will receive.

b. Simplify this expression.

c. Write an expression in simplest form to represent the share of the profits each sales manager will receive.

Simplify each expression. (Lesson 8-1)

49. $\dfrac{\frac{p^3}{2n}}{-\frac{p^2}{4n}}$ **50.** $\dfrac{\frac{m+q}{5}}{\frac{m^2+q^2}{5}}$ **51.** $\dfrac{\frac{x+y}{2x-y}}{\frac{x+y}{2x+y}}$

Graph each function. State the domain and range. (Lesson 7-1)

52. $y = 2(3)^x$ **53.** $y = 5(2)^x$ **54.** $y = 0.5(4)^x$ **55.** $y = 4\left(\frac{1}{3}\right)^x$

Find $(f + g)(x)$, $(f - g)(x)$, $(f \cdot g)(x)$, and $\left(\frac{f}{g}\right)(x)$ for each $f(x)$ and $g(x)$. (Lesson 6-1)

56. $f(x) = x + 9$
$g(x) = x - 9$

57. $f(x) = 2x - 3$
$g(x) = 4x + 9$

58. $f(x) = 2x^2$
$g(x) = 8 - x$

59. GEOMETRY The width of a rectangular prism is w centimeters. The height is 2 centimeters less than the width. The length is 4 centimeters more than the width. If the volume of the prism is 8 times the measure of the length, find the dimensions of the prism. (Lesson 5-5)

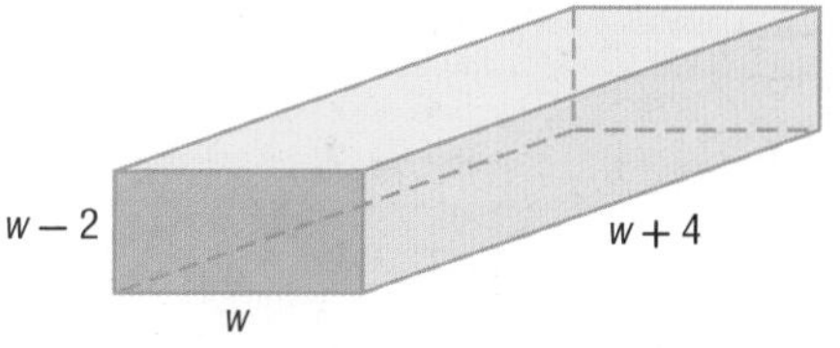

Skills Review

Graph each polynomial function. Estimate the x-coordinates at which the relative maxima and relative minima occur. State the domain and range for each function.

60. $f(x) = x^3 + 2x^2 - 3x - 5$ **61.** $f(x) = x^4 - 8x^2 + 10$

CHAPTER 8

Mid-Chapter Quiz

Lessons 8-1 through 8-3

Simplify each expression. (Lesson 8-1)

1. $\frac{2x^2y^5}{7x^3yz} \cdot \frac{14xyz^2}{18x^4y}$

2. $\frac{24a^4b^6}{35ab^3} \div \frac{12abc}{7a^2c}$

3. $\frac{3x-3}{x^2+x-2} \cdot \frac{4x+8}{6x+18}$

4. $\frac{m^2+3m+2}{9} \div \frac{m+1}{3m+15}$

5. $\frac{\frac{r^2+3r}{r+1}}{\frac{3r}{3r+3}}$

6. $\frac{\frac{2y}{y^2-4}}{\frac{3}{y^2-4y+4}}$

7. MULTIPLE CHOICE For all $r \neq \pm 2$, $\frac{r^2+6r+8}{r^2-4} =$ ___. (Lesson 8-1)

A $\frac{r-2}{r+4}$

B $\frac{r+4}{r-2}$

C $\frac{r+2}{r-4}$

D $\frac{r+4}{r+2}$

8. MULTIPLE CHOICE Identify all values of x for which $\frac{x^2-16}{(x^2-6x-27)(x+1)}$ is undefined. (Lesson 8-1)

F $-3, -1$

G $3, 1, -9$

H $-3, -1, 9$

J -1

9. What is the LCM of $x^2 - x$ and $3 - 3x$? (Lesson 8-2)

Simplify each expression. (Lesson 8-2)

10. $\frac{2x}{4x^2y} + \frac{x}{3xy^3}$

11. $\frac{3}{4m} + \frac{2}{3mn^2} - \frac{4}{n}$

12. $\frac{6}{r^2-3r-18} - \frac{1}{r^2+r-6}$

13. $\frac{3x+6}{x+y} + \frac{6}{-x-y}$

14. $\frac{x-4}{x^2-3x-4} + \frac{x+1}{2x-8}$

15. Determine the perimeter of the rectangle. (Lesson 8-2)

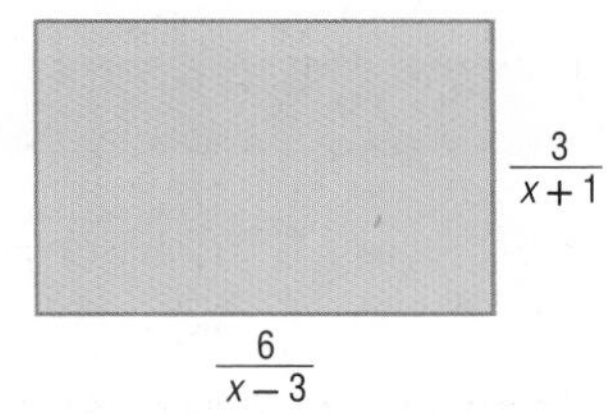

16. TRAVEL Lucita is going to a beach 100 miles away. She travels half the distance at one rate. The rest of the distance, she travels 15 miles per hour slower. (Lesson 8-2)

a. If x represents the faster pace in miles per hour, write an expression that represents the time spent at that pace.

b. Write an expression for the amount of time spent at the slower pace.

c. Write an expression for the amount of time Lucita needs to complete the trip.

Identify the asymptotes, domain, and range of each function. (Lesson 8-3)

17.

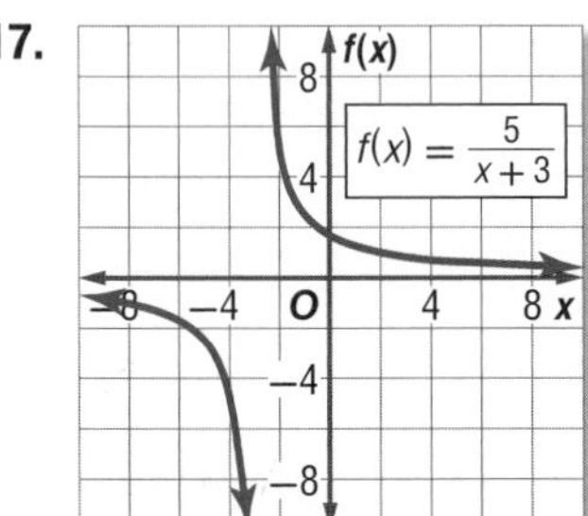

18.

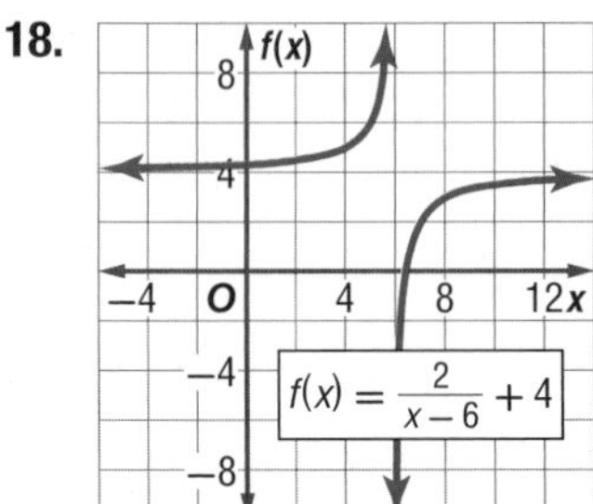

Graph each reciprocal function. State the domain and range. (Lesson 8-3)

19. $f(x) = \frac{4}{x}$

20. $f(x) = \frac{1}{3x}$

21. $f(x) = \frac{6}{x-1}$

22. $f(x) = \frac{-2}{x} + 4$

23. $f(x) = \frac{3}{x+2} - 5$

24. $f(x) = -\frac{1}{x-3} + 2$

25. SANDWICHES A group makes 45 sandwiches to take on a picnic. The number of sandwiches a person can eat depends on how many people go on the trip. (Lesson 8-3)

a. Write a function to represent this situation.

b. Graph the function.

LESSON 8-4

Graphing Rational Functions

Then	Now	Why?
You graphed reciprocal functions.	**1** Graph rational functions with vertical and horizontal asymptotes. **2** Graph rational functions with oblique asymptotes and point discontinuity.	Regina bought a digital SLR camera and a photo printer for \$350. The manufacturer claims that ink and photo paper cost \$0.47 per photo. The rational function $C(p) = \frac{0.47p + 350}{p}$ can be used to determine the average cost $C(p)$ for printing p photos.

NewVocabulary

rational function
vertical asymptote
horizontal asymptote
oblique asymptote
point discontinuity

Common Core State Standards

Content Standards

A.CED.2 Create equations in two or more variables to represent relationships between quantities; graph equations on coordinate axes with labels and scales.

F.IF.9 Compare properties of two functions each represented in a different way (algebraically, graphically, numerically in tables, or by verbal descriptions).

Mathematical Practices

7 Look for and make use of structure.

1 Vertical and Horizontal Asymptotes

A **rational function** has an equation of the form $f(x) = \frac{a(x)}{b(x)}$, where $a(x)$ and $b(x)$ are polynomial functions and $b(x) \neq 0$.

In order to graph a rational function, it is helpful to locate the zeros and asymptotes. A zero of a rational function $f(x) = \frac{a(x)}{b(x)}$ occurs at every value of x for which $a(x) = 0$.

KeyConcept Vertical and Horizontal Asymptotes

Words If $f(x) = \frac{a(x)}{b(x)}$, $a(x)$ and $b(x)$ are polynomial functions with no common factors other than 1, and $b(x) \neq 0$, then:

- $f(x)$ has a **vertical asymptote** whenever $b(x) = 0$.
- $f(x)$ has at most one **horizontal asymptote.**
  - If the degree of $a(x)$ is greater than the degree of $b(x)$, there is no horizontal asymptote.
  - If the degree of $a(x)$ is less than the degree of $b(x)$, the horizontal asymptote is the line $y = 0$.
  - If the degree of $a(x)$ equals the degree of $b(x)$, the horizontal asymptote is the line $y = \frac{\text{leading coefficient of } a(x)}{\text{leading coefficient of } b(x)}$.

Examples

No horizontal asymptote

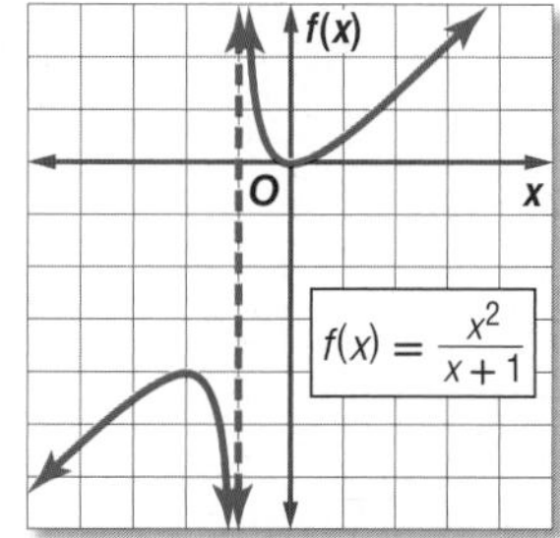

Vertical asymptote: $x = -1$

One horizontal asymptote

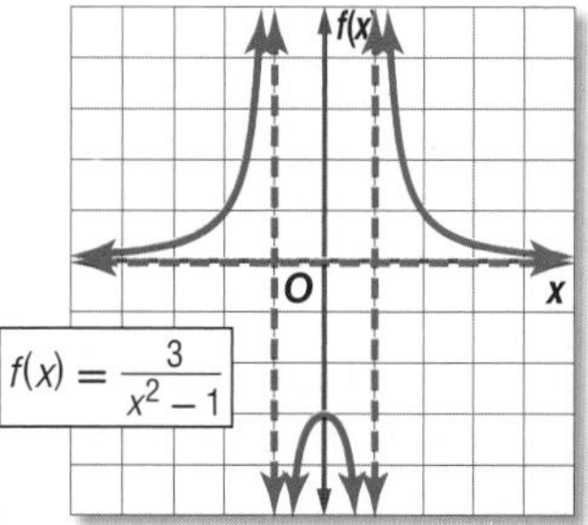

Vertical asymptotes: $x = -1, x = 1$
Horizontal asymptote: $f(x) = 0$

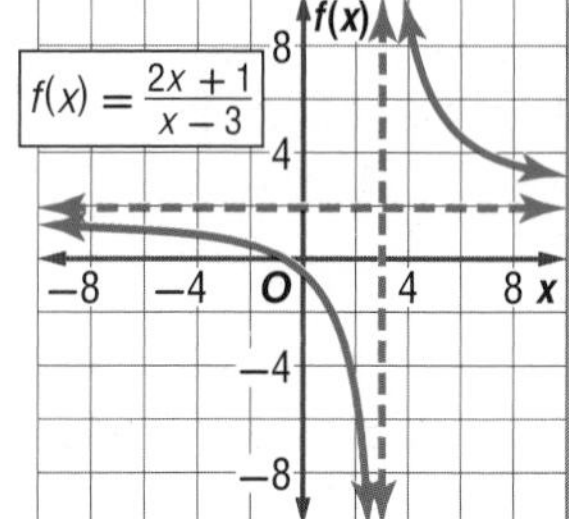

Vertical asymptote: $x = 3$
Horizontal asymptote: $f(x) = 2$

JGI/Blend Images/Getty Images

> **WatchOut!**
> **Zeros vs. Vertical Asymptotes** Zeros of rational functions occur at the values that make the numerator equal to zero. Vertical asymptotes occur at the values that make the denominator equal to zero.

The asymptotes of a rational function can be used to draw the graph of the function. Additionally, the asymptotes can be used to divide a graph into regions to find ordered pairs on the graph.

Example 1 Graph with no Horizontal Asymptote

Graph $f(x) = \frac{x^3}{x - 1}$.

Step 1 Find the zeros.

$x^3 = 0$ Set $a(x) = 0$.

$x = 0$ Take the cube root of each side.

There is a zero at $x = 0$.

Step 2 Draw the asymptotes.

Find the vertical asymptote.

$x - 1 = 0$ Set $b(x) = 0$.

$x = 1$ Add 1 to each side.

There is a vertical asymptote at $x = 1$.

The degree of the numerator is greater than the degree of the denominator. So, there is no horizontal asymptote.

> **StudyTip**
> **Graphing Calculator** The TABLE feature of a graphing calculator can be used to calculate decimal values for x and y.

Step 3 Draw the graph.

Use a table to find ordered pairs on the graph. Then connect the points.

x	$f(x)$
−3	6.75
−2	2.67
−1	0.5
0	0
0.5	−0.25
1.5	6.75
2	8
3	13.5

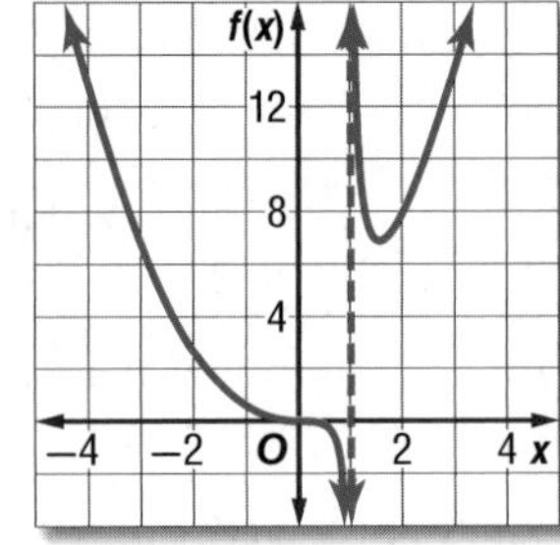

GuidedPractice

Graph each function.

1A. $f(x) = \frac{x^2 - x - 6}{x + 1}$

1B. $f(x) = \frac{(x + 1)^3}{(x + 2)^2}$

In the real world, sometimes values on the graph of a rational function are not meaningful. In the graph at the right, x-values such as time, distance, and number of people cannot be negative in the context of the problem. So, you do not even need to consider that portion of the graph.

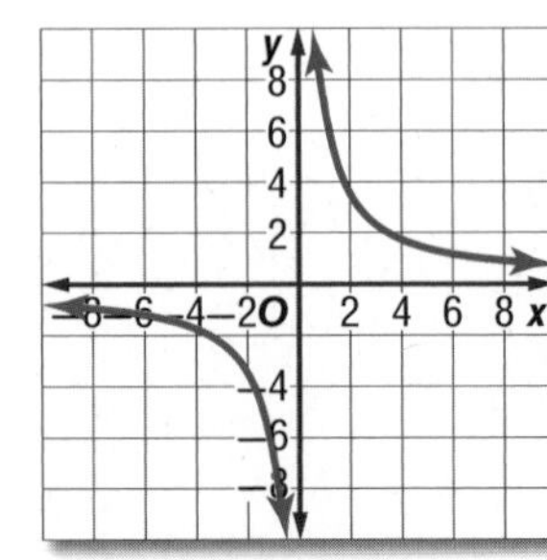

Real-World Example 2 Use Graphs of Rational Functions

AVERAGE SPEED A boat traveled upstream at r_1 miles per hour. During the return trip to its original starting point, the boat traveled at r_2 miles per hour. The average speed for the entire trip R is given by the formula $R = \frac{2r_1r_2}{r_1 + r_2}$.

a. Let r_1 be the independent variable, and let R be the dependent variable. Draw the graph if $r_2 = 10$ miles per hour.

The function is $R = \frac{2r_1(10)}{r_1 + (10)}$ or $R = \frac{20r_1}{r_1 + 10}$.

The vertical asymptote is $r_1 = -10$.

Graph the vertical asymptote and the function.

Notice that the horizontal asymptote is $R = 20$.

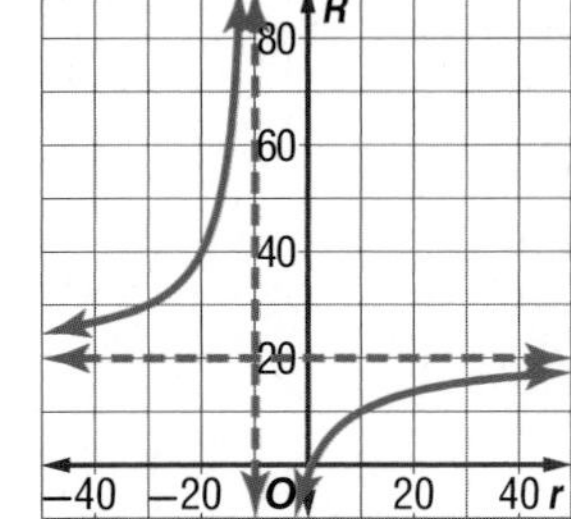

b. What is the R-intercept of the graph?

The R-intercept is 0.

c. What domain and range values are meaningful in the context of the problem?

In the problem context, speeds are nonnegative values. Therefore, only values of r_1 greater than or equal to 0 and values of R between 0 and 20 are meaningful.

Real-World Career

U.S. Coast Guard Boatswain's Mate
The most versatile member of the U.S. Coast Guard's operational team is the boatswain's mate. BMs are capable of performing almost any task. Training for BMs is accomplished through 12 weeks of intensive training.

Guided Practice

2. SALARIES A company uses the formula $S(x) = \frac{45x + 25}{x + 1}$ to determine the salary in thousands of dollars of an employee during his xth year. Graph $S(x)$. What domain and range values are meaningful in the context of the problem? What is the meaning of the horizontal asymptote for the graph?

2 Oblique Asymptotes and Point Discontinuity

An **oblique asymptote**, sometimes called a *slant asymptote*, is an asymptote that is neither horizontal nor vertical.

Key Concept Oblique Asymptotes

Words If $f(x) = \frac{a(x)}{b(x)}$, $a(x)$ and $b(x)$ are polynomial functions with no common factors other than 1 and $b(x) \neq 0$, then $f(x)$ has an oblique asymptote if the degree of $a(x)$ minus the degree of $b(x)$ equals 1. The equation of the asymptote is $f(x) = \frac{a(x)}{b(x)}$ with no remainder.

Example $f(x) = \frac{x^4 + 3x^3}{x^3 - 1}$

Vertical asymptote: $x = 1$

Oblique asymptote: $f(x) = x + 3$

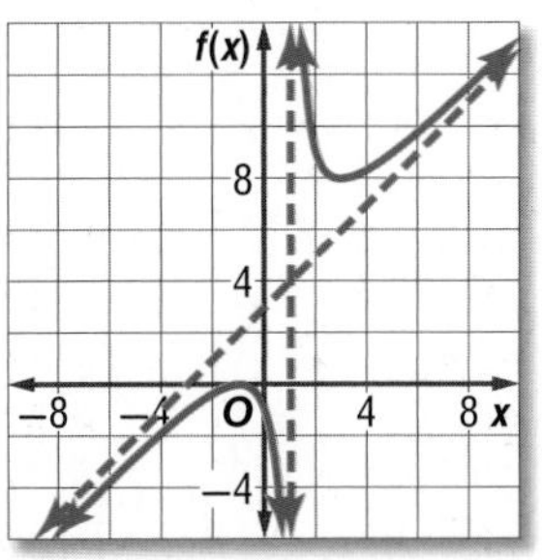

PA1 Tom Sperduto, US Coast Guard

StudyTip

Oblique Asymptotes Oblique asymptotes occur for rational functions that have a numerator polynomial that is one degree higher than the denominator polynomial.

Example 3 Determine Oblique Asymptotes

Graph $f(x) = \dfrac{x^2 + 4x + 4}{2x - 1}$.

Step 1 Find the zeros.

$x^2 + 4x + 4 = 0$	Set $a(x) = 0$.
$(x + 2)^2 = 0$	Factor.
$x + 2 = 0$	Take the square root of each side.
$x = -2$	Subtract 2 from each side.

There is a zero at $x = -2$.

Step 2 Find the asymptotes.

$2x - 1 = 0$	Set $b(x) = 0$.
$2x = 1$	Add 1 to each side.
$x = \frac{1}{2}$	Divide each side by 2.

There is a vertical asymptote at $x = \frac{1}{2}$.

The degree of the numerator is greater than the degree of the denominator, so there is no horizontal asymptote.

The difference between the degree of the numerator and the degree of the denominator is 1, so there is an oblique asymptote.

Divide the numerator by the denominator to determine the equation of the oblique asymptote.

The equation of the asymptote is the quotient excluding any remainder.

$$\begin{array}{r} \frac{1}{2}x + \frac{9}{4} \\ 2x - 1 \overline{)\, x^2 + 4x + 4} \\ (-)x^2 - \frac{1}{2}x \\ \hline \frac{9}{2}x + 4 \\ (-)\frac{9}{2}x - \frac{9}{4} \\ \hline \frac{25}{4} \end{array}$$

Thus, the oblique asymptote is the line $f(x) = \frac{1}{2}x + \frac{9}{4}$.

Step 3 Draw the asymptotes, and then use a table of values to graph the function.

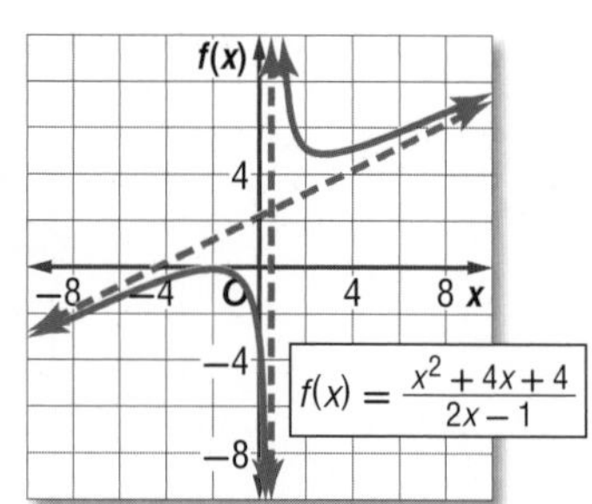

GuidedPractice

Graph each function.

3A. $f(x) = \dfrac{x^2}{x - 2}$

3B. $f(x) = \dfrac{x^3 - 1}{x^2 - 4}$

In some cases, graphs of rational functions may have **point discontinuity**, which looks like a hole in the graph. This is because the function is undefined at that point.

KeyConcept Point Discontinuity

Words If $f(x) = \frac{a(x)}{b(x)}$, $b(x) \neq 0$, and $x - c$ is a factor of both $a(x)$ and $b(x)$, then there is a point discontinuity at $x = c$.

Example $f(x) = \frac{(x+2)(x+1)}{x+1}$
$= x + 2;\ x \neq -1$

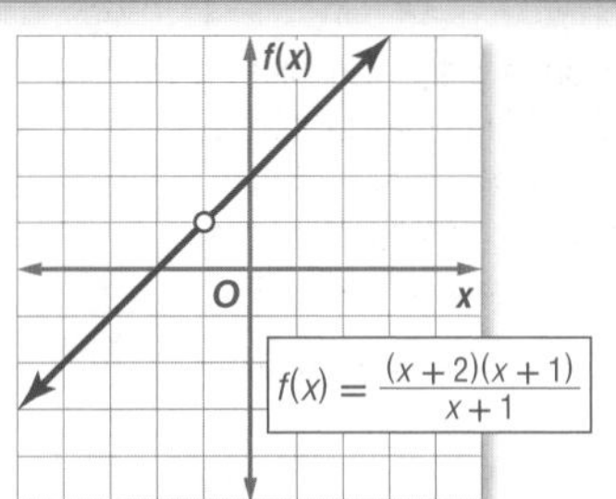

WatchOut!

Holes Remember that a common factor in the numerator and denominator can signal a hole.

Example 4 Graph with Point Discontinuity

Graph $f(x) = \frac{x^2 - 16}{x - 4}$.

Notice that $\frac{x^2 - 16}{x - 4} = \frac{(x + 4)\overset{1}{\cancel{(x - 4)}}}{\underset{1}{\cancel{x - 4}}}$ or $x + 4$.

Therefore, the graph of $f(x) = \frac{x^2 - 16}{x - 4}$ is the graph of $f(x) = x + 4$ with a hole at $x = 4$.

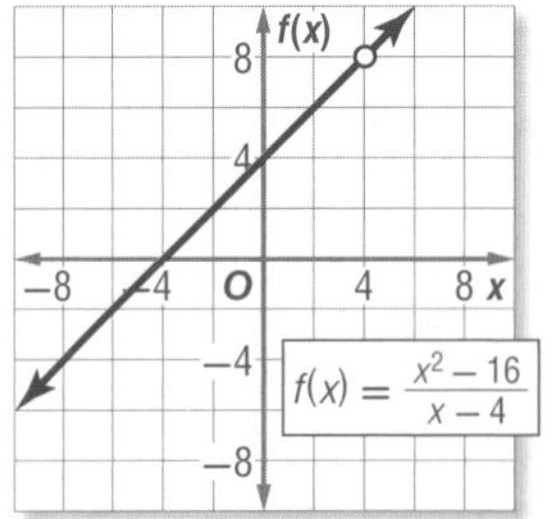

GuidedPractice

Graph each function.

4A. $f(x) = \frac{x^2 + 4x - 5}{x + 5}$

4B. $f(x) = \frac{x^3 + 2x^2 - 9x - 18}{x^2 - 9}$

Check Your Understanding

● = Step-by-Step Solutions begin on page R14.

Example 1 **Graph each function.**

1. $f(x) = \frac{x^4 - 2}{x^2 - 1}$

2. $f(x) = \frac{x^3}{x + 2}$

Example 2

3. CCSS **REASONING** Eduardo is a kicker for his high school football team. So far this season, he has made 7 out of 11 field goals. He would like to improve his field goal percentage. If he can make x consecutive field goals, his field goal percentage can be determined using the function $P(x) = \frac{7 + x}{11 + x}$.

a. Graph the function.

b. What part of the graph is meaningful in the context of this problem?

c. Describe the meaning of the intercept of the vertical axis.

d. What is the equation of the horizontal asymptote? Explain its meaning with respect to Eduardo's field goal percentage.

Examples 3–4 **Graph each function.**

4. $f(x) = \frac{6x^2 - 3x + 2}{x}$

5. $f(x) = \frac{x^2 + 8x + 20}{x + 2}$

6. $f(x) = \frac{x^2 - 4x - 5}{x + 1}$

7. $f(x) = \frac{x^2 + x - 12}{x + 4}$

Practice and Problem Solving

Extra Practice is on page R8.

Example 1 **Graph each function.**

8. $f(x) = \dfrac{x^4}{6x + 12}$

9. $f(x) = \dfrac{x^3}{8x - 4}$

10. $f(x) = \dfrac{x^4 - 16}{x^2 - 1}$

11. $f(x) = \dfrac{x^3 + 64}{16x - 24}$

Example 2

12. SCHOOL SPIRIT As president of Student Council, Brandy is getting T-shirts made for a pep rally. Each T-shirt costs \$9.50, and there is a set-up fee of \$75. The student council plans to sell the shirts, but each of the 15 council members will get one for free.

a. Write a function for the average cost of a T-shirt to be sold. Graph the function.

b. What is the average cost if 200 shirts are ordered? if 500 shirts are ordered?

c. How many T-shirts must be ordered to bring the average cost under \$9.75?

Examples 2–3 **Graph each function.**

13. $f(x) = \dfrac{x}{x + 2}$

14. $f(x) = \dfrac{5}{(x - 1)(x + 4)}$

15. $f(x) = \dfrac{4}{(x - 2)^2}$

16. $f(x) = \dfrac{x - 3}{x + 1}$

17. $f(x) = \dfrac{1}{(x + 4)^2}$

18. $f(x) = \dfrac{2x}{(x + 2)(x - 5)}$

19. $f(x) = \dfrac{(x - 4)^2}{x + 2}$

20. $f(x) = \dfrac{(x + 3)^2}{x - 5}$

21. $f(x) = \dfrac{x^3 + 1}{x^2 - 4}$

22. $f(x) = \dfrac{4x^3}{2x^2 + x - 1}$

23. $f(x) = \dfrac{3x^2 + 8}{2x - 1}$

24. $f(x) = \dfrac{2x^2 + 5}{3x + 4}$

25. $f(x) = \dfrac{x^4 - 2x^2 + 1}{x^3 + 2}$

26. $f(x) = \dfrac{x^4 - x^2 - 12}{x^3 - 6}$

27. CCSS PERSEVERANCE The current in amperes in an electrical circuit with three resistors in a series is given by the equation $I = \dfrac{V}{R_1 + R_2 + R_3}$, where V is the voltage in volts in the circuit and R_1, R_2, and R_3 are the resistances in ohms of the three resistors.

a. Let R_1 be the independent variable, and let I be the dependent variable. Graph the function if $V = 120$ volts, $R_2 = 25$ ohms, and $R_3 = 75$ ohms.

b. Give the equation of the vertical asymptote and the R_1- and I-intercepts of the graph.

c. Find the value of I when the value of R_1 is 140 ohms.

d. What domain and range values are meaningful in the context of the problem?

Example 4 **Graph each function.**

28. $f(x) = \dfrac{x^2 - 2x - 8}{x - 4}$

29. $f(x) = \dfrac{x^2 + 4x - 12}{x - 2}$

30. $f(x) = \dfrac{x^2 - 25}{x + 5}$

31. $f(x) = \dfrac{x^2 - 64}{x - 8}$

32. $f(x) = \dfrac{(x - 4)(x^2 - 4)}{x^2 - 6x + 8}$

33. $f(x) = \dfrac{(x + 5)(x^2 + 2x - 3)}{x^2 + 8x + 15}$

34. $f(x) = \dfrac{3x^4 + 6x^3 + 3x^2}{x^2 + 2x + 1}$

35. $f(x) = \dfrac{2x^4 + 10x^3 + 12x^2}{x^2 + 5x + 6}$

36. BUSINESS Liam purchased a snow plow for \$4500 and plows the parking lots of local businesses. Each time he plows a parking lot, he incurs a cost of \$50 for gas and maintenance.

a. Write and graph the rational function representing his average cost per customer as a function of the number of parking lots.

b. What are the asymptotes of the graph?

c. Why is the first quadrant in the graph the only relevant quadrant?

d. How many total parking lots does Liam need to plow for his average cost per parking lot to be less than \$80?

37. FINANCIAL LITERACY Kristina bought a new cell phone with Internet access. The phone cost \$150, and her monthly usage charge is \$30 plus \$10 for the Internet access.

a. Write and graph the rational function representing her average monthly cost as a function of the number of months Kristina uses the phone.

b. What are the asymptotes of the graph?

c. Why is the first quadrant in the graph the only relevant quadrant?

d. After how many months will the average monthly charge be \$45?

38. CCSS SENSE-MAKING Alana plays softball for Centerville High School. So far this season she has gotten a hit 4 out of 12 times at bat. She is determined to improve her batting average. If she can get x consecutive hits, her batting average can be determined using $B(x) = \frac{4 + x}{12 + x}$.

a. Graph the function.

b. What part of the graph is meaningful in the context of the problem?

c. Describe the meaning of the intercept of the vertical axis.

d. What is the equation of the horizontal asymptote? Explain its meaning with respect to Alana's batting average.

Graph each function.

39. $f(x) = \frac{x + 1}{x^2 + 6x + 5}$ **40.** $f(x) = \frac{x^2 - 10x - 24}{x + 2}$ **41.** $f(x) = \frac{6x^2 + 4x + 2}{x + 2}$

H.O.T. Problems Use Higher-Order Thinking Skills

42. OPEN ENDED Sketch the graph of a rational function with a horizontal asymptote $y = 1$ and a vertical asymptote $x = -2$.

43. CHALLENGE Compare and contrast $g(x) = \frac{x^2 - 1}{x(x^2 - 2)}$ and $f(x)$ shown at the right.

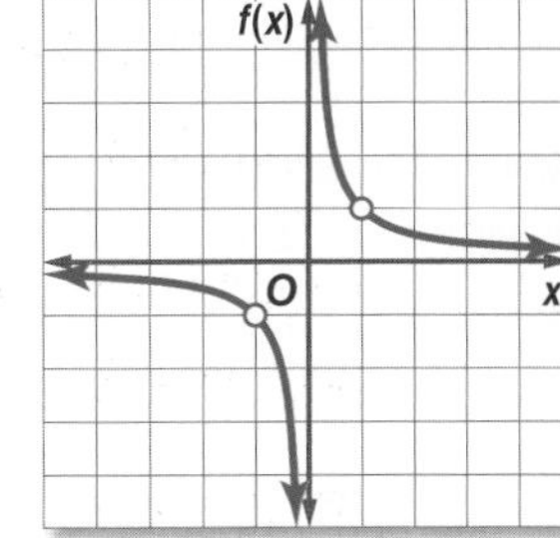

44. REASONING What is the difference between the graphs of $f(x) = x - 2$ and $g(x) = \frac{(x + 3)(x - 2)}{x + 3}$?

45. PROOF A rational function has an equation of the form $f(x) = \frac{a(x)}{b(x)}$, where $a(x)$ and $b(x)$ are polynomial functions and $b(x) \neq 0$. Show that $f(x) = \frac{x}{a - b} + c$ is a rational function.

46. WRITING IN MATH How can factoring be used to determine the vertical asymptotes or point discontinuity of a rational function?

Standardized Test Practice

47. PROBABILITY Of the 6 courses offered by the music department at her school, Kaila must choose exactly 2 of them. How many different combinations of 2 courses are possible for Kaila if there are no restrictions on which 2 courses she can choose?

A 48
B 18
C 15
D 12

48. The projected sales of a game cartridge is given by the function $S(p) = \frac{3000}{2p + a}$, where $S(p)$ is the number of cartridges sold, in thousands, p is the price per cartridge, in dollars, and a is a constant. If 100,000 cartridges are sold at \$10 per cartridge, how many cartridges will be sold at \$20 per cartridge?

F 20,000
G 50,000
H 60,000
J 150,000

49. GRIDDED RESPONSE Five distinct points lie in a plane such that 3 of the points are on line ℓ and 3 of the points are on a different line m. What is the total number of lines that can be drawn so that each line passes through exactly 2 of these 5 points?

50. GEOMETRY In the figure below, what is the value of $w + x + y + z$?

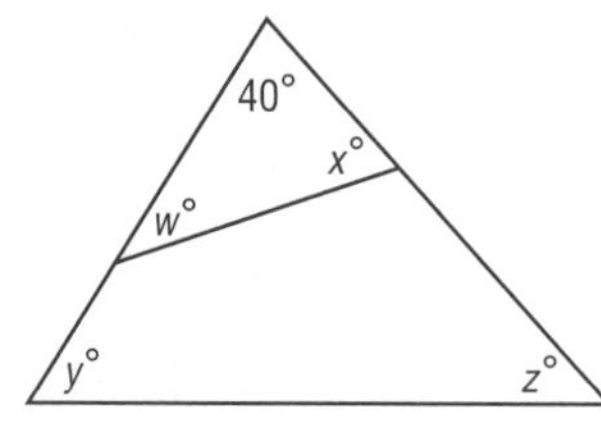

A 140
B 280
C 320
D 360

Spiral Review

Graph each function. State the domain and range. (Lesson 8-3)

51. $f(x) = \frac{-5}{x + 2}$

52. $f(x) = \frac{4}{x - 1} - 3$

53. $f(x) = \frac{1}{x + 6} + 1$

Simplify each expression. (Lesson 8-2)

54. $\frac{m}{m^2 - 4} + \frac{2}{3m + 6}$

55. $\frac{y}{y + 3} - \frac{6y}{y^2 - 9}$

56. $\frac{5}{x^2 - 3x - 28} + \frac{7}{2x - 14}$

57. $\frac{d - 4}{d^2 + 2d - 8} - \frac{d + 2}{d^2 - 16}$

Simplify each expression. (Lesson 6-6)

58. $y^{\frac{5}{3}} \cdot y^{\frac{7}{3}}$

59. $x^{\frac{3}{4}} \cdot x^{\frac{9}{4}}$

60. $\left(b^{\frac{1}{3}}\right)^{\frac{3}{5}}$

61. $\left(a^{-\frac{2}{3}}\right)^{-\frac{1}{6}}$

Skills Review

62. TRAVEL Mr. and Mrs. Wells are taking their daughter to college. The table shows their distances from home after various amounts of time.

a. Find the average rate of change in their distances from home between 1 and 3 hours after leaving home.

b. Find the average rate of change in their distances from home between 0 and 5 hours after leaving home.

Time (h)	Distance (mi)
0	0
1	55
2	110
3	165
4	165
5	225

EXTEND 8-4

Graphing Technology Lab
Graphing Rational Functions

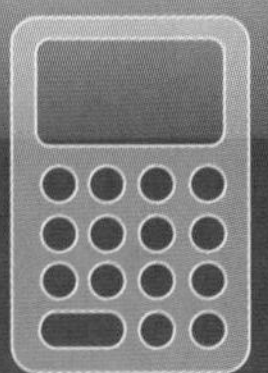

A TI-83/84 Plus graphing calculator can be used to explore graphs of rational functions. These graphs have some features that never appear in the graphs of polynomial functions.

Activity 1 Graph with Asymptotes

Graph $y = \frac{8x - 5}{2x}$ in the standard viewing window. Find the equations of any asymptotes. State the domain and range of the function.

Step 1 Enter the equation in the **Y=** list, and then graph.

KEYSTROKES: [Y=] [(] 8 [X,T,θ,n] [−] 5 [)] [÷] [(] 2 [X,T,θ,n] [)] [ZOOM] 6

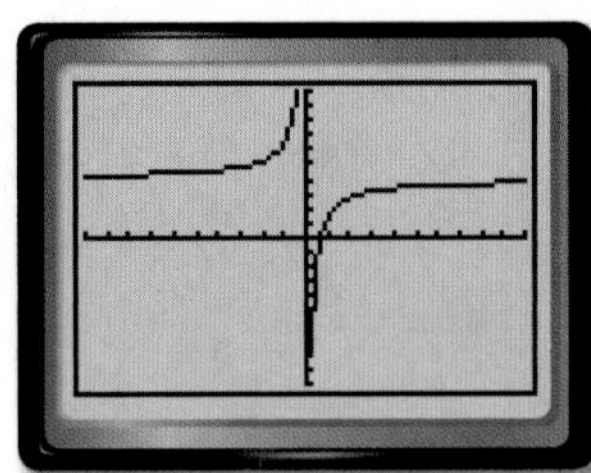

[−10, 10] scl: 1 by [−10, 10] scl: 1

Step 2 Examine the graph.

By looking at the equation, we can determine that if $x = 0$, the function is undefined. The equation of the vertical asymptote is $x = 0$. Notice what happens to the y-values as x grows larger and as x gets smaller. The y-values approach 4. So, the equation for the horizontal asymptote is $y = 4$. The domain is $\{x \mid x \neq 0\}$, and the range is all real numbers.

Activity 2 Graph with Point Discontinuity

Graph $y = \frac{x^2 - 16}{x + 4}$ in the window [−5, 4.4] by [−10, 2] with scale factors of 1.

Step 1 Because the function is not continuous, put the calculator in dot mode.

KEYSTROKES: [MODE] [▼] [▼] [▼] [▼] [▶] [ENTER]

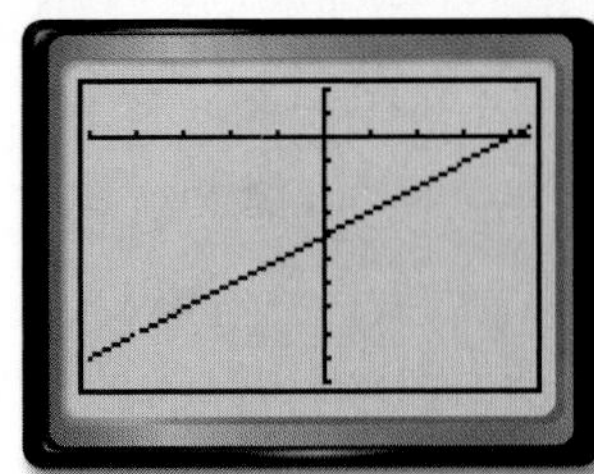

[−5, 4.4] scl: 1 by [−20, 2] scl: 1

Step 2 Examine the graph.

This graph looks like a line with a break in continuity at $x = -4$. This happens because the denominator is 0 when $x = -4$. Therefore, the function is undefined when $x = -4$.

If you **TRACE** along the graph, when you come to $x = -4$, you will see that there is no corresponding y-value.

Exercises

Use a graphing calculator to graph each function. Write the x-coordinates of any points of discontinuity and/or the equations of any asymptotes. State the domain and range.

1. $f(x) = \frac{1}{x}$
2. $f(x) = \frac{x}{x + 2}$
3. $f(x) = \frac{2}{x - 4}$
4. $f(x) = \frac{2x}{3x - 6}$
5. $f(x) = \frac{4x + 2}{x - 1}$
6. $f(x) = \frac{x^2 - 9}{x + 3}$

LESSON 8-5 Variation Functions

Then

- You wrote and graphed linear equations.

Now

1. Recognize and solve direct and joint variation problems.
2. Recognize and solve inverse and combined variation problems.

Why?

- While building skateboard ramps, Yu determined that the best ramps were the ones in which the length of the top of the ramp was 1.5 times as long as the height of the ramp.

As shown in the table, the length of the top of the ramp depends on the height of a ramp. The length increases as the height increases, but the ratio remains the same, or is *constant.*

The equation $\frac{\ell}{h} = 1.5$ can be written as $\ell = 1.5h$.

The length *varies directly* with the height of the ramp.

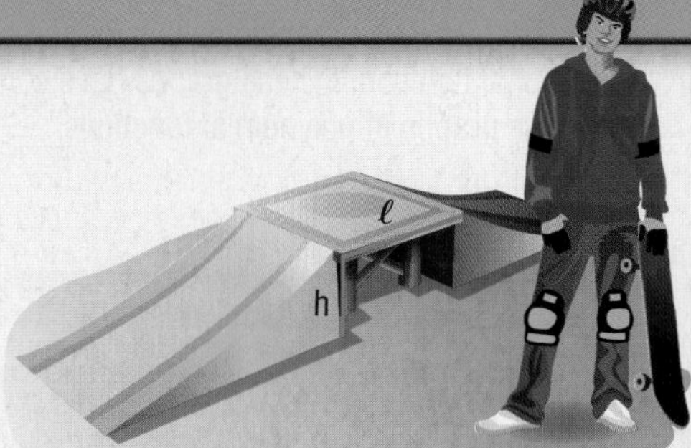

Length (ℓ)	Height (h)	Ratio $\frac{\ell}{h}$
3	2	1.5
6	4	1.5
9	6	1.5
12	8	1.5

NewVocabulary
direct variation
constant of variation
joint variation
inverse variation
combined variation

Common Core State Standards

Content Standards
A.CED.2 Create equations in two or more variables to represent relationships between quantities; graph equations on coordinate axes with labels and scales.

Mathematical Practices
1 Make sense of problems and persevere in solving them.
4 Model with mathematics.

1 Direct Variation and Joint Variation

The relationship given by $\ell = 1.5h$ is an example of direct variation. A **direct variation** can be expressed in the form $y = kx$. In this equation, k is called the **constant of variation**.

Notice that the graph of $\ell = 1.5h$ is a straight line through the origin. A direct variation is a special case of an equation written in slope-intercept form, $y = mx + b$. When $m = k$ and $b = 0$, $y = mx + b$ becomes $y = kx$. So the slope of a direct variation equation is its constant of variation.

To express a direct variation, we say that y varies directly as x. In other words, as x increases, y increases or decreases at a constant rate.

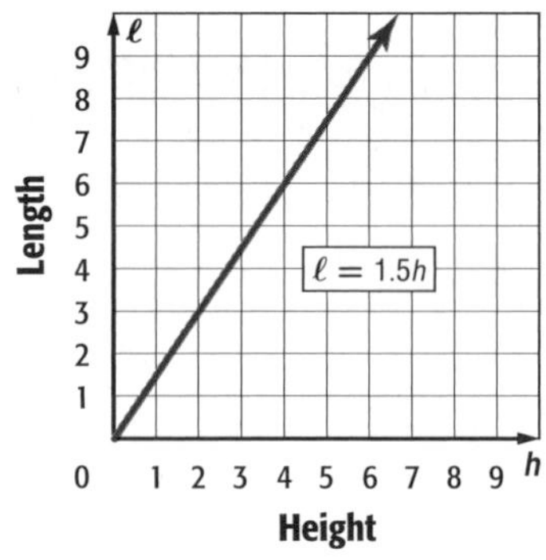

KeyConcept Direct Variation

Words	y varies directly as x if there is some nonzero constant k such that $y = kx$. k is called the *constant of variation.*
Example	If $y = 3x$ and $x = 7$, then $y = 3(7)$ or 21.

If you know that y varies directly as x and one set of values, you can use a proportion to find the other set of corresponding values.

$$y_1 = kx_1 \quad \text{and} \quad y_2 = kx_2$$

$$\frac{y_1}{x_1} = k \qquad \frac{y_2}{x_2} = k \qquad \text{Therefore, } \frac{y_1}{x_1} = \frac{y_2}{x_2}.$$

Using the properties of equality, you can find many other proportions that relate these same x- and y-values.

PT

Example 1 Direct Variation

If y varies directly as x and $y = 15$ when $x = -5$, find y when $x = 7$.

Use a proportion that relates the values.

$\frac{y_1}{x_1} = \frac{y_2}{x_2}$ Direct variation

$\frac{15}{-5} = \frac{y_2}{7}$ $y_1 = 15$, $x_1 = -5$, and $x_2 = 7$

$15(7) = -5(y_2)$ Cross multiply.

$105 = -5y_2$ Simplify.

$-21 = y_2$ Divide each side by -5.

GuidedPractice

1. If r varies directly as t and $r = -20$ when $t = 4$, find r when $t = -6$.

Another type of variation is joint variation. **Joint variation** occurs when one quantity varies directly as the product of two or more other quantities.

StudyTip

Joint Variation Some mathematicians consider joint variation a special type of combined variation.

KeyConcept Joint Variation

Words	y varies jointly as x and z if there is some nonzero constant k such that $y = kxz$.
Example	If $y = 5xz$, $x = 6$, and $z = -2$, then $y = 5(6)(-2)$ or -60.

If you know that y varies jointly as x and z and one set of values, you can use a proportion to find the other set of corresponding values.

$y_1 = kx_1z_1$ and $y_2 = kx_2z_2$

$\frac{y_1}{x_1z_1} = k$ $\frac{y_2}{x_2z_2} = k$ Therefore, $\frac{y_1}{x_1z_1} = \frac{y_2}{x_2z_2}$.

Example 2 Joint Variation

Suppose y varies jointly as x and z. Find y when $x = 9$ and $z = 2$, if $y = 20$ when $z = 3$ and $x = 5$.

Use a proportion that relates the values.

$\frac{y_1}{x_1z_1} = \frac{y_2}{x_2z_2}$ Joint variation

$\frac{20}{5(3)} = \frac{y_2}{9(2)}$ $y_1 = 20$, $x_1 = 5$, $z_1 = 3$, $x_2 = 9$, and $z_2 = 2$

$20(9)(2) = 5(3)(y_2)$ Cross multiply.

$360 = 15y_2$ Simplify.

$24 = y_2$ Divide each side by 15.

GuidedPractice

2. Suppose r varies jointly as v and t. Find r when $v = 2$ and $t = 8$, if $r = 70$ when $v = 10$ and $t = 4$.

2 Inverse Variation and Combined Variation

Another type of variation is inverse variation. If two quantities x and y show **inverse variation**, their product is equal to a constant k.

Inverse variation is often described as one quantity increasing while the other quantity is decreasing. For example, speed and time for a fixed distance vary inversely with each other; the faster you go, the less time it takes you to get there.

KeyConcept Inverse Variation

Words	y varies inversely as x if there is some nonzero constant k such that $xy = k$ or $y = \frac{k}{x}$, where $x \neq 0$ and $y \neq 0$.
Example	If $xy = 2$, and $x = 6$, then $y = \frac{2}{6}$ or $\frac{1}{3}$.

StudyTip

Direct and Inverse Variation You can identify the type of variation by looking at a table of values for x and y. If the quotient $\frac{y}{x}$ has a constant value, y varies directly as x. If the product xy has a constant value, y varies inversely as x.

Suppose y varies inversely as x such that $xy = 6$ or $y = \frac{6}{x}$. The graph of this equation is shown at the right. Since k is a positive value, as the values of x increase, the values of y decrease.

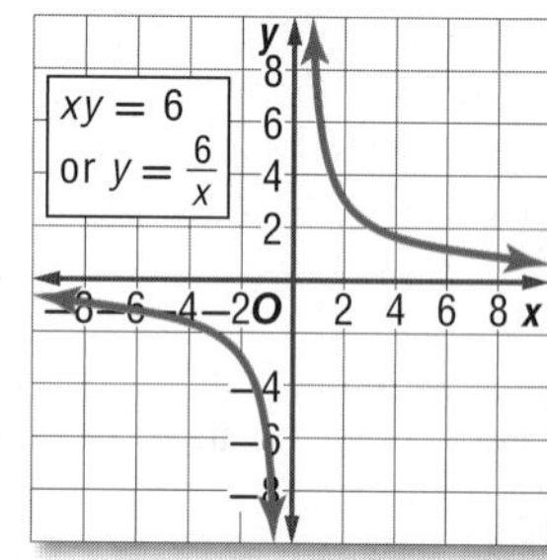

Notice that the graph of an inverse variation is a reciprocal function.

A proportion can be used with inverse variation to solve problems in which some quantities are known. The following proportion is only one of several that can be formed.

$$x_1y_1 = k \text{ and } x_2y_2 = k$$

$$x_1y_1 = x_2y_2 \quad \text{Substitution Property of Equality}$$

$$\frac{x_1}{y_2} = \frac{x_2}{y_1} \quad \text{Divide each side by } y_1y_2.$$

Example 3 Inverse Variation

If a varies inversely as b and $a = 28$ when $b = -2$, find a when $b = -10$.

Use a proportion that relates the values.

$\frac{a_1}{b_2} = \frac{a_2}{b_1}$ — Inverse Variation

$\frac{28}{-10} = \frac{a_2}{-2}$ — $a_1 = 28$, $b_1 = -2$, and $b_2 = -10$

$28(-2) = -10(a_2)$ — Cross multiply.

$-56 = -10(a_2)$ — Simplify.

$5\frac{3}{5} = a_2$ — Divide each side by -10.

GuidedPractice

3. If x varies inversely as y and $x = 24$ when $y = 4$, find x when $y = 12$.

Inverse variation is often used in real-world situations.

Real-World Example 4 Write and Solve an Inverse Variation

MUSIC The length of a violin string varies inversely as the frequency of its vibrations. A violin string 10 inches long vibrates at a frequency of 512 cycles per second. Find the frequency of an 8-inch violin string.

Let $v_1 = 10, f_1 = 512$, and $v_2 = 8$. Solve for f_2.

$v_1 f_1 = v_2 f_2$ — Original equation

$10 \cdot 512 = 8 \cdot f_2$ — $v_1 = 10$, $f_1 = 512$, and $v_2 = 8$

$\frac{5120}{8} = f_2$ — Divide each side by 8.

$640 = f_2$ — Simplify.

The 8-inch violin string vibrates at a frequency of 640 cycles per second.

Real-WorldLink

When you pluck a string, it vibrates back and forth. This causes mechanical energy to travel through the air in waves. The number of times per second these waves hit our ear is called the *frequency*. The more waves per second, the higher the pitch.

GuidedPractice

4. The apparent length of an object is inversely proportional to one's distance from the object. Earth is about 93 million miles from the Sun. Jupiter is about 483.6 million miles from the Sun. Find how many times as large the diameter of the Sun would appear on Earth as on Jupiter.

Another type of variation is combined variation. **Combined variation** occurs when one quantity varies directly and/or inversely as two or more other quantities.

If you know that y varies directly as x, y varies inversely as z, and one set of values, you can use a proportion to find the other set of corresponding values.

$$y_1 = \frac{kx_1}{z_1} \quad \text{and} \quad y_2 = \frac{kx_2}{z_2}$$

$$\frac{y_1 z_1}{x_1} = k \qquad \frac{y_2 z_2}{x_2} = k \qquad \text{Therefore, } \frac{y_1 z_1}{x_1} = \frac{y_2 z_2}{x_2}.$$

Example 5 Combined Variation

StudyTip

Combined Variation Quantities that vary directly appear in the numerator. Quantities that vary inversely appear in the denominator.

Suppose f varies directly as g, and f varies inversely as h. Find g when $f = 18$ and $h = -3$, if $g = 24$ when $h = 2$ and $f = 6$.

First set up a correct proportion for the information given.

$f_1 = \frac{kg_1}{h_1}$ and $f_2 = \frac{kg_2}{h_2}$ — g varies directly as f, so g goes in the numerator. h varies inversely as f, so h goes in the denominator.

$k = \frac{f_1 h_1}{g_1}$ and $k = \frac{f_2 h_2}{g_2}$ — Solve for k.

$\frac{f_1 h_1}{g_1} = \frac{f_2 h_2}{g_2}$ — Set the two proportions equal to each other.

$\frac{6(2)}{24} = \frac{18(-3)}{g_2}$ — $f_1 = 6$, $g_1 = 24$, $h_1 = 2$, $f_2 = 18$, and $h_2 = -3$

$24(18)(-3) = 6(2)(g_2)$ — Cross multiply.

$-1296 = 12g_2$ — Simplify.

$-108 = g_2$ — Divide each side by 12.

When $f = 18$ and $h = -3$, the value of g is -108.

GuidedPractice

5. Suppose p varies directly as r, and p varies inversely as t. Find t when $r = 10$ and $p = -5$, if $t = 20$ when $p = 4$ and $r = 2$.

Check Your Understanding

= Step-by-Step Solutions begin on page R14.

Examples 1–3

1. If y varies directly as x and $y = 12$ when $x = 8$, find y when $x = 14$.

2. Suppose y varies jointly as x and z. Find y when $x = 9$ and $z = -3$, if $y = -50$ when z is 5 and x is -10.

3. If y varies inversely as x and $y = -18$ when $x = 16$, find x when $y = 9$.

Example 4

4. **TRAVEL** A map of Illinois is scaled so that 2 inches represents 15 miles. How far apart are Chicago and Rockford if they are 12 inches apart on the map?

Example 5

5. Suppose a varies directly as b, and a varies inversely as c. Find b when $a = 8$ and $c = -3$, if $b = 16$ when $c = 2$ and $a = 4$.

6. Suppose d varies directly as f, and d varies inversely as g. Find g when $d = 6$ and $f = -7$, if $g = 12$ when $d = 9$ and $f = 3$.

Practice and Problem Solving

Extra Practice is on page R8.

Example 1

If x varies directly as y, find x when $y = 8$.

7. $x = 6$ when $y = 32$
8. $x = 11$ when $y = -3$
9. $x = 14$ when $y = -2$
10. $x = -4$ when $y = 10$

11. **MOON** Astronaut Neil Armstrong, the first man on the Moon, weighed 360 pounds on Earth with all his equipment on, but weighed only 60 pounds on the Moon. Write an equation that relates weight on the Moon m with weight on Earth w.

Example 2

If a varies jointly as b and c, find a when $b = 4$ and $c = -3$.

12. $a = -96$ when $b = 3$ and $c = -8$
13. $a = -60$ when $b = -5$ and $c = 4$
14. $a = -108$ when $b = 2$ and $c = 9$
15. $a = 24$ when $b = 8$ and $c = 12$

16. CCSS **MODELING** According to the A.C. Nielsen Company, the average American watches 4 hours of television a day.
    a. Write an equation to represent the average number of hours spent watching television by m household members during a period of d days.
    b. Assume that members of your household watch the same amount of television each day as the average American. How many hours of television would the members of your household watch in a week?

Example 3

If f varies inversely as g, find f when $g = -6$.

17. $f = 15$ when $g = 9$
18. $f = 4$ when $g = 28$
19. $f = -12$ when $g = 19$
20. $f = 0.6$ when $g = -21$

21. **COMMUNITY SERVICE** Every year students at West High School collect canned goods for a local food pantry. They plan to distribute flyers to homes in the community asking for donations. Last year, 12 students were able to distribute 1000 flyers in four hours.
    a. Write an equation that relates the number of students s to the amount of time t it takes to distribute 1000 flyers.
    b. How long would it take 15 students to hand out the same number of flyers this year?

Example 4

22. **BIRDS** When a group of snow geese migrate, the distance that they fly varies directly with the amount of time they are in the air.

   a. A group of snow geese migrated 375 miles in 7.5 hours. Write a direct variation equation that represents this situation.

   b. Every year, geese migrate 3000 miles from their winter home in the southwest United States to their summer home in the Canadian Arctic. Estimate the number of hours of flying time that it takes for the geese to migrate.

Example 5

23. Suppose a varies directly as b, and a varies inversely as c. Find b when $a = 5$ and $c = -4$, if $b = 12$ when $c = 3$ and $a = 8$.

24. Suppose x varies directly as y, and x varies inversely as z. Find z when $x = 10$ and $y = -7$, if $z = 20$ when $x = 6$ and $y = 14$.

Determine whether each relation shows *direct* or *inverse* variation, or *neither*.

25.

x	y
4	12
8	24
16	48
32	96

26.

x	y
8	2
4	4
−2	−8
−8	−2

27.

x	y
2	4
3	9
4	16
5	25

28. If y varies inversely as x and $y = 6$ when $x = 19$, find y when $x = 2$.

29. If x varies inversely as y and $x = 16$ when $y = 5$, find x when $y = 20$.

30. Suppose a varies directly as b, and a varies inversely as c. Find b when $a = 7$ and $c = -8$, if $b = 15$ when $c = 2$ and $a = 4$.

31. Suppose x varies directly as y, and x varies inversely as z. Find z when $x = 8$ and $y = -6$, if $z = 26$ when $x = 8$ and $y = 13$.

State whether each equation represents a *direct, joint, inverse,* or *combined* variation. Then name the constant of variation.

32. $\frac{x}{y} = 2.75$

33. $fg = -2$

34. $a = 3bc$

35. $10 = \frac{xy^2}{z}$

36. $y = -11x$

37. $\frac{n}{p} = 4$

38. $9n = pr$

39. $-2y = z$

40. $a = 27b$

41. $c = \frac{7}{d}$

42. $-10 = gh$

43. $m = 20cd$

44. **CCSS PRECISION** The volume of a gas v varies inversely as the pressure p and directly as the temperature t.

   a. Write an equation to represent the volume of a gas in terms of pressure and temperature. Is your equation a *direct, joint, inverse,* or *combined* variation?

   b. A certain gas has a volume of 8 liters, a temperature of 275 Kelvin, and a pressure of 1.25 atmospheres. If the gas is compressed to a volume of 6 liters and is heated to 300 Kelvin, what will the new pressure be?

   c. If the volume stays the same, but the pressure drops by half, then what must have happened to the temperature?

45. **VACATION** The time it takes the Levensteins to reach Lake Tahoe varies inversely with their average rate of speed.

   a. If they are 800 miles away, write and graph an equation relating their travel time to their average rate of speed.

   b. What minimum average speed will allow them to arrive within 18 hours?

46. MUSIC The maximum number of songs that a digital audio player can hold depends on the lengths and the quality of the songs that are recorded. A song will take up more space on the player if it is recorded at a higher quality, like from a CD, than at a lower quality, like from the Internet.

a. If a certain player has 5400 megabytes of storage space, write a function that represents the number of songs the player can hold as a function of the average size of the songs.

b. Is your function a *direct, joint, inverse,* or *combined* variation?

c. Suppose the average file size for a high-quality song is 8 megabytes and the average size for a low-quality song is 5 megabytes. Determine how many more songs the player can hold if they are low quality than if they are high quality.

47. GRAVITY According to the Law of Universal Gravitation, the attractive force F in newtons between any two bodies in the universe is directly proportional to the product of the masses m_1 and m_2 in kilograms of the two bodies and inversely proportional to the square of the distance d in meters between the bodies. That is, $F = \frac{Gm_1m_2}{d^2}$. G is the universal gravitational constant. Its value is 6.67×10^{-11} Nm^2/kg^2.

a. The distance between Earth and the Moon is about 3.84×10^8 meters. The mass of the Moon is 7.36×10^{22} kilograms. The mass of Earth is 5.97×10^{24} kilograms. What is the gravitational force that the Moon and Earth exert upon each other?

b. The distance between Earth and the Sun is about 1.5×10^{11} meters. The mass of the Sun is about 1.99×10^{30} kilograms. What is the gravitational force that the Sun and Earth exert upon each other?

c. Find the gravitational force exerted on each other by two 1000-kilogram iron balls at a distance of 0.1 meter apart.

H.O.T. Problems Use Higher-Order Thinking Skills

48. CCSS CRITIQUE Jamil and Savannah are setting up a proportion to begin solving the combined variation in which z varies directly as x and z varies inversely as y. Who has set up the correct proportion? Explain your reasoning.

Jamil

$z_1 = \frac{kx_1}{y_1}$ and $z_2 = \frac{kx_2}{y_2}$

$k = \frac{z_1y_1}{x_1}$ and $k = \frac{z_2y_2}{x_2}$

$\frac{z_1y_1}{x_1} = \frac{z_2y_2}{x_2}$

Savannah

$z_1 = \frac{kx_1}{y_1}$ and $z_2 = \frac{kx_2}{y_2}$

$k = \frac{z_1x_1}{y_1}$ and $k = \frac{z_2x_2}{y_2}$

$\frac{z_1x_1}{y_1} = \frac{z_2x_2}{y_2}$

49. CHALLENGE If a varies inversely as b, c varies jointly as b and f, and f varies directly as g, how are a and g related?

50. REASONING Explain why some mathematicians consider every joint variation a combined variation, but not every combined variation a joint variation.

51. OPEN ENDED Describe three real-life quantities that vary jointly with each other.

52. WRITING IN MATH Determine the type(s) of variation(s) for which 0 cannot be one of the values. Explain your reasoning.

Standardized Test Practice

53. SAT/ACT Rafael left the dorm and drove toward the cabin at an average speed of 40 km/h. Monica left some time later driving in the same direction at an average speed of 48 km/h. After driving for five hours, Monica caught up with Rafael. How long did Rafael drive before Monica caught up?

A 1 hour
B 2 hours
C 4 hours
D 6 hours
E 8 hours

54. 75% of 88 is the same as 60% of what number?

F 100
G 105
H 108
J 110

55. EXTENDED RESPONSE Audrey's hair is 7 inches long and is expected to grow at an average rate of 3 inches per year.

a. Make a table that shows the expected length of Audrey's hair after each of the first 4 years.

b. Write a function that can be used to determine the length of her hair after each year.

c. If she does not get a haircut, determine the length of her hair after 9 years.

56. Which of the following is equal to the sum of two consecutive even integers?

A 144
B 146
C 147
D 148

Spiral Review

Determine any vertical asymptotes and holes in the graph of each rational function. (Lesson 8-4)

57. $f(x) = \frac{1}{x^2 + 5x + 6}$

58. $f(x) = \frac{x + 2}{x^2 + 3x - 4}$

59. $f(x) = \frac{x^2 + 4x + 3}{x + 3}$

60. PHOTOGRAPHY The formula $\frac{1}{q} = \frac{1}{f} - \frac{1}{p}$ can be used to determine how far the film should be placed from the lens of a camera to create a perfect photograph. The variable q represents the distance from the lens to the film, f represents the focal length of the lens, and p represents the distance from the object to the lens. (Lesson 8-3)

a. Solve the formula for $\frac{1}{p}$.

b. Write the expression containing f and q as a single rational expression.

c. If a camera has a focal length of 8 centimeters and the lens is 10 centimeters from the film, how far should an object be from the lens so that the picture will be in focus?

Solve each equation. Check your solutions. (Lesson 7-5)

61. $\log_3 42 - \log_3 n = \log_3 7$

62. $\log_2(3x) + \log_2 5 = \log_2 30$

63. $2 \log_5 x = \log_5 9$

64. $\log_{10} a + \log_{10} (a + 21) = 2$

Given a polynomial and one of its factors, find the remaining factors of the polynomial. Some factors may not be binomials. (Lesson 5-6)

65. $2x^3 - 5x^2 - 28x + 15; x - 5$

66. $3x^3 + 10x^2 - x - 12; x + 3$

Skills Review

Find the LCM of each set of polynomials.

67. $a, 2a, a + 1$

68. $x, 4y, x - y$

69. $8, 24x, 12$

70. $x^4, 3x^2, 2xy$

71. $12a, 15, 4b^2$

72. $x + 2, x - 3, x^2 - x - 6$

LESSON 8-6

Solving Rational Equations and Inequalities

Then

- You simplified rational expressions.

Now

1. Solve rational equations.
2. Solve rational inequalities.

Why?

- A gaming club charges \$20 per month for membership. Members also have to pay \$5 each time they visit the club. If a member visits the club x times in one month, then the charge for that month will be $20 + 5x$. The actual cost per visit will be $\frac{20 + 5x}{x}$. To determine how many visits are needed for the cost per visit to be \$6, you would need to solve the equation $\frac{20 + 5x}{x} = 6$.

NewVocabulary

rational equation
weighted average
rational inequality

Common Core State Standards

Content Standards

A.CED.1 Create equations and inequalities in one variable and use them to solve problems.

A.REI.2 Solve simple rational and radical equations in one variable, and give examples showing how extraneous solutions may arise.

Mathematical Practices

6 Attend to precision.

1 Solve Rational Equations

Equations that contain one or more rational expressions are called **rational equations**. These equations are often easier to solve once the fractions are eliminated. You can eliminate the fractions by multiplying each side by the least common denominator (LCD).

Example 1 Solve a Rational Equation

Solve $\frac{4}{x+3} + \frac{5}{6} = \frac{23}{18}$. Check your solution.

The LCD for the terms is $18(x + 3)$.

$$\frac{4}{x+3} + \frac{5}{6} = \frac{23}{18}$$ Original equation

$$18(x+3)\left(\frac{4}{x+3}\right) + 18(x+3)\left(\frac{5}{6}\right) = 18(x+3)\frac{23}{18}$$ Multiply by LCD.

$$\overset{1}{18(x+3)}\left(\frac{4}{\underset{1}{x+3}}\right) + \overset{3}{18}(x+3)\left(\frac{5}{\underset{1}{6}}\right) = \overset{1}{18}(x+3)\left(\frac{23}{\underset{1}{18}}\right)$$ Divide common factors.

$$72 + 15x + 45 = 23x + 69$$ Multiply.

$$15x + 117 = 23x + 69$$ Simplify.

$$48 = 8x$$ Subtract $15x$ and 69.

$$6 = x$$ Divide.

CHECK

$$\frac{4}{x+3} + \frac{5}{6} = \frac{23}{18}$$ Original equation

$$\frac{4}{6+3} + \frac{5}{6} \stackrel{?}{=} \frac{23}{18}$$ $x = 6$

$$\frac{4}{9} + \frac{5}{6} \stackrel{?}{=} \frac{23}{18}$$ Simplify.

$$\frac{8}{18} + \frac{15}{18} \stackrel{?}{=} \frac{23}{18}$$ Simplify.

$$\frac{23}{18} = \frac{23}{18} \checkmark$$ Add.

GuidedPractice

Solve each equation. Check your solution.

1A. $\frac{2}{x+3} + \frac{3}{2} = \frac{19}{10}$

1B. $\frac{7}{12} + \frac{9}{x-4} = \frac{55}{48}$

Paul Costello/Stone/Getty Images

Math HistoryLink

Brook Taylor (1685–1731) English mathematician Taylor developed a theorem used in calculus known as Taylor's Theorem that relies on the remainders after computations with rational expressions.

Multiplying each side of an equation by the LCD of rational expressions can yield results that are not solutions of the original equation. These are extraneous solutions.

Example 2 Solve a Rational Equation

Solve $\frac{2x}{x+5} - \frac{x^2 - x - 10}{x^2 + 8x + 15} = \frac{3}{x+3}$. Check your solution.

The LCD for the terms is $(x + 3)(x + 5)$.

$$\frac{2x}{x+5} - \frac{x^2 - x - 10}{x^2 + 8x + 15} = \frac{3}{x+3}$$ Original equation

$$\frac{(x+3)(x+5)(2x)}{x+5} - \frac{(x+3)(x+5)(x^2 - x - 10)}{x^2 + 8x + 15} = \frac{(x+3)(x+5)3}{x+3}$$ Multiply by LCD.

$$\frac{(x+3)\overset{1}{\cancel{(x+5)}}(2x)}{\underset{1}{\cancel{x+5}}} - \frac{\overset{1}{\cancel{(x+3)}}\overset{1}{\cancel{(x+5)}}(x^2 - x - 10)}{\underset{1}{\cancel{x^2 + 8x + 15}}} = \frac{(x+5)\overset{1}{\cancel{(x+3)}}3}{\underset{1}{\cancel{x+3}}}$$ Divide common factors.

$$(x+3)(2x) - (x^2 - x - 10) = 3(x+5)$$ Simplify.

$$2x^2 + 6x - x^2 + x + 10 = 3x + 15$$ Distribute.

$$x^2 + 7x + 10 = 3x + 15$$ Simplify.

$$x^2 + 4x - 5 = 0$$ Subtract $3x + 15$.

$$(x+5)(x-1) = 0$$ Factor.

$$x + 5 = 0 \quad \text{or} \quad x - 1 = 0$$ Zero Product Property

$$x = -5 \qquad\qquad x = 1$$

CHECK Try $x = -5$.

$$\frac{2x}{x+5} - \frac{x^2 - x - 10}{x^2 + 8x + 15} = \frac{3}{x+3}$$

$$\frac{2(-5)}{-5+5} - \frac{(-5)^2 - (-5) - 10}{(-5)^2 + 8(-5) + 15} \stackrel{?}{=} \frac{3}{-5+3}$$

$$\frac{-10}{0} - \frac{25 + 5 - 10}{25 - 40 + 15} \neq -\frac{3}{2} \ \times$$

Try $x = 1$.

$$\frac{2x}{x+5} - \frac{x^2 - x - 10}{x^2 + 8x + 15} = \frac{3}{x+3}$$

$$\frac{2(1)}{1+5} - \frac{1^2 - 1 - 10}{1^2 + 8(1) + 15} \stackrel{?}{=} \frac{3}{1+3}$$

$$\frac{2}{6} - \frac{-10}{24} \stackrel{?}{=} \frac{3}{4}$$

$$\frac{8}{24} + \frac{10}{24} \stackrel{?}{=} \frac{3}{4}$$

$$\frac{3}{4} = \frac{3}{4} \ \checkmark$$

When solving a rational equation, any possible solution that results in a zero in the denominator must be excluded from your list of solutions.

Since $x = -5$ results in a zero in the denominator, it is extraneous. Eliminate -5 from the list of solutions. The solution is 1.

ReviewVocabulary

extraneous solutions solutions that do not satisfy the original equation

GuidedPractice

2A. $\frac{5}{y-2} + 2 = \frac{17}{6}$

2B. $\frac{2}{z+1} - \frac{1}{z-1} = \frac{-2}{z^2 - 1}$

2C. $\frac{7n}{3n+3} - \frac{5}{4n-4} = \frac{3n}{2n+2}$

2D. $\frac{1}{p-2} = \frac{2p+1}{p^2 + 2p - 8} + \frac{2}{p+4}$

The **weighted average** is a method for finding the mean of a set of numbers in which some elements of the set carry more importance, or weight, than others. Many real-world problems involving mixtures, work, distance, and interest can be solved by using rational equations.

Real-World Example 3 Mixture Problem

PT

CHEMISTRY Mia adds a 70% acid solution to 12 milliliters of a solution that is 15% acid. How much of the 70% acid solution should be added to create a solution that is 60% acid?

Understand Mia needs to know how much of a solution needs to be added to an original solution to create a new solution.

Plan Each solution has a certain percentage that is acid. The percentage of acid in the final solution must equal the amount of acid divided by the total solution.

	Original	Added	New
Amount of Acid	0.15(12)	0.7(x)	0.15(12) + 0.7x
Total Solution	12	x	12 + x

$$\text{Percentage of acid in solution} = \frac{\text{amount of acid}}{\text{total solution}}$$

Solve

$\frac{\text{percent}}{100} = \frac{\text{amount of acid}}{\text{total solution}}$ Write a proportion.

$\frac{60}{100} = \frac{0.15(12) + 0.7x}{12 + x}$ Substitute.

$\frac{60}{100} = \frac{1.8 + 0.7x}{12 + x}$ Simplify numerator.

$100(12 + x)\frac{60}{100} = 100(12 + x)\frac{1.8 + 0.7x}{12 + x}$ LCD is 100(12 + x). Multiply by LCD.

$\overset{1}{\cancel{100}}(12 + x)\frac{60}{\underset{1}{\cancel{100}}} = 100\overset{1}{\cancel{(12 + x)}}\frac{1.8 + 0.7x}{\underset{1}{\cancel{12 + x}}}$ Divide common factors.

$(12 + x)60 = 100(1.8 + 0.7x)$ Simplify.

$720 + 60x = 180 + 70x$ Distribute.

$540 = 10x$ Subtract 60x and 180.

$54 = x$ Divide by 10.

Check

$\frac{60}{100} = \frac{0.15(12) + 0.7x}{12 + x}$ Original equation

$\frac{60}{100} \stackrel{?}{=} \frac{0.15(12) + 0.7(54)}{12 + 54}$ x = 54

$\frac{60}{100} \stackrel{?}{=} \frac{37.8}{66}$ Simplify.

$0.6 = 0.6$ ✓ Simplify.

Mia needs to add 54 milliliters of the 70% acid solution.

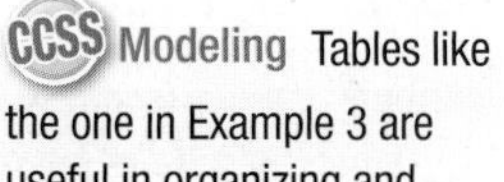

StudyTip

CCSS Modeling Tables like the one in Example 3 are useful in organizing and solving mixture, work, weighted average, and distance problems.

GuidedPractice

3. Jimmy adds a 65% fruit juice solution to 15 milliliters of a drink that is 10% fruit juice. How much of the 65% fruit juice solution must be added to create a fruit punch that is 35% fruit juice?

The formula relating distance, rate, and time can also be used to solve rational equations. The most common use is $d = rt$. However, it can also be represented by $r = \frac{d}{t}$ and $t = \frac{d}{r}$.

PT

Real-World Example 4 Distance Problem

ROWING Sandra is rowing a canoe on Stanhope Lake. Her rate in still water is 6 miles per hour. It takes Sandra 3 hours to travel 10 miles round trip. Assuming that Sandra rowed at a constant rate of speed, determine the rate of the current.

StudyTip

Distance Problems When distances involve round trips, the distance in one direction usually equals the distance in the other direction.

Understand We are given her speed in still water and the time it takes her to travel with the current and against it. We need to determine the speed of the current.

Plan She traveled 5 miles with the current and 5 miles against it. The formula that relates distance, rate, and time is $d = rt$, or $t = \frac{d}{r}$.

Time with the Current	Time Against the Current	Total Time
$\frac{5}{6+r}$	$\frac{5}{6-r}$	3 hours

Solve

$$\frac{5}{6+r} + \frac{5}{6-r} = 3$$ Write the equation.

$$(6+r)(6-r)\frac{5}{6+r} + (6+r)(6-r)\frac{5}{6-r} = (6+r)(6-r)3$$ LCD = (6 + r)(6 − r) Multiply by LCD.

$$\overset{}{\underset{1}{\cancel{(6+r)}}}(6-r)\frac{5}{\underset{1}{\cancel{6+r}}} + (6+r)\underset{1}{\cancel{(6-r)}}\frac{5}{\underset{1}{\cancel{6-r}}} = (6+r)(6-r)3$$ Divide common factors.

$$(6-r)5 + (6+r)5 = (36-r^2)3$$ Simplify.

$$30 - 5r + 30 + 5r = 108 - 3r^2$$ Distribute.

$$60 = 108 - 3r^2$$ Simplify.

$$0 = -3r^2 + 48$$ Subtract 10r.

$$0 = -3(r+4)(r-4)$$ Factor.

$$0 = (r+4)(r-4)$$ Divide each side by −3.

$$r = 4 \text{ or } -4$$ Zero Product Property

Check

$$\frac{5}{6+r} + \frac{5}{6-r} = 3$$ Original equation

$$\frac{5}{6+4} + \frac{5}{6-4} \stackrel{?}{=} 3$$ $r = 4$

$$\frac{5}{10} + \frac{5}{2} \stackrel{?}{=} 3$$ Simplify.

$$\frac{1}{2} + \frac{5}{2} = \frac{6}{2} \checkmark$$ Simplify.

Since speed cannot be negative, the speed of the current is 4 miles per hour.

GuidedPractice

4. **FLYING** The speed of the wind is 20 miles per hour. If it takes a plane 7 hours to fly 2368 miles round trip, determine the plane's speed in still air.

Real-world problems that involve work can often be solved using rational equations.

Real-World Example 5 Work Problems

Real-WorldLink

Since 1997, students from Rock Point School in Burlington, Vermont, spend a week servicing communities throughout the world. While working with Habitat for Humanity, the students spent the time in rural Tennessee, starting and completing the roof of a Habitat home in one week.

Source: Vermont Community Work

COMMUNITY SERVICE Every year, the junior and senior classes at Hillcrest High School build a house for the community. If it takes the senior class 24 days to complete a house and 18 days if they work with the junior class, how long would it take the junior class to complete a house if they worked alone?

Understand We are given how long it takes the senior class working alone and when the classes work together. We need to determine how long it would take the junior class by themselves.

Plan The senior class can complete 1 house in 24 days, so their rate is $\frac{1}{24}$ of a house per day.

The rate for the junior class is $\frac{1}{j}$.

The combined rate for both classes is $\frac{1}{18}$.

Senior Rate	Junior Rate	Combined Rate
$\frac{1}{24}$	$\frac{1}{j}$	$\frac{1}{18}$

Solve

$\frac{1}{24} + \frac{1}{j} = \frac{1}{18}$ Write the equation.

$72j\frac{1}{24} + 72j\frac{1}{j} = 72j\frac{1}{18}$ LCD = $72j$; Multiply by LCD.

$\overset{3}{\cancel{72}}j\frac{1}{\underset{1}{\cancel{24}}} + \overset{1}{\cancel{72}}\cancel{j}\frac{1}{\underset{1}{\cancel{j}}} = \overset{4}{\cancel{72}}j\frac{1}{\underset{1}{\cancel{18}}}$ Divide common factors.

$3j + 72 = 4j$ Distribute.

$72 = j$ Subtract $3j$.

Check Two methods are possible.

Method 1 Substitute values.

$\frac{1}{24} + \frac{1}{j} = \frac{1}{18}$ Original equation

$\frac{1}{24} + \frac{1}{72} \stackrel{?}{=} \frac{1}{18}$ $j = 72$

$\frac{3}{72} + \frac{1}{72} \stackrel{?}{=} \frac{4}{72}$ LCD = 72

$\frac{4}{72} = \frac{4}{72}$ ✓ Simplify.

Method 2 Use a calculator.

It would take the junior class 72 days to complete the house by themselves.

GuidedPractice

5A. It took Anthony and Travis 6 hours to rake the leaves together last year. The previous year it took Travis 10 hours to do it alone. How long will it take Anthony if he rakes them by himself this year?

5B. Noah and Owen paint houses together. If Noah can paint a particular house in 6 days and Owen can paint the same house in 5 days, how long would it take the two of them if they work together?

Paul Burns/Photodisc/Getty Images

2 Solve Rational Inequalities

To solve **rational inequalities**, which are inequalities that contain one or more rational expressions, follow these steps.

KeyConcept Solving Rational Inequalities

Step 1 State the excluded values. These are the values for which the denominator is 0.

Step 2 Solve the related equation.

Step 3 Use the values determined from the previous steps to divide a number line into intervals.

Step 4 Test a value in each interval to determine which intervals contain values that satisfy the inequality.

Example 6 Solve a Rational Inequality

Solve $\frac{x}{3} - \frac{1}{x-2} < \frac{x+1}{4}$.

Step 1 The excluded value for this inequality is 2.

Step 2 Solve the related equation.

$\frac{x}{3} - \frac{1}{x-2} = \frac{x+1}{4}$	Related equation
$\overset{4}{\cancel{12}}(x-2)\frac{x}{\underset{1}{\cancel{3}}} - 12\overset{1}{\cancel{(x-2)}}\frac{1}{\underset{1}{\cancel{x-2}}} = \overset{3}{\cancel{12}}(x-2)\frac{x+1}{\underset{1}{\cancel{4}}}$	LCD is $12(x-2)$. Multiply by LCD.
$4x^2 - 8x - 12 = 3x^2 - 3x - 6$	Distribute.
$x^2 - 5x - 6 = 0$	Subtract $3x^2 - 3x - 6$.
$(x-6)(x+1) = 0$	Factor.
$x = 6$ or -1	Zero Product Property

Step 3 Draw vertical lines at the excluded value and at the solutions to separate the number line into intervals.

Step 4 Now test a sample value in each interval to determine whether the values in the interval satisfy the inequality.

Test $x = -3$.	Test $x = 0$.	Test $x = 4$.	Test $x = 8$.
$\frac{-3}{3} - \frac{1}{-3-2} \overset{?}{<} \frac{-3+1}{4}$	$\frac{0}{3} - \frac{1}{0-2} \overset{?}{<} \frac{0+1}{4}$	$\frac{4}{3} - \frac{1}{4-2} \overset{?}{<} \frac{4+1}{4}$	$\frac{8}{3} - \frac{1}{8-2} \overset{?}{<} \frac{8+1}{4}$
$-1 + \frac{1}{5} \overset{?}{<} -\frac{2}{4}$	$0 + \frac{1}{2} \overset{?}{<} \frac{1}{4}$	$\frac{4}{3} - \frac{1}{2} \overset{?}{<} \frac{5}{4}$	$\frac{32}{12} - \frac{2}{12} \overset{?}{<} \frac{27}{12}$
$-\frac{4}{5} < -\frac{1}{2}$ ✓	$\frac{1}{2} \not< \frac{1}{4}$	$\frac{5}{6} < \frac{5}{4}$ ✓	$\frac{30}{12} \not< \frac{27}{12}$

The statement is true for $x = -3$ and $x = 4$. Therefore, the solution is $x < -1$ or $2 < x < 6$.

StudyTip

Rational Inequalities It is possible that none or all of the intervals will produce a true statement.

GuidedPractice **Solve each inequality.**

6A. $\frac{5}{x} + \frac{6}{5x} > \frac{2}{3}$

6B. $\frac{4}{3x} + \frac{7}{x} < \frac{5}{9}$

Check Your Understanding

= Step-by-Step Solutions begin on page R14.

Examples 1–2 **Solve each equation. Check your solution.**

1. $\frac{4}{7} + \frac{3}{x-3} = \frac{53}{56}$

2. $\frac{7}{3} - \frac{3}{x-5} = \frac{19}{12}$

3. $\frac{10}{2x+1} + \frac{4}{3} = 2$

4. $\frac{11}{4} - \frac{5}{y+3} = \frac{23}{12}$

5. $\frac{8}{x-5} - \frac{9}{x-4} = \frac{5}{x^2-9x+20}$

6. $\frac{14}{x+3} + \frac{10}{x-2} = \frac{122}{x^2+x-6}$

7. $\frac{14}{x-8} - \frac{5}{x-6} = \frac{82}{x^2-14x+48}$

8. $\frac{5}{x+2} - \frac{3}{x-2} = \frac{12}{x^2-4}$

Example 3

9. **CCSS STRUCTURE** Sara has 10 pounds of dried fruit selling for \$6.25 per pound. She wants to know how many pounds of mixed nuts selling for \$4.50 per pound she needs to make a trail mix selling for \$5 per pound.

   a. Let m = the number of pounds of mixed nuts. Complete the following table.

	Pounds	Price per Pound	Total Price
Dried Fruit	10	\$6.25	6.25(10)
Mixed Nuts			
Trail Mix			

   b. Write a rational equation using the last column of the table.

   c. Solve the equation to determine how many pounds of mixed nuts are needed.

Example 4

10. **DISTANCE** Alicia's average speed riding her bike is 11.5 miles per hour. She takes a round trip of 40 miles. It takes her 1 hour and 20 minutes with the wind and 2 hours and 30 minutes against the wind.

    a. Write an expression for Alicia's time with the wind.

    b. Write an expression for Alicia's time against the wind.

    c. How long does it take to complete the trip?

    d. Write and solve the rational equation to determine the speed of the wind.

Example 5

11. **WORK** Kendal and Chandi wax cars. Kendal can wax a particular car in 60 minutes and Chandi can wax the same car in 80 minutes. They plan on waxing the same car together and want to know how long it will take.

    a. How much will Kendal complete in 1 minute?

    b. How much will Kendal complete in x minutes?

    c. How much will Chandi complete in 1 minute?

    d. How much will Chandi complete in x minutes?

    e. Write a rational equation representing Kendal and Chandi working together on the car.

    f. Solve the equation to determine how long it will take them to finish the car.

Example 6 **Solve each inequality. Check your solutions.**

12. $\frac{3}{5x} + \frac{1}{6x} > \frac{2}{3}$

13. $\frac{1}{4c} + \frac{1}{9c} < \frac{1}{2}$

14. $\frac{4}{3y} + \frac{2}{5y} < \frac{3}{2}$

15. $\frac{1}{3b} + \frac{1}{4b} < \frac{1}{5}$

Practice and Problem Solving

Extra Practice is on page R8.

Examples 1–2 **Solve each equation. Check your solutions.**

16. $\frac{9}{x-7} - \frac{7}{x-6} = \frac{13}{x^2 - 13x + 42}$

17. $\frac{13}{y+3} - \frac{12}{y+4} = \frac{18}{y^2 + 7y + 12}$

18. $\frac{14}{x-2} - \frac{18}{x+1} = \frac{22}{x^2 - x - 2}$

19. $\frac{11}{a+2} - \frac{10}{a+5} = \frac{36}{a^2 + 7a + 10}$

20. $\frac{x}{2x-1} + \frac{3}{x+4} = \frac{21}{2x^2 + 7x - 4}$

21. $\frac{2}{y-5} + \frac{y-1}{2y+1} = \frac{2}{2y^2 - 9y - 5}$

Examples 3–5

22. **CHEMISTRY** How many milliliters of a 20% acid solution must be added to 40 milliliters of a 75% acid solution to create a 30% acid solution?

23. **GROCERIES** Ellen bought 3 pounds of bananas for \$0.90 per pound. How many pounds of apples costing \$1.25 per pound must she purchase so that the total cost for fruit is \$1 per pound?

24. **BUILDING** Bryan's volunteer group can build a garage in 12 hours. Sequoia's group can build it in 16 hours. How long would it take them if they worked together?

Example 6 **Solve each inequality. Check your solutions.**

25. $3 - \frac{4}{x} > \frac{5}{4x}$

26. $\frac{5}{3a} - \frac{3}{4a} > \frac{5}{6}$

27. $\frac{x-2}{x+2} + \frac{1}{x-2} > \frac{x-4}{x-2}$

28. $\frac{3}{4} - \frac{1}{x-3} > \frac{x}{x+4}$

29. $\frac{x}{5} + \frac{2}{3} < \frac{3}{x-4}$

30. $\frac{x}{x+2} + \frac{1}{x-1} < \frac{3}{2}$

31. **AIR TRAVEL** It takes a plane 20 hours to fly to its destination against the wind. The return trip takes 16 hours. If the plane's average speed in still air is 500 miles per hour, what is the average speed of the wind during the flight?

32. **FINANCIAL LITERACY** Judie wants to invest \$10,000 in two different accounts. The risky account could earn 9% interest, while the other account earns 5% interest. She wants to earn \$750 interest for the year. Of tables, graphs, or equations, choose the best representation needed and determine how much should be invested in each account.

33. **MULTIPLE REPRESENTATIONS** Consider $\frac{2}{x-3} + \frac{1}{x} = \frac{x-1}{x-3}$.
    a. **Algebraic** Solve the equation for x. Were any values of x extraneous?
    b. **Graphical** Graph $y_1 = \frac{2}{x-3} + \frac{1}{x}$ and $y_2 = \frac{x-1}{x-3}$ on the same graph for $0 < x < 5$.
    c. **Analytical** For what value(s) of x do they intersect? Do they intersect where x is extraneous for the original equation?
    d. **Verbal** Use this knowledge to describe how you can use a graph to determine whether an apparent solution of a rational equation is extraneous.

Solve each equation. Check your solutions.

34. $\frac{2}{y+3} - \frac{3}{4-y} = \frac{2y-2}{y^2 - y - 12}$

35. $\frac{2}{y+2} - \frac{y}{2-y} = \frac{y^2+4}{y^2-4}$

H.O.T. Problems Use Higher-Order Thinking Skills

36. **OPEN ENDED** Give an example of a rational equation that can be solved by multiplying each side of the equation by $4(x+3)(x-4)$.

37. **CHALLENGE** Solve $\frac{1 + \frac{9}{x} + \frac{20}{x^2}}{1 - \frac{25}{x^2}} = \frac{x+4}{x-5}$.

38. **CCSS TOOLS** While using the table feature on the graphing calculator to explore $f(x) = \frac{1}{x^2 - x - 6}$, the values -2 and 3 say "**ERROR.**" Explain its meaning.

39. **WRITING IN MATH** Why should you check solutions of rational equations and inequalities?

Standardized Test Practice

40. Nine pounds of mixed nuts containing 55% peanuts were mixed with 6 pounds of another kind of mixed nuts that contain 40% peanuts. What percent of the new mixture is peanuts?

A 58% **B** 51% **C** 49% **D** 47%

41. Working alone, Dato can dig a 10-foot by 10-foot hole in five hours. Pedro can dig the same hole in six hours. How long would it take them if they worked together?

F 1.5 hours **H** 2.52 hours
G 2.34 hours **J** 2.73 hours

42. An aircraft carrier made a trip to Guam and back. The trip there took three hours and the trip back took four hours. It averaged 6 kilometers per hour on the return trip. Find the average speed of the trip to Guam.

A 6 km/h **C** 10 km/h
B 8 km/h **D** 12 km/h

43. SHORT RESPONSE If a line ℓ is perpendicular to a segment CD at point F and $CF = FD$, how many points on line ℓ are the same distance from point C as from point D?

Spiral Review

Determine whether each relation shows *direct* or *inverse* variation, or *neither*. (Lesson 8-5)

44.

x	y
14	3
28	1.5
56	0.75
112	0.375

45.

x	y
0.2	24
0.6	72
1.8	216
5.4	648

46.

x	y
12	18
24	36
36	18
72	9

Graph each function. (Lesson 8-4)

47. $f(x) = \frac{x + 4}{x^2 + 7x + 12}$

48. $f(x) = \frac{x^2 - 5x - 14}{x - 7}$

49. $f(x) = \frac{x^2 + 3x - 6}{x - 2}$

50. WEATHER The atmospheric pressure P, in bars, of a given height on Earth is given by the formula $P = a \cdot e^{-\frac{k}{H}}$. In the formula, a is the surface pressure on Earth, which is approximately 1 bar, k is the altitude for which you want to find the pressure in kilometers, and H is always 7 kilometers. (Lesson 7-7)

a. Find the pressure for 2, 4, and 7 kilometers.

b. What do you notice about the pressure as altitude increases?

51. COMPUTERS Since computers have been invented, computational speed has multiplied by a factor of 4 about every three years. (Lesson 7-1)

a. If a typical computer operates with a computational speed s today, write an expression for the speed at which you can expect an equivalent computer to operate after x three-year periods.

b. Suppose your computer operates with a processor speed of 2.8 gigahertz and you want a computer that can operate at 5.6 gigahertz. If a computer with that speed is currently unavailable for home use, how long can you expect to wait until you can buy such a computer?

Skills Review

Determine whether the following are possible lengths of the sides of a right triangle.

52. 5, 12, 13 **53.** 60, 80, 100 **54.** 7, 24, 25

EXTEND 8-6

Graphing Technology Lab: Solving Rational Equations and Inequalities

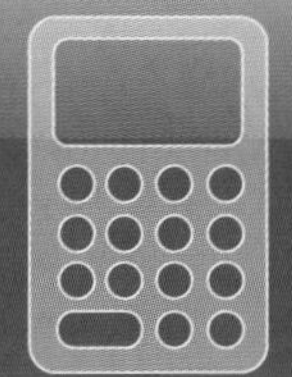

You can use a TI-83/84 Plus graphing calculator to solve rational equations by graphing or by using the table feature. Graph both sides of the equation, and locate the point(s) of intersection.

CCSS Common Core State Standards
Content Standards
A.REI.2 Solve simple rational and radical equations in one variable, and give examples showing how extraneous solutions may arise.
A.REI.11 Explain why the *x*-coordinates of the points where the graphs of the equations $y = f(x)$ and $y = g(x)$ intersect are the solutions of the equation $f(x) = g(x)$; find the solutions approximately, e.g., using technology to graph the functions, make tables of values, or find successive approximations. Include cases where $f(x)$ and/or $g(x)$ are linear, polynomial, rational, absolute value, exponential, and logarithmic functions.
Mathematical Practices
5 Use appropriate tools strategically.

Activity 1 Rational Equation

Solve $\frac{4}{x+1} = \frac{3}{2}$.

Step 1 Graph each side of the equation.

Graph each side of the equation as a separate function. Enter $\frac{4}{x+1}$ as **Y1** and $\frac{3}{2}$ as **Y2**. Then graph the two equations in the standard viewing window.

KEYSTROKES: [Y=] 4 [÷] [(] [X,T,θ,n] [+] 1 [)] [ENTER] 3 [÷] 2 [ZOOM] 6

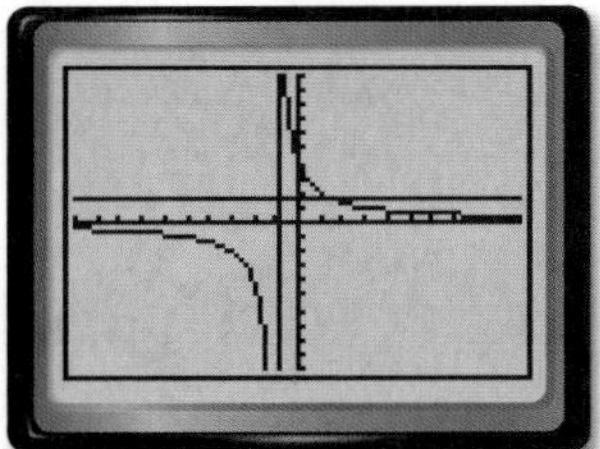

[−10, 10] scl: 1 by [−10, 10] scl: 1

Because the calculator is in connected mode, a vertical line may appear connecting the two branches of the hyperbola. This line is not part of the graph.

Step 2 Use the **intersect** feature.

The **intersect** feature on the **CALC** menu allows you to approximate the ordered pair of the point at which the graphs cross.

KEYSTROKES: [2nd] [CALC] 5

Select one graph and press [ENTER]. Select the other graph, press [ENTER], and press [ENTER] again.

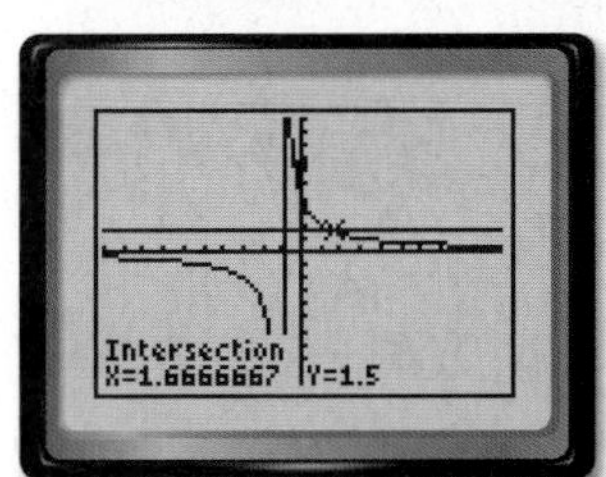

[−10, 10] scl: 1 by [−10, 10] scl: 1

The solution is $1\frac{2}{3}$.

Step 3 Use the **TABLE** feature.

Verify the solution using the **TABLE** feature. Set up the table to show *x*-values in increments of $\frac{1}{3}$.

KEYSTROKES: [2nd] [TBLSET] 0 [ENTER] 1 [÷] 3 [ENTER] [2nd] [TABLE]

X	Y1	Y2
0	4	1.5
.33333	3	1.5
.66667	2.4	1.5
1	2	1.5
1.3333	1.7143	1.5
1.6667	1.5	1.5
2	1.3333	1.5

X=0

The table displays *x*-values and corresponding *y*-values for each graph. At $x = 1\frac{2}{3}$, both functions have a *y*-value of 1.5. Thus, the solution of the equation is $1\frac{2}{3}$.

(continued on the next page)

Graphing Technology Lab
Solving Rational Equations and Inequalities *Continued*

You can use a similar procedure to solve rational inequalities using a graphing calculator.

Activity 2 Rational Inequality

Solve $\frac{3}{x} + \frac{7}{x} > 9$.

Step 1 Enter the inequalities.

Rewrite the problem as a system of inequalities.

The first inequality is $\frac{3}{x} + \frac{7}{x} > y$ or $y < \frac{3}{x} + \frac{7}{x}$. Since this inequality includes the *less than* symbol, shade below the curve. First enter the boundary and then use the arrow and [ENTER] keys to choose the shade below icon, ▙.

The second inequality is $y > 9$. Shade above the curve since this inequality contains *greater than*.

KEYSTROKES: [Y=] [◀] [◀] [ENTER] [ENTER] [ENTER] [▶] [▶] 3 [÷] [X,T,θ,n] [+] 7 [÷] [X,T,θ,n] [ENTER] [◀] [◀] [ENTER] [ENTER] [▶] [▶] 9

Step 2 Graph the system.

KEYSTROKES: [GRAPH]

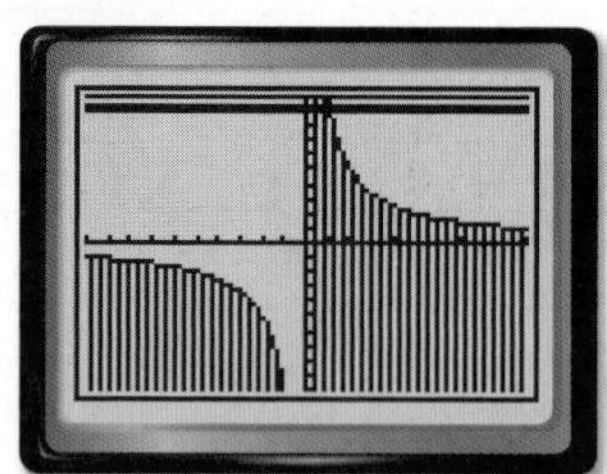

[−10, 10] scl: 1 by [−10, 10] scl: 1

The solution set of the original inequality is the set of x-values of the points in the region where the shadings overlap. Using the calculator's **intersect** feature, you can conclude that the solution set is $\left\{x \middle| 0 < x < 1\frac{1}{9}\right\}$.

Step 3 Use the **TABLE** feature.

Verify using the **TABLE** feature. Set up the table to show x-values in increments of $\frac{1}{9}$.

KEYSTROKES: [2nd] [TBLSET] 0 [ENTER] 1 [÷] 9 [ENTER] [2nd] [TABLE]

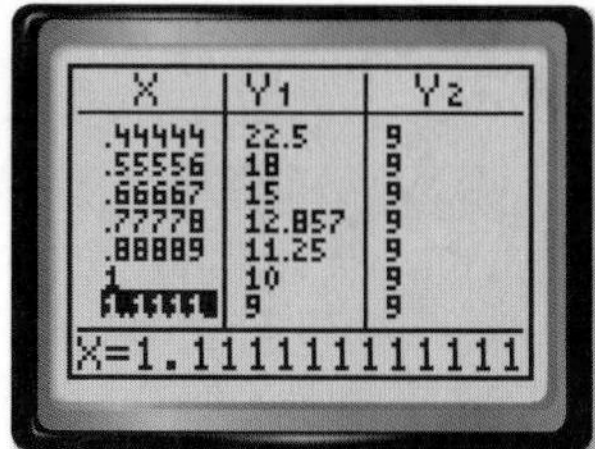

X	Y1	Y2
.44444	22.5	9
.55556	18	9
.66667	15	9
.77778	12.857	9
.88889	11.25	9
1	10	9
1.1111	9	9

X=1.111111111111

Scroll through the table. Notice that for x-values greater than 0 and less than $1\frac{1}{9}$, **Y1** > **Y2**. This confirms that the solution of the inequality is $\left\{x \middle| 0 < x < 1\frac{1}{9}\right\}$.

Exercises

Solve each equation or inequality.

1. $\frac{1}{x} + \frac{1}{2} = \frac{2}{x}$
2. $\frac{1}{x-4} = \frac{2}{x-2}$
3. $\frac{4}{x} = \frac{6}{x^2}$
4. $\frac{1}{1-x} = 1 - \frac{x}{x-1}$
5. $\frac{1}{x+4} = \frac{2}{x^2+3x-4} - \frac{1}{1-x}$
6. $\frac{1}{x} + \frac{1}{2x} > 5$
7. $\frac{1}{x-1} + \frac{2}{x} < 0$
8. $1 + \frac{5}{x-1} \leq 0$
9. $2 + \frac{1}{x-1} \geq 0$

CHAPTER 8 Study Guide and Review

Study Guide

KeyConcepts

Rational Expressions (Lessons 8-1 and 8-2)

- Multiplying and dividing rational expressions is similar to multiplying and dividing fractions.
- To simplify complex fractions, simplify the numerator and the denominator separately, and then simplify the resulting expression.

Reciprocal and Rational Functions (Lessons 8-3 and 8-4)

- A reciprocal function is of the form $f(x) = \frac{1}{a(x)}$, where $a(x)$ is a linear function and $a(x) \neq 0$.
- A rational function is of the form $\frac{a(x)}{b(x)}$, where $a(x)$ and $b(x)$ are polynomial functions and $b(x) \neq 0$.

Direct, Joint, and Inverse Variation (Lesson 8-5)

- Direct Variation: There is a nonzero number k such that $y = kx$.
- Joint Variation: There is a nonzero number k such that $y = kxz$.
- Inverse Variation: There is a nonzero constant k such that $xy = k$ or $y = \frac{k}{x}$, where $x \neq 0$ and $y \neq 0$.

Rational Equations and Inequalities (Lesson 8-6)

- Eliminate fractions in rational equations by multiplying each side of the equation by the LCD.
- Possible solutions of a rational equation must exclude values that result in zero in the denominator.

FOLDABLES StudyOrganizer

Be sure the Key Concepts are noted in your Foldable.

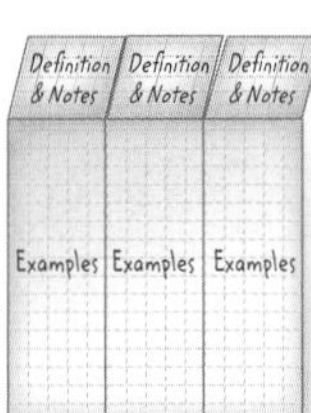

KeyVocabulary

combined variation (p. 565)
complex fraction (p. 532)
constant of variation (p. 562)
direct variation (p. 562)
horizontal asymptote (p. 553)
hyperbola (p. 545)
inverse variation (p. 564)
joint variation (p. 563)
oblique asymptote (p. 555)
point discontinuity (p. 556)
rational equation (p. 570)
rational expression (p. 529)
rational function (p. 553)
rational inequality (p. 575)
reciprocal function (p. 545)
vertical asymptote (p. 553)
weighted average (p. 572)

VocabularyCheck

Choose a term from the list above that best completes each statement or phrase.

1. A(n) ____________ is a rational expression whose numerator and/or denominator contains a rational expression.
2. If two quantities show ____________, their product is equal to a constant k.
3. A(n) ____________ asymptote is a linear asymptote that is neither horizontal nor vertical.
4. A(n) ____________ can be expressed in the form $y = kx$.
5. Equations that contain one or more rational expressions are called ____________.
6. The graph of $y = \frac{x}{x+2}$ has a(n) ____________ at $x = -2$.
7. ____________ occurs when one quantity varies directly as the product of two or more other quantities.
8. A ratio of two polynomial expressions is called a(n) ____________.
9. ____________ looks like a hole in a graph because the graph is undefined at that point.
10. ____________ occurs when one quantity varies directly and/or inversely as two or more other quantities.

Study Guide and Review *Continued*

Lesson-by-Lesson Review

8-1 Multiplying and Dividing Rational Expressions

Simplify each expression.

11. $\frac{-16xy}{27z} \cdot \frac{15z^3}{8x^2}$

12. $\frac{x^2 - 2x - 8}{x^2 + x - 12} \cdot \frac{x^2 + 2x - 15}{x^2 + 7x + 10}$

13. $\frac{x^2 - 1}{x^2 - 4} \cdot \frac{x^2 - 5x - 14}{x^2 - 6x - 7}$

14. $\frac{x + y}{15x} \div \frac{x^2 - y^2}{3x^2}$

15. $\dfrac{\frac{x^2 + 3x - 18}{x + 4}}{\frac{x^2 + 7x + 6}{x + 4}}$

16. **GEOMETRY** A triangle has an area of $3x^2 + 9x - 54$ square centimeters. If the height of the triangle is $x + 6$ centimeters, find the length of the base.

Example 1

Simplify $\frac{4a}{3b} \cdot \frac{9b^4}{2a^2}$.

$$\frac{4a}{3b} \cdot \frac{9b^4}{2a^2} = \frac{2 \cdot 2 \cdot a \cdot 3 \cdot 3 \cdot b \cdot b \cdot b \cdot b}{3 \cdot b \cdot 2 \cdot a \cdot a}$$

$$= \frac{6b^3}{a}$$

Example 2

Simplify $\frac{r^2 + 5r}{2r} \div \frac{r^2 - 25}{6r - 12}$.

$$\frac{r^2 + 5r}{2r} \div \frac{r^2 - 25}{6r - 12} = \frac{r^2 + 5r}{2r} \cdot \frac{6r - 12}{r^2 - 25}$$

$$= \frac{r(r + 5)}{2r} \cdot \frac{6(r - 2)}{(r + 5)(r - 5)}$$

$$= \frac{3(r - 2)}{r - 5}$$

8-2 Adding and Subtracting Rational Expressions

Simplify each expression.

17. $\frac{9}{4ab} + \frac{5a}{6b^2}$

18. $\frac{3}{4x - 8} - \frac{x - 1}{x^2 - 4}$

19. $\frac{y}{2x} + \frac{4y}{3x^2} - \frac{5}{6xy^2}$

20. $\frac{2}{x^2 - 3x - 10} - \frac{6}{x^2 - 8x + 15}$

21. $\frac{3}{3x^2 + 2x - 8} + \frac{4x}{2x^2 + 6x + 4}$

22. $\dfrac{\frac{3}{2x + 3} - \frac{x}{x + 1}}{\frac{2x}{x + 1} + \frac{5}{2x + 3}}$

23. **GEOMETRY** What is the perimeter of the rectangle?

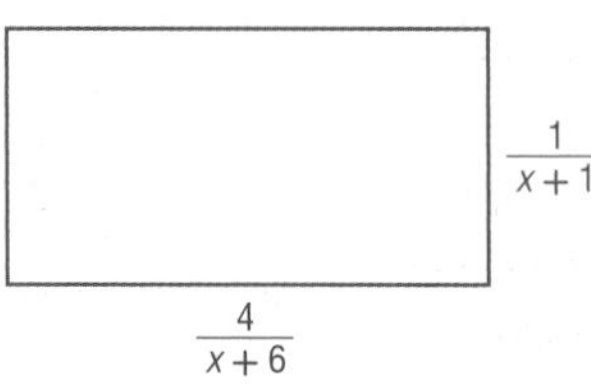

Example 3

Simplify $\frac{3a}{a^2 - 4} - \frac{2}{a - 2}$.

$$\frac{3a}{a^2 - 4} - \frac{2}{a - 2} = \frac{3a}{(a - 2)(a + 2)} - \frac{2}{a - 2}$$

$$= \frac{3a}{(a - 2)(a + 2)} - \frac{2(a + 2)}{(a - 2)(a + 2)}$$

$$= \frac{3a - 2(a + 2)}{(a - 2)(a + 2)}$$ Subtract numerators.

$$= \frac{3a - 2a - 4}{(a - 2)(a + 2)}$$ Distributive Property

$$= \frac{a - 4}{(a - 2)(a + 2)}$$ Simplify.

8-3 Graphing Reciprocal Functions

Graph each function. State the domain and range.

24. $f(x) = \frac{10}{x}$

25. $f(x) = -\frac{12}{x} + 2$

26. $f(x) = \frac{3}{x+5}$

27. $f(x) = \frac{6}{x-9}$

28. $f(x) = \frac{7}{x-2} + 3$

29. $f(x) = -\frac{4}{x+4} - 8$

30. **CONSERVATION** The student council is planting 28 trees for a service project. The number of trees each person plants depends on the number of student council members.

a. Write a function to represent this situation.

b. Graph the function.

Example 4

Graph $f(x) = \frac{3}{x+2} - 1$. State the domain and range.

$a = 3$: The graph is stretched vertically.

$h = -2$: The graph is translated 2 units left. There is an asymptote at $x = -2$.

$k: = -1$: The graph is translated 1 unit down. There is an asymptote at $f(x) = -1$.

Domain: $\{x \mid x \neq -2\}$,
Range: $\{f(x) \mid f(x) \neq -1\}$

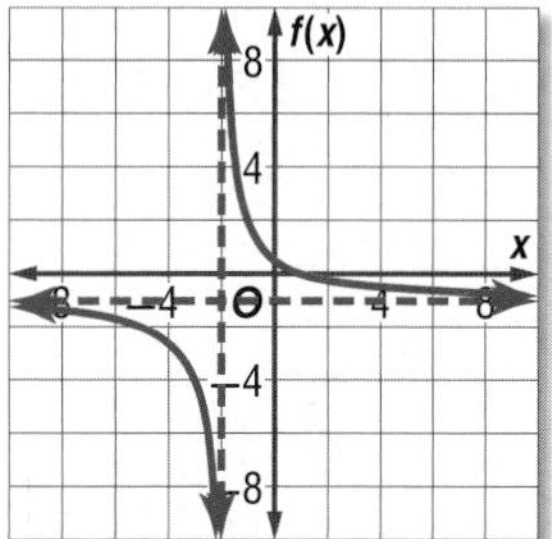

8-4 Graphing Rational Functions

Determine the equations of any vertical asymptotes and the values of x for any holes in the graph of each rational function.

31. $f(x) = \frac{3}{x^2 + 4x}$

32. $f(x) = \frac{x+2}{x^2 + 6x + 8}$

33. $f(x) = \frac{x^2 - 9}{x^2 - 5x - 24}$

Graph each rational function.

34. $f(x) = \frac{x+2}{(x+5)^2}$

35. $f(x) = \frac{x}{x+1}$

36. $f(x) = \frac{x^2 + 4x + 4}{x+2}$

37. $f(x) = \frac{x-1}{x^2 + 5x + 6}$

38. **SALES** Aliyah is selling magazine subscriptions. Out of the first 15 houses, she sold subscriptions to 10 of them. Suppose Aliyah goes to x more houses and sells subscriptions to all of them. The percentage of houses that she sold to out of the total houses can be determined using $P(x) = \frac{10 + x}{15 + x}$.

a. Graph the function.

b. What domain and range values are meaningful in the context of the problem?

Example 5

Determine the equation of any vertical asymptotes and the values of x for any holes in the graph of $f(x) = \frac{x^2 - 1}{x^2 + 2x - 3}$.

$$\frac{x^2 - 1}{x^2 + 2x - 3} = \frac{(x-1)(x+1)}{(x-1)(x+3)}$$

The function is undefined for $x = 1$ and $x = -3$.

Since $\frac{(x-1)(x+1)}{(x-1)(x+3)} = \frac{x+1}{x+3}$, $x = -3$ is a vertical asymptote, and $x = 1$ represents a hole in the graph.

Example 6

Graph $f(x) = \frac{1}{6x(x-1)}$.

The function is undefined for $x = 0$ and $x = 1$. Because $\frac{1}{6x(x-1)}$ is in simplest form, $x = 0$ and $x = 1$ are vertical asymptotes. Draw the two asymptotes and sketch the graph.

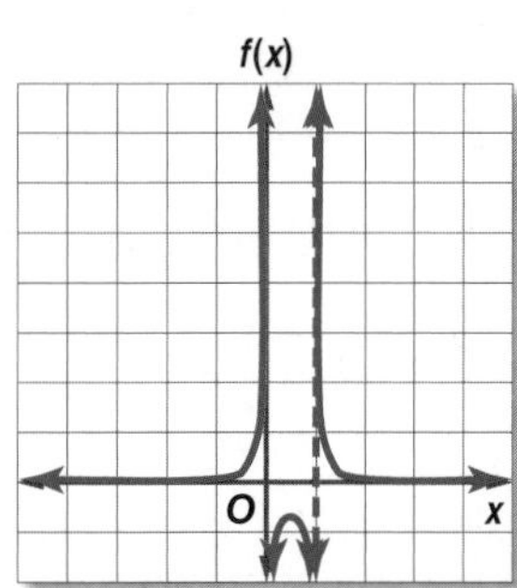

Study Guide and Review *Continued*

8-5 Variation Functions

39. If a varies directly as b and $b = 18$ when $a = 27$, find a when $b = 10$.

40. If y varies inversely as x and $y = 15$ when $x = 3.5$, find y when $x = -5$.

41. If y varies inversely as x and $y = -3$ when $x = 9$, find y when $x = 81$.

42. If y varies jointly as x and z, and $x = 8$ and $z = 3$ when $y = 72$, find y when $x = -2$ and $z = -5$.

43. If y varies jointly as x and z, and $y = 18$ when $x = 6$ and $z = 15$, find y when $x = 12$ and $z = 4$.

44. JOBS Lisa's earnings vary directly with how many hours she babysits. If she earns $68 for 8 hours of babysitting, find her earnings after 5 hours of babysitting.

Example 7

If y varies inversely as x and $x = 24$ when $y = -8$, find x when $y = 15$.

$\frac{x_1}{y_2} = \frac{x_2}{y_1}$ Inverse variation

$\frac{24}{15} = \frac{x_2}{-8}$ $x_1 = 24, y_1 = -8, y_2 = 15$

$24(-8) = 15(x_2)$ Cross multiply.

$-192 = 15x_2$ Simplify.

$-12\frac{4}{5} = x_2$ Divide each side by 15.

When $y = 15$, the value of x is $-12\frac{4}{5}$.

8-6 Solving Rational Equations and Inequalities

Solve each equation or inequality. Check your solutions.

45. $\frac{1}{3} + \frac{4}{x-2} = 6$

46. $\frac{6}{x+5} - \frac{3}{x-3} = \frac{6}{x^2+2x-15}$

47. $\frac{2}{x^2-9} = \frac{3}{x^2-2x-3}$

48. $\frac{4}{2x-3} + \frac{x}{x+1} = \frac{-8x}{2x^2-x-3}$

49. $\frac{x}{x+4} - \frac{28}{x^2+x-12} = \frac{1}{x-3}$

50. $\frac{x}{2} + \frac{1}{x-1} < \frac{x}{4}$

51. $\frac{1}{2x} - \frac{4}{5x} > \frac{1}{3}$

52. YARD WORK Lana can plant a garden in 3 hours. Milo can plant the same garden in 4 hours. How long will it take them if they work together?

Example 8

Solve $\frac{3}{x+2} + \frac{1}{x} = 0$.

The LCD is $x(x + 2)$.

$$\frac{3}{x+2} + \frac{1}{x} = 0$$

$$x(x+2)\left(\frac{3}{x+2} + \frac{1}{x}\right) = x(x+2)(0)$$

$$x(x+2)\left(\frac{3}{x+2}\right) + x(x+2)\left(\frac{1}{x}\right) = 0$$

$$3(x) + 1(x+2) = 0$$

$$3x + x + 2 = 0$$

$$4x + 2 = 0$$

$$4x = -2$$

$$x = -\frac{1}{2}$$

CHAPTER 8 Practice Test

Simplify each expression.

1. $\frac{r^2 + rt}{2r} \div \frac{r + t}{16r^2}$

2. $\frac{m^2 - 4}{3m^2} \cdot \frac{6m}{2 - m}$

3. $\frac{m^2 + m - 6}{n^2 - 9} \div \frac{m - 2}{n + 3}$

4. $\dfrac{\frac{x^2 + 4x + 3}{x^2 - 2x - 15}}{\frac{x^2 - 1}{x^2 - x - 20}}$

5. $\frac{x + 4}{6x + 3} + \frac{1}{2x + 1}$

6. $\frac{x}{x^2 - 1} - \frac{3}{2x + 2}$

7. $\frac{1}{y} + \frac{2}{7} - \frac{3}{2y^2}$

8. $\dfrac{2 + \frac{1}{x}}{5 - \frac{1}{x}}$

9. Identify the asymptotes, domain, and range of the function graphed.

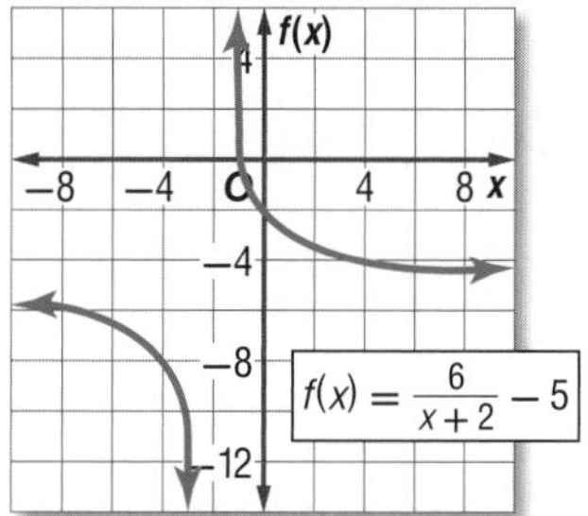

10. **MULTIPLE CHOICE** What is the equation for the vertical asymptote of the rational function $f(x) = \frac{x + 1}{x^2 + 3x + 2}$?

   A $x = -2$

   B $x = -1$

   C $x = 1$

   D $x = 2$

Graph each function.

11. $f(x) = -\frac{8}{x} - 9$

12. $f(x) = \frac{2}{x + 4}$

13. $f(x) = \frac{3}{x - 1} + 8$

14. $f(x) = \frac{5x}{x + 1}$

15. $f(x) = \frac{x}{x - 5}$

16. $f(x) = \frac{x^2 + 5x - 6}{x - 1}$

17. Determine the equations of any vertical asymptotes and the values of x for any holes in the graph of the function $f(x) = \frac{x + 5}{x^2 - 2x - 35}$.

18. Determine the equations of any oblique asymptotes in the graph of the function $f(x) = \frac{x^2 + x - 5}{x + 3}$.

Solve each equation or inequality.

19. $\frac{-1}{x + 4} = 6 - \frac{x}{x + 4}$

20. $\frac{1}{3} = \frac{5}{m + 3} + \frac{8}{21}$

21. $7 + \frac{2}{x} < -\frac{5}{x}$

22. $r + \frac{6}{r} - 5 = 0$

23. $\frac{6}{7} - \frac{3m}{2m - 1} = \frac{11}{7}$

24. $\frac{r + 2}{3r} = \frac{r + 4}{r - 2} - \frac{2}{3}$

25. If y varies inversely as x and $y = 18$ when $x = -\frac{1}{2}$, find x when $y = -10$.

26. If m varies directly as n and $m = 24$ when $n = -3$, find n when $m = 30$.

27. Suppose r varies jointly as s and t. If $s = 20$ when $r = 140$ and $t = -5$, find s when $r = 7$ and $t = 2.5$.

28. **BICYCLING** When Susan rides her bike, the distance that she travels varies directly with the amount of time she is biking. Suppose she bikes 50 miles in 2.5 hours. At this rate, how many hours would it take her to bike 80 miles?

29. **PAINTING** Peter can paint a house in 10 hours. Melanie can paint the same house in 9 hours. How long would it take if they worked together?

30. **MULTIPLE CHOICE** How many liters of a 25% acid solution must be added to 30 liters of an 80% acid solution to create a 50% acid solution?

   F 18

   G 30

   H 36

   J 66

31. What is the volume of the rectangular prism?

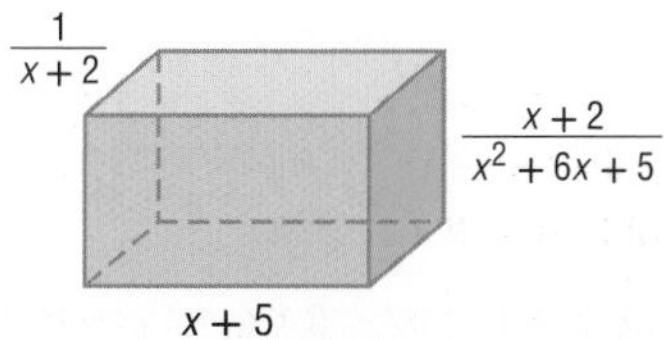

Preparing for Standardized Tests

Guess and Check

It is very important to pace yourself and keep track of how much time you have when taking a standardized test. If time is running short, or if you are unsure how to solve a problem, the guess-and-check strategy may help you determine the correct answer quickly.

Strategies for Guessing and Checking

Step 1

Carefully look over each possible answer choice and evaluate for reasonableness. Eliminate unreasonable answers.

Ask yourself:

- Are there any answer choices that are clearly incorrect?
- Are there any answer choices that are not in the proper format?
- Are there any answer choices that do not have the proper units for the correct answer?

Step 2

For the remaining answer choices, use the guess-and-check method.

- **Equations:** If you are solving an equation, substitute the answer choice for the variable and see if this results in a true number sentence.
- **System of Equations:** For a system of equations, substitute the answer choice for all variables and make sure all equations result in a true number sentence.

Step 3

Choose an answer choice and see if it satisfies the constraints of the problem statement. Identify the correct answer.

- If the answer choice you are testing does not satisfy the problem, move on to the next reasonable guess and check it.
- When you find the correct answer choice, stop.

Standardized Test Example

Read the problem. Identify what you need to know. Then use the information in the problem to solve.

Solve: $\frac{2}{x-3} - \frac{4}{x+3} = \frac{8}{x^2-9}$.

A -1

B 1

C 5

D 7

Radius Images/CORBIS

The solution of the rational equation will be a real number. Since all four answer choices are real numbers, they are all possible correct answers and must be checked. Begin with the first answer choice and check it in the rational equation. Continue until you find the answer choice that results in a true number sentence.

	Check:
Guess: −1	$\frac{2}{(-1)-3} - \frac{4}{(-1)+3} = \frac{8}{(-1)^2-9}$ $-\frac{5}{2} \neq -1$ ✗

	Check:
Guess: 1	$\frac{2}{1-3} - \frac{4}{1+3} = \frac{8}{(1)^2-9}$ $-2 \neq -1$ ✗

	Check:
Guess: 5	$\frac{2}{5-3} - \frac{4}{5+3} = \frac{8}{(5)^2-9}$ $\frac{1}{2} = \frac{1}{2}$ ✓

If $x = 5$, the result is a true number sentence. So, the correct answer is C.

Exercises

Read each problem. Identify what you need to know. Then use the information in the problem to solve.

1. Solve: $\frac{2}{5x} - \frac{1}{2x} = -\frac{1}{2}$.

   A $\frac{1}{10}$

   B $\frac{1}{5}$

   C $\frac{1}{4}$

   D $\frac{1}{2}$

2. The sum of Kevin's, Anna's, and Tia's ages is 40. Anna is 1 year more than twice as old as Tia. Kevin is 3 years older than Anna. How old is Anna?

   F 7 **H** 15

   G 14 **J** 18

3. Determine the point(s) where the following rational function crosses the x-axis.

$$f(x) = \frac{2}{x-1} - \frac{x+4}{3}$$

   A −5

   B 4

   C 2 or 3

   D −5 or 2

4. Rafael's Theatre Company sells tickets for $10. At this price, they sell 400 tickets. Rafael estimates that they would sell 40 fewer tickets for each $2 price increase. What charge would give the most income?

   F 10 **H** 15

   G 13 **J** 20

CHAPTER 8

Standardized Test Practice

Cumulative, Chapters 1 through 8

Multiple Choice

Read each question. Then fill in the correct answer on the answer document provided by your teacher or on a sheet of paper.

1. Greg's father can mow the lawn on his riding mower in 45 minutes. It takes Greg 1 hour 45 minutes to mow the lawn with a push mower. Which of the following rational equations can be solved for the number of minutes t it would take them to mow the lawn working together?

A $\frac{t}{45} + \frac{t}{1.45} = 1$

B $\frac{t}{150} = 1$

C $\frac{t}{45} + \frac{t}{105} = 1$

D $\frac{t + 45}{t + 105} = 1$

2. The total cost of reserving a campsite varies directly as the number of nights the site is rented, as shown in the table.

Days	Total Cost
1	$24
2	$48
3	$72
4	$96

Which equation represents the direct variation?

F $y = x + 24$

G $y = 24x$

H $y = \frac{24}{x}$

J $y = 96x$

3. In which direction must the graph of $y = \frac{1}{x}$ be shifted to produce the graph of $y = \frac{1}{x} + 2$?

A up

B down

C right

D left

Test-Taking Tip

Question 2 Check your answer by substituting 1, 2, 3, and 4 for x and making sure the values in the table are produced.

4. Which of the following is ***not*** an asymptote of the rational function $f(x) = \frac{1}{x^2 - 49}$?

F $f(x) = 0$

G $x = -7$

H $x = 7$

J $f(x) = 1$

5. Simplify the complex fraction.

$$\frac{\frac{(x + 3)^2}{x^2 - 16}}{\frac{x + 3}{x + 4}}$$

A $\frac{x + 3}{x + 4}$

B $\frac{1}{x - 4}$

C $\frac{x + 3}{x - 4}$

D $\frac{x - 4}{x + 3}$

6. A ball was thrown upward with an initial velocity of 16 feet per second from the top of a building 128 feet high. Its height h in feet above the ground t seconds later will be $h = 128 + 16t - 16t^2$.

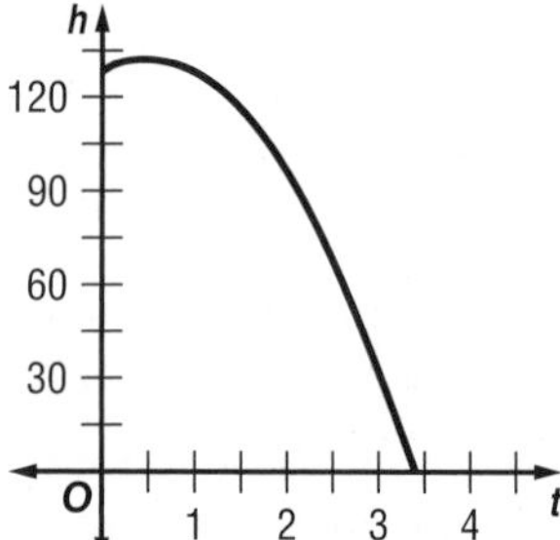

Which is the best conclusion about the ball's action?

F The ball stayed above 128 feet for more than 3 seconds.

G The ball returned to the ground in less than 4 seconds.

H The ball traveled a greater distance going up than it did going down.

J The ball traveled less than 128 feet in 3.4 seconds.

7. Which of these equations describes a relationship in which every negative real number x corresponds to a nonnegative real number y?

A $y = -x$

B $y = x$

C $y = \sqrt{x}$

D $y = x^3$

Short Response/Gridded Response

Record your answers on the answer sheet provided by your teacher or on a sheet of paper.

8. Martha is putting a stone walkway around the garden pictured at the right. About how many feet of stone are needed?

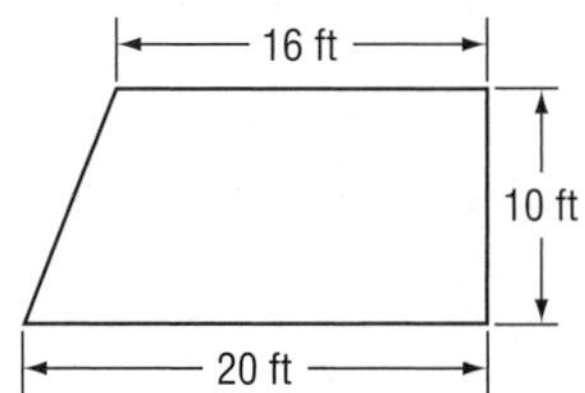

9. GRIDDED RESPONSE Elaine had some money saved for a week-long vacation. The first day of the vacation she spent \$125 on food and a hotel. On the second day, she was given \$80 from her sister for expenses. Elaine then had \$635 left for the rest of the vacation. How much money, in dollars, did she begin the vacation with?

10. Carlos wants to print 800 one-page flyers for his landscaping business. He has a printer that is capable of printing 8 pages per minute. His business partner has another printer that prints 10 pages per minute.

a. How long would it take Carlos' printer to print all the flyers? How long would it take his partner's printer?

b. Set up a rational equation that can be used to find the number of minutes t it would take to print all 800 flyers if both printers are used simultaneously.

c. Solve the equation you wrote in part **b**. How long would it take both printers to print all the flyers if they print simultaneously? Round to the nearest minute.

11. GRIDDED RESPONSE The population of a country can be modeled by the equation $P(t) = 40e^{0.02t}$, where P is the population in millions and t is the number of years since 2005. When will the population be 400 million?

12. What is the area of the shaded region of the rectangle expressed as a polynomial in simplest form?

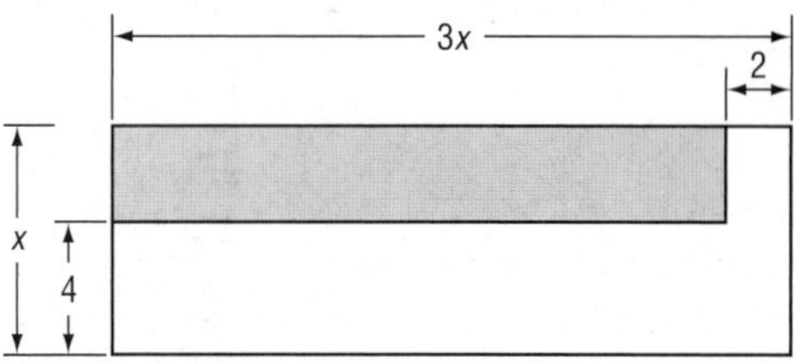

13. GRIDDED RESPONSE Suppose y varies inversely as x and $y = 4$ when $x = 12$. What is y when x is 5? Round to the nearest tenth.

Extended Response

Record your answers on a sheet of paper. Show your work.

14. Use the graph of the rational function at the right to answer each question.

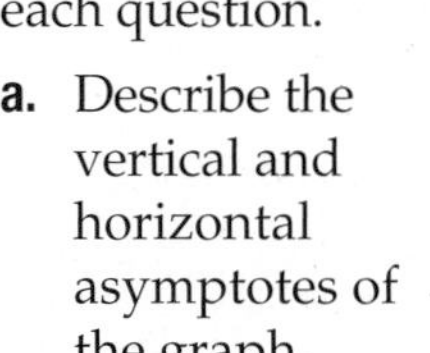

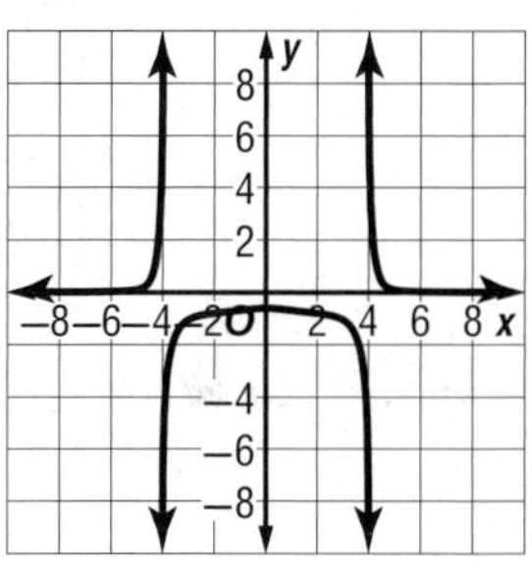

a. Describe the vertical and horizontal asymptotes of the graph.

b. Write the equation of the rational function. Explain how you found your answer.

15. Consider the polynomial function

$$f(x) = 3x^4 + 19x^3 + 7x^2 - 11x - 2.$$

a. What is the degree of the function?

b. What is the leading coefficient of the function?

c. Evaluate $f(1)$, $f(-2)$, and $f(2a)$. Show your work.

Need ExtraHelp?

If you missed Question...	1	2	3	4	5	6	7	8	9	10	11	12	13	14	15
Go to Lesson...	8-5	8-4	8-3	8-3	8-1	8-7	2-4	1-4	3-1	8-5	7-8	5-1	8-4	8-3	6-1

CHAPTER 9

Conic Sections

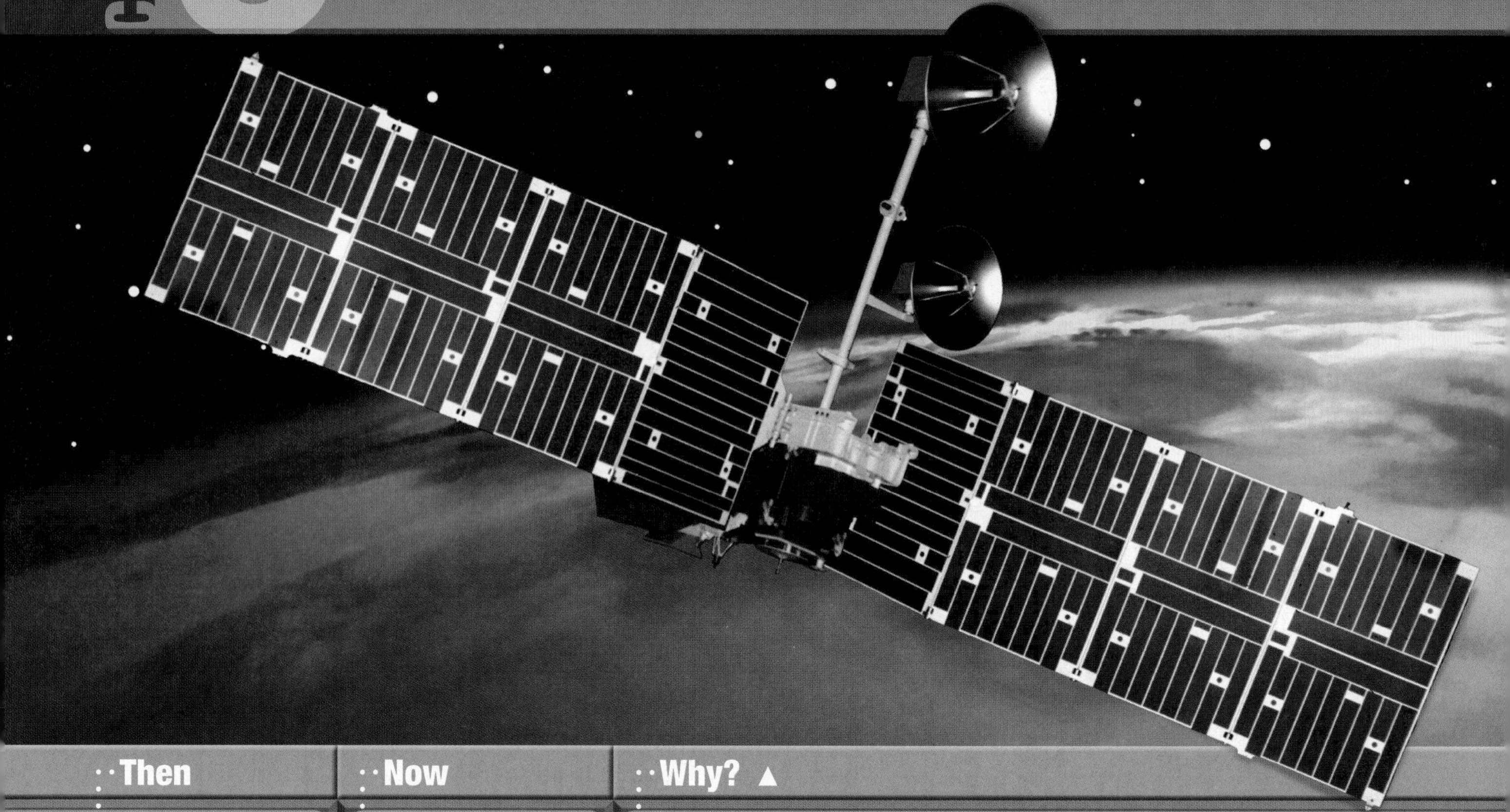

Then

You solved systems of linear equations algebraically and graphically.

Now

You will:

- Use the Midpoint and Distance Formulas.
- Write and graph equations of parabolas, circles, ellipses, and hyperbolas.
- Identify conic sections.
- Solve systems of quadratic equations and inequalities.

Why? ▲

SPACE Conic sections are evident in many aspects of space. Equations of circles are used to pilot spacecraft and satellites in circular orbits around Earth and the Moon. Planets travel in elliptical paths, not circular ones as previously thought. Comets travel along one branch of a hyperbola, which can help us to predict when they will appear again.

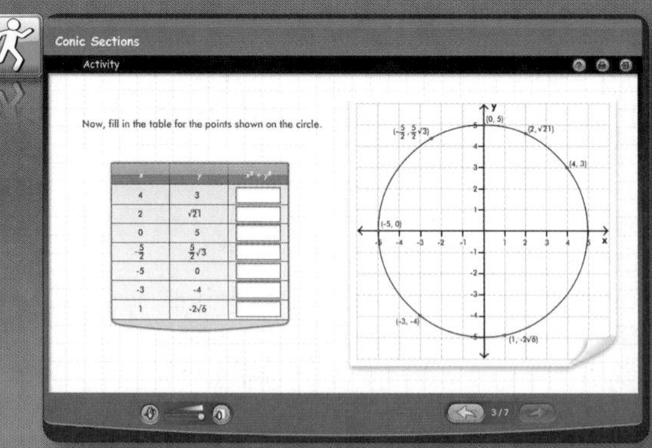

connectED.mcgraw-hill.com Your Digital Math Portal

Animation 
Vocabulary  
eGlossary 
Personal Tutor 
Virtual Manipulatives 
Graphing Calculator 
Audio 
Foldables  
Self-Check Practice 
Worksheets 

Get Ready for the Chapter

Diagnose Readiness | You have two options for checking prerequisite skills.

1 Textbook Option Take the Quick Check below. Refer to the Quick Review for help.

QuickCheck

Solve each equation by completing the square.

1. $x^2 + 8x + 7 = 0$
2. $x^2 + 5x - 6 = 0$
3. $x^2 - 8x + 15 = 0$
4. $x^2 + 2x - 120 = 0$
5. $2x^2 + 7x - 15 = 0$
6. $2x^2 + 3x - 5 = 0$
7. $x^2 - \frac{3}{2}x - \frac{23}{16} = 0$
8. $3x^2 - 4x = 2$

Solve each system of equations by using either substitution or elimination.

9. $y = x + 3$
   $2x - y = -1$
10. $2x - 5y = -18$
    $3x + 4y = 19$
11. $4y + 6x = -6$
    $5y - x = 35$
12. $x = y - 8$
    $4x + 2y = 4$
13. **MONEY** The student council paid \$15 per registration for a conference. They also paid \$10 for T-shirts for a total of \$180. Last year, they spent \$12 per registration and \$9 per T-shirt for a total of \$150 to buy the same number of registrations and T-shirts. Write and solve a system of two equations that represents the number of registrations and T-shirts bought each year.

QuickReview

Example 1

Solve $x^2 + 6x - 16 = 0$ by completing the square.

$$x^2 + 6x = 16$$
$$x^2 + 6x + 9 = 16 + 9$$
$$(x + 3)^2 = 25$$
$$x + 3 = \pm 5$$
$$x + 3 = 5 \quad \text{or} \quad x + 3 = -5$$
$$x = 2 \qquad\qquad x = -8$$

Example 2

Solve the system of equations algebraically.
$3y = x - 9$
$4x + 5y = 2$

Since x has a coefficient of 1 in the first equation, use the substitution method. First, solve that equation for x.

$3y = x - 9 \quad \rightarrow \quad x = 3y + 9$

$4(3y + 9) + 5y = 2$	Substitute $3y + 9$ for x.
$12y + 36 + 5y = 2$	Distributive Property
$17y = -34$	Combine like terms.
$y = -2$	Divide each side by 17.

To find x, use $y = -2$ in the first equation.

$3(-2) = x - 9$	Substitute -2 for y.
$-6 = x - 9$	Multiply.
$3 = x$	Add 9 to each side.

The solution is $(3, -2)$.

2 Online Option Take an online self-check Chapter Readiness Quiz at connectED.mcgraw-hill.com.

Get Started on the Chapter

You will learn several new concepts, skills, and vocabulary terms as you study Chapter 9. To get ready, identify important terms and organize your resources. You may wish to refer to Chapter 0 to review prerequisite skills.

FOLDABLES® StudyOrganizer

Conic Sections Make this Foldable to help you organize your Chapter 9 notes about conic sections. Begin with eight sheets of grid paper.

1. **Staple** the stack of grid paper along the top to form a booklet.

2. **Cut** seven lines from the bottom of the top sheet, six lines from the second sheet, and so on.

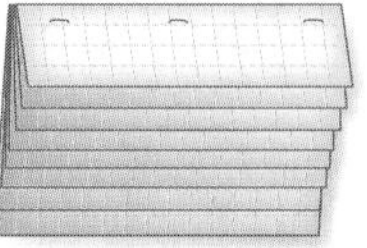

3. **Label** with lesson numbers as shown.

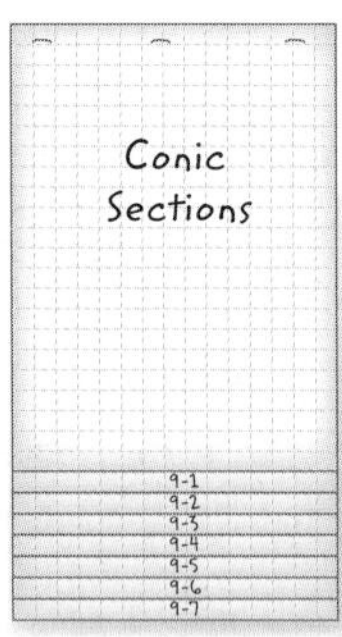

NewVocabulary

English		Español
parabola	p. 599	parábola
focus	p. 599	foco
directrix	p. 599	directriz
circle	p. 607	círculo
center of a circle	p. 607	centro de un círculo
radius	p. 607	radio
ellipse	p. 615	elipse
foci	p. 615	focos
major axis	p. 615	eje mayor
minor axis	p. 615	eje menor
center of an ellipse	p. 615	centro de una elipse
vertices	p. 615	vértices
co-vertices	p. 615	co-vértices
constant sum	p. 616	suma constante
hyperbola	p. 624	hipérbola
transverse axis	p. 624	eje transversal
conjugate axis	p. 624	eje conjugado
constant difference	p. 627	diferencia constante

ReviewVocabulary

quadratic equation ecuación cuadrática an equation of the form $ax^2 + bx + c = 0$, where $a \neq 0$

system of equations sistema de ecuaciones a set of equations with the same variables

x- and y-intercepts intersecciones x y y the x- or y-coordinate of the point at which a graph crosses the x- or y-axis

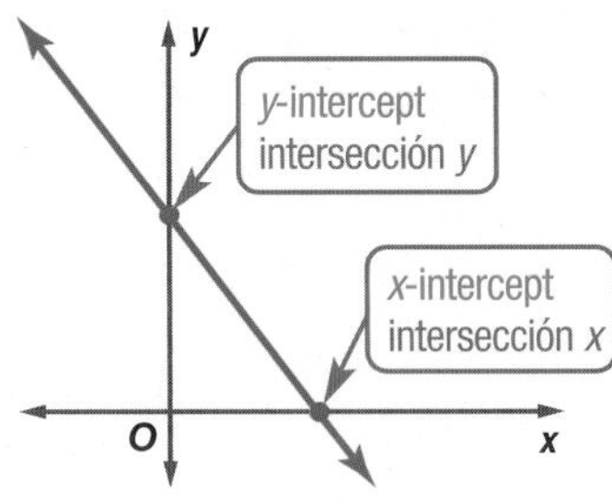

LESSON 9-1 Midpoint and Distance Formulas

Then	Now	Why?
You found the slope of a line passing through two points.	**1** Find the midpoint of a segment on the coordinate plane. **2** Find the distance between two points on the coordinate plane.	The Zero Milestone in Washington, D.C., was established in 1919. It was intended to serve as the origin for all highway measures with highway markers across the United States displaying the distances from the Zero Milestone.

POINT FOR THE MEASUREMENT OF DISTANCES FROM WASHINGTON ON HIGHWAYS OF THE UNITED STATES

CCSS Common Core State Standards

Content Standards

A.CED.4 Rearrange formulas to highlight a quantity of interest, using the same reasoning as in solving equations.

Mathematical Practices

2 Reason abstractly and quantitatively.

1 The Midpoint Formula

Recall that point M is the midpoint of segment PQ if M is between P and Q and $PM = MQ$. There is a formula for the coordinates of the midpoint of a segment in terms of the coordinates of the endpoints.

KeyConcept Midpoint Formula

Words If a line segment has endpoints $P(x_1, y_1)$ and $Q(x_2, y_2)$, then the midpoint of the segment has coordinates

$$M\left(\frac{x_1 + x_2}{2}, \frac{y_1 + y_2}{2}\right).$$

Model

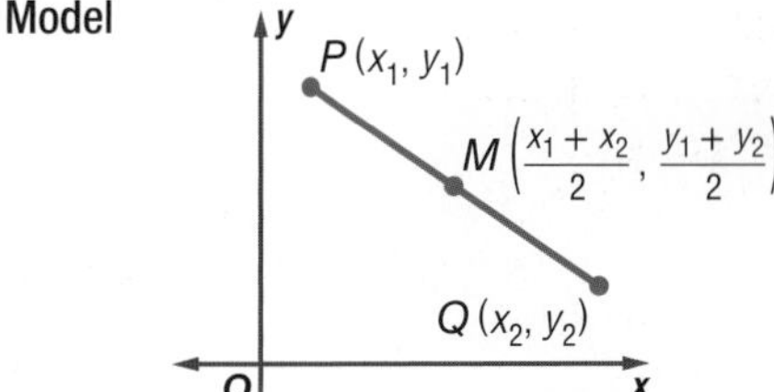

Example 1 Find a Midpoint

Find the coordinates of M, the midpoint of $\overline{JK}$, for $J(-1, 2)$ and $K(6, 1)$.

Let J be (x_1, y_1) and K be (x_2, y_2).

$M\left(\frac{x_1 + x_2}{2}, \frac{y_1 + y_2}{2}\right)$ Midpoint Formula

$= M\left(\frac{-1 + 6}{2}, \frac{2 + 1}{2}\right)$ $(x_1, y_1) = (-1, 2), (x_2, y_2) = (6, 1)$

$= M\left(\frac{5}{2}, \frac{3}{2}\right)$ or $M\left(2\frac{1}{2}, 1\frac{1}{2}\right)$ Simplify.

GuidedPractice

1A. Find the coordinates of the midpoint of $\overline{AB}$ for $A(5, 12)$ and $B(-4, 8)$.

1B. Find the coordinates of the midpoint of $\overline{CD}$ for $C(4, 5)$ and $D(14, 13)$.

2 The Distance Formula

The distance between two points, a and b, on a number line is $|a - b|$ or $|b - a|$. You can use this fact and the Pythagorean Theorem to derive a formula for the distance between two points on a coordinate plane.

U.S. Department of Transportation

StudyTip

Distance In mathematics, just as in real-world situations, distances are always nonnegative.

Let d represent the distance between (x_1, y_1) and (x_2, y_2).

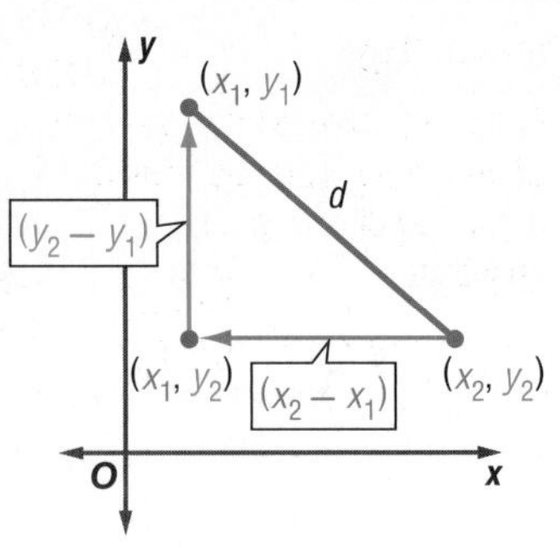

$c^2 = a^2 + b^2$	Pythagorean Theorem
$d^2 = \lvert x_2 - x_1 \rvert^2 + \lvert y_2 - y_1 \rvert^2$	Substitute.
$d^2 = (x_2 - x_1)^2 + (y_2 - y_1)^2$	$\lvert x_2 - x_1 \rvert^2 = (x_2 - x_1)^2$, $\lvert y_2 - y_1 \rvert^2 = (y_2 - y_1)^2$
$d = \sqrt{(x_2 - x_1)^2 + (y_2 - y_1)^2}$	Find the nonnegative square root of each side.

KeyConcept Distance Formula

Words The distance between two points with coordinates (x_1, y_1) and (x_2, y_2) is given by $d = \sqrt{(x_2 - x_1)^2 + (y_2 - y_1)^2}$.

Model

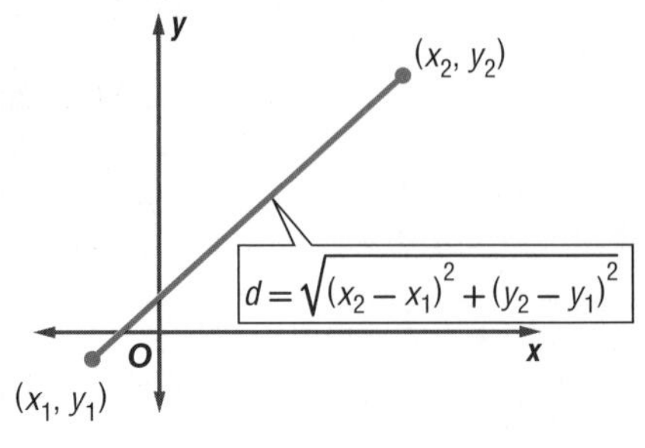

Real-WorldLink

There are more than 700 disc golf courses in the U.S. These courses are permanent installations, usually located in public parks, where players actually "drive" and "putt" with specially-styled discs.

Real-World Example 2 Find the Distance Between Two Points

DISC GOLF Troy's disc is 20 feet short and 8 feet to the right of the basket. On his first putt, the disc lands 2 feet to the left and 3 feet beyond the basket. If the disc went in a straight line, how far did it go?

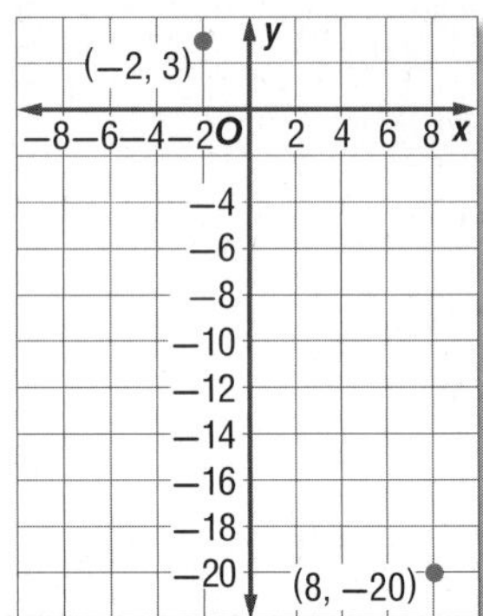

Model the situation. If the basket is at (0, 0), then the location of the disc is (8, −20). The location after the first putt is (−2, 3).

$d = \sqrt{(x_2 - x_1)^2 + (y_2 - y_1)^2}$	Distance Formula
$= \sqrt{(-2 - 8)^2 + [3 - (-20)]^2}$	$(x_1, y_1) = (8, -20)$ and $(x_2, y_2) = (-2, 3)$
$= \sqrt{(-10)^2 + 23^2}$	Simplify.
$= \sqrt{629}$ or about 25	

The disc traveled about 25 feet on his first putt.

GuidedPractice

2. Sharon hits a golf ball 12 feet above the hole and 3 feet to the left. Her first putt traveled to 2 feet above the cup and 1 foot to the right. How far did the ball travel on her first putt?

Rich Reed/National Geographic/Getty Images

StudyTip
Midpoints The coordinates of the midpoint are the means of the coordinates of the endpoints.

There most likely will be problems involving the Midpoint and Distance Formulas on standardized tests you will have to take.

Standardized Test Example 3 Find the Midpoint Between Coordinates

A coordinate grid is placed over a Florida map. St. Augustine is located at (3, 13), and Rockledge is located at (8, −1). If Port Orange is halfway between St. Augustine and Rockledge, which is closest to the distance in coordinate units from St. Augustine to Port Orange?

A 4.75 **B** 7.43 **C** 14.9 **D** 19

Read the Test Item

The question asks you to find the distance between one city and the midpoint. Find the midpoint, and then use the Distance Formula.

Solve the Test Item

Use the Midpoint Formula to find the coordinates of Port Orange.

$$\text{midpoint} = \left(\frac{3+8}{2}, \frac{13+(-1)}{2}\right) \quad \text{Midpoint Formula}$$
$$= (5.5, 6) \quad \text{Simplify.}$$

Use the Distance Formula to find the distance between St. Augustine (3, 13) and Port Orange (5.5, 6).

$$\text{distance} = \sqrt{(3-5.5)^2 + (13-6)^2} \quad \text{Distance Formula}$$
$$= \sqrt{(-2.5)^2 + 7^2} \quad \text{Evaluate exponents and add.}$$
$$= \sqrt{55.25} \text{ or about } 7.43 \quad \text{Simplify.}$$

The answer is B.

Test-TakingTip
Answer Check To check your answer, find the distance between Port Orange and Rockledge. Since Port Orange is at the midpoint, these distances should be equal.

GuidedPractice

3. The coordinates for points *A* and *B* are (−4, −5) and (10, −7), respectively. Find the distance between the midpoint of *A* and *B* and point *B*.

F $\sqrt{10}$ units **G** $5\sqrt{10}$ units **H** $\sqrt{50}$ units **J** $10\sqrt{5}$ units

Check Your Understanding

● = Step-by-Step Solutions begin on page R14.

Example 1 CCSS **PRECISION** **Find the midpoint of the line segment with endpoints at the given coordinates.**

1. (−4, 7), (3, 9)
2. (8, 2), (−1, −5)
3. (11, 6), (18, 13.5)
4. (−12, −2), (−10.5, −6)

Example 2 **Find the distance between each pair of points with the given coordinates.**

5. (3, −5), (13, −11)
6. (8, 1), (−2, 9)
7. (0.25, 1.75), (3.5, 2.5)
8. (−4.5, 10.75), (−6.25, −7)

Example 3 **9. MULTIPLE CHOICE** The map of a mall is overlaid with a numeric grid. The kiosk for the cell phone store is halfway between The Ice Creamery and the See Clearly eyeglass store. If the ice cream store is at (2, 4) and the eyeglass store is at (78, 46), find the distance the kiosk is from the eyeglass store.

A 43.4 units **B** 47.2 units **C** 62.4 units **D** 94.3 units

Practice and Problem Solving

Extra Practice is on page R9.

Example 1 **Find the midpoint of the line segment with endpoints at the given coordinates.**

10. (20, 3), (15, 5) **11.** (−27, 4), (19, −6) **12.** (−0.4, 7), (11, −1.6)

13. (5.4, −8), (9.2, 10) **14.** (−5.3, −8.6), (−18.7, 1) **15.** (−6.4, −8.2), (−9.1, −0.8)

Example 2 **Find the distance between each pair of points with the given coordinates.**

16. (1, 2), (6, 3) **17.** (3, −4), (0, 12)

18. (−6, −7), (11, −12) **19.** (−10, 8), (−8, −8)

20. (4, 0), (5, −6) **21.** (7, 9), (−2, −10)

22. (−4, −5), (15, 17) **23.** (14, −20), (−18, 25)

Example 3 **24. TRACK AND FIELD** A shot put is thrown from the inside of a circle. A coordinate grid is placed over the shot put circle. The toe board is located at the front of the circle at (−4, 1), and the back of the circle is at (5, 2). If the center of the circle is halfway between these two points, what is the distance from the toe board to the center of the circle?

Find the midpoint of the line segment with endpoints at the given coordinates. Then find the distance between the points.

25. (−93, 15), (90, −15) **26.** (−22, 42), (57, 2)

27. (−70, −87), (59, −14) **28.** (−98, 5), (−77, 64)

29. (41, −45), (−25, 75) **30.** (90, 60), (−3, −2)

31. (−1.2, 2.5), (0.34, −7) **32.** (−7.54, 3.89), (4.04, −0.38)

33 $\left(-\frac{5}{12}, -\frac{1}{3}\right), \left(-\frac{17}{2}, -\frac{5}{3}\right)$ **34.** $\left(-\frac{5}{4}, -\frac{13}{2}\right), \left(-\frac{4}{3}, -\frac{5}{6}\right)$

35. $\left(-3\sqrt{2}, -4\sqrt{5}\right), \left(-3\sqrt{3}, 9\right)$ **36.** $\left(\frac{\sqrt{3}}{3}, \frac{\sqrt{2}}{4}\right), \left(\frac{-2\sqrt{3}}{3}, \frac{\sqrt{2}}{4}\right)$

37. SPACE Use the labeled points on the outline of the circular crater on Mars to estimate its diameter in kilometers. Assume each unit on the coordinate system is 1 kilometer.

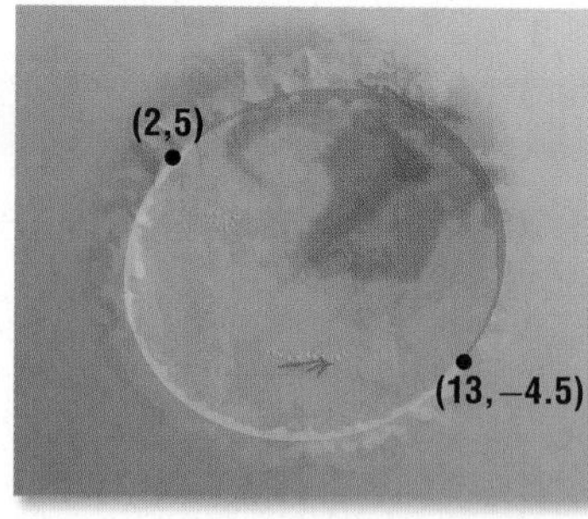

38. CCSS MODELING Triangle *ABC* has vertices *A*(2, 1), *B*(−6, 5), and *C*(−2, −3).

a. An isosceles triangle has two sides with equal length. Is triangle *ABC* isosceles? Explain.

b. An equilateral triangle has three sides of equal length. Is triangle *ABC* equilateral? Explain.

c. Triangle *EFG* is formed by joining the midpoints of the sides of triangle *ABC*. What type of triangle is *EFG*? Explain.

d. Describe any relationship between the lengths of the sides of the two triangles.

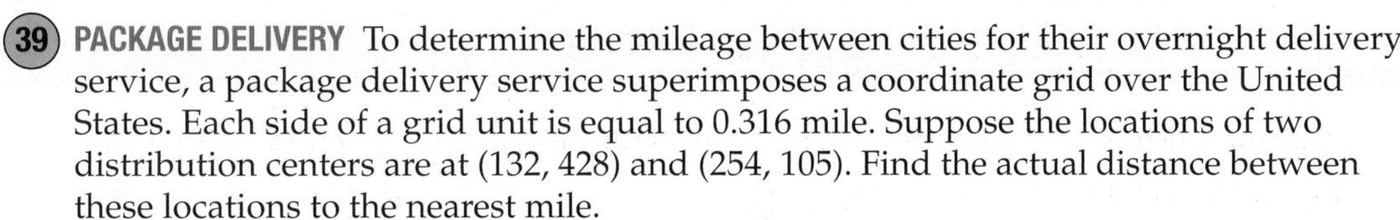

39 **PACKAGE DELIVERY** To determine the mileage between cities for their overnight delivery service, a package delivery service superimposes a coordinate grid over the United States. Each side of a grid unit is equal to 0.316 mile. Suppose the locations of two distribution centers are at (132, 428) and (254, 105). Find the actual distance between these locations to the nearest mile.

40. **HIKING** Orlando wants to hike from his camp to a waterfall. The waterfall is 5 miles south and 8 miles east of his campsite.

a. Use the Distance Formula to determine how far the waterfall is from the campsite.

b. Verify your answer in part a by using the Pythagorean Theorem to determine the distance between the campsite and the waterfall.

c. Orlando wants to stop for lunch halfway to the waterfall. If the camp is at the origin, where should he stop?

41. **MULTIPLE REPRESENTATIONS** Triangle XYZ has vertices $X(4, 9)$, $Y(8, -9)$, and $Z(-6, 5)$.

a. Concrete Draw $\triangle XYZ$ on a coordinate plane.

b. Numerical Find the coordinates of the midpoint of each side of the triangle.

c. Geometric Find the perimeter of $\triangle XYZ$ and the perimeter of the triangle with vertices at the points found in part **b**.

d. Analytical How do the perimeters in part **c** compare?

H.O.T. Problems Use Higher-Order Thinking Skills

42. **CHALLENGE** Find the coordinates of the point that is three fourths of the way from $P(-1, 12)$ to $Q(5, -10)$.

43. **REASONING** Identify all the points in a plane that are 3 units or less from the point (5, 6). What figure does this make?

44. **CCSS ARGUMENTS** Triangle ABC is a right triangle.

a. Find the midpoint of the hypotenuse. Call it point Q.

b. Classify $\triangle BQC$ according to the lengths of its sides. Include sufficient evidence to support your conclusion.

c. Classify $\triangle BQA$ according to its angles.

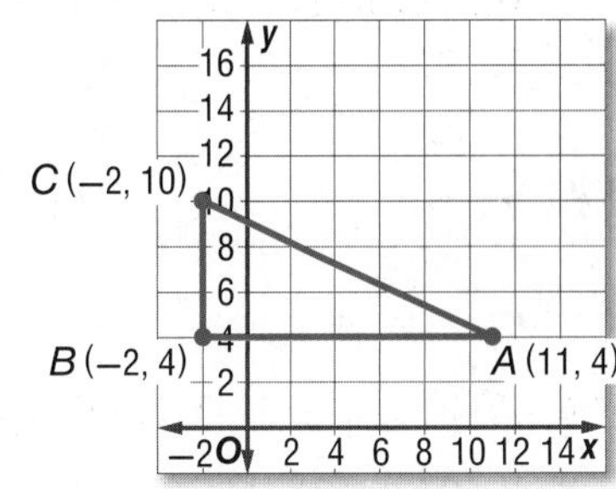

45. **OPEN ENDED** Plot two points, and find the distance between them. Does it matter which ordered pair is first when using the Distance Formula? Explain.

46. **WRITING IN MATH** Explain how the Midpoint Formula can be used to approximate the halfway point between two locations on a map.

Standardized Test Practice

47. SHORT RESPONSE You currently earn \$8.10 per hour and your boss gives you a 10% raise. What is your new hourly wage?

48. SAT/ACT A right circular cylinder has a radius of 3 and a height of 5. Which of the following dimensions of a rectangular solid will have a volume closest to that of the cylinder?

A 5, 5, 6
B 5, 6, 6
C 5, 5, 5
D 4, 5, 6
E 3, 5, 9

49. GEOMETRY If the sum of the lengths of the two legs of a right triangle is 49 inches and the hypotenuse is 41 inches, find the longer of the two legs.

F 9 in.
G 40 in.
H 42 in.
J 49 in.

50. Five more than 3 times a number is 17. Find the number.

A 3
B 4
C 5
D 6

Spiral Review

Solve each equation. Check your solutions. (Lesson 8-6)

51. $\frac{12}{v^2 - 16} - \frac{24}{v - 4} = 3$

52. $\frac{w}{w - 1} + w = \frac{4w - 3}{w - 1}$

53. $\frac{4n^2}{n^2 - 9} - \frac{2n}{n + 3} = \frac{3}{n - 3}$

54. SWIMMING When a person swims under water, the pressure in his or her ears varies directly with the depth at which he or she is swimming. (Lesson 8-5)

a. Write a direct variation equation that represents this situation.

b. Find the pressure at 60 feet.

c. It is unsafe for amateur divers to swim where the water pressure is more than 65 pounds per square inch. How deep can an amateur diver safely swim?

d. Make a table showing the number of pounds of pressure at various depths of water. Use the data to draw a graph of pressure versus depth.

Solve each equation or inequality. Round to the nearest ten-thousandth. (Lesson 7-6)

55. $9^{z - 4} = 6.28$

56. $8.2^{n - 3} = 42.5$

57. $2.1^{t - 5} = 9.32$

58. $8^{2n} > 52^{4n + 3}$

59. $7^{p + 2} \leq 13^{5 - p}$

60. $3^{y + 2} \geq 8^{3y}$

Solve each equation. (Lesson 6-7)

61. $(6n - 5)^{\frac{1}{3}} + 3 = -2$

62. $(5x + 7)^{\frac{1}{5}} + 3 = 5$

63. $(3x - 2)^{\frac{1}{5}} + 6 = 5$

Skills Review

Write each quadratic equation in vertex form. Then identify the vertex, axis of symmetry, and direction of opening.

64. $y = -x^2 - 4x + 8$

65. $y = x^2 - 6x + 1$

66. $y = -2x^2 + 20x - 35$

LESSON 9-2

Parabolas

Then	Now	Why?
• You graphed quadratic functions.	1 Write equations of parabolas in standard form. 2 Graph parabolas.	• Satellite dishes can be used to send and receive signals and can be seen attached to residential homes and businesses. A satellite dish is a type of antenna constructed to receive signals from orbiting satellites. The signals are reflected off of the dish's parabolic surface to a common collection point.

NewVocabulary
parabola
focus
directrix
latus rectum
standard form
general form

Common Core State Standards

Content Standards
A.SSE.1.b Interpret complicated expressions by viewing one or more of their parts as a single entity.
A.CED.2 Create equations in two or more variables to represent relationships between quantities; graph equations on coordinate axes with labels and scales.

Mathematical Practices
1 Make sense of problems and persevere in solving them.

1 Equations of Parabolas

A **parabola** can be defined as the set of all points in a plane that are the same distance from a given point called the **focus** and a given line called the **directrix**.

The line segment through the focus of a parabola and perpendicular to the axis of symmetry is called the **latus rectum**. The endpoints of the latus rectum lie on the parabola.

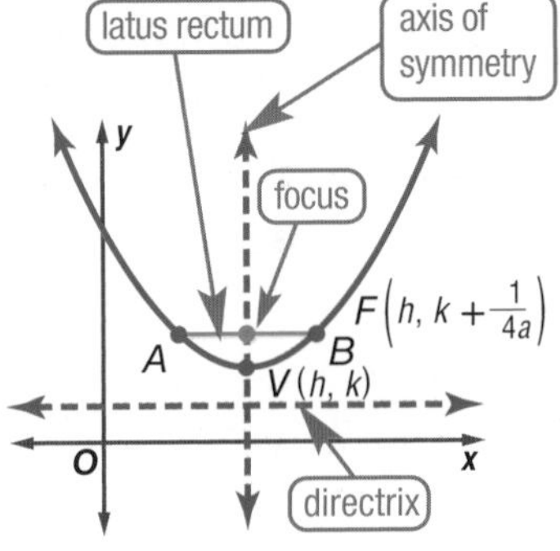

KeyConcept Equations of Parabolas

Form of Equation	$y = a(x - h)^2 + k$	$x = a(y - k)^2 + h$
Direction of Opening	upward if $a > 0$, downward if $a < 0$	right if $a > 0$, left if $a < 0$
Vertex	(h, k)	(h, k)
Axis of Symmetry	$x = h$	$y = k$
Focus	$\left(h, k + \frac{1}{4a}\right)$	$\left(h + \frac{1}{4a}, k\right)$
Directrix	$y = k - \frac{1}{4a}$	$x = h - \frac{1}{4a}$
Length of Latus Rectum	$\left\lvert\frac{1}{a}\right\rvert$ units	$\left\lvert\frac{1}{a}\right\rvert$ units

The **standard form** of the equation of a parabola with vertex (h, k) and axis of symmetry $x = h$ is $y = a(x - h)^2 + k$.

- If $a > 0$, k is the minimum value of the related function and the parabola opens upward.
- If $a < 0$, k is the maximum value of the related function and the parabola opens downward.

An equation of a parabola in the form $y = ax^2 + bx + c$ is the **general form**. Any equation in general form can be written in standard form. The shape of a parabola and the distance between the focus and directrix depend on the value of a in the equation.

Sean Russell/Getty Images

ReviewVocabulary

Completing the Square rewriting a quadratic expression as a perfect square trinomial

Example 1 Analyze the Equation of a Parabola

Write $y = 2x^2 - 12x + 6$ in standard form. Identify the vertex, axis of symmetry, and direction of opening of the parabola.

$y = 2x^2 - 12x + 6$	Original equation
$= 2(x^2 - 6x) + 6$	Factor 2 from the x- and x^2-terms.
$= 2(x^2 - 6x + \blacksquare) + 6 - 2(\blacksquare)$	Complete the square on the right side.
$= 2(x^2 - 6x + 9) + 6 - 2(9)$	The 9 added when you complete the square is multiplied by 2.
$= 2(x - 3)^2 - 12$	Factor.

The vertex of this parabola is located at $(3, -12)$, and the equation of the axis of symmetry is $x = 3$. The parabola opens upward.

GuidedPractice

1. Write $y = 4x^2 + 16x + 34$ in standard form. Identify the vertex, axis of symmetry, and direction of opening of the parabola.

2 Graph Parabolas

In Chapter 4, you learned that the graph of the quadratic equation $y = a(x - h)^2 + k$ is a transformation of the parent graph of $y = x^2$ translated h units horizontally and k units vertically, and reflected and/or dilated depending on the value of a.

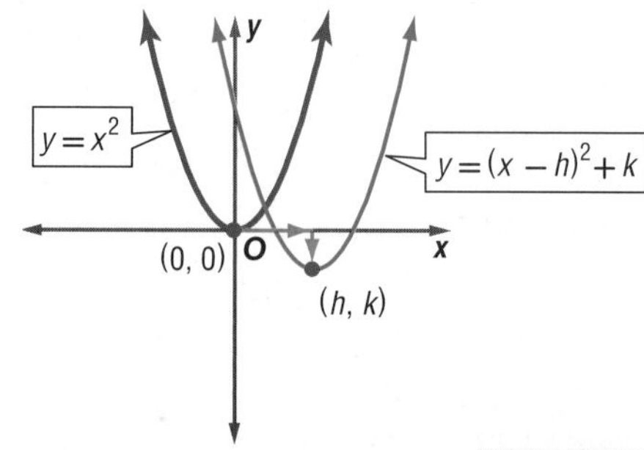

Example 2 Graph Parabolas

WatchOut!

CCSS Structure Carefully examine the values for h and k before beginning to graph an equation.

- If h is positive, translate the graph h units to the right.
- If h is negative, translate the graph $|h|$ units to the left.
- If k is positive, translate the graph k units up.
- If k is negative, translate the graph $|k|$ units down.

Graph each equation.

a. $y = -3x^2$

For this equation, $h = 0$ and $k = 0$. The vertex is at the origin. Since the equation of the axis of symmetry is $x = 0$, substitute some small positive integers for x and find the corresponding y-values.

x	y
1	−3
2	−12
3	−27

Since the graph is symmetric about the y-axis, the points at $(-1, -3)$, $(-2, -12)$, and $(-3, -27)$ are also on the parabola. Use all of these points to draw the graph.

b. $y = -3(x - 4)^2 + 5$

The equation is of the form $y = a(x - h)^2 + k$, where $h = 4$ and $k = 5$. The graph of this equation is the graph of $y = -3x^2$ in part **a** translated 4 units to the right and up 5 units. The vertex is now at $(4, 5)$.

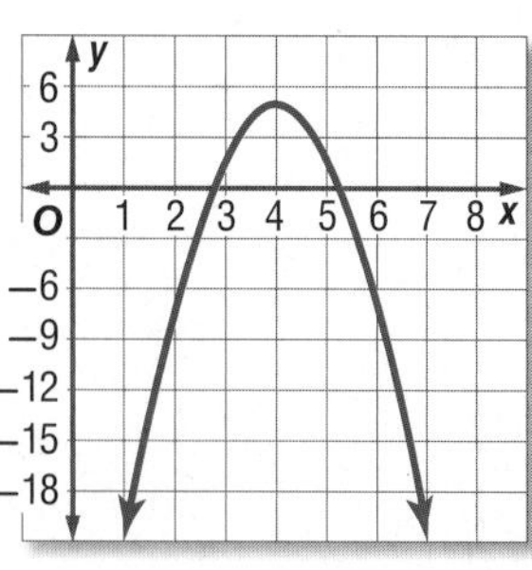

GuidedPractice

2A. $y = 2x^2$

2B. $y = 2(x - 1)^2 - 4$

StudyTip

Graphing When graphing these functions, it may be helpful to sketch the graph of the parent function.

Equations of parabolas with vertical axes of symmetry have the parent function $y = x^2$ and are of the form $y = a(x - h)^2 + k$. These are functions. Equations of parabolas with horizontal axes of symmetry are of the form $x = a(y - k)^2 + h$ and are not functions. The parent graph for these equations is $x = y^2$.

Example 3 Graph an Equation in General Form

Graph each equation.

a. $\mathbf{2x - y^2 = 4y + 10}$

Step 1 Write the equation in the form $x = a(y - k)^2 + h$.

$2x - y^2 = 4y + 10$	Original equation
$2x = y^2 + 4y + 10$	Add y^2 to each side to isolate the x-term.
$2x = (y^2 + 4y + \blacksquare) + 10 - \blacksquare$	Complete the square.
$2x = (y^2 + 4y + 4) + 10 - 4$	Add and subtract 4, since $\left(\frac{4}{2}\right)^2 = 4$.
$2x = (y + 2)^2 + 6$	Factor and subtract.
$x = \frac{1}{2}(y + 2)^2 + 3$	$(h, k) = (3, -2)$

Step 2 Use the equation to find information about the graph. Then draw the graph based on the parent graph, $x = y^2$.

vertex: $(3, -2)$

axis of symmetry: $y = -2$

focus: $\left(3 + \frac{1}{4\left(\frac{1}{2}\right)}, -2\right)$ or $(3.5, -2)$

directrix: $x = 3 - \frac{1}{4\left(\frac{1}{2}\right)}$ or 2.5

direction of opening: right, since $a > 0$

length of latus rectum: $\left|\frac{1}{\left(\frac{1}{2}\right)}\right|$ or 2 units

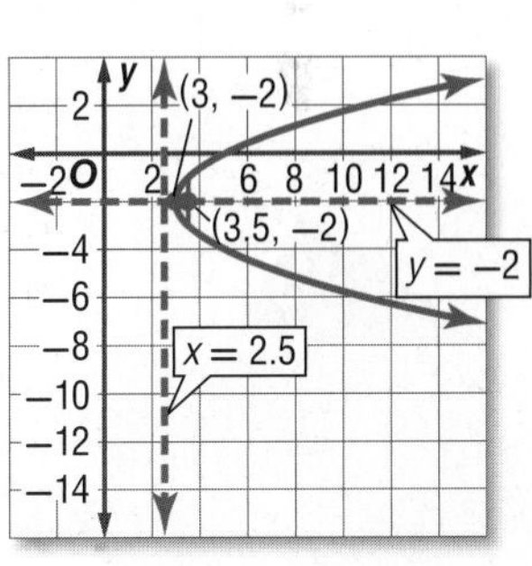

ReadingMath

latus rectum from the Latin *latus*, meaning side, and *rectum*, meaning straight

b. $\mathbf{y + 2x^2 + 32 = -16x - 1}$

Step 1

$y + 2x^2 + 32 = -16x - 1$	Original equation
$y = -2x^2 - 16x - 33$	Solve for y.
$y = -2(x^2 + 8x + \blacksquare) - 33 - \blacksquare$	Complete the square.
$y = -2(x^2 + 8x + 16) - 33 - (-32)$	Add and subtract −32.
$y = -2(x + 4)^2 - 1$	Factor and simplify.

Step 2

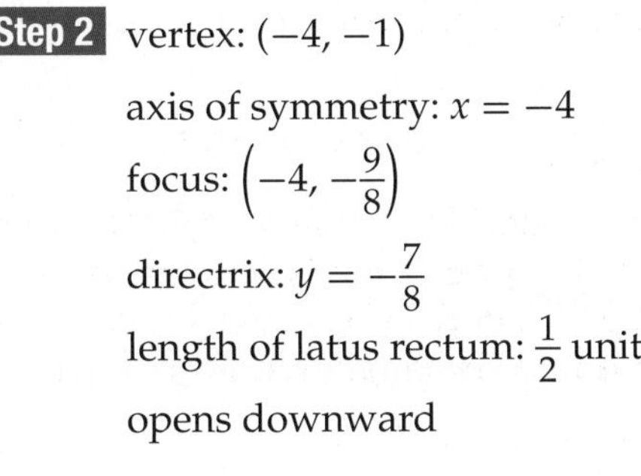

vertex: $(-4, -1)$

axis of symmetry: $x = -4$

focus: $\left(-4, -\frac{9}{8}\right)$

directrix: $y = -\frac{7}{8}$

length of latus rectum: $\frac{1}{2}$ unit

opens downward

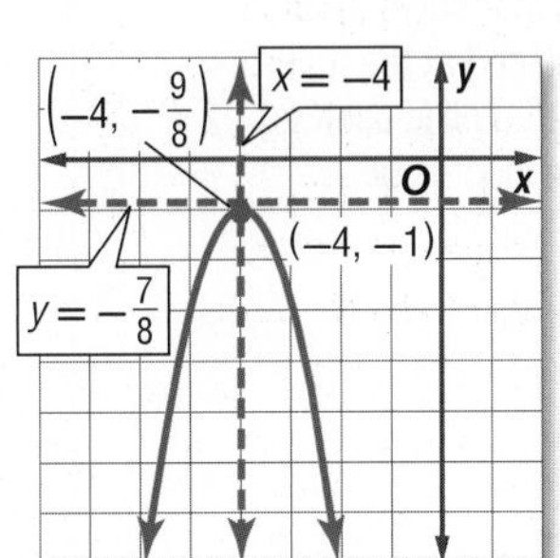

GuidedPractice

3A. $3x - y^2 = 4x + 25$

3B. $y = x^2 + 6x - 4$

You can use specific information about a parabola to write an equation and draw a graph.

Example 4 Write an Equation of a Parabola

Write an equation for a parabola with vertex at (−2, −4) and directrix $y = 1$. Then graph the equation.

The directrix is a horizontal line, so the equation of the parabola is of the form $y = a(x - h)^2 + k$. Find a, h, and k.

- The vertex is at (−2, −4), so $h = -2$ and $k = -4$.
- Use the equation of the directrix to find a.

$y = k - \frac{1}{4a}$	Equation of directrix
$1 = -4 - \frac{1}{4a}$	Replace y with 1 and k with −4.
$5 = -\frac{1}{4a}$	Add 4 to each side.
$20a = -1$	Multiply each side by $4a$.
$a = -\frac{1}{20}$	Divide each side by 20.

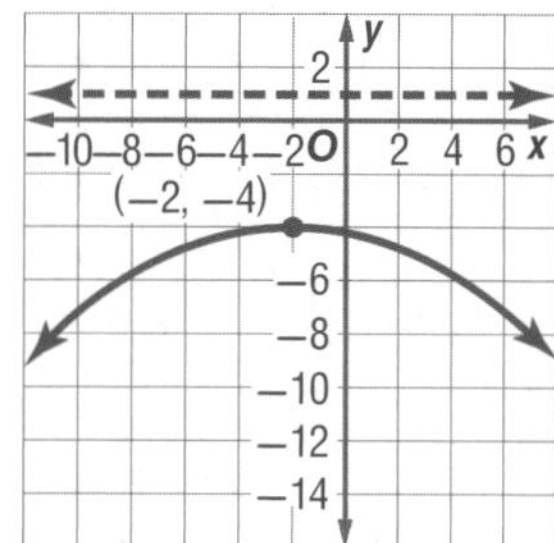

So, the equation of the parabola is $y = -\frac{1}{20}(x + 2)^2 - 4$.

GuidedPractice

Write an equation for each parabola described below. Then graph the equation.

4A. vertex (1, 3), focus (1, 5)

4B. focus (5, 6), directrix $x = -2$

Real-WorldLink

In California's Mojave Desert, parabolic mirrors are used to heat oil that flows through tubes placed at the focus. The heated oil is used to produce electricity.

Source: Solel

Parabolas are often used in the real world.

Real-World Example 5 Write an Equation for a Parabola

ENVIRONMENT Solar energy may be harnessed by using parabolic mirrors. The mirrors reflect the rays from the Sun to the focus of the parabola. The focus of each parabolic mirror at the facility described at the left is 6.25 feet above the vertex. The latus rectum is 25 feet long.

a. Assume that the focus is at the origin. Write an equation for the parabola formed by each mirror.

In order for the mirrors to collect the Sun's energy, the parabola must open upward. Therefore, the vertex must be below the focus.

focus: (0, 0)   vertex: (0, −6.25)

The measure of the latus rectum is 25. So $25 = \left|\frac{1}{a}\right|$, and $a = \frac{1}{25}$.

Using the form $y = a(x - h)^2 + k$, an equation for the parabola formed by each mirror is $y = \frac{1}{25}x^2 - 6.25$.

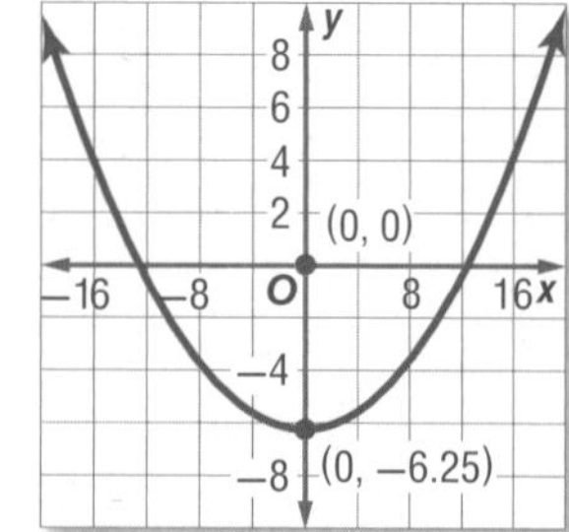

b. Graph the equation.

Now use all of the information to draw a graph.

GuidedPractice

5. Write and graph an equation for a parabolic mirror that has a focus 4.5 feet above the vertex and a latus rectum that is 18 feet long, when the focus is at the origin.

Check Your Understanding

= Step-by-Step Solutions begin on page R14.

Example 1 **Write each equation in standard form. Identify the vertex, axis of symmetry, and direction of opening of the parabola.**

1. $y = 2x^2 - 24x + 40$
2. $y = 3x^2 - 6x - 4$
3. $x = y^2 - 8y - 11$
4. $x + 3y^2 + 12y = 18$

Examples 2–3 **Graph each equation.**

5. $y = (x - 4)^2 - 6$
6. $y = 4(x + 5)^2 + 3$
7. $y = -3x^2 - 4x - 8$
8. $x = 3y^2 - 6y + 9$

Example 4 **Write an equation for each parabola described below. Then graph the equation.**

9. vertex (0, 2), focus (0, 4)
10. vertex (−2, 4), directrix $x = -1$
11. focus (3, 2), directrix $y = 8$
12. vertex (−1, −5), focus (−5, −5)

Example 5

13. **ASTRONOMY** Consider a parabolic mercury mirror like the one described at the beginning of the lesson. The focus is 6 feet above the vertex and the latus rectum is 24 feet long.
    a. Assume that the focus is at the origin. Write an equation for the parabola formed by the parabolic microphone.
    b. Graph the equation.

Practice and Problem Solving

Extra Practice is on page R9.

Example 1 **Write each equation in standard form. Identify the vertex, axis of symmetry, and direction of opening of the parabola.**

14. $y = x^2 - 8x + 13$
15. $y = 3x^2 + 42x + 149$
16. $y = -6x^2 - 36x - 8$
17. $y = -3x^2 - 9x - 6$
18. $x = \frac{1}{3}y^2 - 3y + 4$
19. $x = \frac{2}{3}y^2 - 4y + 12$

Examples 2–3 **Graph each equation.**

20. $y = \frac{1}{3}x^2$
21. $y = -2x^2$
22. $y = -2(x - 2)^2 + 3$
23. $y = 3(x - 3)^2 - 5$
24. $x = \frac{1}{2}y^2$
25. $4x - y^2 = 2y + 13$

Example 4 **Write an equation for each parabola described below. Then graph the equation.**

26. vertex (0, 1), focus (0, 4)
27. vertex (1, 8), directrix $y = 3$
28. focus (−2, −4), directrix $x = -6$
29. focus (2, 4), directrix $x = 10$
30. vertex (−6, 0), directrix $x = 2$
31. vertex (9, 6), focus (9, 5)

Example 5

32. **BASEBALL** When a ball is thrown, the path it travels is a parabola. Suppose a baseball is thrown from ground level, reaches a maximum height of 50 feet, and hits the ground 200 feet from where it was thrown. Assuming this situation could be modeled on a coordinate plane with the focus of the parabola at the origin, find the equation of the parabolic path of the ball. Assume the focus is on ground level.

33. **CCSS PERSEVERANCE** Ground antennas and satellites are used to relay signals between the NASA Mission Operations Center and the spacecraft it controls. One such parabolic dish is 146 feet in diameter. Its focus is 48 feet from the vertex.
    a. Sketch two options for the dish, one that opens up and one that opens left.
    b. Write two equations that model the sketches in part **a**.
    c. If you wanted to find the depth of the dish, does it matter which equation you use? Why or why not?

34. UMBRELLAS A beach umbrella has an arch in the shape of a parabola that opens downward. The umbrella spans 6 feet across and is $1\frac{1}{2}$ feet high. Write an equation of a parabola to model the arch, assuming that the origin is at the point where the pole and umbrella meet at the vertex of the arch.

35) AUTOMOBILES An automobile headlight contains a parabolic reflector. The light coming from the source bounces off the parabolic reflector and shines out the front of the headlight. The equation of the cross section of the reflector is $y = \frac{1}{12}x^2$. How far from the vertex should the filament for the high beams be placed?

36. MULTIPLE REPRESENTATIONS Start with a sheet of wax paper that is about 15 inches long and 12 inches wide.

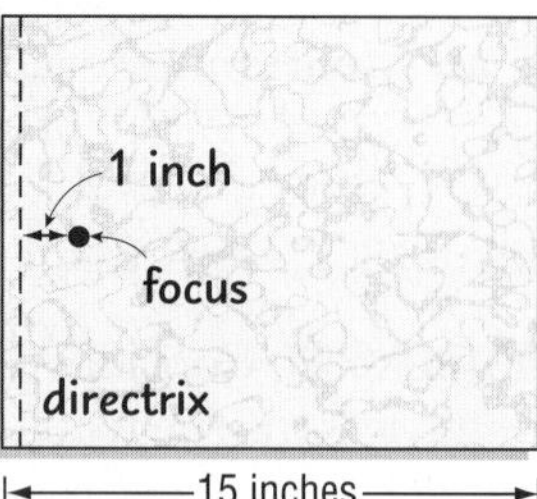

a. Concrete Make a line that is perpendicular to the sides of the sheet by folding the sheet near one end. Open up the paper again. This line is the directrix. Mark a point about midway between the sides of the sheet so that the distance from the directrix is about 1 inch. This is the focus.

b. Concrete Start with a new sheet of wax paper. Form another outline of a parabola with a focus that is about 3 inches from the directrix.

c. Concrete On a new sheet of a wax paper, form a third outline of a parabola with a focus that is about 5 inches from the directrix.

d. Verbal Compare the shapes of the three parabolas. How does the distance between the focus and the directrix affect the shape of a parabola?

H.O.T. Problems Use Higher-Order Thinking Skills

37. REASONING How do you change the equation of the parent function $y = x^2$ to shift the graph to the right?

38. OPEN ENDED Two different parabolas have their vertex at $(-3, 1)$ and contain the point with coordinates $(-1, 0)$. Write two possible equations for these parabolas.

39. CCSS CRITIQUE Brianna and Russell are graphing $\frac{1}{4}y^2 + x = 0$. Is either of them correct? Explain your reasoning.

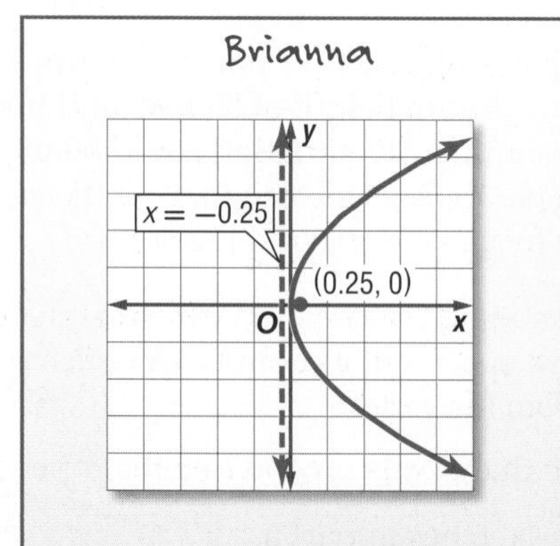

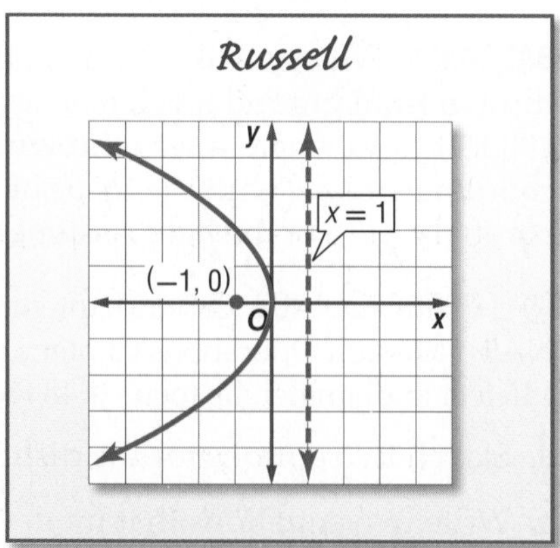

40. WRITING IN MATH Why are parabolic shapes used in the real world?

Standardized Test Practice

41. A gardener is placing a fence around a 1320-square-foot rectangular garden. He ordered 148 feet of fencing. If he uses all the fencing, what is the length of the longer side of the garden?

A 30 ft
B 34 ft
C 44 ft
D 46 ft

42. SAT/ACT When a number is divided by 5, the result is 7 more than the number. Find the number.

F $-\frac{35}{4}$
G $-\frac{35}{6}$
H $\frac{35}{6}$
J $\frac{28}{4}$
K $\frac{35}{4}$

43. GEOMETRY What is the area of the following square, if the length of $\overline{BD}$ is $2\sqrt{2}$?

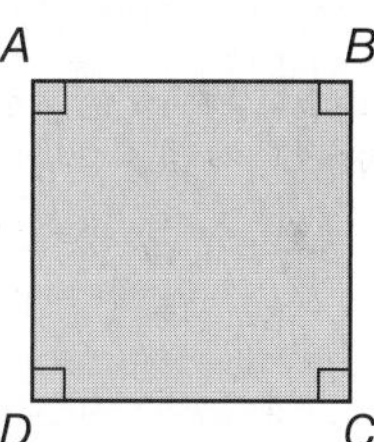

A 1
B 2
C 3
D 4

44. SHORT RESPONSE The measure of the smallest angle of a triangle is two thirds the measure of the middle angle. The measure of the middle angle is three sevenths of the measure of the largest angle. Find the largest angle's measure.

Spiral Review

45. GEOMETRY Find the perimeter of a triangle with vertices at (2, 4), (−1, 3), and (1, −3). (Lesson 9-1)

46. WORK A worker can powerwash a wall of a certain size in 5 hours. Another worker can do the same job in 4 hours. If the workers work together, how long would it take to do the job? Determine whether your answer is reasonable. (Lesson 8-6)

Solve each equation or inequality. Round to the nearest ten-thousandth. (Lesson 7-7)

47. $\ln (x + 1) = 1$
48. $\ln (x - 7) = 2$
49. $e^x > 1.6$
50. $e^{5x} \geq 25$

Simplify. (Lesson 6-4)

51. $\sqrt{0.25}$
52. $\sqrt[3]{-0.064}$
53. $\sqrt[4]{z^8}$
54. $-\sqrt[6]{x^6}$

List all of the possible rational zeros of each function. (Lesson 5-8)

55. $h(x) = x^3 + 8x + 6$
56. $p(x) = 3x^3 - 5x^2 - 11x + 3$
57. $h(x) = 9x^6 - 5x^3 + 27$

Skills Review

Simplify each expression.

58. $\sqrt{24}$
59. $\sqrt{45}$
60. $\sqrt{252}$
61. $\sqrt{512}$

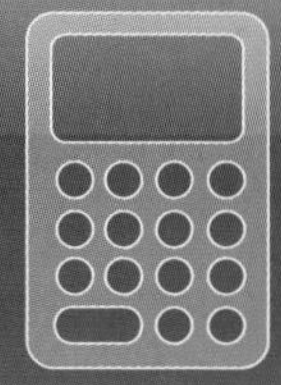

EXPLORE 9-3 Graphing Technology Lab
Equations of Circles

You can use TI-Nspire technology to examine characteristics of circles and the relationship with an equation of the circle.

Activity

Step 1 Draw a circle.

- Add a new **Graphs** page. Select **Window/Zoom** menu and use the **Windows Setting** tool to adjust the window size as shown. From the **View** menu, select **Show Grid**. Then from the **Shapes** menu, select **Circle**. Place the pointer at the point (2, 2) and press **enter** to set the center of the circle. Move the pointer out, creating a circle like the one shown.
- Use the **Point On** tool from the **Points & Line** menu to place a point on the circle.
- Use the **Segment** tool from the **Points & Line** menu to draw the radius.

Step 2 Add labels.

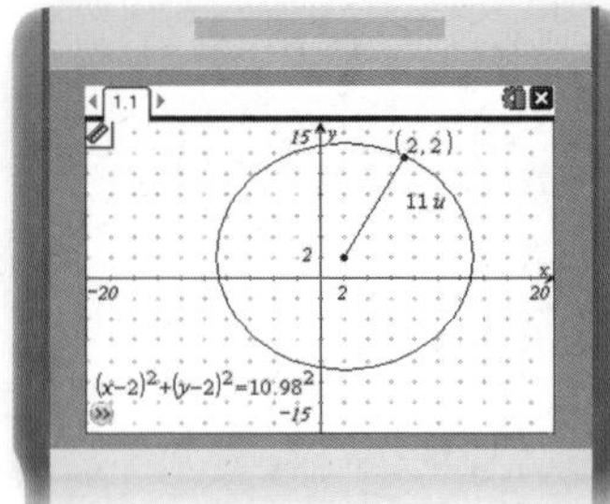

- From the **Actions** menu, select **Coordinates and Equations**. Use the pointer to select the center of the circle to display its coordinate. Then select the circle to display its equation. Move each display outside the circle.
- Use the **Length** tool from the **Measurement** menu to display the length of the radius.

Step 3 Change the radius.

Move the pointer so that a point on the circle is highlighted, then press and hold the center of the touchpad until it is selected. Examine the equation of the circle. Then move the edge of the circle in. Make note of changes in the equation.

Step 4 Move the center of the circle.

Move the pointer so that the center of the circle is highlighted, then press and hold the center of the touchpad until it is selected. Move the center of the circle. Again, examine the equation of the circle.

Analyze the Results

1. How does moving the edge of the circle in or out affect the equation of the circle?
2. What effect does moving the center of the circle have on the equation?
3. Repeat the activity by placing the center of a circle in Quadrant II. Move the center to each of the other two quadrants. How does the equation change?
4. **MAKE A CONJECTURE** Without graphing, write an equation of each circle.
   a. center: (4, 2), radius: 3
   b. center: (−1, 1), radius: 8
   c. center: (−6, −5), radius: 2.5
   d. center: (h, k), radius: r

LESSON

9-3 Circles

Then

- You graphed and wrote equations of parabolas.

Now

1. Write equations of circles.
2. Graph circles.

Why?

- When an object is thrown into water, ripples move out from the center forming concentric circles. If the point where the object entered the water is assigned coordinates, each ripple can be modeled by an equation of a circle.

NewVocabulary

circle
center
radius

Common Core State Standards

Content Standards

A.SSE.1.b Interpret complicated expressions by viewing one or more of their parts as a single entity.

A.CED.4 Rearrange formulas to highlight a quantity of interest, using the same reasoning as in solving equations.

Mathematical Practices

4 Model with mathematics.

1 Equations of Circles

A **circle** is the set of all points in a plane that are equidistant from a given point in the plane, called the **center**. Any segment with endpoints at the center and a point on the circle is a **radius** of the circle.

Assume that (x, y) are the coordinates of a point on the circle at the right. The center is at (h, k), and the radius is r. You can find an equation of the circle by using the Distance Formula.

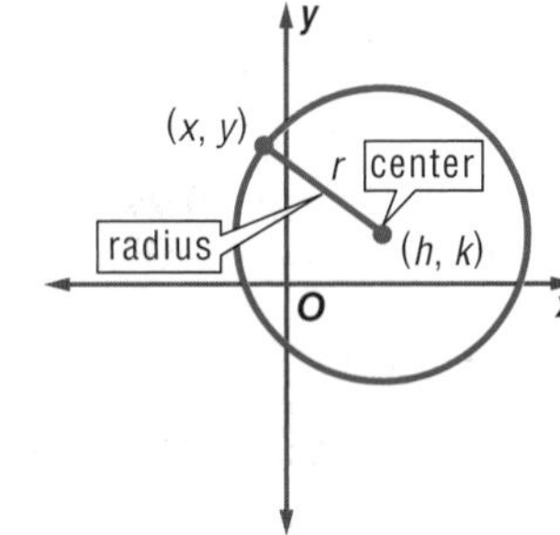

$\sqrt{(x_2 - x_1)^2 + (y_2 - y_1)^2} = d$ Distance Formula

$\sqrt{(x - h)^2 + (y - k)^2} = r$ $(x_1, y_1) = (h, k)$, $(x_2, y_2) = (x, y)$, $d = r$

$(x - h)^2 + (y - k)^2 = r^2$ Square each side.

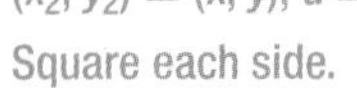

KeyConcept Equations of Circles

Standard Form of Equation	$x^2 + y^2 = r^2$	$(x - h)^2 + (y - k)^2 = r^2$
Center	$(0, 0)$	(h, k)
Radius	r	r

You can use the standard form of the equation of a circle to write an equation for a circle given the center and the radius or diameter.

Real-World Example 1 Write an Equation Given the Radius

DELIVERY Appliances + More offers free delivery within 35 miles of the store. The Jacksonville store is located 100 miles north and 45 miles east of the corporate office. Write an equation to represent the delivery boundary of the Jacksonville store if the origin of the coordinate system is the corporate office.

Since the corporate office is at $(0, 0)$, the Jacksonville store is at $(45, 100)$. The boundary of the delivery region is the circle centered at $(45, 100)$ with radius 35 miles.

$(x - h)^2 + (y - k)^2 = r^2$ Equation of a circle

$(x - 45)^2 + (y - 100)^2 = 35^2$ $(h, k) = (45, 100)$ and $r = 35$

$(x - 45)^2 + (y - 100)^2 = 1225$ Simplify.

GuidedPractice

1. **WI-FI** A certain wi-fi phone has a range of 30 miles in any direction. If the phone is 4 miles south and 3 miles west of headquarters, write an equation to represent the area within which the phone can operate via the Wi-Fi system.

Le Club Symphonie/Ian Nolan/Photodisc/Getty Images

You can write the equation of a circle when you know the location of the center and a point on the circle.

StudyTip

Center-Radius Form Standard form is sometimes referred to as *center-radius form* because the center and radius of the circle are apparent in the equation.

PT

Example 2 Write an Equation from a Graph

Write an equation for the graph.

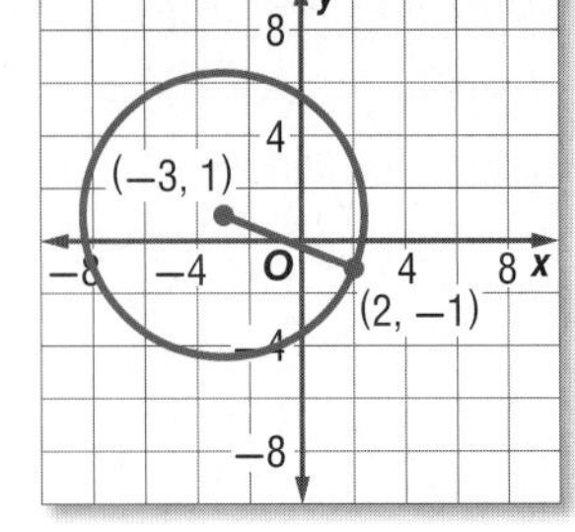

$(x - h)^2 + (y - k)^2 = r^2$ — Standard form

$(2 + 3)^2 + (-1 - 1)^2 = r^2$ — $x = 2, y = -1, h = -3, k = 1$

$(5)^2 + (-2)^2 = r^2$ — Simplify.

$25 + 4 = r^2$ — Evaluate the exponents.

$29 = r^2$ — Add.

So, the equation of the circle is $(x + 3)^2 + (y - 1)^2 = 29$.

GuidedPractice

2A.

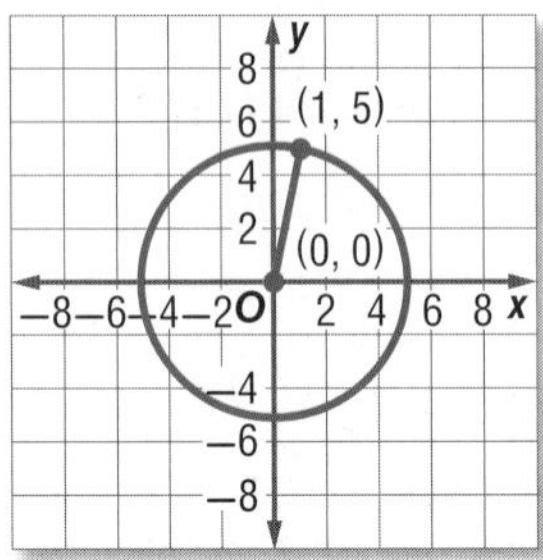

2B.

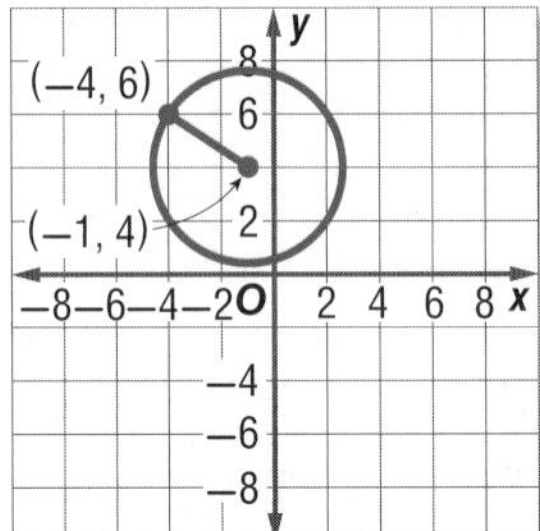

You can use the Midpoint and Distance Formulas when you know the endpoints of the radius or diameter of a circle.

Example 3 Write an Equation Given a Diameter

Write an equation for a circle if the endpoints of a diameter are at (7, 6) and (−1, −8).

Step 1 Find the center.

$(h, k) = \left(\frac{x_1 + x_2}{2}, \frac{y_1 + y_2}{2}\right)$ — Midpoint Formula

$= \left(\frac{7 + (-1)}{2}, \frac{6 + (-8)}{2}\right)$ — $(x_1, y_1) = (7, 6), (x_2, y_2) = (-1, -8)$

$= \left(\frac{6}{2}, \frac{-2}{2}\right)$ — Add.

$= (3, -1)$ — Simplify.

Step 2 Find the radius.

$r = \sqrt{(x_2 - x_1)^2 + (y_2 - y_1)^2}$ — Distance Formula

$= \sqrt{(3 - 7)^2 + (-1 - 6)^2}$ — $(x_1, y_1) = (7, 6), (x_2, y_2) = (3, -1)$

$= \sqrt{(-4)^2 + (-7)^2}$ — Subtract.

$= \sqrt{65}$ — Simplify.

The radius of the circle is $\sqrt{65}$ units, so $r^2 = 65$. Substitute h, k, and r^2 into the standard form of the equation of a circle. An equation of the circle is $(x - 3)^2 + (y + 1)^2 = 65$.

GuidedPractice

3. Write an equation for a circle if the endpoints of a diameter are at (3, −3) and (1, 5).

StudyTip

Axis of Symmetry Every diameter in a circle is an axis of symmetry. There are infinitely many axes of symmetry in a circle.

2 Graph Circles

You can use symmetry to help you graph circles.

Example 4 Graph an Equation in Standard Form

Find the center and radius of the circle with equation $x^2 + y^2 = 100$. Then graph the circle.

- The center of the circle is at (0, 0), and the radius is 10.
- The table lists some integer values for x and y that satisfy the equation.

x	y
0	10
6	8
8	6
10	0

- Because the circle is centered at the origin, it is symmetric about the y-axis. Therefore, the points at (−6, 8), (−8, 6), and (−10, 0) lie on the graph.
- The circle is also symmetric about the x-axis, so the points (−6, −8), (−8, −6), (0, −10), (6, −8), and (8, −6) lie on the graph.
- Plot all of these points and draw the circle that passes through them.

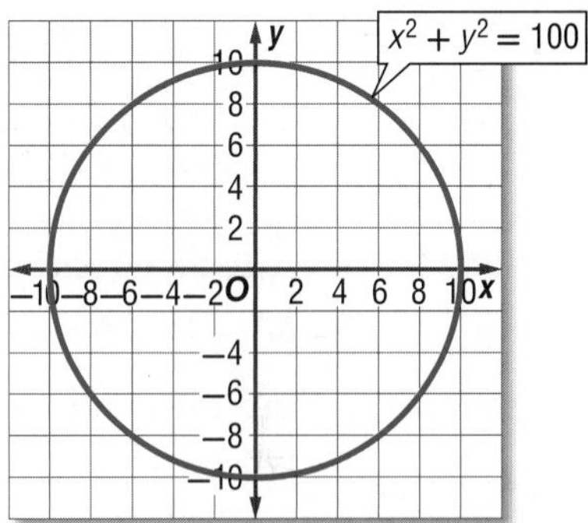

GuidedPractice

4. Find the center and radius of the circle with equation $x^2 + y^2 = 81$. Then graph the circle.

Circles with centers that are not (0, 0) can be graphed by using translations. The graph of $(x - h)^2 + (y - k)^2 = r^2$ is the graph of $x^2 + y^2 = r^2$ translated h units horizontally and k units vertically.

Example 5 Graph an Equation Not in Standard Form

Find the center and radius of the circle with equation $x^2 + y^2 - 8x + 12y - 12 = 0$. Then graph the circle.

Complete the squares.

$$x^2 + y^2 - 8x + 12y - 12 = 0$$
$$x^2 - 8x + \blacksquare + y^2 + 12y + \blacksquare = 12 + \blacksquare + \blacksquare$$
$$x^2 - 8x + 16 + y^2 + 12y + 36 = 12 + 16 + 36$$
$$(x - 4)^2 + (y + 6)^2 = 64$$

The center of the circle is at (4, −6), and the radius is 8. The graph of $(x - 4)^2 + (y + 6)^2 = 64$ is the same as $x^2 + y^2 = 64$ translated 4 units to the right and down 6 units.

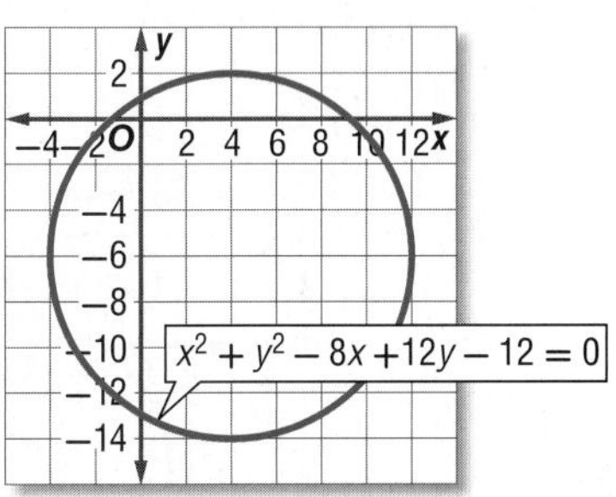

GuidedPractice

5. Find the center and radius of the circle with equation $x^2 + y^2 + 4x - 10y - 7 = 0$. Then graph the circle.

Check Your Understanding

= Step-by-Step Solutions begin on page R14.

Example 1

1. **WEATHER** On average, the eye of a tornado is about 200 feet across. Suppose the center of the eye is at the point (72, 39). Write an equation to represent the boundary of the eye.

Write an equation for each circle given the center and radius.

2. center: $(-2, -6)$, $r = 4$ units
3. center: $(1, -5)$, $r = 3$ units

Example 2

Write an equation for each graph.

4. 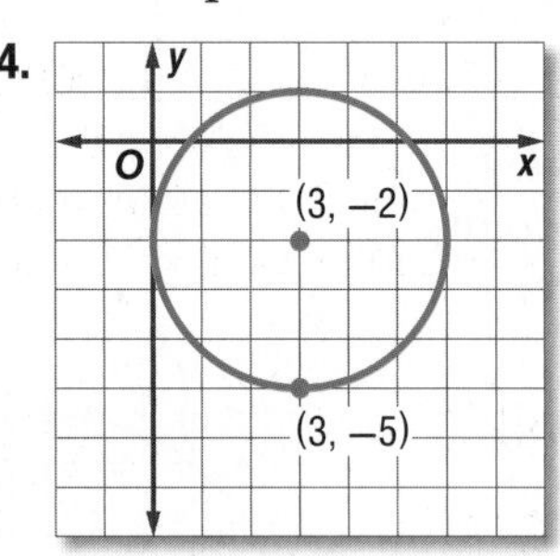

5. 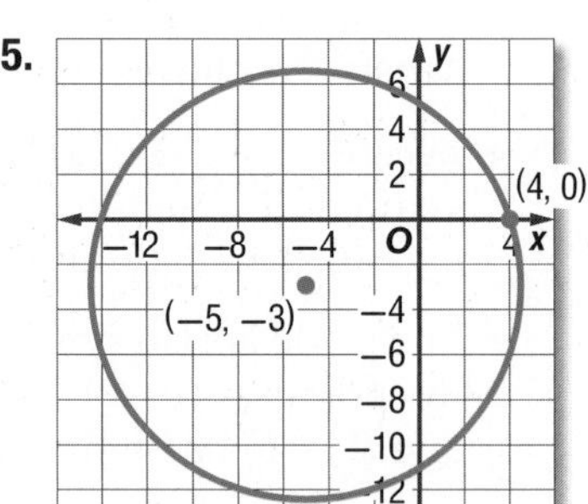

Example 3

Write an equation for each circle given the endpoints of a diameter.

6. $(-1, -7)$ and $(0, 0)$
7. $(4, -2)$ and $(-4, -6)$

Examples 4–5 **Find the center and radius of each circle. Then graph the circle.**

8. $x^2 + y^2 = 16$
9. $x^2 + (y - 7)^2 = 9$
10. $(x - 4)^2 + (y - 4)^2 = 25$
11. $x^2 + y^2 - 4x + 8y - 5 = 0$

Practice and Problem Solving

Extra Practice is on page R9.

Example 1

Write an equation for each circle given the center and radius.

12. center: $(4, 9)$, $r = 6$
13. center: $(-3, 1)$, $r = 4$
14. center: $(-7, -3)$, $r = 13$
15. center: $(-2, -1)$, $r = 9$
16. center: $(1, 0)$, $r = \sqrt{15}$
17. center: $(0, -6)$, $r = \sqrt{35}$
18. **CCSS MODELING** The radar for a county airport control tower is located at (5, 10) on a map. It can detect a plane up to 20 miles away. Write an equation for the outer limits of the detection area.

Example 2

Write an equation for each graph.

19. 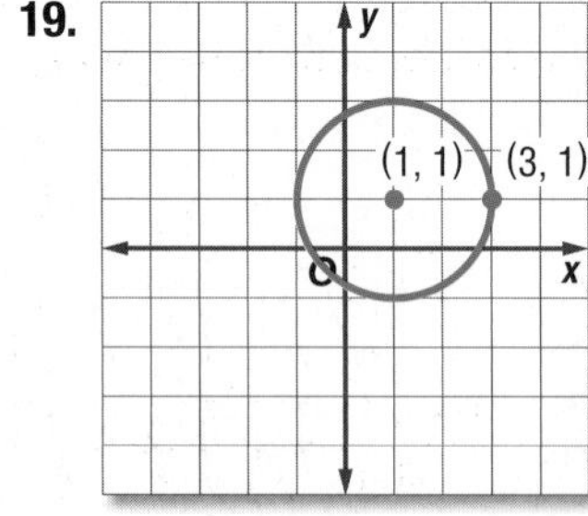

20. 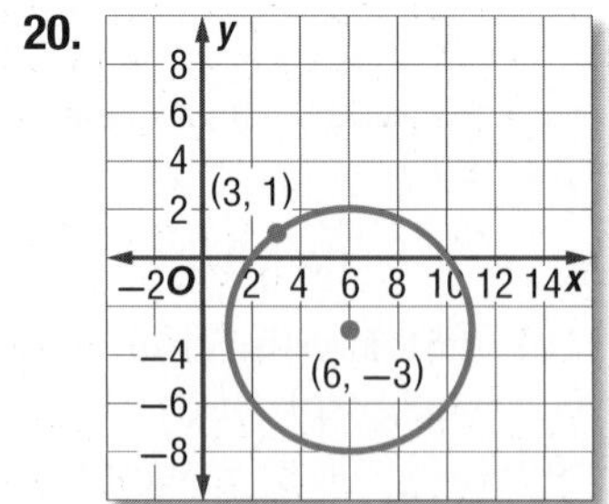

21. 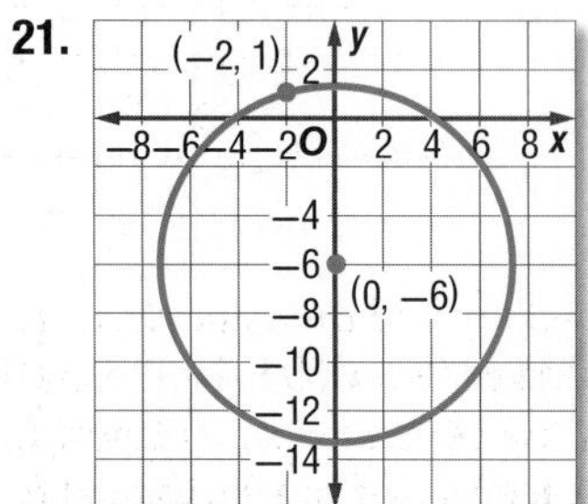

22. 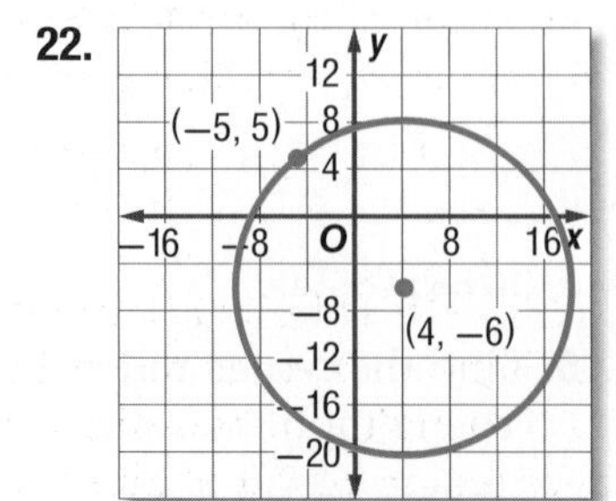

Example 3 **Write an equation for each circle given the endpoints of a diameter.**

23. (2, 1) and (2, −4)
24. (−4, −10) and (4, −10)
25. (5, −7) and (−2, −9)
26. (−6, 4) and (4, 8)
27. (2, −5) and (6, 3)
28. (18, 11) and (−19, −13)

29. **LAWN CARE** A sprinkler waters a circular section of lawn.

a. Write an equation to represent the boundary of the sprinkler area if the endpoints of a diameter are at (−12, 16) and (12, −16).

b. What is the area of the lawn that the sprinkler waters?

30. **SPACE** Apollo 8 was the first manned spacecraft to orbit the Moon at an average altitude of 185 kilometers above the Moon's surface. Write an equation to model a single circular orbit of the command module if the endpoints of a diameter of the Moon are at (1740, 0) and (−1740, 0). Let the center of the Moon be at the origin of the coordinate system measured in kilometers.

Examples 4–5 **Find the center and radius of each circle. Then graph the circle.**

31. $x^2 + y^2 = 75$
32. $(x - 3)^2 + y^2 = 4$
33. $(x - 1)^2 + (y - 4)^2 = 34$
34. $x^2 + (y - 14)^2 = 144$
35. $(x - 5)^2 + (y + 2)^2 = 16$
36. $x^2 + y^2 = 256$
37. $(x - 4)^2 + y^2 = \frac{8}{9}$
38. $\left(x + \frac{2}{3}\right)^2 + \left(y - \frac{1}{2}\right)^2 = \frac{16}{25}$
39. $x^2 + y^2 + 4x = 9$
40. $x^2 + y^2 - 6y + 8x = 0$
41. $x^2 + y^2 + 2x + 4y = 9$
42. $x^2 + y^2 - 3x + 8y = 20$
43. $x^2 + y^2 + 6y = -50 - 14x$
44. $x^2 - 18x + 53 = 18y - y^2$
45. $2x^2 + 2y^2 - 4x + 8y = 32$
46. $3x^2 + 3y^2 - 6y + 12x = 24$

47. **SPACE** A satellite is in a circular orbit 25,000 miles above Earth.

a. Write an equation for the orbit of this satellite if the origin is at the center of Earth. Use 8000 miles as the diameter of Earth.

b. Draw a sketch of Earth and the orbit to scale. Label your sketch.

48. CCSS **SENSE-MAKING** Suppose an unobstructed radio station broadcast could travel 120 miles. Assume the station is centered at the origin.

a. Write an equation to represent the boundary of the broadcast area with the origin as the center.

b. If the transmission tower is relocated 40 miles east and 10 miles south of the current location, and an increased signal will transmit signals an additional 80 miles, what is an equation to represent the new broadcast area?

49. **GEOMETRY** Concentric circles are circles with the same center but different radii. Refer to the graph at the right where $\overline{AB}$ is a diameter of the circle.

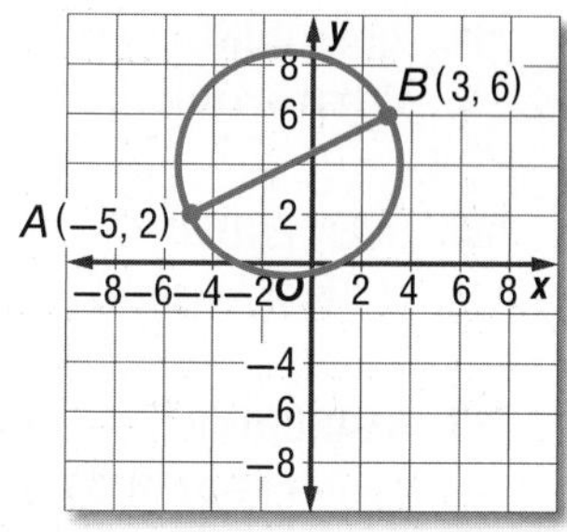

a. Write an equation of the circle concentric with the circle at the right, with radius 4 units greater.

b. Write an equation of the circle concentric with the circle at the right, with radius 2 units less.

c. Graph the circles from parts a and b on the same coordinate plane.

50. **EARTHQUAKES** The Rose Bowl is located about 35 miles west and 40 miles north of downtown Los Angeles. Suppose an earthquake occurs with its epicenter about 55 miles from the stadium. Assume that the origin of a coordinate plane is located at the center of downtown Los Angeles. Write an equation for the set of points that could be the epicenter of the earthquake.

CCSS PRECISION Write an equation for the circle that satisfies each set of conditions.

51. center $(9, -8)$, passes through $(19, 22)$

52. center $\left(-\sqrt{15}, 30\right)$, passes through the origin

53. center at $(8, -9)$, tangent to y-axis

54. center at $(2, 4)$, tangent to x-axis

55. center in the first quadrant; tangent to $x = 5$, the x-axis, and the y-axis

56. center in the second quadrant; tangent to $y = 1$, $y = 5$, and the y-axis

57. **MULTIPLE REPRESENTATIONS** Graph $y = \sqrt{9 - x^2}$ and $y = -\sqrt{9 - x^2}$ on the same graphing calculator screen.

a. **Verbal** Describe the graph formed by the union of these two graphs.

b. **Algebraic** Write an equation for the union of the two graphs.

c. **Verbal** Most graphing calculators cannot graph the equation $x^2 + y^2 = 49$ directly. Describe a way to use a graphing calculator to graph the equation. Then graph the equation.

d. **Analytical** Solve $(x - 2)^2 + (y + 1)^2 = 4$ for y. Why do you need two equations to graph a circle on a graphing calculator?

e. **Verbal** Do you think that it is easier to graph the equation in part **d** using graph paper and a pencil or using a graphing calculator? Explain.

Find the center and radius of each circle. Then graph the circle.

58. $x^2 - 12x + 84 = -y^2 + 16y$

59. $4x^2 + 4y^2 + 36y + 5 = 0$

60. $\left(x + \sqrt{5}\right)^2 + y^2 - 8y = 9$

61. $x^2 + 2\sqrt{7}x + 7 + \left(y - \sqrt{11}\right)^2 = 11$

H.O.T. Problems Use Higher-Order Thinking Skills

62. **ERROR ANALYSIS** Heather says that $(x - 2)^2 + (y + 3)^2 = 36$ and $(x - 2) + (y + 3) = 6$ are equivalent equations. Carlota says that the equations are *not* equivalent. Is either of them correct? Explain your reasoning.

63. **OPEN ENDED** Consider graphs with equations of the form $(x - 3)^2 + (y - a)^2 = 64$. Assign three different values for a, and graph each equation. Describe all graphs with equations of this form.

64. **REASONING** Explain why the phrase "in a plane" is included in the definition of a circle. What would be defined if the phrase were *not* included?

65. **OPEN ENDED** Concentric circles have the same center, but most often, not the same radius. Write equations of two concentric circles. Then graph the circles.

66. **REASONING** Assume that (x, y) are the coordinates of a point on a circle. The center is at (h, k), and the radius is r. Find an equation of the circle by using the Distance Formula.

67. **WRITING IN MATH** The circle with equation $(x - a)^2 + (y - b)^2 = r^2$ lies in the first quadrant and is tangent to both the x-axis and the y-axis. Sketch the circle. Describe the possible values of a, b, and r. Do the same for a circle in Quadrants II, III, and IV. Discuss the similarities among the circles.

Standardized Test Practice

68. GRIDDED RESPONSE Two circles, both with a radius of 6, have exactly one point in common. If A is a point on one circle and B is a point on the other circle, what is the maximum possible length for the line segment $\overline{AB}$?

69. In the senior class, there are 20% more girls than boys. If there are 180 girls, how many more girls than boys are there among the seniors?

A 30

B 36

C 90

D 144

70. A $1000 deposit is made at a bank that pays 2% compounded weekly. How much will you have in your account at the end of 10 years?

F $1200.00

G $1218.99

H $1221.36

J $1224.54

71. The mean of six numbers is 20. If one of the numbers is removed, the average of the remaining numbers is 15. What is the number that was removed?

A 42

B 43

C 45

D 48

Spiral Review

Graph each equation. (Lesson 9-2)

72. $y = -\frac{1}{2}(x - 1)^2 + 4$

73. $4(x - 2) = (y + 3)^2$

74. $(y - 8)^2 = -4(x - 4)$

Find the midpoint of the line segment with endpoints at the given coordinates. Then find the distance between the points. (Lesson 9-1)

75. $\left(-3, -\frac{2}{11}\right), \left(5, \frac{9}{11}\right)$

76. $(2\sqrt{3}, -5), (-3\sqrt{3}, 9)$

77. $(2.5, 4), (-2.5, 2)$

78. If y varies directly as x and $y = 8$ when $x = 6$, find y when $x = 15$. (Lesson 8-5)

79. If y varies jointly as x and z and $y = 80$ when $x = 5$ and $z = 8$, find y when $x = 16$ and $z = 2$. (Lesson 8-5)

80. If y varies inversely as x and $y = 16$ when $x = 5$, find y when $x = 20$. (Lesson 8-5)

Evaluate each expression. (Lesson 7-3)

81. $\log_9 243$

82. $\log_2 \frac{1}{32}$

83. $\log_3 \frac{1}{81}$

84. $\log_{10} 0.001$

85. AMUSEMENT PARKS The velocity v in feet per second of a roller coaster at the bottom of a hill is related to the vertical drop h in feet and the velocity v_0 in feet per second of the coaster at the top of the hill by the formula $v_0 = \sqrt{v^2 - 64h}$. (Lesson 6-5)

a. Explain why $v_0 = v - 8\sqrt{h}$ is not equivalent to the given formula.

b. What velocity must the coaster have at the top of the hill to achieve a velocity of 125 feet per second at the bottom?

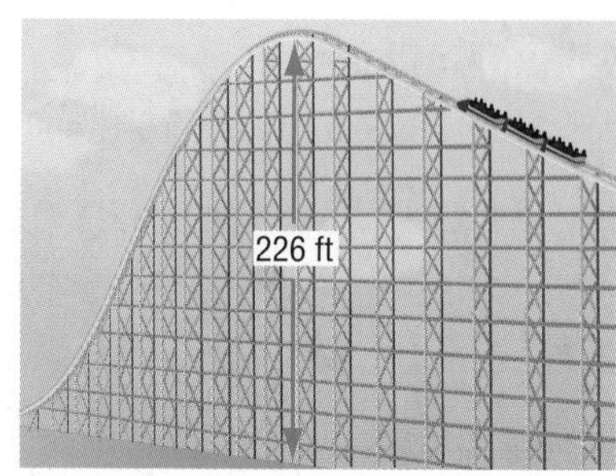

Skills Review

Solve each equation by completing the square.

86. $x^2 + 3x - 18 = 0$

87. $2x^2 - 3x - 3 = 0$

88. $x^2 + 2x + 6 = 0$

EXPLORE 9-4

Algebra Lab
Investigating Ellipses

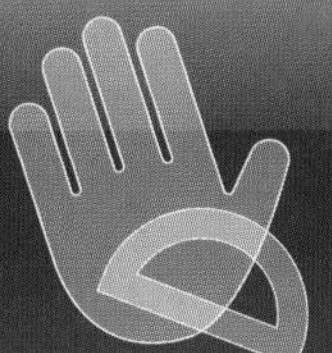

Follow the steps below to construct a type of conic section.

CCSS **Common Core State Standards**
Mathematical Practices
5 Use appropriate tools strategically.

Activity Make an Ellipse

Step 1 Place two thumbtacks in a piece of cardboard, about 1 foot apart.

Step 2 Tie a knot in a piece of string and loop it around the thumbtacks. Place your pencil in the string.

Step 3 Keep the string tight and draw a curve. Continue drawing until you return to your starting point.

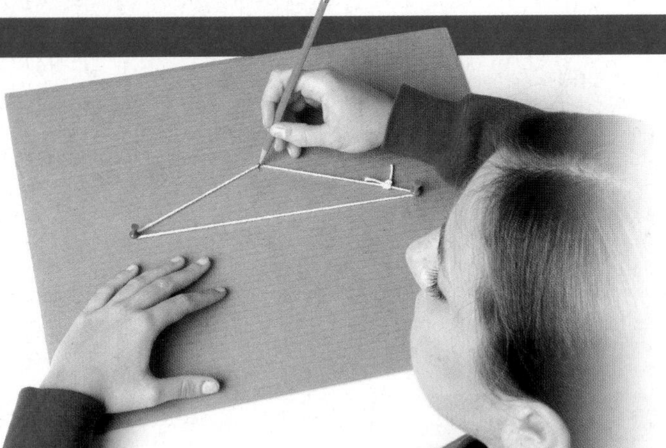

The curve you have drawn is called an **ellipse**. The points where the thumbtacks are located are called the **foci** of the ellipse. *Foci* is the plural of *focus*.

Model and Analyze

Place a large piece of grid paper on a piece of cardboard.

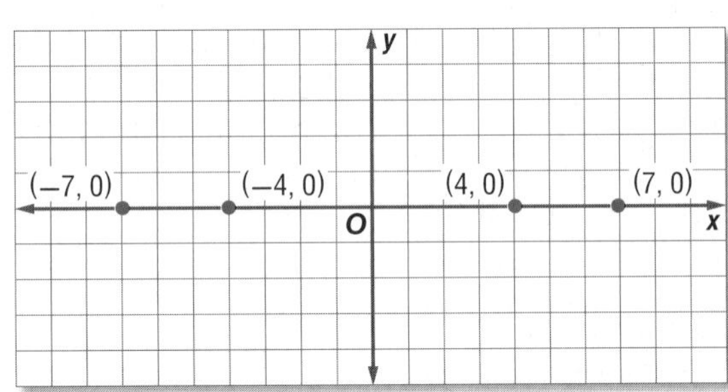

1. Place the thumbtacks at (7, 0) and (−7, 0). Choose a string long enough to loop around both thumbtacks. Draw an ellipse.
2. Repeat Exercise 1, but place the thumbtacks at (4, 0) and (−4, 0). Use the same loop of string and draw an ellipse. How does this ellipse compare to the one in Exercise 1?

Place the thumbtacks at each set of points and draw an ellipse. You may change the length of the loop of string if you like.

3. (11, 0), (−11, 0)
4. (3, 0), (−3, 0)
5. (13, 3), (−9, 3)

Make a Conjecture

Describe what happens to the shape of an ellipse when each change is made.

6. The thumbtacks are moved closer together.
7. The thumbtacks are moved farther apart.
8. The length of the loop of string is increased.
9. The thumbtacks are arranged vertically.
10. One thumbtack is removed, and the string is looped around the remaining thumbtack.
11. Pick a point on one of the ellipses you have drawn. Use a ruler to measure the distances from that point to the points where the thumbtacks were located. Add the distances. Repeat for other points on the same ellipse. What relationship do you notice?
12. Could this activity be done with a rubber band instead of a piece of string? Explain.

LESSON 9-4 Ellipses

Then

- You graphed and wrote equations for circles.

Now

1. Write equations of ellipses.
2. Graph ellipses.

Why?

- Mercury, like all of the planets of our solar system, does not orbit the Sun in a perfect circular path. At its farthest point, Mercury is about 43 million miles from the Sun. At its closest point, it is only about 28.5 million miles from the Sun. This orbit is in the shape of an ellipse with the Sun at a focus.

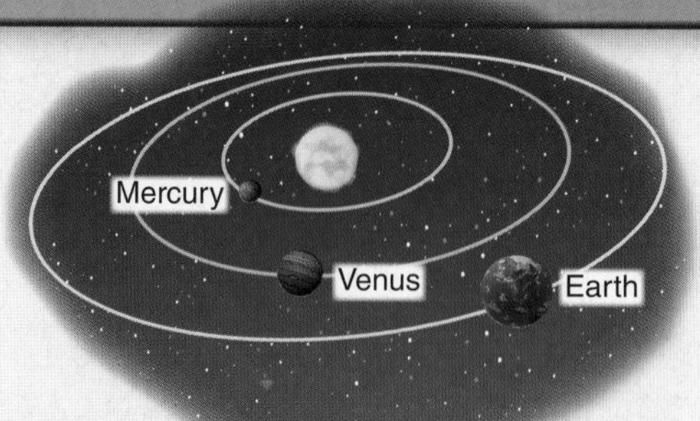

NewVocabulary
ellipse
foci
major axis
minor axis
center
vertices
co-vertices
constant sum

Common Core State Standards

Content Standards

A.SSE.1.b Interpret complicated expressions by viewing one or more of their parts as a single entity.

A.CED.2 Create equations in two or more variables to represent relationships between quantities; graph equations on coordinate axes with labels and scales.

Mathematical Practices

7 Look for and make use of structure.

1 Equations of Ellipses

An **ellipse** is the set of all points in a plane such that the sum of the distances from two fixed points is constant. These two points are called the **foci** of the ellipse.

Every ellipse has two axes of symmetry, the **major axis** and the **minor axis**. The axes are perpendicular at the center of the ellipse.

The foci of an ellipse always lie on the major axis. The endpoints of the major axis are the **vertices** of the ellipse and the endpoints of the minor axis are the **co-vertices** of the ellipse.

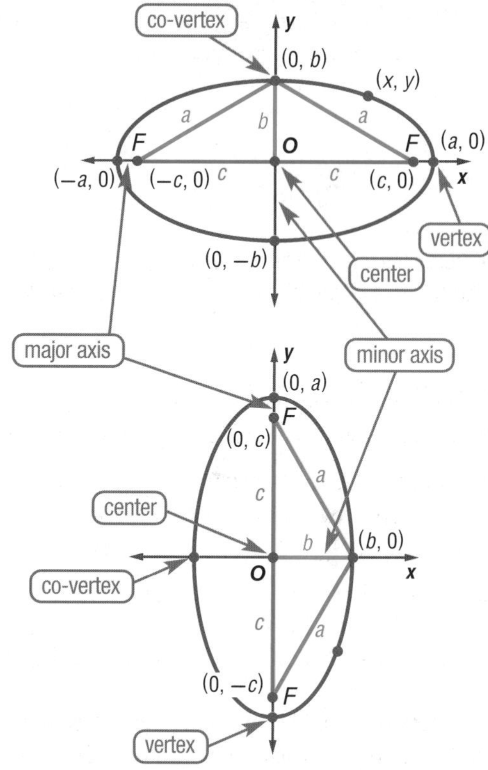

KeyConcept Equations of Ellipses Centered at the Origin

Standard Form	$\frac{x^2}{a^2} + \frac{y^2}{b^2} = 1$	$\frac{y^2}{a^2} + \frac{x^2}{b^2} = 1$
Orientation	horizontal	vertical
Foci	$(c, 0), (-c, 0)$	$(0, c), (0, -c)$
Length of Major Axis	$2a$ units	$2a$ units
Length of Minor Axis	$2b$ units	$2b$ units

There are several important relationships among the many parts of an ellipse.

- The length of the major axis, $2a$ units, equals the sum of the distances from the foci to any point on the ellipse.
- The values of a, b, and c are related by the equation $c^2 = a^2 - b^2$.
- The distance from a focus to either co-vertex is a units.

The sum of the distances from the foci to any point on the ellipse, or the **constant sum**, must be greater than the distance between the foci.

StudyTip

Major Axis In standard form, if the x^2-term has the greater denominator, then the major axis is horizontal. If the y^2-term has the greater denominator, then it is vertical.

Example 1 Write an Equation Given Vertices and Foci

Write an equation for the ellipse.

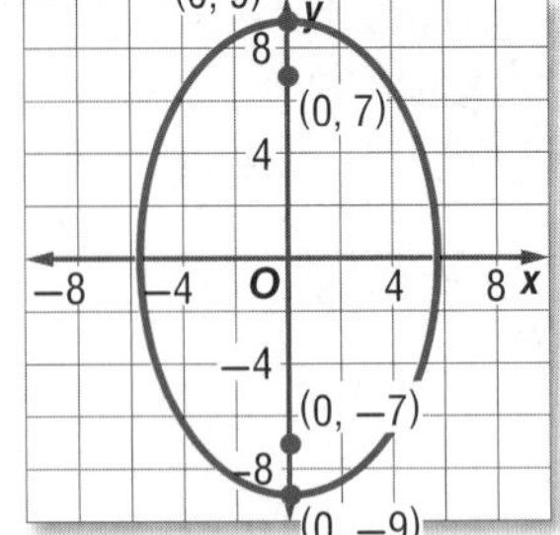

Step 1 Find the center.
The foci are equidistant from the center.
The center is at (0, 0).

Step 2 Find the value of a.
The vertices are (0, 9) and (0, −9), so the length of the major axis is 18.
The value of a is 18 ÷ 2 or 9, and $a^2 = 81$.

Step 3 Find the value of b.
We can use $c^2 = a^2 - b^2$ to find b.
The foci are 7 units from the center, so $c = 7$.

$c^2 = a^2 - b^2$ Equation relating a, b, and c
$49 = 81 - b^2$ $a = 9$ and $c = 7$
$b^2 = 32$ Solve for b^2.

Step 4 Write the equation.
Because the major axis is vertical, a^2 goes with y and b^2 goes with x.
The equation for the ellipse is $\frac{y^2}{81} + \frac{x^2}{32} = 1$.

GuidedPractice

1. Write an equation for an ellipse with vertices at (−4, 0) and (4, 0) and foci at (2, 0) and (−2, 0).

Like other graphs, the graph of an ellipse can be translated. When the graph is translated h units right and k units up, the center of the translation is (h, k). This is equivalent to replacing x with $x - h$ and replacing y with $y - k$ in the parent function.

KeyConcept Equations of Ellipses Centered at (h, k)

Standard Form	$\frac{(x-h)^2}{a^2} + \frac{(y-k)^2}{b^2} = 1$	$\frac{(y-k)^2}{a^2} + \frac{(x-h)^2}{b^2} = 1$
Orientation	horizontal	vertical
Foci	$(h \pm c, k)$	$(h, k \pm c)$
Vertices	$(h \pm a, k)$	$(h, k \pm a)$
Co-vertices	$(h, k \pm b)$	$(h \pm b, k)$

We can use this information to determine the equations for ellipses. The original ellipse at the right is horizontal and has a major axis of 10 units, so $a = 5$.

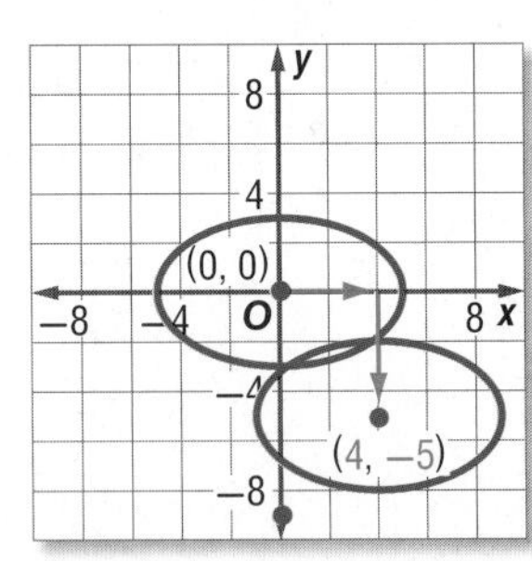

The length of the minor axis is 6 units, so $b = 3$.

The ellipse is translated 4 units right and 5 units down. So, the value of h is 4 and the value of k is −5.

The equation for the original ellipse is $\frac{x^2}{25} + \frac{y^2}{9} = 1$.

The equation for the translation is $\frac{(x-4)^2}{25} + \frac{(y+5)^2}{9} = 1$.

You can also determine the equation for an ellipse if you are given all four vertices.

Example 2 Write an Equation Given the Lengths of the Axes

Write an equation for the ellipse with vertices at (6, −8) and (6, 4) and co-vertices at (3, −2) and (9, −2).

The x-coordinate is the same for both vertices, so the ellipse is vertical.

The center of the ellipse is at $\left(\frac{6+6}{2}, \frac{-8+4}{2}\right)$ or (6, −2).

The length of the major axis is 4 − (−8) or 12 units, so $a = 6$.

The length of the minor axis is 9 − 3 or 6 units, so $b = 3$.

The equation for the ellipse is $\frac{(y+2)^2}{36} + \frac{(x-6)^2}{9} = 1$. $a^2 = 36, b^2 = 9$

GuidedPractice

2. Write an equation for the ellipse with vertices at (−3, 8) and (9, 8) and co-vertices at (3, 12) and (3, 4).

Many real-world phenomena can be represented by ellipses.

Real-WorldCareer

Aerospace Technician Aerospace technicians work for NASA, helping engineers research and develop virtual reality and verbal communication between humans and computer systems. Although a bachelor's degree is desired, on-the-job training is available.

Source: NASA

Real-World Example 3 Write an Equation for an Ellipse

SPACE Refer to the application at the beginning of the lesson. Mercury's greatest distance from the Sun, or *aphelion*, is about 43 million miles. Mercury's closest distance, or *perihelion*, is about 28.5 million miles. The diameter of the Sun is about 870,000 miles. Use this information to determine an equation relating Mercury's elliptical orbit around the Sun in millions of miles.

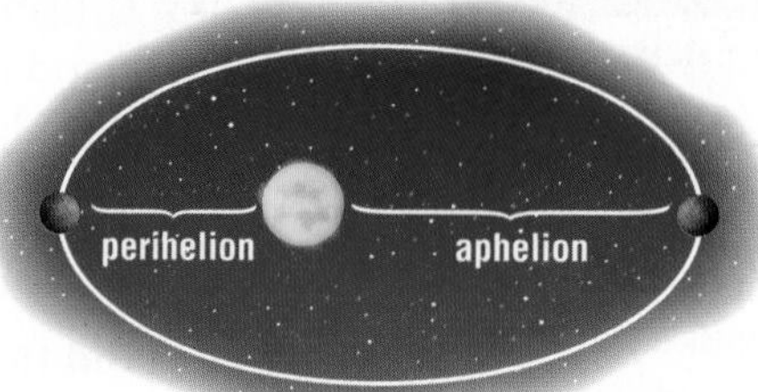

Understand We need to determine an equation representing Mercury's orbit around the Sun.

Plan Including the diameter of the Sun, the sum of the perihelion and aphelion equals the length on the major axis of the ellipse. We can use this information to determine the values of a, b, and c.

Solve Find the value of a.
The value of a is one half the length of the major axis.
$a = 0.5(43 + 28.5 + 0.87)$ or 36.185

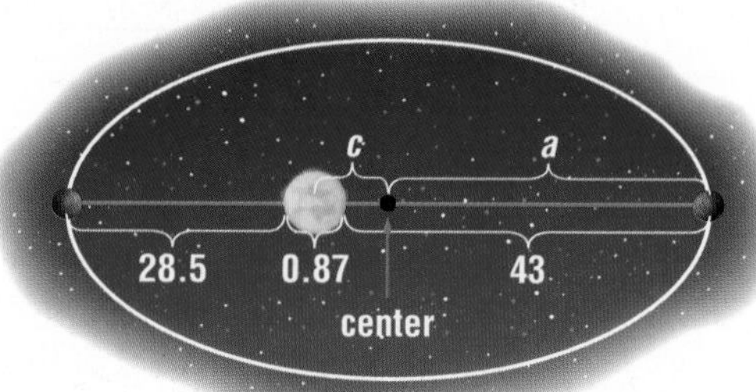

Find the value of c.
The value of c is the distance from the center of the ellipse to the focus. This distance is equal to a minus the perihelion and the radius of the Sun.
$c = 36.185 - 28.5 - 0.435$ or 7.25

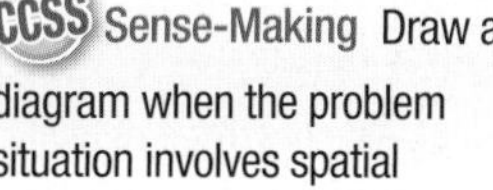

Problem-SolvingTip

CCSS Sense-Making Draw a diagram when the problem situation involves spatial reasoning or geometric figures.

NASA-MSFC

(continued on the next page)

Find the value of b.

$$c^2 = a^2 - b^2$$ Equation relating a, b, and c

$$(7.25)^2 = (36.185)^2 - b^2$$ $c = 7.25$ and $a = 36.185$

$$52.5625 = 1309.3542 - b^2$$ Simplify.

$$b^2 = 1256.828$$ Solve for b^2.

$$b = 35.4518$$ Take the square root of each side.

So, with the center of the orbit at the origin, the equation relating Mercury's orbit around the Sun can be modeled by

$$\frac{x^2}{1309.3542} + \frac{y^2}{1256.792} = 1.$$

Check Use your answer to recalculate a, b, and c. Then determine the aphelion and perihelion based on your answer. Compare to the actual values.

Real-WorldLink

Earth's orbit around the Sun is nearly circular, with only about a 3% difference between perihelion and aphelion.

Source: *The Astronomer*

GuidedPractice

3. SPACE Pluto's distance from the Sun is 2.757 billion miles at perihelion and about 4.583 billion miles at aphelion. Determine an equation relating Pluto's orbit around the Sun in billions of miles with the center of the horizontal ellipse at the origin.

2 Graph Ellipses

When you are given an equation for an ellipse that is not in standard form, you can write it in standard form by completing the square for both x and y. Once the equation is in standard form, you can use it to graph the ellipse.

Example 4 Graph an Ellipse

Find the coordinates of the center and foci, and the lengths of the major and minor axes of an ellipse with equation $25x^2 + 9y^2 + 250x - 36y + 436 = 0$. Then graph the ellipse.

Step 1 Write in standard form. Complete the square for each variable to write this equation in standard form.

$$25x^2 + 9y^2 + 250x - 36y + 436 = 0$$ Original equation

$$25x^2 + 250x + 9y^2 - 36y = -436$$ Associative Property

$$25(x^2 + 10x) + 9(y^2 - 4y) = -436$$ Distributive Property

$$25(x^2 + 10x + \blacksquare) + 9(y^2 - 4y + \blacksquare) = -436 + 25(\blacksquare) + 9(\blacksquare)$$ Complete the squares.

$$25(x^2 + 10x + 25) + 9(y^2 - 4y + 4) = -436 + 25(25) + 9(4)$$ $5^2 = 25$ and $(-2)^2 = 4$

$$25(x + 5)^2 + 9(y - 2)^2 = 225$$ Write as perfect squares.

$$\frac{(x + 5)^2}{9} + \frac{(y - 2)^2}{25} = 1$$ Divide each side by 225.

Step 2 Find the center.
$h = -5$ and $k = 2$, so the center of the ellipse is at $(-5, 2)$.

Step 3 Find the lengths of the axes and graph.
The ellipse is vertical.
$a^2 = 25$, so $a = 5$. $b^2 = 9$, so $b = 3$.
The length of the major axis is $2 \cdot 5$ or 10.
The length of the minor axis is $2 \cdot 3$ or 6.
The vertices are at $(-5, 7)$ and $(-5, -3)$.
The co-vertices are at $(-2, 2)$ and $(-8, 2)$.

StockTrek/Getty Images

Step 4 Find the foci.
$c^2 = 25 - 9$ or 16, so $c = 4$.
The foci are at $(-5, 6)$ and $(-5, -2)$.

Step 5 Graph the ellipse.
Draw the ellipse that passes through the vertices and co-vertices.

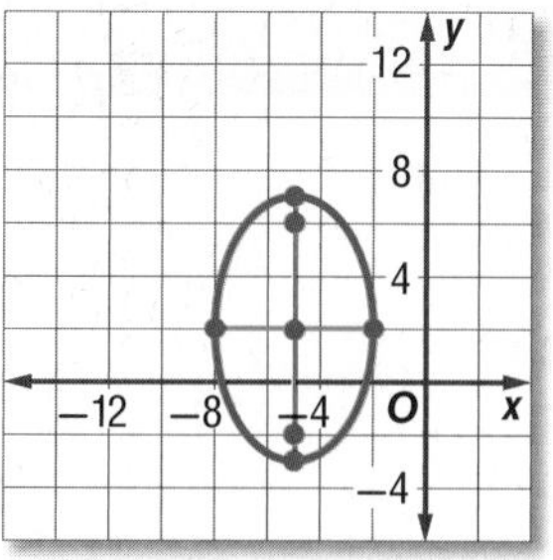

GuidedPractice

4. Find the coordinates of the center and foci and the lengths of the major and minor axes of the ellipse with equation $x^2 + 4y^2 - 2x + 24y + 21 = 0$. Then graph the ellipse.

Check Your Understanding

 = Step-by-Step Solutions begin on page R14.

Example 1 **Write an equation for each ellipse.**

1.

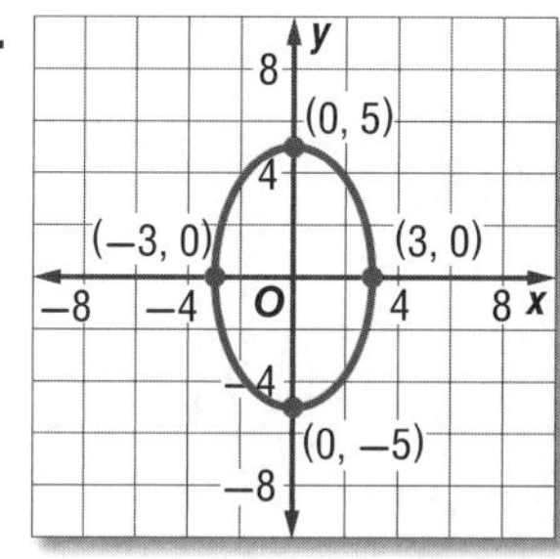

2.

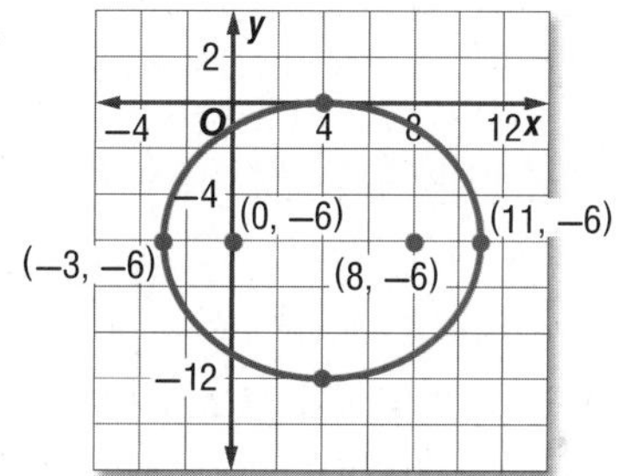

Example 2 **Write an equation for an ellipse that satisfies each set of conditions.**

3. vertices at $(-2, -6)$ and $(-2, 4)$, co-vertices at $(-5, -1)$ and $(1, -1)$

4. vertices at $(-2, 5)$ and $(14, 5)$, co-vertices at $(6, 1)$ and $(6, 9)$

Example 3

5. CCSS **SENSE-MAKING** An architectural firm sent a proposal to a city for building a coliseum, shown at the right.

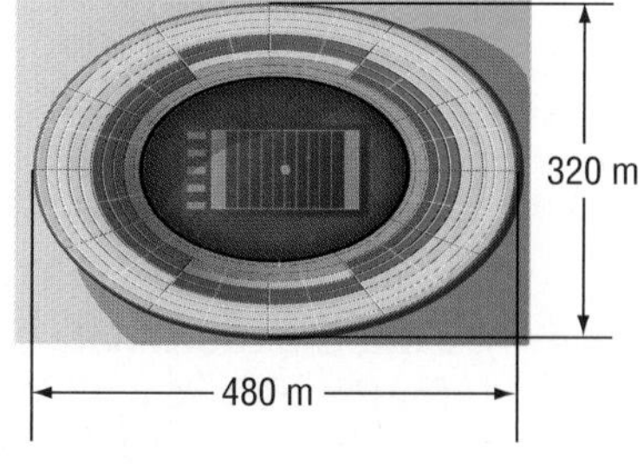

a. Determine the values of a and b.

b. Assuming that the center is at the origin, write an equation to represent the ellipse.

c. Determine the coordinates of the foci.

6. **SPACE** Earth's orbit is about 91.4 million miles at perihelion and about 94.5 million miles at aphelion. Determine an equation relating Earth's orbit around the Sun in millions of miles with the center of the horizontal ellipse at the origin.

Example 4 **Find the coordinates of the center and foci and the lengths of the major and minor axes for the ellipse with the given equation. Then graph the ellipse.**

7. $\frac{(y+1)^2}{64} + \frac{(x-5)^2}{28} = 1$

8. $\frac{(x+2)^2}{48} + \frac{(y-1)^2}{20} = 1$

9. $4x^2 + y^2 - 32x - 4y + 52 = 0$

10. $9x^2 + 25y^2 + 72x - 150y + 144 = 0$

Practice and Problem Solving

Extra Practice is on page R9.

Example 1 **Write an equation for each ellipse.**

11.

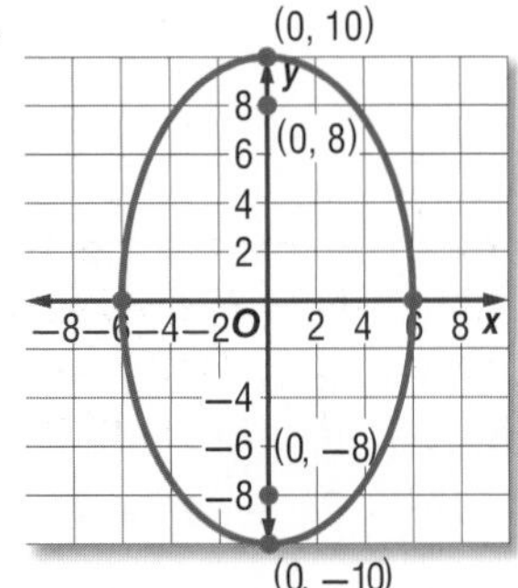

12.

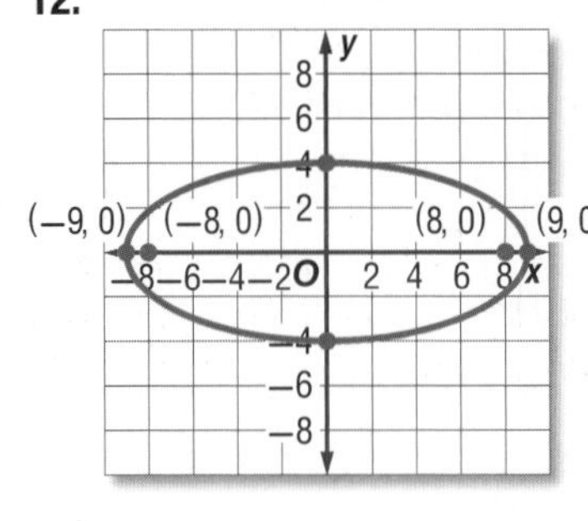

13.

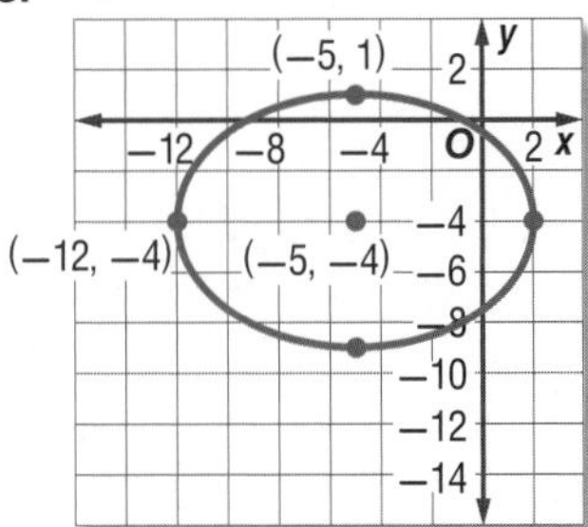

14.

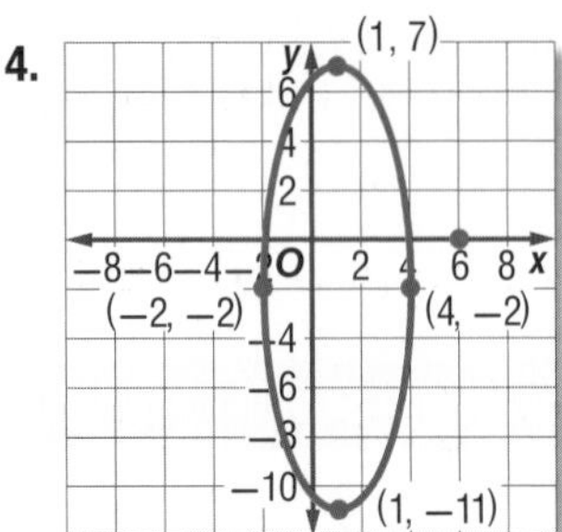

15.

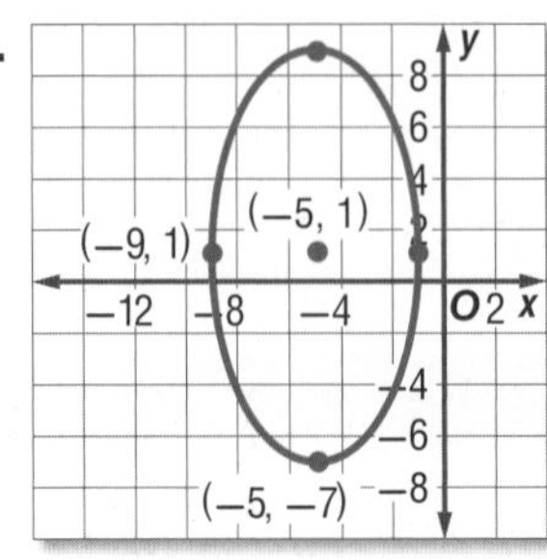

16.

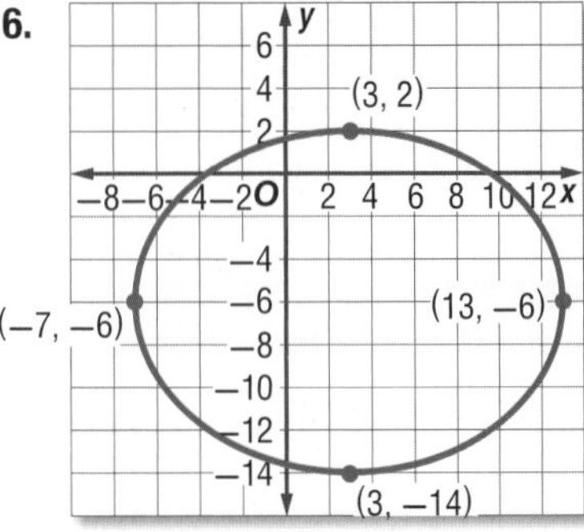

Example 2 **Write an equation for an ellipse that satisfies each set of conditions.**

17. vertices at (−6, 4) and (12, 4), co-vertices at (3, 12) and (3, −4)

18. vertices at (−1, 11) and (−1, 1), co-vertices at (−4, 6) and (2, 6)

19. center at (−2, 6), vertex at (−2, 16), co-vertex at (1, 6)

20. center at (3, −4), vertex at (8, −4), co-vertex at (3, −2)

21. vertices at (4, 12) and (4, −4), co-vertices at (1, 4) and (7, 4)

22. vertices at (−11, 2) and (−1, 2), co-vertices at (−6, 0) and (−6, 4)

Example 3 **23.** **CCSS MODELING** The opening of a tunnel in the mountains can be modeled by semiellipses, or halves of ellipses. If the opening is 14.6 meters wide and 8.6 meters high, determine an equation to represent the opening with the center at the origin.

Example 4 **Find the coordinates of the center and foci and the lengths of the major and minor axes for the ellipse with the given equation. Then graph the ellipse.**

24. $\frac{(x-3)^2}{36} + \frac{(y-2)^2}{128} = 1$

25. $\frac{(x+6)^2}{50} + \frac{(y-3)^2}{72} = 1$

26. $\frac{x^2}{27} + \frac{(y-5)^2}{64} = 1$

27. $\frac{(x+4)^2}{16} + \frac{y^2}{75} = 1$

28. $3x^2 + y^2 - 6x - 8y - 5 = 0$

29. $3x^2 + 4y^2 - 18x + 24y + 3 = 0$

30. $7x^2 + y^2 - 56x + 6y + 93 = 0$

31. $3x^2 + 2y^2 + 12x - 20y + 14 = 0$

32. **SPACE** Like the planets, Halley's Comet travels around the Sun in an elliptical orbit. The aphelion is 3282.9 million miles and the perihelion is 54.87 million miles. Determine an equation relating the comet's orbit around the Sun in millions of miles with the center of the horizontal ellipse at the origin.

Write an equation for an ellipse that satisfies each set of conditions.

33. center at (−5, −2), focus at (−5, 2), co-vertex at (−8, −2)

34. center at (4, −3), focus at (9, −3), co-vertex at (4, −5)

35. foci at (−2, 8) and (6, 8), co-vertex at (2, 10)

36. foci at (4, 4) and (4, 14), co-vertex at (0, 9)

37. **GOVERNMENT** The Oval Office is located in the West Wing of the White House. It is an elliptical shaped room used as the main office by the President of the United States. The long axis is 10.9 meters long and the short axis is 8.8 meters long. Write an equation to represent the outer walls of the Oval Office. Assume that the center of the room is at the origin.

38. **SOUND** A whispering gallery is an elliptical room in which a faint whisper at one focus cannot be heard by other people in the room, but can easily be heard by someone at the other focus. Suppose an ellipse is 400 feet long and 120 feet wide. What is the distance between the foci?

39. **MULTIPLE REPRESENTATIONS** The *eccentricity* of an ellipse measures how circular the ellipse is.

a. Graphical Graph $\frac{x^2}{81} + \frac{y^2}{36} = 1$ and $\frac{x^2}{81} + \frac{y^2}{9} = 1$ on the same graph.

b. Verbal Describe the difference between the two graphs.

c. Algebraic The eccentricity of an ellipse is $\frac{c}{a}$. Find the eccentricity for each.

d. Analytical Make a conjecture about the relationship between the value of an ellipse's eccentricity and the shape of the ellipse as compared to a circle.

H.O.T. Problems Use Higher-Order Thinking Skills

40. **ERROR ANALYSIS** Serena and Karissa are determining the equation for an ellipse with foci at (−4, −11) and (−4, 5) and co-vertices at (2, −3) and (−10, −3). Is either of them correct? Explain your reasoning.

Serena	Karissa
$\frac{(x-4)^2}{64} + \frac{(y+3)^2}{36} = 1$	$\frac{(x+4)^2}{100} + \frac{(y+3)^2}{36} = 1$

41. **OPEN ENDED** Write an equation for an ellipse with a focus at the origin.

42. **CHALLENGE** When the values of a and b are equal, an ellipse is a circle. Use this information and your knowledge of ellipses to determine the formula for the area of an ellipse in terms of a and b.

43. **CHALLENGE** Determine an equation for an ellipse with foci at $(2, \sqrt{6})$ and $(2, -\sqrt{6})$ that passes through $(3, \sqrt{6})$.

44. **CCSS ARGUMENTS** What happens to the location of the foci as an ellipse becomes more circular? Explain your reasoning.

45. **REASONING** An ellipse has foci at (−7, 2) and (18, 2). If (2, 14) is a point on the ellipse, show that (2, −10) is also a point on the ellipse.

46. **WRITING IN MATH** Explain why the domain is $\{x \mid -a \leq x \leq a\}$ and the range is $\{y \mid -b \leq y \leq b\}$ for an ellipse with equation $\frac{x^2}{a^2} + \frac{y^2}{b^2} = 1$.

Standardized Test Practice

47. Multiply.

$$(2 + 3i)(4 + 7i)$$

A $8 + 21i$
B $-13 + 26i$
C $-6 + 10i$
D $13 + 12i$

48. The average lifespan of American women has been tracked, and the model for the data is $y = 0.2t + 73$, where $t = 0$ corresponds to 1960. What is the meaning of the y-intercept?

F In 2007, the average lifespan was 60.
G In 1960, the average lifespan was 58.
H In 1960, the average lifespan was 73.
J The lifespan is increasing 0.2 years every year.

49. GRIDDED RESPONSE If we decrease a number by 6 and then double the result, we get 5 less than the number. What is the number?

50. SAT/ACT The length of a rectangular prism is one inch greater than its width. The height is three times the length. Find the volume of the prism.

A $3x^3 + x^2 + 3x$
B $x^3 + x^2 + x$
C $3x^3 + 6x^2 + 3x$
D $3x^3 + 3x^2 + 3x$
E $3x^3 + 3x^2$

Spiral Review

Write an equation for the circle that satisfies each set of conditions. (Lesson 9-3)

51. center $(8, -9)$, passes through $(21, 22)$

52. center at $(4, 2)$, tangent to x-axis

53. center in the second quadrant; tangent to $y = -1$, $y = 9$, and the y-axis

54. ENERGY A parabolic mirror is used to collect solar energy. The mirrors reflect the rays from the Sun to the focus of the parabola. The focus of a particular mirror is 9.75 feet above the vertex, and the latus rectum is 39 feet long. (Lesson 9-2)

a. Assume that the focus is at the origin. Write an equation for the parabola formed by the mirror.

b. One foot is exactly 0.3048 meter. Rewrite the equation for the mirror in meters.

c. Graph one of the equations for the mirror.

d. Which equation did you choose to graph? Explain why.

Simplify each expression. (Lesson 8-2)

55. $\frac{6}{d^2 + 4d + 4} + \frac{5}{d + 2}$

56. $\frac{a}{a^2 - a - 20} + \frac{2}{a + 4}$

57. $\frac{x}{x + 1} + \frac{3}{x^2 - 4x - 5}$

Solve each equation. (Lesson 7-4)

58. $\log_{10} (x^2 + 1) = 1$

59. $\log_b 64 = 3$

60. $\log_b 121 = 2$

Simplify. (Lesson 5-1)

61. $-5ab^2(-3a^2b + 6a^3b - 3a^4b^4)$

62. $2xy(3xy^3 - 4xy + 2y^4)$

63. $(4x^2 - 3y^2 + 5xy) - (8xy + 3y^2)$

64. $(10x^2 - 3xy + 4y^2) - (3x^2 + 5xy)$

Skills Review

Write an equation of the line passing through each pair of points.

65. $(-2, 5)$ and $(3, 1)$

66. $(7, 1)$ and $(7, 8)$

67. $(-3, 5)$ and $(2, 2)$

CHAPTER 9 Mid-Chapter Quiz

Lessons 9-1 through 9-4

Find the midpoint of the line segment with endpoints at the given coordinates. (Lesson 9-1)

1. (7, 4), (−1, −5)

2. (−2, −9), (−6, 0)

Find the distance between each pair of points with the given coordinates. (Lesson 9-1)

3. (0, 6), (−2, 5)

4. (10, 1), (0, −4)

5. HIKING Carla and Lance left their campsite and hiked 6 miles directly north and then turned and hiked 7 miles east to view a waterfall. (Lesson 9-1)

a. How far is the waterfall from their campsite?

b. Let the campsite be located at the origin on a coordinate grid. At the waterfall they decide to head directly back to the campsite. If they stop halfway between the waterfall and the campsite for lunch, at what coordinates will they stop for lunch?

Write each equation in standard form. Identify the vertex, axis of symmetry, and direction of opening of the parabola. (Lesson 9-2)

6. $y = 3x^2 - 12x + 21$

7. $x - 2y^2 = 4y + 6$

8. $y = \frac{1}{2}x^2 + 12x - 8$

9. $x = 3y^2 + 5y - 9$

10. BRIDGES Write an equation of a parabola to model the shape of the suspension cable of the bridge shown. Assume that the origin is at the lowest point of the cables. (Lesson 9-2)

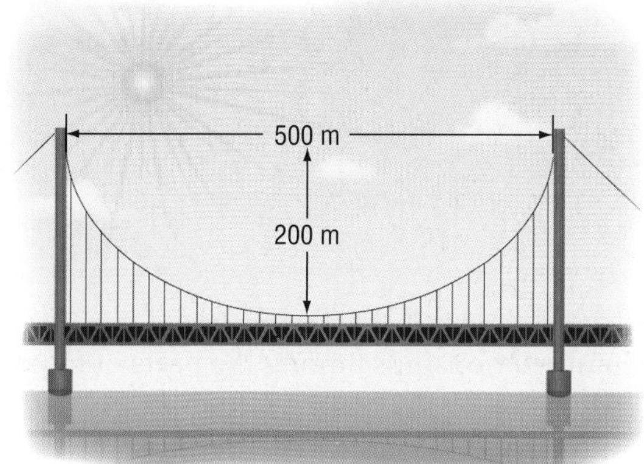

Identify the coordinates of the vertex and focus, the equation of the axis of symmetry and directrix, and the direction of opening of the parabola with the given equation. Then find the length of the latus rectum. (Lesson 9-2)

11. $y = x^2 + 6x + 5$

12. $x = -2y^2 + 4y + 1$

13. Find the center and radius of the circle with equation $(x - 1)^2 + y^2 = 9$. Then graph the circle. (Lesson 9-3)

14. Write an equation for a circle that has center at (3, −2) and passes through (3, 4). (Lesson 9-3)

15. Write an equation for a circle if the endpoints of a diameter are at (8, 31) and (32, 49). (Lesson 9-3)

16. MULTIPLE CHOICE What is the radius of the circle with equation $x^2 + 2x + y^2 + 14y + 34 = 0$? (Lesson 9-3)

A 2

B 4

C 8

D 16

Find the coordinates of the center and foci and the lengths of the major and minor axes of the ellipse with the given equation. Then graph the ellipse. (Lesson 9-4)

17. $\frac{(x + 4)^2}{16} + \frac{(y - 2)^2}{9} = 1$

18. $\frac{(x - 1)^2}{20} + \frac{(y + 2)^2}{4} = 1$

19. $4y^2 + 9x^2 + 16y - 90x + 205 = 0$

20. MULTIPLE CHOICE Which equation represents an ellipse with endpoints of the major axis at (−4, 10) and (−4, −6) and foci at about (−4, 7.3) and (−4, −3.3)? (Lesson 9-4)

F $\frac{(x - 2)^2}{36} + \frac{(y + 4)^2}{64} = 1$

G $\frac{(x + 4)^2}{64} + \frac{(y - 2)^2}{36} = 1$

H $\frac{(y - 2)^2}{64} + \frac{(x + 4)^2}{36} = 1$

J $\frac{(x - 2)^2}{64} + \frac{(y + 4)^2}{36} = 1$

LESSON 9-5 Hyperbolas

Then	Now	Why?
You graphed and analyzed equations of ellipses.	1 Write equations of hyperbolas. 2 Graph hyperbolas.	Because Halley's Comet travels around the Sun in an elliptical path, it reappears in our sky. Other comets pass through our sky only once. Many of these comets travel in paths that resemble hyperbolas.

NewVocabulary
hyperbola
transverse axis
conjugate axis
foci
vertices
co-vertices
constant difference

Common Core State Standards

Content Standards
A.SSE.1.b Interpret complicated expressions by viewing one or more of their parts as a single entity.
A.CED.2 Create equations in two or more variables to represent relationships between quantities; graph equations on coordinate axes with labels and scales.

Mathematical Practices
6 Attend to precision.

1 Equations of Hyperbolas

Similar to an ellipse, a **hyperbola** is the set of all points in a plane such that the absolute value of the differences of the distances from the foci is constant.

Every hyperbola has two axes of symmetry, the **transverse axis** and the **conjugate axis**. The axes are perpendicular at the center of the hyperbola.

The **foci** of a hyperbola always lie on the transverse axis. The **vertices** are the endpoints of the transverse axis. The **co-vertices** are the endpoints of the conjugate axis.

As a hyperbola recedes from the center, both halves approach asymptotes.

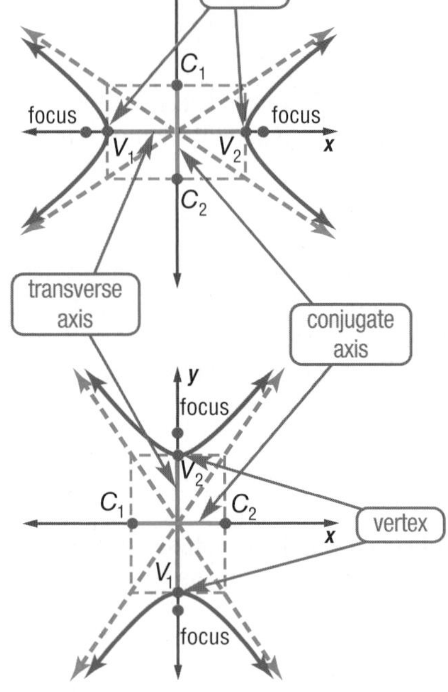

KeyConcept Equations of Hyperbolas Centered at the Origin

Standard Form	$\frac{x^2}{a^2} - \frac{y^2}{b^2} = 1$	$\frac{y^2}{a^2} - \frac{x^2}{b^2} = 1$
Orientation	horizontal	vertical
Foci	$(\pm c, 0)$	$(0, \pm c)$
Length of Transverse Axis	$2a$ units	$2a$ units
Length of Conjugate Axis	$2b$ units	$2b$ units
Equations of Asymptotes	$y = \pm\frac{b}{a}x$	$y = \pm\frac{a}{b}x$

As with ellipses, there are several important relationships among the parts of hyperbolas.

- There are two axes of symmetry.
- The values of a, b, and c are related by the equation $c^2 = a^2 + b^2$.

Example 1 Write an Equation Given Vertices and Foci

Write an equation for the hyperbola shown in the graph.

Step 1 Find the center.

The vertices are equidistant from the center.

The center is at (0, 0).

Step 2 Find the values of a, b, and c.

The value of a is the distance between a vertex and the center, or 4 units.

The value of c is the distance between a focus and the center, or 5 units.

$c^2 = a^2 + b^2$ Equation relating a, b, and c for a hyperbola

$5^2 = 4^2 + b^2$ $c = 5$ and $a = 3$

$9 = b^2$ Subtract 4^2 from each side.

Step 3 Write the equation.

The transverse axis is horizontal, so the equation is $\frac{x^2}{16} - \frac{y^2}{9} = 1$.

GuidedPractice

1. Write an equation for a hyperbola with vertices at (6, 0) and (−6, 0) and foci at (8, 0) and (−8, 0).

Math HistoryLink

Hypatia (415–370 B.C.) Hypatia was a mathematician, scientist, and philosopher in Alexandria, Egypt. She is considered the first woman to write on mathematical topics. Hypatia edited the book *On the Conics of Apollonius*, adding her own problems and examples to clarify the topic for her students. This book developed the ideas of hyperbolas, parabolas, and ellipses.

Hyperbolas can also be determined using the equations of their asymptotes.

Example 2 Write an Equation Given Asymptotes

The asymptotes for a vertical hyperbola are $y = \frac{5}{3}x$ and $y = -\frac{5}{3}x$ and the vertices are at (0, 5) and (0, −5). Write the equation for the hyperbola.

Step 1 Find the center.

The vertices are equidistant from the center.

The center of the hyperbola is at (0, 0).

Step 2 Find the values of a and b.

The hyperbola is vertical, so $a = 5$.

From the asymptotes, $b = 3$.

The value of c is not needed.

Step 3 Write the equation.

The equation for the hyperbola is $\frac{y^2}{25} - \frac{x^2}{9} = 1$.

GuidedPractice

2. The asymptotes for a horizontal hyperbola are $y = \frac{7}{9}x$ and $y = -\frac{7}{9}x$. The vertices are (9, 0) and (−9, 0). Write an equation for the hyperbola.

ReadingMath

Standard Form In the standard form of a hyperbola, the squared terms are subtracted. For an ellipse, they are added.

2 Graphs of Hyperbolas

Hyperbolas can be translated in the same manner as the other conic sections.

KeyConcept Equations of Hyperbolas Centered at (h, k)

Standard Form	$\frac{(x-h)^2}{a^2} - \frac{(y-k)^2}{b^2} = 1$	$\frac{(y-k)^2}{a^2} - \frac{(x-h)^2}{b^2} = 1$
Orientation	horizontal	vertical
Foci	$(h \pm c, k)$	$(h, k \pm c)$
Vertices	$(h \pm a, k)$	$(h, k \pm a)$
Co-vertices	$(h, k \pm b)$	$(h \pm b, k)$
Equations of Asymptotes	$y - k = \pm\frac{b}{a}(x - h)$	$y - k = \pm\frac{a}{b}(x - h)$

StudyTip

Calculator You can graph a hyperbola on a graphing calculator by solving for y, and then graphing the two equations on the same screen.

Example 3 Graph a Hyperbola

Graph $\frac{(x-3)^2}{4} - \frac{(y+2)^2}{16} = 1$. Identify the vertices, foci, and asymptotes.

Step 1 Find the center. The center is at $(3, -2)$.

Step 2 Find a, b, and c. From the equation, $a^2 = 4$ and $b^2 = 16$, so $a = 2$ and $b = 4$.

$c^2 = a^2 + b^2$ — Equation relating a, b, and c for a hyperbola

$c^2 = 2^2 + 4^2$ — $a = 2$, $b = 4$

$c^2 = 20$ — Simplify.

$c = \sqrt{20}$ or about 4.47 — Take the square root of each side.

Step 3 Identify the vertices and foci. The hyperbola is horizontal and the vertices are 2 units from the center, so the vertices are at $(1, -2)$ and $(5, -2)$.
The foci are about 4.47 units from the center.
The foci are at $(-1.47, -2)$ and $(7.47, -2)$.

Step 4 Identify the asymptotes.

$y - k = \pm\frac{b}{a}(x - h)$ — Equation for asymptotes of a horizontal hyperbola

$y - (-2) = \pm\frac{4}{2}(x - 3)$ — $a = 2$, $b = 4$, $h = 3$, and $k = -2$

The equations for the asymptotes are $y = 2x - 8$ and $y = -2x + 4$.

Step 5 Graph the hyperbola. The hyperbola is symmetric about the transverse and conjugate axes. Use this symmetry to plot additional points for the hyperbola.

Use the asymptotes as a guide to draw the hyperbola that passes through the vertices and the other points.

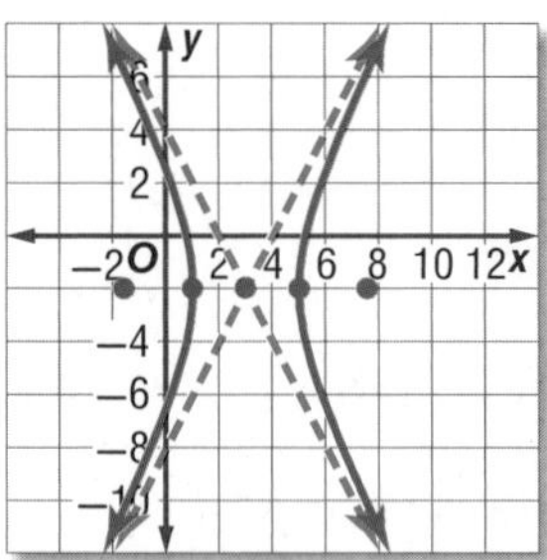

GuidedPractice

3. Graph $\frac{(y-4)^2}{9} - \frac{(x+3)^2}{25} = 1$. Identify the vertices, foci, and asymptotes.

In the equation for any hyperbola, the value of $2a$ represents the **constant difference**. This is the absolute value of the difference between the distances from any point on the hyperbola to the foci of the hyperbola.

Any point on the hyperbola at the right will have the same constant difference, $|y - x|$ or $2a$.

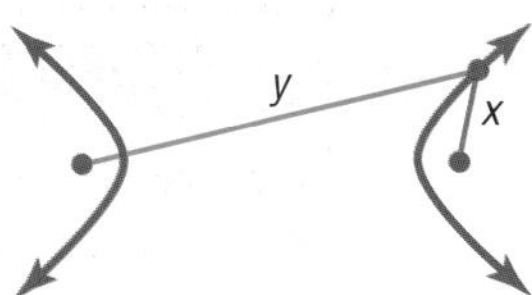

Real-WorldLink

Halley's Comet becomes visible to the unaided eye about every 76 years as it nears the Sun.

Source: NASA

Real-World Example 4 Write an Equation of a Hyperbola

SPACE Earth and the Sun are 146 million kilometers apart. A comet follows a path that is one branch of a hyperbola. Suppose the comet is 30 million miles farther from the Sun than from Earth. Determine the equation of the hyperbola centered at the origin for the path of the comet.

Understand We need to determine the equation for the hyperbola.

Plan Find the center and the values of a and b. Once we have this information, we can determine the equation.

Solve The foci are Earth and the Sun, with the origin between them.

The value of c is $146 \div 2$ or 73.

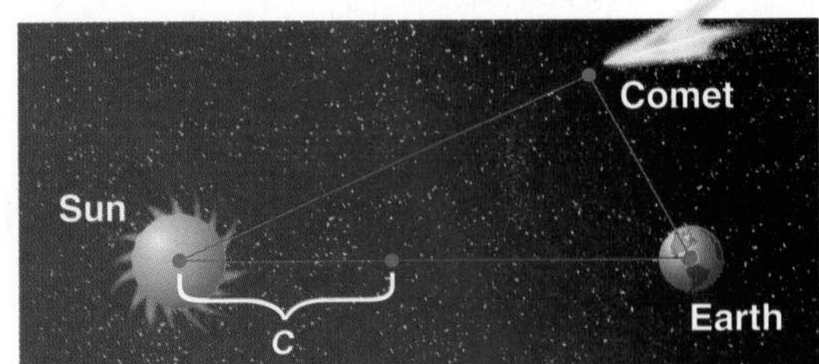

The difference of the distances from the comet to each body is 30. Therefore, a is $30 \div 2$ or 15 million miles.

$c^2 = a^2 + b^2$ Equation relating a, b, and c for a hyperbola

$73^2 = 15^2 + b^2$ $a = 15$ and $c = 73$

$5104 = b^2$ Simplify.

The equation of the hyperbola is $\frac{x^2}{225} - \frac{y^2}{5104} = 1$.

Since the comet is farther from the Sun, it is located on the branch of the hyperbola near Earth.

Check (21, 70) is a point that satisfies the equation.

The distance between this point and the Sun (−73, 0) is

$\sqrt{[21 - (-73)]^2 + (70 - 0)^2}$ or 117.2 million kilometers.

The distance between this point and Earth (73, 0) is

$\sqrt{(21 - 73)^2 + (70 - 0)^2}$ or 87.2 million kilometers.

The difference between these distances is 30. ✓

StockTrek/Photodisc/Getty Images

GuidedPractice

4. SEARCH AND RESCUE Two receiving stations that are 150 miles apart receive a signal from a downed airplane. They determine that the airplane is 80 miles farther from station A than from station B. Determine the equation of the hyperbola centered at the origin on which the plane is located.

StudyTip

Exact Locations A third receiving station is necessary to determine the plane's exact location.

Check Your Understanding

= Step-by-Step Solutions begin on page R14.

Examples 1–2 Write an equation for each hyperbola.

1. 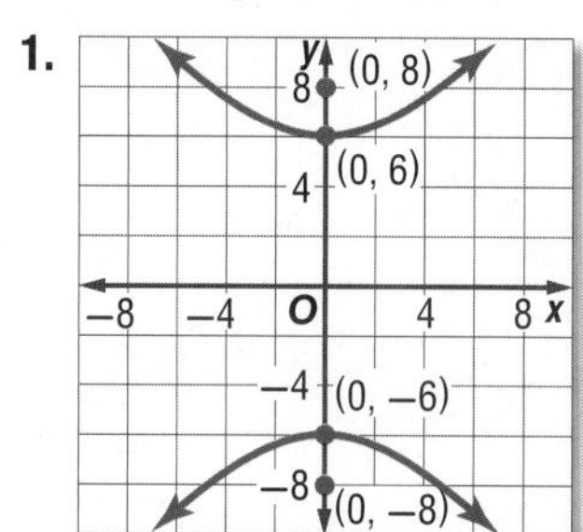

2. 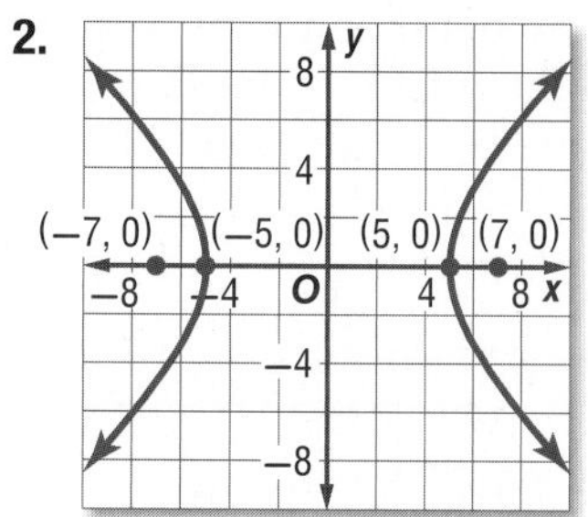

3. 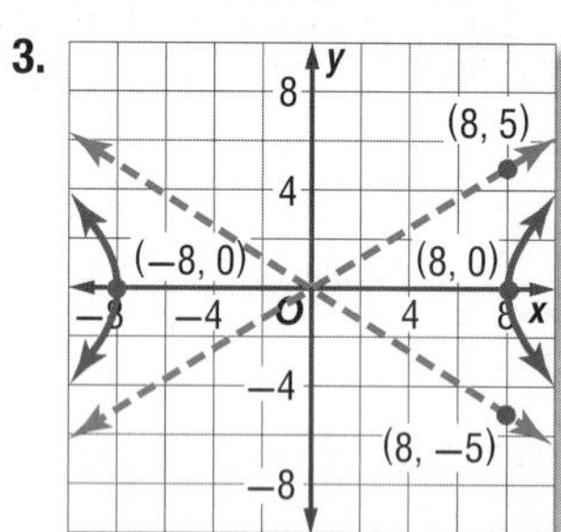

4. 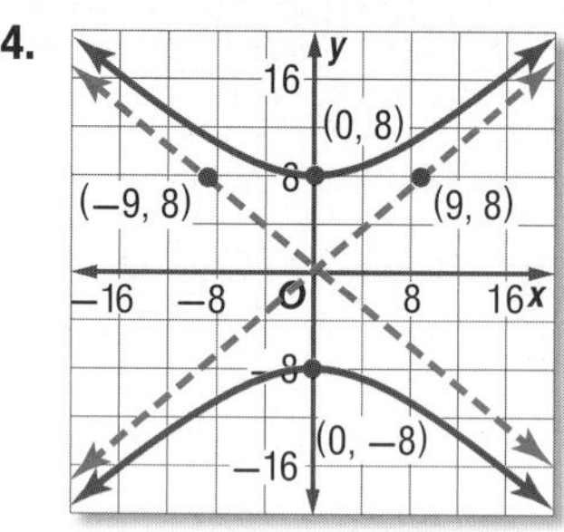

Example 3 CCSS **STRUCTURE** Graph each hyperbola. Identify the vertices, foci, and asymptotes.

5. $\frac{x^2}{64} - \frac{y^2}{49} = 1$

6. $\frac{y^2}{36} - \frac{x^2}{60} = 1$

7. $9y^2 + 18y - 16x^2 + 64x - 199 = 0$

8. $4x^2 + 24x - y^2 + 4y - 4 = 0$

Example 4

9. **NAVIGATION** A ship determines that the difference of its distances from two stations is 60 nautical miles. Write an equation for a hyperbola on which the ship lies if the stations are at $(-80, 0)$ and $(80, 0)$.

Practice and Problem Solving

Extra Practice is on page R9.

Examples 1–2 Write an equation for each hyperbola.

10. 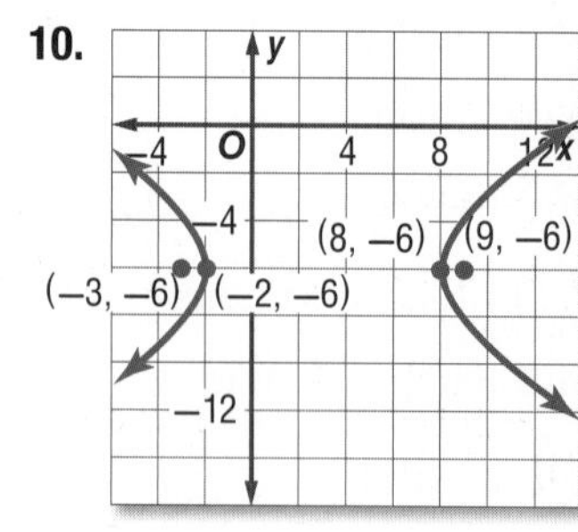

11. 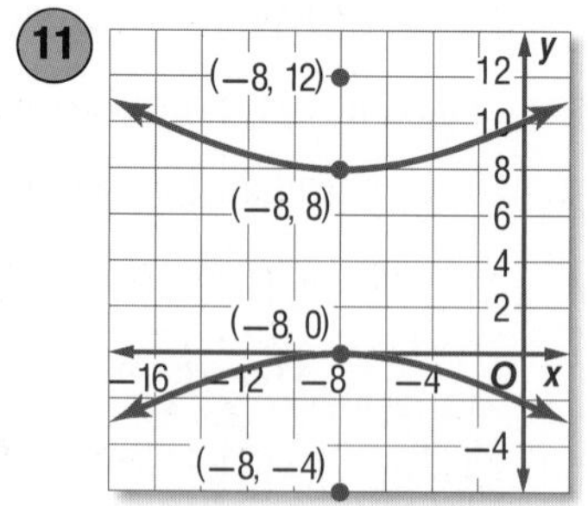

12. 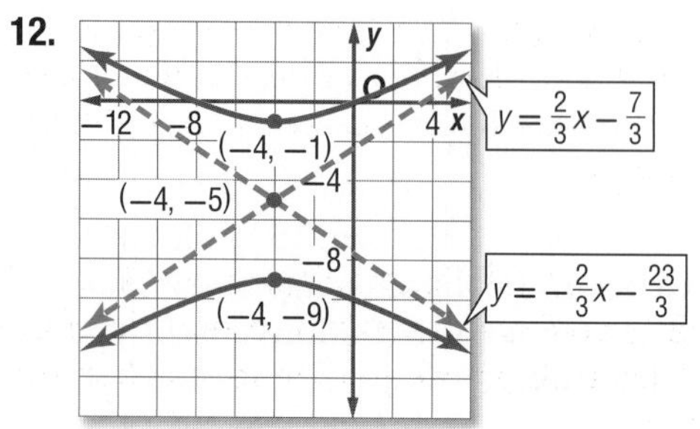

13. 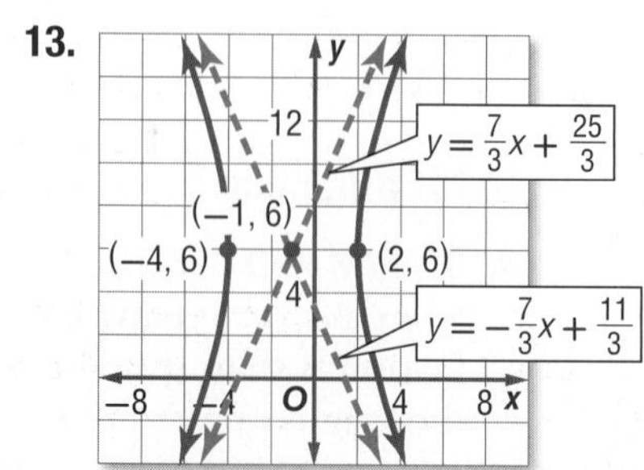

Example 3 **Graph each hyperbola. Identify the vertices, foci, and asymptotes.**

14. $\frac{x^2}{36} - \frac{y^2}{4} = 1$

15. $\frac{y^2}{9} - \frac{x^2}{49} = 1$

16. $\frac{y^2}{36} - \frac{x^2}{25} = 1$

17. $\frac{x^2}{16} - \frac{y^2}{16} = 1$

18. $\frac{(x-3)^2}{16} - \frac{(y+1)^2}{4} = 1$

19. $\frac{(y+5)^2}{16} - \frac{(x+2)^2}{36} = 1$

20. $9y^2 - 4x^2 - 54y + 32x - 19 = 0$

21. $16x^2 - 9y^2 + 128x + 36y + 76 = 0$

22. $25x^2 - 4y^2 - 100x + 48y - 144 = 0$

23. $81y^2 - 16x^2 - 810y + 96x + 585 = 0$

Example 4 24. **NAVIGATION** A ship determines that the difference of its distances from two stations is 80 nautical miles. Write an equation for a hyperbola on which the ship lies if the stations are at $(-100, 0)$ and $(100, 0)$.

Determine whether the following equations represent ellipses or hyperbolas.

25. $4x^2 = 5y^2 + 6$

26. $8x^2 - 2x = 8y - 3y^2$

27. $-5x^2 + 4x = 6y + 3y^2$

28. $7y - 2x^2 = 6x - 2y^2$

29. $6x - 7x^2 - 5y^2 = 2y$

30. $4x + 6y + 2x^2 = -3y^2$

31. **SPACE** Refer to the application at the beginning of the lesson. With the Sun as a focus and the center at the origin, a certain comet's path follows a branch of a hyperbola. If two of the coordinates of the path are (10, 0) and (30, 100) where the units are in millions of miles, determine the equation of the path.

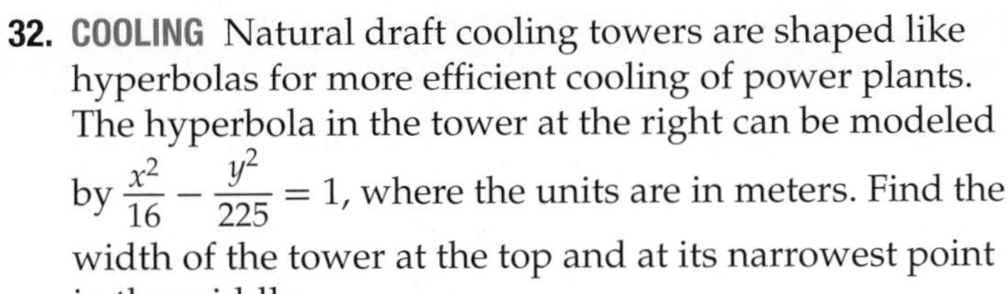

32. **COOLING** Natural draft cooling towers are shaped like hyperbolas for more efficient cooling of power plants. The hyperbola in the tower at the right can be modeled by $\frac{x^2}{16} - \frac{y^2}{225} = 1$, where the units are in meters. Find the width of the tower at the top and at its narrowest point in the middle.

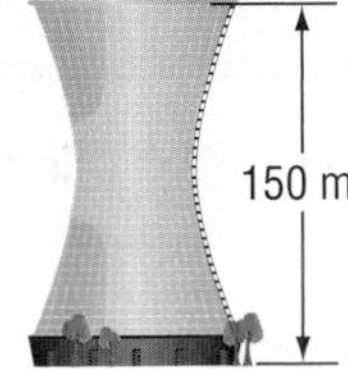

33. **MULTIPLE REPRESENTATIONS** Consider $xy = 16$.

a. **Tabular** Make a table of values for the equation for $-12 \leq x \leq 12$.

b. **Graphical** Graph the hyperbola represented by the equation.

c. **Logical** Determine and graph the asymptotes for the hyperbola.

d. **Analytical** What special property do you notice about the asymptotes? Hyperbolas that represent this property are called *rectangular hyperbolas*.

e. **Analytical** Without any calculations, what do you think will be the coordinates of the vertices for $xy = 25$? for $xy = 36$?

34. **CCSS MODELING** Two receiving stations that are 250 miles apart receive a signal from a downed airplane. They determine that the airplane is 70 miles farther from station B than from station A. Determine the equation of the horizontal hyperbola centered at the origin on which the plane is located.

35. **WEATHER** Luisa and Karl live exactly 4000 feet apart. While on the phone at their homes, Luisa hears thunder out of her window and Karl hears it 3 seconds later out of his. If sound travels 1100 feet per second, determine the equation for the horizontal hyperbola on which the lightning is located.

36. ARCHITECTURE Large pillars with cross sections in the shape of hyperbolas were popular in ancient Greece. The curves can be modeled by the equation $\frac{x^2}{0.16} - \frac{y^2}{4} = 1$, where the units are in feet. If the pillars are 9 feet tall, find the width of the top of each pillar and the width of each pillar at the narrowest point in the middle. Round to the nearest hundredth of a foot.

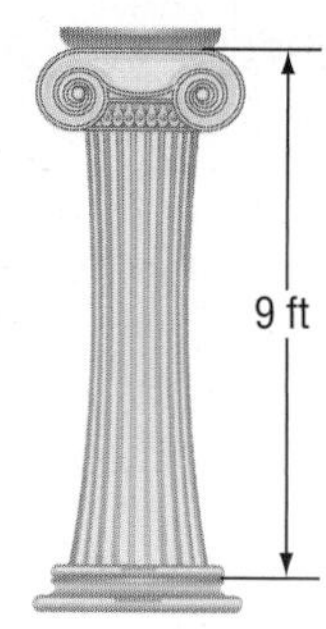

Write an equation for the hyperbola that satisfies each set of conditions.

37. vertices (−8, 0) and (8, 0), conjugate axis of length 20 units

38. vertices (0, −6) and (0, 6), conjugate axis of length 24 units

39. vertices (6, −2) and (−2, −2), foci (10, −2) and (−6, −2)

40. vertices (−3, 4) and (−3, −8), foci (−3, 9) and (−3, −13)

41. centered at the origin with a horizontal transverse axis of length 10 units and a conjugate axis of length 4 units

42. centered at the origin with a vertical transverse axis of length 16 units and a conjugate axis of length 12 units

43. TRIANGULATION While looking for their lost dog in the woods, Lae, Meg, and Cesar hear a bark. Meg hears it 2 seconds after Lae and Cesar hears it 3 seconds after Lae. With Lae at the origin, determine the exact location of their dog if sound travels 1100 feet per second.

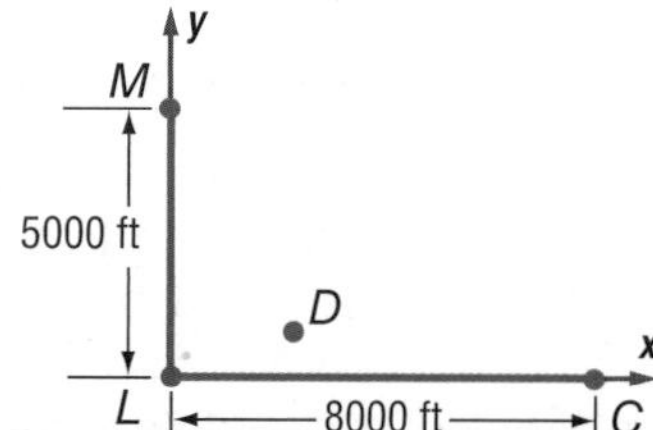

H.O.T. Problems Use Higher-Order Thinking Skills

44. CCSS CRITIQUE Simon and Gabriel are graphing $\frac{y^2}{25} - \frac{x^2}{4} = 1$. Is either of them correct? Explain your reasoning.

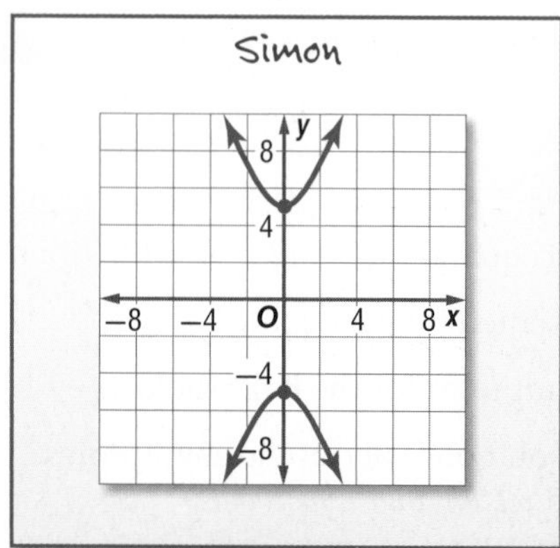

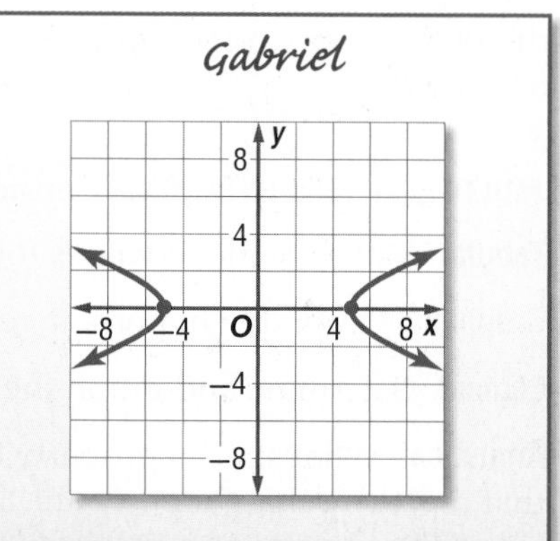

45. CHALLENGE The origin lies on a horizontal hyperbola. The asymptotes for the hyperbola are $y = -x + 1$ and $y = x - 5$. Find the equation for the hyperbola.

46. REASONING What happens to the location of the foci of a hyperbola as the value of a becomes increasingly smaller than the value of b? Explain your reasoning.

47. REASONING Consider $\frac{y^2}{36} - \frac{x^2}{16} = 1$. Describe the change in the shape of the hyperbola and the locations of the vertices and foci if 36 is changed to 9. Explain why this happens.

48. OPEN ENDED Write an equation for a hyperbola with a focus at the origin.

49. WRITING IN MATH Why would you choose a conic section to model a set of data instead of a polynomial function?

Standardized Test Practice

50. You have 6 more dimes than quarters. You have a total of $5.15. How many dimes do you have?

A 13 C 19

B 16 D 25

51. How tall is a tree that is 15 feet shorter than a pole three times as tall as the tree?

F 24.5 ft

G 22.5 ft

H 21.5 ft

J 7.5 ft

52. SHORT RESPONSE A rectangle is 8 feet long and 6 feet wide. If each dimension is increased by the same number of feet, the area of the new rectangle formed is 32 square feet more than the area of the original rectangle. By how many feet was each dimension increased?

53. SAT/ACT When the equation $y = 4x^2 - 5$ is graphed in the coordinate plane, the graph is which of the following?

A line D hyperbola

B circle E parabola

C ellipse

Spiral Review

Write an equation for an ellipse that satisfies each set of conditions. (Lesson 9-4)

54. endpoints of major axis at (2, 2) and (2, −10), endpoints of minor axis at (0, −4) and (4, −4)

55. endpoints of major axis at (0, 10) and (0, −10), foci at (0, 8) and (0, −8)

Find the center and radius of the circle with the given equation. Then graph the circle. (Lesson 9-3)

56. $(x - 3)^2 + y^2 = 16$ **57.** $x^2 + y^2 - 6y - 16 = 0$ **58.** $x^2 + y^2 + 9x - 8y + 4 = 0$

59. BASKETBALL Zonta plays basketball for Centerville High School. So far this season, she has made 6 out of 10 free throws. She is determined to improve her free throw percentage. If she can make x consecutive free throws, her free throw percentage can be determined using $P(x) = \frac{6 + x}{10 + x}$. (Lesson 8-4)

a. Graph the function.

b. What part of the graph is meaningful in the context of the problem?

c. Describe the meaning of the y-intercept.

d. What is the equation of the horizontal asymptote? Explain its meaning with respect to Zonta's shooting percentage.

Solve each equation. (Lesson 7-2)

60. $\left(\frac{1}{7}\right)^{y - 3} = 343$ **61.** $10^{x - 1} = 100^{2x - 3}$ **62.** $36^{2p} = 216^{p - 1}$

Graph each inequality. (Lesson 6-3)

63. $y \geq \sqrt{5x - 8}$ **64.** $y \geq \sqrt{x - 3} + 4$ **65.** $y < \sqrt{6x - 2} + 1$

Skills Review

66. Write an equation for a parabola with vertex at the origin that passes through (2, −8)

67. Write an equation for a parabola with vertex at (−3, −4) that opens up and has y-intercept 8.

LESSON 9-6 Identifying Conic Sections

Then	Now	Why?
You analyzed different conic sections.	**1** Write equations of conic sections in standard form. **2** Identify conic sections from their equations.	Parabolas, circles, ellipses, and hyperbolas are called conic sections because they are the cross sections formed when a double cone is sliced by a plane.

Parabola

Circle and Ellipse

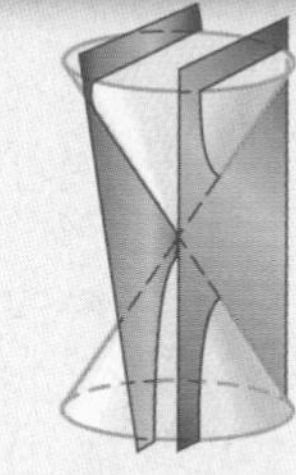
Hyperbola

Common Core State Standards

Content Standards

A.SSE.1.b Interpret complicated expressions by viewing one or more of their parts as a single entity.

F.IF.9 Compare properties of two functions each represented in a different way (algebraically, graphically, numerically in tables, or by verbal descriptions).

Mathematical Practices

3 Construct viable arguments and critique the reasoning of others.

8 Look for and express regularity in repeated reasoning.

1 Conics in Standard Form

The equation for any conic section can be written in the form $Ax^2 + Bxy + Cy^2 + Dx + Ey + F = 0$, where A, B, and C are not all zero. This general form can be converted to the standard forms below by completing the square.

ConceptSummary Standard Forms of Conic Sections

Conic Section	Standard Form of Equation	
Circle	$(x - h)^2 + (y - k) = r^2$	
	Horizontal Axis	**Vertical Axis**
Parabola	$y = a(x - h)^2 + k$	$x = a(y - k)^2 + h$
Ellipse	$\frac{(x-h)^2}{a^2} + \frac{(y-k)^2}{b^2} = 1$	$\frac{(y-k)^2}{a^2} + \frac{(x-h)^2}{b^2} = 1$
Hyperbola	$\frac{(x-h)^2}{a^2} - \frac{(y-k)^2}{b^2} = 1$	$\frac{(y-k)^2}{a^2} - \frac{(x-h)^2}{b^2} = 1$

Example 1 Rewrite an Equation of a Conic Section

Write $16x^2 - 25y^2 - 128x - 144 = 0$ in standard form. State whether the graph of the equation is a *parabola, circle, ellipse,* or *hyperbola*. Then graph the equation.

$16x^2 - 25y^2 - 128x - 144 = 0$ Original equation

$16(x^2 - 8x + ■) - 25y^2 = 144 + 16(■)$ Isolate terms.

$16(x^2 - 8x + 16) - 25y^2 = 144 + 16(16)$ Complete the square.

$16(x - 4)^2 - 25y^2 = 400$ Perfect square

$\frac{(x-4)^2}{25} - \frac{y^2}{16} = 1$ Divide each side by 400.

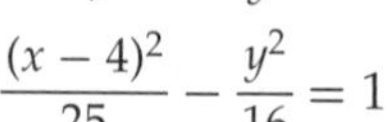

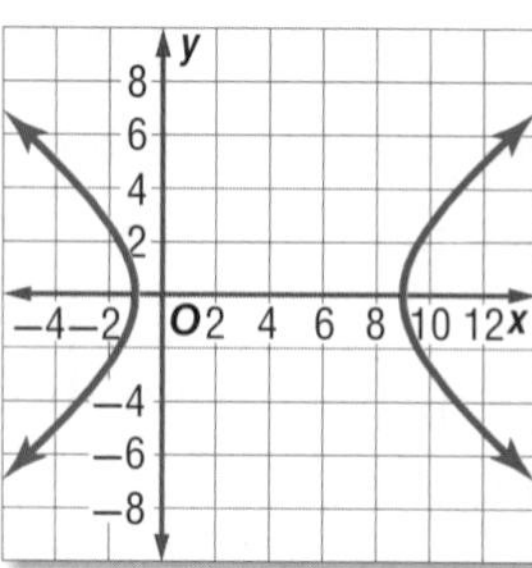

The graph is a hyperbola with its center at (4, 0).

GuidedPractice

1. Write $4x^2 + y^2 - 16x + 8y - 4 = 0$ in standard form. State whether the graph of the equation is a *parabola, circle, ellipse,* or *hyperbola*. Then graph the equation.

ReviewVocabulary

discriminant the expression $b^2 - 4ac$ from the Quadratic Formula

2 Identify Conic Sections

You can determine the type of conic without having to write $Ax^2 + Bxy + Cy^2 + Dx + Ey + F = 0$ in standard form. When there is an xy-term ($B \neq 0$), you can use the discriminant to identify the conic. $B^2 - 4AC$ is the discriminant of $Ax^2 + Bxy + Cy^2 + Dx + Ey + F = 0$.

ConceptSummary Classify Conics with the Discriminant

Discriminant	Conic Section
$B^2 - 4AC < 0$; $B = 0$ and $A = C$	circle
$B^2 - 4AC < 0$; either $B \neq 0$ or $A \neq C$	ellipse
$B^2 - 4AC = 0$	parabola
$B^2 - 4AC > 0$	hyperbola

When $B = 0$, the conic will be either vertical or horizontal. When $B \neq 0$, the conic will be neither vertical nor horizontal.

Horizontal Ellipse: $B = 0$

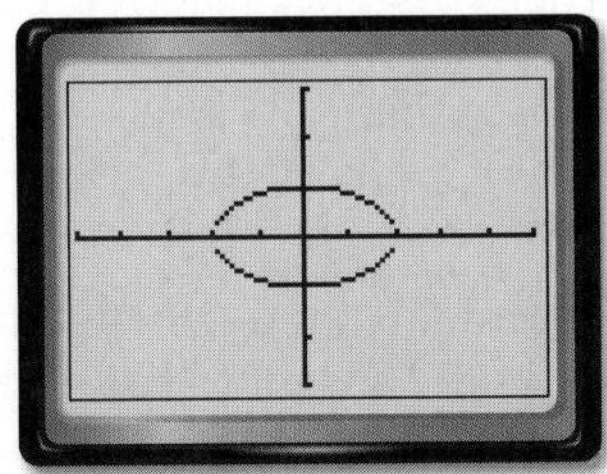

$x^2 + 4y^2 - 4 = 0$

Rotated Ellipse: $B \neq 0$

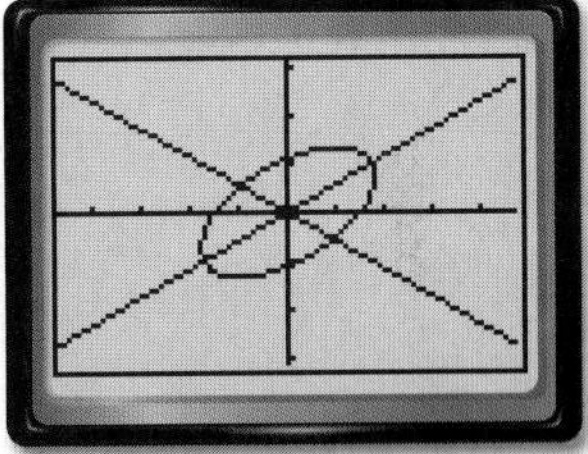

$7x^2 - 6\sqrt{3}xy + 13y^2 - 16 = 0$

StudyTip

Identifying Conics When there is no xy-term ($B = 0$), use A and C.
Parabola: A or $C = 0$ but not both.
Circle: $A = C$
Ellipse: A and C have the same sign but are not equal.
Hyperbola: A and C have opposite signs.

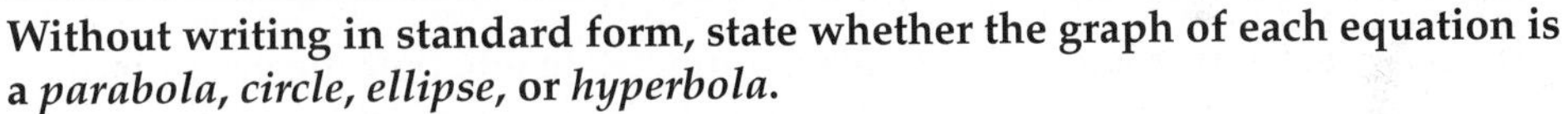

Example 2 Analyze an Equation of a Conic Section

PT

Without writing in standard form, state whether the graph of each equation is a *parabola, circle, ellipse,* or *hyperbola.*

a. $y^2 + 4x^2 - 3xy + 4x - 5y - 8 = 0$

$A = 4$, $B = -3$, and $C = 1$

The discriminant is $(-3)^2 - 4(4)(1)$ or -7.

Because the discriminant is less than 0 and $B \neq 0$, the conic is an ellipse.

b. $3x^2 - 6x + 4y - 5y^2 + 2xy - 4 = 0$

$A = 3$, $B = 2$, and $C = -5$

The discriminant is $2^2 - 4(3)(-5)$ or 64.

Because the discriminant is greater than 0, the conic is a hyperbola.

c. $4y^2 - 8x + 6y - 14 = 0$

$A = 0$, $B = 0$, and $C = 4$

The discriminant is $0^2 - 4(0)(4)$ or 0.

Because the discriminant equals 0, the conic is a parabola.

GuidedPractice

2A. $8y^2 - 6x^2 + 4xy - 6x + 2y - 4 = 0$

2B. $3xy + 4x^2 - 2y + 9x - 3 = 0$

2C. $3x^2 + 16x - 12y + 2y^2 - 6 = 0$

Check Your Understanding

= Step-by-Step Solutions begin on page R14. 

Example 1 **Write each equation in standard form. State whether the graph of the equation is a *parabola, circle, ellipse,* or *hyperbola*. Then graph the equation.**

1. $x^2 + 4y^2 - 6x + 16y - 11 = 0$

2. $x^2 + y^2 + 12x - 8y + 36 = 0$

3. $9y^2 - 16x^2 - 18y - 64x - 199 = 0$

4. $6y^2 - 24y + 28 - x = 0$

Example 2 **Without writing in standard form, state whether the graph of each equation is a *parabola, circle, ellipse,* or *hyperbola*.**

5. $4x^2 + 6y^2 - 3x - 2y = 12$

6. $5y^2 = 2x + 6y - 8 + 3x^2$

7. $8x^2 + 8y^2 + 16x + 24 = 0$

8. $4x^2 - 6y = 8x + 2$

9. $4x^2 - 3y^2 + 8xy - 12 = 2x + 4y$

10. $5xy - 3x^2 + 6y^2 + 12y = 18$

11. $8x^2 + 12xy + 16y^2 + 4y - 3x = 12$

12. $16xy + 8x^2 + 8y^2 - 18x + 8y = 13$

13. **CCSS MODELING** A military jet performs for an air show. The path of the plane during one maneuver can be modeled by a conic section with equation $24x^2 + 1000y - 31{,}680x - 45{,}600 = 0$, where distances are represented in feet.

a. Identify the shape of the curved path of the jet. Write the equation in standard form.

b. If the jet begins its path upward, or ascent, at $x = 0$, what is the horizontal distance traveled by the jet from the beginning of the ascent to the end of the descent?

c. What is the maximum height of the jet?

Practice and Problem Solving

Extra Practice is on page R9.

Example 1 **Write each equation in standard form. State whether the graph of the equation is a *parabola, circle, ellipse,* or *hyperbola*. Then graph the equation.**

14. $3x^2 - 2y^2 + 18x + 8y - 35 = 0$

15. $3x^2 + 24x + 4y^2 - 40y + 52 = 0$

16. $x^2 + y^2 = 16 + 6y$

17. $32x + 28 = y - 8x^2$

18. $7x^2 - 8y = 84x - 2y^2 - 176$

19. $x^2 + 8y = 11 + 6x - y^2$

20. $4y^2 = 24y - x - 31$

21. $112y + 64x = 488 + 7y^2 - 8x^2$

22. $28x^2 + 9y^2 - 188 = 56x - 36y$

23. $25x^2 + 384y - 64y^2 + 200x = 1776$

Example 2 **Without writing in standard form, state whether the graph of each equation is a *parabola, circle, ellipse,* or *hyperbola*.**

24. $4x^2 - 5y = 9x - 12$

25. $4x^2 - 12x = 18y - 4y^2$

26. $9x^2 + 12y = 9y^2 + 18y - 16$

27 $18x^2 - 16y = 12x - 4y^2 + 19$

28. $12y^2 - 4xy + 9x^2 = 18x - 124$

29. $5xy + 12x^2 - 16x = 5y + 3y^2 + 18$

30. $19x^2 + 14y = 6x - 19y^2 - 88$

31. $8x^2 + 20xy + 18 = 4y^2 - 12 + 9x$

32. $5x - 12xy + 6x^2 = 8y^2 - 24y - 9$

33. $18x - 24y + 324xy = 27x^2 + 3y^2 - 5$

34. **LIGHT** A lamp standing near a wall throws an arc of light in the shape of a conic section. Suppose the edge of the light can be represented by the equation $3y^2 - 2y - 4x^2 + 2x - 8 = 0$. Identify the shape of the edge of the light and graph the equation.

Match each graph with its corresponding equation.

a. $x^2 + y^2 - 8x - 4y = -4$ **b.** $9x^2 - 16y^2 - 72x + 64y = 64$ **c.** $9x^2 + 16y^2 = 72x + 64y - 64$

35. 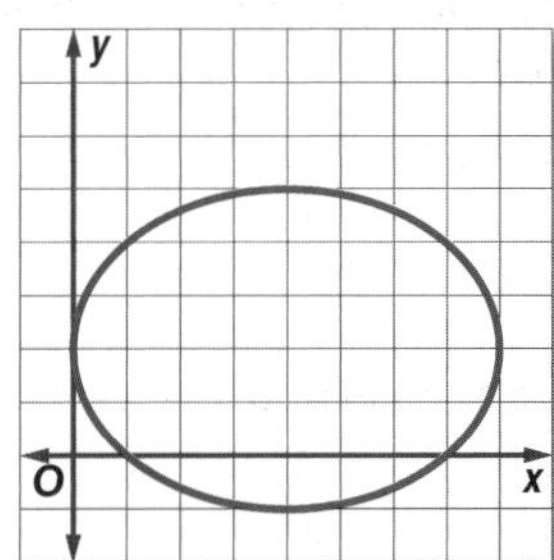

36. 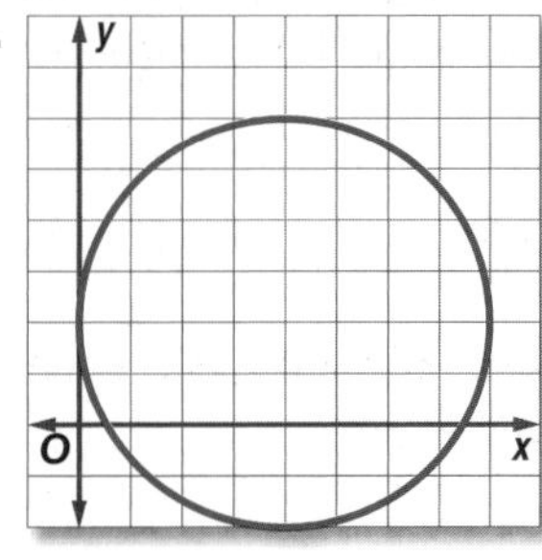

37. 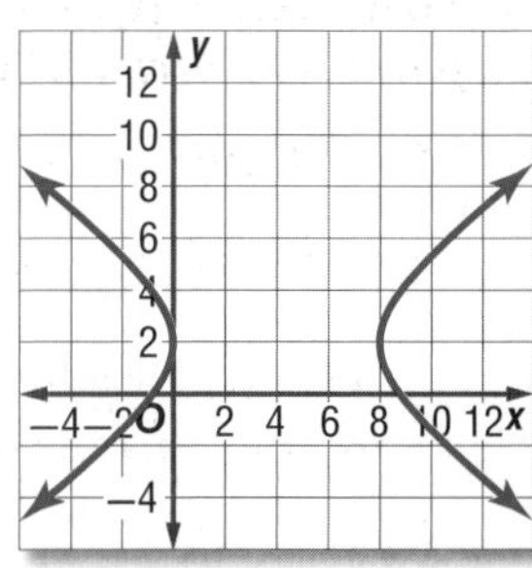

For Exercises 38–41, match each situation with an equation that could be used to represent it.

a. $47.25x^2 - 9y^2 + 18y + 33.525 = 0$ **b.** $25x^2 + 100y^2 - 1900x - 2200y + 45{,}700 = 0$

c. $16x^2 - 90x + y - 0.25 = 0$ **d.** $x^2 + y^2 - 18x - 30y - 14{,}094 = 0$

38. COMPUTERS the boundary of a wireless network with a range of 120 feet

39. FITNESS the oval path of your foot on an exercise machine

40. COMMUNICATIONS the position of a cell phone between two cell towers

41. SPORTS the height of a football above the ground after being kicked

42. CCSS SENSE-MAKING The shape of the cables in a suspension bridge is approximately parabolic. If the towers for a planned bridge are 1000 meters apart and the lowest point of the suspension cables is 200 meters below the top of the towers, write the equation in standard form with the origin at the vertex.

43. MULTIPLE REPRESENTATIONS Consider an ellipse with center (3, −2), vertex $M(-1, -2)$, and co-vertex $N(3, -4)$.

a. Analytical Determine the standard form of the equation of the ellipse.

b. Algebraic Convert part **a** to $Ax^2 + Bxy + Cy^2 + Dx + Ey + F = 0$ form.

c. Graphical Graph the ellipse.

d. Analytical If the ellipse is rotated such that M is moved to (3, −6), determine the location of N and the angle of rotation.

H.O.T. Problems Use Higher-Order Thinking Skills

44. CHALLENGE When a plane passes through the vertex of a cone, a *degenerate* conic is formed.

a. Determine the type of conic represented by $4x^2 + 8y^2 = 0$.

b. Graph the conic.

c. Describe the difference between this degenerate conic and a standard conic of the same type with $A = 4$ and $B = 8$.

45. REASONING Is the following statement *sometimes*, *always*, or *never* true? Explain.

When a conic is vertical and $A = C$, it is a circle.

46. OPEN ENDED Write an equation of the form $Ax^2 + Bxy + Cy^2 + Dx + Ey + F = 0$, where $A = 9C$, that represents a parabola.

47. WRITING IN MATH Compare and contrast the graphs of the four types of conics and their corresponding equations.

Standardized Test Practice

48. SAT/ACT A class of 25 students took a science test. Ten students had a mean score of 80. The other students had an average score of 60. What is the average score of the whole class?

A 66
B 68
C 70
D 72
E 78

49. Six times a number minus 11 is 43. What is the number?

F 12
G 11
H 10
J 9

50. EXTENDED RESPONSE The amount of water remaining in a storage tank as it is drained can be represented by the equation $L = -4t^2 - 10t + 130$, where L represents the number of liters of water remaining and t represents the number of minutes since the drain was opened. How many liters of water were in the tank initially? Determine to the nearest tenth of a minute how long it will take for the tank to drain completely.

51. Ruben has a square piece of paper with sides 4 inches long. He rolled up the paper to form a cylinder. What is the volume of the cylinder?

A $\frac{4}{\pi}$
B $\frac{16}{\pi}$
C 4π
D 16π

Spiral Review

52. ASTRONOMY Suppose a comet's path can be modeled by a branch of the hyperbola with equation $\frac{y^2}{225} - \frac{x^2}{400} = 1$. Find the coordinates of the vertices and foci and the equations of the asymptotes for the hyperbola. Then graph the hyperbola. (Lesson 9-5)

Find the coordinates of the center and foci and the lengths of the major and minor axes for the ellipse with the given equation. Then graph the ellipse. (Lesson 9-4)

53. $\frac{y^2}{18} + \frac{x^2}{9} = 1$

54. $4x^2 + 8y^2 = 32$

55. $x^2 + 25y^2 - 8x + 100y + 91 = 0$

Graph each function. (Lesson 8-3)

56. $f(x) = \frac{3}{x}$

57. $f(x) = \frac{-2}{x + 5}$

58. $f(x) = \frac{6}{x - 2} - 4$

59. SPACE A radioisotope is used as a power source for a satellite. The power output P (in watts) is given by $P = 50e^{-\frac{t}{250}}$, where t is the time in days. (Lesson 7-8)

a. Is the formula for power output an example of exponential *growth* or *decay*? Explain your reasoning.

b. Find the power available after 100 days.

c. Ten watts of power are required to operate the equipment in the satellite. How long can the satellite continue to operate?

Skills Review

Solve each system of equations.

60. $6g - 8h = 50$
$6h = 22 - 4g$

61. $3u + 5v = 6$
$2u - 4v = -7$

62. $10m - 9n = 15$
$5m - 4n = 10$

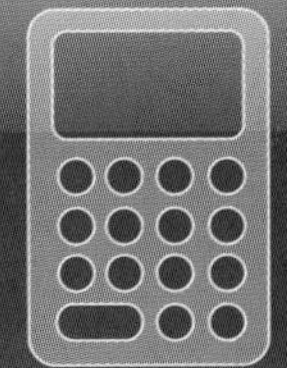

EXTEND 9-6 Graphing Technology Lab
Analyzing Quadratic Relations

You can use TI-Nspire technology to analyze quadratic relations.

Activity 1 Characteristics of a Parabolic Relation

Graph $f(x) = 9x^2 + 1$, $g(x) = -x^2 + 3x - 4$, and $(y - 3)^2 = -4(x - 2)$. Identify the maxima, minima, and axes of symmetry.

Step 1 Add a new **Graphs** and **Calculator** page.

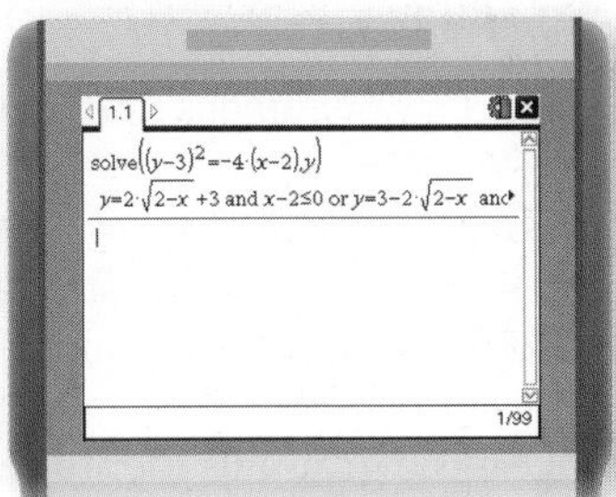

Step 2 Enter $f(x)$ into **f1** and $g(x)$ into **f2**.

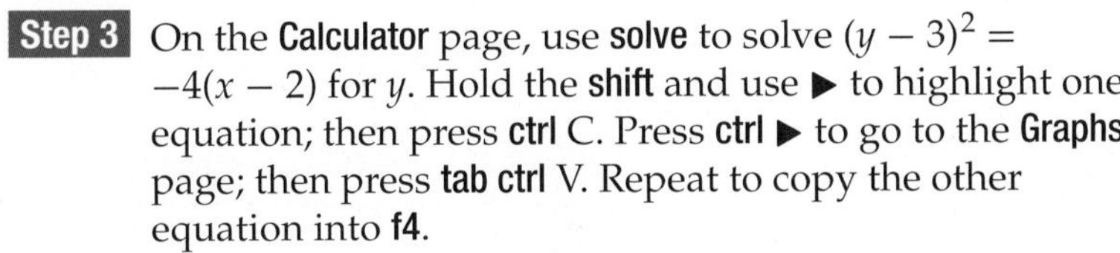

Step 3 On the **Calculator** page, use **solve** to solve $(y - 3)^2 = -4(x - 2)$ for y. Hold the **shift** and use ▶ to highlight one equation; then press **ctrl** C. Press **ctrl** ▶ to go to the **Graphs** page; then press **tab ctrl** V. Repeat to copy the other equation into **f4**.

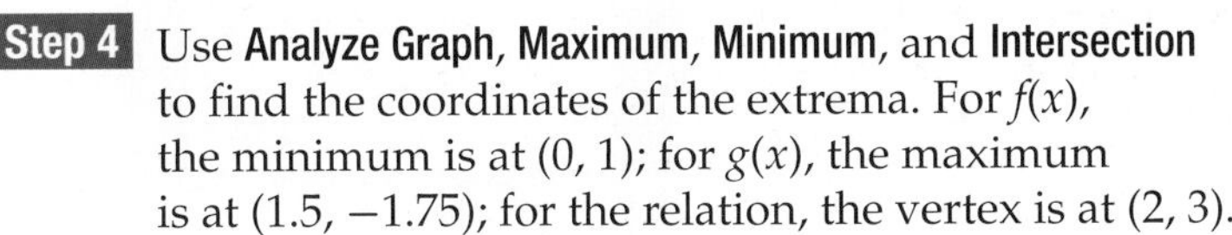

Step 4 Use **Analyze Graph**, **Maximum**, **Minimum**, and **Intersection** to find the coordinates of the extrema. For $f(x)$, the minimum is at (0, 1); for $g(x)$, the maximum is at (1.5, −1.75); for the relation, the vertex is at (2, 3).

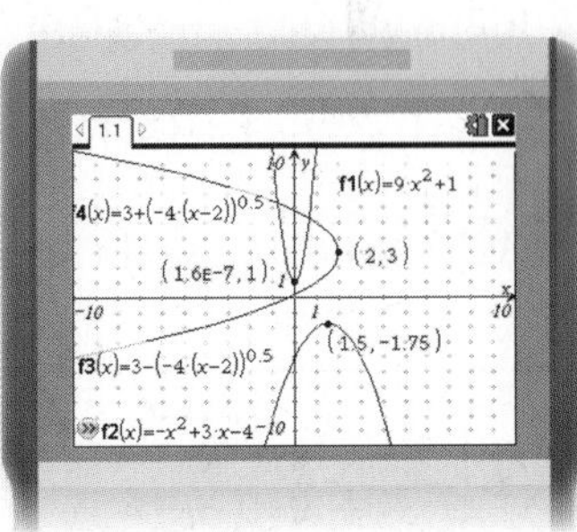

[−10, 10] scl: 1 by [−10, 10] scl: 1

Step 5 You can use the coordinates of the extrema to find the equations of the axes of symmetry. For $f(x)$, the equation of the axis of symmetry is $x = 1.5$; for $g(x)$, the equation of the axis of symmetry is $x = 0$; for the relation, the equation of the axis of symmetry is $y = 3$.

You can also use a graphing calculator to determine the equation of a parabola.

Activity 2 Write an Equation for a Parabola

Given that $f(x)$ has zeros at −2 and 4 and $f(x)$ opens downward, write an equation for the parabola.

Step 1 Add a new **Graphs** and **Calculator** page.

Step 2 Because the zeros are $x = -2$ and $x = 4$, the factors of the quadratic equation are $(x + 2)$ and $(x - 4)$.

Step 3 Use the **expand** command on the **Calculator** page to multiply $(x + 2)$ and $(x - 4)$. So, $y = x^2 - 2x - 8$.

Step 4 On the **Graphs** page, graph $y = x^2 - 2x - 8$. Verify the roots and direction of opening.

Step 5 The function in the graph opens upward, not downward, so multiply $x^2 - 2x - 8$ by −1. Thus, $y = -x^2 + 2x + 8$. Graph this function.

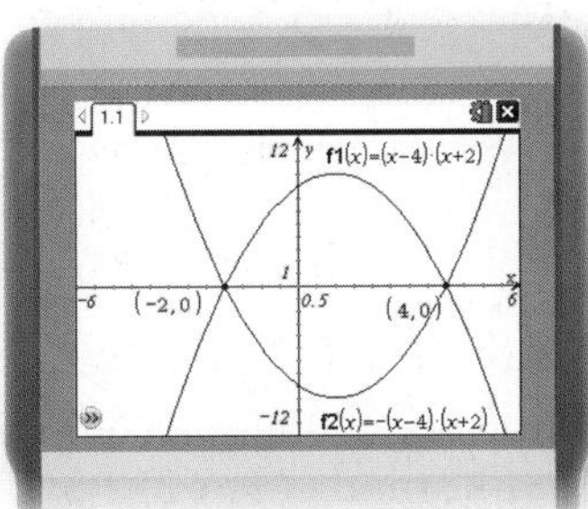

[−6, 6] scl: 0.5 by [−12, 12] scl: 1

An equation for the parabola that has zeros at −2 and 4 and opens downward is $y = -x^2 + 2x + 8$.

(continued on the next page)

Graphing Technology Lab

Analyzing Quadratic Relations *Continued*

You can use a graphing calculator to determine an equation of a quadratic relation.

Activity 3 Write an Equation for an Ellipse

Write an equation for an ellipse that has vertices at (−3, 3) and (−3, −7) and co-vertices at (0, −2) and (−6, −2).

Step 1 Add a new **Graphs** and **Calculator** page.

Step 2 Turn on the grid from **View, Show Axis.** Then use **Points, Points On** to graph the four points. Use **Actions, Coordinates and Equations** to show the coordinates of the points.

Step 3 Use the graph to identify the intersection point of the segment formed by the vertices and the segment formed by the co-vertices. The center is at (−3, −2).

Step 4 Identify other important characteristics. The ellipse is oriented vertically. The length of the major axis is 10, so $a = 5$. The length of the minor axis is 6, so $b = 3$.

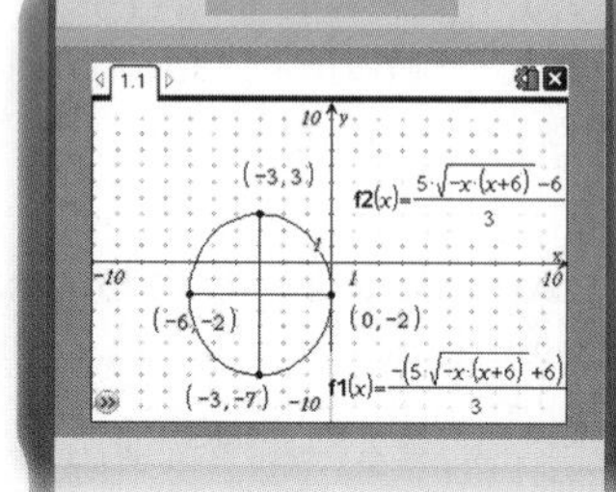

[−10, 10] scl: 1 by [−10, 10] scl: 1

Step 5 Write the equation in standard form.

$$\frac{[y-(-2)]^2}{5^2} + \frac{[x-(-3)]^2}{3^2} = 1 \text{ or}$$

$$\frac{(y+2)^2}{25} + \frac{(x+3)^2}{9} = 1$$

Step 6 Check the equation by using **Solve** under the **Algebra** menu on the **Calculator** page to solve for y. Copy and paste the two equations into the **Graph** page to graph the ellipse.

Exercises

Graph each function and relation. Identify the maxima, minima, and axes of symmetry.

1. $f(x) = -(x-3)^2 + 12$, $g(x) = x^2 + x - 12$, and $(y+5)^2 = -12(x-2)$.

2. $f(x) = -0.25x^2 - 3x - 6$, $g(x) = 2x^2 + 2x + 4$, and $(y+1)^2 = 2(x+6)$.

3. Given that $f(x)$ has zeros at −1 and 3 and $f(x)$ opens upward, write an equation for the parabola.

4. Given that $f(x)$ has zeros at −3 and −1 and $f(x)$ opens downward, write an equation for the parabola.

5. Write an equation for an ellipse that has vertices at (−6, 2) and (−6, −8) and co-vertices at (−3, −3) and (−9, −3).

6. Write an equation for an ellipse that has vertices at (−13, 2) and (1, 2) and co-vertices at (−6, 4) and (−6, 0).

EXPLORE 9-7

Graphing Technology Lab
Linear-Nonlinear Systems

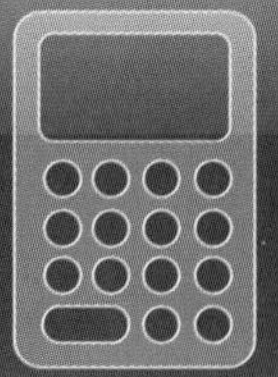

You can use a TI-83/84 Plus application to solve linear-nonlinear systems by using the **Y=** menu to graph each equation on the same set of axes.

Example Linear-Quadratic System

Solve the system of equations.
$3y - 4x = -7$
$4x^2 + 3y^2 = 91$

Step 1 Solve each equation for y.

$$3y - 4x = -7 \qquad 4x^2 + 3y^2 = 91$$
$$3y = 4x - 7 \qquad 3y^2 = 91 - 4x^2$$
$$y = \frac{4}{3}x - \frac{7}{3} \qquad y = \pm\sqrt{\frac{91 - 4x^2}{3}}$$

Step 2 Enter $y = \frac{4}{3}x - \frac{7}{3}$ as **Y1**, $y = \sqrt{\frac{91 - 4x^2}{3}}$ as **Y2**, and $y = -\sqrt{\frac{91 - 4x^2}{3}}$ as **Y3**.
Then graph the equations in a standard viewing window.

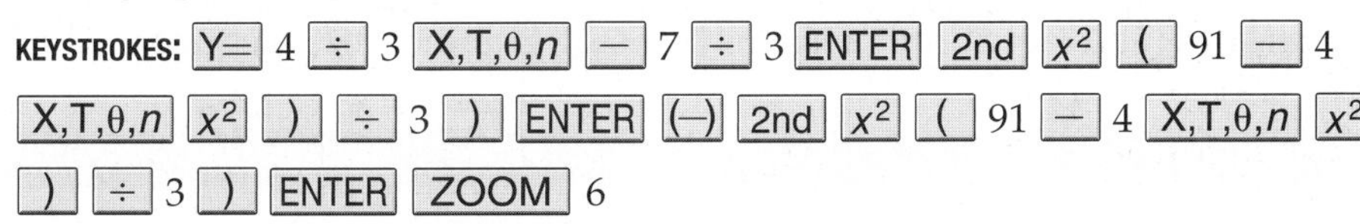

KEYSTROKES: [Y=] 4 [÷] 3 [X,T,θ,n] [−] 7 [÷] 3 [ENTER] [2nd] [x^2] [(] 91 [−] 4 [X,T,θ,n] [x^2] [)] [÷] 3 [)] [ENTER] [(−)] [2nd] [x^2] [(] 91 [−] 4 [X,T,θ,n] [x^2] [)] [÷] 3 [)] [ENTER] [ZOOM] 6

Step 3 Find the intersection of $y = \frac{4}{3}x - \frac{7}{3}$ with $y = \sqrt{\frac{91 - 4x^2}{3}}$.

KEYSTROKES: Press [2nd] [TRACE] 5 [ENTER] [ENTER] [ENTER]. The two graphs intersect at (4, 3).

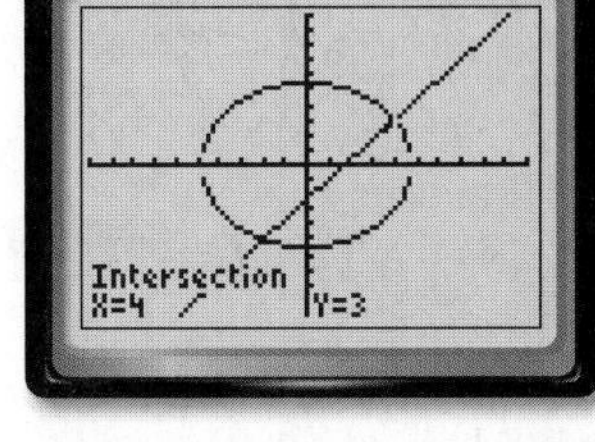

Step 4 Find the intersection of $y = \frac{4}{3}x - \frac{7}{3}$ with $y = -\sqrt{\frac{91 - 4x^2}{3}}$.

KEYSTROKES: Press [2nd] [TRACE] 5 [ENTER] [▼] [ENTER]. Then use [◄] to move the cursor to the second intersection point.

Press [ENTER]. The two graphs intersect at (−2, −5).

The solutions of the system are (4, 3) and (−2, −5).

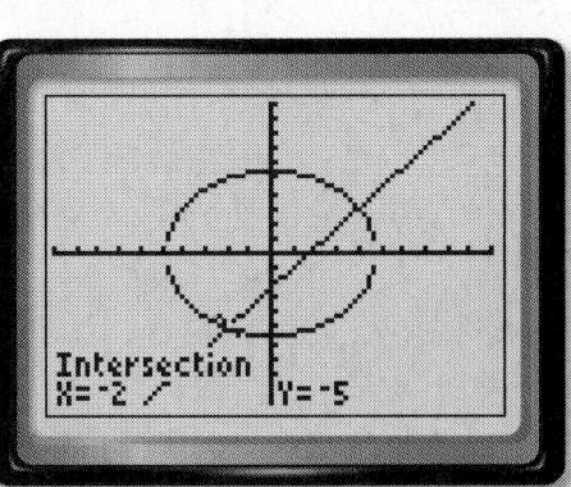

Exercises

Use a graphing calculator to solve each system of equations.

1. $x^2 + y^2 = 100$
   $x + y = 2$
2. $2y - x = 11$
   $5x^2 + 2y^2 = 407$
3. $21x + 9y = -36$
   $7x^2 + 9y^2 = 1152$

LESSON 9-7 Solving Linear-Nonlinear Systems

Then

- You solved systems of linear equations.

Now

1. Solve systems of linear and nonlinear equations algebraically and graphically.
2. Solve systems of linear and nonlinear inequalities graphically.

Why?

- Have you ever wondered how law enforcement agencies can track a cell phone user's location? A person using a cell phone can be located in respect to three cellular towers. The respective coordinates and distances each tower is from the caller are used to pinpoint the caller's location. This is accomplished using a system of quadratic equations.

CCSS Common Core State Standards

Content Standards

A.REI.11 Explain why the x-coordinates of the points where the graphs of the equations $y = f(x)$ and $y = g(x)$ intersect are the solutions of the equation $f(x) = g(x)$; find the solutions approximately, e.g., using technology to graph the functions, make tables of values, or find successive approximations. Include cases where $f(x)$ and/or $g(x)$ are linear, polynomial, rational, absolute value, exponential, and logarithmic functions.

Mathematical Practices

6 Attend to precision.

1 Systems of Equations

When a system of equations consists of a linear and a nonlinear equation, the system may have zero, one, or two solutions. Some of the possible solutions are shown below.

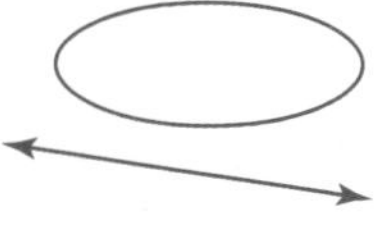
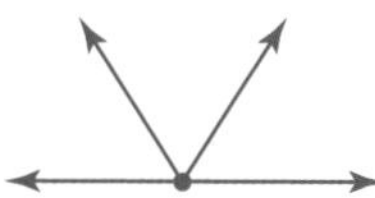

You can solve linear-quadratic systems by using graphical or algebraic methods.

Example 1 Linear-Quadratic System

Solve the system of equations. $9x^2 + 25y^2 = 225$ (1)
$10y + 6x = 6$ (2)

Step 1 Solve the linear equation for y.

$10y + 6x = 6$ Equation (2)
$y = -0.6x + 0.6$ Solve for y.

Step 2 Substitute into the quadratic equation and solve for x.

$9x^2 + 25y^2 = 225$ Quadratic equation
$9x^2 + 25(-0.6x + 0.6)^2 = 225$ Substitute $-0.6x + 0.6$ for y.
$9x^2 + 25(0.36x^2 - 0.72x + 0.36) = 225$ Simplify.
$9x^2 + 9x^2 - 18x + 9 = 225$ Distribute.
$18x^2 - 18x - 216 = 0$ Simplify.
$x^2 - x - 12 = 0$ Divide each side by 18.
$(x - 4)(x + 3) = 0$ Factor.
$x = 4$ or -3 Zero Product Property

Step 3 Substitute x-values into the linear equation and solve for y.

$y = -0.6x + 0.6$	Equation (2)	$y = -0.6x + 0.6$
$= -0.6(4) + 0.6$	Substitute the x-values	$= -0.6(-3) + 0.6$
$= -1.8$	Simplify.	$= 2.4$

The solutions of the system are $(4, -1.8)$ and $(-3, 2.4)$.

GuidedPractice

1A. $3y + x^2 - 4x - 17 = 0$
$3y - 10x + 38 = 0$

1B. $3(y - 4) - 2(x - 3) = -6$
$5x^2 + 2y^2 - 53 = 0$

If a quadratic system contains two conic sections, the system may have anywhere from zero to four solutions. Some graphical representations are shown below.

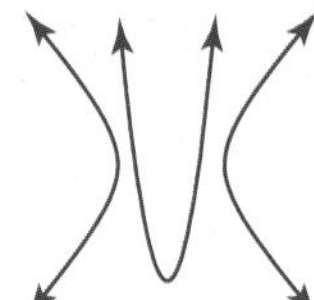  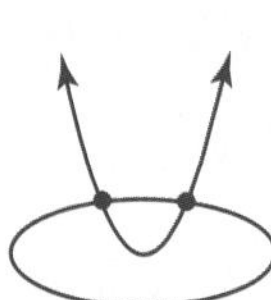 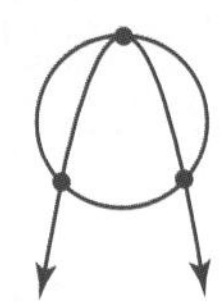 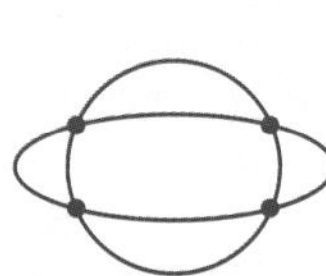

You can use elimination to solve quadratic-quadratic systems.

Example 2 Quadratic-Quadratic System

Solve the system of equations.

$x^2 + y^2 = 45$ (1)

$y^2 - x^2 = 27$ (2)

$y^2 + x^2 = 45$	Equation (1), Commutative Property
$(+)\ y^2 - x^2 = 27$	Equation (2)
$2y^2 = 72$	Add.
$y^2 = 36$	Divide each side by 2.
$y = \pm 6$	Take the square root of each side.

Substitute 6 and −6 into one of the original equations and solve for x.

$x^2 + y^2 = 45$	Equation (1)	$x^2 + y^2 = 45$
$x^2 + 6^2 = 45$	Substitute for y.	$x^2 + (-6)^2 = 45$
$x^2 = 9$	Subtract 36 from each side.	$x^2 = 9$
$x = \pm 3$	Take the square root of each side.	$x = \pm 3$

The solutions are (−3, −6), (−3, 6), (3, −6), and (3, 6).

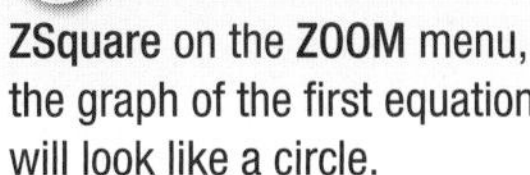

StudyTip

CCSS Tools If you use **ZSquare** on the **ZOOM** menu, the graph of the first equation will look like a circle.

GuidedPractice

2A. $x^2 + y^2 = 8$
$x^2 + 3y = 10$

2B. $3x^2 + 4y^2 = 48$
$2x^2 - y^2 = -1$

2 Systems of Inequalities

Systems of quadratic inequalities can be solved by graphing.

Example 3 Quadratic Inequalities

Solve the system of inequalities by graphing.

$x^2 + y^2 \leq 49$

$x^2 - 4y^2 > 16$

The intersection of the graphs, shaded green, represents the solution of the system.

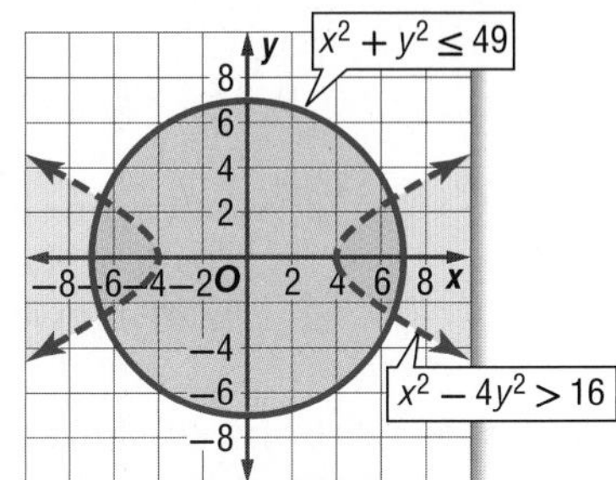

CHECK (6, 0) is in the shaded area. Use this point to check your solution.

$x^2 + y^2 \leq 49$	$x^2 - 4y^2 > 16$
$6^2 + 0^2 \overset{?}{\leq} 49$	$62 - 4(0)^2 \overset{?}{>} 16$
$36 \leq 49$ ✓	$36 > 16$ ✓

GuidedPractice

3A. $5x^2 + 2y^2 \leq 10$
$y \geq x^2 - 2x + 1$

3B. $x^2 - y^2 \leq 8$
$x^2 + y^2 \geq 120$

Systems involving absolute value can also be solved by graphing.

StudyTip

Graphing Calculator Like linear inequalities, systems of quadratic and absolute value inequalities can be checked with a graphing calculator.

Example 4 Quadratics with Absolute Value

Solve the system of inequalities by graphing.

$y \geq |2x - 4|$
$y \leq -x^2 + 4x + 2$

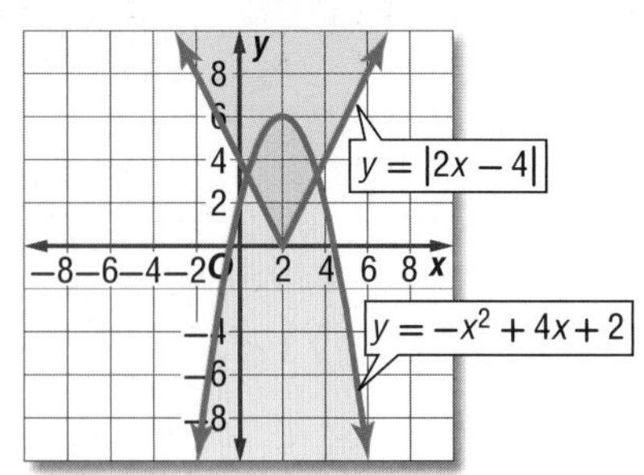

Graph the boundary equations. Then shade appropriately.

The intersection of the graphs, shaded green, represents the solution to the system.

CHECK (2, 4) is in the shaded area. Use the point to check your solution.

$y \geq \|2x - 4\|$	$y \leq -x^2 + 4x + 2$
$4 \overset{?}{\geq} \|2(2) - 4\|$	$4 \overset{?}{\leq} -(2)^2 + 4(2) + 2$
$4 \geq 0$ ✓	$4 \leq 6$ ✓

GuidedPractice

4A. $y > |-0.5x + 2|$
$\frac{x^2}{16} + \frac{y^2}{36} \leq 1$

4B. $x^2 + y^2 \leq 49$
$y \geq |x^2 + 1|$

Check Your Understanding

= Step-by-Step Solutions begin on page R14.

Examples 1–2 **Solve each system of equations.**

1. $8y = -10x$
$y^2 = 2x^2 - 7$

2. $x^2 + y^2 = 68$
$5y = -3x + 34$

3. $y = 12x - 30$
$4x^2 - 3y = 18$

4. $6y^2 - 27 = 3x$
$6y - x = 13$

5. $x^2 + y^2 = 16$
$x^2 - y^2 = 20$

6. $y^2 - 2x^2 = 8$
$3y^2 + x^2 = 52$

7. $x^2 + 2y = 7$
$y^2 - x^2 = 8$

8. $4y^2 - 3x^2 = 11$
$3y^2 + 2x^2 = 21$

9. **CCSS PERSEVERANCE** Refer to the beginning of the lesson. A person using a cell phone can be located with respect to three cellular towers. In a coordinate system where one unit represents one mile, the location of the caller is determined to be 50 miles from the tower at the origin. The person is also 40 miles from a tower at (0, 30) and 13 miles from a tower at (35, 18). Where is the caller?

Examples 3–4 **Solve each system of inequalities by graphing.**

10. $6x^2 + 9(y - 2)^2 \leq 36$
$x^2 + (y + 3)^2 \leq 25$

11. $16x^2 + 4y^2 \leq 64$
$y \geq -x^2 + 2$

12. $4x^2 - 8y^2 \geq 32$
$y \geq |1.5x| - 8$

13. $x^2 + 8y^2 < 32$
$y < -|x - 2| + 2$

Practice and Problem Solving

Extra Practice is on page R9.

Examples 1–2 **Solve each system of equations.**

14. $3x^2 - 2y^2 = -24$
$2y = -3x$

15. $5x^2 + 4y^2 = 20$
$5y = 7x + 35$

16. $x^2 + 3x = -4y - 2$
$y = -2x + 1$

17. $y = 2x$
$4x^2 - 2y^2 = -36$

18. $2y = x + 10$
$y^2 - 4y = 5x + 10$

19. $9y = 8x - 19$
$8x + 11 = 2y^2 + 5y$

20. $2y^2 + 5x^2 = 26$
$2x^2 - y^2 = 5$

21. $x^2 + y^2 = 16$
$x^2 - 4x + y^2 = 12$

22. $x^2 + y^2 = 8$
$5y^2 = 3x^2$

23. $y^2 - x^2 + 3y = 26$
$x^2 + 2y^2 = 34$

24. $x^2 - y^2 = 25$
$x^2 + y^2 + 7 = 0$

25. $x^2 - 10x + 2y^2 = 47$
$y^2 - 2x^2 = -14$

26. FIREWORKS Two fireworks are set off simultaneously but from different altitudes. The height y in feet of one is represented by $y = -16t^2 + 120t + 10$, where t is the time in seconds. The height of the other is represented by $y = -16t^2 + 60t + 310$.

a. After how many seconds are the fireworks the same height?

b. What is that height?

Examples 3–4 CCSS **TOOLS** **Solve each system of inequalities by graphing.**

27. $x^2 + y^2 \geq 36$
$x^2 + 9(y + 6)^2 \leq 36$

28. $-x > y^2$
$4x^2 + 14y^2 \leq 56$

29. $12x^2 - 4y^2 \geq 48$
$16(x - 4)^2 + 25y^2 < 400$

30. $8y^2 - 3x^2 \leq 24$
$2y > x^2 - 8x + 14$

31. $y > x^2 - 6x + 8$
$x \geq y^2 - 6y + 8$

32. $x^2 + y^2 \geq 9$
$25x^2 + 64y^2 \leq 1600$

33. $16(x - 3)^2 + 4y^2 \leq 64$
$y \leq -|x - 2| + 2$

34. $x^2 - 4x + y^2 + 6y \leq 23$
$y > |x - 2| - 6$

35. $2y - 4 \geq |x + 4|$
$12 - 2y > x^2 + 12x + 36$

36. $18y^2 - 3x^2 \leq 54$
$y \geq |2x| - 6$

37. $x^2 + y^2 < 16$
$y \geq |x - 2| + 6$

38. $x^2 < y - 2$
$y \leq |x + 8| - 7$

39. SPACE Two satellites are placed in orbit about Earth. The equations of the two orbits are $\frac{x^2}{(300)^2} + \frac{y^2}{(900)^2} = 1$ and $\frac{x^2}{(600)^2} + \frac{y^2}{(690)^2} = 1$, where distances are in kilometers and Earth is the center of each curve.

a. Solve each equation for y.

b. Use a graphing calculator to estimate the intersection points of the two orbits.

c. Compare the orbits of the two satellites.

40. PETS Taci's dog was missing one day. Fortunately, he was wearing an electronic monitoring device. If the dog is 10 units from the tree, 13 units from the tower, and 20 units from the house, determine the coordinates of his location.

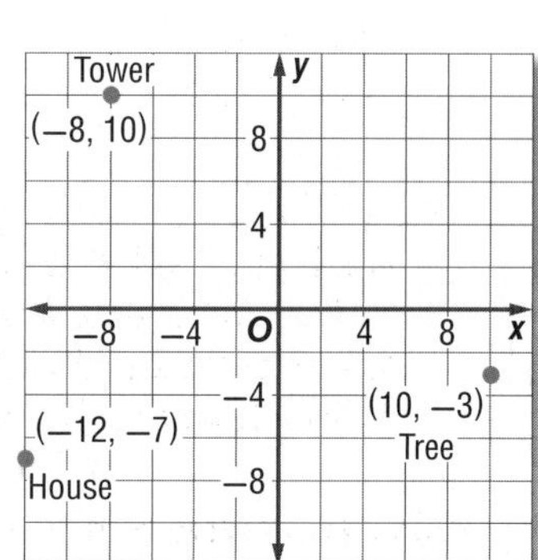

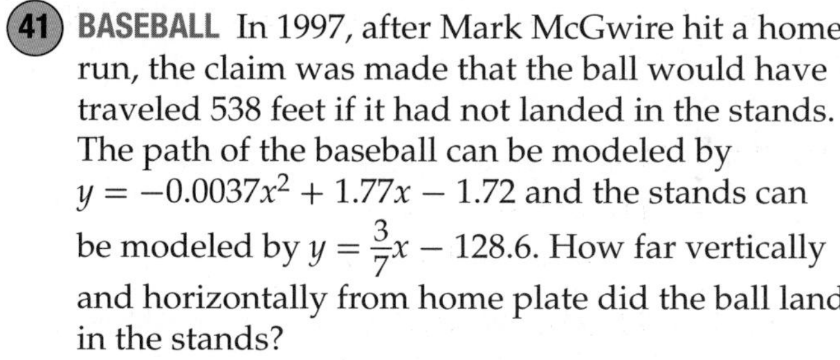

41 BASEBALL In 1997, after Mark McGwire hit a home run, the claim was made that the ball would have traveled 538 feet if it had not landed in the stands. The path of the baseball can be modeled by $y = -0.0037x^2 + 1.77x - 1.72$ and the stands can be modeled by $y = \frac{3}{7}x - 128.6$. How far vertically and horizontally from home plate did the ball land in the stands?

42. ADVERTISING The corporate logo for an automobile manufacturer is shown at the right. Write a system of three equations to model this logo.

Write a system of equations that satisfies each condition.

43. a circle and an ellipse that intersect at one point

44. a parabola and an ellipse that intersect at two points

45. a hyperbola and a circle that do not intersect

46. an ellipse and a parabola that intersect at three points

47. an ellipse and a hyperbola that intersect at four points

48. FINANCIAL LITERACY Prices are often set on an equilibrium curve, where the supply of a certain product equals its corresponding demand by consumers. An economist represents the supply of a product with $y = p^2 + 10p$ and the corresponding demand with $y = -p^2 + 40p$, where p is the price. Determine the equilibrium price.

49. PAINTBALL The shape of a paintball field is modeled by $x^2 + 4y^2 = 10{,}000$ in yards where the center is at the origin. The teams are provided with short-range walkie-talkies with a maximum range of 80 yards. Are the teams capable of hearing each other anywhere on the field? Explain your reasoning graphically.

50. MOVING Lena is moving to a new city and needs for the location of her new home to satisfy the following conditions.

- It must be less than 10 miles from the office where she will work.
- Because of the terrible smell of the local paper mill, it must be at least 15 miles away from the mill.

If the paper mill is located 9.5 miles east and 6 miles north of Lena's office, write and graph a system of inequalities to represent the area(s) were she should look for a home.

H.O.T. Problems Use Higher-Order Thinking Skills

51. CHALLENGE Find all values of k for which the following system of equations has two solutions.

$$\frac{x^2}{a^2} + \frac{y^2}{b^2} = 1 \qquad x^2 + y^2 = k^2$$

52. CCSS ARGUMENTS When the vertex of a parabola lies on an ellipse, how many solutions can the quadratic system represented by the two graphs have? Explain your reasoning using graphs.

53. OPEN ENDED Write a system of equations, one a hyperbola and the other an ellipse, for which a solution is $(-4, 8)$.

54. WRITING IN MATH Explain how sketching the graph of a quadratic system can help you solve it.

Mark Renders/Stringer/Getty Images News/Getty Images

Standardized Test Practice

55. SHORT RESPONSE Solve.

$$4x - 3y = 0$$
$$x^2 + y^2 = 25$$

56. You have 16 stamps. Some are postcard stamps that cost \$0.23, and the rest cost \$0.41. If you spent a total of \$5.30 on the stamps, how many postcard stamps do you have?

A 7
B 8
C 9
D 10

57. Ms. Talbot received a promotion and a 7.2% raise. Her new salary is \$53,600 a year. What was her salary before the raise?

F \$50,000
G \$53,600
H \$55,000
J \$57,500

58. SAT/ACT When a number is multiplied by $\frac{2}{3}$, the result is 188. Find the number.

A 292
B 282
C 272
D 262
E $125\frac{1}{3}$

Spiral Review

Match each equation with the situation that it could represent. (Lesson 9-6)

a. $9x^2 + 4y^2 - 36 = 0$
b. $0.004x^2 - x + y - 3 = 0$
c. $x^2 + y^2 - 20x + 30y - 75 = 0$

59. SPORTS the flight of a baseball

60. PHOTOGRAPHY the oval opening in a picture frame

61. GEOGRAPHY the set of all points 20 miles from a landmark

Find the coordinates of the vertices and foci and the equations of the asymptotes for the hyperbola with the given equation. Then graph the hyperbola. (Lesson 9-5)

62. $\frac{y^2}{16} - \frac{x^2}{25} = 1$

63. $\frac{(y-3)^2}{25} - \frac{(x-2)^2}{16} = 1$

64. $6y^2 = 2x^2 + 12$

Simplify each expression. (Lesson 8-1)

65. $\frac{12p^2 + 6p - 6}{4(p+1)^2} \div \frac{6p - 3}{2p + 10}$

66. $\frac{x^2 + 6x + 9}{x^2 + 7x + 6} \div \frac{4x + 12}{3x + 3}$

67. $\frac{r^2 + 2r - 8}{r^2 + 4r + 3} \div \frac{r - 2}{3r + 3}$

Graph each function. State the domain and range. (Lesson 7-1)

68. $f(x) = -\left(\frac{1}{5}\right)^x$

69. $y = -2.5(5)^x$

70. $f(x) = 2\left(\frac{1}{3}\right)^x$

Skills Review

Solve each equation or formula for the specified variable.

71. $d = rt$, for r

72. $x = \frac{-b}{2a}$, for a

73. $V = \frac{1}{3}\pi r^2 h$, for h

74. $A = \frac{1}{2}h(a + b)$, for b

CHAPTER 9

Study Guide and Review

Study Guide

KeyConcepts

Midpoint and Distance Formulas (Lesson 9-1)

- $M = \left(\frac{x_1 + x_2}{2}, \frac{y_1 + y_2}{2}\right)$
- $d = \sqrt{(x_2 - x_1)^2 + (y_2 - y_1)^2}$

Parabolas (Lesson 9-2)

- Standard Form: $y = a(x - h)^2 + k$
  $x = a(y - k)^2 + h$

Circles (Lesson 9-3)

- The equation of a circle with center (h, k) and radius r can be written in the form $(x - h)^2 + (y - k)^2 = r^2$.

Ellipses (Lesson 9-4)

- Standard Form: horizontal $\frac{(x - h)^2}{a^2} + \frac{(y - k)^2}{b^2} = 1$
  vertical $\frac{(y - k)^2}{a^2} + \frac{(x - h)^2}{b^2} = 1$

Hyperbolas (Lesson 9-5)

- Standard Form: horizontal $\frac{(x - h)^2}{a^2} - \frac{(y - k)^2}{b^2} = 1$
  vertical $\frac{(y - k)^2}{a^2} - \frac{(x - h)^2}{b^2} = 1$

Solving Linear-Nonlinear Systems (Lesson 9-7)

- Systems of quadratic equations can be solved using substitution and elimination.
- A system of quadratic equations can have zero, one, two, three, or four solutions.

FOLDABLES® StudyOrganizer

Be sure the Key Concepts are noted in your Foldable.

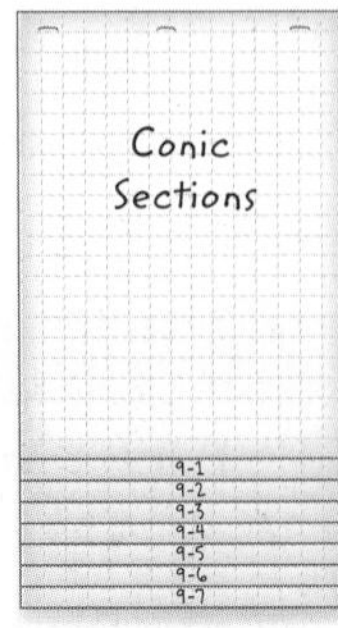

KeyVocabulary

center (of a circle) (p. 607)
center (of an ellipse) (p. 615)
circle (p. 607)
conjugate axis (p. 624)
constant difference (p. 627)
constant sum (p. 616)
co-vertices (of a hyperbola) (p. 624)
co-vertices (of an ellipse) (p. 615)
directrix (p. 599)
ellipse (p. 615)
foci (of a hyperbola) (p. 624)
foci (of an ellipse) (p. 615)
focus (p. 599)
hyperbola (p. 624)
latus rectum (p. 599)
major axis (p. 615)
minor axis (p. 615)
parabola (p. 599)
radius (p. 607)
transverse axis (p. 624)
vertices (of a hyperbola) (p. 624)
vertices (of an ellipse) (p. 615)

VocabularyCheck

State whether each sentence is *true* or *false*. If *false*, replace the underlined term to make a true sentence.

1. The set of all points in a plane that are equidistant from a given point in the plane, called the focus, forms a circle.
2. A(n) ellipse is the set of all points in a plane such that the sum of the distances from the two fixed points is constant.
3. The endpoints of the major axis of an ellipse are the foci of the ellipse.
4. The radius is the distance from the center of a circle to any point on the circle.
5. The line segment with endpoints on a parabola, through the focus of the parabola, and perpendicular to the axis of symmetry is called the latus rectum.
6. Every hyperbola has two axes of symmetry, the transverse axis and the major axis.
7. A directrix is the set of all points in a plane that are equidistant from a given point in the plane, called the center.
8. A hyperbola is the set of all points in a plane such that the absolute value of the sum of the distances from any point on the hyperbola to two given points is constant.
9. A parabola can be defined as the set of all points in a plane that are the same distance from the focus and a given line called the directrix.
10. The major axis is the longer of the two axes of symmetry of an ellipse.

Lesson-by-Lesson Review

9-1 Midpoint and Distance Formulas

Find the midpoint of the line segment with endpoints at the given coordinates.

11. $(-8, 6), (3, 4)$

12. $(-6, 0), (-1, 4)$

13. $\left(\frac{3}{4}, \frac{2}{3}\right), \left(-\frac{1}{3}, \frac{1}{4}\right)$

14. $(15, 20), (18, 21)$

Find the distance between each pair of points with the given coordinates.

15. $(10, -3), (1, -5)$

16. $(0, 6), (-9, 7)$

17. $\left(\frac{1}{4}, \frac{1}{2}\right), \left(\frac{3}{2}, \frac{5}{4}\right)$

18. $(5, -3), (7, -1)$

19. **HIKING** Marc wants to hike from his camp to a waterfall. The waterfall is 5 miles south and 8 miles east of his campsite.

a. How far away is the waterfall?

b. Marc wants to stop for lunch halfway to the waterfall. Where should he stop?

Example 1

Find the midpoint of a line segment whose endpoints are at (−4, 8) and (10, −1).

Let $(x_1, y_1) = (-4, 8)$ and $(x_2, y_2) = (10, -1)$.

$$\left(\frac{x_1 + x_2}{2}, \frac{y_1 + y_2}{2}\right) = \left(\frac{-4 + 10}{2}, \frac{8 + (-1)}{2}\right)$$

$$= \left(\frac{6}{2}, \frac{7}{2}\right) \text{ or } \left(3, \frac{7}{2}\right)$$

Example 2

Find the distance between $P(5, -3)$ and $Q(-1, 5)$.
Let $(x_1, y_1) = (5, -3)$ and $(x_2, y_2) = (-1, 5)$.

$d = \sqrt{(x_2 - x_1)^2 + (y_2 - y_1)^2}$	Distance Formula
$= \sqrt{(-1 - 5)^2 + [5 - (-3)]^2}$	Substitute.
$= \sqrt{36 + 64}$	Subtract.
$= \sqrt{100}$ or 10 units	Simplify.

9-2 Parabolas

Graph each equation.

20. $y = 3x^2 + 24x - 10$

21. $3y - x^2 = 8x - 11$

22. $x = \frac{1}{2}y^2 - 4y + 3$

23. $x = y^2 - 14y + 25$

Write each equation in standard form. Identify the vertex, axis of symmetry, and direction of opening of the parabola.

24. $y = -\frac{1}{2}x^2$

25. $y = 4x^2 - 16x + 9$

26. $x - 6y = y^2 + 4$

27. $x = y^2 + 14y + 20$

28. **SPORTS** When a football is kicked, the path it travels is shaped like a parabola. Suppose a football is kicked from ground level, reaches a maximum height of 50 feet, and lands 200 feet away. Assuming the football was kicked at the origin, write an equation of the parabola that models the flight of the football.

Example 3

Write $3y - x^2 = 4x + 7$ in standard form. Identify the vertex, axis of symmetry, and direction of opening of the parabola.

Write the equation in the form $y = a(x - h)^2 + k$ by completing the square.

$3y = x^2 + 4x + 7$	Isolate the terms with x.
$3y = (x^2 + 4x + ■) + 7 - ■$	Complete the square.
$3y = (x^2 + 4x + 4) + 7 - 4$	$\left(\frac{4}{2}\right)^2 = 4$
$3y = (x + 2)^2 + 3$	$(x^2 + 4x + 4) = (x + 2)$
$y = \frac{1}{3}(x + 2)^2 + 1$	Divide each side by 3.

Vertex: $(-2, 1)$; axis of symmetry: $x = -2$; direction of opening: upward since $a > 0$.

Study Guide and Review *Continued*

9-3 Circles

Write an equation for the circle that satisfies each set of conditions.

29. center $(-1, 6)$, radius 3 units

30. endpoints of a diameter $(2, 5)$ and $(0, 0)$

31. endpoints of a diameter $(4, -2)$ and $(-2, -6)$

Find the center and radius of each circle. Then graph the circle.

32. $(x + 5)^2 + y^2 = 9$

33. $(x - 3)^2 + (y + 1)^2 = 25$

34. $(x + 2)^2 + (y - 8)^2 = 1$

35. $x^2 + 4x + y^2 - 2y - 11 = 0$

36. SOUND A loudspeaker in a school is located at the point (65, 40). The speaker can be heard in a circle with a radius of 100 feet. Write an equation to represent the possible boundary of the loudspeaker sound.

Example 4

Find the center and radius of the circle with equation $x^2 - 2x + y^2 + 6y + 6 = 0$. Then graph the circle.

Complete the squares.

$$x^2 - 2x + y^2 + 6y + 6 = 0$$
$$(x^2 - 2x + \blacksquare) + (y^2 + 6y + \blacksquare) = -6 + \blacksquare + \blacksquare$$
$$(x^2 - 2x + 1) + (y^2 + 6y + 9) = -6 + 1 + 9$$
$$(x - 1)^2 + (y + 3)^2 = 4$$

The center of the circle is at $(1, -3)$ and the radius is 2.

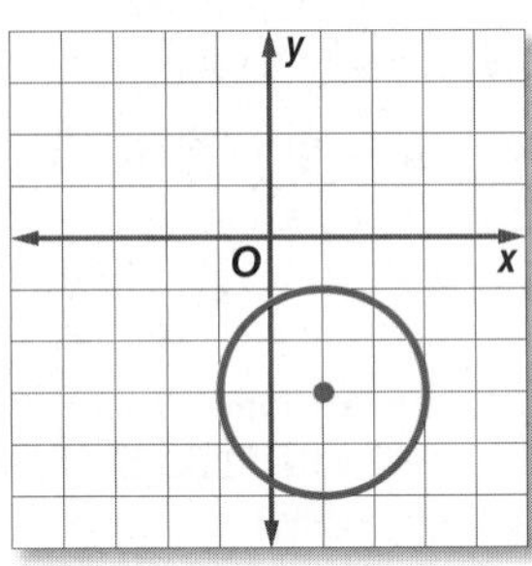

9-4 Ellipses

Find the coordinates of the center and foci and the lengths of the major and minor axes for the ellipse with the given equation. Then graph the ellipse.

37. $\frac{x^2}{9} + \frac{y^2}{36} = 1$

38. $\frac{y^2}{10} + \frac{x^2}{5} = 1$

39. $\frac{x^2}{36} + \frac{(y - 4)^2}{4} = 1$

40. $27x^2 + 9y^2 = 81$

41. $\frac{(x + 1)^2}{25} + \frac{(y - 2)^2}{16} = 1$

42. $9x^2 + 4y^2 + 54x - 8y + 49 = 0$

43. $9x^2 + 25y^2 - 18x + 50y - 191 = 0$

44. $7x^2 + 3y^2 - 28x - 12y = -19$

45. LANDSCAPING The Martins have a garden in their front yard that is shaped like an ellipse. The major axis is 16 feet and the minor axis is 10 feet. Write an equation to model the garden. Assume the origin is at the center of the garden and the major axis is horizontal.

Example 5

Find the coordinates of the center and foci and the lengths of the major and minor axes for the ellipse with equation $9x^2 + 16y^2 - 54x + 32y - 47 = 0$. Then graph the ellipse.

First, convert to standard form.

$$9x^2 + 16y^2 - 54x + 32y - 47 = 0$$
$$9(x^2 - 6x + \blacksquare) + 16(y^2 + 2y + \blacksquare) = 47 + 9(\blacksquare) + 16(\blacksquare)$$
$$9(x^2 - 6x + 9) + 16(y^2 + 2y + 1) = 47 + 9(9) + 16(1)$$
$$9(x - 3)^2 + 16(y + 1)^2 = 144$$
$$\frac{(x - 3)^2}{16} + \frac{(y + 1)^2}{9} = 1$$

The center of the ellipse is $(3, -1)$. The ellipse is horizontal. $a^2 = 16$, so $a = 4$. $b^2 = 9$, so $b = 3$. The length of the major axis is $2 \cdot 4$ or 8. The length of the minor axis is $2 \cdot 3$ or 6. To find the foci: $c^2 = 16 - 9$ or 7, so $c = \sqrt{7}$. The foci are $(3 + \sqrt{7}, -1)$ and $(3 - \sqrt{7}, -1)$.

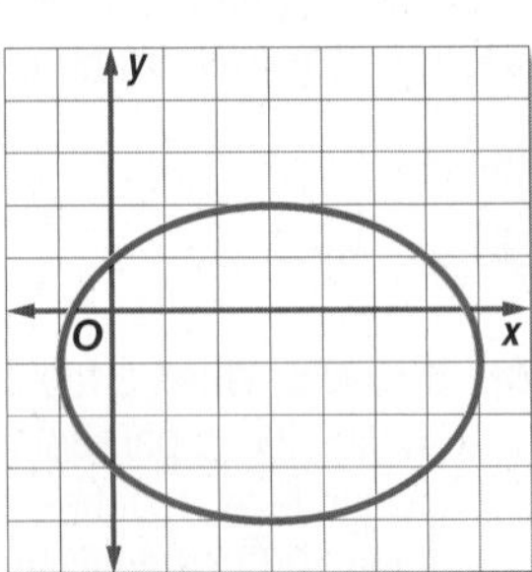

9-5 Hyperbolas

Graph each hyperbola. Identify the vertices, foci, and asymptotes.

46. $\frac{y^2}{9} - \frac{x^2}{4} = 1$

47. $\frac{(x-3)^2}{1} - \frac{(y+2)^2}{4} = 1$

48. $\frac{(y+1)^2}{16} - \frac{(x-4)^2}{9} = 1$

49. $4x^2 - 9y^2 = 36$

50. $9y^2 - x^2 - 4x + 18y + 4 = 0$

51. **MIRRORS** A hyperbolic mirror is shaped like one branch of a hyperbola. It reflects light rays directed at one focus toward the other focus. Suppose a hyperbolic mirror is modeled by the upper branch of the hyperbola $\frac{y^2}{9} - \frac{x^2}{16} = 1$. A light source is located at $(-10, 0)$. Where should the light hit the mirror so that the light will be reflected to $(0, -5)$?

Example 6

Graph $9x^2 - 4y^2 - 36x - 8y - 4 = 0$. Identify the vertices, foci, and asymptotes.

Complete the square.

$$9x^2 - 4y^2 - 36x - 8y - 4 = 0$$
$$9(x^2 - 4x + \blacksquare) - 4(y^2 + 2y + \blacksquare) = 4 + 9(\blacksquare) - 4(\blacksquare)$$
$$9(x^2 - 4x + 4) - 4(y^2 + 2y + 1) = 4 + 9(4) - 4(1)$$
$$9(x-2)^2 - 4(y+1)^2 = 36$$
$$\frac{(x-2)^2}{4} - \frac{(y+1)^2}{9} = 1$$

The center is at $(2, -1)$. The vertices are at $(0, -1)$ and $(4, -1)$. The foci are at $\left(2 + \sqrt{13}, -1\right)$ and $\left(2 - \sqrt{13}, -1\right)$. The equations of the asymptotes are $y + 1 = \pm\frac{3}{2}(x - 2)$

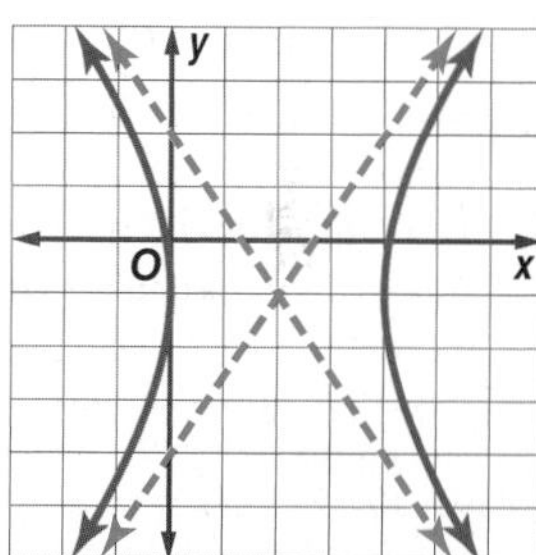

9-6 Identifying Conic Sections

Write each equation in standard form. State whether the graph of the equation is a *parabola, circle, ellipse,* or *hyperbola.* Then graph.

52. $3x^2 + 12x - y + 8 = 0$

53. $9x^2 + 16y^2 = 144$

54. $x^2 + y^2 - 8x - 2y + 8 = 0$

55. $-9x^2 + y^2 + 36x - 45 = 0$

Without writing the equation in standard form, state whether the graph of the equation is a *parabola, circle, ellipse,* or *hyperbola.*

56. $7x^2 + 9y^2 = 63$

57. $5y^2 + 2y + 4x - 13x^2 = 81$

58. $x^2 - 8x + 16 = 6y$

59. $x^2 + 4x + y^2 - 285 = 0$

60. **LIGHT** Suppose the edge of a shadow can be represented by the equation $16x^2 + 25y^2 - 32x - 100y - 284 = 0$.
   a. What is the shape of the shadow?
   b. Graph the equation.

Example 7

Write $3x^2 + 3y^2 - 12x + 30y + 39 = 0$ in standard form. State whether the graph of the equation is a *parabola, circle, ellipse,* or *hyperbola.* Then graph the equation.

$$3x^2 + 3y^2 - 12x + 30y + 39 = 0$$
$$3(x^2 - 4x + \blacksquare) + 3(y^2 + 10y + \blacksquare) = -39 + 3(\blacksquare) + 3(\blacksquare)$$
$$3(x^2 - 4x + 4) + 3(y^2 + 10y + 25) = -39 + 3(4) + 3(25)$$
$$3(x-2)^2 + 3(y+5)^2 = 48$$
$$(x-2)^2 + (y+5)^2 = 16$$

In this equation $A = 3$ and $C = 3$. Since A and C are both positive and $A = C$, the graph is a circle. The center is at $(2, -5)$, and the radius is 4.

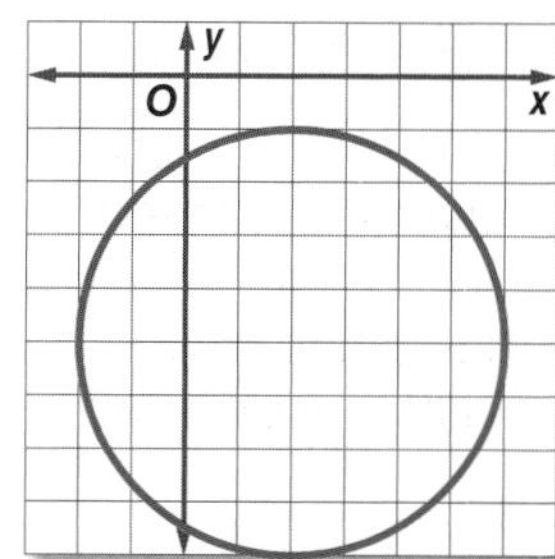

9-7 Solving Linear-Nonlinear Systems

Solve each system of equations.

61. $x^2 + y^2 = 8$
$x + y = 0$

62. $x - 2y = 2$
$y^2 - x^2 = 2x + 4$

63. $y + x^2 = 4x$
$y + 4x = 16$

64. $3x^2 - y^2 = 11$
$x^2 + 4y^2 = 8$

65. $5x^2 + y^2 = 30$
$9x^2 - y^2 = -16$

66. $\frac{x^2}{30} + \frac{y^2}{6} = 1$
$x = y$

67. PHYSICAL SCIENCE Two balls are launched into the air at the same time. The heights they are launched from are different. The height y in feet of one is represented by $y = -16t^2 + 80t + 25$ where t is the time in seconds. The height of the other ball is represented by $y = -16t^2 + 30t + 100$.

a. After how many seconds are the balls at the same height?

b. What is this height?

68. ARCHITECTURE An architect is building the front entrance of a building in the shape of a parabola with the equation $y = -\frac{1}{10}(x - 10)^2 + 20$. While the entrance is being built, the construction team puts in two support beams with equations $y = -x + 10$ and $y = x - 10$. Where do the support beams meet the parabola?

Solve each system of inequalities by graphing.

69. $x^2 + y^2 < 64$
$x^2 + 16(y - 3)^2 < 16$

70. $x^2 + y^2 < 49$
$16x^2 - 9y^2 \geq 144$

71. $x + y < 4$
$9x^2 - 4y^2 \geq 36$

72. $x^2 + y^2 < 25$
$4x^2 - 9y^2 < 36$

73. $x^2 + y^2 < 36$
$4x^2 + 9y^2 > 36$

74. $y^2 < x$
$x^2 - 4y^2 < 16$

Example 8

Solve the system of equations.

$x^2 + y^2 = 100$
$3x - y = 10$

Use substitution to solve the system.
First, rewrite $3x - y = 10$ as $y = 3x - 10$.

$$\begin{aligned} x^2 + y^2 &= 100 \\ x^2 + (3x - 10)^2 &= 100 \\ x^2 + 9x^2 - 60x + 100 &= 100 \\ 10x^2 - 60x + 100 &= 100 \\ 10x^2 - 60x &= 0 \\ 10x(x - 6) &= 0 \end{aligned}$$

$10x = 0$ or $x - 6 = 0$
$x = 0$ $\quad x = 6$

Now solve for y.

$y = 3x - 10 = 3(0) - 10 = -10$ $\quad$ $y = 3x - 10 = 3(6) - 10 = 8$

The solutions of the system are (0, −10) and (6, 8)

Example 9

Solve the system of inequalities by graphing.

$x^2 + y^2 \leq 9$
$2y \geq x^2 + 4$

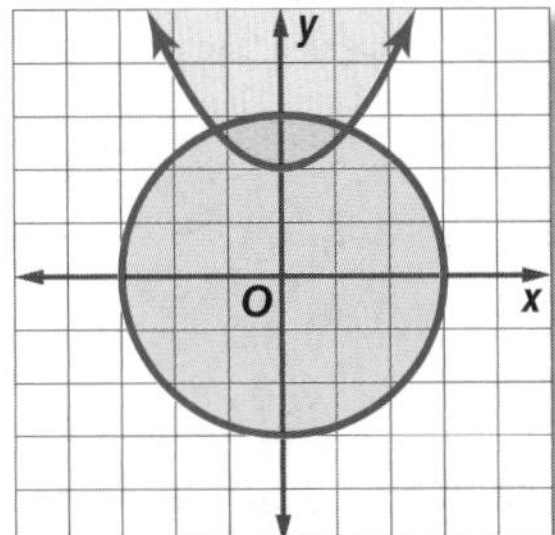

The solution is the green shaded region.

CHAPTER 9 Practice Test

Find the midpoint of the line segment with endpoints at the given coordinates.

1. $(8, 3), (-4, 9)$

2. $\left(\frac{3}{4}, 0\right), \left(\frac{1}{2}, -1\right)$

3. $(-10, 0), (-2, 6)$

Find the distance between each pair of points with the given coordinates.

4. $(-5, 8), (4, 3)$

5. $\left(\frac{1}{3}, \frac{2}{3}\right), \left(-\frac{5}{6}, -\frac{11}{6}\right)$

6. $(4, -5), (4, 9)$

State whether the graph of each equation is a *parabola, circle, ellipse,* or *hyperbola*. Then graph the equation.

7. $y^2 = 64 - x^2$

8. $4x^2 + y^2 = 16$

9. $4x^2 - 9y^2 + 8x + 36y = 68$

10. $\frac{1}{2}x^2 - 3 = y$

11. $y = -2x^2 - 5$

12. $16x^2 + 25y^2 = 400$

13. $x^2 + 6x + y^2 = 16$

14. $\frac{y^2}{4} - \frac{x^2}{16} = 1$

15. $(x + 2)^2 = 3(y - 1)$

16. $4x^2 + 16y^2 + 32x + 63 = 0$

17. **MULTIPLE CHOICE** Which equation represents a hyperbola that has vertices at $(-3, -3)$ and $(5, -3)$ and a conjugate axis of length 6 units?

A $\frac{(y-1)^2}{16} - \frac{(x+3)^2}{9} = 1$

B $\frac{(x-1)^2}{16} - \frac{(y+3)^2}{9} = 1$

C $\frac{(y+1)^2}{16} - \frac{(x-3)^2}{9} = 1$

D $\frac{(x+1)^2}{16} - \frac{(y-3)^2}{9} = 1$

18. **CARPENTRY** Ellis built a window frame shaped like the top half of an ellipse. The window is 40 inches tall at its highest point and 160 inches wide at the bottom. What is the height of the window 20 inches from the center of the base?

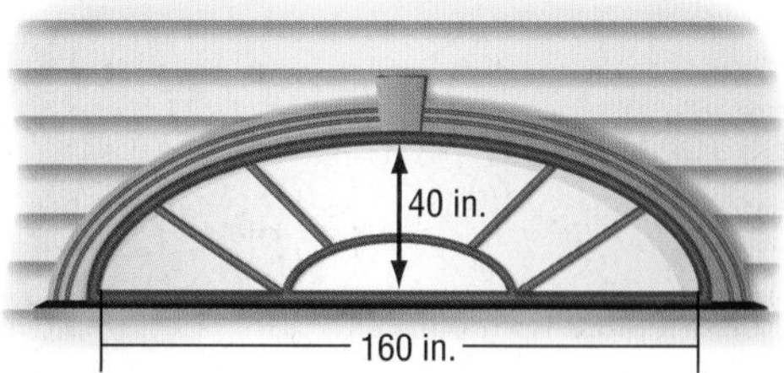

Solve each system of equations.

19. $x^2 + y^2 = 100$
$y = -x - 2$

20. $x^2 + 2y^2 = 11$
$x + y = 2$

21. $x^2 + y^2 = 34$
$y^2 - x^2 = 9$

Solve each system of inequalities.

22. $x^2 + y^2 \leq 9$
$y > -x^2 + 2$

23. $\frac{(x-2)^2}{4} - \frac{(y-4)^2}{9} \geq 1$
$x - 4y < 8$

24. **MULTIPLE CHOICE** Which is NOT the equation of a parabola?

F $y = 3x^2 + 5x - 3$

G $2y + 3x^2 + x - 9 = 0$

H $x = 3(y + 1)^2$

J $x^2 + 2y^2 + 6x = 10$

25. **FORESTRY** A forest ranger at an outpost in the Sam Houston National Forest and another ranger at the primary station both heard an explosion. The outpost and the primary station are 6 kilometers apart.

a. If one ranger heard the explosion 6 seconds before the other, write an equation that describes all the possible locations of the explosion. Place the two ranger stations on the x-axis with the midpoint between the stations at the origin. The transverse axis is horizontal. (*Hint*: The speed of sound is about 0.35 kilometer per second.)

b. Draw a sketch of the possible locations of the explosion. Include the ranger stations in the drawing.

CHAPTER 9

Preparing for Standardized Tests

Use a Formula

Sometimes it is necessary to use a formula to solve problems on standardized tests. In some cases you may even be given a sheet of formulas that you are permitted to reference while taking the test.

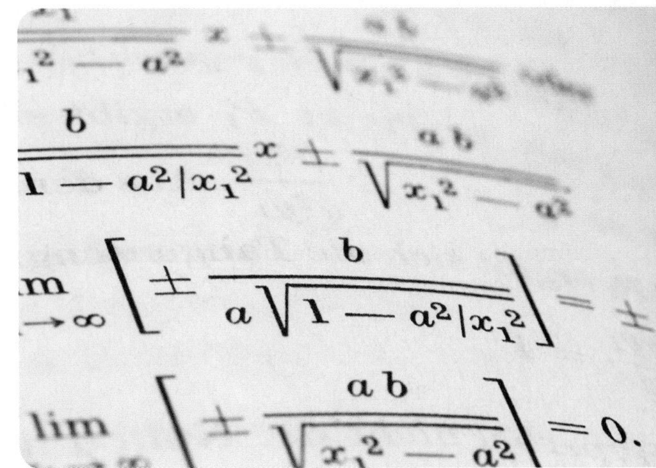

Strategies for Using a Formula

Step 1

Read the problem statement carefully.

Ask yourself:

- What am I being asked to solve?
- What information is given in the problem?
- Are there any formulas that I can use to help me solve the problem?

Step 2

Solve the problem and check your solution.

- Substitute the known quantities that are given in the problem statement into the formula.
- Simplify to solve for the unknown values in the formula.
- Check to make sure your answer makes sense. If time permits, check your answer.

Standardized Test Example

Read the problem. Identify what you need to know. Then use the information in the problem to solve. Show your work.

What is the distance between points *A* and *B* on the coordinate plane? Round your answer to the nearest tenth if necessary.

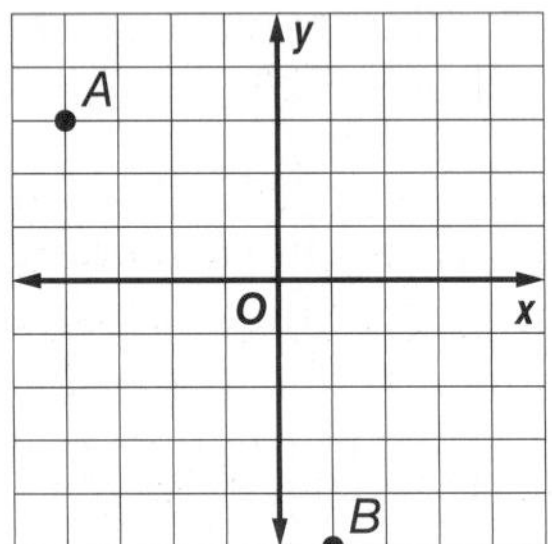

Scoring Rubric	
Criteria	**Score**
Full Credit: The answer is correct and a full explanation is provided that shows each step.	2
Partial Credit: • The answer is correct but the explanation is incomplete. • The answer is incorrect but the explanation is correct.	1
No Credit: Either an answer is not provided or the answer does not make sense.	0

pm/Alamy

Read the problem statement carefully. You are given the coordinates of two points on a coordinate plane and asked to find the distance between them. To solve this problem, you must use the **Distance Formula**.

Example of a 2-point response:

Use the Distance Formula to find the distance between points $A(-4, 3)$ and $B(1, -5)$.

$$d = \sqrt{(x_2 - x_1)^2 + (y_2 - y_1)^2}$$
$$= \sqrt{[1 - (-4)]^2 + [(-5) - 3]^2}$$
$$= \sqrt{5^2 + (-8)^2}$$
$$= \sqrt{25 + 64}$$
$$= \sqrt{89} \text{ or about } 9.4$$

The distance between points A and B is about 9.4 units.

The steps, calculations, and reasoning are clearly stated. The student also arrives at the correct answer. So, this response is worth the full 2 points.

Exercises

Read each problem. Identify what you need to know. Then use the information in the problem to solve. Show your work.

1. What is the midpoint of segment CD with endpoints $C(5, -12)$ and $D(-9, 4)$?

2. Katrina is making a map of her hometown on a coordinate plane. She plots the school at $S(7, 3)$ and the park at $P(-4, 12)$. If the scale of the map is 1 unit = 250 yards, what is the actual distance between the school and the park? Round to the nearest yard.

3. Mr. Washington is making a concrete table for his backyard. The tabletop will be circular with a diameter of 6 feet and a depth of 6 inches. How much concrete will Mr. Washington need to make the top of the table? Round to the nearest cubic foot.

4. What is the equation, in standard form, of the hyperbola graphed below?

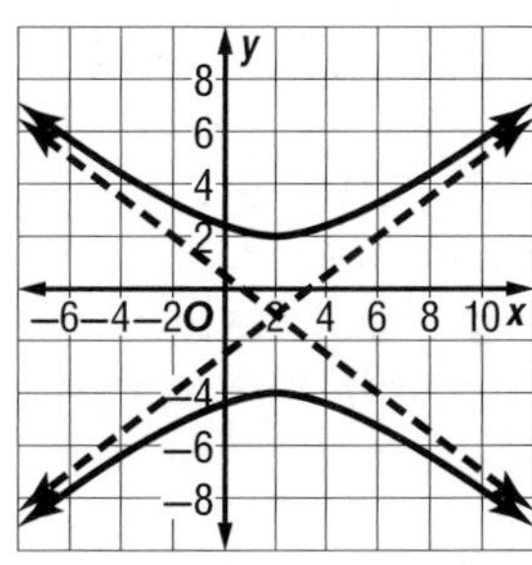

5. If the surface area of a cube is increased by a factor of 9, what is the change in the length of the sides of the cube?

   A The length is 2 times the original length.

   B The length is 3 times the original length.

   C The length is 6 times the original length.

   D The length is 9 times the original length.

Standardized Test Practice

Cumulative, Chapters 1 through 9

Multiple Choice

Read each question. Then fill in the correct answer on the answer document provided by your teacher or on a sheet of paper.

1. Which is the first *incorrect* step in simplifying $\log_3 \frac{3}{48}$?

Step 1: $\log_3 \frac{3}{48} = \log_3 3 - \log_3 48$

Step 2: $= 1 - 16$

Step 3: $= -15$

A Step 1

B Step 2

C Step 3

D Each step is correct.

2. Which is the equation for the parabola that has vertex $(-3, -23)$ and passes through the point $(1, 9)$?

F $y = x^2 + 10x + 7$

G $y = x^2 - 6x + 19$

H $y = 2x^2 + 12x - 5$

J $y = 2x^2 - 3x + 10$

3. What are the vertices of the ellipse with equation $\frac{(x-3)^2}{36} + \frac{(y-2)^2}{144} = 1$?

A $(-3, 2)$ and $(9, 2)$

B $(-2, 3)$ and $(10, 3)$

C $(3, -10)$ and $(3, 14)$

D $(2, -11)$ and $(4, 13)$

4. Hooke's Law states that the force needed to keep a spring stretched x units is directly proportional to x. If a force of 40 N is required to maintain a spring stretched to 5 centimeters, what force is needed to keep the spring stretched 14 centimeters?

F 8 N

G 19 N

H 112 N

J 1600 N

Test-Taking Tip

Question 2 You can check your answer by submitting 1 for x and making sure that the y-value is 9.

5. Angela is making a map of her backyard on a coordinate grid. She plots point $G(-4, -6)$ to represent her mom's garden and point $S(3, 7)$ to represent the rope swing hanging on an oak tree. If the scale of the map is 1 unit = 5 feet, what is the approximate distance between the garden and the rope swing?

A 74 feet

B 79 feet

C 82 feet

D 90 feet

6. If $\sqrt{x+5} + 1 = 4$, what is the value of x?

F 4 G 10 H 11 J 20

7. The area of the base of a rectangular suitcase measures $3x^2 + 5x - 4$ square units. The height of the suitcase measures $2x$ units. Which polynomial expression represents the volume of the suitcase?

A $3x^3 + 5x^2 - 4x$

B $6x^2 + 10x - 8$

C $6x^3 + 10x^2 - 8x$

D $3x^3 + 10x^2 - 4$

8. Malina was given this geoboard to model the slope $-\frac{3}{4}$.

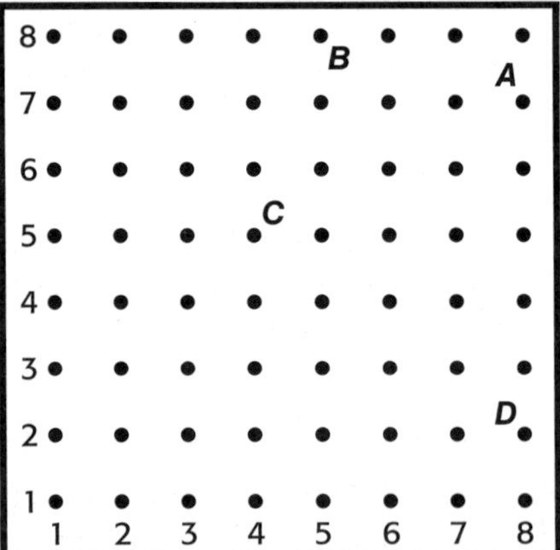

If the peg in the upper right-hand corner represents the origin on a coordinate plane, where could Malina place a rubber band to represent the given slope?

F from peg A to peg B

G from peg A to peg C

H from peg B to peg D

J from peg C to peg D

Short Response/Gridded Response

Record your answers on the answer sheet provided by your teacher or on a sheet of paper.

9. A placekicker kicks a ball upward with a velocity of 32 feet per second. Ignoring the height of the kicking tee, how long after the football is kicked does it hit the ground? Use the formula $h(t) = v_0t - 16t^2$, where $h(t)$ is the height of an object in feet, v_0 is the object's initial velocity in feet per second, and t is the time in seconds.

10. GRIDDED RESPONSE What is the maximum number of solutions of a system of equations that consists of a circle and a hyperbola?

11. Lupe is preparing boxes of assorted chocolates. Chocolate-covered peanuts cost \$7 per pound. Chocolate-covered caramels cost \$6.50 per pound. The boxes of assorted candies contain five more pounds of peanut candies than caramel candies. If the total amount sold was \$575, how many pounds of each candy were needed to make the boxes?

12. GRIDDED RESPONSE What is the y-coordinate of the midpoint of segment AB with endpoints $A(0.8, 5.32)$ and $B(0.44, 2.2)$?

13. Marc went shopping and bought two shirts, three pairs of pants, one belt, and two pairs of shoes. The following matrix shows the prices for each item respectively.

$$\begin{bmatrix} \$20.15 & \$32 & \$15 & \$25.99 \end{bmatrix}$$

Use matrix multiplication to find the total amount of money Marc spent while shopping.

Extended Response

Record your answers on a sheet of paper. Show your work.

14. Clarence graphed the quadratic equation $h(t) = -16t^2 + 128t$ to model the flight of a firework. The parabola shows the height, in feet, of the firework t seconds after it was launched.

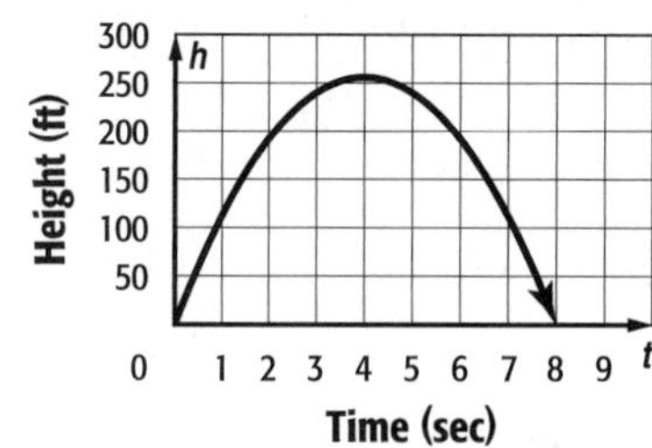

a. What is the vertex of the parabola?

b. What does the vertex of the parabola represent?

c. How long is the firework in the air before it lands?

15. The Colonial High School Yearbook Staff is selling yearbooks and chrome picture frames engraved with the year. The number of yearbooks and frames sold to members of each grade is shown in the table.

Sales for Each Class		
Grade	Yearbooks	Frames
9th	423	256
10th	464	278
11th	546	344
12th	575	497

a. Find the difference in the sales of yearbooks and frames made to the 10th and 11th grade classes.

b. Find the total number of yearbooks and frames sold.

c. A yearbook costs \$48 and a frame costs \$18. Find the sales of yearbooks and frames for each class.

Need ExtraHelp?

If you missed Question...	1	2	3	4	5	6	7	8	9	10	11	12	13	14	15
Go to Lesson...	7-5	9-2	9-4	8-5	9-1	6-7	5-1	2-3	4-2	9-7	3-1	9-1	3-6	9-2	1-3

CHAPTER 10

Sequences and Series

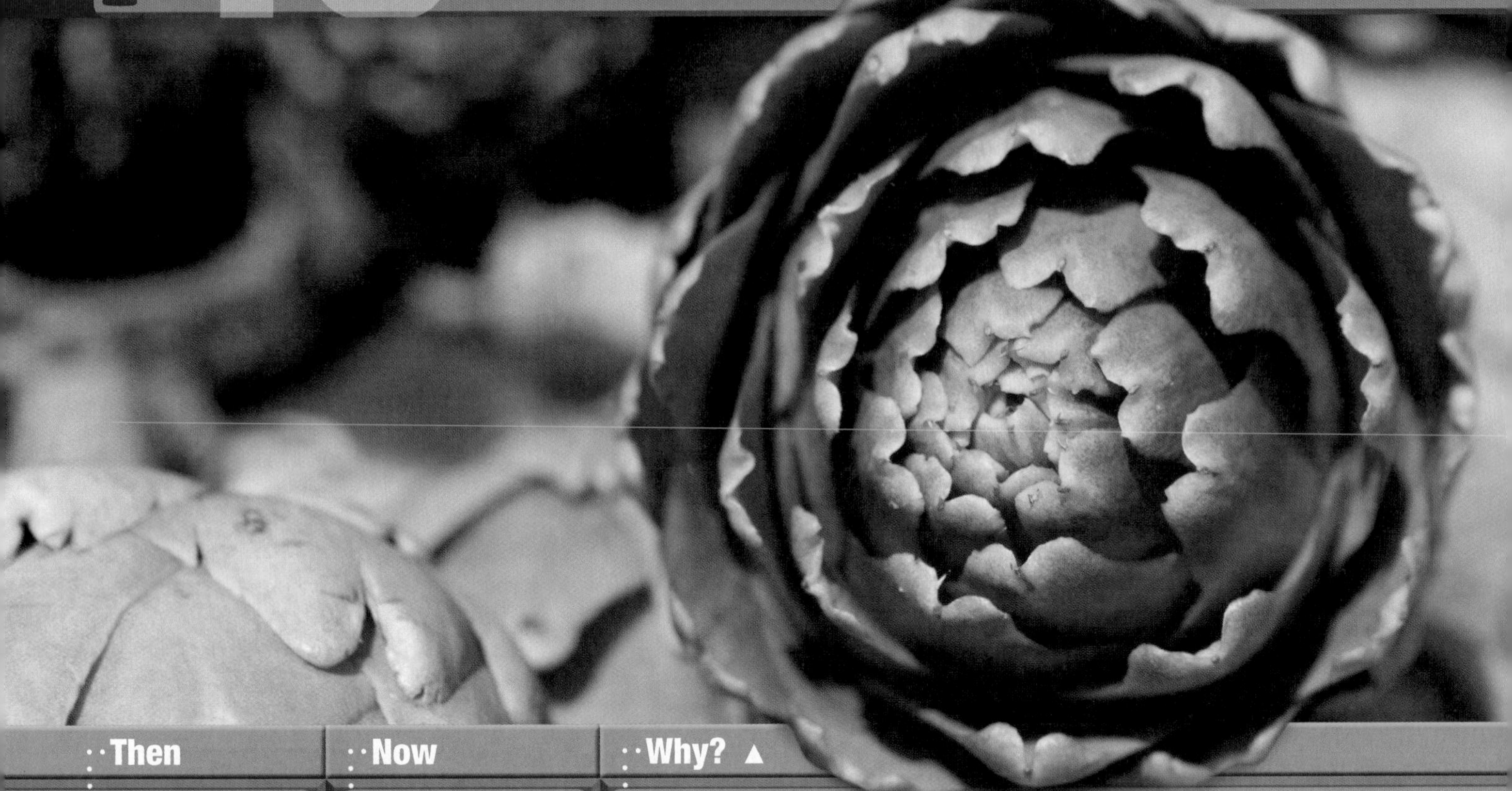

Then

- You simplified and evaluated algebraic expressions.

Now

You will:

- Use arithmetic and geometric sequences and series.
- Use special sequences and iterate functions.
- Expand powers by using the Binomial Theorem.
- Prove statements by using mathematical induction.

Why? ▲

CONSERVATION AND NATURE Mathematics occurs in aspects of nature in astonishing ways. The Fibonacci sequence manifests itself in seeds, flowers, pine cones, fruits, and vegetables. Sequences and series can further help us conserve our natural resources by making water filtration systems more efficient.

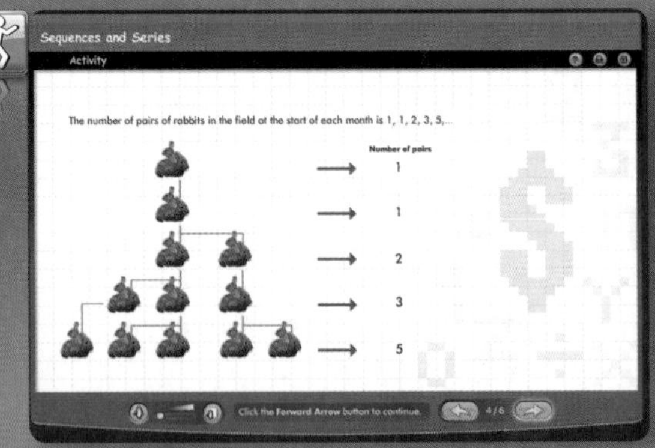

connectED.mcgraw-hill.com Your Digital Math Portal

Animation 
Vocabulary 
eGlossary 

Personal Tutor 

Virtual Manipulatives 
Graphing Calculator 
Audio 
Foldables 

Self-Check Practice 
Worksheets 

Get Ready for the Chapter

Diagnose Readiness | **You have two options for checking prerequisite skills.**

1 Textbook Option Take the Quick Check below. Refer to the Quick Review for help.

QuickCheck

Solve each equation.

1. $-6 = 7x + 78$
2. $768 = 3x^4$
3. $23 - 5x = 8$
4. $2x^3 + 4 = -50$
5. **PLANTS** Lauri has 48 plants for her two gardens. She plants 12 in the small garden. In the other garden she wants 4 plants in each row. How many rows will she have?

QuickReview

Example 1

Solve $25 = 3x^3 + 400$.

$25 = 3x^3 + 400$	Original equation
$-375 = 3x^3$	Subtract 400 from each side.
$-125 = x^3$	Divide each side by 3.
$\sqrt[3]{-125} = \sqrt[3]{x^3}$	Take the cube root of each side.
$-5 = x$	Simplify.

QuickCheck

Graph each function.

6. $\{(1, 3), (2, 5), (3, 7), (4, 9), (5, 11)\}$
7. $\{(1, -15), (2, -12), (3, -9), (4, -6), (5, -3)\}$
8. $\left\{(1, 27), (2, 9), (3, 3), (4, 1), \left(5, \frac{1}{3}\right)\right\}$
9. $\left\{(1, 1), (2, 2), \left(3, \frac{5}{2}\right), \left(4, \frac{11}{4}\right), \left(5, \frac{23}{8}\right)\right\}$
10. **DAY CARE** A child care center has expenses of \$125 per day. They charge \$50 per child per day. The function $P(c) = 50c - 125$ gives the amount of money the center makes when there are c children there. How much will they make if there are 8 children?

QuickReview

Example 2

Graph the function $\{(1, 1), (2, 4), (3, 9), (4, 16), (5, 25)\}$. State the domain and range.

The domain of a function is the set of all possible x-values. So, the domain of the function is $\{1, 2, 3, 4, 5\}$. The range of a function is the set of all possible y-values. So, the range of this function is $\{1, 4, 9, 16, 25\}$.

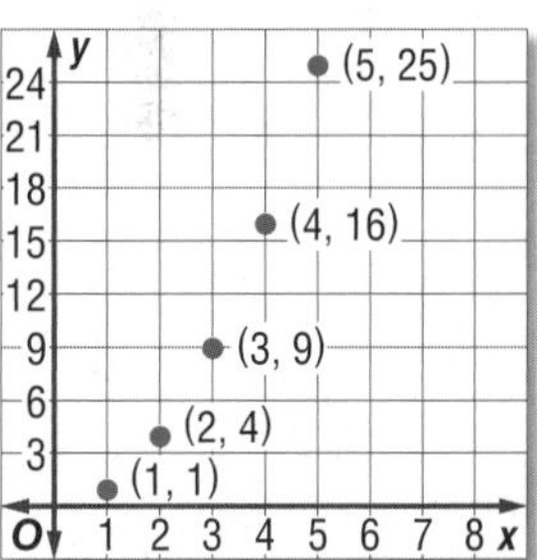

QuickCheck

Evaluate each expression for the given value(s) of the variable(s).

11. $\frac{a}{3}(b + c)$ if $a = 9$, $b = -2$, and $c = -8$
12. $r + (n - 2)t$ if $r = 15$, $n = 5$, and $t = -1$
13. $x \cdot y^{z+1}$ if $x = -2$, $y = \frac{1}{3}$, and $z = 5$
14. $\frac{a(1 - bc)^2}{1 - b}$ if $a = -3$, $b = -4$, and $c = 1$

QuickReview

Example 3

Evaluate $2 \cdot 3^{x+y}$ if $x = -2$ and $y = -3$.

$2 \cdot 3^{x+y} = 2 \cdot 3^{-2+(-3)}$	Substitute.
$= 2 \cdot 3^{-5}$	Simplify.
$= \frac{2}{3^5}$	Rewrite with positive exponent.
$= \frac{2}{243}$	Evaluate the power.

2 Online Option Take an online self-check Chapter Readiness Quiz at connectED.mcgraw-hill.com.

Get Started on the Chapter

You will learn several new concepts, skills, and vocabulary terms as you study Chapter 10. To get ready, identify important terms and organize your resources. You may wish to refer to Chapter 0 to review prerequisite skills.

FOLDABLES StudyOrganizer

Sequences and Series Make this Foldable to help you organize your Chapter 10 notes about sequences and series. Begin with one $8\frac{1}{2}''$ by 11″ sheet of paper.

1 **Fold** in half, matching the short sides.

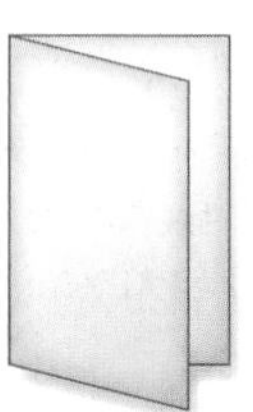

2 **Unfold** and fold the long side up 2 inches to form a pocket.

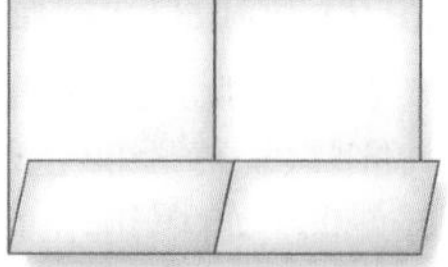

3 **Staple** or glue the outer edges to complete the pocket.

4 **Label** each side as shown. Use index cards to record notes and examples.

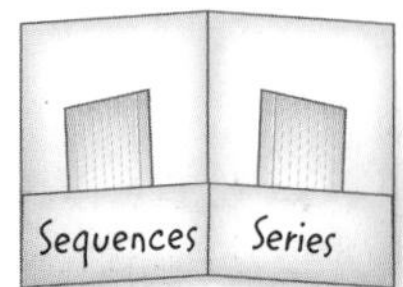

NewVocabulary

English		Español
sequence	p. 659	sucesión
finite sequence	p. 659	sucesión finita
infinite sequence	p. 659	sucesión infinita
arithmetic sequence	p. 659	sucesión aritmética
common difference	p. 659	diferencia común
geometric sequence	p. 661	sucesión geométrica
common ratio	p. 661	razón común
arithmetic means	p. 667	media aritmética
series	p. 668	serie
arithmetic series	p. 668	serie aritmética
partial sum	p. 668	suma parcial
geometric means	p. 675	media geométrica
geometric series	p. 676	serie geométrica
convergent series	p. 683	serie convergente
divergent series	p. 683	serie divergente
recursive sequence	p. 692	sucesión recursiva
iteration	p. 694	iteración
mathematical induction	p. 705	inducción matemática
induction hypothesis	p. 705	hipótesis inductiva

ReviewVocabulary

coefficient coeficiante the numerical factor of a monomial

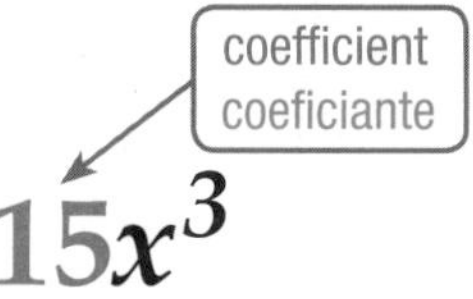

formula fórmula a mathematical sentence that expresses the relationship between certain quantities

function función a relation in which each element of the domain is paired with exactly one element in the range

LESSON 10-1 Sequences as Functions

Then	Now	Why?
You analyzed linear and exponential functions.	**1** Relate arithmetic sequences to linear functions. **2** Relate geometric sequences to exponential functions.	During their routine, a high school marching band marches in rows. There is one performer in the first row, three performers in the next row, and five in the third row. This pattern continues for the rest of the rows.

NewVocabulary
sequence
term
finite sequence
infinite sequence
arithmetic sequence
common difference
geometric sequence
common ratio

Common Core State Standards

Content Standards
F.IF.4 For a function that models a relationship between two quantities, interpret key features of graphs and tables in terms of the quantities, and sketch graphs showing key features given a verbal description of the relationship.

Mathematical Practices
2 Reason abstractly and quantitatively.
7 Look for and make use of structure.

1 Arithmetic Sequences

A **sequence** is a set of numbers in a particular order or pattern. Each number in a sequence is called a **term**. A sequence may be a **finite sequence** containing a limited number of terms, such as $\{-2, 0, 2, 4, 6\}$, or an **infinite sequence** that continues without end, such as $\{0, 1, 2, 3, \ldots\}$. The first term of a sequence is denoted a_1, the second term is denoted a_2, and so on.

KeyConcept Sequences as Functions

Words A sequence is a function in which the domain consists of natural numbers, and the range consists of real numbers.

Symbols

Domain:	1	2	3	...	n	the position of a term
Range:	a_1	a_2	a_3	...	a_n	the terms of the sequence

Examples

Finite Sequence	Infinite Sequence
{3, 6, 9, 12, 15}	{3, 6, 9, 12, 15, ...}
Domain: {1, 2, 3, 4, 5} Range: {3, 6, 9, 12, 15}	Domain: {all natural numbers} Range: $\{y \mid y \text{ is a multiple of } 3, y \geq 3\}$

In an **arithmetic sequence**, each term is determined by adding a constant value to the previous term. This constant value is called the **common difference**.

Consider the sequence 3, 6, 9, 12, 15. This sequence is arithmetic because the terms share a common difference. Each term is 3 more than the previous term.

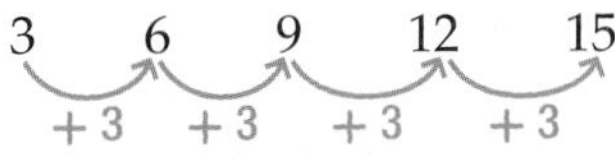

Example 1 Identify Arithmetic Sequences

Determine whether each sequence is arithmetic.

a. 5, −6, −17, −28, ...

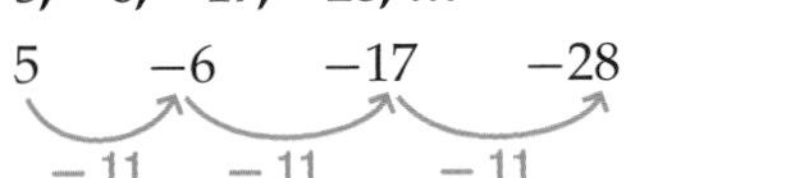

The common difference is −11. The sequence is arithmetic.

b. −4, 12, 28, 42, ...

−4 12 28 42
+ 16 + 16 + 14

There is no common difference. This is not an arithmetic sequence.

GuidedPractice

1A. 7, 12, 16, 20, ...

1B. −6, 3, 12, 21, ...

ThinkStock/SuperStock

You can use the common difference to find terms of an arithmetic sequence.

PT

Example 2 Graph an Arithmetic Sequence

Consider the arithmetic sequence 18, 14, 10, … .

a. Find the next four terms of the sequence.

Step 1 To determine the common difference, subtract any term from the term directly after it. The common difference is $10 - 14$ or -4.

Step 2 To find the next term, add -4 to the last term.

Continue to add -4 to find the following terms.

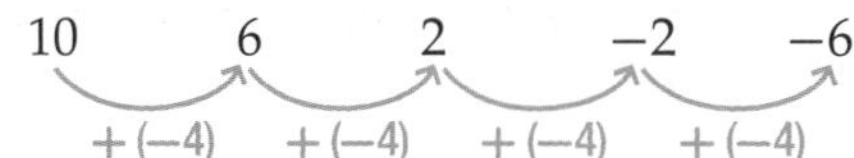

The next four terms are 6, 2, -2, and -6.

b. Graph the first seven terms of the sequence. The domain contains the terms {1, 2, 3, 4, 5, 6, 7} and the range contains the terms {18, 14, 10, 6, 2, -2, -6}. So, graph the corresponding ordered pairs.

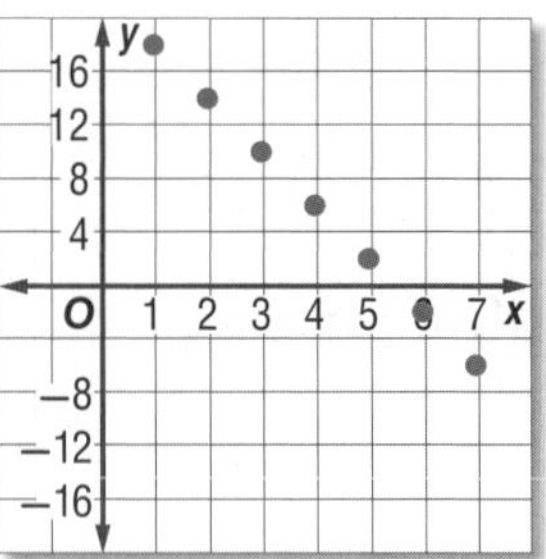

GuidedPractice

2. Find the next four terms of the arithmetic sequence 18, 11, 4, … . Then graph the first seven terms.

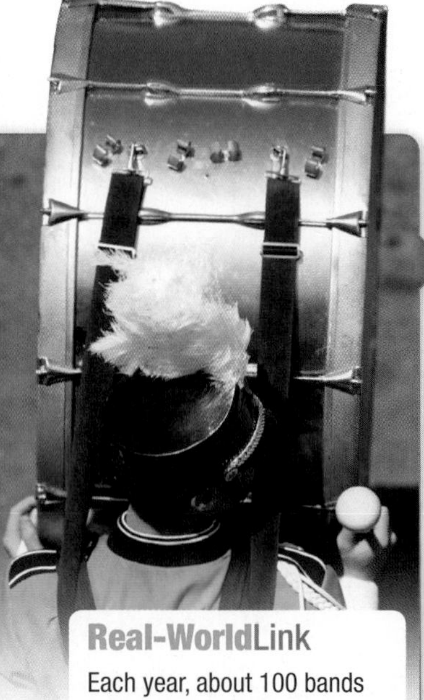

Real-WorldLink

Each year, about 100 bands compete in the Bands of America Grand National Championships.

Source: Bands of America

Notice that the graph of the terms of the arithmetic sequence lie on a line. An arithmetic sequence is a linear function in which the term number n is the independent variable, the term a_n is the dependent variable, and the common difference is the slope.

PT

Real-World Example 3 Find a Term

MARCHING BANDS Refer to the beginning of the lesson. Suppose the director wants to determine how many performers will be in the 14th row during the routine.

Understand Because the difference between any two consecutive rows is 2, the common difference for the sequence is 2.

Plan Use point-slope form to write an equation for the sequence. Let $m = 2$ and $(x_1, y_1) = (3, 5)$. Then solve for $x = 14$.

Solve

$(y - y_1) = m(x - x_1)$	Point-slope form
$(y - 5) = 2(x - 3)$	$m = 2$ and $(x_1, y_1) = (3, 5)$
$y - 5 = 2x - 6$	Multiply.
$y = 2x - 1$	Add 5 to each side.
$y = 2(14) - 1$	Replace x with 14.
$y = 28 - 1$ or 27	Simplify.

Check You can find the terms of the sequence by adding 2, starting with row 1, until you reach row 14.

GuidedPractice

3. MONEY Geraldo's employer offers him a pay rate of \$9 per hour with a \$0.15 raise every three months. How much will Geraldo earn per hour after 3 years?

2 Geometric Sequences

Geometric Sequences Another type of sequence is a geometric sequence. In a **geometric sequence**, each term is determined by multiplying a nonzero constant by the previous term. This constant value is called the **common ratio**.

Consider the sequence $\frac{1}{16}, \frac{1}{4}, 1, 4, 16$. This sequence is geometric because the terms share a common ratio. Each term is 4 times as much as the previous term.

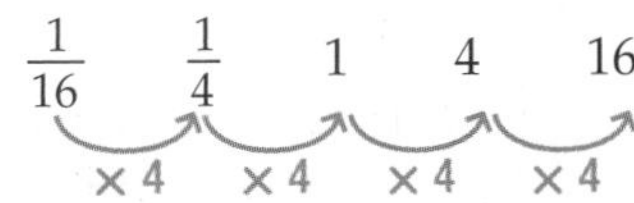

PT

Example 4 Identify Geometric Sequences

Determine whether each sequence is geometric.

a. −2, 6, −18, 54, ...

Find the ratios of the consecutive terms.

$\frac{6}{-2} = -3 \qquad \frac{-18}{6} = -3 \qquad \frac{54}{-18} = -3$

The ratios are the same, so the sequence is geometric.

b. 8, 16, 24, 32, ...

$\frac{16}{8} = 2 \qquad \frac{24}{16} = 1.5 \qquad \frac{32}{24} = 1.\overline{3}$

The ratios are not the same, so the sequence is not geometric.

WatchOut!

Ratios If you find the ratio of a term to the previous term, set up the remaining ratios the same way.

GuidedPractice

4A. −8, 2, −0.5, 0.125, ...

4B. 1, 3, 7, 15, ...

When given a set of information, you can create a problem that relates a story.

PT

Example 5 Graph a Geometric Sequence

Consider the geometric sequence 32, 8, 2, ...

a. Find the next three terms of the sequence.

Step 1 Find the value of the common ratio: $\frac{2}{8}$ or $\frac{1}{4}$.

Step 2 To find the next term, multiply the previous term by $\frac{1}{4}$.

Continue multiplying by $\frac{1}{4}$ to find the following terms.

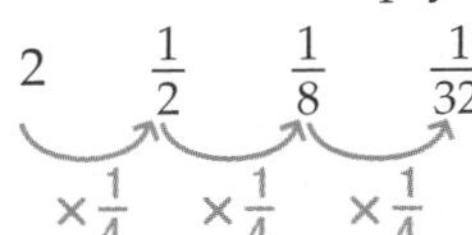

The next three terms are $\frac{1}{2}, \frac{1}{8}$, and $\frac{1}{32}$.

b. Graph the first six terms of the sequence.

Domain: {1, 2, 3, 4, 5, 6}

Range: $\left\{32, 8, 2, \frac{1}{2}, \frac{1}{8}, \frac{1}{32}\right\}$

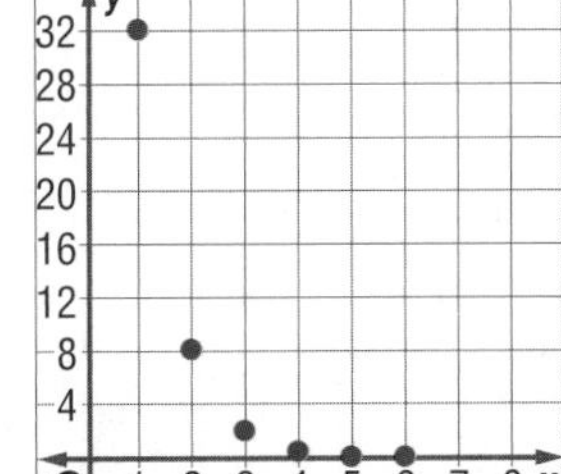

GuidedPractice

5. Find the next two terms of 7, 21, 63, ... Then graph the first five terms.

ReviewVocabulary

exponential function a function of the form $f(x) = b^x$, where $b > 0$ and $b \neq 1$

Examine the graph in Example 5. While the graph of an arithmetic sequence is linear, the graph of a geometric sequence is exponential and can be represented by $f(x) = r^x$, where r is the common ratio, $r > 0$, and $r \neq 1$.

Arithmetic

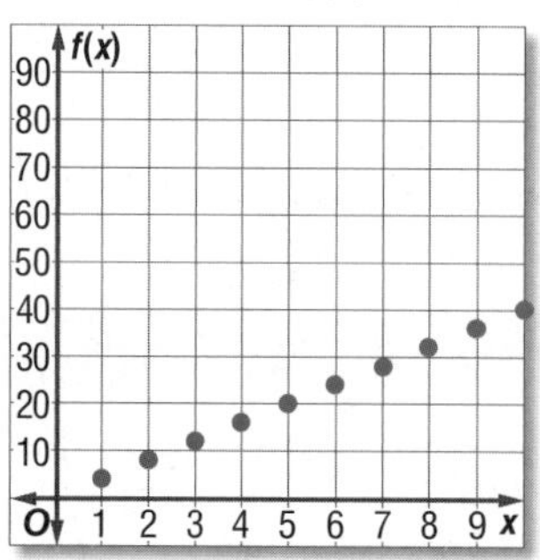

Geometric

x	1	2	3	4	5	6	7	8	9	10
$f(x)$	4	8	12	16	20	24	28	32	36	40

x	1	2	3	4	5	6
$f(x)$	2	4	8	16	32	64

Arithmetic and geometric sequences are functions in which the domain, defined by the term number n, contains the set of or subset of positive integers. The characteristics of arithmetic and geometric sequences can be used to classify sequences.

Example 6 Classify Sequences

Determine whether each sequence is *arithmetic*, *geometric*, or *neither*. Explain your reasoning.

a. 16, 24, 36, 54, …

Check for a common difference.

$54 - 36 = 18$ $\quad$ $36 - 24 = 12$ ✗

Check for a common ratio.

$\frac{54}{36} = \frac{3}{2}$ $\quad$ $\frac{36}{24} = \frac{3}{2}$ $\quad$ $\frac{24}{16} = \frac{3}{2}$ ✓

Because there is a common ratio, the sequence is geometric.

b. 1, 4, 9, 16, …

Check for a common difference.

$16 - 9 = 7$ $\quad$ $9 - 4 = 5$ ✗

Check for a common ratio.

$\frac{16}{9} = 1.\overline{7}$ $\quad$ $\frac{9}{4} = 2.25$ ✗

Because there is no common difference or ratio, the sequence is neither arithmetic nor geometric.

c. 23, 17, 11, 5, …

Check for a common difference.

$5 - 11 = -6$ $\quad$ $11 - 17 = -6$ $\quad$ $17 - 23 = -6$ ✓

Because there is a common difference, the sequence is arithmetic.

GuidedPractice

6A. $\frac{5}{3}, 2, \frac{7}{3}, \frac{8}{3}, \ldots$

6B. $2, -\frac{3}{2}, \frac{9}{8}, -\frac{27}{32}, \ldots$

6C. $-4, 4, 5, -5, \ldots$

Check Your Understanding

● = Step-by-Step Solutions begin on page R14.

Example 1 **Determine whether each sequence is arithmetic. Write *yes* or *no*.**

1. 8, −2, −12, −22,
2. −19, −12, −5, 2, 9
3. 1, 2, 4, 8, 16
4. 0.6, 0.9, 1.2, 1.8, ...

Example 2 **Find the next four terms of each arithmetic sequence. Then graph the sequence.**

5. 6, 18, 30, ...
6. 15, 6, −3, ...
7. −19, −11, −3, ...
8. −26, −33, −40, ...

Example 3

9. **FINANCIAL LITERACY** Kelly is saving her money to buy a car. She has \$250, and she plans to save \$75 per week from her job as a waitress.
   a. How much will Kelly have saved after 8 weeks?
   b. If the car costs \$2000, how long will it take her to save enough money at this rate?

Example 4 **Determine whether each sequence is geometric. Write *yes* or *no*.**

10. −8, −5, −1, 4, ...
11. 4, 12, 36, 108, ...
12. 27, 9, 3, 1, ...
13. 7, 14, 21, 28, ...

Example 5 **Find the next three terms of each geometric sequence. Then graph the sequence.**

14. 8, 12, 18, 27, ...
15. 8, 16, 32, 64, ...
16. 250, 50, 10, 2, ...
17. $9, -3, 1, -\frac{1}{3}, \ldots$

Example 6 **Determine whether each sequence is *arithmetic*, *geometric*, or *neither*. Explain your reasoning.**

18. 5, 1, 7, 3, 9, ...
19. 200, −100, 50, −25, ...
20. 12, 16, 20, 24, ...

Practice and Problem Solving

Extra Practice is on page R10.

Example 1 **Determine whether each sequence is arithmetic. Write *yes* or *no*.**

21. $\frac{1}{2}, \frac{1}{3}, \frac{1}{4}, \frac{1}{5}, \ldots$
22. −9, −3, 0, 3, 9
23. 14, −5, −19, ...
24. $\frac{2}{9}, \frac{5}{9}, \frac{8}{9}, \frac{11}{9}, \ldots$

Example 2 **Find the next four terms of each arithmetic sequence. Then graph the sequence.**

25. −4, −1, 2, 5, ...
26. 10, 2, −6, −14, ...
27. −5, −11, −17, −23, ...
28. −19, −2, 15, ...
29. $\frac{1}{5}, \frac{4}{5}, \frac{7}{5}, \ldots$
30. $\frac{2}{3}, -\frac{1}{3}, -\frac{4}{3}$

Example 3

31. **THEATER** There are 28 seats in the front row of a theater. Each successive row contains two more seats than the previous row. If there are 24 rows, how many seats are in the last row of the theater?

32. **CCSS SENSE-MAKING** Mario began an exercise program to get back in shape. He plans to row 5 minutes on his rowing machine the first day and increase his rowing time by one minute and thirty seconds each day.
    a. How long will he row on the 18th day?
    b. On what day will Mario first row an hour or more?
    c. Is it reasonable for this pattern to continue indefinitely? Explain.

Example 4 **Determine whether each sequence is geometric. Write *yes* or *no*.**

33. 21, 14, 7, ...
34. 124, 186, 248, ...
35. −27, 18, −12, ...
36. 162, 108, 72, ...
37. $\frac{1}{2}, -\frac{1}{4}, 1, -\frac{1}{2}, \ldots$
38. −4, −2, 0, 2, ...

Example 5 **Find the next three terms of the sequence. Then graph the sequence.**

39. 0.125, −0.5, 2, ...
40. 18, 12, 8, ...
41. 64, 48, 36, ...
42. 81, 108, 144, ...
43. $\frac{1}{3}, 1, 3, 9, \ldots$
44. 1, 0.1, 0.01, 0.001, ...

Example 6 **Determine whether each sequence is *arithmetic*, *geometric*, or *neither*. Explain your reasoning.**

45. 3, 12, 27, 48, ...
46. 1, −2, −5, −8, ...
47. 12, 36, 108, 324, ...
48. $-\frac{2}{5}, -\frac{2}{25}, -\frac{2}{125}, -\frac{2}{625}, \ldots$
49. $\frac{5}{2}, 3, \frac{7}{2}, 4, \ldots$
50. 6, 9, 14, 21, ...

51. **READING** Sareeta took an 800-page book on vacation. If she was already on page 112 and is going to be on vacation for 8 days, what is the minimum number of pages she needs to read per day to finish the book by the end of her vacation?

52. **DEPRECIATION** Tammy's car is expected to depreciate at a rate of 15% per year. If her car is currently valued at $24,000, to the nearest dollar, how much will it be worth in 6 years?

53. **CCSS REGULARITY** When a piece of paper is folded onto itself, it doubles in thickness. If a piece of paper that is 0.1 mm thick could be folded 37 times, how thick would it be?

H.O.T. Problems Use Higher-Order Thinking Skills

54. **REASONING** Explain why the sequence 8, 10, 13, 17, 22 is not arithmetic.

55. **OPEN ENDED** Describe a real-life situation that can be represented by an arithmetic sequence with a common difference of 8.

56. **CHALLENGE** The sum of three consecutive terms of an arithmetic sequence is 6. The product of the terms is −42. Find the terms.

57. **ERROR ANALYSIS** Brody and Gen are determining whether the sequence 8, 8, 8, ... is *arithmetic*, *geometric*, *neither*, or *both*. Is either of them correct? Explain your reasoning.

Brody	Gen
The sequence has a common difference of 0. The sequence is arithmetic.	The sequence has a common ratio of 1. The sequence is geometric.

58. **OPEN ENDED** Find a geometric sequence, an arithmetic sequence, and a sequence that is neither geometric nor arithmetic that begins 3, 9, ...

59. **REASONING** If a geometric sequence has a ratio r such that $|r| < 1$, what happens to the terms as n increases? What would happen to the terms if $|r| \geq 1$?

60. **WRITING IN MATH** Describe what happens to the terms of a geometric sequence when the common ratio is doubled. What happens when it is halved? Explain your reasoning.

Standardized Test Practice

61. SHORT RESPONSE Mrs. Aguilar's rectangular bedroom measures 13 feet by 11 feet. She wants to purchase carpet for the bedroom that costs \$2.95 per square foot, including tax. How much will it cost to carpet her bedroom?

62. The pattern of filled circles and white circles below can be described by a relationship between two variables.

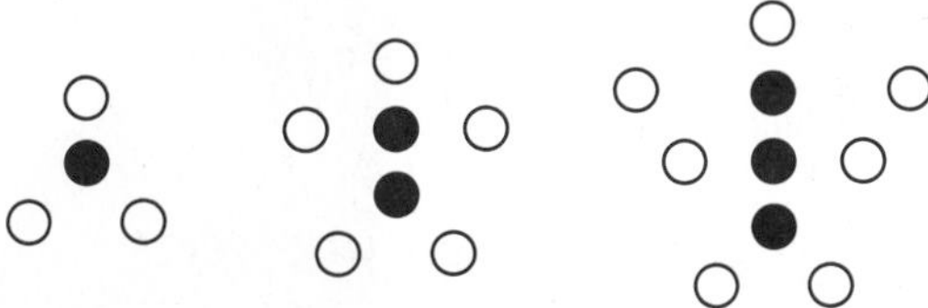

Which rule relates w, the number of white circles, to f, the number of dark circles?

A $w = 3f$

B $f = \frac{1}{2}w - 1$

C $w = 2f + 1$

D $f = \frac{1}{3}w$

63. SAT/ACT Donna wanted to determine the average of her six test scores. She added the scores correctly to get T, but divided by 7 instead of 6. Her average was 12 less than the actual average. Which equation could be used to determine the value of T?

F $6T + 12 = 7T$

G $\frac{T}{7} = \frac{T - 12}{6}$

H $\frac{T}{7} + 12 = \frac{T}{6}$

J $\frac{T}{6} = \frac{T - 12}{7}$

K $\frac{T}{6} = 12 - \frac{T}{7}$

64. Find the next term in the geometric sequence $8, 6, \frac{9}{2}, \frac{27}{8}, \ldots$.

A $\frac{11}{8}$

B $\frac{27}{16}$

C $\frac{9}{4}$

D $\frac{81}{32}$

Spiral Review

Solve each system of equations. (Lesson 9-7)

65. $y = 5$
$y^2 = x^2 + 9$

66. $y - x = 1$
$x^2 + y^2 = 25$

67. $3x = 8y^2$
$8y^2 - 2x^2 = 16$

Write each equation in standard form. State whether the graph of the equation is a *parabola, circle, ellipse,* or *hyperbola*. Then graph the equation. (Lesson 9-6)

68. $6x^2 + 6y^2 = 162$

69. $4y^2 - x^2 + 4 = 0$

70. $x^2 + y^2 + 6y + 13 = 40$

Graph each function. (Lesson 8-4)

71. $f(x) = \frac{6}{(x - 2)(x + 3)}$

72. $f(x) = \frac{-3}{(x - 2)^2}$

73. $f(x) = \frac{x^2 - 36}{x + 6}$

74. HEALTH A certain medication is eliminated from the bloodstream at a steady rate. It decays according to the equation $y = ae^{-0.1625t}$, where t is in hours. Find the half-life of this substance. (Lesson 7-8)

Skills Review

Write an equation of each line.

75. passes through (6, 4), $m = 0.5$

76. passes through $\left(2, \frac{1}{2}\right)$, $m = -\frac{3}{4}$

77. passes through (0, −6), $m = 3$

78. passes through (0, 4), $m = \frac{1}{4}$

79. passes through (1, 3) and $\left(8, -\frac{1}{2}\right)$

80. passes through (−5, 1) and (5, 16)

LESSON 10-2 Arithmetic Sequences and Series

Then

- You determined whether a sequence was arithmetic.

Now

1. Use arithmetic sequences.
2. Find sums of arithmetic series.

Why?

- In the 18th century, a teacher asked his class of elementary students to find the sum of the counting numbers 1 through 100. A pupil named Karl Gauss correctly answered within seconds, astonishing the teacher. Gauss went on to become a great mathematician.

  He solved this problem by using an arithmetic series.

NewVocabulary
arithmetic means
series
arithmetic series
partial sum
sigma notation

Common Core State Standards

Content Standards
A.CED.4 Rearrange formulas to highlight a quantity of interest, using the same reasoning as in solving equations.

Mathematical Practices
8 Look for and express regularity in repeated reasoning.

1 Arithmetic Sequences

In Lesson 10-1, you used the point-slope form to find a specific term of an arithmetic sequence. It is possible to develop an equation for any term of an arithmetic sequence using the same process.

Consider the arithmetic sequence $a_1, a_2, a_3, \ldots, a_n$ in which the common difference is d.

$(y - y_1) = m(x - x_1)$	Point-slope form
$(a_n - a_1) = d(n - 1)$	$(x, y) = (n, a_n)$, $(x_1, y_1) = (1, a_1)$, and $m = d$
$a_n = a_1 + d(n - 1)$	Add a_1 to each side.

You can use this equation to find any term in an arithmetic sequence when you know the first term and the common difference.

KeyConcept *n*th Term of an Arithmetic Sequence

The nth term a_n of an arithmetic sequence in which the first term is a_1 and the common difference is d is given by the following formula, where n is any natural number.

$$a_n = a_1 + (n - 1)d$$

You will prove this formula in Exercise 80.

Example 1 Find the *n*th Term

Find the 12th term of the arithmetic sequence 9, 16, 23, 30, … .

Step 1 Find the common difference.

$16 - 9 = 7 \quad 23 - 16 = 7 \quad 30 - 23 = 7$

So, $d = 7$.

Step 2 Find the 12th term.

$a_n = a_1 + (n - 1)d$	nth term of an arithmetic sequence
$a_{12} = 9 + (12 - 1)(7)$	$a_1 = 9$, $d = 7$, and $n = 12$
$= 9 + 77$ or 86	Simplify.

GuidedPractice

Find the indicated term of each arithmetic sequence.

1A. $a_1 = -4, d = 6, n = 9$

1B. a_{20} for $a_1 = 15, d = -8$

akg-images

If you are given some terms of an arithmetic sequence, you can write an equation for the nth term of the sequence.

Example 2 Write Equations for the nth Term

Write an equation for the nth term of each arithmetic sequence.

a. $5, -13, -31, \ldots$

$d = -13 - 5$ or -18; 5 is the first term.

$a_n = a_1 + (n - 1)d$	nth term of an arithmetic sequence
$a_n = 5 + (n - 1)(-18)$	$a_1 = 5$ and $d = -18$
$a_n = 5 + (-18n + 18)$	Distributive Property
$a_n = -18n + 23$	Simplify.

StudyTip

Checking Solutions Check your solution by using it to determine the first three terms of the sequence.

b. $a_5 = 19, d = 6$

First, find a_1.

$a_n = a_1 + (n - 1)d$	nth term of an arithmetic sequence
$19 = a_1 + (5 - 1)(6)$	$a_5 = 19$, $n = 5$, and $d = 6$
$19 = a_1 + 24$	Multiply.
$-5 = a_1$	Subtract 24 from each side.

Then write the equation.

$a_n = a_1 + (n - 1)d$	nth term of an arithmetic sequence
$a_n = -5 + (n - 1)(6)$	$a_1 = -5$ and $d = 6$
$a_n = -5 + (6n - 6)$	Distributive Property
$a_n = 6n - 11$	Simplify.

GuidedPractice

2A. $12, 3, -6, \ldots$

2B. $a_6 = 12, d = 8$

Sometimes you are given two terms of a sequence, but they are not consecutive terms of the sequence. The terms between any two nonconsecutive terms of an arithmetic sequence, called **arithmetic means**, can be used to find missing terms of a sequence.

Example 3 Find Arithmetic Means

ReadingMath

arithmetic mean the average of two or more numbers

arithmetic means the terms between any two nonconsecutive terms of an arithmetic sequence

Find the arithmetic means in the sequence -8, _?_, _?_, _?_, _?_, 22, ...

Step 1 Since there are four terms between the first and last terms given, there are 4 + 2 or 6 total terms, so n = 6.

Step 2 Find d.

$a_n = a_1 + (n - 1)d$	nth term of an arithmetic sequence
$22 = -8 + (6 - 1)d$	$a_1 = -8$, $a_6 = 22$, and $n = 6$
$30 = 5d$	Distributive Property
$6 = d$	Divide each side by 5.

Step 3 Use d to find the four arithmetic means.

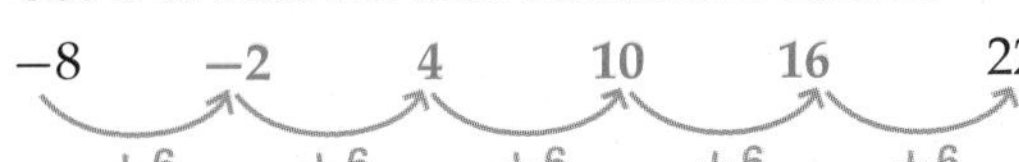

The arithmetic means are -2, 4, 10, and 16.

GuidedPractice

3. Find the five arithmetic means between -18 and 36.

2 Arithmetic Series

A **series** is formed when the terms of a sequence are added. An **arithmetic series** is the sum of the terms of an arithmetic sequence. The sum of the first n terms is called the **partial sum** and is denoted S_n.

KeyConcept Partial Sum of an Arithmetic Series

Formula	Given	The sum S_n of the first n terms is:
General	a_1 and a_n	$S_n = n\left(\frac{a_1 + a_n}{2}\right)$
Alternate	a_1 and d	$S_n = \frac{n}{2}[2a_1 + (n - 1)d]$

Sometimes a_1, a_n, or n must be determined before the sum of an arithmetic series can be found. When this occurs, use the formula for the nth term.

Example 4 Use the Sum Formulas

PT

Find the sum of 12 + 19 + 26 + … + 180.

Step 1 $a_1 = 12$, $a_n = 180$, and $d = 19 - 12$ or 7.

We need to find n before we can use one of the formulas.

$a_n = a_1 + (n - 1)d$ — nth term of an arithmetic sequence

$180 = 12 + (n - 1)(7)$ — $a_n = 180$, $a_1 = 12$, and $d = 7$

$168 = 7n - 7$ — Simplify.

$25 = n$ — Solve for n.

Step 2 Use either formula to find S_n.

$S_n = \frac{n}{2}[2a_1 + (n - 1)d]$ — Sum formula

$S_{25} = \frac{25}{2}[2(12) + (25 - 1)(7)]$ — $n = 25$, $a_1 = 12$, and $d = 7$

$S_{25} = 12.5(192)$ or 2400 — Simplify.

GuidedPractice

Find the sum of each arithmetic series.

4A. $2 + 4 + 6 + \ldots + 100$

4B. $n = 16$, $a_n = 240$, and $d = 8$.

You can use a sum formula to find terms of a series.

WatchOut!

Common Difference

Don't confuse the sign of the common difference in an arithmetic sequence. Check that the rule actually produces the terms of a sequence.

Example 5 Find the First Three Terms

Find the first three terms of the arithmetic series in which $a_1 = 7$, $a_n = 79$, and $S_n = 430$.

Step 1 Find n.

$S_n = n\left(\frac{a_1 + a_n}{2}\right)$ — Sum formula

$430 = n\left(\frac{7 + 79}{2}\right)$ — $S_n = 430$, $a_1 = 7$, and $a_n = 79$

$430 = n(43)$ — Simplify.

$10 = n$ — Divide each side by 43.

Step 2 Find d.

$a_n = a_1 + (n - 1)d$	nth term of an arithmetic sequence
$79 = 7 + (10 - 1)d$	$a_n = 79$, $a_1 = 7$, and $n = 10$
$72 = 9d$	Subtract 7 from each side.
$8 = d$	Divide each side by 9.

Step 3 Use d to determine a_2 and a_3.

$a_2 = 7 + 8$ or 15 $\quad$ $a_3 = 15 + 8$ or 23

The first three terms are 7, 15, and 23.

GuidedPractice

Find the first three terms of each arithmetic series.

5A. $S_n = 120$, $n = 8$, $a_n = 36$ $\quad$ **5B.** $a_1 = -24$, $a_n = 288$, $S_n = 5280$

The sum of a series can be written in shorthand by using **sigma notation**.

ReadingMath

Sigma Notation The name comes from the Greek letter sigma, which is used in the notation.

KeyConcept Sigma Notation

Symbols

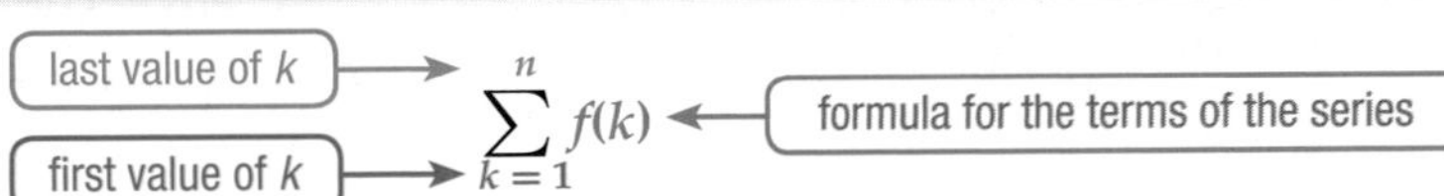

$$\sum_{k=1}^{n} f(k)$$

Example

$$\sum_{k=1}^{12}(4k + 2) = [4(1) + 2] + [4(2) + 2] + [4(3) + 2] + \cdots + [4(12) + 2]$$
$$= 6 + 10 + 14 + \cdots + 50$$

Standardized Test Example 6 Use Sigma Notation

Find $\sum_{k=4}^{18}(6k - 1)$.

A 846 $\quad$ **B** 910 $\quad$ **C** 975 $\quad$ **D** 1008

Read the Test Item

You need to find the sum of the series. Find a_1, a_n, and n.

Solve the Test Item

There are $18 - 4 + 1$ or 15 terms, so $n = 15$.

$a_1 = 6(4) - 1$ or 23 $\quad$ $a_n = 6(18) - 1$ or 107

Find the sum.

$S_n = n\left(\frac{a_1 + a_n}{2}\right)$	Sum formula
$S_{15} = 15\left(\frac{23 + 107}{2}\right)$	$n = 15$, $a_1 = 23$, and $a_n = 107$
$S_{15} = 15(65)$ or 975	The correct answer is C.

Test-TakingTip

CCSS Perseverance Sometimes it is necessary to break a problem into parts, solve each part, then combine the solutions of the parts.

GuidedPractice

6. Find $\sum_{m=9}^{21}(5m + 6)$.

F 972 $\quad$ **G** 1053 $\quad$ **H** 1281 $\quad$ **J** 1701

Check Your Understanding

● = Step-by-Step Solutions begin on page R14.

Example 1 **Find the indicated term of each arithmetic sequence.**

1. $a_1 = 14, d = 9, n = 11$

2. a_{18} for 12, 25, 38, ...

Example 2 **Write an equation for the nth term of each arithmetic sequence.**

3. 13, 19, 25, ...

4. $a_5 = -12, d = -4$

Example 3 **Find the arithmetic means in each sequence.**

5. 6, ?, ?, ?, 42

6. −4, ?, ?, ?, 8

Example 4 **Find the sum of each arithmetic series.**

7. the first 50 natural numbers

8. $4 + 8 + 12 + \cdots + 200$

9. $a_1 = 12, a_n = 188, d = 4$

10. $a_n = 145, d = 5, n = 21$

Example 5 **Find the first three terms of each arithmetic series.**

11. $a_1 = 8, a_n = 100, S_n = 1296$

12. $n = 18, a_n = 112, S_n = 1098$

Example 6 **13. MULTIPLE CHOICE** Find $\sum_{k=1}^{12} (3k + 9)$.

A 45

B 78

C 342

D 410

Practice and Problem Solving

Extra Practice is on page R10.

Example 1 **Find the indicated term of each arithmetic sequence.**

14. $a_1 = -18, d = 12, n = 16$

15. $a_1 = -12, n = 66, d = 4$

16. $a_1 = 9, n = 24, d = -6$

17. a_{15} for −5, −12, −19, ...

18. a_{10} for −1, 1, 3, ...

19. a_{24} for 8.25, 8.5, 8.75, ...

Example 2 **Write an equation for the *n*th term of each arithmetic sequence.**

20. 24, 35, 46, ...

21. 31, 17, 3, ...

22. $a_9 = 45, d = -3$

23. $a_7 = 21, d = 5$

24. $a_4 = 12, d = 0.25$

25. $a_5 = 1.5, d = 4.5$

26. 9, 2, −5, ...

27. $a_6 = 22, d = 9$

28. $a_8 = -8, d = -2$

29. $a_{15} = 7, d = \frac{2}{3}$

30. −12, −17, −22, ...

31. $a_3 = -\frac{4}{5}, d = \frac{1}{2}$

32. **CCSS STRUCTURE** José averaged 123 total pins per game in his bowing league this season. He is taking bowling lessons and hopes to bring his average up by 8 pins each new season.

a. Write an equation to represent the *n*th term of the sequence.

b. If the pattern continues, during what season will José average 187 per game?

c. Is it reasonable for this pattern to continue indefinitely? Explain.

Example 3 **Find the arithmetic means in each sequence.**

33. 24, ?, ?, ?, ?, −1

34. −6, ?, ?, ?, ?, 49

35. −28, ?, ?, ?, ?, 7

36. 84, ?, ?, ?, ?, 39

37. −12, ?, ?, ?, ?, ?, −66

38. 182, ?, ?, ?, ?, ?, 104

Example 4 **Find the sum of each arithmetic series.**

39. the first 100 even natural numbers

40. the first 200 odd natural numbers

41. the first 100 odd natural numbers

42. the first 300 even natural numbers

43. $-18 + (-15) + (-12) + \cdots + 66$

44. $-24 + (-18) + (-12) + \cdots + 72$

45. $a_1 = -16, d = 6, n = 24$

46. $n = 19, a_n = 154, d = 8$

47. **CONTESTS** The prizes in a weekly radio contest began at $150 and increased by $50 for each week that the contest lasted. If the contest lasted for eleven weeks, how much was awarded in total?

Example 5 **Find the first three terms of each arithmetic series.**

48. $n = 32, a_n = -86, S_n = 224$

49. $a_1 = 48, a_n = 180, S_n = 1368$

50. $a_1 = 3, a_n = 66, S_n = 759$

51 $n = 28, a_n = 228, S_n = 2982$

52. $a_1 = -72, a_n = 453, S_n = 6858$

53. $n = 30, a_n = 362, S_n = 4770$

54. $a_1 = 19, n = 44, S_n = 9350$

55. $a_1 = -33, n = 36, S_n = 6372$

56. **PRIZES** A radio station is offering a total of $8500 in prizes over ten hours. Each hour, the prize will increase by $100. Find the amounts of the first and last prize.

Example 6 **Find the sum of each arithmetic series.**

57. $\sum_{k=1}^{16} (4k - 2)$

58. $\sum_{k=4}^{13} (4k + 1)$

59. $\sum_{k=5}^{16} (2k + 6)$

60. $\sum_{k=0}^{12} (-3k + 2)$

61. **FINANCIAL LITERACY** Daniela borrowed some money from her parents. She agreed to pay $50 at the end of the first month and $25 more each additional month for 12 months. How much does she pay in total after the 12 months?

62. **GRAVITY** When an object is in free fall and air resistance is ignored, it falls 16 feet in the first second, an additional 48 feet during the next second, and 80 feet during the third second. How many total feet will the object fall in 10 seconds?

Use the given information to write an equation that represents the *n*th term in each arithmetic sequence

63. The 100th term of the sequence is 245. The common difference is 13.

64. The eleventh term of the sequence is 78. The common difference is −9.

65. The sixth term of the sequence is −34. The 23rd term is 119.

66. The 25th term of the sequence is 121. The 80th term is 506.

67. CCSS **MODELING** The rectangular tables in a reception hall are often placed end-to-end to form one long table. The diagrams below show the number of people who can sit at each of the table arrangements.

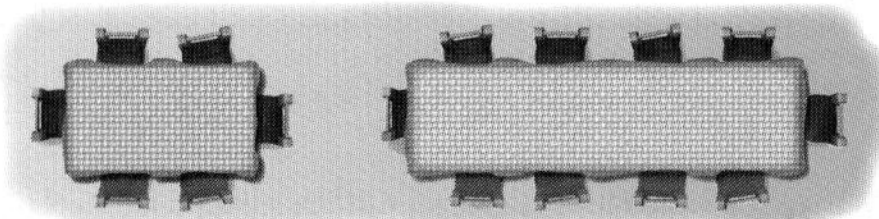

a. Make drawings to find the next three numbers as tables are added one at a time to the arrangement.

b. Write an equation representing the *n*th number in this pattern.

c. Is it possible to have seating for exactly 100 people with such an arrangement? Explain.

68. **PERFORMANCE** A certain company pays its employees according to their performance. Belinda is paid a flat rate of \$200 per week plus \$24 for every unit she completes. If she earned \$512 in one week, how many units did she complete?

69. **SALARY** Terry currently earns \$28,000 per year. If Terry expects a \$4000 increase in salary every year, after how many years will he have a salary of \$100,000 per year?

70. **SPORTS** While training for cross country, Silvia plans to run 3 miles per day for the first week, and then increase the distance by a half mile each of the following weeks.

    a. Write an equation to represent the nth term of the sequence.

    b. If the pattern continues, during which week will she be running 10 miles per day?

    c. Is it reasonable for this pattern to continue indefinitely? Explain.

71. **MULTIPLE REPRESENTATIONS** Consider $\sum_{k=1}^{x}(2k+2)$.

    a. **Tabular** Make a table of the partial sums of the series for $1 \leq k \leq 10$.

    b. **Graphical** Graph (k, partial sum).

    c. **Graphical** Graph $f(x) = x^2 + 3x$ on the same grid.

    d. **Verbal** What do you notice about the two graphs?

    e. **Analytical** What conclusions can you make about the relationship between quadratic functions and the sum of arithmetic series?

    f. **Algebraic** Find the arithmetic series that relates to $g(x) = x^2 + 8x$.

Find the value of x.

72. $\sum_{k=3}^{x}(6k-5) = 928$

73. $\sum_{k=5}^{x}(8k+2) = 1032$

H.O.T. Problems Use Higher-Order Thinking Skills

74. **CCSS CRITIQUE** Eric and Juana are determining the formula for the nth term for the sequence $-11, -2, 7, 16, \ldots$. Is either of them correct? Explain your reasoning.

Eric	Juana
$d = 16 - 7$ or 9, $a_1 = -11$	$d = 16 - 7$ or 9, $a_1 = -11$
$a_n = -11 + (n-1)9$	$a_n = 9n - 11$
$= 9n - 20$	

75. **REASONING** If a is the third term in an arithmetic sequence, b is the fifth term, and c is the eleventh term, express c in terms of a and b.

76. **CHALLENGE** There are three arithmetic means between a and b in an arithmetic sequence. The average of the arithmetic means is 16. What is the average of a and b?

77. **CHALLENGE** Find S_n for $(x + y) + (x + 2y) + (x + 3y) + \ldots$.

78. **OPEN ENDED** Write an arithmetic series with 8 terms and a sum of 324.

79. **WRITING IN MATH** Compare and contrast arithmetic sequences and series.

80. **PROOF** Prove the formula for the nth term of an arithmetic sequence.

81. **PROOF** Derive a sum formula that does not include a_1.

82. **PROOF** Derive the Alternate Sum Formula using the General Sum Formula.

Standardized Test Practice

83. SAT/ACT The measures of the angles of a triangle form an arithmetic sequence. If the measure of the smallest angle is 36°, what is the measure of the largest angle?

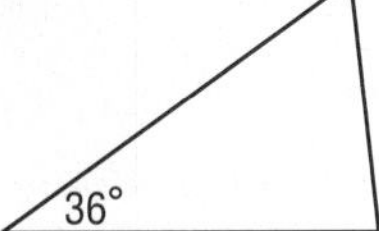

A 54°
B 75°
C 84°
D 90°
E 97°

84. The area of a triangle is $\frac{1}{2}q^2 - 8$ and the height is $q + 4$. Which expression best describes the triangle's base?

F $(q + 1)$
G $(q + 2)$
H $(q - 3)$
J $(q - 4)$.

85. The expression $1 + \sqrt{2} + \sqrt[3]{3}$ is equivalent to

A $\sum_{k=1}^{3} k^{\frac{1}{k}}$

B $\sum_{k=1}^{3} k^k$

C $\sum_{k=1}^{3} k^{-k}$

D $\sum_{k=1}^{3} \sqrt{k}$

86. SHORT RESPONSE Trevor can type a 200-word essay in 6 hours. Minya can type the same essay in $4\frac{1}{2}$ hours. If they work together, how many hours will it take them to type the essay?

Spiral Review

Determine whether each sequence is arithmetic. Write *yes* or *no*. (Lesson 10-1)

87. $-6, 4, 14, 24, \ldots$

88. $2, \frac{7}{5}, \frac{4}{5}, \frac{1}{5}, \ldots$

89. $10, 8, 5, 1, \ldots$

Solve each system of inequalities by graphing. (Lesson 9-7)

90. $x + 2y > 1$
$x^2 + y^2 \leq 25$

91. $x + y \leq 2$
$4x^2 - y^2 \geq 4$

92. $x^2 + y^2 \geq 4$
$4y^2 + 9x^2 \leq 36$

93. PHYSICS The distance a spring stretches is related to the mass attached to the spring. This is represented by $d = km$, where d is the distance, m is the mass, and k is the spring constant. When two springs with spring constants k_1 and k_2 are attached in a series, the resulting spring constant k is found by the equation $\frac{1}{k} = \frac{1}{k_1} + \frac{1}{k_2}$. (Lesson 8-6)

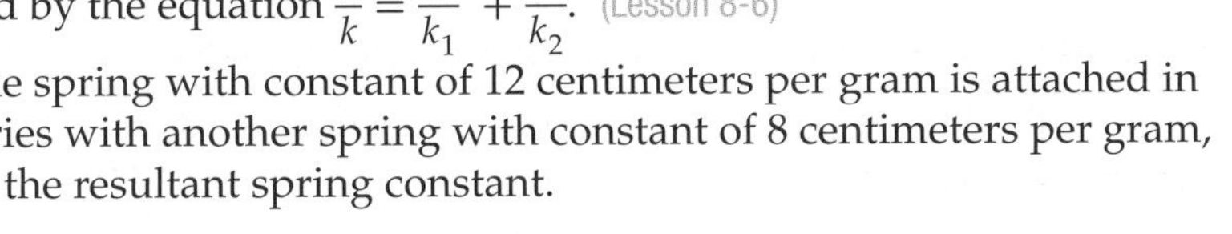

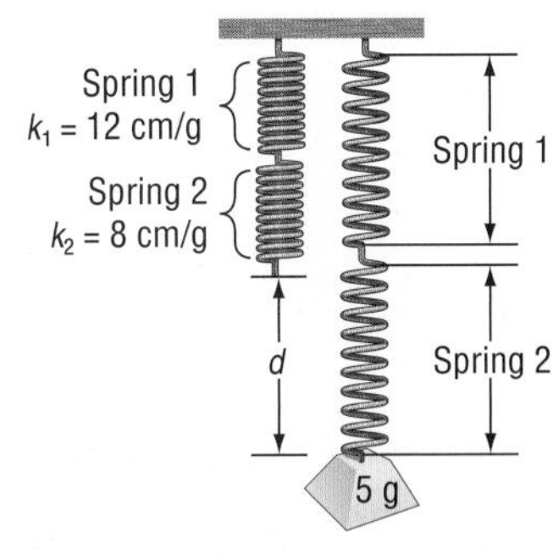

a. If one spring with constant of 12 centimeters per gram is attached in a series with another spring with constant of 8 centimeters per gram, find the resultant spring constant.

b. If a 5-gram object is hung from the series of springs, how far will the springs stretch? Is this answer reasonable in this context?

Graph each function. State the domain and range. (Lesson 7-1)

94. $f(x) = \frac{2}{3}(2^x)$

95. $f(x) = 4^x + 3$

96. $f(x) = 2\left(\frac{1}{3}\right)^x - 1$

Skills Review

Solve each equation. Round to the nearest ten-thousandth.

97. $5^x = 52$

98. $4^{3p} = 10$

99. $3^{n+2} = 14.5$

100. $16^{d-4} = 3^{3-d}$

LESSON 10-3 Geometric Sequences and Series

Then	Now	Why?
You determined whether a sequence was geometric.	**1** Use geometric sequences. **2** Find sums of geometric series.	Julian sees a new band at a concert. He e-mails a link for the band's Web site to five of his friends. They each forward the link to five of their friends. The link is forwarded again following the same pattern. How many people will receive the link on the eighth round of e-mails?.

NewVocabulary
geometric means
geometric series

Common Core State Standards

Content Standards
A.SSE.4 Derive the formula for the sum of a finite geometric series (when the common ratio is not 1), and use the formula to solve problems.

Mathematical Practices
8 Look for and express regularity in repeated reasoning.

1 Geometric Sequences

As with arithmetic sequences, there is a formula for the nth term of a geometric sequence. This formula can be used to determine any term of the sequence.

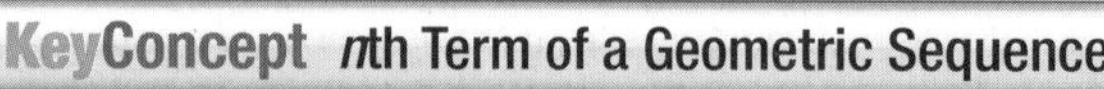

KeyConcept *n*th Term of a Geometric Sequence

The nth term a_n of a geometric sequence in which the first term is a_1 and the common ratio is r is given by the following formula, where n is any natural number.

$$a_n = a_1 r^{n-1}$$

You will prove this formula in Exercise 68.

Real-World Example 1 Find the *n*th Term

MUSIC If the pattern continues, how many e-mails will be sent in the eighth round?

Understand We need to determine the number of forwarded e-mails on the eighth round. Five e-mails were sent on the first round. Each of the five recipients sent five e-mails on the second round, and so on.

Plan This is a geometric sequence, and the common ratio is 5. Use the formula for the nth term of a geometric sequence.

Solve

$a_n = a_1 r^{n-1}$	nth term of a geometric sequence
$a_8 = 5(5)^{8-1}$	$a_1 = 5$, $r = 5$, and $n = 8$
$a_8 = 5(78{,}125)$ or $390{,}625$	$5^7 = 78{,}125$

Check Write out the first eight terms by multiplying by the common ratio.

5, 25, 125, 625, 3125, 15,625, 78,125, 390,625

There will be 390,625 e-mails sent on the 8th round.

GuidedPractice

1. **E-MAILS** Shira receives a joke in an e-mail that asks her to forward it to four of her friends. She forwards it, then each of her friends forwards it to four of their friends, and so on. If the pattern continues, how many people will receive the e-mail on the ninth round of forwarding?

Rubberball/Nicole Hill/Getty Images

Math HistoryLink

Archytas (428–347 B.C.) Geometric sequences, or geometric progressions, were first studied by the Greek mathematician Archytas. His studies of these sequences came from his interest in music and octaves.

If you are given some of the terms of a geometric sequence, you can determine an equation for finding the nth term of the sequence.

PT

Example 2 Write an Equation for the nth Term

Write an equation for the nth term of each geometric sequence.

a. 0.5, 2, 8, 32, ...

$r = 8 \div 2$ or 4; 0.5 is the first term.

$a_n = a_1r^{n-1}$ — nth term of a geometric sequence

$a_n = 0.5(4)^{n-1}$ — $a_1 = 0.5$ and $r = 4$

b. $a_4 = 5$ and $r = 6$

Step 1 Find a_1.

$a_n = a_1r^{n-1}$ — nth term of a geometric sequence

$5 = a_1(6^{4-1})$ — $a_n = 5$, $r = 6$, and $n = 4$

$5 = a_1(216)$ — Evaluate the power.

$\frac{5}{216} = a_1$ — Divide each side by 216.

Step 2 Write the equation.

$a_n = a_1r^{n-1}$ — nth term of a geometric sequence

$a_n = \frac{5}{216}(6)^{n-1}$ — $a_1 = \frac{5}{216}$ and $r = 6$

GuidedPractice

Write an equation for the nth term of each geometric sequence.

2A. $-0.25, 2, -16, 128, \ldots$

2B. $a_3 = 16, r = 4$

Like arithmetic means, **geometric means** are the terms between two nonconsecutive terms of a geometric sequence. The common ratio r can be used to find the geometric means.

ReadingMath

Geometric Means

A geometric mean can also be represented geometrically. In the figure below, h is the geometric mean between x and y.

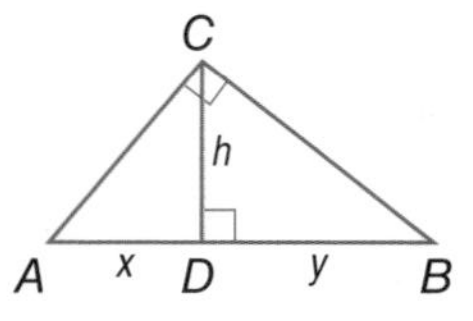

PT

Example 3 Find Geometric Means

Find three geometric means between 2 and 1250.

Step 1 Since there are three terms between the first and last term, there are 3 + 2 or 5 total terms, so $n = 5$.

Step 2 Find r.

$a_n = a_1r^{n-1}$ — nth term of a geometric sequence

$1250 = 2r^{5-1}$ — $a_n = 1250$, $a_1 = 2$, and $n = 5$

$625 = r^4$ — Divide each side by 2.

$\pm 5 = r$ — Take the 4th root of each side.

Step 3 Use r to find the four arithmetic means.

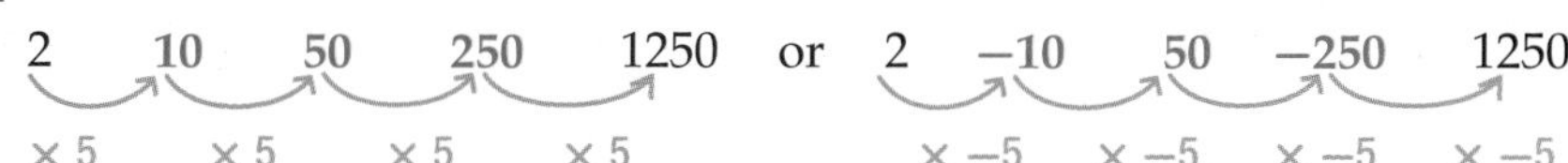

The geometric means are 10, 50, and 250 or −10, 50, and −250.

GuidedPractice

3. Find four geometric means between 0.5 and 512.

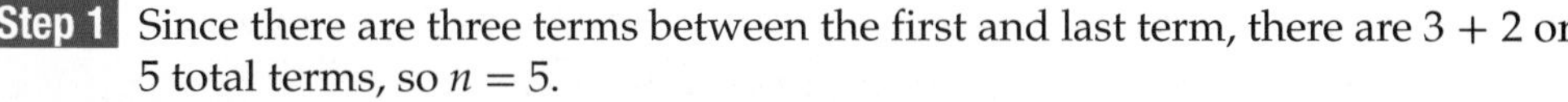

2 Geometric Series A **geometric series** is the sum of the terms of a geometric sequence. The sum of the first n terms of a series is denoted S_n. You can use either of the following formulas to find the partial sum S_n of the first n terms of a geometric series.

KeyConcept Partial Sum of a Geometric Series

Given	The sum S_n of the first n terms is:
a_1 and n	$S_n = \frac{a_1 - a_1 r^n}{1 - r}, r \neq 1$
a_1 and a_n	$S_n = \frac{a_1 - a_n r}{1 - r}, r \neq 1$

Real-World Example 4 Find the Sum of a Geometric Series

MUSIC Refer to the beginning of the lesson. If the pattern continues, what is the total number of e-mails sent in the eight rounds?

Five e-mails are sent in the first round and there are 8 rounds of e-mails. So, $a_1 = 5$, $r = 5$ and $n = 8$.

$S_n = \frac{a_1 - a_1 r^n}{1 - r}$ Sum formula

$S_8 = \frac{5 - 5 \cdot 5^8}{1 - 5}$ $a_1 = 5, r = 5,$ and $n = 8$

$S_8 = \frac{-1{,}953{,}120}{-4}$ Simplify the numerator and denominator.

$S_8 = 488{,}280$ Divide.

There will be 488,280 e-mails sent after 8 rounds.

GuidedPractice

Find the sum of each geometric series.

4A. $a_1 = 2, n = 10, r = 3$

4B. $a_1 = 2000, a_n = 125, r = \frac{1}{2}$

As with arithmetic series, sigma notation can also be used to represent geometric series.

Example 5 Sum in Sigma Notation

WatchOut!

Sigma Notation Notice in Example 5 that you are being asked to evaluate the sum from the 3rd term to the 10th term..

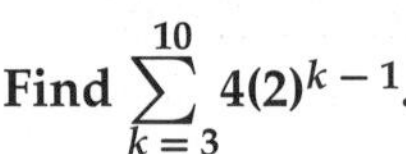

Find $\sum_{k=3}^{10} 4(2)^{k-1}$.

Find a_1, r, and n. In the first term, $k = 3$ and $a_1 = 4 \cdot 2^{3-1}$ or 16. The base of the exponential function is r, so $r = 2$. There are $10 - 3 + 1$ or 8 terms, so $n = 8$.

$S_n = \frac{a_1 - a_1 r^n}{1 - r}$ Sum formula

$= \frac{16 - 16(2)^8}{1 - 2}$ $a_1 = 16, r = 2,$ and $n = 8$

$= 4080$ Use a calculator.

GuidedPractice

Find each sum.

5A. $\sum_{k=4}^{12} \frac{1}{4} \cdot 3^{k-1}$

5B. $\sum_{k=2}^{9} \frac{2}{3} \cdot 4^{k-1}$

You can use the formula for the sum of a geometric series to help find a particular term of the series.

Example 6 Find the First Term of a Series

Find a_1 in a geometric series for which $S_n = 13{,}116$, $n = 7$, and $r = 3$.

$S_n = \frac{a_1 - a_1 r^n}{1 - r}$	Sum formula
$13{,}116 = \frac{a_1 - a_1(3^7)}{1 - 3}$	$S_n = 13{,}116$, $r = 3$, and $n = 7$
$13{,}116 = \frac{a_1(1 - 3^7)}{1 - 3}$	Distributive Property
$13{,}116 = \frac{-2186a_1}{-2}$	Subtract.
$13{,}116 = 1093a_1$	Simplify.
$12 = a_1$	Divide each side by 1093.

GuidedPractice

6. Find a_1 in a geometric series for which $S_n = -26{,}240$, $n = 8$, and $r = -3$.

Check Your Understanding

= Step-by-Step Solutions begin on page R14.

Example 1

1. **CCSS REGULARITY** Dean is making a family tree for his grandfather. He was able to trace many generations. If Dean could trace his family back 10 generations, starting with his parents how many ancestors would there be?

Example 2

Write an equation for the *n*th term of each geometric sequence.

2. 2, 4, 8, …

3. 18, 6, 2, …

4. −4, 16, −64, …

5. $a_2 = 4$, $r = 3$

6. $a_6 = \frac{1}{8}$, $r = \frac{3}{4}$

7. $a_2 = -96$, $r = -8$

Example 3

Find the geometric means of each sequence.

8. 0.25, ?, ?, ?, 64

9. 0.20, ?, ?, ?, 125

Example 4

10. **GAMES** Miranda arranges some rows of dominoes so that after she knocks over the first one, each domino knocks over two more dominoes when it falls. If there are ten rows, how many dominoes does Miranda use?

Example 5

Find the sum of each geometric series.

11. $\sum_{k=1}^{6} 3(4)^{k-1}$

12. $\sum_{k=1}^{8} 4\left(\frac{1}{2}\right)^{k-1}$

Example 6

Find a_1 for each geometric series described.

13. $S_n = 85\frac{5}{16}$, $r = 4$, $n = 6$

14. $S_n = 91\frac{1}{12}$, $r = 3$, $n = 7$

15. $S_n = 1020$, $a_n = 4$, $r = \frac{1}{2}$

16. $S_n = 121\frac{1}{3}$, $a_n = \frac{1}{3}$, $r = \frac{1}{3}$

Practice and Problem Solving

Extra Practice is on page R10.

Example 1

17. **WEATHER** Heavy rain in Brieanne's town caused the river to rise. The river rose three inches the first day, and each day after rose twice as much as the previous day. How much did the river rise in five days?

Find a_n for each geometric sequence.

18. $a_1 = 2400, r = \frac{1}{4}, n = 7$

19. $a_1 = 800, r = \frac{1}{2}, n = 6$

20. $a_1 = \frac{2}{9}, r = 3, n = 7$

21. $a_1 = -4, r = -2, n = 8$

22. **BIOLOGY** A certain bacteria grows at a rate of 3 cells every 2 minutes. If there were 260 cells initially, how many are there after 21 minutes?

Example 2

Write an equation for the *n*th term of each geometric sequence.

23. $-3, 6, -12, \ldots$

24. $288, -96, 32, \ldots$

25. $-1, 1, -1, \ldots$

26. $\frac{1}{3}, \frac{2}{9}, \frac{4}{27}, \ldots$

27. $8, 2, \frac{1}{2}, \ldots$

28. $12, -16, \frac{64}{3}, \ldots$

29. $a_3 = 28, r = 2$

30. $a_4 = -8, r = 0.5$

31. $a_6 = 0.5, r = 6$

32. $a_3 = 8, r = \frac{1}{2}$

33. $a_4 = 24, r = \frac{1}{3}$

34. $a_4 = 80, r = 4$

Example 3

Find the geometric means of each sequence.

35. 810, ?, ?, ?, 10

36. 640, ?, ?, ?, 2.5

37. $\frac{7}{2}$, ?, ?, ?, $\frac{56}{81}$

38. $\frac{729}{64}$, ?, ?, ?, $\frac{324}{9}$

39. Find two geometric means between 3 and 375.

40. Find two geometric means between 16 and −2.

Example 4

41. CCSS **PERSEVERANCE** A certain water filtration system can remove 70% of the contaminants each time a sample of water is passed through it. If the same water is passed through the system four times, what percent of the original contaminants will be removed from the water sample?

Find the sum of each geometric series.

42. $a_1 = 36, r = \frac{1}{3}, n = 8$

43. $a_1 = 16, r = \frac{1}{2}, n = 9$

44. $a_1 = 240, r = \frac{3}{4}, n = 7$

45. $a_1 = 360, r = \frac{4}{3}, n = 8$

46. **VACUUMS** A vacuum claims to pick up 80% of the dirt every time it is run over the carpet. Assuming this is true, what percent of the original amount of dirt is picked up after the seventh time the vacuum is run over the carpet?

Example 5

Find the sum of each geometric series.

47. $\sum_{k=1}^{7} 4(-3)^{k-1}$

48. $\sum_{k=1}^{8} (-3)(-2)^{k-1}$

49. $\sum_{k=1}^{9} (-1)(4)^{k-1}$

50. $\sum_{k=1}^{10} 5(-1)^{k-1}$

Example 6

Find a_1 for each geometric series described.

51. $S_n = -2912, r = 3, n = 6$

52. $S_n = -10{,}922, r = 4, n = 7$

53. $S_n = 1330, a_n = 486, r = \frac{3}{2}$

54. $S_n = 4118, a_n = 128, r = \frac{2}{3}$

55. $a_n = 1024, r = 8, n = 5$

56. $a_n = 1875, r = 5, n = 7$

57 **SCIENCE** One minute after it is released, a gas-filled balloon has risen 100 feet. In each succeeding minute, the balloon rises only 50% as far as it rose in the previous minute. How far will it rise in 5 minutes?

58. **CHEMISTRY** Radon has a half-life of about 4 days. This means that about every 4 days, half of the mass of radon decays into another element. How many grams of radon remain from an initial 60 grams after 4 weeks?

59. **CCSS REASONING** A virus goes through a computer, infecting the files. If one file was infected initially and the total number of files infected doubles every minute, how many files will be infected in 20 minutes?

60. **GEOMETRY** In the figure, the sides of each equilateral triangle are twice the size of the sides of its inscribed triangle. If the pattern continues, find the sum of the perimeters of the first eight triangles.

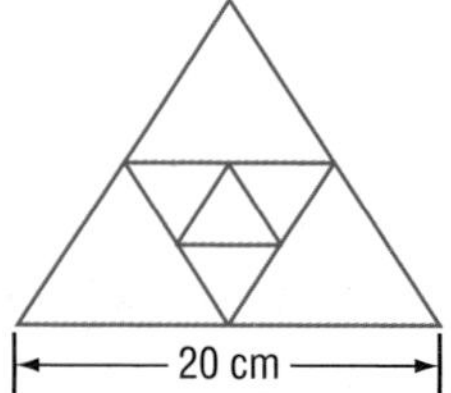

61. **PENDULUMS** The first swing of a pendulum travels 30 centimeters. If each subsequent swing travels 95% as far as the previous swing, find the total distance traveled by the pendulum after the 30th swing.

62. **PHONE CHAINS** A school established a phone chain in which every staff member calls two other staff members to notify them when the school closes due to weather. The first round of calls begins with the superintendent calling both principals. If there are 94 total staff members and employees at the school, how many rounds of calls are there?

63. **TELEVISIONS** High Tech Electronics advertises a weekly installment plan for the purchase of a popular brand of high definition television. The buyer pays $5 at the end of the first week, $5.50 at the end of the second week, $6.05 at the end of the third week, and so on for one year. (Assume that 1 year = 52 weeks.)

a. What will the payments be at the end of the 10th, 20th, and 40th weeks?

b. Find the total cost of the TV.

c. Why is the cost found in part **b** not entirely accurate?

H.O.T. Problems Use Higher-Order Thinking Skills

64. **PROOF** Derive the General Sum Formula using the Alternate Sum Formula.

65. **PROOF** Derive a sum formula that does not include a_1.

66. **OPEN ENDED** Write a geometric series for which $r = \frac{3}{4}$ and $n = 6$.

67. **REASONING** Explain how $\sum_{k=1}^{10} 3(2)^{k-1}$ needs to be altered to refer to the same series if $k = 1$ changes to $k = 0$. Explain your reasoning.

68. **PROOF** Prove the formula for the nth term of a geometric sequence.

69. **CHALLENGE** The fifth term of a geometric sequence is $\frac{1}{27}$th of the eighth term. If the ninth term is 702, what is the eighth term?

70. **CHALLENGE** Use the fact that h is the geometric mean between x and y in the figure at the right to find h^4 in terms of x and y.

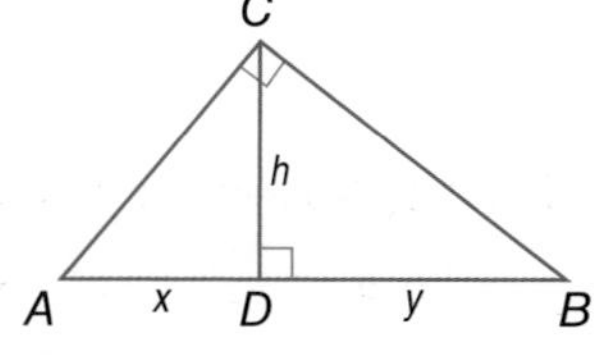

71. **OPEN ENDED** Write a geometric series with 6 terms and a sum of 252.

72. **WRITING IN MATH** How can you classify a sequence? Explain your reasoning.

Standardized Test Practice

73. Which of the following is closest to $\sqrt[3]{7.32}$?

A 1.8
B 1.9
C 2.0
D 2.1

74. The first term of a geometric series is 5, and the common ratio is −2. How many terms are in the series if its sum is −6825?

F 5
G 9
H 10
J 12

75. SHORT RESPONSE Danette has a savings account. She withdraws half of the contents every year. After 4 years, she has $2000 left. How much did she have in the savings account originally?

76. SAT/ ACT The curve below could be part of the graph of which function?

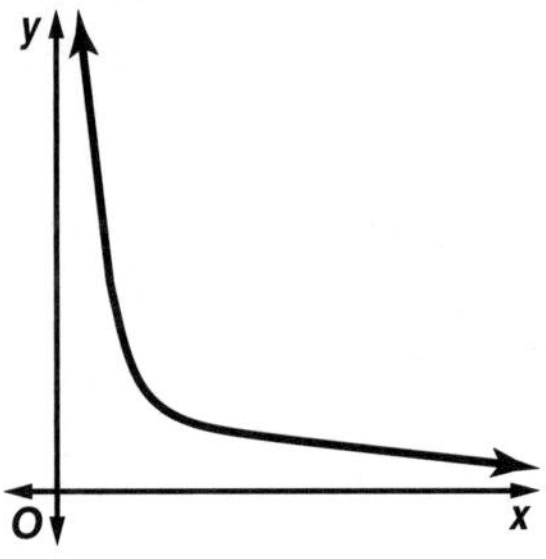

A $y = \sqrt{x}$
B $y = x^2 - 5x + 4$
C $y = -x + 20$
D $y = \log x$
E $xy = 4$

Spiral Review

77. MONEY Elena bought a high-definition LCD television at the electronics store. She paid $200 immediately and $75 each month for a year and a half. How much did Elena pay in total for the TV? (Lesson 10-2)

Determine whether each sequence is *arithmetic*, *geometric*, or *neither*. Explain your reasoning. (Lesson 10-1)

78. $\frac{1}{10}, \frac{3}{5}, \frac{7}{20}, \frac{17}{20}, \ldots$

79. $-\frac{7}{25}, -\frac{13}{50}, -\frac{6}{25}, -\frac{11}{50}, \ldots$

80. $-\frac{22}{3}, -\frac{68}{9}, -\frac{208}{27}, -\frac{632}{81}, \ldots$

Find the center and radius of each circle. Then graph the circle. (Lesson 9-3)

81. $(x - 3)^2 + (y - 1)^2 = 25$

82. $(x + 3)^2 + (y + 7)^2 = 81$

83. $(x - 3)^2 + (y + 7)^2 = 50$

84. Suppose y varies jointly as x and z. Find y when $x = 9$ and $z = -5$, if $y = -90$ when $z = 15$ and $x = -6$. (Lesson 8-5)

85. SHOPPING A certain store found that the number of customers who will attend a sale can be modeled by $N = 125\sqrt[3]{100Pt}$, where N is the number of customers expected, P is the percent of the sale discount, and t is the number of hours the sale will last. Find the number of customers the store should expect for a sale that is 50% off and will last four hours. (Lesson 6-4)

Skills Review

Evaluate each expression if $a = -2$, $b = \frac{1}{3}$, and $c = -12$.

86. $\frac{3ab}{c}$

87. $\frac{a - c}{a + c}$

88. $\frac{a^3 - c}{b^2}$

89. $\frac{c + 3}{ab}$

Algebra Lab
Area Under a Curve

The Morgans are renovating the outside of their house so that there is an archway above the front entrance. Mr. Morgan made a scale drawing of the archway in which each line on the grid paper represents one foot of the actual archway. Mrs. Morgan modeled the shape of the top with the quadratic equation $y = -0.25x^2 + 3x$.

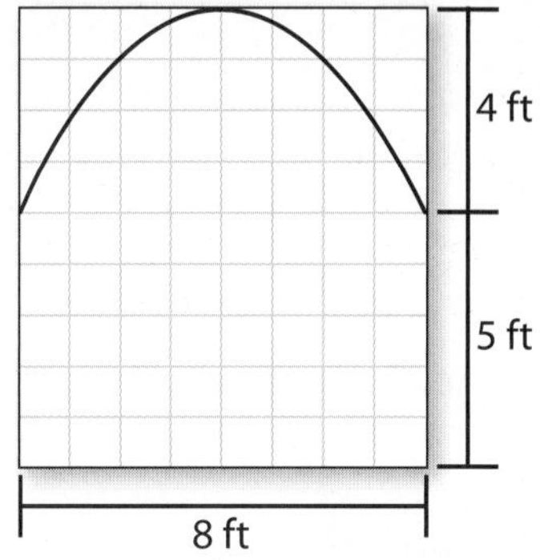

Activity

Find the area of the opening under the archway.

Method 1

Step 1 Make a table of values for $y = -0.25x^2 + 3x$. Then graph the equation.

x	0	1	2	3	4	5	6	7	8	9	10	11	12
y	0	2.75	5	6.75	8	8.75	9	8.75	8	6.75	5	2.75	0

Step 2 Divide the figure into regions.

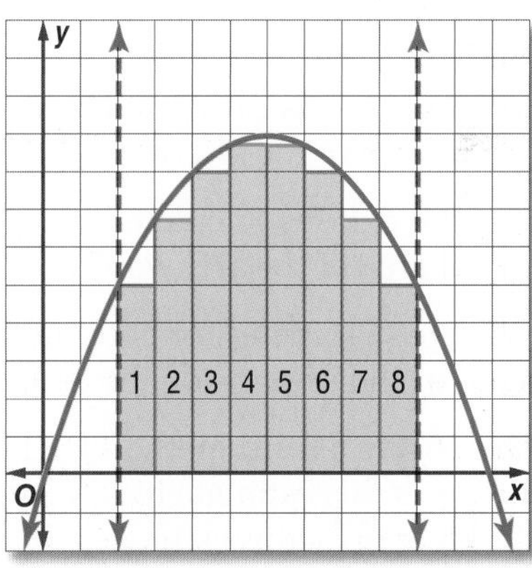

To estimate the area inside the archway, you can divide the archway into rectangles as shown in blue.

Because the left and right sides of the archway are 5 feet high and $y = 5$ when $x = 2$ and when $x = 10$, the opening of the entrance extends from $x = 2$ to $x = 10$.

Step 3 Find the area of the regions.

Rectangle	1	2	3	4	5	6	7	8
Width (ft)	1	1	1	1	1	1	1	1
Height (ft)	5	6.75	8	8.75	8.75	8	6.75	5
Area (ft^2)	5	6.75	8	8.75	8.75	8	6.75	5

The approximate area of the archway is the sum of the areas of the rectangles.

$5 + 6.75 + 8 + 8.75 + 8.75 + 8 + 6.75 + 5 = 57 \text{ ft}^2$

(continued on the next page)

Area Under a Curve *Continued*

Method 2

Step 1 Draw a second graph of the equation and divide into regions. Divide the archway into rectangles as shown in blue.

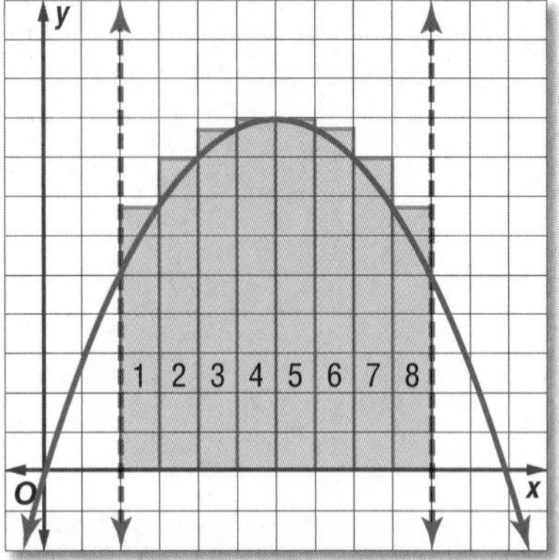

Step 2 Find the area of the regions.

Rectangle	1	2	3	4	5	6	7	8
Width (ft)	1	1	1	1	1	1	1	1
Height (ft)	6.75	8	8.75	9	9	8.75	8	6.75
Area (ft^2)	6.75	8	8.75	9	9	8.75	8	6.75

The approximate area of the archway is the sum of the areas of the rectangles.

$6.75 + 8 + 8.75 + 9 + 9 + 8.75 + 8 + 6.75 = 65 \text{ ft}^2$

Both Method 1 and Method 2 illustrate how to approximate the area under a curve over a specified interval.

Analyze the Results

1. Is the area of the regions calculated using Method 1 greater than or less than the actual area of the archway? Explain your reasoning.
2. Is the area of the regions calculated using Method 2 greater than or less than the actual area of the archway? Explain your reasoning.
3. Compare the area estimates for both methods. How could you find the best estimate for the area inside the archway? Explain your reasoning.
4. The diagram shows a third method for finding an estimate of the area of the archway. Is this estimate for the area greater than or less than the actual area? How does this estimate compare to the other two estimates of the area?

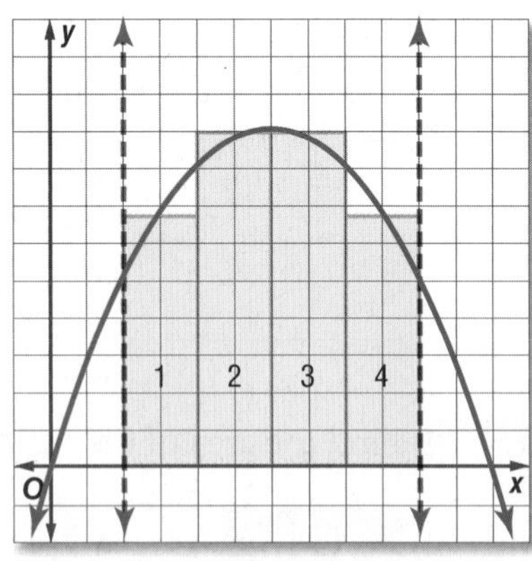

Exercises

Estimate the area described by any method. Make a table of values, draw graphs with rectangles, and make a table for the areas of the rectangles. Compare each estimate to the actual area.

5. the area under the curve for $y = -x^2 + 4$, from $x = -2$ to $x = 2$, and above the x–axis
6. the area under the curve for $y = x^3$, from $x = 0$ to $x = 4$, and above the x–axis
7. the area under the curve for $y = x^2$, from $x = -3$ to $x = 3$, and above the x–axis

LESSON 10-4 Infinite Geometric Series

:: Then	:: Now	:: Why?
● You found sums of finite geometric series.	● **1** Find sums of infinite geometric series. **2** Write repeating decimals as fractions.	● With their opponent on the 10-yard line, the defense is penalized half the distance to the goal, placing the ball on the 5-yard line. If they continue to be penalized in this way, where will the ball eventually be placed? Will they ever reach the goal line? How many total penalty yards will the defense have incurred? These questions can be answered by looking at infinite geometric series.

NewVocabulary
infinite geometric series
convergent series
divergent series
infinity

Common Core State Standards

Mathematical Practices
6 Attend to precision.
8 Look for and express regularity in repeated reasoning.

1 Infinite Geometric Series An **infinite geometric series** has an infinite number of terms. A series that has a sum is a **convergent series**, because its sum converges to a specific value. A series that does not have a sum is a **divergent series**.

When you evaluated the sum S_n of an infinite geometric series for the first n terms, you were finding the partial sum of the series. It is also possible to find the sum of an entire series. In the application above, it seems that the ball will eventually reach the goal line, and the defense will be penalized a total of 10 yards. This value is the actual sum of the infinite series $5 + 2.5 + 1.25 + \dots$. The graph of S_n for $1 \leq n \leq 10$ is shown on the left below. As n increases, S_n approaches 10.

KeyConcept Convergent and Divergent Series

Convergent Series	Divergent Series
Words The sum approaches a finite value.	**Words** The sum does not approach a finite value.
Ratio $\lvert r \rvert < 1$	**Ratio** $\lvert r \rvert \geq 1$
Example $5 + 2.5 + 1.25 + \dots$	**Example** $\frac{1}{16} + \frac{1}{8} + \frac{1}{4} + \dots$

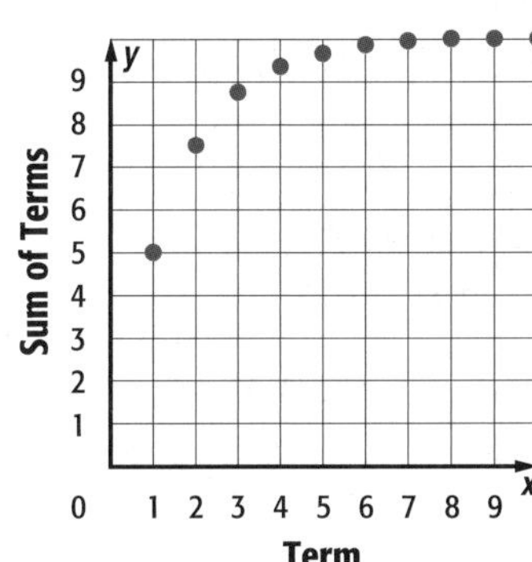

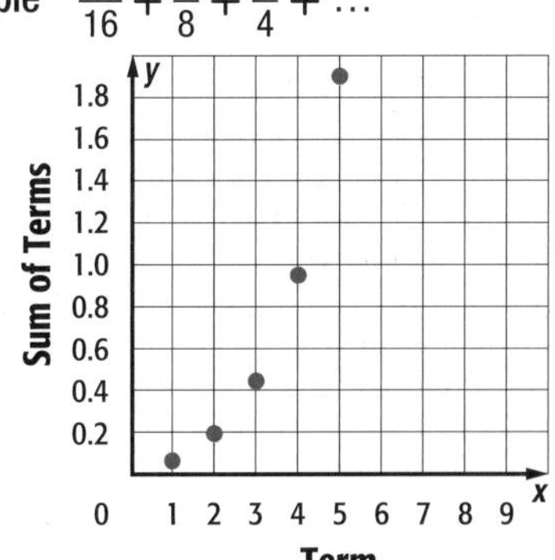

Example 1 Convergent and Divergent Series

Determine whether each infinite geometric series is *convergent* or *divergent*.

a. $54 + 36 + 24 + \dots$

Find the value of r.

$r = \frac{36}{54}$ or $\frac{2}{3}$; since $-1 < \frac{2}{3} < 1$, the series is convergent.

Darrin Klimek/Digital Vision/Getty Images

b. $8 + 12 + 18 + \ldots$

$r = \frac{12}{8}$ or 1.5; since 1.5 > 1, the series is divergent.

GuidedPractice

1A. $2 + 3 + 4.5 + \ldots$

1B. $100 + 50 + 25 + \ldots$

StudyTip

Absolute Value Recall that $|r| < 1$ means $-1 < r < 1$.

When $|r| < 1$, the value of r^n will approach 0 as n increases. Therefore, the partial sums of the infinite geometric series will approach $\frac{a_1 - a_1(0)}{1 - r}$ or $\frac{a_1}{1 - r}$.

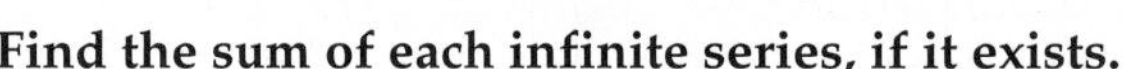

KeyConcept Sum of an Infinite Geometric Series

The sum S of an infinite geometric series with $|r| < 1$ is given by

$$S = \frac{a_1}{1 - r}.$$

If $|r| \geq 1$, the series has no sum.

When an infinite geometric series is divergent, $|r| \geq 1$ and the series has no sum because the absolute value of r^n will increase infinitely as n increases.

The table at the right shows the partial sums for the divergent series $4 + 16 + 64 + \ldots$. As n increases, S_n increases rapidly without limit.

n	S_n
5	1364
10	1,398,100
15	1,431,655,764

PT

Example 2 Sum of an Infinite Series

Find the sum of each infinite series, if it exists.

Determine whether each infinite geometric series is *convergent* or *divergent*.

a. $\frac{2}{3} + \frac{6}{15} + \frac{18}{75} + \ldots$

Step 1 Find the value of r to determine if the sum exists.

$r = \frac{6}{15} \div \frac{2}{3}$ or $\frac{3}{5}$ — Divide consecutive terms.

Since $\left|\frac{3}{5}\right| < 1$, the sum exists.

Step 2 Use the formula to find the sum.

$S = \frac{a_1}{1 - r}$ — Sum formula

$= \frac{\frac{2}{3}}{1 - \frac{3}{5}}$ — $a_1 = \frac{2}{3}$ and $r = \frac{3}{5}$

$= \frac{2}{3} \div \frac{2}{5}$ or $\frac{5}{3}$ — Simplify.

b. $6 + 9 + 13.5 + 20.25 + \ldots$

$r = \frac{9}{6}$ or 1.5; since $|1.5| \geq 1$, the series diverges and the sum does not exist.

StudyTip

Convergence and Divergence A series converges when the absolute value of a term is smaller than the absolute value of the previous term. An infinite arithmetic series will always be divergent.

GuidedPractice

2A. $4 - 2 + 1 - 0.5 + \ldots$

2B. $16 + 20 + 25 + \ldots$

Sigma notation can be used to represent infinite series. If a sequence goes to **infinity**, it continues without end. The infinity symbol ∞ is placed above the $\sum$ to indicate that a series is infinite.

PT

Example 3 Infinite Series in Sigma Notation

Find $\sum_{k=1}^{\infty} 18\left(\frac{4}{5}\right)^{k-1}$.

$$S = \frac{a_1}{1-r} \quad \text{Sum formula}$$

$$= \frac{18}{1-\frac{4}{5}} \quad a_1 = 18 \text{ and } r = \frac{4}{5}$$

$$= \frac{18}{\frac{1}{5}} \text{ or } 90 \quad \text{Simplify.}$$

GuidedPractice

3. Find $\sum_{k=1}^{\infty} 12\left(\frac{3}{4}\right)^{k-1}$.

2 Repeating Decimals

Repeating Decimals A repeating decimal is the sum of an infinite geometric series. For instance, $0.\overline{45} = 0.454545\ldots$ or $0.45 + 0.0045 + 0.000045 + \ldots$. The formula for the sum of an infinite series can be used to convert the decimal to a fraction.

Problem-SolvingTip

CCSS Sense-Making In many cases, it is possible to solve a problem in more than one way. Use the method with which you are most comfortable.

PT

Example 4 Write a Repeating Decimal as a Fraction

Write $0.\overline{63}$ as a fraction.

Method 1 Use the sum of an infinite series.

$$0.\overline{63} = 0.63 + 0.0063 + \ldots$$

$$= \frac{63}{100} + \frac{63}{10{,}000} + \ldots$$

$$S = \frac{a_1}{1-r} \quad \text{Sum formula}$$

$$= \frac{\frac{63}{100}}{1-\frac{1}{100}} \quad a_1 = \frac{63}{100} \text{ and } r = \frac{1}{100}$$

$$= \frac{63}{99} \text{ or } \frac{7}{11} \quad \text{Simplify.}$$

Method 2 Use algebraic properties.

$x = 0.\overline{63}$	Let $x = 0.\overline{63}$.
$x = 0.636363\ldots$	Write as a repeating decimal.
$100x = 63.636363\ldots$	Multiply each side by 100.
$99x = 63$	Subtract x from $100x$ and $0.\overline{63}$ from $63.\overline{63}$.
$x = \frac{63}{99}$ or $\frac{7}{11}$	Divide each side by 99.

StudyTip

Repeating Decimals Every repeating decimal is a rational number and can be written as a fraction.

GuidedPractice

4. Write $0.\overline{21}$ as a fraction.

Check Your Understanding

◯ = Step-by-Step Solutions begin on page R14.

Example 1 **Determine whether each infinite geometric series is *convergent* or *divergent*.**

1. $16 - 8 + 4 - \ldots$

2. $32 - 48 + 72 - \ldots$

3. $0.5 + 0.7 + 0.98 + \ldots$

4. $1 + 1 + 1 + \ldots$

Example 2 **Find the sum of each infinite series, if it exists.**

5. $440 + 220 + 110 + \ldots$

6. $520 + 130 + 32.5 + \ldots$

7. $\frac{1}{4} + \frac{3}{8} + \frac{9}{16} + \ldots$

8. $\frac{32}{9} + \frac{16}{3} + 8 + \ldots$

9. CCSS **SENSE-MAKING** A certain drug has a half-life of 8 hours after it is administered to a patient. What percent of the drug is still in the patient's system after 24 hours?

Example 3 **Find the sum of each infinite series, if it exists.**

10. $\sum_{k=1}^{\infty} 5 \cdot 4^{k-1}$

11. $\sum_{k=1}^{\infty} (-2) \cdot (0.5)^{k-1}$

12. $\sum_{k=1}^{\infty} 3 \cdot \left(\frac{4}{5}\right)^{k-1}$

13. $\sum_{k=1}^{\infty} \frac{1}{2} \cdot \left(\frac{3}{4}\right)^{k-1}$

Example 4 **Write each repeating decimal as a fraction.**

14. $0.\overline{35}$

15. $0.\overline{642}$

Practice and Problem Solving

Extra Practice is on page R10.

Example 1 **Determine whether each infinite geometric series is *convergent* or *divergent*.**

16. $21 + 63 + 189 + \ldots$

17. $480 + 360 + 270 + \ldots$

18. $\frac{3}{4} + \frac{9}{8} + \frac{27}{16} + \ldots$

19. $\frac{5}{6} + \frac{10}{9} + \frac{40}{27} + \ldots$

20. $0.1 + 0.01 + 0.001 + \ldots$

21. $0.008 + 0.08 + 0.8 + \ldots$

Example 2 **Find the sum of each infinite series, if it exists.**

22. $18 + 21.6 + 25.92 + \ldots$

23. $-3 - 4.2 - 5.88 - \ldots$

24. $\frac{1}{2} + \frac{1}{6} + \frac{1}{18} + \ldots$

(25) $\frac{12}{5} + \frac{6}{5} + \frac{3}{5} + \ldots$

26. $21 + 14 + \frac{28}{3} + \ldots$

27. $32 + 40 + 50 + \ldots$

28. **SWINGS** If Kerry does not push any harder after his initial swing, the distance traveled per swing will decrease by 10% with each swing. If his initial swing traveled 6 feet, find the total distance traveled when he comes to rest.

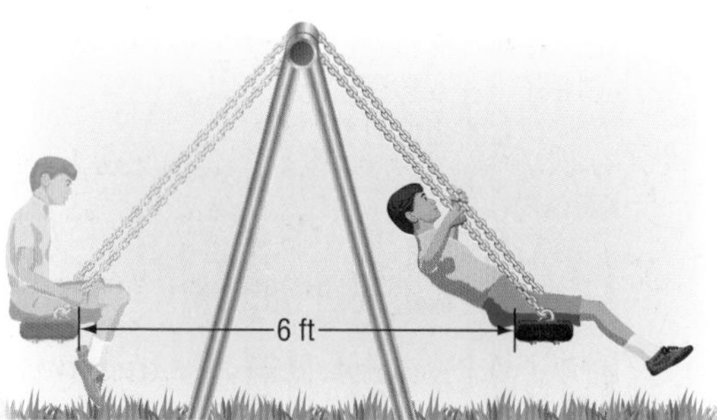

Example 3 **Find the sum of each infinite series, if it exists.**

29. $\sum_{k=1}^{\infty} \frac{4}{3} \cdot \left(\frac{5}{4}\right)^{k-1}$

30. $\sum_{k=1}^{\infty} \frac{1}{4} \cdot 3^{k-1}$

31. $\sum_{k=1}^{\infty} \frac{5}{3} \cdot \left(\frac{3}{7}\right)^{k-1}$

32. $\sum_{k=1}^{\infty} \frac{2}{3} \cdot \left(\frac{4}{3}\right)^{k-1}$

33. $\sum_{k=1}^{\infty} \frac{8}{3} \cdot \left(\frac{5}{6}\right)^{k-1}$

34. $\sum_{k=1}^{\infty} \frac{1}{8} \cdot \left(\frac{1}{12}\right)^{k-1}$

Example 4 **Write each repeating decimal as a fraction.**

35. $0.3\overline{21}$

36. $0.1\overline{45}$

37. $2.\overline{18}$

38. $4.\overline{96}$

39. $0.12\overline{14}$

40. $0.43\overline{36}$

41. **FANS** A fan is running at 10 revolutions per second. After it is turned off, its speed decreases at a rate of 75% per second. Determine the number of revolutions completed by the fan after it is turned off.

42. **CCSS PRECISION** Kamiko deposited $5000 into an account at the beginning of the year. The account earns 8% interest each year.

a. How much money will be in the account after 20 years? (*Hint*: Let $5000(1 + 0.08)^1$ represent the end of the first year.)

b. Is this series *convergent* or *divergent*? Explain.

43. **RECHARGEABLE BATTERIES** A certain rechargeable battery is advertised to recharge back to 99.9% of its previous capacity with every charge. If its initial capacity is 8 hours of life, how many total hours should the battery last?

Find the sum of each infinite series, if it exists.

44. $\frac{7}{5} + \frac{21}{20} + \frac{63}{80} + \ldots$

45. $\frac{15}{4} + \frac{5}{2} + \frac{5}{3} + \ldots$

46. $-\frac{16}{9} + \frac{4}{3} - 1 + \ldots$

47. $\frac{15}{8} + \frac{5}{2} + \frac{10}{3} + \ldots$

48. $\frac{21}{16} + \frac{7}{4} + \frac{7}{3} + \ldots$

49. $-\frac{18}{7} + \frac{12}{7} - \frac{8}{7} + \ldots$

50. **MULTIPLE REPRESENTATIONS** In this problem, you will use a square of paper that is at least 8 inches on a side.

a. **Concrete** Let the square be one unit. Cut away one half of the square. Call this piece Term 1. Next, cut away one half of the remaining sheet of paper. Call this piece Term 2. Continue cutting the remaining paper in half and labeling the pieces with a term number as long as possible. List the fractions represented by the pieces.

b. **Numerical** If you could cut the squares indefinitely, you would have an infinite series. Find the sum of the series.

c. **Verbal** How does the sum of the series relate to the original square of paper?

51. **PHYSICS** In a physics experiment, a steel ball on a flat track is accelerated, and then allowed to roll freely. After the first minute, the ball has rolled 120 feet. Each minute the ball travels only 40% as far as it did during the preceding minute. How far does the ball travel?

52. **PENDULUMS** A pendulum travels 12 centimeters on its first swing and 95% of that distance on each swing thereafter. Find the total distance traveled by the pendulum when it comes to rest.

53. **TOYS** If a rubber ball can bounce back to 95% of its original height, what is the total vertical distance that it will travel if it is dropped from an elevation of 30 feet?

54. **CARS** During a maintenance inspection, a tire is removed from a car and spun on a diagnostic machine. When the machine is turned off, the spinning tire completes 20 revolutions the first second and 98% of the revolutions each additional second. How many revolutions does the tire complete before it stops spinning?

55 **ECONOMICS** A state government decides to stimulate its economy by giving \$500 to every adult. The government assumes that everyone who receives the money will spend 80% on consumer goods and that the producers of these goods will in turn spend 80% on consumer goods. How much money is generated for the economy for every \$500 that the government provides?

56. **SCIENCE MUSEUM** An exhibit at a science museum offers visitors the opportunity to experiment with the motion of an object on a spring. One visitor pulls the object down and lets it go. The object travels 1.2 feet upward before heading back the other way. Each time the object changes direction, it decreases its distance by 20% when compared to the previous direction. Find the total distance traveled by the object.

Match each graph with its corresponding description.

57.

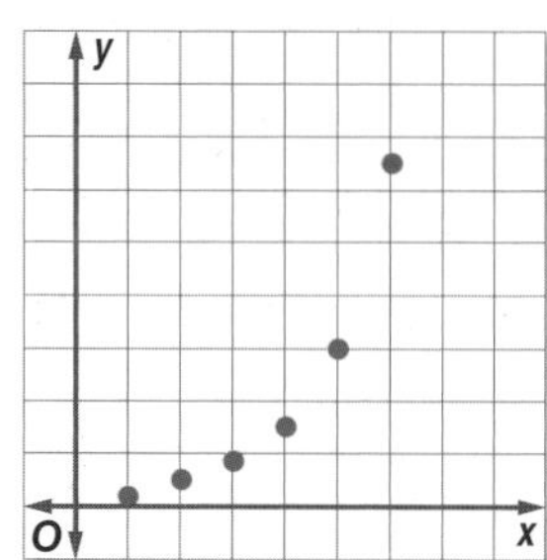

58.

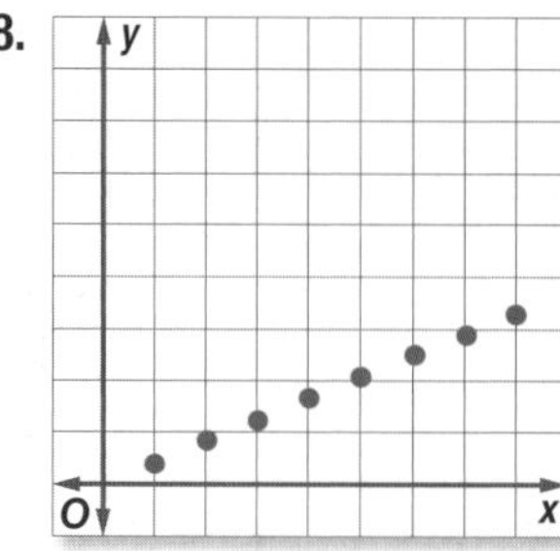

59.

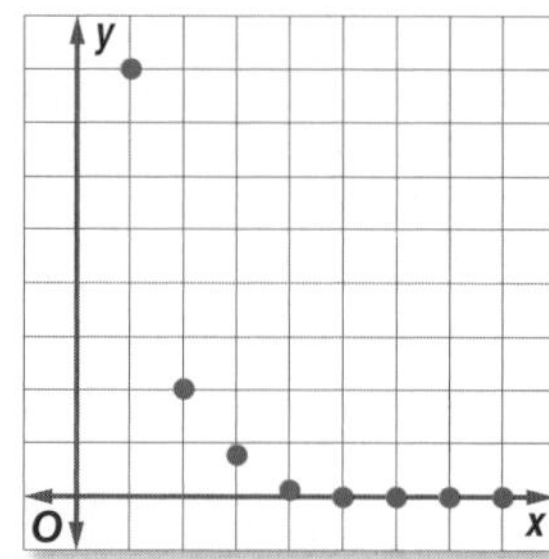

a. converging geometric series

b. diverging geometric series

c. converging arithmetic series

d. diverging arithmetic series

H.O.T. Problems Use Higher-Order Thinking Skills

60. **ERROR ANALYSIS** Emmitt and Austin are asked to find the sum of $1 - 1 + 1 - \ldots$. Is either of them correct? Explain your reasoning.

Emmitt	Austin
The sum is 0 because the sum of each pair of terms in the sequence is 0.	There is no sum because $\lvert r \rvert \geq 1$, and the series diverges.

61. **PROOF** Derive the formula for the sum of an infinite geometric series.

62. **CHALLENGE** For what values of b does $3 + 9b + 27b^2 + 81b^3 + \ldots$ have a sum?

63. **REASONING** When does an infinite geometric series have a sum, and when does it not have a sum? Explain your reasoning.

64. CCSS **ARGUMENTS** Determine whether the following statement is *sometimes*, *always*, or *never* true. Explain your reasoning.

If the absolute value of a term of any geometric series is greater than the absolute value of the previous term, then the series is divergent.

65. **OPEN ENDED** Write an infinite series with a sum that converges to 9.

66. **OPEN ENDED** Write $3 - 6 + 12 - \ldots$ using sigma notation in two different ways.

67. **WRITING IN MATH** Explain why an arithmetic series is always divergent.

Standardized Test Practice

68. SAT/ACT What is the sum of an infinite geometric series with a first term of 27 and a common ratio of $\frac{2}{3}$?

A 18
B 34
C 41
D 65
E 81

69. Adelina, Michelle, Masao, and Brandon each simplified the same expression at the board. Each student's work is shown below. The teacher said that while two of them had a correct answer, only one of them had arrived at the correct conclusion using correct steps.

Adelina's work

$$x^2x^{-5} = \frac{x^2}{x^{-5}}$$
$$= x^7, x \neq 0$$

Masao's work

$$x^2x^{-5} = \frac{x^2}{x^5}$$
$$= \frac{1}{x^3}, x \neq 0$$

Michelle's work

$$x^2x^{-5} = \frac{x^2}{x^{-5}}$$
$$= x^{-3}, x \neq 0$$

Brandon's work

$$x^2x^{-5} = \frac{x^2}{x^5}$$
$$= x^3, x \neq 0$$

Which is a completely accurate simplification?

F Adelina's work
G Michelle's work
H Masao's work
J Brandon's work

70. GRIDDED RESPONSE Evaluate $\log_8 60$ to the nearest hundredth.

71. GEOMETRY The radius of a large sphere was multiplied by a factor of $\frac{1}{3}$ to produce a smaller sphere.

How does the volume of the smaller sphere compare to the volume of the larger sphere?

A The volume of the smaller sphere is $\frac{1}{9}$ as large.
B The volume of the smaller sphere is $\frac{1}{\pi^3}$ as large.
C The volume of the smaller sphere is $\frac{1}{27}$ as large.
D The volume of the smaller sphere is $\frac{1}{3}$ as large.

Spiral Review

72. GAMES An audition is held for a TV game show. At the end of each round, one half of the prospective contestants are eliminated from the competition. On a particular day, 524 contestants begin the audition. (Lesson 10-3)

a. Write an equation for finding the number of contestants who are left after n rounds.

b. Using this method, will the number of contestants who are to be eliminated always be a whole number? Explain.

73. CLUBS A quilting club consists of 9 members. Every week, each member must bring one completed quilt square. (Lesson 10-2)

a. Find the first eight terms of the sequence that describes the total number of squares that have been made after each meeting.

b. One particular quilt measures 72 inches by 84 inches and is being designed with 4-inch squares. After how many meetings will the quilt be complete?

Skills Review

Find each function value.

74. $f(x) = 5x - 9, f(6)$

75. $g(x) = x^2 - x, g(4)$

76. $h(x) = x^2 - 2x - 1, h(3)$

EXTEND 10-4

Graphing Technology Lab
Limits

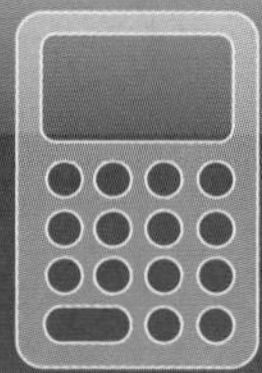

You may have noticed that in some geometric sequences, the later the term in the sequence, the closer the value is to 0. Another way to describe this is that as n increases, a_n approaches 0. The value that the terms of a sequence approach, in this case 0, is called the **limit** of the sequence. Other types of infinite sequences may also have limits. If the terms of a sequence do not approach a unique value, we say that the limit of the sequence does not exist.

You can use a TI-83/84 Plus graphing calculator to help find the limits of infinite sequences.

PT

Activity

Find the limit of the geometric sequence $1, \frac{1}{4}, \frac{1}{16}, \ldots$.

Step 1 Enter the sequence.

The formula for this sequence is $a_n = \left(\frac{1}{4}\right)^{n-1}$.

- Position the cursor on **L1** in the **STAT EDIT 1: Edit...** screen and enter the formula **seq(N,N,1,10,1)**. This generates the values 1, 2, ..., 10 of the index **N**.

  KEYSTROKES: STAT ENTER ▲ 2nd [STAT] ▶ 5 X,T,θ,n , X,T,θ,n , 1 , 10 , 1) ENTER

- Position the cursor on **L2** and enter the formula **seq((1/4)^(N-1),N,1,10,1)**. This generates the first ten terms of the sequence.

  KEYSTROKES: ▶ ▲ 2nd [STAT] ▶ 5 (1 ÷ 4) ^ (X,T,θ,n − 1) , X,T,θ,n , 1 , 10 , 1) ENTER

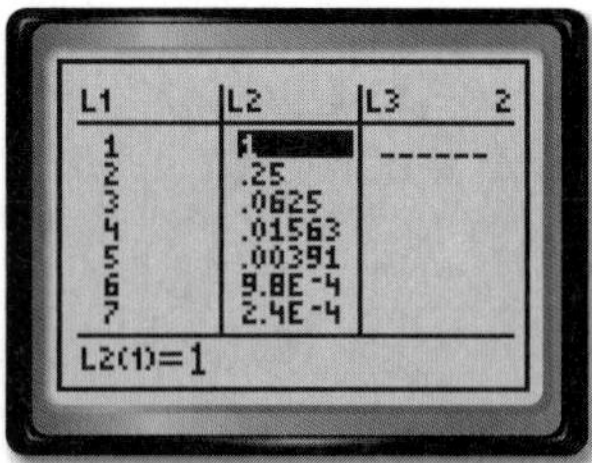

Notice that as n increases, the terms of the given sequence get closer and closer to 0. If you scroll down, you can see that for $n \geq 6$ the terms are so close to 0 that the calculator expresses them in scientific notation. This suggests that the limit of the sequence is 0.

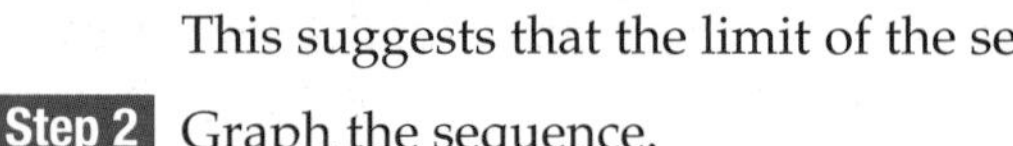

Step 2 Graph the sequence.

Use **STAT PLOT** to graph the sequence. Use **L1** as the **Xlist** and **L2** as the **Ylist**.

The graph also shows that, as n increases, the terms approach 0. In fact, for $n \geq 3$, the marks appear to lie on the horizontal axis. This strongly suggests that the limit of the sequence is 0.

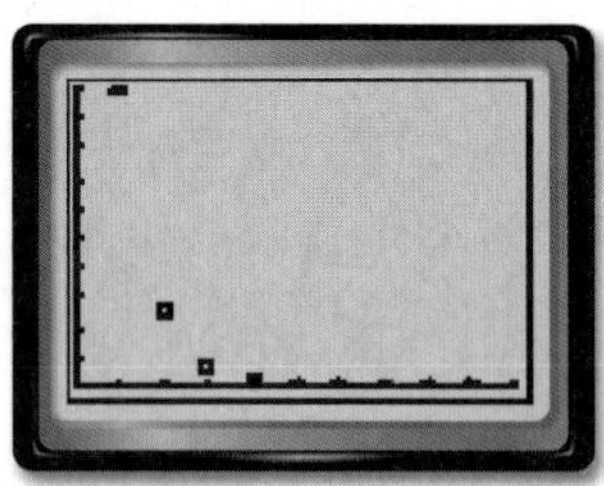

[0, 10] scl: 1 by [0, 1] scl: 0.1

Exercises

Find the limit of each sequence.

1. $a_n = \left(\frac{1}{3}\right)^n$
2. $a_n = \left(-\frac{1}{3}\right)^n$
3. $a_n = 5^n$
4. $a_n = \frac{1}{n^2}$
5. $a_n = \frac{3^n}{3^n + 1}$
6. $a_n = \frac{n^2}{n + 2}$

CHAPTER 10

Mid-Chapter Quiz

Lessons 10-1 through 10-4

Determine whether each sequence is *arithmetic, geometric,* or neither. Explain your reasoning. (Lesson 10-1)

1. 5, −3, −12, −22, −33...

2. $\frac{1}{5}, \frac{7}{10}, \frac{6}{5}, \frac{17}{10}, \frac{11}{5}, \ldots$

3. **HOUSING** Laura is a real estate agent. She needs to sell 15 houses in 6 months. (Lesson 10-1)

   a. By the end of the first 2 months she has sold 4 houses. If she sells 2 houses each month for the rest of the 6 months, will she meet her goal? Explain.

   b. If she has sold 5 houses by the end of the first month, how many will she have to sell on average each month in order to meet her goal?

4. **GEOMETRY** The figures below show a pattern of filled squares and white squares. (Lesson 10-1)

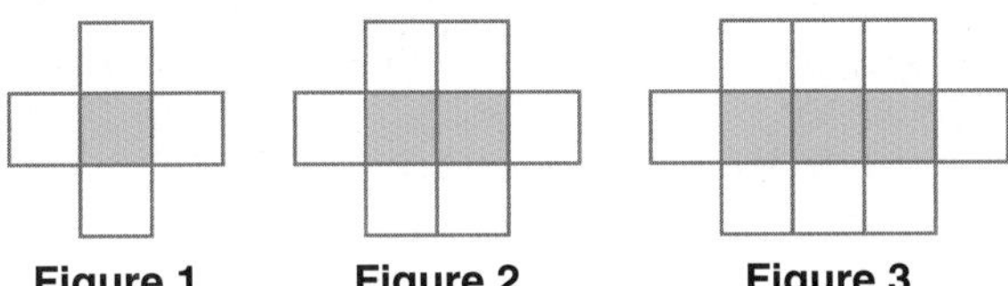

   a. Write an equation representing the *n*th number in this pattern where *n* is the number of white squares.

   b. Is it possible to have exactly 84 white squares in an arrangement? Explain.

Find the indicated term of each arithmetic sequence. (Lesson 10-2)

5. $a_1 = 10, d = -5, n = 9$

6. $a_1 = -8, d = 4, n = 99$

Find the sum of each arithmetic series. (Lesson 10-2)

7. $-15 + (-11) + (-7) + \cdots + 53$

8. $a_1 = -12, d = 8, n = 22$

9. $\sum_{k=11}^{50} (-3k + 1)$

10. **MULTIPLE CHOICE** What is the sum of the first 50 odd numbers? (Lesson 10-2)

   A 2550

   B 2500

   C 2499

   D 2401

Find the indicated term for each geometric sequence. (Lesson 10-3)

11. $a_2 = 8, r = 2, a_8 = ?$

12. $a_3 = 0.5, r = 8, a_{10} = ?$

13. **MULTIPLE CHOICE** What are the geometric means of the sequence below? (Lesson 10-3)

   0.5, ______, ______, ______, 2048

   F 512.375, 1024.25, 1536.125

   G 683, 1365.5, 2048

   H 2, 8, 32

   J 4, 32, 256

14. **INCOME** Peter works for a house building company for 4 months per year. He starts out making $3000 per month. At the end of each month, his salary increases by 5%. How much money will he make in those 4 months? (Lesson 10-3)

Evaluate the sum of each geometric series. (Lesson 10-3)

15. $\sum_{k=1}^{8} 3 \cdot 2^{k-1}$

16. $\sum_{k=1}^{9} 4 \cdot (-1)^{k-1}$

17. $\sum_{k=1}^{20} -2\left(\frac{2}{3}\right)^{k-1}$

Find the sum of each infinite series, if it exists. (Lesson 10-4)

18. $\sum_{n=1}^{\infty} 9 \cdot 2^{n-1}$

19. $\sum_{n=1}^{\infty} (4) \cdot (0.5)^{n-1}$

20. $\sum_{n=1}^{\infty} 12 \cdot \left(\frac{2}{3}\right)^{n-1}$

LESSON 10-5 Recursion and Iteration

Then

- You explored compositions of functions.

Now

1. Recognize and use special sequences.
2. Iterate functions.

Why?

- The female honeybee is produced after the queen mates with a male, so the female has two parents, a male and a female. The male honeybee, however, is produced by the queen's unfertilized eggs and thus has only one parent, a female. The family tree for the honeybee follows a special sequence.

Generation	1	2	3	4	5	6
Ancestors	1	1	2	3	5	8

NewVocabulary
Fibonacci sequence
recursive sequence
explicit formula
recursive formula
iteration

Common Core State Standards

Mathematical Practices
6 Attend to precision.
8 Look for and express regularity in repeated reasoning.

1 Special Sequences

Notice that every term in the list of ancestors is the sum of the previous two terms. This special sequence is called the **Fibonacci sequence**, and it is found in many places in nature. The Fibonacci sequence is an example of a **recursive sequence**. In a recursive sequence, each term is determined by one or more of the previous terms.

The formulas you have used for sequences thus far have been explicit formulas. An **explicit formula** gives a_n as a function of n, such as $a_n = 3n + 1$. The formula that describes the Fibonacci sequence, $a_n = a_{n-2} + a_{n-1}$, is a **recursive formula**, which means that every term will be determined by one or more of the previous terms. An initial term must be given in a recursive formula.

KeyConcept Recursive Formulas for Sequences

Arithmetic Sequence	$a_n = a_{n-1} + d$, where d is the common difference
Geometric Sequence	$a_n = r \cdot a_{n-1}$, where r is the common ratio

Example 1 Use a Recursive Formula

Find the first five terms of the sequence in which $a_1 = -3$ and $a_{n+1} = 4a_n - 2$, if $n \geq 1$.

$a_{n+1} = 4a_n - 2$	Recursive formula
$a_{1+1} = 4a_1 - 2$	$n = 1$
$a_2 = 4(-3) - 2$ or -14	$a_1 = -3$
$a_3 = 4(-14) - 2$ or -58	$a_2 = -14$
$a_4 = 4(-58) - 2$ or -234	$a_3 = -58$
$a_5 = 4(-234) - 2$ or -938	$a_4 = -234$

The first five terms of the sequence are −3, −14, −58, −234, and −938.

GuidedPractice

1. Find the first five terms of the sequence in which $a_1 = 8$ and $a_{n+1} = -3a_n + 6$, if $n \geq 1$.

In order to find a recursive formula, first determine the initial term. Then evaluate the pattern to generate the later terms. The recursive formula that generates a sequence does not include the value of the initial term.

StudyTip

Sequences and Recursive Formulas Like arithmetic and geometric sequences, recursive formulas define functions in which the domain is the set of positive integers, represented by the term number *n*.

PT

Example 2 Write Recursive Formulas

Write a recursive formula for each sequence.

a. 2, 10, 18, 26, 34, ...

Step 1 Determine whether the sequence is arithmetic or geometric.
The sequence is arithmetic because each term after the first can be found by adding a common difference.

Step 2 Find the common difference.

$d = 10 - 2$ or 8

Step 3 Write the recursive formula.

$a_n = a_{n-1} + d$ Recursive formula for arithmetic sequence

$a_n = a_{n-1} + 8$ $d = 8$

A recursive formula for the sequence is $a_n = a_{n-1} + 8, a_1 = 2$.

b. 16, 56, 196, 686, 2401, ...

Step 1 Determine whether the sequence is arithmetic or geometric.
The sequence is geometric because each term after the first can be found after multiplying by a common ratio.

Step 2 Find the common ratio.

$r = \frac{56}{16}$ or 3.5

Step 3 Write the recursive formula.

$a_n = r \cdot a_{n-1}$ Recursive formula for geometric sequence

$a_n = 3.5a_{n-1}$ $r = 3.5$

A recursive formula for the sequence is $a_n = 3.5a_{n-1}, a_1 = 16$.

c. $a_4 = 108$ and $r = 3$

Step 1 Determine whether the sequence is arithmetic or geometric.
Because r is given, the sequence is geometric.

Step 2 Write the recursive formula.

$a_n = r \cdot a_{n-1}$ Recursive formula for geometric sequence

$a_n = 3a_{n-1}$ $r = 3$

A recursive formula for the sequence is $a_n = 3a_{n-1}, a_1 = 4$.

GuidedPractice

Write a recursive formula for each sequence.

2A. 8, 20, 50, 125, 312.5, ... **2B.** 8, 17, 26, 35, 44, ... **2C.** $a_3 = 16$ and $r = 4$

Real-World Example 3 Use a Recursive Formula

Real-WorldLink

In 2008, the average credit card debt for college students was about $3173.

Source: *USA Today*

FINANCIAL LITERACY Nate had $15,000 in credit card debt when he graduated from college. The balance increased by 2% each month due to interest, and Nate could only make payments of $400 per month. Write a recursive formula for the balance of his account each month. Then determine the balance after five months.

Step 1 Write the recursive formula.

Let a_n represent the balance of the account in the nth month. The initial balance a_1 is $15,000. After one month, interest is added and a payment is made.

	initial balance	+	balance times 2%	−	monthly payment
$a_2 =$	a_1	+	$(a_1 \times 0.02)$	−	400

$a_2 = 1.02a_1 - 400$

The formula is $a_n = 1.02a_{n-1} - 400$.

Step 2 Find the next five terms.

$a_n = 1.02a_{n-1} - 400$	Recursive formula
$a_2 = (15{,}000 \times 1.02) - 400$ or 14,900	$a_1 = 15{,}000$
$a_3 = (14{,}900 \times 1.02) - 400$ or 14,798	$a_2 = 14{,}900$
$a_4 = (14{,}798 \times 1.02) - 400$ or 14,693.96	$a_3 = 14{,}798$
$a_5 = (14{,}693.96 \times 1.02) - 400$ or 14,587.84	$a_4 = 14{,}693.96$
$a_6 = (14{,}587.84 \times 1.02) - 400$ or 14,479.60	$a_5 = 14{,}587.84$

After the fifth month, the balance will be $14,479.60.

GuidedPractice

3. Write a recursive formula for a $10,000 debt, at 2.5% interest per month, with a $600 monthly payment. Then find the first five balances.

ReviewVocabulary

Composition of functions
A function is performed, and then a second function is performed on the result of the first function.

2 Iteration

Iteration is the process of repeatedly composing a function with itself. Consider x_0. The first iterate is $f(x_0)$, the second iterate is $f(f(x_0))$, the third iterate is $f(f(f(x_0)))$, and so on.

Iteration can be used to recursively generate a sequence. Start with the initial value x_0. Let $x_1 = f(x_0)$, $x_2 = f(f(x_0))$, and so on.

Example 4 Iterate a Function

Find the first three iterates x_1, x_2, and x_3 of $f(x) = 5x + 4$ for an initial value of $x_0 = 2$.

$x_1 = f(x_0)$	Iterate the function.
$= 5(2) + 4$ or 14	$x_0 = 2$
$x_2 = f(x_1)$	Iterate the function.
$= 5(14) + 4$ or 74	$x_1 = 14$
$x_3 = f(x_2)$	Iterate the function.
$= 5(74) + 4$ or 374	$x_2 = 74$

The first three iterates are 14, 74, and 374.

GuidedPractice

4. Find the first three iterates x_1, x_2, and x_3 of $f(x) = -3x + 8$ for an initial value of $x_0 = 6$.

Corbis Premium RF/Alamy

Check Your Understanding

= Step-by-Step Solutions begin on page R14.

Example 1 **Find the first five terms of each sequence described.**

1. $a_1 = 16, a_{n+1} = a_n + 4$
2. $a_1 = -3, a_{n+1} = a_n + 8$
3. $a_1 = 5, a_{n+1} = 3a_n + 2$
4. $a_1 = -4, a_{n+1} = 2a_n - 6$

Example 2 **Write a recursive formula for each sequence.**

5. 3, 8, 18, 38, 78, ...
6. 5, 14, 41, 122, 365, ...

Example 3

7. **FINANCING** Ben financed a $1500 rowing machine to help him train for the college rowing team. He could only make a $100 payment each month, and his bill increased by 1% due to interest at the end of each month.
   a. Write a recursive formula for the balance owed at the end of each month.
   b. Find the balance owed after the first four months.
   c. How much interest has accumulated after the first six months?

Example 4 **Find the first three iterates of each function for the given initial value.**

8. $f(x) = 5x + 2, x_0 = 8$
9. $f(x) = -4x + 2, x_0 = 5$
10. $f(x) = 6x + 3, x_0 = -4$
11. $f(x) = 8x - 4, x_0 = -6$

Practice and Problem Solving

Extra Practice is on page R10.

Example 1 CCSS **PERSEVERANCE** **Find the first five terms of each sequence described.**

12. $a_1 = 10, a_{n+1} = 4a_n + 1$
13. $a_1 = -9, a_{n+1} = 2a_n + 8$
14. $a_1 = 12, a_{n+1} = a_n + n$
15. $a_1 = -4, a_{n+1} = 2a_n + n$
16. $a_1 = 6, a_{n+1} = 3a_n - n$
17. $a_1 = -2, a_{n+1} = 5a_n + 2n$
18. $a_1 = 7, a_2 = 10, a_{n+2} = 2a_n + a_{n+1}$
19. $a_1 = 4, a_2 = 5, a_{n+2} = 4a_n - 2a_{n+1}$
20. $a_1 = 4, a_2 = 3x, a_n = a_{n-1} + 4a_{n-2}$
21. $a_1 = 3, a_2 = 2x, a_n = 4a_{n-1} - 3a_{n-2}$
22. $a_1 = 2, a_2 = x + 3, a_n = a_{n-1} + 6a_{n-2}$
23. $a_1 = 1, a_2 = x, a_n = 3a_{n-1} + 6a_{n-2}$

Example 2 **Write a recursive formula for each sequence.**

24. 16, 10, 7, 5.5, 4.75, ...
25. 32, 12, 7, 5.75, ...
26. 4, 15, 224, 50,175, ...
27. 1, 2, 9, 730, ...
28. 9, 33, 129, 513, ...
29. 480, 128, 40, 18, ...
30. 393, 132, 45, 16, ...
31. 68, 104, 176, 320, ...

Example 3

32. **FINANCIAL LITERACY** Mr. Edwards and his company deposit $20,000 into his retirement account at the end of each year. The account earns 8% interest before each deposit.
   a. Write a recursive formula for the balance in the account at the end of each year.
   b. Determine how much is in the account at the end of each of the first 8 years.

Example 4 **Find the first three iterates of each function for the given initial value.**

33. $f(x) = 12x + 8, x_0 = 4$
34. $f(x) = -9x + 1, x_0 = -6$
35. $f(x) = -6x + 3, x_0 = 8$
36. $f(x) = 8x + 3, x_0 = -4$
37. $f(x) = -3x^2 + 9, x_0 = 2$
38. $f(x) = 4x^2 + 5, x_0 = -2$
39. $f(x) = 2x^2 - 5x + 1, x_0 = 6$
40. $f(x) = -0.25x^2 + x + 6, x_0 = 8$
41. $f(x) = x^2 + 2x + 3, x_0 = \frac{1}{2}$
42. $f(x) = 2x^2 + x + 1, x_0 = -\frac{1}{2}$

43 **FRACTALS** Consider the figures at the right. The number of blue triangles increases according to a specific pattern.

a. Write a recursive formula for the number of blue triangles in the sequence of figures.

b. How many blue triangles will be in the sixth figure?

44. **FINANCIAL LITERACY** Miguel's monthly car payment is $234.85. The recursive formula $b_n = 1.005b_{n-1} - 234.85$ describes the balance left on the loan after n payments. Find the balance of the $10,000 loan after each of the first eight payments.

45. **CONSERVATION** Suppose a lake is populated with 10,000 fish. A year later, 80% of the fish have died or been caught, and the lake is replenished with 10,000 new fish. If the pattern continues, will the lake eventually run out of fish? If not, will the population of the lake converge to any particular value? Explain.

46. **GEOMETRY** Consider the pattern at the right.

a. Write a sequence of the total number of triangles in the first six figures.

b. Write a recursive formula for the number of triangles.

c. How many triangles will be in the tenth figure?

47. **SPREADSHEETS** Consider the sequence with $x_0 = 20{,}000$ and $f(x) = 0.3x + 5000$.

a. Enter x_0 in cell A1 of your spreadsheet. Enter "= (0.3)*(A1) + 5000" in cell A2. What answer does it provide?

b. Copy cell A2, highlight cells A3 through A70, and paste. What do you notice about the sequence?

c. How do spreadsheets help analyze recursive sequences?

48. **VIDEO GAMES** The final monster in Helena's video game has 100 health points. During the final battle, the monster regains 10% of its health points after every 10 seconds. If Helena can inflict damage to the monster that takes away 10 health points every 10 seconds without getting hurt herself, will she ever kill the monster? If so, when?

H.O.T. Problems Use Higher-Order Thinking Skills

49. CCSS **CRITIQUE** Marcus and Armando are finding the first three iterates of $f(x) = 5x - 3$ for an initial value of $x_0 = 4$. Is either of them correct? Explain.

Marcus	Armando
f(4) = 5(4) − 3 or 17	f(4) = 5(4) − 3 or 17
f(17) = 5(17) − 3 or 82	f(17) = 5(17) − 3 or 82
The first three iterates are 4, 17, and 82.	f(82) = 5(82) − 3 or 407
	The first three iterates are 17, 82, and 407.

50. **CHALLENGE** Find a recursive formula for 5, 23, 98, 401, … .

51. **REASONING** Is the statement *"If the first three terms of a sequence are identical, then the sequence is not recursive"* sometimes, *always*, or *never* true? Explain your reasoning.

52. **OPEN ENDED** Write a function for which the first three iterates are 9, 19, and 39.

53. **WRITING IN MATH** Why is it useful to represent a sequence with an explicit or recursive formula?

Standardized Test Practice

54. GEOMETRY In the figure shown, $a + b + c = ?$

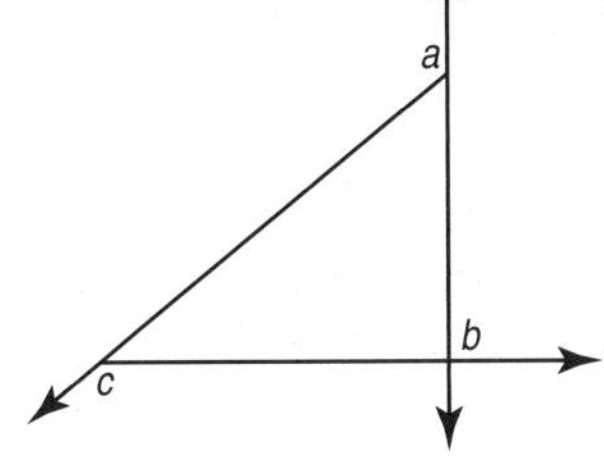

A 180°
B 270°
C 360°
D 450°

55. EXTENDED RESPONSE Bill launches a model rocket from ground level. The rocket's height h in meters is given by the equation $h = -4.9t^2 + 56t$, where t is the time in seconds after the launch.

a. What is the maximum height the rocket will reach?

b. How long after it is launched will the rocket reach its maximum height? Round to the nearest tenth of a second.

c. How long after it is launched will the rocket land? Round to the nearest tenth of a second.

56. Which of the following is true about the graphs of $y = 3(x - 4)^2 + 5$ and $y = 3(x + 4)^2 + 5$?

F Their vertices are maximums.
G The graphs have the same shape with different vertices.
H The graphs have different shapes with different vertices.
J One graph has a vertex that is a maximum, while the other graph has a vertex that is a minimum.

57. Which factors could represent the length times the width?

$A = 16x^4 - 25y^2$

A $(4x - 5y)(4x - 5y)$
B $(4x + 5y)(4x - 5y)$
C $(4x^2 - 5y)(4x^2 + 5y)$
D $(4x^2 + 5y)(4x^2 + 5y)$

Spiral Review

Write each repeating decimal as a fraction. (Lesson 10-4)

58. $0.\overline{7}$ **59.** $5.\overline{126}$ **60.** $6.\overline{259}$

61. SPORTS Adrahan is training for a marathon, about 26 miles. He begins by running 2 miles. Then, when he runs every other day, he runs one and a half times the distance he ran the time before. (Lesson 10-3)

a. Write the first five terms of a sequence describing his training schedule.

b. When will he exceed 26 miles in one run?

c. When will he have run 100 total miles?

State whether the events are *independent* or *dependent*. (Lesson 0-6)

62. tossing a penny and rolling a number cube

63. choosing first and second place in an academic competition

Skills Review

Find each product.

64. $(y + 4)(y + 3)$ **65.** $(x - 2)(x + 6)$ **66.** $(a - 8)(a + 5)$

67. $(4h + 5)(h + 7)$ **68.** $(9p - 1)(3p - 2)$ **69.** $(2g + 7)(5g - 8)$

EXTEND

10-5 Spreadsheet Lab: Amortizing Loans

When a payment is made on a loan, part of the payment is used to cover the interest that has accumulated since the last payment. The rest is used to reduce the *principal*, or original amount of the loan. This process is called *amortization.* You can use a spreadsheet to analyze the payments, interest, and balance on a loan. A table that shows this kind of information is called an *amortization schedule.*

Common Core State Standards
Mathematical Practices
5 Use appropriate tools strategically.

Example

LOANS Gloria just bought a new computer for $695. The store is letting her make monthly payments of $60.78 at an interest rate of 9% for one year. How much will she still owe after six months?

Every month, the interest on the remaining balance will be $\frac{9\%}{12}$ or 0.75%. You can find the balance after a payment by multiplying the balance after the previous payment by 1 + 0.0075 or 1.0075 and then subtracting 60.78.

In a spreadsheet, the column of numbers represents the number of payments, and Column B shows the balance. Enter the interest rate and monthly payment in cells in Column A so that they can be easily updated if the information changes.

The spreadsheet at the right shows the formulas for the balances after each of the first six payments. After six months, Gloria still owes $355.28.

Computer Loan

	A	B	C
1	Interest Rate	=695*(1+A2)–A5	
2	0.0075	=B1*(1+A2)–A5	
3		=B2*(1+A2)–A5	
4	Monthly payment	=B3*(1+A2)–A5	
5	60.78	=B4*(1+A2)–A5	
6		=B5*(1+A2)–A5	
7			

Sheet 1 / Sheet 2 / Sheet 3

Model and Analyze

1. Let b_n be the balance left on Gloria's loan after n months. Write an equation relating b_n and b_{n+1}.
2. Payments at the beginning of a loan go more toward interest than payments at the end. What percent of Gloria's loan remains to be paid after half a year?
3. Extend the spreadsheet to the whole year. What is the balance after 12 payments? Why is it not 0?
4. Suppose Gloria decides to pay $70 every month. How long would it take her to pay off the loan?
5. Suppose that, based on how much she can afford, Gloria will pay a variable amount each month in addition to the $60.78. Explain how the flexibility of a spreadsheet can be used to adapt to this situation.
6. Ethan has a three-year, $12,000 motorcycle loan. The annual interest rate is 6%, and his monthly payment is $365.06. After fifteen months, he receives an inheritance which he wants to use to pay off the loan. How much does he owe at that point?

LESSON

10-6 The Binomial Theorem

Then	Now	Why?
You worked with combinations.	**1** Use Pascal's triangle to expand powers of binomials. **2** Use the Binomial Theorem to expand powers of binomials.	A manager plans to hire 8 new employees. Not wanting to appear biased, the manager wants to hire a combination of males and females that has at least a 10% chance of occurring randomly. If there are an equal number of male and female applicants, is the probability of randomly hiring 6 men and 2 women less than 10%?

NewVocabulary
Pascal's triangle

Common Core State Standards

Content Standards
A.APR.5 Know and apply the Binomial Theorem for the expansion of $(x + y)^n$ in powers of x and y for a positive integer n, where x and y are any numbers, with coefficients determined for example by Pascal's Triangle.

Mathematical Practices
4 Model with mathematics.

1 Pascal's Triangle

In the 13th century, the Chinese discovered a pattern of numbers that would later be referred to as **Pascal's triangle**. This pattern can be used to determine the coefficients of an expanded binomial $(a + b)^n$.

$(a + b)^0$ 1
$(a + b)^1$ 1 1
$(a + b)^2$ 1 2 1
$(a + b)^3$ 1 3 3 1
$(a + b)^4$ 1 4 6 4 1
$(a + b)^5$ 1 5 10 10 5 1

For example, the expanded form of
$(a + b)^5 = 1a^5 + 5a^4b + 10a^3b^2 + 10a^2b^3 + 5ab^4 + 1b^5$.

Real-World Example 1 Use Pascal's Triangle

Find the probability of hiring 6 men and 2 women by expanding $(m + f)^8$.

Write three more rows of Pascal's triangle and use the pattern to write the expansion.

5 1 5 10 10 5 1
6 1 6 15 20 15 6 1
7 1 7 21 35 35 21 7 1
8 1 8 28 56 70 56 28 8 1

$(m + f)^8 = m^8 + 8m^7f + 28m^6f^2 + 56m^5f^3 + 70m^4f^4 + 56m^3f^5 + 28m^2f^6 + 8mf^7 + f^8$

By adding the coefficients of the polynomial, we determine that there are 256 combinations of males and females that could be hired.

$28m^6f^2$ represents the number of combinations with 6 males and 2 females. Therefore, there is a $\frac{28}{256}$ or about an 11% chance of randomly hiring 6 males and 2 females.

GuidedPractice

1. Expand $(c + d)^9$.

2 The Binomial Theorem

Instead of writing out row after row of Pascal's triangle, you can use the **Binomial Theorem** to expand a binomial. Recall that ${}_nC_r = \frac{n!}{r!(n - r)!}$.

Somos/Veer/Getty Images

StudyTip

Combinations Recall that both ${}_nC_0$ and ${}_nC_n$ equal 1.

KeyConcept Binomial Theorem

If n is a natural number, then $(a + b)^n =$

$${}_nC_0\, a^n b^0 + {}_nC_1\, a^{n-1} b^1 + {}_nC_2\, a^{n-2} b^2 + \cdots + {}_nC_n\, a^0 b^n = \sum_{k=0}^{n} \frac{n!}{k!(n-k)!} a^{n-k} b^k.$$

To use the theorem, replace n with the value of the exponent. Notice how the terms will follow the pattern of Pascal's triangle, and the coefficients will be symmetric.

Example 2 Use the Binomial Theorem

Expand $(a + b)^7$.

Method 1 Use combinations.

Replace n with 7 in the Binomial Theorem.

$$(a + b)^7 = {}_7C_0\, a^7 + {}_7C_1\, a^6 b + {}_7C_2\, a^5 b^2 + {}_7C_3\, a^4 b^3 + {}_7C_4\, a^3 b^4 + {}_7C_5\, a^2 b^5 + {}_7C_6\, ab^6 + {}_7C_7\, b^7$$

$$= a^7 + \frac{7!}{6!} a^6 b + \frac{7!}{2!5!} a^5 b^2 + \frac{7!}{3!4!} a^4 b^3 + \frac{7!}{4!3!} a^3 b^4 + \frac{7!}{5!2!} a^2 b^5 + \frac{7!}{6!} ab^6 + b^7$$

$$= a^7 + 7a^6 b + 21a^5 b^2 + 35a^4 b^3 + 35a^3 b^4 + 21a^2 b^5 + 7ab^6 + b^7$$

Method 2 Use Pascal's triangle.

Use the Binomial Theorem to determine exponents, but instead of finding the coefficients by using combinations, look at the seventh row of Pascal's triangle.

6			1		6		15		20		15		6		1			
7		1		7		21		35		35		21		7		1		
8	1		8		28		56		70		56		28		8		1	

$$(a + b)^7 = a^7 + {}_7a^6 b + {}_{21}a^5 b^2 + {}_{35}a^4 b^3 + {}_{35}a^3 b^4 + {}_{21}a^2 b^5 + {}_7ab^6 + b^7$$

GuidedPractice

2. Expand $(x + y)^{10}$.

When the binomial to be expanded has coefficients other than 1, the coefficients will no longer be symmetric. In these cases, you may want to use the Binomial Theorem.

Example 3 Coefficients Other Than 1

StudyTip

Graphing Calculator You can calculate ${}_nC_r$ by using a graphing calculator.

Press MATH and choose PRB 3.

Expand $(5a - 4b)^4$.

$(5a - 4b)^4$

$$= {}_4C_0\,(5a)^4 + {}_4C_1\,(5a)^3(-4b) + {}_4C_2\,(5a)^2(-4b)^2 + {}_4C_3\,(5a)(-4b)^3 + {}_4C_4\,(-4b)^4$$

$$= 625a^4 + \frac{4!}{3!}(125a^3)(-4b) + \frac{4!}{2!2!}(25a^2)(16b^2) + \frac{4!}{3!}(5a)(-64b^3) + 256b^4$$

$$= 625a^4 - 2000a^3 b + 2400a^2 b^2 - 1280ab^3 + 256b^4$$

GuidedPractice

3. Expand $(3x + 2y)^5$.

Sometimes you may need to find only one term in a binomial expansion. To do this, you can use the summation formula for the Binomial Theorem, $\sum_{k=0}^{n} \frac{n!}{k!(n-k)!} a^{n-k} b^k$.

Example 4 Determine a Single Term

Find the fifth term of $(y + z)^{11}$.

Step 1 Use the Binomial Theorem to write the expansion in sigma notation.

$$(y + z)^{11} = \sum_{k=0}^{11} \frac{11!}{k!(11 - k)!} y^{11 - k} z^k$$

Step 2

$$\frac{11!}{k!(11 - k)!} y^{11 - k} z^k = \frac{11!}{4!(11 - 4)!} y^{11 - 4} z^4$$ For the fifth term, $k = 4$.

$$= 330y^7z^4$$ $C(11, 4) = 330$

GuidedPractice

4. Find the sixth term of $(c + d)^{10}$.

ConceptSummary Binomial Expansion

In a binomial expansion of $(a + b)^n$,

- there are $n + 1$ terms.
- n is the exponent of a in the first term and b in the last term.
- in successive terms, the exponent of a decreases by 1, and the exponent of b increases by 1.
- the sum of the exponents in each term is n.
- the coefficients are symmetric.

Check Your Understanding

 = Step-by-Step Solutions begin on page R14.

Examples 1–3 **Expand each binomial.**

1. $(c + d)^5$ **2.** $(g + h)^7$ **3.** $(x - 4)^6$

4. $(2y - z)^5$ **5.** $(x + 3)^5$ **6.** $(y - 4z)^4$

7. **GENETICS** If a woman is equally as likely to have a baby boy or a baby girl, use binomial expansion to determine the probability that 5 of her 6 children are girls. Do not consider identical twins.

Example 4 **Find the indicated term of each expression.**

8. fourth term of $(b + c)^9$ **9.** fifth term of $(x + 3y)^8$ **10.** third term of $(a - 4b)^6$

11. sixth term of $(2c - 3d)^8$ **12.** last term of $(5x + y)^5$ **13.** first term of $(3a + 8b)^5$

14. CCSS **MODELING** The color of a particular flower is determined by the combination of two genes, also called *alleles*. If the flower has two red alleles r, the flower is red. If the flower has two white alleles w, the flower is white. If the flower has one allele of each color, the flower will be pink. In a lab, two pink flowers are mated and eventually produce 1000 offspring. How many of the 1000 offspring will be pink?

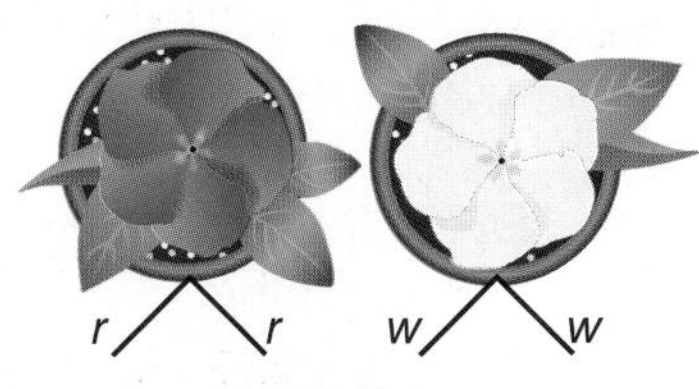

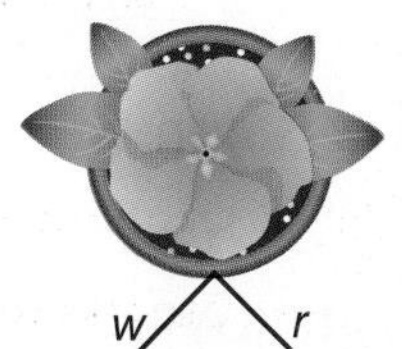

Practice and Problem Solving

Extra Practice is on page R10.

Examples 1–3 **Expand each binomial.**

15. $(a - b)^6$
16. $(c - d)^7$
17. $(x + 6)^6$
18. $(y - 5)^7$
19. $(2a + 4b)^4$
20. $(3a - 4b)^5$

21. **COMMITTEES** If an equal number of men and women applied to be on a community planning committee and the committee needs a total of 10 people, find the probability that 7 of the members will be women. Assume that committee members will be chosen randomly.

22. **BASEBALL** If a pitcher is just as likely to throw a ball as a strike, find the probability that 11 of his first 12 pitches are balls.

Example 4 **Find the indicated term of each expression.**

23. third term of $(x + 2z)^7$
24. fourth term of $(y - 3x)^6$
25. seventh term of $(2a - 2b)^8$
26. sixth term of $(4x + 5y)^6$
27. fifth term of $(x - 4)^9$
28. fourth term of $(c + 6)^8$

Expand each binomial.

29. $\left(x + \frac{1}{2}\right)^5$
30. $\left(x - \frac{1}{3}\right)^4$
31. $\left(2b + \frac{1}{4}\right)^5$
32. $\left(3c + \frac{1}{3}\right)^5$

33. **CCSS SENSE-MAKING** In $\frac{n!}{k!(n-k)!}p^k q^{n-k}$, let p represent the likelihood of a success and q represent the likelihood of a failure.

 a. If a place-kicker makes 70% of his kicks within 40 yards, find the likelihood that he makes 9 of his next 10 attempts from within 40 yards.

 b. If a quarterback completes 60% of his passes, find the likelihood that he completes 8 of his next 10 attempts.

 c. If a team converts 30% of their two-point conversions, find the likelihood that they convert 2 of their next 5 conversions.

H.O.T. Problems Use Higher-Order Thinking Skills

34. **CHALLENGE** Find the sixth term of the expansion of $(\sqrt{a} + \sqrt{b})^{12}$. Explain your reasoning.

35. **REASONING** Explain how the terms of $(x + y)^n$ and $(x - y)^n$ are the same and how they are different.

36. **REASONING** Determine whether the following statement is *true* or *false*. Explain your reasoning.

 The eighth and twelfth terms of $(x + y)^{20}$ have the same coefficients.

37. **OPEN ENDED** Write a power of a binomial for which the second term of the expansion is $6x^4y$.

38. **WRITING IN MATH** Explain how to write out the terms of Pascal's triangle.

Standardized Test Practice

39. PROBABILITY A desk drawer contains 7 sharpened red pencils, 5 sharpened yellow pencils, 3 unsharpened red pencils, and 5 unsharpened yellow pencils. If a pencil is taken from the drawer at random, what is the probability that it is yellow, given that it is one of the sharpened pencils?

A $\frac{5}{12}$

B $\frac{7}{20}$

C $\frac{5}{8}$

D $\frac{1}{5}$

40. GRIDDED RESPONSE Two people are 17.5 miles apart. They begin to walk toward each other along a straight line at the same time. One walks at the rate of 4 miles per hour, and the other walks at the rate of 3 miles per hour. In how many hours will they meet?

41. GEOMETRY Christie has a cylindrical block that she needs to paint for an art project.

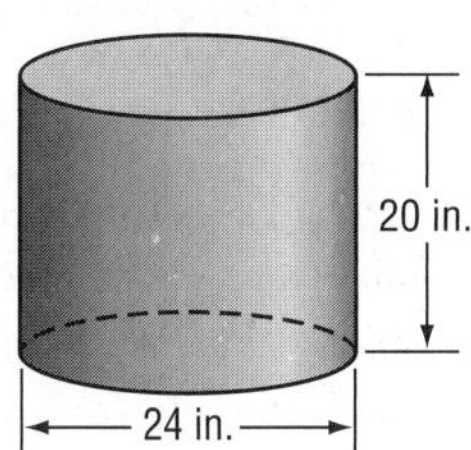

What is the surface area of the cylinder in square inches rounded to the nearest square inch?

F 1960 **H** 5127

G 2413 **J** 6635

42. Which of the following is a linear function?

A $y = \frac{x+3}{x+2}$

B $y = (3x + 2)^2$

C $y = \frac{x+3}{2}$

D $y = |3x| + 2$

Spiral Review

Find the first five terms of each sequence. (Lesson 10-5)

43. $a_1 = -2, a_{n+1} = a_n + 5$

44. $a_1 = 3, a_{n+1} = 4a_n - 10$

45. $a_1 = 4, a_{n+1} = 3a_n - 6$

Find the sum of each infinite geometric series, if it exists. (Lesson 10-4)

46. $-6 + 3 - \frac{3}{2} + \ldots$

47. $\frac{3}{4} + \frac{1}{4} + \frac{1}{12} + \ldots$

48. $\sqrt{3} + 3 + \sqrt{27} + \ldots$

49. TRAVEL A trip between two towns takes 4 hours under ideal conditions. The first 150 miles of the trip is on an interstate, and the last 130 miles is on a highway with a speed limit that is 10 miles per hour less than on the interstate. (Lesson 8-6)

a. If x represents the speed limit on the interstate, write expressions for the time spent at that speed and for the time spent on the other highway.

b. Write and solve an equation to find the speed limits on the two highways.

Skills Review

State whether each statement is *true* or *false* when $n = 1$. Explain.

50. $\frac{(n+1)(n+1)}{2} = 2$

51. $3n + 5$ is even.

52. $n^2 - 1$ is odd.

EXTEND 10-6

Algebra Lab: Combinations and Pascal's Triangle

Recall that an arrangement or selection of objects in which order is not important is called a *combination*. For example, selecting 2 snacks from a choice of 6 is a combination of 6 objects taken 2 at a time and can be written ${}_6C_2$ or $C(6, 2)$.

CCSS Common Core State Standards
Content Standards
A.APR.5 Know and apply the Binomial Theorem for the expansion of $(x + y)^n$ in powers of x and y for a positive integer n, where x and y are any numbers, with coefficients determined for example by Pascal's Triangle.

Activity

A contestant on a game show has the opportunity to win up to five prizes, one for each of five rounds of the game. If the contestant wins a round, he or she may choose one prize. Determine the number of ways that prizes can be chosen.

Step 1 If a contestant does not win any rounds, he or she receives 0 prizes. This represents 5 items taken 0 at a time.

$${}_nC_r = \frac{n!}{(n-r)!\,r!} \quad \text{Definition of combination}$$

$${}_5C_0 = \frac{5!}{(5-0)!\,0!} \quad n = 5 \text{ and } r = 0$$

$$= \frac{120}{120(1)} \text{ or } 1 \quad 5! = 120 \text{ and } 0! = 1$$

There is 1 way to receive 0 prizes.

If a contestant wins one round, any one of the prizes can be selected. If a contestant wins two rounds, two prizes can be chosen. If three rounds are won, three prizes can be chosen, and so on. In how many ways can 1 prize be chosen? 2 prizes? 3, 4, and 5 prizes? We can determine these answers by examining Pascal's triangle.

Step 2 Examine Pascal's triangle.

List Rows 0 through 5 of Pascal's triangle.

Row 0						1					
Row 1					1		1				
Row 2				1		2		1			
Row 3			1		3		3		1		
Row 4		1		4		6		4		1	
Row 5	1		5		10		10		5		1

The number of ways one prize can be chosen from 5 can be determined by looking at Row 5. The first number in Row 5 represents the number of ways to choose 0 prizes, the second number represents the number of ways to choose 1 prize, and so on.

Analyze the Result

1. Make a conjecture about how the numbers in one of the rows can be used to find the number of ways that 0, 1, 2, 3, 4, ..., n objects can be selected from n objects.
2. Suppose the rules of the game are changed so that there are 6 rounds and 6 prizes from which to choose. Find the number of ways that 0, 1, 2, 3, 4, 5, or 6 prizes can be chosen. Which row of Pascal's triangle can be used to find the answers?
3. Use Pascal's triangle to find ${}_8C_0$, ${}_8C_1$, ${}_8C_2$, ${}_8C_3$, ${}_8C_4$, ${}_8C_5$, ${}_8C_6$, ${}_8C_7$, and ${}_8C_8$. State the row number that you used to find the answers.

LESSON 10-7

Proof by Mathematical Induction

Then

- You have proved the sum of an arithmetic series.

Now

1. Prove statements by using mathematical induction.
2. Disprove statements by finding a counterexample.

Why?

- When dominoes are set up closely and the first domino is knocked down, the rest of the dominoes come tumbling down. All that is needed with this setup is for the first domino to fall, and the rest will follow. The same is true with mathematical induction.

NewVocabulary
mathematical induction
induction hypothesis

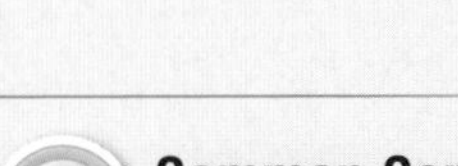

Common Core State Standards

Content Standards
A.SSE.1.b Interpret complicated expressions by viewing one or more of their parts as a single entity.
A.APR.5 Know and apply the Binomial Theorem for the expansion of $(x + y)^n$ in powers of x and y for a positive integer n, where x and y are any numbers, with coefficients determined for example by Pascal's Triangle.

Mathematical Practices
1 Make sense of problems and persevere in solving them.
3 Construct viable arguments and critique the reasoning of others.

1 Mathematical Induction

Mathematical induction is a method of proving statements involving natural numbers.

KeyConcept Mathematical Induction

To prove that a statement is true for all natural numbers n,

Step 1 Show that the statement is true for $n = 1$.

Step 2 Assume that the statement is true for some natural number k. This assumption is called the **induction hypothesis**.

Step 3 Show that the statement is true for the next natural number $k + 1$.

Example 1 Prove Summation

Prove that $1^3 + 2^3 + 3^3 + \cdots + n^3 = \frac{n^2(n+1)^2}{4}$.

Step 1 When $n = 1$, the left side of the equation is 1^3 or 1.
The right side is $\frac{1^2(1+1)^2}{4}$ or 1. Thus, the statement is true for $n = 1$.

Step 2 Assume that $1^3 + 2^3 + 3^3 + \cdots + k^3 = \frac{k^2(k+1)^2}{4}$ for a natural number k.

Step 3 Show that the given statement is true for $n = k + 1$.

$$1^3 + 2^3 + 3^3 + \cdots + k^3 = \frac{k^2(k+1)^2}{4}$$ Inductive hypothesis

$$1^3 + 2^3 + \cdots + k^3 + (k+1)^3 = \frac{k^2(k+1)^2}{4} + (k+1)^3$$ Add $(k + 1)^3$ to each side.

$$= \frac{k^2(k+1)^2 + 4(k+1)^3}{4}$$ The LCD is 4.

$$= \frac{(k+1)^2[k^2 + 4(k+1)]}{4}$$ Factor.

$$= \frac{(k+1)^2(k^2 + 4k + 4)}{4}$$ Simplify.

$$= \frac{(k+1)^2(k+2)^2}{4}$$ Factor.

The last expression is the statement to be proved, where n has been replaced by $k + 1$. This proves the conjecture.

GuidedPractice

1. Prove that $1^2 + 2^2 + 3^2 + \cdots + n^2 = \frac{n(n+1)(2n+1)}{6}$.

Comstock Images/Alamy

Along with summation, mathematical induction can be used to prove divisibility.

Example 2 Prove Divisibility

Prove that $8^n - 1$ is divisible by 7 for all natural numbers n.

Step 1 When $n = 1$, $8^n - 1 = 8^1 - 1$ or 7. Since 7 is divisible by 7, the statement is true for $n = 1$.

Step 2 Assume that $8^k - 1$ is divisible by 7 for some natural number k. This means that there is a natural number r such that $8^k - 1 = 7r$.

Step 3 Show that the statement is true for $n = k + 1$.

$8^k - 1 = 7r$ Inductive hypothesis

$8^k = 7r + 1$ Add 1 to each side.

$8(8^k) = 8(7r + 1)$ Multiply each side by 8.

$8^{k+1} = 56r + 8$ Simplify.

$8^{k+1} - 1 = 56r + 7$ Subtract 1 from each side.

$8^{k+1} - 1 = 7(8r + 1)$ Factor.

Since r is a natural number, $8r + 1$ is a natural number and $7(8r + 1)$ is divisible by 7. Therefore, $8^{k+1} - 1$ is divisible by 7.

This proves that 8^{n-1} is divisible by 7 for all natural numbers n.

StudyTip

Regularity r is a whole number used to show divisibility in proof. When a value equals $4r$, then it must be divisible by 4.

GuidedPractice

2. Prove that $7^n - 1$ is divisible by 6 for all natural numbers n.

2 Counterexamples

Counterexamples Statements can be proved false by using mathematical induction. An easier method is by finding a counterexample, which is a specific case in which the statement is false.

Example 3 Use a Counterexample to Disprove

Find a counterexample to disprove the statement that $2^n + 2n^2$ is divisible by 4 for any natural number n.

Test different values of n.

n	$2^n + 2n^2$	Divisible by 4?
1	$2^1 + 2(1)^2 = 2 + 2$ or 4	yes
2	$2^2 + 2(2)^2 = 4 + 8$ or 12	yes
3	$2^3 + 2(3)^2 = 8 + 18$ or 26	no

The value $n = 3$ is a counterexample for the statement.

ReviewVocabulary

counterexample One of the synonyms of *counter* is to *contradict*, so a counterexample is an example that contradicts a hypothesis.

GuidedPractice

3. Find a counterexample to disprove $1^2 + 2^2 + 3^2 + \cdots + n^2 = \frac{n(3n - 1)}{2}$.

Check Your Understanding

= Step-by-Step Solutions begin on page R14.

Example 1 **Prove that each statement is true for all natural numbers.**

1. $1 + 3 + 5 + \cdots + (2n - 1) = n^2$

2. $1 + 2 + 3 + \cdots + n = \frac{n(n + 1)}{2}$

3. NUMBER THEORY A number is *triangular* if it can be represented visually by a triangular array.

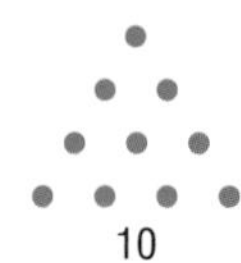

a. The first triangular number is 1. Find the next 5 triangular numbers.

b. Write a formula for the nth triangular number.

c. Prove that the sum of the first n triangular numbers equals $\frac{n(n + 1)(n + 2)}{6}$.

Example 2 **Prove that each statement is true for all natural numbers.**

4. $10^n - 1$ is divisible by 9.

5. $4^n - 1$ is divisible by 3.

Example 3 **Find a counterexample to disprove each statement.**

6. $3^n + 1$ is divisible by 4.

7. $2^n + 3^n$ is divisible by 4.

Practice and Problem Solving

Extra Practice is on page R10.

Example 1 CCSS **ARGUMENTS** **Prove that each statement is true for all natural numbers.**

8. $\frac{1}{2} + \frac{1}{2^2} + \frac{1}{2^3} + \cdots + \frac{1}{2^n} = 1 - \frac{1}{2^n}$

9. $2 + 5 + 8 + \cdots + (3n - 1) = \frac{n(3n + 1)}{2}$

10. $1 + 2 + 4 + \cdots + 2^{n-1} = 2^n - 1$

11. $1 + 5 + 9 + \cdots + (4n - 3) = n(2n - 1)$

12. $1 + 4 + 7 + \cdots + (3n - 2) = \frac{n(3n - 1)}{2}$

13 $3 + 7 + 11 + \cdots + (4n - 1) = 2n^2 + n$

14. $\frac{1}{2} + \frac{1}{6} + \frac{1}{12} + \cdots + \frac{1}{n(n + 1)} = \frac{n}{n + 1}$

15. $1^2 + 3^2 + 5^2 + \cdots + (2n - 1)^2 = \frac{n(2n - 1)(2n + 1)}{3}$

16. GEOMETRY According to the Interior Angle Sum Formula, if a convex polygon has n sides, then the sum of the measures of the interior angles of a polygon equals $180(n - 2)$. Prove this formula for $n \geq 3$ using mathematical induction and geometry.

Example 2 **Prove that each statement is true for all natural numbers.**

17. $5^n + 3$ is divisible by 4.

18. $9^n - 1$ is divisible by 8.

19. $12^n + 10$ is divisible by 11.

20. $13^n + 11$ is divisible by 12.

Example 3 **Find a counterexample to disprove each statement.**

21. $1 + 2 + 3 + \cdots + n = n^2$

22. $1 + 8 + 27 + \cdots + n^3 = (2n + 2)^2$

23. $n^2 - n + 15$ is prime.

24. $n^2 + n + 23$ is prime.

25 **NATURE** The terms of the Fibonacci sequence are found in many places in nature. The number of spirals of seeds in sunflowers is a Fibonacci number, as is the number of spirals of scales on a pinecone. The Fibonacci sequence begins 1, 1, 2, 3, 5, 8, ... Each element after the first two is found by adding the previous two terms. If f_n stands for the nth Fibonacci number, prove that $f_1 + f_2 + \ldots + f_n = f_{n+2} - 1$.

Prove that each statement is true for all natural numbers or find a counterexample.

26. $7^n + 5$ is divisible by 6.

27. $18^n - 1$ is divisible by 17.

28. $n^2 + 21n + 7$ is a prime number.

29. $n^2 + 3n + 3$ is a prime number.

30. $500 + 100 + 20 + \cdots + 4 \cdot 5^{4-n} = 625\left(1 - \frac{1}{5^n}\right)$

31. $\frac{1}{1 \cdot 2 \cdot 3} + \frac{1}{2 \cdot 3 \cdot 4} + \frac{1}{3 \cdot 4 \cdot 5} + \cdots + \frac{1}{n(n+1)(n+2)} = \frac{n(n+3)}{4(n+1)(n+2)}$

32. **CCSS PERSEVERANCE** Refer to the figures below.

Figure 1

Figure 2

Figure 3

a. There is a total of 5 squares in the second figure. How many squares are there in the third figure?

b. Write a sequence for the first five figures.

c. How many squares are there in a standard 8×8 checkerboard?

d. Write a formula to represent the number of squares in an $n \times n$ grid.

H.O.T. Problems Use Higher-Order Thinking Skills

33. **CHALLENGE** Suggest a formula to represent $2 + 4 + 6 + \cdots + 2n$, and prove your hypothesis using mathematical induction.

REASONING **Determine whether the following statements are *true* or *false*. Explain.**

34. If you cannot find a counterexample to a statement, then it is true.

35. If a statement is true for $n = k$ and $n = k + 1$, then it is also true for $n = 1$.

36. **CHALLENGE** Prove $\sum_{k=1}^{n} k^3 = \left(\frac{n(n+1)}{2}\right)^2$.

37. **REASONING** Find a counterexample to $x^3 + 30 > x^2 + 20x$.

38. **OPEN ENDED** Write a sequence, the formula that produces it, and determine the formula for the sum of the terms of the sequence. Then prove the formula with mathematical induction.

39. **WRITING IN MATH** Explain how the concept of dominoes can help you understand the power of mathematical induction.

40. **WRITING IN MATH** Provide a real-world example other than dominoes that describes mathematical induction.

Standardized Test Practice

41. Which of the following is a counterexample to the statement below?

$$n^2 + n - 11 \text{ is prime.}$$

A $n = -6$
B $n = 4$
C $n = 5$
D $n = 6$

42. PROBABILITY Latisha wants to create a 7-character password. She wants to use an arrangement of the first 3 letters of her first name (lat), followed by an arrangement of the 4 digits in 1986, the year she was born. How many possible passwords can she create in this way?

F 72
G 144
H 288
J 576

43. GRIDDED RESPONSE A gear that is 8 inches in diameter turns a smaller gear that is 3 inches in diameter. If the larger gear makes 36 revolutions, how many revolutions does the smaller gear make in that time?

44. SHORT RESPONSE Write an equation for the nth line. Show how it fits the pattern for each given line in the list.

Line 1: $1 \times 0 = 1 - 1$

Line 2: $2 \times 1 = 4 - 2$

Line 3: $3 \times 2 = 9 - 3$

Line 4: $4 \times 3 = 16 - 4$

Line 5: $5 \times 4 = 25 - 5$

Spiral Review

Find the indicated term of each expansion. (Lesson 10-6)

45. fourth term of $(x + 2y)^6$

46. fifth term of $(a + b)^6$

47. fourth term of $(x - y)^9$

48. BIOLOGY In a particular forest, scientists are interested in how the population of wolves will change over the next two years. One model for animal population is the Verhulst population model, $p_{n+1} = p_n + rp_n(1 - p_n)$, where n represents the number of time periods that have passed, p_n represents the percent of the maximum sustainable population that exists at time n, and r is the growth factor. (Lesson 10-5)

a. To find the population of the wolves after one year, evaluate $p_1 = 0.45 + 1.5(0.45)(1 - 0.45)$.

b. Explain what each number in the expression in part **a** represents.

c. The current population of wolves is 165. Find the new population by multiplying 165 by the value in part **a**.

Find the exact solution(s) of each system of equations. (Lesson 9-7)

49. $x^2 + y^2 - 18x + 24y + 200 = 0$
$4x + 3y = 0$

50. $4x^2 + y^2 = 16$
$x^2 + 2y^2 = 4$

Skills Review

Evaluate each expression.

51. $P(8, 2)$

52. $P(9, 1)$

53. $P(12, 6)$

54. $C(5, 2)$

55. $C(8, 4)$

56. $C(20, 17)$

57. $P(12, 2)$

58. $P(7, 2)$

59. $C(8, 6)$

60. $C(9, 4) \cdot C(5, 3)$

61. $C(6, 1) \cdot C(4, 1)$

62. $C(10, 5) \cdot C(8, 4)$

CHAPTER 10

Study Guide and Review

Study Guide

KeyConcepts

Arithmetic Sequences and Series (Lessons 10-1 and 10-2)

- The nth term a_n of an arithmetic sequence with first term a_1 and common difference d is given by $a_n = a_1 + (n - 1)d$, where n is any positive integer.
- The sum S_n of the first n terms of an arithmetic series is given by $S_n = \frac{n}{2}[2a_1 + (n - 1)d]$ or $S_n = \frac{n}{2}(a_1 + a_n)$.

Geometric Sequences and Series (Lessons 10-3 and 10-4)

- The nth term a_n of a geometric sequence with first term a_1 and common ratio r is given by $a_n = a_1 \cdot r^{n-1}$, where n is any positive integer.
- The sum S_n of the first n terms of a geometric series is given by $S_n = \frac{a_1(1 - r^n)}{1 - r}$ or $S_n = \frac{a_1 - a_1 r^n}{1 - r}$, where $r \neq 1$.
- The sum S of an infinite geometric series with $-1 < r < 1$ is given by $S_n = \frac{a_1}{1 - r}$.

Recursion and Iteration (Lesson 10-5)

- In a recursive formula, each term is formulated from one or more previous terms.

The Binomial Theorem (Lesson 10-6)

- The Binomial Theorem:

$$(a + b)^n = \sum_{k=0}^{n} \frac{n!}{(n - k)!\,k!} a^{n-k} b^k$$

Mathematical Induction (Lesson 10-7)

- Mathematical induction is a method of proof used to prove statements about the positive integers.

FOLDABLES® StudyOrganizer

Be sure the Key Concepts are noted in your Foldable.

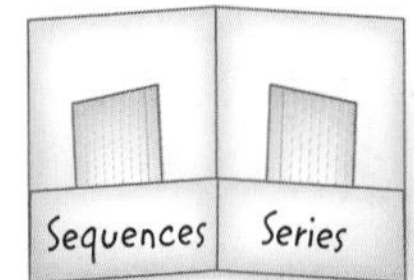

KeyVocabulary

arithmetic means (p. 667)
arithmetic sequence (p. 659)
arithmetic series (p. 668)
common difference (p. 659)
common ratio (p. 661)
convergent series (p. 683)
divergent series (p. 683)
explicit formula (p. 692)
Fibonacci sequence (p. 692)
finite sequence (p. 659)
geometric means (p. 675)
geometric sequence (p. 661)
geometric series (p. 676)
induction hypothesis (p. 705)
infinite geometric series (p. 683)
infinite sequence (p. 659)
infinity (p. 685)
iteration (p. 694)
mathematical induction (p. 705)
partial sum (p. 668)
Pascal's triangle (p. 699)
recursive formula (p. 692)
recursive sequence (p. 692)
sequence (p. 659)
series (p. 668)
sigma notation (p. 669)
term (p. 659)

VocabularyCheck

State whether each sentence is *true* or *false*. If *false*, replace the underlined term to make a true sentence.

1. An infinite geometric series that has a sum is called a <u>convergent series</u>.
2. <u>Mathematical induction</u> is the process of repeatedly composing a function with itself.
3. The <u>arithmetic means</u> of a sequence are the terms between any two non-successive terms of an arithmetic sequence.
4. A <u>term</u> is a list of numbers in a particular order.
5. The sum of the first n terms of a series is called the <u>partial sum</u>.
6. The formula $a_n = a_{n-2} + a_{n-1}$ is a <u>recursive formula</u>.
7. A <u>geometric sequence</u> is a sequence in which every term is determined by adding a constant value to the previous term.
8. An infinite geometric series that does not have a sum is called a <u>partial sum</u>.
9. Eleven and 17 are two <u>geometric means</u> between 5 and 23 in the sequence 5, 11, 17, 23.
10. Using the <u>Binomial Theorem</u>, $(x - 2)^4$ can be expanded to $x^4 - 8x^3 + 24x^2 - 32x + 16$.

Lesson-by-Lesson Review

10-1 Sequences as Functions

Find the indicated term of each arithmetic sequence.

11. $a_1 = 9, d = 3, n = 14$

12. $a_1 = -3, d = 6, n = 22$

13. $a_1 = 10, d = -4, n = 9$

14. $a_1 = -1, d = -5, n = 18$

Example 1

Find the 11th term of an arithmetic sequence if $a_1 = -15$ and $d = 6$.

$a_n = a_1 + (n - 1)d$ Formula for the *n*th term

$a_{11} = -15 + (11 - 1)6$ $n = 11, a_1 = -15, d = 6$

$a_{11} = 45$ Simplify.

10-2 Arithmetic Sequences and Series

Find the arithmetic means in each sequence.

15. -12, __, __, __, 8

16. 15, __, __, 29

17. 12, __, __, __, __, -8

18. 72, __, __, __, 24

19. **BANKING** Carson saves \$40 every 2 months. If he saves at this rate for two years, how much will he have at the end of two years?

Find S_n for each arithmetic series.

20. $a_1 = 16, a_n = 48, n = 6$

21. $a_1 = 8, a_n = 96, n = 20$

22. $9 + 14 + 19 + \cdots + 74$

23. $16 + 7 + -2 + \cdots + -65$

24. **DRAMA** Laura has a drama performance in 12 days. She plans to practice her lines each night. On the first night she rehearses her lines 2 times. The next night she rehearses her lines 4 times. The third night she rehearses her lines 6 times. On the eleventh night, how many times has she rehearsed her lines?

Find the sum of each arithmetic series.

25. $\sum_{k=5}^{21} (3k - 2)$

26. $\sum_{k=0}^{10} (6k - 1)$

27. $\sum_{k=4}^{12} (-2k + 5)$

Example 2

Find the two arithmetic means between 3 and 39.

$a_n = a_1 + (n - 1)d$ Formula for the *n*th term

$a_4 = 3 + (4 - 1)d$ $n = 4, a_1 = 3$

$39 = 3 + 3d$ $a_4 = 39$

$12 = d$ Simplify.

The arithmetic means are $3 + 12$ or 15 and $15 + 12$ or 27.

Example 3

Find S_n for the arithmetic series with $a_1 = 18$, $a_n = 56$, and $n = 8$.

$S_n = \frac{n}{2}(a_1 + a_n)$ Sum formula

$S_8 = \frac{8}{2}(18 + 56)$ $n = 8, a_1 = 18, a_n = 56$

$= 296$ Simplify.

Example 4

Evaluate $\sum_{k=3}^{15} 5k + 1$.

Use the formula $S_n = \frac{n}{2}(a_1 + a_n)$. There are 13 terms, $a_1 = 5(3) + 1$ or 16, and $a_{13} = 5(15) + 1$ or 76.

$S_{13} = \frac{13}{2}(16 + 76)$

$= 598$

Study Guide and Review *Continued*

10-3 Geometric Sequences and Series

Find the indicated term for each geometric sequence.

28. $a_1 = 5, r = 2, n = 7$

29. $a_1 = 11, r = 3, n = 3$

30. $a_1 = 128, r = -\frac{1}{2}, n = 5$

31. a_8 for $\frac{1}{8}, \frac{3}{8}, \frac{9}{8}, \dots$

Find the geometric means in each sequence.

32. 6, __, __, 162

33. 8, __, __, __, 648

34. −4, __, __, 108

35. **SAVINGS** Nolan has a savings account with a current balance of $1500. What would be Nolan's account balance after 4 years if he receives 5% interest annually?

Find S_n for each geometric series.

36. $a_1 = 15, r = 2, n = 4$

37. $a_1 = 9, r = 4, n = 6$

38. $5 - 10 + 20 - \cdots$ to 7 terms

39. $243 + 81 + 27 + \cdots$ to 5 terms

Evaluate the sum of each geometric series.

40. $\sum_{k=1}^{7} 3 \cdot (-2)^{k-1}$

41. $\sum_{k=1}^{8} -1\left(\frac{2}{3}\right)^{k-1}$

42. **ADVERTISING** Natalie is handing out fliers to advertise the next student council meeting. She hands out fliers to 4 people. Then, each of those 4 people hand out 4 fliers to 4 other people. Those 4 then hand out 4 fliers to 4 new people. If Natalie is considered the first round, how many people will have been given fliers after 4 rounds?

Example 5

Find the sixth term of a geometric sequence for which $a_1 = 9$ and $r = 4$.

$a_n = a_1 \cdot r^{n-1}$ Formula for the *n*th term

$a_6 = 9 \cdot 4^{6-1}$ $n = 6, a_1 = 9, r = 4$

$a_6 = 9216$

The sixth term is 9216.

Example 6

Find two geometric means between 1 and 27.

$a_n = a_1 \cdot r^{n-1}$ Formula for the *n*th term

$a_4 = 1 \cdot r^{4-1}$ $n = 4$ and $a_1 = 1$

$27 = r^3$ $a_4 = 27$

$3 = r$ Simplify.

The geometric means are 1(3) or 3 and 3(3) or 9.

Example 7

Find the sum of a geometric series for which $a_1 = 3, r = 5$, and $n = 11$.

$S_n = \frac{a_1 - a_1r^n}{1 - r}$ Sum formula

$S_{11} = \frac{3 - 3 \cdot 5^{11}}{1 - 5}$ $n = 11, a_1 = 3, r = 5$

$S_{11} = 36{,}621{,}093$ Use a calculator.

Example 8

Evaluate $\sum_{k=1}^{6} 2 \cdot (4)^{k-1}$.

$S_6 = \frac{2 - 2 \cdot 4^6}{1 - 4}$ $n = 6, a_1 = 2, r = 4$

$= \frac{-8190}{-3}$ Simplify.

$= 2730$ Simplify.

10-4 Infinite Geometric Series

Find the sum of each infinite series, if it exists.

43. $a_1 = 8, r = \frac{3}{4}$

44. $\frac{5}{6} - \frac{20}{18} + \frac{80}{54} - \frac{320}{162} + \cdots$

45. $\sum_{k=1}^{\infty} 3\left(\frac{1}{2}\right)^{k-1}$

46. **PHYSICAL SCIENCE** Maddy drops a ball off of a building that is 60 feet high. Each time the ball bounces, it bounces back to $\frac{2}{3}$ its previous height. If the ball continues to follow this pattern, what will be the total distance that the ball travels?

Example 9

Find the sum of the infinite geometric series for which $a_1 = 15$ and $r = \frac{1}{3}$.

$S = \frac{a_1}{1 - r}$ Sum formula

$= \frac{15}{1 - \frac{1}{3}}$ $a_1 = 15, r = \frac{1}{3}$

$= \frac{15}{\frac{2}{3}}$ or 22.5 Simplify.

10-5 Recursion and Iteration

Find the first five terms of each sequence.

47. $a_1 = -3, a_{n+1} = a_n + 4$

48. $a_1 = 5, a_{n+1} = 2a_n - 5$

49. $a_1 = 1, a_{n+1} = a_n + 5$

50. **SAVINGS** Sari has a savings account with a $12,000 balance. She has a 5% interest rate that is compounded monthly. Every month Sari adds $500 to the account. The recursive formula $b_n = 1.05b_{n-1} + 500$ describes the balance in Sari's savings account after n months. Find the balance of Sari's account after 3 months. Round your answer to the nearest penny.

Find the first three iterates of each function for the given initial value.

51. $f(x) = 2x + 1, x_0 = 3$

52. $f(x) = 5x - 4, x_0 = 1$

53. $f(x) = 6x - 1, x_0 = 2$

54. $f(x) = 3x + 1, x_0 = 4$

Example 10

Find the first five terms of the sequence in which $a_1 = 1$, $a_{n+1} = 3a_n + 2$.

$a_{n+1} = 3a_n + 2$ Recursive formula

$a_{1+1} = 3a_1 + 2$ $n = 1$

$a_2 = 3(1) + 2$ or 5 $a_1 = 1$

$a_{2+1} = 3a_2 + 2$ $n = 2$

$a_3 = 3(5) + 2$ or 17 $a_2 = 5$

$a_{3+1} = 3a_3 + 2$ $n = 3$

$a_4 = 3(17) + 2$ or 53 $a_3 = 17$

$a_{4+1} = 3a_4 + 2$ $n = 4$

$a_5 = 3(53) + 2$ or 161 $a_4 = 53$

The first five terms of the sequence are 1, 5, 17, 53, and 161.

Example 11

Find the first three iterates of the function $f(x) = 3x - 2$ for the initial value of $x_0 = 2$.

$x_1 = f(x_0)$	$x_2 = f(x_1)$	$x_3 = f(x_2)$
$= f(2)$	$= f(4)$	$= f(10)$
$= 3(2) - 2$	$= 3(4) - 2$	$= 3(10) - 2$
$= 4$	$= 10$	$= 28$

The first three iterates are 4, 10, and 28.

10-6 The Binomial Theorem

Expand each binomial.

55. $(a + b)^3$

56. $(y - 3)^7$

57. $(3 - 2z)^5$

58. $(4a - 3b)^4$

59. $\left(x - \frac{1}{4}\right)^5$

Find the indicated term of each expression.

60. third term of $(a + 2b)^8$

61. sixth term of $(3x + 4y)^7$

62. second term of $(4x - 5)^{10}$

Example 12

Expand $(x - 3y)^4$.

$$\begin{aligned}(x - 3y)^4 &= x^4 + {}_4C_1x^3(-3y) + {}_4C_2x^2(-3y)^2 + {}_4C_3(-3y)^4 \\ &\quad + {}_4C_4(-3y)^4 \\ &= x^4 + \frac{4!}{3!}x^3(-3y) + \frac{4!}{2!2!}x^2(9y^2) + \frac{4!}{3!}x(-27y^3) + 81y^4 \\ &= x^4 + -12x^3y + 54x^2y^2 + -108xy^3 + 81y^4\end{aligned}$$

Example 13

Find the fourth term of $(x + y)^8$.

Use the Binomial Theorem to write the expansion in sigma notation.

$$(x + y)^8 = \sum_{k=0}^{8} \frac{8!}{k!(8 - k)!}x^{8 - k}y^k$$

For the fourth term, $k = 3$.

$$\begin{aligned}\frac{8!}{k!(8 - k)!}x^{8 - k}y^k &= \frac{8!}{3!(8 - 3)!}x^{8 - 3}y^3 \\ &= 56x^5y^3\end{aligned}$$

10-7 Proof by Mathematical Induction

Prove that each statement is true for all positive integers.

63. $2 + 6 + 12 + \cdots + n(n + 1) = \frac{n(n + 1)(n + 2)}{3}$

64. $7^n - 1$ is divisible by 6.

65. $5^n - 1$ is divisible by 4.

Find a counterexample for each statement.

66. $8^n + 3$ is divisible by 11.

67. $6^{n+1} - 2$ is divisible by 17.

68. $n^2 + 2n + 4$ is prime.

69. $n + 19$ is prime.

Example 14

Prove that $9^n + 3$ is divisible by 4.

Step 1 When $n = 1$, $9^n + 3 = 9^1 + 3$ or 12. Since 12 divided by 4 is 3, the statement is true for $n = 1$.

Step 2 Assume that $9^k + 3$ is divisible by 4 for some positive integer k. This means that $9^k + 3 = 4r$ for some whole number r.

Step 3

$$\begin{aligned}9^k + 3 &= 4r \\ 9^k &= 4r - 3 \\ 9^{k+1} &= 36r - 27 \\ 9^{k+1} + 3 &= 36r - 27 + 3 \\ 9^{k+1} + 3 &= 36r - 24 \\ 9^{k+1} + 3 &= 4(9r - 6)\end{aligned}$$

Since r is a whole number, $9r - 6$ is a whole number. Thus, $9^{k+1} + 3$ is divisible by 4, so the statement is true for $n = k + 1$.

Therefore, $9^n + 3$ is divisible by 4 for all positive integers n.

CHAPTER 10 Practice Test

1. Find the next 4 terms of the arithmetic sequence 81, 72, 63, … .

2. Find the 25th term of an arithmetic sequence for which $a_1 = 9$ and $d = 5$.

3. **MULTIPLE CHOICE** What is the eighth term in the arithmetic sequence that begins 18, 20.2, 22.4, 24.6, …?

   A 26.8

   B 29

   C 31.2

   D 33.4

4. Find the four arithmetic means between −9 and 11.

5. Find the sum of the arithmetic series for which $a_1 = 11$, $n = 14$, and $a_n = 22$.

6. **MULTIPLE CHOICE** What is the next term in the geometric sequence below?

$$10, \frac{5}{2}, \frac{5}{8}, \frac{5}{32} \cdots$$

   F $\frac{5}{8}$

   G $\frac{5}{32}$

   H $\frac{5}{128}$

   J $\frac{5}{256}$

7. Find the three geometric means between 6 and 1536.

8. Find the sum of the geometric series for which $a_1 = 15$, $r = \frac{2}{3}$, and $n = 5$.

Find the sum of each series, if it exists.

9. $\sum_{k=2}^{12}(3k - 1)$

10. $\sum_{k=1}^{\infty}\frac{1}{2}(3^k)$

11. $45 + 37 + 29 + \cdots + -11$

12. $\frac{1}{8} + \frac{2}{24} + \frac{4}{72} + \cdots$

13. Write $0.\overline{65}$ as a fraction.

Find the first five terms of each sequence.

14. $a_1 = -1, a_{n+1} = 3a_n + 5$

15. $a_1 = 4, a_{n+1} = a_n + n$

16. **MULTIPLE CHOICE** What are the first 3 iterates of $f(x) = -5x + 4$ for an initial value of $x_0 = 3$?

   A 3, −11, 59

   B −11, 59, −291

   C −1, −6, −11

   D 59, −291, 1459

17. Expand $(2a - 3b)^4$.

18. What is the coefficient of the fifth term of $(m + 3n)^6$?

19. Find the fourth term of the expansion of $(c + d)^9$.

Prove that each statement is true for all positive integers.

20. $1 + 6 + 36 + \cdots + 6^{n-1} = \frac{1}{5}(6^n - 1)$.

21. $11^n - 1$ is divisible by 10.

22. Find a counterexample for the following statement.

$2^n + 4^n$ is divisible by 4.

23. **SCHOOL** There are an equal number of girls and boys in Mr. Marshall's science class. He needs to choose 8 students to represent his class at the science fair. What is the probability that 5 are boys?

24. **PENDULUM** Laurie swings a pendulum. The distance traveled per swing decreases by 15% with each swing. If the pendulum initially traveled 10 inches, find the total distance traveled when the pendulum comes to a rest.

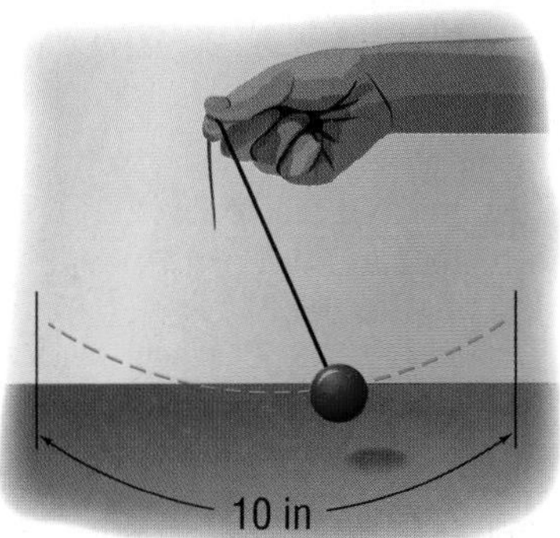

Preparing for Standardized Tests

Look For a Pattern

One of the most common problem-solving strategies is to look for a pattern. The ability to recognize patterns, model them algebraically, and extend them is a valuable problem-solving tool.

Strategies for Looking For a Pattern

Step 1

Identify the pattern.

- Compare the numbers, shapes, or graphs in the pattern.
- **Ask yourself:** How are the terms of the pattern related?
- **Ask yourself:** Are there any common operations that lead from one term to the next?

Step 2

Generalize the pattern.

- Write a rule using words to describe how the terms of the pattern are generated.
- Assign variables and write an algebraic expression to model the pattern if appropriate.

Step 3

Find missing terms, extend the pattern, and solve the problem.

- Use your pattern or your rule to finding missing terms and/or extend the pattern to solve the problem.
- Check your answer to make sure it makes sense.

Standardized Test Example

Read the problem. Identify what you need to know. Then use the information in the problem to solve.

Use the sequence of squares shown. How many squares will be needed to make the ninth figure of the sequence?

A 55 C 74

B 65 D 82

Figure 1 Figure 2 Figure 3

Digital Vision/Getty Images

Read the problem statement carefully. You are given three figures of a sequence and asked to find how many squares will be needed to make the ninth figure.

Look for a pattern in the figures of squares. Count the number of squares in each figure.

Write an expression to model this pattern.

Words	The number of squares is equal to the square of the figure number plus one.
Variable	Let n represent the figure number.
Equation	$a_n = n^2 + 1$

Use your expression to extend the pattern and find the number of squares in the ninth figure.

$a_9 = 9^2 + 1 = 82$

So, the ninth figure will have 82 squares. The correct answer is D.

Exercises

Read each problem. Use a pattern to solve the problem.

1. The numbers below form a famous mathematical sequence of numbers known as the Fibonacci sequence. What is the next Fibonacci number in the sequence?

1, 1, 2, 3, 5, 8, 13, 21, ...

A 36

B 34

C 31

D 29

2. What is the missing number in the table?

n	a_n
1	0
2	2
3	6
4	12
5	??
6	30

F 17

G 18

H 20

J 21

CHAPTER 10

Standardized Test Practice

Cumulative, Chapters 1 through 10

Multiple Choice

Read each question. Then fill in the correct answer on the answer document provided by your teacher or on a sheet of paper.

1. Find the next term of the arithmetic sequence.

$$7, 13, 19, 25, 31, \ldots$$

A 36
B 37
C 38
D 39

2. Lynette gets an enlargement of a 4-inch by 6-inch picture so that the new print has dimensions that are 4 times the dimensions of her original. How does the area of the enlargement compare to the area of the original picture?

F The area is twice as large.

G The area is four times as large.

H The area is eight times as large.

J The area is sixteen times as large.

3. Evaluate $\sum_{k=1}^{15}(8k - 1)$.

A 119
B 826
C 945
D 1072

4. What is the effect on the graph of the equation $y = 3x^2$ when the equation is changed to $y = 2x^2$?

F The graph of $y = 2x^2$ is a reflection of the graph of $y = 3x^2$ across the y-axis.

G The graph is rotated 90 degrees about the origin.

H The graph is narrower.

J The graph is wider.

5. Write the formula for the nth term of the geometric sequence shown in the table.

n	a_n
1	5
2	10
3	20
4	40
5	80

A $a_n = (5)^n$

B $a_n = 5(2)^{n-1}$

C $a_n = 2(5)^{n-1}$

D $a_n = 5(2)^n$

6. The table shows a dimension of a square tent and the number of people that the tent can fit.

Length of Tent (yards)	Number of People
2	7
5	28
6	39
8	67
12	147

Let ℓ represent the length of the tent and n represent the number of people that can fit in the tent. Identify the equation that best represents the relationship between the length of the tent and the number of people that can fit in the tent.

F $\ell = n^2 + 3$
G $n = \ell^2 + 3$
H $\ell = 3n + 1$
J $n = 3\ell + 1$

7. An air filter claims to remove 90% of the contaminants in the air each time air is circulated through the filter. If the same volume of air is circulated through the filter three times, what percent of the original contaminants will be removed from the air?

A 0.01% B 0.1% C 99.0% D 99.9%

8. At the movies, the cost of 2 boxes of popcorn and 1 soft drink is \$11.50. The cost of 3 boxes of popcorn and 4 soft drinks is \$27.25. Which pair of equations can be used to determine p, the cost of a box of popcorn, and s, the cost of a soft drink?

F $2p + s = 27.25$
$3p + 4s = 11.50$

G $2p - s = 11.50$
$3p - 4s = 27.25$

H $2p + s = 11.50$
$3p + 4s = 27.25$

J $p + s = 11.50$
$p + 4 = 27.25$

9. Which of the following geometric series does *not* converge to a sum?

A $\sum_{k=1}^{\infty} 4 \cdot \left(\frac{9}{10}\right)^{k-1}$

B $\sum_{k=1}^{\infty} \frac{1}{5} \cdot \left(\frac{3}{2}\right)^{k-1}$

C $\sum_{k=1}^{\infty} \frac{7}{6} \cdot \left(\frac{1}{3}\right)^{k-1}$

D $\sum_{k=1}^{\infty} (-2) \cdot \left(\frac{5}{6}\right)^{k-1}$

Test-Taking Tip

Question 9 Understand the terms used in Algebra and how to apply them. A geometric series converges to a sum if the common ratio r has an absolute value less than 1.

Short Response/Gridded Response

Record your answers on the answer sheet provided by your teacher or on a sheet of paper.

10. What are the dimensions of the matrix that results from the multiplication shown?

$$\begin{bmatrix} a & b & c \\ d & e & f \\ g & h & i \\ j & k & l \end{bmatrix} \cdot \begin{bmatrix} 7 \\ 4 \\ 6 \end{bmatrix}$$

11. GRIDDED RESPONSE Consider the pattern below. Into how many pieces will the sixth figure of the pattern be divided?

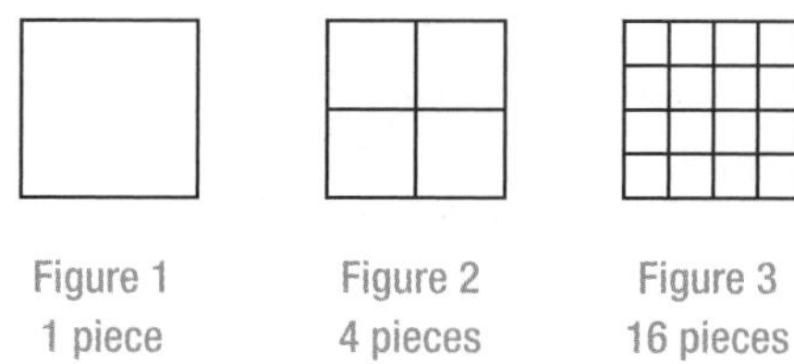

12. Use the Binomial Theorem to expand the expression $(c + d)^6$.

13. GRIDDED RESPONSE Kara has a cylindrical container that she needs to fill with dirt so she can plant some flowers.

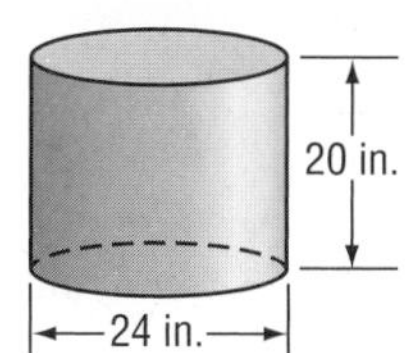

What is the volume of the cylinder in cubic inches rounded to the nearest cubic inch?

14. Bacteria in a culture are growing exponentially with time, as shown in the table.

Hours	Bacteria
0	1000
1	2000
2	4000

Write an equation to express the number of bacteria, y, with respect to time, t.

15. GRIDDED RESPONSE What is the value of $f[g(6)]$ if $f(x) = 2x + 4$ and $g(x) = x^2 + 5$?

Extended Response

Record your answers on a sheet of paper. Show your work.

16. Prove that the sum of any two odd integers is even.

17. The endpoints of a diameter of a circle are at $(-1, 0)$ and $(5, -8)$.

a. What are the coordinates of the center of the circle? Explain your method.

b. Find the radius of the circle. Explain your method.

c. Write an equation of the circle.

18. A cyclist travels from Centerville to Springfield in 2.5 hours. If she increases her speed, she can make the trip in 2 hours.

a. Does this situation represent a direct or inverse variation? Explain your reasoning.

b. If the trip from Centerville to Springfield takes 2.5 hours when traveling at 12 miles per hour, what must the speed be to make the trip in 2 hours?

Need ExtraHelp?

If you missed Question...	1	2	3	4	5	6	7	8	9	10	11	12	13	14	15	16	17	18
Go to Lesson...	10-2	1-1	10-2	4-7	10-3	2-4	10-3	3-1	10-4	3-6	10-5	10-6	5-7	7-8	6-1	10-7	9-3	8-5

CHAPTER 11

Statistics and Probability

Then

You calculated weighted averages.

Now

You will:

- Evaluate surveys, studies, and experiments.
- Create and use graphs of probability distributions.
- Use the Empirical Rule to find probabilities.
- Compare sample statistics and population statistics.

Why? ▲

EDUCATION Probability and statistics are used in all facets of education. Surveys and experiments are done to find out which teaching methods promote the most learning. Statistics are used to determine grades when classes are curved, or when college professors weight their grades.

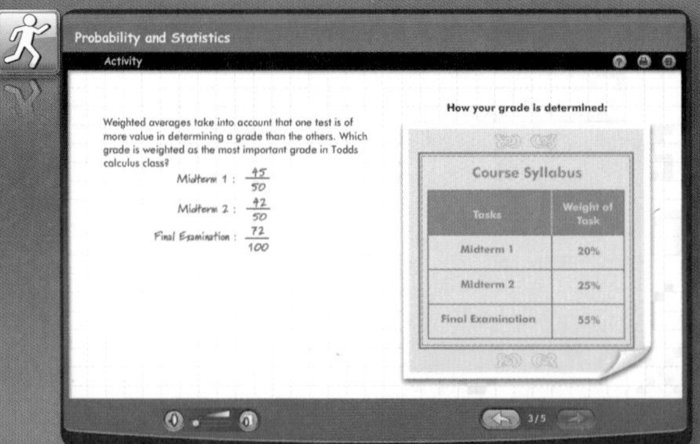

connectED.mcgraw-hill.com Your Digital Math Portal

Animation 
Vocabulary 

eGlossary 
Personal Tutor 
Virtual Manipulatives 
Graphing Calculator 
Audio 
Foldables 
Self-Check Practice 
Worksheets 

Get Ready for the Chapter

Diagnose Readiness | You have two options for checking prerequisite skills.

Textbook Option Take the Quick Check below. Refer to the Quick Review for help.

QuickCheck

Find the mean, median, and mode for each set of data.

1. number of customers at a store each day during the last two weeks:
   78, 80, 101, 66, 73, 92, 97, 125, 110, 76, 89, 90, 82, 87
2. a student's quiz scores for the first grading period:
   88, 70, 85, 92, 88, 77, 98, 88, 70, 82
3. the number of touchdowns scored by a player over the last 10 years:
   7, 5, 10, 12, 4, 10, 11, 6, 9, 3

A die is rolled and a coin is tossed. Find each probability.

4. $P(4, \text{heads})$
5. $P(\text{odds, tails})$
6. $P(2 \text{ or } 4, \text{heads})$

Expand each binomial.

7. $(a - 2)^4$
8. $(m - a)^5$
9. $(2b - x)^4$
10. $(2a + b)^6$
11. $(3x - 2y)^5$
12. $(3x + 2y)^4$
13. $\left(\frac{a}{2} + 2\right)^5$
14. $\left(3 + \frac{m}{3}\right)^5$

QuickReview

Example 1

The number of days it rained in each month over the past year are shown below. Find the mean, median, and mode.

4, 2, 9, 16, 13, 9, 8, 9, 7, 6, 8, 5

Mean $\bar{x} = \frac{4 + 2 + 9 + 16 + 13 + 9 + 8 + 9 + 7 + 6 + 8 + 5}{12}$ or 8 days

Median 2, 4, 5, 6, 7, 8, 8, 9, 9, 9, 13, 16

$\frac{8 + 8}{2}$ or 8 days

Mode The value that occurs most often in the set is 9, so the mode of the data set is 9 days.

Example 2

A die is rolled and a coin is tossed. What is the probability that the die shows a 1 and the coin lands tails up?

$P(1, \text{tails}) = \frac{1}{6} \cdot \frac{1}{2}$ or $\frac{1}{12}$

Example 3

Expand $(a + b)^4$.

Replace n with 4 in the Binomial Theorem.

$(a + b)^4$

$= {}_4C_0a^4 + {}_4C_1a^3b + {}_4C_2a^2b^2 + {}_4C_3ab^3 + {}_4C_4b^4$

$= \frac{4!}{(4-0)! \cdot 0!}a^4 + \frac{4!}{(4-1)! \cdot 1!}a^3b + \frac{4!}{(4-2)! \cdot 2!}a^2b^2 + \frac{4!}{(4-3)! \cdot 3!}ab^3 + \frac{4!}{(4-4)! \cdot 4!}b^4$

$= \frac{24}{24}a^4 + \frac{24}{6 \cdot 1}a^3b + \frac{24}{2 \cdot 2}a^2b^2 + \frac{24}{1 \cdot 6}ab^3 + \frac{24}{1 \cdot 4}b^4$

$= a^4 + 4a^3b + 6a^2b^2 + 4ab^3 + b^4$

2 Online Option Take an online self-check Chapter Readiness Quiz at connectED.mcgraw-hill.com.

Get Started on the Chapter

You will learn several new concepts, skills, and vocabulary terms as you study Chapter 11. To get ready, identify important terms and organize your resources. You may wish to refer to Chapter 0 to review prerequisite skills.

FOLDABLES® StudyOrganizer

Probability and Statistics Make this Foldable to help you organize your Chapter 11 notes about probability and statistics. Begin with a sheet of $8\frac{1}{2}$" by 11" paper.

1 **Fold** in half lengthwise.

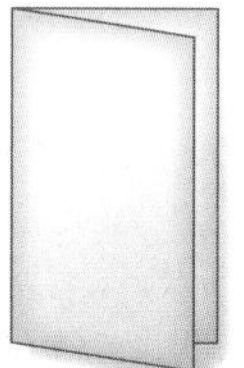

2 **Fold** the top to the bottom.

3 **Open.** Cut along the second fold to make two tabs.

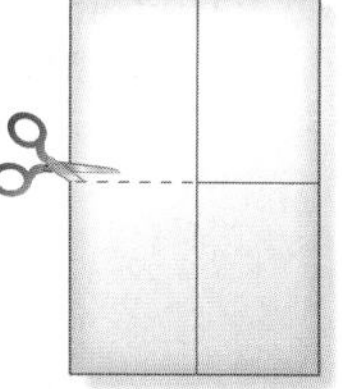

4 **Label** each tab as shown.

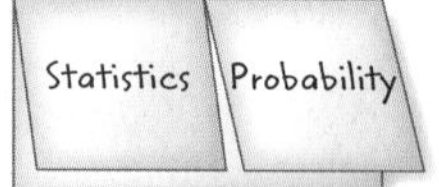

NewVocabulary

English		Español
parameter	p. 723	parámetro
statistic	p. 723	estadística
survey	p. 723	exámenes
experiment	p. 723	experimento
observational study	p. 723	estudio de observación
random variable	p. 742	variable aleatoria
probability distribution	p. 743	distribución de probabilidad
expected value	p. 745	valor previsto
binomial experiment	p. 752	experimento binomio
binomial distribution	p. 754	distribución binomial
normal distribution	p. 760	distribución normal
z-value	p. 762	valor de z
confidence interval	p. 769	intervalo de la confianza
inferential statistics	p. 769	estadística deductiva
statistical inference	p. 769	inferencia estadística
hypothesis test	p. 771	prueba de hipótesis
null hypothesis	p. 771	hipótesis nula
alternative hypothesis	p. 771	hipótesis alternativa

ReviewVocabulary

combination combinación an arrangement or selection of objects in which order is not important

permutation permutación a group of objects or people arranged in a certain order

random arbitrario Unpredictable, or not based on any predetermined characteristics of the population; when a die is tossed, a coin is flipped, or a spinner is spun, the outcome is a random event.

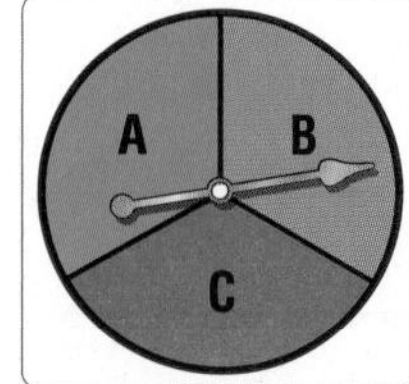

LESSON 11-1

Designing a Study

:: Then	:: Now	:: Why?
You identified various sampling techniques.	**1** Classify study types. **2** Design statistical studies.	According to a recent study, 88% of teen cell phone users in the U.S. send text messages, and one in three teens sends more than 100 texts per day.

NewVocabulary

parameter
statistic
bias
random sample
survey
experiment
observational study

Common Core State Standards

Content Standards

S.IC.3 Recognize the purposes of and differences among sample surveys, experiments, and observational studies; explain how randomization relates to each.

S.IC.5 Use data from a randomized experiment to compare two treatments; use simulations to decide if differences between parameters are significant.

Mathematical Practices

3 Construct viable arguments and critique the reasoning of others.

1 Classifying Studies

In a statistical study, data are collected and used to answer questions about a population characteristic or **parameter**. Due to time and money constraints, it may be impractical or impossible to collect data from each member of a population. Therefore, in many studies, a sample of the population is taken, and a measure called a **statistic** is calculated using the data. The sample statistic, such as the sample mean or sample standard deviation, is then used to make inferences about the population parameter.

The steps in a typical statistical study are shown below.

Identify the objective of the study. → Choose a sample, and collect data. → Organize the data, and calculate sample statistics. → Make inferences, and draw conclusions about the population.

To obtain good information and draw accurate conclusions about a population, it is important to select an *unbiased* sample. A **bias** is an error that results in a misrepresentation of members of a population. A poorly chosen sample can cause biased results. To reduce the possibility of selecting a biased sample, a **random sample** can be taken, in which members of the population are selected entirely by chance.
You will review other sampling methods in Exercise 32.

The following study types can be used to collect sample information.

KeyConcept Study Types

Definition	Example
In a **survey**, data are collected from responses given by members of a population regarding their characteristics, behaviors, or opinions.	To determine whether the student body likes the new cafeteria menu, the student council asks a random sample of students for their opinion.
In an **experiment**, the sample is divided into two groups: • an *experimental group* that undergoes a change, and • a *control group* that does not undergo the change. The effect on the experimental group is then compared to the control group.	A restaurant is considering creating meals with chicken instead of beef. They randomly give half of a group of participants meals with chicken and the other half meals with beef. Then they ask how they like the meals.
In an **observational study**, members of a sample are measured or observed without being affected by the study.	Researchers at an electronics company observe a group of teenagers using different laptops and note their reactions.

Radius Images/Jupiterimages

Example 1 Classify Study Types

Determine whether each situation describes a *survey*, an *experiment*, or an *observational study*. Then identify the sample, and suggest a population from which it may have been selected.

a. MUSIC A record label wants to test three designs for an album cover. They randomly select 50 teenagers from local high schools to view the covers while they watch and record their reactions.

This is an observational study, because the company is going to observe the teens without them being affected by the study. The sample is the 50 teenagers selected, and the population is all potential purchasers of this album.

StudyTip

Census

A *census* is a survey in which each member of a population is questioned. Therefore, when a census is conducted, there is no sample.

b. RECYCLING The city council wants to start a recycling program. They send out a questionnaire to 200 random citizens asking what items they would recycle.

This is a survey, because the data are collected from participants' responses in the questionnaire. The sample is the 200 people who received the questionnaire, and the population is all of the citizens of the city.

GuidedPractice

1A. **RESEARCH** Scientists study the behavior of one group of dogs given a new heartworm treatment and another group of dogs given a false treatment or *placebo*.

1B. **YEARBOOKS** The yearbook committee conducts a study to determine whether students would prefer to have a print yearbook or both print and digital yearbooks.

To determine when to use a survey, experiment, or observational study, think about how the data will be obtained and whether or not the participants will be affected by the study.

Example 2 Choose a Study Type

Determine whether each situation calls for a *survey*, an *experiment*, or an *observational study*. Explain your reasoning.

a. MEDICINE A pharmaceutical company wants to test whether a new medicine is effective.

The treatment will need to be tested on a sample group, which means that the members of the sample will be affected by the study. Therefore, this situation calls for an experiment.

b. ELECTIONS A news organization wants to randomly call citizens to gauge opinions on a presidential election.

This situation calls for a survey because members of the sample population are asked for their opinion.

GuidedPractice

2A. **RESEARCH** A research company wants to study smokers and nonsmokers to determine whether 10 years of smoking affects lung capacity.

2B. **PETS** A national pet chain wants to know whether customers would pay a small annual fee to participate in a rewards program. They randomly select 200 customers and send them questionnaires.

2 Designing Studies

Designing Studies The questions chosen for a survey or procedures used in an experiment can also introduce bias, and thus, affect the results of the study.

A survey question that is poorly written may result in a response that does not accurately reflect the opinion of the participant. Therefore, it is important to write questions that are clear and precise. Avoid survey questions that:

- are confusing or wordy
- cause a strong reaction
- encourage a certain response
- address more than one issue

Questions can also introduce bias if there is not enough information given for the participant to give an accurate response.

Example 3 Identify Bias in Survey Questions

Determine whether each survey question is *biased* or *unbiased*. If biased, explain your reasoning.

a. Don't you agree that the cafeteria should serve healthier food?

This question is biased because it encourages a certain response. The phrase "don't you agree" encourages you to agree that the cafeteria should serve healthier food.

b. How often do you exercise?

This question is unbiased because it is clearly stated and does not encourage a certain response.

GuidedPractice

3A. How many glasses of water do you drink a day?

3B. Do you prefer watching exciting action movies or boring documentaries?

When designing a survey, clearly state the objective, identify the population, and carefully choose unbiased survey questions.

Real-World Example 4 Design a Survey

TECHNOLOGY Jim is writing an article for his school newspaper about online courses. He wants to conduct a survey to determine how many students at his school would be interested in taking an online course from home. State the objective of the survey, suggest a population, and write two unbiased survey questions.

Step 1 State the objective of the survey.

The objective of the survey is to determine students' interest in taking an online course from home.

Step 2 Identify the population.

The population is the student body.

Step 3 Write unbiased survey questions.

Possible survey questions:

- "Do you have Internet access at home?"
- "If offered, would you take an online course?"

GuidedPractice

4. TECHNOLOGY In a follow-up article, Jim decides to conduct a survey to determine how many teachers from his school with at least five years of experience would be interested in teaching an online course. State the objective of the survey, suggest a population, and write two unbiased survey questions.

Real-WorldLink

Online Courses In 2009, about 1.2 million students took at least one online course.

Source: International Association for K-12 Online Learning

Ephraim Ben-Shimon/Corbis

To avoid introducing bias in experiments, the experimental and control groups should be randomly selected and the experiment should be designed so that everything about the two groups is alike (except for the treatment or procedure).

Example 5 Identify Flaws in Experiments

Identify any flaws in the design of the experiment, and describe how they could be corrected.

Experiment: An electronics company wants to test whether using a new graphing calculator increases students' test scores. A random sample is taken. Calculus students in the experimental group are given the new calculator to use, and Algebra 2 students in the control group are asked to use their own calculator.

Results: When given the same test, the experimental group scored higher than the control group. The company concludes that the use of this calculator increases test scores.

Calculus students are more likely to score higher when given the same test as Algebra 2 students. Therefore, the flaw is that the experimental group consists of Calculus students and the control group consists of Algebra 2 students. This flaw could be corrected by selecting a random sample of all Calculus or all Algebra 2 students.

StudyTip

Bias in Experiments An experiment is biased when the participants know which group they are in.

GuidedPractice

5. Experiment: A research firm tests the effectiveness of a de-icer on car locks. They use a random sample of drivers in California and Minnesota for the control and experimental groups.

Results: They concluded that the de-icer is effective.

When designing an experiment, clearly state the objective, identify the population, determine the experimental and control groups, and define the procedure.

Real-World Example 6 Design an Experiment

PLANTS A research company wants to test the claim of the advertisement shown at the right. State the objective of the experiment, suggest a population, determine the experimental and control groups, and describe a sample procedure.

Taller tomato plants in just 3 weeks!

Thrive

Step 1 State the objective, and identify the population.

The objective of the experiment is to determine whether tomato plants given the plant food grow taller in three weeks than tomato plants not given the food. The population is all tomato plants.

Step 2 Determine the experimental and control groups.

The experimental group is the tomato plants given the food, and the control group is the tomato plants not given the food.

Step 3 Describe a sample procedure.

Measure the heights of the plants in each group, and give the experimental group the plant food. Then, wait three weeks, measure the heights of the plants again, and compare the heights for each group to see if the claim was valid.

GuidedPractice

6. SPORTS A company wants to determine whether wearing a new tennis shoe improves jogging time. State the objective of the experiment, suggest a population, determine the experimental and control groups, and describe a sample procedure.

Check Your Understanding

= Step-by-Step Solutions begin on page R14.

Example 1 **Determine whether each situation describes a *survey*, an *experiment*, or an *observational study*. Then identify the sample, and suggest a population from which it may have been selected.**

1. **SCHOOL** A group of high school students is randomly selected and asked to complete the form shown.

Do you agree with the new lunch rules?
- ☐ agree
- ☐ disagree
- ☐ don't care

2. **DESIGN** An advertising company wants to test a new logo design. They randomly select 20 participants and watch them discuss the logo.

Example 2 **CCSS ARGUMENTS Determine whether each situation calls for a *survey*, an *experiment*, or an *observational study*. Explain your reasoning.**

3. **LITERACY** A literacy group wants to determine whether high school students that participated in a recent national reading program had higher standardized test scores than high school students that did not participate in the program.

4. **RETAIL** The research department of a retail company plans to conduct a study to determine whether a dye used on a new T-shirt will begin fading before 50 washes.

Example 3 **Determine whether each survey question is *biased* or *unbiased*. If biased, explain your reasoning.**

5. Which student council candidate's platform do you support?

6. How long have you lived at your current address?

Example 4

7. **HYBRIDS** A car manufacturer wants to determine what the demand in the U.S. is for hybrid vehicles. State the objective of the survey, suggest a population, and write two unbiased survey questions.

Example 5

8. Identify any flaws in the experiment design, and describe how they could be corrected.

Experiment: A research company wants to determine whether a new vitamin boosts energy levels and decides to test the vitamin at a college campus. A random sample is taken. The experimental group consists of students who are given the vitamin, and the control group consists of instructors who are given a placebo.

Results: When given a physical test, the experimental group outperformed the control group. The company concludes that the vitamin is effective.

Example 6

9. **SPORTS** A research company wants to conduct an experiment to test the claim of the protein shake shown. State the objective of the experiment, suggest a population, determine the experimental and control groups, and describe a sample procedure.

Practice and Problem Solving

Extra Practice is on page R11.

Example 1 **Determine whether each situation describes a *survey*, an *experiment*, or an *observational study*. Then identify the sample, and suggest a population from which it may have been selected.**

10. **FOOD** A grocery store conducts an online study in which customers are randomly selected and asked to provide feedback on their shopping experience.

11. **GRADES** A research group randomly selects 80 college students, half of whom took a physics course in high school, and compares their grades in a college physics course.

12. **HEALTH** A research group randomly chooses 100 people to participate in a study to determine whether eating blueberries reduces the risk of heart disease for adults.

13. **TELEVISION** A television network mails a questionnaire to randomly selected people across the country to determine whether they prefer watching sitcoms or dramas.

Example 2 **Determine whether each situation calls for a *survey*, an *experiment*, or an *observational study*. Explain your reasoning.**

14. **FASHION** A fashion magazine plans to poll 100 people in the U.S. to determine whether they would be more likely to buy a subscription if given a free issue.

15. **TRAVEL** A travel agency randomly calls 250 U.S. citizens and asks them what their favorite vacation destination is.

16. **FOOD** Chee wants to examine the eating habits of 100 random students at lunch to determine how many students eat in the cafeteria.

17. **ENGINEERING** An engineer is planning to test 50 metal samples to determine whether a new titanium alloy has a higher strength than a different alloy.

Example 3 **Determine whether each survey question is *biased* or *unbiased*. If biased, explain your reasoning.**

18. Do you think that the school needs a new gym and football field?

19. Which is your favorite football team, the Dallas Cowboys or the Pittsburgh Steelers?

20. Do you play any extracurricular sports?

21. Don't you agree that students should carpool to school?

Example 4 22. **COLLEGE** A school district wants to conduct a survey to determine the number of juniors in the district who are planning to attend college after high school. State the objective of the survey, suggest a population, and write two unbiased survey questions.

Example 5 23. Identify any flaws in the experiment design, and describe how they could be corrected.

Experiment: A supermarket chain wants to determine whether shoppers are more likely to buy sunscreen if it is located near the checkout line. The experimental group consists of a group of stores in the midwest in which the sunscreen was moved next to the checkout line, and the control group consists of stores in Arizona in which the sunscreen was not moved.

Results: The Arizona stores sold more sunscreen than the midwest stores. The company concluded that moving the sunscreen closer to the checkout line did not increase sales.

Example 6 24. **CCSS ARGUMENTS** In chemistry class, Pedro learned that copper objects become dull over time because the copper reacts with air to form a layer of copper oxide. He plans to use the supplies shown below to determine whether a mixture of lemon juice and salt will remove copper oxide from pennies.

2 lemons

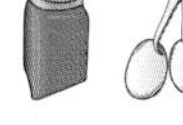
1 teaspoon

30 dull pennies

plastic bowl

a. State the objective of the experiment, suggest a population, determine the experimental and control groups, and describe a sample procedure.

b. What factors do you think should be considered when selecting pennies for the experiment? Explain your reasoning.

25. **REPORTS** The graph shown is from a report on the average number of minutes 8- to 18-year-olds in the U.S. spend on cell phones each day.

a. Describe the sample and suggest a population.

b. What type of sample statistic do you think was calculated for this report?

c. Describe the results of the study for each age group.

d. Who do you think would be interested in this type of report? Explain your reasoning.

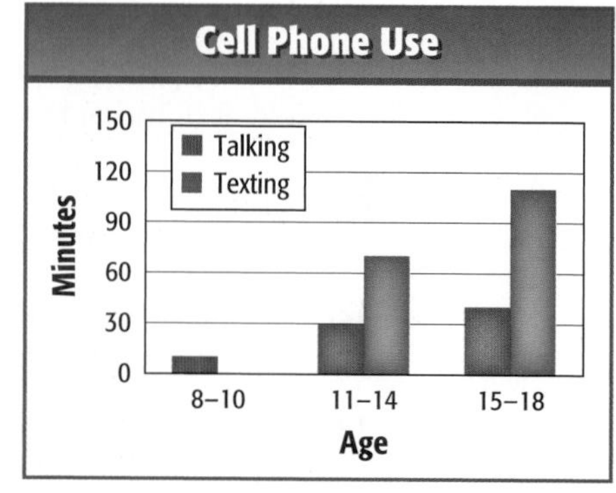

26. CCSS **PERSEVERANCE** In 1936, the *Literary Digest* reported the results of a statistical study used to predict whether Alf Landon or Franklin D. Roosevelt would win the presidential election that year. The sample consisted of 2.4 million Americans, including subscribers to the magazine, registered automobile owners, and telephone users. The results concluded that Landon would win 57% of the popular vote. The actual election results are shown.

ELECTORAL VOTE

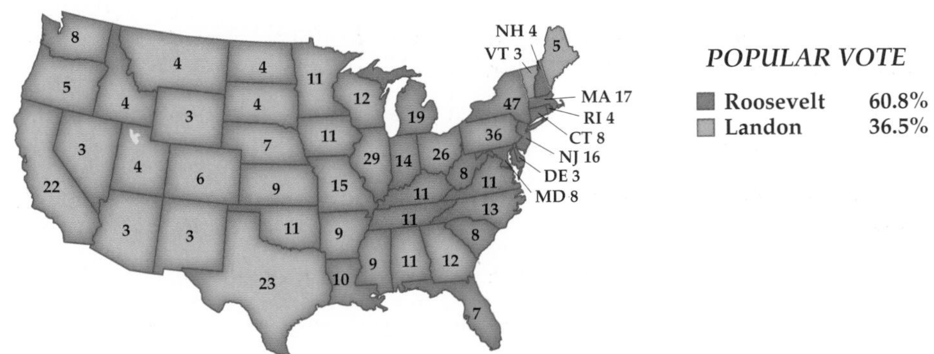

a. Describe the type of study performed, the sample taken, and the population.

b. How do the predicted and actual results compare?

c. Do you think that the survey was biased? Explain your reasoning.

27. **MULTIPLE REPRESENTATIONS** The results of two experiments concluded that Product A is 70% effective and Product B is 80% effective.

```
randInt(0,9)
```

a. **NUMERICAL** To simulate the experiment for Product A, use the random number generator on a graphing calculator to generate 30 integers between 0 and 9. Let 0–6 represent an effective outcome and 7–9 represent an ineffective outcome.

b. **TABULAR** Copy and complete the frequency table shown using the results from part **a.** Then use the data to calculate the probability that Product A was effective. Repeat to find the probability for Product B.

Product A	
Number	**Frequency**
0–6	
7–9	

c. **ANALYTICAL** Compare the probabilities that you found in part **b.** Do you think that the difference in the effectiveness of each product is significant enough to justify selecting one product over the other? Explain.

d. **LOGICAL** Suppose Product B costs twice as much as Product A. Do you think the probability of the product's effectiveness justifies the price difference to a consumer? Explain.

H.O.T. Problems Use Higher-Order Thinking Skills

REASONING Determine whether each statement is *true* or *false*. If false, explain.

28. To save time and money, population parameters are used to estimate sample statistics.

29. Observational studies and experiments can both be used to study cause-and-effect relationships.

30. **OPEN ENDED** Design an observational study. Identify the objective of the study, define the population and sample, collect and organize the data, and calculate a sample statistic.

31. **CHALLENGE** What factors should be considered when determining whether a given statistical study is reliable?

32. **WRITING IN MATH** Research each of the following sampling methods. Then describe each method and discuss whether using the method could result in bias.

a. convenience sample

b. self-selected sample

c. stratified sample

d. systematic sample

Standardized Test Practice

33. GEOMETRY In $\triangle ABC$, $BC > AB$. Which of the following must be true?

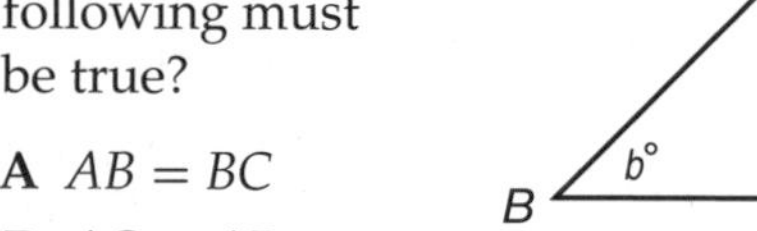

A $AB = BC$

B $AC < AB$

C $a > 60$

D $a = b$

34. SHORT RESPONSE What is the solution set of $4^{4x^2 - 2x - 4} = 4^{-2}$?

35. SAT/ACT A pie is divided evenly between 3 boys and a girl. If one boy gives one half of his share to the girl and a second boy keeps two thirds of his share and gives the rest to the girl, what portion will the girl have in all?

F $\frac{5}{24}$ H $\frac{1}{2}$ K $\frac{13}{12}$

G $\frac{11}{24}$ J $\frac{13}{24}$

36. Which equation represents a hyperbola?

A $y^2 = 49 - x^2$ C $y = 49x^2$

B $y = 49 - x^2$ D $y = \frac{49}{x}$

Spiral Review

37. Prove that the statement *$9^n - 1$ is divisible by 8* is true for all natural numbers. (Lesson 10-7)

38. INTRAMURALS Ofelia is taking ten shots in the intramural free-throw shooting competition. How many sequences of hits and misses are there that result in her making eight shots and missing two? (Lesson 10-6)

Solve each system of equations. (Lesson 9-7)

39. $y = x + 3$
$y = 2x^2$

40. $x^2 + y^2 = 36$
$y = x + 2$

41. $y^2 + x^2 = 9$
$y = 7 - x$

42. $y + x^2 = 3$
$x^2 + 4y^2 = 36$

43. $x^2 + y^2 = 64$
$x^2 + 64y^2 = 64$

44. $y^2 = x^2 - 25$
$x^2 - y^2 = 7$

Find the distance between each pair of points with the given coordinates. (Lesson 9-1)

45. (9, −2), (12, −14)

46. (−4, −10), (−3, −11)

47. (1, −14), (−6, 10)

48. (−4, 9), (1, −3)

49. (2.3, −1.2), (−4.5, 3.7)

50. (0.23, 0.4), (0.68, −0.2)

Simplify. Assume that no variable equals 0. (Lesson 5-1)

51. $(5cd^2)(-c^4d)$

52. $(7x^3y^{-5})(4xy^3)$

53. $\frac{a^2n^6}{an^5}$

54. $(n^4)^4$

55. $\frac{-y^5z^7}{y^2z^5}$

56. $(-2r^2t)^3(3rt^2)$

Write a quadratic equation with the given root(s). Write the equation in the form $ax^2 + bx + c = 0$, where a, b, and c are integers. (Lesson 4-3)

57. −3, 9

58. $-\frac{1}{3}, -\frac{3}{4}$

59. 4, −5

Skills Review

60. TESTS Ms. Bonilla's class of 30 students took a biology test. If 20 of her students had an average of 83 on the test and the other students had an average score of 74, what was the average score of the whole class?

61. DRIVING During a 10-hour trip, Kwan drove 4 hours at 60 miles per hour and 6 hours at 65 miles per hour. What was her average rate, in miles per hour, for the entire trip?

EXTEND 11-1

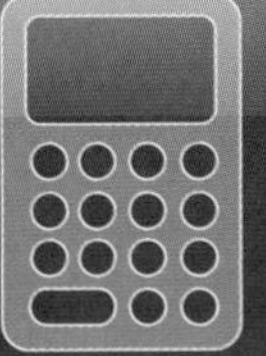

Graphing Calculator Lab
Simulations and Margin of Error

Common Core State Standards
Content Standards
S.IC.4 Use data from a sample survey to estimate a population mean or proportion; develop a margin of error through the use of simulation models for random sampling.
S.IC.6 Evaluate reports based on data.
Mathematical Practices
5 Use appropriate tools strategically.

The Pew Research Center conducted a survey of a random sample of teens and concluded that 43% of all teens who take their cell phones to school text in class on a daily basis. How accurately did their random sample represent all teens?

As you learned in the previous lesson, a survey of a random sample is a valuable tool for generalizing information about a larger population. The program in the following activity makes use of a random number generator (randInt(a, b)) to simulate the results of a random sampling survey.

Activity 1 Random Sampling Simulation

Use the following program that simulates the texting survey to measure the percent of teens who text in class for random sample sizes of 20, 50, and 100 students.

Step 1 Input the following program into a graphing calculator.

```
Program:SIMTEST
:Input "SAMPLE SIZE ",S
:0→A
:0→B
:Lbl Z
:A + 1→A
:randInt(1, 100) →C
:If C ≤ 43
:B + 1→B
:If A<S
:Goto Z
:100B/S→P
:Disp "PERCENT WHO TEXT",P
:Stop
```

Step 2 Run 10 trials of the program for each sample size of 20, 50, and 100. Press ENTER to run the program again each time.

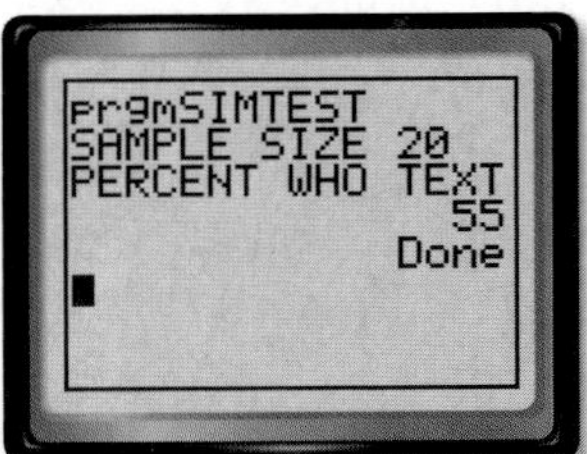

Step 3 Record the percent who text for each trial in the table below.

Sample Size	1	2	3	4	5	6	7	8	9	10
20										
50										
100										

Analyze the Results

1. Discard the percent that is farthest from the Pew survey result of 43% for each sample size. What is the range of the remaining nine percents for each sample size?

2. What is the farthest any of these remaining trials is from the 43% for each sample size?

3. The positive or negative of the result found in Exercise 2 is known as the **margin of error**. For your results, which sample size had the smallest margin of error?

4. What would you expect to happen to the margin of error if we used a sample size of 500?

(continued on the next page)

Simulations and Margin of Error *Continued*

Statisticians have found that for large populations, the margin of error for a random sample of size n can be approximated by the following formula.

KeyConcept **Margin of Error Formula**

$$\text{Margin of error} = \pm\frac{1}{\sqrt{n}}(100)$$

Since n is in the denominator, the margin of error will decrease as the size of the random sample increases. This expression can also be used to determine the size of a random sample necessary to achieve a desired level of reliability.

Activity 2 Margin of Error and Sample Size

PT

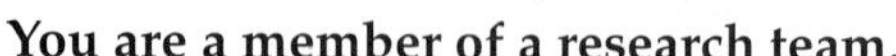

You are a member of a research team.

a. You need to decide whether to conduct a survey with a margin of error of ±3% or ±2%. What sample size would be needed to achieve each goal?

Set each percent equal to the margin of error formula and solve for n.

$\pm 3\% = \pm\frac{1}{\sqrt{n}}(100)$	Margin of error formula	$\pm 2\% = \pm\frac{1}{\sqrt{n}}(100)$
$0.03\sqrt{n} = 1$	Multiply by $\frac{\sqrt{n}}{100}$.	$0.02\sqrt{n} = 1$
$\sqrt{n} = 33.333$	Divide.	$\sqrt{n} = 50$
$n = 1111.11$	Square each side.	$n = 2500$

A random sample of about 1100 would have a margin of error of about ±3%, while a random sample of 2500 would have a margin of error of ±2%.

b. Suppose the finance director would like to reduce the cost of the survey by using a random sample of 100. What would be the margin of error for this sample size?

Substitute 100 for n in the margin of error formula.

$$\text{margin of error} = \pm\frac{1}{\sqrt{n}}(100) \quad \text{Margin of error formula}$$

$$= \pm\frac{1}{\sqrt{100}}(100) \text{ or } \pm 10\% \quad n = 100$$

A random sample of 100 would have a margin of error of ±10%.

Exercises

5. What random sample size would produce a margin of error of ±1%?
6. What margin of error can be expected when using a sample size of 500?
7. What are some reasons that a research center might decide that a survey with a margin of error of ±3% would be more desirable than one with a margin of error of ±2%?
8. What is the range for the percent of students that text in class that the research center can expect from any random survey they conduct with a sample size of 2500?
9. If a survey with a random sample of 2500 students is conducted, is it possible that only 20% of the students could respond that they text during class? If so, how could this be possible?
10. If Step 2 from Activity 1 is repeated using a sample size of 2500 and the range for the percents is found to be 19%–23%, would this result cause you to question the model?

LESSON 11-2

Distributions of Data

∴ Then	∴ Now	∴ Why?
● You calculated measures of central tendency and variation.	● 1 Use the shapes of distributions to select appropriate statistics. 2 Use the shapes of distributions to compare data.	● After four games as a reserve player, Craig joined the starting lineup and averaged 18 points per game over the *remaining* games. Craig's scoring average for the *entire* season was less than 18 points per game as a result of the lack of playing time in the first four games.

NewVocabulary

distribution
negatively skewed distribution
symmetric distribution
positively skewed distribution

Common Core State Standards

Content Standards

S.IC.1 Understand statistics as a process for making inferences about population parameters based on a random sample from that population.

Mathematical Practices

1 Make sense of problems and persevere in solving them.

1 Analyzing Distributions

A **distribution** of data shows the observed or theoretical frequency of each possible data value. In Lesson 0-9, you described distributions of sample data using statistics. You used the mean or median to describe a distribution's center and standard deviation or quartiles to describe its spread. Analyzing the shape of a distribution can help you decide which measure of center or spread best describes a set of data.

The shape of the distribution for a set of data can be seen by drawing a curve over its histogram.

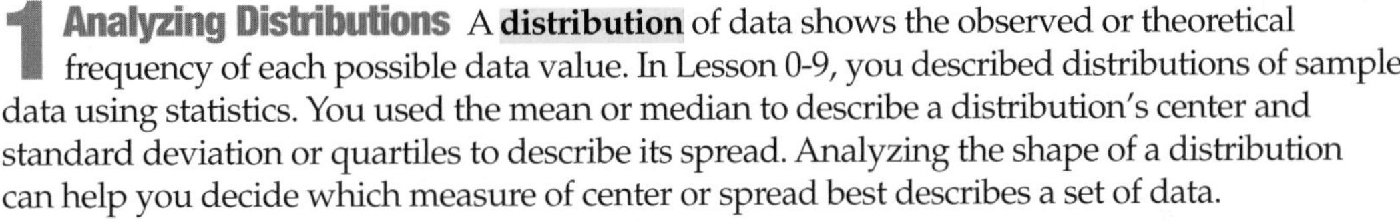

KeyConcept Symmetric and Skewed Distributions

Negatively Skewed Distribution

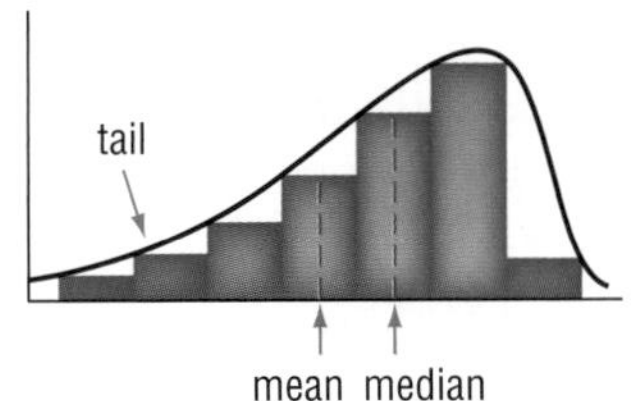

- The mean is less than the median.
- The majority of the data are on the right of the mean.

Symmetric Distribution

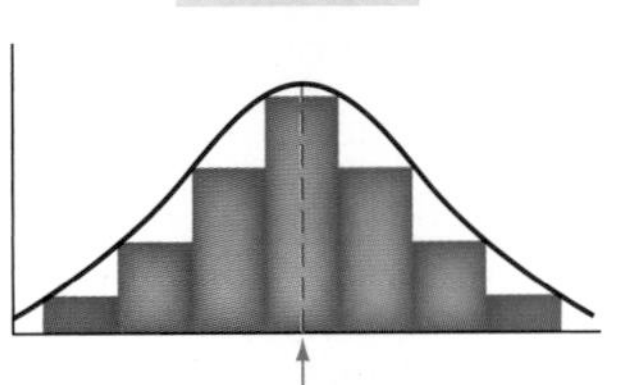

- The mean and median are approximately equal.
- The data are evenly distributed on both sides of the mean.

Positively Skewed Distribution

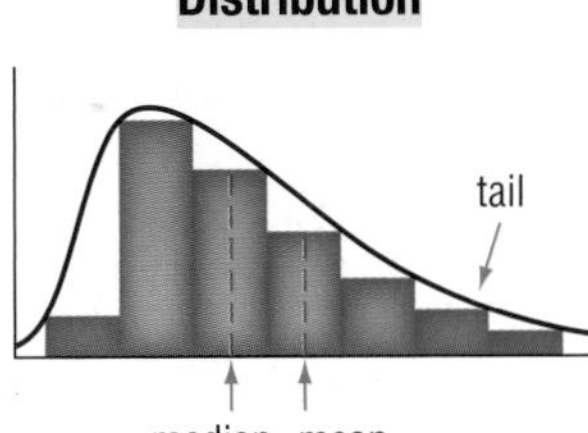

- The mean is greater than the median.
- The majority of the data are on the left of the mean.

When a distribution is symmetric, the mean and standard deviation accurately reflect the center and spread of the data. However, when a distribution is skewed, these statistics are not as reliable. Recall that outliers have a strong effect on the mean of a data set, while the median is less affected. Similarly, when a distribution is skewed, the mean lies away from the majority of the data toward the tail. The median is less affected, so it stays near the majority of the data.

When choosing appropriate statistics to represent a set of data, first determine the skewness of the distribution.

- If the distribution is relatively symmetric, the mean and standard deviation can be used.
- If the distribution is skewed or has outliers, use the five-number summary to describe the center and spread of the data.

Real-WorldLink

The first portable computer, the Osborne I, was available for sale in 1981 for $1795. The computer weighed 24 pounds and included a 5-inch display. Laptops can now be purchased for as little as $250 and can weigh as little as 3 pounds.

Source: Computer History Museum

Real-World Example 1 Describe a Distribution Using a Histogram

COMPUTERS The prices for a random sample of personal computers are shown.

Price (dollars)							
723	605	847	410	440	386	572	523
374	915	734	472	420	508	613	659
706	463	470	752	671	618	538	425
811	502	490	552	390	512	389	621

a. Use a graphing calculator to create a histogram. Then describe the shape of the distribution.

First, press STAT ENTER and enter each data value. Then, press 2nd [STAT PLOT] ENTER ENTER and choose the histogram icon. Finally, adjust the window to the dimensions shown.

The majority of the computers cost between $400 and $700. Some of the computers are priced significantly higher, forming a tail for the distribution on the right. Therefore, the distribution is positively skewed.

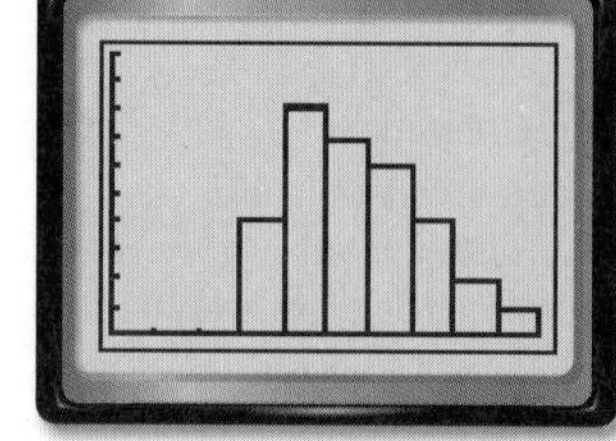

[0, 1000] scl: 100 by [0, 10] scl: 1

b. Describe the center and spread of the data using either the mean and standard deviation or the five-number summary. Justify your choice.

The distribution is skewed, so use the five-number summary to describe the center and spread. Press STAT ▶ ENTER ENTER and scroll down to view the five-number summary.

The prices for this sample range from $374 to $915. The median price is $530.50, and half of the computers are priced between $451.50 and $665.

GuidedPractice

1. RAINFALL The annual rainfall for a region over a 24-year period is shown below.

A. Use a graphing calculator to create a histogram. Then describe the shape of the distribution.

B. Describe the center and spread of the data using either the mean and standard deviation or the five-number summary. Justify your choice.

Annual Rainfall (in.)					
27.2	30.2	35.8	26.1	39.3	20.6
28.9	23.0	32.7	26.8	22.7	25.4
29.6	36.8	33.4	28.4	21.9	20.8
24.7	30.6	27.7	31.4	34.9	37.1

A box-and-whisker plot can also be used to identify the shape of a distribution. The position of the line representing the median indicates the center of the data. The "whiskers" show the spread of the data. If one whisker is considerably longer than the other and the median is closer to the shorter whisker, then the distribution is skewed.

KeyConcept Box-and-Whisker Plots as Distributions

Negatively Skewed

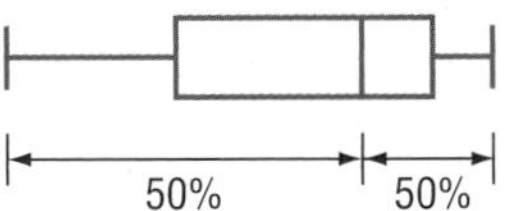

The data to the left of the median are distributed over a wider range than the data to the right. The data have a tail to the left.

Symmetric

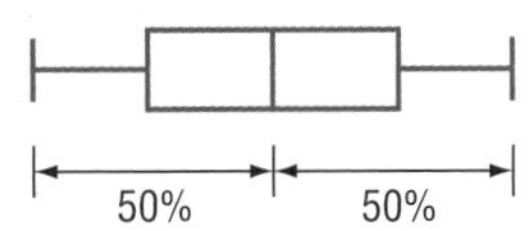

The data are equally distributed to the left and right of the median.

Positively Skewed

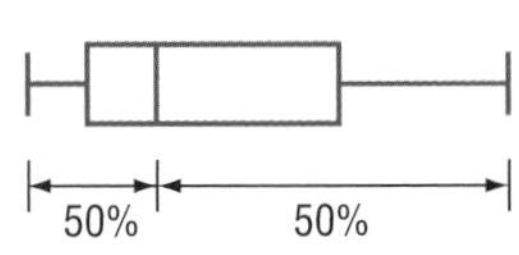

The data to the right of the median are distributed over a wider range than the data to the left. The data have a tail to the right.

Example 2 Describe a Distribution Using a Box-and-Whisker Plot

HOMEWORK The students in Mr. Fejis' language arts class found the average number of minutes that they each spent on homework each night.

Minutes per Night					
62	53	46	66	38	45
52	46	73	39	42	56
64	54	48	59	70	60
49	54	48	57	70	33

a. Use a graphing calculator to create a box-and-whisker plot. Then describe the shape of the distribution.

Enter the data as **L1**. Press 2nd [STAT PLOT] ENTER ENTER and choose the box plot icon. Adjust the window to the dimensions shown.

The lengths of the whiskers are approximately equal, and the median is in the middle of the data. This indicates that the data are equally distributed to the left and right of the median. Thus, the distribution is symmetric.

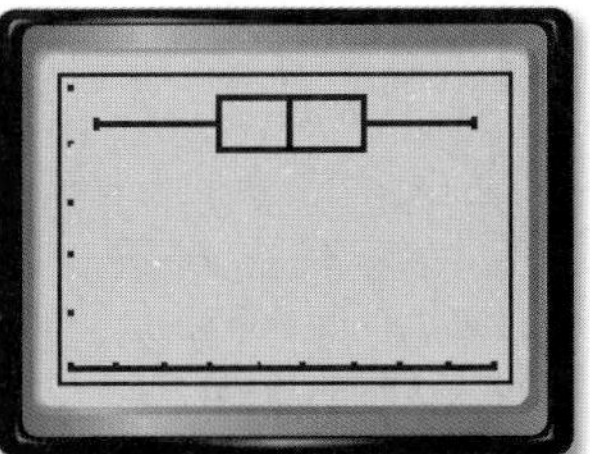

[30, 75] scl: 5 by [0, 5] scl: 1

b. Describe the center and spread of the data using either the mean and standard deviation or the five-number summary. Justify your choice.

The distribution is symmetric, so use the mean and standard deviation to describe the center and spread. The average number of minutes that a student spent on homework each night was 53.5 with standard deviation of about 10.5.

WatchOut!

Standard Deviation Recall from Lesson 0-9 that the formulas for standard deviation for a population σ and for a sample s are slightly different. In Example 2, times for all of the students in Mr. Fejis' class are being analyzed, so use the population standard deviation.

GuidedPractice

2. CELL PHONE Janet's parents have given her a prepaid cell phone. The number of minutes she used each month for the last two years are shown in the table.

Minutes Used per Month			
582	608	670	620
667	598	671	613
537	511	674	627
638	661	642	641
668	673	680	695
658	653	670	688

A. Use a graphing calculator to create a box-and-whisker plot. Then describe the shape of the distribution.

B. Describe the center and spread of the data using either the mean and standard deviation or the five-number summary. Justify your choice.

2 Comparing Distributions

Comparing Distributions To compare two sets of data, first analyze the shape of each distribution. Use the mean and standard deviation to compare two symmetric distributions. Use the five-number summaries to compare two skewed distributions or a symmetric distribution and a skewed distribution.

Example 3 Compare Data Using Histograms

PT

TEST SCORES **Test scores from Mrs. Morash's class are shown below.**

Chapter 3 Test Scores
81, 81, 92, 99, 61, 67, 86, 82, 76, 73, 62, 97, 97, 72, 72, 84, 77, 88, 92, 93, 76, 74, 66, 78, 76, 69, 84, 87, 83, 87, 92, 87, 82

Chapter 4 Test Scores
87, 73, 69, 83, 74, 86, 74, 69, 79, 84, 79, 74, 83, 74, 86, 69, 91, 73, 79, 83, 69, 79, 83, 74, 86, 79, 79, 78, 83, 79, 86, 79, 84

a. Use a graphing calculator to create a histogram for each data set. Then describe the shape of each distribution.

Chapter 3 Test Scores

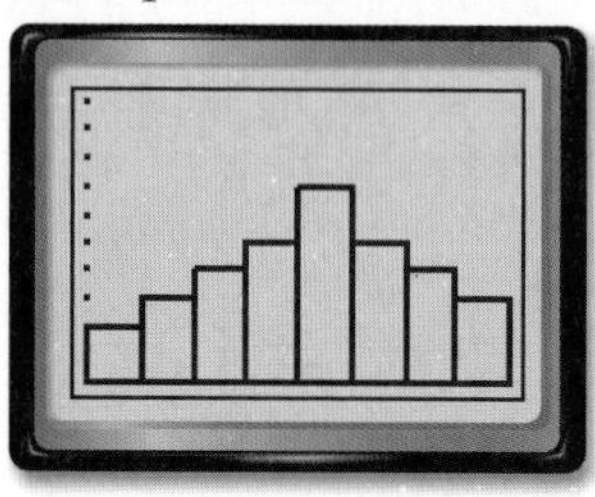

[60, 100] scl: 5 by [0, 10] scl: 1

Chapter 4 Test Scores

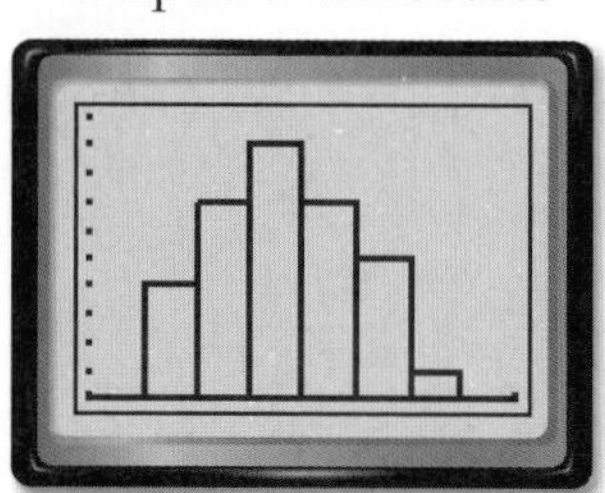

[60, 100] scl: 5 by [0, 10] scl: 1

Both distributions are symmetric.

b. Compare the distributions using either the means and standard deviations or the five-number summaries. Justify your choice.

The distributions are symmetric, so use the means and standard deviations.

Chapter 3 Test Scores

Chapter 4 Test Scores

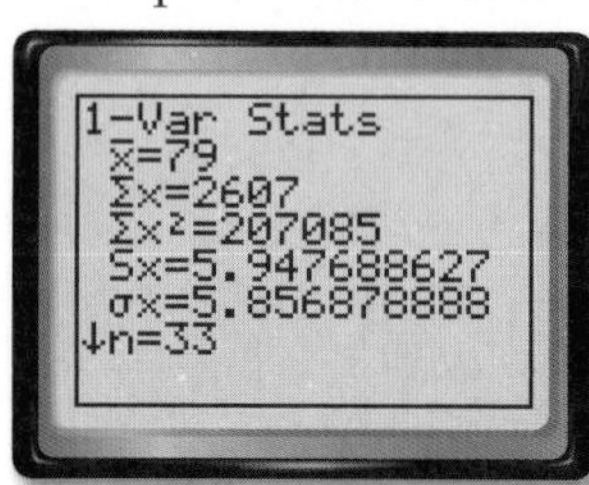

The Chapter 4 test scores, while lower in average, have a much smaller standard deviation, indicating that the scores are more closely grouped about the mean. Therefore, the mean for the Chapter 4 test scores is a better representation of the data than the mean for the Chapter 3 test scores.

StudyTip

CCSS Tools To compare two sets of data, enter one set as **L1** and the other as **L2**. In order to calculate statistics for a set of data in **L2**, press [STAT] [▶] [ENTER] [2nd] [L2] [ENTER].

GuidedPractice

3. TYPING The typing speeds of the students in two classes are shown below.

A. Use a graphing calculator to create a histogram for each data set. Then describe the shape of each distribution.

B. Compare the distributions using either the means and standard deviations or the five-number summaries. Justify your choice.

3rd Period (wpm)
23, 38, 27, 28, 40, 45, 32, 33, 34, 27, 40, 22, 26, 34, 29, 31, 35, 33, 37, 38, 28, 29, 39, 42

6th Period (wpm)
38, 26, 43, 46, 23, 24, 27, 36, 22, 21, 26, 27, 31, 32, 27, 25, 23, 22, 28, 29, 28, 33, 23, 24

Box-and-whisker plots can be displayed alongside one another, making them useful for side-by-side comparisons of data.

Example 4 Compare Data Using Box-and-Whisker Plots

POINTS The points scored per game by a professional football team for the 2008 and 2009 football seasons are shown.

2008							
7	51	24	27	17	35	27	33
28	30	27	21	24	30	14	20

2009							
20	9	3	10	6	14	3	10
3	37	7	21	13	41	20	23

a. Use a graphing calculator to create a box-and-whisker plot for each data set. Then describe the shape of each distribution.

Enter the 2008 scores as **L1**. Graph these data as **Plot1** by pressing 2nd [STAT PLOT] ENTER ENTER and choosing ⊏⊐···. Enter the 2009 scores as **L2**. Graph these data as **Plot2** by pressing 2nd [STAT PLOT] ▼ ENTER ENTER and choosing ⊏⊐···. For **Xlist**, enter **L2**. Adjust the window to the dimensions shown.

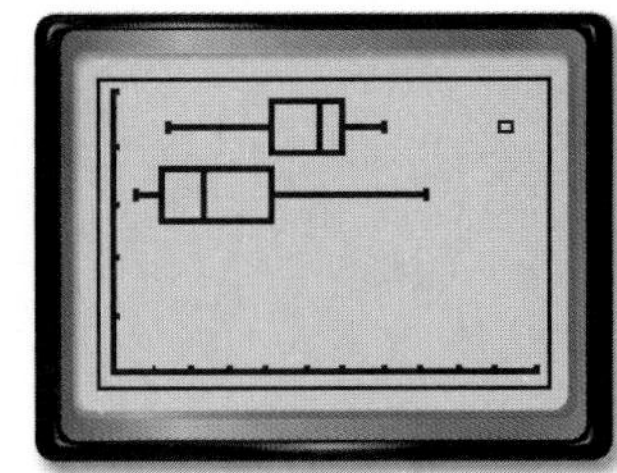

[0, 55] scl: 5 by [0, 5] scl: 1

For the 2008 scores, the left whisker is longer than the right and the median is closer to the right whisker. The distribution is negatively skewed.

For the 2009 scores, the right whisker is longer than the left and the median is closer to the left whisker. The distribution is positively skewed.

b. Compare the distributions using either the means and standard deviations or the five-number summaries. Justify your choice.

The distributions are skewed, so use the five-number summaries to compare the data.

The lower quartile for the 2008 season and the upper quartile for the 2009 season are both 20.5. This means that 75% of the scores from the 2008 season were greater than 20.5 and 75% of the scores from the 2009 season were less than 20.5.

The minimum of the 2008 season is approximately equal to the lower quartile for the 2009 season. This means that 25% of the scores from the 2009 season are lower than any score achieved in the 2008 season. Therefore, we can conclude that the team scored a significantly higher amount of points during the 2008 season than the 2009 season.

> **StudyTip**
> **Outliers** Recall from Lesson 0-9 that outliers are data that are more than 1.5 times the interquartile range beyond the upper or lower quartile. All outliers should be plotted, but the whiskers should be drawn to the least and greatest values that are not outliers.

GuidedPractice

4. GOLF Robert recorded his golf scores for his sophomore and junior seasons.

A. Use a graphing calculator to create a box-and-whisker plot for each data set. Then describe the shape of each distribution.

B. Compare the distributions using either the means and standard deviations or the five-number summaries. Justify your choice.

Sophomore Season
42, 47, 43, 46, 50, 47, 52, 45, 53, 55, 48, 39, 40, 49, 47, 50

Junior Season
44, 38, 46, 48, 42, 41, 42, 46, 43, 40, 43, 43, 44, 45, 39, 44

Check Your Understanding

● = Step-by-Step Solutions begin on page R14.

Example 1

1. **EXERCISE** The amount of time that James ran on a treadmill for the first 24 days of his workout is shown.

Time (minutes)											
23	10	18	24	13	27	19	7	25	30	15	22
10	28	23	16	29	26	26	22	12	23	16	27

a. Use a graphing calculator to create a histogram. Then describe the shape of the distribution.

b. Describe the center and spread of the data using either the mean and standard deviation or the five-number summary. Justify your choice.

Example 2

2. **RESTAURANTS** The total number of times that 20 random people either ate at a restaurant or bought fast food in a month are shown.

Restaurants or Fast Food									
4	7	5	13	3	22	13	6	5	10
7	18	4	16	8	5	15	3	12	6

a. Use a graphing calculator to create a box-and-whisker plot. Then describe the shape of the distribution.

b. Describe the center and spread of the data using either the mean and standard deviation or the five-number summary. Justify your choice.

Example 3

3. CCSS **TOOLS** The total fundraiser sales for the students in two classes at Cantonville High School are shown.

Mrs. Johnson's Class (dollars)					
6	14	17	12	38	15
11	12	23	6	14	28
16	13	27	34	25	32
21	24	21	17	16	

Mr. Edmunds' Class (dollars)					
29	38	21	28	24	33
14	19	28	15	30	6
31	23	33	12	38	28
18	34	26	34	24	37

a. Use a graphing calculator to create a histogram for each data set. Then describe the shape of each distribution.

b. Compare the distributions using either the means and standard deviations or the five-number summaries. Justify your choice.

Example 4

4. **RECYCLING** The weekly totals of recycled paper for the junior and senior classes are shown.

Junior Class (pounds)					
14	24	8	26	19	38
12	15	12	18	9	24
12	21	9	15	13	28

Senior Class (pounds)					
25	31	35	20	37	27
22	32	24	28	18	32
25	32	22	29	26	35

a. Use a graphing calculator to create a box-and-whisker plot for each data set. Then describe the shape of each distribution.

b. Compare the distributions using either the means and standard deviations or the five-number summaries. Justify your choice.

Practice and Problem Solving

Extra Practice is on page R11.

Examples 1–2 **For Exercises 5 and 6, complete each step.**

a. Use a graphing calculator to create a histogram and a box-and-whisker plot. Then describe the shape of the distribution.

b. Describe the center and spread of the data using either the mean and standard deviation or the five-number summary. Justify your choice.

5. **FANTASY** The weekly total points of Kevin's fantasy football team are shown.

Total Points							
165	140	88	158	101	137	112	127
53	151	120	156	142	179	162	79

6. **MOVIES** The students in one of Mr. Peterson's classes recorded the number of movies they saw over the past month.

Movies Seen											
14	11	17	9	6	11	7	8	12	13	10	9
5	11	7	13	9	12	10	9	15	11	13	15

Example 3 CCSS **MODELING** **For Exercises 7 and 8, complete each step.**

a. Use a graphing calculator to create a histogram for each data set. Then describe the shape of each distribution.

b. Compare the distributions using either the means and standard deviations or the five-number summaries. Justify your choice.

7. **SAT** A group of students took the SAT their sophomore year and again their junior year. Their scores are shown.

Sophomore Year Scores					
1327	1663	1708	1583	1406	1563
1637	1521	1282	1752	1628	1453
1368	1681	1506	1843	1472	1560

Junior Year Scores					
1728	1523	1857	1789	1668	1913
1834	1769	1655	1432	1885	1955
1569	1704	1833	2093	1608	1753

8. **INCOME** The total incomes for 18 households in two neighboring cities are shown.

Yorkshire (thousands of dollars)					
68	59	61	78	58	66
56	72	86	58	63	53
68	58	74	60	103	64

Applewood (thousands of dollars)					
52	55	60	61	55	65
65	60	45	37	41	71
50	61	65	66	87	55

Example 4 9. **TUITION** The annual tuitions for a sample of public colleges and a sample of private colleges are shown. Complete each step.

a. Use a graphing calculator to create a box-and-whisker plot for each data set. Then describe the shape of each distribution.

b. Compare the distributions using either the means and standard deviations or the five-number summaries. Justify your choice.

Public Colleges (dollars)					
3773	3992	3004	4223	4821	3880
3163	4416	5063	4937	3321	4308
4006	3508	4498	3471	4679	3612

Private Colleges (dollars)					
10,766	13,322	12,995	15,377	16,792	9147
15,976	11,084	17,868	7909	12,824	10,377
14,304	10,055	12,930	16,920	10,004	11,806

10. DANCE The total amount of money that a random sample of seniors spent on prom is shown. Complete each step.

a. Use a graphing calculator to create a box-and-whisker plot for each data set. Then describe the shape of each distribution.

b. Compare the distributions using either the means and standard deviations or the five-number summaries. Justify your choice.

Boys (dollars)					
253	288	304	283	348	276
322	368	247	404	450	341
291	260	394	302	297	272

Girls (dollars)					
682	533	602	504	635	541
489	703	453	521	472	368
562	426	382	668	352	587

11. BASKETBALL Refer to the beginning of the lesson. The points that Craig scored in the remaining games are shown.

Points Scored			
18	10	18	21
9	25	13	17
17	12	24	19
20	17	27	21

a. Use a graphing calculator to create a box-and-whisker plot. Describe the center and spread of the data.

b. Craig scored 0, 2, 1, and 0 points in the first four games. Use a graphing calculator to create a box-and-whisker plot that includes the new data. Then find the mean and median of the new data set.

c. What effect does adding the scores from the first four games have on the shape of the distribution and on how you should describe the center and spread?

12. SCORES Allison's quiz scores are shown.

Math Quiz Scores					
83	76	86	82	84	57
86	62	90	96	76	89
76	88	86	86	92	94

a. Use a graphing calculator to create a box-and-whisker plot. Describe the center and spread.

b. Allison's teacher allows students to drop their two lowest quiz scores. Use a graphing calculator to create a box-and-whisker plot that reflects this change. Then describe the center and spread of the new data set.

H.O.T. Problems Use Higher-Order Thinking Skills

13 CHALLENGE Approximate the mean and median for each distribution of data.

a.

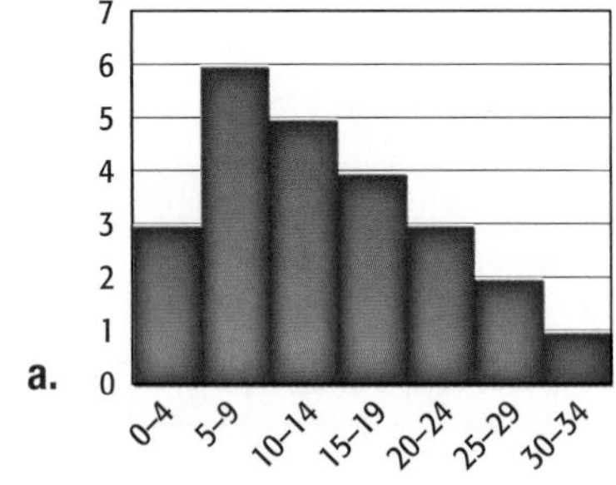

b.

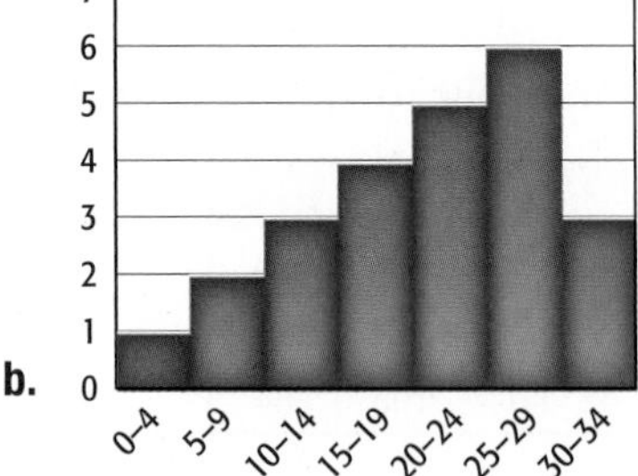

c.

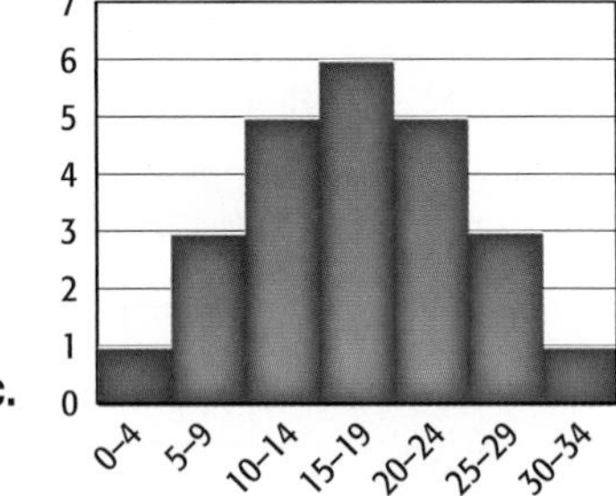

14. CCSS ARGUMENTS Distributions of data are not always symmetric or skewed. If a distribution has a gap in the middle, like the one shown, two separate clusters of data may result, forming a *bimodal distribution.* How can the center and spread of a bimodal distribution be described?

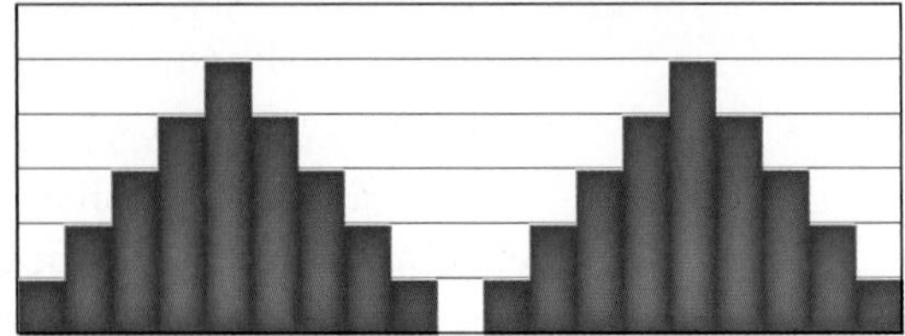

15. OPEN ENDED Find a real-world data set that appears to represent a symmetric distribution and one that does not. Describe each distribution. Create a visual representation of each set of data.

16. WRITING IN MATH Explain the difference between positively skewed, negatively skewed, and symmetric sets of data, and give an example of each.

Standardized Test Practice

17. DISTRIBUTIONS Which of the following is a characteristic of a negatively skewed distribution?

A The majority of the data are on the left of the mean.

B The mean and median are approximately equal.

C The mean is greater than the median.

D The mean is less than the median.

18. SHORT RESPONSE The average test score of a class of c students is 80, and the average test score of a class of d students is 85. When the scores of both classes are combined, the average score is 82. What is the value of $\frac{c}{d}$?

19. SAT/ACT What is the multiplicative inverse of $2i$?

F $-2i$

G -2

H $-\frac{i}{2}$

J $\frac{1}{2}$

K $\frac{i}{2}$

20. Which equation best represents the graph?

A $y = 4x$

B $y = x^2 + 4$

C $y = 4^{-x}$

D $y = -4^x$

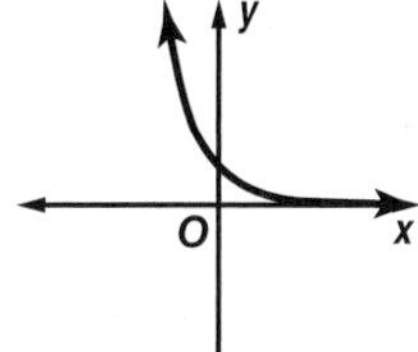

Spiral Review

Determine whether each survey question is biased. Explain your reasoning. (Lesson 11-1)

21. What toppings do you prefer on your pizza?

22. What is your favorite class, and what teacher gives the easiest homework?

23. Don't you hate how high gas prices are?

24. PARTIES Suppose each time a new guest arrives at a party, he or she shakes hands with each person already at the party. Prove that after n guests have arrived, a total of $\frac{n(n-1)}{2}$ handshakes have taken place. (Lesson 10-7)

25. ASTRONOMY The orbit of Pluto can be modeled by $\frac{x^2}{39.5^2} + \frac{y^2}{38.3^2} = 1$, where the units are astronomical units. Suppose a comet is following a path modeled by $x = y^2 + 20$. (Lesson 9-7)

a. Find the point(s) of intersection of the orbits of Pluto and the comet.

b. Will the comet necessarily hit Pluto? Explain.

c. Where do the graphs of $y = 2x + 1$ and $2x^2 + y^2 = 11$ intersect?

d. What are the coordinates of the points that lie on the graphs of both $x^2 + y^2 = 25$ and $2x^2 + 3y^2 = 66$?

Skills Review

Determine whether each situation involves a *permutation* or a *combination*. Then find the number of possibilities.

26. the winner of the first, second, and third prizes in a contest with 8 finalists

27. selecting two of eight employees to attend a business seminar

28. an arrangement of the letters in the word *MATH*

29. placing an algebra book, a geometry book, a chemistry book, an English book, and a health book on a shelf

LESSON 11-3 Probability Distributions

Then

- You used statistics to describe symmetrical and skewed distributions of data.

Now

1. Construct a probability distribution.
2. Analyze a probability distribution and its summary statistics.

Why?

- Mutual funds are professionally managed investments that offer diversity to investors. An accurate analysis of the fund's current and expected performance can help an investor determine if the fund suits their needs.

NewVocabulary
random variable
discrete random variable
continuous random variable
probability distribution
theoretical probability distribution
experimental probability distribution
Law of Large Numbers
expected value

Common Core State Standards

Content Standards
S.MD.7 Analyze decisions and strategies using probability concepts (e.g., product testing, medical testing, pulling a hockey goalie at the end of a game).

Mathematical Practices
2 Reason abstractly and quantitatively.

1 Construct a Probability Distribution

A sample space is the set of all possible outcomes in a distribution. Consider a distribution of values represented by the sum of the values on two dice and a distribution of the miles per gallon for a sample of cars.

Sum of Two Dice

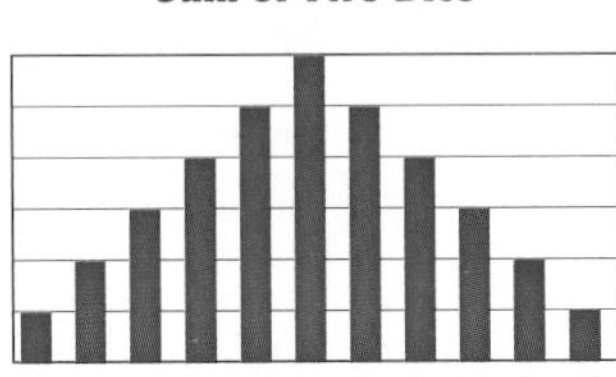

Miles Per Gallon

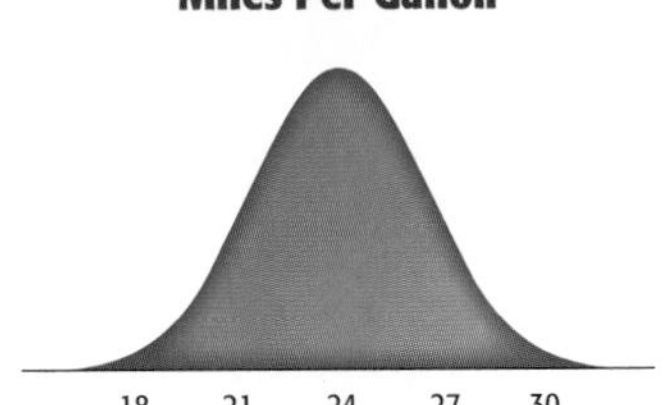

The sum of the values on the dice can be any integer from 2 to 12. So, the sample space is [2, 3, ..., 11, 12]. This distribution is *discrete* because the number of possible values in the sample space can be counted.

The distribution of miles per gallon is *continuous*. While the sample space includes any positive value less than a certain maximum (around 100), the data can take on an infinite number of values within this range.

The value of a **random variable** is the numerical outcome of a random event. A random variable can be discrete or continuous. **Discrete random variables** represent countable values. **Continuous random variables** can take on any value.

Example 1 Identify and Classify Random Variables

PT

Identify the random variable in each distribution, and classify it as *discrete* or *continuous*. Explain your reasoning.

a. the number of songs found on a random selection of mp3 players

The random variable X is the number of songs on any mp3 player in the random selection of players. The number of songs is countable, so X is discrete.

b. the weights of football helmets sent by a manufacturer

The random variable X is the weight of any particular helmet. The weight of any particular helmet can be anywhere within a certain range, typically 6 to 8 pounds. Therefore, X is continuous.

GuidedPractice

1A. the exact distances of a sample of discus throws

1B. the ages of counselors at a summer camp

Jack Hollingsworth/CORBIS

Study Tip

Discrete vs. Continuous Variables representing height, weight, and capacity will always be continuous variables because they can take on any positive value.

A **probability distribution** for a particular random variable is a function that maps the sample space to the probabilities of the outcomes in the sample space. Probability distributions can be represented by tables, equations, or graphs. In this lesson, we will focus on discrete probability distributions.

A probability distribution has the following properties.

KeyConcept Probability Distribution

- A probability distribution can be determined theoretically or experimentally.
- A probability distribution can be discrete or continuous.
- The probability of each value of X must be at least 0 and not greater than 1.
- The sum of all the probabilities for all of the possible values of X must equal 1. That is, $\Sigma P(X) = 1$.

Review Vocabulary

Theoretical and Experimental Probability Theoretical probability is based on assumptions, and experimental probability is based on experiments. (Lesson 0-5)

A **theoretical probability distribution** is based on what is expected to happen. For example, the distribution for flipping a fair coin is $P(\text{heads}) = 0.5$, $P(\text{tails}) = 0.5$.

Example 2 Construct a Theoretical Probability Distribution

X represents the sum of the values on two dice.

a. Construct a relative-frequency table.

The theoretical probabilities associated with rolling two dice can be described using a relative-frequency table. When two dice are rolled, 36 total outcomes are possible. To determine the relative frequency, or theoretical probability, of each outcome, divide the frequency by 36.

Sum	2	3	4	5	6	7	8	9	10	11	12	
Frequency	1	2	3	4	5	6	5	4	3	2	1	Sum: 36
Relative Frequency	$\frac{1}{36}$	$\frac{1}{18}$	$\frac{1}{12}$	$\frac{1}{9}$	$\frac{5}{36}$	$\frac{1}{6}$	$\frac{5}{36}$	$\frac{1}{9}$	$\frac{1}{12}$	$\frac{1}{18}$	$\frac{1}{36}$	

b. Graph the theoretical probability distribution.

The graph shows the probability distribution for the sum of the values on two dice X. The bars are separated on the graph because the distribution is discrete (no other values of X are possible).

Each unique outcome of X is indicated on the horizontal axis, and the probability of each outcome occurring $P(X)$ is indicated on the vertical axis.

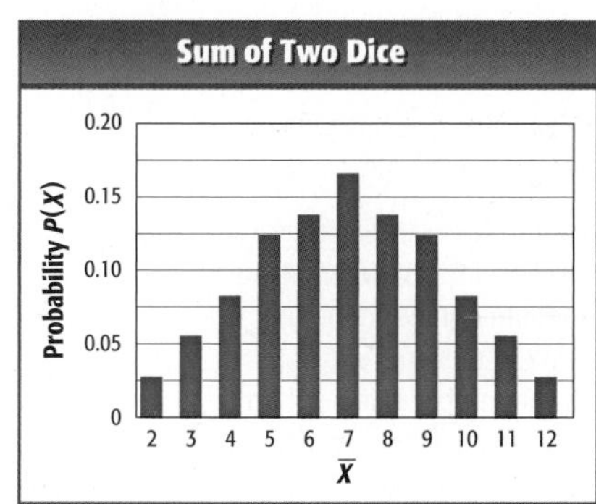

Guided Practice

2. X represents the sum of the values of two spins of the wheel.

A. Construct a relative-frequency table.

B. Graph the theoretical probability distribution.

An **experimental probability distribution** is a distribution of probabilities estimated from experiments. Simulations can be used to construct an experimental probability distribution. When constructing this type of distribution, use the frequency of occurrences of each observed value to compute its probability.

Example 3 Construct an Experimental Probability Distribution

X represents the sum of the values found by rolling two dice.

a. Construct a relative-frequency table.

Roll two dice 100 times or use a random number generator to complete the simulation and create a simulation tally sheet.

ReviewVocabulary

Simulations and Random Number Generators For more practice on simulations and random number generators, see Extend 12-1.

Sum	Tally	Frequency	Sum	Tally	Frequency
2	||	2	8	卌 卌 ||	12
3	卌	5	9	卌 卌 |	11
4	卌 |	6	10	卌 |||	8
5	卌 卌 |||	13	11	|||	3
6	卌 卌 ||||	14	12	||||	4
7	卌 卌 卌 卌 ||	22			

Calculate the experimental probability of each value by dividing its frequency by the total number of trials, 100.

Sum	2	3	4	5	6	7	8	9	10	11	12
Relative Frequency	0.02	0.05	0.06	0.13	0.14	0.22	0.12	0.11	0.08	0.03	0.04

b. Graph the experimental probability distribution.

The graph shows the discrete probability distribution for the sum of the values shown on two dice X.

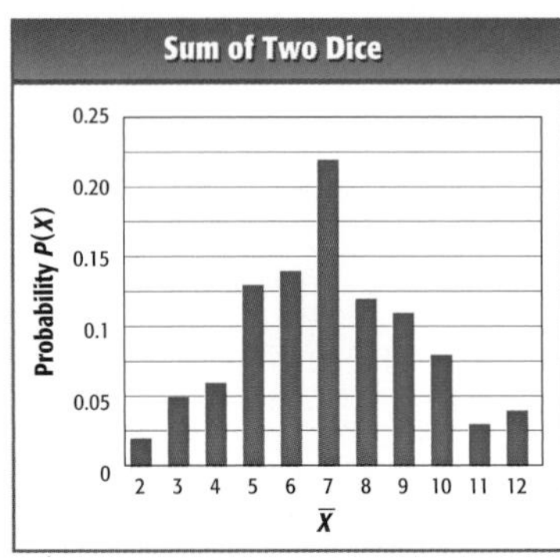

StudyTip

Random Number Generators and Proportions When using a random number generator to simulate events with different probabilities, set up a proportion. For example, suppose there are 3 possible outcomes with probabilities of A: 0.25, B:0.35, and C:0.40. Random numbers 1–25 can represent A, 26–60 represent B, and 61–100 represent C.

GuidedPractice

3. X represents the sum of the values of two spins of the wheel.

   A. Construct a relative-frequency table for 100 trials.

   B. Graph the experimental probability distribution.

Notice that this graph is different from the theoretical graph in Example 2. With small sample sizes, experimental distributions may vary greatly from their associated theoretical distributions. However, as the sample size increases, experimental probabilities will more closely resemble their associated theoretical probabilities. This is due to the **Law of Large Numbers**, which states that the variation in a data set decreases as the sample size increases.

Math HistoryLink

Christian Huygens (1629–1695) This Dutchman was the first to discuss games of chance. "Although in a pure game of chance the results are uncertain, the chance that one player has to win or to lose depends on a determined value." This became known as the *expected value.*

2 Analyze a Probability Distribution

Probability distributions are often used to analyze financial data. The two most common statistics used to analyze a discrete probability distribution are the mean, or expected value, and the standard deviation. The **expected value** $E(X)$ of a discrete random variable of a probability distribution is the weighted average of the variable.

KeyConcept Expected Value of a Discrete Random Variable

Words The expected value of a discrete random variable is the weighted average of the values of the variable. It is calculated by finding the sum of the products of every possible value of X and its associated probability $P(X)$.

Symbols $E(X) = \Sigma\, [X \cdot P(X)]$

Real-World Example 4 Expected Value

GAME SHOWS A game-show contestant has won one spin of the wheel at the right. Find the expected value of his winnings.

Each prize value represents a value of X and each percent represents the corresponding probability $P(X)$. Find $E(X)$.

$$\begin{aligned} E(X) &= \Sigma[X \cdot P(X)] \\ &= 0(0.20) + 25{,}000(0.08) + 15{,}000(0.14) + 10{,}000(0.22) + 5000(0.36) \\ &= 0 + 2000 + 2100 + 2200 + 1800 \\ &= 8100 \end{aligned}$$

The expected value of the contestant's winnings is $8100.

WatchOut!

Expected Value The expected value is what you *expect* to happen in the long run, not necessarily what *will* happen.

GuidedPractice

4. PRIZES Curt won a ticket for a prize. The distribution of the values of the tickets and their relative frequencies are shown. Find the expected value of his winnings.

Value ($)	1	10	100	1000	5000	25,000
Frequency	5000	100	25	5	1	1

Sometimes the expected value does not provide enough information to fully analyze a probability distribution. For example, suppose two wheels had roughly the same expected value. Which one would you choose? Which one is *riskier*? The standard deviation can provide more insight into the expected value of a probability distribution.

The formula for calculating the standard deviation of a probability distribution is similar to the one used for a set of data.

KeyConcept Standard Deviation of a Probability Distribution

Words For each value of X, subtract the mean from X and square the difference. Then multiply by the probability of X. The sum of each of these products is the variance. The standard deviation is the square root of the variance.

Symbols Variance: $\sigma^2 = \Sigma[[X - E(X)]^2 \cdot P(X)]$

Standard Deviation: $\sigma = \sqrt{\sigma^2}$

Real-WorldCareer

Mutual Fund Manager Mutual fund managers buy and sell fund investments according to the investment objective of the fund. Investment management includes financial statement analysis, asset and stock selection, and monitoring of investments. Certification beyond a bachelor's degree is required.

Source: International Financial Services, London

Real-World Example 5 Standard Deviation of a Distribution

DECISION MAKING Jimmy is thinking about investing $10,000 in two different investment funds. The expected rates of return and the corresponding probabilities for each fund are listed below.

Fund A	Fund B
50% chance of an $800 profit	30% chance of a $2400 profit
20% chance of a $1200 profit	10% chance of a $1900 profit
20% chance of a $600 profit	40% chance of a $200 loss
10% chance of a $100 loss	20% chance of a $400 loss

a. Find the expected value of each investment.

Fund A: $E(X) = 0.50(800) + 0.20(1200) + 0.20(600) + 0.10(-100)$ or 750

Fund B: $E(X) = 0.30(2400) + 0.10(1900) + 0.40(-200) + 0.20(-400)$ or 750

An investment of $10,000 in Fund A or Fund B will expect to yield $750.

b. Find each standard deviation.

Fund A:

Profit, X	$P(X)$	$[X - E(X)]^2$	$[X - E(X)]^2 \cdot P(X)$
800	0.50	$(800 - 750)^2 = 2500$	$2500 \cdot 0.50 = 1250$
1200	0.20	$(1200 - 750)^2 = 202{,}500$	$202{,}500 \cdot 0.20 = 40{,}500$
600	0.20	$(600 - 750)^2 = 22{,}500$	$22{,}500 \cdot 0.20 = 4500$
−100	0.10	$(-100 - 750)^2 = 722{,}500$	$722{,}500 \cdot 0.10 = 72{,}250$
			$\Sigma[[(X - E(X)]^2 \cdot P(X)] = 118{,}500$
			$\sqrt{118{,}500} \approx 344.2$

Fund B:

Profit, X	$P(X)$	$[X - E(X)]^2$	$[X - E(X)]^2 \cdot P(X)$
2400	0.30	$(2400 - 750)^2 = 2{,}722{,}500$	$2{,}722{,}500 \cdot 0.30 = 816{,}750$
1900	0.10	$(1900 - 750)^2 = 1{,}322{,}500$	$1{,}322{,}500 \cdot 0.10 = 132{,}250$
−200	0.40	$(-200 - 750)^2 = 902{,}500$	$902{,}500 \cdot 0.40 = 361{,}000$
−400	0.20	$(-400 - 750)^2 = 1{,}322{,}500$	$1{,}322{,}500 \cdot 0.20 = 264{,}500$
			$\Sigma[[(X - E(X)]^2 \cdot P(X)] = 1{,}574{,}500$
			$\sqrt{1{,}574{,}500} \approx 1254.8$

c. Which investment would you advise Jimmy to choose, and why?

Jimmy should choose Fund A. While the funds have identical expected values, the standard deviation of Fund B is almost four times the standard deviation for Fund A. This means that the expected value for Fund B will have about four times the variability than Fund A and will be riskier with a greater chance for gains and losses.

StudyTip

Return on Investment When investing $1000 in a product that has a 6% expected return, the investor can expect a 0.06(1000) or $60 profit.

GuidedPractice

5. DECISION MAKING Compare a $10,000 investment in the two funds. Which investment would you recommend, and why?

Fund C	Fund D
30% chance of a $1000 profit	40% chance of a $1000 profit
40% chance of a $500 profit	30% chance of a $600 profit
20% chance of a $100 loss	15% chance of a $100 profit
10% chance of a $300 loss	15% chance of a $200 loss

Tetra Images/Getty Images

Check Your Understanding

 = Step-by-Step Solutions begin on page R14.

Example 1 **Identify the random variable in each distribution, and classify it as *discrete* or *continuous*. Explain your reasoning.**

1. the number of pages linked to a Web page
2. the number of stations in a cable package
3. the amount of precipitation in a city per month
4. the number of cars passing through an intersection in a given time interval

Examples 2–5 5. X represents the sum of the values of two spins of the wheel.

a. Construct a relative-frequency table showing the theoretical probabilities.
b. Graph the theoretical probability distribution.
c. Construct a relative-frequency table for 100 trials.
d. Graph the experimental probability distribution.
e. Find the expected value for the sum of two spins of the wheel.
f. Find the standard deviation for the sum of two spins of the wheel.

Practice and Problem Solving

Extra Practice is on page R11.

Example 1 **Identify the random variable in each distribution, and classify it as *discrete* or *continuous*. Explain your reasoning.**

6. the number of texts received per week
7. the number of diggs (or "likes") for a Web page
8. the height of a plant after a specific amount of time
9. the number of files infected by a computer virus

Examples 2–5 10. CCSS **PERSEVERANCE** A contestant has won a prize on a game show. The frequency table at the right shows the number of winners for 3200 hypothetical players.

a. Construct a relative-frequency table showing the theoretical probability.
b. Graph the theoretical probability distribution.
c. Construct a relative-frequency table for 50 trials.
d. Graph the experimental probability distribution.
e. Find the expected value.
f. Find the standard deviation.

Prize, X	Winners
\$100	1120
\$250	800
\$500	480
\$1000	320
\$2500	256
\$5000	128
\$7500	64
\$10,000	32

11. **SNOW DAYS** The following probability distribution lists the probable number of snow days per school year at North High School. Use this information to determine the expected number of snow days per year.

Number of Snow Days Per Year									
Days	0	1	2	3	4	5	6	7	8
Probability	0.1	0.1	0.15	0.15	0.25	0.1	0.08	0.05	0.02

12. **CARDS** In a standard deck of 52 cards, there are 4 different suits.

a. If jacks = 11, queens = 12, kings = 13, and aces = 1, what is the expected value of a card that is drawn from a standard deck?
b. If you are dealt 7 cards with replacement, what is the expected number of spades?

13. **RAFFLES** The table shows the probability distribution for a raffle if 100 tickets are sold for $1 each. There is 1 prize for $20, 5 prizes for $10, and 10 prizes for $5.

Distribution of Prizes				
Prize	no prize	$20	$10	$5
Probability	0.84	0.01	0.05	0.10

a. Graph the theoretical probability distribution.

b. Find the expected value.

c. Interpret the results you found in part **b**. What can you conclude about the raffle?

14. **CCSS TOOLS** Based on previous data, the probability distribution of the number of students running for class president is shown.

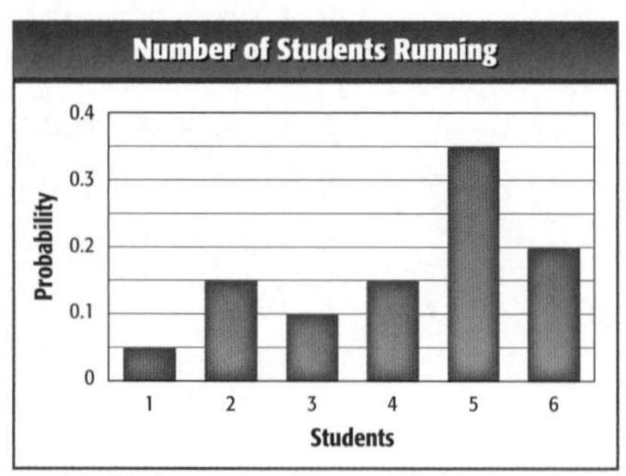

a. Determine the expected number of students who will run. Interpret your results.

b. Construct a relative-frequency table for 50 trials.

c. Graph the experimental probability distribution.

15. **BASKETBALL** The distribution below lists the probability of the number of major upsets in the first round of a basketball tournament each year.

Number of Upsets Per Year									
Upsets	0	1	2	3	4	5	6	7	8
Probability	$\frac{1}{32}$	$\frac{1}{16}$	$\frac{3}{32}$	$\frac{1}{8}$	$\frac{1}{8}$	$\frac{5}{16}$	$\frac{1}{8}$	$\frac{3}{32}$	$\frac{1}{32}$

a. Determine the expected number of upsets. Interpret your results.

b. Find the standard deviation.

c. Construct a relative-frequency table for 50 trials.

d. Graph the experimental probability distribution.

16. **RAFFLES** The French Club sold 500 raffle tickets for $1 each. The first prize ticket will win $100, 2 second prize tickets will each win $10, and 5 third prize tickets each win $5.

a. What is the expected value of a single ticket?

b. Calculate the standard deviation of the probability distribution.

c. **DECISION MAKING** The Glee Club is offering a raffle with a similar expected value and a standard deviation of 2.2. In which raffle should you participate? Explain your reasoning.

17. **DECISION MAKING** Carmen is thinking about investing $10,000 in two different investment funds. The expected rates of return and the corresponding probabilities for each fund are listed below. Compare the two investments using the expected value and standard deviation. Which investment would you advise Carmen to choose, and why?

Fund A
30% chance of a $1900 profit
30% chance of a $600 profit
15% chance of a $200 loss
25% chance of a $500 loss

Fund B
40% chance of a $1600 profit
10% chance of a $900 profit
10% chance of a $300 loss
40% chance of a $400 loss

18. **MULTIPLE REPRESENTATIONS** In this problem, you will investigate geometric probability.

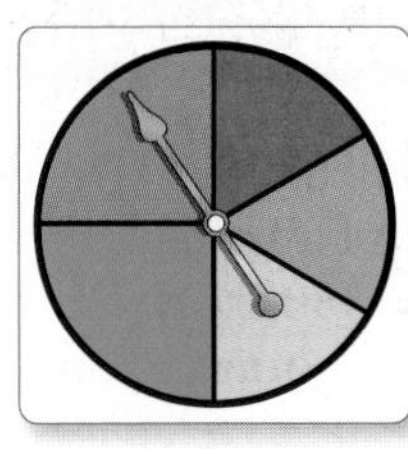

a. **Tabular** The spinner shown has a radius of 2.5 inches. Copy and complete the table below.

Color	Probability	Sector Area	Total Area	$\frac{\text{Sector Area}}{\text{Total Area}}$
red				
orange				
yellow				
green				
blue				

b. **Verbal** Make a conjecture about the relationship between the ratio of the area of the sector to the total area and the probability of the spinner landing on each color.

c. **Analytical** Consider the dartboard shown. Predict the probability of a dart landing in each area of the board. Assume that any dart thrown will land on the board and is equally likely to land at any point on the board.

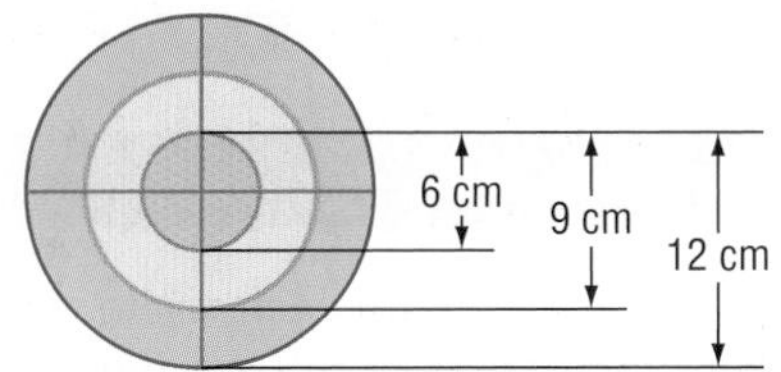

d. **Tabular** Construct a relative-frequency table for throwing 100 darts.

e. **Graphical** Graph the experimental probability distribution.

H.O.T. Problems Use Higher-Order Thinking Skills

19. **CCSS CRITIQUE** Liana and Shannon each created a probability distribution for the sum of two spins on the spinner at the right. Is either of them correct? Explain your reasoning.

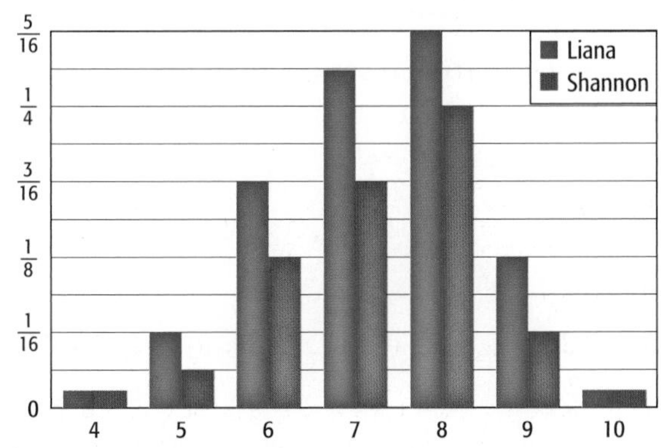

20. **REASONING** Determine whether the following statement is *true* or *false*. Explain.

If you roll a die 10 times, you will roll the expected value at least twice.

21. **OPEN ENDED** Create a discrete probability distribution that shows five different outcomes and their associated probabilities.

22. **REASONING** Determine whether the following statement is *true* or *false*. Explain.

Random variables that can take on an infinite number of values are continuous.

23. **OPEN ENDED** Provide examples of a discrete probability distribution and a continuous probability distribution. Describe the differences between them.

24. **WRITING IN MATH** Compare and contrast two investments that have identical expected values and significantly different standard deviations.

Standardized Test Practice

25. GRIDDED RESPONSE The height $f(x)$ of a bouncing ball after x bounces is represented by $f(x) = 140(0.8)^x$. How many times higher is the first bounce than the fifth bounce?

26. PROBABILITY Andres has a bag that contains 4 red, 6 yellow, 2 blue, and 4 green marbles. If he reaches into the bag and removes a marble without looking, what is the probability that it will not be yellow?

A $\frac{1}{8}$

B $\frac{1}{4}$

C $\frac{3}{8}$

D $\frac{5}{8}$

27. GEOMETRY Find the area of the shaded portion of the figure to the nearest square inch.

60°
15 in.

F 79

G 94

H 589

J 707

28. SAT/ACT If x and y are positive integers, which of the following expressions is equivalent to $\frac{(5^x)^y}{5^x}$?

A 1^y

B ± 1

C 5^y

D 5^{xy-1}

E 5^{xy-x}

Spiral Review

29. ARTICLES Peter and Paul each write articles for an online magazine. Their employer keeps track of the number of *likes* received by each article. (Lesson 11-2)

a. Use a graphing calculator to create a histogram for each data set. Then describe the shape of each distribution.

b. Compare the distributions using either the means and standard deviations or the five-number summaries. Justify your choice.

Peter's Articles
16, 22, 19, 31, 24, 8, 40, 19, 33, 18, 36, 21, 55, 3, 16, 44, 22, 39, 12, 18, 13, 20, 67, 31, 13, 38, 31, 22, 26, 28

Paul's Articles
41, 38, 29, 33, 36, 55, 51, 19, 49, 56, 28, 52, 49, 19, 38, 33, 42, 61, 72, 55, 48, 39, 37, 43, 48, 45, 52, 43, 34, 29

Determine whether the situation calls for a *survey*, an *observational study*, or an *experiment*. Explain your reasoning. (Lesson 11-1)

30. You want to test a drug that reverses male pattern baldness.

31. You want to find voters' opinions on recent legislation.

Find the first five terms of each geometric sequence described. (Lesson 10-3)

32. $a_1 = 0.125, r = 1.5$

33. $a_1 = 0.5, r = 2.5$

34. $a_1 = 4, r = 0.5$

35. $a_1 = 12, r = \frac{1}{3}$

36. $a_1 = 21, r = \frac{2}{3}$

37. $a_1 = 80, r = \frac{5}{4}$

38. COMMUNICATION A microphone is placed at the focus of a parabolic reflector to collect sound for the television broadcast of a football game. Write an equation for the cross section, assuming that the focus is at the origin, the focus is 6 inches from the vertex, and the parabola opens to the right. (Lesson 9-2)

Solve each equation. Check your solutions. (Lesson 7-4)

39. $\log_9 x = \frac{3}{2}$

40. $\log_{\frac{1}{10}} x = -3$

41. $\log_b 9 = 2$

Skills Review

Expand each power.

42. $(a - b)^3$

43. $(m + n)^4$

44. $(r + n)^8$

CHAPTER 11

Mid-Chapter Quiz

Lessons 11-1 through 11-3

Determine whether each situation describes a *survey*, an *experiment*, or an *observational study*. Then identify the sample, and suggest a population from which it may have been selected. (Lesson 11-1)

1. A high school principal wants to test five ideas for a new school mascot. He randomly selects 15 high school students to view pictures of the ideas while he watches and records their reactions.

2. Half of the employees of a grocery store are randomly chosen for an extra hour lunch break. The managers then compare their attitudes with their co-workers.

3. Students want to create a school yearbook. They send out a questionnaire to 100 students asking what they would like to showcase in the yearbook.

4. The producers of a sitcom want to determine if a new character that they are planning to introduce will be well received. They show a clip of the show with the new character to 50 randomly chosen participants and then record the participants' reactions.

5. **MULTIPLE CHOICE** Which survey question is *unbiased*? (Lesson 11-1)

   A Do you like days like today?
   B Which is your favorite theme park, Park A or Park B?
   C Don't you think that carrots taste better than celery?
   D How often do you go to the movies?

6. **PARENTS** The table below shows the ages of parents who volunteered to assist in a neighborhood bake sale. (Lesson 11-2)

Ages of Parents (years)				
28	34	33	45	31
33	41	34	36	42
30	29	32	40	36
29	33	29	28	44
47	31	28	27	29

   a. Use a graphing calculator to create a box-and-whisker plot. Then describe the shape of the distribution.

   b. Describe the center and spread of the data using either the mean and standard deviation or the five-number summary. Justify your choice.

7. **TRAINING** Aiden and Mark's training times for the 40-meter dash are shown. (Lesson 11-2)

Aiden's 40-Meter Dash Times (seconds)					
4.84	4.94	4.87	4.78	5.04	4.98
4.83	5.03	4.74	5.15	4.82	4.91
4.62	4.83	4.76	4.93	4.85	4.82
4.76	4.98	4.94	5.05	4.94	5.04
4.86	4.85	4.71	4.66	4.91	4.82

Mark's 40-Meter Dash Times (seconds)					
5.03	4.76	4.69	4.52	4.81	4.78
4.65	4.66	4.83	4.95	4.64	4.76
4.43	4.64	4.50	4.58	4.68	4.65
4.83	4.78	4.71	4.81	4.76	4.84
4.61	4.63	4.33	4.46	4.74	4.63

   a. Use a graphing calculator to create a histogram for each data set. Then describe the shape of each distribution.

   b. Compare the distributions using either the means and standard deviations or the five-number summaries. Justify your choice.

8. **MULTIPLE CHOICE** Find the expected value of winning one of the following prizes. (Lesson 11-3)

   F $1950
   G $2100
   H $3000
   J $3450

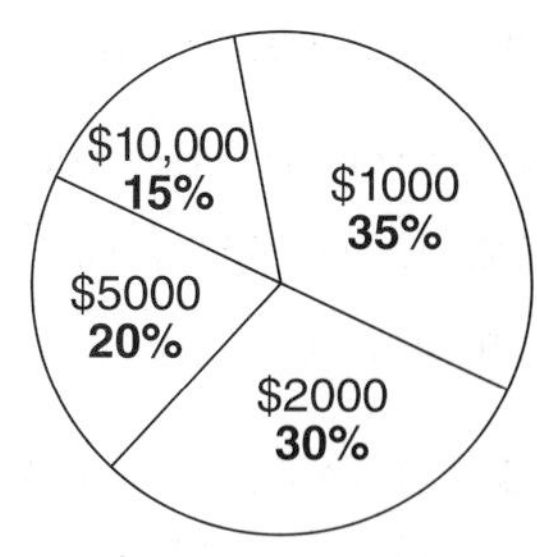

Identify the random variable in each distribution, and classify it as *discrete* or *continuous*. Explain your reasoning. (Lesson 11-3)

9. the number of calls received by an operator

10. the number of books sold at a yard sale

11. the height of students in a gym class

12. the weight of animals on a farm

LESSON 11-4 The Binomial Distribution

Then	Now	Why?
You used the Binomial Theorem.	**1** Identify and conduct a binomial experiment. **2** Find probabilities using binomial distributions.	Jessica forgot to study for her civics quiz. The quiz consists of five multiple-choice questions with each question having four answer choices. Jessica randomly circles an answer for each question. In order to pass, she needs to answer at least four questions correctly.

NewVocabulary
binomial experiment
binomial distribution

Common Core State Standards

Content Standards

S.MD.6 Use probabilities to make fair decisions (e.g., drawing by lots, using a random number generator).

S.MD.7 Analyze decisions and strategies using probability concepts (e.g., product testing, medical testing, pulling a hockey goalie at the end of a game).

Mathematical Practices

4 Model with mathematics.

1 Binomial Experiments

Each question on a multiple-choice quiz, like the one described above, can be thought of as a trial with two possible outcomes, correct or incorrect. If Jessica guesses on each question, the probability that she answers a question correctly is the same for all five questions.

Jessica's guessing on each question is an example of a binomial experiment. A **binomial experiment** is a probability experiment that satisfies the following conditions.

KeyConcept Binomial Experiments

- There is a fixed number of independent trials n.
- Each trial has only two possible outcomes, success or failure.
- The probability of success p is the same in every trial. The probability of failure q is $1 - p$.
- The random variable X is the number of successes in n trials.

Many probability experiments are or can be reduced to binomial experiments.

Example 1 Identify a Binomial Experiment

Determine whether each experiment is a binomial experiment or can be reduced to a binomial experiment. If so, describe a trial, determine the random variable, and state n, p, and q.

a. The spinner at the right is spun 20 times to see how many times it lands on red.

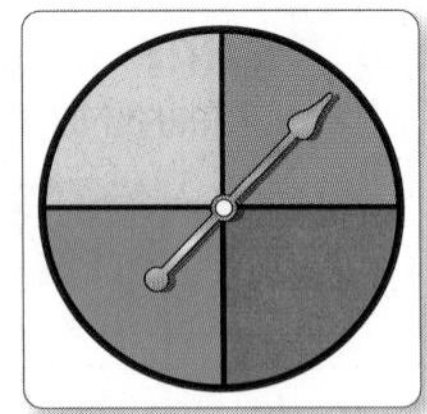

This experiment can be reduced to a binomial experiment with success being that the spinner lands on red and failure being any other outcome. Thus, a trial is a spin, and the random variable X represents the number of reds spun. The number of trials n is 20, the probability of success p is $\frac{1}{4}$ or 0.25, and the probability of failure q is $1 - 0.25$ or 0.75.

b. One hundred students are randomly asked their favorite food.

This is not a binomial experiment because there are many possible outcomes.

GuidedPractice

1A. Seventy-five students are randomly asked if they own a car.

1B. Four cards are removed from a deck to see how many aces are selected.

Fancy Photography/Veer

Use the following guidelines when conducting a binomial experiment.

KeyConcept Conducting Binomial Experiments

Step 1 Describe a trial for the situation, and determine the number of trials to be conducted.

Step 2 Define a success, and calculate the theoretical probabilities of success and failure.

Step 3 Describe the random variable X.

Step 4 Design and conduct a simulation to determine the experimental probability.

A binomial experiment can be conducted to compare experimental and theoretical probabilities.

PT

Example 2 Design a Binomial Experiment

Conduct a binomial experiment to determine the probability of drawing an odd-numbered card from a deck of cards. Then compare the experimental and theoretical probabilities of the experiment.

Step 1 A trial is drawing a card from a deck. The number of trials conducted can be any number greater than 0. We will use 52.

Step 2 A success is drawing an odd-numbered card. The odd-numbered cards in a deck are 3, 5, 7, and 9, and they occur once in each of the four suits. Therefore, there are $4 \cdot 4$ or 16 odd-numbered cards in the deck. The probability of drawing an odd-numbered card, or the probability of success, is $\frac{16}{52}$ or $\frac{4}{13}$. The probability of failure is $1 - \frac{4}{13}$ or $\frac{9}{13}$.

Step 3 The random variable X represents the number of odd-numbered cards drawn in 52 trials.

Step 4 Use the random number generator on a calculator to create a simulation. Assign the integers 0–12 to accurately represent the probability data.

Odd-numbered cards 0, 1, 2, 3
Other cards 4, 5, 6, 7, 8, 9, 10, 11, 12

Make a frequency table and record the results as you run the generator.

Outcome	Tally	Frequency
Odd-Numbered Card	𝍸 𝍸 II	12
Other Cards	𝍸 𝍸 𝍸 𝍸 𝍸 𝍸 𝍸 𝍸	40

An odd-numbered card was drawn 12 times, so the experimental probability is $\frac{12}{52}$ or about 23.1%. This is less than the theoretical probability of $\frac{16}{52}$ or about 30.8%.

StudyTip

Random Number Generator To generate all 52 random numbers using a graphing calculator, press MATH ◀ 5 and then enter the desired range followed by the number of trials. For example, type (0, 12, 52) for Example 2.

GuidedPractice

2. Conduct a binomial experiment to determine the probability of drawing an even-numbered card from a deck of cards. Then compare the experimental and theoretical probabilities of the experiment.

2 Binomial Distribution

Binomial Distribution In the binomial experiment in Example 2, there were 12 successes in 52 trials. If you conducted that same experiment again, there may be any number of successes from 0 to 52. This situation can be represented by a binomial distribution. A **binomial distribution** is a frequency distribution of the probability of each value of X, where the random variable X represents the number of successes in n trials. Because X is a discrete random variable, a binomial distribution is a *discrete probability distribution.*

The probabilities in a binomial distribution can be calculated using the following formula.

> **StudyTip**
>
> **Binomial Probability Formula** In the Binomial Probability Formula, X represents the number of successes in n trials. Thus, the exponent for q, $n - X$, represents the number of failures in n trials.

KeyConcept Binomial Probability Formula

The probability of X successes in n independent trials is

$$P(X) = {}_nC_X\, p^X q^{n-X},$$

where p is the probability of success of an individual trial and q is the probability of failure on that same individual trial ($q = 1 - p$).

Notice that the Binomial Probability Formula is an adaptation of the Binomial Theorem you have already studied. The expression ${}_nC_X\, p^X q^{n-X}$ represents the $p^X q^{n-X}$ term in the binomial expansion of $(p + q)^n$.

PT

Standardized Test Example 3 Find a Probability

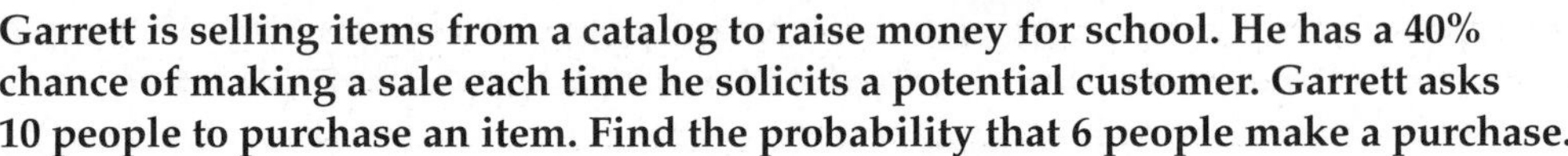

Garrett is selling items from a catalog to raise money for school. He has a 40% chance of making a sale each time he solicits a potential customer. Garrett asks 10 people to purchase an item. Find the probability that 6 people make a purchase.

A 8.6% **B** 11.2% **C** 24% **D** 40%

Read the Test Item

We need to find the probability that 6 people purchase an item. A success is making a sale, so $p = 0.4$, $q = 1 - 0.4$ or 0.6, and $n = 10$.

Solve the Test Item

$P(X) = {}_nC_X\, p^X q^{n-X}$	Binomial Probability Formula
$P(6) = {}_{10}C_6\, (0.4)^6\, (0.6)^{10-6}$	$n = 10$, $X = 6$, $p = 0.4$, and $q = 0.6$
≈ 0.111	Simplify.

The probability of Garrett making six sales is about 0.111 or 11.1%. So, the correct answer is B.

> **Test-TakingTip**
>
> **CCSS Precision** A common error when using the Binomial Probability Formula is to focus on only the successes and to forget the failures. Notice that $P(6) \neq (0.4)^6$ and $P(6) \neq {}_{10}C_6(0.4)^6$.

GuidedPractice

3. **TELEMARKETING** At Jenny's telemarketing job, 15% of the calls that she makes to potential customers result in a sale. She makes 20 calls in a given hour. What is the probability that 5 calls result in a sale?

   F 6.7% **G** 8.3% **H** 10.3% **J** 11.9%

If, on average, 40% of the people Garrett solicits make a purchase and he solicits 10 people, he can probably expect to make 10(0.40) or 4 sales. This value represents the mean of the binomial distribution. In general, the mean of a binomial distribution can be calculated by the following formula.

KeyConcept Mean of a Binomial Distribution

The mean μ of a binomial distribution is given by $\mu = np$, where n is the number of trials and p is the probability of success.

You can find the probability distribution for a binomial experiment by fully expanding the binomial $(p + q)^n$. A probability distribution can be helpful when solving for problems that allow multiple numbers of successes.

Real-World Example 4 Full Probability Distribution

TEST TAKING Refer to the beginning of the lesson.

a. Determine the probabilities associated with the number of questions Jessica answered correctly by calculating the probability distribution.

If there are four answer choices for each question, then the probability that Jessica guesses and answers a question correctly is $\frac{1}{4}$ or 0.25. In this binomial experiment, $n = 5$, $p = 0.25$, and $q = 1 - 0.25$ or 0.75. Expand the binomial $(p + q)^n$.

$$(p + q)^n$$
$$= 1p^5 + 5p^4q + 10p^3q^2 + 10p^2q^3 + 5pq^4 + 1q^5$$
$$= (0.25)^5 + 5(0.25)^4(0.75) + 10(0.25)^3(0.75)^2 + 10(0.25)^2(0.75)^3 + 5(0.25)(0.75)^4 + (0.75)^5$$
$$\approx 0.001 + 0.015 + 0.089 + 0.264 + 0.396 + 0.237$$

0.1%	1.5%	8.9%	26.4%	39.6%	23.7%
5 correct	4 correct	3 correct	2 correct	1 correct	0 correct

The graph shows the binomial probability distribution for the number of questions that Jessica answered correctly.

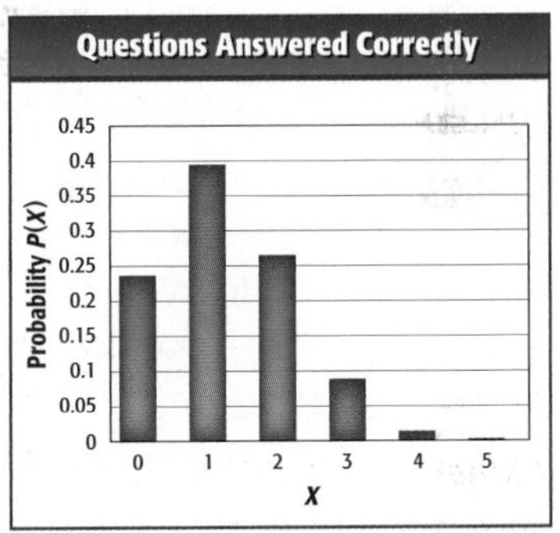

b. What is the probability that Jessica passes the quiz?

Jessica must answer at least four questions correctly to pass the quiz. The probability that Jessica answers *at least* four correct is the sum of the probabilities that she answers four or five correct and is about 1.5% + 0.1% or 1.6%. So, Jessica has about a 1.6% chance of passing, which is not likely.

c. How many questions should Jessica expect to answer correctly?

Find the mean.

$\mu = np$ — Mean of a Binomial Distribution

$= 5(0.25)$ or 1.25 — $n = 5$ and $p = 0.25$

The mean of the distribution is 1.25. On average, Jessica should expect to answer one question correctly when she guesses on five.

Real-WorldLink

ACT The math portion of the ACT college entrance exam includes 60 multiple-choice questions that each have five answer choices.

Source: ACT

StudyTip

Mean and Expected Value The mean of a binomial distribution can be any positive rational number. The expected value of a binomial distribution, however, should be rounded to the nearest whole number since a fraction of a success is not possible.

GuidedPractice

4. TEST TAKING Suppose Jessica's civics quiz consisted of five true-or-false questions instead of multiple-choice questions.

A. Determine the probabilities associated with the number of answers Jessica answered correctly by calculating the probability distribution.

B. What is the probability that Jessica passes the quiz?

C. How many questions should Jessica expect to answer correctly?

Check Your Understanding

= Step-by-Step Solutions begin on page R14.

Example 1

Determine whether each experiment is a binomial experiment or can be reduced to a binomial experiment. If so, describe a trial, determine the random variable, and state n, p, and q.

1. A study finds that 58% of people have pets. You ask 100 people how many pets they have.
2. You roll a die 15 times and find the sum of all of the rolls.
3. A poll found that 72% of students plan on going to the homecoming dance. You ask 30 students if they are going to the homecoming dance.

Example 2

4. Conduct a binomial experiment to determine the probability of drawing an ace or a king from a deck of cards. Then compare the experimental and theoretical probabilities of the experiment.

Example 3

5. **GAMES** Aiden has earned five spins of the wheel on the right. He will receive a prize each time the spinner lands on WIN. What is the probability that he receives three prizes?

A 4.2% **C** 7.1%
B 5.8% **D** 8.8%

Example 4

6. **CCSS PRECISION** A poll at Steve's high school was taken to see if students are in favor of spending class money to expand the junior-senior parking lot. Steve surveyed 6 random students from the population.

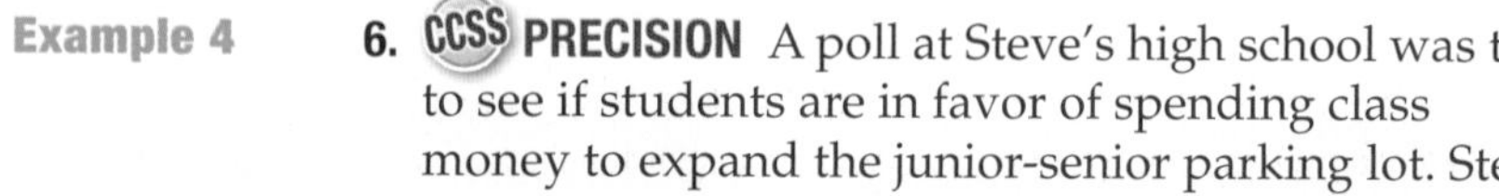

Expand the Parking Lot	
favor	85%
oppose	15%

a. Determine the probabilities associated with the number of students that Steve asked who are in favor of expanding the parking lot by calculating the probability distribution.

b. What is the probability that no more than 2 people are in favor of expanding the parking lot?

c. How many students should Steve expect to find who are in favor of expanding the parking lot?

Practice and Problem Solving

Extra Practice is on page R11.

Example 1

Determine whether each experiment is a binomial experiment or can be reduced to a binomial experiment. If so, describe a trial, determine the random variable, and state n, p, and q.

7. There is a 35% chance that it rains each day in a given month. You record the number of days that it rains for that month.
8. A survey found that on a scale of 1 to 10, a movie received a 7.8 rating. A movie theater employee asks 200 patrons to rate the movie on a scale of 1 to 10.
9. A ball is hidden under one of the hats shown below. A hat is chosen, one at a time, until the ball is found.

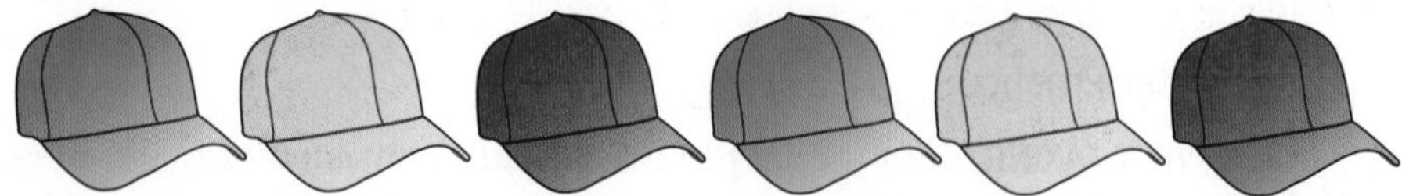

Example 2

10. **DICE** Conduct a binomial experiment to determine the probability of rolling a 7 with two dice. Then compare the experimental and theoretical probabilities of the experiment.
11. **MARBLES** Conduct a binomial experiment to determine the probability of pulling a red marble from the bag. Then compare the experimental and theoretical probabilities of the experiment.

12. **SPINNER** Conduct a binomial experiment to determine the probability of the spinner stopping on an even number. Then compare the experimental and theoretical probabilities of the experiment.

13. **CARDS** Conduct a binomial experiment to determine the probability of drawing a face card out of a standard deck of cards. Then compare the experimental and theoretical probabilities of the experiment.

Example 3

14. **PERSONAL MEDIA PLAYERS** According to a recent survey, 85% of high school students own a personal media player. What is the probability that 6 out of 10 random high school students own a personal media player?

15. **CARS** According to a recent survey, 92% of high school seniors drive their own car. What is the probability that 10 out of 12 random high school students drive their own car?

16. **SENIOR PROM** According to a recent survey, 25% of high school upperclassmen think that the junior-senior prom is the most important event of the school year. What is the probability that 3 out of 15 random high school upperclassmen think this way?

17. **FOOTBALL** A certain football team has won 75.7% of their games. Find the probability that they win 7 of their next 12 games.

18. **GARDENING** Peter is planting 24 irises in his front yard. The flowers he bought were a combination of two varieties, blue and white. The flowers are not blooming yet, but Peter knows that the probability of having a blue flower is 75%. What is the probability that 20 of the flowers will be blue?

19. **FOOTBALL** A field goal kicker is accurate 75% of the time from within 35 yards. What is the probability that he makes exactly 7 of his next 10 kicks from within 35 yards?

Range (yd)	Accuracy (%)
0–35	75
35–45	62
45+	20

20. **BABIES** Mr. and Mrs. Davis are planning to have 3 children. The probability of each child being a boy is 50%. What is the probability that they will have 2 boys?

Example 4

21. **CCSS SENSE-MAKING** According to a recent survey, 52% of high school students own a laptop. Ten random students are chosen.

    a. Determine the probabilities associated with the number of students that own a laptop by calculating the probability distribution.

    b. What is the probability that at least 8 of the 10 students own a laptop?

    c. How many students should you expect to own a laptop?

22. **ATHLETICS** A survey was taken to see the percent of students that participate in sports for their school. Six random students are chosen.

Student Athletics	
0 sports	20%
1 sport	55%
2 sports	20%
3+ sports	5%

    a. Determine the probabilities associated with the number of students playing in at least one sport by calculating the probability distribution.

    b. What is the probability that no more than 2 of the students participated in a sport?

    c. How many students should you expect to have participated in at least one sport?

23. CCSS **MODELING** An online poll showed that 57% of adults still own vinyl records. Moe surveyed 8 random adults from the population.

a. Determine the probabilities associated with the number of adults that still own vinyl records by calculating the probability distribution.

b. What is the probability that no less than 6 of the people surveyed still own vinyl records?

c. How many people should Moe expect to still own vinyl records?

A binomial distribution has a 60% rate of success. There are 18 trials.

24. What is the probability that there will be at least 12 successes?

25. What is the probability that there will be 12 failures?

26. What is the expected number of successes?

27. **DECISION MAKING** Six roommates randomly select someone to wash the dishes each day.

a. What is the probability that the same person has to wash the dishes 3 times in a given week?

b. What method can the roommates use to select who washes the dishes each day?

28. **DECISION MAKING** A committee of five people randomly selects someone to take the notes of each meeting.

a. What is the probability that a person takes notes less than twice in 10 meetings?

b. What method can the committee use to select the notetaker each meeting?

c. If the method described in part **b** results in the same person being notetaker for nine straight meetings, would this result cause you to question the method?

Each binomial distribution has n trials and p probability of success. Determine the most likely number of successes.

29. $n = 8, p = 0.6$
30. $n = 10, p = 0.4$
31. $n = 6, p = 0.8$
32. $n = 12, p = 0.55$
33. $n = 9, p = 0.75$
34. $n = 11, p = 0.35$

35. **SWEEPSTAKES** A beverage company is having a sweepstakes. The probabilities of winning selected prizes are shown at the right. If Ernesto purchases 8 beverages, what is the probability that he wins at least one prize?

Odds of Winning	
beverage	1 in 10
CD	1 in 200
hat	1 in 250
MP3 player	1 in 20,000
car	1 in 25,000,000

Each binomial distribution has n trials and p probability of success. Determine the probability of s successes.

36. $n = 8, p = 0.3, s \geq 2$
37. $n = 10, p = 0.2, s > 2$
38. $n = 6, p = 0.6, s \leq 4$
39. $n = 9, p = 0.25, s \leq 5$
40. $n = 10, p = 0.75, s \geq 8$
41. $n = 12, p = 0.1, s < 3$

H.O.T. Problems Use Higher-Order Thinking Skills

42. **CHALLENGE** A poll of students determined that 88% wanted to go to college. Eight random students are chosen. The probability that at least x students want to go to college is about 0.752 or 75.2%. Solve for x.

43. **WRITING IN MATH** What should you consider when using a binomial distribution to make a decision?

44. **OPEN ENDED** Describe a real-world setting within your school or community activities that seems to fit a binomial distribution. Identify the key components of your setting that connect to binomial distributions.

45. **WRITING IN MATH** Describe how binomial distributions are connected to Pascal's triangle.

46. **WRITING IN MATH** Explain the relationship between a binomial experiment and a binomial distribution.

Standardized Test Practice

47. EXTENDED RESPONSE Carly is taking a 10 question multiple-choice test, in which each question has four choices. If she guesses on each question, what is the probability that she will get

a. 7 questions correct?

b. 9 questions correct?

c. 0 questions correct?

d. 3 questions correct?

48. What is the maximum point of the graph of the equation $y = -2x^2 + 16x + 5$?

A $(-4, -59)$ C $(4, 37)$

B $(-4, -91)$ D $(4, 101)$

49. GEOMETRY On a number line, point X has coordinate -8 and point Y has coordinate 4. Point P is $\frac{2}{3}$ of the way from X to Y. What is the coordinate of P?

F -4 H 0

G -2 J 2

50. SAT/ACT The cost of 4 CDs is d dollars. At this rate, what is the cost, in dollars, of 36 CDs?

A $9d$ D $\frac{d}{36}$

B $144d$ E $\frac{36}{d}$

C $\frac{9d}{4}$

Spiral Review

Identify the random variable in each distribution, and classify it as *discrete* or *continuous*. Explain your reasoning. (Lesson 11-3)

51. the number of customers at an amusement park

52. the running time of a movie

53. the number of hot dogs sold at a sporting event

54. the distance between two cities

55. FINANCIAL LITERACY The prices of entrees offered by a restaurant are shown. (Lesson 11-2)

Prices (dollars)			
11.25	14.75	9.00	17.25
19.75	9.75	20.25	15.50
16.50	21.50	10.25	22.75
12.75	18.50	23.00	13.50

a. Use a graphing calculator to create a box-and-whisker plot. Then describe the shape of the distribution.

b. Describe the center and spread of the data using either the mean and standard deviation or the five-number summary. Justify your choice.

Find the missing value for each arithmetic sequence. (Lesson 10-1)

56. $a_5 = 12, a_{16} = 133, d = ?$

57. $a_9 = -34, a_{22} = 44, d = ?$

58. $a_4 = 18, a_n = 95, d = 7, n = ?$

59. $a_8 = ?, a_{19} = 31, d = 8$

60. $a_6 = ?, a_{20} = 64, d = 7$

61. $a_7 = -28, a_n = 76, d = 8, n = ?$

62. ASTRONOMY The table at the right shows the closest and farthest distances of Venus and Jupiter from the Sun in millions of miles. (Lesson 9-4)

Planet	Closest	Farthest
Venus	66.8	67.7
Jupiter	460.1	507.4

a. Write an equation for the orbit of each planet. Assume that the center of the orbit is the origin and the center of the Sun is a focus that lies on the x-axis.

b. Which planet has an orbit that is closer to a circle?

Write an equivalent exponential or logarithmic function. (Lesson 7-7)

63. $e^{-x} = 5$

64. $e^2 = 6x$

65. $\ln e = 1$

66. $\ln 5.2 = x$

67. $e^{x+1} = 9$

68. $e^{-1} = x^2$

69. $\ln \frac{7}{3} = 2x$

70. $\ln e^x = 3$

Skills Review

71. MUSIC Tina owns 11 pop, 6 country, 16 rock, and 7 rap CDs. Find each probability if she randomly selects 4 CDs.

a. P(2 rock)

b. P(1 rap)

c. P(1 rock and 2 country)

LESSON 11-5 The Normal Distribution

Then

- You constructed and analyzed discrete probability distributions.

Now

1. Use the Empirical Rule to analyze normally distributed variables.
2. Apply the standard normal distribution and *z*-values.

Why?

- Extensive observations of Swiss cherry trees found that the mean flowering date is April 21 with a standard deviation of about 10 days. Therefore, 95% of the time, a Swiss cherry tree will have a flowering date between April 1 and May 3.

NewVocabulary
normal distribution
Empirical Rule
z-value
standard normal distribution

Common Core State Standards

Content Standards
S.ID.4 Use the mean and standard deviation of a data set to fit it to a normal distribution and to estimate population percentages. Recognize that there are data sets for which such a procedure is not appropriate. Use calculators, spreadsheets, and tables to estimate areas under the normal curve.

Mathematical Practices
6 Attend to precision.
8 Look for and express regularity in repeated reasoning.

1 The Normal Distribution

Distributions of mileages of different sample sizes of cars are shown below. As the sample size increases, the distributions become more and more symmetrical and resemble the curve at the right, due to the Law of Large Numbers.

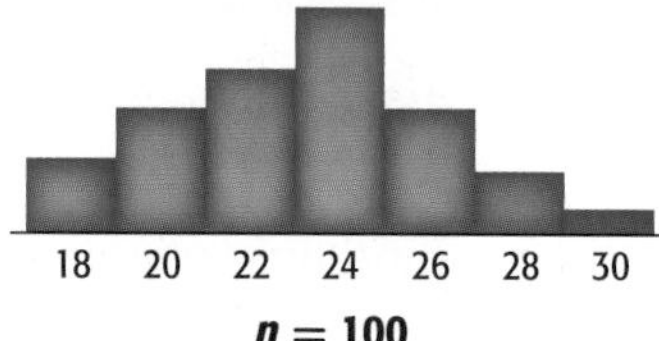

$n = 100$

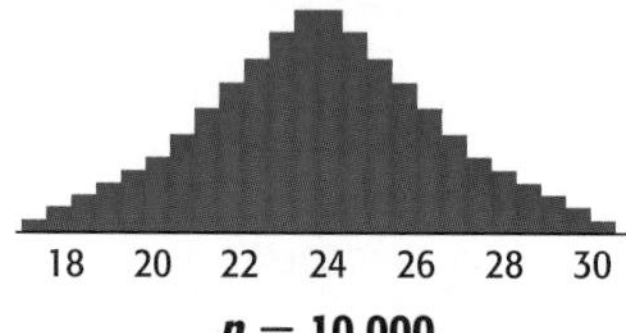

$n = 10{,}000$

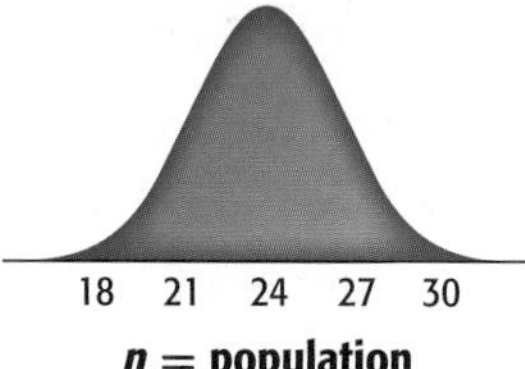

$n =$ population

The curve at the right is a **normal distribution**, a continuous, symmetric, bell-shaped distribution of a random variable. It is the most common *continuous probability distribution*. The characteristics of the normal distribution are as follows.

KeyConcept The Normal Distribution

- The graph of the curve is continuous, bell-shaped, and symmetric with respect to the mean.
- The mean, median, and mode are equal and located at the center.
- The curve approaches, but never touches, the *x*-axis.
- The total area under the curve is equal to 1 or 100%.

The area under the normal curve represents the amount of data within a certain interval or the probability that a random data value falls within that interval. The **Empirical Rule** can be used to determine the area under the normal curve at specific intervals.

KeyConcept The Empirical Rule

In a normal distribution with mean μ and standard deviation σ,

- approximately 68% of the data fall within 1σ of the mean,
- approximately 95% of the data fall within 2σ of the mean, and
- approximately 99.7% of the data fall within 3σ of the mean.

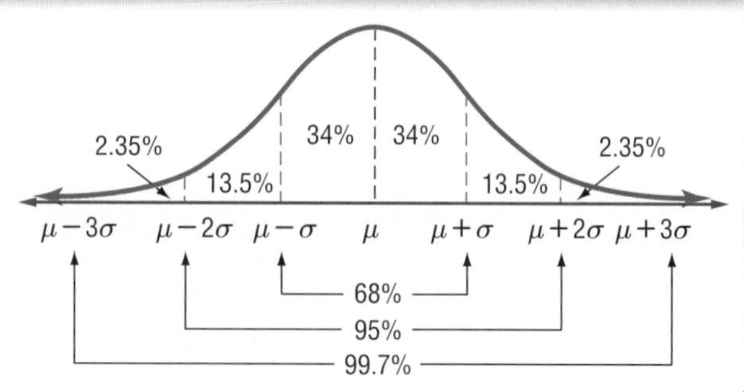

Frank Krahmer/Photographer's Choice RF/Getty Images

StudyTip

Normal Distributions In all of these cases, the number of data values must be large for the distribution to be approximately normal.

Example 1 Use the Empirical Rule to Analyze Data

A normal distribution has a mean of 21 and a standard deviation of 4.

a. Find the range of values that represent the middle 68% of the distribution.

The middle 68% of data in a normal distribution is the range from $\mu - \sigma$ to $\mu + \sigma$. Therefore, the range of values in the middle 68% is $17 < X < 25$.

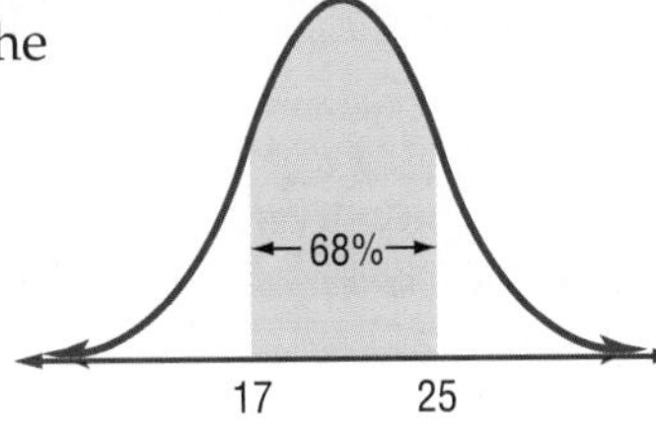

b. What percent of the data will be greater than 29?

29 is 2σ more than μ. 95% of the data fall between $\mu - 2\sigma$ and $\mu + 2\sigma$, so the remaining data values represented by the two tails covers 5% of the distribution. We are only concerned with the upper tail, so 2.5% of the data will be greater than 29.

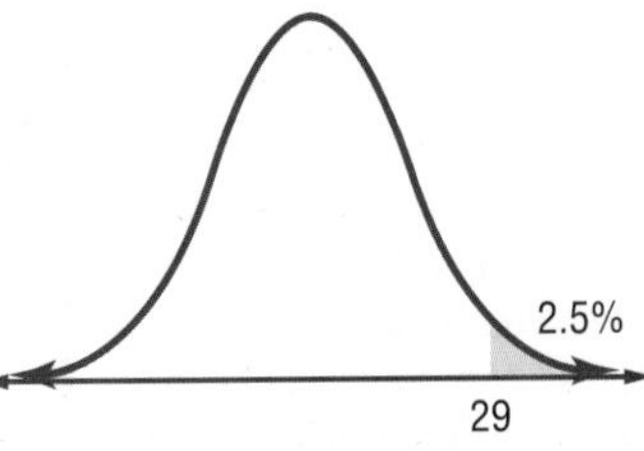

GuidedPractice

1. A normal distribution has a mean of 8.2 and a standard deviation of 1.3.

A. Find the range of values that represent the middle 95% of the distribution.

B. What percent of the data will be less than 4.3?

Real-WorldLink

While the average adult American male is 5 feet 10 inches, the average height of adult males in the Netherlands is the highest worldwide, at almost 6 feet 1 inch.

Source: Eurostats Statistical Yearbook

Real-World Example 2 Use the Empirical Rule to Analyze a Distribution

HEIGHTS The heights of 1800 adults are normally distributed with a mean of 70 inches and a standard deviation of 2 inches.

a. About how many adults are between 66 and 74 inches?

66 and 74 are 2σ away from the mean. Therefore, about 95% of the data are between 66 and 74.

Since $1800 \times 95\% = 1710$, we know that about 1710 of the adults are between 66 and 74 inches tall.

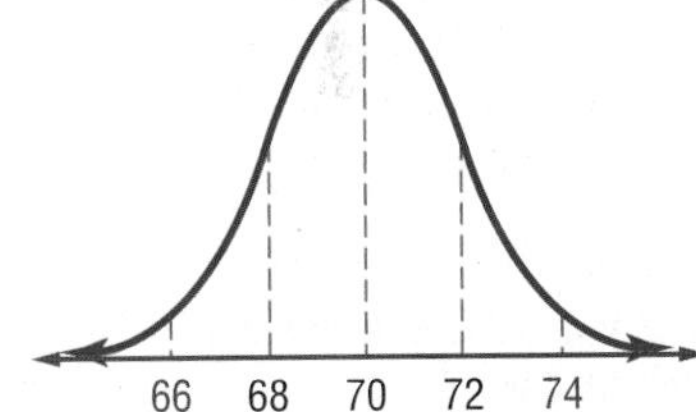

b. What is the probability that a random adult is more than 72 inches tall?

From the curve, values greater than 72 are more than 1σ from the mean. 13.5% are between 1σ and 2σ, 2.35% are between 2σ and 3σ, and 0.15% are greater than 3σ.

So, the probability that an adult selected at random has a height greater than 72 inches is $13.5 + 2.35 + 0.15$ or 16%.

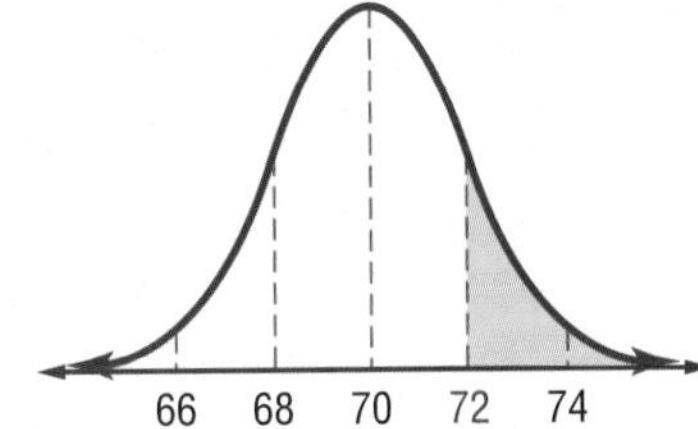

GuidedPractice

2. NETWORKING SITES The number of friends per member in a sample of 820 members is normally distributed with a mean of 38 and a standard deviation of 12.

A. About how many members have between 26 and 50 friends?

B. What is the probability that a random member will have more than 14 friends?

2 Standard Normal Distribution

Standard Normal Distribution The Empirical Rule is only useful for evaluating specific values, such as $\mu + \sigma$. Once the data set is *standardized*, however, any data value can be evaluated. Data are standardized by converting them to z-values, also known as z-scores. The **z-value** represents the number of standard deviations that a given data value is from the mean. Therefore, z-values can be used to determine the position of any data value within a set of data.

KeyConcept Formula for z-Values

The z-value for a data value X in a set of normally distributed data is given by $z = \frac{X - \mu}{\sigma}$, where μ is the mean and σ is the standard deviation.

StudyTip

Symmetry The normal distribution is symmetrical, so when you are asked for the middle or outside set of data, the z-values will be opposites.

Example 3 Use z-Values to Locate Position

PT

Find z if $X = 18$, $\mu = 22$, and $\sigma = 3.1$. Indicate the position of X in the distribution.

$z = \frac{X - \mu}{\sigma}$ Formula for z-values

$= \frac{18 - 22}{3.1}$ $X = 18, \mu = 22, \sigma = 3.1$

≈ -1.29 Simplify.

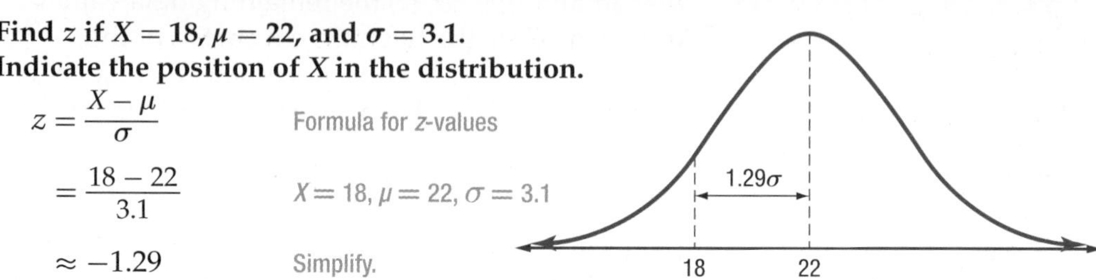

The z-value that corresponds to $X = 18$ is approximately -1.29. Therefore, 18 is about 1.29 standard deviations less than the mean of the distribution.

GuidedPractice

3. Find X if $\mu = 39$, $\sigma = 8.2$, and $z = 0.73$. Indicate the position of X in the distribution.

Any combination of mean and standard deviation is possible for a normally distributed set of data. As a result, there are infinitely many normal probability distributions. This makes comparing two individual distributions difficult. Different distributions *can* be compared, however, once they are standardized using z-values. The **standard normal distribution** is a normal distribution with a mean of 0 and a standard deviation of 1.

StudyTip

Standard Normal Distribution The standard normal distribution is the set of all z-values.

KeyConcept Characteristics of the Standard Normal Distribution

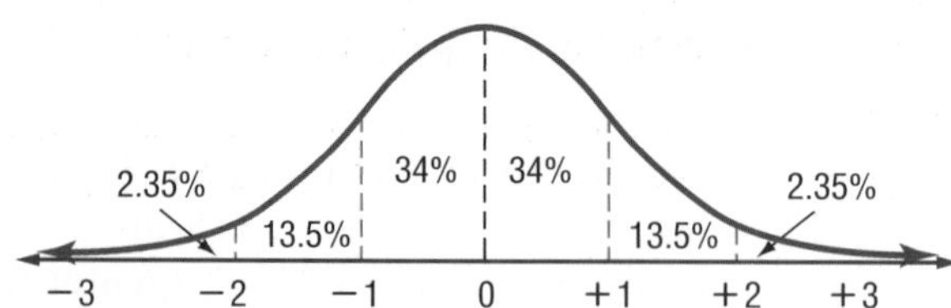

- The total area under the curve is equal to 1 or 100%.
- Almost all of the area is between $z = -3$ and $z = 3$.
- The distribution is symmetric.
- The mean is 0, and the standard deviation is 1.

The standard normal distribution allows us to assign actual areas to the intervals created by z-values. The area under the normal curve corresponds to the proportion of data values in an interval as well as the probability of a random data value falling within the interval. For example, the area between $z = 0$ and $z = 1$ is 0.34. Therefore, the probability of a z-value being in this interval is 34%.

Real-World Example 4 Find Probabilities

Real-WorldLink

Video Uploading According to a recent study, 52% of people who said they upload videos to the Web do it through sites such as Facebook and MySpace. The rest use video-sharing sites like YouTube and Google Video.

Source: *Pew Internet and American Life Project*

VIDEOS The number of videos uploaded daily to a video sharing site is normally distributed with $\mu = 181{,}099$ videos and $\sigma = 35{,}644$ videos. Find each probability. Then use a graphing calculator to sketch the corresponding area under the curve.

a. $P(180{,}000 < X < 200{,}000)$

The question is asking for the percent of days when between 180,000 and 200,000 videos are uploaded. First, find the corresponding z-values for $X = 180{,}000$ and $X = 200{,}000$.

$$z = \frac{X - \mu}{\sigma}$$ Formula for z-values

$$= \frac{180{,}000 - 181{,}099}{35{,}644} \text{ or about } -0.03$$ $X = 180{,}000$, $\mu = 181{,}099$, and $\sigma = 35{,}644$

Use 200,000 to find the other z-value.

$$z = \frac{X - \mu}{\sigma}$$ Formula for z-values

$$= \frac{200{,}000 - 181{,}099}{35{,}644} \text{ or about } 0.53$$ $X = 200{,}000$, $\mu = 181{,}099$, and $\sigma = 35{,}644$

The range of z-values that corresponds to $180{,}000 < X < 200{,}000$ is $-0.03 < z < 0.53$. Find the area under the normal curve within this interval.

You can use a graphing calculator to display the area that corresponds to any z-value by selecting 2nd [DISTR]. Then, under the **DRAW** menu, select **ShadeNorm(lower *z* value, upper *z* value)**. The area between $z = -0.03$ and $z = 0.53$ is about 0.21 as shown in the graph.

[−4, 4] scl: 1 by [0, 0.5] scl: 0.125

Therefore, about 21% of the time, there will be between 180,000 and 200,000 video uploads on a given day.

b. $P(X > 250{,}000)$

$$z = \frac{X - \mu}{\sigma}$$ Formula for z-values

$$= \frac{250{,}000 - 181{,}099}{35{,}644} \text{ or about } 1.93$$ $X = 250{,}000$, $\mu = 181{,}099$, and $\sigma = 35{,}644$

StudyTip

Range of *z*-values The majority of data values are within $\pm 4\sigma$ of the mean, so setting the maximum value of z equal to 4 is sufficient in part **b**. Use the window [−4, 4] by [0, 0.5] when using **ShadeNorm.**

Using a graphing calculator, you can find the area between $z = 1.93$ and $z = 4$ to be about 0.027.

Therefore, the probability that more than 250,000 videos will be uploaded is about 2.7%.

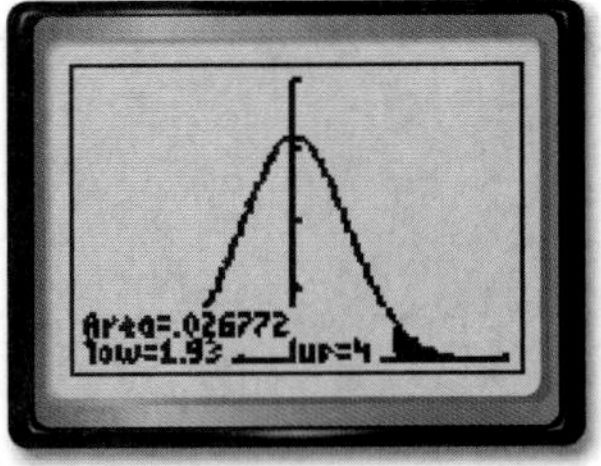

[−4, 4] scl: 1 by [0, 0.5] scl: 0.125

GuidedPractice

4. TIRES The life spans of a certain tread of tire are normally distributed with $\mu = 31{,}066$ miles and $\sigma = 1644$ miles. Find each probability. Then use a graphing calculator to sketch the corresponding area under the curve.

A. $P(30{,}000 < X < 32{,}000)$ **B.** $P(X > 35{,}000)$

Another method for calculating the area between two z-values is 2nd [DISTR] **normalcdf**(*lower z value, upper z value*).

Check Your Understanding

= Step-by-Step Solutions begin on page R14.

Example 1 **A normal distribution has a mean of 416 and a standard deviation of 55.**

1. Find the range of values that represent the middle 99.7% of the distribution.
2. What percent of the data will be less than 361?

Example 2

3. **CCSS TOOLS** The number of texts sent per day by a sample of 811 teens is normally distributed with a mean of 38 and a standard deviation of 7.
   a. About how many teens sent between 24 and 38 texts?
   b. What is the probability that a teen selected at random sent less than 45 texts?

Example 3 **Find the missing variable. Indicate the position of X in the distribution.**

4. z if $\mu = 89$, $X = 81$, and $\sigma = 11.5$
5. z if $\mu = 13.3$, $X = 17.2$, and $\sigma = 1.9$
6. X if $z = -1.38$, $\mu = 68.9$, and $\sigma = 6.6$
7. σ if $\mu = 21.1$, $X = 13.7$, and $z = -2.40$

Example 4

8. **CONCERTS** The number of concerts attended per year by a sample of 925 teens is normally distributed with a mean of 1.8 and a standard deviation of 0.5. Find each probability. Then use a graphing calculator to sketch the area under each curve.
   a. $P(X < 2)$
   b. $P(1 < X < 3)$

Practice and Problem Solving

Extra Practice is on page R11.

Example 1 **A normal distribution has a mean of 29.3 and a standard deviation of 6.7.**

9. Find the range of values that represent the outside 5% of the distribution.
10. What percent of the data will be between 22.6 and 42.7?

Example 2

11. **GYMS** The number of visits to a gym per year by a sample of 522 members is normally distributed with a mean of 88 and a standard deviation of 19.
    a. About how many members went to the gym at least 50 times?
    b. What is the probability that a member selected at random went to the gym more than 145 times?

Example 3 **Find the missing variable. Indicate the position of X in the distribution.**

12. z if $\mu = 3.3$, $X = 3.8$, and $\sigma = 0.2$
13. z if $\mu = 19.9$, $X = 18.7$, and $\sigma = 0.9$
14. μ if $z = -0.92$, $X = 44.2$, and $\sigma = 8.3$
15. X if $\mu = 138.8$, $\sigma = 22.5$, and $z = 1.73$

Example 4

16. **VENDING** A vending machine dispenses about 8.2 ounces of coffee. The amount varies and is normally distributed with a standard deviation of 0.3 ounce. Find each probability. Then use a graphing calculator to sketch the corresponding area under the curve.
    a. $P(X < 8)$
    b. $P(X > 7.5)$

17. **CAR BATTERIES** The useful life of a certain car battery is normally distributed with a mean of 113,627 miles and a standard deviation of 14,266 miles. The company makes 20,000 batteries a month.
    a. About how many batteries will last between 90,000 and 110,000 miles?
    b. About how many batteries will last more than 125,000 miles?
    c. What is the probability that if you buy a car battery at random, it will last less than 100,000 miles?

18. **FOOD** The shelf life of a particular snack chip is normally distributed with a mean of 173.3 days and a standard deviation of 23.6 days.
    a. About what percent of the product lasts between 150 and 200 days?
    b. About what percent of the product lasts more than 225 days?
    c. What range of values represents the outside 5% of the distribution?

19 **FINANCIAL LITERACY** The insurance industry uses various factors including age, type of car driven, and driving record to determine an individual's insurance rate. Suppose insurance rates for a sample population are normally distributed.

a. If the mean annual cost per person is \$829 and the standard deviation is \$115, what is the range of rates you would expect the middle 68% of the population to pay annually?

b. If 900 people were sampled, how many would you expect to pay more than \$1000 annually?

c. Where on the distribution would you expect a person with several traffic citations to lie? Explain your reasoning.

d. How do you think auto insurance companies use each factor to calculate an individual's insurance rate?

20. **STANDARDIZED TESTS** Nikki took three national standardized tests and scored an 86 on all three. The table shows the mean and standard deviation of each test.

	Math	Science	Social Studies
μ	76	81	72
σ	9.7	6.2	11.6

a. Calculate the z-values that correspond to her score on each test.

b. What is the probability of a student scoring an 86 or *lower* on each test?

c. On which test was Nikki's standardized score the highest? Explain your reasoning.

H.O.T. Problems Use Higher-Order Thinking Skills

21. **CCSS CRITIQUE** A set of normally distributed tree diameters have mean 11.5 centimeters, standard deviation 2.5, and range from 3.6 to 19.8. Monica and Hiroko are to find the range that represents the middle 68% of the data. Is either of them correct? Explain.

Monica	Hiroko
The data span 16.2 cm. 68% of 16.2 is about 11 cm. Center this 11-cm range around the mean of 11.5 cm. This 68% group will range from about 6 cm to about 17 cm.	The middle 68% span from $\mu + \sigma$ to $\mu - \sigma$. So we move 2.5 cm below 11.5 and then 2.5 cm above 11.5. The 68% group will range from 9 cm to 14 cm.

22. **CHALLENGE** A case of portable media players has an average battery life of 8.2 hours with a standard deviation of 0.7 hour. Eight of the players have a battery life greater than 9.3 hours. If the sample is normally distributed, how many portable media players are in the case?

23. **REASONING** The term *six sigma process* comes from the notion that if one has six standard deviations between the mean of a process and the nearest specification limit, there will be practically no items that fail to meet the specifications. Is this a true assumption? Explain.

24. **REASONING** *True* or *false*: According to the Empirical Rule, in a normal distribution, most of the data will fall within one standard deviation of the mean. Explain.

25. **OPEN ENDED** Find a set of real-world data that appears to be normally distributed. Calculate the range of values that represent the middle 68%, the middle 95%, and the middle 99.7% of the distribution.

26. **WRITING IN MATH** Describe the relationship between the z-value, the position of an interval of X in the normal distribution, the area under the normal curve, and the probability of the interval occurring. Use an example to explain your reasoning.

Standardized Test Practice

27. The lifetimes of 10,000 light bulbs are normally distributed. The mean lifetime is 300 days, and the standard deviation is 40 days. How many light bulbs will last between 260 and 340 days?

A 2500 **C** 5000
B 3400 **D** 6800

28. Which description best represents the graph?

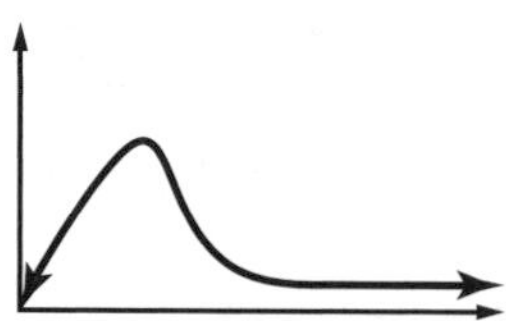

F negatively skewed **H** normal distribution
G no correlation **J** positively skewed

29. SHORT RESPONSE In the figure below, $RT = TS$ and $QR = QT$. What is the value of x?

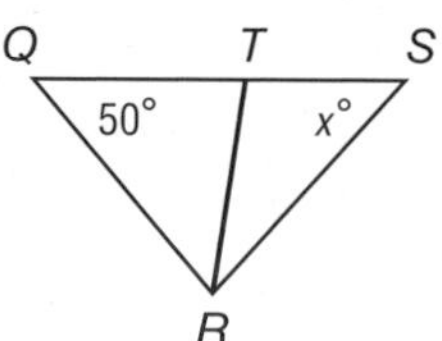

30. SAT/ACT The integer 99 can be expressed as a sum of n consecutive positive integers. The value of n could be which of the following?

I. 2
II. 3
III. 6

A I only **D** I and II
B II only **E** I, II, and III
C III

Spiral Review

31. SNOW There is a 25% chance that it snows each day during a given week. Find the probability that it snows 3 out of the next 7 days. (Lesson 11-4)

Identify the random variable in each distribution, and classify it as *discrete* or *continuous*. Explain your reasoning. (Lesson 11-3)

32. the number of pages in a newspaper

33. the amount of precipitation in a city per month

34. BRIDGES The Sydney Harbour Bridge connects the central business district to northern metropolitan Sydney. It has an arch in the shape of a parabola that opens downward. Write an equation of a parabola to model the arch, assuming that the origin is at the surface of the water, beneath the vertex of the arch. (Lesson 9-2)

Identify the type of function represented by each graph. (Lesson 2-7)

35.

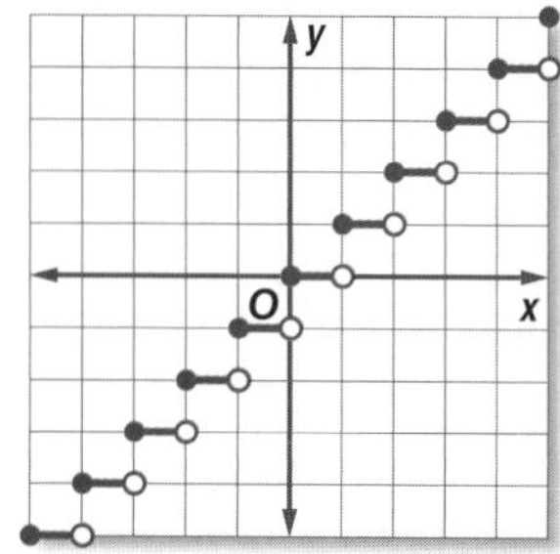

36.

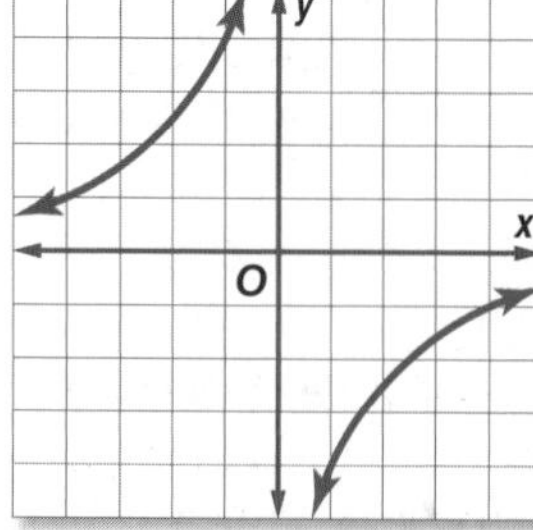

37.

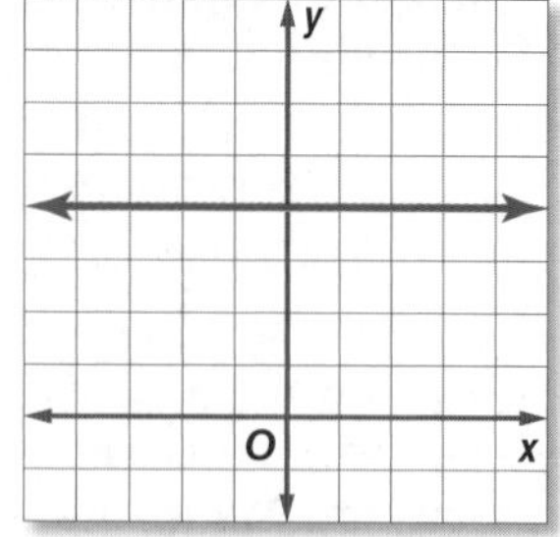

Alberto Coto/Photodisc/Getty Images

Skills Review

38. Calculate the standard deviation of the population of data.

13	18	17	21	16	9	11	28	8	10
7	19	16	16	12	19	21	11	8	13

EXTEND 11-5

Spreadsheet Lab: Normal Approximation of Binomial Distributions

In Lesson 11-4, you used a binomial expansion to find a full probability distribution. A spreadsheet can be used to quickly find and graph a full distribution for any number of trials.

Common Core State Standards
Content Standards
S.ID.4 Use the mean and standard deviation of a data set to fit it to a normal distribution and to estimate population percentages. Recognize that there are data sets for which such a procedure is not appropriate. Use calculators, spreadsheets, and tables to estimate areas under the normal curve.

Activity 1 Full Probability Distribution

PLAYING CARDS **Tom randomly selects a card from a deck of 52 playing cards, records its suit, and replaces it. Use a spreadsheet to construct and graph a full probability distribution for *X*, the number of hearts that Tom selects if he chooses 4, 20, or 100 cards.**

Since one fourth of the cards in a standard deck are hearts, the probability of success is 25% or 0.25, and the probability of failure is 75% or 0.75.

Step 1 Enter the numbers 0 to 4 in column A. In B1, enter the binomial probability formula as =COMBIN(4,A1)*(0.25)^A1*(0.75)^(4−A1). Copy and paste this formula in cells B2:B5.

Step 2 Select cells B1:B5, and insert a clustered column bar graph. Use the values in column A for Category *(X)* axis labels.

Step 3 Select cells A1:A5, and autofill the through A21. In C1, enter the formula =COMBIN(20,A1)*(0.25)^A1* (0.75)^(20−A1). Copy this formula in cells C2:C21. Repeat Step 2 to graph this distribution.

Step 4 Autofill column A through A101. In D1, enter the formula =COMBIN(100,A1)*(0.25)^A1* (0.75)^(100−A1). Copy this formula in cells D2:C101. Repeat Step 2 to graph this distribution.

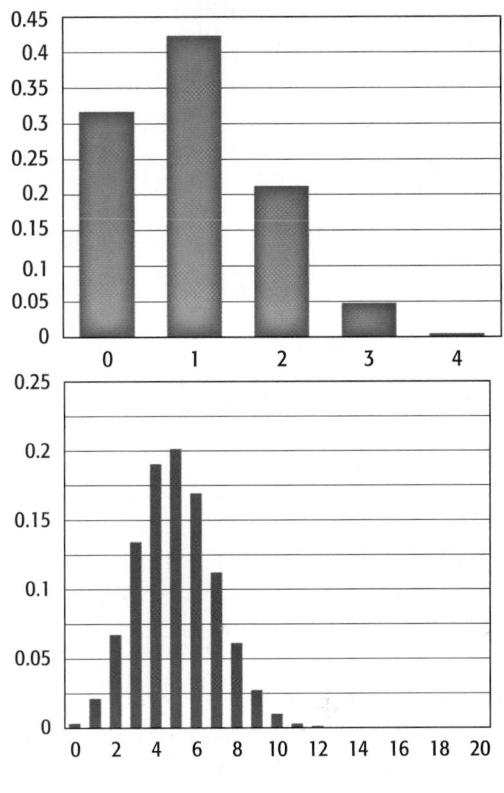

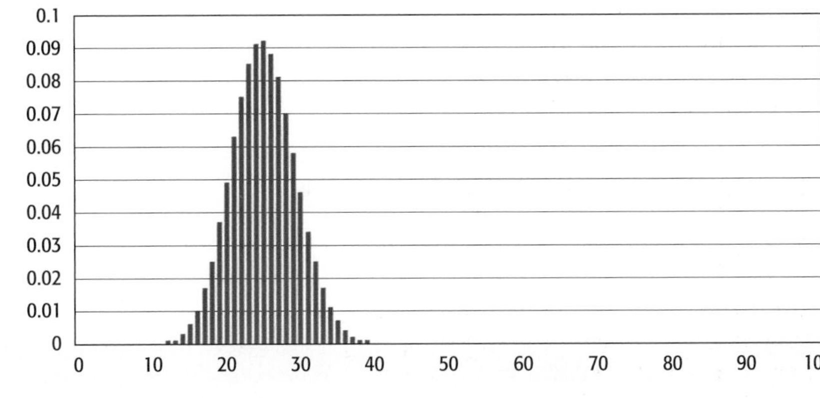

Notice that the number of trials increases, the shape of the graph of the distribution becomes more symmetric with its center at the mean, $\mu = np$. If the number of trials becomes large enough, the shape of the distribution approaches a normal curve. Therefore, a normal distribution can be used to approximate the binomial distribution.

KeyConcept Normal Approximation of a Binomial Distribution

In a binomial distribution with n trials, a probability of success p, and a probability of failure q, such that $np \geq 5$ and $nq \geq 5$, a binomial distribution can be approximated by a normal distribution with $\mu = np$ and $\sigma = \sqrt{npq}$.

(continued on the next page)

Normal Approximation of Binomial Distributions *Continued*

Once the mean and standard deviation are calculated, a z-value can be determined, and the corresponding probability can be found as demonstrated in the next activity.

Activity 2 Normal Approximation of a Binomial Distribution

JURY DUTY According to a poll, 60% of the registered voters in a city have never been called for jury duty. Mariah conducts a random survey of 300 registered voters. What is the probability that at least 170 of those voters have never been called for jury duty?

This is a binomial experiment with $n = 300$, $p = 0.6$, and $q = 0.4$. Since $np = 300(0.6)$ or 180 and $nq = 300(0.4)$ or 120 are both greater than 5, the normal distribution can be used to approximate the binomial distribution.

Step 1 The mean μ is np or 180. Find the standard deviation σ.

$\sigma = \sqrt{npq}$ — Standard deviation of a binomial distribution

$= \sqrt{300(0.6)(0.4)}$ — $n = 300$, $p = 0.6$, and $q = 0.4$

≈ 8.49 — Simplify.

Step 2 Write the problem in probability notation using X. The probability that at least 170 people have never been called for jury duty is $P(X \geq 170)$.

Step 3 Find the corresponding z-value for X.

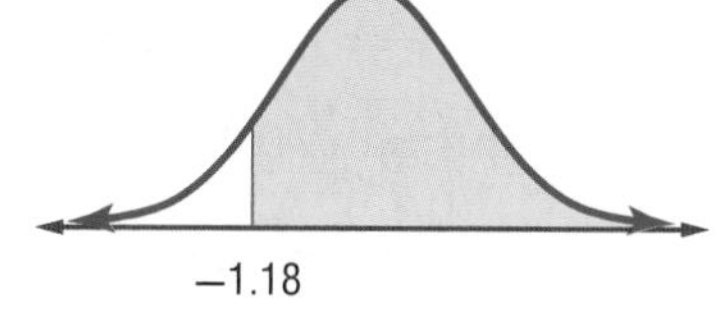

$z = \frac{X - \mu}{\sigma}$ — Formula for z-values

$= \frac{170 - 180}{8.49}$ — $X = 170$, $\mu = 180$, and $\sigma = 8.49$

≈ -1.18 — Simplify.

Step 4 Use a calculator to find the area under the normal curve to the right of z.

KEYSTROKES: [2nd] [DISTR] 2 [(−)] 1.18 [,] 4 [)] [ENTER]

The approximate area to the right of z is 0.881. Therefore, the probability that at least 170 of the registered voters have never been called for jury duty is about 88.1%.

Exercises

1. Use a spreadsheet to construct the graph of the full probability distribution for X, the number of times a 3 is rolled from rolling a die 25 times.

2. **SUMMER JOBS** According to an online poll, 80% of high school upperclassmen have summer jobs. Tadeo thinks the number should be lower so he conducts a survey of 480 random upperclassmen. What is the probability that no more than 380 of the surveyed upperclassmen have summer jobs?

3. **WORK** According to an online poll, 28% of adults feel that the standard 40-hour work week should be increased. Sheila interviews 300 adults at the mall. What is the probability that more than 80 but fewer than 100 of those surveyed will say that the work week should be increased?

Confidence Intervals and Hypothesis Testing

:: Then	:: Now	:: Why?
• You applied the standard normal distribution and *z*-values.	• **1** Find confidence intervals for normally distributed data. **2** Perform hypothesis tests on normally distributed data.	• In a recent Gallup Poll, 1514 teens who owned a portable media player had an average of 1033 songs. The poll had the following disclaimer: "For results based on the total sample of national teens, one can say with 95% confidence that the margin of sampling error is ±31 songs."

NewVocabulary
inferential statistics
statistical inference
confidence interval
maximum error of estimate
hypothesis test
null hypothesis
alternative hypothesis
critical region
left-tailed test
two-tailed test
right-tailed test

Common Core State Standards

Content Standards
S.IC.1 Understand statistics as a process for making inferences about population parameters based on a random sample from that population.
S.IC.4 Use data from a sample survey to estimate a population mean or proportion; develop a margin of error through the use of simulation models for random sampling.

Mathematical Practices
7 Look for and make use of structure.

1 Confidence Intervals

Inferential statistics are used to draw conclusions or **statistical inferences** about a population using a sample. For example, the sample mean of 1033 songs per portable media player can be used to estimate the population mean.

A **confidence interval** is an estimate of a population parameter stated as a range with a specific degree of certainty. Typically, statisticians use 90%, 95%, and 99% confidence intervals, but any percentage can be considered. In the opening example, we are 95% confident that the population mean is within 31 songs of 1033.

The confidence interval for a normal distribution is equivalent to the area under the standard normal curve between $-z$ and z, as shown. A 95% confidence interval for a population mean implies that we are 95% sure that the mean will fall within the range of z-values.

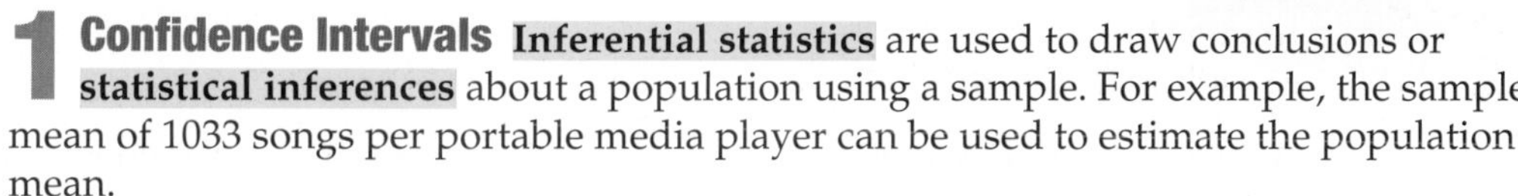

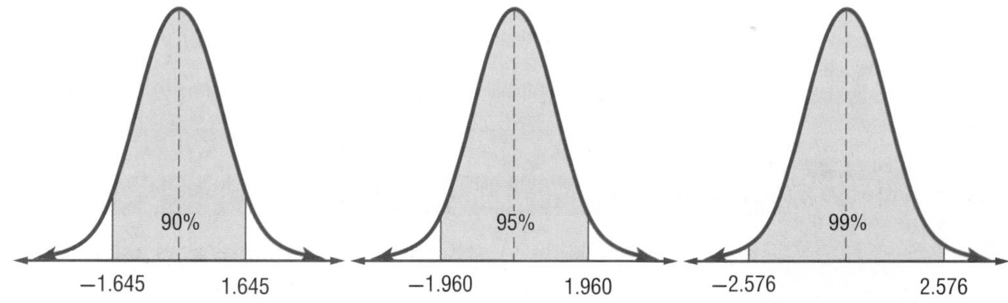

Suppose you want 95% confidence when conducting an experiment. The corresponding z-value is 1.960, where 2.5% of the area lies to the left of $-z$ and 2.5% lies to the right of z.

To find the confidence interval, use the maximum error of estimate and the sample mean. The **maximum error of estimate** E is the maximum difference between the estimate of the population mean μ and its actual value.

KeyConcept Maximum Error of Estimate

The maximum error of estimate E for a population mean is given by

$$E = z \cdot \frac{s}{\sqrt{n}},$$

where z is the z-value that corresponds to a particular confidence level, s is the standard deviation of the sample, and n is the sample size; $n \geq 30$.

Fancy Photography/Veer

Example 1 Maximum Error of Estimate

SOCIAL NETWORKING A poll of 218 randomly selected members of a social networking Web site showed that they spent an average of 14 minutes per day on the site with a standard deviation of 3.1 minutes. Use a 95% confidence interval to find the maximum error of estimate for the time spent on the site.

In a 95% confidence interval, 2.5% of the area lies in each tail. The corresponding z-value is 1.960.

$E = z \cdot \frac{s}{\sqrt{n}}$	Maximum Error of Estimate
$= 1.960 \cdot \frac{3.1}{\sqrt{218}}$	$z = 1.960$, $s = 3.1$, and $n = 218$
≈ 0.41	Simplify.

This means that you can be 95% confident that the population mean time spent on the site will be within 0.41 minute of the sample mean of 14 minutes.

StudyTip

CCSS Perseverance The *z*-value that corresponds to a particular confidence level is known as the *critical value*. The most commonly used levels and their corresponding *z*-value are shown below.

Confidence Level	*z*-Value
90%	1.645
95%	1.960
99%	2.576

GuidedPractice

1. **TEACHING** A poll of 184 randomly selected high school teachers showed that they spend an average of 16.8 hours per week grading, planning, and preparing for class. The standard deviation is 2.9 hours. Use a 90% confidence interval to find the maximum error of estimate for the amount of time spent per week.

Once the maximum error of estimate E is found, a confidence interval CI for the population mean can be determined by adding $\pm E$ to the sample mean.

KeyConcept Confidence Interval for the Population Mean

A confidence interval for a population mean is given by

$$CI = \overline{x} \pm E \text{ or } \overline{x} \pm z \cdot \frac{s}{\sqrt{n}},$$

where $\overline{x}$ is the sample mean and E is the maximum error of estimate.

Real-World Example 2 Confidence Interval

SCHOOL WORK A sample of 200 students was asked the average time they spent on homework each weeknight. The mean time was 52.5 minutes with a standard deviation of 5.1 minutes. Determine a 99% confidence interval for the population mean.

$CI = \overline{x} \pm z \cdot \frac{s}{\sqrt{n}}$	Confidence Interval for Population Mean
$= 52.5 \pm 2.576 \cdot \frac{5.1}{\sqrt{200}}$	$\overline{x} = 52.5$, $z = 2.576$, $s = 5.1$, and $n = 200$
$\approx 52.5 \pm 0.93$	Use a calculator.

The 99% confidence interval is $51.57 \leq \mu \leq 53.43$. Therefore, we are 99% confident that the population mean time is between 51.57 and 53.43 minutes.

Real-WorldLink

On average, students sleep 8.1 hours per weekday and spend 7.5 hours on educational activities, such as attending class or doing homework.

Source: Bureau of Labor Statistics

GuidedPractice

2. **SCHOOL ATHLETICS** A sample of 224 students showed that they attend an average of 2.6 school athletic events per year with a standard deviation of 0.8. Determine a 90% confidence interval for the population mean.

2 Hypothesis Testing

While a confidence interval provides an estimate of the population mean, a **hypothesis test** is used to assess a specific claim about the mean. Typical claims are that the mean is equal to, is greater than, or is less than a specific value.

There are two parts to a hypothesis test: the null hypothesis and the alternative hypothesis. The **null hypothesis** H_0 is a statement of equality to be tested. The **alternative hypothesis** H_a is a statement of inequality that is the complement of the null hypothesis. A *claim* can be part of the null or alternative hypothesis.

PT

Example 3 Claims and Hypotheses

Identify the null and alternative hypotheses for each statement. Then identify the statement that represents the claim.

a. A school administrator thinks it takes less than 3 minutes to evacuate the entire building for a fire drill.

less than 3 minutes — not less than three minutes

$\mu < 3$ — $\mu \geq 3$

The claim is $\mu < 3$, and it is the alternative hypothesis because it does not include equality. The null hypothesis is $\mu \geq 3$, which is the complement of $\mu < 3$.

H_0: $\mu \geq 3$ — H_a: $\mu < 3$ (claim)

b. The owner of a deli says that there are 2 ounces of ham in a sandwich.

H_0: $\mu = 2$ (claim); H_a: $\mu \neq 2$

The claim is $\mu = 2$, and it is the null hypothesis because it is an equality. The alternative hypothesis is $\mu \neq 2$, which is the complement of $\mu = 2$.

GuidedPractice

3A. Jill thinks it takes longer than 10 minutes for the school bus to reach school from her stop.

3B. Mindy thinks there aren't 35 peanuts in every package.

ReadingMath

Significance Level
A significance level of 5% means that we need the data to provide evidence against H_0 so strong that the data would occur no more than 5% of the time when H_0 is true.

Once a claim is made, a sample is collected and analyzed, and the null hypothesis is either *rejected* or *not rejected.* From this information, the claim is then rejected or not rejected. This is determined by whether the sample mean falls within the critical region. The **critical region** is the range of values that suggests a significant enough difference to reject the null hypothesis. The critical region is determined by the significance level α, most commonly 1%, 5%, or 10%, and the inequality sign of the alternate hypothesis.

ConceptSummary Tests of Significance

If H_a: $\mu < k$, the hypothesis test is a **left-tailed test**.	If H_a: $\mu \neq k$, the hypothesis test is a **two-tailed test**.	If H_a: $\mu > k$, the hypothesis test is a **right-tailed test**.

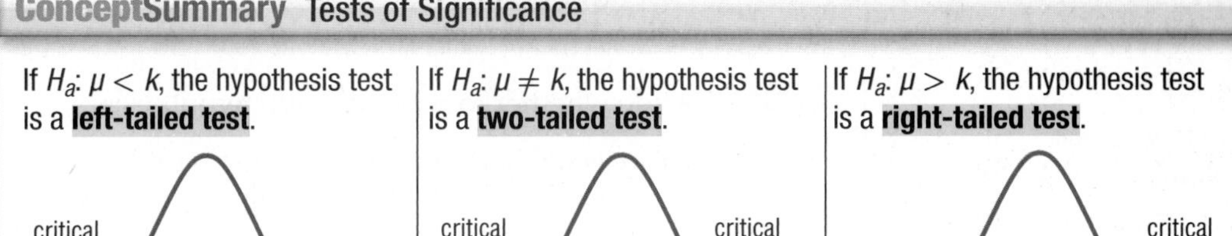

The value k represents the claim about the population mean μ.

StudyTip

z-Values and *z*-Statistics In hypothesis testing, a *z*-value is used to determine the critical region. A *z*-statistic is calculated to see if the sample mean falls within that critical region, and thus determines whether or not to reject the null hypothesis.

The type of hypothesis test that we are using in this lesson is known as a *z*-test. Once the area corresponding to the significance level is determined, a *z*-statistic for the sample mean is calculated. The *z*-statistic is the *z*-value for the sample. When testing at 5% significance, it is only 5% likely for the *z*-statistic to fall within the critical region *and* for H_0 to be true. Therefore, when the *z*-statistic falls within the critical region, H_0 is rejected.

Use these steps to test a hypothesis.

KeyConcept Steps for Hypothesis Testing

Step 1 State the hypotheses and identify the claim.

Step 2 Determine the critical value(s) and critical region.

Step 3 Calculate the z-statistic using $z = \dfrac{\overline{x} - \mu}{\frac{s}{\sqrt{n}}}$.

Step 4 Reject or fail to reject the null hypothesis.

Step 5 Make a conclusion about the claim.

Real-WorldCareer

Politician Politicians are elected officials in the federal, state, and local governments. They influence public decision making and make laws that affect everyone. Many politicians have business, teaching, or legal experience prior to getting involved in politics.

Real-World Example 4 One-Sided Hypothesis Test

STUDENT COUNCIL Lindsey, the president of the student council, has heard complaints that the cafeteria lunch line moves too slowly. The dining services coordinator assures her that the average wait time is 6 minutes or less. Using a sample of 65 customers, Lindsey calculated a mean wait time of 6.3 minutes and a standard deviation of 1.1 minutes. Test the hypothesis at 5% significance.

Step 1 State the claim and hypotheses. H_0: $\mu \leq 6$ (claim) and H_a: $\mu > 6$.

Step 2 Determine the critical region.

The alternative hypothesis is $\mu > 6$, so this is a right-tailed test. We are testing at 5% significance, so we need to identify the z-value that corresponds with the upper 5% of the distribution.

KEYSTROKES: 2nd [DISTR] 3 .95 ENTER

Step 3 Calculate the z-statistic for the sample data.

$$z = \frac{\overline{x} - \mu}{\frac{s}{\sqrt{n}}} \qquad \text{Formula for } z\text{-statistic}$$

$$= \frac{6.3 - 6}{\frac{1.1}{\sqrt{65}}} \qquad \overline{x} = 6.3, \mu = 6, s = 1.1, \text{ and } n = 65$$

$$\approx 2.20 \qquad \text{Simplify.}$$

StudyTip

Rejecting the Null Hypothesis When the null hypothesis is rejected, it is implied that the alternative hypothesis is accepted.

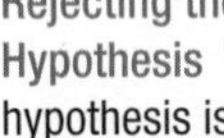

Step 4 Decide whether to reject the null hypothesis.

H_0 is rejected because the z-statistic for the sample falls within the critical region.

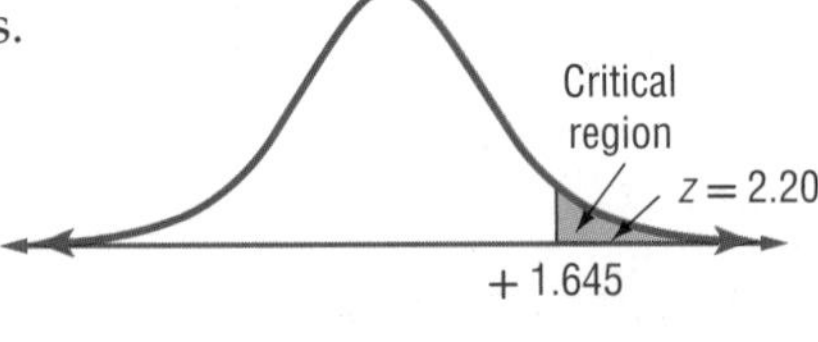

Step 5 Make a conclusion about the claim.

Therefore, there is enough evidence to reject the claim that the average wait time is 6 minutes or less.

GuidedPractice

4. **EXERCISE** Julian thinks it takes less than 60 minutes for people to get in a full workout at the gym. Using a sample of 128 members, he calculated a mean workout time of 58.4 minutes with a standard deviation of 8.9 minutes. Test the hypothesis at 10% significance.

For a two-sided test, the significance level α must be divided by 2 in order to determine the critical value at each tail.

PT

Example 5 Two-Sided Hypothesis Test

ADVERTISEMENTS **Jerrod wants to determine if the advertisement shown is accurate. Using a sample of 42 pizzas, Jerrod calculates a mean of 99.8 pieces and a standard deviation of 0.8. Test the hypothesis at 5% significance.**

Step 1 State the claim and hypotheses.

H_0: $\mu = 100$ (claim) and H_a: $\mu \neq 100$.

Step 2 Determine the critical region.

The alternative hypothesis is $\mu \neq 100$, so this is a two-tailed test. We are testing at 5% significance, so we need to identify the z-value that corresponds with upper and lower 2.5% of the distribution.

KEYSTROKES: [2nd] [DISTR] 3 0.25 [ENTER] and [2nd] [DISTR] 3 .975 [ENTER].

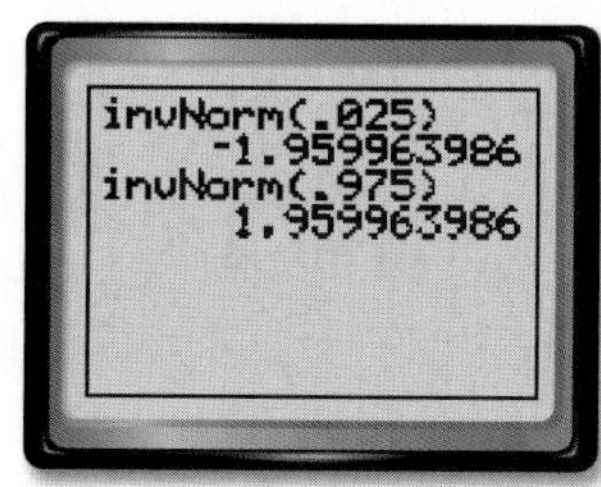

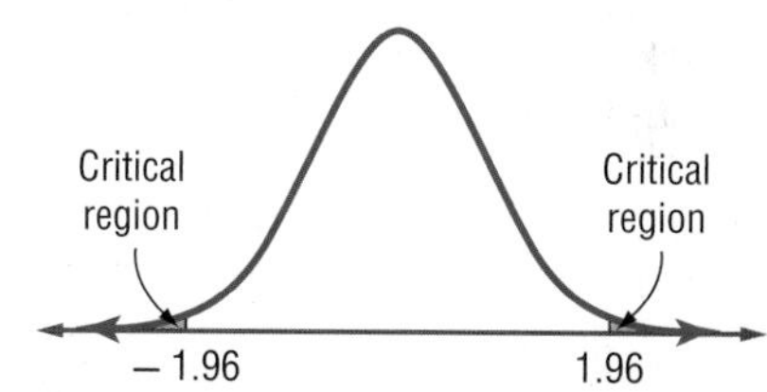

Step 3 Calculate the z-statistic for the sample data.

$$z = \frac{\overline{x} - \mu}{\frac{s}{\sqrt{n}}}$$ Formula for z-statistic

$$= \frac{99.8 - 100}{\frac{0.8}{\sqrt{42}}}$$ $\overline{x} = 99.8$, $\mu = 100$, $s = 0.8$, and $n = 42$

$$\approx -1.62$$ Simplify.

Step 4 Reject or fail to reject the null hypothesis.

H_0 is not rejected because the z-value for the sample does not fall within the critical region.

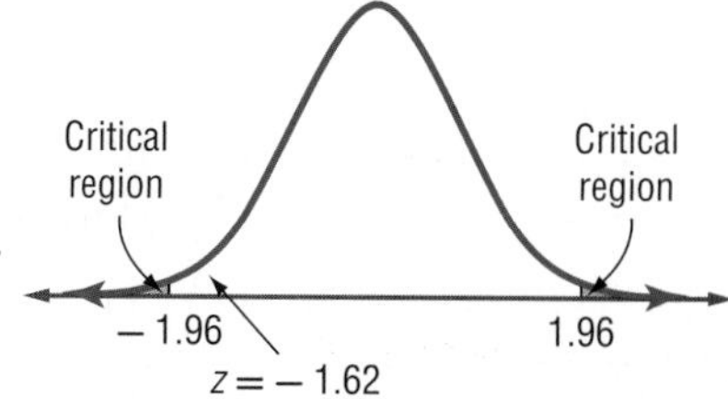

Step 5 Make a conclusion about the claim.

Therefore, there is not enough evidence to reject the claim that there are 100 pieces of pepperoni on every large pizza.

GuidedPractice

5. **JUICE** A manufacturer claims there are 12 ounces of juice in every can. Using a sample of 68 cans, Susan calculated a mean of 11.95 ounces with a standard deviation of 0.15 ounce. Test the hypothesis at 1% significance.

StudyTip

z-Test The z-test can be used when σ is known or when $n \geq 30$.

WatchOut!

Inferential Statistics Inferential statistics are not designed to definitely prove a hypothesis. They depend on probability statements and therefore, may not be correct.

Check Your Understanding

 = Step-by-Step Solutions begin on page R14.

Example 1

1. **LUNCH** A sample of 145 high school seniors was asked how many times they go out for lunch per week. The mean number of times was 2.4 with a standard deviation of 0.7. Use a 90% confidence level to calculate the maximum error of estimate.

Example 2

2. **PRACTICE** A poll of 233 randomly chosen high school athletes showed that they spend an average of 1.6 hours practicing their sport during the off-season. The standard deviation is 0.5 hour. Determine a 99% confidence interval for the population mean.

Example 3

Identify the null and alternative hypotheses for each statement. Then identify the statement that represents the claim.

3. Lori thinks it takes a fast-food restaurant less than 2 minutes to serve her meal after she orders it.
4. A snack label states that one serving contains one gram of fat.
5. Mrs. Hart's review game takes at least 20 minutes to complete.
6. The tellers at a bank can complete no more than 18 transactions per hour.

CCSS **REASONING** **Identify the hypotheses and claim, decide whether to reject the null hypothesis, and make a conclusion about the claim.**

Example 4

7. **COMPACT DISCS** A manufacturer of blank compact discs claims that each disc can hold at least 84 minutes of music. Using a sample of 219 compact discs, Cayla calculated a mean time of 84.1 minutes per disc with a standard deviation of 1.9 minutes. Test the hypothesis at 5% significance.

Example 5

8. **GOLF TEES** A company claims that each golf tee they produce is 5 centimeters in length. Using a sample of 168 tees, Angelene calculated a mean of 5.1 centimeters with a standard deviation of 0.3. Test the hypothesis at 10% significance.

Practice and Problem Solving

Extra Practice is on page R11.

Example 1

9. **MUSIC** A sample of 76 albums had a mean run time of 61.3 minutes with a standard deviation of 5.2 minutes. Use a 95% confidence level to calculate the maximum error of estimate.

Example 2

10. **COLLEGE** A poll of 218 students at a university showed that they spend 11.8 hours per week studying. The standard deviation is 3.7 hours. Determine a 90% confidence interval for the population mean.

Example 3

Identify the null and alternative hypotheses for each statement. Then identify the statement that represents the claim.

11. Julian sends at least six text messages to his best friend every day.
12. A car company states that one of their vehicles gets 27 miles per gallon.
13. A company advertisement states that it takes no more than 2 hours to paint a 200-square-foot room.
14. A singer plays at least 18 songs at every concert.

Identify the hypotheses and claim, decide whether to reject the null hypothesis, and make a conclusion about the claim.

Example 4

15. **PIZZA** A pizza chain promises a delivery time of less than 30 minutes. Using a sample of 38 deliveries, Chelsea calculated a mean delivery time of 29.6 minutes with a standard deviation of 3.9 minutes. Test the hypothesis at 1% significance.

Example 5

16. **CHEESE** A company claims that each package of cheese contains exactly 24 slices. Using a sample of 93 packages, Mr. Matthews calculated a mean of 24.1 slices with a standard deviation of 0.5. Test the hypothesis at 5% significance.

17. **DECISION MAKING** The number of peaches in 40 random cans is shown below. Should the manufacturer place a label on the can promising exactly 12 peaches in every can? Explain your reasoning.

13, 14, 13, 14, 12, 12, 12, 11, 15, 12, 13, 13, 14, 13, 14, 12, 15, 11, 11, 14,
13, 14, 14, 13, 12, 12, 12, 12, 13, 13, 11, 14, 14, 13, 14, 13, 13, 14, 12, 12

18. **CCSS ARGUMENTS** The number of chocolate chips in 40 random cookies is shown below. Should the manufacturer place a label on the package promising exactly 20 chips on every cookie? Explain your reasoning.

21, 19, 20, 20, 19, 19, 18, 21, 19, 17, 19, 18, 18, 20, 20, 19, 18, 20, 19, 20,
21, 21, 19, 17, 17, 18, 19, 19, 20, 17, 22, 21, 21, 20, 19, 18, 19, 17, 17, 21

19. **MULTIPLE REPRESENTATIONS** In this problem, you will explore how the confidence interval is affected by the sample size and the confidence level. Consider a sample of data where $\overline{x} = 25$ and $s = 3$.

   a. **GRAPHICAL** Graph the 90% confidence interval for $n = 50$, 100, and 250 on a number line.

   b. **ANALYTICAL** How does the sample size affect the confidence interval?

   c. **GRAPHICAL** Graph the 90%, 95%, and 99% confidence intervals for $n = 150$.

   d. **ANALYTICAL** How does the confidence level affect the confidence interval?

   e. **ANALYTICAL** How does decreasing the size of the confidence interval affect the accuracy of the confidence interval?

H.O.T. Problems Use Higher-Order Thinking Skills

20. **ERROR ANALYSIS** Tim and Judie want to test whether a delivery service meets their promised time of 45 minutes or less. Their hypotheses are shown below. Is either of them correct?

Tim	Judie
$H_0: \mu < 45$ (claim)	$H_0: \mu \leq 45$
$H_a: \mu \geq 45$	$H_a: \mu > 45$ (claim)

21. **CHALLENGE** A 95% confidence interval for the mean weight of a 20-ounce box of cereal was $19.932 \leq \mu \leq 20.008$ with a sample standard deviation of 0.128 ounces. Determine the sample size that led to this interval.

22. **REASONING** Determine whether the following statement is *sometimes*, *always*, or *never* true. Explain your reasoning.

If a confidence interval contains the H_0 value of μ, then it is not rejected.

23. **WRITING IN MATH** How can a statistical test be used in a decision-making process?

24. **OPEN ENDED** Design and conduct your own research study, and draw conclusions based on the results of a hypothesis test. Write a brief summary of your findings.

Standardized Test Practice

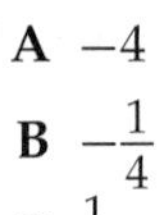

25. **GEOMETRY** In the graph below, line ℓ passes through the origin. What is the value of $\frac{a}{b}$?

A -4

B $-\frac{1}{4}$

C $\frac{1}{4}$

D 4

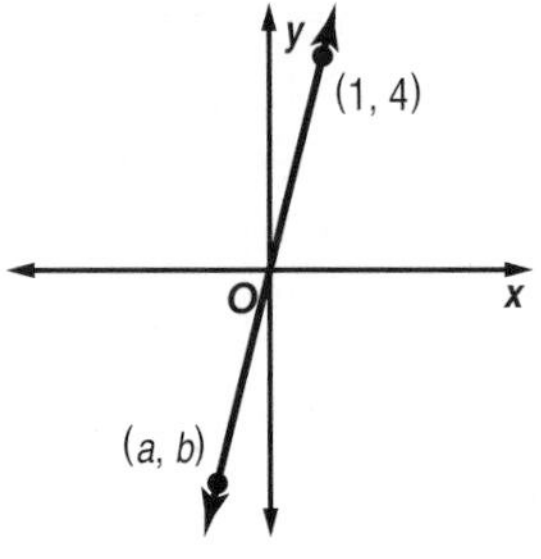

26. **SAT/ACT** If $5 + i$ and $5 - i$ are the roots of the equation $x^2 - 10x + c = 0$, what is the value of c?

F -26 G -25 H 25 J 26

27. The Service Club at Jake's school was founded 8 years ago. The number of members of the club by year is shown in the table. Which linear equation best models the data?

A $y = 1.4x$

B $y = 1.4x + 10.4$

C $y = 1.6x$

D $y = 1.6x + 11.1$

Year	Participation
0	11
2	13
4	15
6	19
8	22

28. **SHORT RESPONSE** Solve for x: $\log_2 (x - 6) = 3$.

Spiral Review

29. **HEALTH** The heights of students at Madison High School are normally distributed with a mean of 66 inches and a standard deviation of 2 inches. Of the 1080 students in the school, how many would you expect to be less than 62 inches tall? (Lesson 11-5)

30. **CAR WASH** The Spanish Club is washing cars to raise money. They have determined that 65% of the customers donate more than the minimum amount for the car wash. What is the probability that at least 4 of the next 5 customers will donate more than the minimum? (Lesson 11-4)

Find a_n for each geometric sequence. (Lesson 10-3)

31. $a_1 = \frac{1}{3}, r = 3, n = 8$

32. $a_1 = \frac{1}{64}, r = 4, n = 9$

33. $a_4 = 16, r = 0.5, n = 8$

Write each equation in standard form. State whether the graph of the equation is a *parabola, circle, ellipse,* or *hyperbola*. Then graph the equation. (Lesson 9-6)

34. $4x^2 + 2y^2 = 8$

35. $x^2 = 8y$

36. $(x - 1)^2 - 9(y - 4)^2 = 36$

Write an equation in slope-intercept form for each graph. (Lesson 2-4)

37.

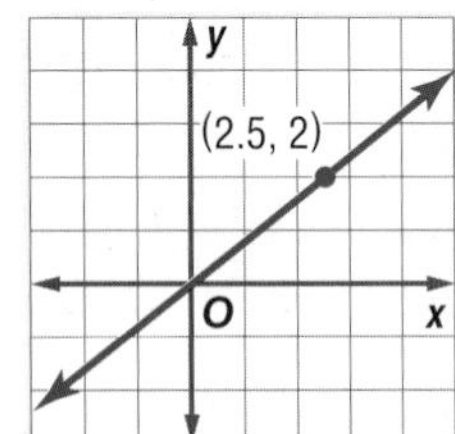

38.

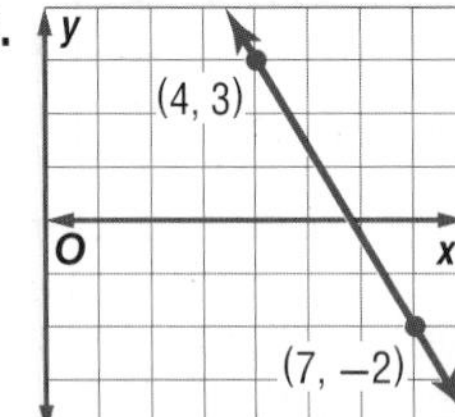

39.

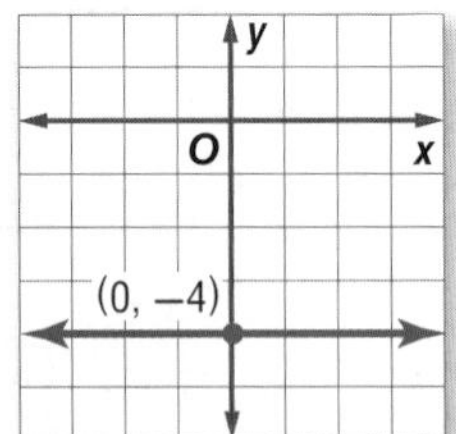

Skills Review

Find each missing measure. Round to the nearest tenth, if necessary.

40.

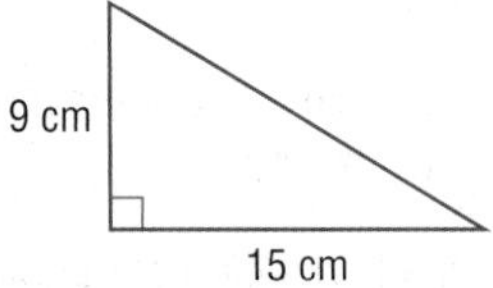

41.

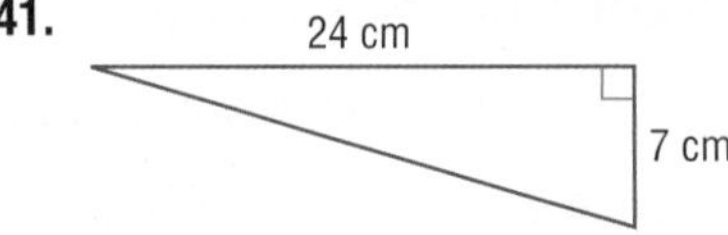

42.

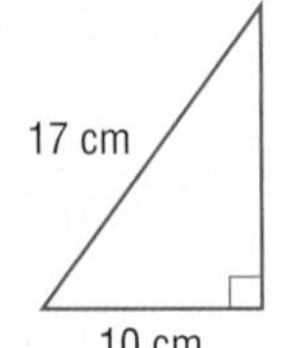

CHAPTER 11

Study Guide and Review

Study Guide

KeyConcepts

Designing a Study (Lesson 11-1)

- A survey, an experiment, or an observational study can be used to collect information.
- A bias is an error that results in a misrepresentation of members of a population.

Distributions of Data (Lesson 11-2)

- Use the mean and standard deviation to describe a symmetric distribution.
- Use the five-number summary to describe a skewed distribution.

Probability Distributions (Lesson 11-3)

- A theoretical probability distribution is based on what is expected to happen. An experimental probability distribution is a distribution of probabilities estimated from experiments.
- The expected value of a discrete random variable is the weighted average of the values of the variable.

Binomial Distributions (Lesson 11-4)

- A binomial experiment has a fixed number of independent trials, only two possible outcomes for each trial, and the same probability of success for every trial.
- A binomial distribution is a frequency distribution of the probability of each value of a random variable.

Normal Distributions (Lesson 11-5)

- The graph of a normal distribution is bell-shaped.
- The z-value represents the number of standard deviations that a given data value is from the mean.

Hypothesis Testing (Lesson 11-6)

- A confidence interval for a population mean μ is given by $CI = \bar{x} \pm E$ or $\bar{x} \pm z \cdot \frac{s}{\sqrt{n}}$, where $\bar{x}$ is the sample mean and E is the maximum error of estimate.
- A hypothesis test is used to assess a specific claim.

FOLDABLES StudyOrganizer

Be sure the following Key Concepts are noted in your Foldable.

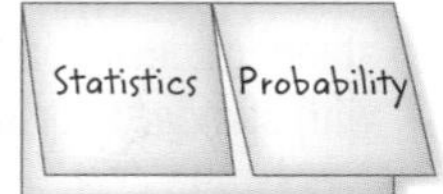

KeyVocabulary

alternative hypothesis H_a (p. 771)
bias (p. 723)
binomial distribution (p. 754)
confidence interval (p. 769)
continuous random variable (p. 742)
discrete random variable (p. 742)
Empirical Rule (p. 760)
expected value $E(X)$ (p. 745)
experiment (p. 723)
experimental probability distribution (p. 744)
hypothesis test (p. 771)
inferential statistics (p. 769)
maximum error of estimate (p. 769)
negatively skewed distribution (p. 733)
normal distribution (p. 760)
null hypothesis H_0 (p. 771)
observational study (p. 723)
parameter (p. 723)
positively skewed distribution (p. 733)
probability distribution (p. 743)
random variable (p. 742)
standard normal distribution (p. 762)
statistic (p. 723)
statistical inference (p. 769)
survey (p. 723)
symmetric distribution (p. 733)
theoretical probability distribution (p. 743)
z-value (p. 762)

VocabularyCheck

Choose a term from the list above that best completes each statement.

1. A(n) ____________ for a particular random variable is a function that maps the sample space to the probabilities of the outcomes of the sample space.
2. A(n) ____________ is an error that results in a misrepresentation of members of a population.
3. In a statistical study, data are collected and used to answer questions about a population characteristic or ____________.
4. The ____________ can be used to determine the area under the normal curve at specific intervals.
5. In a(n) ____________, members of a sample are measured or observed without being affected by the study.
6. A ____________ is an estimate of a population parameter stated as a range with a specific degree of certainty.

Study Guide and Review *Continued*

Lesson-by-Lesson Review

11-1 Designing a Study

Determine whether each situation describes a *survey,* an *experiment,* or an *observational study.* Then identify the sample, and suggest a population from which it may have been selected.

7. **SHOPPING** Every tenth shopper coming out of a store is asked questions about his or her satisfaction with the store.
8. **MILK SHAKE** A fast food restaurant gives 25 of their customers a sample of a new milk shake and employees monitor their reactions as they taste it.
9. **SCHOOL** Every fifth person coming out of a high school is asked what their favorite class is.

Example 1

A dealership wants to test different promotions. They randomly select 100 customers and ask them which promotion they prefer. Does this situation describe a *survey*, an *experiment*, or an *observational study*? Identify the sample, and suggest a population from which it may have been selected.

This is a survey, because the data are collected from participants' responses. The sample is the 100 customers that were selected, and the population is all potential customers.

11-2 Distributions of Data

10. **DOGSLED** The Iditarod is a race across Alaska. The table shows the winning times, in days, for recent years.

    Iditarod Winning Times
    9.1, 9.4, 10.3, 9.3, 9.6, 8.7, 9.5, 9.4, 9.2, 17.3, 15.4, 15.5, 14.2, 12.0, 16.6, 13.5, 13.0, 18.1, 12.4, 11.6, 11.5, 11.3, 11.3, 13.1, 11.2, 11.6, 11.6, 9.7

    a. Use a graphing calculator to create a histogram. Then describe the shape of the distribution.
    b. Describe the center and spread of the data using either the mean and standard deviation or the five-number summary. Justify your choice.

11. **SWIMMING** Kelly's practice times in the 400-meter individual medley are shown in the table.

    Times in Seconds
    301, 311, 320, 308, 312, 307, 303, 305, 309, 308, 304, 302, 311, 313, 313, 316, 314, 306, 329, 326, 319, 310, 306, 309, 320, 318, 315, 318, 314, 309

    a. Use a graphing calculator to create a box-and-whisker plot. Then describe the shape of the distribution.
    b. Describe the center and spread of the data using either the mean and standard deviation or the five-number summary. Justify your choice.

Example 2

Data collected from a group of third graders is shown.

Number of Years Playing an Instrument
2.5, 2.4, 3.1, 2.9, 4.2, 1.3, 2.6, 2.4, 3.3, 1.9, 3.4, 4.8, 2.3, 1.7, 3.2, 2.3, 3.5, 2.2, 3.6, 1.2, 4.4, 2.1, 3.4, 4.5, 1.9, 1.5, 1.4, 0.7, 1.2, 2.5, 1.9, 2.0, 2.4, 2.5, 3.4

a. Use a graphing calculator to create a histogram. Then describe the shape of the distribution.

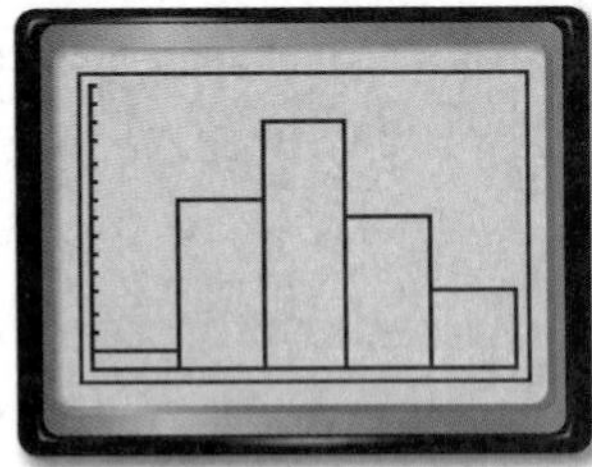

[0, 5] scl: 1 by [0, 15] scl: 1

The distribution is symmetric.

b. Describe the center and spread of the data using either the mean and standard deviation or the five-number summary. Justify your choice.

The distribution is symmetric, so use the mean and standard deviation. The mean number of years is about 2.6 with standard deviation of about 1 year.

11-3 Probability Distributions

Identify the random variable in each distribution, and classify it as *discrete* or *continuous*. Explain your reasoning.

12. the number of ice cream sandwiches sold at an ice cream shop

13. the time it takes to run a 5-kilometer race

14. **DANCE RECITALS** The probability distribution lists the probable number of dance recitals per year at Rena's Dance School. Determine the expected number of dance recitals per year.

Number of Dance Recitals Per Year					
Recitals	0	1	2	3	4
Probability	0.3	0.3	0.13	0.13	0.14

15. **SNOW DAYS** The distribution lists the number of snow days per year at Washington Elementary over the past 26 years. Determine the expected number of snow days this year.

Number of Snow Days Per Year					
Snow Days	0	1	2	3	4
Frequency	4	8	6	3	5

Example 3

Identify the random variable in a distribution of the number of DVDs on display at a store. Classify it as *discrete* or *continuous*. Explain your reasoning.

The random variable X is the number of DVDs on display. The number of DVDs is countable, so X is discrete.

Example 4

MEDICINE The probability distribution lists the probable number of drops of medicine that a veterinarian administers to her sick patients. Find the expected number of drops of medicine.

Number of Drops of Medicine				
Drops	1	2	3	4
Probability	0.5	0.3	0.1	0.1

Each drop of medicine represents a value of X, and each decimal represents the corresponding probability $P(X)$.

$$E(X) = \Sigma[X \cdot P(X)]$$
$$= 1(0.5) + 2(0.3) + 3(0.1) + 4(0.1)$$
$$= 0.5 + 0.6 + 0.3 + 0.4 \text{ or } 1.8$$

The expected number of drops of medicine is 1.8.

11-4 The Binomial Distribution

Determine whether each experiment is a binomial experiment or can be reduced to a binomial experiment. If so, describe a trial, determine the random variable, and state n, p, and q.

16. A survey found that 30% of adults like chocolate ice cream more than any other flavor. You ask 35 adults if they prefer chocolate ice cream more than any other flavor.

17. Thirty random guests from Sheila's birthday party are asked their favorite song.

18. **WATCHES** According to an online poll, 74% of adults wear watches. Timmy surveyed 25 random adults. What is the probability that 20 of the adults surveyed wear a watch?

19. **SEASONS** Of 1108 people surveyed, 68% say that summer is their favorite season. What is the probability that at least 15 of 20 randomly selected people will prefer summer?

Example 5

WORK According to an online poll, 40% of adults feel that the standard 40-hour workweek should be increased. Sheila conducts a survey of 10 random adults. What is the probability that 3 of the surveyed adults feel that the standard 40-hour workweek should be increased?

A success is an adult agreeing that the 40-hour workweek should be increased, so $p = 0.4$, $X = 3$, $q = 1 - 0.4$ or 0.6, and $n = 10$.

$$P(X) = {}_nC_x p^X q^{n-X}$$
$$P(6) = {}_{10}C_3(0.4)^3(0.6)^{10-3}$$
$$\approx 0.215$$

The probability that three surveyed adults will feel that the standard 40-hour workweek should be increased is about 0.215 or 21.5%.

Study Guide and Review *Continued*

11-5 The Normal Distribution

20. **RUNNING TIMES** The times in the 40-meter dash for a select group of professional football players are normally distributed with a mean of 4.74 seconds and a standard deviation of 0.13 second.
    a. About what percent of players have times between 4.6 and 4.8 seconds?
    b. About how many of a sample of 800 players will have times below 4.5 seconds?
21. **ATTENDANCE** The number of tickets sold at high school basketball games in a particular conference are normally distributed with a mean of 68.7 and a standard deviation of 13.1.
    a. About what percent of the games sell fewer than 75 tickets?
    b. About how many of a sample of 200 games will sell more than 100 tickets?
22. **COMMUTING** The number of minutes it takes Phil to commute to work each day are normally distributed with a mean of 18.6 and a standard deviation of 3.5.
    a. About what percent of the time will it take Phil more than 20 minutes to commute to work?
    b. About how many of a sample of 50 days will it take Phil less than 15 minutes to commute to work?

Example 6

TEST SCORES The midterm test scores for the students in Mrs. Hendrix's classes are normally distributed with a mean of 73.2 and standard deviation of 7.8. About how many test scores are between 70 and 80?

Find the corresponding z-values for $X = 70$ and 80.

$$z = \frac{X - \mu}{\sigma} = \frac{70 - 73.2}{7.8} \approx -0.41 \qquad z = \frac{X - \mu}{\sigma} = \frac{80 - 73.2}{7.8} \approx 0.87$$

Using a graphing calculator, you can find the area between $-0.41 < z < 0.87$ to be about 0.47.

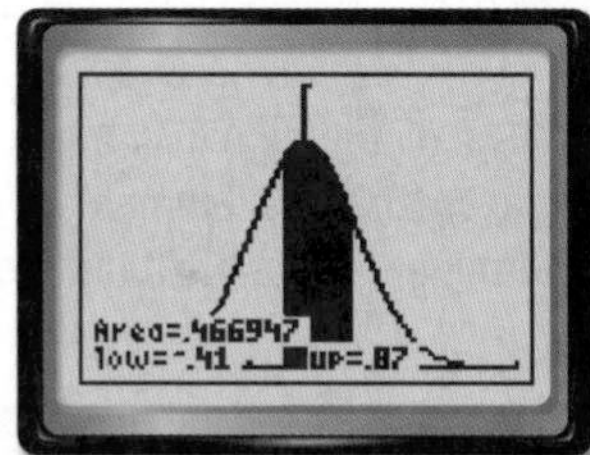

[−4, 4] scl: 1 by [0, 0.5] scl: 0.125

11-6 Confidence Intervals and Hypothesis Testing

23. **INTERNET** A sample of 300 students was asked the average amount of time they spend online during a weeknight. The mean time was 64.3 minutes with a standard deviation of 7.3 minutes. Determine a 95% confidence interval.

Identify the null and alternative hypotheses for each statement. Then identify the statement that represents the claim.

24. A cupcake shop owner says that they sell 200 cupcakes everyday.
25. A technician says that it takes at least 45 minutes to diagnose a computer problem.
26. A member of a swim team says that she practices no more than 35 minutes on Mondays.

Example 7

TELEVISION WATCHING A sample of 100 teenagers was asked the average amount of time they spent watching television on the weekends. The mean time was 65 minutes with a standard deviation of 1.6 minutes. Determine a 95% confidence interval for the population.

$CI = \bar{x} \pm 1.96 \cdot \frac{s}{\sqrt{n}}$ Confidence Interval Formula

$= 65 \pm 1.96 \cdot \frac{1.6}{\sqrt{100}}$ $\bar{x} = 65, s = 1.6, n = 100$

$= 65 \pm 0.31$ Use a calculator.

The 95% confidence interval is $64.69 \leq \mu \leq 65.31$. Therefore, we are 95% confident that the population mean time is between 64.69 and 65.31 minutes.

Practice Test

1. **BUTTERFLIES** Students in a biology class are learning about the monarch butterfly's life cycle. Each student is given a caterpillar. When a caterpillar turns into chrysalis, it is placed in a glass enclosure with food and a heat lamp and examined.

   a. Determine whether the situation describes a *survey*, an *experiment* or an *observational study.*

   b. Identify the sample, and suggest a population from which it was selected.

2. **HEIGHTS The heights of Ms. Joy's dance students are shown.**

Height (Inches)				
60	64	62	69	64
63	65	64	66	73
74	63	62	65	64
68	70	66	63	61

   a. Use a graphing calculator to create a box-and-whisker plot. Then describe the shape of the distribution.

   b. Describe the center and spread of the data using either the mean and standard deviation or the five-number summary. Justify your choice.

3. A binomial distribution has a 65% rate of success. There are 15 trials.

   a. What is the probability that there will be exactly 12 successes?

   b. What is the probability that there will be at least 10 successes?

Determine whether each experiment is a binomial experiment or can be reduced to a binomial experiment. If so, describe a trial, determine the random variable, and state n, p, and q.

4. A poll found that 65% of high school teachers own a pet. You ask 15 high school teachers if they own a pet.

5. A study finds that 20% of families in a town have a land line telephone. You ask 55 families how many land line telephones they have.

6. A survey found that on a scale of 1 to 5, a sandwich received a 3.5 rating. A restaurant manager asks 150 customers to rate the sandwich on a scale of 1 to 5.

Identify the random variable in each distribution, and classify it as *discrete* or *continuous.* Explain your reasoning.

7. the number of laps that Jesse swims

8. the body temperatures of patients in a hospital

9. the weights of pets in a pet store

10. **MULTIPLE CHOICE** The table shows the number of gift cards previously won in a mall contest. What is the expected value of the gift card that is won?

    A $250.00

    B $223.15

    C $143.25

    D $100.23

Amount, X	Winners
$100	495
$125	405
$150	285
$200	180
$250	90
$300	45

11. **WEIGHTS** The weights of 1500 bodybuilders are normally distributed with a mean of 190.6 pounds and a standard deviation of 5.8 pounds.

    a. About how many bodybuilders are between 180 and 190 pounds?

    b. What is the probability that a bodybuilder selected at random has a weight greater than 195 pounds?

A normal distribution has a mean of 16.4 and a standard deviation of 2.6.

12. Find the range of values that represent the middle 95% of the distribution.

13. What percent of the data will be less than 19?

14. **MULTIPLE CHOICE** A survey of 300 randomly selected members of a fitness club showed that they spent an average of 25.3 minutes per visit in the gym with a standard deviation of 8.6 minutes. Which is the maximum error of estimate for the time spent in the gym using a 95% confidence interval?

    F 0.82 G 0.97 H 1.28 J 2.86

Identify the null and alternative hypotheses for each statement. Then identify the statement that represents the claim.

15. A restaurant owner says that each milk shake contains 5 strawberries.

16. A valet says that no more than 56 cars can be parked in the parking lot at one time.

CHAPTER 11 Preparing for Standardized Tests

Solve Multi-Step Problems

Some problems that you will encounter on standardized tests require you to solve multiple parts in order to come up with the final solution. Use this lesson to practice these types of problems.

Comstock/Comstock Images/Getty

Strategies for Solving Multi-Step Problems

Step 1

Read the problem statement carefully.

Ask yourself:

- What am I being asked to solve? What information is given?
- Are there any intermediate steps that need to be completed before I can solve the problem?

Step 2

Organize your approach.

- List the steps you will need to complete in order to solve the problem.
- Remember that there may be more than one possible way to solve the problem.

Step 3

Solve and check.

- Work as efficiently as possible to complete each step and solve.
- If time permits, check your answer.

Standardized Test Example

Read the problem. Identify what you need to know. Then use the information in the problem to solve.

There are 15 boys and 12 girls in Mrs. Lawrence's homeroom. Suppose a committee is to be made up of 6 randomly selected students. What is the probability that the committee will contain 3 boys and 3 girls? Round your answer to the nearest tenth of a percent.

A 27.2% **C** 31.5%

B 29.6% **D** 33.8%

Read the problem statement carefully. You are asked to find the probability that a committee will be made up of 3 boys and 3 girls. Finding this probability involves successfully completing several steps.

Step 1 Find the number of possible successes.

There are C(15, 3) ways to choose 3 boys from 15, and there are C(12, 3) to choose 3 girls from 12. Use the Fundamental Counting Principle to find s, the number of possible successes.

$s = C(15, 3) \times C(12, 3) = \frac{15!}{12!3!} \times \frac{12!}{9!3!}$ or 100,100

Step 2 Find the total number of possible outcomes.

Compute the number of ways 6 people can be chosen from a group of 27 students.

$C(27, 6) = 296{,}010$

Step 3 Compute the probability.

Find the probability by comparing the number of successes to the number of possible outcomes.

$P(3 \text{ boys}, 3 \text{ girls}) = \frac{100{,}100}{296{,}010} \approx 0.33816$

So, there is about a 33.8% chance of selecting 3 boys and 3 girls for the committee. The answer is D.

Exercises

Read the problem. Identify what you need to know. Then use the information in the problem to solve.

1. There are 52 cards in a standard deck. Of these, 4 of the cards are aces. What is the probability of a randomly dealt 5-card hand containing a pair of aces? Round your answer to the nearest whole percent.

   A 4%

   B 5%

   C 6%

   D 7%

2. According to the table, what is the probability that a randomly selected camper went on the horse ride, given that the camper is an 8th grader?

Camp Activities			
Grade	Canoe Trip	Horse Ride	Nature Hike
6th	8	6	3
7th	5	4	7
8th	11	9	6

   F 0.731

   G 0.441

   H 0.346

   J 0.153

CHAPTER 11

Standardized Test Practice

Cumulative, Chapters 1 through 11

Multiple Choice

Read each question. Then fill in the correct answer on the answer document provided by your teacher or on a sheet of paper.

1. Suppose the test scores on a final exam are normally distributed with a mean of 74 and a standard deviation of 3. What is the probability that a randomly selected test has a score higher than 77?

A 2.5%
B 13.5%
C 16%
D 34%

2. A diameter of a circle has endpoints $A(4, 6)$ and $B(-3, -1)$. Find the approximate length of the radius.

F 2.5 units
G 4.9 units
H 5.1 units
J 9.9 units

3. An equation can be used to find the total cost of a pizza with a certain diameter. Using the table below, find the equation that best represents y, the total cost, as a function of x, the diameter in inches.

Diameter, x (in.)	Total Cost, y
9	\$10.80
12	\$14.40
20	\$24.00

A $y = 1.2x$
B $x = 1.2y$
C $y = 0.83x$
D $y = x + 1.80$

4. Which shows the functions correctly listed in order from widest to narrowest graph?

F $y = 8x^2, y = 2x^2, y = \frac{1}{2}x^2, y = -\frac{4}{5}x^2$
G $y = -\frac{4}{5}x^2, y = \frac{1}{2}x^2, y = 2x^2, y = 8x^2$
H $y = \frac{1}{2}x^2, y = -\frac{4}{5}x^2, y = 2x^2, y = 8x^2$
J $y = 8x^2, y = 2x^2, y = -\frac{4}{5}x^2, y = \frac{1}{2}x^2$

Test-Taking Tip

Question 5 You can use a scientific calculator to find the standard deviation. Enter the data values as a list and calculate the 1-Var statistics.

5. The table at the right shows the grades earned by students on a science test. Calculate the standard deviation of the test scores.

76	84	91	75	83
82	65	94	90	71
92	84	83	88	80
78	84	89	95	93

A 7.82 B 8.03 C 8.23 D 8.75

6. Which expression is equivalent to $(6a - 2b) - \frac{1}{4}(4a + 12b)$?

F $5a + 10b$
G $10a + 10b$
H $5a + b$
J $5a - 5b$

7. Simplify $\sqrt[6]{27x^3}$.

A $3x^2$ B $3x$ C $\sqrt{3}x$ D $\sqrt{3x}$

8. In the figure below, lines a and b are parallel. What are the measures of the angles in the triangle?

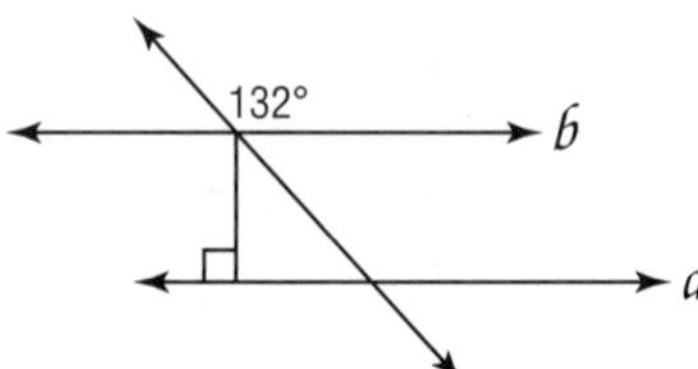

F 42, 48, 90
G 42, 90, 132
H 48, 52, 90
J 48, 90, 132

9. Which of the following functions represents exponential decay?

A $y = 0.2(7)^x$
B $y = (0.5)^x$
C $y = 4(9)^x$
D $y = 5\left(\frac{4}{3}\right)^x$

10. Using the table below, which expression can be used to determine the nth term of the sequence?

n	1	2	3	4
y	6	10	14	18

F $y = 6n$
G $y = n + 5$
H $y = 2n + 1$
J $y = 2(2n + 1)$

Short Response/Gridded Response

Record your answers on the answer sheet provided by your teacher or on a sheet of paper.

11. Determine whether each of the following situations calls for a *survey*, an *observational study*, or an *experiment*. Explain the process.

a. Caroline wants to find out if a particular plant food helps plants to grow faster than just water.

b. Allison wants to find opinions on favorite candidates in the upcoming student counsel elections.

c. Manuel wants to find out if people who get regular exercise sleep better at night.

12. GRIDDED RESPONSE Carla received a map of some walking paths through her college campus. Paths A, B, and C are parallel. What is the length x to the nearest tenth of a foot?

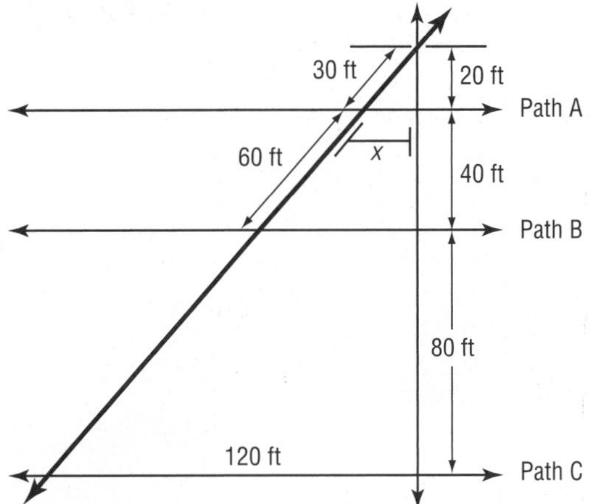

13. Alex wants to find the area of a triangle. He draws the triangle on a coordinate plane and finds that it has vertices at (2, 1), (3, 4), and (1, 4). Find the area of the triangle using determinants.

14. GRIDDED RESPONSE Perry drove to the gym at an average rate of 30 miles per hour. It took him 45 minutes. Going home, he took the same route, but drove at a rate of 45 miles per hour. How many miles is it to his house from the gym?

Extended Response

Record your answers on a sheet of paper. Show your work.

15. Martin is taking a multiple choice test that has 8 questions. Each question has four possible answers: A, B, C, or D. Martin forgot to study for the test, so he must guess at each answer.

a. What is the probability of guessing a correct answer on the test?

b. What is the expected number of correct answers if Martin guesses at each question?

c. What is the probability that Martin will get at least half of the questions correct? Round your answer to the nearest tenth of a percent.

16. Christine had one dress and three sweaters cleaned at the dry cleaner and the charge was $19.50. The next week, she had two dresses and two sweaters cleaned for a total charge of $23.00.

a. Let d represent the price of cleaning a dress and s represent the price of cleaning a sweater. Write a system of linear equations to represent the prices of cleaning each item.

b. Solve the system of equations using substitution or elimination. Explain your choice of method.

c. What will the charge be if Christine takes two dresses and four sweaters to be cleaned?

Need ExtraHelp?

If you missed Question...	1	2	3	4	5	6	7	8	9	10	11	12	13	14	15	16
Go to Lesson...	11-5	9-1	2-4	4-7	0-9	5-1	6-6	3-1	7-1	10-2	11-1	1-4	3-8	8-5	11-4	3-1

CHAPTER 12

Trigonometric Functions

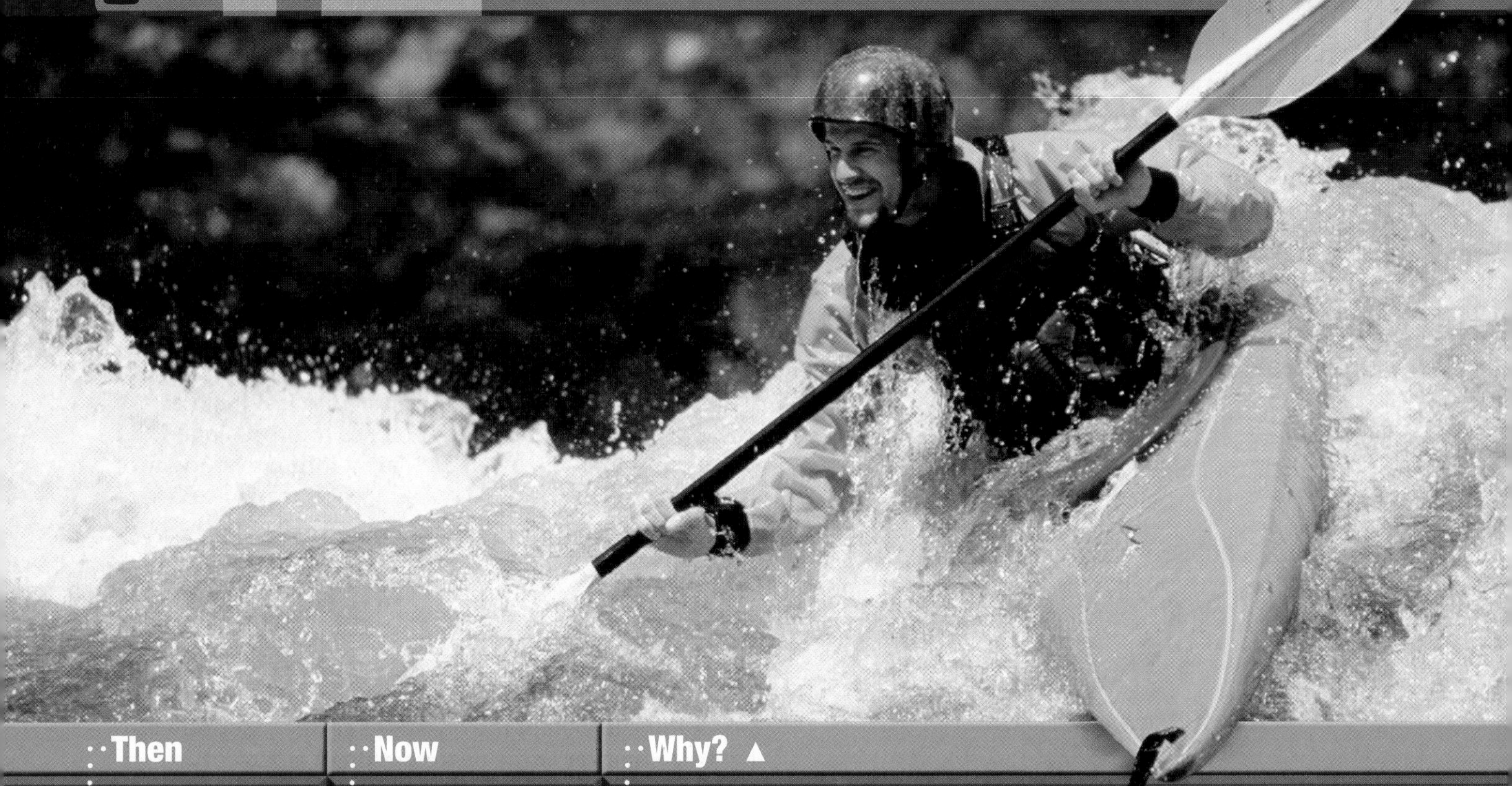

Then

You have graphed and analyzed functions.

Now

You will:

- Find values of trigonometric functions.
- Solve problems by using right triangle trigonometry.
- Solve triangles by using the Law of Sines and Law of Cosines.
- Graph trigonometric functions.

Why? ▲

WATER SPORTS Knowing trigonometric functions has practical applications in water sports. For instance, you can use right triangle trigonometry to find the distance a kayak has traveled down river. If you are familiar with angles and angle measures, then you have a better understanding of how impressive it is to be able to do a 540° rotation on a wakeboard.

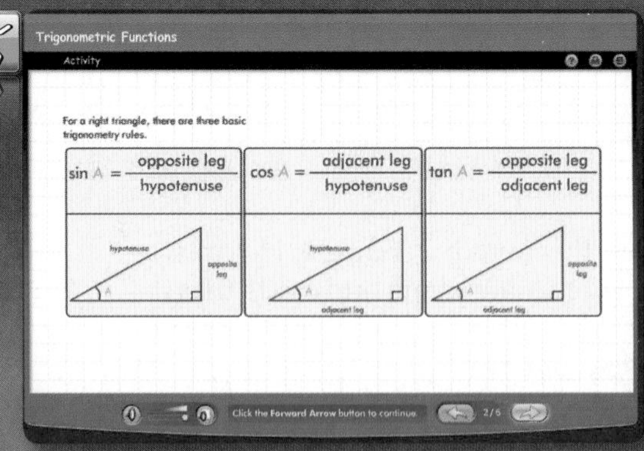

connectED.mcgraw-hill.com Your Digital Math Portal

Animation  Vocabulary  eGlossary  Personal Tutor  Virtual Manipulatives  Graphing Calculator  Audio  Foldables  Self-Check Practice  Worksheets 

Get Ready for the Chapter

Diagnose Readiness | **You have two options for checking prerequisite skills.**

1 Textbook Option Take the Quick Check below. Refer to the Quick Review for help.

QuickCheck

Find the value of x to the nearest tenth.

1. 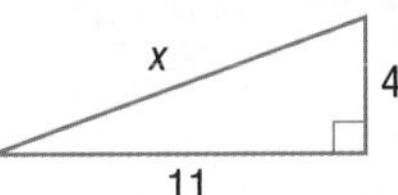

2. 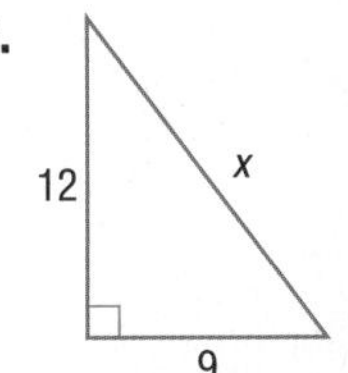

3. 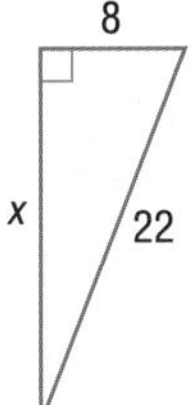

4. Laura has a rectangular garden in her backyard that measures 12 feet by 15 feet. She wants to put a rock walkway on the diagonal. How long will the walkway be? Round to the nearest tenth of a foot.

Find each missing measure. Write all radicals in simplest form.

5. 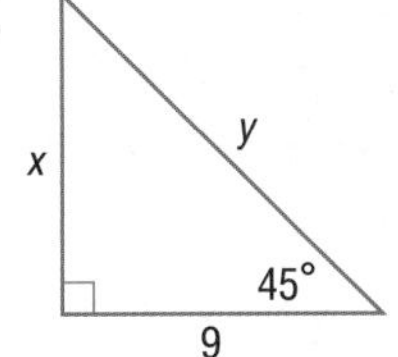

6. 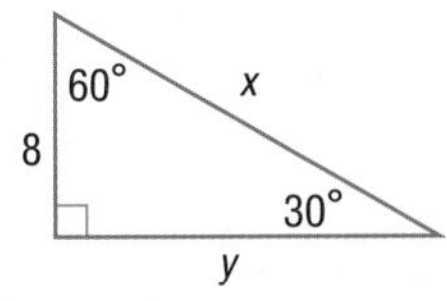

7. 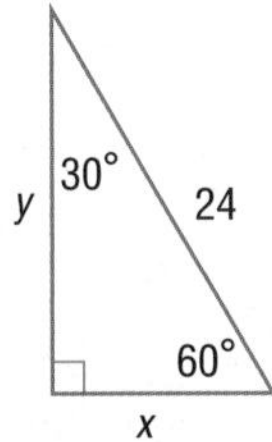

8. A ladder leans against a wall at a 45° angle. If the ladder is 12 feet long, how far up the wall does the ladder reach?

QuickReview

Example 1

Find the missing measure of the right triangle.

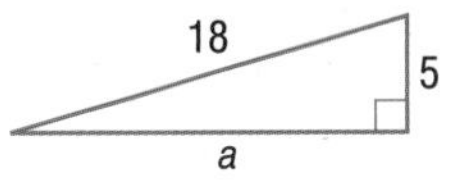

$c^2 = a^2 + b^2$	Pythagorean Theorem
$18^2 = a^2 + 5^2$	Replace c with 18 and b with 5.
$324 = a^2 + 25$	Simplify.
$299 = a^2$	Subtract 25 from each side.
$17.3 \approx a$	Take the positive square root of each side.

Example 2

Find the missing measures. Write all radicals in simplest form.

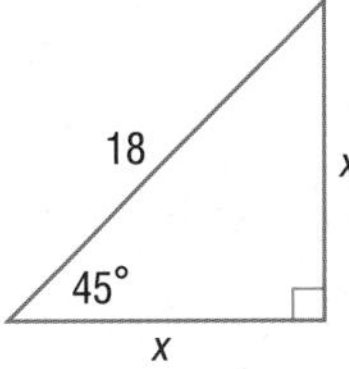

$x^2 + x^2 = 18^2$	Pythagorean Theorem
$2x^2 = 18^2$	Combine like terms.
$2x^2 = 324$	Simplify.
$x^2 = 162$	Divide each side by 2.
$x = \sqrt{162}$	Take the positive square root of each side.
$x = 9\sqrt{2}$	Simplify.

2 Online Option Take an online self-check Chapter Readiness Quiz at connectED.mcgraw-hill.com.

Get Started on the Chapter

You will learn several new concepts, skills, and vocabulary terms as you study Chapter 12. To get ready, identify important terms and organize your resources. You may wish to refer to Chapter 0 to review prerequisite skills.

FOLDABLES® StudyOrganizer

Trigonometric Functions Make this Foldable to help you organize your Chapter 12 notes about trigonometric functions. Begin with four pieces of grid paper.

1 **Stack** paper together and measure 2.5 inches from the bottom.

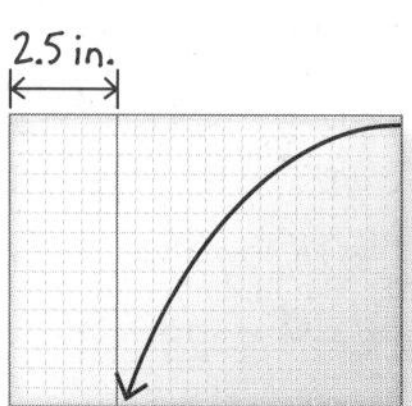

2 **Fold** on the diagonal.

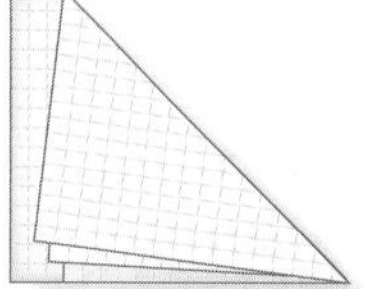

3 **Staple** along the diagonal to form a book.

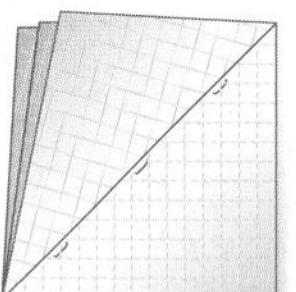

4 **Label** the edge as Trigonometric Functions.

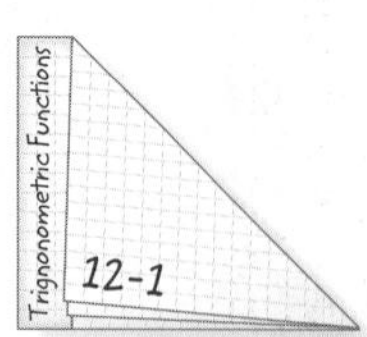

NewVocabulary

English		Español
trigonometry	p. 790	trigonometría
sine	p. 790	seno
cosine	p. 790	coseno
tangent	p. 790	tangente
cosecant	p. 790	cosecante
secant	p. 790	secante
cotangent	p. 790	cotangente
angle of elevation	p. 794	ángulo de depresión
angle of depression	p. 794	ángulo de elevación
standard position	p. 799	posición estándar
radian	p. 801	radián
Law of Sines	p. 815	Ley de los senos
ambiguous case	p. 816	caso ambiguo
Law of Cosines	p. 823	Ley de los cosenos
unit circle	p. 830	círculo unitario
circular function	p. 830	funciones circulares
periodic function	p. 831	función periódica
cycle	p. 831	ciclo
period	p. 831	período
amplitude	p. 837	amplitud
frequency	p. 838	frecuencia

ReviewVocabulary

function función a relation in which each element of the domain is paired with exactly one element in the range

inverse function función inversa two functions *f* and *g* are inverse functions if and only if both of their compositions are the identity function

EXPLORE 12-1

Spreadsheet Lab
Investigating Special Right Triangles

You can use a spreadsheet to investigate side measures of special right triangles.

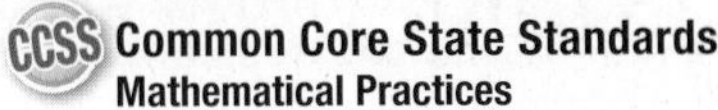

CCSS Common Core State Standards
Mathematical Practices
8 Look for and express regularity in repeated reasoning.

Activity 45°-45°-90° Triangle

The legs of a 45°-45°-90° triangle, *a* and *b*, are equal in measure. What patterns do you observe in the ratios of the side measures of these triangles?

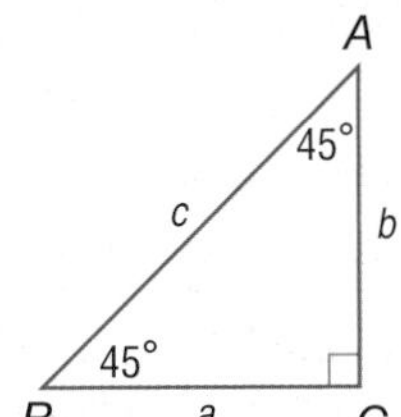

Step 1 Enter the indicated formulas in the spreadsheet. The formula uses the Pythagorean Theorem in the form $c = \sqrt{a^2 + b^2}$.

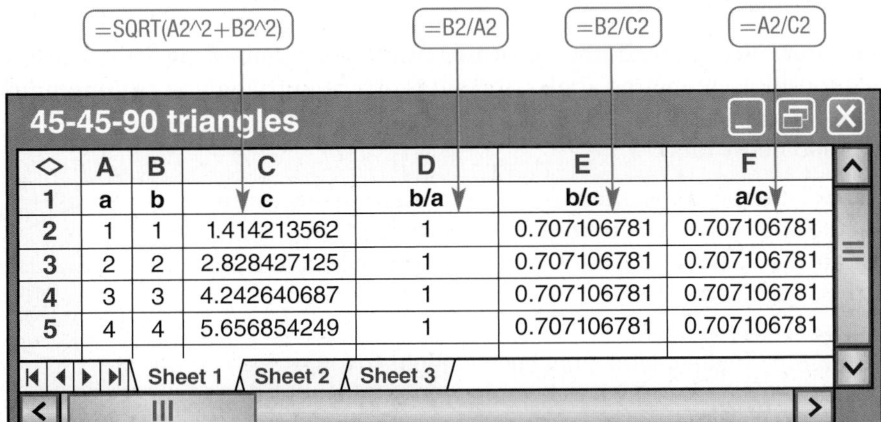

◇	A	B	C	D	E	F
1	a	b	c	b/a	b/c	a/c
2	1	1	1.414213562	1	0.707106781	0.707106781
3	2	2	2.828427125	1	0.707106781	0.707106781
4	3	3	4.242640687	1	0.707106781	0.707106781
5	4	4	5.656854249	1	0.707106781	0.707106781

Step 2 Examine the results. Because 45°-45°-90° triangles share the same angle measures, these triangles are all similar. The ratios of the sides of these triangles are all the same. The ratios of side *b* to side *a* are 1. The ratios of side *b* to side *c* and of side *a* to side *c* are approximately 0.71.

Model and Analyze

Use the spreadsheet below for 30°-60°-90° triangles.

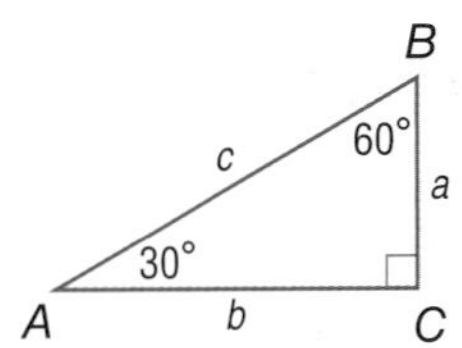

30-60-90 triangles

◇	A	B	C	D	E	F
1	a	b	c	b/a	b/c	a/c
2	1		2			
3	2		4			
4	3		6			
5	4		8			

Sheet 1 / Sheet 2 / Sheet 3

1. Copy and complete the spreadsheet above.
2. Describe the relationship among the 30°-60°-90° triangles with the dimensions given.
3. What patterns do you observe in the ratios of the side measures of these triangles?

LESSON 12-1 Trigonometric Functions in Right Triangles

Then

- You used the Pythagorean Theorem to find side lengths of right triangles.

Now

1. Find values of trigonometric functions for acute angles.
2. Use trigonometric functions to find side lengths and angle measures of right triangles.

Why?

- The altitude of a person parasailing depends on the length of the tow rope ℓ and the angle the rope makes with the horizontal $x°$. If you know these two values, you can use a ratio to find the altitude of the person parasailing.

NewVocabulary
trigonometry
trigonometric ratio
trigonometric function
sine
cosine
tangent
cosecant
secant
cotangent
reciprocal functions
inverse sine
inverse cosine
inverse tangent
angle of elevation
angle of depression

Common Core State Standards

Mathematical Practices
6 Attend to precision.

1 Trigonometric Functions for Acute Angles

Trigonometry is the study of relationships among the angles and sides of a right triangle. A **trigonometric ratio** compares the side lengths of a right triangle. A **trigonometric function** has a rule given by a trigonometric ratio.

The Greek letter *theta* θ is often used to represent the measure of an acute angle in a right triangle. The *hypotenuse*, the *leg opposite* θ, and the *leg adjacent to* θ are used to define the six trigonometric functions.

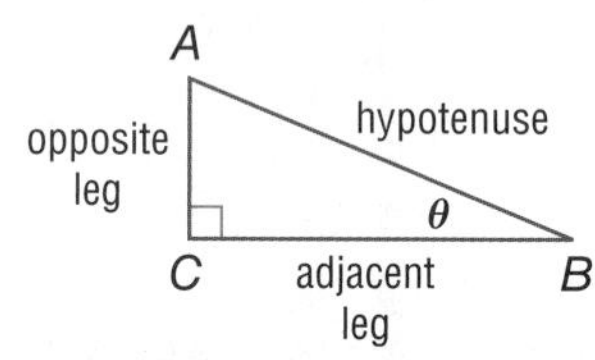

KeyConcept Trigonometric Functions in Right Triangles

Words If θ is the measure of an acute angle of a right triangle, then the following trigonometric functions involving the opposite side *opp*, the adjacent side *adj*, and the hypotenuse *hyp* are true.

Symbols

$\sin$ (**sine**) $\theta = \frac{\text{opp}}{\text{hyp}}$ $\csc$ (**cosecant**) $\theta = \frac{\text{hyp}}{\text{opp}}$

$\cos$ (**cosine**) $\theta = \frac{\text{adj}}{\text{hyp}}$ $\sec$ (**secant**) $\theta = \frac{\text{hyp}}{\text{adj}}$

$\tan$ (**tangent**) $\theta = \frac{\text{opp}}{\text{adj}}$ $\cot$ (**cotangent**) $\theta = \frac{\text{adj}}{\text{opp}}$

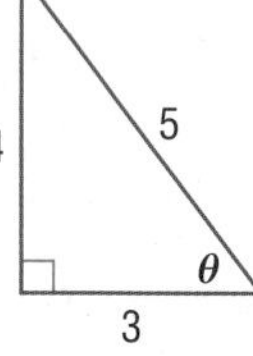

Examples

$\sin \theta = \frac{4}{5}$ $\cos \theta = \frac{3}{5}$ $\tan \theta = \frac{4}{3}$

$\csc \theta = \frac{5}{4}$ $\sec \theta = \frac{5}{3}$ $\cot \theta = \frac{3}{4}$

Example 1 Evaluate Trigonometric Functions

Find the values of the six trigonometric functions for angle θ.

leg opposite θ: $BC = 8$ leg adjacent θ: $AC = 15$ hypotenuse: $AB = 17$

$\sin \theta = \frac{\text{opp}}{\text{hyp}} = \frac{8}{17}$ $\cos \theta = \frac{\text{adj}}{\text{hyp}} = \frac{15}{17}$ $\tan \theta = \frac{\text{opp}}{\text{adj}} = \frac{8}{15}$

$\csc \theta = \frac{\text{hyp}}{\text{opp}} = \frac{17}{8}$ $\sec \theta = \frac{\text{hyp}}{\text{adj}} = \frac{17}{15}$ $\cot \theta = \frac{\text{adj}}{\text{opp}} = \frac{15}{8}$

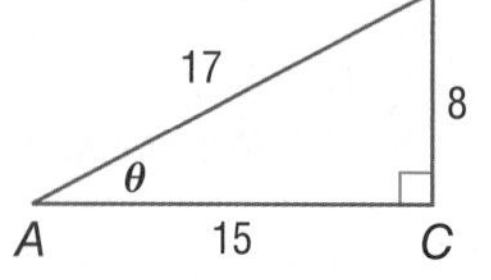

GuidedPractice

1. Find the values of the six trigonometric functions for angle B.

isifa Image Service s.r.o./Alamy

StudyTip

Memorize Trigonometric Ratios SOH-CAH-TOA is a mnemonic device for remembering the first letter of each word in the ratios for sine, cosine, and tangent.

$\sin \theta = \frac{\text{opp}}{\text{hyp}}$

$\cos \theta = \frac{\text{adj}}{\text{hyp}}$

$\tan \theta = \frac{\text{opp}}{\text{adj}}$

Notice that the cosecant, secant, and cotangent ratios are reciprocals of the sine, cosine, and tangent ratios, respectively. These are called the **reciprocal functions**.

$$\csc \theta = \frac{1}{\sin \theta} \qquad \sec \theta = \frac{1}{\cos \theta} \qquad \cot \theta = \frac{1}{\tan \theta}$$

The domain of any trigonometric function is the set of all acute angles θ of a right triangle. So, trigonometric functions depend only on the measures of the acute angles, not on the side lengths of a right triangle.

Example 2 Find Trigonometric Ratios

If $\sin B = \frac{5}{8}$, find the exact values of the five remaining trigonometric functions for B.

Step 1 Draw a right triangle and label one acute angle B. Label the opposite side 5 and the hypotenuse 8.

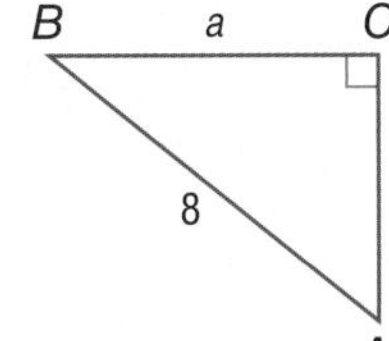

Step 2 Use the Pythagorean Theorem to find a.

$a^2 + b^2 = c^2$	Pythagorean Theorem
$a^2 + 5^2 = 8^2$	$b = 5$ and $c = 8$
$a^2 + 25 = 64$	Simplify.
$a^2 = 39$	Subtract 25 from each side.
$a = \pm\sqrt{39}$	Take the square root of each side.
$a = \sqrt{39}$	Length cannot be negative.

Step 3 Find the other values.

Since $\sin B = \frac{5}{8}$, $\csc B = \frac{\text{hyp}}{\text{opp}}$ or $\frac{8}{5}$.

$$\cos B = \frac{\text{adj}}{\text{hyp}} = \frac{\sqrt{39}}{8} \qquad \sec B = \frac{\text{hyp}}{\text{adj}} = \frac{8}{\sqrt{39}} \text{ or } \frac{8\sqrt{39}}{39}$$

$$\tan B = \frac{\text{opp}}{\text{adj}} = \frac{5}{\sqrt{39}} \text{ or } \frac{5\sqrt{39}}{39} \qquad \cot B = \frac{\text{adj}}{\text{opp}} = \frac{\sqrt{39}}{5}$$

GuidedPractice

2. If $\tan B = \frac{3}{7}$, find exact values of the remaining trigonometric fuctions for B.

ReadingMath

Labeling Triangles Throughout this chapter, a capital letter is used to represent both a vertex of a triangle and the measure of the angle at that vertex. The same letter in lowercase is used to represent both the side opposite that angle and the length of the side.

Angles that measure 30°, 45°, and 60° occur frequently in trigonometry.

KeyConcept Trigonometric Values for Special Angles

30°-60°-90°

$\sin 30° = \frac{1}{2}$ $\quad \cos 30° = \frac{\sqrt{3}}{2}$ $\quad \tan 30° = \frac{\sqrt{3}}{3}$

$\sin 60° = \frac{\sqrt{3}}{2}$ $\quad \cos 60° = \frac{1}{2}$ $\quad \tan 60° = \sqrt{3}$

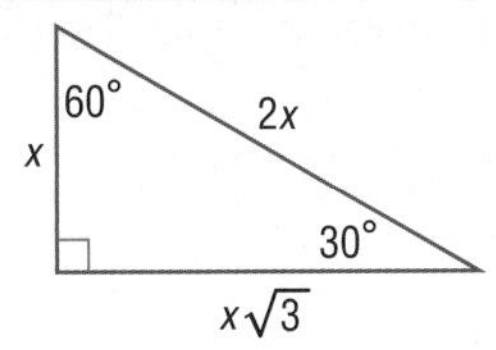

45°-45°-90°

$\sin 45° = \frac{\sqrt{2}}{2}$ $\quad \cos 45° = \frac{\sqrt{2}}{2}$ $\quad \tan 45° = 1$

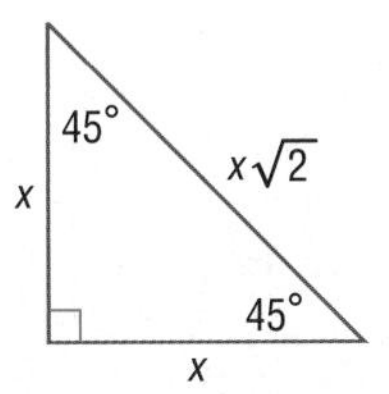

2 Use Trigonometric Functions

Use Trigonometric Functions You can use trigonometric functions to find missing side lengths and missing angle measures of right triangles.

Example 3 Find a Missing Side Length

Use a trigonometric function to find the value of x. Round to the nearest tenth if necessary.

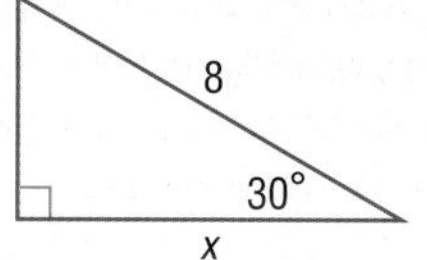

The length of the hypotenuse is 8. The missing measure is for the side adjacent to the 30° angle. Use the cosine function to find x.

$\cos \theta = \frac{\text{adj}}{\text{hyp}}$	Cosine function
$\cos 30° = \frac{x}{8}$	Replace θ with 30°, *adj* with x, and *hyp* with 8.
$\frac{\sqrt{3}}{2} = \frac{x}{8}$	$\cos 30° = \frac{\sqrt{3}}{2}$
$\frac{8\sqrt{3}}{2} = x$	Multiply each side by 8.
$6.9 \approx x$	Use a calculator.

StudyTip

Choose a Function If the length of the hypotenuse is unknown, then either the sine or cosine function must be used to find the missing measure.

GuidedPractice

3A.

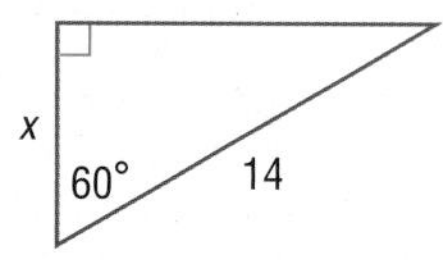

3B.

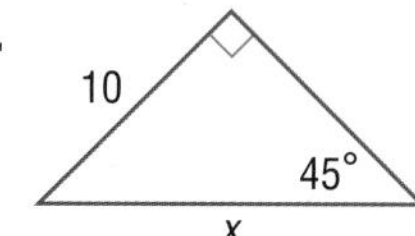

You can use a calculator to find the missing side lengths of triangles that do not have 30°, 45°, or 60° angles.

Real-World Example 4 Find a Missing Side Length

BUILDINGS To calculate the height of a building, Joel walked 200 feet from the base of the building and used an inclinometer to measure the angle from his eye to the top of the building. If his eye level is at 6 feet, how tall is the building?

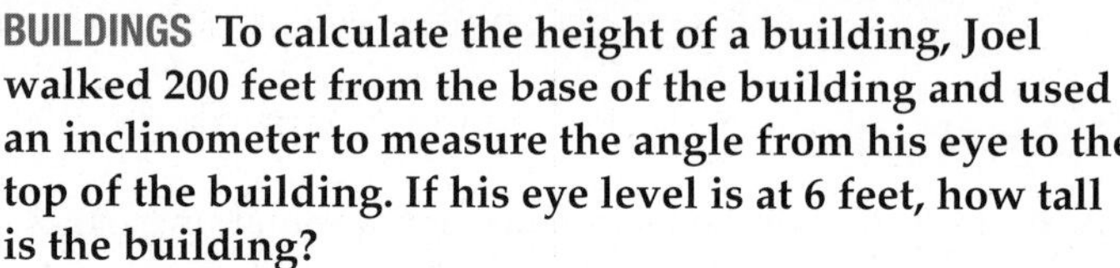

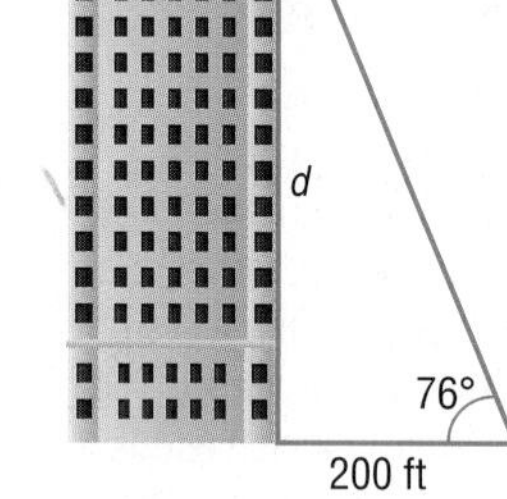

The measured angle is 76°. The side adjacent to the angle is 200 feet. The missing measure is the side opposite the angle. Use the tangent function to find d.

$\tan \theta = \frac{\text{opp}}{\text{adj}}$	Tangent function
$\tan 76° = \frac{d}{200}$	Replace θ with 76°, *opp* with d, and *adj* with 200.
$200 \tan 76° = d$	Multiply each side by 200.
$802 \approx d$	Use a calculator to simplify: 200 [TAN] 76 [ENTER].

Because the inclinometer was 6 feet above the ground, the height of the building is approximately 808 feet.

Real-WorldLink

Inclinometers measure the angle of Earth's magnetic field as well as the pitch and roll of vehicles, sailboats, and airplanes. They are also used for monitoring volcanoes and well drilling.

Source: Science Magazine

GuidedPractice

4. Use a trigonometric function to find the value of x. Round to the nearest tenth if necessary.

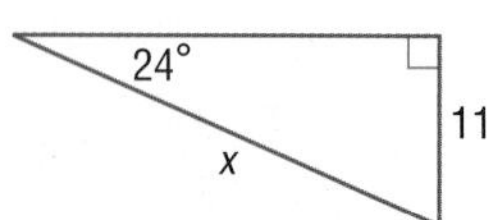

When solving equations like $3x = -27$, you use the inverse of multiplication to find x. You also can find angle measures by using the inverse of sine, cosine, or tangent.

ReadingMath

CCSS Precision The expression $\sin^{-1} x$ is read *the inverse sine of x* and is interpreted as the angle whose sine is x. Be careful not to confuse this notation with the notation for negative exponents; $\sin^{-1} x \neq \frac{1}{\sin x}$. Instead, this notation is similar to the notation for an inverse function, $f^{-1}(x)$.

KeyConcept Inverse Trigonometric Ratios

Words	If $\angle A$ is an acute angle and the sine of A is x, then the **inverse sine** of x is the measure of $\angle A$.
Symbols	If $\sin A = x$, then $\sin^{-1} x = m\angle A$.
Example	$\sin A = \frac{1}{2} \rightarrow \sin^{-1} \frac{1}{2} = m\angle A \rightarrow m\angle A = 30°$
Words	If $\angle A$ is an acute angle and the cosine of A is x, then the **inverse cosine** of x is the measure of $\angle A$.
Symbols	If $\cos A = x$, then $\cos^{-1} x = m\angle A$.
Example	$\cos A = \frac{\sqrt{2}}{2} \rightarrow \cos^{-1} \frac{\sqrt{2}}{2} = m\angle A \rightarrow m\angle A = 45°$
Words	If $\angle A$ is an acute angle and the tangent of A is x, then the **inverse tangent** of x is the measure of $\angle A$.
Symbols	If $\tan A = x$, then $\tan^{-1} x = m\angle A$.
Example	$\tan A = \sqrt{3} \rightarrow \tan^{-1} \sqrt{3} = m\angle A \rightarrow m\angle A = 60°$

If you know the sine, cosine, or tangent of an acute angle, you can use a calculator to find the measure of the angle, which is the inverse of the trigonometric ratio.

Example 5 Find a Missing Angle Measure

Find the measure of each angle. Round to the nearest tenth if necessary.

a. $\angle N$

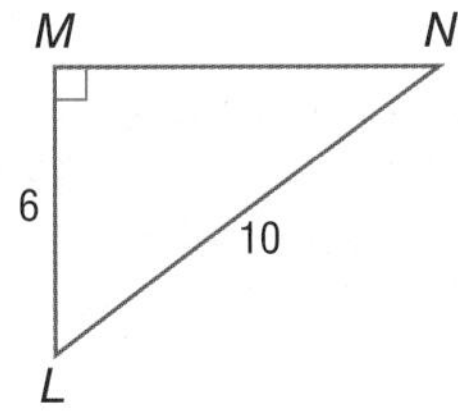

You know the measure of the side opposite $\angle N$ and the measure of the hypotenuse. Use the sine function.

$\sin N = \frac{6}{10}$ — $\sin \theta = \frac{\text{opp}}{\text{hyp}}$

$\sin^{-1} \frac{6}{10} = m\angle N$ — Inverse sine

$36.9° \approx m\angle N$ — Use a calculator.

b. $\angle B$

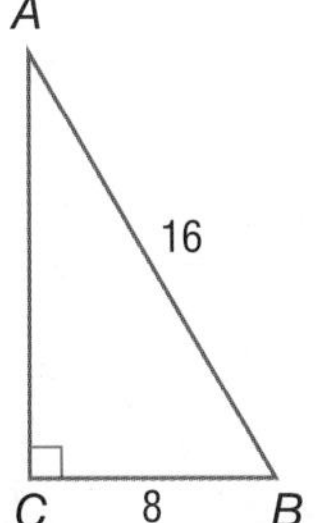

Use the cosine function.

$\cos B = \frac{8}{16}$ — $\cos \theta = \frac{\text{adj}}{\text{hyp}}$

$\cos^{-1} \frac{8}{16} = m\angle B$ — Inverse cosine

$60° = m\angle B$ — Use a calculator.

GuidedPractice **Find x. Round to the nearest tenth if necessary.**

5A.

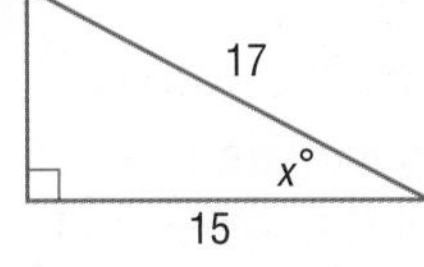

5B.

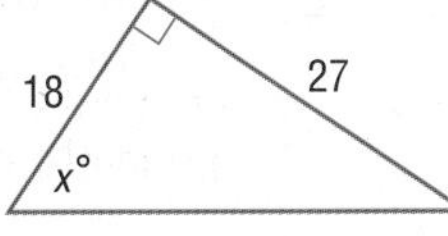

StudyTip

Angles of Elevation and Depression The angle of elevation and the angle of depression are congruent since they are alternate interior angles of parallel lines.

In the figure at the right, the angle formed by the line of sight from the swimmer and a line parallel to the horizon is called the **angle of elevation**. The angle formed by the line of sight from the lifeguard and a line parallel to the horizon is called the **angle of depression**.

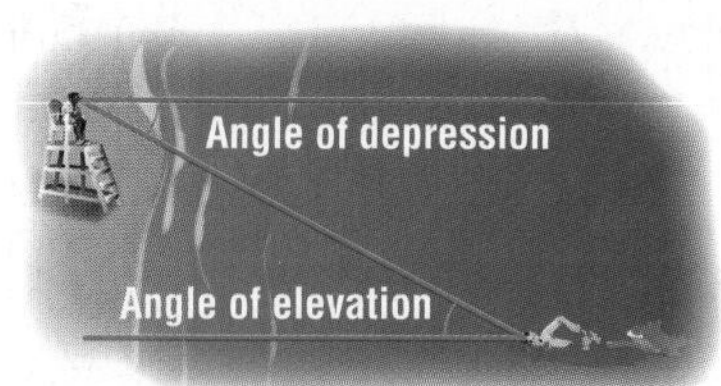

Real-World Example 6 Use Angles of Elevation and Depression

a. GOLF A golfer is standing at the tee, looking up to the green on a hill. If the tee is 36 yards lower than the green and the angle of elevation from the tee to the hole is 12°, find the distance from the tee to the hole.

Write an equation using a trigonometric function that involves the ratio of the vertical rise (side opposite the 12° angle) and the distance from the tee to the hole (hypotenuse).

$\sin 12° = \frac{36}{x}$ $\sin \theta = \frac{\text{opp}}{\text{hyp}}$

$x \sin 12° = 36$ Multiply each side by x.

$x = \frac{36}{\sin 12°}$ Divide each side by sin 12°.

$x \approx 173.2$ Use a calculator.

So, the distance from the tee to the hole is about 173.2 yards.

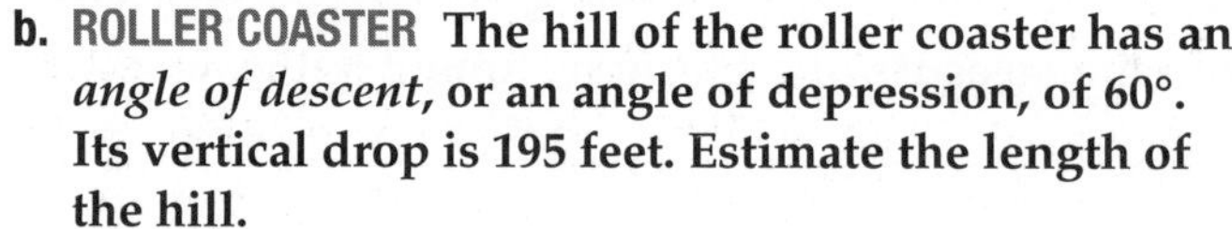

b. ROLLER COASTER The hill of the roller coaster has an *angle of descent*, or an angle of depression, of 60°. Its vertical drop is 195 feet. Estimate the length of the hill.

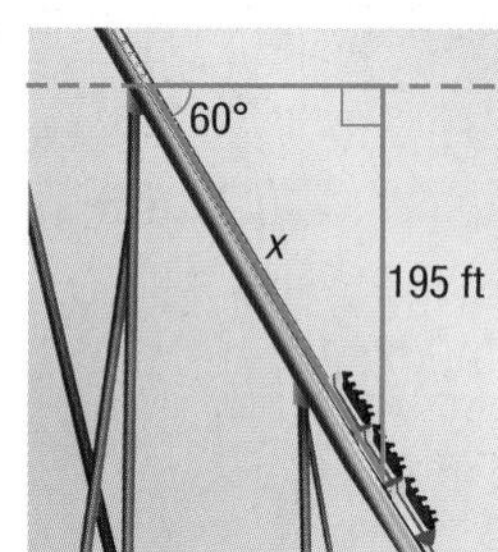

Write an equation using a trigonometric function that involves the ratio of the vertical drop (side opposite the 60° angle) and the length of the hill (hypotenuse).

$\sin 60° = \frac{195}{x}$ $\sin \theta = \frac{\text{opp}}{\text{hyp}}$

$x \sin 60° = 195$ Multiply each side by x.

$x = \frac{195}{\sin 60°}$ Divide each side by sin 60°.

$x \approx 225.2$ Use a calculator.

So, the length of the hill is about 225.2 feet.

Real-WorldLink

The steepest roller coasters in the world have angles of descent that are close to 90°.

Source: Ultimate Roller Coaster

GuidedPractice

6A. MOVING A ramp for unloading a moving truck has an angle of elevation of 32°. If the top of the ramp is 4 feet above the ground, estimate the length of the ramp.

6B. LADDERS A 14-ft long ladder is placed against a house at an angle of elevation of 72°. How high above the ground is the top of the ladder?

Check Your Understanding

= Step-by-Step Solutions begin on page R14.

Example 1 **Find the values of the six trigonometric functions for angle θ.**

1. 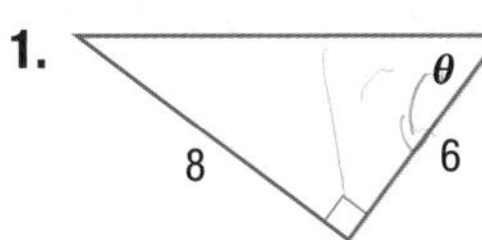

2. 16, 12, θ

Example 2 **In a right triangle, $\angle A$ is acute. Find the values of the five remaining trigonometric funtions.**

3. $\cos A = \frac{4}{7}$

4. $\tan A = \frac{20}{21}$

Examples 3–4 **Use a trigonometric function to find the value of x. Round to the nearest tenth.**

5. 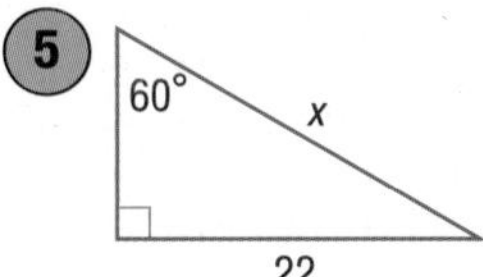

6. 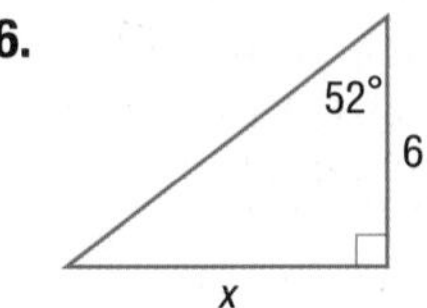

7. 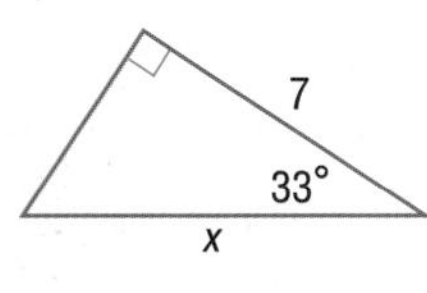

Example 5 **Find the value of x. Round to the nearest tenth.**

8. 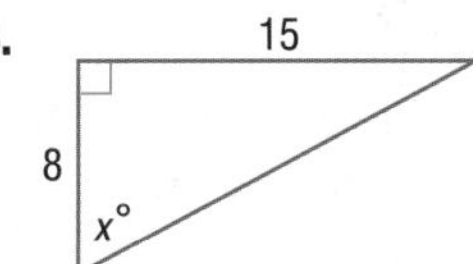

9. 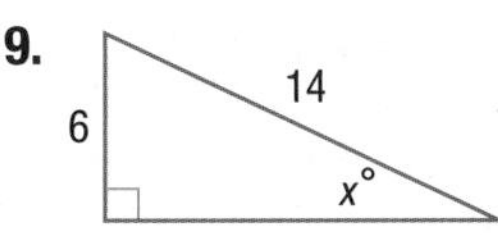

10. 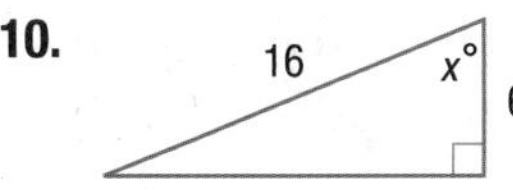

Example 6

11. CCSS **SENSE-MAKING** Christian found two trees directly across from each other in a canyon. When he moved 100 feet from the tree on his side (parallel to the edge of the canyon), the angle formed by the tree on his side and the tree on the other side was 70°. Find the distance across the canyon.

12. **LADDERS** The recommended angle of elevation for a ladder used in fire fighting is 75°. At what height on a building does a 21-foot ladder reach if the recommended angle of elevation is used? Round to the nearest tenth.

Practice and Problem Solving

Extra Practice is on page R12.

Example 1 **Find the values of the six trigonometric functions for angle θ.**

13. 12, 13, θ

14. 9, 40, θ

15. 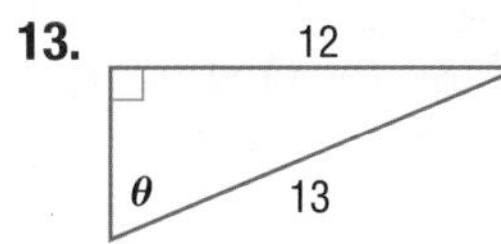

16. 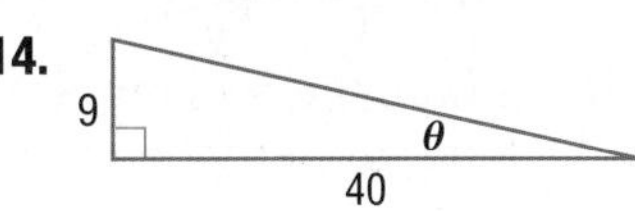

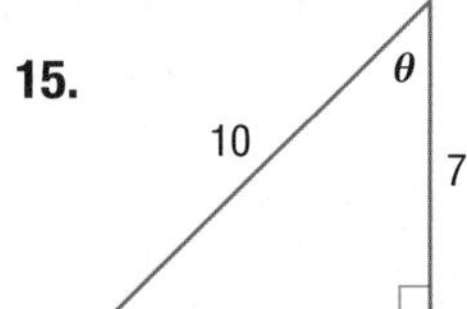

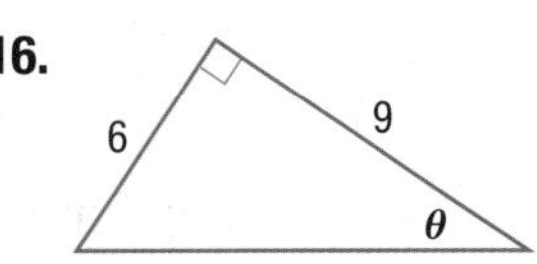

Example 2 **In a right triangle, $\angle A$ and $\angle B$ are acute. Find the values of the five remaining trigonometric funtions.**

17. $\tan A = \frac{8}{15}$

18. $\cos A = \frac{3}{10}$

19. $\tan B = 3$

20. $\sin B = \frac{4}{9}$

Examples 3–4 **Use a trigonometric function to find each value of x. Round to the nearest tenth.**

21. 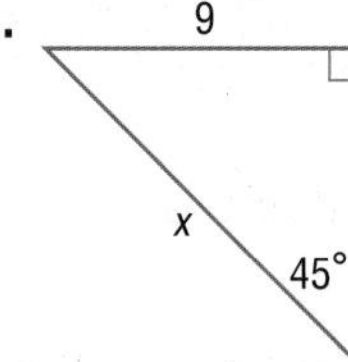

22. 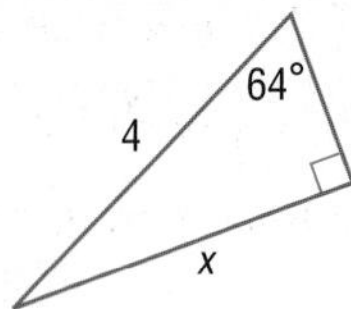

23. 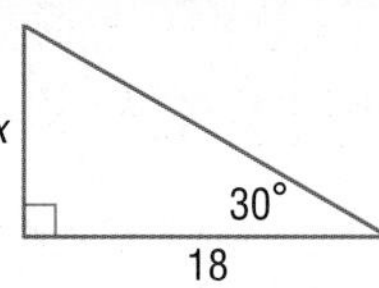

24. 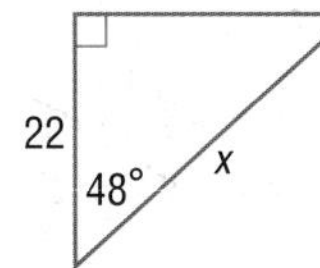

25. 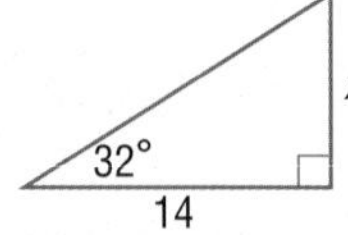

26. 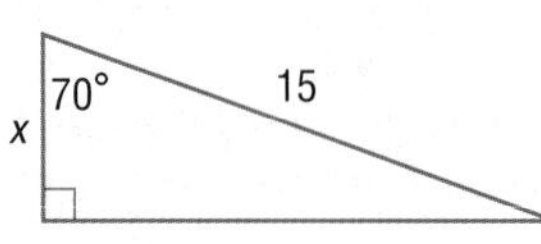

27. **PARASAILING** Refer to the beginning of the lesson and the figure at the right. Find a, the altitude of a person parasailing, if the tow rope is 250 feet long and the angle formed is 32°. Round to the nearest tenth.

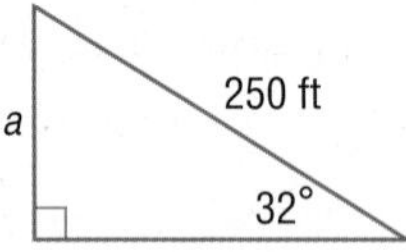

28. **CCSS MODELING** Devon wants to build a rope bridge between his treehouse and Cheng's treehouse. Suppose Devon's treehouse is directly behind Cheng's treehouse. At a distance of 20 meters to the left of Devon's treehouse, an angle of 52° is measured between the two treehouses. Find the length of the rope.

Example 5 **Find the value of x. Round to the nearest tenth.**

29. 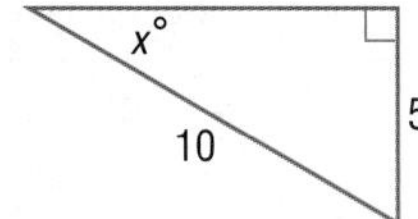

30. 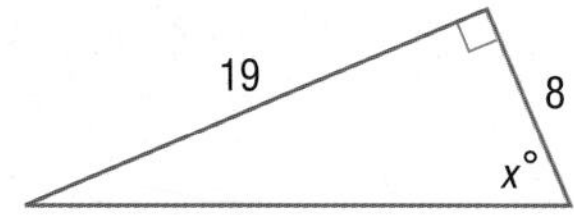

31. 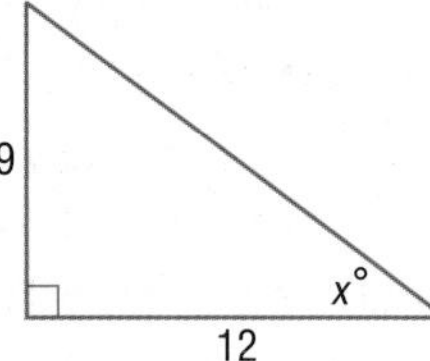

32. 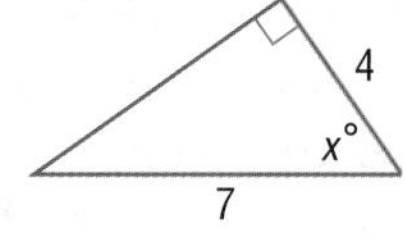

33. 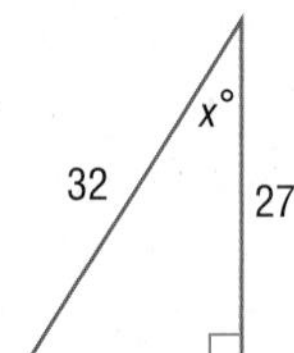

34. 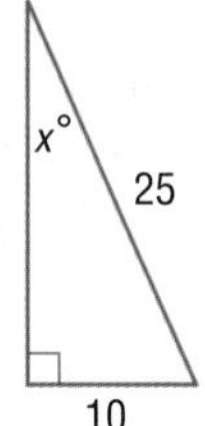

Example 6

35. **SQUIRRELS** Adult flying squirrels can make glides of up to 160 feet. If a flying squirrel glides a horizontal distance of 160 feet and the angle of descent is 9°, find its change in height.

36. **HANG GLIDING** A hang glider climbs at a 20° angle of elevation. Find the change in altitude of the hang glider when it has flown a horizontal distance of 60 feet.

Use trigonometric functions to find the values of x and y. Round to the nearest tenth.

37. 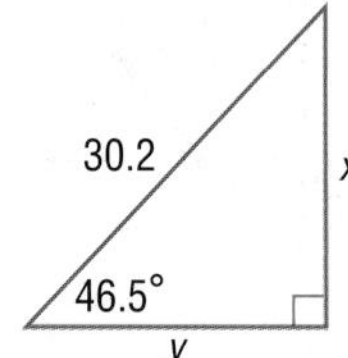

38. 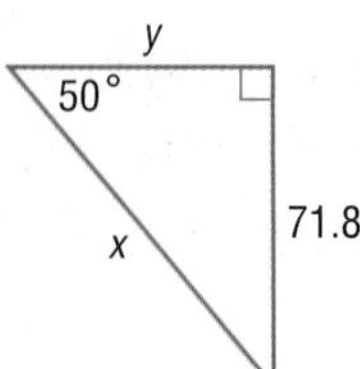

39. 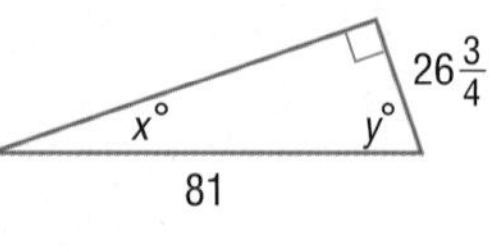

Solve each equation.

40. $\cos A = \frac{3}{19}$

41. $\sin N = \frac{9}{11}$

42. $\tan X = 15$

43. $\sin T = 0.35$

44. $\tan G = 0.125$

45. $\cos Z = 0.98$

46. MONUMENTS A monument casts a shadow 24 feet long. The angle of elevation from the end of the shadow to the top of the monument is 50°.

a. Draw and label a right triangle to represent this situation.

b. Write a trigonometric function that can be used to find the height of the monument.

c. Find the value of the function to determine the height of the monument to the nearest tenth.

47. NESTS Tabitha's eyes are 5 feet above the ground as she looks up to a bird's nest in a tree. If the angle of elevation is 74.5° and she is standing 12 feet from the tree's base, what is the height of the bird's nest? Round to the nearest foot.

48. RAMPS Two bicycle ramps each cover a horizontal distance of 8 feet. One ramp has a 20° angle of elevation, and the other ramp has a 35° angle of elevation, as shown at the right.

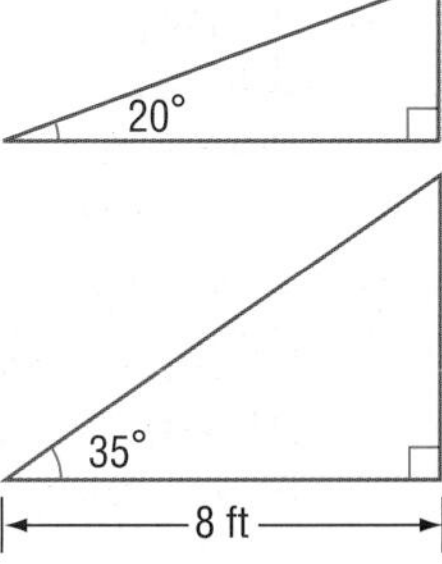

a. How much taller is the second ramp than the first? Round to the nearest tenth.

b. How much longer is the second ramp than the first? Round to the nearest tenth.

49. FALCONS A falcon at a height of 200 feet sees two mice A and B, as shown in the diagram.

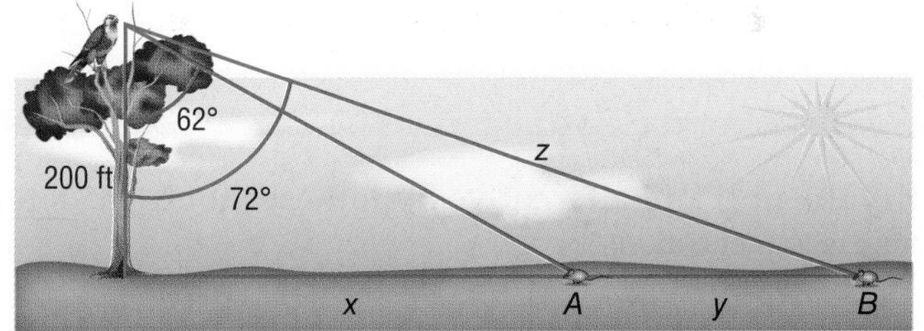

a. What is the approximate distance z between the falcon and mouse B?

b. How far apart are the two mice?

In $\triangle ABC$, $\angle C$ is a right angle. Use the given measurements to find the missing side lengths and missing angle measures of $\triangle ABC$. Round to the nearest tenth if necessary.

50. $m\angle A = 36°, a = 12$

51. $m\angle B = 31°, b = 19$

52. $a = 8, c = 17$

53. $\tan A = \frac{4}{5}, a = 6$

H.O.T. Problems Use Higher-Order Thinking Skills

54. CHALLENGE A line segment has endpoints $A(2, 0)$ and $B(6, 5)$, as shown in the figure at the right. What is the measure of the acute angle θ formed by the line segment and the x-axis? Explain how you found the measure.

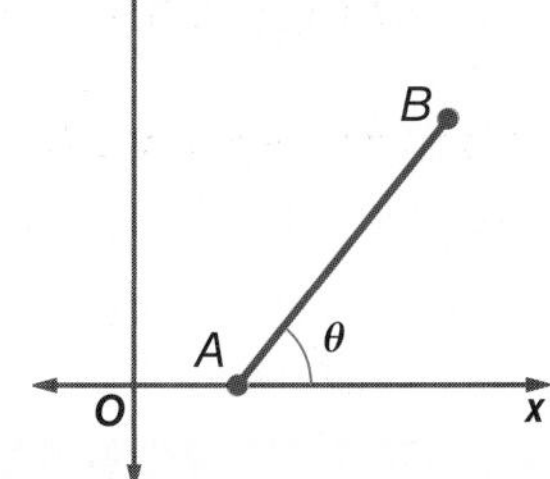

55. CCSS ARGUMENTS Determine whether the following statement is *true* or *false*. Explain your reasoning.

For any acute angle, the sine function will never have a negative value.

56. OPEN ENDED In right triangle ABC, $\sin A = \sin C$. What can you conclude about $\triangle ABC$? Justify your reasoning.

57. WRITING IN MATH A roof has a slope of $\frac{2}{3}$. Describe the connection between the slope and the angle of elevation θ that the roof makes with the horizontal. Then use an inverse trigonometric function to find θ.

Standardized Test Practice

58. EXTENDED RESPONSE Your school needs 5 cases of yearbooks. Neighborhood Yearbooks lists a case of yearbooks at $153.85 with a 10% discount on an order of 5 cases. Yearbooks R Us lists a case of yearbooks at $157.36 with a 15% discount on 5 cases.

a. Which company would you choose?

b. What is the least amount that you would have to spend for the yearbooks?

59. SHORT RESPONSE As a fundraiser, the marching band sold T-shirts and hats. They sold a total of 105 items and raised $1170. If the cost of a hat was $10 and the cost of a T-shirt was $15, how many T-shirts were sold?

60. A hot dog stand charges price x for a hot dog and price y for a drink. Two hot dogs and one drink cost $4.50. Three hot dogs and two drinks cost $7.25. Which matrix could be multiplied by $\begin{bmatrix} 4.50 \\ 7.25 \end{bmatrix}$ to find x and y?

A $\begin{bmatrix} -1 & 1 \\ 2 & -1 \end{bmatrix}$

B $\begin{bmatrix} 2 & -1 \\ -3 & 2 \end{bmatrix}$

C $\begin{bmatrix} 1 & 2 \\ -1 & 3 \end{bmatrix}$

D $\begin{bmatrix} 1 & -1 \\ -1 & 2 \end{bmatrix}$

61. SAT/ACT The length and width of a rectangle are in the ratio of 5:12. If the rectangle has an area of 240 square centimeters, what is the length, in centimeters, of its diagonal?

F 24

G 26

H 28

J 30

K 32

Spiral Review

Identify the null and alternative hypotheses for each statement. Then identify the statement that represents the claim. (Lesson 11-6)

62. Jack thinks that it takes less than 10 minutes to ride his bike from his home to the store.

63. A deli sign says that one 12-inch turkey sandwich contains three ounces of meat.

64. Mrs. Thomas takes at least 15 minutes to prepare a cake.

65. SWIMMING POOL The number of visits to a community swimming pool per year by a sample of 425 members is normally distributed with a mean of 90 and a standard deviation of 15. (Lesson 11-5)

a. About what percent of the members went to the pool at least 45 times?

b. What is the probability that a member selected at random went to the pool more than 120 times?

c. What percent of the members went to the pool between 75 and 105 times?

66. POLLS A polling company wants to estimate how many people are in favor of a new environmental law. The polling company polls 20 people. The probability that a person is in favor of the law is 0.5. (Lesson 11-4)

a. What is the probability that exactly 12 people are in favor of the new law?

b. What is the expected number of people in favor of the law?

Skills Review

Find each product. Include the appropriate units with your answer.

67. $4.3 \text{ miles}\left(\frac{5280 \text{ feet}}{1 \text{ mile}}\right)$

68. $8 \text{ gallons}\left(\frac{8 \text{ pints}}{1 \text{ gallon}}\right)$

69. $\left(\frac{5 \text{ dollars}}{3 \text{ meters}}\right)21 \text{ meters}$

70. $\left(\frac{18 \text{ cubic inches}}{5 \text{ seconds}}\right)24 \text{ seconds}$

71. $65 \text{ degrees}\left(\frac{10 \text{ centimeters}}{3 \text{ degrees}}\right)$

72. $\left(\frac{7 \text{ liters}}{30 \text{ minutes}}\right)10 \text{ minutes}$

LESSON

12-2 Angles and Angle Measure

Then	Now	Why?
You used angles with degree measures.	1 Draw and find angles in standard position. 2 Convert between degree measures and radian measures.	A sundial is an instrument that indicates the time of day by the shadow that it casts on a surface marked to show hours or fractions of hours. The shadow moves around the dial 15° every hour.

NewVocabulary
standard position
initial side
terminal side
coterminal angles
radian
central angle
arc length

Common Core State Standards

Content Standards
F.TF.1 Understand radian measure of an angle as the length of the arc on the unit circle subtended by the angle.

Mathematical Practices
2 Reason abstractly and quantitatively.

1 Angles in Standard Position

An angle on the coordinate plane is in **standard position** if the vertex is at the origin and one ray is on the positive x-axis.

- The ray on the x-axis is called the **initial side** of the angle.
- The ray that rotates about the center is called the **terminal side**.

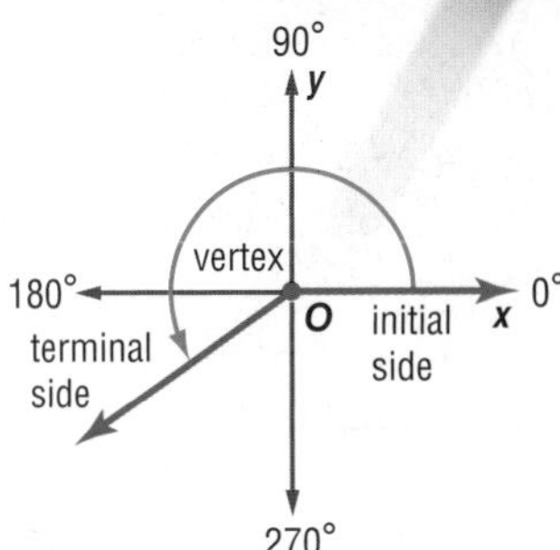

KeyConcept Angle Measures

If the measure of an angle is positive, the terminal side is rotated counterclockwise.

If the measure of an angle is negative, the terminal side is rotated clockwise.

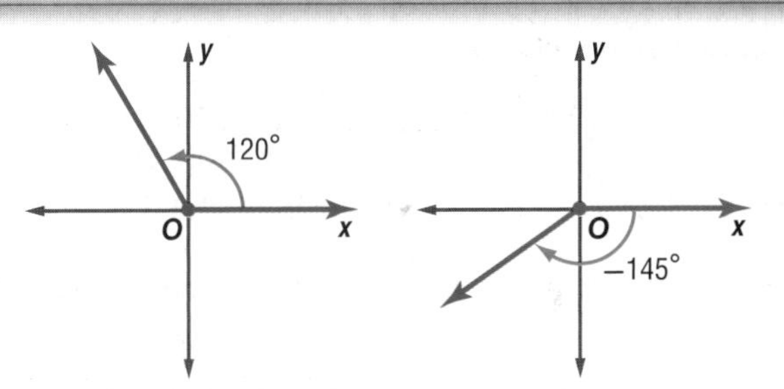

Example 1 Draw an Angle in Standard Position

Draw an angle with the given measure in standard position.

a. 215° $215° = 180° + 35°$

Draw the terminal side of the angle 35° counterclockwise past the negative x-axis.

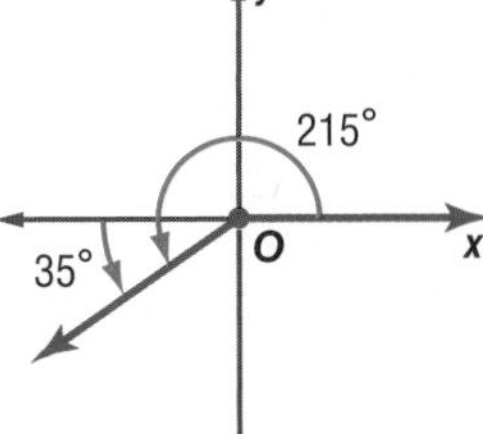

b. −40°

The angle is negative. Draw the terminal side of the angle 40° clockwise from the positive x-axis.

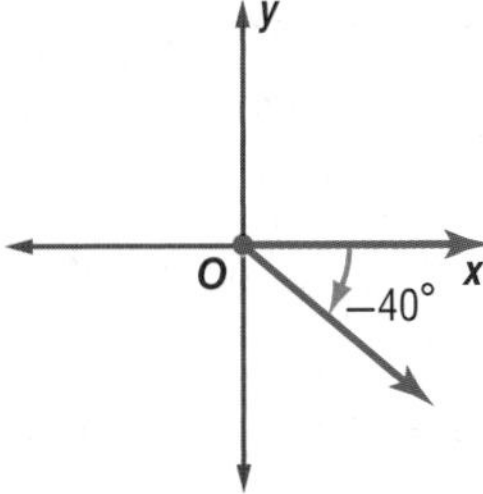

GuidedPractice

1A. 80°

1B. −105°

The terminal side of an angle can make more than one complete rotation. For example, a complete rotation of 360° plus a rotation of 120° forms an angle that measures 360° + 120° or 480°.

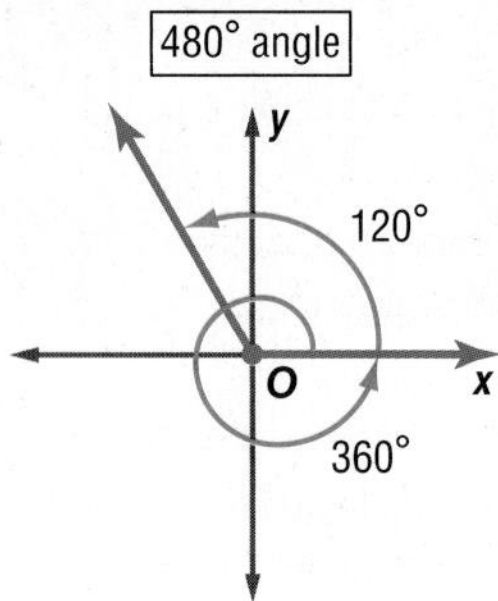

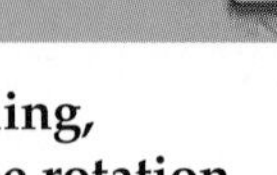

Real-World Example 2 Draw an Angle in Standard Position

WAKEBOARDING ***Wakeboarding*** **is a combination of surfing, skateboarding, snowboarding, and water skiing. One maneuver involves a 540-degree rotation in the air. Draw an angle in standard position that measures 540°.**

540° = 360° + 180°

Draw the terminal side of the angle 180° past the positive x-axis.

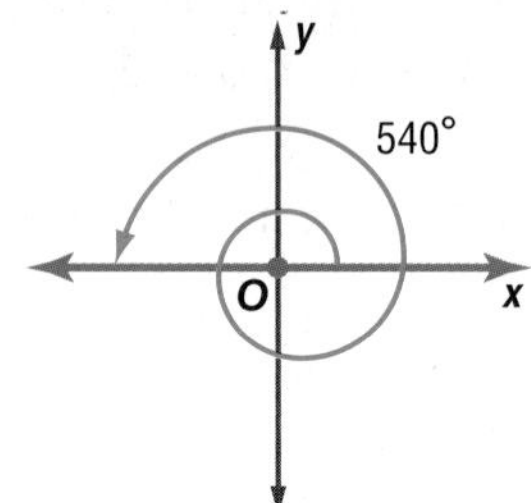

Real-WorldLink

Wakeboarding is one of the fastest-growing water sports in the United States. Participation has increased more than 100% in recent years.

Source: King of Wake

GuidedPractice

2. Draw an angle in standard position that measures 600°.

Two or more angles in standard position with the same terminal side are called **coterminal angles**. For example, angles that measure 60°, 420°, and −300° are coterminal, as shown in the figure at the right.

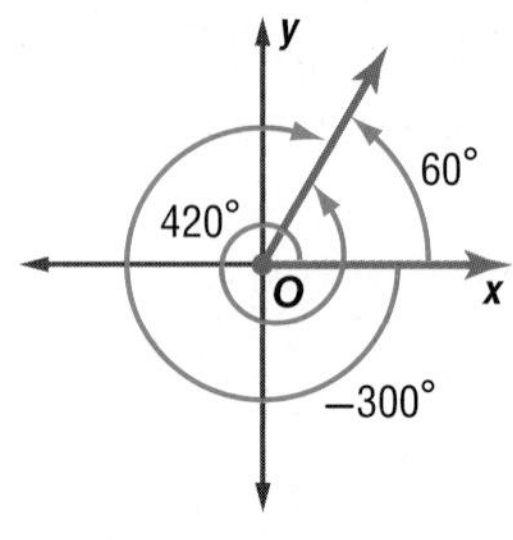

An angle that is coterminal with another angle can be found by adding or subtracting a multiple of 360°.

- 60° + 360° = 420°
- 60° − 360° = −300°

ReadingMath

Angle of Rotation
In trigonometry, an angle is sometimes referred to as an *angle of rotation.*

Example 3 Find Coterminal Angles

Find an angle with a positive measure and an angle with a negative measure that are coterminal with each angle.

a. 130°

positive angle: 130° + 360° = 490° Add 360°.
negative angle: 130° − 360° = −230° Subtract 360°.

b. −200°

positive angle: −200° + 360° = 160° Add 360°.
negative angle: −200° − 360° = −560° Subtract 360°.

GuidedPractice

3A. 15° **3B.** −45°

Paul Bradbury/Digital Vision/Getty Images

StudyTip

CCSS Structure As with degrees, radians measure the amount of rotation from the initial side to the terminal side.

- The measure of an angle in radians is positive if its rotation is counterclockwise.
- The measure is negative if the rotation is clockwise.

2 Convert Between Degrees and Radians

Angles can also be measured in units that are based on arc length. One **radian** is the measure of an angle θ in standard position with a terminal side that intercepts an arc with the same length as the radius of the circle.

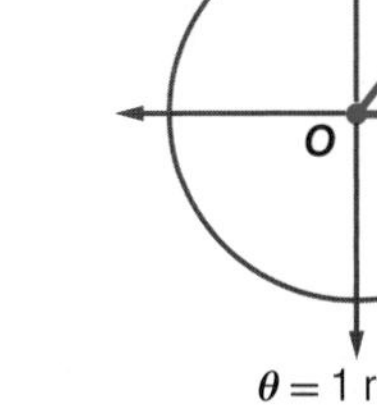

$\theta = 1$ radian

The circumference of a circle is $2\pi r$. So, one complete revolution around a circle equals 2π radians. Since 2π radians $= 360°$, degree measure and radian measure are related by the following equations.

$$2\pi \text{ radians} = 360° \qquad \pi \text{ radians} = 180°$$

KeyConcept Convert Between Degrees and Radians

Degrees to Radians	Radians to Degrees
To convert from degrees to radians, multiply the number of degrees by $\frac{\pi \text{ radians}}{180°}$.	To convert from radians to degrees, multiply the number of radians by $\frac{180°}{\pi \text{ radians}}$.

ReadingMath

Radian Measures

The word *radian* is usually omitted when angles are expressed in radian measure. Thus, when no units are given for an angle measure, radian measure is implied.

Example 4 Convert Between Degrees and Radians

Rewrite the degree measure in radians and the radian measure in degrees.

a. $-30°$

$$-30° = -30° \cdot \frac{\pi \text{ radians}}{180°}$$

$$= \frac{-30\pi}{180} \text{ or } -\frac{\pi}{6} \text{ radians}$$

b. $\frac{5\pi}{2}$

$$\frac{5\pi}{2} = \frac{5\pi}{2} \text{ radians} \cdot \frac{180°}{\pi \text{ radians}}$$

$$= \frac{900°}{2} \text{ or } 450°$$

GuidedPractice

4A. $120°$

4B. $-\frac{3\pi}{8}$

ConceptSummary Degrees and Radians

The diagram shows equivalent degree and radian measures for special angles.

You may find it helpful to memorize the following equivalent degree and radian measures. The other special angles are multiples of these angles.

$30° = \frac{\pi}{6}$ $\qquad 45° = \frac{\pi}{4}$

$60° = \frac{\pi}{3}$ $\qquad 90° = \frac{\pi}{2}$

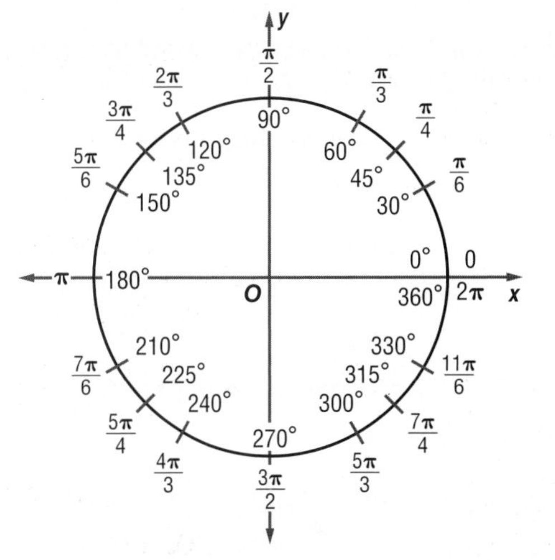

A **central angle** of a circle is an angle with a vertex at the center of the circle. If you know the measure of a central angle and the radius of the circle, you can find the length of the arc that is intercepted by the angle.

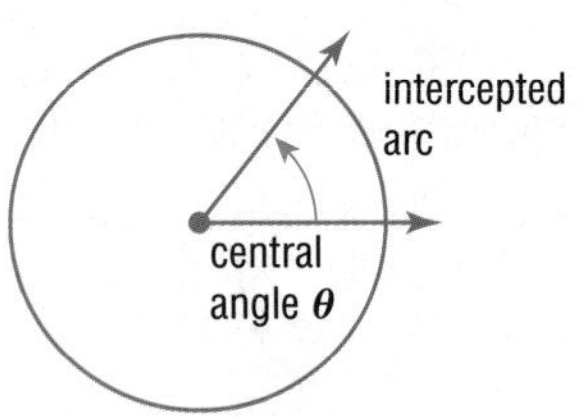

KeyConcept Arc Length

Words For a circle with radius r and central angle θ (in radians), the **arc length** s equals the product of r and θ.

Model

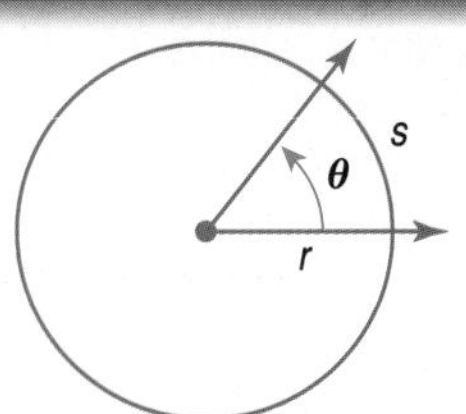

Symbols $s = r\theta$

You will justify this formula in Exercise 52

Real-World Example 5 Find Arc Length

TRUCKS Monster truck tires have a radius of 33 inches. How far does a monster truck travel in feet after just three fourths of a tire rotation?

Step 1 Find the central angle in radians.

$\theta = \frac{3}{4} \cdot 2\pi$ or $\frac{3\pi}{2}$ — The angle is $\frac{3}{4}$ of a complete rotation.

Step 2 Use the radius and central angle to find the arc length.

$s = r\theta$ — Write the formula for arc length.

$= 33 \cdot \frac{3\pi}{2}$ — Replace r with 33 and θ with $\frac{3\pi}{2}$.

≈ 155.5 in. — Use a calculator to simplify.

≈ 13.0 ft — Divide by 12 to convert to feet.

So, the truck travels about 13 feet after three fourths of a tire rotation.

WatchOut!

Arc Length Remember to write the angle measure in radians, not degrees, when finding arc length. Also, recall that the number of radians in a complete rotation is 2π.

GuidedPractice

5. A circle has a diameter of 9 centimeters. Find the arc length if the central angle is 60°. Round to the nearest tenth.

Check Your Understanding

Examples 1–2 **Draw an angle with the given measure in standard position.**

1. 140° **2.** −60° **3.** 390°

Example 3 **Find an angle with a positive measure and an angle with a negative measure that are coterminal with each angle.**

4. 25° **5.** 175° **6.** −100°

Example 4 **Rewrite each degree measure in radians and each radian measure in degrees.**

7. $\frac{\pi}{4}$ **8.** 225° **9.** −40°

Example 5 **10. CCSS REASONING** A tennis player's swing moves along the path of an arc. If the radius of the arc's circle is 4 feet and the angle of rotation is 100°, what is the length of the arc? Round to the nearest tenth.

Practice and Problem Solving

Extra Practice is on page R12.

Examples 1–2 **Draw an angle with the given measure in standard position.**

11. 75°
12. 160°
13. −90°
14. −120°
15. 295°
16. 510°

17. GYMNASTICS A gymnast on the uneven bars swings to make a 240° angle of rotation.

18. FOOD The lid on a jar of pasta sauce is turned 420° before it comes off.

Example 3 **Find an angle with a positive measure and an angle with a negative measure that are coterminal with each angle.**

19. 50°
20. 95°
21. 205°
22. 350°
23. −80°
24. −195°

Example 4 **Rewrite each degree measure in radians and each radian measure in degrees.**

25. 330°
26. $\frac{5\pi}{6}$
27. $-\frac{\pi}{3}$
28. −50°
29. 190°
30. $-\frac{7\pi}{3}$

Example 5 **31. SKATEBOARDING** The skateboard ramp at the right is called a *quarter pipe*. The curved surface is determined by the radius of a circle. Find the length of the curved part of the ramp.

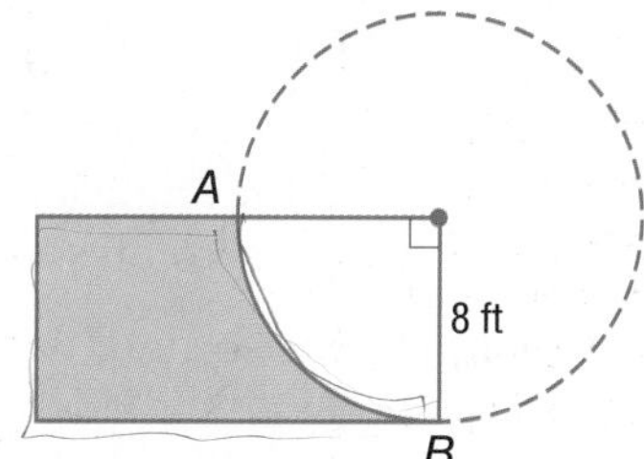

32. RIVERBOATS The paddlewheel of a riverboat has a diameter of 24 feet. Find the arc length of the circle made when the paddlewheel rotates 300°.

Find the length of each arc. Round to the nearest tenth.

33.

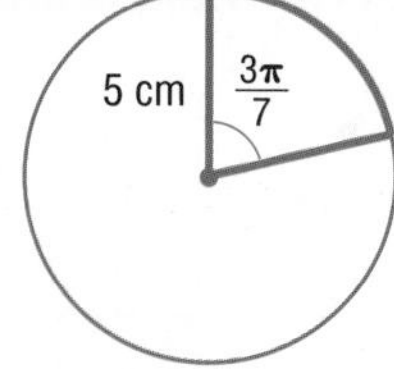

34.

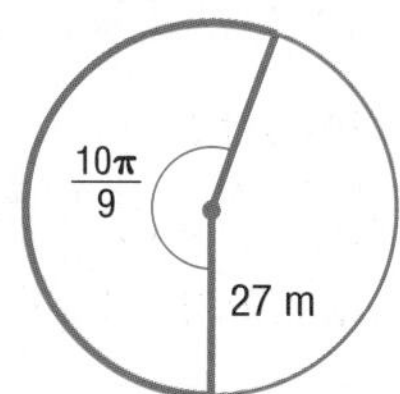

35. CLOCKS How long does it take for the minute hand on a clock to pass through 2.5π radians?

36. CCSS PERSEVERANCE Refer to the beginning of the lesson. A shadow moves around a sundial 15° every hour.

a. After how many hours is the angle of rotation of the shadow $\frac{8\pi}{5}$ radians?

b. What is the angle of rotation in radians after 5 hours?

c. A sundial has a radius of 8 inches. What is the arc formed by a shadow after 14 hours? Round to the nearest tenth.

Find an angle with a positive measure and an angle with a negative measure that are coterminal with each angle.

37. 620°
38. −400°
39. $-\frac{3\pi}{4}$
40. $\frac{19\pi}{6}$

41 **SWINGS** A swing has a 165° angle of rotation.

a. Draw the angle in standard position.

b. Write the angle measure in radians.

c. If the chains of the swing are $6\frac{1}{2}$ feet long, what is the length of the arc that the swing makes? Round to the nearest tenth.

d. Describe how the arc length would change if the lengths of the chains of the swing were doubled.

42. **MULTIPLE REPRESENTATIONS** Consider $A(-4, 0)$, $B(-4, 6)$, $C(6, 0)$, and $D(6, 8)$.

a. **Geometric** Draw $\triangle EAB$ and $\triangle ECD$ with E at the origin.

b. **Algebraic** Find the values of the tangent of $\angle BEA$ and the tangent of $\angle DEC$.

c. **Algebraic** Find the slope of $\overline{BE}$ and $\overline{ED}$.

d. **Verbal** What conclusions can you make about the relationship between slope and tangent?

Rewrite each degree measure in radians and each radian measure in degrees.

43. $\frac{21\pi}{8}$ **44.** 124° **45.** −200° **46.** 5

47. **CAROUSELS** A carousel makes 5 revolutions per minute. The circle formed by riders sitting in the outside row has a radius of 17.2 feet. The circle formed by riders sitting in the inside row has a radius of 13.1 feet.

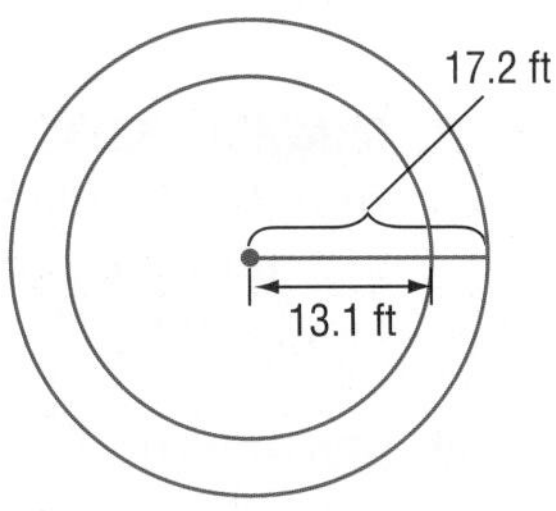

a. Find the angle θ in radians through which the carousel rotates in one second.

b. In one second, what is the difference in arc lengths between the riders sitting in the outside row and the riders sitting in the inside row?

H.O.T. Problems Use Higher-Order Thinking Skills

48. **CCSS CRITIQUE** Tarshia and Alan are writing an expression for the measure of an angle coterminal with the angle shown at the right. Is either of them correct? Explain your reasoning.

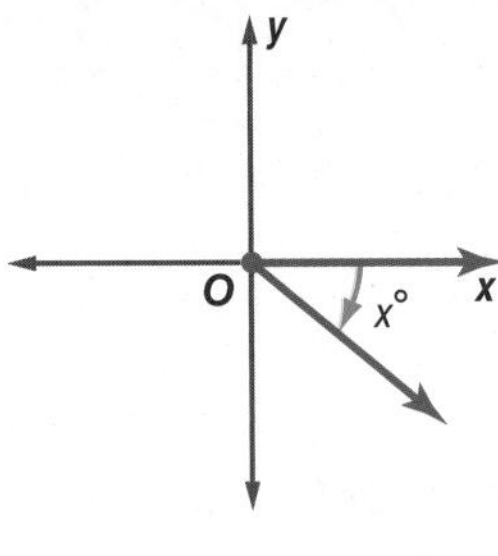

Tarshia	Alan
The measure of a coterminal angle is $(x - 360)°$.	The measure of a coterminal angle is $(360 - x)°$.

49. **CHALLENGE** A line makes an angle of $\frac{\pi}{2}$ radians with the positive x-axis at the point (2, 0). Find an equation for this line.

50. **REASONING** Express $\frac{1}{8}$ of a revolution in degrees and in radians. Explain your reasoning.

51. **OPEN ENDED** Draw and label an acute angle in standard position. Find two angles, one positive and one negative, that are coterminal with the angle.

52. **REASONING** Justify the formula for the length of an arc.

53. **WRITING IN MATH** Use a circle with radius r to describe what one degree and one radian represent. Then explain how to convert between the measures.

Standardized Test Practice

54. SHORT RESPONSE If $(x + 6)(x + 8) - (x - 7)(x - 5) = 0$, find x.

55. Which of the following represents an inverse variation?

A

x	2	5	10	20	25	50
y	50	20	10	5	4	2

B

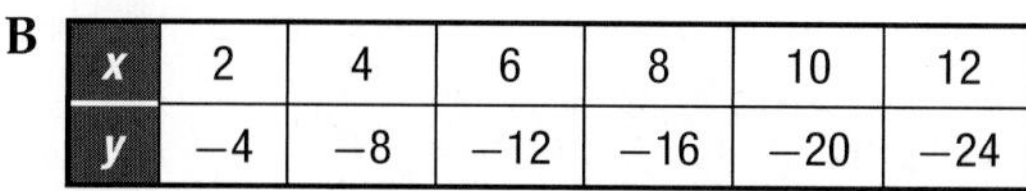

x	2	4	6	8	10	12
y	−4	−8	−12	−16	−20	−24

C

x	1	2	3	4	5	6
y	5	10	15	20	25	30

D

x	10	9	8	7	6	5
y	5	6	7	8	9	10

56. GEOMETRY If the area of the figure is 60 square units, what is the length of side $\overline{XZ}$?

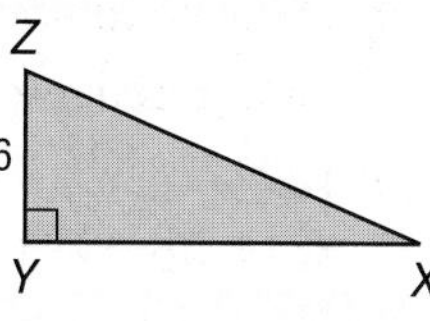

F $2\sqrt{34}$

G $2\sqrt{109}$

H $4\sqrt{34}$

J $4\sqrt{109}$

57. SAT/ACT The first term of a sequence is −6, and every term after the first is 8 more than the term immediately preceding it. What is the value of the 101st term?

A 788

B 794

C 802

D 806

E 814

Spiral Review

Find the values of the six trigonometric functions for angle θ. (Lesson 12-1)

58.

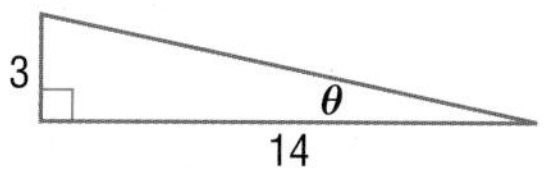

59.

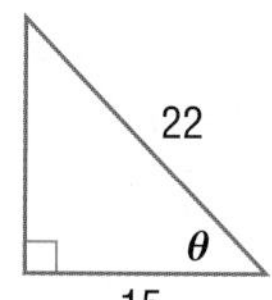

60.

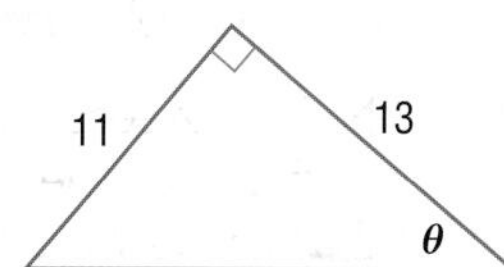

Identify the null and alternative hypotheses for each statement. Then identify the statement that represents the claim. (Lesson 11-6)

61. Tom drinks at least eight glasses of water every day.

62. Juanita says that she has two umbrellas in her car.

63. MANUFACTURING The sizes of CDs made by a company are normally distributed with a standard deviation of 1 millimeter. The CDs are supposed to be 120 millimeters in diameter, and they are made for drives that are 122 millimeters wide. (Lesson 11-5)

a. What percent of the CDs would you expect to be greater than 120 millimeters?

b. If the company manufactures 1000 CDs per hour, how many of the CDs made in one hour would you expect to be between 119 and 122 millimeters?

c. About how many CDs per hour will be too large to fit in the drives?

64. FINANCIAL LITERACY If the rate of inflation is 2%, the cost of an item in future years can be found by iterating the function $c(x) = 1.02x$. Find the cost of a $70 digital audio player in four years if the rate of inflation remains constant. (Lesson 10-5)

Skills Review

Use the Pythagorean Theorem to find the length of the hypotenuse for each right triangle with the given side lengths.

65. $a = 12, b = 15$

66. $a = 8, b = 17$

67. $a = 14, b = 11$

EXTEND 12-2

Geometry Lab
Areas of Parallelograms

The area of any triangle can be found using the sine ratios in the triangle. A similar process can be used to find the area of a parallelogram.

Activity

Find the area of parallelogram $ABCD$.

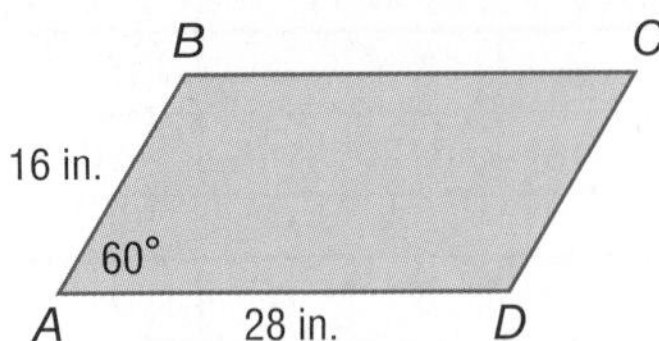

Step 1 Draw diagonal $\overline{BD}$.

$\overline{BD}$ divides the parallelogram into two congruent triangles, $\triangle ABD$ and $\triangle CDB$.

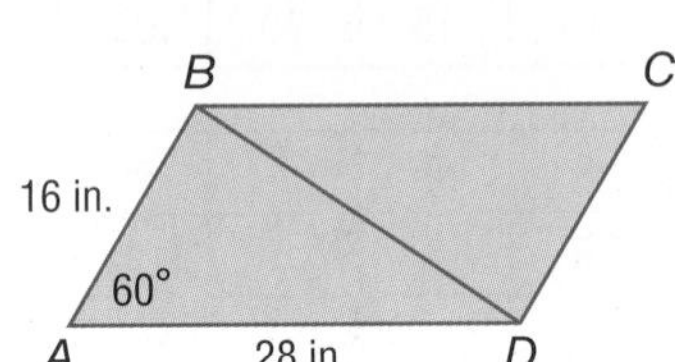

Step 2 Find the area of $\triangle ABD$.

Use the sine ratio to determine the height h from B to $\overline{AD}$.

$\sin \theta = \frac{\text{opp}}{\text{hyp}}$ — Definition of sine

$\sin \theta = \frac{h}{AB}$ — $h = \text{opp}$, $AB = \text{hyp}$

$AB \sin \theta = h$ — Solve for h.

So, $h = AB \sin \theta$.

$\text{Area} = \frac{1}{2}bh$ — Area of a triangle

$= \frac{1}{2}(AD)(AB) \sin A$ — $b = AD$, $h = AB \sin A$

$= \frac{1}{2}(28)(16) \sin 60°$ — $AD = 28$, $AB = 16$, $A = 60°$

$= 224\left(\frac{\sqrt{3}}{2}\right)$ — Multiply and evaluate sin 60°.

$= 112\sqrt{3}$ — Simplify.

Step 3 Find the area of $\square ABCD$.

The area of $\square ABCD$ is equal to the sum of the areas of $\triangle ABD$ and $\triangle CDB$. Because $\triangle ABD \cong \triangle CDB$, the areas of $\triangle ABD$ and $\triangle CDB$ are equal. So, the area of $\square ABCD$ equals twice the area of $\triangle ABD$.

$2 \cdot 112\sqrt{3} = 224\sqrt{3}$ or about 387.98 square inches.

Exercises

For each of the following,

a. find the area of each parallelogram.

b. find the area of each parallelogram when the included angle is half the given measure.

c. find the area of each parallelogram when the included angle is twice the given measure.

1.

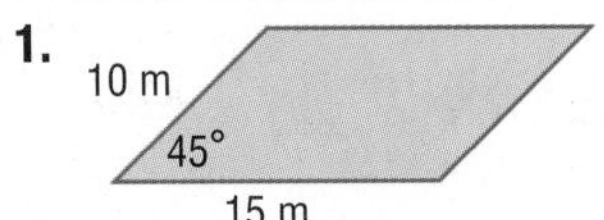

2.

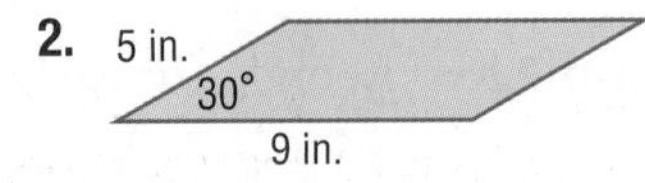

3.

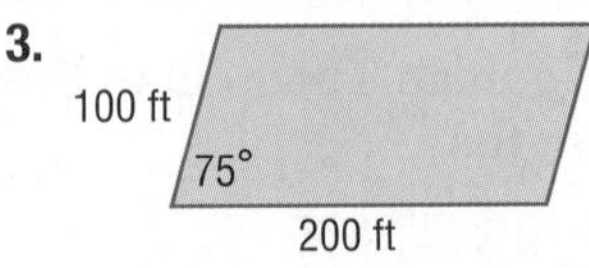

LESSON 12-3

Trigonometric Functions of General Angles

Then

- You found values of trigonometric functions for acute angles.

Now

1. Find values of trigonometric functions for general angles.
2. Find values of trigonometric functions by using reference angles.

Why?

- In the ride at the right, the cars rotate back and forth about a central point. The positions of the arms supporting the cars can be described using trigonometric angles in standard position, with the central point of the ride at the origin of a coordinate plane.

NewVocabulary
quadrantal angle
reference angle

Common Core State Standards

Mathematical Practices
6 Attend to precision.

1 Trigonometric Functions for General Angles

You can find values of trigonometric functions for angles greater than 90° or less than 0°.

KeyConcept Trigonometric Functions of General Angles

Let θ be an angle in standard position and let $P(x, y)$ be a point on its terminal side. Using the Pythagorean Theorem, $r = \sqrt{x^2 + y^2}$. The six trigonometric functions of θ are defined below.

$\sin \theta = \frac{y}{r}$ $\quad \cos \theta = \frac{x}{r}$ $\quad \tan \theta = \frac{y}{x}, x \neq 0$

$\csc \theta = \frac{r}{y}, y \neq 0$ $\quad \sec \theta = \frac{r}{x}, x \neq 0$ $\quad \cot \theta = \frac{x}{y}, y \neq 0$

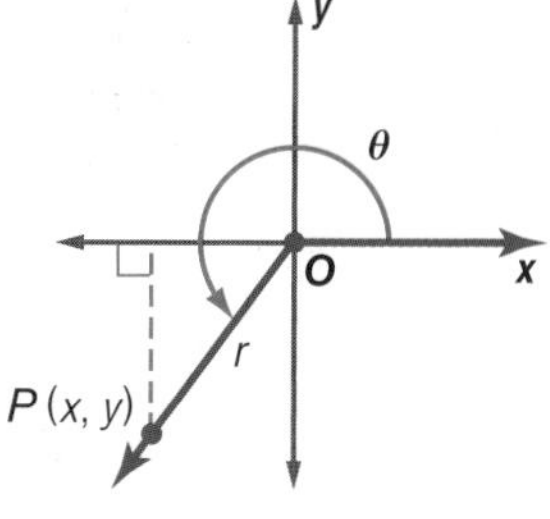

PT

Example 1 Evaluate Trigonometric Functions Given a Point

The terminal side of θ in standard position contains the point at (−3, −4). Find the exact values of the six trigonometric functions of θ.

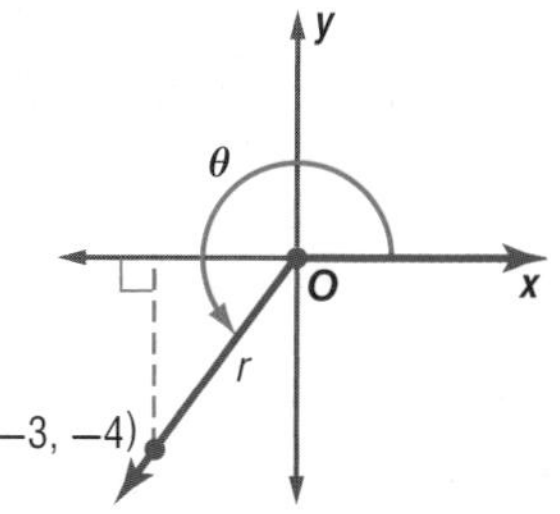

Step 1 Draw the angle, and find the value of r.

$$r = \sqrt{x^2 + y^2}$$
$$= \sqrt{(-3)^2 + (-4)^2}$$
$$= \sqrt{25} \text{ or } 5$$

Step 2 Use $x = -3$, $y = -4$, and $r = 5$ to write the six trigonometric ratios.

$\sin \theta = \frac{y}{r} = \frac{-4}{5}$ or $-\frac{4}{5}$ $\quad \cos \theta = \frac{x}{r} = \frac{-3}{5}$ or $-\frac{3}{5}$ $\quad \tan \theta = \frac{y}{x} = \frac{-4}{-3}$ or $\frac{4}{3}$

$\csc \theta = \frac{r}{y} = \frac{5}{-4}$ or $-\frac{5}{4}$ $\quad \sec \theta = \frac{r}{x} = \frac{5}{-3}$ or $-\frac{5}{3}$ $\quad \cot \theta = \frac{x}{y} = \frac{-3}{-4}$ or $\frac{3}{4}$

GuidedPractice

1. The terminal side of θ in standard position contains the point at (−6, 2). Find the exact values of the six trigonometric functions of θ.

Joseph Sohm/drr.net

If the terminal side of angle θ in standard position lies on the x- or y-axis, the angle is called a **quadrantal angle**.

StudyTip

Quadrantal Angles The measure of a quadrantal angle is a multiple of $90°$ or $\frac{\pi}{2}$.

KeyConcept Quadrantal Angles

$\theta = 0°$ or 0 radians	$\theta = 90°$ or $\frac{\pi}{2}$ radians	$\theta = 180°$ or π radians	$\theta = 270°$ or $\frac{3\pi}{2}$ radians
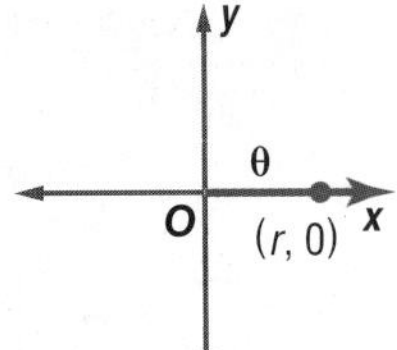	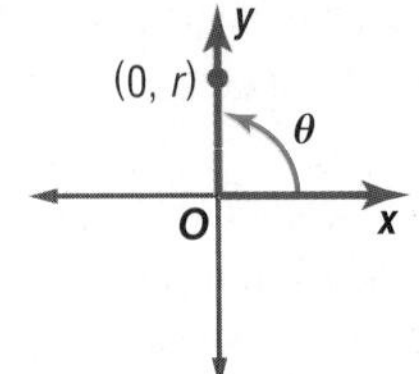	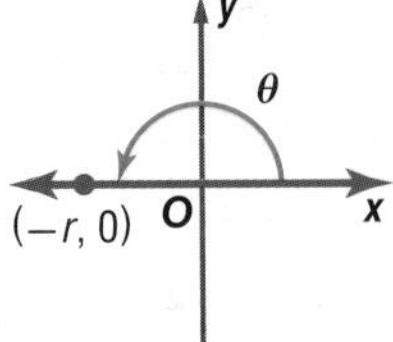	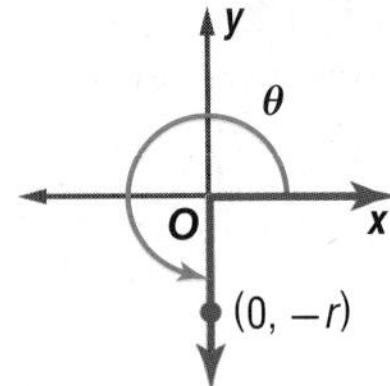

Example 2 Quadrantal Angles

The terminal side of θ in standard position contains the point at (0, 6). Find the values of the six trigonometric functions of θ.

The point at (0, 6) lies on the positive y-axis, so the quadrantal angle θ is 90°. Use $x = 0$, $y = 6$, and $r = 6$ to write the trigonometric functions.

$\sin\theta = \frac{y}{r} = \frac{6}{6}$ or 1 $\quad \cos\theta = \frac{x}{r} = \frac{0}{6}$ or 0 $\quad \tan\theta = \frac{y}{x} = \frac{6}{0}$ undefined

$\csc\theta = \frac{r}{y} = \frac{6}{6}$ or 1 $\quad \sec\theta = \frac{r}{x} = \frac{6}{0}$ undefined $\quad \cot\theta = \frac{x}{y} = \frac{0}{6}$ or 0

GuidedPractice

2. The terminal side of θ in standard position contains the point at (−2, 0). Find the values of the six trigonometric functions of θ.

2 Trigonometric Functions with Reference Angles

If θ is a nonquadrantal angle in standard position, its **reference angle** θ' is the acute angle formed by the terminal side of θ and the x-axis. The rules for finding the measures of reference angles for $0° < \theta < 360°$ or $0° < \theta < 2\pi$ are shown below.

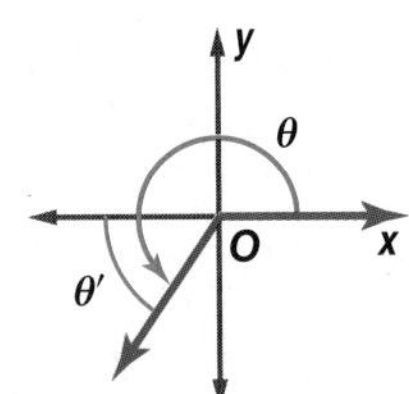

ReadingMath

Theta Prime θ' is read *theta prime.*

KeyConcept Reference Angles

Quadrant I	Quadrant II	Quadrant III	Quadrant IV
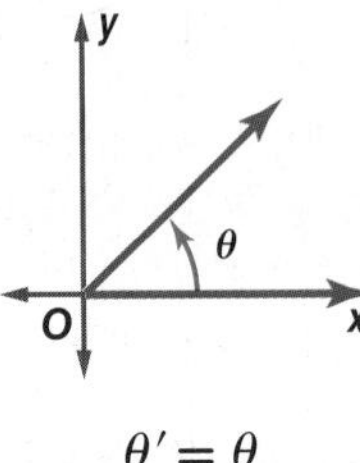	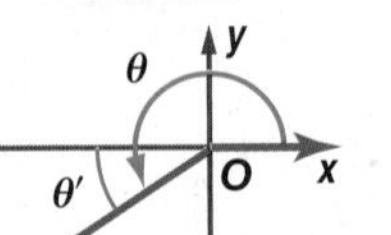	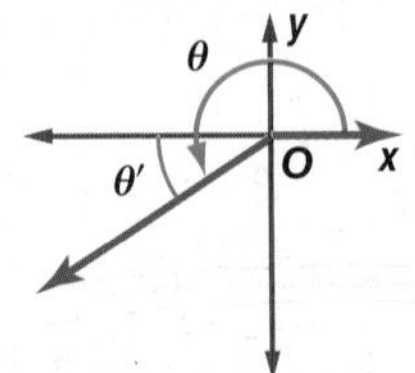	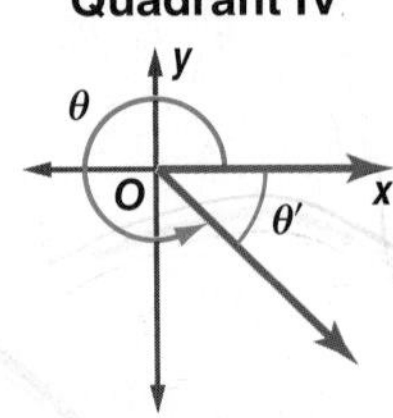
$\theta' = \theta$	$\theta' = 180° - \theta$ $\theta' = \pi - \theta$	$\theta' = \theta - 180°$ $\theta' = \theta - \pi$	$\theta' = 360° - \theta$ $\theta' = 2\pi - \theta$

If the measure of θ is greater than 360° or less than 0°, then use a coterminal angle with a positive measure between 0° and 360° to find the reference angle.

StudyTip

Graphing Angles You can refer to the diagram in the Lesson 12-2 Concept Summary to help you sketch angles.

Example 3 Find Reference Angles

PT

Sketch each angle. Then find its reference angle.

a. 210°

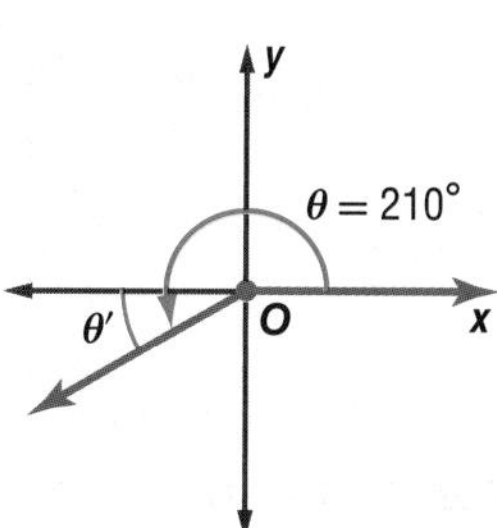

The terminal side of 210° lies in Quadrant III.

$\theta' = \theta - 180°$

$= 210° - 180°$ or $30°$

b. $-\frac{5\pi}{4}$

coterminal angle: $-\frac{5\pi}{4} + 2\pi = \frac{3\pi}{4}$

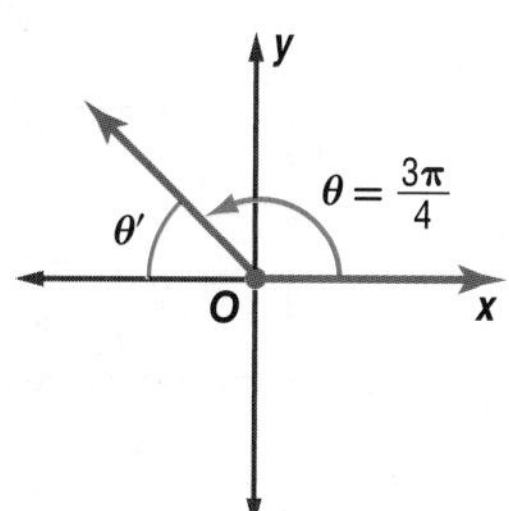

The terminal side of $\frac{3\pi}{4}$ lies in Quadrant II.

$\theta' = \pi - \theta$

$= \pi - \frac{3\pi}{4}$ or $\frac{\pi}{4}$

GuidedPractice

3A. $-110°$

3B. $\frac{2\pi}{3}$

You can use reference angles to evaluate trigonometric functions for any angle θ. The sign of a function is determined by the quadrant in which the terminal side of θ lies. Use these steps to evaluate a trigonometric function for any angle θ.

KeyConcept Evaluate Trigonometric Functions

Step 1 Find the measure of the reference angle θ'.

Step 2 Evaluate the trigonometric function for θ'.

Step 3 Determine the sign of the trigonometric function value. Use the quadrant in which the terminal side of θ lies.

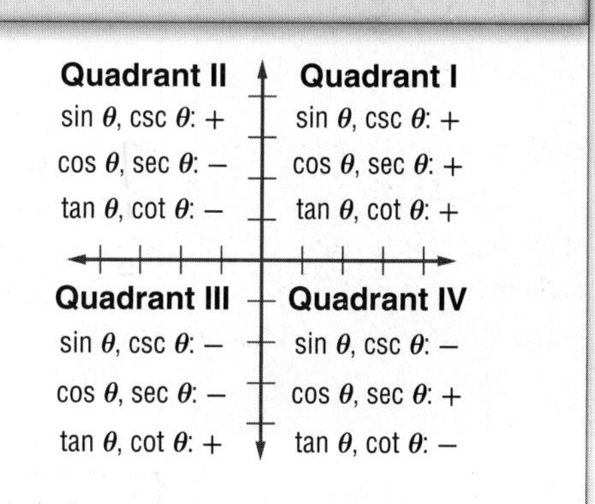

You can use the trigonometric values of angles measuring 30°, 45°, and 60° that you learned in Lesson 12-1.

Trigonometric Values for Special Angles

Sine	Cosine	Tangent	Cosecant	Secant	Cotangent
$\sin 30° = \frac{1}{2}$	$\cos 30° = \frac{\sqrt{3}}{2}$	$\tan 30° = \frac{\sqrt{3}}{3}$	$\csc 30° = 2$	$\sec 30° = \frac{2\sqrt{3}}{3}$	$\cot 30° = \sqrt{3}$
$\sin 45° = \frac{\sqrt{2}}{2}$	$\cos 45° = \frac{\sqrt{2}}{2}$	$\tan 45° = 1$	$\csc 45° = \sqrt{2}$	$\sec 45° = \sqrt{2}$	$\cot 45° = 1$
$\sin 60° = \frac{\sqrt{3}}{2}$	$\cos 60° = \frac{1}{2}$	$\tan 60° = \sqrt{3}$	$\csc 60° = \frac{2\sqrt{3}}{3}$	$\sec 60° = 2$	$\cot 60° = \frac{\sqrt{3}}{3}$

Example 4 Use a Reference Angle to Find a Trigonometric Value

Find the exact value of each trigonometric function.

a. cos 240°

The terminal side of 240° lies in Quadrant III.

$\theta' = \theta - 180°$ — Find the measure of the reference angle.

$= 240° - 180°$ or $60°$ — $\theta = 240°$

$\cos 240° = -\cos 60°$ or $-\frac{1}{2}$ — The cosine function is negative in Quadrant III.

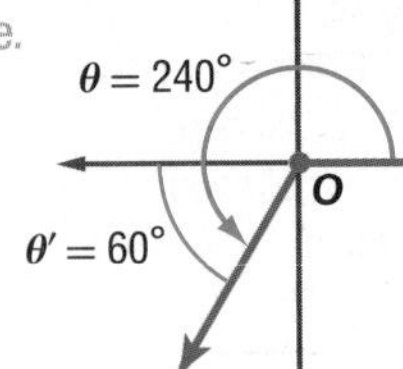

b. $\csc \frac{5\pi}{6}$

The terminal side of $\frac{5\pi}{6}$ lies in Quadrant II.

$\theta' = \pi - \theta$ — Find the measure of the reference angle.

$= \pi - \frac{5\pi}{6}$ or $\frac{\pi}{6}$ — $\theta = \frac{5\pi}{6}$

$\csc \frac{5\pi}{6} = \csc \frac{\pi}{6}$ — The cosecant function is positive in Quadrant II.

$= \csc 30°$ — $\frac{\pi}{6}$ radians $= 30°$

$= 2$ — $\csc 30° = \frac{1}{\sin 30}$

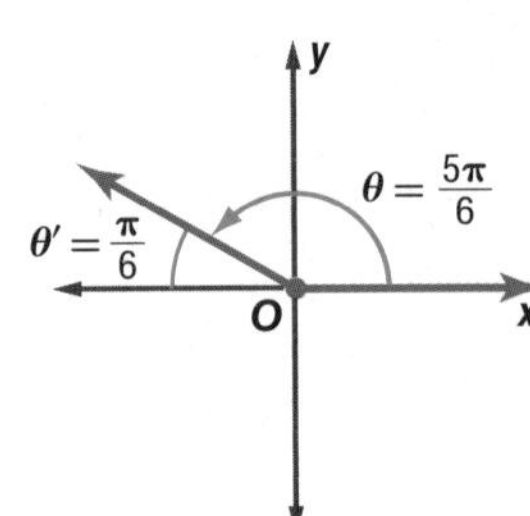

GuidedPractice

4A. cos 135°

4B. $\tan \frac{5\pi}{6}$

Real-WorldLink

On a swing ride, riders experience weightlessness just like the drop side of a roller coaster. The ride lasts one minute and reaches speeds of 60 miles per hour in both directions.

Source: Cedar Point

Real-World Example 5 Use Trigonometric Functions

RIDES The swing arms of the ride at the right are 84 feet long and the height of the axis from which the arms swing is 97 feet. What is the total height of the ride at the peak of the arc?

coterminal angle: $-200° + 360° = 160°$

reference angle: $180° - 160° = 20°$

$\sin \theta = \frac{y}{r}$ — Sine function

$\sin 20° = \frac{y}{84}$ — $\theta = 20°$ and $r = 84$

$84 \sin 20° = y$ — Multiply each side by 84.

$28.7 \approx y$ — Use a calculator to solve for y.

Since y is approximately 28.7 feet, the total height of the ride at its peak is $28.7 + 97$ or about 125.7 feet.

GuidedPractice

5. RIDES A similar ride that is smaller has swing arms that are 72 feet long. The height of the axis from which the arms swing is 88 feet, and the angle of rotation from the standard position is −195°. What is the total height of the ride at the peak of the arc?

Check Your Understanding

= Step-by-Step Solutions begin on page R14.

Examples 1–2 **The terminal side of θ in standard position contains each point. Find the exact values of the six trigonometric functions of θ.**

1. (1, 2) **2.** (−8, −15) **3.** (0, −4)

Example 3 **Sketch each angle. Then find its reference angle.**

4. 300° **5.** 115° **6.** $-\frac{3\pi}{4}$

Example 4 **Find the exact value of each trigonometric function.**

7. $\sin \frac{3\pi}{4}$ **8.** $\tan \frac{5\pi}{3}$ **9.** sec 120° **10.** sin 300°

Example 5 **11. ENTERTAINMENT** Alejandra opens her portable DVD player so that it forms a 125° angle. The screen is $5\frac{1}{2}$ inches long.

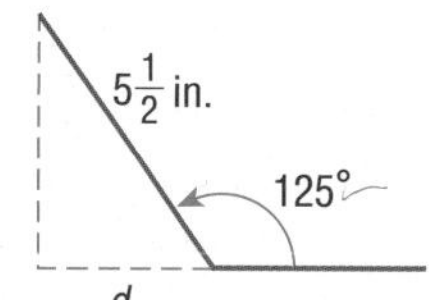

a. Redraw the diagram so that the angle is in standard position on the coordinate plane.

b. Find the reference angle. Then write a trigonometric function that can be used to find the distance to the wall d that she can place the DVD player.

c. Use the function to find the distance. Round to the nearest tenth.

Practice and Problem Solving

Extra Practice is on page R12.

Examples 1–2 **The terminal side of θ in standard position contains each point. Find the exact values of the six trigonometric functions of θ.**

12. (5, 12) **13.** (−6, 8) **14.** (3, 0)

15. (0, −7) **16.** (4, −2) **17.** (−9, −3)

Example 3 **Sketch each angle. Then find its reference angle.**

18. 195° **19.** 285° **20.** −250°

21. $\frac{7\pi}{4}$ **22.** $-\frac{\pi}{4}$ **23.** 400°

Example 4 **Find the exact value of each trigonometric function.**

24. sin 210° **25.** tan 315° **26.** cos 150° **27.** csc 225°

28. $\sin \frac{4\pi}{3}$ **29.** $\cos \frac{5\pi}{3}$ **30.** $\cot \frac{5\pi}{4}$ **31.** $\sec \frac{11\pi}{6}$

Example 5 **32. CCSS REASONING** A soccer player x feet from the goalie kicks the ball toward the goal, as shown in the figure. The goalie jumps up and catches the ball 7 feet in the air.

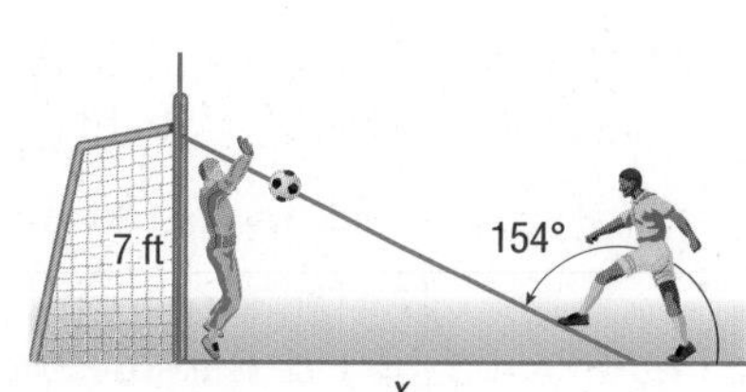

a. Find the reference angle. Then write a trigonometric function that can be used to find how far from the goalie the soccer player was when he kicked the ball.

b. About how far away from the goalie was the soccer player?

33 **SPRINKLER** A sprinkler rotating back and forth shoots water out a distance of 10 feet. From the horizontal position, it rotates 145° before reversing its direction.At a 145° angle, about how far to the left of the sprinklerdoes the water reach?

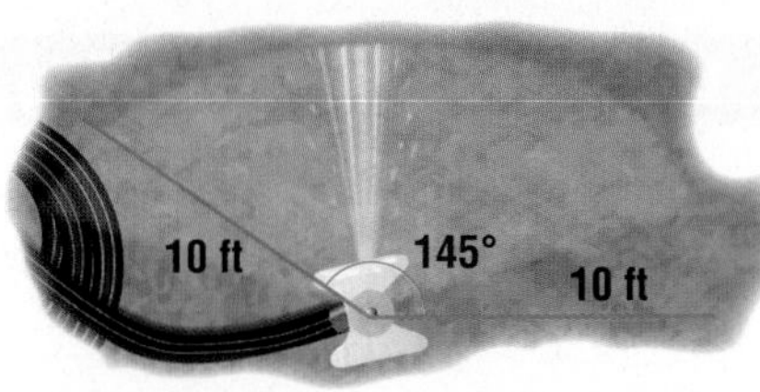

34. **BASKETBALL** The formula $R = \frac{v_0{}^2 \sin 2\theta}{32}$ gives the distance of a basketball shot with an initial velocity of V_0 feet per second at an angle θ with the ground.

 a. If the basketball was shot with an initial velocity of 24 feet per second at an angle of 75°, how far will the basketball travel?

 b. If the basketball was shot at an angle of 65° and traveled 10 feet, what was its initial velocity?

 c. If the basketball was shot with an initial velocity of 30 feet per second and traveled 12 feet, at what angle was it shot?

35. **PHYSICS** A rock is shot off the edge of a ravine with a slingshot at an angle of 65° and with an initial velocity of 6 meters per second. The equation that represents the horizontal distance of the rock x is $x = v_0 (\cos \theta)t$, where v_0 is the initial velocity, θ is the angle at which it is shot, and t is the time in seconds. About how far does the rock travel after 4 seconds?

36. **FERRIS WHEELS** The Wonder Wheel Ferris wheel at Coney Island has a radius of about 68 feet and is 15 feet off the ground. After a person gets on the bottom car, the Ferris wheel rotates 202.5° counterclockwise before stopping. How high above the ground is this car when it has stopped?

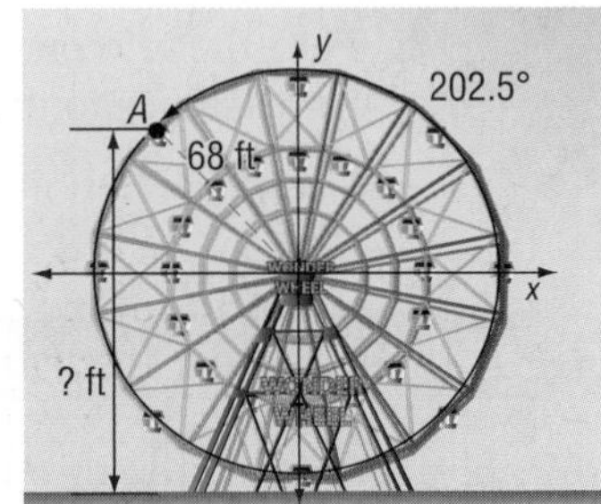

Suppose θ is an angle in standard position whose terminal side is in the given quadrant. For each function, find the exact values of the remaining five trigonometric functions of θ.

37. $\sin \theta = \frac{4}{5}$, Quadrant II

38. $\tan \theta = -\frac{2}{3}$, Quadrant IV

39. $\cos \theta = -\frac{8}{17}$, Quadrant III

40. $\cot \theta = -\frac{12}{5}$, Quadrant IV

Find the exact value of each trigonometric function.

41. $\cot 270°$

42. $\csc 180°$

43. $\sin 570°$

44. $\tan \left(-\frac{7\pi}{6}\right)$

45. $\cos \left(-\frac{11\pi}{6}\right)$

46. $\cot \frac{9\pi}{4}$

H.O.T. Problems Use Higher-Order Thinking Skills

47. **CHALLENGE** For an angle θ in standard position, $\sin \theta = \frac{\sqrt{2}}{2}$ and $\tan \theta = -1$. Can the value of θ be 225°? Justify your reasoning.

48. **CCSS ARGUMENTS** Determine whether $3 \sin 60° = \sin 180°$ is *true* or *false*. Explain your reasoning.

49. **REASONING** Use the sine and cosine functions to explain why $\cot 180°$ is undefined.

50. **OPEN ENDED** Give an example of a negative angle θ for which $\sin \theta > 0$ and $\cos \theta < 0$.

51. **WRITING IN MATH** Describe the steps for evaluating a trigonometric function for an angle θ that is greater than 90°. Include a description of a reference angle.

Standardized Test Practice

52. **GRIDDED RESPONSE** If the sum of two numbers is 21 and their difference is 3, what is their product?

53. **GEOMETRY** D is the midpoint of $\overline{BC}$, and A and E are the midpoints of $\overline{BD}$ and $\overline{DC}$, respectively. If the length of $\overline{AE}$ is 12, what is the length of $\overline{BC}$?

A 6 **C** 24
B 12 **D** 48

54. The expression $(-6 + i)^2$ is equivalent to which of the following expressions?

F $-12i$ **H** $36 - 12i$
G $36 - i$ **J** $35 - 12i$

55. **SAT/ACT** Of the following, which is least?

A $1 + \frac{1}{4}$ **D** $1 \times \frac{1}{4}$
B $1 - \frac{1}{4}$ **E** $\frac{1}{4} - 1$
C $1 \div \frac{1}{4}$

Spiral Review

Rewrite each radian measure in degrees. (Lesson 12-2)

56. $\frac{4}{3}\pi$ 57. $\frac{11}{6}\pi$ 58. $-\frac{17}{4}\pi$

Solve each equation. (Lesson 12-1)

59. $\cos a = \frac{13}{17}$ 60. $\sin 30 = \frac{b}{6}$ 61. $\tan c = \frac{9}{4}$

62. **ARCHITECTURE** A memorial being constructed in a city park will be a brick wall, with a top row of six gold-plated bricks engraved with the names of six local war veterans. Each row has two more bricks than the row above it. Prove that the number of bricks in the top n rows is $n^2 + 5n$. (Lesson 10-7)

63. **LEGENDS** There is a legend of a king who wanted to reward a boy for a good deed. The king gave the boy a choice. He could have $1,000,000 at once, or he could be rewarded daily for a 30-day month, with one penny on the first day, two pennies on the second day, and so on, receiving twice as many pennies each day as the previous day. How much would the second option be worth? (Lesson 10-3)

Write an equation for each circle given the endpoints of a diameter. (Lesson 9-3)

64. (2, −4), (10, 2) 65. (−1, −10), (−7, 6) 66. (9, 0), (4, −7)

Simplify each expression. (Lesson 8-2)

67. $\frac{5}{x^2 + 6x + 8} + \frac{x}{x^2 - 3x - 28}$ 68. $\frac{3x}{x^2 + 8x - 20} - \frac{6}{x^2 + 7x - 18}$ 69. $\frac{4}{3x^2 + 12x} + \frac{2x}{x^2 - 2x - 24}$

Solve each equation or inequality. Round to the nearest ten-thousandth. (Lesson 7-6)

70. $8^x = 30$ 71. $5^x = 64$ 72. $3^{x+2} = 41$

Evaluate each expression. (Lesson 6-6)

73. $16^{-\frac{1}{4}}$ 74. $27^{\frac{4}{3}}$ 75. $25^{-\frac{5}{2}}$

Skills Review

Solve for x.

76. $\frac{x + 2}{18} = \frac{x - 2}{9}$ 77. $\frac{x + 5}{x - 1} = \frac{7}{4}$ 78. $\frac{5}{x + 8} = \frac{15}{2x + 20}$

LESSON 12-4 Law of Sines

Then	Now	Why?
You found side lengths and angle measures of right triangles.	1 Find the area of a triangle using two sides and an included angle. 2 Use the Law of Sines to solve triangles.	Mars has hundreds of thousands of craters. These craters are named after famous scientists, science fiction authors, and towns on Earth. The craters named Wahoo, Wabash, and Naukan are shown in the figure. You can use trigonometry to find the distance between Wahoo and Naukan.

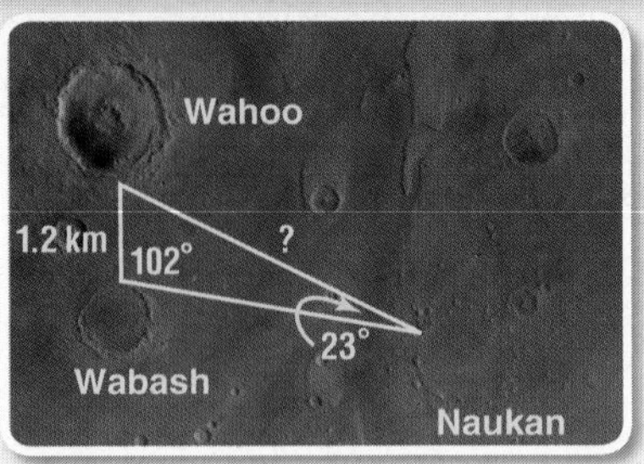

NewVocabulary
Law of Sines
solving a triangle
ambiguous case

Common Core State Standards

Mathematical Practices
1 Make sense of problems and persevere in solving them.
3 Construct viable arguments and critique the reasoning of others.

1 Find the Area of a Triangle

In the triangle at the right, $\sin A = \frac{h}{c}$, or $h = c \sin A$.

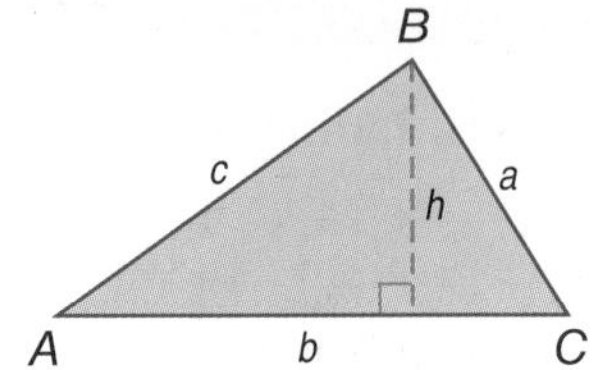

$\text{Area} = \frac{1}{2}bh$ Formula for area of a triangle

$\text{Area} = \frac{1}{2}b(c \sin A)$ Replace h with $c \sin A$.

$\text{Area} = \frac{1}{2}bc \sin A$ Simplify.

You can use this formula or two other formulas to find the area of a triangle if you know the lengths of two sides and the measure of the included angle.

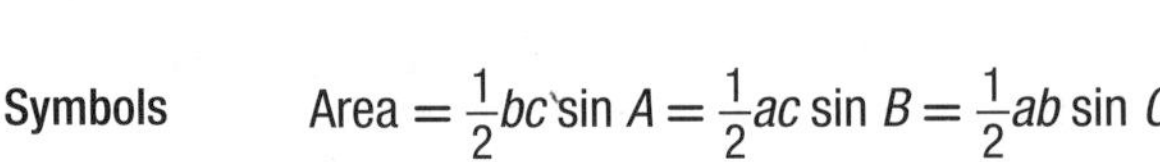

KeyConcept Area of a Triangle

Words The area of a triangle is one half the product of the lengths of two sides and the sine of their included angle.

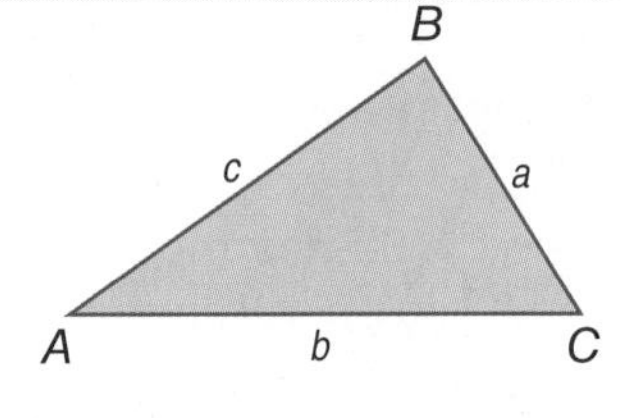

Symbols $\text{Area} = \frac{1}{2}bc \sin A = \frac{1}{2}ac \sin B = \frac{1}{2}ab \sin C$

Example 1 Find the Area of a Triangle

Find the area of $\triangle ABC$ to the nearest tenth.

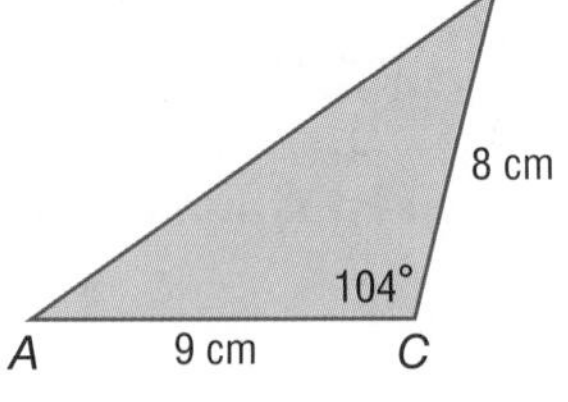

In $\triangle ABC$, $a = 8$, $b = 9$, and $C = 104°$.

$\text{Area} = \frac{1}{2}ab \sin C$ Based on the known measures, use the third area formula.

$= \frac{1}{2}(8)(9) \sin 104°$ Substitution

$\approx 34.9 \text{ cm}^2$ Simplify.

MENTAL CHECK Round the sin 104° to sin 90° because the sin of 90° is 1.

$\frac{1}{2}(8)(9)\sin 90° = \frac{1}{2}(8)(9)(1) = 36$

This is close to the answer of 34.9 square centimeters.

GuidedPractice

1. Find the area of $\triangle ABC$ to the nearest tenth if $A = 31°$, $b = 18$ meters, and $c = 22$ meters.

JPL/NASA

Math HistoryLink

Pauline Sperry (1885–1967) During the 1920s, Pauline Sperry wrote two textbooks, *Short Course in Spherical Trigonometry* and *Plane Trigonometry*. In 1923, she became the first woman to be promoted to assistant professor in the mathematics department at the University of California, Berkeley.

2 Use the Law of Sines to Solve Triangles

You can use the area formulas to derive the **Law of Sines**, which shows the relationships between side lengths of a triangle and the sines of the angles opposite them.

$\frac{1}{2}bc \sin A = \frac{1}{2}ac \sin B = \frac{1}{2}ab \sin C$ — Set the area formulas equal to each other.

$bc \sin A = ac \sin B = ab \sin C$ — Multiply each expression by 2.

$\frac{bc \sin A}{abc} = \frac{ac \sin B}{abc} = \frac{ab \sin C}{abc}$ — Divide each expression by *abc*.

$\frac{\sin A}{a} = \frac{\sin B}{b} = \frac{\sin C}{c}$ — Simplify.

KeyConcept Law of Sines

In $\triangle ABC$, if sides with lengths *a*, *b*, and *c* are opposite angles with measures *A*, *B*, and *C*, respectively, then the following is true.

$$\frac{\sin A}{a} = \frac{\sin B}{b} = \frac{\sin C}{c}$$

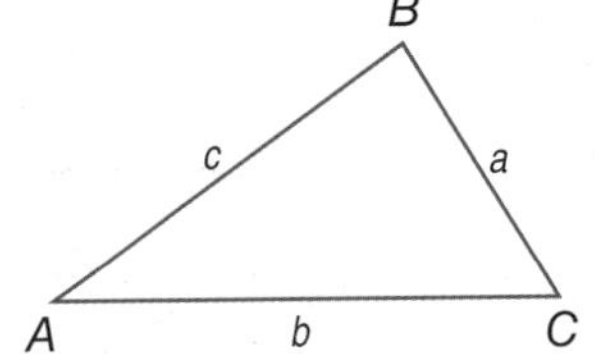

You can use the Law of Sines to solve a triangle if you know either one of the following.

- the measures of two angles and any side (angle-angle-side AAS or angle-side-angle ASA cases)
- the measures of two sides and the angle opposite one of the sides (side-side-angle SSA case)

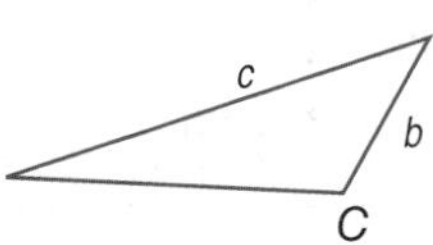

Using given measures to find all unknown side lengths and angle measures of a triangle is called **solving a triangle**.

StudyTip

CCSS Reasoning The Law of Sines may also be written as $\frac{a}{\sin A} = \frac{b}{\sin B} = \frac{c}{\sin C}$. So, the expressions below could also be used to solve the triangle in Example 2.

- $\frac{a}{\sin 55°} = \frac{3}{\sin 80°}$
- $\frac{b}{\sin 45°} = \frac{3}{\sin 80°}$

Example 2 Solve a Triangle Given Two Angles and a Side

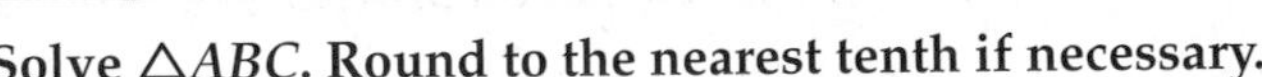

Solve $\triangle ABC$. Round to the nearest tenth if necessary.

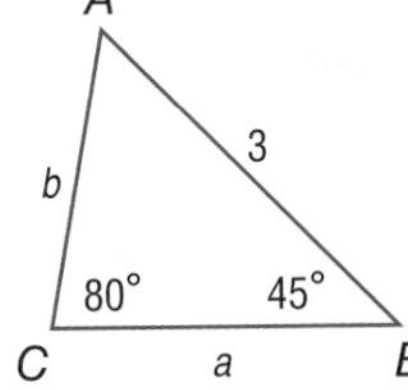

Step 1 Find the measure of the third angle.
$m\angle A = 180 - (80 + 45)$ or $55°$

Step 2 Use the Law of Sines to find side lengths *a* and *b*. Write an equation to find each variable.

$\frac{\sin A}{a} = \frac{\sin C}{c}$	Law of Sines	$\frac{\sin B}{b} = \frac{\sin C}{c}$
$\frac{\sin 55°}{a} = \frac{\sin 80°}{3}$	Substitution	$\frac{\sin 45°}{b} = \frac{\sin 80°}{3}$
$a = \frac{3 \sin 55°}{\sin 80°}$	Solve for each variable.	$b = \frac{3 \sin 45°}{\sin 80°}$
$a \approx 2.5$	Use a calculator.	$b \approx 2.2$

So, $A = 55°$, $a \approx 2.5$, and $b \approx 2.2$.

GuidedPractice

2. Solve $\triangle NPQ$ if $P = 42°$, $Q = 65°$, and $n = 5$.

If you are given the measures of two angles and a side, exactly one triangle is possible. However, if you are given the measures of two sides and the angle opposite one of them, zero, one, or two triangles may be possible. This is known as the **ambiguous case**. So, when solving a triangle using the SSA case, zero, one, or two solutions are possible.

StudyTip

***A* is Acute** In the figures at the right, the altitude h is compared to a because h is the minimum distance from C to $\overline{AB}$ when A is acute.

$\sin A = \frac{\text{opp}}{\text{hyp}}$

$\sin A = \frac{h}{b}$

KeyConcept Possible Triangles in SSA Case

Consider a triangle in which a, b, and $m\angle A$ are given.

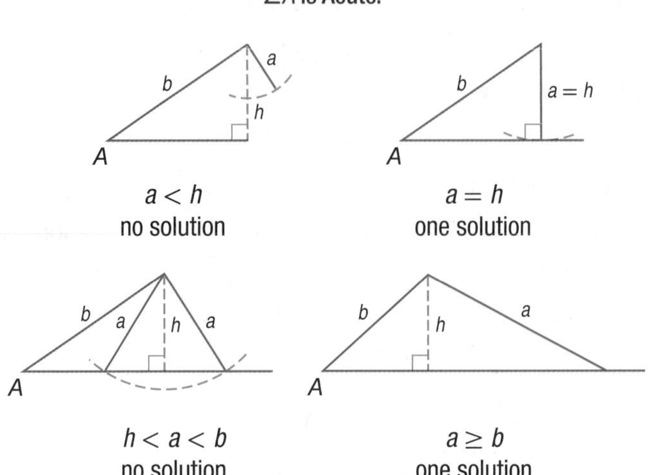

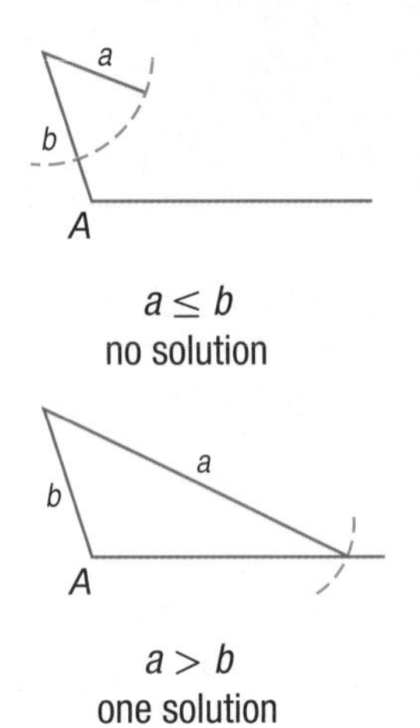

Since $\sin A = \frac{h}{b}$, you can use $h = b \sin A$ to find h in acute triangles.

Example 3 Solve a Triangle Given Two Sides and an Angle

Determine whether each triangle has *no* solution, *one* solution, or *two* solutions. Then solve the triangle. Round side lengths to the nearest tenth and angle measures to the nearest degree.

a. In $\triangle RST$, $R = 105°$, $r = 9$, and $s = 6$.

Because $\angle R$ is obtuse and $9 > 6$, you know that one solution exists.

S, t, 9, 105°, R, 6, T

Step 1 Use the Law of Sines to find $m\angle S$.

$\frac{\sin S}{6} = \frac{\sin 105°}{9}$ Law of Sines

$\sin S = \frac{6 \sin 105°}{9}$ Multiply each side by 6.

$\sin S \approx 0.6440$ Use a calculator.

$S \approx 40°$ Use the $\sin^{-1}$ function.

Step 2 Find $m\angle T$.

$m\angle T \approx 180 - (105 + 40)$ or $35°$

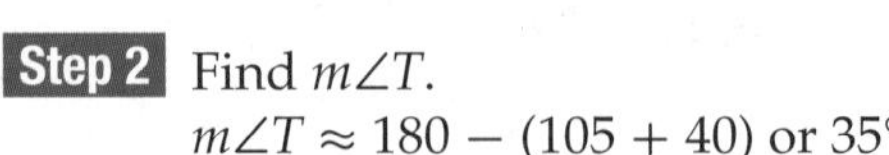

Step 3 Use the Law of Sines to find t.

$\frac{\sin 35°}{t} \approx \frac{\sin 105°}{9}$ Law of Sines

$t \approx \frac{9 \sin 35°}{\sin 105°}$ Solve for t.

$t \approx 5.3$ Use a calculator.

So, $S \approx 40°$, $T \approx 35°$, and $t \approx 5.3$.

b. In $\triangle ABC$, $A = 54°$, $a = 6$, and $b = 8$.

Since $\angle A$ is acute and $6 < 8$, find h and compare it to a.

$b \sin A = 8 \sin 54°$ $b = 8$ and $A = 54°$

≈ 6.5 Use a calculator.

Since $6 \leq 6.5$ or $a \leq h$, there is no solution.

c. In $\triangle ABC$, $A = 35°$, $a = 17$, and $b = 20$.

Since $\angle A$ is acute and $17 < 20$, find h and compare it to a.

$b \sin A = 20 \sin 35°$ $b = 20$ and $A = 35°$

≈ 11.5 Use a calculator.

Since $11.5 < 17 < 20$ or $h < a < b$, there are two solutions. So, there are two triangles to be solved

Case 1 $\angle B$ **is acute.**

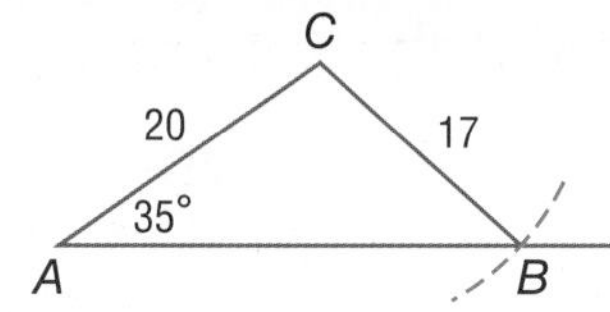

Step 1 Find $m\angle B$.

$\frac{\sin B}{20} = \frac{\sin 35°}{17}$ Law of Sines

$\sin B = \frac{20 \sin 35°}{17}$ Solve for sin B.

$\sin B \approx 0.6748$ Use a calculator.

$B \approx 42°$ Find $\sin^{-1} 0.6748$.

Step 2 Find $m\angle C$.

$m\angle C \approx 180 - (35 + 42)$ or $103°$

Step 3 Find c.

$\frac{\sin 103°}{c} = \frac{\sin 35°}{17}$ Law of Sines

$c = \frac{17 \sin 103°}{\sin 35°}$ Solve for c.

$c \approx 28.9$ Simplify.

Case 2 $\angle B$ **is obtuse.**

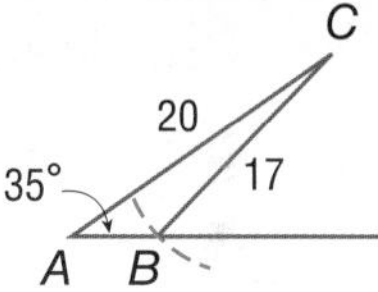

Step 1 Find $m\angle B$.

The sine function also has a positive value in Quadrant II. So, find an obtuse angle B for which $\sin B \approx 0.6748$.

$m\angle B \approx 180° - 42°$ or $138°$

Step 2 Find $m\angle C$.

$m\angle C \approx 180 - (35 + 138)$ or $7°$

Step 3 Find c.

$\frac{\sin 7°}{c} \approx \frac{\sin 35°}{17}$ Law of Sines

$c \approx \frac{17 \sin 7°}{\sin 35°}$ Solve for c.

$c \approx 3.6$ Simplify.

So, one solution is $B \approx 42°$, $C \approx 103°$, and $c \approx 28.9$, and another solution is $B \approx 138°$, $C \approx 7°$, and $c \approx 3.6$.

StudyTip

Reference Angle In the triangle in Case 2, you are using the reference angle 42° to find the other value of B.

GuidedPractice

Determine whether each triangle has *no* solution, *one* solution, or *two* solutions. Then solve the triangle. Round side lengths to the nearest tenth and angle measures to the nearest degree.

3A. In $\triangle RST$, $R = 95°$, $r = 10$, and $s = 12$.

3B. In $\triangle MNP$, $N = 32°$, $n = 7$, and $p = 4$.

3C. In $\triangle ABC$, $A = 47°$, $a = 15$, and $b = 18$.

Real-World Example 4 Use the Law of Sines to Solve a Problem

BASEBALL A baseball is hit between second and third bases and is caught at point *B*, as shown in the figure. How far away from second base was the ball caught?

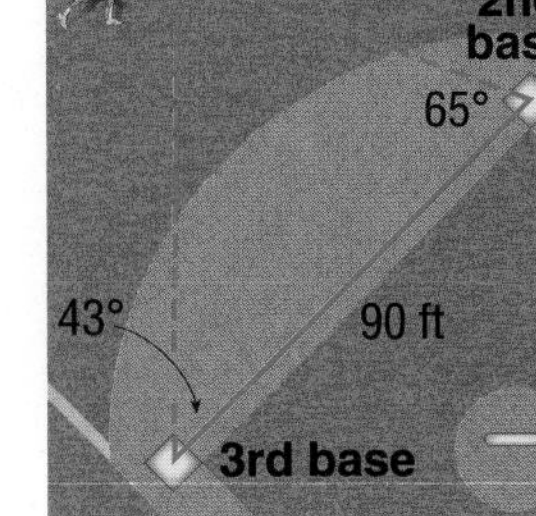

$\frac{\sin 72°}{90} = \frac{\sin 43°}{x}$ Law of Sines

$x \sin 72° = 90 \sin 43°$ Cross products

$x = \frac{90 \sin 43°}{\sin 72°}$ Solve for x.

$x \approx 64.5$ Use a calculator.

So, the distance is about 64.5 feet.

Guided Practice

4. How far away from third base was the ball caught?

Real-World Link

High school and college baseball fields share the same infield dimensions as professional baseball fields. The outfield dimensions vary greatly.

Source: *Baseball Digest Magazine*

Check Your Understanding

Example 1 **Find the area of $\triangle ABC$ to the nearest tenth, if necessary.**

1.

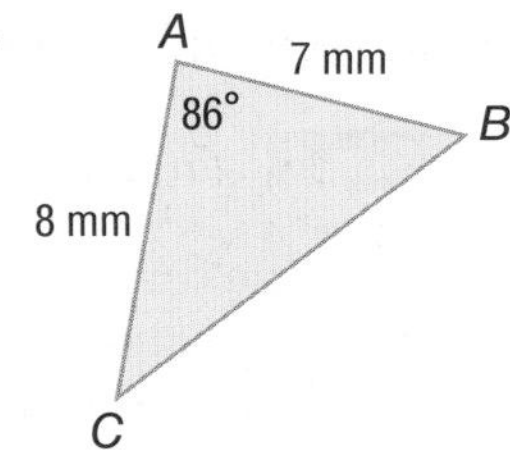

2.

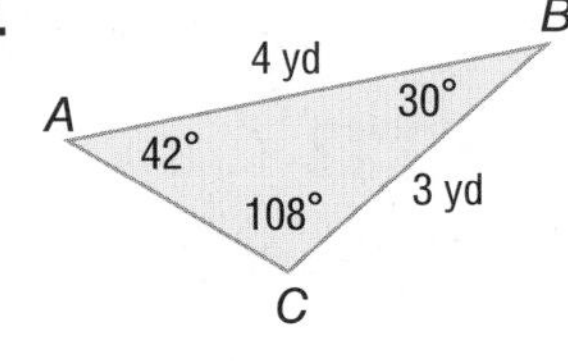

3. $A = 40°, b = 11$ cm, $c = 6$ cm

4. $B = 103°, a = 20$ in., $c = 18$ in.

Example 2 **Solve each triangle. Round side lengths to the nearest tenth and angle measures to the nearest degree.**

5.

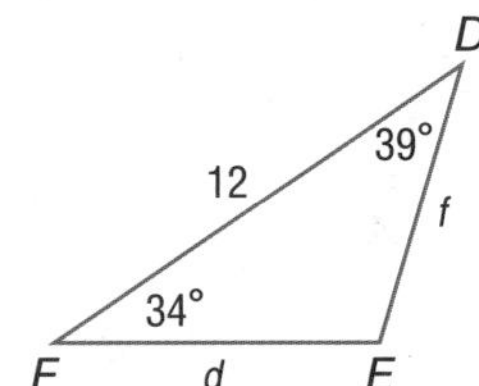

6.

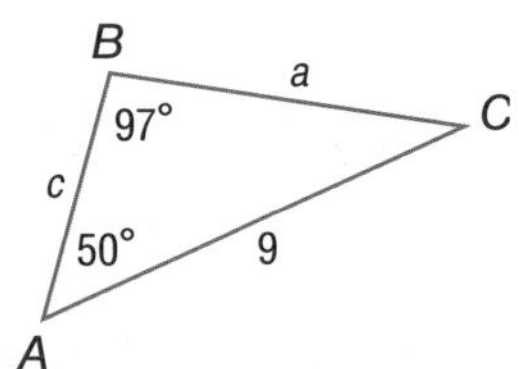

7. Solve $\triangle FGH$ if $G = 80°, H = 40°$, and $g = 14$.

Example 3 **CCSS PERSEVERANCE Determine whether each $\triangle ABC$ has *no* solution, *one* solution, or *two* solutions. Then solve the triangle. Round side lengths to the nearest tenth and angle measures to the nearest degree.**

8. $A = 95°, a = 19, b = 12$

9. $A = 60°, a = 15, b = 24$

10. $A = 34°, a = 8, b = 13$

11. $A = 30°, a = 3, b = 6$

Example 4 **12. SPACE** Refer to the beginning of the lesson. Find the distance between the Wahoo Crater and the Naukan Crater on Mars.

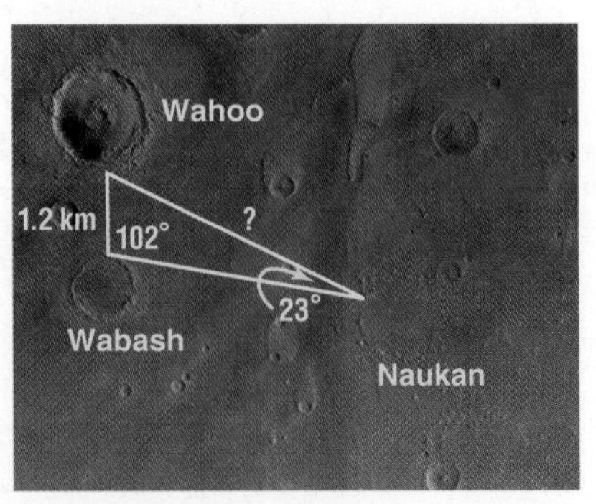

Practice and Problem Solving

Extra Practice is on page R12.

Example 1 **Find the area of $\triangle ABC$ to the nearest tenth.**

13.

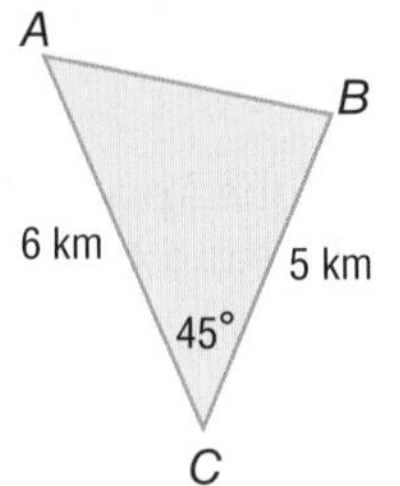

14.

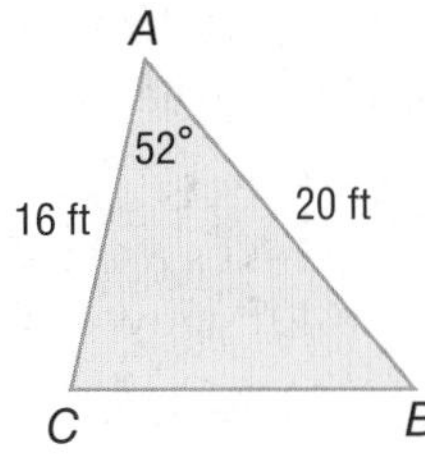

15.

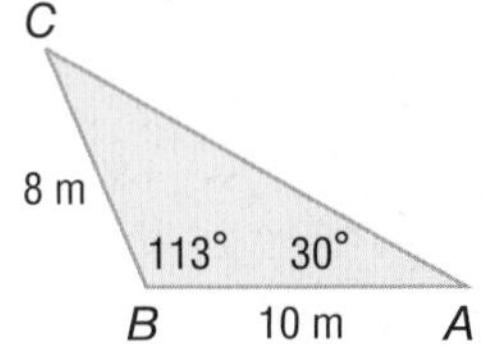

16.

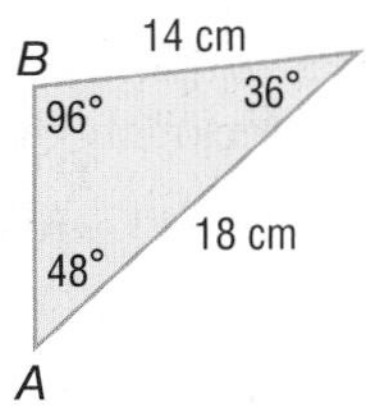

17. $C = 25°, a = 4$ ft, $b = 7$ ft

18. $A = 138°, b = 10$ in., $c = 20$ in.

19. $B = 92°, a = 14.5$ m, $c = 9$ m

20. $C = 116°, a = 2.7$ cm, $b = 4.6$ cm

Example 2 **CCSS REASONING** **Solve each triangle. Round side lengths to the nearest tenth and angle measures to the nearest degree.**

21.

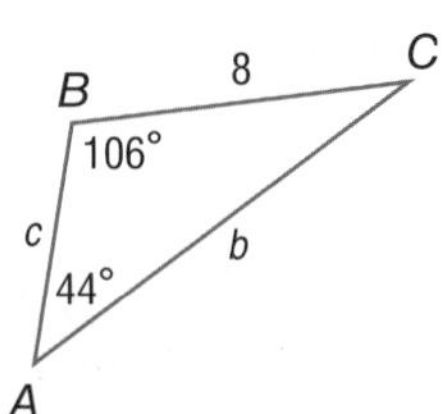

22.

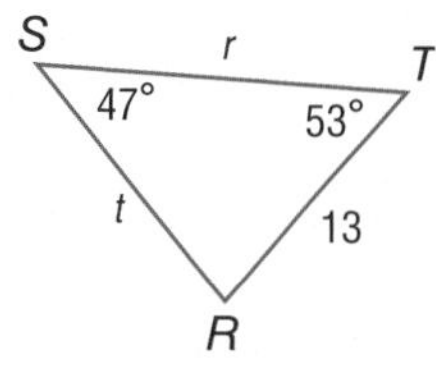

23.

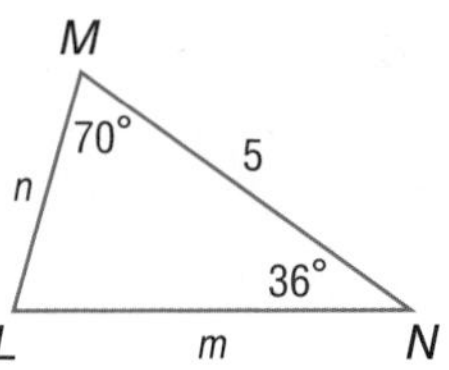

24.

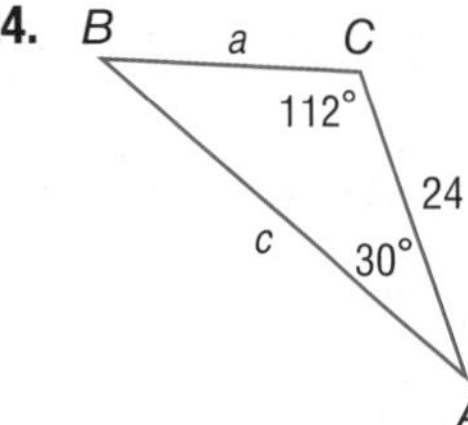

25 Solve $\triangle HJK$ if $H = 53°$, $J = 20°$, and $h = 31$.

26. Solve $\triangle NPQ$ if $P = 109°$, $Q = 57°$, and $n = 22$.

27. Solve $\triangle ABC$ if $A = 50°$, $a = 2.5$, and $C = 67°$.

28. Solve $\triangle ABC$ if $B = 18°$, $C = 142°$, and $b = 20$.

Example 3 **Determine whether each $\triangle ABC$ has *no* solution, *one* solution, or *two* solutions. Then solve the triangle. Round side lengths to the nearest tenth and angle measures to the nearest degree.**

29. $A = 100°, a = 7, b = 3$

30. $A = 75°, a = 14, b = 11$

31. $A = 38°, a = 21, b = 18$

32. $A = 52°, a = 9, b = 20$

33. $A = 42°, a = 5, b = 6$

34. $A = 44°, a = 14, b = 19$

35. $A = 131°, a = 15, b = 32$

36. $A = 30°, a = 17, b = 34$

Example 4 **GEOGRAPHY** **In Hawaii, the distance from Hilo to Kailua is 57 miles, and the distance from Hilo to Captain Cook is 55 miles.**

Kailua 57 mi Hilo
80° 55 mi
Captain Cook

37. What is the measure of the angle formed at Hilo?

38. What is the distance between Kailua and Captain Cook?

39 **TORNADOES** Tornado sirens A, B, and C form a triangular region in one area of a city. Sirens A and B are 8 miles apart. The angle formed at siren A is 112°, and the angle formed at siren B is 40°. How far apart are sirens B and C?

40. **MYSTERIES** The Bermuda Triangle is a region of the Atlantic Ocean between Bermuda, Miami, Florida, and San Juan, Puerto Rico. It is an area where ships and airplanes have been rumored to mysteriously disappear.

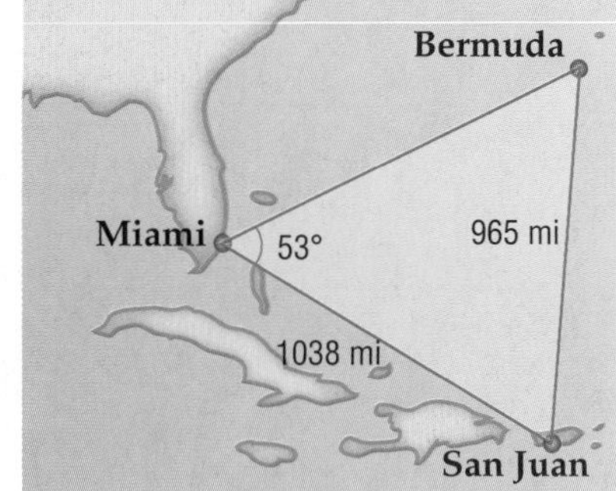

a. What is the distance between Miami and Bermuda?

b. What is the approximate area of the Bermuda Triangle?

41. **BICYCLING** One side of a triangular cycling path is 4 miles long. The angle opposite this side is 64°. Another angle formed by the triangular path measures 66°.

a. Sketch a drawing of the situation. Label the missing sides a and b.

b. Write equations that could be used to find the lengths of the missing sides.

c. What is the perimeter of the path?

42. **ROCK CLIMBING** Savannah S and Leon L are standing 8 feet apart in front of a rock climbing wall, as shown at the right. What is the height of the wall? Round to the nearest tenth.

H.O.T. Problems Use Higher-Order Thinking Skills

43. **CCSS CRITIQUE** In $\triangle RST$, $R = 56°$, $r = 24$, and $t = 12$. Cameron and Gabriela are using the Law of Sines to find T. Is either of them correct? Explain your reasoning.

Cameron

$$\frac{\sin T}{12} = \frac{\sin 56°}{24}$$

$$\sin T \approx 0.4145$$

$$T \approx 24.5°$$

Gabriela

Since $r > t$, there is no solution.

44. **OPEN ENDED** Create an application problem involving right triangles and the Law of Sines. Then solve your problem, drawing diagrams if necessary.

45. **CHALLENGE** Using the figure at the right, derive the formula Area $= \frac{1}{2}bc \sin A$.

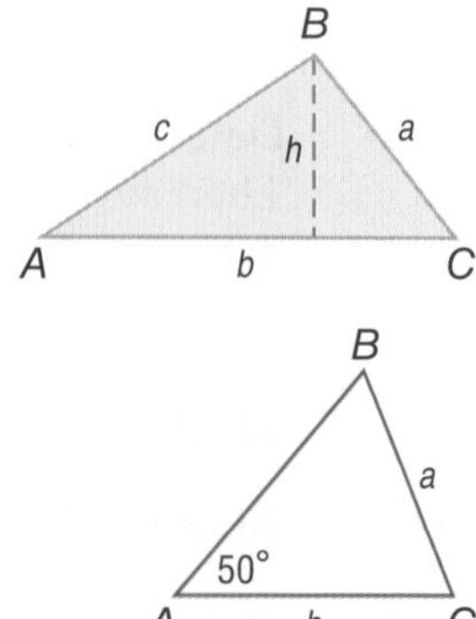

46. **REASONING** Find the side lengths of two different triangles ABC that can be formed if $A = 55°$ and $C = 20°$.

47. **WRITING IN MATH** Use the Law of Sines to explain why a and b do not have unique values in the figure shown.

48. **OPEN ENDED** Given that $E = 62°$ and $d = 38$, find a value for e such that no triangle DEF can exist. Explain your reasoning.

Standardized Test Practice

49. SHORT RESPONSE Given the graphs of $f(x)$ and $g(x)$, what is the value of $f(g(4))$?

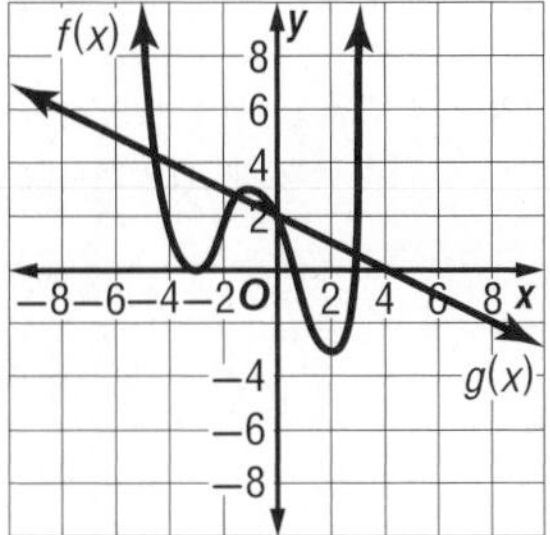

50. STATISTICS If the average of seven consecutive odd integers is n, what is the median of these seven integers?

A 0

B 7

C n

D $n - 2$

51. One zero of $f(x) = x^3 - 7x^2 - 6x + 72$ is 4. What is the factored form of the expression $x^3 - 7x^2 - 6x + 72$?

F $(x - 6)(x + 3)(x + 4)$

G $(x - 6)(x + 3)(x - 4)$

H $(x + 6)(x + 3)(x - 4)$

J $(x + 12)(x - 1)(x - 4)$

52. SAT/ACT Three people are splitting \$48,000 using the ratio 5 : 4 : 3. What is the amount of the greatest share?

A \$12,000

B \$16,000

C \$20,000

D \$24,000

E \$30,000

Spiral Review

Find the exact value of each trigonometric function. (Lesson 12-3)

53. $\sin 210°$

54. $\cos \frac{3}{4}\pi$

55. $\cot 60°$

Find an angle with a positive measure and an angle with a negative measure that are coterminal with each angle. (Lesson 12-2)

56. $125°$

57. $-32°$

58. $\frac{2}{3}\pi$

59. CLOCKS Jun's grandfather clock is broken. When she sets the pendulum in motion by holding it against the side of the clock and letting it go, it swings 24 centimeters to the other side, then 18 centimeters back, then 13.5 centimeters, and so on. What is the total distance that the pendulum swings before it stops? (Lesson 10-5)

Find the sum of each infinite series, if it exists. (Lesson 10-4)

60. $64 + 48 + 36 + \ldots$

61. $27 + 36 + 48 + \ldots$

62. $\sum_{n=1}^{\infty} 0.5(1.1)^n$

63. ASTRONOMY At its closest point, Earth is 91.8 million miles from the center of the Sun. At its farthest point, Earth is 94.9 million miles from the center of the Sun. Write an equation for the orbit of Earth, assuming that the center of the orbit is the origin and the Sun lies on the x-axis. (Lesson 9-4)

Simplify. (Lesson 6-4)

64. $\sqrt{(x - 4)^2}$

65. $\sqrt{(y + 2)^4}$

66. $\sqrt[3]{(a - b)^6}$

Skills Review

Evaluate each expression if $w = 6$, $x = -4$, $y = 1.5$, and $z = \frac{3}{4}$.

67. $w^2 + y^2 - 6xz$

68. $x^2 + z^2 + 5wy$

69. $wy + xz + w^2 - x^2$

EXTEND

12-4 Geometry Lab Regular Polygons

You can use central angles of circles to investigate characteristics of regular polygons inscribed in a circle. Recall that a regular polygon is inscribed in a circle if each of its vertices lies on the circle.

Activity Collect the Data

Step 1 Use a compass to draw a circle with a radius of one inch.

Step 2 Inscribe an equilateral triangle inside the circle. To do this, use a protractor to measure three angles of 120° at the center of the circle, since $\frac{360°}{3} = 120°$. Then connect the points where the sides of the angles intersect the circle using a straightedge.

Step 3 The **apothem** of a regular polygon is a segment that is drawn from the center of the polygon perpendicular to a side of the polygon. Use the cosine of angle θ to find the length of an apothem, labeled a in the diagram.

Model and Analyze

1. Make a table like the one shown below and record the length of the apothem of the equilateral triangle. Inscribe each regular polygon named in the table in a circle with radius one inch. Copy and complete the table.

Number of Sides, n	θ	a	Number of Sides, n	θ	a
3	60		7		
4	45		8		
5			9		
6			10		

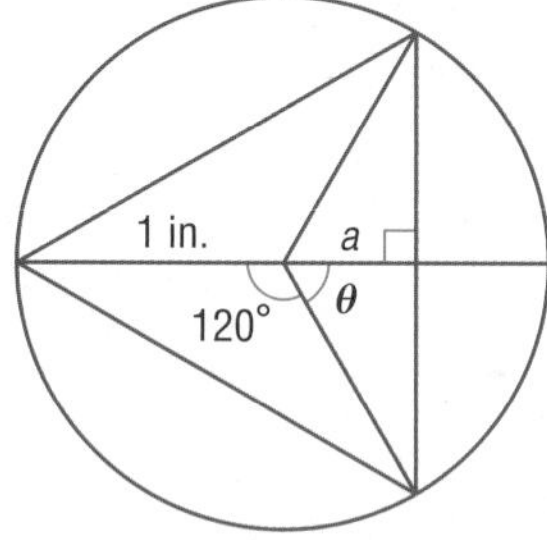

2. What do you notice about the measure of θ as the number of sides of the inscribed polygon increases?
3. What do you notice about the value of a?
4. **MAKE A CONJECTURE** Suppose you inscribe a 30-sided regular polygon inside a circle. Find the measure of angle θ.
5. Write a formula that gives the measure of angle θ for a polygon with n sides.
6. Write a formula that gives the length of the apothem of a regular polygon inscribed in a circle with radius one inch.
7. How would the formula you wrote in Exercise 5 change if the apothem of the circle was not one inch?

Ed-Imaging

LESSON 12-5

Law of Cosines

Then

- You solved triangles by using the Law of Sines.

Now

1. Use the Law of Cosines to solve triangles.
2. Choose methods to solve triangles.

Why?

- A *submersible* is an underwater vessel used for exploring the depths of the ocean. You can use trigonometry to find the distance from a ship used to lower a submersible into the ocean and a shipwreck spotted by the submersible on the ocean floor.

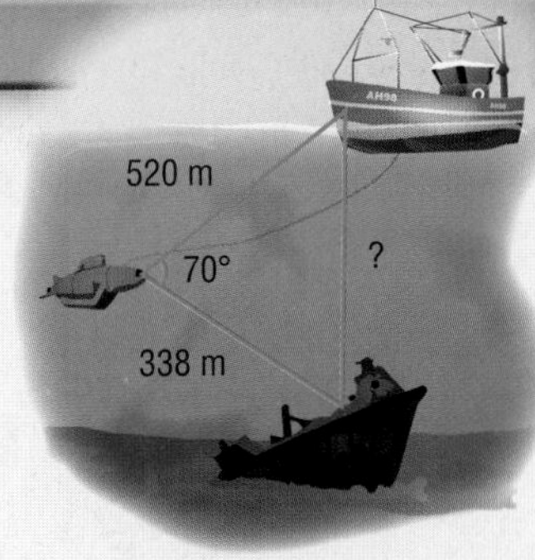

NewVocabulary
Law of Cosines

Common Core State Standards

Mathematical Practices
1 Make sense of problems and persevere in solving them.
3 Construct viable arguments and critique the reasoning of others.

1 Use Law of Cosines to Solve Triangles

You cannot use the Law of Sines to solve a triangle like the one shown above. You can use the **Law of Cosines** if:

- the measures of two sides and the included angle are known (side-angle-side case).
- the measures of three sides are known (side-side-side case).

KeyConcept Law of Cosines

In $\triangle ABC$, if sides with lengths a, b, and c are opposite angles with measures A, B, and C, respectively, then the following are true.

$$a^2 = b^2 + c^2 - 2bc \cos A$$
$$b^2 = a^2 + c^2 - 2ac \cos B$$
$$c^2 = a^2 + b^2 - 2ab \cos C$$

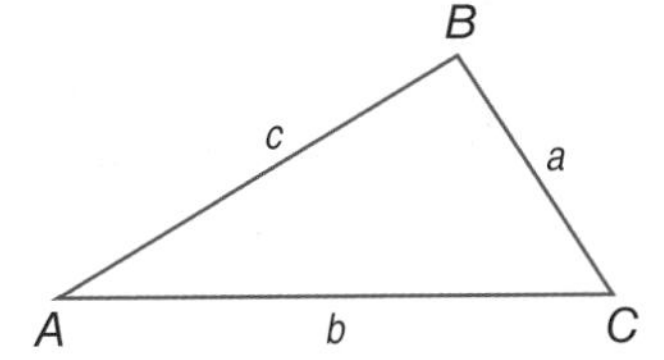

Example 1 Solve a Triangle Given Two Sides and the Included Angle

Solve $\triangle ABC$.

A, b, 5, C, 7, 36°, B

Step 1 Use the Law of Cosines to find the missing side length.

$b^2 = a^2 + c^2 - 2ac \cos B$	Law of Cosines
$b^2 = 7^2 + 5^2 - 2(7)(5) \cos 36°$	$a = 7, c = 5, B = 36°$
$b^2 \approx 17.4$	Use a calculator to simplify.
$b \approx 4.2$	Take the positive square root of each side.

Step 2 Use the Law of Sines to find a missing angle measure.

$\frac{\sin A}{7} \approx \frac{\sin 36°}{4.2}$	$\frac{\sin A}{a} = \frac{\sin B}{b}$
$\sin A \approx \frac{7 \sin 36°}{4.2}$	Multiply each side by 7.
$A \approx 78°$	Use the $\sin^{-1}$ function.

Step 3 Find the measure of the other angle.
$m\angle C \approx 180° - (36° + 78°)$ or $66°$

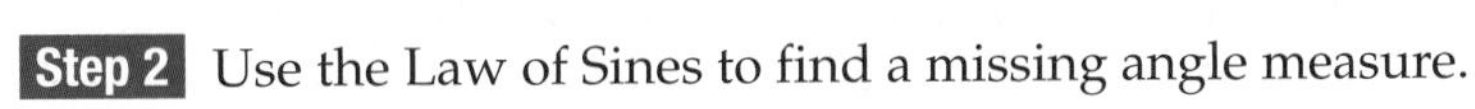

So, $b \approx 4.2$, $A \approx 78°$, and $C \approx 66°$.

GuidedPractice

1. Solve $\triangle FGH$ if $G = 82°$, $f = 6$, and $h = 4$.

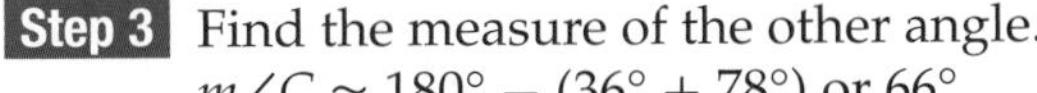

When you are only given the three side lengths of a triangle, you can solve it by using the Law of Cosines. The first step is to find the measure of the largest angle. This is done to ensure the other two angles are acute when using the Law of Sines.

PT

Example 2 Solve a Triangle Given Three Sides

Solve $\triangle ABC$.

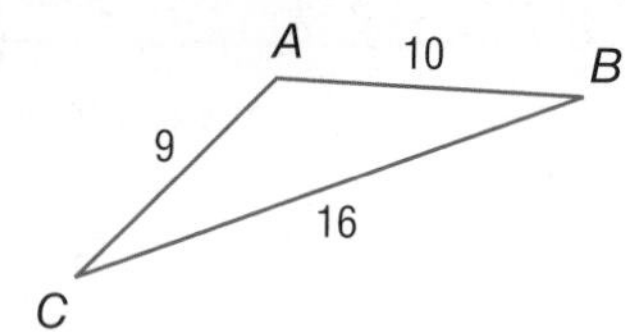

Step 1 Use the Law of Cosines to find the measure of the largest angle, $\angle A$.

$a^2 = b^2 + c^2 - 2bc \cos A$	Law of Cosines
$16^2 = 9^2 + 10^2 - 2(9)(10) \cos A$	$a = 16$, $b = 9$, and $c = 10$
$16^2 - 9^2 - 10^2 = -2(9)(10) \cos A$	Subtract 9^2 and 10^2 from each side.
$\frac{16^2 - 9^2 - 10^2}{-2(9)(10)} = \cos A$	Divide each side by $-2(9)(10)$.
$-0.4167 \approx \cos A$	Use a calculator to simplify.
$115° \approx A$	Use the $\cos^{-1}$ function.

Step 2 Use the Law of Sines to find the measure of $\angle B$.

$\frac{\sin B}{9} \approx \frac{\sin 115°}{16}$	$\frac{\sin B}{b} = \frac{\sin A}{a}$
$\sin B \approx \frac{9 \sin 115°}{16}$	Multiply each side by 9.
$\sin B \approx 0.5098$	Use a calculator.
$B \approx 31°$	Use the $\sin^{-1}$ function.

Step 3 Find the measure of $\angle C$.

$m\angle C \approx 180° - (115° + 31°)$ or about $34°$

So, $A \approx 115°$, $B \approx 31°$, and $C \approx 34°$.

GuidedPractice

2. Solve $\triangle ABC$ if $a = 5$, $b = 11$, and $c = 8$.

StudyTip

Alternative Method After finding $m\angle A$ in Step 1, the Law of Cosines could be used again to find the measure of a second angle.

ReviewVocabulary

oblique a triangle that has no right angle

2 Choose a Method to Solve Triangles

You can use the Law of Sines and the Law of Cosines to solve problems involving oblique triangles. You need to know the measure of at least one side and any two other parts. If the triangle has a solution, you must decide whether to use the Law of Sines or the Law of Cosines to begin solving it.

ConceptSummary Solving Oblique Triangles

Given	Begin by Using
two angles and any sides	Law of Sines
two sides and an angle opposite one of them	Law of Sines
two sides and their included angle	Law of Cosines
three sides	Law of Cosines

Real-World Example 3 Use the Law of Cosines

SCUBA DIVING A scuba diver looks up 20° and sees a turtle 9 feet away. She looks down 40° and sees a blue parrotfish 12 feet away. How far apart are the turtle and the blue parrotfish?

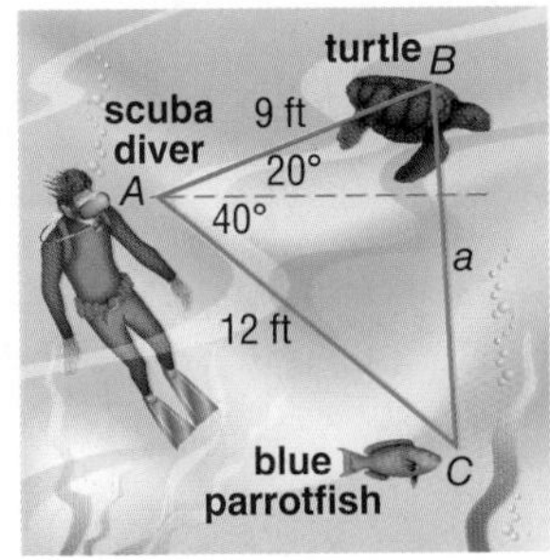

Understand You know the angles formed when the scuba diver looks up and when she looks down. You also know how far away the turtle and the blue parrotfish are from the scuba diver.

Plan Use the information to draw and label a diagram. Since two sides and the included angle of a triangle are given, you can use the Law of Cosines to solve the problem.

Solve

$a^2 = b^2 + c^2 - 2bc \cos A$	Law of Cosines
$a^2 = 12^2 + 9^2 - 2(12)(9) \cos 60$	$b = 12$, $c = 9$, and $A = 60$
$a^2 = 117$	Use a calculator.
$a \approx 10.8$	Find the positive value of a.

So, the turtle and the blue parrotfish are about 10.8 feet apart.

Check Using the Law of Sines, you can find that $B \approx 74°$ and $C \approx 46°$. Since $C < A < B$ and $c < a < b$, the solution is reasonable.

Real-WorldLink

The record for the deepest seawater scuba dive was 1044 feet, made by a scuba diver in the Red Sea.

Source: *Guinness Book of World Records*

GuidedPractice

3. MARATHONS Amelia ran 6 miles in one direction. She then turned 79° and ran 7 miles. At the end of the run, how far was Amelia from her starting point?

Check Your Understanding

= Step-by-Step Solutions begin on page R14.

Examples 1–2 **Solve each triangle. Round side lengths to the nearest tenth and angle measures to the nearest degree.**

1.

2.

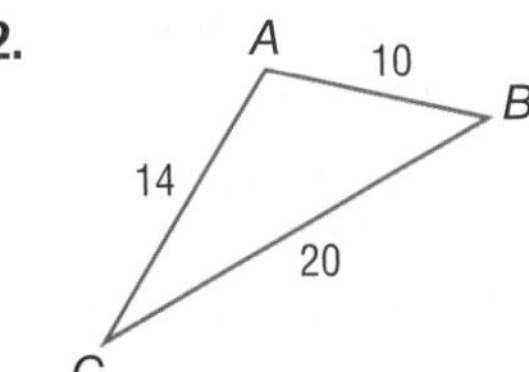

3. $a = 5, b = 8, c = 12$

4. $B = 110°, a = 6, c = 3$

Example 3 **CCSS PRECISION Determine whether each triangle should be solved by beginning with the Law of *Sines* or the Law of *Cosines*. Then solve the triangle.**

5.

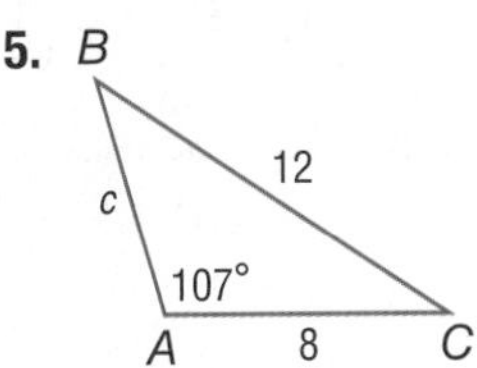

6.

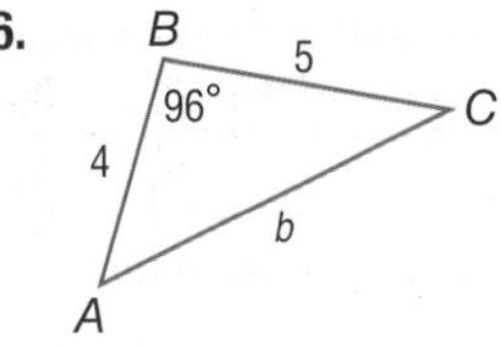

7. In $\triangle RST$, $R = 35°$, $s = 16$, and $t = 9$.

8. FOOTBALL In a football game, the quarterback is 20 yards from Receiver A. He turns 40° to see Receiver B, who is 16 yards away. How far apart are the two receivers?

Practice and Problem Solving

Extra Practice is on page R12.

Examples 1–2 **Solve each triangle. Round side lengths to the nearest tenth and angle measures to the nearest degree.**

9.

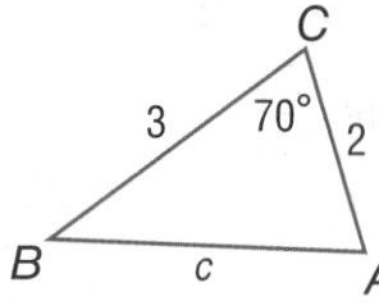

10.

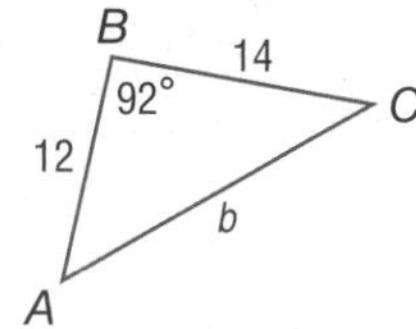

11.

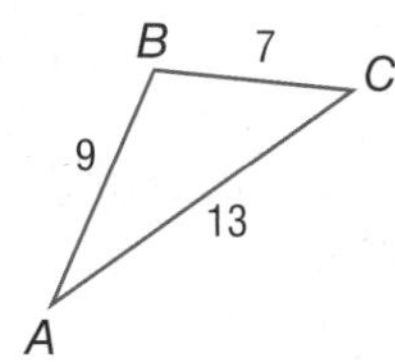

12.

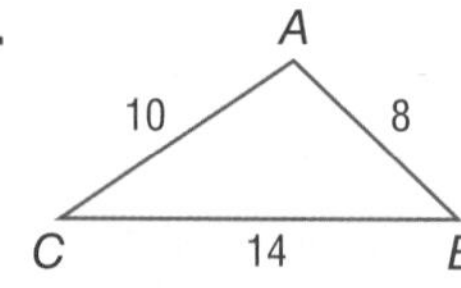

13. $A = 116°, b = 5, c = 3$

14. $C = 80°, a = 9, b = 2$

15. $f = 10, g = 11, h = 4$

16. $w = 20, x = 13, y = 12$

Example 3 **Determine whether each triangle should be solved by beginning with the Law of *Sines* or the Law of *Cosines*. Then solve the triangle.**

17.

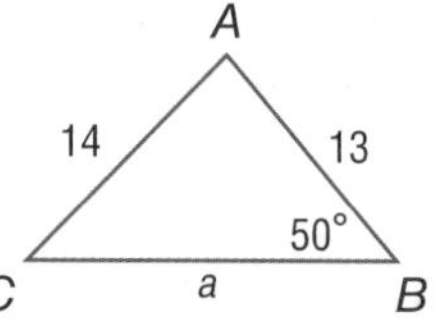

18.

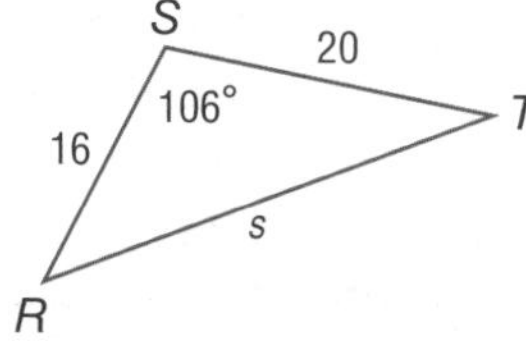

19.

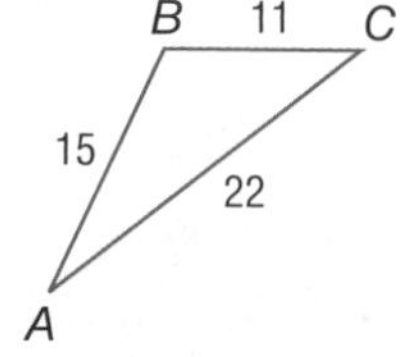

20.

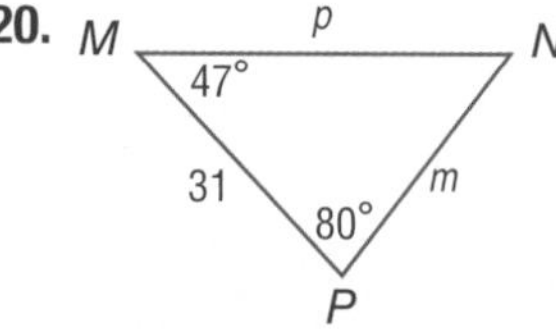

21. In $\triangle ABC$, $C = 84°$, $c = 7$, and $a = 2$.

22. In $\triangle HJK$, $h = 18$, $j = 10$, and $k = 23$.

23 **EXPLORATION** Find the distance between the ship and the shipwreck shown in the diagram. Round to the nearest tenth.

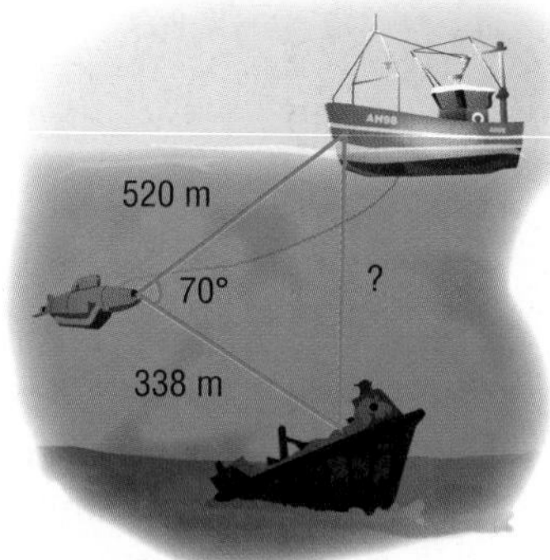

24. **GEOMETRY** A parallelogram has side lengths 8 centimeters and 12 centimeters. One angle between them measures 42°. To the nearest tenth, what is the length of the shorter diagonal?

25. **RACING** A triangular cross-country course has side lengths 1.8 kilometers, 2 kilometers, and 1.2 kilometers. What are the angles formed between each pair of sides?

26. CCSS **MODELING** A triangular plot of farm land measures 0.9 by 0.5 by 1.25 miles.

a. If the plot of land is fenced on the border, what will be the angles at which the fences of the three sides meet? Round to the nearest degree.

b. What is the area of the plot of land?

27. **LAND** Some land is in the shape of a triangle. The distances between each vertex of the triangle are 140 yd, 210 yd and 300 yd, respectively. Use the Law of Cosines to find the area of the land to the nearest square yard.

28. RIDES Two bumper cars at an amusement park ride collide as shown below.

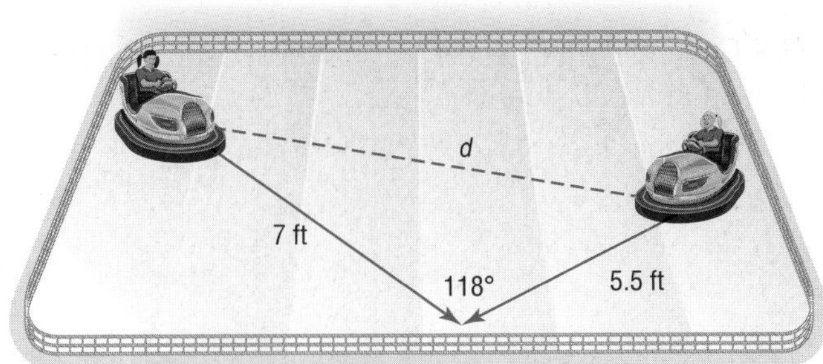

a. How far apart d were the two cars before they collided?

b. Before the collision, a third car was 10 feet from car 1 and 13 feet from car 2. Describe the angles formed by cars 1, 2, and 3 before the collision.

29. PICNICS A triangular picnic area is 11 yards by 14 yards by 10 yards.

a. Sketch and label a drawing to represent the picnic area.

b. Describe how you could find the area of the picnic area.

c. What is the area? Round to the nearest tenth.

30. WATERSPORTS A person on a personal watercraft makes a trip from point A to point B to point C traveling 28 miles per hour. She then returns from point C back to her starting point traveling 35 miles per hour. How many minutes did the entire trip take? Round to the nearest tenth.

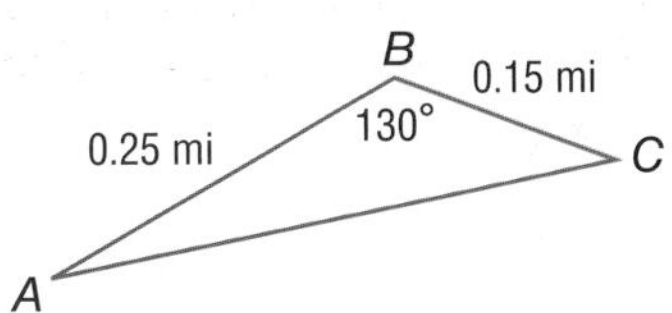

Solve each triangle. Round side lengths to the nearest tenth and angle measures to the nearest degree.

31

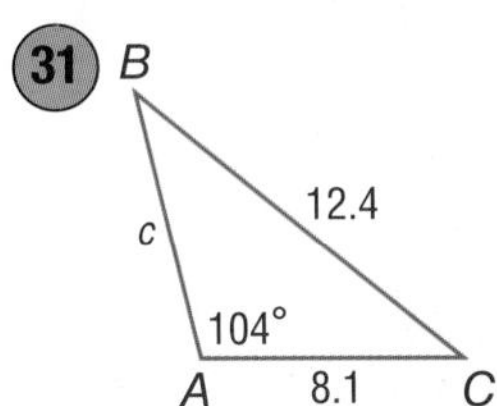

32.

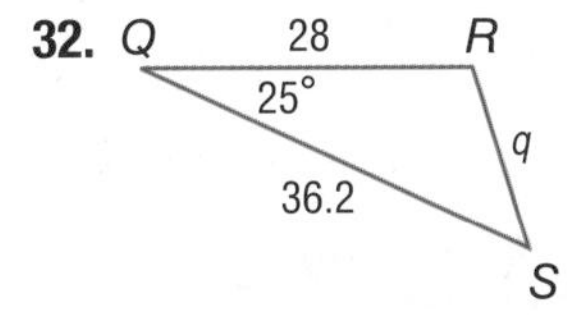

33.

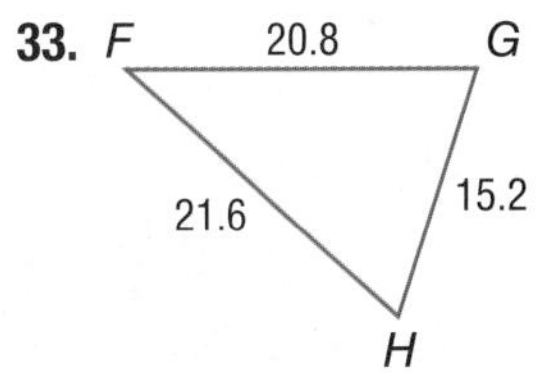

H.O.T. Problems Use Higher-Order Thinking Skills

34. CHALLENGE Use the figure and the Pythagorean Theorem to derive the Law of Cosines. Use the hints below.

- First, use the Pythagorean Theorem for $\triangle DBC$.
- In $\triangle ADB$, $c^2 = x^2 + h^2$.
- $\cos A = \frac{x}{c}$

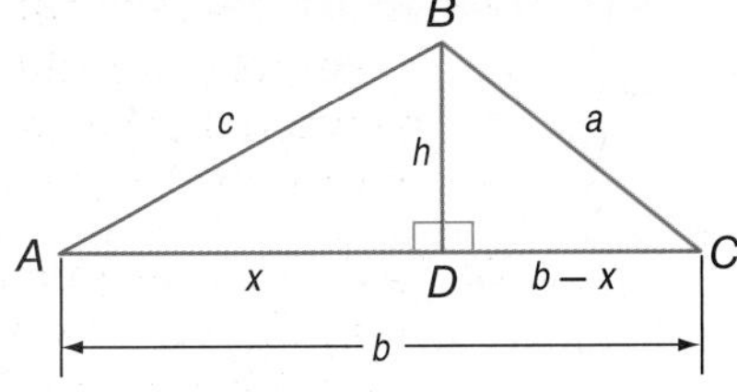

35. CCSS ARGUMENTS Three sides of a triangle measure 10.6 centimeters, 8 centimeters, and 14.5 centimeters. Explain how to find the measure of the largest angle. Then find the measure of the angle to the nearest degree.

36. OPEN ENDED Create an application problem involving right triangles and the Law of Cosines. Then solve your problem, drawing diagrams if necessary.

37. WRITING IN MATH How do you know which method to use when solving a triangle?

Standardized Test Practice

38. SAT/ACT If c and d are different positive integers and $4c + d = 26$, what is the sum of all possible values of c?

A 6
B 10
C 15
D 21
E 28

39. If $6^y = 21$, what is y?

F $\log 12 - \log 6$
G $\dfrac{\log 21}{\log 6}$
H $\dfrac{\log 6}{\log 21}$
J $\log\left(\dfrac{6}{21}\right)$

40. GEOMETRY Find the perimeter of the figure.

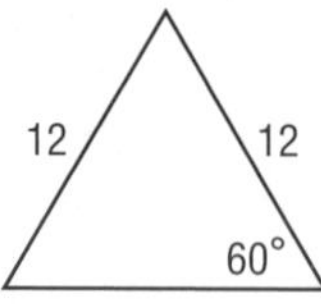

A 24 B 30 C 36 D 48

41. SHORT RESPONSE Solve the equation below for x.

$$\frac{1}{x-1} + \frac{5}{8} = \frac{23}{6x}$$

Spiral Review

Find the area of $\triangle ABC$ to the nearest tenth. (Lesson 12-4)

42.

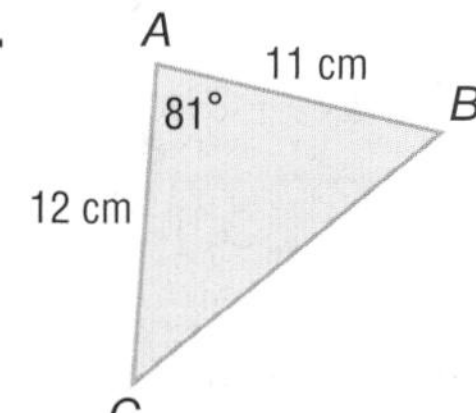

43.

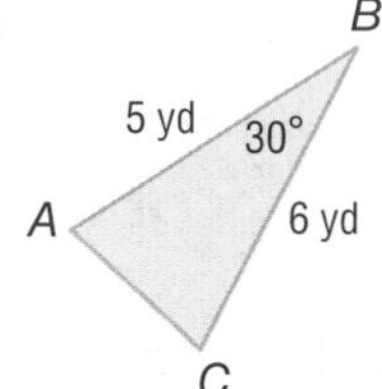

44.

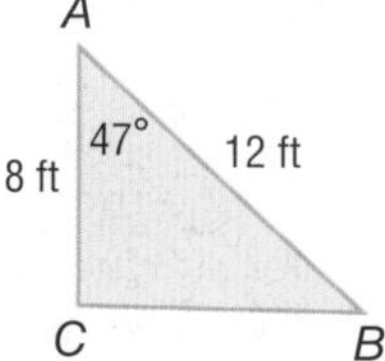

The terminal side of θ in standard position contains each point. Find the exact values of the six trigonometric functions of θ. (Lesson 12-3)

45. $(8, 5)$ **46.** $(-4, -2)$ **47.** $(6, -9)$

48. ATHLETIC SHOES The prices for a random sample of athletic shoes are shown. (Lesson 11-2)

Price (dollars)				
70	300	400	250	250
150	120	250	100	70
150	160	200	170	300

a. Use a graphing calculator to create a box-and-whisker plot. Then describe the shape of the distribution.

b. Describe the center and spread of the data using either the mean and standard deviation or the five-number summary. Justify your choice.

49. BUSINESS During the month of June, MediaWorld had revenue of \$2700 from sales of a certain DVD box set. During the July Blowout Sale, the set was on sale for \$10 off. Revenue from the set was \$3750 in July with 30 more sets sold than were sold in June. Find the price of the DVD set for June and the price for July. (Lesson 9-7)

Without writing the equation in standard form, state whether the graph of each equation is a *parabola, circle, ellipse,* or *hyperbola*. (Lesson 9-6)

50. $x^2 + y^2 - 8x - 6y + 5 = 0$ **51.** $3x^2 - 2y^2 + 32y - 134 = 0$ **52.** $y^2 + 18y - 2x = -84$

Skills Review

Sketch each angle. Then find its reference angle.

53. $245°$ **54.** $-15°$ **55.** $\frac{5}{4}\pi$

CHAPTER 12

Mid-Chapter Quiz

Lessons 12-1 through 12-5

Solve $\triangle XYZ$ by using the given measurements. Round measures of sides to the nearest tenth and measures of angles to the nearest degree. (Lesson 12-1)

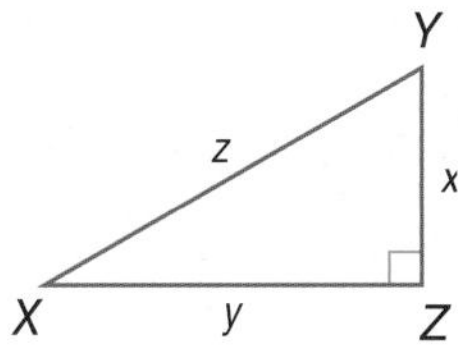

1. $Y = 65°, x = 16$

2. $X = 25°, x = 8$

3. Find the values of the six trigonometric functions for angle θ. (Lesson 12-1)

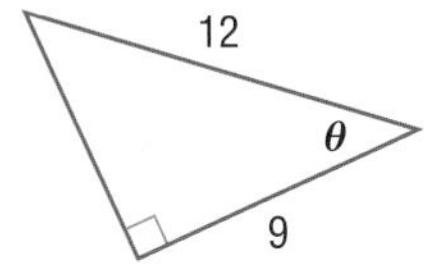

4. Draw an angle measuring $-80°$ in standard position. (Lesson 12-2)

Rewrite each degree measure in radians and each radian measure in degrees. (Lesson 12-2)

5. $215°$

6. $-350°$

7. $\frac{8\pi}{5}$

8. $\frac{9\pi}{2}$

9. **MULTIPLE CHOICE** What is the length of the arc below rounded to the nearest tenth? (Lesson 12-2)

   A 4.2 cm

   B 17.1 cm

   C 53.9 cm

   D 2638.9 cm

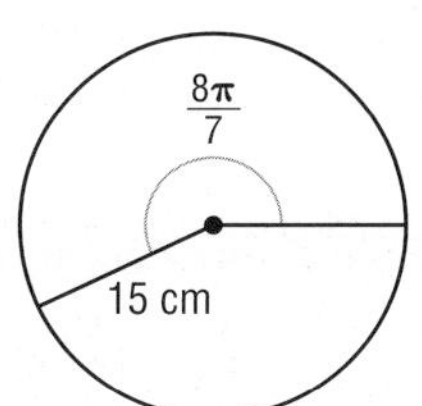

Find the exact value of each trigonometric function. (Lesson 12-3)

10. $\tan \pi$

11. $\cos \frac{3\pi}{4}$

The terminal side of θ in standard position contains each point. Find the exact values of the six trigonometric functions of θ. (Lesson 12-3)

12. $(0, -5)$

13. $(6, 8)$

14. **MULTIPLE CHOICE** Suppose θ is an angle in standard position with $\cos \theta > 0$. In which quadrant(s) does the terminal side of θ lie? (Lesson 12-3)

    F I

    G II

    H III

    J I and IV

15. **GARDEN** Lana has a garden in the shape of a triangle as pictured below. She wants to fill the garden with top soil. What is the area of the triangle? (Lesson 12-4)

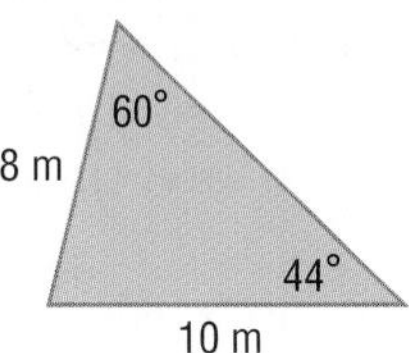

Determine whether each triangle has *no* solution, *one* solution, or *two* solutions. Then solve the triangle. Round side lengths to the nearest tenth and angle measures to the nearest degree. (Lesson 12-4)

16. $A = 38°, a = 18, c = 25$

17. $A = 65°, a = 5, b = 7$

18. $A = 115°, a = 12, b = 8$

Solve each triangle. Round side lengths to the nearest tenth and angle measures to the nearest degree. (Lesson 12-5)

19.

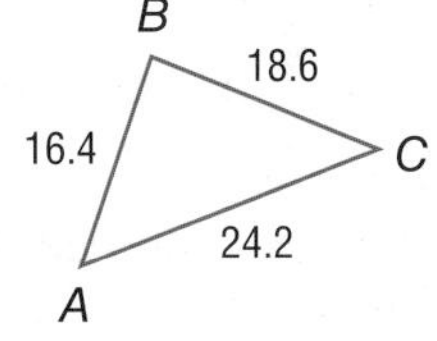

20.

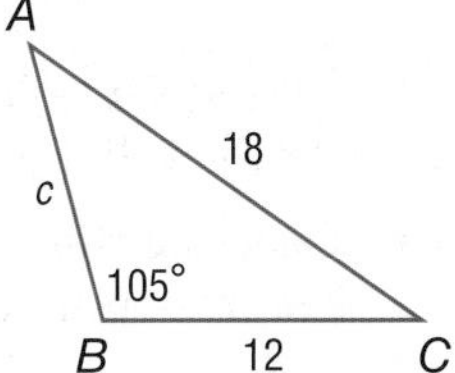

21. Eric and Zach are camping. Erik leaves Zach at the campsite and walks 4.5 miles. He then turns at a 120° angle and walks another 2.5 miles. If Eric were to walk directly back to Zach, how far would he walk? (Lesson 12-5)

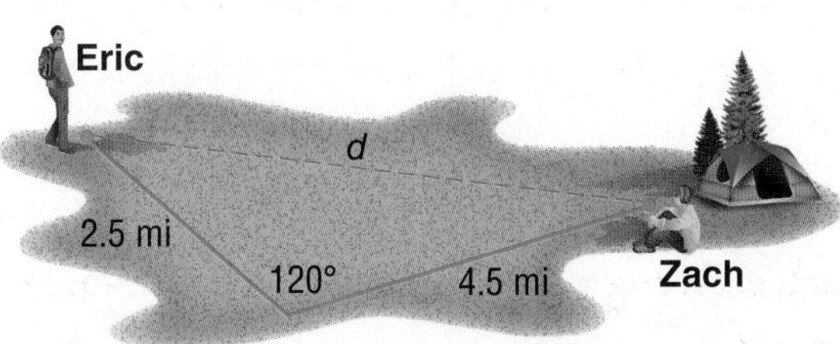

LESSON 12-6

Circular and Periodic Functions

Then	Now	Why?
• You evaluated trigonometric functions using reference angles.	1 Find values of trigonometric functions based on the unit circle. 2 Use the properties of periodic functions to evaluate trigonometric functions.	• The pedals on a bicycle rotate as the bike is being ridden. The height of a pedal is a function of time, as shown in the figure at the right. Notice that the pedal makes one complete rotation every two seconds.

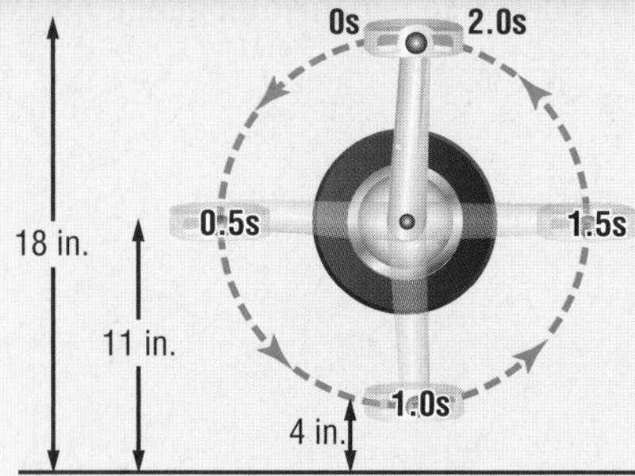

NewVocabulary

unit circle
circular function
periodic function
cycle
period

Common Core State Standards

Content Standards

F.TF.1 Understand radian measure of an angle as the length of the arc on the unit circle subtended by the angle.

F.TF.2 Explain how the unit circle in the coordinate plane enables the extension of trigonometric functions to all real numbers, interpreted as radian measures of angles traversed counterclockwise around the unit circle.

Mathematical Practices

7 Look for and make use of structure.

1 Circular Functions

A **unit circle** is a circle with a radius of 1 unit centered at the origin on the coordinate plane. You can use a point P on the unit circle to generalize sine and cosine functions.

$$\sin \theta = \frac{y}{r} = \frac{y}{1} \text{ or } y \qquad \cos \theta = \frac{x}{r} = \frac{x}{1} \text{ or } x$$

So, the values of $\sin \theta$ and $\cos \theta$ are the y-coordinate and x-coordinate, respectively, of the point where the terminal side of θ intersects the unit circle.

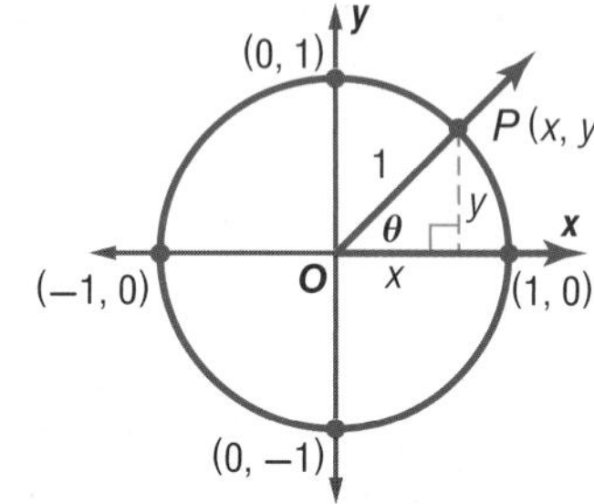

KeyConcept Functions on a Unit Circle

Words If the terminal side of an angle θ in standard position intersects the unit circle at $P(x, y)$, then $\cos \theta = x$ and $\sin \theta = y$.

Symbols $P(x, y) = P(\cos \theta, \sin \theta)$

Example If $\theta = 120°$, $P(x, y) = P(\cos 120°, \sin 120°)$.

Model

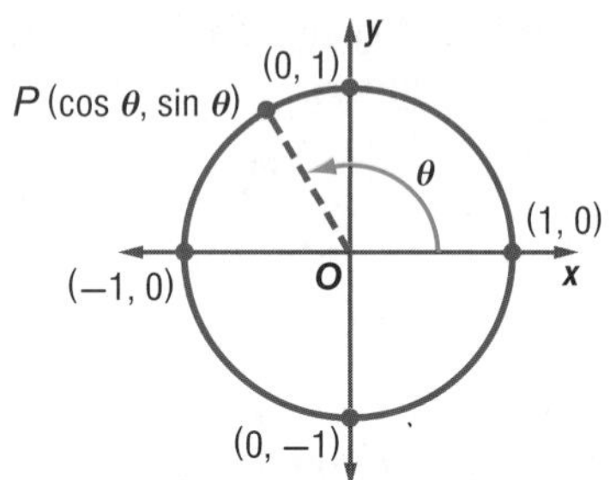

Both $\cos \theta = x$ and $\sin \theta = y$ are functions of θ. Because they are defined using a unit circle, they are called **circular functions**.

PT

Example 1 Find Sine and Cosine Given a Point on the Unit Circle

The terminal side of angle θ in standard position intersects the unit circle at $P\left(\frac{1}{2}, \frac{\sqrt{3}}{2}\right)$. Find $\cos \theta$ and $\sin \theta$.

$P\left(\frac{1}{2}, \frac{\sqrt{3}}{2}\right) = P(\cos \theta, \sin \theta)$

$\cos \theta = \frac{1}{2} \qquad \sin \theta = \frac{\sqrt{3}}{2}$

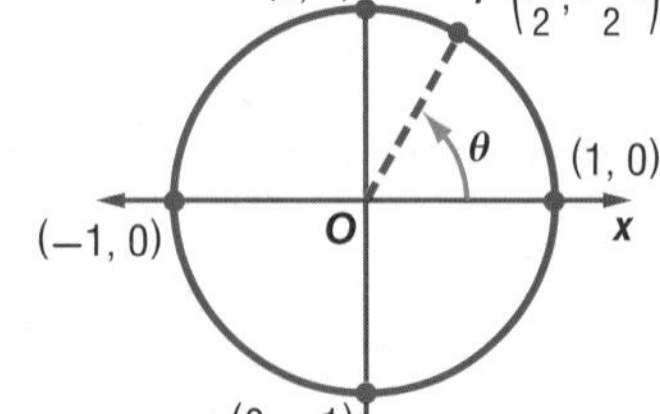

GuidedPractice

1. The terminal side of angle θ in standard position intersects the unit circle at $P\left(\frac{3}{5}, -\frac{4}{5}\right)$. Find $\cos \theta$ and $\sin \theta$.

StudyTip

Cycles A cycle can begin at any point on the graph of a periodic function. In Example 2, if the beginning of the cycle is at $\frac{\pi}{2}$, then the pattern repeats at $\frac{3\pi}{2}$. The period is $\frac{3\pi}{2} - \frac{\pi}{2}$ or π.

2 Periodic Functions

A **periodic function** has y-values that repeat at regular intervals. One complete pattern is a **cycle**, and the horizontal length of one cycle is a **period**.

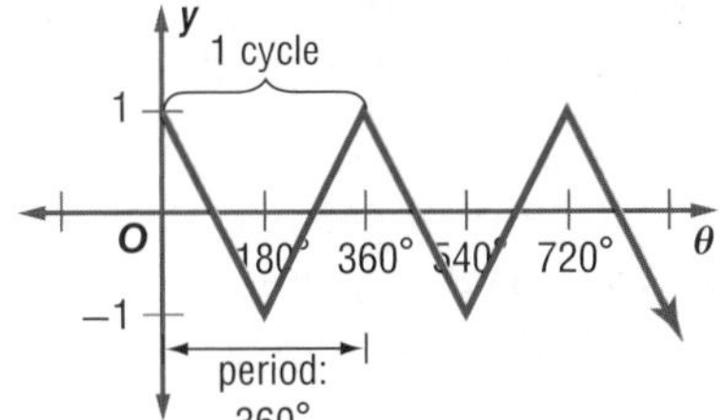

θ	y
0°	1
180°	−1
360°	1
540°	−1
720°	1

The cycle repeats every 360°.

Example 2 Identify the Period

Determine the period of the function.

The pattern repeats at π, 2π, and so on. So, the period is π.

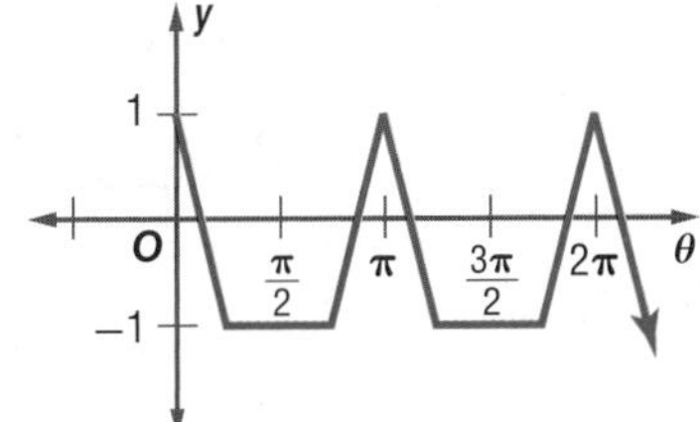

Guided Practice

2. Graph a function with a period of 4.

The rotations of wheels, pedals, carousels, and objects in space are all periodic.

Real-World Example 3 Use Trigonometric Functions

CYCLING Refer to the beginning of the lesson. The height of a bicycle pedal varies periodically as a function of time, as shown in the figure.

a. Make a table showing the height of a bicycle pedal at 0, 0.5, 1.0, 1.5, 2.0, 2.5, and 3.0 seconds.

At 0 seconds, the pedal is 18 inches high. At 0.5 second, the pedal is 11 inches high. At 1.0 second, the pedal is 4 inches high, and so on.

Time (s)	Height (in.)
0	18
0.5	11
1.0	4
1.5	11
2.0	18
2.5	11
3.0	4

b. Identify the period of the function.

The period is the time it takes to complete one rotation. So, the period is 2 seconds.

c. Graph the function. Let the horizontal axis represent the time t and the vertical axis represent the height h in inches that the pedal is from the ground.

The maximum height of the pedal is 18 inches, and the minimum height is 4 inches. Because the period of the function is 2 seconds, the pattern of the graph repeats in intervals of 2 seconds.

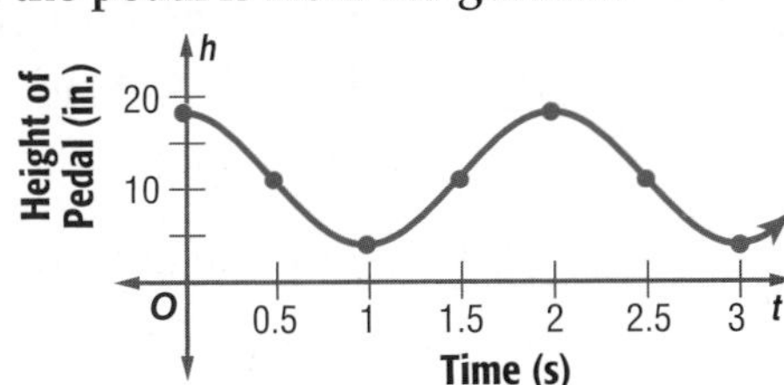

Guided Practice

3. CYCLING Another cyclist pedals the same bike at a rate of 1 revolution per second.

A. Make a table showing the height of a bicycle pedal at times 0, 0.5, 1.0, 1.5, 2.0, 2.5, and 3.0 seconds.

B. Identify the period and graph the function.

Real-WorldLink

Most competitive cyclists pedal at rates of more than 200 rotations per minute. Most other people pedal at between 90 and 120 rotations per minute.

Source: SpringerLink

StudyTip

Sine and Cosine To help you remember that for (x, y) on a unit circle, $x = \cos\theta$ and $y = \sin\theta$, notice that alphabetically x comes before y and *cosine* comes before *sine*.

The exact values of $\cos\theta$ and $\sin\theta$ for special angles are shown on the unit circle at the right. The cosine values are the x-coordinates of the points on the unit circle, and the sine values are the y-coordinates.

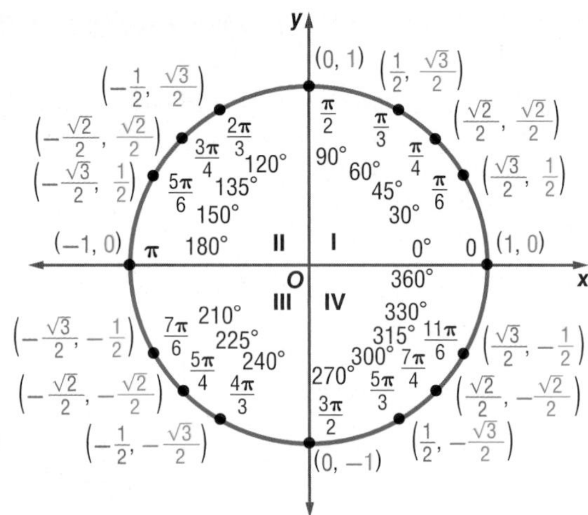

You can use this information to graph the sine and cosine functions. Let the horizontal axis represent the values of θ and the vertical axis represent the values of $\sin\theta$ or $\cos\theta$.

The cycles of the sine and cosine functions repeat every 360°. So, they are periodic functions. The period of each function is 360° or 2π.

Consider the points on the unit circle for $\theta = 45°$, $\theta = 150°$, and $\theta = 270°$.

$$(\cos 45°, \sin 45°) = \left(\frac{\sqrt{2}}{2}, \frac{\sqrt{2}}{2}\right)$$

$$(\cos 150°, \sin 150°) = \left(-\frac{\sqrt{3}}{2}, \frac{1}{2}\right)$$

$$(\cos 270°, \sin 270°) = (0, -1)$$

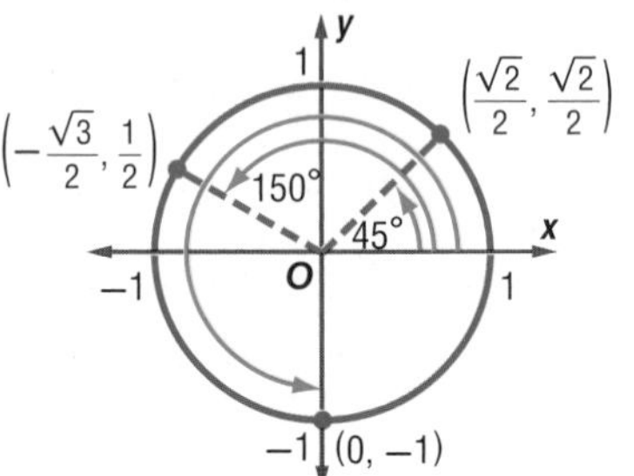

StudyTip

Radians The sine and cosine functions can also be graphed using radians as the units on the θ-axis.

These points can also be shown on the graphs of the sine and cosine functions.

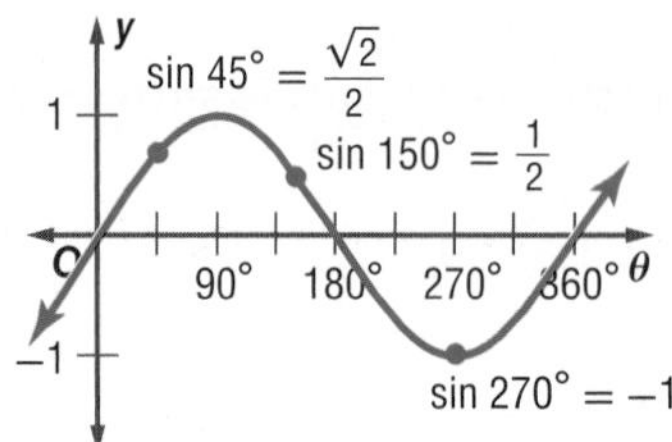

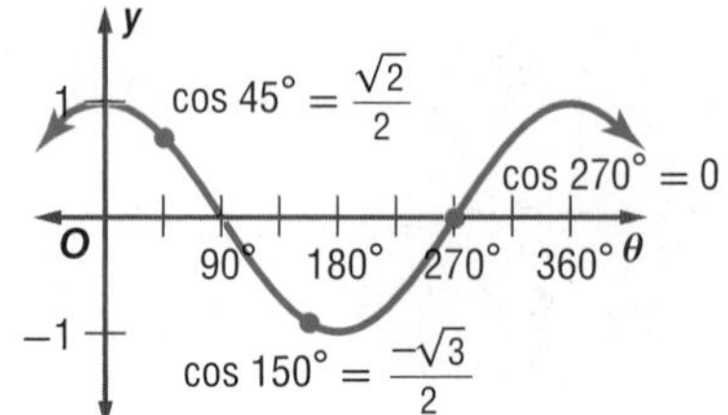

Since the period of the sine and cosine functions is 360°, the values repeat every 360°. So, $\sin(x + 360°) = \sin x$, and $\cos(x + 360°) = \cos x$.

Example 4 Evaluate Trigonometric Expressions

Find the exact value of each expression.

a. $\cos 480°$

$$\cos 480° = \cos(120° + 360°)$$
$$= \cos 120°$$
$$= -\frac{1}{2}$$

b. $\sin \frac{11\pi}{4}$

$$\sin\frac{11\pi}{4} = \sin\left(\frac{3\pi}{4} + \frac{8\pi}{4}\right)$$
$$= \sin\frac{3\pi}{4}$$
$$= \frac{\sqrt{2}}{2}$$

GuidedPractice

4A. $\cos\left(-\frac{3\pi}{4}\right)$

4B. $\sin 420°$

Check Your Understanding

= Step-by-Step Solutions begin on page R14.

Example 1 **CCSS STRUCTURE** **The terminal side of angle θ in standard position intersects the unit circle at each point P. Find $\cos \theta$ and $\sin \theta$.**

1. $P\left(\frac{15}{17}, \frac{8}{17}\right)$

2. $P\left(-\frac{\sqrt{2}}{2}, \frac{\sqrt{2}}{2}\right)$

Example 2 **Determine the period of each function.**

3.

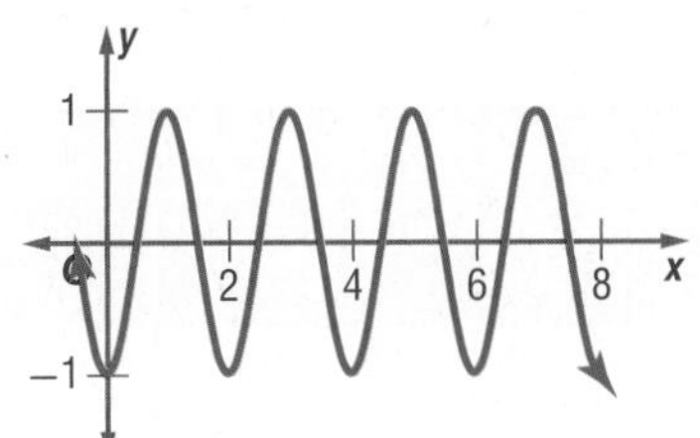

4.

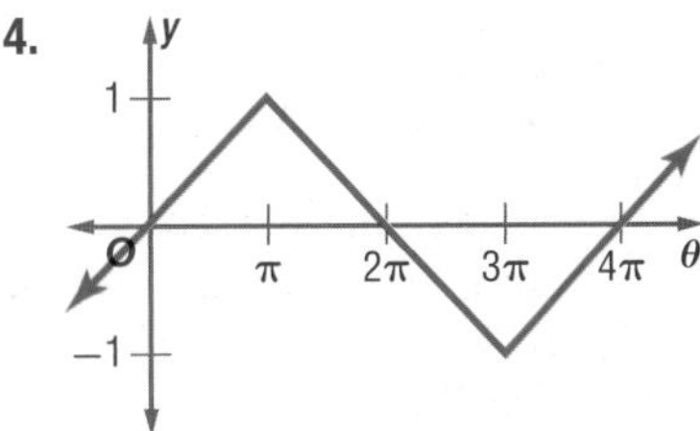

Example 3 **5. SWINGS** The height of a swing varies periodically as the function of time. The swing goes forward and reaches its high point of 6 feet. It then goes backward and reaches 6 feet again. Its lowest point is 2 feet. The time it takes to swing from its high point to its low point is 1 second.

a. How long does it take for the swing to go forward and back one time?

b. Graph the height of the swing h as a function of time t.

Example 4 **Find the exact value of each expression.**

6. $\sin \frac{13\pi}{6}$

7. $\sin(-60°)$

8. $\cos 540°$

Practice and Problem Solving

Extra Practice is on page R12.

Example 1 **The terminal side of angle θ in standard position intersects the unit circle at each point P. Find $\cos \theta$ and $\sin \theta$.**

9. $P\left(\frac{6}{10}, -\frac{8}{10}\right)$

10. $P\left(-\frac{10}{26}, -\frac{24}{26}\right)$

11. $P\left(\frac{\sqrt{3}}{2}, \frac{1}{2}\right)$

12. $P\left(\frac{\sqrt{6}}{5}, \frac{\sqrt{19}}{5}\right)$

Example 2 **Determine the period of each function.**

13.

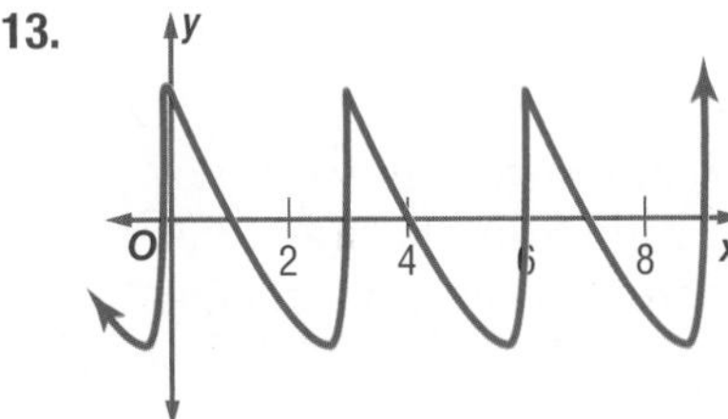

14.

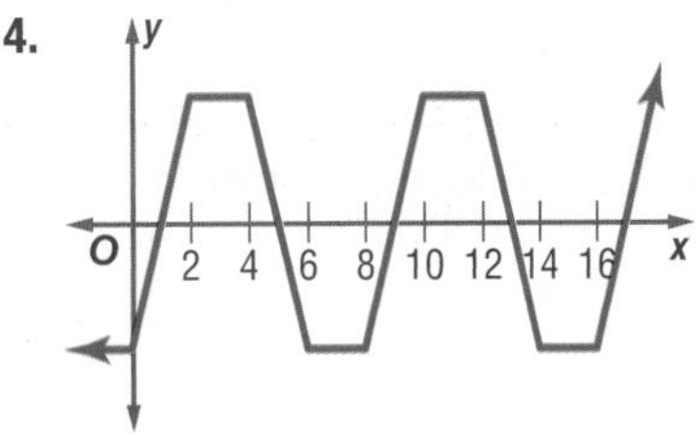

15.

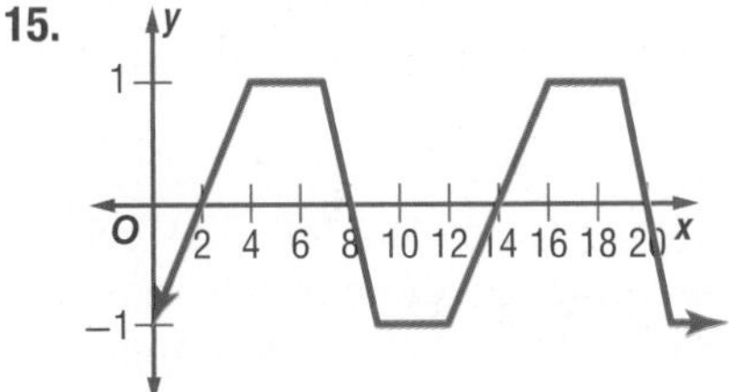

16.

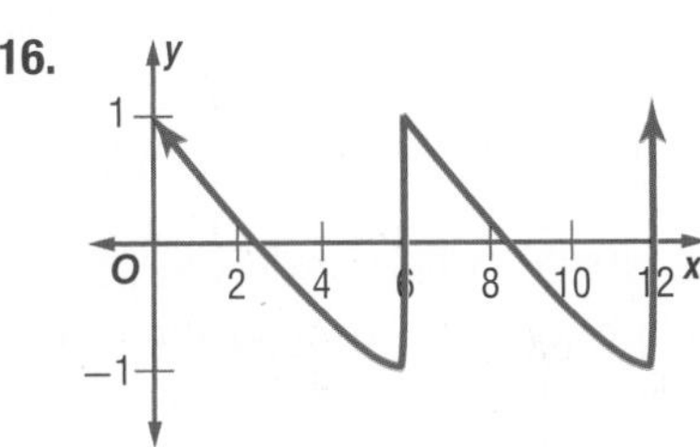

Determine the period of each function.

17.

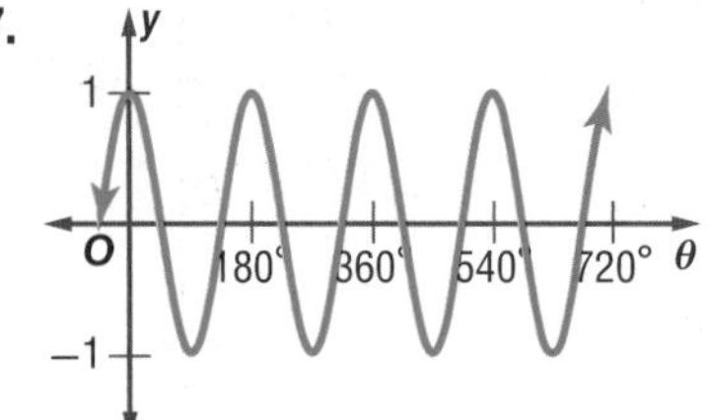

18.

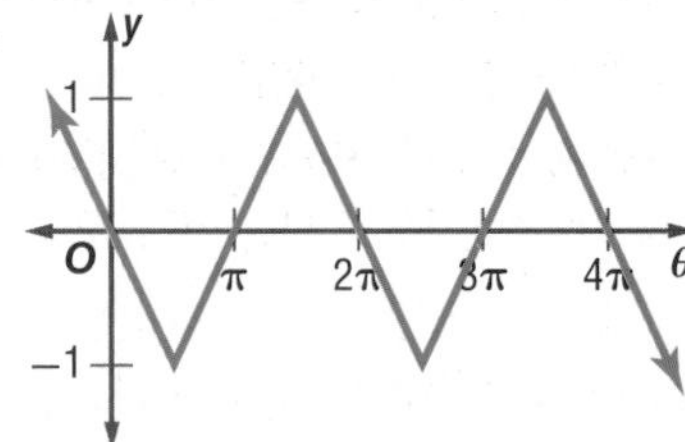

Example 3

19. WEATHER In a city, the average high temperature for each month is shown in the table.

a. Sketch a graph of the function representing this situation.

b. Describe the period of the function.

Average High Temperatures			
Month	**Temperature (°F)**	**Month**	**Temperature (°F)**
Jan	36	July	85
Feb.	41	Aug.	84
Mar.	52	Sept.	78
Apr.	64	Oct.	66
May	74	Nov.	52
Jun.	82	Dec.	41

Source: The Weather Channel

Example 4

Find the exact value of each expression.

20. $\sin \frac{7\pi}{3}$

21. $\cos (-60°)$

22. $\cos 450°$

23. $\sin \frac{11\pi}{4}$

24. $\sin (-45°)$

25. $\cos 570°$

26. CCSS **SENSE-MAKING** In the engine at the right, the distance d from the piston to the center of the circle, called the *crankshaft*, is a function of the speed of the piston rod. Point R on the piston rod rotates 150 times per second.

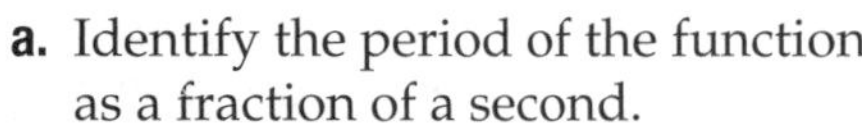

a. Identify the period of the function as a fraction of a second.

b. The shortest distance d is 0.5 inch, and the longest distance is 3.5 inches. Sketch a graph of the function. Let the horizontal axis represent the time t. Let the vertical axis represent the distance d.

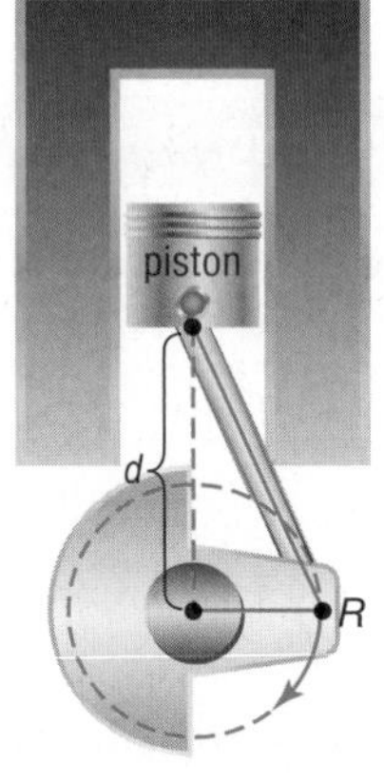

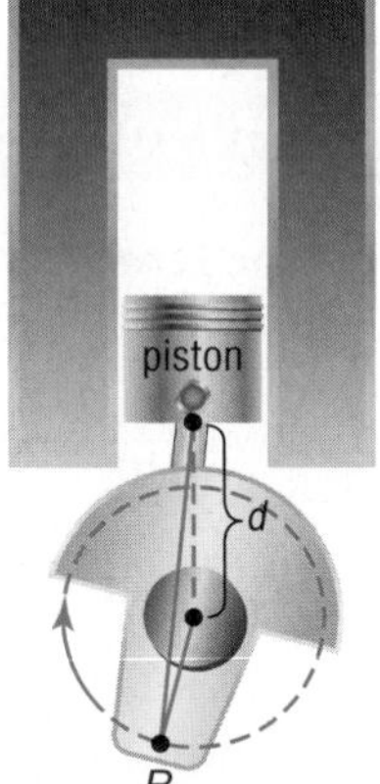

27 **TORNADOES** A tornado siren makes 2.5 rotations per minute and the beam of sound has a radius of 1 mile. Ms. Miller's house is 1 mile from the siren. The distance of the sound beam from her house varies periodically as a function of time.

a. Identify the period of the function in seconds.

b. Sketch a graph of the function. Let the horizontal axis represent the time t from 0 seconds to 60 seconds. Let the vertical axis represent the distance d the sound beam is from Ms. Miller's house at time t.

28. FERRIS WHEEL A Ferris wheel in China has a diameter of approximately 520 feet. The height of a compartment h is a function of time t. It takes about 30 seconds to make one complete revolution. Let the height at the center of the wheel represent the height at time 0. Sketch a graph of the function.

29. 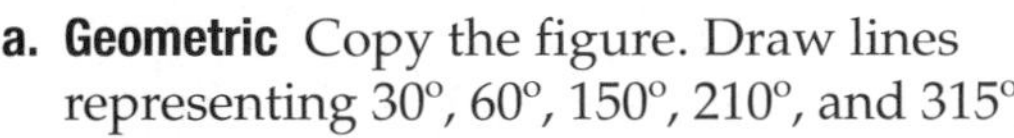 **MULTIPLE REPRESENTATIONS** The terminal side of an angle in standard position intersects the unit circle at P, as shown in the figure.

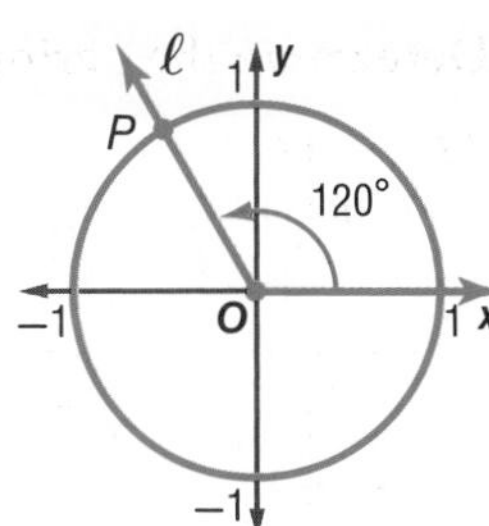

a. Geometric Copy the figure. Draw lines representing 30°, 60°, 150°, 210°, and 315°.

b. Tabular Use a table of values to show the slope of each line to the nearest tenth.

c. Analytical What conclusions can you make about the relationship between the terminal side of the angle and the slope? Explain your reasoning.

30. **POGO STICK** A person is jumping up and down on a pogo stick at a constant rate. The difference between his highest and lowest points is 2 feet. He jumps 50 times per minute.

a. Describe the independent variable and dependent variable of the periodic function that represents this situation. Then state the period of the function in seconds.

b. Sketch a graph of the jumper's change in height in relation to his starting point. Assume that his starting point is halfway between his highest and lowest points. Let the horizontal axis represent the time t in seconds. Let the vertical axis represent the height h.

Find the exact value of each expression.

31. $\cos 45° - \cos 30°$

32. $6(\sin 30°)(\sin 60°)$

33. $2 \sin \frac{4\pi}{3} - 3 \cos \frac{11\pi}{6}$

34. $\cos\left(-\frac{2\pi}{3}\right) + \frac{1}{3} \sin 3\pi$

35. $(\sin 45°)^2 + (\cos 45°)^2$

36. $\frac{(\cos 30°)(\cos 150°)}{\sin 315°}$

H.O.T. Problems Use Higher-Order Thinking Skills

37. **CCSS CRITIQUE** Francis and Benita are finding the exact value of $\cos \frac{-\pi}{3}$. Is either of them correct? Explain your reasoning.

Francis	Benita
$\cos \frac{-\pi}{3} = -\cos \frac{\pi}{3}$ $= -0.5$	$\cos \frac{-\pi}{3} = \cos\left(-\frac{\pi}{3} + 2\pi\right)$ $= \cos \frac{5\pi}{3}$ $= 0.5$

38. **CHALLENGE** A ray has its endpoint at the origin of the coordinate plane, and point $P\left(\frac{1}{2}, -\frac{\sqrt{3}}{2}\right)$ lies on the ray. Find the angle θ formed by the positive x-axis and the ray.

39. **REASONING** Is the period of a sine curve *sometimes*, *always*, or *never* a multiple of π? Justify your reasoning.

40. **OPEN ENDED** Draw the graph of a periodic function that has a maximum value of 10 and a minimum value of −10. Describe the period of the function.

41. **WRITING IN MATH** Explain how to determine the period of a periodic function from its graph. Include a description of a cycle.

Standardized Test Practice

42. SHORT RESPONSE Describe the translation of the graph of $f(x) = x^2$ to the graph of $g(x) = (x + 4)^2 - 3$.

43. The rate of population decline of Hampton Cove is modeled by $P(t) = 24{,}000e^{-0.0064t}$, where t is time in years from this year and 24,000 is the current population. In how many years will the population be 10,000?

A 14 B 104 C 137 D 375

44. SAT/ACT If $d^2 + 8 = 21$, then $d^2 - 8 =$

F 0 H 13 K 161

G 5 J 31

45. STATISTICS If the average of three different positive integers is 65, what is the greatest possible value of one of the integers?

A 192 B 193 C 194 D 195

46. GRIDDED RESPONSE If $8xy + 3 = 3$, what is the value of xy?

Spiral Review

Solve each triangle. Round side lengths to the nearest tenth and angle measures to the nearest degree. (Lesson 12-5)

47.

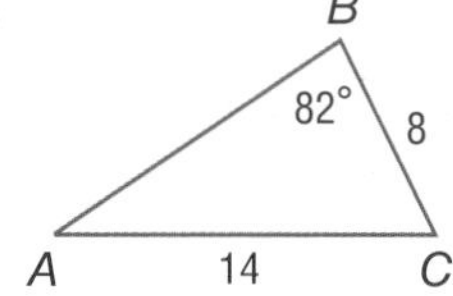

48.

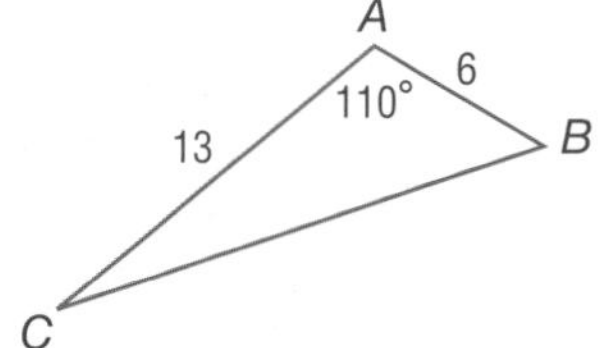

49.

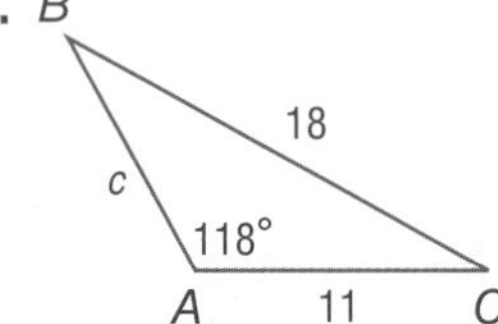

Determine whether each triangle has *no* solution, *one* solution, or *two* solutions. Then solve the triangle. Round side lengths to the nearest tenth and angle measures to the nearest degree. (Lesson 12-4)

50. $A = 72°, a = 6, b = 11$

51. $A = 46°, a = 10, b = 8$

52. $A = 110°, a = 9, b = 5$

A binomial distribution has a 70% rate of success. There are 10 trials. (Lesson 11-4)

53. What is the probability that there will be 3 failures?

54. What is the probability that there will be at least 7 successes?

55. What is the expected number of successes?

56. GAMES The diagram shows the board for a game in which spheres are dropped down a chute. A pattern of nails and dividers causes the spheres to take various paths to the sections at the bottom. For each section, how many paths through the board lead to that section? (Lesson 10-6)

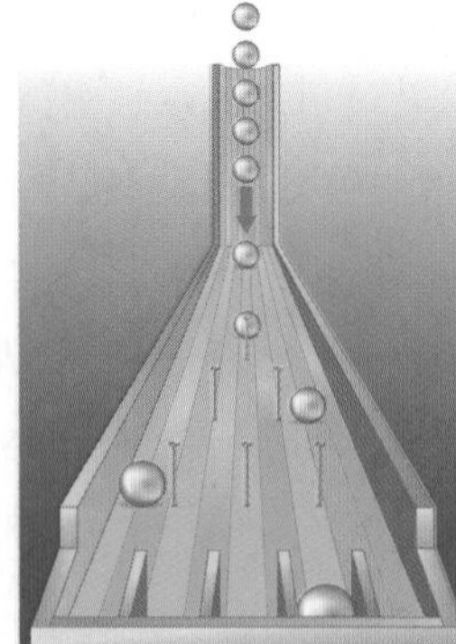

57. SALARIES Phillip's current salary is \$40,000 per year. His annual pay raise is always a percent of his salary at the time. What would his salary be if he got four consecutive 4% increases? (Lesson 10-2)

Find the exact solution(s) of each system of equations. (Lesson 9-7)

58. $y = x + 2$
$y = x^2$

59. $4x + y^2 = 20$
$4x^2 + y^2 = 100$

Skills Review

Simplify each expression.

60. $\dfrac{240}{\left|1 - \frac{5}{4}\right|}$

61. $\dfrac{180}{\left|2 - \frac{1}{3}\right|}$

62. $\dfrac{90}{\left|2 - \frac{11}{4}\right|}$

LESSON 12-7 Graphing Trigonometric Functions

Then

- You examined periodic functions.

Now

1. Describe and graph the sine, cosine, and tangent functions.
2. Describe and graph other trigonometric functions.

Why?

- Visible light waves have different wavelengths or periods. Red has the longest wavelength and violet has the shortest wavelength.

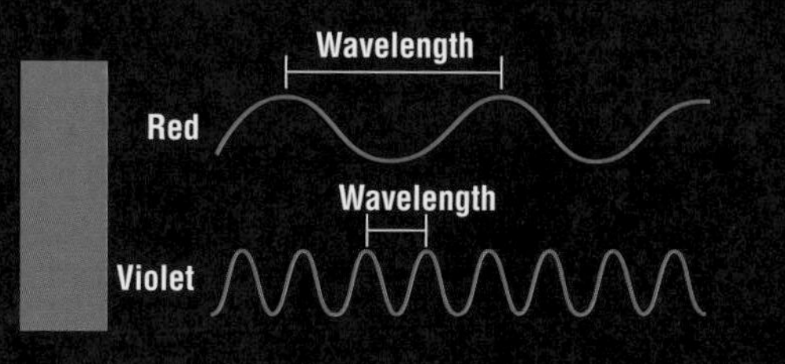

NewVocabulary
amplitude
frequency

Common Core State Standards

Content Standards
F.IF.7.e Graph exponential and logarithmic functions, showing intercepts and end behavior, and trigonometric functions, showing period, midline, and amplitude.
F.TF.5 Choose trigonometric functions to model periodic phenomena with specified amplitude, frequency, and midline.

Mathematical Practices
1 Make sense of problems and persevere in solving them.

1 Sine, Cosine, and Tangent Functions

Trigonometric functions can also be graphed on the coordinate plane. Recall that graphs of periodic functions have repeating patterns, or *cycles*. The horizontal length of each cycle is the *period*. The **amplitude** of the graph of a sine or cosine function equals half the difference between the maximum and minimum values of the function.

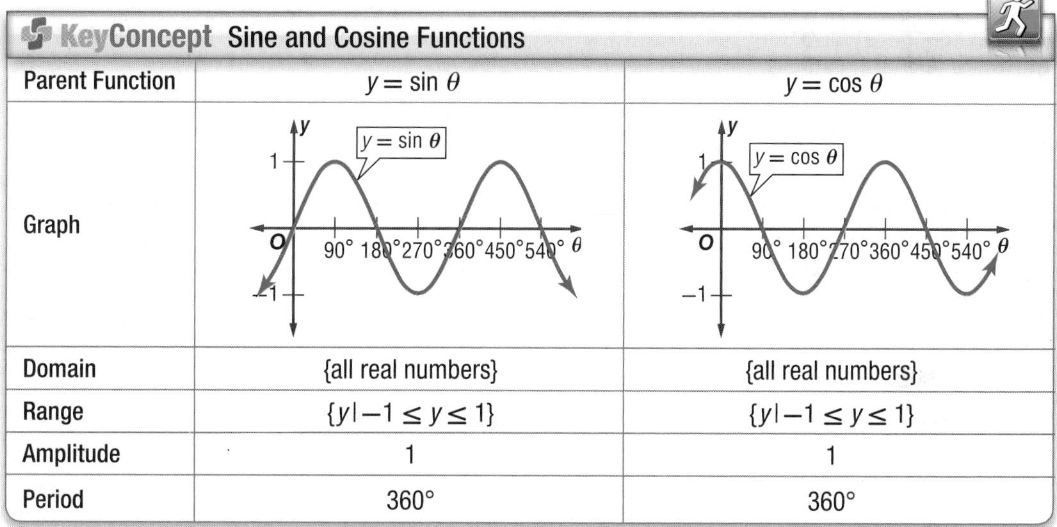

KeyConcept Sine and Cosine Functions

Parent Function	$y = \sin \theta$	$y = \cos \theta$
Graph		
Domain	{all real numbers}	{all real numbers}
Range	$\{y \mid -1 \leq y \leq 1\}$	$\{y \mid -1 \leq y \leq 1\}$
Amplitude	1	1
Period	360°	360°

As with other functions, trigonometric functions can be transformed. For the graphs of $y = a \sin b\theta$ and $y = a \cos b\theta$, the amplitude $= |a|$ and the period $= \frac{360°}{|b|}$.

Example 1 Find Amplitude and Period

Find the amplitude and period of $y = 4 \cos 3\theta$.

amplitude: $|a| = |4|$ or 4

period: $\frac{360°}{|b|} = \frac{360°}{|3|}$ or 120°

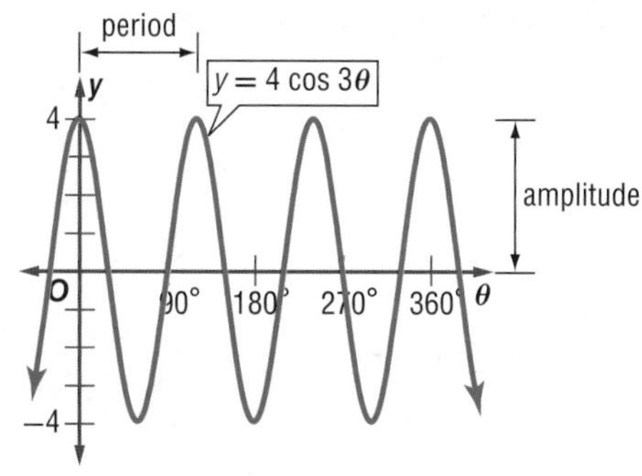

GuidedPractice

Find the amplitude and period of each function.

1A. $y = \cos \frac{1}{2}\theta$

1B. $y = 3 \sin 5\theta$

StudyTip

Periods In $y = a \sin b\theta$ and $y = a \cos b\theta$, b represents the number of cycles in 360°. In Example 1, the 3 in $y = 4 \cos 3\theta$ indicates that there are three cycles in 360°. So, there is one cycle in 120°.

Use the graphs of the parent functions to graph $y = a \sin b\theta$ and $y = a \cos b\theta$. Then use the amplitude and period to draw the appropriate sine and cosine curves. You can also use θ-intercepts to help you graph the functions.

The θ-intercepts of $y = a \sin b\theta$ and $y = a \cos b\theta$ in one cycle are as follows.

$y = a \sin b\theta$	$y = a \cos b\theta$
$(0, 0), \left(\frac{1}{2} \cdot \frac{360°}{b}, 0\right) \left(\frac{360°}{b}, 0\right)$	$\left(\frac{1}{4} \cdot \frac{360°}{b}, 0\right), \left(\frac{3}{4} \cdot \frac{360°}{b}, 0\right)$

StudyTip

Amplitude The graphs of $y = a \sin b\theta$ and $y = a \cos b\theta$ with amplitude of $|a|$ have maxima at $y = a$ and minima at $y = -a$.

Example 2 Graph Sine and Cosine Functions

Graph each function.

a. $y = 2 \sin \theta$

Find the amplitude, the period, and the x-intercepts: $a = 2$ and $b = 1$.

amplitude: $|a| = |2|$ or 2 → The graph is stretched vertically so that the maximum value is 2 and the minimum value is −2.

period: $\frac{360°}{|b|} = \frac{360°}{|1|}$ or 360° → One cycle has a length of 360°.

x-intercepts: $(0, 0)$

$\left(\frac{1}{2} \cdot \frac{360°}{b}, 0\right) = (180°, 0)$

$\left(\frac{360°}{b}, 0\right) = (360°, 0)$

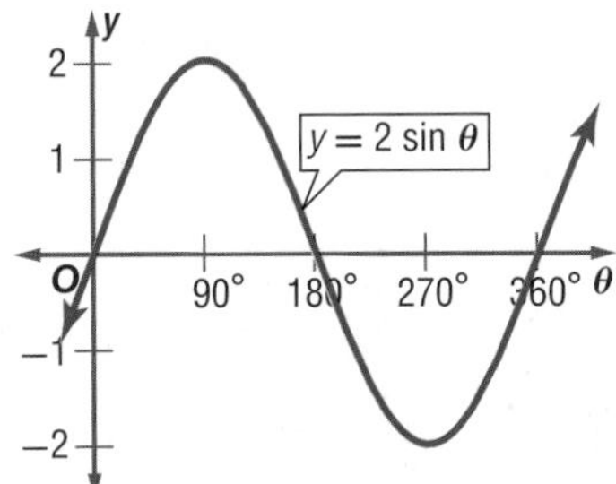

b. $y = \cos 4\theta$

amplitude: $|a| = |1|$ or 1

period: $\frac{360°}{|b|} = \frac{360°}{|4|}$ or 90°

x-intercepts: $\left(\frac{1}{4} \cdot \frac{360°}{b}, 0\right) = (22.5°, 0)$

$\left(\frac{3}{4} \cdot \frac{360°}{b}, 0\right) = (67.5°, 0)$

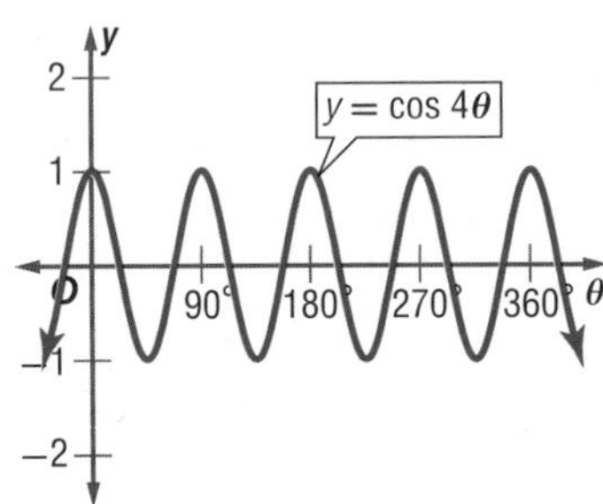

GuidedPractice

2A. $y = 3 \cos \theta$

2B. $y = \frac{1}{2} \sin 2\theta$

Trigonometric functions are useful for modeling real-world periodic motion such as electromagnetic waves or sound waves. Often these waves are described using *frequency*. **Frequency** is the number of cycles in a given unit of time.

The frequency of the graph of a function is the reciprocal of the period of the function. So, if the period of a function is $\frac{1}{100}$ second, then the frequency is 100 cycles per second.

Real-World Example 3 Model Periodic Situations

SOUND Sound that has a frequency below the human range is known as *infrasound*. Elephants can hear sounds in the infrasound range, with frequencies as low as 5 hertz (Hz), or 5 cycles per second.

a. Find the period of the function that models the sound waves.

There are 5 cycles per second, and the period is the time it takes for one cycle. So, the period is $\frac{1}{5}$ or 0.2 second.

b. Let the amplitude equal 1 unit. Write a sine equation to represent the sound wave y as a function of time t. Then graph the equation.

$\text{period} = \frac{2\pi}{\lvert b \rvert}$	Write the relationship between the period and b.
$0.2 = \frac{2\pi}{\lvert b \rvert}$	Substitution
$0.2\lvert b \rvert = 2\pi$	Multiply each side by $\lvert b \rvert$.
$b = 10\pi$	Multiply each side by 5; b is positive.
$y = a \sin b\theta$	Write the general equation for the sine function.
$y = 1 \sin 10\pi t$	$a = 1$, $b = 10\pi$, and $\theta = t$
$y = \sin 10\pi t$	Simplify.

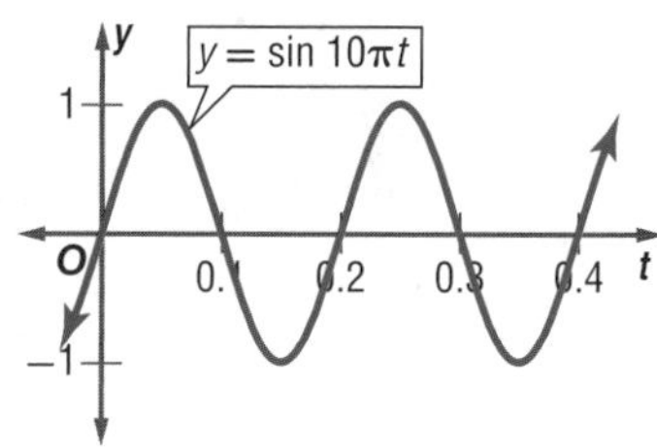

Real-WorldLink

Elephants are able to hear sound coming from up to 5 miles away. Humans can hear sounds with frequencies between 20 Hz and 20,000 Hz.

Source: School for Champions

GuidedPractice

3. SOUND Humans can hear sounds with frequencies as low as 20 hertz.

A. Find the period of the function.

B. Let the amplitude equal 1 unit. Write a cosine equation to model the sound waves. Then graph the equation.

StudyTip

Amplitude and Period Note that the amplitude affects the graph along the vertical axis, and the period affects it along the horizontal axis.

The tangent function is one of the trigonometric functions whose graphs have asymptotes.

KeyConcept Tangent Functions

Parent Function	$y = \tan \theta$	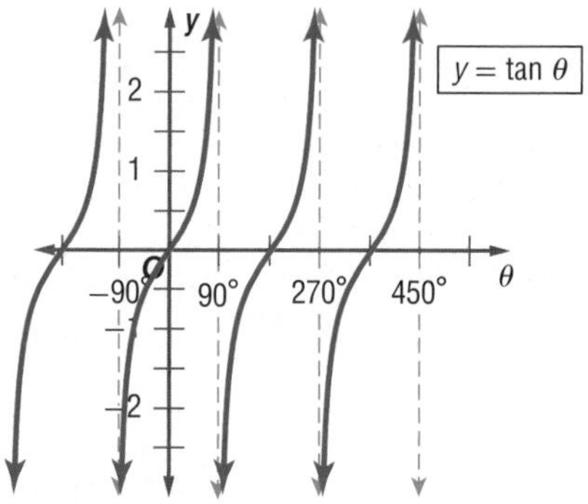 Graph
Domain	$\{\theta \mid \theta \neq 90 + 180n$, n is an integer$\}$	
Range	{all real numbers}	
Amplitude	undefined	
Period	180°	
θ intercepts in one cycle	$(0, 0)$, $\left(\frac{1}{2} \cdot \frac{360°}{b}, 0\right)$, $\left(\frac{360°}{b}, 0\right)$	

For the graph of $y = a \tan b\theta$, the period $= \frac{180°}{\lvert b \rvert}$, there is no amplitude, and the asymptotes are odd multiples of $\frac{180°}{2\lvert b \rvert}$.

StudyTip

Tangent The tangent function does not have an amplitude because it has no maximum or minimum values.

Example 4 Graph Tangent Functions

Find the period of $y = \tan 2\theta$. Then graph the function.

period: $\frac{180^\circ}{|b|} = \frac{180^\circ}{|2|}$ or 90°

asymptotes: $\frac{180^\circ}{2|b|} = \frac{180^\circ}{2|2|}$ or 45°

Sketch asymptotes at $-1 \cdot 45^\circ$ or -45°, $1 \cdot 45^\circ$ or 45°, $3 \cdot 45^\circ$ or 135°, and so on.

Use $y = \tan \theta$, but draw one cycle every 90°.

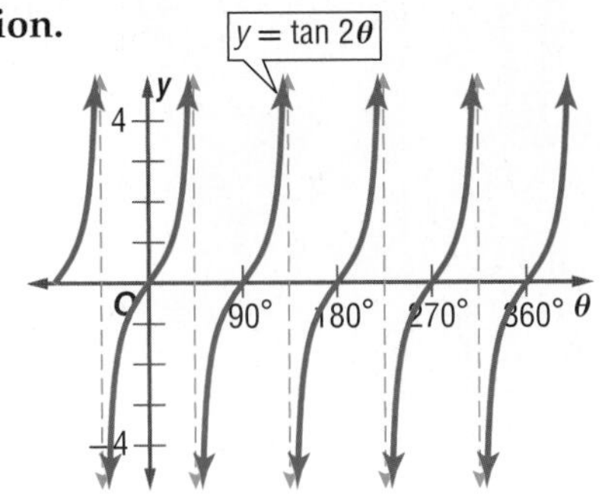

GuidedPractice

4. Find the period of $y = \frac{1}{2} \tan \theta$. Then graph the function.

2 Graphs of Other Trigonometric Functions

The graphs of the cosecant, secant, and cotangent functions are related to the graphs of the sine, cosine, and tangent functions.

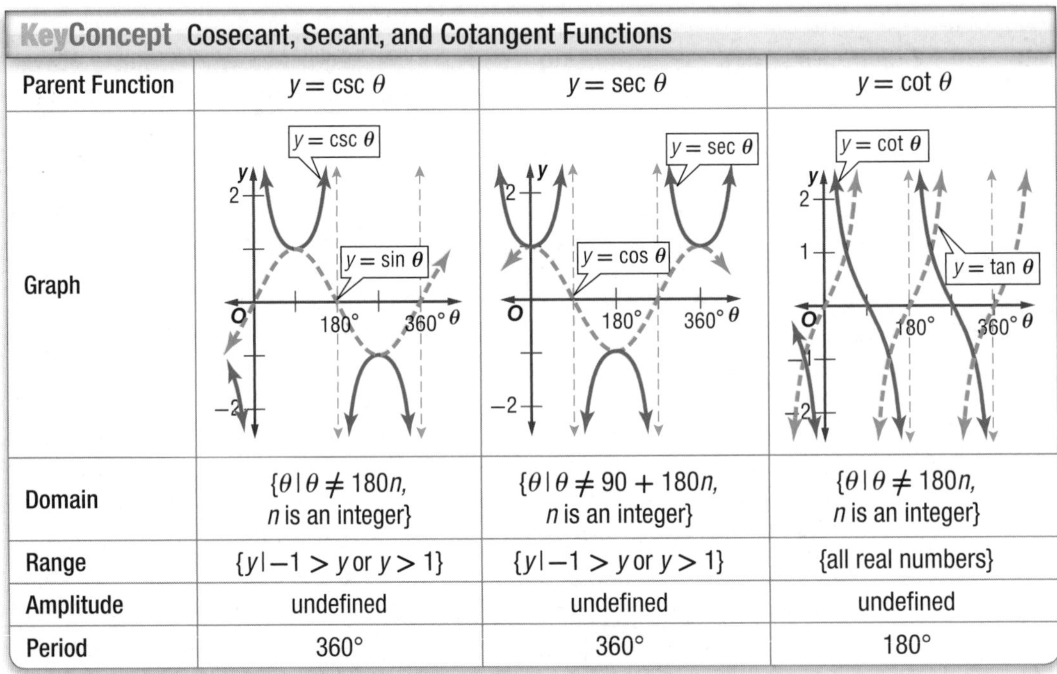

KeyConcept Cosecant, Secant, and Cotangent Functions

Parent Function	$y = \csc \theta$	$y = \sec \theta$	$y = \cot \theta$
Graph			
Domain	$\{\theta \mid \theta \neq 180n$, n is an integer$\}$	$\{\theta \mid \theta \neq 90 + 180n$, n is an integer$\}$	$\{\theta \mid \theta \neq 180n$, n is an integer$\}$
Range	$\{y \mid -1 > y$ or $y > 1\}$	$\{y \mid -1 > y$ or $y > 1\}$	{all real numbers}
Amplitude	undefined	undefined	undefined
Period	360°	360°	180°

StudyTip

Reciprocal Functions You can use the graphs of $y = \sin \theta$, $y = \cos \theta$, and $y = \tan \theta$ to graph the reciprocal functions, but these graphs are not part of the graphs of the cosecant, secant, and cotangent functions.

Example 5 Graph Other Trigonometric Functions

Find the period of $y = 2 \sec \theta$. Then graph the function.

Since $2 \sec \theta$ is a reciprocal of $2 \cos \theta$, the graphs have the same period, 360°. The vertical asymptotes occur at the points where $2 \cos \theta = 0$. So, the asymptotes are at $\theta = 90^\circ$ and $\theta = 270^\circ$.

Sketch $y = 2 \cos \theta$ and use it to graph $y = 2 \sec \theta$.

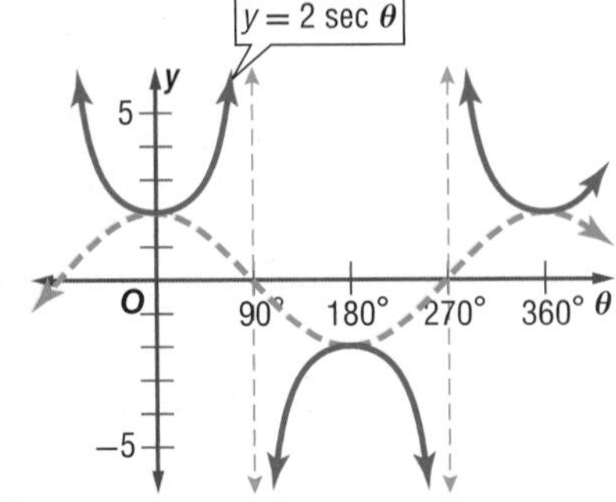

GuidedPractice

5. Find the period of $y = \csc 2\theta$. Then graph the function.

Check Your Understanding

= Step-by-Step Solutions begin on page R14.

Examples 1–2 **Find the amplitude and period of each function. Then graph the function.**

1. $y = 4 \sin \theta$

2. $y = \sin 3\theta$

3. $y = \cos 2\theta$

4. $y = \frac{1}{2} \cos 3\theta$

Example 3

5. SPIDERS When an insect gets caught in a spider web, the web vibrates with a frequency of 14 hertz.

a. Find the period of the function.

b. Let the amplitude equal 1 unit. Write a sine equation to represent the vibration of the web y as a function of time t. Then graph the equation.

Examples 4–5 **Find the period of each function. Then graph the function.**

6. $y = 3 \tan \theta$

7. $y = 2 \csc \theta$

8. $y = \cot 2\theta$

Practice and Problem Solving

Extra Practice is on page R12.

Examples 1–2 **Find the amplitude and period of each function. Then graph the function.**

9. $y = 2 \cos \theta$

10. $y = 3 \sin \theta$

11. $y = \sin 2\theta$

12. $y = \cos 3\theta$

13. $y = \cos \frac{1}{2}\theta$

14. $y = \sin 4\theta$

15. $y = \frac{3}{4} \cos \theta$

16. $y = \frac{3}{2} \sin \theta$

17 $y = \frac{1}{2} \sin 2\theta$

18. $y = 4 \cos 2\theta$

19. $y = 3 \cos 2\theta$

20. $y = 5 \sin \frac{2}{3}\theta$

Example 3

21. CCSS REASONING A boat on a lake bobs up and down with the waves. The difference between the lowest and highest points of the boat is 8 inches. The boat is at *equilibrium* when it is halfway between the lowest and highest points. Each cycle of the periodic motion lasts 3 seconds.

a. Write an equation for the motion of the boat. Let h represent the height in inches and let t represent the time in seconds. Assume that the boat is at equilibrium at $t = 0$ seconds.

b. Draw a graph showing the height of the boat as a function of time.

22. ELECTRICITY The voltage supplied by an electrical outlet is a periodic function that *oscillates*, or goes up and down, between −165 volts and 165 volts with a frequency of 50 cycles per second.

a. Write an equation for the voltage V as a function of time t. Assume that at $t = 0$ seconds, the current is 165 volts.

b. Graph the function.

Examples 4–5 **Find the period of each function. Then graph the function.**

23. $y = \tan \frac{1}{2}\theta$

24. $y = 3 \sec \theta$

25. $y = 2 \cot \theta$

26. $y = \csc \frac{1}{2}\theta$

27. $y = 2 \tan \theta$

28. $y = \sec \frac{1}{3}\theta$

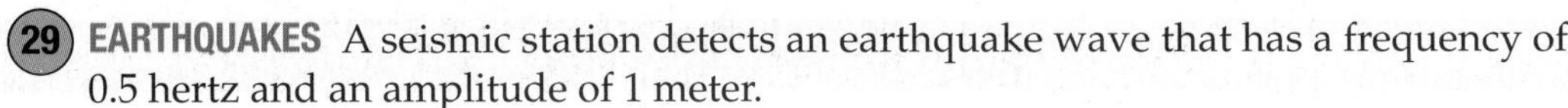

29 **EARTHQUAKES** A seismic station detects an earthquake wave that has a frequency of 0.5 hertz and an amplitude of 1 meter.

a. Write an equation involving sine to represent the height of the wave h as a function of time t. Assume that the equilibrium point of the wave, $h = 0$, is halfway between the lowest and highest points.

b. Graph the function. Then determine the height of the wave after 20.5 seconds.

30. CCSS **PERSEVERANCE** An object is attached to a spring as shown at the right. It oscillates according to the equation $y = 20 \cos \pi t$, where y is the distance in centimeters from its equilibrium position at time t.

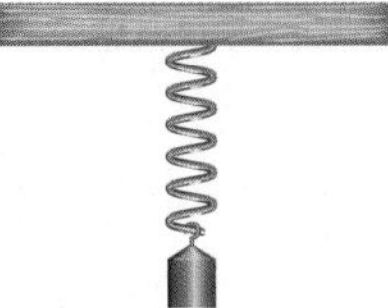

a. Describe the motion of the object by finding the following: the amplitude in centimeters, the frequency in vibrations per second, and the period in seconds.

b. Find the distance of the object from its equilibrium position at $t = \frac{1}{4}$ second.

c. The equation $v = (-20 \text{ cm})(\pi \text{ rad/s}) \cdot \sin(\pi \text{ rad/s} \cdot t)$ represents the velocity v of the object at time t. Find the velocity at $t = \frac{1}{4}$ second.

31. **PIANOS** A piano string vibrates at a frequency of 130 hertz.

a. Write and graph an equation using cosine to model the vibration of the string y as a function of time t. Let the amplitude equal 1 unit.

b. Suppose the frequency of the vibration doubles. Do the amplitude and period increase, decrease, or remain the same? Explain.

Find the amplitude, if it exists, and period of each function. Then graph the function.

32. $y = 3 \sin \frac{2}{3}\theta$ **33.** $y = \frac{1}{2} \cos \frac{3}{4}\theta$ **34.** $y = 2 \tan \frac{1}{2}\theta$

35. $y = 2 \sec \frac{4}{5}\theta$ **36.** $y = 5 \csc 3\theta$ **37.** $y = 2 \cot 6\theta$

Identify the period of the graph and write an equation for each function.

38.

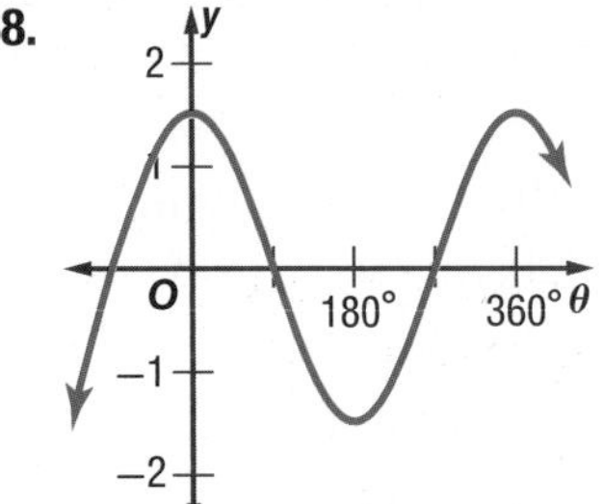

39.

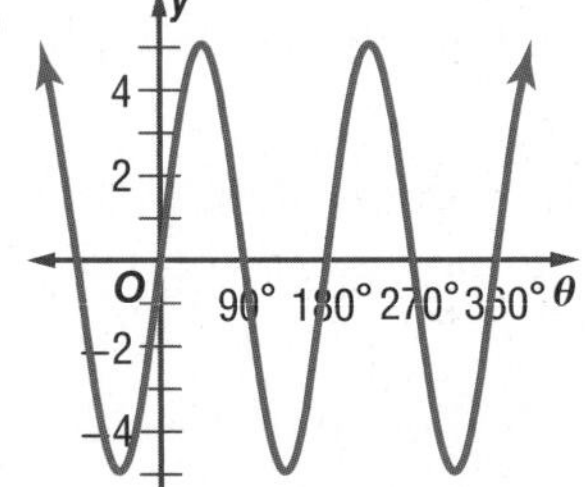

40.

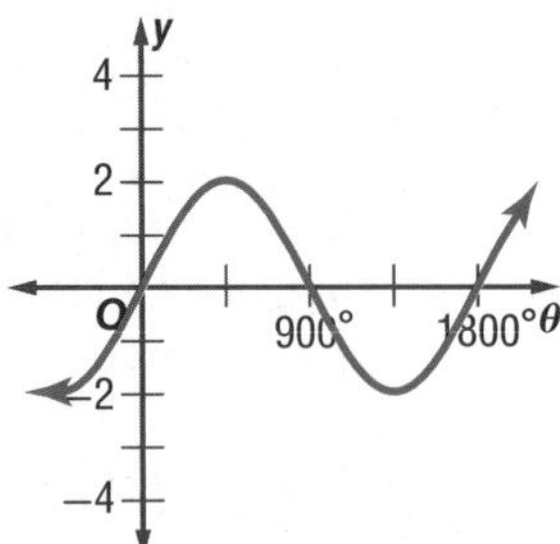

H.O.T. Problems Use Higher-Order Thinking Skills

41. **CHALLENGE** Describe the domain and range of $y = a \cos \theta$ and $y = a \sec \theta$, where a is any positive real number.

42. **REASONING** Compare and contrast the graphs of $y = \frac{1}{2} \sin \theta$ and $y = \sin \frac{1}{2}\theta$.

43. **OPEN ENDED** Write a trigonometric function that has an amplitude of 3 and a period of 180°. Then graph the function.

44. **WRITING IN MATH** How can you use the characteristics of a trigonometric function to sketch its graph?

Standardized Test Practice

45. SHORT RESPONSE Find the 100,001st term of the sequence.

13, 20, 27, 34, 41, …

46. STATISTICS You bowled five games and had the following scores: 143, 171, 167, 133, and 156. What was your average?

A 147 B 153 C 154 D 156

47. Your city had a population of 312,430 ten years ago. If its current population is 418,270, by what percentage has it grown over the past 10 years?

F 25% G 34% H 66% J 75%

48. SAT/ACT If $h + 4 = b - 3$, then $(h - 2)^2 =$

A $h^2 + 4$

B $b^2 - 6b + 3$

C $b^2 - 18b + 81$

D $b^2 - 14b + 49$

E $b^2 - 10b + 25$

Spiral Review

Find the exact value of each expression. (Lesson 12-6)

49. $\cos 120° - \sin 30°$

50. $3(\sin 45°)(\sin 60°)$

51. $4 \sin \frac{4\pi}{3} - 2 \cos \frac{\pi}{6}$

Solve each triangle. Round side lengths to the nearest tenth and angle measures to the nearest degree. (Lesson 12-5)

52. 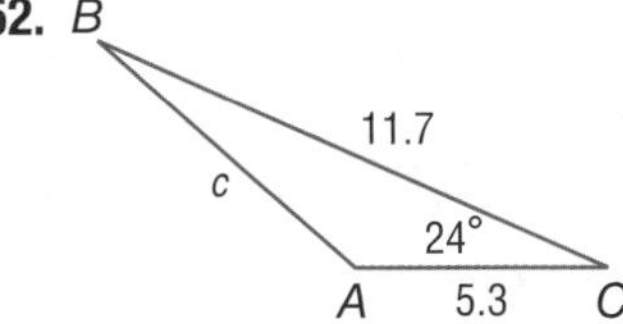

53. 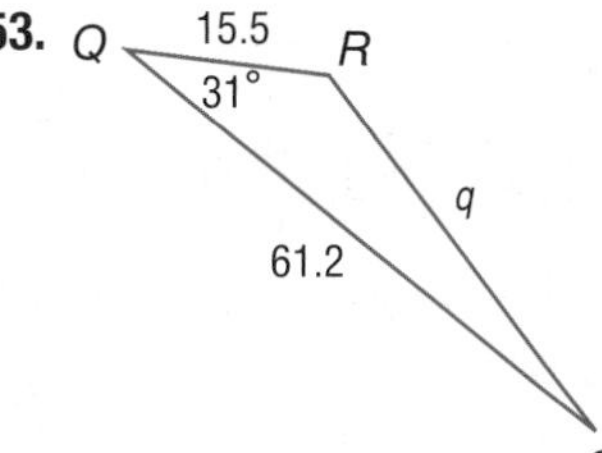

54. 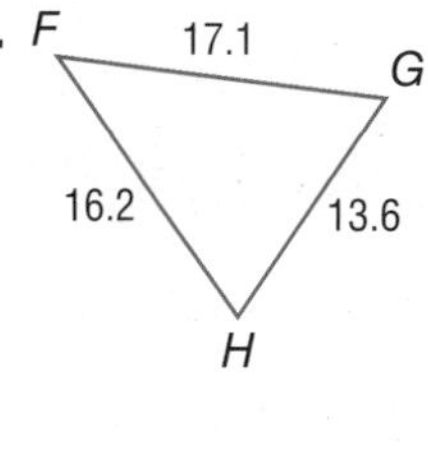

A binomial distribution has a 40% rate of success. There are 12 trials. (Lesson 11-4)

55. What is the probability that there will be exactly 5 failures?

56. What is the probability that there will be at least 8 successes?

57. What is the expected number of successes?

58. BANKING Rita has deposited $1000 in a bank account. At the end of each year, the bank posts interest to her account in the amount of 3% of the balance, but then takes out a $10 annual fee. (Lesson 10-6)

a. Let b_0 be the amount Rita deposited. Write a recursive equation for the balance b_n in her account at the end of n years.

b. Find the balance in the account after four years.

Write an equation for an ellipse that satisfies each set of conditions. (Lesson 9-4)

59. center at (6, 3), focus at (2, 3), co-vertex at (6, 1)

60. foci at (2, 1) and (2, 13), co-vertex at (5, 7)

Skills Review

Graph each function.

61. $y = 2(x - 3)^2 - 4$

62. $y = \frac{1}{3}(x + 5)^2 + 2$

63. $y = -3(x + 6)^2 + 7$

EXPLORE

12-8 Graphing Technology Lab
Trigonometric Graphs

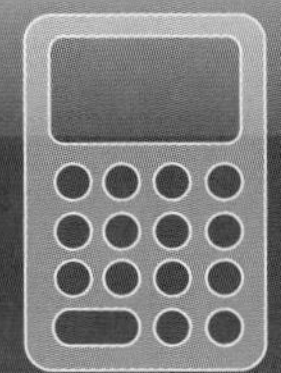

You can use a TI-83/84 Plus graphing calculator to explore transformations of the graphs of trigonometric functions.

CCSS Common Core State Standards
Content Standards
F.BF.3 Identify the effect on the graph of replacing $f(x)$ by $f(x) + k$, $k\,f(x)$, $f(kx)$, and $f(x + k)$ for specific values of k (both positive and negative); find the value of k given the graphs. Experiment with cases and illustrate an explanation of the effects on the graph using technology.
Mathematical Practices
5 Use appropriate tools strategically.

Activity 1 k in $y = \sin \theta + k$

Graph $y = \sin \theta$, $y = \sin \theta + 2$, and $y = \sin \theta - 3$ on the same coordinate plane. Describe any similarities and differences among the graphs.

Set the viewing window to match the window shown at the right. Let **Y1** $= \sin \theta$, **Y2** $= \sin \theta + 2$, and **Y3** $= \sin \theta - 3$.

KEYSTROKES: [Y=] [SIN] [X,T,θ,n] [)] [ENTER]
[SIN] [X,T,θ,n] [)] [+] 2 [ENTER]
[SIN] [X,T,θ,n] [)] [−] 3 [GRAPH]

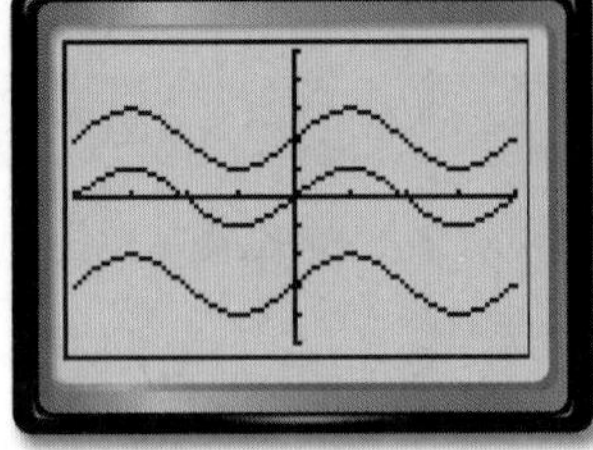

[−360, 360] scl: 90 by [−5, 5] scl: 1

The graphs have the same shape, but different vertical positions.

Activity 2 h in $y = \sin (\theta - h)$

Graph $y = \sin \theta$, $y = \sin (\theta + 45°)$, and $y = \sin (\theta - 90°)$ on the same coordinate plane. Describe any similarities and differences among the graphs.

Let **Y1** $= \sin \theta$, **Y2** $= \sin (\theta + 45)$, and **Y3** $= \sin (\theta - 90)$.
Be sure to clear the entries from Activity 1.

KEYSTROKES: [Y=] [SIN] [X,T,θ,n] [)] [ENTER]
[SIN] [X,T,θ,n] [+] 45 [)] [ENTER]
[SIN] [X,T,θ,n] [−] 90 [)] [GRAPH]

[−360, 360] scl: 90 by [−5, 5] scl: 1

The graphs have the same shape, but different horizontal positions.

Model and Analyze

Repeat the activities for the cosine and tangent functions.

1. What are the domain and range of the functions in Activities 1 and 2?
2. What is the effect of adding a constant to a trigonometric function?
3. What is the effect of adding a constant to θ in a trigonometric function?

Repeat the activities for each of the following. Describe the relationship between each pair of graphs.

4. $y = \sin \theta + 4$
   $y = \sin (2\theta) + 4$
5. $y = \cos \left(\frac{1}{2}\theta\right)$
   $y = \cos \frac{1}{2}(\theta + 45°)$
6. $y = 2 \sin \theta$
   $y = 2 \sin \theta - 1$
7. $y = \cos \theta - 3$
   $y = \cos (\theta - 90°) - 3$
8. Write a general equation for the sine, cosine, and tangent functions after changes in amplitude a, period b, horizontal position h, and vertical position k.

LESSON 12-8

Translations of Trigonometric Graphs

::Then	::Now	::Why?
● You translated exponential functions.	● 1 Graph horizontal translations of trigonometric graphs and find phase shifts. 2 Graph vertical translations of trigonometric graphs.	● The graphs at the right represent the waves in a bay during high and low tides. Notice that the shape of the waves does not change.

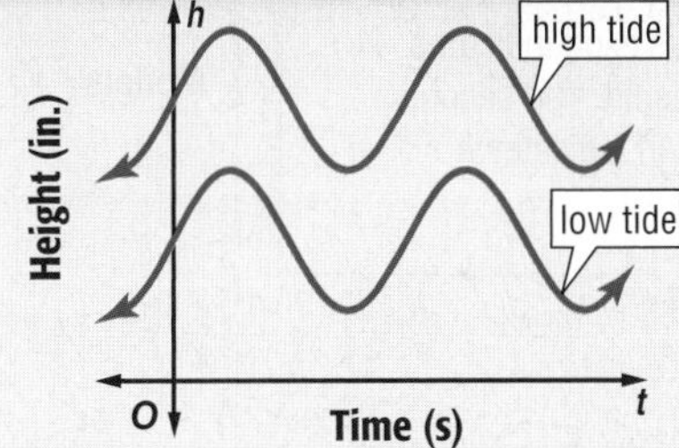

NewVocabulary
phase shift
vertical shift
midline

Common Core State Standards

Content Standards

F.IF.7.e Graph exponential and logarithmic functions, showing intercepts and end behavior, and trigonometric functions, showing period, midline, and amplitude.

F.BF.3 Identify the effect on the graph of replacing $f(x)$ by $f(x) + k$, $k\,f(x)$, $f(kx)$, and $f(x + k)$ for specific values of k (both positive and negative); find the value of k given the graphs. Experiment with cases and illustrate an explanation of the effects on the graph using technology.

Mathematical Practices

4 Model with mathematics.

1 Horizontal Translations

Recall that a *translation* occurs when a figure is moved from one location to another on the coordinate plane without changing its orientation. A horizontal translation of a periodic function is called a **phase shift**.

KeyConcept Phase Shift

Words The phase shift of the functions $y = a \sin b(\theta - h)$, $y = a \cos b(\theta - h)$, and $y = a \tan b(\theta - h)$ is h, where $b > 0$.

Models

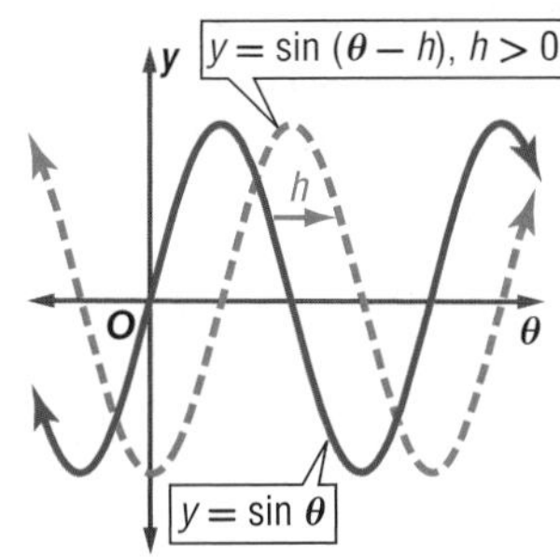

If $h > 0$, the shift is h units to the right.

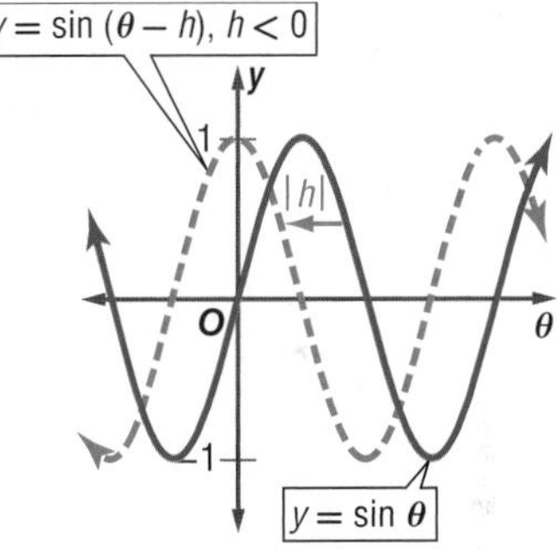

If $h < 0$, the shift is $|h|$ units to the left.

Examples

$y = \cos(\theta - 90°)$ The phase shift is 90° to the right.

$y = \tan(\theta + 30°)$ The phase shift is 30° to the left.

The secant, cosecant, and cotangent can be graphed using the same rules.

Example 1 Graph Phase Shifts

State the amplitude, period, and phase shift for $y = \sin(\theta - 90°)$. Then graph the function.

amplitude: $a = 1$

period: $\frac{360°}{|b|} = \frac{360°}{1}$ or 360°

phase shift: $h = 90°$

Graph $y = \sin\theta$ shifted 90° to the right.

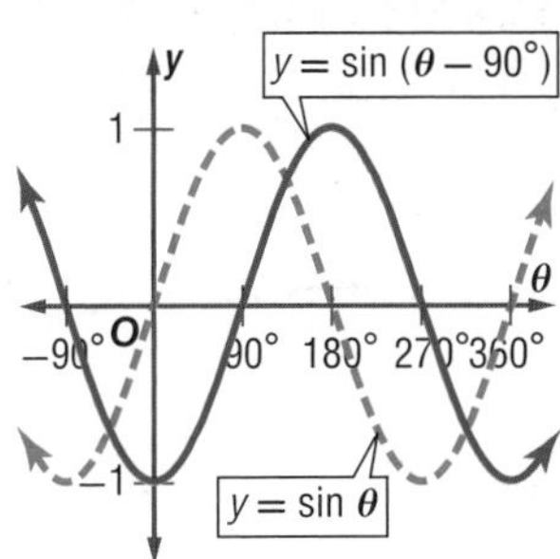

GuidedPractice

1. State the amplitude, period, and phase shift for $y = 2\cos(\theta + 45°)$. Then graph the function.

2 Vertical Translations

Recall that the graph of $y = x^2 + 5$ is the graph of the parent function $y = x^2$ shifted up 5 units. Similarly, graphs of trigonometric functions can be translated vertically through a **vertical shift**.

StudyTip

Notation Note that $\sin(\theta + x) \neq \sin \theta + x$. The first expression indicates a phase shift. The second expression indicates a vertical shift.

KeyConcept Vertical Shift

Words The vertical shift of the functions $y = a \sin b\theta + k$, $y = a \cos b\theta + k$, and $y = a \tan b\theta + k$ is k.

Models

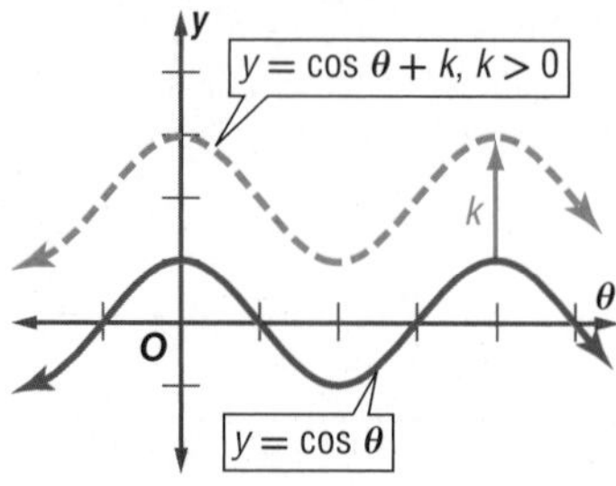

If $k > 0$, the shift is k units up.

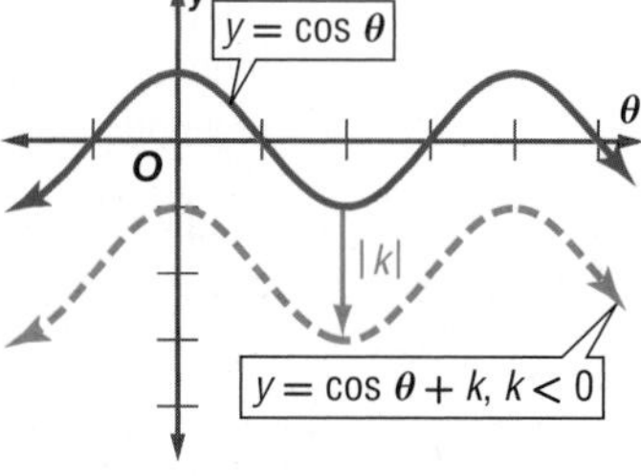

If $k < 0$, the shift is $|k|$ units down.

Examples

$y = \sin \theta + 4$ The vertical shift is 4 units up.

$y = \tan \theta - 3$ The vertical shift is 3 units down.

The secant, cosecant, and cotangent can be graphed using the same rules.

When a trigonometric function is shifted vertically k units, the line $y = k$ is the new horizontal axis about which the graph oscillates. This line is called the **midline**, and it can be used to help draw vertical translations.

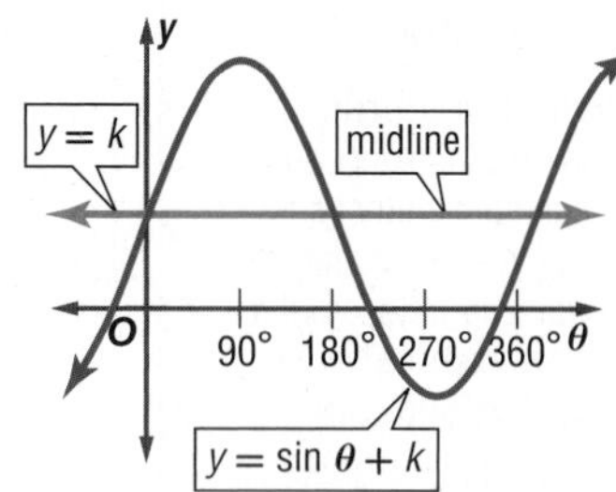

StudyTip

Using Color It may be helpful to first graph the parent function in one color. Next, apply the vertical shift and graph the function in another color. Then apply the change in amplitude and graph the function in the final color.

Example 2 Graph Vertical Translations

State the amplitude, period, vertical shift, and equation of the midline for $y = \frac{1}{2} \cos \theta - 2$. Then graph the function.

amplitude: $|a| = \frac{1}{2}$

period: $\frac{2\pi}{|b|} = \frac{2\pi}{|1|}$ or 2π

vertical shift: $k = -2$

midline: $y = -2$

To graph $y = \frac{1}{2} \cos \theta - 2$, first draw the midline. Then use it to graph $y = \frac{1}{2} \cos \theta$ shifted 2 units down.

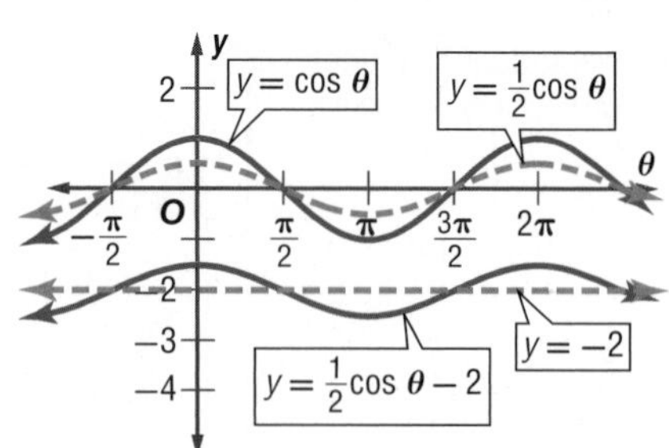

GuidedPractice

2. State the amplitude, period, vertical shift, and equation of the midline for $y = \tan \theta + 3$. Then graph the function.

You can use the following steps to graph trigonometric functions involving phase shifts and vertical shifts.

ConceptSummary Graph Trigonometric Functions

$$y = a \sin b(\theta - h) + k$$

amplitude: a; period: b; phase shift: h; vertical shift: k

Step 1 Determine the vertical shift, and graph the midline.

Step 2 Determine the amplitude, if it exists. Use dashed lines to indicate the maximum and minimum values of the function.

Step 3 Determine the period of the function, and graph the appropriate function.

Step 4 Determine the phase shift, and translate the graph accordingly.

Example 3 Graph Transformations

State the amplitude, period, phase shift, and vertical shift for $y = 3 \sin \frac{2}{3}(\theta - \pi) + 4$. Then graph the function.

amplitude: $|a| = 3$

period: $\frac{2\pi}{|b|} = \frac{2\pi}{\left|\frac{2}{3}\right|}$ or 3π — The period indicates that the graph will be stretched.

phase shift: $h = \pi$ — The graph will shift π to the right.

vertical shift: $k = 4$ — The graph will shift 4 units up.

midline: $y = 4$ — The graph will oscillate around the line $y = 4$.

Step 1 Graph the midline.

Step 2 Since the amplitude is 3, draw dashed lines 3 units above and 3 units below the midline.

Step 3 Graph $y = 3 \sin \frac{2}{3}\theta + 4$ using the midline as a reference.

Step 4 Shift the graph π units to the right.

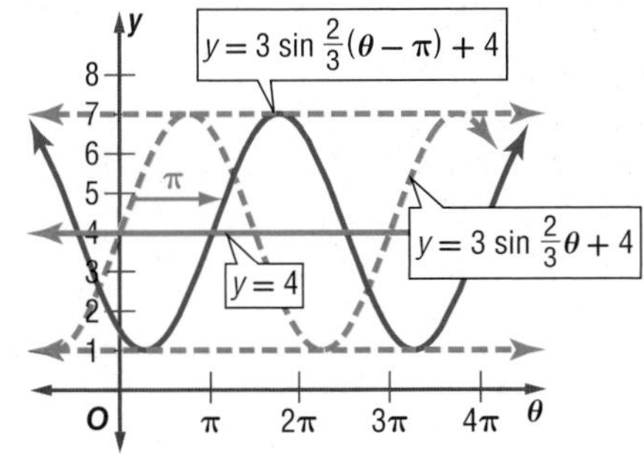

CHECK You can check the accuracy of your transformation by evaluating the function for various values of θ and confirming their location on the graph.

StudyTip

Verifying a Graph After drawing the graph of a trigonometric function, select values of θ and evaluate them in the equation to verify your graph.

GuidedPractice

3. State the amplitude, period, phase shift, and vertical shift for $y = 2 \cos \frac{1}{2}\left(\theta + \frac{\pi}{2}\right) - 2$. Then graph the function.

The sine wave occurs often in physics, signal processing, music, electrical engineering, and many other fields.

Real-WorldLink

In some wave pools, surfers can ride waves up to 70 meters.

Source: Orlando Wave Pool

Real-World Example 4 Represent Periodic Functions

WAVE POOL The height of water in a wave pool oscillates between a maximum of 13 feet and a minimum of 5 feet. The wave generator pumps 6 waves per minute. Write a sine function that represents the height of the water at time *t* seconds. Then graph the function.

Step 1 Write the equation for the midline, and determine the vertical shift.

$y = \frac{13 + 5}{2}$ or 9 — The midline lies halfway between the maximum and minimum values.

Since the midline is $y = 9$, the vertical shift is $k = 9$.

Step 2 Find the amplitude.

$|a| = |13 - 9|$ or 4 — Find the difference between the midline value and the maximum value.

So, $a = 4$.

Step 3 Find the period.

Since there are 6 waves per minute, there is 1 wave every 10 seconds. So, the period is 10 seconds.

$10 = \frac{2\pi}{|b|}$ — period $= \frac{2\pi}{|b|}$

$|b| = \frac{2\pi}{10}$ — Solve for $|b|$.

$b = \pm\frac{\pi}{5}$ — Simplify.

WatchOut!

Parent Functions Often the graph of a trigonometric function can be represented by more than one equation. For example, the graphs of $y = \cos \theta$ and $y = \sin (\theta + 90°)$ are the same.

Step 4 Write an equation for the function.

$h = a \sin b(t - h) + k$ — Write the equation for sine relating height h and time t.

$= 4 \sin \frac{\pi}{5}(t - 0) + 9$ — Substitution: $a = 4$, $b = \frac{\pi}{5}$, $h = 0$, $k = 9$

$= 4 \sin \frac{\pi}{5}t + 9$ — Simplify.

Then graph the function.

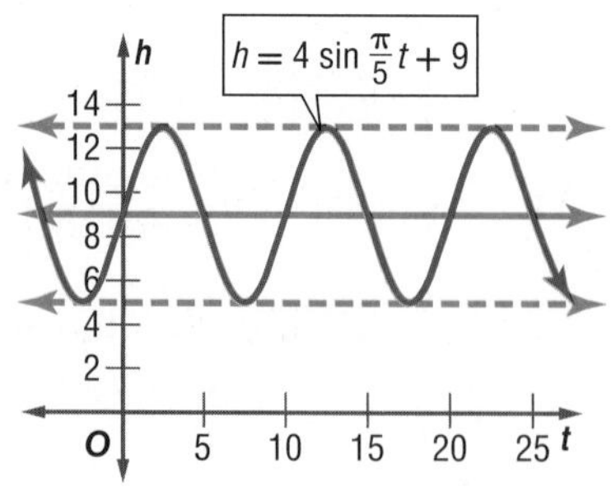

GuidedPractice

4. **WAVE POOL** The height of water in a wave pool oscillates between a maximum of 14 feet and a minimum of 6 feet. The wave generator pumps 5 waves per minute. Write a cosine function that represents the height of water at time *t* seconds. Then graph the function.

Stephen Frink/Photographer's Choice RF/Getty Images

Check Your Understanding

= Step-by-Step Solutions begin on page R14.

Example 1 **State the amplitude, period, and phase shift for each function. Then graph the function.**

1. $y = \sin(\theta - 180°)$
2. $y = \tan\left(\theta - \frac{\pi}{4}\right)$
3. $y = \sin\left(\theta - \frac{\pi}{2}\right)$
4. $y = \frac{1}{2}\cos(\theta + 90°)$

Example 2 **State the amplitude, period, vertical shift, and equation of the midline for each function. Then graph the function.**

5. $y = \cos\theta + 4$
6. $y = \sin\theta - 2$
7. $y = \frac{1}{2}\tan\theta + 1$
8. $y = \sec\theta - 5$

Example 3 **CCSS REGULARITY** **State the amplitude, period, phase shift, and vertical shift for each function. Then graph the function.**

9. $y = 2\sin(\theta + 45°) + 1$
10. $y = \cos 3(\theta - \pi) - 4$
11. $y = \frac{1}{4}\tan 2(\theta + 30°) + 3$
12. $y = 4\sin\frac{1}{2}\left(\theta - \frac{\pi}{2}\right) + 5$

Example 4

13. **EXERCISE** While doing some moderate physical activity, a person's blood pressure oscillates between a maximum of 130 and a minimum of 90. The person's heart rate is 90 beats per minute. Write a sine function that represents the person's blood pressure P at time t seconds. Then graph the function.

Practice and Problem Solving

Extra Practice is on page R12.

Example 1 **State the amplitude, period, and phase shift for each function. Then graph the function.**

14. $y = \cos(\theta + 180°)$
15. $y = \tan(\theta - 90°)$
16. $y = \sin(\theta + \pi)$
17. $y = 2\sin\left(\theta + \frac{\pi}{2}\right)$
18. $y = \tan\frac{1}{2}(\theta + 30°)$
19. $y = 3\cos\left(\theta - \frac{\pi}{3}\right)$

Example 2 **State the amplitude, period, vertical shift, and equation of the midline for each function. Then graph the function.**

20. $y = \cos\theta + 3$
21. $y = \tan\theta - 1$
22. $y = \tan\theta + \frac{1}{2}$
23. $y = 2\cos\theta - 5$
24. $y = 2\sin\theta - 4$
25. $y = \frac{1}{3}\sin\theta + 7$

Example 3 **State the amplitude, period, phase shift, and vertical shift for each function. Then graph the function.**

26. $y = 4\sin(\theta - 60°) - 1$
27. $y = \cos\frac{1}{2}(\theta - 90°) + 2$
28. $y = \tan(\theta + 30°) - 2$
29. $y = 2\tan 2\left(\theta + \frac{\pi}{4}\right) - 5$
30. $y = \frac{1}{2}\sin\left(\theta - \frac{\pi}{2}\right) + 4$
31. $y = \cos 3(\theta - 45°) + \frac{1}{2}$
32. $y = 3 + 5\sin 2(\theta - \pi)$
33. $y = -2 + 3\sin\frac{1}{3}\left(\theta - \frac{\pi}{2}\right)$

Example 4

34. **TIDES** The height of the water in a harbor rose to a maximum height of 15 feet at 6:00 P.M. and then dropped to a minimum level of 3 feet by 3:00 A.M. The water level can be modeled by the sine function. Write an equation that represents the height h of the water t hours after noon on the first day.

35. LAKES A buoy marking the swimming area in a lake oscillates each time a speed boat goes by. Its distance d in feet from the bottom of the lake is given by $d = 1.8 \sin \frac{3\pi}{4}t + 12$, where t is the time in seconds. Graph the function. Describe the minimum and maximum distances of the buoy from the bottom of the lake when a boat passes by.

36. FERRIS WHEEL Suppose a Ferris wheel has a diameter of approximately 520 feet and makes one complete revolution in 30 minutes. Suppose the lowest car on the Ferris wheel is 5 feet from the ground. Let the height at the top of the wheel represent the height at time 0. Write an equation for the height of a car h as a function of time t minutes. Then graph the function.

Write an equation for each translation.

37. $y = \sin x$, 4 units to the right and 3 units up

38. $y = \cos x$, 5 units to the left and 2 units down

39. $y = \tan x$, π units to the right and 2.5 units up

40. JUMP ROPE The graph at the right approximates the height of a jump rope h in inches as a function of time t in seconds. A maximum point on the graph is (1.25, 68), and a minimum point is (2.75, 2).

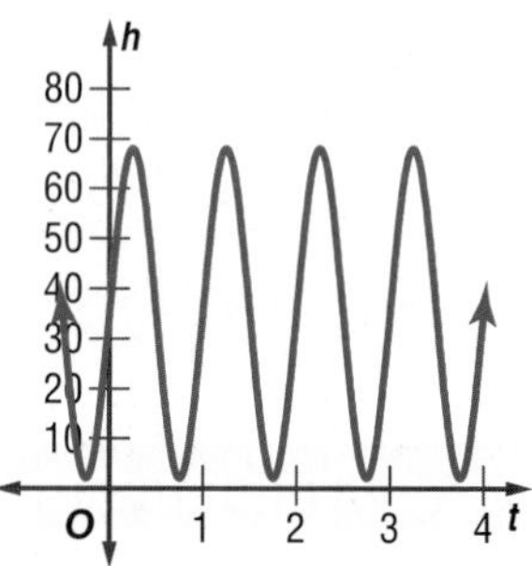

a. Describe what the maximum and minimum points mean in the context of the situation.

b. What is the equation for the midline, the amplitude, and the period of the function?

c. Write an equation for the function.

41. CAROUSEL A horse on a carousel goes up and down 3 times as the carousel makes one complete rotation. The maximum height of the horse is 55 inches, and the minimum height is 37 inches. The carousel rotates once every 21 seconds. Assume that the horse starts and stops at its median height.

a. Write an equation to represent the height of the horse h as a function of time t seconds.

b. Graph the function.

c. Use your graph to estimate the height of the horse after 8 seconds. Then use a calculator to find the height to the nearest tenth.

42. CCSS REASONING During one month, the outside temperature fluctuates between 40°F and 50°F. A cosine curve approximates the change in temperature, with a high of 50°F being reached every four days.

a. Describe the amplitude, period, and midline of the function that approximates the temperature y on day d.

b. Write a cosine function to estimate the temperature y on day d.

c. Sketch a graph of the function.

d. Estimate the temperature on the 7th day of the month.

Find a coordinate that represents a maximum for each graph.

43. $y = -2 \cos \left(x - \frac{\pi}{2}\right)$

44. $y = 4 \sin \left(x + \frac{\pi}{3}\right)$

45. $y = 3 \tan \left(x + \frac{\pi}{2}\right) + 2$

46. $y = -3 \sin \left(x - \frac{\pi}{4}\right) - 4$

Compare each pair of graphs.

47. $y = -\cos 3\theta$ and $y = \sin 3(\theta - 90°)$

48. $y = 2 + 0.5 \tan \theta$ and $y = 2 + 0.5 \tan (\theta + \pi)$

49. $y = 2 \sin \left(\theta - \frac{\pi}{6}\right)$ and $y = -2 \sin \left(\theta + \frac{5\pi}{6}\right)$

Identify the period of each function. Then write an equation for the graph using the given trigonometric function.

50. sine

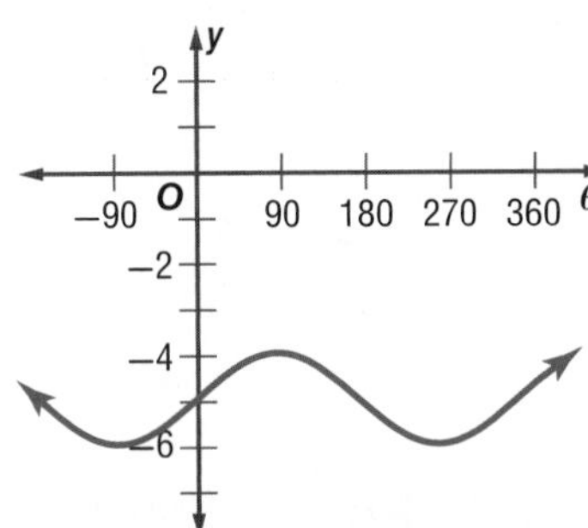

51. cosine

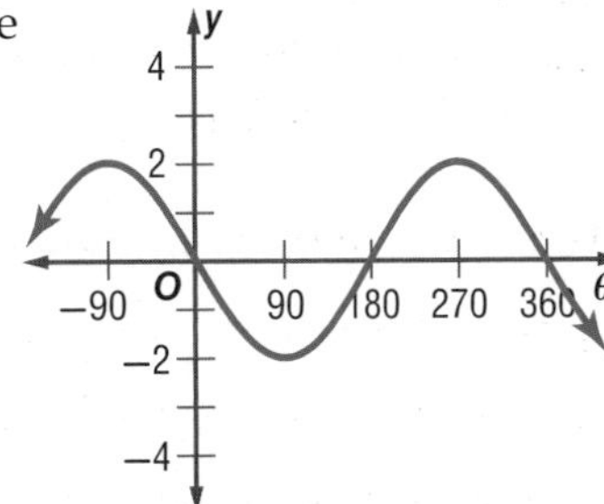

52. cosine

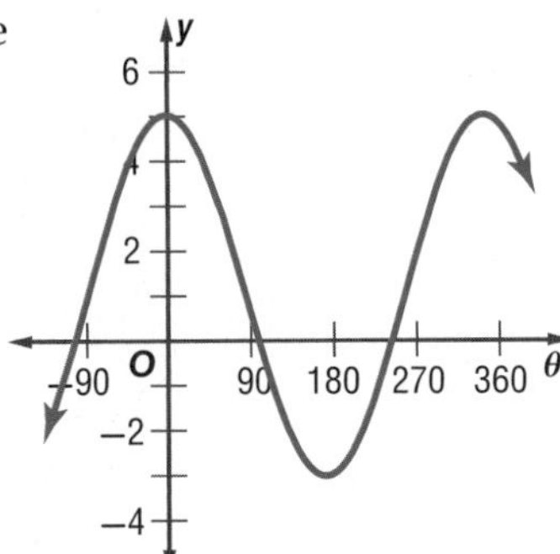

53 sine

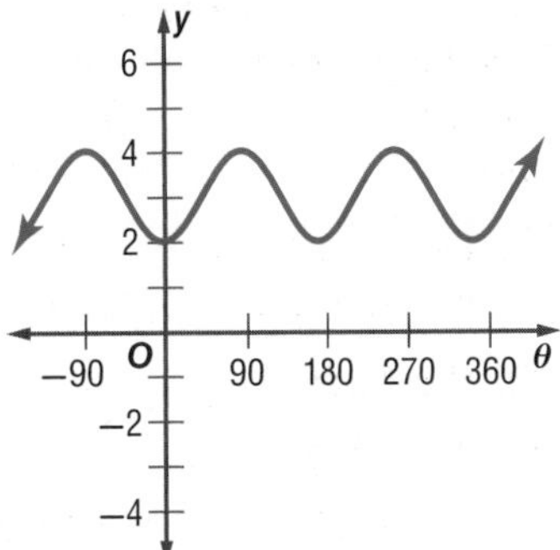

State the period, phase shift, and vertical shift. Then graph the function.

54. $y = \csc (\theta + \pi)$

55. $y = \cot \theta + 6$

56. $y = \cot \left(\theta - \frac{\pi}{6}\right) - 2$

57. $y = \frac{1}{2} \csc 3(\theta - 45°) + 1$

58. $y = 2 \sec \frac{1}{2}(\theta - 90°)$

59. $y = 4 \sec 2\left(\theta + \frac{\pi}{2}\right) - 3$

H.O.T. Problems Use Higher-Order Thinking Skills

60. CCSS **ARGUMENTS** If you are given the amplitude and period of a cosine function, is it *sometimes*, *always*, or *never* possible to find the maximum and minimum values of the function? Explain your reasoning.

61. **REASONING** Describe how the graph of $y = 3 \sin 2\theta + 1$ is different from $y = \sin \theta$.

62. **WRITING IN MATH** Describe two different phase shifts that will translate the sine curve onto the cosine curve shown at the right. Then write an equation for the new sine curve using each phase shift.

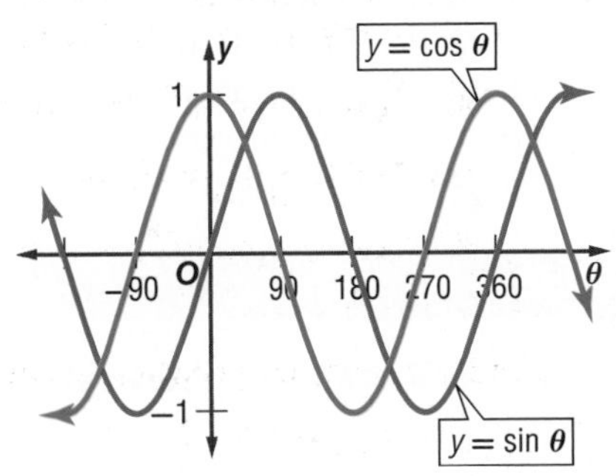

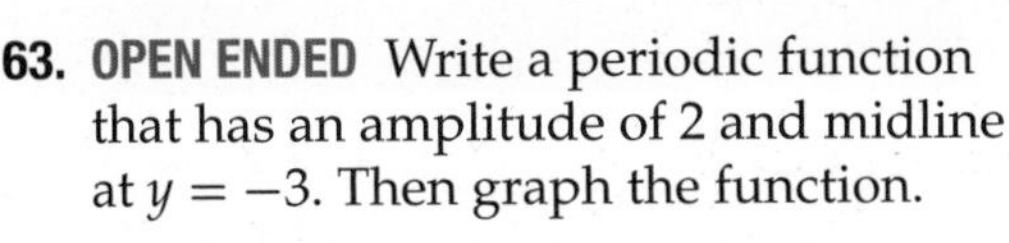

63. **OPEN ENDED** Write a periodic function that has an amplitude of 2 and midline at $y = -3$. Then graph the function.

64. **REASONING** How many different sine graphs pass through the origin $(n\pi, 0)$? Explain your reasoning.

Standardized Test Practice

65. GRIDDED RESPONSE The expression $\frac{3x-1}{4} + \frac{x+6}{4}$ is how much greater than x?

66. Expand $(a - b)^4$.

A $a^4 - b^4$

B $a^4 - 4ab + b^4$

C $a^4 + 4a^3b + 6a^2b^2 + 4ab^3 + b^4$

D $a^4 - 4a^3b + 6a^2b^2 - 4ab^3 + b^4$

67. Solve $\sqrt{x-3} + \sqrt{x+2} = 5$.

F 7

G 0, 7

H 7, 13

J no solution

68. GEOMETRY Using the figures below, what is the average of a, b, c, d, and f?

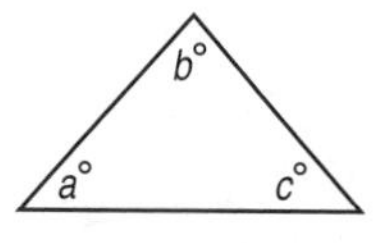

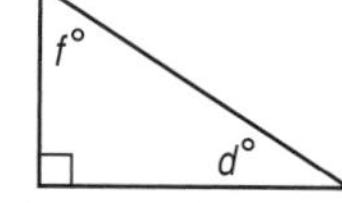

A 21 B 45 C 50 D 54

Spiral Review

Find the amplitude and period of each function. Then graph the function. (Lesson 12-7)

69. $y = 2 \cos \theta$

70. $y = 3 \sin \theta$

71. $y = \sin 2\theta$

Find the exact value of each expression. (Lesson 12-6)

72. $\sin \frac{4\pi}{3}$

73. $\sin (-30°)$

74. $\cos 405°$

Determine whether each situation describes a *survey*, an *experiment*, or an *observational study*. Then identify the sample, and suggest a population from which it may have been selected. (Lesson 11-1)

75. A group of 220 adults is randomly split into two groups. One group exercises for an hour a day and the other group does not. The body mass indexes are then compared.

76. A soccer coach randomly selects some of his players and gives them a questionnaire asking about their daily sleeping habits.

77. A teacher randomly selects 100 students who have part-time jobs and compares their grades.

78. GEOMETRY Equilateral triangle ABC has a perimeter of 39 centimeters. If the midpoints of the sides are connected, a smaller equilateral triangle results. Suppose the process of connecting midpoints of sides and drawing new triangles is continued indefinitely. (Lesson 10-4)

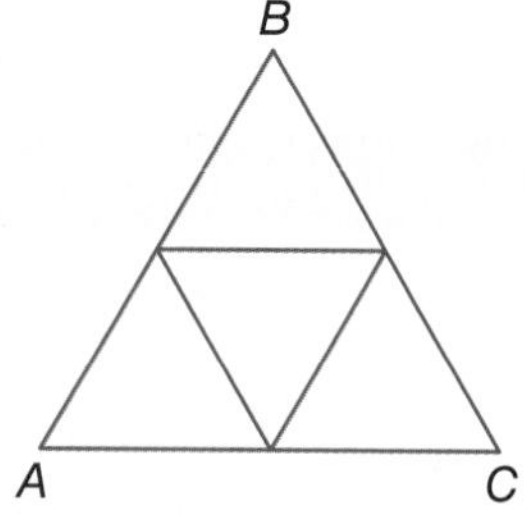

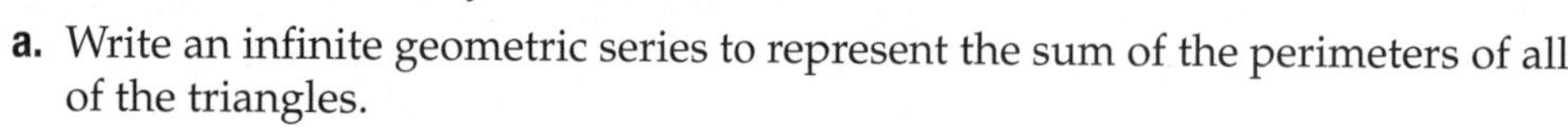

a. Write an infinite geometric series to represent the sum of the perimeters of all of the triangles.

b. Find the sum of the perimeters of all of the triangles.

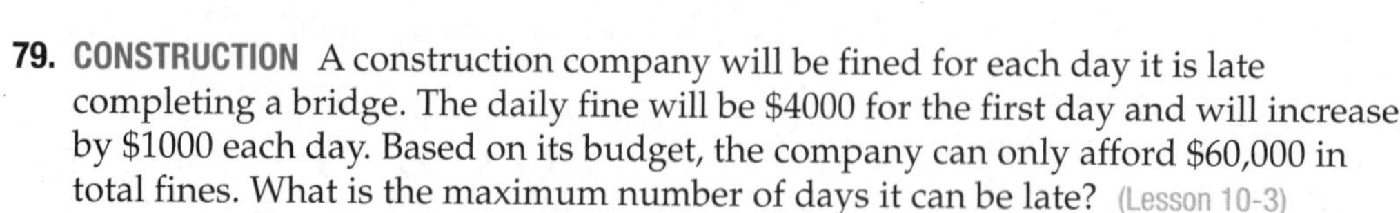

79. CONSTRUCTION A construction company will be fined for each day it is late completing a bridge. The daily fine will be $4000 for the first day and will increase by $1000 each day. Based on its budget, the company can only afford $60,000 in total fines. What is the maximum number of days it can be late? (Lesson 10-3)

Skills Review

Find each value of θ. Round to the nearest degree.

80. $\sin \theta = \frac{7}{8}$

81. $\tan \theta = \frac{9}{10}$

82. $\cos \theta = \frac{1}{4}$

83. $\cos \theta = \frac{4}{5}$

84. $\sin \theta = \frac{5}{6}$

85. $\tan \theta = \frac{2}{7}$

LESSON

12-9 Inverse Trigonometric Functions

Then	Now	Why?
You graphed trigonometric functions.	**1** Find values of inverse trigonometric functions. **2** Solve equations by using inverse trigonometric functions.	The leaning bookshelf at the right is 15 inches from the wall and reaches a height of 75 inches. In Lesson 13-1, you learned how to use the inverse of a trigonometric function to find the measure of acute angle θ. $\tan \theta = \frac{15}{75}$ or 0.2 Use the tangent function. Find an angle that has a tangent of 0.2. 2nd [TAN^{-1}] .2 ENTER 11.30993247 So, the measure of θ is about 11°.

NewVocabulary
principal values
Arcsine function
Arccosine function
Arctangent function

Common Core State Standards

Content Standards
A.CED.2 Create equations in two or more variables to represent relationships between quantities; graph equations on coordinate axes with labels and scales.

Mathematical Practices
7 Look for and make use of structure.

1 Inverse Trigonometric Functions

If you know the value of a trigonometric function for an angle, you can use the *inverse* to find the angle. Recall that an inverse function is the relation in which all values of x and y are reversed. The inverse of $y = \sin x$, $x = \sin y$, is graphed at the right.

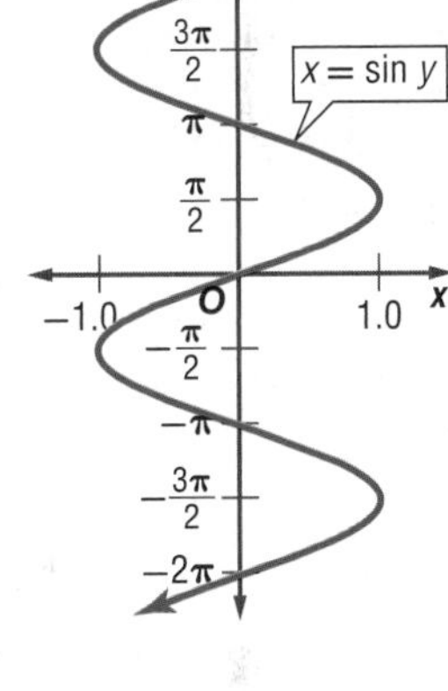

Notice that the inverse is not a function because there are many values of y for each value of x. If you restrict the domain of the sine function so that $-\frac{\pi}{2} \leq x \leq \frac{\pi}{2}$, then the inverse is a function.

The values in this restricted domain are called **principal values**. Trigonometric functions with restricted domains are indicated with capital letters.

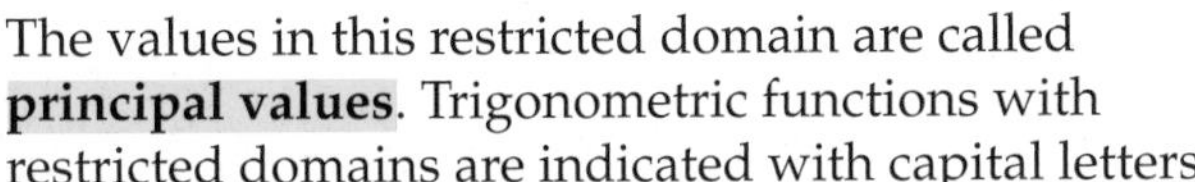

- $y = \text{Sin } x$ if and only if $y = \sin x$ and $-\frac{\pi}{2} \leq x \leq \frac{\pi}{2}$.
- $y = \text{Cos } x$ if and only if $y = \cos x$ and $0 \leq x \leq \pi$.

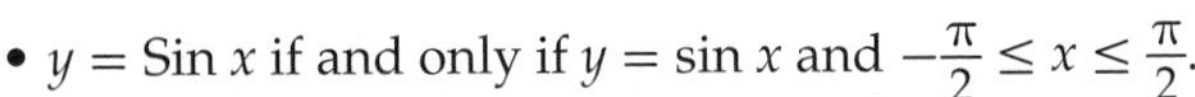

- $y = \text{Tan } x$ if and only if $y = \tan x$ and $-\frac{\pi}{2} \leq x \leq \frac{\pi}{2}$.

You can use functions with restricted domains to define inverse trigonometric functions. The inverses of the sine, cosine, and tangent functions are the **Arcsine**, **Arccosine**, and **Arctangent** functions, respectively.

KeyConcept Inverse Trigonometric Functions

Inverse Function	Symbols	Domain	Range	Model
Arcsine	$y = \text{Arcsin } x$ $y = \text{Sin}^{-1} x$	$-1 \leq x \leq 1$	$-\frac{\pi}{2} \leq y \leq \frac{\pi}{2}$ $-90° \leq y \leq 90°$	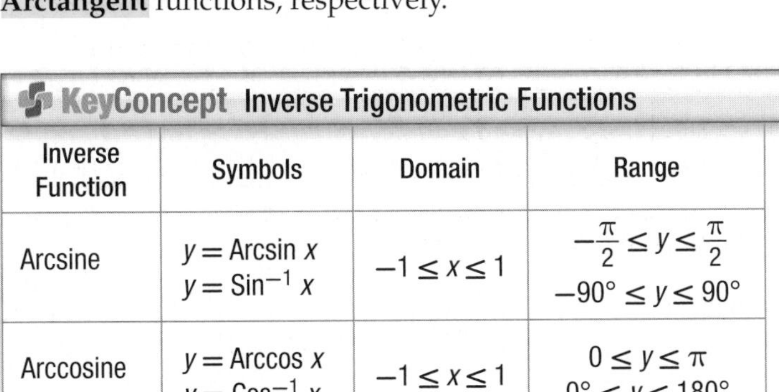
Arccosine	$y = \text{Arccos } x$ $y = \text{Cos}^{-1} x$	$-1 \leq x \leq 1$	$0 \leq y \leq \pi$ $0° \leq y \leq 180°$	
Arctangent	$y = \text{Arctan } x$ $y = \text{Tan}^{-1} x$	all real numbers	$-\frac{\pi}{2} \leq y \leq \frac{\pi}{2}$ $-90° \leq y \leq 90°$	

ReviewVocabulary

inverse functions If f and f^{-1} are inverse functions, then $f(a) = b$ if and only if $f^{-1}(b) = a$.

In the relation $y = \cos^{-1} x$, if $x = \frac{1}{2}$, $y = 60°$, $300°$, and all angles that are coterminal with those angles. In the function $y = \text{Cos}^{-1} x$, if $x = \frac{1}{2}$, $y = 60°$ only.

Example 1 Evaluate Inverse Trigonometric Functions

Find each value. Write angle measures in degrees and radians.

a. $\text{Cos}^{-1}\left(-\frac{1}{2}\right)$

Find the angle θ for $0° \leq \theta \leq 180°$ that has a cosine value of $-\frac{1}{2}$.

Method 1 Use a unit circle.

Find a point on the unit circle that has an x-coordinate of $-\frac{1}{2}$.

When $\theta = 120°$, $\cos \theta = -\frac{1}{2}$.

So, $\text{Cos}^{-1}\left(-\frac{1}{2}\right) = 120°$ or $\frac{2\pi}{3}$.

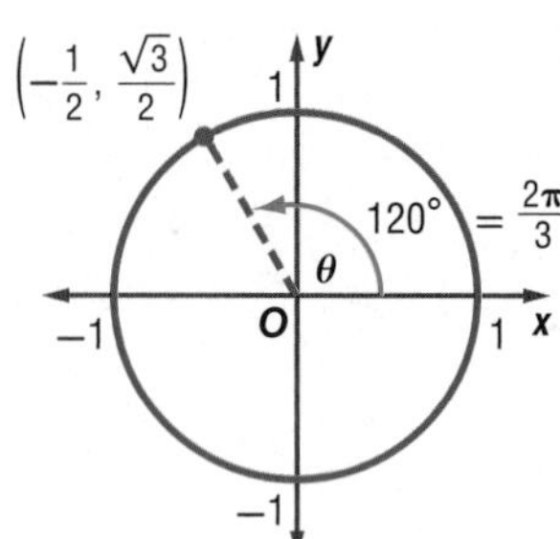

Method 2 Use a calculator.

KEYSTROKES: [2nd] [COS^{-1}] [(−)] 1 [÷] 2 [)] [ENTER] 120

Therefore, $\text{Cos}^{-1}\left(-\frac{1}{2}\right) = 120°$ or $\frac{2\pi}{3}$.

StudyTip

Angle Measure Remember that when evaluating an inverse trigonometric function, the result is an angle measure.

b. Arctan 1

Find the angle θ for $-90° \leq \theta \leq 90°$ that has a tangent value of 1.

KEYSTROKES: [2nd] [TAN^{-1}] 1 [ENTER] 45 Therefore, Arctan $1 = 45°$ or $\frac{\pi}{4}$.

GuidedPractice

1A. $\text{Cos}^{-1} 0$

1B. Arcsin $\left(-\frac{\sqrt{2}}{2}\right)$

When finding a value when there are multiple trigonometric functions involved, use the order of operations to solve.

Example 2 Find a Trigonometric Value

Find $\tan\left(\text{Cos}^{-1} \frac{1}{2}\right)$. Round to the nearest hundredth.

Use a calculator.

KEYSTROKES: [TAN] [2nd] [COS^{-1}] 1 [÷] 2 [)] [)] [ENTER] 1.732050808

So, $\tan\left(\text{Cos}^{-1} \frac{1}{2}\right) \approx 1.73$.

CHECK $\text{Cos}^{-1} \frac{1}{2} = 60°$ and $\tan 60° \approx 1.73$. So, the answer is correct.

GuidedPractice

Find each value. Round to the nearest hundredth.

2A. $\sin\left(\text{Tan}^{-1} \frac{3}{8}\right)$

2B. $\cos\left(\text{Arccos} -\frac{\sqrt{2}}{2}\right)$

2 Solve Equations by Using Inverses

You can rewrite trigonometric equations to solve for the measure of an angle.

Standardized Test Example 3 Find an Angle Measure

If Sin $\theta = -0.35$, find θ.

A $-20.5°$ **B** $-0.6°$ **C** $0.6°$ **D** $20.5°$

Read the Test Item

The sine of angle θ is -0.35. This can be written as Arcsin $(-0.35) = \theta$.

Solve the Test Item

Use a calculator.

KEYSTROKES: 2nd [SIN^{-1}] (–) .35 ENTER -20.48731511

So, $\theta \approx -20.5°$. The answer is A.

Test-Taking Tip

Eliminate Possibilities The function Sin restricts the possible angle measures to Quadrants I or IV. Because -0.35 is negative, look for an angle measure in Quadrant IV.

Guided Practice

3. If Tan $\theta = 1.8$, find θ.

F $0.03°$ **G** $29.1°$ **H** $60.9°$ **J** no solution

Inverse trigonometric functions can be used to determine angles of inclination, depression, and elevation.

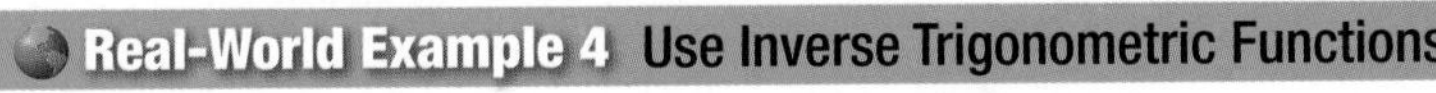

Real-World Example 4 Use Inverse Trigonometric Functions

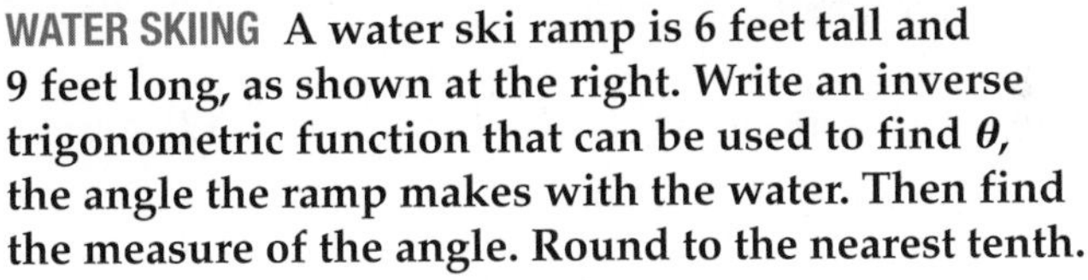

WATER SKIING A water ski ramp is 6 feet tall and 9 feet long, as shown at the right. Write an inverse trigonometric function that can be used to find θ, the angle the ramp makes with the water. Then find the measure of the angle. Round to the nearest tenth.

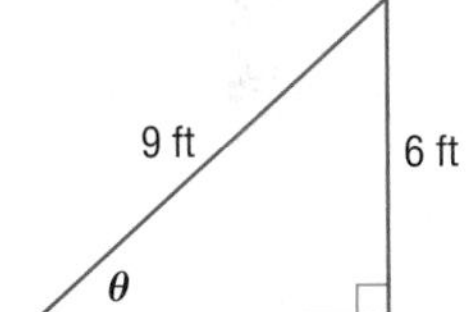

Because the measures of the opposite side and the hypotenuse are known, you can use the sine function.

$\sin \theta = \frac{6}{9}$ Sine function

$\theta = \text{Sin}^{-1} \frac{6}{9}$ Inverse sine function

$\theta \approx 41.8°$ Use a calculator.

So, the angle of the ramp is about $41.8°$.

CHECK Using your calculator, $\sin 41.8 \approx 0.66653 \approx \frac{6}{9}$. So, the answer is correct.

Guided Practice

4. SKIING A ski trail is shown at the right. Write an inverse trigonometric function that can be used to find θ, the angle the trail makes with the ground in the valley. Then find the angle. Round to the nearest tenth.

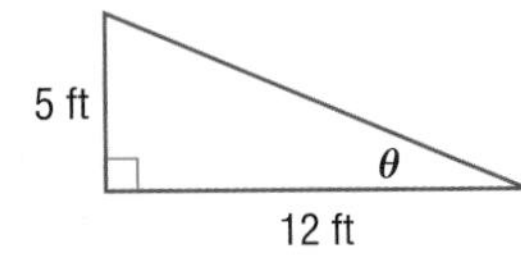

Real-World Career

Sport Science Administrator A sport science administrator provides sport science information to players, coaches, and parents. He or she implements testing, training, and treatment programs for athletes. A master's degree in sport science or a related area is recommended.

Aaron Josefczyk/Icon SMI/Corbis

Check Your Understanding

= Step-by-Step Solutions begin on page R14.

Example 1 **Find each value. Write angle measures in degrees and radians.**

1. $\text{Sin}^{-1} \frac{1}{2}$
2. $\text{Arctan} (-\sqrt{3})$
3. $\text{Arccos} (-1)$

Example 2 **Find each value. Round to the nearest hundredth if necessary.**

4. $\cos \left(\text{Arcsin} \frac{4}{5}\right)$
5. $\tan (\text{Cos}^{-1} 1)$
6. $\sin \left(\text{Sin}^{-1} \frac{\sqrt{3}}{2}\right)$

Example 3

7. **MULTIPLE CHOICE** If $\text{Sin}\ \theta = 0.422$, find θ.

A 25°　　B 42°　　C 48°　　D 65°

Solve each equation. Round to the nearest tenth if necessary.

8. $\text{Cos}\ \theta = 0.9$
9. $\text{Sin}\ \theta = -0.46$
10. $\text{Tan}\ \theta = 2.1$

Example 4

11. **SNOWBOARDING** A cross section of a superpipe for snowboarders is shown at the right. Write an inverse trigonometric function that can be used to find θ, the angle that describes the steepness of the superpipe. Then find the angle to the nearest degree.

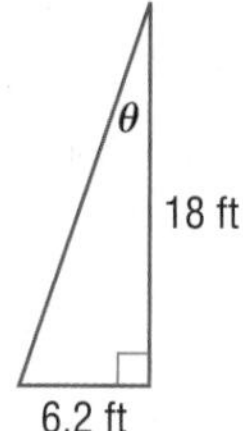

Practice and Problem Solving

Extra Practice is on page R12.

Example 1 **Find each value. Write angle measures in degrees and radians.**

12. $\text{Arcsin} \left(\frac{\sqrt{3}}{2}\right)$
13. $\text{Arccos} \left(\frac{\sqrt{3}}{2}\right)$
14. $\text{Sin}^{-1} (-1)$
15. $\text{Tan}^{-1} \sqrt{3}$
16. $\text{Cos}^{-1} \left(-\frac{\sqrt{3}}{2}\right)$
17. $\text{Arctan} \left(-\frac{\sqrt{3}}{3}\right)$

Example 2 **Find each value. Round to the nearest hundredth if necessary.**

18. $\tan (\text{Cos}^{-1} 1)$
19. $\tan \left[\text{Arcsin} \left(-\frac{1}{2}\right)\right]$
20. $\cos \left(\text{Tan}^{-1} \frac{3}{5}\right)$
21. $\sin \left(\text{Arctan} \sqrt{3}\right)$
22. $\cos \left(\text{Sin}^{-1} \frac{4}{9}\right)$
23. $\sin \left[\text{Cos}^{-1} \left(-\frac{\sqrt{2}}{2}\right)\right]$

Example 3 **Solve each equation. Round to the nearest tenth if necessary.**

24. $\text{Tan}\ \theta = 3.8$
25. $\text{Sin}\ \theta = 0.9$
26. $\text{Sin}\ \theta = -2.5$
27. $\text{Cos}\ \theta = -0.25$
28. $\text{Cos}\ \theta = 0.56$
29. $\text{Tan}\ \theta = -0.2$

Example 4

30. CCSS **SENSE-MAKING** A boat is traveling west to cross a river that is 190 meters wide. Because of the current, the boat lands at point Q, which is 59 meters from its original destination point P. Write an inverse trigonometric function that can be used to find θ, the angle at which the boat veered south of the horizontal line. Then find the measure of the angle to the nearest tenth.

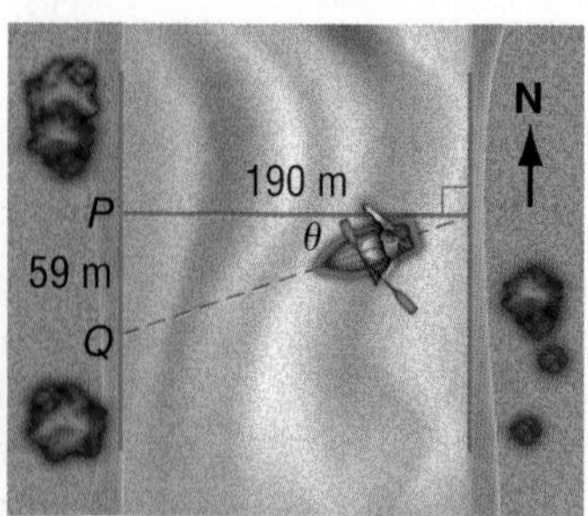

31. TREES A 24-foot tree is leaning 2.5 feet left of vertical, as shown in the figure. Write an inverse trigonometric function that can be used to find θ, the angle at which the tree is leaning. Then find the measure of the angle to the nearest degree.

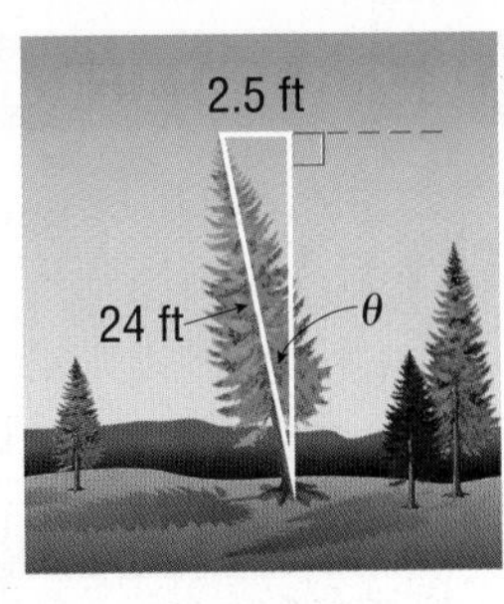

32. DRIVING An expressway off-ramp curve has a radius of 52 meters and is designed for vehicles to safely travel at speeds up to 45 kilometers per hour (or 12.5 meters per second). The equation below represents the angle θ of the curve. What is the measure of the angle to the nearest degree?

$$\tan \theta = \frac{(12.5 \text{ m/s})^2}{(52 \text{ m})(9.8 \text{ m/s}^2)}$$

33. TRACK AND FIELD A shot-putter throws the shot with an initial speed of 15 meters per second. The expression $\frac{15 \text{ m/s} (\sin x)}{9.8 \text{ m/s}^2}$ represents the time in seconds at which the shot reached its maximum height. In the expression, x is the angle at which the shot was thrown. If the maximum height of the shot was reached in 1.0 second, at what angle was it thrown? Round to the nearest tenth.

Solve each equation for $0 \leq \theta \leq 2\pi$.

34. $\csc \theta = 1$ **35.** $\sec \theta = -1$ **36.** $\sec \theta = 1$

37. $\csc \theta = \frac{1}{2}$ **38.** $\cot \theta = 1$ **39.** $\sec \theta = 2$

40. MULTIPLE REPRESENTATIONS Consider $y = \text{Cos}^{-1} x$.

a. **Graphical** Sketch a graph of the function. Describe the domain and the range.

b. **Symbolic** Write the function using different notation.

c. **Numerical** Choose a value for x between -1 and 0. Then evaluate the inverse cosine function. Round to the nearest tenth.

d. **Analytical** Compare the graphs of $y = \cos x$ and $y = \text{Cos}^{-1} x$.

H.O.T. Problems Use Higher-Order Thinking Skills

41. CHALLENGE Determine whether *cos (Arccos x) = x for all values of x is true* or *false*. If false, give a counterexample.

42. CCSS CRITIQUE Desiree and Oscar are solving $\cos \theta = 0.3$ where $90 < \theta < 180$. Is either of them correct? Explain your reasoning.

Desiree	Oscar
$\cos \theta = 0.3$	$\cos \theta = 0.3$
$\cos^{-1} 0.3 = 162.5°$	$\cos^{-1} 0.3 = 72.5°$

43. REASONING Explain how the domain of $y = \text{Sin}^{-1} x$ is related to the range of $y = \text{Sin}\ x$.

44. OPEN ENDED Write an equation with an Arcsine function and an equation with a Sine function that both involve the same angle measure.

45. WRITING IN MATH Compare and contrast the relations $y = \tan^{-1} x$ and $y = \text{Tan}^{-1} x$. Include information about the domains and ranges.

46. REASONING Explain how $\text{Sin}^{-1} 8$ and $\text{Cos}^{-1} 8$ are undefined while $\text{Tan}^{-1} 8$ is defined.

Standardized Test Practice

47. Simplify $\dfrac{\frac{2}{x}+2}{\frac{2}{x}-2}$.

A $\dfrac{1+x}{1-x}$

B $\dfrac{2}{x}$

C $\dfrac{1-x}{1+x}$

D $-x$

48. SHORT RESPONSE What is the equation of the graph below?

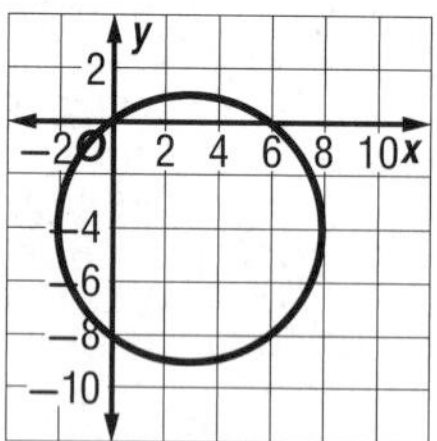

49. If $f(x) = 2x^2 - 3x$ and $g(x) = 4 - 2x$, what is $g[f(x)]$?

F $g[f(x)] = 4 + 6x - 8x^2$

G $g[f(x)] = 4 + 6x - 4x^2$

H $g[f(x)] = 20 - 26x + 8x^2$

J $g[f(x)] = 44 - 38x + 8x^2$

50. If g is a positive number, which of the following is equal to $12g$?

A $\sqrt{144g}$

B $\sqrt{12g^2}$

C $\sqrt{24g^2}$

D $6\sqrt{4g^2}$

Spiral Review

51. RIDES The Cosmoclock 21 is a huge Ferris wheel in Japan. The diameter is 328 feet. Suppose a rider enters the ride at 0 feet, and then rotates in 90° increments counterclockwise. The table shows the angle measures of rotation and the height in feet above the ground of the rider. (Lesson 12-8)

a. A function that models the data is $y = 164 \cdot [\sin (x - 90°)] + 164$. Identify the vertical shift, amplitude, period, and phase shift of the graph.

b. Write an equation using the sine that models the position of a rider on the Vienna Giant Ferris Wheel in Austria, with a diameter of 200 feet. Check by plotting the points and your equation with a graphing calculator.

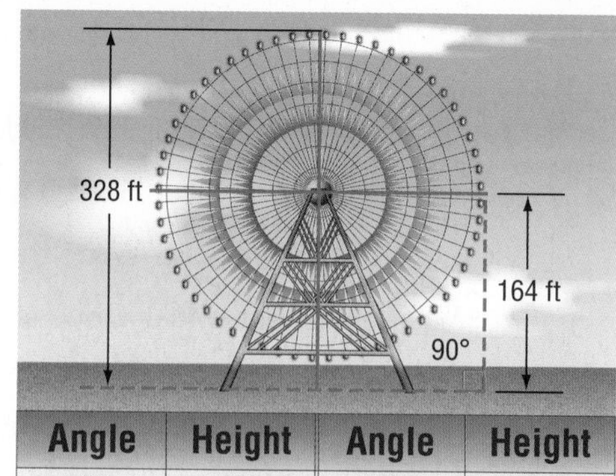

Angle	Height	Angle	Height
0°	0	450°	164
90°	164	540°	328
180°	328	630°	164
270°	164	720°	0
360°	0		

52. TIDES The world's record for the highest tide is held by the Minas Basin in Nova Scotia, Canada, with a tidal range of 54.6 feet. A tide is at equilibrium when it is at its normal level halfway between its highest and lowest points. Write an equation to represent the height h of the tide. Assume that the tide is at equilibrium at $t = 0$, that the high tide is beginning, and that the tide completes one cycle in 12 hours. (Lesson 12-7)

Solve each equation. (Lesson 7-4)

53. $\log_3 5 + \log_3 x = \log_3 10$

54. $\log_4 a + \log_4 9 = \log_4 27$

55. $\log_{10} 16 - \log_{10} 2t = \log_{10} 2$

56. $\log_7 24 - \log_7 (y + 5) = \log_3 8$

Skills Review

Find the exact value of each trigonometric function.

57. $\cos 3\pi$

58. $\tan 120°$

59. $\sin 300°$

60. $\sec \frac{7\pi}{6}$

CHAPTER 12

Study Guide and Review

Study Guide

KeyConcepts

Right Triangle Trigonometry (Lesson 12-1)

- $\sin \theta = \frac{\text{opp}}{\text{hyp}}$, $\cos \theta = \frac{\text{adj}}{\text{hyp}}$, $\tan \theta = \frac{\text{opp}}{\text{adj}}$,

  $\csc \theta = \frac{\text{hyp}}{\text{opp}}$, $\sec \theta = \frac{\text{hyp}}{\text{adj}}$, $\cot \theta = \frac{\text{adj}}{\text{opp}}$

Angle Measures and Trigonometric Functions of General Angles (Lessons 12-2 and 12-3)

- The measure of an angle is determined by the amount of rotation from the initial side to the terminal side.
- You can find the exact values of the six trigonometric functions of θ, given the coordinates of a point $P(x, y)$ on the terminal side of the angle.

Law of Sines and Law of Cosines (Lessons 12-4 and 12-5)

- $\frac{\sin A}{a} = \frac{\sin B}{b} = \frac{\sin C}{c}$
- $a^2 = b^2 + c^2 - 2bc \cos A$

  $b^2 = a^2 + c^2 - 2ac \cos B$

  $c^2 = a^2 + b^2 - 2ab \cos C$

Circular and Inverse Trigonometric Functions (Lessons 12-6 and 12-9)

- If the terminal side of an angle θ in standard position intersects the unit circle at $P(x, y)$, then $\cos \theta = x$ and $\sin \theta = y$.
- $y = \text{Sin } x$ if $y = \sin x$ and $-\frac{\pi}{2} \leq x \leq \frac{\pi}{2}$

Graphing Trigonometric Functions (Lesson 12-7)

- For trigonometric functions of the form $y = a \sin b\theta$ and $y = a \cos b\theta$, the amplitude is $|a|$, and the period is $\frac{360°}{|b|}$ or $\frac{2\pi}{b}$.
- The period of $y = a \tan b\theta$ is $\frac{180°}{|b|}$ or $\frac{\pi}{|b|}$.

FOLDABLES StudyOrganizer

Be sure the Key Concepts are noted in your Foldable.

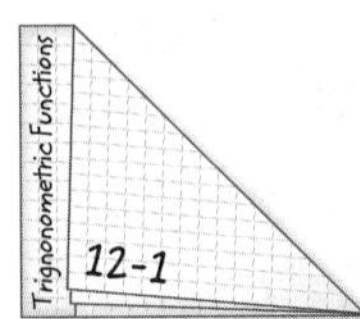

KeyVocabulary

ambiguous case (p. 816)
amplitude (p. 837)
angle of depression (p. 794)
angle of elevation (p. 794)
Arccosine function (p. 853)
Arcsine function (p. 853)
Arctangent function (p. 853)
central angle (p. 802)
circular function (p. 830)
cosecant (p. 790)
cosine (p. 790)
cotangent (p. 790)
coterminal angles (p. 800)
cycle (p. 831)
frequency (p. 838)
initial side (p. 799)
Law of Cosines (p. 823)
Law of Sines (p. 815)
midline (p. 846)
period (p. 831)
periodic function (p. 831)
phase shift (p. 845)
principal values (p. 853)
quadrantal angle (p. 808)
radian (p. 801)
reference angle (p. 808)
secant (p. 790)
sine (p. 790)
solving a triangle (p. 815)
standard position (p. 799)
tangent (p. 790)
terminal side (p. 799)
trigonometric function (p. 790)
trigonometric ratio (p. 790)
trigonometry (p. 790)
unit circle (p. 830)
vertical shift (p. 846)

VocabularyCheck

State whether each sentence is *true* or *false*. If *false*, replace the underlined term to make a true sentence.

1. The <u>Law of Cosines</u> is used to solve a triangle when two angles and any sides are known.
2. An angle on the coordinate plane is in <u>standard position</u> if the vertex is at the origin and one ray is on the positive x-axis.
3. <u>Coterminal angles</u> are angles in standard position that have the same terminal side.
4. A horizontal translation of a periodic function is called a <u>phase shift</u>.
5. The inverse of the sine function is the <u>cosecant function</u>.
6. The <u>cycle</u> of the graph of a sine or cosine function equals half the difference between the maximum and minimum values of the function.

Lesson-by-Lesson Review

12-1 Trigonometric Functions in Right Triangles

Solve $\triangle ABC$ by using the given measurements. Round measures of sides to the nearest tenth and measures of angles to the nearest degree.

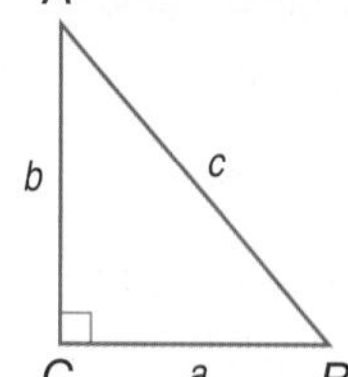

7. $c = 12, b = 5$
8. $a = 10, B = 55°$
9. $B = 75°, b = 15$
10. $B = 45°, c = 16$
11. $A = 35°, c = 22$
12. $\sin A = \frac{2}{3}, a = 6$

13. **TRUCK** The back of a moving truck is 3 feet off of the ground. What length does a ramp off the back of the truck need to be in order for the angle of elevation of the ramp to be 20°?

Example 1

Solve $\triangle ABC$ by using the given measurements. Round measures of sides to the nearest tenth and measures of angles to the nearest degree.

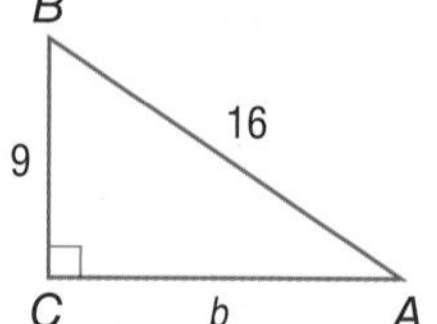

Find b. $a^2 + b^2 = c^2$

$9^2 + b^2 = 16^2$

$b = \sqrt{16^2 - 9^2}$

$b \approx 13.2$

Find A. $\sin A = \frac{9}{16}$

Use a calculator.

To the nearest degree, $A = 34°$.

Find B. $34° + B \approx 90°$

$B \approx 56°$

Therefore, $b \approx 13.2$, $A \approx 34°$, and $B \approx 56°$.

12-2 Angles and Angle Measure

Rewrite each degree measure in radians and each radian measure in degrees.

14. 215°
15. $\frac{5\pi}{2}$
16. -3π
17. $-315°$

Find one angle with positive measure and one angle with negative measure coterminal with each angle.

18. 265°
19. $-65°$
20. $\frac{7\pi}{2}$

21. **BICYCLE** A bicycle tire makes 8 revolutions in one second. The tire has a radius of 15 inches. Find the angle θ in radians through which the tire rotates in one second.

Example 2

Rewrite 160° in radians.

$160° = 160°\left(\frac{\pi \text{ radians}}{180°}\right)$

$= \frac{160\pi}{180}$ radians or $\frac{8\pi}{9}$

Example 3

Find one angle with positive measure and one angle with negative measure coterminal with 150°.

positive angle:

$150° + 360° = 510°$ Add 360°.

negative angle:

$150° - 360° = -210°$ Subtract 360°.

12-3 Trigonometric Functions of General Angles

Find the exact value of each trigonometric function.

22. cos 135°

23. tan 150°

24. sin 2π

25. $\cos \frac{3\pi}{2}$

The terminal side of θ in standard position contains each point. Find the exact values of the six trigonometric functions of θ.

26. $P(-4, 3)$

27. $P(5, 12)$

28. $P(16, -12)$

29. **BALL** A ball is thrown off the edge of a building at an angle of 70° and with an initial velocity of 5 meters per second. The equation that represents the horizontal distance of the ball x is $x = v_0 (\cos \theta)t$, where v_0 is the initial velocity, θ is the angle at which it is thrown, and t is the time in seconds. About how far will the ball travel in 10 seconds?

Example 4

Find the exact value of sin 120°.

Because the terminal side of 120° lies in Quadrant II, the reference angle θ' is 180° − 120° or 60°. The sine function is positive in Quadrant II, so $\sin 120° = \sin 60°$ or $\frac{\sqrt{3}}{2}$.

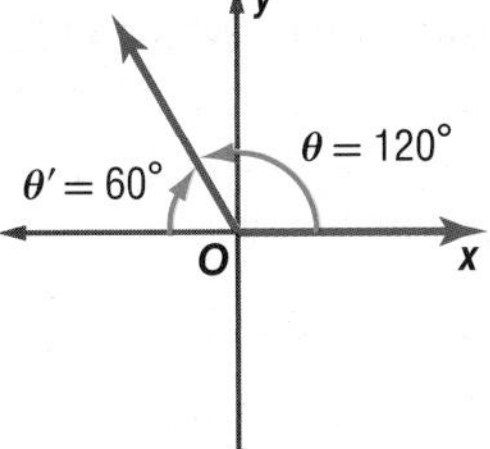

Example 5

The terminal side of θ in standard position contains the point (6, 5). Find the exact values of the six trigonometric functions of θ.

$\sin \theta = \frac{y}{r}$ or $\frac{5\sqrt{61}}{61}$

$\cos \theta = \frac{x}{r}$ or $\frac{6\sqrt{61}}{61}$

$\tan \theta = \frac{y}{x}$ or $\frac{5}{6}$

$\csc \theta = \frac{r}{y}$ or $\frac{\sqrt{61}}{5}$

$\sec \theta = \frac{r}{x}$ or $\frac{\sqrt{61}}{6}$

$\cot \theta = \frac{x}{y}$ or $\frac{6}{5}$

12-4 Law of Sines

Determine whether each triangle has *no* solution, *one* solution, or *two* solutions. Then solve each triangle. Round measures of sides to the nearest tenth and measures of angles to the nearest degree.

30. $C = 118°, c = 10, a = 4$

31. $A = 25°, a = 15, c = 18$

32. $A = 70°, a = 5, c = 16$

33. **BOAT** Kira and Mallory are standing on opposite sides of a river. How far is Kira from the boat? Round to the nearest tenth if necessary.

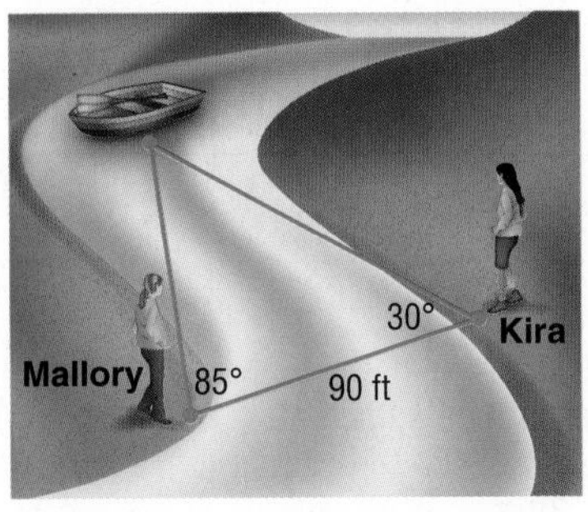

Example 6

Solve $\triangle ABC$.

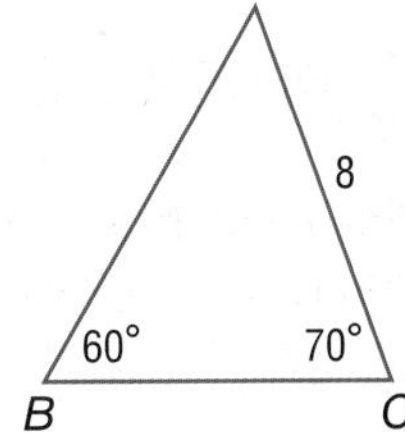

First, find the measure of the third angle.

$$60° + 70° + a = 180°$$
$$A = 50°$$

Now use the Law of Sines to find a and c. Write two equations, each with one variable.

$$\frac{\sin B}{b} = \frac{\sin C}{c} \qquad \frac{\sin B}{b} = \frac{\sin A}{a}$$

$$\frac{\sin 60°}{8} = \frac{\sin 70°}{c} \qquad \frac{\sin 60°}{8} = \frac{\sin 50°}{a}$$

$$c = \frac{8 \sin 70°}{\sin 60°} \qquad a = \frac{8 \sin 50°}{\sin 60°}$$

$$c \approx 8.7 \qquad a \approx 7.1$$

Therefore, $A = 50°$, $c \approx 8.7$, and $a \approx 7.1$.

CHAPTER 12 Study Guide and Review *Continued*

12-5 Law of Cosines

Determine whether each triangle should be solved by beginning with the Law of *Sines* or Law of *Cosines*. Then solve each triangle. Round measures of sides to the nearest tenth and measures of angles to the nearest degree.

34.

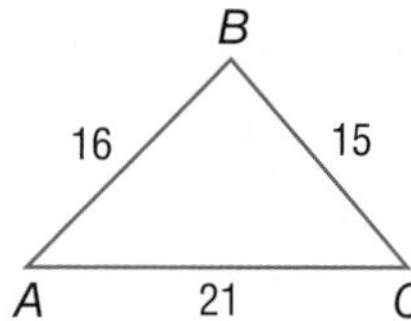

35.

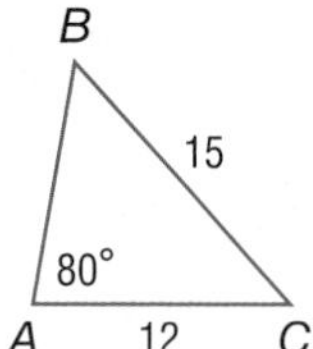

36. $C = 75°$, $a = 5$, $b = 7$

37. $A = 42°$, $a = 9$, $b = 13$

38. $b = 8.2$, $c = 15.4$, $A = 35°$

39. **FARMING** A farmer wants to fence a piece of his land. Two sides of the triangular field have lengths of 120 feet and 325 feet. The measure of the angle between those sides is 70°. How much fencing will the farmer need?

Example 7

Solve $\triangle ABC$ for $C = 55°$, $b = 11$, and $a = 18$.

You are given the measure of two sides and the included angle. Begin by drawing a diagram and using the Law of Cosines to determine c.

B
18
55°
A
11
C

$c^2 = a^2 + b^2 - 2ab \cos C$

$c^2 = 182 + 112 - 2(18)(11) \cos 55°$

$c^2 \approx 217.9$

$c \approx 14.8$

Next, you can use the Law of Sines to find the measure of angle A.

$\frac{\sin A}{18} \approx \frac{\sin 55°}{14.8}$

$\sin A \approx \frac{18 \sin 55°}{14.8}$ or A is about 85.0°

The measure of the angle B is approximately $180 - (85.0 + 55)$ or 40.0°.

Therefore, $c \approx 14.8$, $A \approx 85.0°$, and $B \approx 40.0°$.

12-6 Circular and Periodic Functions

Find the exact value of each function.

40. $\cos(-210°)$

41. $(\cos 45°)(\cos 210°)$

42. $\sin -\frac{7\pi}{4}$

43. $\left(\cos \frac{\pi}{2}\right)\left(\sin \frac{\pi}{2}\right)$

44. Determine the period of the function.

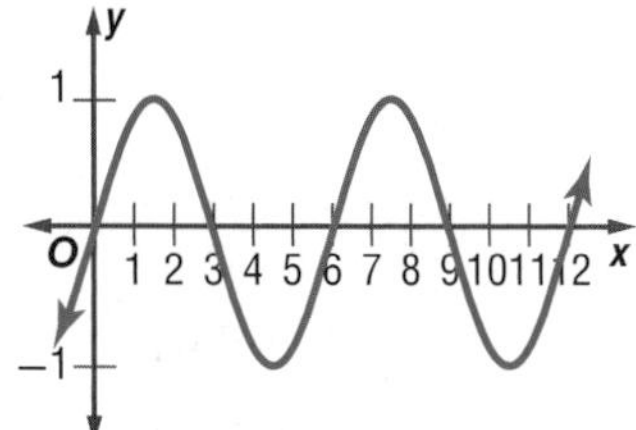

45. A wheel with a diameter of 18 inches completes 4 revolutions in 1 minute. What is the period of the function that describes the height of one spot on the outside edge of the wheel as a function of time?

Example 8

Find the exact value of sin 510°.

$\sin 510° = \sin(360° + 150°)$

$= \sin 150°$

$= \frac{1}{2}$

Example 9

Determine the period of the function below.

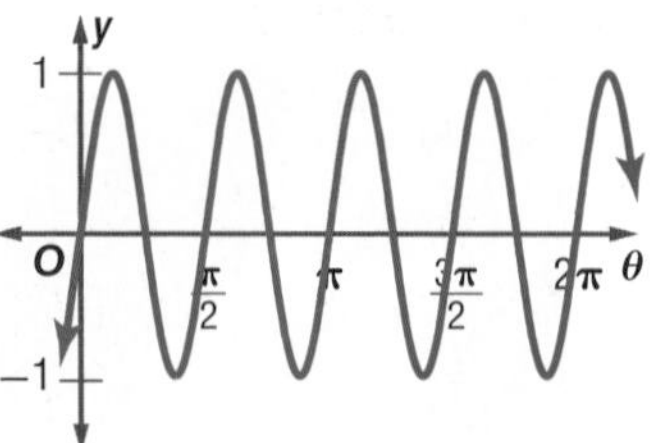

The pattern repeats itself at $\frac{\pi}{2}$, π, and so on. So, the period is $\frac{\pi}{2}$.

12-7 Graphing Trigonometric Functions

Find the amplitude, if it exists, and period of each function. Then graph the function.

46. $y = 4 \sin 2\theta$

47. $y = \cos \frac{1}{2}\theta$

48. $y = 3 \csc \theta$

49. $y = 3 \sec \theta$

50. $y = \tan 2\theta$

51. $y = 2 \csc \frac{1}{2}\theta$

52. When Lauren jumps on a trampoline it vibrates with a frequency of 10 hertz. Let the amplitude equal 5 feet. Write a sine equation to represent the vibration of the trampoline y as a function of time t.

Example 10

Find the amplitude and period of $y = 2 \cos 4\theta$. Then graph the function.

amplitude: $|a| = |2|$ or 2. The graph is stretched vertically so that the maximum value is 2 and the minimum value is −2.

period:
$\frac{360°}{|b|} = \frac{360°}{|4|}$ or 90°

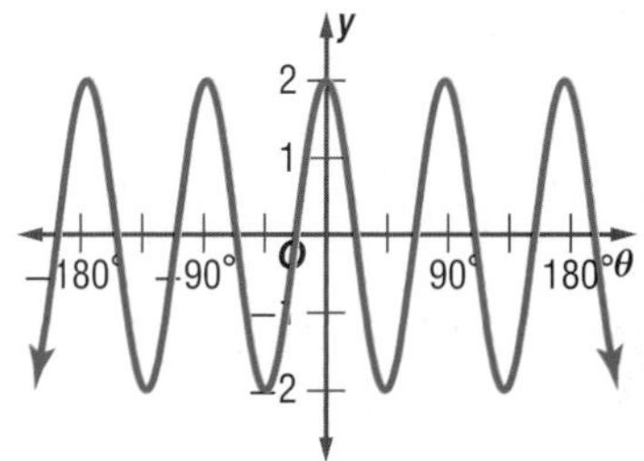

12-8 Translations of Trigonometric Graphs

State the vertical shift, amplitude, period, and phase shift of each function. Then graph the function.

53. $y = 3 \sin [2(\theta - 90°)] + 1$

54. $y = \frac{1}{2} \tan [2(\theta - 30°)] - 3$

55. $y = 2 \sec \left[3\left(\theta - \frac{\pi}{2}\right)\right] + 2$

56. $y = \frac{1}{2} \cos \left[\frac{1}{4}\left(\theta + \frac{\pi}{4}\right)\right] - 1$

57. $y = \frac{1}{3} \sin \left[\frac{1}{3}(\theta - 90°)\right] + 2$

58. The graph below approximates the height y of a rope that two people are twirling as a function of time t in seconds. Write an equation for the function.

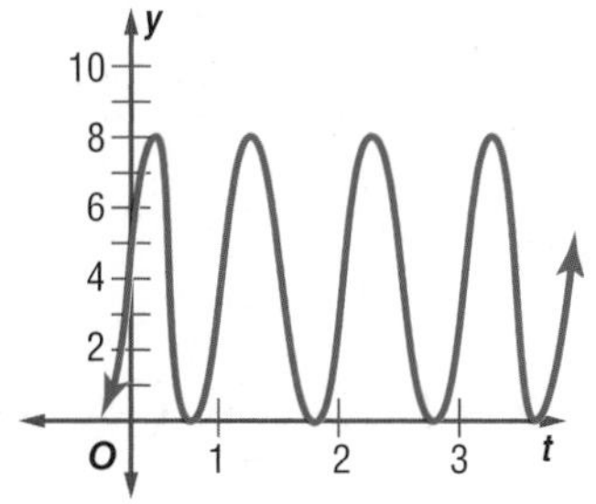

Example 11

State the vertical shift, amplitude, period, and phase shift of $y = 2 \sin \left[3\left(\theta + \frac{\pi}{2}\right)\right] + 4$. Then graph the function.

Identify the values of k, a, b, and h.

$k = 4$, so the vertical shift is 4.

$a = 2$, so the amplitude is 2.

$b = 3$, so the period is $\frac{2\pi}{|3|}$ or $\frac{2\pi}{3}$.

$h = -\frac{\pi}{2}$, so the phase shift is $\frac{\pi}{2}$ to the left.

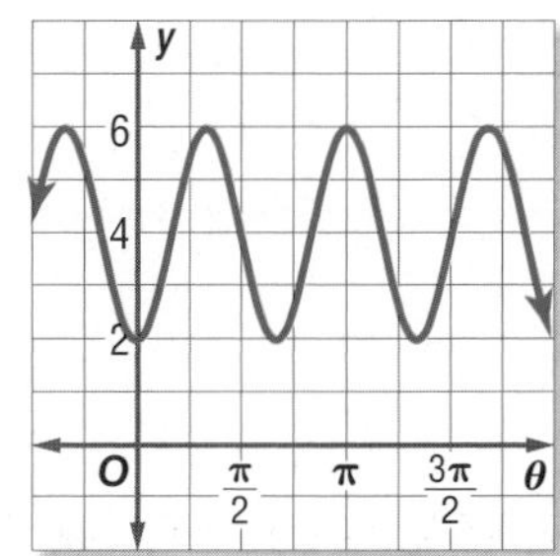

12-9 Inverse Trigonometric Functions

Evaluate each inverse trigonometric function. Write angle measures in degrees and radians.

59. $\text{Sin}^{-1}(1)$

60. Arctan (0)

61. Arcsin $\frac{\sqrt{3}}{2}$

62. $\text{Cos}^{-1}\frac{\sqrt{2}}{2}$

63. $\text{Tan}^{-1} 1$

64. Arccos 0

65. **RAMPS** A bicycle ramp is 5 feet tall and 10 feet long, as shown below. Write an inverse trigonometric function that can be used to find θ, the angle the ramp makes with the ground. Then find the angle.

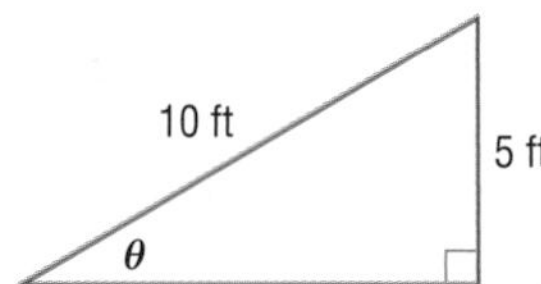

Evaluate each inverse trigonometric function. Round to the nearest hundredth if necessary.

66. $\tan\left(\text{Cos}^{-1}\frac{1}{3}\right)$

67. $\sin\left(\text{Arcsin} -\frac{\sqrt{2}}{2}\right)$

68. $\sin(\text{Tan}^{-1} 0)$

Solve each equation. Round to the nearest tenth if necessary.

69. $\text{Tan}\ \theta = -1.43$

70. $\text{Sin}\ \theta = 0.8$

71. $\text{Cos}\ \theta = 0.41$

Example 12

Evaluate $\text{Cos}^{-1}\frac{1}{2}$. Write angle measures in degrees and radians.

Find the angle θ for $0° \leq \theta \leq 180°$ that has a cosine value of $\frac{1}{2}$.

Use a unit circle.

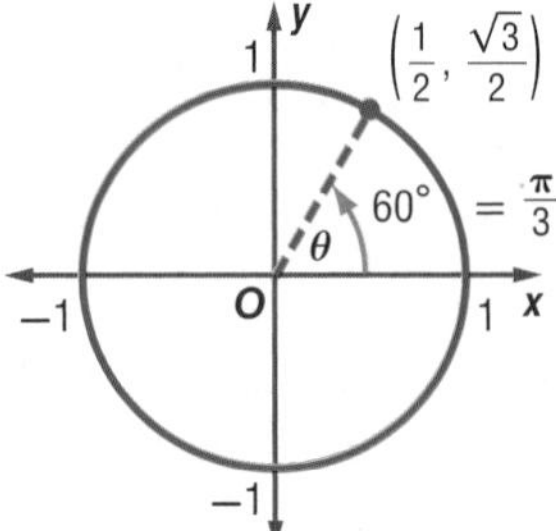

Find a point on the unit circle that has an x-coordinate of $\frac{1}{2}$.

When $\theta = 60°$, $\text{Cos}\ \theta = \frac{1}{2}$.

So, $\text{Cos}^{-1} = 60°$ or $\frac{\pi}{3}$.

Example 13

Evaluate $\sin\left(\text{Tan}^{-1}\frac{1}{2}\right)$. Round to the nearest hundredth.

Use a calculator.

KEYSTROKES: SIN 2nd [TAN^{-1}] 1 ÷ 2)) ENTER 0.4472135955

So, $\sin\left(\text{Tan}^{-1}\frac{1}{2}\right) \approx 0.45$.

Example 14

If $\text{Cos}\ \theta = 0.72$, find θ.

Use a calculator.

KEYSTROKES: 2nd [COS^{-1}] .72) ENTER 43.9455195623

So, $\theta \approx 43.9°$.

CHAPTER 12

Practice Test

Solve $\triangle ABC$ by using the given measurements. Round measures of sides to the nearest tenth and measures of angles to the nearest degree.

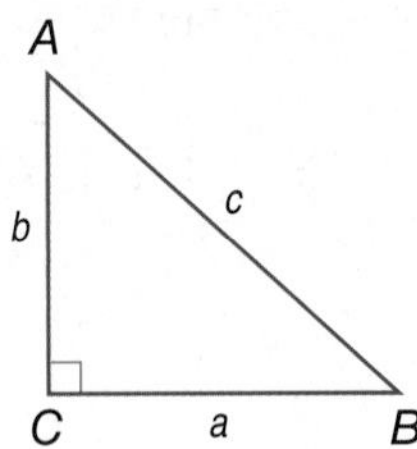

1. $A = 36°, c = 9$
2. $a = 12, A = 58°$
3. $B = 85°, b = 8$
4. $a = 9, c = 12$

Rewrite each degree measure in radians and each radian measure in degrees.

5. 325°
6. −175°
7. $\frac{9\pi}{4}$
8. $-\frac{5\pi}{6}$
9. Determine whether $\triangle ABC$, with $A = 110°$, $a = 16$, and $b = 21$, has *no* solution, *one* solution, or *two* solutions. Then solve the triangle, if possible. Round measures of sides to the nearest tenth and measures of angles to the nearest degree.

Find the exact value of each function. Write angle measures in degrees.

10. $\cos(-90°)$
11. $\sin 585°$
12. $\cot \frac{4\pi}{3}$
13. $\sec\left(-\frac{9\pi}{4}\right)$
14. $\tan\left(\text{Cos}^{-1}\frac{4}{5}\right)$
15. $\text{Arccos}\,\frac{1}{2}$
16. The terminal side of angle θ in standard position intersects the unit circle at point $P\left(\frac{1}{2}, \frac{\sqrt{3}}{2}\right)$. Find $\cos\theta$ and $\sin\theta$.
17. **MULTIPLE CHOICE** What angle has a tangent and sine that are both negative?

    A 65°

    B 120°

    C 265°

    D 310°

18. **NAVIGATION** Airplanes and ships measure distance in nautical miles. The formula 1 nautical mile = $6077 - 31 \cos 2\theta$ feet, where θ is the latitude in degrees, can be used to find the approximate length of a nautical mile at a certain latitude. Find the length of a nautical mile when the latitude is 120°.

Find the amplitude and period of each function. Then graph the function.

19. $y = 2 \sin 3\theta$
20. $y = \frac{1}{2} \cos 2\theta$
21. **MULTIPLE CHOICE** What is the period of the function $y = 3 \cot \theta$?

    F 120°

    G 180°

    H 360°

    J 1080°

22. Determine whether $\triangle XYZ$, with $y = 15$, $z = 9$, and $X = 105°$, should be solved by beginning with the Law of *Sines* or Law of *Cosines*. Then solve the triangle. Round measures of sides to the nearest tenth and measures of angles to the nearest degree.

State the amplitude, period, and phase shift for each function. Then graph the function.

23. $y = \cos(\theta + 180)$
24. $y = \frac{1}{2} \tan\left(\theta - \frac{\pi}{2}\right)$
25. **WHEELS** A water wheel has a diameter of 20 feet. It makes one complete revolution in 45 seconds. Let the height at the top of the wheel represent the height at time 0. Write an equation for the height of point h in the diagram below as a function of time t. Then graph the function.

Preparing for Standardized Tests

Using a Scientific Calculator

Scientific calculators and graphing calculators are powerful problem-solving tools. As you have likely seen, some test problems that you encounter have steps or computations that require the use of a scientific calculator.

Strategies for Using a Scientific Calculator

Step 1

Familiarize yourself with the various functions of a scientific calculator as well as when they should be used.

- **Scientific notation**—for calculating large numbers
- **Logarithmic and exponential functions**—growth and decay problems, compound interest
- **Trigonometric functions**—problems involving angles, triangle problems, indirect measurement problems
- **Square roots and *n*th roots**—distance on a coordinate plane, Pythagorean Theorem

Step 2

Use your scientific or graphing calculator to solve the problem.

- Remember to work as efficiently as possible. Some steps may be done mentally or by hand, while others must be done using your calculator.
- If time permits, check your answer.

Standardized Test Example

Read the problem. Identify what you need to know. Then use the information in the problem to solve.

When Molly stands at a distance of 18 feet from the base of a tree, she forms an angle of 57° with the top of the tree. What is the height of the tree to the nearest tenth?

A 27.7 ft

B 28.5 ft

C 29.2 ft

D 30.1 ft

Read the problem carefully. You are given some measurements and asked to find the height of a tree. It may be helpful to first sketch a model of the problem.

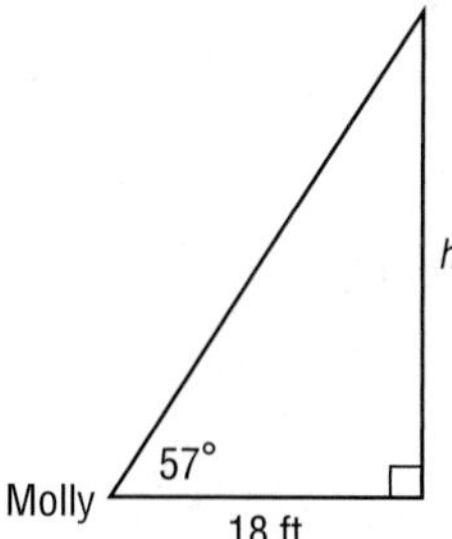

Use a trigonometric function to relate the lengths and the angle measure in the right triangle.

$\text{tangent } \theta = \frac{\text{opposite}}{\text{adjacent}}$ Definition of tangent ratio

$\tan 57° = \frac{h}{18}$ Substitute.

You need to evaluate tan 57° to solve for the height of the tree h. Use a scientific calculator.

$1.53986 \approx \frac{h}{18}$ Use a calculator.

$27.71748 \approx h$ Multiply each side by 18.

The height of the tree is about 27.7 feet. The correct answer is A.

Exercises

Read each problem. Identify what you need to know. Then use the information in the problem to solve.

1. An airplane takes off and climbs at a constant rate. After traveling 800 yards horizontally, the plane has climbed 285 yards vertically. What is the plane's angle of elevation during the takeoff and initial climb?

   A 15.6°

   B 18.4°

   C 19.6°

   D 22.3°

2. What is the angle of the bike ramp below?

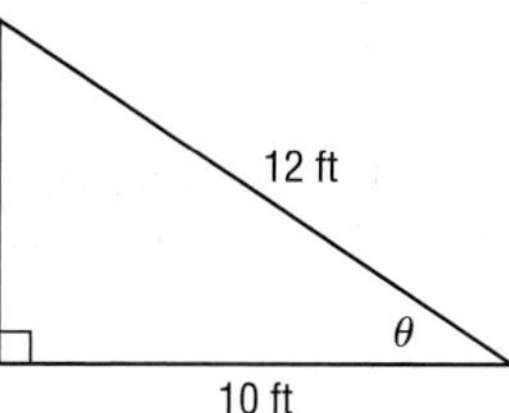

   F 26.3°

   G 28.5°

   H 30.4°

   J 33.6°

CHAPTER 12

Standardized Test Practice

Cumulative, Chapters 1 through 12

Multiple Choice

Read each question. Then fill in the correct answer on the answer document provided by your teacher or on a sheet of paper.

1. What is the value of x? Round to the nearest tenth if necessary.

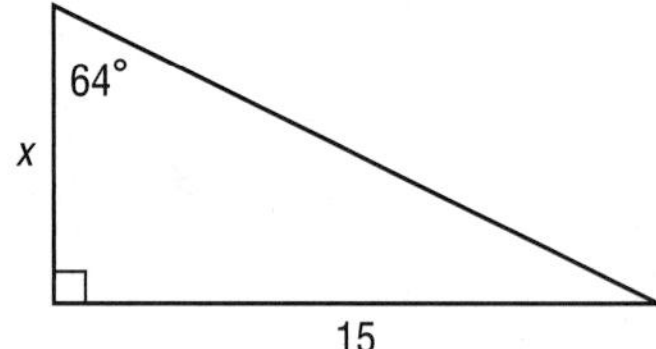

A 6.5

B 6.9

C 7.1

D 7.3

2. Marvin rides his bike at a speed of 21 miles per hour and can ride his training loop 10 times in the time that it takes his younger brother to complete the training loop 8 times. Which is a reasonable estimate for Marvin's younger brother's speed?

F between 14 mph and 15 mph

G between 15 mph and 16 mph

H between 16 mph and 17 mph

J between 17 mph and 18 mph

3. Suppose a Ferris wheel has a diameter of 68 feet. The wheel rotates 12° each time a new passenger is picked up. How far would you travel when the wheel rotates 12°? Round to the nearest tenth if necessary.

A 7.1 ft

B 7.5 ft

C 7.8 ft

D 14.2 ft

4. What is the slope of a line parallel to $y - 2 = 4(x + 1)$?

F -4

G $-\frac{1}{4}$

H $\frac{1}{4}$

J 4

5. What is the exact value of sin 240°?

A $-\frac{\sqrt{3}}{2}$

B $-\frac{1}{2}$

C $\frac{\sqrt{2}}{3}$

D $\frac{\sqrt{3}}{2}$

6. What is the solution of the system of equations shown below?

$$\begin{cases} x - y + z = 0 \\ -5x + 3y - 2z = -1 \\ 2x - y + 4z = 11 \end{cases}$$

F (0, 3, 3)

G (2, 5, 3)

H no solution

J infinitely many solutions

7. Find m in triangle MNO if $n = 12.4$ centimeters, $M = 35°$, and $N = 74°$. Round to the nearest tenth.

A 7.4 cm

B 8.5 cm

C 14.6 cm

D 35.9 cm

8. The results of a recent poll are organized in the matrix.

	For	Against
Proposition 1	1553	771
Proposition 2	689	1633
Proposition 3	2088	229

Based on these results, which conclusion is NOT valid?

F There were 771 votes cast against Proposition 1.

G More people voted against Proposition 1 than voted for Proposition 2.

H Proposition 2 has little chance of passing.

J More people voted for Proposition 1 than for Proposition 3.

9. The graph of which of the following equations is symmetrical about the y-axis?

A $y = x^2 + 3x - 1$

B $y = -x^2 + x$

C $y = 6x^2 + 9$

D $y = 3x^2 - 3x + 1$

10. What is the remainder when $x^3 - 7x + 5$ is divided by $x + 3$?

F -11

G -1

H 1

J 11

Test-Taking Tip

Question 7 Use the Law of Sines to solve the triangle.

Short Response/Gridded Response

Record your answers on the answer sheet provided by your teacher or on a sheet of paper.

11. The speed a tsunami, or tidal wave, can travel is modeled by the equation $s = 356\sqrt{d}$, where s is the speed in kilometers per hour and d is the average depth of the water in kilometers. A tsunami is found to be traveling at 145 kilometers per hour. What is the average depth of the water? Round to the nearest hundredth.

12. GRIDDED RESPONSE Suppose you deposit \$500 in an account paying 4.5% interest compounded semiannually. Find the dollar value of the account rounded to the nearest penny after 10 years.

13. In order to remain healthy, a horse requires 10 pounds of hay per day.

a. Write an equation to represent the amount of hay needed to sustain x horses for d days.

b. Is your equation a direct, joint, or inverse variation? Explain.

c. How much hay do three horses need for the month of July?

14. GRIDDED RESPONSE What is the radius of the circle with equation $x^2 + y^2 + 8x + 8y + 28 = 0$?

15. Anna is training to run a 10-kilometer race. The table below lists the times she received in several 1-kilometer races. The times are listed in minutes. Describe the center and spread of the data using either the mean and standard deviation or the five-number summary. Justify your choice.

7.25	8.10
7.40	6.75
7.20	7.35
7.10	7.25
8.00	7.45

16. GRIDDED RESPONSE The pattern of squares below continues infinitely, with more squares being added at each step. How many squares are in the tenth step?

Step 1 Step 2 Step 3

Extended Response

Record your answers on a sheet of paper. Show your work.

17. Amanda's hours at her summer job for one week are listed in the table below. She earns \$6 per hour.

Amanda's Work Hours	
Sunday	0
Monday	6
Tuesday	4
Wednesday	0
Thursday	2
Friday	6
Saturday	8

a. Write an expression for Amanda's total weekly earnings.

b. Evaluate the expression from part **a** by using the Distributive Property.

c. Michael works with Amanda and also earns \$6 per hour. If Michael's earnings were \$192 this week, write and solve an equation to find how many more hours Michael worked than Amanda.

Need ExtraHelp?

If you missed Question...	1	2	3	4	5	6	7	8	9	10	11	12	13	14	15	16	17
Go to Lesson...	12-1	1-3	12-2	2-4	12-3	3-4	12-4	3-5	4-1	5-2	6-3	7-2	8-5	9-3	11-2	10-2	1-3

CHAPTER 13

Trigonometric Identities and Equations

Then

- You graphed trigonometric functions and determined the period, amplitude, phase shifts, and vertical shifts.

Now

You will:

- Use and verify trigonometric identities.
- Use the sum and difference of angles identities.
- Use the double- and half-angle identities.
- Solve trigonometric equations.

Why? ▲

ELECTRONICS Many aspects of electronics can be modeled by trigonometric functions. Radio, television, cellular telephones, and wireless Internet all communicate through radio waves that are modeled by trigonometric functions. The amount of power in an electronic gadget can be found by using a trigonometric equation.

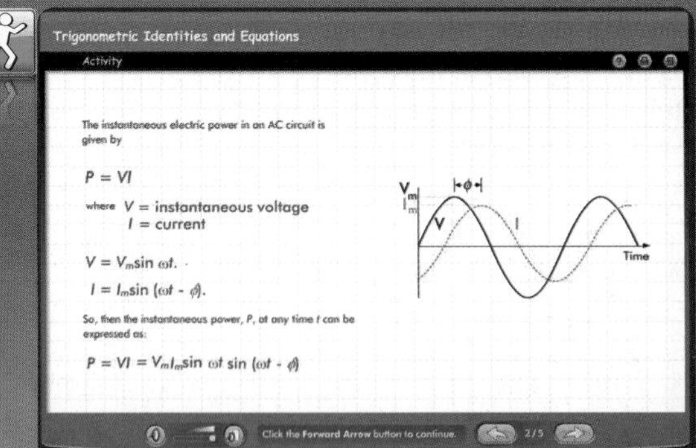

connectED.mcgraw-hill.com Your Digital Math Portal

Animation 
Vocabulary  
eGlossary 
Personal Tutor 
Virtual Manipulatives 
Graphing Calculator 
Audio 
Foldables 
Self-Check Practice 
Worksheets 

Maria Teijeiro/Photodisc/Getty Images

Get Ready for the Chapter

Diagnose Readiness | You have two options for checking prerequisite skills.

Textbook Option Take the Quick Check below. Refer to the Quick Review for help.

QuickCheck

Factor completely. If the polynomial is not factorable, write *prime*.

1. $-16a^2 + 4a$

2. $5x^2 - 20$

3. $x^3 + 9$

4. $2y^2 - y - 15$

5. **GEOMETRY** The area of a rectangular piece of cardboard is $x^2 + 6x + 8$ square inches. If the cardboard has a length of $(x + 4)$ inches, what is the width?

Solve each equation by factoring.

6. $x^2 + 6x = 0$

7. $x^2 + 2x - 35 = 0$

8. $x^2 - 9 = 0$

9. $x^2 - 7x + 12 = 0$

10. **GARDENING** Peyton is building a flower bed in her back yard. The area of the flower bed will be 42 square feet. Find the possible values for x.

Find the exact value of each trigonometric function.

11. sin 45°

12. cos 225°

13. tan 150°

14. sin 120°

15. **RIDES** The distance from the highest point of a Ferris wheel to the ground can be found by multiplying 90 feet by sin 90°. What is the height of the Ferris wheel when it is halfway between the tallest point and the ground?

QuickReview

Example 1

Factor $x^3 + 2x^2 - 24x$ completely.

$x^3 + 2x^2 - 24x = x(x^2 + 2x - 24)$

The product of the coefficients of the x terms must be -24, and their sum must be 2. The product of 6 and -4 is -24 and their sum is 2.

$x(x^2 + 2x - 24) = x(x + 6)(x - 4)$

Example 2

Solve $x^2 + 6x + 5 = 0$ by factoring.

$x^2 + 6x + 5 = 0$ Original equation

$(x + 5)(x + 1) = 0$ Factor.

$x + 5 = 0$ or $x + 1 = 0$

$x = -5$ $\quad x = -1$

The solution set is $\{-5, -1\}$.

Example 3

Find the exact value of cos 135°.

The reference angle is 180° − 135° or 45°.

cos 45° is $\frac{\sqrt{2}}{2}$. Since 135° is in the second quadrant, $\cos 135° = -\frac{\sqrt{2}}{2}$.

2 Online Option Take an online self-check Chapter Readiness Quiz at connectED.mcgraw-hill.com.

Get Started on the Chapter

You will learn several new concepts, skills, and vocabulary terms as you study Chapter 13. To get ready, identify important terms and organize your resources. You may wish to refer to Chapter 0 to review prerequisite skills.

FOLDABLES StudyOrganizer

Trigonometric Identities and Equations Make this Foldable to help you organize your Chapter 13 notes about trigonometric identities and equations. Begin with one sheet of 11″ × 17″ paper and four sheets of grid paper.

1 **Fold** the short sides of the 11″ × 17″ paper to meet in the middle.

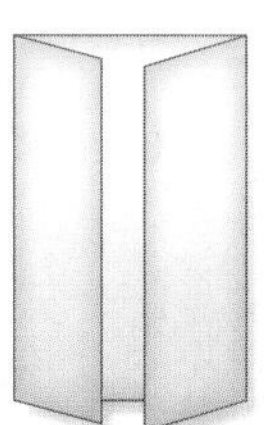

2 **Cut** each tab in half as shown.

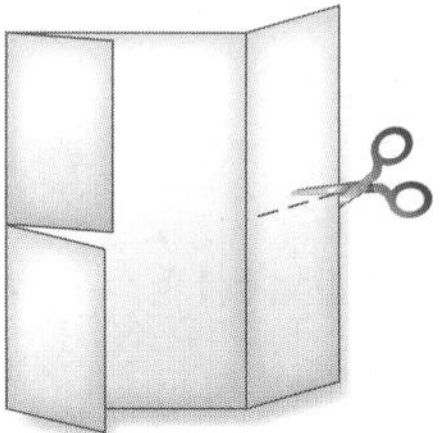

3 **Cut** four sheets of grid paper in half and fold the half-sheets in half.

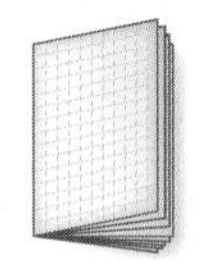

4 **Insert** two folded half-sheets under each of the four tabs and staple along the fold. Label each tab as shown.

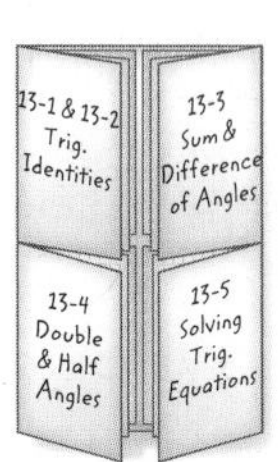

NewVocabulary

English		Español
trigonometric identity	p. 873	identidad trigonométrica
quotient identity	p. 873	identidad de cociente
reciprocal identity	p. 873	identidad recíproca
Pythagorean identity	p. 873	identidad Pitagórica
cofunction identity	p. 873	identidad de función conjunta
negative angle identity	p. 873	identidad negativa de ángulo
trigonometric equation	p. 901	ecuación trigonométrica

ReviewVocabulary

formula fórmula a mathematical sentence that expresses the relationship between certain quantities

identity identidad an equality that remains true regardless of the values of any variables that are in it

trigonometric functions funciones rigonométricas For any angle, with measure θ, a point $P(x, y)$ on its terminal side, $r = \sqrt{x^2 + y^2}$, the trigonometric functions of θ are as follows.

$\sin \theta = \frac{y}{r}$ $\cos \theta = \frac{x}{r}$ $\tan \theta = \frac{y}{x}$

$\csc \theta = \frac{r}{y}$ $\sec \theta = \frac{r}{x}$ $\cot \theta = \frac{x}{y}$

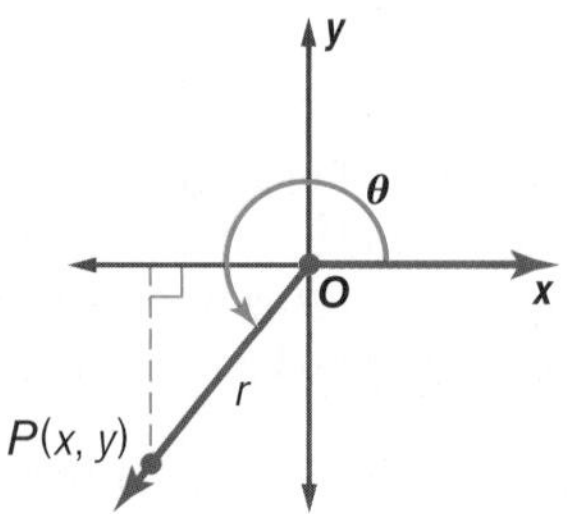

LESSON 13-1 Trigonometric Identities

Then	Now	Why?
You evaluated trigonometric functions.	**1** Use trigonometric identities to find trigonometric values. **2** Use trigonometric identities to simplify expressions.	The amount of light that a source provides to a surface is called the *illuminance*. The illuminance E in foot candles on a surface is related to the distance R in feet from the light source. The formula $\sec \theta = \frac{I}{ER^2}$, where I is the intensity of the light source measured in candles and θ is the angle between the light beam and a line perpendicular to the surface, can be used in situations in which lighting is important, as in photography.

NewVocabulary
trigonometric identity

Common Core State Standards

Content Standards
F.TF.8 Prove the Pythagorean identity $\sin^2(\theta) + \cos^2(\theta) = 1$ and use it to find $\sin(\theta)$, $\cos(\theta)$, or $\tan(\theta)$ given $\sin(\theta)$, $\cos(\theta)$, or $\tan(\theta)$ and the quadrant of the angle.

Mathematical Practices
2 Reason abstractly and quantitatively.
7 Look for and make use of structure.

1 Find Trigonometric Values

The equation above can also be written as $E = \frac{I \cos \theta}{R^2}$. This is an example of a trigonometric identity. A **trigonometric identity** is an equation involving trigonometric functions that is true for all values for which every expression in the equation is defined.

If you can show that a specific value of the variable in an equation makes the equation false, then you have produced a *counterexample*. It only takes one counterexample to prove that an equation is not an identity.

KeyConcept Basic Trigonometric Identities

Quotient Identities		
$\tan \theta = \frac{\sin \theta}{\cos \theta}$, $\cos \theta \neq 0$		$\cot \theta = \frac{\cos \theta}{\sin \theta}$, $\sin \theta \neq 0$
Reciprocal Identities		
$\sin \theta = \frac{1}{\csc \theta}$, $\csc \theta \neq 0$		$\csc \theta = \frac{1}{\sin \theta}$, $\sin \theta \neq 0$
$\cos \theta = \frac{1}{\sec \theta}$, $\sec \theta \neq 0$		$\sec \theta = \frac{1}{\cos \theta}$, $\cos \theta \neq 0$
$\tan \theta = \frac{1}{\cot \theta}$, $\cot \theta \neq 0$		$\cot \theta = \frac{1}{\tan \theta}$, $\tan \theta \neq 0$
Pythagorean Identities		
$\cos^2 \theta + \sin^2 \theta = 1$	$\tan^2 \theta + 1 = \sec^2 \theta$	$\cot^2 \theta + 1 = \csc^2 \theta$
Cofunction Identities		
$\sin\left(\frac{\pi}{2} - \theta\right) = \cos \theta$	$\cos\left(\frac{\pi}{2} - \theta\right) = \sin \theta$	$\tan\left(\frac{\pi}{2} - \theta\right) = \cot \theta$
Negative Angle Identities		
$\sin(-\theta) = -\sin \theta$	$\cos(-\theta) = \cos \theta$	$\tan(-\theta) = -\tan \theta$

The negative angle identities are sometimes called *odd-even* identities.

The identity $\tan \theta = \frac{\sin \theta}{\cos \theta}$ is true except for angle measures such as 90°, 270°, … , 90° + k180°, where k is an integer. The cosine of each of these angle measures is 0, so $\tan \theta$ is not defined when $\cos \theta = 0$. An identity similar to this is $\cot \theta = \frac{\cos \theta}{\sin \theta}$.

Ryan McVay/Stone/Getty images

You can use trigonometric identities to find exact values of trigonometric functions. You can find approximate values by using a graphing calculator.

Example 1 Use Trigonometric Identities

a. Find the exact value of $\cos \theta$ if $\sin \theta = \frac{1}{4}$ and $90° < \theta < 180°$.

$\cos^2 \theta + \sin^2 \theta = 1$	Pythagorean identity
$\cos^2 \theta = 1 - \sin^2 \theta$	Subtract $\sin^2 \theta$ from each side.
$\cos^2 \theta = 1 - \left(\frac{1}{4}\right)^2$	Substitute $\frac{1}{4}$ for $\sin \theta$.
$\cos^2 \theta = 1 - \frac{1}{16}$	Square $\frac{1}{4}$.
$\cos^2 \theta = \frac{15}{16}$	Subtract: $\frac{16}{16} - \frac{1}{16} = \frac{15}{16}$.
$\cos \theta = \pm\frac{\sqrt{15}}{4}$	Take the square root of each side.

Since θ is in the second quadrant, $\cos \theta$ is negative. Thus, $\cos \theta = -\frac{\sqrt{15}}{4}$.

CHECK Use a calculator to find an approximate answer.

Step 1 Find Arcsin $\frac{1}{4}$.

$\sin^{-1} \frac{1}{4} \approx 14.48°$ Use a calculator.

Because $90° < \theta < 180°$, $\theta \approx 180° - 14.48°$ or about 165.52°.

Step 2 Find $\cos \theta$.

Replace θ with 165.52°.

$\cos 165.52° \approx -0.97$

Step 3 Compare with the exact value.

$-\frac{\sqrt{15}}{4} \stackrel{?}{\approx} 0.97$

$-0.968 \approx 0.97$ ✓

b. Find the exact value of $\csc \theta$ if $\cot \theta = -\frac{3}{5}$ and $270° < \theta < 360°$.

$\cot^2 \theta + 1 = \csc^2 \theta$	Pythagorean identity
$\left(-\frac{3}{5}\right)^2 + 1 = \csc^2 \theta$	Substitute $-\frac{3}{5}$ for $\cot \theta$.
$\frac{9}{25} + 1 = \csc^2 \theta$	Square $-\frac{3}{5}$.
$\frac{34}{25} = \csc^2 \theta$	Add: $\frac{9}{25} + \frac{25}{25} = \frac{34}{25}$.
$\pm\frac{\sqrt{34}}{5} = \csc \theta$	Take the square root of each side.

Since θ is in the fourth quadrant, $\csc \theta$ is negative. Thus, $\csc \theta = -\frac{\sqrt{34}}{5}$.

StudyTip

Quadrants Here is a table to help you remember which ratios are positive and which are negative in each quadrant.

Function	+	−
$\sin \theta$	1, 2	3, 4
$\cos \theta$	1, 4	2, 3
$\tan \theta$	1, 3	2, 4
$\csc \theta$	1, 2	3, 4
$\sec \theta$	1, 4	2, 3
$\cot \theta$	1, 3	2, 4

GuidedPractice

1A. Find $\sin \theta$ if $\cos \theta = \frac{1}{3}$ and $270° < \theta < 360°$.

1B. Find $\sec \theta$ if $\sin \theta = -\frac{2}{7}$ and $180° < \theta < 270°$.

2 Simplify Expressions

Simplify Expressions Simplifying an expression that contains trigonometric functions means that the expression is written as a numerical value or in terms of a single trigonometric function, if possible.

StudyTip

Simplifying It is often easiest to write all expressions in terms of sine and/or cosine.

Example 2 Simplify an Expression

Simplify $\frac{\sin\theta\csc\theta}{\cot\theta}$.

$$\frac{\sin\theta\csc\theta}{\cot\theta} = \frac{\sin\theta\left(\frac{1}{\sin\theta}\right)}{\frac{1}{\tan\theta}} \qquad \csc\theta = \frac{1}{\sin\theta} \text{ and } \cot\theta = \frac{1}{\tan\theta}$$

$$= \frac{1}{\frac{1}{\tan\theta}} \qquad \frac{\sin\theta}{\sin\theta} = 1$$

$$= \frac{1}{1}\cdot\frac{\tan\theta}{1} \text{ or } \tan\theta \qquad \frac{a}{b}\div\frac{c}{d} = \frac{a}{b}\cdot\frac{d}{c}$$

GuidedPractice

Simplify each expression.

2A. $\frac{\tan^2\theta\csc^2\theta - 1}{\sec^2\theta}$

2B. $\frac{\sec\theta}{\sin\theta}(1 - \cos^2\theta)$

Simplifying trigonometric expressions can be helpful when solving real-world problems.

Real-World Example 3 Simplify and Use an Expression

LIGHTING Refer to the beginning of the lesson.

a. Solve the formula in terms of E.

$$\sec\theta = \frac{I}{ER^2} \qquad \text{Original equation}$$

$$ER^2\sec\theta = I \qquad \text{Multiply each side by } ER^2.$$

$$ER^2\frac{1}{\cos\theta} = I \qquad \frac{1}{\cos\theta} = \sec\theta$$

$$\frac{E}{\cos\theta} = \frac{I}{R^2} \qquad \text{Divide each side by } R^2.$$

$$E = \frac{I\cos\theta}{R^2} \qquad \text{Multiply each side by } \cos\theta.$$

Math HistoryLink

Aryabhatta (476–550 A.D.) Among Indian mathematicians, Aryabhatta is probably the most famous. His name is closely associated with trigonometry. He was the first to introduce inverse trigonometric functions and spherical trigonometry. Aryabhatta also calculated approximations for pi and trigonometric functions.

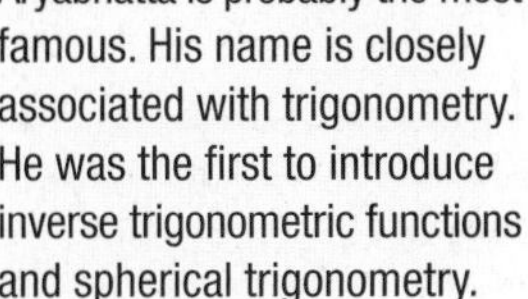

b. Is the equation in part a equivalent to $R^2 = \frac{I\tan\theta\cos\theta}{E}$? Explain.

$$R^2 = \frac{I\tan\theta\cos\theta}{E} \qquad \text{Original equation}$$

$$ER^2 = I\tan\theta\cos\theta \qquad \text{Multiply each side by } E.$$

$$E = \frac{I\tan\theta\cos\theta}{R^2} \qquad \text{Divide each side by } R^2.$$

$$E = \frac{I\left(\frac{\sin\theta}{\cos\theta}\right)\cos\theta}{R^2} \qquad \tan\theta = \frac{\sin\theta}{\cos\theta}$$

$$E = \frac{I\sin\theta}{R^2} \qquad \text{Simplify.}$$

No; the equations are not equivalent. $R^2 = \frac{I\tan\theta\cos\theta}{E}$ simplifies to $E = \frac{I\sin\theta}{R^2}$.

GuidedPractice

3. Rewrite $\cot^2\theta - \tan^2\theta$ in terms of $\sin\theta$.

Check Your Understanding

● = Step-by-Step Solutions begin on page R14.

Example 1 **Find the exact value of each expression if $0° < \theta < 90°$.**

1. If $\cot \theta = 2$, find $\tan \theta$.
2. If $\sin \theta = \frac{4}{5}$, find $\cos \theta$.
3. If $\cos \theta = \frac{2}{3}$, find $\sin \theta$.
4. If $\cos \theta = \frac{2}{3}$, find $\csc \theta$.

Example 2 **Simplify each expression.**

5. $\tan \theta \cos^2 \theta$
6. $\csc^2 \theta - \cot^2 \theta$
7. $\frac{\cos \theta \csc \theta}{\tan \theta}$

Example 3

8. **CCSS PERSEVERANCE** When unpolarized light passes through polarized sunglass lenses, the intensity of the light is cut in half. If the light then passes through another polarized lens with its axis at an angle of θ to the first, the intensity of the light is again diminished. The intensity of the emerging light can be found by using the formula $I = I_0 - \frac{I_0}{\csc^2 \theta}$, where I_0 is the intensity of the light incoming to the second polarized lens, I is the intensity of the emerging light, and θ is the angle between the axes of polarization.

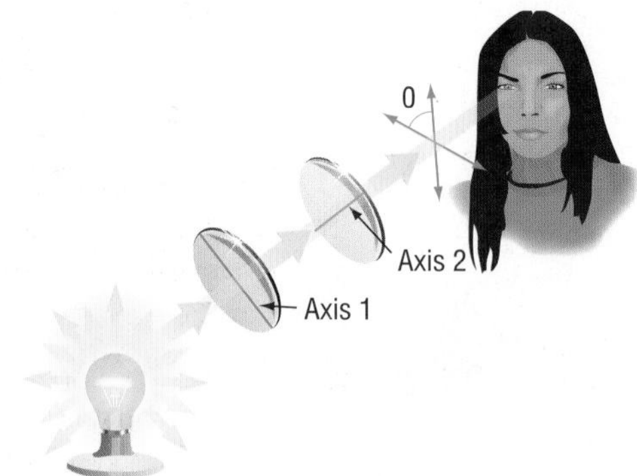

   a. Simplify the formula in terms of $\cos \theta$.

   b. Use the simplified formula to determine the intensity of light that passes through a second polarizing lens with axis at 30° to the original.

Practice and Problem Solving

Extra Practice is on page R13.

Example 1 **Find the exact value of each expression if $0° < \theta < 90°$.**

9. If $\cos \theta = \frac{3}{5}$, find $\csc \theta$.
10. If $\sin \theta = \frac{1}{2}$, find $\tan \theta$.
11. If $\sin \theta = \frac{3}{5}$, find $\cos \theta$.
12. If $\tan \theta = 2$, find $\sec \theta$.

Find the exact value of each expression if $180° < \theta < 270°$.

13. If $\cos \theta = -\frac{3}{5}$, find $\csc \theta$.
14. If $\sec \theta = -3$, find $\tan \theta$.
15. If $\cot \theta = \frac{1}{4}$, find $\csc \theta$.
16. If $\sin \theta = -\frac{1}{2}$, find $\cos \theta$.

Find the exact value of each expression if $270° < \theta < 360°$.

17. If $\cos \theta = \frac{5}{13}$, find $\sin \theta$.
18. If $\tan \theta = -1$, find $\sec \theta$.
19. If $\sec \theta = \frac{5}{3}$, find $\cos \theta$.
20. If $\csc \theta = -\frac{5}{3}$, find $\cos \theta$.

Example 2 **Simplify each expression.**

21. $\sec \theta \tan^2 \theta + \sec \theta$
22. $\cos\left(\frac{\pi}{2} - \theta\right)\cot \theta$
23. $\cot \theta \sec \theta$
24. $\sin \theta (1 + \cot^2 \theta)$
25. $\sin\left(\frac{\pi}{2} - \theta\right)\sec \theta$
26. $\frac{\cos(-\theta)}{\sin(-\theta)}$

Example 3

27 **ELECTRONICS** When there is a current in a wire in a magnetic field, such as in a hairdryer, a force acts on the wire. The strength of the magnetic field can be determined using the formula $B = \frac{F \csc \theta}{I\ell}$, where F is the force on the wire, I is the current in the wire, ℓ is the length of the wire, and θ is the angle the wire makes with the magnetic field. Rewrite the equation in terms of $\sin \theta$. (*Hint:* Solve for F.)

Simplify each expression.

28. $\frac{1 - \sin^2 \theta}{\sin^2 \theta}$

29. $\tan \theta \csc \theta$

30. $\frac{1}{\sin^2 \theta} - \frac{\cos^2 \theta}{\sin^2 \theta}$

31. $2(\csc^2 \theta - \cot^2 \theta)$

32. $(1 + \sin \theta)(1 - \sin \theta)$

33. $2 - 2 \sin^2 \theta$

34. **SUN** The ability of an object to absorb energy is related to a factor called the emissivity e of the object. The emissivity can be calculated by using the formula $e = \frac{W \sec \theta}{AS}$, where W is the rate at which a person's skin absorbs energy from the Sun, S is the energy from the Sun in watts per square meter, A is the surface area exposed to the Sun, and θ is the angle between the Sun's rays and a line perpendicular to the body.

a. Solve the equation for W. Write your answer using only $\sin \theta$ or $\cos \theta$.

b. Find W if $e = 0.80$, $\theta = 40°$, $A = 0.75$ m^2, and $S = 1000$ W/m^2. Round to the nearest hundredth.

35. **CCSS MODELING** The map shows some of the buildings in Maria's neighborhood that she visits on a regular basis. The sine of the angle θ formed by the roads connecting the dance studio, the school, and Maria's house is $\frac{4}{9}$.

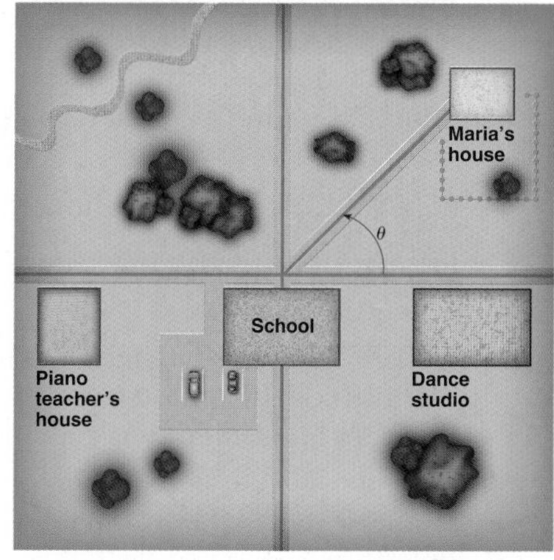

a. What is the cosine of the angle?

b. What is the tangent of the angle?

c. What are the sine, cosine, and tangent of the angle formed by the roads connecting the piano teacher's house, the school, and Maria's house?

36. **MULTIPLE REPRESENTATIONS** In this problem, you will use a graphing calculator to determine whether an equation may be a trigonometric identity. Consider the trigonometric identity $\tan^2 \theta - \sin^2 \theta = \tan^2 \theta \sin^2 \theta$.

a. Tabular Copy and complete the table below.

θ	0°	30°	45°	60°
$\tan^2 \theta - \sin^2 \theta$				
$\tan^2 \theta \sin^2 \theta$				

b. Graphical Use a graphing calculator to graph $\tan^2 \theta - \sin^2 \theta = \tan^2 \theta \sin^2 \theta$ as two separate functions. Sketch the graph.

c. Analytical If the graphs of the two functions do not match, then the equation is not an identity. Do the graphs coincide?

d. Analytical Use a graphing calculator to determine whether the equation $\sec^2 x - 1 = \sin^2 x \sec^2 x$ may be an identity. (Be sure your calculator is in degree mode.)

37. **SKIING** A skier of mass m descends a θ-degree hill at a constant speed. When Newton's laws are applied to the situation, the following system of equations is produced: $F_n - mg \cos \theta = 0$ and $mg \sin \theta - \mu_k F_n = 0$, where g is the acceleration due to gravity, F_n is the normal force exerted on the skier, and μ_k is the coefficient of friction. Use the system to define μ_k as a function of θ.

Simplify each expression.

38. $\dfrac{\tan\left(\frac{\pi}{2} - \theta\right)\sec \theta}{1 - \csc^2 \theta}$

39. $\dfrac{\cos\left(\frac{\pi}{2} - \theta\right) - 1}{1 + \sin(-\theta)}$

40. $\dfrac{\sec \theta \sin \theta + \cos\left(\frac{\pi}{2} - \theta\right)}{1 + \sec \theta}$

41. $\dfrac{\cot \theta \cos \theta}{\tan(-\theta) \sin\left(\frac{\pi}{2} - \theta\right)}$

H.O.T. Problems Use Higher-Order Thinking Skills

42. **CCSS CRITIQUE** Clyde and Rosalina are debating whether an equation from their homework assignment is an identity. Clyde says that since he has tried ten specific values for the variable and all of them worked, it must be an identity. Rosalina argues that specific values could only be used as counterexamples to prove that an equation is not an identity. Is either of them correct? Explain your reasoning.

43. **CHALLENGE** Find a counterexample to show that $1 - \sin x = \cos x$ is *not* an identity.

44. **REASONING** Demonstrate how the formula about illuminance from the beginning of the lesson can be rewritten to show that $\cos \theta = \frac{ER^2}{I}$.

45. **WRITING IN MATH** Pythagoras is most famous for the Pythagorean Theorem. The identity $\cos^2 \theta + \sin^2 \theta = 1$ is an example of a Pythagorean identity. Why do you think that this identity is classified in this way?

46. **PROOF** Prove that $\tan(-a) = -\tan a$ by using the quotient and negative angle identities.

47. **OPEN ENDED** Write two expressions that are equivalent to $\tan \theta \sin \theta$.

48. **REASONING** Explain how you can use division to rewrite $\sin^2 \theta + \cos^2 \theta = 1$ as $1 + \cot^2 \theta = \csc^2 \theta$.

49. **CHALLENGE** Find $\cot \theta$ if $\sin \theta = \frac{3}{5}$ and $90° \leq \theta < 180°$.

50. **ERROR ANALYSIS** Jordan and Ebony are simplifying $\frac{\sin^2 \theta}{\cos^2 \theta + \sin^2 \theta}$. Is either of them correct? Explain your reasoning.

Jordan

$$\frac{\sin^2 \theta}{\cos^2 \theta + \sin^2 \theta} = \frac{\sin^2 \theta}{\cos^2 \theta} + \frac{\sin^2 \theta}{\sin^2 \theta}$$
$$= \tan^2 \theta + 1$$
$$= \sec^2 \theta$$

Ebony

$$\frac{\sin^2 \theta}{\cos^2 \theta + \sin^2 \theta} = \frac{\sin^2 \theta}{1}$$
$$= \sin^2 \theta$$

Standardized Test Practice

51. Refer to the figure below. If $\cos D = 0.8$, what is the length of $\overline{DF}$?

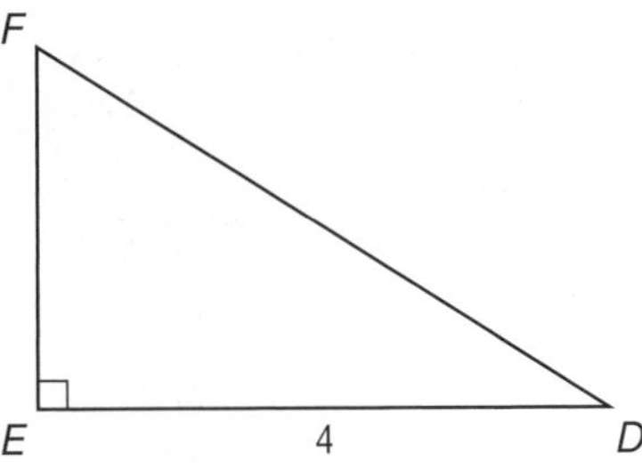

A 5
B 4
C 3.2
D $\frac{4}{5}$

52. PROBABILITY There are 16 green marbles, 2 red marbles, and 6 yellow marbles in a jar. How many yellow marbles need to be added to the jar in order to double the probability of selecting a yellow marble?

F 4
G 6
H 8
J 12

53. SAT/ACT Ella is 6 years younger than Amanda. Zoe is twice as old as Amanda. The total of their ages is 54. Which equation can be used to find Amanda's age?

A $x + (x - 6) + 2(x - 6) = 54$
B $x - 6x + (x + 2) = 54$
C $x - 6 + 2x = 54$
D $x + (x - 6) + 2x = 54$
E $2(x + 6) + (x + 6) + x = 54$

54. Which of the following functions represents exponential growth?

F $y = (0.3)^x$
G $y = (1.3)^x$
H $y = x^3$
J $y = x^{\frac{1}{3}}$

Spiral Review

Find each value. Write angle measures in radians. Round to the nearest hundredth. (Lesson 12-9)

55. $\text{Cos}^{-1}\left(-\frac{1}{2}\right)$

56. $\text{Sin}^{-1}\frac{\pi}{2}$

57. $\text{Arctan}\,\frac{\sqrt{3}}{3}$

58. $\tan\left(\text{Cos}^{-1}\frac{6}{7}\right)$

59. $\sin\left(\text{Arctan}\,\frac{\sqrt{3}}{3}\right)$

60. $\cos\left(\text{Arcsin}\,\frac{3}{5}\right)$

61. PHYSICS A weight is attached to a spring and suspended from the ceiling. At equilibrium, the weight is located 4 feet above the floor. The weight is pulled down 1 foot and released. Write the equation for the height h of the weight above the floor as a function of time t seconds assuming the weight returns to its lowest position every 4 seconds. (Lesson 12-8)

Evaluate the sum of each geometric series. (Lesson 10-3)

62. $\sum_{k=1}^{5} \frac{1}{4} \cdot 2^{k-1}$

63. $\sum_{k=1}^{7} 81\left(\frac{1}{3}\right)^{k-1}$

64. $\sum_{k=1}^{8} \frac{1}{3} \cdot 5^{k-1}$

Skills Review

Solve each equation.

65. $a + 1 = \frac{6}{a}$

66. $\frac{9}{t-3} = \frac{t-4}{t-3} + \frac{1}{4}$

67. $\frac{5}{x+1} - \frac{1}{3} = \frac{x+2}{x+1}$

LESSON 13-2 Verifying Trigonometric Identities

Then	Now	Why?
You used identities to find trigonometric values and simplify expressions.	**1** Verify trigonometric identities by transforming one side of an equation into the form of the other side. **2** Verify trigonometric identities by transforming each side of the equation into the same form.	While running on a circular track, Lamont notices that his body is not perpendicular to the ground. Instead, it leans away from a vertical position. The nonnegative acute angle θ that Lamont's body makes with the vertical is called the *angle of incline* and is described by the equation $\tan \theta = \frac{v^2}{gR}$. This is not the only equation that describes the angle of incline in terms of trigonometric functions. Another such equation is $\sin \theta = \cos \frac{v^2}{gR}\theta$, where $0 \leq \theta \leq 90°$. Are these two equations completely independent of one another or are they merely different versions of the same relationship?

Common Core State Standards

Content Standards

F.TF.8 Prove the Pythagorean identity $\sin^2(\theta) + \cos^2(\theta) = 1$ and use it to find $\sin(\theta)$, $\cos(\theta)$, or $\tan(\theta)$ given $\sin(\theta)$, $\cos(\theta)$, or $\tan(\theta)$ and the quadrant of the angle.

Mathematical Practices

1 Make sense of problems and persevere in solving them.

8 Look for and express regularity in repeated reasoning.

1 Transform One Side of an Equation

You can use the basic trigonometric identities along with the definitions of the trigonometric functions to verify identities. If you wish to show an identity, you need to show that it is true for all values of θ.

KeyConcept Verifying Identities by Transforming One Side

Step 1 Simplify one side of an equation until the two sides of the equation are the same. It is often easier to work with the more complicated side of the equation.

Step 2 Transform that expression into the form of the simpler side.

Example 1 Transform One Side of an Equation

Verify that $\frac{\sin^2 \theta}{1 - \cos \theta} = 1 + \cos \theta$ is an identity.

$\frac{\sin^2 \theta}{1 - \cos \theta} \stackrel{?}{=} 1 + \cos \theta$ — Original equation

$\frac{1 + \cos \theta}{1 + \cos \theta} \cdot \frac{\sin^2 \theta}{1 - \cos \theta} \stackrel{?}{=} 1 + \cos \theta$ — Multiply the numerator and denominator by $1 + \cos \theta$.

$\frac{\sin^2 \theta(1 + \cos \theta)}{1 - \cos^2 \theta} \stackrel{?}{=} 1 + \cos \theta$ — $(1 + \cos \theta)(1 - \cos \theta) = 1 - \cos^2 \theta$

$\frac{\sin^2 \theta(1 + \cos \theta)}{\sin^2 \theta} \stackrel{?}{=} 1 + \cos \theta$ — $\sin^2 \theta = 1 - \cos^2 \theta$

$1 + \cos \theta = 1 + \cos \theta$ ✓ — Divide the numerator and denominator by $\sin^2 \theta$.

GuidedPractice

1. Verify that $\cot^2 \theta - \cos^2 \theta = \cot^2 \theta \cos^2 \theta$ is an identity.

Nick Wilson/AUS /Allsport/Getty Images

When you verify a trigonometric identity, you are really working backward. In Example 1, consider the last step $1 + \cos\theta = 1 + \cos\theta$. Since that step is clearly true, you can conclude that the next-to-last step is also true, and so on, all the way back to the original equation.

Standardized Test Example 2 Simplify an Expression

$\dfrac{\cos\theta \csc\theta}{\tan\theta} =$

A $\cot\theta$ **B** $\csc\theta$ **C** $\cot^2\theta$ **D** $\csc^2\theta$

Read the Test Item

Find an expression that is always equal to the given expression. Notice that all of the answer choices involve either $\cot\theta$ or $\csc\theta$. So work toward eliminating the other trigonometric functions.

Solve the Test Item

Transform the given expression to match one of the choices.

$$\frac{\cos\theta\csc\theta}{\tan\theta} = \frac{\cos\theta\left(\frac{1}{\sin\theta}\right)}{\frac{\sin\theta}{\cos\theta}}$$ $\csc\theta = \frac{1}{\sin\theta}$ and $\tan\theta = \frac{\sin\theta}{\cos\theta}$

$$= \frac{\frac{\cos\theta}{\sin\theta}}{\frac{\sin\theta}{\cos\theta}}$$ Multiply.

$$= \frac{\cos\theta}{\sin\theta} \cdot \frac{\cos\theta}{\sin\theta}$$ Invert the denominator and multiply.

$$= \cot\theta \cdot \cot\theta$$ $\cot\theta = \frac{\cos\theta}{\sin\theta}$

$$= \cot^2\theta$$ Multiply.

The answer is C.

WatchOut!

CCSS Perseverance Verifying an identity is like checking the solution of an equation. You must simplify one or both sides separately until they are the same.

Test-TakingTip

Checking Answers Verify your answer by choosing values for θ. Then evaluate the original expression and compare to your answer choice.

GuidedPractice

2. $\tan^2\theta(\cot^2\theta - \cos^2\theta) =$

F $\cot^2\theta$ **G** $\tan^2\theta$ **H** $\cos^2\theta$ **J** $\sin^2\theta$

2 Transform Each Side of an Equation

Sometimes it is easier to transform each side of an equation separately into a common form. The following suggestions may be helpful as you verify trigonometric identities.

KeyConcept Suggestions for Verifying Identities

- Substitute one or more basic trigonometric identities to simplify the expression.
- Factor or multiply as necessary. You may have to multiply both the numerator and denominator by the same trigonometric expression.
- Write each side of the identity in terms of sine and cosine only. Then simplify each side as much as possible.
- The properties of equality do not apply to identities as with equations. Do not perform operations to the quantities on each side of an unverified identity.

Example 3 Verify by Transforming Each Side

Verify that $1 - \tan^4 \theta = 2\sec^2 \theta - \sec^4 \theta$ is an identity.

$1 - \tan^4 \theta \stackrel{?}{=} 2\sec^2 \theta - \sec^4 \theta$	Original equation
$(1 - \tan^2 \theta)(1 + \tan^2 \theta) \stackrel{?}{=} \sec^2 \theta\,(2 - \sec^2 \theta)$	Factor each side.
$[1 - (\sec^2 \theta - 1)]\sec^2 \theta \stackrel{?}{=} (2 - \sec^2 \theta)\sec^2 \theta$	$1 + \tan^2 \theta = \sec^2 \theta$
$(2 - \sec^2 \theta)\sec^2 \theta = (2 - \sec^2 \theta)\sec^2 \theta$ ✓	Simplify.

GuidedPractice

3. Verify that $\csc^2 \theta - \cot^2 \theta = \cot \theta \tan \theta$ is an identity.

Check Your Understanding

◯ = Step-by-Step Solutions begin on page R14.

Examples 1–3 **CCSS PRECISION** **Verify that each equation is an identity.**

1. $\cot \theta + \tan \theta = \dfrac{\sec^2 \theta}{\tan \theta}$
2. $\cos^2 \theta = (1 + \sin \theta)(1 - \sin \theta)$
3. $\sin \theta = \dfrac{\sec \theta}{\tan \theta + \cot \theta}$
4. $\tan^2 \theta = \dfrac{1 - \cos^2 \theta}{\cos^2 \theta}$
5. $\tan^2 \theta \csc^2 \theta = 1 + \tan^2 \theta$
6. $\tan^2 \theta = (\sec \theta + 1)(\sec \theta - 1)$

Example 2

7. **MULTIPLE CHOICE** Which expression can be used to form an identity with $\dfrac{\tan^2 \theta + 1}{\tan^2 \theta}$?

   A $\sin^2 \theta$ **B** $\cos^2 \theta$ **C** $\tan^2 \theta$ **D** $\csc^2 \theta$

Practice and Problem Solving

Extra Practice is on page R13.

Example 1 **Verify that each equation is an identity.**

8. $\cos^2 \theta + \tan^2 \theta \cos^2 \theta = 1$
9. $\cot \theta\,(\cot \theta + \tan \theta) = \csc^2 \theta$
10. $1 + \sec^2 \theta \sin^2 \theta = \sec^2 \theta$
11. $\sin \theta \sec \theta \cot \theta = 1$
12. $\dfrac{1 - \cos \theta}{1 + \cos \theta} = (\csc \theta - \cot \theta)^2$
13. $\dfrac{1 - 2\cos^2 \theta}{\sin \theta \cos \theta} = \tan \theta - \cot \theta$
14. $\tan \theta = \dfrac{\sec \theta}{\csc \theta}$
15. $\cos \theta = \sin \theta \cot \theta$
16. $(\sin \theta - 1)(\tan \theta + \sec \theta) = -\cos \theta$
17. $\cos \theta \cos(-\theta) - \sin \theta \sin(-\theta) = 1$

Example 2

18. **LADDER** Some students derived an expression for the length of a ladder that, when carried flat, could fit around a corner from a 5-foot-wide hallway into a 7-foot-wide hallway, as shown. They determined that the maximum length ℓ of a ladder that would fit was given by $\ell(\theta) = \dfrac{7\sin \theta + 5\cos \theta}{\sin \theta \cos \theta}$. When their teacher worked the problem, she concluded that $\ell(\theta) = 7\sec \theta + 5\csc \theta$. Are the two expressions equivalent?

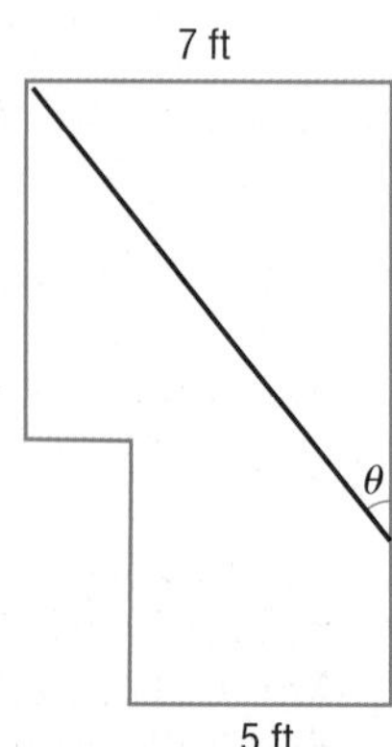

Example 3 **Verify that each equation is an identity.**

19. $\sec \theta - \tan \theta = \frac{1 - \sin \theta}{\cos \theta}$

20. $\frac{1 + \tan \theta}{\sin \theta + \cos \theta} = \sec \theta$

21. $\sec \theta \csc \theta = \tan \theta + \cot \theta$

22. $\sin \theta + \cos \theta = \frac{2 \sin^2 \theta - 1}{\sin \theta - \cos \theta}$

23. $(\sin \theta + \cos \theta)^2 = \frac{2 + \sec \theta \csc \theta}{\sec \theta \csc \theta}$

24. $\frac{\cos \theta}{1 - \sin \theta} = \frac{1 + \sin \theta}{\cos \theta}$

25. $\csc \theta - 1 = \frac{\cot^2 \theta}{\csc \theta + 1}$

26. $\cos \theta \cot \theta = \csc \theta - \sin \theta$

27. $\sin \theta \cos \theta \tan \theta + \cos^2 \theta = 1$

28. $(\csc \theta - \cot \theta)^2 = \frac{1 - \cos \theta}{1 + \cos \theta}$

29. $\csc^2 \theta = \cot^2 \theta + \sin \theta \csc \theta$

30. $\frac{\sec \theta - \csc \theta}{\csc \theta \sec \theta} = \sin \theta - \cos \theta$

31. $\sin^2 \theta + \cos^2 \theta = \sec^2 \theta - \tan^2 \theta$

32. $\sec \theta - \cos \theta = \tan \theta \sin \theta$

33. **CCSS SENSE-MAKING** The diagram at the right represents a game of tetherball. As the ball rotates around the pole, a conical surface is swept out by the line segment $\overline{SP}$. A formula for the relationship between the length L of the string and the angle θ that the string makes with the pole is given by the equation $L = \frac{g \sec \theta}{\omega^2}$. Is $L = \frac{g \tan \theta}{\omega^2 \sin \theta}$ also an equation for the relationship between L and θ?

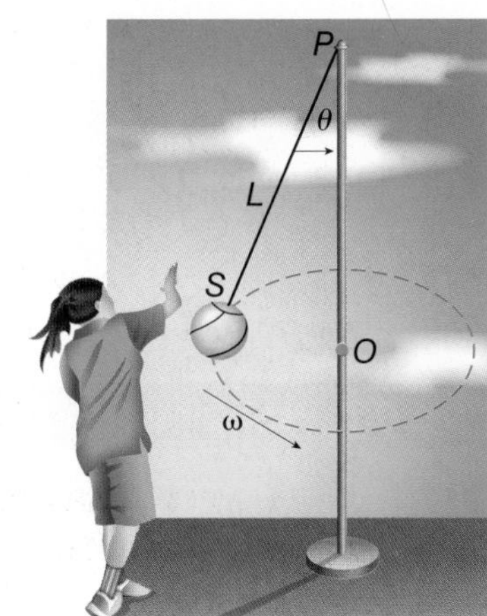

34. **RUNNING** A portion of a racetrack has the shape of a circular arc with a radius of 16.7 meters. As a runner races along the arc, the sine of her angle of incline θ is found to be $\frac{1}{4}$. Find the speed of the runner. Use the Angle of Incline Formula given at the beginning of the lesson, $\tan \theta = \frac{v^2}{gR}$, where $g = 9.8$ and R is the radius. (*Hint*: Find $\cos \theta$ first.)

When simplified, would the expression be equal to 1 or −1?

35. $\cot(-\theta) \tan(-\theta)$

36. $\sin \theta \csc(-\theta)$

37. $\sin^2(-\theta) + \cos^2(-\theta)$

38. $\sec(-\theta) \cos(-\theta)$

39. $\sec^2(-\theta) - \tan^2(-\theta)$

40. $\cot(-\theta) \cot\left(\frac{\pi}{2} - \theta\right)$

Simplify the expression to either a constant or a basic trigonometric function.

41. $\frac{\tan\left(\frac{\pi}{2} - \theta\right)\csc \theta}{\csc^2 \theta}$

42. $\frac{1 + \tan \theta}{1 + \cot \theta}$

43. $(\sec^2 \theta + \csc^2 \theta) - (\tan^2 \theta + \cot^2 \theta)$

44. $\frac{\sec^2 \theta - \tan^2 \theta}{\cos^2 x + \sin^2 x}$

45. $\tan \theta \cos \theta$

46. $\cot \theta \tan \theta$

47. $\sec \theta \sin\left(\frac{\pi}{2} - \theta\right)$

48. $\frac{1 + \tan^2 \theta}{\csc^2 \theta}$

49. **PHYSICS** When a firework is fired from the ground, its height y and horizontal displacement x are related by the equation $y = \frac{-gx^2}{2v_0^2 \cos^2 \theta} + \frac{x \sin \theta}{\cos \theta}$, where v_0 is the initial velocity of the projectile, θ is the angle at which it was fired, and g is the acceleration due to gravity. Rewrite this equation so that $\tan \theta$ is the only trigonometric function that appears in the equation.

50. ELECTRONICS When an alternating current of frequency f and peak current I_0 passes through a resistance R, the power delivered to the resistance at time t seconds is $P = I_0^2 R \sin^2 2\pi ft$.

a. Write an expression for the power in terms of $\cos^2 2\pi ft$.

b. Write an expression for the power in terms of $\csc^2 2\pi ft$.

51. THROWING A BALL In this problem, you will investigate the path of a ball represented by the equation $h = \frac{v_0^2 \sin^2 \theta}{2g}$, where θ is the measure of the angle between the ground and the path of the ball, v_0 is its initial velocity in meters per second, and g is the acceleration due to gravity. The value of g is 9.8 m/s^2.

a. If the initial velocity of the ball is 47 meters per second, find the height of the ball at 30°, 45°, 60°, and 90°. Round to the nearest tenth.

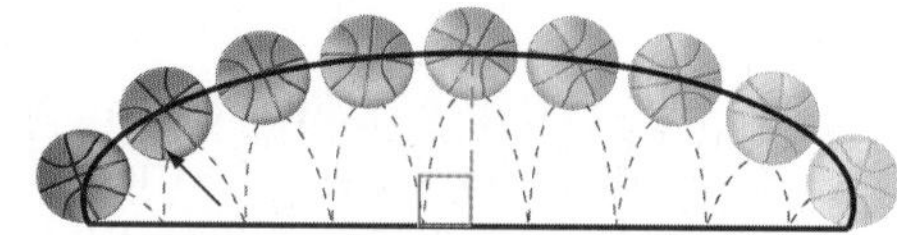

b. Graph the equation on a graphing calculator.

c. Show that the formula $h = \frac{v_0^2 \tan^2 \theta}{2g \sec^2 \theta}$ is equivalent to the one given above.

H.O.T. Problems Use Higher-Order Thinking Skills

52. CCSS ARGUMENTS Identify the equation that does not belong with the other three. Explain your reasoning.

$\sin^2 \theta + \cos^2 \theta = 1$	$1 + \cot^2 \theta = \csc^2 \theta$
$\sin^2 \theta - \cos^2 \theta = 2 \sin^2 \theta$	$\tan^2 \theta + 1 = \sec^2 \theta$

53. CHALLENGE Transform the right side of $\tan^2 \theta = \frac{\sin^2 \theta}{\cos^2 \theta}$ to show that $\tan^2 \theta = \sec^2 \theta - 1$.

54. WRITING IN MATH Explain why you cannot square each side of an equation when verifying a trigonometric identity.

55. REASONING Explain why $\sin^2 \theta + \cos^2 \theta = 1$ is an identity, but $\sin \theta = \sqrt{1 - \cos \theta}$ is not.

56. WRITE A QUESTION A classmate is having trouble trying to verify a trigonometric identity involving multiple trigonometric functions to multiple degrees. Write a question to help her work through the problem.

57. WRITING IN MATH Why do you think expressions in trigonometric identities are often rewritten in terms of sine and cosine?

58. CHALLENGE Let $x = \frac{1}{2} \tan \theta$, where $-\frac{\pi}{2} < \theta < \frac{\pi}{2}$. Write $f(x) = \frac{x}{\sqrt{1 + 4x^2}}$ in terms of a single trigonometric function of θ.

59. REASONING Justify the three basic Pythagorean identities.

Standardized Test Practice

60. SAT/ACT A small business owner must hire seasonal workers as the need arises. The following list shows the number of employees hired monthly for a 5-month period.

5, 14, 6, 8, 12

If the mean of these data is 9, what is the population standard deviation for these data? (Round your answer to the nearest tenth.)

A 3.5
B 3.9
C 5.7
D 8.6
E 12.3

61. Find the center and radius of the circle with equation $(x - 4)^2 + y^2 - 16 = 0$.

F $C(-4, 0)$; $r = 4$ units
G $C(-4, 0)$; $r = 16$ units
H $C(4, 0)$; $r = 4$ units
J $C(4, 0)$; $r = 16$ units

62. GEOMETRY The perimeter of a right triangle is 36 inches. Twice the length of the longer leg minus twice the length of the shorter leg is 6 inches. What are the lengths of all three sides?

A 3 in., 4 in., 5 in.
B 6 in., 8 in., 10 in.
C 9 in., 12 in., 15 in.
D 12 in., 16 in., 20 in.

63. Simplify $128^{\frac{1}{4}}$.

F $2\sqrt[4]{2}$
G $2\sqrt[4]{8}$
H 4
J $4\sqrt[4]{2}$

Spiral Review

Find the exact value of each expression. (Lesson 13-1)

64. $\tan \theta$, if $\cot \theta = 2$; $0° \leq \theta < 90°$

65. $\sin \theta$, if $\cos \theta = \frac{2}{3}$; $0° \leq \theta < 90°$

66. $\csc \theta$, if $\cos \theta = -\frac{3}{5}$; $90° < \theta < 180°$

67. $\cos \theta$, if $\sec \theta = \frac{5}{3}$; $270° < \theta < 360°$

68. ARCHITECTURE The support for a roof is shaped like two right triangles, as shown at the right. Find θ. (Lesson 12-9)

69. FAST FOOD The table shows the probability distribution for value meals ordered at a fast food restaurant on Saturday mornings. Use this information to determine the expected value of the meals ordered. (Lesson 11-3)

Value Meals Ordered				
Meals	\$3	\$4	\$5	\$6
Probability	0.5	0.2	0.1	0.2

Find the coordinates of the vertices and foci and the equations of the asymptotes for the hyperbolas with the given equations. Then graph the hyperbola. (Lesson 9-5)

70. $\frac{y^2}{18} - \frac{x^2}{20} = 1$

71. $\frac{(y + 6)^2}{20} - \frac{(x - 1)^2}{25} = 1$

72. $x^2 - 36y^2 = 36$

Skills Review

Simplify.

73. $\frac{2 + \sqrt{2}}{5 - \sqrt{2}}$

74. $\frac{x + 1}{\sqrt{x^2 - 1}}$

75. $\frac{x - 1}{\sqrt{x} - 1}$

76. $\frac{-2 - \sqrt{3}}{1 + \sqrt{3}}$

LESSON 13-3

Sum and Difference of Angles Identities

::Then	::Now	::Why?
You found values of trigonometric functions for general angles.	**1** Find values of sine and cosine by using sum and difference identities. **2** Verify trigonometric identities by using sum and difference identities.	Have you ever been using a wireless Internet provider and temporarily lost the signal? Waves that pass through the same place at the same time cause interference. Interference occurs when two waves combine to have a greater, or smaller, amplitude than either of the component waves.

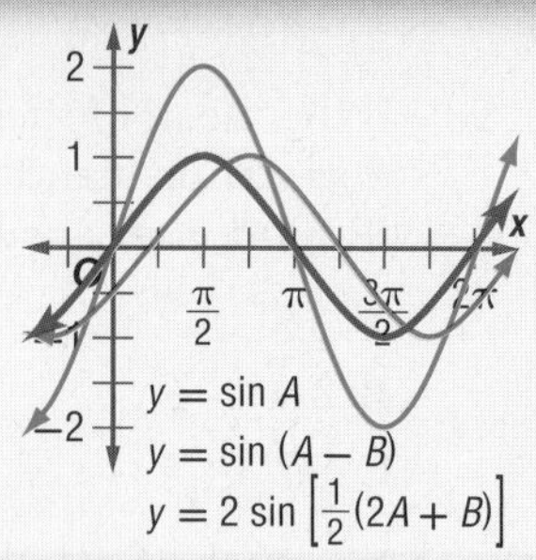

Common Core State Standards

Content Standards

F.TF.8 Prove the Pythagorean identity $\sin^2(\theta) + \cos^2(\theta) = 1$ and use it to find $\sin(\theta)$, $\cos(\theta)$, or $\tan(\theta)$ given $\sin(\theta)$, $\cos(\theta)$, or $\tan(\theta)$ and the quadrant of the angle.

Mathematical Practices

3 Construct viable arguments and critique the reasoning of others.

6 Attend to precision.

1 Sum and Difference Identities

Notice that the third equation shown above involves the sum of A and B. It is often helpful to use formulas for the trigonometric values of the difference or sum of two angles. For example, you could find the exact value of sin 15° by evaluating sin (60° − 45°). Formulas exist that can be used to evaluate expressions like $\sin (A - B)$ or $\cos (A + B)$.

KeyConcept Sum and Difference Identities

Sum Identities	Difference Identities
$\sin (A + B) = \sin A \cos B + \cos A \sin B$	$\sin (A - B) = \sin A \cos B - \cos A \sin B$
$\cos (A + B) = \cos A \cos B - \sin A \sin B$	$\cos (A - B) = \cos A \cos B + \sin A \sin B$
$\tan (A + B) = \dfrac{\tan A + \tan B}{1 - \tan A \tan B}$	$\tan (A - B) = \dfrac{\tan A - \tan B}{1 + \tan A \tan B}$

Example 1 Find Trigonometric Values

Find the exact value of each expression.

a. sin 105°

Use the identity $\sin (A + B) = \sin A \cos B + \cos A \sin B$.

$\sin 105° = \sin (60° + 45°)$ — $A = 60°$ and $B = 45°$

$= \sin 60° \cos 45° + \cos 60° \sin 45°$ — Sum identity

$= \left(\frac{\sqrt{3}}{2} \cdot \frac{\sqrt{2}}{2}\right) + \left(\frac{1}{2} \cdot \frac{\sqrt{2}}{2}\right)$ — Evaluate each expression.

$= \frac{\sqrt{6}}{4} + \frac{\sqrt{2}}{4}$ or $\frac{\sqrt{6} + \sqrt{2}}{4}$ — Multiply.

b. cos (−120°)

Use the identity $\cos (A - B) = \cos A \cos B + \sin A \sin B$.

$\cos (-120) = \cos (60° - 180°)$ — $A = 60°$ and $B = 180°$

$= \cos 60° \cos 180° + \sin 60° \sin 180°$ — Difference identity

$= \frac{1}{2} \cdot (-1) + \frac{\sqrt{3}}{2} \cdot 0$ — Evaluate each expression.

$= -\frac{1}{2}$ — Multiply.

GuidedPractice

1A. sin 15° **1B.** cos (−15°)

You can use the sum and difference of angles identities to solve real-world applications.

Real-World Example 2 Sum and Difference of Angles Identities

A geologist measures the angle between one side of a rectangular lot and the line from her position to the opposite corner of the lot as 30°. She then measures the angle between that line and the line to the point on the property where a river crosses as 45°. She stands 100 yards from the opposite corner of the property. How far is she from the point at which the river crosses the property line?

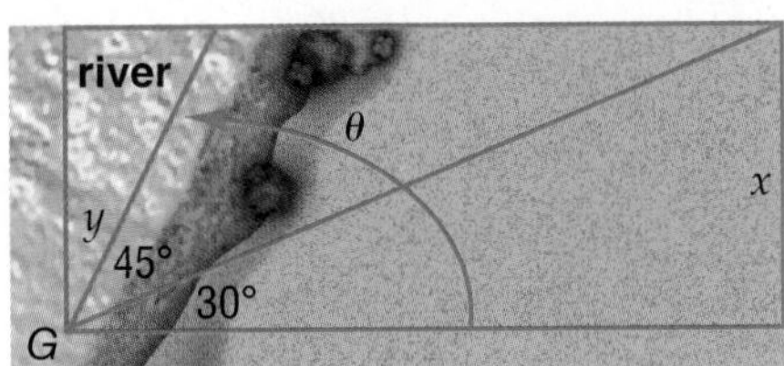

Problem-Solving Tip

Make a Model Make a model to visualize a problem situation. A model can be a drawing or a figure made of different objects, such as algebra tiles or folded paper.

Understand The question asks for the distance between the geologist and the point where the river crosses the property line, or y.

Plan Draw a picture that labels all the things that you know from the information given.

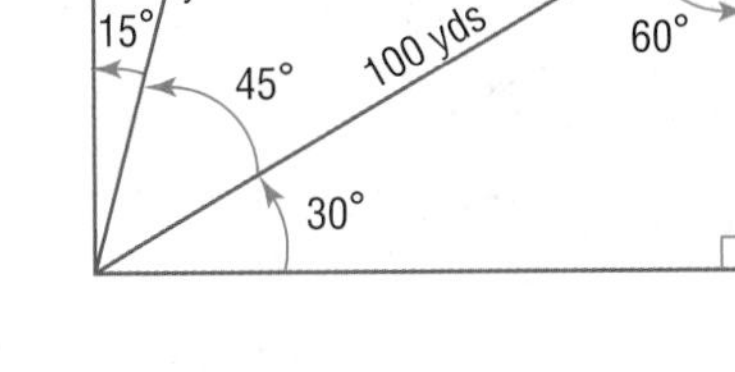

Solve Solve for x.

$\sin 30° = \frac{x}{100}$ Definition of sine

$x = 100 \sin 30°$

$x = 50$ Since the lot is rectangular, opposite sides are equal.

Now look at the triangle on the far left and solve for y.

$\cos 15° = \frac{50}{y}$ Definition of cosine

$\cos (45° - 30°) = \frac{50}{y}$ $15 = 45 - 30$

$\cos 45° \cos 30° + \sin 45° \sin 30° = \frac{50}{y}$ Difference identity

$\frac{\sqrt{2}}{2} \cdot \frac{\sqrt{3}}{2} + \frac{\sqrt{2}}{2} \cdot \frac{1}{2} = \frac{50}{y}$ Evaluate.

$\frac{\sqrt{6} + \sqrt{2}}{4} = \frac{50}{y}$ Simplify.

$(\sqrt{6} + \sqrt{2})y = 200$ Cross products

$$y = \frac{200}{(\sqrt{6} + \sqrt{2})} \cdot \frac{(\sqrt{6} - \sqrt{2})}{(\sqrt{6} - \sqrt{2})}$$

$$y = 50(\sqrt{6} - \sqrt{2})$$

$y = 50\sqrt{6} - 50\sqrt{2}$ or about 51.8

The geologist is about 51.8 yards from the point where the river crosses the property line.

Check Use a calculator to find the Arccos of $\frac{50}{51.8} \approx 15°$. ✓

Guided Practice

2. The harmonic motion of an object can be described by $x = 4 \cos \left(2\pi t - \frac{\pi}{4}\right)$, where x is distance from the equilibrium point in inches and t is time in minutes. Find the exact distance from the equilibrium point at 45 seconds.

StudyTip

CCSS Sense-Making Make a list of the trigonometric values for the angles between 0° and 360° for which the sum and difference identities can be easily used. Use your list as a reference.

2 Verify Trigonometric Identities

You can also use the sum and difference identities to verify identities.

Example 3 Verify Trigonometric Identities

Verify that each equation is an identity.

a. $\cos(90° - \theta) = \sin\theta$

$\cos(90° - \theta) \stackrel{?}{=} \sin\theta$	Original equation
$\cos 90° \cos\theta + \sin 90° \sin\theta \stackrel{?}{=} \sin\theta$	Sum identity
$0 \cdot \cos\theta + 1 \cdot \sin\theta \stackrel{?}{=} \sin\theta$	Evaluate each expression.
$\sin\theta = \sin\theta$ ✓	Simplify.

b. $\sin\left(\theta + \frac{\pi}{2}\right) = \cos\theta$

$\sin\left(\theta + \frac{\pi}{2}\right) \stackrel{?}{=} \cos\theta$	Original equation
$\sin\theta\cos\frac{\pi}{2} + \cos\theta\sin\frac{\pi}{2} \stackrel{?}{=} \cos\theta$	Sum identity
$\sin\theta \cdot 0 + \cos\theta \cdot 1 \stackrel{?}{=} \cos\theta$	Evaluate each expression.
$\cos\theta = \cos\theta$ ✓	Simplify.

GuidedPractice

3A. $\sin(90° - \theta) = \cos\theta$

3B. $\cos(90° + \theta) = -\sin\theta$

Check Your Understanding

 = Step-by-Step Solutions begin on page R14.

Example 1 **Find the exact value of each expression.**

1. $\cos 165°$

2. $\cos 105°$

3. $\cos 75°$

4. $\sin(-30°)$

5. $\sin 135°$

6. $\sin(-210°)$

Example 2

7. **CCSS MODELING** Refer to the beginning of the lesson. *Constructive interference* occurs when two waves combine to have a greater amplitude than either of the component waves. *Destructive interference* occurs when the component waves combine to have a smaller amplitude. The first signal can be modeled by the equation $y = 20\sin(3\theta + 45°)$. The second signal can be modeled by the equation $y = 20\sin(3\theta + 225°)$.

a. Find the sum of the two functions.

b. What type of interference results when signals modeled by the two equations are combined?

Example 3 **Verify that each equation is an identity.**

8. $\sin(90° + \theta) = \cos\theta$

9. $\cos\left(\frac{3\pi}{2} - \theta\right) = -\sin\theta$

10. $\tan\left(\theta + \frac{\pi}{2}\right) = -\cot\theta$

11. $\sin(\theta + \pi) = -\sin\theta$

Practice and Problem Solving

Extra Practice is on page R13.

Example 1 **Find the exact value of each expression.**

12. $\sin 165°$

13. $\cos 135°$

14. $\cos \frac{7\pi}{12}$

15. $\sin \frac{\pi}{12}$

16. $\tan 195°$

17. $\cos \left(-\frac{\pi}{12}\right)$

Example 2

18. ELECTRONICS In a certain circuit carrying alternating current, the formula $c = 2 \sin (120t)$ can be used to find the current c in amperes after t seconds.

a. Rewrite the formula using the sum of two angles.

b. Use the sum of angles formula to find the exact current at $t = 1$ second.

Example 3 **Verify that each equation is an identity.**

19. $\cos \left(\frac{\pi}{2} + \theta\right) = -\sin \theta$

20. $\cos (60° + \theta) = \sin (30° - \theta)$

21. $\cos (180° + \theta) = -\cos \theta$

22. $\tan (\theta + 45°) = \frac{1 + \tan \theta}{1 - \tan \theta}$

23. CCSS REASONING The monthly high temperatures for Minneapolis, Minnesota, can be modeled by the equation $y = 31.65 \sin \left(\frac{\pi}{6}x - 2.09\right) + 52.35$, where the months x are represented by January = 1, February = 2, and so on. The monthly low temperatures for Minneapolis can be modeled by the equation $y = 30.15 \sin \left(\frac{\pi}{6}x - 2.09\right) + 32.95$.

a. Write a new function by adding the expressions on the right side of each equation and dividing the result by 2.

b. What is the meaning of the function you wrote in part **a**?

Find the exact value of each expression.

24. $\tan 165°$

25. $\sec 1275°$

26. $\sin 735°$

27. $\tan \frac{23\pi}{12}$

28. $\csc \frac{5\pi}{12}$

29. $\cot \frac{113\pi}{12}$

30. FORCE In the figure at the right, the effort F necessary to hold a safe in position on a ramp is given by $F = \frac{W(\sin A + \mu \cos A)}{\cos A - \mu \sin A}$, where W is the weight of the safe and $\mu = \tan \theta$. Show that $F = W \tan (A + \theta)$.

31. QUILTING As part of a quilt that is being made, the quilter places two right triangular swatches together to make a new triangular piece. One swatch has sides 6 inches, 8 inches, and 10 inches long. The other swatch has sides 8 inches, $8\sqrt{3}$ inches, and 16 inches long. The pieces are placed with the sides of eight inches against each other, as shown in the figure, to form triangle ABC.

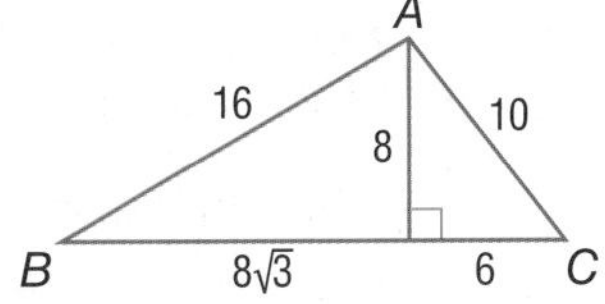

a. What is the exact value of the sine of angle BAC?

b. What is the exact value of the cosine of angle BAC?

c. What is the measure of angle BAC?

d. Is the new triangle formed from the two triangles also a right triangle?

32. **OPTICS** When light passes symmetrically through a prism, the index of refraction n of the glass with respect to air is $n = \dfrac{\sin\left[\frac{1}{2}(a+b)\right]}{\sin\frac{b}{2}}$, where a is the measure of the deviation angle and b is the measure of the prism apex angle.

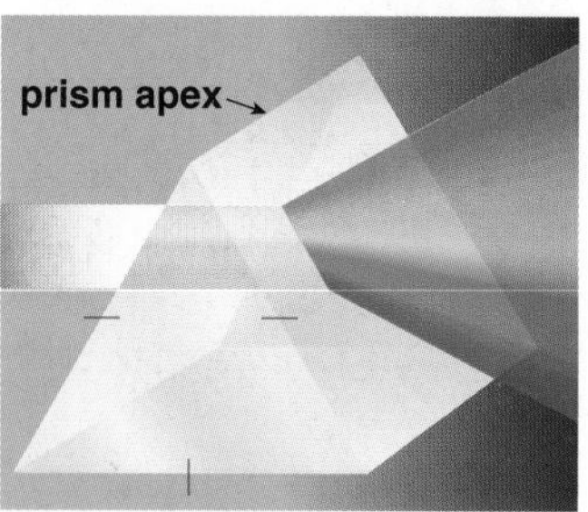

a. Show that for the prism shown, $n = \sqrt{3}\sin\frac{a}{2} + \cos\frac{a}{2}$.

b. Find n for the prism shown.

33. **MULTIPLE REPRESENTATIONS** In this problem, you will disprove the hypothesis that $\sin(A + B) = \sin A + \sin B$.

a. Tabular Copy and complete the table.

A	B	sin A	sin B	sin (A + B)	sin A + sin B
30°	90°				
45°	60°				
60°	45°				
90°	30°				

b. Graphical Assume that B is always 15° less than A. Use a graphing calculator to graph $y = \sin(x + x - 15)$ and $y = \sin x + \sin(x - 15)$ on the same screen.

c. Analytical Determine whether $\cos(A + B) = \cos A + \cos B$ is an identity. Explain your reasoning.

Verify that each equation is an identity.

34. $\sin(A + B) = \dfrac{\tan A + \tan B}{\sec A \sec B}$

35. $\cos(A + B) = \dfrac{1 - \tan A \tan B}{\sec A \sec B}$

36. $\sec(A - B) = \dfrac{\sec A \sec B}{1 + \tan A \tan B}$

37. $\sin(A + B)\sin(A - B) = \sin^2 A - \sin^2 B$

H.O.T. Problems Use Higher-Order Thinking Skills

38. **REASONING** Simplify the following expression without expanding any of the sums or differences.

$$\sin\left(\frac{\pi}{3} - \theta\right)\cos\left(\frac{\pi}{3} + \theta\right) - \cos\left(\frac{\pi}{3} - \theta\right)\sin\left(\frac{\pi}{3} + \theta\right)$$

39. **WRITING IN MATH** Use the information at the beginning of the lesson and in Exercise 7 to explain how the sum and difference identities are used to describe wireless Internet interference. Include an explanation of the difference between constructive and destructive interference.

40. **CHALLENGE** Derive an identity for $\cot(A + B)$ in terms of $\cot A$ and $\cot B$.

41. **CCSS ARGUMENTS** The figure shows two angles A and B in standard position on the unit circle. Use the Distance Formula to find d, where $(x_1, y_1) = (\cos B, \sin B)$ and $(x_2, y_2) = (\cos A, \sin A)$.

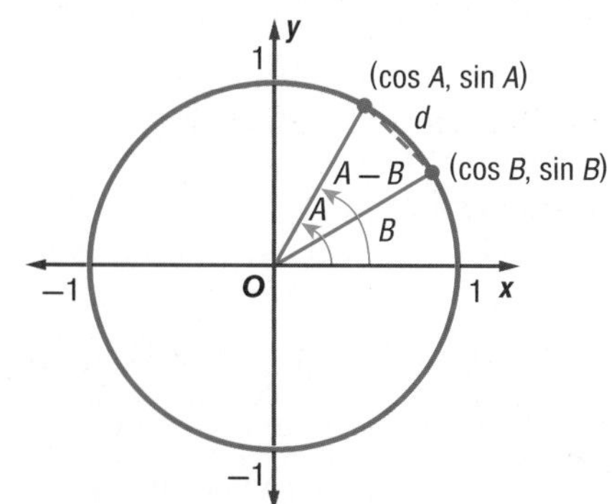

42. **OPEN ENDED** Consider the following theorem. *If A, B, and C are the angles of an oblique triangle, then tan A + tan B + tan C = tan A tan B tan C.* Choose values for A, B, and C. Verify that the conclusion is true for your specific values.

Standardized Test Practice

43. GRIDDED RESPONSE The mean of seven numbers is 0. The sum of three of the numbers is −9. What is the sum of the remaining four numbers?

44. The variables a, b, c, d, and f are integers in a sequence, where $a = 2$ and $b = 12$. To find the next term, double the last term and add that result to one less than the next-to-last term. For example, $c = 25$, because $2(12) = 24$, $2 - 1 = 1$, and $24 + 1 = 25$. What is the value of f?

A 74

B 144

C 146

D 256

45. SAT/ACT Solve $x^2 - 5x < 14$.

F $\{x \mid -7 < x < 2\}$

G $\{x \mid x < -7 \text{ or } x > 2\}$

H $\{x \mid -2 < x < 7\}$

J $\{x \mid x < -2 \text{ or } x > 7\}$

K $\{x \mid x > -2 \text{ and } x < 7\}$

46. PROBABILITY A math teacher is randomly distributing 15 yellow pencils and 10 green pencils. What is the probability that the first pencil she hands out will be yellow and the second pencil will be green?

A $\frac{1}{24}$

B $\frac{1}{4}$

C $\frac{2}{5}$

D $\frac{23}{25}$

Spiral Review

Verify that each equation is an identity. (Lesson 13-2)

47. $\frac{\sin \theta}{\tan \theta} + \frac{\cos \theta}{\cot \theta} = \cos \theta + \sin \theta$

48. $\sec \theta (\sec \theta - \cos \theta) = \tan^2 \theta$

Simplify each expression. (Lesson 13-1)

49. $\sin \theta \csc \theta - \cos^2 \theta$

50. $\cos^2 \theta \sec \theta \csc \theta$

51. $\cos \theta + \sin \theta \tan \theta$

52. GUITAR When a guitar string is plucked, it is displaced from a fixed point in the middle of the string and vibrates back and forth, producing a musical tone. The exact tone depends on the frequency, or number of cycles per second, that the string vibrates. To produce an A, the frequency is 440 cycles per second, or 440 hertz (Hz). (Lesson 12-6)

a. Find the period of this function.

b. Graph the height of the fixed point on the string from its resting position as a function of time. Let the maximum distance above the resting position have a value of 1 unit, and let the minimum distance below this position have a value of 1 unit.

Prove that each statement is true for all positive integers. (Lesson 10-7)

53. $4^n - 1$ is divisible by 3.

54. $5^n + 3$ is divisible by 4.

Skills Review

Solve each equation.

55. $7 + \sqrt{4x + 8} = 9$

56. $\sqrt{y + 21} - 1 = \sqrt{y + 12}$

57. $\sqrt{4z + 1} = 3 + \sqrt{4z - 2}$

CHAPTER 13

Mid-Chapter Quiz

Lessons 13-1 through 13-3

Simplify each expression. (Lesson 13-1)

1. $\cot \theta \sec \theta$
2. $\dfrac{1 - \cos^2 \theta}{\sin^2 \theta}$
3. $\dfrac{1}{\cos \theta} - \dfrac{\sin^2 \theta}{\cos \theta}$
4. $\cos\left(\dfrac{\pi}{2} - \theta\right) \csc \theta$

5. **HISTORY** In 1861, the United States 34-star flag was adopted. For this flag, $\tan \theta = \dfrac{31.5}{51}$. Find $\sin \theta$.

Find the value of each expression. (Lesson 13-1)

6. $\sin \theta$, if $\cos \theta = \dfrac{3}{5}$; $0° < \theta < 90°$

7. $\csc \theta$, if $\cot \theta = \dfrac{1}{2}$; $270° < \theta < 360°$

8. $\tan \theta$, if $\sec \theta = \dfrac{4}{3}$; $0° < \theta < 90°$

9. **MULTIPLE CHOICE** Which of the following is equivalent to $\dfrac{\cos \theta}{1 - \sin^2 \theta}$? (Lesson 13-1)

   A $\cos \theta$

   B $\csc \theta$

   C $\tan \theta$

   D $\sec \theta$

10. **AMUSEMENT PARKS** Suppose a child on a merry-go-round is seated on an outside horse. The diameter of the merry-go-round is 16 meters. The angle of inclination is represented by the equation $\tan \theta = \dfrac{v^2}{gR}$, where R is the radius of the circular path, v is the speed in meters per second, and g is 9.8 meters per second squared. (Lesson 13-1)

    a. If the sine of the angle of inclination of the child is $\dfrac{1}{5}$, what is the angle of inclination made by the child?

    b. What is the velocity of the merry-go-round?

    c. If the speed of the merry-go-round is 3.6 meters per second, what is the value of the angle of inclination of a rider?

Verify that each of the following is an identity. (Lesson 13-2)

11. $\cot^2 \theta + 1 = \dfrac{\cot \theta}{\cos \theta \cdot \sin \theta}$

12. $\dfrac{\cos \theta \csc \theta}{\cot \theta} = 1$

13. $\dfrac{\sin \theta \tan \theta}{1 - \cos \theta} = (1 + \cos \theta) \sec \theta$

14. $\tan \theta(1 - \sin \theta) = \dfrac{\cos \theta \sin \theta}{1 + \sin \theta}$

15. **COMPUTER** The front of a computer monitor is usually measured along the diagonal of the screen as shown below. (Lesson 13-2)

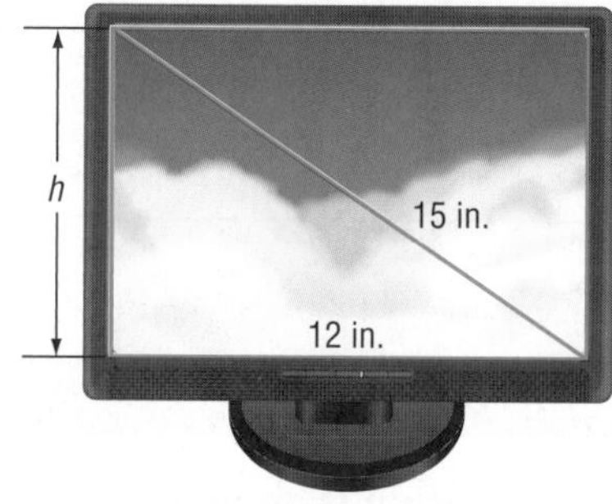

    a. Find h.

    b. Using the diagram shown, show that $\cot \theta = \dfrac{\cos \theta}{\sin \theta}$.

Verify that each of the following is an identity. (Lesson 13-2)

16. $\tan^2 \theta + 1 = \dfrac{\tan \theta}{\cos \theta \cdot \sin \theta}$

17. $\dfrac{\sin \theta \cdot \sec \theta}{\sec \theta - 1} = (\sec \theta + 1) \cot \theta$

18. $\sin^2 \theta \cdot \tan^2 \theta = \tan^2 \theta - \sin^2 \theta$

19. $\cot \theta(1 - \cos \theta) = \dfrac{\cos \theta \cdot \sin \theta}{1 + \cos \theta}$

Find the exact value of each expression. (Lesson 13-3)

20. $\cos 105°$

21. $\sin (-135°)$

22. $\tan 15°$

23. $\cot 75°$

24. **MULTIPLE CHOICE** What is the exact value of $\cos \dfrac{5\pi}{12}$? (Lesson 13-3)

    F $\sqrt{2}$

    G $\dfrac{\sqrt{6} + \sqrt{2}}{2}$

    H $\dfrac{\sqrt{6} - \sqrt{2}}{4}$

    J $\dfrac{\sqrt{6} + \sqrt{2}}{4}$

25. Verify that $\cos 30° \cos \theta + \sin 30° \sin \theta = \sin 60° \cos \theta + \cos 60° \sin \theta$ is an identity. (Lesson 13-3)

LESSON

13-4 Double-Angle and Half-Angle Identities

Then

- You found values of sine and cosine by using sum and difference identities.

Now

1. Find values of sine and cosine by using double-angle identities.
2. Find values of sine and cosine by using half-angle identities.

Why?

- Chicago's Buckingham Fountain contains jets placed at specific angles that shoot water into the air to create arcs. When a stream of water shoots into the air with velocity v at an angle of θ with the horizontal, the model predicts that the water will travel a horizontal distance of $D = \frac{v^2}{g}\sin 2\theta$ and reach a maximum height of $H = \frac{v^2}{2g}\sin^2\theta$. The ratio of H to D helps determine the total height and width of the fountain. Express $\frac{H}{D}$ as a function of θ.

Common Core State Standards

Content Standards

F.TF.8 Prove the Pythagorean identity $\sin^2(\theta) + \cos^2(\theta) = 1$ and use it to find $\sin(\theta)$, $\cos(\theta)$, or $\tan(\theta)$ given $\sin(\theta)$, $\cos(\theta)$, or $\tan(\theta)$ and the quadrant of the angle.

Mathematical Practices

3 Construct viable arguments and critique the reasoning of others.

6 Attend to precision.

1 Double-Angle Identities

It is sometimes useful to have identities to find the value of a function of twice an angle or half an angle.

KeyConcept Double-Angle Identities

The following identities hold true for all values of θ.

$$\sin 2\theta = 2\sin\theta\cos\theta$$

$$\cos 2\theta = \cos^2\theta - \sin^2\theta$$

$$\cos 2\theta = 2\cos^2\theta - 1$$

$$\cos 2\theta = 1 - 2\sin^2\theta$$

$$\tan 2\theta = \frac{2\tan\theta}{1-\tan^2\theta}$$

Example 1 Double-Angle Identities

Find the exact value of $\sin 2\theta$ if $\sin\theta = \frac{2}{3}$ and θ is between 0° and 90°.

Step 1 Use the identity $\sin 2\theta = 2\sin\theta\cos\theta$ to find the value of $\cos\theta$.

$\cos^2\theta = 1 - \sin^2\theta$ — $\cos^2\theta + \sin^2\theta = 1$

$\cos^2\theta = 1 - \left(\frac{2}{3}\right)^2$ — $\sin\theta = \frac{2}{3}$

$\cos^2\theta = \frac{5}{9}$ — Subtract.

$\cos\theta = \pm\frac{\sqrt{5}}{3}$ — Take the square root of each side.

Since θ is in the first quadrant, cosine is positive. Thus, $\cos\theta = \frac{\sqrt{5}}{3}$.

Step 2 Find $\sin 2\theta$.

$\sin 2\theta = 2\sin\theta\cos\theta$ — Double-angle identity

$= 2\left(\frac{2}{3}\right)\left(\frac{\sqrt{5}}{3}\right)$ — $\sin\theta = \frac{2}{3}$ and $\cos\theta = \frac{\sqrt{5}}{3}$

$= \frac{4\sqrt{5}}{9}$ — Multiply.

GuidedPractice

1. Find the exact value of $\sin 2\theta$ if $\cos\theta = -\frac{1}{3}$ and $90° < \theta < 180°$.

Example 2 Double-Angle Identities

Find the exact value of each expression if $\sin \theta = \frac{2}{3}$ and θ is between 0° and 90°.

a. $\cos 2\theta$

Since we know the values of $\cos \theta$ and $\sin \theta$, we can use any of the double-angle identities for cosine. We will use the identity $\cos 2\theta = 1 - 2\sin^2 \theta$.

$\cos 2\theta = 1 - 2\sin^2 \theta$ — Double-angle identity

$= 1 - 2\left(\frac{2}{3}\right)^2$ or $\frac{1}{9}$ — $\sin \theta = \frac{2}{3}$

b. $\tan 2\theta$

Step 1 Find $\tan \theta$ to use the double-angle identity for $\tan 2\theta$.

$\tan \theta = \frac{\sin \theta}{\cos \theta}$ — Definition of tangent

$= \frac{\frac{2}{3}}{\frac{\sqrt{5}}{3}}$ — $\sin \theta = \frac{2}{3}$ and $\cos \theta = \frac{\sqrt{5}}{3}$

$= \frac{2}{\sqrt{5}}$ or $\frac{2\sqrt{5}}{5}$ — Rationalize the denominator.

Step 2 Find $\tan 2\theta$.

$\tan 2\theta = \frac{2\tan \theta}{1 - \tan^2 \theta}$ — Double-angle identity

$= \frac{2\left(\frac{2\sqrt{5}}{5}\right)}{1 - \left(\frac{2\sqrt{5}}{5}\right)^2}$ — $\tan \theta = \frac{2\sqrt{5}}{5}$

$= \frac{2\left(\frac{2\sqrt{5}}{5}\right)}{\frac{25}{25} - \frac{20}{25}}$ — Square the denominator.

$= \frac{\frac{4\sqrt{5}}{5}}{\frac{1}{5}}$ — Simplify.

$= \frac{4\sqrt{5}}{5} \cdot \frac{5}{1}$ or $4\sqrt{5}$ — $\frac{a}{b} \div \frac{c}{d} = \frac{a}{b} \cdot \frac{d}{c}$

GuidedPractice

Find the exact value of each expression if $\cos \theta = -\frac{1}{3}$ and $90° < \theta < 180°$.

2A. $\cos 2\theta$

2B. $\tan 2\theta$

StudyTip

Deriving Formulas You can use the identity for $\sin(A + B)$ to find the sine of twice an angle θ, $\sin 2\theta$, and the identity for $\cos(A + B)$ to find the cosine of twice an angle θ, $\cos 2\theta$.

Real-WorldCareer

Electrician An electrician specializes in the wiring of electrical components. Electricians serve an apprenticeship lasting 3–5 years. Schooling in electrical theory and building codes is required. Certification requires work experience and a passing score on a written test.

2 Half-Angle Identities

It is sometimes useful to have identities to find the value of a function of half an angle.

KeyConcept Half-Angle Identities

The following identities hold true for all values of θ.

$\sin \frac{\theta}{2} = \pm\sqrt{\frac{1 - \cos \theta}{2}}$ $\quad \cos \frac{\theta}{2} = \pm\sqrt{\frac{1 + \cos \theta}{2}}$ $\quad \tan \frac{\theta}{2} = \pm\sqrt{\frac{1 - \cos \theta}{1 + \cos \theta}}, \cos \theta \neq -1$

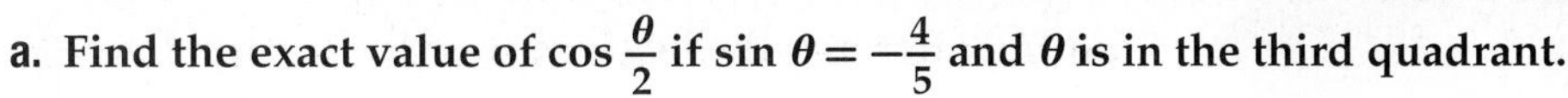

Example 3 Half-Angle Identities

a. Find the exact value of $\cos \frac{\theta}{2}$ if $\sin \theta = -\frac{4}{5}$ and θ is in the third quadrant.

$\cos^2 \theta = 1 - \sin^2 \theta$ — Use a Pythagorean identity to find $\cos \theta$.

$\cos^2 \theta = 1 - \left(-\frac{4}{5}\right)^2$ — $\sin \theta = -\frac{4}{5}$

$\cos^2 \theta = 1 - \frac{16}{25}$ — Evaluate exponent.

$\cos^2 \theta = \frac{9}{25}$ — Subtract.

$\cos \theta = \pm\frac{3}{5}$ — Take the square root of each side.

Since θ is in the third quadrant, $\cos \theta = -\frac{3}{5}$.

$\cos \frac{\theta}{2} = \pm\sqrt{\frac{1 + \cos \theta}{2}}$ — Half-angle identity

$= \pm\sqrt{\frac{1 - \frac{3}{5}}{2}}$ — $\cos \theta = -\frac{3}{5}$

$= \pm\sqrt{\frac{1}{5}}$ — Simplify.

$= \pm\frac{1}{\sqrt{5}} \cdot \frac{\sqrt{5}}{\sqrt{5}}$ or $\pm\frac{\sqrt{5}}{5}$ — Rationalize the denominator.

If θ is between 180° and 270°, $\frac{\theta}{2}$ is between 90° and 135°. So, $\cos \frac{\theta}{2}$ is $-\frac{\sqrt{5}}{5}$.

b. Find the exact value of cos 67.5°.

$\cos 67.5° = \cos \frac{135°}{2}$ — $67.5° = \frac{135°}{2}$

$= \sqrt{\frac{1 + \cos 135°}{2}}$ — $\cos \frac{\theta}{2} = \pm\sqrt{\frac{1 + \cos \theta}{2}}$

$= \sqrt{\frac{1 - \frac{\sqrt{2}}{2}}{2}}$ — 67.5° is in Quadrant I; the value is positive.

$= \sqrt{\frac{\frac{2}{2} - \frac{\sqrt{2}}{2}}{2}}$ — $1 = \frac{2}{2}$

$= \sqrt{\frac{\frac{2 - \sqrt{2}}{2}}{2}}$ — Subtract fractions.

$= \sqrt{\frac{2 - \sqrt{2}}{2} \cdot \frac{1}{2}}$ — $\frac{a}{b} \div \frac{c}{d} = \frac{a}{b} \cdot \frac{d}{c}$

$= \sqrt{\frac{2 - \sqrt{2}}{4}}$ — Multiply.

$= \frac{\sqrt{2 - \sqrt{2}}}{\sqrt{4}}$ — $\sqrt{\frac{a}{b}} = \frac{\sqrt{a}}{\sqrt{b}}$

$= \frac{\sqrt{2 - \sqrt{2}}}{2}$ — Simplify.

StudyTip

Choosing the Sign In the first step of the solution, you may want to determine the quadrant in which the terminal side of $\frac{\theta}{2}$ will lie. Then you can use the correct sign from that point on.

ReadingMath

Plus or Minus The first sign of the half-angle identity is read *plus or minus*. Unlike with the double-angle identities, you must determine the sign.

GuidedPractice

3. Find the exact value of $\sin \frac{\theta}{2}$ if $\sin \theta = \frac{2}{3}$ and θ is in the second quadrant.

Real-WorldLink

The City Hall Park Fountain in New York City is located in the heart of Manhattan in front of City Hall.

Source: Fodor's

Real-World Example 4 Simplify Using Double-Angle Identities

FOUNTAIN Refer to the beginning of the lesson. Find $\frac{H}{D}$.

$$\frac{H}{D} = \frac{\frac{v^2}{2g}\sin^2\theta}{\frac{v^2}{g}\sin 2\theta} \qquad \text{Original equation}$$

$$= \frac{\frac{v^2\sin^2\theta}{2g}}{\frac{v^2\sin 2\theta}{g}} \qquad \text{Simplify the numerator and denominator.}$$

$$= \frac{v^2\sin^2\theta}{2g} \cdot \frac{g}{v^2\sin 2\theta} \qquad \frac{a}{b} \div \frac{c}{d} = \frac{a}{b} \cdot \frac{d}{c}$$

$$= \frac{\sin^2\theta}{2\sin 2\theta} \qquad \text{Simplify.}$$

$$= \frac{\sin^2\theta}{4\sin\theta\cos\theta} \qquad \sin 2\theta = 2\sin\theta\cos\theta$$

$$= \frac{1}{4} \cdot \frac{\sin\theta}{\cos\theta} \qquad \text{Simplify.}$$

$$= \frac{1}{4}\tan\theta \qquad \frac{\sin\theta}{\cos\theta} = \tan\theta$$

GuidedPractice

Find each value.

4A. $\sin 135°$

4B. $\cos \frac{7\pi}{8}$

Recall that you can use the sum and difference identities to verify identities. Double- and half-angle identities can also be used to verify identities.

Example 5 Verify Identities

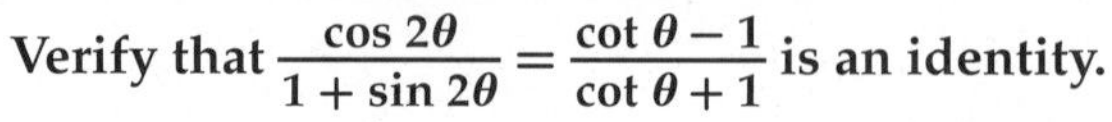

Verify that $\frac{\cos 2\theta}{1 + \sin 2\theta} = \frac{\cot\theta - 1}{\cot\theta + 1}$ is an identity.

$$\frac{\cos 2\theta}{1+\sin 2\theta} \stackrel{?}{=} \frac{\cot\theta - 1}{\cot\theta + 1} \qquad \text{Original equation}$$

$$\frac{\cos 2\theta}{1+\sin 2\theta} \stackrel{?}{=} \frac{\frac{\cos\theta}{\sin\theta} - 1}{\frac{\cos\theta}{\sin\theta} + 1} \qquad \cot\theta = \frac{\cos\theta}{\sin\theta}$$

$$\frac{\cos 2\theta}{1+\sin 2\theta} \stackrel{?}{=} \frac{\cos\theta - \sin\theta}{\cos\theta + \sin\theta} \qquad \text{Multiply numerator and denominator by } \sin\theta.$$

$$\frac{\cos 2\theta}{1+\sin 2\theta} \stackrel{?}{=} \frac{\cos\theta - \sin\theta}{\cos\theta + \sin\theta} \cdot \frac{\cos\theta + \sin\theta}{\cos\theta + \sin\theta} \qquad \text{Multiply the right side by 1.}$$

$$\frac{\cos 2\theta}{1+\sin 2\theta} \stackrel{?}{=} \frac{\cos^2\theta - \sin^2\theta}{\cos^2\theta + 2\cos\theta\sin\theta + \sin^2\theta} \qquad \text{Multiply.}$$

$$\frac{\cos 2\theta}{1+\sin 2\theta} \stackrel{?}{=} \frac{\cos^2\theta - \sin^2\theta}{1 + 2\cos\theta\sin\theta} \qquad \text{Simplify.}$$

$$\frac{\cos 2\theta}{1+\sin 2\theta} = \frac{\cos 2\theta}{1+\sin 2\theta} \checkmark \qquad \cos^2\theta - \sin^2\theta = \cos 2\theta;\ 2\cos\theta\sin\theta = \sin 2\theta$$

GuidedPractice

5. Verify that $4\cos^2 x - \sin^2 2x = 4\cos^4 x$.

Paul Coco/Painet Inc.

Check Your Understanding

= Step-by-Step Solutions begin on page R14.

Examples 1–3 **CCSS PRECISION Find the exact values of $\sin 2\theta$, $\cos 2\theta$, $\sin \frac{\theta}{2}$, and $\cos \frac{\theta}{2}$.**

1. $\sin \theta = \frac{1}{4}$; $0° < \theta < 90°$

2. $\sin \theta = \frac{4}{5}$; $90° < \theta < 180°$

3. $\cos \theta = -\frac{5}{13}$; $\frac{\pi}{2} < \theta < \pi$

4. $\cos \theta = \frac{3}{5}$; $270° < \theta < 360°$

5. $\tan \theta = -\frac{8}{15}$; $90° < \theta < 180°$

6. $\tan \theta = \frac{5}{12}$; $\pi < \theta < \frac{3\pi}{2}$

Find the exact value of each expression.

7. $\sin \frac{\pi}{8}$

8. $\cos 15°$

Example 4

9. SOCCER A soccer player kicks a ball at an angle of 37° with the ground with an initial velocity of 52 feet per second. The distance d that the ball will go in the air if it is not blocked is given by $d = \frac{2v^2 \sin \theta \cos \theta}{g}$. In this formula, g is the acceleration due to gravity and is equal to 32 feet per second squared, and v is the initial velocity.

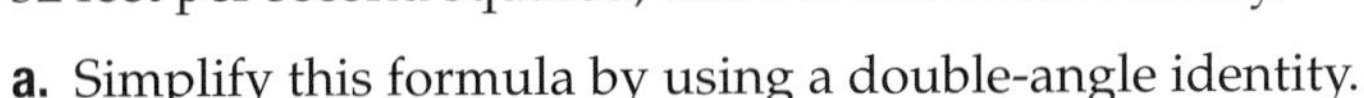

a. Simplify this formula by using a double-angle identity.

b. Using the simplified formula, how far will this ball go?

Example 5 **Verify that each equation is an identity.**

10. $\tan \theta = \frac{1 - \cos 2\theta}{\sin 2\theta}$

11. $(\sin \theta + \cos \theta)^2 = 1 + 2 \sin \theta \cos \theta$

Practice and Problem Solving

Extra Practice is on page R13.

Examples 1–3 **Find the exact values of $\sin 2\theta$, $\cos 2\theta$, $\sin \frac{\theta}{2}$, and $\cos \frac{\theta}{2}$.**

12. $\sin \theta = \frac{2}{3}$; $90° < \theta < 180°$

13. $\sin \theta = -\frac{15}{17}$; $\pi < \theta < \frac{3\pi}{2}$

14. $\cos \theta = \frac{3}{5}$; $\frac{3\pi}{2} < \theta < 2\pi$

15. $\cos \theta = \frac{1}{5}$; $270° < \theta < 360°$

16. $\tan \theta = \frac{4}{3}$; $180° < \theta < 270°$

17. $\tan \theta = -2$; $\frac{\pi}{2} < \theta < \pi$

Find the exact value of each expression.

18. $\sin 75°$

19. $\sin \frac{3\pi}{8}$

20. $\cos \frac{7\pi}{12}$

21. $\tan 165°$

22. $\tan \frac{5\pi}{12}$

23. $\tan 22.5°$

24. GEOGRAPHY The Mercator projection of the globe is a projection on which the distance between the lines of latitude increases with their distance from the equator. The calculation of the location of a point on this projection involves the expression $\tan\left(45° + \frac{L}{2}\right)$, where L is the latitude of the point.

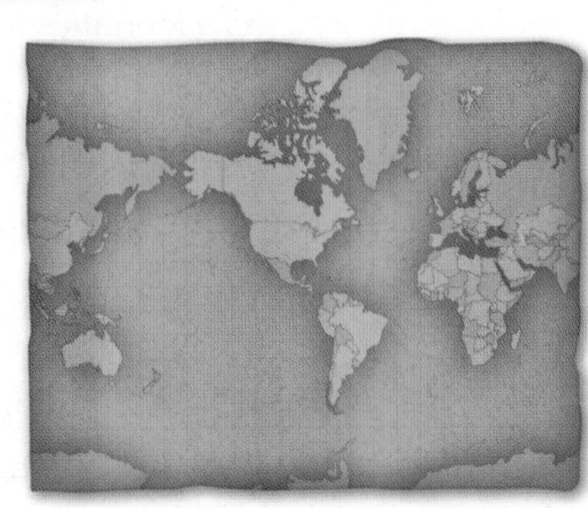

a. Write this expression in terms of a trigonometric function of L.

b. The latitude of Tallahassee, Florida, is 30° north. Find the value of the expression if $L = 30°$.

Example 4

25. **ELECTRONICS** Consider an AC circuit consisting of a power supply and a resistor. If the current I_0 in the circuit at time t is $I_0 \sin t\theta$, then the power delivered to the resistor is $P = I_0^2 R \sin^2 t\theta$, where R is the resistance. Express the power in terms of $\cos 2t\theta$.

Example 5

Verify that each equation is an identity.

26. $\tan 2\theta = \dfrac{2}{\cot \theta - \tan \theta}$

27. $1 + \dfrac{1}{2} \sin 2\theta = \dfrac{\sec \theta + \sin \theta}{\sec \theta}$

28. $\sin \dfrac{\theta}{2} \cos \dfrac{\theta}{2} = \dfrac{\sin \theta}{2}$

29. $\tan \dfrac{\theta}{2} = \dfrac{\sin \theta}{1 + \cos \theta}$

30. **FOOTBALL** Suppose a place kicker consistently kicks a football with an initial velocity of 95 feet per second. Prove that the horizontal distance the ball travels in the air will be the same for $\theta = 45° + A$ as for $\theta = 45° - A$. Use the formula given in Exercise 9.

Find the exact values of $\sin 2\theta$, $\cos 2\theta$, and $\tan 2\theta$.

31. $\cos \theta = \dfrac{4}{5}$; $0° < \theta < 90°$

32. $\sin \theta = \dfrac{1}{3}$; $0 < \theta < \dfrac{\pi}{2}$

33. $\tan \theta = -3$; $90° < \theta < 180°$

34. $\sec \theta = -\dfrac{4}{3}$; $90° < \theta < 180°$

35. $\csc \theta = -\dfrac{5}{2}$; $\dfrac{3\pi}{2} < \theta < 2\pi$

36. $\cot \theta = \dfrac{3}{2}$; $180° < \theta < 270°$

H.O.T. Problems Use Higher-Order Thinking Skills

37. CCSS **CRITIQUE** Teresa and Nathan are calculating the exact value of sin 15°. Is either of them correct? Explain your reasoning.

Teresa

$\sin (A - B) = \sin A \cos B - \cos A \sin B$

$\sin (45 - 30) = \sin 45 \cos 30 - \cos 45 \sin 30$

$= \dfrac{\sqrt{2}}{2} \cdot \dfrac{\sqrt{3}}{2} - \dfrac{\sqrt{2}}{2} \cdot \dfrac{1}{2}$

$= \dfrac{\sqrt{4}}{4}$

Nathan

$\sin \dfrac{A}{2} = \pm\sqrt{\dfrac{1 - \cos A}{2}}$

$\sin \dfrac{30}{2} = \pm\sqrt{\dfrac{1 - \frac{1}{2}}{2}}$

$= 0.5$

38. **CHALLENGE** Circle O is a unit circle. Use the figure to prove that $\tan \dfrac{1}{2}\theta = \dfrac{\sin \theta}{1 + \cos \theta}$.

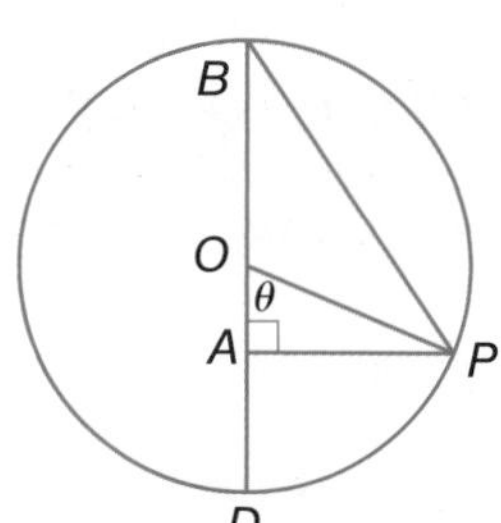

39. **WRITING IN MATH** Write a short paragraph about the conditions under which you would use each of the three identities for $\cos 2\theta$.

40. **PROOF** Use the formula for $\sin (A + B)$ to derive the formula for $\sin 2\theta$, and use the formula for $\cos (A + B)$ to derive the formula for $\cos 2\theta$.

41. **REASONING** Derive the half-angle identities from the double-angle identities.

42. **OPEN ENDED** Suppose a golfer consistently hits the ball so that it leaves the tee with an initial velocity of 115 feet per second and $d = \dfrac{2v^2 \sin \theta \cos \theta}{g}$. Explain why the maximum distance is attained when $\theta = 45°$.

Standardized Test Practice

43. SHORT RESPONSE Angles C and D are supplementary. The measure of angle C is seven times the measure of angle D. Find the measure of angle D in degrees.

44. SAT/ACT Ms. Romero has a list of the yearly salaries of the staff members in her department. Which measure of data describes the middle income value of the salaries?

A mean
B median
C mode
D range
E standard deviation

45. Identify the domain and range of the function $f(x) = |4x + 1| - 8$.

F $D = \{x \mid -3 \leq x \leq 1\}$, $R = \{y \mid y \geq -8\}$
G $D = \{\text{all real numbers}\}$, $R = \{y \mid y \geq -8\}$
H $D = \{x \mid -3 \leq x \leq 1\}$, $R = \{\text{all real numbers}\}$
J $D = \{\text{all real numbers}\}$, $R = \{\text{all real numbers}\}$

46. GEOMETRY Angel is putting a stone walkway around a circular pond. He has enough stones to make a walkway 144 feet long. If he uses all of the stones to surround the pond, what is the radius of the pond?

A $\frac{12}{\pi}$ ft
B $\frac{72}{\pi}$ ft
C 72π ft
D 144π ft

Spiral Review

Find the exact value of each expression. (Lesson 13-3)

47. $\sin 135°$
48. $\cos 105°$
49. $\sin 285°$
50. $\cos(-30°)$
51. $\sin(-240°)$
52. $\cos(-120°)$

Verify that each equation is an identity. (Lesson 13-2)

53. $\cot\theta + \sec\theta = \frac{\cos^2\theta + \sin\theta}{\sin\theta\cos\theta}$

54. $\sin^2\theta + \tan^2\theta = (1 - \cos^2\theta) + \frac{\sec^2\theta}{\csc^2\theta}$

Determine whether each triangle should be solved by beginning with the Law of *Sines* or Law of *Cosines*. Then solve each triangle. Round measures of sides to the nearest tenth and measures of angles to the nearest degree. (Lesson 12-5)

55.
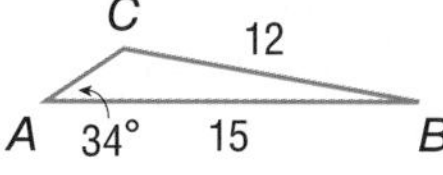

56.
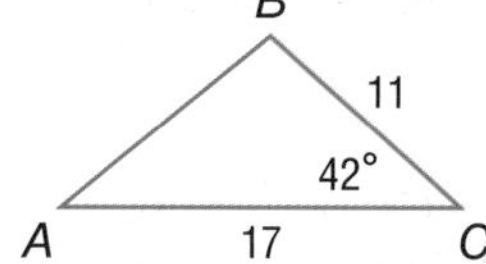

57.
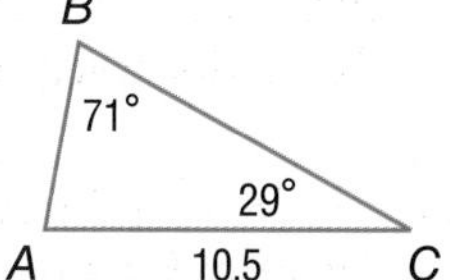

Skills Review

Solve each equation by factoring.

58. $x^2 + 5x - 24 = 0$
59. $x^2 - 3x - 28 = 0$
60. $x^2 - 4x = 21$

EXPLORE 13-5

Graphing Technology Lab
Solving Trigonometric Equations

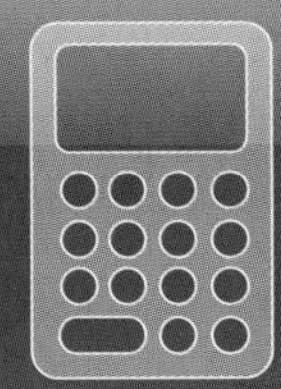

The graph of a trigonometric function is made up of points that represent all values that satisfy the function. To solve a trigonometric equation, you need to find all values of the variable that satisfy the equation. You can use a TI-83/84 Plus graphing calculator to solve trigonometric equations by graphing each side of the equation as a function and then locating the points of intersection.

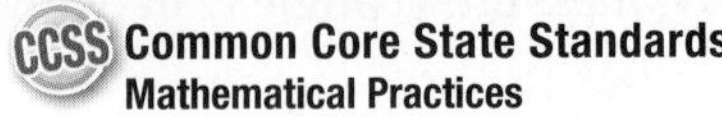

CCSS Common Core State Standards
Mathematical Practices
5 Use appropriate tools strategically.

Activity 1 Real Solutions

Use a graphing calculator to solve $\sin x = 0.4$ if $0° \leq x < 360°$.

Step 1 Enter and graph related equations. Rewrite the equation as two equations, **Y1** $= \sin x$ and **Y2** $= 0.4$. Then graph the two equations. Because the interval is in degrees, set your calculator to degree mode.

KEYSTROKES: [MODE] [▼] [▼] [▶] [ENTER]
[Y=] [SIN] [X,T,θ,n] [)]
[ENTER] 0.4 [ENTER] [GRAPH]

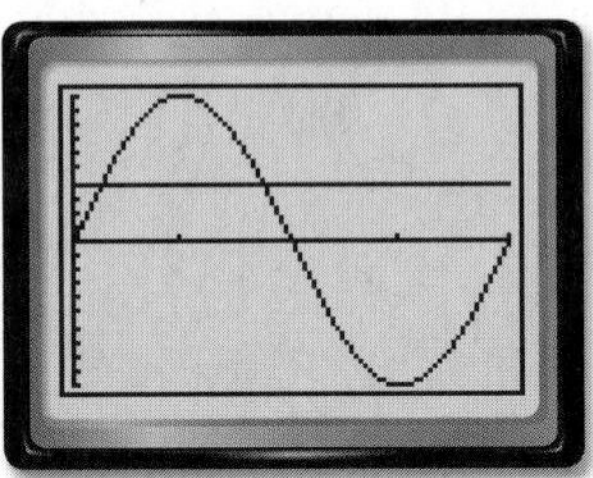

[0, 360] scl: 90 by [−1, 1] scl: 0.1

Step 2 Approximate the solutions. Based on the graph, you can see that there are two points of intersection in the interval $0° \leq x < 360°$. Use the **CALC** feature to determine the x-values at which the two graphs intersect.

The solutions are $x \approx 23.57°$ and $x \approx 156.4°$.

Activity 2 No Real Solutions

Use a graphing calculator to solve $\tan^2 x \cos x + 3 \cos x = 0$ if $0° \leq x < 360°$.

Step 1 Enter and graph related equations. The related equations to be graphed are $y_1 = \tan^2 x \cos x + 3 \cos x$ and $y_2 = 0$.

KEYSTROKES: [Y=] [TAN] [X,T,θ,n] [)] [x^2] [COS]
[X,T,θ,n] [)] [+] 3 [COS] [X,T,θ,n]
[)] [ENTER] 0 [ENTER]

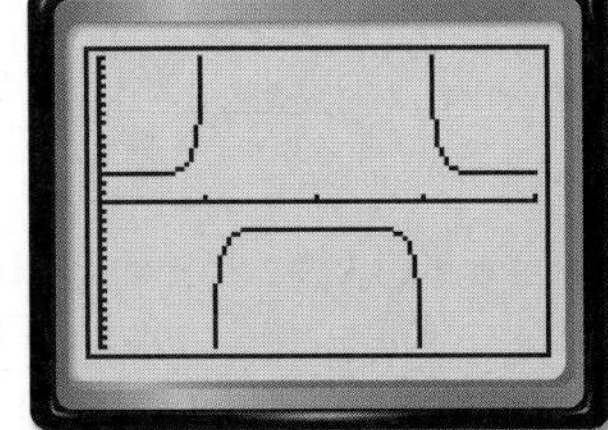

[0, 360] scl: 90 by [−15, 15] scl: 1

Step 2 These two functions do not intersect.

Therefore, the equation $\tan^2 x \cos x + 3 \cos x = 0$ has no real solutions.

Exercises

Use a graphing calculator to solve each equation for the values of x indicated.

1. $\sin x = 0.7$; $0° \leq x < 360°$
2. $\tan x = \cos x$; $0° \leq x < 360°$
3. $3 \cos x + 4 = 0.5$; $0° \leq x < 360°$
4. $0.25 \cos x = 3.4$; $-720° \leq x < 720°$
5. $\sin 2x = \sin x$; $0° \leq x < 360°$
6. $\sin 2x - 3 \sin x = 0$ if $-360° \leq x < 360°$

LESSON

13-5 Solving Trigonometric Equations

:: Then	:: Now	:: Why?
● You verified trigonometric identities.	● **1** Solve trigonometric equations. **2** Find extraneous solutions from trigonometric equations.	● When you ride a Ferris wheel that has a diameter of 40 meters and turns at a rate of 1.5 revolutions per minute, the height above the ground, in meters, of your seat after t minutes can be modeled by the equation $$h = 21 - 20 \cos 3\pi t.$$ After the ride begins, how long is it before your seat is 31 meters above the ground for the first time?

NewVocabulary
trigonometric equations

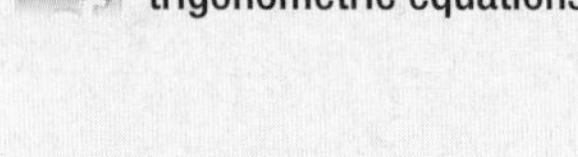

Common Core State Standards

Content Standards
F.TF.8 Prove the Pythagorean identity $\sin^2(\theta) + \cos^2(\theta) = 1$ and use it to find $\sin(\theta)$, $\cos(\theta)$, or $\tan(\theta)$ given $\sin(\theta)$, $\cos(\theta)$, or $\tan(\theta)$ and the quadrant of the angle.

Mathematical Practices
4 Model with mathematics.
6 Attend to precision.

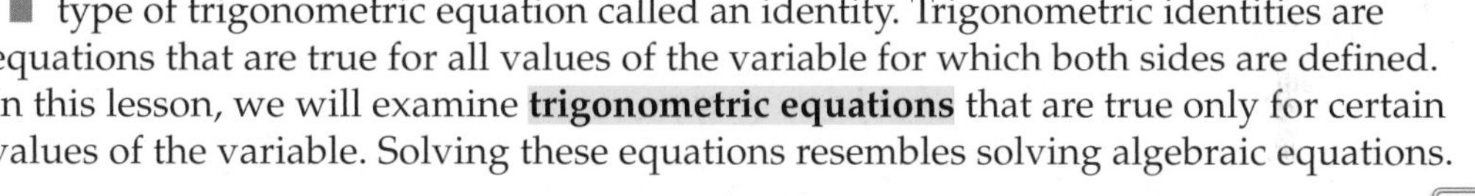

1 Solve Trigonometric Equations

So far in this chapter, we have studied a special type of trigonometric equation called an identity. Trigonometric identities are equations that are true for all values of the variable for which both sides are defined. In this lesson, we will examine **trigonometric equations** that are true only for certain values of the variable. Solving these equations resembles solving algebraic equations.

Example 1 Solve Equations for a Given Interval

PT

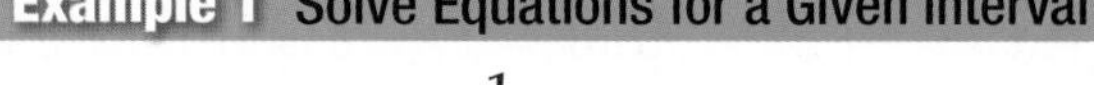

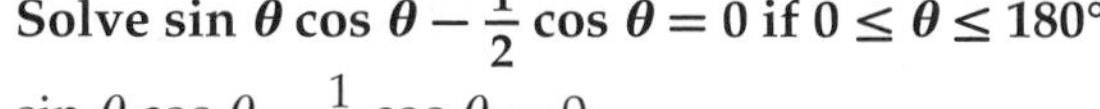

Solve $\sin\theta\cos\theta - \frac{1}{2}\cos\theta = 0$ if $0 \le \theta \le 180°$.

$\sin\theta\cos\theta - \frac{1}{2}\cos\theta = 0$ — Original equation

$\cos\theta\left(\sin\theta - \frac{1}{2}\right) = 0$ — Factor.

$\cos\theta = 0$ or $\sin\theta - \frac{1}{2} = 0$ — Zero Product Property

$\theta = 90°$ $\qquad \sin\theta = \frac{1}{2}$

$\theta = 30°$ or $150°$

The solutions are 30°, 90°, and 150°.

CHECK You can check the answer by graphing $y = \sin\theta\cos\theta$ and $y = \frac{1}{2}\cos\theta$ in the same coordinate plane on a graphing calculator. Then find the points where the graphs intersect. You can see that there are infinitely many such points, but we are only interested in the points between 0° and 180°.

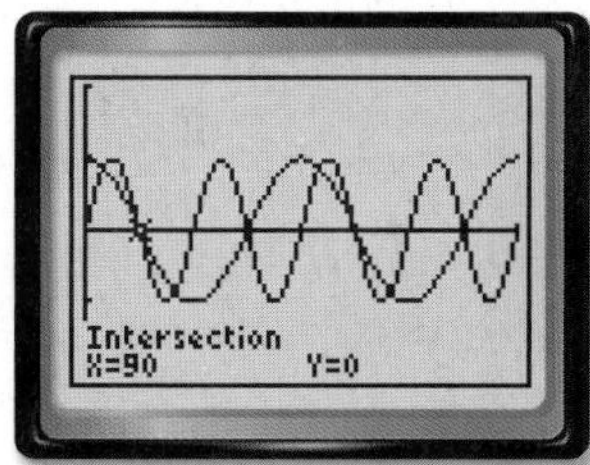

[0, 720] scl: 90 by [−1, 1] scl: 0.5

GuidedPractice

1. Find all solutions of $\sin 2\theta = \cos\theta$ if $0 \le \theta \le 2\pi$.

Trigonometric equations are usually solved for values of the variable between 0° and 360° or between 0 radians and 2π radians. There are solutions outside that interval. These other solutions differ by integral multiples of the period of the function.

Hans-jürgen Hermann/age fotostock

Example 2 Infinitely Many Solutions

Solve $\cos \theta + 1 = 0$ for all values of θ if θ is measured in radians.

$$\cos \theta + 1 = 0$$
$$\cos \theta = -1$$

Look at the graph of $y = \cos \theta$ to find solutions of $\cos \theta = -1$.

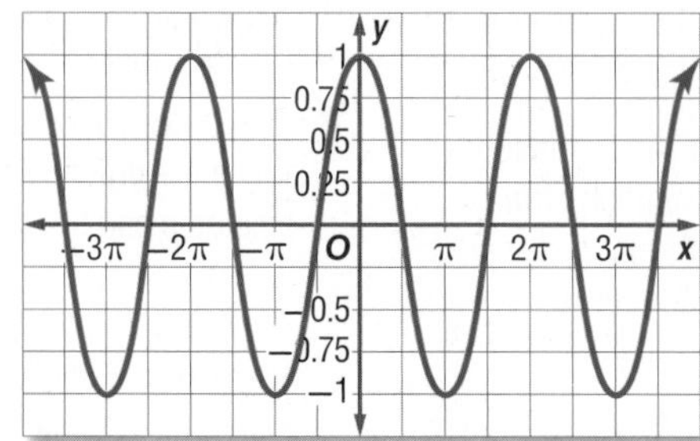

StudyTip

Expressing Solutions as Multiples The expression $\pi + 2k\pi$ includes 3π and its multiples, so it is not necessary to list them separately.

The solutions are π, 3π, 5π, and so on, and $-\pi$, -3π, -5π, and so on. The only solution in the interval 0 radians to 2π radians is π. The period of the cosine function is 2π radians. So the solutions can be written as $\pi + 2k\pi$, where k is any integer.

GuidedPractice

2A. Solve $\cos 2\theta + \cos \theta + 1 = 0$ for all values of θ if θ is measured in degrees.

2B. Solve $2 \sin \theta = -1$ for all values of θ if θ is measured in radians.

Trigonometric equations are often used to solve real-world problems.

Real-World Example 3 Solve Trigonometric Equations

AMUSEMENT PARKS Refer to the beginning of the lesson. How long after the Ferris wheel starts will your seat first be 31 meters above the ground?

$h = 21 - 20 \cos 3\pi t$	Original equation
$31 = 21 - 20 \cos 3\pi t$	Replace h with 31.
$10 = -20 \cos 3\pi t$	Subtract 21 from each side.
$-\frac{1}{2} = \cos 3\pi t$	Divide each side by −20.
$\cos^{-1}\left(-\frac{1}{2}\right) = 3\pi t$	Take the Arccosine.
$\frac{2\pi}{3} = 3\pi t$ or $\frac{4\pi}{3} = 3\pi t$	The Arccosine of $-\frac{1}{2}$ is $\frac{2\pi}{3}$ or $\frac{4\pi}{3}$.
$\frac{2\pi}{3} + 2\pi k = 3\pi t$ or $\frac{4\pi}{3} + 2\pi k = 3\pi t$	k is any integer.
$\frac{2}{9} + \frac{2}{3}k = t$ $\frac{4}{9} + \frac{2}{3}k = t$	Divide each term by 3π.

The least positive value for t is obtained by letting $k = 0$ in the first expression.

Therefore, $t = \frac{2}{9}$ of a minute or about 13 seconds.

GuidedPractice

3. How long after the Ferris wheel starts will your seat first be 41 meters above the ground?

2 Extraneous Solutions

Some trigonometric equations have no solution. For example, the equation $\cos \theta = 4$ has no solution because all values of $\cos \theta$ are between -1 and 1, inclusive. Thus, the solution set for $\cos \theta = 4$ is empty.

Example 4 Determine Whether a Solution Exists

Solve each equation.

a. $\mathbf{2 \sin^2 \theta - 3 \sin \theta - 2 = 0}$ **if** $\mathbf{0 \leq \theta \leq 2\pi}$

$2 \sin^2 \theta - 3 \sin \theta - 2 = 0$ — Original equation

$(\sin \theta - 2)(2 \sin \theta + 1) = 0$ — Factor.

$\sin \theta - 2 = 0$ or $2 \sin \theta + 1 = 0$ — Zero Product Property

$\sin \theta = 2$ $\qquad$ $2 \sin \theta = -1$

This is not a solution since all values of $\sin \theta$ are between -1 and 1, inclusive. $\qquad$ $\sin \theta = -\frac{1}{2}$

$\theta = \frac{7\pi}{6}$ or $\frac{11\pi}{6}$

The solutions are $\frac{7\pi}{6}$ or $\frac{11\pi}{6}$.

> **Problem-Solving Tip**
>
> **CCSS Regularity** Look for patterns in your solutions. Look for pairs of solutions that differ by exactly π or 2π and write your solutions with the simplest possible pattern.

CHECK $2 \sin \theta - 3 \sin \theta - 2 = 0$

$$2 \sin^2 \left(\frac{7\pi}{6}\right) - 3 \sin \left(\frac{7\pi}{6}\right) - 2 \stackrel{?}{=} 0$$
$$2\left(\frac{1}{4}\right) - 3\left(-\frac{1}{2}\right) - 2 \stackrel{?}{=} 0$$
$$\frac{1}{2} + \frac{3}{2} - 2 \stackrel{?}{=} 0$$
$$0 = 0 \checkmark$$

$2 \sin^2 \theta - 3 \sin \theta - 2 = 0$

$$2 \sin^2 \left(\frac{11\pi}{6}\right) - 3 \sin \left(\frac{11\pi}{6}\right) - 2 \stackrel{?}{=} 0$$
$$2\left(\frac{1}{4}\right) - 3\left(-\frac{1}{2}\right) - 2 \stackrel{?}{=} 0$$
$$\frac{1}{2} + \frac{3}{2} - 2 \stackrel{?}{=} 0$$
$$0 = 0 \checkmark$$

b. $\mathbf{\sin \theta = 1 + \cos \theta}$ **if** $\mathbf{0° \leq \theta < 360°}$

$\sin \theta = 1 + \cos \theta$ — Original equation

$\sin^2 \theta = (1 + \cos \theta)^2$ — Square each side.

$1 - \cos^2 \theta = 1 + 2 \cos \theta + \cos^2 \theta$ — $\sin^2 \theta = 1 - \cos^2 \theta$

$0 = 2 \cos \theta + 2 \cos^2 \theta$ — Set the left side equal to 0.

$0 = 2 \cos \theta (1 + \cos \theta)$ — Factor.

$1 + \cos \theta = 0$ or $2 \cos \theta = 0$ — Zero Product Property

$\cos \theta = -1$ $\qquad$ $\cos \theta = 0$

$\theta = 180$ $\qquad$ $\theta = 90°$ or $270°$

CHECK $\sin \theta = 1 + \cos \theta$

$$\sin 90° \stackrel{?}{=} 1 + \cos 90°$$
$$1 \stackrel{?}{=} 1 + 0$$
$$1 = 1 \checkmark$$

$\sin \theta = 1 + \cos \theta$

$$\sin 180° \stackrel{?}{=} 1 + \cos 180°$$
$$0 \stackrel{?}{=} 1 + (-1)$$
$$0 = 0 \checkmark$$

$\sin \theta = 1 + \cos \theta$

$$\sin 270° \stackrel{?}{=} 1 + \cos 270°$$
$$-1 \stackrel{?}{=} 1 + 0$$
$$-1 \neq 1 \quad X$$

The solutions are 90° and 180°.

Guided Practice

4A. $\sin^2 \theta + 2 \cos^2 \theta = 4$

4B. $\cos^2 \theta + 3 = 4 - \sin^2 \theta$

If an equation cannot be solved easily by factoring, try rewriting the expression using trigonometric identities. However, using identities and some algebraic operations, such as squaring, may result in extraneous solutions. So, it is necessary to check your solutions using the original equation.

StudyTip

Solving Trigonometric Equations Remember that *solving a trigonometric equation* means solving for all values of the variable.

PT

Example 5 Solve Trigonometric Equations by Using Identities

Solve $2 \sec^2 \theta - \tan^4 \theta = -1$ for all values of θ if θ is measured in degrees.

$2 \sec^2 \theta - \tan^4 \theta = -1$ Original equation

$2(1 + \tan^2 \theta) - \tan^4 \theta = -1$ $\sec^2 \theta = 1 + \tan^2 \theta$

$2 + 2 \tan^2 \theta - \tan^4 \theta = -1$ Distributive Property

$\tan^4 \theta - 2 \tan^2 \theta - 3 = 0$ Set one side of the equation equal to 0.

$(\tan^2 \theta - 3)(\tan^2 \theta + 1) = 0$ Factor.

$\tan^2 \theta - 3 = 0$ or $\tan^2 \theta + 1 = 0$ Zero Product Property

$\tan^2 \theta = 3$ $\tan^2 \theta = -1$

$\tan \theta = \pm\sqrt{3}$ This part gives no solutions since $\tan^2 \theta$ is never negative.

$\theta = 60° + 180°k$ and $\theta = -60° + 180°k$, where k is any integer. The solutions are $60° + 180°k$ and $-60° + 180°k$.

GuidedPractice

Solve each equation.

5A. $\sin \theta \cot \theta - \cos^2 \theta = 0$

5B. $\dfrac{\cos \theta}{\cot \theta} + 2 \sin^2 \theta = 0$

Check Your Understanding

 = Step-by-Step Solutions begin on page R14. 

Example 1 **CCSS REGULARITY** **Solve each equation if $0° \leq \theta \leq 360°$.**

1. $2 \sin \theta + 1 = 0$

2. $\cos^2 \theta + 2 \cos \theta + 1 = 0$

3. $\cos 2\theta + \cos \theta = 0$

4. $2 \cos \theta = 1$

5. $\cos \theta = -\dfrac{\sqrt{3}}{2}$

6. $\sin 2\theta = -\dfrac{\sqrt{3}}{2}$

7. $\cos 2\theta = 8 - 15 \sin \theta$

8. $\sin \theta + \cos \theta = 1$

Example 2 **Solve each equation for all values of θ if θ is measured in radians.**

9. $4 \sin^2 \theta - 1 = 0$

10. $2 \cos^2 \theta = 1$

11. $\cos 2\theta \sin \theta = 1$

12. $\sin \dfrac{\theta}{2} + \cos \dfrac{\theta}{2} = \sqrt{2}$

13. $\cos 2\theta + 4 \cos \theta = -3$

14. $\sin \dfrac{\theta}{2} + \cos \theta = 1$

Solve each equation for all values of θ if θ is measured in degrees.

15. $\cos 2\theta - \sin^2 \theta + 2 = 0$

16. $\sin^2 \theta - \sin \theta = 0$

17. $2 \sin^2 \theta - 1 = 0$

18. $\cos \theta - 2 \cos \theta \sin \theta = 0$

19. $\cos 2\theta \sin \theta = 1$

20. $\sin \theta \tan \theta - \tan \theta = 0$

Example 3

21. **LIGHT** The number of hours of daylight d in Hartford, Connecticut, may be approximated by the equation $d = 3 \sin \dfrac{2\pi}{365}t + 12$, where t is the number of days after March 21.

a. On what days will Hartford have exactly $10\frac{1}{2}$ hours of daylight?

b. Using the results in part **a**, tell what days of the year have at least $10\frac{1}{2}$ hours of daylight. Explain how you know.

Examples 4–5 **Solve each equation.**

22. $\sin^2 2\theta + \cos^2 \theta = 0$

23. $\tan^2 \theta + 2 \tan \theta + 1 = 0$

24. $\cos^2 \theta + 3 \cos \theta = -2$

25. $\sin 2\theta - \cos \theta = 0$

26. $\tan \theta = 1$

27. $\cos 8\theta = 1$

28. $\sin \theta + 1 = \cos 2\theta$

29. $2 \cos^2 \theta = \cos \theta$

Practice and Problem Solving

Extra Practice is on page R13.

Example 1 **Solve each equation for the given interval.**

30. $\cos^2 \theta = \frac{1}{4}; 0° \leq \theta \leq 360°$

31. $2 \sin^2 \theta = 1; 90° < \theta < 270°$

32. $\sin 2\theta - \cos \theta = 0; 0 \leq \theta \leq 2\pi$

33. $3 \sin^2 \theta = \cos^2 \theta; 0 \leq \theta \leq \frac{\pi}{2}$

34. $2 \sin \theta + \sqrt{3} = 0; 180° < \theta < 360°$

35. $4 \sin^2 \theta - 1 = 0; 180° < \theta < 360°$

Example 2 **Solve each equation for all values of θ if θ is measured in radians.**

36. $\cos 2\theta + 3 \cos \theta = 1$

37. $2 \sin^2 \theta = \cos \theta + 1$

38. $\cos^2 \theta - \frac{3}{2} = \frac{5}{2} \cos \theta$

39. $3 \cos \theta - \cos \theta = 2$

Solve each equation for all values of θ if θ is measured in degrees.

40. $\sin \theta - \cos \theta = 0$

41. $\tan \theta - \sin \theta = 0$

42. $\sin^2 \theta = 2 \sin \theta + 3$

43. $4 \sin^2 \theta = 4 \sin \theta - 1$

Example 3 44. **ELECTRONICS** One of the tallest structures in the world is a television transmitting tower located near Fargo, North Dakota, with a height of 2064 feet. What is the measure of θ if the length of the shadow is 1 mile?

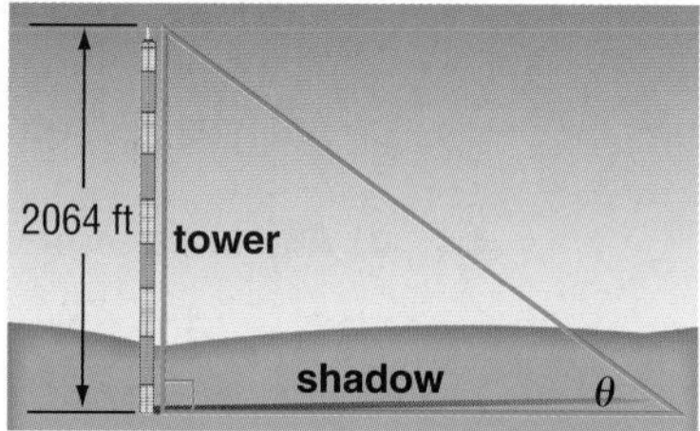

Examples 4–5 **Solve each equation.**

45. $2 \sin^2 \theta = 3 \sin \theta + 2$

46. $2 \cos^2 \theta + 3 \sin \theta = 3$

47. $\sin^2 \theta + \cos 2\theta = \cos \theta$

48. $2 \cos^2 \theta = -\cos \theta$

49. **CCSS SENSE-MAKING** Due to ocean tides, the depth y in meters of the River Thames in London varies as a sine function of x, the hour of the day. On a certain day that function was $y = 3 \sin\left[\frac{\pi}{6}(x - 4)\right] + 8$, where $x = 0, 1, 2, \ldots, 24$ corresponds to 12:00 midnight, 1:00 A.M., 2:00 A.M., …, 12:00 midnight the next night.

   a. What is the maximum depth of the River Thames on that day?

   b. At what times does the maximum depth occur?

Solve each equation if θ is measured in radians.

50. $(\cos \theta)(\sin 2\theta) - 2 \sin \theta + 2 = 0$

51. $2 \sin^2 \theta + (\sqrt{2} - 1) \sin \theta = \frac{\sqrt{2}}{2}$

Solve each equation if θ is measured in degrees.

52. $\sin 2\theta + \frac{\sqrt{3}}{2} = \sqrt{3} \sin \theta + \cos \theta$

53. $1 - \sin^2 \theta - \cos \theta = \frac{3}{4}$

Solve each equation.

54. $2 \sin \theta = \sin 2\theta$

55. $\cos \theta \tan \theta - 2 \cos^2 \theta = -1$

56. DIAMONDS According to Snell's Law, $n_1 \sin \boldsymbol{i} = n_2 \sin r$, where n_1 is the index of refraction of the medium the light is exiting, n_2 is the index of refraction of the medium the light is entering, $\boldsymbol{i}$ is the degree measure of the angle of incidence, and r is the degree measure of the angle of refraction.

a. The index of refraction of a diamond is 2.42, and the index of refraction of air is 1.00. If a beam of light strikes a diamond at an angle of 35°, what is the angle of refraction?

b. Explain how a gemologist might use Snell's Law to determine whether a diamond is genuine.

57 **CCSS PERSEVERANCE** A wave traveling in a guitar string can be modeled by the equation $D = 0.5 \sin (6.5x) \sin (2500t)$, where D is the displacement in millimeters at the position x millimeters from the left end of the string at time t seconds. Find the first positive time when the point 0.5 meter from the left end has a displacement of 0.01 millimeter.

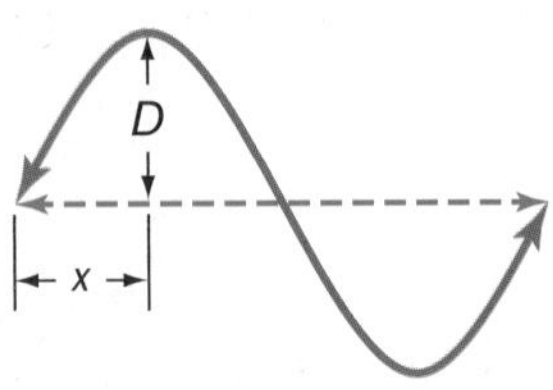

58. MULTIPLE REPRESENTATIONS Consider the trigonometric inequality $\sin \theta \geq \frac{1}{2}$.

a. Tabular Construct a table of values for $0° \leq \theta \leq 360°$. For what values of θ is $\sin \theta \geq \frac{1}{2}$?

b. Graphical Graph $y = \sin \theta$ and $y = \frac{1}{2}$ on the same graph for $0° \leq \theta \leq 360°$. For what values of θ is the graph of $y = \sin \theta$ above the graph of $y = \frac{1}{2}$?

c. Analytic Based on your answers for parts **a** and **b**, solve $\sin \theta \geq \frac{1}{2}$ for all values of θ.

d. Algebraic Solve each inequality if $0 \leq \theta \leq 360°$. Then solve each for all values of θ.

i. $\cos \theta \geq \frac{\sqrt{2}}{2}$

ii. $2 \sin \theta \leq \sqrt{3}$

iii. $-\sin \theta \geq 0$

iv. $\cos \theta - 1 < -\frac{1}{2}$

H.O.T. Problems Use Higher-Order Thinking Skills

59. CHALLENGE Solve $\sin 2x < \sin x$ for $0 \leq x \leq 2\pi$ without a calculator.

60. REASONING Compare and contrast solving trigonometric equations with solving linear and quadratic equations. What techniques are the same? What techniques are different? How many solutions do you expect?

61. WRITING IN MATH Why do trigonometric equations often have infinitely many solutions?

62. OPEN ENDED Write an example of a trigonometric equation that has exactly two solutions if $0° \leq \theta \leq 360°$.

63. CHALLENGE How many solutions in the interval $0° \leq \theta \leq 360°$ should you expect for $a \sin (b\theta + c) = d$, if $a \neq 0$ and b is a positive integer?

Standardized Test Practice

64. EXTENDED RESPONSE Charles received \$2500 for a graduation gift. He put it into a savings account in which the interest rate was 5.5% per year.

a. How much did he have in his savings account after 5 years if he made no deposits or withdrawals?

b. After how many years will the amount in his savings account have doubled?

65. PROBABILITY Find the probability of rolling three 3s if a number cube is rolled three times.

A $\frac{1}{216}$ C $\frac{1}{6}$

B $\frac{1}{36}$ D $\frac{1}{4}$

66. Use synthetic substitution to find $f(-2)$ for the function below.

$$f(x) = x^4 + 10x^2 + x + 8$$

F 62 H 30

G 38 J 8

67. SAT/ACT The pattern of dots below continues infinitely, with more dots being added at each step.

Step 1 Step 2 Step 3

Which expression can be used to determine the number of dots in the nth step?

A $2n$ D $2(n + 2)$

B $n(n + 2)$ E $2(n + 1)$

C $n(n + 1)$

Spiral Review

Find the exact value of each expression. (Lesson 13-4)

68. $\cos 165°$

69. $\sin 22\frac{1}{2}°$

70. $\sin \frac{7\pi}{8}$

71. $\cos \frac{7\pi}{12}$

Verify that each equation is an identity. (Lesson 13-3)

72. $\sin (270° - \theta) = -\cos \theta$

73. $\cos (90° + \theta) = -\sin \theta$

74. $\cos (90° - \theta) = \sin \theta$

75. $\sin (90° - \theta) = \cos \theta$

76. WATER SAFETY A harbor buoy bobs up and down with the waves. The distance between the highest and lowest points is 4 feet. The buoy moves from its highest point to its lowest point and back to its highest point every 10 seconds. (Lesson 12-7)

a. Write an equation for the motion of the buoy. Assume that it is at equilibrium at $t = 0$ and that it is on the way up from the normal water level.

b. Draw a graph showing the height of the buoy as a function of time.

c. What is the height of the buoy after 12 seconds?

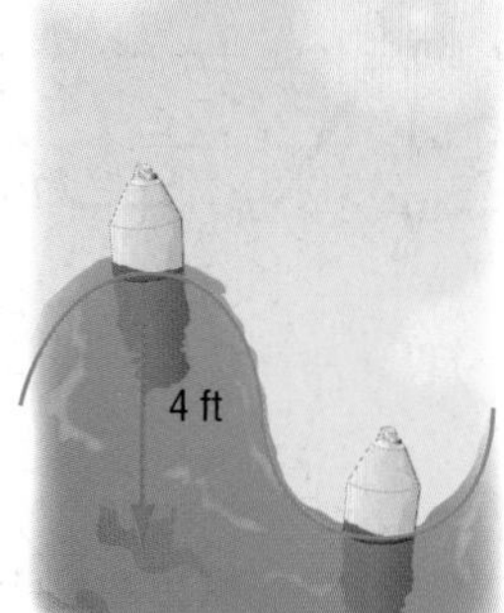

Find the first three terms of each arithmetic series described. (Lesson 10-2)

77. $a_1 = 17, a_n = 197, S_n = 2247$

78. $a_1 = -13, a_n = 427, S_n = 18{,}423$

79. $n = 31, a_n = 78, S_n = 1023$

80. $n = 19, a_n = 103, S_n = 1102$

Skills Review

Graph each rational function.

81. $f(x) = \frac{1}{(x + 3)^2}$

82. $f(x) = \frac{x + 4}{x - 1}$

83. $f(x) = \frac{x + 2}{x^2 - x - 6}$

CHAPTER 13

Study Guide and Review

Study Guide

Key Concepts

Trigonometric Identities (Lessons 13-1, 13-2, and 13-5)

- Trigonometric identities describe the relationships between trigonometric functions.
- Trigonometric identities can be used to simplify, verify, and solve trigonometric equations and expressions.

Sum and Difference of Angles Identities (Lesson 13-3)

- For all values of A and B:

$\cos(A \pm B) = \cos A \cos B \mp \sin A \sin B$

$\sin(A \pm B) = \sin A \cos B \pm \cos A \sin B$

$\tan(A \pm B) = \frac{\tan A \pm \tan B}{1 \mp \tan A \tan B}$

Double-Angle and Half-Angle Identities (Lesson 13-4)

- Double-angle identities:

$\sin 2\theta = 2 \sin \theta \cos \theta$

$\cos 2\theta = \cos^2 \theta - \sin^2 \theta$

$\cos 2\theta = 1 - 2 \sin^2 \theta$

$\cos 2\theta = 2 \cos^2 \theta - 1$

$\tan 2\theta = \frac{2 \tan \theta}{1 - \tan^2 \theta}$

- Half-angle identities:

$\sin \frac{\theta}{2} = \pm\sqrt{\frac{1 - \cos \theta}{2}}$

$\cos \frac{\theta}{2} = \pm\sqrt{\frac{1 + \cos \theta}{2}}$

$\tan \frac{\theta}{2} = \pm\sqrt{\frac{1 - \cos \theta}{1 + \cos \theta}}, \cos \theta \neq -1$

FOLDABLES Study Organizer

Be sure the Key Concepts are noted in your Foldable.

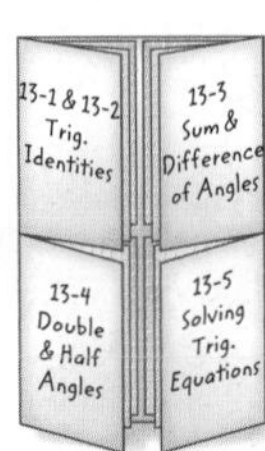

Key Vocabulary

cofunction identity (p. 873)

negative angle identity (p. 873)

Pythagorean identity (p. 873)

quotient identity (p. 873)

reciprocal identity (p. 873)

trigonometric equation (p. 901)

trigonometric identity (p. 873)

Vocabulary Check

Choose the correct term to complete each sentence.

1. The ____________ can be used to find the sine or cosine of 75° if the sine and cosine of 90° and 15° are known.
2. The identities $\tan \theta = \frac{\sin \theta}{\cos \theta}$ and $\cot \theta = \frac{\cos \theta}{\sin \theta}$ are examples of ____________.
3. A ____________ is an equation involving trigonometric functions that is true for all values for which every expression in the equation is defined.
4. The ____________ can be used to find sin 60° using 30° as a reference.
5. A ____________ is true for only certain values of the variable.
6. The ____________ formula can be used to find $\cos 22\frac{1}{2}°$.
7. The identities $\csc \theta = \frac{1}{\sin \theta}$ and $\sec \theta = \frac{1}{\cos \theta}$ are examples of ____________.
8. The ____________ can be used to find the sine or cosine of 120° if the sine and cosine of 90° and 30° are known.
9. $\cos^2 \theta + \sin^2 \theta = 1$ is an example of a ____________.

Lesson-by-Lesson Review

13-1 Trigonometric Identities

Find the value of each expression.

10. $\sin \theta$, if $\cos \theta = \frac{\sqrt{2}}{2}$ and $270° < \theta < 360°$

11. $\sec \theta$, if $\cot \theta = \frac{\sqrt{2}}{2}$ and $90° < \theta < 180°$

12. $\tan \theta$, if $\cot \theta = 2$ and $0° < \theta < 90°$

13. $\cos \theta$, if $\sin \theta = -\frac{3}{5}$ and $180° < \theta < 270°$

14. $\csc \theta$, if $\cot \theta = -\frac{4}{5}$ and $270° < \theta < 360°$

15. SOCCER For international matches, the maximum dimensions of a soccer field are 110 meters by 75 meters. Find $\sin \theta$.

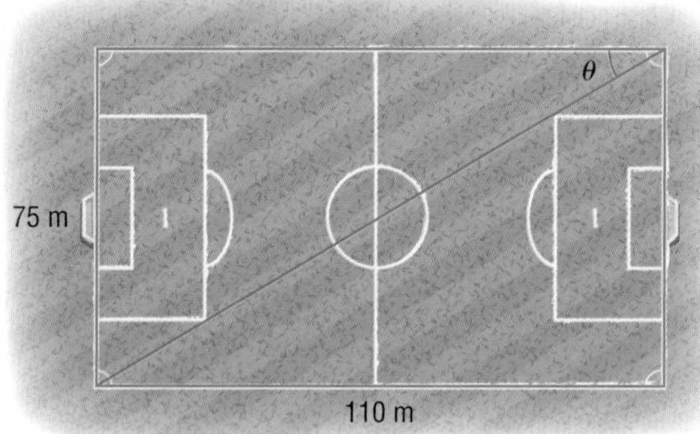

Simplify each expression.

16. $1 - \tan \theta \sin \theta \cos \theta$

17. $\tan \theta \csc \theta$

18. $\sin \theta + \cos \theta \cot \theta$

19. $\cos \theta (1 + \tan^2 \theta)$

Example 1

Find $\sin \theta$ if $\cos \theta = \frac{3}{4}$ and $0° < \theta < 90°$.

$\cos^2 \theta + \sin^2 \theta = 1$ — Trigonometric identity

$\sin^2 \theta = 1 - \cos^2 \theta$ — Subtract $\cos^2 \theta$ from each side.

$\sin^2 \theta = 1 - \left(\frac{3}{4}\right)^2$ — Substitute $\frac{3}{4}$ for $\cos \theta$.

$\sin^2 \theta = 1 - \frac{9}{16}$ — Square $\frac{3}{4}$.

$\sin^2 \theta = \frac{7}{16}$ — Subtract.

$\sin \theta = \pm\frac{\sqrt{7}}{4}$ — Take the square root of each side.

Because θ is in the first quadrant, $\sin \theta$ is positive. Thus, $\sin \theta = \frac{\sqrt{7}}{4}$.

Example 2

Simplify $\cos \theta \sec \theta \cot \theta$.

$$\cos \theta \sec \theta \cot \theta = \cos \theta \left(\frac{1}{\cos \theta}\right)\left(\frac{\cos \theta}{\sin \theta}\right)$$
$$= \cot \theta$$

13-2 Verifying Trigonometric Identities

Verify that each of the following is an identity.

20. $\tan \theta \cos \theta + \cot \theta \sin \theta = \sin \theta + \cos \theta$

21. $\frac{\cos \theta}{\cot \theta} + \frac{\sin \theta}{\tan \theta} = \sin \theta + \cos \theta$

22. $\sec^2 \theta - 1 = \frac{\sin^2 \theta}{1 - \sin^2 \theta}$

23. GEOMETRY The right triangle shown at the right is used in a special quilt. Use the measures of the sides of the triangle to show that $\tan^2 \theta + 1 = \sec^2 \theta$.

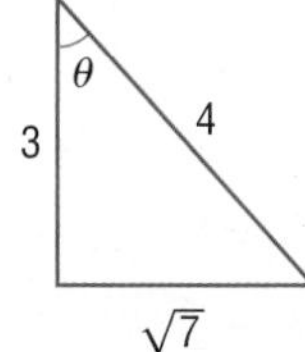

Example 3

Verify that $\frac{\cos \theta + 1}{\sin \theta} = \cot \theta + \csc \theta$ is an identity.

$\frac{\cos \theta + 1}{\sin \theta} \stackrel{?}{=} \cot \theta + \csc \theta$ — Original equation

$\frac{\cos \theta}{\sin \theta} + \frac{1}{\sin \theta} \stackrel{?}{=} \cot \theta + \csc \theta$ — Simplify.

$\cot \theta + \csc \theta = \cot \theta + \csc \theta$ ✓ — Simplify.

Study Guide and Review *Continued*

13-3 Sum and Difference of Angles Identities

Find the exact value of each expression.

24. $\cos(-135°)$

25. $\cos 15°$

26. $\sin 210°$

27. $\sin 105°$

28. $\tan 75°$

29. $\cos 105°$

Verify that each of the following is an identity.

30. $\sin(\theta + 90) = \cos\theta$

31. $\sin\left(\frac{3\pi}{2} - \theta\right) = -\cos\theta$

32. $\tan(\theta - \pi) = \tan\theta$

Example 4

Find the exact value of sin 75°.

Use $\sin(A + B) = \sin A\cos B + \cos A\sin B$.

$$\begin{aligned}\sin 75° &= \sin(30° + 45°)\\ &= \sin 30°\cos 45° + \cos 30°\sin 45°\\ &= \left(\frac{1}{2}\right)\left(\frac{\sqrt{2}}{2}\right) + \left(\frac{\sqrt{3}}{2}\right)\left(\frac{\sqrt{2}}{2}\right)\\ &= \frac{\sqrt{2}}{4} + \frac{\sqrt{6}}{4} \text{ or } \frac{\sqrt{2}+\sqrt{6}}{4}\end{aligned}$$

13-4 Double-Angle and Half-Angle Identities

Find the exact values of $\sin 2\theta$, $\cos 2\theta$, $\sin\frac{\theta}{2}$, and $\cos\frac{\theta}{2}$ for each of the following.

33. $\cos\theta = \frac{4}{5}$; $0° < \theta < 90°$

34. $\sin\theta = -\frac{1}{4}$; $180° < \theta < 270°$

35. $\cos\theta = -\frac{2}{3}$; $\frac{\pi}{2} < \theta < \pi$

36. **BASEBALL** The infield of a baseball diamond is a square with side length 90 feet.

- **a.** Find the length of the diagonal.
- **b.** Write the ratio for sin 45° using the lengths of the baseball diamond.
- **c.** Use the formula $\sin\frac{\theta}{2} = \pm\sqrt{\frac{1-\cos\theta}{2}}$ to verify the ratio you wrote in part **b**.

Example 5

Find the exact value of $\sin\frac{\theta}{2}$ if $\cos\theta = -\frac{3}{5}$ and θ is in the second quadrant.

$\sin\frac{\theta}{2} = \pm\sqrt{\frac{1-\cos\theta}{2}}$ — Half-angle identity

$= \pm\sqrt{\frac{1-\left(-\frac{3}{5}\right)}{2}}$ — $\cos\theta = -\frac{3}{5}$

$= \pm\sqrt{\frac{\frac{8}{5}}{2}}$ — Subtract.

$= \pm\sqrt{\frac{4}{5}}$ — Divide.

$= \pm\frac{2\sqrt{5}}{5}$ — Simplify.

Since θ is in the second quadrant, $\sin\frac{\theta}{2} = \frac{2\sqrt{5}}{5}$.

13-5 Solving Trigonometric Equations

Find all solutions of each equation for the given interval.

37. $2\cos\theta - 1 = 0$; $0° \le \theta < 360°$

38. $4\cos^2\theta - 1 = 0$; $0 \le \theta < 2\pi$

39. $\sin 2\theta + \cos\theta = 0$; $0° \le \theta < 360°$

40. $\sin^2\theta = 2\sin\theta + 3$; $0° \le \theta < 360°$

41. $4\cos^2\theta - 4\cos\theta + 1 = 0$; $0 \le \theta < 2\pi$

Example 6

Find all solutions of $\sin 2\theta - \cos\theta = 0$ if $0 \le \theta < 2\pi$.

$\sin 2\theta - \cos\theta = 0$ — Original equation

$2\sin\theta\cos\theta - \cos\theta = 0$ — Double-angle identity

$\cos\theta(2\sin\theta - 1) = 0$ — Factor.

$\cos\theta = 0$ or $2\sin\theta - 1 = 0$

$\theta = \frac{\pi}{2}, \frac{3\pi}{2}$ $\quad$ $\sin\theta = \frac{1}{2}$; $\theta = \frac{\pi}{6}, \frac{5\pi}{6}$

CHAPTER 13

Practice Test

1. **MULTIPLE CHOICE** Which expression is equivalent to $\sin \theta + \cos \theta \cot \theta$?

   A $\cot \theta$ C $\sec \theta$

   B $\tan \theta$ D $\csc \theta$

2. Verify that $\cos (30° - \theta) = \sin (60° + \theta)$ is an identity.

3. Verify that $\cos (\theta - \pi) = -\cos \theta$.

4. **MULTIPLE CHOICE** What is the exact value of $\sin \theta$, if $\cos \theta = -\frac{3}{5}$ and $90° < \theta < 180°$?

   F $\frac{5}{3}$

   G $\frac{\sqrt{34}}{8}$

   H $-\frac{4}{5}$

   J $\frac{4}{5}$

Find the value of each expression.

5. $\cot \theta$, if $\sec \theta = \frac{4}{3}$; $270° < \theta < 360°$

6. $\tan \theta$, if $\cos \theta = -\frac{1}{2}$; $90° < \theta < 180°$

7. $\sec \theta$, if $\csc \theta = -2$; $180° < \theta < 270°$

8. $\cot \theta$, if $\csc \theta = -\frac{5}{3}$; $270° < \theta < 360°$

9. $\sec \theta$, if $\sin \theta = \frac{1}{2}$; $0° \leq \theta < 90°$

Verify that each of the following is an identity.

10. $\sin \theta (\cot \theta + \tan \theta) = \sec \theta$

11. $\frac{\cos^2 \theta}{1 - \sin \theta} = \frac{\cos \theta}{\sec \theta - \tan \theta}$

12. $(\tan \theta + \cot \theta)^2 = \csc^2 \theta \sec^2 \theta$

13. $\frac{1 + \sec \theta}{\sec \theta} = \frac{\sin^2 \theta}{1 - \cos \theta}$

14. $\frac{\sin \theta}{1 - \cos \theta} = \csc \theta + \cot \theta$

15. **MULTIPLE CHOICE** What is the exact value of $\tan \frac{\pi}{8}$?

    A $\frac{\sqrt{2 - \sqrt{3}}}{2}$

    B $\sqrt{2} - 1$

    C $1 - \sqrt{2}$

    D $-\frac{\sqrt{2 - \sqrt{3}}}{2}$

16. **HISTORY** Some researchers believe that the builders of ancient pyramids, such as the Great Pyramid of Khufu, may have tried to build the faces as equilateral triangles. Later they had to change to other types of triangles. Suppose a pyramid is built such that a face is an equilateral triangle of side length 18 feet.

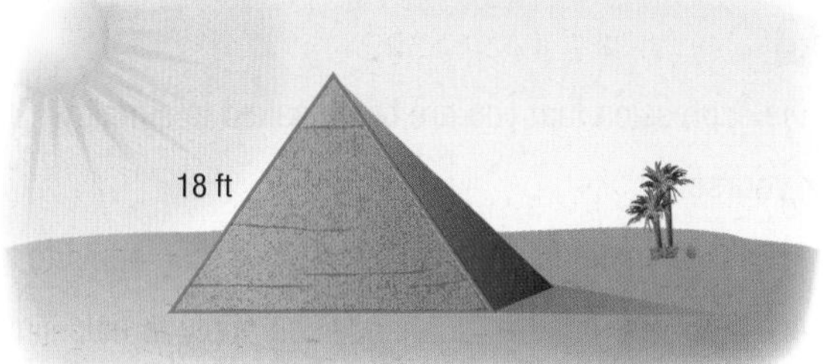

    a. Find the height of the equilateral triangle.

    b. Use the formula $\sin 2\theta = 2 \sin \theta \cos \theta$ and the measures of the equilateral triangle and its height to show that $\sin 2(30°) = \sin 60°$. Find the exact values.

Find the exact value of each expression.

17. $\cos (-225°)$

18. $\sin 480°$

19. $\cos 75°$

20. $\sin 165°$

21. **ROCKETS** A model rocket is launched with an initial velocity of 20 meters per second. The range of a projectile is given by the formula $R = \frac{v^2}{g} \sin 2\theta$, where R is the range, v is the initial velocity, g is acceleration due to gravity or 9.8 meters per second squared, and θ is the launch angle. What angle is needed in order for the rocket to reach a range of 25 meters?

Solve each equation for all values of θ if θ is measured in radians.

22. $2 \cos^2 \theta - 3 \cos \theta - 2 = 0$

23. $2 \sin 3\theta - 1 = 0$

Solve each equation for $0° \leq \theta \leq 360°$ if θ is measured in degrees.

24. $\cos 2\theta + \cos \theta = 2$

25. $\sin \theta \cos \theta - \frac{1}{2} \sin \theta = 0$

CHAPTER 13

Preparing for Standardized Tests

Simplify Expressions

Some standardized test questions will require you to use the properties of algebra to simplify expressions. Follow the steps below to help prepare to solve these kinds of problems.

Royalty-Free/CORBIS

Strategies for Simplifying Expressions

Step 1

Study the expression that you are being asked to simplify.

Ask yourself:

- Are there any mathematical operations I can apply to help simplify the expression?
- Are there any laws or identities I can apply to help simplify the expression?

Step 2

Solve the problem and check your solution.

- Use the order of operations.
- Combine terms and factor as appropriate.
- Apply laws and identities.

Step 3

Check your solution if time permits.

- Retrace the steps in your work to make sure you answered the question thoroughly and accurately.
- If needed, sometimes you can use your scientific calculator to help you check your solution. Evaluate the original expression and your answer for some value and make sure they are the same.

Standardized Test Example

Solve the problem below. Responses will be graded using the short-response scoring rubric shown.

Simplify the trigonometric expression shown below by writing it in terms of sin θ. Show your work to receive full credit.

$$\frac{\cos \theta}{\sec \theta + \tan \theta}$$

Scoring Rubric	
Criteria	**Score**
Full Credit: The answer is correct and a full explanation is provided that shows each step.	2
Partial Credit: • The answer is correct, but the explanation is incomplete. • The answer is incorrect, but the explanation is correct.	1
No Credit: Either an answer is not provided or the answer does not make sense.	0

Read the problem statement carefully. You are given a trigonometric expression and asked to simplify it by writing it in terms of $\sin \theta$. So, your final answer must contain only numbers and terms involving $\sin \theta$. Show your work to receive full credit.

Example of a 2-point response:

Use trigonometric identities to simplify the expression.

$$\frac{\cos \theta}{\sec \theta + \tan \theta} = \frac{\cos \theta}{\frac{1}{\cos \theta} + \frac{\sin \theta}{\cos \theta}}$$ Definition of sec θ and tan θ

$$= \frac{\cos \theta}{\frac{1 + \sin \theta}{\cos \theta}}$$ Simplify the denominator.

$$= \frac{\cos^2 \theta}{1 + \sin \theta}$$ Simplify the complex fraction.

$$= \frac{1 - \sin^2 \theta}{1 + \sin \theta}$$ Pythagorean identity

$$= \frac{(1 + \sin \theta)(1 - \sin \theta)}{1 + \sin \theta}$$ Factor.

$$= 1 - \sin \theta$$ Simplify.

The simplified expression is $1 - \sin \theta$.

The steps, calculations, and reasoning are clearly stated. The student also arrives at the correct answer. So this response is worth the full 2 points.

Exercises

Solve each problem. Show your work. Responses will be graded using the short-response scoring rubric given at the beginning of the lesson.

1. Simplify $\frac{\sec \theta}{\cot \theta + \tan \theta}$ by writing it in terms of $\sin \theta$.

2. What is $\frac{10a^{-3}}{29b^4} \div \frac{5a^{-5}}{16b^{-7}}$?

3. Write $\frac{y + 1}{y - 1} + \frac{y + 2}{y - 2} + \frac{y}{y^2 - 3y + 2}$ in simplest form.

4. Simplify $\frac{\cot^2 \theta - \csc^2 \theta}{\tan^2 \theta - \sec^2 \theta}$ by writing it as a constant.

5. Multiply $(-5 + 2i)(6 - i)(4 + 3i)$.

6. Simplify $(\cot \theta + 1)^2 - 2 \cot \theta$ by writing it in terms of $\csc \theta$.

7. Express $\frac{4 - \sqrt{7}}{3 + \sqrt{7}}$ in simplest form.

CHAPTER 13

Standardized Test Practice

Cumulative, Chapters 1 through 13

Multiple Choice

Read each question. Then fill in the correct answer on the answer document provided by your teacher or on a sheet of paper.

1. The profit p that Selena's Shirt Store makes in a day can be represented by the inequality $10t + 200 < p < 15t + 250$, where t represents the number of shirts sold. If the store sold 45 shirts on Friday, which of the following is a reasonable amount that the store made?

A $200 B $625 C $850 D $950

2. Use a sum or difference of angles identity to find the exact value of cos 75°.

F $\frac{\sqrt{2} - \sqrt{6}}{4}$

H $\frac{\sqrt{6} - \sqrt{2}}{4}$

G $\frac{\sqrt{2} + \sqrt{6}}{2}$

J $\frac{\sqrt{6} + \sqrt{2}}{4}$

3. Use the table to determine the expression that best represents the degree measure of an interior angle of a regular polygon with n sides.

Polygon	Number of Sides	Angle Measure
triangle	3	60
quadrilateral	4	90
pentagon	5	108
hexagon	6	120
heptagon	7	128.5
octagon	8	135

A $(180 + n) \div n$

B $\frac{180}{n}$

C $[180(n - 2)] \div n$

D $30(n - 1)$

4. Which of the following best describes the graphs of $y = 3x - 5$ and $4y = 12x + 16$?

F The lines have the same y-intercept.

G The lines have the same x-intercept.

H The lines are perpendicular.

J The lines are parallel.

Test-TakingTip

Question 2 You can check your answer using a scientific calculator. Find cos 75° and compare it to the value of your answer.

5. What is the product of $[5 \quad -2 \quad 3]$ and $\begin{bmatrix} 1 & -2 \\ 0 & 3 \\ 2 & 5 \end{bmatrix}$?

A $\begin{bmatrix} 11 \\ -1 \end{bmatrix}$

C $\begin{bmatrix} 5 & -10 \\ 0 & -6 \\ 6 & -15 \end{bmatrix}$

B $[11 \quad -1]$

D undefined

6. Which quadratic equation has roots $\frac{1}{2}$ and $\frac{1}{3}$?

F $5x^2 - 5x - 2 = 0$

G $5x^2 - 5x + 1 = 0$

H $6x^2 + 5x - 1 = 0$

J $6x^2 - 5x + 1 = 0$

7. How can you express $\cos \theta \csc \theta \cot \theta$ in terms of $\sin \theta$?

A $\frac{1 - \sin^2 \theta}{\sin^2 \theta}$

C $\frac{\sin^2 \theta}{2}$

B $\frac{1 + \sin^2 \theta}{\sin^2 \theta}$

D $\frac{1 - \sin^2 \theta}{\sin \theta}$

8. The area of a rectangle is $25a^4 - 16b^2$. Which factors could represent the length times width?

F $(5a^2 + 4b)(5a^2 + 4b)$

H $(5a - 4b)(5a - 4b)$

G $(5a^2 + 4b)(5a^2 - 4b)$

J $(5a + 4b)(5a - 4b)$

9. What is the domain of $f(x) = \sqrt{5x - 3}$?

A $\left\{x \middle| x > \frac{3}{5}\right\}$

C $\left\{x \middle| x \geq \frac{3}{5}\right\}$

B $\left\{x \middle| x > -\frac{3}{5}\right\}$

D $\left\{x \middle| x \geq -\frac{3}{5}\right\}$

10. If the equation $y = 3^x$ is graphed, which of the following values of x would produce a point closest to the x-axis?

F $\frac{3}{4}$

H 0

G $\frac{1}{4}$

J $-\frac{3}{4}$

Short Response/Gridded Response

Record your answers on the answer sheet provided by your teacher or on a sheet of paper.

11. Use right triangle *LMN* at the right to show that $\sin 2N = \frac{2n\ell}{m^2}$.

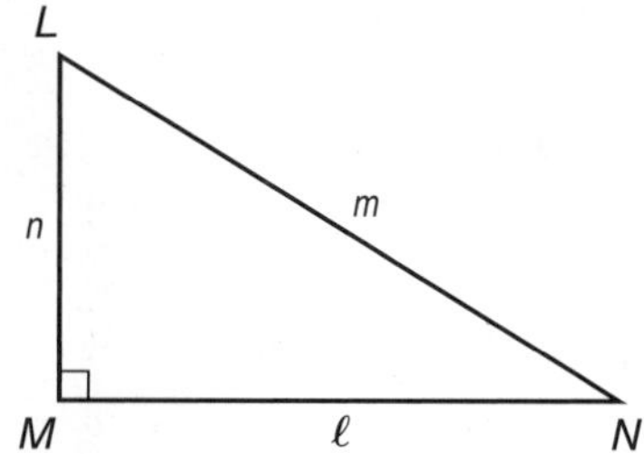

12. GRIDDED RESPONSE Solve the trigonometric equation below in the interval from 0 to 2π. Round your answer to the nearest hundredth if necessary.

$$3 \cos \frac{t}{3} = 2$$

13. Kelly is designing a 12-inch by 12-inch scrapbook page. She cuts one picture that is 4 inches by 6 inches. She decides that she wants the next picture to be 75% as large as the first picture and the third picture to be 150% as large as the second picture. What are the approximate dimensions of the third picture?

14. Identify the amplitude and period of the function graphed below. Then write an equation for the function.

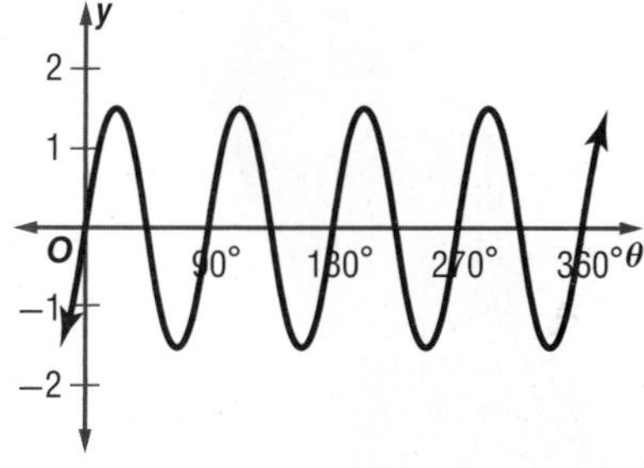

15. GRIDDED RESPONSE A coordinate grid is placed over a map. Emilee's house is located at (−3, −2), and Oliver's house is located at (2, 5). A side of each square represents one block. What is the approximate distance between Emilee's house and Oliver's house?

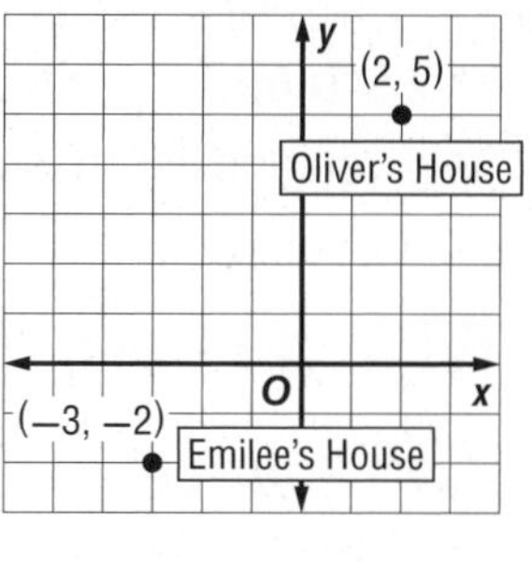

Extended Response

Record your answers on a sheet of paper. Show your work.

16. Kyla's annual salary is \$50,000. Each year she gets a 6% raise.

a. To the nearest dollar, what will her salary be in four years?

b. To the nearest dollar, what will her salary be in 10 years?

17. COMPACT DISCS According to a recent survey, 91% of high school students do not buy compact discs. 8 random students are chosen.

a. Determine the probabilities associated with the number of students that do not buy compact discs by calculating the probability distribution.

b. What is the probability that at least 7 of the 8 students do not buy compact discs?

c. How many students should you expect to not buy compact discs?

Need ExtraHelp?

If you missed Question...	1	2	3	4	5	6	7	8	9	10	11	12	13	14	15	16	17
Go to Lesson...	1-5	13-3	2-4	3-1	3-6	4-3	13-1	5-2	6-3	7-1	13-4	13-5	8-5	12-8	9-1	10-2	11-4

Student Handbook

This **Student Handbook** can help you answer these questions.

What if I Need More Practice?

The **Extra Practice** section provides additional problems for each lesson so you have ample opportunity to practice new skills.

What if I Need to Check a Homework Answer?

The answers to odd-numbered problems are included in **Selected Answers and Solutions**.

What if I Forget a Vocabulary Word?

The **English-Spanish Glossary** provides definitions and page numbers of important or difficult words used throughout the textbook.

What if I Need to Find Something Quickly?

The **Index** alphabetically lists the subjects covered throughout the entire textbook and the pages on which each subject can be found.

What if I Forget a Formula?

Inside the back cover of your math book is a list of **Formulas and Symbols** that are used in the book.

Rubberball/Getty Images

Extra Practice

CHAPTER 1 Equations and Inequalities

Evaluate each expression if $x = 2.2$, $y = -4$, and $z = 5$. (Lesson 1-1)

1. $5x + z(2y - 3)$
2. $(6 - y)^2 - 3$
3. $z^3 - 7x(y + 6)$
4. $3y(2z + 5x) - 7$

Evaluate each expression if $a = \frac{3}{5}$, $b = -1.4$, and $c = 3$. (Lesson 1-1)

5. $\frac{a - b}{a + b}$
6. $\frac{b^2 + a}{ac}$

7. **INTEREST** The formula $A = P(1 + r)^t$ can be used to calculate the value of an investment A after t years that starts with a principal p and has an interest rate r compounded annually. If Jack deposits $3500 into an investment with an interest rate of 5%, how much interest will he earn in 3 years? (Lesson 1-1)

Simplify each expression. (Lesson 1-2)

8. $7(3x - 4y) + 2(4x + 3y)$
9. $-2(5a + 3b) - 8(2a - b)$
10. $3(-6c - 5d) - 4(9c - 8d)$

11. **FOOD** The menu for a fast food restaurant is shown in the table below.

Item	Cost
cheeseburger	$1.19
French fries	$1.79
milkshake	$2.09

Meiko buys 2 cheeseburgers, 2 fries, and 2 milkshakes. (Lesson 1-2)

a. Illustrate the Distributive Property by writing two expressions to represent the cost of the food.

b. Use the Distributive Property to find out how much the restaurant receives from selling this food.

Solve each equation. Check your solution. (Lesson 1-3)

12. $a - 21 = 42$
13. $2x + 5 = -13$
14. $4(c - 5) - 2(3c + 5) = 6$

Solve each equation. Check your solutions. (Lesson 1-4)

15. $|x + 11| = 21$
16. $4|y - 7| = -3$
17. $-3|2a - 5| = -9$

Solve each inequality. Then graph the solution set on a number line. (Lesson 1-5)

18. $d - 6 > -3$
19. $2c + 7 \geq 15$
20. $3x + 8 \leq 5$
21. $21 < -3(2y + 1)$

22. **TAXI** A taxi's rates are shown below. (Lesson 1-5)

Initial fare	$2.00
Mileage charge	$1.75

a. Shannon wants to keep his fare under $35. How many miles can he travel in the taxi?

b. Samantha has $21 for her taxi fare. If she plans on traveling 20 miles in the taxi, does she have enough money to cover the fare? Explain your answer.

Solve each inequality. Graph the solution set on a number line. (Lesson 1-6)

23. $-1 < d + 9 < 7$
24. $|y| < -5$

25. **PARTY** Mrs. Jackson is buying pizzas for her son's birthday party. She wants to spend at least $40 but no more than $50 for the pizzas. The pizza prices are shown below. (Lesson 1-6)

Size	Cost
medium	$8.50
large	$11.50

If she buys 2 large pizzas, how many medium pizzas can she buy to stay within her range?

State the domain and range of each relation. Then determine whether each relation is a *function*. If it is a function, determine if it is *one-to-one, onto, both,* or *neither.* (Lesson 2-1)

1.

x	y
−2	4
−1	1
2	4
2	6

2. 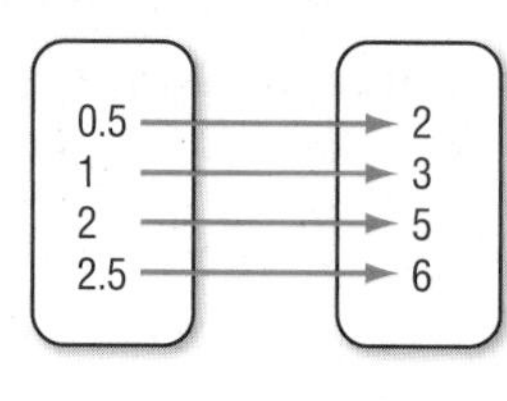

Find the x-intercept and y-intercept of the graph of each equation. Then graph the equation using the intercepts. (Lesson 2-2)

3. $y = 3x - 9$

4. $2y - 5x = 10$

5. $4x + 5y = -20$

6. $-2y + 3x = -12$

7. **REPAIR** An auto mechanic charges an initial fee of $25 plus an hourly fee of $35. (Lesson 2-2)

   a. Write an equation to represent the situation.

   b. How much did it cost Stacy if the mechanic fixed her car in 3.5 hours?

Find the slope of the line that passes through each pair of points. (Lesson 2-3)

8. (1, −7), (−3, 5)

9. (−1, 8), (3, −12)

10. (−2, −15), (4, 9)

11. **ANIMALS** The table shows the weight in pounds of a Chihuahua for the first four weeks after its birth. (Lesson 2-3)

Time (wk)	Weight (lb)
0	2.0
1	2.5
2	3.25
3	4.7
4	5.6

   a. Find the average rate of change in the weight of the puppy between weeks 1 and 3.

   b. Find the average rate of change in the weight of the puppy between weeks 0 and 4.

Write an equation of the line passing through each pair of points. (Lesson 2-4)

12. (−3, −14), (1, −2)

13. (−2, 11), (2, 3)

14. (−4, 2), (2, 5)

15. **TELEVISION** The table shows the number of LCD televisions sold in recent years at an electronics store. (Lesson 2-5)

Year	Number of TVs Sold
2006	733
2007	838
2008	943
2009	1048
2015	?

   a. Using x as the number of years since 2000, make a scatter plot, graph a line of fit, and describe the correlation.

   b. Use two ordered pairs to write a prediction equation.

   c. Use your prediction equation to predict the missing value.

Graph each function. Identify the domain and range. (Lesson 2-6)

16. $g(x) = |-2x|$

17. $f(x) = [[x - 3]]$

18. $h(x) = 3|x| + 2$

Describe the translation in each function. Then graph the function. (Lesson 2-7)

19. $y = x^2 - 3$

20. $y = |x| + 5$

21. $y = (x + 2)^2$

22. $y = |x - 4|$

Graph each inequality. (Lesson 2-8)

23. $y > -3$

24. $y < -2$

25. $2x - 5y \geq 10$

Solve each system of equations by using substitution or elimination. (Lesson 3-1)

1. $y = 3x + 4$
$y = -2x - 6$

2. $y = -4x - 5$
$y - 5x = -14$

3. $2a - 3b = -5$
$2a + 3b = 13$

4. $-5a + 2b = 16$
$4a - 5b = -6$

Solve each system of inequalities by graphing. (Lesson 3-2)

5. $y > 3x + 1$
$y < -2x - 1$

6. $y \leq -x - 2$
$y \leq 4x + 1$

7. COLLEGE The total score on a college entrance exam for the math and verbal sections is 2000. The college requires a math score of at least 875 and a verbal score of at least 800. (Lesson 3-2)

a. Write and graph a system of inequalities to represent this situation.

b. Give two examples of acceptable math and verbal scores for entrance to the college.

Graph each system of inequalities. Name the coordinates of the vertices of the feasible region. Find the maximum and minimum value of the given function for this region. (Lesson 3-3)

8. $x \geq -2$
$y \leq 4$
$y \geq x - 1$
$f(x, y) = 2x - y$

9. $y \geq -3$
$x \leq 5$
$y \leq 2x + 4$
$f(x, y) = -4x + 5y$

Solve each system of equations. (Lesson 3-4)

10. $2x - 3y + z = -4$
$x + 2y - 3z = -9$
$4x - y + 2z = -3$

11. $-3x + 4y - 2z = -14$
$2x - 3y - 4z = 6$
$-5x - 2y + 3z = 35$

12. BUSINESS A car dealership is combining two of their lots. Each entry in the matrices represents the number of new car models of each color. Express the total number of new car models of each color after the lots are combined as a matrix. (Lesson 3-5)

$$\text{Lot 1} = \begin{bmatrix} 1 & 5 & 7 \\ 8 & 4 & 3 \\ 6 & 5 & 2 \end{bmatrix}, \text{Lot 2} = \begin{bmatrix} 7 & 5 & 3 \\ 6 & 2 & 5 \\ 6 & 7 & 5 \end{bmatrix}$$

Find each product, if possible. (Lesson 3-6)

$$A = \begin{bmatrix} -2 & 4 \\ 3 & -5 \end{bmatrix} \quad B = \begin{bmatrix} -1 & 2 & 6 \\ -4 & -2 & 5 \end{bmatrix}$$

$$C = \begin{bmatrix} -7 & 8 & 4 \\ -2 & -5 & 6 \\ 4 & 9 & 1 \end{bmatrix} \quad D = \begin{bmatrix} 1 & 6 \\ -7 & 5 \\ -5 & 4 \\ 2 & -2 \end{bmatrix}$$

13. AD

14. CD

15. AB

16. BC

Evaluate each determinant. (Lesson 3-7)

17. $\begin{vmatrix} 3 & 4 \\ -5 & -3 \end{vmatrix}$

18. $\begin{vmatrix} -2 & 6 & 0 \\ -1 & 3 & 2 \\ 4 & -1 & 5 \end{vmatrix}$

Use Cramer's Rule to solve each system of equations. (Lesson 3-7)

19. $2x - 3y = 18$
$5x + 7y = -13$

20. $-3x - 2y = -5$
$-4x + 6y = 28$

21. $x + 5y - z = 14$
$2x - y + z = -8$
$3x + 2y + z = -1$

22. $x + y + z = -3$
$3x - 2y = 16$
$2x - y + 3z = -5$

23. EVENT At a high school football game, the concession stand sells hot chocolate for \$1.75 and sodas for \$1.25. At the last game, 175 drinks were sold for a total of \$243.75. Use Cramer's Rule to determine how many of each type of drink were sold. (Lesson 3-7)

Use a matrix equation to solve each system of equations. (Lesson 3-8)

24. $5x - 3y = -9$
$-4x - 6y = 24$

25. $-2x + 4y = 4$
$6x - 5y = -19$

26. $2x + 3y = 6$
$6x + 9y = -10$

Extra Practice

Complete parts a–c for each quadratic function.
a. Find the y-intercept, the equation of the axis of symmetry, and the x-coordinate of the vertex.
b. Make a table of values that includes the vertex.
c. Use this information to graph the function.
(Lesson 4-1)

1. $f(x) = x^2 - 3$
2. $f(x) = 2x^2 - 5x$

Use the related graph of each equation to determine its solutions. (Lesson 4-2)

3. $x^2 + x - 6 = 0$

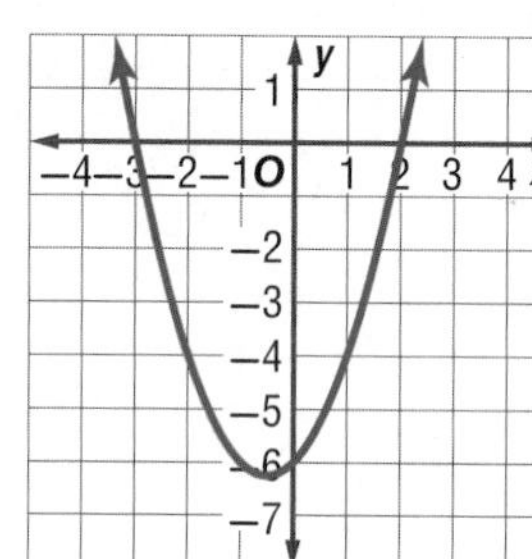

4. $x^2 - 2x - 3 = 0$

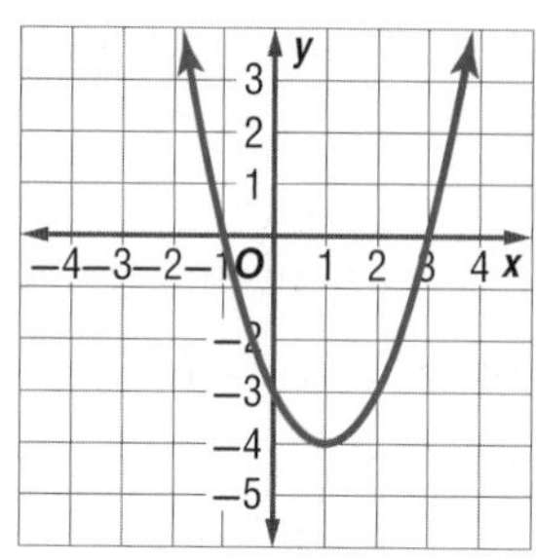

5. **ROCKET** A science teacher is conducting an experiment with a model rocket. The function $h = -16t^2 + 64t$ models the height of the rocket, where h is the height of the rocket in feet and t is the time in seconds after it is launched. Determine how many seconds the rocket is in the air. (Lesson 4-2)

Solve each equation by factoring. (Lesson 4-3)

6. $2x^2 + x - 10 = 0$
7. $2x^2 + x = 28$

8. **GAMES** Julio constructed a platform for a bean bag toss game. The plans for the original platform had dimensions of 3 feet by 5 feet. He made his platform larger by adding x feet to each side. The area of the new platform is 35 square feet. (Lesson 4-3)

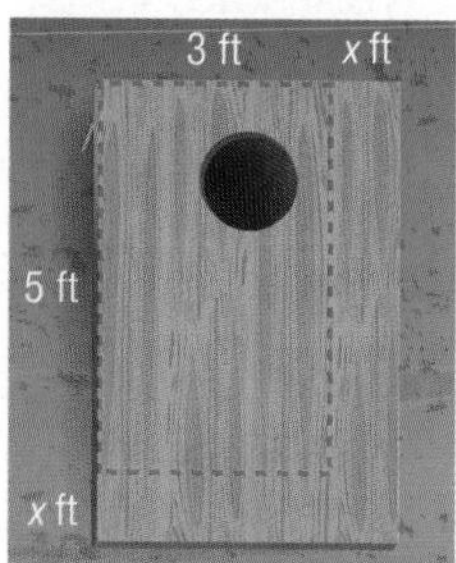

a. Write a quadratic equation that represents the area of his platform.

b. Find the dimensions of the platform Julio made.

Simplify. (Lesson 4-4)

9. $(10 + 3i) + (3 - 7i)$
10. $(2 + i)(2 - i)$
11. $\frac{5}{1 + 3i}$

Solve each equation by completing the square. (Lesson 4-5)

12. $x^2 - 2x - 24 = 0$
13. $x^2 - 2x = 35$
14. $5x^2 - 4x + 1 = 0$
15. $3x^2 + 6x + 10 = 0$

Solve each equation by using the Quadratic Formula. (Lesson 4-6)

16. $2x^2 - 3x - 9 = 0$
17. $4x^2 + 2x - 1 = 0$
18. $3x^2 + 4x - 2 = 0$
19. $3x^2 - 5x - 10 = 0$

Write an equation in vertex form for each parabola. (Lesson 4-7)

20.

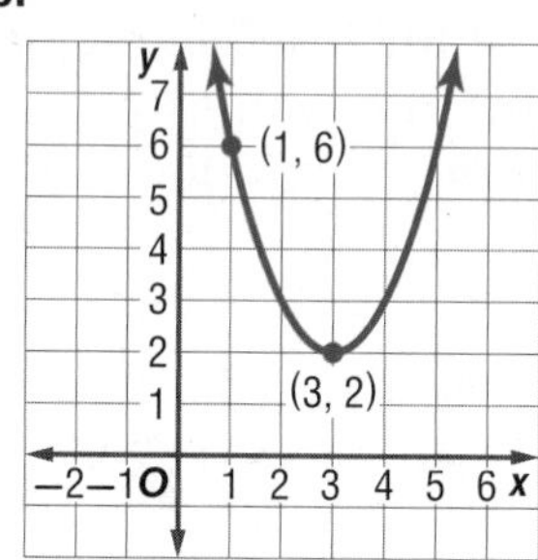

21.

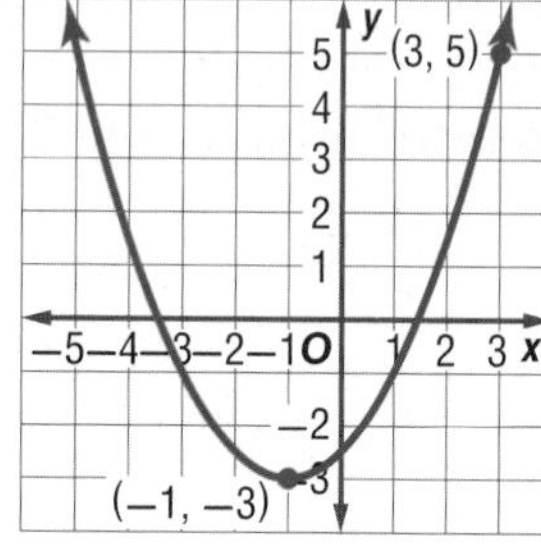

22. **VIDEO GAMES** The daily sales of a video game company can be modeled by the function $S(p) = -2p^2 + 80p + 1250$, where $S(p)$ is the amount of sales and p is the price per game. (Lesson 4-7)

    a. Write the function in vertex form.

    b. What is the maximum amount of daily sales the company can expect to make?

Solve each inequality algebraically. (Lesson 4-8)

23. $x^2 + 3x < 18$
24. $x^2 - 3x \leq 28$
25. $2x^2 - 13x \geq -20$

Simplify. (Lesson 5-1)

1. $(5x^2 + 3x - 7) - (3x^2 - x + 4)$
2. $(2y^2 - 4y - 5) + (3y^2 - 2y + 1)$
3. $2cd(3c - 4d) + 4d(c + 2d)$
4. $(r - s)(r + s)(3r - 2s)$

Simplify. (Lesson 5-2)

5. $\dfrac{4x^3y^4 - 10xy^5}{2xy}$
6. $(x^3 + 4x^2 - 4x - 7) \div (x + 1)$
7. **SHOES** The number of shoes sold by a store can be modeled by $-s^2 + 12s$, where s is the number of employees. Find the average number of shoes sold by each worker. (Lesson 5-2)

Find $p(-2)$ and $p(4)$ for each function. (Lesson 5-3)

8. $p(x) = -2x^2 + 5x - 3$
9. $p(x) = -3x^3 - x^2 + 2x - 8$
10. $p(x) = x^4 - 5x^3 + 4x + 6$
11. **CELL PHONES** The yearly sales of cell phones can be modeled by the function $s(t) = 0.45t^4 + 5.4t^3 - 15t^2 - 300t + 1000$, where $s(t)$ is the annual sales in millions of dollars and t is the number of years after 2005. (Lesson 5-4)
    a. Graph the function for $0 \le t \le 10$.
    b. Describe the turning points of the graph and its end behavior.
    c. What trends in cell phone sales does this graph suggest?
    d. Is it reasonable that the trend will continue indefinitely? Explain.

Factor completely. If the polynomial is not factorable, write *prime*. (Lesson 5-5)

12. $6x^4 - 5y^7$
13. $27a^3 - 125d^3$
14. $2ax^2 - 3bx^2 + cx^2 - 2ay^2 + 3by^2 - cy^2$

Given a polynomial and one of its factors, find the remaining factors of the polynomial. (Lesson 5-6)

15. $x^3 - 2x^2 - 5x + 6$; $x - 1$
16. $2x^3 - x^2 - 25x - 12$; $x + 3$
17. $6x^3 + 11x^2 + x - 4$; $2x - 1$

Find all the zeros of each function. (Lesson 5-7)

18. $f(x) = x^3 - 4x^2 - 7x + 10$
19. $f(x) = x^4 - 8x^2 - 9$
20. $f(x) = x^4 + 3x^2 - 4$
21. **BUSINESS** The profit in hundreds of dollars for selling c calculators per day can be modeled by $R(c) = -0.005c^4 + 0.25c^3 + 0.01c^2 - 2.5c + 100$. (Lesson 5-7)
    a. How many positive real zeros, negative real zeros, and imaginary zeros exist?
    b. What is the meaning of the zeros in this situation?

Find all the rational zeros of each function. (Lesson 5-8)

22. $f(x) = 20x^3 - 29x^2 - 25x + 6$
23. $f(x) = 16x^3 - 44x^2 + 4x + 15$
24. $f(x) = 4x^4 - 81x^2 - 110x + 75$
25. Refer to the graph below. Find all the zeros of $f(x) = x^3 - 5x^2 - 12x + 36$. (Lesson 5-8)

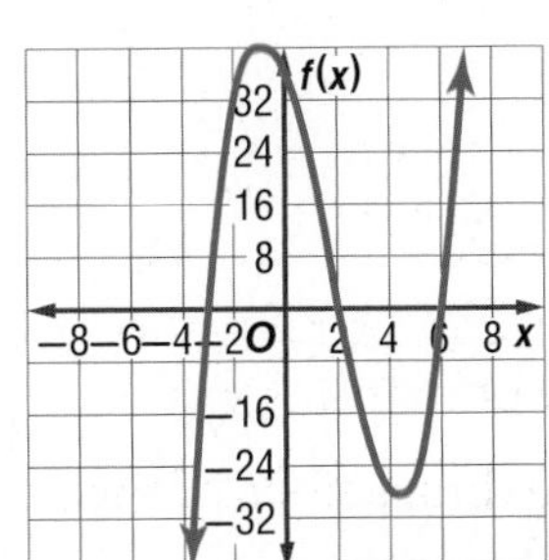

Extra Practice

Find $(f + g)(x)$, $(f - g)(x)$, $(f \cdot g)(x)$, and $\left(\frac{f}{g}\right)(x)$ for each $f(x)$ and $g(x)$. (Lesson 6-1)

1. $f(x) = x - 4$
   $g(x) = 2x + 3$

2. $f(x) = -x + 3$
   $g(x) = x^2 - 4$

Find $[f \circ g](x)$ and $[g \circ f](x)$, if they exist.

3. $f(x) = 3x + 2$
   $g(x) = x - 3$

4. $f(x) = 3x^2$
   $g(x) = x^2 + 5$

Determine whether each pair of functions are inverse functions. Write *yes* or *no*. (Lesson 6-2)

5. $f(x) = 2x - 8$
   $g(x) = \frac{1}{2}x + 4$

6. $f(x) = 3x - 4$
   $g(x) = 3x + 7$

7. **GEOMETRY** The formula for the volume of a sphere is $V = \frac{4}{3}\pi r^3$, where r is the radius of the sphere. (Lesson 6-2)
   a. Find the inverse of the function.
   b. Explain what purpose $V^{-1}(x)$ serves.

Identify the domain and range of each function. (Lesson 6-3)

8. $f(x) = \sqrt{2x} + 4$

9. $f(x) = \sqrt{x - 1} - 2$

10. $f(x) = -3\sqrt{x + 2}$

Simplify. (Lesson 6-4)

11. $\sqrt{36a^2b^4}$

12. $\sqrt[3]{343x^6y^9}$

13. $\sqrt[5]{s^{10}t^{15}}$

Simplify. (Lesson 6-5)

14. $4\sqrt{3x} \cdot 5\sqrt{12x}$

15. $\sqrt{36} + 2\sqrt{32} - \sqrt{50}$

16. $\frac{2}{\sqrt{5} - 3}$

17. **GEOMETRY** Find the exact perimeter or circumference of each shape. (Lesson 6-5)

a.

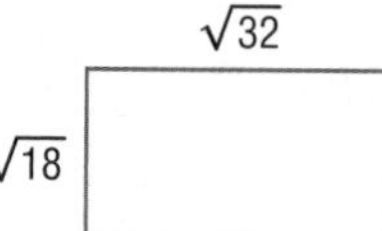

b.

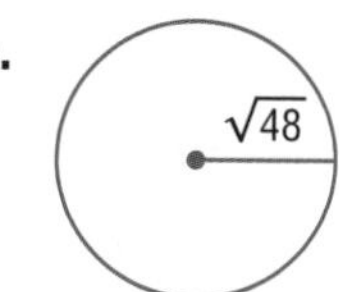

Simplify each expression. (Lesson 6-6)

18. $a^{\frac{1}{4}} \cdot a^{\frac{2}{3}}$

19. $\sqrt[6]{125x^9}$

20. $\frac{a^{\frac{1}{2}} - 3}{a^{\frac{1}{2}} + 3}$

21. **HERON'S FORMULA** The area A of a triangle with sides a, b, and c is given by $A = \sqrt{s(s - a)(s - b)(s - c)}$ where s is the semiperimeter of the triangle. Determine the exact area of the triangle below. (Lesson 6-6)

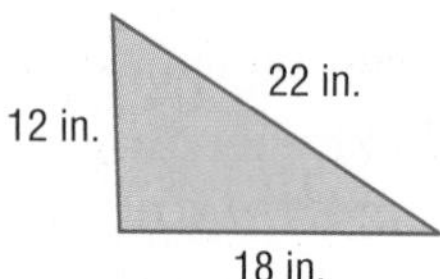

Solve each equation. (Lesson 6-7)

22. $\sqrt{5x + 14} - 2 = 6$

23. $\sqrt{4x + 15} = \sqrt{2x + 9}$

24. $\sqrt{x + 7} = \sqrt{x} - 1$

25. **WIND** The wind speed s in miles per hour can be modeled by the formula $s = 150\sqrt{\frac{p}{a}}$, where a is the area of a sail in square feet and p is the pressure of the wind on the sail in pounds per square feet. Determine the area of a sail if the wind speed is 75 mph and the pressure is 20 pounds per square feet. (Lesson 6-7)

Graph each function. State the domain and range. (Lesson 7-1)

1. $f(x) = 3(4)^x$
2. $f(x) = 2^{3x} - 3$

Solve each equation. (Lesson 7-2)

3. $2^{x-1} = 8^{x+3}$
4. $5^{2x+12} = 25^{10x-12}$
5. **BIOLOGY** During an experiment, the number of cells of a virus can be modeled by $f(t) = 2^{t-2}$, where t is the time in days and $f(t)$ is the number of cells. Determine how many days have passed if there are 64 virus cells. (Lesson 7-2)

Evaluate each expression. (Lesson 7-3)

6. $\log_8 64$
7. $\log_7 1$
8. $\log_5 \frac{1}{25}$
9. **DOLPHINS** The number of dolphins living in an ocean region after t months can be approximated by $n(t) = 920 \log_{10} (t - 1)$. (Lesson 7-3)
   a. How many dolphins are in the ocean region after 4 months?
   b. How many dolphins are in the ocean region after 2 years?

Solve each equation. (Lesson 7-4)

10. $\log_{25} x = \frac{3}{2}$
11. $\log_9 81 = x$
12. $\log_6 (5x - 3) = \log_6 (x + 9)$
13. $\log_8 (x^2 + 6) = \log_8 (5x)$

Solve each equation. Check your solution. (Lesson 7-5)

14. $\log_4 3 + \log_4 x = \log_4 12$
15. $\log_7 6 - \log_7 2 = \log_7 x$
16. $4 \log_2 x = \log_2 81$
17. $\log_6 x + \log_6 (x + 5) = 2$
18. **POPULATION** The wolf population per year at a national park is listed in the table below. (Lesson 7-5)

Year	Population
2005	420
2006	399
2007	379
2008	360
2009	342
2020	?

   a. Determine the annual percent of decrease of the population.
   b. Write a logarithmic function for the time in years based upon population from 2005.
   c. How many wolves will there be in the national park in 2020?

Solve each equation. Round to the nearest ten-thousandth. (Lesson 7-6)

19. $5^x = 60$
20. $4^{x^2} = 21$
21. $3^{x-2} = 4^x$

Solve each equation. Round to the nearest ten-thousandth. (Lesson 7-7)

22. $4e^x - 6 = 11$
23. $3e^{-x} + 4 = 17$
24. $2e^{3x} - 8 = 21$
25. **INTEREST** Stefanie deposited $2000 into a savings account paying 2.5% interest compounded continuously. (Lesson 7-8)
   a. Determine how long it will take Stefanie to have a balance of $2400.
   b. How long will it take Stefanie to double her investment?
   c. If Stefanie wanted to have $5000 after 10 years, how much should she have invested?

Extra Practice

Simplify each expression. (Lesson 8-1)

1. $\dfrac{(x-2)(x+5)}{x^2+2x-8}$

2. $\dfrac{(x+3)(x-4)}{x^2+x-20}$

3. $\dfrac{x^2-16}{x^2-8x+12} \cdot \dfrac{x^2+3x-18}{x^2-7x+12}$

4. $\dfrac{x^2+10x+21}{x^2-5x+6} \div \dfrac{x^2+3x-28}{x^2-8x+15}$

5. **BUSINESS** The average hourly income I of a college student x years after graduation can be modeled by the function $I(x) = \dfrac{11x^2+33x+22}{x+2}$. Simplify the function. (Lesson 8-2)

Simplify each expression. (Lesson 8-2)

6. $\dfrac{2}{x^2-7x+12} + \dfrac{6}{x^2+x-20}$

7. $\dfrac{4}{x^2-11x+30} - \dfrac{3}{x^2-13x+40}$

Identify the asymptotes, domain, and range of each function. (Lesson 8-3)

8.

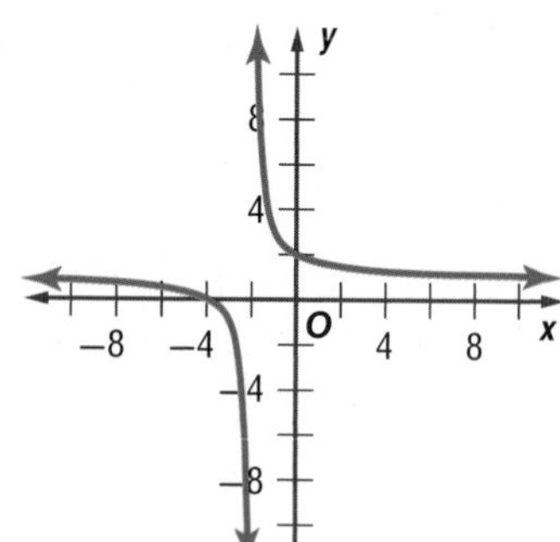

9.

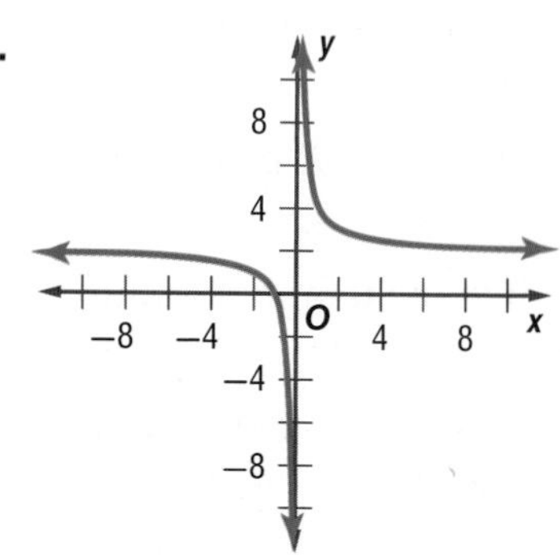

Identify the asymptotes, domain, and range of each function. (Lesson 8-3)

10.

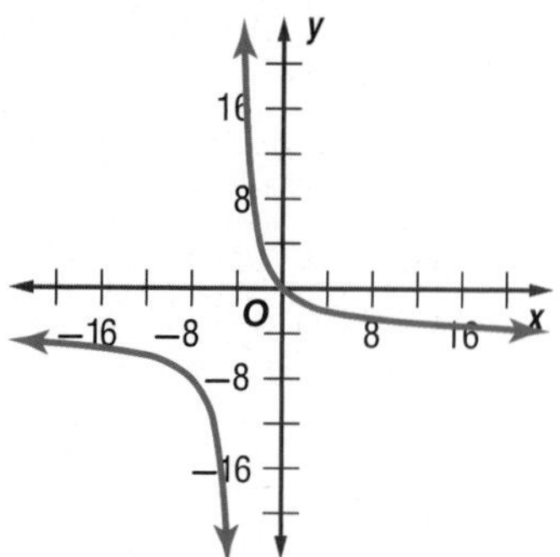

11. **FIELD TRIP** The American history class is planning a field trip to Washington, D.C. It costs $300 to rent the bus, and each ticket to the museum costs $12. (Lesson 8-3)

    a. If C represents the cost for each student and n represents the number of students, write an equation to represent the cost to each student as a function of how many students go on the field trip.

    b. Are there any limitations to the range or domain in this situation? Explain.

Graph each function. (Lesson 8-4)

12. $f(x) = \dfrac{2x}{x-3}$

13. $f(x) = \dfrac{x-4}{x+2}$

If a varies jointly as b and c, find a when $b = 2$ and $c = -3$, given each of the following. (Lesson 8-5)

14. $a = -12$ when $b = 4$ and $c = 6$

15. $a = -36$ when $b = 12$ and $c = -4$

Solve each equation. Check your solution. (Lesson 8-6)

16. $\dfrac{3}{x+4} + \dfrac{1}{x-1} = \dfrac{17}{24}$

17. $\dfrac{12}{x+4} - \dfrac{3}{x-3} = \dfrac{24}{x^2+x-12}$

18. $\dfrac{14}{x-5} - \dfrac{7}{x+5} = \dfrac{-7}{x^2-25}$

19. **EXERCISE** Natalie can bike 10 kilometers per hour faster than Aaron. By the time Natalie travels 60 kilometers, Aaron has gone 40 kilometers. (Lesson 8-6)

    a. Write an expression for Natalie's time.

    b. Write an expression for Aaron's time.

    c. Write and solve the rational equation to determine the speed of each biker.

Find the distance between each pair of points with the given coordinates. (Lesson 9-1)

1. (1, 3), (−4, 5)

2. (−2, 4), (6, 1)

3. (−2, −2), (−6, 8)

4. (−3, −3), (4, −5)

Write each equation in standard form. Identify the vertex, axis of symmetry, and direction of opening of the parabola. (Lesson 9-2)

5. $y = x^2 + 6x + 7$

6. $y = 2x^2 - 8x + 12$

7. TRACK A long jumper makes a leap of 24 feet, and his feet reach a maximum height of 6 feet. A parabola on the coordinate plane with the focus at the origin models the event. What is the equation of the parabolic path of the jump? (Lesson 9-2)

Write an equation for each circle given the endpoints of a diameter. (Lesson 9-3)

8. (1, 3), (−4, 5)

9. (−2, 4), (6, 1)

10. GEARS The endpoints of the diameter of a gear of a bicycle are (−2, 5) and (2, −5). Write an equation to represent a two-dimensional sketch of the gear. (Lesson 9-3)

Write an equation for each ellipse. (Lesson 9-4)

11.

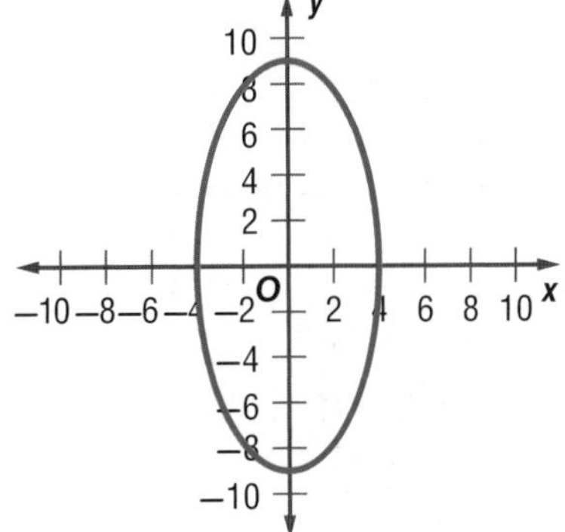

12.

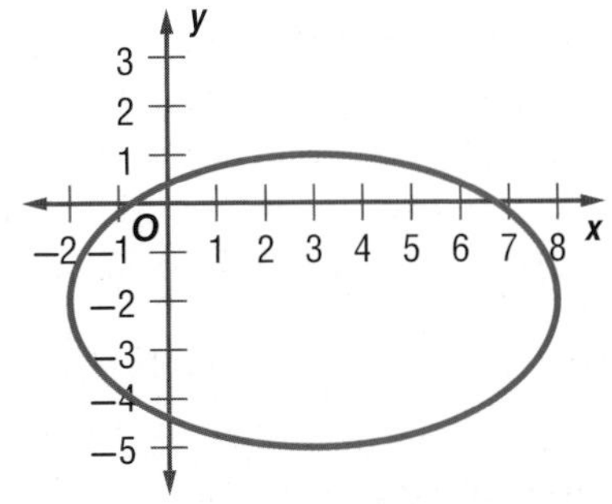

Write an equation for each hyperbola. (Lesson 9-5)

13.

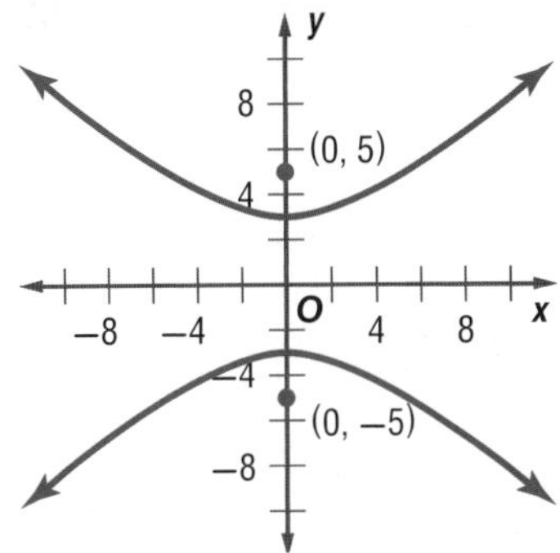

14.

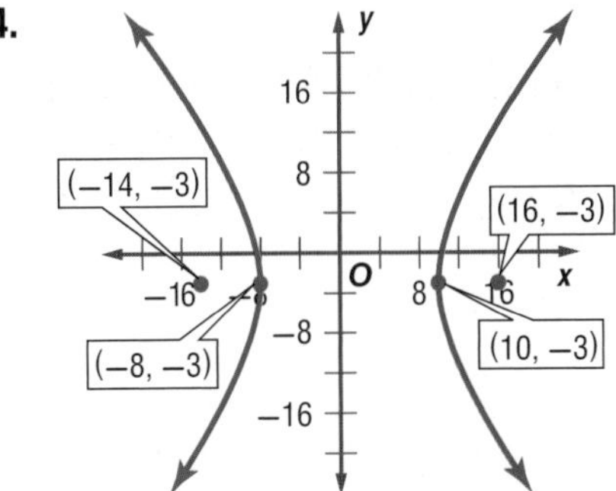

Without writing in standard form, state whether the graph of each equation is a *parabola, circle, ellipse,* or *hyperbola.* (Lesson 9-6)

15. $8y^2 - 16x + 12y - 20 = 0$

16. $9x^2 - 18x + 12y - 15y^2 + 10xy + 12 = 0$

17. $2y^2 + 8x^2 - 6xy + 8x - 10y - 12 = 0$

18. TUNNEL A tunnel through a mountain has an elliptical opening that is 100 feet wide and 48 feet high. Write the equation of the ellipse in standard form. (Lesson 9-6)

Solve each system of equations. (Lesson 9-7)

19. $x^2 - y^2 = 21$
$y^2 + x^2 = 29$

20. $y = -3x + 1$
$y - x^2 = 23 - 8x$

21. $y = x^2 - 2x - 3$
$y = 2x - 3$

Extra Practice

Determine whether each sequence is *arithmetic, geometric,* or *neither.* Explain your reasoning. (Lesson 10-1)

1. 3, 6, 9, 12, …

2. 4, 12, 36, 108, …

3. −1, 5, −25, 125, …

4. 2, 5, 9, 12, …

Write an equation for the *n*th term of each arithmetic sequence. (Lesson 10-2)

5. 3, −12, −27, …

6. 6, 13, 20, …

7. **GAMES** Susie is playing a game in which she has to stack cups as shown in the diagram below. (Lesson 10-2)

How many cups does she need to be able to stack the cups 8 rows high?

Write an equation for the *n*th term of each geometric sequence. (Lesson 10-3)

8. 0.75, 3, 12, …

9. 0.4, −2, 10, …

Find the sum of each infinite series, if it exists. (Lesson 10-4)

10. 8 + 14 + 24.5 + …

11. $\frac{1}{3} + \frac{1}{9} + \frac{1}{27} + \dots$

12. 20 + 15 + 11.25 + …

13. −5 + (−12) + (−28.8) + …

14. **MONEY** Dustin is saving money from his summer job to buy a gaming system. He saved one dollar the first week, and each following week he saved twice as much as the week before. Dustin saved enough money to buy the gaming system after 8 weeks. How much did the system cost? (Lesson 10-3)

Find the first five terms of each sequence described. (Lesson 10-5)

15. $a_1 = 6, a_{n+1} = 4a_n + 3$

16. $a_1 = 11, a_{n+1} = 3a_n - n$

17. $a_1 = 4, a_2 = x, a_n = 2a_{n-2} - 3a_{n-1}$

Write a recursive formula for each sequence. (Lesson 10-5)

18. 3, 7, 11, 15, …

19. $a_3 = 96$ and $r = 2$

20. 8, 16, 32, 64, …

21. $a_5 = -4$ and $d = -2$

Find the indicated term of each expression. (Lesson 10-6)

22. third term of $(a + 2b)^6$

23. fourth term of $(y - 4x)^7$

24. **GEOMETRY** Suppose each of the dimensions of the cube shown below is increased by 0.25 centimeters. (Lesson 10-6)

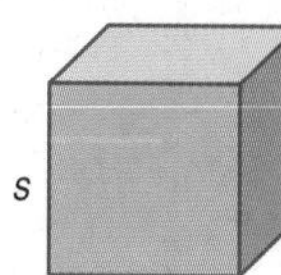

a. Write a binomial expression to represent the volume of the new cube.

b. Expand the binomial expression.

Find a counterexample to disprove each statement. (Lesson 10-7)

25. $n^2 + 4n - 3$ is prime.

26. $2n^2 + 3n - 3$ is prime.

Determine whether each situation describes a *survey*, an *experiment*, or an *observational study*. Then identify the sample, and suggest a population from which it may have been selected. (Lesson 11-1)

1. Every fifth person coming out of a football stadium is asked how often they go to a football game.

2. A company posts a controversial advertisement in their store and records the reactions of the customers.

3. A randomly selected group of people are chosen to test the effects of a new drug on energy levels.

4. A randomly selected group of people are asked their opinions on political topics.

5. **COLLEGE** The SAT scores of Mr. Williams' calculus class are given below. (Lesson 11-2)

SAT Scores				
2020	1250	1500	1700	1600
1830	1230	1330	1450	1250
1560	1450	1350	1430	1640
2210	1550	2200	2050	1750
2400	1900	2140	1280	1800

   a. Use a graphing calculator to create a histogram. Then describe the shape of the distribution.

   b. Describe the center and spread of the data using either the mean and standard deviation or the five-number summary. Justify your choice.

Identify the random variable in each distribution, and classify it as *discrete* or *continuous*. Explain your reasoning. (Lesson 11-3)

6. the number of golf balls hit into a lake

7. the life of a battery

8. the amount of gasoline in a gas tank

9. the number of stamps in a collection

10. **DOGS** According to a recent survey, 34% of homeowners own a dog. What is the probability that out of 10 randomly selected homeowners, more than 5 own a dog? (Lesson 11-4)

11. **ASSEMBLIES** The number of assemblies at West High School each year is normally distributed with a mean of 12.4 and a standard deviation of 1.6. (Lesson 11-5)

   a. What is the probability that there will be more than 10 assemblies in a given year?

   b. If the school has existed for 30 years, in how many of those years were there between 11 and 13 assemblies?

12. **STUDENT COUNCIL** The number of students that run for student council each year is normally distributed with a mean of 16.8 students and a standard deviation of 3.7. (Lesson 11-5)

   a. What is the probability that fewer than 10 students run in a given year?

   b. If the school has kept records for 20 years, in how many of those years were there between 15 and 20 students who ran for student council?

13. **CLOTHES** A sample of 225 high school students spent a mean of $125 on clothes for the new school year. Calculate the maximum error of estimate for a 90% confidence level if the standard deviation was $10.40. (Lesson 11-6)

14. **CEREAL** A cereal manufacturer claims that every box of cereal contains 100 raisins. Jackie thought there were fewer than 100, so she tested 30 random boxes of cereal. Test the hypothesis at 5% significance. Identify the hypotheses and claim, decide whether to reject the null hypothesis, and make a conclusion about the claim. (Lesson 11-6)

Number of Raisins					
98	97	99	95	98	99
100	98	99	99	99	98
102	97	100	96	98	99
100	98	99	97	99	100
101	97	100	101	99	99

Extra Practice

Use a trigonometric function to find the value of x. Round to the nearest tenth. (Lesson 12-1)

1.

2.

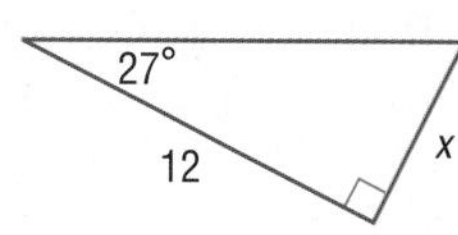

3. FLAGPOLE Tina is standing away from a flagpole. The angle of elevation from the ground to the top of the flagpole is 15°. The flagpole is 50 feet high. Determine how far Tina is standing from the flagpole. (Lesson 12-1)

Rewrite each degree measure in radians and each radian measure in degrees. (Lesson 12-2)

4. 300°

5. $-80°$

6. $-\frac{2\pi}{3}$

7. $-\frac{\pi}{6}$

Find the exact value of each trigonometric function. (Lesson 12-3)

8. $\sin 120°$

9. $\csc \frac{3\pi}{4}$

Solve each triangle. Round side lengths to the nearest tenth and angle measures to the nearest degree. (Lesson 12-4)

10.

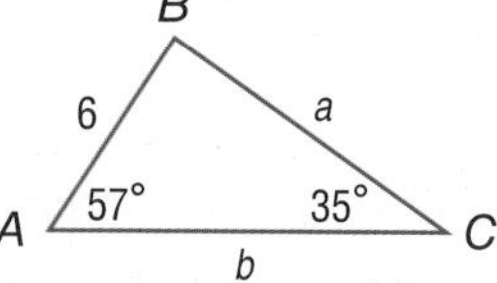

11.

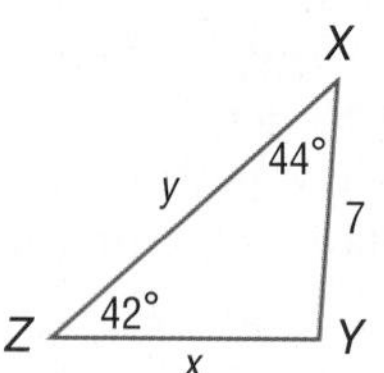

12. SAILS One side of a triangular sail of a sailboat is 25 feet long. The angle opposite this side is 40°. Another angle formed by the sail measures 55°. What is the perimeter of the sail to the nearest tenth? (Lesson 12-4)

Solve each triangle. Round side lengths to the nearest tenth and angle measures to the nearest degree. (Lesson 12-5)

13.

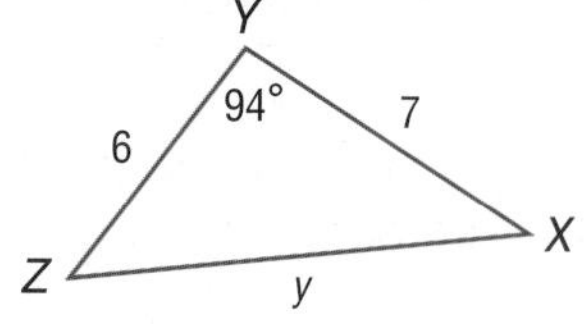

14.

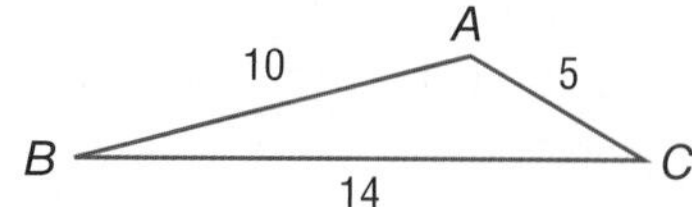

15. GARDENING A triangular flower bed has sides measuring 4.5 feet, 6 feet, and 7.5 feet. Find the measure of the smallest angle. (Lesson 12-5)

Find the exact value of each function. (Lesson 12-6)

16. $\cos(-60)°$

17. $\sin\left(-\frac{3\pi}{4}\right)$

Find the amplitude and period of each function. Then graph the function. (Lesson 12-7)

18. $y = 3\cos\theta$

19. $y = \frac{1}{2}\sin 3\theta$

State the amplitude, period, phase shift, and vertical shift for each function. Then graph the function. (Lesson 12-8)

20. $y = 2\tan(\theta + 30°) + 3$

21. $y = 4\cos\left(\theta - \frac{\pi}{2}\right) - 2$

Solve each equation. Round to the nearest tenth if necessary. (Lesson 12-9)

22. $\text{Sin}\ \theta = -0.58$

23. $\text{Cos}\ \theta = 0.32$

24. $\text{Tan}\ \theta = 2.7$

Find the exact value of each expression if $0° < \theta < 90°$. (Lesson 13-1)

1. If $\cos \theta = \frac{3}{5}$, find $\sin \theta$.

2. If $\tan \theta = 2$, find $\cot \theta$.

3. If $\sin \theta = \frac{\sqrt{5}}{3}$, find $\cos \theta$.

4. If $\csc \theta = \frac{3\sqrt{5}}{5}$, find $\tan \theta$.

Verify that each equation is an identity. (Lesson 13-2)

5. $\frac{\sin^2 \theta}{\cos^2 \theta} \cdot \csc^2 \theta = 1 + \tan^2 \theta$

6. $\sec \theta \cot \theta = \csc \theta$

7. $\sin \theta \cot \theta = \cos \theta$

8. $\frac{\sec^2 \theta}{\csc^2 \theta} = \tan^2 \theta$

9. **CONSTRUCTION** A window has the dimensions shown below. Use the measures of the sides of the triangle to show that $\sin^2\theta + \cos^2\theta = 1$. (Lesson 13-2)

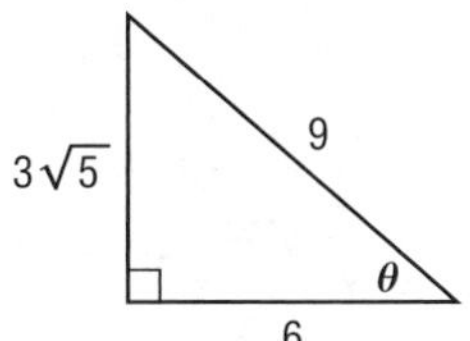

Find the exact value of each expression. (Lesson 13-3)

10. $\cos (-30°)$

11. $\sin (-120°)$

12. $\sin 135°$

13. $\cos 225°$

Find the exact value of each expression. (Lesson 13-4)

14. $\sin 15°$

15. $\cos \frac{\pi}{12}$

16. $\tan 67.5°$

17. $\sin 165°$

Solve each equation if $0° \leq \theta \leq 360°$. (Lesson 13-5)

18. $2 \sin \theta = 1$

19. $2 \cos \theta + 1 = 0$

20. $4 \cos^2 \theta - 1 = 0$

21. **TOYS** A toy's height h in inches can be modeled by the equation $h = 4 \sin \frac{11\pi}{12} t$, where t represents time in seconds. (Lesson 13-5)
   a. At what point will the toy be 4 inches above its resting position for the first time?
   b. At what point will the toy be 2 inches above its resting position for the first time?

22. **BUOYS** As waves move past a buoy, the height in feet of the buoy oscillates according to the equation $y = 6 \cos 30t$, where t is the time in seconds. If the buoy is at the top of the wave at 0 seconds, during what time intervals will the buoy be 3 feet below the midline? (Lesson 13-5)

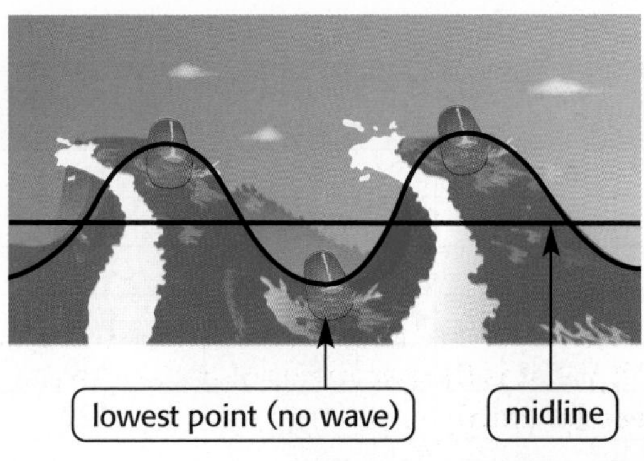

Selected Answers and Solutions

Go to Hotmath.com for step-by-step solutions of most odd-numbered exercises free of charge.

CHAPTER 0

Preparing for Advanced Algebra

Lesson 0-1

1. D = {1, 2, 3}, R = {6, 7, 10}; yes **3.** D = {1, 2}, R = {5, 7, 9}; no **5.** D = {−2, −1, 0, 3}, R = {−3, −2, 2}; yes **7.** D = {−1, 0, 1, 2, 3}, R = {−3, −2, −1, 2, 3, 4}; no **9.** I **11.** none

Lesson 0-2

1. $a^2 + 6a + 8$ **3.** $h^2 - 16$ **5.** $b^2 + b - 12$ **7.** $r^2 - 5r - 24$ **9.** $p^2 + 16p + 64$ **11.** $2c^2 - 9c - 5$ **13.** $6m^2 - 7m - 20$ **15.** $2q^2 - 13q - 34$ **17a.** $n - 7$, $n + 2$ **17b.** $n^2 - 5n - 14$

Lesson 0-3

1. $4x(3x + 1)$ **3.** $4ab(2b - 3)$ **5.** $(y + 3)(y + 9)$ **7.** $(3y + 1)(y + 4)$ **9.** $(3x + 4)(x + 8)$ **11.** $(y - 4)(y - 1)$ **13.** $2(3a - b)(a - 8b)$ **15.** $(2x - 3y)(9x - 2y)$ **17.** $(3x - 4)^2$ **19.** $(x + 12)(x - 12)$ **21.** $(4y + 1)(4y - 1)$ **23.** $4(3y + 2)(3y - 2)$

Lesson 0-4

1. 60 **3.** 18 **5.** 120 **7.** 6 **9.** 6 **11.** permutations, 720 **13.** permutations, 5040 **15.** combinations, 15 **17.** permutations, 3024 **19a.** 2,176,782,336; 1,402,410,240 **19b.** 308,915,776; 712,882,560; The password with one digit is more secure, because the chance of someone guessing this password at random is $\frac{1}{712{,}882{,}560}$, which is less than the chance of someone guessing a 6-character password that contains only letters, $\frac{1}{308{,}915{,}776}$.

Lesson 0-5

1a.

Color	Frequency	Experimental Probability
red	6	0.12
blue	7	0.14
yellow	9	0.18
orange	12	0.24
purple	6	0.10
green	11	0.22

1b.

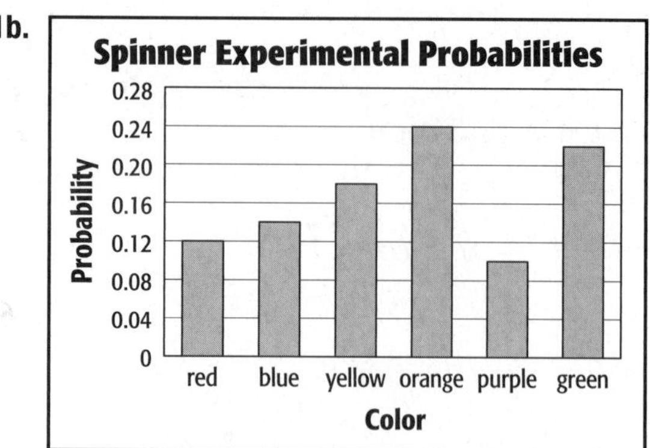

1c.

Color	Frequency	Experimental Probability	Theoretical Probability
red	6	0.12	$0.1\overline{6}$
blue	7	0.14	$0.1\overline{6}$
yellow	9	0.18	$0.1\overline{6}$
orange	12	0.24	$0.1\overline{6}$
purple	6	0.10	$0.1\overline{6}$
green	11	0.22	$0.1\overline{6}$

1d.

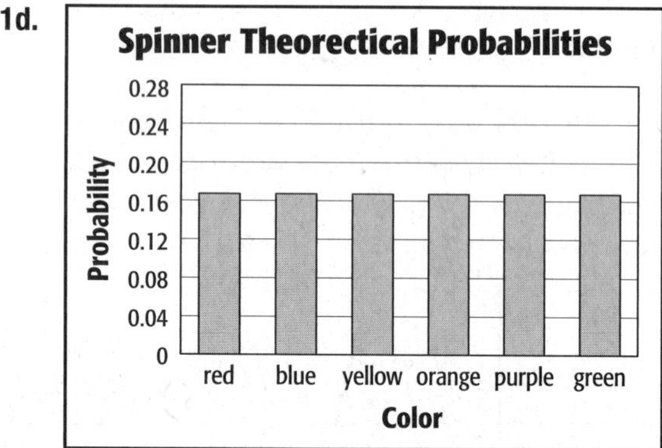

1e. Sample answer: Since all bars in the graph of the theoretical probabilities are the same height, the graph represents a uniform distribution. This means that in theory, the chance of landing on any one of the colors is equally likely. The graph of the experimental probabilities indicates that in practice, it is more likely that the spinner will land on orange or green than on any of the other colors, since the heights of those bars are taller than any others in the graph. **3a.** mutually exclusive, $\frac{1}{2}$ **3b.** not mutually exclusive, $\frac{4}{13}$ **3c.** not mutually exclusive, $\frac{7}{13}$ **5a.** mutually exclusive, $\frac{11}{20}$ **5b.** not mutually exclusive, $\frac{29}{40}$ **5c.** not mutually exclusive, $\frac{2}{5}$

7. When events are mutually exclusive, $P(A \text{ and } B)$ will always equal 0, so the probability will simplify to $P(A \text{ or } B) = P(A) + P(B)$. **9a.** 1 to 3 **9b.** 1 to 1

Lesson 0-6

1. independent; $\frac{1}{36}$ **3.** $\frac{1}{36}$ **5.** $\frac{25}{36}$ **7.** $\frac{5}{18}$ **9.** $\frac{5}{17}$ **11.** $\frac{1}{8}$ **13.** 16% **15a.** $\frac{32}{41}$ **15b.** $\frac{2}{5}$ **15c.** $\frac{3}{7}$ **17a.** $\frac{91}{115}$ or about 79.1% **17b.** $\frac{239}{308}$ or about 77.6%

Lesson 0-7

1. similar **3.** neither **5.** similar **7.** 8; 21 **9.** 10.2; 13.6 **11.** $4\frac{1}{2}$ in.

Lesson 0-8

1. 39 ft **3.** 8.3 cm **5.** 5 **7.** 9.2 **9.** 8.5 **11.** yes **13.** no **15.** yes **17.** about 2.66 m

Lesson 0-9

1. 451.8 min, 399 min, no mode **3.** ≈34.4 text messages, 35 text messages, 35 text messages **5.** Sample; Walk A: 47, ≈242.0, ≈15.6; Walk B: 92, ≈1115.4, ≈33.4; since the sample standard deviation of Walk B is greater than that of Walk A, there is more variability in the number of sponsors obtained by participants in Walk B than in Walk A. **7.** 18, 23, 25, 27, 29; Sample answer: There are 18 students in the smallest math class at Central High and 29 students in the largest class. 25% of the classes have less than 23 students, 50% of the classes have less than 25 students, and 75% of the classes have less than 27 students. **9.** 13; Sample answer: The interval beyond which any outliers would lie is $14.25 < X < 60.25$. Since $13 < 14.25$, it is an outlier.

Data Set	Mean	Median	Mode	Range	Standard Deviation
with outlier	≈35.8	36	36	38	≈9.3
without outlier	≈37.3	36	36	29	≈7.5

Removing the outlier did not affect the mode. However, the removal did affect the mean, median, standard deviation, and range. The mean and median increased, and the standard deviation and range decreased.
11a. Yes; sample answer: The interval beyond which any outliers would lie is $15.35 < X < 17.35$. Since $14.9 < 15.35$, it is an outlier. **11b.** No; sample answer: The new interval beyond which any outliers would lie would be $15.4 < X < 17.4$. Since $17.4 > 17.35$, it would not be an outlier. **11c.** Sample answers: data recording errors, manufacturing errors

CHAPTER 1

Equations and Inequalities

Chapter 1 Get Ready

1. 12.25 **3.** −66.15 **5.** $1\frac{13}{15}$ **7.** $-1\frac{1}{3}$ **9.** $10\frac{1}{2}$ yd **11.** −64 **13.** 15.625 **15.** $\frac{2401}{81}$ **17.** $-\frac{3375}{8}$ **19.** true **21.** false **23.** yes

Lesson 1-1

1. 4.6 **3.** 18.4 **5.** 11.6 **7.** 0.96875 **9.** 0.6 **11.** $6\frac{4}{15}$ **13.** 28 **15.** −13.4 **17a.** 1524.6 mi **17b.** 720 mi **19.** 20

21
$$\frac{b^2c^2}{ad} = \frac{(-0.8)^2(5)^2}{(-4)\left(\frac{1}{5}\right)} \quad a=-4,\ b=-0.8,\ c=5,\ d=\frac{1}{5}$$
$$= \frac{(0.64)(25)}{(-0.8)} \quad \text{Evaluate the numerator and denominator separately.}$$
$$= \frac{16}{-0.8} \quad \text{Simplify the numerator.}$$
$$= -20 \quad \text{Simplify the fraction.}$$

23. 3.71 **25.** $\frac{1}{2}(x + 7)(2x)$ **27a.** 584,336,233.6 mi **27b.** 8761 h **27c.** yes; $\frac{8761}{24} = 365$ days or 1 year **29.** 544 **31.** 13.8 **33.** 131.25 **35.** $6\pi x^3$

37
$$t = 50 + \frac{n-40}{4} \quad \text{Write the formula.}$$
$$= 50 + \frac{120-40}{4} \quad n = 120$$
$$= 50 + \frac{80}{4} \quad \text{Evaluate the numerator of the fraction.}$$
$$= 50 + 20 \quad \text{Simplify the fraction.}$$
$$= 70 \quad \text{Simplify.}$$
If the number of chirps is 120, then the temperature is 70°F.

39a. \$3.91; \$5.36; \$7.31 **39b.** \$4.42; \$6.62; \$11.62; Sample answer: the average prices found in part **b** become increasingly higher with time.

41
$$y = \sqrt{b^2\left(1 - \frac{x^2}{a^2}\right)} \quad \text{Write the equation.}$$
$$= \sqrt{8^2\left(1 - \frac{3^2}{6^2}\right)} \quad a = 6,\ b = 8,\ x = 3$$
$$= \sqrt{64\left(1 - \frac{9}{36}\right)} \quad \text{Evaluate the powers.}$$
$$= \sqrt{64\left(\frac{27}{36}\right)} \quad \text{Simplify inside the parentheses.}$$
$$= \sqrt{48} \quad \text{Simplify.}$$
$$\approx 6.9 \quad \text{Use a calculator.}$$

43. Lauren; −12 − 20 = −32. **45.** Sample answer: b; Since t is time, it must be nonnegative. So $-2t(2t + 1)$ will be negative for all values of t other than 0. The maximum value of $-2t(2t + 1)$ is 0, which occurs when $t = 0$. Thus, the maximum value of $-2t(2t + 1) + 6$ is 6. **47.** Sample answer: $y\left(\frac{-4z}{x^2} - x\right) + z$ **49.** A table of on-base percentages is limited to those situations listed, while a formula can be used to find any on-base percentage. **51.** 9 mo **53.** B **55.** 10 cm **57.** $6x(x + 2)$ **59.** 3 and 11 **61.** 5 **63.** 11 **65.** −4 **67.** $\frac{5}{8}$

Lesson 1-2

1. N, W, Z, Q, R **3.** I, R **5.** Associative Property (×) **7.** Commutative Property (+) **9.** 7; $-\frac{1}{7}$ **11.** −3.8; $\frac{1}{3.8}$ **13a.** $22(2 + 4 + 3 + 1 + 5 + 6 + 7)$ or $22(2) + 22(4) + 22(3) + 22(1) + 22(5) + 22(6) + 22(7)$ **13b.** \$616 **13c.** If she continues to mow the same number of lawns, at the end of next week she will have the money. This may not be reasonable because not all the lawns she mowed this week may need to be mowed again next week.
15. $24a + 9b$ **17.** $-16x + 22y$ **19.** Q, R **21.** Q, R

23 $-\sqrt{144} = -12$ belongs to the set of integers (Z), the set of rationals (Q), and the set of reals (R).

25. I, R **27.** Distributive Property **29.** Inverse Property (×) **31.** −12.1; $\frac{1}{12.1}$ **33.** $-\frac{6}{13}$; $\frac{13}{6}$ **35.** $-\sqrt{15}$; $\frac{1}{\sqrt{15}}$ **37.** $12b + 6c$ **39.** $40x - 20y$
41. $28g - 48k$ **43.** 53(60 + 60); 53(60) + 53(60); 6360 yd^2

45 **a.** $(2 + 1)4.50 = (3)4.50$ Distributive Property
$= \$13.50$ Simplify.

b. The amount left over is \$20 − \$13.50 or \$6.50.
\$6.50 ÷ 2 = 3.25
Because Billie cannot buy part of a sandwich, she can buy 3 cold sandwiches.

c. In two weeks, or ten school days, Billie buys a hot lunch 3 times and buys a cold sandwich 3 times. She has to pack lunch 10 − (3 + 3) = 10 − 6 or 4 times.

47. $\frac{27}{5}c - \frac{199}{20}d$ **49.** $-42x - 72y - 30z$

51 **a.** $-\sqrt{6}$ is an irrational number because the square root of 6 is not a perfect square.
3, or $\frac{3}{1}$, is a rational number, integer, whole number, and natural number.
$\frac{-15}{3}$, or −5, is a rational number and integer.
4.1, or $\frac{41}{10}$, is a rational number.
π is an irrational number.
0, or $\frac{0}{1}, \frac{0}{2}, \ldots$, is a rational number, integer, and whole number.
$\frac{3}{8}$ is a rational number.
$\sqrt{36}$, or 6, is a rational number, integer, whole number, and natural number.

irrational	rational	integer	whole	natural
$-\sqrt{6}$, π	3, $\frac{-15}{3}$, 4.1, 0, $\frac{3}{8}$, $\sqrt{36}$	3, $\frac{-15}{3}$, 0, $\sqrt{36}$	3, 0, $\sqrt{36}$	3, $\sqrt{36}$

b. Use a calculator to find the decimal form of $-\sqrt{6}$ and π; $-\sqrt{6} \approx -2.449$, 3 = 3.0, $\frac{-15}{3} = -5$, 4.1 = 4.1, π ≈ 3.14, 0 = 0, $\frac{3}{8} = 0.375$, $\sqrt{36} = 6$. Since $-5 < -2.449 < 0 < 0.375 < 3.14 < 4.1 < 6$, the numbers from least to greatest are $\frac{-15}{3}$, $-\sqrt{6}$, 0, $\frac{3}{8}$, π, 4.1, $\sqrt{36}$.

c. Draw a number line with tick marks at integers from −6 to 6. Then use the decimal forms in part b to graph each number.

$-\frac{15}{3}$ $-\sqrt{6}$ 0 $\frac{3}{8}$ 3π 4.1 $\sqrt{36}$

−6 −5 −4 −3 −2 −1 0 1 2 3 4 5 6

d. Sample answer: By converting the real numbers into decimal form, the decimal points can be easily lined up and the numbers compared.

53. $\sqrt{81}$; It is a rational number, while the other three are irrational numbers. **55.** No; Luna did not distribute the negative sign to the second term and Sophia switched the a and b terms because usually a comes first. The correct answer is $32a - 46b$.
57. Sample answer: $\sqrt{5} \cdot \sqrt{5} = \sqrt{25}$ or 5, which is not irrational **59.** Sample answer: (a) 3.2 and (b) $\sqrt{10}$
61. Sample answer: The Commutative Property does not hold for subtraction or division because order matters with these two operations. In addition or multiplication, the order does not matter. For example, $2 + 4 = 4 + 2$ and $2 \cdot 4 = 4 \cdot 2$. However, with subtraction, $2 - 4 \neq 4 - 2$, and with division, $2 \div 4 \neq 4 \div 2$. **63.** C **65.** B **67.** 24 **69.** about 2.66 m
71. $3(3x^2 - x + 6)$ **73.** $10x(x - 2)$ **75.** $6(2x^2 - 3x - 4)$
77. $y^2 + y - 2$ **79.** $b^2 - 10b + 21$ **81.** $p^2 - 8p - 9$
83. $\frac{10}{9}$ **85.** 8 **87.** ≈−1.176 **89.** −1.7

Lesson 1-3

1. $12[x + (-3)]$ **3.** The sum of five times a number and 7 equals 18. **5.** The difference between five times a number and the cube of that number is 12.
7. Reflexive Property **9.** 53 **11.** −8 **13.** −6 **15.** 3
17. 4 **19.** $q = \frac{8r - 3}{5}$ **21.** B **23.** $8x^2$ **25.** $\frac{x}{4} + 5$
27. The quotient of the sum of 3 and a number and 4 is 5.

29 Let n = the number of home runs that Jacobs hit. Then $n + 6$ = the number of home runs that Cabrera hit.

$n + (n + 6) = 46$ Cabrera and Jacobs hit a combined total of 46 home runs.
$2n + 6 = 46$ Simplify.
$2n = 40$ Subtract 6 from each side.
$n = 20$ Divide each side by 2.

So, Jacobs hit 20 home runs and Cabrera hit $n + 6 = 20 + 6$ or 26 home runs.

31. Substitution **33.** Multiplication (=) **35.** 5 **37.** −3

39 $5(-2x - 4) - 3(4x + 5) = 97$ Original equation
$-10x - 20 - 12x - 15 = 97$ Apply the Distributive Property.
$-22x - 35 = 97$ Simplify the left side.
$-22x = 132$ Add 35 to each side.
$x = -6$ Divide each side by −22.

41. −3 **43.** s = length of a side; $5s = 100$; 20 in.
45. $m = \frac{E}{c^2}$ **47.** $h = \frac{z}{\pi q^3}$ **49.** $a = \frac{y - bx - c}{x^2}$
51a. $V = \pi \cdot r \cdot r \cdot h$ **51b.** $h = \frac{V}{\pi r^2}$ **53.** −2 **55.** −4

57. $-\frac{117}{11}$ **59.** x = the cost of rent each month; $622 + 428 + 240 + 144 + 12x = 10{,}734$; \$775 per month

61 **a.** The integers from −5 to 5 are −5, −4, −3, −2, −1, 0, 1, 2, 3, 4, and 5. Draw a number line and plot a point at each integer.

−5 −4 −3 −2 −1 0 1 2 3 4 5

b. −5 and 5 are 5 units from zero, −4 and 4 are 4 units from zero, and so on.

Integer	Distance from Zero
−5	5
−4	4
−3	3
−2	2
−1	1
0	0
1	1
2	2
3	3
4	4
5	5

c. The points (x, y) = (integer, distance from zero) are (−5, 5), (−4, 4), (−3, 3), (−2, 2), (−1, 1), (0, 0), (1, 1), (2, 2), (3, 3), (4, 4), and (5, 5).

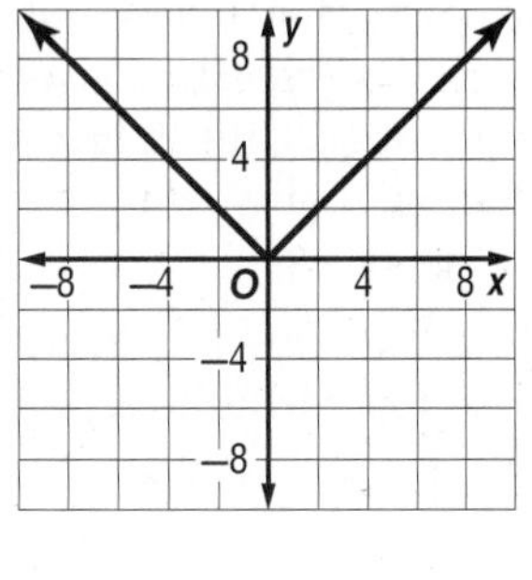

d. For positive integers, the distance from zero is the same as the integer. For negative integers, the distance is the integer with the opposite sign because distance is always positive.

63. $y_1 = y_2 - \sqrt{d^2 - (x_2 - x_1)^2}$ **65.** Sample answer: $3(x - 4) = 3x + 5$; $2(3x - 1) = 6x - 2$ **67.** D **69.** A **71.** $-3x + 6y + 6z$ **73.** 605 ft **75.** $4\frac{1}{5}$ **77.** $2x$ **79.** $-3\frac{2}{3}$ **81.** $-5x$

Lesson 1-4

1. 12 **3.** −108 **5a.** $|x - 78| = 2$ **5b.** least: 76°F; greatest: 80°F **5c.** 77°F; This would ensure a minimum temperature of 76°F. **7.** {15, −7} **9.** ∅ **11.** $\left\{\frac{6}{5}, -\frac{4}{5}\right\}$ **13.** {2} **15.** 25 **17.** 9.2 **19.** 49.2 **21.** −63 **23.** {34, −8} **25.** {4, −14} **27.** {−2, −10} **29.** {2}

31

$2\|3x - 4\| + 8 = 6$	Original equation
$2\|3x - 4\| + 8 - 8 = 6 - 8$	Subtract 8 from each side.
$2\|3x - 4\| = -2$	Simplify.
$\|3x - 4\| = -1$	Divide each side by 2.

Because the absolute value of a number is always positive or zero, this sentence is never true. So, there is no solution.

33. ∅ **35.** $|x - 5.67| = 0.02$; heaviest: 5.69 g; lightest: 5.65 g **37.** 28 **39.** $\left\{1, \frac{1}{5}\right\}$ **41.** $\left\{-\frac{8}{3}\right\}$

43 The average altitude c is 100 ft. Since the altitude can be plus or minus 245 feet, the range r is 245.

$\|x - c\| = r$	Absolute value equation
$\|x - 100\| = 245$	$c = 100$ and $r = 245$

$|x - 100| = 245$ means $x - 100 = 245$ or $x - 100 = -245$.

Case 1	Case 2
$x - 100 = 245$	$x - 100 = -245$
$x - 100 + 100 = 245 + 100$	$x - 100 + 100 = -245 + 100$
$x = 345$	$x = -145$

The solutions are 345 and −145. This means the maximum is 345 ft above sea level; the minimum is −145 ft or 145 ft below sea level. The maximum is reasonable, but the minimum is not. Florida's lowest point should be at sea level where Florida meets the Atlantic Ocean and the Gulf of Mexico.

45. Ling; Ana included an extraneous solution. She would have caught this error if she had checked to see if her answers were correct by substituting the values into the original equation. **47.** Sometimes; this is only true for certain values of a. For example, it is true for $a = 8$; if $8 > 7$, then $11 > 10$. However, it is not true for $a = -8$; if $8 > 7$, then $5 \not> 10$. **49.** Always; starting with numbers between 1 and 5 and subtracting 3 will produce numbers between −2 and 2. These all have an absolute value less than or equal to 2. **51.** Sample answer: First, isolate the absolute value symbol by subtracting each side by c, and then dividing each side by a. You then have $|x - b|$ equals a mathematical expression. Take away the absolute value symbol, and form two new equations by setting $x - b$ equal to both the positive and negative values of the expression. Solve each equation for x. Then substitute each solution into the original equation, and confirm whether they are correct. **53.** $\frac{5}{8}$ **55.** E **57.** −2 **59a.** \$6800 **59b.** \$535.83 **59c.** 1 mo **61.** Distributive **63.** $10x + 2y$ **65.** $11m + 10a$ **67.** $32c - 46d$ **69.** 2 **71.** −8 **73.** $-\frac{4}{7}$

Lesson 1-5

1. $b < 8$

3. $x \le -6$

5. $w < 2$

7

$s \ge \frac{s+6}{5}$	Original inequality
$5s \ge s + 6$	Multiply each side by 5.
$4s \ge 6$	Subtract s from each side.
$s \ge 1.5$	Divide each side by 4.

The solution set is $\{s \mid s \ge 1.5\}$.

9. 40 bags

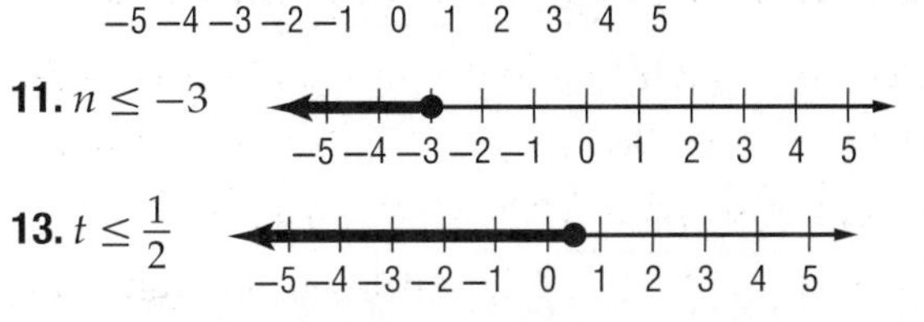

11. $n \le -3$

13. $t \le \frac{1}{2}$

15. $k < 27$

17. $z < 3$

−5 −4 −3 −2 −1 0 1 2 3 4 5

(19) $12 < -4(3c - 6)$ Original inequality

$-3 > 3c - 6$ Divide each side by −4, reversing the inequality symbol.

$3 > 3c$ Add 6 to each side.

$1 > c$ Divide each side by 3.

The solution set is $\{c \mid c < 1\}$.

21. $z < 3$

−5 −4 −3 −2 −1 0 1 2 3 4 5

23. $3x - 12 < 21; x < 11$ **25.** $5x - 6 > x; x > 1.5$
27. 8 hours
29. $x > -\frac{3}{4}$

−5 −4 −3 −2 −1 0 1 2 3 4 5

31. $y > 18.75$

15 16 17 18 19 20 21 22 23 24 25

33. $v > -4.5$

−5 −4 −3 −2 −1 0 1 2 3 4 5

35. $r > -\frac{3}{4}$

−5 −4 −3 −2 −1 0 1 2 3 4 5

37a. $250 + 0.03(500a) \geq 700$ **37b.** $a \geq 30$; He must sell at least 30 advertisements. **39.** $\frac{x}{3} + 4 \leq 2x + 12$; $x \geq -4.8$

(41) a. Let d = the number of miles by which Jamie should increase her average daily run. Then $5 + d$ = her average daily distance after the increase.

3 times	average daily distance	is at least	length of a marathon
$3 \cdot$	$(5 + d)$	$\geq$	26.2

So, the inequality is $3(5 + d) \geq 26.2$.

b. $3(5 + d) \geq 26.2$ Original inequality

$5 + d \geq 8.73$ Divide each side by 3. Round to the nearest hundredth.

$d \geq 3.73$ Subtract 5 from each side.

In order to have enough endurance to run a marathon, Jamie should increase the distance of her average daily run by at least 3.73 miles.

43a. Sample answer:

Point	Resulting Statement	True or False
(0, 0)	$0 \geq 3$	False
(1, 1)	$1 \geq \frac{5}{2}$	False
(2, 2)	$2 \geq 2$	True
(3, 3)	$3 \geq \frac{3}{2}$	True
(4, 4)	$4 \geq 1$	True

43b. Sample answer:

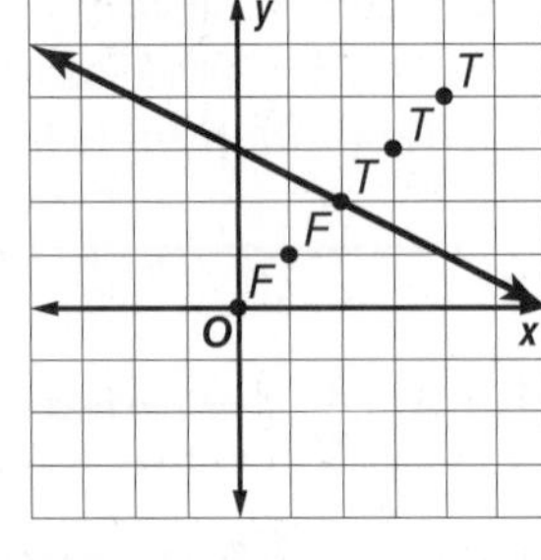

43c. Sample answer: The points on or above the line result in true statements, and the points below the line result in false statements. This is true for all points on the coordinate plane. **45.** No; Sample answer: Madlynn reversed the inequality sign when she added 1 to each side. Emilie did not reverse the inequality sign at all. **47.** Using the Triangle Inequality Theorem, we know that the sum of the lengths of any 2 sides of a triangle must be greater than the length of the remaining side. This generates 3 inequalities to examine.

$3x + 4 + 2x + 5 > 4x$, $x > -9$

$3x + 4 + 4x > 2x + 5$, $x > 0.2$

$2x + 5 + 4x > 3x + 4$, $x > -\frac{1}{3}$

In order for all 3 conditions to be true, x must be greater than 0.2. **49.** Sample answer: When one number is greater than another number, it is either more positive or less negative than that number. When these numbers are multiplied by a negative value, their roles are reversed. That is, the number that was more positive is now more negative than the other number. Thus, it is now *less than* that number and the inequality symbol needs to be reversed.
51. A **53.** D **55.** $\left\{-\frac{1}{3}, 3\right\}$ **57.** $|t - 3647.5| = 891.5$
59a. $SA = 2\pi r(r + h)$ **59b.** 78π cm^2 **59c.** Sample answer: The formula in part b is quicker. **61.** $\{-9, 9\}$
63. $\left\{\frac{1}{2}, 7\right\}$ **65.** $\{-6, 2\}$

Lesson 1-6

1. $\{g \mid -12 < g < -2\}$

−15 −12 −9 −6 −3 0 3 6 9 12 15

3. $\{z \mid z > -3 \text{ or } z < -6\}$

−10 −8 −6 −4 −2 0 2 4 6 8 10

5. $\{c \mid c \geq 8 \text{ or } c \leq -8\}$

−10 −8 −6 −4 −2 0 2 4 6 8 10

7. $\{z \mid -6 < z < 6\}$

−10 −8 −6 −4 −2 0 2 4 6 8 10

9. $\left\{v \,\middle|\, v > 3 \text{ or } v < -\frac{19}{3}\right\}$

−10 −8 −6 −4 −2 0 2 4 6 8 10

11. $43.96 \leq c \leq 77.94$; between \$43.96 and \$77.94

13. $\{d \mid -1 \leq d \leq 0.5\}$

−5 −4 −3 −2 −1 0 1 2 3 4 5

15. $\{y \mid y < -4 \text{ or } y > 7\}$

17. $\{k \mid -4 > k \text{ or } k > 4\}$

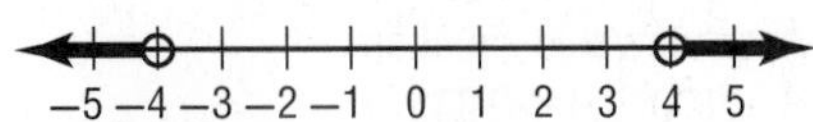

19 $|8t + 3| \leq 4$ is equivalent to $-4 \leq 8t + 3 \leq 4$.

$$\begin{aligned} -4 \leq \quad 8t + 3 \quad &\leq 4 \\ -4 - 3 \leq 8t + 3 - 3 &\leq 4 - 3 \\ -7 \leq \quad 8t \quad &\leq 1 \\ -\frac{7}{8} \leq \quad \frac{8t}{8} \quad &\leq \frac{1}{8} \\ -\frac{7}{8} \leq \quad t \quad &\leq \frac{1}{8} \end{aligned}$$

The solution set is $\left\{t \,\middle|\, -\frac{7}{8} \leq t \leq \frac{1}{8}\right\}$ or $\left[-\frac{7}{8}, \frac{1}{8}\right]$.

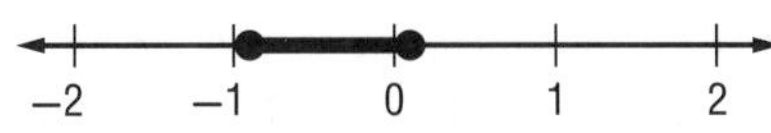

21. $\left\{j \,\middle|\, j \geq \frac{8}{5} \text{ or } j \leq -\frac{16}{5}\right\}$

23. $|x - 1| \leq 5$ **25.** $|x + 9| \leq 3$ **27.** $|x - 2| \geq 10$ **29.** $|x + 3| > 1$

31 A healthy weight w for a fully grown female Labrador retriever is 55 pounds to 70 pounds. This can be represented by $55 \leq w \leq 70$.

33. $\{k \mid 2 < k < 4\}$

35. $\{h \mid h < -15 \text{ or } h > 15\}$

37. $\{z \mid z < -1 \text{ or } z > 5\}$

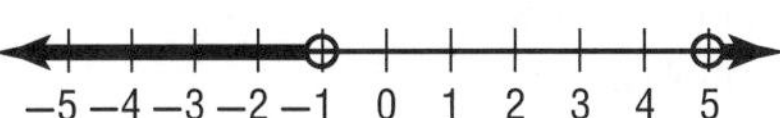

39. $\left\{f \,\middle|\, f > \frac{26}{5} \text{ or } f < -\frac{22}{5}\right\}$

41. $|x + 5| \geq 4$ **43.** $6 \leq |x - 2| \leq 10$

45. $\{n \mid -7 < n < 1\}$

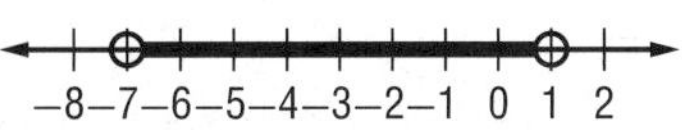

47. $\varnothing$

49. $\left\{g \,\middle|\, g \geq -\frac{2}{3}\right\}$

51. $|s - 88| > 38$; $\{s \mid s > 126 \text{ or } s < 50\}$ **53.** Sample answer: David; when Sarah converted the absolute value into two inequalities, she mistakenly switched the inequality symbols. **55.** False; sample answer: the graph of $x > 2$ and $x > 5$ is a ray bounded only on one end. **57.** true **59.** Sample answer: The graph on the left indicates a solution set from -3 to 5 for the inequality $|x - 1| \leq 4$. The graph on the right indicates a solution set of all numbers less than or equal to -3 or greater than or equal to 5 for the inequality $|x - 1| \geq 4$. **61.** Each of these has a non-empty solution set except for $x > 5$ and $x < 1$. There are no values of x that are simultaneously greater than 5 and less than 1. **63.** C **65.** 60 **67a.** $750 \leq x \leq 990$ **67b.** 110 g **69.** $\{0, 10\}$ **71.** $\varnothing$ **73.** Transitive (=)

Chapter 1 Study Guide and Review

1. false; nonnegative **3.** true **5.** true **7.** false; or **9.** true **11.** 3 **13.** 10 **15.** 21 **17.** ≈ 169.65 in^3 **19.** N, W, Z, Q, R **21.** $11x + 2y$ **23.** $5m + 41n$ **25.** -7 **27.** $\frac{3}{2}$ **29.** \$8 **31.** $m = \frac{r + 5}{pn}$ **33.** about 8 in. **35.** $\{2, 10\}$ **37.** $\{14\}$

39. $a \geq -6$

41. $x \leq -\frac{2}{9}$

43. 3 or fewer slices each

45. $\{x \mid -2 < x < 4\}$

47. $\left\{m \,\middle|\, \frac{13}{5} \leq m < \frac{24}{5}\right\}$

49. $\{p \mid -5 \leq p \leq 33\}$

51. $\varnothing$ **53.** $20 \leq 2.50(3) + 1.25b \leq 30$; $10 \leq b \leq 18$

CHAPTER 2 Linear Relations and Functions

Chapter 2 Get Ready

1. (4, 1); I **3.** (0, 0); origin **5.** (−4, −4); III **7.** -15 **9.** 10 **11.** \$40 **13.** $b = \frac{a}{3} - 3$ **15.** $x = \frac{8}{3} + \frac{4}{3}y$

Lesson 2-1

1. D = {5, 6, −2}, R = {3, −8, 1}; function; both

3 The domain is the set of x-values: {−2, 1, 4, 8}; the range is the set of y-values: {−4, −2, 6}. Since each element of the domain is paired with exactly one element of the range, the relation is a function. Since each element of the range corresponds to an element of the domain, the relation is an onto function.

5.

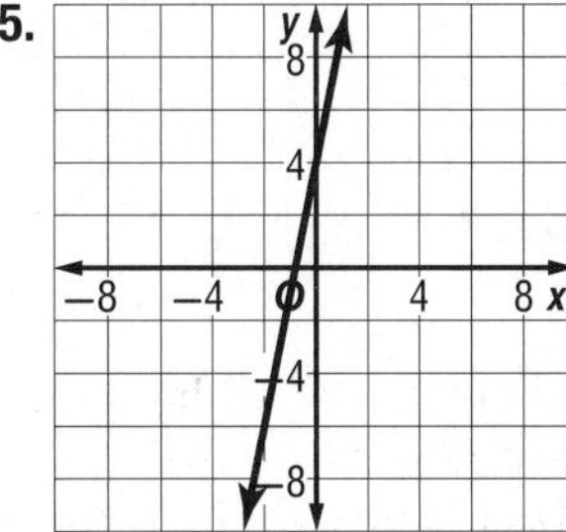

D = {all real numbers}, R = {all real numbers}; function; both; continuous

Selected Answers and Solutions

7. 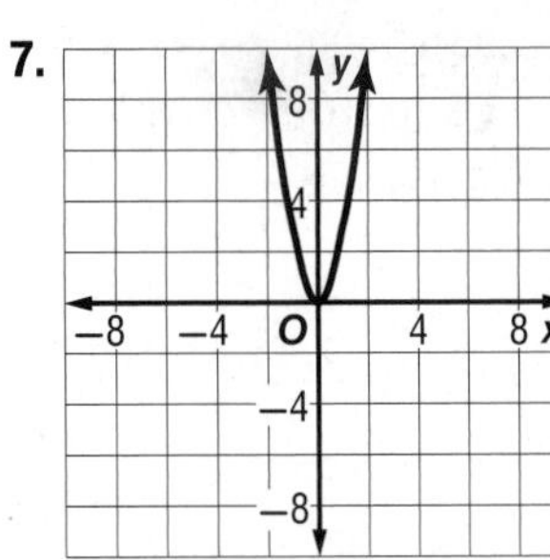

D = {all real numbers}, R = {y| $y \geq 0$}; function; onto; continuous **9.** 4 **11.** D = {−0.3, 0.4, 1.2}, R = {−6, −3, −1, 4}; not a function **13.** D = {−3, −1, 3, 5}, R = {−4, 0, 3}; function; onto

15. 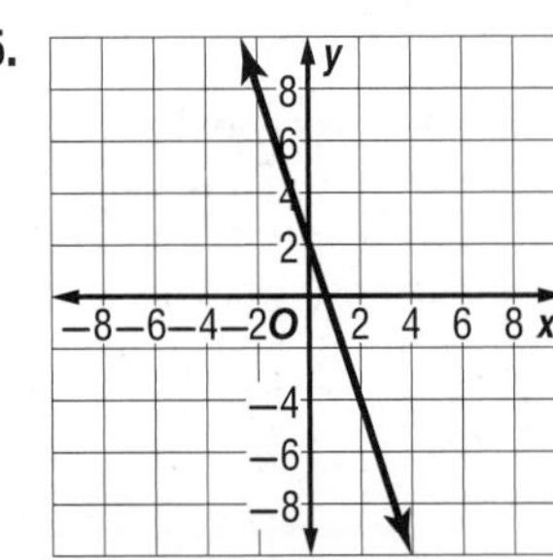

D = {all real numbers}, R = {all real numbers}; function; both; continuous

17. 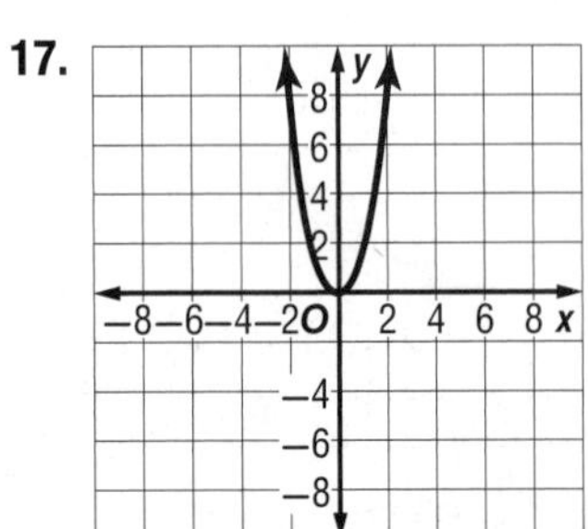

D = {all real numbers}, R = {$y \mid y \geq 0$}; function; onto; continuous

19. 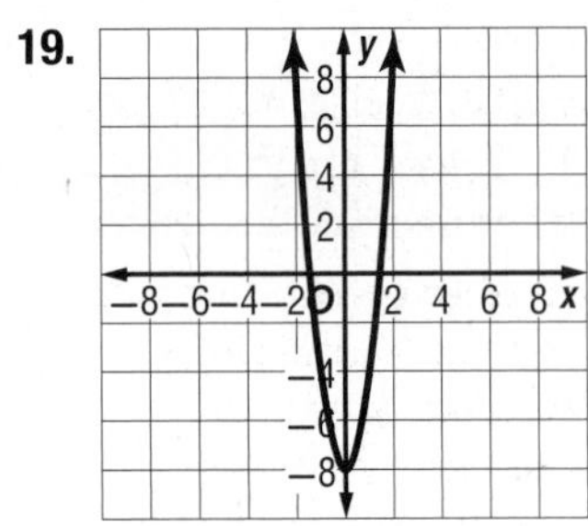

D = {all real numbers}, R = {$y \mid y \geq -8$}; function; onto; continuous

21

$f(x) = 5x^3 + 1$	Original function
$f(-8) = 5(-8)^3 + 1$	Substitute −8 for each x.
$= 5(-512) + 1$	Evaluate $(-8)^3$.
$= -2560 + 1$	Multiply.
$= -2559$	Simplify.

23a. {(0, 1), (20, 1.6), (40, 2.2), (60, 2.8), (80, 3.4), (100, 4)}

23b.

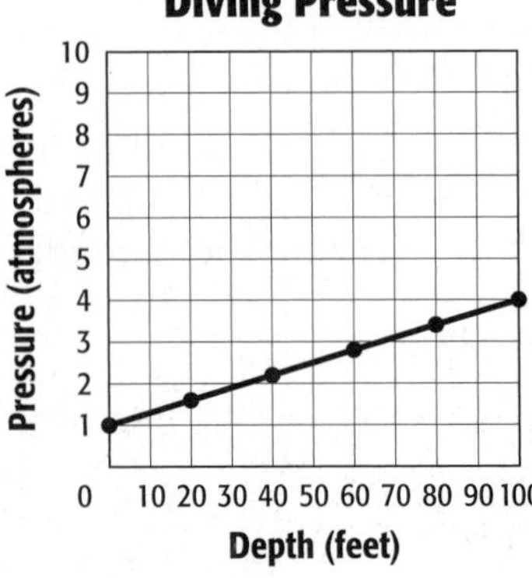

23c. D = {$x \mid x \geq 0$}, R = {$y \mid y \geq 1$}; continuous **23d.** Yes; each domain value is paired with only one range value so the relation is a function. **25.** 29 **27.** −72 **29.** −267 **31.** −4.5

33

$P(t) = 15 + 3t$	Original function
$P(8) = 15 + 3(8)$	Substitute 8 for each t.
$= 15 + 24$	Multiply.
$= 39$	Simplify.

After 8 months, Chaz will have 39 podcasts.

35. Sample answer: Omar; Madison did not square the 3 before multiplying by −4. **37.** Never; if the graph crosses the y-axis twice, then there will be two separate y-values that correspond to $x = 0$, which violates the vertical line test. **39.** Sample answer: False; a function is onto and not one-to-one if all of the elements of the domain correspond to an element of the range, but more than one element of the domain corresponds to the same element of the range. **41.** A **43.** J **45.** $6 > y > 2$ **47.** $x > \frac{7}{4}$ or $x < -\frac{11}{4}$ **49.** $15x \leq 120$; She can buy up to 8 shirts. **51.** $\frac{3}{4}$ or $\frac{7}{4}$ **53.** 33a **55.** $10c + 36d$ **57.** 4 **59.** −4 **61.** −4 **63.** −6

Lesson 2-2

1. Yes; it can be written as $f(x) = \frac{x}{5} + \frac{12}{5}$. **3.** No; x has an exponent that is not 1.

5 **a.**

$m(x) = 0.75x$	Original function
$m(4) = 0.75(4)$	Substitute 4 for x.
$= 3$	Simplify.

If you have 4 CDs, you have 3 hours of music.

b.

$m(x) = 0.75x$	Original function
$6 = 0.75x$	Substitute 6 for $m(x)$.
$8 = x$	Divide each side by 0.75.

If the trip is 6 hours long, you should bring 8 CDs.

7. $6x - y = -5$; $A = 6, B = -1, C = -5$ **9.** $8x + 9y = 6$; $A = 8, B = 9, C = 6$ **11.** $2x - 3y = 12$; $A = 2, B = -3, C = 12$

13. $\frac{5}{2}$; −10

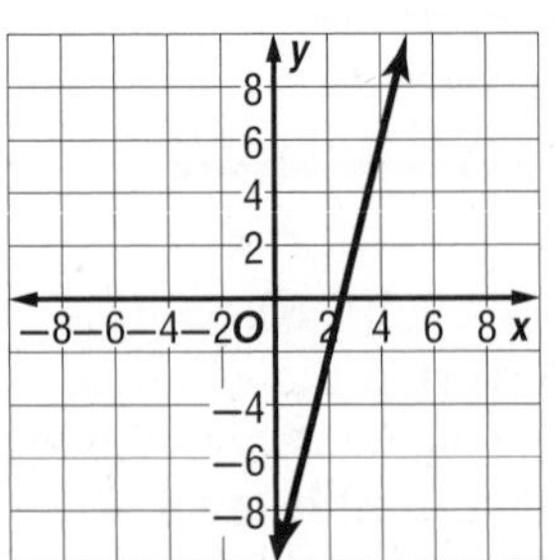

15. 7; $-\frac{21}{4}$

17. No; x has an exponent other than 1. **19.** No; x has an exponent other than 1. **21.** No; it cannot be written in $f(x) = mx + b$ form. **23.** No; it cannot be written in $f(x) = mx + b$ form; There is an xy term. **25a.** 260 m **25b.** Kingda Ka; Sample answer: The Kingda Ka travels 847.5 meters in 25 seconds, so it travels a greater distance in the same amount of time. **27.** $8x + 3y = -6$; $A = 8$,

$B = 3, C = -6$ **29.** $2x + y = -11; A = 2, B = 1, C = -11$

31
$$\begin{aligned} 2.4y &= -14.4x && \text{Original equation} \\ 14.4x + 2.4y &= 0 && \text{Add 14.4x to each side.} \\ 144x + 24y &= 0 && \text{Multiply each side by 10.} \\ 6x + y &= 0 && \text{Divide each side by 24.} \end{aligned}$$
$A = 6, B = 1, C = 0.$

33. $5x + 32y = 160; A = 5, B = 32, C = 160$

35. $-0.5; -4$

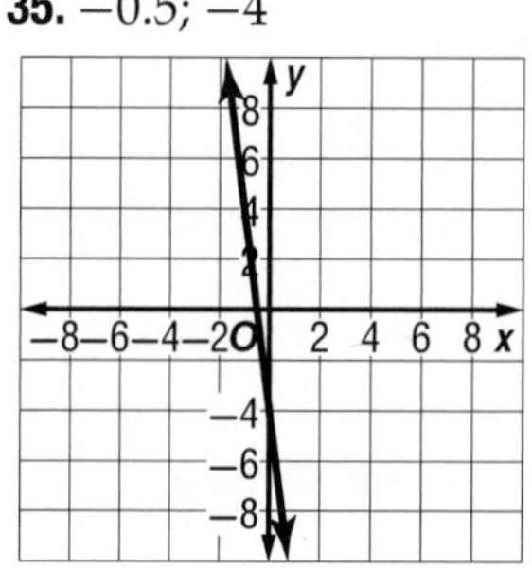

37. $-7; 10.5$

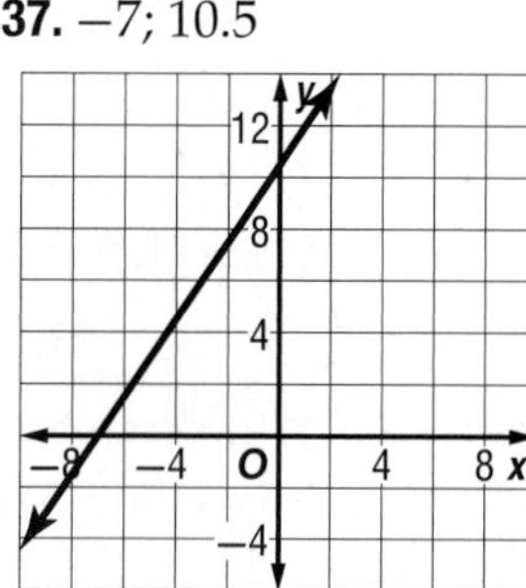

39. $12; -18$

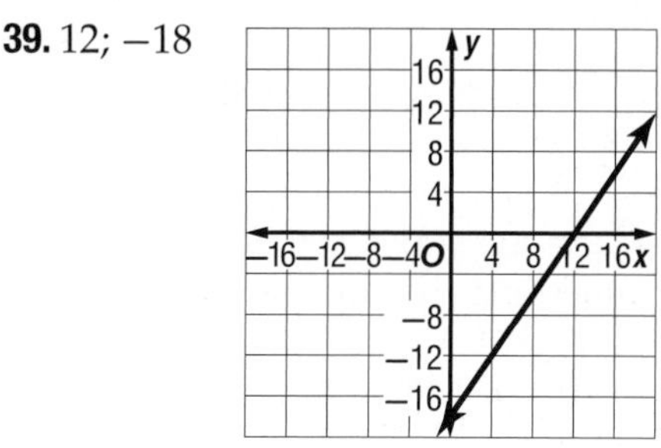

41a. $1.75m + 1.5n = 525$

41b.

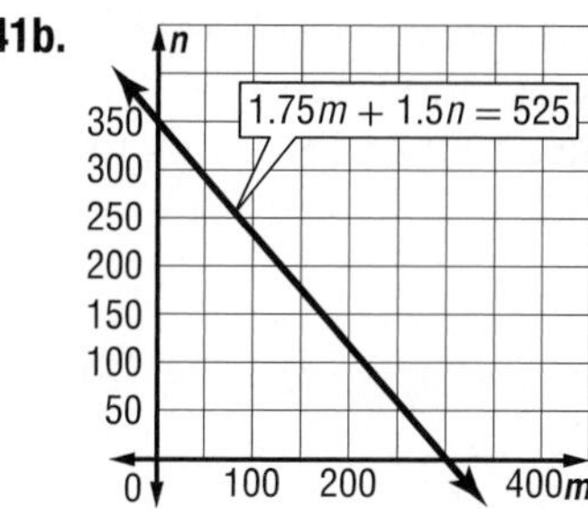

Yes; the graph passes the vertical line test.

41c. No; the amount that Latonya will sell is $1.75 \cdot 100 + 1.5 \cdot 200$, which is \$475.
43a. $y = 3x + 13$ **43b.** \$31 **45.** $4x - 40y = -59; A = 4, B = -40, C = -59$

47 The x-intercept is the value of x when $y = 0$.
$$\begin{aligned} \frac{6x+15}{4} &= 3y - 12 && \text{Original equation} \\ \frac{6x+15}{4} &= 3(0) - 12 && \text{Substitute 0 for } y. \\ \frac{6x+15}{4} &= -12 && \text{Simplify.} \\ 6x + 15 &= -48 && \text{Multiply each side by 4.} \\ 6x &= -63 && \text{Subtract 15 from each side.} \\ x &= -\frac{63}{6} && \text{Divide each side by 6.} \\ x &= -10.5 && \text{Simplify.} \end{aligned}$$
The x-intercept is -10.5.
The y-intercept is the value of y when $x = 0$.
$$\begin{aligned} \frac{6x+15}{4} &= 3y - 12 && \text{Original equation} \\ \frac{6(0)+15}{4} &= 3y - 12 && \text{Substitute 0 for } x. \\ \frac{15}{4} &= 3y - 12 && \text{Simplify.} \\ \frac{63}{4} &= 3y && \text{Add 12 to each side.} \\ 5.25 &= y && \text{Divide each side by 3.} \end{aligned}$$
The y-intercept is 5.25.

49. $-1\frac{1}{75}; 6\frac{1}{3}$ **51a.**

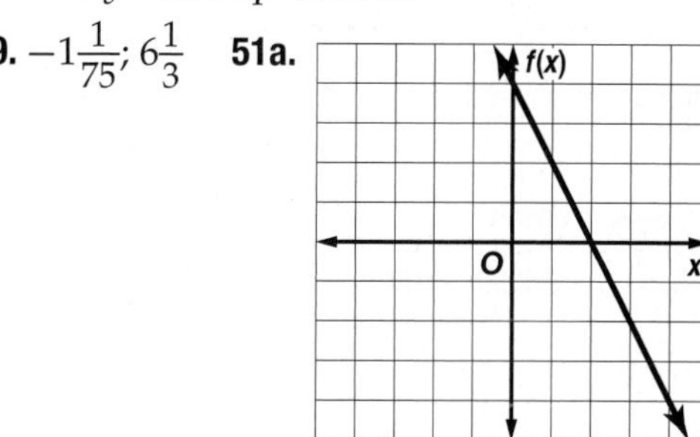

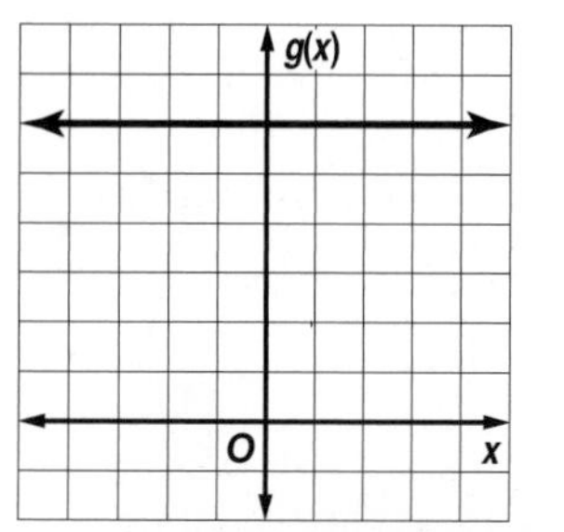

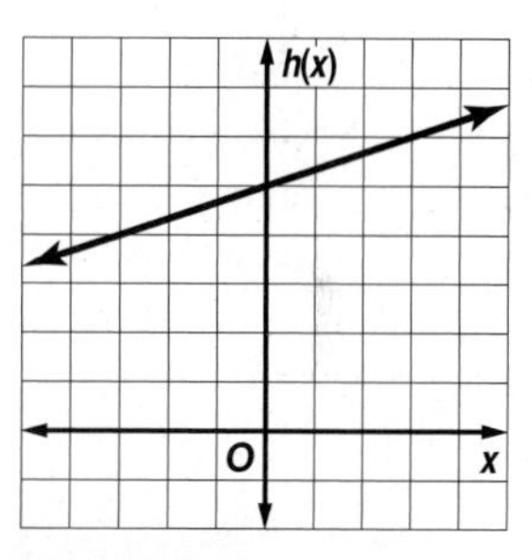

51b.

Function	One-to-One	Onto
$f(x) = -2x + 4$	yes	yes
$g(x) = 6$	no	no
$h(x) = \frac{1}{3}x + 5$	yes	yes

51c. No; horizontal lines are neither one-to-one nor onto because only one y-value is used and it is repeated for every x-value. Every other linear function is one-to-one and onto because every x-value has one unique y-value that is not used by any other x-element and every possible y-value is used. **53.** Sample answer: $f(x) = 2(x - 3)$ **55.** $y = 2xy$; Sample answer: $y = 2xy$ is not a linear function. **57.** C **59.** J **61.** D = {8, −4, −1}, R = {6, 3, 9}; not a function **63.** D = {−3, −4, 7}, R = {−1, −2, 9}; function; both **65.** 0.78 **67.** about −0.583 **69.** $\frac{2}{3}$ **71.** $-\frac{5}{4}$ **73.** $-\frac{1}{3}$ **75.** 9

Lesson 2-3

1. 6 feet/min **3a.** about 11,000 per year **3b.** about −5000 per year **3c.** The positive rate in part **a** represents an increase in the sales of digital cameras. The negative rate in part **b** represents a decrease in sales of film cameras. **5.** −3 **7.** $\frac{3}{5}$

9 Use the ordered pairs (3, 20) and (6, 40).

$$\text{rate of change} = \frac{\text{change in } y}{\text{change in } x}$$

$$= \frac{\text{change in height}}{\text{change in time}} \begin{matrix} \leftarrow \text{mm} \\ \leftarrow \text{days} \end{matrix}$$
$$= \frac{40 - 20}{6 - 3}$$
$$= \frac{20}{3}$$

The rate of change is $\frac{20}{3}$ mm/day.

11a. 0.15°/h **11b.** –0.125°/h; Yes; the number should be negative because her temperature is dropping. **11c.** Monday 8:00 A.M.–Monday 8:00 P.M. **13.** $\frac{14}{15}$ **15.** –2 **17.** $\frac{5}{3}$ **19.** 5

21 The line passes through (0, 16) and (10, 8).

$m = \frac{y_2 - y_1}{x_2 - x_1}$ Slope Formula

$= \frac{8 - 16}{10 - 0}$ $(x_1, y_1) = (0, 16), (x_2, y_2) = (10, 8)$

$= -\frac{8}{10}$ or -0.8 Simplify.

23. $\frac{4}{3}$ **25.** 3 **27.** $\frac{6}{5}$

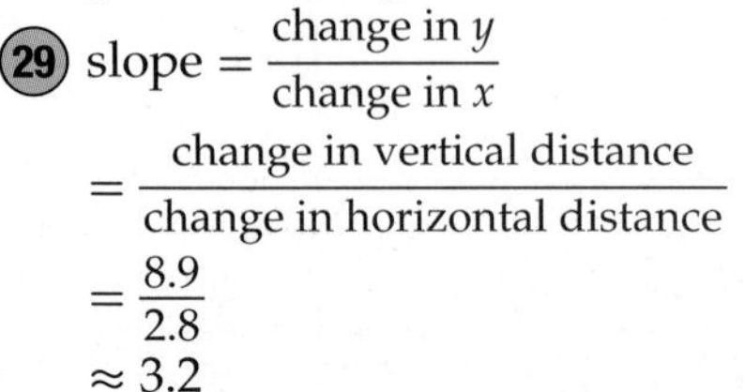

29 slope $= \frac{\text{change in } y}{\text{change in } x}$

$= \frac{\text{change in vertical distance}}{\text{change in horizontal distance}}$

$= \frac{8.9}{2.8}$

≈ 3.2

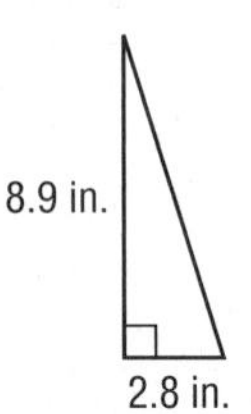

31. 9 **33.** 5 **35a.**

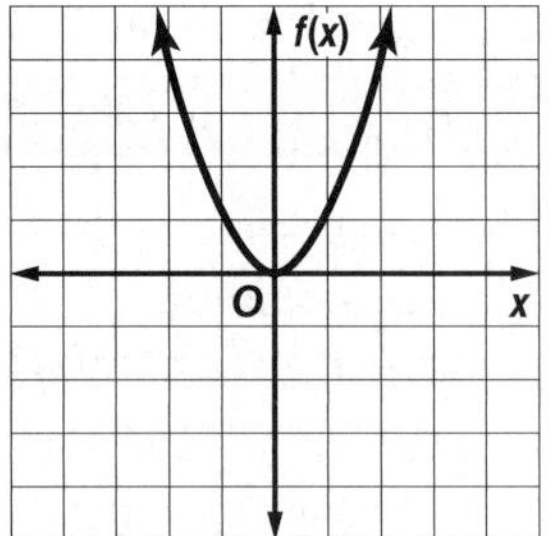

35b.

x	–4	–3	–2	–1	0	1	2	3	4
f(x)	16	9	4	1	0	1	4	9	16
slope		–7	–5	–3	–1	1	3	5	7

35c. Sample answer: The rate of change is not constant. The rate of change decreases as x approaches zero and then increases as x approaches infinity. **37.** Sample answer: Because the slope from (2, 3) to (5, 8) is the same as the slope from (5, 8) to (11, y), find the slope between each pair of points and set them equal to each other. Then solve for y.

$$\frac{8 - 3}{5 - 2} = \frac{y - 8}{11 - 5}$$
$$\frac{5}{3} = \frac{y - 8}{6}$$
$$30 = 3(y - 8)$$
$$10 = y - 8$$
$$18 = y$$

39. Sometimes; the slope of a vertical line is undefined. **41.** $\frac{3}{2}$ **43.** H **45.** Yes; it can be written in $f(x) = mx + b$ form. **47.** No; it cannot be written in $f(x) = mx + b$ form. **49.** –46 **51.** 336 **53.** II **55.** 3.5 **57.** $\frac{7}{3}$

Lesson 2-4

1. $y = 1.5x + 5$ **3.** $y = -2x + 11$ **5.** A

7 The slope of the given line is $\frac{7}{8}$. Lines that are parallel have the same slope, so the slope of the line parallel to the given line is $\frac{7}{8}$.

$y - y_1 = m(x - x_1)$ Point-slope form

$y - (-10) = \frac{7}{8}(x - 4)$ $(x_1, y_1) = (4, -10)$ and $m = \frac{7}{8}$

$y + 10 = \frac{7}{8}x - \frac{7}{2}$ Distributive Property

$y = \frac{7}{8}x - \frac{27}{2}$ Subtract 10 from each side.

9. $y = -\frac{1}{2}x + 5$ **11.** $y = 4.5x - 6.5$ **13.** $y = 4x - 15$ **15.** $y = -\frac{1}{4}x - 1$ **17.** $y = 2x - 2$ **19.** $y = -8x - 20$ **21.** $y = -0.5x + 3.35$ **23.** $y = \frac{1}{2}x$ **25.** $y = -\frac{1}{2}x + 6$ **27.** $y = 180x + 5900$ **29.** $y = -25x + 300$

31 First, find the slope. The line passes through (–6, 2) and (0, 6).

$m = \frac{y_2 - y_1}{x_2 - x_1}$ Slope Formula

$= \frac{6 - 2}{0 - (-6)}$ $(x_1, y_1) = (-6, 2), (x_2, y_2) = (0, 6)$

$= \frac{4}{6}$ or $\frac{2}{3}$ Simplify.

The graph intersects the y-axis at 6. So, $b = 6$. Substitute the values into the slope-intercept equation.

$y = mx + b$ Slope-intercept form

$y = \frac{2}{3}x + 6$ $m = \frac{2}{3}, b = 6$

33. 10 mi

35 **a.** Let x be the number of people Ms. Cooper recruits and let y be the amount of money she earns. Use the points (10, 100) and (14, 120) to represent this situation.

$m = \frac{y_2 - y_1}{x_2 - x_1}$ Slope Formula

$= \frac{120 - 100}{14 - 10}$ $(x_1, y_1) = (10, 100), (x_2, y_2) = (14, 120)$

$= \frac{20}{4}$ or 5 Simplify.

Use the slope and either of the given points with the point-slope form to write the equation.

$y - y_1 = m(x - x_1)$ Point-slope form

$y - 100 = 5(x - 10)$ $(x_1, y_1) = (10, 100), m = 5$

$y - 100 = 5x - 50$ Distributive Property

$y = 5x + 50$ Add 100 to each side.

b. The y-intercept of the graph of $y = 5x + 50$ is 50. This represents the money Ms. Cooper would make if she had no recruits. So, $50 is her daily salary.

c. Find the value of y when $x = 20$.

$y = 5x + 50$ Use the equation you found in part a.

$y = 5(20) + 50$ Replace x with 20.

$y = 150$ Simplify.

So, Ms. Cooper would earn $150 in a day if she recruits 20 people.

37. Sample answer: Sometimes; while the two sets of parallel and perpendicular lines will always

form a quadrilateral with four 90° angles, that figure will always be a rectangle, but not necessarily a square. **39.** Sample answer: $y - 0 = a\left(x + \frac{b}{a}\right)$
41. Sample answer: $y - d = -\frac{d}{c}(x - 0)$ **43.** A
45. G **47.** $-\frac{5}{3}$ **49.** $\frac{1}{5}$ **51.** $x \geq -4$ **53.** $x \geq -\frac{21}{13}$
55. yes **57.** $8c^2 + 8c - 30$ **59.** $-6a^2 + 7a + 20$
61. $\frac{3}{2}$ **63.** $\frac{1}{5}$ **65.** $-\frac{1}{9}$

Lesson 2-5

1 **a.** Graph the data as ordered pairs with the depth on the horizontal axis and the temperature on the vertical axis. Draw a line through two points that appear to represent the data well, such as (0, 22) and (2000, 6).

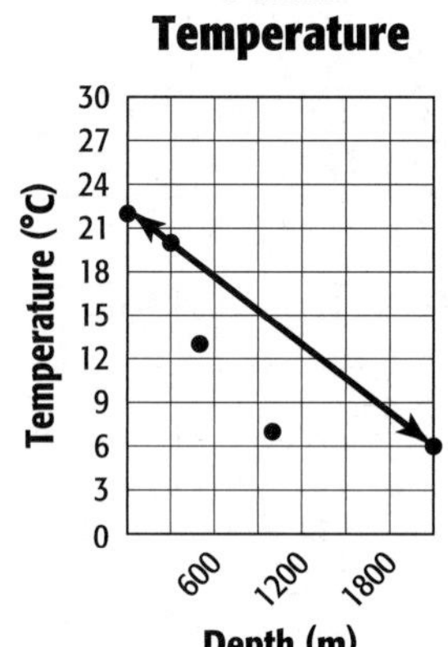

The data show a weak negative correlation.

b. Sample answer: Use (0, 22) and (2000, 6) to find an equation. First, find the slope of the line through (0, 22) and (2000, 6).

$m = \frac{y_2 - y_1}{x_2 - x_1}$ Slope Formula

$= \frac{6 - 22}{2000 - 0}$ $(x_1, y_1) = (0, 22), (x_2, y_2) = (2000, 6)$

≈ -0.008 Simplify.

Then write the equation.

$y - y_1 = m(x - x_1)$ Point-slope form

$y - 22 = -0.008(x - 0)$ $(x_1, y_1) = (0, 22), m = -0.008$

$y - 22 = -0.008x$ Simplify.

$y = -0.008x + 22$ Add 22 to each side.

c. Sample answer: Find y when $x = 2500$.

$y = -0.008x + 22$ Prediction equation

$= -0.008(2500) + 22$ $x = 2500$

$= -20 + 22$ or 2 Simplify.

So, at a depth of 2500 m, the temperature in the ocean is about 2°C.

3a.

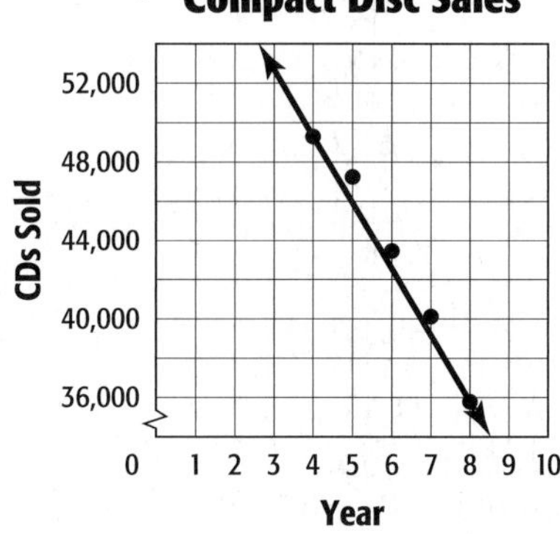

strong negative correlation

3b. Sample answer, using (4, 49,300) and (8, 20,193): $y = -7276.75x + 78{,}407$ **3c.** Sample answer: 12,916 CDs

5 **a.** Graph the data as ordered pairs with the month on the horizontal axis and the number of gallons sold on the vertical axis. Let 1 represent January, 2 represent February, and so on. Draw a line through two points that appear to represent the data well, such as (1, 37) and (8, 131).

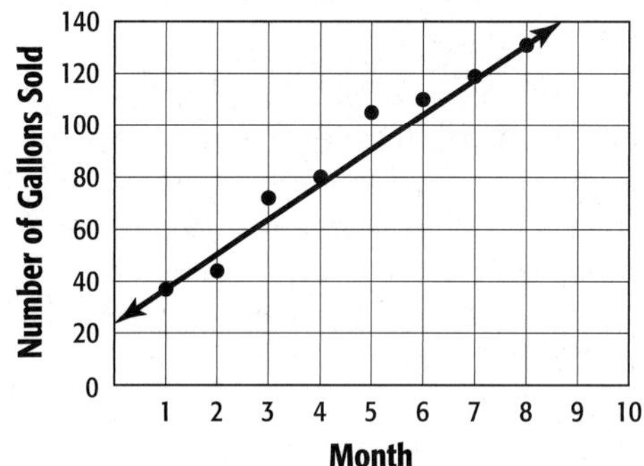

The data show a strong positive correlation.

b. Sample answer: Use (1, 37) and (8, 131) to find an equation. First, find the slope of the line through (1, 37) and (8, 131).

$m = \frac{y_2 - y_1}{x_2 - x_1}$ Slope Formula

$= \frac{131 - 37}{8 - 1}$ $(x_1, y_1) = (1, 37), (x_2, y_2) = (8, 131)$

$= \frac{94}{7}$ Simplify.

Then write the equation.

$y - y_1 = m(x - x_1)$ Point-slope form

$y - 37 = \frac{94}{7}(x - 1)$ $(x_1, y_1) = (1, 37), m = \frac{94}{7}$

$y - 37 = \frac{94}{7}x - \frac{94}{7}$ Distributive Property

$y = \frac{94}{7}x + \frac{165}{7}$ Add 37 to each side.

c. Sample answer: Find y when $x = 9$.

$y = \frac{94}{7}x + \frac{165}{7}$ Prediction equation

$= \frac{94}{7}(9) + \frac{165}{7}$ $x = 9$

$= \frac{1011}{7}$ Simplify.

≈ 144.4 Use a calculator.

So, about 144 gallons of ice cream are sold in September.

7.

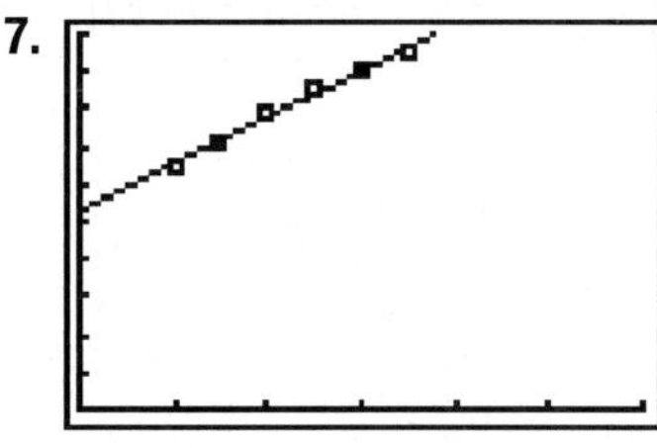

[0, 12] scl: 1 by [0, 1000] scl: 100

$y = 61.9x + 530.2$ (x is the number of years after 2002); $1.149 million in sales

9 **a.** Graph the data as ordered pairs with the year on the horizontal axis and the attendance on the vertical axis.

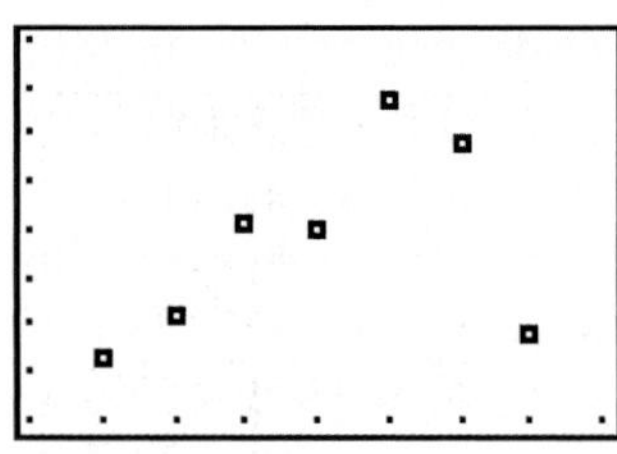

[2002, 2010] scl: 1 by [1,600,000, 2,400,000] scl: 100,000

b. Use a graphing calculator to find a regression equation for the data. Enter the years in **L1** and the attendance in **L2**. Then select **LinReg(ax + b)** on the **STAT CALC** menu. The regression equation is approximately $y = 40{,}634.9x - 79{,}543{,}385.8$. **c.** The slope of the regression line indicates that the attendance increases by about 40,634.9 each year. **d.** The correlation is stronger between wins and attendance. The correlation coefficient between years and attendance is 0.42; the correlation coefficient between wins and attendance is about 0.79. The greater correlation coefficient implies a stronger correlation.

11a. $y = 3.1x - 6177$; $r = 0.63$ **11b.** about \$64.2 million **11c.** $y = 2.6x - 5170$; $r = 0.986$ **11d.** about \$62.4 million **11e.** Sample answer: The new equation has a correlation coefficient, 0.986, that is extremely close to 1, so this equation should accurately represent the data. **13.** Sample answer: If a and b have a positive correlation, then they are both increasing. If b and c have a negative correlation and b is increasing, then c must be decreasing. If c and d have a positive correlation and c is decreasing, then d must be decreasing. If a is increasing and d is decreasing, then they must have a negative correlation.

15. 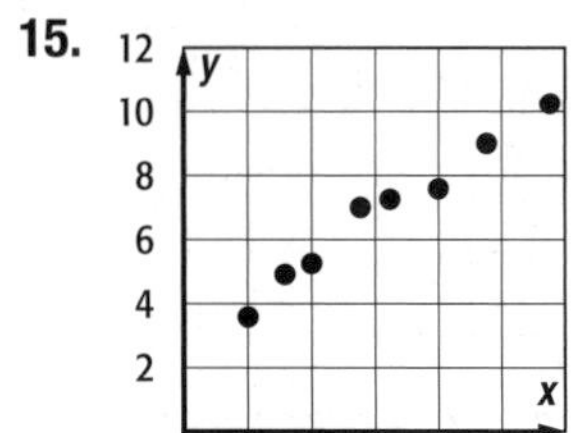

a; Sample answer: The data show a strong positive correlation which means that the correlation coefficient r should be close to 1.

17. 2 **19.** J **21.** $y = 2.5x - 6$ **23.** $y = -3x - 6$ **25.** 17.5 mi/hr **27.** 1.5 J/N **29.** 120 **31.** $\frac{17}{3}, -\frac{25}{3}$

Lesson 2-6

1. 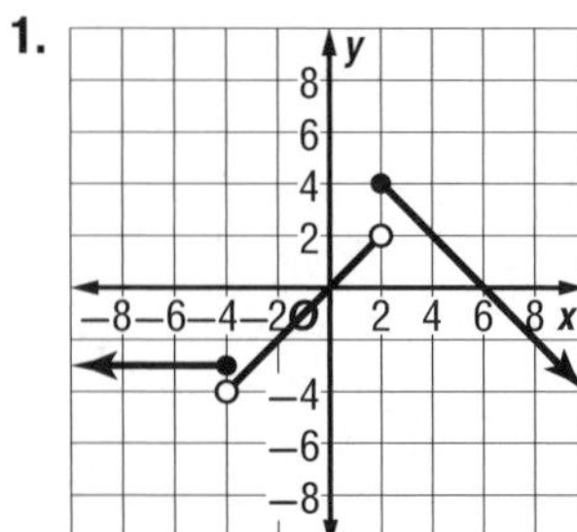

$D = \{\text{all real numbers}\}$; $R = \{y \mid y \leq 4\}$

3. $g(x) = \begin{cases} x + 4 \text{ if } x < -2 \\ -3 \text{ if } -2 \leq x \leq 3 \\ -2x + 12 \text{ if } x > 3 \end{cases}$

5 If the number of tickets sold is greater than 0 but less than or equal to 250, then the drama club must do 1 performance. If the number of tickets sold is greater than 250 but less than or equal to 500, then the drama club must do 2 performances, and so on. You can use the pattern to make a table, where x is the number of tickets sold and $P(x)$ is the number of performances. Then graph.

x	$P(x)$
$0 < x \leq 250$	1
$250 < x \leq 500$	2
$500 < x \leq 750$	3
$750 < x \leq 1000$	4
$1000 < x \leq 1250$	5

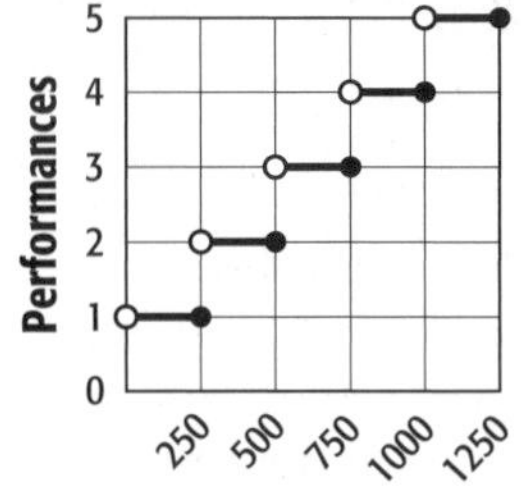

7. 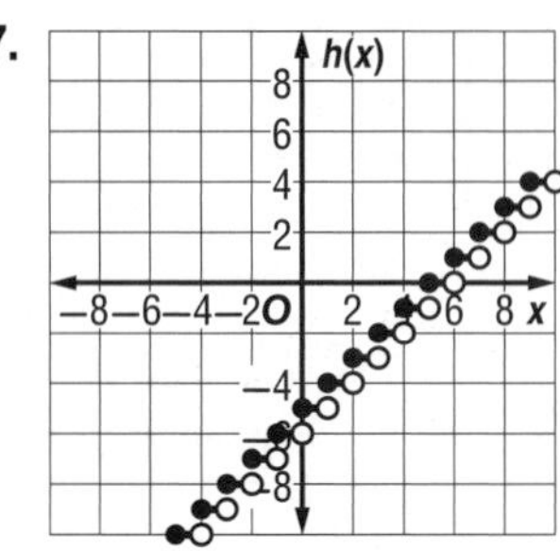
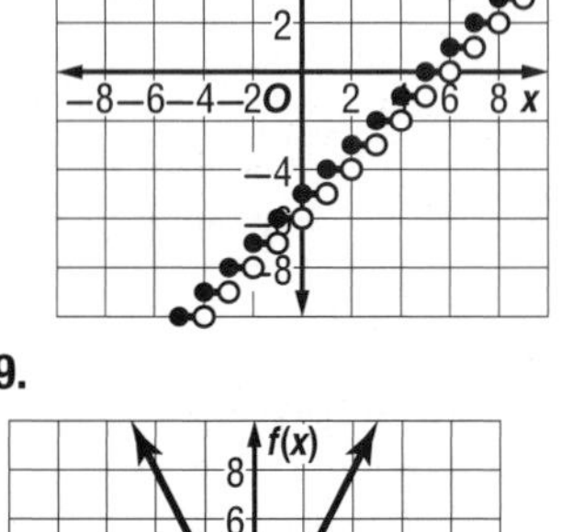

$D = \{\text{all real numbers}\}$; $R = \{\text{all integers}\}$

9. 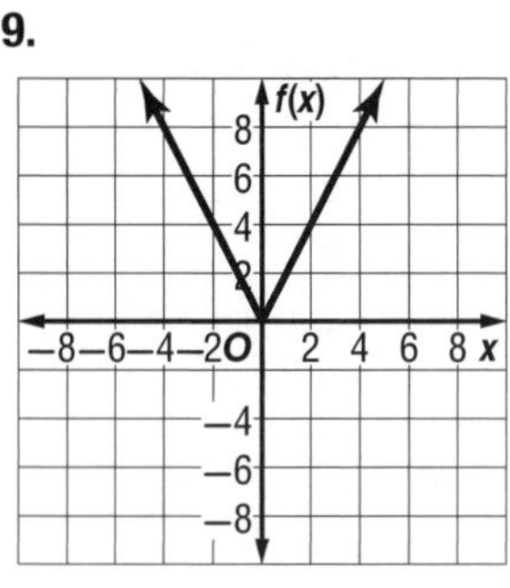

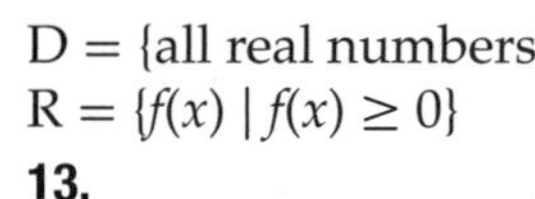

$D = \{\text{all real numbers}\}$; $R = \{f(x) \mid f(x) \geq 0\}$

11. 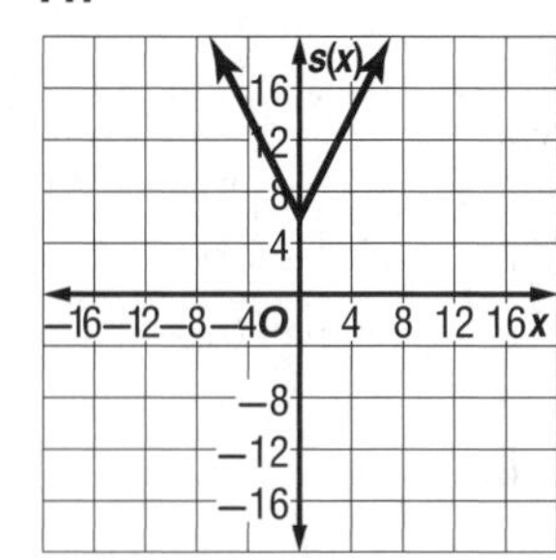

$D = \{\text{all real numbers}\}$; $R = \{s(x) \mid s(x) \geq 6\}$

13. 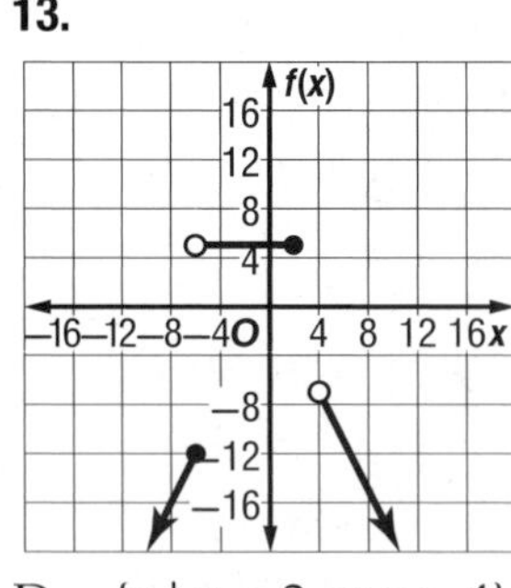

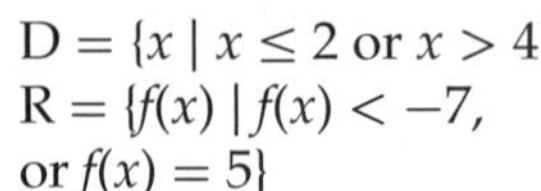

$D = \{x \mid x \leq 2 \text{ or } x > 4\}$; $R = \{f(x) \mid f(x) < -7, \text{ or } f(x) = 5\}$

15. 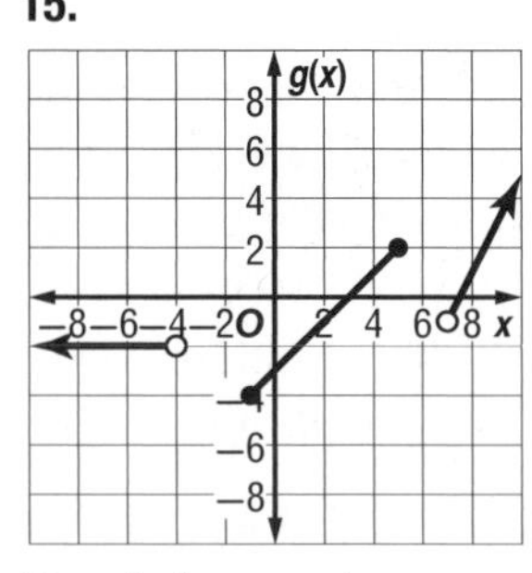

$D = \{x \mid x < -4, -1 \leq x \leq 5, \text{ or } x > 7\}$; $R = \{g(x) \mid g(x) \geq -4\}$

17 The left portion of the graph is the graph of $g(x) = -x - 4$. There is a circle at $(-3, -1)$, so the linear function is defined for $\{x \mid x < -3\}$. The middle portion of the graph is the graph of $g(x) = x + 1$. There are dots at $(-3, -2)$ and $(1, 2)$, so the linear function is defined for $\{x \mid -3 \leq x \leq 1\}$. The right portion of the graph is the graph of

$g(x) = -6$. There is a circle at $(4, -6)$, so the linear function is defined for $\{x \mid x > 4\}$. Write the piecewise-defined function.

$$g(x) = \begin{cases} -x - 4 \text{ if } x < -3 \\ x + 1 \text{ if } -3 \leq x \leq 1 \\ -6 \text{ if } x > 4 \end{cases}$$

19. $g(x) = \begin{cases} 8 \text{ if } x \leq -1 \\ 2x \text{ if } 4 \leq x \leq 6 \\ 2x - 15 \text{ if } x > 7 \end{cases}$

21.

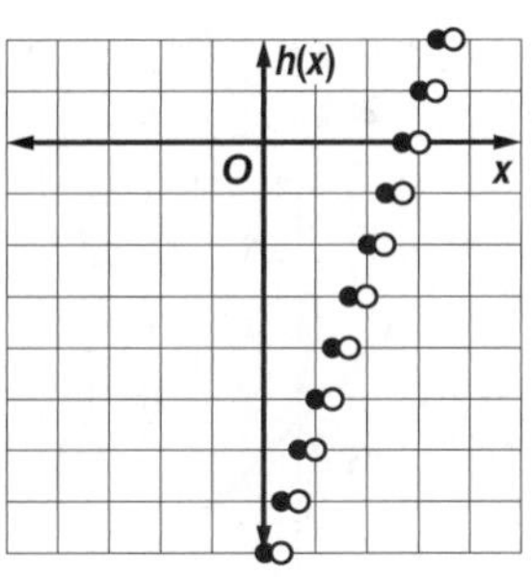

D = {all real numbers};
R = {all integers}

23.

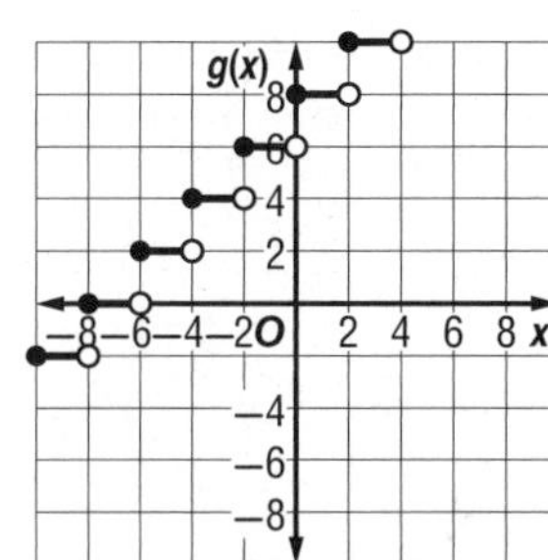

D = {all real numbers};
R = {all even integers}

25.

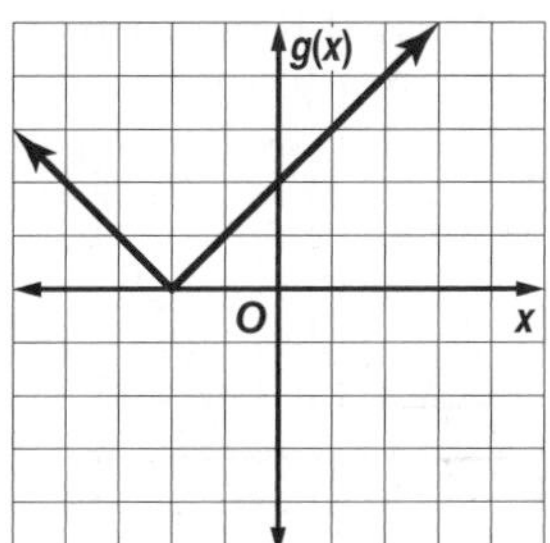

D = {all real numbers};
R = $\{g(x) \mid g(x) \geq 0\}$

27.

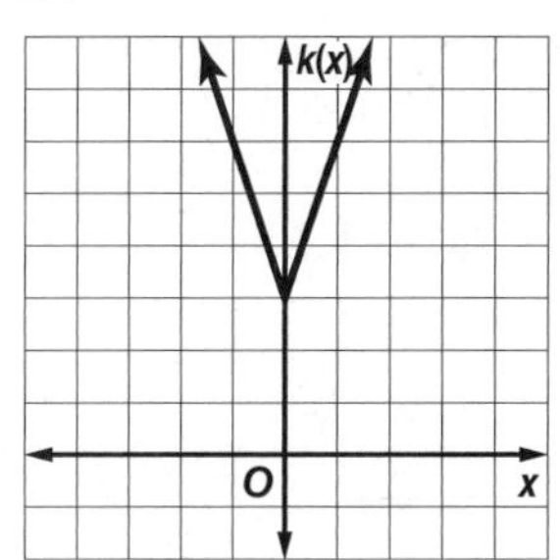

D = {all real numbers};
R = $\{k(x) \mid k(x) \geq 3\}$

29.

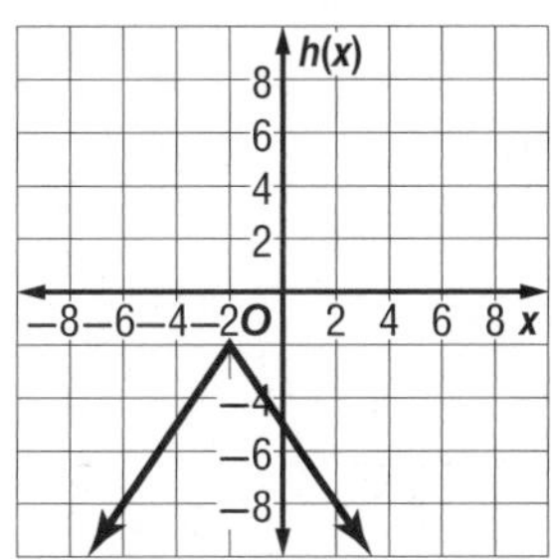

D = {all real numbers};
R = $\{h(x) \mid h(x) \leq -2\}$

31a. $f(a) = |a - 60|$

31b. $\{a \mid a \geq 0\}$

31c.

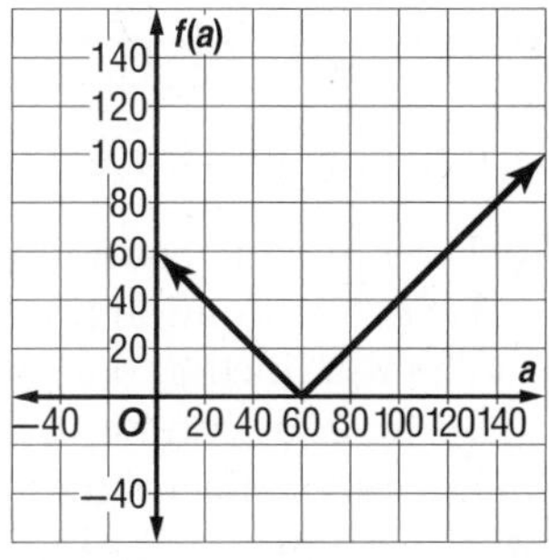

33 $f(x) = 0.5x$ if $x > 0$, $f(x) = 0$ if $x = 0$, and $f(x) = -0.5x$ if $x < 0$. So, according to the definition of absolute value, $f(x) = |0.5x|$.

35.

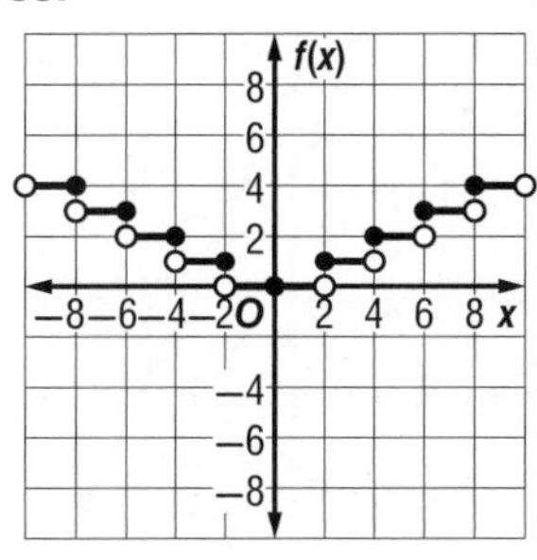

D = {all real numbers};
R = {all whole numbers}

37.

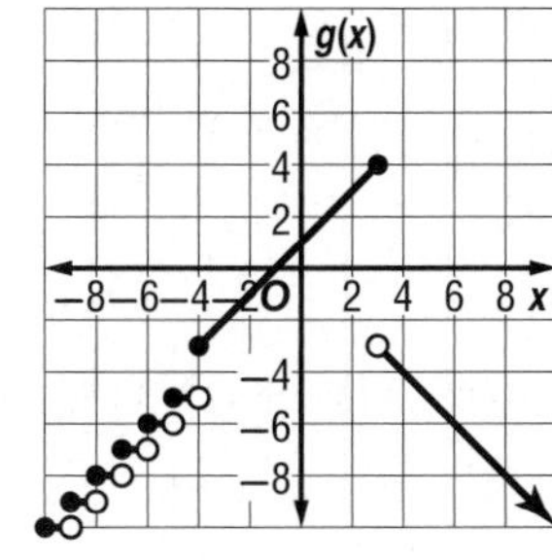

D = {all real numbers};
R = $\{g(x) \mid g(x) \leq 4\}$

39a.

x	−4	−3	−2	−1	0	1	2	3	4
f(x)	0	−1	−2	−3	−4	−3	−2	−1	0

x	−4	−3	−2	−1	0	1	2	3	4
g(x)	12	9	6	3	0	3	6	9	12

39b.

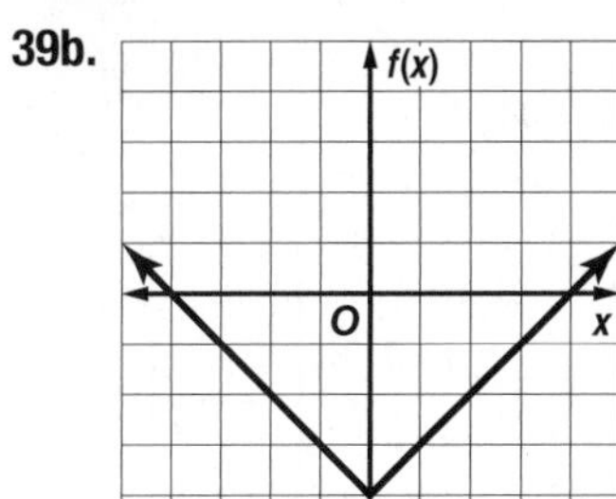

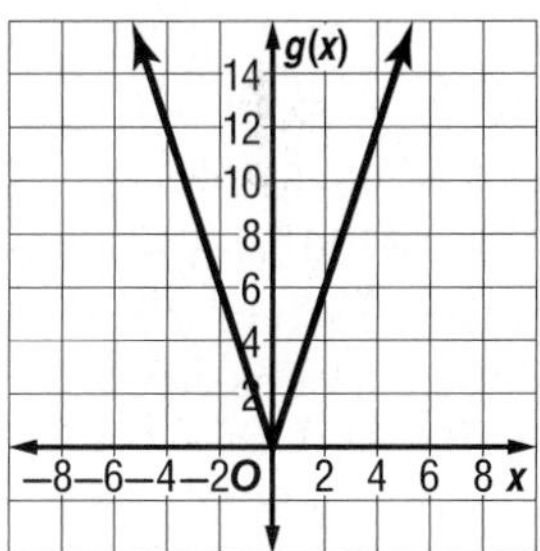

39c.

x	−4	−3	−2	−1	0	1	2	3	4
f(x)	0	−1	−2	−3	−4	−3	−2	−1	0
slope		−1	−1	−1	−1	1	1	1	1

x	−4	−3	−2	−1	0	1	2	3	4
g(x)	12	9	6	3	0	3	6	9	12
slope		−3	−3	−3	−3	3	3	3	3

39d. The two sections of an absolute value graph have opposite slopes. The slope is constant for each section of the graph.

41.

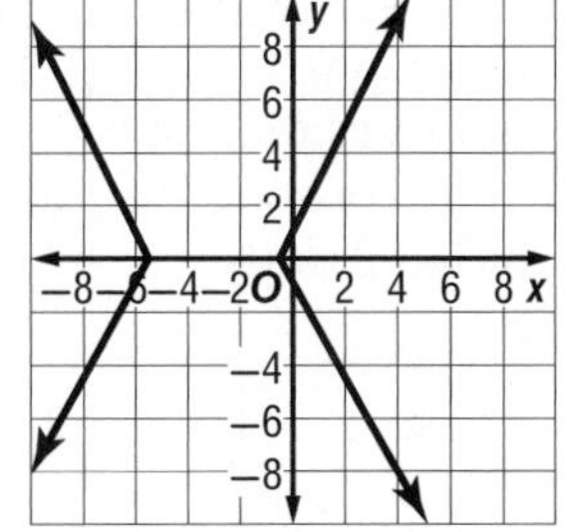

43. Sample answer: $f(x) = -|x - 2|$ **45.** $3n + 1$
47. J **49a.** $y = 0.10x + 30.34$ **49b.** $r = 0.987$
49c. about 110 **51.** $y = -\frac{3}{2}x + 6$ **53.** $-8c + 6$
55. −99

57.

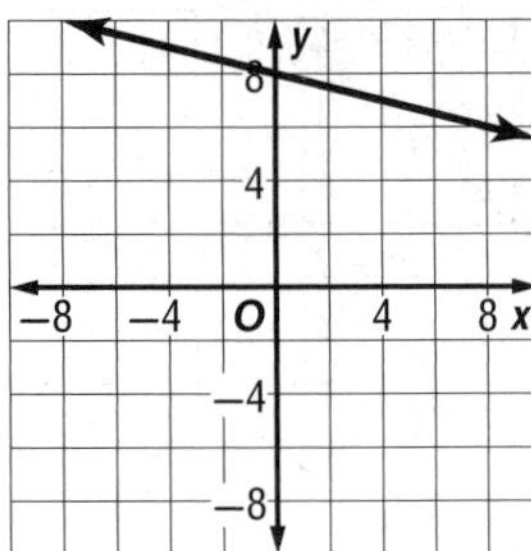

59.

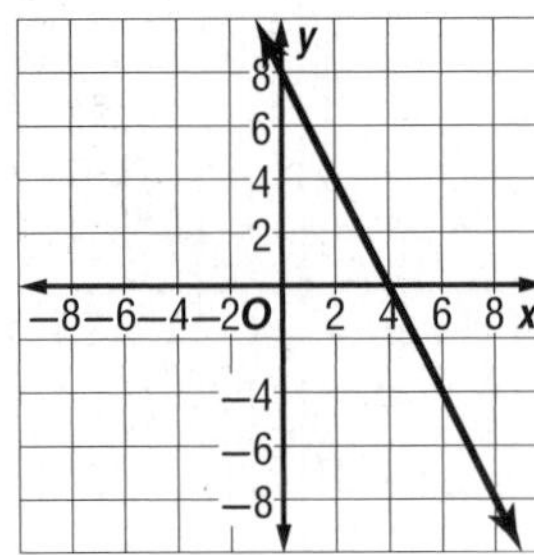

Lesson 2-7

1. linear

3. translation of the graph of $y = x^2$ down 4 units

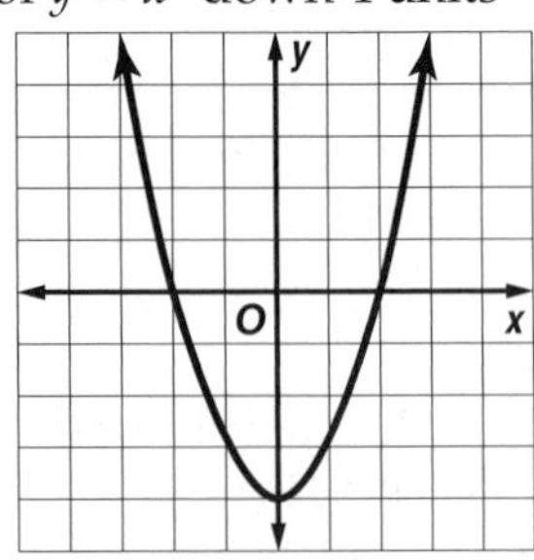

5. reflection of the graph of $y = |x|$ in the x-axis

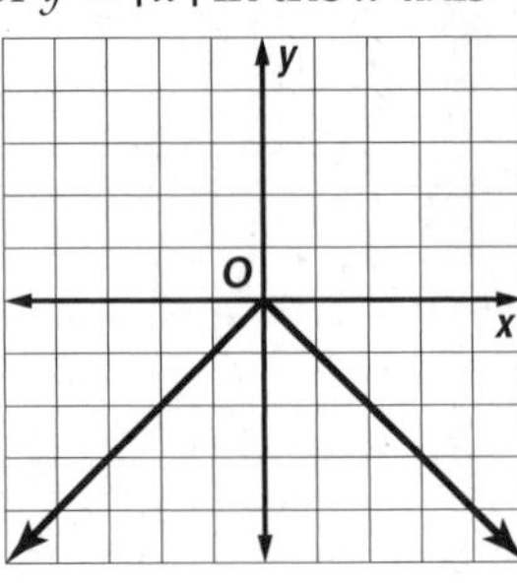

7. A vertical compression of the graph of $y = x$; the slope is not as steep as that of $y = x$.

9. The function is a dilation and translation. The graph of $f(x) = \frac{1}{2}|x - 12|$ compresses the graph $f(x) = |x|$ vertically and translates it 12 units to the right.

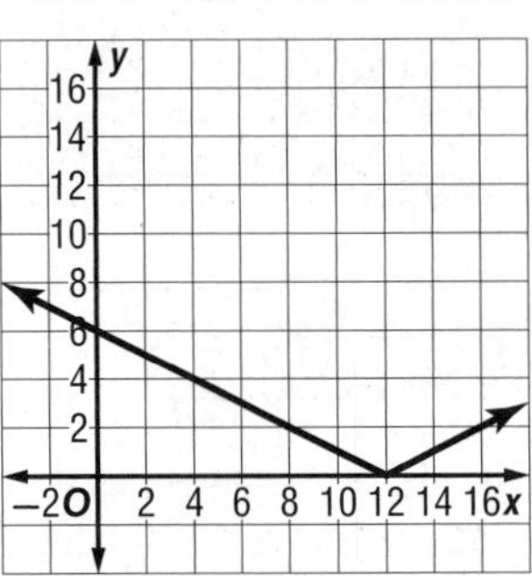

11 The graph is a curve that appears symmetrical. The graph represents a quadratic function.

13. linear

15. translation of the graph of $y = |x|$ down 3 units

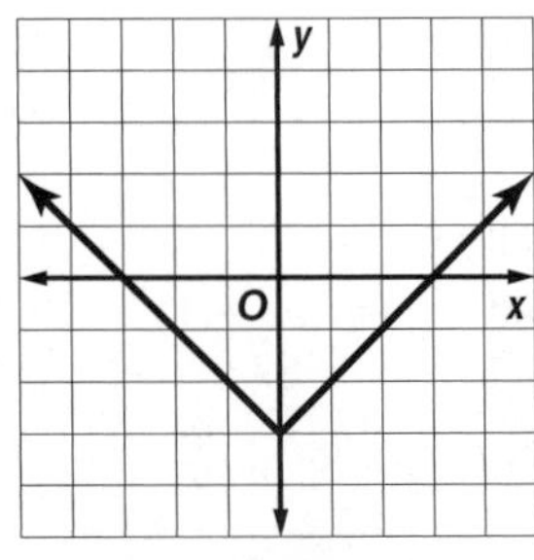

17. translation of the graph of $y = x$ up 2 units or left 2 units

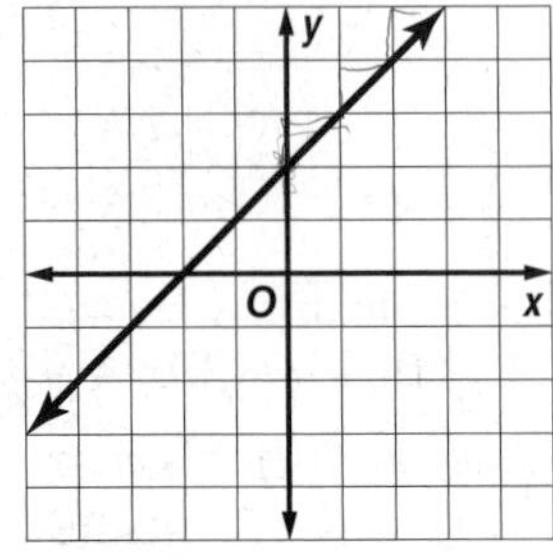

19. translation of the graph of $y = |x|$ left 6 units

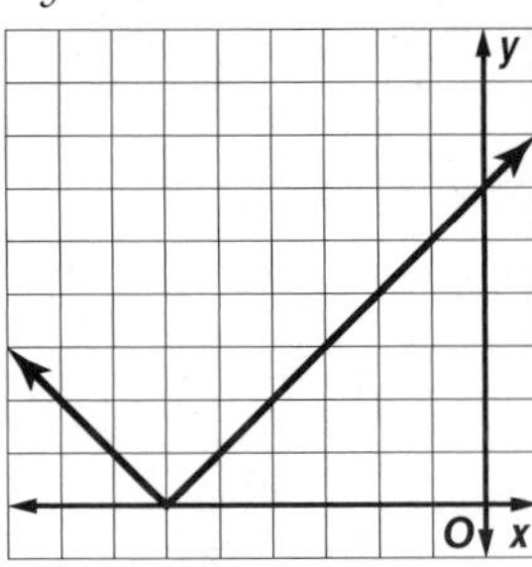

21. reflection of the graph of $y = x^2$ in the x-axis

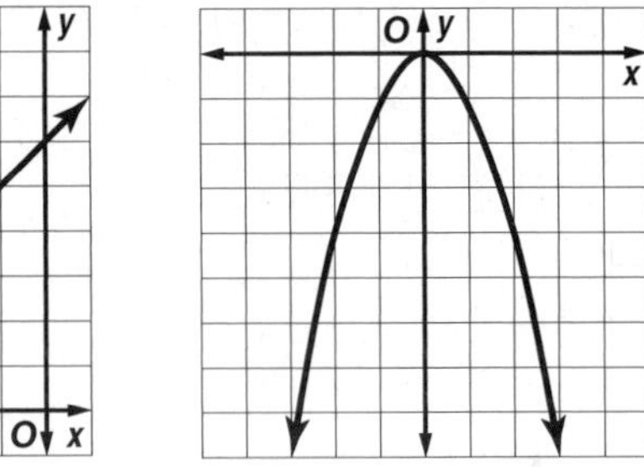

23. reflection of the graph of $y = |x|$ in the y-axis

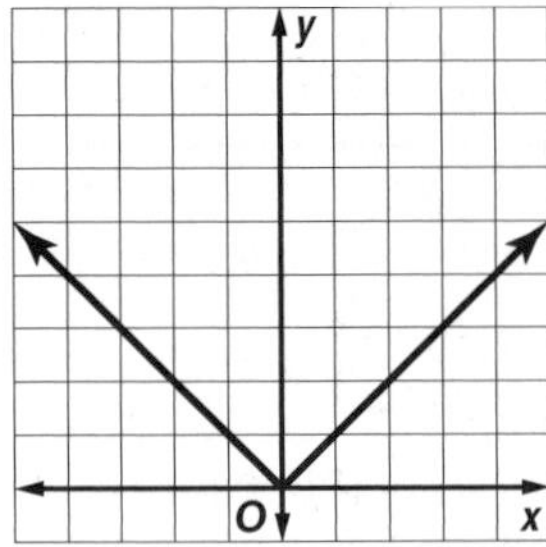

25. reflection of the graph of $y = x$ in the y-axis

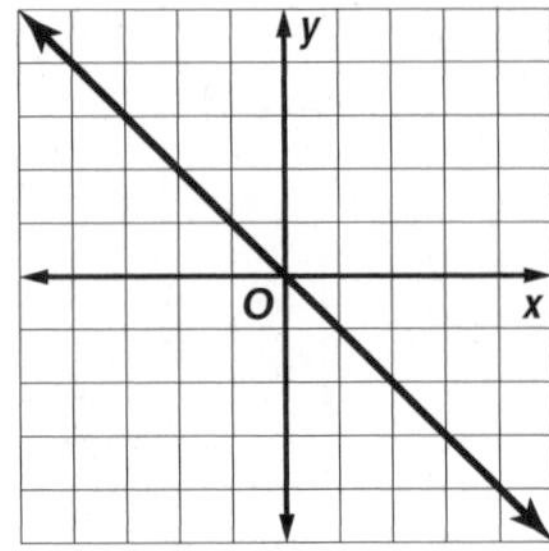

27. vertical stretch of the graph of $y = x$; The slope is steeper than that of $y = x$.

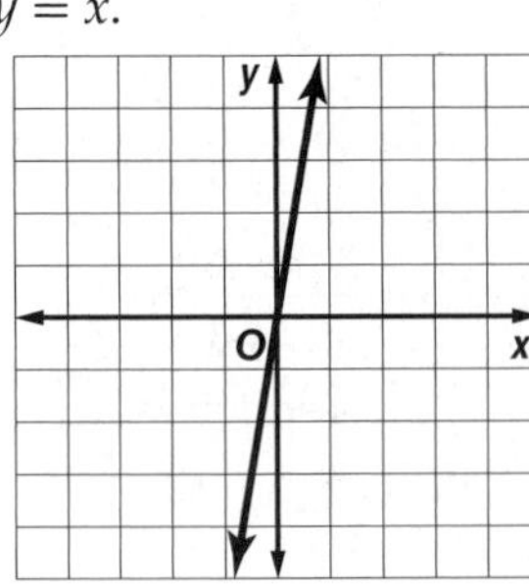

29. The dilation compressed the graph of $y = |x|$ horizontally.

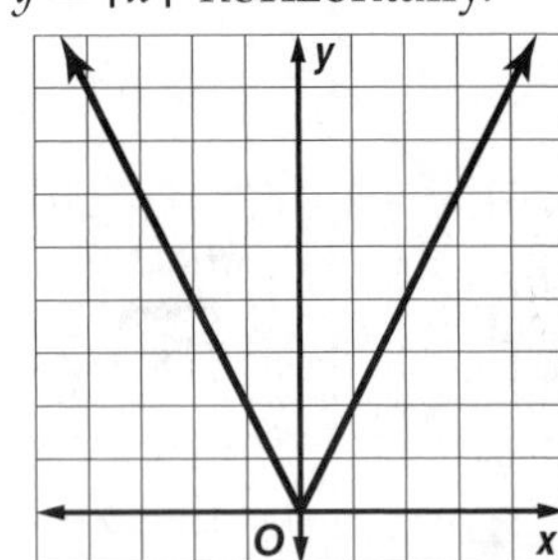

31. vertical compression of the graph of $y = x^2$

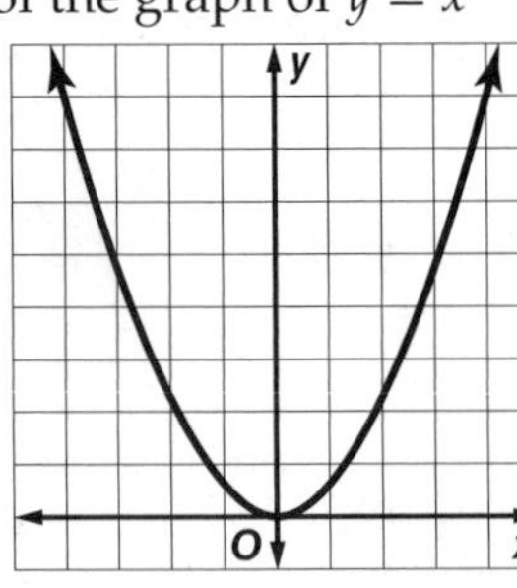

33. $y = x^2 + 1$ **35.** $y = x - 5$ **37.** $y = (x - 2)^2$

39 The blue line has a y-intercept of 4 and a slope of 1. So, an equation for the blue line is $y = x + 4$. The red line has a y-intercept of 2 and a slope of 1. So, an equation for the red line is $y = x + 2$. The red line is a translation of the blue line 2 units down.

41 $y = (x + 4)^2 - 6$ **43.** Sample answer: Since a vertical translation concerns only y-values and a horizontal translation concerns only x-values, order is irrelevant.

45. Sample graph:

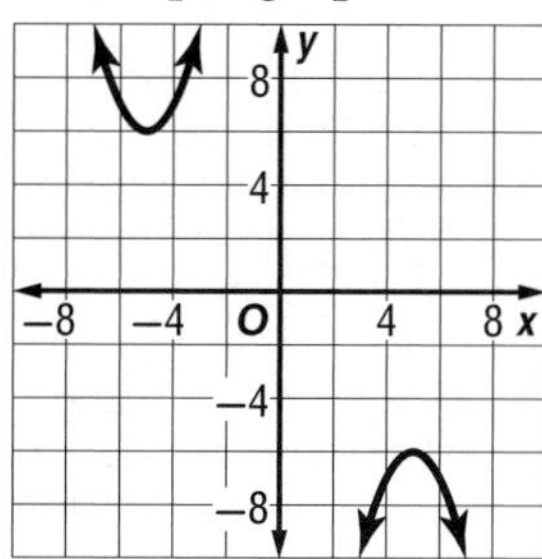

Sample answer: The figure in Quadrant II has been reflected and moved left 10 units.

47. Sample answer: It is not always true. When the axis of symmetry of the parabola is not along the y-axis, the graphs of the preimage and image will be different. **49.** G **51.** A

53.

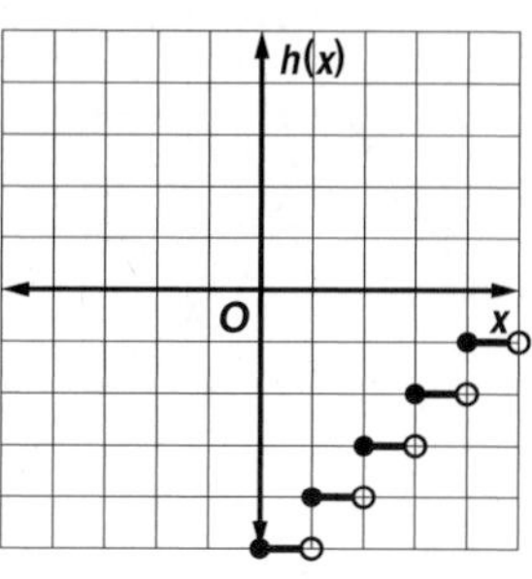

D = {all real numbers}, R = {all integers}

55a. $y = 7.83x - 15{,}620.70$

55b. $r = 0.953$

55c. about 118 people

57. $2 > y > -3$

59. 20 **61.** yes

63. −52 **65.** 84

Lesson 2-8

1.

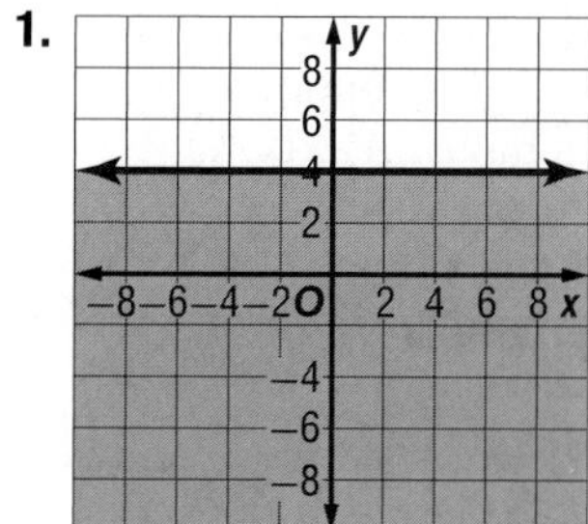

3.

5a. $3.45g + 2.41q \leq 50$

5b.

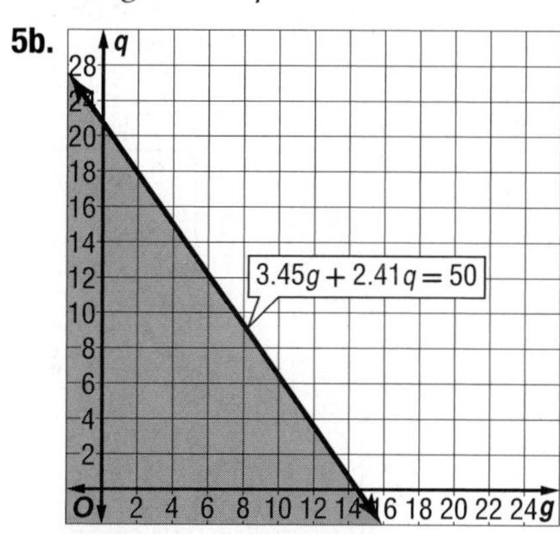

5c. No; (10, 8) is not in the shaded region.

7.

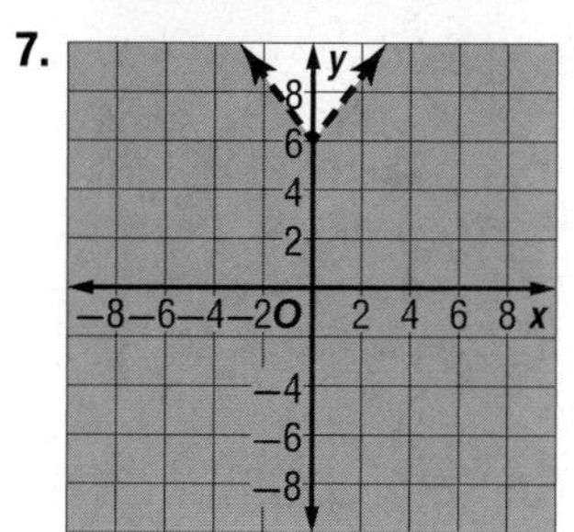

9.

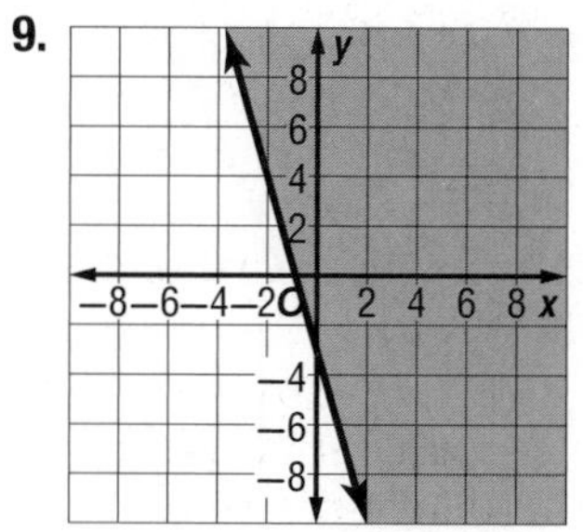

11.

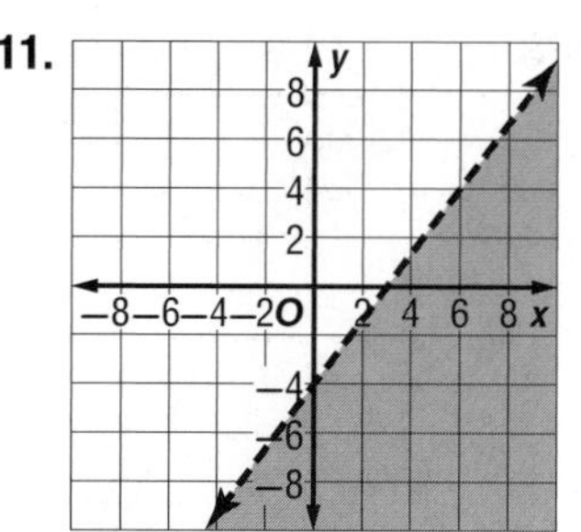

13.

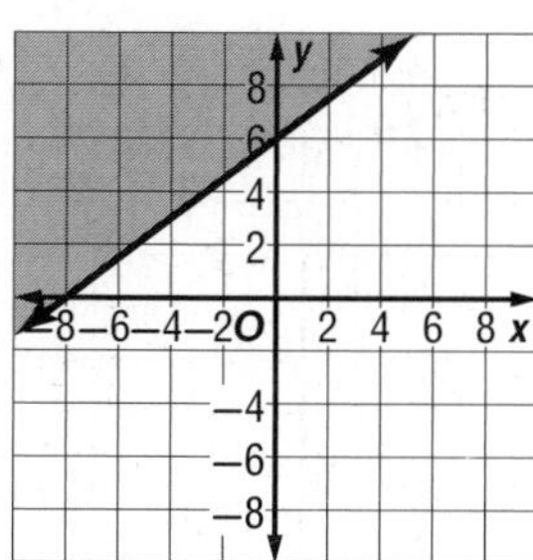

15.

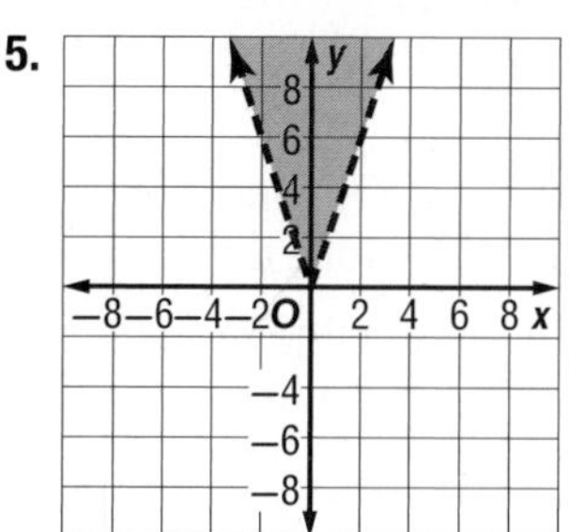

17 $y - 6 < |-2x|$ Original equation

$y < |-2x| + 6$ Add 6 to each side.

Since the inequality symbol is <, the boundary is dashed. Graph $y = |-2x| + 6$. Then test (0, 0).

$y < |-2x| + 6$ Write the inequality.

$0 \stackrel{?}{<} |-2(0)| + 6$ $(x, y) = (0, 0)$

$0 < 6$ ✓ True

The region that contains (0, 0) is shaded.

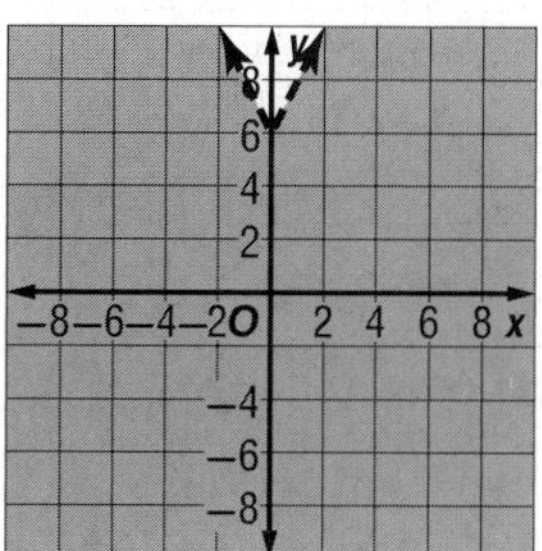

19.

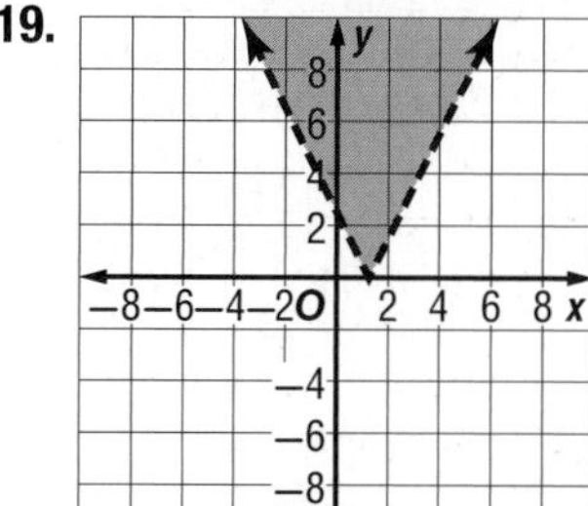

21a. $8a + 6b \geq 700$

21b.

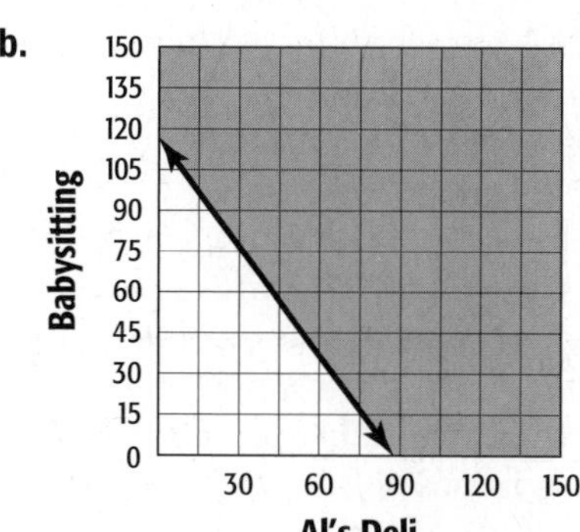

21c. yes

23.

25.

27. all real numbers (The graph would be shaded everywhere.)

29 a. $20d$ represents the cost of d DVDs and $15c$ represents the cost of c CDs. Because the sum can equal the maximum \$400, the inequality symbol is $\leq$. The inequality is $20d + 15c \leq 400$.
b. Graph the boundary $20d + 15c = 400$. Since the inequality symbol is $\leq$, the boundary is solid. Then test the point (0, 0).

$20d + 15c \leq 400$	Original inequality
$20(0) + 15(0) \overset{?}{\leq} 400$	$(c, d) = (0, 0)$
$0 \leq 400$ ✓	True

The region that contains (0, 0) is shaded.

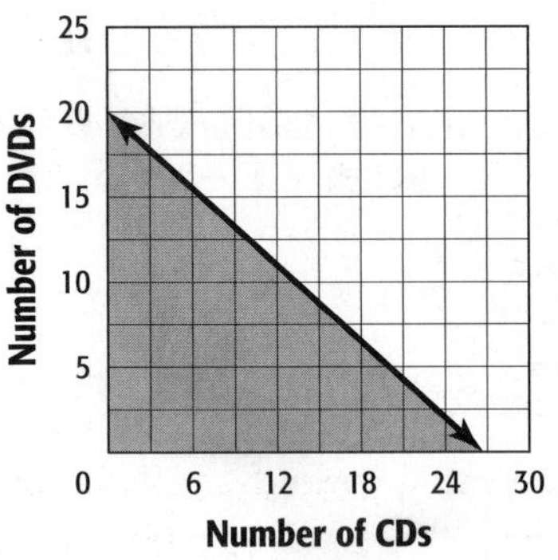

c. Sample answer: Since the points (18, 5), (12, 10), and (6, 15) lie inside the shaded region, they satisfy the inequality. This means that Susan can buy 18 CDs and 5 DVDs, 12 CDs and 10 DVDs, or 6 CDs and 15 DVDs.

31.

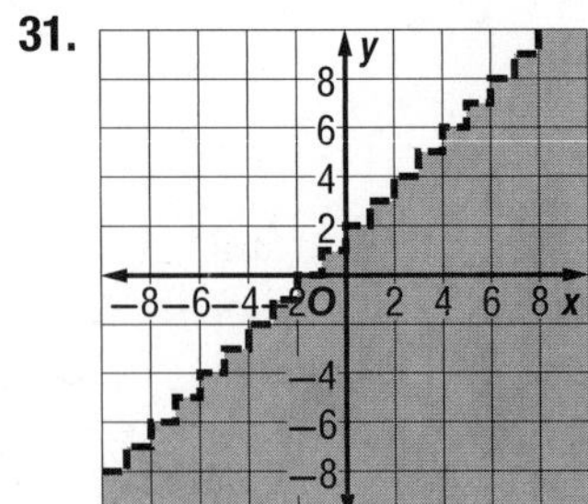

33. Sample answer: $|y| < x$ **35.** Paulo; $x - y \geq 2$ can be written as $y \leq x - 2$. **37.** Sample answer: One possibility is when $|y| < 0$. In order for there to be a solution, the absolute value of y will need to be less than 0, and, by definition of absolute value, this is impossible. **39.** C **41.** K **43.** $y = |x + 4| - 5$

45.

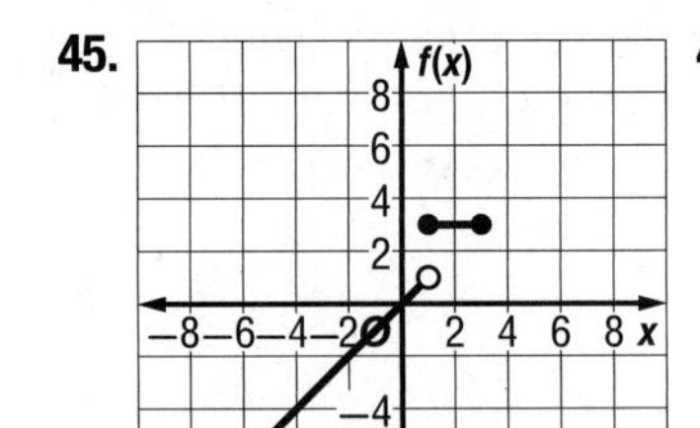

47.

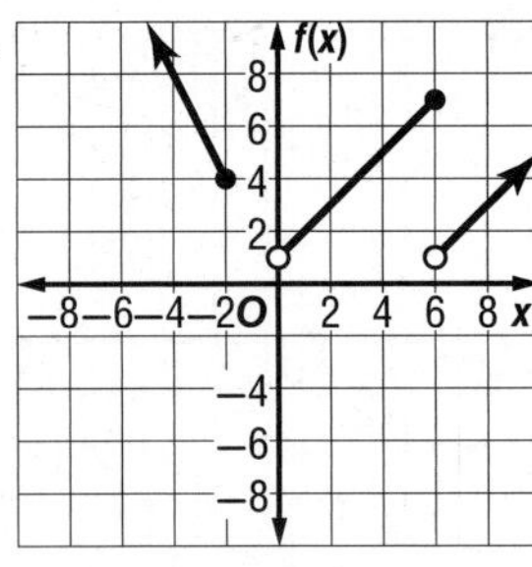

49. $4x - 15y = 6$; $A = 4$, $B = -15$, $C = 6$ **51.** 1000 **53.** $-6x^2 - 23x - 15$

55.

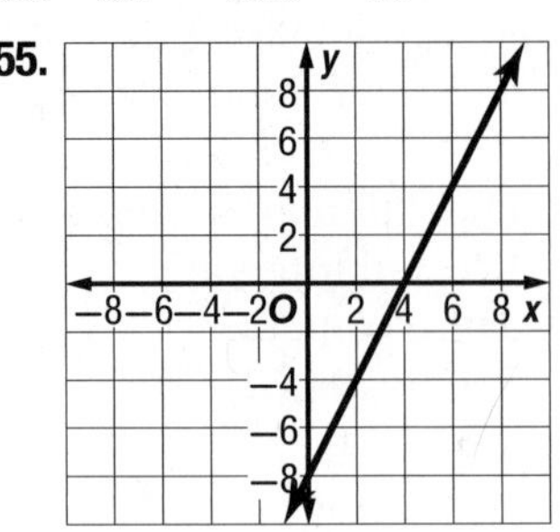

57.

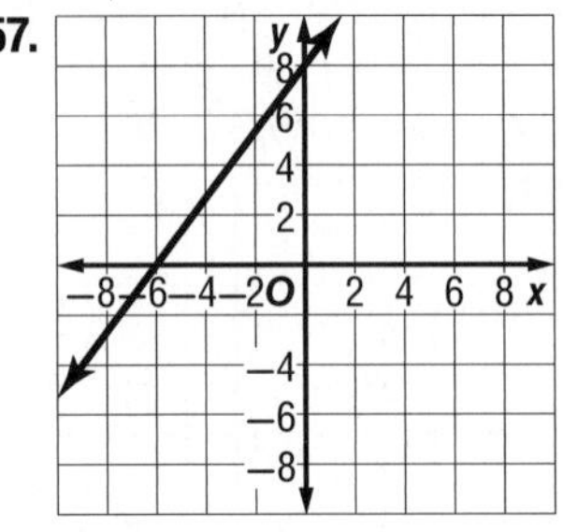

Chapter 2 Study Guide and Review

1. one-to-one **3.** identity **5.** scatter plot **7.** D = {1, 3, 5, 7}, R = {2, 4, 6, 8}; function; both **9.** D = {−4, −2, 1, 3}, R = {−4, 1, 3, 5}; not a function **11.** −10 **13.** 2 **15.** $3a + 2$ **17.** D = {1, 2, 3, 4, 5, ...}, R = {5.75, 9, 12.25, 15.5, 18.75, ...}; a function; discrete **19.** No; the variables have an exponent other than 1. **21.** yes **23.** No; x appears in a denominator. **25.** $12x - y = 0$; 12, −1, 0 **27.** $x - 2y = -3$; 1, −2, −3 **29.** 18 **31.** −1 **33.** $y = -2x - 11$ **35.** $y = 2x + 8$ **37.** $y = -4x + 25$ **39.** $y = 4x - 2$ **41.** $y = 32x + 250$ **43.** Sample answer using (1, 44) and (5, 24): $y = -5x + 49$

45.

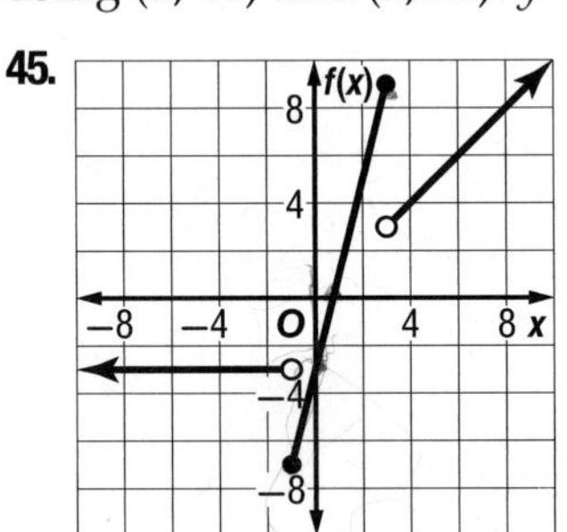

D = {all real numbers}, R = $\{f(x) \mid f(x) \geq -7\}$

47.

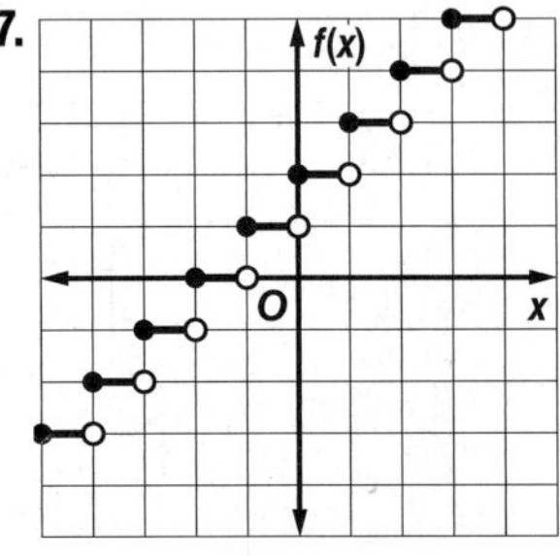

D = {all real numbers}, R = {all integers}

49. quadratic **51.** $y = x^2$ shifted down 3 units **53.** parabola

55.

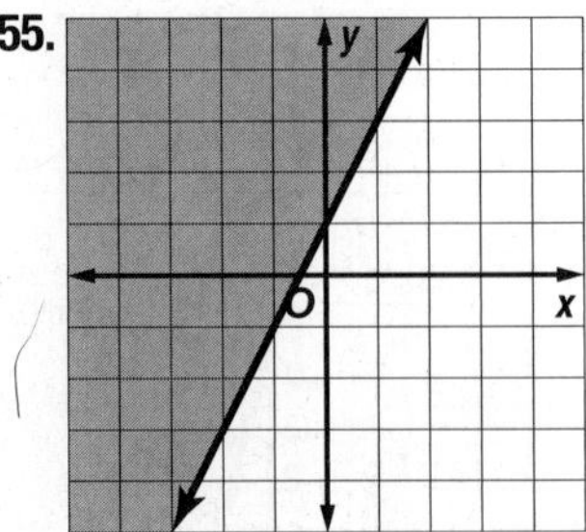

57.

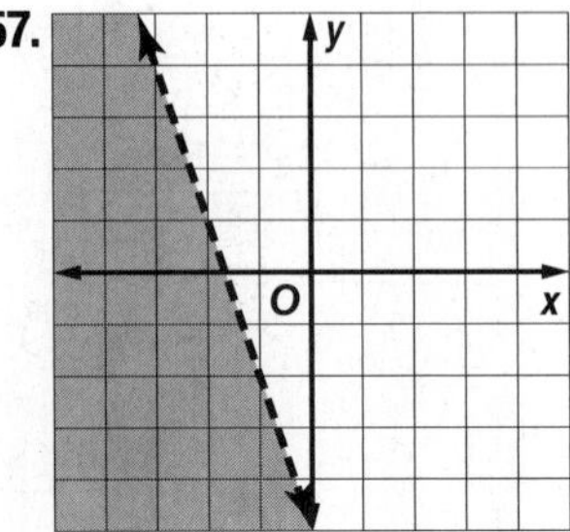

59.

61.

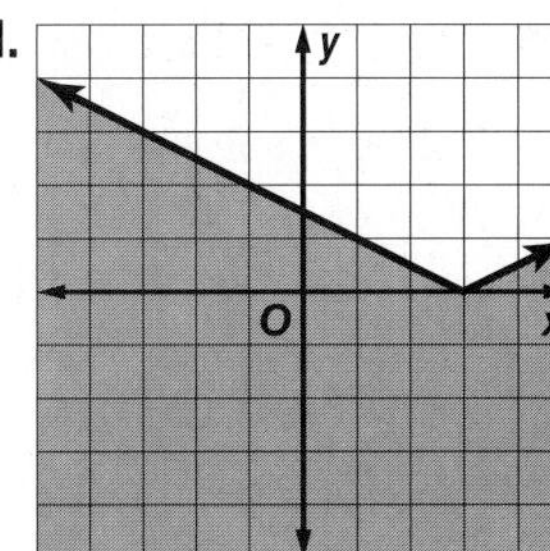

13.

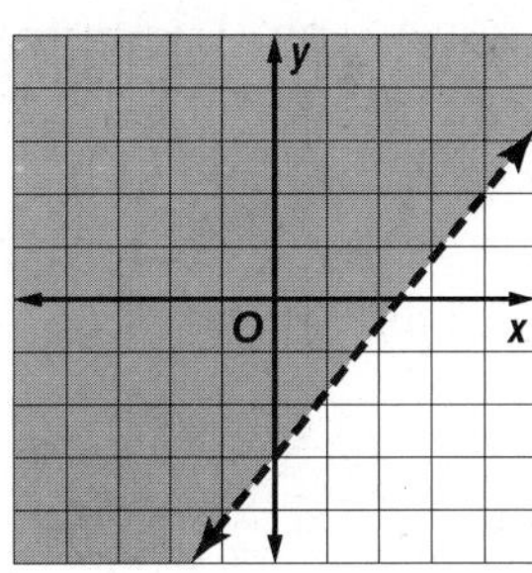

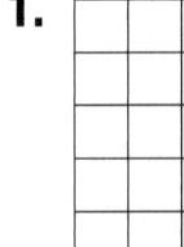

CHAPTER 3

Systems of Equations and Inequalities

Chapter 3 Get Ready

1.

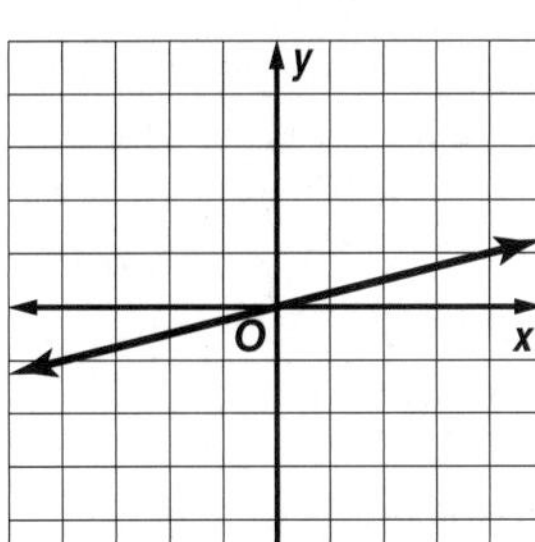

3.

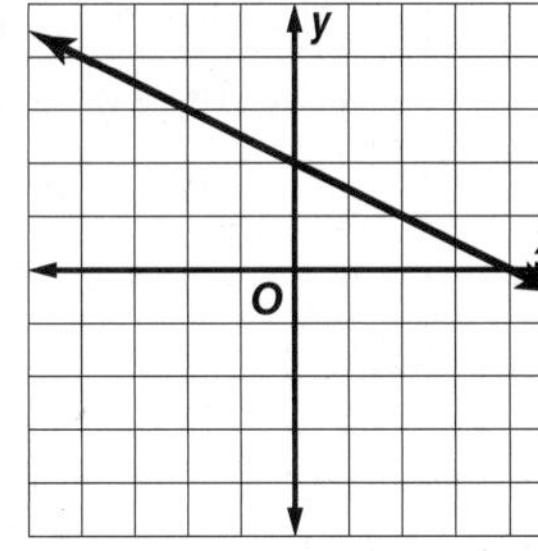

5.

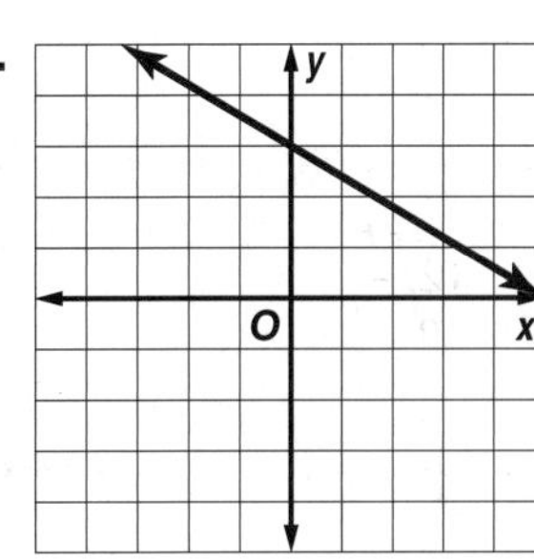

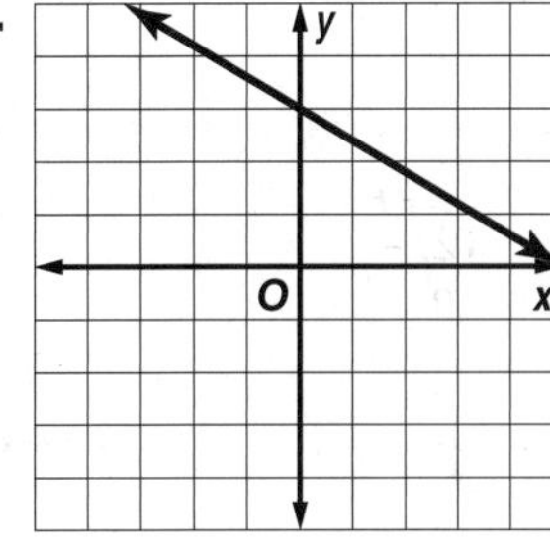

7a. $8.50a + 5.25c = 650$

7b.

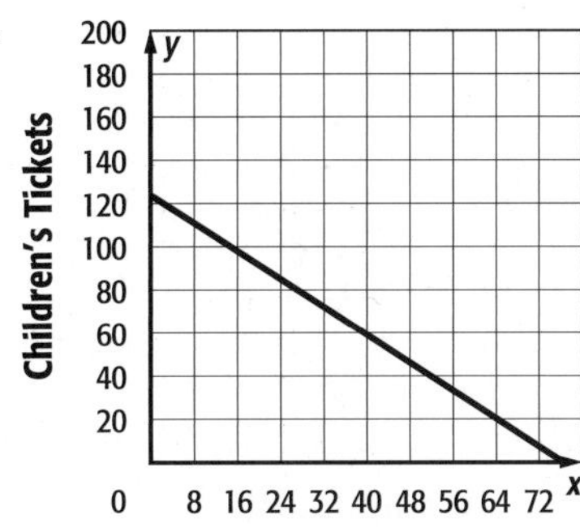

9.

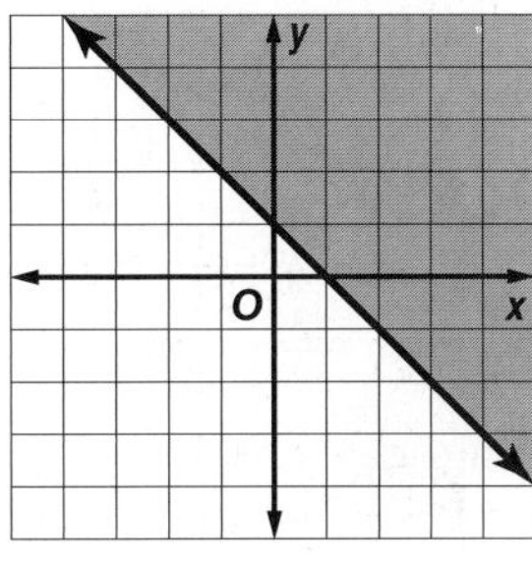

11.

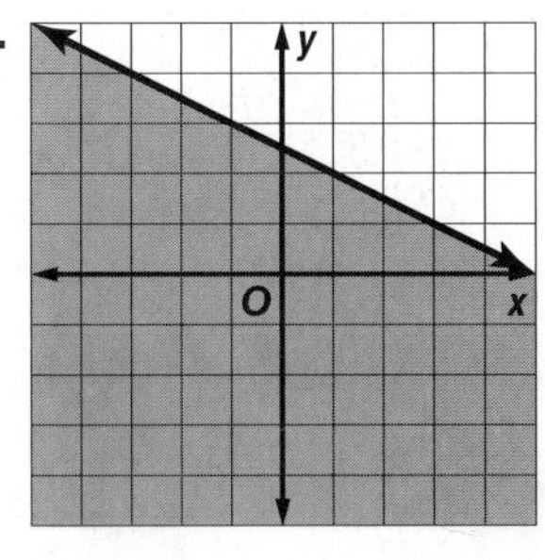

Lesson 3-1

1. (3, 5) **3.** (3, −3) **5.** (6, 7)

7 Write each equation in slope-intercept form. Then graph.

$4x + 5y = -41 \quad \rightarrow \quad y = -\frac{4}{5}x - \frac{41}{5}$

$3y - 5x = 5 \quad \rightarrow \quad y = \frac{5}{3}x + \frac{5}{3}$

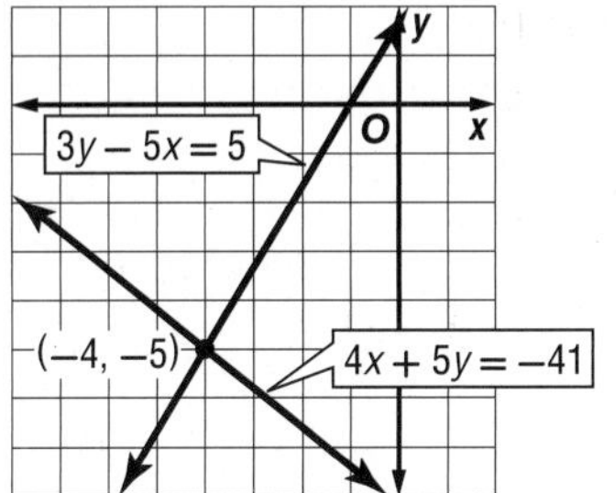

The graphs appear to intersect at (−4, −5). So, the solution is (−4, −5).

9a. $y = 0.15x + 2.70$, $y = 0.25x$

9b. \$6.75 for 27 photos

9c. You should use EZ Online photos if you are printing more than 27 digital photos, and the local pharmacy if you are printing fewer than 27 photos.

11.

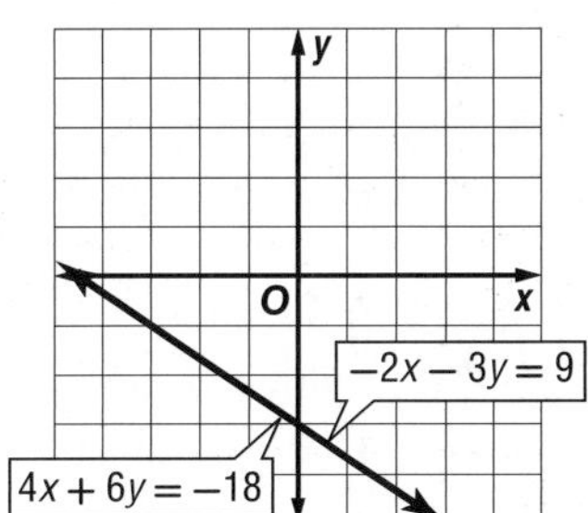

consistent and dependent

13. (−2, 1)

15 $2a + 8b = -8$ Multiply by 5. $\rightarrow$ $10a + 40b = -40$

$3a - 5b = 22$ Multiply by 8. $\rightarrow$ $24a - 40b = 176$

$10a + 40b = -40$ Equation 1 × 5

$(+)\ 24a - 40b = 176$ Equation 2 × 8

$34a = 136$ Add the equations.

$a = 4$ Divide each side by 34.

Substitute 4 for a into either original equation.

$2a + 8b = -8$ Equation 1

$2(4) + 8b = -8$ $a = 4$

$8 + 8b = -8$ Multiply.

$8b = -16$ Subtract 8 from each side.

$b = -2$ Divide each side by 8.

The solution is (4, −2).

17. (5, 1) **19.** (−2, 7) **21.** (−4, −3) **23.** no solution **25.** B **27.** (4, −1) **29.** 250 T-shirts **31.** (−3, −4) **33.** infinite solutions **35.** (−1.5, −2)

37. consistent and independent

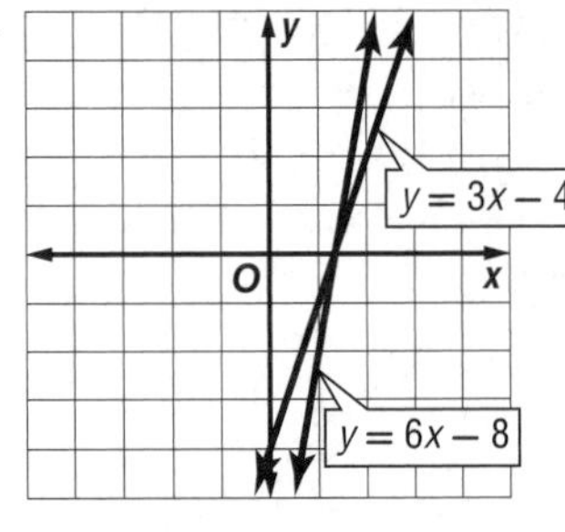

39. inconsistent

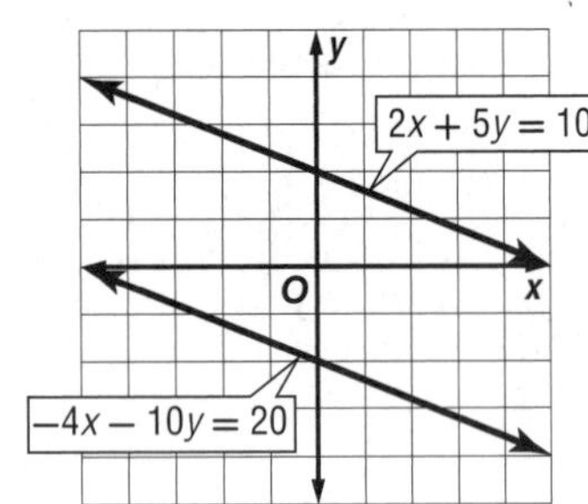

41 Write each equation in slope-intercept form. Then graph.

$-5x - 6y = 13 \quad \rightarrow \quad y = -\frac{5}{6}x - \frac{13}{6}$

$12y + 10x = -26 \quad \rightarrow \quad y = -\frac{5}{6}x - \frac{13}{6}$

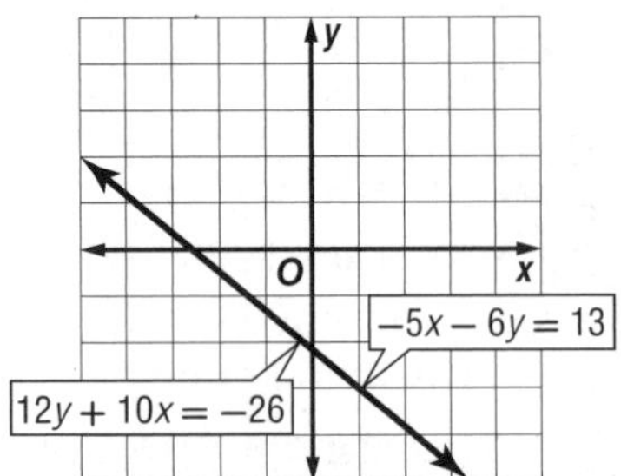

Because the equations are equivalent, their graphs are the same line. The system is consistent and dependent.

43. infinite solutions **45.** (8, 4) **47.** (−3, −1) **49a.** $x + y = 13$ and $4x + 2y = 38$ **49b.** 6 doubles matches and 7 singles matches **51.** (0, 4) **53.** no solution **55.** (8, −6) **57.** (5, 4) **59.** (2.07, −0.39) **61.** (15.03, 10.98) **63.** (−5, 4) **65.** infinite solutions **67.** (16, −8)

69 **a.** Sample answer: For men, you can use (0, 10) and (44, 9.69) to write an equation.

$m = \frac{y_2 - y_1}{x_2 - x_1}$ Slope Formula

$= \frac{9.69 - 10}{44 - 0}$ $(x_1, y_1) = (0, 10), (x_2, y_2) = (44, 9.69)$

$= -0.00705$ Simplify.

At $x = 0$, $y = 10.0$. So, the y-intercept is 10.

$y = mx + b$ Slope-intercept form

$y_m = -0.00705x + 10$ $m = -0.00705, b = 10$

For women, you can use (0, 11.4) and (44, 10.78) to write an equation.

$m = \frac{y_2 - y_1}{x_2 - x_1}$ Slope Formula

$= \frac{10.78 - 11.4}{44 - 0}$ $(x_1, y_1) = (0, 11.4), (x_2, y_2) = (44, 10.78)$

$= -0.01409$ Simplify.

At $x = 0$, $y = 11.4$. So, the y-intercept is 11.4.

$y = mx + b$ Slope-intercept form

$y_w = -0.01409x + 11.4$ $m = -0.01409, b = 11.4$

b.

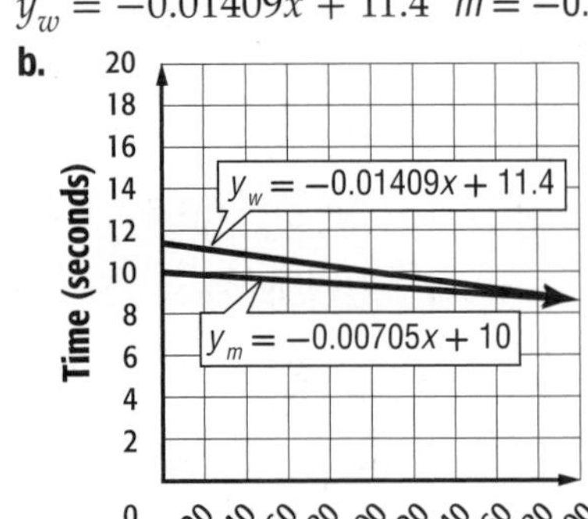

The graphs appear to intersect at (198, 8.6). This means that 198 years after 1964, the winning time for both men and women will be about 8.6 seconds. So, based on these data, the women's performance will catch up to the men's performance 198 years after 1964, or in the year 2162. The next Olympic year would be 2164. This prediction is not reasonable. It is unlikely that women's times will ever catch up to men's times because the times cannot continue to increase and decrease infinitely.

71 Let x represent the cost of an adult and y represent the cost of a student.

Van A: $2x + 5y = 77$

Van B: $2x + 7y = 95$

$2x + 5y = 77$ Multiply by −1. $\rightarrow -2x - 5y = -77$

$-2x - 5y = -77$ Equation 1 × (−1)

$(+)\ 2x + 7y = 95$ Equation 2

$2y = 18$ Add the equations.

$y = 9$ Divide each side by 2.

$2x + 7y = 95$ Equation 2

$2x + 7(9) = 95$ $y = 9$

$2x + 63 = 95$ Multiply.

$2x = 32$ Subtract 63 from each side.

$x = 16$ Divide each side by 2.

The solution is (16, 9). So, the cost of an adult is \$16 and the cost of a student is \$9.

73 Find an equation for the diagonal that goes through (6, 3) and (2, 9).

Slope:

$m = \frac{y_2 - y_1}{x_2 - x_1}$

$= \frac{9 - 3}{2 - 6}$

$= \frac{6}{-4}$ or $-\frac{3}{2}$

Equation:

$y - y_1 = m(x - x_1)$

$y - 3 = -\frac{3}{2}(x - 6)$

$y - 3 = -\frac{3}{2}x + 9$

$y = -\frac{3}{2}x + 12$

Find an equation for the diagonal that goes through (3, 4) and (11, 18).

Slope:

$m = \frac{y_2 - y_1}{x_2 - x_1}$

$= \frac{18 - 4}{11 - 3}$

$= \frac{14}{8}$ or $\frac{7}{4}$

Equation:

$y - y_1 = m(x - x_1)$

$y - 4 = \frac{7}{4}(x - 3)$

$y - 4 = \frac{7}{4}x - \frac{21}{4}$

$y = \frac{7}{4}x - \frac{5}{4}$

Find the point of intersection of the diagonals.

$y = -\frac{3}{2}x + 12$ Equation 1

$\frac{7}{4}x - \frac{5}{4} = -\frac{3}{2}x + 12$ Substitute $\frac{7}{4}x - \frac{5}{4}$ for y.

$\frac{13}{4}x - \frac{5}{4} = 12$ Add $\frac{3}{2}x$ to each side.

$\frac{13}{4}x = \frac{53}{4}$ Add $\frac{5}{4}$ to each side.

$x = \frac{53}{13}$ Multiply each side by $\frac{4}{13}$.

$y = -\frac{3}{2}x + 12$ Equation 1

$y = -\frac{3}{2}\left(\frac{53}{13}\right) + 12$ $x = \frac{53}{13}$

$y = -\frac{159}{26} + 12$ Multiply.

$y = \frac{153}{26}$ Simplify.

The diagonals intersect at $\left(\frac{53}{13}, \frac{153}{26}\right)$.

75a.

Equation 1	
x	y
0	$\frac{16}{3}$
1	5
2	$\frac{14}{3}$
3	$\frac{13}{3}$
4	4

Equation 2	
x	y
0	−4
1	−2
2	0
3	2
4	4

Equation 3	
x	y
0	10
1	5
2	0
3	−5
4	−10

75b. Equations 1 and 2 intersect at (4, 4), equations 2 and 3 intersect at (2, 0), and equations 1 and 3 intersect at (1, 5); there is no solution that satisfies all three equations.

75c.

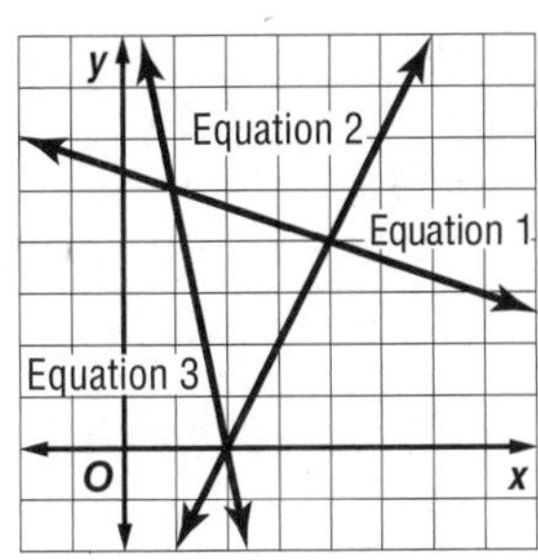

75d. If all three lines intersect at the same point, then the system has a solution. The system has no solution if the lines intersect at 3 different points, or if two or three lines are parallel.

77. $a \neq 0, b = 3$

79. Sample answer:

$4x + 5y = 21 \rightarrow 3(4x + 5y = 21)$

$3x - 2y = 10 \rightarrow 4(3x - 2y = 10)$

$12x + 15y = 63$

$(-)\ 12x - 8y = 40$

$23y = 23$

$y = 1$

$4x + 5(1) = 21$

$4x + 5 = 21$

$4x = 16$

$x = 4$

The solution is (4, 1).

81. $12xy + 18y^2 - 15y$ **83.** J **85a.** $10s + 15\ell \geq 350$

85b.

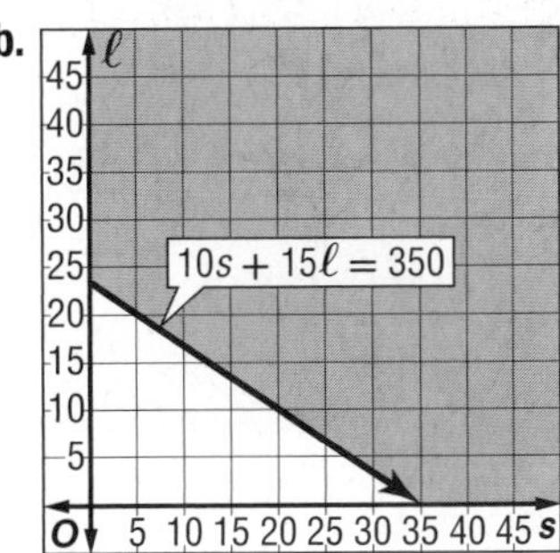

85c. no **87.** $-|x - 3|$ **89.** 7 **91.** 3.2 **93.** −8

95. yes **97.** no

Lesson 3-2

1.

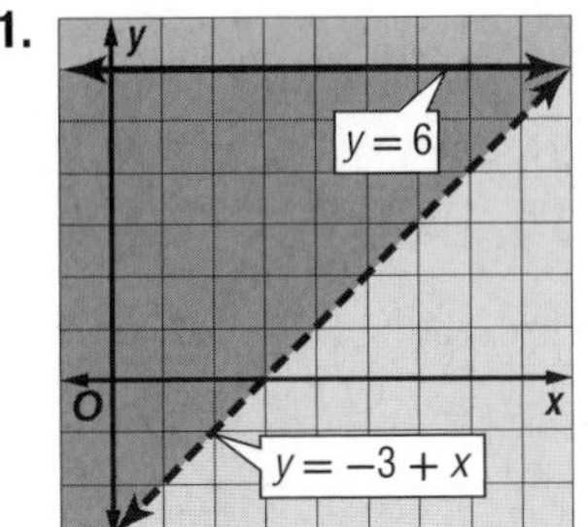

3.

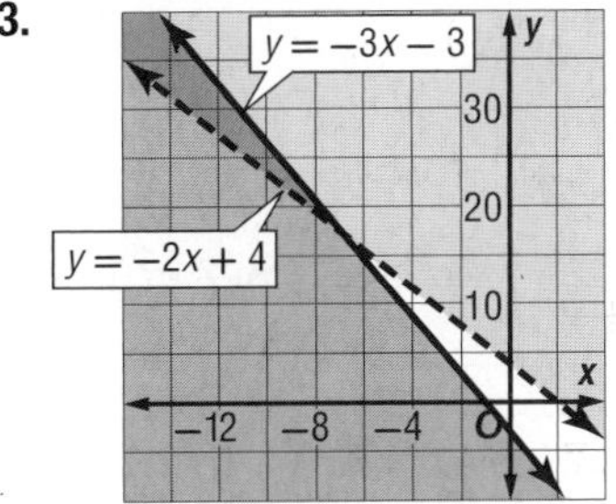

5. **7.**

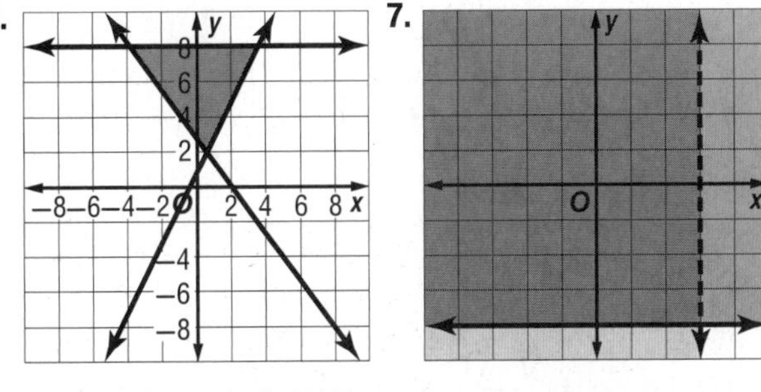

(3.5, 8), (−4, 8), (0.5, 2)

9 The solution of $y < -3x + 4$ is the region to the left of the boundary. The solution of $3y + x > -6$ is the region above the boundary. The intersection of the two regions is the solution of the system.

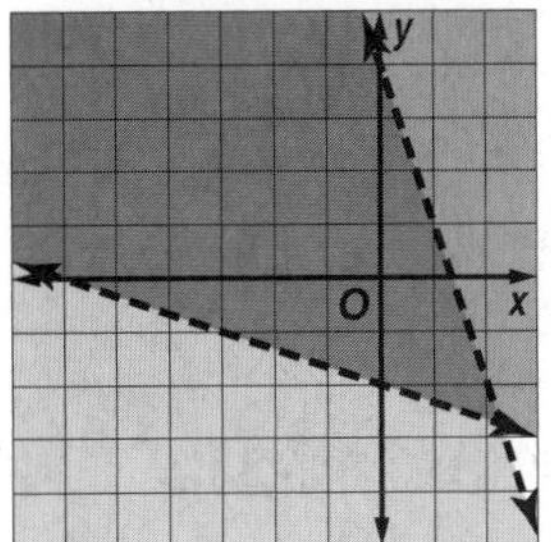

11.

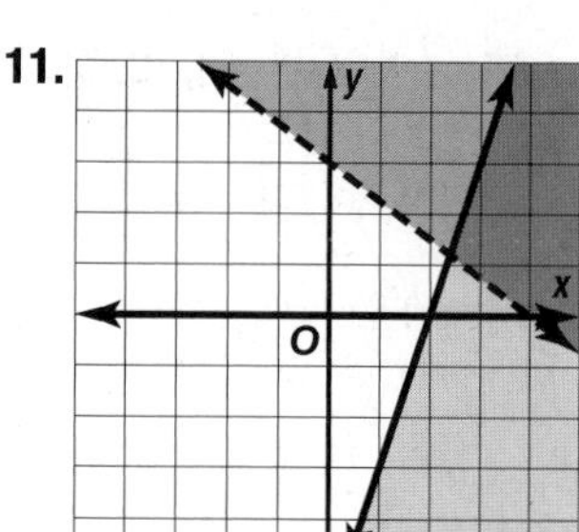

13.

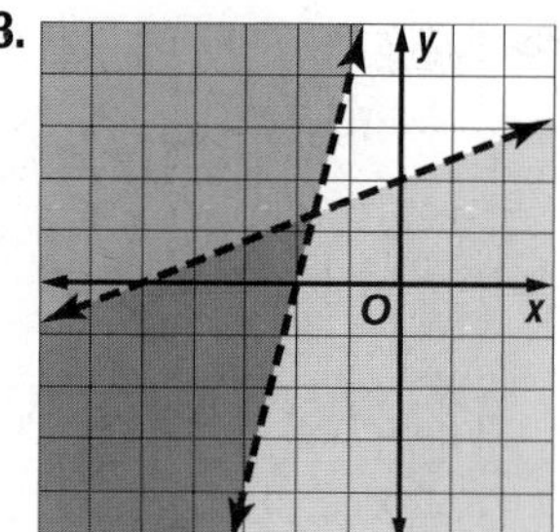

15.

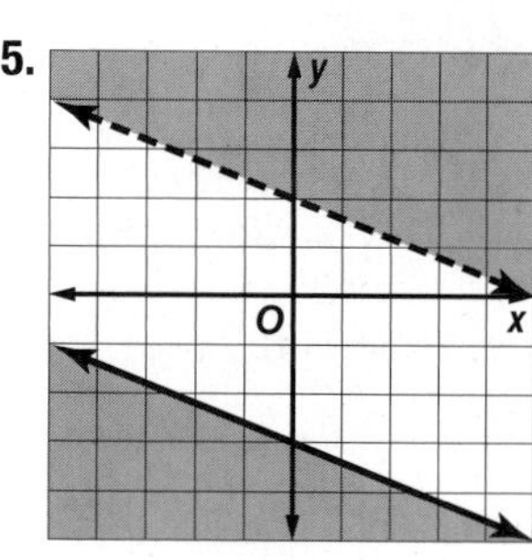

17.

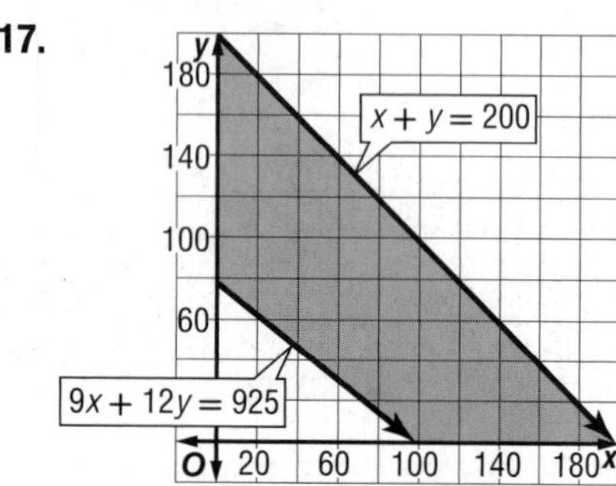

19.

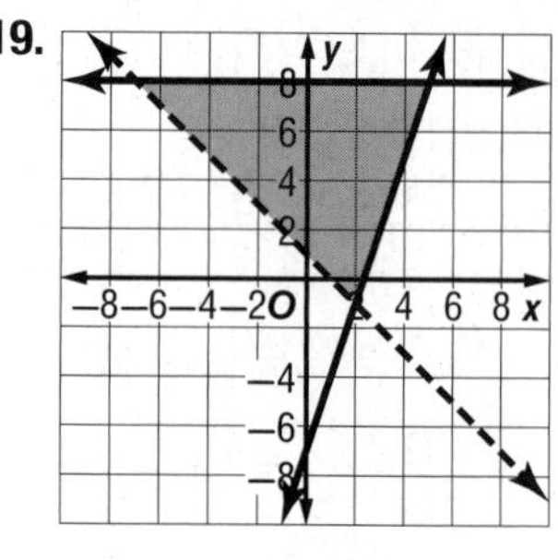

(2, −1), (5, 8), (−7, 8)

21.

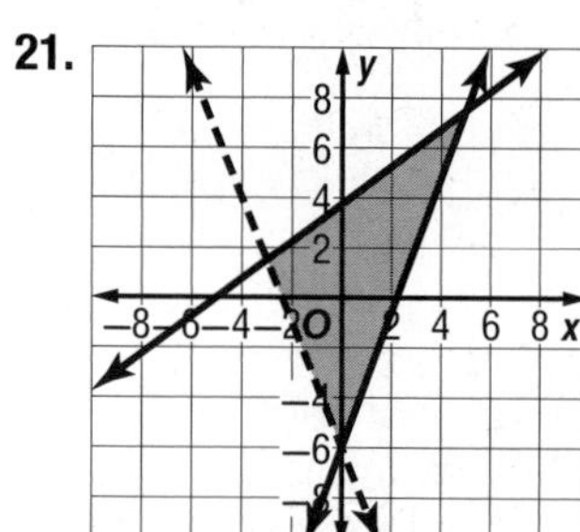

(−3, 1.5), (5, 7.5), (0, −6)

23.

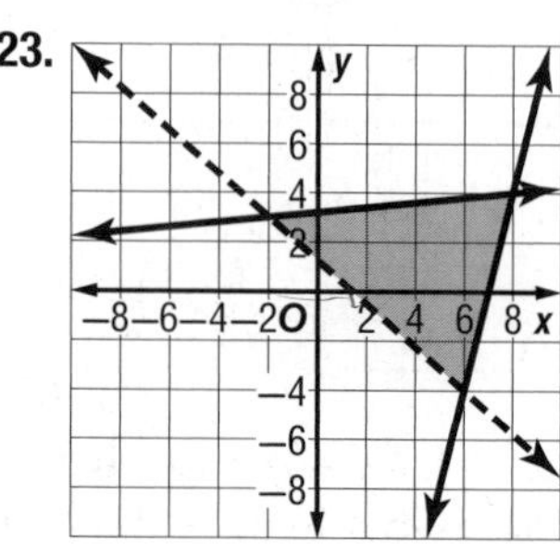

(8, 4), (6, −4), (−2, 3)

25 Let d represent the number of daytime minutes and n represent the number of nighttime minutes. Write a system of inequalities and then graph.

$d + n \leq 800$ Maximum number of minutes is 800.

$d \geq 2n$ At least twice as many daytime minutes as nighttime minutes

$n \geq 200$ At least 200 nighttime minutes

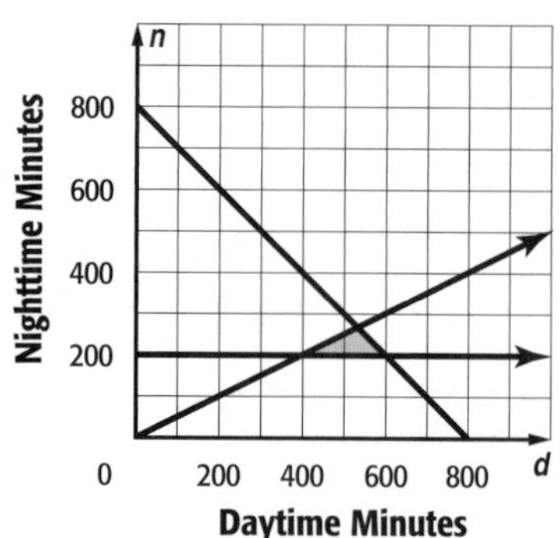

The intersection points are at (400, 200), (533.3, 266.7), and (600, 200).

$0.15d + 0.1n = 0.15(400) + 0.1(200)$ 400 daytime min., 200 nighttime min.

$= 60 + 20$ or \$80 Simplify.

$0.15d + 0.1n = 0.15(533.3) + 0.1(266.7)$ 533.3 daytime min., 266.7 nighttime min.

$\approx 80.0 + 26.7$ or \$106.70 Simplify.

$0.15d + 0.1n = 0.15(600) + 0.1(200)$ 600 daytime min., 200 nighttime min.

$= 90 + 20$ or \$110 Simplify.

So, his maximum bill is \$110 and his minimum bill is \$80.

27a.

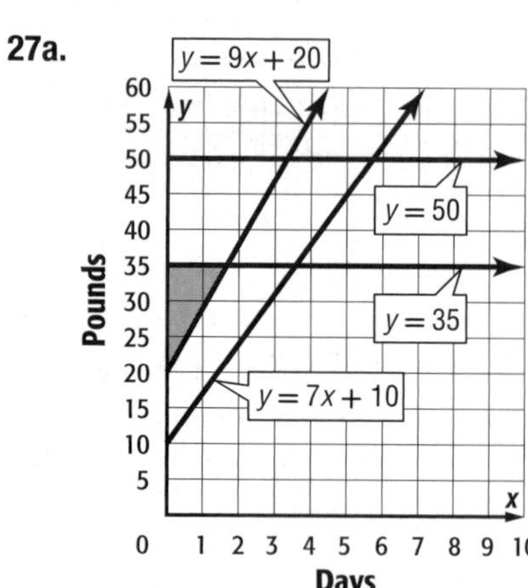

27b. $3\frac{1}{3}$ days **27c.** Marc; Jessica could last about a quarter of a day longer than Marc.

29.

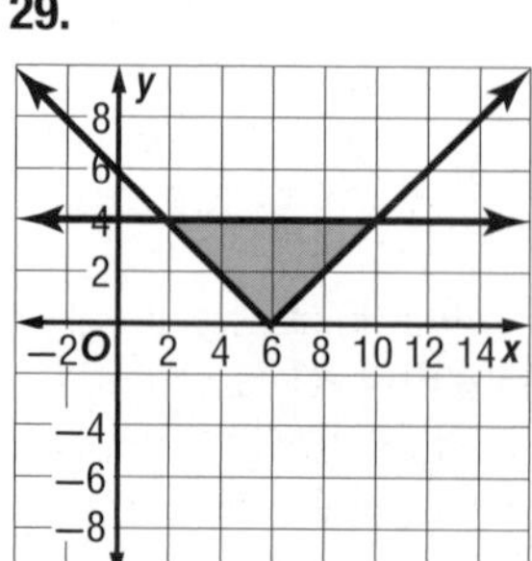

31.

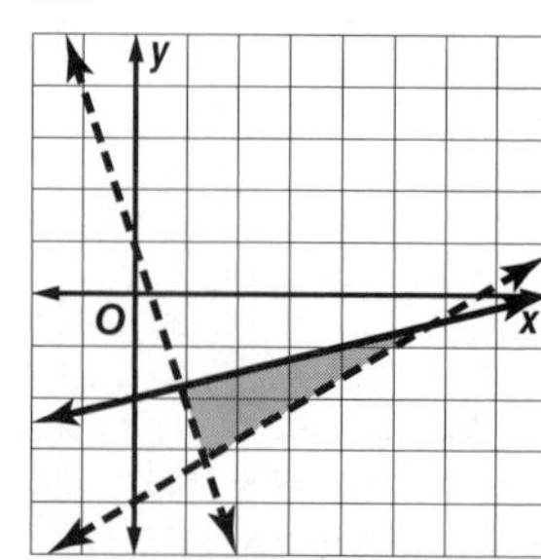

33.

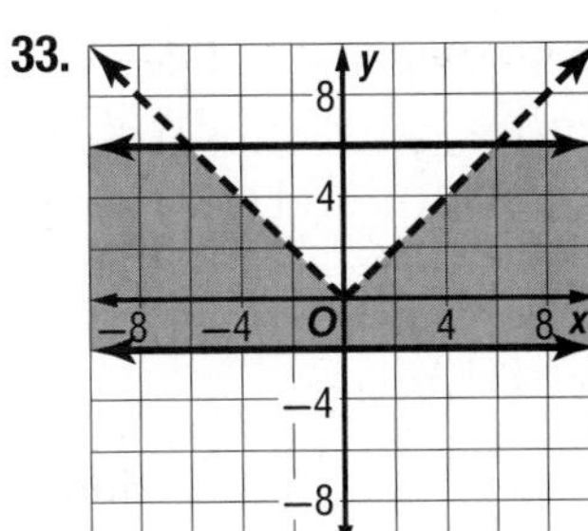

35.

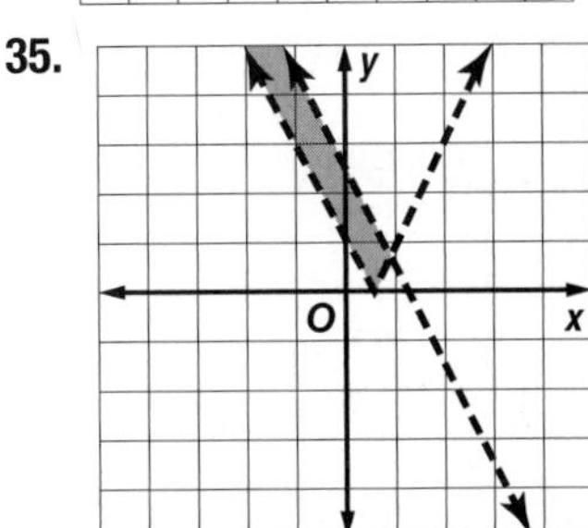

37.

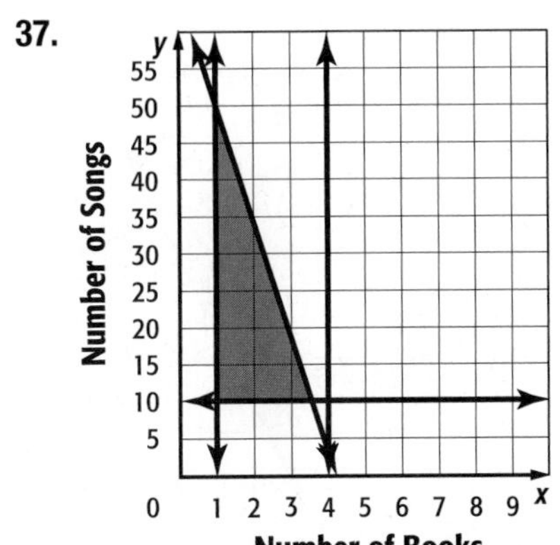

39. Let $w =$ the number of hours writing, and let $e =$ the number of hours exercising.
$w + e \leq 35$
$7 \leq e \leq 15$
$20 \leq w \leq 25$

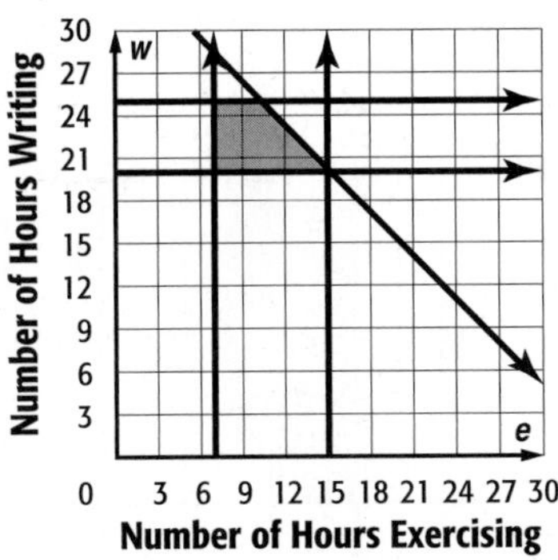

41. $(-6, -2)$, $\left(-3\frac{13}{17}, 6\frac{16}{17}\right)$, $\left(9\frac{1}{7}, 3\frac{5}{7}\right)$, $(0.8, -8.8)$

43 Let x represent the amount in the fund that pays 6% interest and y represent the amount in the fund that pays 10% interest. Write a system of inequalities and then graph.

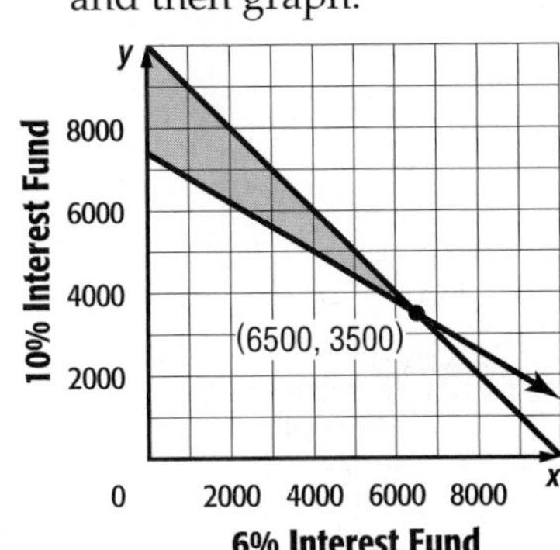

$x + y \leq 10{,}000$ Total amount invested is up to \$10,000.
$0.06x + 0.10y \geq 740$ Total amount earned is at least \$740.
The least amount Mr. Hoffman can invest in the risky fund, or the 10% interest fund, is \$3500.

45.

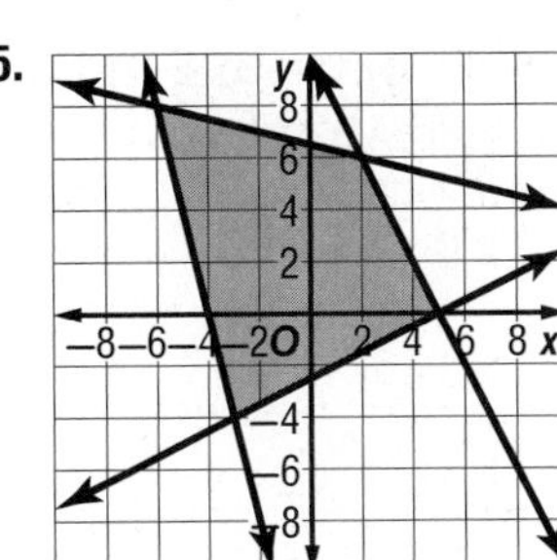

$(-3, -4)$, $(-6, 8)$, $(2, 6)$, $(5, 0)$; 75 units^2

47. Sample answer:
$y \geq 2x - 6$,
$y \leq -0.5x + 4$,
$y \geq -3x - 6$; 47

49. Sample answer: Shade each inequality in their standard way, by shading above the line if $y >$ and shading below the line if $y <$ (or you can use test points). Once you determine where to shade for each inequality, the area where *every* inequality needs to be shaded is the actual solution. This is only the shaded area. **51.** A **53.** $\frac{4}{5}z$

55.

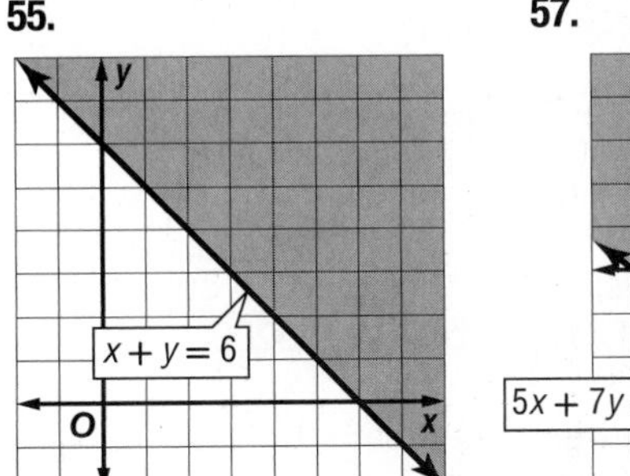

57.

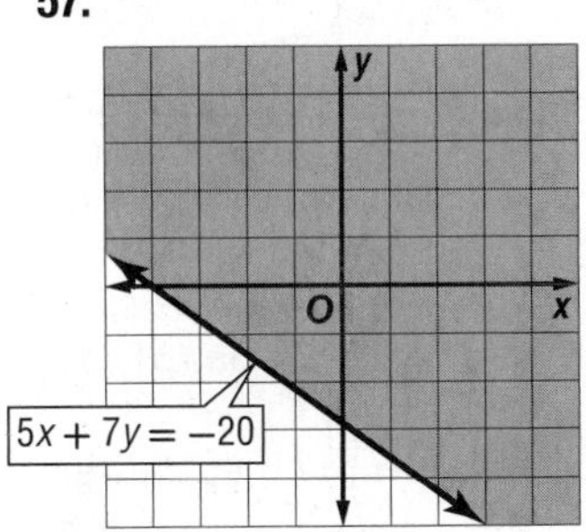

59.

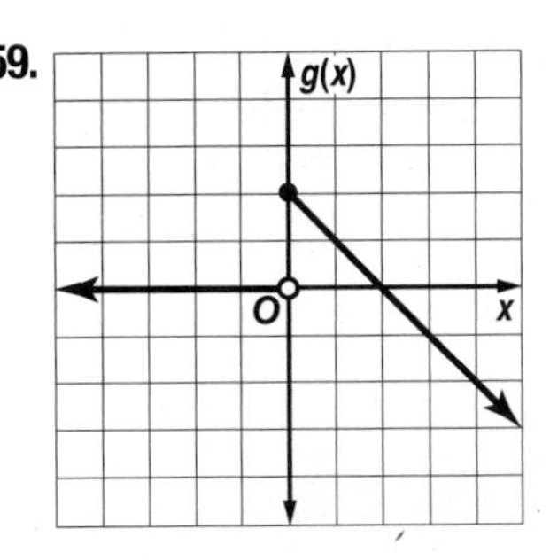

D = {all real numbers}, R = $\{g(x) \mid g(x) \leq 2\}$

61.

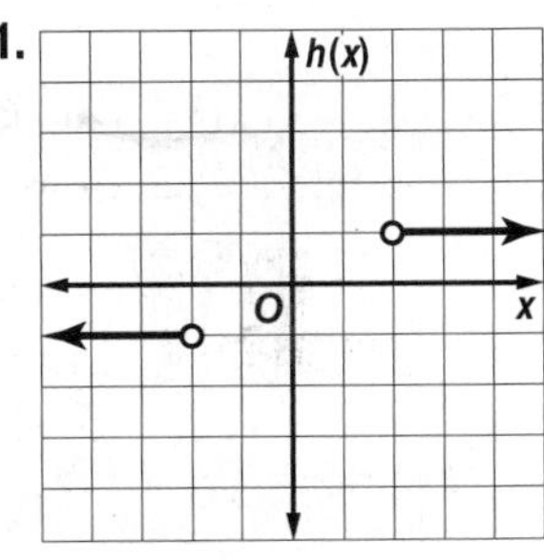

D = $\{x \mid x < -2$ or $x > 2\}$, R = $\{-1, 1\}$

63. -1 **65.** 3 **67.** 4.5

Lesson 3-3

1.

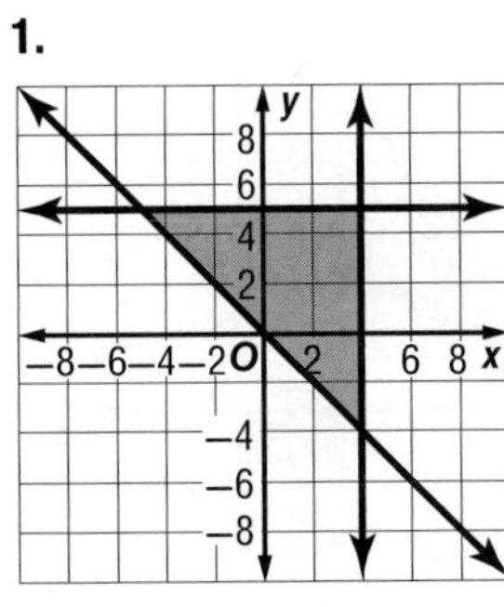

$(4, 5)$, $(4, -4)$, $(-5, 5)$; max = 28, min = -35

3.

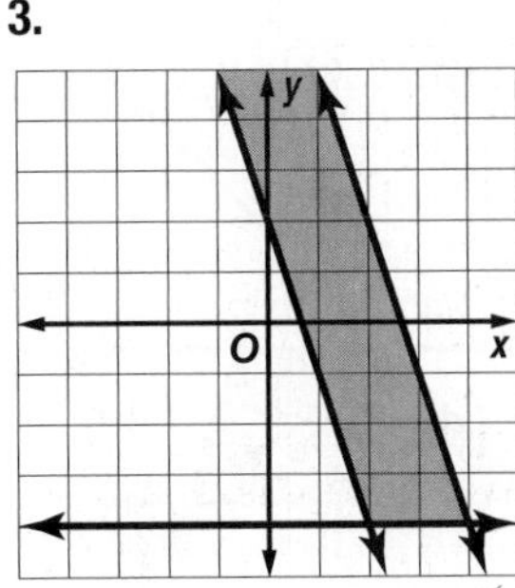

$(2, -4)$, $(4, -4)$; max does not exist, min = -52

5. 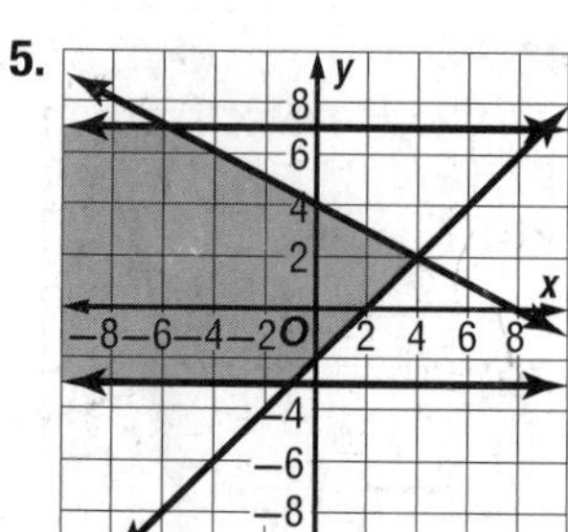

(4, 2), (−1, −3), (−6, 7); max does not exist; min = −30

7a. $g \geq 0, c \geq 0, 1.5g + c \leq 85, 2g + 0.5c \leq 40$

7b. 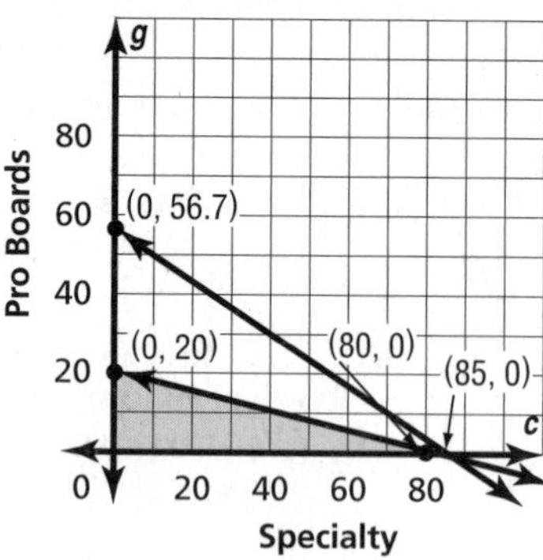

7c. (0, 0), (0, 20), (80, 0) **7d.** $f(c, g) = 65c + 50g$
7e. 80 specialty boards, 0 pro boards; $5200

9 Graph the inequalities and locate the vertices.

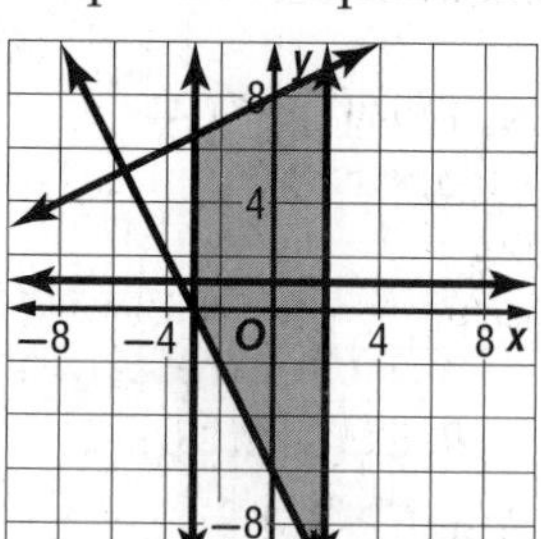

The vertices are at (2, −10), (−3, 0), (−3, 6.5), and (2, 9). Evaluate the function at each vertex.

(x, y)	−4x − 9y	f(x, y)
(2, −10)	−4(2) − 9(−10)	82
(−3, 0)	−4(−3) − 9(−0)	12
(−3, 6.5)	−4(−3) − 9(6.5)	−46.5
(2, 9)	−4(2) − 9(9)	−89

The maximum value is 82 at (2, −10). The minimum value is −89 at (2, 9).

11. 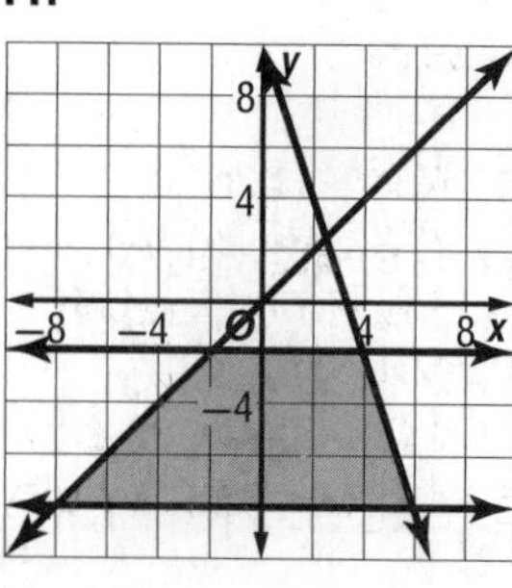

(6, −8), (4, −2), (−2, −2), (−8, −8); max = −8, min = −152

13. 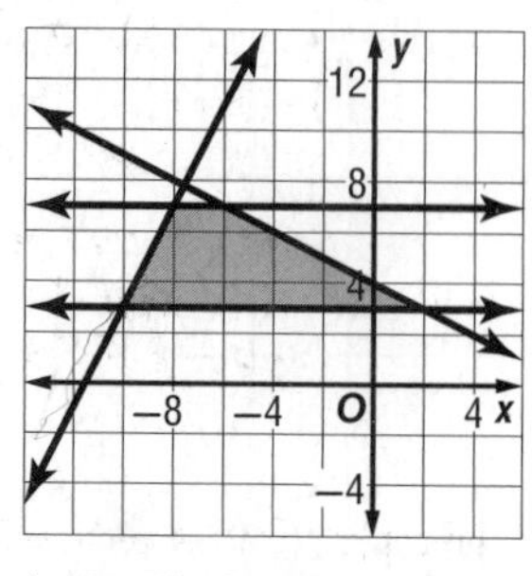

(−10, 3), (2, 3), (−6, 7), (−8, 7); max = 59, min = 9

15 Graph the inequalities and locate the vertices.

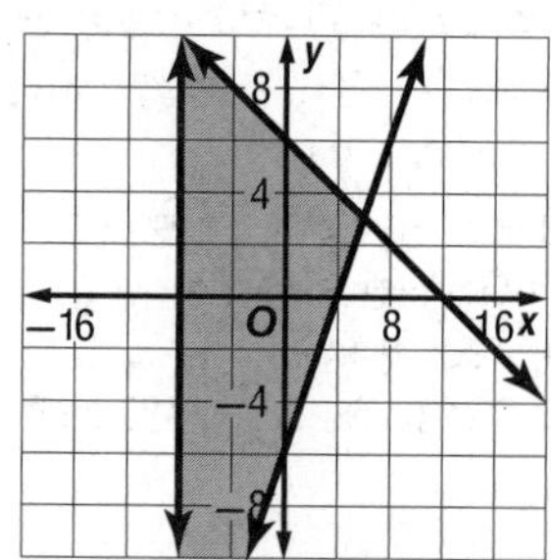

The vertices are at (6, 3), (−8, 10), and (−8, −18). Evaluate the function at each vertex.

(x, y)	10x − 6y	f(x, y)
(6, 3)	10(6) − 6(3)	42
(−8, 10)	10(−8) − 6(10)	−140
(−8, −18)	10(−8) − 6(−18)	28

The maximum value is 42 at (6, 3). The minimum value is −140 at (−8, 10).

17. 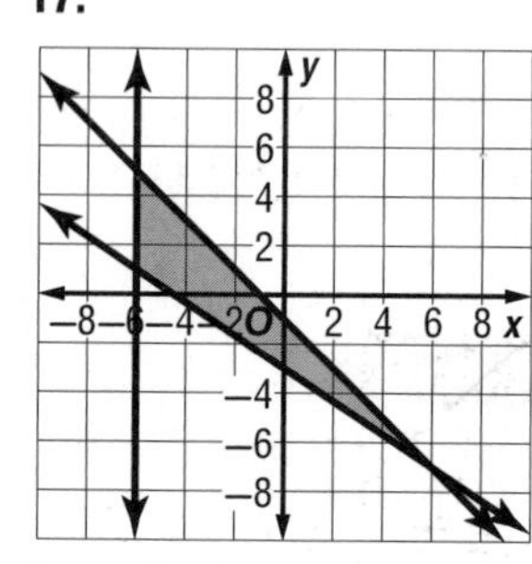

(−6, 1), (6, −7), (−6, 5); max = 48, min = 0

19. 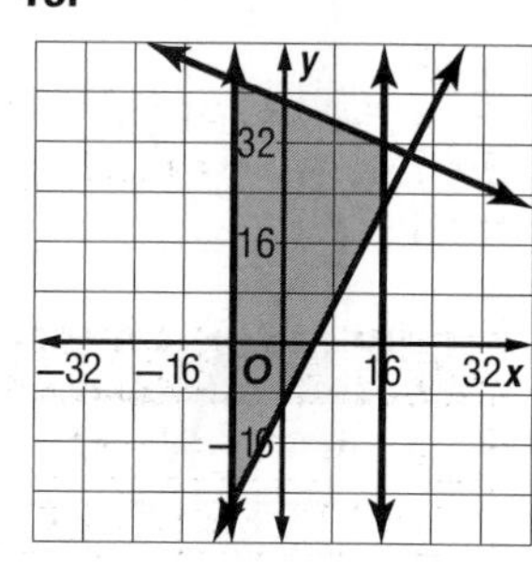

(−8, 44), (16, 32), (−8, −26), (16, 22); max = 672, min = −486

21. 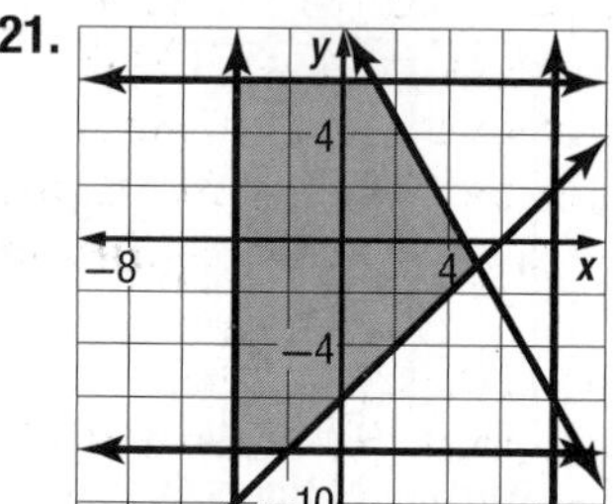

(5, −1), (1, 6), (−2, −8), (−4, −8), (−4, 6), max = 60, min = −112

23. 225 yellow cakes, 0 strawberry cakes
25a. $a \geq 0, b \geq 0, a + b \leq 45, 4a + 5b \leq 200$

25b. 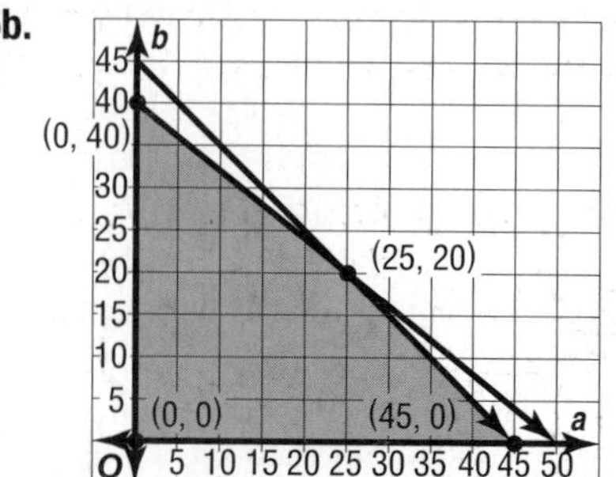

(0, 0), (0, 40), (25, 20), (45, 0)

25c. 25 sheds, 20 play houses
25d. $1250

27 **a.** Let x represent the number of small packages and y represent the number of large packages. Write a system of inequalities. Then graph the inequalities and locate the vertices.

$x \geq 0$	number of small packages ≥ 0
$y \geq 0$	number of large packages ≥ 0
$25x + 50y \leq 4200$	weight of packages ≤ 4200 lb
$3x + 5y \leq 480$	capacity of packages ≤ 480 cu ft

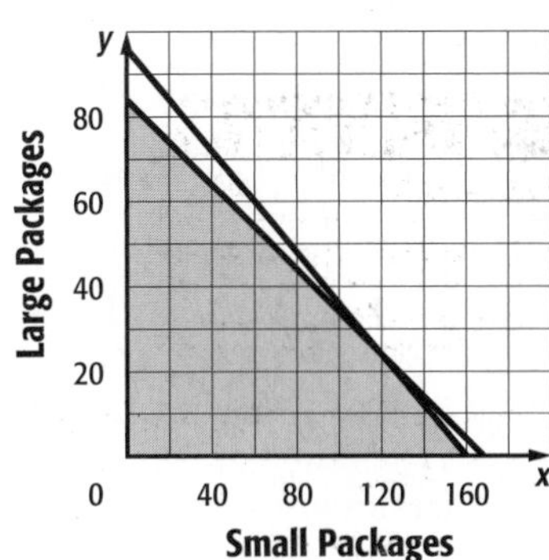

The vertices are at (0, 84), (120, 24), and (160, 0). Evaluate the function $f(x, y) = 5x + 8y$ at each vertex.

(x, y)	5x + 8y	f(x, y)
(0, 84)	5(0) + 8(84)	672
(120, 124)	5(120) + 8(24)	792
(160, 0)	5(160) + 8(0)	800

To maximize revenue, 160 small packages and 0 large packages should be placed on a train car. **b.** The maximum revenue per train car is \$800. **c.** No; if revenue is maximized, the company will not deliver any large packages, and customers with large packages to ship will probably choose another carrier.

29. Sample answer: $-2 \geq y \geq -6$, $4 \leq x \leq 9$
31. b; The feasible region of Graph b is unbounded while the other three are bounded.
33. Sample answer: Even though the region is bounded, multiple maximums occur at A and B and all of the points on the boundary of the feasible region containing both A and B. This happened because that boundary of the region has the same slope as the function. **35.** \$70.20 **37.** D

39.

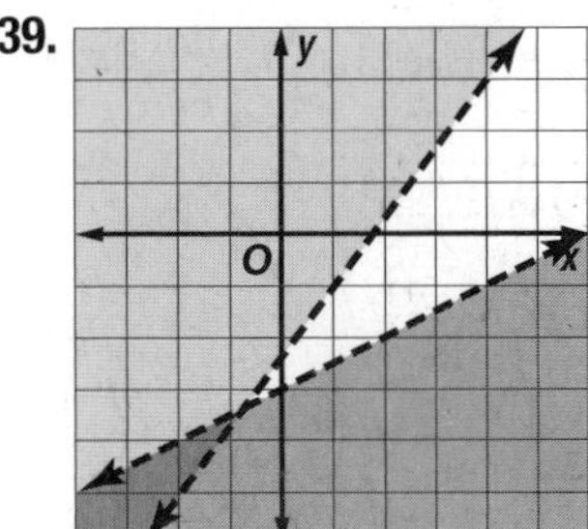

41. $7x + 15y = 330$, $8x + 16y = 360$; hats: 15, shirts: 15
43. $y = -\frac{1}{2}x + \frac{7}{2}$

45. 6; −2

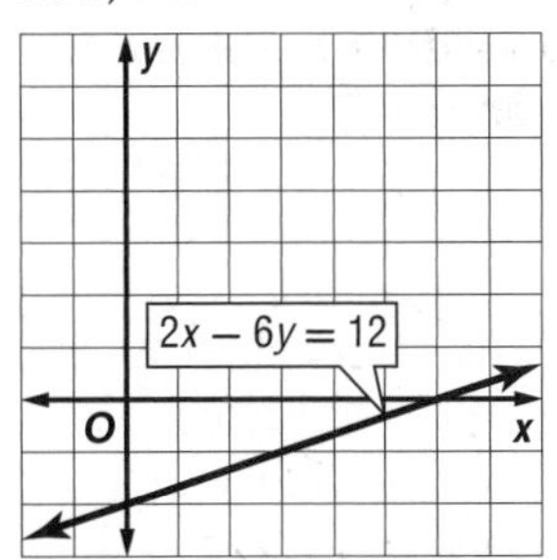

47. 5; 2

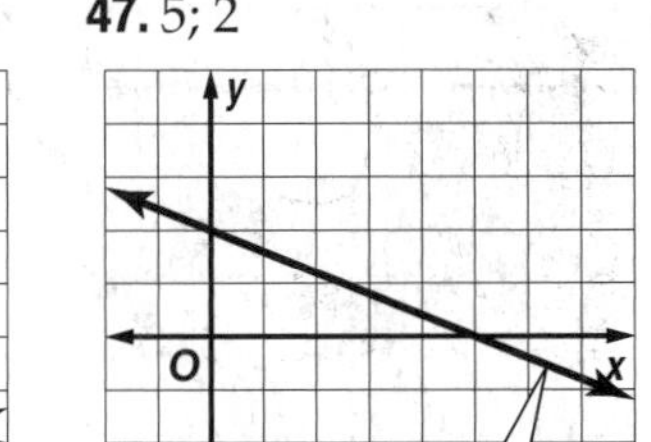

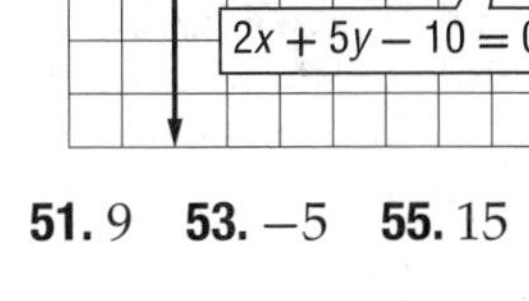

49. $\frac{1}{2}$; −2

y
O
x
y = 4x − 2

51. 9 **53.** −5 **55.** 15

Lesson 3-4

1. (−2, −3, 5) **3.** (−4, 3, 6) **5.** infinite solutions
7a. $s + d + t = 7$, $d = 2s$, $0.3s + 0.6d + 0.6t = 3.6$
7b. 2 sitcoms, 4 dramas, 1 talk show

9 $-a + 4b + 2c = -13$ Multiply by 4. → $-4a + 16b + 8c = -52$

$-4a + 16b + 8c = -52$ Equation 3 (× 4)
$(+)\ 4a + 5b - 6c = 2$ Equation 1
$21b + 2c = -50$

$-a + 4b + 2c = -13$ Multiply by −3. → $3a - 12b - 6c = 39$

$3a - 12b - 6c = 39$ Equation 3 × (−3)
$(+)\ -3a - 2b + 7c = -15$ Equation 2
$-14b + c = 24$

The resulting system of two equations and two variables is shown below.

$21b + 2c = -50$ New Equation 1
$-14b + c = 24$ New Equation 2

$-14b + c = 24$ Multiply by −2. → $28b - 2c = -48$

$21b + 2c = -50$ New Equation 1
$(+)\ 28b - 2c = -48$ New Equation 2 × (−2)
$49b = -98$ Add the equations.
$b = -2$ Divide each side by 49.

$-14b + c = 24$ New Equation 2
$-14(-2) + c = 24$ Replace b with −2.
$28 + c = 24$ Multiply.
$c = -4$ Subtract 28 from each side.

$-a + 4b + 2c = -13$ Original Equation 3
$-a + 4(-2) + 2(-4) = -13$ $b = -2$ and $c = -4$
$-a - 8 - 8 = -13$ Multiply.
$-a = 3$ Add 16 to each side.
$a = -3$ Multiply each side by −1.

The solution is (−3, −2, −4).

11. (−2, −1, 4) **13.** infinite solutions **15.** (−4, −1, 6) **17.** no solution **19.** infinite solutions **21.** roller coasters: 5; bumper cars: 1; water slides: 4

23 a = the amount invested in account A
b = the amount invested in account B
c = the amount invested in account C

$a + b + c = 100{,}000$ She invested a total of \$100,000.
$a = c + 30{,}000$ She invested \$30,000 more in account A than account C.
$0.04a + 0.08b + 0.1c = 6300$ The expected interest earned is \$6300.

Substitute $a = c + 30{,}000$ in Equations 1 and 3.

$$\begin{aligned} a + b + c &= 100{,}000 && \text{Equation 1} \\ c + 30{,}000 + b + c &= 100{,}000 && a = c + 30{,}000 \\ 30{,}000 + b + 2c &= 100{,}000 && \text{Add.} \\ b + 2c &= 70{,}000 && \text{Simplify.} \end{aligned}$$

$$\begin{aligned} 0.04a + 0.08b + 0.1c &= 6300 && \text{Equation 3} \\ 0.04(c + 30{,}000) + 0.08b + 0.1c &= 6300 && a = c + 30{,}000 \\ 0.04c + 1200 + 0.08b + 0.1c &= 6300 && \text{Distribute.} \\ 1200 + 0.08b + 0.14c &= 6300 && \text{Add.} \\ 0.08b + 0.14c &= 5100 && \text{Simplify.} \end{aligned}$$

Solve the system of two equations in two variables.
$b + 2c = 70{,}000$ Multiply by -0.08. $\rightarrow$ $0.08b + 0.14c = 5100$

$$\begin{aligned} -0.08b - 0.16c &= -5600 \\ (+)\ 0.08b + 0.14c &= 5100 \\ \hline -0.02c &= -500 \\ c &= 25{,}000 \end{aligned}$$

Substitute to find b.

$$\begin{aligned} b + 2c &= 70{,}000 && \text{Remaining equation in two variables} \\ b + 2(25{,}000) &= 70{,}000 && c = 25{,}000 \\ b + 50{,}000 &= 70{,}000 && \text{Distribute.} \\ b &= 20{,}000 && \text{Simplify.} \end{aligned}$$

Substitute to find a.

$$\begin{aligned} a + b + c &= 100{,}000 && \text{Equation 1} \\ a + 20{,}000 + 25{,}000 &= 100{,}000 && b = 20{,}000,\ c = 25{,}000 \\ a + 45{,}000 &= 100{,}000 && \text{Add.} \\ a &= 55{,}000 && \text{Simplify.} \end{aligned}$$

The solution is (55,000, 20,000, 25,000). She invested \$55,000 in account A, \$20,000 in account B, and \$25,000 in account C.

25. $y = -3x^2 + 4x - 6$; $a = -3$, $b = 4$, $c = -6$

27. Sample answer:

$$\begin{aligned} 3x + 4y + z &= -17 \\ 2x - 5y - 3z &= -18; \\ -x + 3y + 8z &= 47 \end{aligned}$$

$$\begin{aligned} 3x + 4y + z &= -17 \\ 3(-5) + 4(-2) + 6 &= -17 \\ -15 + (-8) + 6 &= -17 \\ &= -17 \checkmark \end{aligned}$$

$$\begin{aligned} 2x - 5y - 3z &= -18 \\ 2(-5) - 5(-2) - 3(6) &= -18 \\ -10 + 10 - 18 &= -18 \\ -18 &= -18 \checkmark \end{aligned}$$

$$\begin{aligned} -x + 3y + 8z &= 47 \\ -(-5) + 3(-2) + 8(6) &= 47 \\ 5 - 6 + 48 &= 47 \\ 47 &= 47 \checkmark \end{aligned}$$

29. Sample answer: First, combine two of the original equations using elimination to form a new equation with three variables. Next, combine a different pair of the original equations using elimination to eliminate the same variable and form a second equation with three variables. Do the same thing with a third pair of the original equations. You now have a system of three equations with three variables. Follow the same procedure you learned in this section. Once you find the three variables, you need to use them to find the eliminated variable. **31.** J **33.** A **35.** 16; –8 **37.** 9; –8 **39.** (6, 1) **41.** (8, –5)

Lesson 3-5

1a. $\begin{bmatrix} 23 & 21 & 21 & 42 & 61 \\ 25 & 24 & 32 & 49 & 70 \end{bmatrix}$ **1b.** Sample answer: City: The sum is 168. However, this value is irrelevant since it is the sum of 5 different types of data. Highway: The sum is 200. However, this value is irrelevant since it is the sum of 5 different types of data. **1c.** Sample answer: The sums are 48, 45, 53, 91, and 131. These values are irrelevant since they are the sums of 2 different types of data. **3.** impossible

5. $\begin{bmatrix} 7 & 31 & -14 \\ 1 & -6 & 2 \end{bmatrix}$ **7.** $\begin{bmatrix} -90 & 54 & -12 & -18 \\ -36 & 66 & -84 & 12 \\ -24 & 48 & 60 & -162 \end{bmatrix}$

9. $\begin{bmatrix} 50 & 36 \\ -87 & 41 \end{bmatrix}$ **11.** $\begin{bmatrix} -24 & 29 \\ -38 & -7 \end{bmatrix}$

13a. $\begin{bmatrix} 3 & 2 & 2 & 1 \\ 4 & 3 & 2 & 3 \\ 5 & 5 & 4 & 4 \\ 1 & 5 & 5 & 2 \end{bmatrix}$ **13b.** Sample answer: Brand C; it was given the highest rating possible for cost and comfort, and a high rating for looks, and it will last a fairly long time. **13c.** Sample answer: Yes; finding the sum of the rows and then calculating the average will provide an easy way to compare the data.

15 To find the sum of two matrices, they must have the same dimensions. Since the dimensions of the matrices are 2×2 and 2×3, it is impossible to find the sum.

17. $\begin{bmatrix} -33 & -26 \\ 25 & 75 \end{bmatrix}$ **19.** impossible

21a. Library A: $\begin{bmatrix} 10{,}000 \\ 5000 \\ 5000 \end{bmatrix}$; Library B: $\begin{bmatrix} 15{,}000 \\ 10{,}000 \\ 2500 \end{bmatrix}$; Library C: $\begin{bmatrix} 4000 \\ 700 \\ 800 \end{bmatrix}$ **21b.** $\begin{bmatrix} 29{,}000 \\ 15{,}700 \\ 8300 \end{bmatrix}$ **21c.** $\begin{bmatrix} 6000 \\ 4300 \\ 4200 \end{bmatrix}$

21d. $\begin{bmatrix} 25{,}000 \\ 15{,}000 \\ 7500 \end{bmatrix}$; Sample answer: The sum represents the combined size of the two libraries.

23. $\begin{bmatrix} 5 & 24 \\ -\frac{167}{12} & -10 \end{bmatrix}$

25. $\begin{bmatrix} -8a & 32b & 8c - 8b \\ -104 & 80 & -40c \end{bmatrix}$

27 $-5\left(\begin{bmatrix} 4 & -8 \\ 8 & -9 \end{bmatrix} + \begin{bmatrix} 4 & -2 \\ -3 & -6 \end{bmatrix}\right)$

$= -5\begin{bmatrix} 4+4 & -8+(-2) \\ 8+(-3) & -9+(-6) \end{bmatrix}$ Add corresponding elements.

$= -5\begin{bmatrix} 8 & -10 \\ 5 & -15 \end{bmatrix}$ Simplify.

$= \begin{bmatrix} -5(8) & -5(-10) \\ -5(5) & -5(-15) \end{bmatrix}$ Distribute the scalar.

$= \begin{bmatrix} -40 & 50 \\ -25 & 75 \end{bmatrix}$ Multiply.

29 **a.** Write a matrix to show the American records and a matrix to show the world records. Then subtract.

$\begin{bmatrix} 24.63 \text{ s} \\ 53.99 \text{ s} \\ 1{:}57.41 \text{ min} \\ 8{:}16.22 \text{ min} \end{bmatrix} - \begin{bmatrix} 24.13 \text{ s} \\ 53.52 \text{ s} \\ 1{:}56.54 \text{ min} \\ 8{:}16.22 \text{ min} \end{bmatrix}$

$= \begin{bmatrix} 24.63 \text{ s} - 24.13 \text{ s} \\ 53.99 \text{ s} - 53.52 \text{ s} \\ 1{:}57.41 \text{ min} - 1{:}56.54 \text{ min} \\ 8{:}16.22 \text{ min} - 8{:}16.22 \text{ min} \end{bmatrix}$ Subtract corresponding elements.

$= \begin{bmatrix} 0.5 \text{ s} \\ 0.47 \text{ s} \\ 0.87 \text{ s} \\ 0 \text{ s} \end{bmatrix}$ Simplify.

b. In the 50-meter, the fastest American time is 0.5 second behind the world record. In the 100 m, the fastest American time is 0.47 second behind the world record. In the 200 m, the fastest American time is 0.87 second behind the world record. In the 800 m, the American and world records are the same. So, it was an American who set the world record.

c. In the 50-meter and 100-meter events, the fastest times were set at the Olympics. These times became world records.

31. To show that the Commutative Property of Matrix Addition is true for 2 × 2 matrices, let $A = \begin{bmatrix} a & b \\ c & d \end{bmatrix}$ and $B = \begin{bmatrix} e & f \\ g & h \end{bmatrix}$. Show that $A + B = B + A$.

$A + B = \begin{bmatrix} a & b \\ c & d \end{bmatrix} + \begin{bmatrix} e & f \\ g & h \end{bmatrix}$ Substitution

$= \begin{bmatrix} a+e & b+f \\ c+g & d+h \end{bmatrix}$ Definition of matrix addition

$= \begin{bmatrix} e+a & f+b \\ g+c & h+d \end{bmatrix}$ Commutative Property of Addition for Real Numbers

$= \begin{bmatrix} e & f \\ g & h \end{bmatrix} + \begin{bmatrix} a & b \\ c & d \end{bmatrix}$ Definition of matrix addition

$= B + A$ Substitution

33. $\begin{bmatrix} 7 & 5 \\ -1 & -5 \end{bmatrix}$ **35.** Sample answer: $A = \begin{bmatrix} 6 & 1 \\ 6 & 3 \end{bmatrix}$ and $B = \begin{bmatrix} 3 & 2 \\ 4 & 2 \end{bmatrix}$ **37.** C **39.** F **41.** (3, −1, 4)

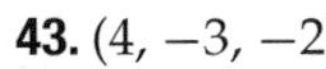

43. (4, −3, −2)

45.

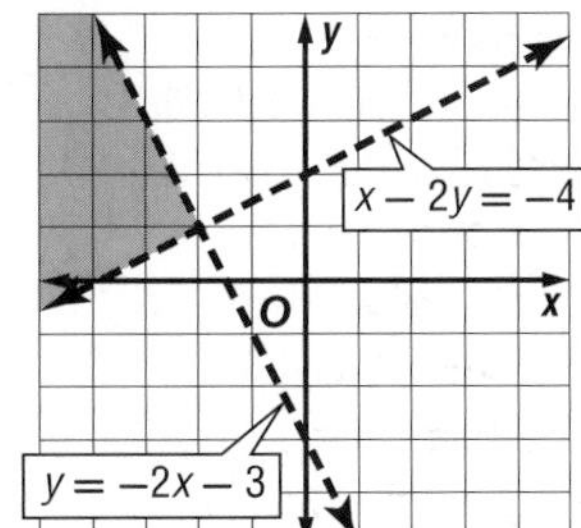

47.

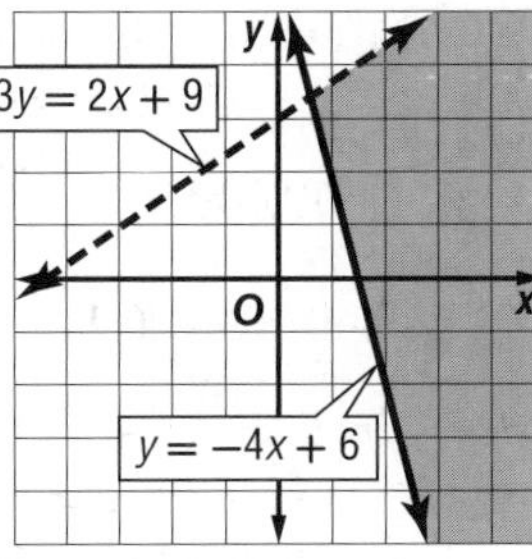

49. 350,349 − x = 15,991; 334,358 **51.** $-10a - b$

Lesson 3-6

1. 2 × 3 **3.** 8 × 10 **5.** $\begin{bmatrix} 0 & 44 \\ 8 & -34 \end{bmatrix}$

7 The product of a 2 × 1 matrix and a 1 × 3 matrix is a 2 × 3 matrix.

$\begin{bmatrix} -9 \\ 6 \end{bmatrix} \cdot \begin{bmatrix} -1 & -10 & 1 \end{bmatrix}$

$= \begin{bmatrix} -9(-1) & -9(-10) & -9(1) \\ 6(-1) & 6(-10) & 6(1) \end{bmatrix}$ Multiply the column by the row.

$= \begin{bmatrix} 9 & 90 & -9 \\ -6 & -60 & 6 \end{bmatrix}$ Simplify.

9. $\begin{bmatrix} -44 \\ 25 \end{bmatrix}$ **11.** $\begin{bmatrix} -16 & -1 \\ -6 & 10 \end{bmatrix}$

13. No; $\begin{bmatrix} 53 & -87 \\ -2 & -60 \end{bmatrix} \neq \begin{bmatrix} 62 & -33 \\ 28 & -69 \end{bmatrix}$.

15. 2 × 4 **17.** undefined **19.** undefined **21.** [26]

23 $\begin{bmatrix} -3 & -7 \\ -2 & -1 \end{bmatrix} \cdot \begin{bmatrix} 4 & 4 \\ 9 & -3 \end{bmatrix}$

$= \begin{bmatrix} -3(4) + (-7)(9) & -3(4) + (-7)(-3) \\ & \end{bmatrix}$ Multiply the 1st row in the 1st matrix by each column in the 2nd matrix.

$\begin{bmatrix} -3 & -7 \\ -2 & -1 \end{bmatrix} \cdot \begin{bmatrix} 4 & 4 \\ 9 & -3 \end{bmatrix}$

$= \begin{bmatrix} -3(4) + (-7)(9) & -3(4) + (-7)(-3) \\ -2(4) + (-1)(9) & -2(4) + (-1)(-3) \end{bmatrix}$ Multiply the 2nd row in the 1st matrix by each column in the 2nd matrix.

$= \begin{bmatrix} -12 + (-63) & -12 + 21 \\ -8 + (-9) & -8 + 3 \end{bmatrix}$ Multiply.

$= \begin{bmatrix} -75 & 9 \\ -17 & -5 \end{bmatrix}$ Add.

25. undefined **27.** $\begin{bmatrix} -40 & 64 \\ 22 & 1 \end{bmatrix}$

29a. $I = \begin{bmatrix} 3 & 2 & 2 \\ 2 & 3 & 1 \\ 4 & 3 & 0 \end{bmatrix}$, $C = \begin{bmatrix} 220 \\ 250 \\ 360 \end{bmatrix}$ **29b.** $\begin{bmatrix} \$1880 \\ \$1550 \\ \$1630 \end{bmatrix}$

29c. \$5060 **31.** $PQR = \begin{bmatrix} -22 & 240 \\ 44 & -12 \end{bmatrix}$ and $RQP = \begin{bmatrix} 34 & -40 \\ -220 & -44 \end{bmatrix}$

33. No; $R(P + Q) = \begin{bmatrix} 34 & -6 \\ -64 & -30 \end{bmatrix}$ and $PR + QR = \begin{bmatrix} 22 & 72 \\ 14 & -18 \end{bmatrix}$.

35 a. The bonuses can be found by multiplying the number of cars sold by the amount of bonus given for each new and used car sold.

Cars **Bonus**

$$C = \begin{bmatrix} 27 & 49 \\ 35 & 36 \\ 9 & 56 \\ 15 & 62 \end{bmatrix} \qquad B = \begin{bmatrix} 1000 \\ 500 \end{bmatrix}$$

$$CB = \begin{bmatrix} 27 & 49 \\ 35 & 36 \\ 9 & 56 \\ 15 & 62 \end{bmatrix} \cdot \begin{bmatrix} 1000 \\ 500 \end{bmatrix}$$ Write an equation.

$$= \begin{bmatrix} 27(1000) + 49(500) \\ 35(1000) + 36(500) \\ 9(1000) + 56(500) \\ 15(1000) + 62(500) \end{bmatrix}$$ Multiply columns by rows.

$$= \begin{bmatrix} 51{,}500 \\ 53{,}000 \\ 37{,}000 \\ 46{,}000 \end{bmatrix}$$ Simplify.

Westin earned the most, \$53,000.

b. total amount on bonuses = 51,500 + 53,000 + 37,000 + 46,000 or \$187,500

37. $\begin{bmatrix} -10 - 4.5y & 36.75 \\ 2x + 4 + 3y^2 & -6x - 4.5y - 12 \\ 3.6y + 26 & -83.4 \end{bmatrix}$

39. $\begin{bmatrix} -1.5x - 1.5y + 15 \\ y^2 + xy - 3x - 6 \\ 1.2x + 1.2y - 39 \end{bmatrix}$

41. undefined

43. $\begin{bmatrix} 15x + 69y - 12 \\ -18y^2 + 42.75y - 18xy + 20.25x \end{bmatrix}$

45a. A: \$421; B: \$274; C: \$150; D: \$68

45b. A: \$357.85; B: \$232.90; C: \$127.50; D: \$57.80

47a. $c(A + B)$

$= c\left(\begin{bmatrix} a & b \\ d & e \end{bmatrix} + \begin{bmatrix} w & x \\ y & z \end{bmatrix}\right)$ Substitution

$= c\begin{bmatrix} a + w & b + x \\ d + y & e + z \end{bmatrix}$ Definition of matrix addition

$= \begin{bmatrix} ca + cw & cb + cx \\ cd + cy & ce + cz \end{bmatrix}$ Definition of scalar multiplication

$= \begin{bmatrix} ca & cb \\ cd & ce \end{bmatrix} + \begin{bmatrix} cw & cx \\ cy & cz \end{bmatrix}$ Definition of matrix addition

$= cA + cB$ Substitution

47b. $C(A + B) = \begin{bmatrix} a & b \\ c & d \end{bmatrix}\left(\begin{bmatrix} e & f \\ g & h \end{bmatrix} + \begin{bmatrix} j & k \\ m & n \end{bmatrix}\right)$ Substitution

$= \begin{bmatrix} a & b \\ c & d \end{bmatrix}\begin{bmatrix} e + j & f + k \\ g + m & h + n \end{bmatrix}$ Definition of matrix addition

$= \begin{bmatrix} a(e + j) + b(g + m) & a(f + k) + b(h + n) \\ c(e + j) + d(g + m) & c(f + k) + d(h + n) \end{bmatrix}$ Definition of matrix multiplication

$= \begin{bmatrix} ea + ja + gb + mb & fa + ka + hb + nb \\ ec + jc + gd + md & fc + kc + hd + nd \end{bmatrix}$ Distributive Property

$= \begin{bmatrix} ea + gb + ja + mb & fa + hb + ka + nb \\ ec + gd + jc + md & fc + hd + kc + nd \end{bmatrix}$ Commutative Property of Addition

$= \begin{bmatrix} ea + gb & fa + hb \\ ec + gd & fc + hd \end{bmatrix} + \begin{bmatrix} ja + mb & ka + nb \\ jc + md & kc + nd \end{bmatrix}$ Definition of matrix addition

$= CA + CB$ Definition of matrix multiplication

$(A + B)C = \left(\begin{bmatrix} a_{11} & a_{12} \\ a_{21} & a_{22} \end{bmatrix} + \begin{bmatrix} b_{11} & b_{12} \\ b_{21} & b_{22} \end{bmatrix}\right)\begin{bmatrix} c_{11} & c_{12} \\ c_{21} & c_{22} \end{bmatrix}$ Substitution

$= \begin{bmatrix} a_{11} + b_{11} & a_{12} + b_{12} \\ a_{21} + b_{21} & a_{22} + b_{22} \end{bmatrix}\begin{bmatrix} c_{11} & c_{12} \\ c_{21} & c_{22} \end{bmatrix}$ Definition of matrix addition

$= \begin{bmatrix} (a_{11} + b_{11})c_{11} + (a_{12} + b_{12})c_{21} & (a_{11} + b_{11})c_{12} + (a_{12} + b_{12})c_{22} \\ (a_{11} + b_{11})c_{11} + (a_{12} + b_{12})c_{21} & (a_{21} + b_{21})c_{12} + (a_{22} + b_{22})c_{22} \end{bmatrix}$ Definition of matrix multiplication

$= \begin{bmatrix} a_{11}c_{11} + b_{11}c_{11} + a_{12}c_{21} + b_{12}c_{21} & a_{11}c_{12} + b_{11}c_{12} + a_{12}c_{22} + b_{12}c_{22} \\ a_{21}c_{11} + b_{21}c_{11} + a_{22}c_{21} + b_{22}c_{21} & a_{21}c_{12} + b_{21}c_{12} + a_{22}c_{22} + b_{22}c_{22} \end{bmatrix}$ Distributive Property

$= \begin{bmatrix} a_{11}c_{11} + a_{12}c_{21} + b_{11}c_{11} + b_{12}c_{21} & a_{11}c_{12} + a_{12}c_{22} + b_{11}c_{12} + b_{12}c_{22} \\ a_{21}c_{11} + a_{22}c_{21} + b_{21}c_{11} + b_{22}c_{21} & a_{21}c_{12} + a_{22}c_{22} + b_{21}c_{12} + b_{22}c_{22} \end{bmatrix}$ Commutative Property of Addition

$= \begin{bmatrix} a_{11}c_{11} + a_{12}c_{21} & a_{11}c_{12} + a_{12}c_{22} \\ a_{21}c_{11} + a_{22}c_{21} & a_{21}c_{12} + a_{22}c_{22} \end{bmatrix} + \begin{bmatrix} b_{11}c_{11} + b_{12}c_{21} & b_{11}c_{12} + b_{12}c_{22} \\ b_{21}c_{11} + b_{22}c_{21} & b_{21}c_{12} + b_{22}c_{22} \end{bmatrix}$ Definition of matrix addition

$= AC + BC$ Definition of matrix multiplication

47c. $$(AB)C = \left(\begin{bmatrix} a_{11} & a_{12} \\ a_{21} & a_{22} \end{bmatrix}\begin{bmatrix} b_{11} & b_{12} \\ b_{21} & b_{22} \end{bmatrix}\right)\begin{bmatrix} c_{11} & c_{12} \\ c_{21} & c_{22} \end{bmatrix}$$ Substitution

$$= \begin{bmatrix} a_{11}b_{11} + a_{12}b_{21} & a_{11}b_{12} + a_{12}b_{22} \\ a_{21}b_{11} + a_{22}b_{21} & a_{21}b_{12} + a_{22}b_{22} \end{bmatrix}\begin{bmatrix} c_{11} & c_{12} \\ c_{21} & c_{22} \end{bmatrix}$$ Definition of matrix multiplication

$$= \begin{bmatrix} (a_{11}b_{11} + a_{12}b_{21})c_{11} + (a_{11}b_{12} + a_{12}b_{22})c_{21} & (a_{11}b_{11} + a_{12}b_{21})c_{12} + (a_{11}b_{12} + a_{12}b_{22})c_{22} \\ (a_{21}b_{11} + a_{22}b_{21})c_{11} + (a_{21}b_{12} + a_{22}b_{22})c_{21} & (a_{21}b_{11} + a_{22}b_{21})c_{12} + (a_{21}b_{12} + a_{22}b_{22})c_{22} \end{bmatrix}$$ Definition of matrix multiplication

$$= \begin{bmatrix} a_{11}b_{11}c_{11} + a_{12}b_{21}c_{11} + a_{11}b_{12}c_{21} + a_{12}b_{22}c_{21} & a_{11}b_{11}c_{12} + a_{12}b_{21}c_{12} + a_{11}b_{12}c_{22} + a_{12}b_{22}c_{22} \\ a_{21}b_{11}c_{11} + a_{22}b_{21}c_{11} + a_{21}b_{12}c_{21} + a_{22}b_{22}c_{21} & a_{21}b_{11}c_{12} + a_{22}b_{21}c_{12} + a_{21}b_{12}c_{22} + a_{22}b_{22}c_{22} \end{bmatrix}$$ Distributive Property

$$= \begin{bmatrix} a_{11}b_{11}c_{11} + a_{11}b_{12}c_{21} + a_{12}b_{21}c_{11} + a_{12}b_{22}c_{21} & a_{11}b_{11}c_{12} + a_{11}b_{12}c_{22} + a_{12}b_{21}c_{12} + a_{12}b_{22}c_{22} \\ a_{21}b_{11}c_{11} + a_{21}b_{12}c_{21} + a_{22}b_{21}c_{11} + a_{22}b_{22}c_{21} & a_{21}b_{11}c_{12} + a_{21}b_{12}c_{22} + a_{22}b_{21}c_{12} + a_{22}b_{22}c_{22} \end{bmatrix}$$ Commutative Property of Addition

$$= \begin{bmatrix} a_{11}(b_{11}c_{11} + b_{12}c_{21}) + a_{12}(b_{21}c_{11} + b_{22}c_{21}) & a_{11}(b_{11}c_{12} + b_{12}c_{22}) + a_{12}(b_{21}c_{12} + b_{22}c_{22}) \\ a_{11}(b_{11}c_{11} + b_{12}c_{21}) + a_{22}(b_{21}c_{11} + b_{22}b_{21}) & a_{21}(b_{11}c_{12} + b_{12}c_{22}) + a_{22}(b_{21}c_{12} + b_{22}c_{22}) \end{bmatrix}$$ Distributive Property

$$= \begin{bmatrix} a_{11} & a_{12} \\ a_{21} & a_{22} \end{bmatrix}\begin{bmatrix} b_{11}c_{11} + b_{12}c_{21} & b_{11}c_{12} + b_{12}c_{22} \\ b_{21}c_{11} + b_{22}c_{21} & b_{21}c_{12} + b_{22}c_{22} \end{bmatrix}$$ Definition of matrix multiplication

$$= \begin{bmatrix} a_{11} & a_{12} \\ a_{21} & a_{22} \end{bmatrix}\left(\begin{bmatrix} b_{11} & b_{12} \\ b_{21} & b_{22} \end{bmatrix}\begin{bmatrix} c_{11} & c_{12} \\ c_{21} & c_{22} \end{bmatrix}\right)$$ Definition of matrix multiplication

$$= A(BC)$$ Substitution

47d. $$c(AB) = c\left(\begin{bmatrix} a_{11} & a_{12} \\ a_{21} & a_{22} \end{bmatrix}\begin{bmatrix} b_{11} & b_{12} \\ b_{21} & b_{22} \end{bmatrix}\right)$$ Substitution

$$= c\begin{bmatrix} a_{11}b_{11} + a_{12}b_{21} & a_{11}b_{12} + a_{12}b_{22} \\ a_{21}b_{11} + a_{22}b_{21} & a_{21}b_{12} + a_{22}b_{22} \end{bmatrix}$$ Definition of matrix addition

$$= \begin{bmatrix} c(a_{11}b_{11} + a_{12}b_{21}) & c(a_{11}b_{12} + a_{12}b_{22}) \\ c(a_{21}b_{11} + a_{22}b_{21}) & c(a_{21}b_{12} + a_{22}b_{22}) \end{bmatrix}$$ Definition of scalar multiplication

$$= \begin{bmatrix} ca_{11}b_{11} + ca_{12}b_{21} & ca_{11}b_{12} + ca_{12}b_{22} \\ ca_{21}b_{11} + ca_{22}b_{21} & ca_{21}b_{12} + ca_{22}b_{22} \end{bmatrix}$$ Distributive Property

$$= \begin{bmatrix} ca_{11} & ca_{12} \\ ca_{21} & ca_{22} \end{bmatrix}\begin{bmatrix} b_{11} & b_{12} \\ b_{21} & b_{22} \end{bmatrix}$$ Definition of matrix multiplication

$$= c\begin{bmatrix} a_{11} & a_{12} \\ a_{21} & a_{22} \end{bmatrix}\begin{bmatrix} b_{11} & b_{12} \\ b_{21} & b_{22} \end{bmatrix}$$ Definition of scalar multiplication

$$= (cA)B$$ Substitution

$$c(AB) = c\left(\begin{bmatrix} a_{11} & a_{12} \\ a_{21} & a_{22} \end{bmatrix}\begin{bmatrix} b_{11} & b_{12} \\ b_{21} & b_{22} \end{bmatrix}\right)$$ Substitution

$$= c\begin{bmatrix} a_{11}b_{11} + a_{12}b_{21} & a_{11}b_{12} + a_{12}b_{22} \\ a_{21}b_{11} + a_{22}b_{21} & a_{21}b_{12} + a_{22}b_{22} \end{bmatrix}$$ Definition of matrix multiplication

$$= \begin{bmatrix} c(a_{11}b_{11} + a_{12}b_{21}) & c(a_{11}b_{12} + a_{12}b_{22}) \\ c(a_{21}b_{11} + a_{22}b_{21}) & c(a_{21}b_{12} + a_{22}b_{22}) \end{bmatrix}$$ Definition of scalar multiplication

$$= \begin{bmatrix} ca_{11}b_{11} + ca_{12}b_{21} & ca_{11}b_{12} + ca_{12}b_{22} \\ ca_{21}b_{11} + ca_{22}b_{21} & ca_{21}b_{12} + ca_{22}b_{22} \end{bmatrix}$$ Distributive Property

$$= \begin{bmatrix} a_{11}cb_{11} + a_{12}cb_{21} & a_{11}cb_{12} + a_{12}cb_{22} \\ a_{21}cb_{11} + a_{22}cb_{21} & a_{21}cb_{12} + a_{22}cb_{22} \end{bmatrix}$$ Commutative Property

$$= \begin{bmatrix} a_{11} & a_{12} \\ a_{21} & a_{22} \end{bmatrix}\begin{bmatrix} cb_{11} & cb_{12} \\ cb_{21} & cb_{22} \end{bmatrix}$$ Definition of matrix multiplication

$$= A(cB)$$ Substitution

---

49. $a = 2, b = 1, c = 3, d = 4$ **51.** 12 **53.** H

55. $\begin{bmatrix} 42 & -24 \\ -42 & -31 \end{bmatrix}$ **57.** $\begin{bmatrix} -80 & 44 \\ 68 & -4 \end{bmatrix}$ **59.** (8, 3, 1)

61a. Sample answer: $y = 116.25x - 231{,}176.97$ **61b.** Sample answer: $2950.53 **61c.** The value predicted by the equation is significantly lower than the one given in the graph.

63. translation 4 units right and 3 units up

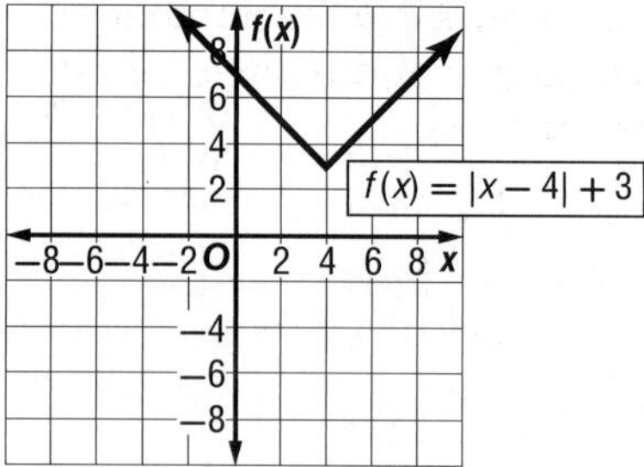

65. translation 2 units left and 6 units down

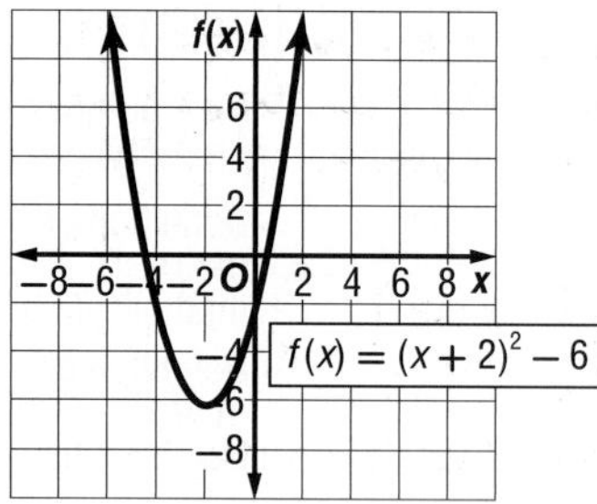

Lesson 3-7

1. 26 **3.** −128

5 Rewrite the first two columns to the right of the determinant. Then find the products of the elements of the diagonals.

$$\left|\begin{array}{ccc}3 & -2 & 2\\ -4 & 2 & -5\\ -3 & 1 & 4\end{array}\right|\begin{array}{cc}3 & -2\\ -4 & 2\\ -3 & 1\end{array}\qquad \begin{array}{l}3(2)(4) = 24\\ -2(-5)(-3) = -30\\ 2(-4)(1) = -8\end{array}$$

$$\left|\begin{array}{ccc}3 & -2 & 2\\ -4 & 2 & -5\\ -3 & 1 & 4\end{array}\right|\begin{array}{cc}3 & -2\\ -4 & 2\\ -3 & 1\end{array}\qquad \begin{array}{l}-3(2)(2) = -12\\ 1(-5)(3) = -15\\ 4(-4)(-2) = 32\end{array}$$

The sum of the first group is 24 + (−30) + (−8) or −14. The sum of the second group is −12 + (−15) + 32 or 5. The first sum minus the second sum is −14 − 5 or −19.

7. −284 **9.** 72 **11.** 182 **13.** (6, −3) **15.** (4, −1) **17a.** 15.75 units^2 **17b.** 482,343.75 mi^2

19. $\left(\frac{66}{7}, -\frac{116}{7}, -\frac{41}{7}\right)$ **21.** (−4, −2, 8)

23. (−1, −3, 7) **25.** (4, 0, 8) **27.** 3 **29.** −135 **31.** −459 **33.** 0 **35.** 728 **37.** −952 **39.** (8, −5)

41 $x = \dfrac{\begin{vmatrix} m & b\\ n & g\end{vmatrix}}{|C|} = \dfrac{\begin{vmatrix} -39 & -5\\ 54 & 8\end{vmatrix}}{\begin{vmatrix} -4 & -5\\ 5 & 8\end{vmatrix}}$

$x = \dfrac{-39(8) - 54(-5)}{-4(8) - 5(-5)}$

$= \dfrac{-312 + 270}{-32 + 25}$

$= \dfrac{-42}{-7}$ or 6

$y = \dfrac{\begin{vmatrix} a & m\\ f & n\end{vmatrix}}{|C|} = \dfrac{\begin{vmatrix} -4 & -39\\ 5 & 54\end{vmatrix}}{\begin{vmatrix} -4 & -5\\ 5 & 8\end{vmatrix}}$

$y = \dfrac{-4(54) - 5(-39)}{-4(8) - 5(-5)}$

$= \dfrac{-216 + 195}{-32 + 25}$

$= \dfrac{-21}{-7}$ or 3

The solution of the system is (6, 3).

43. (−3, −7) **45.** (4, −2, 5) **47.** 6 **49.** 2 m^2

51. (4, 8, −5) **53.** $\left(-\frac{6187}{701}, -\frac{2904}{701}, -\frac{4212}{701}\right)$

55 **a.** Let x = the number of medium drinks. Then $2x$ = the number of small drinks. Let y = the number of large drinks.

number of drinks: $x + 2x + y = 1385 \rightarrow 3x + y = 1385$

total sales: $1.75x + 1.15(2x) + 2.25y = 2238.75$

$\rightarrow 4.05x + 2.25y = 2238.75$

$x = \dfrac{\begin{vmatrix} m & b\\ n & g\end{vmatrix}}{|C|} = \dfrac{\begin{vmatrix} 1385 & 1\\ 2238.75 & 2.25\end{vmatrix}}{\begin{vmatrix} 3 & 1\\ 4.05 & 2.25\end{vmatrix}}$

$x = \dfrac{1385(2.25) - 2238.75(1)}{3(2.25) - 4.05(1)}$

$= \dfrac{3116.25 - 2238.75}{6.75 - 4.05}$

$= \dfrac{877.5}{2.7}$ or 325

$y = \dfrac{\begin{vmatrix} a & m\\ f & n\end{vmatrix}}{|C|} = \dfrac{\begin{vmatrix} 3 & 1385\\ 4.05 & 2238.75\end{vmatrix}}{\begin{vmatrix} 3 & 1\\ 4.05 & 2.25\end{vmatrix}}$

$y = \dfrac{3(2238.75) - 4.05(1385)}{3(2.25) - 4.05(1)}$

$= \dfrac{6716.25 - 5609.25}{6.75 - 4.05}$

$= \dfrac{1107}{2.7}$ or 410

He sold 325 medium drinks, 2(325) or 650 small drinks, and 410 large drinks.

b. total sales = sales of small + sales of medium + sales of large

= 1.25(650 − 140) + 1.75(325 + 125) + 2.25(410 + 35)

= 637.50 + 787.50 + 1001.25

= $2426.25

c. It seems like it was a good move for the vendor. Although he sold 140 fewer small drinks, he sold 125 more medium drinks and 35 more large drinks. On the whole, he made $2,426.25 − $2,238.75 or $187.50 more this week than in the previous week.

57. Sample answer: There is no unique solution of the system. There are either infinite or no solutions.

59. 0 **61.** Sample answer: Given a 2 × 2 system of linear equations, if the determinant of the matrix of coefficients is 0, then the system does not have a unique solution. The system may have no solution and the graphical representation shows two parallel lines. The system may have infinitely many solutions in which the graphical representation will be the same line. **63.** H **65.** D

67. no **69a.** $\begin{bmatrix} \$4.50 & \$6.75 & \$9.50\\ \$4.50 & \$6.75 & \$9.50\\ \$4.00 & \$6.25 & \$8.75\\ \$4.75 & \$7.50 & \$10.25\end{bmatrix}$

69b. $\begin{bmatrix} \$3.60 & \$5.40 & \$7.60 \\ \$3.60 & \$5.40 & \$7.60 \\ \$3.20 & \$5.00 & \$7.00 \\ \$3.80 & \$6.00 & \$8.20 \end{bmatrix}$

69c. $\begin{bmatrix} \$0.90 & \$1.35 & \$1.90 \\ \$0.90 & \$1.35 & \$1.90 \\ \$0.80 & \$1.25 & \$1.75 \\ \$0.95 & \$1.50 & \$2.05 \end{bmatrix}$

71.

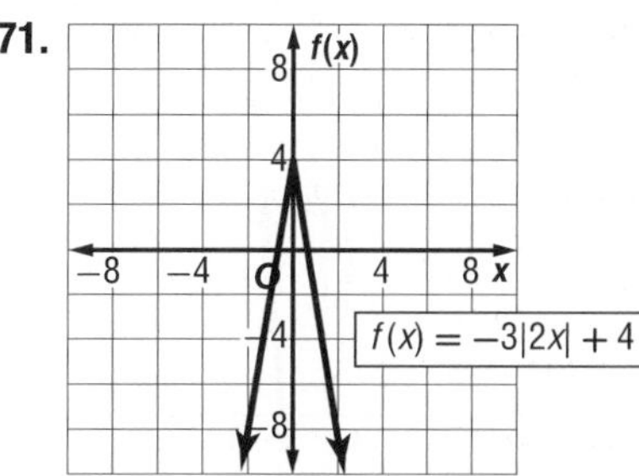

73. (−8, 2) **75.** (−3, −4)

Lesson 3-8

1. no **3.** yes **5.** $\begin{bmatrix} 0 & -1 \\ -\frac{1}{3} & -2 \end{bmatrix}$ **7.** $\begin{bmatrix} -\frac{1}{3} & 0 \\ \frac{5}{6} & \frac{1}{2} \end{bmatrix}$

9. (−2, 5) **11.** (1, −2) **13.** no **15.** no **17.** $\begin{bmatrix} \frac{1}{3} & 0 \\ 0 & \frac{1}{2} \end{bmatrix}$

19. $\begin{bmatrix} \frac{1}{3} & 0 \\ -\frac{5}{3} & 1 \end{bmatrix}$

21 $\begin{vmatrix} -5 & -4 \\ 4 & 2 \end{vmatrix} = -5(2) - 4(-4)$ or 6 Find the determinant.

Since the determinant $\neq 0$, the inverse exists.

$A^{-1} = \frac{1}{ad - bc}\begin{bmatrix} d & -b \\ -c & a \end{bmatrix}$ Definition of inverse

$= \frac{1}{-5(2) - (-4)(4)}\begin{bmatrix} 2 & 4 \\ -4 & -5 \end{bmatrix}$ $a = -5, b = -4, c = 4, d = 2$

$= \frac{1}{6}\begin{bmatrix} 2 & 4 \\ -4 & -5 \end{bmatrix}$ or $\begin{bmatrix} \frac{1}{3} & \frac{2}{3} \\ -\frac{2}{3} & -\frac{5}{6} \end{bmatrix}$ Simplify.

23. $\begin{bmatrix} \frac{9}{74} & \frac{5}{74} \\ -\frac{2}{37} & \frac{3}{37} \end{bmatrix}$ **25.** $\begin{bmatrix} \frac{7}{22} & \frac{4}{11} \\ \frac{4}{11} & \frac{3}{11} \end{bmatrix}$ **27.** no solution

29. (−1, 5) **31.** no solution **33.** (−5, 0)

35. $\left(\frac{3}{4}, 3\right)$

37 **a.** Let x = the number of people who own CD players and let y = the number of people who own digital audio players. The following expressions represent the change in player ownership.
Keeping or switching to CD players: $0.35x + 0.12y$
Keeping or switching to digital audio players: $0.65x + 0.88y$

		From	
		CD	DAP
To	CD	0.35	0.12
	DAP	0.65	0.88

b. Let x = the number of people who will own CD players next year and let y = the number of people who will own digital audio players next year. Write a matrix equation.

$$\begin{bmatrix} 0.35 & 0.12 \\ 0.65 & 0.88 \end{bmatrix} \cdot \begin{bmatrix} 7748 \\ 17{,}252 \end{bmatrix} = \begin{bmatrix} x \\ y \end{bmatrix}$$

$$\begin{bmatrix} 4782 \\ 20{,}218 \end{bmatrix} = \begin{bmatrix} x \\ y \end{bmatrix}$$

So, about 20,218 people will own digital audio players next year.

c. Let x = the number of people who owned CD players last year and let y = the number of people who owned digital audio players last year. Write a matrix equation.

$$\begin{bmatrix} 0.35 & 0.12 \\ 0.65 & 0.88 \end{bmatrix} \cdot \begin{bmatrix} x \\ y \end{bmatrix} = \begin{bmatrix} 7748 \\ 17{,}252 \end{bmatrix}$$

Find the inverse of the coefficient matrix.

A^{-1}

$= \frac{1}{0.35(0.88) - 0.12(0.65)}\begin{bmatrix} 0.88 & -0.12 \\ -0.65 & 0.35 \end{bmatrix}$ $a = 0.35, b = 0.12, c = 0.65, d = 0.88$

$= \frac{1}{0.23}\begin{bmatrix} 0.88 & -0.12 \\ -0.65 & 0.35 \end{bmatrix}$ Simplify.

Multiply each side of the matrix equation by the inverse matrix.

$$\frac{1}{0.23}\begin{bmatrix} 0.88 & -0.12 \\ -0.65 & 0.35 \end{bmatrix} \cdot \begin{bmatrix} 0.35 & 0.12 \\ 0.65 & 0.88 \end{bmatrix} \cdot \begin{bmatrix} x \\ y \end{bmatrix} =$$

$$\frac{1}{0.23}\begin{bmatrix} 0.88 & -0.12 \\ -0.65 & 0.35 \end{bmatrix} \cdot \begin{bmatrix} 7748 \\ 17{,}252 \end{bmatrix}$$

$$\begin{bmatrix} 1 & 0 \\ 0 & 1 \end{bmatrix} \cdot \begin{bmatrix} x \\ y \end{bmatrix} = \frac{1}{0.23}\begin{bmatrix} 4748 \\ 1002 \end{bmatrix}$$

$$\begin{bmatrix} x \\ y \end{bmatrix} = \begin{bmatrix} 20{,}643 \\ 4357 \end{bmatrix}$$

So, about 4357 people owned digital audio players last year.

39. The system would have to consist of two equations that are the same or one equation that is a multiple of the other.

41. Sample answer: $\begin{bmatrix} 2 & 3 \\ 4 & 6 \end{bmatrix} \cdot \begin{bmatrix} x \\ y \end{bmatrix} = \begin{bmatrix} 9 \\ 10 \end{bmatrix}$; any matrix that has a determinant equal to 0, such as $\begin{bmatrix} 1 & 0 \\ 0 & 1 \end{bmatrix}$ **43.** C **45.** $\left(\frac{1}{2}, \frac{1}{4}\right)$ **47.** −54 **49.** 551

51. $\begin{bmatrix} -22 \\ -7 \end{bmatrix}$ **53.** 179 gal of skim and 21 gal of whole milk **55.** absolute value

Chapter 3 Study Guide and Review

1. unbounded **3.** constant matrix **5.** dimensions **7.** identity **9.** inconsistent **11.** (0, 2) **13.** (−3, 4) **15.** 4 h **17.** (−2, −7) **19.** (3, 5)

21.

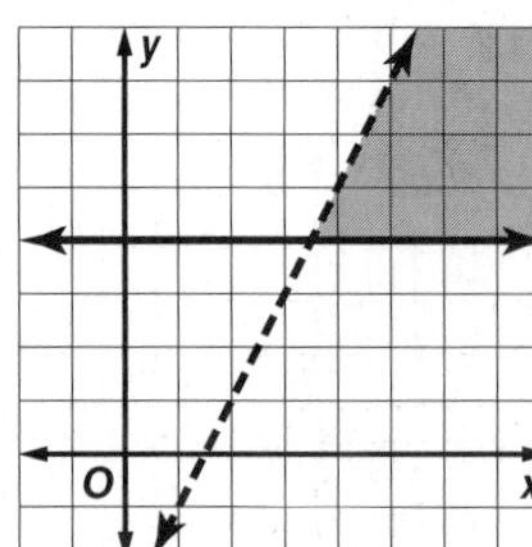

23.

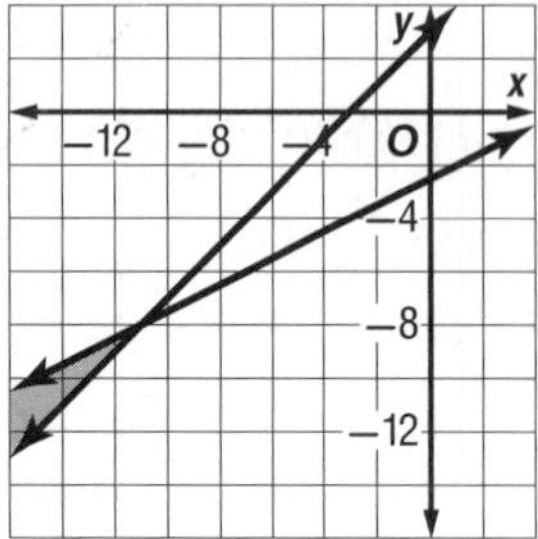

25.

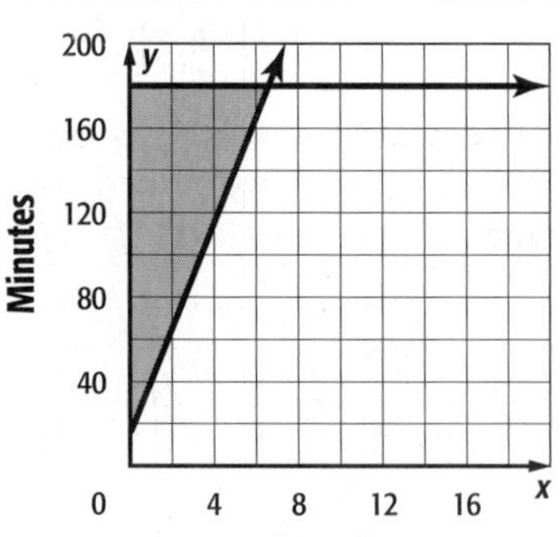

27. \$480; 12 outdoor, 16 indoor **29.** (5, −5, −4)

31. $\begin{bmatrix} -3 & 27 \\ 9 & 12 \end{bmatrix}$ **33a.** buying price: $\begin{bmatrix} 15 \\ 25 \\ 30 \end{bmatrix}$;

33b. selling price: $\begin{bmatrix} 35 \\ 55 \\ 85 \end{bmatrix}$; **33c.** $\begin{bmatrix} 35 \\ 55 \\ 85 \end{bmatrix} - \begin{bmatrix} 15 \\ 25 \\ 30 \end{bmatrix} = \begin{bmatrix} 20 \\ 30 \\ 55 \end{bmatrix}$

35. $\begin{bmatrix} -12 & 17 \\ 82 & 26 \end{bmatrix}$ **37.** \$28.56 **39.** −44 **41.** (2, −3, 6)

43. $\frac{1}{2}\begin{bmatrix} 2 & -4 \\ -3 & 7 \end{bmatrix}$ **45.** does not exist **47.** (2, 1)

CHAPTER 4

Quadratic Functions and Relations

Chapter 4 Get Ready

1. 6 **3.** 4 **5.** 3 **7a.** $f(x) = 9x$ **7b.** 157,680 mi **9.** $(x + 8)(x + 5)$ **11.** $(2x - 1)(x + 4)$ **13.** prime **15.** $(x + 8)$ feet

Lesson 4-1

1a. y-int = 0; axis of symmetry: $x = 0$; x-coordinate = 0

1b.

x	$f(x)$
−2	12
−1	3
0	0
1	3
2	12

1c.

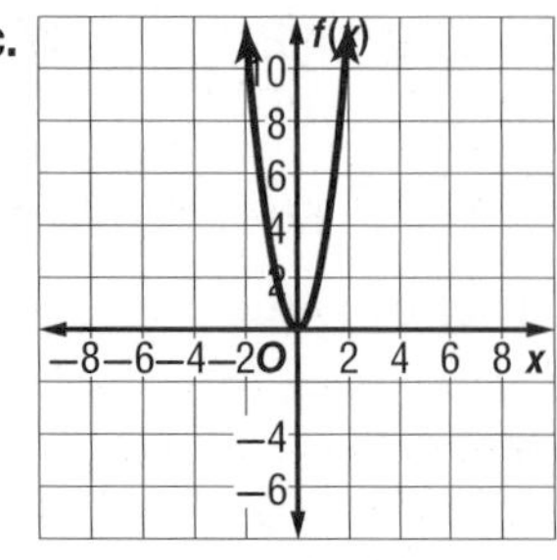

3a. y-int = 0; axis of symmetry: $x = 2$; x-coordinate = 2

3b.

x	$f(x)$
0	0
1	−3
2	−4
3	−3
4	0

3c.

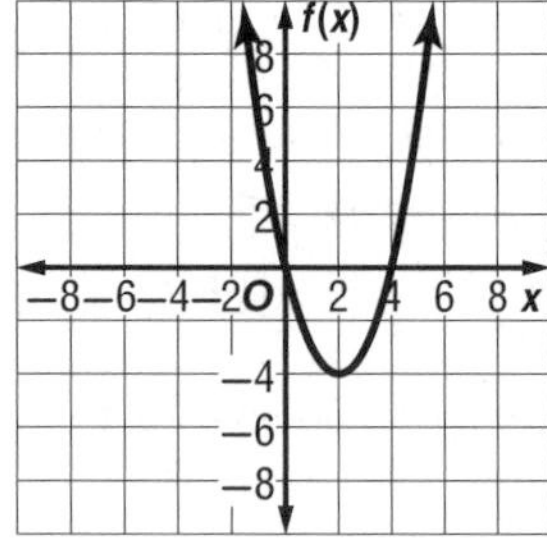

5a. y-int = −3; axis of symmetry: $x = 0.75$; x-coordinate = 0.75

5b.

x	$f(x)$
−1	7
0	−3
0.75	−5.25
1.5	−3
2.5	7

5c.

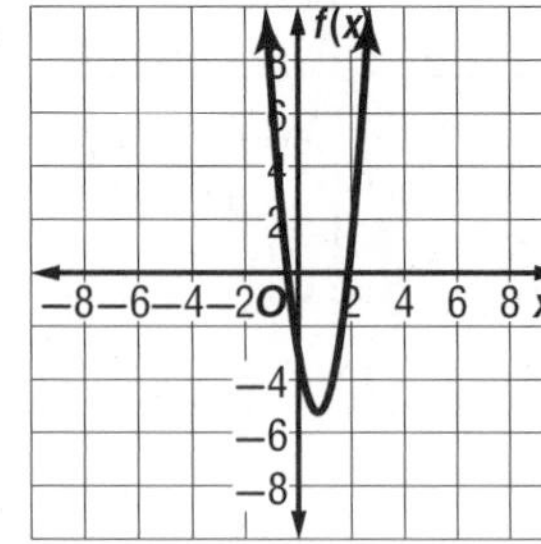

7. max = 8; D = {all real numbers}, R = $\{f(x) \mid f(x) \leq 8\}$

9. min = $-\frac{1}{3}$; D = {all real numbers}, R = $\left\{f(x) \,\middle|\, f(x) \geq -\frac{1}{3}\right\}$ **11.** \$2.88 **13a.** y-int = 0; axis of symmetry: $x = 0$; x-coordinate = 0

13b.

x	$f(x)$
−2	−8
−1	−2
0	0
1	−2
2	−8

13c.

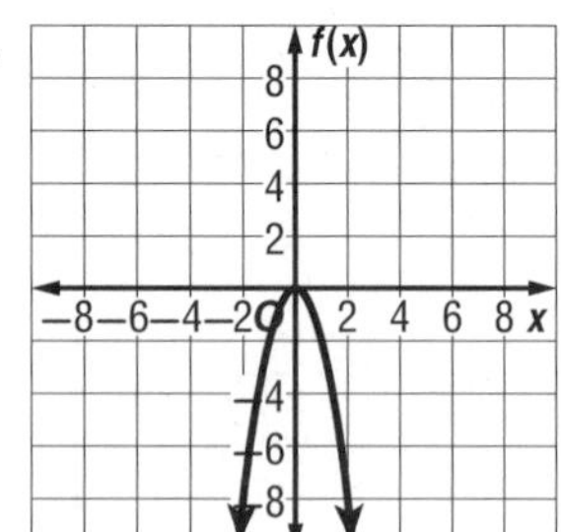

15a. y-int = 3; axis of symmetry: $x = 0$; x-coordinate = 0

15b.

x	$f(x)$
−2	7
−1	4
0	3
1	4
2	7

15c.

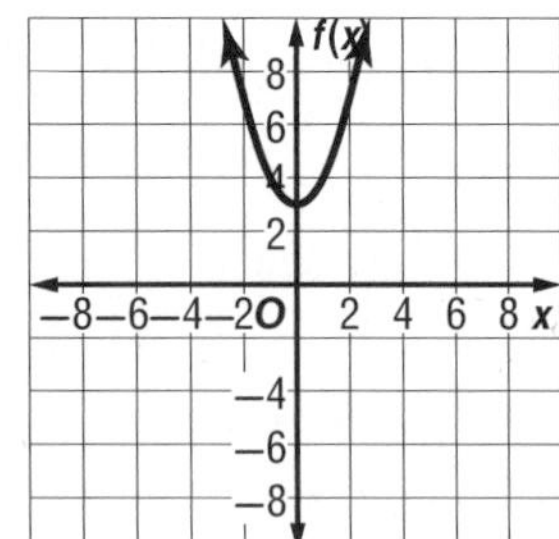

17a. y-int = 5; axis of symmetry: $x = 0$; x-coordinate = 0

17b.

x	$f(x)$
−2	−7
−1	2
0	5
1	2
2	−7

17c.

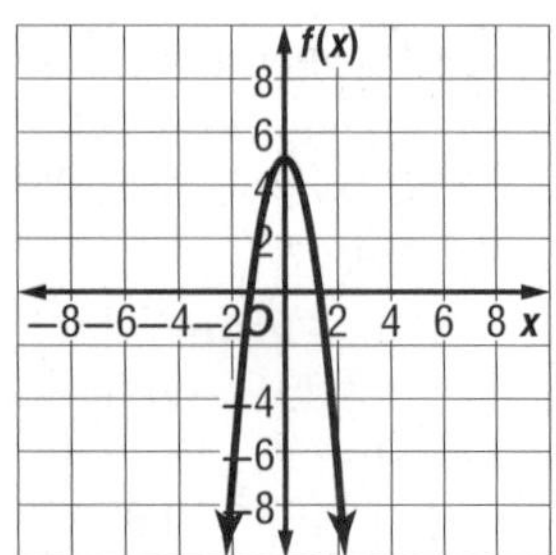

19 a. $f(x) = ax^2 + bx + c$

$\downarrow \quad \downarrow \quad \downarrow$

$f(x) = 1x^2 - 3x - 10 \quad a = 1, b = -3, c = -10$

The y-intercept is $c = -10$.

$x = -\frac{b}{2a}$ Equation of the axis of symmetry

$= -\frac{(-3)}{2(1)}$ $a = 1$ and $b = -3$

$= \frac{3}{2}$ or 1.5 Simplify.

The equation of the axis of symmetry is $x = 1.5$. So, the x-coordinate of the vertex is 1.5.

b. Select five points, with the vertex in the middle and two points on either side of the vertex, including the y-intercept and its reflection.

x	$f(x)$	
0	−10	←reflection of y-intercept
1	−12	
1.5	−12.25	←vertex
2	−12	
3	−10	←y-intercept

c. Graph the five points from the table, connecting them with a smooth curve.

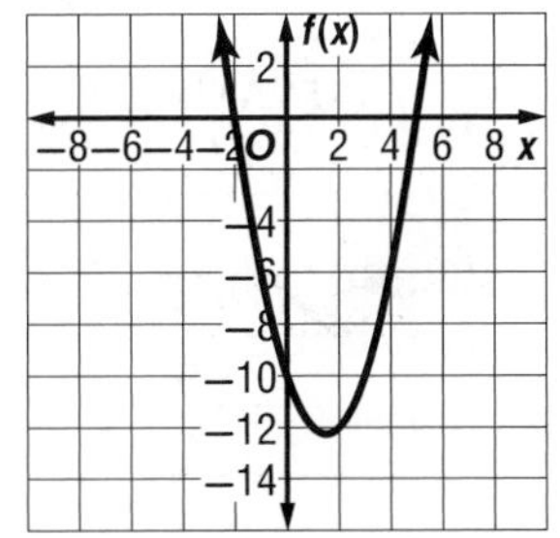

21a. y-int = 9; axis of symmetry: $x = 0.75$; x-coordinate = 0.75

21b.

x	$f(x)$
−1	4
0	9
0.75	10.125
1.5	9
2.5	4

21c.

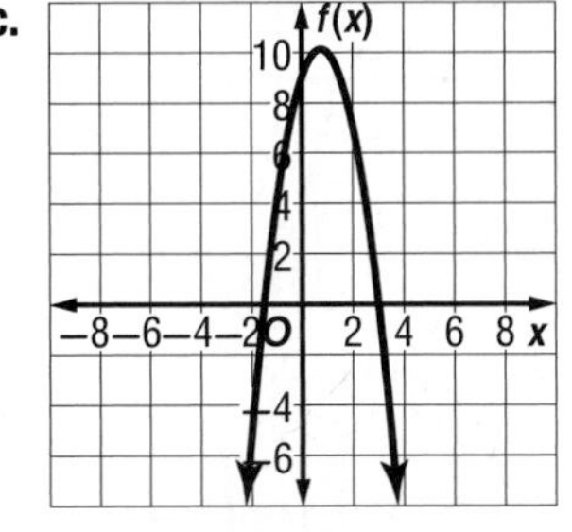

23. max = −12; D = {all real numbers}, R = $\{f(x) \mid f(x) \le -12\}$ **25.** max = 13.25; D = {all real numbers}, R = $\{f(x) \mid f(x) \le 13.25\}$ **27.** max = 7; D = {all real numbers}, R = $\{f(x) \mid f(x) \le 7\}$ **29.** min = −9; D = {all real numbers}, R = $\{f(x) \mid f(x) \ge -9\}$ **31.** min = −74; D = {all real numbers}, R = $\{f(x) \mid f(x) \ge -74\}$ **33a.** y-int = −9; axis of symmetry: $x = 1.5$; x-coordinate of vertex = 1.5

33b.

x	$f(x)$
0	−9
1	−13
1.5	−13.5
2	−13
3	−9

33c.

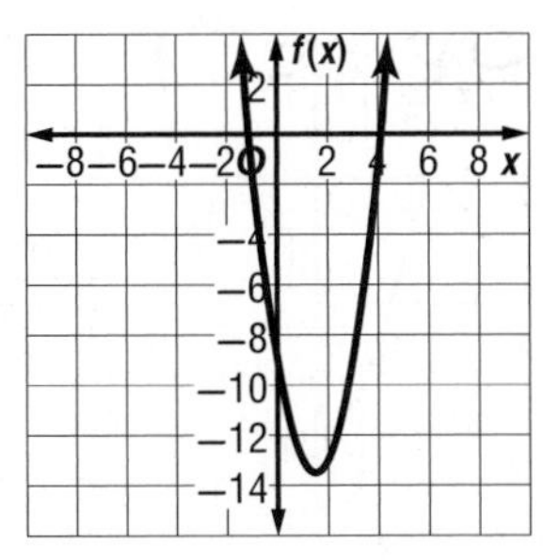

35a. y-int = 0; axis of symmetry: $x = \frac{5}{8}$; x-coordinate of vertex = $\frac{5}{8}$

35b.

x	$f(x)$
$-\frac{3}{4}$	−6
$\frac{1}{4}$	1
$\frac{5}{8}$	1.5625
1	1
2	−6

35c.

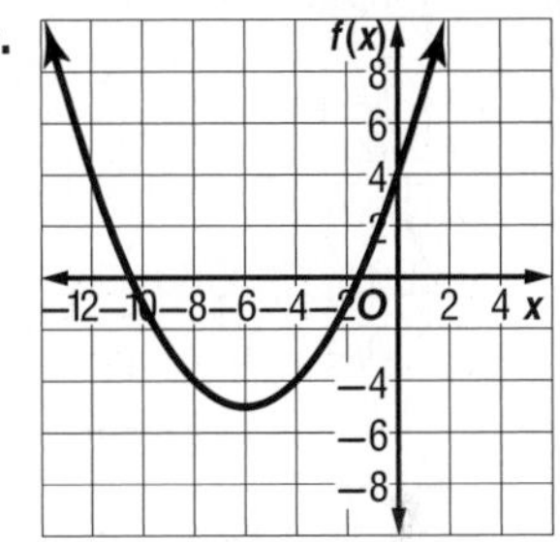

37a. y-int = 4; axis of symmetry: $x = -6$; x-coordinate of vertex = −6

37b.

x	$f(x)$
−10	−1
−8	−4
−6	−5
−4	−4
−2	−1

33c.

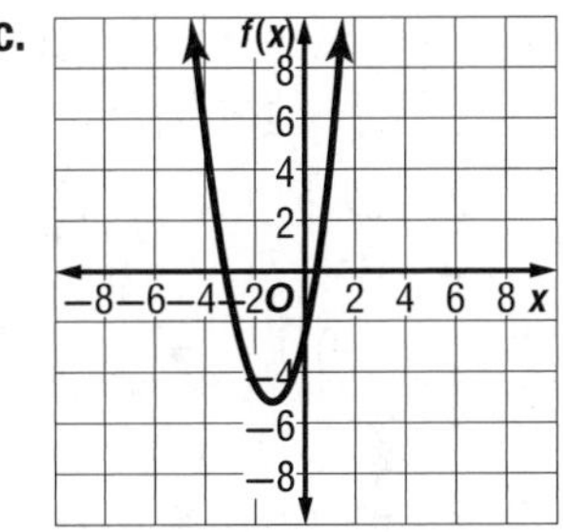

39a. y-int = −2.5; axis of symmetry: $x = -\frac{4}{3}$; x-coordinate of vertex = $-\frac{4}{3}$

39b.

x	$f(x)$
$-\frac{11}{3}$	3
$-\frac{8}{3}$	−2.5
$-\frac{4}{3}$	$-5\frac{1}{6}$
0	−2.5
1	3

39c.

41a. $I(x) = -x^2 + 6x + 475$ **41b.** D = $\{x \mid 0 \le x \le 25\}$, R = $\{y \mid 0 \le y \le 484\}$ **41c.** \$11; Because the function has a maximum at $x = 3$, it is in the domain. Therefore, three \$0.50 increases is reasonable. **41d.** \$484 **43.** max = 23 **45.** max = −0.10 **47.** max = −4.11

49 $a = -5$, so the graph opens down and has a maximum value. The maximum value is the y-coordinate of the vertex.

$x = -\frac{b}{2a}$ Equation of the axis of symmetry

$= -\frac{4}{2(-5)}$ or 0.4 $a = -5$ and $b = 4$

The x-coordinate of the vertex is 0.4. Find the y-coordinate of the vertex by evaluating the function for $x = 0.4$.

$f(x) = -5x^2 + 4x - 8$ Original function

$= -5(0.4)^2 + 4(0.4) - 8$ $x = 0.4$

$= -7.2$ The maximum value of the function is −7.2.

The domain is all real numbers. The range is all real numbers less than or equal to the maximum value, or $\{f(x) \mid f(x) \le -7.2\}$.

51. min = −9.375; D = {all real numbers}, R = $\{f(x) \mid f(x) \ge -9.375\}$ **53.** min = −23.5; D = {all real numbers}, R = $\{f(x) \mid f(x) \ge -23.5\}$

55. $f(x) = x^2 - 4x - 5$ **57.** $f(x) = x^2 - 6x + 8$

59 **a.** Let x = the number of 5-cent increases. Then $65 + 5x$ = price per can in cents and $600 - 100x$ = number of cans. Let $f(x)$ = income as a function of x.

Income = number of cans times the price per can

$f(x) \quad = \quad (600 - 100x) \quad \cdot \quad (65 + 5x)$

Solve for x in the equation.

$f(x) = (600 - 100x) \cdot (65 + 5x)$

$= 600(65) + 600(5x) + (-100x)(65) + (-100x)(5x)$ Distribute.

$= 39{,}000 + 3000x - 6500x - 500x^2$ Multiply.

$= 39{,}000 - 3500x - 500x^2$ Simplify.

So, the equation is $f(x) = 39{,}000 - 3500x - 500x^2$ or $f(x) = -500x^2 - 3500x + 39{,}000$.

$x = -\frac{b}{2a}$ Equation of the axis of symmetry

$x = -\frac{(-3500)}{2(-500)}$ or -3.5 $a = -500$ and $b = -3500$

$f(x) = -500x^2 - 3500x + 39{,}000$ Original function

$= -500(-3.5)^2 - 3500(-3.5) + 39{,}000$ $x = -3.5$

$= 45{,}125$ Simplify.

The coordinates of the vertex are $(-3.5, 45{,}125)$. The y-intercept is 39,000.

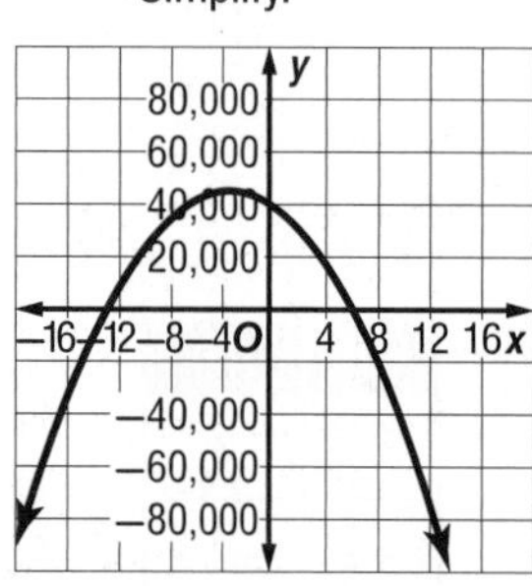

b. Let x = the number of 5-cent decreases. Then $65 - 5x$ = price per can and $600 + 100x$ = number of cans. Let $f(x)$ = income as a function of x.

Income = number of cans times the price per can

$f(x) \quad = \quad (600 + 100x) \quad \cdot \quad (65 - 5x)$

Solve for x in the equation.

$f(x) = (600 + 100x) \cdot (65 - 5x)$

$= 600(65) + 600(-5x) + (100x)(65) + (100x)(-5x)$ Distribute.

$= 39{,}000 - 3000x + 6500x - 500x^2$ Multiply.

$= 39{,}000 + 3500x - 500x^2$ Simplify.

So, the equation is $f(x) = 39{,}000 + 3500x - 500x^2$ or $f(x) = -500x^2 + 3500x + 39{,}000$.

$x = -\frac{b}{2a}$ Equation of the axis of symmetry

$x = -\frac{(3500)}{2(-500)}$ or 3.5 $a = -500$ and $b = 3500$

$f(x) = -500x^2 + 3500x + 39{,}000$ Original function

$= -500(-3.5)^2 + 3500(3.5) + 39{,}000$ $x = -3.5$

$= 45{,}125$ Simplify.

The coordinates of the vertex are (3.5, 45,125). So, Omar should have 3 price decreases and charge $65 - 3(5)$ or 50 cents. Or, he should have 4 price decreases and charge $65 - 4(5)$ or 45 cents.

c. $f(x) = -500x^2 + 3500x + 39{,}000$ Income function

$= -500(3)^2 + 3500(3) + 39{,}000$ Replace x with 3.

$= 45{,}000$ Simplify.

So, his income would be 45,000 cents or $450 per week.

61. Madison; sample answer: $f(x)$ has a maximum of -2. $g(x)$ has a maximum of 1. **63a.** $a = 22$; $b = 26$; $c = -6$; $d = 2$ **63b.** 0 **63c.** maximum **65.** Sample answer: A function is quadratic if it has no other terms than a quadratic term, linear term, and constant term. The function has a maximum if the coefficient of the quadratic term is negative and has a minimum if the coefficient of the quadratic term is positive. **67.** K **69.** C

71. $\begin{bmatrix} -\frac{1}{4} & -\frac{1}{24} \\ 0 & \frac{1}{6} \end{bmatrix}$ **73.** 45 **75.** 0 **77.** No; it cannot be written as $y = mx + b$. **79.** Yes; it is written in $y = mx + b$ form, $m = 0$. **81.** -13

Lesson 4-2

1. no real solution

3. -4

5 Graph the related function $f(x) = x^2 - 3x - 18$. The equation of the axis of symmetry is $x = -\frac{(-3)}{2(1)}$ or 1.5. Make a table using x-values around 1.5. Then graph each point.

x	−3	0	1.5	3	6
f(x)	0	−18	−20.25	−18	0

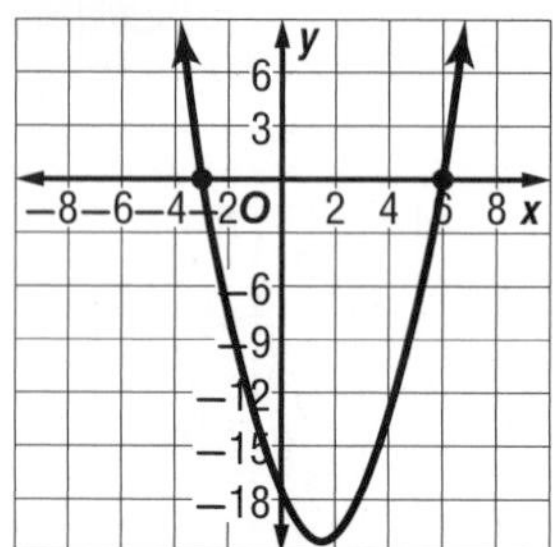

The zeros of the function are −3 and 6. So, the solutions of the equation are −3 and 6.

7.

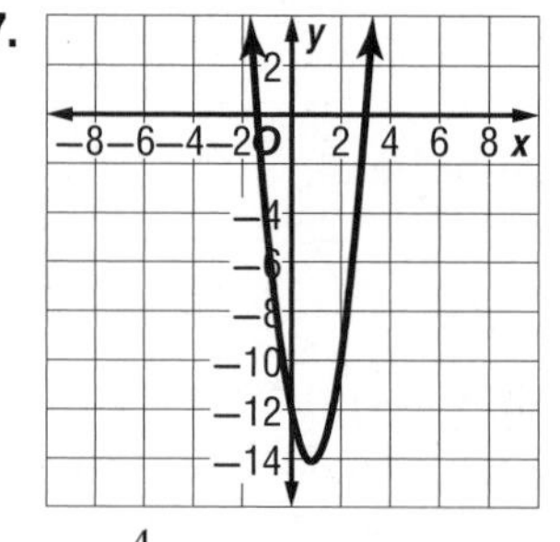

3, $-\frac{4}{3}$

9.

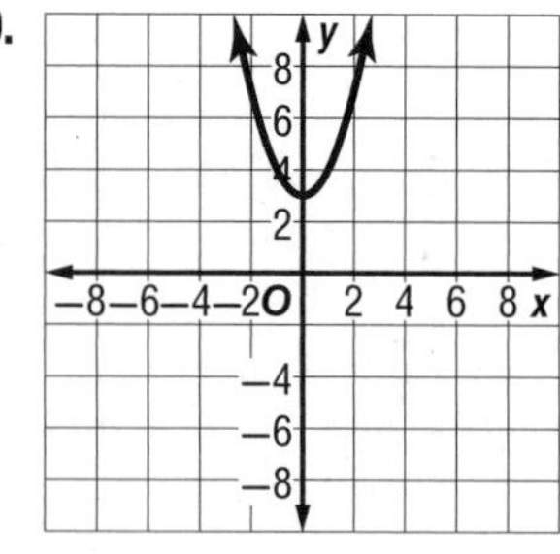

no real solution

11.

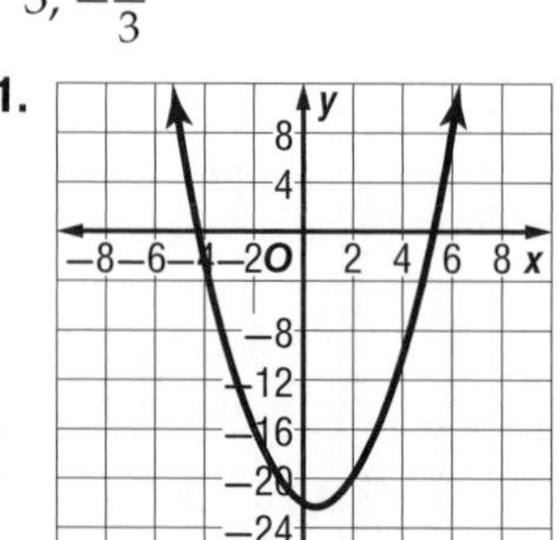

between −5 and −4, between 5 and 6

13. 5 seconds **15.** no real solution **17.** −2 **19.** −3, 4

21.

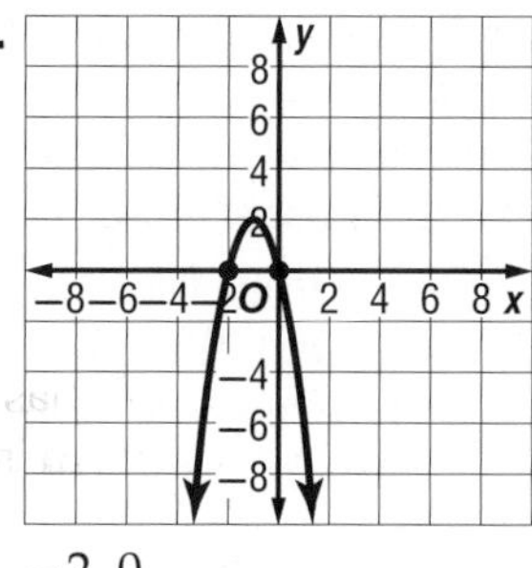

−2, 0

23.

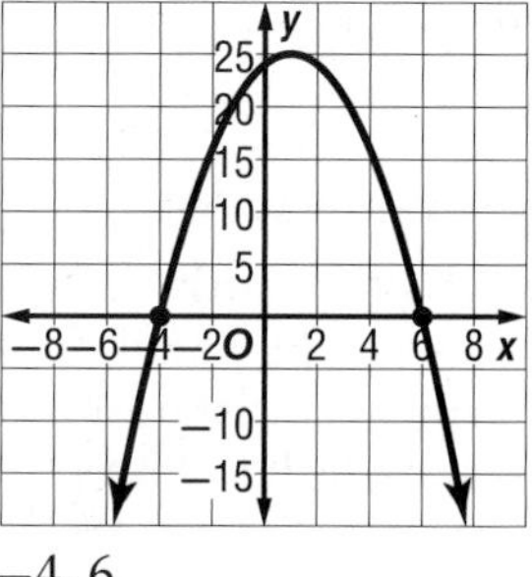

−4, 6

25.

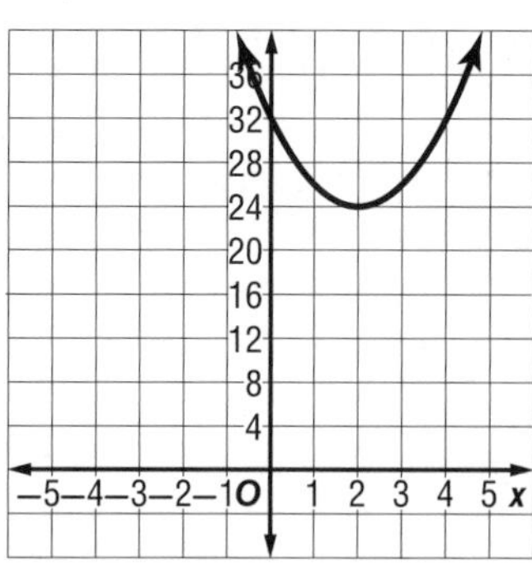

no real solution

27.

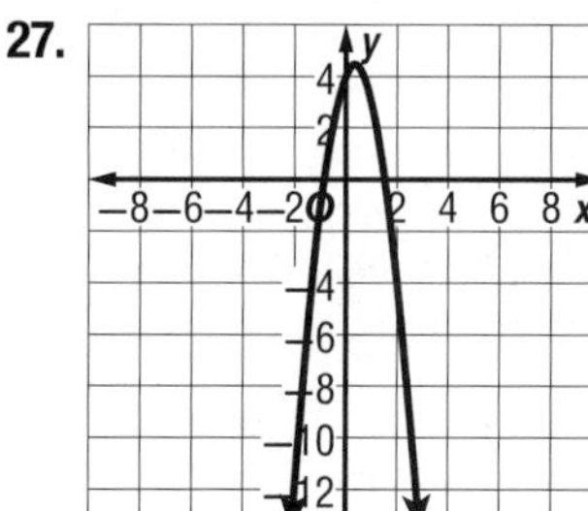

between −1 and 0, between 1 and 2

29.

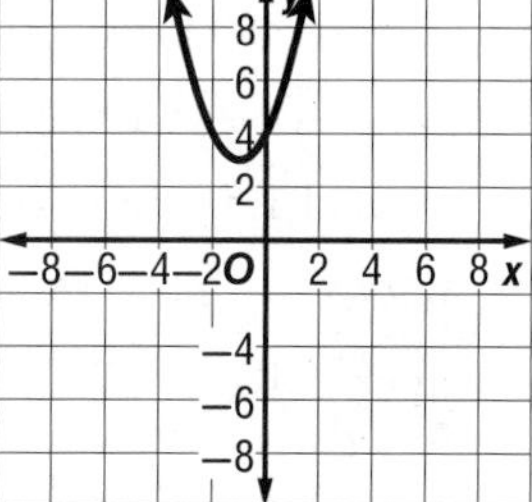

no real solution

31. between 0 and 1; between 2 and 3

33 Let x = one of the numbers. Then $-15 - x$ = the other number.

$x(-x - 15) = -54$ The product is −54.
$-x^2 - 15x = -54$ Distributive Property
$-x^2 - 15x + 54 = 0$ Add 54 to each side.

Graph the related function $f(x) = -x^2 - 15x + 54$. The equation of the axis of symmetry is $x = -\frac{(-15)}{2(-1)}$ or −7.5. Make a table using x-values around −7.5. Then graph each point.

x	−20	−15	−10	−7.5	−5	0	5
$f(x)$	−46	54	104	110.25	104	54	−46

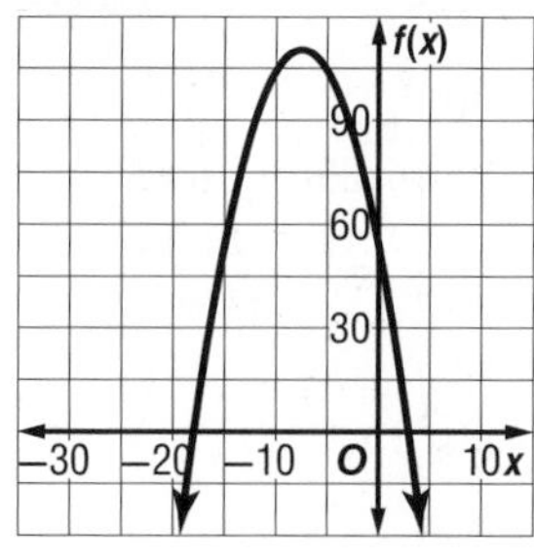

The zeros of the function are −18 and 3. So, the numbers are −18 and 3.

35. about −5.0 and 17.0 **37.** 11 and −19
39. about 3.4375 seconds

41.

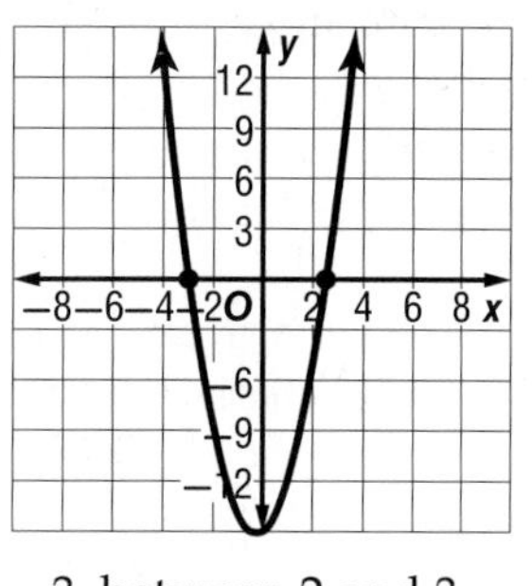

−3, between 2 and 3

43.

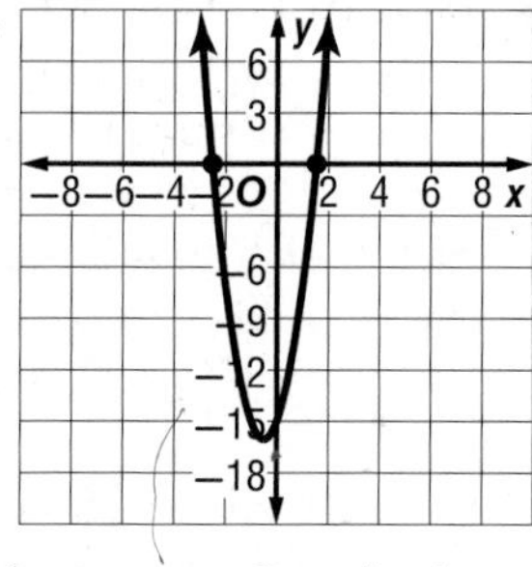

between −3 and −2, between 1 and 2

45.

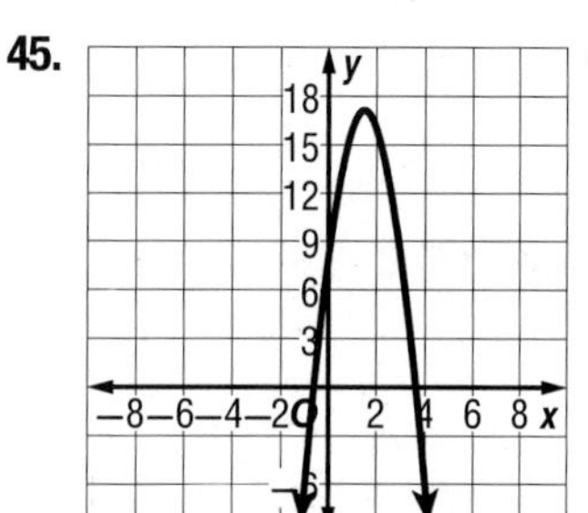

between −1 and 0, between 4 and 5

47.

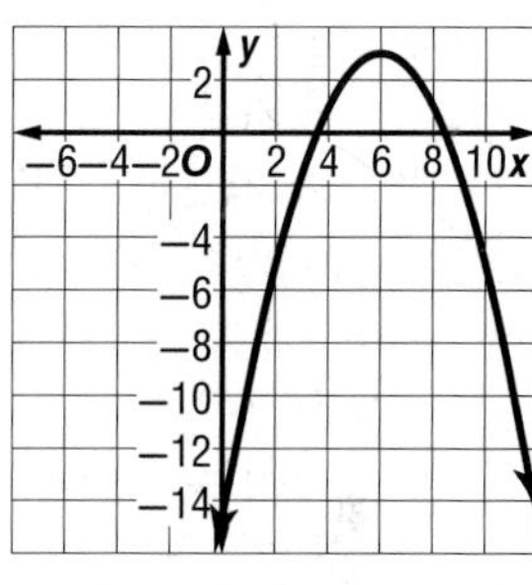

between 3 and 4, between 8 and 9

49 Find t when $h_0 = 60$ and $h(t) = 0$.

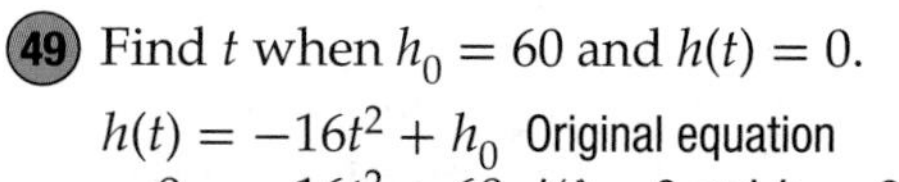

$h(t) = -16t^2 + h_0$ Original equation
$0 = -16t^2 + 60$ $h(t) = 0$ and $h_0 = 60$

Graph the related function $f(t) = -16t^2 + 60$.

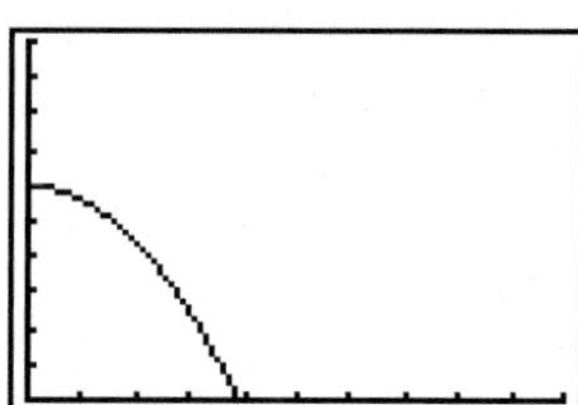

[0, 5] scl: 0.5 by [0, 100] scl: 10

Use the **Zero** feature in the **CALC** menu to find the positive zero of the function, since time cannot be negative. $x \approx 1.94$, so it would take the balloon about 1.94 seconds to hit the ground. Tony's brother should be 4.4 ft/s • 1.94 s or about 8.5 feet from the target when Tony lets go of the balloon.

51. 25 seconds **53.** $k = 8$ **55.** $f(x) = -5x^2 + 30x + 80$

57. $\frac{1}{20}$ **59.** H **61.** maximum, −12; D = {all real numbers}, R = $\{f(x) \mid f(x) \leq -12\}$

63. maximum, 15; D = {all real numbers}, R = $\{f(x) \mid f(x) \leq 15\}$ **65.** no **67a.** $\begin{bmatrix} 10 & 10 & 15 \\ 25 & 35 & 45 \end{bmatrix}$

67b. 2; $\begin{bmatrix} 20 & 20 & 30 \\ 50 & 70 & 90 \end{bmatrix}$ **67c.** $\begin{bmatrix} 10 & 10 & 15 \\ 25 & 35 & 45 \end{bmatrix}$; the number of additional shirts that he needs to stock

69. (−3, 4) **71.** $x \leq -\frac{8}{3}$ **73.** $x \leq \frac{19}{9}$ **75.** 1

$= 0$ **3.** $6x^2 - 11x - 10 = 0$
$+ 4)$ **7.** $(x - 7)(x + 3)$ **9.** $(4x - 3)(4x - 1)$
3. 0, 9 **15.** 6 **17.** $x^2 - 14x + 49 = 0$
$- 31x + 6 = 0$ **21.** $17c(3c^2 - 2)$
$+ 2)(x - 2)$

25 $48cg + 36cf - 4dg - 3df$ Original expression
$= (48cg + 36cf) + (-4dg - 3df)$ Group terms with common factors.
$= 12c(4g + 3f) + (-d)(4g + 3f)$ Factor the GCF from each group.
$= (12c - d)(4g + 3f)$ Distributive Property

27. $(x - 11)(x + 2)$ **29.** $(5x - 1)(3x + 2)$
31. $3(2x - 1)(3x + 4)$ **33.** $(3x + 5)(3x - 5)$
35. $-\frac{2}{5}, 6$ **37.** 0, 9 **39.** 8, −3
41. 2, −2 **43.** $5, \frac{3}{4}$ **45.** 24 and 26 or −24 and −26
47. $x = 20$; 24 in. by 18 in. **49.** $-\frac{1}{2}, \frac{5}{6}$ **51.** $-\frac{3}{2}$
53. 6, −6

55 To find the number of movie screens that produces a profit, first find the number of movie screens in which the profit is zero. Solve $-x^2 + 48x - 512 = 0$.
$ac = -1(-512)$ or 512
$m = 16$; $p = 32$ $mp = 512$ and $ac = 512$; $m + p = 48$ and $b = 48$
$-x^2 + 16x + 32x - 512 = 0$ Write the pattern.
$(-x^2 + 16x) + (32x - 512) = 0$ Group terms with common factors.
$-x(x - 16) + 32(x - 16) = 0$ Group terms with common factors.
$(-x + 32)(x - 16) = 0$ Distributive Property
$-x + 32 = 0$ or $x - 16 = 0$ Zero Product Property
$x = 32$ $x = 16$ Solve each equation.
The solutions are 16 and 32. When there are 16 or 32 movie screens, the profit is zero. Since a is negative, the graph of the function opens down and has a maximum value. So, $P(x)$ is nonnegative for $16 \leq x \leq 32$. When there are 16 to 32 movie screens, the company will not lose money.

57. $25x^2 - 100x + 51 = 0$ **59.** $-3, \frac{1}{2}$ **61.** $1, -\frac{5}{4}$
63. $-\frac{3}{2}, \frac{5}{6}$ **65.** $x^2 - 6^2$; $(x + 6)(x - 6)$ **67.** 20 in. by 15 in. **69.** 13 cm **71.** $2(3 - 4y)(3a + 8b)$

73 $6a^2b^2 - 12ab^2 - 18b^3$ Original expression
$= 6b^2(a^2 - 2a - 3b)$ Factor the GCF, $6b^2$.

75. $2(2x - 3y)(8a + 3b)$ **77.** $(x + y)(x - y)(5a + 2b)$
79. Sample answer: Neither is correct; Gwen didn't have like terms in the parentheses in the third line; Morgan made a sign error in the fourth line.
81. $5x^2(2x - 3y)(4x^2 + 6xy + 9y^2)$
83. Sample answer:
$(x - p)(x - q) = 0$ Original equation
$x^2 - px - qx + pq = 0$ Multiply.
$x^2 - (p + q)x + pq = 0$ Simplify.
$x = -\frac{b}{2a}$ Formula for axis of symmetry
$x = -\frac{-(p + q)}{2(1)}$ $a = 1$ and $b = -(p + q)$
$x = \frac{p + q}{2}$ Simplify.
x is midway between p and q. Definition of midpoint

85. Sample answer: Always; in order to factor using perfect square trinomials, the coefficient of the linear term, bx, must be a multiple of 2, or even. **87.** 192 square units **89.** H **91.** −2, 4 **93.** −2

95.

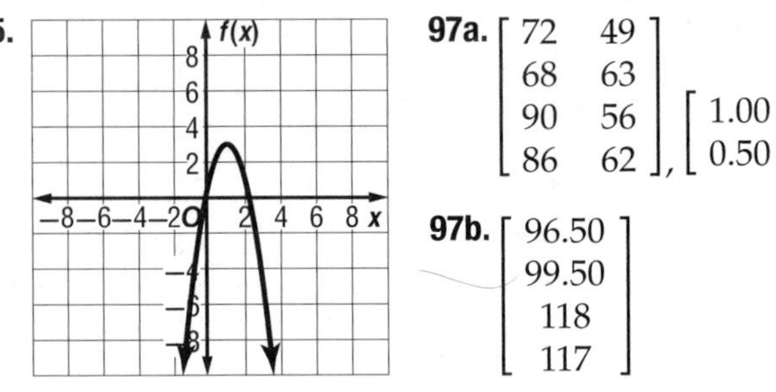

97a. $\begin{bmatrix} 72 & 49 \\ 68 & 63 \\ 90 & 56 \\ 86 & 62 \end{bmatrix}, \begin{bmatrix} 1.00 \\ 0.50 \end{bmatrix}$

97b. $\begin{bmatrix} 96.50 \\ 99.50 \\ 118 \\ 117 \end{bmatrix}$

97c. juniors **97d.** $431 **99.** 16

Lesson 4-4

1. $9i$ **3.** 12 **5.** 1 **7.** $\pm 2i\sqrt{2}$ **9.** 3, −2 **11.** $-3 + 2i$
13. $70 - 60i$ **15.** $\frac{1}{2} - \frac{1}{2}i$ **17.** $12 + 6j$ amps **19.** $13i$
21. $9i$ **23.** $-144i$ **25.** i **27.** −7 **29.** 9 **31.** $30 + 16i$
33. $1 + i$ **35.** $\frac{1}{3} - \frac{5}{3}i$

37 $3x^2 + 48 = 0$ Original equation
$3x^2 = -48$ Subtract 48 from each side.
$x^2 = -16$ Divide each side by 3.
$x = \pm\sqrt{-16}$ Square Root Property
$x = \pm 4i$ $\sqrt{-16} = \sqrt{16} \cdot \sqrt{-1}$ or $4i$

39. $\pm i\sqrt{5}$ **41.** $\pm 4i$ **43.** 2, −3 **45.** $\frac{4}{3}, 4$ **47.** 25, −2
49. $4i$ **51.** 8 **53.** $-21 + 15i$ **55.** $\frac{15}{13} + \frac{16}{13}i$
57. $11 + 23i$ **59.** $\frac{1}{7} - \frac{4\sqrt{3}}{7}i$

61 $V = C \cdot I$ Electricity formula
$= (3 + 6j) \cdot (5 - j)$ $C = 3 + 6j$ and $I = 5 - j$
$= 3(5) + 3(-j) + 6j(5) + 6j(-j)$ FOIL Method
$= 15 - 3j + 30j - 6j^2$ Multiply.
$= 15 - 3j + 30j - 6(-1)$ $j^2 = -1$
$= 21 + 27j$ Simplify.
The voltage is $21 + 27j$ volts.

63. $(3 + i)x^2 + (-2 + i)x - 8i + 7$
65a. Sample answer: $x^2 + 9 = 0$
65b. $f(x) = x^2 + 9$
65c. Sample answer: $x^2 - 4x + 5 = 0$

65d.

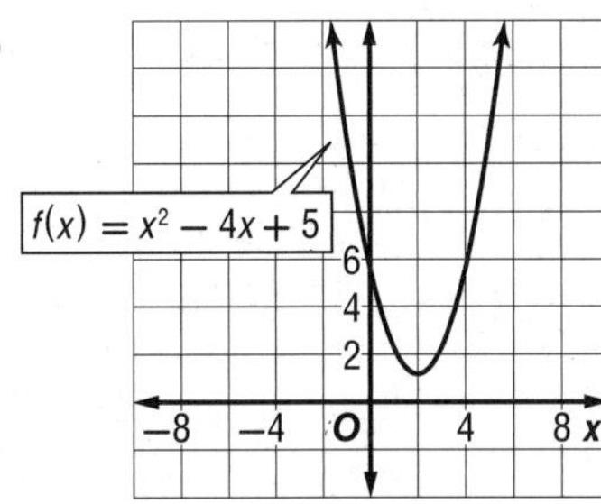

65e. Sample answer: A quadratic equation will have only complex solutions when the graph of the related function has no x-intercepts.
67. $-11 - 2i$ **69.** Sample answer: $(4 + 2i)(4 - 2i)$
71a. $\triangle CBE \cong \triangle ADE$ **71b.** $\angle AED \cong \angle CEB$ (Vertical angles) $\overline{DE} \cong \overline{BE}$ (Both have length x.) $\angle ADE \cong \angle CBE$ (Given) Consecutive angles and the included side are all congruent, so the triangles are congruent by the ASA Property. **71c.** $\overline{EC} \cong \overline{EA}$ by CPCTC (corresponding parts of congruent triangles are congruent.) $EA = 7$, so $EC = 7$. **73.** J **75.** $-5, \frac{3}{2}$
77. $-\frac{1}{2}, \frac{4}{3}$ **79.** 3, 16 **81.** −9, −12
83.

Hours Raking Leaves
$x + y = 15$
$10x + 12y = 120$
Hours Cutting Grass

85. yes
87. no
89. yes

Lesson 4-5

1. $\{-8.45, -3.55\}$ **3.** $\{-12.87, -5.13\}$ **5.** 25 ft
7. 6.25; $(x - 2.5)^2$ **9.** $\{2 - i\sqrt{5}, 2 + i\sqrt{5}\}$
11. $\{-4.37, 1.37\}$ **13.** $\{-6.45, -1.55\}$ **15.** $\{-1.47, 7.47\}$
17. $\{-7.65, -2.35\}$ **19.** $\{-1, 3\}$ **21.** $\{4.67, 10.33\}$
23. $\{-0.95, 3.95\}$ **25.** $\{4, 5\}$ **27.** 64; $(x + 8)^2$
29. 20.25; $(x + 4.5)^2$ **31.** $\{-4.61, 2.61\}$ **33.** $\{1, 3\}$
35. $\left\{\frac{3 - i\sqrt{31}}{4}, \frac{3 + i\sqrt{31}}{4}\right\}$

37 $3x^2 - 6x - 9 = 0$ Original equation
$x^2 - 2x - 3 = 0$ Divide by the coefficient of the quadratic term, 3.
$x^2 - 2x = 3$ Add 3 to each side.
$x^2 - 2x + 1 = 3 + 1$ Since $\left(\frac{-2}{2}\right)^2 = 1$, add 1 to each side.
$(x - 1)^2 = 4$ Write the left side as a perfect square.
$x - 1 = \pm 2$ Square Root Property
$x = \pm 2 + 1$ Add 1 to each side.
$x = 2 + 1$ or $x = -2 + 1$ Write as two equations.
$= 3$ $= -1$ Simplify.
The solution set is $\{-1, 3\}$ or $\{x \mid x = -1, 3\}$.
39. $\{-2 - i\sqrt{7}, -2 + i\sqrt{7}\}$ **41.** $\{5 - 2i, 5 + 2i\}$
43. $\left\{\frac{7 - i\sqrt{47}}{4}, \frac{7 + i\sqrt{47}}{4}\right\}$ **45.** $\{2.65 - i\sqrt{1.5775}, 2.65 + i\sqrt{1.5775}\}$ **47.** $\{-0.89, 5.39\}$ **49.** $\{2.38, 4.62\}$

51. $\{-1.26, 0.26\}$
53 **a.** The time in which the firework explodes is the t-coordinate of the vertex.
$x = -\frac{b}{2a}$ Equation of the axis of symmetry
$x = -\frac{(25)}{2(-1.5)}$ or $8\frac{1}{3}$ $a = -1.5$ and $b = 25$
So, the firework explodes after $8\frac{1}{3}$ seconds.
b. The time in which the firework explodes is the d-coordinate of the vertex.
$d = -1.5t^2 + 25t$ Original function
$d = -1.5\left(8\frac{1}{3}\right)^2 + 25\left(8\frac{1}{3}\right)$ $t = 8\frac{1}{3}$
$d \approx 104.2$ Simplify.
So, the firework explodes at a height of about 104.2 feet.
55. 2.56; $(x - 1.6)^2$
57a.

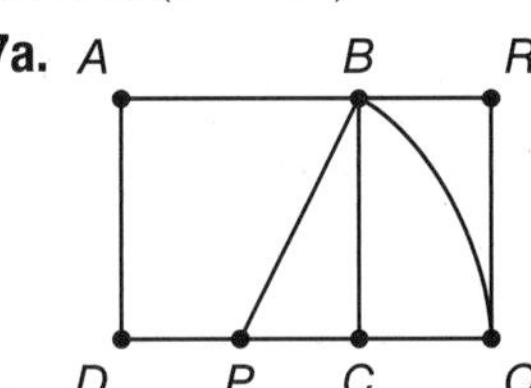

57b. $x = \frac{1 + \sqrt{5}}{2}$
57c.

CQ	x
2	$1 + \sqrt{5}$
3	$\frac{3 + 3\sqrt{5}}{2}$
4	$2 + 2\sqrt{5}$

57d. Sample answer: the x-values are multiples of $\frac{1 + \sqrt{5}}{2}$; $x = \frac{n(1 + \sqrt{5})}{2}$. **59.** $x = \frac{-b}{2} \pm \sqrt{\frac{b^2}{4} - c}$
61. Sample answer: $x^2 - \frac{2}{3}x + \frac{1}{9} = \frac{1}{4}$; $\left\{\frac{5}{6}, -\frac{1}{6}\right\}$
63. E **65.** 125 **67.** $39 + 80i$ **69.** $\frac{4}{39} - \frac{19}{39}i$
71. $5x^2 - 28x - 12 = 0$ **73.** 1 \$100, 3 \$50, and 6 \$20 checks
75.

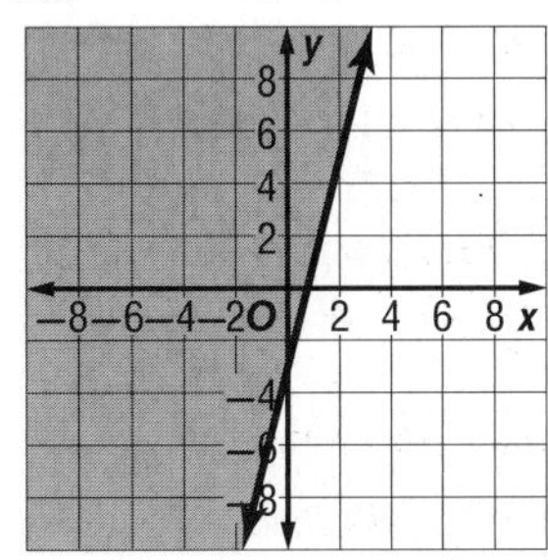

77.

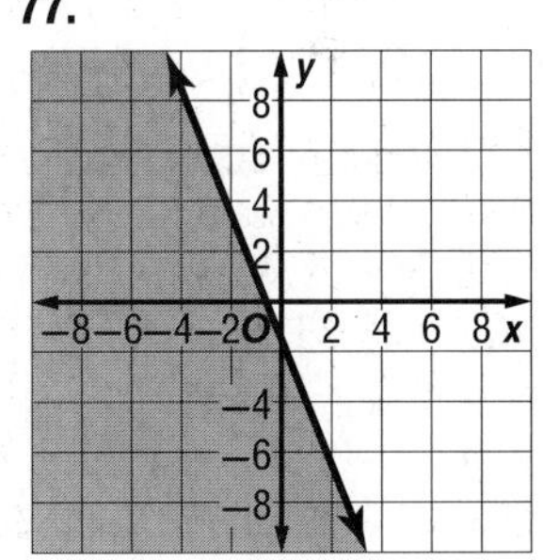

79. $f(x) = \begin{cases} -8 \text{ if } x \leq -2 \\ 2x - 6 \text{ if } -2 < x \leq 5 \\ 2x - 18 \text{ if } x > 5 \end{cases}$ **81.** −4 **83.** −136

Lesson 4-6

1 $x = \frac{-b \pm \sqrt{b^2 - 4ac}}{2a}$ Quadratic Formula
$= \frac{-12 \pm \sqrt{12^2 - 4(1)(-9)}}{2(1)}$ $a = 1$, $b = 12$, and $c = -9$
$= \frac{-12 \pm \sqrt{144 + 36}}{2}$ Multiply.
$= \frac{12 \pm \sqrt{180}}{2}$ Simplify.
$= \frac{-12 \pm 6\sqrt{5}}{2}$ $\sqrt{180} = 6\sqrt{5}$

$x = \frac{-12 + 6\sqrt{5}}{2}$ or $x = \frac{-12 - 6\sqrt{5}}{2}$ Write as two equations.

$= -6 + 3\sqrt{5}$ $= -6 - 3\sqrt{5}$ Simplify.

The solutions are $-6 + 3\sqrt{5}$ and $-6 - 3\sqrt{5}$.

3. $\left(\frac{5 + \sqrt{57}}{8}, \frac{5 - \sqrt{57}}{8}\right)$ **5.** (1.5, −0.2)

7. $\left(\frac{2 + 2\sqrt{7}}{3}, \frac{2 - 2\sqrt{7}}{3}\right)$ **9.** about 0.78 second

11a. −36 **11b.** 2 complex roots **13a.** −76

13b. 2 complex roots **15.** $\frac{-3 \pm \sqrt{15}}{2}$ **17.** $\frac{-7 \pm \sqrt{129}}{8}$

19. $\frac{-3 \pm i\sqrt{71}}{8}$

21 **a.** $b^2 - 4ac = 3^2 - 4(2)(-3)$ $a = 2, b = 3, c = -3$

$= 9 + 24$ or 33 Simplify.

b. The discriminate is positive and not a perfect square. So, there are 2 irrational roots.

c. $x = \frac{-b \pm \sqrt{b^2 - 4ac}}{2a}$ Quadratic Formula

$= \frac{-3 \pm \sqrt{3^2 - 4(2)(-3)}}{2(2)}$ $a = 2$, $b = 3$, and $c = -3$

$= \frac{-3 \pm \sqrt{9 + 24}}{4}$ Multiply.

$= \frac{-3 \pm \sqrt{33}}{4}$ Simplify.

23a. 49 **23b.** 2 rational **23c.** $\frac{1}{6}$, −1 **25a.** −87

25b. 2 complex **25c.** $\frac{3 \pm i\sqrt{87}}{6}$ **27a.** 36

27b. 2 rational **27c.** 1, $-\frac{1}{5}$ **29a.** 1 **29b.** 2 rational

29c. −1, $-\frac{4}{3}$ **31a.** −16 **31b.** 2 complex **31c.** $-1 \pm 2i$

33a. 0 **33b.** about 2.3 seconds **35a.** 64

35b. 2 rational **35c.** 0, $-\frac{8}{5}$ **37a.** 160 **37b.** 2 irrational

37c. $\frac{-1 \pm \sqrt{10}}{6}$ **39a.** 13.48 **39b.** 2 irrational

39c. $\frac{-0.7 \pm \sqrt{3.37}}{0.6}$

41 **a.** $y = -0.26x^2 - 0.55x + 91.81$ Original equation

$= -0.26(10)^2 - 0.55(10) + 91.81$ Replace x with 10.

$= 60.31$ Simplify.

$y = -0.26x^2 - 0.55x + 91.81$ Original equation

$= -0.26(15)^2 - 0.55(15) + 91.81$ Replace x with 15.

$= 25.06$ Simplify.

For 2010, the number of deaths per 100,000 is 60.31. For 2015, the number is 25.06.

b. $y = -0.26x^2 - 0.55x + 91.81$ Original equation

$50 = -0.26x^2 - 0.55x + 91.81$ Replace y with 50.

$0 = -0.26x^2 - 0.55x + 41.81$ Subtract 50 from each side.

$x = \frac{-b \pm \sqrt{b^2 - 4ac}}{2a}$ Quadratic Formula

$= \frac{-(-0.55) \pm \sqrt{(-0.55)^2 - 4(-0.26)(41.81)}}{2(-0.26)}$ $a = -0.26$, $b = -0.55$, and $c = 41.81$

$= \frac{0.55 \pm \sqrt{43.7894}}{-0.52}$ Simplify.

$x \approx -13.8$ or $x \approx 11.7$

Since the number of years after 2000 cannot be negative, the solution is 11.58. So, 11.7 years after 2000, or in 2011, the death rate will be 50 per 100,000.

c. $y = -0.26x^2 - 0.55x + 91.81$ Original equation

$0 = -0.26x^2 - 0.55x + 91.81$ Replace y with 0.

$x = \frac{-b \pm \sqrt{b^2 - 4ac}}{2a}$ Quadratic Formula

$= \frac{-(-0.55) \pm \sqrt{(-0.55)^2 - 4(-0.26)(91.81)}}{2(-0.26)}$ $a = -0.26$, $b = -0.55$, and $c = 91.81$

$= \frac{0.55 \pm \sqrt{95.7849}}{-0.52}$ Simplify.

$x \approx -19.88$ or $x \approx 17.76$

Since the number of years after 2000 cannot be negative, the solution is 17.76. So, 17.76 years after 2000, or in 2017, the death rate will be 0 per 100,000. Sample answer: This prediction is not reasonable because the death rate from cancer will never be 0 unless a cure is found. If and when a cure will be found cannot be predicted.

43. Jonathan is correct; you must first write the equation in the form $ax^2 + bx + c = 0$ to determine the values of a, b, and c. Therefore, the value of c is −7, not 7. **45a.** Sample answer: Always; when a and c are opposite signs, then ac will always be negative and $-4ac$ will always be positive. Since b^2 will also always be positive, then $b^2 - 4ac$ represents the addition of two positive values, which will never be negative. Hence, the discriminant can never be negative and the solutions can never be imaginary. **45b.** Sample answer: Sometimes; the roots will only be irrational if $b^2 - 4ac$ is not a perfect square.

47. −0.75 **49.** B **51.** 112.5 in^2 **53.** 42.25; $(x + 6.5)^2$

55. $\frac{4}{25}$; $\left(x + \frac{2}{5}\right)^2$ **57.** $4i$ **59.** 27 hours of flight instruction and 23 hours in the simulator

61. $y = x^2 + 1$ **63.** $y = |x + 3|$

Lesson 4-7

1. $y = (x + 3)^2 - 7$ **3.** $y = 4(x + 3)^2 - 12$

5.

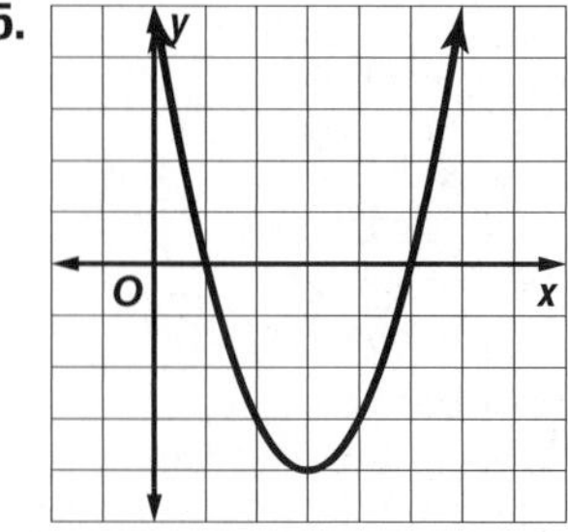

7.

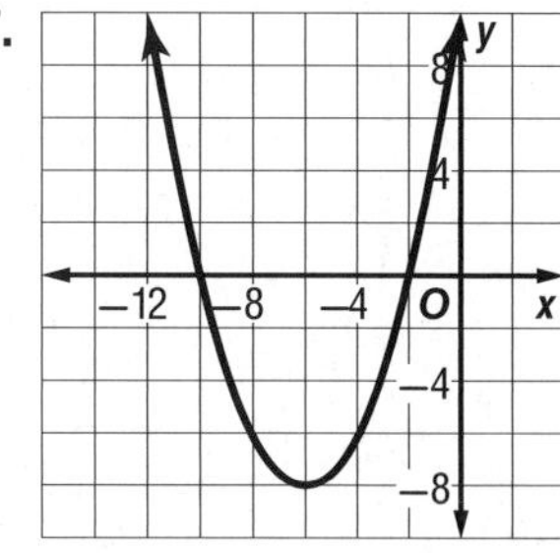

9. $y = (x - 3)^2 - 6$

11 $y = x^2 + 2x + 7$ Original equation.

$y = (x^2 + 2x + 1) + 7 - 1$ Complete the square by adding $\left(\frac{2}{2}\right)^2$ or 1. Balance the equation by subtracting 1.

$y = (x + 1)^2 + 6$ Write $x^2 + 2x + 1$ as a perfect square.

13. $y = (x + 4)^2$ **15.** $y = 3\left(x + \frac{5}{3}\right)^2 - \frac{25}{3}$

17. $y = -4(x + 3)^2 + 21$ **19.** $y = -(x + 2)^2 + 3$

21. $y = -15(x - 8.5)^2 + 4083.75$

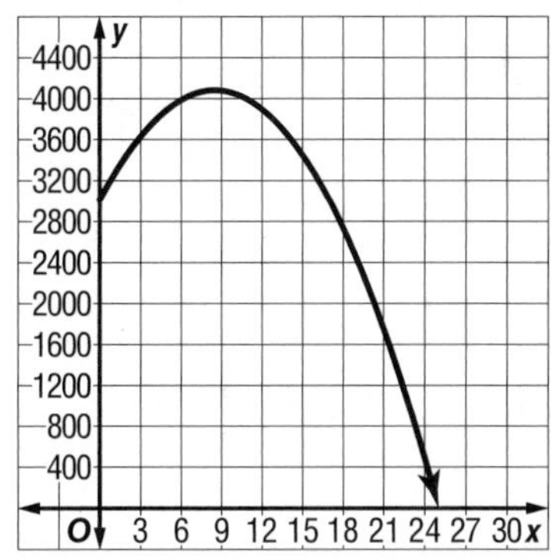

23.

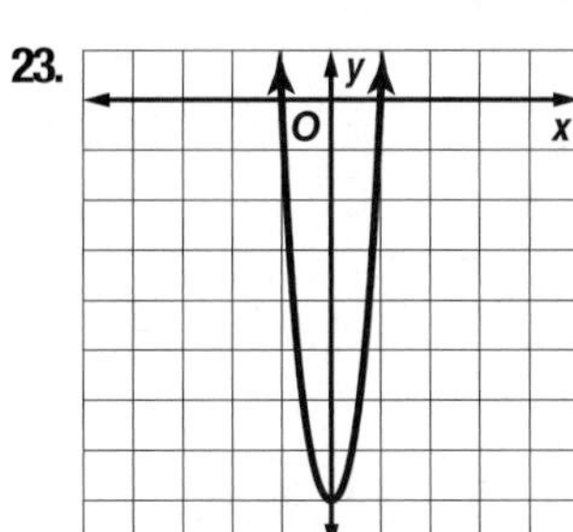

25.

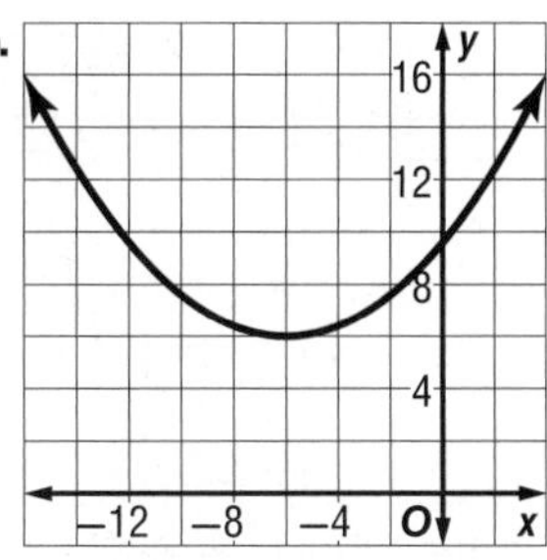

27.

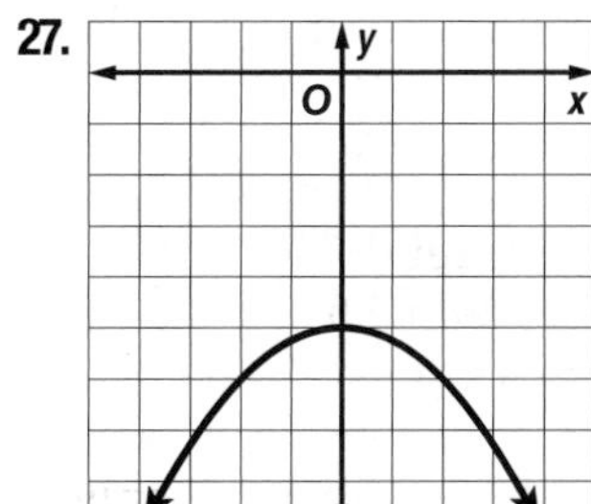

29.

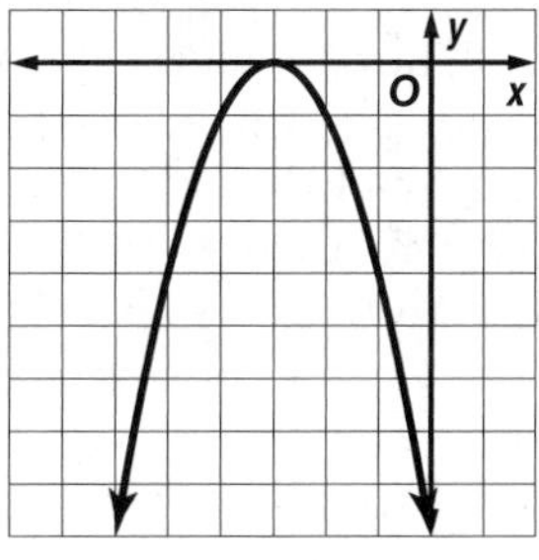

31.

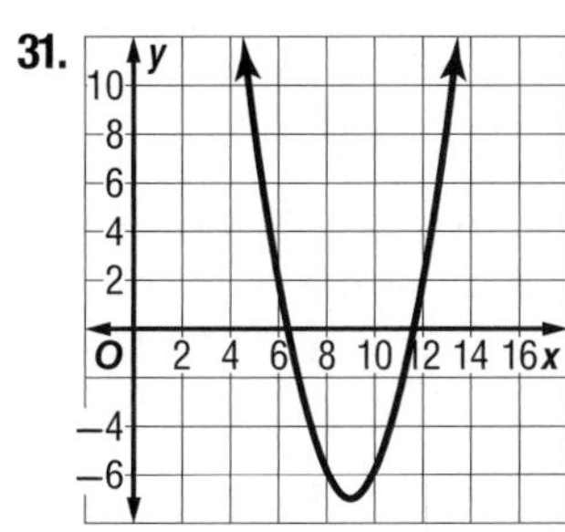

33.

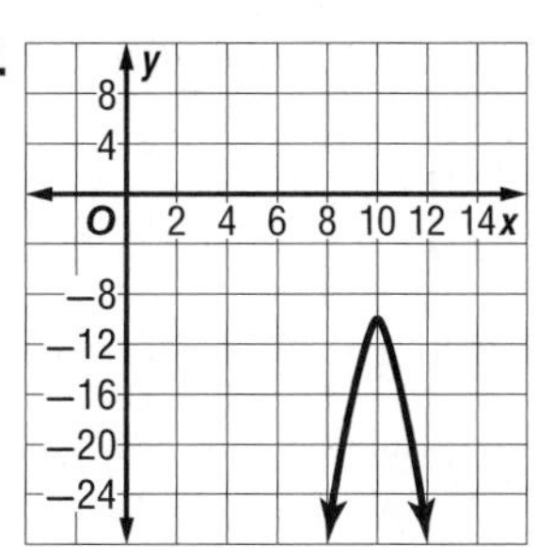

35. $y = 9(x - 6)^2 + 1$ **37.** $y = -\frac{2}{3}(x - 3)^2$

39. $y = \frac{1}{3}x^2 + 5$ **41.** $y = 3\left(x - \frac{2}{3}\right)^2 - \frac{10}{3}$; $\left(\frac{2}{3}, -\frac{10}{3}\right)$, $x = \frac{2}{3}$, opens up **43.** $y = -(x + 2.35)^2 + 8.3225$; $(-2.35, 8.3225)$, $x = -2.35$, opens down

45. $y = \left(x - \frac{1}{3}\right)^2 - 3$; $\left(\frac{1}{3}, -3\right)$, $x = \frac{1}{3}$, opens up

47 **a.** $S(t) = \frac{1}{2}at^2 + v_0t$ Original equation

$S(t) = \frac{1}{2}(0.002)t^2 + 0.0097t$ $a = 0.002$ and $v_0 = 0.0097$ mi/s

$S(t) = 0.001t^2 + 0.0097t$ Simplify.

$S(t) = 0.001(t^2 + 9.7t)$ Group $ax^2 + bx$ and factor, dividing by a.

$S(t) = 0.001(t^2 + 9.7t + 4.86^2) - 0.024$

Complete the square by adding 4.86^2 inside the parentheses. This is an overall addition of 0.024. Balance the equation by subtracting 0.024.

$S(t) = 0.001(t + 4.86)^2 - 0.024$ Write as a perfect square.

b. $a \cdot t = v$ acceleration • time = velocity

$0.002\ \text{mi/s}^2 \cdot t = (68 - 35)\ \text{mi/h}$ Substitution

$0.002\ \text{mi/s}^2 \cdot t = 33\ \text{m/h}$ Simplify.

$0.002\ \text{mi/s}^2 \cdot t = \frac{33\ \text{mi}}{1\ \text{h}} \cdot \frac{1\ \text{h}}{3600\ \text{s}}$ Convert hours to seconds.

$0.002\ \text{mi/s}^2 \cdot t = 0.009\ \text{mi/s}$ Simplify.

$t = \frac{0.009\ \text{mi}}{\text{s}} \div \frac{0.002\ \text{mi}}{\text{s}^2}$ Divide each side by $0.002\ \text{mi/s}^2$.

$t = \frac{0.009\ \text{mi}}{\text{s}} \cdot \frac{\text{s}^2}{0.002\ \text{mi}}$ Multiply by the reciprocal.

$t = 4.58\ \text{s}$ Simplify.

It will take about 4.58 seconds for Valerie to accelerate from 35 mi/h to 68 mi/h.

c. Yes; if we substitute $\frac{1}{8}$ for $S(t)$ and solve for t, we get 7.346 seconds. This is how long Valerie will be on the ramp. Since it will take her 4.58 seconds to accelerate to 68 mph, she will be on the ramp long enough to accelerate to match the average expressway speed.

49. The equation of a parabola can be written in the form $y = ax^2 + bx + c$ with $a \neq 0$. For each of the three points, substitute the value of the x-coordinate for x in the equation and substitute the value of the y-coordinate for y in the equation. This will produce three equations in three variables a, b, and c. Solve the system of equations to find the values of a, b, and c. These values determine the quadratic equation.

51. Sample answer: The variable a represents different values for these functions, so making $a = 0$ will have a different effect on each function. For $f(x)$, when $a = 0$, the graph will be a horizontal line, $f(x) = k$. For $g(x)$, when $a = 0$, the graph will be linear, but not necessarily horizontal, $g(x) = bx + c$.

53. B **55.** D **57.** $\frac{-15 \pm \sqrt{561}}{8}$ **59.** $\frac{3 \pm \sqrt{39}}{5}$

61. 0.0025 **63.** minimum, $9\frac{1}{3}$ **65.** minimum, −12

67.

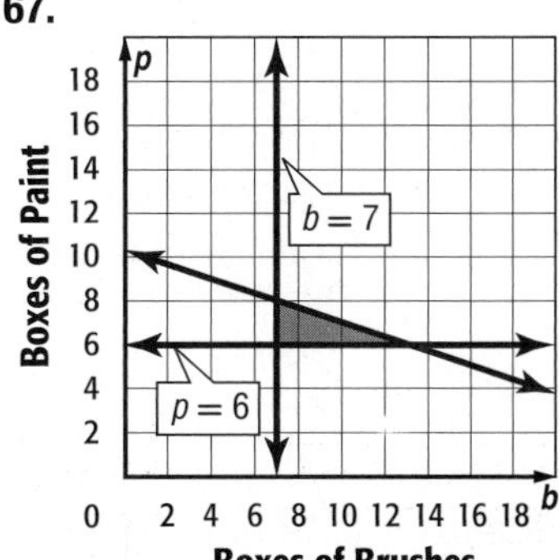

69.

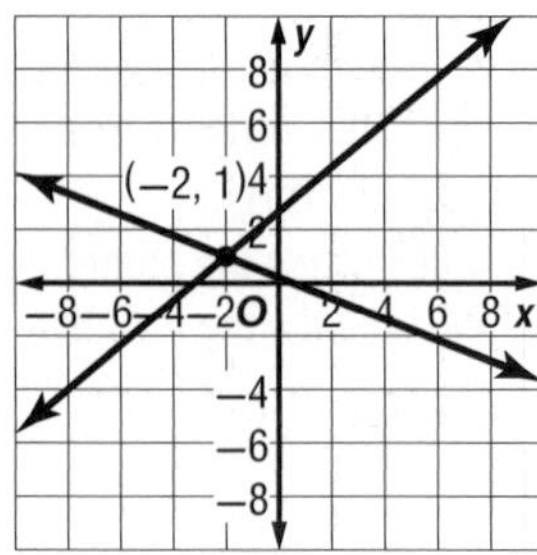

71. 9 **73.** 52 **75.** yes

Lesson 4-8

1.

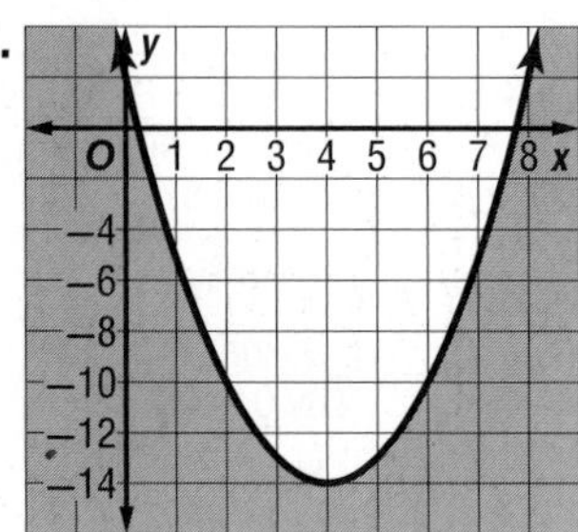

3.

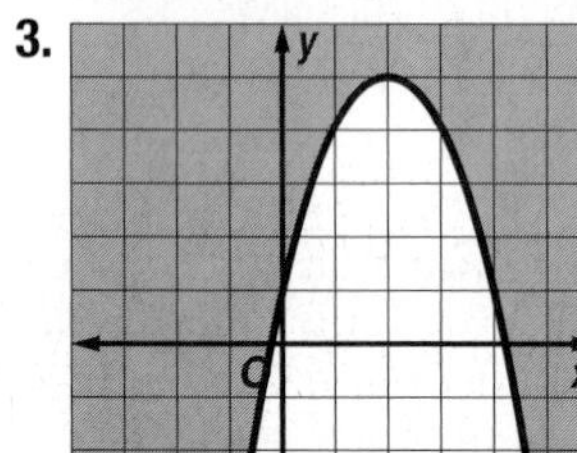

5. $\{x \mid -5 < x < -3\}$
7. $\{x \mid 0.29 \leq x \leq 1.71\}$
9. $\{x \mid -8 < x < 2\}$

11

$-x^2 + 12x = 28$ Related quadratic equation

$-x^2 + 12x - 28 = 0$ Subtract 28 from each side.

$x = \dfrac{-b \pm \sqrt{b^2 - 4ac}}{2a}$ Quadratic Formula

$= \dfrac{-12 \pm \sqrt{122 - 4(-1)(-28)}}{2(-1)}$ $a = -1$, $b = 12$, and $c = -28$

$x = \dfrac{-12 + \sqrt{32}}{-2}$ or $x = \dfrac{-12 - \sqrt{32}}{-2}$ Simplify and write as two equations.

≈ 3.17 ≈ 8.83 Simplify.

Plot 3.17 and 8.83 on a number line. Use dots since these values are solutions of the original inequality.

$x \leq 3.17$ $3.17 \leq x \leq 8.83$ $x \geq 8.83$

1 2 3 4 5 6 7 8 9 10

Test a value from each of the three intervals to see if it satisfies the original inequality.

$x \leq 3.17$
Test $x = 0$.
$-x^2 + 12x \geq 28$
$-(0)^2 + 12(0) \geq 28$
$0 \not\geq 28$

$3.17 \leq x \leq 8.83$
Test $x = 5$.
$-x^2 + 12x \geq 28$
$-(5)^2 + 12(5) \geq 28$
$35 \geq 28$

$x \geq 8.83$
Test $x = 10$.
$-x^2 + 12x \geq 28$
$-(10)^2 + 12(10) \geq 28$
$20 \not\geq 28$

The solution set is $\{x \mid 3.17 \leq x \leq 8.83\}$ or $[3.17, 8.83]$.

13.

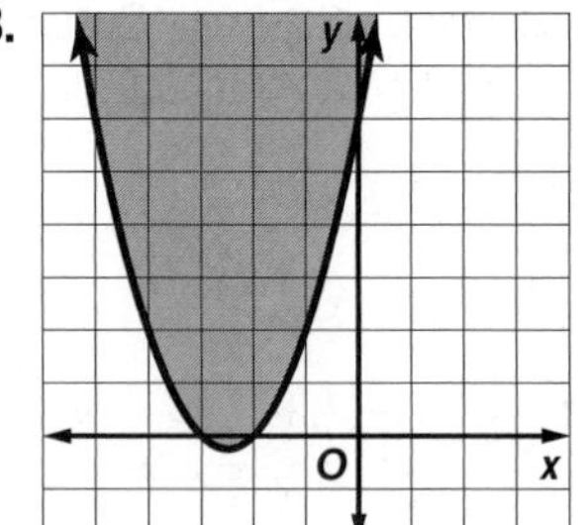

15.

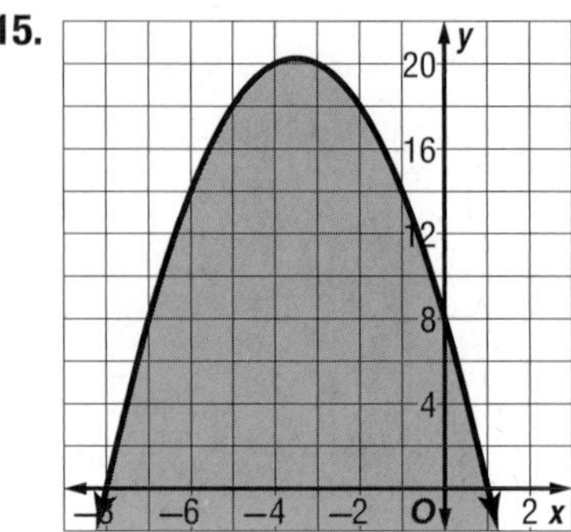

17.

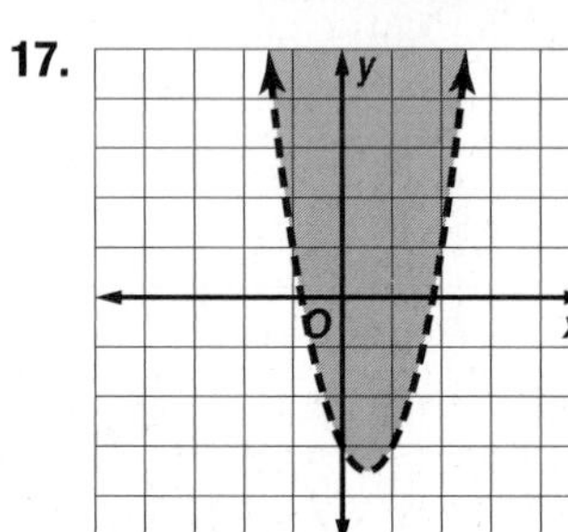

19. $\{x \mid 1.1 < x < 7.9\}$
21. {all real numbers}
23. $\{x \mid x < -1.42$ or $x > 8.42\}$ **25.** $\varnothing$
27. $\{x \mid x < -0.73$ or $x > 2.73\}$
29. $\{x \mid -0.5 \leq x \leq 2.5\}$

31 The function describes the height of the arch. You want to find the values of x for which $f(x) \geq 7$.

$f(x) \geq 7$ Original inequality

$-x^2 + 6x + 1 \geq 7$ $f(x) = -x^2 + 6x + 1$

$-x^2 + 6x - 6 \geq 0$ Subtract 7 from each side.

Graph the related function $y = -x^2 + 6x - 6$ using a graphing calculator.

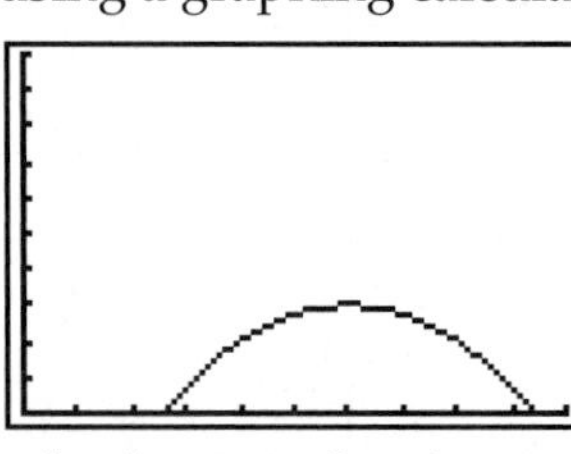

[0, 5] scl: 0.5 by [0, 10] scl: 1

At $x \approx 1.26$ and $x \approx 4.73$, $f(x) \geq 7$. So, at about 1.26 ft to 4.73 ft from the sides of the arch, the height is at least 7 ft.

33. $\{x \mid 4 < x < 5\}$ **35.** $\{x \mid -1 < x < 2\}$ **37.** $\{x \mid x \leq -2.32$ or $x \geq 4.32\}$ **39.** $\{x \mid x \leq -1.58$ or $x \geq 1.58\}$ **41.** {all real numbers} **43.** $\{x \mid -2.84 < x < 0.84\}$

45a.

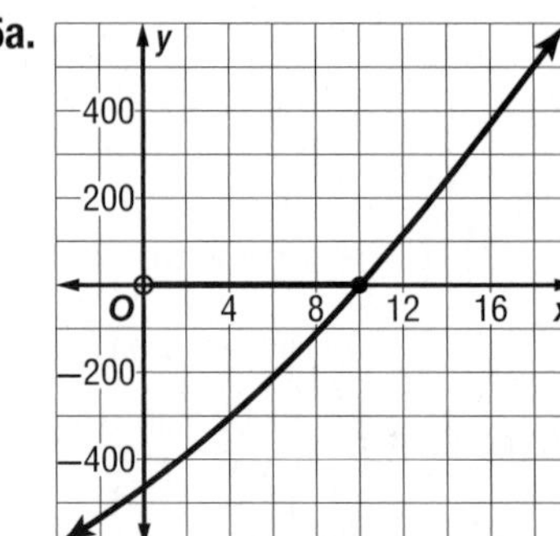

45b. greater than 0 ft but no more than 10.04 ft
47. $y \leq -x^2 + 2x + 6$
49. $\{x \mid x < -1.06$ or $x > 7.06\}$

51 $11 = 4x^2 + 7x$ Related quadratic equation

$0 = 4x^2 + 7x - 11$ Subtract 11 from each side.

$x = \dfrac{-b \pm \sqrt{b^2 - 4ac}}{2a}$ Quadratic Formula

$x = \dfrac{-7 \pm \sqrt{7^2 - 4(4)(-11)}}{2(4)}$ $a = 4$, $b = 7$, and $c = -11$

$x = \dfrac{-7 + \sqrt{255}}{8}$ or $x = \dfrac{-7 - \sqrt{255}}{8}$ Simplify and write as two equations.

$= 1$ $= -2.75$ Simplify.

Plot −2.75 and 1 on a number line. Use dots since these values are solutions of the original inequality.

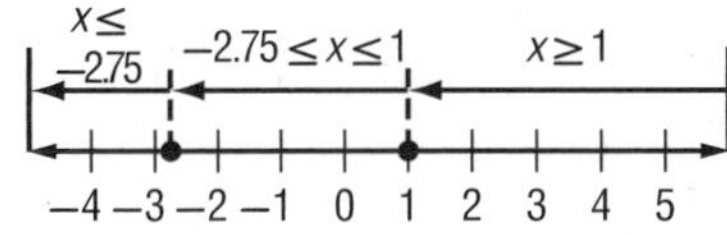

Test a value from each of the three intervals to see if it satisfies the original inequality.

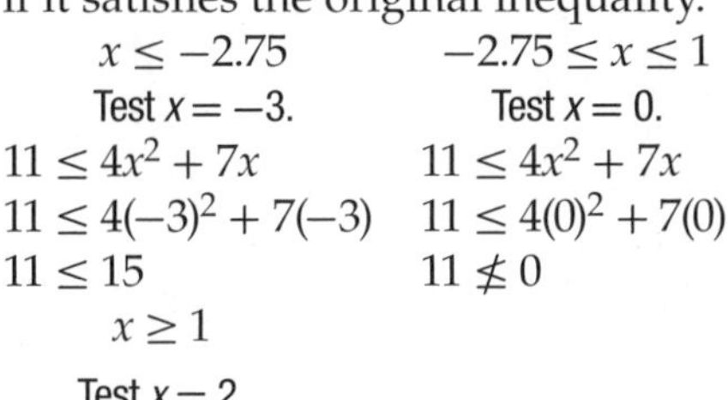

$x \le -2.75$ $-2.75 \le x \le 1$

Test $x = -3$. Test $x = 0$.

$11 \le 4x^2 + 7x$ $11 \le 4x^2 + 7x$

$11 \le 4(-3)^2 + 7(-3)$ $11 \le 4(0)^2 + 7(0)$

$11 \le 15$ $11 \not\le 0$

$x \ge 1$

Test $x = 2$.

$11 \le 4x^2 + 7x$

$11 \le 4(2)^2 + 7(2)$

$11 \le 30$

The solution set is $\{x \mid x \le -2.75 \text{ or } x \ge 1\}$.

53. $\{x \mid x < 0.61 \text{ or } x > 2.72\}$

55a.

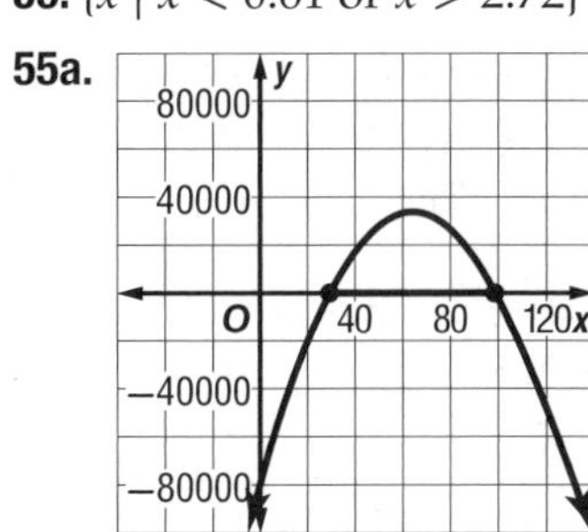

55b. from 30,000 to 98,000 digital audio players **55c.** The graph is shifted down 25,000 units. The manufacturer must sell from 48,000 to 81,000 digital audio players.

57a. Sample answer: $x^2 + 2x + 1 \ge 0$ **57b.** Sample answer: $x^2 - 4x + 6 < 0$ **59.** No; the graphs of the inequalities intersect the x-axis at the same points.

61.

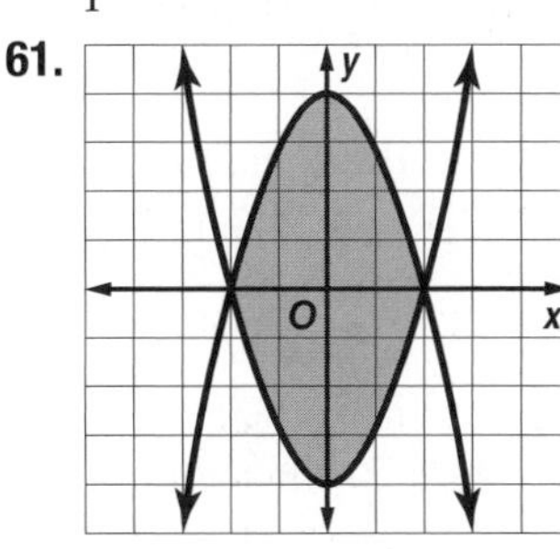

63. 15 **65.** G
67. $y = 2(x - 3)^2 - 4$
69. $y = 0.25(x + 4)^2 + 3$
71. −152; 2 complex roots **73.** $\begin{bmatrix} 0 & -21 \\ -14 & -16 \end{bmatrix}$

75. $\begin{bmatrix} 10 & -38 & -22 \\ -14 & 40 & 20 \end{bmatrix}$

77. $-6x + 24$

79. $8y - 12z$ **81.** $2.5x + 3y$

Chapter 4 Study Guide and Review

1. false, standard form **3.** false, factored form
5. false, completing the square **7.** true

9a. y-int: 12; $x = -\frac{5}{2}$; $-\frac{5}{2}$

9b.

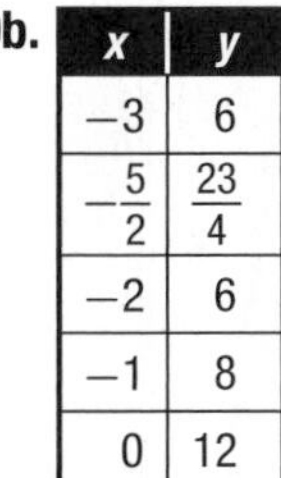

x	y
−3	6
$-\frac{5}{2}$	$\frac{23}{4}$
−2	6
−1	8
0	12

9c.

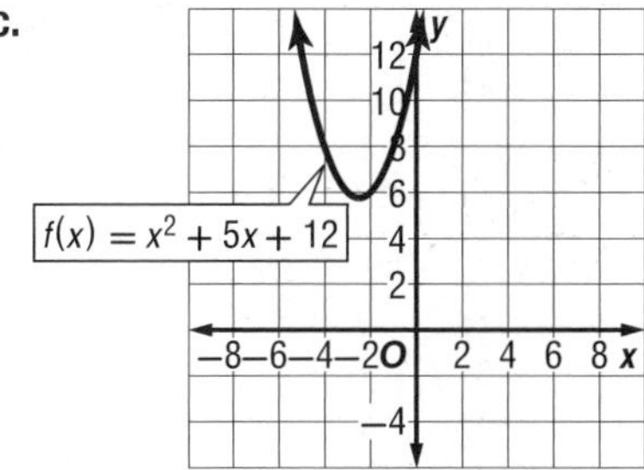

11a. y-int: −5; $x = \frac{9}{4}$; $\frac{9}{4}$

11b.

x	y
1	2
2	5
$\frac{9}{4}$	$\frac{41}{8}$
3	4
4	−1

11c.

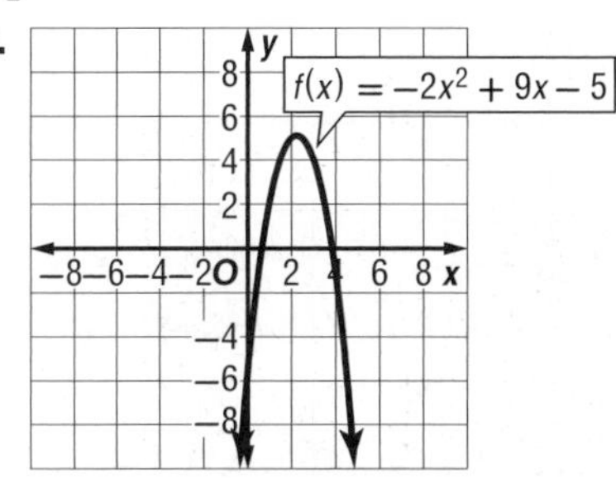

13. max; 1.25; D = {all real numbers}; R = $\{y \mid y \le 1.25\}$
15. 75 T-shirts at $15 each **17.** $\left(-1, \frac{3}{2}\right)$ **19.** 7.5 seconds
21. $x^2 + 10x + 21 = 0$ **23.** $3x^2 - x - 2 = 0$
25. $4x^2 + 5x + 1 = 0$ **27.** $\left\{-\frac{1}{2}, 3\right\}$ **29.** $x = 12$; 9 ft by 14 ft **31.** $15 + 3i$ **33.** $28 + 3i$ **35.** $x = \pm 5i$
37. $x = \pm i\sqrt{5}$ **39.** $x = \pm \frac{1}{2}i$ **41.** 4; $(x - 2)^2$ **43.** 1.44; $(x + 1.2)^2$ **45.** $\frac{9}{25}$; $\left(x + \frac{3}{5}\right)^2$ **47.** $\{1 \pm i\sqrt{7}\}$
49. $\left\{1, -\frac{5}{2}\right\}$ **51a.** 0 **51b.** 1 real rational root **51c.** {5}
53a. 153 **53b.** 2 irrational real roots
53c. $\left\{\frac{-3 \pm 3\sqrt{17}}{4}\right\}$ **55a.** −32 **55b.** 2 complex roots
55c. $\{1 \pm 2i\sqrt{2}\}$ **57a.** −47 **57b.** 2 complex roots
57c. $\left\{\frac{-5 \pm i\sqrt{47}}{4}\right\}$

59. $y = -3(x - 1)^2 + 5$; (1, 5); $x = 1$; opens down

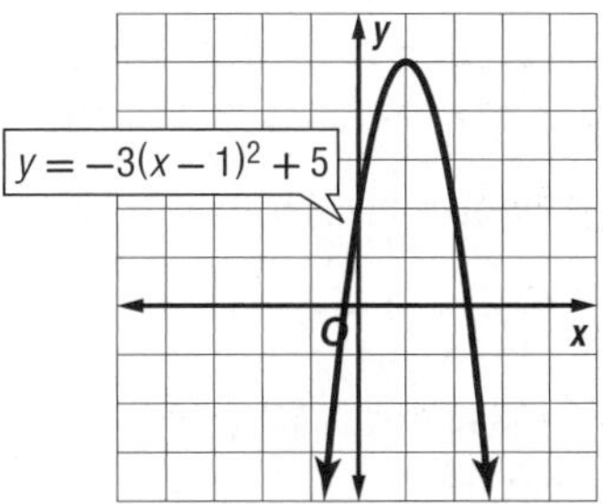

61. $y = -\frac{1}{2}(x + 2)^2 + 14$; (−2, 14); $x = -2$; opens down

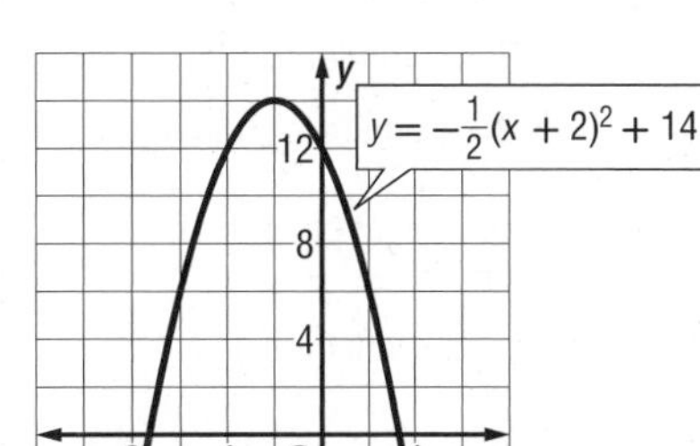

63. $f(x) = -x^2 + 10x$; 5 and 5

65.

$y < -x^2 + 5x - 6$

67.

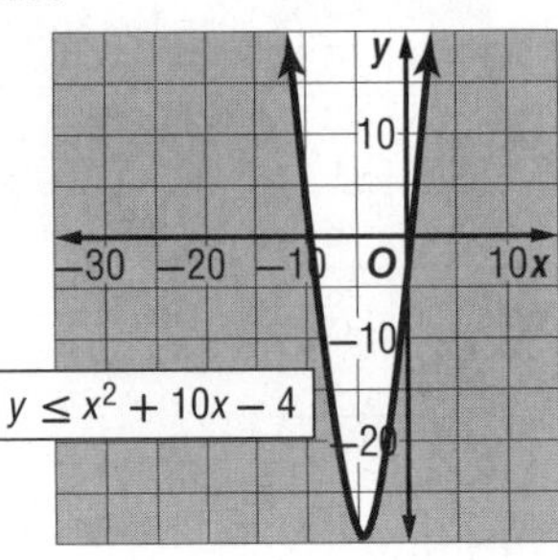

69. $\{x \mid x < -6 \text{ or } x > -2\}$ **71.** $\left\{x \,\middle|\, x < -4 \text{ or } x > \frac{5}{2}\right\}$

73. $\left\{x \,\middle|\, x < \frac{2}{3} \text{ or } x > 2\right\}$

CHAPTER 5
Polynomials and Polynomial Functions

Chapter 5 Get Ready

1. $-5 + (-13)$ **3.** $5mr + (-7mp)$ **5.** $20 + (-2x)$ **7.** $-3b^2 - 2b + 1$ **9.** $-\frac{9}{4}z - \frac{15}{4}$ **11.** $-4, 2$ **13.** $-\frac{4}{3}, \frac{1}{2}$ **15.** about 1.77 seconds

Lesson 5-1

1. $-8a^5b^2$ **3.** $\frac{8a^6}{27b^3}$ **5.** yes, 1 **7.** no **9.** $-2x^2 - 6x + 3$ **11.** $8ab + 10a$ **13.** $n^2 - 2n - 63$ **15.** $750 - 2.5x$ **17.** $-8b^5c^3$ **19.** $-yz^2$ **21.** $\frac{a^2c^2}{2b^4}$ **23.** z^{18} **25.** yes; 3 **27.** no **29.** $3b^2 + 6b - 5$ **31.** $8x^3 + 4xy$

33 $(a + b)(a^3 - 3ab - b^2)$
$= a(a^3 - 3ab - b^2) + b(a^3 - 3ab - b^2)$ Distributive Property
$= a(a^3) - a(3ab) - a(b^2) + b(a^3) - b(3ab) - b(b^2)$ Distributive Property
$= a^4 - 3a^2b - ab^2 + a^3b - 3ab^2 - b^3$ Multiply.
$= a^4 + a^3b - 3a^2b - 4ab^2 - b^3$ Simplify.

35. $10c^3 - c^2 + 4c$ **37.** $12a^2b + 8a^2b^2 - 15ab^2 + 4b^2$ **39.** $4a^2x - 2a^2y + 10abx - 5aby + 6b^2x - 3b^2y$ **41.** $\frac{y^4}{81x^4}$ **43.** $\frac{x^6}{16y^{14}}$ **45.** b **47.** $\frac{2}{5}cd^4$ **49.** $\frac{1}{2}x^6y^3$

51 **a.** $d = rt$ distance = rate • time
$t = \frac{d}{r}$ Solve the formula for time.
$= \frac{2.367 \times 10^{21}\text{ m}}{3 \times 10^8 \text{ m/s}}$ ← Distance from Andromeda to Earth ← Speed of light
$= \frac{2.367}{3} \cdot \frac{10^{21}}{10^8} \cdot \frac{\text{m}}{\text{m/s}}$ Separate to get powers of the same base.
$\approx 0.789 \cdot 10^{21-8}$ s Subtract exponents.
$\approx 0.789 \times 10^{13}$ s Simplify.
It takes about 7.89×10^{12} seconds or about 250,190.26 years.

b. $t = \frac{d}{r}$ Write the formula.
$= \frac{2.28 \times 10^{11}\text{ m}}{3 \times 10^8 \text{ m/s}}$ ← Distance from the Sun to Mars ← Speed of light
$= \frac{2.28}{3} \cdot \frac{10^{11}}{10^8} \cdot \frac{\text{m}}{\text{m/s}}$ Separate to get powers of the same base.
$= 0.76 \times 10^{11-8}$ s Subtract exponents.
$= 0.76 \times 10^3$ s Simplify.
$= 760$ s Simplify.
It takes 760 seconds or about 12.67 minutes.

53. $2n^4 - 3n^3p + 6n^4p^4$ **55.** $b^3 + \frac{b}{a} + \frac{1}{a^2}$ **57.** $2n^5 - 14n^3 + 4n^2 - 28$ **59.** $64n^3 - 240n^2 + 300n - 125$ **61a.** $0.155x^2 + 8.818x + 835.8$ **61b.** $0.061x^2 - 10.57x + 112.4$ **63.** 9 **65.** $\frac{1}{a^n} = \frac{a^0}{a^n} = a^{0-n} = a^{-n}$ **67.** Sample answer: We would have a 0 in the denominator, which makes the expression undefined. **69.** Sample answer: Astronomy deals with very large numbers that are sometimes difficult to work with because they contain so many digits. Properties of exponents make very large or very small numbers more manageable. As long as you know how far away a planet is from a light source, you can divide that distance by the speed of light to obtain how long it will take light to reach that planet. **71.** D **73.** D **75.** $x > 5$ or $x < -8$

77.

(0, 8) (4, 8) (2, −4)

79.

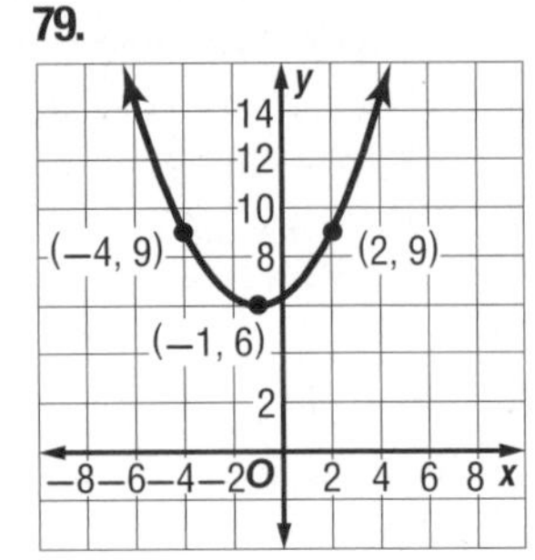

81. 42 **83.** 28 **85.** $\frac{7}{8}$ **87.** $\frac{1}{2}$ **89.** $\frac{5}{9}$ **91.** $4x(3ax^2 + 5bx + 8c)$ **93.** $(3y + 2)(4y + 3)$ **95.** $(2x - 3)(4a - 3)$

Lesson 5-2

1. $4y + 2x - 2$ **3.** $x - 8 - \frac{4}{x+2}$ **5.** $3z^3 - 15z^2 + 36z - 105 + \frac{309}{z+3}$ **7.** A **9.** $6a + 6 + \frac{21}{3a-2}$ **11.** $3y + 5$ **13.** $x + 3y - 2$ **15.** $2a^2 + b - 3$ **17.** $3np - 6 + 7p$ **19.** $-w + 16 + \frac{1000}{w}$

21

$$\begin{array}{r} b^2 - 5b + 6 \\ b + 1 \overline{) b^3 - 4b^2 + b - 2} \\ (-)\ b^3 + b^2 \quad\quad\quad\quad \\ -5b^2 + b \quad\quad \\ (-)\ -5b^2 - 5b \quad\quad \\ 6b - 2 \\ (-)\ 6b + 6 \\ -8 \end{array}$$

The quotient is $b^2 - 5b + 6$, and the remainder is -8. So, the expression equals $b^2 - 5b + 6 - \frac{8}{b+1}$.

23. $x^4 + 4x^3 + 12x^2 + 52x + 208 + \frac{832}{x-4}$ **25.** $g^3 + 2g^2 + g + 2 - \frac{14}{g-2}$ **27.** $2x^4 + x^3 - x + \frac{2}{3} - \frac{2}{9x+3}$ **29.** $b^2 - 4b + 8 - \frac{8}{b+1}$ **31.** $2y^5 - y^4 + y^3 + y^2 - y - 3$ **33.** $V(t) = t^2 + 5t + 6$

35 **a.**

$$\begin{array}{r} 3500 \\ a^2 + 100 \overline{) 3500a^2} \quad\quad\quad \\ (-)\ 3500a^2 + 350{,}000 \\ -350{,}000 \end{array}$$

The quotient is 3500, and the remainder is −350,000. So, the expression equals $3500 - \frac{350{,}000}{a^2 + 100}$.

b. $n = 3500 - \frac{350{,}000}{a^2 + 100}$ Write the equation.

$= 3500 - \frac{350{,}000}{15^2 + 100}$ $a = 15$

$= 3500 - \frac{350{,}000}{325}$ Simplify.

≈ 2423 subscriptions

37. $\frac{4c^2d - 3d}{2}$ **39.** $n^2 - n - 1$

41. $3z^4 - z^3 + 2z^2 - 4z + 9 - \frac{13}{z + 2}$ **43.** Sample answer: Sharon; Jamal actually divided by $x + 3$. **45.** Sample answer: The degree of the quotient plus the degree of the divisor equals the degree of the dividend. **47.** $\frac{5}{x^2}$ does not belong with the other three. The other three expressions are polynomials. Since the denominator of $\frac{5}{x^2}$ contains a variable, it is not a polynomial. **49.** A **51.** 360 **53.** $3x^3 + 2x^2 + x + 4$ **55.** $23a^2 - 24a$ **57.** $8x^5y^8z^3$ **59.** 0 to 10 ft or 24 to 34 ft **61.** $4 \pm \sqrt{19}$ **63.** between -6 and -5; between -3 and -2 **65.** between -1 and 0; between 2 and 3 **67.** -21 **69.** -20 **71.** $-9d^2$

Lesson 5-3

1. degree = 6, leading coefficient = 11 **3.** not in one variable because there are two variables, x and y **5.** $w(5) = -247$; $w(-4) = 104$ **7.** $4y^9 - 5y^6 + 2$ **9.** $1536a^3 - 426a^2 - 144a + 82$ **11a.** $f(x) \to -\infty$ as $x \to -\infty$. $f(x) \to +\infty$ as $x \to +\infty$. **11b.** Since the end behavior is in opposite directions, it is an odd-degree function. **11c.** The graph intersects the x-axis at three points, so there are three real zeros. **13.** not in one variable because there are two variables, x and y **15.** degree = 6, leading coefficient = -12 **17.** degree = 4, leading coefficient = -5 **19.** degree = 2, leading coefficient = 3 **21.** degree = 9, leading coefficient = 2 **23.** $p(-6) = 1227$; $p(3) = 66$ **25.** $p(-6) = -156$; $p(3) = 78$

27 $p(x) = -x^3 + 3x^2 - 5$ Original function

$p(-6) = -(-6)^3 + 3(-6)^2 - 5$ Replace x with -6.

$= 216 + 108 - 5$ Simplify.

$= 319$ Simplify.

$p(x) = -x^3 + 3x^2 - 5$ Original function

$p(-6) = -(3)^3 + 3(3)^2 - 5$ Replace x with 3.

$= -27 + 27 - 5$ Simplify.

$= -5$ Simplify.

29. $18a^2 - 12a + 3$ **31.** $2b^4 - 4b^2 + 3$ **33.** $-64y^3 + 144y^2 - 104y + 25$ **35a.** $f(x) \to +\infty$ as $x \to -\infty$. $f(x) \to +\infty$ as $x \to +\infty$. **35b.** Since the end behavior is in the same direction, it is an even-degree function. **35c.** The graph intersects the x-axis at four points, so there are four real zeros. **37a.** $f(x) \to -\infty$ as $x \to -\infty$. $f(x) \to +\infty$ as $x \to +\infty$. **37b.** Since the end behavior is in opposite directions, it is an odd-degree function. **37c.** The graph intersects the x-axis at one point, so there is one real zero. **39a.** $f(x) \to -\infty$ as $x \to -\infty$. $f(x) \to -\infty$ as $x \to +\infty$. **39b.** Since the end behavior is in the same direction, it is an even-degree function. **39c.** The graph intersects the x-axis at two points, so there are two real zeros.

41 $KE(v) = 0.5mv^2$ Original function

$= 0.5(171)(11)^2$ Replace m with 171 and v with 11.

$= 10{,}345.5$ Simplify.

The kinetic energy is 10,345.5 kg-m/s or 10,345 joules. **43.** $p(-2) = -16$; $p(8) = 1024$ **45.** $p(-2) = -0.5$; $p(8) = 3112$ **47.** D **49.** A **51.** $3a^3 - 24a^2 + 240a + 66$ **53.** $5a^6 - 298a^2 + 1008a - 928$

55a.

x	$p(x)$
−7	−585
−6	0
−4	240
−3	135
−2	0
0	−144
1	−105
2	0
4	240
6	0
7	−585

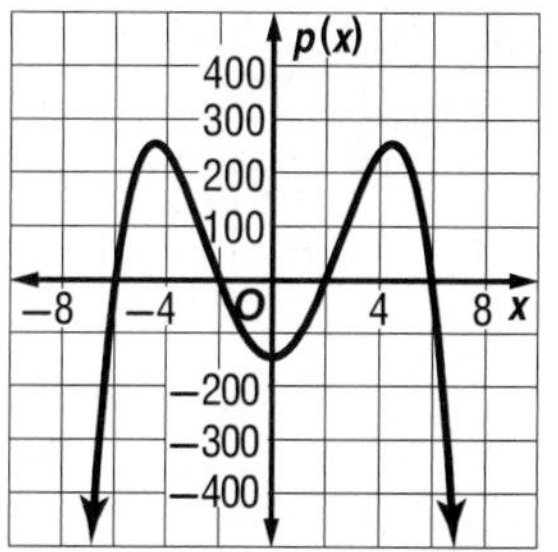

55b. $-6, -2, 2, 6$ **55c.** 2000 and 6000 items **55d.** Sample answer: The negative values should not be considered because the company will not produce negative items.

57 The degree, 4, is even and the leading coefficient, -5, is negative. So, $f(x) \to -\infty$ as $x \to -\infty$ and $f(x) \to -\infty$ as $x \to +\infty$.

59. $h(x) \to +\infty$ as $x \to -\infty$; $h(x) \to -\infty$ as $x \to +\infty$ **61.** $g(x) \to -\infty$ as $x \to -\infty$; $g(x) \to +\infty$ as $x \to +\infty$ **63.** Sample answer: Virginia is correct; an even function will have an even number of zeros and the double root represents 2 zeros. **65.** Sample answer: $f(x) \to +\infty$ as $x \to -\infty$; $f(x) \to +\infty$ as $x \to +\infty$; $\frac{f(x)}{g(x)}$ will become a 2-degree function with a positive leading coefficient. **67.** Sometimes; a polynomial function with four real roots may be a sixth-degree polynomial function with two imaginary roots. A polynomial function that has four real roots is at least a fourth-degree polynomial. **69.** Student A

71a.

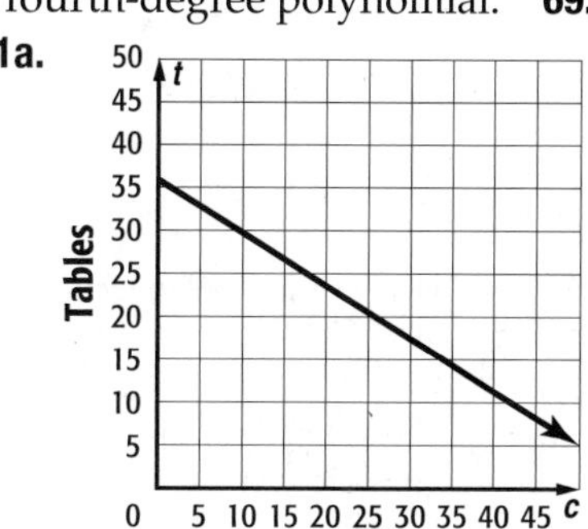

71b. $t = 0.5c$

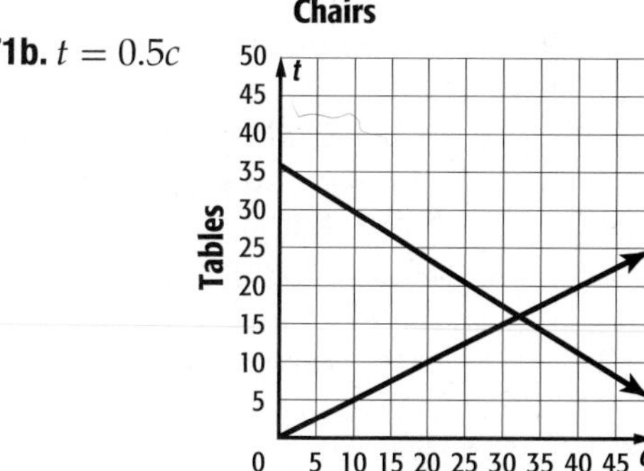

71c. 16 tables and 32 chairs **71d.** Sample answer: This can be determined by the intersection of the graphs. This point of intersection is the optimal amount of tables and chairs manufactured.
73. $2x^2y^2 + 4x^4y^4z^2$ **75.** $6c^3 - 1 + 4a^5cd^2$ **77.** yes; 6
79a. $h(d) = -2d^2 + 4d + 6$; The graph opens downward and is narrower than the parent graph, and the vertex is at (1, 8).
79b. $h(d) = -2(d - 1.25)^2 + 12.5$; it shifted the graph up 4.5 ft and to the right 3 in. **81.** $x \leq -\frac{2}{3}$ or $x \geq 2$
83. minimum; $-\frac{4}{3}$ **85.** maximum; 11

Lesson 5-4

1.

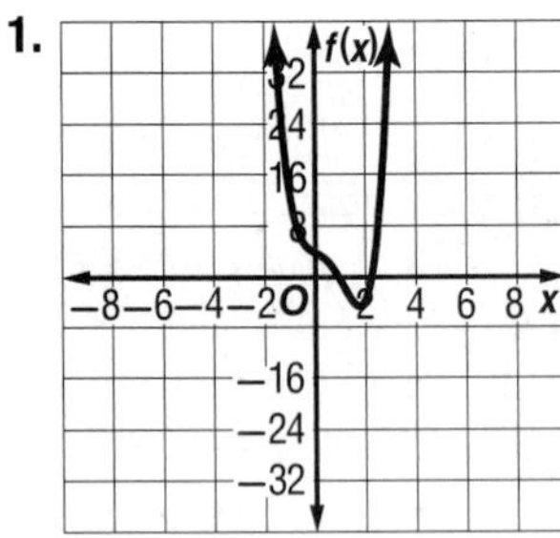

3.

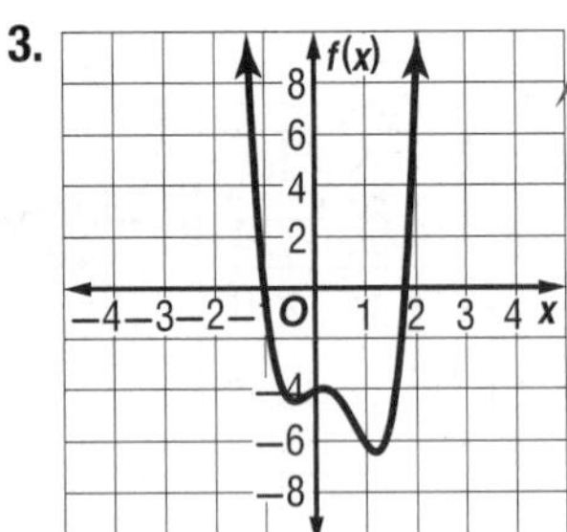

5. between −2 and −1

7. between 0 and 1 and between 2 and 3

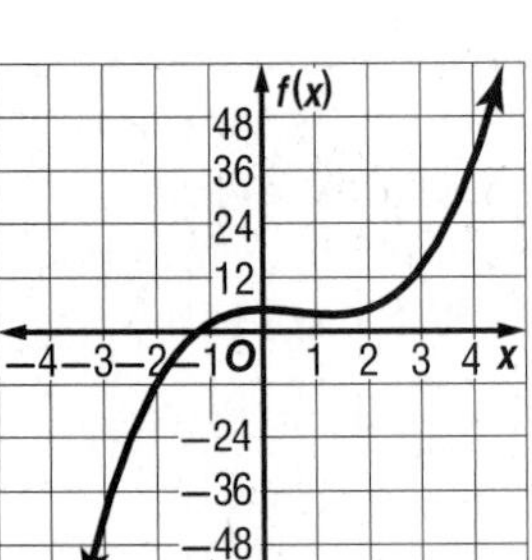

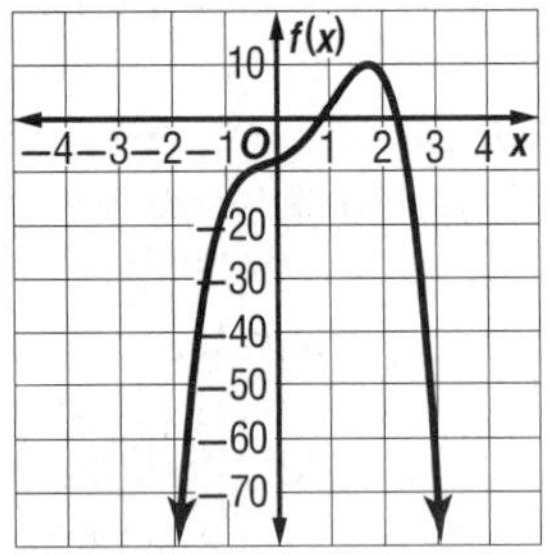

9. rel. max at $x \approx -1.8$; rel. min at $x \approx 1.1$; D = {all real numbers}, R = {all real numbers}

11. rel. max at $x \approx 2.4$; rel. min at $x \approx 0.3$; D = {all real numbers}, R = {all real numbers}

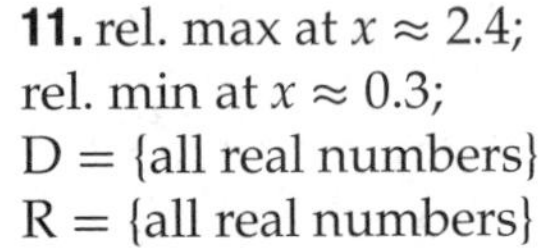

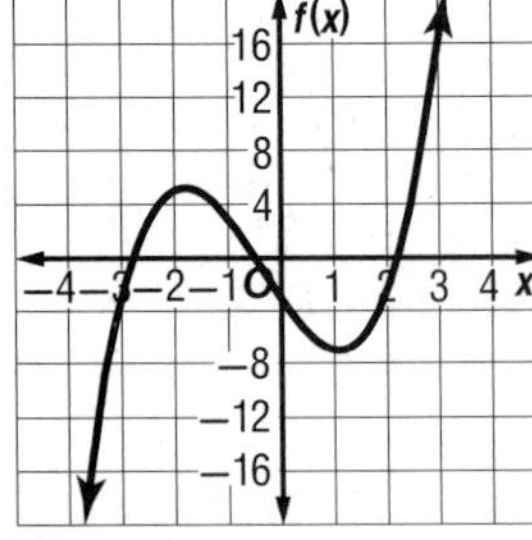

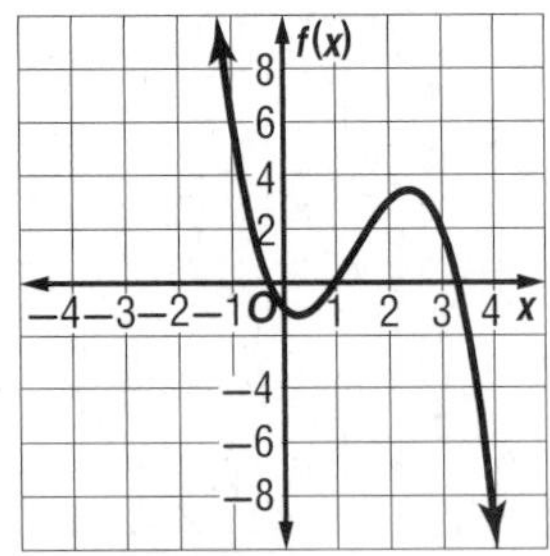

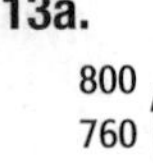

13a.

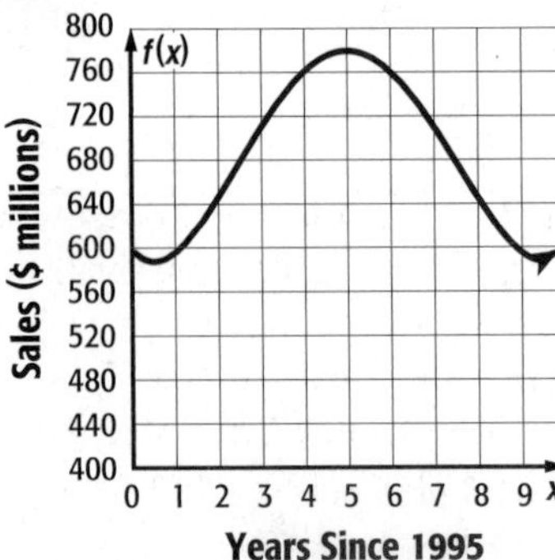

13b. Sample answer: Relative maximum at $x = 5$ and relative minimum at $x \approx 9.5$. $f(x) \to \infty$ as $x \to -\infty$ and $f(x) \to \infty$ as $x \to \infty$. The graph increases when $x < 5$ and $x > 9.5$ and decreases when $5 < x < 9.5$.

13c.

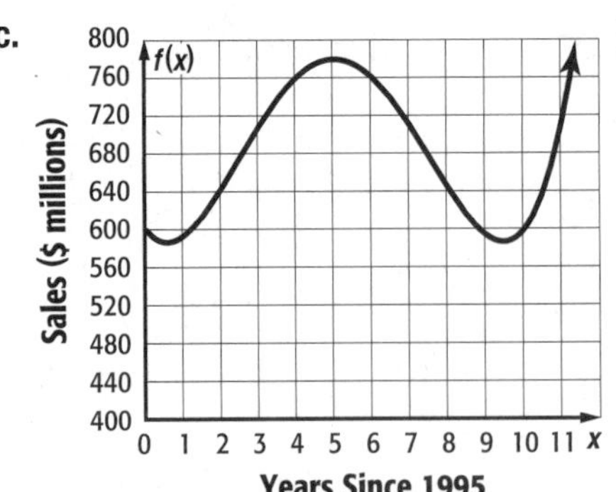

Sample answer: This suggests a dramatic increase in sales.

13d. Sample answer: No; with so many other forms of media on the market today, CD sales will not increase dramatically. In fact, the sales will probably decrease. The function appears to be accurate only until about 2005.

15 **a.** Since $f(x)$ is a third-degree polynomial function, it will have either 3 or 1 real zeros. Look at the values of $f(x)$ to locate the zeros. Then use the points to sketch the graph.

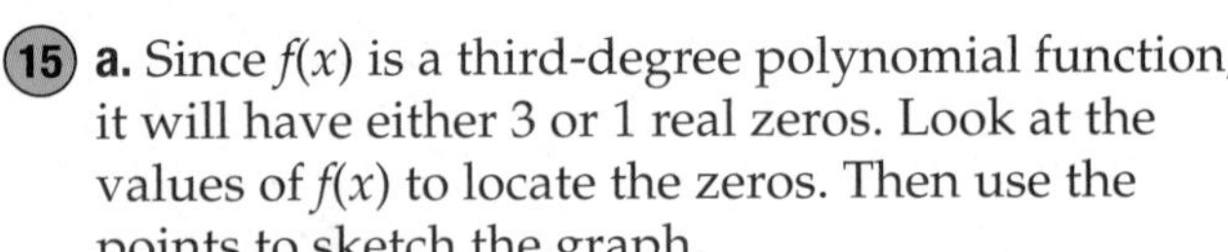

x	$f(x)$
−4	92
−3	41
−2	12
−1	−1
0	−4
1	−3
2	−4
3	−13
4	−36

← change in sign

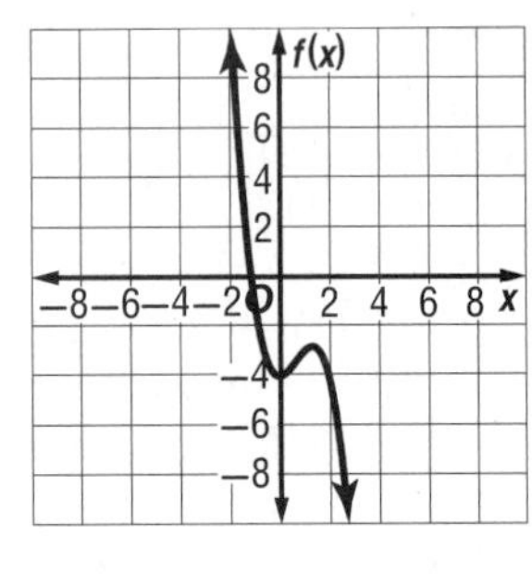

b. The value of $f(x)$ changes signs between $x = -2$ and $x = 1$. So, there is a zero between −2 and −1.
c. The value of $f(x)$ at $x = 0$ is less than the surrounding points, so there must be a relative minimum near $x = 0$. The value of $f(x)$ near $x = 1$ is greater than the surrounding points, so there must be a relative maximum near $x = 1$.

17a.

x	$f(x)$
−4	−155
−3	−80
−2	−33
−1	−8
0	1
1	0
2	−5
3	−8
4	−3
5	16

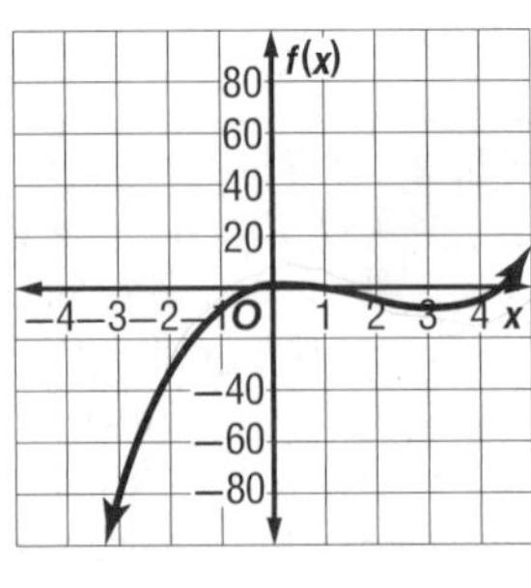

17b. at $x = 1$, between −1 and 0, and between $x = 4$ and $x = 5$ **17c.** rel. max: $x \approx \frac{1}{3}$, rel. min: $x = 3$

Selected Answers and Solutions

19a.

x	$f(x)$
−4	−176
−3	−77
−2	−22
−1	1
0	4
1	−1
2	−2
3	13
4	56

19b. between $x = -2$ and $x = -1$, between $x = 0$ and $x = 1$, and between $x = 2$ and $x = 3$

19c. rel. max: near $x = -0.3$; rel. min: near $x = 1.6$

21a.

x	$f(x)$
−4	372
−3	141
−2	36
−1	−3
0	−12
1	−3
2	36
3	141
4	372

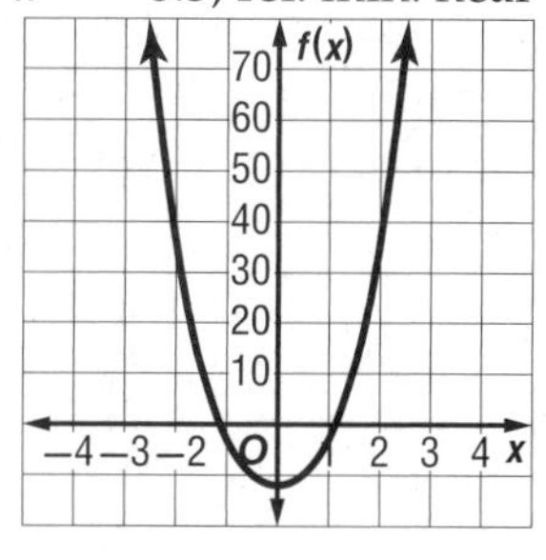

21b. between $x = -2$ and $x = -1$ and between $x = 1$ and $x = 2$ **21c.** min: near $x = 0$ **23.** rel. max: $x = -2.73$; rel. min: $x = 0.73$ **25.** rel. max: $x = 1.34$, no rel. min

27. Sample answer:

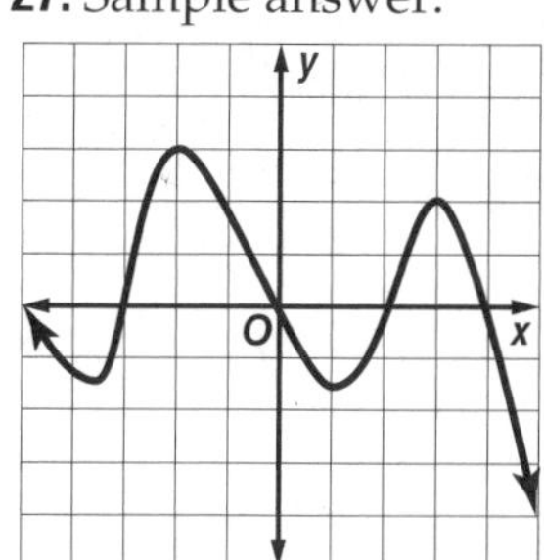

29. Sample answer:

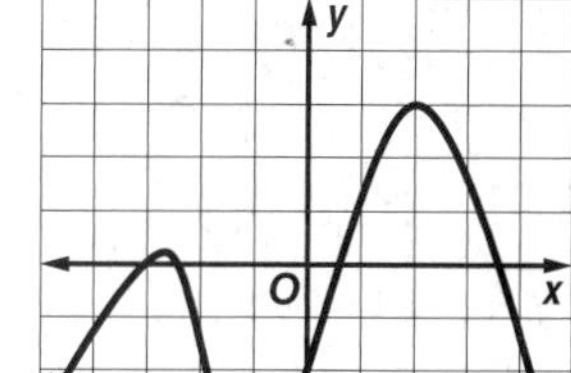

31. Sample answer:

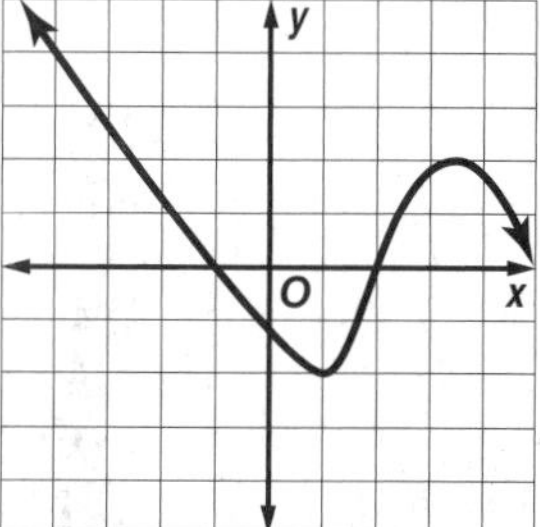

33 a.

x	$d(x)$
0	0
1	0.0145
2	0.056
3	0.1215
4	0.208
5	0.3125
6	0.432
7	0.5635
8	0.704
9	0.8505
10	1

b. Plot the points in the table and connect with a smooth curve.

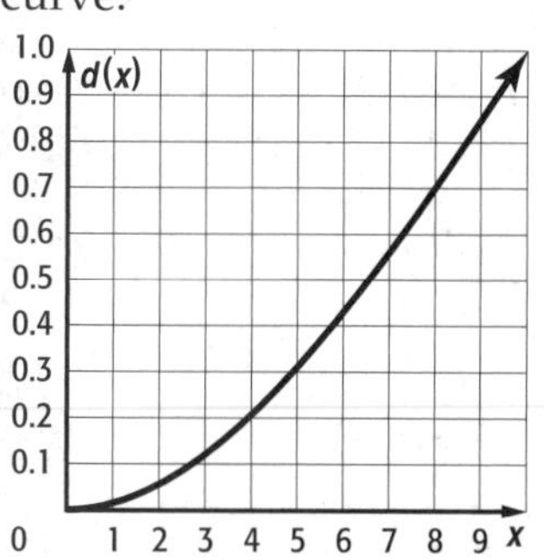

c. $d(x) \to +\infty$ as $x \to +\infty$; as x increases, $d(x)$ increases.

d. Sample answer: Since the diving board is only 10 feet long, x cannot be greater than 10. So, this trend cannot continue indefinitely.

35a. −2.5 (min), −0.5 (max), 1.5 (min) **35b.** −3.5, −1, 0, 3 **35c.** 4 **35d.** D = {all real numbers}; R = $\{y \mid y \geq -3.1\}$ **37a.** −3.5 (min), −2.5(max), −2 (min), −1 (max), 1 (min) **37b.** −3.75, −3.25, −2, −1.75, −0.25, 2.9 **37c.** 6 **37d.** D = {all real numbers}; R = $\{y \mid y \geq -5\}$ **39a.** −2 (max), 1 (min) **39b.** −3, −0.5, 2 **39c.** 3 **39d.** D = {all real numbers}; R = {all real numbers}

41. $1.25

43 a. Make a table of values and graph the function.

x	$f(x)$
−2	−16
−1.5	−4.5
−1	−4
−0.5	−4.6
0	−4
0.5	−3.3
1	−4
1.5	−3.5
2	8

← change in sign

The value of $f(x)$ changes signs between $x = 1.5$ and $x = 2$. So, there is a zero between these values, at approximately 1.75. The x-intercept ≈ 1.75. The y-intercept is at −4. The value of $f(x)$ at $x \approx -1.25$ and at $x \approx 0.5$ is greater than the surrounding points, so $x \approx -1.25$ and $x \approx 0.5$ are turning points. The value of $f(x)$ at $x \approx -0.5$ and at $x \approx 1.25$ is less than the surrounding points, so $x \approx -0.5$ and $x \approx 1.25$ are turning points.

b. The function does not have an axis of symmetry because the graph is not a parabola.

c. The function is increasing in the intervals $x \leq -1.25$, $-0.5 \leq x \leq 0.5$, and $x \geq 1.25$. The function is decreasing in the intervals $-1.25 \leq x \leq -0.5$ and $0.5 \leq x \leq 1.25$.

45a. no zeros, no x-intercepts, y-intercept: 5; no turning points **45b.** no axis of symmetry **45c.** decreasing: $x \leq -4$; constant: $-4 < x \leq 0$; increasing: $x > 0$

47. As the x-values approach large positive or negative numbers, the term with the largest degree becomes more and more dominant in determining the value of $f(x)$.

Selected Answers and Solutions

49. Sample answer:

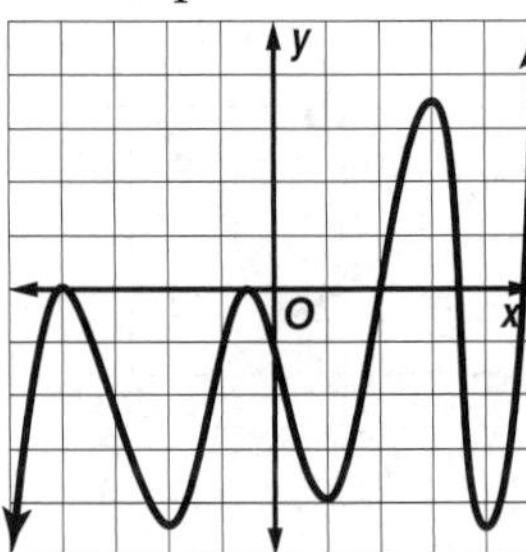

51. Sample answer: No; $f(x) = x^2 + x$ is an even degree, but $f(1) \neq f(-1)$. **53.** Sample answer: The degree will help determine whether the graph is even or odd and the maximum number of zeros and turning points for the graph. The leading coefficient determines the end behavior of the graph, and, along with the degree, builds the shape of the graph. The zeros and turning points allow for the plotting of specific points in the center of the graph. All of these things combine for an accurate sketch of the graph of a polynomial function. **55.** 95 **57.** C **59.** $f(x) \to -\infty$ as $x \to -\infty$. $f(x) \to -\infty$ as $x \to +\infty$. Since the end behavior is in the same direction, it is an even-degree function. The graph intersects the x-axis at six points, so there are six real zeros. **61.** $(x - 2)(x + 3)$ **63.** $2a^2 + a - 3 - \frac{3}{a - 1}$ **65.** $(x + 6)(x + 3)$ **67.** $(a + 8)(a - 2)$ **69.** $(3x - 4)(2x + 1)$

Lesson 5-5

1. $(a + b)(3x + 2y - z)$ **3.** prime
5. $12q(w - q)(w^2 + qw + q^2)$
7. $x^2(a - b)(a^2 + ab + b^2)(a + b)(a^2 - ab + b^2)$
9. $(2c - 5d)(4c^2 + 12cd + 25d^2)$ **11.** $4, -4, \pm\sqrt{3}$
13. $-3, \frac{3 \pm 3i\sqrt{3}}{2}$ **15.** 5 ft **17.** not possible
19. $\sqrt{6}, -\sqrt{6}, 2\sqrt{3}, -2\sqrt{3}$ **21.** $x(4x + y)(16x^2 - 4xy + y^2)$
23. $y^3(x^2 + y^2)(x^4 - x^2y^2 + y^2)$ **25.** prime
27. $(6x^2 - 5y^2)(2a - 3b + 4c)$

29 $8x^5 - 25y^3 + 80x^4 - x^2y^3 + 200x^3 - 10xy^3$ Original expression

$= (8x^5 + 80x^4 + 200x^3) + (-25y^3 - x^2y^3 - 10xy^3)$ Group to find a GCF.

$= 8x^3(x^2 + 10x + 25) - y^3(25 + x^2 + 10x)$ Factor the GCF.

$= 8x^3(x + 5)^2 - y^3(x + 5)^2$ Perfect squares

$= (8x^3 - y^3)(x + 5)^2$ Distributive Property

$= (2x - y)(4x^2 + 2xy + y^2)(x + 5)^2$ Difference of cubes

31. $6, -6, \pm 2i\sqrt{5}$ **33.** $\pm\sqrt{7}, \pm i\sqrt{13}$ **35.** $-\frac{1}{4}, \frac{1 \pm i\sqrt{3}}{8}$
37. $-15(x^2)^2 + 18(x^2) - 4$ **39.** not possible
41. $4(2x^5)^2 + 1(2x^5) + 6$ **43.** $\pm\sqrt{5}, \pm i\sqrt{2}$
45. $\pm\frac{2\sqrt{3}}{3}, \pm\frac{\sqrt{15}}{3}$ **47.** $\pm\frac{\sqrt{6}}{6}, \pm i\frac{\sqrt{3}}{2}$
49. $(x^2 + 25)(x + 5)(x - 5)$ **51.** $x(x + 2)(x - 2)(x^2 + 4)$
53. $(5x + 4y + 5z)(3a - 2b + c)$
55. $x(x + 3)(x - 3)(3x + 2)(2x - 5)$ **57.** $x = 8$; 5, 8, 11
59. $\pm\frac{2\sqrt{3}}{3}, \pm i\frac{\sqrt{2}}{2}$ **61.** $\pm\frac{1}{3}, \pm i\frac{\sqrt{10}}{2}$ **63.** $3, -3, \pm i\frac{\sqrt{15}}{3}$

65

$x^6 - 26x^3 - 27 = 0$ Original equation

$(x^3)^2 - 26(x^3) - 27 = 0$ $(x^3)^2 = x^6$

$u^2 - 26u - 27 = 0$ Let $u = x^3$.

$(u - 27)(u + 1) = 0$ Factor.

$u - 27 = 0$ or $u + 1 = 0$ Zero Product Property

$x^3 - 27 = 0$ or $x^3 + 1 = 0$ Replace u with x^3.

$x^3 - 27 = 0$

$(x - 3)(x^2 + 3x + 9) = 0$ Difference of Two Cubes

$x - 3 = 0$ or $x^2 + 3x + 9 = 0$ Zero Product Property

$x = 3$ $\quad x = \frac{-3 \pm \sqrt{3^2 - 4(1)(9)}}{2(1)}$

$= \frac{-3 \pm 3i\sqrt{3}}{2}$

$x^3 + 1 = 0$

$(x + 1)(x^2 - x + 1) = 0$ Sum of Two Cubes

$x + 1 = 0$ or $x^2 - x + 1 = 0$ Zero Product Property

$x = -1$ $\quad x = \frac{-(-1) \pm \sqrt{(-1)^2 - 4(1)(1)}}{2(1)}$

$= \frac{1 \pm i\sqrt{3}}{2}$

The solutions are $-1, 3, \frac{-3 \pm 3i\sqrt{3}}{2}$, and $\frac{1 \pm i\sqrt{3}}{2}$.

67. $-1, 1, \pm\frac{1}{2}$ **69.** $\pm i\sqrt{5}, \pm i\sqrt{3}$ **71a.** 2 ft **71b.** 176 ft^2 **71c.** 428 ft^2

73 **a.** $f(x) = (x + 6)[x + x + (x + 2) + (x + 2)] + x[x + (x + 2) + (x + 2)] + x(x + 2)$

$= (x + 6)(4x + 4) + x(3x + 4) + x(x + 2)$

$= 4x^2 + 24x + 4x + 24 + 3x^2 + 4x + x^2 + 2x$

$= 8x^2 + 34x + 24$

b. $f(x) = 8x^2 + 34x + 24$ Original function

$1366 = 8x^2 + 34x + 24$ Replace $f(x)$ with 1366.

$0 = 8x^2 + 34x - 1342$ Subtract 1366 from each side.

$0 = 2(4x^2 + 17x - 671)$ Factor.

$0 = 2(4x + 61)(x - 11)$ Perfect squares

$4x + 61 = 0$ or $x - 11 = 0$ Zero Product Property

$x = -15.25$ $\quad x = 11$ Simplify.

Since distance cannot be negative, $x = 11$ ft.

75. $(x + 2)^3(x - 2)^3$ **77.** $(x + y)^3(x - y)^3$
79. $(6x^n + 1)^2$ **81.** Sample answer: $a = 1, b = -1$
83. Sample answer: The factors can be determined by the x-intercepts of the graph. An x-intercept of 5 represents a factor of $(x - 5)$. **85.** D **87.** D
89. rel. max. at $x \approx 1.5$, rel. min. at $x \approx 0.1$;

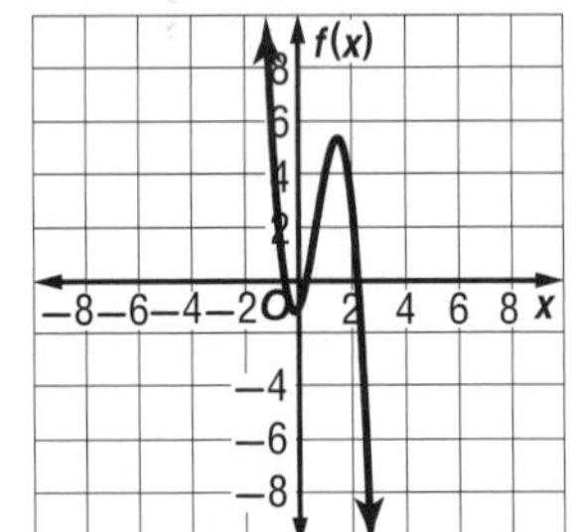

91. degree = 4; leading coefficient = 5 **93.** degree = 7; leading coefficient = −1 **95.** 18 skis and 10 snowboards **97.** $x + 2 - \frac{10}{x + 4}$ **99.** $8x^2 - 12x + 24 - \frac{42}{x + 2}$

Lesson 5-6

1. 58; −20 **3.** 12,526 **5.** $x + 4, x - 4$ **7.** $x - 5, 2x - 1$
9. 71; −6 **11.** −435; −15 **13.** −4150; 85 **15.** 647; −4
17. $(x - 1)^2$ **19.** $x - 4, x + 1$ **21.** $x + 6, 2x + 7$
23. $x + 1, x^2 + 2x + 3$ **25.** $x - 4, 3x - 2$

27 **a.**

1	−0.04	0.8	0.5	−1	0
		−0.04	0.76	1.26	0.26
	−0.04	0.76	1.26	0.26	0.26

$f(1) = 0.26$ ft/s; At 1 second, the speed of the boat is 0.26 ft/s.

2	−0.04	0.8	0.5	−1	0
		−0.08	1.44	3.88	5.76
	−0.04	0.72	1.94	2.88	5.76

$f(2) = 5.76$ ft/s; At 2 seconds, the speed of the boat is 5.76 ft/s.

3	−0.04	0.8	0.5	−1	0
		−0.12	2.04	7.62	19.86
	−0.04	0.68	2.54	6.62	19.86

$f(3) = 19.86$ ft/s; At 3 seconds, the speed of the boat
is 19.86 ft/s.

b.

6	−0.04	0.8	0.5	−1	0
		−0.24	3.36	23.16	132.96
	−0.04	0.56	3.86	22.16	132.96

$f(6) = 132.96$ ft/s; This means the boat is traveling at 132.96 ft/s when it passes the second buoy.

29. $x + 2, x - 3, x^2 - x + 4$ **31a.** $g(x) = 9x^4 + 50x^3 + 51x^2 - 150x - 72$ **31b.**

x	$g(x)$
−5	−9922
−4	−4160
−3	−1242
−2	−112
−1	70
0	−72
1	−130
2	88
3	558
4	1040
5	1078
6	0

31c. There is a zero between $x = -2$ and $x = -1$ because $g(x)$ changes sign between the two values. There are also zeros between −1 and 0 and between $x = 1$ and $x = 2$ because $g(x)$ changes sign between the two values. There is also a zero at $x = 6$.

31d.

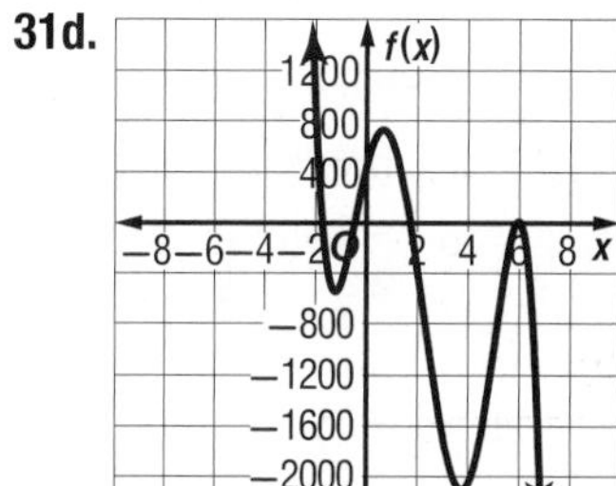

33

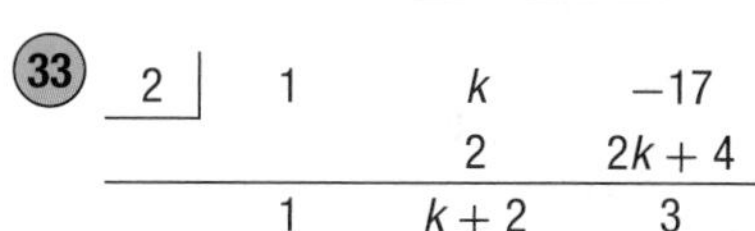

2	1	k	−17
		2	$2k + 4$
	1	$k + 2$	3

$-17 + 2k + 4 = 3$ Write an equation.
$-13 + 2k = 3$ Simplify.
$2k = 16$ Add 13 to each side.
$k = 8$ Divide each side by 2.

35. −3 **37.** $\pm\sqrt{6}, \pm\sqrt{3}$ **39a.** $x - c$ is a factor of $f(x)$. **39b.** $x - c$ is not a factor of $f(x)$. **39c.** $f(x) = x - c$
41. Sample answer: $f(x) = -x^3 + x^2 + x + 10$
43. Sample answer: A zero can be located using the Remainder Theorem and a table of values by determining when the output, or remainder, is equal to zero. For instance, if $f(6)$ leaves a remainder if 2 and $f(7)$ leaves a remainder of −1, then you know that there is a zero between $x = 6$ and $x = 7$. **45.** 4 **47.** B
49. $\pm 3, \pm i\sqrt{3}$ **51a.** −1.5 (max.), 0.5 (min.), 2.5 (max.) **51b.** −3.5, 3.75 **51c.** 4 **51d.** D = {all real numbers}; R = $\{y \mid y \leq 4.5\}$ **53a.** −3 (min.), −1 (max.), 1 (min.) **53b.** −0.25, 3 **53c.** 4 **53d.** D = {all real numbers}; R = $\{y \mid y \geq -4.5\}$

55.

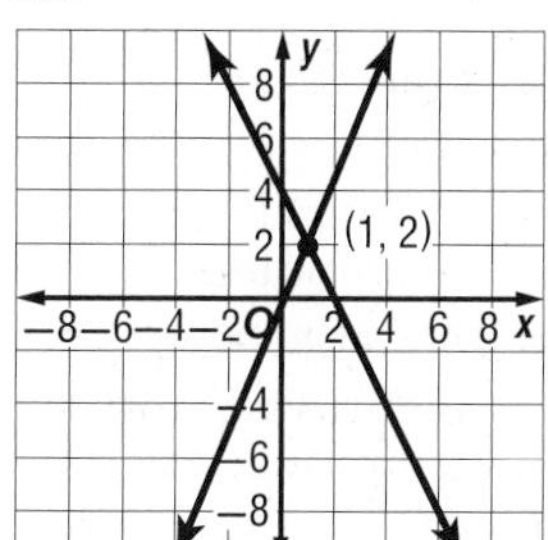

57.

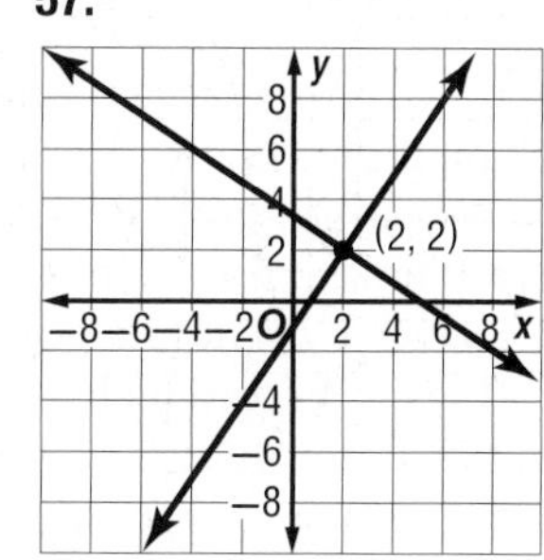

59. $4a^2 - 8a + 16$ **61.** $79a^2 - 58a + 12$
63. $-4a^4 - 24a^2 - 48a - 22$

Lesson 5-7

1. −2, 5; 2 real **3.** $-\frac{3}{2}, \frac{3}{2}, -\frac{3}{2}i, \frac{3}{2}i$; 2 real, 2 imaginary
5. 3 or 1; 0; 0 or 2 **7.** 1 or 3; 0 or 2; 0, 2, or 4
9. −8, −2, 1 **11.** $-4, 6, -4i, 4i$
13. $f(x) = x^3 - 9x^2 + 14x + 24$
15. $f(x) = x^4 - 3x^3 - x^2 - 27x - 90$
17. $-2, \frac{3}{2}$; 2 real **19.** $-1, \frac{1 \pm i\sqrt{3}}{2}$; 1 real, 2 imaginary
21. $-\frac{8}{3}, 1$; 2 real **23.** $-\frac{5}{2}, \frac{5}{2}, -\frac{5}{2}i, \frac{5}{2}i$; 2 real, 2 imaginary
25. −2, −2, 0, 2, 2; 5 real

27 Find the number of sign changes for $f(x)$ and $f(-x)$.

$f(x) = x^4 - 5x^3 + 2x^2 + 5x + 7$
yes yes no no

Since there are 2 sign changes, the function has 0 or 2 positive real zeros.

$f(-x) = x^4 + 5x^3 + 2x^2 - 5x + 7$
no no yes yes

Since there are 2 sign changes, the function has 0 or 2 negative real zeros.
Since f(x) has degree 4, the function has 4 zeros.
So, the function could have the following.
2 positive real zeros, 2 negative real zeros, 0 imaginary zeros
2 positive real zeros, 0 negative real zeros, 2 imaginary zeros
0 positive real zeros, 0 negative real zeros, 4 imaginary zeros

29. 0 or 2; 1, 2 or 4 **31.** 0 or 2; 0 or 2; 2, 4, or 6
33. −6, −2, 1 **35.** $-4, 7, -5i, 5i$ **37.** $4, 4, -2i, 2i$
39. $f(x) = x^3 - 2x^2 - 13x - 10$ **41.** $f(x) = x^4 + 2x^3 + 5x^2 + 8x + 4$ **43.** $f(x) = x^4 - x^3 - 20x^2 + 50x$

45 **a.** Find the number of sign changes for $P(x)$ and $P(-x)$.

$P(x) = -0.006x^4 + 0.15x^3 - 0.05x^2 - 1.8x$
yes yes no

Since there are 2 sign changes, the function has 0 or 2 positive real zeros.

$P(-x) = -0.006x^4 - 0.15x^3 - 0.05x^2 + 1.8x$
no no yes

Since there is 1 sign change, the function has 1 negative real zero.
Since $f(x)$ has degree 4, the function has 4 zeros.

Selected Answers and Solutions

So, the function could have the following.
2 positive real zeros, 1 negative real zero, 1 imaginary zero
0 positive real zeros, 1 negative real zero, 3 imaginary zeros
b. Negative zeros do not make sense in the problem because the number of computers produced cannot be negative. Since the zeros are values of x for which $P(x) = 0$, nonnegative zeros represent numbers of computers produced per hour which lead to no profit for the manufacturer.

47 Because $f(x)$ is of degree 3, there are at most 3 zeros. For $f(x)$, there is one sign change, so there are 1 or 0 positive real zeros. For $f(-x)$, there are two sign changes, so there are 2 or 0 negative real zeros. You can find the zeros by factoring, dividing, or using synthetic division.

$$f(x) = 4x^3 + 2x^2 - 4x - 2$$
$$= 2x^2(2x + 1) - 2(2x + 1)$$
$$= (2x^2 - 2)(2x + 1)$$
$$= 2(x^2 - 1)(2x + 1)$$

The zeros are -1, $-\frac{1}{2}$, and 1. Plot all three zeros on a coordinate plane. You need to determine what happens to the graph between the zeros and at its extremes. Use a table to analyze values close to the zeros. Use these values to complete the graph.

x	-2	$-\frac{3}{4}$	0	3
y	-18	0.4375	-2	112

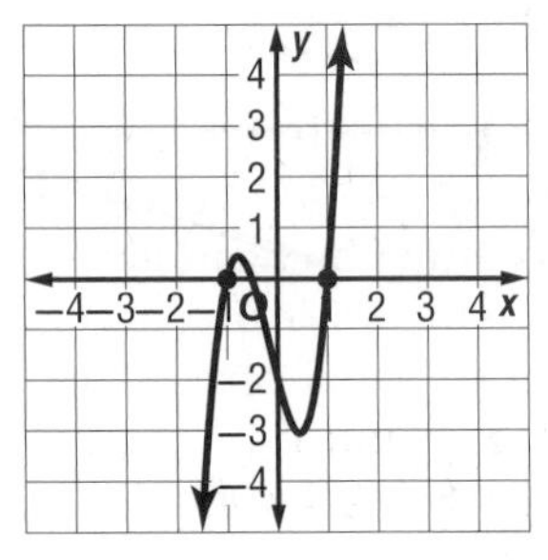

49.

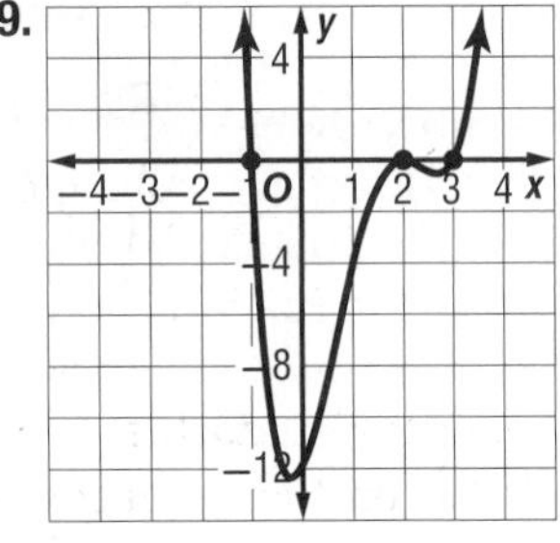

51. b **53a.** 3 or 1, 0, 2 or 0

53b.

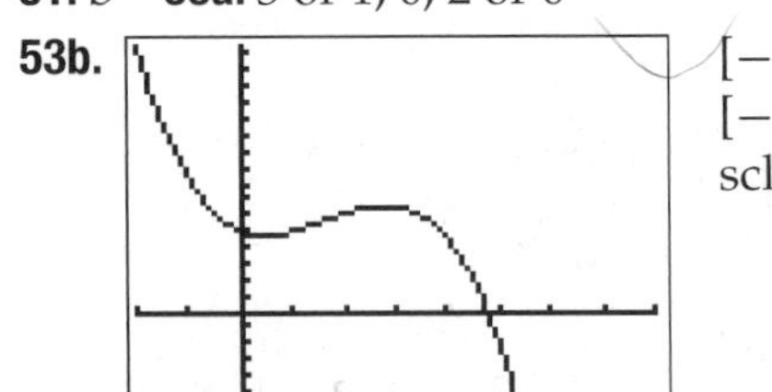

[−10, 40] scl: 5 by [−4000, 13,200] scl: 100

53c. 23.8; Sample answer: According to the model, the music hall will not earn any money after 2026.
55. 1 positive, 2 negative, 2 imaginary; Sample answer: The graph crosses the positive x-axis once, and crosses the negative x-axis twice. Because the degree of the polynomial is 5, there are 5 − 3 or 2 imaginary zeros. **57.** Sample answer: $f(x) = (x + 2i)(x - 2i)(3x + 5)(x + \sqrt{5})(x - \sqrt{5})$; Use conjugates for the imaginary and irrational values.
59a. Sample answer: $f(x) = x^4 + 4x^2 + 4$ **59b.** Sample answer: $f(x) = x^3 + 6x^2 + 9x$ **61.** C **63.** H **65.** $f(-8) = -1638$; $f(4) = 342$ **67.** $f(-8) = -63{,}940$; $f(4) = 1868$
69. $(a^2 + b^2)(a^4 - a^2b^2 + b^4)$ **71.** $(a - 4)(a - 2)(5a + 2b)$
73. 0.25 s **75.** $\pm1, \pm2, \pm4, \pm8, \pm\frac{1}{5}, \pm\frac{2}{5}, \pm\frac{4}{5}, \pm\frac{8}{5}$

Lesson 5-8

1. $\pm1, \pm2, \pm3, \pm4, \pm6, \pm8, \pm12, \pm24$
3. 5 in. × 9 in. × 28 in.

5 If $\frac{p}{q}$ is a rational zero, then p is a factor of 12 and q is a factor of 2.
p: $\pm1, \pm2, \pm3, \pm4, \pm6, \pm12$ q: $\pm1, \pm2$
possible rational zeros: $\frac{p}{q} = \pm1, \pm2, \pm3, \pm4, \pm6, \pm12, \pm\frac{1}{2}, \pm\frac{3}{2}$
There are no changes of signs for $f(x)$, so there are no positive real zeros. There are 4 changes of signs for $f(-x)$, so there are 0, 2, or 4 negative real zeros. Make a table for synthetic division and test the possible negative values.

$\frac{p}{q}$	2	11	26	29	12
−1	2	9	17	12	0
−2	2	7	12	5	2
−3	2	5	11	−4	24
−4	2	3	14	−27	120
−6	2	−1	32	−163	990
−12	2	−13	182	−2155	25,872
$-\frac{1}{2}$	2	10	21	18.5	2.75
$-\frac{3}{2}$	2	8	14	8	0

Because $f(-1) = 0$ and $f\left(-\frac{3}{2}\right) = 0$, there are zeros at $-\frac{3}{2}$ and -1. **7.** $-\frac{1}{2}, \frac{-5 \pm i\sqrt{23}}{8}$ **9.** $-\frac{1}{2}, \frac{3}{2}, 1 + 2i, 1 - 2i$
11. $\pm1, \pm2, \pm4, \pm7, \pm8, \pm14, \pm28, \pm56$

13 If $\frac{p}{q}$ is a rational zero, then p is a factor of 35 and q is a factor of 3.
p: $\pm1, \pm5, \pm7, \pm35$ q: $\pm1, \pm3$
possible rational zeros: $\frac{p}{q} = \pm1, \pm5, \pm7, \pm35, \pm\frac{1}{3}, \pm\frac{5}{3}, \pm\frac{7}{3}, \pm\frac{35}{3}$

15. $\pm1, \pm2, \pm3, \pm6, \pm7, \pm14, \pm21, \pm42, \pm\frac{1}{2}, \pm\frac{3}{2}, \pm\frac{7}{2}, \pm\frac{21}{2}, \pm\frac{1}{4}, \pm\frac{3}{4}, \pm\frac{7}{4}, \pm\frac{21}{4}, \pm\frac{1}{8}, \pm\frac{3}{8}, \pm\frac{7}{8}, \pm\frac{21}{8}$
17. $\pm1, \pm2, \pm4, \pm8, \pm16, \pm32, \pm64, \pm128, \pm\frac{1}{2}, \pm\frac{1}{4}, \pm\frac{1}{8}, \pm\frac{1}{16}$ **19.** −5, −3, −2 **21.** $-5, \frac{3}{4}, 5$ **23.** −1, 2
25. $-\frac{1}{4}$ **27.** −7, 1, 3 **29.** 2, −1, i, $-i$ **31.** 0, 3, $-i$, i
33. $-2, \frac{4}{3}, \frac{-3 \pm i}{2}$ **35.** $3, \frac{2}{3}, -\frac{2}{3}, \frac{-3 \pm \sqrt{13}}{2}$
37. $-\frac{1}{2}, \frac{1}{3}, \frac{1}{2}, \frac{3}{4}$ **39a.** $V(x) = 324x^3 + 54x^2 - 19x - 2$
39b. 3, $1.05i$, $-4.22i$; 3 is the only reasonable value for x. The other two values are imaginary.

41 **a.**
$V = \pi r^2 h$ Volume of a cylinder
$V = \pi r^2(r + 6)$ $h = r + 6$
$V = \pi r^2(r) + \pi r^2(6)$ Distributive Property

$V = \pi r^3 + 6\pi r^2$ Simplify.

b. $V = \pi r^3 + 6\pi r^2$ Volume of a cylinder

$160\pi = \pi r^3 + 6\pi r^2$ Substitute.

$0 = \pi r^3 + 6\pi r^2 - 160\pi$ Subtract 160π from each side.

$0 = r^3 + 6r^2 - 160$ Divide each side by π.

p: ±1, ±2, ±4, ±5, ±8, ±20, ±32, ±40, ±80, ±160

q: ±1

possible rational zeros: $\frac{p}{q}$ = ±1, ±2, ±4, ±5, ±8, ±20, ±32, ±40, ±80, ±160

Make a table and test some possible rational zeros.

r	1	6	0	−160
1	1	7	7	−153
2	1	8	16	−128
3	1	9	27	−79
4	1	10	40	0

$V(r) = 0$, so there is a zero at $x = 4$. Use the Quadratic Formula to find zeros of the depressed polynomial $r^2 + 10r + 40$.

$x = \frac{-b \pm \sqrt{b^2 - 4ac}}{2a}$ Quadratic Formula

$= \frac{-10 \pm \sqrt{10^2 - 4(1)(40)}}{2(1)}$ $a = 1$, $b = 10$, and $c = 40$

$= \frac{-10 \pm \sqrt{100 - 160}}{2}$ Multiply.

$= \frac{-10 \pm \sqrt{-60}}{2}$ Simplify.

$= \frac{-10 \pm 2i\sqrt{15}}{2}$ $\sqrt{-60} = 2i\sqrt{15}$

$= -5 \pm i\sqrt{15}$ Simplify.

Since the radius cannot be an imaginary number, the only reasonable value is $r = 4$ in.

c. Since $r = 4$ in., $h = r + 6$ or 10 in.

43a. $30x^3 - 478x^2 + 1758x + 1092 = 0$ **43b.** 1, 2, 3, 4, 6, 7, 12, 13, 14, 21, 26, 28, 39, 42, 52, 78, 84, 91, 156, 182, 273, 364, 546, 1092 **43c.** 2013 **43d.** No; Sample answer: Music sales fluctuate from 2005 to 2015, then increase indefinitely. It is not reasonable to expect sales to increase forever. **45.** 2, 3, 3, −3, −4

47. Sample answer: $f(x) = x^4 - 12x^3 + 47x^2 - 38x - 58$

49. Sample answer: $f(x) = 4x^5 + 3x^3 + 8x + 18$

51. Sample answer: For any polynomial function, the constant term represents p and the leading coefficient represents q. The possible zeros of the function can be found with $\pm\frac{p}{q}$ where the fraction is every combination of factors of p and q. For example, if p is 4 and q is 3, then ±4, ±2, ±1, $\pm\frac{4}{3}$, $\pm\frac{2}{3}$, and $\pm\frac{1}{3}$ are all possible zeros.

53. J **55.** 6 **57.** $f(x) = x^4 - 4x^3 + 11x^2 - 64x - 80$

59. $(x - 1)(x + 2)(x + 1)$ **61.** $(x - 3)(x + 4)(x - i)$

63a. about 3.5 s **63b.** 430 ft **65.** $3x^3 + 12x$ **67.** 32

69. $18c + 2$

Chapter 5 Study Guide and Review

1. true **3.** false; depressed polynomial **5.** true

7. true **9.** true **11.** $\frac{7x}{y^4}$ **13.** $r^2 + 8r - 5$

15. $m^3 - m^2p - mp^2 + p^3$ **17.** $3x^3 + 2x^2y^2 - 4xy$

19. $a^3 + 3a^2 - 4a + 2$ **21.** $x^2 + 3x - 40$ units^2

23. This is not a polynomial in one variable. It has two variables, x and y. **25.** $p(-2) = -3$; $p(x + h) = x^2 + 2xh + h^2 + 2x + 2h - 3$

27. $p(-2) = -25$; $p(x + h) = 3 - 5x^2 - 10xh - 5h^2 + x^3 + 3hx^2 + 3h^2x + h^3$

29a.

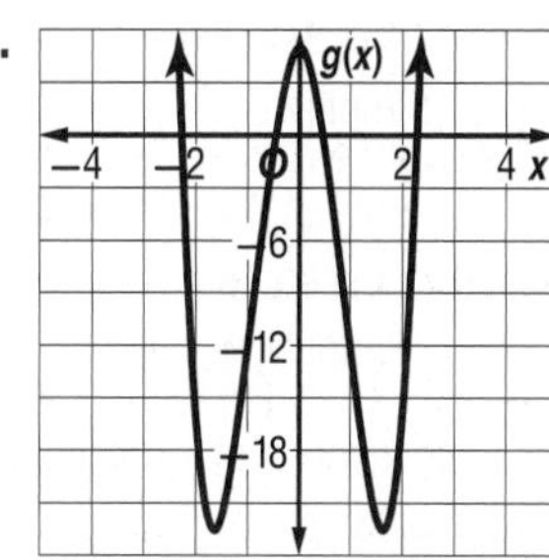

29b. between −3 and −2, between −1 and 0, between 0 and 1, between 2 and 3

29c. rel. max: $x \approx 0$; rel. min: $x \approx 1.62$ and $x \approx -1.62$

31a.

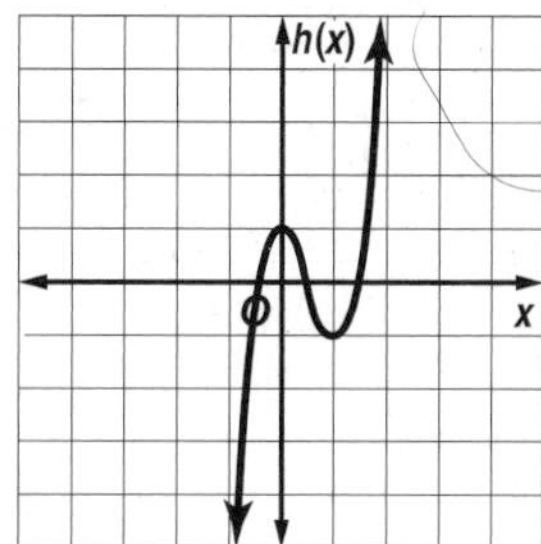

31b. between −1 and 0, between 0 and 1, and between 1 and 2

31c. rel. max: $x \approx 0$; rel. min: $x \approx 1$ **33.** 2 relative maxima and 1 relative minima

35. prime

37. $(2y + z)(3a + 2b - c)$ **39.** $\pm\frac{\sqrt{3}}{2}$, $\pm\frac{\sqrt{2}}{2}$

41. $f(-2) = 1$; $f(4) = 13$ **43.** $f(-2) = 16$; $f(4) = 118$

45. $x + 2$ and $3x - 1$ **47.** $x + 3$, $x + 4$

49. positive real zeros: 0
negative real zeros: 4, 2, or 0
imaginary zeros: 4, 2, or 0

51. positive real zeros: 2 or 0
negative real zeros: 1
imaginary zeros: 4 or 2

53. $-2, -1 \pm \sqrt{2}$ **55.** $-2, \pm 2i$

CHAPTER 6
Inverses and Radical Functions and Relations

Chapter 6 Get Ready

1. between 0 and 1, and between 3 and 4

3. between 1 and 2 seconds **5.** $3x + 2 - \frac{20}{x + 4}$

7. $3x^3 - 4x^2 + 5x - 3 + \frac{6}{x - 3}$

Lesson 6-1

1. $(f + g)(x) = 4x + 1$; $(f - g)(x) = -2x + 3$; $(f \cdot g)(x) = 3x^2 + 5x - 2$; $\left(\frac{f}{g}\right)(x) = \frac{x + 2}{3x - 1}$, $x \neq \frac{1}{3}$

3. $f \circ g$ is undefined, D = ∅, R = ∅; $g \circ f$ = {(2, 8), (6, 13), (12, 11), (7, 15)}, D = {2, 6, 7, 12}, R = {8, 11, 13, 15}. **5.** $[f \circ g](x) = -15x + 18$, R = {all multiples of 3}; $[g \circ f](x) = -15x - 6$, R = $\{y \mid y = 5x - 6 \text{ for } -\infty \leq x \leq \infty\}$. **7.** Either way, she will have $228.95 taken from her paycheck. If she takes the college savings plan deduction before taxes, $76 will go to her college plan and $152.95 will go to taxes. If she takes the college savings plan deduction after taxes, only $62.70 will go to her college plan and $166.25 will go to taxes.

9. $(f + g)(x) = 6x - 3$; $(f - g)(x) = -4x + 1$; $(f \cdot g)(x) = 5x^2 - 7x + 2$; $\left(\frac{f}{g}\right)(x) = \frac{x - 1}{5x - 2}$, $x \neq \frac{2}{5}$

11 $(f+g)(x) = f(x) + g(x)$ Addition of functions
$= (3x) + (-2x + 6)$ $f(x) = 3x$ and $g(x) = -2x + 6$
$= x + 6$ Simplify.

$(f-g)(x) = f(x) - g(x)$ Subtraction of functions
$= (3x) - (-2x + 6)$ Substitution
$= 5x - 6$ Simplify.

$(f \cdot g)(x) = f(x) \cdot g(x)$ Multiplication of functions
$= (3x)(-2x + 6)$ Substitution
$= -6x^2 + 18x$ Simplify.

$\left(\frac{f}{g}\right)(x) = \frac{f(x)}{g(x)}$ Division of functions
$= \frac{3x}{-2x+6}, x \neq 3$ Substitution

13. $(f+g)(x) = x^2 + x - 5$; $(f-g)(x) = x^2 - x + 5$; $(f \cdot g)(x) = x^3 - 5x^2$; $\left(\frac{f}{g}\right)(x) = \frac{x^2}{x-5}, x \neq 5$ **15.** $(f+g)(x) = 4x^2 - 8x$; $(f-g)(x) = 2x^2 + 8x - 8$; $(f \cdot g)(x) = 3x^4 - 24x^3 + 8x^2 + 32x - 16$; $\left(\frac{f}{g}\right)(x) = \frac{3x^2-4}{x^2-8x+4}, x \neq 4 \pm 2\sqrt{3}$ **17.** $f \circ g = \{(-4, 4)\}$, $D = \{-4\}$, $R = \{4\}$; $g \circ f = \{(-8, 0), (0, -4), (2, -5), (-6, -1)\}$, $D = \{-6, 0, 2\}$, $R = \{-5, -4, -1, 0\}$ **19.** $f \circ g$ is undefined, $D = \varnothing$, $R = \varnothing$; $g \circ f$ is undefined, $D = \varnothing$, $R = \varnothing$. **21.** $f \circ g$ is undefined, $D = \varnothing$, $R = \varnothing$; $g \circ f = \{(-4, 0), (1, 2)\}$, $D = \{-4, 1\}$, $R = \{0, 2\}$. **23.** $f \circ g = \{(4, 6), (3, -8)\}$, $D = \{3, 4\}$, $R = \{-8, 6\}$; $g \circ f$ is undefined, $D = \varnothing$, $R = \varnothing$. **25.** $f \circ g = \{(3, -1), (6, 11)\}$, $D = \{3, 6\}$, $R = \{-1, 11\}$; $g \circ f = \{(-4, 5), (-2, 4), (-1, 8)\}$, $D = \{-4, -2, -1\}$, $R = \{4, 5, 8\}$.

27 $[f \circ g](x) = f[g(x)]$ Composition of functions
$= f(x + 5)$ Replace $g(x)$ with $x + 5$.
$= 2(x + 5)$ Substitute $x + 5$ for x in $f(x)$.
$= 2x + 10$ Distributive Property

For $[f \circ g](x)$, $D = \{\text{all real numbers}\}$ and $R = \{\text{all even numbers}\}$.
$[g \circ f](x) = g[f(x)]$ Composition of functions
$= g(2x)$ Replace $f(x)$ with $2x$.
$= 2x + 5$ Substitute $2x$ for x in $g(x)$.

For $[g \circ f](x)$, $D = \{\text{all real numbers}\}$ and $R = \{\text{all odd numbers}\}$.

29. $[f \circ g](x) = 3x - 2$, $D = \{\text{all real numbers}\}$, $R = \{y \mid y = 2x - 2 \text{ for } -\infty \leq x \leq \infty\}$; $[g \circ f](x) = 3x + 8$, $D = \{\text{all real numbers}\}$, $R = \{y \mid y = 3x + 8 \text{ for } -\infty \leq x \leq \infty\}$ **31.** $[f \circ g](x) = x^2 - 6x - 2$, $D = \{\text{all real numbers}\}$, $R = \{y \mid y = x^2 - 6x - 2 \text{ for } -\infty \leq x \leq \infty\}$; $[g \circ f](x) = x^2 + 6x - 8$, $D = \{\text{all real numbers}\}$, $R = \{y \mid y = x^2 + 6x - 8 \text{ for } -\infty \leq x \leq \infty\}$ **33.** $[f \circ g](x) = 4x^3 + 7$, $D = \{\text{all real numbers}\}$, $R = \{y \mid y = 4x^3 + 7 \text{ for } -\infty \leq x \leq \infty\}$; $[g \circ f](x) = 64x^3 - 48x^2 + 12x + 1$, $D = \{\text{all real numbers}\}$, $R = \{y \mid y = 64x^3 - 48x^2 + 12x + 1 \text{ for } -\infty \leq x \leq \infty\}$ **35.** $[f \circ g](x) = 128x^4 + 96x^3 + 18x^2$, $D = \{\text{all real numbers}\}$, $R = \{y \mid y = 128x^4 + 96x^3 + 18x^2 \text{ for } -\infty \leq x \leq \infty\}$; $[g \circ f](x) = 32x^4 + 6x^2$, $D = \{\text{all real numbers}\}$, $R = \{y \mid y = 32x^4 + 6x^2 \text{ for } -\infty \leq x \leq \infty\}$ **37a.** $p(x) = 0.65x$; $t(x) = 1.0625x$ **37b.** Since $[p \circ t](x) = [t \circ p](x)$, either function represents the price. **37c.** $1587.75 **39.** $2(g \cdot f)(x) = 2x^3 - 4x^2 - 30x + 72$; $D = \{\text{all real numbers}\}$ **41.** 25 **43.** 483 **45.** -5 **47.** $-30a + 5$ **49.** $-10a^2 + 10a + 1$

51 **a.** Let $w(x)$ represent the function for women and $m(x)$ represent the function for men.
$(w + m)(x) = w(x) + m(x)$ Addition of functions
$= (1086.4x + 56{,}610) + (999.2x + 66{,}450)$ Substitution
$= 2085.6x + 123{,}060$ Simplify.

The equation $y = 2085.6x + 123{,}060$ models the total number.
b. $(f - g)(x)$ = the number of men employed in the U.S. − the number of women employed in the U.S. So, the function represents the difference in the number of men and women employed in the U.S.

53. 0 **55.** 1 **57.** 256 **59.** Sample answer: $f(x) = x - 9$, $g(x) = x + 5$ **61a.** $D = \{\text{all real numbers}\}$ **61b.** $D = \{x \mid x \geq 0\}$ **63.** Sample answer: Many situations in the real world involve complex calculations in which multiple functions are used. In order to solve some problems, a composition of those functions may need to be used. For example, the product of a manufacturing plant may have to go through several processes in a particular order, in which each process is described by a function. By finding the composition, only one calculation must be made to find the solution to the problem. **65.** G **67.** D **69.** $-3, 2, 4$ **71.** $-3, 5, \frac{1}{2}$ **73.** 1; 1; 2 **75.** 2 or 0; 2 or 0; 4, 2, or 0 **77.** (1, 2, 3) **79.** (3, −1, 5) **81.** $x = \frac{12 + 7y}{5}$ **83.** $x = \frac{15 - 8yz}{4}$ **85.** $k = \pm\sqrt{A - b}$

Lesson 6-2

1. $\{(10, -9), (-3, 1), (-5, 8)\}$

3. $f^{-1}(x) = -\frac{1}{3}x$

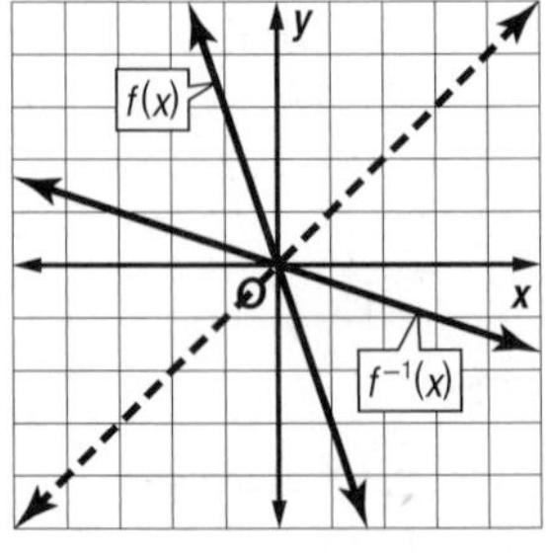

5. $h^{-1}(x) = \pm\sqrt{x + 3}$

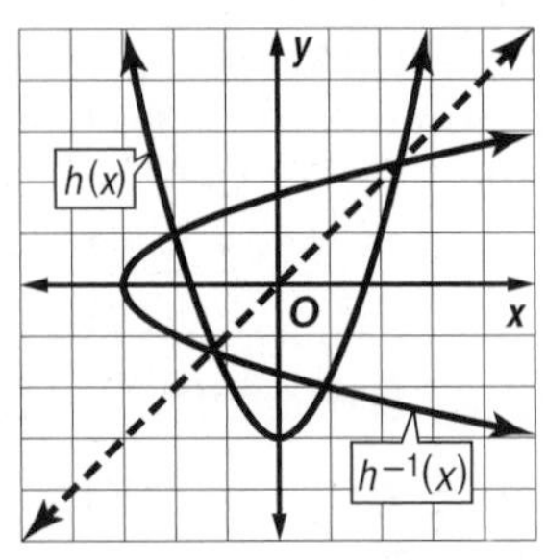

7. no **9.** $\{(6, -8), (-2, 6), (-3, 7)\}$ **11.** $\{(-1, 8), (-1, -8), (-8, -2), (8, 2)\}$ **13.** $\{(-5, 1), (6, 2), (-7, 3), (8, 4), (-9, 5)\}$

15. $f^{-1}(x) = x - 2$

17. $f^{-1}(x) = -\frac{x}{2} + \frac{1}{2}$

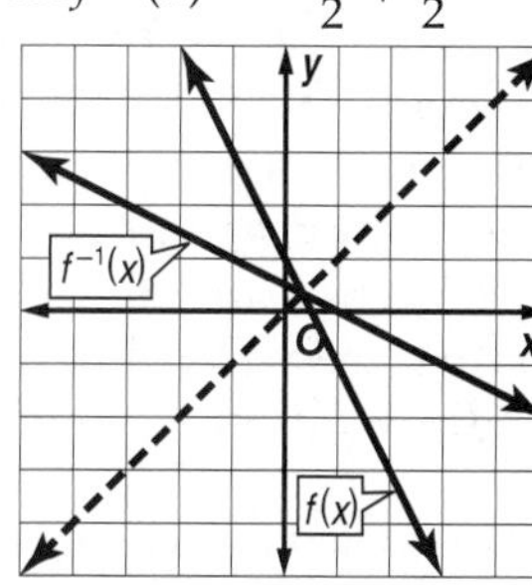

19. $f^{-1}(x) = -\frac{3}{5}(x + 8)$

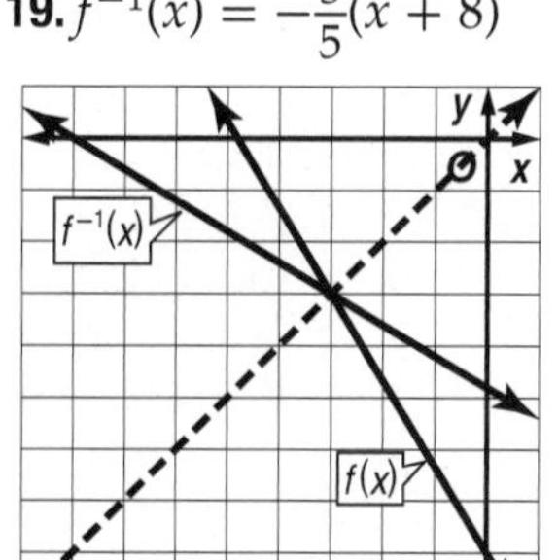

21. $f^{-1}(x) = \frac{1}{4}x$

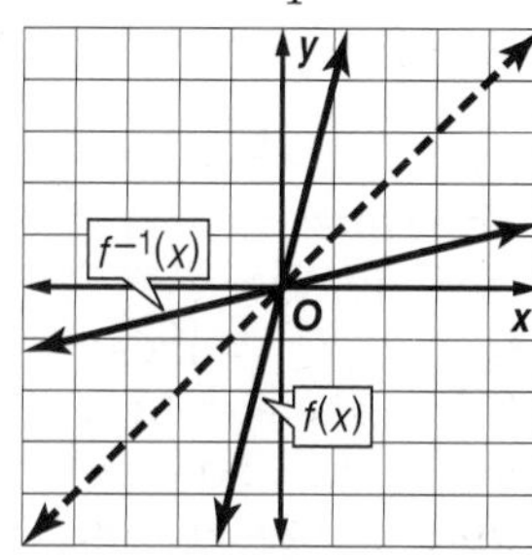

23. $f^{-1}(x) = \pm\sqrt{\frac{1}{5}x}$

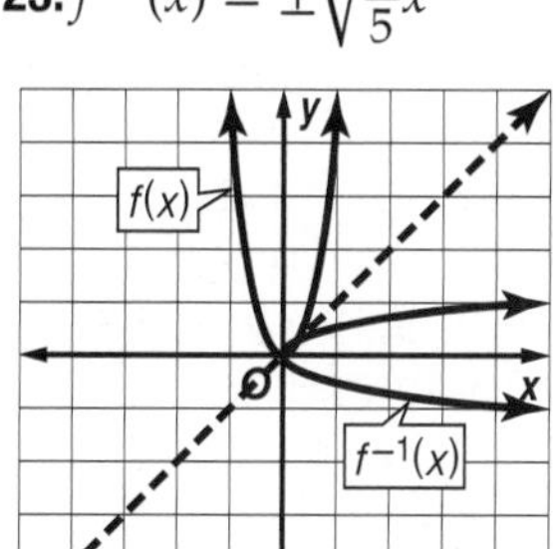

25. $f^{-1}(x) = \pm\sqrt{2x + 2}$

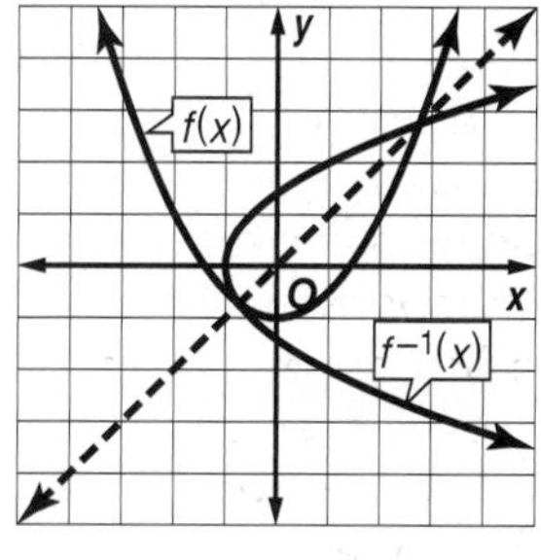

27. no **29.** yes **31.** yes **33.** yes

35

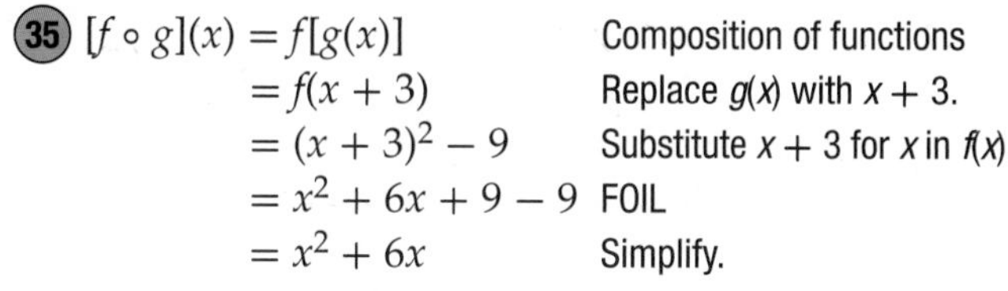

$[f \circ g](x) = f[g(x)]$	Composition of functions
$= f(x + 3)$	Replace $g(x)$ with $x + 3$.
$= (x + 3)^2 - 9$	Substitute $x + 3$ for x in $f(x)$.
$= x^2 + 6x + 9 - 9$	FOIL
$= x^2 + 6x$	Simplify.

$[g \circ f](x) = g[f(x)]$	Composition of functions
$= g(x^2 - 9)$	Replace $f(x)$ with $x^2 - 9$.
$= (x^2 - 9) + 3$	Substitute $x^2 - 9$ for x in $g(x)$.
$= x^2 - 6$	Simplify.

The functions are not inverses because $[f \circ g](x) \neq [g \circ f](x)$.

37. yes **39a.** $c(g) = 2.95g$ **39b.** $c\ (m) \approx 0.105m$

41 **a.**

$A = \pi r^2$	Write the formula
$y = \pi x^2$	Write the formula using x and y.
$x = \pi y^2$	Exchange x and y in the equation.
$\frac{x}{\pi} = y^2$	Divide each side by π.
$\sqrt{\frac{x}{\pi}} = y$	Take the positive square root of each side.

So, replacing y with r and x with A, the inverse is $r = \sqrt{\frac{A}{\pi}}$.

b. $r = \sqrt{\frac{A}{\pi}}$ Write the inverse of the function.

$= \sqrt{\frac{36}{\pi}}$ $A = 36$

≈ 3.39 cm Simplify.

43. yes **45.** no **47.** no **49a.** $F^{-1}(x) = \frac{5}{9}(x - 32)$; $F[F^{-1}(x)] = \frac{9}{5}\left[\frac{5}{9}(x - 32)\right] + 32 = x - 32 + 32 = x$; $F^{-1}[F(x)] = \frac{5}{9}\left(\frac{9}{5}x + 32 - 32\right) = \frac{5}{9}\left(\frac{9}{5}x + 0\right) = x.$

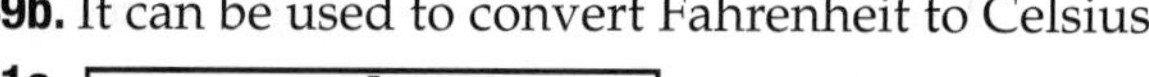

49b. It can be used to convert Fahrenheit to Celsius.

51a.

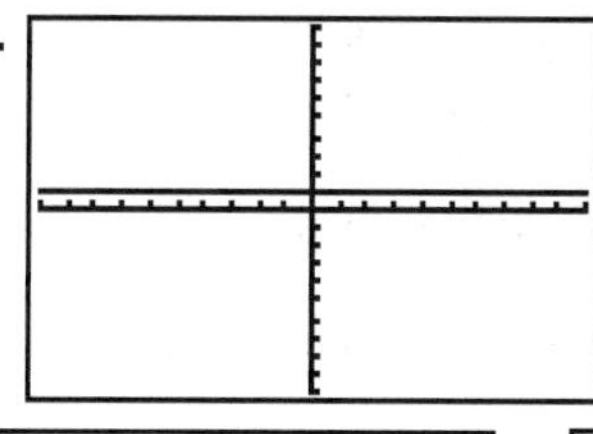

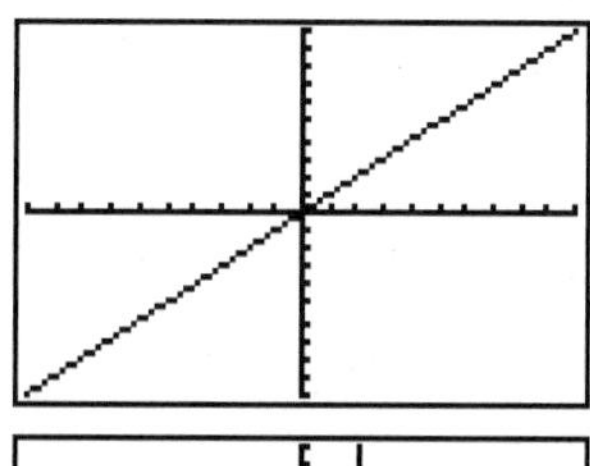

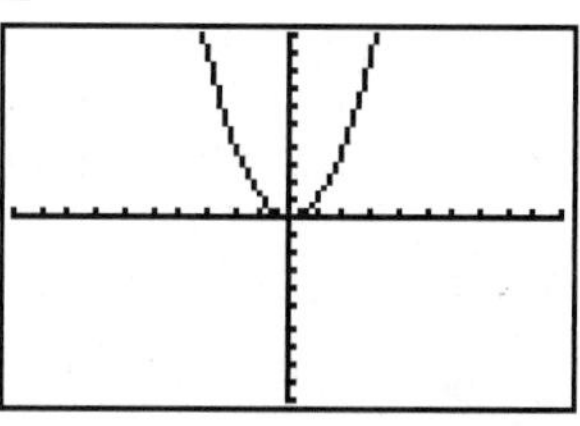

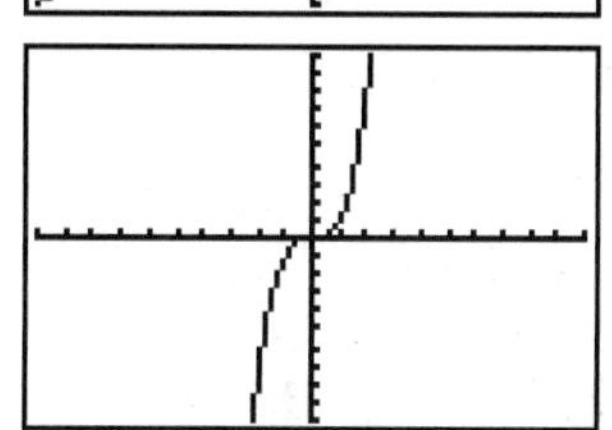

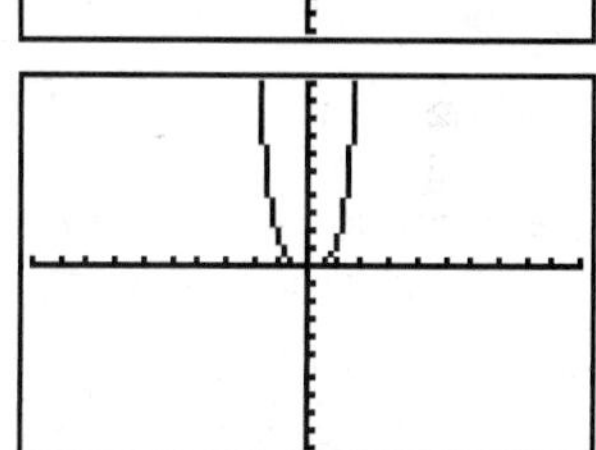

51b.

Function	Inverse a function?
$y = x^0$ or $y = 1$	no
$y = x^1$ or $y = x$	yes
$y = x^2$	no
$y = x^3$	yes
$y = x^4$	no

51c. n is odd. **53.** Sample answer: $f(x) = 2x$, $f^{-1}(x) = 0.5x$; $f[f^{-1}(x)] = f^{-1}[f(x)] = x$

55. $y = \frac{1}{m}x - \frac{b}{m}$

57. 30 in.

59. J **61.** 12 **63.** 0

65. 289; $(x + 17)^2$

67. $23 + 14i$ **69.** i

71. $\frac{1}{2}$ **73.** $\frac{3}{2}$

75.

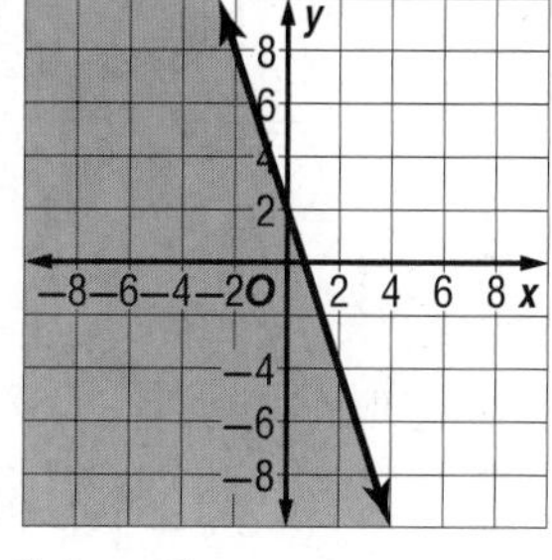

Lesson 6-3

1. D = $\{x \mid x \geq 0\}$; R = $\{f(x) \mid f(x) \geq 0\}$

3. D = $\{x \mid x \geq -8\}$; R = $\{f(x) \mid f(x) \geq -2\}$

5.

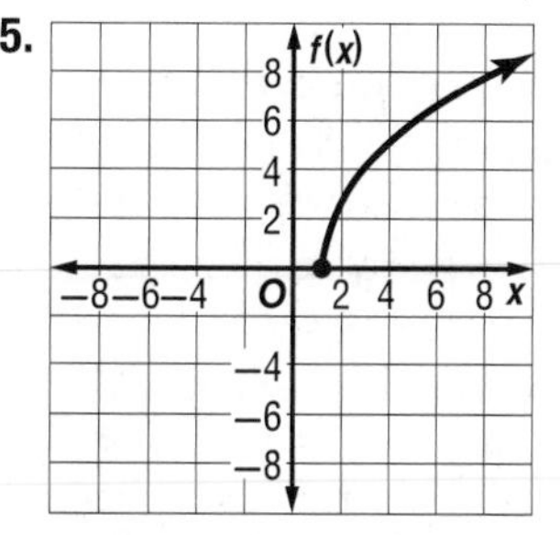

D = $\{x \mid x \geq 1\}$;

R = $\{f(x) \mid f(x) \geq 0\}$

7.

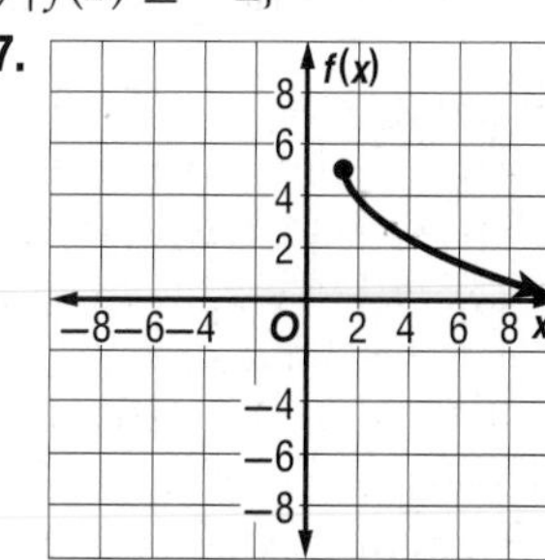

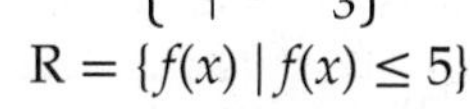

D = $\left\{x \mid x \geq \frac{5}{3}\right\}$;

R = $\{f(x) \mid f(x) \leq 5\}$

9. [graph]

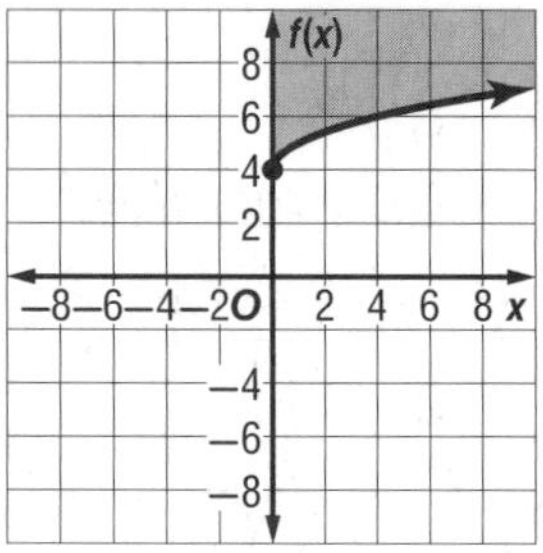

11. [graph]

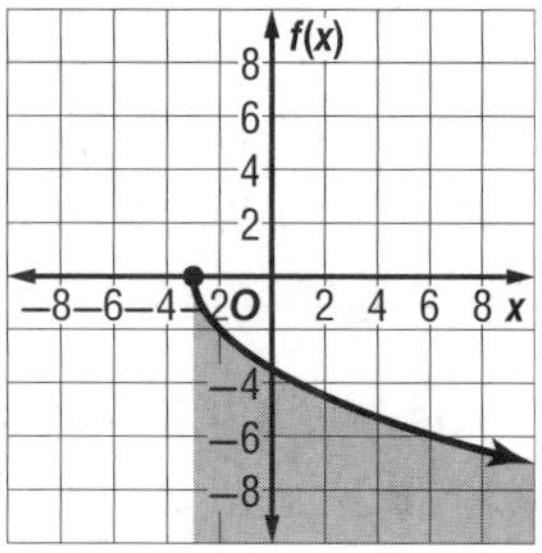

13. D = $\{x \mid x \geq 0\}$; R = $\{f(x) \mid f(x) \leq 2\}$

15 The domain only includes values for which the radicand is nonnegative.

$x - 2 \geq 0$ Write an inequality.

$x \geq 2$ Add 2 to each side.

The domain is $\{x \mid x \geq 2\}$. Find $f(2)$ to find the lower limit of the range.

$f(2) = 4\sqrt{2 - 2} - 8$

$= 0 - 8$ or -8

The range is $\{f(x) \mid f(x) \geq -8\}$.

17. D = $\{x \mid x \geq 4\}$; R = $\{f(x) \mid f(x) \geq -6\}$

19.

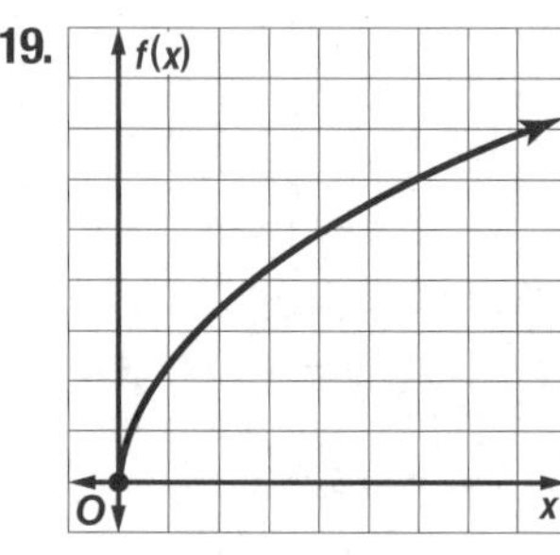

D = $\{x \mid x \geq 0\}$;
R = $\{f(x) \mid f(x) \geq 0\}$

21.

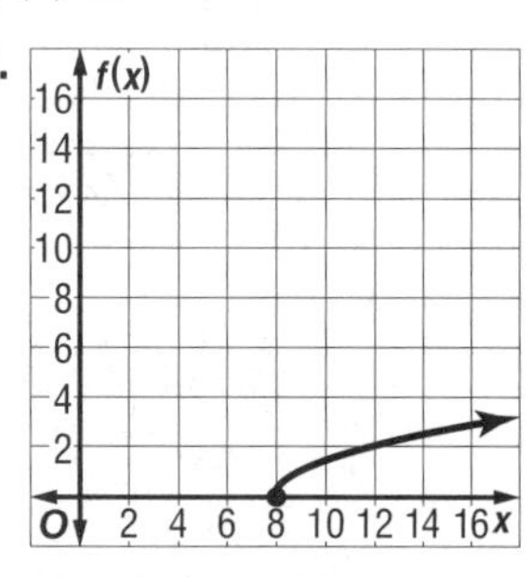

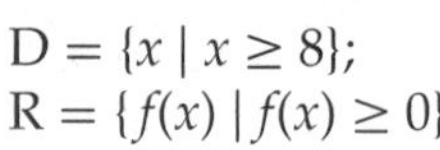

D = $\{x \mid x \geq 8\}$;
R = $\{f(x) \mid f(x) \geq 0\}$

23.

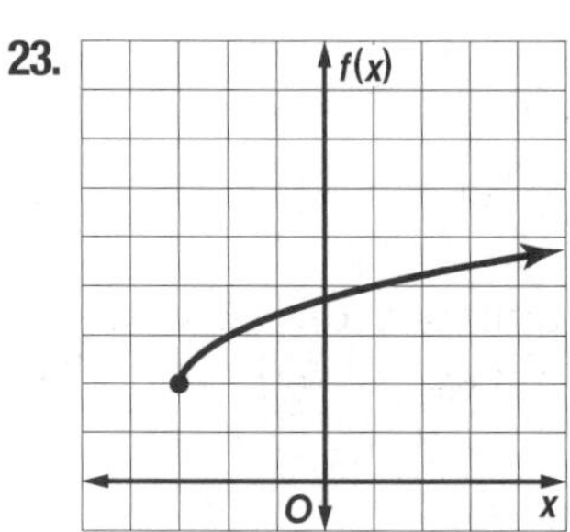

D = $\{x \mid x \geq -3\}$;
R = $\{f(x) \mid f(x) \geq 2\}$

25.

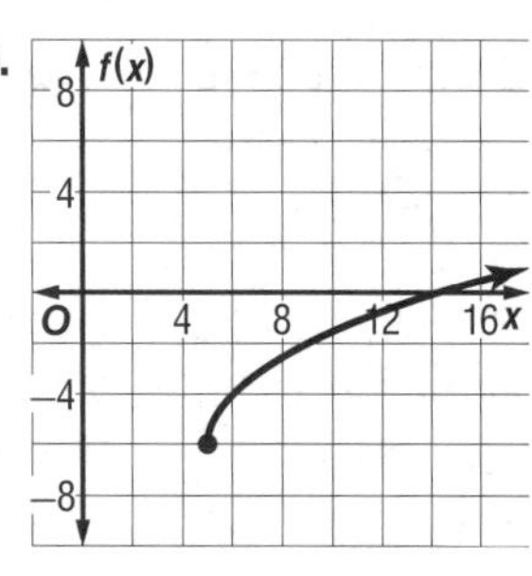

D = $\{x \mid x \geq 5\}$;
R = $\{f(x) \mid f(x) \geq -6\}$

27.

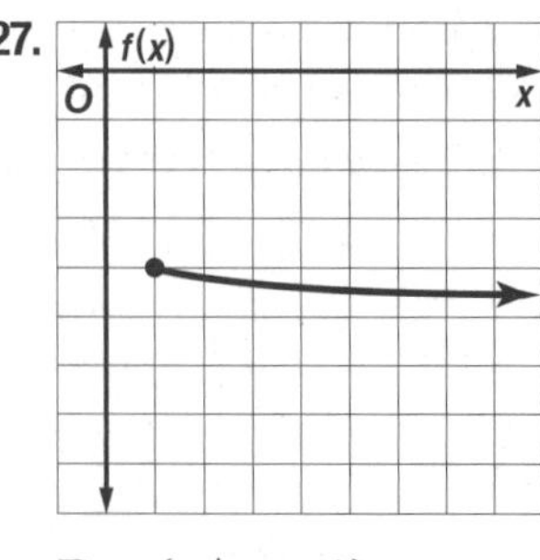

D = $\{x \mid x \geq 1\}$;
R = $\{f(x) \mid f(x) \leq -4\}$

29. 1936 ft

31.

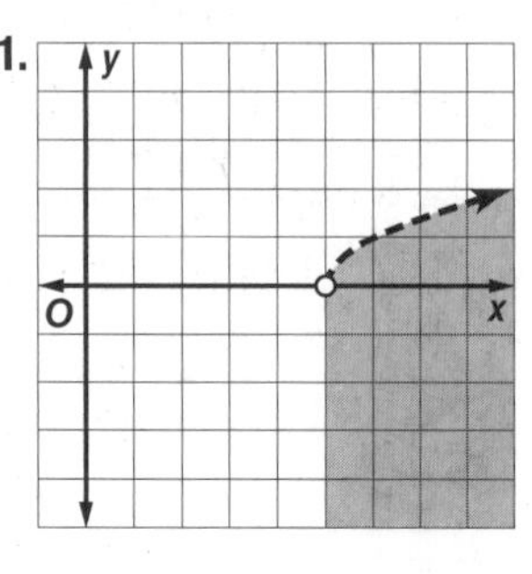

33

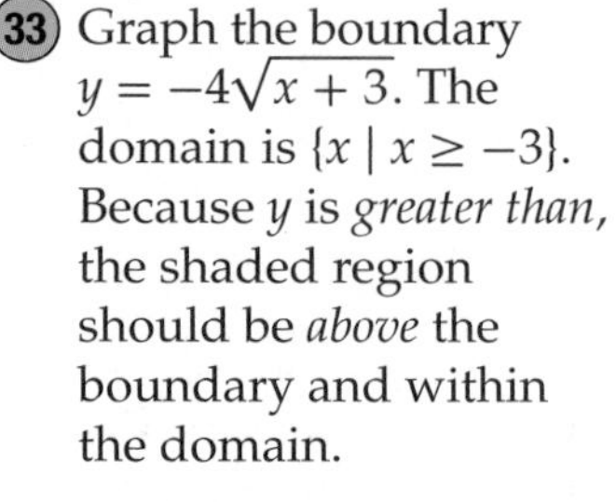

Graph the boundary $y = -4\sqrt{x + 3}$. The domain is $\{x \mid x \geq -3\}$. Because y is *greater than*, the shaded region should be *above* the boundary and within the domain.

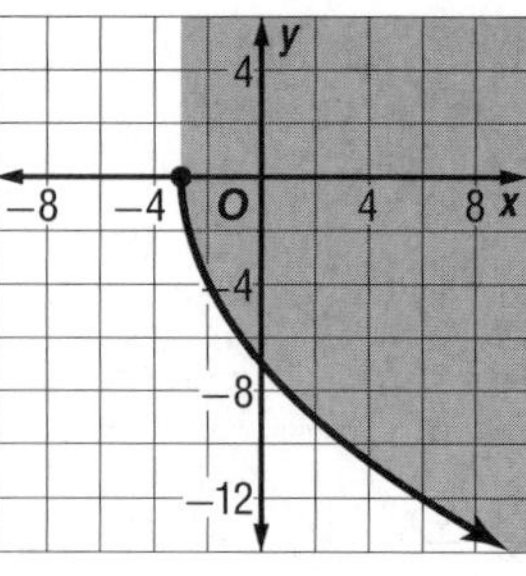

35. [graph]

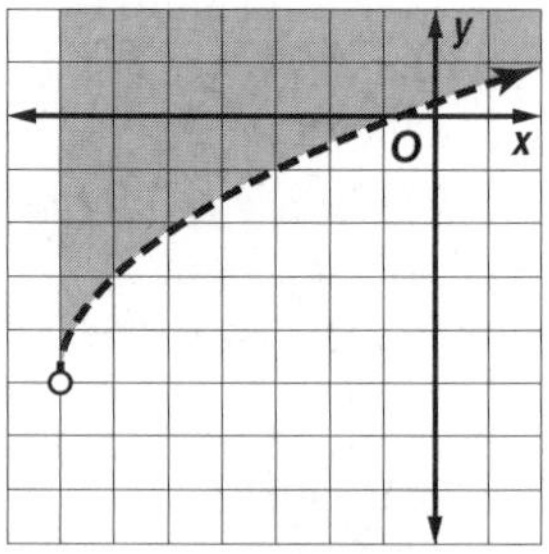

37.

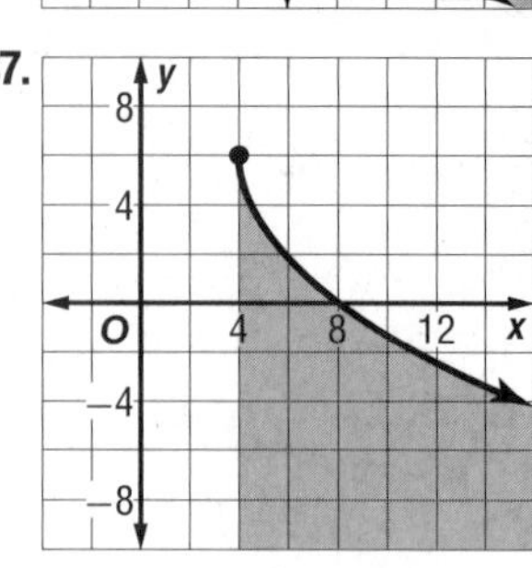

39a. $v = \sqrt{\frac{2E}{m}}$ **39b.** about 36.5 m/s **39c.** about 85,135 m/s **41.** $f(x) = \sqrt{x - 4} - 6$

43. $f(x) = -\sqrt{x + 6} - 6$

45 **a.** $T = 2\pi\sqrt{\frac{L}{g}}$ Original function

$T = 2\pi\sqrt{\frac{L}{32}}$ Replace g with 32.

Make a table of values for $0 \leq L \leq 10$. Graph the points and connect with a smooth curve.

L	T
0	0
1	1.11
2	1.57
3	1.92
4	2.22
5	2.48
6	2.72
7	2.94
8	3.14
9	3.33
10	3.51

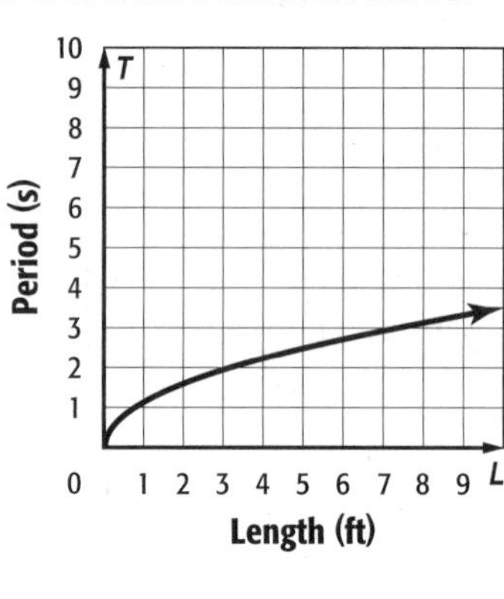

b. Use the table that you made in part a. (2, 1.57) means the period for a pendulum 2 feet long is about 1.57 seconds. (5, 2.48) means the period for a pendulum 5 feet long is about 2.48 seconds. (8, 3.14) means the period for a pendulum 8 feet long is about 3.14 seconds.

47. Sample answer: $y = -\sqrt{x + 4} + 6$

49. Sample answer: $y = -\sqrt{x - 8} + 14$

51. Molly; $y = \sqrt{5x + 10}$ has an x-intercept of -2 and would be at the right of the given graph.

53a. Sample answer: The original is $y = x^2 + 2$ and inverse is $y = \pm\sqrt{x - 2}$. **53b.** Sample answer: The original is $y = \pm\sqrt{x} + 4$ and inverse is $y = (x - 4)^2$. **55.** G **57.** E **59.** no **61.** $[d \circ h](m) = \frac{m}{1440}$ **63.** rational **65.** rational

Lesson 6-4

1. $\pm 10y^4$

③ $\sqrt{(y-6)^8} = \sqrt{[(y-6)^4]^2}$
$= (y-6)^4$

5. $\pm 4iy^2$ **7.** 7.616 **9.** -2.122
11. about 4.088×10^8 m

⑬ $\pm\sqrt{225a^{16}b^{36}} = \pm\sqrt{(15a^8b^{18})^2}$
$= \pm 15a^8b^{18}$

15. $-4c^2|d|$ **17.** $-20x^{16}y^{20}$ **19.** $(x^2+6)^8$ **21.** $2a^2b^4$
23. $3b^6c^4$ **25.** $\pm i(x+2)^4$ **27.** $|x^3|$ **29.** a^4
31. $(4x-7)^8$ **33.** $4|(5x-2)^3|$ **35.** $2a^3b^2$ **37.** 8 cm
39. -12.247 **41.** 0.787 **43.** -5.350 **45.** 29.573
47. $14|c^3|d^2$ **49.** $-3a^5b^3$ **51.** $20x^8|y^3|$ **53.** $4(x+y)^2$
55. about 141 million mi

㊼57 bald eagle:
$P = 73.3\sqrt[4]{m^3}$
$= 73.3\sqrt[4]{4.5^3}$
≈ 226.5 Cal/d

golden retriever:
$P = 73.3\sqrt[4]{m^3}$
$= 73.3\sqrt[4]{30^3}$
≈ 939.6 Cal/d

komodo dragon:
$P = 73.3\sqrt[4]{m^3}$
$= 73.3\sqrt[4]{72^3}$
≈ 1811.8 Cal/d

bottlenose dolphin:
$P = 73.3\sqrt[4]{m^3}$
$= 73.3\sqrt[4]{156^3}$
≈ 3235.5 Cal/d

Asian elephant:
$P = 73.3\sqrt[4]{m^3}$
$= 73.3\sqrt[4]{2300^3}$
$\approx 24{,}344.4$ Cal/d

59. Kimi is correct; Ashley's error was keeping the y^2 inside the absolute value symbol. **61.** Sample answer: Sometimes; when $x = -3$, $\sqrt[4]{(-x)^4} = |(-x)|$ or 3. When $x = 3$, $\sqrt[4]{(-x)^4} = |3|$ or 3.
63. Sample answers: 1, 64 **65.** $2\sqrt[3]{2xy}$ **67.** -0.1
69. $\frac{1}{625}$ **71.** G **73.** B

75.
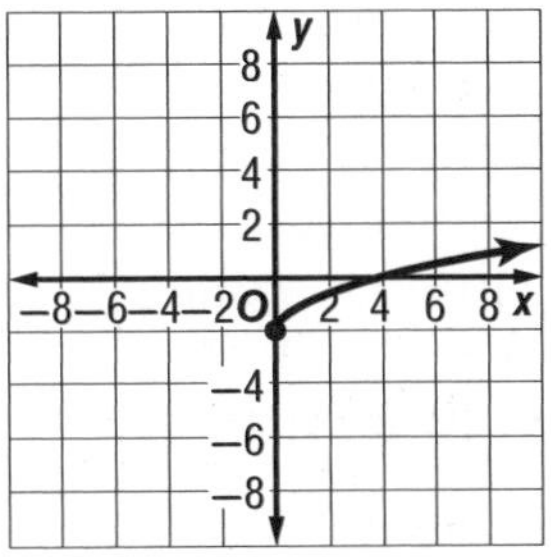

77. 3.41 kg and 49.53 cm
79. $4x^2 + 22x - 34$
81. $4a^4 + 24a^2 + 36$

83. $(-2, 3)$; $x = -2$; down
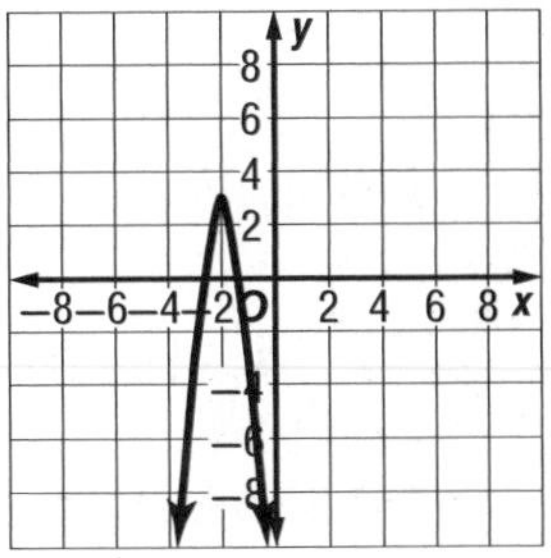

85. $(2, -2)$; $x = 2$; up
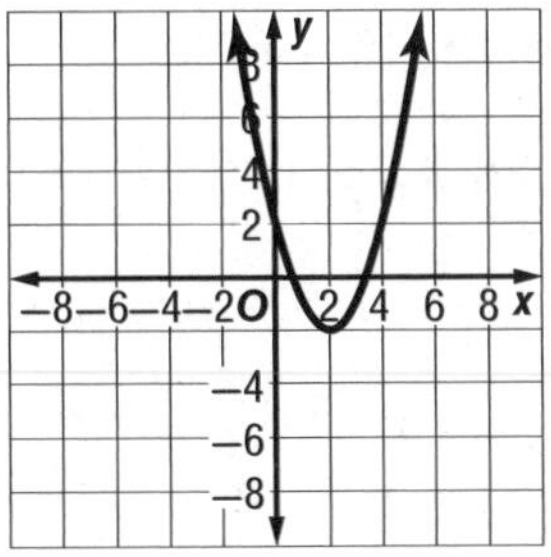

87. $x^2 + 9x + 20$ **89.** $a^2 - 7a - 18$ **91.** $x^2 + xy - 2y^2$

Lesson 6-5

1. $6b^2c^2\sqrt{ac}$ **3.** $\frac{c^2\sqrt{cd}}{d^5}$ **5.** $60x$ **7.** $36xy$
9. $20\sqrt{2} + 13\sqrt{3}$ **11.** $12\sqrt{3} + 16\sqrt{5} + 40 + 6\sqrt{15}$
13. $\frac{15 - 5\sqrt{2}}{7}$ **15.** $-2 - \sqrt{2}$ **17.** $32 - 2\sqrt{3}$ cm
19. $3a^7b\sqrt{ab}$ **21.** $3|a^3|bc^2\sqrt{2bc}$ **23.** $\frac{\sqrt{70xy}}{10y^2}$
25. $\frac{\sqrt[4]{28b^2x^3}}{2|b|}$ **27.** $32a^5b^3\sqrt{b}$ **29.** $25x^6y^3\sqrt{2xy}$
31. $18\sqrt{3} + 14\sqrt{2}$ **33.** $9\sqrt{6} + 72\sqrt{2} - 7\sqrt{3}$

㉟35
$A = \ell w$ — Area of a rectangle
$= (8 + \sqrt{3})(\sqrt{6})$ — $\ell = 8 + \sqrt{3}$ and $w = \sqrt{6}$
$= 8 \cdot \sqrt{6} + \sqrt{3} \cdot \sqrt{6}$ — Distributive Property
$= 8\sqrt{6} + \sqrt{18}$ — Product Property
$= 8\sqrt{6} + 3\sqrt{2}$ ft^2 — Simplify.

37. $56\sqrt{3} + 42\sqrt{6} - 36\sqrt{2} - 54$ **39.** 1260
41. $6\sqrt{3} + 6\sqrt{2}$ **43.** $\frac{20 - 7\sqrt{3}}{11}$ **45.** $2yz^4\sqrt[3]{2y}$
47. $3|a|b^3\sqrt[4]{2a^2bc}$ **49.** $\frac{\sqrt[4]{1500a^2b^3x^3y^2}}{5|a|b}$

㊿51
$\frac{x+1}{\sqrt{x}-1} = \frac{x+1}{\sqrt{x}-1} \cdot \frac{\sqrt{x}+1}{\sqrt{x}+1}$ — $\sqrt{x} + 1$ is the conjugate of $\sqrt{x} - 1$.
$= \frac{x(\sqrt{x}) + 1(x) + 1(\sqrt{x}) + 1(1)}{\sqrt{x}(\sqrt{x}) + 1(\sqrt{x}) + -1(\sqrt{x}) + (-1)(1)}$ — Multiply.
$= \frac{x\sqrt{x} + x + \sqrt{x} + 1}{x + \sqrt{x} - \sqrt{x} - 1}$ — Simplify.
$= \frac{x(\sqrt{x} + 1) + (\sqrt{x} + 1)}{x - 1}$ — Simplify.
$= \frac{(x+1)(\sqrt{x}+1)}{x-1}$ — Simplify.

53. $\frac{\sqrt{x^3 - x}}{x^2 - 1}$ **55.** $|a|$ **57.** a^2
59a. $a^2 + b^2 = c^2$
$1^2 + 1^2 = c^2$
$2 = c^2$
$c = \sqrt{2}$

59b.
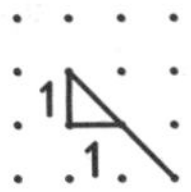

59c. $\sqrt{2} + \sqrt{2}$ units is the length of the hypotenuse of an isosceles right triangle with legs of length 2 units. Therefore, $\sqrt{2} + \sqrt{2} > 2$.

59d.
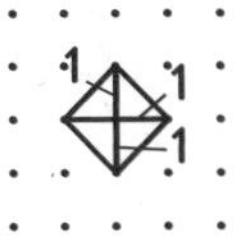

59e. The square creates 4 triangles with a base of 1 and a height of 1. Therefore the area of each triangle is $\frac{1}{2}bh = \frac{1}{2}(1)(1)$ or $\frac{1}{2}$. $4\left(\frac{1}{2}\right) = 2$. The area of the square is 2, so $\sqrt{2} \cdot \sqrt{2} = 2$.

61. $\left(\frac{-1 - i\sqrt{3}}{2}\right)^3 = \left(\frac{-1 - i\sqrt{3}}{2}\right) \cdot \left(\frac{-1 - i\sqrt{3}}{2}\right) \cdot \left(\frac{-1 - i\sqrt{3}}{2}\right)$
$= \frac{(-1 - i\sqrt{3})(-1 - i\sqrt{3})(-1 - i\sqrt{3})}{8}$

Selected Answers and Solutions

$$= \frac{(1 + i\sqrt{3} + i\sqrt{3} + 3i^2)(-1 - i\sqrt{3})}{8}$$
$$= \frac{(2i\sqrt{3} - 2)(-1 - i\sqrt{3})}{8}$$
$$= \frac{-2i\sqrt{3} - 6i^2 + 2 + 2i\sqrt{3}}{8}$$
$$= \frac{-6i^2 + 2}{8} = \frac{8}{8} \text{ or } 1$$

63. $a = 1, b = 256; a = 2, b = 16; a = 4, b = 4; a = 8, b = 2$ **65.** Sample answer: It is only necessary to use absolute values when it is possible that n could be odd or even and still be defined.
It is when the radicand must be nonnegative in order for the root to be defined that the absolute values are not necessary. **67.** G **69.** C **71.** $9ab^3$

73.

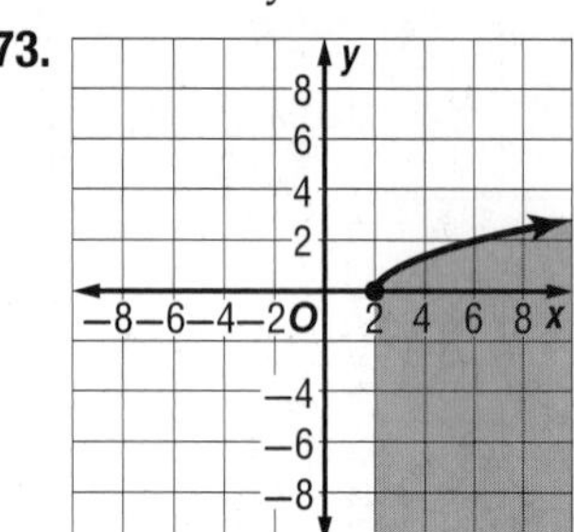

75. $-4, 4, -i, i$ **77.** $-4, 2 + 2i\sqrt{3}, 2 - 2i\sqrt{3}$
79. $\frac{3}{2}, \frac{-3 + 3i\sqrt{3}}{4}, \frac{-3 - 3i\sqrt{3}}{4}$ **81.** 9 small, 4 large **83.** $\frac{1}{3}$ **85.** $\frac{7}{12}$ **87.** $\frac{5}{12}$

Lesson 6-6

1. $\sqrt[4]{10}$ **3.** $15^{\frac{1}{3}}$ **5.** 7 **7.** 25

9 $\ell = A^{\frac{1}{2}}$ Write the formula.
$= 169^{\frac{1}{2}}$ $A = 169$
$= \sqrt{169}$ Write in radical form.
$= 13$ ft Simplify.

11. $x^{\frac{3}{5}}$ **13.** $\sqrt{3g}$ **15.** $\frac{g - 2g^{\frac{1}{2}} + 1}{g - 1}$ **17.** $\sqrt[7]{16}$ **19.** $\sqrt{x^9}$
21. $63^{\frac{1}{4}}$ **23.** $5x^{\frac{1}{2}}$ **25.** 4 **27.** $\frac{1}{3}$ **29.** about 2.64 cm
31. $a^{\frac{25}{36}}$ **33.** $\frac{y^{\frac{1}{5}}}{y}$

35 $\frac{\sqrt[4]{27}}{\sqrt[4]{3}} = \frac{27^{\frac{1}{4}}}{3^{\frac{1}{4}}}$ Rational exponents
$= \frac{(3^3)^{\frac{1}{4}}}{3^{\frac{1}{4}}}$ $27 = 3^3$
$= \frac{3^{\frac{3}{4}}}{3^{\frac{1}{4}}}$ Power of a power
$= 3^{\frac{3}{4} - \frac{1}{4}}$ Quotient of powers
$= 3^{\frac{2}{4}}$ Simplify.
$= 3^{\frac{1}{2}}$ Simplify.
$= \sqrt{3}$ Write in radical form.

37. $\sqrt[3]{9} \cdot \sqrt{g}$ **39.** $\frac{x + 4x^{\frac{3}{4}} + 8x^{\frac{1}{2}} + 16x^{\frac{1}{4}} + 16}{x - 16}$
41. $28.27x^{\frac{4}{3}}y^{\frac{2}{5}}z^4$ units^2 **43.** $6 - 2 \cdot 4^{\frac{1}{3}}$ **45.** $x^{\frac{10}{3}}$ **47.** $y^{\frac{3}{20}}$
49. $\sqrt{6}$ **51.** $\frac{w^{\frac{1}{8}}}{w}$

53 $\frac{f^{-\frac{1}{4}}}{4f^{\frac{1}{2}} \cdot f^{-\frac{1}{3}}} = \frac{f^{\frac{1}{3}}}{4f^{\frac{1}{2}} \cdot f^{\frac{1}{4}}}$ Defn. of negative exponents
$= \frac{f^{\frac{1}{3}}}{4f^{\frac{3}{4}}}$ Add powers.
$= \frac{f^{\frac{1}{3}}}{4f^{\frac{3}{4}}} \cdot \frac{f^{\frac{1}{4}}}{f^{\frac{1}{4}}}$ Multiply to rationalize the denominator.
$= \frac{f^{\frac{7}{12}}}{4f}$ $f^{\frac{1}{3}} \cdot f^{\frac{1}{4}} = f^{\frac{1}{3} + \frac{1}{4}}$

55. $c^{\frac{1}{2}}$ **57.** $23\sqrt[6]{23}$ **59.** 3 **61.** $\frac{ab\sqrt{c}}{c}$ **63.** $2\sqrt{6} - 5$

65a.

x	$f(x)$	$g(x)$
−2	−8	−1.26
−1	−1	−1
0	0	0
1	1	1
2	8	1.26

65b.

65c. It is a reflection in the line $y = x$. **67a.** Sample answer: $\sqrt[4]{(-16)^3} = \sqrt[4]{-4096}$; there is no real number that when raised to the fourth power results in a negative number. **67b.** Sample answer: $\sqrt[4]{-1}$
69. Sample answer: It may be easier to simplify an expression when it has rational exponents because all the properties of exponents apply. We do not have as many properties dealing directly with radicals. However, we can convert all radicals to rational exponents, and then use the properties of exponents to simplify. **71.** B **73.** C **75.** $9\sqrt{3}$ **77.** $6y^2z\sqrt[3]{7}$
79. $-6; x + h - 2$ **81.** $-21; 6x + 6h + 3$
83. $20; x^2 + 2xh + h^2 - x - h$ **85.** $\{0, 11\}$ **87.** $\{-\frac{3}{4}, 4\}$
89. $\{3\}$ **91.** 5 in. by 4 in. **93.** $3x - 4$ **95.** $x - 8\sqrt{x} + 16$
97. $9x + 6\sqrt{x} + 1$

Lesson 6-7

1. 20 **3.** 13

5 $\sqrt[3]{x - 2} = 3$ Original equation
$(\sqrt[3]{x - 2})^3 = 3^3$ Raise each side to the third power.
$x - 2 = 27$ Evaluate each side.
$x = 29$ Add 2 to each side.

7. 2 **9.** 49 **11.** $\frac{27}{2}$ **13a.** about 9.5 s **13b.** about 324 feet **15.** $-\frac{4}{3} \le x \le \frac{77}{3}$ **17.** $1 \le y \le 5$ **19.** $x > 1$
21. $x \le -11$ **23.** 22 **25.** 3 **27.** no real solution **29.** $\frac{1}{4}$

Selected Answers and Solutions

31. 9 **33.** $\frac{81}{16}$ **35.** 1 m **37.** 3 **39.** 83 **41.** 61 **43.** 3
45. 18 **47.** 2 **49.** F **51.** $x \geq 43$ **53.** no real solution

55

$\sqrt{d+3} + \sqrt{d+7} > 4$	Original inequality
$\sqrt{d+3} > 4 - \sqrt{d+7}$	Subtract $\sqrt{d+7}$ from each side.
$d + 3 > 16 - 8\sqrt{d+7} + d + 7$	Square each side.
$\frac{5}{2} < \sqrt{d+7}$	Simplify.
$\frac{25}{4} < d + 7$	Square each side.
$-\frac{3}{4} < d$	Subtract 7 from each side.
$d > -\frac{3}{4}$	Rewrite inequality.

57. $-\frac{5}{2} \leq y \leq 2$ **59.** $a > 8$ **61.** $0 \leq c < 3$
63. $M = \left(\frac{L}{0.46}\right)^3$

65

$A = \pi r^2$	Original formula
$250{,}000 = \pi r^2$	Replace A with 250,000.
$79{,}577.5 \approx r^2$	Divide each side by π.
$\sqrt{79{,}577.5} \approx r$	Take the square root of each side.
$282.1 \approx r$	Use a calculator.

The radius is about 282 ft.

67. $\sqrt{x+2} - 7 = -10$
69. never

$$\frac{\sqrt{(x^2)^2}}{-x} = x$$
$$\frac{x^2}{-x} = x$$
$$x^2 = (x)(-x)$$
$$x^2 \neq -x^2$$

71. They are the same number. **73.** 3
75. Sometimes; Sample answer: When the radicand is negative, then there will be extraneous roots. **77.** G
79. A **81.** 81 **83.** $4x^2y^2\sqrt{5}$ **85.** $y^{-1} = \frac{-x-3}{2}$
87. $y^{-1} = \pm\frac{1}{2}\sqrt{x} - \frac{3}{2}$ **89a.** $f(x) \to +\infty$ as $x \to +\infty$, $f(x) \to +\infty$ as $x \to -\infty$ **89b.** even **89c.** 0 **91.** $\frac{1}{6}$
93. $\frac{5}{8}$ **95.** 16 **97.** $2\frac{1}{2}$

Chapter 6 Study Guide and Review

1. identity function **3.** composition of functions
5. rationalizing the denominator **7.** inverse relations
9. radical function
11. $[f \circ g](x) = x^2 - 14x + 50$; $[g \circ f](x) = x^2 - 6$
13. $[f \circ g](x) = 20x - 4$; $[g \circ f](x) = 20x - 1$
15. $[f \circ g](x) = x^2 + 4x$; $[g \circ f](x) = x^2 + 2x - 2$
17. $f^{-1}(x) = \frac{x+6}{5}$

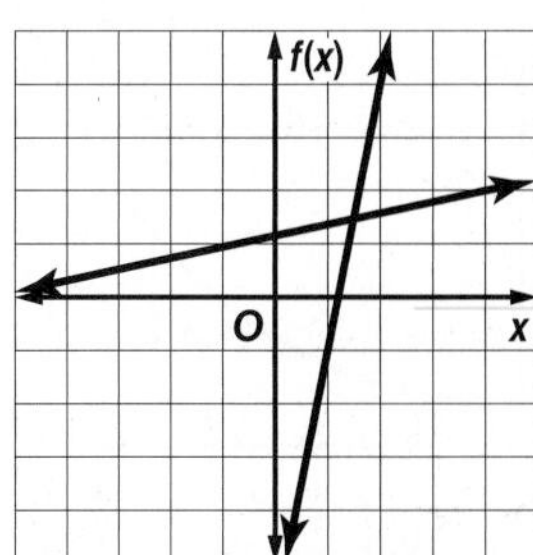

19. $f^{-1}(x) = 2x - 6$

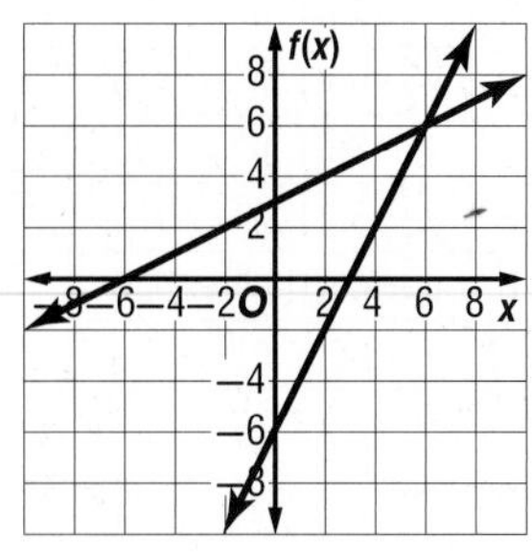

21. $f^{-1}(x) = \pm\sqrt{x}$

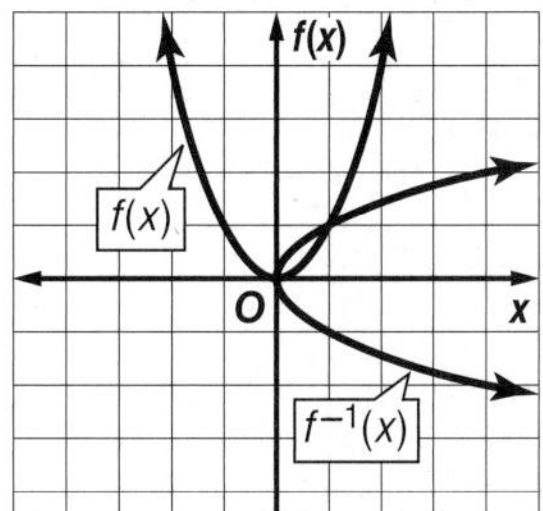

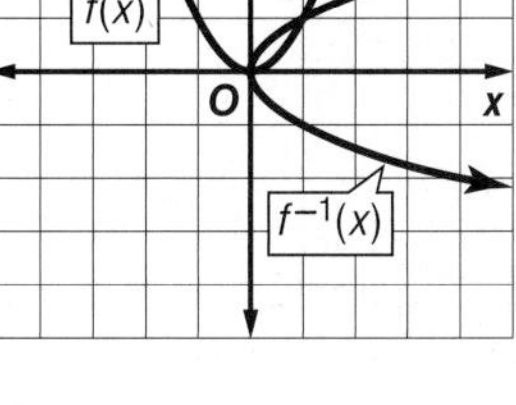

23. $1200 **25.** yes
27. no **29.** no

31.

D = $\{x \mid x \geq 0\}$;
R = $\{f(x) \mid f(x) \geq 0\}$

33.

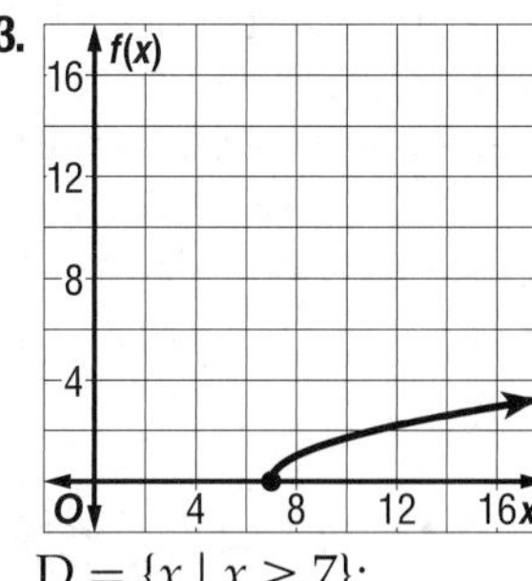

D = $\{x \mid x \geq 7\}$;
R = $\{f(x) \mid f(x) \geq 0\}$

35.

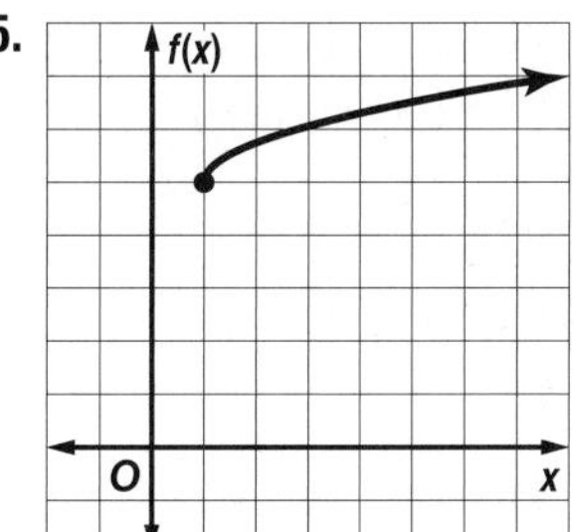

D = $\{x \mid x \geq 1\}$;
R = $\{f(x) \mid f(x) \geq 5\}$
37. about 9.8 in.

39.

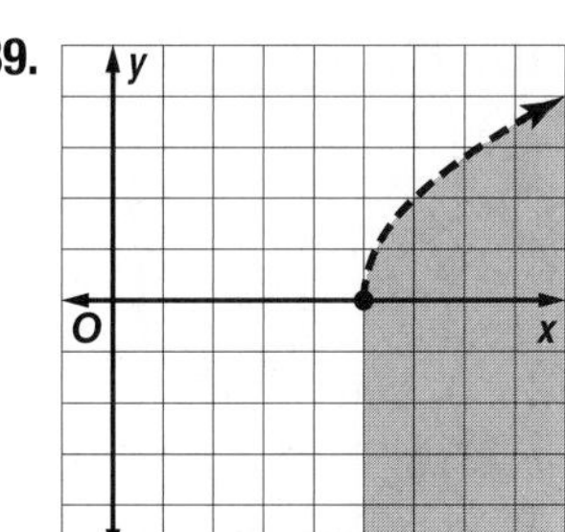

41. ± 11 **43.** 6 **45.** $(x^2 + 2)^3$ **47.** $a^2|b^3|$ **49.** 10 m/s
51. $12ab^2\sqrt{ab}$ **53.** $80\sqrt{2}$ **55.** $\frac{m^2\sqrt{6mp}}{p^6}$
57. $-\sqrt{15} - 3\sqrt{2}$ **59.** $x^{\frac{7}{6}}$
61. $\frac{d^{\frac{5}{12}}}{d}$ **63.** 3 **65.** $4a^{\frac{2}{3}}b^{\frac{4}{5}}c^2\pi$ units^2 **67.** $\frac{100}{9}$
69. 2 **71.** no solution **73.** 3 **75.** $\frac{1}{3} \leq x < \frac{10}{3}$
77. $x \geq \frac{4}{3}$ **79.** no solution **81.** $x > \frac{5}{2}$

CHAPTER 7 Exponential and Logarithmic Functions and Relations

Chapter 7 Get Ready

1. a^{12} **3.** $\frac{-3x^6}{2y^3z^5}$ **5.** 5 g/cm^3

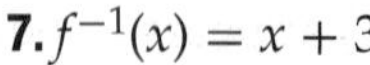
7. $f^{-1}(x) = x + 3$

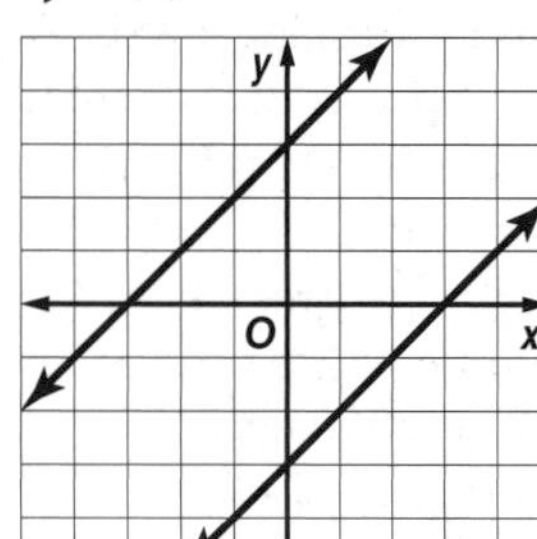

9. $f^{-1}(x) = 4x + 12$

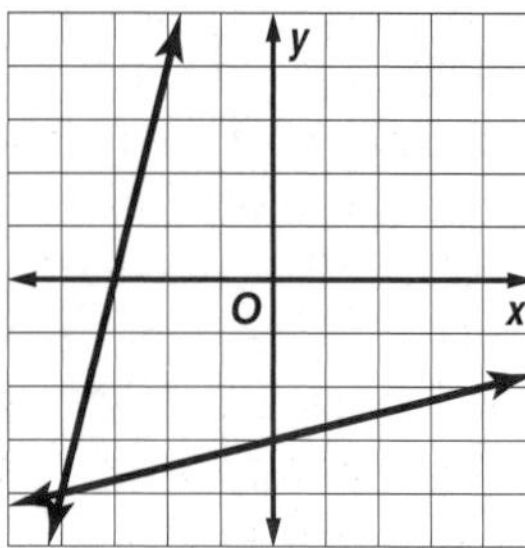

11. $f^{-1}(x) = 3x - 12$

13. no

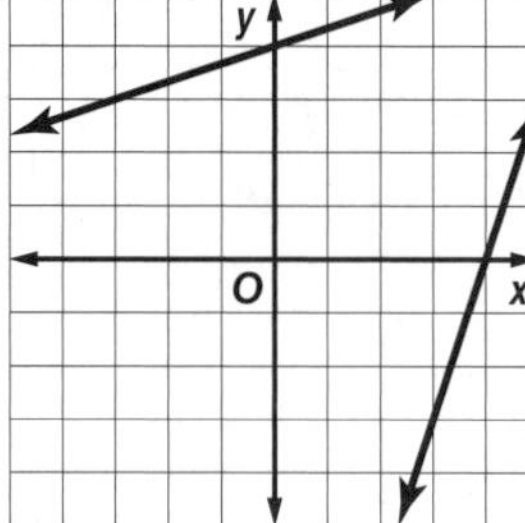

Lesson 7-1

1.

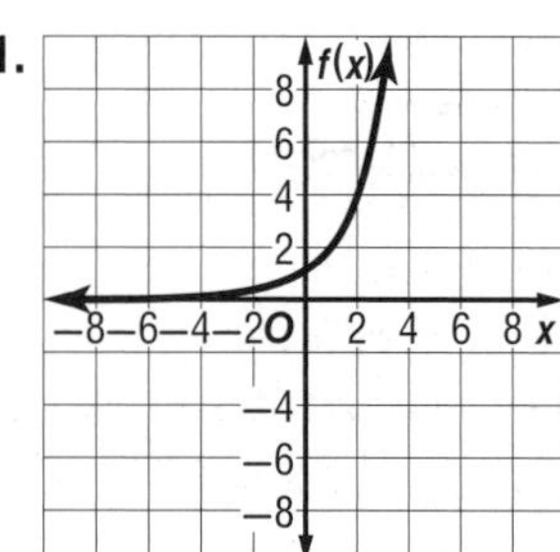

D = {all real numbers}; R = $\{f(x) \mid f(x) > 0\}$

3 Make a table of values. Then plot the points, and sketch the graph.

x	$f(x) = 3^{x-2} + 4$
−2	$3^{-2-2} + 4 = 4\frac{1}{81}$
−1	$3^{-1-2} + 4 = 4\frac{1}{27}$
0	$3^{0-2} + 4 = 4\frac{1}{9}$
1	$3^{1-2} + 4 = 4\frac{1}{3}$
2	$3^{2-2} + 4 = 5$
3	$3^{3-2} + 4 = 7$
4	$3^{4-2} + 4 = 13$

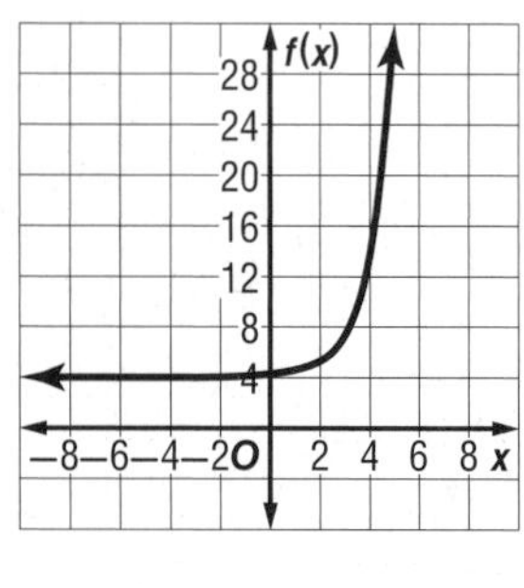

The domain is all real numbers and the range is all real numbers greater than 4.
D = {all real numbers}; R = $\{f(x) \mid f(x) > 4\}$

5.

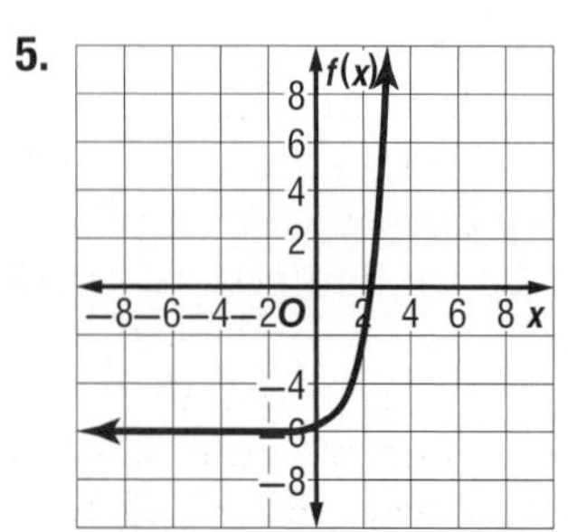

D = {all real numbers}; R = $\{f(x) \mid f(x) > -6\}$

7.

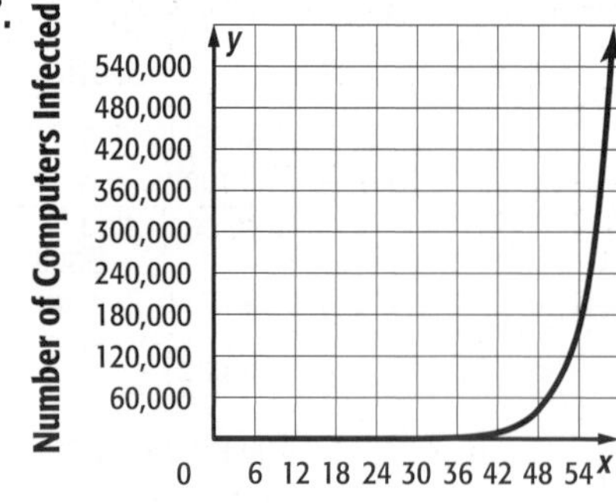

Minutes

9.

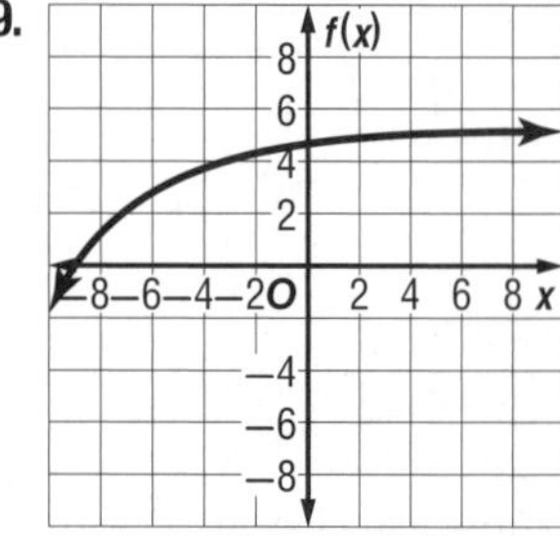

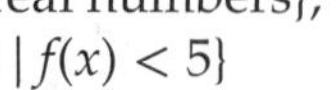
D = {all real numbers}; R = $\{f(x) \mid f(x) < 5\}$

11.

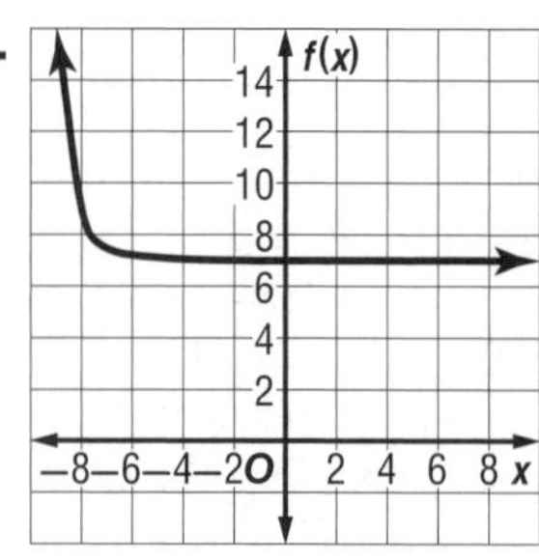

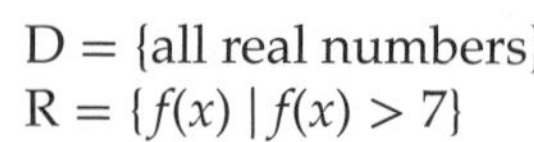
D = {all real numbers}; R = $\{f(x) \mid f(x) > 7\}$

13.

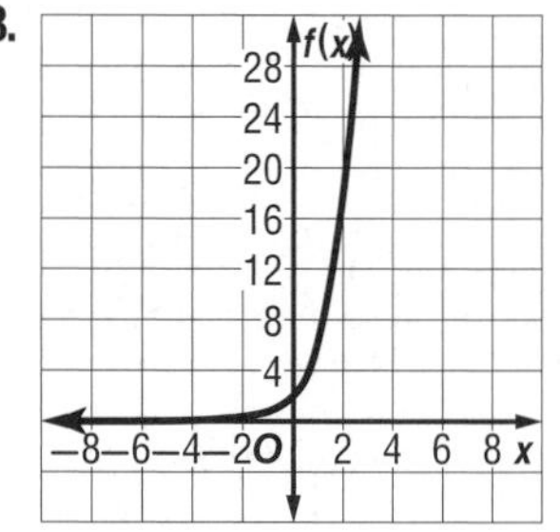

D = {all real numbers}; R = $\{f(x) \mid f(x) > 0\}$

15.

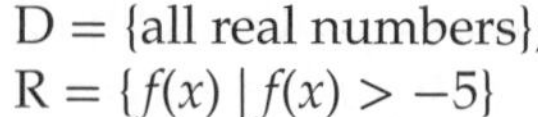
D = {all real numbers}; R = $\{f(x) \mid f(x) > -5\}$

17.

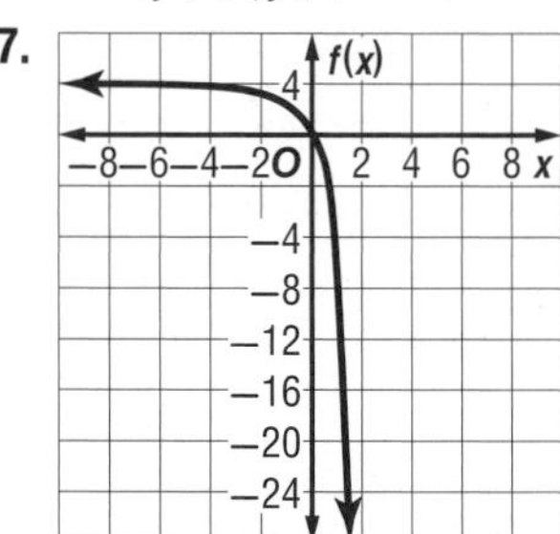

D = {all real numbers}; R = $\{f(x) \mid f(x) < 4\}$

19

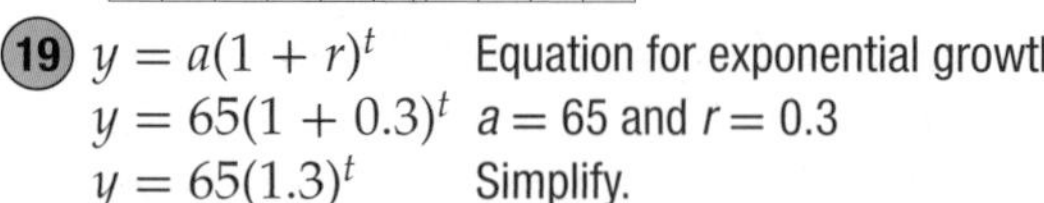

$y = a(1 + r)^t$	Equation for exponential growth
$y = 65(1 + 0.3)^t$	$a = 65$ and $r = 0.3$
$y = 65(1.3)^t$	Simplify.

Make a table of values. Then plot the points, and sketch the graph.

t	$y = 65(0.3)^t$
0	$y = 65(1.3)^0 = 65$
2	$y = 65(1.3)^2 \approx 110$
4	$y = 65(1.3)^4 \approx 186$
6	$y = 65(1.3)^6 \approx 314$
8	$y = 65(1.3)^8 \approx 530$
10	$y = 65(1.3)^{10} \approx 896$

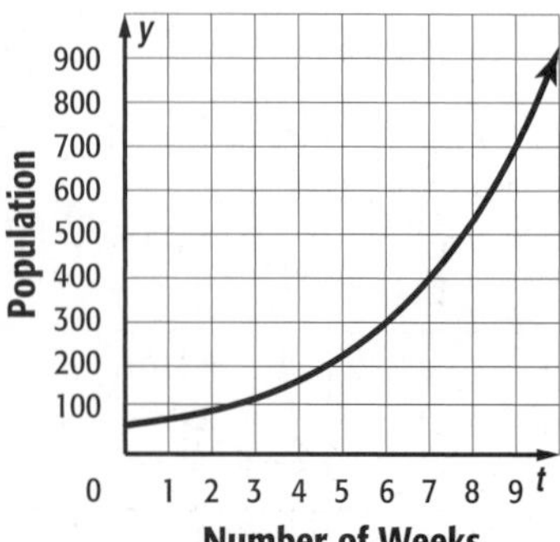

21.

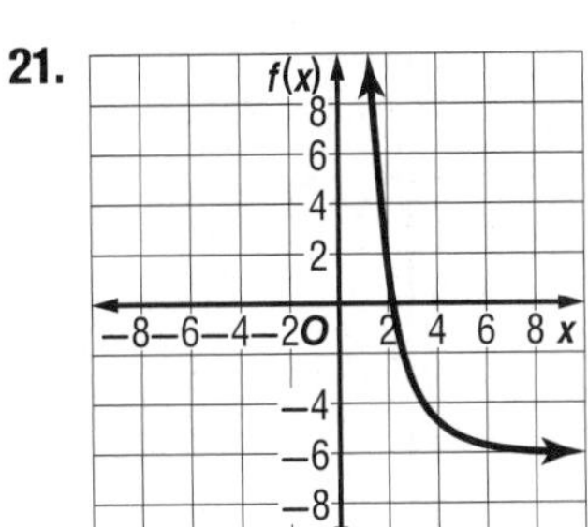

D = {all real numbers}; R = $\{f(x) \mid f(x) > -6\}$

23.

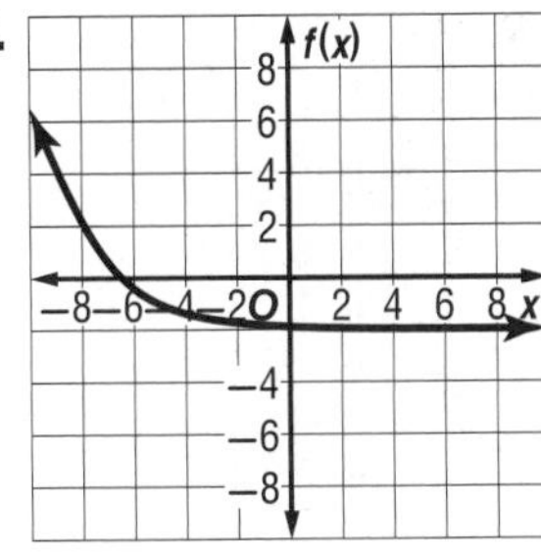

D = {all real numbers}; R = $\{f(x) \mid f(x) > -2\}$

25.

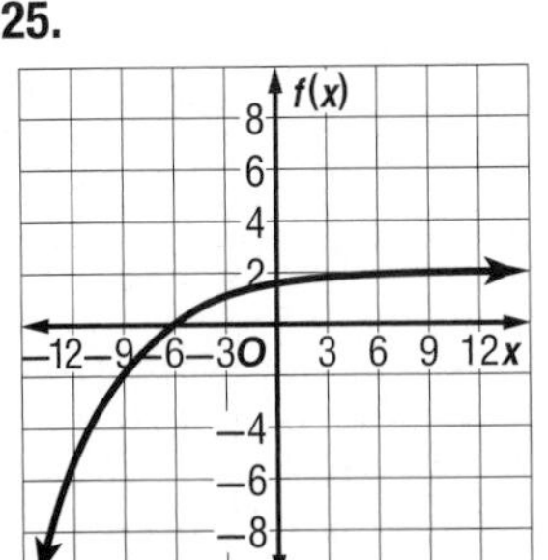

D = {all real numbers}; R = $\{f(x) \mid f(x) < 2\}$

27a. decay; 0.9

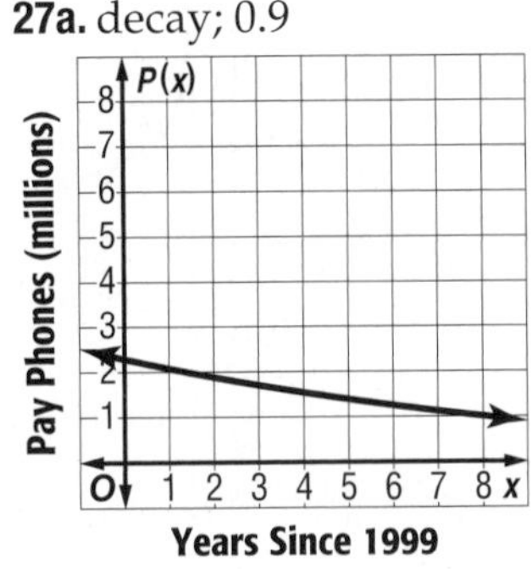

27b. The y-intercept represents the number of pay phones in 1999. The asymptote is the x-axis. The number of pay phones can approach 0, but will never equal 0. This makes sense as there will probably always be a need for some pay phones. **29a.** $f(x) = 18(1.25)^{x-1}$

29b. growth; 1.25

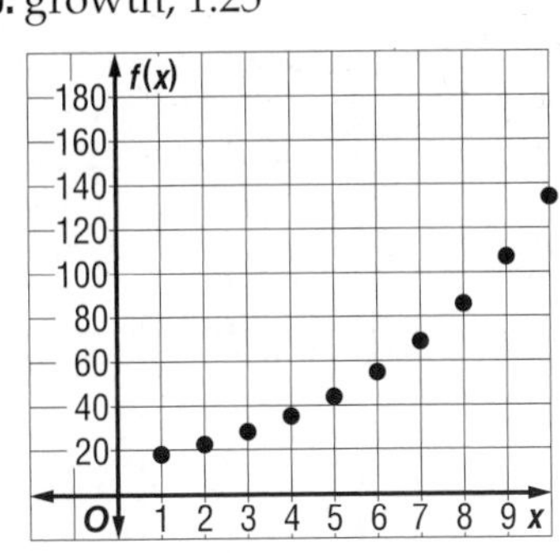

29c. 134
31. $g(x) = 4(2)^{x-3}$ or $g(x) = \frac{1}{2}(2^x)$

33 **a.**

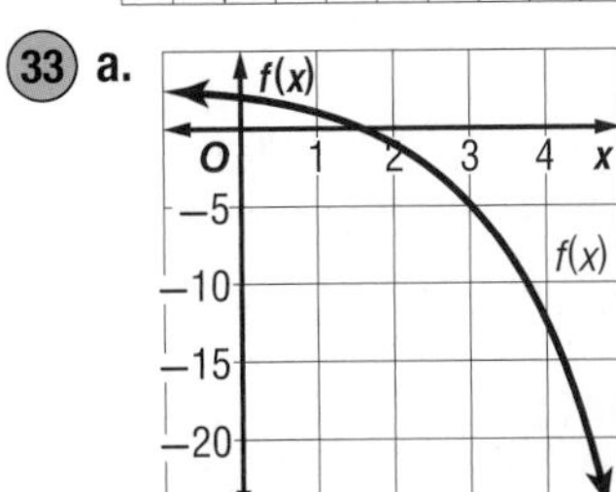

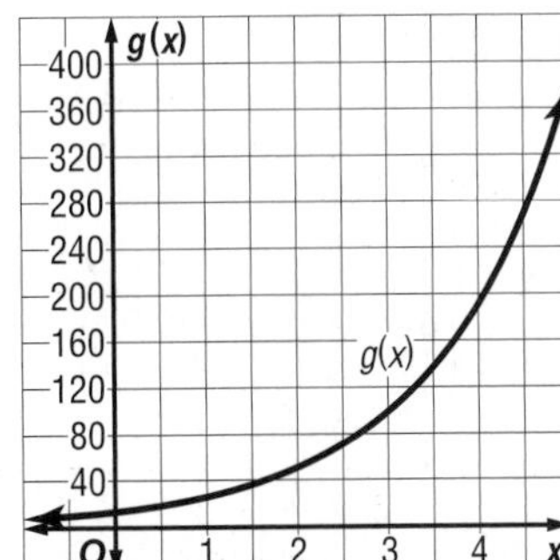

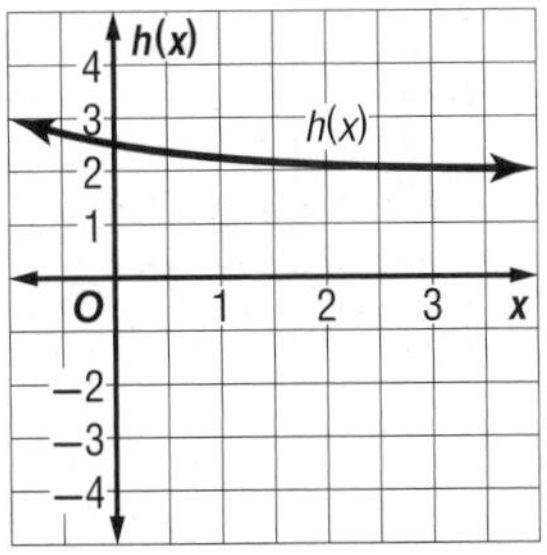

b. Sample answer: The graph of $f(x)$ appears to be the graph of $f(x) = b^x$ reflected across the x-axis. As the values of x increase, the output values decrease.
c. Sample answer: The graphs of $g(x)$ and $h(x)$ appear to be translated to the left.
d. Sample answer: $f(x)$ and $g(x)$ are growth and $h(x)$ is decay; the absolute value of the output is increasing for the growth functions and decreasing for the decay function.
35. Vince; the graphs of the function would be the same. **37.** Sample answer: 10 **39.** 12 **41.** H

43. 8 **45.** $x \geq 21$ **47.** $d > -\frac{3}{4}$ **49.** $\frac{y^{\frac{3}{5}}}{y}$ **51.** $3x^{\frac{5}{3}} + 4x^{\frac{8}{3}}$

53. $\sqrt{3}$ **55.** about 204.88 ft **57.** $\frac{1}{f^3}$ **59.** $8xy^4$

Lesson 7-2

1. 12 **3.** −10 **5a.** $c = 2^{\frac{t}{15}}$ **5b.** 16 cells **7.** $x \geq 4.5$ **9.** 0

11

$81^{a+2} = 3^{3a+1}$	Original equation
$(3^4)^{a+2} = 3^{3a+1}$	Rewrite 81 as 3^4.
$3^{4(a+2)} = 3^{3a+1}$	Power of a Power
$3^{4a+8} = 3^{3a+1}$	Distributive Property
$4a + 8 = 3a + 1$	Property of Equality for Exponential Functions
$a + 8 = 1$	Subtract $3a$ from each side.
$a = -7$	Subtract 8 from each side.

13. $\frac{5}{3}$ **15a.** $y = 10{,}000(1.045)^x$ **15b.** about \$26,336.52 **17.** $y = 256(0.75)^x$ **19.** $y = 144(3.5)^x$ **21.** \$16,755.63 **23.** \$97,362.61 **25.** $b > \frac{1}{5}$ **27.** $d \geq -1$ **29.** $w < \frac{2}{5}$

31 **a.**

$\frac{a}{w^{1.31}} = \frac{170}{45^{1.31}}$	Write a proportion
$a \cdot 45^{1.31} = w^{1.31} \cdot 170$	Cross Products Property
$a = \frac{w^{1.31} \cdot 170}{45^{1.31}}$	Divide each side by $45^{1.31}$.
$a = 1.16w^{1.31}$	Use a calculator.

b.

$a = 1.16w^{1.31}$	Write the equation.
$= 1.16(430)^{1.31}$	$w = 430$
≈ 3268	Use a calculator.

33. $\frac{1}{7}$ **35.** $-\frac{4}{13}$ **37.** 1 **39a.** $d = 1.30h^{\frac{3}{2}}$ **39b.** about 1001 cm **41a.** 2, 4, 8, 16

41b.

Cuts	Pieces
1	2
2	4
3	8
4	16

41c. $y = 2^x$ **41d.** $y = 0.003(2)^x$ **41e.** about 3,221,225.47 in.

43. Sample answer: Beth; Liz added the exponents instead of multiplying them when taking the power of a power. **45.** Reducing the term will be more beneficial. The multiplier is 1.3756 for the 4-year and 1.3828 for the 6.5%. **47.** Sample answer: $4^x \leq 4^2$ **49.** Sample answer: Divide the final amount by the initial amount. If n is the number of time intervals that pass, take the nth root of the answer. **51.** F **53.** E

55.

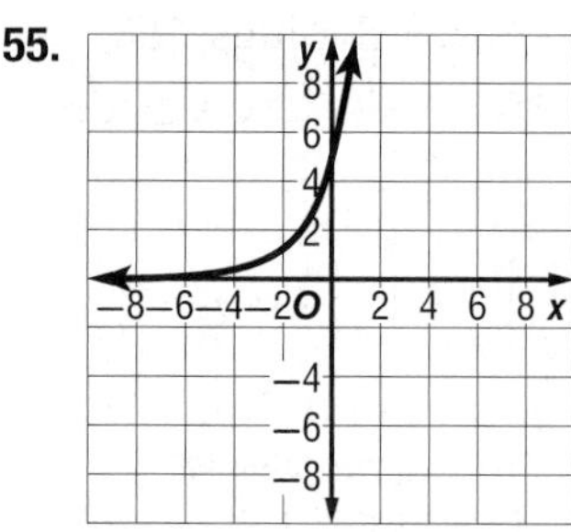

57. 4 **59.** 8.5 **61.** 5 **63.** 5 **65.** −1 **67.** booth 1, 190 lb; booth 2, 150 lb; booth 3, 100 lb **69.** $x^2 - 1$; $x^2 - 6x + 11$ **71.** $-15x - 5$; $-15x + 25$ **73.** $|x + 4|$; $|x| + 4$

Lesson 7-3

1. $8^3 = 512$ **3.** $\log_{11} 1331 = 3$ **5.** 2 **7.** 0

9.

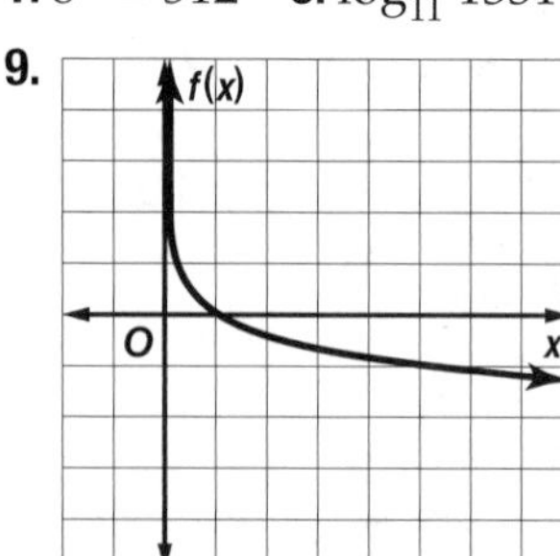

11.

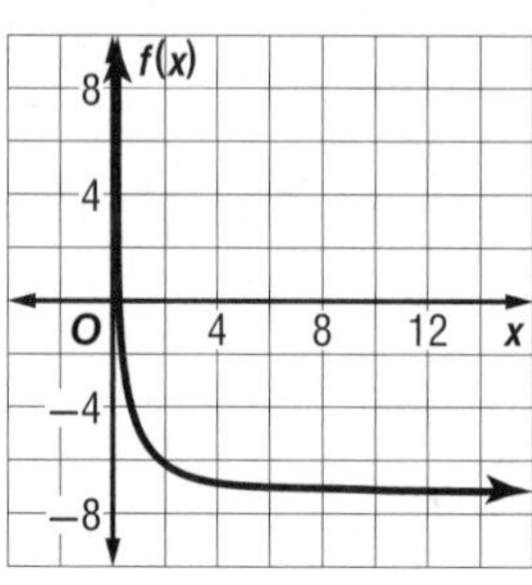

13. $2^4 = 16$ **15.** $9^{-2} = \frac{1}{81}$ **17.** $12^2 = 144$
19. $\log_9 \frac{1}{9} = -1$ **21.** $\log_2 256 = 8$ **23.** $\log_{27} 9 = \frac{2}{3}$
25. −2 **27.** 3 **29.** $\frac{1}{3}$ **31.** $\frac{1}{2}$

33

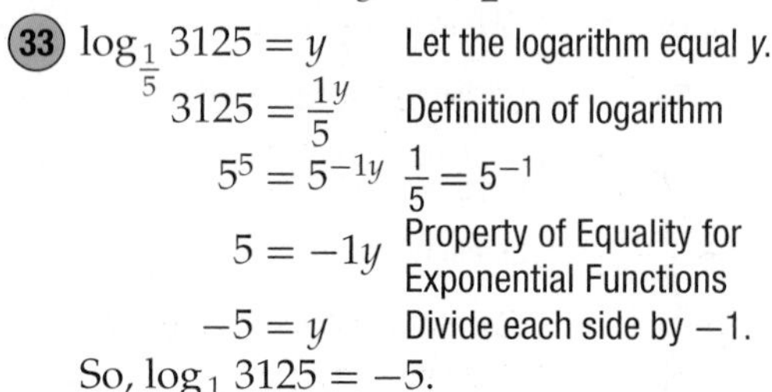

$\log_{\frac{1}{5}} 3125 = y$ Let the logarithm equal y.

$3125 = \frac{1}{5}^y$ Definition of logarithm

$5^5 = 5^{-1y}$ $\frac{1}{5} = 5^{-1}$

$5 = -1y$ Property of Equality for Exponential Functions

$-5 = y$ Divide each side by −1.

So, $\log_{\frac{1}{5}} 3125 = -5$.

35. 4

37.

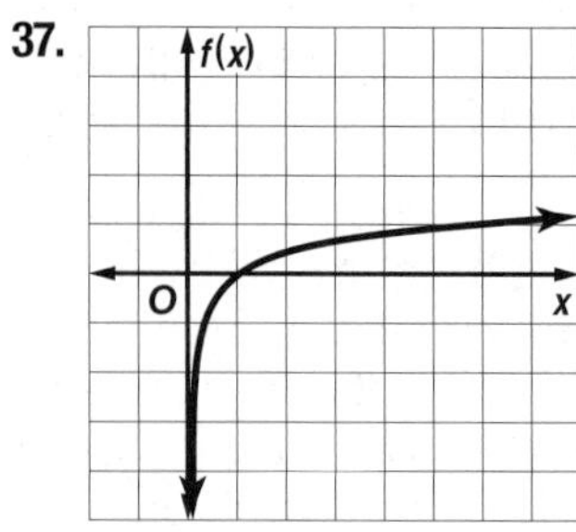

39.

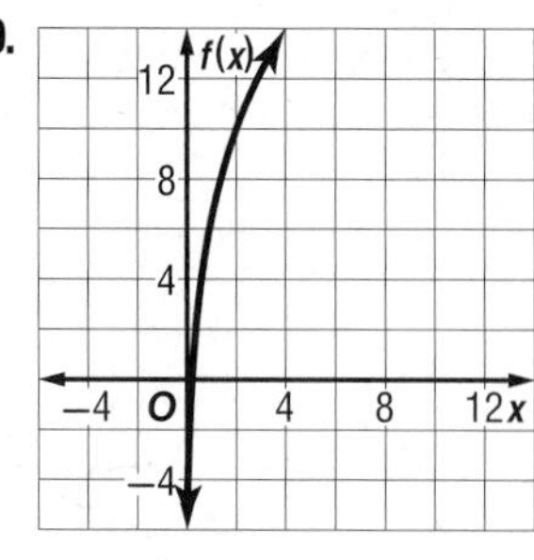

41.

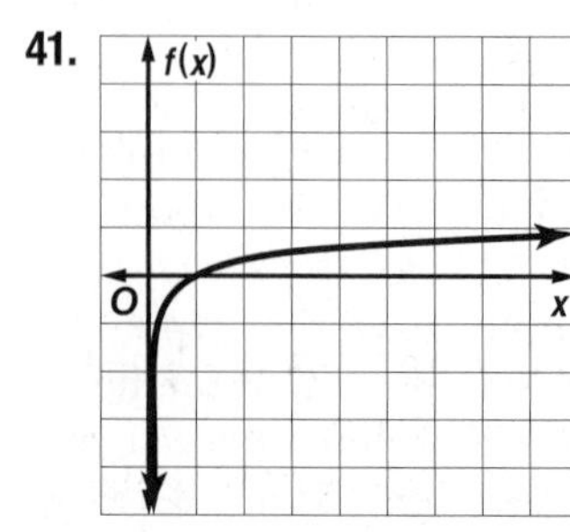

43.

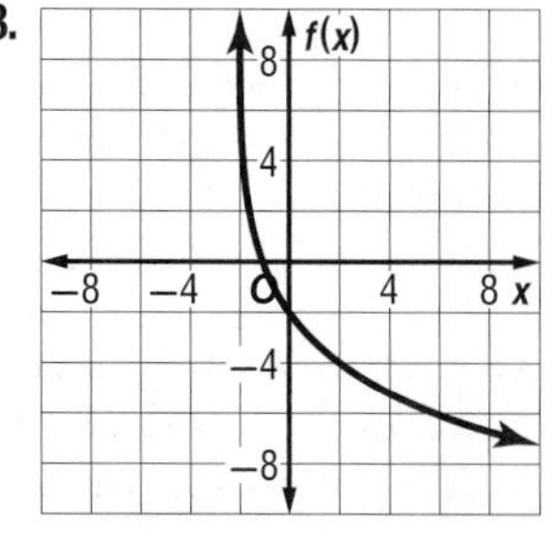

45.

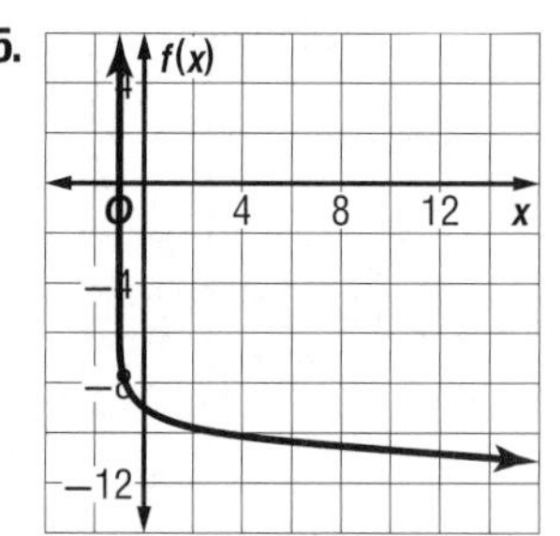

47.

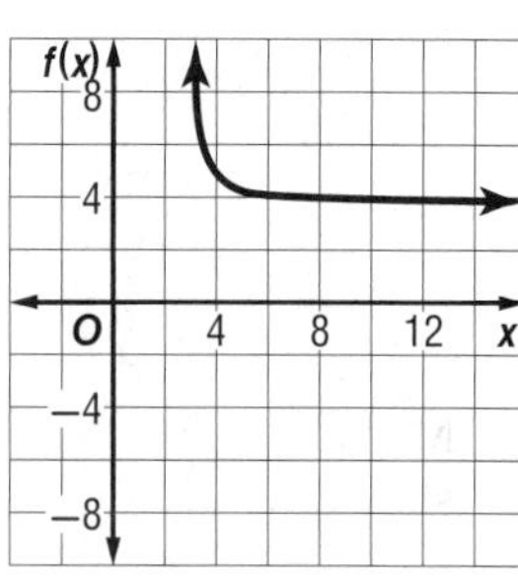

49a. 2 **49b.**

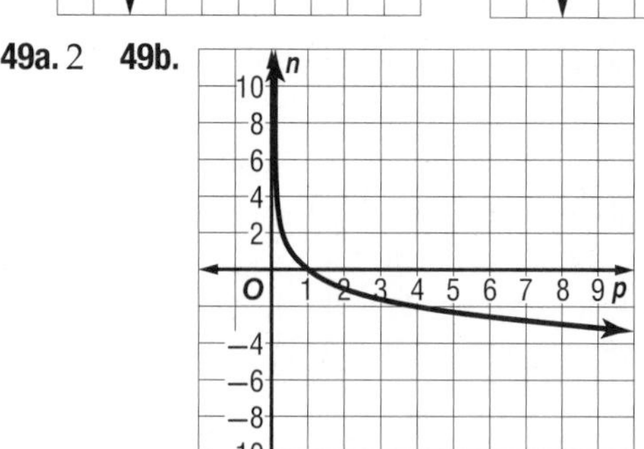

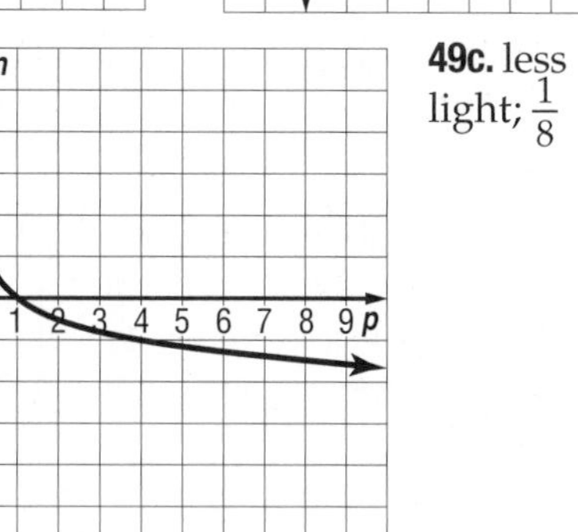

49c. less light; $\frac{1}{8}$

51 This represents a transformation of the graph of $f(x) = \log_2 x$.
$|a| = 4$: The graph expands vertically.
$h = 4$: The graph is translated 4 units to the right.
$k = 6$: The graph is translated 6 units up.

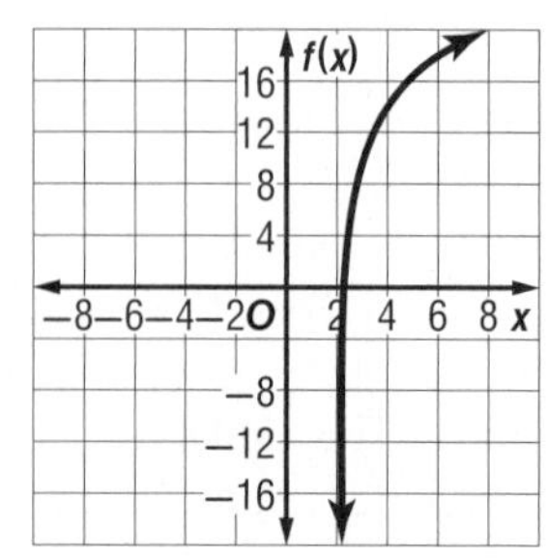

53. **55.**

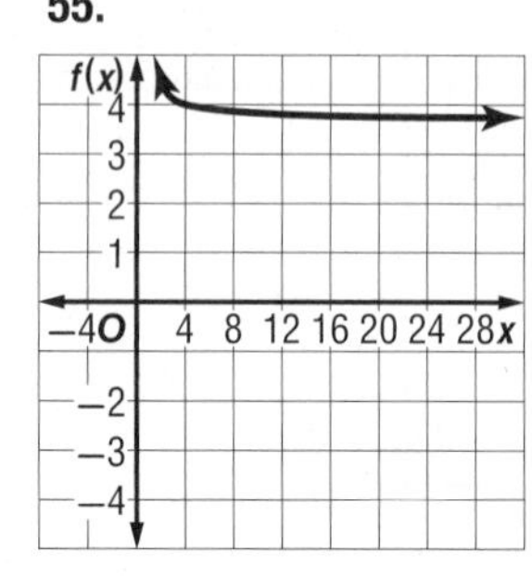

57a. $S(3) \approx 30$, $S(15) = 50$, $S(63) = 70$
57b. If \$3000 is spent on advertising, \$30,000 is returned in sales. If \$15,000 is spent on advertising, \$50,000 is returned in sales. If \$63,000 is spent on advertising, \$70,000 is returned in sales.

57c.

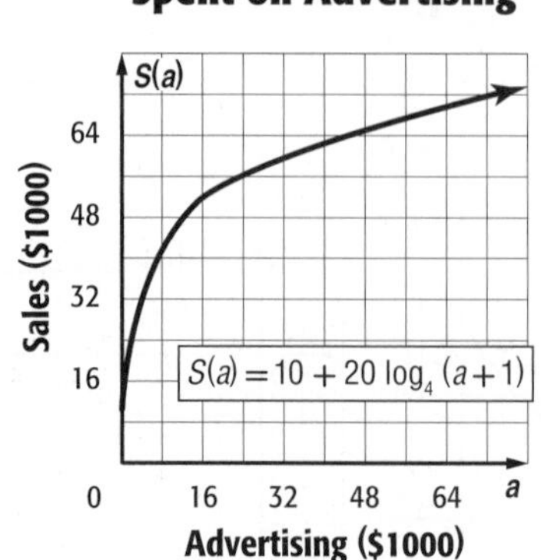

57d. Sample answer: Because eventually the graph plateaus, and no matter how much money you spend you are still returning about the same in sales.

Selected Answers and Solutions

59 **a.** $\log_{\left(1+\frac{0.24}{12}\right)} \frac{A}{2000} = 12t$ Original formula

$\log_{1.02} \frac{A}{2000} = 12t$ Simplify.

$\frac{A}{2000} = 1.02^{12t}$ Definition of logarithm

$A = 2000 \cdot 1.02^{12t}$ Multiply each side by 2000.

Make a table of values. Then plot the points, and sketch the graph.

t	$A = 2000 \cdot 1.02^{12t}$
0	$A = 2000 \cdot 1.02^{12(0)} = 2000$
2	$A = 2000 \cdot 1.02^{12(2)} \approx 3217$
4	$A = 2000 \cdot 1.02^{12(4)} \approx 5174$
6	$A = 2000 \cdot 1.02^{12(6)} \approx 8322$
8	$A = 2000 \cdot 1.02^{12(8)} \approx 13{,}386$
10	$A = 2000 \cdot 1.02^{12(10)} \approx 21{,}530$

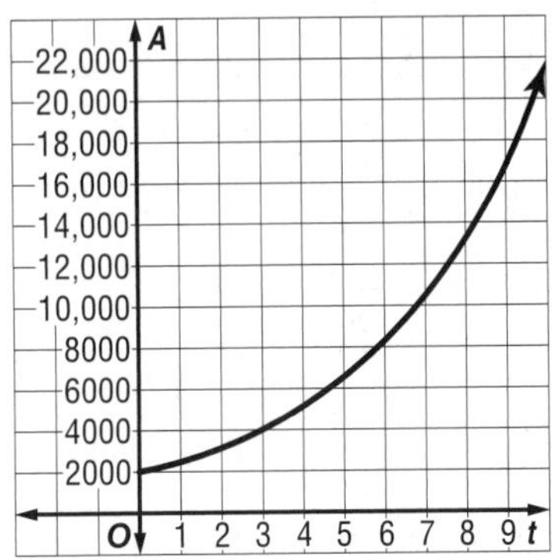

b. From the graph, $A = 4000$ at about $t = 3$. So, it will take approximately 3 years for the debt to double.

c. From the graph, $A = 6000$ at about $t = 4.5$. So, it will take approximately 4.5 years for the debt to triple.

61. Never; if zero were in the domain, the equation would be $y = \log_b 0$.
Then $b^y = 0$.
However, for any real number b, there is no real power that would let $b^y = 0$.

63. $\log_7 51$; Sample answer: $\log_7 51$ equals a little more than 2. $\log_8 61$ equals a little less than 2. $\log_9 71$ equals a little less than 2. Therefore, $\log_7 51$ is the greatest.

65. No; Elisa was closer. She should have $-y = 2$ or $y = -2$ instead of $y = 2$. Matthew used the definition of logarithms incorrectly. **67.** D

69. 80 **71.** $n > 5$ **73.** $n < 3$

75.

77.

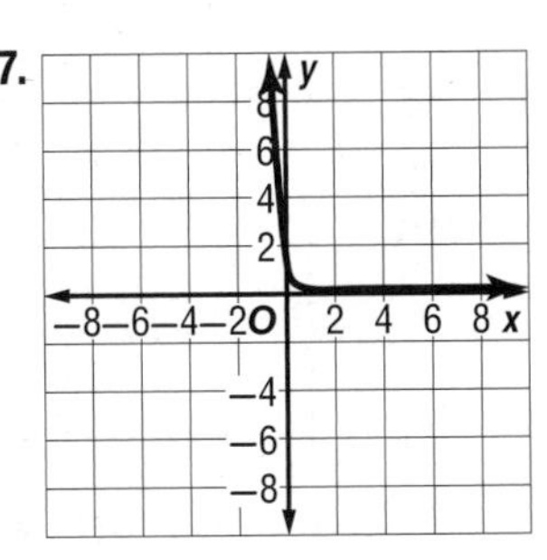

79. $\frac{27\sqrt{15}}{4}$ ft^2 **81.** batteries, \$74; spark plugs, \$58; wiper blades, \$48 **83.** −1 **85.** $x \le -\sqrt{6}$ or $x \ge \sqrt{6}$

Lesson 7-4

1. 16 **3.** C **5.** $\left\{x \,\middle|\, 0 < x \le \frac{1}{64}\right\}$ **7.** $\left\{x \,\middle|\, 2 > x > \frac{4}{3}\right\}$

9. 3125 **11.** −2 **13.** 9

15 $\log_{12}(x^2 - 7) = \log_{12}(x + 5)$ Original equation

$x^2 - 7 = x + 5$ Property of Equality for Exponential Functions

$x^2 - x - 7 = 5$ Subtract x from each side.

$x^2 - x - 12 = 0$ Subtract 5 from each side.

$(x - 4)(x + 3) = 0$ Factor.

$x - 4 = 0$ or $x + 3 = 0$ Zero Product Property

$x = 4$ $\quad x = -3$ Solve each equation.

17. 5 **19.** −3 **21.** 318 mph **23.** $\{x \mid x \ge 256\}$

25 $\log_2 x \le -2$ Original inequality

$0 < x \le 2^{-2}$ Property of Inequality for Exponential Functions

$0 < x \le \frac{1}{4}$ Simplify.

The solution is $\left\{x \,\middle|\, 0 < x \le \frac{1}{4}\right\}$ or $\left(0, \frac{1}{4}\right]$.

27. $\left\{x \,\middle|\, 0 < x < \frac{1}{7}\right\}$ **29.** $\left\{x \,\middle|\, \frac{1}{2} < x \le 1\right\}$

31. $\left\{x \,\middle|\, -\frac{5}{12} < x \le 1\right\}$ **33.** $\{x \mid x \ge 8\}$

35a. 37 **35b.** 61

37 **a.** $\beta = 10 \log_{10}\left(\frac{I}{10^{-12}}\right)$ Original equation

$= 10 \log_{10}\left(\frac{1}{10^{-12}}\right)$ $I = 10$

$= 10 \log_{10} 10^{12}$ Write $\frac{1}{10^{-12}}$ as 10^{12}.

$= 10(12)$ or 120 Definition of logarithm

b. $\beta = 10 \log_{10}\left(\frac{I}{10^{-12}}\right)$ Original equation

$= 10 \log_{10}\left(\frac{10^{-2}}{10^{-12}}\right)$ $I = 10^{-2}$

$= 10 \log_{10} 10^{10}$ Quotient of Powers Property

$= 10(10)$ or 100 Definition of logarithm

c. Sample answer: The power of the logarithm only changes by 2. The power is the answer to the logarithm. That 2 is multiplied by the 10 before the logarithm. So we expect the decibels to change by 20.

39. $6\frac{17}{20}$ **41.** The logarithmic function of the form $y = \log_b x$ is the inverse of the exponential function of the form $y = b^x$. The domain of one of the two inverse functions is the range of the other. The range of one of the two inverse functions is the domain of the other.

43a. less than **43b.** less than **43c.** no **43d.** infinitely many **45.** C **47.** B **49.** 4 **51.** 3 **53.** −3 **55.** $x \le 0$

57. $a \le -3$ **59.** −8 **61.** 5 ft **63.** x^8 **65.** $8p^6n^3$

67. x^3y^4

Lesson 7-5

1. 2.085 **3.** 0.3685 **5.** Mt. Everest: 26,855.44 pascals; Mt. Trisuli: 34,963.34 pascals; Mt. Bonete: 36,028.42 pascals; Mt. McKinley: 39,846.22 pascals;

Mt. Logan: 41,261.82 pascals **7.** 2.4182 **9.** 2
11. 13.4403 **13.** 2.1610

15 $\log_4 \frac{4}{3} = \log_4 4 - \log_4 3$ Quotient Property
$= 1 - \log_4 3$ Inverse Property of Exponents and Logarithms
$\approx 1 - 0.7925$ Replace $\log_4 3$ with 0.7925.
≈ 0.2075 Simplify.

17. 1.5 **19.** 2.1606 **21.** 3.4818 **23.** 8 **25.** 2

27 **a.** $P = \log_{10}\left(1 + \frac{1}{d}\right)$ Original equation
$10^P = 1 + \frac{1}{d}$ Definition of logarithm
$10^P - 1 = \frac{1}{d}$ Subtract 1 from each side.
$d(10^P - 1) = 1$ Multiply each side by d.
$d = \frac{1}{10^P - 1}$ Divide each side by $10^P - 1$.

b. $d = \frac{1}{10^P - 1}$ Write the formula.
$= \frac{1}{10^{0.097} - 1}$ $P = 0.097$
≈ 4 Use a calculator.

c. $P = \log_{10}\left(1 + \frac{1}{d}\right)$ Original equation
$= \log_{10}\left(1 + \frac{1}{1}\right)$ $d = 1$
$= \log_{10} 2$ Simplify.
≈ 0.30103 Replace $\log_{10} 2$ with 0.30103.

The probability is about 30.1%.

29. 2.1133 **31.** 0.1788 **33.** 1.7228 **35.** 2.0478 **37.** 3 **39.** 5 **41.** $85\frac{1}{3}$ **43.** $\left(\frac{x-2}{256}\right)^{\frac{1}{6}}$ **45.** $\sqrt{6}, -\sqrt{6}$ **47.** 5 **49.** 12 **51.** false **53.** false **55.** true **57.** false **59a.** 10^{12} **59b.** 10^4 or about 10,000 times **61a.** Sample answer: $\log_b \frac{xz}{5} = \log_b x + \log_b z - \log_b 5$ **61b.** Sample answer: $\log_b m^4p^6 = 4\log_b m + 6\log_b p$

61c. Sample answer: $\log_b \frac{j^8k}{h^5} = 8\log_b j + \log_b k - 5\log_b h$

63a. $\log_b 1 = 0$, because $b^0 = 1$. **63b.** $\log_b b = 1$, because $b^1 = b$. **63c.** $\log_b b^x = x$, because $b^x = b^x$.
65. $\log_b 24 \neq \log_b 20 + \log_b 4$; all other choices are equal to $\log_b 24$.

67. $x^{3\log_x 2 - \log_x 5} = x^{\log_x 2^3 - \log_x 5}$
$= x^{\log_x 8 - \log_x 5}$
$= x^{\log_x \frac{8}{5}}$
$= \frac{8}{5}$

69. D **71.** growing exponentially **73.** $\frac{1}{2}$, 1 **75.** no solution **77.** $2x$ **79.** 6.3 amps **81.** no **83.** $\frac{3}{5}$ **85.** 10 **87.** $x > 26$

Lesson 7-6

1. 0.6990 **3.** −0.3979 **5.** 3.55×10^{24} ergs **7.** 0.8442

9 $11^{b-3} = 5^b$ Original equation
$\log 11^{b-3} = \log 5^b$ Property of Equality for Logarithmic Functions
$(b-3)\log 11 = b\log 5$ Power Property of Logarithms
$b\log 11 - 3\log 11 = b\log 5$ Distributive Property
$-3\log 11 = b\log 5 - b\log 11$ Subtract $b\log 11$ from each side.
$-3\log 11 = b(\log 5 - \log 11)$ Distributive Property
$\frac{-3\log 11}{\log 5 - \log 11} = b$ Divide each side by $\log 5 - \log 11$.
$9.1237 \approx b$ Use a calculator.

11. $\{p \mid p \leq 4.4190\}$ **13.** $\frac{\log 23}{\log 4} \approx 2.2618$ **15.** $\frac{\log 5}{\log 2} \approx 2.3219$ **17.** 1.0414 **19.** 0.9138 **21.** −1.3979 **23.** 1.7740 **25.** 5.9647 **27.** ±1.1691 **29.** $\{n \mid n > 0.6667\}$ **31.** $\{y \mid y \geq -3.8188\}$ **33.** $\frac{\log 18}{\log 7} \approx 1.4854$ **35.** $\frac{\log 16}{\log 2} = 4$ **37.** $\frac{\log 11}{\log 3} \approx 2.1827$

39 **a.** $n = 35[\log_4(t+2)]$ Original equation
$= 35[\log_4(10+2)]$ $t = 10$
$= 35[\log_4(12)]$ Simplify.
$= 35 \cdot \frac{\log_{10} 12}{\log_{10} 4}$ Change of Base Formula
≈ 62.737 Use a calculator.

In 2010, there are about 62.737 thousand, or 62,737 pet owners.

b. $n = 35[\log_4(t+2)]$ Original equation
$80 = 35[\log_4(t+2)]$ $n = 80$
$2.2857 \approx \log_4(t+2)$ Divide each side by 35.
$4^{2.2857} \approx t + 2$ Definition of logarithm
$22 \approx t$

In 22 years after 2000, or in 2022, there will be 80,000 pet owners.

41. 3.3578 **43.** −0.0710 **45.** 4.7393 **47.** $\{x \mid x \geq 2.3223\}$ **49.** $\{x \mid x \leq 0.9732\}$ **51.** $\{p \mid p \leq 2.9437\}$

53. $\frac{\log 12}{\log 4} \approx 1.7925$ **55.** $\frac{\log 2}{\log 8} = 0.3333$

57. $\frac{\log 7.29}{\log 5} \approx 1.2343$ **59a.** 113.03 cents **59b.** about 218 Hz

61 $4^{x^2-3} = 16$ Original equation
$4^{x^2-3} = 4^2$ Rewrite 16 as 4^2.
$x^2 - 3 = 2$ Property of Equality for Exponential Functions
$x^2 = 5$ Add 3 to each side.
$x = \pm\sqrt{5}$ Take the square root of each side.
$\approx \pm 2.2361$ Use a calculator.

63. 3.5 **65.** −3.8188 **67a.** The solution is between 1.8 and 1.9. **67b.** (1.85, 13) **67c.** Yes; all methods produce the solution of 1.85. They all should produce the same result because you are starting with the same equation. If they do not, then an error was made.

69. $\log_{\sqrt{a}} 3 = \log_a x$ Original equation
$\frac{\log_a 3}{\log_a \sqrt{a}} = \log_a x$ Change of Base Formula
$\frac{\log_a 3}{\frac{1}{2}} = \log_a x$ $\sqrt{a} = a^{\frac{1}{2}}$
$2\log_a 3 = \log_a x$ Multiply numerator and denominator by 2.
$\log_a 3^2 = \log_a x$ Power Property of Logarithms

$3^2 = x$	Property of Equality for Logarithmic Functions
$9 = x$	

71. $\log_3 27 = 3$ and $\log_{27} 3 = \frac{1}{3}$; Conjecture: $\log_a b = \frac{1}{\log_b a}$

Proof: $\log_a b \stackrel{?}{=} \frac{1}{\log_b a}$	Original statement
$\frac{\log_b b}{\log_b a} \stackrel{?}{=} \frac{1}{\log_b a}$	Change of Base Formula
$\frac{1}{\log_b a} = \frac{1}{\log_b a}$	Inverse Property of Exponents and Logarithms

73. B **75.** G **77.** 14 **79.** 15 **81.** 2 **83.** −4, 3 **85.** $32x^3 + 8x^2 - 24x + 16$ **87.** $2^x = 5$ **89.** $5^2 = 25$ **91.** $6^4 = x$

Lesson 7-7

1. $\ln 30 = x$ **3.** $\ln x = 3$ **5.** $7 \ln 2$ **7.** $\ln 17496$ **9.** 2.0794 **11.** 0.1352 **13.** 993.6527 **15.** $\{x \mid -25.0855 < x < 15.0855, x \neq -5\}$ **17.** $\{x \mid x > 3.3673\}$ **19.** about 58 min **21.** $\ln 0.1 = -5x$ **23.** $5.4 = e^x$ **25.** $e^{36} = x + 4$ **27.** $e^7 = e^x$ **29.** $7 \ln 10$

31 $7 \ln \frac{1}{2} + 5 \ln 2 = 7 \ln 2^{-1} + 5 \ln 2$	Rewrite $\frac{1}{2}$ as 2^{-1}.
$= \ln 2^{-7} + \ln 2^5$	Power Property of Logarithms
$= \ln (2^{-7})(2^5)$	Product Property of Logarithms
$= \ln 2^{-2}$	Simplify.
$= -2 \ln 2$	Power Property of Logarithms

33. $\ln 81x^6$ **35.** 3.7955 **37.** 0.6931 **39.** −0.5596 **41.** $\{x \mid x \leq 2.1633\}$ **43.** $\{x \mid x > 8.0105\}$ **45.** $\{x \mid x < -239.8802$ or $x > 239.8802\}$

47

a. $A = Pe^{rt}$	Continuous Compounding Formula
$= 800e^{(0.045)(5)}$	$P = 800, r = 0.045, t = 5$
$= 800e^{0.225}$	Simplify.
≈ 1001.86	Use a calculator.

About \$1001.86 will be in the account.

b. $A = Pe^{rt}$	Continuous Compounding Formula
$1600 = 800e^{(0.045)t}$	$A = 2 \cdot 800$ or $1600, P = 800, r = 0.045$
$2 = e^{0.045t}$	Divide each side by 800.
$\ln 2 = \ln e^{0.045t}$	Property of Equality of Logarithms
$\ln 2 = 0.045t$	$\ln e^x = x$
$\frac{\ln 2}{0.045} = t$	Divide each side by 0.045.
$15.4 \approx t$	Use a calculator.

It would take about 15.4 years to double your money.

c. $A = Pe^{rt}$	Continuous Compounding Formula
$1600 = 800e^{r(9)}$	$A = 1600, P = 800, t = 9$
$2 = e^{9r}$	Divide each side by 800.
$\ln 2 = \ln e^{9r}$	Property of Equality of Logarithms
$\ln 2 = 9r$	$\ln e^x = x$
$\frac{\ln 2}{9} = r$	Divide each side by 9.
$0.077 \approx r$	Use a calculator.

You would need a rate of about 7.7%.

d. $A = Pe^{rt}$	Continuous Compounding Formula
$10{,}000 = Pe^{(0.0475)(12)}$	$A = 10{,}000, r = 0.0475, t = 12$
$10{,}000 = Pe^{0.57}$	Simplify.
$\frac{10{,}000}{e^{0.57}} = P$	Divide each side by $e^{0.57}$.
$5655.25 \approx P$	Use a calculator.

You would need to deposit about \$5655.25.

49. $4 \ln 2 - 3 \ln 5$ **51.** $\ln x + 4 \ln y - 3 \ln z$ **53.** −0.8340 **55.** 1.1301

57a.

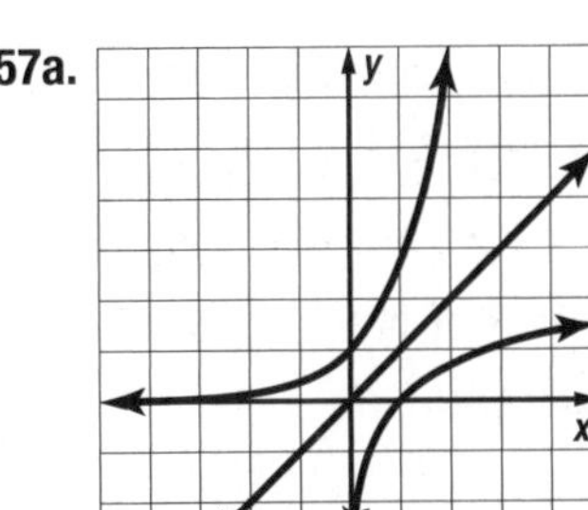

57b. y-axis; $a(x) = -e^x$

57c. $\ln(-x)$ is a reflection across the y-axis. $-\ln x$ is a reflection across the x-axis.

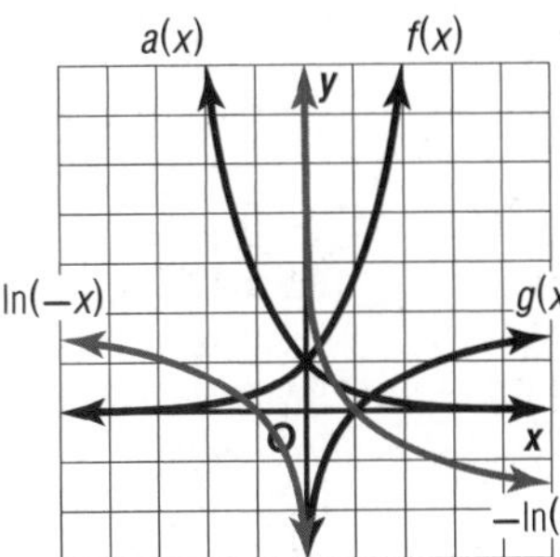

57d. Sample answer: No; these functions are reflections along $y = -x$, which indicates that they are not inverses.

59. Let $p = \ln a$ and $q = \ln b$. That means that $e^p = a$ and $e^q = b$.

$ab = e^p \times e^q$

$ab = e^{p+q}$

$\ln (ab) = (p + q)$

$\ln (ab) = \ln a + \ln b$

61. Sample answer: $e^{\ln 3}$ **63.** B **65.** G **67.** 5.7279 **69.** $x < 7.3059$ **71.** $x \geq 5.8983$ **73.** 10 decibels **75.** $x^2 + 2x + 3$ **77.** $\frac{2}{3}$ **79.** $-\frac{8}{3}$ **81.** $\frac{5}{3}$

Lesson 7-8

1a. 5.545×10^{-10} **1b.** 1,578,843,530 yr **1c.** about 30.48 mg **1d.** 3,750, 120,003 yr

3a.

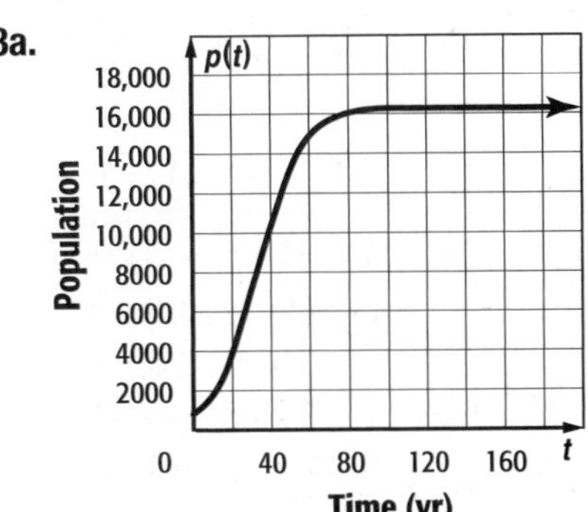

3b. $P(t) = 16{,}500$

3c. 16,500

3d. about 102 years

Selected Answers and Solutions

5 **a.**

$y = 80e^{kt}$	Original formula
$675 = 80e^{k(30)}$	$y = 675$, $t = 30$
$8.4375 = e^{30k}$	Divide each side by 80.
$\ln 8.4375 = \ln e^{30k}$	Property of Equality for Logarithmic Functions
$\ln 8.4375 = 30k$	$\ln e^x = x$
$\frac{\ln 8.4375}{30} = k$	Divide each side by 30.
$0.071 \approx k$	Use a calculator.

b.

$y = 80e^{kt}$	Original formula
$6000 = 80e^{(0.071)t}$	$y = 6000$, $k \approx 0.071$
$75 = e^{0.071t}$	Divide each side by 80.
$\ln 75 = \ln e^{0.071t}$	Property of Equality for Logarithmic Functions
$\ln 75 = 0.071t$	$\ln e^x = x$
$\frac{\ln 75}{0.071} = t$	Divide each side by 0.071.
$60.8 \approx t$	Use a calculator.

The bacteria will reach a population of 6000 cells in about 60.8 minutes.

c.

$35e^{0.0978t} > 80e^{0.071t}$	Formula for exponential growth
$\ln 35e^{0.0978t} > \ln 80e^{0.071t}$	Property of Inequality for Logarithms
$\ln 35 + \ln e^{0.0978t} > \ln 80 + \ln e^{0.071t}$	Product Property of Logarithms
$\ln 35 + 0.0978t > \ln 80 + 0.071t$	$\ln e^x = x$
$0.0268t > \ln 80 - \ln 35$	Subtract $(0.071t + \ln 35)$ from each side.
$t > \frac{\ln 80 - \ln 35}{0.0268}$	Divide each side by 0.0268.
$t > 30.85$	Use a calculator.

The number of cells of this bacteria exceed the number of cells in the other bacteria in about 30.85 minutes.

7

$y = ae^{-0.00012t}$	Equation for the decay of Carbon-14
$0.85a = ae^{-0.00012t}$	$y = 0.85a$
$0.85 = e^{-0.00012t}$	Divide each side by a.
$\ln 0.85 = \ln e^{-0.00012t}$	Property of Equality for Logarithmic Functions
$\ln 0.85 = -0.00012t$	$\ln e^x = x$
$\frac{\ln 0.85}{-0.00012} = t$	Divide each side by −0.00012.
$1354 \approx t$	Use a calculator.

The bone is about 1354 years old.

9. about 14.85 billion yr **11.** about 20.1 yr

13a.

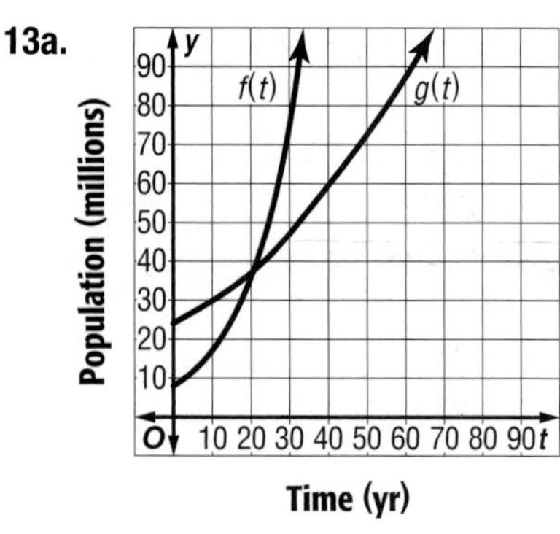

13b. The graphs intersect at $t = 20.79$. Sample answer: This intersection indicates the point at which both functions determine the same population at the same time.

13c. Sample answer: The logistic function $g(t)$ is a more accurate estimate of the country's population since $f(t)$ will continue to grow exponentially and $g(t)$ considers limitations on population growth such as food supply. **15.** $t \approx 113.45$ **17.** Sample answer: The spread of the flu throughout a small town. The growth of this is limited to the population of the town itself. **19.** C **21.** D **23.** $\ln y = 7$ **25.** $5x^4 = e^9$ **27.** $\frac{1}{6}$ **29.** $\frac{5}{8}$ **31.** 16 **33.** $2\frac{1}{2}$

Chapter 7 Study Guide and Review

1. exponential growth **3.** common logarithms **5.** change of base formula **7.** logarithmic function **9.** natural logarithm

11.

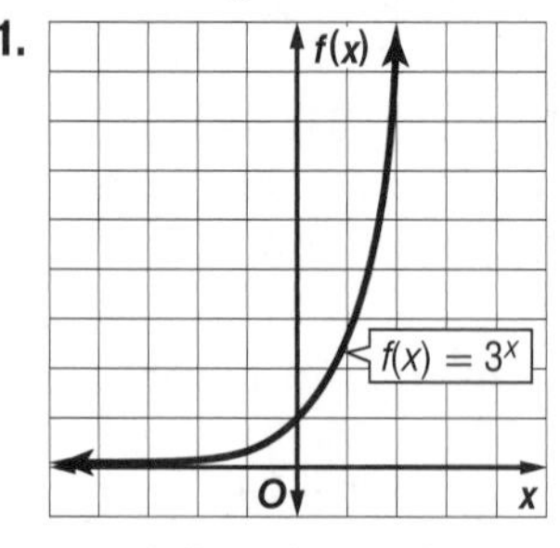

D = {all real numbers}
R = $\{f(x) \mid f(x) > 0\}$

13.

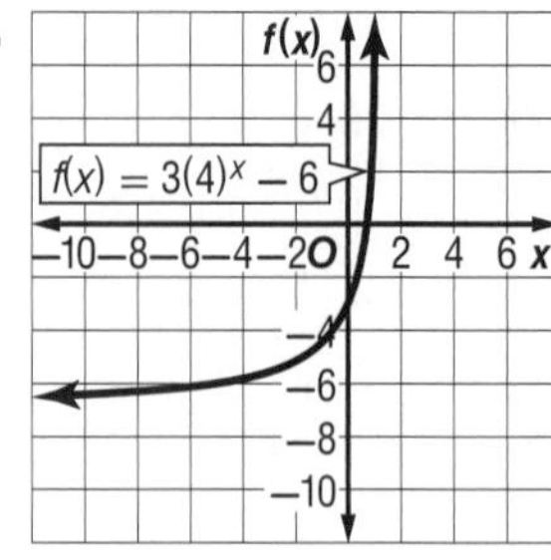

D = {all real numbers}
R = $\{f(x) \mid f(x) > -6\}$

15.

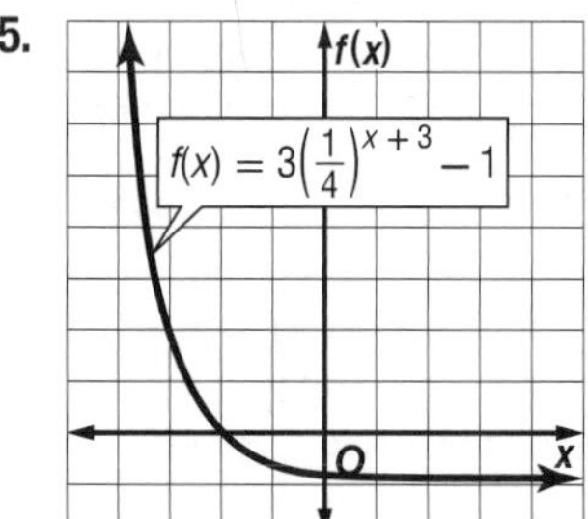

D = all real numbers
R = $\{f(x) \mid f(x) > -1\}$

17a. $f(x) = 120{,}000(0.97)^x$
17b. about 88,491
19. −7 **21.** $\frac{9}{41}$
23. $x \leq -\frac{2}{5}$
25. $2^{-4} = \frac{1}{16}$
27. 4

29.

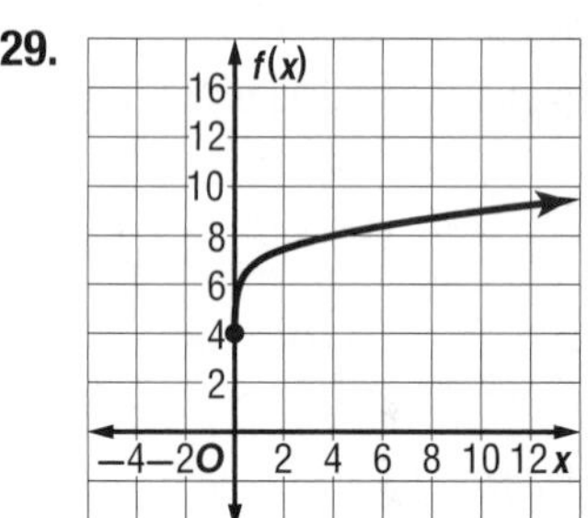

31. $x = 64$ **33.** $0 < x < 64$
35. no solution
37. no solution
39. 1.2920 **41.** 0.8614
43. −0.4307 **45.** $a = 3$
47. $n = 5$ **49.** $x \approx 2.4650$
51. $m \approx 0.6356$
53. $n > 0.5786$
55a. about 8.2 years

55b. about 13.9 years **57.** $x \approx -1.9459$ **59.** $x > 1.9459$ **61.** $x < -2.8904$ **63.** $1054.69 **65.** about 4.1%

CHAPTER 8
Rational Functions and Relations

Chapter 8 Get Ready

1. $x = \frac{15}{14}$ **3.** $k = \frac{32}{5}$ **5.** 27 gallons **7.** $\frac{1}{18}$ **9.** $\frac{43}{6}$
11. $p = 27$ **13.** $k = 17.5$

Lesson 8-1

1. $\frac{x+3}{x+8}$ **3.** D

5

$\frac{a^2x - b^2x}{by - ay} = \frac{x(a^2 - b^2)}{y(b - a)}$	Factor.
$= \frac{x(a - b)(a + b)}{y(b - a)}$	Factor.
$= \frac{-x(b - a)(a + b)}{y(b - a)}$	$a - b = -1(b - a)$

$= \dfrac{-x(b-a)(a+b)}{y(b-a)}$ Eliminate common factors.

$= \dfrac{-x(a+b)}{y}$ Simplify.

7. $\dfrac{2x^2}{3aby^2}$ **9.** $\dfrac{(a-b)(a+1)}{12(a-1)}$ **11.** 4 **13.** $\dfrac{x(x+6)}{x+4}$ **15.** $\dfrac{(x+3)(x-z)}{4}$ **17.** $\dfrac{x(x+2)}{6(x+5)}$ **19.** J **21.** $\dfrac{x^2}{x+6}$ **23.** $-\dfrac{c+4}{c+5}$ **25.** $\dfrac{c}{4ab^2f^2}$ **27.** $\dfrac{32b}{3ac^3f^2}$ **29.** $\dfrac{5a^4c}{3b}$

31 $\dfrac{y^2+8y+15}{y-6} \cdot \dfrac{y^2-9y+18}{y^2-9}$

$= \dfrac{(y+3)(y+5)}{y-6} \cdot \dfrac{(y-3)(y-6)}{(y-3)(y+3)}$ Factor.

$= \dfrac{(y+3)(y+5)}{y-6} \cdot \dfrac{(y-3)(y-6)}{(y-3)(y+3)}$

$= y+5$ Simplify.

33. $\dfrac{(x+4)(x+2)}{2(x-5)}$ **35.** $\dfrac{(x-3)(x+1)}{6(x+7)}$ **37.** $\dfrac{-a^2(a+b)}{b^4}$ **39a.** $\dfrac{33}{121}$ **39b.** $\dfrac{33+m}{121+a}$ **41a.** $T(x) = \dfrac{0.4}{x+3}$ **41b.** about 3.9 mm thick **43.** $\dfrac{1}{4}$ **45.** $\dfrac{x(x+2)(x-1)}{(x+3)(x-7)}$

47 $\dfrac{20x^2y^6z^{-2}}{3a^3c^2} \cdot \left(\dfrac{16x^3y^3}{9acz}\right)^{-1}$

$= \dfrac{20x^2y^6z^{-2}}{3a^3c^2} \cdot \dfrac{9acz}{16x^3y^3}$ $\left(\dfrac{16x^3y^3}{9acz}\right)^{-1} = \dfrac{9acz}{16x^3y^3}$

$= \dfrac{20x^2y^6}{3a^3c^2z^2} \cdot \dfrac{9acz}{16x^3y^3}$ $z^{-2} = \dfrac{1}{z^2}$

$= \dfrac{2 \cdot 2 \cdot 5 \cdot x \cdot x \cdot y \cdot y \cdot y \cdot y \cdot y \cdot y \cdot 3 \cdot 3 \cdot a \cdot c \cdot z}{3 \cdot a \cdot a \cdot a \cdot c \cdot c \cdot z \cdot z \cdot 2 \cdot 2 \cdot 2 \cdot 2 \cdot x \cdot x \cdot x \cdot y \cdot y \cdot y}$ Factor.

$= \dfrac{2 \cdot 2 \cdot 5 \cdot x \cdot x \cdot y \cdot y \cdot y \cdot y \cdot y \cdot y \cdot 3 \cdot 3 \cdot a \cdot c \cdot z}{3 \cdot a \cdot a \cdot a \cdot c \cdot c \cdot z \cdot z \cdot 2 \cdot 2 \cdot 2 \cdot 2 \cdot x \cdot x \cdot x \cdot y \cdot y \cdot y}$ Eliminate common factors.

$= \dfrac{5 \cdot y \cdot y \cdot y \cdot 3}{a \cdot a \cdot c \cdot z \cdot 2 \cdot 2 \cdot x}$ Simplify.

$= \dfrac{15y^3}{4a^2cxz}$ Simplify.

49. $\dfrac{2(4x+1)(2x+1)}{5(2x-1)(x+2)}$ **51.** $\dfrac{2x+1}{-9x(x+2)}$ **53.** $\dfrac{x(x-2)(x+8)}{2(2x-1)(3x+1)}$ **55.** $\dfrac{-2(x-8)(x+4)(x-2)(x+1)}{(2x+1)^2(x^2+2x-6)}$

57 **a.** 5 tracks $\cdot \dfrac{2 \text{ miles}}{1 \text{ track}} \cdot \dfrac{5280 \text{ feet}}{1 \text{ mile}} \cdot \dfrac{1 \text{ car}}{75 \text{ feet}}$

b. 5 tracks $\cdot \dfrac{2 \text{ miles}}{1 \text{ track}} \cdot \dfrac{5280 \text{ feet}}{1 \text{ mile}} \cdot \dfrac{1 \text{ car}}{75 \text{ feet}}$

$= 5 \text{ tracks} \cdot \dfrac{2 \text{ miles}}{1 \text{ track}} \cdot \dfrac{15 \cdot 352 \text{ feet}}{1 \text{ mile}} \cdot \dfrac{1 \text{ car}}{5 \cdot 15 \text{ feet}}$

$= \dfrac{1 \cdot 2 \cdot 352 \cdot 1 \text{ car}}{1 \cdot 1 \cdot 1 \cdot 1}$

$= 704$ cars

c. 704 cars $\cdot \dfrac{8 \text{ attendants}}{1 \text{ car}} \cdot \dfrac{45 \text{ s}}{1 \text{ attendant}} \cdot \dfrac{1 \text{ min}}{60 \text{ s}} \cdot \dfrac{1 \text{ h}}{60 \text{ min}}$

$= 704 \text{ cars} \cdot \dfrac{8 \text{ attendants}}{1 \text{ car}} \cdot \dfrac{45 \text{ s}}{1 \text{ attendant}} \cdot \dfrac{1 \text{ min}}{60 \text{ s}} \cdot \dfrac{1 \text{ h}}{60 \text{ min}}$

$= \dfrac{704 \cdot 8 \cdot 45 \cdot 1 \cdot 1 \text{ h}}{1 \cdot 1 \cdot 60 \cdot 60}$

$= 70.4$ hours

59. Sample answer: The two expressions are equivalent except that the rational expression is undefined at $x = 3$. **61.** $x^2 + x - 6$ **63.** Sample answer: Sometimes; with a denominator like $x^2 + 2$, in which the denominator cannot equal 0, the rational expression can be defined for all values of x. **65.** Sample answer: When the original expression was simplified, a factor of x was taken out of the denominator. If x were to equal 0, then this expression would be undefined. So, the simplified expression is also undefined for x. **67.** J **69.** 4π **71.** 0.2877 **73.** 0.2747 **75.** $10^{1.7}$ or about 50 times **77.** $2y\sqrt[3]{2}$ **79.** $2ab^2\sqrt{10a}$ **81.** $10a - 2b$ **83.** $-3y - 3y^2$ **85.** $x^2 + 9x + 18$

Lesson 8-2

1. $80x^3y^3$ **3.** $3y(y-3)(y-5)$ **5.** $\dfrac{48y^4 + 25x^2}{20xy^3}$ **7.** $\dfrac{21b^4 - 2}{36ab^3}$ **9.** $\dfrac{9x+15}{(x+3)(x+6)}$ **11.** $\dfrac{x-11}{3(x+2)(x-2)}$ **13.** $\dfrac{14x-10}{(x+1)(x-2)}$ **15.** $\dfrac{3y+2}{y+3}$ **17.** $\dfrac{2a+5b}{3b-8a}$ **19.** $180x^2y^4z^2$ **21.** $6(x+4)(2x-1)(2x+3)$ **23.** $\dfrac{28by^2z - 9bx}{105x^3y^4z}$ **25.** $\dfrac{20x^2y + 120y + 6x^2}{15x^3y}$ **27.** $\dfrac{15b^3 + 100ab^2 - 216a}{240ab^3}$

29 $\dfrac{6}{y^2-2y-35} + \dfrac{4}{y^2+9y+20}$

$= \dfrac{6}{(y-7)(y+5)} + \dfrac{4}{(y+4)(y+5)}$ Factor denominators.

$= \dfrac{6(y+4)}{(y-7)(y+5)(y+4)} + \dfrac{4(y-7)}{(y+4)(y+5)(y-7)}$ Multiply by missing factors.

$= \dfrac{6y + 24 + 4y - 28}{(y-7)(y+5)(y+4)}$ Add the numerators.

$= \dfrac{10y - 4}{(y-7)(y+5)(y+4)}$ Simplify.

31. $\dfrac{-10x-10}{(2x-1)(x+6)(x-3)}$ **33.** $\dfrac{2x^2+32x}{3(x-2)(x+3)(2x+5)}$ **35.** $\dfrac{1000x + 800y}{x(x+2y)}$

37 $\dfrac{\frac{4}{x+5} + \frac{9}{x-6}}{\frac{5}{x-6} - \frac{8}{x+5}} = \dfrac{\frac{4(x-6)}{(x+5)(x-6)} + \frac{9(x+5)}{(x+5)(x-6)}}{\frac{5(x+5)}{(x+5)(x-6)} - \frac{8(x-6)}{(x+5)(x-6)}}$

$= \dfrac{\frac{4x - 24 + 9x + 45}{(x+5)(x-6)}}{\frac{5x + 25 - 8x + 48}{(x+5)(x-6)}}$ Simplify the numerator and denominator.

$= \dfrac{\frac{13x+21}{(x+5)(x-6)}}{\frac{-3x+73}{(x+5)(x-6)}}$ Combine like terms.

$= \dfrac{13x+21}{(x+5)(x-6)} \div \dfrac{-3x+73}{(x+5)(x-6)}$ Write as a division expression.

Selected Answers and Solutions

$= \dfrac{13x + 21}{(x + 5)(x - 6)} \cdot \dfrac{(x + 5)(x - 6)}{-3x + 73}$ Multiply by the reciprocal of the divisor.

$= \dfrac{13x + 21}{-3x + 73}$ Simplify.

39. $\dfrac{-x^2 + 33x + 16}{12x^2 + 11x - 27}$ **41.** $420x^5y^4z^3$

43. $(x + 4)(x - 4)(2x + 1)(x - 7)$ **45.** $\dfrac{360a^2 + 5a - 36}{60a^2}$

47. $\dfrac{42x + 41}{6(3x - 1)(x + 8)(2x + 3)}$ **49.** 0 **51.** $\dfrac{5a - 11}{6}$

53. $(x - 3)(x + 2)$ **55.** $-\dfrac{3}{2}$ **57.** -1

59a. $y = \dfrac{70x}{x - 70}$ **59b.** Sample answer: When the object is 70 mm away, y needs to be 0, which is impossible.

61 a. $P_0\left(\dfrac{s_0}{s_0 - x}\right) - P_0\left(\dfrac{s_0}{s_0 - y}\right) = \dfrac{P_0s_0}{s_0 - x} - \dfrac{P_0s_0}{s_0 - y}$

$= \dfrac{P_0s_0(s_0 - y)}{(s_0 - x)(s_0 - y)} - \dfrac{P_0s_0(s_0 - x)}{(s_0 - x)(s_0 - y)}$

$= \dfrac{P_0s_0(s_0 - y) - P_0s_0(s_0 - x)}{(s_0 - x)(s_0 - y)}$

$= \dfrac{P_0s_0s_0 - P_0s_0y - P_0s_0s_0 - P_0s_0x}{(s_0 - x)(s_0 - y)}$

$= \dfrac{P_0s_0x - P_0s_0y}{(s_0 - x)(s_0 - y)}$

b. $\dfrac{P_0s_0x - P_0s_0y}{(s_0 - x)(s_0 - y)} = \dfrac{(500)(332)(70) - (500)(332)(45)}{(332 - 70)(332 - 45)}$

$P_0 = 500$, $s_0 = 332$, $x = 70$, $y = 45$

$= \dfrac{4{,}150{,}000}{75{,}194}$ Simplify.

≈ 55.2 Hz Simplify.

63. $\dfrac{-3x^3 - 2x^2 + 16x - 5}{4x^3 + 18x^2 - 6x}$ **65.** Sample answer: $20a^4b^2c$, $15ab^6$, $9abc$ **67.** D **69.** F **71.** $-\dfrac{4bc}{33a}$ **73.** $(n + 3)(n - 6)$

75. $D = \{x \mid x \geq -0.5\}$, $R = \{y \mid y \leq 0\}$

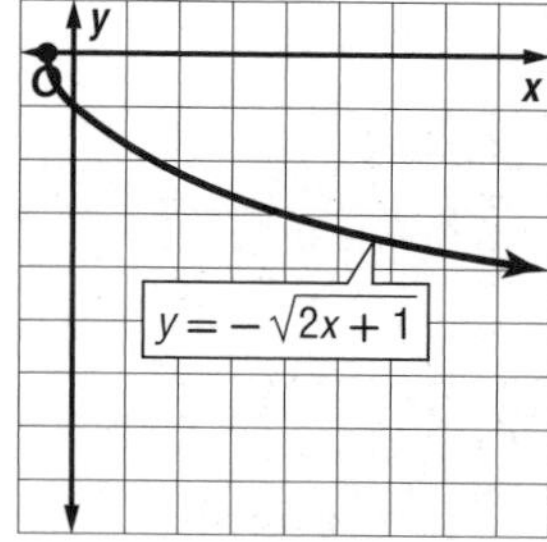

77. $D = \{x \mid x \geq -6\}$, $R = \{y \mid y \geq -3\}$

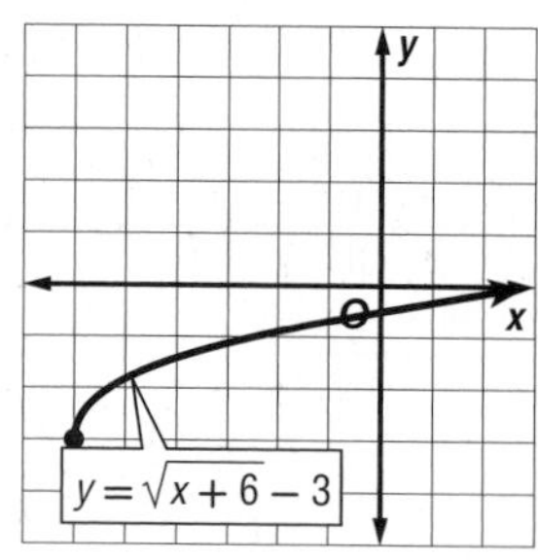

79. $D = \{x \mid x \geq 2\}$, $R = \{y \mid y \geq 4\}$

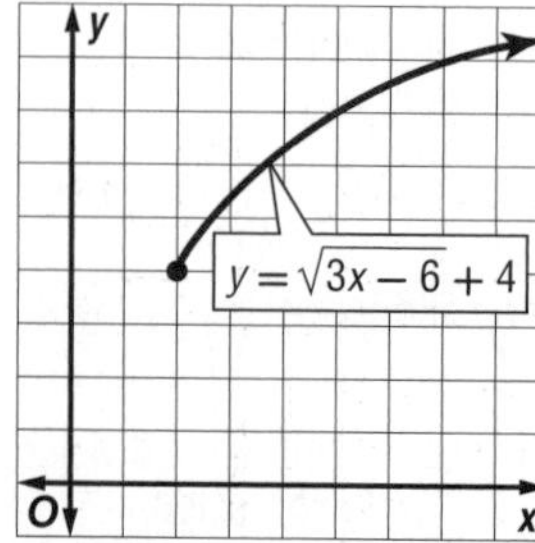

81. $-\dfrac{8}{3}$; 1 real **83.** 0, $3i$, $-3i$; 1 real, 2 imaginary

85.

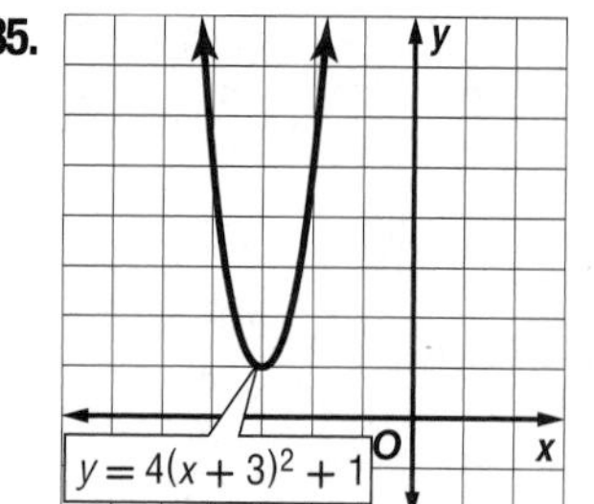

87.

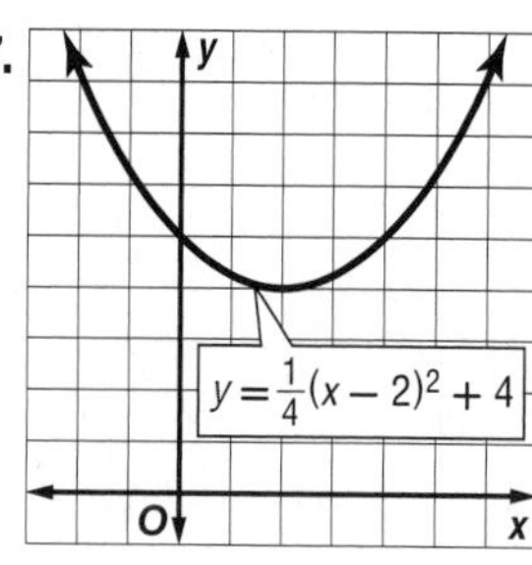

89.

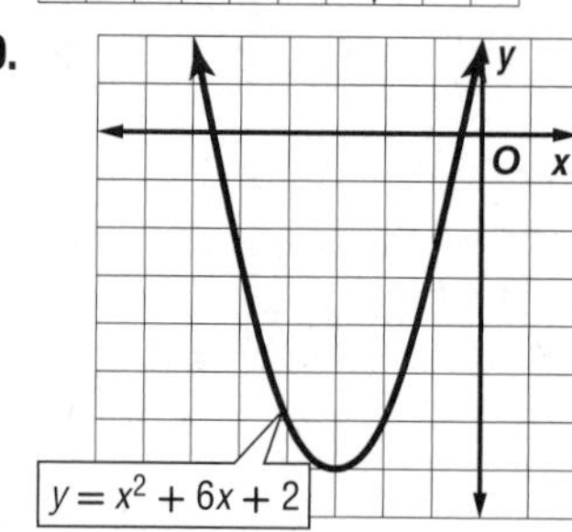

Lesson 8-3

1 $x - 1 = 0$

$x = 1$

$f(x)$ is not defined when $x = 1$. So, there is a vertical asymptote at $x = 1$. From $x = 1$, as x-values decrease, $f(x)$ values approach 0, and as x-values increase, $f(x)$ values approach 0. So there is a horizontal asymptote at $f(x) = 0$. The domain is all real numbers not equal to 1 or $D = \{x \mid x \neq 1\}$. The range is all real numbers not equal to 0 or $R = \{f(x) \mid f(x) \neq 0\}$.

3.

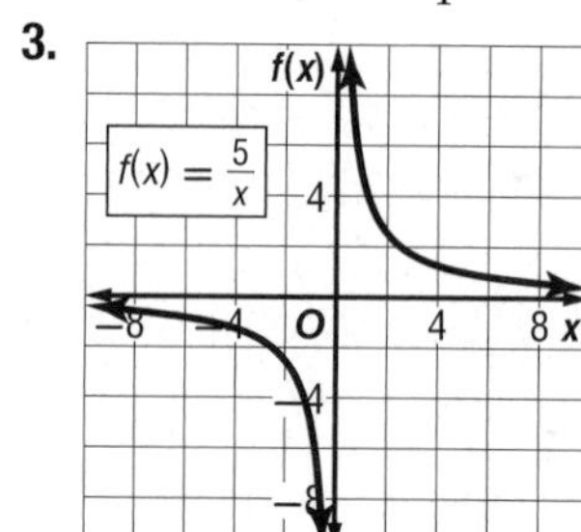

$D = \{x \mid x \neq 0\}$; $R = \{f(x) \mid f(x) \neq 0\}$

5.

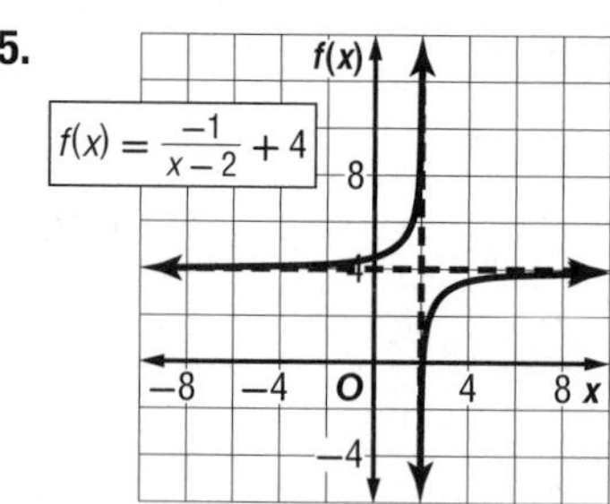

$D = \{x \mid x \neq 2\}$; $R = \{f(x) \mid f(x) \neq 4\}$

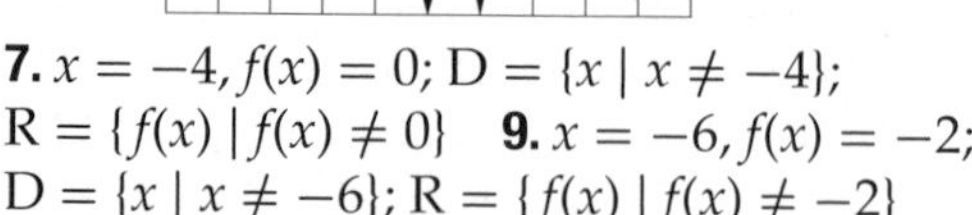

7. $x = -4$, $f(x) = 0$; $D = \{x \mid x \neq -4\}$; $R = \{f(x) \mid f(x) \neq 0\}$ **9.** $x = -6$, $f(x) = -2$; $D = \{x \mid x \neq -6\}$; $R = \{f(x) \mid f(x) \neq -2\}$

11.

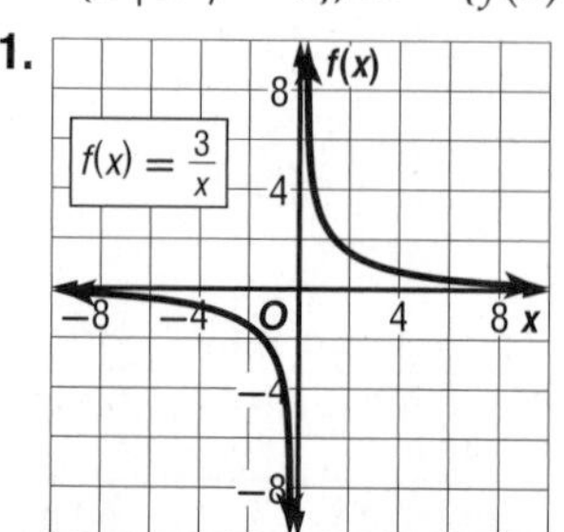

$D = \{x \mid x \neq 0\}$; $R = \{f(x) \mid f(x) \neq 0\}$

13. 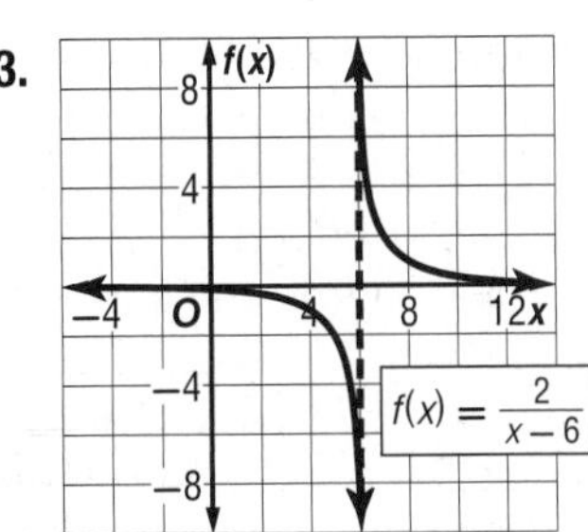

D = {x | x ≠ 6};
R = {f(x) | f(x) ≠ 0}

15 This represents a transformation of the graph of $f(x) = \frac{1}{x}$.
$a = 2$: The graph is stretched vertically.
$k = 3$: The graph is translated 3 units up. There is a horizontal asymptote at $f(x) = 3$. Domain: D = {x | x ≠ 0}. Range: R = {f(x) | f(x) ≠ 3}.

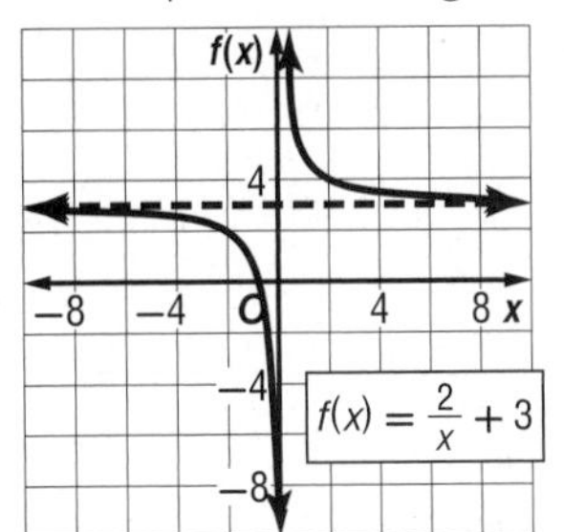

17. D = {x | x ≠ 5};
R = {f(x) | f(x) ≠ 0}

19. 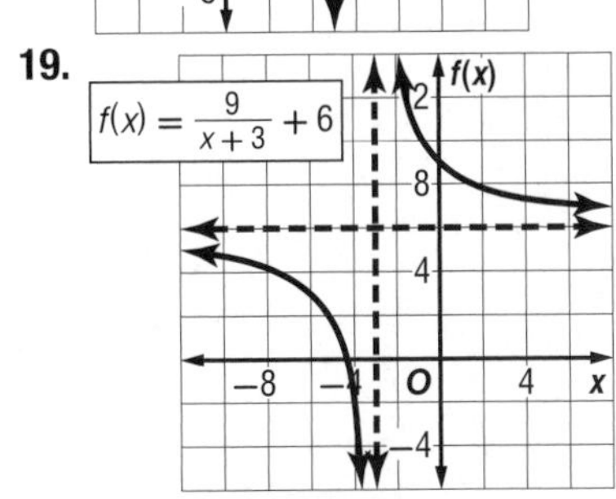

D = {x | x ≠ −3};
R = {f(x) | f(x) ≠ 6}

21. 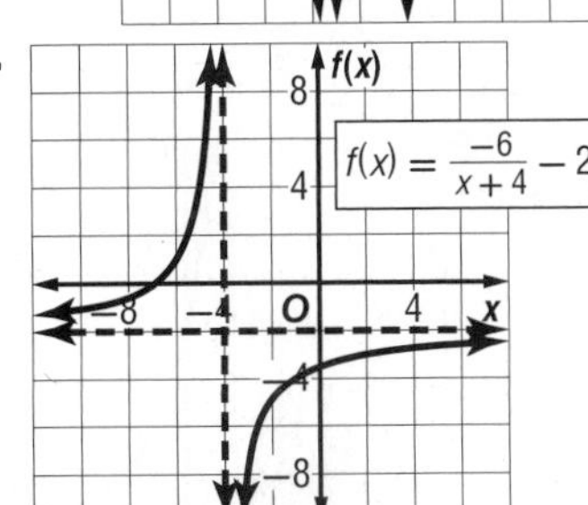

D = {x | x ≠ −4};
R = {f(x) | f(x) ≠ −2}

23a. $m = \frac{5000}{d}$

23b. 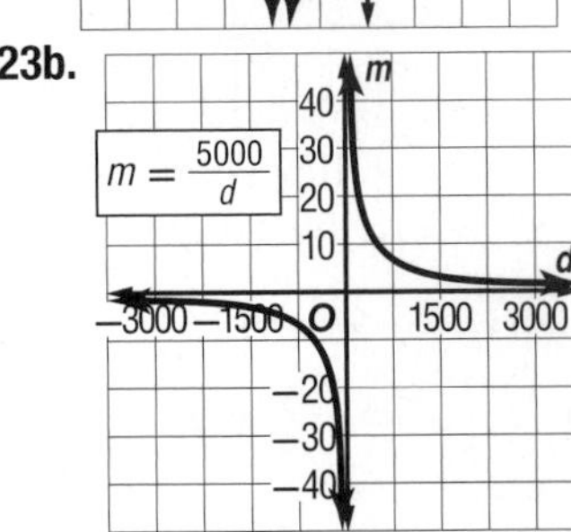

23c. 13.7 mi

25. 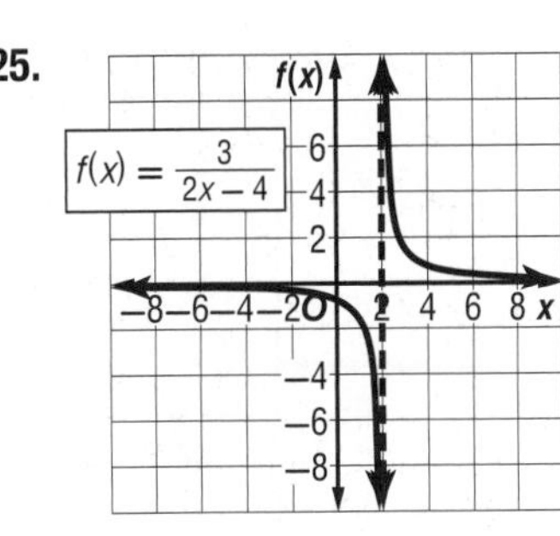

D = {x | x ≠ 2};
R = {f(x) | f(x) ≠ 0}

27. 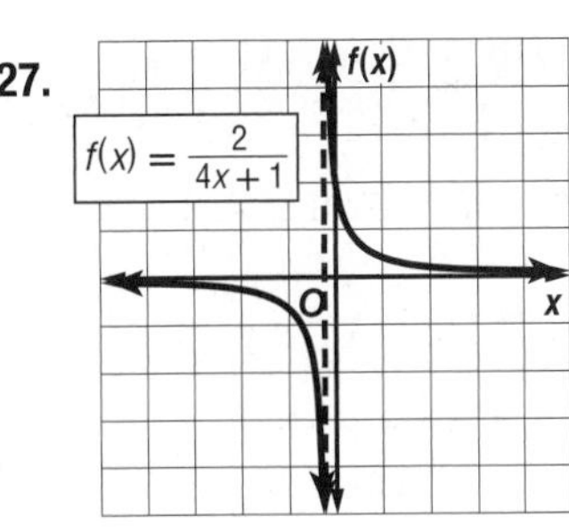

$D = \left\{x \,\middle|\, x \neq \frac{1}{4}\right\}$;
R = {f(x) | f(x) ≠ 0}

29 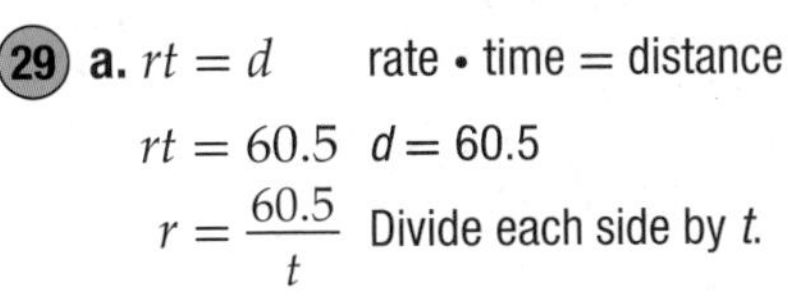
a. $rt = d$ rate • time = distance
$rt = 60.5$ $d = 60.5$
$r = \frac{60.5}{t}$ Divide each side by t.

b. This represents a transformation of the graph of $f(x) = \frac{1}{x}$. There are asymptotes at $t = 0$ and $r = 0$. Since $a = 60.5$, the graph is stretched vertically.

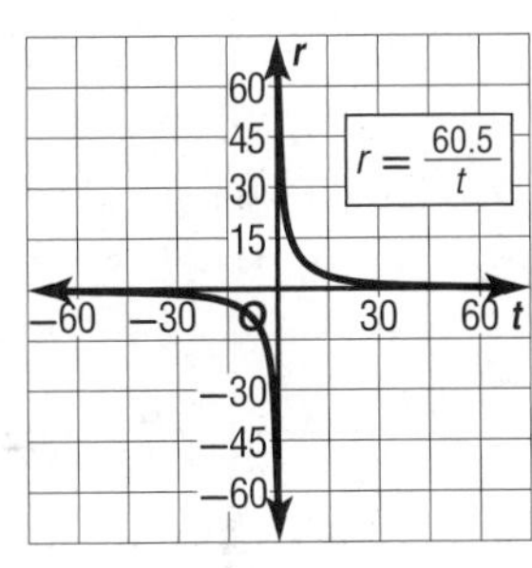

c. $r = \frac{60.5}{t}$ Write the equation.
$= \frac{60.5}{0.48}$ $t = 0.48$
≈ 126 ft/s Use a calculator.

31. D = {x | x ≠ −2};
R = {f(x) | f(x) ≠ −5};
$x = -2, f(x) = -5$

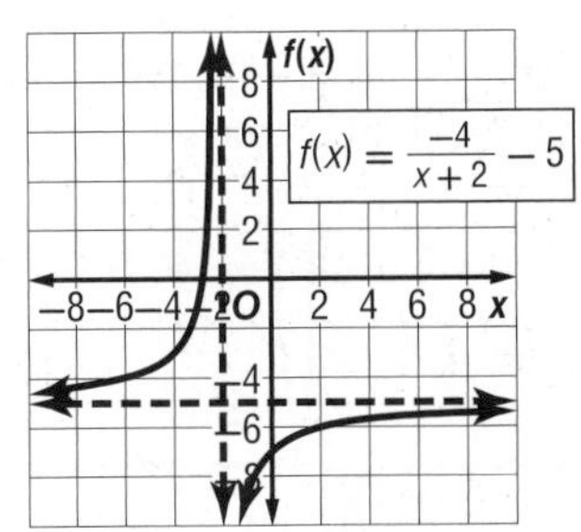

33. D = {x | x ≠ 4};
R = {f(x) | f(x) ≠ 3};
$x = 4, f(x) = 3$

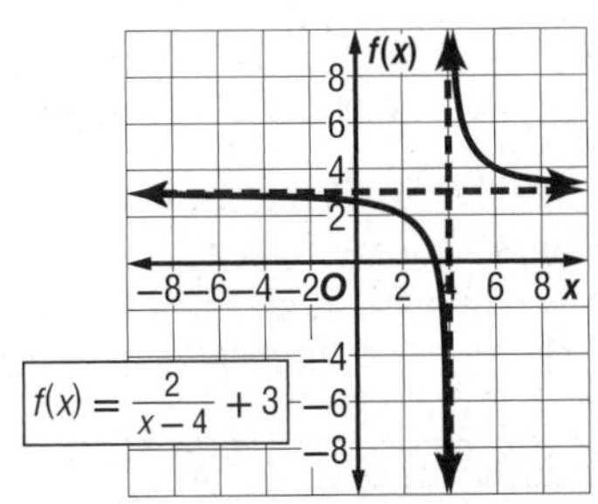

Selected Answers and Solutions

35. $D = \{x \mid x \neq 7\}$; $R = \{f(x) \mid f(x) \neq -8\}$; $x = 7, f(x) = -8$

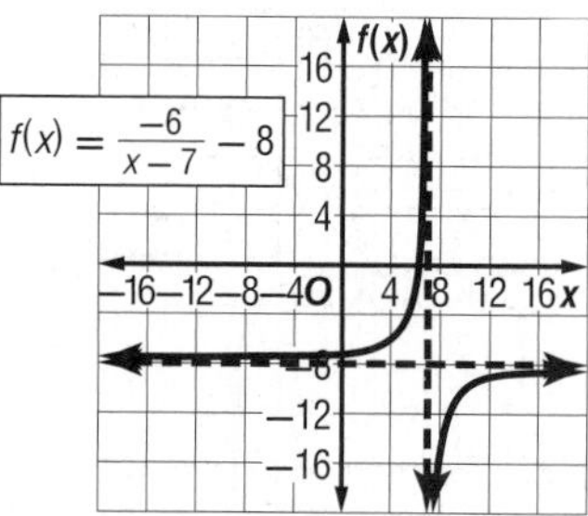

37a.

x	$a(x) = x^2$	$b(x) = x^{-2}$	$c(x) = x^3$	$d(x) = x^{-3}$
−4	16	$\frac{1}{16}$	−64	$-\frac{1}{64}$
−3	9	$\frac{1}{9}$	−27	$-\frac{1}{27}$
−2	4	$\frac{1}{4}$	−8	$-\frac{1}{8}$
−1	1	1	−1	−1
0	0	undefined	0	undefined
1	1	1	1	1
2	4	$\frac{1}{4}$	8	$\frac{1}{8}$
3	9	$\frac{1}{9}$	27	$\frac{1}{27}$
4	16	$\frac{1}{16}$	64	$\frac{1}{64}$

37b.

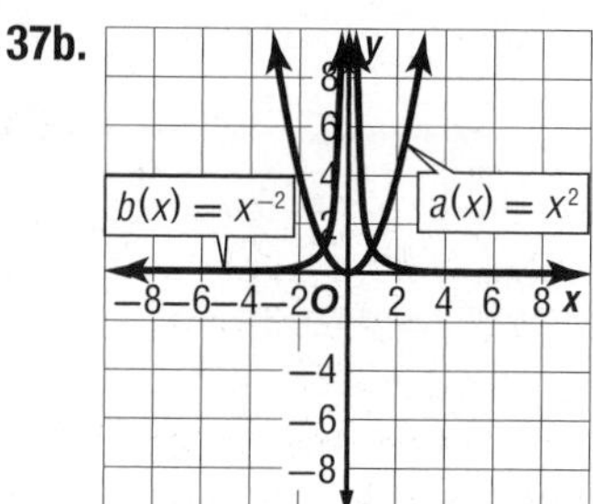

37c. $a(x)$: D = {all real numbers}, $R = \{a(x) \mid a(x) \geq 0\}$; $x \to -\infty, a(x) \to \infty, x \to \infty, a(x) \to \infty$; At $x = 0, a(x) = 0$, so there is a zero at $x = 0$.
$b(x)$: $D = \{x \mid x \neq 0\}$, $R \{b(x) \mid b(x) > 0\}$; $x \to -\infty, a(x) \to 0, x \to \infty, a(x) \to 0$; At $x = 0$, $b(x)$ is undefined, so there is an asymptote at $x = 0$.

37d.

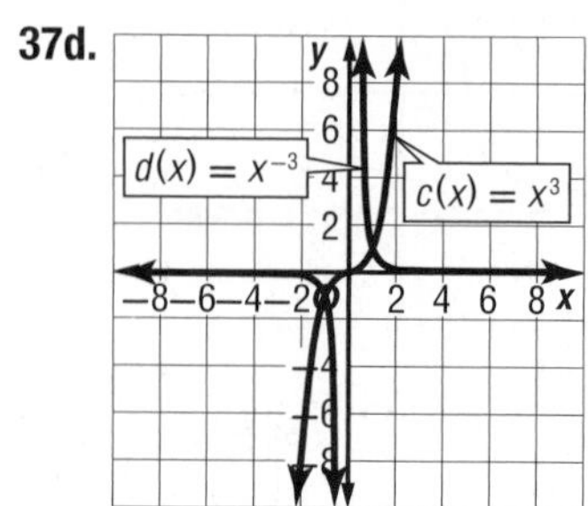

37e. $c(x)$: D = {all real numbers}, R = {all real numbers}; $x \to -\infty, a(x) \to -\infty, x \to \infty, a(x) \to \infty$; At $x = 0$, a(x) = 0, so there is a zero at $x = 0$.
$d(x)$: $D = \{x \mid x \neq 0\}$, $R \{b(x) \mid b(x) \neq 0\}$; $x \to -\infty, a(x) \to 0, x \to \infty, a(x) \to 0$; At $x = 0$, $b(x)$ is undefined, so there is an asymptote at $x = 0$.

37f. For two power functions $f(x) = ax^n$ and $g(x) = ax^{-n}$, for every x, $f(x)$ and $g(x)$ are reciprocals. The domains are similar except that for $g(x)$, $x \neq 0$. Additionally, wherever $f(x)$ has a zero, $g(x)$ is undefined.
39a. The first graph has a vertical asymptote at $x = 0$ and a horizontal asymptote at $y = 0$. The second graph is translated 7 units up and has a vertical asymptote at $x = 0$ and a horizontal asymptote at $y = 7$. **39b.** Both graphs have a vertical asymptote at $x = 0$ and a horizontal asymptote at $y = 0$. The second graph is stretched by a factor of 4. **39c.** The first graph has a vertical asymptote at $x = 0$ and a horizontal asymptote at $y = 0$. The second graph is translated 5 units to the left and has a vertical asymptote at $x = -5$ and a horizontal asymptote at $y = 0$.

39d.

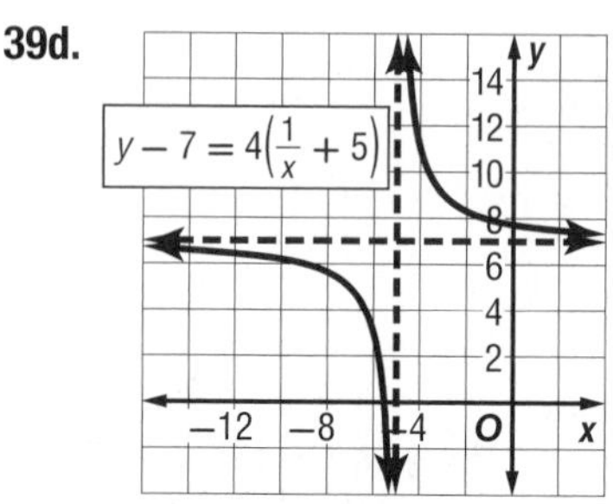

41. Sample answer: $f(x) = \frac{2}{x-3} + 4$ and $g(x) = \frac{5}{x-3} + 4$

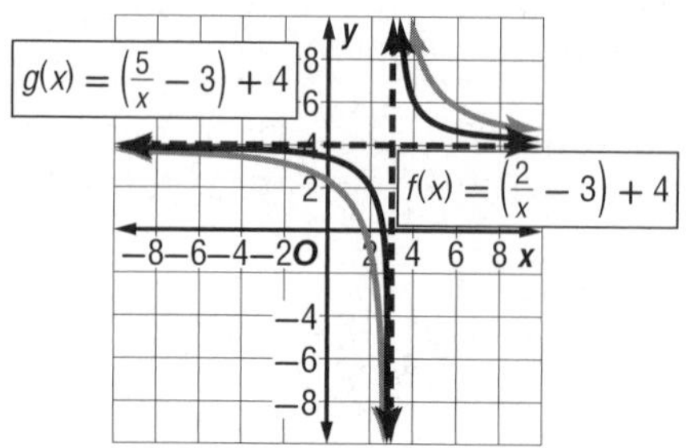

43. 4 **45.** B **47.** B **49.** $-2p$ **51.** $\frac{2x + y}{2x - y}$

53.

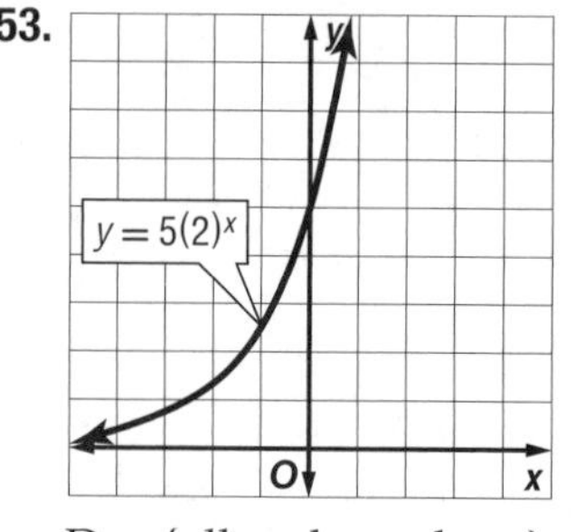

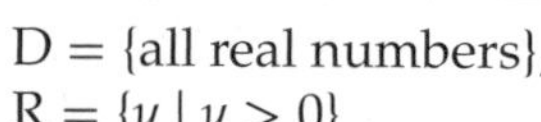

D = {all real numbers}, $R = \{y \mid y > 0\}$

55.

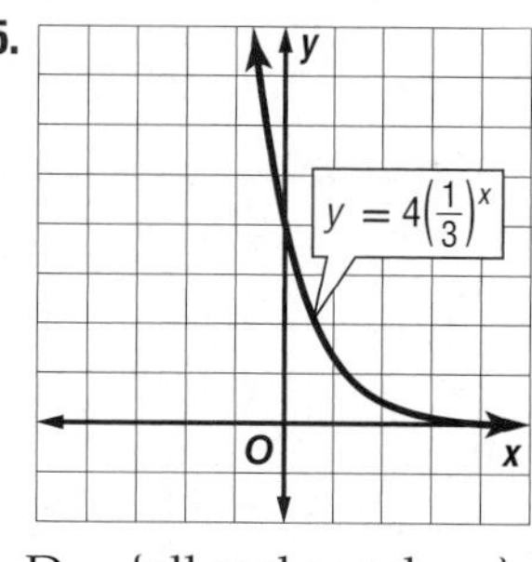

D = {all real numbers}, $R = \{y \mid y > 0\}$

57. $(f + g)(x) = 6x + 6$; $(f - g)(x) = -2x - 12$; $(f \cdot g)(x) = 8x^2 + 6x - 27$; $\left(\frac{f}{g}\right)(x) = \frac{2x - 3}{4x + 9}, x \neq -\frac{9}{4}$
59. $w = 4$ cm, $\ell = 8$ cm, $h = 2$ cm

61. rel. max at $x = 0$, rel. min at $x = -2$ and at $x = 2$; D = {all real numbers}, R = $\{f(x) \mid f(x) \geq -6\}$

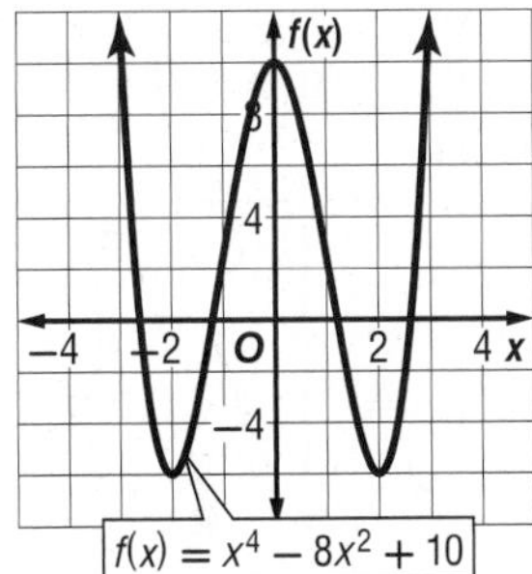

Lesson 8-4

1.

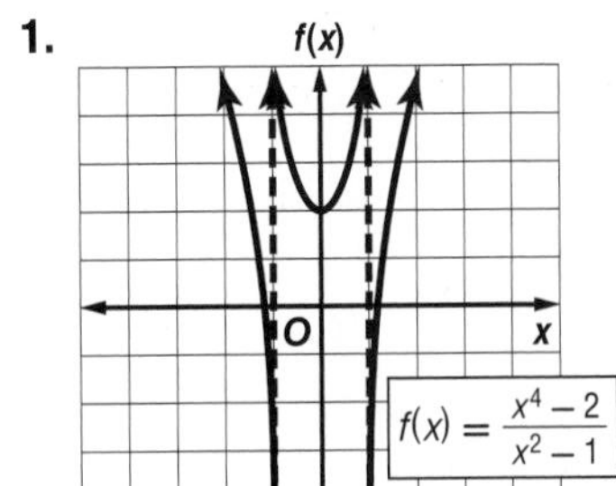

3a.

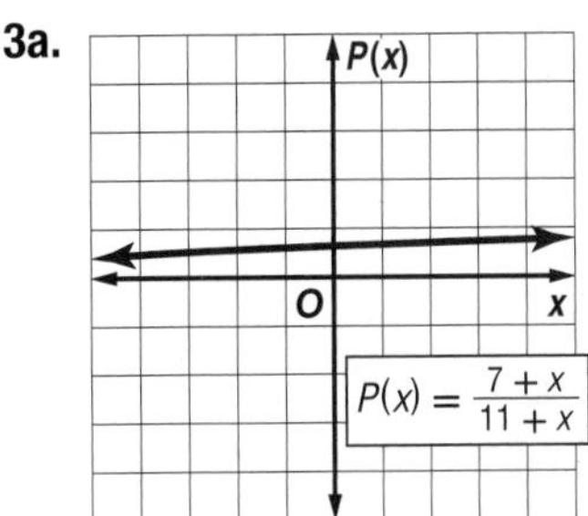

3b. The part in the first quadrant. **3c.** It represents his original field goal percentage of 63.6%. **3d.** $y = 1$; this represents 100% which he cannot achieve because he has already missed 4 field goals.

5 $x^2 + 8x + 20 = 0$ Set $a(x) = 0$.

Since $b^2 - 4ac = 8^2 - 4(1)(20)$ or -16, there are no real roots. So, there are no zeros.

$x + 2 = 0$ Set $b(x) = 0$.

$x = -2$ Subtract 2 from each side.

There is a vertical asymptote at $x = -2$. The degree of the numerator is greater than the degree of the denominator, so there is no horizontal asymptote. The difference between the degree of the numerator and the degree of the denominator is 1, so there is an oblique asymptote.

$$\begin{array}{r} x + 6 \\ x + 2\overline{)x^2 + 8x + 20} \\ (-)\ x^2 + 2x \\ \hline 6x + 10 \\ (-)\ 6x + 12 \\ \hline -2 \end{array}$$

The oblique asymptote is $y = x + 6$.

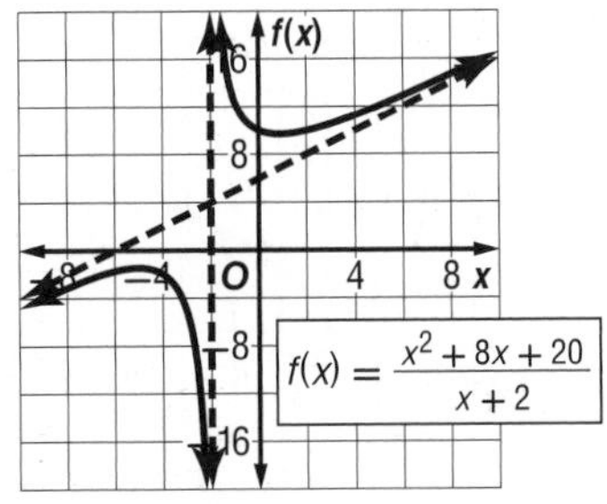

7.

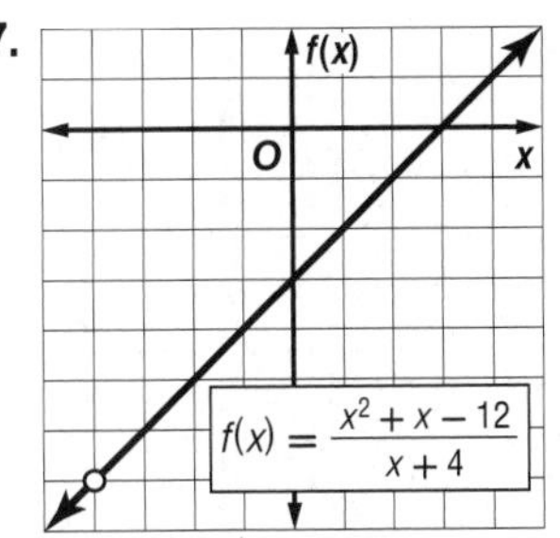

9.

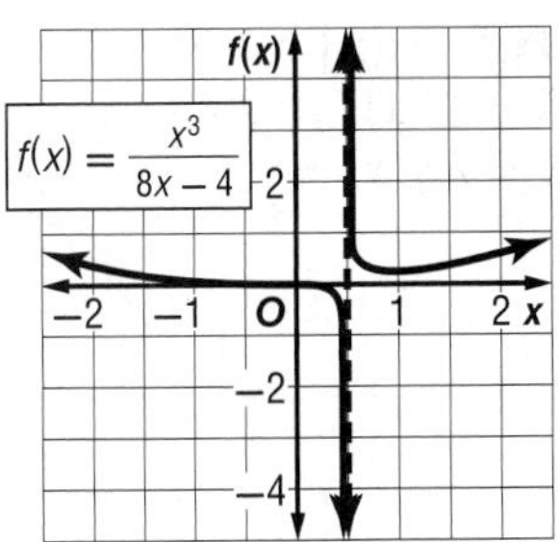

11.

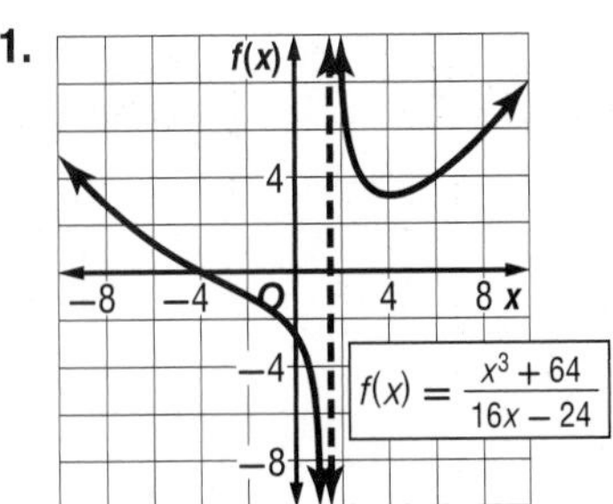

13.

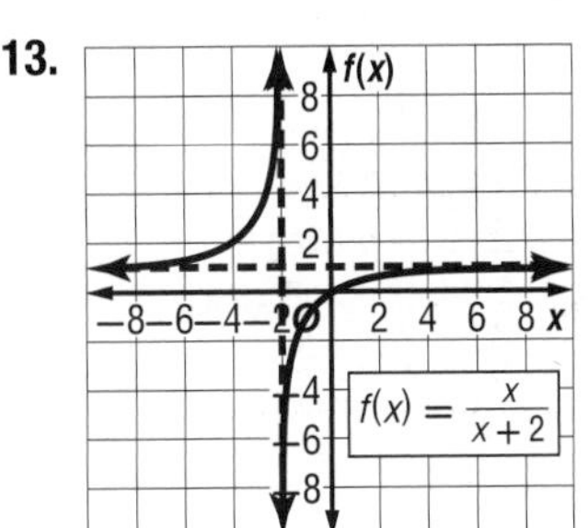

15 Since $a(x) = 4$, there are no zeros. The function is undefined for $x = 2$, so there is a vertical asymptote at $x = 2$. Since the degree of the numerator is less than the degree of the denominator, there is a horizontal asymptote at $f(x) = 0$. The difference between the degree of the numerator and the degree of the denominator is 2, so there is no oblique asymptote.

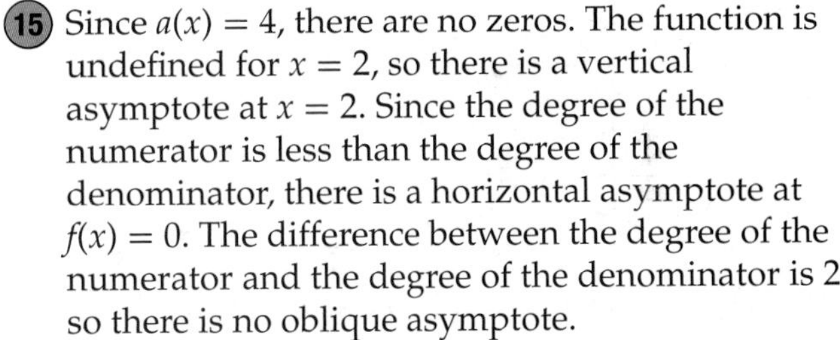

17.

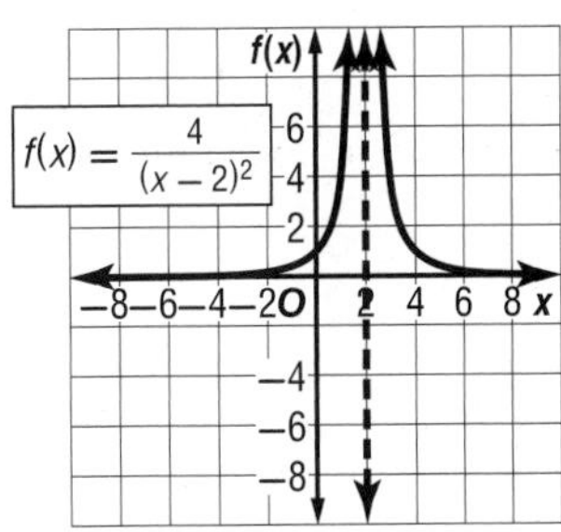

19.

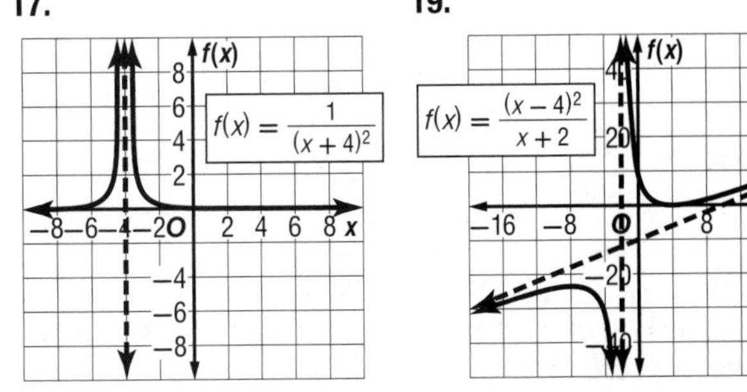

Selected Answers and Solutions

21.

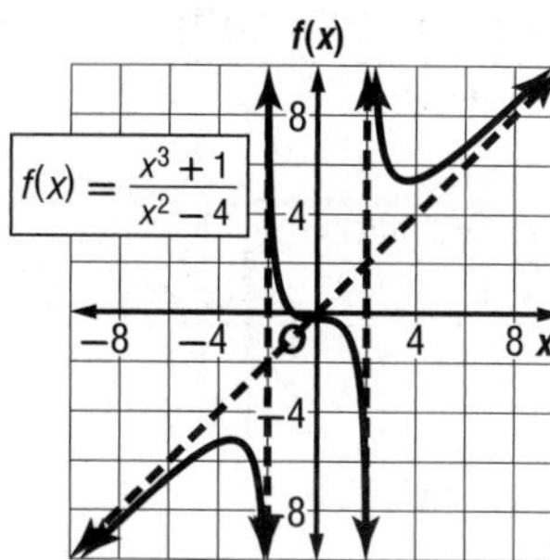

23.

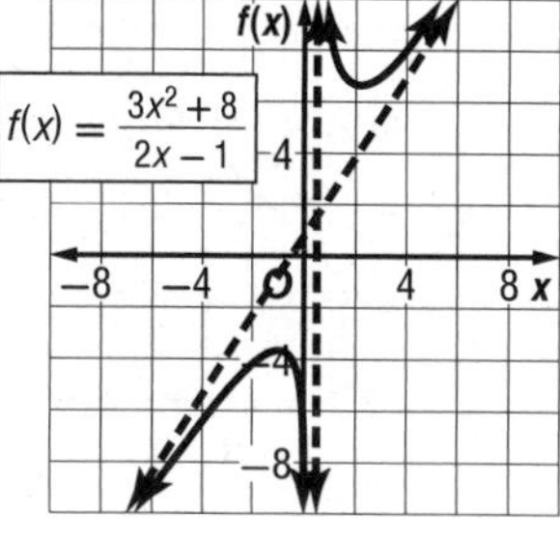

25.

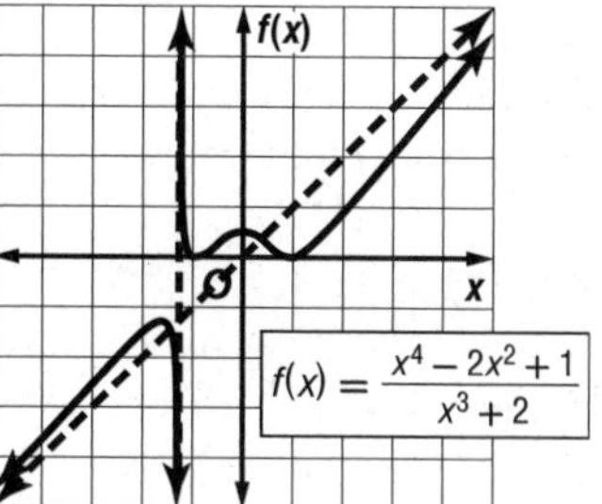

27a.

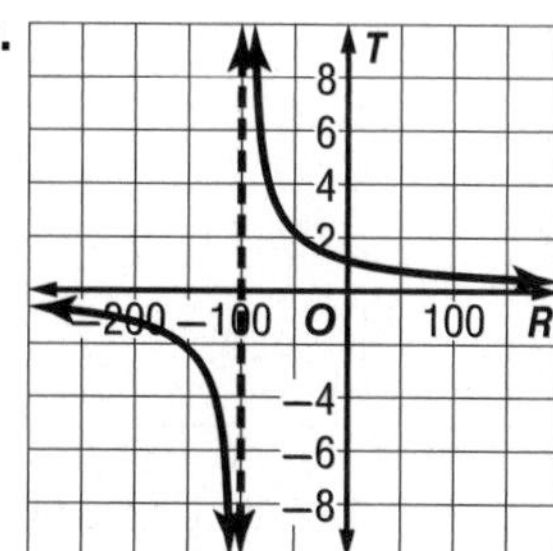

27b. $R_1 = -100$; no R_1-intercept; 1.2
27c. 0.5 amperes
27d. $R_1 \geq 0$ and $0 < I \leq 1.2$

29.

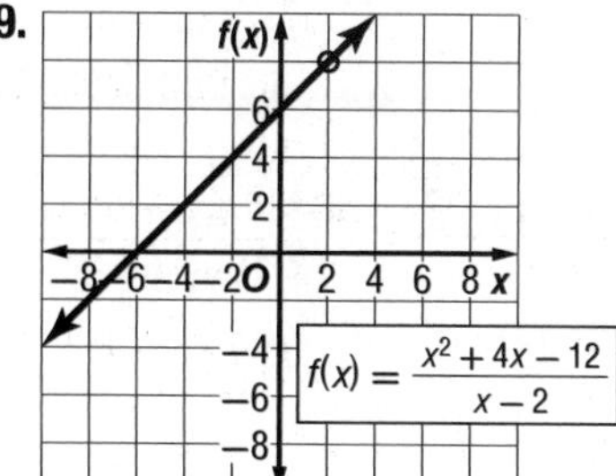

31.

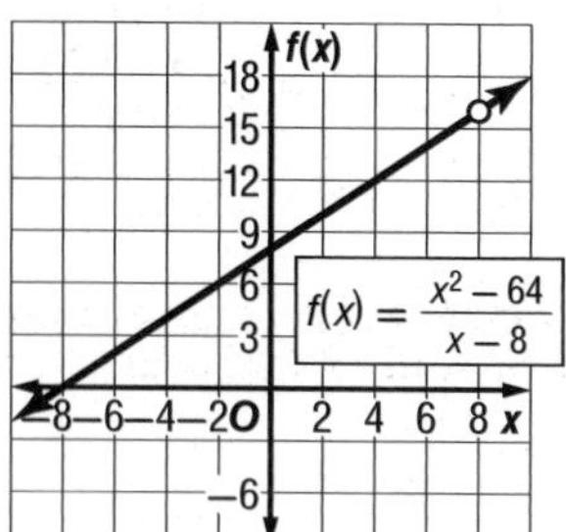

33.

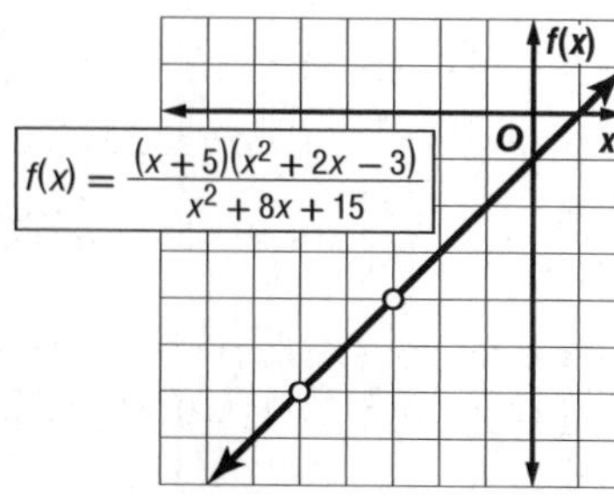

35.

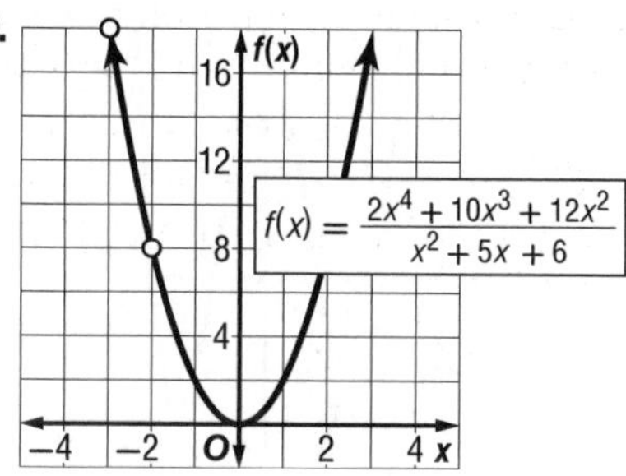

37 **a.** total cost = phone cost + monthly usage charge
= 150 + 40x

average monthly cost = $\frac{\text{total cost}}{\text{number of months}}$

$f(x) = \frac{150 + 40x}{x}$

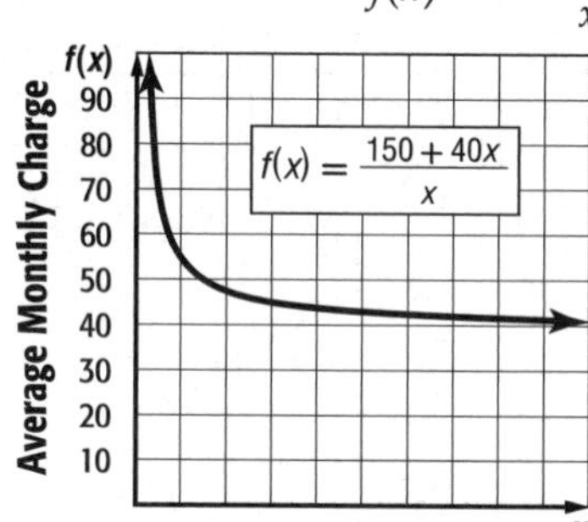

b. The vertical asymptote is $x = 0$. Since the degree of the numerator equals the degree of the denominator, the horizontal asymptote is at $f(x) = \frac{40}{1}$ or $f(x) = 40$.

c. Sample answer: The number of months and the average cost cannot have negative values.

d.

$f(x) = \frac{150 + 40x}{x}$	Write the equation.
$45 = \frac{150 + 40x}{x}$	$f(x) = 45$
$45x = 150 + 40x$	Multiply each side by x.
$5x = 150$	Subtract $40x$ from each side.
$x = 30$	Divide each side by 5.

After 30 months, the average monthly charge will be $45.

39.

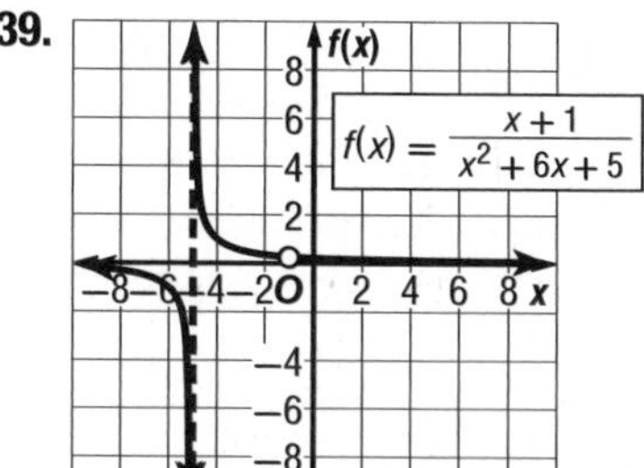

41.

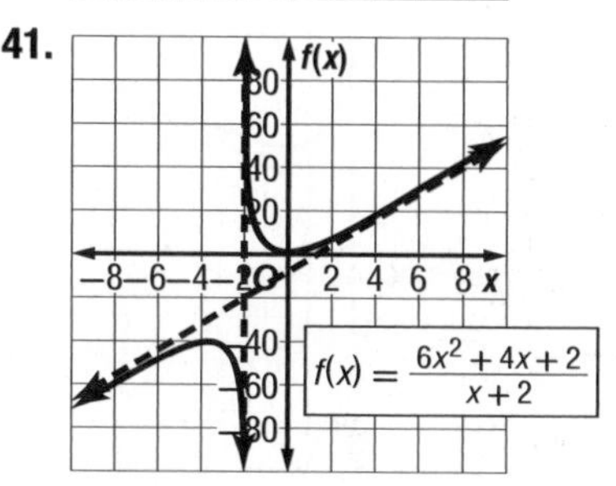

43. Similarities: Both have vertical asymptotes at $x = 0$. Both approach 0 as x approaches $-\infty$ and approach 0 as x approaches ∞. Differences: $f(x)$ has holes at $x = 1$ and $x = -1$, while $g(x)$ has vertical asymptotes at $x = \sqrt{2}$ and $x = -\sqrt{2}$. $f(x)$ has no zeros, but $g(x)$ has zeros at at $x = 1$ and $x = -1$.

45. $f(x) = \frac{x}{a-b} + \frac{c(a-b)}{a-b} = \frac{x + ca - cb}{a-b}$ **47.** C **49.** 4

51.

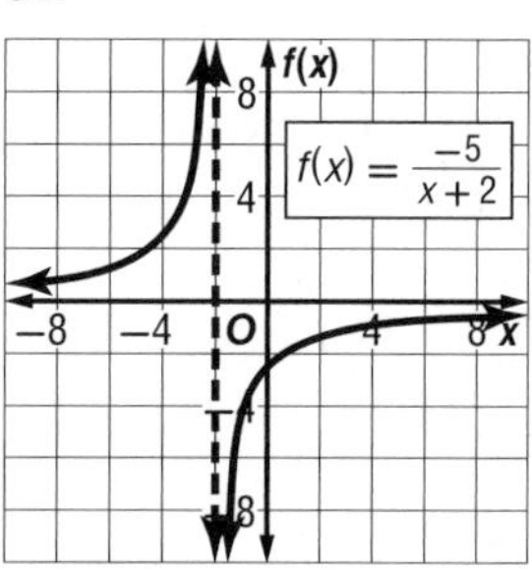

$D = \{x \mid x \neq -2\}$, $R = \{f(x) \mid f(x) \neq 0\}$

53.

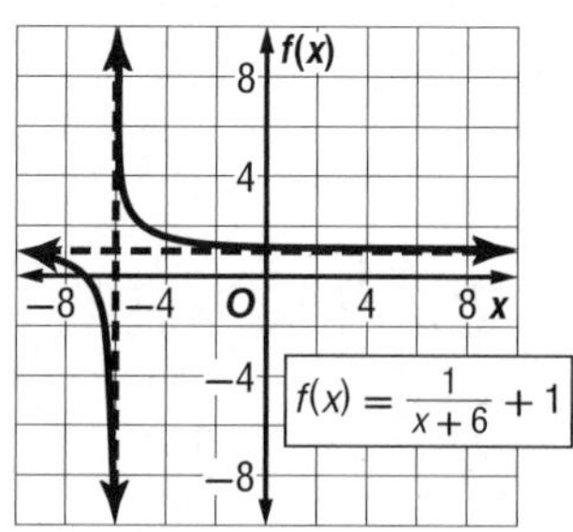

$D = \{x \mid x \neq -6\}$, $R = \{f(x) \mid f(x) \neq 1\}$

55. $\frac{y(y-9)}{(y+3)(y-3)}$ **57.** $\frac{-8d+20}{(d-4)(d+4)(d-2)}$ **59.** x^3 **61.** $a^{\frac{1}{9}}$

Lesson 8-5

1. 21 **3.** −32 **5.** −48 **7.** 1.5 **9.** −56 **11.** $m = \frac{1}{6}w$

13

$\frac{a_1}{b_1c_1} = \frac{a_2}{b_2c_2}$ Joint variation

$\frac{-60}{-5(4)} = \frac{a_2}{4(-3)}$ $a_1 = -60, b_1 = -5, c_1 = 4, b_2 = 4, c_2 = -3$

$-60(4)(-3) = -5(4)(a_2)$ Cross multiply.

$720 = -20a_2$ Simplify.

$-36 = a_2$ Divide each side by −20.

15. −3 **17.** −22.5 **19.** 38 **21a.** $s = \frac{48}{t}$ **21b.** 3.2 hours **23.** −10 **25.** direct **27.** neither

29

$\frac{x_1}{y_2} = \frac{x_2}{y_1}$ Inverse variation

$\frac{16}{20} = \frac{x_2}{5}$ $x_1 = 16, y_1 = 5, y_2 = 20$

$16(5) = 20(x_2)$ Cross multiply.

$80 = 20x_2$ Simplify.

$4 = x_2$ Divide each side by 20.

31. −12 **33.** inverse; −2 **35.** combined; 10 **37.** direct; 4 **39.** direct; −2 **41.** inverse; 7 **43.** joint; 20

45a. $800 = rt$

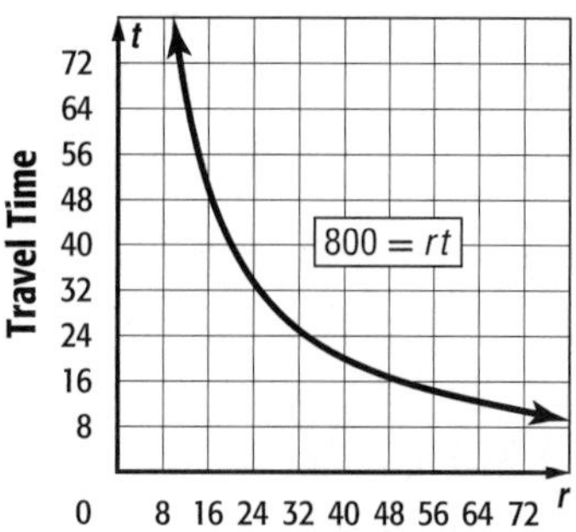

45b. $44.\overline{4}$ mph

47 **a.** $F = G\frac{m_1m_2}{d^2}$ Law of Universal Gravitation

$= (6.67 \times 10^{-11})\frac{(7.36 \times 10^{22})(5.97 \times 10^{24})}{(3.84 \times 10^8)^2}$

$\approx 2 \times 10^{20}$ newtons

b. $F = G\frac{m_1m_2}{d^2}$ Law of Universal Gravitation

$= (6.67 \times 10^{-11})\frac{(1.99 \times 10^{30})(5.97 \times 10^{24})}{(1.5 \times 10^{11})^2}$

$\approx 3.5 \times 10^{22}$ newtons

c. $F = G\frac{m_1m_2}{d^2}$ Law of Universal Gravitation

$= (6.67 \times 10^{-11})\frac{(1000)(1000)}{(0.1)^2}$

$= 6.67 \times 10^{-3}$ newtons

49. a and c are directly related. **51.** Sample answer: The force of an object varies jointly as its mass and acceleration. **53.** D

55a.

Year	Length
1	10
2	13
3	16
4	19

55b. $f(x) = 3x + 7$ **55c.** 34 in. **57.** asymptotes: $x = -2$, $x = -3$ **59.** hole: $x = -3$ **61.** 6 **63.** 3 **65.** $x + 3$, $x - \frac{1}{2}$ or $2x - 1$ **67.** $2a(a + 1)$ **69.** $24x$ **71.** $60ab^2$

Lesson 8-6

1. 11 **3.** 7

5 The LCD for the terms is $(x - 5)(x - 4)$.

$\frac{8}{x-5} - \frac{9}{x-4} = \frac{5}{x^2 - 9x + 20}$ Original equation

$\frac{(x-5)(x-4)(8)}{x-5} - \frac{(x-5)(x-4)(9)}{x-4} = \frac{(x-5)(x-4)(5)}{x^2-9x+20}$ Multiply by the LCD.

$\frac{\cancel{(x-5)}^1(x-4)(8)}{\cancel{x-5}_1} - \frac{(x-5)\cancel{(x-4)}^1(9)}{\cancel{x-4}_1} = \frac{\cancel{(x-5)}^1\cancel{(x-4)}^1(5)}{\cancel{x^2-9x+20}_1}$ Divide common factors.

$(x-4)(8) - (x-5)(9) = 5$ Simplify.

$8x - 32 - 9x + 45 = 5$ Distribute.

$-x + 13 = 5$ Simplify.

$-x = -8$ Subtract 13 from each side.

$x = 8$ Divide each side by −1.

7. 14

9a.

	pounds	price per pound	total price
dried fruit	10	$6.25	6.25(10)
mixed nuts	m	$4.50	$4.5m$
trail mix	$10 + m$	$5.00	$5(10 + m)$

9b. $62.5 + 4.5m = 50 + 5m$ **9c.** 25 **11a.** $\frac{1}{60}$ **11b.** $\frac{x}{60}$ **11c.** $\frac{1}{80}$ **11d.** $\frac{x}{80}$ **11e.** $\frac{x}{60} + \frac{x}{80} = 1$ **11f.** about 34.3 min **13.** $c < 0$, or $\frac{13}{18} < c$ **15.** $b < 0$, or $\frac{35}{12} < b$ **17.** 2 **19.** 1 **21.** ∅

23 cost of 3 pounds of bananas for \$0.90/pound = 0.9(3)
cost of x pounds of apples for \$1.25/pound = $1.25x$
total weight = $3 + x$

$$\frac{\text{total cost}}{\text{total weight}} = 1 \quad \text{Write an equation.}$$

$$\frac{0.9(3) + 1.25x}{3 + x} = 1 \quad \text{Substitute.}$$

$$\frac{2.7 + 1.25x}{3 + x} = 1 \quad \text{Simplify the numerator.}$$

$$\frac{(3 + x)(2.7 + 1.25x)}{3 + x} = (3 + x)(1) \quad \text{LCD is } (3 + x)\text{. Multiply by the LCD.}$$

$$\frac{\overset{1}{\cancel{(3 + x)}}(2.7 + 1.25x)}{\underset{1}{\cancel{3 + x}}} = (3 + x)(1) \quad \text{Divide out common factors.}$$

$$2.7 + 1.25x = 3 + x \quad \text{Simplify.}$$

$$2.7 + 0.25x = 3 \quad \text{Subtract } x \text{ from each side.}$$

$$0.25x = 0.3 \quad \text{Subtract 2.7 from each side.}$$

$$x = 1.2 \quad \text{Divide each side by 0.25.}$$

She must purchase 1.2 pounds of apples.

25. $x < 0$ or $x > 1.75$ **27.** $x < -2$, or $2 < x < 14$

29. $x < -5$ or $4 < x < \frac{17}{3}$ **31.** 55.56 mph **33a.** 1; yes; 3

33b.

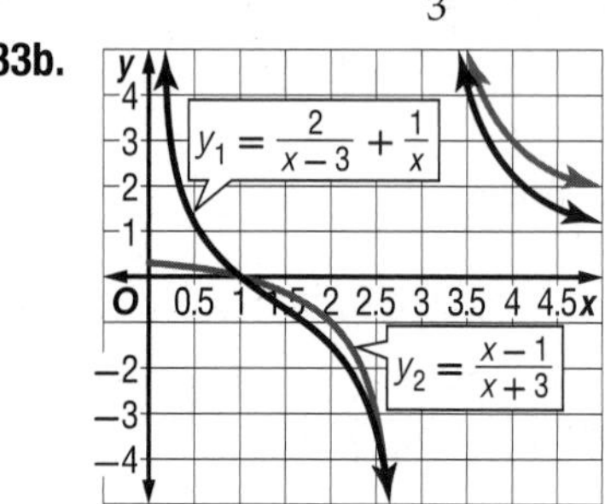

33c. 1; no

33d. Graph both sides of the equation. Where the graphs intersect, there is a solution. If they do not, then the possible solution is extraneous.

35. ∅ **37.** all real numbers except 5, −5, 0

39. Sample answer: Multiplying both sides of a rational inequality can produce extraneous solutions.

41. J **43.** all of the points **45.** direct

47.

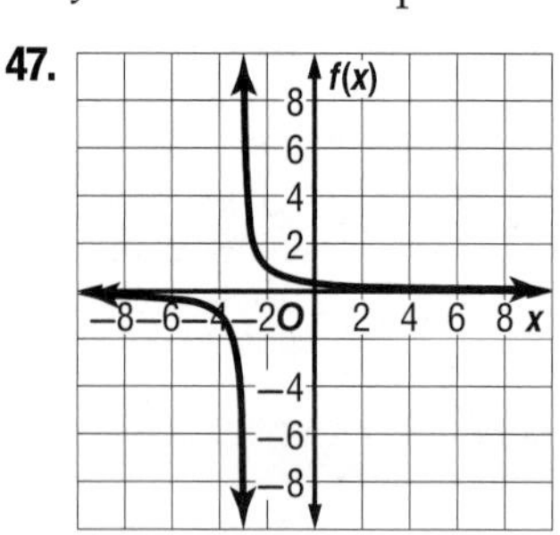

49.

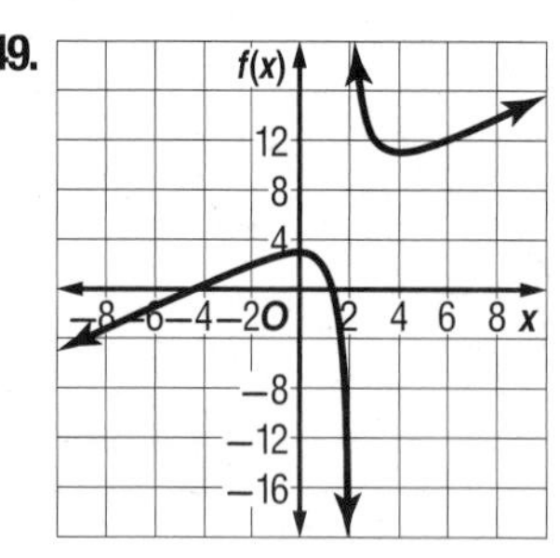

51a. $s \cdot 4^x$ **51b.** 0.5 three-yr periods or 1.5 yr **53.** yes

Chapter 8 Study Guide and Review

1. complex fraction **3.** oblique **5.** rational equations **7.** Joint variation **9.** Point discontinuity **11.** $-\frac{10yz^2}{9x}$

13. $\frac{x-1}{x-2}$ **15.** $\frac{x-3}{x+1}$ **17.** $\frac{27b + 10a^2}{12ab^2}$ **19.** $\frac{3xy^3 + 8y^3 - 5x}{6x^2y^2}$

21. $\frac{12x^2 - 10x + 6}{2(x+2)(3x-4)(x+1)}$ **23.** $\frac{10x + 20}{(x+6)(x+1)}$

25.

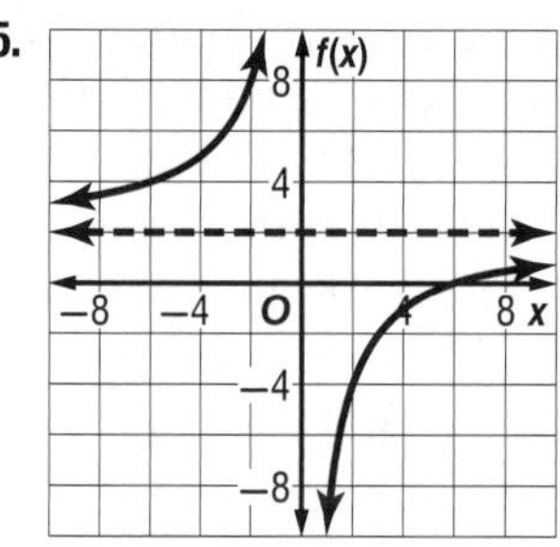

$D = \{x \mid x \neq 0\}$, $R = \{f(x) \mid f(x) \neq 2\}$

27.

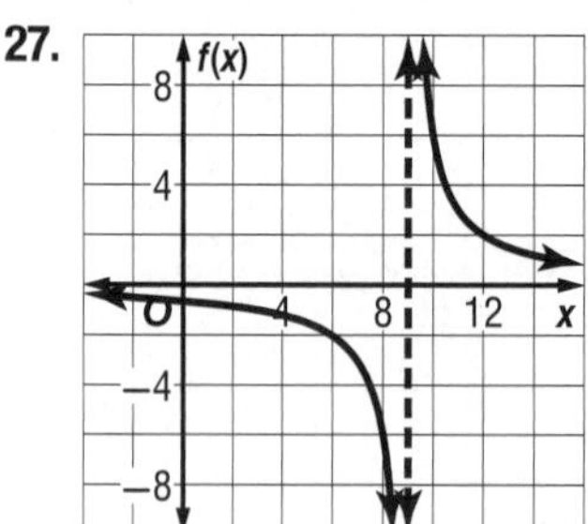

$D = \{x \mid x \neq 9\}$, $R = \{f(x) \mid f(x) \neq 0\}$

29.

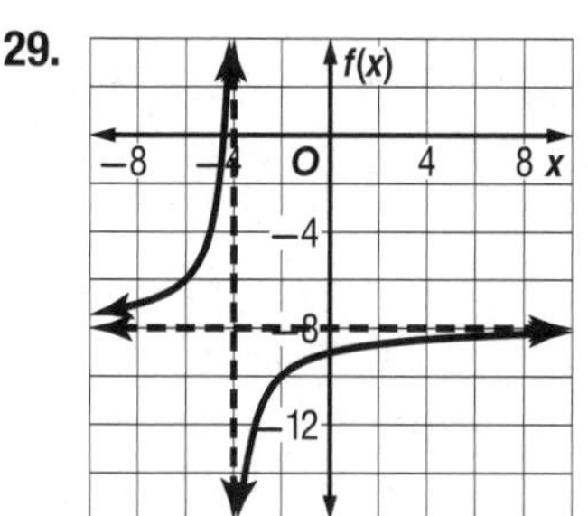

$D = \{x \mid x \neq -4\}$, $R = \{f(x) \mid f(x) \neq -8\}$

31. $x = -4$, $x = 0$ **33.** $x = 8$; hole: $x = -3$

35.

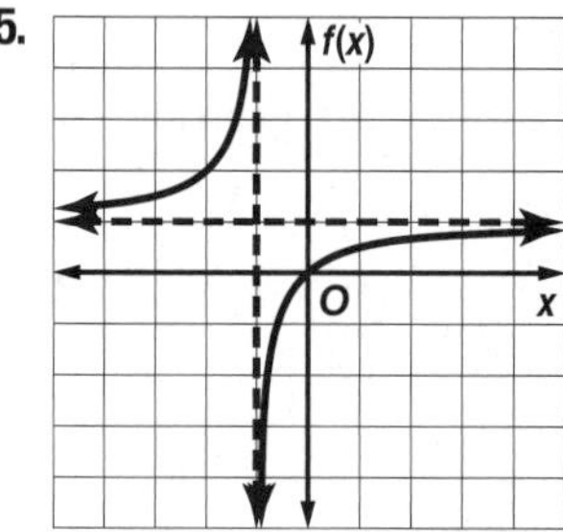

37.

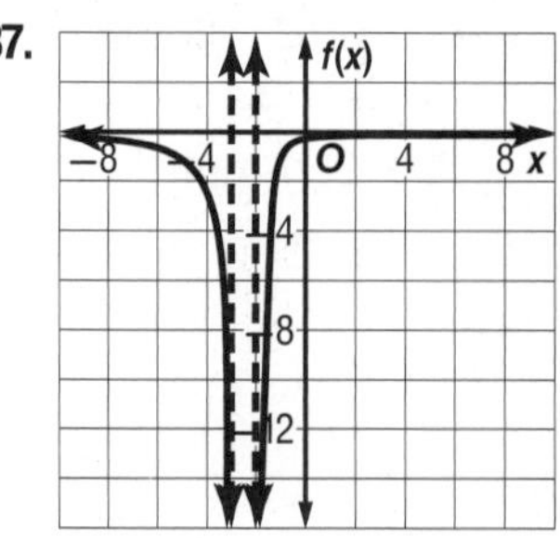

39. $a = 15$ **41.** $y = -\frac{1}{3}$ **43.** $y = \frac{48}{5}$ **45.** $x = \frac{46}{17}$

47. $x = -7$ **49.** $x = 8$ **51.** $-\frac{9}{10} < x < 0$

Selected Answers and Solutions

CHAPTER 9

Conic Sections

Chapter 9 Get Ready

1. $\{-7, -1\}$ **3.** $\{3, 5\}$ **5.** $\left\{-5, \frac{3}{2}\right\}$ **7.** $\left\{\frac{3}{4} \pm \sqrt{2}\right\}$

9. $A'(-2, -5)$, $B'(-1, -1)$, $C'(5, -1)$, $D'(4, -5)$ **11.** $J'(-3, 6)$, $K'(-2, -3)$, $L'(-10, -4)$

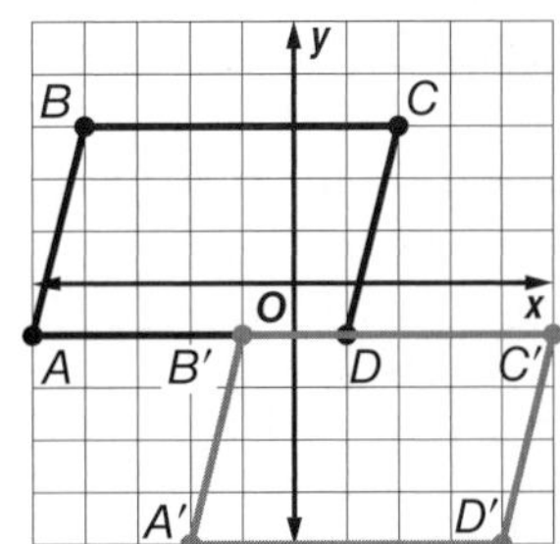

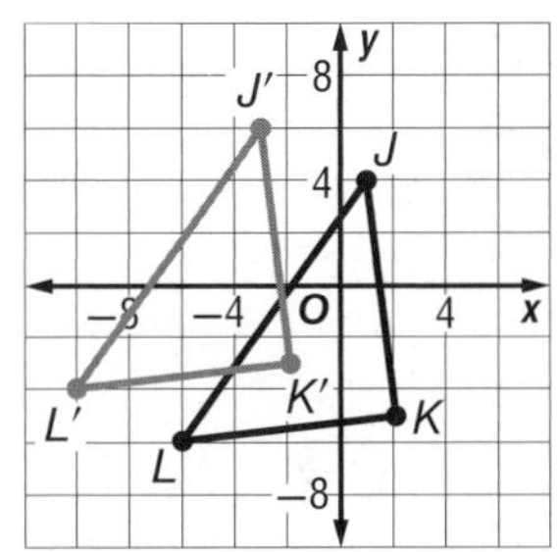

13. (115, 60), (125, 60), (125, 50), and (115, 50)

Lesson 9-1

1. $\left(-\frac{1}{2}, 8\right)$ **3.** (14.5, 9.75) **5.** 11.662 units

7
$$d = \sqrt{(x_2 - x_1)^2 + (y_2 - y_1)^2} \quad \text{Distance Formula}$$
$$= \sqrt{(3.5 - 0.25)^2 + (2.5 - 1.75)^2} \quad (x_1, y_1) = (0.25, 1.75) \text{ and } (x_2, y_2) = (3.5, 2.5)$$
$$= \sqrt{3.25^2 + 0.75^2} \quad \text{Simplify.}$$
$$= \sqrt{11.125} \text{ or about } 3.335 \text{ units} \quad \text{Simplify.}$$

9. A **11.** (−4, −1) **13.** (7.3, 1) **15.** (−7.75, −4.5) **17.** 16.279 units **19.** 16.125 units **21.** 21.024 units **23.** 55.218 units **25.** (−1.5, 0); 185.443 units **27.** (−5.5, −50.5); 148.223 units **29.** (8, 15); 136.953 units **31.** (−0.43, −2.25); 9.624 units

33
$$\left(\frac{x_1 + x_2}{2}, \frac{y_1 + y_2}{2}\right) \quad \text{Midpoint Formula}$$
$$= \left(\frac{-\frac{5}{12} + \left(-\frac{17}{2}\right)}{2}, \frac{-\frac{1}{3} + \left(-\frac{5}{3}\right)}{2}\right) \quad (x_1, y_1) = \left(-\frac{5}{12}, -\frac{1}{3}\right) \text{ and } (x_2, y_2) = \left(-\frac{17}{2}, -\frac{5}{3}\right)$$
$$= \left(\frac{-\frac{107}{12}}{2}, \frac{-2}{2}\right) \quad \text{Simplify.}$$
$$\approx (-4.458, -1) \quad \text{Simplify.}$$
$$d = \sqrt{(x_2 - x_1)^2 + (y_2 - y_1)^2} \quad \text{Distance Formula}$$
$$= \sqrt{\left(-\frac{17}{2} - \left(-\frac{5}{12}\right)\right)^2 + \left(-\frac{5}{3} - \left(-\frac{1}{3}\right)\right)^2} \quad (x_1, y_1) = \left(-\frac{5}{12}, -\frac{1}{3}\right) \text{ and } (x_2, y_2) = \left(-\frac{17}{2}, -\frac{5}{3}\right)$$
$$= \sqrt{\left(-\frac{97}{12}\right)^2 + \left(-\frac{4}{3}\right)^2} \quad \text{Simplify.}$$
$$\approx 8.193 \text{ units} \quad \text{Use a calculator.}$$

35. (−4.719, 0.028); 17.97 units **37.** 14.53 km

39
$$d = \sqrt{(x_2 - x_1)^2 + (y_2 - y_1)^2} \quad \text{Distance Formula}$$
$$= \sqrt{(254 - 132)^2 + (105 - 428)^2} \quad (x_1, y_1) = (132, 428) \text{ and } (x_2, y_2) = (254, 105)$$
$$= \sqrt{122^2 + (-323)^2} \quad \text{Simplify.}$$
$$= \sqrt{119{,}213} \text{ or about } 345 \text{ units} \quad \text{Simplify.}$$

345 units • 0.316 mi/unit ≈ 109 mi

41a.

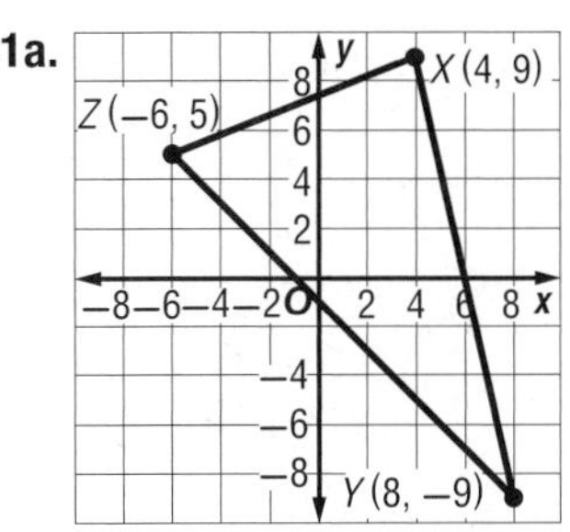

41b. midpoint of $\overline{XY} = (6, 0)$; midpoint of $\overline{YZ} = (1, -2)$; midpoint of $\overline{XZ} = (-1, 7)$ **41c.** The perimeter of $\triangle XYZ$ is $2\sqrt{29} + 14\sqrt{2} + 2\sqrt{85}$ units. perimeter = $\sqrt{29} + 7\sqrt{2} + \sqrt{85}$

41d. The perimeter of $\triangle XYZ$ is twice the perimeter of the smaller triangle. **43.** a circle and its interior with center at (5, 6) and radius 3 units **45.** The distance from A to B equals the distance from B to A. Using the Distance Formula, the solution is the same no matter which ordered pair is used first. **47.** 8.91

49. G **51.** −6, −2 **53.** $\frac{3}{2}$ **55.** 4.8362 **57.** 8.0086 **59.** $\{p \mid p \leq 1.9803\}$ **61.** −20 **63.** $\frac{1}{3}$ **65.** $y = (x - 3)^2 - 8$; (3, −8); $x = 3$; up

Lesson 9-2

1. $y = 2(x - 6)^2 - 32$; vertex (6, −32); axis of symmetry: $x = 6$; opens upward **3.** $x = (y - 4)^2 - 27$; vertex (−27, 4); axis of symmetry: $y = 4$; opens right

5.

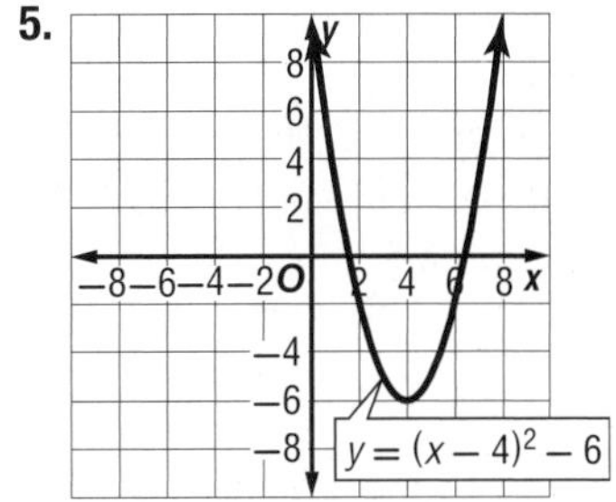

7.

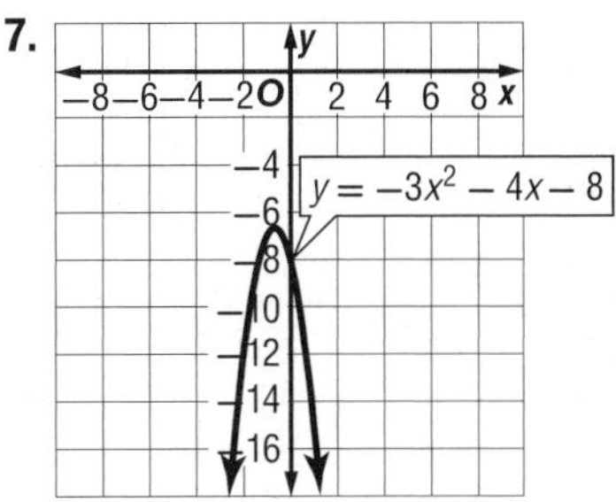

9. $y = \frac{1}{8}x^2 + 2$

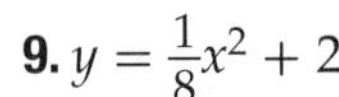

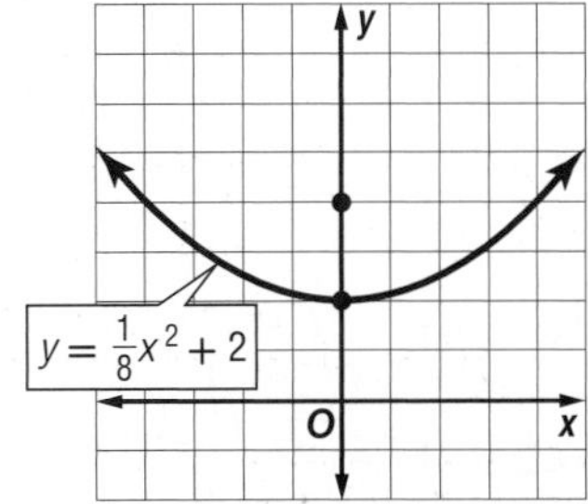

11. $y = -\frac{1}{12}(x-3)^2 + 5$

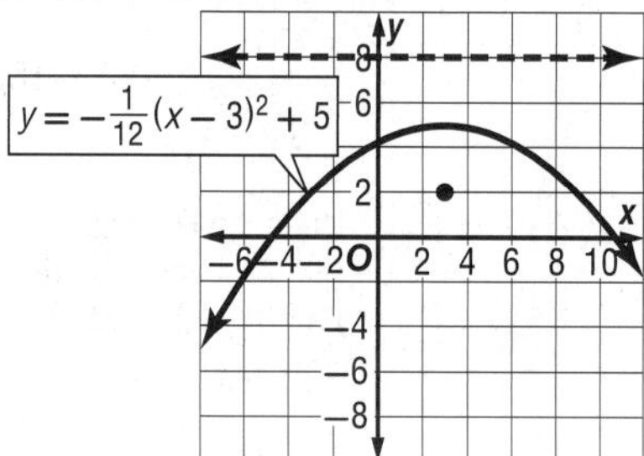

13a. $y = \frac{1}{24}x^2 - 6$

13b.

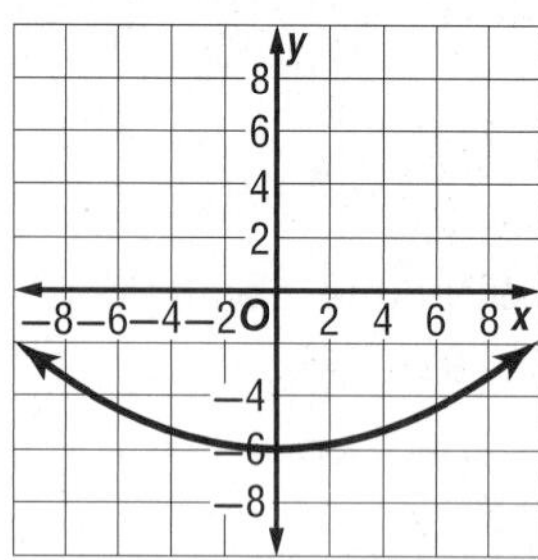

15. $y = 3(x+7)^2 + 2$; vertex $= (-7, 2)$; axis of symmetry: $x = -7$; opens upward

17. $y = -3\left(x + \frac{3}{2}\right)^2 + \frac{3}{4}$; vertex $= \left(-\frac{3}{2}, \frac{3}{4}\right)$; axis of symmetry: $x = -\frac{3}{2}$; opens downward

19. $x = \frac{2}{3}(y-3)^2 + 6$; vertex $= (6, 3)$; axis of symmetry: $y = 3$; opens right

21.

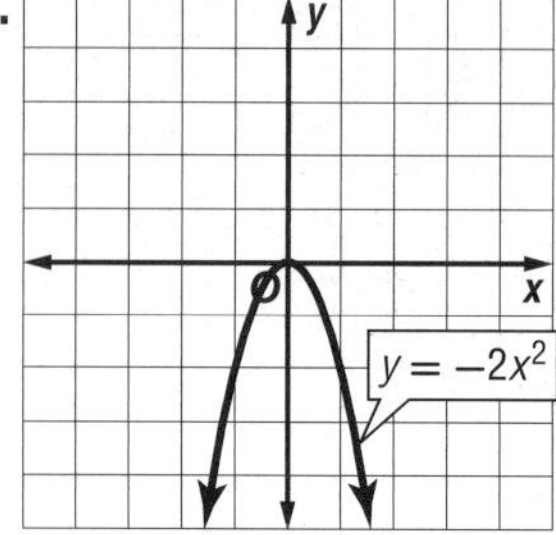

23.

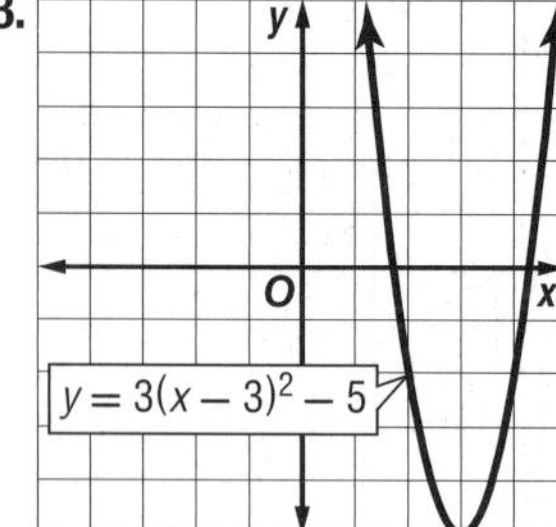

25.

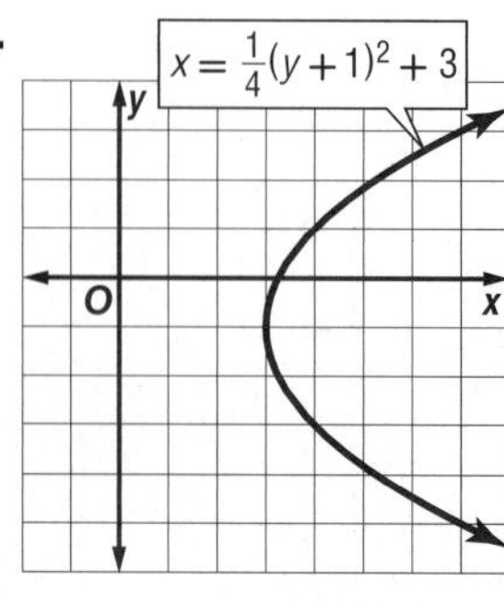

27. $y = \frac{1}{20}(x-1)^2 + 8$

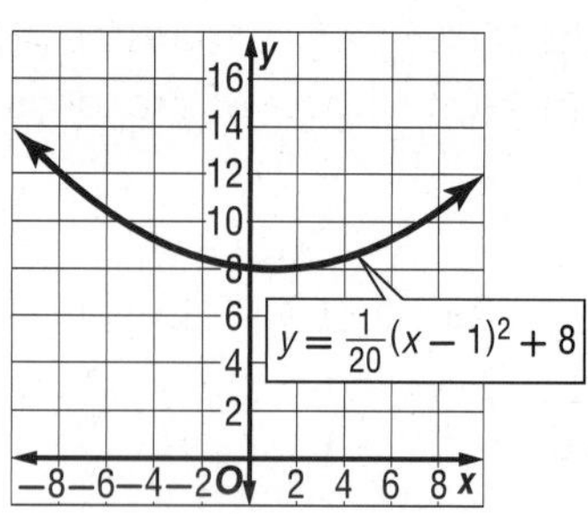

29 The directrix is a vertical line, so the equation of the parabola is of the form $x = a(y-k)^2 + h$. Since the vertex is equidistant from the focus and the directrix, the vertex is at (6, 4).

$x = h - \frac{1}{4a}$	Equation of directrix
$10 = 6 - \frac{1}{4a}$	$x = 10$ and $h = 6$
$4 = -\frac{1}{4a}$	Subtract 6 from each side.
$16a = -1$	Multiply each side by $4a$.
$a = -\frac{1}{16}$	Divide each side by 16.
$x = a(y-k)^2 + h$	Equation of a parabola
$x = -\frac{1}{16}(y-4)^2 + 6$	$a = -\frac{1}{6}$, $(h, k) = (6, 4)$

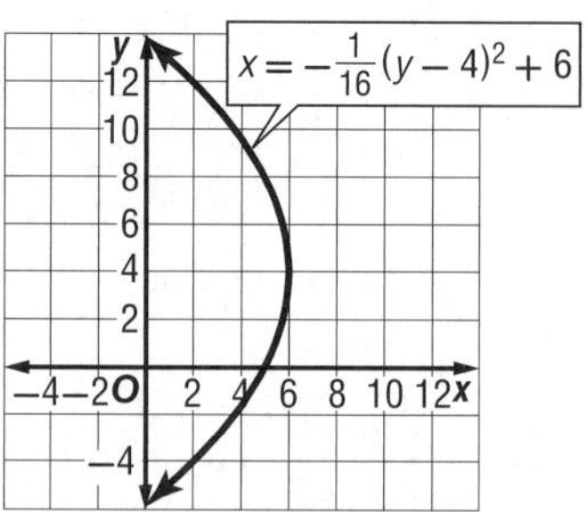

31. $y = -\frac{1}{4}(x-9)^2 + 6$

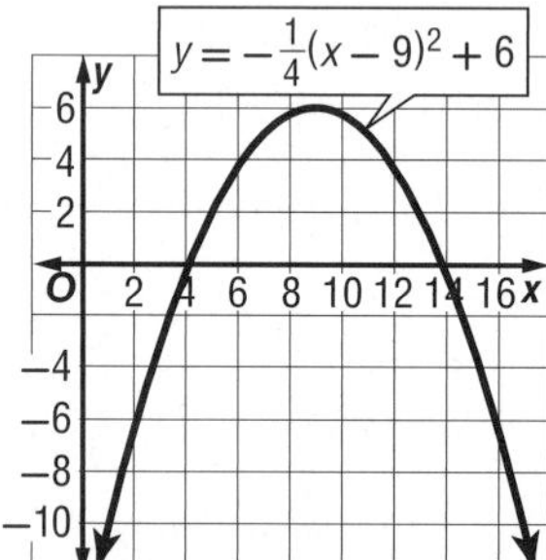

33a.

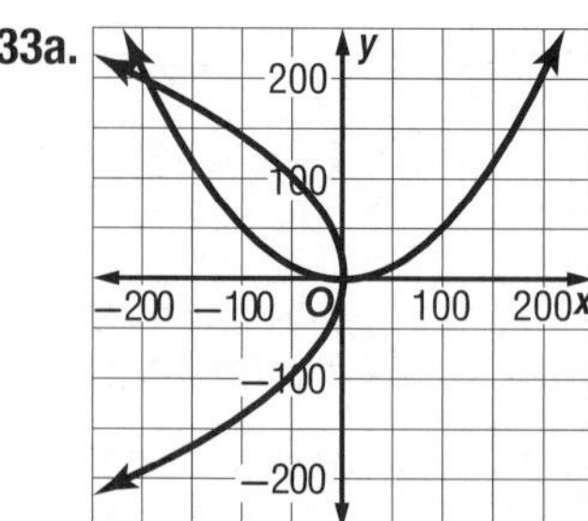

33b. $y = \frac{x^2}{192}$ and $x = \frac{y^2}{-192}$ **33c.** Sample answer: No; except for the direction in which they open, the graphs are identical.

35 The high beams should be placed at the focus. The y-coordinate of the focus is $k + \frac{1}{4a}$.

$k + \frac{1}{4a} = 0 + \frac{1}{4\left(\frac{1}{12}\right)}$ $\quad k = 0, a = \frac{1}{2}$

$= \frac{12}{4}$ or 3 $\quad$ Simplify.

The filament for the high beams should be placed 3 units above the vertex.

37. Rewrite it as $y = (x - h)^2$, where $h > 0$. **39.** Russell; the parabola should open to the left rather than to the right. **41.** C **43.** D **45.** $5\sqrt{2} + 3\sqrt{10}$ units **47.** 1.7183 **49.** $x > 0.4700$ **51.** 0.5 **53.** z^2 **55.** $\pm 1, \pm 2, \pm 3, \pm 6$ **57.** $\pm 1, \pm\frac{1}{3}, \pm\frac{1}{9}, \pm 3, \pm 9, \pm 27$ **59.** $3\sqrt{5}$ **61.** $16\sqrt{2}$

Lesson 9-3

1. $(x - 72)^2 + (y - 39)^2 = 10{,}000$

3. $(x - 1)^2 + (y + 5)^2 = 9$ **5.** $(x + 5)^2 + (y + 3)^2 = 90$

7. $x^2 + (y + 4)^2 = 20$

9. center: (0, 7); radius: 3

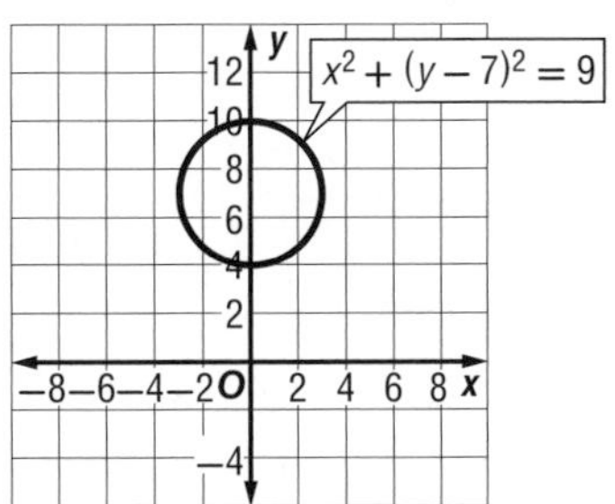

11. center: (2, –4); radius: 5

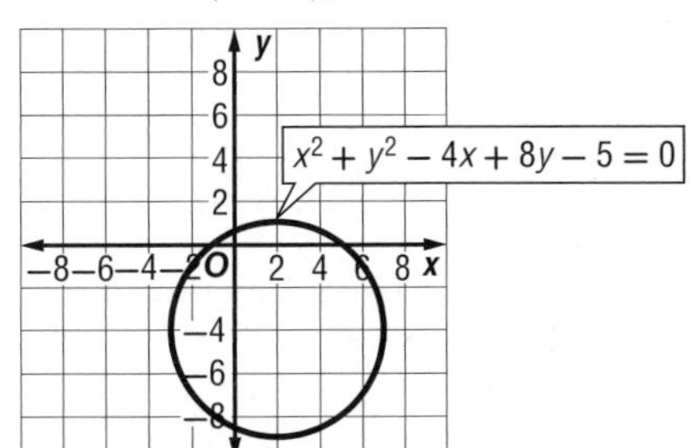

13. $(x + 3)^2 + (y - 1)^2 = 16$ **15.** $(x + 2)^2 + (y + 1)^2 = 81$

17 $(x - h)^2 + (y - k)^2 = r^2$ $\quad$ Equation of a circle

$(x - 0)^2 + [y - (-6)]^2 = (\sqrt{35})^2$ $\quad$ $(h, k) = (0, -6)$ and $r = \sqrt{35}$

$x^2 + (y + 6)^2 = 35$ $\quad$ Simplify.

19. $(x - 1)^2 + (y - 1)^2 = 4$ **21.** $x^2 + (y + 6) = 53$ **23.** $(x - 2)^2 + \left(y + \frac{3}{2}\right)^2 = \frac{25}{4}$ **25.** $\left(x - \frac{3}{2}\right)^2 + (y + 8)^2 = \frac{53}{4}$ **27.** $(x - 4)^2 + (y + 1)^2 = 20$

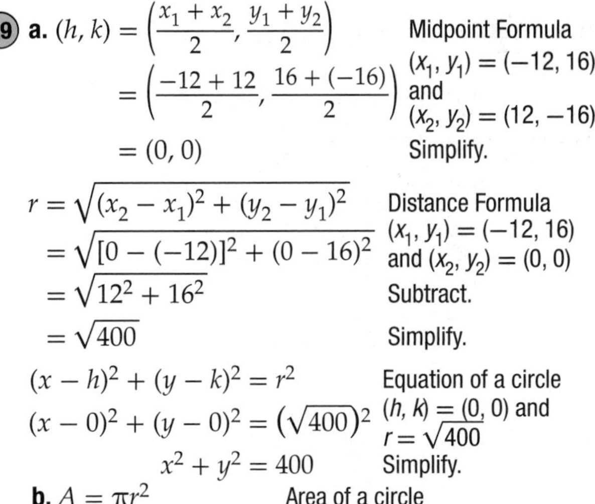

29 **a.** $(h, k) = \left(\frac{x_1 + x_2}{2}, \frac{y_1 + y_2}{2}\right)$ $\quad$ Midpoint Formula

$= \left(\frac{-12 + 12}{2}, \frac{16 + (-16)}{2}\right)$ $\quad$ $(x_1, y_1) = (-12, 16)$ and $(x_2, y_2) = (12, -16)$

$= (0, 0)$ $\quad$ Simplify.

$r = \sqrt{(x_2 - x_1)^2 + (y_2 - y_1)^2}$ $\quad$ Distance Formula

$= \sqrt{[0 - (-12)]^2 + (0 - 16)^2}$ $\quad$ $(x_1, y_1) = (-12, 16)$ and $(x_2, y_2) = (0, 0)$

$= \sqrt{12^2 + 16^2}$ $\quad$ Subtract.

$= \sqrt{400}$ $\quad$ Simplify.

$(x - h)^2 + (y - k)^2 = r^2$ $\quad$ Equation of a circle

$(x - 0)^2 + (y - 0)^2 = (\sqrt{400})^2$ $\quad$ $(h, k) = (0, 0)$ and $r = \sqrt{400}$

$x^2 + y^2 = 400$ $\quad$ Simplify.

b. $A = \pi r^2$ $\quad$ Area of a circle

$= \pi(\sqrt{400})^2$ $\quad$ $r = \sqrt{400}$

$= 400\pi$ $\quad$ Simplify.

≈ 1256.64 units^2 $\quad$ Use a calculator.

31. center: (0, 0); radius: $5\sqrt{3}$

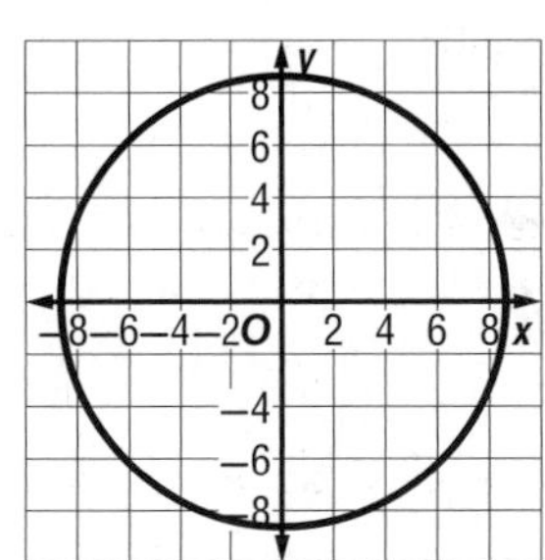

33. center: (1, 4); radius: $\sqrt{34}$

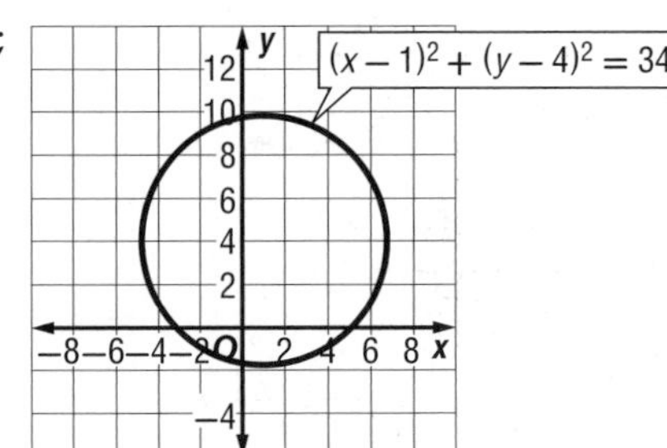

35. center: (5, –2); radius: 4

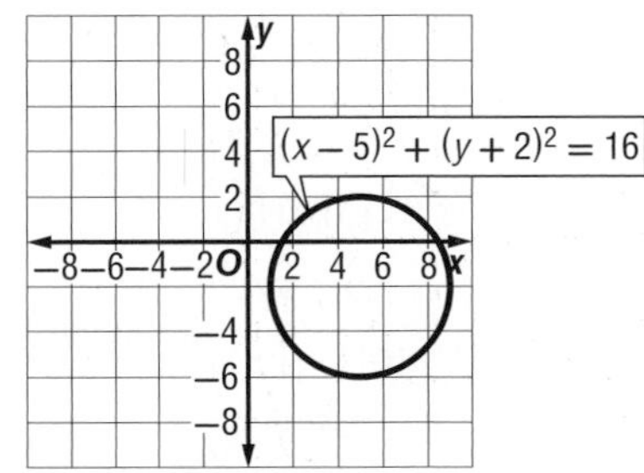

37. center: (4, 0); radius: $\frac{\sqrt{8}}{3}$

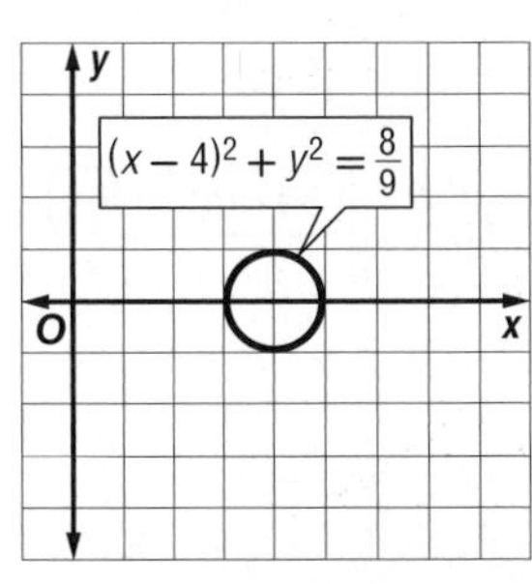

Selected Answers and Solutions

39. center: $(-2, 0)$; radius: $\sqrt{13}$

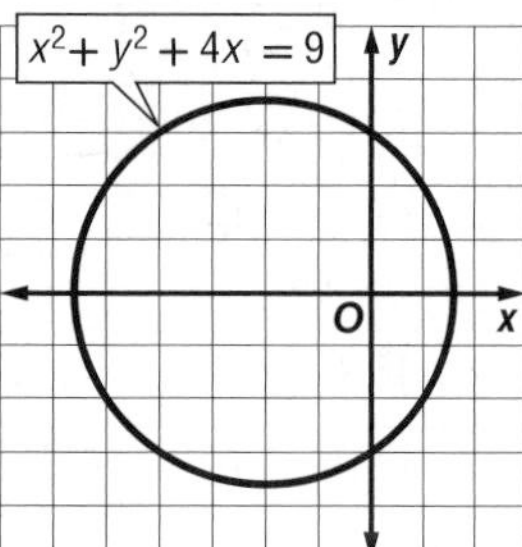

41. center: $(-1, -2)$; radius: $\sqrt{14}$

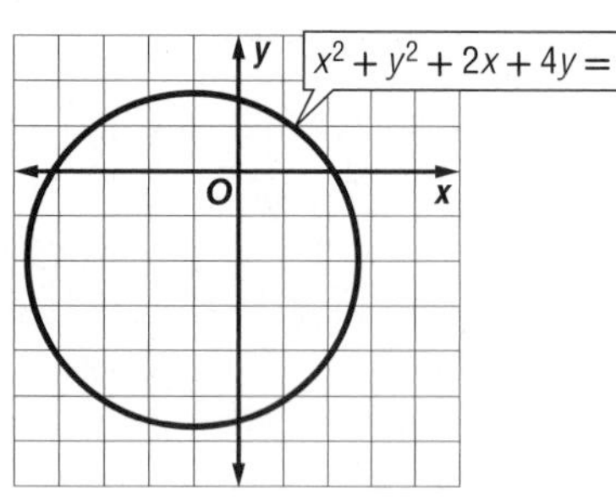

43. center: $(-7, -3)$; radius: $2\sqrt{2}$ units

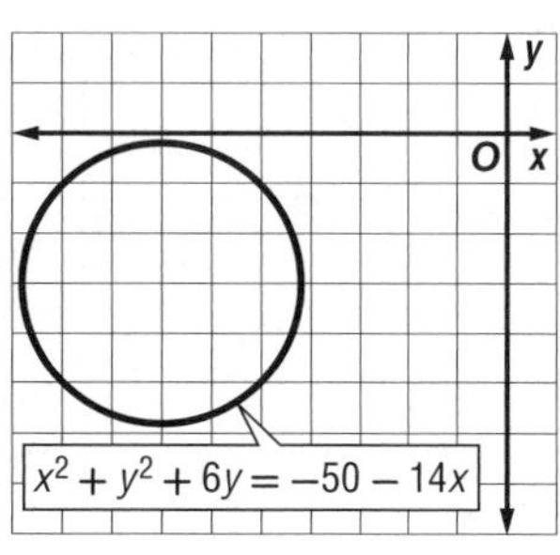

45. center: $(1, -2)$; radius: $\sqrt{21}$

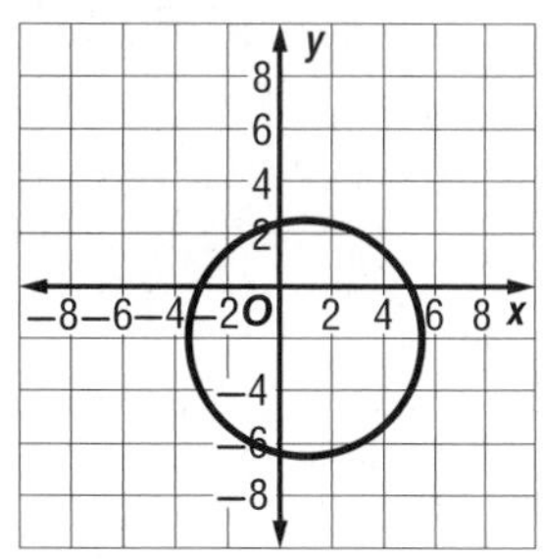

47a. $x^2 + y^2 = 841{,}000{,}000$

47b.

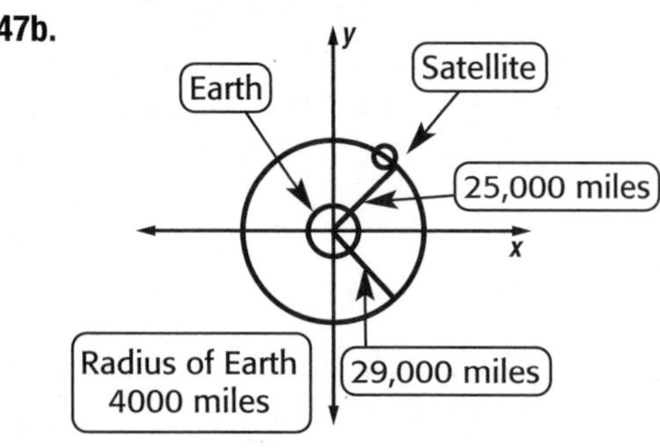

49a. $(x + 1)^2 + (y - 4)^2 = 36 + 16\sqrt{5}$

49b. $(x + 1)^2 + (y - 4)^2 = 24 - 8\sqrt{5}$

49c.

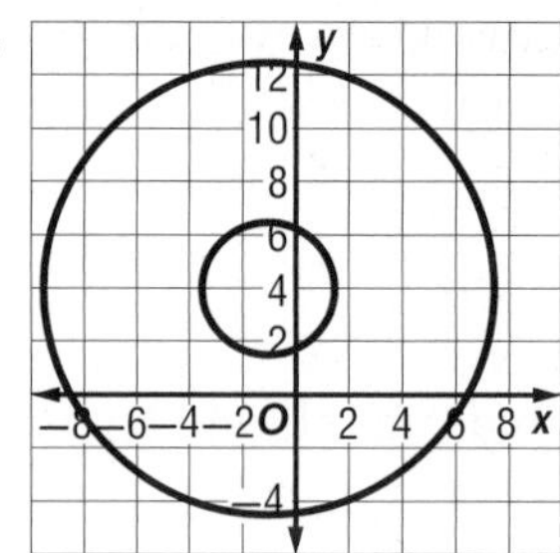

51

$$r = \sqrt{(x_2 - x_1)^2 + (y_2 - y_1)^2} \quad \text{Distance Formula}$$

$$= \sqrt{(19 - 9)^2 + [22 - (-8)]^2} \quad (x_1, y_1) = (9, -8) \text{ and } (x_2, y_2) = (19, 22)$$

$$= \sqrt{10^2 + 30^2} \quad \text{Subtract.}$$

$$= \sqrt{1000} \quad \text{Simplify.}$$

$$(x - h)^2 + (y - k)^2 = r^2 \quad \text{Equation of a circle}$$

$$(x - 9)^2 + [y - (-8)]^2 = (\sqrt{1000})^2 \quad (h, k) = (9, -8) \text{ and } r = \sqrt{1000}$$

$$(x - 9)^2 + (y + 8)^2 = 1000 \quad \text{Simplify.}$$

53. $(x - 8)^2 + (y + 9)^2 = 64$ **55.** $(x - 2.5)^2 + (y - 2.5)^2 = 6.25$ **57a.** circle **57b.** $x^2 + y^2 = 9$

57c. Solve the equation for y: $y = \pm\sqrt{49 - x^2}$. Then graph the positive and negative answers.

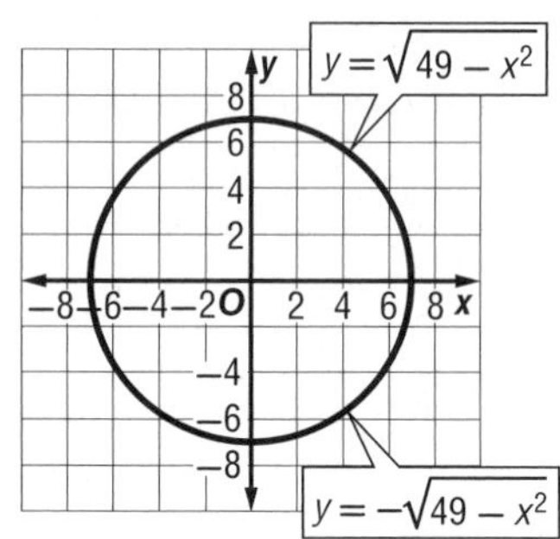

57d. $y = \pm\sqrt{4 - (x - 2)^2} - 1$; because when you solve for y you must take the square root resulting in both a positive and negative answer, so you have to enter the positive equation as Y1 and the negative equation as Y2.

59. center: $\left(0, -\frac{9}{2}\right)$; radius: $\sqrt{19}$

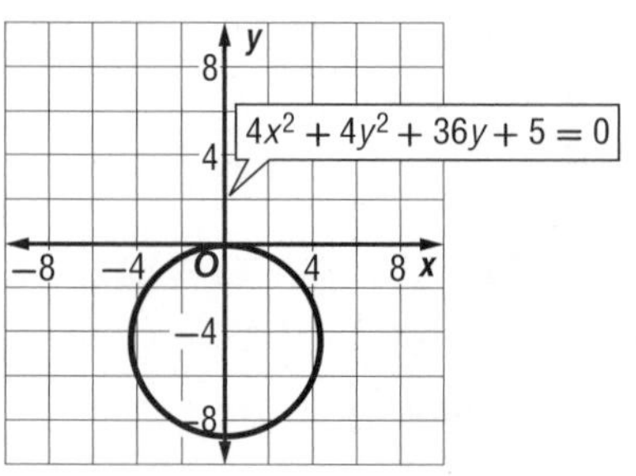

61. center: $(-\sqrt{7}, \sqrt{11})$; radius: $\sqrt{11}$

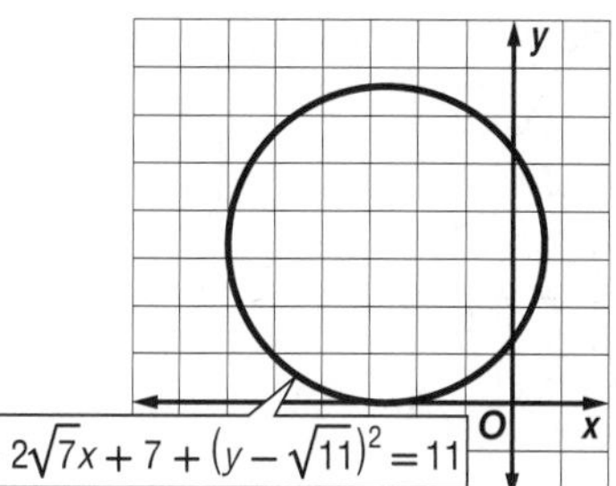

63. Circles with a radius of 8 and centers on the graph of $x = 3$.

65. Sample answer: $(x - 2)^2 + (y - 3)^2 = 25$ and $(x - 2)^2 + (y - 3)^2 = 36$

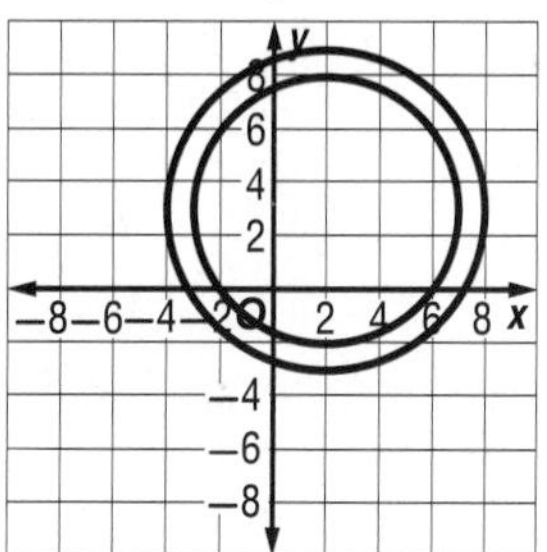

67. Quadrant I

$a > 0, b > 0, a = b, r > 0$

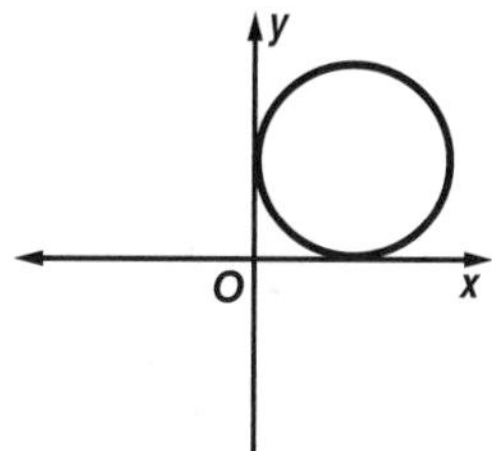

Quadrant II

$a < 0, b > 0, a = -b, r > 0$

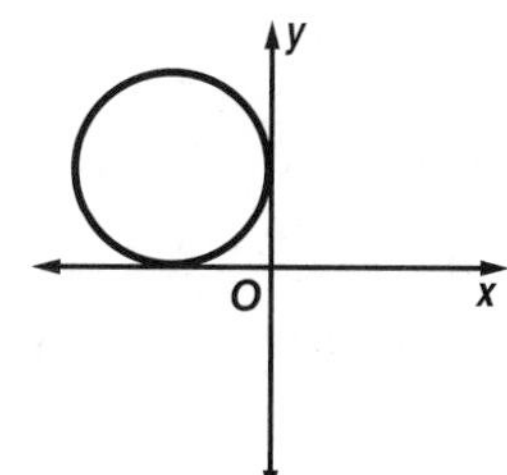

Quadrant III

$a < 0, b < 0, a = b, r > 0$

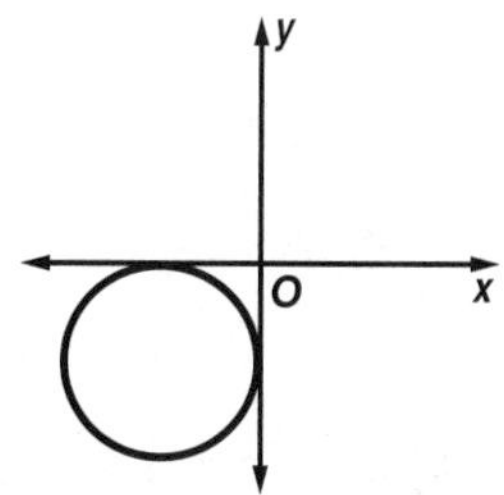

Quadrant IV

$a > 0, b < 0, a = -b, r > 0$

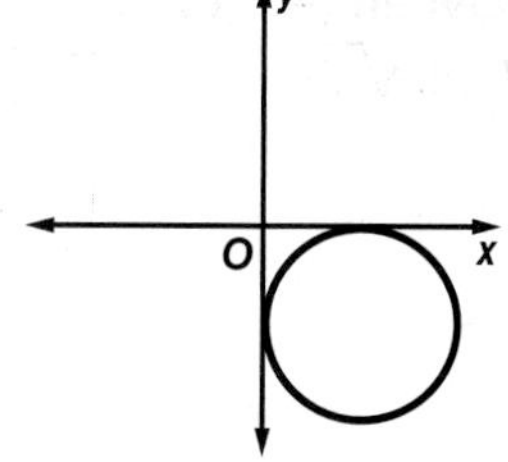

Sample answer: The circle is rotated 90° about the origin from one quadrant to the next.

69. A **71.** C

73.

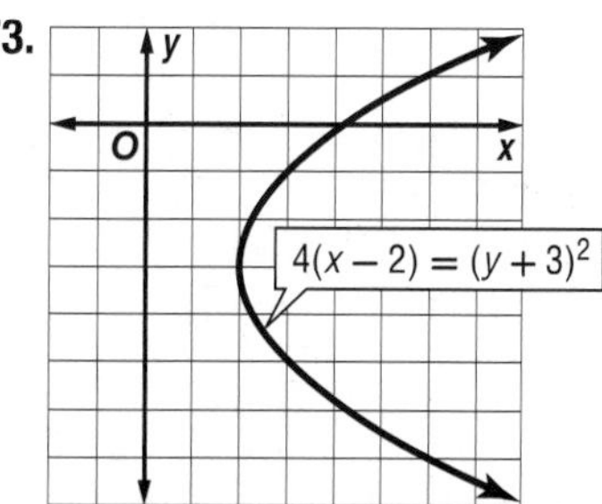

75. $\left(1, \frac{7}{22}\right)$; $\sqrt{65}$ units **77.** (0, 3); $\sqrt{29}$ units **79.** 64

81. $\frac{5}{2}$ **83.** -4 **85a.** The square root of a difference is not the difference of the square roots. **85b.** 34.1 ft/s

87. $\left\{\frac{3 \pm \sqrt{33}}{4}\right\}$

Lesson 9-4

1. $\frac{y^2}{25} + \frac{x^2}{9} = 1$ **3.** $\frac{(y + 1)^2}{25} + \frac{(x + 2)^2}{9} = 1$

5a. $a = 240, b = 160$ **5b.** $\frac{x^2}{57{,}600} + \frac{y^2}{25{,}600} = 1$

5c. about (179, 0) and (−179, 0)

7. center (5, −1); foci (5, 5) and (5, −7); major axis: 16; minor axis: ≈ 10.58

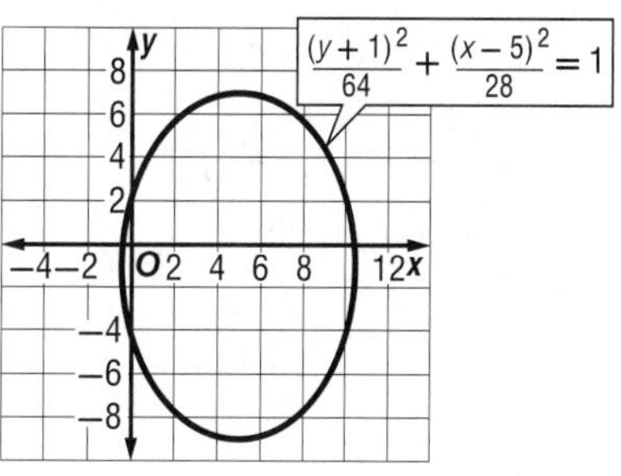

9

$4x^2 + y^2 - 32x - 4y + 52 = 0$	Original equation
$4x^2 - 32x + y^2 - 4y = -52$	Commutative Property
$4(x^2 - 8x) + y^2 - 4y = -52$	Distributive Property
$4(x^2 - 8x + ■) + (y^2 - 4y + ■) = -52 + 4(■) + (■)$	Complete the squares.
$4(x^2 - 8x + 16) + (y^2 - 4y + 4) = -52 + 4(16) + (4)$	$(-4)^2 = 16$ and $(-2)^2 = 4$
$4(x - 4)^2 + (y - 2)^2 = 16$	Write as perfect squares.
$\frac{(x - 4)^2}{4} + \frac{(y - 2)^2}{16} = 1$	Divide each side by 16.

$h = 4$ and $k = 2$, so the center is at (4, 2).
The ellipse is vertical. $a^2 = 16$, so $a = 4$, and $b^2 = 4$, so $b = 2$.
$c^2 = 16 - 4$ or 12, so $c \approx 3.46$.

foci: $(4, 2 + 3.46)$ or $(4, 5.46)$; $(4, 2 - 3.46)$ or $(4, -1.46)$
major axis: $2 \cdot 4$ or 8
minor axis: $2 \cdot 2$ or 4

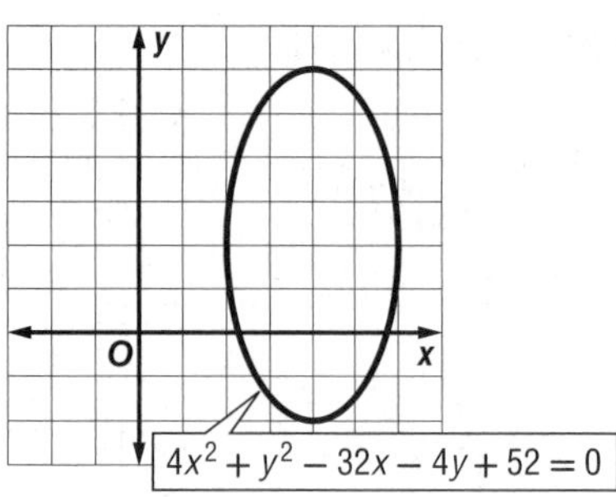

11. $\frac{y^2}{100} + \frac{x^2}{36} = 1$ **13.** $\frac{(x + 5)^2}{49} + \frac{(y + 4)^2}{25} = 1$

15. $\frac{(y - 1)^2}{64} + \frac{(x + 5)^2}{16} = 1$ **17.** $\frac{(x - 3)^2}{81} + \frac{(y - 4)^2}{64} = 1$

19 The x-coordinate is the same for both vertices, so the ellipse is vertical.
length of major axis: 16 − 6 or 10 units, so $a = 10$
length of minor axis: 1 − (−2) or 3 units, so $b = 3$

$\frac{(y - k)^2}{a^2} + \frac{(x - h)^2}{b^2} = 1$ Equation of a vertical ellipse

$\frac{(y - 6)^2}{10^2} + \frac{[x - (-2)]^2}{3^2} = 1$ $(h, k) = (-2, 6)$, $a = 10$, $b = 3$

$\frac{(y - 6)^2}{100} + \frac{(x + 2)^2}{9} = 1$ Simplify.

21. $\frac{(y - 4)^2}{64} + \frac{(x - 4)^2}{9} = 1$ **23.** $\frac{y^2}{73.96} + \frac{x^2}{53.29} = 1$

25. center $(-6, 3)$; foci $(-6, 7.69)$ and $(-6, -1.69)$; major axis: ≈ 16.97; minor axis: ≈ 14.14

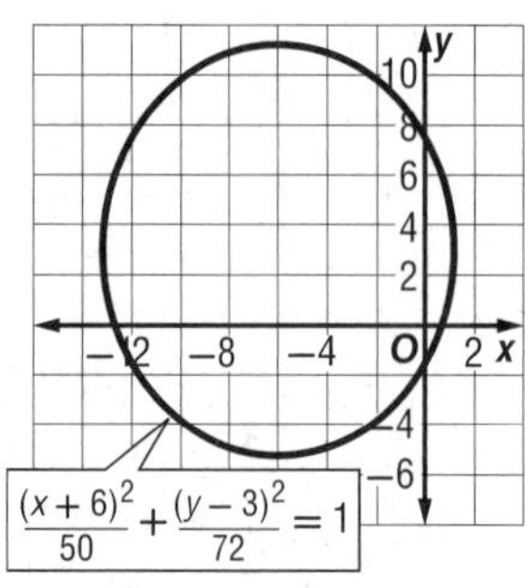

27. center $(-4, 0)$; foci $(-4, 7.68)$ and $(-4, -7.68)$; major axis: ≈ 17.32; minor axis: 8

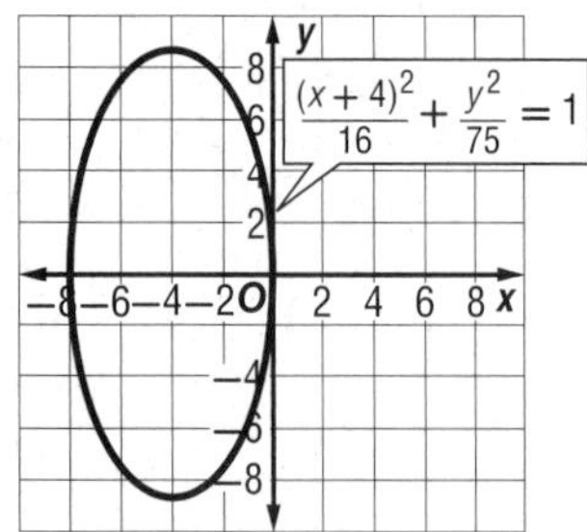

29. center $(3, -3)$; foci $(5.24, -3)$ and $(0.76, -3)$; major axis: ≈ 8.94; minor axis: ≈ 7.75

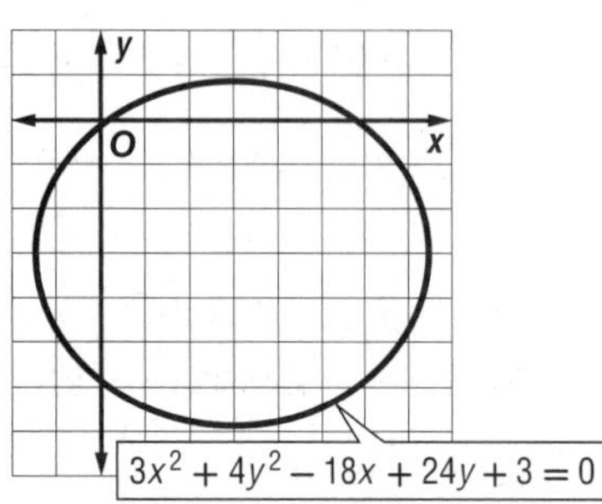

31. center $(-2, 5)$; foci $(-2, 7.83)$ and $(-2, 2.17)$; major axis: ≈ 9.80; minor axis: 8

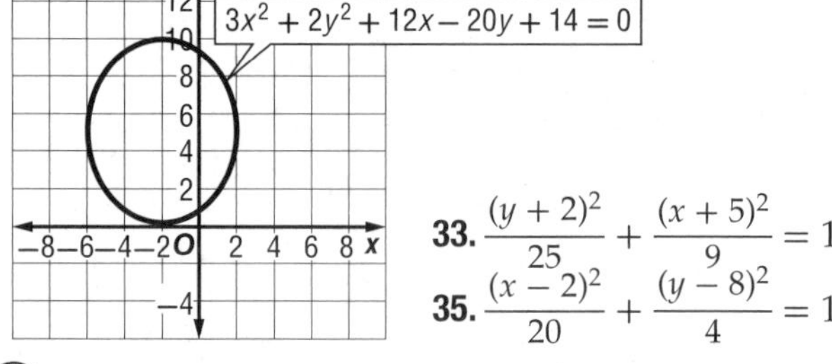

33. $\frac{(y + 2)^2}{25} + \frac{(x + 5)^2}{9} = 1$

35. $\frac{(x - 2)^2}{20} + \frac{(y - 8)^2}{4} = 1$

37 length of major axis $= 2a$
$10.9 = 2a$
$5.45 = a$
length of minor axis $= 2b$
$8.8 = 2b$
$4.4 = b$

$\frac{x^2}{a^2} + \frac{y^2}{b^2} = 1$ Equation of an ellipse

$\frac{x^2}{5.45^2} + \frac{y^2}{4.4^2} = 1$ Substitute.

$\frac{x^2}{29.7025} + \frac{y^2}{19.36} = 1$ Simplify.

39a.

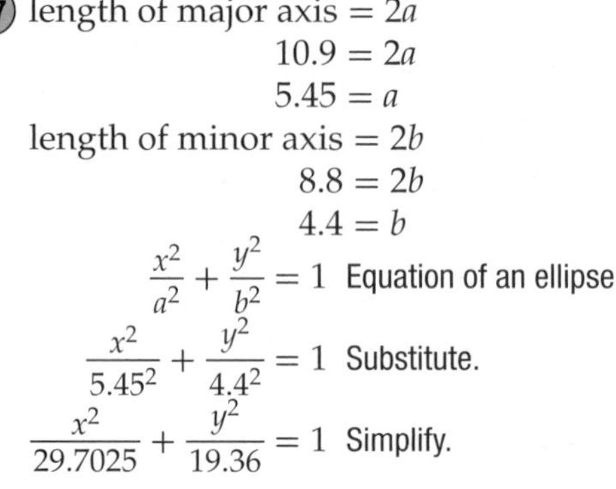

39b. Sample answer: The first graph is more circular than the second graph.
39c. first graph: 0.745; second graph: 0.943
39d. Sample answer: The closer the eccentricity is to 0, the more circular the ellipse.

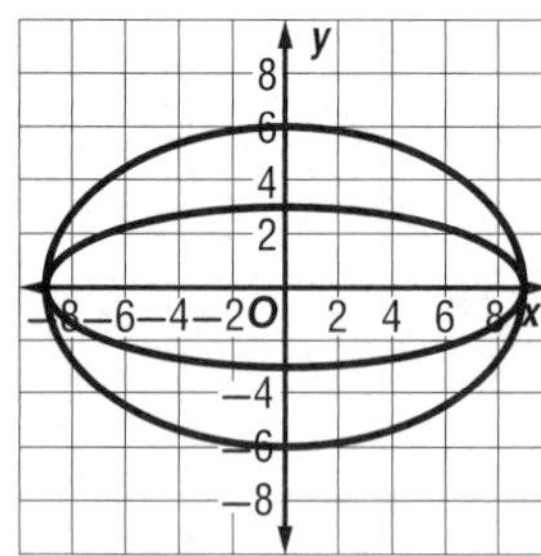

41. Sample answer: $\frac{(x + 4)^2}{40} + \frac{y^2}{24} = 1$

43. $\frac{y^2}{9} + \frac{(x - 2)^2}{3} = 1$

45. For any point on an ellipse, the sum of the distances from that point to the foci is constant by the definition of an ellipse. So, if (2, 14) is on the ellipse, then the sum of the distances from it to the foci will be a certain value consistent with every other point on the ellipse. The distance between $(-7, 2)$ and $(2, 14)$ is $\sqrt{(-7 - 2)^2 + (2 - 14)^2}$ or 15. The distance between (18, 2) and (2, 14) is $\sqrt{(18 - 2)^2 + (2 - 14)^2}$ or 20. The sum of these two distances is 35.
The distance between $(-7, 2)$ and $(2, -10)$ is $\sqrt{(-7 - 2)^2 + [2 - (-10)]^2}$ or 15. The distance between (18, 2) and $(2, -10)$ is $\sqrt{(18 - 2)^2 + [2 - (-10)]^2}$ or 15. The sum of these distances is also 35. Thus, $(2, -10)$ also lies on the ellipse.

47. B **49.** 7 **51.** $(x - 8)^2 + (y + 9)^2 = 1130$

53. $(x + 5)^2 + (y - 4)^2 = 25$ **55.** $\frac{5d + 16}{(d + 2)^2}$

57. $\frac{x^2 - 5x + 3}{(x - 5)(x + 1)}$ **59.** 4 **61.** $15a^3b^3 - 30a^4b^3 + 15a^5b^6$

Selected Answers and Solutions

63. $4x^2 - 3xy - 6y^2$ **65.** $y = -\frac{4}{5}x + \frac{17}{5}$

67. $y = -\frac{3}{5}x + \frac{16}{5}$

Lesson 9-5

1. $\frac{y^2}{36} - \frac{x^2}{28} = 1$ **3.** $\frac{x^2}{64} - \frac{y^2}{25} = 1$

5.

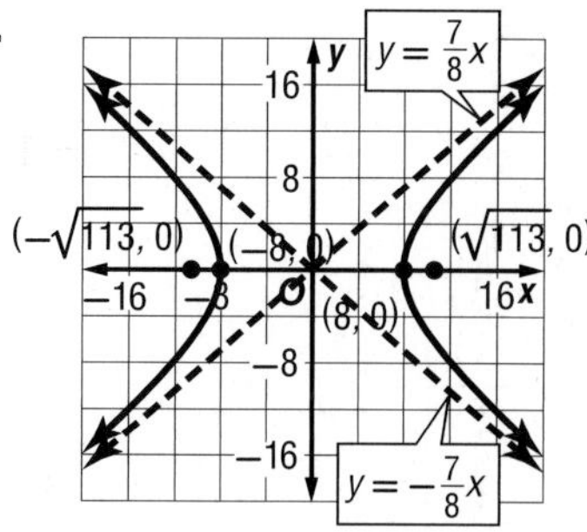

7.

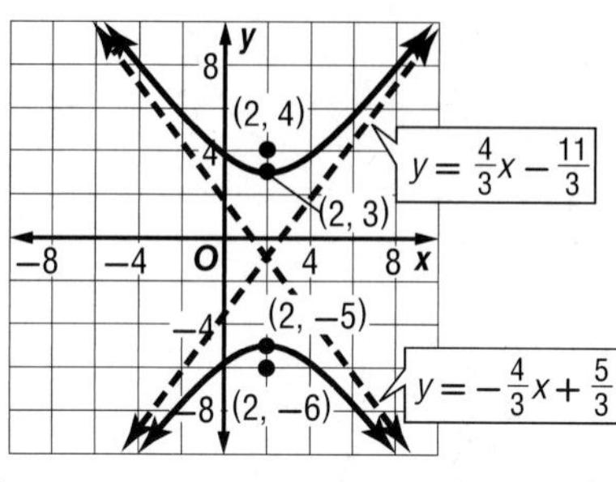

9. $\frac{x^2}{900} - \frac{y^2}{5500} = 1$

11 Since the vertices are equidistant from the center, the center is at (−8, 4). The value of a is the distance between a vertex and the center, or 4 units. The value of c is the distance between a focus and the center, or 8 units.

$c^2 = a^2 + b^2$	Equation relating a, b, and c for a hyperbola
$8^2 = 4^2 + b^2$	$c = 8$ and $a = 4$
$48 = b^2$	Subtract 4^2 from each side.
$\frac{(y-k)^2}{a^2} - \frac{(x-h)^2}{b^2} = 1$	Equation of a vertical hyperbola
$\frac{(y-4)^2}{4^2} - \frac{[x-(-8)]^2}{48} = 1$	$(h, k) = (-8, 4)$, $a = 4$, $b^2 = 48$
$\frac{(y-4)^2}{16} - \frac{(x+8)^2}{48} = 1$	Simplify.

13. $\frac{(x+1)^2}{9} - \frac{(y-6)^2}{49} = 1$

15.

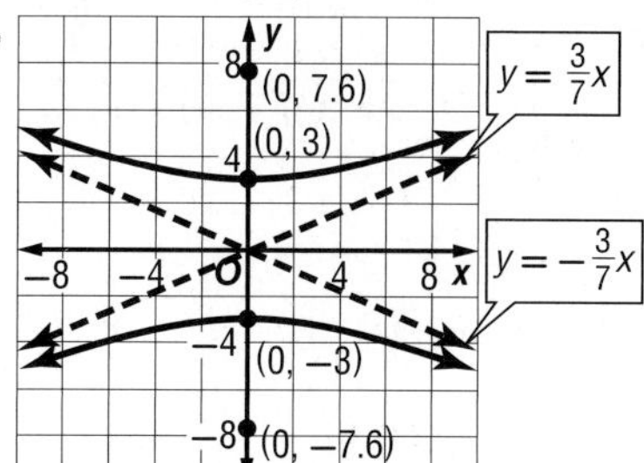

17.

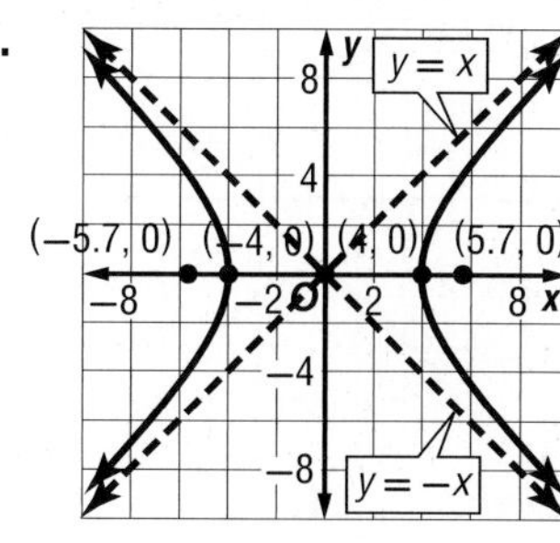

19.

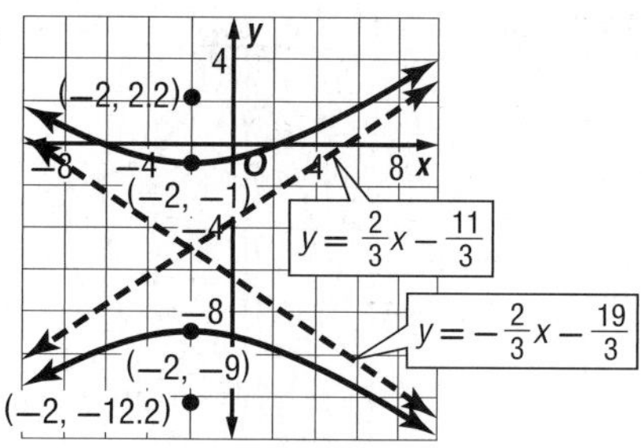

21.

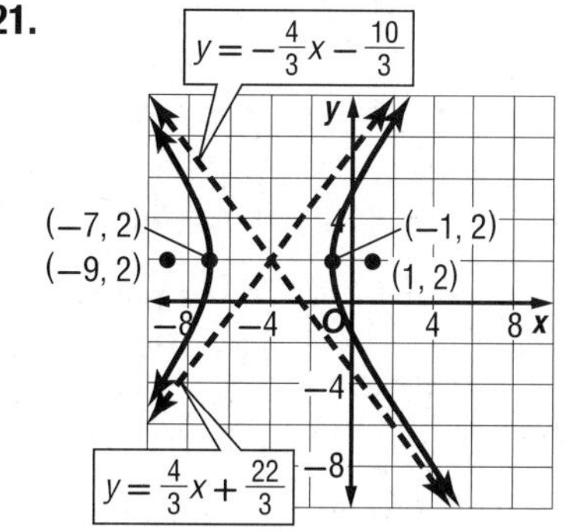

23.

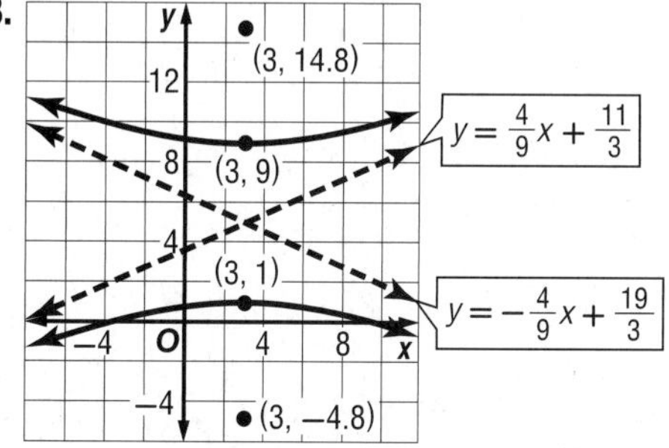

25. hyperbola **27.** ellipse **29.** ellipse

31 Because the center is at the origin, use the equation $\frac{x^2}{a^2} - \frac{y^2}{b^2} = 1$. The hyperbola intersects the x-axis at (10, 0), and one of the vertices is (10, 0). So, $a = 10$. The hyperbola also passes through (30, 100). Use $a = 10$, $x = 30$, and $y = 100$ to solve for b^2.

$\frac{x^2}{a^2} - \frac{y^2}{b^2} = 1$	Equation of an ellipse
$\frac{30^2}{10^2} - \frac{100^2}{b^2} = 1$	$a = 10$, $x = 30$, $y = 100$
$\frac{900}{100} - \frac{10{,}000}{b^2} = 1$	Evaluate exponents.
$9 - \frac{10{,}000}{b^2} = 1$	Simplify.
$-\frac{10{,}000}{b^2} = -8$	Subtract 9 from each side.
$-10{,}000 = -8b^2$	Multiply each side by b^2.
$\frac{-10{,}000}{-8} = b^2$	Divide each side by −8.
$1250 = b^2$	Simplify.

Substitute 1250 for b^2 in the equation $\frac{x^2}{a^2} - \frac{y^2}{b^2} = 1$. So, the equation of the path of the comet is $\frac{x^2}{100} - \frac{y^2}{1250} = 1$.

Selected Answers and Solutions

33a.

x	y
−12	−1.33
−10	−1.6
−8	−2
−6	−2.67
−4	−4
−2	−8
0	undef
2	8
4	4
6	2.67
8	2
10	1.6
12	1.33

33b.

33c. The asymptotes are $y = 0$ and $x = 0$.

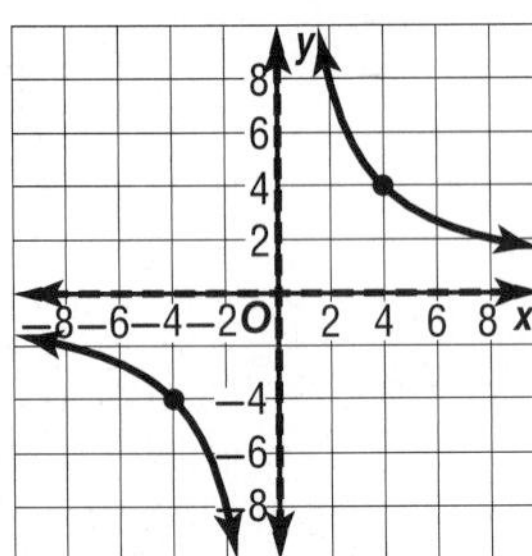

33d. They are perpendicular. **33e.** For $xy = 25$, the vertices will be at (5, 5) and (−5, −5), and for $xy = 36$, they will be at (−6, −6) and (6, 6).

35. $\frac{x^2}{2{,}722{,}500} - \frac{y^2}{1{,}277{,}500} = 1$ **37.** $\frac{x^2}{64} - \frac{y^2}{100} = 1$

39 The vertices are equidistant from the center. The center is at (2, −2). The value of a is the distance between a vertex and the center, or 4 units. The value of c is the distance between a focus and the center, or 8 units.

$c^2 = a^2 + b^2$ Equation relating a, b, and c for a hyperbola

$8^2 = 4^2 + b^2$ $c = 8$ and $a = 4$

$48 = b^2$ Subtract 4^2 from each side.

$\frac{(x-h)^2}{a^2} - \frac{(y-k)^2}{b^2} = 1$ Equation of a horizontal hyperbola

$\frac{(x-2)^2}{4^2} - \frac{[y-(-2)]^2}{48} = 1$ $(h, k) = (2, -2)$, $a = 4$, $b^2 = 48$

$\frac{(x-2)^2}{16} - \frac{(y+2)^2}{48} = 1$ Simplify.

41. $\frac{x^2}{25} - \frac{y^2}{4} = 1$ **43.** (2308, 826) **45.** $\frac{(x-3)^2}{5} - \frac{(y+2)^2}{5} = 1$ **47.** Sample answer: When 36 changes to 9, the vertical hyperbola widens (splits out from the y-axis faster). This is due to a smaller value of y being needed to produce the same value of x. The vertices are moved closer together due to the value of a decreasing from 6 to 3. The foci move farther from the vertices because the difference between c and a increased. **49.** Sample answer: The graphs of ellipses are closed in, while the branches of the hyperbolas extend without bound. There is always an upper and lower limit to the values of the coordinates of an ellipse, while maximum x- and y-values for the coordinates of hyperbolas are infinite. When both of the x^2 and y^2 terms are on the same side of the equation, the equation is for an ellipse if the signs of their coefficients are the same. Otherwise it is a hyperbola. (This is true only for conics that are not rotated.)

51. J **53.** E **55.** $\frac{y^2}{100} + \frac{x^2}{36} = 1$

57. (0, 3), 5 units

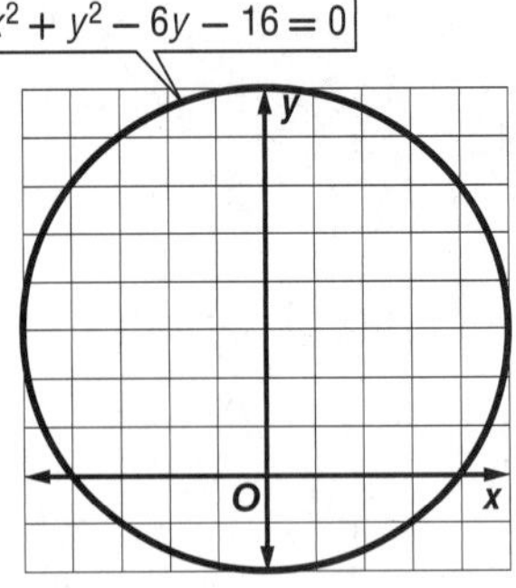

59a.

59b. the part in the first quadrant **59c.** It represents her original free-throw percentage of 60%. **59d.** $P(x) = 1$; this represents 100%, which she cannot achieve because she has already missed 4 free throws. **61.** $\frac{5}{3}$

63.

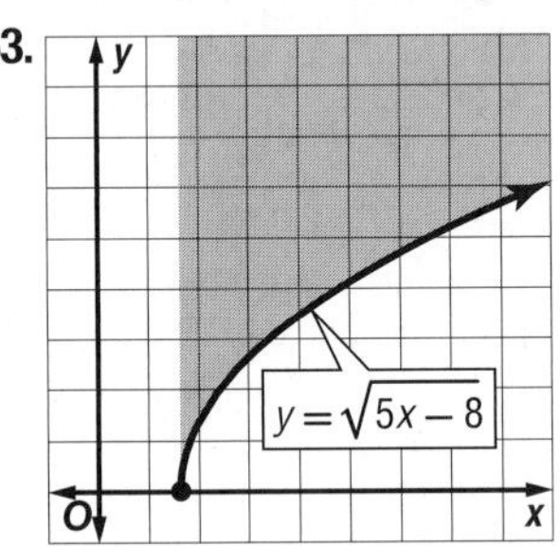

65.

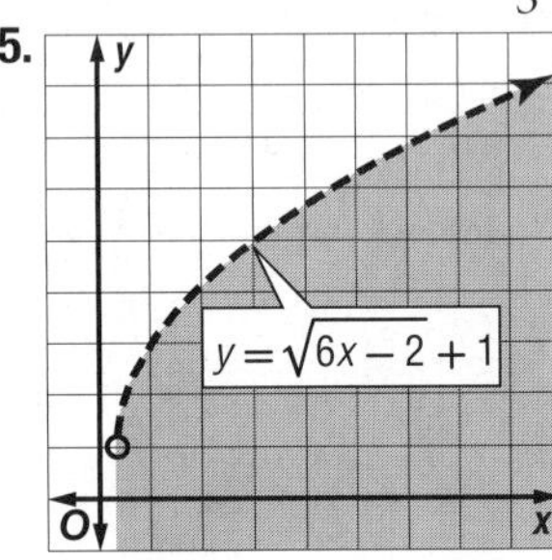

67. $y = \frac{4}{3}(x+3)^2 - 4$

Lesson 9-6

1. $\frac{(x-3)^2}{36} + \frac{(y+2)^2}{9} = 1$; ellipse

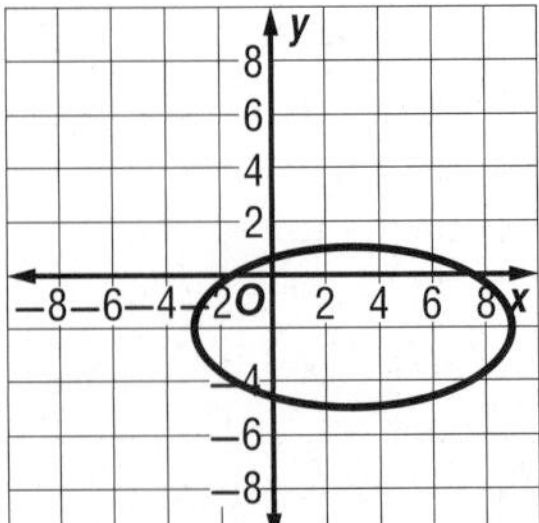

3. $\frac{(y-1)^2}{16} - \frac{(x+2)^2}{9} = 1$; hyperbola

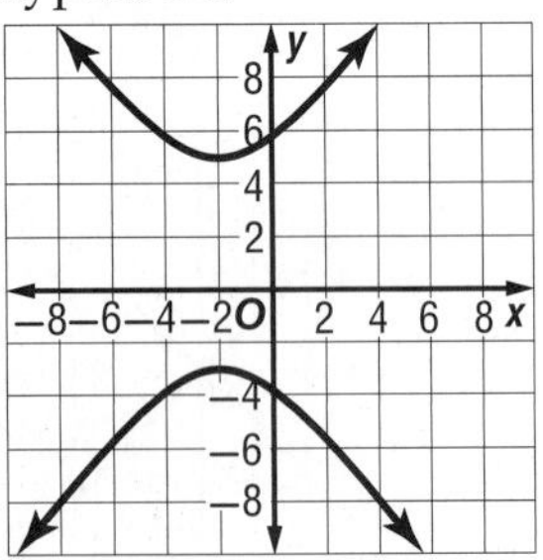

5. ellipse **7.** circle **9.** hyperbola **11.** ellipse

13a. parabola; $y = -0.024(x - 660)^2 + 10{,}500$
13b. about 1320 ft **13c.** 10,500 ft

15. $\frac{(x+4)^2}{32} + \frac{(y-5)^2}{24} = 1$; ellipse

17. $y = 8(x+2)^2 - 4$; parabola

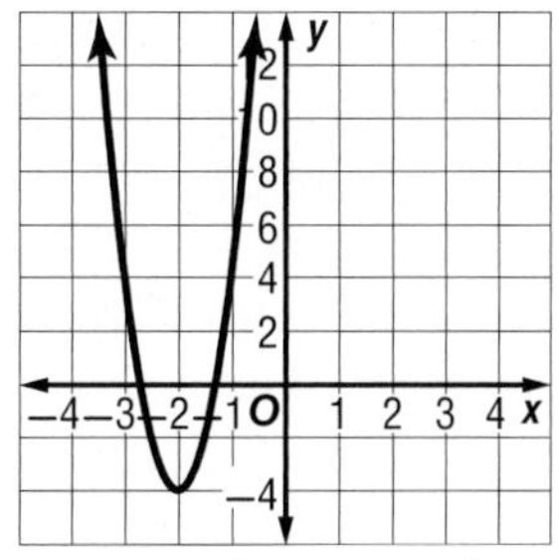

19. $(x-3)^2 + (y+4)^2 = 36$; circle

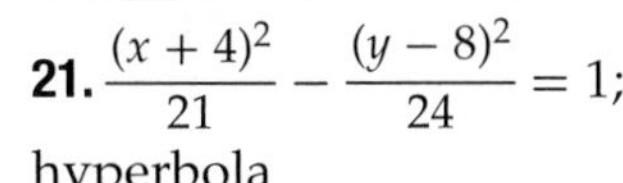

21. $\frac{(x+4)^2}{21} - \frac{(y-8)^2}{24} = 1$; hyperbola

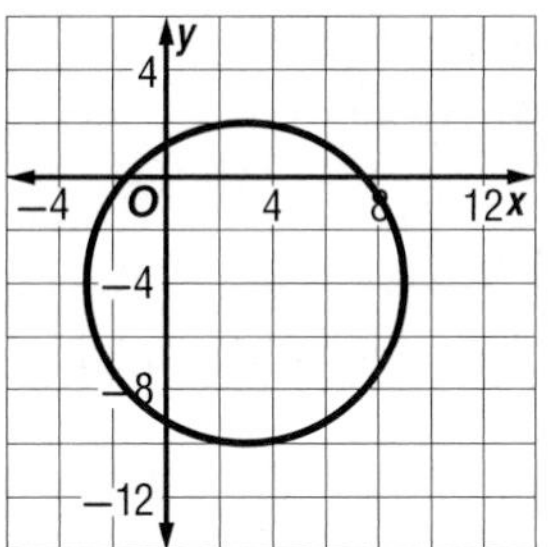

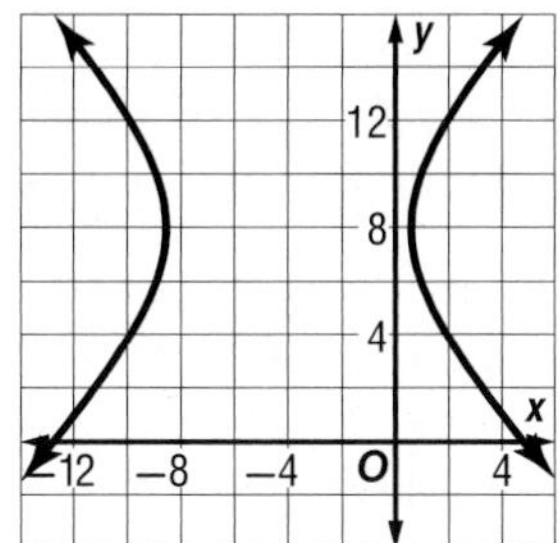

23. $\frac{(x+4)^2}{64} - \frac{(y-3)^2}{25} = 1$; hyperbola

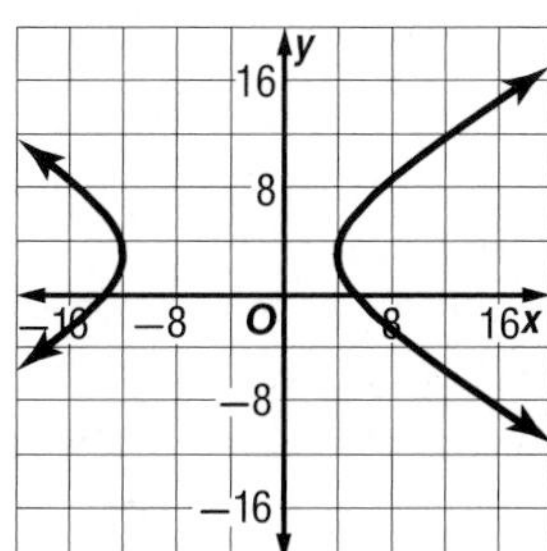

25. circle

(27) $18x^2 - 16y = 12x - 4y^2 + 19$ Original equation
$18x^2 + 4y^2 - 12x - 16y - 19 = 0$ Standard form
$A = 18, B = 0$, and $C = 4$
$B^2 - 4AC = 0^2 - 4(18)(4)$ or -288
Since the discriminant is less than 0 and $A \neq C$, the conic is an ellipse.

29. hyperbola **31.** hyperbola **33.** hyperbola **35.** c **37.** b **39.** b

(41) Equation *c* can be written as $y = -16x^2 + 90x + 0.25$. Since this is an equation of a parabola that opens downward, it could be used to represent the height of a football above the ground after being kicked.

43a. $\frac{(x-3)^2}{16} + \frac{(y+2)^2}{4} = 1$

43b. $x^2 + 4y^2 - 6x + 16y + 9 = 0$

43c.

43d. $N(5, -2)$; 90° counterclockwise

45. Sample answer: Always; when a conic is vertical, $B = 0$. When this is true and $A = C$, the conic is a circle.

47. Sample answer: An ellipse is a flattened circle. Both circles and ellipses are enclosed regions, while hyperbolas and parabolas are not. A parabola has one branch, which is a smooth curve that never ends, and a hyperbola has two such branches that are reflections of each other. In standard form and when there is no xy-term: an equation for a parabola consists of only one squared term, an equation for a circle has values for A and C that are equal, an equation for an ellipse has values for A and C that are the same sign but not equal, and an equation for a hyperbola has values of A and C that have opposite signs. **49.** J **51.** B

53. $(0, 0)$; $(0, \pm 3)$; $6\sqrt{2}$; 6

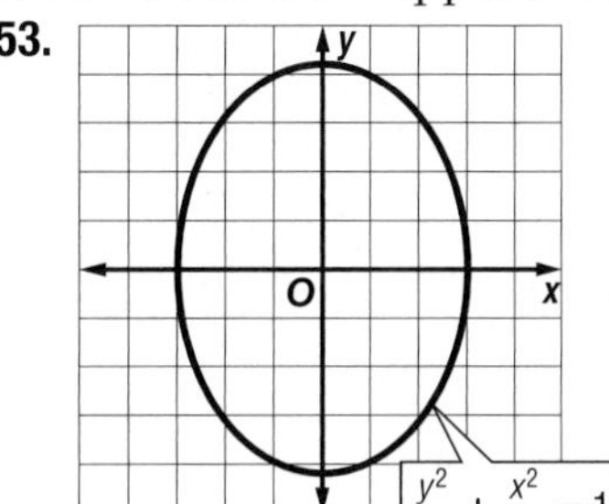

55. $(4, -2)$; $(4 \pm 2\sqrt{6}, -2)$; 10; 2

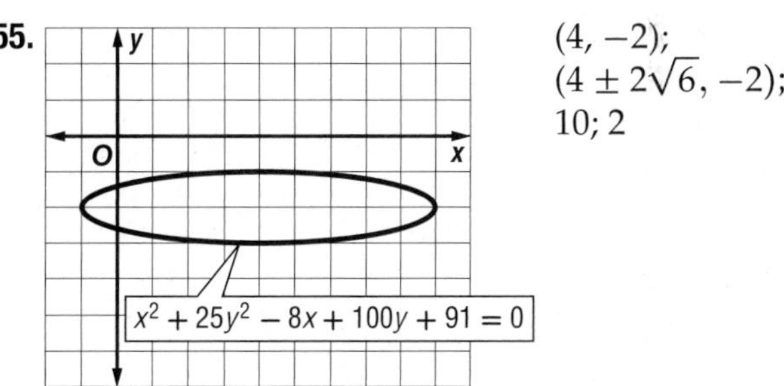

57.

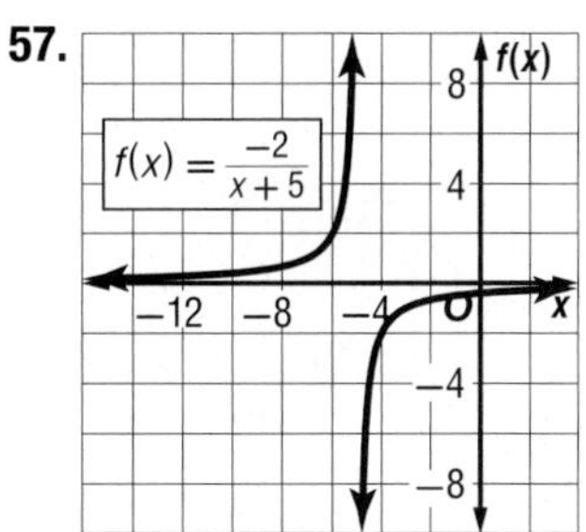

59a. Decay; the exponent is negative. **59b.** about 33.5 watts **59c.** about 402 days **61.** $\left(-\frac{1}{2}, \frac{3}{2}\right)$

Lesson 9-7

1. $(4, -5), (-4, 5)$ **3.** $(3, 6), (6, 42)$ **5.** no solution

(7) Solve the second equation for x^2.
$y^2 - x^2 = 8 \rightarrow y^2 - 8 = x^2$
$x^2 + 2y = 7$ Write the first equation.
$y^2 - 8 + 2y = 7$ Substitute $y^2 - 8$ for x^2.
$y^2 + 2y - 15 = 0$ Subtract 7 from each side.
$(y-3)(y+5) = 0$ Factor.
$y - 3 = 0$ or $y + 5 = 0$ Zero Product Property
$y = 3$ $y = -5$ Solve each equation.
Substitute 3 and -5 into one of the original equations and solve for x.

$x^2 + 2y = 7$	$x^2 + 2y = 7$
$x^2 + 2(3) = 7$	$x^2 + 2(-5) = 7$
$x^2 + 6 = 7$	$x^2 - 10 = 7$
$x^2 = 1$	$x^2 = 17$
$x = \pm 1$	$x = \pm\sqrt{17}$

Selected Answers and Solutions

The solutions are $(-1, 3)$, $(1, 3)$, $(-\sqrt{17}, -5)$, $(\sqrt{17}, -5)$.

9. $(40, 30)$ **11.**

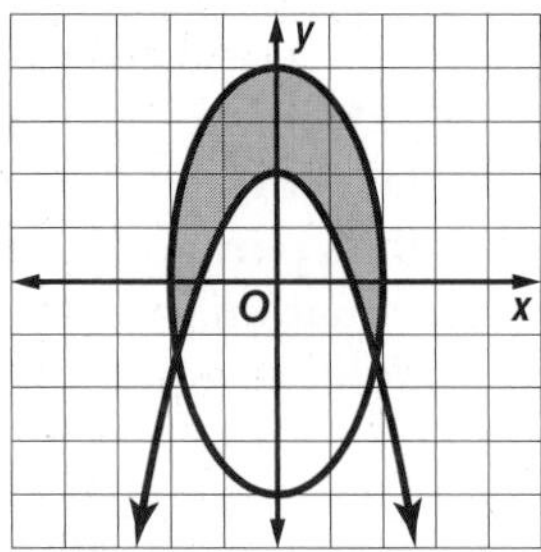

13.

15. no solution
17. $(-3, -6)$, $(3, 6)$
19. $(-1, -3)$, $(8, 5)$
21. $(1, -\sqrt{15})$, $(1, \sqrt{15})$
23. $(-\sqrt{2}, 4)$, $(\sqrt{2}, 4)$
25. $(5, -6)$, $(5, 6)$, $(-3, -2)$, $(-3, 2)$

27.

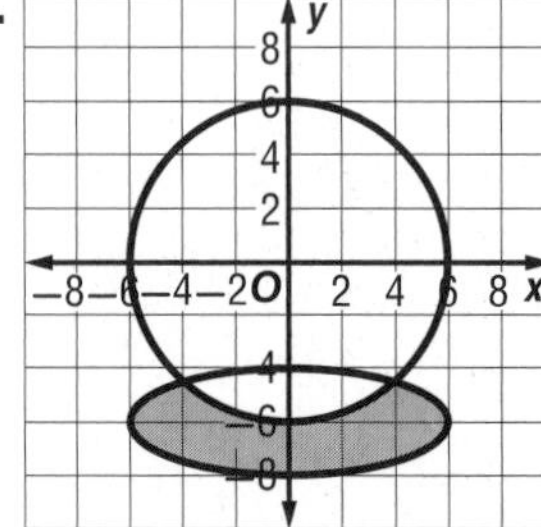

29.

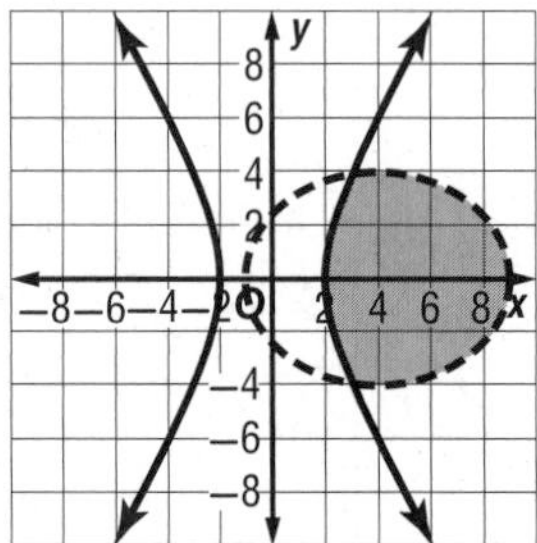

31.

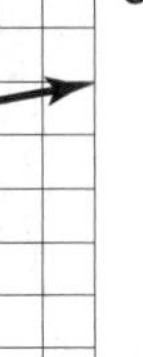

33.

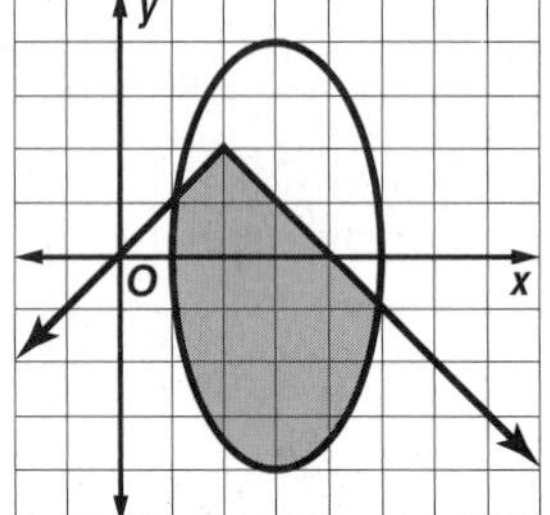

35.

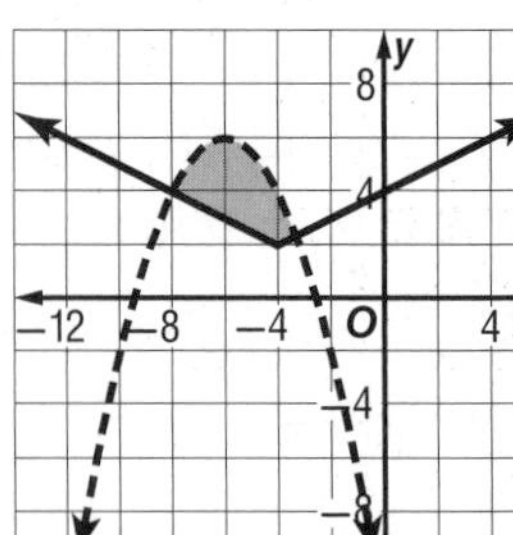

37. no solution

39a. $y = \pm 900\sqrt{1 - \dfrac{x^2}{(300)^2}}$; $y = \pm 690\sqrt{1 - \dfrac{x^2}{(600)^2}}$

39b. Sample answer: $(209, 647)$, $(-209, 647)$, $(-209, -647)$, $(209, -647)$

39c. Sample answer: The orbit of the satellite modeled by the second equation is closer to a circle than the other orbit. The distance on the x-axis is twice as great for one satellite as for the other.

41 $y = -0.0037x^2 + 1.77x - 1.72$ Path of baseball equation

$\frac{3}{7}x - 128.6 = -0.0037x^2 + 1.77x - 1.72$ Substitute $\frac{3}{7}x - 128.6$ for y.

$-128.6 \approx -0.0037x^2 + 1.34x - 1.72$ Substitute $\frac{3}{7}x$ from each side.

$0 \approx -0.0037x^2 + 1.34x + 126.88$ Add 128.6 to each side.

$x = \dfrac{-b \pm \sqrt{b^2 - 4ac}}{2a}$ Quadratic Formula

$x = \dfrac{-1.34 \pm \sqrt{1.34^2 - 4(-0.0037)(126.88)}}{2(-0.0037)}$ $a = -0.0037$, $b = 1.34$, $c = 126.88$

$x \approx \dfrac{-1.34 \pm \sqrt{3.67}}{-0.0074}$ Simplify.

$x \approx -78$ or $x \approx 440$ Use a calculator.

Since the vertical distance cannot be negative, the ball landed about 440 ft from home plate.

$y = \frac{3}{7}x - 128.6$ Original equation

$y = \frac{3}{7}(440) - 128.6$ Replace x with 440.

$y \approx 60$ Use a calculator.

The ball was 60 ft above the playing surface.

43 Sample answer: A circle with the equation $(x + 10)^2 + y^2 = 36$ has its center at $(-10, 0)$ and contains the point $(-4, 0)$. An ellipse with the equation $\frac{x^2}{16} + \frac{y^2}{36} = 1$ has its center at $(0, 0)$ and contains the point $(-4, 0)$. The circle and ellipse intersect only $(-4, 0)$.

45. Sample answer: $x^2 + y^2 = 1$ and $\frac{x^2}{16} - \frac{y^2}{36} = 1$

47. Sample answer: $\frac{x^2}{64} + \frac{y^2}{100} = 1$ and $x^2 - y^2 = 1$

49. Sample answer: No; if one player is in one of the shaded areas and the other player is in the other shaded area, they will not be able to hear each other.

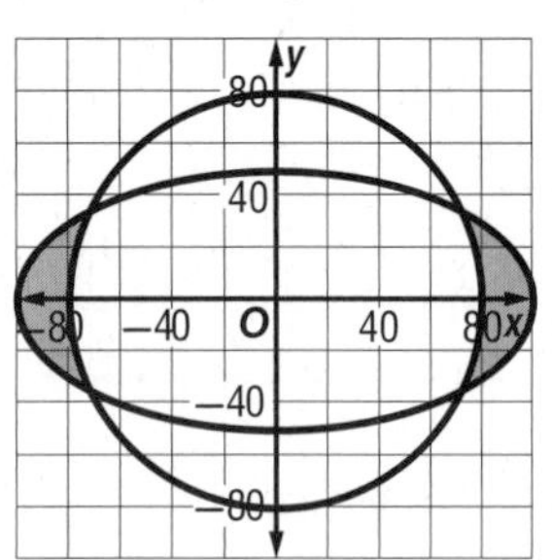

51. $k = a$ or $k = b$ **53.** Sample answer: $\frac{y^2}{128} + \frac{x^2}{32} = 1$ and $\frac{x^2}{8} - \frac{y^2}{64} = 1$ **55.** $(-3, -4)$, $(3, 4)$ **57.** F **59.** b **61.** c

63. $(2, -2)$, $(2, 8)$; $(2, 3 \pm \sqrt{41})$; $y - 3 = \pm\frac{5}{4}(x - 2)$

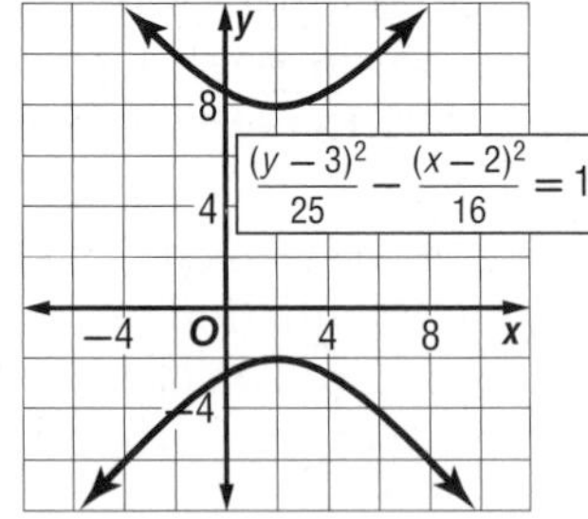

65. $\dfrac{p + 5}{p + 1}$ **67.** $\dfrac{3(r + 4)}{r + 3}$

69.

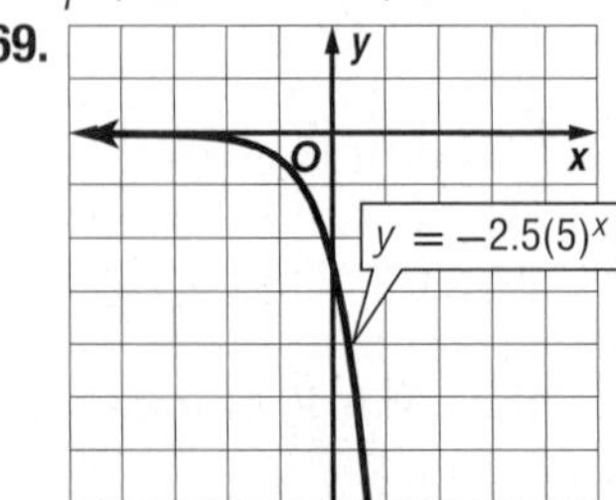

D = {all real numbers}, R = $\{y \mid y < 0\}$

71. $\frac{d}{t} = r$ **73.** $\frac{3V}{\pi r^2} = h$

1. false, center **3.** false, vertices **5.** true **7.** false, circle **9.** true **11.** $\left(-\frac{5}{2}, 5\right)$ **13.** $\left(\frac{5}{24}, \frac{5}{24}\right)$ **15.** $\sqrt{85}$ **17.** $\frac{\sqrt{34}}{4}$ **19a.** $\sqrt{89} \approx 9.4$ miles **19b.** $\left(4, -\frac{5}{2}\right)$

21.

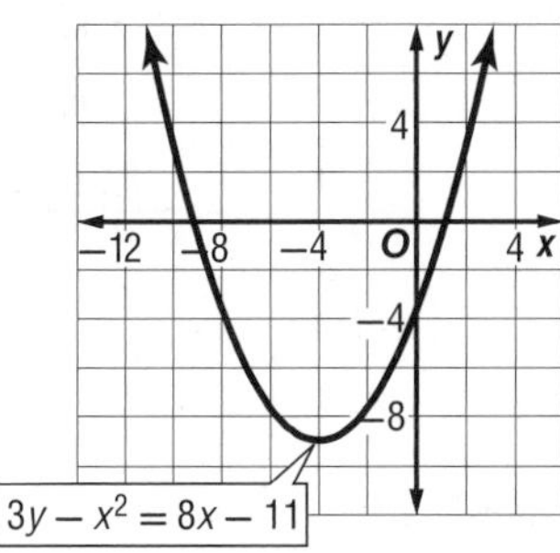

23.

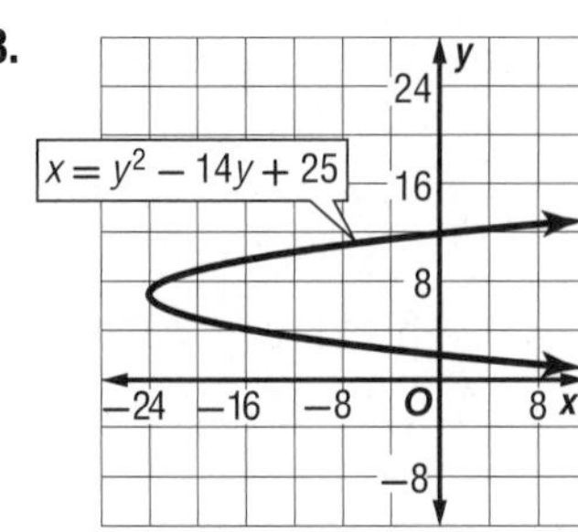

25. $y = 4(x - 2)^2 - 7$; vertex: $(2, -7)$; axis of symmetry: $x = 2$; opens upward **27.** $x = (y + 7)^2 - 29$; vertex $= (-29, -7)$; axis of symmetry: $y = -7$; opens to the right **29.** $(x + 1) + (y - 6)^2 = 9$
31. $(x - 1)^2 + (y + 4)^2 = 13$
33. $(3, -1)$; $r = 5$

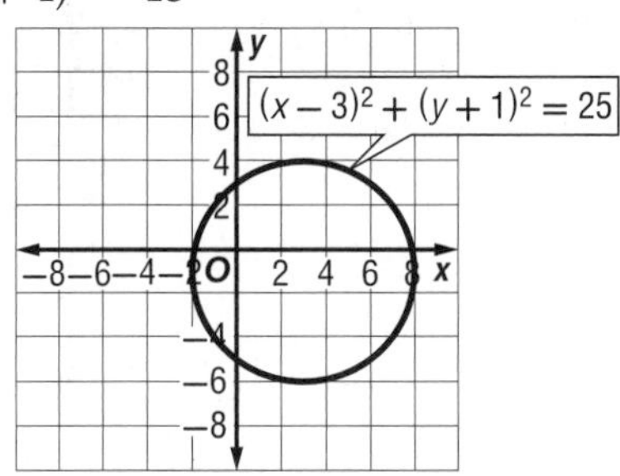

35. $(-2, 1)$; $r = 4$

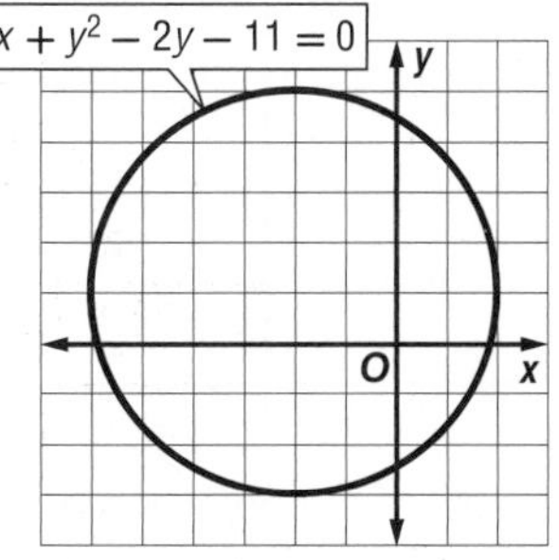

37. $(0, 0)$; $(0, \pm 3\sqrt{3})$; 12; 6

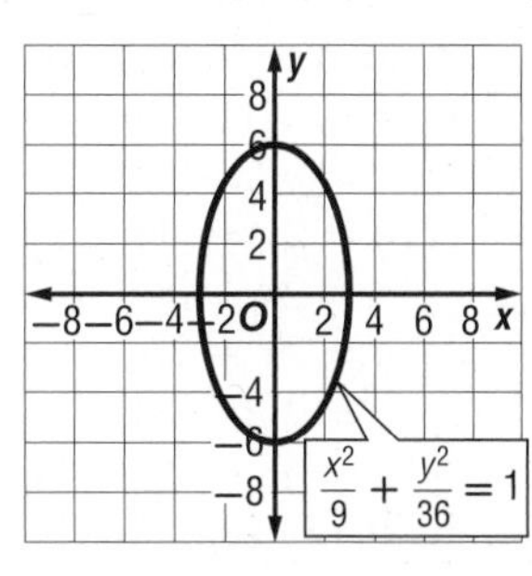

39. $(0, 4)$; $(\pm 4\sqrt{2}, 4)$; 12; 4

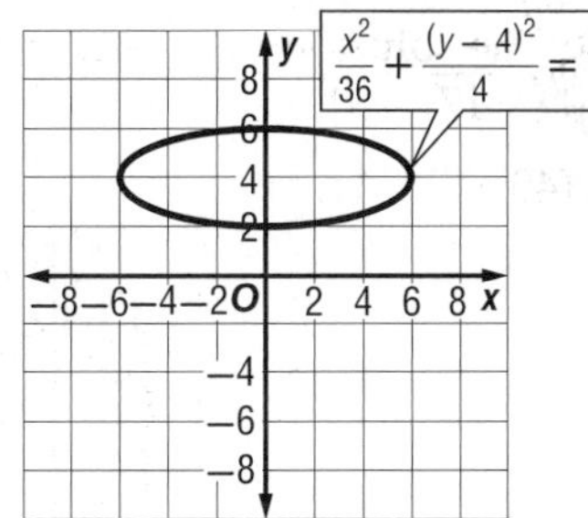

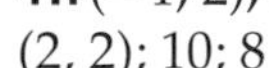
41. $(-1, 2)$; $(-4, 2)$; $(2, 2)$; 10; 8

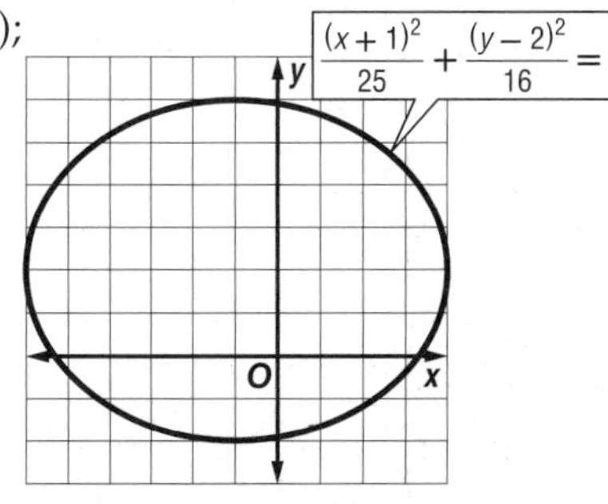

43. $(1, -1)$; $(-3, -1)$, $(5, -1)$; 10; 6

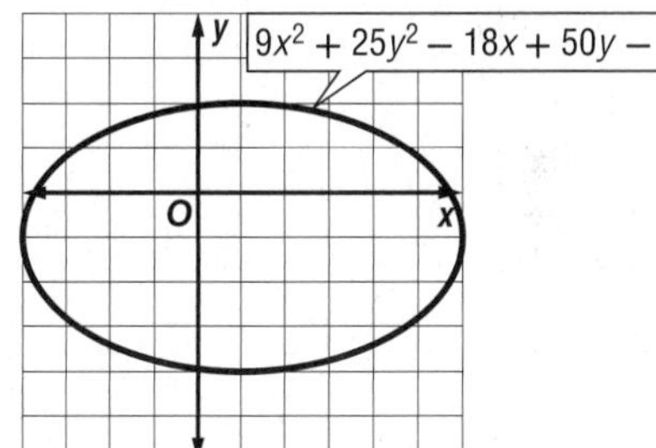

45. $\frac{x^2}{64} + \frac{y^2}{25} = 1$
47. $(2, -2)$, $(4, -2)$; $(3 \pm \sqrt{5}; -2)$; $y + 2 = \pm 2(x - 3)$

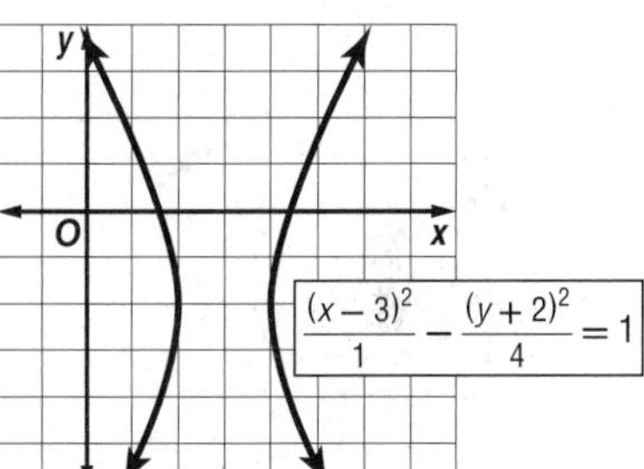

49. $(\pm 3, 0)$; $(\pm\sqrt{13}, 0)$; $y = \pm\frac{2}{3}x$

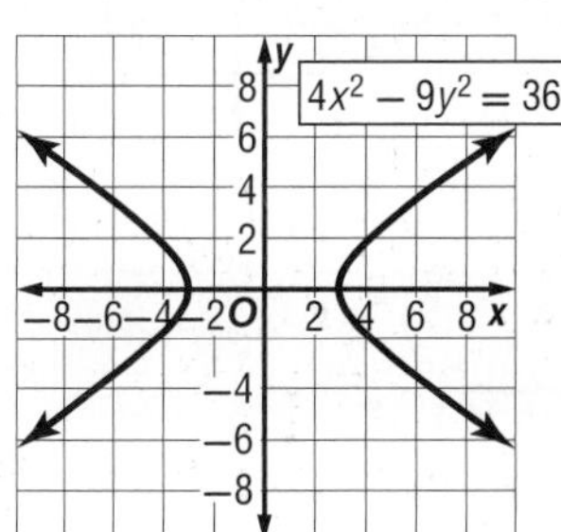

51. $\left(\frac{40 - 24\sqrt{5}}{5}, \frac{45 - 12\sqrt{5}}{5}\right)$
53. $\frac{x^2}{16} + \frac{y^2}{9} = 1$; ellipse

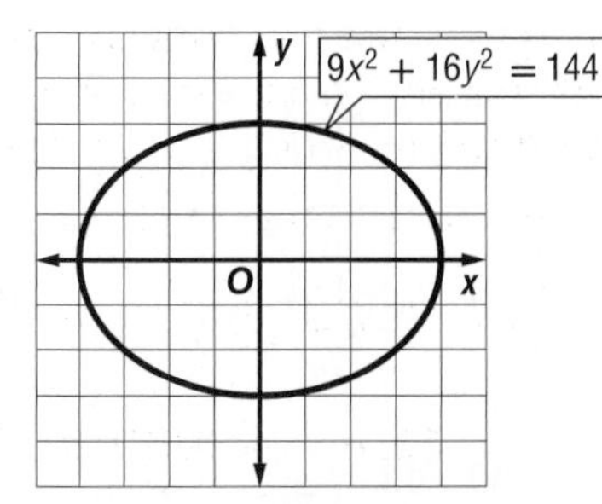

55. $\frac{y^2}{9} - \frac{(x-2)^2}{1} = 1$; hyperbola

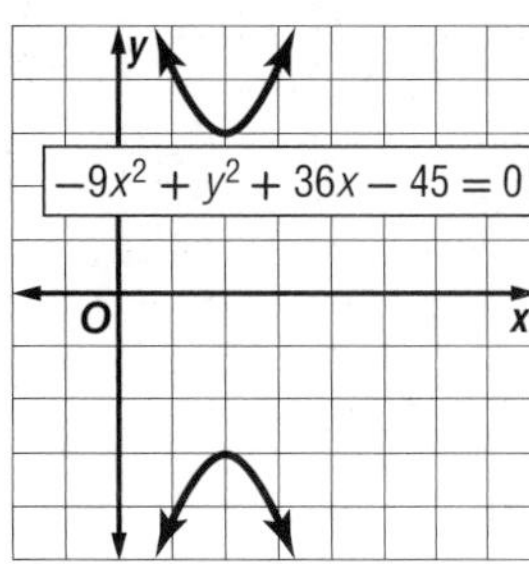

57. hyperbola **59.** circle **61.** (2, −2), (−2, 2)

63. (4, 0) **65.** (1, ±5), (−1, ±5)

67a. 1.5 seconds

67b. 109 feet

69.

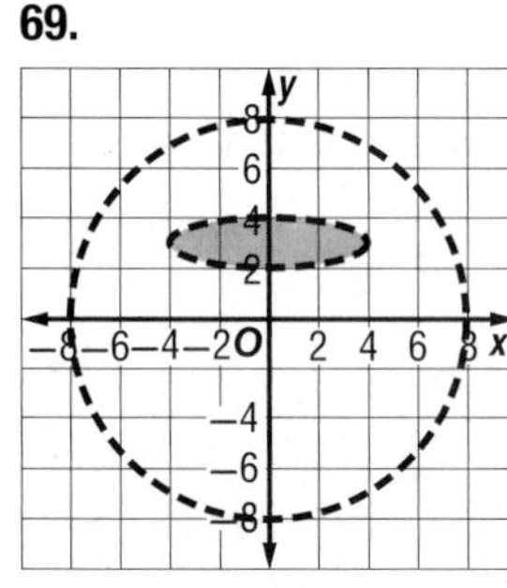

71.

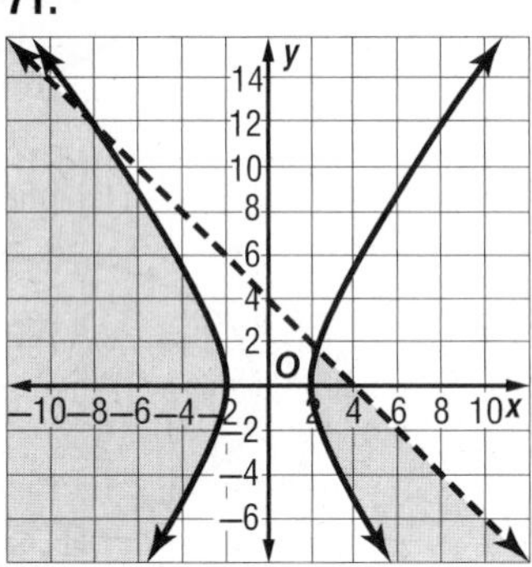

73.

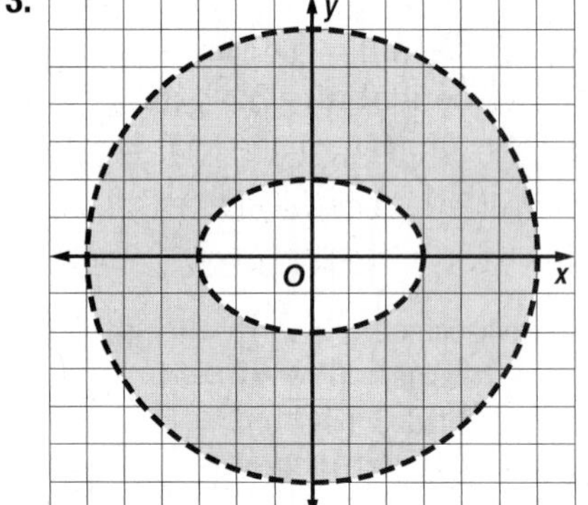

CHAPTER 10
Sequences and Series

Chapter 10 Get Ready

1. $x = -12$ **3.** $x = 3$ **5.** 9 rows

7.

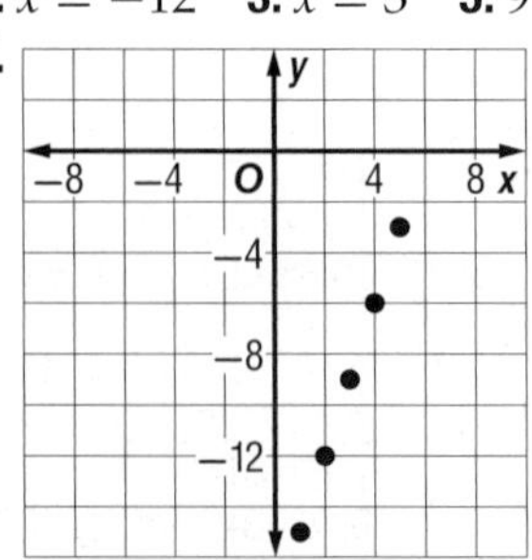

9.

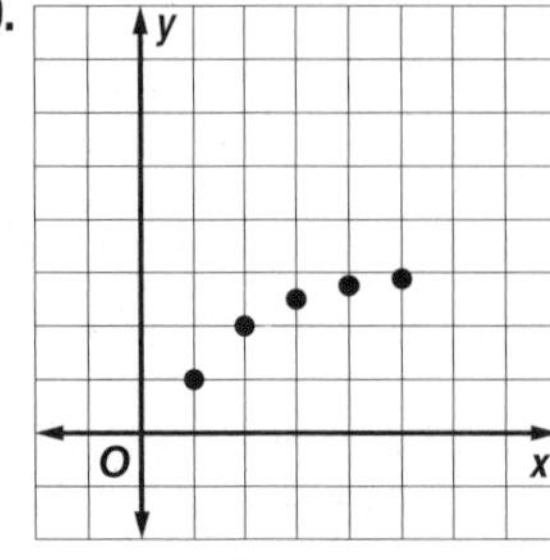

11. −30 **13.** $-\frac{2}{729}$

Lesson 10-1

1. yes **3.** no

5. 42, 54, 66, 78

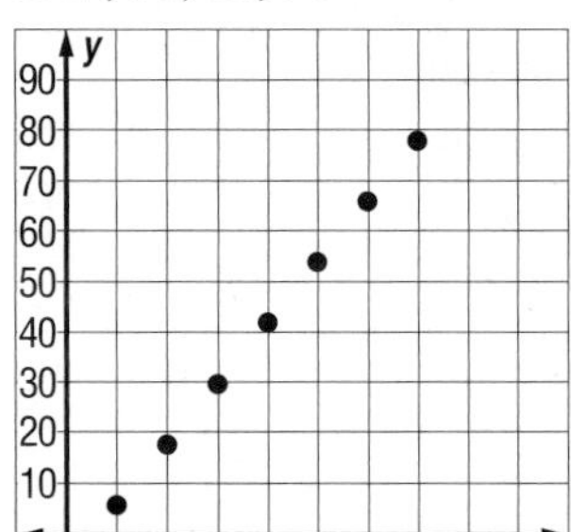

7. 5, 13, 21, 29

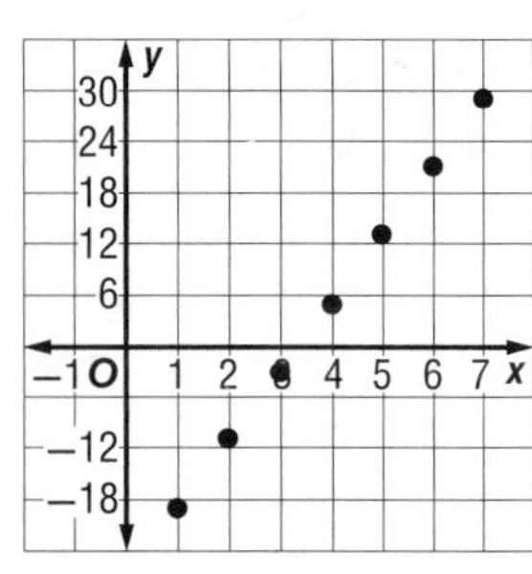

9a. $850 **9b.** 24 wk **11.** yes **13.** no

15. 128, 256, 512

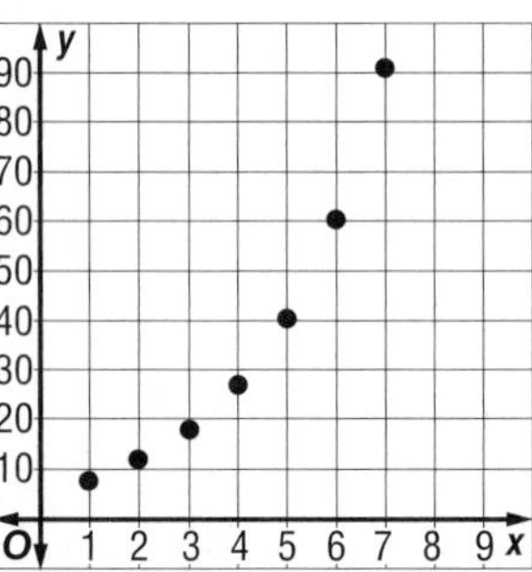

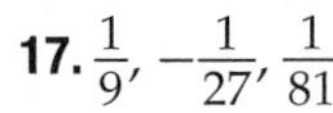

17. $\frac{1}{9}, -\frac{1}{27}, \frac{1}{81}$

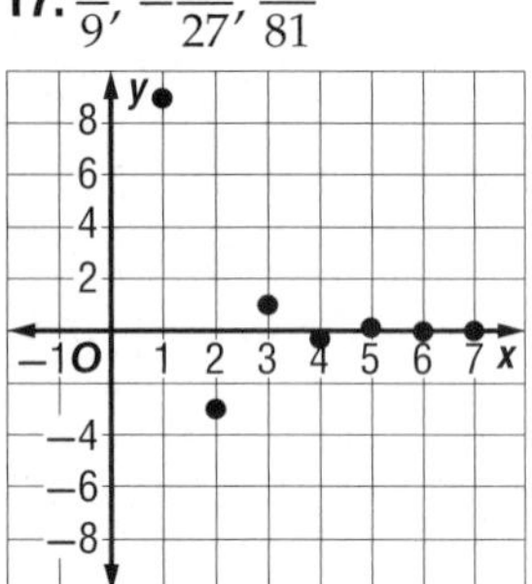

19. Geometric; the common ratio is $-\frac{1}{2}$. **21.** no **23.** no

25. 8, 11, 14, 17

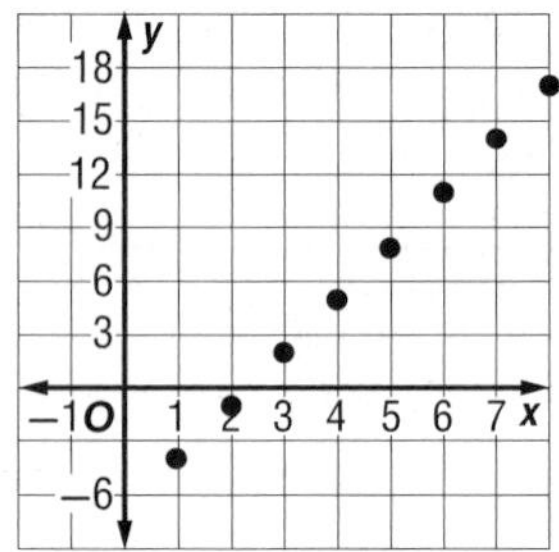

27. −29, −35, −41, −47

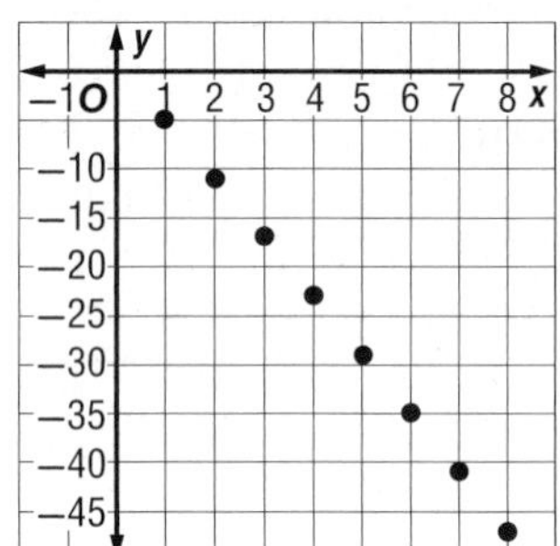

29. $2, \frac{13}{5}, \frac{16}{5}, \frac{19}{5}$

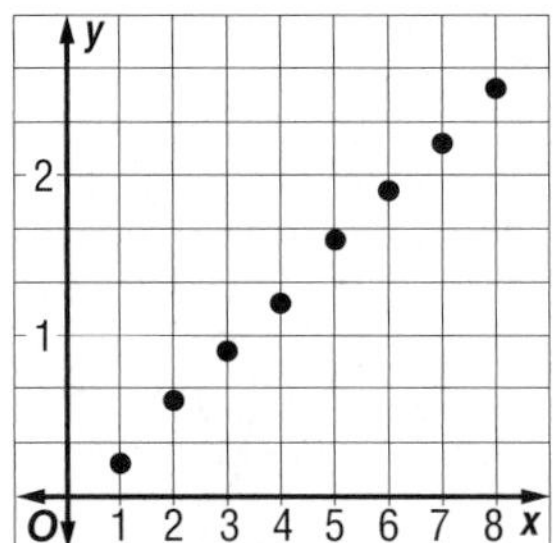

31 The difference between any two consecutive rows is 2, so the common difference is 2. Use point-slope form to write an equation for the sequence. Let $m = 2$ and $(x_1, y_1) = (1, 28)$. Solve for $x = 24$.

$(y - y_1) = m(x - x_1)$	Point-slope form
$(y - 28) = 2(x - 1)$	$m = 2, (x_1, y_1) = (1, 28)$
$y - 28 = 2x - 2$	Multiply.
$y = 2x + 26$	Add 28 to each side.
$y = 2(24) + 26$	Replace x with 24.
$y = 48 + 26$ or 74	Simplify.

There are 74 seats in the last row of the theater.

33. no

35 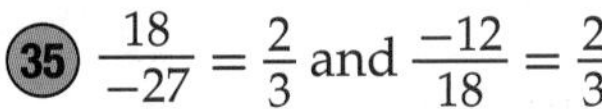
$\frac{18}{-27} = \frac{2}{3}$ and $\frac{-12}{18} = \frac{2}{3}$

Since the ratios of the consecutive terms are the same, the sequence is geometric.

37. no **39.** $-8, 32, -128$

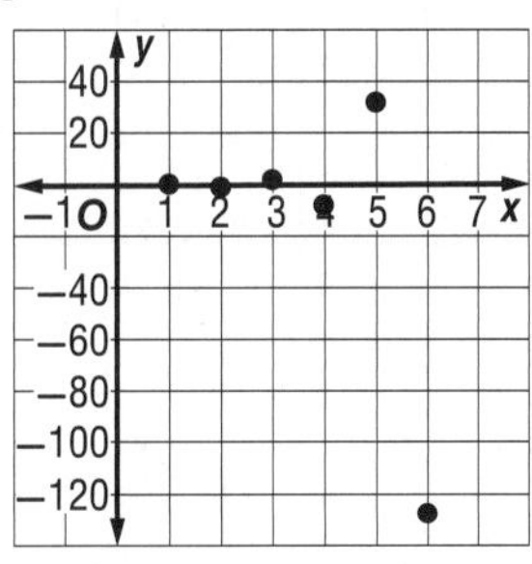

41. $27, \frac{81}{4}, \frac{243}{16}$

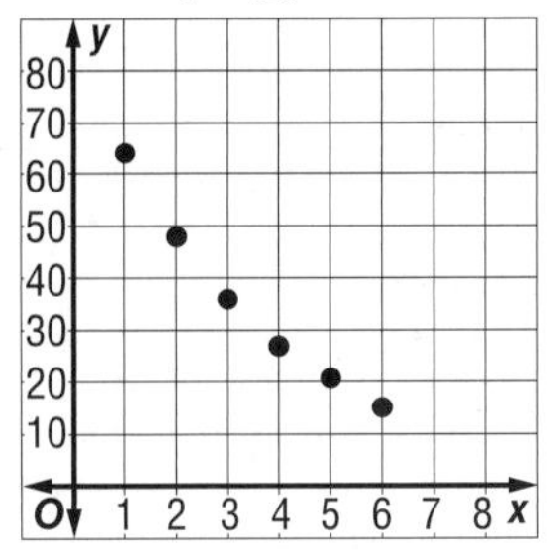

43. $27, 81, 243$

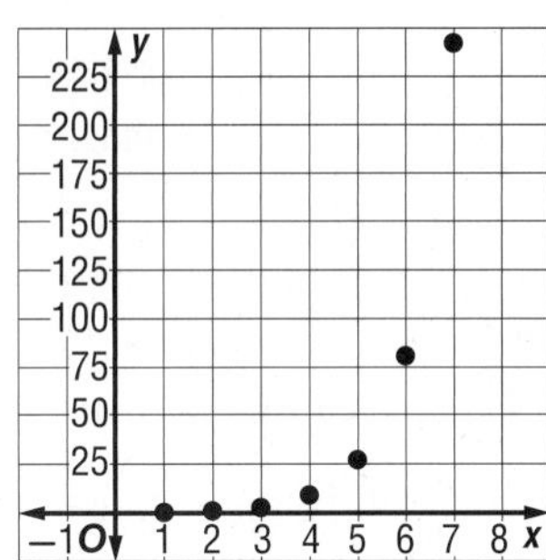

45. Neither; there is no common difference or ratio. **47.** Geometric; the common ratio is 3. **49.** Arithmetic; the common difference is $\frac{1}{2}$. **51.** 86 pg/day **53.** about 13,744 km **55.** Sample answer: A babysitter earns \$20 for cleaning the house and \$8 extra for every hour she watches the children. **57.** Sample answer: Neither; the sequence is both arithmetic and geometric. **59.** Sample answer: If a geometric sequence has a ratio r such that $|r| < 1$, as n increases, the absolute value of the terms will decrease and approach zero because they are continuously being multiplied by a fraction. When $|r| \geq 1$, the absolute value of the terms will increase and approach infinity because they are continuously being multiplied by a value greater than 1. **61.** \$421.85 **63.** H **65.** $(\pm 4, 5)$ **67.** no solution

69. hyperbola; $\frac{x^2}{4} - \frac{y^2}{1} = 1$

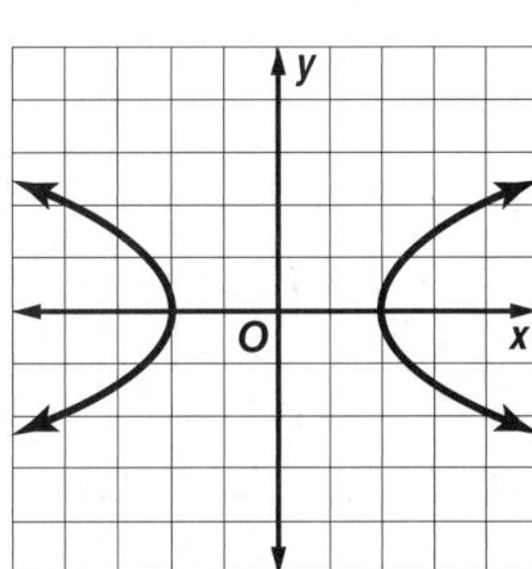

71.

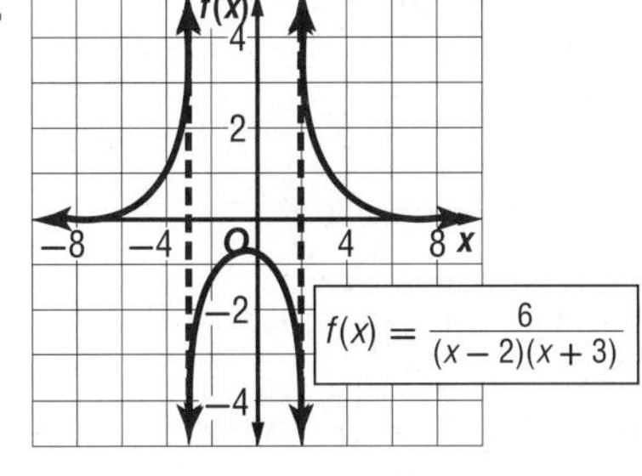

73.

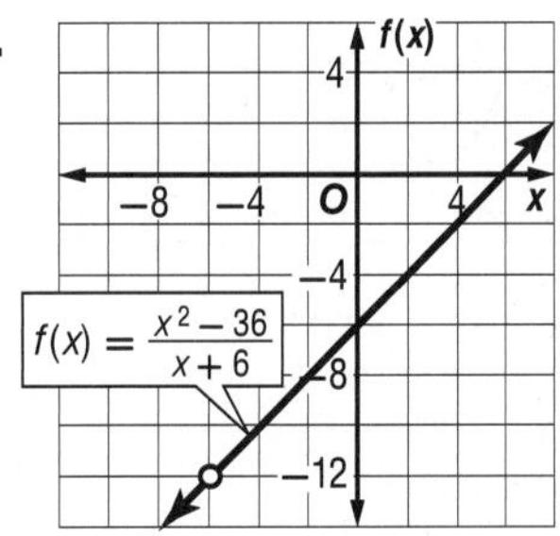

75. $y = 0.5x + 1$
77. $y = 3x - 6$
79. $y = -\frac{1}{2}x + \frac{7}{2}$

Lesson 10-2

1. 104 **3.** $a_n = 6n + 7$ **5.** 15, 24, 33 **7.** 1275 **9.** 4500 **11.** 8, 12, 16 **13.** C **15.** 248 **17.** -103 **19.** 14 **21.** $a_n = -14n + 45$

23

$a_n = a_1 + (n-1)d$	nth term of an arithmetic sequence
$21 = a_1 + (7-1)5$	$a_7 = 21$, $n = 7$, and $d = 5$
$21 = a_1 + 30$	Multiply.
$-9 = a_1$	Subtract 30 from each side.
$a_n = a_1 + (n-1)d$	nth term of an arithmetic sequence
$a_n = -9 + (n-1)5$	$a_1 = -9$ and $d = 5$
$a_n = -9 + (5n - 5)$	Distributive Property
$a_n = 5n - 14$	Simplify.

25. $a_n = 4.5n - 21$ **27.** $a_n = 9n - 32$ **29.** $a_n = \frac{2}{3}n - 3$ **31.** $a_n = \frac{1}{2}n - \frac{23}{10}$ **33.** 19, 14, 9, 4 **35.** $-21, -14, -7, 0$ **37.** $-21, -30, -39, -48, -57$ **39.** 10,100 **41.** 10,000 **43.** 696 **45.** 1272 **47.** \$4400 **49.** 48, 60, 72

51

$S_n = n\left(\frac{a_1 + a_n}{2}\right)$	Sum Formula
$2982 = 28\left(\frac{a_1 + 228}{2}\right)$	$S_n = 2982$, $n = 28$, $a_n = 228$
$2982 = 14a_1 + 3192$	Simplify.
$-210 = 14a_1$	Subtract 3192 from each side.
$-15 = a_1$	Divide each side by 14.
$a_n = a_1 + (n-1)d$	nth term of an arithmetic sequence
$228 = -15 + (28-1)d$	$a_n = 228$, $a_1 = -15$, $n = 28$
$243 = 27d$	Add 15 to each side.
$9 = d$	Divide each side by 27.

$a_1 = -15$, $a_2 = -15 + 9$ or -6, $a_3 = -6 + 9$ or 3

The first three terms are $-15, -6, 3$.

53. $-44, -30, -16$ **55.** $-33, -21, -9$ **57.** 512 **59.** 324 **61.** \$2250 **63.** $a_n = 13n - 1055$ **65.** $a_n = 9n - 88$

67a. 14, 18, 22

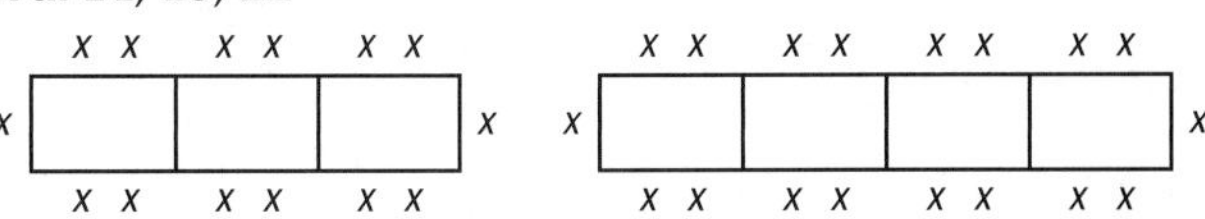

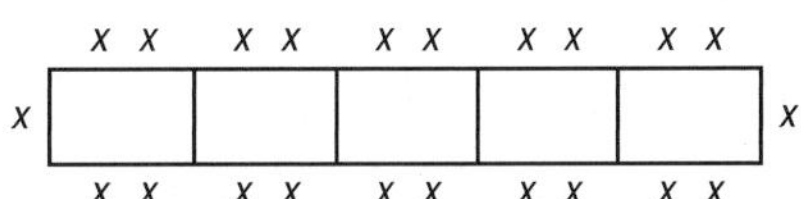

67b. $p_n = 4n + 2$ **67c.** No; there is no whole number n for which $4n + 2 = 100$.

69

$a_n = a_1 + (n-1)d$	nth term of an arithmetic sequence
$100{,}000 = 28{,}000 + (n-1)4000$	$a_n = 100{,}000$, $a_1 = 28{,}000$, $d = 4000$
$72{,}000 = (n-1)4000$	Subtract 28,000 from each side.
$18 = n - 1$	Divide each side by 4000.
$19 = n$	Add 1 to each side.

He will have a salary of \$100,000 in the 19th year.

71a.

n	S_n
1	4
2	10
3	18
4	28
5	40
6	54
7	70
8	88
9	108
10	130

71b.

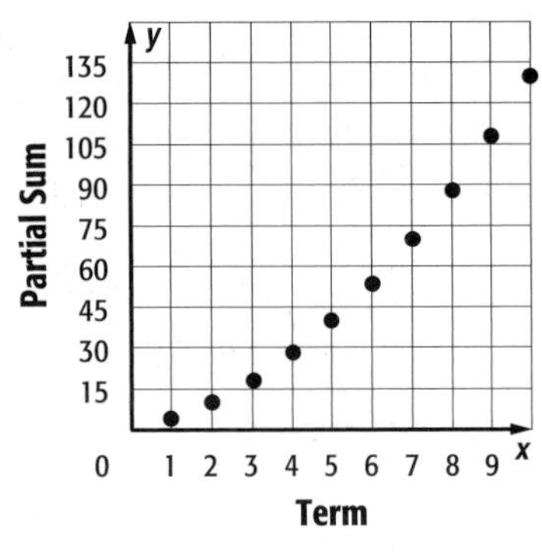

71c.

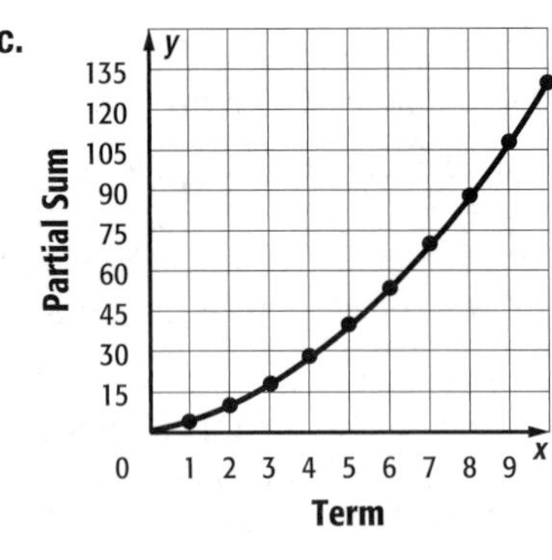

71d. Sample answer: The graphs cover the same range. The domain of the series is the natural numbers, while the domain of the quadratic function is all real numbers, $0 \le x \le 10$.

71e. Sample answer: For every partial sum of an arithmetic series, there is a corresponding quadratic function that shares the same range. **71f.** $\sum_{k=1}^{x} 2k + 7$

73. 16 **75.** $4b - 3a$ **77.** $S_n = nx + y\left(\frac{n^2 + n}{2}\right)$

79. Sample answer: An arithmetic sequence is a list of terms such that any pair of successive terms has a common difference. An arithmetic series is the sum of the terms of an arithmetic sequence.

81.

$S_n = (a_1 + a_n) \cdot \left(\frac{n}{2}\right)$ General sum formula

$a_n = a_1 + (n - 1)d$ Formula for nth term

$a_n - (n - 1)d = a_1$ Subtract $(n - 1)d$ from both sides.

$S_n = [a_n - (n - 1)d + a_n] \cdot \left(\frac{n}{2}\right)$ Substitution

$S_n = [2a_n - (n - 1)d] \cdot \left(\frac{n}{2}\right)$ Simplify.

83. C **85.** A **87.** yes **89.** no

91.

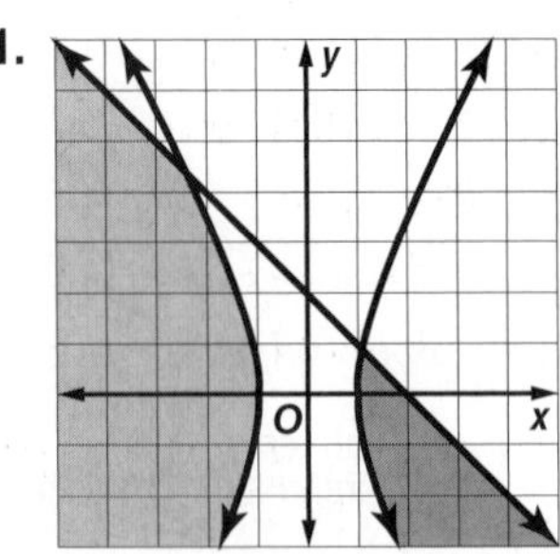

93a. 4.8 cm/g **93b.** 24 cm; The answer is reasonable. The object would stretch the first spring 60 cm and would stretch the second spring 40 cm. The object would have to stretch the combined springs less than it would stretch either of the springs individually.

95.

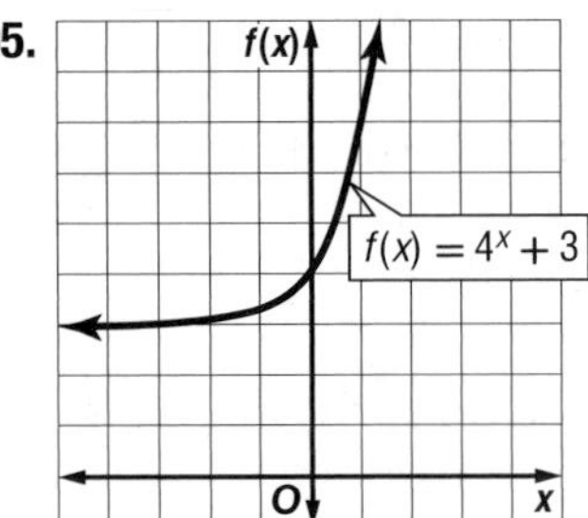

D = {all real numbers}, R = $\{f(x) \mid f(x) > 3\}$ **97.** 2.4550 **99.** 0.4341

Lesson 10-3

1. 2046 **3.** $a_n = 18 \cdot \left(\frac{1}{3}\right)^{n-1}$

5

$a_n = a_1 r^{n-1}$ nth term of a geometric sequence

$4 = a_1(3^{2-1})$ $a_n = 4$, $r = 3$, and $n = 2$

$4 = a_1(3)$ Evaluate the power.

$\frac{4}{3} = a_1$ Divide each side by 3.

$a_n = a_1 r^{n-1}$ nth term of a geometric sequence

$a_n = \frac{4}{3}(3)^{n-1}$ $a_1 = \frac{4}{3}$, $r = 3$

7. $a_n = 12(-8)^{n-1}$ **9.** 1, 5, 25 or −1, 5, −25 **11.** 4095 **13.** $\frac{1}{16}$ **15.** 512 **17.** 93 in. **19.** 25 **21.** 512 **23.** $a_n = (-3)(-2)^{n-1}$ **25.** $a_n = (-1)(-1)^{n-1}$ **27.** $a_n = 8 \cdot \left(\frac{1}{4}\right)^{n-1}$ **29.** $a_n = 7(2)^{n-1}$ **31.** $a_n = \frac{1}{15{,}552}(6)^{n-1}$ **33.** $a_n = 648\left(\frac{1}{3}\right)^{n-1}$ **35.** 270, 90, 30 or −270, 90, −30 **37.** $\frac{7}{3}, \frac{14}{9}, \frac{28}{27}$ or $-\frac{7}{3}, \frac{14}{9}, -\frac{28}{27}$ **39.** 15 and 75 **41.** 99.19% **43.** 31.9375 **45.** 9707.82 **47.** 2188 **49.** −87,381

51

$S_n = \frac{a_1 - a_1 r^n}{1 - r}$ Sum formula

$-2912 = \frac{a_1 - a_1(3^6)}{1 - 3}$ $S_n = -2912$, $r = 3$, and $n = 6$

$-2912 = \frac{a_1(1 - 3^6)}{1 - 3}$ Distributive Property

$-2912 = \frac{-728a_1}{-2}$ Subtract.

$-2912 = 364a_1$ Simplify.

$-8 = a_1$ Divide each side by 364.

53. 64 **55.** 0.25

57

$S_n = \frac{a_1 - a_1 r^n}{1 - r}$ Alternate sum formula

$= \frac{100 - 100(0.5)^5}{1 - (0.5)}$ Substitution

$= 193.75$ ft Use a calculator.

59. 524, 288 **61.** about 471 cm **63a.** \$11.79, \$30.58, \$205.72 **63b.** \$7052.15 **63c.** Each payment made is rounded to the nearest penny, so the sum of the payments will actually be more than the sum found in part b.

65.

$S_n = \frac{a_1 - a_n r}{1 - r}$ Alternate sum formula

$a_n = a_1 \cdot r^{n-1}$ Formula for nth term

$\frac{a_n}{r^{n-1}} = a_1$ Divide both sides by r^{n-1}.

$S_n = \frac{\frac{a_n}{r^{n-1}} - a_n r}{1 - r}$ Substitution.

$= \frac{\frac{a_n}{r^{n-1}} - \frac{a_n r \cdot r^{n-1}}{r^{n-1}}}{1 - r}$ Multiply by $\frac{r^{n-1}}{r^{n-1}}$.

$$= \frac{\frac{a_n(1-r^n)}{r^{n-1}}}{1-r} \quad \text{Simplify.}$$

$$= \frac{a_n(1-r^n)}{r^{n-1}(1-r)} \quad \text{Divide by } (1-r).$$

$$= \frac{a_n(1-r^n)}{r^{n-1}-r^n} \quad \text{Simplify.}$$

67. Sample answer: $n - 1$ needs to change to n, and the 10 needs to change to a 9. When this happens, the terms for both series will be identical (a_1 in the first series will equal a_0 in the second series, and so on), and the series will be equal to each other. **69.** 234 **71.** Sample answer: $4 + 8 + 16 + 32 + 64 + 128$ **73.** B **75.** $32,000 **77.** $1550 **79.** Arithmetic; the common difference is $\frac{1}{50}$.

81. (3, 1), 5 units

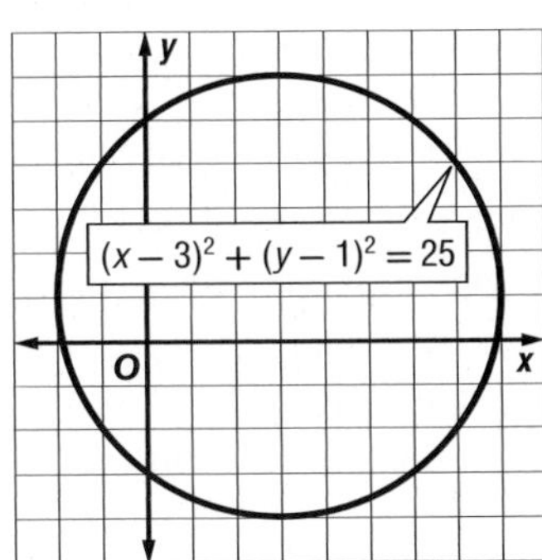

83. $(3, -7)$, $5\sqrt{2}$ units

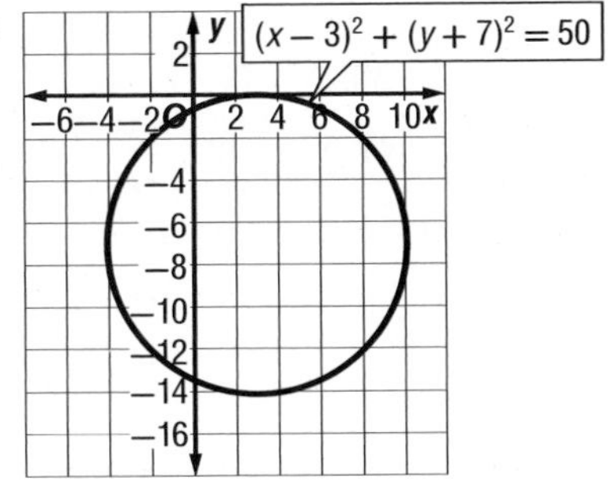

85. 731 customers **87.** $-\frac{5}{7}$ **89.** $\frac{27}{2}$

Lesson 10-4

1. convergent **3.** divergent **5.** 880 **7.** No sum exists. **9.** 12.5% **11.** -4 **13.** 2 **15.** $\frac{214}{333}$ **17.** convergent **19.** divergent **21.** divergent **23.** No sum exists.

25 $r = \frac{6}{5} \div \frac{12}{5}$ or $\frac{1}{2}$ Divide consecutive terms.

Since $\left|\frac{1}{2}\right| < 1$, the sum exists.

$$S = \frac{a_1}{1-r} \quad \text{Sum formula}$$

$$= \frac{\frac{12}{5}}{1-\frac{1}{2}} \quad a_1 = \frac{12}{5} \text{ and } r = \frac{1}{2}$$

$$= \frac{12}{5} \div \frac{1}{2} \text{ or } \frac{24}{5} \quad \text{Simplify.}$$

27. No sum exists. **29.** No sum exists. **31.** $\frac{35}{12}$ **33.** 16

35 $0.3\overline{21} = 0.3 + 0.021 + 0.00021 + \ldots$

$$= \frac{3}{10} + \frac{21}{1000} + \frac{21}{100{,}000} + \ldots$$

$$S = \frac{a_1}{1-r} + \frac{3}{10} \quad \text{Sum formula}$$

$$= \frac{\frac{21}{1000}}{1-\frac{1}{100}} + \frac{3}{10} \quad a_1 = \frac{21}{1000} \text{ and } r = \frac{1}{100}$$

$$= \frac{2100}{99{,}000} + \frac{3}{10} \quad \text{Simplify.}$$

$$= \frac{107}{330} + \frac{3}{10} \text{ or } \frac{53}{165} \quad \text{Simplify.}$$

37. $\frac{24}{11}$ **39.** $\frac{601}{4950}$ **41.** $\frac{40}{3}$ **43.** 8000 hrs **45.** $\frac{45}{4}$ **47.** No sum exists. **49.** $-\frac{54}{35}$ **51.** 200 ft **53.** 1170 ft

55 $S = \frac{a_1}{1-r}$ Sum formula

$$= \frac{500}{1-0.8} \quad a_1 = 500 \text{ and } r = 0.8$$

$$= \frac{500}{0.2} \text{ or } \$2500 \quad \text{Simplify.}$$

57. b **59.** a **61.** Sample answer: The sum of a geometric series is $S_n = \frac{a_1 - a_1r^n}{1-r}$. For an infinite series with $|r| < 1$, $r^n \to 0$ as $n \to \infty$. Thus, $S = \frac{a_1 - a_1(0)}{1-r}$ or $\frac{a_1}{1-r}$. **63.** Sample answer: An infinite geometric series has a sum when the common ratio has an absolute value less than 1. When this occurs, the terms will approach 0 as n approaches infinity. With the future terms almost 0, the sum of the series will approach a limit. When the common ratio is 1 or greater, the terms will keep increasing and approach infinity as n approaches infinity and the sum of the series will have no limit. **65.** Sample answer: $3 + 2 + \frac{4}{3} + \ldots$ **67.** An arithmetic series has a common difference, so each term will eventually become more positive or more negative, but never approach 0. With the terms not approaching 0, the sum will never reach a limit and the series cannot converge. **69.** H **71.** C **73a.** 9, 18, 27, 36, 45, 54, 63, 72 **73b.** 42 meetings **75.** 12

Lesson 10-5

1. 16, 20, 24, 28, 32 **3.** 5, 17, 53, 161, 485 **5.** $a_{n+1} = 2a_n + 2$; $a_1 = 3$ **7a.** $a_n = 1.01a_{n-1} - 100$, $a_1 = 1500$ **7b.** $1415, $1329.15, $1242.44, $1154.87 **7c.** $77.08 **9.** $-18, 74, -294$ **11.** $-52, -420, -3364$ **13.** $-9, -10, -12, -16, -24$ **15.** $-4, -7, -12, -21, -38$

17 $a_{n+1} = 5a_n + 2n$ Recursive formula

$a_{1+1} = 5a_1 + 2(1)$ $n = 1$

$a_2 = 5(-2) + 2(1)$ or -8 $a_1 = -2$

$a_3 = 5(-8) + 2(2)$ or -36 $a_2 = -8, n = 2$

$a_4 = 5(-36) + 2(3)$ or -174 $a_3 = -36, n = 3$

$a_5 = 5(-174) + 2(4)$ or -862 $a_4 = -36, n = 4$

The first five terms are $-2, -8, -36, -174$, and -862.

19. 4, 5, 6, 8, 8 **21.** $3, 2x, 8x - 9, 26x - 36, 80x - 117$ **23.** $1, x, 3x + 6, 15x + 18, 63x + 90$ **25.** $a_{n+1} = 0.25a_n + 4$; $a_1 = 32$ **27.** $a_{n+1} = (a_n)^3 + 1$; $a_1 = 1$ **29.** $a_{n+1} = 0.25a_n + 8$; $a_1 = 480$ **31.** $a_{n+1} = 2a_n - 32$; $a_1 = 84$ **33.** 56, 680, 8168 **35.** $-45, 273, -1635$ **37.** $-3, -18, -963$ **39.** 43, 3484, 24,259,093 **41.** 4.25, 29.5625, 936.0664

43 **a.** The number of blue triangles is 1, 3, and 9. The sequence is geometric because each term after the first can be found after multiplying by a common ratio, 3.

$a_n = r \cdot a_{n-1}$ Recursive formula for geometric sequence

$a_n = 3a_{n-1}, a_1 = 1$ $r = 3$

b. $a_1 = 1, a_2 = 3, a_3 = 9$

$a_n = 3a_{n-1}$ Recursive formula

$a_4 = 3a_3$ $n = 4$

$= 3(9)$ or 27 $a_3 = 9$

$a_5 = 3a_4$ $n = 5$

$= 3(27)$ or 81 $a_4 = 27$

$a_6 = 3a_5$ $n = 6$

$= 3(81)$ or 243 $a_5 = 27$

There will be 243 blue triangles in the sixth figure.

45. No; the population of fish will reach 12,500. Each year, 20% of 12,500 or 2500 fish plus 10,000 additional fish yields 12,500 fish.

47a. 11,000 **47b.** It converges to 7142.857.

47c. Sample answer: They make it easier to analyze recursive sequences because they can produce the first 100 terms instantaneously; it would take a long time to calculate the terms by hand. **49.** Armando; Marcus included x_0 with the iterates and only showed the first 2 iterates. **51.** Sample answer: Sometimes; the recursive formula could involve the first three terms. For example, 2, 2, 2, 8, 20, … is recursive with $a_{n+3} = a_n + a_{n+1} + 2a_{n+2}$.

53. Sample answer: In a recursive sequence, each term is determined by one or more of the previous terms. A recursive formula is used to produce the terms of the recursive sequence. **55a.** 160 m

55b. 5.7 s **55c.** 11.4 s **57.** C **59.** $5\frac{14}{111}$

61a. 2, 3, 4.5, 6.75, 10.125 **61b.** the eighth session

61c. during the ninth session **63.** dependent

65. $x^2 + 4x - 12$ **67.** $4h^2 + 33h + 35$

69. $10g^2 + 19g - 56$

Lesson 10-6

1. $c^5 + 5c^4d + 10c^3d^2 + 10c^2d^3 + 5cd^4 + d^5$

3. $x^6 - 24x^5 + 240x^4 - 1280x^3 + 3840x^2 - 6144x + 4096$

5. $x^5 + 15x^4 + 90x^3 + 270x^2 + 405x + 243$

7. $\frac{3}{32}$ or 0.09375

9 $(x + 3y)^8 = \sum_{k=0}^{8} \frac{8!}{k!(8-k)!} x^{8-k}(3y)^k$

$\frac{8!}{k!(8-k)!} x^{8-k}(3y)^k = \frac{8!}{4!(8-4)!} x^{8-4}(3y)^4$ For the fifth term, $k = 4$.

$= 70x^4(81y^4)$ $C(8, 4) = 70$, $(3y)^4 = 81y^4$

$= 5670x^4y^4$ Simplify.

11. $-108{,}864c^3d^5$ **13.** $243a^5$ **15.** $a^6 - 6a^5b + 15a^4b^2 - 20a^3b^3 + 15a^2b^4 - 6ab^5 + b^6$ **17.** $x^6 + 36x^5 + 540x^4 + 4320x^3 + 19{,}440x^2 + 46{,}656x + 46{,}656$ **19.** $16a^4 + 128a^3b + 384a^2b^2 + 512ab^3 + 256b^4$

21 Let w represent the number of women and m represent the number of men.

$(w + m)^{10} = \sum_{k=0}^{10} \frac{10!}{k!(10-k)!} w^{10-k}m^k$

To find the probability that 7 members are women, find the term in which the exponent of w is 7. Since $10 - 3 = 7$, find the term in which $k = 3$, the fourth term.

$\frac{10!}{k!(10-k)!} w^{10-k}m^k = \frac{10!}{3!(10-3)!} w^{10-3}m^3$ For the fourth term, $k = 3$.

$= 120w^7m^3$ $C(10, 3) = 120$

The probability of choosing a woman is $\frac{1}{2}$ and the probability of choosing a man is $\frac{1}{2}$.

$120w^7m^3 = 120\left(\frac{1}{2}\right)^7\left(\frac{1}{2}\right)^3$ $w = \frac{1}{2}$ and $m = \frac{1}{2}$

$= 120\left(\frac{1}{2}\right)^{10}$ Product of Powers Property

$= \frac{120}{1024}$ $\left(\frac{1}{2}\right)^{10} = \frac{1}{1024}$

$= \frac{15}{128}$ Simplify.

The probability that 7 of the members will be women is $\frac{15}{128}$, or about 0.117.

23. $84x^5z^2$ **25.** $7168a^2b^6$ **27.** $32{,}256x^5$ **29.** $x^5 + \frac{5}{2}x^4 + \frac{5}{2}x^3 + \frac{5}{4}x^2 + \frac{5}{16}x + \frac{1}{32}$ **31.** $32b^5 + 20b^4 + 5b^3 + \frac{5}{8}b^2 + \frac{5}{128}b + \frac{1}{1024}$ **33a.** 0.121 **33b.** 0.121 **33c.** 0.309

35. Sample answer: While they have the same terms, the signs for $(x + y)^n$ will all be positive, while the signs for $(x - y)^n$ will alternate. **37.** Sample answer: $\left(x + \frac{6}{5}y\right)^5$ **39.** A **41.** G **43.** −2, 3, 8, 13, 18 **45.** 4, 6, 12, 30, 84 **47.** $1\frac{1}{8}$ **49a.** $\frac{150}{x}; \frac{130}{x-10}$ **49b.** $\frac{150}{x} + \frac{130}{x-10} = 4$; 75 mph, 65 mph **51.** true; $3(1) + 5 = 8$, which is even

Lesson 10-7

1. Step 1: When $n = 1$, the left side of the given equation is 1. The right side is 1^2 or 1, so the equation is true for $n = 1$.

Step 2: Assume that $1 + 3 + 5 + \cdots + (2k - 1) = k^2$ for some natural number k.

Step 3: $1 + 3 + 5 + \cdots + (2k - 1) + (2(k + 1) - 1)$

$= k^2 + (2(k + 1) - 1)$

$= k^2 + (2k + 2 - 1)$

$= k^2 + 2k + 1$

$= (k + 1)^2$

The last expression is the right side of the equation to be proved, where $n = k + 1$. Thus, the equation is true for $n = k + 1$. Therefore, $1 + 3 + 5 + \cdots + (2n - 1) = n^2$ for all natural numbers n.

3a. 3, 6, 10, 15, 21 **3b.** $a_n = \frac{n(n+1)}{2}$

3c. Step 1: When $n = 1$, the left side of the given equation is $\frac{1(1+1)}{2}$ or 1. The right side is $\frac{1(1+1)(1+2)}{6}$ or 1, so the equation is true for $n = 1$.

Step 2: Assume that $1 + 3 + 6 + \cdots + \frac{k(k+1)}{2} = \frac{k(k+1)(k+2)}{6}$ for some natural number k.

Step 3: $1 + 3 + 6 + \cdots + \frac{k(k+1)}{2} + \frac{(k+1)(k+1+1)}{2}$

$= \frac{k(k+1)(k+2)}{6} + \frac{(k+1)(k+1+1)}{2}$
$= \frac{k(k+1)(k+2)}{6} + \frac{3(k+1)(k+2)}{6}$
$= \frac{(k+1)(k+2)(k+3)}{6}$
$= \frac{(k+1)[(k+1)+1][(k+1)+2]}{6}$

The last expression is the right side of the equation to be proved, where $n = k + 1$. Thus, the equation is true for $n = k + 1$. Therefore, $1 + 3 + 6 + \cdots + \frac{n(n+1)}{2} = \frac{n(n+1)(n+2)}{6}$ for all natural numbers n.

5. Step 1: $4^1 - 1 = 3$, which is divisible by 3. The statement is true for $n = 1$.
Step 2: Assume that $4^k - 1$ is divisible by 3 for some natural number k. This means that $4^k - 1 = 3r$ for some whole number r.
Step 3: $4^k - 1 = 3r$
$4^k = 3r + 1$
$4^{k+1} = 12r + 4$
$4^{k+1} - 1 = 12r + 3$
$4^{k+1} - 1 = 3(4r + 1)$
Since r is a whole number, $4r + 1$ is a whole number. Thus, $4^{k+1} - 1$ is divisible by 3, so the statement is true for $n = k + 1$. Therefore, $4^n - 1$ is divisible by 3 for all natural numbers n. **7.** $n = 1$

9. Step 1: When $n = 1$, the left side of the given equation is 2. The right side is $\frac{1[3(1)+1]}{2}$ or 2, so the equation is true for $n = 1$.
Step 2: Assume that $2 + 5 + 8 + \cdots + (3k - 1) = \frac{k(3k+1)}{2}$ for some natural number k.
Step 3: $2 + 5 + 8 + \cdots + (3k - 1) + [3(k+1) - 1]$
$= \frac{k(3k+1)}{2} + [3(k+1) - 1]$
$= \frac{k(3k+1) + 2[3(k+1) - 1]}{2}$
$= \frac{3k^2 + k + 6k + 6 - 2}{2}$
$= \frac{3k^2 + 7k + 4}{2}$
$= \frac{(k+1)(3k+4)}{2}$
$= \frac{(k+1)[3(k+1)+1]}{2}$
The last expression is the right side of the equation to be proved, where $n = k + 1$. Thus, the equation is true for $n = k + 1$. Therefore, $2 + 5 + 8 + \cdots + (3n - 1) = \frac{n(3n+1)}{2}$ for all natural numbers n.

11. Step 1: When $n = 1$, the left side of the given equation is 1. The right side is $1[2(1) - 1]$ or 1, so the equation is true for $n = 1$.
Step 2: Assume that $1 + 5 + 9 + \cdots + (4k - 3) = k(2k - 1)$ for some natural number k.
Step 3: $1 + 5 + 9 + \cdots + (4k - 3) + [4(k+1) - 3]$
$= k(2k - 1) + [4(k+1) - 3]$
$= 2k^2 - k + 4k + 4 - 3$
$= 2k^2 + 3k + 1$
$= (k+1)(2k+1)$
$= (k+1)[2(k+1) - 1]$
The last expression is the right side of the equation to be proved, where $n = k + 1$. Thus, the equation is true for $n = k + 1$. Therefore, $1 + 5 + 9 + \cdots + (4n - 3) = n(2n - 1)$ for all natural numbers n.

13 Step 1: When $n = 1$, the left side of the given equation is $4(1) - 1$ or 3. The right side is $2(1)^2 + 1$ or 3, so the equation is true for $n = 1$.
Step 2: Assume that $3 + 7 + 11 + \cdots + (4k - 1) = 2k^2 + k$ for some natural number k.
Step 3: $3 + 7 + 11 + \cdots + (4k - 1) + [4(k+1) - 1]$
$= 2k^2 + k + [4(k+1) - 1]$
$= 2k^2 + k + 4k + 3$
$= 2k^2 + 5k + 3$
$= 2k^2 + 4k + 2 + k + 1$
$= [2(k+1)^2] + (k+1)$
The last expression is the right side of the equation to be proved, where $n = k + 1$. Thus, the equation is true for $n = k + 1$. Therefore, $3 + 7 + 11 + \cdots + (4n - 1) = 2n^2 + n$ for all natural numbers n.

15. Step 1: When $n = 1$, the left side of the given equation is 1^2 or 1. The right side is $\frac{1[2(1)-1][2(1)+1]}{3}$ or 1, so the equation is true for $n = 1$.
Step 2: Assume that $1^2 + 3^2 + 5^2 + \cdots + (2k - 1)^2 = \frac{k(2k-1)(2k+1)}{3}$ for some natural number k.
Step 3: $1^2 + 3^2 + 5^2 + \cdots + (2k-1)^2 + [2(k+1) - 1]^2$
$= \frac{k(2k-1)(2k+1)}{3} + [2(k+1) - 1]^2$
$= \frac{k(2k-1)(2k+1) + 3(2k+1)^2}{3}$
$= \frac{(2k+1)[k(2k-1) + 3(2k+1)]}{3}$
$= \frac{(2k+1)(2k^2 - k + 6k + 3)}{3}$
$= \frac{(2k+1)(2k^2 + 5k + 3)}{3}$
$= \frac{(2k+1)(k+1)(2k+3)}{3}$
$= \frac{(k+1)[2(k+1) - 1][2(k+1) + 1]}{3}$
The last expression is the right side of the equation to be proved, where $n = k + 1$. Thus, the equation is true for $n = k + 1$. Therefore, $1^2 + 3^2 + 5^2 + \cdots + (2n - 1)^2 = \frac{n(2n-1)(2n+1)}{3}$ for all natural numbers n.

17. Step 1: $5^1 + 3 = 8$, which is divisible by 4. The statement is true for $n = 1$.
Step 2: Assume $5^k + 3$ is divisible by 4 for some natural number k. This means that $5^k + 3 = 4r$ for some natural number r.
Step 3: $5^k + 3 = 4r$
$5^k = 4r - 3$
$5^{k+1} = 20r - 15$
$5^{k+1} + 3 = 20r - 12$
$5^{k+1} + 3 = 4(5r - 3)$
Since r is a natural number, $5r - 3$ is a natural number. Thus, $5^{k+1} + 3$ is divisible by 4, so the statement is true for $n = k + 1$. Therefore, $5^n + 3$ is divisible by 4 for all natural numbers n.

19. Step 1: $12^1 + 10 = 22$, which is divisible by 11. The statement is true for $n = 1$.
Step 2: Assume that $12^k + 10$ is divisible by 11 for some natural number k. This means that $12^k + 10 = 11r$ for some natural number r.
Step 3: $12^k + 10 = 11r$
$$12^k = 11r - 10$$
$$12^{k+1} = 132r - 120$$
$$12^{k+1} + 10 = 132r - 110$$
$$12^{k+1} + 10 = 11(12r - 10)$$
Since r is a natural number, $12r - 10$ is a natural number. Thus, $12^{k+1} + 10$ is divisible by 11, so the statement is true for $n = k + 1$. Therefore, $12^n + 10$ is divisible by 11 for all natural numbers n. **21.** $n = 2$ **23.** $n = 1$

25 In the sequence 1, 1, 2, 3, 5, 8, …, $f_1 = 1, f_2 = 1, f_3 = 2, f_4 = 3, f_5 = 5, f_6 = 8, \ldots$.
$f_1 + f_2 + \cdots + f_n = f_{n+2} - 1$ Original equation
$f_1 = f_{1+2} - 1$ Let $n = 1$.
$f_1 = f_3 - 1$ Simplify.
Step 1: When $n = 1$, the left side of the given equation is f_1. The right side is $f_3 - 1$. Since $f_1 = 1$ and $f_3 = 2$, the equation becomes $1 = 2 - 1$ and is true for $n = 1$.
Step 2: Assume that $f_1 + f_2 + \cdots + f_k = f_{k+2} - 1$ for some natural number k.
Step 3: $f_1 + f_2 + \cdots + f_k + f_{k+1} = f_{k+2} - 1 + f_{k+1}$
$= f_{k+1} + f_{k+2} - 1$
$= f_{k+3} - 1$, since Fibonacci numbers are produced by adding the two previous Fibonacci numbers.
The last expression is the right side of the equation to be proved, where $n = k + 1$. Thus, the equation is true for $n = k + 1$. Therefore, $f_1 + f_2 + \cdots + f_n = f_{n+2} - 1$ for all natural numbers n.

27. Step 1: $18^1 - 1 = 17$, which is divisible by 17. The statement is true for $n = 1$.
Step 2: Assume that $18^k - 1$ is divisible by 17 for some natural number k. This means that $18^k - 1 = 17r$ for some natural number r.
Step 3: $18^k - 1 = 17r$
$$18^k = 17r + 1$$
$$18^{k+1} = 18(17r + 1)$$
$$18^{k+1} = 306r + 18$$
$$18^{k+1} - 1 = 306r + 17$$
$$18^{k+1} - 1 = 17(18r + 1)$$
Since r is a natural number, $18r + 1$ is a natural number. Thus, $18^{k+1} - 1$ is divisible by 17, so the statement is true for $n = k + 1$. Therefore, $18^n - 1$ is divisible by 17 for all natural numbers n. **29.** $n = 3$

31. Step 1: When $n = 1$, the left side of the given equation is $\frac{1}{1(1+1)(1+2)}$ or $\frac{1}{6}$. The right side is $\frac{1(1+3)}{4(1+1)(1+2)}$ or $\frac{1}{6}$, so the equation is true for $n = 1$.
Step 2: Assume that $\frac{1}{1 \cdot 2 \cdot 3} + \frac{1}{2 \cdot 3 \cdot 4} + \frac{1}{3 \cdot 4 \cdot 5} + \cdots + \frac{1}{k(k+1)(k+2)} = \frac{k(k+3)}{4(k+1)(k+2)}$ for some natural number k.
Step 3: $\frac{1}{1 \cdot 2 \cdot 3} + \frac{1}{2 \cdot 3 \cdot 4} + \cdots + \frac{1}{k(k+1)(k+2)} + \frac{1}{(k+1)(k+2)(k+3)}$
$$= \frac{k(k+3)}{4(k+1)(k+2)} + \frac{1}{(k+1)(k+2)(k+3)}$$
$$= \frac{k(k+3)(k+3)}{4(k+1)(k+2)(k+3)} + \frac{4}{4(k+1)(k+2)(k+3)}$$
$$= \frac{k^3 + 6k^2 + 9k + 4}{4(k+1)(k+2)(k+3)}$$
$$= \frac{(k+1)(k^2 + 5k + 4)}{4(k+1)(k+2)(k+3)}$$
$$= \frac{(k+1)(k+4)}{4(k+2)(k+3)}$$
$$= \frac{(k+1)[(k+1)+3]}{4[(k+1)+1][(k+1)+2]}$$
The last expression is the right side of the equation to be proved, where $n = k + 1$. Thus, the equation is true for $n = k + 1$. Therefore, $\frac{1}{1 \cdot 2 \cdot 3} + \frac{1}{2 \cdot 3 \cdot 4} + \frac{1}{3 \cdot 4 \cdot 5} + \cdots + \frac{1}{n(n+1)(n+2)} = \frac{n(n+3)}{4(n+1)(n+2)}$ for all natural numbers n. **33.** $n(n + 1)$
Step 1: When $n = 1$, the left side of the given equation is 2(1) or 2. The right side is $1(1 + 1)$ or 2, so the equation is true for $n = 1$.
Step 2: Assume that $2 + 4 + 6 + \cdots + 2k = k(k + 1)$ for some natural number k.
Step 3: $2 + 4 + 6 + \cdots + 2k + 2(k + 1)$
$= k(k + 1) + 2(k + 1)$
$= (k + 1)(k + 2)$
$= (k + 1)[(k + 1) + 1]$
The last expression is the right side of the equation to be proved, where $n = k + 1$. Thus, the equation is true for $n = k + 1$. Therefore, $2 + 4 + 6 + \cdots + n^2 = n(n + 1)$ for all natural numbers n. **35.** Sample answer: False; assume $k = 2$, just because a statement is true for $n = 2$ and $n = 3$ does not mean that it is true for $n = 1$.
37. $x = 3$ **39.** Sample answer: When dominoes are set up, after the first domino falls, the rest will fall as well. With induction, once it is proved that the statement is true for $n = 1$ (the first domino), $n = k$ (the second domino), and $n = k + 1$ (the next domino), it will be true for any integer value (any domino).
41. B **43.** 96 **45.** $160x^3y^3$ **47.** $-84x^6y^3$
49. (6, −8), (12, −16) **51.** 56 **53.** 665,280 **55.** 70
57. 132 **59.** 28 **61.** 24

Chapter 10 Study Guide and Review

1. true **3.** true **5.** true **7.** false, arithmetic sequence **9.** false, arithmetic means **11.** 48 **13.** −22 **15.** −7, −2, 3 **17.** 8, 4, 0, −4 **19.** $480 **21.** 1040 **23.** −245 **25.** 629 **27.** −99 **29.** 99 **31.** $\frac{2187}{8}$ **33.** ±24, 72, ±216 **35.** $1823.26 **37.** 12,285 **39.** 363 **41.** $-\frac{6305}{45{,}927}$ **43.** 32 **45.** 6 **47.** −3, 1, 5, 9, 13 **49.** 1, 6, 11, 16, 21 **51.** 7, 15, 31 **53.** 11, 65, 389 **55.** $a^3 + 3a^2b + 3ab^2 + b^3$
57. $-32z^5 + 240z^4 - 720z^3 + 1080z^2 - 810z + 243$
59. $x^5 - \frac{5}{4}x^4 + \frac{5}{8}x^3 - \frac{5}{32}x^2 + \frac{5}{256}x - \frac{1}{1024}$

61. $193{,}536x^2y^5$

63. Step 1: When $n = 1$, the left side of the equation is equal to 2. The right side of the equation is also equal to 2. So the equation is true for $n = 1$.

Step 2: Assume that $2 + 6 + 12 + \cdots + k(k + 1) = \frac{k(k + 1)(k + 2)}{3}$ for some positive integer k.

Step 3: $1 \cdot 2 + 2 \cdot 3 + \cdots + k(k + 1) + (k + 1)(k + 2) =$

$$\frac{k(k + 1)(k + 2)}{3} + (k + 1)(k + 2)$$
$$= \frac{k(k + 1)(k + 2)}{3} + \frac{3(k + 1)(k + 2)}{3}$$
$$= \frac{(k + 1)[k(k + 2) + 3(k + 2)]}{3}$$
$$= \frac{(k + 1)(k + 2)(k + 3)}{3} = \frac{(k + 1)[(k + 1) + 1][(k + 1) + 2]}{3}$$

The last expression is the right side of the equation to be proved, where $n = k + 1$. Thus, the equation is true for $n = k + 1$.

Therefore, $2 + 6 + 12 + \cdots + n(n + 1) = \frac{n(n + 1)(n + 2)}{3}$ for all positive integers n.

65. Step 1: When $n = 1$, $5^1 - 1 = 5$ or 4. Since 4 divided by 4 is 1, the statement is true for $n = 1$.

Step 2: Assume that $5^k - 1$ is divisible by 4 for some positive integer k. This means that $5^k - 1 = 4r$ for some whole number r.

Step 3:
$$5^k - 1 = 4r$$
$$5^k = 4r + 1$$
$$5^{k+1} = 20r + 5$$
$$5^{k+1} - 1 = 20r + 5 - 1$$
$$5^{k+1} - 1 = 20r + 4$$
$$5^{k+1} - 1 = 4(5r + 1)$$

Since r is a whole number, $5r + 1$ is a whole number. Thus, $5^{k+1} - 1$ is divisible by 4, so the statement is true for $n = k + 1$.

Therefore, $5^n - 1$ is divisible by 4 for all positive integers n. **67.** $n = 2$ **69.** $n = 1$

CHAPTER 11
Statistics and Probability

Chapter 11 Get Ready

1. 89 customers, 88 customers, no mode
3. 7.7 touchdowns, 8 touchdowns, 10 touchdowns
5. $\frac{1}{4}$ **7.** $a^4 - 8a^3 + 24a^2 - 32a + 16$
9. $16b^4 - 32b^3 + 24b^2x^2 - 8bx^3 + x^4$ **11.** $243x^5 - 810x^4y + 1080x^3y^2 - 720x^2y^3 + 240xy^4 - 32y^5$
13. $\frac{a^5}{32} + \frac{5a^4}{8} + 5a^3 + 20a^2 + 40a + 32$

Lesson 11-1

1. survey; sample: the students in the study; population: the student body **3.** Observation study; sample answer: The scores of the participants are observed and compared without them being affected by the study. **5.** unbiased **7.** objective: to determine how many people in the U.S. are interested in purchasing a hybrid vehicle; population: the people surveyed; sample survey questions: Do you currently own a hybrid vehicle? Are you planning on purchasing a hybrid vehicle? **9.** objective: to determine whether the protein shake helps athletes recover from exercise; population: all athletes; experiment group: athletes given the protein shake; control group: athletes given a placebo; sample procedure: The researchers could randomly divide the athletes into two groups: an experimental group given the protein shake and a control group given the placebo. Next, they could have the athletes exercise and then drink the protein shake or placebo. Later, the researchers could interview the athletes to see how they feel.

11 This is an observational study because the study group is going to observe the students' performance without directly affecting the students. The sample is the 80 physics students because they are the ones being observed. Both halves of the selected students are included because one half is going to be compared to the other half. The population is all college students that take a physics course.

13. survey; sample: people that receive the questionnaire; population: all viewers **15.** Survey; sample answer: The data will be obtained from opinions given by members of the sample population. **17.** Experiment; sample answer: Metal samples will need to be tested, which means that the members of the sample will be affected by the study. **19.** Biased; sample answer: The question only gives two options, and thus encourages a certain response. **21.** Biased; sample answer: The question encourages a certain response. The phrase "don't you agree" suggests that the students should agree.
23. Sample answer: The flaw is that the experimental group consists of stores in the midwest, and the control group consists of stores on the west coast. On average, the temperature is higher on the west coast than in the midwest, and people use more sunscreen. Therefore, the sunscreen sales in stores located in those regions would likely be different and should not be compared in an experiment.

25 **a.** The sample group is the 8- to 18-year-olds who were actually surveyed. The population is represented by the sample, so the population is all 8- to 18-year-olds in the U.S.

b. average time

c. Interpret the bar graph for each group. The red bar represents talking and the blue bar represents texting. Sample answer: The 8- to 10-year-old group talked for about 10 minutes a day and did not text at all. The 11- to 14-year-old group talked for about 30 minutes a day and texted for about 70 minutes a day. The 15- to 18-year-old group

talked for about 40 minutes a day and texted for about 110 minutes a day.
d. Sample answer: A cell phone company might use a report like this to determine which age group to target in their ads.
27a. See students' work.
27b. Sample answer for Product A: ≈63.3%

Product A	
Number	**Frequency**
0–6	𝍸 𝍸 𝍸 ||||
7–9	𝍸 𝍸 |

Sample answer for Product B: ≈76.7%

Product B	
Number	**Frequency**
0–7	𝍸 𝍸 𝍸 𝍸 |||
8–9	𝍸 ||

27c. Sample answer: Yes; the probability that Product B is effective is 13.4% higher than that of Product A.
27d. Sample answer: It depends on what the product is and how it is being used. For example, if the product is a pencil sharpener, then the lower price may be more important than the effectiveness, and therefore, might not justify the price difference. However, if the product is a life-saving medicine, the effectiveness may be more important than the price, and therefore, might justify the price difference. **29.** true
31 An invalid sampling method and type of sample can produce bias. For example, if a sample is not random, the person conducting the study can influence the results by selecting a specific sample of people. Also, if an experiment is used when an observational study is the more logical type of study to be used, the study can be unreliable. For example, if someone wants to analyze the speeds of vehicles on a specific stretch of highway and decides to place an empty police car on the side of the road, the data will be affected by the police car. The results of this study will show lower speeds than are normally driven on the highway. Biased survey questions and incorrect procedures can affect the reliability of a study as well. A survey question that is poorly written may result in a response that does not accurately reflect the opinion of the participant.
33. C **35.** G **37.** Step 1: $9^1 - 1 = 8$, which is divisible by 8. The statement is true for $n = 1$.
Step 2: Assume that $9^k - 1$ is divisible by 8 for some positive integer k. This means that $9^k - 1 = 8r$ for some whole number r.
Step 3:
$$9^k - 1 = 8r$$
$$9^k = 8r + 1$$
$$9^{k+1} = 72r + 9$$
$$9^{k+1} - 1 = 72r + 8$$
$$9^{k+1} - 1 = 8(9r + 1)$$
Since r is a whole number, $9r + 1$ is a whole number. Thus, $9^{k+1} - 1$ is divisible by 8, so the statement is true for $n = k + 1$. Therefore, $9^n - 1$ is divisible by 8 for all positive integers n. **39.** $\left(\frac{3}{2}, \frac{9}{2}\right)$, $(-1, 2)$
41. no solution **43.** $(\pm 8, 0)$ **45.** $3\sqrt{17}$ units
47. 25 units **49.** $\sqrt{70.25}$ units **51.** $-5c^5d^3$ **53.** an
55. $-y^3z^2$ **57.** $x^2 - 6x - 27 = 0$ **59.** $x^2 + x - 20 = 0$
61. 63

Lesson 11-2

1 **a.** First, press STAT ENTER and enter each data value. Then, press 2nd [STAT PLOT] ENTER ENTER and choose the histogram icon. Finally, adjust the window to dimensions appropriate for the data.

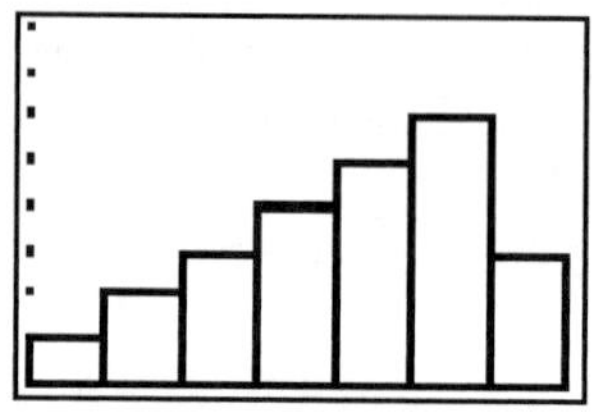

[4, 32] scl: 4 by [0, 8] scl: 1

Since the majority of the data is on the right and there is a tail on the left, the distribution is negatively skewed.
b. Sample answer: The distribution is skewed, so use the five-number summary. Press STAT ▶ ENTER ENTER and scroll down to display the statistics for the data set.

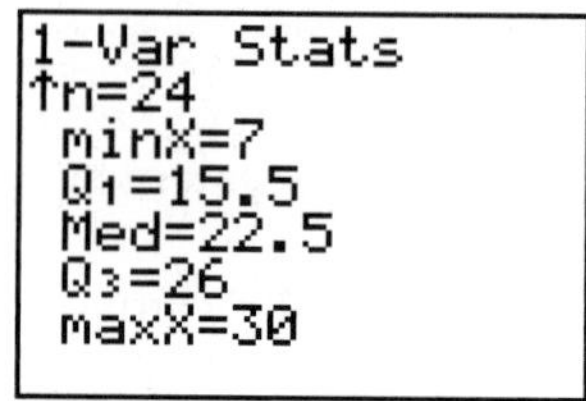

The range is 7 to 30 minutes. The median is 22.5 minutes, and half of the data are between 15.5 and 26 minutes.

3a.

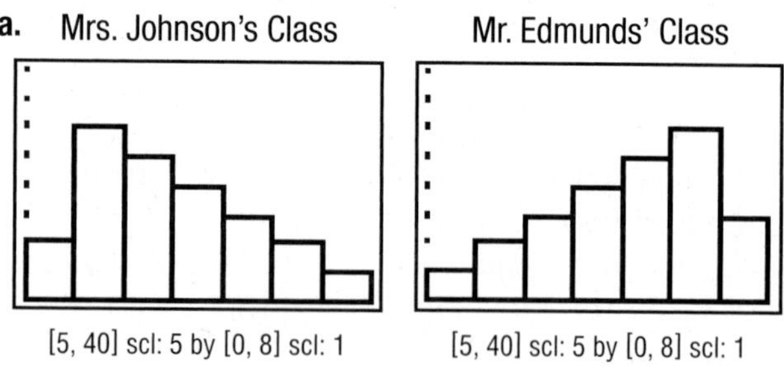

[5, 40] scl: 5 by [0, 8] scl: 1 [5, 40] scl: 5 by [0, 8] scl: 1

Mrs. Johnson's class, positively skewed; Mr. Edmunds' class, negatively skewed. **3b.** Sample answer: The distributions are skewed, so use the five-number summaries. The range for both classes is the same. However, the median for Mrs. Johnson's class is 17 and the median for Mr. Edmunds' class is 28. The lower quartile for Mr. Edmunds' class is 20. Since this is greater than the median for Mrs. Johnson's class, this means that 75% of the data from Mr. Edmunds' class is greater than 50% of the data from Mrs. Johnson's class. Therefore, we can conclude that the students in Mr. Edmunds' class had slightly higher sales overall than the students in Mrs. Johnson's class.

5a.

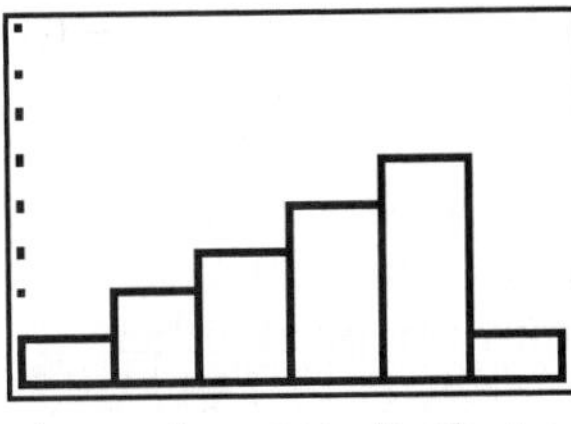

[50, 200] scl: 25 by [0, 8] scl: 1

[50, 200] scl: 25 by [0, 5] scl: 1

negatively skewed

5b. Sample answer: The distribution is skewed, so use the five-number summary. The range is 53 to 179 points. The median is 138.5 points, and half of the data are between 106.5 and 157 points.

7a. Sophomore Year

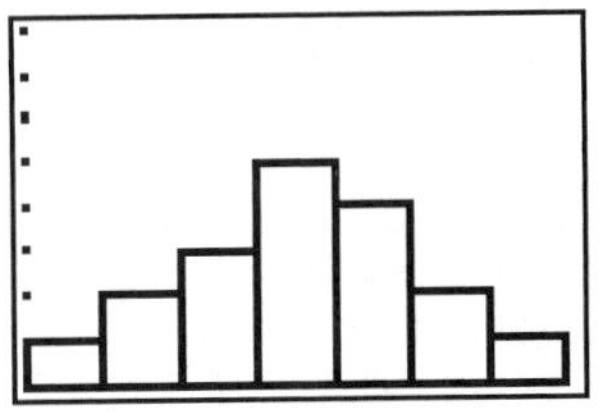

[1200, 1900] scl: 100 by [0, 8] scl: 1

Junior Year

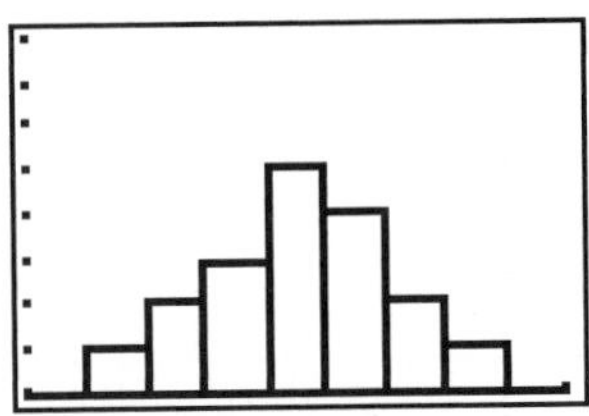

[1300, 2200] scl: 100 by [0, 8] scl: 1

both symmetric

7b. Sample answer: The distributions are symmetric, so use the means and standard deviations. The mean score for sophomore year is about 1552.9 with standard deviation of about 147.2. The mean score for junior year is about 1753.8 with standard deviation of about 159.1. We can conclude that the scores and the variation of the scores from the mean both increased from sophomore year to junior year.

9 **a.** Enter the tuitions for the public colleges as **L1**. Graph these data as **Plot1** by pressing 2nd **[STAT PLOT]** ENTER ENTER and choosing the box plot icon. Enter the tuitions for the private colleges as **L2**. Graph these data as **Plot2** by pressing 2nd **[STAT PLOT]** ▼ ENTER ENTER and choosing the box plot icon. For **Xlist**, enter **L2**. Adjust the window to dimensions appropriate for the data.

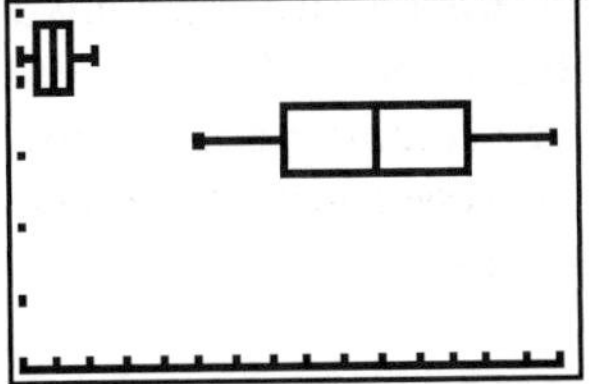

[3000, 18,000] scl: 1000 by [0, 5] scl: 1

For both sets of data, the whiskers are approximately equal, and the median is in the middle of the data. The distributions are symmetric.

b. Sample answer: The distributions are symmetric, so use the means and standard deviations. Press STAT ▶ ENTER ENTER to display the statistics for the public colleges.

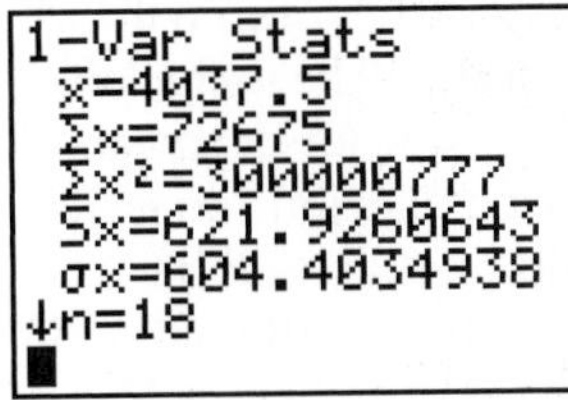

Press STAT ▶ ENTER 2nd **[L1]** ENTER to display the statistics for the private colleges

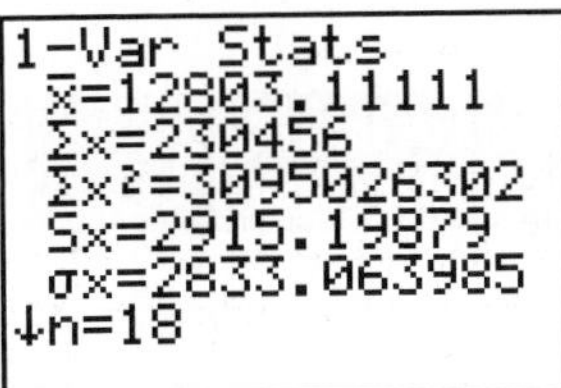

The mean for the public colleges is $4037.50 with standard deviation of about $621.93. The mean for private colleges is about $12,803.11 with standard deviation of about $2915.20. We can conclude that not only is the average cost of private schools far greater than the average cost of public schools, but the variation of the costs from the mean is also much greater.

11a.

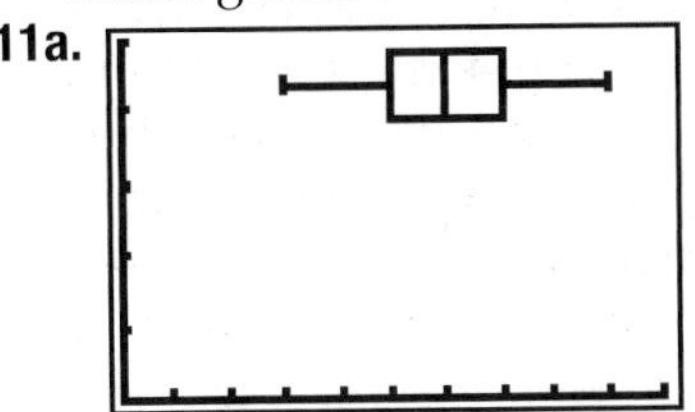

[0, 30] sc: 3 by [0, 5] scl: 1

Sample answer: The distribution is symmetric, so use the mean and standard deviation. The mean of the data is 18 with standard deviation of about 5.2 points.

11b.

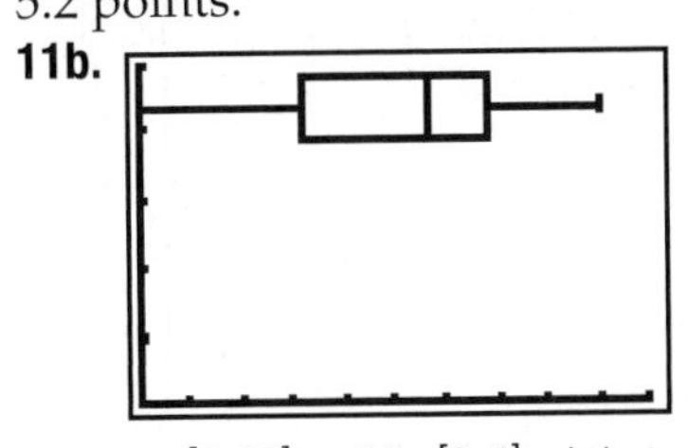

[0, 30] sc: 3 by [0, 5] scl: 1

mean: 14.6; median: 17

11c. Sample answer: Adding the scores from the first four games causes the shape of the distribution to go from being symmetric to being negatively skewed. Therefore, the center and spread should be described using the five-number summary.

13 a. Sample answer: Since the distribution is positively skewed, the median will be to the left of the mean closer to the majority of the data. An estimate for the median is 10. The mean will be more affected by the tail, and will be to the right of the majority of the data. An estimate for the mean is 14.

b. Sample answer: Since the distribution is negatively skewed, the median will be to the right of the mean closer to the majority of the data. An estimate for the median is 24. The mean will be more affected by the tail, and will be to the left of the majority of the data. An estimate for the mean is 20.

c. Sample answer: Since the distribution is symmetric, the mean and median will be approximately equal near the middle of the data. An estimate for the mean and median is 17.

15. Sample answer: The heights of the players on the Pittsburgh Steelers roster appear to represent a normal distribution.

Heights of the Players on the 2009 Pittsburgh Steelers Roster (inches)							
75	74	71	70	74	75	77	72
71	72	70	70	75	78	71	75
77	71	69	70	77	75	74	73
77	71	73	76	76	74	72	75
75	70	70	74	73	76	79	73
71	69	70	77	77	80	75	77
67	74	69	76	77	76		

The mean of the data is about 73.61 in. or 6 ft 1.61 in. The standard deviation is about 2.97 in.

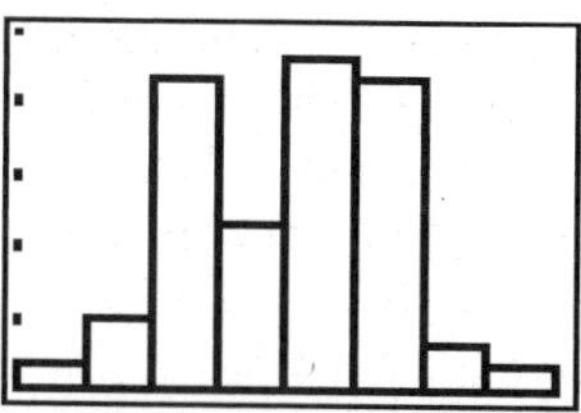

[66, 82] scl: 2 by [0, 15] scl: 3

The birth months of the players do not display central tendency.

Birth Months of the Players on the 2009 Pittsburgh Steelers Roster							
1	12	10	3	11	1	10	5
4	8	9	11	1	1	11	5
8	6	11	4	3	4	8	5
3	7	2	1	11	4	3	2
1	1	6	1	6	8	11	9
3	3	1	6	9	1	9	9
6	5	10	11	11	12		

[0, 12] scl: 2 by [0, 15] scl: 3

17. D **19.** H **21.** unbiased **23.** Biased; sample answer: The question encourages a certain response. The phrase "Don't you hate" encourages you to agree that gas prices are too high. **25a.** (39.2, ±4.4) **25b.** No; the comet and Pluto may not be at either point of intersection at the same time. **25c.** $\left(-\frac{5}{3}, -\frac{7}{3}\right)$, (1, 3) **25d.** (3, ±4), (−3, ±4) **27.** combination; 28 **29.** permutation; 120

Lesson 11-3

1. The random variable X is the number of pages linked to a Web page. The pages are finite and countable, so X is discrete. **3.** The random variable X is the amount of precipitation in a city per month. Precipitation can be anywhere within a certain range. Therefore, X is continuous.

5a.

Sum	Frequency	Relative Frequency
4	1	$\frac{1}{64}$
6	2	$\frac{1}{32}$
7	2	$\frac{1}{32}$
8	3	$\frac{3}{64}$
9	4	$\frac{1}{16}$
10	5	$\frac{5}{64}$
11	4	$\frac{1}{16}$
12	7	$\frac{7}{64}$
13	4	$\frac{1}{16}$
14	7	$\frac{7}{64}$
15	4	$\frac{1}{16}$
16	5	$\frac{5}{64}$
17	4	$\frac{1}{16}$
18	4	$\frac{1}{16}$
19	2	$\frac{1}{32}$
20	3	$\frac{3}{64}$
22	2	$\frac{1}{32}$
24	1	$\frac{1}{64}$

5b.

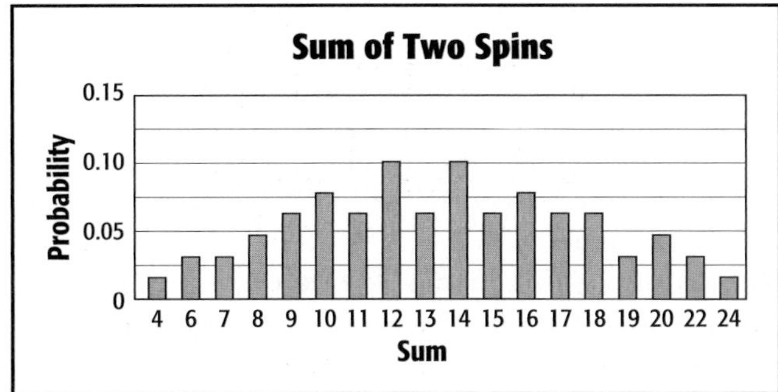

5c.

Sum	Frequency	Relative Frequency
4	1	$\frac{1}{64}$
6	1	$\frac{1}{64}$
7	3	$\frac{3}{64}$
8	3	$\frac{3}{64}$
9	2	$\frac{1}{32}$
10	6	$\frac{3}{32}$
11	5	$\frac{5}{64}$
12	7	$\frac{7}{64}$
13	5	$\frac{5}{64}$
14	8	$\frac{1}{8}$
15	3	$\frac{3}{64}$
16	3	$\frac{3}{64}$
17	4	$\frac{1}{16}$
18	5	$\frac{5}{64}$
19	2	$\frac{1}{32}$
20	3	$\frac{3}{64}$
22	2	$\frac{1}{32}$
24	1	$\frac{1}{64}$

5d.

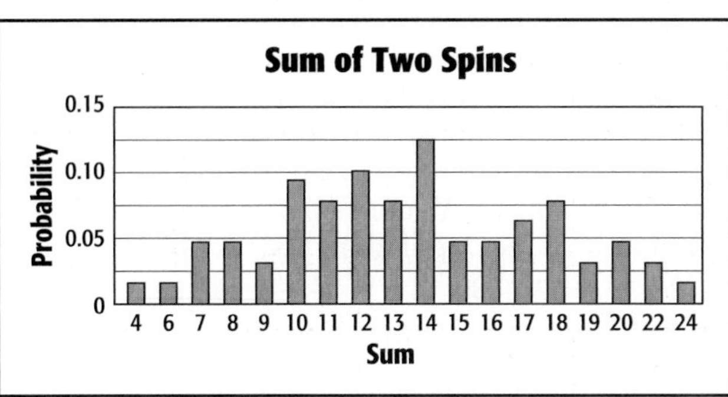

5e. 13.5 **5f.** 4.29 **7.** The random variable X is the number of diggs for a web page. The diggs are finite and countable, so X is discrete. **9.** The random variable X is the number of files infected by a computer virus. The files are finite and countable, so X is discrete.

11 Find the sum of the weighted values of each variable.

$$\begin{aligned}0(0.1) &= 0.00\\ 1(0.1) &= 0.10\\ 2(0.15) &= 0.30\\ 3(0.15) &= 0.45\\ 4(0.25) &= 1.00\\ 5(0.1) &= 0.50\\ 6(0.08) &= 0.48\\ 7(0.05) &= 0.35\\ +\ 8(0.02) &= 0.16\\ \hline &= 3.34\end{aligned}$$

$$\begin{aligned}E(X) &= \Sigma[X \cdot P(X)]\\ &= 3.34\end{aligned}$$

13a.

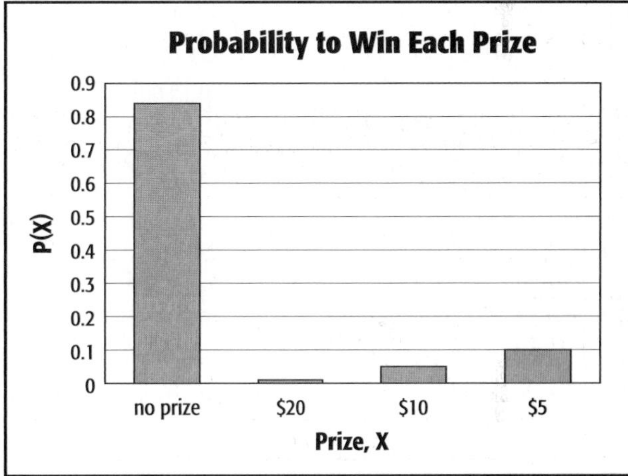

13b. \$1.20 **13c.** Sample answer: The expected value is positive, so a person buying a ticket can expect to win \$0.20 even after the cost of the ticket is considered. Thus, a person would want to participate in this raffle. On the other hand, this raffle is guaranteed to lose money for the organizers and they should change the distribution of prizes or not do the raffle. **15a.** 4.34; Sample answer: The expected number is 4.34, so we can expect there to be 4 upsets. We cannot have 0.34 upsets, so we round to the nearest whole number. **15b.** 1.90

15c.

Number of Upsets, X	Frequency	Relative Frequency
0	1	0.02
1	2	0.04
2	7	0.14
3	4	0.08
4	9	0.18

Number of Upsets, X	Frequency	Relative Frequency
5	17	0.34
6	4	0.08
7	6	0.12
8	0	0

15d.

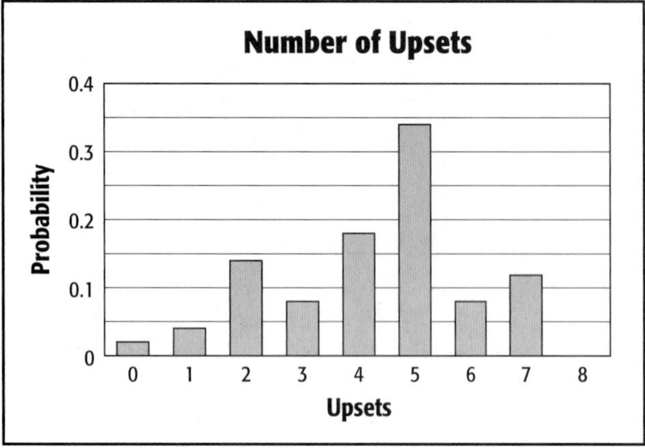

17 Find the expected value of each investment. Multiply the possible value of each fund by the associated probability. A profit is a positive value while a loss is a negative value.

Fund A:

$$\begin{aligned} 0.30(1900) &= 570 \\ 0.30(600) &= 180 \\ 0.15(-200) &= (-30) \\ +\ 0.25(-500) &= (-125) \\ &= 595 \end{aligned}$$

Fund B:

$$\begin{aligned} 0.40(1600) &= 640 \\ 0.10(900) &= 90 \\ 0.10(-300) &= (-30) \\ +\ 0.40(-400) &= (-160) \\ &= 540 \end{aligned}$$

The expected values of Funds A and B are \$595 and \$540, respectively. Calculate each standard deviation.

Fund A:

$$\begin{aligned} (1900 - 595)^2 \cdot 0.30 &= 510{,}907.50 \\ (600 - 595)^2 \cdot 0.30 &= 7.50 \\ (-200 - 595)^2 \cdot 0.15 &= 94{,}803.75 \\ +(-500 - 595)^2 \cdot 0.25 &= 299{,}756.25 \\ \Sigma[[(X - E(X)]^2 \cdot P(X)] &= 905{,}475.00 \\ \sqrt{905{,}475} &\approx 951.6 \end{aligned}$$

Fund B:

$$\begin{aligned} (1600 - 540)^2 \cdot 0.40 &= 449{,}440.00 \\ (900 - 540)^2 \cdot 0.10 &= 12{,}960.00 \\ (-300 - 540)^2 \cdot 0.10 &= 70{,}560.00 \\ +(-400 - 540)^2 \cdot 0.40 &= 353{,}440.00 \\ \Sigma[[(X - E(X)]^2 \cdot P(X)] &= 886{,}400.00 \\ \sqrt{886{,}400} &\approx 941.5 \end{aligned}$$

The expected value of Funds A and B is \$595 and \$540, respectively. The standard deviation for Fund A is about 951.6, while the standard deviation for Fund B is about 941.5. Since the standard deviations are about the same, the funds have about the same amount of risk. Therefore, with a higher expected value, Fund A is the better investment.

19. Sample answer: Liana; Shannon didn't consider every scenario in determining the total probability. For example, in calculating the probability of a sum of 5, she considered spinning a 3 then a 2, but not a 2 then a 3.

21 Ensure that each outcome is independent of the others and that the sum of the probabilities of the outcomes is one. Sample answer: A spinner with 5 equal-sided areas shaded red, blue, yellow, green, and brown.

Color	red	blue	yellow	green	brown
Probability	0.2	0.2	0.2	0.2	0.2

Color	red	blue	yellow	green	brown
Probability	0.2	0.2	0.2	0.2	0.2

23. Sample answer: A discrete probability distribution can be the uniform distribution of the roll of a die. In this type of distribution, there are only a finite number of possibilities. A continuous probability distribution can be the distribution of the lives of 400 batteries. In this distribution, there are an infinite number of possibilities. **25.** 2.4 **27.** H

29a.

Peter's Articles

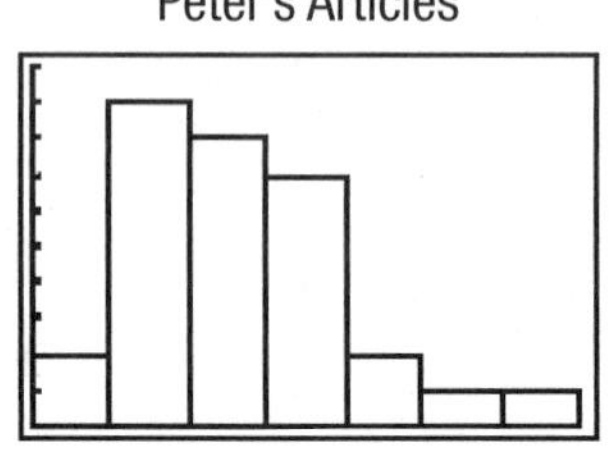

[0, 70] scl: 5 by [0, 10] scl: 1

Paul's Articles

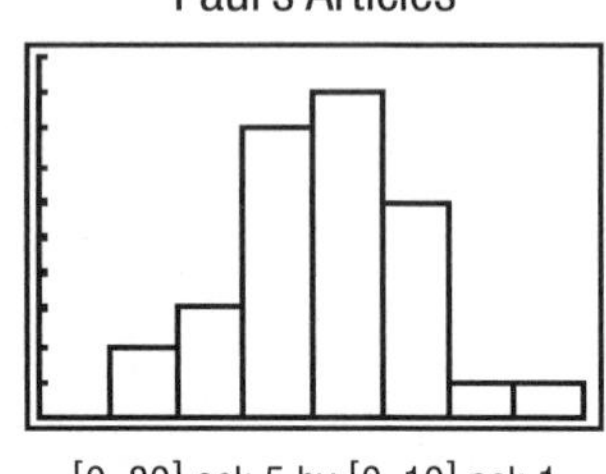

[0, 80] scl: 5 by [0, 10] scl: 1

Peter's articles, positively skewed; Paul's articles, symmetric

29b. Sample answer: One of the distributions is symmetric and the other is skewed, so use the five-number summaries. The range for Peter's articles is 64, and the range for Paul's articles is 53. However, the upper quartile for Peter's is 33, while the lower quartile for Paul's is 34. This means that 75% of Paul's articles have more likes (and are more popular) than 75% of Peter's articles. Therefore, we can conclude that Paul's articles are more popular overall. **31.** Sample answer: This situation calls for a survey because the data will be collected from responses from members of a sample of the population. **33.** 0.5, 1.25, 3.125, 7.8125, 19.53125 **35.** 12, 4, $\frac{4}{3}$, $\frac{4}{9}$, $\frac{4}{27}$ **37.** 80, 100, 125, $\frac{625}{4}$, $\frac{3125}{16}$ **39.** 27 **41.** 3

43. $m^4 + 4m^3n + 6m^2n^2 + 4mn^3 + n^4$

1. This experiment cannot be reduced to a binomial experiment because there are more than two possible outcomes. **3.** This experiment can be reduced to a binomial experiment. Success is yes, failure is no, a trial is asking a student, and the random variable is the number of yeses; $n = 30$, $p = 0.72$, $q = 0.28$.

5 Find the probability that the spinner lands on WIN three times out of five spins. Since the spinner is divided into eight equal spaces and WIN is on two of them, $p = \frac{2}{8}$ or 0.25. So, $q = 1 - 0.25$ or 0.75 and $n = 5$.

$P(X) = {}_nC_X p^X q^{n-X}$

$P(3) = {}_5C_3(0.25)^3(0.75)^{5-3}$

≈ 0.0879

The probability of Aiden receiving three prizes is about 0.088 or 8.8%. So, the correct answer is D.

7. This experiment can be reduced to a binomial experiment. Success is a day that it rains, failure is a day it does not rain, a trial is a day, and the random variable is the number of days it rains; $n =$ the number of days in the month, $p = 0.35$, $q = 0.65$.

9. This experiment cannot be reduced to a binomial experiment because the events are not independent. The probability of choosing the hat that covers the ball changes after each selection.

11. Sample answer:

Step 1 A trial is pulling out a marble. The simulation will consist of 20 trials.

Step 2 A success is pulling out a red marble. The probability of success is $\frac{5}{12}$ and the probability of failure is $\frac{7}{12}$.

Step 3 The random variable X represents the number of red marbles pulled out in 20 trials.

Step 4 Use a random number generator. Let 0–4 represent pulling out a red marble. Let 5–11 represent all other outcomes. Make a frequency table and record the results as you run the generator.

Outcome	Tally	Frequency
Red Marble	𝍸 𝍸	10
Other Outcomes	𝍸 𝍸	10

The experimental probability is $\frac{10}{20}$ or 50%. This is greater than the theoretical probability of $\frac{5}{12}$ or about 41.7%.

13. Sample answer:

Step 1 A trial is drawing a card from a deck. The simulation will consist of 20 trials.

Step 2 A success is drawing a face card. The probability of success is $\frac{3}{13}$ and the probability of failure is $\frac{10}{13}$.

Step 3 The random variable X represents the number of face cards drawn in 20 trials.

Step 4 Use a random number generator. Let 0–2 represent drawing a face card. Let 3–12 represent all other outcomes. Make a frequency table and record the results as you run the generator.

Outcome	Tally	Frequency
Face Card	||	2
Other Cards	𝍸 𝍸 𝍸 |||	18

The experimental probability is $\frac{2}{20}$ or 10%. This is less than the theoretical probability of $\frac{3}{13}$ or about 23.1%.

15. 0.183 or 18.3% **17.** 0.096 or 9.6%

19 Find the probability that the kicker makes 7 of his next 10 kicks from within 35 yards. A success is making a field goal, so $p = 0.75$, $q = 1 - 0.75$ or 0.25, and $n = 10$.

$P(X) = {}_nC_X p^X q^{n-X}$

$P(7) = {}_{10}C_7(0.75)^7(0.25)^{10-7}$

≈ 0.25

The probability of the kicker making 7 of his next 10 kicks from within 35 yards is about 0.25 or 25%.

21a. 0 own a laptop, 0.0006 or 0.06%; 1 owns a laptop, 0.007 or 0.7%, 2 own a laptop, 0.0343 or 3.43%; 3 own a laptop, 0.0991 or 9.91%; 4 own a laptop, 0.1878 or 18.78%; 5 own a laptop, 0.2441 or 24.41%; 6 own a laptop, 0.2204 or 22.04%; 7 own a laptop, 0.1364 or 13.64%; 8 own a laptop, 0.0554 or 5.54%; 9 own a laptop, 0.0133 or 1.33%; 10 own a laptop, 0.0014 or 0.14% **21b.** 0.0701 or 7.01% **21c.** 5 **23a.** 0 own vinyl records, 0.001 or 0.1%; 1 owns vinyl records, 0.012 or 1.2%; 2 own vinyl records, 0.058 or 5.8%; 3 own vinyl records, 0.152 or 15.2%; 4 own vinyl records, 0.253 or 25.3%; 5 own vinyl records, 0.268 or 26.8%; 6 own vinyl records, 0.178 or 17.8%; 7 own vinyl records, 0.067 or 6.7%; 8 own vinyl records, 0.011 or 1.1% **23b.** 0.256 or 25.6% **23c.** 5

25. 0.015 or 1.5% **27a.** about 7.8%

27b. Sample answer: They can roll a six-sided die.

29 Find the mean when $n = 8$ and $p = 0.6$.

$\mu = np$

$= 8(0.6)$ or 4.8

Since there cannot be a fraction of a success, the most likely number of successes is 5.

31. 5 **33.** 7 **35.** 0.603 or 60.3% **37.** 0.322 or 32.2% **39.** 0.99 or 99% **41.** 0.889 or 88.9% **43.** Sample answer: You should consider the type of situation for which the binomial distribution is being used. For example, if a binomial distribution is being used to predict outcomes regarding an athletic event, the probabilities of sucess and failure could change due to other variables such as weather conditions or player health. So, binomial distributions should be used cautiously when making decisions involving events that are not completely random. **45.** Sample answer: A full binomial distribution can be determined by expanding the binomial, which itself utilizes Pascal's triangle. **47a.** 0.003 or 0.3%

48b. 0.00003 or 0.003% **48c.** 0.056 or 5.6%

48d. 0.25 or 25% **49.** H **51.** The random variable X

is the number of customers at an amusement park. The customers are finite and countable, so X is discrete. **53.** The random variable X is the number of hot dogs sold at a sporting event. The hot dogs are finite and countable, so X is discrete.

55a. 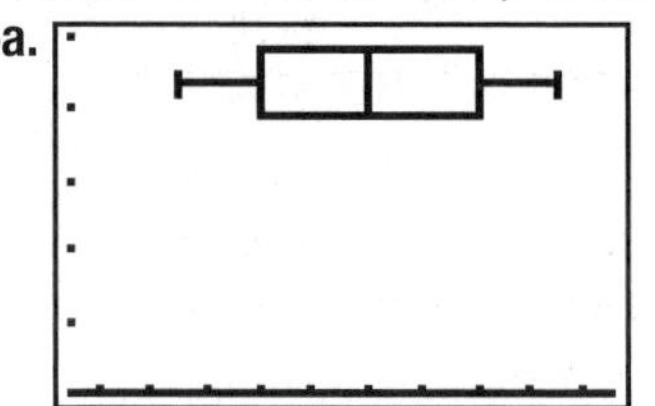 symmetric

[5, 25] scl: 2 by [0, 5] scl: 1

55b. Sample answer: The distribution is symmetric, so use the mean and standard deviation. The mean is about \$16.02 with standard deviation of about \$4.52. **57.** 6 **59.** −57 **61.** 20 **63.** $-x = \ln 5$ **65.** $e^1 = e$ **67.** $x + 1 = \ln 9$ **69.** $e^{2x} = \frac{7}{3}$ **71a.** 0.36 **71b.** 0.42 **71c.** 0.05

Lesson 11-5

1. $251 < X < 581$ **3a.** about 386 teens **3b.** 84% **5.** 2.05 **7.** 3.08 **9.** $X < 15.9$ or $X > 42.7$ **11a.** 509 **11b.** 0.15% **13.** −1.33 **15.** 177.7

17 **a.** Find the z-values associated with 90,000 and 110,000. The mean μ is 113,627 and the standard deviation σ is 14,266.

$$z = \frac{X - \mu}{\sigma} = \frac{90{,}000 - 113{,}627}{14{,}266} \approx -1.656$$

$$z = \frac{X - \mu}{\sigma} = \frac{110{,}000 - 113{,}627}{14{,}266} \approx -0.254$$

Use a graphing calculator to find the area between the z-values.

```
normalcdf(-1.656
,-.254)
          .3508869415
```

The value 0.35 is the percentage of batteries that will last between 90,000 and 110,000 miles. The total number of batteries in this group is 0.35 × 20,000 or 7000.

b. Find the z-value associated with 125,000.

$$z = \frac{X - \mu}{\sigma} = \frac{125{,}000 - 113{,}627}{14{,}266} \approx 0.797$$

We are looking for values greater than 125,000, so we can use a graphing calculator to find the area between $z = 0.797$ and $z = 4$.

```
normalcdf(.797,4
)
          .2126937641
```

The value 0.21 is the percentage of batteries that will last between more than 125,000 miles. The total number of batteries in this group is 0.21 × 20,000 or 4200.

c. Find the z-value associated with 100,000.

$$z = \frac{X - \mu}{\sigma} = \frac{100{,}000 - 113{,}627}{14{,}266} \approx -0.955$$

We are looking for values less than 100,000, so we can use a graphing calculator to find the area between $z = -4$ and $z = -0.955$.

```
normalcdf(-4,-.9
55)
          .1697571502
```

The probability that if you buy a car battery at random, it will last less than 100,000 miles is about 17.0%.

19 **a.** The middle 68% represents all data values within one standard deviation of the mean. Add ±\$115 to \$829. The range of rates is \$714 to \$944.

b. Find the z-value associated with 1000.

$$z = \frac{X - \mu}{\sigma} = \frac{1000 - 829}{115} \approx 1.487$$

We are looking for values more than 1000, so we can use a graphing calculator to find the area between $z = 1.487$ and $z = 4$.

```
normalcdf(1.487,
4)
          .0684757505
```

The probability that a customer selected at random will pay more than \$1000 is about 6.8%. Out of 900 people, about 0.068 • 900 or 62 people will pay more than \$1000.

c. Sample answer: I would expect people with several traffic citations to lie to the far right of the distribution where insurance costs are highest, because I think insurance companies would charge them more.

d. Sample answer: I think auto insurance companies would charge younger people more than older people because they have not been driving as long. I think they would charge more for expensive cars and sports cars and less for cars that have good safety ratings. I think they would charge a person less if they have a good driving record and more if they have had tickets and accidents.

21. Sample answer: Hiroko; Monica's solution would work with a uniform distribution. **23.** Sample answer: True; according to the Empirical Rule, 99% of the data lie within 3 standard deviations of the mean. Therefore, only 1% will fall outside of three sigma. An infinitely small amount will fall outside of six-sigma. **25.** Sample answer: The scores per team in each game of the first round of the 2010 NBA playoffs. The mean is 96.56 and the standard deviation is 11.06. The middle 68% of the distribution is $85.50 < X < 107.62$. The middle 95% is $74.44 < X < 118.68$. The middle 99.7% is $63.38 < X < 129.74$. **27.** D **29.** 32.5 **31.** 17.3% **33.** The random variable X is the amount of precipitation in a city per month. Precipitation can be anywhere within a certain range. Therefore, X is continuous. **35.** greatest integer **37.** constant

Lesson 11-6

1. 0.096 **3.** H_0: $\mu \geq 2$; H_a: $\mu < 2$ (claim) **5.** H_0: $\mu \geq 20$ (claim); H_a: $\mu < 20$ **7.** H_0: $\mu \geq 84$ (claim), H_a: $\mu < 84$; Do not reject H_0; The manufacturer's claim that the discs can hold at least 84 minutes cannot be rejected.

9 The sample size n is 76 and the standard deviation s is 5.2. The mean is not needed to calculate the maximum error of estimate. The z-values which correspond to 95% significance are ±1.96.

$$E = z \cdot \frac{s}{\sqrt{n}}$$
$$= 1.96 \cdot \frac{5.2}{\sqrt{76}}$$
$$\approx 1.17$$

The maximum error of estimate is about 1.17.

11. H_0: $\mu \geq 6$ (claim); H_a: $\mu < 6$ **13.** H_0: $\mu \leq 2$ (claim); H_a: $\mu > 2$ **15.** H_0: $\mu \geq 30$; H_a: $\mu < 30$ (claim); Do not reject H_0; There is not enough evidence to support the pizza chain's claim of a delivery time of less than 30 minutes. **17.** H_0: $\mu = 12$; H_a: $\mu \neq 12$; The mean of the sample data is 12.9 with a standard deviation of about 1.08. The z-statistic is about 5.27, which falls in the critical region at 1% significance. Therefore, the null hypothesis is rejected and the company should not make the claim on the label.

19a.

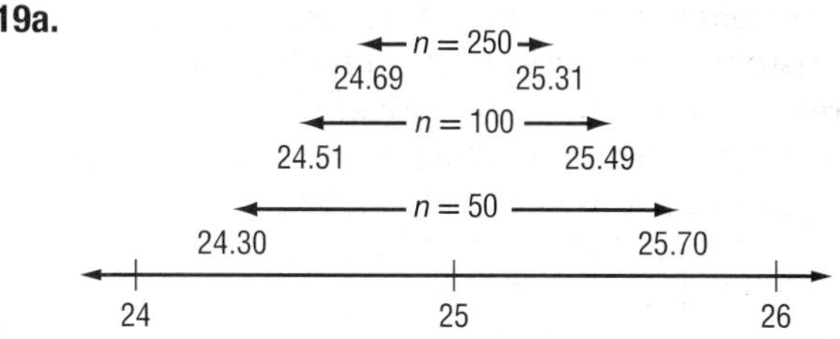

19b. Sample answer: With everything else held constant, increasing the same size will decrease the size of the confidence interval.

19c.

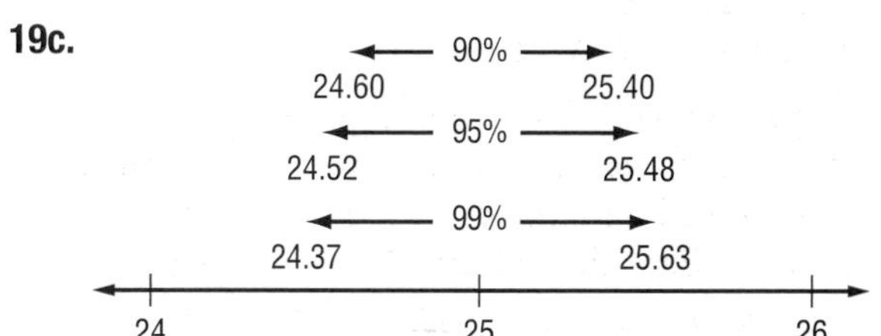

19d. Sample answer: With everything else held constant, increasing the confidence level will decrease the size of the confidence interval.

19e. Sample answer: Expanding the confidence interval reduces the accuracy of the estimate.

21 The range of values for the mean weight is 19.932 to 20.008. Half of this range is the maximum error of estimate.

$$\frac{20.008 - 19.932}{2} = 0.038.$$

Use the formula for calculating the maximum error of estimate to determine the same size. The z-values which correspond to 95% significance are ±1.96.

$$E = z \cdot \frac{s}{\sqrt{n}}$$
$$0.038 = 1.96 \cdot \frac{0.128}{\sqrt{n}}$$
$$\frac{0.038}{1.96} = \frac{0.128}{\sqrt{n}}$$
$$\sqrt{n} \cdot \frac{0.038}{1.96} = 0.128$$
$$\sqrt{n} = 0.128 \div \frac{0.038}{1.96}$$
$$(\sqrt{n})^2 = \left(0.128 \div \frac{0.038}{1.96}\right)^2$$
$$n \approx 44$$

23. Sample answer: You can use a statistical test to help you determine the strength of your decision. **25.** C **27.** B **29.** 27 students **31.** 729 **33.** 1

35. parabola

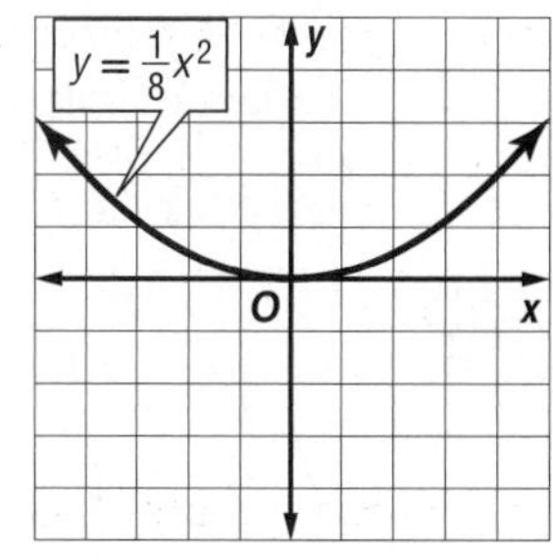

37. $y = 0.8x$ **39.** $y = -4$ **41.** 25

Selected Answers and Solutions

Chapter 11 Study Guide and Review

1. probability distribution **3.** parameter **5.** observational study **7.** survey; sample: every tenth shopper; population: all potential shoppers **9.** survey; sample: every fifth person; population: student body

11a. 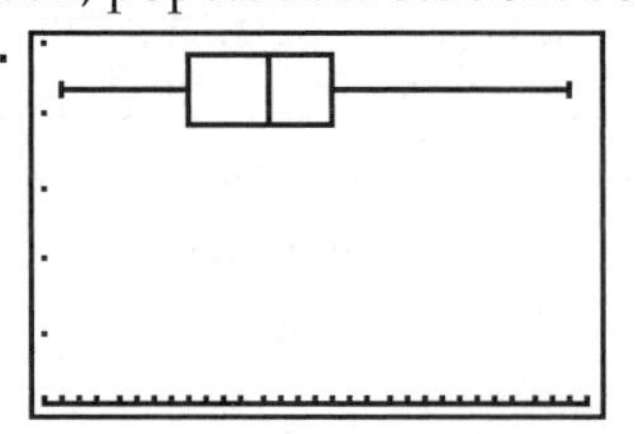 positively skewed

[300, 330] scl: 1 by [0, 5] scl: 1

11b. Sample answer: The distribution is skewed, so use the five-number summary. Kelly's times range from 301 to 329 seconds. The median is 311, and half of the data are between 307 and 316 seconds. **13.** The random variable X is the time it takes to run the race. Time can be anywhere within a certain range. Therefore, X is continuous. **15.** 1.88 snow days **17.** This experiment cannot be reduced to a binomial experiment because there are more than two possible outcomes. **19.** 34.3% **21a.** 68.5% **21b.** about 2 games **23.** $63.5 \leq \mu \leq 65.1$ **25.** H_0: $\mu \geq 45$ (claim); H_a: $\mu < 45$

CHAPTER 12

Trigonometric Functions

Chapter 12 Get Ready

1. 11.7 **3.** 20.5 **5.** $x = 9, y = 9\sqrt{2}$ **7.** $x = 12, y = 12\sqrt{3}$

Lesson 12-1

1. $\sin B = \frac{4}{5}$; $\cos B = \frac{3}{5}$; $\tan B = \frac{4}{3}$; $\csc B = \frac{5}{4}$; $\sec B = \frac{5}{3}$; $\cot B = \frac{3}{4}$ **3.** $\sin A = \frac{\sqrt{33}}{7}$, $\csc A = \frac{7\sqrt{33}}{33}$, $\cot A = \frac{4\sqrt{33}}{33}$, $\sec A = \frac{7}{4}$, $\tan A = \frac{\sqrt{33}}{4}$

5 The side opposite the 60° angle is given. The missing measure is the hypotenuse. Use the sine function to find x.

$\sin \theta = \frac{\text{opp}}{\text{hyp}}$	Sine function
$\sin 60° = \frac{22}{x}$	$\theta = 60°$ and opp = 22
$\frac{\sqrt{3}}{2} = \frac{22}{x}$	$\sin 60° = \frac{\sqrt{3}}{2}$
$\sqrt{3}x = 44$	Cross multiply.
$x = \frac{44}{\sqrt{3}}$	Divide each side by $\sqrt{3}$.
$x \approx 25.4$	Use a calculator.

7. 8.3 **9.** 25.4 **11.** about 274.7 ft **13.** $\sin \theta = \frac{12}{13}$; $\cos \theta = \frac{5}{13}$; $\tan \theta = \frac{12}{5}$; $\csc \theta = \frac{13}{12}$; $\sec \theta = \frac{13}{5}$; $\cot \theta = \frac{5}{12}$ **15.** $\sin \theta = \frac{\sqrt{51}}{10}$; $\cos \theta = \frac{7}{10}$; $\tan \theta = \frac{\sqrt{51}}{7}$; $\csc \theta = \frac{10\sqrt{51}}{51}$; $\sec \theta = \frac{10}{7}$; $\cot \theta = \frac{7\sqrt{51}}{51}$

17. $\sin A = \frac{8}{17}$, $\cos A = \frac{15}{17}$, $\csc A = \frac{17}{8}$, $\sec A = \frac{17}{15}$, $\cot A = \frac{15}{8}$ **19.** $\sin B = \frac{3\sqrt{10}}{10}$, $\cos B = \frac{\sqrt{10}}{10}$, $\csc B = \frac{\sqrt{10}}{3}$, $\sec B = \sqrt{10}$, $\cot B = \frac{1}{3}$ **21.** 12.7

23

$\tan \theta = \frac{\text{opp}}{\text{adj}}$	Tangent function
$\tan 30° = \frac{x}{18}$	Replace θ with 30°, opp with x, and adj with 18.
$\frac{\sqrt{3}}{3} = \frac{x}{18}$	$\tan 30° = \frac{\sqrt{3}}{3}$
$\frac{18\sqrt{3}}{3} = x$	Multiply each side by 18.
$10.4 \approx x$	Use a calculator.

25. 8.7 **27.** 132.5 ft **29.** 30 **31.** 36.9 **33.** 32.5 **35.** 25.3 ft higher **37.** $x = 21.9, y = 20.8$ **39.** $x = 19.3, y = 70.7$ **41.** 54.9 **43.** 20.5 **45.** 11.5

47

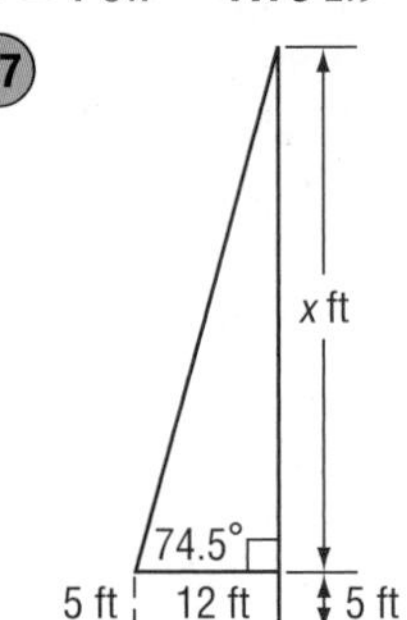

$\tan \theta = \frac{\text{opp}}{\text{adj}}$	Tangent function
$\tan 74.5° = \frac{x}{12}$	Replace θ with 74.5°, opp with x, and adj with 12.
$12 \cdot \tan 74.5° = x$	Multiply each side by 12.
$43 \approx x$	Use a calculator.

So, the height of the bird's nest is 43 + 5 or 48 feet.

49a. about 647.2 ft **49b.** about 239.4 ft **51.** $m\angle A = 59°, a = 31.6, c = 36.9$ **53.** $m\angle A = 38.7°$, $m\angle B = 51.3°, b = 7.5, c = 9.6$ **55.** True; $\sin \theta = \frac{\text{opp}}{\text{hyp}}$ and the values of the opposite side and the hypotenuse of an acute triangle are positive, so the value of the sine function is positive. **57.** Sample answer: The slope describes the ratio of the vertical rise to the horizontal run of the roof. The vertical rise is opposite the angle that the roof makes with the horizontal. The horizontal run is the adjacent side. So, the tangent of the angle of elevation equals the ratio of the rise to the run, or the slope of the roof; $\theta = 33.7°$. **59.** 24 **61.** G **63.** H_0: $\mu = 3$ (claim); H_a: $\mu \neq 3$ **65a.** 99.85% **65b.** 2.5% **65c.** 64% **67.** 22,704 feet **69.** 35 dollars **71.** $216\frac{2}{3}$ centimeters

Lesson 12-2

1. **3.**

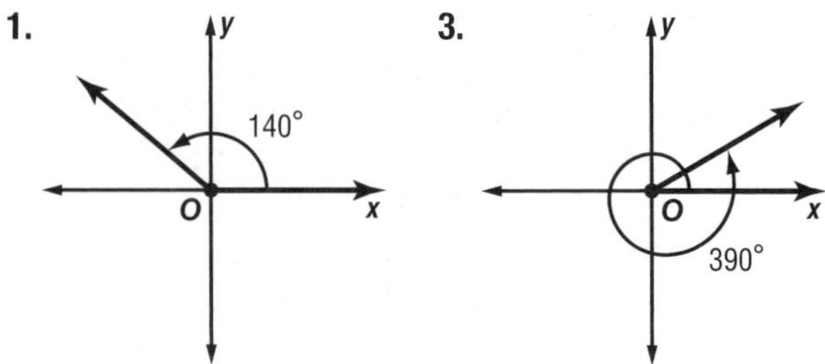

Selected Answers and Solutions

5 Sample answer:
positive angle: $175° + 360° = 535°$
negative angle: $175° - 360° = -185°$

7. $45°$ **9.** $-\frac{2\pi}{9}$

11. **13.**

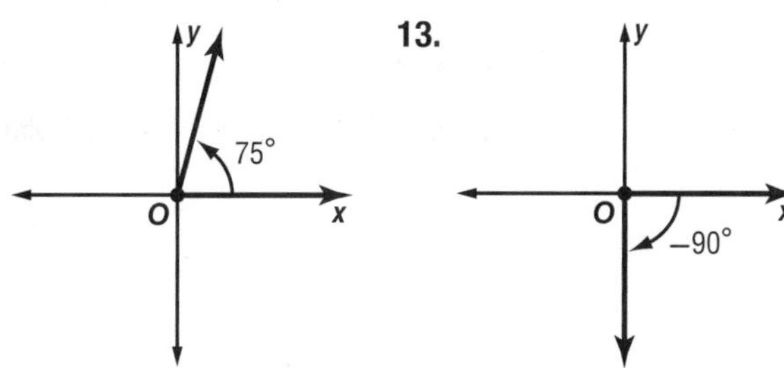

15. **17.**

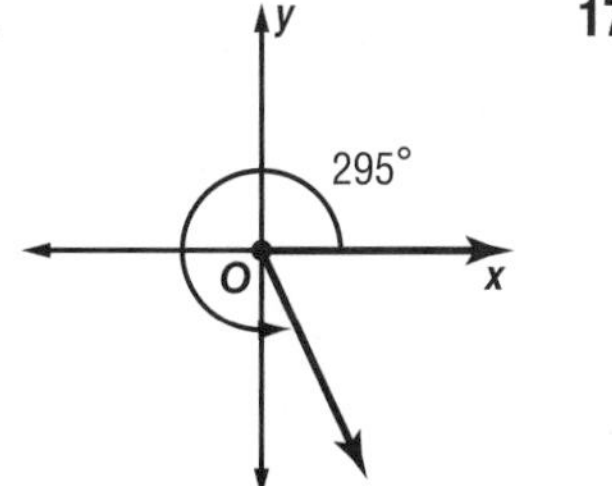

240°

19. Sample answer: $410°, -310°$ **21.** Sample answer: $565°, -155°$ **23.** Sample answer: $280°, -440°$

25 $330° = 330 \cdot \frac{\pi \text{ radians}}{180°}$
$= \frac{330\pi}{180}$ or $\frac{11\pi}{6}$ radians

27. $-60°$ **29.** $\frac{19\pi}{18}$ **31.** about 12.6 ft **33.** 6.7 cm
35. 1 h 15 min **37.** Sample answer: $260°, -100°$
39. Sample answer: $\frac{5\pi}{4}, -\frac{11\pi}{4}$

41 **a.**

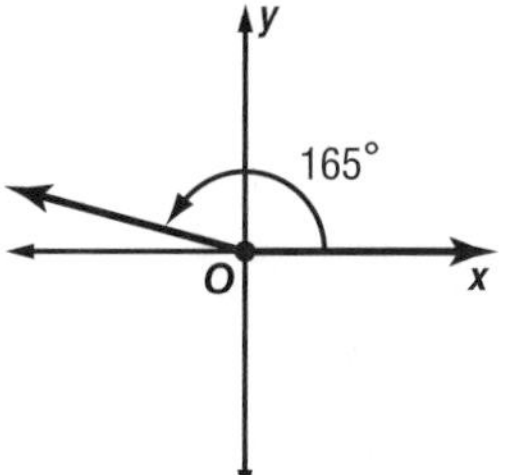

b. $165° = 165 \cdot \frac{\pi \text{ radians}}{180°}$
$= \frac{165\pi}{180}$ or $\frac{11\pi}{12}$ radians

c. $s = r\theta$ Formula for arc length
$= 6.5 \cdot \frac{11\pi}{12}$ $r = 6.5$ and $\theta = \frac{11\pi}{12}$
≈ 18.7 ft Use a calculator.

d. The arc length would double. Since $s = r\theta$, if r is doubled and θ remains unchanged, then the value of s is also doubled.

43. $472.5°$ **45.** $-\frac{10\pi}{9}$ **47a.** $\frac{\pi}{6}$ **47b.** 2.1 ft **49.** $x = 2$

51. Sample answer: $440°$ and $-280°$

80°

53. $\frac{\theta}{2\pi} = \frac{s}{2\pi r}$ Substitute.
$2\pi r\theta = 2\pi s$ Find the cross products.
$r\theta = s$ Divide each side by 2π.

One degree represents an angle measure that equals $\frac{1}{360}$ rotation around a circle. One radian represents the measure of an angle in standard position that intercepts an arc of length r. To change from degrees to radians, multiply the number of degrees by $\frac{\pi \text{ radians}}{180°}$. To change from radians to degrees, multiply the number of radians by $\frac{180°}{\pi \text{ radians}}$.

55. A **57.** B **59.** $\sin \theta = \frac{\sqrt{259}}{22}$, $\cos \theta = \frac{15}{22}$, $\tan \theta = \frac{\sqrt{259}}{15}$, $\csc \theta = \frac{22}{\sqrt{259}}$ or $\frac{22\sqrt{259}}{259}$, $\sec \theta = \frac{22}{15}$, $\cot \theta = \frac{15}{\sqrt{259}}$ or $\frac{15\sqrt{259}}{259}$ **61.** $H_0: \mu \geq 8$ (claim); $H_a: \mu < 8$ **63a.** 50% **63b.** 815 **63c.** 25
65. $3\sqrt{41}$ **67.** $\sqrt{317}$

Lesson 12-3

1. $\sin \theta = \frac{2\sqrt{5}}{5}$, $\cos \theta = \frac{\sqrt{5}}{5}$, $\tan \theta = 2$, $\csc \theta = \frac{\sqrt{5}}{2}$, $\sec \theta = \sqrt{5}$, $\cot \theta = \frac{1}{2}$ **3.** $\sin \theta = -1$, $\cos \theta = 0$, $\tan \theta =$ undefined, $\csc \theta = -1$, $\sec \theta =$ undefined, $\cot \theta = 0$

5. $65°$

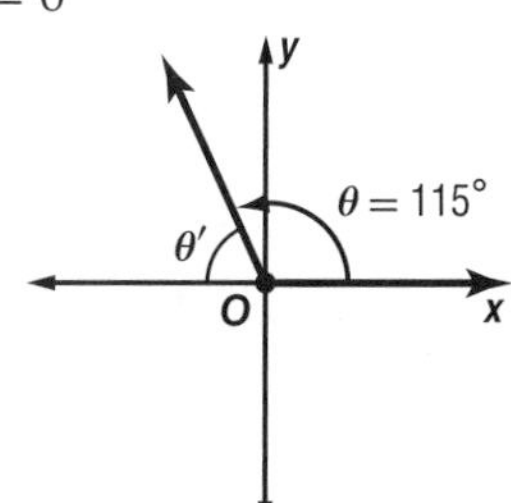

7. $\frac{\sqrt{2}}{2}$ **9.** -2

11a.

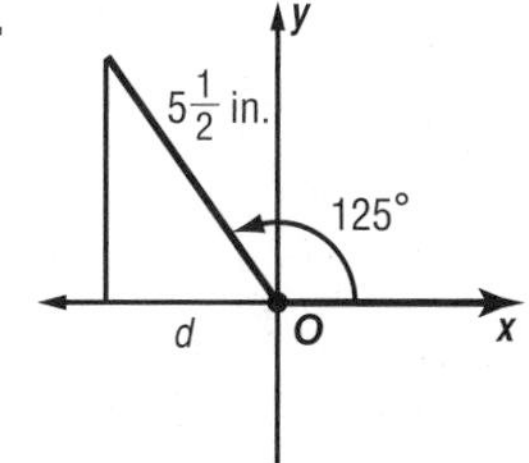

11b. $55°$; $\cos 55° = \frac{d}{5\frac{1}{2}}$
11c. 3.2 in.

13

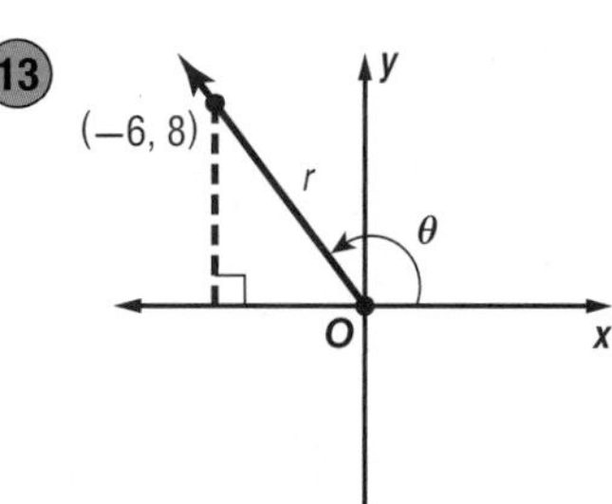

$r = \sqrt{x^2 + y^2}$
$= \sqrt{(-6)^2 + 8^2}$
$= \sqrt{100}$ or 10

Use $x = -6$, $y = 8$, and $r = 10$.
$\sin \theta = \frac{y}{r}$ $\cos \theta = \frac{x}{y}$

$= \frac{8}{10}$ or $\frac{4}{5}$ $\qquad$ $= \frac{-6}{10}$ or $-\frac{3}{5}$

$\tan\theta = \frac{y}{x}$ $\qquad$ $\csc\theta = \frac{r}{y}$

$= \frac{8}{-6}$ or $-\frac{4}{3}$ $\qquad$ $= \frac{10}{8}$ or $\frac{5}{4}$

$\sec\theta = \frac{r}{x}$ $\qquad$ $\cot\theta = \frac{x}{y}$

$= \frac{10}{-6}$ or $-\frac{5}{3}$ $\qquad$ $= \frac{-6}{8}$ or $-\frac{3}{4}$

15. $\sin\theta = -1$, $\cos\theta = 0$, $\tan\theta =$ undefined, $\csc\theta = -1$, $\sec\theta =$ undefined, $\cot\theta = 0$

17. $\sin\theta = -\frac{\sqrt{10}}{10}$, $\cos\theta = -\frac{3\sqrt{10}}{10}$, $\tan\theta = \frac{1}{3}$, $\csc\theta = -\sqrt{10}$, $\sec\theta = -\frac{\sqrt{10}}{3}$, $\cot\theta = 3$

19. 75° **21.** $\frac{\pi}{4}$

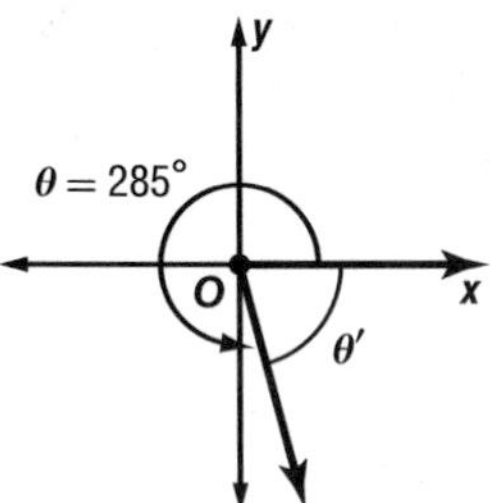

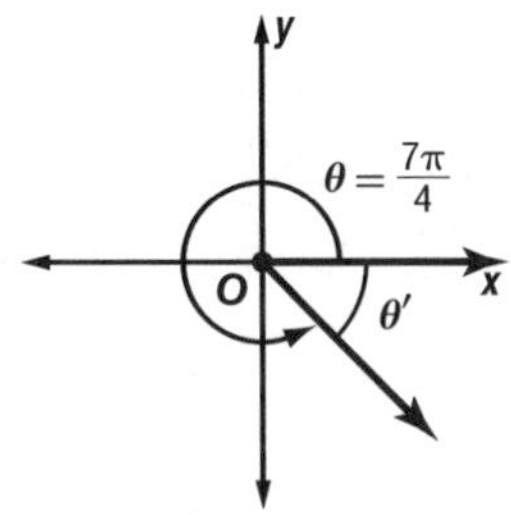

23. 40°

25. −1 **27.** $-\sqrt{2}$ **29.** $\frac{1}{2}$ **31.** $\frac{2\sqrt{3}}{3}$

33

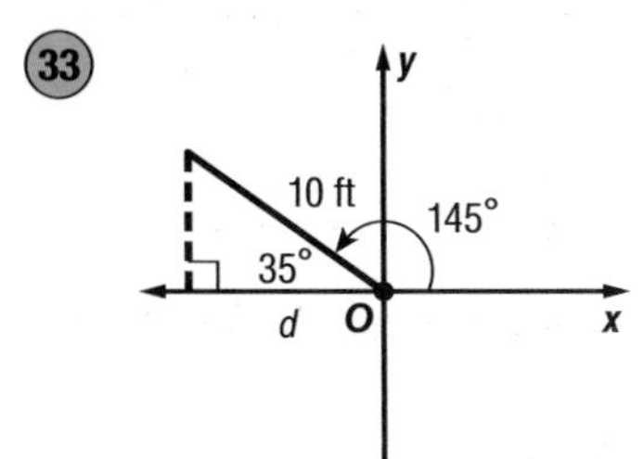

$\cos 35° = \frac{d}{10}$

$10 \cdot \cos 35° = d$

$8.2 \approx d$

The water reaches about 8.2 feet to the left of the sprinkler.

35. about 10.1 m **37.** $\cos\theta = -\frac{3}{5}$, $\tan\theta = -\frac{4}{3}$, $\csc\theta = \frac{5}{4}$, $\sec\theta = -\frac{5}{3}$, $\cot\theta = -\frac{3}{4}$ **39.** $\sin\theta = -\frac{15}{17}$, $\tan\theta = \frac{15}{8}$, $\csc\theta = -\frac{17}{15}$, $\sec\theta = -\frac{17}{8}$, $\cot\theta = \frac{8}{15}$

41. 0 **43.** $-\frac{1}{2}$ **45.** $\frac{\sqrt{3}}{2}$ **47.** No; for $\sin\theta = \frac{\sqrt{2}}{2}$ and $\tan\theta = -1$, the reference angle is 45°. However, for $\sin\theta$ to be positive and $\tan\theta$ to be negative, the reference angle must be in the second quadrant. So, the value of θ must be 135° or an angle coterminal with 135°. **49.** Sample answer: We know that $\cot\theta = \frac{x}{y}$, $\sin\theta = \frac{y}{r}$, and $\cos\theta = \frac{x}{r}$. Since $\sin 180 = 0$, it must be true that $y = 0$. Thus, $\cot\theta = \frac{x}{0}$, which is undefined. **51.** Sample answer: First, sketch the angle and determine in which quadrant it is located. Then use the appropriate rule for finding its reference angle θ'. A reference angle is the acute angle formed by the terminal side of θ and the x-axis. Next, find the value of the trigonometric function for θ'. Finally, use the quadrant location to determine the sign of the trigonometric function value of θ. **53.** C **55.** E **57.** 330° **59.** 40.1° **61.** 66.0° **63.** $10,737,418.23

65. $(x + 4)^2 + (y + 2)^2 = 73$ **67.** $\frac{x^2 + 7x - 35}{(x + 2)(x + 4)(x - 7)}$

69. $\frac{2(3x^2 + 2x - 12)}{3x(x + 4)(x - 6)}$ **71.** 2.5841 **73.** $\frac{1}{2}$ **75.** $\frac{1}{3125}$ **77.** 9

Lesson 12-4

1. 27.9 mm^2

3 Area $= \frac{1}{2}bc \sin A$ — Area Formula

$= \frac{1}{2}(11)(6) \sin 40°$ — Substitution

$\approx 21.2 \text{ cm}^2$ — Simplify.

5. $E = 107°$, $d \approx 7.9$, $f \approx 7.0$ **7.** $F = 60°$, $f \approx 12.3$, $h \approx 9.1$ **9.** no solution **11.** one; $B = 90°$, $C = 60°$, $c \approx 5.2$ **13.** 10.6 km^2 **15.** 36.8 m^2 **17.** 5.9 ft^2 **19.** 65.2 m^2 **21.** $C = 30°$, $b \approx 11.1$, $c \approx 5.8$ **23.** $L = 74°$, $m \approx 4.9$, $n \approx 3.1$

25 $m\angle K = 180 - (53 + 20)$ or 107°

$\frac{\sin H}{h} = \frac{\sin J}{j}$ — Law of Sines

$\frac{\sin 53°}{31} = \frac{\sin 20°}{j}$ — Substitution

$j = \frac{31 \sin 20°}{\sin 53°}$ — Solve for j.

$j \approx 13.3$ — Use a calculator.

$\frac{\sin H}{h} = \frac{\sin K}{k}$ — Law of Sines

$\frac{\sin 53°}{31} = \frac{\sin 107°}{k}$ — Substitution

$k = \frac{31 \sin 107°}{\sin 53°}$ — Solve for k.

$k \approx 37.1$ — Use a calculator.

27. $B = 63°$, $b \approx 2.9$, $c \approx 3.0$ **29.** one; $B \approx 25°$, $C \approx 55°$, $c \approx 5.8$ **31.** one; $B \approx 32°$, $C \approx 110°$, $c \approx 32.1$ **33.** two; $B \approx 53°$, $C \approx 85°$, $c \approx 7.4$; $B \approx 127°$, $C \approx 11°$, $c \approx 1.4$ **35.** no solution **37.** about 28°

39

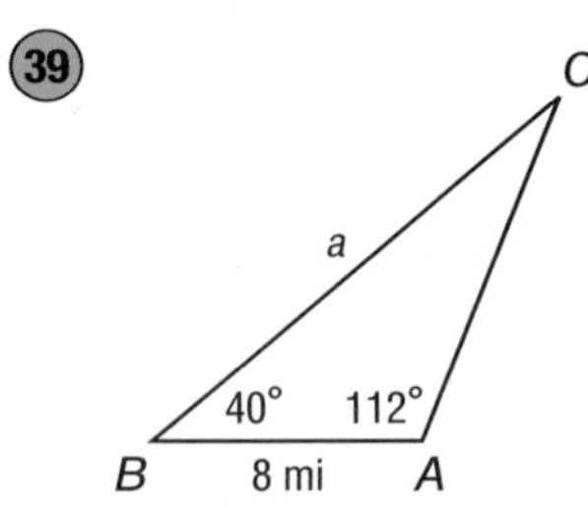

$m\angle C = 180 - (40 + 112)$ or 28°

$\frac{\sin A}{a} = \frac{\sin C}{c}$ — Law of Sines

$\frac{\sin 112°}{a} = \frac{\sin 28°}{8}$ — Substitution

$a = \frac{8 \sin 112°}{\sin 28°}$ — Solve for a.

$a \approx 15.8$ — Use a calculator.

Sirens B and C are about 15.8 miles apart.

41a. Sample answer: 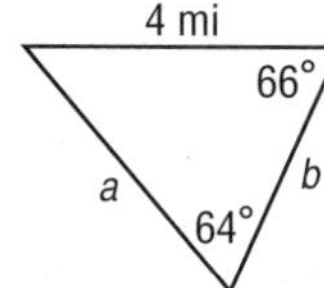

41b. Sample answer: $\frac{\sin 66°}{a} = \frac{\sin 64°}{4}$; $\frac{\sin 50°}{b} = \frac{\sin 64°}{4}$
41c. about 11.5 mi **43.** Cameron; R is acute and $r > t$, so there is one solution.
45. Sample answer:

$\text{Sin } A = \frac{\text{opposite}}{\text{hypotenuse}}$	Definition of sine
$\text{Sin } A = \frac{h}{c}$	h = opposite side, c = hypotenuse
$c \sin A = h$	Multiply both sides by c.
$\text{Area} = \frac{1}{2} \cdot \text{base} \cdot \text{height}$	Area of a triangle
$\text{Area} = \frac{1}{2}bh$	b = base, h = height
$\text{Area} = \frac{1}{2}bc \sin A$	Substitution

47. Sample answer: In the triangle, $B = 115°$. Using the Law of Sines, $\frac{\sin 50°}{a} = \frac{\sin 115°}{b}$. This equation cannot be solved because there are two unknowns. To solve a triangle using the Law of Sines, two sides and an angle must be given or two angles and a side opposite one of the angles must be given.

49. 2 **51.** G **53.** $-\frac{1}{2}$ **55.** $\frac{\sqrt{3}}{3}$ **57.** 328°, −392° **59.** 96 cm **61.** No sum exists. **63.** $\frac{x^2}{8.714 \times 10^{15}} + \frac{y^2}{8.710 \times 10^{15}} = 1$ **65.** $(y + 2)^2$ **67.** 56.25 **69.** 26

Lesson 12-5

1. $A \approx 36°, C \approx 52°, b \approx 5.1$ **3.** $A \approx 18°, B \approx 29°, C \approx 133°$ **5.** Sines; $B \approx 40°, C \approx 33°, c \approx 6.8$

7 Since the lengths of two sides and the measure of the included angle are known, first use the Law of Cosines to find the missing side length.

$r^2 = s^2 + t^2 - 2st \cos R$	Law of Cosines
$r^2 = 16^2 + 9^2 - 2(16)(9) \cos 35°$	$s = 16, t = 9, R = 35°$
$r^2 \approx 101.1$	Use a calculator.
$r \approx 10.1$	Take the positive square root of each side.
$\frac{\sin R}{r} = \frac{\sin T}{t}$	Law of Sines
$\frac{\sin 35°}{10.1} = \frac{\sin T}{9}$	Substitution
$\frac{9 \sin 35°}{10.1} = \sin T$	Multiply each side by 9.
$31° \approx T$	Use the $\sin^{-1}$ function.

$m\angle S = 180 - (35° + 31°)$ or 114°

9. $A \approx 70°, B \approx 40°, c \approx 3.0$ **11.** $A \approx 31°, B \approx 108°, C \approx 41°$ **13.** $a \approx 6.9, B \approx 41°, C \approx 23°$ **15.** $F \approx 65°, G \approx 94°, H \approx 21°$ **17.** Sines; $C \approx 45°, A \approx 85°, a \approx 18.2$
19. Cosines; $A \approx 27°, B \approx 115°, C \approx 38°$
21. Sines; $A \approx 17°, B \approx 79°, b \approx 6.9$

23 $d^2 = 338^2 + 520^2 - 2(338)(520) \cos 70°$
$d^2 \approx 264{,}417.1$
$d \approx 514.2$ m

25. 81°, 36°, 63° **27.** about 13,148 yd²

29a. Sample answer: 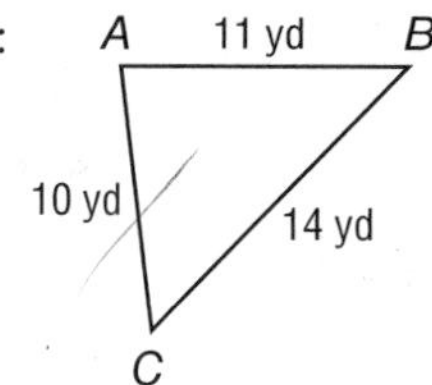

29b. Sample answer: Use the Law of Cosines to find the measure of $\angle A$. Then use the formula $\text{Area} = \frac{1}{2}bc \sin A$. **29c.** 54.6 yd²

31

$\frac{\sin A}{a} = \frac{\sin B}{b}$	Law of Sines
$\frac{\sin 104°}{12.4} = \frac{\sin B}{8.1}$	Substitution
$\frac{8.1 \sin 104°}{12.4} = \sin B$	Multiply each side by 8.1.
$39° \approx B$	Use the $\sin^{-1}$ function.
$m\angle C \approx 180 - (39° + 104°)$ or 37°	
$\frac{\sin A}{a} = \frac{\sin C}{c}$	Law of Sines
$\frac{\sin 104°}{12.4} = \frac{\sin 37°}{c}$	Substitution
$c = \frac{12.4 \sin 37°}{\sin 104°}$	Solve for c.
$c \approx 7.7$	Use a calculator.

33. $F \approx 42°, G \approx 72°, H \approx 66°$ **35.** The longest side is 14.5 centimeters. Use the Law of Cosines to find the measure of the angle opposite the longest side; 102°.
37. When two angles and a side are given or when two sides and an angle opposite one of the sides are given, you can use the Law of Sines to solve a triangle. When two sides and an included angle are given or when three sides are given, you can use the Law of Cosines to solve a triangle. **39.** G **41.** 4, $\frac{23}{15}$
43. 7.5 yd² **45.** $\sin \theta = \frac{5\sqrt{89}}{89}$, $\cos \theta = \frac{8\sqrt{89}}{89}$, $\tan \theta = \frac{5}{8}$, $\csc \theta = \frac{\sqrt{89}}{5}$, $\sec \theta = \frac{\sqrt{89}}{8}$, $\cot \theta = \frac{8}{5}$
47. $\sin \theta = -\frac{3\sqrt{13}}{13}$, $\cos \theta = \frac{2\sqrt{13}}{13}$, $\tan \theta = -1.5$, $\csc \theta = \frac{-\sqrt{13}}{3}$, $\sec \theta = \frac{\sqrt{13}}{2}$, $\cot \theta = -\frac{2}{3}$ **49.** \$60, \$50
51. hyperbola
53. 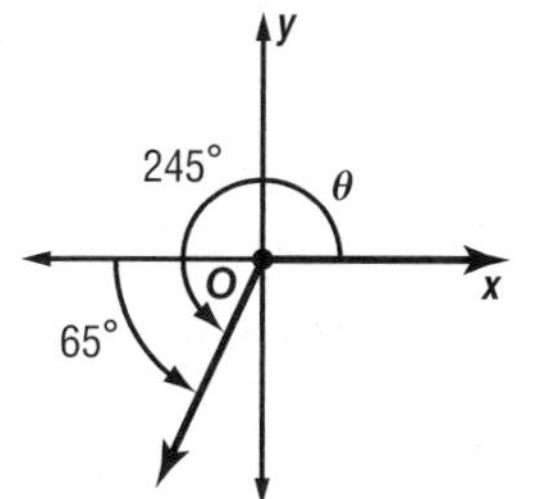

55. 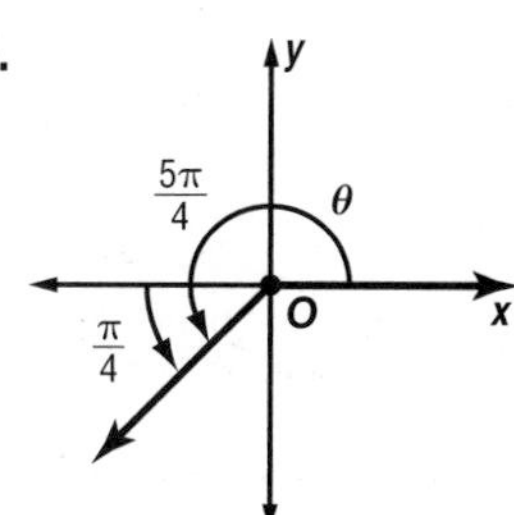

Lesson 12-6

1. $\cos \theta = \frac{15}{17}$, $\sin \theta = \frac{8}{17}$ **3.** 2 **5a.** 4 seconds
5b. Sample answer: 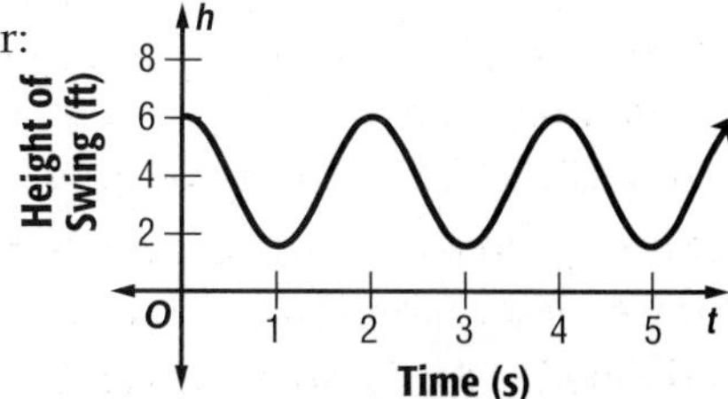

7. $-\frac{\sqrt{3}}{2}$ **9.** $\cos\theta = \frac{3}{5}, \sin\theta = -\frac{4}{5}$

11 $P\left(\frac{\sqrt{3}}{2}, \frac{1}{2}\right) = P(\cos\theta, \sin\theta)$

$\cos\theta = \frac{\sqrt{3}}{2}$ $\sin\theta = \frac{1}{2}$

13. 3 **15.** 12 **17.** 180°

19a. **Average High Temperatures**

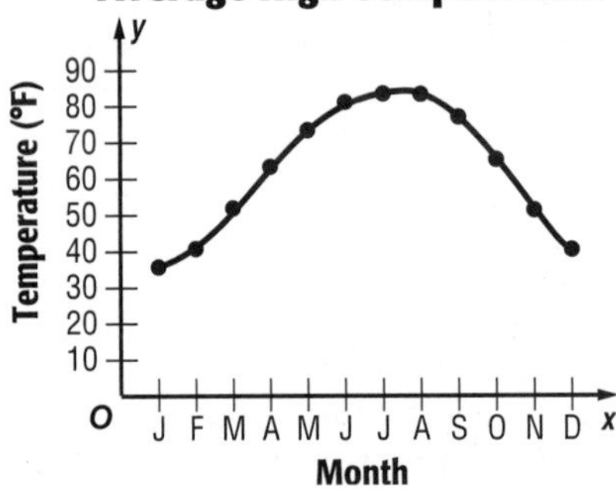

19b. 12 mo or 1 yr **21.** $\frac{1}{2}$ **23.** $\frac{\sqrt{2}}{2}$ **25.** $-\frac{\sqrt{3}}{2}$

27 **a.** The period is the time it takes to complete one rotation. So, the period is 60 seconds ÷ 2.5 or 24 seconds.

b. Since the siren is 1 mile from Ms. Miller's house and the beam of sound has a radius of 1 mile, then the minimum distance of the sound beam from the house is 0 miles and the maximum distance is 2 miles. Draw a sine curve with the pattern repeating every 24 seconds. Sample answer:

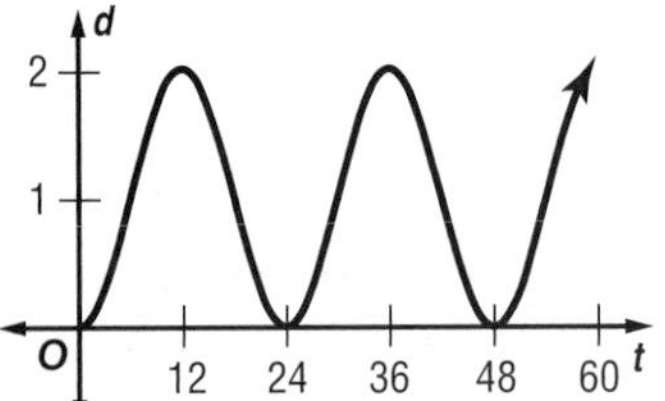

29a.

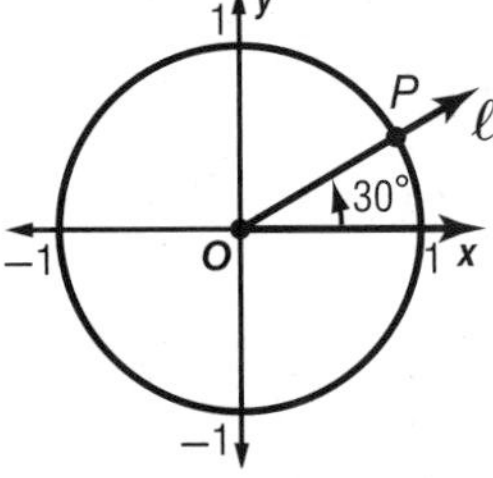

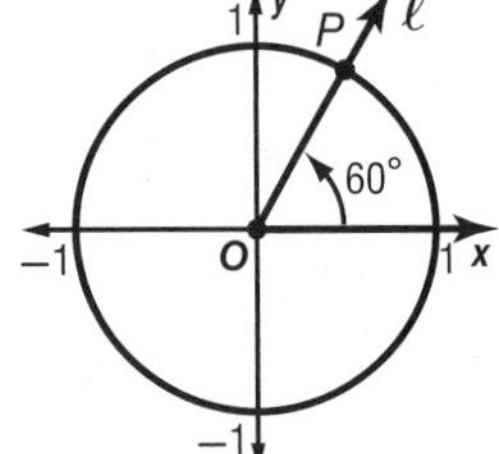

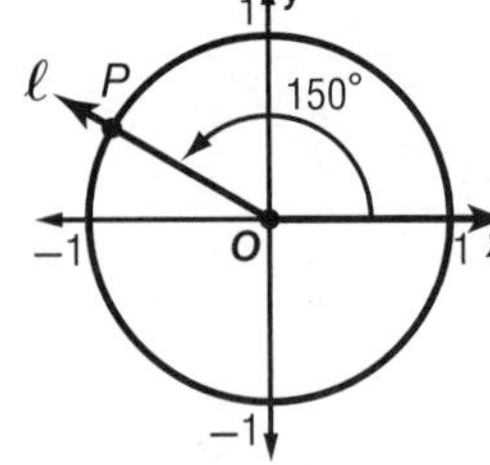

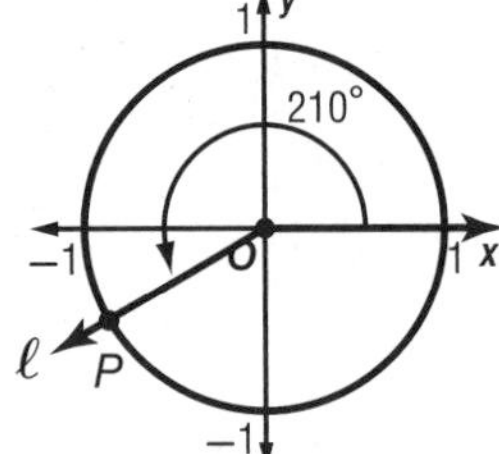

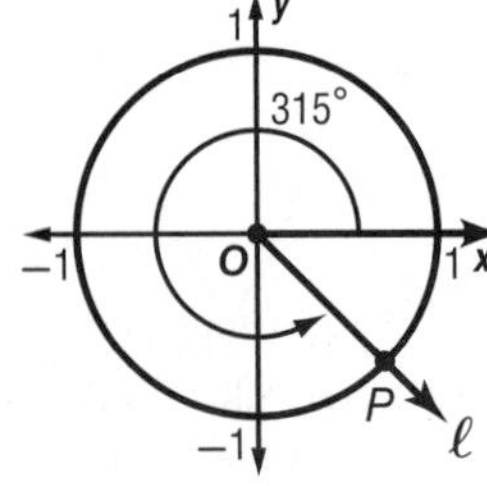

29b.

Angle	Slope
30	0.6
60	1.7
120	−1.7
150	−0.6
210	0.6
315	−1

29c. Sample answer: The slope corresponds to the tangent of the angle. For $\theta = 120°$, the x-coordinate of P is $-\frac{1}{2}$ and the y-coordinate is $\frac{\sqrt{3}}{2}$; slope = $\frac{\text{change in } y}{\text{change in } x}$. Since change in $x = -\frac{1}{2}$ and change in $y = \frac{\sqrt{3}}{2}$, slope $= \frac{\sqrt{3}}{2} \div \left(-\frac{1}{2}\right) = -\sqrt{3}$ or about −1.7.

31 $\cos 45° - \cos 30° = \frac{\sqrt{2}}{2} - \frac{\sqrt{3}}{2}$
$= \frac{\sqrt{2} - \sqrt{3}}{2}$

33. $-\frac{5\sqrt{3}}{2}$ **35.** 1 **37.** Benita; Francis incorrectly wrote $\cos\frac{-\pi}{3} = -\cos\frac{\pi}{3}$. **39.** Sometimes; the period of a sine curve could be $\frac{\pi}{2}$, which is not a multiple of π.
41. The period of a periodic function is the horizontal distance of the part of the graph that is nonrepeating. Each nonrepeating part of the graph is one cycle.
43. C **45.** A **47.** $A \approx 34°, C \approx 64°, c \approx 12.7$
49. $B \approx 33°, C \approx 29°, c \approx 9.9$ **51.** one solution; $B \approx 35°$, $C \approx 99°, c \approx 13.7$ **53.** 0.267 **55.** 7 **57.** $46,794.34
59. (5, 0), (−4, ±6) **61.** 108

Lesson 12-7

1. amplitude: 4; period: 360°

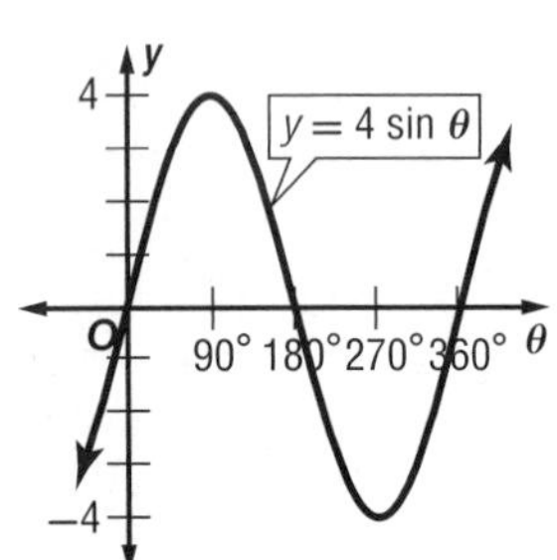

3. amplitude: 1; period: 180°

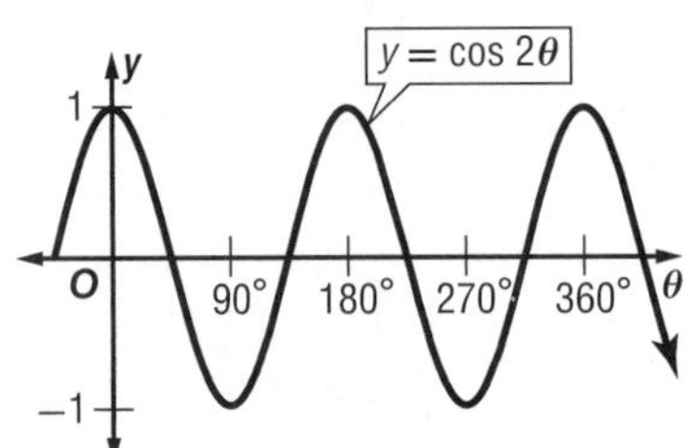

5a. $\frac{1}{14}$ or about 0.07 second
5b. $y = \sin 28\pi t$

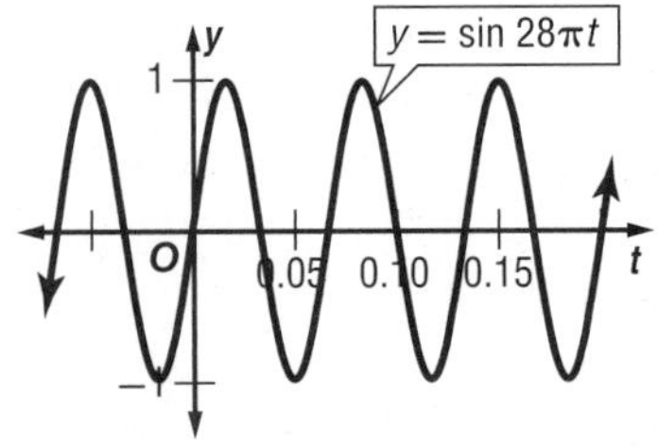

Selected Answers and Solutions

7. period: 360°

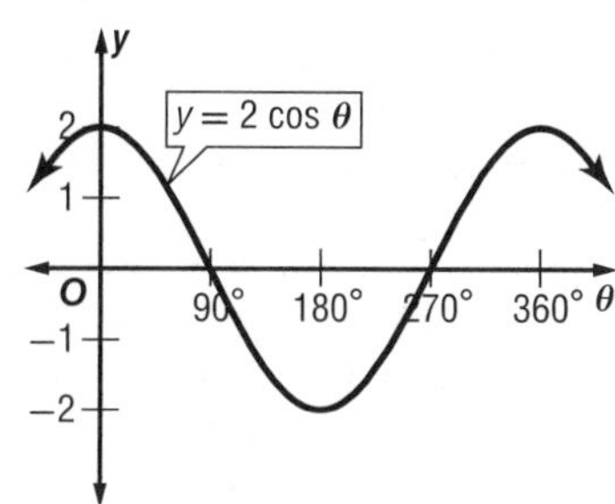

9. amplitude: 2; period: 360°

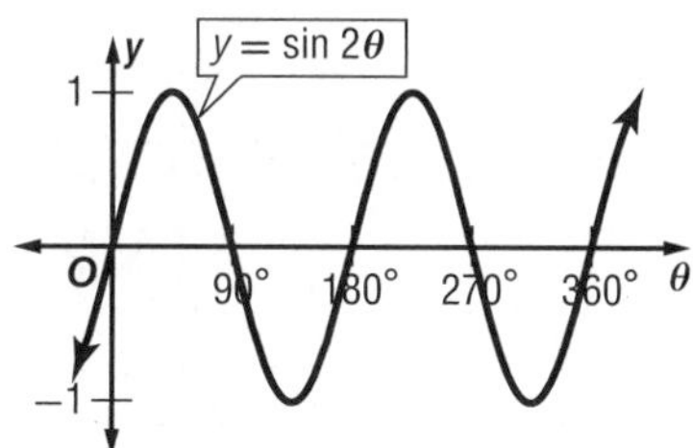

11. amplitude: 1; period: 180°

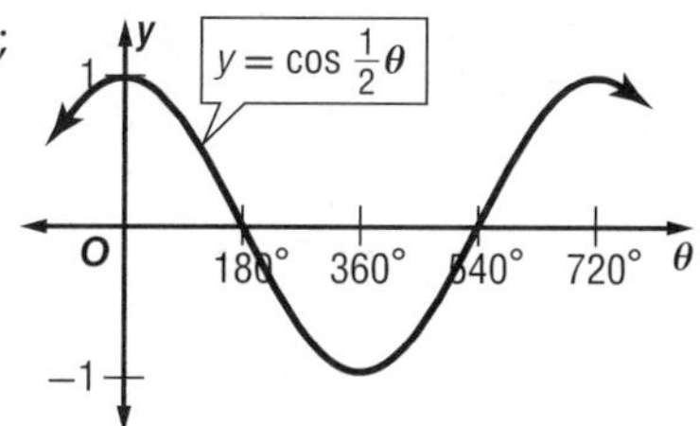

13. amplitude: 1; period: 720°

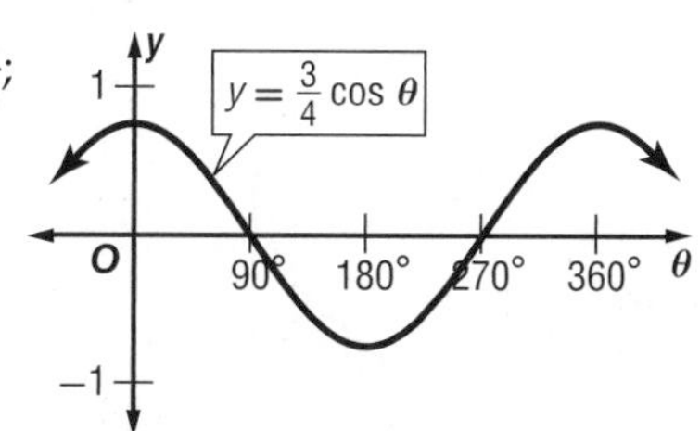

15. amplitude: $\frac{3}{4}$; period: 360°

17 amplitude: $|a| = \left|\frac{1}{2}\right|$ or $\frac{1}{2}$

period: $\frac{360°}{|b|} = \frac{360°}{|2|}$
$= 180°$

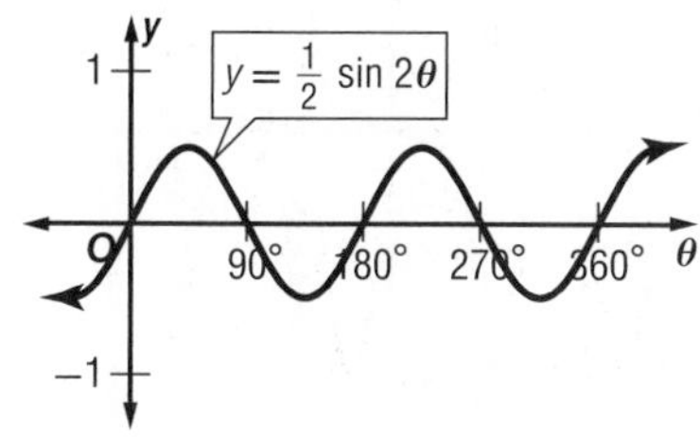

19. amplitude: 3; period: 180°

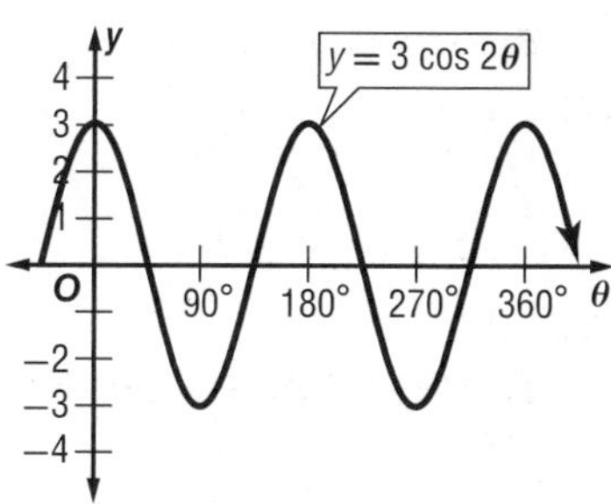

21a. $h = 4 \sin \frac{2}{3}\pi t$

21b.

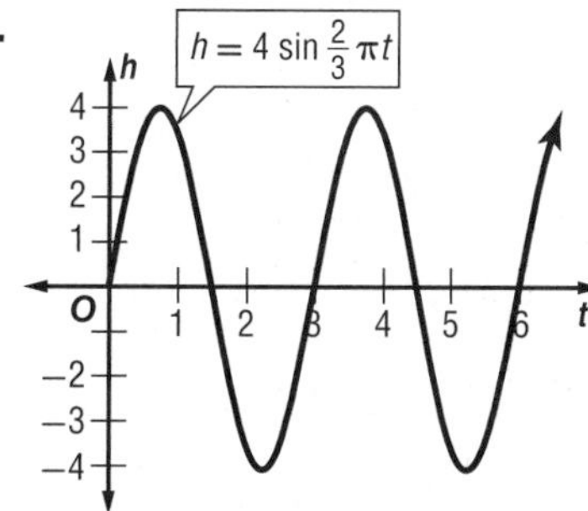

23. period: 360°

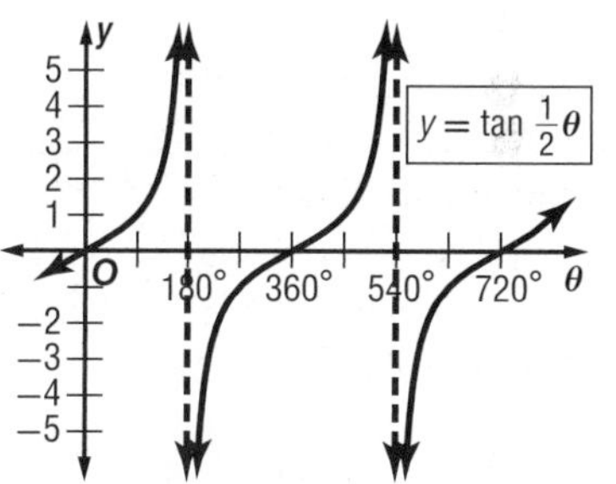

25. period: 180°

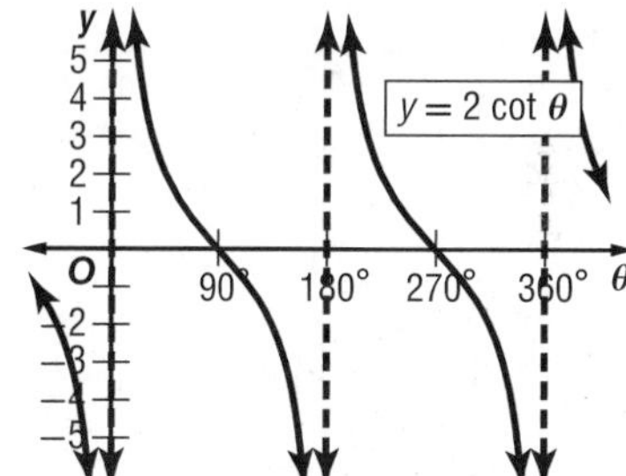

27. period: 180°

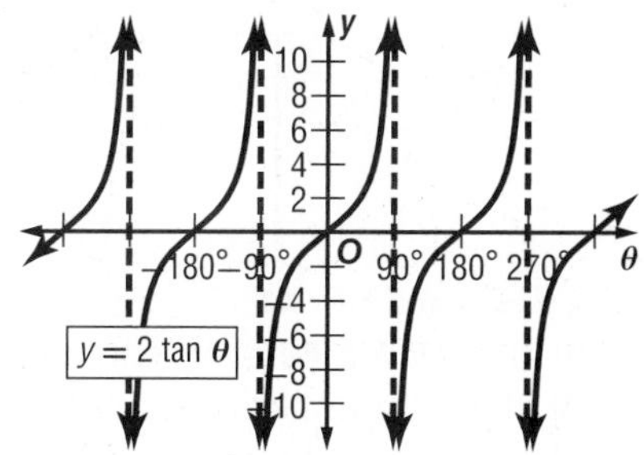

29 **a.** Since the frequency is 0.5, the period is $\frac{1}{0.5}$ or 2.

$\text{period} = \frac{2\pi}{\lvert b \rvert}$	Write the relationship between the period and *b*.
$2 = \frac{2\pi}{\lvert b \rvert}$	Substitution
$b = \pi$	Solve for *b*.
$y = a \sin b\theta$	General equation for the sine function
$h = 1 \sin \pi t$	Replace *y* with *h*, *a* with 1, *b* with π, and θ with *t*.
$h = \sin \pi t$	Simplify.

Selected Answers and Solutions

b.

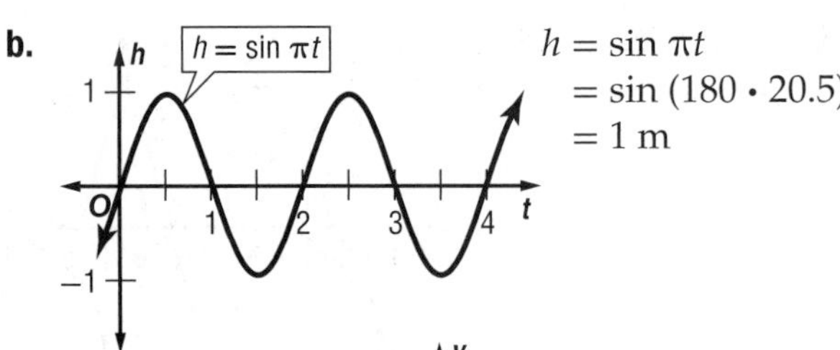

$h = \sin \pi t$
$= \sin (180 \cdot 20.5)$
$= 1$ m

31a. $y = \cos 260\pi t$ The amplitude remains the same. The period decreases because it is the reciprocal of the frequency. **31b.** The amplitude remains the same. The period decreases because it is the reciprocal of the frequency.

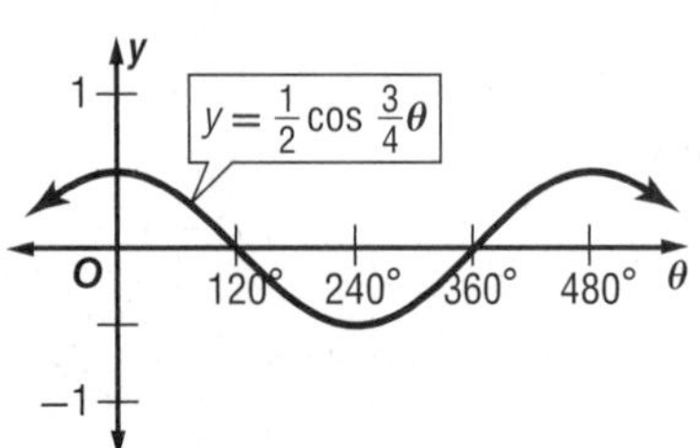

33. amplitude: $\frac{1}{2}$; period: 480°

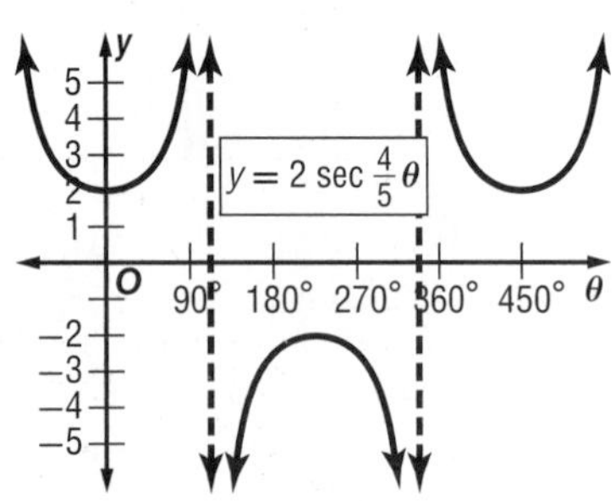

35. amplitude: does not exist; period: 450°

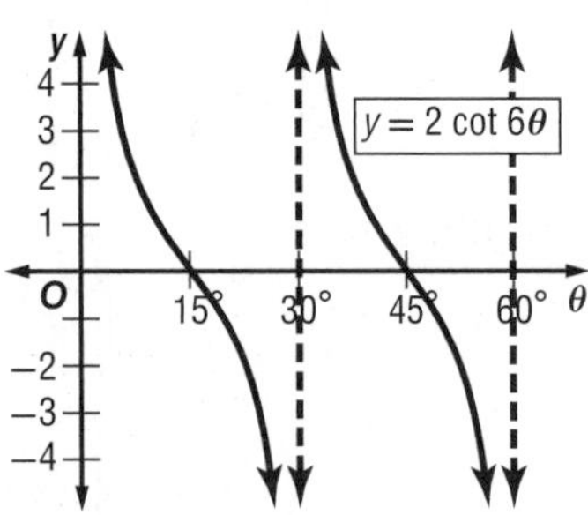

37. amplitude: does not exist; period: 30°

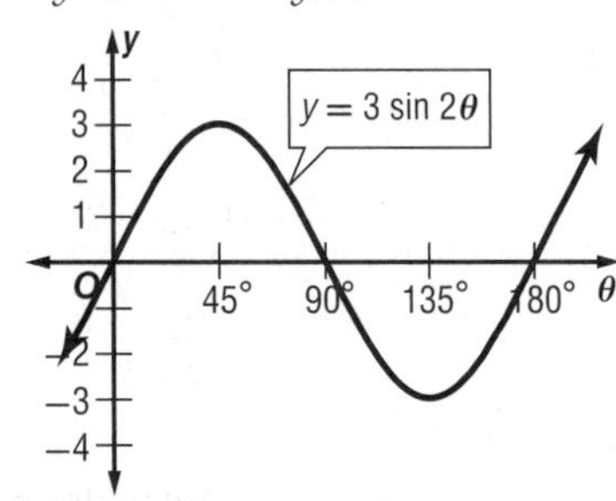

39. 180°; $y = 5 \sin 2\theta$ **41.** The domain of $y = a \cos \theta$ is the set of all real numbers. The domain of $y = a \sec \theta$ is the set of all real numbers except the values for which $\cos \theta = 0$. The range of $y = a \cos \theta$ is $-a \leq y \leq a$. The range of $y = a \sec \theta$ is $y \leq -a$ and $y \geq a$.

43. Sample answer: $y = 3 \sin 2\theta$

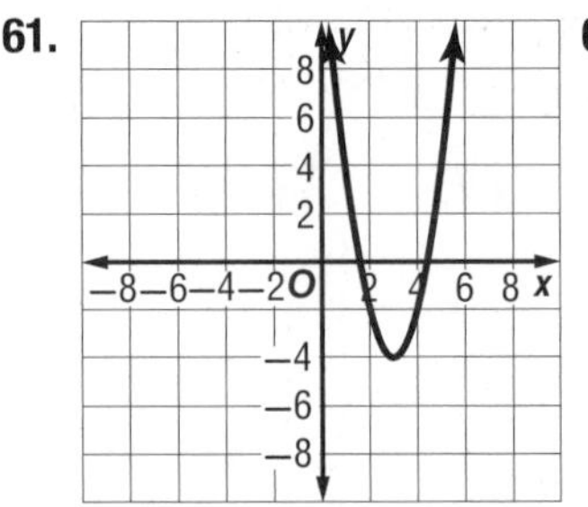

45. 700,013 **47.** G **49.** −1 **51.** $-3\sqrt{3}$ **53.** $R \approx 140°$, $S \approx 9°$, $q \approx 48.6$ **55.** 10.1% **57.** 5 **59.** $\frac{(x-6)^2}{20} + \frac{(y-3)^2}{4} = 1$

61.

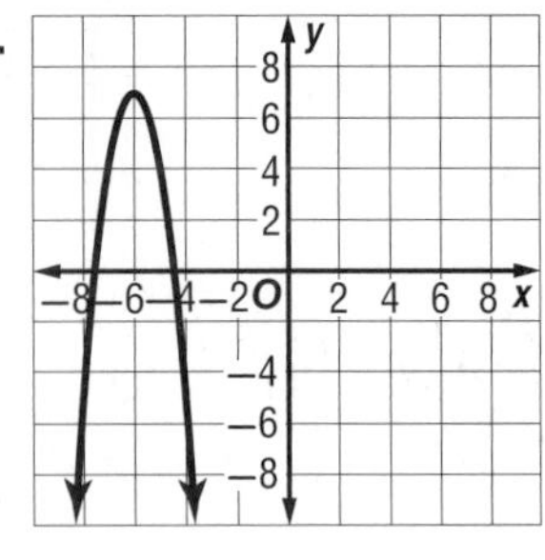

63.

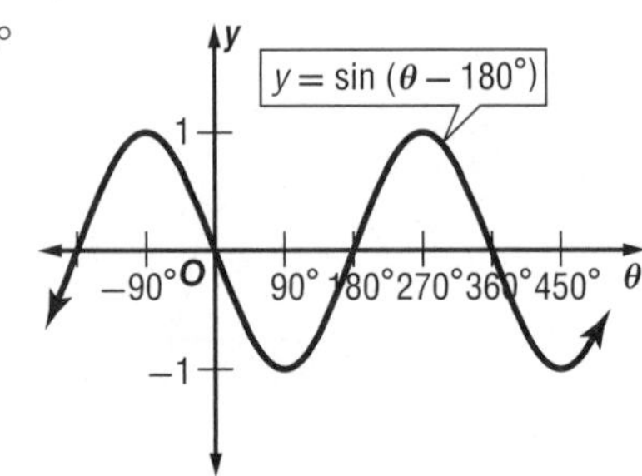

Lesson 12-8

1. 1; 360°; $h = 180°$

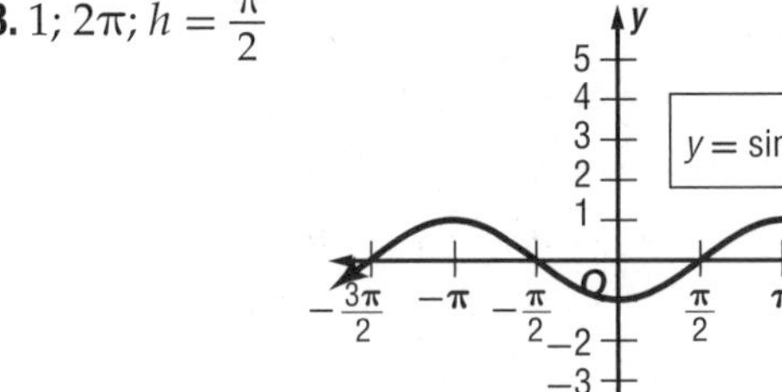

3. 1; 2π; $h = \frac{\pi}{2}$

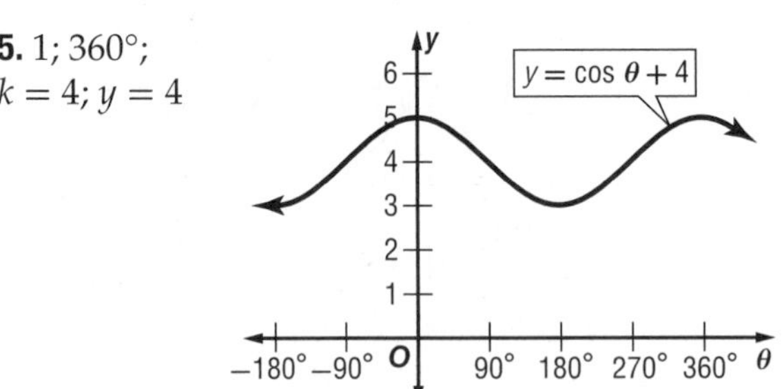

5. 1; 360°; $k = 4$; $y = 4$

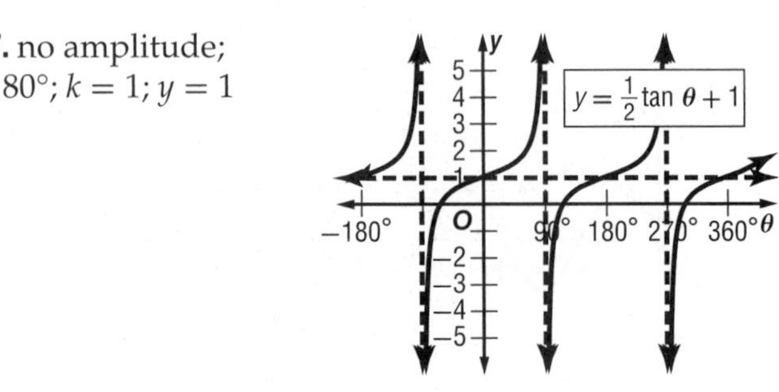

7. no amplitude; 180°; $k = 1$; $y = 1$

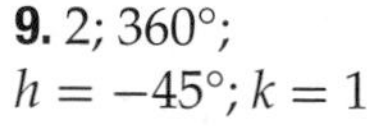

9. 2; 360°; $h = -45°$; $k = 1$

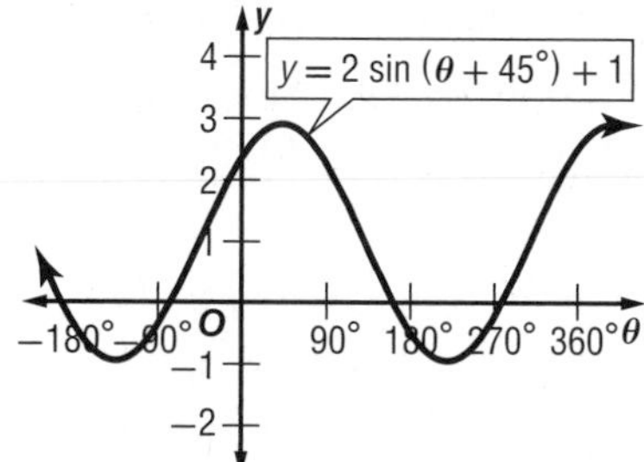

11. no amplitude; 90°; $h = -30°$; $k = 3$

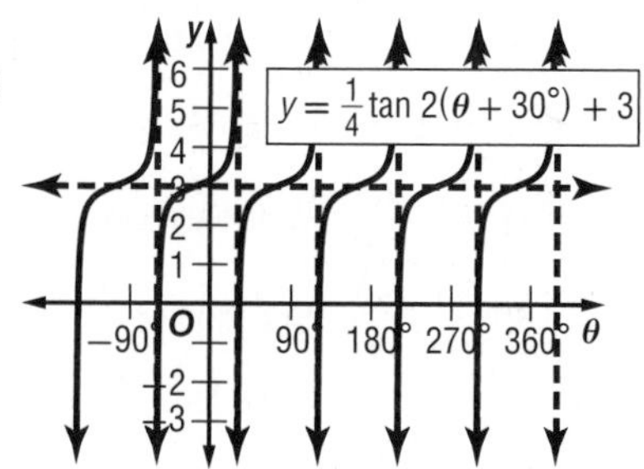

13. $P = 20 \sin 3\pi t + 110$

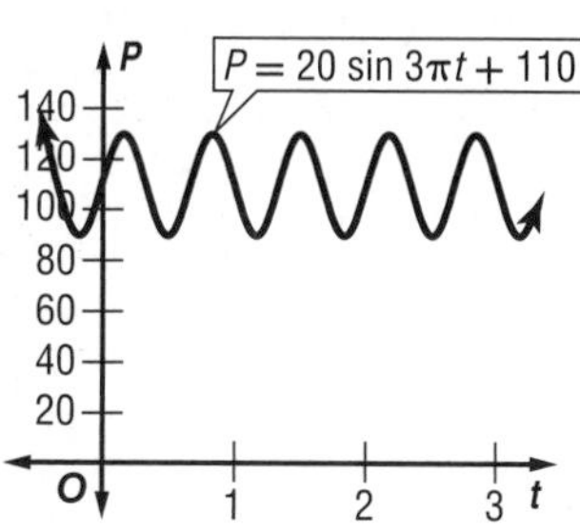

15. no amplitude; 180°; $h = 90°$

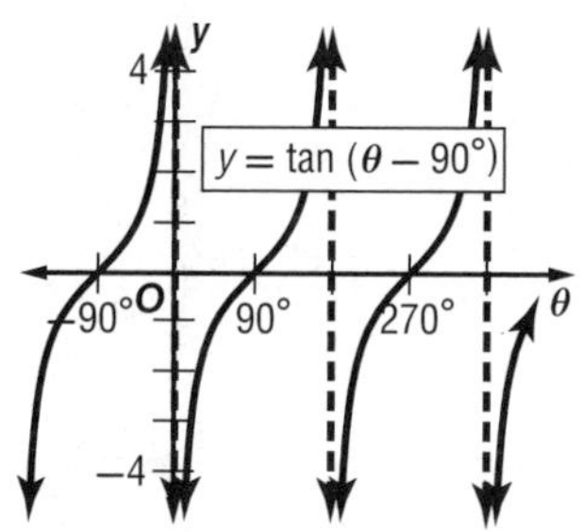

17. 2; 2π; $h = -\frac{\pi}{2}$

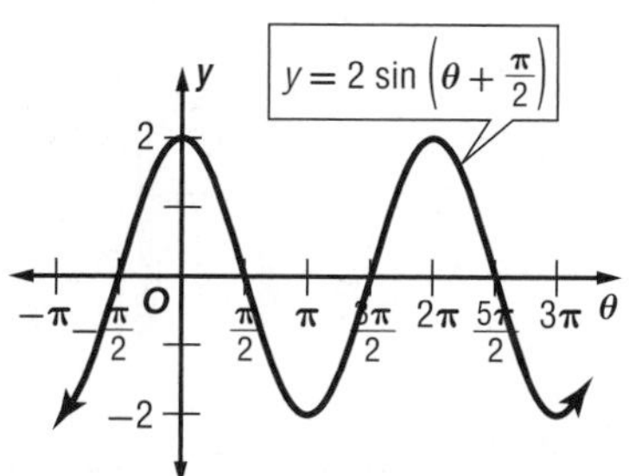

19. 3; 2π; $h = \frac{\pi}{3}$

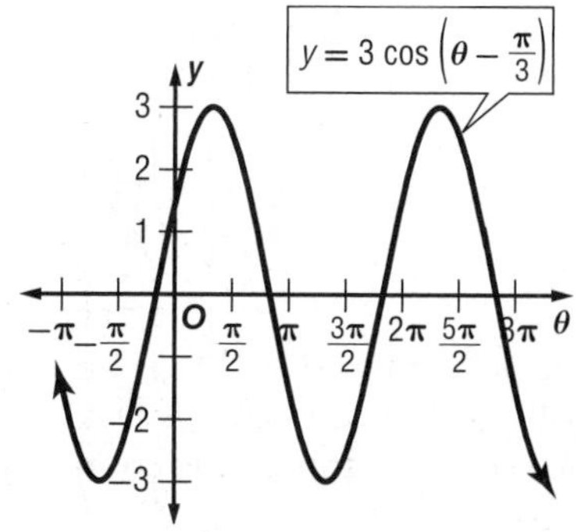

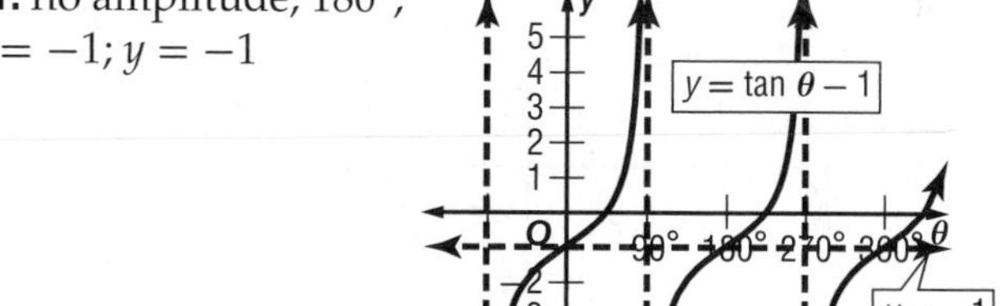

21. no amplitude; 180°; $k = -1$; $y = -1$

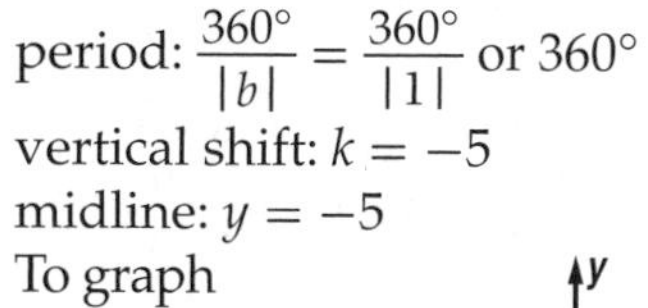

23 amplitude: $|a| = 2$

period: $\frac{360°}{|b|} = \frac{360°}{|1|}$ or 360°

vertical shift: $k = -5$

midline: $y = -5$

To graph $y = 2\cos\theta - 5$, first draw the midline. Then use it to graph $y = 2\cos\theta$ shifted 5 units down.

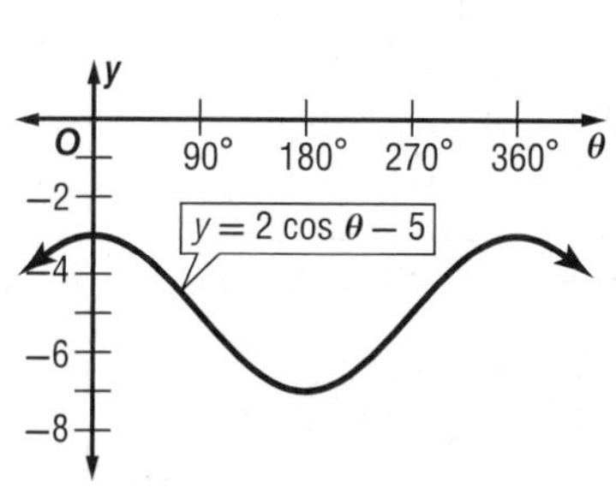

25. $\frac{1}{3}$; 360°; $k = 7$; $y = 7$

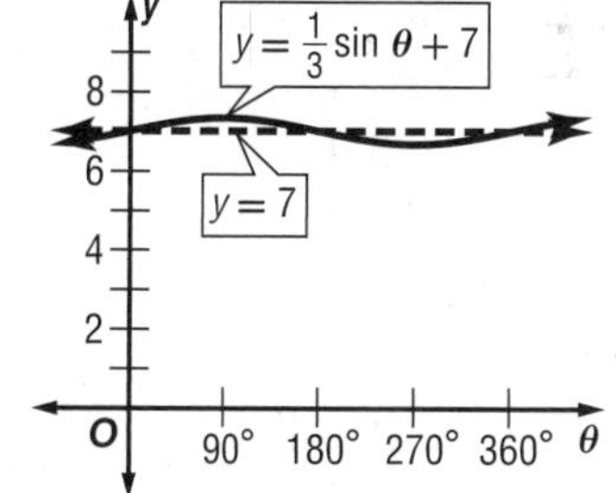

27. 1; 720°; $h = 90°$; $k = 2$

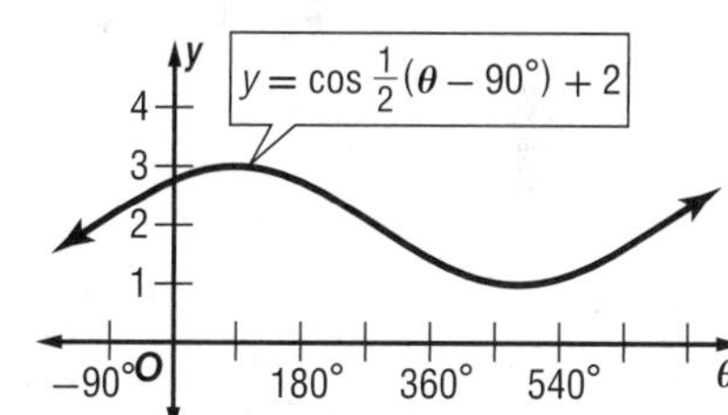

29. no amplitude; $\frac{\pi}{2}$; $h = -\frac{\pi}{4}$; $k = -5$

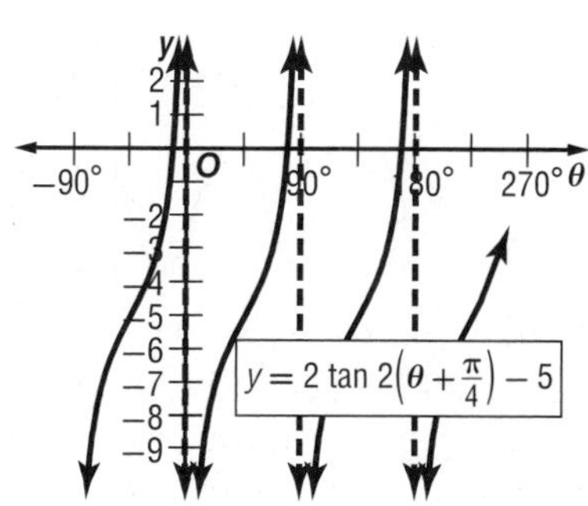

31. 1; 120°; $h = 45°$; $k = \frac{1}{2}$

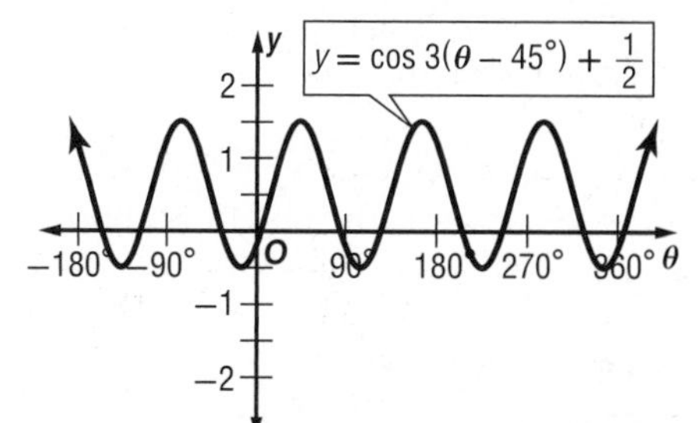

33. 3; 6π; $h = \frac{\pi}{2}$; $k = -2$

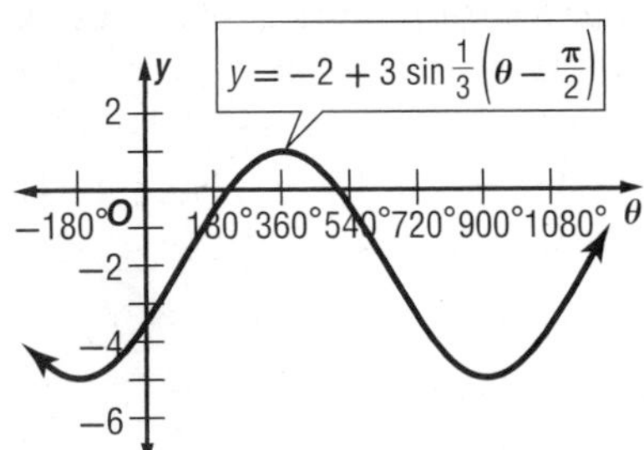

35.

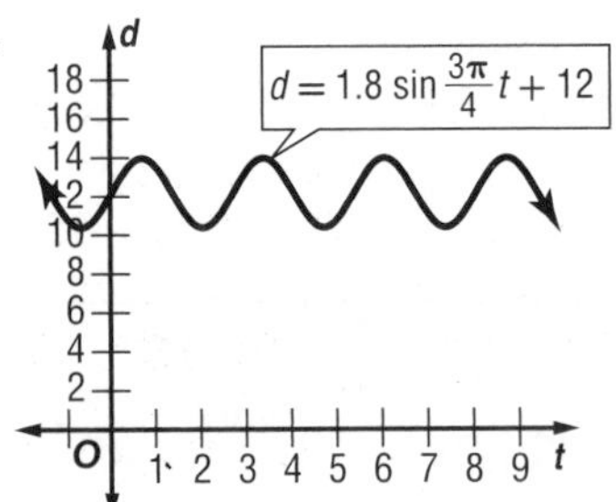

min: 10.2 ft; max: 13.8 ft **37.** $y = \sin (x - 4) + 3$ **39.** $y = \tan (x - \pi) + 2.5$

41 **a.** The midline lies halfway between the maximum and minimum values. So, $y = \frac{55 + 37}{2}$ or 46. Since the midline is $y = 46$, the vertical shift is $k = 46$.

The amplitude is the difference between the midline value and the maximum value. So, $|a| = |55 - 46|$ or 9.

Since the carousel rotates once every 21 seconds and a horse on the carousel goes up and down 3 times per rotation, it goes up and down every $21 \div 3$ or 7 seconds. So, the period is 7 seconds.

$\text{period} = \frac{2\pi}{|b|}$ Write the relationship between the period and b.

$7 = \frac{2\pi}{|b|}$ Substitution

$b = \frac{2\pi}{7}$ Solve for b.

$y = a \sin b\theta + k$ General equation for the sine function

$h = 9 \sin \frac{2\pi}{7} t + 46$ Replace y with h, a with 9, b with $\frac{2\pi}{7}$, θ with t, and k with 46.

b.

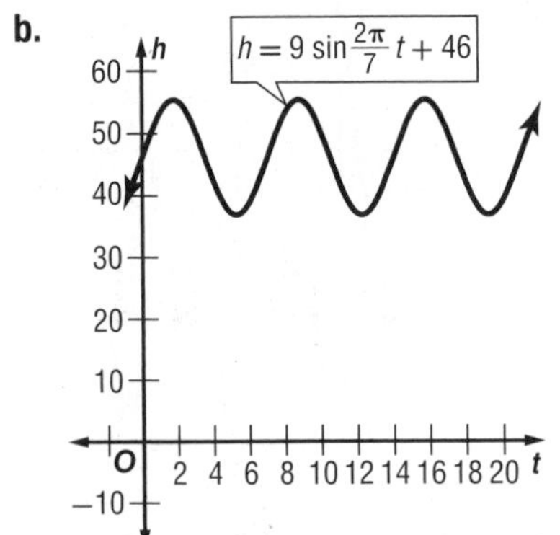

c. Sample answer: On the graph, when $t = 8$, $d \approx 53$. So after 8 seconds, the height is about 53 inches.

$h = 9 \sin \frac{2\pi}{7} t + 46$

$= 9 \sin \frac{2\pi}{7}(8) + 46$

≈ 53.0 in.

43. $\left(\frac{3\pi}{2}, 2\right)$ **45.** no maximum values **47.** The graphs are reflections of each other over the x-axis. **49.** The graphs are identical. **51.** 360°; Sample answer: $y = 2 \cos (\theta + 90°)$

53 The midline lies halfway between the maximum and minimum values. So $y = \frac{4 + 2}{2}$ or 3. Since the midline is $y = 3$, the vertical shift is $k = 3$.

The amplitude is the difference between the midline value and the maximum value. So $|a| = |4 - 3|$ or 1.

Since the cycle repeats every 180°, the period is 180°.

$\text{period} = \frac{360°}{|b|}$ Write the relationship between the period and b.

$180° = \frac{360°}{|b|}$ Substitution

$b = 2$ Solve for b.

The graph is the sine curve shifted 45° to the right. So the phase shift is $h = 45°$.

$y = a \sin b(\theta - h) + k$ General equation for the sine function

$y = 1 \sin 2(\theta - 45°) + 3$ Replace a with 1, b with 2, h with 45°, and k with 3.

$y = \sin 2(\theta - 45°) + 3$ Simplify.

55. 180°; no phase shift; $k = 6$

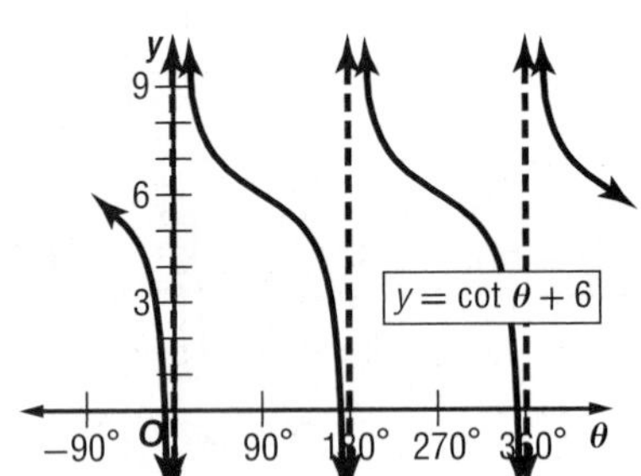

57. 120°; $h = 45°$; $k = 1$

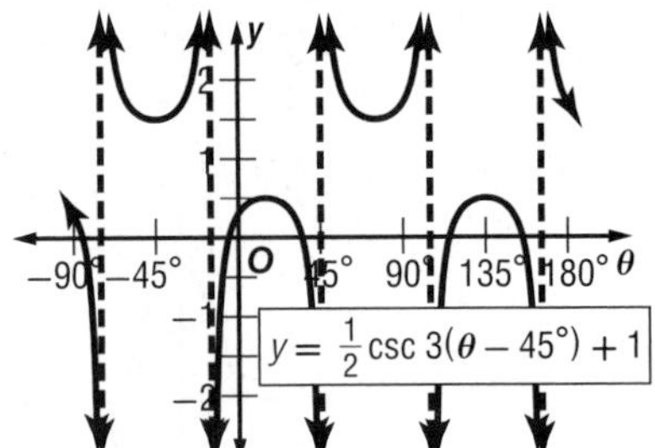

59. π; $h = -\frac{\pi}{2}$; $k = -3$

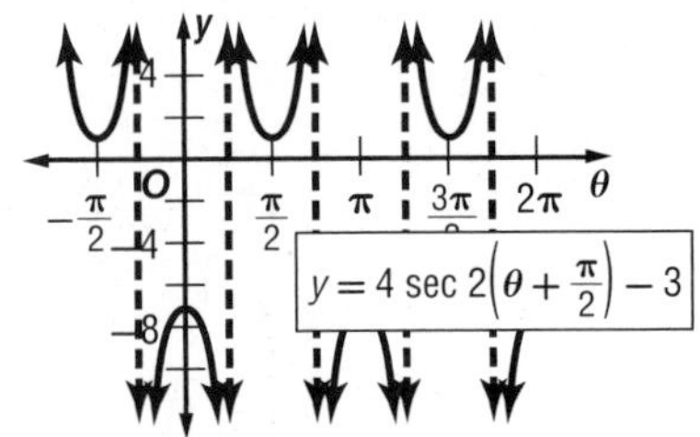

61. The graph of $y = 3 \sin 2\theta + 1$ has an amplitude of 3 rather than an amplitude of 1. It is shifted up 1 unit from the parent graph and is compressed so that it has a period of 180°.

63. Sample answer: $y = 2 \sin \theta - 3$

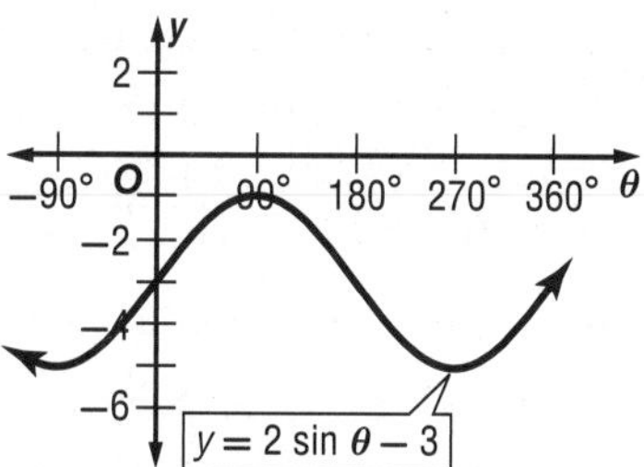

65. 1.25 **67.** F
69. amplitude: 2; period: 360°

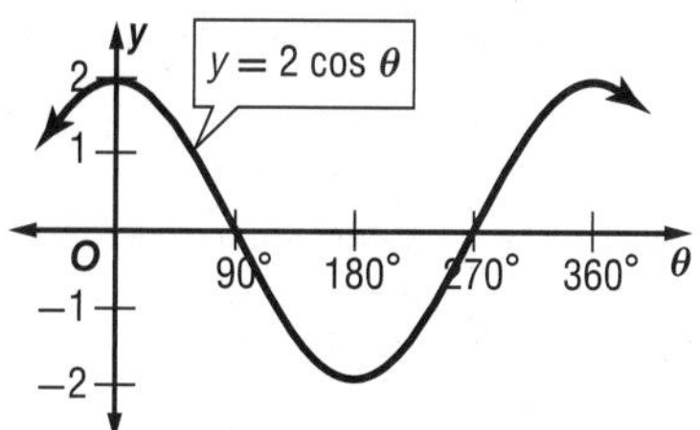

71. amplitude: 1; period: 180°

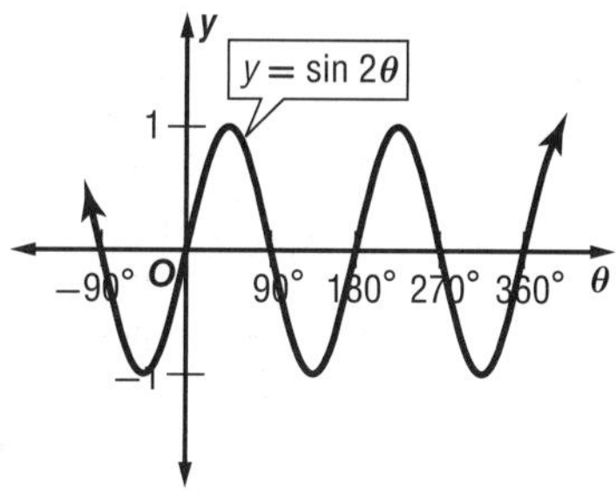

73. $-\frac{1}{2}$ **75.** experiment; sample: people that exercise for an hour a day; population: all adults **77.** observational study; sample: 100 students selected; population: all students that have part-time jobs **79.** 8 days **81.** 42° **83.** 37° **85.** 16°

Lesson 12-9

1. 30°; $\frac{\pi}{6}$ **3.** 180°; π **5.** 0 **7.** A **9.** −27.4°
11. Arctan $\frac{6.2}{18}$; 19°

13 KEYSTROKES: 2nd ENTER [COS^{-1}] 2nd [$\sqrt{\ }$] 3) ÷ 2) ENTER 30

So, $\text{Arccos}\left(\frac{\sqrt{3}}{2}\right) = 30°$ or $\frac{\pi}{6}$.

15. 60°; $\frac{\pi}{3}$ **17.** −30°; $-\frac{\pi}{6}$ **19.** −0.58 **21.** 0.87 **23.** 0.71 **25.** 64.2° **27.** 104.5° **29.** −11.3°
31. Arcsin $\frac{2.5}{24}$; 6°

33

$\frac{15 \text{ m/s} (\sin x)}{9.8 \text{ m/s}^2} = 1 \text{ s}$	Maximum height equals 1.
$\sin x = \frac{9.8}{15}$	Solve for sin *x*.
$x = \text{Sin}^{-1} \frac{9.8}{15}$	Inverse sine function
$x \approx 40.8°$	Use a calculator.

35. π **37.** no solution **39.** $\frac{\pi}{3}, \frac{5\pi}{3}$ **41.** false; $x = 2\pi$
43. The domain of $y = \text{Sin}^{-1} x$ is $-1 \leq x \leq 1$. This is the same as the range of $y = \text{Sin } x$. **45.** Sample answer: $y = \tan^{-1} x$ is a relation that has a domain of all real numbers and a range of all real numbers except odd multiples of $\frac{\pi}{2}$. The relation is not a function. $y = \text{Tan}^{-1} x$ is a function that has a domain of all real numbers and a range of $-\frac{\pi}{2} \leq y \leq \frac{\pi}{2}$.
47. A **49.** G **51a.** 164; 164; 360°, 90°
51b. $y = 100[\sin (x - 90°)] + 100$ **53.** 2 **55.** 4 **57.** −1
59. $-\frac{\sqrt{3}}{2}$

Chapter 12 Study Guide and Review

1. false, Law of Sines **3.** true **5.** false, Arcsine function **7.** $a = 10.9$; $A = 65°$; $B = 25°$ **9.** $A = 15°$; $a = 4.0$; $c = 15.5$ **11.** $B = 55°$; $a = 12.6$; $b = 18.0$ **13.** about 8.8 feet **15.** 450° **17.** $-\frac{7\pi}{4}$ **19.** 295°, −425° **21.** $\frac{4\pi}{15}$ **23.** $-\frac{\sqrt{3}}{3}$ **25.** 0 **27.** $\sin \theta = \frac{12}{13}$, $\cos \theta = \frac{5}{13}$, $\tan \theta = \frac{12}{5}$, $\csc \theta = \frac{13}{12}$, $\sec \theta = \frac{13}{5}$, $\cot \theta = \frac{5}{12}$ **29.** about 17.1 meters **31.** two solutions; First solution: $C = 30°$, $B = 125°$, $b = 29.1$; second solution: $C = 150°$, $B = 5°$, $b = 3.1$ **33.** 98.9 ft **35.** Law of Sines; $B \approx 52°$, $C \approx 48°$, $c \approx 11.3$ **37.** Law of Sines; $B \approx 75°$, $C \approx 63°$, $c \approx 12.0$ or $B \approx 105°$, $C \approx 33°$, $c \approx 7.3$ **39.** 750.5 ft **41.** $\frac{-\sqrt{6}}{4}$ **43.** 0 **45.** 15 seconds

47. amplitude: 1, period: 720°

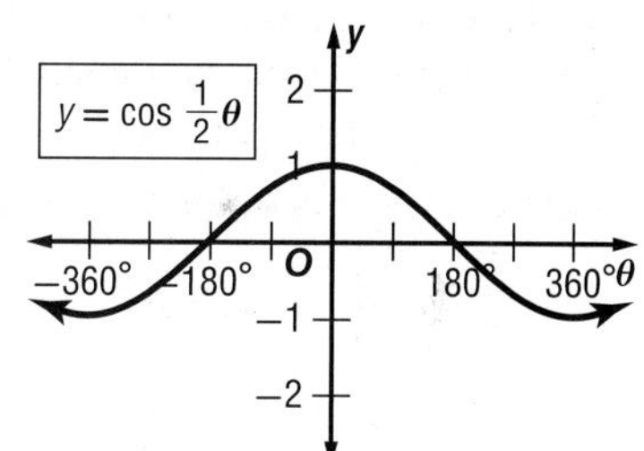

49. amplitude: not defined, period: 360°

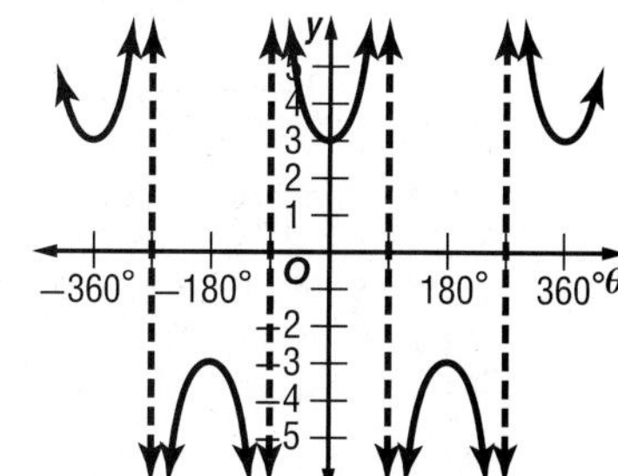

51. amplitude: not defined, period: 720°

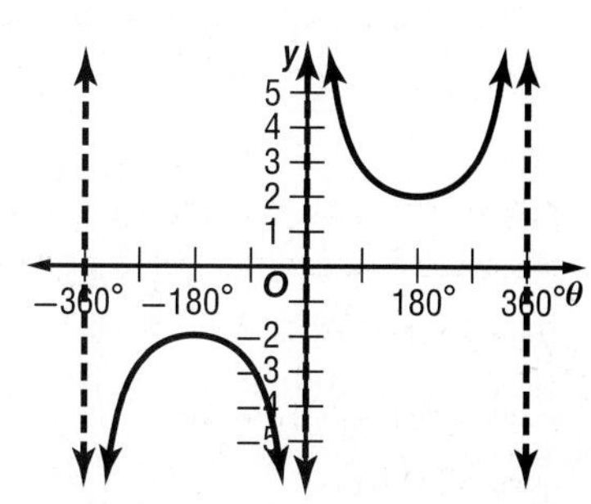

Selected Answers and Solutions

53. vertical shift: up 1; amplitude: 3; period 180°; phase shift: 90° right

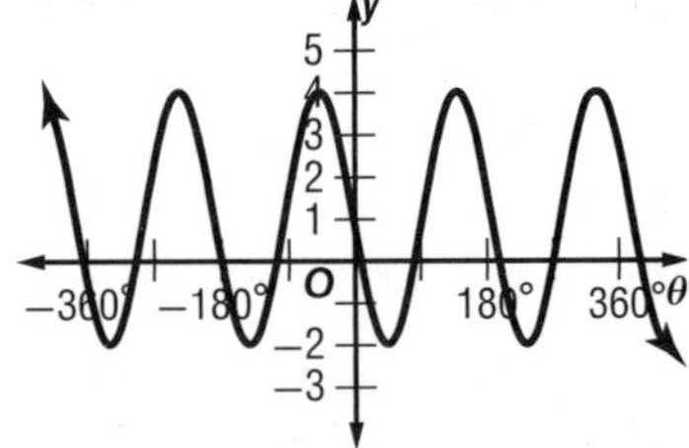

55. vertical shift: up 2 amplitude: not defined period: $\frac{2\pi}{3}$ phase shift: $\frac{\pi}{2}$° right

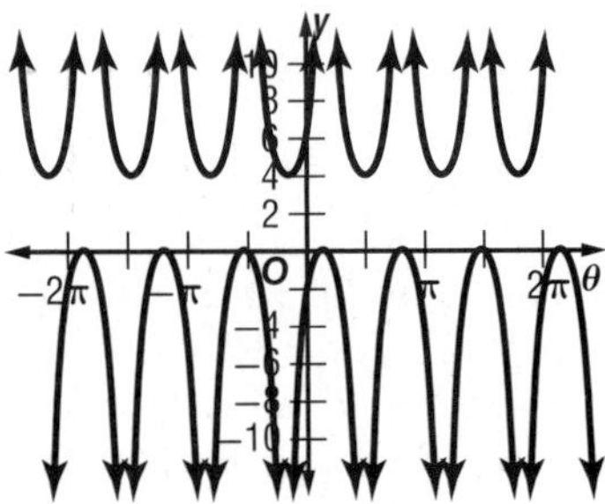

57. vertical shift: up 2 amplitude: $\frac{1}{3}$ period: 1080° phase shift: 90° right

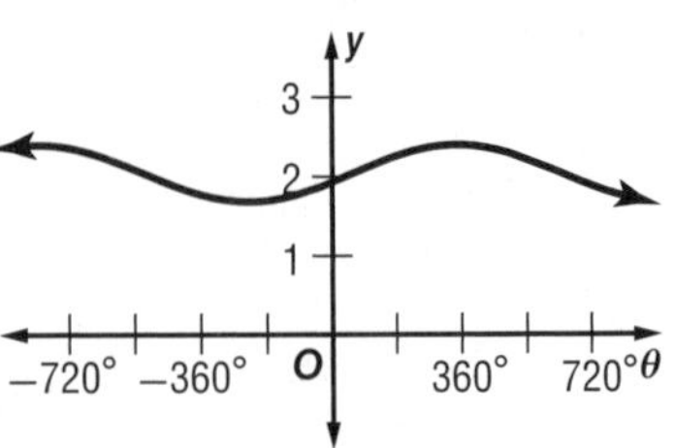

59. $90°, \frac{\pi}{2}$ **61.** $60°, \frac{\pi}{3}$ **63.** $45°, \frac{\pi}{4}$ **65.** $\sin^{-1}\frac{5}{10} = \theta$; 30° **67.** −0.71 **69.** −55.0° **71.** 65.8°

CHAPTER 13

Trigonometric Identities and Equations

Chapter 13 Get Ready

1. $-4a(a-1)$ **3.** prime **5.** $(x+2)$ in. **7.** $\{-7, 5\}$ **9.** $\{3, 4\}$ **11.** $\frac{\sqrt{2}}{2}$ **13.** $-\frac{\sqrt{3}}{3}$ **15.** 45 ft

Lesson 13-1

1. $\frac{1}{2}$ **3.** $\frac{\sqrt{5}}{3}$ **5.** $\sin\theta\cos\theta$ **7.** $\cot^2\theta$ **9.** $\frac{5}{4}$ **11.** $\frac{4}{5}$ **13.** $-\frac{5}{4}$

15
$\cot^2\theta + 1 = \csc^2\theta$ Pythagorean Identity
$\left(\frac{1}{4}\right)^2 + 1 = \csc^2\theta$ Substitute $\frac{1}{4}$ for $\cot\theta$.
$\frac{1}{16} + 1 = \csc^2\theta$ Square $\frac{1}{4}$.
$\frac{17}{16} = \csc^2\theta$ Add.
$\pm\frac{\sqrt{17}}{4} = \csc\theta$ Take the square root of each side.
Since θ is in the third quadrant, $\csc\theta$ is negative. So, $\csc\theta = -\frac{\sqrt{17}}{4}$.

17. $-\frac{12}{13}$ **19.** $\frac{3}{5}$ **21.** $\sec^3\theta$ **23.** $\csc\theta$ **25.** 1

27
$B = \frac{F\csc\theta}{I\ell}$ Original equation
$I\ell \cdot B = F\csc\theta$ Multiply each side by $I\ell$.
$I\ell B = F \cdot \frac{1}{\sin\theta}$ $\frac{1}{\sin\theta} = \csc\theta$
$I\ell B\sin\theta = F$ Multiply each side by $\sin\theta$.
The equation can be written as $F = I\ell B\sin\theta$.

29. $\sec\theta$ **31.** 2 **33.** $2\cos^2\theta$ **35a.** $\frac{\sqrt{65}}{9}$ **35b.** $\frac{4\sqrt{65}}{65}$ **35c.** $\frac{4}{9}, -\frac{\sqrt{65}}{9}, -\frac{4\sqrt{65}}{65}$ **37.** $\mu_k = \tan\theta$

39
$\frac{\cos\left(\frac{\pi}{2}-\theta\right)-1}{1+\sin(-\theta)} = \frac{\sin\theta - 1}{1+\sin(-\theta)}$ $\cos\left(\frac{\pi}{2}-\theta\right) = \sin\theta$
$= \frac{\sin\theta - 1}{1 - \sin\theta}$ $\sin(-\theta) = -\sin\theta$
$= \frac{\sin\theta - 1}{-1(\sin\theta - 1)}$ $1 - \sin\theta = -1(\sin\theta - 1)$
$= \frac{1}{-1}$ or -1 Simplify.

41. $-\cot^2\theta$ **43.** Sample answer: $x = 45°$ **45.** The functions $\cos\theta$ and $\sin\theta$ can be thought of as the lengths of the legs of a right triangle, and the number 1 can be thought of as the measure of the corresponding hypotenuse. **47.** Sample answer: $\frac{\sin\theta}{\cos\theta} \cdot \sin\theta$ and $\frac{\sin^2\theta}{\cos\theta}$ **49.** $-\frac{4}{3}$ **51.** A **53.** D **55.** 2.09 **57.** 0.52 **59.** 0.5 **61.** $d = 4 - \cos\frac{\pi}{2}t$ or $d = 4 - \cos 90°t$ **63.** $\frac{1093}{9}$ **65.** −3, 2 **67.** 2

Lesson 13-2

1.
$\cot\theta + \tan\theta \stackrel{?}{=} \frac{\sec^2\theta}{\tan\theta}$
$\cot\theta + \tan\theta \stackrel{?}{=} \frac{\tan^2\theta + 1}{\tan\theta}$
$\cot\theta + \tan\theta \stackrel{?}{=} \frac{\tan^2\theta}{\tan\theta} + \frac{1}{\tan\theta}$
$\cot\theta + \tan\theta = \tan\theta + \cot\theta$ ✓

3.
$\sin\theta \stackrel{?}{=} \frac{\sec\theta}{\tan\theta + \cot\theta}$
$\sin\theta \stackrel{?}{=} \frac{\frac{1}{\cos\theta}}{\frac{\sin\theta}{\cos\theta} + \frac{\cos\theta}{\sin\theta}}$
$\sin\theta \stackrel{?}{=} \frac{\frac{1}{\cos\theta}}{\frac{\sin^2\theta + \cos^2\theta}{\cos\theta\sin\theta}}$
$\sin\theta \stackrel{?}{=} \frac{\frac{1}{\cos\theta}}{\frac{1}{\cos\theta\sin\theta}}$
$\sin\theta \stackrel{?}{=} \frac{1}{\cos\theta} \cdot \frac{\cos\theta\sin\theta}{1}$
$\sin\theta = \sin\theta$ ✓

5.
$\tan^2\theta\csc^2\theta \stackrel{?}{=} 1 + \tan^2\theta$
$\frac{\sin^2\theta}{\cos^2\theta} \cdot \frac{1}{\sin^2\theta} \stackrel{?}{=} \sec^2\theta$
$\frac{1}{\cos^2\theta} \stackrel{?}{=} \sec^2\theta$
$\sec^2\theta = \sec^2\theta$ ✓

7
$\frac{\tan^2\theta + 1}{\tan^2\theta} = \frac{\sec^2\theta}{\tan^2\theta}$ $\tan^2\theta + 1 = \sec^2\theta$
$= \frac{\frac{1}{\cos^2\theta}}{\frac{\sin^2\theta}{\cos^2\theta}}$ $\sec^2\theta = \frac{1}{\cos^2\theta}$ and $\tan^2\theta = \frac{\sin^2\theta}{\cos^2\theta}$
$= \frac{1}{\cos^2\theta} \cdot \frac{\cos^2\theta}{\sin^2\theta}$ Invert the denominator and multiply.
$= \frac{1}{\sin^2\theta}$ Simplify.
$= \csc^2\theta$ $\csc^2\theta = \frac{1}{\sin^2\theta}$
The answer is D.

9.
$\cot\theta(\cot\theta + \tan\theta) \stackrel{?}{=} \csc^2\theta$
$\cot^2\theta + \cot\theta\tan\theta \stackrel{?}{=} \csc^2\theta$
$\cot^2\theta + \frac{\sin\theta}{\cos\theta} \cdot \frac{\cos\theta}{\sin\theta} \stackrel{?}{=} \csc^2\theta$
$\cot^2\theta + 1 \stackrel{?}{=} \csc^2\theta$
$\csc^2\theta = \csc^2\theta$ ✓

11. $\sin\theta\sec\theta\cot\theta \stackrel{?}{=} 1$

$\sin\theta \cdot \frac{1}{\cos\theta} \cdot \frac{\cos\theta}{\sin\theta} \stackrel{?}{=} 1$

$1 = 1$ ✓

13. $\frac{1 - 2\cos^2\theta}{\sin\theta\cos\theta} \stackrel{?}{=} \tan\theta - \cot\theta$

$\frac{(1 - \cos^2\theta) - \cos^2\theta}{\sin\theta\cos\theta} \stackrel{?}{=} \tan\theta - \cot\theta$

$\frac{\sin^2\theta - \cos^2\theta}{\sin\theta\cos\theta} \stackrel{?}{=} \tan\theta - \cot\theta$

$\frac{\sin^2\theta}{\sin\theta\cos\theta} - \frac{\cos^2\theta}{\sin\theta\cos\theta} \stackrel{?}{=} \tan\theta - \cot\theta$

$\frac{\sin\theta}{\cos\theta} - \frac{\cos\theta}{\sin\theta} \stackrel{?}{=} \tan\theta - \cot\theta$

$\tan\theta - \cot\theta = \tan\theta - \cot\theta$ ✓

15. $\cos\theta \stackrel{?}{=} \sin\theta\cot\theta$

$\cos\theta \stackrel{?}{=} \sin\theta\frac{\cos\theta}{\sin\theta}$

$\cos\theta = \cos\theta$ ✓

17. $\cos\theta\cos(-\theta) - \sin\theta\sin(-\theta) \stackrel{?}{=} 1$

$\cos\theta\cos\theta - \sin\theta(-\sin\theta) \stackrel{?}{=} 1$

$\cos^2\theta + \sin^2\theta \stackrel{?}{=} 1$

$1 = 1$ ✓

19. $\sec\theta - \tan\theta \stackrel{?}{=} \frac{1 - \sin\theta}{\cos\theta}$

$\frac{1}{\cos\theta} - \frac{\sin\theta}{\cos\theta} \stackrel{?}{=} \frac{1 - \sin\theta}{\cos\theta}$

$\frac{1 - \sin\theta}{\cos\theta} = \frac{1 - \sin\theta}{\cos\theta}$ ✓

21. $\sec\theta\csc\theta \stackrel{?}{=} \tan\theta + \cot\theta$

$\frac{1}{\cos\theta} \cdot \frac{1}{\sin\theta} \stackrel{?}{=} \frac{\sin\theta}{\cos\theta} + \frac{\cos\theta}{\sin\theta}$

$\frac{1}{\cos\theta\sin\theta} \stackrel{?}{=} \frac{\sin^2\theta}{\sin\theta\cos\theta} + \frac{\cos^2\theta}{\sin\theta\cos\theta}$

$\frac{1}{\cos\theta\sin\theta} \stackrel{?}{=} \frac{\sin^2\theta + \cos^2\theta}{\sin\theta\cos\theta}$

$\frac{1}{\cos\theta\sin\theta} = \frac{1}{\cos\theta\sin\theta}$ ✓

23. $(\sin\theta + \cos\theta)^2 \stackrel{?}{=} \frac{2 + \sec\theta\csc\theta}{\sec\theta\csc\theta}$

$(\sin\theta + \cos\theta)^2 \stackrel{?}{=} \frac{2 + \frac{1}{\cos\theta} \cdot \frac{1}{\sin\theta}}{\frac{1}{\cos\theta} \cdot \frac{1}{\sin\theta}}$

$(\sin\theta + \cos\theta)^2 \stackrel{?}{=} \left(2 + \frac{1}{\cos\theta\sin\theta}\right) \cdot \frac{\cos\theta\sin\theta}{1}$

$(\sin\theta + \cos\theta)^2 \stackrel{?}{=} 2\cos\theta\sin\theta + 1$

$(\sin\theta + \cos\theta)^2 \stackrel{?}{=} 2\cos\theta\sin\theta + \cos^2\theta + \sin^2\theta$

$(\sin\theta + \cos\theta)^2 = (\sin\theta + \cos\theta)^2$ ✓

25. $\csc\theta - 1 \stackrel{?}{=} \frac{\cot^2\theta}{\csc\theta + 1}$

$\csc\theta - 1 \stackrel{?}{=} \frac{\csc^2\theta - 1}{\csc\theta + 1}$

$\csc\theta - 1 \stackrel{?}{=} \frac{(\csc\theta - 1)(\csc\theta + 1)}{\csc\theta + 1}$

$\csc\theta - 1 = \csc\theta - 1$ ✓

27. $\sin\theta\cos\theta\tan\theta + \cos^2\theta \stackrel{?}{=} 1$

$\sin\theta\cos\theta \cdot \frac{\sin\theta}{\cos\theta} + \cos^2\theta \stackrel{?}{=} 1$

$\sin^2\theta + \cos^2\theta \stackrel{?}{=} 1$

$1 = 1$ ✓

29. $\csc^2\theta \stackrel{?}{=} \cot^2\theta + \sin\theta\csc\theta$

$\csc^2\theta \stackrel{?}{=} \cot^2\theta + \sin\theta \cdot \frac{1}{\sin\theta}$

$\csc^2\theta \stackrel{?}{=} \cot^2\theta + 1$

$\csc^2\theta = \csc^2\theta$ ✓

31. $\sin^2\theta + \cos^2\theta \stackrel{?}{=} \sec^2\theta - \tan^2\theta$

$1 \stackrel{?}{=} \tan^2\theta + 1 - \tan^2\theta$

$1 = 1$ ✓

33. yes

35 $\cot(-\theta)\tan(-\theta) = \frac{1}{\tan(-\theta)} \cdot \tan(-\theta)$ $\quad \frac{1}{\tan(-\theta)} = \cot(-\theta)$

$= 1$ $\quad$ Simplify.

37. 1 **39.** 1 **41.** $\cos\theta$ **43.** 2 **45.** $\sin\theta$ **47.** 1

49. $y = -\frac{gx^2}{2v_0^2}(1 + \tan^2\theta) + x\tan\theta$

51 **a.** $h = \frac{v_0^2\sin^2\theta}{2g} = \frac{47^2\sin^2\theta}{2(9.8)}$ $\quad$ Replace v_0 with 47 and g with 9.8.

$= \frac{2209\sin^2\theta}{19.6}$ $\quad$ Simplify.

$\frac{2209\sin^2 30°}{19.6} \approx 28.2$ m $\quad \theta = 30°$

$\frac{2209\sin^2 45°}{19.6} \approx 56.4$ m $\quad \theta = 45°$

$\frac{2209\sin^2 60°}{19.6} \approx 84.5$ m $\quad \theta = 60°$

$\frac{2209\sin^2 90°}{19.6} \approx 112.7$ m $\quad \theta = 90°$

b. Sample answer: Enter the equation $y = \frac{2209(\sin\theta)^2}{19.6}$. Use the window Xmin = 0, Xmax = 180, Xscl = 10, Ymin = 0, Ymax = 150, Yscl = 10, Xres = 1.

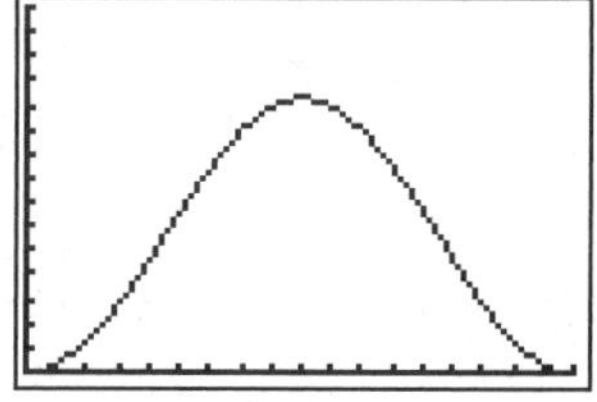

[0, 180] scl: 10 by [0, 150] scl: 10

c. $\frac{v_0^2\tan^2\theta}{2g\sec^2\theta} \stackrel{?}{=} \frac{v_0^2\sin^2\theta}{2g}$

$\frac{v_0^2\left(\frac{\sin^2\theta}{\cos^2\theta}\right)}{2g\left(\frac{1}{\cos^2\theta}\right)} \stackrel{?}{=} \frac{v_0^2\sin^2\theta}{2g}$ $\quad \tan^2\theta = \frac{\sin^2\theta}{\cos^2\theta}$ and $\sec^2\theta = \frac{1}{\cos^2\theta}$

$\frac{v_0^2\sin^2\theta}{2g} = \frac{v_0^2\sin^2\theta}{2g}$ ✓ $\quad$ Simplify.

53. $\tan^2\theta = \frac{\sin^2\theta}{\cos^2\theta} = \frac{1 - \cos^2\theta}{\cos^2\theta} = \frac{1}{\cos^2\theta} - \frac{\cos^2\theta}{\cos^2\theta}$

$= \sec^2\theta - 1$

55. Sample answer: counterexample 45°, 30°

57. Sample answer: Sine and cosine are the trigonometric functions with which most people are familiar, and all trigonometric expressions can be written in terms of sine and cosine. Also, by rewriting complex trigonometric expressions in terms of sine and cosine it may be easier to perform operations and to apply trigonometric properties.

59. Using the unit circle and the Pythagorean Theorem, we can justify $\cos^2\theta + \sin^2\theta = 1$.

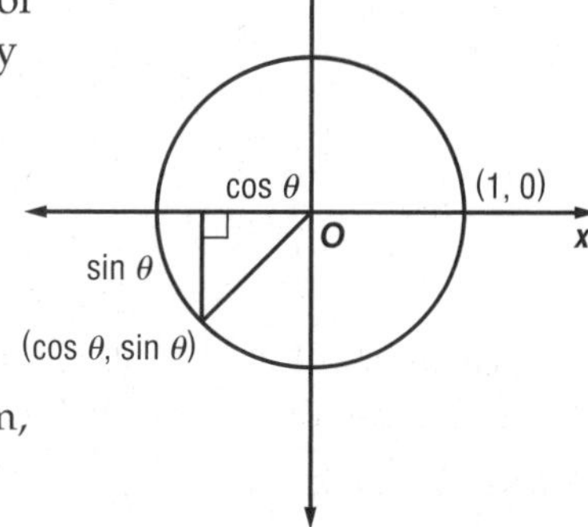

Selected Answers and Solutions

If we divide each term of the identity $\cos^2 \theta + \sin^2 \theta = 1$ by $\cos^2 \theta$, we can justify $1 + \tan^2 \theta = \sec^2 \theta$.

$$\frac{\cos^2 \theta}{\cos^2 \theta} + \frac{\sin^2 \theta}{\cos^2 \theta} = \frac{1}{\cos^2 \theta}$$

$$1 + \tan^2 \theta = \sec^2 \theta$$

If we divide each term of the identity $\cos^2 \theta + \sin^2 \theta = 1$ by $\sin^2 \theta$, we can justify $\cot^2 \theta + 1 = \csc^2 \theta$.

$$\frac{\cos^2 \theta}{\sin^2 \theta} + \frac{\sin^2 \theta}{\sin^2 \theta} = \frac{1}{\sin^2 \theta}$$

$$\cot^2 \theta + 1 = \csc^2 \theta$$

61. H **63.** G **65.** $\frac{\sqrt{5}}{3}$ **67.** $\frac{3}{5}$ **69.** \$4

71. $(1, -6 \pm 2\sqrt{5})$; $(1, -6 \pm 3\sqrt{5})$; $y + 6 = \pm\frac{2\sqrt{5}}{5}(x - 1)$

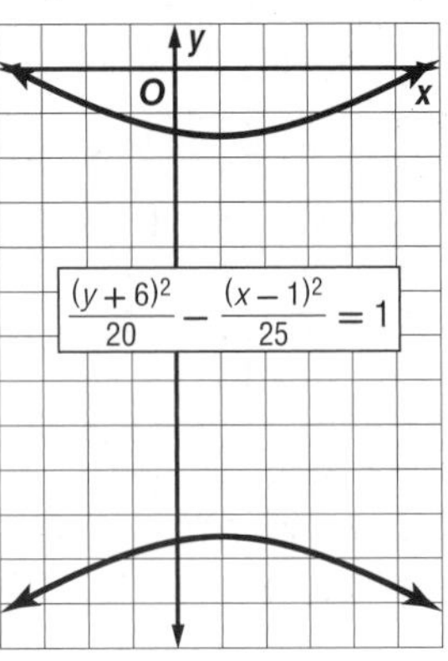

73. $\frac{12 + 7\sqrt{2}}{23}$

75. $\sqrt{x} + 1$

Lesson 13-3

1 $\cos 165° = \cos (120° + 45°)$

$$= \cos 120° \cos 45° - \sin 120° \sin 45°$$

$$= \left(-\frac{1}{2} \cdot \frac{\sqrt{2}}{2}\right) - \left(\frac{\sqrt{3}}{2} \cdot \frac{\sqrt{2}}{2}\right)$$

$$= -\frac{\sqrt{2}}{4} - \frac{\sqrt{6}}{4}$$

$$= -\frac{\sqrt{2} + \sqrt{6}}{4}$$

3. $\frac{\sqrt{6} - \sqrt{2}}{4}$ **5.** $\frac{\sqrt{2}}{2}$ **7a.** 0 **7b.** The interference is destructive. The signals cancel each other completely.

9. $\cos\left(\frac{3\pi}{2} - \theta\right) \stackrel{?}{=} -\sin \theta$

$\cos \frac{3\pi}{2} \cos \theta + \sin \frac{3\pi}{2} \sin \theta \stackrel{?}{=} -\sin \theta$

$0 \cdot \cos \theta - 1 \cdot \sin \theta \stackrel{?}{=} -\sin \theta$

$-\sin \theta = -\sin \theta$ ✓

11. $\sin(\theta + \pi) \stackrel{?}{=} -\sin \theta$

$\sin \theta \cos \pi + \cos \theta \sin \pi \stackrel{?}{=} -\sin \theta$

$(\sin \theta)(-1) + (\cos \theta)(0) \stackrel{?}{=} -\sin \theta$

$-\sin \theta = -\sin \theta$ ✓

13. $-\frac{\sqrt{2}}{2}$ **15.** $\frac{\sqrt{6} - \sqrt{2}}{4}$ **17.** $\frac{\sqrt{2} + \sqrt{6}}{4}$

19. $\cos\left(\frac{\pi}{2} + \theta\right) \stackrel{?}{=} -\sin \theta$

$\cos \frac{\pi}{2} \cos \theta - \sin \frac{\pi}{2} \sin \theta \stackrel{?}{=} -\sin \theta$

$(0)(\cos \theta) - (1)(\sin \theta) \stackrel{?}{=} -\sin \theta$

$-\sin \theta = -\sin \theta$ ✓

21. $\cos (180° + \theta) \stackrel{?}{=} -\cos \theta$

$\cos 180° \cos \theta - \sin 180° \sin \theta \stackrel{?}{=} -\cos \theta$

$-1 \cdot \cos \theta - 0 \cdot \sin \theta \stackrel{?}{=} -\cos \theta$

$-\cos \theta = -\cos \theta$ ✓

23a. $y = 30.9 \sin\left(\frac{\pi}{6}x - 2.09\right) + 42.65$ **23b.** The new function represents the average of the high and low temperatures for each month. **25.** $\sqrt{2} - \sqrt{6}$ **27.** $-2 + \sqrt{3}$ **29.** $2 - \sqrt{3}$

31 **a.** Let X be the endpoint of the segment that is 8 inches long.

$\sin (m\angle BAC)$

$= \sin (m\angle BAX + m\angle XAC)$

$= \sin (m\angle BAX) \cos (m\angle XAC) + \cos (m\angle BAX) \sin (m\angle XAC)$

$= \frac{8\sqrt{3}}{16} \cdot \frac{8}{10} + \frac{8}{16} \cdot \frac{6}{10}$ $\sin = \frac{\text{opp}}{\text{hyp}}$ and $\cos = \frac{\text{adj}}{\text{hyp}}$

$= \frac{4\sqrt{3}}{10} + \frac{3}{10}$ Multiply.

$= \frac{3 + 4\sqrt{3}}{10}$ Add.

b. Let X be the endpoint of the segment that is 8 inches long.

$\cos (m\angle BAC)$

$= \cos (m\angle BAX + m\angle XAC)$

$= \cos (m\angle BAX) \cos (m\angle XAC) - \sin (m\angle BAX) \sin (m\angle XAC)$

$= \frac{8}{16} \cdot \frac{8}{10} - \frac{8\sqrt{3}}{16} \cdot \frac{6}{10}$ $\sin = \frac{\text{opp}}{\text{hyp}}$ and $\cos = \frac{\text{adj}}{\text{hyp}}$

$= \frac{4}{10} - \frac{3\sqrt{3}}{10}$ Multiply.

$\approx \frac{4 - 3\sqrt{3}}{10}$ Add.

c. $\cos (m\angle BAC) = \frac{4 - 3\sqrt{3}}{10}$

$m\angle BAC = \text{Cos}^{-1}\left(\frac{3 + 4\sqrt{3}}{10}\right)$

$\approx 96.9°$

d. Since $m\angle BAC \neq 90$, the triangle formed is not a right triangle.

33a.

A	B	$\sin A$	$\sin B$	$\sin (A + B)$	$\sin A + \sin B$
30°	90°	$\frac{1}{2}$	1	$\frac{\sqrt{3}}{2}$	$\frac{3}{2}$
45°	60°	$\frac{\sqrt{2}}{2}$	$\frac{\sqrt{3}}{2}$	$\frac{\sqrt{2} + \sqrt{6}}{4}$	$\frac{\sqrt{2} + \sqrt{3}}{2}$
60°	45°	$\frac{\sqrt{3}}{2}$	$\frac{\sqrt{2}}{2}$	$\frac{\sqrt{2} + \sqrt{6}}{4}$	$\frac{\sqrt{2} + \sqrt{3}}{2}$
90°	30°	1	$\frac{1}{2}$	$\frac{\sqrt{3}}{2}$	$\frac{3}{2}$

33b.

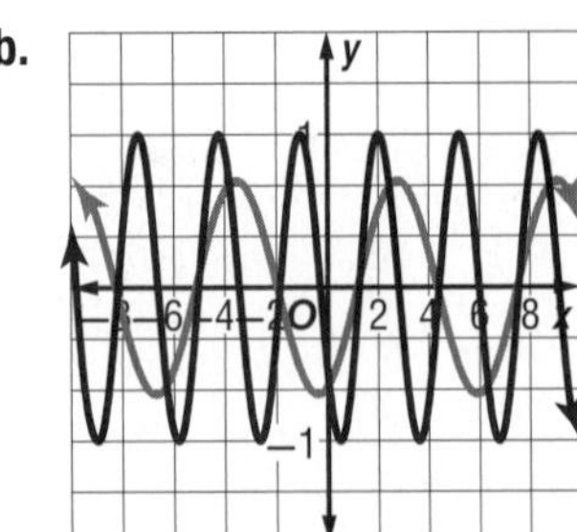

33c. No; a counterexample is: $\cos (30° + 45°) = \cos 30° + \cos 45°$, which equals $\frac{\sqrt{3}}{2} + \frac{\sqrt{2}}{2}$ or about 1.5731. Since a cosine value cannot be greater than 1, this statement must be false.

35 $\cos(A+B) \stackrel{?}{=} \dfrac{1-\tan A \tan B}{\sec A \sec B}$

$\cos(A+B) \stackrel{?}{=} \dfrac{1-\frac{\sin A}{\cos A}\cdot\frac{\sin B}{\cos B}}{\frac{1}{\cos A}\cdot\frac{1}{\cos B}}$

$\cos(A+B) \stackrel{?}{=} \dfrac{1-\frac{\sin A}{\cos A}\cdot\frac{\sin B}{\cos B}}{\frac{1}{\cos A}\cdot\frac{1}{\cos B}}\cdot\dfrac{\cos A\cos B}{\cos A\cos B}$ $\dfrac{\cos A\cos B}{\cos A\cos B}=1$

$\cos(A+B) \stackrel{?}{=} \left(1-\frac{\sin A}{\cos A}\cdot\frac{\sin B}{\cos B}\right)\left(\frac{\cos A\cos B}{1}\right)\left(\frac{\cos A\cos B}{\cos A\cos B}\right)$

$\cos(A+B) \stackrel{?}{=} \dfrac{\cos A\cos B-\sin A\sin B}{1}$ Simplify.

$\cos(A+B) = \cos(A+B)$ ✓ Difference Identity

37. $\sin(A+B)\sin(A-B) \stackrel{?}{=} \sin^2 A-\sin^2 B$
$(\sin A\cos B+\cos A\sin B)(\sin A\cos B-\cos A\sin B) \stackrel{?}{=} \sin^2 A-\sin^2 B$
$(\sin A\cos B)^2-(\cos A\sin B)^2 \stackrel{?}{=} \sin^2 A-\sin^2 B$
$\sin^2 B\cos^2 B-\cos^2 A\sin^2 B \stackrel{?}{=} \sin^2 A-\sin^2 B$
$\sin^2 A\cos^2 B+\sin^2 A\sin^2 B-\sin^2 A\sin^2 B-\cos^2 A\sin^2 B \stackrel{?}{=} \sin^2 A-\sin^2 B$
$\sin^2 A(\cos^2 B+\sin^2 B)-\sin^2 B(\sin^2 A+\cos^2 A) \stackrel{?}{=} \sin^2 A-\sin^2 B$
$(\sin^2 A)(1)-(\sin^2 B)(1) \stackrel{?}{=} \sin^2 A-\sin^2 B$
$\sin^2 A-\sin^2 B=\sin^2 A-\sin^2 B$ ✓

39. Sample answer: To determine wireless Internet interference, you need to determine the sine or cosine of the sum or difference of two angles. Interference occurs when waves pass through the same space at the same time. When the combined waves have a greater amplitude, constructive interference results. When the combined waves have a smaller amplitude, destructive interference results.

41. $d=\sqrt{(\cos A-\cos B)^2+(\sin A-\sin B)^2}$
$d^2=(\cos A-\cos B)^2+(\sin A-\sin B)^2$
$d^2=(\cos^2 A-2\cos A\cos B+\cos^2 B)+(\sin^2 A-2\sin A\sin B+\sin^2 B)$
$d^2=\cos^2 A+\sin^2 A+\cos^2 B+\sin^2 B-2\cos A\cos B-2\sin A\sin B$
$d^2=1+1-2\cos A\cos B-2\sin A\sin B$ $\sin^2 A+\cos^2 A=1$ and $\sin^2 B+\cos^2 B=1$
$d^2=2-2\cos A\cos B-2\sin A\sin B$

Now find the value of d^2 when the angle having measure $A-B$ is in standard position on the unit circle, as shown in the figure below.

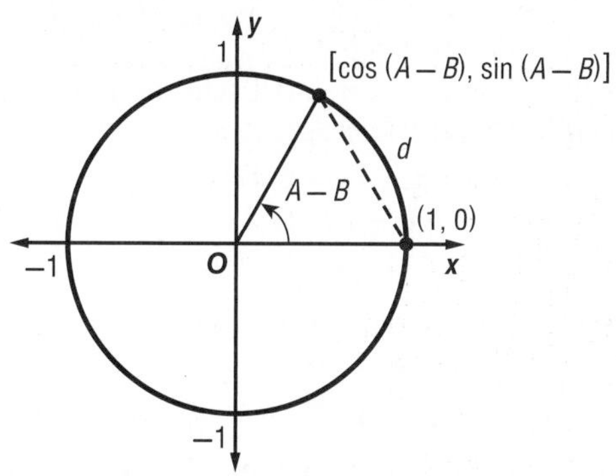

$d=\sqrt{[\cos(A-B)-1]^2+[\sin(A-B)-0]^2}$
$d^2=[\cos(A-B)-1]^2+[\sin(A-B)-0]^2$
$=[\cos^2(A-B)-2\cos(A-B)+1]+\sin^2(A-B)$
$=\cos^2(A-B)+\sin^2(A-B)-2\cos(A-B)+1$
$=1-2\cos(A-B)+1$
$=2-2\cos(A-B)$

43. 9 **45.** H

47. $\dfrac{\sin\theta}{\tan\theta}+\dfrac{\cos\theta}{\cot\theta} \stackrel{?}{=} \cos\theta+\sin\theta$

$\dfrac{\sin\theta}{\frac{\sin\theta}{\cos\theta}}+\dfrac{\cos\theta}{\frac{\cos\theta}{\sin\theta}} \stackrel{?}{=} \cos\theta+\sin\theta$

$\sin\theta\cdot\dfrac{\cos\theta}{\sin\theta}+\cos\theta\cdot\dfrac{\sin\theta}{\cos\theta} \stackrel{?}{=} \cos\theta+\sin\theta$

$\cos\theta+\sin\theta=\cos\theta+\sin\theta$ ✓

49. $\sin^2\theta$ **51.** $\sec\theta$

53. Step 1: $4^1-1=3$, which is divisible by 3. The statement is true for $n=1$.
Step 2: Assume that 4^k-1 is divisible by 3 for some positive integer k. This means that $4^k-1=3r$ for some whole number r.
Step 3: $4^k-1=3r$
$4^k=3r+1$
$4^{k+1}=12r+4$
$4^{k+1}-1=12r+3$
$4^{k+1}-1=3(4r+1)$
Since r is a whole number, $4r+1$ is a whole number. Thus, $4^{k+1}-1$ is divisible by 3, so the statement is true for $n=k+1$. Therefore, 4^n-1 is divisible by 3 for all positive integers n. **55.** -1 **57.** no solution

Lesson 13-4

1. $\frac{\sqrt{15}}{8}, \frac{7}{8}, \frac{\sqrt{8-2\sqrt{15}}}{4}, \frac{\sqrt{8+2\sqrt{15}}}{4}$ **3.** $-\frac{120}{169}, -\frac{119}{169}, \frac{3\sqrt{13}}{13}, \frac{2\sqrt{13}}{13}$ **5.** $-\frac{240}{289}, \frac{161}{289}, \frac{4\sqrt{17}}{17}, \frac{\sqrt{17}}{17}$ **7.** $\frac{\sqrt{2-\sqrt{2}}}{2}$

9a. $d=\dfrac{v^2\sin 2\theta}{g}$ **9b.** ≈ 81 ft

11. $(\sin\theta+\cos\theta)^2 \stackrel{?}{=} 1+2\sin\theta\cos\theta$
$(\sin\theta+\cos\theta)(\sin\theta+\cos\theta) \stackrel{?}{=} 1+2\sin\theta\cos\theta$
$\sin^2\theta+2\sin\theta\cos\theta+\cos^2\theta \stackrel{?}{=} 1+2\sin\theta\cos\theta$
$1+2\sin\theta\cos\theta=1+2\sin\theta\cos\theta$ ✓

13. $\frac{240}{289}, -\frac{161}{289}, \frac{5\sqrt{34}}{34}, -\frac{3\sqrt{34}}{34}$

15 $\sin^2\theta=1-\cos^2\theta$ $\sin^2\theta+\cos^2\theta=1$
$\sin^2\theta=1-\left(\frac{1}{5}\right)^2$ $\cos\theta=\frac{1}{5}$
$\sin^2\theta=\frac{24}{25}$ Subtract.
$\sin\theta=\pm\frac{2\sqrt{6}}{5}$ Take the square root of each side.
Since θ is in the fourth quadrant, sine is negative. So, $\sin\theta=-\frac{2\sqrt{6}}{5}$.
$\sin 2\theta=2\sin\theta\cos\theta$ Double-angle identity
$=2\left(-\frac{2\sqrt{6}}{5}\right)\left(\frac{1}{5}\right)$ $\sin\theta=-\frac{2\sqrt{6}}{5}$ and $\cos\theta=\frac{1}{5}$
$=-\frac{4\sqrt{6}}{25}$ Simplify.
$\cos 2\theta=1-2\sin^2\theta$ Double-angle identity
$=1-2\left(\frac{24}{25}\right)$ $\sin^2\theta=\frac{24}{25}$
$=-\frac{23}{25}$ Simplify.
$\sin\frac{\theta}{2}=\pm\sqrt{\frac{1-\cos\theta}{2}}$ Half-angle identity
$=\pm\sqrt{\frac{1-\frac{1}{5}}{2}}$ $\cos\theta=\frac{1}{5}$

$= \pm\sqrt{\frac{2}{5}}$ Simplify.

$= \pm\frac{\sqrt{2}}{\sqrt{5}} \cdot \frac{\sqrt{5}}{\sqrt{5}}$ or $\pm\frac{\sqrt{10}}{5}$ Rationalize the denominator.

If θ is between 270° and 360°, $\frac{\theta}{2}$ is between 135° and 180°. So, $\sin\frac{\theta}{2}$ is $\frac{\sqrt{10}}{5}$.

$\cos\frac{\theta}{2} = \pm\sqrt{\frac{1+\cos\theta}{2}}$ Half-angle identity

$= \pm\sqrt{\frac{1+\frac{1}{5}}{2}}$ $\cos\theta = \frac{1}{5}$

$= \pm\sqrt{\frac{3}{5}}$ Simplify.

$= \pm\sqrt{\frac{3}{5}} \cdot \frac{\sqrt{5}}{\sqrt{5}}$ or $\pm\frac{\sqrt{15}}{5}$ Rationalize the denominator.

If θ is between 270° and 360°, $\frac{\theta}{2}$ is between 135° and 180°. So, $\cos\frac{\theta}{2}$ is $-\frac{\sqrt{15}}{5}$.

17. $-\frac{4}{5}, -\frac{3}{5}, \sqrt{\frac{\sqrt{5}+1}{2\sqrt{5}}}, \sqrt{\frac{\sqrt{5}-1}{2\sqrt{5}}}$ **19.** $\frac{\sqrt{2+\sqrt{2}}}{2}$

21. $\sqrt{3} - 2$ **23.** $\sqrt{2} - 1$

25 $P = I_0^2 R \sin^2 \theta t$ Original equation

$P = I_0^2 R (\cos^2 \theta t - \cos 2\theta t)$ $\sin^2 \theta t = \cos^2 \theta t - \cos 2\theta t$

$P = I_0^2 R\left(\frac{1}{2}\cos 2\theta t + \frac{1}{2} - \cos 2\theta t\right)$ $\cos^2 \theta t = \frac{1}{2}\cos 2\theta t + \frac{1}{2}$

$P = I_0^2 R\left(\frac{1}{2} - \frac{1}{2}\cos 2\theta t\right)$ Simplify.

$P = \frac{1}{2} I_0^2 R - \frac{1}{2} I_0^2 R \cos 2\theta t$ Distributive Property

27. $1 + \frac{1}{2}\sin 2\theta \stackrel{?}{=} \frac{\sec\theta + \sin\theta}{\sec\theta}$

$\stackrel{?}{=} \frac{\frac{1}{\cos\theta} + \sin\theta}{\frac{1}{\cos\theta}}$

$\stackrel{?}{=} \frac{\frac{1}{\cos\theta} + \sin\theta}{\frac{1}{\cos\theta}} \cdot \frac{\cos\theta}{\cos\theta}$

$\stackrel{?}{=} 1 + \frac{1}{2} \cdot 2\sin\theta\cos\theta$

$\stackrel{?}{=} 1 + \frac{1}{2}\sin 2\theta$ ✓

29. $\tan\frac{\theta}{2} \stackrel{?}{=} \frac{\sin\theta}{1+\cos\theta}$

$\tan\frac{\theta}{2} \stackrel{?}{=} \frac{\sin 2\left(\frac{\theta}{2}\right)}{1+\cos 2\left(\frac{\theta}{2}\right)}$

$\tan\frac{\theta}{2} \stackrel{?}{=} \frac{2\sin\frac{\theta}{2}\cos\frac{\theta}{2}}{1 + 2\cos^2\frac{\theta}{2} - 1}$

$\tan\frac{\theta}{2} \stackrel{?}{=} \frac{2\sin\frac{\theta}{2}\cos\frac{\theta}{2}}{2\cos^2\frac{\theta}{2}}$

$\tan\frac{\theta}{2} \stackrel{?}{=} \frac{\sin\frac{\theta}{2}}{\cos\frac{\theta}{2}}$

$\tan\frac{\theta}{2} = \tan\frac{\theta}{2}$ ✓

31. $\frac{24}{25}, \frac{7}{25}, \frac{24}{7}$ **33.** $-\frac{3}{5}, -\frac{4}{5}, \frac{3}{4}$ **35.** $-\frac{4\sqrt{21}}{25}, \frac{17}{25}, -\frac{4\sqrt{21}}{17}$

37. No; Teresa incorrectly added the square roots, and Nathan used the half-angle identity incorrectly. He used sin 30° in the formula instead of first finding the cosine. **39.** If you are only given the value of $\cos\theta$, then $\cos 2\theta = 2\cos^2\theta - 1$ is the best identity to use. If you are only given the value of $\sin\theta$, then $\cos 2\theta = 1 - 2\sin^2\theta$ is the best identity to use. If you are given the values of both $\cos\theta$ and $\sin\theta$, then $\cos 2\theta = \cos^2\theta - \sin^2\theta$ works just as well as the other two.

41. $1 - 2\sin^2\theta = \cos 2\theta$ Double-angle identity

$1 - 2\sin^2\frac{A}{2} = \cos A$ Substitute $\frac{A}{2}$ for θ and A for 2θ.

$\sin^2\frac{A}{2} = \frac{1-\cos A}{2}$ Solve for $\sin^2\frac{A}{2}$.

$\sin\frac{A}{2} = \pm\sqrt{\frac{1-\cos A}{2}}$ Take the square root of each side.

Find $\cos\frac{A}{2}$.

$2\cos^2\theta - 1 = \cos 2\theta$ Double-angle identity

$2\cos^2\frac{A}{2} - 1 = \cos A$ Substitute $\frac{A}{2}$ for θ and A for 2θ.

$\cos^2\frac{A}{2} = \frac{1+\cos A}{2}$ Solve for $\cos^2\frac{A}{2}$.

$\cos\frac{A}{2} = \pm\sqrt{\frac{1+\cos A}{2}}$ Take the square root of each side.

43. 22.5 **45.** G **47.** $\frac{\sqrt{2}}{2}$ **49.** $\frac{-\sqrt{6}-\sqrt{2}}{4}$ **51.** $\frac{\sqrt{3}}{2}$

53. $\cot\theta + \sec\theta \stackrel{?}{=} \frac{\cos^2\theta + \sin\theta}{\sin\theta\cos\theta}$

$\cot\theta + \sec\theta \stackrel{?}{=} \frac{\cos^2\theta}{\sin\theta\cos\theta} + \frac{\sin\theta}{\sin\theta\cos\theta}$

$\cot\theta + \sec\theta \stackrel{?}{=} \frac{\cos\theta}{\sin\theta} + \frac{1}{\cos\theta}$

$\cot\theta + \sec\theta = \cot\theta + \sec\theta$

55. sines; $B \approx 102°$, $C \approx 44°$, $b \approx 21.0$ or $B \approx 10°$, $C \approx 136°$, $b \approx 3.7$ **57.** sines; $A = 80°$, $a \approx 10.9$, $c \approx 5.4$ **59.** $\{-4, 7\}$

Lesson 13-5

1. 210°, 330° **3.** 60°, 180°, or 300° **5.** 150°, 210°

7 $\cos 2\theta = 8 - 15\sin\theta$ Original equation

$\cos 2\theta + 15\sin\theta - 8 = 0$ Add $15\sin\theta - 8$ to each side.

$1 - 2\sin^2\theta + 15\sin\theta - 8 = 0$ Double-angle identity

$-2\sin^2\theta + 15\sin\theta - 7 = 0$ Simplify.

$(-2\sin\theta + 1)(\sin\theta - 7) = 0$ Factor.

$-2\sin\theta + 1 = 0$ or $\sin\theta - 7 = 0$ Zero Product Property

$\sin\theta = \frac{1}{2}$ $\sin\theta = 7$

$\theta = 30°$ or $150°$ no solution since $0 \le \sin\theta \le 1$

9. $\pm\frac{\pi}{6} + 2k\pi$ or $\pm\frac{5\pi}{6} + 2k\pi$ **11.** $\frac{3\pi}{2} + 2k\pi$ **13.** $\pi + 2k\pi$ **15.** $90° + k \cdot 180°$ **17.** $45° + k \cdot 90°$ **19.** $270° + k \cdot 360°$

21a. There will be $10\frac{1}{2}$ hours of daylight 213 and 335 days after March 21; that is, on October 20 and February 19. **21b.** Every day from February 19 to October 20; sample explanation: Since the longest

day of the year occurs around June 22, the days between February 19 and October 20 must increase in length until June 22 and then decrease in length until October 20. **23.** $\frac{3\pi}{4} + \pi k$ **25.** $\frac{\pi}{2} + \pi k, \frac{\pi}{6} + 2\pi k, \frac{5\pi}{6} + 2\pi k$ **27.** $0° + k \cdot 45°$ or $0 + k \cdot \frac{\pi}{4}$
29. $90° + k \cdot 180°, 60° + k \cdot 360°, 300° + k \cdot 360°$
31. 135°, 225° **33.** $\frac{\pi}{6}$ **35.** 210°, 330°

37

$2 \sin^2 \theta = \cos \theta + 1$	Original equation
$2(1 - \cos^2 \theta) = \cos \theta + 1$	$\sin^2 \theta = 1 - \cos^2 \theta$
$2 - 2\cos^2 \theta = \cos \theta + 1$	Simplify.
$-2\cos^2 \theta - \cos \theta + 1 = 0$	Subtract $\cos \theta + 1$ from each side and simplify.
$(-2\cos \theta + 1)(\cos \theta + 1) = 0$	Factor.
$-2\cos \theta + 1 = 0$ or $\cos \theta + 1 = 0$	Zero Product Property
$\cos \theta = \frac{1}{2}$ $\quad$ $\cos \theta = -1$	

The solutions of $\cos \theta = \frac{1}{2}$ are $\frac{\pi}{3} + 2k\pi$ and $\frac{5\pi}{3} + 2k\pi$.
The solution of $\cos \theta = -1$ is $\pi + 2k\pi$.

39. $0 + 2k\pi$ **41.** $0° + k \cdot 180°$ **43.** $30° + k \cdot 360°, 150° + k \cdot 360°$ **45.** $\frac{7\pi}{6} + 2k\pi, \frac{11\pi}{6} + 2k\pi$ or $210° + k \cdot 360°, 330° + k \cdot 360°$ **47.** $0 + 2k\pi, \frac{\pi}{2} + k\pi$ or $0° + k \cdot 360°, 90° + k \cdot 180°$ **49a.** 11 m **49b.** 7:00 A.M. and 7:00 P.M. **51.** $\frac{\pi}{6} + 2\pi k, \frac{5\pi}{6} + 2\pi k, \frac{5\pi}{4} + 2\pi k, \frac{7\pi}{4} + 2\pi k$ **53.** $120° + 360°k, 240° + 360°k$
55. $\frac{\pi}{6} + 2\pi k, \frac{5\pi}{6} + 2\pi k$

57

$D = 0.5 \sin(6.5x) \sin(2500t)$	Original equation
$0.01 = 0.5 \sin(6.5 \cdot 500) \sin(2500t)$	$D = 0.01$ mm and $x = 0.5 \cdot 1000$ or 500 mm
$0.01 = 0.5 \sin(3250) \sin(2500t)$	Simplify.
$0.1152 \approx \sin(2500t)$	Divide each side by 0.5 sin (3250).
$\text{Sin}^{-1}(0.1152) \approx 2500t$	Use the $\sin^{-1}$ function.
$6.6152 \approx 2500t$	Use a calculator.
$0.0026 \approx t$	Divide each side by 2500.

The time is about 0.0026 second.

59. $\frac{\pi}{3} < x < \pi$ or $\frac{5\pi}{3} < x < 2\pi$ **61.** Sample answer: All trigonometric functions are periodic. Therefore, once one or more solutions are found for a certain interval, there will be additional solutions that can be found by adding integral multiples of the period of the function to those solutions. **63.** 0, b, $2b$ **65.** A **67.** E

69. $\frac{\sqrt{2 - \sqrt{2}}}{2}$ **71.** $-\frac{\sqrt{2 - \sqrt{3}}}{2}$

73.
$$\cos(90° + \theta) \stackrel{?}{=} -\sin \theta$$
$$\cos 90° \cos \theta - \sin 90° \sin \theta \stackrel{?}{=} -\sin \theta$$
$$0 - 1 \sin \theta \stackrel{?}{=} -\sin \theta$$
$$-\sin \theta = -\sin \theta \checkmark$$

75.
$$\sin(90° - \theta) \stackrel{?}{=} \cos \theta$$
$$\sin 90° \cos \theta - \cos 90° \sin \theta \stackrel{?}{=} \cos \theta$$
$$1 \cdot \cos \theta - 0 \cdot \sin \theta \stackrel{?}{=} \cos \theta$$
$$\cos \theta - 0 \stackrel{?}{=} \cos \theta$$
$$\cos \theta = \cos \theta \checkmark$$

77. 17, 26, 35
79. −12, −9, −6
81.

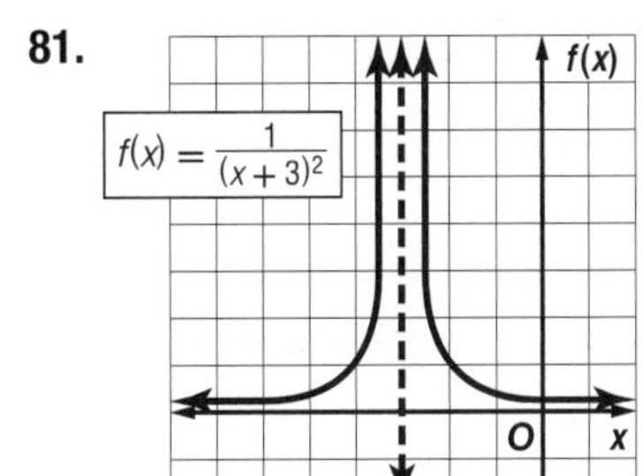

83.

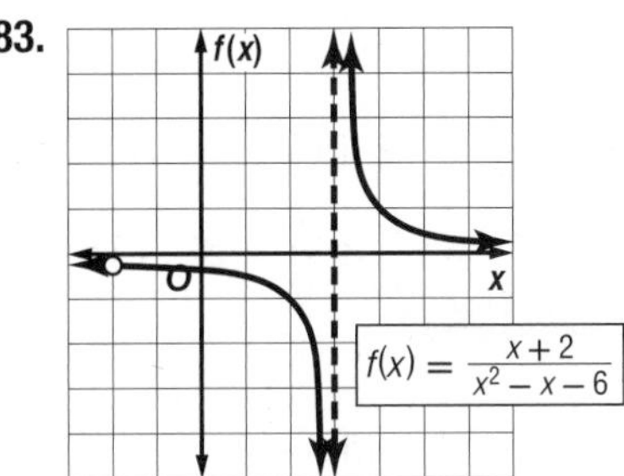

Chapter 13 Study Guide and Review

1. difference of angles identity **3.** trigonometric identity **5.** trigonometric equation **7.** reciprocal identities **9.** Pythagorean identity **11.** $-\sqrt{3}$ **13.** $-\frac{4}{5}$
15. First find the length of the diagonal: $75^2 + 110^2 = c^2$; $5625 + 12{,}100 = c^2$; $17{,}725 = c^2$; $c = 5\sqrt{709}$; $\sin \theta = \frac{75}{5\sqrt{709}} = \frac{15\sqrt{709}}{709}$ **17.** sec θ **19.** sec θ

21.
$$\frac{\cos \theta}{\cot \theta} + \frac{\sin \theta}{\tan \theta} \stackrel{?}{=} \sin \theta + \cos \theta$$
$$\cos \theta \div \frac{\cos \theta}{\sin \theta} + \sin \theta \div \frac{\sin \theta}{\cos \theta} \stackrel{?}{=} \sin \theta + \cos \theta$$
$$\cos \theta \cdot \frac{\sin \theta}{\cos \theta} + \sin \theta \cdot \frac{\cos \theta}{\sin \theta} \stackrel{?}{=} \sin \theta + \cos \theta$$
$$\sin \theta + \cos \theta = \sin \theta + \cos \theta \checkmark$$

23. $\tan^2 \theta + 1 = \left(\frac{\sqrt{7}}{3}\right)^2 + 1 = \frac{7}{9} + 1 = \frac{7}{9} + \frac{9}{9} = \frac{16}{9}$; $\sec^2 \theta = \left(\frac{4}{3}\right)^2 = \frac{16}{9}$

25. $\frac{\sqrt{6} + \sqrt{2}}{4}$ **27.** $\frac{\sqrt{6} + \sqrt{2}}{4}$ **29.** $\frac{-\sqrt{6} + \sqrt{2}}{4}$

31.
$$\sin\left(\frac{3\pi}{2} - \theta\right) \stackrel{?}{=} -\cos \theta$$
$$\sin \frac{3\pi}{2} \cos \theta - \cos \frac{3\pi}{2} \sin \theta \stackrel{?}{=} -\cos \theta$$
$$(-1)\cos \theta - (0)\sin \theta \stackrel{?}{=} -\cos \theta$$
$$-\cos \theta = -\cos \theta \checkmark$$

33. $\sin 2\theta = \frac{24}{25}, \cos 2\theta = \frac{7}{25}, \sin \frac{\theta}{2} = \frac{\sqrt{10}}{10}$, and $\cos \frac{\theta}{2} = \frac{3\sqrt{10}}{10}$ **35.** $\sin 2\theta = -\frac{4\sqrt{5}}{9}, \cos 2\theta = -\frac{1}{9}, \sin \frac{\theta}{2} = \frac{\sqrt{30}}{6}$, and $\cos \frac{\theta}{2} = \frac{\sqrt{6}}{6}$ **37.** 60°, 300°
39. 90°, 210°, 270°, 330° **41.** $\frac{\pi}{3}, \frac{5\pi}{3}$

Selected Answers and Solutions

Glossary/Glosario

MultilingualeGlossary

Go to **connectED.mcgraw-hill.com** for a glossary of terms in these additional languages:

Arabic	Chinese	Hmong	Spanish	Vietnamese
Bengali	English	Korean	Tagalog	
Brazilian Portugese	Haitian Creole	Russian	Urdu	

English

Español

A

absolute value (p. 27) A number's distance from zero on the number line, represented by $|x|$.

valor absoluto Distancia entre un número y cero en una recta numérica; se denota con $|x|$.

absolute value function (p. 103) A function written as $f(x) = |x|$, where $f(x) = \begin{cases} x \text{ if } x > 0 \\ 0 \text{ if } x = 0 \\ -x \text{ if } x < 0 \end{cases}$, for all values of x.

función del valor absoluto Una función que se escribe $f(x) = |x|$, donde $f(x) = \begin{cases} x \text{ si } x > 0 \\ 0 \text{ si } x = 0 \\ -x \text{ si } x < 0 \end{cases}$, para todos los valores de x.

algebraic expression (p. 5) An expression that contains at least one variable.

expresión algebraica Expresión que contiene al menos una variable.

alternative hypothesis (p. 771) Mutually exclusive to the null hypothesis. It is stated as an inequality using $\neq, <, \leq, >$, or $\geq$.

hipótesis alternativa Mutuamente exclusiva a la hipótesis nula. Se indica como usar de la desigualdad $\neq, <, \leq, >$, o $\geq$.

amplitude (p. 837) For functions in the form $y = a \sin b\theta$ or $y = a \cos b\theta$, the amplitude is $|a|$.

amplitud Para funciones de la forma $y = a$ sen $b\theta$ o $y = a \cos b\theta$, la amplitud es $|a|$.

angle of depression (p. 794) The angle between a horizontal line and the line of sight from the observer to an object at a lower level.

ángulo de depresión Ángulo entre una recta horizontal y la línea visual de un observador a una figura en un nivel inferior.

angle of elevation (p. 794) The angle between a horizontal line and the line of sight from the observer to an object at a higher level.

ángulo de elevación Ángulo entre una recta horizontal y la línea visual de un observador a una figura en un nivel superior.

Arccosine (p. 853) The inverse of $y = \cos x$, written as $x = \text{Arccos } y$.

arcocoseno La inversa de $y = \cos x$, que se escribe como $x = \text{arccos } y$.

Arcsine (p. 853) The inverse of $y = \sin x$, written as $x = \text{Arcsin } y$.

arcoseno La inversa de $y = \text{sen } x$, que se escribe como $x = \text{arcsen } y$.

Arctangent (p. 853) The inverse of $y = \tan x$ written as $x = \text{Arctan } y$.

arcotangente La inversa de $y = \tan x$ que se escribe como $x = \text{arctan } y$.

arithmetic mean (p. 667) The terms between any two nonconsecutive terms of an arithmetic sequence.

media aritmética Cualquier término entre dos términos no consecutivos de una sucesión aritmética.

arithmetic sequence (p. 659) A sequence in which each term after the first is found by adding a constant, the common difference d, to the previous term.

sucesión aritmética Sucesión en que cualquier término después del primero puede hallarse sumando una constante, la diferencia común d, al término anterior.

arithmetic series (p. 668) The indicated sum of the terms of an arithmetic sequence.

serie aritmética Suma específica de los términos de una sucesión aritmética.

asymptote (p. 451) A line that a graph approaches.

asíntota Recta a la que se aproxima una gráfica.

axis of symmetry (p. 220) A line about which a figure is symmetric.

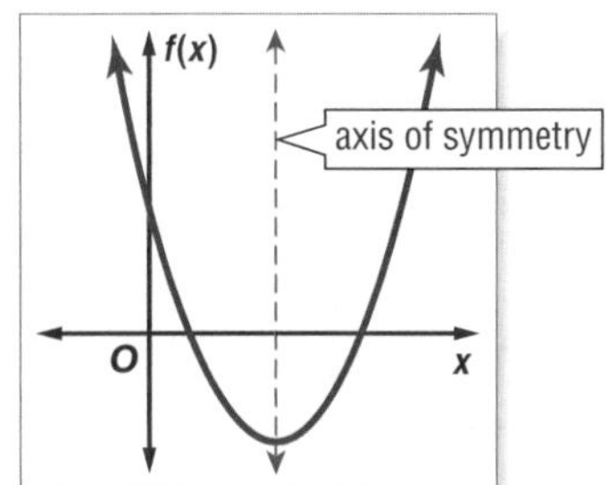

eje de simetría Recta respecto a la cual una figura es simétrica.

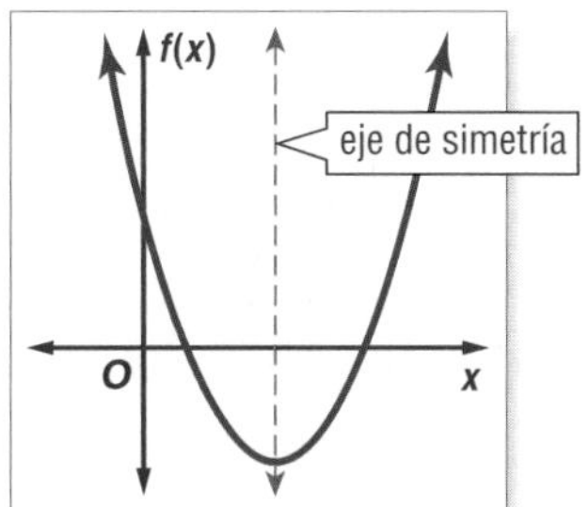

B

$b^{\frac{1}{n}}$ (p. 422) For any real number b and for any positive integer n, $b^{\frac{1}{n}} = \sqrt[n]{b}$, except when $b < 0$ and n is even.

$b^{\frac{1}{n}}$ Para cualquier número real b y para cualquier entero positivo n, $b^{\frac{1}{n}} = \sqrt[n]{b}$, excepto cuando $b < 0$ y n es par.

bias (p. 723) An error that results in a misrepresentation of members of a population.

sesgo Error que resulta en la representación errónea de los miembros de una población.

binomial distribution (p. 754) A distribution that shows the probabilities of the outcomes of a binomial experiment.

distribución binómica Distribución que muestra las probabilidades de los resultados de un experimento binómico.

binomial experiment (p. 752) An experiment in which there are exactly two possible outcomes for each trial, a fixed number of independent trials, and the probabilities for each trial are the same.

experimento binomial Experimento con exactamente dos resultados posibles para cada prueba, un número fijo de pruebas independientes y en el cual cada prueba tiene igual probabilidad.

Binomial Theorem (p. 699) If n is a nonnegative integer, then $(a + b)^n = 1a^nb^0 + \frac{n}{1}a^{n-1}b^1 + \frac{n(n+1)}{1 \cdot 2}a^{n-2}b^2 + \cdots + 1a^0b^n$.

teorema del binomio Si n es un entero no negativo, entonces $(a + b)^n = 1a^nb^0 + \frac{n}{1}a^{n-1}b^1 + \frac{n(n+1)}{1 \cdot 2}a^{n-2}b^2 + \cdots + 1a^0b^n$.

bivariate data (p. 92) Data consisting of pairs of values.

datos bivariados Datos que consiste en pares de valores.

boundary (p. 117) A line or curve that separates the coordinate plane into two regions.

frontera Recta o curva que divide un plano de coordenadas en dos regiones.

bounded (p. 154) A region is bounded when the graph of a system of constraints is a polygonal region.

acotada Una región está acotada cuando la gráfica de un sistema de restricciones es una región poligonal.

box-and-whisker plot (p. 735) A diagram that divides a set of data into four parts using the median and quartiles. A box is drawn around the quartile values and whiskers extend from each quartile to the extreme data points.

diagrama de caja y bigotes Diagrama que divide un conjunto de datos en cuatro partes usando la mediana y los cuartiles. Se dibuja una caja alrededor de los cuartiles y se extienden patillas de cada uno de ellos a los valores extremos.

break-even point (p. 136) The point at which the income equals the cost.

el punto de equilibrio Cuando el punto la renta iguala la causalidad del coste.

C

Cartesian coordinate plane (p. P4) A plane divided into four quadrants by the intersection of the *x*-axis and the *y*-axis at the origin.

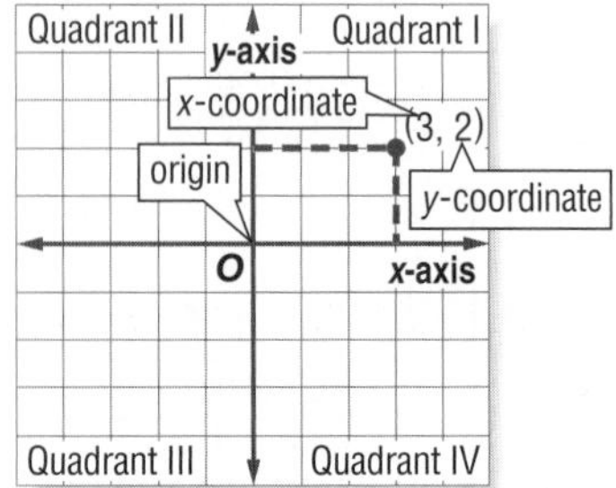

plano de coordenadas cartesiano Plano dividido en cuatro cuadrantes mediante la intersección en el origen de los ejes *x* y *y*.

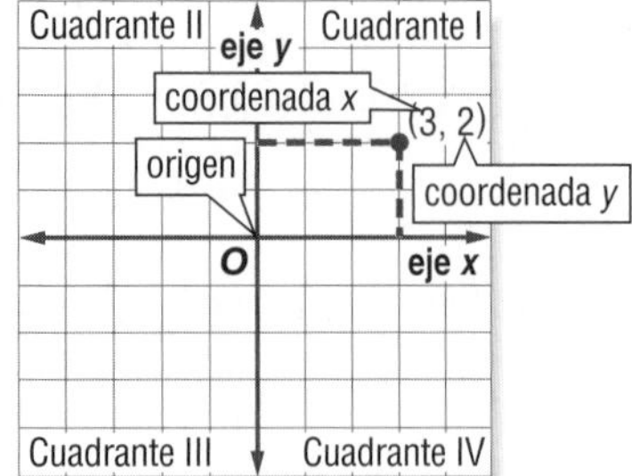

census (p. 724) A survey in which every member of the population is polled.

censo Examen en el cual voten a cada miembro de la población.

center of a circle (p. 607) The point from which all points on a circle are equidistant.

centro de un círculo El punto desde el cual todos los puntos de un círculo están equidistantes.

center of a hyperbola (p. 624) The midpoint of the segment whose endpoints are the foci.

centro de una hipérbola Punto medio del segmento cuyos extremos son los focos.

center of an ellipse (p. 615) The point at which the major axis and minor axis of an ellipse intersect.

centro de una elipse Punto de intersección de los ejes mayor y menor de una elipse.

central angle (p. 802) An angle with a vertex at the center of the circle and sides that are radii.

ángulo central Ángulo cuyo vértice es el centro del círculo y cuyos lados son radios.

Change of Base Formula (p. 494) For all positive numbers *a*, *b*, and *n*, where $a \neq 1$ and $b \neq 1$,

$$\log_a n = \frac{\log_b n}{\log_b a}.$$

fórmula del cambio de base Para todo número positivo *a*, *b* y *n*, donde $a \neq 1$ y $b \neq 1$,

$$\log_a n = \frac{\log_b n}{\log_b a}.$$

circle (p. 607) The set of all points in a plane that are equidistant from a given point in the plane, called the center.

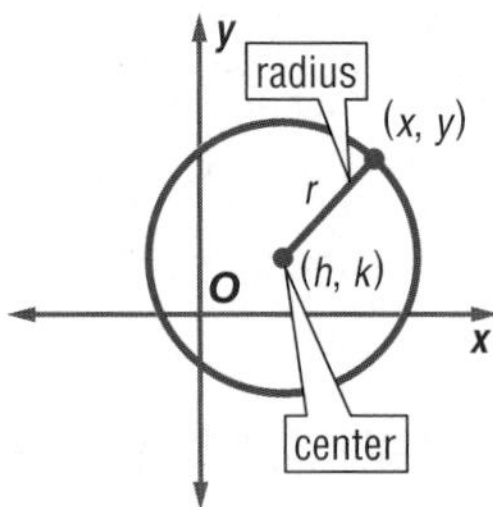

círculo Conjunto de todos los puntos en un plano que equidistan de un punto dado del plano llamado centro.

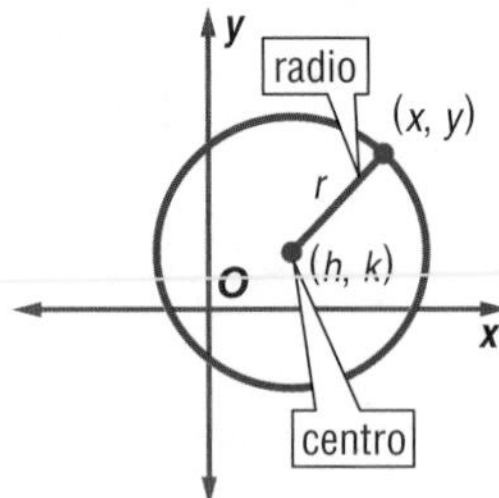

circular function (p. 830) A function defined using a unit circle.

funciones circulares Funciones definidas en un círculo unitario.

coefficient matrix (p. 192) A matrix that contains only the coefficients of a system of equations.

matriz coeficiente Una matriz que contiene solamente los coeficientes de un sistema de ecuaciones.

combination (p. P11) An arrangement of objects in which order is not important.

combinación Arreglo de elementos en que el orden no es importante.

combined variation (p. 565) When one quantity varies directly and/or inversely as two or more other quantities.

variación combinada Cuando una cantidad varía directamente e inverso como dos o más otras cantidades.

common difference (p. 659) The difference between the successive terms of an arithmetic sequence.

diferencia común Diferencia entre términos consecutivos de una sucesión aritmética.

common logarithms (p. 492) Logarithms that use 10 as the base.

logaritmos comunes El logaritmo de base 10.

common ratio (p. 661) The ratio of successive terms of a geometric sequence.

razón común Razón entre términos consecutivos de una sucesión geométrica.

completing the square (p. 257) A process used to make a quadratic expression into a perfect square trinomial.

completar el cuadrado Proceso mediante el cual una expresión cuadrática se transforma en un trinomio cuadrado perfecto.

complex conjugates (p. 249) Two complex numbers of the form $a + bi$ and $a - bi$.

conjugados complejos Dos números complejos de la forma $a + bi$ y $a - bi$.

complex fraction (p. 532) A rational expression whose numerator and/or denominator contains a rational expression.

fracción compleja Expresión racional cuyo numerador o denominador contiene una expresión racional.

complex number (p. 247) Any number that can be written in the form $a + bi$, where a and b are real numbers and i is the imaginary unit.

número complejo Cualquier número que puede escribirse de la forma $a + bi$, donde a y b son números reales e i es la unidad imaginaria.

composition of functions (p. 387) A function is performed, and then a second function is performed on the result of the first function. The composition of f and g is denoted by $f \circ g$, and $[f \circ g](x) = f[g(x)]$.

composición de funciones Se evalúa una función y luego se evalúa una segunda función en el resultado de la primera función. La composición de f y g se define con $f \circ g$, y $[f \circ g](x) = f[g(x)]$.

compound event (p. P14) Two or more simple events.

evento compuesto Dos o más eventos simples.

compound inequality (p. 41) Two inequalities joined by the word *and* or *or*.

desigualdad compuesta Dos desigualdades unidas por las palabras *y* u *o*.

compound interest (p. 462) Interest paid on the principal of an investment and any previously earned interest.

interés compuesto Interés obtenido tanto sobre la inversion inicial como sobre el interes conseguido.

conditional probability (p. P16) The probability of an event occurring given that another event has already occurred.

probabilidad condicional Probabilidad de un evento dado que otro evento ya ha ocurrido.

confidence interval (p. 769) An estimate of a population parameter stated as a range with a specific degree of certainty.

intervalo de la confianza Una estimación de un parámetro de la población indicado como gama con un grado específico de la certeza.

conic section (p. 632) Any figure that can be obtained by slicing a double cone.

sección cónica Cualquier figura obtenida mediante el corte de un cono doble.

conjugate axis (p. 624) The segment of length $2b$ units that is perpendicular to the transverse axis at the center.

eje conjugado El segmento de $2b$ unidades de longitud que es perpendicular al eje transversal en el centro.

conjugates (p. 418) Binomials of the form $a\sqrt{b} + c\sqrt{d}$ and $a\sqrt{b} - c\sqrt{d}$, where a, b, c, and d are rational numbers.

conjugados Binomios de la forma $a\sqrt{b} + c\sqrt{d}$ y $a\sqrt{b} - c\sqrt{d}$, donde a, b, c, y d son números racionales.

consistent (p. 137) A system of equations that has at least one ordered pair that satisfies both equations.

consistente Sistema de ecuaciones para el cual existe al menos un par ordenado que satisface ambas ecuaciones.

constant difference (p. 627) The absolute value of the difference between the distances from any point on the hyperbola to the foci of the hyperbola.

diferencia constante El valor absoluto de la diferencia entre las distancias de cualquier punto en la hipérbola a los focos de la hipérbola.

constant function (p. 109) A linear function of the form $f(x) = b$.

constant matrix (p. 200) A matrix that contains the constants of a system.

constant of variation (pp. 90, 562) The constant k used with direct or inverse variation.

constant sum (p. 616) The sum of the distances from the foci to any point on the ellipse.

constant term (p. 219) In $f(x) = ax^2 + bx + c$, c is the constant term.

constraints (p. 29) Conditions given to variables, often expressed as linear inequalities.

continuous random variable (p. 742) The numerical outcome of a random event that can take on any value.

continuous relation (p. 62) A relation that can be graphed with a line or smooth curve.

convergent series (p. 683) An infinite series with a sum.

correlation coefficient (p. 94) A measure that shows how well data are modeled by a linear equation.

cosecant (p. 790) For any angle, with measure α, a point $P(x, y)$ on its terminal side, $r = \sqrt{x^2 + y^2}$, $\csc \alpha = \frac{r}{y}$.

cosine (p. 790) For any angle, with measure α, a point $P(x, y)$ on its terminal side, $r = \sqrt{x^2 + y^2}$, $\cos \alpha = \frac{x}{r}$.

cotangent (p. 790) For any angle, with measure α, a point $P(x, y)$ on its terminal side, $r = \sqrt{x^2 + y^2}$, $\cot \alpha = \frac{x}{y}$.

coterminal angles (p. 800) Two angles in standard position that have the same terminal side.

co-vertices (pp. 615, 624)
ellipse—The endpoints of the minor axis.
hyperbola—The endpoints of the conjugate axis.

Cramer's Rule (p. 192) A method that uses determinants to solve a system of linear equations.

critical region (p. 771) The range of values that suggests a significant enough difference to reject the null hypothesis.

cycle (p. 831) One complete pattern of a periodic function.

función constante Función lineal de la forma $f(x) = b$.

matriz constante Una matriz que contiene las constantes de un sistema

constante de variación La constante k que se usa en variación directa o inversa.

suma constante La suma de las distancias de los focos a cualquier punto en la elipse.

término constante En $f(x) = ax^2 + bx + c$, c es el término constante.

restricciones Condiciones a que están sujetas las variables, a menudo escritas como desigualdades lineales.

variable aleatoria continua Resultado numérico de un evento aleatorio que puede tomar cualquier valor.

relación continua Relación cuya gráfica puede ser una recta o una curva suave.

serie convergente Serie infinita con una suma.

coeficiente de correlación Una medida que demuestra cómo los datos bien son modelados por una ecuación linear.

cosecante Para cualquier ángulo de medida α, un punto $P(x, y)$ en su lado terminal, $r = \sqrt{x^2 + y^2}$, $\csc \alpha = \frac{r}{y}$.

coseno Para cualquier ángulo de medida α, un punto $P(x, y)$ en su lado terminal, $r = \sqrt{x^2 + y^2}$, $\cos \alpha = \frac{x}{r}$.

cotangente Para cualquier ángulo de medida α, un punto $P(x, y)$ en su lado terminal, $r = \sqrt{x^2 + y^2}$, $\cot \alpha = \frac{x}{y}$.

ángulos coterminales Dos ángulos en posición estándar que tienen el mismo lado terminal.

co-cimas
(elipse)—Puntos finales del eje de menor importancia.
(hipérbola)—Puntos finales del eje conyugal.

regla de Crámer Método que usa determinantes para resolver un sistema de ecuaciones lineales.

región crítica Rango de valores que sugiere una diferencia suficientemente significativa para rechazar la hipótesis nula.

ciclo Un patrón completo de una función periódica.

D

decay factor (p. 454) In exponential decay, the base of the exponential expression, $1 - r$.

factor de decaimiento En decaimiento exponencial, la base de la expresión exponencial, $1 - r$.

degree of a polynomial (p. 305) The greatest degree of any term in the polynomial.

grado de un polinomio Grado máximo de cualquier término del polinomio.

dependent (p. 137) A system of equations that has an infinite number of solutions.

dependiente Sistema de ecuaciones que posee un número infinito de soluciones.

dependent events (p. P16) Events in which the outcome of one event affects the outcome of another event.

evento dependiente Eventos en que el resultado de uno de los eventos afecta el resultado de otro de los eventos.

dependent variable (p. 64) The other variable in a function, usually y, whose values depend on x.

variable dependiente La otra variable de una función, por lo general y, cuyo valor depende de x.

depressed polynomial (p. 354) The quotient when a polynomial is divided by one of its binomial factors.

polinomio reducido El cociente cuando se divide un polinomio entre uno de sus factores binomiales.

descriptive statistics (p. P24) The branch of statistics that focuses on collecting, summarizing, and displaying data.

estadística descriptiva Rama de la estadística cuyo enfoque es la recopilación, resumen y demostración de los datos.

determinant (p. 189) A square array of numbers or variables enclosed between two parallel lines.

determinante Arreglo cuadrado de números o variábles encerrados entre dos rectas paralelas.

diagonal rule (p. 190) A method for finding the determinant of a third-order matrix.

regla diagonal Método para encontrar el determinante de una matriz third-order.

dilation (p. 111) A transformation in which a geometric figure is enlarged or reduced.

homotecia Transformación en que se amplía o se reduce un figura geométrica.

dimensional analysis (p. 310) Performing operations with units.

análisis dimensional Realizar operaciones con unidades.

dimensions of a matrix (p. 169) The number of rows, m, and the number of columns, n, of the matrix written as $m \times n$.

tamaño de una matriz El número de filas, m, y columnas, n, de una matriz, lo que se escribe $m \times n$.

directrix (p. 599) *See* parabola.

directriz Véase parábola.

direct variation (pp. 90, 562) y varies directly as x if there is some nonzero constant k such that $y = kx$. k is called the constant of variation.

variación directa y varía directamente con x si hay una constante no nula k tal que $y = kx$. k se llama la constante de variación.

discrete random variable (p. 742) The numerical outcome of a random event that takes on countable values.

variable aleatoria discreta Resultado numérico de un evento aleatorio que toma valores contables.

discrete relation (p. 62) A relation in which the domain is a set of individual points.

relación discreta Relación en la cual el dominio es un conjunto de puntos individuales.

discriminant (p. 267) In the Quadratic Formula, the expression $b^2 - 4ac$.

discriminante En la fórmula cuadrática, la expresión $b^2 - 4ac$.

Distance Formula (p. 594) The distance between two points with coordinates (x_1, y_1) and (x_2, y_2) is given by

$$d = \sqrt{(x_2 - x_1)^2 + (y_2 - y_1)^2}.$$

fórmula de la distancia La distancia entre dos puntos (x_1, y_1) and (x_2, y_2) viene dada por

$$d = \sqrt{(x_2 - x_1)^2 + (y_2 - y_1)^2}.$$

divergent series (p. 683) An infinite geometric series that does *not* have a sum.

serie divergente Serie geométrica infinita que no tiene suma.

domain (p. P4) The set of all x-coordinates of the ordered pairs of a relation.

dominio El conjunto de todas las coordenadas x de los pares ordenados de una relación.

dot plot (p. 92) Two sets of data plotted as ordered pairs in a coordinate plane.

diagrama del punto Dos conjuntos de datos gráficados como pares ordenados en un plano de coordenadas.

E

e (p. 501) The irrational number 2.71828.... e is the base of the natural logarithms.

e El número irracional 2.71828.... e es la base de los logaritmos naturales.

element (p. 169) Each value in a matrix.

elemento Cada valor de una matriz.

elimination method (p. 139) Eliminate one of the variables in a system of equations by adding or subtracting the equations.

método de eliminación Eliminar una de las variables de un sistema de ecuaciones sumando o restando las ecuaciones.

ellipse (p. 615) The set of all points in a plane such that the sum of the distances from two given points in the plane, called foci, is constant.

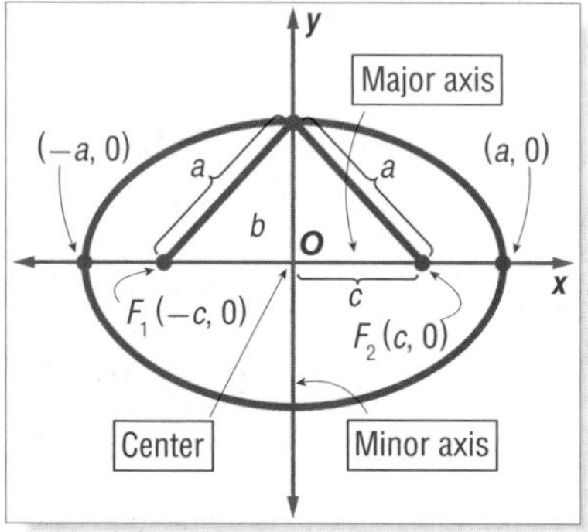

elipse Conjunto de todos los puntos de un plano en los que la suma de sus distancias a dos puntos dados del plano, llamados focos, es constante.

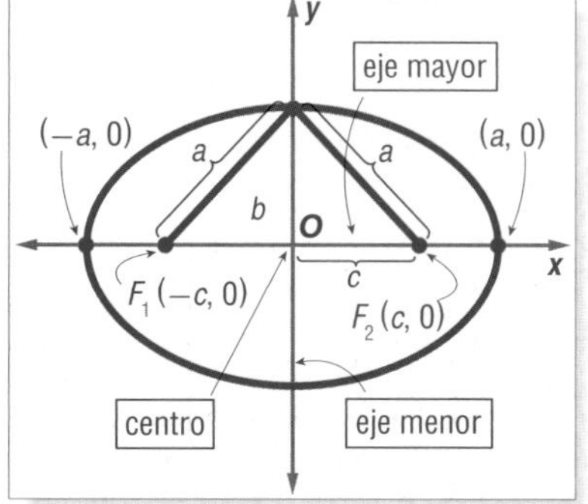

empty set (p. 28) The solution set for an equation that has no solution, symbolized by { } or ø.

conjunto vacío Conjunto solución de una ecuación que no tiene solución, denotado por { } o ø.

end behavior (p. 324) The behavior of the graph as x approaches positive infinity ($+\infty$) or negative infinity ($-\infty$).

comportamiento final El comportamiento de una gráfica a medida que x tiende a más infinito ($+\infty$) o menos infinito ($-\infty$).

equation (p. 18) A mathematical sentence stating that two mathematical expressions are equal.

ecuación Enunciado matemático que afirma la igualdad de dos expresiones matemáticas.

event (p. P9) One or more outcomes of a trial.

evento Uno o más resultados de una prueba.

expected value (p. 745) The expected value of a discrete random variable is the weighted average of the values of the variable.

valor esperado El valor esperado de una variable aleatoria discreta es el promedio ponderado de los valores de la variable.

experiment (p. 723) Something that is intentionally done to people, animals, or objects, and then the response is observed.

experimento Algo se hace intencionalmente poblar, los animales, o los objetos, y entonces la respuesta se observa.

experimental probability (p. P13) What is estimated from observed simulations or experiments.

probabilidad experimental Qué se estima de simulaciones o de experimentos observados.

experimental probability distribution (p. 744) A distribution of probabilities estimated from experiments.

distribución de probabilidad experimental Distribución de probabilidades estimadas a partir de experimentos.

explicit formula (p. 692) Gives a_n as a function of n, such as $a_n = 3n + 1$.

un fórmula explícito Da en función de n, tal como $a_n = 3n + 1$.

exponential decay (p. 453) Exponential decay occurs when a quantity decreases exponentially over time.

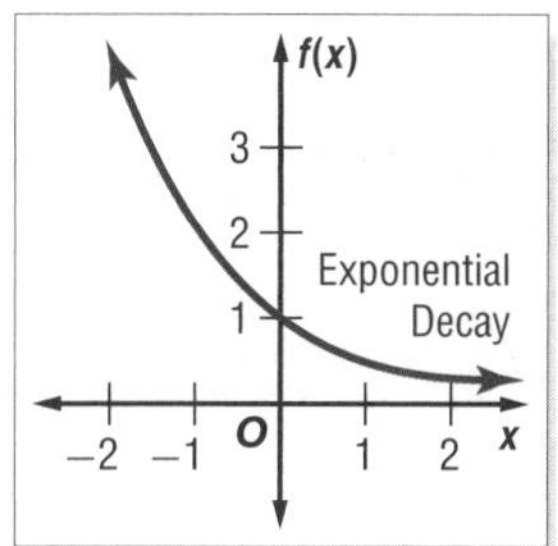

desintegración exponencial Ocurre cuando una cantidad disminuye exponencialmente con el tiempo.

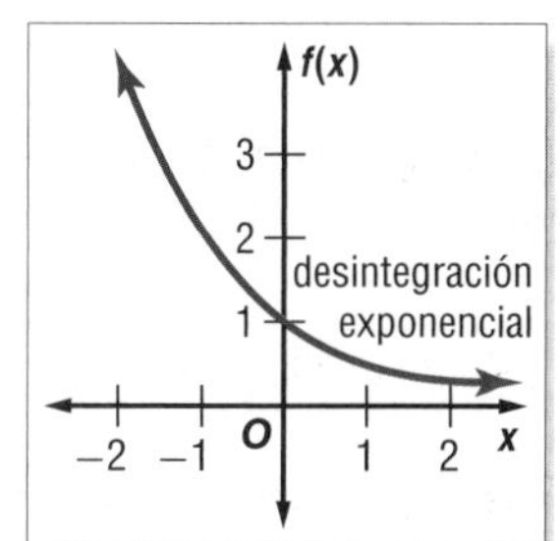

exponential equation (p. 461) An equation in which the variables occur as exponents.

ecuación exponencial Ecuación en que las variables aparecen en los exponentes.

exponential function (p. 451) A function of the form $y = ab^x$, where $a \neq 0$, $b > 0$, and $b \neq 1$.

función exponencial Una función de la forma $y = ab^x$, donde $a \neq 0$, $b > 0$, y $b \neq 1$.

exponential growth (p. 451) Exponential growth occurs when a quantity increases exponentially over time.

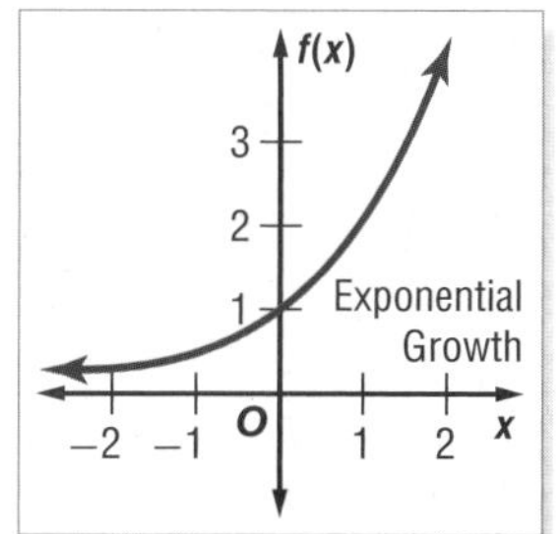

crecimiento exponencial El que ocurre cuando una cantidad aumenta exponencialmente con el tiempo.

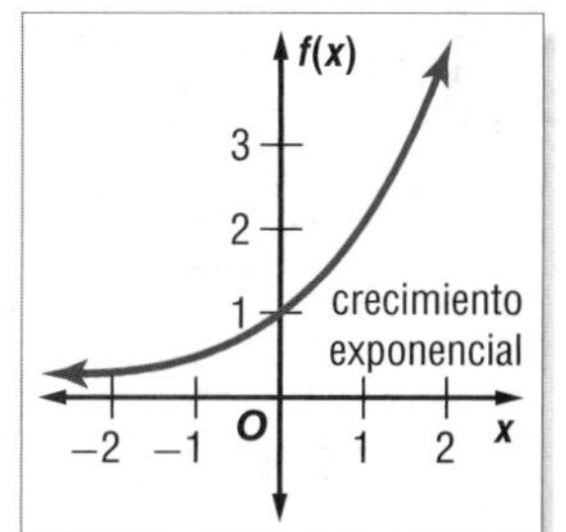

exponential inequality (p. 463) An inequality involving exponential functions.

desigualdad exponencial Desigualdad que contiene funciones exponenciales.

extraneous solution (pp. 29, 429) A number that does not satisfy the original equation.

solución extraña Número que no satisface la ecuación original.

extrema (p. 331) The maximum and minimum values of a function.

extrema Son los valores máximos y mínimos de una función.

F

factored form (p. 238) The form of a polynomial showing all of its factors. $y = a(x - p)(x - q)$ is the factored form of a quadratic equation.

forma reducida La forma de un polinomio que demuestra todos sus factores. $y = a(x - p)(x - q)$ es la forma descompuesta en factores de una ecuación cuadrática.

Factor Theorem (p. 354) The binomial $x - r$ is a factor of the polynomial $P(x)$ if and only if $P(r) = 0$.

teorema factor El binomio $x - r$ es un factor del polinomio $P(x)$ si $P(r) = 0$.

family of graphs (p. 109) A group of graphs that displays one or more similar characteristics.

familia de gráficas Grupo de gráficas que presentan una o más características similares.

feasible region (p. 154) The intersection of the graphs in a system of constraints.

región viable Intersección de las gráficas de un sistema de restricciones.

Fibonacci sequence (p. 692) A sequence in which the first two terms are 1 and each of the additional terms is the sum of the two previous terms.

sucesión de Fibonacci Sucesión en que los dos primeros términos son iguales a 1 y cada término que sigue es igual a la suma de los dos anteriores.

finite sequence (p. 659) A sequence containing a limited number of terms.

secuencia finitas Una secuencia que contiene un número limitado de términos.

five-number summary (p. P26) The three quartiles and the maximum and minimum values in a set of data.

resumen de cinco números Los tres cuartiles y los valores máximo y mínimo de un conjunto de datos.

foci (pp. 615, 624)
ellipse—The two fixed points from which the sum of the distances from a set of all points in a plane is constant.
hyperbola—The two fixed points from which the difference of the distances from a set of all points in a plane is constant.

focos
(elipse) Los dos puntos fijos de los cuales la suma de las distancias de un sistema de todos los puntos en un plano es constante.
(hipérbola) Los dos puntos fijos de los cuales la diferencia de las distancias de un sistema de todos los puntos en un plano es constante.

focus (p. 599) *See* parabola, ellipse, hyperbola.

foco Véase parábola, elipse, hipérbola.

FOIL method (pp. P6, 238) The product of two binomials is the sum of the products of **F** the *first* terms, **O** the *outer* terms, **I** the *inner* terms, and **L** the *last* terms.

método FOIL El producto de dos binomios es la suma de los productos de los primeros (***F**irst*) términos, los términos exteriores (***O**uter*), los términos interiores (***I**nner*) y los últimos (***L**ast*) términos.

formula (p. 6) A mathematical sentence that expresses the relationship between certain quantities.

fórmula Enunciado matemático que describe la relación entre ciertas cantidades.

frequency (p. 838) The number of cycles in a given unit of time.

frecuencia El número de ciclos en una unidad del tiempo dada.

function (p. P4) A relation in which each element of the domain is paired with exactly one element in the range.

función Relación en que a cada elemento del dominio le corresponde un solo elemento del rango.

function notation (p. 64) An equation of y in terms of x can be rewritten so that $y = f(x)$. For example, $y = 2x + 1$ can be written as $f(x) = 2x + 1$.

notación funcional Una ecuación de y en términos de x puede escribirse en la forma $y = f(x)$. Por ejemplo, $y = 2x + 1$ puede escribirse como $f(x) = 2x + 1$.

G

general form (p. 599) An equation of a parabola in the form $y = ax^2 + bx + c$.

forma general Una ecuación de una parábola en la forma $y = ax^2 + bx + c$.

geometric mean (p. 675) The terms between any two nonsuccessive terms of a geometric sequence.

media geométrica Cualquier término entre dos términos no consecutivos de una sucesión geométrica.

geometric sequence (p. 661) A sequence in which each term after the first is found by multiplying the previous term by a constant r, called the common ratio.

sucesión geométrica Sucesión en que cualquier término después del primero puede hallarse multiplicando el término anterior por una constante r, llamada razón común.

geometric series (p. 676) The sum of the terms of a geometric sequence.

serie geométrica La suma de los términos de una sucesión geométrica.

greatest integer function (p. 102) A step function, written as $f(x) = [\![x]\!]$, where $f(x)$ is the greatest integer less than or equal to x.

función del máximo entero Una función etapa que se escribe $f(x) = [\![x]\!]$, donde $f(x)$ es el meaximo entero que es menor que o igual a x.

growth factor (p. 453) In exponential growth, the base of the exponential expression, $1 + r$.

factor del crecimiento En el crecimiento exponencial, la base de la expresión exponencial, $1 + r$.

H

histogram (p. 733) A histogram uses bars to display numerical data that have been organized into equal intervals.

histograma Un histograma usa barras para exhibir datos numéricos que han sido organizados en intervalos iguales.

horizontal asymptote (p. 553) A horizontal line which a graph approaches.

asíntota horizontal Una linea horizontal a que un gráfico acerca.

hyperbola (pp. 545, 624) The set of all points in the plane such that the absolute value of the difference of the distances from two given points in the plane, called foci, is constant.

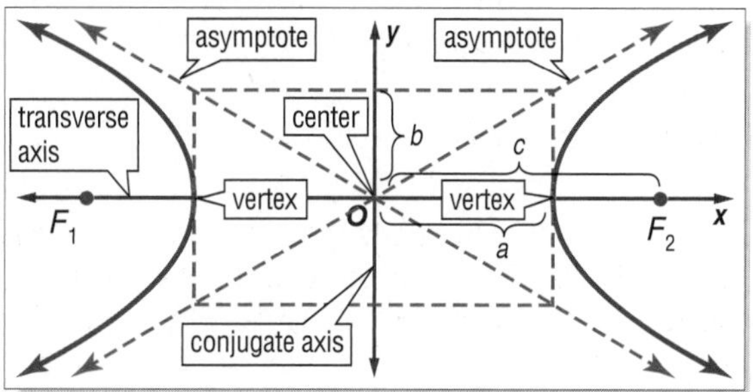

hipérbola Conjunto de todos los puntos de un plano en los que el valor absoluto de la diferencia de sus distancias a dos puntos dados del plano, llamados focos, es constante.

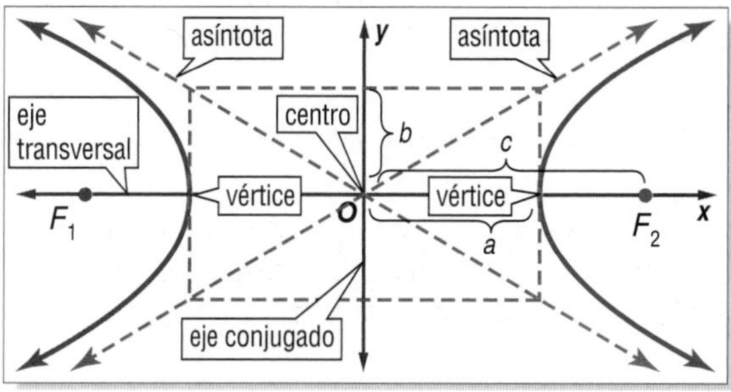

hypothesis test (p. 771) A test used to assess a specific claim about the mean.

prueba de hipótesis Prueba que se usa para evaluar una conjetura específica acerca de la media.

I

identity (p. 350) An equation in which the left side is equal to the right side for all values.

identidad Ecuación en la cual su lado izquierdo es igual a su lado derecho para todos los valores de la variable para los cuales ambos lados están definidos.

identity function (p. 109) The function $I(x) = x$.

unción identidad La función $I(x) = x$.

identity matrix (p. 198) A square matrix that, when multiplied by another matrix, equals that same matrix. If A is any $n \times n$ matrix and I is the $n \times n$ identity matrix, then $A \cdot I = A$ and $I \cdot A = A$.

matriz identidad Matriz cuadrada que al multiplicarse por otra matriz, es igual a la misma matriz. Si A es una matriz de $n \times n$ e I es la matriz identidad de $n \times n$, entonces $A \cdot I = A$ y $I \cdot A = A$.

imaginary unit (p. 246) $\boldsymbol{i}$, or the principal square root of -1.

unidad imaginaria $\boldsymbol{i}$, o la raíz cuadrada principal de -1.

inconsistent (p. 335) A system of equations with no ordered pair that satisfy both equations.

inconsistente Un sistema de ecuaciones para el cual no existe par ordenado alguno que satisfaga ambas ecuaciones.

independent (p. 335) A system of equations with exactly one solution.

independiente Un sistema de ecuaciones que posee una única solución.

independent events (p. P16) Events in which the outcome of one event does not affect the outcome of another event.

eventos independientes Eventos en que el resultado de uno de los eventos no afecta el resultado de otro evento.

independent variable (p. 64) In a function, the variable, usually x, whose values make up the domain.

variable independiente En una función, la variable, por lo general x, cuyos valores forman el dominio.

index (p. 407) In nth roots, the value of n in the symbol $\sqrt[n]{\quad}$. Indicates to what root the value under the radicand is being taken.

indice (de un radical) En las nth raíces, el valor de n en el símbolo $\sqrt[n]{\quad}$. Indica qué raíz se está llevando el valor bajo radicand.

induction hypothesis (p. 705) The assumption that a statement is true for some positive integer k, where $k \geq n$.

hipótesis inductiva El suponer que un enunciado es verdadero para algún entero positivo k, donde $k \geq n$.

inferential statistics (p. 769) Statistics like predictions and hypothesis testing are used to draw conclusions about a population by using a sample.

estadística deductiva La estadística como predicciones y la prueba de la hipótesis es utilizada para dibujar conclusiones sobre una población usando una muestra.

infinite geometric series (p. 683) A geometric series with an infinite number of terms.

serie geométrica infinita Serie geométrica con un número infinito de términos.

infinite sequence (p. 659) A sequence that continues without end.

secuencia infinita Una secuencia que continúa sin extremo.

infinity (pp. 40, 685) Without bound, or continues without end.

infinito Sin límite, o continúa sin extremo.

initial side (p. 799) The fixed ray of an angle.

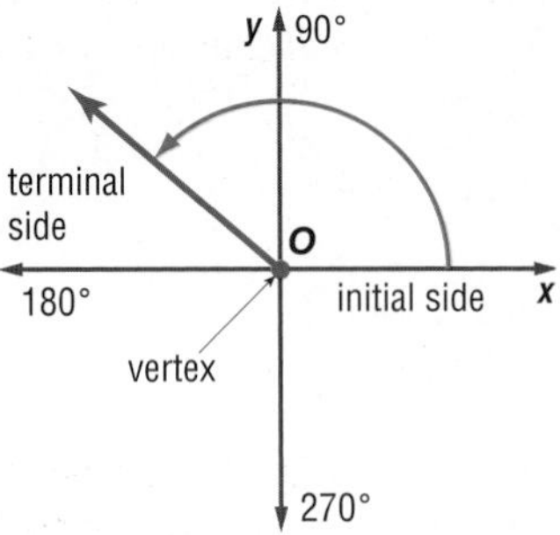

lado inicial de un ángulo El rayo fijo de un ángulo.

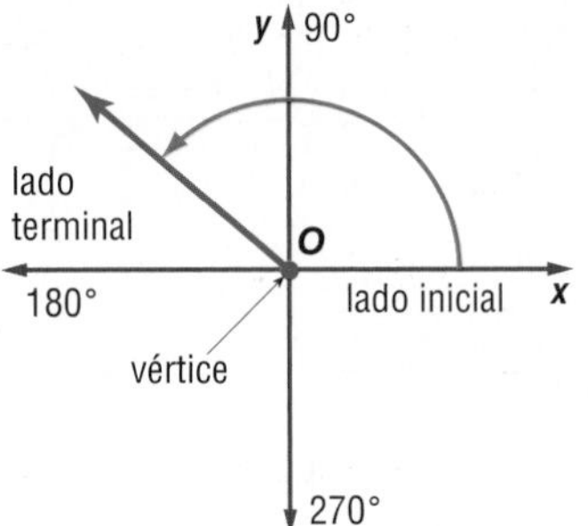

integer (p. 11) $\{\ldots, -3, -2, -1, 0, 1, 2, 3, \ldots\}$

número entero $\{\ldots, -3, -2, -1, 0, 1, 2, 3, \ldots\}$

interquartile range (IQR) (p. P27) The range of the middle half of a set of data. It is the difference between the upper quartile and the lower quartile.

amplitud intercuartílica Amplitud de la mitad central de un conjunto de datos. Es la diferencia entre el cuartil superior y el inferior.

intersection (p. 41) The graph of a compound inequality containing *and.*

intersección Gráfica de una desigualdad compuesta que contiene la palabra *y.*

interval notation (p. 40) A way to describe the solution set of an inequality.

notación del intervalo Una manera de describir el sistema de la solución de una desigualdad.

inverse function (p. 393) Two functions f and g are inverse functions if and only if both of their compositions are the identity function.

función inversa Dos funciones f y g son inversas mutuas si y sólo si las composiciones de ambas son la función identidad.

inverse matrices (p. 198) Two $n \times n$ matrices are inverses of each other if their product is the identity matrix.

matrices inversas Dos matrices de $n \times n$ son inversas mutuas si su producto es la matriz identidad.

inverse of a trigonometric function (p. 853) The arccosine, arcsine, and arctangent relations.

inversa de una función trigonométrica Las relaciones arcocoseno, arcoseno y arcotangente.

inverse relation (p. 393) Two relations are inverse relations if and only if whenever one relation contains the element (a, b) the other relation contains the element (b, a).

relaciones inversas Dos relaciones son relaciones inversas mutuas si y sólo si cada vez que una de las relaciones contiene el elemento (a, b), la otra contiene el elemento (b, a).

inverse variation (p. 564) y varies inversely as x if there is some nonzero constant k such that $xy = k$ or $y = \frac{k}{x}$, where $x \neq 0$ and $y \neq 0$.

variación inversa y varía inversamente con x si hay una constante no nula k tal que $xy = k$ o $y = \frac{k}{x}$, donde $x \neq 0$ y $y \neq 0$.

irrational number (p. 11) A real number that is not rational. The decimal form neither terminates nor repeats.

número irracional Número que no es racional. Su expansión decimal no es ni terminal ni periódica.

iteration (p. 694) The process of composing a function with itself repeatedly.

iteración Proceso de componer una función consigo misma repetidamente.

J

joint variation (p. 563) y varies jointly as x and z if there is some nonzero constant k such that $y = kxz$.

variación conjunta y varía conjuntamente con x y z si hay una constante no nula k tal que $y = kxz$.

L

latus rectum (p. 599) The line segment through the focus of a parabola and perpendicular to the axis of symmetry.

latus rectum El segmento de recta que pasa por el foco de una parábola y que es perpendicular a su eje de simetría.

Law of Cosines (p. 823) Let $\triangle ABC$ be any triangle with a, b, and c representing the measures of sides, and opposite angles with measures A, B, and C, respectively. Then the following equations are true.

$a^2 = b^2 + c^2 - 2bc \cos A$

$b^2 = a^2 + c^2 - 2ac \cos B$

$c^2 = a^2 + b^2 - 2ab \cos C$

Ley de los cosenos Sea $\triangle ABC$ un triángulo cualquiera, con a, b, y c las longitudes de los lados y con ángulos opuestos de medidas A, B y C, respectivamente. Entonces se cumplen las siguientes ecuaciones.

$a^2 = b^2 + c^2 - 2bc \cos A$

$b^2 = a^2 + c^2 - 2ac \cos B$

$c^2 = a^2 + b^2 - 2ab \cos C$

Law of Large Numbers (p. 744) The variation in a data set decreases as the sample size increases.

ley de los grandes números La variación en un conjunto de datos disminuye a medida que aumenta el tamaño de la muestra.

Law of Sines (p. 815) Let $\triangle ABC$ be any triangle with a, b, and c representing the measures of sides opposite angles with measurements A, B, and C, respectively. Then

$$\frac{\sin A}{a} = \frac{\sin B}{b} = \frac{\sin C}{c}.$$

Ley de los senos Sea $\triangle ABC$ cualquier triángulo con a, b y c las longitudes de los lados y con ángulos opuestos de medidas A, B y C, respectivamente. Entonces

$$\frac{\sin A}{a} = \frac{\sin B}{b} = \frac{\sin C}{c}.$$

leading coefficient (p. 322) The coefficient of the term with the highest degree.

coeficiente líder Coeficiente del término de mayor grado.

like radical expressions (p. 417) Two radical expressions in which both the radicands and indices are alike.

expresiones radicales semejantes Dos expresiones radicales en que tanto los radicandos como los índices son semejantes.

limit (p. 690) The value that the terms of a sequence approach.

límite El valor al que tienden los términos de una sucesión.

linear correlation coefficient (p. 94) A value that shows how close data points are to a line.

coeficiente de correlación lineal Valor que muestra la cercanía de los datos a una recta.

linear equation (p. 69) An equation that has no operations other than addition, subtraction, and multiplication of a variable by a constant.

ecuación lineal Ecuación sin otras operaciones que las de adición, sustracción y multiplicación de una variable por una constante.

linear function (p. 69) A function whose ordered pairs satisfy a linear equation.

función lineal Función cuyos pares ordenados satisfacen una ecuación lineal.

linear inequality (p. 117) An inequality that describes a half-plane with a boundary that is a straight line.

desigualdad lineal Desigualdad lineal es una desigualdad que describe un semiplano con un límite que es una línea recta.

linear programming (p. 154) The process of finding the maximum or minimum values of a function for a region defined by inequalities.

programación lineal Proceso de hallar los valores máximo o mínimo de una función lineal en una región definida por las desigualdades.

linear relation (p. 69) A relation that has straight line graphs.

notación del intervalo Una manera de describir el sistema de la solución de una desigualdad.

linear term (p. 219) In the equation $f(x) = ax^2 + bx + c$, bx is the linear term.

término lineal En la ecuación $f(x) = ax^2 + bx + c$, el término lineal es bx.

line of fit (p. 92) A line that closely approximates a set of data.

recta de ajuste Recta que se aproxima estrechamente a un conjunto de datos.

line of reflection (p. 111) The line over which a reflection flips a figure.

línea de la reflexión La línea excedente que una reflexión mueve de un tirón una figura.

Location Principle (p. 330) Suppose $y = f(x)$ represents a polynomial function and a and b are two numbers such that $f(a) < 0$ and $f(b) > 0$. Then the function has at least one real zero between a and b.

principio de ubicación Sea $y = f(x)$ una función polinómica con a y b dos números tales que $f(a) < 0$ y $f(b) > 0$. Entonces la función tiene por lo menos un resultado real entre a y b.

logarithm (p. 468) In the function $x = b^y$, y is called the logarithm, base b, of x. Usually written as $y = \log_b x$ and is read "y equals log base b of x."

logaritmo En la función $x = b^y$, y es el logaritmo en base b, de x. Generalmente escrito como $y = \log_b x$ y se lee "y es igual al logaritmo en base b de x."

logarithmic equation (p. 478) An equation that contains one or more logarithms.

ecuación logarítmica Ecuación que contiene uno o más logaritmos.

logarithmic function (p. 469) The function $y = \log_b x$, where $b > 0$ and $b \neq 1$, which is the inverse of the exponential function $y = b^x$.

función logarítmica La función $y = \log_b x$, donde $b > 0$ y $b \neq 1$, inversa de la función exponencial $y = b^x$.

logarithmic inequality (p. 479) An inequality that contains one or more logarithms.

desigualdad logarítmica Desigualdad que contiene uno o más logaritmos.

logistic growth model (p. 512) A growth model that represents growth that has a limiting factor. Logistic models are the most accurate models for representing population growth.

modelo logístico del crecimiento Un modelo del crecimiento que representa el crecimiento que tiene un factor limitador. Los modelos logísticos son los modelos más exactos para representar crecimiento de la población.

lower quartile (p. P26) The median of the lower half of a set of data, indicated by LQ.

cuartil inferior Mediana de la mitad inferior de un conjunto de datos, se denota con CI.

M

major axis (p. 615) The longer of the two line segments that form the axes of symmetry of an ellipse.

eje mayor El más largo de dos segmentos de recta que forman los ejes de simetría de una elipse.

mapping (p. P4) How each member of the domain is paired with each member of the range.

transformaciones La correspondencia entre cada miembro del dominio con cada miembro del rango.

margin of error (p. 731) The limit on the difference between how a sample responds and how the total population would respond.

margen de error muestral Límite en la diferencia entre las respuestas obtenidas con una muestra y cómo pudiera responder la población entera.

mathematical induction (p. 705) A method of proof used to prove statements about positive integers.

inducción matemática Método de demostrar enunciados sobre los enteros positivos.

matrix (p. 169) Any rectangular array of variables or constants in horizontal rows and vertical columns.

matriz Arreglo rectangular de variables o constantes en filas horizontales y columnas verticales.

matrix equation (p. 200) A matrix form used to represent a system of equations.

ecuación matriz Forma de matriz que se usa para representar un sistema de ecuaciones.

maximum error of estimate (p. 769) The maximum difference between the estimate of the population mean and its actual value.

error máximo de estimación Diferencia máxima entre la estimación de la media de una población y su valor real.

maximum value (p. 222) The y-coordinate of the vertex of the quadratic function $f(x) = ax^2 + bx + c$, where $a < 0$.

valor máximo La coordenada y del vértice de la función cuadrática $f(x) = ax^2 + bx + c$, where $a < 0$.

mean (p. P24) The sum of the values in a set of data divided by the total number of values in the set.

media Suma de los valores en un conjunto de datos dividida entre el número total de valores en el conjunto.

measure of central tendency (p. P24) A number that represents the center or middle of a set of data.

medida de tendencia central Número que representa el centro o medio de un conjunto de datos.

measure of variation (p. P25) A representation of how spread out or scattered a set of data is.

medida de variación Número que representa la dispersión de un conjunto de datos.

median (p. P24) The middle value or the mean of the middle values in a set of data when the data are arranged in numerical order.

mediana Valor del medio o la media de los valores del medio en un conjunto de datos, cuando los datos están ordenados numéricamente.

midline (p. 846) A horizontal axis used as the reference line about which the graph of a periodic function oscillates.

linea media Eje horizontal que se usa como recta de referencia alrededor de la cual oscila la gráfica de una función periódica.

minimum value (p. 222) The y-coordinate of the vertex of the quadratic function $f(x) = ax^2 + bx + c$, where $a > 0$.

valor mínimo La coordenada y del vértice de la función cuadrática $f(x) = ax^2 + bx + c$, donde $a > 0$.

minor axis (p. 615) The shorter of the two line segments that form the axes of symmetry of an ellipse.

eje menor El más corto de los dos segmentos de recta de los ejes de simetría de una elipse.

mode (p. P24) The value or values that appear most often in a set of data.

moda Valor o valores que aparecen con más frecuencia en un conjunto de datos.

mutually exclusive (p. P14) Two events that cannot occur at the same time.

mutuamente exclusivos Dos eventos que no pueden ocurrir simultáneamente.

N

***n*th root** (p. 407) For any real numbers a and b, and any positive integer n, if $a^n = b$, then a is an nth root of b.

raíz enésima Para cualquier número real a y b y cualquier entero positivo n, si $a^n = b$, entonces a se llama una raíz *enésima* de b.

natural base, e (p. 501) An irrational number approximately equal to 2.71828... .

base natural, e Número irracional aproximadamente igual a 2.71828... .

natural base exponential function (p. 501) An exponential function with base e, $y = e^x$.

función exponencial natural La función exponencial de base e, $y = e^x$.

natural logarithm (p. 501) Logarithms with base e, written $\ln x$.

logaritmo natural Logaritmo de base e, el que se escribe $\ln x$.

natural number (p. 11) $\{1, 2, 3, 4, 5, \ldots\}$

número natural $\{1, 2, 3, 4, 5, \ldots\}$

negative correlation (p. 92) When the values in a scatter plot are closely linked in a negative manner.

corelación negativo Cuando los valores en un diagrama de dispersión se ligan de cerca de una manera negativa.

negative exponent (p. 303) For any real number $a \neq 0$ and any integer n, $a^{-n} = \frac{1}{a^n}$ and $\frac{1}{a^{-n}} = a^n$.

exponente negativo Para cualquier número real $a \neq 0$ cualquier entero positivo n, $a^{-n} = \frac{1}{a^n}$ y $\frac{1}{a^{-n}} = a^n$.

normal distribution (p. 760) A continuous, symmetric, bell-shaped distribution of a random variable.

Normal Distribution

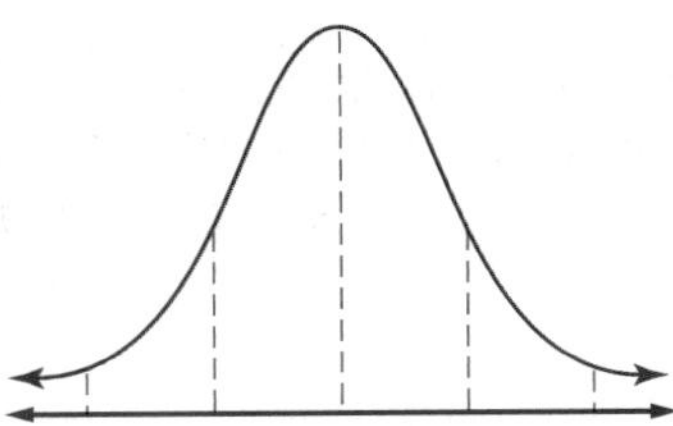

distribución normal Distribución con forma de campana, simétrica y continua de una variable aleatoria.

Distribución normal

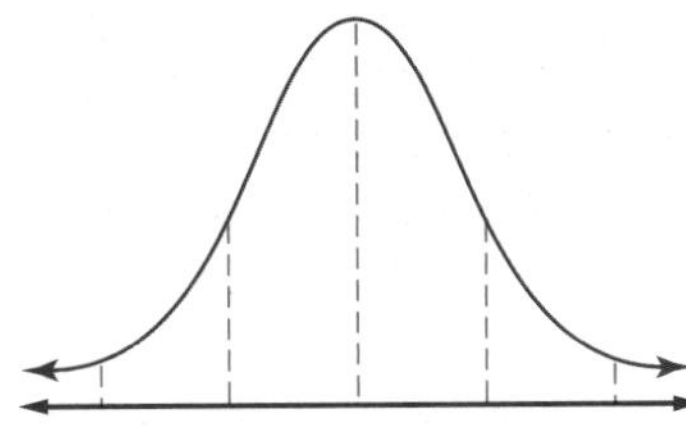

null hypothesis (p. 771) A specific hypothesis to be tested. It is expressed as an equality and is considered true until evidence indicates otherwise.

hipótesis nula Es una hipótesis específica que se probará. Se expresa como igualdad y se considera verdad hasta que la evidencia indica de otra manera.

O

oblique asymptote (p. 555) An asymptote that is neither horizontal nor vertical and is sometimes called a *slant asymptote.*

asíntota oblicuo Una asíntota que es ni horizontal ni la vertical y a veces se llama una *asíntota inclinada.*

observational study (p. 723) Individuals are observed and no attempt is made to influence the results.

estudio de observación Observan a los individuos y no se hace ninguna tentativa de influenciar los resultados.

odds (p. P15) A ratio that compares the number of ways an event can occur to the number of ways that it cannot occur.

posibilidad Razón que compara el número de maneras en que puede ocurrir un evento al número de maneras en que no puede ocurrir dicho evento.

one-to-one function (p. 61) **1.** A function where each element of the range is paired with exactly one element of the domain **2.** A function whose inverse is a function.

función biunívoca **1.** Función en la que a cada elemento del rango le corresponde sólo un elemento del dominio. **2.** Función cuya inversa es una función.

onto function (p. 61) Each element of the range corresponds to an element of the domain.

sobre la función Cada elemento de la gama corresponde a un elemento del dominio. Centro y un punto en el círculo. Índice del crecimiento continuo—la tarifa en la cual algo crece continuamente. El valor de *k* en la función exponencial del crecimiento, $f(x) = ae$.

open sentence (p. 18) A mathematical sentence containing one or more variables.

enunciado abierto Enunciado matemático que contiene una o más variables.

optimize (p. 156) To seek the optimal price or amount that is desired to minimize costs or maximize profits.

optimice Buscar el precio óptimo o ascender que se desea para reducir al mínimo costes o para maximizar de los beneficios.

ordered triple (p. 161) **1.** The coordinates of a point in space **2.** The solution of a system of equations in three variables *x*, *y*, and *z*.

triple ordenado **1.** Las coordenadas de un punto en el espacio **2.** Solución de un sistema de ecuaciones en tres variables *x*, *y* y *z*.

Order of Operations (p. 5)
- **Step 1** Evaluate expressions inside grouping symbols.
- **Step 2** Evaluate all powers.
- **Step 3** Do all multiplications and/or divisions from left to right.
- **Step 4** Do all additions and subtractions from left to right.

orden de las operaciones
- **Paso 1** Evalúa las expresiones dentro de símbolos de agrupamiento.
- **Paso 2** Evalúa todas las potencias.
- **Paso 3** Ejecuta todas las multiplicaciones y divisiones de izquierda a derecha.
- **Paso 4** Ejecuta todas las adiciones y sustracciones de izquierda a derecha.

outcome (p. P9) The results of a probability experiment or an event.

resultados Lo que produce un experimento o evento probabilístico.

outlier (pp. P27, 93) A data point that does not appear to belong to the rest of the set.

valor atípico Dato que no parece pertenecer al resto el conjunto.

P

parabola (pp. 219, 599) The graph of a quadratic function. The set of all points in a plane that are the same distance from a given point, called the focus, and a given line, called the directrix.

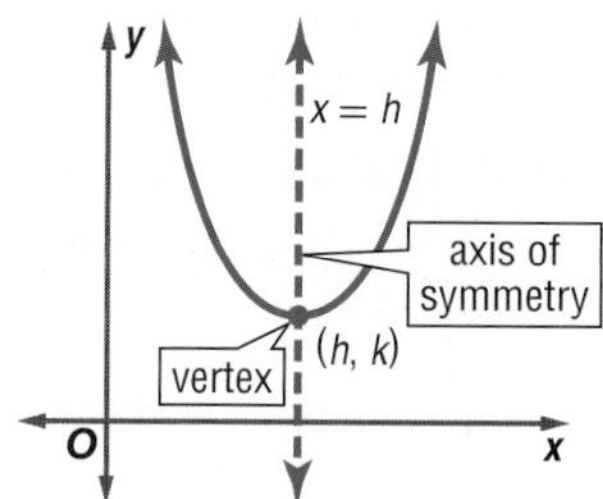

parábola La gráfica de una función cuadrática. Conjunto de todos los puntos de un plano que están a la misma distancia de un punto dado, llamado foco, y de una recta dada, llamada directriz.

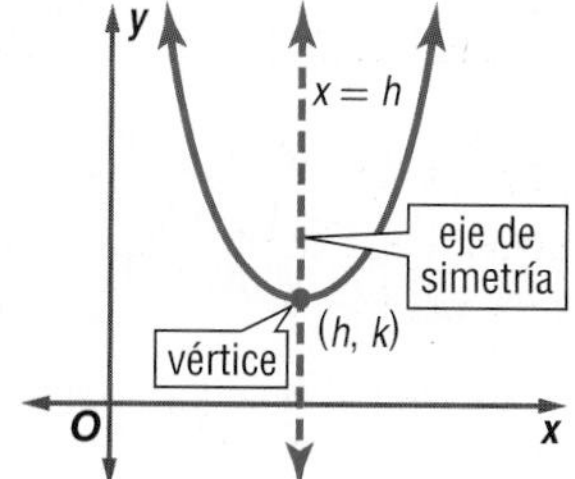

parallel lines (p. 85) Nonvertical coplanar lines with the same slope.

rectas paralelas Rectas coplanares no verticales con la misma pendiente.

parameter (p. 723) A measure that describes a characteristic of a population.

parámetro Una medida que describe una característica de una población.

parent function (p. 109) The simplest, most general function in a family of functions.

función del padre El más simple, la mayoría de la función general en una familia de funciones.

parent graph (p. 109) The simplest of graphs in a family.

gráfica madre La gráfica más sencilla en una familia de gráficas.

partial sum (p. 668) The sum of the first n terms of a series.

suma parcial La suma de los primeros n términos de una serie.

Pascal's triangle (p. 699) A triangular array of numbers such that the $(n + 1)th$ row is the coefficient of the terms of the expansion $(x + y)^n$ for $n = 0, 1, 2 \ldots$

triángulo de Pascal Arreglo triangular de números tal que la fila de $(n + 1)$ está compuesta de proporciona los coeficientes de los términos de la expansión de $(x + y)^n$ para $n = 0, 1, 2 \ldots$

period (p. 831) The least possible value of a for which $f(x) = f(x + a)$.

período El menor valor positivo posible para a, para el cual $f(x) = f(x + a)$.

periodic function (p. 831) **1.** A function with y-values that repeat at regular intervals. **2.** A function is called periodic if there is a number a such that $f(x) = f(x + a)$ for all x in the domain of the function.

función periódica **1.** Una función con y-valores aquella repetición con regularidad. **2.** Función para la cual hay un número a tal que $f(x) = f(x + a)$ para todo x en el dominio de la función.

permutation (p. P9) An arrangement of objects in which order is important.

permutación Arreglo de elementos en que el orden es importante.

perpendicular lines (p. 85) In a plane, any two oblique lines, the product of whose slopes is -1.

rectas perpendiculares En un plano, dos rectas oblicuas cualesquiera cuyas pendientes tienen un producto igual a -1.

phase shift (p. 845) A horizontal translation of a trigonometric function.

desplazamiento de fase Traslación horizontal de una función trigonométrica.

piecewise-defined function (p. 101) A function that is written using two or more expressions.

función por trozos-definida Una función se escribe que usando dos o más expresiones.

piecewise-linear function (p. 102) A function in which the equation for each interval is linear.

función lineal por partes Una función en que la ecuación para cada intervalo es lineal.

point discontinuity (p. 556) If the original function is undefined for $x = a$ but the related rational expression of the function in simplest form is defined for $x = a$, then there is a hole in the graph at $x = a$.

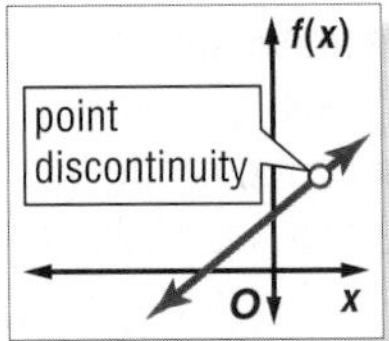

discontinuidad evitable Si la función original no está definida en $x = a$ pero la expresión racional reducida correspondiente de la función está definida en $x = a$, entonces la gráfica tiene una ruptura o corte en $x = a$.

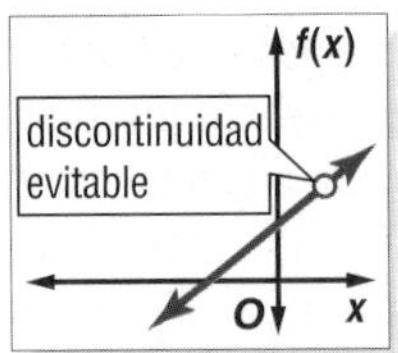

point-slope form (p. 84) An equation in the form $y - y_1 = m(x - x_1)$ where (x_1, y_1) are the coordinates of a point on the line and m is the slope of the line.

forma punto-pendiente Ecuación de la forma $y - y_1 = m(x - x_1)$ donde (x_1, y_1) es un punto en la recta y m es la pendiente de la recta.

polynomial function (p. 323) A function that is represented by a polynomial equation.

función polinomial Función representada por una ecuación polinomial.

polynomial identity (p. 350) A polynomial equation that is true for any values that are substituted for the variables.

identidad polinómica Ecuación polinómica que es verdadera para cualquier valor por el que se reemplazan las variables.

polynomial in one variable (p. 322) $a_nx^n + a_{n-1}x^{n-1} + \cdots + a_2x^2 + a_1x + a_0$, where the coefficients $a_n, a_{n-1}, \ldots, a_0$ represent real numbers, and a_n is not zero and n is a nonnegative integer.

polinomio de una variable $a_nx^n + a_{n-1}x^{n-1} + \cdots + a_2x^2 + a_1x + a_0$, donde los coeficientes $a_n, a_{n-1}, \ldots, a_0$ son números reales, a_n no es nulo y n es un entero no negativo.

population (p. P24) An entire group of living things or objects.

población Un grupo entero de cosas o de objetos vivos.

positive correlation (p. 92) When the values in a scatter plot are closely linked in a positive manner.

correlación positivo Cuando los valores en un diagrama de la dispersión se ligan de cerca de una manera positiva.

power function (p. 320) An equation in the form $f(x) = ax^b$, where a and b are real numbers.

función potencia Ecuación de la forma $f(x) = ax^b$, donde a y b son números reales.

prediction equation (p. 92) An equation suggested by the points of a scatter plot that is used to predict other points.

ecuación de predicción Ecuación sugerida por los puntos de una gráfica de dispersión y que se usa para predecir otros puntos.

prime polynomial (p. 342) A polynomial that cannot be factored.

polinomio primero Un polinomio que no puede ser descompuesto en factores.

principal root (p. 407) The nonnegative root.

raíz principal La raíz no negativa.

principal values (p. 853) The values in the restricted domains of trigonometric functions.

valores principales Valores en los dominios restringidos de las funciones trigonométricas.

probability (p. P13) A measure of the chance that a given event will occur.

probabilidad Medida de la posibilidad de que ocurra un evento dado.

probability distribution (p. 743) A function that maps the sample space to the probabilities of the outcomes in the sample space for a particular random variable.

distribución de probabilidad Función que aplica el espacio muestral a las probabilidades de los resultados en el espacio muestral obtenidos para una variable aleatoria particular.

probability model (p. P13) A mathematical model used to represent the outcomes of an experiment.

modelo de probabilidad Modelo matemático que se usa para representar los resultados de un experimento.

pure imaginary number (p. 246) The square roots of negative real numbers. For any positive real number b,

$$\sqrt{-b^2} = \sqrt{b^2} \cdot \sqrt{-1}, \text{ or } bi.$$

número imaginario puro Raíz cuadrada de un número real negativo. Para cualquier número real positivo b,

$$\sqrt{-b^2} = \sqrt{b^2} \cdot \sqrt{-1}, \text{ ó } bi.$$

Q

quadrantal angle (p. 808) An angle in standard position whose terminal side coincides with one of the axes.

ángulo de cuadrante Ángulo en posición estándar cuyo lado terminal coincide con uno de los ejes.

quadrants (p. P4) The four areas of a Cartesian coordinate plane.

cuadrantes Las cuatro regiones de un plano de coordenadas Cartesiano.

quadratic equation (p. 229) A quadratic function set equal to a value, in the form $ax^2 + bx + c = 0$, where $a \neq 0$.

ecuación cuadrática Función cuadrática igual a un valor, de la forma $ax^2 + bx + c = 0$, donde $a \neq 0$.

quadratic form (p. 345) For any numbers a, b, and c, except for $a = 0$, an equation that can be written in the form $u^2 + u + c = 0$, where u is some expression in x.

forma de ecuación cuadrática Para cualquier número a, b, y c, excepto $a = 0$, una ecuación que puede escribirse de la forma $u^2 + u + c = 0$, donde u es una expresión en x.

Quadratic Formula (p. 264) The solutions of a quadratic equation of the form $ax^2 + bx + c = 0$, where $a \neq 0$, are given by the Quadratic Formula, which is

$$x = \frac{-b \pm \sqrt{b^2 - 4ac}}{2a}.$$

fórmula cuadrática Las soluciones de una ecuación cuadrática de la forma $ax^2 + bx + c = 0$, donde $a \neq 0$, se dan por la fórmula cuadrática, que es

$$x = \frac{-b \pm \sqrt{b^2 - 4ac}}{2a}.$$

quadratic function (pp. 109, 219) A function described by the equation $f(x) = ax^2 + bx + c$, where $a \neq 0$.

función cuadrática Función descrita por la ecuación $f(x) = ax^2 + bx + c$, donde $a \neq 0$.

quadratic inequality (p. 282) An inequality of the form $y > ax^2 + bx + c$, $y \geq ax^2 + bx + c$, $y < ax^2 + bx + c$, or $y \leq ax^2 + bx + c$.

desigualdad cuadrática Desigualdad cuadrática de la forma $y > ax^2 + bx + c$, $y \geq ax^2 + bx + c$, $y < ax^2 + bx + c$, o $y \leq ax^2 + bx + c$.

quadratic term (p. 219) In the equation $f(x) = ax^2 + bx + c$, ax^2 is the quadratic term.

término cuadrático En la ecuación $f(x) = ax^2 + bx + c$, ax^2 el término cuadrático es ax^2.

quartic function (p. 324) A fourth-degree function.

función quartic Una función del cuarto-grado.

quartiles (p. P26) The values that divide a set of data into four equal parts.

cuartiles Valores que dividen un conjunto de datos en cuatro partes iguales.

quintic function (p. 322) A fifth-degree function.

función quintic Una función del quinto-grado.

R

radian (p. 801) The measure of an angle θ in standard position whose rays intercept an arc of length 1 unit on the unit circle.

radián Medida de un ángulo θ en posición normal cuyos rayos intersecan un arco de 1 unidad de longitud en el círculo unitario.

radical equation (p. 429) An equation with radicals that have variables in the radicands.

ecuación radical Ecuación con radicales que tienen variables en el radicando.

radical function (p. 400) A function that contains the root of a variable.

función radical Una función que contiene la raíz de una variable.

radical inequality (p. 431) An inequality that has a variable in the radicand.

desigualdad radical Desigualdad que tiene una variable en el radicando.

radical sign (p. 407) In nth roots, the symbol $\sqrt[n]{\quad}$.

signo radical El símbolo $\sqrt[n]{\quad}$, que se usa par indicar la raíz cuadrada no negative o el símbolo por raíz enésima.

radicand (p. 407) In nth roots, the value inside in the symbol $\sqrt[n]{\quad}$. Indicates the value that is being taken to the nth root.

radicando El número o la expressión que aparece debajo del signo radical.

radius (p. 607) Any segment whose endpoints are the center and a point on the circle.

radio Un segmento cuyos extremos son el centro y un punto del círculo.

random sample (p. 723) A sample in which every member of the population has an equal chance of being selected.

muestra aleatoria Muestra en la cual cada miembro de la población tiene igual posibilidad de ser elegido.

random variable (p. 742) The outcome of a random process that has a numerical value.

variable aleatoria El resultado de un proceso aleatorio que tiene un valor numérico.

range (pp. P4, P25) **1.** The set of all *y*-coordinates of a relation. **2.** The difference between the greatest and least values in a set of data.

rango **1.** Conjunto de todas las coordenadas *y* de una relación. **2.** Diferencia entre el valor mayor y el menor en un conjunto de datos.

rate of change (p. 76) How much a quantity changes on average, relative to the change in another quantity, over time.

tasa de cambio Lo que cambia una cantidad en promedio, respecto al cambio en otra cantidad, por lo general el tiempo.

rate of continuous decay (p. 509) The rate at which something decays continuously. Represented by a constant k in the exponential decay function $f(x) = ae^{-kt}$, where a is the initial value, and t is time in years.

índice de desintegración continúa Ritmo al cual algo se desintegra continuamente. Representado por la constante k en la función de desintegración exponencial $f(x) = ae^{-kt}$, donde a es el valor inicial y t es el tiempo en años.

rate of continuous growth (p. 509) The rate at which something grows continuously. The value of k in the exponential growth function, $f(x) = ae^{kt}$.

el índice del crecimiento continuo Es la tasa en la cual algo crece continuamente. El valor de k en la función exponencial del crecimiento, $f(x) = ae^{kt}$. La tasa en la cual algo crece continuamente.

rational equation (p. 570) Any equation that contains one or more rational expressions.

ecuación racional Cualquier ecuación que contiene una o más expresiones racionales.

rational exponent (p. 423) For any nonzero real number b, and any integers m and n, with $n > 1$, $b^{\frac{m}{n}} = \sqrt[n]{b^m} = \left(\sqrt[n]{b}\right)^m$, except when $b < 0$ and n is even.

exponent racional Para cualquier número real no nulo b y cualquier entero m y n, con $n > 1$, $b^{\frac{m}{n}} = \sqrt[n]{b^m} = \left(\sqrt[n]{b}\right)^m$, excepto cuando $b < 0$ y n es par.

rational expression (p. 529) A ratio of two polynomial expressions.

expresión racional Razón de dos expresiones polinomiales.

rational function (p. 553) An equation of the form $f(x) = \frac{p(x)}{q(x)}$, where $p(x)$ and $q(x)$ are polynomial functions, and $q(x) \neq 0$.

función racional Ecuación de la forma $f(x) = \frac{p(x)}{q(x)}$, donde $p(x)$ y $q(x)$ son funciones polinomiales y $q(x) \neq 0$.

rational inequality (p. 575) Any inequality that contains one or more rational expressions.

desigualdad racional Cualquier desigualdad que contiene una o más expresiones racionales.

rationalizing the denominator (p. 416) To eliminate radicals from a denominator or fractions from a radicand.

racionalizar el denominador La eliminación de radicales de un denominador o de fracciones de un radicando.

rational number (p. 11) Any number $\frac{m}{n}$, where m and n are integers and n is not zero. The decimal form is either a terminating or repeating decimal.

número racional Cualquier número $\frac{m}{n}$, donde m y n son enteros y n no es cero. Su expansión decimal es o terminal o periódica.

Rational Zero Theorem (p. 367) Helps you choose some possible zeros of a polynomial function to test.

El teorema cero racional Ayudas usted elige algunos ceros posibles de una función polinómica para probar.

real numbers (p. 11) All numbers used in everyday life; the set of all rational and irrational numbers.

números reales Todos los números que se usan en la vida cotidiana; el conjunto de los todos los números racionales e irracionales.

reciprocal function (pp. 545, 791) **1.** A function of the form $f(x) = \frac{1}{a(x)}$, where $a(x)$ is a linear function and $a(x) \neq 0$. **2.** Trigonometric functions that are reciprocals of each other.

funciones recíprocas **1.** Una función de la forma $f(x) = \frac{1}{a(x)}$, donde $a(x)$ es una función linear y $a(x) \neq 0$. **2.** Funciones trigonométricas de la función que son reciprocals de uno a.

recursive formula (p. 692) Each term is formulated from one or more previous terms.

fórmula recursiva Cada término proviene de uno o más términos anteriores.

recursive sequence (p. 692) A sequence in which each term is determined by one or more of the previous terms.

sucesión recursiva Una secuencia en la cual cada término es determinado por uno o más de los términos anteriores.

reference angle (p. 808) The acute angle formed by the terminal side of an angle in standard position and the *x*-axis.

ángulo de referencia El ángulo agudo formado por el lado terminal de un ángulo en posición estándar y el eje *x*.

reflection (p. 111) A transformation in which every point of a figure is mapped to a corresponding image across a line of symmetry.

reflexión Transformación en que cada punto de una figura se aplica a través de una recta de simetría a su imagen correspondiente.

regression line (p. 94) A line of best fit.

reca de regresión Una recta de óptimo ajuste.

relation (p. P4) A set of ordered pairs.

relación Conjunto de pares ordenados.

relative frequency (pp. P18, 743) The ratio of the number of observations in a category to the total number of observations.

frecuencia relativa Razón del número de observaciones en una categoría al número total de observaciones.

relative maximum (p. 331) A point on the graph of a function where no other nearby points have a greater *y*-coordinate.

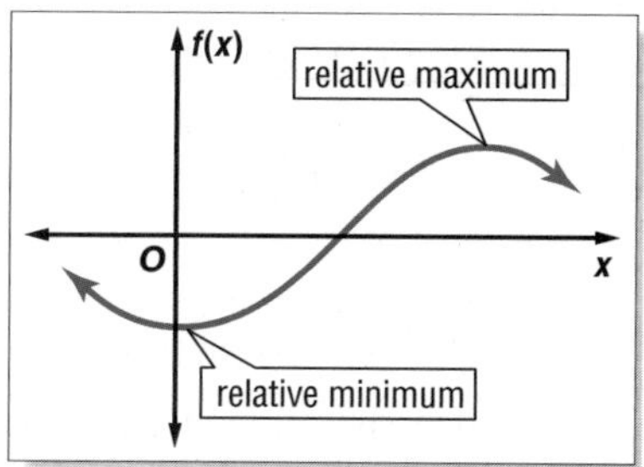

máximo relativo Punto en la gráfica de una función en donde ningún otro punto cercano tiene una coordenada *y* mayor.

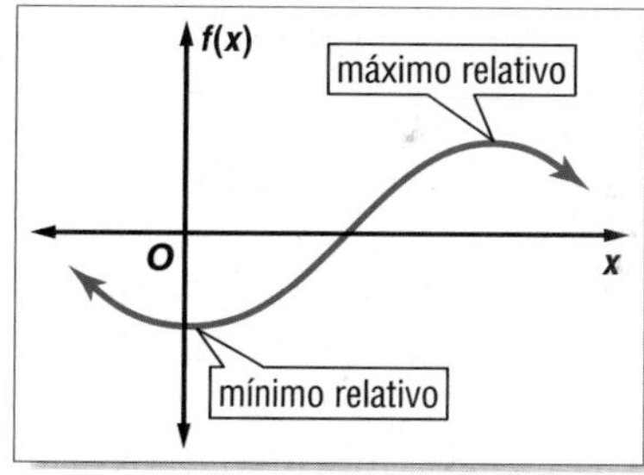

relative minimum (p. 331) A point on the graph of a function where no other nearby points have a lesser *y*-coordinate.

mínimo relativo Punto en la gráfica de una función en donde ningún otro punto cercano tiene una coordenada *y* menor.

root (p. 229) The solutions of a quadratic equation.

raíz Las soluciones de una ecuación cuadrática.

row matrix (p. 169) A matrix that has only one row.

matriz fila Matriz que sólo tiene una fila.

S

sample (pp. P24, 723) A part of a population.

muestra Parte de una población.

sample space (p. P9) The set of all possible outcomes of an experiment.

espacio muestral Conjunto de todos los resultados posibles de un experimento probabilístico.

scalar (p. 173) A constant.

escalar Una constante.

scalar multiplication (p. 173) Multiplying any matrix by a constant called a scalar; the product of a scalar *k* and an $m \times n$ matrix.

multiplicación por escalares Multiplicación de una matriz por una constante llamada escalar; producto de un escalar *k* y una matriz de $m \times n$.

scatter plot (p. 92) A set of data graphed as ordered pairs in a coordinate plane.

gráfica de dispersión Conjuntos de datos graficados como pares ordenados en un plano de coordenadas.

secant (p. 790) For any angle, with measure α, a point $P(x, y)$ on its terminal side, $r = \sqrt{x^2 + y^2}$, $\sec \alpha = \frac{r}{x}$.

secante Para cualquier ángulo de medida α, un punto $P(x, y)$ en su lado terminal, $r = \sqrt{x^2 + y^2}$, $\sec \alpha = \frac{r}{x}$.

second-order determinant (p. 189) The determinant of a 2×2 matrix.

determinante de segundo orden El determinante de una matriz de 2×2.

sequence (p. 659) A list of numbers in a particular order.

sucesión Lista de números en un orden particular.

series (p. 668) The sum of the terms of a sequence.

serie Suma específica de los términos de una sucesión.

set-builder notation (p. 35) The expression of the solution set of an inequality, for example $\{x \mid x > 9\}$.

notación de construcción de conjuntos Escritura del conjunto solucion de una desigualdad, por ejemplo, $\{x \mid x > 9\}$.

sigma notation (p. 669) For any sequence $a_1, a_2, a_3, \ldots$, the sum of the first k terms may be written $\sum_{n=1}^{k} a_n$, which is read "the summation from $n = 1$ to k of a_n." Thus, $\sum_{n=1}^{k} a_n = a_1 + a_2 + a_3 + \cdots + a_k$, where k is an integer value.

notación de suma Para cualquier sucesión $a_1, a_2, a_3, \ldots$, la suma de los k primeros términos puede escribirse $\sum_{n=1}^{k} a_n$, lo que se lee "la suma de $n = 1$ a k de los a_n." Así, $\sum_{n=1}^{k} a_n = a_1 + a_2 + a_3 + \cdots + a_k$, donde k es un valor entero.

simple event (p. P14) One event.

evento simple Un solo evento.

simplify (p. 303) To rewrite an expression without parentheses or negative exponents.

reducir Escribir una expresión sin paréntesis o exponentes negativos.

simulation (p. 731) The use of a probability experiment to mimic a real-life situation.

simulación Uso de un experimento probabilístico para imitar una situación de la vida real.

sine (p. 790) For any angle, with measure α, a point $P(x, y)$ on its terminal side, $r = \sqrt{x^2 + y^2}$, $\sin \alpha = \frac{y}{r}$.

seno Para cualquier ángulo de medida α, un punto $P(x, y)$ en su lado terminal, $r = \sqrt{x^2 + y^2}$, $\sin \alpha = \frac{y}{r}$.

slope (p. 77) The ratio of the change in y-coordinates to the change in x-coordinates.

pendiente La razón del cambio en coordenadas y al cambio en coordenadas x.

slope-intercept form (p. 83) The equation of a line in the form $y = mx + b$, where m is the slope and b is the y-intercept.

forma pendiente-intersección Ecuación de una recta de la forma $y = mx + b$, donde m es la pendiente y b la intersección.

solution (p. 18) A replacement for the variable in an open sentence that results in a true sentence.

solución Sustitución de la variable de un enunciado abierto que resulta en un enunciado verdadero.

solving a right triangle (p. 815) The process of finding the measures of all of the sides and angles of a right triangle.

resolver un triángulo rectángulo Proceso de hallar las medidas de todos los lados y ángulos de un triángulo rectángulo.

solving a triangle (p. 815) Using given measures to find all unknown side lengths and angle measures of a triangle.

resolver un triángulo Usar medidas dadas de hallar todas las medidas de las longitudes y laterales desconocidas del ángulo de un triángulo.

square matrix (p. 198) A matrix with the same number of rows and columns.

matriz cuadrada Matriz con el mismo número de filas y columnas.

square root function (p. 400) A function that contains a square root of a variable.

función radical Función que contiene la raíz cuadrada de una variable.

square root inequality (p. 402) An inequality involving the square root of a variable expression.

desigualdad radical Desigualdad que presenta raíces cuadradas de un expresión con variables.

Square Root Property (p. 247) For any real number n, if $x^2 = n$, then $x = \pm\sqrt{n}$.

Propiedad de la raíz cuadrada Para cualquier número real n, si $x^2 = n$, entonces $x = \pm\sqrt{n}$.

standard deviation (p. P25) The square root of the variance.

desviación estándar La raíz cuadrada de la varianza.

standard form (pp. 70, 229, 599) **1.** A linear equation written in the form $Ax + By = C$, where A, B, and C are integers whose greatest common factor is 1, $A \geq 0$, and A and B are not both zero. **2.** A quadratic equation written in the form $ax^2 + bx + c = 0$, where a, b, and c are integers, and $a \neq 0$.

forma estándar **1.** Ecuación lineal escrita de la forma $Ax + By = C$, donde A, B, y C son enteros cuyo máximo común divisores 1, $A \geq 0$, y A y B no son cero simultáneamente. **2.** Una ecuación cuadrática escrita en la forma $ax^2 + bx + c = 0$, donde a, b, y c son números enteros, y $a \neq 0$.

standard normal distribution (p. 762) A normal distribution with a mean of 0 and a standard deviation of 1.

distribución normal estándar Distribución normal con una media de 0 y una desviación estándar de 1.

standard position (p. 799) An angle positioned so that its vertex is at the origin and its initial side is along the positive x-axis.

posición estándar Ángulo en posición tal que su vértice está en el origen y su lado inicial está a lo largo del eje x positivo.

statistic (p. 723) A measure that describes a characteristic of a sample.

estadística Una medida que describe una característica de una muestra.

statistical inference (p. 769) Use information from a sample to draw conclusions about a population.

inferencia estadística Utiliza la estadística deductiva como las predicciones y la hipótesis que prueban información de uso de una muestra para dibujar las conclusiones acerca de una población.

step function (p. 102) A function whose graph is a series of line segments.

función etapa Función cuya gráfica es una serie de segmentos de recta.

substitution method (p. 138) A method of solving a system of equations in which one equation is solved for one variable in terms of the other.

método de sustitución Método para resolver un sistema de ecuaciones en que una de las ecuaciones se resuelve en una de las variables en términos de la otra.

survey (p. 723) Used to collect information about a population.

encuesta Reunía información acerca de una población.

synthetic division (p. 312) A method used to divide a polynomial by a binomial.

división sintética Método que se usa para dividir un polinomio entre un binomio.

synthetic substitution (p. 352) The use of synthetic division to evaluate a function.

sustitución sintética Uso de la división sintética para evaluar una función polinomial.

system of equations (p. 136) A set of equations with the same variables.

sistema de ecuaciones Conjunto de ecuaciones con las mismas variables.

system of inequalities (p. 146) A set of inequalities with the same variables.

sistema de desigualdades Conjunto de desigualdades con las mismas variables.

tangent (p. 790) **1.** A line that intersects a circle at exactly one point. **2.** For any angle, with measure α, a point $P(x, y)$ on its terminal side, $r = \sqrt{x^2 + y^2}$, $\tan \alpha = \frac{y}{x}$.

tangente **1.** Recta que interseca un círculo en un solo punto. **2.** Para cualquier ángulo, de medida α, un punto $P(x, y)$ en su lado terminal, $r = \sqrt{x^2 + y^2}$, $\tan \alpha = \frac{y}{x}$.

term (p. 659) **1.** The monomials that make up a polynomial. **2.** Each number in a sequence or series.

término **1.** Los monomios que constituyen un polinomio. **2.** Cada número de una sucesión o serie.

terminal side (p. 799) A ray of an angle that rotates about the center.

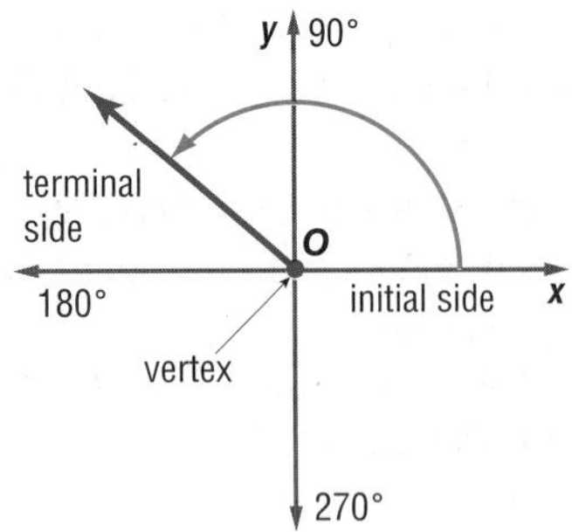

lado terminal Rayo de un ángulo que gira alrededor de un centro.

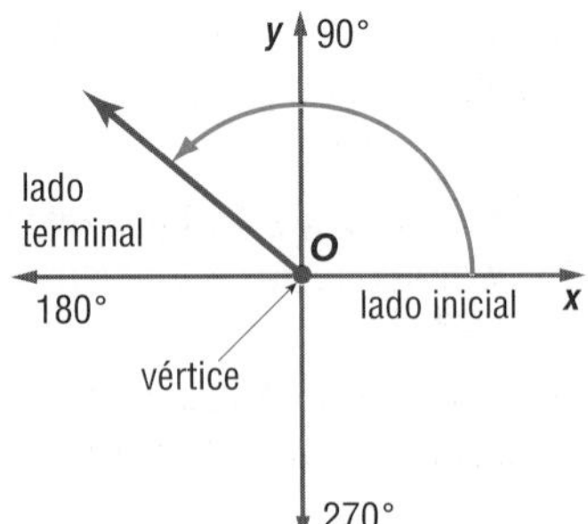

theoretical probability (p. P13) What should occur in a probability experiment.

probabilidad teórica Lo que debería ocurrir en un experimento probabilístico.

theoretical probability distribution (p. 743) A distribution of probabilities based on what is expected to happen.

distribución de probabilidad teórica Distribución de probabilidades basada en lo que se espera que ocurra.

third-order determinant (p. 190) Determinant of a 3×3 matrix.

determinante de tercer orden Determinante de una matriz de 3×3.

translation (p. 110) A figure is moved from one location to another on the coordinate plane without changing its size, shape, or orientation.

traslación Se mueve una figura de un lugar a otro en un plano de coordenadas sin cambiar su tamaño, forma u orientación.

transverse axis (p. 624) The segment of length $2a$ whose endpoints are the vertices of a hyperbola.

eje transversal El segmento de longitud $2a$ cuyos extremos son los vértices de una hipérbola.

tree diagram (p. P9) A diagram that shows all possible outcomes of an event.

diagrama de árbol Diagrama que muestra todos los posibles resultados de un evento.

trigonometric equation (p. 901) An equation containing at least one trigonometric function that is true for some but not all values of the variable.

ecuación trigonométrica Ecuación que contiene por lo menos una función trigonométrica y que sólo se cumple para algunos valores de la variable.

trigonometric functions (p. 790) For any angle, with measure α, a point $P(x, y)$ on its terminal side, $r = \sqrt{x^2 + y^2}$, the trigonometric functions of α are as follows.

$\sin \alpha = \frac{y}{r}$ $\cos \alpha = \frac{x}{r}$ $\tan \alpha = \frac{y}{x}$

$\csc \alpha = \frac{r}{y}$ $\sec \alpha = \frac{r}{x}$ $\cot \alpha = \frac{x}{y}$

funciones trigonométricas Para cualquier ángulo, de medida α, un punto $P(x, y)$ en su lado terminal, $r = \sqrt{x^2 + y^2}$, las funciones trigonométricas de α son las siguientes.

$\sin \alpha = \frac{y}{r}$ $\cos \alpha = \frac{x}{r}$ $\tan \alpha = \frac{y}{x}$

$\csc \alpha = \frac{r}{y}$ $\sec \alpha = \frac{r}{x}$ $\cot \alpha = \frac{x}{y}$

trigonometric identity (p. 873) An equation involving a trigonometric function that is true for all values of the variable for which the function is defined.

identidad trigonométrica Ecuación que involucra una o más funciones trigonométricas y que se cumple para todos los valores de la variable en que el función es definido.

trigonometric ratio (p. 790) Compares the side lengths of a right triangle.

razón trigonometric Compara las longitudes laterales de un triángulo derecho.

trigonometry (p. 790) The study of the relationships between the angles and sides of a right triangle.

trigonometría Estudio de las relaciones entre los lados y ángulos de un triángulo rectángulo.

turning point (p. 331) Point at which a graph turns. The location of relative maxima or minima.

momento crucial Un punto en el cual un gráfico da vuelta. La localización de máximos o de mínimos relativos.

U

unbounded (p. 154) A system of inequalities that forms a region that is open.

no acotado Sistema de desigualdades que forma una región abierta.

uniform probability model (p. P13) An experiment for which all outcomes are equally likely.

modelo de probabilidad uniforme Experimento cuyos resultados son todos equiprobables.

union (p. 42) The graph of a compound inequality containing *or*.

unión Gráfica de una desigualdad compuesta que contiene la palabra *o*.

unit analysis (p. 310) The process of including unit measurement when computing.

análisis de la unidad Proceso de incluir unidades de medida al computar.

unit circle (p. 830) A circle of radius 1 unit whose center is at the origin of a coordinate system.

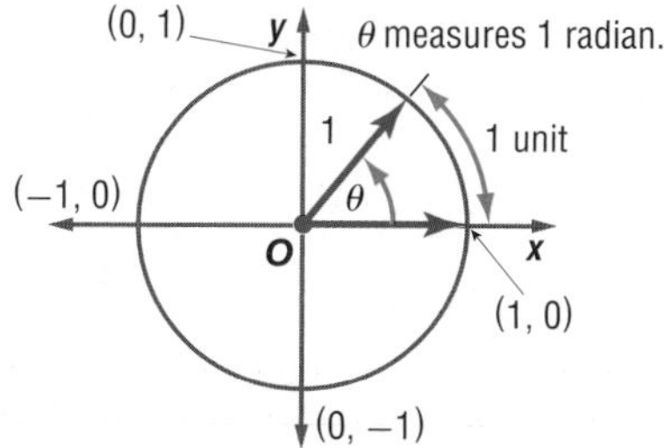

círculo unitario Círculo de radio 1 cuyo centro es el origen de un sistema de coordenadas.

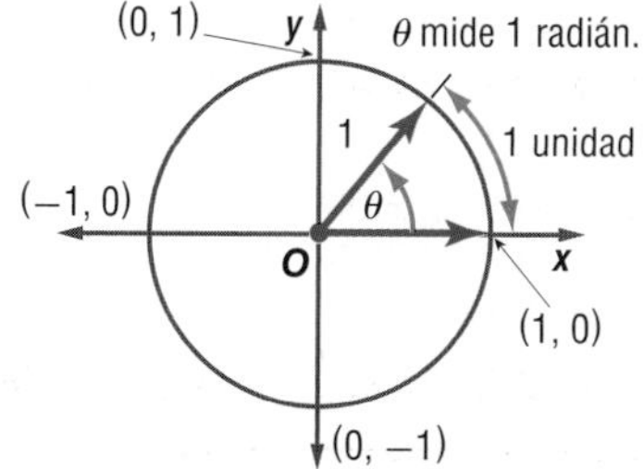

univariate data (p. P24) Data with one variable.

datos univariados Datos con una variable.

V

variable (pp. P24, 5) **1.** A characteristic of a population that can assume different values called *data.* **2.** A symbol, usually a letter, used to represent an unknown quantity.

variable **1.** Característica de una población que puede tomar diferentes valores llamados *datos.* **2.** Símbolo, generalmente una letra, que se usa para representar una cantidad desconocida.

variable matrix (p. 200) A matrix that only contains the variables of a system of equations.

matriz variable Una matriz que contiene solamente las variables de un sistema de ecuaciones.

variance (p. P25) The mean of the squares of the deviations from the arithmetic mean.

varianza Media de los cuadrados de las desviaciones de la media aritmética.

vertex (pp. 154, 220) **1.** Any of the points of intersection of the graphs of the constraints that determine a feasible region. **2.** The point at which the axis of symmetry intersects a parabola. **3.** The point on each branch nearest the center of a hyperbola.

vértice **1.** Cualqeiera de los puntos de intersección de las gráficas que los contienen y que determinan una región viable. **2.** Punto en el que el eje de simetría interseca una parábola. **3.** El punto en cada rama más cercano al centro de una hipérbola.

vertex form (p. 275) A quadratic function in the form $y = a(x - h)^2 + k$, where (h, k) is the vertex of the parabola and $x = h$ is its axis of symmetry.

forma de vértice Función cuadrática de la forma $y = a(x - h)^2 + k$, donde (h, k) es el vértice de la parábola y $x = h$ es su eje de simetría.

vertical asymptote (p. 553) If the related rational expression of a function is written in simplest form and is undefined for $x = a$, then $x = a$ is a vertical asymptote.

asíntota vertical Si la expresión racional que corresponde a una función racional se reduce y está no definida en $x = a$, entonces $x = a$ es una asíntota vertical.

vertical line test (p. 62) If no vertical line intersects a graph in more than one point, then the graph represents a function.

prudba de la recta vertical Si ninguna recta vertical interseca una gráfica en más de un punto, entonces la gráfica representa una función.

vertical shift (p. 846) When graphs of trigonometric functions are translated vertically.

cambio vertical Cuando los gráficos de funciones trigonométricas son translado verticales.

vertices (pp. 615, 624) **ellipse**—The endpoints of the major axis. **hyperbola**—The endpoints of the transverse axis.

vértices **(elipse)** Son las puntos finales del eje principal. **(hipérbola)** Son las puntos finales del eje transversal.

W

weighted average (p. 572) A method for finding the mean of a set of numbers in which some elements of the set carry more importance, or weight, than others.

promedio ponderado Un método para encontrar el medio de un sistema de los números en los cuales algunos elementos del sistema llevan más importancia, o peso, que otros.

whole numbers (p. 11) {0, 1, 2, 3, 4, ...}

números naturales {0, 1, 2, 3, 4, ...}

X

***x*-intercept** (p. 71) The x-coordinate of the point at which a graph crosses the x-axis.

intersección *x* La coordenada x del punto o puntos en que una gráfica interseca o cruza el eje x.

Y

***y*-intercept** (p. 71) The y-coordinate of the point at which a graph crosses the y-axis.

intersección *y* La coordenada y del punto o puntos en que una gráfica interseca o cruza el eje y.

Z

zero matrix (p. 169) A matrix in which every element is zero.

matriz nula matriz cuyos elementos son todos igual a cero.

zeros (p. 229) The x-intercepts of the graph of a function; the points for which $f(x) = 0$.

ceros Las intersecciones x de la gráfica de una función; los puntos x para los que $f(x) = 0$.

z-value (p. 762) The number of standard deviations that a given data value is from the mean.

valor Z Número de variaciones estándar que separa un valor dado de la media.

Index

A

B

C

D

E

G

H

I

J

K

L

M

N

O

Q

R

Index